BIOREGENERATIVE ENGINEERING

The Wiley Bicentennial–Knowledge for Generations

Each generation has its unique needs and aspirations. When Charles Wiley first opened his small printing shop in lower Manhattan in 1807, it was a generation of boundless potential searching for an identity. And we were there, helping to define a new American literary tradition. Over half a century later, in the midst of the Second Industrial Revolution, it was a generation focused on building the future. Once again, we were there, supplying the critical scientific, technical, and engineering knowledge that helped frame the world. Throughout the 20th Century, and into the new millennium, nations began to reach out beyond their own borders and a new international community was born. Wiley was there, expanding its operations around the world to enable a global exchange of ideas, opinions, and know-how.

For 200 years, Wiley has been an integral part of each generation's journey, enabling the flow of information and understanding necessary to meet their needs and fulfill their aspirations. Today, bold new technologies are changing the way we live and learn. Wiley will be there, providing you the must-have knowledge you need to imagine new worlds, new possibilities, and new opportunities.

Generations come and go, but you can always count on Wiley to provide you the knowledge you need, when and where you need it!

William J. Pesce
President and Chief Executive Officer

Peter Booth Wiley
Chairman of the Board

BIOREGENERATIVE ENGINEERING: PRINCIPLES AND APPLICATIONS

SHU Q. LIU

WILEY-INTERSCIENCE
A JOHN WILEY & SONS, INC., PUBLICATION

Published by John Wiley & Sons, Inc., Hoboken, New Jersey
Published simultaneously in Canada

For general information on our other products and services or for technical support, please contact our Customer Care Department within the United States at (800) 762-2974, outside the United States at (317) 572-3993 or fax (317) 572-4002.

Wiley also publishes its books in a variety of electronic formats. Some content that appears in print may not be available in electronic formats. For more information about Wiley products, visit our web site at www.wiley.com.

Wiley Bicentennial Logo: Richard J. Pacifico

Library of Congress Cataloging-in-Publication Data

Liu, Shu Q.
Bioregenerative engineering : principles and applications / Shu Q. Liu.
p. ; cm.
Includes bibliographical references and index.
ISBN 978-0-471-70907-7 (alk. paper)
1. Biomedical engineering. 2. Regeneration (Biology) I. Title.
[DNLM: 1. Biomedical Engineering. 2. Regenerative Medicine. QT 36 L783f 2007]
R856.L57 2007
610.28—dc22

2006027165

Printed in the United States of America

10 9 8 7 6 5 4 3 2 1

CHAPTER SUMMARIES

CONTENTS

PREFACE

Nature has created numerous elegant living systems, including the human, based on the hierarchical functional units—molecule, cell, tissue, and organ. A living system develops through a long evolutionary process, during which the system undergoes genotypic and phenotypic changes in response to environmental simuli. Whereas the environmental and genetic factors play critical roles in evolutionary development, they may induce disorders and injuries of the cell, tissue, or organ, resulting in impairment or destruction of the functional units and preventing the living system from functioning and survival. Since these disorders and injuries are inevitable events during the evolutionary process, Nature has designed various mechanisms for the repair or replacement of injured and disordered cells, tissues, or organs, leading to partial or complete restoration of the structure and function of the living system. Among these mechanisms is cell, tissue, and organ regeneration.

Regeneration is a natural process by which a mature living system repairs or replaces its lost cells, tissues, and organs by activating specific renewal mechanisms, resulting in the restoration of the structure and function of the system. The application of regeneration principles to the treatment of human disease is known as *regenerative medicine*. During the past decade (since the mid-1990s), extensive investigations have been conducted to elucidate the mechanisms of regeneration, leading to the development of regenerative technologies such as stem cell identification, expansion, and transplantation. It is hoped that the transplanted stem cells can engraft to target tissues or organs, differentiate to specified cell types, replace malfunctioned or lost cells, and thus restore the natural structure and function of involved tissues and organs. Preliminary investigations have demonstrated the potential of stem cell transplantation for the treatment of degenerative disorders and cell injury in experimental tests and clinical trials. However, a simple transplantation of stem cells may not solve all the problems in regenerative medicine, since the selected stem cells may not be designed for the therapy of a specified target tissue or may not be able to differentiate into the desired cell types in an environment that is not established for the stem cells. Thus, fundamental issues in regenerative medicine are how to induce stem cells to differentiate into specified functional cell types under given environmental conditions and how to integrate the stem cell-derived cells into the natural system.

Nature has established numerous barriers that prevent the transformation of stem and progenitor cells to specified cell types in developed adult systems, especially in the vital organs such as the brain, heart, and kidney, and thus hinder the regeneration of disordered or lost cells. To resolve such a problem, it is necessary to establish engineering strategies and technologies that alter the expression of specified genes and modulate the phenotypes of target cells, including stem and progenitor cells, and thus to break Nature's barriers and induce appropriate regeneration of disordered or lost cells. Bioregenerative engineering is a discipline established for addressing these issues.

In definition, *bioregenerative engineering* is to induce, modulate, enhance, and/or control regenerative processes by using engineering approaches and thus to improve the restoration of the structure and function of disordered or lost cells, tissues, and organs. Although the term *bioregenerative engineering* has seldom been used, the concept of bioregenerative engineering has long been applied to regenerative medicine. Typical examples include the enhancement of stem cell proliferation and differentiation by transfecting cells with selected mitogenic genes, the elimination of an undesired function by knocking down or knocking out a selected gene, and the improvement of stem cell engraftment, migration, and differentiation by modulating the content, distribution, and pattern of extracellular matrix in a tissue or organ substitute. Given the nature of the discipline, bioregenerative engineering can be considered the engineering aspect of regenerative medicine.

For the past decade, extensive studies have been conducted and a large amount of information has been accumulated in the area of bioregenerative engineering. A reference that systematically summarizes the bioregenerative engineering literature may assist the readers to understand the principles of and design therapeutic strategies in bioregenerative engineering. It was the hope of the author that this book would serve as such a reference.

The author would like to dedicate this book to his mother Jing-zhen Li, father Ding-an Liu, in-laws Tong Wu and Pei-lan Hou, wife Yu-hua Wu, daughter Diana Liu, and son Charley Liu for their sincere support for the work.

S. Q. Liu

March 9, 2006
Evanston, Illinois

INTRODUCTION TO BIOREGENERATIVE ENGINEERING

Bioregenerative engineering is to induce, modulate, and/or control regenerative processes by using molecular, cellular, and tissue engineering approaches and thus improving the restoration of the structure and function of disordered or lost cells, tissues, and/or organs. Bioregenerative engineering is an emerging discipline established by integrating engineering principles and technologies into regenerative medicine. Although the term *bioregenerative engineering* is rarely used, bioregenerative engineering research has been conducted extensively for the past several decades. As we will see throughout the book, this research has elicited significant impacts in essentially all biomedical fields.

Bioregenerative engineering stems from several scientific disciplines, including molecular engineering, cellular engineering, and tissue engineering and may be considered the engineering aspect of regenerative medicine. *Regenerative medicine* is an emerging discipline that addresses restoration of the structure and function of disordered or lost cells, tissues, and organs on the basis of stem cell biology. Strategies for regenerative medicine are to identify and prepare stem and/or progenitor cells and transplant and/or stimulate the identified cells to or in a target tissue, where the stem and/or progenitor cells can differentiate into specified cell types in an appropriate regional environment and thus restore the structure and function of the injured or lost cells. Compared to regenerative medicine, bioregenerative engineering emphasizes the engineering modulation of the regenerative processes at the molecular, cellular, and tissue levels (Fig. I.1).

Regenerative engineering at the molecular level, which may be referred to as *molecular regenerative engineering*, addresses the promotion and control of molecular and cellular activities (e.g., cell signaling, gene expression, cell division, differentiation, migration, adhesion, secretion, and contraction/relaxation); the activation and control of residential stem and progenitor cells; the mobilization and recruitment of remote stem and progenitor cells; and the formation of functional structures by controlled administration of proteins, genes, antisense oligonucleotides, siRNA, and pharmacological substances. Examples of molecular regenerative engineering include the control of a target signaling pathway, the regulation of specific gene expression, and the enhancement or reduction in the proliferation and differentiation of a specified cell type by transfecting target cells with growth regulator genes.

Regenerative engineering at the cellular level, which may be referred to as *cellular regenerative engineering*, addresses the preparation, modulation, and transplantation of autogenous and/or allogenic stem/progenitor cells in a controlled manner, resulting in enhanced regeneration of functional cells and structures. Examples of cellular regenerative engineering include the transplantation of hematopoietic stem and progenitor cells to repopulate impaired leukocytes due to leukemia, the transplantation of embryonic and bone marrow-derived stem cells to the heart to differentiate into cardiomyocytes in cardiac infarction, and the transplantation of neuronal stem cells to the brain to alleviate the symptoms of Alzheimer's and Parkinson's diseases.

Regenerative engineering at the tissue level, which may be referred to as *tissue regenerative engineering*, addresses the construction of tissue-mimicking scaffolds integrated with mature, stem, or progenitor cells, and the implantation of the tissue scaffolds into target organs, thus inducing, enhancing, and/or controlling the regeneration of functional cells and tissues. An artificial scaffold may either function as a tissue substitute or serve as a framework for the regeneration of lost tissues. Examples of tissue regenerative engineering include the construction and implantation of artificial tissues and organs, such as joints, heart valves, and blood vessels. Other approaches, such as reduction of stretch-induced vascular bypass graft injury by structural reinforcement and stimulation of intestinal expansion by mechanical stretching, can also be used to engineer the regeneration at the tissue level. The overall goal of the three regenerative engineering approaches is to improve the therapeutic effects of regenerative medicine (Fig. I.1).

An important basis for bioregenerative engineering is that the cell is capable of conducting natural regenerative processes in response to cell injury and death. Examples of cell regeneration include the renewal of blood cells, epithelial cells in the gastrointestinal system and the skin, endothelial and smooth muscle cells in the vascular system, and hepatocytes. While certain cell types, such as the blood cell and epithelial cell, conduct rapid and intensive regeneration even under physiological conditions, other cell types, such as the neuron and cardiomyocyte, experience very limited regeneration even in response to cell injury and death. These cell-specific characteristics are evolved based on the intrinsic regenerative mechanisms unique to distinct cell types. The clarification of the control mechanisms of cell regeneration is an important task for regenerative engineering research.

The human body is an integrated system composed of a hierarchy of structures, including molecules, cells, tissues, and organs. Although Nature has designed and created these structures with nearly perfect functionality and protective mechanisms, unnecessary or even harmful alterations do occur as a result of gene mutation and environmental stimulation by chemical, biological, and physical pathogens, resulting in pathogenic disorders that may harm or destroy the physiological systems. In response to these changes, the molecules, cells, tissues, and organs are capable of detecting and repairing pathogenic disorders to a certain extent. However, the repairing capability is limited and dependent on a number of factors, including the state of the human protective systems, the nature of gene mutation, and the type and strength of environmental pathogens. In the case of defect or impairment of the protective mechanisms and/or exposure to an unusual pathogen, the human systems may not be able to conduct self-repair or regeneration processes. In severe cases, death is the ultimate consequence. Bioregenerative engineering is established to enhance and improve the repair and regeneration processes and thus to help the human systems recover from pathological disorders.

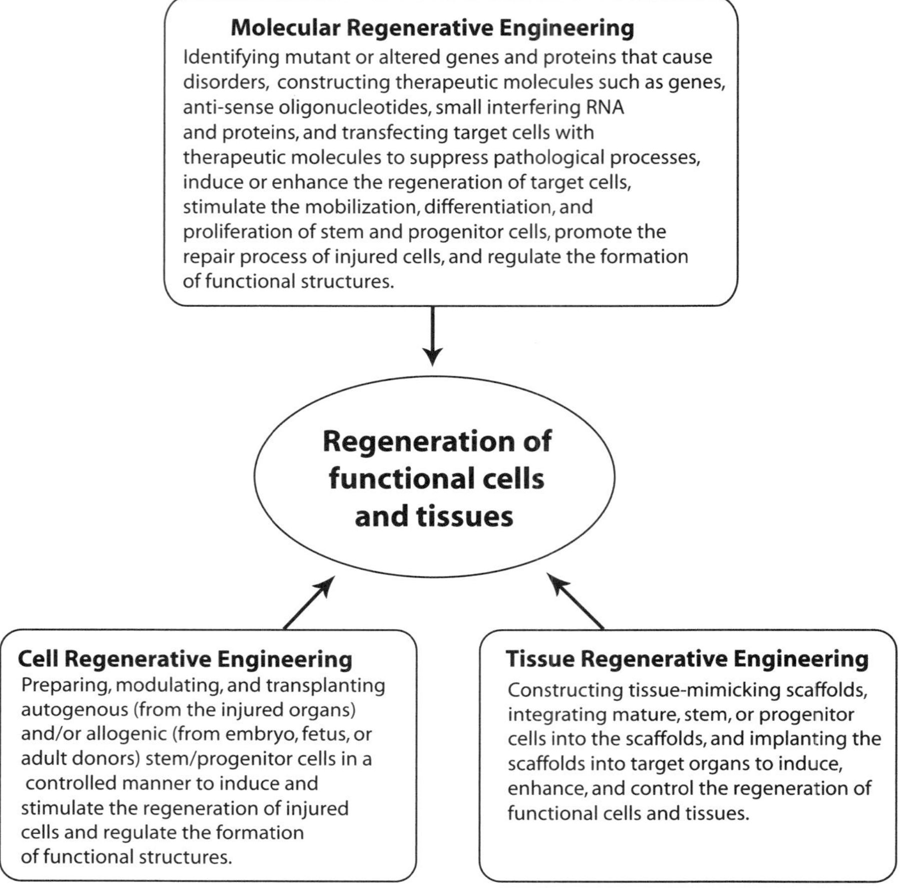

Figure I.1. Regenerative engineering at the molecular, cellular, and tissue levels.

During the past decade, regenerative medicine has become a popular research topic. However, current work relies primarily on simple engineering approaches, such as cell collection, expansion, and transplantation. These approaches may not change the fundamental course of natural processes and thus may not be sufficient to achieve optimal therapeutic effects. For certain types of vital organ, such as the brain and heart, Nature does not develop sophisticated regenerative mechanisms, presumably because these organs are well protected from environmental hazards and are not subject to frequent injury. However, injury and disorder do occur in these vital organs, often with deadly consequences. Thus, simple engineering approaches that do not alter the natural process may not be effective in inducing and enhancing the regeneration of these organs. A sophisticated engineering strategy and technology may be necessary to overcome Nature's barriers and to achieve the goal of regenerative therapies for these vital organs. Although it is a challenging task, regenerative therapies can be significantly improved by incorporating engineering principles and technologies into regenerative medicine.

The primary goal of this book is to introduce to the principles and technologies of bioregenerative engineering. Since bioregenerative engineering is built on the basis of various biomedical disciplines, including molecular biology, cell biology, developmental biology, physiology, pathology and bioengineering, the book will also address these fundamental disciplines. The book consists of two parts: the foundations of bioregenerative engineering, and the principles and applications of bioregenerative engineering. The first part covers the molecular, cellular, and developmental foundations of bioregenerative engineering. The second part covers general mechanisms and technologies of bioregenerative engineering, as well as the application of bioregenerative engineering to selected organ systems. For each organ system, the engineering tests and therapies are discussed at the molecular, cellular, and tissue levels, if applicable.

For the past decade, bioregenerative engineering has undergone rapid development, and engineering-based therapeutic approaches have been extensively tested in experimental models and clinical trials. A large amount of information has been accumulated in the literature. Although it is difficult to cover the information in all aspects in a single book, it was the hope of the author that this book would introduce to the readers the fundamental concepts, experimental approaches, and potential applications of bioregenerative engineering.

PART I

FOUNDATIONS OF BIOREGENERATIVE ENGINEERING

SECTION 1

MOLECULAR BASIS FOR BIOREGENERATIVE ENGINEERING

1

STRUCTURE AND FUNCTION OF MACROMOLECULES

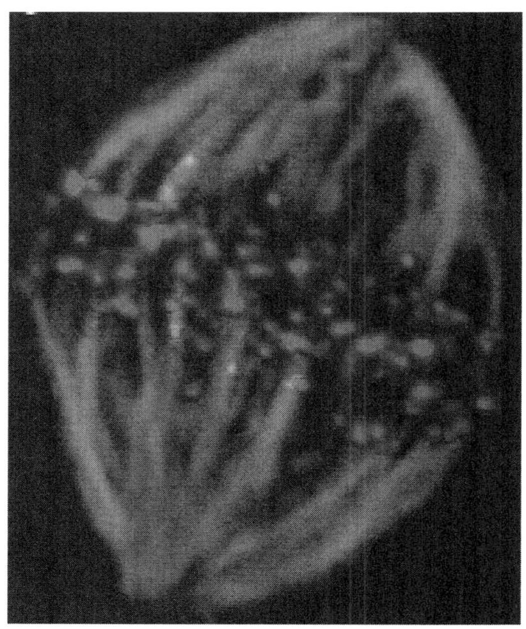

Organization of chromosomes and microtubules in an epithelial cell in the metaphase. Green and red: immunochemically labeled tubulin and kinetochores, respectively. (Reprinted by permission from Kapoor TM et al: Chromosomes can congress to the metaphase plate before biorientation, *Science* 311:388–91, 2006. Copyright 2006, AAAS.) See color insert.

Bioregenerative Engineering: Principles and Applications, by Shu Q. Liu

A living organism is composed of several basic elements: water, electrolytes, nucleotides, amino acids, sugars, and lipids. Water is the most abundant substance in a living system, occupying about 70–85% of the total volume in most cells. All living organisms were originally developed in water, and all biochemical and enzymatic reactions in a cell take place in an aqueous environment. Thus, water is the most important element in a living organism.

A cell consists of a number of electrolytes, including sodium, potassium, calcium, magnesium, chloride, phosphate, sulfate, and bicarbonate. These electrolytes participate in fundamental processes, such as the establishment and maintenance of cell membrane and action potentials, regulation of biochemical and enzymatic reactions, control of muscular contraction and relaxation, formation of mechanical supporting and protection systems, and maintenance of the internal environment. The functions of these electrolytes will be discussed throughout the book where applicable.

Other elements, including nucleotides, sugars, amino acids, and lipids, participate in the formation of macromolecules, including deoxyribose and ribose nucleic acids (nucleotides and sugars), proteins (amino acids), phospholipids (lipids), and polysaccharides (sugars). These macromolecules are essential to the formation, survival, function, and regeneration of living organisms. This chapter focuses on the structure and function of these macromolecules.

DEOXYRIBONUCLEIC ACIDS (DNA)

DNA is the molecule for the transmission, processing, and storage of hereditary information. A DNA molecule is capable of replicating itself, a fundamental mechanism for the transmission of hereditary information from the mother to the daughter generation. A DNA molecule can be transcribed into various types of RNA, including messenger RNA (mRNA), ribosomal RNA (rRNA), and transfer RNA (tRNA). These RNA molecules participate in the translation of proteins. The translated proteins are transported to various compartments of a cell, serving as not only structural constituents, which provide the cell with shape and strength, but also enzymes and signaling molecules, which regulate cellular activities and functions.

Composition and Structure of DNA [1.1]

A DNA molecule is constituted by joining together a large number of nucleotides. A *nucleotide* is composed of three elements, including a base, a β-D-2-deoxyribose, and a phosphate group. There exist four types of base, which are nitrogen-containing ring compounds, including two pyrimidines—cytosine and thymine, denoted as C and T, respectively—and two purines—adenine and guanine, denoted as A and G, respectively (Fig. 1.1). A base is capable of forming a complex with a deoxyribose (Fig. 1.2), giving rise to a molecule known as *nucleoside*. Collectively, there are four types of nucleoside based on the four bases, including cytidine, thymidine, adenosine, and guanosine. With the addition of 1, 2, or 3 phosphate groups, a nucleoside is converted to a nucleotide, known as *nucleoside monophosphate*, *diphosphate*, or *triphosphate*, respectively (Fig. 1.3). The nomenclature for various individual nucleosides and nucleotides are listed in Table 1.1.

A complete DNA molecule is a double-stranded helical polymer of nucleotides. Each stand is composed of a pentose–phosphate backbone and bases positioned on the side of

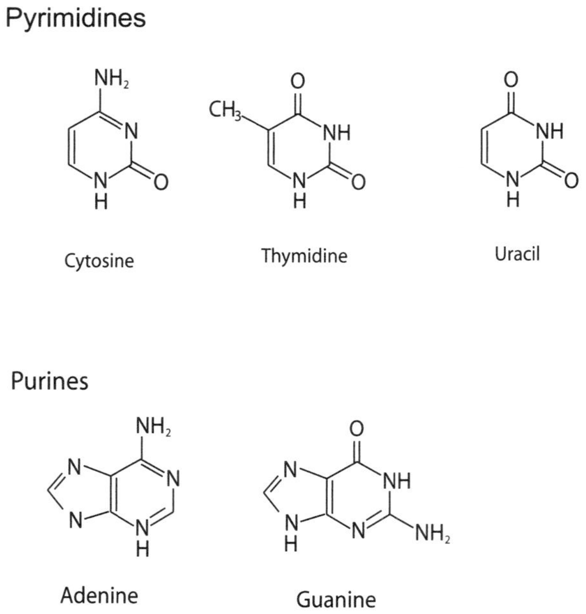

Pyrimidines

Cytosine Thymidine Uracil

Purines

Adenine Guanine

Figure 1.1. Chemical structure of pyrimidines (cytosine, thymine, and uracil) and purines (adenine and guanine) that constitute DNA and RNA.

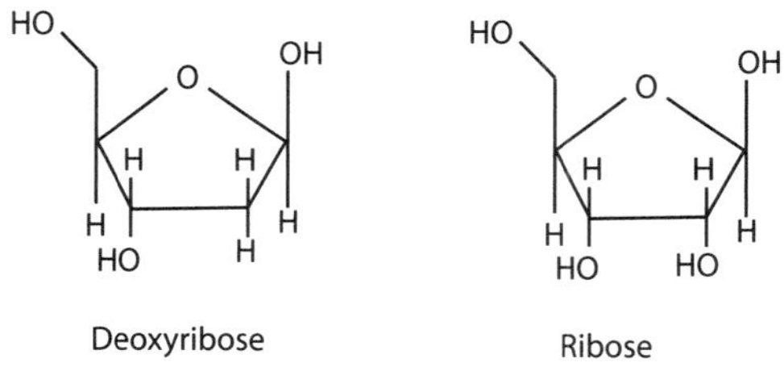

Deoxyribose Ribose

Figure 1.2. Chemical structure of deoxyribose and ribose.

TABLE 1.1. Nomenclature of Nucleosides and Nucleotides for DNA

Types of Base	Cytosine (C)	Thymine (T)	Adenine (A)	Guanine (G)
Nucleosides	Deoxycytidine (dC)	Deoxythymidine (dT)	Deoxyadenosine (dA)	Deoxyguanosine (dG)
Nucleotides	Deoxycytidine 5′ monophosphate (dCMP)	Deoxythymidine 5′ monophosphate (dTMP)	Deoxyadenosine 5′ monophosphate (dAMP)	Deoxyguanosine 5′ monophosphate (dGMP)
	Deoxycytidine 5′ diphosphate (dCDP)	Deoxythymidine 5′ diphosphate (dTDP)	Deoxyadenosine 5′ diphosphate (dADP)	Deoxyguanosine 5′ diphosphate (dGDP)
	Deoxycytidine 5′ triphosphate (dCTP)	Deoxythymidine 5′ triphosphate (dTTP)	Deoxyadenosine 5′ triphosphate (dATP)	Deoxyguanosine 5′ triphosphate (dGTP)

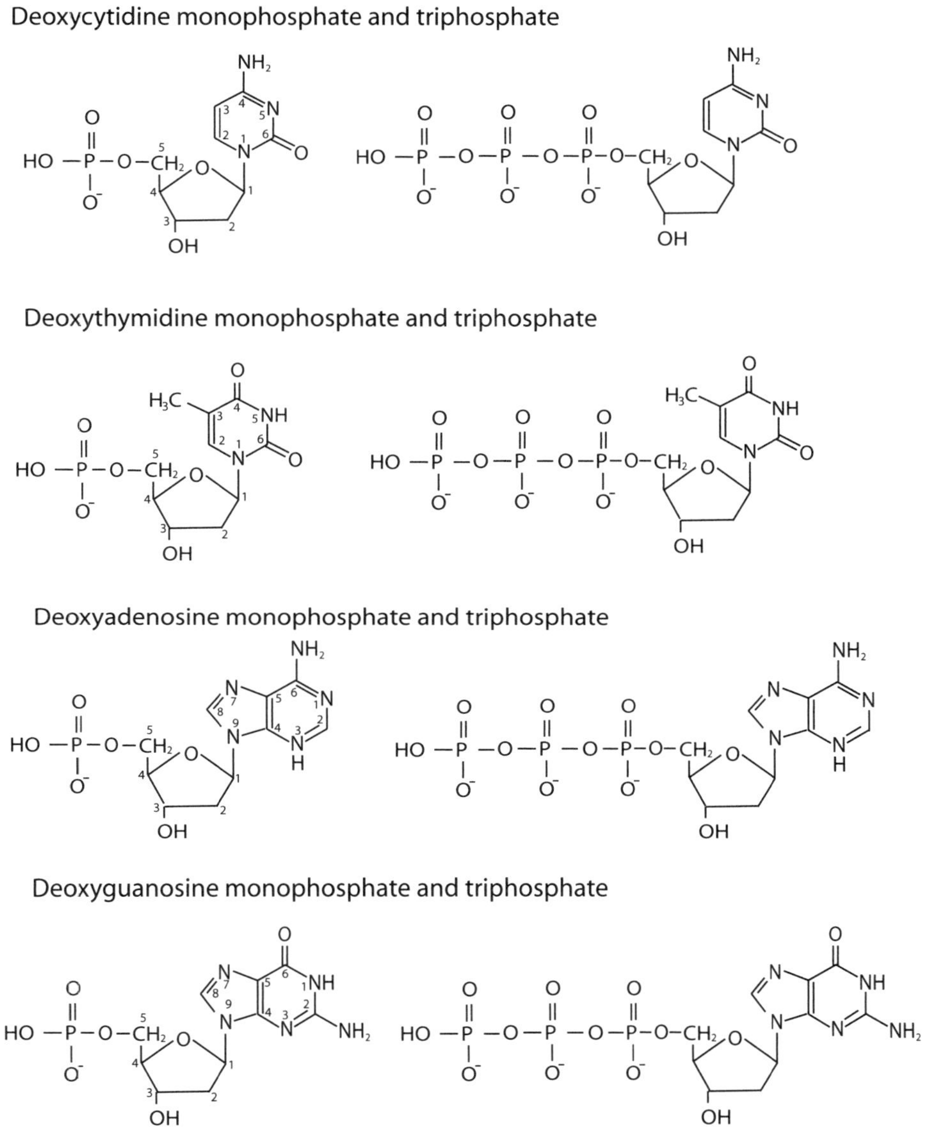

Deoxycytidine monophosphate and triphosphate

Deoxythymidine monophosphate and triphosphate

Deoxyadenosine monophosphate and triphosphate

Deoxyguanosine monophosphate and triphosphate

Figure 1.3. Chemical structure of the nucleotides that constitute DNA.

the backbone. The backbone is formed by joining the pentose molecules via covalent phosphodiester bonds. A phosphate group joins one pentose at the 5′ carbon and to the next pentose at the 3′ carbon (Fig. 1.4). The four bases are attached to the pentose–phosphate backbone and aligned on the same side of each DNA chain. The two strands of DNA are attached to each other via hydrogen bonds between the base pairs on the basis of the complementary principle, specifica, A with T and C with G. Note that a double-ringed purine base (A or G) is always paired with a single-ringed pyrimidine (T or C) (Fig. 1.5). A hydrogen bond is established between a positively charged H and a negatively

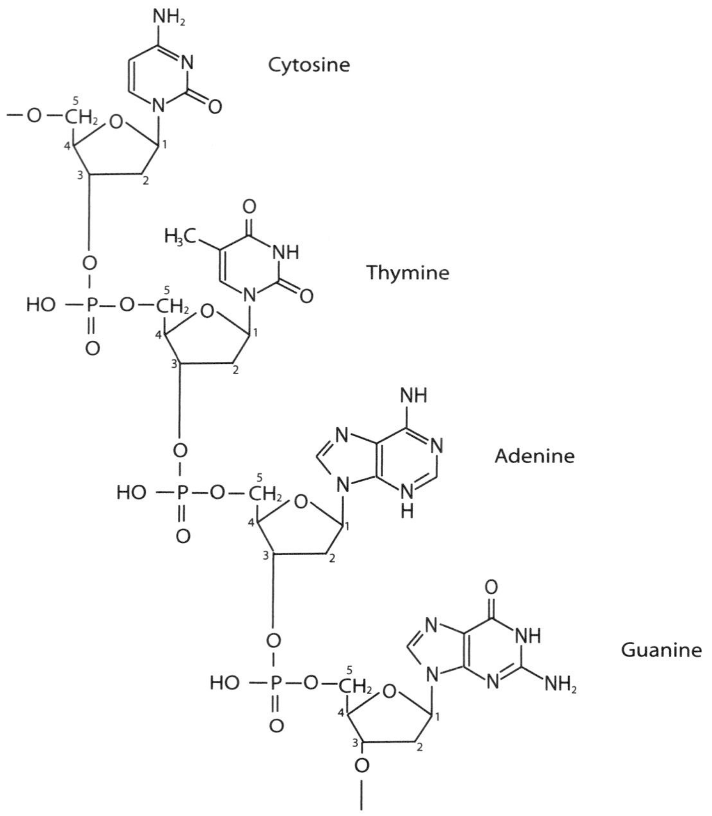

Figure 1.4. Formation of phosphodiester bonds between pentose molecules, constituting the backbone of DNA (based on bibliography 1.1).

charged *acceptor*, such as an O and N. A hydrogen bond is relatively weak, about 3% of the strength of a covalent bond. However, an array of hydrogen bonds, as found in the double-stranded helical DNA molecule, can be very strong if all hydrogen bonds are aligned on the same side of a DNA strand. For a double-stranded DNA molecule, the two sugar-phosphate backbones run in opposite directions, defined as an *antiparallel* arrangement. One strand is defined as the $5' \rightarrow 3'$ strand; the other, the $3' \rightarrow 5'$ strand.

The four bases, A, G, C, and T, are organized into a series of a large number of distinct sequences, each forming an independent functional unit known as the *gene*. In eukaryotic cells, genes are composed of coding sequences called exons and noncoding sequences called *introns*. The *exons* contain codons for specific proteins with each codon composed of three nucleotides that specify an amino acid. In contrast, the *introns* are sequences for the regulation of gene transcription. The intron sequences do not contain protein-coding sequences. The different sequences or structures of DNA are defined as *genotypes*, which determine the chemical, physical, and functional characteristics, or *phenotypes*, of various living organisms.

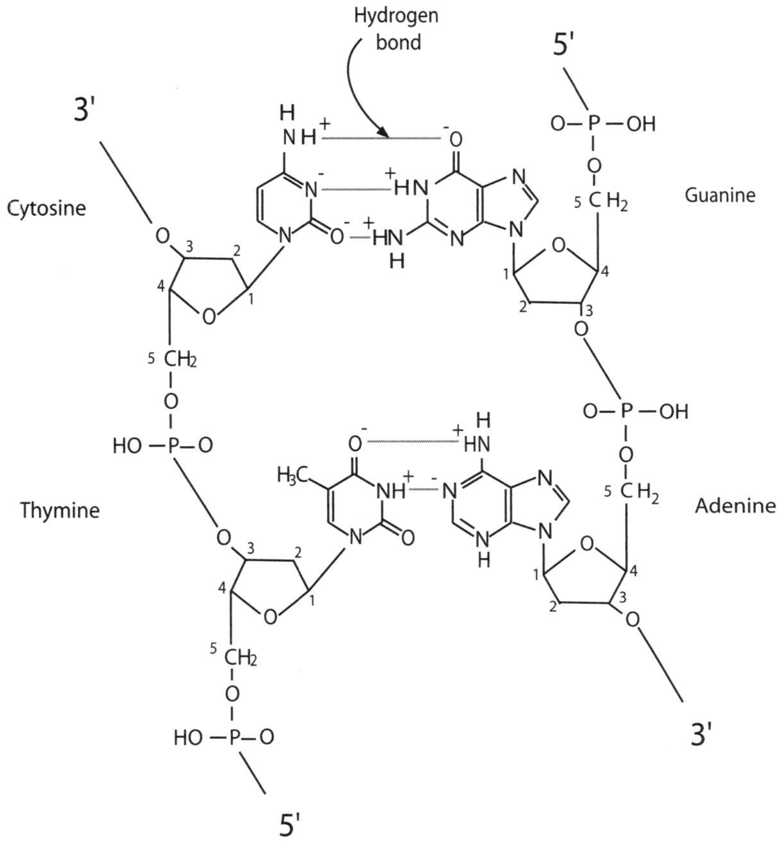

Figure 1.5. Formation of hydrogen bonds that link two single DNA strands into a double DNA helical structure (based on bibliography 1.1).

A gene carries genetic information in a particular form that can be stored, processed, copied, transcribed to generate messenger RNA (mRNA), and transmitted from the mother cells to the progeny. The double-stranded helical structure is ideal for such purposes. Since both nucleotide strands of a DNA molecule are complementary to each other, both strands carry identical genetic information. Such an organization ensures precise information transfer during DNA replication. When a DNA molecule is ready for replication, the two strands separate. Each strand can serve as a template for the synthesis of a new strand. A newly synthesized strand is identical to the complementary counterpart of the template strand. Thus, each daughter DNA is a pair of an original template and a newly synthesized strand. The daughter DNA molecules can in turn serve as templates for further DNA replication. In such a way, the genetic information can be stored, copied, and transmitted from generation to generation endlessly.

Organization of Chromosomes [1.2]

Each DNA molecule is packaged in a nuclear structure, known as *chromosome.* In humans, each cell contains 46 chromosomes, including a pair of sex chromosomes (XX for female or XY for male), which are arranged in 23 pairs. One of each paired chromosomes is derived from each parent. Thus, each individual offspring inherits two copies of the same chromosome. The total chromosomes contain about 6.6×10^9 DNA base pairs. The spacing distance is about 3.4 Å per base pair. Each human cell contains 46 DNA molecules (one DNA molecule for each chromosome) with a total length of about 2 m. Given the small size of a cell nucleus (about 5–10 μm), the DNA molecules must be folded to fit in the nucleus. The folding and packaging of DNA are accomplished with the assistance of packaging proteins (primarily histones). DNA and the packaging proteins together form a structure called *chromatin*, which is organized to form a chromosome. Chromatin exists in two states: heterochromatin and euchromatin. *Heterochromatin* is a highly condensed form of chromatin and is not involved in RNA transcription, whereas *euchromatin* is less condensed and is ready for RNA transcription.

There are several levels of DNA packaging, including (1) formation of nucleosomes, which are DNA coils around protein cores; (2) folding of the nucleosome DNA into a fiber structure (~30 nm in diameter); (3) additional folding of the fiber DNA into thicker bundles (100–300 nm); and (4) formation of loop domains, each containing 15–100 kilo–base pairs (kb) (i.e., 15,000–100,000 base pairs). At the first level, a string-like DNA molecule is coiled around a series of core complexes of proteins known as *histones* to form nucleosomes. Each nucleosome contains about 170 base pairs (bp) of DNA. Uncoiled DNA fragment between two adjacent nucleosomes is referred to as *linker DNA*, which is about 30 bp in length. There are several types of histones (H), including H2A, H2B, H3, and H4, which constitute the core complex of nucleosomes. These histones contain positive charges, which neutralize the negative charges of the DNA phosphate groups, thus stabilizing the DNA–histone complex structures. Each human cell nucleus contains about 3×10^7 nucleosomes.

At higher levels, a string of DNA with coiled nucleosomes is folded and condensed into chromatin fibers about 30 nm in diameter. These fibers are further folded into thicker chromatin bundles. The folded chromatins form large loops, each containing thousands to millions of base pairs of DNA. The chromatin loops are organized into a chromosomal structure by nuclear matrix proteins, also known as *nonhistone chromosomal proteins*, which form chromosomal scaffolds. A well-characterized complex of nuclear matrix proteins is the condensin, which can be phosphorylated by the cyclin-dependent kinase-1/cyclin B complex and controls the final level of chromosomal condensation. With various levels of folding and confinement by nuclear matrix proteins, a DNA molecule is greatly reduced in length and well organized, allowing the fit of the molecule into a chromosome.

Functional Units of DNA [1.3]

All DNA molecules in the 23 pairs of chromosomes constitute the *genome*. Each DNA molecule within a chromosome is composed of several types of functional units, including the *genes*, a *centromere*, two *telomeres*, and numbers of *replication origins* (approximately 1 per 100,000 bp). A gene is a functional unit for the process and transmission of genetic information and for coding a specific protein. In total, there are more than 50,000 genes in the human genome. Each offspring individual inherits two copies of the same gene, one from the mother and the other from the father. Several terms, such as genetic locus and alleles, are often used in genetic analysis. Genetic *locus* is defined as the chromosomal location of the two copies of each gene. *Alleles* are the forms of a gene at a genetic locus. Some genes exist in two or more alternate forms. Each gene is located at a specified locus of a chromosome. When the two copies of the gene at the same locus are identical, the individual who carries the gene is defined as a *homozygote*. When the two gene copies are different at the same locus, the individual is said a *heterozygote*. In humans, about 80% genetic loci contain homozygous genes and about 20% loci contain heterozygous genes.

Each gene encodes specific information for the transcription of an mRNA, which can be translated to a specific protein. The processes of mRNA transcription and protein translation are referred to as *gene expression*. At a given time, only a fraction of genes is expressed in the genome. The regulation of gene expression is a complicated process, involving a variety of signaling pathways and regulatory factors. In addition to the genes, there exist a large number of DNA sequences, which contain no information for protein coding in each DNA molecule. These noncoding sequences may participate in the regulation of gene stability and function, although the exact function remains poorly understood.

The *centromere* is a chromosomal structure that mediates the separation of a chromosome during mitosis and meiosis. In each DNA molecule or chromosome, the centromere is located at the point where a chromosome is attached to the microtubule-based spindle (Fig. 1.6). During mitosis, the centromeric regions of the two sister chromatids separate and are pulled by microtubules toward opposite poles. Each centromere region is composed of heterochromatin, which does not contain coding genes. A centromere contains

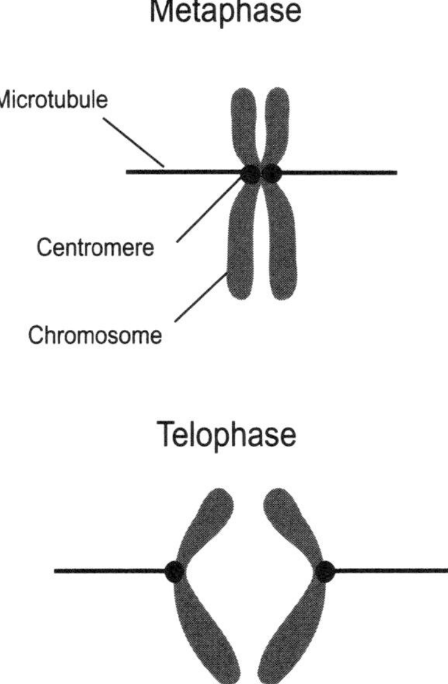

Figure 1.6. Location of centromere and separation of chromatids during mitosis (from metaphase to telophase). Based on bibliography 1.3.

a substructure called *kinetochore*, which binds microtubules and directs the movement of chromosome during mitosis. The DNA sequence of a centromere forms complexes with proteins, known as *centromere proteins*, including centromere protein (CENP)A, B, and C. These proteins regulate the function of the centromere. Centromere protein A [which has a molecular weight of 17 kilodaltons (17 kDa)] possibly mediates the formation of kinetochore and assists the attachment of centromere protein C to the kinetochore. Centromere protein B (65 kDa) may regulate the formation and organization of the centromeric heterochromatin. Centromere protein C (107 kDa) plays a critical role in the assembly of the kinetochore. Centromere proteins A and C are required for mitosis. The blockade of centromere protein C with neutralizing antibody, which is injected into to the cell nuclei, induces alterations in the kinetochore and cell arrest in mitosis.

Telomeres are DNA sequences found at the two ends of a DNA molecule. A telomere possesses several basic functions: (1) controlling the integrity of the DNA ends, (2) guiding the DNA replication machinery during DNA replication, and (3) providing signals that allow the DNA replication machinery to recognize the DNA ends without joining two DNA molecules mistakenly. In mammals, telomeres contain numerous repeats of the sequence 5′-TTAGGG-3′. Such an organization results in a unique structure at the two ends of each DNA molecule, rich in G in one strand while rich in C in the other strand. In mammalian cells, telomeres form complexes with nuclear proteins, such as telomere repeat factors (TRFs) 1 and 2, which play a critical role in regulating the stability and function of telomeres. Alterations in the binding of telomere repeat factor 2 to the telomeres cause an increase in the possibility of chromosome-to-chromosome fusion.

The origins of replication are sites along a DNA molecule where DNA replication is initiated. The DNA sequences of the replication origins have been characterized in lower levels of organisms, such as bacteria, yeast, and viruses. A replication origin sequence is about 300 bp in length. Such a structure allows the binding of initiator proteins and helicase, initiating the formation of a replication bubble and two replication forks. At the replication fork, the synthesis and annealing of a RNA primer activate a DNA polymerase, initiating DNA replication.

DNA Replication [1.4]

DNA replication is a process that synthesizes a copy of the entire genome based on the mother template during cell mitosis, ensuring the transmission of an identical genome from the mother to the daughter cell during mitosis. DNA replication is accomplished through several steps, including replication initiation, DNA extension, and sequence proofreading. Since *Escherichia coli* have been used for investigating the mechanisms of DNA replication, the process of DNA replication is described here on the basis of an *E. coli* model. The mechanisms of *E. coli* DNA replication are similar to those in eukaryotic cells.

Initiation. The replication of *E. coli* DNA is initiated at a specific site known as the *replication origin*, which is composed of several elements including a consensus 13-bp sequence and several binding sites for regulatory proteins including dnaA protein and helicase. On the binding of dnaA protein to the replication origin, a helicase binds to the replication origin and unwinds the double-stranded DNA, resulting in the formation of regionally separated single DNA strands with free bases. The two separation points flanking the replication origin are known as *replication forks*. With continuous separation of the DNA double strands, the replication forks are dynamically moving away from the replication origin. The separation of the replication origin and the formation of the replication forks prepare the synthesis of DNA.

The initiation of DNA synthesis requires the presence of several components in *E. coli*: RNA primers (~30 bps in length) and DNA polymerases. A RNA primer is specific to the sequence of a replication origin and is synthesized by a RNA polymerase or primase. On the separation of a replication origin, a primase forms a complex with the template DNA as well as with several regulatory proteins, including dnaB, dnaT, priA, priB, and priC, leading to the synthesis of a specific RNA primer. The synthesis of a RNA primer induces the binding of a DNA polymerase. The synthesized RNA primer anneals to the replication origin, initiating DNA synthesis.

DNA Extension. DNA synthesis is a process by which the annealed RNA primer is elongated on the template DNA strand according to the base-pairing principle. Such a process requires the presence of several types of DNA polymerase and a DNA ligase. A bound, activated DNA polymerase is capable of selecting deoxynucleotides complementary to that of the template DNA and adding these deoxynucleotides to the RNA primer one at a time, resulting in the elongation of the daughter DNA molecule. The elongation of DNA occurs always in the $5' \rightarrow 3'$ direction. At each replication origin, there goes bidirectional DNA synthesis. At each replication fork, DNA synthesis is conducted simultaneously along the two separated DNA strands in opposite directions, because of different polarities of the two DNA strands (Fig. 1.7). As one DNA template directs DNA elongation toward the dynamically moving replication fork, a process defined as the *leading strand DNA elongation*, the other DNA template directs DNA synthesis in a direction away from the fork, which is defined as *lagging strand elongation*. The leading strand DNA elongation is continuous, whereas the lagging strand elongation is discontinuous; thus, DNA is synthesized segment by segment, due to the constraints of the fork moving direction and the DNA synthesis direction (Fig. 1.7).

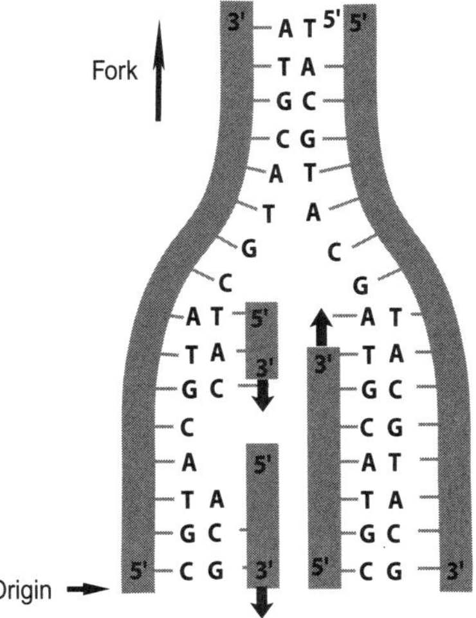

Figure 1.7. DNA replication along the two template DNA strands (based on bibliography 1.4).

Three types of DNA polymerase are found in *E. coli*: polymerases I, II, and III. DNA polymerase I possesses three functions: (1) catalyzing DNA extension in the 5′–3′ direction following a RNA primer on a DNA template, (2) serving as an exonuclease that eliminates mismatched deoxynucleotides and removes RNA primers on the lagging template following DNA extension, and (3) degrading double-stranded DNA in the 5′ → 3′ direction. DNA polymerase II is involved in the repair of damaged DNA. DNA polymerase III possesses functions similar to those of DNA polymerase I, but with different target DNA strands. DNA polymerase III can act on both DNA strands, whereas DNA polymerase I works primarily on the lagging strand, completing DNA replication based on the template segments that have not been duplicated. DNA polymerase I also removes RNA primers following the completion of DNA extension. In addition, a DNA ligase is needed to join all newly synthesized DNA segments on the lagging template by catalyzing the formation of a phosphodiester bond between the 5′ phosphate of a nucleotide and the OH group of an adjacent nucleotide.

Proofreading. During DNA synthesis, incorrect nucleotides could be mistakenly inserted into the daughter DNA. These incorrect nucleotides are removed by an enzymatic process known as *DNA proofreading* or *DNA proofediting*. The enzymes that catalyze DNA extension, including DNA polymerases I and III, can serve as nucleases, which are responsible for the removal of incorrect nucleotides. These nucleases can recognize and excise mismatched bases by hydrolysis at the 5′ end of the mismatched nucleotide. Since these enzymes act in the 3′ → 5′ direction, the excision of a nucleotide at the 5′ end will leave a free 3′ OH group in the preceding base, allowing the insertion of a correct nucleotide.

DNA Replication in Prokaryotic and Eukaryotic Cells. The processes described above are observed in prokaryotic *E. coli*. DNA replication is similar between prokaryotic and eukaryotic cells in many aspects, but there are differences. First, the time required for DNA replication differs between the two types of cell. A DNA replication–cell division cycle for *E. coli* is about 40 min, whereas that for eukaryotic cells is much longer. For instance, the cell division cycle is about 1.5 h in yeast and 24 h in mammalian cells. Second, eukaryotic cells contain usually multiple chromosomes. These cells must conduct DNA replication simultaneously in all chromosomes in a coordinated manner. An effective approach is to initiate DNA replication at multiple replication origins. Such a mechanism has been demonstrated by the existence of multiple DNA extension locations after a "pulse" exposure to radioactively labeled thymidine. In yeast, there exist about 400 replication origins in the 17 chromosomes. Third, the types of DNA polymerase are different between prokaryotic and eukaryotic cells. In eukaryotic cells, four types of DNA polymerases have been found: α, β, γ, and δ. The eukaryotic DNA polymerases α and δ are similar to prokaryotic polymerase II. DNA polymerase β may be responsible in DNA repair. The γ type may be involved in the replication of mitochondrial DNA.

RIBONUCLEIC ACID (RNA)

RNA Composition and Structure [1.5]

Ribonucleic acid (RNA) is a molecule that transmits genetic information from DNA to proteins. Similar to a DNA molecule, RNA is composed of linearly joined nucleotides,

TABLE 1.2. Nomenclature of Nucleosides and Nucleotides for RNA

Types of Base	Cytosine (C)	Uracil (U)	Adenine (A)	Guanine (G)
Nucleosides	Cytidine	Uridine	Adenosine	Guanosine
Nucleotides	Cytidine monophosphate (CMP)	Uridine monophosphate (UMP)	Adenosine monophosphate (AMP)	Guanosine monophosphate (GMP)
	Cytidine diphosphate (CDP)	Uridine diphosphate (UDP)	Adenosine diphosphate (ADP)	Guanosine diphosphate (GDP)
	Cytidine triphosphate (CTP)	Uridine triphosphate (UTP)	Adenosine triphosphate (ATP)	Guanosine triphosphate (GTP)

each including a nitrogenous base, a pentose, and a phosphate group. Unlike DNA, the pentose for RNA is β-D-ribose instead of β-D-2-deoxyribose (Fig. 1.2) and the four bases are cytosine, uracil, adenine, and guanine with uracil in place of thymine (Fig. 1.1). Furthermore, RNA is a single-stranded molecule and is relatively short compared to DNA molecules. The nomenclatures for various RNA nucleosides and nucleotides are listed in Table 1.2.

RNA is synthesized via DNA transcription, a process similar to DNA replication in certain aspects. To transcribe a RNA molecule, a DNA molecule opens locally into single-stranded forms. One of the two strands serves as a template for RNA synthesis according to the base-pairing principle. That is, for a deoxynucleotide A on the template, a ribonucleotide U (but not T) is added; and for a C on the template, a G is added. The RNA molecule is elongated by adding ribonucleotides one by one. These ribonucleotides are joined together via covalent bonds. The size and type of RNA transcribed from a region of DNA is controlled by proteins called *gene regulatory factors*. These proteins bind to specific sites of DNA and regulate the process of RNA transcription. For any given time, some genes are activated for RNA transcription, while others are not. The selection of gene activation is controlled by gene regulatory proteins, which are activated by upstream signaling pathways. The process of RNA synthesis stops when a DNA stop codon appears.

There exist three types of RNA molecules: messenger RNA (mRNA), transfer RNA (tRNA), and ribosome RNA (rRNA). An mRNA molecule is a copy of a specific gene that carries the information or codon for a protein. Thus, mRNA directs the translation or synthesis of a specific protein. An rRNA molecule serves as a machine for protein synthesis or translation with a specific mRNA as a template. A tRNA molecule is responsible for the transfer of amino acids to an rRNA that synthesizes proteins based on an mRNA transcript. The three types of RNA work coordinately in protein translation.

RNA Transcription [1.5]

RNA transcription is a process that transfers genetic information from a gene to an mRNA, which is then translated to a protein. This sequence of processes is also known as *gene expression*. RNA transcription is similar to DNA replication except that (1) a different enzyme, the RNA polymerase, is used for RNA synthesis; (2) a uridine is used instead of thymidine; and (3) a single RNA strand is synthesized. Studies based on the *E. coli* system have established that RNA transcription is accomplished through several steps, including initiation, elongation, and termination. Similar processes are found in eukaryotic RNA transcription with several exceptions (see below). Here, the E. coli system is used to describe the mechanisms of RNA transcription.

Initiation. A RNA molecule is transcribed from a specific gene and encodes information for the sequence of amino acids of a protein. RNA transcription is initiated at a DNA site next to a special DNA sequence known as a *promoter*, which is found in each gene and activated in response to the binding of transcription factors to specific DNA sequences known as enhancers. There are two types of promoter in the *E. coli* genome with consensus sequences TTGACAT and TATAAT, which are located at the –35 and –10 sites relative to the first base pair (+1) to be transcribed. On the binding of transcription factors, a RNA polymerase binds to a gene at the promoter region and unwinds the double-stranded DNA at the –10 site, forming an open promoter complex or "transcription bubble" (Fig. 1.8). Such a process initiates RNA transcription.

Elongation. The elongation of RNA is catalyzed by RNA polymerase, a complex enzyme that contains four subunits: α, β, β', and σ subunits. The subunits can be separated from each another and bound to the other units. A complete complex of RNA polymerase subunits is required for the initiation of RNA transcription. RNA polymerase can recruit nucleotides and synthesize RNA based on a template gene according to the base-pairing principle. The enzyme can catalyze the addition of a nucleoside monophosphate to an existing RNA fragment while a diphosphate group is released from a recruited nucleoside triphosphate. The release of the diphosphate provides energy for the synthetic process. The "transcription bubble" continues to extend when RNA is synthesized (Fig. 1.8). RNA synthesis occurs in the $5' \rightarrow 3'$ direction.

Termination. RNA synthesis is terminated when the RNA polymerase recognizes a termination sequence on the DNA template strand. This sequence contains a GC-rich segment followed by a stretch of six or more adenosine nucleotides. The GC-rich segment of the synthesized RNA can fold regionally, forming a short, hairpin-shaped, double-stranded RNA segment. This segment is followed by U-rich segment. These structures serve as a termination signal. The RNA polymerase is then released and RNA synthesis terminated when the enzyme recognizes these structures.

RNA Transcription and Processing in Eukaryotes. While RNA transcription in eukaryotes is similar to that in prokaryotes in principle, there are several differences. Unlike prokaryotes, eukaryotes express at least three RNA polymerases, including polymerase I,

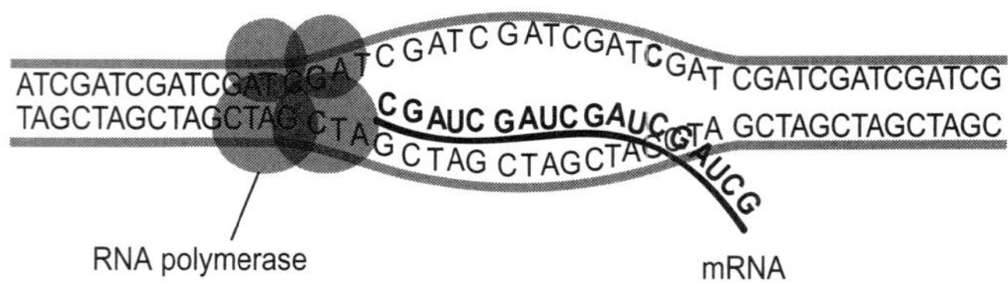

Figure 1.8. Schematic representation of RNA transcription (based on bibliography 1.5).

II, and III. These polymerases possess distinct functions. Polymerase I is for the synthesis of rRNA, polymerase II is for mRNA, whereas polymerase III is for tRNA.

In addition, transcribed RNAs in the nucleus are processed in eukaryotes before being delivered to the cytoplasm for protein translation. Immediately after the transcription of an mRNA molecule, an enzyme called *guanyltransferase* is activated, adding a 7-methylguanosine residue to the 5′ end of the mRNA transcript. A sequence at the 3′ end of the transcript, usually AAUAAA, activates an endonuclease, which cleaves the mRNA molecule (~20 bp) at the 3′ end. This step is followed by the activation of a poly(A) polymerase, inducing the addition of a poly(A) tail of 150–200 adenosine residues to the 3′ end of the mRNA. These processes generate a complete primary structure of an mRNA molecule.

Unlike prokaryotes, eukaryotes contain primary mRNAs that are composed of two types of sequence known as *exons* and *introns*. The exons contain protein coding regions and are used for protein translation, whereas the introns are sequences that interrupt the exons and do not contain protein coding information. An mRNA transcript must be spliced to remove the introns before protein translation occurs. A mature mRNA molecule is produced after the removal of the introns and is ready for protein translation. In eukaryotes, mRNA splicing occurs in a structure called *splicesome*, which is composed of splicing enzymes and associated factors. In addition to mRNAs, tRNAs, and rRNAs are also processed by splicing to remove introns.

PROTEINS

Protein Composition and Structure [1.6]

Proteins are molecules constituted with amino acids and are major components of living cells, constituting about 50% of the cell dry weight. There are two types of proteins in the cell: structural and regulatory proteins. Structural proteins participate in cell construction, giving a cell the shape, strength, and elasticity, whereas regulatory proteins control biological processes, such as cell-to-cell and cell-to-matrix communication, intracellular enzyme activation, signal transduction, control of gene expression, transport of necessary compounds for cell metabolism, cell division, cell differentiation, cell migration, and cell apoptosis. It is important to note that structural proteins are also involved in the regulation of cellular activities, which become clear in the following examples. Structural proteins include, for example, actin filaments, microtubules, and intermediate filaments, which form an integrated structure called *cytoskeleton*. While these proteins provide the cell with shape and strength, they play a critical role in the regulation of cell mitosis, migration, and adhesion. Regulatory proteins are found in the cell membrane, cytoplasm, and nucleus. Cell membrane proteins, such as growth factor receptors, adhesion molecules, and integrins, are responsible for cell-to-cell and cell-to-matrix interactions. Cytoplasmic proteins are mostly enzymes. Nuclear proteins are involved in the regulation of chromosomal organization and gene expression.

A protein consists of one or more peptides, which are constituted with 20 types of amino acid with distinct structure and chemical properties (Table 1.3). The combination of these amino acids gives rise to a variety of different peptides. The sequence of amino acids is specified by a corresponding gene. The length of a peptide varies widely, ranging from several amino acid residues, such as oxytocin, to about 25,000 residues, including titin. Most peptides are composed of 100–1000 amino acids.

TABLE 1.3. Amino Acids Found in Humans and Animals*

Amino Acids	Chemical Forms	Three-Letter Abbreviation	One-Letter Abbreviation
Alanine	$CH_3CHCOOH$ | NH_2	Ala	A
Arginine	NH ‖ $H_2NCNHCH_2CH_2CH_2CHCOOH$ | NH_2	Arg	R
Asparagine	O ‖ $N_2NCCH_2CHCOOH$ | NH_2	Asn	N
Aspartic acid	$HOOCCH_2CHCOOH$ | NH_2	Asp	D
Cysteine	$HSCH_2CHCOOH$ | NH_2	Cys	C
Glutamic acid	$HOOCCH_2CH_2CHCOOH$ | NH_2	Glu	E
Glutamine	O ‖ $N_2NCCH_2CH_2CHCOOH$ | NH_2	Gln	Q
Glycine	$HCHCOOH$ | NH_3^+	Gly	G
Histidine	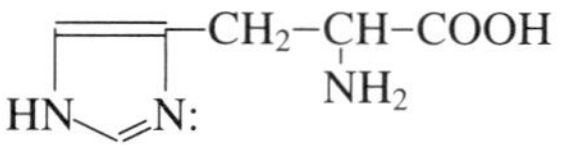	His	H
Isoleucine	CH_3 | $CH_3CH_2CHCHCOOH$ | NH_2	Ile	I
Leucine	$CH_2CHCH_2CHCOOH$ | | CH_3 NH_2	Leu	L

TABLE 1.3. *Continued*

Amino Acids	Chemical Forms	Three-Letter Abbreviation	One-Letter Abbreviation
Lysine	$H_2NCH_2CH_2CH_2CH_2CH(NH_2)COOH$	Lys	K
Methionine	$CH_3SCH_2CH_2CH(NH_2)COOH$	Met	M
Phenylalanine	$C_6H_5-CH_2-CH(NH_2)-COOH$	Phe	F
Proline	pyrrolidine ring, $\overset{+}{N}H_2$, COOH	Pro	P
Serine	$HOCH_2CH(NH_2)COOH$	Ser	S
Threonine	$CH_3CH(OH)CH(NH_2)COOH$	Thr	T
Tryptophan	indole (N–H) $-CH_2-CH(NH_2)-COOH$	Trp	W
Tyrosine	$HO-C_6H_4-CH_2-CH(NH_2)-COOH$	Tyr	Y
Valine	$(CH_3)_2CHCH(NH_2)COOH$	Val	V

*Based on bibliography 1.6.

With the exception of one amino acid, proline, all amino acids are formed on the basis of a common structure, containing a central α-carbon atom bound with an amino group, a carboxyl group, a hydrogen atom, and a sidechain, the R group. Each amino acid has a distinct sidechain (Table 1.3). It is the sidechain that determines the unique chemical and physical properties of an amino acid. Proline has a cyclic sidechain linking the α carbon and the nitrogen, forming an imino group. Based on the structure and properties of the sidechains, amino acids can be classified into two major groups: charged and uncharged amino acids.

Charged amino acids include those with positive and negative charges. Amino acids with positive charges include arginine, histidine, and lysine, which are also known as *basic amino acids*. Those with negative charges include aspartic acid and glutamic acid, also known as *acidic amino acids*. The charged amino acids are all hydrophilic in nature, capable of interacting with water, which determines the hydrophilic features of proteins.

The remaining 15 amino acids are all *uncharged*. Among these amino acids, five are polar amino acids, including serine, threonine, tyrosine, asparagines, and glutamine. The sidechains of these amino acids contain either polar hydrogen bond donors or acceptors, capable of interacting with water. These are considered hydrophilic amino acids. The remaining 10 uncharged amino acids possess nonpolar sidechains and interact poorly with water. These amino acids include glycine, alanine, valine, leucine, cysteine, methionine, praline, isoleucine, phenylalanine, and tryptophan. These are considered hydrophobic amino acids. Because of the versatility of the amino acids, a large number of proteins can be constructed.

Several amino acids can be modified by the addition of various chemical groups under the action of specific enzymes. For instance, a phosphate group can be added to serine, threonine, tyrosine, and histidine by a family of enzymes known as *protein kinases*, resulting in amino acid phosphorylation. Such a process plays a critical role in the regulation of a variety of cellular activities, such as cell division, adhesion, and migration. Other types of amino acid modification include the addition of the methyl group to lysine and the hydroxyl group to proline, as well as the formation of disulfate bonds between adjacent cysteine residues. These modifications are essential to the function and stability of proteins.

Protein Translation [1.7]

Protein translation is a process that synthesizes protein. The synthesis of proteins requires three basic elements: ribosomes, messenger RNA (mRNA), and transfer RNA (tRNA). Ribosomes are composed of rRNA components and regulatory proteins. There are three rRNA modules in *E. coli* ribosomes with distinct molecular sizes: 5*S*, 16*S*, and 23*S* (*S*: sedimentation velocity, a measure of molecular size). These molecules are transcribed from specific DNA sequences at specified locations of a DNA molecule. rRNA components and ribosomal proteins form two complex subunits. In prokaryotes, the two subunits are 50*S* and 30*S* in molecular size, whereas in eukaryotes they are 60*S* and 40*S*. These ribosomal structures serve as machineries for protein synthesis.

A messenger RNA molecule is a sequence of linear nucleotides, carrying genetic codons that dictate the sequence of amino acids for a specific protein. The genetic codons are stored in a DNA molecule. DNA transcription transmits the genetic codons to mRNA molecules. Each amino acid is represented by a codon of three nucleotides. In other words, each 3-nucleotide codon determines a type of amino acid to be incorporated into a peptide at a specified site. Genetic codes for all amino acids found in humans are listed in Table 1.4. In addition to the codons for amino acids, each mRNA molecule contains stop codons, including UAG, UGA, and UAA, which signal the termination of peptide translation when recognized by a ribosome.

A tRNA molecule is responsible for the transport of an amino acid to a ribosome during protein synthesis. Transfer RNA is able to not only identify a specific amino acid but also recognize a corresponding codon on a mRNA molecule, ensuring the placement of the amino acid to an appropriate position. There are two functional domains in each tRNA molecule: an anticodon and an amino acid attachment site. The anticodon is a seven-nucleotide sequence that recognizes and binds to a mRNA site according to the complementary rule. The amino acid attachment site binds to an amino acid. Each amino acid

TABLE 1.4. Genetic Codes for Amino Acids*

Amino Acids	Genetic Codes					
Ala	GCU	GCC	GCA	GCG		
Arg	CGU	CGC	CGA	CGG	AGA	AGG
Asn	AAU	AAC				
Asp	GAU	GAC				
Cys	UGU	UGC				
Gln	CAA	CAG				
Glu	GAA	GAG				
Gly	GGU	GGC	GGA	GGG		
His	CAU	CAC				
Ile	AUU	AUC	AUA			
Leu	UUA	UUG	CUU	CUC	CUA	CUG
Lys	AAA	AAG				
Met	AUG					
Phe	UUU	UUC				
Pro	CCU	CCC	CCA	CCG		
Ser	UCU	UCC	UCA	UCG	AGU	AGC
Thr	ACU	ACC	ACA	ACG		
Trp	UGG					
Tyr	UAU	UAC				
Val	GUU	GUC	GUA	GUG		
Initiation codes	AUG	GUG				
Stop codes	UAA	UAG	UGA			

*Based on bibliography 1.7.

is carried by a specific tRNA molecule. Like rRNA, a tRNA molecule is coded by a specific gene at a specified location in a DNA molecule.

Protein translation is accomplished via three steps: initiation, elongation, and termination. These processes have been well understood in prokaryotes. Here, the prokaryote protein translation system is used as an example.

Initiation. The initiation of protein translation requires RNA molecules, including mRNA, rRNA, and tRNA, and rRNA-constituted ribosomes. In addition, three regulatory factors, termed *initiation factors* (IFs) 1, 2, and 3, are necessary. Several steps are involved for the initiation of protein translation. First, the translation of protein starts with the activation of IF3, which stimulates the binding of mRNA to the 30*S* subunit of ribosome. mRNA binding induces the attachment of the 50*S* subunit to the 30*S* subunit, forming a complete ribosome (note that the two ribosomal subunits are present in separate forms without protein synthesis when they are not activated). Second, IF2 is activated to bind GTP and fMet-tRNA, which is a translation initiator tRNA carrying *N*-formylmethionine (fMet). This combination stimulates the attachment of fMet-tRNA to a specific initiation codon (AUG or GUG) on the mRNA molecule localized to an rRNA site, known as the P site. An initiation codon is preceded by a sequence that can hybridize to rRNA. The GTP molecule provides energy for the ribosomal assembly process. When a phosphate group is removed from the GTP molecule, IF2 and IF3 are released from the ribosomal complex.

Elongation. *Elongation* is a process by which amino acids are added to the initiation fMet-tRNA one at a time. Such process is regulated by several protein elongation factors (EFs), including EF-Tu, EF-Ts, and EF-G. The elongation factor EF-Tu regulates the attachment of aminoacyl-tRNAs to a specific mRNA codon localized to an rRNA site adjacent to the P site, known as the *A site*. The binding of GTP to the aminoacyl-tRNA is required in this process for supplying energy. When GTP is hydrolyzed, elongation factor EF-Ts attaches to the ribosome, regulating the release of the EF-Tu-GDP complex, leaving tRNA at the A site. Note that at this state the first amino acid or a synthesized partial peptide is attached to the P site. An enzyme called *peptidyltransferase* catalyzes a process that transfer the amino acid or partial peptide from the P to the A site. At the same time, the elongation factor EF-G initiates a process that moves the mRNA molecule by three base pairs in the $5' \rightarrow 3'$ direction, which is associated with the release of the tRNA at the P site and the transfer of the peptide together with the associated tRNA from the A to the P site (Fig. 1.9).

Termination. *Termination* of protein translation is initiated when one of the three stop codons of the mRNA, specifically UAG, UAA, and UGA, appears at the A site. At least three regulatory proteins, known as *release factors* (RFs) 1, 2, and 3, regulate translation termination. These release factors can recognize and bind to a stop codon at the A site of the ribosome, inducing the release of synthesized peptide from the P site. The ribosome is then dissociated into two free subunits. Synthesized peptides undergo protein splicing and folding processes, eventually forming proteins with a three-dimensional structure.

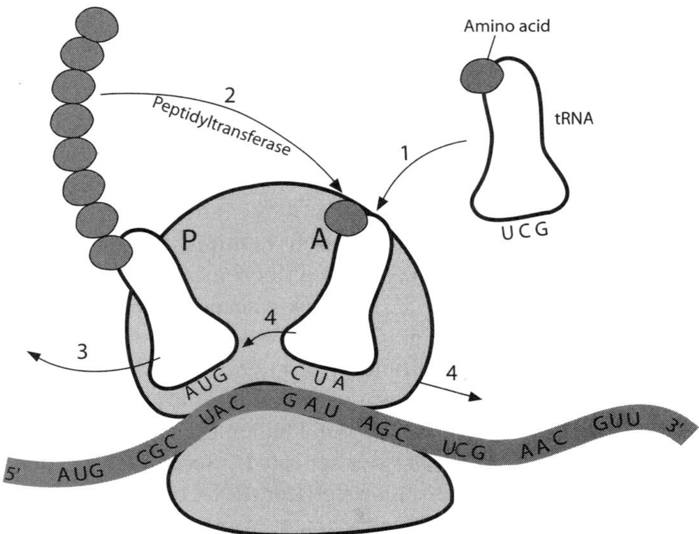

Figure 1.9. Schematic representation of protein translation. Several steps are involved in protein elongation: (1) recruitment of an aminoacyl-tRNA molecule to site A; (2) transfer of a partial peptide from site P to site A; (3) removal of the tRNA at site P; (4) movement of the peptide–tRNA complex from site A to site P; (5) movement of the rRNA complex toward the 3′ direction by three base pairs (based on bibliography 1.7).

Protein Folding and Architecture [1.8]

All proteins are folded into a three-dimensional structure after being translated. A protein is composed of one or more peptides, which are sequences of linearly jointed amino acids via peptide bonds. Three atoms from each amino acid, including the nitrogen from the amino group, the α-carbon, and the carbon from the carboxyl group, join together to form a central structure for each peptide called the *polypeptide backbone*. A peptide chain is synthesized in a ribosome based on the codons of an mRNA. The peptide end with a free amino group is referred to as the *N-terminus*, where peptide synthesis begins, and the end with a carboxyl group is the *C-terminus*. The counting of the number of amino acids starts at the *N*-terminus. Most peptide bonds can rotate freely, giving the flexibility of forming a variety of different conformations for proteins.

A peptide is usually folded into a three-dimensional (3D) structure, resulting in a final conformation that exhibits maximal stability and functionality. A denatured protein molecule under harsh conditions, such as extreme pH and a high concentration of urea, lose not only its conformation but also function. However, proteins are able to refold back to their original conformation and regain their functions under restored physiological conditions. Various chemical and physical features of the amino acids in a peptide chain determine the process of protein folding and the 3D protein conformation. The hydrophobic and hydrophilic features of amino acids play a critical role in controlling protein folding and conformation. The uncharged hydrophobic nonpolar sidechains of the amino acids intend to pack themselves in the core of a protein to minimize exposure to water molecules. The core of a protein contains the most conservative amino acids. In contrast, most charged and polar sidechains of amino acids are localized to the surface of a protein and capable of interacting with water molecules.

By X-ray diffraction and nuclear magnetic resonance spectroscopy, the structure of proteins can be determined to atomic accuracy. Protein structural analysis has demonstrated that proteins usually contain two types of secondary structure: α-helix and β-sheet. These structures are common in most proteins, although the overall conformation varies from protein to protein. An *α-helix* is a right-handed coiled structure with 3.6 residues per turn (Fig. 1.10). The helical structure is formed on the basis of hydrogen bonding between adjacent polar groups of the peptide backbone. A *β-sheet* is a structure containing several antiparallel segments of a single peptide, which is formed as a result

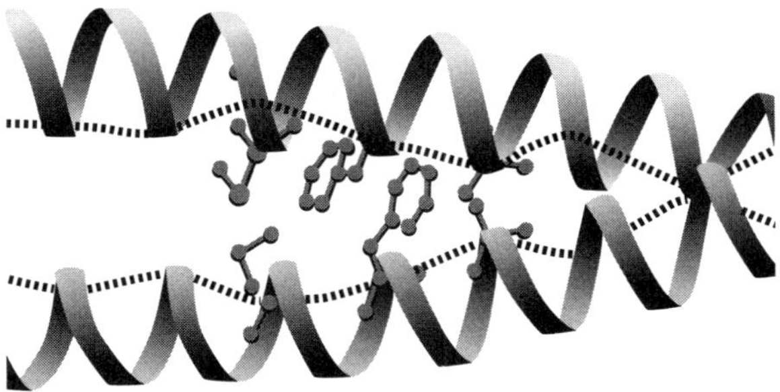

Figure 1.10. Crystallographic structure of a parallel α-helical coiled-coil dimer of the intermediate filament protein vimentin. (Reprinted from Burkhard P et al: *Trends Cell Biol* 11:82–8, 2001 by permission of Elsevier.)

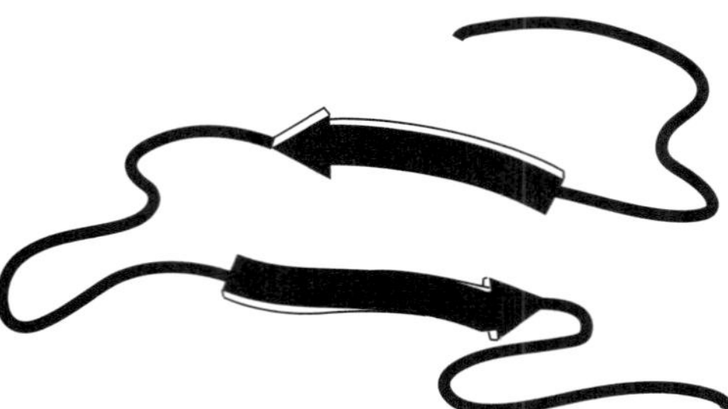

Figure 1.11. Schematic representation of the structure of a β-sheet-containing protein (based on bibliography 1.8).

of the back-and-forth turning of the peptide (Fig. 1.11). The antiparallel segments are linked together by lateral hydrogen bonds between the polar groups of the peptide backbone.

There exist peptide fragments that are not organized into regular structures like α-helices and β-sheets. These peptides exhibit disordered structures and can move more freely than a α-helix and β-sheet. Such structures are often found at the *N*- and *C*-termini. Collectively, α-helices, β-sheets, and irregular fragments are referred to as *secondary structures* (note that the amino acid sequences are known as the *primary structure*). These structures can be further organized into higher levels of 3D conformation. Examples of 3D protein structures include the coiled-coil structure, which is a complex of paired α-helices interacting laterally via hydrophobic bonds (Fig. 1.10), and the β-barrel structure, which is a cylindrical structure formed by a number of β-sheets. The formation of these superstructures enhances the stability of proteins. Furthermore, most proteins are composed of multiple peptides, which are integrated into a protein molecule. Proteins also form complexes such as dimmers and trimers. All these forms are essential for the functionality of proteins.

Changes in Protein Conformation [1.8]

Proteins undergo dynamic changes in their conformation. The atoms of a protein may move in a very fast speed in the order of 100 m/s within a nanometer range. Protein–protein interactions may induce conformational changes on the molecular scale. Such conformational changes play a critical role in the regulation of molecular activities. For instance, the binding of a growth factor to its receptor may cause a conformational change in the receptor, initiating autophosphorylation of the cytoplasmic receptor tyrosine kinase, which is associated with most growth factors. Conformational changes and autophosphorylation are critical processes for the activation of mitogenic signaling pathways. In the contractile apparatus of muscular cells, a conformational change in the myosin head, on the activation of a myosin molecule, is an essential process that initiates the sliding of actin filaments and the generation of forces. Conformational changes in protein kinases and phosphatases, such as the Src protein tyrosine kinase and the Src homology 2 domain-containing protein tyrosine phosphatase, control the state of molecular activation. It is commonly received that a protein conformational change is an essential process for the regulation of protein functions.

LIPIDS

There are various types of lipid molecules in mammalian cells, including phospholipids, glycolipids, steroids, and triglycerides. These lipid molecules play important roles in the construction of cellular structures and in the regulation of cellular functions. Lipids are the basic building blocks for cell and subcellular membranes, serve as hormones and intracellular signaling molecules, and contribute to the production of energy. The structure and function of common lipids are briefly described in the following sections.

Phospholipids [1.9]

Phospholipids are the primary constituents of cell membrane. A phospholipid contains several basic elements, including an alcohol, two fatty acid chains, and a phosphate group, which can be bonded with another alcohol group. Based on the alcohol structure, phospholipids can be further divided into two subgroups: phosphoglycerides and sphingolipids. A *phosphoglyceride* contains a glycerol as an alcohol group, whereas a *sphingolipid* contains a sphingosine.

Phosphoglycerides. In a phosphoglyceride molecule, two of the three —OH groups of the alcohol (glycerol) molecule are bonded with fatty acids via ester bonds, and the remaining —OH group is esterified by a phosphate group. The phosphate group can be bonded with another chemical groups, which can be either inositol, serine, ethanolamine, or choline (Fig. 1.12). Each combination gives rise to a distinct phosphoglyceride, including phosphatidylinositol, phosphatidylserine, phosphatidylethanolamine, or phosphatidylcholine, respectively. It is important to note that some of these phosphoglycerides not only contribute to the construction of cell membrane, but also serve as signaling molecules. For instance, phosphatidylinositol (Fig. 1.13), when phosphorylated into phosphatidylinositol 3,4 biphosphates, play a critical role in the regulation of G-protein receptor-initiated signal transduction (see Chapter 5).

A phosphoglyceride molecule contains a polar alcohol head and nonpolar fatty acid tails, which render the molecule amphipathic in nature, that is, hydrophilic at the head and hydrophobic at the tail. Such a modular arrangement determines the form of phosphoglyceride aggregation while they are mixed in an aqueous solution. The hydrophilic head of phosphoglyceride interacts with the water molecules, while the hydrophobic tail intends to interact with the hydrophobic tails from different phosphoglyceride molecules, resulting in the spontaneous formation of a lipid bilayer with two water-contacting surfaces composed of hydrophilic heads and an internal layer composed of hydrophobic tails (Fig. 1.14). Indeed, all cell membranes are established on the basis of such a principle.

Sphingolipids. *Sphingolipids* are composed of a sphingosine, two fatty acid chains, and a phosphate group (Fig. 1.15). These molecules possess properties similar to those of phosphoglycerides and can be found in the membrane of many cell types. Sphingolipids can aggregate into microdomains in cell membranes, which may play a role in targeting specific proteins to the plasma membrane and in organizing membrane-associated signaling pathways. For instance, K^+ and other ion channels are localized to lipid microdomains on the cell surface. Sphingolipids can interact with ion channels and mediate the localization of the ion channels. Sphingolipids are also involved in the regulation of cellular

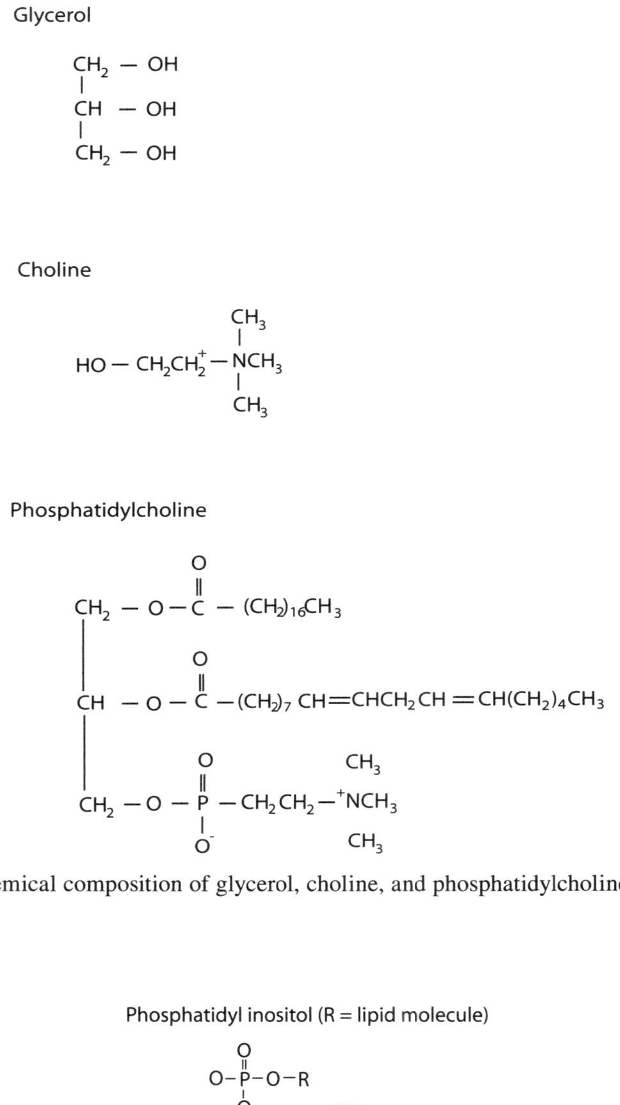

Glycerol

CH_2-OH
$CH-OH$
CH_2-OH

Choline

$HO-CH_2CH_2^{+}-N(CH_3)_3$

Phosphatidylcholine

$CH_2-O-C(=O)-(CH_2)_{16}CH_3$
$CH-O-C(=O)-(CH_2)_7CH{=}CHCH_2CH{=}CH(CH_2)_4CH_3$
$CH_2-O-P(=O)(O^-)-CH_2CH_2-{}^{+}N(CH_3)_3$

Figure 1.12. Chemical composition of glycerol, choline, and phosphatidylcholine (based on bibliography 1.9).

Phosphatidyl inositol (R = lipid molecule)

O–P(=O)–O–R
OH OH OH HO OH

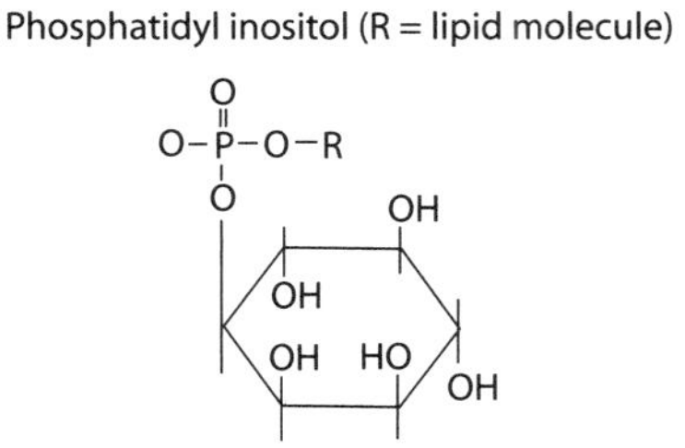

Figure 1.13. Chemical composition of phosphatidyl inositols (PI). Based on bibliography 1.9.

activities, such as cell apoptosis and proliferation. Some sphingolipid molecules such as ceramide and sphingosine promote cell apoptosis, whereas others such as sphingosine-1-phosphate induce cell proliferation. These observations demonstrate the involvement of sphingolipids in the regulation of signal transduction.

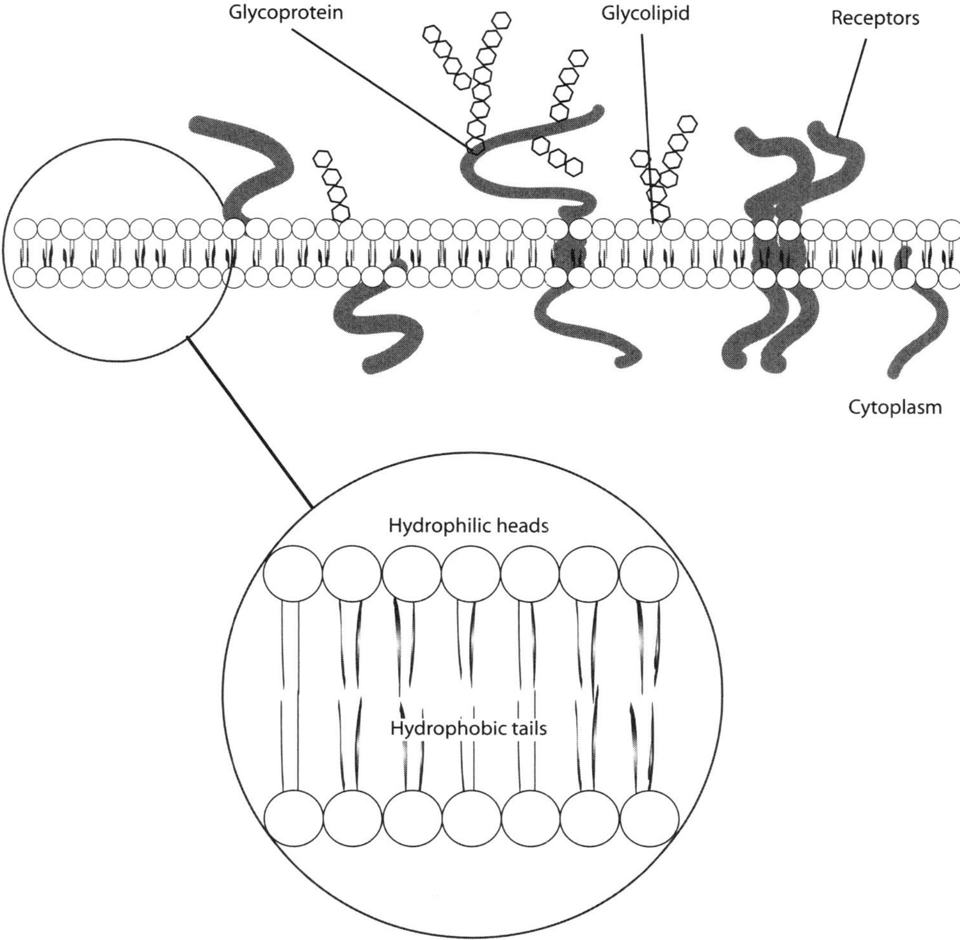

Figure 1.14. Constituents of a lipid bilayer.

Sphingolipids

$$CH_3(CH_2)_{12}CH{=}CH{-}\underset{\displaystyle OH}{\underset{|}{CH}}{-}\underset{\displaystyle NH_2}{\underset{|}{CH}}{-}CH_2OH$$

Figure 1.15. Chemical composition of the sphingolipid molecule sphingosine(based on bibliography 1.9).

Glycolipids [1.10]

Glycolipids are lipid molecules that contain carbohydrates or sugar residues, including glucose and galactose. Glycolipids are found in the membrane of all cell types and primarily distributed in the extracellular half of the lipid bilayer, constituting about 5% of the lipid molecules. These molecules often form aggregates via self-assembly. Given the unique location of the glycolipids, it has been speculated that these molecules may play roles in regulating the interaction of cells with extracellular factors and may also serve to protect the cell from extreme chemical conditions.

Steroids [1.11]

Steroids are lipids that contain a structure with four fused rings, including three cyclohexane rings and a cyclopentane ring. Steroid compounds found in mammalian cells include cholesterol, vitamin D, progesterone, adrenocortical hormones, gonadal hormones, and bile salts. Cholesterol is the most abundant steroid in mammals and exists in a free form or a form esterified with fatty acids (Fig. 1.16). It participates in the construction of cell membrane structures, and is a basic molecule for the derivation of other steroids, such as vitamin D, adrenocortical hormones, gonadal hormones, and bile salts.

Because of the influence of blood cholesterol on atherogenesis, cholesterol has received much attention. A high level of blood cholesterol is referred to as *hypercholesterolemia*, a condition facilitating the development of atherogenesis. Cholesterol can be taken up from digested foods and also synthesized by the liver. The rate of cholesterol synthesis is dependent on the blood level of cholesterol, which controls cholesterol synthesis based on a negative feedback mechanism. Namely, a high level of blood cholesterol inhibits cholesterol synthesis. Dietary cholesterol can be absorbed into blood from the small intestine in a form known as *chylomicron*, which contains cholesterol, triglycerides, and apoproteins. Under the action of lipoprotein lipase, triglycerides are released from the chylomicrons, forming chylomicron remnants that consist of cholesteryl esters and apoproteins. The chylomicron remnants can be taken up by hepatocytes in the liver, where cholesterol is cleaved and released into the blood. Free cholesterol molecules can be taken up by cells for various purposes, including the construction of cell membrane, formation of bile, synthesis of hormones, and generation of endogenous lipoproteins.

In the liver, cholesterol can be used for generating endogenous low-density lipoproteins (LDL), which is circulated in the blood for 1–2 days and constitutes the major pool of plasma cholesterol (60–70% of total cholesterol). Circulating LDL is a major form that delivers cholesterol to needy cells. The release of cholesterol from a LDL molecule results in the formation of high-density lipoproteins (HDL), consisting of apoproteins and residual cholesterol. HDL can be reused in the liver to form LDL.

Clinically, hypercholesterolemia can be divided into two groups: primary and secondary. The primary hypercholesterolemia is an inherited disease and is induced by genetic defects. In some patients, the cholesterol metabolic disorder is due to a single gene defect, which is called monogenic hypercholesterolemia. This type of disorder can be predicted on the basis of the Mendelian genetic mechanism; some members of a family inherit the

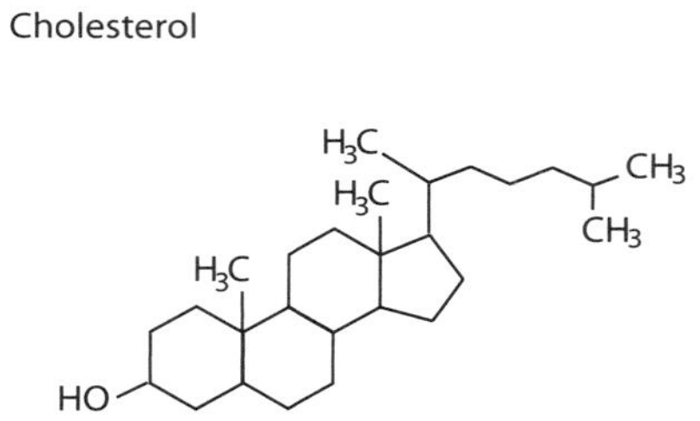

Cholesterol

H_3C H_3C CH_3 CH_3 H_3C HO

Figure 1.16. Chemical composition of cholesterol.

disease whereas others do not. In other patients, it is caused by a combination of multiple gene mutations and environmental stimulation. This type of cholesterol metabolic disorder is called polygenic hypercholesterolemia. In this case, the plasma cholesterol level of all family members may increase at some time during their lifespans, if the intake of saturated fats and cholesterol is high. Of all patients with hypercholesterolemia, the vast majority belong to the polygenic type. Secondary hypercholesterolemia is a complication of metabolic disorders, such as diabetes.

BIBLIOGRAPHY

1.1. Composition and Structure of DNA

Berg JM, Tymoczko JL, Stryer L: *Biochemistry*, 5th ed, Freeman, New York, 2002.

Jones ME: Pyrimidine nucleotide biosynthesis in animals: Genes, enzymes, and regulation of UMP biosynthesis, *Annu Rev Biochem* 49:253–79, 1980.

Kornberg A: Biologic synthesis of deoxyribonucleic acid, *Science* 131:1503–8, 1960.

Meselson M, Stahl FW: The replication of DNA in Escherichia coli, *Proc Natl Acad Sci USA* 44:671–82, 1958.

Watson JD, Crick FHC: Genetical implications of the structure of deoxyribonucleic acid, *Nature* 171:964–7, 1953.

Watson JD, Crick FHC: Molecular structure of nucleic acids: A structure for deoxyribose nucleic acid, *Nature* 171:737–8, 1953.

Wilkins MHF, Stokes AR, Wilson HR: Molecular structure of deoxypentose nucleic acids, *Nature* 171:738–40, 1953.

1.2. Organization of Chromosomes

Grunstein M: Histones as regulators of genes, *Sci Am* 267(4):68–74B, 1992.

Heck MMS: Condensins, cohesins, and chromosome architecture: How to make and break a mitotic chromosome, *Cell* 91:5–8, 1997.

Kornberg RD: Chromatin structure: A repeating unit of histones and DNA, *Science* 184:868–71, 1974.

Koshland D, Strunnikov A: Mitotic chromosome condensation, *Annu Rev Cell Biol* 12:305–33, 1996.

Luger K, Mader AW, Richmond RK, Sargent DE, Richmond TJ: Crystal structure of the nucleosome core particle at 2.8 A resolution, *Nature* 389:251–60, 1997.

Paulson JR, Laemmli UK: The structure of histone-depleted metaphase chromosomes, *Cell* 12:817–28, 1977.

Richmond TJ, Finch JT, Rushton B, Rhodes D, Klug A: Structure of the nucleosome core particle at 7A resolution, *Nature* 311:532–7, 1984.

Saitoh Y, Laemmli UK: Metaphase chromosome structure: Bands arise from a differential folding path of the highly AT-rich scaffold, *Cell* 76:609–22, 1994.

Van Holde KE, Zlatanovai J: Chromatin higher order structure: Chasing a mirage, *J Biol Chem* 270:8373–6, 1995.

Huebert DJ, Bernstein BE: Genomic views of chromatin, *Curr Opin Genet Dev* 15(5):476–81, Oct 2005.

Zhao H, Dean A: Organizing the genome: Enhancers and insulators, *Biochem Cell Biol* 83(4):516–24, Aug 2005.

Luger K, Hansen JC: Nucleosome and chromatin fiber dynamics, *Curr Opin Struct Biol* 15(2):188–96, April 2005.

Gruenbaum Y, Margalit A, Goldman RD, Shumaker DK, Wilson KL: The nuclear lamina comes of age, *Nat Rev Mol Cell Biol* 6(1):21–31, Jan 2005.

Becskei A, Mattaj IW: Quantitative models of nuclear transport, *Curr Opin Cell Biol* 17(1):27–34, Feb 2005.

Gregory RI, Shiekhattar R: Chromatin modifiers and carcinogenesis, *Trends Cell Biol* 14(12):695–702, Dec 2004.

Wienberg J: The evolution of eutherian chromosomes, *Curr Opin Genet Dev* 14(6):657–66, Dec 2004.

1.3. Functional Units of DNA

Blackburn EH: Switching and signaling at the telomere, *Cell* 106:661–73, 2001.

Blattner ER, Plunkett G, Bloch CA, Perna NT, Burland V et al: The complete genome sequence of Escherichia coli K-12, *Science* 277:1453–62, 1997.

Carbon J: Yeast centromeres: Structure and function, *Cell* 37:351–3, 1984.

Clarke L: Centromeres of budding and fission yeasts, *Trends Genet* 6:150–4, 1990.

Greider CW: Telomeres do D-loop-Tloop, *Cell* 97:419–2, 1999.

Ingram VM: Gene mutations in human hemoglobin: The chemical difference between normal and sickle cell hemoglobin, *Nature* 180:326–8, 1957.

Fritsch EF, Lawn RM, Maniatis T: Molecular cloning and characterization of the human beta-like globin gene cluster, *Cell* 19:959–72, 1980.

Gilbert W, de Souza SJ, Long M: Origin of genes, *Proc Natl Acad Sci USA* 94:7698–703, 1997.

Henikoff S, Greene EA, Pietrokovski S, Bork P, Attwood TK, Hood L: Gene families: The taxonomy of protein paralogs and chimeras, *Science* 278:609–14, 1997.

International Human Genome Sequencing Consortium: Initial sequencing and analysis of the human genome, *Nature* 409:860–921, 2001.

Mouse Genome Sequencing Consortium: Initial sequence and comparative analysis of the mouse genome, *Nature* 420:520–62, 2002.

Pluta AE, MacKay AM, Ainsztein AM, Goldberg AG, Earnshaw WC: The centromere: Hub of chromosomal activities, *Science* 270:1591–4, 1995.

Rubin GM et al (and 54 others): Comparative genomics of the eukaryotes, *Science* 287:2204–215, 2000.

Schueler MG, Higgins AW, Rudd MK, Gustashaw K, Willard HE: Genomic and genetic definition of a functional human centromere, *Science* 294:109–15, 2001.

Sun X, Wahlstrom J, Karpen G: Molecular structure of a functional Drosophila centromere, *Cell* 91:1007–19, 1997.

Szostak JW, Blackburn EH: Cloning yeast telomeres on linear plasmid vectors, *Cell* 29:245–55, 1982.

Stoltzfus A, Spencer DE, Zuker M, Logsdon JM Jr, Doolittle WE: Testing the exon theory of genes: The evidence from protein structure, *Science* 265:202–7, 1994.

Venter JC et al (and 273 colleagues): The sequence of the human genome, *Science* 291:1304–51, 2001.

Wiens GR, Sorger PK: Centromeric chromatin and epigenetic effects in kinetochore assembly, *Cell* 93:313–6, 1998.

Zakian VA: Telomeres: Beginning to understand the end, *Science* 270:1601–7, 1995.

1.4. DNA Replication

Kornberg A, Baker TA: *DNA Replication*, 2nd ed, Freeman, New York, 1991.

Baltimore D: RNA-dependent DNA polymerase in virions of RNA tumour viruses, *Nature* 226:1209–11, 1970.

Temin HM, Mizutani S: RNA-dependent DNA polymerase in virions of Rous sarcoma virus, *Nature* 226:1211–3, 1970.

Baker TA, Bell SP: Polymerases and the replisome: Machines within machines, *Cell* 92:296–305, 1998.

Bell SP, Dutta A: DNA replication in eukaryotic cells, *Annu Rev Biochem* 71:333–74, 2002.

Benkovic SJ, Valentine AM, Salinas F: Replisome-mediated DNA replication, *Annu Rev Biochem* 70:181–208, 2001.

Frick DN, Richardson CC: DNA primases, *Annu Rev Biochem* 70:39–80, 2001.

Gilbert DM: Making sense of eukaryotic DNA replication origins, *Science* 294:96–100, 2001.

Hubscher U, Maga G, Spadari S: Eukaryotic DNA polymerases, *Annu Rev Biochem* 71:133–63, 2002.

Kunkel TA, Bebenek K: DNA replication fidelity, *Annu Rev Biochem* 69:497–529, 2000.

McEachern MJ, Krauskopf A, Blackburn EH: Telomeres and their control, *Annu Rev Genet* 34:331–58, 2000.

Ogawa T, Okazaki T: Discontinuous DNA replication, *Annu Rev Biochem* 49:421–57, 1980.

Stinthcomb D, Struhl K, Davis RW: Isolation and characterization of a yeast chromosomal replicator, *Nature* 282:39–43, 1979.

Waga S, Stillman B: Anatomy of a DNA replication fork revealed by reconstitution of SV 40 DNA replication in vitro, *Nature* 369:207–12, 1994.

Waga S, Stillman B: The DNA replication fork in eukaryotic cells, *Annu Rev Biochem* 67:721–51, 1998.

West SC: DNA helicases: New breeds of translocating motors and molecular pumps, *Cell* 86:177–80, 1996.

Batty DP, Wood RD: Damage recognition in nucleotide excision repair of DNA, *Gene* 241:193–204, 2000.

De Laat WL, Jaspers NGJ, Hoeijmakers JHJ: Molecular mechanism of nucleotide excision repair, *Genes Dev* 13:768–85, 1999.

Goodman MF: Error-prone repair DNA polymerases in prokaryotes and eukaryotes, *Annu Rev Biochem* 71:17–50, 2002.

Harfe BD, Jinks-Robertson S: DNA mismatch repair and genetic instability, *Annu Rev Genet* 34:359–99, 2000.

Hoeijmakers JHJ: Genome maintenance mechanisms for preventing cancer, *Nature* 411:366–74, 2001.

Khanna KK, Jackson SP: DNA double strand breaks: Signaling, repair and the cancer connection, *Nature Genet* 27:247–54, 2001.

Livneh Z: DNA damage control by novel DNA polymerases: translesion replication and mutagenesis, *J Biol Chem* 276:25639–42, 2001.

Modrich P: Strand-specific mismatch repair in mammalian cells, *J Biol Chem* 272:24727–30, 1997.

Sancar A: DNA excision repair, *Annu Rev Biochem* 65:43–81, 1996.

1.5. RNA Composition, Structure, and Transcription

Brenner S, Jacob F, Meselson M: An unstable intermediate carrying information from genes to ribosomes for protein synthesis, *Nature* 190:576–81, 1961.

Butler JE, Kadonaga JT: The RNA polymerase II core promoter: A key component in the regulation of gene expression, *Genes Dev* 16(20):2583–92, 2002.

Shilatifard A, Conaway RC, Conaway JW: The RNA polymerase II elongation complex, *Annu Rev Biochem* 72:693–715, 2003.

Bushnell DA, Westover KD, Davis RE, Kornberg RD: Structural basis of transcription: An RNA polymerase II-TFIIB cocrystal at 4.5 Angstroms, *Science* 303:983–8, 2004.

Westover KD, Bushnell DA, Kornberg RD: Structural basis of transcription: Separation of RNA from DNA by RNA polymerase II, *Science* 303:1014–6, 2004.

Boyer LA, Lee TI, Cole MF, Johnstone SE, Levine SS, Zucker JP, Guenther MG, Kumar RM, Murray HL, Jenner RG, Gifford DK, Melton DA, Jaenisch R, Young RA: Core transcriptional regulatory circuitry in human embryonic stem cells, *Cell* 122:947–56, 2005.

Myers LC, Kornberg RD: Mediator of transcriptional regulation, *Annu Rev Biochem* 69:729–49, 2000.

Naar AM, Lemon BD, Tjian R: Transcriptional coactivator complexes, *Annu Rev Biochem* 70:475–501, 2001.

Orphanides G, Reinberg D: A unified theory of gene expression, *Cell* 108:439–51, 2002.

Atchison ML: Enhancers: Mechanisms of action and cell specificity, *Annu Rev Cell Biol* 4:127–53, 1988.

Berger SL: Histone modifications in transcriptional regulation, *Curr Opin Genet Dev* 12:142–8, 2002.

Dvir A, Conaway JW, Conaway RC: Mechanism of transcription initiation and promoter escape by RNA polymerase II, *Curr Opin Genet Dev* 11:209–14, 2001.

Conaway JW, Shilatifard A, Dvir A, Conaway RC: Control of elongation by RNA polymerase II, *Trends Biochem Sci* 25:375–80, 2000.

Horn PJ, Peterson CL: Molecular biology. Chromatin higher order folding—wrapping up transcription, *Science* 297:1824–7, 2002.

McKenna NJ, O'Malley BW: Combinatorial control of gene expression by nuclear receptors and coregulators, *Cell* 108:465–74, 2002.

Naar AM, Lemon BD, Tjian R: Transcriptional coactivator complexes, *Annu Rev Biochem* 70:475–501, 2001.

Narlikar GJ, Fan HY, Kingston RE: Cooperation between complexes that regulate chromatin structure and transcription, *Cell* 108:475–87, 2002.

Pabo CO, Sauer RT: Transcription factors: Structural families and principles of DNA recognition, *Annu Rev Biochem* 61:1053–95, 1992.

Abelson J, Trotta CR, Li H: tRNA splicing, *J Biol Chem* 273:12685–8, 1998.

Blanc V, Davidson NO: C-to-U RNA editing: Mechanisms leading to genetic diversity, *J Biol Chem* 278:1395–8, 2003.

Maas S, Rich A, Nishikura K: A-to-I RNA editing: Recent news and residual mysteries, *J Biol Chem* 278:1391–4, 2003.

Madison-Antenucci S, Grams J, Hajduk SL: Editing machines: The complexities of trypanosome RNA editing, *Cell* 108:435–8, 2002.

Maniatis T, Tasic B: Alternative pre-mRNA splicing and proteome expansion in metazoans, *Nature* 418:236–43, 2002.

Moore MJ: Nuclear RNA turnover, *Cell* 108:431–4, 2002.

Proudfoot NJ, Furger A, Dye MJ: Integrating mRNA processing with transcription, *Cell* 108:501–12, 2002.

Ruskin B, Krainer AR, Maniatis T, Green MR: Excision of an intact intron as a novel lariat structure during pre-mRNA splicing in vitro, *Cell* 38:317–31, 1984.

Smith CW, Valcarcel J: Alternative pre-mRNA splicing: The logic of combinatorial control, *Trends Biochem Sci* 25:381–8, 2000.

1.6. Protein Composition and Structure

Branden C, Tooze J: *Introduction to Protein Structure*, 2nd ed, Garland, New York, 1999.

Chothia C, Finkelstein AV: The classification and origins of protein folding patterns, *Annu Rev Biochem* 59:1007–39, 1990.

Holm L, Sander C: Mapping the protein universe, *Science* 273:595–602, 1996.

Kendrew JC: The three-dimensional structure of a protein molecule, *Sci Am* 205:96–111, 1961.

Levitt M, Gerstein M, Huang E, Subbiah S, Tsai J: Protein folding: The endgame, *Annu Rev Biochem* 66:549–79, 1997.

Richardson JS: The anatomy and taxonomy of protein structure, *Adv Protein Chem* 34:167–339, 1981.

Sanger F: Sequences, sequences, and sequences, *Annu Rev Biochem* 57:1–28, 1988.

Umbarger HE: Amino acid biosynthesis and its regulation, *Annu Rev Biochem* 47:533–606, 1978.

1.7. Protein Translation

Ban N, Nissen P, Hansen J, Moore PB, Steitz TA: The complete atomic structure of the large ribosomal subunit at 2.4 A resolution, *Science* 289:905–20, 2000.

Dever TE: Gene-specific regulation by general translation factors, *Cell* 108:545–56, 2002.

Gray NK, Wickens M: Control of translation initiation in animals, *Annu Rev Cell Dev Biol* 14:399–458, 1998.

Moore PB, Steitz TA: The structural basis of large ribosomal subunit function, *Annu Rev Biochem* 72:813–50, 2003.

Moore PB, Steitz TA: The involvement of RNA in ribosome function, *Nature* 418:229–35, 2002.

Ogle JM, Ramakrishnan V: Structural insights into translational fidelity, *Annu Rev Biochem* 74:129–77, 2005.

Sachs AB: Cell cycle-dependent translation initiation: IRES elements prevail, *Cell* 101(3):243–5, 2000.

1.8. Protein Folding and Architecture

Burkhard P, Stetefeld J, Strelkov SV: Coiled coils: A highly versatile protein folding motif, *Trends Cell Biol* 11:82–8, 2001.

Crick FHC, Barnett L, Brenner S, Watts-Tobin RJ: General nature of the genetic code for proteins, *Nature* 192:1227–32, 1961.

Nirenberg M, Leder P: RNA codewords and protein synthesis, *Science* 145:1399–407, 1964.

Nirenberg MW, Matthaei JH: The dependence of cell-free protein synthesis in *E. coli* upon naturally occurring or synthetic polyribonucleotides, *Proc Natl Acad Sci USA* 47:1588–602, 1961.

Gahmberg CG, Tolvanen M: Why mammalian cell surface proteins are glycoproteins, *Trends Biochem Sci* 21:308–11, 1996.

Gething MJ, Sambrook J: Protein folding in the cell, *Nature* 355:33–45, 1992.

Schiene C, Fischer G: Enzymes that catalyse the restructuring of proteins, *Curr Opin Struct Biol* 10:40–5, 2000.

Sigler PB, Xu Z, Rye HS, Burston SG, Fenton WA, Horwich AL: Structure and function in GroEL-mediated protein folding, *Annu Rev Biochem* 67:581–608, 1998.

Udenfriend S, Kodukula K: How glycosylphosphatidylinositol-anchored membrane proteins are made, *Annu Rev Biochem* 64:563–91, 1995.

Zhang FL, Casey PJ: Protein prenylation: Molecular mechanisms and functional consequences, *Annu Rev Biochem* 65:241–69, 1996.

1.9. Phospholipids

Deguchi H, Yegneswaran S, Griffin JH: Sphingolipids as bioactive regulators of thrombin generation, *J Biol Chem* 279:12036–12042, 2004.

Martens JR, O'Connell K, Tamkun M: Targeting of ion channels to membrane microdomains: Localization of KV channels to lipid rafts, *Trends Pharmacol Sci* 25:16–21, 2004.

van Meer G, Burger KN: Sphingolipid trafficking—sorted out? *Trends Cell Biol* 2:332–7, 1992.

Wakil SJ, Stoops JK, Joshi VC: Fatty acid synthesis and its regulation, *Annu Rev Biochem* 52:537–79, 1983.

Lee A: Membrane structure, *Curr Biol* 11:R811–4, 2001.

McLaughlin S, Murray D: Plasma membrane phosphoinositide organization by protein electrostatics, *Nature* 438:605–11, 2005.

Behnia R, Munro S: Organelle identity and the signposts for membrane traffic, *Nature* 438:597–604, 2005.

Chalfant CE, Spiegel S: Sphingosine 1-phosphate and ceramide 1-phosphate: Expanding roles in cell signaling, *J Cell Sci* 118:4605–12, 2005.

Niggli V: Regulation of protein activities by phosphoinositide phosphates, *Annu Rev Cell Dev Biol* 21:57–79, 2005.

Rosen H, Goetzl EJ: Sphingosine 1-phosphate and its receptors: An autocrine and paracrine network, *Nature Rev Immunol* 5:560–70, 2005.

Branton D, Cohen CM, Tyler J: Interaction of cytoskeletal proteins on the human erythrocyte membrane, *Cell* 24:24–32, 1981.

Thompson TE, Tillack TW: Organization of glycosphingolipids in bilayers and plasma membranes of mammalian cells, *Annu Rev Biophys Biophys Chem* 14:361–86, 1985.

1.10. Glycolipids

Ramstedt B, Slotte JP: Membrane properties of sphingomyelins, *FEBS Lett* 531:33–7, 2002.

Cremesti AE, Goni FM, Kolesnick R: Role of sphingomyelinase and ceramide in modulating rafts: Do biophysical properties determine biologic outcome? *FEBS Lett* 531:47–53, 2002.

Lee A: Membrane structure, *Curr Biol* 11:R811–4, 2001.

Bush CA, Martin-Pastor M, Imberty A: Structure and conformation of complex carbohydrates of glycoproteins, glycolipids, and bacterial polysaccharides, *Annu Rev Biophys Biomol Struct* 28:269–93, 1999.

Stoffel W, Bosio A: Myelin glycolipids and their functions, *Curr Opin Neurobiol* 7:654–61, 1997.

Udenfriend S, Kodukula K: How glycosylphosphatidylinositol-anchored membrane proteins are made, *Annu Rev Biochem* 64:563–91, 1995.

Englund PT: The structure and biosynthesis of glycosyl phosphatidylinositol protein anchors, *Annu Rev Biochem* 62:121–38, 1993.

Paulson JC, Colley KJ: Glycosyltransferases. Structure, localization, and control of cell type-specific glycosylation, *J Biol Chem* 264:17615–8, 1989.

Low MG: Glycosyl-phosphatidylinositol: A versatile anchor for cell surface proteins, *FASEB J* 3:1600–8, 1989.

Thompson TE, Tillack TW: Organization of glycosphingolipids in bilayers and plasma membranes of mammalian cells, *Annu Rev Biophys Biophys Chem* 14:361–86, 1985.

Rauvala H, Finne J: Structural similarity of the terminal carbohydrate sequences of glycoproteins and glycolipids, *FEBS Lett* 97:1–8, 1979.

1.11. Steroids

Simons K, Toomre D: Lipid rafts and signal transduction, *Nature Rev Mol Cell Biol* 1:31–9, 2000.

Incardona JP, Eaton S: Cholesterol in signal transduction, *Curr Opin Cell Biol* 12:193–203, 2000.

Bruce C, Chouinard RA Jr, Tall AR: Plasma lipid transfer proteins, high-density lipoproteins, and reverse cholesterol transport, *Annu Rev Nutr* 18:297–330, 1998.

Lagrost L, Desrumaux C, Masson D, Deckert V, Gambert P: Structure and function of the plasma phospholipid transfer protein, *Curr Opin Lipidol* 9:203–9, 1998.

Simons K, Ikonen E: Functional rafts in cell membranes, *Nature* 387:569–72, 1997.

Yeagle PL: Lipid regulation of cell membrane structure and function, *FASEB J* 3:1833–42, 1989.

Kummerow FA: Modification of cell membrane composition by dietary lipids and its implications for atherosclerosis, *Ann NY Acad Sci* 414:29–43, 1983.

Griffiths AJF: *An Introduction to Genetic Analysis*, 6th ed, Freeman, New York, 1996.

Bretscher MS, Munro S: Cholesterol and the Golgi apparatus, *Science* 261:1280–1, 1993.

2

REGULATION OF GENE EXPRESSION

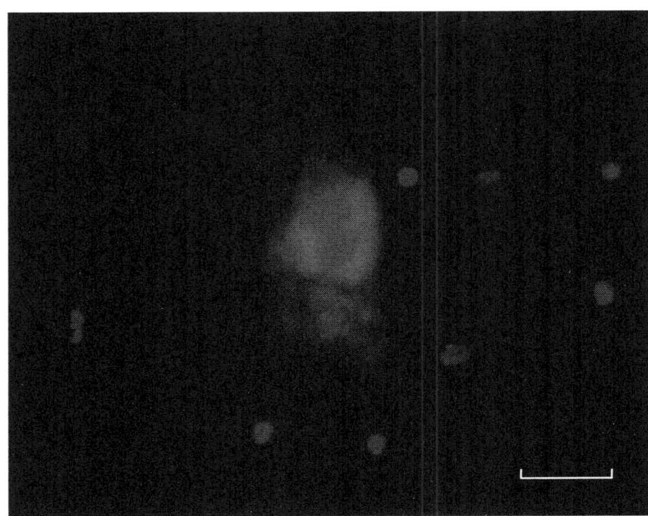

Mobilization of albumin-positive hepatocyte-like cells into the blood in response to the upregulation of interleukin-6 in ischemic cardiac injury. The mobilized hepatocyte-like cells can be recruited to injured cardiac tissue, promoting the proliferation of cardiomyocytes in the zone of cardiac injury. Red: albumin. Blue: cell nuclei. Sclae: 10 μm. See color insert.

Bioregenerative Engineering: Principles and Applications, by Shu Q. Liu

Gene expression is the transcription of mRNA (transfer of genetic information from DNA to mRNA) and the translation of protein from a specific mRNA (transfer of genetic information from mRNA to protein). Gene expression is regulated at several key levels, including the transcription of pre-mRNA, the conversion of pre-mRNA to mature mRNA, and the synthesis of protein. The regulation of gene expression involves basically two types of factors: the *cis*-acting regulatory elements and *trans*-acting regulatory factors. *Cis-acting regulatory elements* are DNA-sequences that are protein-binding sequences and are found near the initiation site of gene transcription. *Trans-acting regulatory factors* are proteins that bind to specific *cis*-acting elements and control gene transcription. In addition, as we see below, the components of chromatin also participate in the regulation of gene expression.

There are several common features for the regulation of gene expression:

1. *Trans*-acting factors can bind only to specific *cis*-acting elements.
2. The binding of *trans*-acting factors to *cis* elements can induce conformational changes in the chromatin, a critical process for regulating gene transcription.
3. The binding of *trans*-acting factors may induce positive or negative regulation of gene transcription. Positive regulation is a process by which the binding of *trans*-acting factors stimulates gene transcription, whereas negative regulation exerts an opposite effect.
4. A transcription regulatory signal can be effective for only a limited time. The signal is turned off after a mission is accomplished.

The principles of gene expression regulation are briefly discussed in this section.

BASIC DNA ELEMENTS FOR REGULATING GENE EXPRESSION [2.1]

The expression of genes is controlled at the level of mRNA transcription. For each gene, there are specialized short sequences, known as *cis*-acting regulatory elements, which mediate the attachment and activation of RNA polymerases, necessary processes for initiating mRNA transcription. In mammalian cells, *cis*-acting regulatory elements include promoters, promoter-proximal elements, and enhancers. Each *cis*-acting element can be recognized and activated by *trans*-acting regulatory factors. These factors can bind to the promoter and initiate the attachment of RNA polymerase II, inducing RNA transcription. A typical promoter is composed of TATA nucleotides, known as the *TATA box*. The TATA box is usually found at a site about 30 bp upstream the first base pair of the transcription site. On the action of *trans*-acting factors, a promoter mediates the initiation of RNA transcription in coordination with promoter-proximal elements and enhancers. Typical promoter-proximal elements, which often include the CCAAT box and GC-rich segments, are located about 100–200 bp away from the TATA box. Experiments with point mutation have demonstrated that a mutation in the TATA or the promoter-proximal elements induces a decrease in the rate of RNA transcription, whereas mutation in other regions exhibits little influence (Fig. 2.1). Each gene contains several sequence motifs of enhancers, which are located at a distance, some times up to 50 kb, from the TATA box. Different motifs serve as binding sits for specific *trans*-acting factors. The function of promoter-proximal elements and enhancers is to facilitate the activity of the promoters and enhance gene transcription.

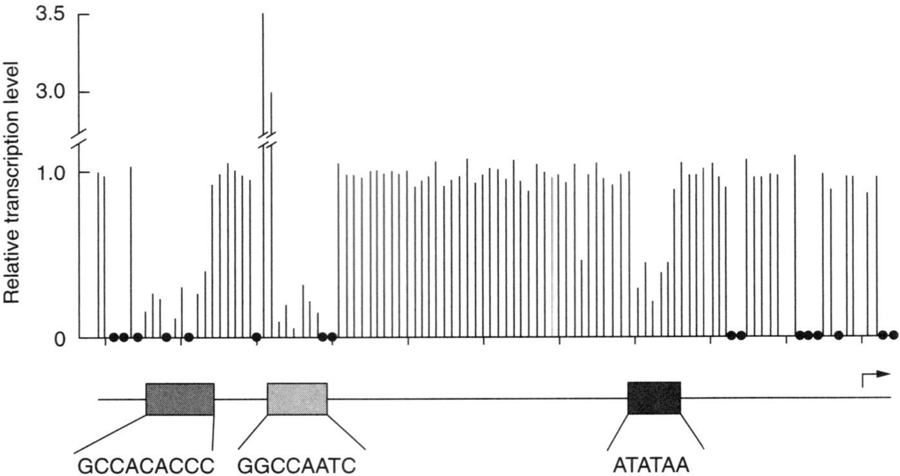

Figure 2.1. Identification of the promoter and promoter-proximal elements that control the expression of the β-globin gene. Point mutations were induced on most nucleotides through the promoter region of the β-globin gene, and the effect of each point mutation was tested. Mutations within the TATA promoter and the promoter-proximal elements caused a significant decrease in the level of transcription, suggesting that these *cis*-elements are critical to the transcription of the β-globin gene. (Reprinted by permission from Maniatis T, Goodbourn S, Fischer JA: *Science* 236:1237–45, 1987. Copyright 1987 AAAS.)

TRANS-ACTING REGULATORY FACTORS [2.1]

Trans-acting regulatory factors are proteins that bind to and activate target genes and regulate RNA transcription. In mammalian cells, there exist a large number of *trans*-acting factors. These factors, while in their inactive state, are present in the cytoplasm. Once activated in response to the stimulation of signaling molecules (see Chapter 5), a *trans*-acting factor can be translocated from the cytoplasm to the nucleus and bind to specific *cis*-acting elements, or promoters and enhancers, of target genes. A *trans*-acting factor, once bound to a *cis*-acting element, can form a complex with RNA polymerase II and other necessary regulatory factors (See Table 2.1), assisting the polymerase to initiate mRNA transcription. Each *trans*-acting factor contains two domains. The first domain is responsible for binding to *cis*-acting elements, whereas the second is for activating mRNA transcription. A typical example of *trans*-acting factors is the complex of *t*ranscription *f*actors for RNA polymerase *II*, or TFII. Each TFII is composed of a TATA box-binding protein (TBP) and several associated proteins (Fig. 2.2). These proteins are referred to as *basal trans-acting factors*. Examples of other *trans*-acting factors include activator protein 1 and nuclear factor κ B. The functions of these factors are discussed in Chapter 5.

REGULATION OF GENE TRANSCRIPTION

At the transcription level, several mechanisms are involved in the regulation of gene expression, including the activation of *trans*-acting factors, modification of the chromatin structure, and DNA modulation. These mechanisms are briefly described as follows.

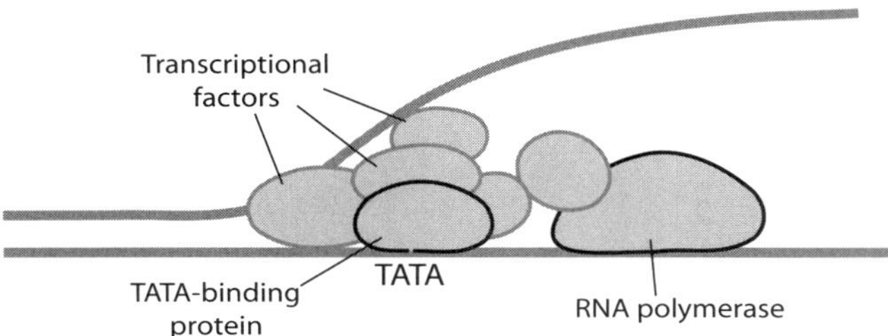

Figure 2.2. Schematic representation of controlling gene transcription by *trans*-acting factors (based on bibliography 2.1).

Control of the Activity of *Trans*-Acting factors [2.2]

In eukaryotes, there are two types of *trans*-acting factor: transcription-activating proteins (transcriptional activators) and transcription-inhibiting proteins (transcriptional repressors). The two types of factor regulate gene transcription via controlling the function of the RNA polymerase holoenzyme and the structure of chromatin. These factors dynamically counterbalance each other, ensuring an appropriate level of gene expression.

Trans-acting factors interact with DNA transcription apparatus, which is composed of *cis*-acting elements and protein coactivators. While *trans*-acting proteins act on specific *cis* elements, protein coactivators enhance the regulation of gene transcription, which is independent of specific *cis* elements. The binding of *trans*-acting factors to *cis*-acting DNA elements induces activation and promotes reorientation of coactivator proteins, which are necessary conditions for the initiation of gene transcription.

The activity of *trans*-acting proteins is controlled by three basic approaches: modification of preexisting proteins, de novo synthesis of proteins, and targeted degradation. These processes may coordinate in a time- and space-dependent manner, providing synergistic inputs for the regulation of gene transcription. Among these regulatory processes, the modification of preexisting proteins is the most important process that causes immediate activation of signaling events, resulting in prompt cellular activities. Common types of modification include phosphorylation and dimerization of proteins. Protein *phosphorylation* occurs primarily on the serine, threonine, and tyrosine residues, and is regulated by protein kinases and protein phosphatases, which phosphorylate and dephosphorylate target proteins, respectively. The phosphorylation of signaling molecules may lead to activation or deactivation of *trans*-acting factors, depending on the nature of specific signaling pathways. In most cases, phosphorylation induces activation of *trans*-acting factors. Such a process enhances the interaction of *trans*-acting factors with nuclear export machineries, which facilitate the translocation of the *trans*-acting factors from the cytoplasm to the nucleus. In addition, the phosphorylation of the DNA-binding domain of a *trans*-acting factor mediates the binding of the trans-acting factor to the target DNA. The involvement of protein phosphorylation in specific signaling pathways is described on Chapter 5.

Dimerization is another form of modification of transcriptional factors. Most *trans*-acting factors bind to DNA in the form of homodimer or heterodimer. The dimerization of *trans*-acting factors enhances their binding to target DNA and provides a mechanism for the specificity and selectivity of gene transcription. For instance, the members of the activating protein-1 family, including Jun, Fos, and activating transcription factor (ATF), form homodimers (Jun-Jun, Fos-Fos, ATF-ATF) and heterodimers (Fos-Jun, Jun-ATF,

TABLE 2.1. Characteristics of Selected Gene Transcription Regulators*

Proteins	Alternative Names	Amino Acids	Molecular Weight (kDa)	Expression	Functions
RNA polymerase II, subunit 1	RNA polymerase II 14.5-kDa subunit, polymerase RNA II DNA-directed polypeptide I 14.5 kDa, DNA-directed RNA polymerase II polypeptide I	125	14	Ubiquitous	Forming a complex with subunits 2 and 3 and regulating mRNA synthesis in eukaryotes
RNA polymerase II, subunit 2	RNA polymerase II subunit B, RNA polymerase II 140-kDa subunit, DNA-directed RNA polymerase II 140-kDa polypeptide	1174	134	Ubiquitous	Forming a complex with subunits 1 and 3 and regulating mRNA synthesis
RNA polymerase II, subunit 3	RNA polymerase II subunit C, DNA-directed RNA polymerase II 33-kDa polypeptide	275	31	Ubiquitous	Forming a complex with subunits 1 and 2 and regulating mRNA synthesis
TATA box-binding protein	GTF2D, SCA17, spinocerebellar ataxia 17	339	38	Ubiquitous	Together with TATA box-binding protein-associated factors, forming transcription factor IID (TFIID), which positions the RNA polymerase II to a target gene promoter and regulates mRNA transcription

*Based on bibliography 2.1.

Fos-ATF). Each type of dimer targets specific genes, thus increasing the specificity and also the capacity of transcriptional regulation. repress

In contrast to gene activation by *trans*-acting proteins, gene transcription can be inhibited by transcriptional protein repressors. The mechanisms of transcriptional repression are diverse. Some repressors can bind to transcription activators, reducing their activities. The TATA-box-binding protein (TBP) is a major target of transcription repressors. For instance, the binding of the repressor NC2 to the TATA-box-binding protein prevents the assembly of the RNA polymerase II holoenzyme into the initiation complex of gene transcription, thus repressing gene transcription. Another repression mechanism involves the competition of transcriptional repressors with transcriptional activators for DNA binding sites. The outcome, activation, or repression of gene transcription is determined on the basis of the relative activity of transcriptional activators and repressors as well as their affinity to target DNA binding sites.

In addition to the modification of existing *trans*-acting factors, de novo synthesis of *trans*-acting proteins contributes to the regulation of gene transcription. The synthesis of a *trans*-acting factor is induced by upregulating the expression of a specific gene that encodes the *trans*-acting factor. An increase in the level of a *trans*-acting factor enhances gene transcription. However, this approach is relatively slow and is not as effective as the modification of existing *trans*-acting factors. A *trans*-acting factor is usually degraded by enzymes when its task is accomplished. The degradation of *trans*-acting factors is an effective approach ensuring that gene transcription can be stopped promptly.

Chromatin Modification [2.3]

In addition to the regulation of gene transcription by transcriptional activators and repressors, dynamic changes in chromatin components contribute significantly to the regulation of gene transcription. Chromatin is composed of nucleosomes, the basic chromatin units. In each nucleosome, DNA is wrapped around a cylinder-like histone core. The nucleosomes are further arranged into three-dimensional structures. The structure of chromatin components can be modified via different processes, including histone phosphorylation on the serine and threonine residues, histone ubiquitination, histone acetylation on the lysine residue, and histone methylation on the lysine and arginine residues. These processes significantly influence the activities of gene transcription.

Histone phosphorylation can be induced by certain members of the mitogen-activated protein kinase family. Such a process is often a result of growth factor action and facilitates the transcription of mitogenic genes. *Histone ubiquitination* is a process that regulates gene transcription via degrading histone (see Chapter 5). Such a process influences the stability of chromatin and thus the activity of gene transcription. A typical example is the ubiquitination of histone H1. This type of histone is involved in the packaging of chromatin nucleosomes into a higher-ordered structure and is referred to as a *linker histone*. The formation of a higher-order chromatin structure usually results in the repression of gene transcription. The degradation of histone H1 thus promotes activation of gene transcription.

The *acetylation of histone* occurs primarily on the lysine residues at the *N*-terminus. This process is catalyzed by the histone acetyl transferase and can be reversed by the histone deacetylase. The acetylation of histones is thought to weaken the interaction of histone cores with DNA, increase the mobility of DNA and accessibility of *trans*-acting factors to *cis*-acting elements, and thus facilitate transcriptional activation. Furthermore,

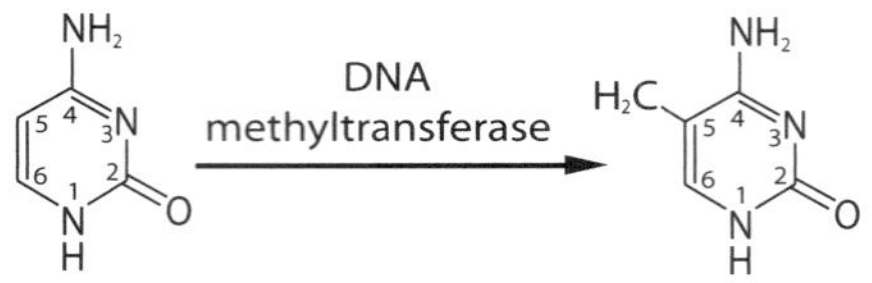

Figure 2.3. Methylation of cytosine, catalyzed by the DNA methyltransferase (DNMTs). Based on bibliography 2.4.

histone acetylation may enhance the association of *trans*-acting factors and *cis*-acting elements and facilitates gene transcription. Another means of nucleosome modification is *histone methylation*, which occurs usually on the lysine and arginine residues. A well-characterized example is the methylation of histone H3. In this type of histone, methylation is catalyzed on lysine 4 (the 4th lysine residue) and lysine 9. The outcome of methylation is dependent on the location of methylation. Methylation on the 4th lysine residue of histone H3 is associated with transcriptional activation, whereas methylation on lysine 9 suppresses gene transcription.

DNA Modification [2.4]

In addition to the modification of histones, the structure of DNA can be modified, which influences gene transcription. A well-characterized form of modification is *DNA methylation* at cytosines located prior to guanosines, which are referred to as the *CpG dinucleotides*. These dinucleotides are rare, but can be found in certain regions of a gene in the form of cluster. A cluster of CpGs is termed a *CpG island*, which is often found in the promoter region of a gene. DNA methylation is found in mammalian cells and is catalyzed by an enzyme known as *DNA methyltransferase* (DNMT), which transfers a methyl group from the donor molecule *S*-adenosyl-methionine to a cytosine base (Fig. 2.3). DNA methylation is often associated with inactive genes, whereas active genes are relatively demethylated, suggesting that DNA methylation suppresses gene transcription. In particular, DNA methylation in a promoter region causes gene silencing. While the mechanisms remain poorly understood, it has been hypothesized that DNA methylation may enhance the binding of methyl-CpG-binding proteins (MBPs), which are capable of recognizing and binding to methylated cytosine–guanosine groups. These methyl-CpG-binding proteins have been found to recruit histone deacetylase to chromatin, enhancing histone deacetylation. As a result, the accessibility of *trans*-acting factors to *cis*-acting elements is reduced and gene transcription is repressed. A physiological aspect of DNA methylation is that DNA methylation may help maintain certain genes in an inactive state. Such a process is highly selective and physiologically active genes may not be the target of DNA methylation.

REGULATION OF PRE-mRNA CONVERSION TO MATURE mRNA

Gene transcription is a process that produces a primary transcript known as pre-mRNA. This form of mRNA cannot be directly processed for protein translation. To translate a protein, it is necessary to convert the pre-mRNA to mature mRNA. Such a conversion is accomplished via several post-transcriptional processes, including capping at the 5′ terminus, polyadenylation at the 3′ terminus, pre-mRNA splicing, and pre-mRNA transport.

5′-Terminal Capping and Decapping of pre-mRNA [2.5]

5′-Terminal capping is a process by which a 7-methyl guanosine is linked to the first nucleoside of the transcript with a 5′-triphosphate (Fig. 2.4). This process is necessary for efficient splicing and export of pre-mRNA as well as for the initiation of protein translation. Several enzymes are required for 5′-terminal capping, including RNA triphosphatase, RNA guanylyltransferase, and RNA methyltransferase. The RNA triphosphatase catalyzes the formation of a diphosphate from a triphosphate at the 5′-terminus of the mRNA transcript. The RNA guanylyltransferase induces the capping of the 5′-diphosphate with a GMP. The RNA methyltransferase catalyzes the methylation at the N7 position of the 5′-guanine base. These three steps are necessary for the 5′-terminal capping and take place during transcription when the transcript is about 25–30 nucleotides in length. The capping structure is required for the arrangement of mRNA in the ribosome for protein translation. The 5′ capping also stabilizes the structure of mRNA and protects mRNA from degradation by 5′-exoribonucleases.

The 5′-terminal cap of mRNA can be removed by decapping enzymes, a process leading to mRNA degradation. There are two types of decapping enzyme, including the

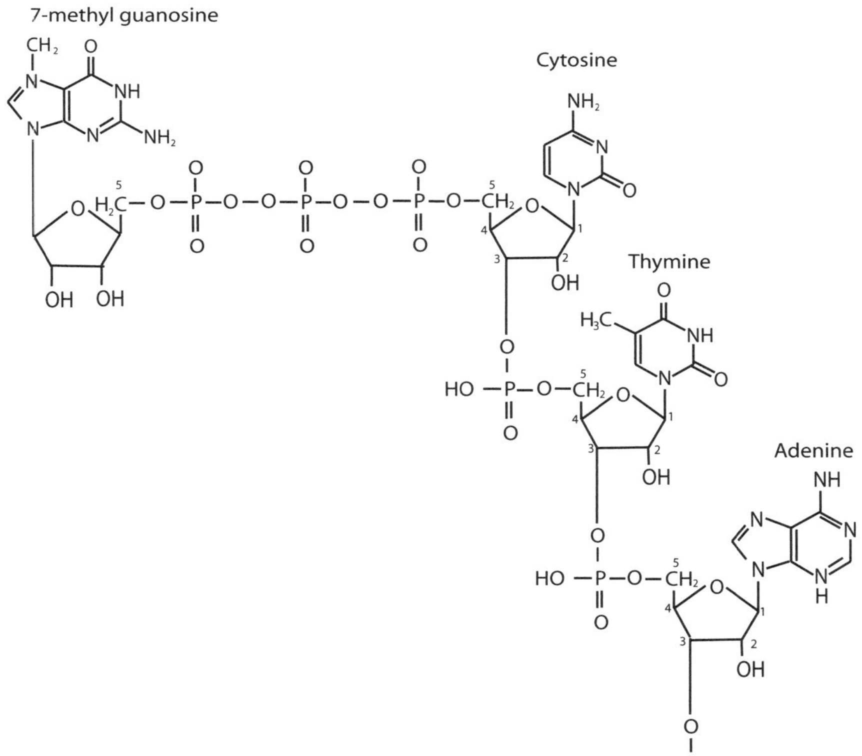

Figure 2.4. Chemical structure of the mRNA cap. The cap consists of N7-methyl guanosine linked by an inverted 5′-5′ triphosphate bridge to the first nucleoside of the mRNA chain (base N can be adenine, guanine, cytosine, or uracil). (Reprinted by permission from Gu M, Lima CD: Processing the message: Structural insights into capping and decapping mRNA, *Curr Opin Struct Biol* 15:99–106, 2005.)

Dcpl/Dcp2 decapping complex and DcpS. The Dcpl/Dcp2 complex regulates mRNA cap degradation in the 5′→3′ mRNA decay pathway. Within the Dcpl/Dcp2 complex, Dcp2 can catalyze the hydrolysis of the 5′-cap, resulting in the release of GDP from the mRNA, and Dcpl serves as a regulator that stimulates the activity of Dcp2. In contrast, DcpS regulates mRNA cap degradation in the 3′→5′ mRNA decay pathway and catalyzes residual mRNA cap hydrolysis.

Polyadenylation [2.6]

Polyadenylation is a process by which a poly-A sequence about 200 bases is added to a pre-mRNA molecule, a process necessary for the maturation of pre-mRNA. There are two basic steps for polyadenylation: cleavage of the nascent pre-mRNA transcript and addition of a poly-A sequence to the 3′ terminus. Several proteins are involved in the regulation of pre-mRNA cleavage and polyadenylation, including the cleavage and polyadenylation stimulation factor (CPSF), cleavage stimulation factor (CstF), cleavage factor I (CFI), and cleavage factor II (CFII). These factors form a complex and bind to a pre-mRNA site that is downstream to a ubiquitous sequence AAUAAA and immediately upstream to a GU-rich sequence. The AAUAAA sequence serves as an initiation signal for the binding of the protein complex. A CPSF molecule, CPSF-160, is responsible for recognizing the AAUAAA signal. The binding of the regulatory factors induces the cleavage of the pre-mRNA sequence at the binding site. The resulting 3′-terminus by cleavage is immediately polyadenylated by a poly(A) polymerase (Fig. 2.5). The polyadenylation of pre-mRNA contributes to the stabilization of mRNA.

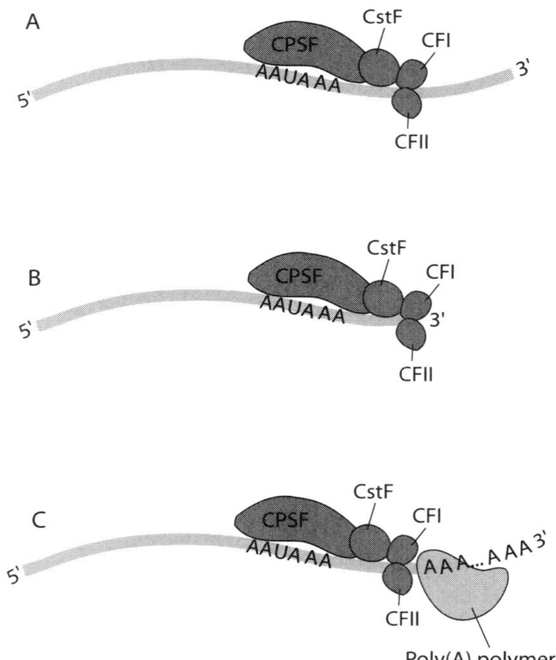

Figure 2.5. Schematic representation of polyadenylation of pre-mRNA (based on bibliography 2.6).

Pre-mRNA Splicing [2.7]

Pre-mRNA splicing is a process by which a transcribed pre-mRNA molecule is modified by removing the introns and rejoining the exons. This process is critical to the maturation of mRNA and the selection of various specific mRNA molecules from the same pre-mRNA. A pre-mRNA molecule contains protein-coding exons and noncoding introns. To generate mature mRNA, the introns must be removed and the exons must be rejoined in an appropriate order. Pre-mRNA splicing takes place in a structure known as *spliceosome*, which consists of a number of small nuclear ribonucleoproteins (SNRNPs) and several small nuclear RNA (snRNA) molecules (note that snRNAs are involved in not only RNA splicing, but also maintenance of the telomeres or the chromosome ends). These components are thought to mediate the cleaving and rejoining of nucleotides (Fig. 2.6). In the spliceosome, multiple mature mRNA molecules with different functions can be formed by alternative splicing and selective combination of exons, which are regulated by specific regulatory small nuclear ribonucleoproteins and snRNAs. This is a mechanism for producing cell- or tissue-specific proteins based on the same pre-mRNA molecule. The mechanisms of selective splicing remain poorly understood.

mRNA Transport [2.8]

mRNA transport is a process by which transcribed and processed mRNA is moved from the cell nucleus to the cytoplasm. The transcription of mRNA takes place in the cell nucleus, whereas protein translation occurs in the cytoplasm. Spliced mRNA must be transported to the cytoplasm and localized to designated sites for appropriate protein

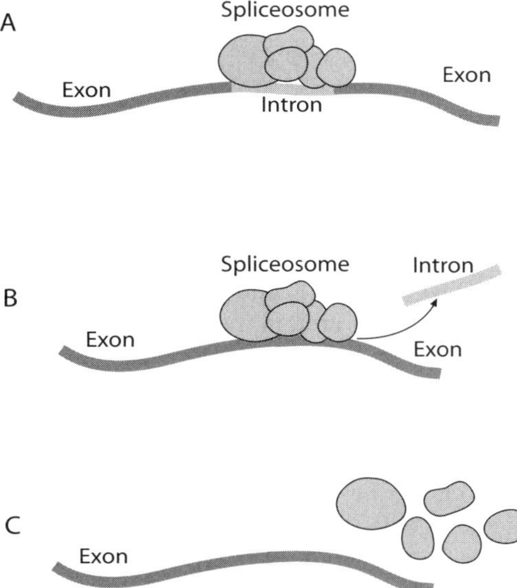

Figure 2.6. Schematic representation of pre-mRNA splicing (based on bibliography 2.7).

translation. The regulation of these processes involves specified RNA sequences, which are termed *cis-acting RNA elements*, and regulatory RNA-binding proteins, which are termed *trans-acting RNA-binding* factors. The *trans*-acting RNA-binding factors can recognize and interact with *cis*-acting RNA elements to form ribonucleotide protein (RNP) complexes. These complexes can move from the nucleus to the cytoplasm and interact with motor proteins, such as kinesin and dynein molecules, which further move the ribonucleotide protein complexes along the microtubules to the destination sites. The ribonucleotide protein complexes can also interact with myosin molecules, which move the complexes along the actin filaments. Once transported to destination sites, *trans*-acting RNA-anchoring proteins recognize and interact with the ribonucleotide protein complexes, anchoring the complexes to the specified sites. At these sites, the ribonucleotide protein complexes interact with ribosomes, initiating protein translation. It is interesting to note that unspliced mRNA cannot be transported. The exact mechanisms for selective mRNA transport remain poorly understood.

In addition to mRNA capping, splicing, and transport, the posttranscriptional stability of mRNA, which determines the concentration of mRNA, is also a critical factor for the regulation of gene expression. The mRNA concentration is controlled by the rates of RNA transcription, splicing, and transport as well as the rate of mRNA degradation. The stability of mRNA varies widely with half-life ranging from minutes to hours. The mRNAs of cytokines and growth factors, which are involved in short-term signaling events, are usually short-lived. In contrast, the mRNAs of constitutive proteins possess a relatively long-term lifespan. The stability of mRNAs is controlled by enzymes known as *nucleases*. These enzymes are often coupled to protein translation and can be activated when excessive proteins are produced, inducing degradation of specific mRNAs and reduction in protein translation.

REGULATION OF PROTEIN TRANSLATION [2.9]

In addition to the regulation of transcription and conversion of pre-mRNA to mature mRNA, gene expression is also controlled at the protein translation level. There are two known regulatory mechanisms at this level: modification at the 5′-terminus of mRNA and modification of initiation factors. The modification of mRNA at the 5′-terminus is usually a negative mechanism of translation regulation by RNA-binding proteins. These proteins serve as translation repressors, bind to mRNA at the 5′-terminus, and block the assembly of initiation factors and the $40S$ subunit of ribosomes. Such a process results in the suppression of protein translation. The modification of translation initiation factors is another powerful means for the regulation of protein translation. A major form of modification is phosphorylation, which controls the activity of translation initiation factors and thus regulates the rate of protein translation. For example, eukaryotic initiation factor 2 (eIF2) is a protein translation initiation factor. The α subunit of this protein can be phosphorylated on the Ser 51 residue by a serine/threonine protein kinase known as *RNA-dependent protein kinase* (PKR), which can be activated in response to the stimulation of various factors, such as viruses and double-stranded RNA (dsRNA). The phosphorylation of the eukaryotic initiation factor 2 reduces its activity and thus results in the suppression of protein synthesis. In the case of virus infection, the suppression of protein synthesis helps to reduce viral propagation.

BIBLIOGRAPHY

2.1. DNA Regulatory Elements and Transcriptional Factors

Cramer P, Bushnell DA, Fu J, Gnatt AL, Maier-Davis B et al: Architecture of RNA polymerase II and implications for the transcription mechanism, *Science* 288:640–8, 2000.

Acker J, Mattei MG, Wintzerith M, Roeckel N, Depetris D et al: Chromosomal localization of human RNA polymerase II subunit genes, *Genomics* 20:496–9, 1994.

Fanciulli M, Bruno T, Di Padova M, De Angelis R, Lovari S et al: The interacting RNA polymerase II subunits, hRPB11 and hRPB3, are coordinately expressed in adult human tissues and down-regulated by doxorubicin, *FEBS Lett* 427:236–40, 1998.

Pati UK, Weissman SM: The amino acid sequence of the human RNA polymerase II 33-kDa subunit hRPB 33 is highly conserved among eukaryotes, *J Biol Chem* 265:8400–3, 1990.

Hobbs NK, Bondareva AA, Barnett S, Capecchi MR et al: Removing the vertebrate-specific TBP N terminus disrupts placental beta-2M-dependent interactions with the maternal immune system, *Cell* 110:43–54, 2002.

Human protein reference data base, Johns Hopkins University and the Institute of Bioinformatics, at http://www.hprd.org/protein.

Imbert G, Trottier Y, Beckmann J, Mandel JL: The gene for the TATA binding protein (TBP) that contains a highly polymorphic protein coding CAG repeat maps to 6q27, *Genomics* 21:667–8, 1994.

Kao CC, Lieberman PM, Schmidt MC, Zhou Q, Pei R et al: Cloning of a transcriptionally active human TATA binding factor, *Science* 248:1646–50, 1990.

Martianov I, Viville S, Davidson I: RNA polymerase II transcription in murine cells lacking the TATA binding protein, *Science* 298:1036–9, 2002.

Nikolov DB, Hu SH, Lin J, Gasch A, Hoffmann A et al: Crystal structure of TFIID TATA-box binding protein, *Nature* 360:40–6, 1992.

Nikolov DB, Chen H, Halay ED, Hoffman A, Roeder RG et al: Crystal structure of a human TATA box-binding protein/TATA element complex, *Proc Natl Acad Sci USA*, 93(10):4862–7, May 14, 1996.

Peterson MG, Tanese N, Pugh BF, Tjian R: Functional domains and upstream activation properties of cloned human TATA binding protein, *Science* 248:1625–30, 1990.

Rosen DR, Trofatter JA, Brown RH, Jr: Mapping of the human TATA-binding protein gene (TBP) to chromosome 6qter, *Cytogenet Cell Genet* 69:279–80, 1995.

Veenstra GJC, Weeks DL, Wolffe AP: Distinct roles for TBP and TBP-like factor in early embryonic gene transcription in Xenopus, *Science* 290:2312–4, 2000.

2.2. Control of the Activity of *Trans*-Acting Factors

Holmberg CI, Tran SE, Eriksson JE, Sistonen L: Multisite phosphorylation provides sophisticated regulation of transcription factors, *Trends Biochem Sci* 27:619–27, 2002.

Firulli AB: A HANDful of questions: The molecular biology of the heart and neural crest derivatives (HAND)-subclass of basic helix-loop-helix transcription factors, *Gene* 312:27–40, 2003.

Falke D, Juliano RL: Selective gene regulation with designed transcription factors: Implications for therapy, *Curr Opin Mol Ther* 5:161–6, 2003.

Roeder RG: Transcriptional regulation and the role of diverse coactivators in animal cells, *FEBS Lett* 579:909–15, 2005.

McDonnell DP, Norris JD: Connections and regulation of the human estrogen receptor, *Science* 296:1642–4, 2002.

Della Fazia MA, Servillo G, Sassone-Corsi P: Cyclic AMP signalling and cellular proliferation: Regulation of CREB and CREM, *FEBS Lett* 410:22–4, 1997.

Sassone-Corsi P: Transcription factors responsive to cAMP, *Annu Rev Cell Dev Biol* 11:355–77, 1995.

2.3. Chromatin Modification

van Attikum H, Gasser SM: The histone code at DNA breaks: a guide to repair? *Nature Rev Mol Cell Biol* 6:757–65, 2005.

Herman JG, Baylin SB: Gene silencing in cancer in association with promoter hypermethylation, *N Engl J Med* 349:2042–54, 2003.

Thiriet C, Hayes JJ: Chromatin in need of a fix: Phosphorylation of H2AX connects chromatin to DNA repair, *Mol Cell* 18:617–22, 2005.

Martin C, Zhang Y: The diverse functions of histone lysine methylation, *Nature Rev Mol Cell Biol* 6:838–49, 2005.

Henikoff S, Ahmad K: Assembly of variant histones into chromatin, *Annu Rev Cell Dev Biol* 21:133–53, 2005.

2.4. DNA Modification

Baylin SB: DNA methylation and gene silencing in cancer, *Nat Clin Pract Oncol* 2 (Suppl 1):S4–11, 2005.

Herman JG, Baylin SB: Gene silencing in cancer in association with promoter hypermethylation, *N Engl J Med* 349:2042–54, 2003.

Jones PA, Laird PW: Cancer epigenetics comes of age, *Nature Genet* 21:163–7, 1999.

Jones PA, Baylin SB: The fundamental role of epigenetic events in cancer, *Nature Rev Genet* 3:415–28, 2002.

Herman JG: Hypermethylation of tumor suppressor genes in cancer, *Semin Cancer Biol* 9:359–67, 1999.

Holliday R: Epigenetic inheritance based on DNA methylation, *EXS* 64:452–68, 1993.

2.5. 5′-Terminal Capping of Pre-mRNA

Parker R, Song H: The enzymes and control of eukaryotic mRNA turnover, *Nat Struct Mol Biol* 11:121–7, 2004.

Dunckley T, Parker R: The DCP2 protein is required for mRNA decapping in Saccharomyces cerevisiae and contains a functional MutT motif, *EMBO J* 18:5411–22, 1999.

Wang Z, Jiao X, Carr-Schmid A, Kiledjian M: The hDcp2 protein is a mammalian mRNA decapping enzyme, *Proc Natl Acad Sci USA* 99:12663–8, 2002.

Lykke-Andersen J: Identification of a human decapping complex associated with hUpf proteins in nonsense-mediated decay, *Mol Cell Biol* 22:8114–21, 2002.

van Dijk E, Cougot N, Meyer S, Babajko S, Wahle E et al: Seraphin, Human Dcp2: A catalytically active mRNA decapping enzyme located in specific cytoplasmic structures, *EMBO J* 21:6915–24, 2002.

Wang Z, Kiledjian M: Functional link between the mammalian exosome and mRNA decapping, *Cell* 107:751–62, 2001.

Liu H, Rodgers ND, Jiao X, Kiledjian M: The scavenger mRNA decapping enzyme DcpS is a member of the HIT family of pyrophosphatases, *EMBO J* 21:4699–708, 2002.

Lima CD, Wang LK, Shuman S: Structure and mechanism of yeast RNA triphosphatase: An essential component of the mRNA capping apparatus, *Cell* 99:533–43, 1999.

Changela A, Ho CK, Martins A, Shuman S, Mondragon A: Structure and mechanism of the RNA triphosphatase component of mammalian mRNA capping enzyme, *EMBO J* 20:2575–86, 2001.

Håkansson K, Doherty AJ, Shuman S, Wigley DB: X-ray crystallography reveals a large conformational change during guanyl transfer by mRNA capping enzymes, *Cell* 89:545–53, 1997.

Shuman S, Lima CD: The polynucleotide ligase/RNA capping enzyme superfamily of covalent nucleotidyltransferases, *Curr Opin Struct Biol* 14:757–64, 2004.

Håkansson K, Wigley DB: Structure of a complex between a cap analogue and mRNA guanylyl transferase demonstrates the structural chemistry of RNA capping, *Proc Natl Acad Sci USA* 95:1505–10, 1998.

Ho CK, Sriskanda V, McCracken S, Bentley D, Schwer B et al: The gaunylyltransferase domain of mammalian mRNA capping enzyme binds to the phosphorylated carboxy-terminal domain of RNA polymerase II, *J Biol Chem* 273:9577–85, 1998.

Bentley D: The mRNA assembly line: transcription and processing machines in the same factory, *Curr Opin Cell Biol* 14:336–42, 2002.

She M, Decker CJ, Sundramurthy K, Liu Y, Chen N et al: Crystal structure of Dcp1p and its functional implications in mRNA decapping, *Nat Struct Mol Biol* 11:249–56, 2004.

van Dijk E, Le Hir H, Seraphin B: DcpS can act in the 5′-3′ mRNA decay pathway in addition to the 3′-5′ pathway, *Proc Natl Acad Sci USA* 100:12081–6, 2003.

Gu M, Lima CD: Processing the message: Structural insights into capping and decapping mRNA, *Curr Opin Struct Biol* 15:99–106, 2005.

2.6. Polyadenylation

Colgan DF, Manley JL: Mechanism and regulation of mRNA polyadenylation, *Genes Dev* 11:2755–66, 1997.

Zhao J, Hyman L, Moore C: Formation of mRNA 3′ ends in eukaryotes: Mechanism, regulation, and interrelationships with other steps in mRNA synthesis, *Microbiol Mol Biol Rev* 63:405–45, 1999.

Shatkin AJ, Manley JL: The ends of the affair: Capping and polyadenylation, *Nat Struct Biol* 7:838–42, 2000.

de Vries H, Ruegsegger U, Hubner W, Friedlein A, Langen H et al: Human pre-mRNA cleavage factor IIm contains homologs of yeast proteins and bridges two other cleavage factors, *EMBO J* 19:5895–904, 2000.

Dichtl B, Keller W: Recognition of polyadenylation sites in yeast pre-mRNAs by cleavage and polyadenylation factor, *EMBO J* 20:3197–209, 2001.

Zhao J, Hyman L, Moore C: Formation of mRNA 3′ ends in eukaryotes: Mechanism, regulation, and interrelationships with other steps in mRNA synthesis, *Microbiol Mol Biol Rev* 63:405–45, 1999.

Buratowski S: Connections between mRNA 3′ end processing and transcription termination, *Curr Opin Cell Biol* 17:257–61, 2005.

Proudfoot N: New perspectives on connecting messenger RNA 3′ end formation to transcription, *Curr Opin Cell Biol* 16:272–8, 2004.

Proudfoot N, O'Sullivan J: Polyadenylation: A tail of two complexes, *Curr Biol* 12:R855–7, 2002.

2.7. Pre-mRNA Splicing

Black DL: Mechanisms of alternative pre-messenger RNA splicing, *Annu Rev Biochem* 72:291–336, 2003.

Chan SP, Kao DI, Tsai WY, Cheng SC: The Prp19p-associated complex in spliceosome activation, *Science* 302:279–82, 2003.

Fairbrother WG, Yeh RF, Sharp PA, Burge CB: Predictive identification of exonic splicing enhancers in human genes, *Science* 297:1007–13, 2002.

Konarska MM, Query CC: Insights into the mechanisms of splicing: More lessons from the ribosome, *Genes Dev* 19:2255–60, 2005.

Makarov EM, Makarova OV, Urlaub H, Gentzel M, Will CL et al: Small nuclear ribonucleoprotein remodeling during catalytic activation of the spliceosome, *Science* 298:2205–8, 2002.

Maniatis T, Tasic B: Alternative pre-mRNA splicing and proteome expansion in metazoans, *Nature* 418:236–43, 2002.

Moore MJ, Sharp PA: Evidence for two active sites in the spliceosome provided by stereochemistry of pre-mRNA splicing, *Nature* 365:364–8, 1993.

Nelson KK, Green MR: Mechanism for cryptic splice site activation during pre-mRNA splicing, *Proc Natl Acad Sci* 87:6253–7, 1990.

Ogle JM, Ramakrishnan V: Structural insights into translational fidelity, *Annu Rev Biochem* 74:129–77, 2005.

Zhou Z, Licklider LJ, Gygi SP, Reed R: Comprehensive proteomic analysis of the human spliceosome, *Nature* 419:182–5, 2002.

Jurica MS, Moore MJ: Pre-mRNA splicing: Awash in a sea of proteins, *Mol Cell* 12:5–14, 2003.

Staley JP, Guthrie C: Mechanical devices of the spliceosome: Motors, clocks, springs, and things, *Cell* 92:315–26, 1998.

2.8. mRNA Transport

Kindler S, Wang H, Richter D, Tiedge H: RNA transport and local control of translation, *Annu Rev Cell Dev Biol* 21:223–45, 2005.

Ainger K, Avossa D, Diana AS, Barry C, Barbarese E et al: Transport and localization elements in myelin basic protein mRNA, *J Cell Biol* 138:1077–87, 1997.

Chartrand P, Singer RH, Long RM: RNP localization and transport in yeast, *Annu Rev Cell Dev Biol* 17:297–310, 2001.

Chennathukuzhi V, Morales CR, El-Alfy M, Hecht NB: The kinesin KIF17b and RNA-binding protein TB-RBP transport specific cAMP-responsive element modulator-regulated mRNAs in male germ cells, *Proc Natl Acad Sci USA* 100:15566–71, 2003.

Farina KL, Singer RH: The nuclear connection in RNA transport and localization, *Trends Cell Biol* 12:466–72, 2002.

Goldstein LS, Yang Z: Microtubule-based transport systems in neurons: The roles of kinesins and dyneins, *Annu Rev Neurosci* 23:39–71, 2000.

Hachet O, Ephrussi A: Splicing of oskar RNA in the nucleus is coupled to its cytoplasmic localization, *Nature* 428:959–63, 2004.

Han JR, Yiu GK, Hecht NB: Testis/brain RNA-binding protein attaches translationally repressed and transported mRNAs to microtubules, *Proc Natl Acad Sci USA* 92:9550–54, 1995.

Job C, Eberwine J: Localization and translation of mRNA in dendrites and axons, *Nature Rev Neurosci* 2:889–98, 2001.

Johnstone O, Lasko P: Translational regulation and RNA localization in Drosophila oocytes and embryos, *Annu Rev Genet* 35:365–406, 2001.

Kapp LD, Lorsch JR: The molecular mechanics of eukaryotic translation, *Annu Rev Biochem* 73:657–704, 2004.

Kloc M, Zearfoss NR, Etkin LD: Mechanisms of subcellular mRNA localization, *Cell* 108:533–44, 2002.

Kress TL, Yoon YJ, Mowry KL: Nuclear RNP complex assembly initiates cytoplasmic RNA localization, *J Cell Biol* 165:203–11, 2004.

2.9. Regulation of Protein Translation

Dean KA, Aggarwal AK, Wharton RP: Translational repressors in Drosophila, *Trends Genet* 18:572–7, 2002.

Gray NK, Wickens M: Control of translation initiation in animals, *Annu Rev Cell Dev Biol* 14:399–458, 1998.

Dey M, Cao C, Dar AC, Tamura T, Ozato K et al: Mechanistic link between PKR dimerization, autophosphorylation, and eIF2alpha substrate recognition, *Cell* 122:901–13, 2005.

Johnstone O, Lasko P: Translational regulation and RNA localization in Drosophila oocytes and embryos, *Annu Rev Genet* 35:365–406, 2001.

Okabe M et al: Translational repression determines a neuronal potential in Drosophila asymmetric cell division, *Nature* 411:94–8, 2001.

Hershey JWB: Translational control in mammalian cells, *Annu Rev Biochem* 60:717–55, 1991

Holcik M, Sonenberg N: Translational control in stress and apoptosis, *Nat Rev Mol Cell Biol* 6:318–27, 2005.

Dar AC, Dever TE, Sicheri F: Higher-order substrate recognition of eIF2alpha by the RNA-dependent protein kinase PKR, *Cell* 122:887–900, 2005.

Su Q, Wang S, Baltzis D, Qu LK, Wong AH et al: Tyrosine phosphorylation acts as a molecular switch to full-scale activation of the eIF2 alpha RNA-dependent protein kinase, *Proc Natl Acad Sci USA* 103:63–8, 2006.

3

STRUCTURE AND FUNCTION OF CELLULAR COMPONENTS

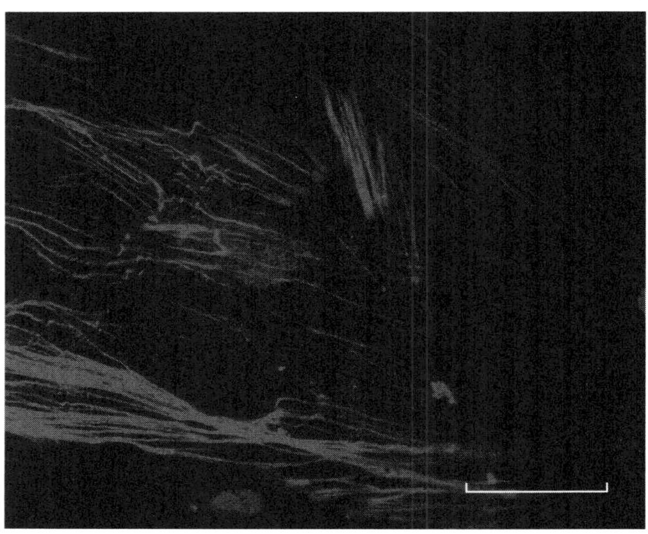

α-Actin filaments in vascular smooth muscle cells derived from the mouse aorta. Smooth muscle cells were collected from the medial layer of the mouse aorta and cultured for 10 days. The α-actin filaments were labeled with an anti-smooth-muscle α actin antibody (red in color) and observed by fluorescence microscopy. Cell nuclei were labeled with Hoechst 33258 (blue in color). Scale bar: 5 μm. See color insert.

Bioregenerative Engineering: Principles and Applications, by Shu Q. Liu

A mammalian cell is composed of numbers of subcellular organelles, including the cell membrane, cytoskeleton, smooth and rough endoplasmic reticulum, Golgi apparatus, lysosomes, peroxisomes, mitochondria, and nucleus. A *cell membrane* is a phospholipid bilayer, which encloses cell contents and separates a cell into different compartments. The *cytoskeleton* is constituted with three distinct elements, including actin filaments, microtubules, and intermediate filaments, which not only give a cell shape, strength, and elasticity but also regulate various cellular functions. *Endoplasmic reticulum* is the site where proteins and phospholipids are synthesized. The *Golgi apparatus* is an organelle in which proteins are processed and modulated. *Lysosomes* contain digestive enzymes, participating in the degradation of engulfed molecules or microorganisms. *Peroxisomes* contain enzymes for the mediation of oxidative reactions. *Mitochondria* are machineries that generate and store energy in the form of ATP. The *nucleus* contains chromosomes and is the center for the storage and processing of genetic information. It becomes clear that each cellular organelle possesses distinct structure and function, yet all cell organelles work together in a highly coordinated manner, ensuring appropriate regulation of cellular activities and functions. In this chapter, the structure, organization, and function of major cellular organelles and compartments are briefly reviewed.

CELL MEMBRANE [3.1]

The cell membrane is composed of lipids and proteins. As discussed in Chapter 1, lipids are amphipathic in nature (i.e., each molecule contains a polar hydrophilic and a nonpolar hydrophobic end) and can spontaneously form bilayers when mixed with an aqueous solution. The most abundant lipids are phospholipids in the cell membrane. Each phospholipid molecule contains a polar hydrophilic head and two nonpolar hydrophobic tails. In addition, cholesterol molecules can be found in a cell membrane. The membrane of a mammalian cell contains about 1×10^9 lipid molecules. Lipid molecules constitute about half of the membrane mass, while the remaining half is primarily proteins. The lipid composition is asymmetric between the two lipid layers of the cell membrane. For instance, glycosphingolipids are found primarily in the external layer, whereas phosphatidylserine is in the internal layer. The primary functions of cell membranes are to separate cellular contents from the extracellular space, create a suitable internal environment for intracellular activities, and establish subcellular compartments for various metabolic and signaling processes.

A lipid bilayer is a fluid-like structure. Lipid molecules can move laterally or diffuse within a lipid monolayer, but cannot change the molecular polarity or flip from one lipid layer to the other. The fluid-like feature of lipid bilayers is dependent on the composition of the cell membrane. For instance, cholesterol molecules reduce the fluidity of cell membranes, and thus enhance the membrane rigidity. The fluidity of a cell membrane ensures dynamic movement of membrane components, including not only lipids but also proteins. The movement of membrane molecules is critical to the function of these molecules as well as the cell. For instance, integrins move toward the leading edge of cell migration and participate in the construction of focal adhesion contacts, regulating cell attachment to the substrate (Fig. 3.1). Growth factor receptors move dynamically, resulting in the redistribution of the receptors to regions that require increased signal inputs from growth factors.

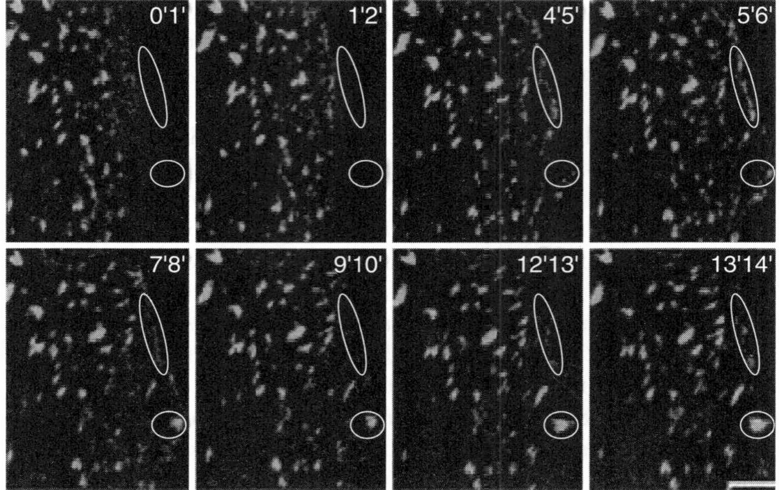

Figure 3.1. Dynamic formation of β3 integrin complexes in porcine arterial endothelial cells. Endothelial cells were transfected with a GFP-β3 integrin gene and cultured to confluence. Cell wound was created by mechanical scraping, which induces cell migration. The images were taken from migrating endothelial cells. Note that new integrin aggregates form at the leading edge of the migrating cells (within the ovals). The times of the sequential images are indicated at the upper right corners. Scale bar: 5 μm. (Reprinted from Zaidel-Bar R et al: *J Cell Sci* 116:4605–13, 2003 by permission of The Company of Biologists Ltd.)

The cell membrane contains various types and amounts of proteins, depending on the type and function of the cell. For instance, a myelin membrane, which encloses and protects the nerve axon, contains proteins about 25% of the membrane mass, whereas a cell membrane that is involved in extensive molecular transport and ligand–receptor interaction may contain up to 75% proteins. Cell membrane proteins may serve as ligand receptors, ion pumps, water and ion channels, or molecule carriers. Membrane proteins can be divided into several classes based on the structure and relationship with the lipid bilayer. One type is transmembrane proteins, which pass through the cell membrane and consist of three domains: the extracellular, transmembrane, and intracellular domains. The extracellular and intracellular domains are usually hydrophilic, whereas the transmembrane domain is hydrophobic. The hydrophilic domains can interact with water-soluble proteins, while the hydrophobic domain interacts with the fatty acid tails of membrane lipids via covalent bonds, serving as an anchoring structure for the protein. The second type of membrane protein is found at the external surface of a cell membrane. These proteins attach to the lipid bilayer via the linkage of oligosaccharides. The third type of protein attaches to the intracellular side of the cell membrane via covalent bonds with fatty acids. In addition, some proteins attach to membrane proteins via noncovalent bonds. The structural relationship between a protein and the cell membrane usually determines the protein function. For instance, transmembrane proteins are responsible for molecular transport across the cell membrane and signal transduction from extracellular ligands to intracellular signaling pathways. Proteins attached to the cytosolic side of the cell membrane usually serve as signaling molecules, which relay signals from transmembrane protein receptors.

CYTOSKELETON

The cell contains a filamentous framework, known as the *cytoskeleton*. There are three cytoskeletal elements: the actin filaments, intermediate filaments, and microtubules. These filaments not only determine the shape and mechanical strength but also participate in the regulation of cellular activities, such as cell adhesion, division, migration, and apoptosis. The structure and function of these filaments are briefly discussed here.

Actin Filaments

Structure and Organization of Actin Filaments [3.2]. An *actin filament* is a helical structure of 8 nm in diameter and is established via polymerization of actin monomers. Each actin monomer contains about 375 amino acid residues with a molecular size about 43 kDa. In mammalian cells, there exist several isoforms of actin (see examples listed in Table 3.1), including the α and β isoforms in muscular cells and β and γ isoforms in nonmuscular cells. The α type of actin constitutes the contractile actin filaments in skeletal, cardiac, and smooth muscle cells. The β and γ types of actin participate in the constitution of the cytoskeleton. Actin filaments with various actin isoforms are localized to different compartments in both muscular and nonmuscular cells. For instance, in nonmuscular cells, β-actin is primarily found near the edge of the cell membrane, whereas γ-actin constitutes stress fibers, which are distributed more uniformly. An actin filament is a polarized structure. When an actin filament is bound with myosin molecules, an array of asymmetric arrowhead-like structures appears under an electron microscope. The end of an actin filament consistent with the arrowhead is defined as the pointed end, whereas the other end is defined the barbed end.

Actin monomers can be self-assembled or polymerized into actin filaments through biochemical reactions (Fig. 3.2). *Actin polymerization* is accomplished in several steps, including actin nucleation, filament growth, and ATP hydrolysis. *Actin nucleation* is a process that induces the formation of actin trimers. These trimeric actin structures, known as *actin nuclei*, serve as initiators for actin polymerization or filament growth. In addition, actin polymerization can be initiated from the barbed end of grown actin filaments or random sites along the side of actin filaments (Fig. 3.2). The addition of an ATP-actin to an actin nucleus or an actin filament triggers hydrolysis of ATP into ADP and phosphate. The phosphate group dissociates from the actin, leaving a newly added actin molecule with a tightly bound ADP.

An actin filament can be simultaneously polymerized and depolymerized at both ends. Under a steady physiological condition, the addition of actin subunits to the barbed end of an actin filament is counterbalanced by the dissociation of actin subunits from the pointed end, resulting in a relatively constant density for actin monomers and filaments. However, the rate of polymerization and depolymerization may change in response to environmental alterations. For instance, an increase in the concentration of ATP and the presence of cations lower the critical level of actin monomers, enhancing actin polymerization. Actin monomers above a critical concentration can be all assembled into actin filaments.

Actin-Binding Proteins. Actin polymerization and depolymerization are regulated by numbers of actin-binding proteins. These proteins are classified into various groups on the basis of their functions, including actin monomer-binding proteins, actin filament-

TABLE 3.1. Characteristics of Selected Actin Isoforms*

Proteins	Alternative Names	Amino Acids	Molecular Weight (kDa)	Expression	Functions
Actin α, cardiac	Smooth muscle actin, cardiac actin α, actin α	377	42	Cardiomyocytes and smooth muscle cells	Forming contractile actin filaments in cardiomyocytes and smooth muscle cells
Actin α, skeletal 1	Actin α1	377	42	Skeletal muscle	Forming contractile actin filaments in skeletal muscle cells
Actin α2	Vascular smooth muscle actin, vascular smooth muscle actin α, vascular smooth muscle actin α2, actin 2α	377	42	Vascular smooth muscle cells	Forming actin contractile filaments in vascular smooth muscle cells
Actin β	Cytoskeletal actin β	375	42	Primarily nonmuscular cells	Constituting the cytoskeleton of nonmuscular cells, regulating the motility of nonmuscular cells
Actin γl	Cytoskeletal actin γ	375	42	Primarily nonmuscular cells	Cytoplasmic actin found in nonmuscular cells, constituting cytoskeleton, and mediating cell motility
Actin γ2 enteric smooth muscle	Actin α3, smooth muscle actin γ	376	42	Intestinal smooth muscle cells	Constituting the cytoskeleton of intestinal smooth muscle cells

*Based on bibliography 3.2.

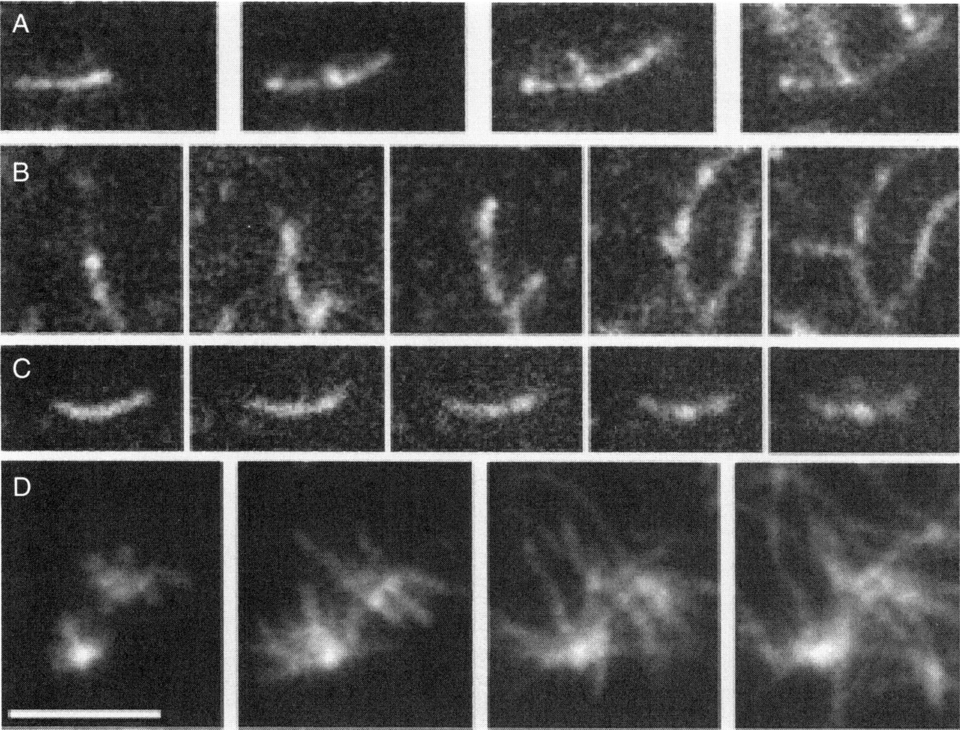

Figure 3.2. Actin filament polymerization and branching. Monomer actin molecules were prepared from rabbit skeletal muscle and labeled on Cys-374 with rhodamine. Actin polymerization was induced in the presence of 20% rhodamine-actin and observed by total internal reflection fluorescence microscopy (TIRFM). The images were subsequently captured at 100, 130, 170, and 210 s after initiating actin polymerization. Scale bar: 4 μm. (Reprinted by permission from Amann KJ et al: *Proc Natl Acad Sci USA* 98:15009–13, copyright 2001, National Academy of Sciences, USA.)

capping proteins, actin filament-binding proteins, actin filament-severing proteins, and actin filament crosslinking proteins.

Actin Monomer-Binding Proteins [3.3]. The family of *actin monomer-binding proteins* (see Table 3.2) includes several molecules, including β-thymosins, cofilins, profilins, and formins, which bind actin monomers and regulate the activities of the actin molecules. *β-Thymosins* are molecules that primarily bind to and sequester ATP-actin monomers, and thus inhibit actin polymerization. *Cofilins* bind ADP-actin with high affinity and destabilize actin filaments. However, a controversial role of cofilins has been observed. *Profilins* bind to ADP- and ATP-free actin monomers and play a role in sequestration of actin monomers. Profilins also inhibit nucleation and elongation at the pointed end of an actin filament, but do not influence the nucleation and elongation at the barbed end. *Formin* is a homodimer composed of formin homology 1 (FH1) and formin homology 2 (FH2) domains. The FH2 domain can bind to monomer actin and induce the nucleation and polymerization of actin filaments. Furthermore, The FH2 domain can bind to the barbed

TABLE 3.2. Characteristics of Selected Actin Monomer-Binding Proteins*

Proteins	Alternative Names	Amino Acids	Molecular Weight (kDa)	Expression	Functions
Cofilin 1	CFL1	166	19	Ubiquitous	Binding and depolymerizing filamentous F-actin, inhibiting the polymerization of monomeric G-actin; however, the role of cofilins is controversial
Cofilin 2	CFL2, muscle cofilin	166	19	Skeletal muscle, heart, brain, lung, liver, pancreas, kidney	Same as those for cofilins 1
Profilin	PFN1	140	15	Ubiquitous	Binding to ADP- and ATP-free actin monomers, and sequestering actin monomers
β-Thymosin	Thymosin β4 X chromosome	44	5	Ubiquitous	Sequestering actin monomers, inhibiting actin polymerization, enhancing cardiac cell survival, migration, and regeneration
Formin		844	95	Ubiquitous	Promoting nucleation and polymerization of actin filaments

*Based on bibliography 3.3.

end of actin filaments and promote the elongation of the filaments. The FH1 domain can bind to the actin-binding protein profilin. This process enhances the elongation of actin filaments.

Actin Filament-Capping Proteins [3.4]. Actin filament-capping proteins (see Table 3.3) are molecules that bind either the pointed or the barbed end of actin filaments and prevent actin polymerization or depolymerization. This family of proteins includes gelsolins, heterodimeric capping proteins, the actin-related protein (Arp)2/3 complex, tropomyosin, nebulin, and tropomodulin. *Gelsolins* are capable of binding to the barbed end and the side of actin filaments and inhibiting actin polymerization. *Heterodimeric capping proteins* bind and cap the barbed end of actin filaments, and impose effects similar to those of gelsolins. *Arp 2/3* is a complex of Arp 2 and Arp 3, which binds and caps the pointed end of an actin filament and promotes the attachment of the capped end to a different actin filament and the formation of actin filament branches. It has been shown that this process is regulated by the ρ family GTPases. ρ GTPases activate a protein known as the *Wiskott–Aldrich syndrome protein* (WASP), which in turn activates the Arp2/3 complex. Other actin filament-binding proteins, including tropomyosin, nebulin, capZ, and tropomodulin, bind to the side or ends of actin filaments and contribute to the stability of the filaments. *Tropomyosin* binds the side of actin filaments, induces an increase in the stiffness of the filaments, and stimulates the interaction of actin filaments with myosin. *Nebulin* is found in skeletal muscle cells and plays a role in the control of the length of actin filaments. *Tropomodulin* binds to the pointed end and enhances the stability of actin filaments.

Actin Filament-Severing Proteins. Actin filament-severing proteins include gelsolins, fragmin/severin, and cofilins. These molecules are able to sever actin filaments into short fragments and promote actin filament depolymerization. *Gelsolins* are also capping molecules for the barbed end of actin filaments. *Cofilins* can also bind to actin monomers.

Actin Filament-Crosslinking Proteins [3.5]. Actin filament crosslinking proteins (Table 3.4) include α-actinin, fimbrin, villin, and filamin. These molecules can bind simultaneously to multiple actin filaments and induce crosslink of actin filaments. *α-Actinin* is associated with actin stress fibers and the Z-disk of striated muscular actin fibers. In addition, α-actinin is a constituent of focal adhesion contacts, structures that mediate cell attachment and migration. This molecule possesses multiple functions. *Fimbrin* can bind and crosslink actin filaments in microvilli. *Villin* has a similar function as fimbrin. *Filamin* can not only crosslink actin filaments but also anchor actin filaments to integrins, major constituents of focal adhesion contacts. All these actin filament crosslinking molecules enhance the stability of actin filaments.

Regulation of Actin Assembly and Disassembly [3.6]. In mammalian cells, actin filaments undergo a dynamic turnover process, or simultaneous assembly and disassembly, under physiological conditions. The rate of turnover is dependent on cell types. Nonmuscular cells exhibit actin filament turnover at a timescale of minutes, while muscular cells demonstrate actin filament turnover at a scale of days. Actin polymerization (assembly) and depolymerization (disassembly) can be observed in living cells with fluorescent marker-tagged actin monomers. The fluorescent markers can be incorporated into actin filaments. Following photobleaching of fluorescent actin filaments, the bleached region

TABLE 3.3. Characteristics of Selected Actin Filament-Capping Proteins*

Proteins	Alternative Names	Amino Acids	Molecular Weight (kDa)	Expression	Functions
Gelsolin	Actin depolymerizing factor (ADF), Brevin	782	86	Ubiquitous	Inhibiting actin polymerization, severing actin filaments, and promoting actin filament depolymerization
Actin capping protein α1	Muscle Z-line actin filament capping protein α1, CAPZA1, F-actin capping protein α1 subunit	286	33	Skeletal muscle, red blood cells, placenta	Found at the Z line of muscular cells, binding to barbed end of actin filaments, and inhibiting actin polymerization
Actin-related protein 2	ARP2, actin-like protein 2	394	45	Ubiquitous	Constituting the ARP2/3 complex and participating in regulation of cell shape and motility via actin assembly and protrusion
Actin-related protein 3	ARP3, actin-like protein 3	418	47	Ubiquitous	Constituting the ARP2/3 complex and regulating actin assembly
Tropomyosin 1	Tropomyosin skeletal muscle α, tropomyosin 1α chain, α tropomyosin	284	33	Skeletal muscle	Binding to actin filaments in striated muscle cells, stabilizing actin filaments, and regulating calcium-dependent interaction of actin filaments with myosin molecules during muscle contraction
Nebulin	NEB	6669	773	Skeletal muscle cells	Coexisting with thick and thin filaments within sarcomeres of skeletal muscle and playing a critical role in both integrity and stability of contractile filaments
Tropomodulin	Tropomodulin 1, erythrocyte tropomodulin, E-tropomodulin	359	41	Skeletal muscle, heart, brain, lung, liver, kidney, pancreas	Binding to the pointed end and enhancing stability of actin filaments

*Based on bibliography 3.4.

TABLE 3.4. Characteristics of Selected Actin Filament Crosslinking Proteins*

Proteins	Alternative Names	Amino Acids	Molecular Weight (kDa)	Expression	Functions
Actinin α1	ACTN1	892	103	Nonmuscular cells	Interacting with actin filaments and regulating the assembly of actin filaments and focal adhesion contacts
Actinin α2	ACTN2, α actinin skeletal muscle isoform 2, F-actin-crosslinking protein	894	104	Skeletal muscle, cardiomyocytes	A muscle-specific α actinin that anchors actin filaments to the Z disks
Fimbrin	Intestine-specific plastin, I-plastin, plastin 1, accumentin	629	70	Intestine, lung, kidney, leukocytes	Binding to and crosslinking actin filaments
Villin	Villin 1	827	93	Intestine, kidney	Inducing crosslinking of actin filaments
Filamin A	Filamin α, filamin 1 (FLN1), actin-binding protein 280 (ABP280), nonmuscle filamin, α-filamin, endothelial actin-binding protein	2647	281	Primarily nonmuscular cells	Inducing crosslinking of actin filaments and regulating the organization and remodeling of actin cytoskeleton by interacting with integrins and transmembrane receptors
Filamin B	β filamin, filamin 1 (actin-binding protein-280)-like, actin-binding-like protein, truncated actin-binding protein Actin-binding protein 276/278, ABP276/278 truncated actin-binding protein	2602	278	Heart, skeletal muscle, brain, lung, liver, kidney, pancreas, uterus, ovary	Inducing actin filament crosslink in muscular and nonmuscular cells and regulating the organization of actin filaments

*Based on bibliography 3.5.

can be replaced with fluorescent actin filaments, suggesting dynamic reassembly of actin filaments. In nonmuscular cells, there exists a relatively high concentration of unpolymerized actin monomers (50–100 μM). Such a concentration allows rapid actin polymerization in response to stimulations that initiate cell adhesion and migration. Indeed, the concentration of actin monomers is a critical factor that controls actin filament assembly and disassembly.

The dynamics of actin assembly–disassembly is regulated by actin regulatory and binding proteins. Sequestration of actin monomers and the capping of actin filaments at the ends are two mechanisms that control the rate of actin filament assembly. As discussed above, profilin and thymosin can bind and sequester actin monomers and reduce the concentration of free actin monomers, suppressing the polymerization of actin filaments. Profilin- or thymosin-bound actin monomers have reduced capability of initiating nucleation. An increase in the activity of actin filament-capping proteins promotes actin polymerization.

Actin filaments are found in all mammalian cells and are organized into various patterns and structures. For instance, actin filaments form a network in the cortical region of the cell, while forming fiber bundles within filopodia or microvilli. The pattern formation of actin filaments is a process that may be regulated by the Rho family of GTPases, which includes Rho, Rac, and Cdc42 (see Table 3.5). These molecules have been shown to regulate distinct processes of actin assembly. Activated *Cdc42* stimulates the formation of filopodia, *Rho* enhances the formation of actin "stress fibers," while *Rac* promotes the formation of cortical network of actin filaments (Fig. 3.3). Although the signaling path-

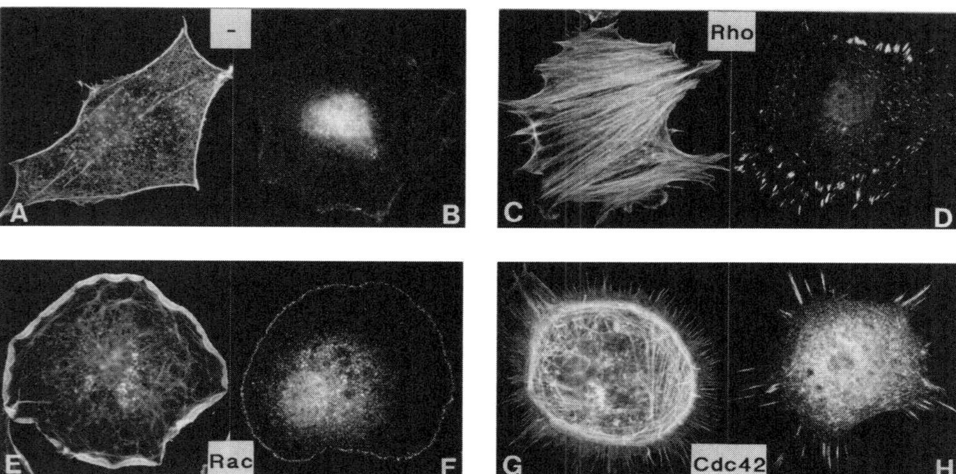

Figure 3.3. Influence of Rho, Rac, and Cdc42 on the organization of actin filaments and morphology of cells: (A,B) quiescent serum-starved Swiss 3T3 fibroblasts labeled for actin filaments and vinculin; (C,D) treatment of cells with lysophosphatidic acid, a growth stimulator, which activates Rho, leading to the formation of organized actin filaments or stress fibers (C) and focal adhesion contacts (D); (E,F) microinjection of Rac induces the formation of lamellipodia (E) and focal adhesion contacts (F); (G,H) microinjection of FGD1, an exchange factor for Cdc42, leads to formation of filopodia (G) and the focal adhesion contacts (H). (Reprinted by permission from Hall A: *Science* 279:509–14, 1998.)

TABLE 3.5. Characteristics of Selected Factors that Regulate the Formation of Actin Filaments*

Proteins	Alternative names	Amino Acids	Molecular Weight (kDa)	Expression	Functions
RhoA	RHOA, aplysia Ras-related homolog 12, oncogene RhoH12, RHOH12, RHO12, transforming protein RhoA, Ras homolog gene family member A	193	22	Leukocytes, platelets	Enhancing the formation of actin stress fibers, regulating cell migration, polarity, and protrusion
RhoB	Oncogene RhoH6, RhoH6, aplysia Ras-related homolog 6	196	22	Nervous system, macrophage, lung	Regulating the formation of actin filaments and assembly of focal adhesion contacts, promoting cell adhesion, vesicle trafficking, MAPK signaling, and immunity
RhoC	Ras homolog gene family member C, aplysia Ras-related homolog 9, oncogene RhoH9, transforming protein RhoC	193	22	Leukocytes, lung, breast, carcinoma cells	Enhancing the formation of actin filaments, regulating cell motility, and mediating tumorigenesis
RAC1	p21-Rac1, Ras-related C3 botulinum toxin substrate 1, Rho family small GTP-binding protein RAC1, Ras-like protein TC25, TC-25	211	23	Ubiquitous	A small Ras GTP-binding protein that regulates cell survival, growth, cytoskeletal reorganization, and the activation of protein kinases
Cdc42	Cell division cycle 42, G25K, GTP-binding protein 25 kDa	191	21		A small ρ GTPase that regulates cell morphology, migration, endocytosis, polarity, and cell cycle progression; also regulates actin polymerization via interaction with neural Wiskott–Aldrich syndrome protein (N-WASP), which subsequently activates the Arp2/3 protein complex

*Based on bibliography 3.6.

ways for these molecules remain poorly understood. these observations provide insights into the mechanisms by which actin filaments form distinct patterns.

Actin assembly and disassembly are regulated by extracellular factors. For instance, *growth factors* and *cytokines* stimulate cell attachment and migration, which are associated with increased actin assembly. These observations suggest a role for growth factors and cytokines in the regulation of actin polymerization or depolymerization. However, exact mechanisms remain poorly understood. In addition, fluid shear stress has been shown to influence actin assembly in vascular endothelial cells. In cell culture models, the introduction of fluid shear stress to endothelial cells enhances actin filament assembly, forming actin "stress fibers." Shear stress-induced deformation of cell membrane receptors or other cell structures may play a role in the initiation of such a process. However, the signaling pathways that transduce shear stress signals remain to be identified.

Function of Actin Filaments [3.7]. Actin filaments participate in a number of functions, including cell contraction, migration, and division. In contractile cells, including skeletal, cardiac, and smooth muscle cells, actin filaments interact with myosin molecules, causing filament sliding and cell contraction, a fundamental process for force generation. In noncontractile cells, directed actin polymerization contributes to regional extension of cell membrane, a primary step in cell migration. The interaction of actin filaments and myosin molecules provide forces that induce cell traction and movement. During cell division, actin filaments form a ring-shaped structure between two premature daughter cells, known as the *contractile ring*, underneath the plasma membrane. Contraction of the ring is initiated following cell mitosis. Such an activity separates the mother cytoplasm into two daughter compartments. While chromosome separation is defined as *mitosis*, cytoplasmic separation is defined as *cytokenesis*.

BIBLIOGRAPHY

3.1. Cell Membrane

Mineo C, Gill GN, Anderson RG: Regulated migration of epidermal growth factor receptor from caveolae, *J Biol Chem* 274:30636–43, 1999.

Bretscher M: The molecules of the cell membrane, *Sci Am* 253:100–8, 1985.

Dowham W: Molecular basis for membrane phospholipid diversity: Why are there so many lipids? *Annu Rev Biochem* 66:199–212, 1997.

Englund PT: The structure and biosynthesis of glycosyl phosphatidylinositol protein anchors, *Annu Rev Biochem* 62:121–38, 1993.

Farazi TA, Waksman G, Gordon J: The biology and enzymology of protein N-myristoylation, *Annu Rev Biochem* 276:39501–64, 2001.

Petty HR: *Molewlar Biology of Membrmles: Structure and Function*, Plenum Press, New York, 1993.

Singer SJ: The structure and insertion of integral proteins in membranes, *Annu Rev Cell Biol* 6:247–96, 1990.

Singer SJ, Nicolson GL: The fluid mosaic model of the structure of cell membranes, *Science* 175:720–31, 1972.

Tamm LK, Arora A, Kleinschmidt JH: Structure and assembly of beta-barrel membrane proteins, *J Biol Chem* 276:32399–402, 2001.

Towler DA, Gordon J, Adams SP, Glaser L: The biology and enzymology of eukaryotic protein acylation, *Annu Rev Biochem* 57:69–99, 1988.

White SH, Ladokhin AS, Jayasinghe S, Hristoya K: How membranes shape protein structure, *J Biol Chem* 276:32395–8, 2001.

Yeagle PL: *The Membranes of Cells*, 2nd ed, Academic Press, San Diego, 1993.

Zhang FL, Casey PJ: Protein prenylation: Molecular mechanisms and functional consequences, *Annu Rev Biochem* 65:241–69, 1996.

Simons K, Vaz WL: Model systems, lipid rafts, and cell membranes, *Annu Rev Biophys Biomol Struct* 33:269–95, 2004.

Vereb G, Szollosi J, Matko J, Nagy P, Farkas T et al: Dynamic, yet structured: The cell membrane three decades after the Singer-Nicolson model, *Proc Natl Acad Sci USA* 100:8053–8, 2003.

Edidin M: The state of lipid rafts: From model membranes to cells, *Annu Rev Biophys Biomol Struct* 32:257–83, 2003.

Lipowsky R: The conformation of membranes, *Nature* 349:475–81, 1991.

3.2. Structure and Organization of Actin Filaments

α-Actin, Cardiac

Gunning P, Ponte P, Kedes L, Eddy R, Shows T: Chromosomal location of the co-expressed human skeletal and cardiac actin genes, *Proc Natl Acad Sci USA* 81:1813–7, 1984.

Hamada H, Petrino MG, Kakunaga T: Molecular structure and evolutionary origin of human cardiac muscle actin gene, *Proc Natl Acad Sci USA* 79:5901–5, 1982.

Humphries SE, Whittall R, Minty A, Buckingham M, Williamson R: There are approximately 20 actin genes in the human genome, *Nucleic Acids Res* 9:4895–908, 1981.

Mogensen J, Klausen IC, Pedersen AK, Egeblad H, Bross P et al: Alpha-cardiac actin is a novel disease gene in familial hypertrophic cardiomyopathy, *J Clin Invest* 103:R39–43, 1999.

Olson TM, Michels VV, Thibodeau SN, Tai YS, Keating MT: Actin mutations in dilated cardiomyopathy, a heritable form of heart failure, *Science* 280:750–2, 1998.

Schwartz K, de la Bastie D, Bouveret P, Oliviero P, Alonso S et al: Alpha-skeletal muscle actin mRNAs accumulate in hypertrophied adult rat hearts, *Circ Res* 59:551–5, 1986.

Takai E, Akita H, Shiga N, Kanazawa K, Yamada S et al: Mutational analysis of the cardiac actin gene in familial and sporadic dilated cardiomyopathy, *Am J Med Genet* 86:325–7, 1999.

Dunwoodie SL, Joya JE, Arkell RM, Hardeman EC: Multiple regions of the human cardiac actin gene are necessary for maturation-based expression in striated muscle, *J Biol Chem* 269:12212–9, 1994.

Skeletal α-Actin

Agrawal PB, Strickland CD, Midgett C, Morales A, Newburger DE et al: Heterogeneity of nemaline myopathy cases with skeletal muscle alpha-actin gene mutations, *Annu Neurol* 56:86–96, 2004.

Akkari PA, Eyre HJ, Wilton SD, Callen DF, Lane SA et al: Assignment of the human skeletal muscle alpha actin gene (ACTA1) to 1q42 by fluorescence in situ hybridization, *Cytogenet Cell Genet* 65:265–7, 1994.

Crawford K, Flick R, Close L, Shelly D, Paul R et al: Mice lacking skeletal muscle actin show reduced muscle strength and growth deficits and die during the neonatal period, *Mol Cell Biol* 22:5887–96, 2002.

Gunning P, Ponte P, Kedes L, Eddy R, Shows T: Chromosomal location of the co-expressed human skeletal and cardiac actin genes, *Proc Natl Acad Sci USA* 81:1813–7, 1984.

Gunning P, Ponte P, Okayama H, Engel J, Blau H et al: Isolation and characterization of full-length cDNA clones for human alpha-, beta-, and gamma-actin mRNAs: Skeletal but not cytoplasmic actins have an amino-terminal cysteine that is subsequently removed, *Mol Cell Biol* 3:787–95, 1983.

Hanauer A, Levin M, Heilig R, Daegelen D, Kahn A et al: Isolation and characterization of cDNA clones for human skeletal muscle alpha actin, *Nucleic Acids Res* 11:3503–16, 1983.

lkovski B, Cooper ST, Nowak K, Ryan MM, Yang N et al: Nemaline myopathy caused by mutations in the muscle alpha-skeletal-actin gene, *Am J Hum Genet* 68:1333–43, 2001.

Laing NG, Clarke NF, Dye DE, Liyanage K, Walker KR et al: Actin mutations are one cause of congenital fibre type disproportion, *Annu Neurol* 56:689–94, 2004.

Nowak KJ, Wattanasirichaigoon D, Goebel HH, Wilce M, Pelin K et al (and 20 others): Mutations in the skeletal muscle alpha-actin gene in patients with actin myopathy and nemaline myopathy, *Nature Genet* 23:208–12, 1999.

Taylor A, Erba HP, Muscat GEO, Kedes L: Nucleotide sequence and expression of the human skeletal alpha-actin gene: evolution of functional regulatory domains, *Genomics* 3:323–36, 1988.

Vascular SMC α-Actin

Kumar MS, Hendrix JA, Johnson AD, Owens GK: Smooth muscle alpha-actin gene requires two E-boxes for proper expression in vivo and is a target of class I basic helix-loop-helix proteins, *Circ Res* 92:840–7, 2003.

Ueyama H, Bruns G, Kanda N: Assignment of the vascular smooth muscle actin gene ACTSA to human chromosome 10, *Jpn J Hum Genet* 35:145–50, 1990.

Ueyama H, Hamada H, Battula N, Kakunaga T: Structure of a human smooth muscle actin gene (aortic type) with a unique intron site, *Mol Cell Biol* 4:1073–8, 1984.

β-Actin

Erba HP, Eddy R, Shows T, Kedes L, Gunning P: Structure, chromosome location, and expression of the human gamma-actin gene: Differential evolution, location, and expression of the cytoskeletal beta- and gamma-actin genes, *Mol Cell Biol* 8:1775–89, 1988.

Habets GGM, van der Kammen RA, Willemsen V, Balemans M, Wiegant J et al: Sublocalization of an invasion-inducing locus and other genes on human chromosome 7, *Cytogenet Cell Genet* 60:200–5, 1992.

Kedes L, Ng SY, Lin CS, Gunning P, Eddy R et al: The human beta-actin multigene family, *Trans Assoc Am Phys* 98:42–6, 1985.

Leavitt J, Bushar G, Kakunaga T, Hamada H, Hirakawa T et al: Variations in expression of mutant beta-actin accompanying incremental increases in human fibroblast tumorigenicity, *Cell* 28:259–68, 1982.

Nakajima-Iijima S, Hamada H, Reddy P, Kakunaga T: Molecular structure of the human cytoplasmic beta-actin gene; interspecies homology of sequences in the introns, *Proc Natl Acad Sci* 82:6133–7, 1985.

Ng SY, Gunning P, Eddy R, Ponte P, Leavitt J et al: Evolution of the functional human beta-actin gene and its multi-pseudogene family: Conservation of the noncoding regions and chromosomal dispersion of pseudogenes, *Mol Cell Biol* 5:2720–32, 1985.

Toyama S, Toyama S: A variant form of beta-actin in a mutant of KB cells resistant to cytochalasin B, *Cell* 37:609–14, 1984.

Ueyama H, Inazawa J, Nishino H, Ohkubo I, Miwa T: FISH localization of human cytoplasmic actin genes ACTB to 7p22 and ACTGl to 17q25 and characterization of related pseudogenes, *Cytogenet Cell Genet* 74:221–4, 1996.

γ-Actin

Erba HP, Eddy R, Shows T, Kedes L, Gunning P: Structure, chromosome location, and expression of the human gamma-actin gene: differential evolution, location, and expression of the cytoskeletal beta- and gamma-actin genes, *Mol Cell Biol* 8:1775–89, 1988.

Erba HP, Gunning P, Kedes L: Nucleotide sequence of the human gamma cytoskeletal actin mRNA: Anomalous evolution of vertebrate non-muscle actin genes, *Nucleic Acids Res* 14:5275–94, 1986.

Leisel TP, Boujemaa R, Pantaloni D, Carlier MF: Reconstitution of actin-based motility of Listeria and Shigella using pure proteins, *Nature* 401:613–6, 1999.

Otterbein LR, Graceffa P, Dominguez R: The crystal structure of uncomplexed actin in the ADP state, *Science* 293:708–11, 2001.

Ueyama H, Inazawa J, Nishino H, Ohkubo I, Miwa T: FISH localization of human cytoplasmic actin genes ACTB to 7p22 and ACTG1 to 17q25 and characterization of related pseudogenes, *Cytogenet Cell Genet* 74:221–4, 1996.

van Wijk E, Krieger E, Kemperman MH, De Leenheer EMR, Huygen PLM et al: A mutation in the gamma actin 1 (ACTG1) gene causes autosomal dominant hearing loss (DFNA20/26), *J Med Genet* 40:879–84, 2003.

Zhu M, Yang T, Wei S, DeWan AT, Morell RJ et al: Mutations in the gamma-actin gene (ACTG1) are associated with dominant progressive deafness (DFNA20/26), *Am J Hum Genet* 73:1082–91, 2003.

Miwa T, Manabe Y, Kurokawa K, Kamada S, Kanda N et al: Structure, chromosome location, and expression of the human smooth muscle (enteric type) gamma-actin gene: Evolution of six human actin genes, *Mol Cell Biol* 11:3296–306, 1991.

Szucsik JC, Lessard JL: Cloning and sequence analysis of the mouse smooth muscle gamma-enteric actin gene, *Genomics* 28:154–62, 1995.

Ueyama H, Inazawa J, Nishino H, Han-Xiang D, Ochiai Y et al: Chromosomal mapping of the human smooth muscle actin gene (enteric type, ACTA3) to 2p13.1 and molecular nature of the HindIII polymorphism, *Genomics* 25:720–3, 1995.

Human protein reference data base, Johns Hopkins University and the Institute of Bioinformatics, at http://www.hprd.org/protein.

3.3. Actin Monomer-Binding Proteins

Cofilin

Ghosh M, Song X, Mouneimne G, Sidani M, Lawrence DS et al: Cofilin promotes actin polymerization and defines the direction of cell motility, *Science* 304:743–6, 2004.

Gillett GT, Fox MF, Rowe PSN, Casimir CM, Povey S: Mapping of human non-muscle type cofilin (CFL1) to chromosome 11q13 and muscle-type cofilin (CFL2) to chromosome 14, *Annu Hum Genet* 60:201–11, 1996.

Kuhn TB, Meberg PJ, Brown MD, Bernstein BW, Minamide LS et al: Regulating actin dynamics in neuronal growth cones by ADF/cofilin and Rho family GTPases, *J Neurobiol* 44:126–44, 2000.

Maekawa M, Ishizaki T, Boku S, Watanabe N, Fujita A et al: Signaling from Rho to the actin cytoskeleton through protein kinases ROCK and LIM-kinase, *Science* 285:895–8, 1999.

Ono S, Minami N, Abe H, Obinata T: Characterization of a novel cofilin isoform that is predominantly expressed in mammalian skeletal muscle, *J Biol Chem* 269:15280–6, 1994.

Profilin

Kovar DR, Harris ES, Mahaffy R, Higgs HN, Pollard TD: Control of the assembly of ATP- and ADP-actin by formins and profilin, *Cell* 124:423–35, 2006.

Ampe C, Markey F, Lindberg U, Vandekerckhove J: The primary structure of human platelet profilin: reinvestigation of the calf spleen profilin sequence, *FEBS Lett* 228:17–21, 1988.

Goldschmidt-Clermont PJ, Janmey PA: Profilin, a weak CAP for actin and RAS, *Cell* 66:419–21, 1991.

Kwiatkowski DJ, Aklog L, Ledbetter DH, Morton CC: Identification of the functional profilin gene, its localization to chromosome subband 17p13.3, and demonstration of its deletion in some patients with Miller-Dieker syndrome, *Am J Hum Genet* 46:559–67, 1990.

Kwiatkowski DJ, Bruns GAP: Human profilin: Molecular cloning, sequence comparison, and chromosomal analysis, *J Biol Chem* 263:5910–5, 1988.

Theriot JA, Mitchison TJ: The three faces of profilin, *Cell* 75:835–8, 1993.

Vojtek A, Haarer B, Field J, Gerst J, Pollard TD et al: Evidence for a functional link between profilin and CAP in the yeast S. cerevisiae, *Cell* 66:497–505, 1991.

Witke W, Sutherland JD, Sharpe A, Arai M, Kwiatkowski DJ: Profilin I is essential for cell survival and cell division in early mouse development, *Proc Natl Acad Sci USA* 98:3832–6, 2001.

β-Thymosin

Bock-Marquette I, Saxena A, White MD, DiMaio JM, Srivastava D: Thymosin beta-4 activates integrin-linked kinase and promotes cardiac cell migration, survival and cardiac repair, *Nature* 432:466–72, 2004.

Clark EA, Golub TR, Lander ES, Hynes RO: Genomic analysis of metastasis reveals an essential role for RhoC, *Nature* 406:532–5, 2000.

Clauss IM, Wathelet MG, Szpirer J, Islam MQ, Levan G et al: Human thymosin-beta-4/6-26 gene is part of a multigene family composed of seven members located on seven different chromosomes, *Genomics* 9:174–80, 1991.

Gondo H, Kudo J, White JW, Barr C, Selvanayagam P et al: Differential expression of the human thymosin-beta(4) gene in lymphocytes, macrophages, and granulocytes, *J Immunol* 139:3840–8, 1987.

Li X, Zimmerman A, Copeland NG, Gilbert DJ, Jenkins NA et al: The mouse thymosin beta-4 gene: structure, promoter identification, and chromosome localization, *Genomics* 32:388–94, 1996.

Formin

Chang F, Drubin D, Nurse P: cdc12p, a protein required for cytokinesis in fission yeast, is a component of the cell division ring and interacts with profilin, *J Cell Biol* 137:169–82, 1997.

Kovar DR, Kuhn JR, Tichy AL, Pollard TD: The fission yeast cytokinesis formin Cdc12p is a barbed end actin filament capping protein gated by profilin, *J Cell Biol* 161:875–87, 2003.

Kovar DR, Pollard TD: Insertional assembly of actin filament barbed ends in association with formins produces piconewton forces, *Proc Natl Acad Sci USA* 101:14725–30, 2004.

Pruyne D, Evangelista M, Yang C, Bi E, Zigmond S et al: Role of formins in actin assembly: nucleation and barbed-end association, *Science* 297:612–5, 2002.

Romero S, Le Clainche C, Didry D, Egile C, Pantaloni D et al: Formin is a processive motor that requires profilin to accelerate actin assembly and associated ATP hydrolysis, *Cell* 119:419–29, 2004.

Sagot I, Rodal AA, Moseley J, Goode BL, Pellman D: An actin nucleation mechanism mediated by Bni1 and profilin, *Nature Cell Biol* 4:626–31, 2002.

Otomo T, Tomchick DR, Otomo C, Panchal SC, Machius M et al: Structural basis of actin filament nucleation and processive capping by a formin homology 2 domain, *Nature* 433:488–94, 2005.

Li F, Higgs HN: The mouse Formin mDia1 is a potent actin nucleation factor regulated by autoinhibition, *Curr Biol* 13:1335–40, 2003.

Kovar DR, Wu JQ, Pollard TD: Profilin-mediated competition between capping protein and formin Cdc12p during cytokinesis in fission yeast, *Mol Biol Cell* 16:2313–24, 2005.

Kovar DR, Pollard TD: Insertional assembly of actin filament barbed ends in association with formins produces piconewton forces, *Proc Natl Acad Sci USA* 101:14725–30, 2004.

Kobielak A, Pasolli HA, Fuchs E: Mammalian formin-1 participates in adherens junctions and polymerization of linear actin cables, *Nature Cell Biol* 6:21–30, 2004.

Human protein reference data base, Johns Hopkins University and the Institute of Bioinformatics, at http://www.hprd.org/protein.

3.4. Actin Filament-Capping Proteins

Gelsolin

Kwiatkowski DJ, Ozelius L, Schuback D, Gusella J, Breakefield XO: The gelsolin (GSN) cDNA clone, from 9q32-34, identifies BclI and StuI RFLPs, *Nucleic Acids Res* 17:4425 (only), 1989.

Kwiatkowski DJ, Stossel TP, Orkin SH, Mole JE, Colten HR et al: Plasma and cytoplasmic gelsolins are encoded by a single gene and contain a duplicated actin-binding domain, *Nature* 323:455–8, 1986.

Lee WM, Galbraith RM: The extracellular actin-scavenger system and actin toxicity, *N Engl J Med* 326:1335–41, 1992.

Vasconcellos CA, Allen PG, Wohl ME, Drazen JM, Janmey PA et al: Reduction in viscosity of cystic fibrosis sputum in vitro by gelsolin, *Science* 263:969–71, 1994.

Witke W, Sharpe AH, Hartwig JH, Azuma T, Stossel TP, Kwiatkowski DJ: Hemostatic, inflammatory, and fibroblast responses are blunted in mice lacking gelsolin, *Cell* 81:41–51, 1995.

Capping Protein α1

Barron-Casella EA, Torres MA, Scherer SW, Heng HHQ, Tsui, LC et al: Sequence analysis and chromosomal localization of human Cap Z: Conserved residues within the actin-binding domain may link Cap Z to gelsolin/severin and profilin protein families, *J Biol Chem* 270:21472–9, 1995.

Hart MC, Korshunova YO, Cooper JA: Mapping of the mouse actin capping protein alpha subunit genes and pseudogenes, *Genomics* 39:264–70, 1997.

ARP2

Leisel TP, Boujemaa R, Pantaloni D, Carlier, MF: Reconstitution of actin-based motility of Listeria and Shigella using pure proteins, *Nature* 401:613–6, 1999.

Machesky LM, Reeves E, Wientjes F, Mattheyse FJ, Grogan A et al: Mammalian actin-related protein 2/3 complex localizes to regions of lamellipodial protrusion and is composed of evolutionarily conserved proteins, *Biochem J* 328:105–12, 1997.

Marchand JB, Kaiser DA, Pollard TD, Higgs HN: Interaction of WASP/Scar proteins with actin and vertebrate Arp2/3 complex, *Nature Cell Biol* 3:76–82, 2001.

Prehoda KE, Scott JA, Mullins RD, Lim WA: Integration of multiple signals through cooperative regulation of the N-WASP-Arp2/3 complex, *Science* 290:801–6, 2000.

Robinson RC, Turbedsky K, Kaiser DA, Marchand JB, Higgs HN et al: Crystal structure of Arp2/3 complex, *Science* 294:1679–84, 2001.

Volkmann N, Amann KJ, Stoilova-McPhie S, Egile C, Winter DC et al: Structure of Arp2/3 complex in its activated state and in actin filament branch junctions, *Science* 293:2456–9, 2001.

Welch MD, DePace AH, Verma S, Iwamatsu A, Mitchison TJ: The human Arp2/3 complex is composed of evolutionarily conserved subunits and is localized to cellular regions of dynamic actin filament assembly, *J Cell Biol* 138:375–84, 1997.

Welch MD, Iwamatsu A, Mitchison TJ: Actin polymerization is induced by Arp2/3 protein complex at the surface of Listeria monocytogenes, *Nature* 385:265–9, 1997.

ARP3

Machesky LM, Reeves E, Wientjes F, Mattheyse FJ, Grogan A et al: Mammalian actin-related protein 2/3 complex localizes to regions of lamellipodial protrusion and is composed of evolutionarily conserved proteins, *Biochem J* 328:105–12, 1997.

Robinson RC, Turbedsky K, Kaiser DA, Marchand JB, Higgs HN et al: Crystal structure of Arp2/3 complex, *Science* 294:1679–84, 2001.

Volkmann N, Amann KJ, Stoilova-McPhie S, Egile C, Winter DC et al: Structure of Arp2/3 complex in its activated state and in actin filament branch junctions, *Science* 293:2456–9, 2001.

Welch MD, DePace AH, Verma S, Iwamatsu A, Mitchison TJ: The human Arp2/3 complex is composed of evolutionarily conserved subunits and is localized to cellular regions of dynamic actin filament assembly, *J Cell Biol* 138:375–84, 1997.

Welch MD, Iwamatsu A, Mitchison TJ: Actin polymerization is induced by Arp2/3 protein complex at the surface of Listeria monocytogenes, *Nature* 385:265–9, 1997.

Tropomyosin 1

Brown JH, Kim KH, Jun G, Greenfield NJ, Dominguez R et al: Deciphering the design of the tropomyosin molecule, *Proc Natl Acad Sci USA* 98:8496–501, 2001.

Eyre H, Akkari PA, Wilton SD, Callen DC, Baker E et al: Assignment of the human skeletal muscle alpha-tropomyosin gene (TPM1) to band 15q22 by fluorescence in situ hybridization, *Cytogenet Cell Genet* 69:15–7, 1995.

Lees-Miller JP, Helfman DM: The molecular basis for tropomyosin isoform diversity, *BioEssays* 13:429–37, 1991.

Schleef M, Werner K, Satzger U, Kaupmann K, Jockusch H: Chromosomal localization and genomic cloning of the mouse alpha-tropomyosin gene Tpm-1, *Genomics* 17:519–21, 1993.

Thierfelder L, Watkins H, MacRae C, Lamas R, McKenna W et al: Alpha-tropomyosin and cardiac troponin T mutations cause familial hypertrophic cardiomyopathy: A disease of the sarcomere, *Cell* 77:701–12, 1994.

Tiso N, Rampoldi L, Pallavicini A, Zimbello R, Pandolfo D et al: Fine mapping of five human skeletal muscle genes: Alpha-tropomyosin, beta-tropomyosin, troponin-I slow-twitch, troponin-I fast-twitch, and troponin-C fast, *Biochem Biophys Res Commun* 230:347–50, 1997.

Watkins H, McKenna WJ, Thierfelder L, Suk HJ, Anan R et al: Mutations in the genes for cardiac troponin T and alpha-tropomyosin in hypertrophic cardiomyopathy, *N Engl J Med* 332:1058–64, 1995.

Nebulin

Donner K, Sandbacka M, Lehtokari VL, Wallgren-Pettersson C et al: Complete genomic structure of the human nebulin gene and identification of alternatively spliced transcripts, *Eur J Hum Genet* 12:744–51, 2004.

Labeit S, Kolmerer B: The complete primary structure of human nebulin and its correlation to muscle structure, *J Mol Biol* 248:308–15, 1995.

Limongi MZ, Pelliccia F, Rocchi A: Assignment of the human nebulin gene (NEB) to chromosome band 2q24.2 and the alpha-1 (III) collagen gene (COL3A1) to chromosome band 2q32.2 by in situ hybridization: the FRA2G common fragile site lies between the two genes in the 2q31 band, *Cytogenet Cell Genet* 77:259–60, 1997.

Pelin K, Hilpela P, Donner K, Sewry C, Akkari PA et al: Mutations in the nebulin gene associated with autosomal recessive nemaline myopathy, *Proc Natl Acad Sci USA* 96:2305–10, 1999.

Schurr E, Skamene E, Gros P: Mapping of the gene coding for the muscle protein nebulin (Neb) to the proximal region of mouse chromosome 2, *Cytogenet Cell Genet* 57:214–6, 1991.

Stedman H, Browning K, Oliver N, Oronzi-Scott M, Fischbeck K et al: Nebulin cDNAs detect a 25-kilobase transcript in skeletal muscle and localize to human chromosome 2, *Genomics* 2:1–7, 1988.

Wang K, Knipfer M, Huang QQ, van Heerden A, Hsu LCL et al: Human skeletal muscle nebulin sequence encodes a blueprint for thin filament architecture: Sequence motifs and affinity profiles of tandem repeats and terminal SH3, *J Biol Chem* 271:4304–14, 1996.

Tropomodulin

Chu X, Thompson D, Yee LJ, Sung LA: Genomic organization of mouse and human erythrocyte tropomodulin genes encoding the pointed end capping protein for the actin filaments, *Gene* 256:271–81, 2000.

Conley CA: Leiomodin and tropomodulin in smooth muscle, *Am J Physiol Cell Physiol* 280: C1645–56, 2001.

Fowler VM, Sussmann MA, Miller PG, Flucher BE, Daniels MP: Tropomodulin is associated with the free (pointed) ends of the thin filaments in rat skeletal muscle, *J Cell Biol* 120:411–20, 1993.

Lench NJ, Telford EA, Andersen SE, Moynihan TP, Robinson PA et al: An EST and STS-based YAC contig map of human chromosome 9q22.3, *Genomics* 38:199–205, 1996.

Sung LA, Fan YS, Lin CC: Gene assignment, expression, and homology of human tropomodulin, *Genomics* 34:92–6, 1996.

Sung LA, Fowler VM, Lambert K, Sussman MA, Karr D et al: Molecular cloning and characterization of human fetal liver tropomodulin: a tropomyosin-binding protein, *J Biol Chem* 267:2616–21, 1992

Human protein reference data base, Johns Hopkins University and the Institute of Bioinformatics, at http://www.hprd.org/protein.

3.5. Actin Filament Crosslinking Proteins

α-Actinin

Youssoufian H, McAfee M, Kwiatkowski DJ: Cloning and chromosomal localization of the human cytoskeletal alpha-actinin gene reveals linkage to the beta-spectrin gene, *Am J Hum Genet* 47:62–72, 1990.

Beggs AH, Byers TJ, Knoll JHM, Boyce FM, Bruns GAP et al: Cloning and characterization of two human skeletal muscle alpha-actinin genes located on chromosomes 1 and 11, *J Biol Chem* 267:9281–8, 1992.

Beggs AH, Phillips HA, Kozman H, Mulley JC, Wilton SD et al: A (CA)n repeat polymorphism for the human skeletal muscle alpha-actinin gene ACTN2 and its localization on the linkage map of chromosome 1, *Genomics* 13:1314–5, 1992.

Harper SQ, Crawford RW, DellRusso C, Chamberlain JS: Spectrin-like repeats from dystrophin and alpha-actinin-2 are not functionally interchangeable, *Hum Mol Genet* 11:1807–15, 2002.

Mills MA, Yang N, Weinberger RP, Vander Woude DL, Beggs AH et al: Differential expression of the actin-binding proteins, alpha-actinin-2 and -3, in different species: implications for the evolution of functional redundancy, *Hum Mol Genet* 10:1335–46, 2001.

Fimbrin

Lin CS, Shen W, Chen ZP, Tu YH, Matsudaira P: Identification of I-plastin, a human fimbrin isoform expressed in intestine and kidney, *Mol Cell Biol* 14:2457–67, 1994.

Villin

Phillips MJ, Azuma T, Meredith SLM, Squire JA, Ackerley CA et al: Abnormalities in villin gene expression and canalicular microvillus structure in progressive cholestatic liver disease of childhood, *Lancet* 362:1112–9, 2003.

Pringault E, Arpin M, Garcia A, Finidori J, Louvard D: A human villin cDNA clone to investigate the differentiation of intestinal and kidney cells in vivo and in culture, *EMBO J* 5:3119–24, 1986.

Pringault E, Robine S, Louvard D: Structure of the human villin gene, *Proc Natl Acad Sci USA* 88:10811–5, 1991.

Rousseau-Merck MF, Simon-Chazottes D, Arpin M, Pringault E, Louvard D et al: Localization of the villin gene on human chromosome 2q35-q36 and on mouse chromosome 1, *Hum Genet* 78:130–3, 1988.

Schurr E, Skamene E, Morgan K, Chu ML, Gros P: Mapping of Col3a1 and Col6a3 to proximal murine chromosome 1 identifies conserved linkage of structural protein genes between murine chromosome 1 and human chromosome 2q, *Genomics* 8:477–86, 1990.

Filamin A

Chakarova C, Wehnert MS, Uhl K, Sakthivel S, Vosberg HP et al: Genomic structure and fine mapping of the two human filamin gene paralogues FLNB and FLNC and comparative analysis of the filamin gene family, *Hum Genet* 107:597–611, 2000.

Gariboldi M, Maestrini E, Canzian F, Manenti G, De Gregorio L et al: Comparative mapping of the actin-binding protein 280 genes in human and mouse, *Genomics* 21:428–30, 1994.

Gorlin JB, Yamin R, Egan S, Stewart M, Stossel TP et al: Human endothelial actin-binding protein (ABP-280, nonmuscle filamin): A molecular leaf spring, *J Cell Biol* 111:1089–105, 1990.

Loy CJ, Sim KS, Yong EL: Filamin—a fragment localizes to the nucleus to regulate androgen receptor and coactivator functions, *Proc Natl Acad Sci USA* 100:4562–7, 2003.

Maestrini E, Patrosso C, Mancini M, Rivella S, Rocchi M et al: Mapping of two genes encoding isoforms of the actin binding protein ABP-280, a dystrophin like protein, to Xq28 and to chromosome 7, *Hum Mol Genet* 2:761–6, 1993.

Robertson SP, Twigg SRF, Sutherland-Smith AJ, Biancalana V, Gorlin RJ et al: Localized mutations in the gene encoding the cytoskeletal protein filamin A cause diverse malformations in humans, *Nature Genet* 33:487–91, 2003.

Sheen VL, Feng Y, Graham D, Takafuta T, Shapiro SS et al: Filamin A and filamin B are co-expressed within neurons during periods of neuronal migration and can physically interact, *Hum Mol Genet* 11:2845–54, 2002.

Vadlamudi RK, Li F, Adam L, Nguyen D, Ohta Y, Stossel TP et al: Filamin is essential in actin cytoskeletal assembly mediated by p21-activated kinase 1, *Nature Cell Biol* 4:681–90, 2002.

Filamin B

Brocker F, Bardenheuer W, Vieten L, Julicher K, Werner N et al: Assignment of human filamin gene FLNB to human chromosome band 3p14.3 and identification of YACs containing the complete FLNB transcribed region, *Cytogenet Cell Genet* 85:267–8, 1999.

Chakarova C, Wehnert MS, Uhl K, Sakthivel S, Vosberg HP et al: Genomic structure and fine mapping of the two human filamin gene paralogues FLNB and FLNC and comparative analysis of the filamin gene family, *Hum Genet* 107:597–611, 2000.

Krakow D, Robertson SP, King LM, Morgan T, Sebald ET et al (and 20 others): Mutations in the gene encoding filamin B disrupt vertebral segmentation, joint formation and skeletogenesis, *Nature Genet* 36:405–10, 2004.

Sheen VL, Feng Y, Graham D, Takafuta T, Shapiro SS et al: Filamin A and filamin B are co-expressed within neurons during periods of neuronal migration and can physically interact, *Hum Mol Genet* 11:2845–54, 2002.

Takafuta T, Wu G, Murphy GF, Shapiro SS: Human beta-filamin is a new protein that interacts with the cytoplasmic tail of glycoprotein Ib-alpha, *J Biol Chem* 273:17531–8, 1998.

Human protein reference data base, Johns Hopkins University and the Institute of Bioinformatics, at http://www.hprd.org/protein.

3.6. Regulation of Actin Assembly and Disassembly

RhoA

Cannizzaro LA, Madaule P, Hecht F, Axel R, Croce CM et al: Chromosome localization of human ARH genes, a ras-related gene family, *Genomics* 6:197–203, 1990.

Kiss C, Li J, Szeles A, Gizatullin RZ, Kashuba VI, Lushnikova T et al: Assignment of the ARHA and GPX1 genes to human chromosome bands 3p21.3 by in situ hybridization and with somatic cell hybrids, *Cytogenet Cell Genet* 79:228–30, 1997.

Maekawa M, Ishizaki T, Boku S, Watanabe N, Fujita A et al: Signaling from Rho to the actin cytoskeleton through protein kinases ROCK and LIM-kinase, *Science* 285:895–8, 1999.

Nakamura M, Nagano T, Chikama T, Nishida T: Role of the small GTP-binding protein Rho in epithelial cell migration in the rabbit cornea, *Invest Ophthalm Vis Sci* 42:941–7, 2001.

Sin WC, Haas K, Ruthazer ES, Cline HT: Dendrite growth increased by visual activity requires NMDA receptor and Rho GTPases, *Nature* 419:475–80, 2002.

Wang HR, Zhang Y, Ozdamar B, Ogunjimi AA, Alexandrova E et al: Regulation of cell polarity and protrusion formation by targeting RhoA for degradation, *Science* 302:1775–9, 2003.

Wu KY, Hengst U, Cox LJ, Macosko EZ, Jeromin A et al: Local translation of RhoA regulates growth cone collapse [letter], *Nature* 436:1020–4, 2005.

RhoB

Liu AX, Cerniglia GJ, Bernhard EJ, Prendergast GC: RhoB is required to mediate apoptosis in neoplastically transformed cells after DNA damage, *Proc Natl Acad Sci USA* 98:6192–7, 2001.

Zhang J, Zhu J, Bu X, Cushion M, Kinane TB et al: Cdc42 and RhoB activation are required for mannose receptor-mediated phagocytosis by human alveolar macrophages, *Mol Biol Cell* 16:824–34, 2005.

Cannizzaro LA, Madaule P, Hecht F, Axel R, Croce CM et al: Chromosome localization of human ARH genes, a ras-related gene family, *Genomics* 6:197–203, 1990.

Chardin P, Madaule P, Tavitian A: Coding sequence of human rho cDNAs clone 6 and clone 9, *Nucleic Acid Res* 16:2717 (only), 1988.

Madaule P, Axel R: A novel ras-related gene family, *Cell* 41:31–40, 1985.

Maekawa M, Ishizaki T, Boku S, Watanabe N, Fujita A et al: Signaling from Rho to the actin cytoskeleton through protein kinases ROCK and LIM-kinase, *Science* 285:895–8, 1999.

Ridley AJ, Hall A: The small GTP-binding protein rho regulates the assembly of focal adhesions and actin stress fibers in response to growth factors, *Cell* 70:389–99, 1992.

Sandilands E, Cans C, Fincham VJ, Brunton VG, Mellor H et al: RhoB and actin polymerization coordinate Src activation with endosome-mediated delivery to the membrane, *Dev Cell* 7:855–69, 2004.

RhoC

Cannizzaro LA, Madaule P, Hecht F, Axel R, Croce CM, Huebner K: Chromosome localization of human ARH genes, a ras-related gene family, *Genomics* 6:197–203, 1990.

Chardin P, Madaule P, Tavitian A: Coding sequence of human rho cDNAs clone 6 and clone 9, *Nucleic Acid Res* 16:2717 (only), 1988.

Clark EA, Golub TR, Lander ES, Hynes RO: Genomic analysis of metastasis reveals an essential role for RhoC, *Nature* 406:532–5, 2000.

Morris SW, Valentine MB, Kirstein MN, Huebner K: Reassignment of the human ARH9 RAS-related gene to chromosome 1p13-p21, *Genomics* 15:677–9, 1993.

Rose R, Weyand M, Lammers M, Ishizaki T, Ahmadian MR et al: Structural and mechanistic insights into the interaction between Rho and mammalian Dia [letter], *Nature* 435:513–8, 2005.

Rac1

Benvenuti F, Hugues S, Walmsley M, Ruf S, Fetler L et al: Requirement of Rac1 and Rac2 expression by mature dendritic cells for T cell priming, *Science* 305:1150–3, 2004.

Chang HY, Ready DF: Rescue of photoreceptor degeneration in rhodopsin-null Drosophila mutants by activated Rac1, *Science* 290:1978–80, 2000.

Eden S, Rohatgi R, Podtelejnikov AV, Mann M, Kirschner MW: Mechanism of regulation of WAVE1-induced actin nucleation by Rac1 and Nck, *Nature* 418:790–3, 2002.

Gu Y, Filippi MD, Cancelas JA, Siefring JE, Williams EP et al: Hematopoietic cell regulation by Rac1 and Rac2 guanosine triphosphatases, *Science* 302:445–9, 2003.

Joneson T, McDonough M, Bar-Sagi D, Van Aelst L: RAC regulation of actin polymerization and proliferation by a pathway distinct from Jun kinase, *Science* 274:1374–6, 1996.

Jordan P, Brazao R, Boavida MG, Gespach C, Chastre E: Cloning of a novel human Rac1b splice variant with increased expression in colorectal tumors, *Oncogene* 18:6835–9, 1999.

Katoh H, Negishi M: RhoG activates Rac1 by direct interaction with the Dock180-binding protein Elmo, *Nature* 424:461–4, 2003.

Kheradmand F, Werner E, Tremble P, Symons M, Werb Z: Role of Rac1 and oxygen radicals in collagenase-1 expression induced by cell shape change, *Science* 280:898–902, 1998.

Kissil JL, Johnson KC, Eckman MS, Jacks T: Merlin phosphorylation by p21-activated kinase 2 and effects of phosphorylation on merlin localization, *J Biol Chem* 277:10394–9, 2002.

Lanzetti L, Rybin V, Malabarba MG, Christoforidis S, Scita G et al: The Eps8 protein coordinates EGF receptor signalling through Rac and trafficking through Rab5, *Nature* 408:374–7, 2000.

Malecz N, McCabe PC, Spaargaren C, Qiu RG, Chuang Y et al: Synaptojanin 2, a novel Rac1 effector that regulates clathrin-mediated endocytosis, *Curr Biol* 10:1383–6, 2000.

Manser E, Leung T, Salihuddin H, Zhao ZS, Lim L: A brain serine/threonine protein kinase activated by Cdc42 and Rac1, *Nature* 367:40–6, 1994.

Miki H, Yamaguchi H, Suetsugu S, Takenawa T: IRSp53 is an essential intermediate between Rac and WAVE in the regulation of membrane ruffling, *Nature* 408:732–5, 2000.

Nakaya Y, Kuroda S, Katagiri YT, Kaibuchi K, Takahashi Y: Mesenchymal-epithelial transition during somitic segmentation is regulated by differential roles of Cdc42 and Rac1, *Dev Cell* 7:425–38, 2004.

Radisky DC, Levy DD, Littlepage LE, Liu H, Nelson CM et al: Rac1b and reactive oxygen species mediate MMP-3-induced EMT and genomic instability, *Nature* 436:123–7, 2005.

Simon AR, Vikis HG, Stewart S, Fanburg BL, Cochran BH et al: Regulation of STAT3 by direct binding to the Rac1 GTPase, *Science* 290:144–7, 2000.

Sin WC, Haas K, Ruthazer ES, Cline HT: Dendrite growth increased by visual activity requires NMDA receptor and Rho GTPases, *Nature* 419:475–80, 2002.

Walmsley MJ, Ooi SKT, Reynolds LF, Smith SH, Ruf S et al: Critical roles for Rac1 and Rac2 GTPases in B cell development and signaling, *Science* 302:459–62, 2003.

Xiao GH, Beeser A, Chernoff J, Testa JR: p21-activated kinase links Rac/Cdc42 signaling to merlin, *J Biol Chem* 277:883–6, 2002.

Cdc42

Deloukas P, Schuler GD, Gyapay G, Beasley EM, Soderlund C et al (and 60 others): A physical map of 30,000 human genes, *Science* 282:744–6, 1998.

Erickson JW, Zhang C, Kahn RA, Evans T, Cerione RA: Mammalian cdc42 is a brefeldin A-sensitive component of the Golgi apparatus, *J Biol Chem* 271:26850–4, 1996.

Etienne-Manneville S, Hall A: Cdc42 regulates GSK-3-beta and adenomatous polyposis coli to control cell polarity, *Nature* 421:753–6, 2003.

Garrett WS, Chen LM, Kroschewski R, Ebersold M et al: Developmental control of endocytosis in dendritic cells by Cdc42, *Cell* 102:325–34, 2000.

Irie F, Yamaguchi Y: EphB receptors regulate dendritic spine development via intersectin, Cdc42 and N-WASP, *Nature Neurosci* 5:1117–8, 2002.

Kim AS, Kakalis LT, Abdul-Manan N, Liu GA, Rosen MK: Autoinhibition and activation mechanisms of the Wiskott-Aldrich syndrome protein, *Nature* 404:151–8, 2000.

Manser E, Leung T, Salihuddin H, Tan L, Lim L: A non-receptor tyrosine kinase that inhibits the GTPase activity of p21cdc42, *Nature* 363:364–7, 1993.

Munemitsu S, Innis MA, Clark R, McCormick F, Ullrich A et al: Molecular cloning and expression of a G25K cDNA, the human homolog of the yeast cell cycle gene CDC42, *Mol Cell Biol* 10:5977–82, 1990.

Nakaya Y, Kuroda S, Katagiri YT, Kaibuchi K, Takahashi Y: Mesenchymal-epithelial transition during somitic segmentation is regulated by differential roles of Cdc42 and Rac1, *Dev Cell* 7:425–38, 2004.

Nalbant P, Hodgson L, Kraynov V, Toutchkine A, Hahn KM: Activation of endogenous Cdc42 visualized in living cells, *Science* 305:1615–9, 2004.

Nicole S, White PS, Topaloglu H, Beigthon P, Salih M et al: The human CDC42 gene: Genomic organization, evidence for the existence of a putative pseudogene and exclusion as a SJS1 candidate gene, *Hum Genet* 105:98–103, 1999.

Shinjo K, Koland JG, Hart MJ, Narasimhan V, Johnson DI et al: Molecular cloning of the gene for the human placental GTP-binding protein G(p) (G25K): identification of this GTP-binding protein as the human homolog of the yeast cell-division-cycle protein CDC42, *Proc Natl Acad Sci USA* 87:9853–7, 1990.

Sin WC, Haas K, Ruthazer ES, Cline HT: Dendrite growth increased by visual activity requires NMDA receptor and Rho GTPases, *Nature* 419:475–80, 2002.

Wu WJ, Erickson JW, Lin R, Cerione RA: The gamma-subunit of the coatomer complex binds Cdc42 to mediate transformation, *Nature* 405:800–4, 2000.

Yasuda S, Oceguera-Yanez F, Kato T, Okamoto M et al: Cdc42 and mDia3 regulate microtubule attachment to kinetochores, *Nature* 428:767–71, 2004.

Zheng Y, Fischer DJ, Santos MF, Tigyi G, Pasteris NG et al: The faciogenital dysplasia gene product FGD1 functions as a Cdc42Hs-specific guanine-nucleotide exchange factor, *J Biol Chem* 271:33169–72, 1996.

Human protein reference data base, Johns Hopkins University and the Institute of Bioinformatics, at http://www.hprd.org/protein.

3.7. Function of Actin Filaments

Bamburg JR, Wiggan OP: ADF/cofilin and actin dynamics in disease, *Trends Cell Biol* 12:598–605, 2002.

Campbell KP: Three muscular dystrophies: Loss of cytoskeleton-extracellular matrix linkage, *Cell* 80:675–9, 1995.

Holmes KC, Popp D, Gebhard W, Kabsch W: Atomic model of the actin filament, *Nature* 347:44–9, 1990.

Luna EJ, Hitt AL: Cytoskeleton—plasma membrane interactions, *Science* 258:955–64, 1992.

Schmidt A, Hall MN: Signaling to the actin cytoskeleton, *Annu Rev Cell Dev Biol* 14:305–38, 1998.

Winder SJ: Structural insights into actin-binding, branching and bundling proteins, *Curr Opin Cell Biol* 15:14–22, 2003.

Finer JT, Simmons RM, Spudich JA: Single myosin molecule mechanics: piconewton forces and nanometre steps, *Nature* 368:113–9, 1994.

Geeves MA, Holmes KC: Structural mechanism of muscle contraction, *Annu Rev Biochem* 68:687–728, 1999.

Goldman YE: Wag the tail: Structural dynamics of actomyosin, *Cell* 93:1–4, 1998.

Huxley HE: The mechanism of muscular contraction, *Science* 164:1356–65, 1969.

Pantaloni D, Le Clainche C, Carlier MF: Mechanism of actin-based motility, *Science* 292:1502–6, 2001.

Rayment I, Smith C, Yount RG: The active site of myosin, *Annu Rev Physiol* 58:671–702, 1996.

Schroder RR, Manstein DJ, Jahn W, Holden H, Rayment I et al: Three-dimensional atomic model of F-actin decorated with Dictyostelium myosin S1, *Nature* 364:171–4, 1993.

Welch MD, Mallavarapu A, Rosenblatt J, Mitchison TJ: Actin dynamics in vivo, *Curr Opin Cell Biol* 9:54–61, 1997.

Rayment I, Holden HM, Whittaker M, Yohn CB, Lorenz M et al: Structure of the actin-myosin complex and its implications for muscle contraction, *Science* 261:58–65, 1993.

Rayment I, Rypniewski WR, Schmidt-Base K, Smith R, Tomchick DR et al: Three-dimensional structure of myosin subfragment-1: A molecular motor, *Science* 261:50–58, 1993.

Ruppel KM, Spudich JA: Structure-function analysis of the motor domain of myosin, *Annu Rev Cell Dev Biol* 12:543–73, 1996.

Small JV, Glotzer M: Cell structure and dynamics, *Curr Opin Cell Biol* 18:1–3, 2006.

Tan JL, Ravid S, Spudich JA: Control of nonmuscle myosins by phosphorylation, *Annu Rev Biochem* 61:721–59, 1992.

Cramer LP, Mitchison TJ, Theriot JA: Actin-dependent motile forces and cell motility, *Curr Opin Cell Biol* 6:82–6, 1994.

Theriot JA, Mitchison TJ: Actin microfilament dynamics in locomoting cells, *Nature* 352:126–31, 1991.

Microtubules

Structure and Organization of Microtubules [3.8]. *Microtubules* are hollow polymeric microcylinders about 20 nm in diameter and up to 20 μm in length. Microtubules are composed of dimeric tubulins. There are three types of tubulin: α, β, and γ (see Table 3.6). The α- and β-tubulins are the primary constituents of microtubules, whereas the γ-tubulin regulates the nucleation of microtubule assembly. Each tubulin molecule used for constructing the microtubules is a heterodimer of α- and β-tubulin. In mammalian cells, there are several isoforms for α- as well as for β-tubulin. These isoforms have similar structures, but are originated from different genes. All tubulin isoforms can be polymerized into microtubules. Tubulin can be found in all mammalian cells. However, the distribution of tubulin varies in different cell types. For instance, the nerve cells exhibit a higher concentration of tubulin than do other cell types. The tubulin genes are highly conserved among different species.

In microtubules, α- and β-tubulin dimers are uniformly aligned along the axis of the microtubule, forming parallel protofilaments. In each protofilament, the α- or β-tubulin subunits are always arranged in the same direction, giving a polarity to microtubules with a plus and minus end. Each microtubule is composed of 13 protofilaments (Fig. 3.4). In

TABLE 3.6. Characteristics of Selected Tubulin Isoforms*

Proteins	Alternative Names	Amino Acids	Molecular Weight (kDa)	Expression	Functions
Tubulin $\alpha 1$	TUBA1, Tubulin α testis-specific	448	50	Nervous system, testis	Constituting microtubules
Tubulin $\alpha 2$	TUBA2	450	50	Thymus, leukocytes, intestine, overy, testis	Constituting microtubules
Tubulin $\alpha 3$	Tubulin α brain-specific, B $\alpha 1$	451	50	Brain	Constituting microtubules in central nervous system
Tubulin α, ubiquitous	K-$\alpha 1$	451	50	Ubiquitous	Constituting microtubules
Tubulin β	TUBB	445	50	Brain	Constituting microtubules
Tubulin γ	TUBG1, TUBG, tubulin $\gamma 1$ chain, $\gamma 1$ tubulin, γ tubulin complex component 1, tubulin γ polypeptide	451	51	Heart, lung, liver, kidney, intestine, ovary, skeletal muscle	Regulating the nucleation of microtubule assembly

*Based on bibliography 3.8.

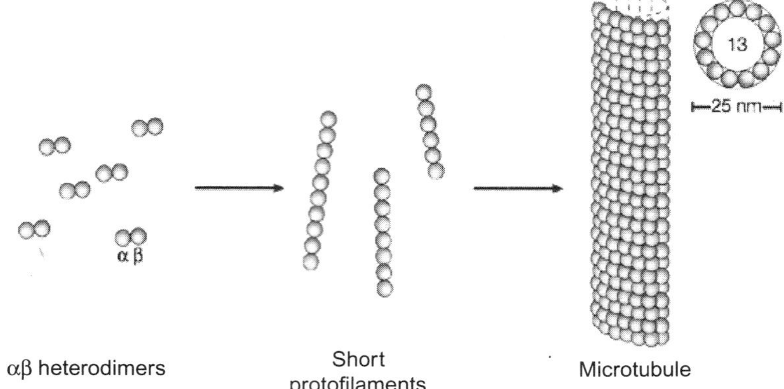

Figure 3.4. Formation of a microtubule from tubulin molecules. A microtubule is formed via several steps: (1) an α, β-tubulin monomer aggregates to form a tubulin heterodimer; (2) the tubulin heterodimers form short linear protofilaments; (3) 13 protofilaments are joined together laterally to organize into a microtubule. (Adapted by permission from Macmillan Publishers Ltd.: Westermann S, Weber K: *Nature Rev Mol Cell Biol* 4:938–48, copyright 2003.)

an interphase cell, microtubules are distributed in the radial direction with the minus end attached to the centrosome and the plus end toward the cell periphery.

It is important to note that several substances, including colchicine, colcemid, and taxol, are commonly used to modulate the assembly, structure, stability, and function of microtubules. *Colchicine* is an alkaloid extracted from meadow saffron. Colchicine can bind to tubulin and suppress tubulin polymerization or microtubule assembly. Since microtubules undergo continuously depolymerization, a treatment with colchicine facilitates the disassembly of microtubules. Once tubulin molecules are polymerized into microtubules, colchicine can no longer bind to tubulin. *Colcemid* is a substance similar to colchicine in function. Since the disassemble of microtubules interrupt cell mitosis, colchicine and colcemid are used to treat cancer. *Taxol* is derived from yew trees and can bind to polymerized microtubules. The binding of taxol enhances the stability of microtubules, inhibiting tubulin depolymerization. Such an effect induces cell arrest during mitosis. Taxol is also used as a drug for the treatment of cancer.

Microtubule Assembly and Disassembly [3.9]. Microtubule assembly is accomplished via tubulin polymerization, whereas its disassembly is via tubulin depolymerization. There are two critical processes, which are involved in microtubule assembly: nucleation and elongation. *Nucleation* is the formation of short tubulin protofilaments or oligomers, which further form a short initiating microtubule (Fig. 3.4). *Elongation* is the growth of microtubules based on the initial microtubule segment. Microtubule assembly can be simulated in vitro with tubulins in the presence of Mg^{2+} and GTP. The initial nucleation from tubulin heteodimers is a more difficult process than elongation. Thus nucleation is usually a slower process than elongation. While a microtubule is elongating via tubulin polymerization, there also exists simultaneous tubulin depolymerization. The rate of tubulin polymerization and depolymerization is dependent on the concentration of free tubulins. At a critical concentration of free tubulin, the rate of tubulin polymerization is counterbalanced by that of depolymerization, and microtubules cease growing.

Microtubules are connected at their minus end to a central structure within the cell, known as the *centrosome*, which is located in the nucleus during the interphase. The centrosome is considered the origin where microtubules grow from. The relationship of microtubules with the centrosome can be verified by observing the growth of degraded microtubules. A treatment with colcemid induces the degradation of microtubules. In the presence of fluorescent marker-tagged tubulins, it can be found that new microtubules grow from the centrosome following the removal of colcemid. These microtubules continuously elongate toward the cell periphery until a complete microtubule network is reestablished. Each centrosome contains two cylindrical structures perpendicular to each other, known as *centrioles*. During the interphase, the centrosome can be split into two daughter centrosomes, which move to opposite sides of the nucleus during the early stage of cell mitosis, serving as two poles for anchoring microtubule spindles (Fig. 3.5).

A microtubule undergoes rapid assembly and disassembly. The tubulins within a microtubule could be completely replaced with new tubulins within a period as short as 20 min. Such a process can be detected by injecting fluorescent marker-tagged tubulins into a living cell and observed by fluorescence microscopy. It is interesting to note that microtubules undergo alternating growth and retraction, resulting in a dynamic change in the length of microtubules. These dynamic changes are critical for the redistribution of microtubules within a cell.

Microtubule dynamics requires the presence of GTPs, which produce energy by hydrolysis. Each α- and β-tubulin is bound with a GTP molecule, which is required for tubulin polymerization. On the polymerization of a tubulin heterodimer to a microtubule, the GTP molecule associated with the β-tubulin can be hydrolyzed to produce energy, whereas the GTP molecule associated with the α-tubulin serves as a constituent of the tubulin and cannot be hydrolyzed. The energy produced by the hydrolysis of the β-tubulin-associated GTP is used for microtubule depolymerization, but not for polymerization. This can be

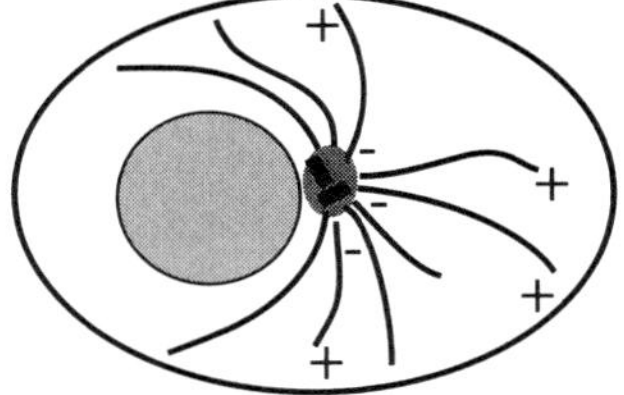

Dividing cell

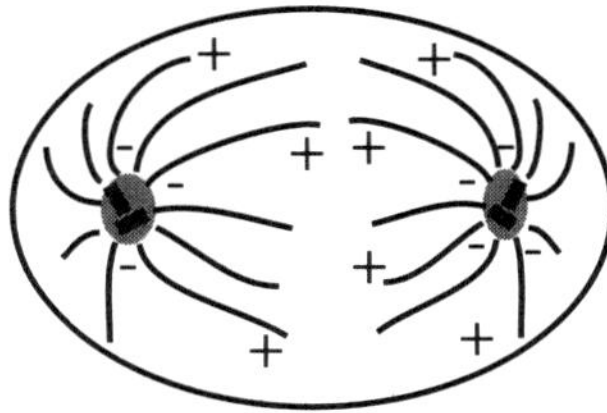

Figure 3.5. Centrosomes and microtubules in nondividing and dividing cells (based on bibliography 3.9).

verified by using GTP analogs that cannot be hydrolyzed. Tubulins associated with GTP analogs can be polymerized. However, once incorporated into a microtubule, these tubulin molecules cannot be depolymerized, suggesting that GTP hydrolysis is critical to the depolymerization of microtubules.

A microtubule can be assembled at the plus and minus ends, but exhibits different assembly rate at these ends under a given condition. The assembly of microtubules can be observed by using in vitro experiments. Isolated microtubules from cells can grow in the presence of free tubulins. The plus end of a microtubule grows about 3 times faster than the minus end. Since microtubules are aligned in the radial direction of a cell with the plus ends pointing at the periphery, microtubules often grow from the cell center to the periphery.

Regulation of Microtubule Dynamics [3.9]. The assembly and disassembly of microtubules are processes regulated by microtubule-associated proteins (Table 3.7). Two major types of microtubule-associated proteins have been identified in nerve cells: the high-molecular-weight proteins and the τ proteins. The *high-molecular-weight proteins* include microtubule-associated proteins 1 and 2 with molecular weights 200 and 300 kDa, respectively. The *τ proteins* have molecular weights ranging from 55 to 62 kDa. Each of these microtubule-associated proteins contains two domains; the first domain is capable of binding to microtubules, and the second domain binds to other types of intracellular structures. The binding of microtubule-associated proteins to microtubules prevents microtubules from depolymerization and enhances the stability of the microtubules. The exact regulatory mechanisms, however, remain to be investigated.

Function of Microtubules [3.10]. One of the primary functions of microtubules is the control of cell polarity. Microtubules exhibit nonuniform tubulin polymerization and depolymerization through the cell. Such a nonuniform feature is critical to the controlled distribution of microtubules, potentially contributing to cell polarization. At a given time, some microtubules may undergo predominant polymerization, while others may experience depolymerization. Fast-growing microtubules may be capped or protected by capping molecules, yielding stabilized microtubules in a specified direction. Meanwhile, uncapped microtubules are not stable and cannot grow as rapidly as the capped microtubules. The rapid growth of the capped microtubules causes regional extension of the cell membrane, leading to the formation of cell polarity.

Microtubules play a critical role in the transport of intracellular organelles and vesicles, which are required for a variety of metabolic and signaling activities. The transport function is accomplished by coordinated interactions of motor proteins, including kinesin and dynein, with microtubules. Each motor molecule is composed of two heavy chains and several light chains. Each heavy chain contains a globular head and a tail. The head interacts directly with microtubules and induces the sliding of the motor protein along a microtubule, a process dependent on ATPs, whereas the tail binds to an intracellular component to be moved. The light chains also play a role in the regulation of motor protein movement.

The motor proteins *kinesin* and *dynein* (Table 3.8) are both involved in the transport of intracellular organelles and chromosome separation during mitosis. However, kinesin and dynein move in opposite directions along a microtubule. Kinesin can only move intracellular organelles from the centrosome or the minus end of the microtubule to the cell periphery or the plus end of the microtubule, whereas dynein moves toward the cen-

TABLE 3.7. Characteristics of Selected Microtubule-Associated Proteins*

Proteins	Alternative Names	Amino Acids	Molecular Weight (kDa)	Expression	Functions
Microtubule -associated protein 1A	Microtubule-associated protein 1-like, MTAP1A	2805	306	Central nerveous system	Regulating microtubule assembly
Microtubule -associated protein 1B	MAP1B, MAP1 light chain LC1	2468	271	Central nerveous system	Regulating microtubule assembly
Microtubule-associated protein-2	MAP2, dendrite-specific MAP	1858	203	Central nerveous system	Regulating microtubule assembly
τ	Microtubule-associated protein τ, MAPT, MTBT1, neurofibrillary tangle protein, paired helical filament τ	758	79	Central nerveous system	Regulating microtubule assembly, playing a critical role in both integrity and functionality of neurons[a]

[a]Note that τ mutation induces the formation of neurofibrillary tangles and causes neurodegenerative disorders such as Alzheimer's disease, Parkinson's disease, frontotemporal dementia, and supranuclear palsy.
*Based on bibliography 3.9.

TABLE 3.8. Characteristics of the Motor Proteins Kinesin and Dynein*

Proteins	Alternative Names	Amino Acids	Molecular Weight (kDa)	Expression	Functions
Kinesin heavy chain 2	HK2	679	77	Nervous system, lung, spleen, stomach, testis, placenta	A constituent for the microtubule-associated motor protein kinesin, which mediates movement of cytoplasmic structures such as chromosomes and vesicles[a]
Kinesin light chain	Kinesin 2	569	65	Nervous system	A constituent for the microtubule-associated motor protein kinesin (see above for function)
Dynein cytoplasmic heavy chain 1	Dynein cytoplasmic heavy polypeptide 1	4646	532	Brain, heart, lung, pancreas, liver, testis	A constituent for the microtubule-associated motor protein dynein, which move organelles and vesicles from cell periphery or plus end of microtubule to centrosome or minus end of microtubule
Dynein cytoplasmic intermediate chain 1	Cytoplasmic dynein intermediate chain 1	645	73	Brain heart, lung, pancreas, liver, testis	A constituent for the microtubule-associated motor protein dynein (see above for dynein function)
Dynein cytoplasmic light chain 1	8-kDa dynein light chain, cytoplasmic dynein light polypeptide	89	10	Brain, heart, lung, pancreas, liver, testis	A constituent for the microtubule-associated motor protein dynein (see above for dynein function)

[a]Note that kinesins move organelles from the centrosome or the minus end of the microtubule to the cell periphery or the plus end of the microtubule.
*Based on bibliography 3.10.

trosome. A *kinesin* molecule is a tetramer composed of two heavy and two light chains. The heavy chain is located at the *N*-terminus of the molecule, and the light chains are at the *C*-terminus. The *N*-terminal heavy chains form the motor domains with the microtubule-binding regions, which mediate the sliding motion of the kinesin molecule along the microtubule. The *N*-terminal heavy chain also possesses an ATP-binding site, which serves as an ATPase and interacts with ATP molecules to provide energy for kinesin movement. The movement caused by kinesin molecules can be readily verified by using in vitro assays with purified motor proteins and microtubules. When microtubules are mixed with kinesin-coated polystyrene beads, the beads move toward the plus end of microtubules.

Dyneins are a family of motor proteins that are divided into two groups: the axonal and cytoplasmic dyneins. The axonal dynein molecules are responsible for organelle transport within the neuronal axon. Cytoplasmic dyneins mediate intracellular motility, protein sorting, and movement of intracellular organelles such as endosomes and lysosomes. A dynein molecule is comprised of two force-generating heavy chains and several intermediate and light chains. The heavy chains contain ATPases, which interact with ATP molecules and generate energy for mechanical movement. The motility of dynein molecules can be observed by using in vitro assays with purified dynein molecules and microtubules. Dynein molecules can move intracellular organelles from the cell periphery or the plus end of the microtubule to the centrosome or the minus end of the microtubule.

Microtubules are well known for their role in regulating cell mitosis or the segregation of chromosomes. Microtubules and associated proteins constitute a key structure for cell mitosis, known as the *mitotic spindle*, which plays a critical role in the alignment and separation of chromosomes. During the early stage of mitosis or prophase, the centrosome is separated into two daughter centrosomes, which move toward the two opposite poles. A mitotic spindle is initiated from the two centrosomes and gradually forms a polar structure. During metaphase, chromosomes are attached to the spindle microtubules. The shortening of the microtubules induces the movement of separated daughter chromosomes from the cell center toward the two centrosome poles. The destruction of microtubules by a treatment with colchicine interrupts cell mitosis.

Intermediate Filaments

Structure and Organization of Intermediate Filaments [3.11]. *Intermediate filaments* are one of the three types of filamentous structures that constitute the cytoskeleton. The term "intermediate" is derived from the fact that the diameter of intermediate filaments (~10 nm) is between the other two types of cytoskeletal filaments (Fig. 3.6), specifically, actin filaments (~8 nm) and microtubules (~25 nm). Intermediate filaments are composed of various molecules, including keratin, vimentin, neurofilament protein, and nuclear lamin. The constituent molecules of intermediate filaments are fibrous in shape. To form an intermediate filament, two molecules are organized into a parallel dimer with the amino termini at one end and the carboxyl termini at the other end. For most types of intermediate filaments, the two dimers in turn form an antiparallel tetramer bundle with the amino termini of one dimer arranged with the carboxyl termini of the other dimer at each end of the tetramer bundle (Fig. 3.7). The tetramers are the basic units that are assembled into helical intermediate filaments via bundle–lateral interactions. Because of the antiparallel feature of the tetramer bundles, intermediate filaments do not exhibit polarity.

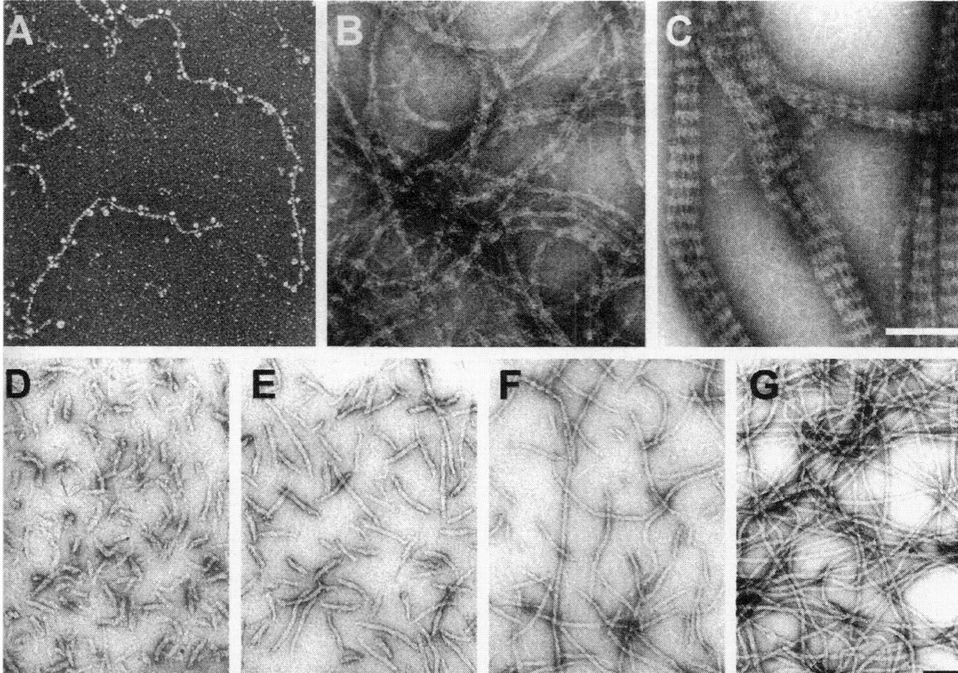

Figure 3.6. Electron micrographs of intermediate filaments at different assembly stages. (A–C) Lamin A/C. Lamin filaments can be dialyzed in pH 6.5/150 mM NaCl buffer, generating linear head-to-tail fibers (panel A). In the presence of Ca^{2+}, lamin filaments can be dialyzed into beaded long filaments (panel B). Panel C shows assembled lamin filaments. (D–G) Assembly of recombinant human vimentin. Vimentin filament assembly was initiated by adding filament buffer and fixed with 0.1% glutaraldehyde at 10 s (panel D), 1 min (panel E), 5 min, (panel F), and 1 h (panel G). Scale bar: 100 nm. (Reprinted by permission from Herrmann H, Aebi U: *Annu Rev Biochem*, 73:749–89, copyright 2004 by *Annual Reviews*, www.annualreviews.org.)

On the basis of constituents, intermediate filaments are classified into several subtypes, including keratin filaments, vimentin filaments, neurofilaments, and lamin filaments (see list in Table 3.9), which are found in different cell types. *Keratin filaments* are composed of various types of keratin and are present in epithelial cells, the hair, and the nails. Individual keratin molecules are different in structure and can be grouped into to subfamilies, including types I and II keratins, based on the properties of amino acids. Type I keratins are acidic with a molecular weight 40–70 kDa, whereas type II keratins are basic or neutral with a similar molecular weight. Both type I and type II keratins are required for the constitution of keratin filaments. In a typical epithelial cell, keratin filaments are connected at the end to *desmosomes*, a cell junction structure that joins two neighboring cells. In addition, keratin filaments anchor to hemidesmosomes, a structure that mediates cell attachment to the basal lamina.

Vimentin filaments are present in fibroblasts, endothelial cells, and leukocytes, and contain a single type of molecule: vimentin. In addition, there exist vimentin-related filaments, which exhibit structure and properties similar to those of vimentin filaments. One type is *desmin filaments*, which are composed of desmin and are present primarily in

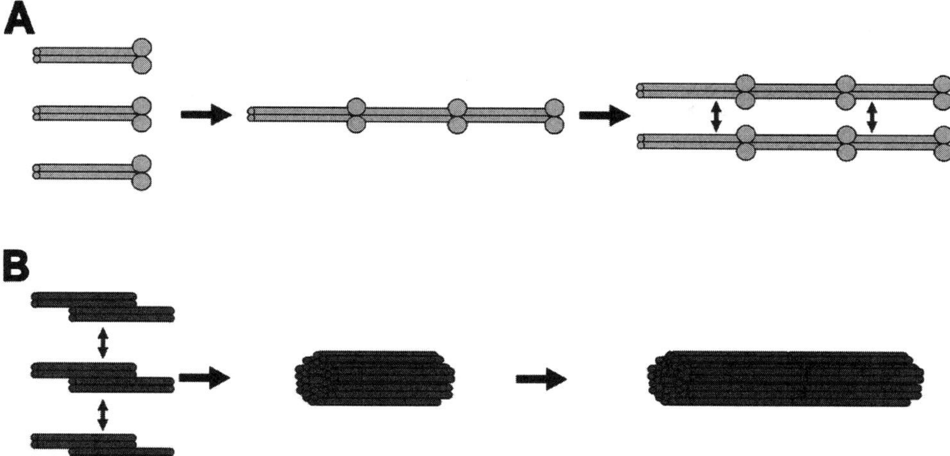

Figure 3.7. Schematic representation of intermediate filament assembly. (A) Lamin filament assembly. Lamin dimers are first associated into head-to-tail filaments, which are further associated laterally into complete filaments. (B) Vimentin filament assembly. Vimentin molecules first form antiparallel half-staggered double dimers (or tetramers), which form complete vimentin filaments. (Reprinted by permission from Herrmann H, Aebi U: *Annu Rev Biochem*, 73:749–89, copyright 2004 by *Annual Reviews*, www.annualreviews.org.)

muscle cells, including smooth, skeletal, and cardiac muscle cells. Desmin filaments often anchor to cell junctions. Another type is *glial filaments* composed of glial fibrillary acidic proteins. This type of intermediate filament is found in astrocytes of the central nervous system and Schwann cells of the peripheral nervous system. It is important to note that vimentin and vimentin-related proteins can be crosslinked together, but these proteins cannot be crosslinked with keratin-based intermediate filaments.

Neurofilaments are present in neurons, arranged primarily along the axon. There are three types of neurofilament proteins, including neurofilament-L, -M, and -H, based on low, medium, and high molecular weights, respectively. These molecular types can be found within all neurofilaments. In a typical neuronal axon, neurofilaments are uniformly spaced with a high density. These filaments are laterally crosslinked, providing mechanical strength to the axon.

Lamin filaments are found in the nuclear lamina, which is a ~20-nm membrane lining the internal surface of the nuclear membrane. Lamin filaments are composed of two types of lamin: lamin A (or A/C) and lamin B. In structure, lamin is similar to other intermediate filament proteins. However, lamin contains signaling structures that direct lamin transport from the cytosol to nucleus. The lamin filaments undergo dynamic disassembly during early mitosis and reassembly during the late mitosis in coordination with chromosome reorganization and separation. In interphase cells, lamin filaments are organized into a dense lattice network. The network is interrupted at nuclear pores, which allow the transport of molecules from and to the nucleus.

Function of Intermediate Filaments [3.12]. A major function of intermediate filaments is to provide mechanical strength to cells and tissues. Such a function is supported by observations from transgenic keratin-deficient animal models. In transgenic mice with a

TABLE 3.9. Characteristics of Selected Molecules that Constitute Intermediate Filaments*

Proteins	Alternative Names	Amino Acids	Molecular Weight (kDa)	Expression	Functions
Keratin	Cytokeratin, cytokeratin 1, keratin type II cytoskeletal 1, hair α protein, 67-kDa cytokeratin	644	66	Keratinocytes	Constituting keratin filaments in epithelial cells, hair, and nails
Vimentin	Vim	467	57	Epithelial cells, fibroblasts, muscular cells	Constituting vimentin intermediate filaments
Neurofilament protein heavy polypeptide	Neurofilament heavy polypeptide, neurofilament triplet H protein, 200-kDa neurofilament protein	1026	112	Nervous system	Constituting neurofilaments in neurons
Neurofilament protein light polypeptide	Neurofilament protein light chain, neurofilament triplet L protein, 68-kDa neurofilament protein, neurofilament protein	544	62	Nervous system	Constituting neurofilaments in neurons
Lamin A/C	Lamin A, lamin C, 70-kDa lamin	702	79	Ubiquitous	A constituent for nuclear lamina that regulates nuclear stability, chromatin structure, and gene expression
Lamin B	LMNB1, LMNB	586	66	Ubiquitous	A constituent for nuclear lamina

*Based on bibliography 3.11.

mutant keratin gene that lacks the amino/carboxyl-terminal domains, the mechanical strength of epidermis reduces significantly, resulting in cell injury in response to mechanical impacts that are harmless to normal cells. In human genetic diseases with mutation in the keratin gene, epidermal cells and tissues demonstrate a similar phenomenon, leading to skin blistering. In the human or animal skin, there exists a layer of keratin filaments that are highly crosslinked. Such a keratin layer serves as a protective structure for internal tissues.

The function of intermediate filaments is not limited to the enhancement of mechanical strength. Various types of intermediate filaments are bound to other cytoskeletal filaments. For instance, desmin filaments are linked to actin filaments in muscular cells, suggesting a role for the desmin filaments in regulating the interaction of contractile filaments. In addition, desmin filaments are attached to cell junctions, suggesting a role for these filaments in regulating cell-to-cell interactions.

ENDOPLASMIC RETICULUM [3.13]

Endoplasmic reticulum (ER) is a cytosolic membrane system consisting of lipid bilayers and is involved in the synthesis of proteins and lipids as well as in the sequestration and release of calcium. There is a rich network of interconnected tubular branches or sheets in the ER, forming a continuous membrane system in each cell. The ER membrane constitutes about 50% of the total cell lipid membrane. The ER tubular structures occupy about 10% of the total volume of the cell. There are two types of ER: rough and smooth. *Rough ER* is defined as ER with attached ribosomes on the cytosolic surface, whereas *smooth ER* is that without ribosomes.

ER is involved in the synthesis of proteins as well as lipids. Ribosomes bound to the ER are sites for protein translation. Proteins translated by ribosomes are transported to the rough ER for further processing before being released into the cytosol. In the lumen of the rough ER, proteins are modified by ER resident protein enzymes, a process critical in protein folding and assembly. An important enzyme for protein modification is protein disulfide isomerase in the rough ER. This enzyme catalyzes the formation of disulfide (S—S) bonds between cysteines, a proces critical in the formation of a three-dimensional protein structure. Another function of rough ER is to add sugar residues to proteins, a process known as *glycosylation*, which results in the formation of glycoproteins. The addition of sugar residues to proteins is catalyzed by enzymes present in the rough ER. A typical enzyme is oligosaccharyl transferase, which is localized to the ER membrane. This enzyme catalyzes the addition of a preformed oligosaccharide, composed of *N*-acetylglucosamine, mannose, and glucose, to the side NH_2 group of asparagines. The original oligosaccharide chain is trimmed or processed to remove certain sugar residues while the glycoproteins are still in the ER. Glycoproteins will be further processed when the molecules are transported into the Golgi apparatus (see the following section). Glycoproteins serve as cell membrane receptors. The sugar residues play a critical role in the recognition of and interaction with extracellular ligands.

The *smooth ER* constitutes a small fraction of the ER system in most cells and is connected to the rough ER. Rough ER segments are often found in smooth ER-dominant regions. A primary function of the smooth ER is to transport proteins from the ER to the Golgi apparatus. In addition, smooth ER is involved in the synthesis of lipids. Almost all lipid bilayers are assembled within the ER system. The ER system of hepatocytes is involved in the synthesis of lipoproteins. These molecules are released into blood and serve as lipid carriers between various tissues and organs. Cells for the synthesis of steroid hormones are rich in smooth ER.

The ER system also plays a critical role in the storage and controlled release of calcium. In an inactive state, calcium is stored in the ER, where calcium-binding proteins sequester calcium. In response to stimulation for intracellular signaling processes that require calcium, the calcium channels of the ER are open, resulting in the release of calcium. Calcium mediates a variety of molecular processes, ranging from actin–myosin interaction to activation of signaling protein kinases.

GOLGI APPARATUS [3.14]

The *Golgi apparatus* is a stack of lipid membrane cisternae and tubular networks and is involved in the synthesis of carbohydrates and in the modification and sorting of proteins transported from the ER. The Golgi apparatus is located near the cell nucleus and centrosome. There exist several subsystems in the Golgi apparatus, including the *cis*-Golgi network, *cis*-cisterna, medial cisterna, *trans*-Golgi cisterna, and *trans*-Golgi network (Fig. 3.8). The *cis*-Golgi network is a membrane tubular network, which is connected to the *cis-cis*terna and serves as the entrance for protein-containing vesicles transported from the ER. Proteins are transported from the *cis*-Golgi network to the *cis*-cisterna. The *cis*-cisterna is adjacent, but not connected to the medial cisterna. Proteins are transported from the *cis*-cisterna to the medial cisterna via vesicular carriers. Similarly, the medial cisterna is not connected to the *trans*-cisterna. Vesicular transport is required for the movement of proteins from the medial cisterna to the *trans*-cisterna. The *trans*-cisterna is connected to the *trans* network, which serves as an exit for processed proteins. The exiting proteins are carried by vesicles to cellular compartments, including cell membranes, secretary vesicles, and lysosomes, where proteins are used for various purposes.

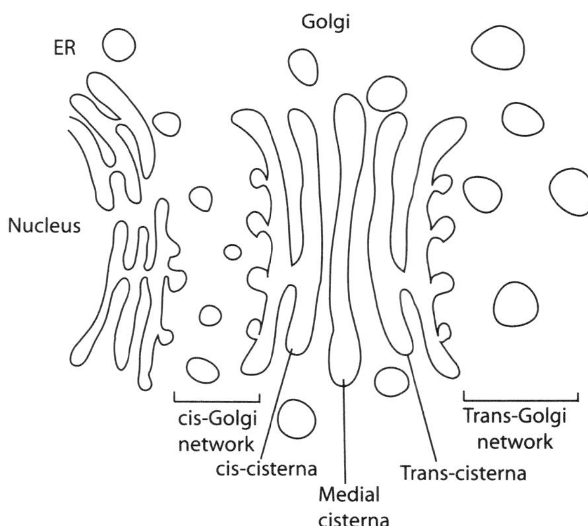

Figure 3.8. Schematic representation of Golgi apparatus (based on bibliography 3.14).

Major functions of the Golgi apparatus are to *modify proteins* and *synthesize carbohydrates*. Proteins are preliminarily modified in the ER by the addition of oligosaccharides. When transported to the Golgi apparatus, the proteins are further processed by glycosylation, or the addition of complex oligosaccharides and high-mannose-content oligosaccharides. The glycosylation process, which occurs through the Golgi cisternae, is critical to the formation of glycolproteins. In addition, the Golgi apparatus assembles *proteoglycans*, a process involving the polymerization of glycosaminoglycans (GAG) and the linkage of GAG chains to core proteins. Proteoglycans are deployed to the extracellular space and serve as ground substance. It is important to note that lipid vesicles can bud from the Golgi network and cisternae. These vesicles play a critical role for the transport of proteins between the Golgi subsystems and from the Golgi apparatus to destination compartments.

ENDOSOMES AND LYSOSOMES [3.15]

Endosomes are lipid vesicles that form by budding from cell membranes during *endocytosis*, a process by which cells ingest macromolecules and cell debris. Endocytosis is initiated when a stimulating macromolecule contacts the cell membrane. In response to such a contact, the stimulated region of the cell membrane invaginates, pinches off from the cell membrane, encloses the stimulating macromolecule, and forms an endosome. Most cells are capable of ingesting fluids, solutes, and small molecules, while phagocytic cells, such as macrophages and neutrophils, can take up large particles with a diameter in the order of μm (micrometers), such as bacteria and cell debris. Endosomes in phagocytic cells are also known as *phagosomes*. Endocytosis in phagocytic cells plays a critical role in protecting cells from bacterial infection and in scavenging debris from damaged and dead cells. Endosomes or phagosomes are eventually transformed to lysosomes, where ingested contents are degraded by enzymes.

Lysosomes are lipid membrane vesicles in which ingested molecules or particles are digested or degraded. All mammalian cells contain lysosomes. A typical lysosome contains numbers of hydrolytic enzymes, including proteases, lipases, phospholipases, and glycosidases, which degrade a variety of molecules. These digestive enzymes are synthesized by ribosomes in the rough ER, processed in the ER and Golgi apparatus, and delivered to lysosomes by Golgi vesicles. The internal environment of lysosomes is highly acidic with a pH value of ~5, which is advantageous for the activation of the hydrolytic enzymes. The internal H^+ concentration is maintained by H^+ pumps in the lysosomal membrane at the expense of energy from ATP molecules. The final products of the digestion, including saccharides, amino acids, and nucleotides, are transported across the lysosomal membrane to the cytosol, where these products are recycled.

In addition to the endosomes formed by endocytosis, there is another route that delivers materials to lysosomes for digestion. This route is used for the destruction and disposal of intracellular obsolete structures and organelles, a process known as *autophagy*. An obsolete organelle is usually enclosed by an ER membrane, forming an autophagosome. The autophagosome is then fused with a lysosome or endosome, where the enclosed organelle is degraded and disposed. Thus, endosomes and lysosomes play a critical role in the destruction and clearance of externally ingested materials as well as internally obsolete subcellular organelles.

MITOCHONDRIA [3.16]

Structure and Organization

Mitochondria are intracellular lipid membrane organelles that generate, store, and dispatch energy necessary for molecular activities. There are two types of specialized membrane for each mitochondrion: the internal and external membrane. These membranes divide a mitochondrion into two compartments: the *internal matrix space* and the *intermembrane space*. While the external membrane appears smooth, the internal membrane forms numbers of protrusions into the internal matrix space, known as *cristae*. The protrusions greatly increase the surface area of the internal membrane, which is necessary for membrane-related energy-generating processes. Each mitochondrial compartment and membrane contains distinct proteins that are developed for specialized functions as discussed below.

The external layer of mitochondria is composed of a large number of porins, proteins that form channels across the membrane. The porin channels allow the transport of water, salts, small proteins, and other molecules with a molecular weight $<\sim 5$ kDa. Most of these molecules, however, cannot pass through the internal membrane. Because of the high permeability of the external membrane, electrolytes, water, and small molecules are equilibrated between the intermembrane space and the cytosol.

The internal membrane of the mitochondria is different from the external membrane. It is composed of a high density of cadiolipin, a phospholipid molecule containing four fatty acids. The presence of this lipid molecule renders the internal membrane highly impermeable to ions. The internal membrane contains a variety of specialized transport proteins, which exhibit selective permeability to molecules necessary for intramitochondrial activities. Because of the selective permeability of the internal membrane, the environment in the internal matrix space is different from that of the intermembrane space. Most importantly, the internal membrane consists of enzymes of the intracellular respiratory chain, forming an enzymatic cascade responsible for oxidation reactions and energy generation. One enzyme, known as *ATP synthase*, catalyzes the formation of ATP molecules.

The internal matrix space of mitochondria contains enzymes that metabolize pyruvate and fatty acids, generating acetyl CoA. This space also contains enzymes that oxidize acetyl CoA. The end products of these enzymatic reactions include nicotine adenine dinucleotide hydride (NADH) and CO_2. NADH is a form of nicotine adenine dinucleotide (NAD) with the addition of two electrons and is a major carrier and source of electrons for energy generation in the mitochondria. CO_2 is a waste product, which is released into the blood and removed from the lung and kidney. The internal matrix also contains mitochondrial DNA, ribosomes, tRNA, and enzymes necessary for regulating the expression of mitochondrial genes.

ATP Generation

The primary function of mitochondria is generation of energy in the form of ATP for molecular and cellular activities. Sources for mitochondrial energy generation are fatty acids and glycogens, or glucose polymers. Fatty acids are a more efficient form than glycogen for energy generation. The oxidation of fatty acids can generate energy 6 times as much as that of an equal amount of glycogen. Fatty acids are mainly stored in fat cells, whereas glycogens are stored in liver and muscle cells. It is important to note that glucose can be converted to fatty acids, but fatty acids cannot be converted to glucose.

For fatty acid oxidation, fatty acid molecules are transported through the external and internal membranes of the mitochondria to the internal matrix. Each fatty acid is processed through a four-enzyme oxidation cycle, which catalyzes the oxidation of fatty acids. Each cycle reduces a fatty acid by two carbons, giving an acetyl CoA and two distinct high-energy electron carriers: NADH and $FADH_2$ (flavin adenine dinucleotide hydride). The acetyl CoA molecule is further oxidized in the citric cycle, and NADH and $FADH_2$ are used for electron transfer in energy generation.

For glycogen metabolism, cells first break down glycogen into glucose 1-phosphate, which occurs in the cytosol. Each glucose 1-phosphate is further catalyzed into two pyruvate molecules, which are transported from the cytosol into the mitochondrial internal matrix. The pyruvate molecules are catalyzed by a complex of enzymes and coenzymes into acetyl CoA and CO_2. The acetyl CoA molecule is further oxidized for energy generation through the citric cycle.

The *citric cycle*, also known as the Krebs cycle or tricarboxylic acid cycle, is the principal process that oxidizes fatty acids and pyruvates. About 60% of carbohydrates are processed by the citric cycle. Such a process produces CO_2 as a waste and high-energy electrons, which are carried by NADH and $FADH_2$ and used for the generation of ATP molecules. The citric cycle is a sequence of enzymatic events, starting with the formation of citric acid from acetyl CoA or pyruvate. Each cycle produces 2 CO_2, 2 H_2O, 1 $FADH_2$, 3 NADH with 3 H^+, and 1 GTP. The GTP molecule is converted to ATP by direct transfer of a high-energy phosphate group.

In the citric cycle, most energy from the oxidation of carbohydrates is saved in the form of high-energy electrons, which are carried by NADH and $FADH_2$. These electrons are transferred through the respiratory chain to oxygen, providing energy for the formation of ATP molecules. Such a process is referred to as *oxidative phosphorylation*. It has been hypothesized that oxidative phosphorylation is dependent on a chemiosmotic process. In such a process, chemically generated high-energy electrons from the hydrogen of NADH and $FADH_2$ are transported through the electron-carrying molecules of the respiratory chain localized to the mitochondrial internal membrane (note that each hydrogen atom gives a proton H^+ and an electron e^-). The energy released from the electron transfer is used to pump H^+ from the matrix side to the intermembrane side of the internal membrane, establishing a proton gradient across the internal membrane. This gradient drives H^+ flow in the opposite direction, providing energy for the synthesis of ATPs from ADPs and phosphates by ATP synthase.

CELL NUCLEI [3.17]

The cell nucleus is an organelle that contains the hereditary molecules—DNAs. The nucleus is enclosed with a nuclear envelope, which contains two lipid membranes: the outer and inner membranes. The outer membrane is a continuation of the adjacent ER membrane, and the intermembrane space is connected to the ER. The nucleus membranes are supported by an internal layer and an external layer of intermediate filaments. The internal supporting layer is a relatively dense structure composed of nuclear lamin and is defined as the *nuclear lamina*. The external supporting layer is composed of loosely organized intermediate filaments. These intermediate filament-containing layers protect the nucleus from mechanical impacts and injury. Across the nucleus membrane and dense nuclear lamina, there exist pores, which allow the transport of selected molecules between the cytosol and nucleus. The nucleus contains chromosomes. The structure and function of chromosomes are discussed in Chapter 1.

BIBLIOGRAPHY

3.8. Structure and Organization of Microtubules

α-Tubulin

Ravelli RBG, Gigant B, Curmi PA, Jourdain I, Lachkar S et al: Insight into tubulin regulation from a complex with colchicine and a stathmin-like domain, *Nature* 428:198–202, 2004.

Villasante A, Wang D, Dobner P, Dolph P, Lewis SA et al: Six mouse alpha-tubulin mRNAs encode five distinct isotypes: Testis-specific expression of two sister genes, *Mol Cell Biol* 6:2409–19, 1986.

Wilde CD, Chow LT, Wefald FC, Cowan NJ. Structure of two human alpha-tubulin genes, *Proc Natl Acad Sci USA* 79:96–100, 1982.

Dode C, Weil D, Levilliers J, Crozet F, Chaib H et al: Sequence characterization of a newly identified human alpha-tubulin gene (TUBA2), *Genomics* 47:125–30, 1998.

Hall JL, Cowan NJ: Structural features and restricted expression of a human alpha-tubulin gene, *Nucleic Acids Res* 13:207–23, 1985.

Miller FD, Naus CCG, Durand M, Bloom FE, Milner RJ: Isotypes of alpha-tubulin are differentially regulated during neuronal maturation, *J Cell Biol* 105:3065–73, 1987.

Watts NR, Sackett DL, Ward RD, Miller MW, Wingfield PT et al: HIV-1 rev depolymerizes microtubules to form stable bilayered rings, *J Cell Biol* 150:349–60, 2000.

Cowan NJ, Dobner PR, Fuchs EV, Cleveland DW: Expression of human alpha-tubulin genes: Interspecies conservation of 3-prime untranslated regions, *Mol Cell Biol* 3:1738–45, 1983.

β-Tubulin

Cleveland DW, Lopata MA, MacDonald RJ, Cowan NJ, Rutter WJ et al: Number and evolutionary conservation of alpha- and beta-tubulin and cytoplasmic beta- and gamma-actin genes using specific cloned cDNA probes, *Cell* 20:95–105, 1980.

Cleveland DW, Sullivan KF: Molecular biology and genetics of tubulin, *Annu Rev Biochem* 54:331–65, 1985.

Ravelli RBG, Gigant B, Curmi PA, Jourdain I, Lachkar S et al: Insight into tubulin regulation from a complex with colchicine and a stathmin-like domain, *Nature* 428:198–202, 2004.

Wang HW, Nogales E: Nucleotide-dependent bending flexibility of tubulin regulates microtubule assembly, *Nature* 435:911–5, 2005.

Yen TJ, Machlin PS, Cleveland DW: Autoregulated instability of beta-tubulin mRNAs by recognition of the nascent amino terminus of beta-tubulin, *Nature* 334:580–5, 1988.

γ-Tubulin

Aldaz H, Rice LM, Stearns T, Agard DA: Insights into microtubule nucleation from the crystal structure of human gamma-tubulin, *Nature* 435:523–7, 2005.

Oakley CE, Oakley BR: Identification of gamma-tubulin, a new member of the tubulin superfamily encoded by mipA gene of Aspergillus nidulans, *Nature* 338:662–4, 1989.

Rommens JM, Durocher F, McArthur J, Tonin P, LeBlanc, JF et al: Generation of a transcription map at the HSD17B locus centromeric to BRCA1 at 17q21, *Genomics* 28:530–42, 1995.

Stearns T, Evans L, Kirschner M: Gamma-tubulin is a highly conserved component of the centrosome, *Cell* 65:825–36, 1991.

Wise DO, Krahe R, Oakley BR: The gamma-tubulin gene family in humans, *Genomics* 67:164–70, 2000.

Human protein reference data base, Johns Hopkins University and the Institute of Bioinformatics, at http://www.hprd.org/protein.

3.9. Microtubule Assembly and Disassembly

Microtubule-Associated Protein 1A

Fink JK, Jones SM, Esposito C, Wilkowski J: Human microtubule-associated protein 1a (MAP1A) gene: genomic organization, cDNA sequence, and developmental- and tissue-specific expression, *Genomics* 35:577–85, 1996.

Hammarback JA, Obar RA, Hughes SM, Vallee RB: MAP1B is encoded as a polyprotein that is processed to form a complex N-terminal microtubule-binding domain, *Neuron* 7:129–39, 1991.

Ikeda A, Zheng QY, Zuberi AR, Johnson KR et al: Microtubule-associated protein 1A is a modifier of tubby hearing (moth1), *Nature Genet* 30:401–5, 2002.

Langkopf A, Hammarback JA, Muller R, Vallee RB, Garner CC et al: Microtubule-associated proteins 1A and LC2: Two proteins encoded in one messenger RNA, *J Biol Chem* 267:16561–6, 1992.

Lien LL, Feener CA, Fischbach N, Kunkel LM: Cloning of human microtubule-associated protein 1B and the identification of a related gene on chromosome 15, *Genomics* 22:273–80, 1994.

Microtubule-Associated Protein 1B

Allen E, Ding J, Wang W, Pramanik S, Chou J et al: Gigaxonin-controlled degradation of MAP1B light chain is critical to neuronal survival, *Nature* 438:224–8, 2005.

Edelmann W, Zervas M, Costello P, Roback L, Fischer I et al: Neuronal abnormalities in microtubule-associated protein 1B mutant mice, *Proc Natl Acad Sci USA* 93:1270–5, 1996.

Lien LL, Boyce FM, Kleyn P, Brzustowicz LM, Menninger J et al: Mapping of human microtubule-associated protein 1B in proximity to the spinal muscular atrophy locus at 5q13, *Proc Natl Acad Sci USA* 88:7873–6, 1991.

Lien LL, Feener CA, Fischbach N, Kunkel LM: Cloning of human microtubule-associated protein 1B and the identification of a related gene on chromosome 15, *Genomics* 22:273–80, 1994.

Wirth B, Voosen B, Rohrig D, Knapp M, Piechaczek B et al: Fine mapping and narrowing of the genetic interval of the spinal muscular atrophy region by linkage studies, *Genomics* 15:113–8, 1993.

Zhang YQ, Bailey AM, Matthies HJG, Renden RB, Smith MA et al: Drosophila fragile X-related gene regulates the MAP1B homolog Futsch to control synaptic structure and function, *Cell* 107:591–603, 2001.

MAP2

Garner CC, Tucker RP, Matus A: Selective localization of messenger RNA for cytoskeletal protein MAP2 in dendrites, *Nature* 336:674–7, 1988.

Kalcheva N, Albala J, O'Guin K, Rubino H, Garner C et al: Genomic structure of human microtubule-associated protein 2 (MAP-2) and characterization of additional MAP-2 isoforms, *Proc Natl Acad Sci USA* 92:10894–8, 1995.

Marsden KM, Doll T, Ferralli J, Botteri F, Matus A: Transgenic expression of embryonic MAP2 in adult mouse brain: Implications for neuronal polarization, *J Neurosci* 16:3265–73, 1996.

τ Proteins

Abel KJ, Boehnke M, Prahalad M, Ho P, Flejter WL et al: A radiation hybrid map of the BRCA1 region of chromosome 17q12-q21, *Genomics* 17:632–41, 1993.

Alonso ADC, Grundke-Iqbal I, Iqbal K: Alzheimer's disease hyperphosphorylated tau sequesters normal tau into tangles of filaments and disassembles microtubules, *Nature Med* 2:783–7, 1996.

Andreadis A, Brown WM, Kosik KS: Structure and novel exons of the human tau gene, *Biochemistry* 31:10626–33, 1992.

Clark LN, Poorkaj P, Wszolek Z, Geschwind DH, Nasreddine ZS et al: Pathogenic implications of mutations in the tau gene in pallido-ponto-nigral degeneration and related neurodegenerative disorders linked to chromosome 17, *Proc Natl Acad Sci* 95:13103–7, 1998.

Conrad C, Andreadis A, Trojanowski JQ, Dickson DW, Kang D et al: Genetic evidence for the involvement of tau in progressive supranuclear palsy, *Ann Neurol* 41:277–81, 1997.

Giasson BI, Forman MS, Higuchi M, Golbe LI, Graves CL, Kotzbauer PT et al: Initiation and synergistic fibrillization of tau and alpha-synuclein, *Science* 300:636–40, 2003.

Goedert M, Spillantini MG, Potier MC, Ulrich J, Crowther RA: Cloning and sequencing of the cDNA encoding an isoform of microtubule-associated protein tau containing four tandem repeats: differential expression of tau protein mRNAs in human brain, *EMBO J* 8:393–9, 1989.

Goedert M, Wischik CM, Crowther RA, Walker JE, Klug A: Cloning and sequencing of the cDNA encoding a core protein of the paired helical filament of Alzheimer disease: Identification as the microtubule-associated protein tau, *Proc Natl Acad Sci USA* 85:4051–5, 1988.

Gotz J, Chen F, van Dorpe J, Nitsch RM: Formation of neurofibrillary tangles in P301L tau transgenic mice induced by A-beta42 fibrils, *Science* 293:1491–5, 2001.

Holzer M, Craxton M, Jakes R, Arendt T, Goedert M: Tau gene (MAPT) sequence variation among primates, *Gene* 341:313–22, 2004.

Hong M, Zhukareva V, Vogelsberg-Ragaglia V, Wszolek Z, Reed L et al: Mutation-specific functional impairments in distinct tau isoforms of hereditary FTDP-17, *Science* 282:1914–7, 1998.

Hutton M, Lendon CL, Rizzu P, Baker M, Froelich S et al: Association of missense and 5-prime-splice-site mutations in tau with the inherited dementia FTDP-17, *Nature* 393:702–5, 1998.

Lewis J, Dickson DW, Lin WL, Chisholm L, Corral A et al: Enhanced neurofibrillary degeneration in transgenic mice expressing mutant tau and APP, *Science* 293:1487–91, 2001.

Lewis J, McGowan E, Rockwood J, Melrose H, Nacharaju P et al: Neurofibrillary tangles, amyotrophy and progressive motor disturbance in mice expressing mutant (P301L) tau protein, *Nature Genet* 25:402–5, 2000.

Human protein reference data base, Johns Hopkins University and the Institute of Bioinformatics, at http://www.hprd.org/protein.

3.10. Function of Microtubules

Kinesin Heavy Chain 2

Debernardi S, Fontanella E, De Gregorio L, Pierotti MA, Delia D: Identification of a novel human kinesin-related gene (HK2) by the cDNA differential display technique, *Genomics* 42:67–73, 1997.

Homma N, Takei Y, Tanaka Y, Nakata T, Terada S et al: Kinesin superfamily protein 2A (KIF2A) functions in suppression of collateral branch extension, *Cell* 114:229–39, 2003.

Kinesin Light Chain

Cabeza-Arvelaiz Y, Shih LCN, Hardman N, Asselbergs F, Bilbe G et al: Cloning and genetic organization of the human kinesin light-chain (KLC) gene, *DNA Cell Biol* 12:881–92, 1993.

Chernajovsky Y, Brown A, Clark J: Human kinesin light (beta) chain gene: DNA sequence and functional characterization of its promoter and first exon, *DNA Cell Biol* 15:965–74, 1996.

Goedert M, Marsh S, Carter N: Localization of the human kinesin light chain gene (KNS2) to chromosome 14q32.3 by fluorescence in situ hybridization, *Genomics* 32:173–5, 1996.

Kamal A, Stokin GB, Yang Z, Xia C, Goldstein LS: Axonal transport of amyloid precursor protein is mediated by direct binding to the kinesin light chain subunit of kinesin-I, *Neuron* 28:449–59, 2000.

Dynein

Criswell PS, Ostrowski LE, Asai DJ: A novel cytoplasmic dynein heavy chain: Expression of DHC1b in mammalian ciliated epithelial cells, *J Cell Sci* 109:1891–8, 1996.

Hafezparast M, Klocke R, Ruhrberg C, Marquardt A, Ahmad-Annuar A et al: Mutations in dynein link motor neuron degeneration to defects in retrograde transport, *Science* 300:808–12, 2003.

Harada A, Takei Y, Kanai Y, Tanaka Y, Nonaka S et al: Golgi vesiculation and lysosome dispersion in cells lacking cytoplasmic dynein, *J Cell Biol* 141:51–9, 1998.

Kural C, Kim H, Syed S, Goshima G, Gelfand VI et al: Kinesin and dynein move a peroxisome in vivo: A tug-of-war or coordinated movement? *Science* 308:1469–72, 2005.

Mikami A, Paschal BM, Vallee RB: Molecular cloning of retrograde transport motor cytoplasmic dynein, *Neuron* 10:788–96, 1993.

Narayan D, Desai T, Banks A, Patanjali SR, Ravikumar TS et al: Localization of the human cytoplasmic dynein heavy chain (DNECL) to 14qter by fluorescence in situ hybridization, *Genomics* 22:660–1, 1994.

Vaisberg EA, Grissom PM, McIntosh JR: Mammalian cells express three distinct dynein heavy chains that are localized to different cytoplasmic organelles, *J Cell Biol* 133:831–42, 1996.

Vaisberg EA, Koonce MP, McIntosh JR: Cytoplasmic dynein plays a role in mammalian mitotic spindle formation, *J Cell Biol* 123:849–58, 1993.

Cytoplasmic Dynein Intermediate Chain 1

Crackower MA, Sinasac DS, Xia J, Motoyama J, Prochazka M et al: Cloning and characterization of two cytoplasmic dynein intermediate chain genes in mouse and human, *Genomics* 55:257–67, 1999.

Horikawa I, Parker ES, Solomon GG, Barrett JC: Upregulation of the gene encoding a cytoplasmic dynein intermediate chain in senescent human cells, *J Cell Biochem* 82:415–21, 2001.

Dynein Light Chain

Dick T, Ray K, Salz HK, Chia W: Cytoplasmic dynein (ddlc1) mutations cause morphogenetic defects and apoptotic cell death in Drosophila melanogaster, *Mol Cell Biol* 16:1966–77, 1996.

Fuhrmann JC, Kins S, Rostaing P, El Far O, Kirsch J et al: Gephyrin interacts with dynein light chains 1 and 2, components of motor protein complexes, *J Neurosci* 22:5393–402, 2002.

Jaffrey SR, Snyder SH: PIN: An associated protein inhibitor of neuronal nitric oxide synthase, *Science* 274:774–7, 1996.

Kaiser FJ, Tavassoli K, Van den Bemd GJ, Chang GTG, Horsthemke B et al: Nuclear interaction of the dynein light chain LC8a with the TRPS1 transcription factor suppresses the transcriptional repression activity of TRPS1, *Hum Mol Genet* 12:1349–58, 2003.

Bornens M: Centrosome composition and microtubule anchoring mechanisms, *Curr Opin Cell Biol* 14:25–34, 2002.

Desai A, Mitchison TJ: Microtubule polymerization dynamics, *Annu Rev Cell Dev Biol* 13:83–117, 1997.

Hays T, Li M: Kinesin transport: Driving kinesin in the neuron, *Curr Biol* 11:R136–9, 2001.

Job D, Valiron O, Oakley B: Microtubule nucleation, *Curr Opin Cell Biol* 15:111–7, 2003.

Karsenti E, Vernos I: The mitotic spindle: A self-made machine, *Science* 294:543–7, 2001.

Surrey T, Nedelec F, Leibler S, Karsenti E: Physical properties determining self-organization of motors and microtubules, *Science* 292:1167–71, 2001.

Carazo-Salas RE, Gruss OJ, Mattaj IW, Karsenti E: Ran-GTP coordinates regulation of microtubule nucleation and dynamics during mitotic-spindle assembly, *Nature Cell Biol* 3:228–34, 2001.

Mitchison T, Kirschner M: Dynamic instability of microtubule growth, *Nature* 312:237–42, 1984.

Nogales E, Whittaker M, Milligan RA, Downing KH: High-resolution model of the microtubule, *Cell* 96:79–88, 1999.

Rieder CL, Salmon ED: The vertebrate cell kinetochore and its roles during mitosis, *Trends Cell Biol* 8:310–8, 1998.

Yildiz A, Tomishige M, Vale RD, Selvin PR: Kinesin walks hand-over-hand, *Science* 303:676–8, 2004.

Rogers GC, Rogers SL, Schwimmer TA, Ems-McClung SC, Walczak CE et al: Two mitotic kinesins cooperate to drive sister chromatid separation during anaphase, *Nature* 427:364–70, 2004.

Vale RD: The molecular motor toolbox for intracellular transport, *Cell* 112:467–80, 2003.

Vallee RB, Stehman SA: How dynein helps the cell find its center: A servomechanical model, *Trends Cell Biol* 15:288–94, 2005.

Vallee RB, Sheetz MP: Targeting of motor proteins, *Science* 271:1539–44, 1996.

Human protein reference data base, Johns Hopkins University and the Institute of Bioinformatics, at http://www.hprd.org/protein.

3.11. Structure and Organization of Intermediate Filaments

Keratin

Chipev CC, Korge BP, Markova N, Bale SJ, DiGiovanna JJ et al: A leucine-to-proline mutation in the H1 subdomain of keratin 1 causes epidermolytic hyperkeratosis, *Cell* 70:821–8, 1992.

Compton JG: Epidermal disease: Faulty keratin filaments take their toll, *Nature Genet* 6:6–7, 1994.

Compton JG, DiGiovanna JJ, Santucci SK, Kearns KS, Amos CI et al: Linkage of epidermolytic hyperkeratosis to the type II keratin gene cluster on chromosome 12q, *Nature Genet* 1:301–5, 1992.

Rothnagel JA, Dominey AM, Dempsey LD, Longley MA, Greenhalgh DA et al: Mutations in the rod domains of keratins 1 and 10 in epidermolytic hyperkeratosis, *Science* 257:1128–30, 1992.

Schimkat M, Baur MP, Henke J: Inheritance of some electrophoretic phenotypes of human hair, *Hum Genet* 85:311–4, 1990.

Sprecher E, Yosipovitch G, Bergman R, Ciubutaro D, Indelman M et al: Epidermolytic hyperkeratosis and epidermolysis bullosa simplex caused by frameshift mutations altering the V2 tail domains of keratin 1 and keratin 5, *J Invest Dermatol* 120:623–6, 2003.

Syder AJ, Yu QC, Paller AS, Giudice G, Pearson R et al: Genetic mutations in the K1 and K10 genes of patients with epidermolytic hyperkeratosis: Correlation between location and disease severity, *J Clin Invest* 93:1533–42, 1994.

Neurofilament Heavy Chain

Bucan M, Gatalica B, Nolan P, Chung A, Leroux A et al: Comparative mapping of 9 human chromosome 22q loci in the laboratory mouse, *Hum Mol Genet* 2:1245–52, 1993.

Collard JF, Cote F, Julien JP: Defective axonal transport in a transgenic mouse model of amyotrophic lateral sclerosis, *Nature* 375:61–4, 1995.

Elder GA, Friedrich VL Jr, Kang C, Bosco P, Gourov A et al: Requirement of heavy neurofilament subunit in the development of axons with large calibers, *J Cell Biol* 143:195–205, 1998.

Hirokawa N, Takeda S: Gene targeting studies begin to reveal the function of neurofilament proteins, *J Cell Biol* 143:1–4, 1998.

Lees JF, Shneidman PS, Skuntz SF, Carden MJ, Lazzarini RA: The structure and organization of the human heavy neurofilament subunit (NF-H) and the gene encoding it, *EMBO J* 7:1947–55, 1988.

Mattei MG, Dautigny A, Pham-Dinh D, Passage E, Mattei JF et al: The gene encoding the large human neurofilament subunit (NF-H) maps to the q121-q131 region on human chromosome 22, *Hum Genet* 80:293–5, 1988.

Rao MV, Houseweart MK, Williamson TL, Crawford TO, Folmer J et al: Neurofilament-dependent radial growth of motor axons and axonal organization of neurofilaments does not require the neurofilament heavy subunit (NF-H) or its phosphorylation, *J Cell Biol* 143:171–81, 1998.

Rouleau GA, Merel P, Lutchman M, Sanson M, Zucman J et al: Alteration in a new gene encoding a putative membrane-organizing protein causes neuro-fibromatosis type 2, *Nature* 363:515–21, 1993.

Watson CJ, Gaunt L, Evans G, Patel K, Harris R et al: A disease-associated germline deletion maps the type 2 neurofibromatosis (NF2) gene between the Ewing sarcoma region and the leukaemia inhibitory factor locus, *Hum Mol Genet* 2:701–4, 1993.

Zhu Q, Lindenbaum M, Levavasseur F, Jacomy H, Julien JP: Disruption of the NF-H gene increases axonal microtubule content and velocity of neurofilament transport: Relief of axonopathy resulting from the toxin beta,beta-prime-iminodipropionitrile, *J Cell Biol* 143:183–93, 1998.

Vimentin

Geisler N, Plessmann U, Weber K: Amino acid sequence characterization of mammalian vimentin, the mesenchymal intermediate filament protein, *FEBS Lett* 163:22–4, 1983.

Colucci-Guyon E, Portier MM, Dunia I, Paulin D, Pournin S et al: Mice lacking vimentin develop and reproduce without an obvious phenotype, *Cell* 79:679–94, 1994.

Ferrari S, Battini R, Kaczmarek L, Rittling S, Calabretta B et al: Coding sequence and growth regulation of the human vimentin gene, *Mol Cell Biol* 6:3614–20, 1986.

Ferrari S, Cannizzaro LA, Battini R, Huebner K, Baserga R: The gene encoding human vimentin is located on the short arm of chromosome 10, *Am J Hum Genet* 41:616–26, 1987.

Gieser L, Swaroop A: Expressed sequence tags and chromosomal localization of cDNA clones from a subtracted retinal pigment epithelium library, *Genomics* 13:873–6, 1992.

Mathew CG, Wakeling W, Jones E, Easton D, Fisher R et al: Regional localization of polymorphic markers on chromosome 10 by physical and genetic mapping, *Ann Hum Genet* 54:121–9, 1990.

Mor-Vaknin N, Punturieri A, Sitwala K, Markovitz DM: Vimentin is secreted by activated macrophages, *Nature Cell Biol* 5:59–63, 2003.

Perreau J, Lilienbaum A, Vasseur M, Paulin D: Nucleotide sequence of the human vimentin gene and regulation of its transcription in tissues and cultured cells, *Gene* 62:7–16, 1988.

Zhang X, Diab IH, Zehner ZE: ZBP-89 represses vimentin gene transcription by interacting with the transcriptional activator Sp1, *Nucleic Acids Res* 31:2900–14, 2003.

Neurofilament Light Chain

Fabrizi GM, Cavallaro T, Angiari C, Bertolasi L et al: Giant axon and neurofilament accumulation in Charcot-Marie-Tooth disease type 2E, *Neurology* 62:1429–31, 2004.

Hirokawa N, Takeda S: Gene targeting studies begin to reveal the function of neurofilament proteins, *J Cell Biol* 143:1–4, 1998.

Hurst J, Flavell D, Julien JP, Meijer D, Mushynski W et al: The human neurofilament gene (NEFL) is located on the short arm of chromosome 8, *Cytogenet Cell Genet* 45:30–2, 1987.

Nguyen MD, Lariviere RC, Julien JP: Deregulation of Cdk5 in a mouse model of ALS: toxicity alleviated by perikaryal neurofilament inclusions, *Neuron* 30:135–47, 2001.

Previtali SC, Zerega B, Sherman DL, Brophy PJ, Dina G et al: Myotubularin-related 2 protein phosphatase and neurofilament light chain protein, both mutated in CMT neuropathies, interact in peripheral nerve, *Hum Mol Genet* 12:1713–23, 2003.

Zhu Q, Couillard-Despres S, Julien JP: Delayed maturation of regenerating myelinated axons in mice lacking neurofilaments, *Exp Neurol* 148:299–316, 1997.

Lamin A/C

Bonne G, Di Barletta MR, Varnous S, Becane HM, Hammouda EH et al: Mutations in the gene encoding lamin A/C cause autosomal dominant Emery-Dreifuss muscular dystrophy, *Nature Genet* 21:285–8, 1999.

De Sandre-Giovannoli A, Bernard R, Cau P, Navarro C, Amiel J et al: Lamin A truncation in Hutchinson-Gilford progeria, *Science* 300:2055, 2003.

Eriksson M, Brown WT, Gordon LB, Glynn MW, Singer J et al: Recurrent de novo point mutations in lamin A cause Hutchinson-Gilford progeria syndrome, *Nature* 423:293–8, 2003.

Fatkin D, MacRae C, Sasaki T, Wolff MR, Porcu M, et al: Missense mutations in the rod domain of the lamin A/C gene as causes of dilated cardiomyopathy and conduction-system disease, *New Engl J Med* 341:1715–24, 1999.

Fisher DZ, Chaudhary N, Blobel G: cDNA sequencing of nuclear lamins A and C reveals primary and secondary structural homology to intermediate filament proteins, *Proc Natl Acad Sci USA* 83:6450–4, 1986.

Goldman RD, Shumaker DK, Erdos MR, Eriksson M, Goldman AE et al: Accumulation of mutant lamin A causes progressive changes in nuclear architecture in Hutchinson-Gilford progeria syndrome, *Proc Natl Acad Sci USA* 101:8963–8, 2004.

Lin F, Worman HJ: Structural organization of the human gene encoding nuclear lamin A and nuclear lamin C, *J Biol Chem* 268:16321–6, 1993.

Lloyd DJ, Trembath RC, Shackleton S: A novel interaction between lamin A and SREBP1: Implications for partial lipodystrophy and other laminopathies, *Hum Mol Genet* 11:769–77, 2002.

McKeon FD, Kirschner MW, Caput D: Homologies in both primary and secondary structure between nuclear envelope and intermediate filament proteins, *Nature* 319:463–8, 1986.

Mounkes LC, Kozlov S, Hernandez L, Sullivan T, Stewart CL: A progeroid syndrome in mice is caused by defects in A-type lamins, *Nature* 423:298–301, 2003.

Scaffidi P, Misteli T: Reversal of the cellular phenotype in the premature aging disease Hutchinson-Gilford progeria syndrome, *Nature Med* 11:440–5, 2005.

Shackleton S, Lloyd DJ, Jackson SNJ, Evans R, Niermeijer MF et al: LMNA, encoding lamin A/C, is mutated in partial lipodystrophy, *Nature Genet* 24:153–6, 2000.

Lamin B

Furukawa K, Hotta Y: cDNA cloning of a germ cell specific lamin B3 from mouse spermatocytes and analysis of its function by ectopic expression in somatic cells, *EMBO J* 12:97–106, 1993.

Justice MJ, Gilbert DJ, Kinzler KW, Vogelstein B, Buchberg AM et al: A molecular genetic linkage map of mouse chromosome 18 reveals extensive linkage conservation with human chromosomes 5 and 18, *Genomics* 13:1281–8, 1992.

Lin F, Worman HJ: Structural organization of the human gene (LMNB1) encoding nuclear lamin B1, *Genomics* 27:230–6, 1995.

Maeno H, Sugimoto K, Nakajima N: Genomic structure of the mouse gene (Lmnbl) encoding nuclear lamin B1, *Genomics* 30:342–6, 1995.

Vergnes L, Peterfy M, Bergo MO, Young SG, Reue K: Lamin B1 is required for mouse development and nuclear integrity, *Proc Natl Acad Sci USA* 101:10428–33, 2004.

Wydner KL, McNeil JA, Lin F, Worman HJ, Lawrence JB: Chromosomal assignment of human nuclear envelope protein genes LMNA, LMNB1, and LBR by fluorescence in situ hybridization, *Genomics* 32:474–8, 1996.

Human protein reference data base, Johns Hopkins University and the Institute of Bioinformatics, at http://www.hprd.org/protein.

3.12. Function of Intermediate Filaments

Fuchs E, Cleveland DW: A structural scaffolding of intermediate filaments in health and disease, *Science* 279:514–9, 1998.

Getsios S, Huen AC, Green KJ: Working out the strength and flexibility of desmosomes, *Nat Rev Mol Cell Biol* 5:271–81, 2004.

Helfand BT, Chang L, Goldman RD: The dynamic and motile properties of intermediate filaments, *Annu Rev Cell Dev Biol* 19:445–67, 2003.

Worman HJ, Courvalin JC: The nuclear lamina and inherited disease, *Trends Cell Biol* 12:591–8, 2002.

Hutchison CJ: Lamins: Building blocks or regulators of gene expression? *Nat Rev Mol Cell Biol* 3:848–58, 2002.

Clarke EJ, Allan V: Intermediate filaments: Vimentin moves in, *Curr Biol* 12:R596–8, 2002.

Green KJ, Gaudry CA: Are desmosomes more than tethers for intermediate filaments? *Nat Rev Mol Cell Biol* 1:208–16, 2000.

Chou YH, Helfand BT, Goldman RD: New horizons in cytoskeletal dynamics: Transport of intermediate filaments along microtubule tracks, *Curr Opin Cell Biol* 13:106–9, 2001.

Coulombe PA, Bousquet O, Ma L, Yamada S, Wirtz D: The "ins" and "outs" of intermediate filament organization, *Trends Cell Biol* 10:420–8, 2000.

Herrmann H, Aebi U: Intermediate filaments and their associates: Multi-talented structural elements specifying cytoarchitecture and cytodynamics, *Curr Opin Cell Biol* 12:79–90, 2000.

Stuurman N, Heins S, Aebi U: Nuclear lamins: Their structure, assembly, and interactions, *J Struct Biol* 122:42–66, 1998.

Herrmann H, Aebi U: Intermediate filament assembly: Fibrillogenesis is driven by decisive dimer-dimer interactions, *Curr Opin Struct Biol* 8:177–85, 1998.

3.13. Endoplasmic Reticulum

Voelker DR: Bridging gaps in phospholipid transport, *Trends Biochem Sci* 30:396–404, 2005.

Hebert DN, Garman SC, Molinari M: The glycan code of the endoplasmic reticulum: Asparagine-linked carbohydrates as protein maturation and quality-control tags, *Trends Cell Biol* 15:364–70, 2005.

Rapoport TA, Goder V, Heinrich SU, Matlack KE: Membrane-protein integration and the role of the translocation channel, *Trends Cell Biol* 14:568–75, 2004.

Watanabe R, Riezman H: Differential ER exit in yeast and mammalian cells, *Curr Opin Cell Biol* 16:350–5, 2004.

Toyoshima C, Inesi G: Structural basis of ion pumping by Ca2+-ATPase of the sarcoplasmic reticulum, *Annu Rev Biochem* 73:269–92, 2004.

Pfeffer S: Membrane domains in the secretory and endocytic pathways, *Cell* 112:507–17, 2003.

Venkatachalam K, van Rossum DB, Patterson RL, Ma HT, Gill DL: The cellular and molecular basis of store-operated calcium entry, *Nature Cell Biol* 4:E263–72, 2002.

Holthuis JC, Pomorski T, Raggers RJ, Sprong H, Van Meer G: The organizing potential of sphingolipids in intracellular membrane transport, *Physiol Rev* 81:1689–723, 2001.

Hirschberg CB, Robbins PW, Abeijon C: Transporters of nucleotide sugars, ATP, and nucleotide sulfate in the endoplasmic reticulum and Golgi apparatus, *Annu Rev Biochem* 67:49–69, 1998.

3.14. Golgi Apparatus

Machesky LM, Bornens M: Cell structure and dynamics, *Curr Opin Cell Biol* 15:2–5, 2003.

Puthenveedu MA, Linstedt AD: Subcompartmentalizing the Golgi apparatus, *Curr Opin Cell Biol* 17:369–75, 2005.

Rios RM, Bornens M: The Golgi apparatus at the cell centre, *Curr Opin Cell Biol* 15:60–6, 2003.

Pfeffer SR: Constructing a Golgi complex, *J Cell Biol* 155:873–5, 2001.

Glick BS: Organization of the Golgi apparatus, *Curr Opin Cell Biol* 12:450–6, 2000.

Bannykh SI, Nishimura N, Balch WE: Getting into the Golgi, *Trends Cell Biol* 8:21–5, 1998.

Munro S: The Golgi apparatus: defining the identity of Golgi membranes, *Curr Opin Cell Biol* 17:395–401, 2005.

Lee MC, Miller EA, Goldberg J, Orci L, Schekman R: Bi-directional protein transport between the ER and Golgi, *Annu Rev Cell Dev Biol* 20:87–123, 2004.

Altan-Bonnet N, Sougrat R, Lippincott-Schwartz J: Molecular basis for Golgi maintenance and biogenesis, *Curr Opin Cell Biol* 16:364–72, 2004.

de Graffenried CL, Bertozzi CR: The roles of enzyme localisation and complex formation in glycan assembly within the Golgi apparatus, *Curr Opin Cell Biol* 16:356–63, 2004.

Graham TR: Membrane targeting: Getting Arl to the Golgi, *Curr Biol* 14:R483–5, 2004.

Palmer KJ, Stephens DJ: Biogenesis of ER-to-Golgi transport carriers: Complex roles of COPII in ER export, *Trends Cell Biol* 14:57–61, 2004.

3.15. Endosomes and Lysosomes

Piper RC, Luzio JP: CUPpling calcium to lysosomal biogenesis, *Trends Cell Biol* 14:471–3, 2004.

Fevrier B, Raposo G: Exosomes: Endosomal-derived vesicles shipping extracellular messages, *Curr Opin Cell Biol* 16:415–21, 2004.

Miaczynska M, Pelkmans L, Zerial M: Not just a sink: Endosomes in control of signal transduction, *Curr Opin Cell Biol* 16:400–6, 2004.

Gruenberg J, Stenmark H: The biogenesis of multivesicular endosomes, *Nat Rev Mol Cell Biol* 5:317–23, 2004.

Piddini E, Vincent JP: Modulation of developmental signals by endocytosis: Different means and many ends, *Curr Opin Cell Biol* 15:474–81, 2003.

Raiborg C, Rusten TE, Stenmark H: Protein sorting into multivesicular endosomes, *Curr Opin Cell Biol* 15:446–55, 2003.

Bonifacino JS, Traub LM: Signals for sorting of transmembrane proteins to endosomes and lysosomes, *Annu Rev Biochem* 72:395–447, 2003.

Sandvig K, van Deurs B: Endocytosis, intracellular transport, and cytotoxic action of Shiga toxin and ricin, *Physiol Rev* 76:949–66, 1996.

Nixon RA, Cataldo AM: The endosomal-lysosomal system of neurons: New roles, *Trends Neurosci* 18:489–96, 1995.

Bomsel M, Mostov K: Sorting of plasma membrane proteins in epithelial cells, *Curr Opin Cell Biol* 3:647–53, 1991.

3.16. Mitochondria

Chen XJ, Butow RA: The organization and inheritance of the mitochondrial genome, *Nature Rev Genet* 6:815–25, 2005.

Frey TG, Mannella CA: The internal structure of mitochondria, *Trends Biochem Sci* 25:319–24, 2000.

Bonen L: The mitochondrial genome: So simple yet so complex, *Curr Opin Genet Dev* 1:515–22, 1991.

3.17. Cell Nuclei

Grunstein M: Histones as regulators of genes, *Sci Am* 267(4): 68–74B, 1992.

Heck MMS: Condensins, cohesins, and chromosome architecture: How to make and break a mitotic chromosome, *Cell* 91:5–8, 1997.

Kornberg RD: Chromatin structure: A repeating unit of histones and DNA, *Science* 184:868–71, 1974.

Koshland D, Strunnikov A: Mitotic chromosome condensation, *Annu Rev Cell Biol* 12:305–33, 1996.

Luger K, Mader AW, Richmond RK, Sargent DE, Richmond TJ: Crystal structure of the nucleosome core particle at 2.8 A resolution, *Nature* 389:251–60, 1997.

Paulson JR, Laemmli UK: The structure of histone-depleted metaphase chromosomes, *Cell* 12:817–28, 1977.

Richmond TJ, Finch JT, Rushton B, Rhodes D, Klug A: Structure of the nucleosome core particle at 7A resolution, *Nature* 311:532–7, 1984.

Saitoh Y, Laemmli UK: Metaphase chromosome structure: Bands arise from a differential folding path of the highly AT-rich scaffold, *Cell* 76:609–22, 1994.

Van Holde KE, Zlatanovai J: Chromatin higher order structure: Chasing a mirage, *J Biol Chem* 270:8373–6, 1995.

4

EXTRACELLULAR MATRIX

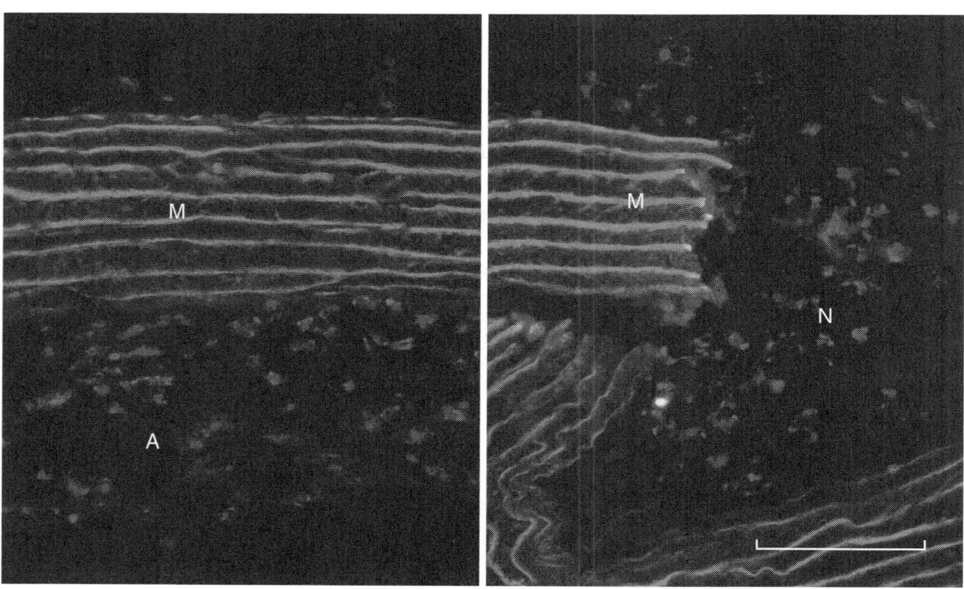

Transverse fluorescent micrographs showing the distribution of CD 11b/c-positive leukocytes in the media and adventitia of matrix-based aortic substitutes. The density of leukocytes in the elastic lamina-containing media was significantly lower than that in the collagen-containing adventitia. Note that leukocytes did not migrate into the gaps between the elastic laminae at the end of the aortic matrix substitutes (right). Red: antibody-labeled CD 11 b/c. Green: elastic laminae. Blue: Hoechst 33258-labeled cell nuclei. M, media. A, adventitia. N, neointima. Scale: 100 μm. (Reprinted from Liu SQ et al: *J Biolo Chem* 280:39294–301, 2005 with permission from the American Society for Biochemistry and Molecular Biology). See color insert.

Bioregenerative Engineering: Principles and Applications, by Shu Q. Liu

The *extracellular matrix* is the noncellular structure found in the extracellular space. This structure is composed of collagen fibers, elastic fibers or laminae, and proteoglycans. All components of extracellular matrix are produced and released by cells residing in the same tissue. Extracellular matrix plays critical roles in several aspects: (1) constituting a matrix framework that supports and organizes cells, tissues, and organs; (2) contributing to the morphogenesis and shape formation of tissues and organs; (3) providing mechanical strength to and protecting tissues and organs from injury; and (4) participating in the regulation of cell adhesion, proliferation, migration, and apoptosis. These aspects are outlined in this chapter.

The extracellular matrix can be used as biological materials for the regeneration of lost tissues and organs. Compared with synthetic polymer materials, extracellular matrix components are naturally occurring polymeric materials that are nontoxic and compatible to host cells and tissues and participate in the maintenance and regulation of cellular functions as described above. In particular, the collagen matrix has been used to construct scaffolds in experimental models for the reconstruction of a variety of tissues, such as the liver, pancreas, bones, and blood vessels. Since collagen matrix promotes cell adhesion, proliferation, and migration, collagen-based scaffolds enhance the regeneration of impaired tissues and organs. As other polymeric materials, extracellular matrix can be engineered and fabricated into various shapes and forms as desired. Thus, extracellular matrix components are preferred materials for the repair, regeneration, and engineering of malfunctioned tissues and organs.

COLLAGEN MATRIX

Composition and Formation of Collagen Matrix [4.1]

The *collagen matrix* is the most abundant type of extracellular matrix that is found primarily in connective tissues, such as the subcutaneous tissue, bone, and the adventitia of tubular organs, including blood vessels, airways, esophagus, stomach, and intestines. In mammalian tissues, there exist more than 20 types of collagen matrix, classified as collagen types I, II, III, and so on. Among these types of collagen, types I, II, III, IV, V, IX, XI, and XII are commonly found in connective tissue. Each type of collagen matrix is formed with one or more types of collagen molecule. A typical collagen molecule is a helical fibrillar structure composed of three peptide chains, termed α *chains*. A large number of collagen genes have been identified; each encodes a distinct collagen α chain. Combinations of various α chains give rise to different types of collagen fibril. Table 4.1 lists 22 types of representative collagen peptide chains.

Collagens are synthesized first as procollagen molecules in the cytoplasm of several cell types, including fibroblast, osteoblast, smooth muscle cell, and endothelial cell. Procollagen molecules are released to the extracellular space, cleaved by proteinases to remove procollagen peptides, and self-assembled into various forms of matrix structure. Collagen types I, II, III, V, and XI are organized into filamentous structures, known as *collagen fibrils*, with a diameter of ~10–100 nm. These fibrils usually form larger collagen bundles as found in the subcutaneous tissue and the adventitia of tubular organs. Collagen types I and V are often found in the bone, skin, cornea, tendon, ligament, and internal organs, such as the lung, liver, pancreas, and kidney. Mutation of the collagen type I genes causes several disorders, including osteogenesis imperfecta, idiopathic osteoporosis, and

TABLE 4.1. Characteristics of Selected Collagen Molecules*

Proteins	Alternative Names	Amino Acids	Molecular Weight (kDa)	Expression	Functions
Collagen type I α1	COL1A1, collagen α1 chain	1464	139	Bone, cartilage, connective tissue, skin, cornea, tendon, ligament, internal organs	Constituting the collagen matrix
Collagen type I α2	COL1A2, collagen I α2 polypeptide	1366	129	Bone, cartilage, connective tissue, skin, cornea, tendon, ligament, internal organs	Constituting the collagen matrix
Collagen type II α1	COL2A1, chondrocalcin, collagen type XI α3 (COL11A3), cartilage collagen	1487	142	Cartilage, notochord, intervertebral disks, vitreous humor of the eye	Constituting the collagen matrix
Collagen type III α1	COL3A1	1466	139	Connective tissues, skin, lung, blood vessels	Constituting the collagen matrix
Collagen type IV α1	COL4A1, collagen of basement membrane, α1 chain	1669	161	Brain, heart, blood vessel, liver, pancreas, kidney, placenta, eye	Constituting basal lamina or basement membrane
Collagen type IV α2	COL4A2, collagen of basement membrane α2 chain	1712	167	Brain, heart, blood vessel, liver, pancreas, kidney, placenta, eye	Constituting the basal lamina or basement membrane
Collagen type V α1	COL5A1	1838	184	Ubiquitous	Present in tissues containing type I collagen, regulating the assembly of type I collagen fibers
Collagen type V α2	COL5A2	1496	145	Bone, skin, cornea, tendon, ligament, lung, liver, pancreas, kidney	Constituting the extracellular matrix and regulating the assembly of type I collagen fibers

Collagen type VI α1	COL6A1	1028	109	Ubiquitous	Constituting the extracellular matrix and regulating the integrity of tissues
Collagen type VII α1	COL7A1	2944	295	Skin, mouth	Found near the basement membrane of stratified squamous epithelia, forming fibrils that contribute to anchoring of epithelia to underlying stroma
Collagen type VIII α1	Endothelial collagen	744	73	Endothelial cells, skin, kidney, cornea, leukocytes	Constituting the extracellular matrix, a major component of the basement membrane of corneal endothelium
Collagen type IX α1	COL9A1, cartilage specific short collagen	921	92	Cartilage, ear, eye (usually found in tissues containing type II collagen)	Constituting the collagen matrix
Collagen type X α1	COL10A1	680	66	Cartilage	Constituting the cartilage matrix
Collagen type XI α1	COL11A1	1818	183	Cartilage, cornea	Constituting the extracellular matrix
Collagen type XII α1	COL12A1	3063	333	Skin, bone	Often found in association with type I collagen and regulating the interaction of collagen I fibrils with other matrix components
Collagen type XIII α1	COL13A1	717	70	Eye, placenta	Containing a transmembrane domain, often localized to the cell membrane, and possibly regulating cell–cell interaction and angiogenesis

TABLE 4.1. *Continued*

Proteins	Alternative Names	Amino Acids	Molecular Weight (kDa)	Expression	Functions
Collagen type XIV α1	COL14A1, undulin	1796	194	Heart, blood vessels, brain, skeletal muscle, liver, uterus, tendon, skin	Constituting the matrix of connective tissues (note that this type of collagen is often associated with mature collagen fibrils)
Collagen type XV α1	COL15A1	1388	142	Ubiquitous	Often found near basal lamina or basement membrane and regulating interaction of basal lamina with underlying connective tissue
Collagen type XVI α1	COL16A1	1603	158	Skin, cartilage, skeletal muscle, placenta	Often found in association with collagen type I and II fibrils, regulating the integrity and organization of collagen matrix
Collagen type XVII	COL17A1	1497	180	Skin, cornea, intestine, esophagus, testis, spleen	A transmembrane protein that regulates the adhesion of epithelial cells to underlying basal lamina
Collagen type XVIII α1	COL18A1	1516	154	Brain, heart, blood vessel, liver, pancreas, kidney, intestine, ovary, skeletal muscle	Constituting the collagen matrix and generating endostatin, an antiangiogenic protein, by proteolytic cleavage of the *C*-terminal fragment of the molecule
Collagen type XIX α1	COL19A1	1143	115	Skin	Function remains to be determined

*Based on bibliography 4.1.

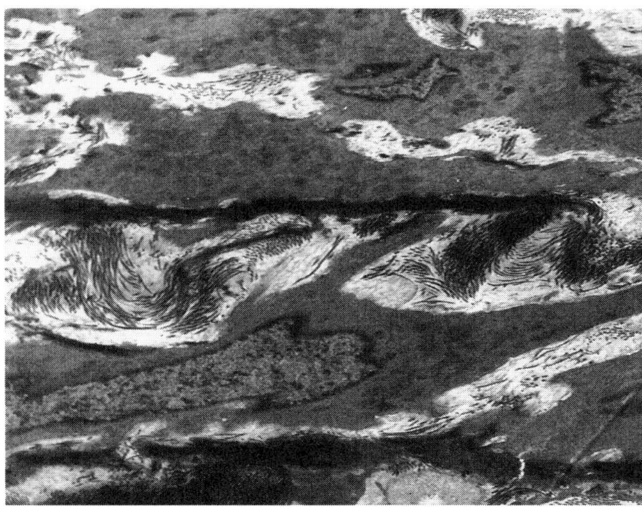

Figure 4.1. Electron micrograph showing collagen fibrils in the wall of a rat mesenteric artery. Scale bar: 1 μm.

atypical Marfan syndrome. Collagen types II and XI are found in the cartilage, notochord, and intervertebral disks. Collagen type III is found in blood vessels (Fig. 4.1), skin, and internal organs. These collagen fibrous structures play critical roles in the support and protection of cells and in the regulation of cellular functions, such as cell adhesion, proliferation, and migration. Collagen types IX and XII are molecules that link other types of collagen fibrils and are known as *fibril-associated collagens*. These types are found in the cartilage, tendon, and ligament. In contrast to the filamentous collagen molecules, collagen type IV participates in the construction of a membrane-like structure, known as the basal lamina or basement membrane, which underlies epithelial and endothelial cells (note that other components of a basal lamina include laminin, entactin, perlecan, nidogen, and heparan sulfate proteoglycans; see Fig. 4.2). Additional types of collagen are listed in Table 4.1.

Function of Collagen Matrix [4.2]

The collagen matrix plays several roles in a mammalian tissue or organ. The collagen matrix serves as a structural material that supports cells, helps organize cells into various forms of tissues and organs, and protects cells from mechanical injury. In addition, the collagen matrix participates in the regulation of cellular activities such as cell survival, adhesion, proliferation, and migration. Collagen molecules can directly interact with cells via the cell membrane collagen receptors, or indirectly via the mediation of *fibronectin*, a matrix component that binds collagen molecules at one side and cell membrane matrix receptors, known as *integrins*, at the other side. The binding of collagen and fibronectin molecules to the matrix receptors initiate the activation of intracellular signaling pathways that stimulate or activate mitogenic processes, including cell adhesion, survival, proliferation, and migration.

Given the structural and functional features, collagen matrix has long been used for constructing drug delivery devices and scaffolds for tissue regeneration. Collagen matrix

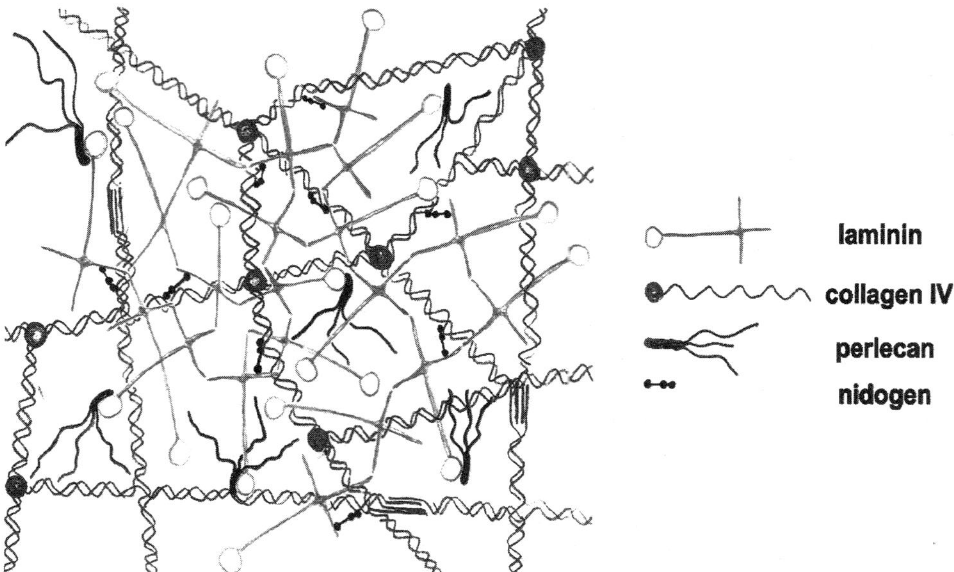

Figure 4.2. Schematic representation of endothelial cell basement membrane. Major components of endothelial cell basement membrane include laminin 8 and 10 isoforms, collagen type IV, and nidogen 1 and 2. (Reprinted from Hallmann R et al: *Physiol Rev* 85:979–1000, 2005 with permission from the American Physiological Society.)

has been used for constructing tissue scaffolds in several forms: collagen gel, collagen mesh, composite structures with other types of extracellular matrix molecules such as elastic fibers and proteoglycans, and decellularized natural collagen matrix scaffolds. The constructed structures can be used for various purposes of regenerative medicine. Collagen gels and meshes are suitable for drug delivery, whereas the cell-free natural collagen matrix can be used as scaffolds or grafts for the repair or regeneration of various tissues and organs, such as blood vessels, airways, intestines, stomach, and bladder.

To prepare collagen gels, natural collagen-containing tissues can be collected and degraded (note that collagen fibers are insoluble), and soluble collagen molecules can be extracted. Collagen molecules can be crosslinked into a gel structure with appropriate pH, temperature, and ionic strength. A therapeutic substance can be blended with the collagen molecules during gele formation. The collagen gel can be implanted or injected into a target tissue, and the therapeutic substance can be released at the rate of collagen gel degradation. In addition, collagen gel can be mixed with selected cells and used to deliver cells into a target tissue to replace malfunctioned cells. The delivered cells can be integrated into and restore the function of the target tissue. Sponge-like collagen matrix can be prepared in vitro and used as a framework for tissue regeneration and as a scaffold for repairing traumatized tissues.

Whereas a native collagen matrix is mechanical tough and strong, an in vitro crosslinked collagen gel exhibits low mechanical strength. Several methods have been developed and used to strengthen collagen gels. One method is to treat collagen gels with glutaraldehyde, which induces collagen crosslink and increases the strength of collagen gels. However, glutaraldehyde is toxic to cells and significantly influences cellular activities and functions. Another method is to facilitate collagen crosslink by introducing glycation. This method enhances the strength of collagen gels without significantly compromising the cell functions.

Native collagen matrix is a suitable material for the construction of tissue scaffolds. Such a material maintains the natural biological and mechanical characteristics and exhibits superior biocompatibility compared to in vitro crossliked collagen gels or matrix. To prepare a native collagen matrix, mammalian tissue specimens can be collected from selected structures, such as the submucosa of intestines, the adventitia of blood vessels, and the subcutaneous tissue. Cells in these specimens can be removed by various enzymatic and hydrolytic methods. Such treatments eliminate the cellular immunogenicity of allogenic tissues (note that extracellular matrix molecules exhibit little immunogenicity). The resulting cell-free collagen matrix can be tailored into a scaffold with a desired form and used for tissue repair or regeneration.

ELASTIC FIBERS AND LAMINAE

Composition and Structure of Elastic Laminae [4.3]

Elastic fibers and laminae are major extracellular matrix components found in mammalian tissues and organs. Elastic fibers are present in the lung, connective tissue, the submucosa of intestines, and the wall of veins, whereas elastic laminae are found primarily in the media of large and medium arteries (Fig. 4.3). Elastic fibers and laminae are composed of several proteins, including elastin, microfibrils, and microfibril-associated proteins. Elastin is the most abundant protein in elastic fibers and laminae. In this section, arterial elastic laminae are used as an example to describe the composition, structure, and function of elastin-based extracellular matrix.

Elastic laminae (see Table 4.2) are concentrically organized layers composed of tightly organized elastic fibers. These elastic fibers are composed of microfibrils and amorphous elastin and are arranged predominantly in the circumferential direction of arteries. Elastin

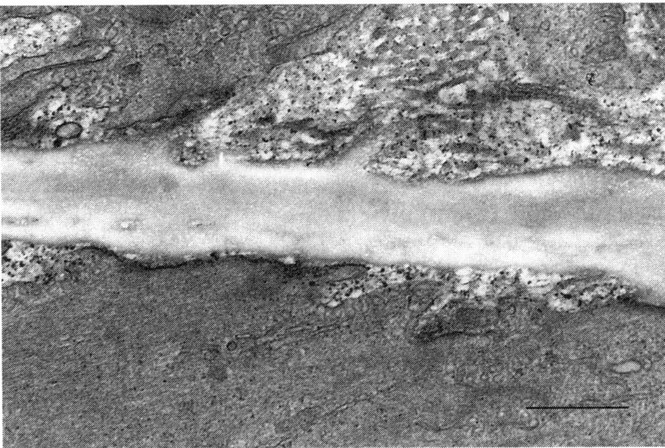

Figure 4.3. Electron micrograph of elastic laminae in the wall of a rat pulmonary artery. Scale bar: 1 μm.

TABLE 4.2. Characteristics of Selected Molecules Constituting the Elastic Laminae*

Proteins	Alternative Names	Amino Acids	Molecular Weight (kDa)	Expression	Functions
Tropoelastin	ELN	757	66	Blood vessels, skin, lung, kidney, cartilage	Constituting elastin, a major component of elastic fibers and laminae
Fibrillin 1	FBN1	2871	312	Blood vessels, skin, lung, kidney, cartilage	A constitutive component of microfibrils, which are organized into scaffolds for deposition of elastin and assembly of elastic fibers, and also a component of nonelastic matrix, causing Marfan syndrome when mutated
Fibrillin 2	FBN2	2911	314	Blood vessel	Same as fibrillin 1
Fibrillin 3	FBN3	2809	300	Skin, lung, kidney, skeletal muscle	Same as fibrillin 1
Microfibril-associated glycoprotein	MAGP, microfibril-associated protein 2	183	21	Skin, lung, kidney, skeletal muscle	Regulating the assembly and stability of microfibrils and elastic fibers

*Based on bibliography 4.3.

is the most abundant protein found in large arteries and contributes to approximately half the dry mass of the arterial wall. Mature elastin is a highly insoluble and hydrophobic protein, and is formed by the crosslinking of the 72-kDa elastin precursor, known as *tropoelastin*. In mammals, approximately 75% of tropoelastin is composed of four amino acids, including glycine, valine, alanine, and proline. Tropoelastin is produced by several cell types, including the smooth muscle cell (SMC) and endothelial cell (EC), and is released into the extracellular space where crosslinking and elastin formation take place. A mature elastin molecule contains two types of domain: the hydrophobic and crosslinking domains. The hydrophobic domains are rich in nonpolar amino acids, including glycine, valine, proline, and alanine, which are often arranged in repeats of three to six amino acid peptides, such as GVGVP, GGVP, and GVGVAP. The crosslinking domains are rich in alanine and lysine; the latter is subject to enzymatic crosslinking by lysyl oxidase. The lysine-containing crosslinking domains appear to be well conserved through evolution, whereas the hydrophobic domains display considerable variability. The structural conservation in the cross-linking domains renders elastin a highly inert and nonimmunogenic protein.

In the extracellular space, tropoelastin molecules are aligned and assembled into elastin based on a nonelastin microfibril mesh. Microfibrils are filaments of 8–16 nm in diameter and are composed of glycoproteins known as *fibrillins* and several microfibril-associated glycoproteins (MAGPs), including MAGP1 and MAGP2. It is thought that microfibrils are established prior to elastin assembly, providing a scaffold for the deposition, alignment, and crosslinking of tropoelastin. The MAGPs have been proposed to mediate the interaction of microfibrils with tropoelastin. One possible role of the MAGPs is to bind to the *C*-terminus of tropoelastin and stabilize tropoelastin prior to enzymatic crosslinking. The *C*-terminus of tropoelastin is critical to the formation of elastin. The lack of the *C*-terminus reduces the assembly of elastic laminae.

Following organized deposition to and alignment along the microfibrils, tropoelastin molecules are crosslinked into elastin via enzymatic reactions mediated by lysyl oxidase. This enzyme catalyzes oxidative deamination of the lysine residues, converting lysine to allysine (α-amino adipic δ-semialdehyde). Most lysine residues in tropoelastin are involved in such an enzymatic reaction. Lysine and allysine residues are then condensed spontaneously, resulting in the formation of elastin-specific crosslinks known as *desmosines* and *isodesmosines*. These crosslinks play a critical role in the assembly of elastin fibers. Since desmosines and isodesmosines are very stable in structure, elastic fibers are considered one of the toughest materials found in mammalian systems.

In addition to lysyl oxidase, elastin assembly may be regulated by other factors. For instance, negatively charged extracellular glycosaminoglycans may interact with the positively charged lysine residues of tropoelastin to promote elastin assembly. Conversely, glycosaminoglycans containing galactose derivatives, such as dermatan and chondroitin sulfate, have been linked to impaired elastogenesis, promoting the degradation of elastic fibers. The overexpression of a chondroitin sulfate-deficient proteoglycan known as *versican* (variant V3) increased tropoelastin expression and elastic fiber formation in vitro, and resulted in elastic lamina formation in balloon-injured carotid arteries in vivo. Another protein, latent transforming growth factor β-binding protein 2 (LTBP2), is coexpressed with tropoelastin and may contribute to elastic fiber formation. These examples demonstrate that various reactions are possible for the formation of elastic fibers and laminae, due to the participation of different extracellular components, although the regulatory mechanisms of elastin assembly remain to be clarified.

In mammals, arteries contain concentric elastic laminae with circumferentially aligned elastic fibers, whereas veins consist of a network of elastic fiber bundles aligned predominantly in the axial direction of the vessel. When observed by optical and electron microscopy, elastic laminae or fibers appear amorphous under physiological conditions. Historically, it has been thought that elastic laminae and fibers are stable structures that undergo little turnover and remodeling through the lifespan. However, recent studies have demonstrated that mechanical stretch in hypoxia-induced pulmonary hypertension can induce swelling and reorganization of elastic laminae within several hours. These observations suggest that elastic laminae and fibers may undergo dynamic remodeling in response to environmental stimuli.

Function of Elastic Fibers and Laminae [4.4]

Large arteries are composed of multiple layers of elastic laminae. These laminae have long been known to contribute to the structural stability, mechanical strength, and elasticity of the arterial wall. Arteries are subject to extensive mechanical stress induced by arterial blood pressure. Without the support of the elastic laminae, vascular cells may be overstretched under arterial blood pressure. The mechanical stretch may induce structural change in or degradation of elastic fibers or laminae. The degradation of elastic laminae has long been considered a major factor for reducing the strength of the arterial wall and inducing arterial aneurysms. The importance of the elastic laminae has been demonstrated in experimental arterial reconstruction with vein grafts. Veins have only loosely organized elastic fibers instead of elastic laminae, although veins and arteries both possess a strong collagen-containing adventitia. When a vein is used as an arterial substitute and exposed to arterial blood pressure, about 60% of endothelial cells and SMCs die within 12 h of implantation due to mechanical stretch. The lack of multilayer elastic laminae reduces the strength of the vein graft wall, contributing to the injury and death of vascualr cells. Thus, elastic laminae are a critical structure for the stability and mechanical strength of the arterial wall.

In addition to structural support, elastic laminae contribute to the elasticity of arteries. The recoil of the arterial wall is a critical mechanism for the continuation of bloodflow during diastole when cardiac ejection is ceased. The unique amino acid organization and crosslinking patterns of elastin are commonly regarded as important determinants for the elasticity of elastic fibers and laminae. Investigations by nuclear magnetic resonance have demonstrated that the backbones of elastin amino acid chains are highly mobile and individual amino acid residues are able to move freely. The crosslinks help organize the tropoelastin peptide chains into a filamentous network, which is an effective structure for the storage of recoiling energy under mechanical stretch. Observations by electron microscopy suggest the presence of ordered filamentous structures in elastic fibers under extensive mechanical stretch (in a range of strain or degree of stretch of ~150–200% with respect to the unstretched state), while amorphous appearance is observed without mechanical stretch. The structure and organization of elastin provide a basis for the elastic properties of elastic fibers.

Elastic laminae have also been shown to serve as a signaling structure and play a role in regulating arterial morphogenesis and pathogenesis. An important contribution of elastic laminae is to confine smooth muscle cells to the arterial media by inhibiting smooth muscle cell proliferation and migration, thus preventing intimal hyperplasia under physiological conditions. In addition, elastic laminae exhibit antiinflammatory effects relative

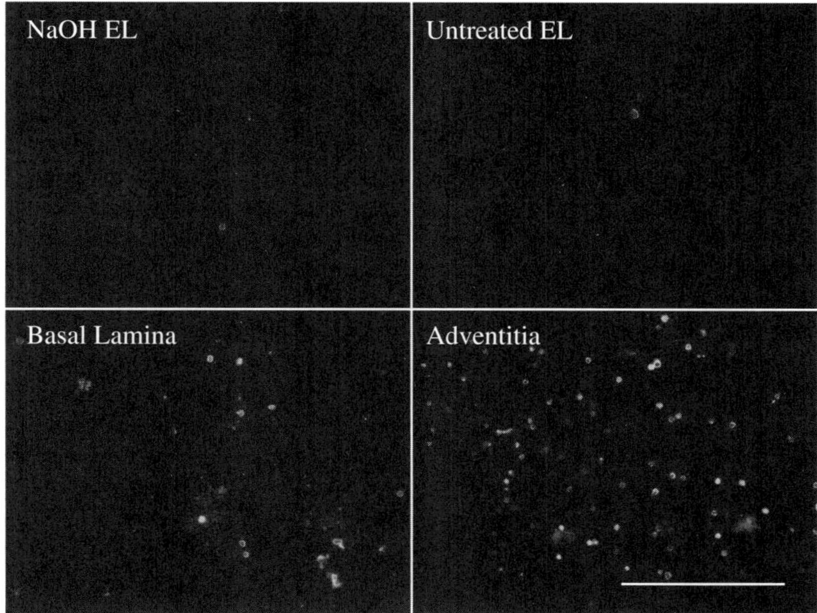

Figure 4.4. En face fluorescent micrographs showing monocytes adhered to NaOH-treated and untreated elastic lamina, basal lamina, and adventitia. EL: elastic lamina. Scale: 100 μm. (Reprinted from Liu SQ et al: *J Biol Chem* 280:39294–301, 2005 with permission from the American Society for Biochemistry and Molecular Biology.)

to collagen matrix. In particular, elastic laminae are capable of inhibiting leukocyte adhesion to and transmigration through the arterial media (Fig. 4.4 and chapter-opening Figure, above). Such inhibitory effects are potentially mediated by an inhibitory receptor known as signal-regulatory protein (SIRP) α. Elastic lamina degradation peptides extracted from arterial specimens bind to and activate SIRP α in monocytes, and induce the recruitment and phosphorylation of a protein tyrosine phosphatase known as SH2 domain-containing protein tyrosine phosphatase (SHP)-1. SHP-1 dephosphorylates mitogenic protein tyrosine kinases (see Chapter 5), resulting in the suppression of monocyte adhesion and activation. These anti-inflammatory effects render elastic laminae a potential material for vascular reconstruction. This issue is discussed in detail in Chapter 15.

PROTEOGLYCANS

Composition and Structure of Proteoglycans [4.5]

Proteoglycan is a complex molecule composed of a core protein and a large number of glycosaminoglycans (GAGs). The core protein is a 10–600-kDa chain-shaped protein. The protein chain can link to GAGs via covalent bonds, forming proteoglycan (Fig. 4.5). A proteoglycan molecule is different in structure and form from a glycoprotein, another type of protein with sugar residues. A proteoglycan is defined as a molecule with long unbranched GAG sidechains and is found primarily in the extracellular space. A glyco-

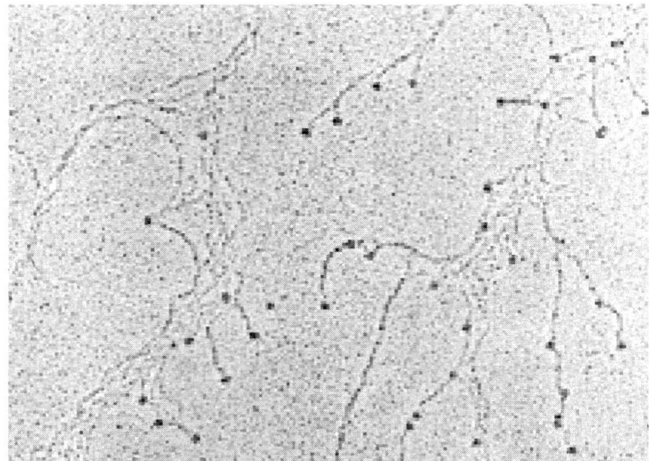

Figure 4.5. Electron micrograph showing interaction of neurocan with hyaluronan. The fiber-like structures are hyaluronan aggregates. Neurocan molecules often interact with the hyaluronan fibers at the end. (Reprinted from Retzler C et al: *J Biol Chem* 271:17107–13, 1996 with permission from the American Society for Biochemistry and Molecular Biology.)

protein usually contains short branched oligosaccharide chains and is found primarily in the cell membrane. Glycoproteins often serve as receptors.

Glycosaminoglycans are polysaccharide chains constituted with repeating disaccharides. Each disaccharide unit is composed of an amino sugar, either an *N*-acetylglucosamine or *N*-acetylgalactosamine. The other molecule is a uronic acid, which is a sugar acid generated by oxidation of the terminal —CH_2OH group of a sugar molecule to a carboxyl (=COOH) group. A large number of disaccharides are sulfated in a GAG molecule. The presence of the carboxyl and sulfate groups renders the GAGs negatively charged. Based on the type and arrangement of the sugar molecules as well as the location and number of sulfate bonds, GAGs can be classified into several groups: chondroitin sulfate and dermatan sulfate, heparin and heparan sulfate, keratan sulfate, and hyaluronic acid.

Chondroitin sulfate and *dermatan sulfate* are GAGs composed of about 60 repeating disaccharide units, each containing a D-glucuronic acid residue and an *N*-acetyl-D-galactosamine residue linked by glycosidic bonds (Fig. 4.6). These GAGs are sulfated at the C4 and C6 locations of the galactosamine residue. Chondroitin sulfate is a GAG with the C4 sulfate bond, whereas dermatan sulfate is a GAG with the C6 sulfate bond. These GAGs are found in cartilage, bone, connective tissue, and blood vessels and serve as ground substance, which supports and protects cells from injury.

Heparin is a highly sulfated GAG composed of repeating disaccharides of D-glucuronic acid and *N*-acetyl-D-glucosamine (Fig. 4.6). *Heparan sulfate* is similar to heparin in structure, but contains fewer *N*- and *O*-sulfate bonds. Heparin and heparan sulfate are generated in hepatocytes and vascular endothelial cells. These GAGs possess potent anticoagulant and antithrombogenic properties. Heparan sulfate is present on the surface of endothelial cells and plays a critical role in the maintenance of blood fluidity.

Keratan sulfate is a GAG consisting of repeating disaccharide units containing D-galactose and *N*-acetyl-D-glucosamine-6-sulfate. This type of GAG is found in the cornea

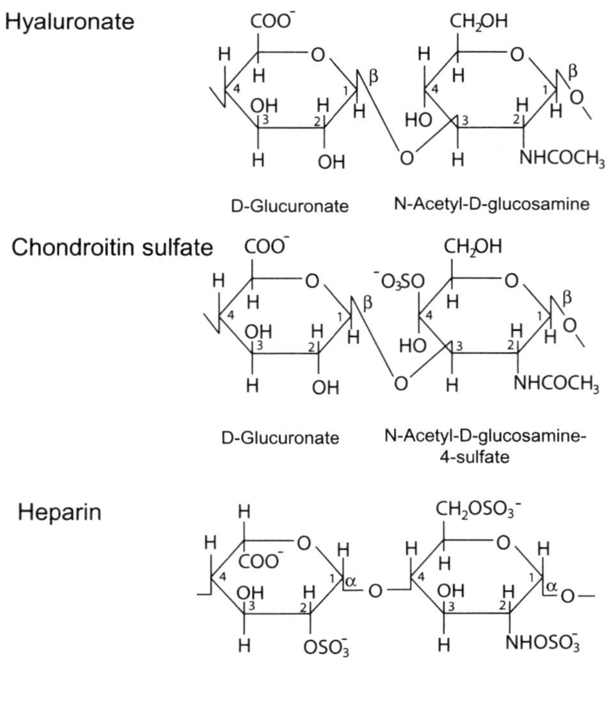

Figure 4.6. Chemical composition of the basic units of hyaluronate, chondroitin sulfate, dermatan sulfate, heparin, and keratan sulfate. Based on bibliography 4.5.

and cartilage. *Hyaluronic acid* is composed of more than 2000 disaccharide units, each containing a D-glucuronic acid and an *N*-acetyl-D-glucosamine residue linked by glycosidic bonds. This GAG is found in the vitreous humor, synovial fluids, cartilage, and blood vessels.

The GAGs described above can form various types of proteoglycan, including aggrecan, β-glycan, decorin, perlecan, syndecans, and versican. *Aggrecan* is a proteoglycan with a molecular weight of ~210 kDa, and is composed of about 130 chondroitin sulfate and keratan sulfate chains. This type of proteoglycan is found primarily in cartilage, forms complexes with hyaluronic acids, and serves as a ground substance in which cells reside. β-*Glycan* is a molecule with a molecular weight of ~36 kDa. It contains a single GAG chain constituted by chondroitin sulfate and dermatan sulfate. β-Glycan is present in extracellular matrix and cell membrane and play a role in mediating the activity of transforming growth factor β.

Decorin is a ~40-kDa proteoglycan with a single chondroitin sulfate and dermatan sulfate GAG chain. It is present in connective tissues and can bind to collagen type I, regulating the organization of the collagen matrix. It also binds to transforming growth factor β and mediates the activity of this growth factor. *Perlecan* is about 500 kDa in molecular weight and is composed of 2–15 heparan sulfate GAG chains. It is found primarily in the basal laminae of various organs and plays a role in the mechanical support of the basal lamina and mediating cellular activities (see next section). *Syndecans* are a family of proteoglycans, which include four members: syndecan-1, -2, -3, and -4; each member is encoded by a distinct gene. These are cell-associated proteoglycans and their structure and function are discussed in the next section.

Versicans are another family of proteoglycans, including versican-0, -1, -2, -3, and 4. These isoforms are generated by alternative splicing of the mRNA transcript for the versican core protein. A versican proteoglycan contains primarily chondroitin sulfate GAGs. Versicans are found in the extracellular matrix of blood vessels and synthesized by vascular smooth muscle cells. Versicans can bind to growth factors, enzymes, and other extracellular matrix components, and play a critical role in mediating the proliferation and migration of smooth muscle cells. The level of versicans is increased in response to vascular injury, promoting inflammatory reactions, lipid accumulation, mitogenic activity of smooth muscle cells, and intimal hyperplasia. See Table 4.3 for further information on these proteoglycans and additional proteoglycans.

Function of Proteoglycans [4.6]

There are several functions for proteoglycans in general. The most important function of proteoglycans is probably to serve as ground substances that support and protect cells from mechanical injury. Proteoglycans are found primarily in extracellular space and are highly hydrophilic. These molecules are negatively charged and can attract positively charged ions such as Na^+ and K^+. These ions create an osmotic gradient, resulting in the accumulation of water in proteoglycan molecules. Given their hydrophilic nature, these molecules can absorb a large amount of water and form a gel-like structure even at a very low concentration. Such a structure can resist a high level of compressive stress induced by mechanical impacts. The gel-like structure of proteoglycans also helps to organize cells within a tissue and organ.

Proteoglycans play a role in lubricating joint surfaces and preventing blood coagulation. Hyaluronic acids and hyaluronic acid-containing proteoglycans are present in the joint fluid and serve as lubricants, which reduce friction between the joint surfaces. Heparin and heparan sulfate are molecules that prevent blood coagulation and thrombogenesis. These molecules can inhibit the conversion of prothrombin to thrombin, a protease that cleaves soluble fibrinogen and catalyzes the formation of insoluble fibrin. The insoluble fibrin forms a solid meshwork at the site of endothelial injury and stimulates activation and adhesion of leukocytes and platelets. The fibrin meshwork serves as a soil for thrombogenesis and atherogenesis. The inhibition of thrombin formation by heparin or heparan sulfate prevents blood coagulation, thrombogenesis, and atherogenesis.

Proteoglycans are also involved in regulating the activity of signaling molecules. Proteoglycans can form complexes with growth factors, such as fibroblast growth factor and transforming growth factor. Such a process may activate or inhibit the activity of a growth factor, depending on the nature of the proteoglycans and growth factors. For instance, the interaction of fibroblast growth factor with heparan sulfate-containing proteoglycans can promote the activation of the growth factor. In contrast, the binding of transforming growth factor to proteoglycans inhibits the activity of the growth factor.

Proteoglycans participate directly in the regulation of cellular activities and functions. A heparan sulfate proteoglycan molecule found in the basal lamina, known as *perlecan*, has been shown to serve as an inhibitor for vascular smooth muscle cells. At the site of vascular injury, smooth muscle cells are activated to proliferate and migrate from the media to the intima of blood vessels, processes contributing to intimal hyperplasia and

TABLE 4.3. Characteristics of Selected Proteoglycans*

Proteins	Alternative Names	Amino Acids	Molecular Weight (kDa)	Expression	Functions
Versican	Chondroitin sulfate proteoglycan 2, glial hyaluronate-binding protein	3396	373	Blood vessel, liver, lung, uterus, kidney, prostate gland	Regulating cell proliferation and migration
Decorin	Proteoglycan II (PGII), dermatan sulfate proteoglycans II, bone proteoglycan II	359	40	Lung, kidney, skin, skeletal muscle, bone, cartilage, ligament	Constituting the matrix of connective tissues, binding to type I collagen fibrils, regulating matrix assembly, and suppressing tumor cell growth
Perlecan	Heparan sulfate proteoglycan of basement membrane, heparan sulfate proteoglycan 2 (HSPG2)	4393	469	Blood vessel, intestine, cartilage, kidney	Constituting the basement membrane, contributing to stabilization of matrix molecules, regulating glomerular permeability to macromolecules, and regulating cell adhesion
Aggrecan 1	AGC1, chondroitin sulfate proteoglycan core protein 1 (CSPCP1), cartilage-specific proteoglycan core protein 1	2415	250	Cartilage, brain	Constituting the extracellular matrix of cartilage, protecting cartilage from compression injury, and causing skeletal dysplasia and spinal degeneration when mutated

TABLE 4.3. *Continued*

Proteins	Alternative Names	Amino Acids	Molecular Weight (kDa)	Expression	Functions
Biglycan	BGN, proteoglycan I (PG-I), bone/cartilage proteoglycan I, dermatan sulfate proteoglycan I (DSPG-1)	368	42	Bone, cartilage, skin, ligament, brain, lung	Binding to collagen fibrils, regulating both assembly and integrity of extracellular matrix, promoting neuronal survival, and mediating macrophage-related inflammatory reactions
β-Glycan	Transforming growth factor β receptor type 3	849	93	Heart	Found at cell surface and in extracellular matrix, interacting with transforming growth factor, and participating in regulation of cell proliferation and differentiation
Syndecan 2	SYND2, heparan sulfate proteoglycan (HSPG), fibroglycan	201	22	Bone, liver, skin	A transmembrane heparan sulfate proteoglycan; mediating cell binding, proliferation, and migration; regulating cytoskeletal integrity and organization; and mediating HIV transmission to T lymphocytes
Neurocan	Chondroitin sulfate proteoglycan 3	1321	143	Nervous system	A chondroitin sulfate proteoglycan that mediates the adhesion and migration of neural cells
Keratocan	KERA, KTN, keratan sulfate proteoglycan	352	41	Cornea	A keratan sulfate proteoglycan found in cornea and a critical component in corneal transparency

*Based on bibliography 4.5.

atherogenesis. The perlecan molecules in the basal lamina, which resides beneath the endothelium, inhibit the proliferation and migration of smooth muscle cells and thus suppress intimal hyperplasia and atherogenesis.

In general, proteoglycans may regulate the activity of signaling molecules and cells via several approaches: (1) immobilizing signaling molecules and thus confining the molecules to a specified location, (2) blocking or stimulating the activity of signaling molecules via binding interactions, and (3) protecting signaling molecules from enzymatic degradation. Various types of proteoglycan may elect to use different mediating approaches.

While most proteoglycans are present in extracellular space, there exist cell-associated proteoglycans. A typical example is the proteoglycan family of *syndecans*. These proteoglycans are transmembrane receptor type of molecules. Each syndecan molecule is composed of an extracellular domain, a single-span transmembrane domain, and a cytoplasmic domain. The extracellular domain of syndecans contains GAGs, such as chondroitin sulfate and heparan sulfate. The intracellular domain of syndecans interacts with actin filaments. Syndecans are found in fibroblasts and epithelial cells, and serve as receptors for extracellular matrix components, including fibronectin and collagen. These proteoglycan molecules can bind to growth factors, such as fibroblast growth factor, and mediate the interaction of growth factors with their receptors. Such an activity contributes to the regulation of embryonic development, angiogenesis, and tumorigenesis.

Another type of cell-associated proteoglycan is heparan sulfate-containing proteoglycans. In addition to the role in regulating blood coagulation, heparan sulfate proteoglycans can mediate the activity of several signaling pathways involving Wnt, hedgehog, transforming growth factor, and fibroblast growth factor. Such a mediating process is critical to embryonic development and pathogenic remodeling. Furthermore, heparan sulfate proteoglycans are involved in the regulation of tumorigenesis. These molecules may promote tumor growth and metastasis in at least two types of tumor: myeloma and breast cancer. Understanding the role of proteoglycans in regulating signaling processes may lead to the development of new therapeutic approaches for tumors and other pathological disorders.

MATRIX METALLOPROTEINASES

Matrix metalloproteinases (MMPs) are enzymes that induce degradation of extracellular matrix components. There are more than 20 types of MMPs, which are found in mammalian tissues and produced by different cell types. Each type of MMP can target one or more extracellular matrix components, although the activity of MMPs is not highly specific. Almost all MMPs are synthesized in cells as preproenzymes and released as inactive forms known as pro-MMPs. The inactive forms of MMPs can be activated by tissue and plasma proteinases or membrane-type MMPs (MT-MMPs), which cleave pro-MMPs. The production and activation of MMPs are highly regulated processes, which are critical to a number of physiological processes, including embryonic morphogenesis, neurite outgrowth, ovulation, bone growth, angiogenesis, apoptosis, and wound healing. In addition, MMPs are involved in the pathogenesis of a number of disorders, including cancer metastasis, atherosclerosis, skin ulceration, gastric ulcer, corneal ulceration, liver fibrosis, and emphysema. MMPs mediated these physiological and pathological processes via inducing matrix degradation, which promotes two fundamental cellular activities: cell migration and proliferation.

The expression of MMPs can be induced by several types of stimulating factors, such as growth factors, cytokines, phorbol esters, and mechanical stress. In general, factors that mediate inflammatory and growth reactions likely stimulate the expression of MMPs. These growth and inflammatory factors induce activation of cell signaling pathways involving ERK1/2, stress-activated protein kinase (SAPK)/JNK, and p38, resulting in the upregulation of the MMP genes (see Chapter 5 for the signaling pathways). The physiological significance is that MMP-induced matrix degradation, in association with the upregulation of growth factors (e.g., epidermal growth factor and platelet-derived growth factor) and inflammatory cytokines (e.g., tumor necrosis factor α and interleukin-1), facilitates cell migration, an essential process for tissue regeneration in wound healing and leukocyte infiltration in inflammatory reactions. Several factors, such as transforming growth factor β, retinoic acids, glucocorticoids, exert an inhibitory effect on the activity and expression of MMPs. The effects of stimulatory and inhibitory factors are coordinately regulated under physiological and pathological conditions. In a quiescent state, the activity and expression of MMPs are inhibited. The activity and expression of MMPs are usually upregulated in pathological disorders such as mechanical and chemical trauma, atherosclerosis, and carcinogenesis.

Structural Features of MMPs [4.7]

A MMP is composed of several common domains, including a signal peptide domain, a propeptide domain, and a catalytic domain (Fig. 4.7). There are other MMP domains,

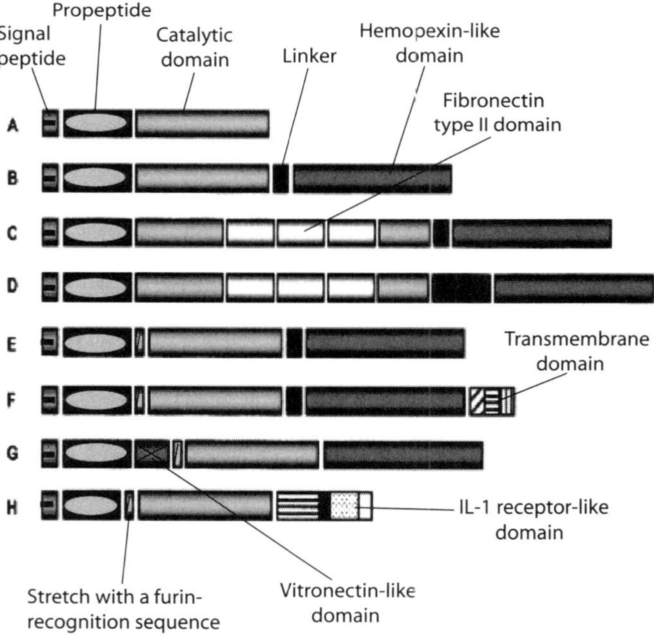

Figure 4.7. Schematic representation of the structure of matrix metalloproteinases (MMPs). (Reprinted from Nagase H, Woessner JF Jr: *J Biol Chem* 274:21491–4, 1999 with permission from the American Society for Biochemistry and Molecular Biology.)

such as hemopexin-like domain, fibronectin type II domain, vitronectin-like domain, furin recognition sequence, linker, and transmembrane-cytoplasmic domain, but these domains are not present in all MMPs. Figure 4.7 shows the common and specific domains for MMPs. There are several structural features that are important for the function of MMPs. The propeptide domain contains a PRCG(V/N)PD sequence, which inhibits the activity of the zinc-dependent catalytic domain of MMPs and renders the enzymes inactive. The removal of the propeptide by an enzyme or chemical compound induces the activation of the MMPs. The catalytic domain can bind to zinc and calcium ions, which are necessary for the stability and activity of MMPs. Several MMPs, including MMP2 and MMP9, contain repeated fibronectin type II-like sequences in the catalytic domain. These sequences mediate the interaction of MMPs with substrate molecules, such as collagen. MMPs that possess a collagenase activity contain a *C*-terminal hemopexin-like domain. This domain is essential for cleaving helical collagen fibrils. The transmembrane domain found in MT-MMPs mediates the integration of the enzymes to the cell membrane.

Activation of MMPs [4.8]

Matrix metalloproteinases (see Table 4.4) are released from cells as inactive pro-MMPs. The inactive forms can be activated by a number of factors, including proteinases, mercurial compounds, reactive-oxygen species, and protein-denaturing reagents, under experimental conditions in vitro. These factors can degrade or remove the propeptide domain, which inhibits the catalytic activity of MMPs. Under physiological conditions in an in vivo system, the propeptide domain of pro-MMPs is cleaved by proteinases, resulting in the activation of MMPs. Plasmin is a typical proteinase that activates pro-MMPs. Furthermore, certain types of pro-MMP, such as pro-MMP2, are activated by a group of MMPs, known as *membrane-type MMPs* (MT-MMPs), which are anchored to the cell membrane. For instance, MT1-MMP can cleave and activate pro-MMP2 on the cell surface. MT-MMP-induced activation of MMP is critical to several biological processes, including angiogenesis, cell migration, and cancer metastasis. In these processes, activated MMPs on the cell surface induce matrix degradation, creating a channel that allows cell migration.

The activity of MMPs can be suppressed by a family of molecules, known as tissue inhibitors of metalloproteinases (TIMPs). This family includes four known members: TIMP1, TIMP2, TIMP3, and TIMP4, with molecular weight ranging from 21 to 30 kDa. TIMPs can inhibit cell migration, tumor cell invasion, and angiogenesis via their negative influence on MMPs. Thus, TIMPs participate in the regulation of the activity of MMPs together with growth factors and cytokines. In addition, TIMPs exert other activities. These activities are dependent on the type of TIMPs and the type of target cells. TIMP1 and TIMP2 have been shown to stimulate cell proliferation and prevent cell apoptosis. However, TIMP2 has also been found to inhibit the proliferation of vascualr endothelial cells and angiogenesis. TIMP3 can induce apoptosis in carcinoma cells and melanoma cells. These diverse activities of TIMPs play important roels in the regulation of not only matrix degradation but also cellular activities.

TABLE 4.4. Characteristics of Selected MMPs*

Proteins	Alternative Names	Amino Acids	Molecular Weight (kDa)	Expression	Functions
MMP1	Fibroblast collagenase, interstitial collagenase	469	54	Connective tissue, bone	An enzyme that degrades collagen types I, II, and III
MMP2	Collagenase type 4, collagenase type 4A, 72-kDa gelatinase, gelatinase A, neutrophil gelatinase	660	74	Connective tissue, skin, bone, blood vessel	An enzyme that degrades type IV collagen and gelatin; also regulates vascularization and inflammatory reactions
MMP3	Stromelysin I, STMY1, STR1, progelatinase, transin, proteoglycanase	477	54	Connective tissue, cartilage, skin, blood vessel	An enzyme that degrades fibronectin, laminin, collagens (types III, IV, IX, and X), and proteoglycans; also plays a role in regulation of wound repair, atherogenesis, and tumor invasion
MMP7	Uterine matrilysin, putative metalloproteinase 1 (PUMP1), Matrin	267	30	Ubiquitous	A proteinase that degrades proteoglycans, fibronectin, elastin and casein; also participates in regulation of wound healing
MMP8	Neutrophil collagenase, collagenase 2	467	52	Leukocytes, bone marrow	A proteinase that degrades collagen types I, II, and III
MMP9	Gelatinase B, 92-kDa gelatinase, collagenase type IV B, 92 KD collagenase type IV, collagenase type V, macrophage gelatinase	707	78	Leukocytes, bone marrow, skin, intestine, liver, kidney, lung, blood vessel	A proteinase that degrades collagen types IV and V, mediates IL8-induced mobilization of hematopoietic progenitor cells from the bone marrow
MMP10	Stromelysin II (STMY2), Transin-2	476	54	Skin, leukocytes, heart, blood vessel, kidney, lung, liver	A proteinase that degrades proteoglycans and fibronectin

MMP11	Stromelysin 3 (STMY3)	488	55	Connective tissue	A proteinase that degrades α1 proteinase inhibitor, fibronectin, laminin
MMP12	Macrophage metalloelastase	470	54	macrophages	A proteinase that degrades elastin and contributes to the development of aneurysm and emphysema
MMP13	Collagenase 3	471	54	Cartilage, skin, blood vessel	A proteinase that degrades collagen types I, II, and III contributes to cartilage turnover and osteoarthritis
MMP14	Membrane type matrix metalloproteinase 1 (MT1 MMP)	582	66	Cartilage, skin, connective tissue	A membrane-type proteinase that activates MMP2 and participates in the regulation of tumor invasion
MMP15	Membrane type matrix metalloproteinase 2 (MT2-MMP)	669	76	Connective tissue	A membrane-type MMP that activates progelatinase A
TIMP1	Metalloproteinase inhibitor 1, fibroblast collagenase inhibitor, collagenase inhibitor, erythroid-potentiating activity	207	23	Bone marrow, connective tissue, bone	A natural inhibitor of the matrix metalloproteinases (MMPs), promoting cell proliferation and preventing cell apoptosis
TIMP2	Metalloproteinase inhibitor 2	220	24	Liver, retina, placenta	Inhibitory MMPs, inhibiting the activity of MMPs and also inhibiting proliferation of endothelial cells and angiogenesis
TIMP3	Metalloproteinase inhibitor 3	211	24	Brain, heart, lung, kidney, liver, pancreas, blood vessel	Inhibitory MMPs
TIMP4	Metalloproteinase inhibitor 4	224	26	Heart, blood vessel, kidney, pancreas, and intestine	Inhibitory matrix metalloproteinases, participating in regulation of platelet aggregation and recruitment

*Based on bibliography 4.7.

BIBLIOGRAPHY

4.1. Composition and Formation of Collagen Matrix

Collagen Type I

Bateman JF, Lamande SR, Dahl HHM, Chan D, Mascara T et al: A frameshift mutation results in a truncated nonfunctional carboxyl-terminal pro-alpha-1(I) propeptide of type I collagen in osteogenesis imperfecta, *J Biol Chem* 264:10960–4, 1989.

Bonadio J, Ramirez F, Barr M: An intron mutation in the human alpha-1(I) collagen gene alters the efficiency of pre-mRNA splicing and is associated with osteogenesis imperfecta type II, *J Biol Chem* 265:2262–8, 1990.

Bonaventure J, Cohen-Solal L, Lasselin C, Maroteaux P: A dominant mutation in the COL1A1 gene that substitutes glycine for valine causes recurrent lethal osteogenesis imperfecta, *Hum Genet* 89:640–6, 1992.

Cabral WA, Mertts MV, Makareeva E, Colige A, Tekin M et al: Type I collagen triplet duplication mutation in lethal osteogenesis imperfecta shifts register of alpha chains throughout the helix and disrupts incorporation of mutant helices into fibrils and extracellular matrix, *J Biol Chem* 278:10006–12, 2003.

Chu ML, de Wet W, Bernard M, Ding JF, Morabito M et al: Human pro-alpha-1(I) collagen gene structure reveals evolutionary conservation of a pattern of introns and exons, *Nature* 310:337–40, 1984.

Chu ML, de Wet W, Bernard M, Ramirez F: Fine structural analysis of the human pro-alpha-1(I) collagen gene: Promoter structure, AluI repeats, and polymorphic transcripts, *J Biol Chem* 260:2315–20, 1985.

Dalgleish R: The human type I collagen mutation database, *Nucleic Acids Res* 25:181–7, 1997.

Deak SB, Scholz PM, Amenta PS, Constantinou CD, Levi-Minzi SA et al: The substitution of arginine for glycine 85 of the alpha-1(I) procollagen chain results in mild osteogenesis imperfecta: The mutation provides direct evidence for three discrete domains of cooperative melting of intact type I collagen, *J Biol Chem* 266:21827–32, 1991.

Di Lullo GA, Sweeney SM, Korkko J, Ala-Kokko L, San Antonio JD: Mapping the ligand-binding sites and disease-associated mutations on the most abundant protein in the human, type I collagen, *J Biol Chem* 277:4223–31, 2002.

Grant SFA, Reid DM, Blake G, Herd R, Fogelman I et al: Reduced bone density and osteoporosis associated with a polymorphic Sp1 binding site in the collagen type I-alpha 1 gene, *Nature Genet* 14:203–5, 1996.

Lohmueller KE, Pearce CL, Pike M, Lander ES, Hirschhorn JN: Meta-analysis of genetic association studies supports a contribution of common variants to susceptibility to common disease, *Nature Genet* 33:177–82, 2003.

Schnieke A, Harbers K, Jaenisch R: Embryonic lethal mutation in mice induced by retrovirus insertion into the alpha-1 (I) collagen gene, *Nature* 304:315–20, 1983.

Uitterlinden AG, Burger H, Huang Q, Yue F, McGuigan FEA et al: Relation of alleles of the collagen type I-alpha-1 gene to bone density and the risk of osteoporotic fractures in postmenopausal women, *New Eng J Med* 338:1016–21, 1998

Collagen Type II

Ala-Kokko L, Baldwin CT, Moskowitz RW, Prockop DJ: Single base mutation in the type II procollagen gene (COL2A1) as a cause of primary osteoarthritis associated with a mild chondrodysplasia, *Proc Natl Acad Sci USA* 87:6565–8, 1990.

Bogaert R, Tiller GE, Weis MA, Gruber HE, Rimoin DL et al: An amino acid substitution (gly853-to-glu) in the collagen alpha-1(II) chain produces hypochondrogenesis, *J Biol Chem* 267:22522–6, 1992.

Chan D, Cole WG, Chow CW, Mundlos S, Bateman JF: A COL2A1 mutation in achondrogenesis type II results in the replacement of type II collagen by type I and III collagens in cartilage, *J Biol Chem* 270:1747–53, 1995.

Cheah KSE, Stoker NG, Griffin JR, Grosveld FG, Solomon E: Identification and characterization of the human type II collagen gene (COL2A1), *Proc Natl Acad Sci USA* 82:2555–9, 1985.

Horton WA, Machado MA, Ellard J, Campbell D, Bartley J et al: Characterization of a type II collagen gene (COL2A1) mutation identified in cultured chondrocytes from human hypochondrogenesis, *Proc Natl Acad Sci USA* 89:4583–7, 1992.

Knowlton RG, Katzenstein PL, Moskowitz RW, Weaver EJ, Malemud CJ et al: Genetic linkage of a polymorphism in the type II procollagen gene (COL2A1) to primary osteoarthritis associated with mild chondrodysplasia, *New Engl J Med* 322:526–30, 1990.

Liu YF, Chen WM, Lin YF, Yang RC, Lin MW et al: Type II collagen variants and inherited osteonecrosis of the femoral head, *New Engl J Med* 352:2294–301, 2005.

Solomon E, Hiorns LR, Spurr N, Kurkinen M, Barlow D et al: Chromosomal assignments of the genes coding for human types II, III and IV collagen: A dispersed gene family, *Proc Natl Acad Sci USA* 82:3330–4, 1985.

Tiller GE, Polumbo PA, Weis MA, Bogaert R, Lachman RS et al: Dominant mutations in the type II collagen gene, COL2A1, produce spondyloepimetaphyseal dysplasia, Strudwick type, *Nature Genet* 11:87–9, 1995.

Collagen Type III

Janeczko RA, Ramirez F: Nucleotide and amino acid sequences of the entire human alpha-1(III) collagen, *Nucleic Acids Res* 17:6742, 1989.

Kontusaari S, Tromp G, Kuivaniemi H, Romanic AM, Prockop DJ: A mutation in the gene for type III procollagen (COL3A1) in a family with aortic aneurysms, *J Clin Invest* 86:1465–73, 1990.

Liu X, Wu H, Byrne M, Krane S, Jaenisch R: Type III collagen is crucial for collagen I fibrillogenesis and for normal cardiovascular development, *Proc Natl Acad Sci USA* 94:1852–6, 1997.

Solomon E, Hiorns LR, Spurr N, Kurkinen M, Barlow D et al: Chromosomal assignments of the genes coding for human types II, III and IV collagen: A dispersed gene family, *Proc Natl Acad Sci USA* 82:3330–4, 1985.

Tromp G, Wu Y, Prockop DJ, Madhatheri SL, Kleinert C et al: Sequencing of cDNA from 50 unrelated patients reveals that mutations in the triple-helical domain of type III procollagen are an infrequent cause of aortic aneurysms, *J Clin Invest* 91:2539–45, 1993.

Collagen Type IV α 1

Bowcock AM, Hebert JM, Christiano AM, Wijsman E, Cavalli-Sforza LL et al: The pro alpha 1 (IV) collagen gene is linked to the D13S3 locus at the distal end of human chromosome 13q, *Cytogenet Cell Genet* 45:234–6, 1987.

Bowcock AM, Hebert JM, Wijsman E, Gadi I, Cavalli-Sforza LL, Boyd CD: High recombination between two physically close human basement membrane collagen genes at the distal end of chromosome 13q, *Proc Natl Acad Sci USA* 85:2701–5, 1988.

Brinker JM, Gudas LJ, Loidl HR, Wang SY, Rosenbloom J et al: Restricted homology between human alpha-1 type IV and other procollagen chains, *Proc Natl Acad Sci USA* 82:3649–53, 1985.

Emanuel BS, Sellinger BT, Gudas LJ, Myers JC: Localization of the human procollagen alpha-1(IV) gene to chromosome 13q34 by in situ hybridization, *Am J Hum Genet* 38:38–44, 1986.

Gould DB, Phalan FC, Breedveld GJ, van Mil SE, Smith RS et al: Mutations in Col4a1 cause perinatal cerebral hemorrhage and porencephaly, *Science* 308:1167–71, 2005.

Maeshima Y, Colorado PC, Torre A, Holthaus KA, Grunkemeyer JA et al: Distinct antitumor properties of a type IV collagen domain derived from basement membrane, *J Biol Chem* 275:21340–8, 2000.

Collagen Type IV α 2

Boyd CD, Toth-Fejel S, Gadi IK, Litt M, Condon MR et al: The genes coding for human pro alpha-1(IV) collagen and pro alpha-2(IV) collagen are both located at the end of the long arm of chromosome 13, *Am J Hum Genet* 42:309–14, 1988.

Griffin CA, Emanuel BS, Hansen JR, Cavenee WK, Myers JC: Human collagen genes encoding basement membrane alpha-1(IV) and alpha-2(IV) chains map to the distal long arm of chromosome 13, *Proc Natl Acad Sci UAS* 84:512–16, 1987.

Haniel A, Welge-Lussen U, Kuhn K, Poschl E: Identification and characterization of a novel transcriptional silencer in the human collagen type IV gene COL4A2, *J Biol Chem* 270:11209–15, 1995.

Killen PD, Francomano CA, Yamada Y, Modi WS, O'Brien SJ: Partial structure of the human alpha-2(IV) collagen chain and chromosomal localization of the gene (COL4A2), *Hum Genet* 77:318–24, 1987.

Solomon E, Hall V, Kurkinen M: The human alpha-2(IV) collagen gene, COL4A2, is syntenic with the alpha-1(IV) gene, COL4A1, on chromosome 13, *Ann Hum Genet* 51:125–7, 1987.

Collagen V α 1

Caridi G, Pezzolo A, Bertelli R, Gimelli G, Di Donato A et al: Mapping of the human COL5A1 gene to chromosome 9q34.3, *Hum Genet* 90:174–6, 1992.

Fichard A, Kleman JP, Ruggiero F: Another look at collagen V and XI molecules, *Matrix Biol* 14:515–31, 1994.

Greenspan DS, Byers MG, Eddy RL, Cheng W, Jani-Sait S et al: Human collagen gene COL5A1 maps to the q34.2-q34.3 region of chromosome 9, near the locus for nail-patella syndrome, *Genomics* 12:836–7, 1992.

Greenspan DS, Northrup H, Au KS, McAllister KA, Francomano CA et al: COL5A1: Fine genetic mapping and exclusion as candidate gene in families with nail-patella syndrome, tuberous sclerosis 1, hereditary hemorrhagic telangiectasia, and Ehlers-Danlos syndrome type II, *Genomics* 25:737–9, 1995.

Takahara K, Hoffman GG, Greenspan DS: Complete structural organization of the human alpha-1(V) collagen gene (COL5A1): Divergence from the conserved organization of other characterized fibrillar collagen genes, *Genomics* 29:588–97, 1995.

Takahara K, Sato Y, Okazawa K, Okamoto N, Noda A et al: Complete primary structure of human collagen alpha-1(V) chain, *J Biol Chem* 266:13124–9, 1991.

Collagen Type V α 2

Andrikopoulos K, Liu X, Keene DR, Jaenisch R, Ramirez F: Targeted mutation in the col5a2 gene reveals a regulatory role for type V collagen during matrix assembly, *Nature Genet* 9:31–6, 1995.

Emanuel BS, Cannizzaro LA, Seyer JM, Myers JC: Human alpha-1(III) and alpha-2(V) procollagen genes are located on the long arm of chromosome 2, *Proc Natl Acad Sci USA* 82:3385–9, 1985.

Grond-Ginsbach C, Wigger F, Morcher M, von Pein F, Grau A et al: Sequence analysis of the COL5A2 gene in patients with spontaneous cervical artery dissections, *Neurology* 58:1103–5, 2002.

Huerre-Jeanpierre C, Henry I, Bernard M, Gallano P, Weil D et al: The pro-alpha-2(V) collagen gene (COL5A2) maps to 2q14–2q32, syntenic to the pro-alpha-1(III) collagen locus (COL3A1), *Hum Genet* 73:64–7, 1986.

Collagen Type VI α 1

Bonaldo P, Braghetta P, Zanetti M, Piccolo S, Volpin D et al: Collagen VI deficiency induces early onset myopathy in the mouse: An animal model for Bethlem myopathy, *Hum Mol Genet* 7:2135–40, 1998.

Davies GE, Howard CM, Farrer MJ, Coleman MM, Bennett LB et al: Genetic variation in the COL6A1 region is associated with congenital heart defects in trisomy 21 (Down's syndrome), *Ann Hum Genet* 59:253–69, 1995.

Hessle H, Engvall E: Type VI collagen: Studies on its localization, structure, and biosynthetic form with monoclonal antibodies, *J Biol Chem* 259:3955–61, 1984.

Irwin WA, Bergamin N, Sabatelli P, Reggiani C, Megighian A et al: Mitochondrial dysfunction and apoptosis in myopathic mice with collagen VI deficiency, *Nature Genet* 35:367–71, 2003.

Lamande SR, Shields KA, Kornberg AJ, Shield LK, Bateman JF: Bethlem myopathy and engineered collagen VI triple helical deletions prevent intracellular multimer assembly and protein secretion, *J Biol Chem* 274:21817–22, 1999.

Trueb B, Winterhalter KH: Type VI collagen is composed of a 200kD subunit and two 140kD subunits, *EMBO J* 5:2815–19, 1986.

Collagen Type VII

Bentz H, Morris NP, Murray LW, Sakai LY, Hollister DW et al: Isolation and partial characterization of a new human collagen with an extended triple-helical structural domain, *Proc Natl Acad Sci USA* 80:3168–72, 1983.

Chen M, Kasahara N, Keene DR, Chan L, Hoeffler WK et al: Restoration of type VII collagen expression and function in dystrophic epidermolysis bullosa, *Nature Genet* 32:578–83, 2002.

Christiano AM, Greenspan DS, Lee S, Uitto J: Cloning of human type VII collagen: Complete primary sequence of the alpha-1(VII) chain and identification of intragenic polymorphisms, *J Biol Chem* 269:20256–62, 1994.

Greenspan DS, Byers MG, Eddy RL, Hoffman GG, Shows TB: Localization of the human collagen gene COL7A1 to 3p21.3 by fluorescence in situ hybridization, *Cytogenet Cell Genet* 62:35–6, 1993.

Parente MG, Chung LC, Ryynanen J, Woodley DT, Wynn KC et al: Human type VII collagen: cDNA cloning and chromosomal mapping of the gene, *Proc Natl Acad Sci USA* 88:6931–5, 1991.

Ryynanen J, Sollberg S, Parente MG, Chung LC, Christiano AM et al: Type VII collagen gene expression by cultured human cells and in fetal skin: Abundant mRNA and protein levels in epidermal keratinocytes, *J Clin Invest* 89:163–8, 1992.

Collagen Type VIII

Muragaki Y, Mattei MG, Yamaguchi N, Olsen BR, Ninomiya Y: The complete primary structure of the human alpha-1(VIII) chain and assignment of its gene (COL8A1) to chromosome 3, *Eur J Biochem* 197:615–22, 1991.

Muragaki Y, Shiota C, Inoue M, Ooshima A, Olsen BR et al: Alpha-1(VIII)-collagen gene transcripts encode a short-chain collagen polypeptide and are expressed by various epithelial, endothelial and mesenchymal cells in newborn mouse tissues, *Eur J Biochem* 207:895–902, 1992.

Yamaguchi N, Benya PD, van der Rest M, Ninomiya Y: The cloning and sequencing of alpha-1(VIII) collagen cDNAs demonstrate that type VIII collagen is a short chain collagen and contains triple-helical and carboxyl-terminal non-triple-helical domains similar to those of type X collagen, *J Biol Chem* 264:16022–9, 1989

Collagen Type IX

Eyre DR, Apon S, Wu JJ, Ericsson LH, Walsh KA: Collagen type IX: Evidence for covalent linkages to type II collagen in cartilage, *FEBS Lett* 220:337–41, 1987.

Fassler R, Schnegelsberg PNJ, Dausman J, Shinya T, Muragaki Y et al: Mice lacking alpha 1 (IX) collagen develop noninflammatory degenerative joint disease, *Proc Natl Acad Sci* 91:5070–4, 1994.

Hagg R, Hedbom E, Mollers U, Aszodi A, Fassler R et al: Absence of the alpha-1(IX) chain leads to a functional knock-out of the entire collagen IX protein in mice, *J Biol Chem* 272:20650–4, 1997.

Kimura T, Mattei MG, Stevens JW, Goldring MB, Ninomiya Y et al: Molecular cloning of rat and human type IX collagen cDNA and localization of the alpha1(IX) gene on the human chromosome 6, *Eur J Biochem* 179:71–8, 1989.

Mayne R, van der Rest M, Ninomiya Y, Olsen BR: The structure of type IX collagen, *Ann NY Acad Sci* 460:38–46, 1985.

McCormick D, van der Rest M, Goodship J, Lozano G, Ninomiya T, Olsen BR: Structure of the glycosaminoglycan domain in the type IX collagen-proteoglycan, *Proc Natl Acad Sci USA* 84:4044–8, 1987.

Ninomiya Y, Olsen BR: Synthesis and characterization of cDNA encoding a cartilage-specific short collagen, *Proc Natl Acad Sci USA* 81:3014–8, 1984.

Collagen Type X

Apte S, Mattei MG, Olsen BR: Cloning of human alpha-1(X) collagen DNA and localization of the COL10A1 gene to the q21-q22 region of human chromosome 6, *FEBS Lett* 282:393–6, 1991.

Gress CJ, Jacenko O: Growth plate compressions and altered hematopoiesis in collagen X null mice, *J Cell Biol* 149:983–93, 2000.

Rosati R, Horan GSB, Pinero GJ, Garofalo S, Keene DR et al: Normal long bone growth and development in type X collagen-null mice, *Nature Genet* 8:129–35, 1994.

Zheng Q, Zhou G, Chen Y, Garcia-Rojas X, Lee B: Type X collagen gene regulation by Runx2 contributes directly to its hypertrophic chondrocyte-specific expression in vivo, *J Cell Biol* 162:833–42, 2003.

Collagen Type XI

Bernard M, Yoshioka H, Rodriguez E, van der Rest M, Kimura T et al: Cloning and sequencing of pro-alpha1(XI) collagen cDNA demonstrated that type XI belongs to the fibrillar class of collagens and reveals that the expression of the gene is not restricted to cartilaginous tissue, *J Biol Chem* 263:17159–66, 1988.

Li Y, Lacerda DA, Warman ML, Beier DR, Yoshioka H et al: A fibrillar collagen gene, Col11a1, is essential for skeletal morphogenesis, *Cell* 80:423–30, 1995.

Morris NP, Bachinger HP: Type XI collagen is a heterotrimer with the composition (1alpha,2alpha,3alpha) retaining non-triple-helical domains, *J Biol Chem* 262:11345–50, 1987.

Wu JJ, Eyre DR: Structural analysis of cross-linking domains in cartilage type XI collagen: Insights on polymeric assembly, *J Biol Chem* 270:18865–70, 1995.

Zhidkova NI, Justice SK, Mayne R: Alternative mRNA processing occurs in the variable region of the pro-alpha-1(XI) and pro-alpha-2(XI) collagen chains, *J Biol Chem* 270:9486–93, 1995.

Collagen Type XII

Gerecke DR, Olson PF, Koch M, Knoll JHM, Taylor R et al: Complete primary structure of two splice variants of collagen XII, and assignment of alpha-1(XII) collagen (COL12A1), alpha-1(IX) collagen (COL9A1), and alpha-1(XIX) collagen (COL19A1) to human chromosome 6q12-q13, *Genomics* 41:236–42, 1997.

Gordon MK, Gerecke DR, Olsen BR: Type XII collagen: Distinct extracellular matrix component discovered by cDNA cloning, *Proc Natl Acad Sci USA* 84:6040–4, 1987.

Collagen Type XIII

Horelli-Kuitunen N, Kvist AP, Helaakoski T, Kivirikko K, Pihlajaniemi T et al: The order and transcriptional orientation of the human COL13A1 and P4HA genes on chromosome 10 long arm determined by high-resolution FISH, *Genomics* 46:299–302, 1997.

Shows TB, Tikka L, Byers MG, Eddy RL, Haley LL et al: Assignment of the human collagen alpha-1(XIII) chain gene (COL13A1) to the q22 region of chromosome 10, *Genomics* 5:128–33, 1989.

Sund M, Ylonen R, Tuomisto A, Sormunen R, Tahkola J et al: Abnormal adherence junctions in the heart and reduced angiogenesis in transgenic mice overexpressing mutant type XIII collagen, *EMBO J* 20:5153–64, 2001.

Tikka L, Pihlajaniemi T, Henttu P, Prockop DJ, Tryggvason K: Gene structure for the alpha-1 chain of a human short-chain collagen (type XIII) with alternatively spliced transcripts and translation termination codon at the 5-prime end of the last exon, *Proc Natl Acad Sci USA* 85:7491–5, 1988.

Collagen Type XIV

Bauer M, Dieterich W, Ehnis T, Schuppan D: Complete primary structure of human collagen type XIV (undulin), *Biochim Biophys Acta* 1354:183–8, 1997.

Dublet B, van der Rest M: Type XIV collagen, a new homotrimeric molecule extracted from fetal bovine skin and tendon, with a triple helical disulfide-bonded domain homologous to type IX and type XII collagens, *J Biol Chem* 266:6853–8, 1991.

Just M, Herbst H, Hummel M, Durkop H, Tripier D et al: Undulin is a novel member of the fibronectin-tenascin family of extracellular matrix glycoproteins, *J Biol Chem* 266:17326–32, 1991.

Schnittger S, Herbst H, Schuppan D, Dannenberg C, Bauer M et al: Localization of the undulin gene (UND) to human chromosome band 8q23, *Cytogenet Cell Genet* 68:233–4, 1995.

Schuppan D, Cantaluppi MC, Becker J, Veit A, Bunte T et al: Undulin, an extracellular matrix glycoprotein associated with collagen fibrils, *J Biol Chem* 265:8823–32, 1990.

Collagen Type XV

Myers JC, Kivirikko S, Gordon MK, Dion AS, Pihlajaniemi T: Identification of a previously unknown human collagen chain, alpha-1(XV), characterized by extensive interruptions in the triple-helical region, *Proc Natl Acad Sci USA* 89:10144–8, 1992.

Eklund L, Piuhola J, Komulainen J, Sormunen R, Ongvarrasopone C et al: Lack of type XV collagen causes a skeletal myopathy and cardiovascular defects in mice, *Proc Natl Acad Sci USA* 98:1194–9, 2001.

Hagg PM, Muona A, Lietard J, Kivirikko S, Pihlajaniemi T: Complete exon-intron organization of the human gene for the alpha-1 chain of type XV collagen (COL15A1) and comparison with the homologous Col18a1 gene, *J Biol Chem* 273:17824–31, 1998.

Huebner K, Cannizzaro LA, Jabs EW, Kivirikko S, Manzone H et al: Chromosomal assignment of a gene encoding a new collagen type (COL15A1) to 9q21-q22, *Genomics* 14:220–4, 1992.

Kivirikko S, Heinamaki P, Rehn M, Honkanen N, Myers JC et al: Primary structure of the alpha-1 chain of human type XV collagen and exon-intron organization in the 3-prime region of the corresponding gene, *J Biol Chem* 269:4773–9, 1994.

Muragaki Y, Abe N, Ninomiya Y, Olsen BR, Ooshima A: The human alpha-1(XV) collagen chain contains a large amino-terminal non-triple helical domain with a tandem repeat structure and homology to alpha-1(XVIII) collagen, *J Biol Chem* 269:4042–6, 1994.

Rehn M, Hintikka E, Pihlajaniemi T: Primary structure of the alpha 1 chain of mouse type XVIII collagen, partial structure of the corresponding gene, and comparison of the alpha 1(XVIII) chain with its homologue, the alpha 1(XV) collagen chain, *J Biol Chem* 269:13929–35, 1994.

Collagen Type XVI

Pan TC, Zhang RZ, Mattei MG, Timpl R, Chu ML: Cloning and chromosomal location of human alpha-1(XVI) collagen, *Proc Natl Acad Sci USA* 89:6565–9, 1992.

Yamaguchi N, Kimura S, McBride OW, Hori H, Yamada Y et al: Molecular cloning and partial characterization of a novel collagen chain, alpha-1(XVI), consisting of repetitive collagenous domains and cysteine-containing non-collagenous segments, *J Biochem* 112:856–63, 1992.

Collagen Type XVII

Li K, Sawamura D, Giudice GJ, Diaz LA, Mattei MG et al: Genomic organization of collagenous domains and chromosomal assignment of human 180-kDa bullous pemphigoid antigen-2, a novel collagen of stratified squamous epithelium, *J Biol Chem* 266:24064–9, 1991.

Li K, Tamai K, Tan EML, Uitto J: Cloning of type XVII collagen: Complementary and genomic sequences of mouse 180-kDa bullous pemphigoid antigen (BPAG2) predict an interrupted collagenous domain, a transmembranous segment, and unusual features in the 5-prime end of the gene and the 3-prime-untranslated region of the mRNA, *J Biol Chem* 268:8825–34, 1993.

McGrath JA, Gatalica B, Christiano AM, Li K, Owaribe K et al: Mutations in the 180-kD bullous pemphigoid antigen (BPAG2), a hemidesmosomal transmembrane collagen (COL17A1), in generalized atrophic benign epidermolysis bullosa, *Nature Genet* 11:83–6, 1995.

Schacke H, Schumann H, Hammami-Hauasli N, Raghunath M, Bruckner-Tuderman L: Two forms of collagen XVII in keratinocytes. A full-length transmembrane protein and a soluble ectodomain, *J Biol Chem* 273:25937–43, 1998.

Collagen Type XVIII

O'Reilly MS, Boehm T, Shing Y, Fukai N, Vasios G et al: Endostatin: An endogenous inhibitor of angiogenesis and tumor growth, *Cell* 88:277–85, 1997.

Oh SP, Warman ML, Seldin MF, Cheng SD, Knoll JHM et al: Cloning of cDNA and genomic DNA encoding human type XVIII collagen and localization of the alpha-1(XVIII) collagen gene to mouse chromosome 10 and human chromosome 21, *Genomics* 19:494–9, 1994.

Rehn M, Hintikka E, Pihlajaniemi T: Characterization of the mouse gene for the alpha-1 chain of type XVIII collagen (Col18a1) reveals that the three variant N-terminal polypeptide forms are transcribed from two widely separated promoters, *Genomics* 32:436–46, 1996.

Rehn M, Pihlajaniemi T: Alpha-1(XVIII), a collagen chain with frequent interruptions in the collagenous sequence, a distinct tissue distribution, and homology with type XV collagen, *Proc Natl Acad Sci USA* 91:4234–8, 1994.

Saarela J, Ylikarppa R, Rehn M, Purmonen S, Pihlajaniemi T: Complete primary structure of two variant forms of human type XVIII collagen and tissue-specific differences in the expression of the corresponding transcripts, *Matrix Biol* 16:319–28, 1998.

Collagen Type XIX

Gerecke DR, Olson PF, Koch M, Knoll JHM, Taylor R et al: Complete primary structure of two splice variants of collagen XII, and assignment of alpha-1(XII) collagen (COL12A1), alpha-1(IX) collagen (COL9A1), and alpha-1(XIX) collagen (COL19A1) to human chromosome 6q12-q13, *Genomics* 41:236–42, 1997.

Inoguchi K, Yoshioka H, Khaleduzzaman M, Ninomiya Y: The mRNA for alpha-1(XIX) collagen chain, a new member of FACITs, contains a long unusual 3-prime untranslated region and displays many unique splicing variants, *J Biochem* 117:137–46, 1995.

Khaleduzzaman M, Sumiyoshi H, Ueki Y, Inoguchi K, Ninomiya Y et al: Structure of the human type XIX collagen (COL19A1) gene, which suggests it has arisen from an ancestor gene of the FACIT family, *Genomics* 45:304–12, 1997.

Myers JC, Sun MJ, D'Ippolito JA, Jabs EW, Neilson EG et al: Human cDNA clones transcribed from an unusually high molecular weight RNA encode a new collagen chain. *Gene* 123:211–17, 1993.

Yoshioka H, Zhang H, Ramirez F, Mattei MG, Moradi-Ameli M et al: Synteny between the loci for a novel FACIT-like collagen (D6S228E) and alpha 1(IX) collagen (COL9A1) on 6q12-q14 in humans, *Genomics* 13:884–6, 1992.

Human protein reference data base, Johns Hopkins University and the Institute of Bioinformatics, at http://www.hprd.org/protein.

4.2. Function of Collagen Matrix

Hafemann B, Ensslen S, Erdmann C, Niedballa R, Zuhlke A et al: Use of a collagen/elastin-membrane for the engineering of dermis, *Burns* 25:373–84, 1999.

Tay BK, Le AX, Heilman M, Lotz J, Bradford DS: Use of a collagen-hydroxyapatite matrix in spinal fusion. A rabbit model, *Spine* 23:2276–81, 1998.

Hakim S, Merguerian PA, Chavez DR: Use of biodegradadble mesh as a transport for a cultured uroepitheial graft: An improved method using collagen gel, *Urology* 44:139–42, 1994.

Iozzo RV: Basement membrane proteoglycans: From cellar to ceiling, *Nat Rev Mol Cell Biol* 6:646–56, 2005.

Franzke CW, Bruckner P, Bruckner-Tuderman L: Collagenous transmembrane proteins: Recent insights into biology and pathology. *J Biol Chem* 280:4005–8, 2005.

Kjaer M: Role of extracellular matrix in adaptation of tendon and skeletal muscle to mechanical loading, *Physiol Rev* 84:649–98, 2004.

Myllyharju J, Kivirikko KI: Collagens, modifying enzymes and their mutations in humans, flies and worms, *Trends Genet* 20:33–43, 2004.

Kalluri R: Basement membranes: Structure, assembly and role in tumour angiogenesis, *Nature Rev Cancer* 3:422–33, 2003.

Marneros AG, Olsen BR: The role of collagen-derived proteolytic fragments in angiogenesis, *Matrix Biol* 20:337–45, 2001.

4.3. Composition and Structure of Elastic Laminae

Elastin

Hinek A: Biological roles of the non-integrin elastin/laminin receptor, *Biol Chem* 377:471–80, 1996.

Mecham RP, Broekelmann T, Davis EC, Gibson MA, Brown-Augsburger P: Elastic fibre assembly: Macromolecular interactions, *Ciba Found Symp* 192:172–81, 1995.

Robert L: Interaction between cells and elastin, the elastin-receptor, *Connect Tissue Res* 40:75–82, 1999.

Urry DW, Hugel T, Seitz M, Gaub HE, Sheiba L et al: Elastin: A representative ideal protein elastomer, *Phi Trans R Soc Lond B Biol Sci* 357:169–84, 2002.

Vrhovski B, Weiss AS: Biochemistry of tropoelastin, *Eur J Biochem* 258:1–18, 1998.

Curran ME, Atkinson DL, Ewart AK, Morris CA, Leppert MF et al: The elastin gene is disrupted by a translocation associated with supravalvular aortic stenosis, *Cell* 73:159–68, 1993.

Emanuel BS, Cannizzaro L, Ornstein-Goldstein N, Indik ZK, Yoon K et al: Chromosomal localization of the human elastin gene, *Am J Hum Genet* 37:873–82, 1985.

Ewart AK, Jin W, Atkinson D, Morris CA, Keating MT: Supravalvular aortic stenosis associated with a deletion disrupting the elastin gene, *J Clin Invest* 93:1071–7, 1994.

Ewart AK, Morris CA, Atkinson D, Jin W, Sternes K et al: Hemizygosity at the elastin locus in a developmental disorder, Williams syndrome, *Nature Genet* 5:11–16, 1993.

Faury G, Pezet M, Knutsen RH, Boyle WA, Heximer SP et al: Developmental adaptation of the mouse cardiovascular system to elastin haploinsufficiency, *J Clin Invest* 112:1419–28, 2003.

Fazio MJ, Mattei MG, Passage E, Chu ML et al: Human elastin gene: new evidence for localization to the long arm of chromosome 7, *Am J Hum Genet* 48:696–703, 1991.

Li DY, Brooke B, Davis EC, Mecham RP, Sorensen LK et al: Elastin is an essential determinant of arterial morphogenesis, *Nature* 393:276–80, 1998.

Olson TM, Michels VV, Urban Z, Csiszar K, Christiano AM et al: A 30 kb deletion within the elastin gene results in familial supravalvular aortic stenosis, *Hum Molec Genet* 4:1677–9, 1995.

Urban Z, Riazi S, Seidl TL, Katahira J, Smoot LB et al: Connection between elastin haploinsufficiency and increased cell proliferation in patients with supravalvular aortic stenosis and Williams-Beuren syndrome, *Am J Hum Genet* 71:30–44, 2002.

Urban Z, Zhang J, Davis EC, Maeda GK, Kumar A et al: Supravalvular aortic stenosis: Genetic and molecular dissection of a complex mutation in the elastin gene, *Hum Genet* 109:512–20, 2001.

Fibrillin 1

Dietz HC, Cutting GR, Pyeritz RE, Maslen CL, Sakai LY et al: Marfan syndrome caused by a recurrent de novo missense mutation in the fibrillin gene, *Nature* 352:337–9, 1991.

Dietz HC, McIntosh I, Sakai LY, Corson GM, Chalberg SC et al: Four novel FBN1 mutations: Significance for mutant transcript level and EGF-like domain calcium binding in the pathogenesis of Marfan syndrome, *Genomics* 17:468–75, 1993.

Dietz HC, Pyeritz RE: Mutations in the human gene for fibrillin-1 (FBN1) in the Marfan syndrome and related disorders, *Hum Molec Genet* 4:1799–809, 1995.

Gayraud B, Keene DR, Sakai LY, Ramirez F: New insights into the assembly of extracellular microfibrils from the analysis of the fibrillin 1 mutation in the tight skin mouse, *J Cell Biol* 150:667–79, 2000.

Hollister DW, Godfrey M, Sakai LY, Pyeritz RE: Immunohistologic abnormalities of the microfibrillar-fiber system in the Marfan syndrome, *New Engl J Med* 323:152–9, 1990.

Kainulainen K, Karttunen L, Puhakka L, Sakai L, Peltonen L: Mutations in the fibrillin gene responsible for dominant ectopia lentis and neonatal Marfan syndrome, *Nature Genet* 6:64–9, 1994.

Kielty CM, Raghunath M, Siracusa LD, Sherratt MJ, Peters R et al: The tight skin mouse: Demonstration of mutant fibrillin-1 production and assembly into abnormal microfibrils, *J Cell Biol* 140:1159–66, 1998.

Lee B, Godfrey M, Vitale E, Hori H, Mattei MG et al: Linkage of Marfan syndrome and a phenotypically related disorder to two different fibrillin genes, *Nature* 352:330–4, 1991.

Pereira L, Andrikopoulos K, Tian J, Lee SY, Keene DR et al: Targetting of the gene encoding fibrillin-1 recapitulates the vascular aspect of Marfan syndrome, *Nature Genet* 17:218–22, 1997.

Reinhardt DP, Ono RN, Notbohm H, Muller PK, Bachinger HP et al: Mutations in calcium-binding epidermal growth factor modules render fibrillin-1 susceptible to proteolysis: A potential disease-causing mechanism in Marfan syndrome, *J Biol Chem* 275:12339–45, 2000.

Sakai LY, Keene DR, Engvall E: Fibrillin, a new 350-kD glycoprotein, is a component of extracellular microfibrils, *J Cell Biol* 103:2499–509, 1986.

Sakai LY, Keene DR, Glanville RW, Bachinger HP: Purification and partial characterization of fibrillin, a cysteine-rich structural component of connective tissue microfibrils, *J Biol Chem* 266:14763–70, 1991.

Fibrillin 2

Belleh S, Zhou G, Wang M, Der Kaloustian VM, Pagon RA et al: Two novel fibrillin-2 mutations in congenital contractural arachnodactyly, *Am J Med Genet* 92:7–12, 2000.

Maslen C, Babcock D, Raghunath M, Steinmann B: A rare branch-point mutation is associated with missplicing of fibrillin-2 in a large family with congenital contractural arachnodactyly, *A J Hum Genet* 60:1389–98, 1997.

Putnam EA, Zhang H, Ramirez F, Milewicz DM: Fibrillin-2 (FBN2) mutations result in the Marfan-like disorder, congenital contractural arachnodactyly, *Nature Genet* 11:456–8, 1995.

Zhang H, Apfelroth SD, Hu W, Davis EC, Sanguineti C et al: Structure and expression of fibrillin-2, a novel microfibrillar component preferentially located in elastic matrices, *J Cell Biol* 124:855–63, 1994.

Zhang H, Hu W, Ramirez F: Developmental expression of fibrillin genes suggests heterogeneity of extracellular microfibrils, *J Cell Biol* 129:1165–76, 1995.

Fibrillin 3

Corson GM, Charbonneau NL, Keene DR, Sakai LY: Differential expression of fibrillin-3 adds to microfibril variety in human and avian, but not rodent, connective tissues, *Genomics* 83:461–72, 2004.

Nagase T, Nakayama M, Nakajima D, Kikuno R, Ohara O: Prediction of the coding sequences of unidentified human genes. XX. The complete sequences of 100 new cDNA clones from brain which code for large proteins in vitro, *DNA Res* 8:85–95, 2001.

MAGP

Chen Y, Faraco J, Yin W, Germiller J, Francke U et al: Structure, chromosomal localization, and expression pattern of the murine Magp gene, *J Biol Chem* 268:27381–9, 1993.

Faraco J, Bashir M, Rosenbloom J, Francke U: Characterization of the human gene for microfibril-associated glycoprotein (MFAP2), assignment to chromosome 1p36.1-p35, and linkage to D1S170, *Genomics* 25:630–7, 1995.

Human protein reference data base, Johns Hopkins University and the Institute of Bioinformatics, at http://www.hprd.org/protein.

4.4. Function of Elastic Fibers and Laminae

Brooke BS, Karnik SK, Li DY: Extracellular matrix in vascular morphogenesis and disease: structure versus signal, *Trends Cell Biol* 13:51–6, 2003.

Karnik SK, Brooke BS, Bayes-Genis A, Sorensen L, Wythe JD et al: A critical role for elastin signaling in vascular morphogenesis and disease, *Development* 130:411–23, 2003.

Li DY, Brooke B, Davis EC, Mecham RP, Sorensen LK et al: Elastin is an essential determinant of arterial morphogenesis, *Nature* 393:276–80, 1998.

Liu SQ, Tieche C, Alkema PK: Neointima formation on elastic lamina and collagen matrix scaffolds implanted in the rat aorta, *Biomaterials* 25:1869–82, 2004.

Keating MT: Genetic approaches to disease and regeneration, *Phil Trans R Soc Lond B Biol Sci* 359:795–8, 2004.

Urry DW, Hugel T, Seitz M, Gaub HE, Sheiba L et al: Elastin: A representative ideal protein elastomer, *Phil Trans R Soc Lond B Biol Sci* 357:169–84, 2002.

Gosline J, Lillie M, Carrington E, Guerette P, Ortlepp C et al: Elastic proteins: Biological roles and mechanical properties, *Phil Trans R Soc Lond B Biol Sci* 357:121–32, 2002.

Morris CA, Mervis CB: Williams syndrome and related disorders, *Annu Rev Genom Hum Genet* 1:461–84, 2000.

Tatham AS, Shewry PR: Elastomeric proteins: Biological roles, structures and mechanisms, *Trends Biochem Sci* 25:567–71, 2000.

Dietz HC, Mecham RP: Mouse models of genetic diseases resulting from mutations in elastic fiber proteins, *Matrix Biol* 19:481–8, 2000.

Liu SQ, Alkema PK, Tieche C, Tefft BJ, Liu DZ et al: Negative regulation of monocyte adhesion to arterial elastic laminae by signal-regulatory protein alpha and SH2 domain-containing protein tyrosine phosphatase-1, *J Biol Chem* 280:39294–301, 2005.

Tieche C, Alkema PK, Liu SQ: Arterial elastic laminae: Anti-inflammatory effects and potential application to arterial reconstruction, *Front Biosci* 9:2205–17, 2004.

Milewicz DM, Urban Z, Boyd C: Genetic disorders of the elastic fiber system, *Matrix Biol* 19:471–80, 2000.

Ramirez F: Pathophysiology of the microfibril/elastic fiber system: Introduction, *Matrix Biol* 19:455–6, 2000.

Robert L: Interaction between cells and elastin, the elastin-receptor, *Connect Tissue Res* 40:75–82, 1999.

Ye C, Rabinovitch M: New developments in the pulmonary circulation in children, *Curr Opin Cardiol* 7:124–33, 1992.

4.5. Composition and Structure of Proteoglycans

Versican

Bode-Lesniewska B, Dours-Zimmermann MT, Odermatt BF, Briner J, Heitz PU et al: Distribution of the large aggregating proteoglycan versican in adult human tissues, *J Histochem Cytochem* 44:303–12, 1996.

Dours-Zimmermann MT, Zimmermann DR: A novel glycosaminoglycan attachment domain identified in two alternative splice variants of human versican, *J Biol Chem* 269:32992–8, 1994.

Iozzo RV, Naso MF, Cannizzaro LA, Wasmuth JJ, McPherson JD: Mapping of the versican proteoglycan gene (CSPG2) to the long arm of human chromosome 5 (5q12-5q14), *Genomics* 14:845–51, 1992.

Kjellen L, Lindahl U: Proteoglycans: Structures and interactions, *Annu Rev Biochem* 60:443–75, 1991.

Naso MF, Morgan JL, Buchberg AM, Siracusa LD, Iozzo RV: Expression pattern and mapping of the murine versican gene (Cspg2) to chromosome 13, *Genomics* 29:297–300, 1995.

Naso MF, Zimmermann DR, Iozzo RV: Characterization of the complete genomic structure of the human versican gene and functional analysis of its promoter, *J Biol Chem* 269:32999–3008, 1994.

Yoon H, Liyanarachchi S, Wright FA, Davuluri R, Lockman JC et al: Gene expression profiling of isogenic cells with different TP53 gene dosage reveals numerous genes that are affected by TP53 dosage and identifies CSPG2 as a direct target of p53, *Proc Natl Acad Sci USA* 99:15632–7, 2002.

Zako M, Shinomura T, Ujita M, Ito K, Kimata K: Expression of PG-M(V3), an alternatively spliced form of PG-M without a chondroitin sulfate attachment region in mouse and human tissues, *J Biol Chem* 270:3914–8, 1995.

Zimmermann DR, Ruoslahti E: Multiple domains of the large fibroblast proteoglycan, versican, *EMBO J* 8:2975–81, 1989.

Decorin

McBride OW, Fisher LW, Young MF: Localization of PGI (biglycan, BGN) and PGII (decorin, DCN, PG-40) genes on human chromosomes Xq13-qter and 12q, respectively, *Genomics* 6:219–55, 1990.

Moscatello DK, Santra M, Mann DM, McQuillan DJ, Wong AJ et al: Decorin suppresses tumor cell growth by activating the epidermal growth factor receptor, *J Clin Invest* 101:406–12, 1998.

Pulkkinen L, Alitalo T, Krusius T, Peltonen L: Expression of decorin in human tissues and cell lines and defined chromosomal assignment of the gene locus (DCN), *Cytogenet Cell Genet* 60:107–11, 1992.

Pulkkinen L, Kainulainen K, Krusius T, Makinen P, Schollin J et al: Deficient expression of the gene coding for decorin in a lethal form of Marfan syndrome, *J Biol Chem* 265:17780–5, 1990.

Scholzen T, Solursh M, Suzuki S, Reiter R, Morgan JL et al: The murine decorin: Complete cDNA cloning, genomic organization, chromosomal assignment, and expression during organogenesis and tissue differentiation, *J Biol Chem* 269:28270–81, 1994.

Perlecan

Arikawa-Hirasawa E, Watanabe H, Takami H, Hassell JR, Yamada Y: Perlecan is essential for cartilage and cephalic development, *Nature Genet* 23:354–8, 1999.

Chakravarti S, Phillips SL, Hassell JR: Assignment of the perlecan (heparan sulfate proteoglycan) gene to mouse chromosome 4, *Mam Genome* 1:270–2, 1991.

Cohen IR, Grassel S, Murdoch AD, Iozzo RV: Structural characterization of the complete human perlecan gene and its promoter, *Proc Natl Acad Sci USA* 90:10404–8, 1993.

Dodge GR, Kovalszky I, Chu ML, Hassell JR, McBride OW et al: Heparan sulfate proteoglycan of human colon: Partial molecular cloning, cellular expression, and mapping of the gene (HSPG2) to the short arm of human chromosome 1, *Genomics* 10:673–80, 1991.

Iozzo RV, Pillarisetti J, Sharma B, Murdoch AD, Danielson KG et al: Structural and functional characterization of the human perlecan gene promoter: transcriptional activation by transforming growth factor-beta via a nuclear factor 1-binding element, *J Biol Chem* 272:5219–28, 1997.

Kallunki P, Eddy RL, Byers MG, Kestila M, Shows TB et al: Cloning of human heparan sulfate proteoglycan core protein, assignment of the gene (HSPG2) to 1p36.1-p35 and identification of a BamHI restriction fragment length polymorphism, *Genomics* 11:389–96, 1991.

Mongiat M, Otto J, Oldershaw R, Ferrer F, Sato JD et al: Fibroblast growth factor-binding protein is a novel partner for perlecan protein core, *J Biol Chem* 276:10263–71, 2001.

Nicole S, Davoine CS, Topaloglu H, Cattolico L, Barral D et al: Perlecan, the major proteoglycan of basement membranes, is altered in patients with Schwartz-Jampel syndrome (chondrodystrophic myotonia), *Nature Genet* 26:480–3, 2000.

Sharma B, Handler M, Eichstetter I, Whitelock JM, Nugent MA et al: Antisense targeting of perlecan blocks tumor growth and angiogenesis in vivo, *J Clin Invest* 102:1599–608, 1998.

Wintle RF, Kisilevsky R, Noonan D, Duncan AMV: In situ hybridization to human chromosome 1 of a cDNA probe for the gene encoding the basement membrane heparan sulfate proteoglycan (HSPG), *Cytogenet Cell Genet* 54:60–1, 1990.

Aggrecan-1

Finkelstein JE, Doege K, Yamada Y, Pyeritz RE, Graham JM Jr et al: Analysis of the chondroitin sulfate proteoglycan core protein (CSPGCP) gene in achondroplasia and pseudoachondroplasia, *Am J Hum Genet* 48:97–102, 1991.

Gleghorn L, Ramesar R, Beighton P, Wallis G: A mutation in the variable repeat region of the aggrecan gene (AGC1) causes a form of spondyloepiphyseal dysplasia associated with severe, premature osteoarthritis, *Am J Hum Genet* 77:484–90, 2005.

Just W, Klett C, Vetter U, Vogel W: Assignment of the human aggrecan gene AGC1 to 15q25-q26.2 by in situ hybridization, *Hum Genet* 92:516–18, 1993.

Korenberg JR, Chen XN, Doege K, Grover J, Roughley PJ: Assignment of the human aggrecan gene (AGC1) to 15q26 using fluorescence in situ hybridization analysis, *Genomics* 16:546–8, 1993.

Valhmu WB, Palmer GD, Rivers PA, Ebara S, Cheng JF et al: Structure of the human aggrecan gene: Exon-intron organization and association with the protein domains, *Biochem J* 309:535–42, 1995.

Watanabe H, Kimata K, Line S, Strong D, Gao L et al: Mouse cartilage matrix deficiency (cmd) caused by a 7 bp deletion in the aggrecan gene, *Nature Genet* 7:154–7, 1994.

Watanabe H, Nakata K, Kimata K, Nakanishi I, Yamada Y: Dwarfism and age-associated spinal degeneration of heterozygote cmd mice defective in aggrecan, *Proc Natl Acad Sci USA* 94:6943–7, 1997.

Biglycan

Chatterjee A, Faust CJ, Herman GE: Genetic and physical mapping of the biglycan gene on the mouse X chromosome, *Mam Genome* 4:33–6, 1993.

Fisher LW, Heegaard AM, Vetter U, Vogel W, Just W et al: Human biglycan gene: Putative promoter, intron-exon junctions, and chromosomal localization, *J Biol Chem* 266:14371–7, 1991.

Fisher LW, Termine JD, Young MF: Deduced-protein sequence of bone small proteoglycan I (biglycan) shows homology with proteoglycan II (decorin) and several nonconnective tissue proteins in a variety of species, *J Biol Chem* 264:4571–6, 1989.

Geerkens C, Vetter U, Just W, Fedarko NS, Fisher LW et al: The X-chromosomal human biglycan gene BGN is subject to X inactivation but is transcribed like an X-Y homologous gene, *Hum Genet* 96:44–52, 1995.

Schaefer L, Babelova A, Kiss E, Hausser HJ, Baliova M et al: The matrix component biglycan is proinflammatory and signals through toll-like receptors 4 and 2 in macrophages, *J Clin Invest* 115:2223–33, 2005.

Traupe H, van den Ouweland AMW, van Oost BA, Vogel W, Vetter U et al: Fine mapping of the human biglycan (BGN) gene within the Xq28 region employing a hybrid cell panel, *Genomics* 13:481–3, 1992.

Wegrowski Y, Pillarisetti J, Danielson KG, Suzuki S, Iozzo RV: The murine biglycan: Complete cDNA cloning, genomic organization, promoter function, and expression, *Genomics* 30:8–17, 1995.

Xu T, Bianco P, Fisher LW, Longenecker G, Smith E et al: Targeted disruption of the biglycan gene leads to an osteoporosis-like phenotype in mice, *Nature Genet* 20:78–82, 1998.

β-glycan

Brown CB, Boyer AS, Runyan RB, Barnett JV: Requirement of type III TGF-beta receptor for endocardial cell transformation in the heart, *Science* 283:2080–2, 1999.

Chen W, Kirkbride KC, How T, Nelson CD, Mo J et al: Beta-arrestin 2 mediates endocytosis of type III TGF-beta receptor and down-regulation of its signaling, *Science* 301:1394–7, 2003.

Johnson DW, Qumsiyeh M, Benkhalifa M, Marchuk DA: Assignment of human transforming growth factor-beta type I and type III receptor genes (TGFBR1 and TGFBR3) to 9q33-q34 and 1p32-p33, respectively, *Genomics* 28:356–7, 1995.

Lewis KA, Gray PC, Blount AL, MacConell LA, Wiater E et al: Betaglycan binds inhibin and can mediate functional antagonism of activin signalling, *Nature* 404:411–4, 2000.

Syndecan 2

Bobardt MD, Saphire ACS, Hung HC, Yu X, van der Schueren B et al: Syndecan captures, protects, and transmits HIV to T lymphocytes, *Immunity* 18:27–39, 2003.

David G, van der Schueren B, Marynen P, Cassiman JJ, van den Berghe H: Molecular cloning of amphiglycan, a novel integral membrane heparan sulfate proteoglycan expressed by epithelial and fibroblastic cells, *J Cell Biol* 118:961–9, 1992.

Lories V, De Broeck H, David G, Cassiman JJ, Van den Berghe H: Heparan sulfate proteoglycans of human lung fibroblasts: Structural heterogeneity of the core proteins of the hydrophobic cell-associated forms, *J Biol Chem* 262:854–9, 1987.

Marynen P, Zhang J, Cassiman JJ, Van den Berghe H, David G: Partial primary structure of the 48- and 90-kilodalton core proteins of cell surface-associated heparan sulfate proteoglycans of lung fibroblasts: Prediction of an integral membrane domain and evidence for multiple distinct core proteins at the cell surface of human lung fibroblasts, *J Biol Chem* 264:7017–24, 1989.

Spring J, Goldberger OA, Jenkins NA, Gilbert DJ, Copeland NG et al: Mapping of the syndecan genes in the mouse: Linkage with members of the Myc gene family, *Genomics* 21:597–601, 1994.

Neurocan

Prange CK, Pennacchio LA, Lieuallen K, Fan W, Lennon GG: Characterization of the human neurocan gene, CSPG3, *Gene* 221:199–205, 1998.

Rauch U, Grimpe B, Kulbe G, Arnold-Ammer I, Beier DR, Fassler R: Structure and chromosomal localization of the mouse neurocan gene, *Genomics* 28:405–10, 1995.

Rauch U, Karthikeyan L, Maurel P, Margolis RU, Margolis RK: Cloning and primary structure of neurocan, a developmentally regulated, aggregating chondroitin sulfate proteoglycan of brain, *J Biol Chem* 267:19536–47, 1992.

Zhang Y, Rauch U, Perez MTR: Accumulation of neurocan, a brain chondroitin sulfate proteoglycan, in association with the retinal vasculature in RCS rats, *Invest Ophthal Vis Sci* 44:1252–61, 2003.

Keratocan

Khan AO, Kambouris M: A novel KERA mutation associated with autosomal recessive cornea plana, *Ophthalm Genet* 25:147–52, 2004.

Lehmann OJ, El-Ashry MF, Ebenezer ND, Ocaka L, Francis PJ et al: A novel keratocan mutation causing autosomal recessive cornea plana, *Invest Ophthalm Vis Sci* 42:3118–22, 2001.

Liu CY, Shiraishi A, Kao CWC, Converse RL, Funderburgh JL et al: The cloning of mouse keratocan cDNA and genomic DNA and the characterization of its expression during eye development, *J Biol Chem* 273:22584–8, 1998.

Pellegata NS, Dieguez-Lucena JL, Joensuu T, Lau S, Montgomery KT et al: Mutations in KERA, encoding keratocan, cause cornea plana, *Nature Genet* 25:91–5, 2000.

Tasheva ES, Funderburgh JL, Funderburgh ML, Corpuz LM, Conrad GW: Structure and sequence of the gene encoding human keratocan, *DNA Seq* 10:67–74, 1999.

Tasheva ES, Pettenati M, Von Kap-Her C, Conrad GW: Assignment of keratocan gene (KERA) to human chromosome band 12q22 by in situ hybridization, *Cytogenet Cell Genet* 88:244–5, 2000.

Human protein reference data base, Johns Hopkins University and the Institute of Bioinformatics, at http://www.hprd.org/protein.

4.6. Function of Proteoglycans

Lin X: Functions of heparan sulfate proteoglycans in cell signaling during development, *Development* 131:6009–21, 2004.

Iozzo RV: Basement membrane proteoglycans: From cellar to ceiling, *Nat Rev Mol Cell Biol* 6:646–56, 2005.

Handel TM, Johnson Z, Crown SE, Lau EK, Proudfoot AE: Regulation of protein function by glycosaminoglycans—as exemplified by chemokines. *Annu Rev Biochem* 74:385–410, 2005.

Sugahara K, Mikami T, Uyama T, Mizuguchi S, Nomura K et al: Recent advances in the structural biology of chondroitin sulfate and dermatan sulfate. *Curr Opin Struct Biol* 13:612–20, 2003.

Trowbridge JM, Gallo RL: Dermatan sulfate: New functions from an old glycosaminoglycan. *Glycobiology* 12:117R–25R, 2002.

Sasisekharan R, Shriver Z, Venkataraman G, Narayanasami U: Roles of heparan-sulphate glycosaminoglycans in cancer, *Nature Rev Cancer* 2:521–8, 2002.

Woods A, Couchman JR: Syndecan-4 and focal adhesion function, *Curr Opin Cell Biol* 13:578–83, 2001.

Chait A, Wight TN: Interaction of native and modified low-density lipoproteins with extracellular matrix, *Curr Opin Lipidol* 11:457–63, 2000.

Funderburgh JL: Keratan sulfate: structure, biosynthesis, and function, *Glycobiology* 10:951–8, 2000.

Fransson LA, Belting M, Jonsson M, Mani K, Moses J et al: Biosynthesis of decorin and glypican, *Matrix Biol* 19:367–76, 2000.

Ornitz DM: FGFs, heparan sulfate and FGFRs: Complex interactions essential for development, *Bioessays* 22:108–12, 2000.

Prydz K, Dalen KT: Synthesis and sorting of proteoglycans, *J Cell Sci* 113:193–205, 2000.

Iozzo RV: The biology of the small leucine-rich proteoglycans. Functional network of interactive proteins, *J Biol Chem* 274:18843–6, 1999.

Iozzo RV: Matrix proteoglycans: From molecular design to cellular function, *Annu Rev Biochem* 67:609–52, 1998.

Rosenberg RD, Shworak NW, Liu J, Schwartz JJ, Zhang L: Heparan sulfate proteoglycans of the cardiovascular system. Specific structures emerge but how is synthesis regulated? *J Clin Invest* 99:2062–70, 1997.

Wight TN, Merrilees MJ: Proteoglycans in atherosclerosis and restenosis. Key roles for versican, *Circ Res* 94:1158–67, 2004.

Segev A, Nili N, Strauss BH: The role of perlecan in arterial injury and angiogenesis, *Cardiovasc Res* 63:603–10, 2004.

4.7. Structural Features of MMPs

MMP1

Boire A, Covic L, Agarwal A, Jacques S, Sherifi S et al: PAR1 is a matrix metalloprotease-1 receptor that promotes invasion and tumorigenesis of breast cancer cells, *Cell* 120:303–31, 2005.

Goldberg GI, Wilhelm SM, Kronberger A, Bauer EA, Grant GA et al: Human fibroblast collagenase: Complete primary structure and homology to an oncogene transformation-induced rat protein, *J Biol Chem* 261:6600–5, 1986.

Joos L, He JQ, Shepherdson MB, Connett JE, Anthonisen NR et al: The role of matrix metalloproteinase polymorphisms in the rate of decline in lung function, *Hum Molec Genet* 11:569–76, 2002.

Minn AJ, Gupta GP, Siegel PM, Bos PD, Shu W et al: Genes that mediate breast cancer metastasis to lung, *Nature* 436:518–24, 2005.

Pendas AM, Santamaria I, Alvarez MV, Pritchard M, Lopez-Otin C: Fine physical mapping of the human matrix metalloproteinase genes clustered on chromosome 11q22.3, *Genomics* 37:266–9, 1996.

Saffarian S, Collier IE, Marmer BL, Elson EL, Goldberg G: Interstitial collagenase is a Brownian ratchet driven by proteolysis of collagen, *Science* 306:108–11, 2004.

MMP2

Bigg HF, Shi YE, Liu YE, Steffensen B, Overall CM: Specific, high affinity binding of tissue inhibitor of metalloproteinases-4 (TIMP-4) to the COOH-terminal hemopexin-like domain of human gelatinase A, *J Biol Chem* 272:15496–500, 1997.

Brooks PC, Silletti S, von Schalscha TL, Friedlander M, Cheresh DA: Disruption of angiogenesis by PEX, a noncatalytic metalloproteinase fragment with integrin binding activity, *Cell* 92:391–400, 1998.

Collier IE, Bruns GAP, Goldberg GI, Gerhard DS: On the structure and chromosome location of the 72- and 92-kDa human type IV collagenase genes, *Genomics* 9:429–34, 1991.

Corry DB, Rishi K, Kanellis J, Kiss A, Song L et al: Decreased allergic lung inflammatory cell egression and increased susceptibility to asphyxiation in MMP2-deficiency, *Nature Immun* 3:347–353, 2002.

Devarajan P, Johnston JJ, Ginsberg SS, Van Wart HE, Berliner N: Structure and expression of neutrophil gelatinase cDNA: Identity with type IV collagenase from HT1080 cells, *J Biol Chem* 267:25228–32, 1992.

Huhtala P, Chow LT, Tryggvason K: Structure of the human type IV collagenase gene, *J Biol Chem* 265:11077–82, 1990.

Huhtala P, Eddy RL, Fan YS, Byers MG, Shows TB et al: Completion of the primary structure of the human type IV collagenase preproenzyme and assignment of the gene (CLG4) to the q21 region of chromosome 16, *Genomics* 6:554–9, 1990.

Irwin JC, Kirk D, Gwatkin RBL, Navre M, Cannon P et al: Human endometrial matrix metalloproteinase-2, a putative menstrual proteinase: Hormonal regulation in cultured stromal cells and messenger RNA expression during the menstrual cycle, *J Clin Invest* 97:438–47, 1996.

Kato T, Kure T, Chang JH, Gabison EE, Itoh T et al: Diminished corneal angiogenesis in gelatinase A-deficient mice, *FEBS Lett* 508:187–90, 2001.

Martignetti JA, Al Aqeel A, Al Sewairi W, Boumah CE, Kambouris M et al: Mutation of the matrix metalloproteinase 2 gene (MMP2) causes a multicentric osteolysis and arthritis syndrome, *Nature Genet* 28:261–5, 2001.

Matsumura S, Iwanaga S, Mochizuki S, Okamoto H, Ogawa S et al: Targeted deletion or pharmacological inhibition of MMP-2 prevents cardiac rupture after myocardial infarction in mice, *J Clin Invest* 115:599–609, 2005.

McQuibban GA, Butler GS, Gong JH, Bendall L, Power C et al: Matrix metalloproteinase activity inactivates the CXC chemokine stromal cell-derived factor-1, *J Biol Chem* 276:43503–8, 2001.

McQuibban GA, Gong JH, Tam EM, McCulloch CAG, Clark-Lewis I et al: Inflammation dampened by gelatinase A cleavage of monocyte chemoattractant protein-3, *Science* 289:1202–6, 2000.

Minn AJ, Gupta GP, Siegel PM, Bos PD, Shu W et al: Genes that mediate breast cancer metastasis to lung, *Nature* 436:518–24, 2005.

Morgunova E, Tuuttila A, Bergmann U, Isupov M, Lindqvist Y et al: Structure of human pro-matrix metalloproteinase-2: Activation mechanism revealed, *Science* 284:1667–70, 1999.

Zhang K, McQuibban GA, Silva C, Butler GS, Johnston JB et al: HIV-induced metalloproteinase processing of the chemokine stromal cell derived factor-1 causes neurodegeneration, *Nature Neurosci* 6:1064–71, 2003.

MMP3

D'Souza CA, Mak B, Moscarello MA: The up-regulation of stromelysin-1 (MMP-3) in a spontaneously demyelinating transgenic mouse precedes onset of disease, *J Biol Chem* 277:13589–96, 2002.

Formstone CJ, Byrd PJ, Ambrose HJ, Riley JH, Hernandez D et al: The order and orientation of a cluster of metalloproteinase genes, stromelysin 2, collagenase, and stromelysin, together with D11S385, on chromosome 11q22-q23, *Genomics* 16:289–91, 1993.

Pendas AM, Santamaria I, Alvarez MV, Pritchard M, Lopez-Otin C: Fine physical mapping of the human matrix metalloproteinase genes clustered on chromosome 11q22.3, *Genomics* 37:266–9, 1996.

Quinones S, Saus J, Otani Y, Harris ED Jr, Kurkinen M: Transcriptional regulation of human stromelysin, *J Biol Chem* 264:8339–44, 1989.

Radisky DC, Levy DD, Littlepage LE, Liu H, Nelson CM et al: Rac1b and reactive oxygen species mediate MMP-3-induced EMT and genomic instability, *Nature* 436:123–7, 2005.

Saus J, Quinones S, Otani Y, Nagase H, Harris ED Jr, Kurkinen M: The complete primary structure of human matrix metalloproteinase-3: Identity with stromelysin, *J Biol Chem* 263:6742–5, 1988.

Sternlicht MD, Lochter A, Sympson CJ, Huey B, Rougier JP et al: The stromal proteinase MMP3/ stromelysin-1 promotes mammary carcinogenesis, *Cell* 98:137–46, 1999.

Ye S, Eriksson P, Hamsten A, Kurkinen M, Humphries SE et al: Progression of coronary atherosclerosis is associated with a common genetic variant of the human stromelysin-1 promoter which results in reduced gene expression, *J Biol Chem* 271:13055–60, 1996.

MMP7

Crawford HC, Scoggins CR, Washington MK, Matrisian LM, Leach SD: Matrix metalloproteinase-7 is expressed by pancreatic cancer precursors and regulates acinar-to-ductal metaplasia in exocrine pancreas, *J Clin Invest* 109:1437–44, 2002.

Gaire M, Magbanua Z, McDonnell S, McNeil L, Lovett DH et al: Structure and expression of the human gene for the matrix metalloproteinase matrilysin, *J Biol Chem* 269:2032–40, 1994.

Knox JD, Boreham DR, Walker JA, Morrison DP, Matrisian LM, Nagle RB, Bowden GT: Mapping of the metalloproteinase gene matrilysin (MMP7) to human chromosome 11q21-q22, *Cytogenet Cell Genet* 72:179–82, 1996.

Li Q, Park PW, Wilson CL, Parks WC: Matrilysin shedding of syndecan-1 regulates chemokine mobilization and transepithelial efflux of neutrophils in acute lung injury, *Cell* 111:635–46, 2002.

Pendas AM, Santamaria I, Alvarez MV, Pritchard M, Lopez-Otin C: Fine physical mapping of the human matrix metalloproteinase genes clustered on chromosome 11q22.3, *Genomics* 37:266–9, 1996.

Wilson CL, Ouellette AJ, Satchell DP, Ayabe T, Lopez-Boado YS et al: Regulation of intestinal alpha-defensin activation by the metalloproteinase matrilysin in innate host defense, *Science* 286:113–7, 1999.

Zuo F, Kaminski N, Eugui E, Allard J, Yakhini Z et al: Gene expression analysis reveals matrilysin as a key regulator of pulmonary fibrosis in mice and humans, *Proc Natl Acad Sci USA* 99:6292–7, 2002.

MMP8

Balbin M, Fueyo A, Tester AM, Pendas AM, Pitiot AS et al: Loss of collagenase-2 confers increased skin tumor susceptibility to male mice, *Nature Genet* 35:252–7, 2003.

Devarajan P, Mookhtiar K, Van Wart H, Berliner N: Structure and expression of the cDNA encoding human neutrophil collagenase, *Blood* 77:2731–8, 1991.

Hasty KA, Pourmotabbed TF, Goldberg GI, Thompson JP, Spinella DG et al: Human neutrophil collagenase: A distinct gene product with homology to other matrix metalloproteinases, *J Biol Chem* 265:11421–4, 1990.

Nagase H, Barrett AJ, Woessner JF Jr: Nomenclature and glossary of the matrix metalloproteinases, *Matrix Suppl* 1:421–4, 1992.

Pendas AM, Santamaria I, Alvarez MV, Pritchard M, Lopez-Otin C: Fine physical mapping of the human matrix metalloproteinase genes clustered on chromosome 11q22.3, *Genomics* 37:266–9, 1996.

MMP9

Collier IE, Bruns GAP, Goldberg GI, Gerhard DS: On the structure and chromosome location of the 72- and 92-kDa human type IV collagenase genes, *Genomics* 9:429–34, 1991.

Coussens LM, Tinkle CL, Hanahan D, Werb Z: MMP-9 supplied by bone mar row-derived cells contributes to skin carcinogenesis, *Cell* 103:481–90, 2000.

Dubois B, Masure S, Hurtenbach U, Paemen L, Heremans H et al: Resistance of young gelatinase B-deficient mice to experimental autoimmune encephalomyelitis and necrotizing tail lesions, *J Clin Invest* 104:1507–15, 1999.

Gu Z, Kaul M, Yan B, Kridel SJ, Cui J et al: S-nitrosylation of matrix metalloproteinases: Signaling pathway to neuronal cell death, *Science* 297:1186–90, 2002.

Heissig B, Hattori K, Dias S, Friedrich M, Ferris B et al: Recruitment of stem and progenitor cells from the bone marrow niche requires MMP-9 mediated release of Kit-ligand, *Cell* 109:625–37, 2002.

Huhtala P, Tuuttila A, Chow LT, Lohi J, Keski-Oja J et al: Complete structure of the human gene for 92-kDa type IV collagenase: Divergent regulation of expression for the 92- and 72-kilodalton enzyme genes in HT-1080 cells, *J Biol Chem* 266:16485–90, 1991.

Linn R, DuPont BR, Knight CB, Plaetke R, Leach RJ: Reassignment of the 92-kDa type IV collagenase gene (CLG4B) to human chromosome 20, *Cytogenet Cell Genet* 72:159–61, 1996.

Pruijt JFM, Fibbe WE, Laterveer L, Pieters RA, Lindley IJD et al: Prevention of interleukin-8-induced mobilization of hematopoietic progenitor cells in rhesus monkeys by inhibitory antibodies against the metalloproteinase gelatinase B (MMP-9), *Proc Natl Acad Sci USA* 96:10863–8, 1999.

Vu TH, Shipley JM, Bergers G, Berger JE, Helms JA et al: MMP-9/gelatinase B is a key regulator of growth plate angiogenesis and apoptosis of hypertrophic chondrocytes, *Cell* 93:411–22, 1998.

Wang X, Lee SR, Arai K, Lee SR, Tsuji K et al: Lipoprotein receptor-mediated induction of matrix metalloproteinase by tissue plasminogen activator, *Nature Med* 9:1313–17, 2003.

Yan C, Wang H, Toh Y, Boyd DD: Repression of 92-kDa type IV collagenase expression by MTA1 is mediated through direct interactions with the promoter via a mechanism, which is both dependent on and independent of histone deacetylation, *J Biol Chem* 278:2309–16, 2003.

Yu Q, Stamenkovic I: Cell surface-localized matrix metalloproteinase-9 proteolytically activates TGF-beta and promotes tumor invasion and angiogenesis, *Genes Dev* 14:163–76, 2000.

Zhang B, Henney A, Eriksson P, Hamsten A, Watkins H et al: Genetic variation at the matrix metalloproteinase-9 locus on chromosome 20q12.2-13.1, *Hum Genet* 105:418–23, 1999.

MMP10

Jung JY, Warter S, Rumpler Y: Localization of stromelysin 2 gene to the q22.3-23 region of chromosome 11 by in situ hybridization, *Ann Genet* 33:21–3, 1990.

Pendas AM, Santamaria I, Alvarez MV, Pritchard M, Lopez-Otin C: Fine physical mapping of the human matrix metalloproteinase genes clustered on chromosome 11q22.3, *Genomics* 37:266–9, 1996.

MMP11

Levy A, Zucman J, Delattre O, Mattei MG, Rio MC et al: Assignment of the human stromelysin 3 (STMY3) gene to the q11.2 region of chromosome 22, *Genomics* 13:881–3, 1992.

Matrisian LM: Metalloproteinases and their inhibitors in matrix remodeling, *Trends Genet* 6:121–5, 1990.

Pei D, Weiss SJ: Furin-dependent intracellular activation of the human stromelysin-3 zymogen, *Nature* 375:244–7, 1995.

Tryggvason K, Huhtala P, Tuuttila A, Chow L, Keski-Oja J et al: Structure and expression of type IV collagenase genes, *Cell Differ Dev* 32:307–12, 1990.

MMP12

Belaaouaj A, Shipley JM, Kobayashi DK, Zimonjic DB, Popescu N et al: Human macrophage metalloelastase: Genomic organization, chromosomal location, gene linkage, and tissue-specific expression, *J Biol Chem* 270:14568–75, 1995.

Curci JA, Liao S, Huffman MD, Shapiro SD, Thompson RW: Expression and localization of macrophage elastase (matrix metalloproteinase-12) in abdominal aortic aneurysms, *J Clin Invest* 102:1900–10, 1998.

Hautamaki RD, Kobayashi DK, Senior RM, Shapiro SD: Requirement for macrophage elastase for cigarette smoke-induced emphysema in mice, *Science* 277:2002–4, 1997.

Kaminski N, Allard JD, Pittet JF, Zuo F, Griffiths MJ et al: Global analysis of gene expression in pulmonary fibrosis reveals distinct programs regulating lung inflammation and fibrosis, *Proc Natl Acad Sci USA* 97:1778–83, 2000.

Morris DG, Huang X, Kaminski N, Wang Y, Shapiro SD et al: Loss of integrin alpha-v-beta-6-mediated TGF-beta activation causes Mmp12-dependent emphysema, *Nature* 422:169–73, 2003.

Pendas AM, Santamaria I, Alvarez MV, Pritchard M, Lopez-Otin C: Fine physical mapping of the human matrix metalloproteinase genes clustered on chromosome 11q22.3, *Genomics* 37:266–9, 1996.

Shapiro SD, Griffin GL, Gilbert DJ, Jenkins NA, Copeland NG et al: Molecular cloning, chromosomal localization, and bacterial expression of a murine macrophage metalloelastase, *J Biol Chem* 267:4664–71, 1992.

Shapiro SD, Kobayashi DK, Ley TJ: Cloning and characterization of a unique elastolytic metalloproteinase produced by human alveolar macrophages, *J Biol Chem* 268:23824–9, 1993.

Shipley JM, Wesselschmidt RL, Kobayashi DK, Ley TJ, Shapiro SD: Metalloelastase is required for macrophage-mediated proteolysis and matrix invasion in mice, *Proc Natl Acad Sci USA* 93:3942–6, 1996.

MMP13

Freije JMP, Diez-Itza I, Balbin M, Sanchez LM, Blasco R et al: Molecular cloning and expression of collagenase-3, a novel human matrix metalloproteinase produced by breast carcinomas, *J Biol Chem* 269:16766–73, 1994.

Inada M, Wang Y, Byrne MH, Rahman MU, Miyaura C et al: Critical roles for collagenase-3 (Mmp13) in development of growth plate cartilage and in endochondral ossification, *Proc Natl Acad Sci USA* 101:17192–7, 2004.

Kennedy AM, Inada M, Krane SM, Christie PT, Harding B et al: MMP13 mutation causes spondyloepimetaphyseal dysplasia, Missouri type (SEMD(MO)), *J Clin Invest* 115:2832–42, 2005.

Mitchell PG, Magna HA, Reeves LM, Lopresti-Morrow LL, Yocum SA et al: Cloning, expression, and type II collagenolytic activity of matrix metalloproteinase-13 from human osteoarthritic cartilage, *J Clin Invest* 97:761–8, 1996.

Pendas AM, Matilla T, Estivill X, Lopez-Otin C: The human collagenase-3 (CLG3) gene is located on chromosome 11q22.3 clustered to other members of the matrix metalloproteinase gene family, *Genomics* 26:615–18, 1995.

Pendas AM, Santamaria I, Alvarez MV, Pritchard M, Lopez-Otin C: Fine physical mapping of the human matrix metalloproteinase genes clustered on chromosome 11q22.3, *Genomics* 37:266–9, 1996.

Reboul P, Pelletier JP, Tardif G, Cloutier JM, Martel-Pelletier J: The new collagenase, collagenase-3, is expressed and synthesized by human chondrocytes but not by synoviocytes: A role in osteoarthritis, *J Clin Invest* 97:2011–9, 1996.

MMP14

Holmbeck K, Bianco P, Caterina J, Yamada S, Kromer M et al: MT1-MMP-deficient mice develop dwarfism, osteopenia, arthritis, and connective tissue disease due to inadequate collagen turnover, *Cell* 99:81–92, 1999.

Mignon C, Okada A, Mattei MG, Basset P: Assignment of the human membrane-type matrix metalloproteinase (MMP14) gene to 14q11-q12 by in situ hybridization, *Genomics* 28:360–1, 1995.

Oh J, Takahashi R, Adachi E, Kondo S, Kuratomi S et al: Mutations in two matrix metalloproteinase genes, MMP-2, and MT1-MMP, are synthetic lethal in mice, *Oncogene* 23:5041–8, 2004.

Sato H, Takino T, Okada Y, Cao J, Shinagawa A et al: A matrix metalloproteinase expressed on the surface of invasive tumor cells, *Nature* 370:61–5, 1994.

Takino T, Sato H, Yamamoto E, Seiki M: Cloning of a human gene potentially encoding a novel matrix metalloproteinase having a C-terminal transmembrane domain, *Gene* 155:293–8, 1995.

MMP15

Mattei MG, Roeckel N, Olsen BR, Apte SS: Genes of the membrane-type matrix metalloproteinase (MT-MMP) gene family, MMP14, MMP15, and MMP16, localize to human chromosomes 14, 16, and 8, respectively, *Genomics* 40:168–9, 1997.

Sato H, Tanaka M, Takino T, Inoue M, Seiki M: Assignment of the human genes for membrane-type-1, -2, and -3 matrix metalloproteinases (MMP14, MMP15, and MMP16) to 14q12.2, 16q12.2-q21, and 8q21, respectively, by in situ hybridization, *Genomics* 39:412–3, 1997.

Takino T, Sato H, Shinagawa A, Seiki M: Identification of the second membrane-type matrix metalloproteinase (MT-MMP-2) gene from a human placenta cDNA library: MT-MMPs form a unique membrane-type subclass in the MMP family, *J Biol Chem* 270:23013–20, 1995.

TIMP1

Carmichael DF, Sommer A, Thompson RC, Anderson DC, Smith CG et al: Primary structure and cDNA cloning of human fibroblast collagenase inhibitor, *Proc Natl Acad Sci USA* 83:2407–11, 1986.

Carrel L, Cottle AA, Goglin KC, Willard HF: A first-generation X-inactivation profile of the human X chromosome, *Proc Natl Acad Sci USA* 96:14440–4, 1999.

Derry JMJ, Barnard PJ: Physical linkage of the A-raf-1, properdin, synapsin I, and TIMP genes on the human and mouse X chromosomes, *Genomics* 12:632–8, 1992.

Docherty AJP, Lyons A, Smith BJ, Wright EM, Stephens PE et al: Sequence of human tissue inhibitor of metalloproteinases and its identity to erythroid-potentiating activity, *Nature* 318:66–9, 1985.

Gasson JC, Golde DW, Kaufman SE, Westbrook CA, Hewick RM et al: Molecular characterization and expression of the gene encoding human erythroid-potentiating activity, *Nature* 315:768–71, 1985.

Huebner K, Isobe M, Gasson JC, Golde DW, Croce CM: Localization of the gene encoding human erythroid-potentiating activity to chromosome region Xp11.1-Xp11.4, *Am J Hum Genet* 38:819–26, 1986.

Spurr NK, Goodfellow PN, Docherty AJP: Chromosomal assignment of the gene encoding the human tissue inhibitor of metalloproteinases to Xp11.1-p11.4, *Ann Hum Genet* 51:189–94, 1987.

Zeiss CJ, Acland GM, Aguirre GD, Ray K: TIMP-1 expression is increased in X-linked progressive retinal atrophy despite its exclusion as a candidate gene, *Gene* 225:67–75, 1998.

TIMP2

De Clerck Y, Szpirer C, Aly MS, Cassiman JJ, Eeckhout Y et al: The gene for tissue inhibitor of metalloproteinases-2 is localized on human chromosome arm 17q25, *Genomics* 14:782–4, 1992.

Hammani K, Blakis A, Morsette D, Bowcock AM, Schmutte, C et al: Structure and characterization of the human tissue inhibitor of metalloproteinases-2 gene, *J Biol Chem* 271:25498–505, 1996.

Noda K, Ishida S, Inoue M, Obata K, Oguchi Y et al: Production and activation of matrix metalloproteinase-2 in proliferative diabetic retinopathy, *Invest Ophthalm Vis Sci* 44:2163–70, 2003.

Seo DW, Li H, Guedez L, Wingfield PT, Diaz T et al: TIMP-2 mediated inhibition of angiogenesis: An MMP-independent mechanism, *Cell* 114:171–80, 2003.

Stetler-Stevenson WG, Krutzsch HC, Liotta LA: Tissue inhibitor of metalloproteinase (TIMP-2): A new member of the metalloproteinase inhibitor family, *J Biol Chem* 264:17374–8, 1989.

TIMP3

Apte SS, Mattei MG, Olsen BR: Cloning of the cDNA encoding human tissue inhibitor of metalloproteinases-3 (TIMP-3) and mapping of the TIMP3 gene to chromosome 22, *Genomics* 19:86–90, 1994.

Langton KP, Barker MD, McKie N: Localization of the functional domains of human tissue inhibitor of metalloproteinases-3 and the effects of a Sorsby's fundus dystrophy mutation, *J Biol Chem* 273:16778–81, 1998.

Langton KP, McKie N, Curtis A, Goodship JA, Bond PM et al: A novel tissue inhibitor of metalloproteinases-3 mutation reveals a common molecular phenotype in Sorsby's fundus dystrophy, *J Biol Chem* 275:27027–31, 2000.

Mohammed FF, Smookler DS, Taylor SEM, Fingleton B, Kassiri Z et al: Abnormal TNF activity in Timp3-/- mice leads to chronic hepatic inflammation and failure of liver regeneration, *Nature Genet* 36:969–77, 2004.

Qi JH, Ebrahem Q, Moore N, Murphy G, Claesson-Welsh L et al: A novel function for tissue inhibitor of metalloproteinases-3 (TIMP3): inhibition of angiogenesis by blockage of VEGF binding to VEGF receptor-2, *Nature Med* 9:407–15, 2003.

Stohr H, Roomp K, Felbor U, Weber BHF: Genomic organization of the human tissue inhibitor of metalloproteinases-3 (TIMP3): *Genome Res* 5:483–7, 1995.

Weber BHF, Vogt G, Pruett RC, Stohr H, Felbor U: Mutations in the tissue inhibitor metalloproteinases-3 (TIMP3) in patients with Sorsby's fundus dystrophy, *Nature Genet* 8:352–6, 1994.

TIMP4

Bigg HF, Shi YE, Liu YE, Steffensen B, Overall CM: Specific, high affinity binding of tissue inhibitor of metalloproteinases-4 (TIMP-4) to the COOH-terminal hemopexin-like domain of human gelatinase A, *J Biol Chem* 272:15496–500, 1997.

Greene J, Wang M, Liu YE, Raymond LA, Rosen C et al: Molecular cloning and characterization of human tissue inhibitor of metalloproteinase 4, *J Biol Chem* 271:30375–80, 1996.

Leco KJ, Apte SS, Taniguchi GT, Hawkes SP, Khokha R et al: Murine tissue inhibitor of metalloproteinases-4 (Timp-4): cDNA isolation and expression in adult mouse tissues, *FEBS Lett* 401:213–7, 1997.

Olson TM, Hirohata S, Ye J, Leco K, Seldin MF et al: Cloning of the human tissue inhibitor of metalloproteinase-4 gene (TIMP4) and localization of the TIMP4 and Timp4 gene to human chromosome 3p25 and mouse chromosome 6, respectively, *Genomics* 51:148–51, 1998.

Wang M, Liu YE, Greene J, Sheng S, Fuchs A et al: Inhibition of tumor growth and metastasis of human breast cancer cells transfected with tissue inhibitor of metalloproteinase 4, *Oncogene* 14:2767–74, 1997.

Human protein reference data base, Johns Hopkins University and the Institute of Bioinformatics, at http://www.hprd.org/protein.

4.8. Activation of MMPs

Nagase H, Woessner JF Jr: Matrix metalloproteinases, *J Biol Chem* 274:21491–4, 1999.

Massova I, Kotra LP, Fridman R, Mobashery S: Matrix metalloproteinases: Structures, evolution, and diversification, *FASEB J* 12:1075–95, 1998.

Sternlicht MD, Werb Z: How matrix metalloproteinases regulate cell behavior, *Annu Rev Cell Dev Biol* 17:463–516, 2001.

Visse R, Nagase H: Matrix metalloproteinases and tissue inhibitors of metalloproteinases: Structure, function, and biochemistry, *Circ Res* 92:827–39, 2003.

SECTION 2

REGULATORY MECHANISMS OF REGENERATION

5

CELL SIGNALING PATHWAYS AND MECHANISMS

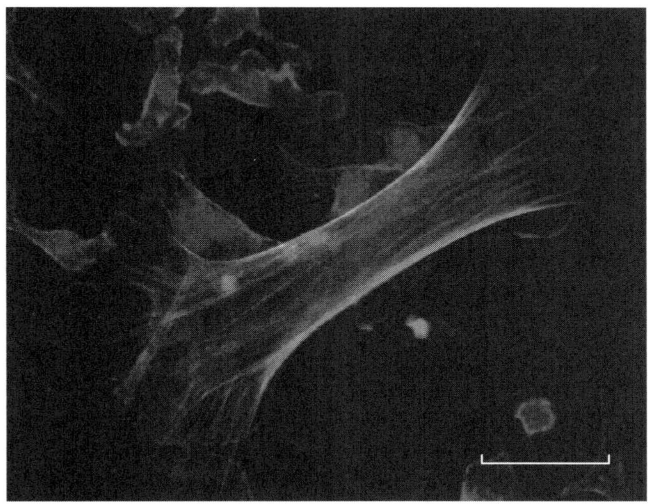

Interaction of bone marrow-derived smooth muscle α actin+ cells with CD11b+ cells in culture. Mouse bone marrow cells were collected and cultured in DMEM with 10% FBS. Smooth muscle α actin+ cells (green) were often found on top of CD11b+ cells (red). When the CD11b+ cells were selectively removed, the number of smooth muscle α actin+ cells was reduced, suggesting that the CD11b+ bone marrow cells play a role in regulating the development of smooth muscle cells from bone marrow progenitor cells. Blue: cell nuclei. Scale: 10 μm. See color insert.

Bioregenerative Engineering: Principles and Applications, by Shu Q. Liu

PRINCIPLES OF CELL SIGNALING [5.1]

In a multicellular system, there exist signaling activities at the molecular and cellular levels for cell-to-cell and cell-to-matrix communications, which are essential for the function of cells, tissues, and organs. In order to conduct physiological function, a cell must communicate with other cells to achieve coordinated cellular activities. Often, synchronized molecular and cellular activities are required for tissue and organ functions. A typical example is the control of the contractile activity of skeletal muscle cells in response to an electrical stimulation from a nerve axon. In such a regulatory process, an electrical action potential is generated from a motor neuron center and transmitted through the nerve axon to the muscle cell. At the junction, or synapse, between the nerve axon and the muscle cell, the electrical signal is converted to a chemical signal, which induces the activation of an intracellular signaling cascade in the muscle cells, resulting in cell contraction. Without neuron-to-muscle cell signal transduction, the skeletal muscle cells cannot contract or relax in a controlled manner. Thus, coordinated cell signaling is a fundamental process for accomplishing complicated functional tasks. Here, the principles of cell signaling and common cell signaling pathways are discussed.

Factors Serving as Signals

There are two basic types of factors, which serve as "signals" for regulating molecular and cellular activities as well as cell-to-cell and cell-to-matrix communications: extracellular factors and intracellular factors (Fig. 5.1). *Extracellular facotrs* are present in extracellular space and serve as signals that initiate and control molecular and cellular activities. *Intracellular factors* are present in the cytoplasm, serve as elements for signaling pathways, and regulate cellular activities. In general, an extracellular signal must cooperate with intracellular signals to initiate and control a cellular activity. The extracellular signal may initiate the activation of an intracellular signaling pathway via interacting with a cell membrane or cytoplasmic signaling factor, whereas the intracellular signaling pathway relays the extracellular signal to target subcellular organelles, initiating a specified activity, such as gene expression, cell adhesion, cell proliferation, or cell migration.

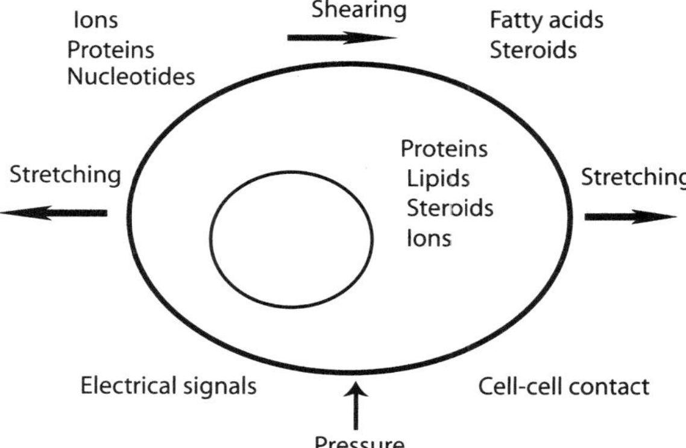

Figure 5.1. Extracellular and intracellular signals that regulate cell activities and functions (based on bibliography 5.1).

Cellular activities are often initiated in response to environmental or extracellular stimulations or cues. Extracellular factors that serve as signaling factors and induce cellular activities may include biochemical molecules and substances, electrical signals, and physical factors. *Biochemical molecules and substances* include cell-secreted factors and extracellular matrix factors, such as ions, proteins, nucleotides, fatty acids, and steroids. Some of these factors, such as proteins (growth factors and extracellular matrix components) and nucleotides, can act on cell membrane receptors, while others, such as steroids, can diffuse through the cell membrane and interact directly with intracellular receptors, inducing activation of signaling pathways. *Electrical signals* are action potentials, or changes in membrane potentials. An electrical signal initiates the activation of nerve and muscular cells. *Physical factors*, such as mechanical forces and deformation, can serve as signaling factors. For instance, mechanical shearing and stretching forces due to bloodflow and pressure, respectively, induce various cellular activities, ranging from cell migration, proliferation to apoptosis, in the vascular system. These biochemical, electrical, and physical signals initiate and control specified cellular activities.

Intracellular signaling factors include proteins, lipids, steroids, and ions. These factors are also referred to as *regulatory factors*. In each cell, there exist a large number of proteins. Most of these protiens serve as signaling molecules. Common types of signaling proteins include cell membrane receptors, ion channels, protein kinases, protein phosphatases, and enzymes. Certain types of structural proteins, such as actin filaments, also play a role in signal transduction. A variety of lipid molecules in the cell membrane and endoplasmic reticulum serve as signaling factors. Ions, especially, calcium, play a critical role in signal transduction and the regulation of cellular activities. The roles of these signaling factors are discussed in the following sections.

Types of Cell Signaling

There are various cell signaling mechanisms. Based on the range of signal transduction, cell signaling events can be divided into long- and short-range events (Fig. 5.2). *Long-range signaling events* involve cells from different systems. These events are mediated by either hormones or synapses. Hormones are biochemical factors, which are synthesized by endocrine gland cells and released into blood. These factors regulate the activities of remote cells. This type of signaling is also known as *endocrine* signaling. Since a hormone is delivered via the bloodflow, it can reach almost all cells in the body. The specificity of hormones is dependent on the receptors on the target cells. *Synapses* are subcellular structures found at the junctions between different neurons and between neurons and muscular cells. In a synapse-related signaling event, signals from a presynaptic neuron are conducted to another neuron or a peripheral muscular cell via the propagation of electrical action potentials, which stimulate the release of neurotransmitters at the terminal of the presynaptic cell. The synapse in turn delivers the neurotransmitters to the postsynaptic target cell, initiating postsynaptic cell activation. Synaptic signaling events influence target cells with great precision. In contrast to long-range signaling, *short-range signaling events* involve cells within a local neighborhood. These events are mediated by regional chemical factors and mechanical forces. Changes in mechanical forces often induce the activation and release of chemical factors, which in turn transmit mechanical signals to the cell. Signaling events mediated by regional chemical factors are often referred to as *paracrine* signaling. Chemical factors for paracrine signaling are usually

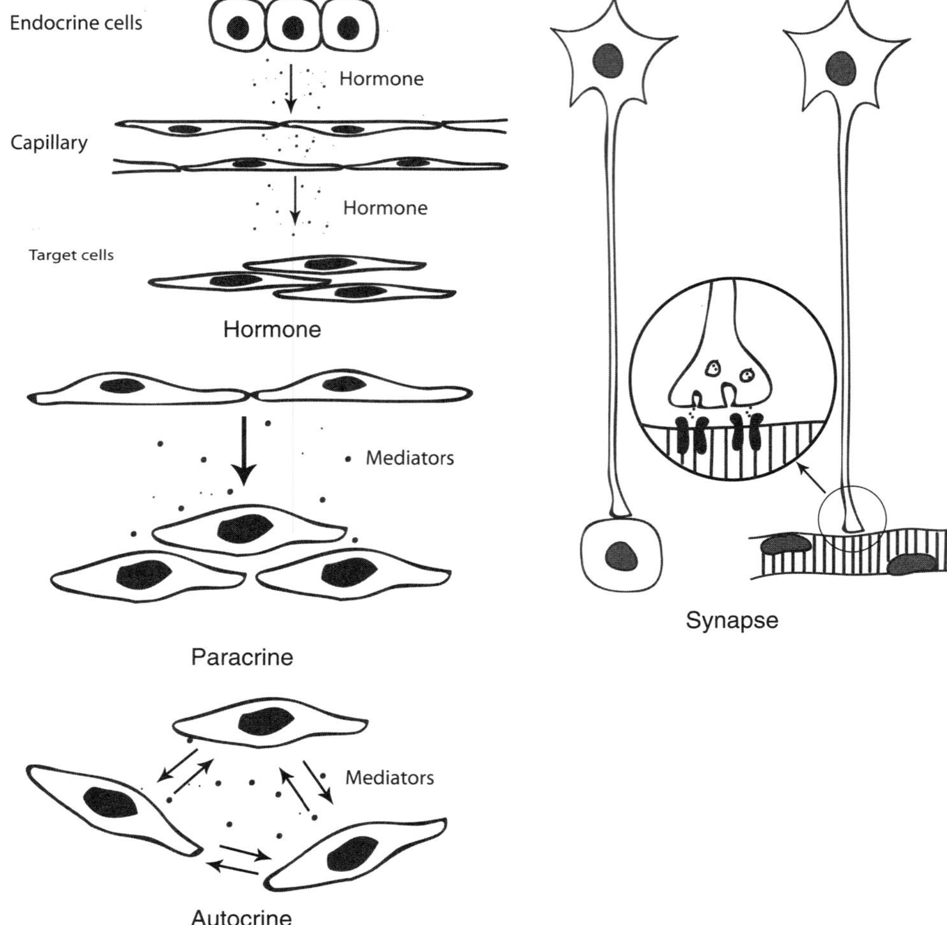

Figure 5.2. Long-range (hormone- and synapse-mediated) and short-range (paracrine and autocrine) cell signaling (based on bibliography 5.1).

endocytosed by neighboring cells or immobilized by extracellular matrix, ensuring a local influence without spreading to remote cells.

General Mechanisms of Cell Signaling

On the stimulation of an extracellular signal, a cell responds by activating its intracellular signaling pathways, leading to specific cellular activities, such as gene expression, cell division, cell migration, cell adhesion, and cell apoptosis. Several steps are involved in the activation of the intracellular signaling pathways: interaction of an extracellular signaling factor with a corresponding receptor or factor in the target cell, transduction of the extracellular signal to the target cell, activation of intracellular biochemical and/or electrical reactions, and termination of intracellular reactions.

A large number of molecules, including proteins, lipids, and ions, can serve as intracellular signaling molecules and participate in signal transduction. Proteins, including receptors, enzymes, and adapters, are among the common signaling molecules. Receptors are distributed in the cell membrane or cytoplasm, and are responsible for receiving extracellular signals and transmitting the signals into intracellular signaling pathways. Enzymes involved in cell signaling primarily include protein kinases, protein phosphatases, and GTPases. Protein kinases and phosphatases are responsible for protein phosphorylation and dephosphorylation, respectively, which are critical processes for cell signaling. GTPases serve as switches for signal transduction to downstream signaling elements. Adapter proteins help to target, recruit, and co-localize proteins. Some adapter proteins serve as scaffolds or docking sites for signaling proteins.

Signaling molecules exist in two modes: inactive and active. In unstimulated cells, signaling molecules are present mostly in the inactive mode. In response to a stimulation, specified signaling molecules can be activated by various mechanisms, including chemical modification, enzymatic activation, conformational changes in signaling molecules, alterations in molecular concentration, and col/localization and clustering of molecules. In most cases, multiple mechanisms are involved for the same or various signaling molecules at a given time and location.

There are several common features for cell signaling. These include signaling specificity, the involvement of signaling cascades, and crosstalk between signaling pathways. In general, each signaling molecule reacts with specific upstream and downstream molecules, ensuring the induction of specific activities. Various signaling pathways are designed and developed for specific cellular activities. For the regulation of each cellular activity, multiple signaling molecules and pathways may be involved. In addition, each signaling molecule may exhibit multiple functions. The abundance and multifunctionality of signaling molecules may be a mechanism that ensures the accomplishment of cellular activities and functions.

Each signaling pathway is composed of a number of signaling molecules, which relay signals from extracellular space to the cytoplasm and nucleus. Such a cascade is also referred to as a *signaling cascade*. In each cell, there exist multiple signaling pathways. These pathways often communicate or crosstalk with each other via branching pathways, forming signaling networks. Through these crosstalk pathways, a signaling molecule can be activated by different upstream signaling molecules and can act on different downstream effectors. With such an approach, various upstream signals can be converged to a single signaling pathway, and one activated signaling molecule can initiate multiple downstream activities. In addition, signals in one pathway can influence signals in other pathways. Thus crosstalk is an effective approach for amplifying and controlling signals and facilitating signal transduction. In the following sections, common signaling pathways in mammalian cells are briefly reviewed.

PROTEIN TYROSINE KINASE-MEDIATED CELL SIGNALING [5.2]

Protein tyrosine kinases belong to the superfamily of protein kinases. *Protein kinases* are enzymes that catalyze the phosphorylation of, or the addition of a phosphate group to, a substrate protein. Protein phosphorylation is a major form of molecular modification, which mediates a variety of molecular activities, including enzymatic activation, cell signaling, gene transcription, activation of ion channels, and reorganization of cytoskeletal proteins. Protein kinases represent a large family of signaling molecules. The genes encoding protein kinases constitute about 1.7% of total genes in the human. Based on the types of target amino acid, protein kinases are classified into four types: serine/threonine-, tyrosine-, histidine-, and aspartate- (or glutamate-) specific protein kinases. These protein kinases catalyze the phosphorylation of substrate proteins on serine/threonine, tyrosine, histidine, and aspartate (or glutamate), respectively. Among these protein kinases, serine/threonine and tyrosine protein kinases can phosphorylate signaling proteins, play critical roles in signal transduction. Thus, these two types of protein kinase are the focus of this book. The protein tyrosine kinases are covered in this section. The protein serine/threonine kinases are covered later in this chapter.

Structure and Function

Protein tyrosine kinases are enzymes that catalyze the phosphorylation of the tyrosine residues of substrate proteins. A protein tyrosine kinase is often found as an integral part of growth factor receptors in the cell membrane and is referred to as *receptor protein tyrosine kinase.* A receptor with a protein tyrosine kinase is known as *protein tyrosine kinase receptor.* Such a receptor is composed of an extracellular domain for ligand binding, a transmembrane domain for anchoring the receptor to the cell membrane, and an intracellular domain for interaction with cytoplasmic signaling molecules. The receptor protein tyrosine kinase is localized to the cytoplasmic domain of the growth factor receptor. The receptor tyrosine kinase assists the growth factor receptor in transducing signals that induce essential cell activities, such as proliferation, differentiation, and migration. There are also nonreceptor protein tyrosine kinases such as the Src family of protein tyrosine kinases. These kinases are discussed on page 180.

A signaling pathway mediated by a protein tyrosine kinase receptor is composed of several components, including the extracellular ligands, cell membrane receptors, cytoplasmic signaling elements, and transcriptional factors. Extracellular ligands are usually growth factors produced and released by cells. There are a number of growth factors that interact and activate protein tyrosine kinase receptors. These include epidermal growth factor (EGF), platelet-derived growth factors (PDGFs) A and B, nerve growth factor (NGF), basic fibroblast growth factor (bFGF), vascular endothelial growth factor (VEGF), insulin-like growth factor (ILGF), nerve growth factor (NGF), hepatocyte growth factor (HGF), and ephrins. Based on the type of ligand growth factors, growth factor receptors are classified as EGF receptor, PDGF receptor, NGF receptor, FGF receptor, VEGF receptor, insulin receptor, HGF receptor, and Eph receptor groups, respectively (see Fig. 5.3 for schematic representation of growth factor receptors). The ligand–receptor interaction is highly specific. Each growth factor only interacts with and activates a specific protein tyrosine kinase receptor.

The EGFR group is composed of several receptors, including EGFR (epidermal growth factor receptor), ERBB2 (v-erb-b2 erythroblastic leukemia viral oncogene homolog 2 or neuro/glioblastoma-derived oncogene homolog), ERBB3 (v-erb-b2 erythroblastic leukemia viral oncogene homolog 3), and ERBB4 (v-erb-a erythroblastic leukemia viral oncogene homolog 4). These receptors are expressed in various cell types, including epithelial, mesenchymal, and neural cells. The EGFR contains two extracellular cysteine-rich domains, a transmembrane domain, and a cytoplasmic tyrosine kinase domain (Fig. 5.3). This receptor interacts with EGF and similar ligands, regulating cell development, morphogenesis, regeneration, and tumorigenesis.

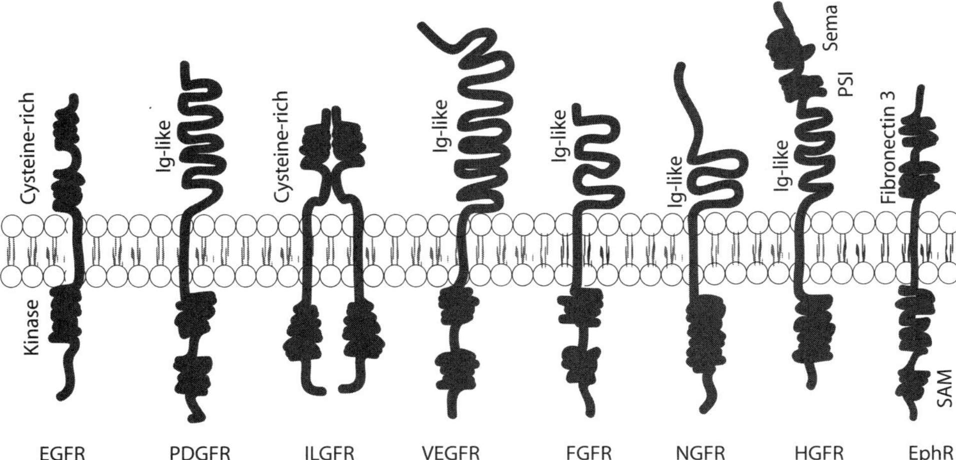

Figure 5.3. Schematic representation of growth factor receptors. Sema: semaphorin domain. Ig: immunoglobulin domain. PSI: plexin, semaphorin, and integrin domain. SAM: sterile alpha motif (based on bibliography 5.2).

The PDGFR group consists of several molecules, including PDGFRα, PDGFRβ, CSF1R, KIT/SCFR, FLK2, and FLT3. Each of these receptors is composed of five IgD-like domains in the extracellular domains (Fig. 5.3). The intrinsic protein tyrosine kinase domain is split into two parts by an intervening segment. These receptors play a critical role in regulating the development of connective tissue cells and vascular smooth muscle cells.

The insulin receptor group includes three members: insulin receptor, insulin growth factor (IGF)1 receptor, and IRR, which are present in the form of dimers linked by disulfide bonds (Fig. 5.3). These receptors regulate processes related to cell survival.

The VEGFR group includes three members, including VEGFR1, VEGFR2, and VEGFR3, which are characterized by the presence of seven Ig-like domains in the extracellular domain (Fig. 5.3). These receptors are expressed primarily in vascular endothelial cells and regulate the development of endothelial cells, as well as angiogenesis and vasculogenesis.

The FGFR group consists of FGFR1, FGFR2, FGFR3, and FGFR4. The extracellular region contains three Ig-like domains (Fig. 5.3). These receptors mediate the development and morphogenesis of a variety of cell types in connective tissues and the cardiovascular system.

The nerve growth factor receptor group is composed of three members, including TrkA, TrkB, and TrkC. The extracellular domain of these receptors contains a LRD domain and two Ig-like domains, and the intracellular region contains a single protein tyrosine kinase domain (Fig. 5.3). These receptors are expressed in neurons and neural glial cells and play important roles in regulating the development and morphogenesis of nerve cells.

The HGFR group is composed of two members: Met and Ron. These receptors are expressed primarily in hepatocytes and also in other epithelial cells. The extracellular domain of the receptor is composed of a Sema (semaphorin) domain, a PSI (plexin, semaphorin, and integrin) domain, and four Ig-like domains and the intracellular domain contains a single-protein tyrosine kinase domain (Fig. 5.3). HGF receptors regulate the development and regeneration of the liver and other epithelial tissues.

The Eph receptor group includes at least 14 members, EPHA1 (8 members) and EPHB1 (6 members), and interacts with various types of ephrin. A typical Eph receptor contains two fibronectin 3 domains in the extracellular region and a tyrosine kinase domain in the cytoplasmic region. The cytoplasmic region also contains a SAM (sterile α motif) (Fig. 5.3). These receptors are expressed primarily in nerve cells and vascular endothelial cells, and play a critical role in the regulation of cell migration and morphogenesis.

Most receptors listed above interact with intracellular signaling molecules, including Src homology 2 domain-containing molecules and adapter proteins, leading to activation of a mitogenic signaling cascade composed of mitogen-activated protein kinase kinase kinases (MAPKKKs), mitogen-activated protein kinase kinases (MAPKKs), and mitogen-activated protein kinases (MAPKs). These molecules regulate cell proliferation, differentiation, and migration.

Signaling Mechanisms

The activation of protein tyrosine kinase receptor signaling pathways begins with the binding of extracellular ligands to the protein tyrosine kinase receptors (Fig. 5.4). These receptors undergo a dimerization process on the interaction with ligands, a common mechanism for activating the protein tyrosine kinase receptor signaling pathways. For various groups of protein tyrosine kinase receptors, there exist different forms of dimerization. For example, PDGF receptors are dimerized into a symmetric structure on the binding of a disulfide-bonded PDGF dimeric complex (Fig. 5.4). In contrast, EGF receptors undergo conformational changes in response to EGF binding, forming receptor–receptor complexes. The dimerization of receptors brings the intracellular domains of the two receptors together, a critical process for initiating autophosphorylation of the protein tyrosine kinase domains on tyrosine residues.

The autophosphorylation of a receptor protein tyrosine kinase activates the tyrosine kinase domain, leading to the activation of downstream signaling molecules, sucs as Src homology (SH)2 domain-containing protein tyrosine kinases and adapter proteins.

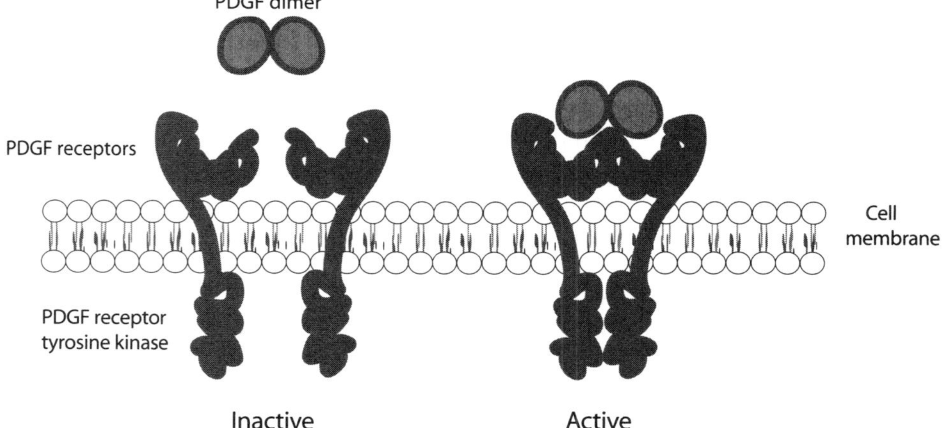

Figure 5.4. Schematic representation of the interaction of platelet-derived growth factor (PDGF) with PDGF receptor. A PDGF molecule forms a dimer with another PDGF molecule (e.g., PDGF-AA and PDGF-BB). A PDGF dimer can interact with two PDGF receptors, causing crosslink between the two PDGF receptors and autophosphorylation of the receptor protein tyrosine kinase in the cytoplasmic domain (based on bibliography 5.2).

A typical example of SH2 domain-containing kinases is the Src protein tyrosine kinase. The molecule possesses intrinsic enzymatic kinase activities and can induce phosphorylation on substrate proteins. *Adapter proteins* are proteins that serve as linkers between receptor protein kinases and downstream signaling molecules. Typical examples of adapter proteins include Grb2/sos, Crk, Nck, and Shc (these and other examples are listed in Table 5.1). These molecules do not possess intrinsic enzymatic kinase activities.

The interaction of an autophosphorylated receptor tyrosine kinase with a SH2 domain-containing kinase or an adapter protein is a critical process for signal transduction. A phosphorylated tyrosine residue in the receptor tyrosine kinase can serve as a docking site for the SH2 domain of signaling molecules, which is organized into a pocket-like structure specific for interaction with a phosphorylated tyrosine. Each receptor tyrosine kinase is capable of interacting with multiple SH2 domain-containing molecules at various tyrosine sites. For example, the PDGF receptor tyrosine kinase contains about 12 tyrosine residues that can be phosphorylated within the intracellular domain (11 are outside and 1 is within the tyrosine kinase). Some of these phosphorylated tyrosine residues serve as docking sites for SH2 domain-containing tyrosine kinases, such as Src, GAP, SHP2, and PLCγ, and others serve as docking sites for adapter signaling proteins, such as Shc, Nck, and Grb2/Sos. The phosphorylated tyrosine at site 581 (pY581) can interact with Src and Shc, pY740 can interact with Shc and PI3-kinase, pY751 can interact with PI3-kinase and Nck, and pY771 can interact with GAP and Shc. The interaction of the tyrosine kinase domain with downstream SH2 domain-containing molecules activates downstream signaling molecules, leading to activation of mitogenic cellular activities, such as cell proliferation, differentiation, and migration. (See Table 5.2.)

Although different tyrosine kinase receptors are present in the cell membrane and responsible for transducing distinct extracellular signals into the cell, there are common mechanisms of action. Here, the PDGF receptor is used as an example to demonstrate how the protein tyrosine kinase transduces PDGF signals into intracellular signaling pathways (Fig. 5.5). On the binding of PDGF ligands, PDGF receptors undergo dimerization, inducing autophosphorylation of the PDGF receptor tyrosine kinase domain. The activated tyrosine kinase domain recruits the adapter protein complex Grb2/Sos to the pY 716 site of the protein tyrosine kinase. Grb2/Sos activates the Ras protein by stimulating the substitution of GTP for GDP in the Ras protein. Activated Ras induces the activation of at least two cascades of signaling molecules, including the extracellular signal-regulated protein kinase (ERK)1/2 cascade and the c-Jun *N*-terminal kinase/stress-activated protein kinase (JNK/SAPK) cascade. Both pathways are collectively known as the *mitogen-activated protein kinase* (MAPK) pathways.

For the ERK1/2 pathway, Ras activates several protein kinases including Raf-1, A-Raf, and B-Raf, members of the mitogen-activated protein kinase kinase kinase (MAPKKK) family. The Raf kinases phosphorylate MAPK/ERK kinase (MEK)1/2, members of the MAPKK family. MEK1/2 in turn phosphorylates the tyrosine and threonine residues of ERK1/2, which is a protein complex belonging to the MAPK family. Activated ERK1/2 can translocate from the cytoplasm to the nucleus, where it activates transcriptional factors such as c-Fos, cAMP response element binding protein (CREB), early growth response (Egr)1, and Elk1, initiating the expression of mitogenic genes.

For the JNK/SAPK pathway, Ras can activate MEK kinase (MEKK) 1, 2, and 3, which are members of the MAPKKK family, possibly via the mediation of Rac/Cdc42 and p21-activated protein kinase (PAK). Activated MEKK 1, 2, and 3 phosphorylate MEK4, a member of the MAPKK family. MEK4 in turn phosphorylates JNK/SAPKs, which belong

TABLE 5.1. Characteristics of Selected Signaling Molecules*

Proteins	Alternative Names	Amino Acids	Molecular Weight (kDa)	Expression	Functions
Src	Avian sarcoma virus protein (ASV), p60-Src, tyrosine kinase pp60c-src, tyrosine protein kinase SRC-1, protooncogene SRC	536	60	Ubiquitous	A protein tyrosine kinase that regulates cell survival, proliferation, migration, and cell–cell communication
Grb2	Growth factor receptor-bound protein 2, abundant Src homology protein	217	25	Ubiquitous	Containing a SH2 domain and two SH3 domains[a] interacting with growth factor receptors, serving as an adaptor protein, and mediating the transduction of mitogenic signals, such as growth factors
Sos	"Son of sevenless" *Drosophila* homolog 1, SOS1 guanine nucleotide exchange factor	1333	152	Ubiquitous	A guanine nucleotide exchange factor that binds to Grb2 and activates the Ras protein
Crk	v-crk sarcoma virus CT10 oncogene homolog, oncogene Crk	304	34	Lung, kidney	An adapter protein that binds to tyrosine-phosphorylated proteins via interaction with phosphotyrosine residues, and mediating intracellular signal transduction
Nck	Noncatalytic region of tyrosine kinase, NCK1, NCK α, NCK adaptor protein 1, cytoplasmic protein NCK1, SH2/SH3 adaptor protein NCK α	377	43	Ubiquitous	An adaptor protein that transduces signals from receptor tyrosine kinases to downstream signal recipients, such as the Ras protein
Shc	p66, Src homology 2 domain containing (SHC) transforming protein 1	583	63	Ubiquitous	Serving as an adaptor protein that mediates the activation of Ras proteins in response to stimulation of mitogenic factors

[a]Note that the SH2 domain binds tyrosine-phosphorylated sequences and the two SH3 domains bind to proline-rich regions of substrate proteins.

*Based on bibliography 5.2.

TABLE 5.2. Characteristics of Selected Signaling Molecules*

Proteins	Alternative Names	Amino Acids	Molecular Weight (kDa)	Expression	Functions
GAP	GTPase-activating protein (GAP), guanosine triphosphatase-activating protein, Ras p21 protein activator 1 (RASA1), p120GAP, RasGAP	1047	116	Ubiquitous	A GTPase-activating protein that activates the GTPase of the p21 Ras protein, induces conversion of GTP to GDP on Ras protein, and thus suppresses Ras activity, resulting in inhibition of cell mitogenic activities
SHP2	SH2 containing protein tyrosine phosphatase 2, protein tyrosine phosphatase nonreceptor type 11 (PTPN11), protein tyrosine phosphatase 2C (PTP2C), tyrosine phosphatase SHP2, PTP-1D, SHPTP2	593	68	Ubiquitous	A molecule of protein tyrosine phosphatase family, regulating cell proliferation and migration
Phospholipase Cγ	PLC γ1, PLC1, PLCG1, PLC148	1290	149	Ubiquitous	Catalyzing the formation of inositol 1,4,5-trisphosphate and diacylglycerol from phosphatidylinositol 4,5-bisphosphate, an important process for G-protein receptor- and Ras-mediated cell signaling

TABLE 5.2. *Continued*

Proteins	Alternative Names	Amino Acids	Molecular Weight (kDa)	Expression	Functions
Phosphatidylinositol 3-kinase catalytic subunit α	PI3 kinase α, PI3K α, PIK3CA, p110 α, PI3 kinase p110 subunit α	1068	124	Ubiquitous	Serving as catalytic subunit of PI3 kinase, which is composed of a regulatory subunit and a catalytic subunit, inducing phosphorylation of phosphatidylinositol (PtdIns), PtdIns4P, and PtdIns(4,5)P2 on the 3-hydroxyl group of the inositol ring, regulating cell proliferation and differentiation, and contributing to development of cancers
Phosphatidylinositol 3-kinase catalytic subunit β	PI3K β, phosphatidylinositol-4, 5-bisphosphate 3-kinase catalytic subunit β isoform, phosphoinositide 3-kinase catalytic β polypeptide, phosphatidylinositol 3-kinase catalytic 110-kDa β, p110-β, PI3-kinase p110 subunit β	1070	123	Ubiquitous	Same as PI3 kinase α
Phosphatidylinositol 3-kinase catalytic subunit γ	PI3K γ, phosphatidylinositol 3 kinase catalytic subunit γ isoform, phosphatidylinositol 3 kinase catalytic 110 kDa γ, p110 γ, PI3 kinase p110 subunit γ	1102	126	Heart, liver, skeletal muscle, pancreas, leukocytes	In addition to functions as described for PI3 kinase α, PI3 kinase γ regulates inflammatory reactions

*Based on bibliography 5.2.

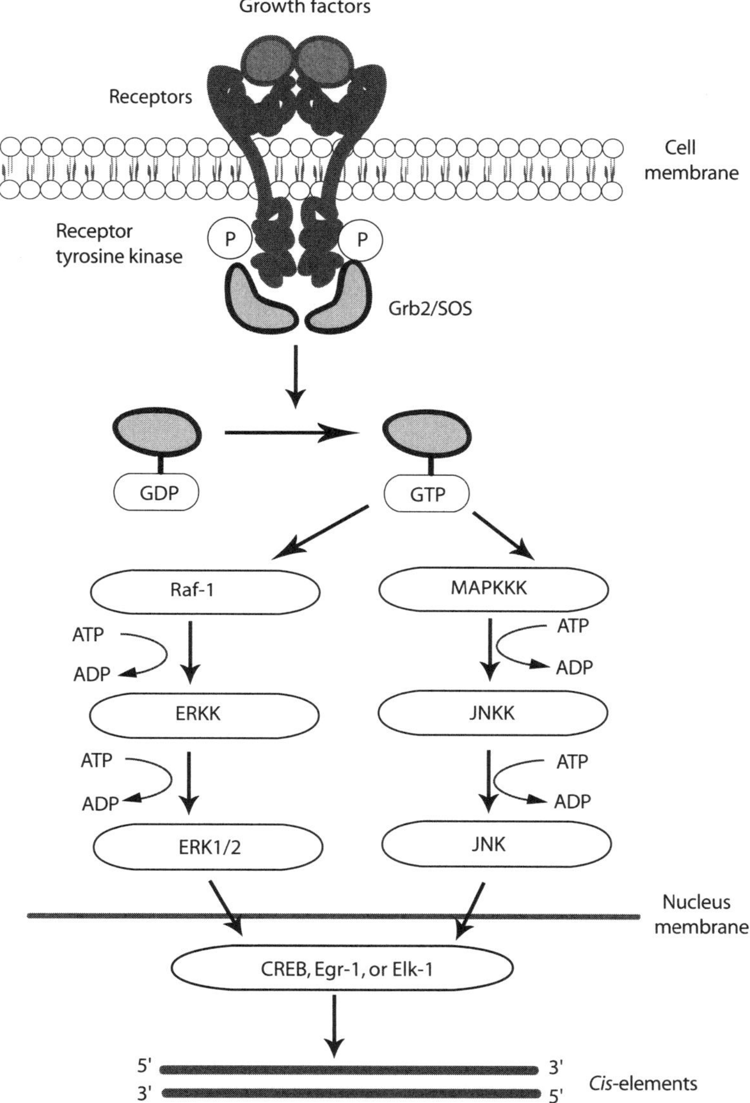

Figure 5.5. Schematic representation of protein tyrosine kinase receptor-mediated cell signaling (based on bibliography 5.2).

to the MAPK family. Activated JNK/SAPKs can translocate from the cytoplasm to the cell nucleus and activate transcriptional factors c-Jun, activating transcription factor (ATF)2, and Elk1. These transcriptional factors interact with corresponding cis-elements in target genes, initiating mitogenic mRNA transcription.

It is important to note that, in addition to the stimulatory effect on transcriptional factors as described above, c-Fos activated by ERK1/2 and c-Jun activated by JNK/SAPKs

TABLE 5.3. Characteristics of Selected Signaling Molecules*

Proteins	Alternative Names	Amino Acids	Molecular Weight (kDa)	Expression	Functions
PDGF receptor β	Platelet-derived growth factor receptor β, PDGFRB	1106	124	Blood vessel, kidney, pancreas, bone marrow	A cell membrane tyrosine kinase receptor that interacts with platelet-derived growth factors and regulates cell survival, proliferation, differentiation, and migration
H-Ras	Harvey murine sarcoma virus oncogene, HRAS, RASH1, HRSP, C-H-RAS	189	21	Lung, intestine, stomach, and thyroid gland	A protein homologous to the product of transforming gene of Harvey murine sarcoma virus, mediating mitogenic signal transduction, and contributing to carcinogenesis when overexpressed
K-Ras	Kirsten murine sarcoma virus 2, RASK2, oncogene KRAS2, V KI RAS2 Kirsten rat sarcoma 2 viral oncogene homolog, CK RAS, cK Ras protein, K Ras p21 protein, c-Kirsten Ras protein, transforming protein p21	189	21	Lung, liver, intestine, leukocytes	A protein homologous to product of transforming gene of Kirsten murine sarcoma virus, mediating mitogenic signal transduction, and contributing to carcinogenesis when mutated
Raf-1	Oncogene RAF1, c-raf, v-raf-1 murine leukemia viral oncogene homolog 1	648	73	Widely expressed	A MAP kinase kinase kinase phosphorylating MEK1 and MEK2, which in turn phosphorylates the serine/threonine-specific protein kinases ERK1/2 and regulates cell survival, proliferation, and differentiation

MEK1	MAPK/ERK kinase 1, mitogen-activated protein kinase kinase 1 (MAPKK1, MAP2K1, MKK1, MAP kinase kinase 1), dual-specificity mitogen-activated protein kinase kinase 1, ERK activator kinase 1, mitogen-activated protein kinase (MAPK)/ extracellular signal-regulated kinase (ERK) kinase 1, MAPK/ERK kinase 1 (MEK1)	393	43	Brain, heart, lung, blood vessel, liver, kidney, spleen, intestine, thymus, skeletal muscle	A serine/threonine kinase serving as an element for the ERK1/2 signaling pathway, activating mitogen-activated protein kinases including ERK1/2, and regulating cell survival, development, proliferation, and differentiation
MEK2	MAPK/ERK kinase 2, MAP kinase/Erk kinase 2, MAP2K2, MKK2, MAPKK2, MAPK/ERK kinase 2, ERK activator kinase 2, mitogen-activated protein kinase kinase 2	400	44	Neutrophils	Similar to function of MEK1
MEK3	MAPK/ERK kinase 3, mitogen-activated protein kinase kinase 3, MAPKK3, MAP2K3, MKK3, and protein kinase mitogen-activated kinase 3	352	40	Heart, blood vessel, lung, liver, kidney, spleen, pancreas. Intestine, thymus, prostate gland, ovary	Similar to function of MEK1, and activating MAPK14/p38-MAPK
MEK4	MAPK/ERK kinase 4, mitogen-activated protein kinase kinase 4, SAPK/ERK kinase 1, MAPKK4, MAP2K4, MKK4, JNK-activated kinase 1, and JNKK1	399	44	Skeletal muscle	Similar to the function of MEK1, activating MAPK8/JNK1, MAPK9/JNK2, and MAPK14/p38
Extracellular signal-regulated kinasel	ERK1/2, mitogen-activated protein 3 (MAPK3), p44ERK1, p44MAPK	379	44	Ubiquitous	A serine, threonine, and tyrosine kinase that forms a complex with ERK2 and regulates cell survival, proliferation, and differentiation

TABLE 5.3. *Continued*

Proteins	Alternative Names	Amino Acids	Molecular Weight (kDa)	Expression	Functions
Extracellular signal-regulated kinase 2	ERK2, mitogen-activated protein kinase 1, mitogen-activated protein kinase 2 (MAPK2), protein kinase mitogen-activated 1, protein kinase mitogen-activated 2, protein tyrosine kinase ERK2, p42MAPK, p41MAPK	360	42	Ubiquitous	A serine, threonine, and tyrosine kinase that forms a complex with ERK1 and regulates cell survival, proliferation, and differentiation
c-Fos	Oncogene FOS, FOS, and v-fos FBJ murine osteosarcoma viral oncogene homolog	380	41	Ubiquitous	Forming heterodimeric transcriptional factor complexes, known as activating protein (AP)-1, with proteins of the c-Jun family, and regulating cell survival, development, proliferation, differentiation, and transformation
cAMP response element-binding protein	CREB, CREB1, cAMP response element-binding protein 1, transactivator protein	341	37	Brain, heart, thymus, adrenal gland	A transcriptional factor belonging to the leucine zipper family of DNA-binding proteins, binding to the cAMP-responsive element of target genes, regulating cell survival, proliferation, and differentiation, and regulating hepatic gluconeogenesis
early growth response 1	Egr-1, nerve growth factor-induced protein A, zinc finger protein 225	543	58	Nervous system, cartilage, blood vessel, liver, stomach, leukocytes	Serving as a transcriptional factor that binds to mitogenic genes, regulating cell proliferation, differentiation, and apoptosis
Elk-1	Transforming protein elk-1	428	45	Ubiquitous	Acting as a transcriptional factor that is a target of ERK1/2 in the cell nucleus, and binding to and activating the serum response element in the promoter region of the c-fos gene

MEK kinase 1	MEKK1				
MEK kinase 2	MEKK2, mitogen-activated protein kinase kinase kinase 2 (MAPKKK2), MAPK/ERK kinase kinase 2, MEK kinase 2	620	70	Ubiquitous	Acting as a serine/threonine protein kinase, phosphorylating MAPK7 and MAP2K4, and regulating cell proliferation, differentiation, and apoptosis
MEK kinase 3	MEKK3, mitogen-activated kinase kinase kinase 3, MAP/ERK kinase kinase 3, MAPKKK3, and MAP3K3	626	71	Ubiquitous	Activating the stress-activated protein kinase (SAPK) and extracellular signal-regulated protein kinase (ERK) pathways; also regulating cell development and growth
Rac1	Ras-related C3 botulinum toxin substrate 1, ρ family small GTP-binding protein Rac1, p21-Rac1	192	21	Ubiquitous	Acting as a Ras superfamily GTPase, and regulating cell proliferation, differentiation, and cytoskeletal reorganization
Cdc42	Cell division cycle 42, GTP-binding protein, 25 kDa, G25K	191	21	Ubiquitous	A small GTPase of the ρ subfamily, regulating cell development, morphology, polarity, migration, transformation, and endocytosis
p21-activated protein kinase	PAK, p21/CDC42/RAC1-activated kinase 1, p21-activated kinase 1, serine/threonine kinase PAK 1, P65 PAK, and α PAK	553	62	Ubiquitous	Serving as a target for the small GTP-binding proteins Cdc42 and Rac; also regulating cytoskeleton reorganization and cell motility

TABLE 5.3. *Continued*

Proteins	Alternative Names	Amino Acids	Molecular Weight (kDa)	Expression	Functions
JNK1/SAPK	c-Jun kinase 1, c-Jun *N*-terminal kinase 1, JNK1 α protein kinase, protein kinase JNK1, mitogen-activated protein kinase 8 (MAPK8), stress-activated protein kinase (SAPK)	427	48	Ubiquitous	Acting as a serine/threonine kinase and key element for MAP kinase signaling pathways, mediating the expression of immediate–early genes in response to cell stimulation, regulating cell development, proliferation, differentiation, migration, and transcription; also regulating UV radiation- and TNF α-induced apoptosis
c-Jun	Jun, protooncogene c Jun, v-jun avian sarcoma virus 17 oncogene homolog	331	36	Ubiquitous	Forming heterodimeric transcriptional factor complexes, known as activating protein (AP)-1, with proteins of the c-Fos family; also regulating cell survival, development, proliferation, differentiation, and apoptosis
Activating transcription factor 2	ATF2, cAMP response element-binding protein 2 (CREB2), cyclic AMP-dependent transcription factor ATF 2	505	55	Brain, uterus, lymphocytes	Acting as a transcriptional factor, forming a homodimer or heterodimer with c-Jun, binding to the cAMP-responsive element (CRE) of target genes, stimulating CRE-dependent transcription, and serving as a histone acetyltransferase (HAT) that acetylates histones H2B and H4 and activating transcription by modulating chromatin components

*Based on bibliography 5.2.

can form heterodimers and homodimers, known as activating proteins (AP)1. AP1 serves as a transcriptional factor, which interacts with AP1-specific *cis* elements and regulates the expression of mitogen genes. In addition to the ERK1/2 and JNK/SAPK pathways, there are several other pathways, which transduce signals from protein tyrosine kinase receptors. These include the p38 MAPK, ERK3, ERK5, and ERK6 pathways. Although different signaling molecules are involved, these pathways follow hierarchical orders similar to the ERK1/2 and JNK/SAPK pathways. (See Table 5.3.)

After extracellular ligand signals are transduced into the cell via corresponding protein tyrosine kinase receptors, the ligand–receptor complexes are clustered and internalized via endocytosis, resulting in the formation of endosomes. Within the endosomes, the ligands are dissociated from the receptors. The dimeric receptors are also split into monomers. The receptor tyrosine kinases are dephosphorylated by phosphatases. Monomeric receptors are recycled back to the cell membrane for further use.

BIBLIOGRAPHY

5.1. Principles of Cell Signaling

Marshall MS: Ras target proteins in eukaryotic cells, *FASEB J* 9:1311–8, 1995.

Johnson GL, Vaillancourt RR: Sequential protein kinase reactions controlling cell growth and differentiation, *Curr Opin Cell Biol* 6:230–8, 1994.

Heldin CH: Protein tyrosine kinase receptor signaling overview, in *Handbook of Cell Signaling*, Bradshaw RA, and Dennis EA, eds, Academic Press, Amsterdam, Vol. 1, 2004, pp 391–396.

Krauss G: *Biochemistry of Signal Transduction and Regulation*, 3rd ed, Wiley-VCH GmbH, 2003.

Mor A, Philips MR: Compartmentalized Ras/MAPK signaling, *Annu Rev Immunol* (in press).

Liu YC, Penninger J, Karin M: Immunity by ubiquitylation: A reversible process of modification, *Nat Rev Immunol* 5:941–52, 2005.

Massague J, Seoane J, Wotton D: Smad transcription factors, *Genes Dev* 19:2783–810, 2005.

Songyang Z, Cantley LC: Recognition and specificity in protein tyrosine kinase-mediated signaling, *Trends Biochem Sci* 20:470–5, 1995.

Kolch W: Coordinating ERK/MAPK signalling through scaffolds and inhibitors, *Nat Rev Mol Cell Biol* 6:827–37, 2005.

Chen ZJ: Ubiquitin signalling in the NF-kappaB pathway, *Nat Cell Biol* 7:758–65, 2005.

Pasquale EB: Eph receptor signalling casts a wide net on cell behaviour, *Nat Rev Mol Cell Biol* 6:462–75, 2005.

Berton G, Mocsai A, Lowell CA: Src and Syk kinases: Key regulators of phagocytic cell activation, *Trends Immunol* 26:208–14, 2005.

Jumaa H, Hendriks RW, Reth M: B cell signaling and tumorigenesis, *Annu Rev Immunol* 23:415–45, 2005.

Blobel CP: ADAMs: Key components in EGFR signalling and development, *Nat Rev Mol Cell Biol* 6:32–43, 2005.

5.2. Protein Tyrosine Kinase-Mediated Cell Signaling

Src

Avizienyte E, Wyke AW, Jones RJ, McLean GW, Westhoff MA, Brunton VG et al: Src-induced de-regulation of E-cadherin in colon cancer cells requires integrin signaling, *Nature Cell Biol* 4:632–8, 2002.

Czernilofsky AP, Levinson AD, Varmus HE, Bishop JM, Tischer E et al: Correction to the nucleotide sequence of the src gene of Rous sarcoma virus, *Nature* 301:736–8, 1983.

Irby RB, Mao W, Coppola D, Kang J, Loubeau JM et al: Activating SRC mutation in a subset of advanced human colon cancers, *Nature Genet* 21:187–90, 1999.

Le Beau MM, Westbrook CA, Diaz MO, Rowley JD: Evidence for two distinct c-src loci on human chromosomes 1 and 20, *Nature* 312:70–1, 1984.

Lowe C, Yoneda T, Boyce BF, Chen H, Mundy GR, Soriano P: Osteopetrosis in Src-deficient mice is due to an autonomous defect of osteoclasts, *Proc Nat Acad Sci USA* 90:4485–9, 1993.

Morris CM, Honeybone LM, Hollings PE, Fitzgerald PH: Localization of the SRC oncogene to chromosome band 20q11.2 and loss of this gene with deletion (20q) in two leukemic patients, *Blood* 74:1768–73, 1989.

Parker RC, Mardon G, Lebo RV, Varmus HE, Bishop JM: Isolation of duplicated human c-src genes located on chromosomes 1 and 20, *Mol Cell Biol* 5:831–8, 1985.

Sakaguchi AY, Mohandas T, Naylor SL: A human c-src gene resides on the proximal long arm of chromosome 20 (cen-q13.1), *Cancer Genet Cytogenet* 18:123–9, 1985.

Soriano P, Montgomery C, Geske R, Bradley A: Targeted disruption of the c-src proto-oncogene leads to osteopetrosis in mice, *Cell* 64:693–702, 1991.

Xing L, Venegas AM, Chen A, Garrett-Beal L, Boyce BF et al: Genetic evidence for a role for Src family kinases in TNF family receptor signaling and cell survival, *Genes Dev* 15:241–53, 2001.

Grb2

Bochmann H, Gehrisch S, Jaross W: The gene structure of the human growth factor bound protein GRB2, *Genomics* 56:203–7, 1999.

Cheng AM, Saxton TM, Sakai R, Kulkarni S, Mbamalu G et al: Mammalian Grb2 regulates multiple steps in embryonic development and malignant transformation, *Cell* 95:793–803, 1998.

Huebner K, Kastury K, Druck T, Salcini Lanfrancone L et al: Chromosome locations of genes encoding human signal transduction adapter proteins, Nck (NCK), Shc (SHC1), and Grb2 (GRB2), *Genomics* 22:281–7, 1994.

Lowenstein EJ, Daly RJ, Batzer AG, Li W, Margolis B et al: The SH2 and SH3 domain-containing protein GRB2 links receptor tyrosine kinases to ras signaling, *Cell* 70:431–42, 1992.

Matuoka K, Shibata M, Yamakawa A, Takenawa T: Cloning of ASH, a ubiquitous protein composed of one Src homology region (SH) 2 and two SH3 domains, from human and rat cDNA libraries, *Proc Natl Acad Sci USA* 89:9015–9, 1992.

Yulug IG, Egan SE, See CG, Fisher EMC: Mapping GRB2, a signal transduction gene in the human and the mouse, *Genomics* 22:313–8, 1994.

Zhang S, Weinheimer C, Courtois M, Kovacs A, Zhang CE et al: The role of the Grb2-p38 MAPK signaling pathway in cardiac hypertrophy and fibrosis, *J Clin Invest* 111:833–41, 2003.

Sos

Bowtell D, Fu P, Simon MA, Senior P: Identification of murine homologues of the Drosophila Son of sevenless gene: Potential activators of ras, *Proc Natl Acad Sci USA* 89:6511–5, 1992.

Chardin P, Camonis JH, Gale NW, Van Aelst L, Schlessinger J et al: Human Sos1: A guanine nucleotide exchange factor for Ras that binds to GRB2, *Science* 260:1338–43, 1993.

Sibilia M, Fleischmann A, Behrens A, Stingl L, Carroll J, Watt FM et al: The EGF receptor provides an essential survival signal for SOS-dependent skin tumor development, *Cell* 102:211–20, 2000.

Soisson SM, Nimnual AS, Uy M, Bar-Sagi D, Kuriyan J: Crystal structure of the Dbl and pleckstrin homology domains from the human Son of sevenless protein, *Cell* 95:259–68, 1998.

Webb GC, Jenkins NA, Largaespada DA, Copeland NG, Fernandez CS et al: Mammalian homologues of the Drosophila Son of sevenless gene map to murine chromosomes 17 and 12 and to human chromosomes 2 and 14, respectively, *Genomics* 18:14–9, 1993.

Crk

Cardoso C, Leventer RJ, Ward HL, Toyo-oka K, Chung J et al: Refinement of a 400-kb critical region allows genotypic differentiation between isolated lissencephaly, Miller-Dieker syndrome, and other phenotypes secondary to deletions of 17p13.3, *Am J Hum Genet* 72:918–30, 2003.

Feller SM, Ren RB, Hanafusa H, Baltimore D: SH2 and SH3 domains as molecular adhesives: The interactions of crk and abl, *Trends Biochem Sci* 19:453–8, 1994.

Fioretos T, Heisterkamp N, Groffen J, Benjes S, Morris C: CRK proto-oncogene maps to human chromosome band 17p13, *Oncogene* 8:2853–5, 1993.

Matsuda M, Tanaka S, Nagata S, Kojima A, Kurata T et al: Two species of human CRK cDNA encode proteins with distinct biological activities, *Mol Cell Biol* 12:3482–9, 1992.

Reichman CT, Mayer BJ, Keshav S, Hanafusa H: The product of the cellular crk gene consists primarily of SH2 and SH3 regions, *Cell Growth Differ* 3:451–60, 1992.

Nck

Chen M, She H, Davis EM, Spicer CM, Kim L et al: Identification of Nck family genes, chromosomal localization, expression, and signaling specificity, *J Biol Chem* 273:25171–8, 1998.

Eden S, Rohatgi R, Podtelejnikov AV, Mann M, Kirschner MW: Mechanism of regulation of WAVE1-induced actin nucleation by Rac1 and Nck, *Nature* 418:790–3, 2002.

Huebner K, Kastury K, Druck T, Salcini AE, Lanfrancone L et al: Chromosome locations of genes encoding human signal transduction adapter proteins, Nck (NCK), Shc (SHC1), and Grb2 (GRB2), *Genomics* 22:281–7, 1994.

Vorobieva N, Protopopov A, Protopopov M, Kashuba V, Allikmets RL et al: Localization of human ARF2 and NCK genes and 13 other NotI-linking clones to chromosome 3 by fluorescence in situ hybridization, *Cytogenet Cell Genet* 68:91–4, 1995.

Shc

Harun RB, Smith KK, Leek JP, Markham AF, Norris A et al: Characterization of human SHC p66 cDNA and its processed pseudogene mapping to Xq12-q13.1, *Genomics* 42:349–52, 1997.

Huebner K, Kastury K, Druck T, Salcini AE, Lanfrancone L et al: Chromosome locations of genes encoding human signal transduction adapter proteins, Nck (NCK), Shc (SHC1), and Grb2 (GRB2), *Genomics* 22:281–7, 1994.

McGlade J, Cheng A, Pelicci G, Pelicci PG, Pawson T: Shc proteins are phosphorylated and regulated by the v-src and v-fps protein-tyrosine-kinases, *Proc Natl Acad Sci USA* 89:8869–73, 1992.

Migliaccio E, Giorgio M, Mele S, Pelicci G, Reboldi P et al: The p66(shc) adaptor protein controls oxidative stress response and life span in mammals, *Nature* 402:309–13, 1999.

Nemoto S, Finkel T: Redox regulation of forkhead proteins through a p66shc-dependent signaling pathway, *Science* 295:2450–2, 2002.

Pelicci G, Lanfrancone L, Grignani F, McGlade J, Cavallo F et al: A novel transforming protein (SHC) with an SH2 domain is implicated in mitogenic signal transduction, *Cell* 70:93–104, 1992.

Yulug IG, Egan SE, See CG, Fisher EMC: A human SHC-related sequence maps to chromosome 17, the SHC gene maps to chromosome 1, *Hum Genet* 96:245–8, 1995.

Zhang L, Camerini V, Bender TP, Ravichandran KS: A nonredundant role for the adapter protein Shc in thymic T cell development, *Nature Immunol* 3:749–55, 2002.

GAP

Friedman E, Gejman PV, Martin GA, McCormick F: Nonsense mutations in the C-terminal SH2 region of the GTPase activating protein (GAP) gene in human tumours, *Nature Genet* 5:242–7, 1993.

Hsieh CL, Vogel US, Dixon RA, Francke U: Chromosome localization and cDNA sequence of murine and human genes for ras p21 GTPase activating protein (GAP), *Somat Cell Mol Genet* 15:579–90, 1989.

Lemons RS, Espinosa R III, Rebentisch M, McCormick F, Ladner M et al: Chromosomal localization of the gene encoding GTPase-activating protein (RASA) to human chromosome 5, bands q13-q15, *Genomics* 6:383–5, 1990.

Mitsudomi T, Friedman E, Gejman PV, McCormick F, Gazdar AF: Genetic analysis of the catalytic domain of the GAP gene in human lung cancer cell lines, *Hum Genet* 93:27–31, 1994.

Trahey M, Wong G, Halenbeck R, Rubinfeld B, Martin GA et al: Molecular cloning of two types of GAP complementary DNA from human placenta, *Science* 242:1697–1700, 1988.

SHP2

Ahmad S, Banville D, Zhao Z, Fischer EH, Shen SH: A widely expressed human protein-tyrosine phosphatase containing src homology 2 domains, *Proc Natl Acad Sci USA* 90:2197–201, 1993.

Dechert U, Duncan AMV, Bastien L, Duff C, Adam M et al: Protein-tyrosine phosphatase SH-PTP2 (PTPN11) is localized to 12q24.1–24.3, *Hum Genet* 96:609–15, 1995.

Hof P, Pluskey S, Dhe-Paganon S, Eck MJ, Shoelson SE: Crystal structure of the tyrosine phosphatase SHP-2, *Cell* 92:441–50, 1998.

Isobe M, Hinoda Y, Imai K, Adachi M: Chromosomal localization of an SH2 containing tyrosine phosphatase (SH-PTP3) gene to chromosome 12q24.1, *Oncogene* 9:1751–3, 1994.

Qu CK, Yu WM, Azzarelli B, Cooper S, Broxmeyer HE et al: Biased suppression of hematopoiesis and multiple developmental defects in chimeric mice containing Shp-2 mutant cells, *Mol Cell Biol* 18:6075–82, 1998.

Saxton TM, Ciruna BG, Holmyard D, Kulkarni S, Harpal K et al: The SH2 tyrosine phosphatase Shp2 is required for mammalian limb development, *Nature Genet* 24:420–3, 2000.

Saxton TM, Henkemeyer M, Gasca S, Shen R, Rossi DJ et al: Abnormal mesoderm patterning in mouse embryos mutant for the SH2 tyrosine phosphatase Shp-2, *EMBO J* 16:2352–64, 1997.

Tartaglia M, Mehler EL, Goldberg R, Zampino G, Brunner HG et al: Mutations in PTPN11, encoding the protein tyrosine phosphatase SHP-2, cause Noonan syndrome, *Nature Genet* 29:465–8, 2001.

PLC γ

Bivona TG, Perez de Castro I, Ahearn IM, Grana TM, Chiu VK et al: Phospholipase C-gamma activates Ras on the Golgi apparatus by means of RasGRP1, *Nature* 424:694–8, 2003.

Bristol A, Hall SM, Kriz RW, Stahl ML, Fan YS et al: Phospholipase C-148: chromosomal location and deletion mapping of functional domains, *Cold Spring Harbor Symp Quant Biol* 53:915–20, 1988.

Nelson KK, Knopf JL, Siracusa LD: Localization of phospholipase C-gamma 1 to mouse chromosome 2, *Mam Genome* 3:597–600, 1992.

Patterson RL, van Rossum DB, Ford DL, Hurt KJ, Bae SS et al: Phospholipase C-gamma is required for agonist-induced Ca(2+) entry, *Cell* 111:529–41, 2002.

Ye K, Aghdasi B, Luo HR, Moriarity JL, Wu FY et al: Phospholipase C-gamma-1 is a physiological guanine nucleotide exchange factor for the nuclear GTPase PIKE, *Nature* 415:541–4, 2002.

PI3 Kinase α

Campbell IG, Russell SE, Choong DYH, Montgomery KG, Ciavarella ML et al: Mutation of the PIK3CA gene in ovarian and breast cancer, *Cancer Res* 64:7678–81, 2004.

Drakas R, Tu X, Baserga R: Control of cell size through phosphorylation of upstream binding factor 1 by nuclear phosphatidylinositol 3-kinase, *Proc Natl Acad Sci USA* 101:9272–6, 2004.

Hiles ID, Otsu M, Volinia S, Fry MJ, Gout I et al: Phosphatidylinositol 3-kinase: Structure and expression of the 110kd catalytic subunit, *Cell* 70:419–29, 1992.

Ma YY, Wei SJ, Lin YC, Lung JC, Chang TC et al: PIK3CA as an oncogene in cervical cancer, *Oncogene* 19:2739–44, 2000.

Samuels Y, Wang Z, Bardelli A, Silliman N, Ptak J et al: High frequency of mutations of the PIK3CA gene in human cancers, *Science* 304:554, 2004.

Shayesteh L, Lu Y, Kuo WL, Baldocchi R et al: PIK3CA is implicated as an oncogene in ovarian cancer, *Nature Genet* 21:99–102, 1999.

Shi SH, Jan LY, Jan YN: Hippocampal neuronal polarity specified by spatially localized mPar3/mPar6 and PI 3-kinase activity, *Cell* 112:63–75, 2003.

Volinia S, Hiles I, Ormondroyd E, Nizetic D, Antonacci R et al: Molecular cloning, cDNA sequence and chromosomal localization of the human phosphatidylinositol 3-kinase p110-alpha (PIK3CA) gene, *Genomics* 24:472–7, 1994.

PI3 Kinase β

Hu P, Mondino A, Skolnik EY, Schlessinger J: Cloning of a novel, ubiquitously expressed human phosphatidylinositol 3-kinase and identification of its binding site on p85, *Mol Cell Biol* 13:7677–88, 1993.

Jackson SP, Schoenwaelder SM, Goncalves I, Nesbitt WS, Yap CL et al: PI 3-kinase p110-beta: A new target for antithrombotic therapy, *Nature Med* 11:507–14, 2005.

Kossila M, Sinkovic M, Karkkainen P, Laukkanen MO, Miettinen R et al: Gene encoding the catalytic subunit p110-beta of human phosphatidylinositol 3-kinase: Cloning, genomic structure, and screening for variants in patients with type 2 diabetes, *Diabetes* 49:1740–3, 2000.

PI3 Kinase γ

Barber DF, Bartolome A, Hernandez C, Flores JM, Redondo C et al: PI3K-gamma inhibition blocks glomerulonephritis and extends lifespan in a mouse model of systemic lupus, *Nature Med* 11:933–5, 2005.

Camps M, Ruckle T, Ji H, Ardissone V, Rintelen F, Shaw J et al: Blockade of PI3K-gamma suppresses joint inflammation and damage in mouse models of rheumatoid arthritis, *Nature Med* 11:936–43, 2005.

Crackower MA, Oudit GY, Kozieradzki I, Sarao R, Sun H et al: Regulation of myocardial contractility and cell size by distinct PI3K-PTEN signaling pathways, *Cell* 110:737–49, 2002.

Hirsch E, Katanaev VL, Garlanda C, Azzolino O, Pirola L et al: Central role for G protein-coupled phosphoinositide 3-kinase gamma in inflammation, *Science* 287:1049–53, 2000.

Jiang K, Zhong B, Gilvary DL, Corliss BC, Hong-Geller E et al: Pivotal role of phosphoinositide-3 kinase in regulation of cytotoxicity in natural killer cells, *Nature Immun* 1:419–25, 2000.

Kratz CP, Emerling BM, Bonifas J, Wang W, Green ED et al: Genomic structure of the PIK3CG gene on chromosome band 7q22 and evaluation as a candidate myeloid tumor suppressor, *Blood* 99:372–4, 2002.

Li Z, Jiang H, Xie W, Zhang Z, Smrcka AV et al: Roles of PLC-beta-2 and -beta-3 and PI3K-gamma in chemoattractant-mediated signal transduction, *Science* 287:1046–9, 2000.

Patrucco E, Notte A, Barberis L, Selvetella G, Maffei A et al: PI3K-gamma modulates the cardiac response to chronic pressure overload by distinct kinase-dependent and -independent effects, *Cell* 118:375–87, 2004.

Sasaki T, Irie-Sasaki J, Horie Y, Bachmaier K, Fata JE et al: Colorectal carcinomas in mice lacking the catalytic subunit of PI(3)K-gamma, *Nature* 406:897–902, 2000.

Sasaki T, Irie-Sasaki J, Jones RG, Oliveira-dos-Santos AJ, Stanford WL et al: Function of PI3K-gamma in thymocyte development, T cell activation, and neutrophil migration, *Science* 287:1040–6, 2000.

Stoyanov B, Volinia S, Hanck T, Rubio I, Loubtchenkov M et al: Cloning and characterization of a G protein-activated human phosphoinositide-3 kinase, *Science* 269:690–3, 1995.

PDGFR β

Francke U, Yang-Feng TL, Brissenden JE, Ullrich A: Chromosomal mapping of genes involved in growth control, *Cold Spring Harbor Symp Quant Biol* 51:855–66, 1986.

Gronwald RGK, Grant FJ, Haldeman BA, Hart CE, O'Hara PJ et al: Cloning and expression of a cDNA coding for the human platelet-derived growth factor receptor: Evidence for more than one receptor class, *Proc Natl Acad Sci USA* 85:3435–9, 1988.

Klinghoffer RA, Mueting-Nelsen PF, Faerman A, Shani M, Soriano P: The two PDGF receptors maintain conserved signaling in vivo despite divergent embryological functions, *Mol Cell* 7:343–54, 2001.

Matsui T, Heidaran M, Miki T, Popescu N, La Rochelle W et al: Isolation of a novel receptor cDNA establishes the existence of two PDGF receptor genes, *Science* 243:800–4, 1989.

Roberts WM, Look AT, Roussel MF, Sherr CJ: Tandem linkage of human CSF-1 receptor (c-fms) and PDGF receptor genes, *Cell* 55:655–61, 1988.

Yarden Y, Escobedo JA, Kuang WJ, Yang-Feng TL, Daniel TO et al: Structure of the receptor for platelet-derived growth factor helps define a family of closely related growth factor receptors, *Nature* 323:226–32, 1986.

H-Ras

Bos JL: ras oncogenes in human cancer: A review, *Cancer Res* 49:4682–9, 1989.

Chang EH, Gonda MA, Ellis RW, Scolnick EM, Lowy DR: Human genome contains four genes homologous to transforming genes of Harvey and Kirsten murine sarcoma viruses, *Proc Natl Acad Sci USA* 79:4848–52, 1982.

Colby WW, Hayflick JS, Clark SG, Levinson AD: Biochemical characterization of polypeptides encoded by mutated human Ha-rasl genes, *Mol Cell Biol* 6:730–4, 1986.

Dajee M, Lazarov M, Zhang JY, Cai T, Green CL et al: NF-kappa-B blockade and oncogenic Ras trigger invasive human epidermal neoplasia, *Nature* 421:639–43, 2003.

Krontiris TG, Devlin B, Karp DD, Robert NJ, Risch N: An association between the risk of cancer and mutations in the HRAS1 minisatellite locus, *New Engl J Med* 329:517–23, 1993.

Krontiris TG, DiMartino NA, Colb M, Parkinson DR: Unique allelic restriction fragments of the human Ha-ras locus in leukocyte and tumour DNAs of cancer patients, *Nature* 313:369–74, 1985.

Mochizuki N, Yamashita S, Kurokawa K, Ohba Y, Nagai T et al: Spatio-temporal images of growth-factor-induced activation of Ras and Rapl, *Nature* 411:1065–8, 2001.

Oft M, Akhurst RJ, Balmain A: Metastasis is driven by sequential elevation of H-ras and Smad2 levels, *Nature Cell Biol* 4:487–94, 2002.

Sears R, Leone G, DeGregori J, Nevins JR: Ras enhances Myc protein stability, *Mol Cell* 3:169–79, 1999.

Seeburg PH, Colby WW, Capon DJ, Goeddel DV, Levinson AD: Biological properties of human c-Ha-rasl genes mutated at codon 12, *Nature* 312:71–5, 1984.

Stallings RL, Crawford BD, Black RJ, Chang EH: Assignment of RAS proto-oncogenes in Chinese hamsters: Implications for mammalian gene linkage conservation and neoplasia, *Cytogenet Cell Genet* 43:2–5, 1986.

Tong L, de Vos AM, Milburn MV, Jancarik J, Noguchi S et al: Structural differences between a RAS oncogene protein and the normal protein, *Nature* 337:90–3, 1989.

Weijzen S, Rizzo P, Braid M, Vaishnav R, Jonkheer SM et al: Activation of Notch-1 signaling maintains the neoplastic phenotype in human Ras-transformed cells, *Nature Med* 8:979–86, 2002.

K-Ras

Almoguera C, Shibata D, Forrester K, Martin J, Arnheim N et al: Most human carcinomas of the exocrine pancreas contain mutant c-K-ras genes, *Cell* 53:549–54, 1988.

Bollag G, Adler F, elMasry N, McCabe PC, Connor E Jr et al: Biochemical characterization of a novel KRAS insertion mutation from a human leukemia, *J Biol Chem* 271:32491–4, 1996.

Burmer GC, Loeb LA: Mutations in the KRAS2 oncogene during progressive stages of human colon carcinoma, *Proc Natl Acad Sci USA* 86:2403–7, 1989.

Capon DJ, Seeburg PH, McGrath JP, Hayflick JS, Edman U et al: Activation of Ki-ras2 gene in human colon and lung carcinomas by two different point mutations, *Nature* 304:507–13, 1983.

Dinulescu DM, Ince TA, Quade BJ, Shafer SA, Crowley D et al: Role of K-ras and Pten in the development of mouse models of endometriosis and endometrioid ovarian cancer, *Nature Med* 11:63–70, 2005.

Feig LA, Bast RC Jr, Knapp RC, Cooper GM: Somatic activation of ras-K gene in a human ovarian carcinoma, *Science* 223:698–701, 1984.

Holland EC, Celestino J, Dai C, Schaefer L, Sawaya RE et al: Combined activation of Ras and Akt in neural progenitors induces glioblastoma formation in mice, *Nature Genet* 25:55–7, 2000.

Johnson L, Mercer K, Greenbaum D, Bronson RT, Crowley D et al: Somatic activation of the K-ras oncogene causes early onset lung cancer in mice, *Nature* 410:1111–6, 2001.

Johnson SM, Grosshans H, Shingara J, Byrom M, Jarvis R et al: RAS is regulated by the let-7 microRNA family, *Cell* 120:635–47, 2005.

Liu E, Hjelle B, Morgan R, Hecht F, Bishop JM: Mutations of the Kirsten-ras proto-oncogene in human preleukaemia, *Nature* 330:186–8, 1987.

McGrath JP, Capon DJ, Smith DH, Chen EY, Seeburg PH et al: Structure and organization of the human Ki-ras proto-oncogene and a related processed pseudogene, *Nature* 304:501–6, 1983.

Porta M, Malats N, Jariod M, Grimalt JO, Rifa J et al: Serum concentrations of organochlorine compounds and K-ras mutations in exocrine pancreatic cancer, *Lancet* 354:2125–9, 1999.

Santos E, Martin-Zanca D, Reddy EP, Pierotti MA, Della Porta G et al: Malignant activation of a K-ras oncogene in lung carcinoma but not in normal tissue of the same patient, *Science* 223:661–4, 1984.

Sidransky D, Tokino T, Hamilton SR, Kinzler KW, Levin B et al: Identification of RAS oncogene mutations in the stool of patients with curable colorectal tumors, *Science* 256:102–5, 1992.

Zhang Z, Wang Y, Vikis HG, Johnson L, Liu G et al: Wildtype Kras2 can inhibit lung carcinogenesis in mice, *Nature Genet* 29:25–33, 2001.

Raf1

Alavi A, Hood JD, Frausto R, Stupack DG, Cheresh DA: Role of Raf in vascular protection from distinct apoptotic stimuli, *Science* 301:94–6, 2003.

Bonner T, O'Brien SJ, Nash WG, Rapp UR, Morton CC et al: The human homologs of the raf (mil) oncogene are located on human chromosomes 3 and 4, *Science* 223:71–4, 1984.

Bonner TI, Kerby SB, Sutrave P, Gunnell MA, Mark G et al: Structure and biological activity of human homologs of the raf/mil oncogene, *Mol Cell Biol* 5:1400–7, 1985.

Bonner TI, Oppermann H, Seeburg P, Kerby SB, Gunnell MA et al: The complete coding sequence of the human raf oncogene and the corresponding structure of the c-raf-1 gene, *Nucleic Acids Res* 14:1009–15, 1986.

Kasid U, Pfeifer A, Weichselbaum RR, Dritschilo A, Mark GE: The raf oncogene is associated with a radiation-resistant human laryngeal cancer, *Science* 237:1039–41, 1987.

O'Neill E, Rushworth L, Baccarini M, Kolch W: Role of the kinase MST2 in suppression of apoptosis by the proto-oncogene product Raf-1, *Science* 306:2267–70, 2004.

Shimizu K, Nakatsu Y, Sekiguchi M, Hokamura K, Tanaka K et al: Molecular cloning of an activated human oncogene, homologous to v-raf, from primary stomach cancer, *Proc Natl Acad Sci USA* 82:5641–5, 1985.

Wang HG, Rapp UR, Reed JC: Bcl-2 targets the protein kinase Raf-1 to mitochondria, *Cell* 87:629–38, 1996.

Yamaguchi O, Watanabe T, Nishida K, Kashiwase K, Higuchi Y et al: Cardiac-specific disruption of the c-raf-1 gene induces cardiac dysfunction and apoptosis, *J Clin Invest* 114:937–43, 2004.

MEK1

Crews CM, Alessandrini A, Erikson RL: The primary structure of MEK, a protein kinase that phosphorylates the ERK gene product, *Science* 258:478–80, 1992.

Favata MF, Horiuchi KY, Manos EJ, Daulerio AJ, Stradley DA et al: Identification of a novel inhibitor of mitogen-activated protein kinase kinase, *J Biol Chem* 273:18623–32, 1998.

Giroux S, Tremblay M, Bernard D, Cadrin-Girard JF, Aubry S et al: Embryonic death of Mek1-deficient mice reveals a role for this kinase in angiogenesis in the labyrinthine region of the placenta, *Curr Biol* 9:369–72, 1999.

Meloche S, Gopalbhai K, Beatty BG, Scherer SW, Pellerin J: Chromosome mapping of the human genes encoding the MAP kinase kinase MEK1 (MAP2K1) to 15q21 and MEK2 (MAP2K2) to 7q32, *Cytogenet Cell Genet* 88:249–52, 2000.

Orth K, Palmer LE, Bao ZQ, Stewart S, Rudolph AE et al: Inhibition of the mitogen-activated protein kinase kinase superfamily by a Yersinia effector, *Science* 285:1920–3, 1999.

Perry RLS, Parker MH, Rudnicki MA: Activated MEK1 binds the nuclear MyoD transcriptional complex to repress transactivation, *Mol Cell* 8:291–301, 2001.

Rampoldi L, Zimbello R, Bortoluzzi S, Tiso N, Valle G et al: Chromosomal localization of four MAPK signaling cascade genes: MEK1, MEK3, MEK4 and MEKK5, *Cytogenet Cell Genet* 78:301–3, 1997.

Sebolt-Leopold JS, Dudley DT, Herrera R, Van Becelaere K, Wiland A et al: Blockade of the MAP kinase pathway suppresses growth of colon tumors in vivo, *Nature Med* 5:810–6, 1999.

Seger R, Krebs EG: The MAPK signaling cascade, *FASEB J* 9:726–35, 1995.

Seger R, Seger D, Lozeman FJ, Ahn NG, Graves LM et al: Human T-cell mitogen-activated protein kinase kinases are related to yeast signal transduction kinases, *J Biol Chem* 267:25628–31, 1992.

Takekawa M, Tatebayashi K, Saito H: Conserved docking site is essential for activation of mammalian MAP kinase kinases by specific MAP kinase kinase kinases, *Mol Cell* 18:295–306, 2005.

Zheng CF, Guan KL: Cloning and characterization of two distinct human extracellular signal-regulated kinase activator kinases, MEK1 and MEK2, *J Biol Chem* 268:11435–9, 1993.

MEK2

Belanger LF, Roy S, Tremblay M, Brott B, Steff AM et al: Mek2 is dispensable for mouse growth and development, *Mol Cell Biol* 23:4778–87, 2003.

Brott BK, Alessandrini A, Largaespada DA, Copeland NG, Jenkins NA et al: MEK2 is a kinase related to MEK1 and is differentially expressed in murine tissues, *Cell Growth Differ* 4:921–9, 1993.

Favata MF, Horiuchi KY, Manos EJ, Daulerio AJ, Stradley DA et al: Identification of a novel inhibitor of mitogen-activated protein kinase kinase, *J Biol Chem* 273:18623–32, 1998.

Meloche S, Gopalbhai K, Beatty BG, Scherer SW, Pellerin J: Chromosome mapping of the human genes encoding the MAP kinase kinase MEK1 (MAP2K1) to 15q21 and MEK2 (MAP2K2) to 7q32, *Cytogenet Cell Genet* 88:249–52, 2000.

Orth K, Palmer LE, Bao ZQ, Stewart S, Rudolph AE et al: Inhibition of the mitogen-activated protein kinase kinase superfamily by a Yersinia effector, *Science* 285:1920–3, 1999.

Zheng CF, Guan KL: Cloning and characterization of two distinct human extracellular signal-regulated kinase activator kinases, MEK1 and MEK2, *J Biol Chem* 268:11435–9, 1993.

MEK3

Derijard B, Raingeaud J, Barrett T, Wu IH, Han J et al: Independent human MAP kinase signal transduction pathways defined by MEK AND MKK isoforms, *Science* 267:682–5, 1995.

Han J, Wang X, Jiang Y, Ulevitch RJ, Lin S: Identification and characterization of a predominant isoform of human MKK3, *FEBS Lett* 403:19–22, 1997.

Lu HT, Yang DD, Wysk M, Gatti E, Mellman I et al: Defective IL-12 production in mitogen-activated protein (MAP) kinase kinase 3 (Mkk3)-deficient mice, *EMBO J* 18:1845–57, 1999.

Orth K, Palmer LE, Bao ZQ, Stewart S, Rudolph AE et al: Inhibition of the mitogen-activated protein kinase kinase superfamily by a Yersinia effector, *Science* 285:1920–3, 1999.

Rampoldi L, Zimbello R, Bortoluzzi S, Tiso N, Valle G et al: Chromosomal localization of four MAPK signaling cascade genes: MEK1, MEK3, MEK4 and MEKK5, *Cytogenet Cell Genet* 78:301–3, 1997.

MEK4

Ganiatsas S, Kwee L, Fujiwara Y, Perkins A, Ikeda T et al: SEK1 deficiency reveals mitogen-activated protein kinase cascade crossregulation and leads to abnormal hepatogenesis, *Proc Natl Acad Sci USA* 95:6881–6, 1998.

Lin A, Minden A, Martinetto H, Claret FX, Lange-Carter C et al: Identification of a dual specificity kinase that activates the Jun kinases and p38-Mpk2, *Science* 268:286–90, 1995.

Rampoldi L, Zimbello R, Bortoluzzi S, Tiso N, Valle G et al: Chromosomal localization of four MAPK signaling cascade genes: MEK1, MEK3, MEK4 and MEKK5, *Cytogenet Cell Genet* 78:301–3, 1997.

White RA, Hughes RT, Adkison LR, Bruns G, Zon LI: The gene encoding protein kinase SEK1 maps to mouse chromosome 11 and human chromosome 17, *Genomics* 34:430–2, 1996.

Wu Z, Wu J, Jacinto E, Karin M: Molecular cloning and characterization of human JNKK2, a novel jun NH(2)-terminal kinase-specific kinase, *Mol Cell Biol* 17:7407–16, 1997.

ERK1/2

Seger R, Krebs EG: The MAPK signaling cascade, *FASEB J* 9:726–35, 1995.

Charest DL, Mordret G, Harder KW, Jirik F, Pelech SL: Molecular cloning, expression, and characterization of the human mitogen-activated protein kinase p44erk1, *Mol Cell Biol* 13:4679–90, 1993.

Forcet C, Stein E, Pays L, Corset V, Llambi F et al: Netrin-1-mediated axon outgrowth requires deleted in colorectal cancer-dependent MAPK activation, *Nature* 417:443–7, 2002.

Li L, Wysk M, Gonzalez FA, Davis RJ: Genomic loci of human mitogen-activated protein kinases, *Oncogene* 9:647–9, 1994.

Mazzucchelli C, Vantaggiato C, Ciamei A, Fasano S, Pakhotin P et al: Knockout of ERK1 MAP kinase enhances synaptic plasticity in the striatum and facilitates striatal-mediated learning and memory, *Neuron* 34:807–20, 2002.

Pages G, Guerin S, Grall D, Bonino F, Smith A et al: Defective thymocyte maturation in p44 MAP kinase (Erk 1) knockout mice, *Science* 286:1374–8, 1999.

Saba-El-Leil MK, Malo D, Meloche S: Chromosomal localization of the mouse genes encoding the ERK1 and ERK2 isoforms of MAP kinases, *Mam Genome* 8:141–2, 1997.

Waskiewicz AJ, Flynn A, Proud CG, Cooper JA: Mitogen-activated protein kinases activate the serine/threonine kinases Mnk1 and Mnk2, *EMBO J* 16:1909–20, 1997.

ERK2

Bhalla US, Ram PT, Iyengar R: MAP kinase phosphatase as a locus of flexibility in a mitogen-activated protein kinase signaling network, *Science* 297:1018–23, 2002.

Boulton TG, Nye SH, Robbins DJ, Ip NY, Radziejewska E et al: ERKs: A family of protein-serine/threonine kinases that are activated and tyrosine phosphorylated in response to insulin and NGF, *Cell* 65:663–75, 1991.

Khokhlatchev AV, Canagarajah B, Wilsbacher J, Robinson M, Atkinson M et al: Phosphorylation of the MAP kinase ERK2 promotes its homodimerization and nuclear translocation, *Cell* 93:605–15, 1998.

Li L, Wysk M, Gonzalez FA, Davis RJ: Genomic loci of human mitogen-activated protein kinases, *Oncogene* 9:647–9, 1994.

Saba-El-Leil MK, Malo D, Meloche S: Chromosomal localization of the mouse genes encoding the ERK1 and ERK2 isoforms of MAP kinases, *Mam Genome* 8:141–2, 1997.

Seger R, Krebs EG: The MAPK signaling cascade, *FASEB J* 9:726–35, 1995.

Stefanovsky VY, Pelletier G, Hannan R, Gagnon-Kugler T, Rothblum LI et al: An immediate response of ribosomal transcription to growth factor stimulation in mammals is mediated by ERK phosphorylation of UBF, *Mol Cell* 8:1063–73, 2001.

c-Fos

Bakin AV, Curran T: Role of DNA 5-methylcytosine transferase in cell transformation by fos, *Science* 283:387–90, 1999.

Barker PE, Rabin M, Watson M, Breg WR, Ruddle FH et al: Human c-fos oncogene mapped within chromosomal region 14q21-q31, *Proc Natl Acad Sci USA* 81:5826–30, 1984.

David JP, Mehic D, Bakiri L, Schilling AF, Mandic V et al: Essential role of RSK2 in c-Fos-dependent osteosarcoma development, *J Clin Invest* 115:664–72, 2005.

Dony C, Gruss P: Proto-oncogene c-fos expression in growth regions of fetal bone and mesodermal web tissue, *Nature* 328:711–4, 1987.

Glover JNM, Harrison SC: Crystal structure of the heterodimeric bZIP transcription factor c-Fos–c-Jun bound to DNA, *Nature* 373:257–61, 1995.

Grigoriadis AE, Wang ZQ, Cecchini MG, Hofstetter W, Felix R et al: c-Fos: A key regulator of osteoclast-macrophage lineage determination and bone remodeling, *Science* 266:443–8, 1994.

Grigoriadis AE, Wang ZQ, Wagner EF: Fos and bone cell development: Lessons from a nuclear oncogene, *Trends Genet* 11:436–41, 1995.

Ruther U, Garber C, Komitowski D, Muller R, Wagner EF: Deregulated c-fos expression interferes with normal bone development in transgenic mice, *Nature* 325:412–6, 1987.

Saez E, Rutberg SE, Mueller E, Oppenheim H, Smoluk J et al: c-fos is required for malignant progression of skin tumors, *Cell* 82:721–32, 1995.

van Straaten F, Muller R, Curran T, Van Beveren C, Verma IM: Complete nucleotide sequence of a human c-onc gene: Deduced amino acid sequence of the human c-fos protein, *Proc Natl Acad Sci USA* 80:3183–7, 1983.

Wang ZQ, Ovitt C, Grigoriadis AE, Mohle-Steinlein U, Ruther U et al: Bone and haematopoietic defects in mice lacking c-fos, *Nature* 360:741–5, 1992.

Zhang J, Zhang D, McQuade JS, Behbehani M, Tsien JZ et al: c-fos regulates neuronal excitability and survival, *Nature Genet* 30:416–20, 2002.

CREB

Barco A, Alarcon JM, Kandel ER: Expression of constitutively active CREB protein facilitates the late phase of long-term potentiation by enhancing synaptic capture, *Cell* 108:689–703, 2002.

Barton K, Muthusamy N, Chanyangam M, Fischer C, Clendenin C et al: Defective thymocyte proliferation and IL-2 production in transgenic mice expressing a dominant-negative form of CREB, *Nature* 379:81–5, 1996.

Chen AE, Ginty DD, Fan CM: Protein kinase A signalling via CREB controls myogenesis induced by Wnt proteins, *Nature* 433:317–22, 2005.

Cole TJ, Copeland NG, Gilbert DJ, Jenkins NA, Schutz G, Ruppert S: The mouse CREB (cAMP responsive element binding protein) gene: Structure promoter analysis, and chromosomal localization, *Genomics* 13:974–82, 1992.

Euskirchen G, Royce TE, Bertone P, Martone R, Rinn JL et al: CREB binds to multiple loci on human chromosome 22, *Mol Cell Biol* 24:3804–14, 2004.

Fentzke RC, Korcarz CE, Lang RM, Lin H, Leiden JM: Dilated cardiomyopathy in transgenic mice expressing a dominant-negative CREB transcription factor in the heart, *J Clin Invest* 101:2415–26, 1998.

Herzig S, Hedrick S, Morantte I, Koo SH, Galimi F et al: CREB controls hepatic lipid metabolism through nuclear hormone receptor PPAR-gamma, *Nature* 426:190–3, 2003.

Herzig S, Long F, Jhala US, Hedrick S, Quinn R et al: CREB regulates hepatic gluconeogenesis through the coactivator PGC-1, *Nature* 413:179–83, 2001.

Hoeffler JP, Meyer TE, Yun Y, Jameson JL, Habener JF: Cyclic AMP-responsive DNA-binding protein: Structure based on a cloned placental cDNA, *Science* 242:1430–3, 1988.

Mantamadiotis T, Lemberger T, Bleckmann SC, Kern H, Kretz O et al: Disruption of CREB function in brain leads to neurodegeneration, *Nature Genet* 31:47–54, 2002.

Parker D, Ferreri K, Nakajima T, LaMorte VJ, Evans R et al: Phosphorylation of CREB at ser-133 induces complex formation with CREB-binding protein via a direct mechanism, *Mol Cell Biol* 16:694–703, 1996.

Radhakrishnan I, Perez-Alvarado GC, Parker D, Dyson HJ, Montminy MR et al: Solution structure of the KIX domain of CBP bound to the transactivation domain of CREB: A model for activator: coactivator interactions, *Cell* 91:741–52, 1997.

Rudolph D, Tafuri A, Gass P, Hammerling GJ, Arnold B et al: Impaired fetal T cell development and perinatal lethality in mice lacking the cAMP response element binding protein, *Proc Natl Acad Sci USA* 95:4481–6, 1998.

Taylor AK, Klisak I, Mohandas T, Sparkes RS, Li C et al: Assignment of the human gene for CREB1 to chromosome 2q32.3-q34, *Genomics* 7:416–21, 1990.

Egr1

Ayadi A, Zheng H, Sobieszczuk P, Buchwalter G, Moerman P et al: Net-targeted mutant mice develop a vascular phenotype and up-regulate egr-1, *EMBO J* 20:5139–52, 2001.

de Belle I, Huang RP, Fan Y, Liu C, Mercola D et al: p53 and Egr-1 additively suppress transformed growth in HT1080 cells but Egr-1 counteracts p53-dependent apoptosis, *Oncogene* 18:3633–42, 1999.

Fahmy RG, Dass CR, Sun LQ, Chesterman CN, Khachigian LM: Transcription factor Egr-1 supports FGF-dependent angiogenesis during neovascularization and tumor growth, *Nature Med* 9:1026–32, 2003.

Liao Y, Shikapwashya ON, Shteyer E, Dieckgraefe BK, Hruz PW et al: Delayed hepatocellular mitotic progression and impaired liver regeneration in early growth response-1-deficient mice, *J Biol Chem* 279:43107–16, 2004.

Liu C, Adamson E, Mercola D: Transcription factor EGR-1 suppresses the growth and transformation of human HT-1080 fibrosarcoma cells by induction of transforming growth factor beta-1, *Proc Natl Acad Sci USA* 93:11831–6, 1996.

Liu C, Yao J, de Belle I, Huang RP, Adamson E et al: The transcription factor EGR-1 suppresses transformation of human fibrosarcoma HT1080 cells by coordinated induction of transforming growth factor-beta-1, fibronectin, and plasminogen activator inhibitor-1, *J Biol Chem* 274:4400–11, 1999.

Virolle T, Krones-Herzig A, Baron V, De Gregorio G, Adamson ED et al: Ergl promotes growth and survival of prostate cancer cells: Identification of novel Egrl target genes, *J Biol Chem* 278:11802–10, 2003.

Elk1

Giovane A, Sobieszczuk P, Mignon C, Mattei MG, Wasylyk B: Locations of the ets subfamily members net, elk1, and sapl (ELK3, ELK1, and ELK4) on three homologous regions of the mouse and human genomes, *Genomics* 29:769–72, 1995.

Janz M, Lehmann U, Olde Weghuis D, de Leeuw B, Geurts van Kessel A et al: Refined mapping of the human Ets-related gene Elk-1 to Xp11.2-p11.4, distal to the OATL1 region, *Hum Genet* 94:442–4, 1994.

Rao VN, Huebner K, Isobe M, ar-Rushdi A, Croce CM et al: Elk, tissue-specific ets-related genes on chromosomes X and 14 near translocation breakpoints, *Science* 244:66–70, 1989.

Tamai Y, Taketo M, Nozaki M, Seldin MF: Mouse Elk oncogene maps to chromosome X and a novel Elk oncogene (Elk3) maps to chromosome 10, *Genomics* 26:414–6, 1995.

Wang Z, Wang DZ, Hockemeyer D, McAnally J et al: Myocardin and ternary complex factors compete for SRF to control smooth muscle gene expression, *Nature* 428:185–9, 2004.

MEKK2

Cheng J, Zhang D, Kim K, Zhao Y, Zhao Y, Su B: Mipl, an MEKK2-interacting protein, controls MEKK2 dimerization and activation, *Mol Cell Biol* 25:5955–64, 2005.

Raviv Z, Kalie E, Seger R: MEK5 and ERK5 are localized in the nuclei of resting as well as stimulated cells, while MEKK2 translocates from the cytosol to the nucleus upon stimulation, *J Cell Sci* 117:1773–1784, 2004.

Hammaker DR, Boyle DL, Chabaud-Riou M, Firestein GS: Regulation of c-Jun N-terminal kinase by MEKK-2 and mitogen-activated protein kinase kinase kinases in rheumatoid arthritis, *J Immunol* 172:1612–8, 2004.

MEKK3

Ellinger-Ziegelbauer H, Brown K, Kelly K, Siebenlist U: Direct activation of the stress-activated protein kinase (SAPK) and extracellular signal-regulated protein kinase (ERK) pathways by an inducible mitogen-activated protein kinase/ERK kinase kinase 3 (MEKK) derivative, *J Biol Chem* 272:2668–74, 1997.

Yang J, Boerm M, McCarty M, Bucana C, Fidler IJ et al: Mekk3 is essential for early embryonic cardiovascular development, *Nature Genet* 24:309–13, 2000.

Rac1

Chan AY, Coniglio SJ, Chuang YY, Michaelson D, Knaus UG et al: Roles of the Rac1 and Rac3 GTPases in human tumor cell invasion, *Oncogene* 24:7821–9, 2005.

Zhang B, Yang L, Zheng Y: Novel intermediate of Rac GTPase activation by guanine nucleotide exchange factor, *Biochem Biophys Res Commun* 331:413–21, 2005.

Papaharalambus C, Sajjad W, Syed A, Zhang C, Bergo MO et al: Tumor necrosis factor alpha stimulation of Racl activity. Role of isoprenylcysteine carboxylmethyltransferase, *J Biol Chem* 280:18790–6, 2005.

Kanekura K, Hashimoto Y, Kita Y, Sasabe J, Aiso S et al: A Racl/phosphatidylinositol 3-kinase/Akt3 anti-apoptotic pathway, triggered by AlsinLF, the product of the ALS2 gene, antagonizes Cu/Zn-superoxide dismutase (SOD1) mutant-induced motoneuronal cell death, *J Biol Chem* 280:4532–43, 2005.

Fera E, O'Neil C, Lee W, Li S, Pickering JG: Fibroblast growth factor-2 and remodeled type I collagen control membrane protrusion in human vascular smooth muscle cells: Biphasic activation of Racl, *J Biol Chem* 279:35573–82, 2004.

Meriane M, Charrasse S, Comunale F, Mery A, Fort P et al: Participation of small GTPases Racl and Cdc42Hs in myoblast transformation, *Oncogene* 21:2901–7, 2002.

Cdc42

Erickson JW, Zhang C, Kahn RA, Evans T, Cerione RA: Mammalian cdc42 is a brefeldin A-sensitive component of the Golgi apparatus, *J Biol Chem* 271:26850–4, 1996.

Etienne-Manneville S, Hall A: Cdc42 regulates GSK-3-beta and adenomatous polyposis coli to control cell polarity, *Nature* 421:753–6, 2003.

Garrett WS, Chen LM, Kroschewski R, Ebersold M, Turley S et al: Developmental control of endocytosis in dendritic cells by Cdc42, *Cell* 102:325–34, 2000.

Irie F, Yamaguchi Y: EphB receptors regulate dendritic spine development via intersectin, Cdc42 and N-WASP, *Nature Neurosci* 5:1117–18, 2002.

Jensen SJ, Sulman EP, Maris JM, Matise TC, Vojta PJ et al: An integrated transcript map of human chromosome 1p35-p36, *Genomics* 42:126–36, 1997.

Manser E, Leung T, Salihuddin H, Tan L, Lim L: A non-receptor tyrosine kinase that inhibits the GTPase activity of p21cdc42, *Nature* 363:364–7, 1993.

Marks PW, Kwiatkowski DJ: Genomic organization and chromosomal location of murine Cdc42, *Genomics* 38:13–8, 1996.

Munemitsu S, Innis MA, Clark R, McCormick F, Ullrich A et al: Molecular cloning and expression of a G25K cDNA, the human homolog of the yeast cell cycle gene CDC42, *Mol Cell Biol* 10:5977–82, 1990.

Nakaya Y, Kuroda S, Katagiri YT, Kaibuchi K, Takahashi Y: Mesenchymal-epithelial transition during somitic segmentation is regulated by differential roles of Cdc42 and Racl, *Dev Cell* 7:425–38, 2004.

Nalbant P, Hodgson L, Kraynov V, Toutchkine A, Hahn KM: Activation of endogenous Cdc42 visualized in living cells, *Science* 305:1615–9, 2004.

Nicole S, White PS, Topaloglu H, Beigthon P, Salih M et al: The human CDC42 gene: Genomic organization, evidence for the existence of a putative pseudogene and exclusion as a SJS1 candidate gene, *Hum Genet* 105:98–103, 1999.

Schuler GD, Boguski MS, Stewart EA, Stein LD, Gyapay G et al: A gene map of the human genome, *Science* 274:540–6, 1996.

Sin WC, Haas K, Ruthazer ES, Cline HT: Dendrite growth increased by visual activity requires NMDA receptor and Rho GTPases, *Nature* 419:475–80, 2002.

Wu WJ, Erickson JW, Lin R, Cerione RA: The gamma-subunit of the coatomer complex binds Cdc42 to mediate transformation, *Nature* 405:800–4, 2000.

Yasuda S, Oceguera-Yanez F, Kato T, Okamoto M, Yonemura S et al: Cdc42 and mDia3 regulate microtubule attachment to kinetochores, *Nature* 428:767–71, 2004.

PAK

Bekri S, Adelaide J, Merscher S, Grosgeorge J, Caroli-Bosc F et al: Detailed map of a region commonly amplified at 11q13-q14 in human breast carcinoma, *Cytogenet Cell Genet* 79:125–31, 1997.

Brown JL, Stowers L, Baer M, Trejo J, Coughlin S et al: Human Ste20 homologue hPAK1 links GTPases to the JNK MAP kinase pathway, *Curr Biol* 6:598–605, 1996.

Knaus UG, Morris S, Dong HJ, Chernoff J, Bokoch GM: Regulation of human leukocyte p21-activated kinases through G protein-coupled receptors, *Science* 269:221–3, 1995.

Lei M, Lu W, Meng W, Parrini MC, Eck MJ et al: Structure of PAK1 in an autoinhibited conformation reveals a multistage activation switch, *Cell* 102:387–97, 2000.

Parrini MC, Lei M, Harrison SC, Mayer BJ: Pak1 kinase homodimers are autoinhibited in trans and dissociated upon activation by Cdc42 and Rac1, *Mol Cell* 9:73–83, 2002.

Sanders LC, Matsumura F, Bokoch GM, de Lanerolle P: Inhibition of myosin light chain kinase by p21-activated kinase, *Science* 283:2083–5, 1999.

Vadlamudi RK, Li F, Adam L, Nguyen D, Ohta Y et al: Filamin is essential in actin cytoskeletal assembly mediated by p21-activated kinase 1, *Nature Cell Biol* 4:681–90, 2002.

JNK/SAPK

Chang NS, Doherty J, Ensign A: JNK1 physically interacts with WW domain-containing oxidoreductase (WOX1) and inhibits WOX1-mediated apoptosis, *J Biol Chem* 278:9195–202, 2003.

Derijard B, Hibi M, Wu IH, Barrett T, Su B et al: JNK1: A protein kinase stimulated by UV light and Ha-Ras that binds and phosphorylates the c-Jun activation domain, *Cell* 76:1025–37, 1994.

De Smaele E, Zazzeroni F, Papa S, Nguyen DU, Jin R et al: Induction of gadd45-beta by NF-kappa-B downregulates pro-apoptotic JNK signaling, *Nature* 414:308–13, 2001.

Dong C, Yang DD, Tournier C, Whitmarsh AJ, Xu J et al: JNK is required for effector T-cell function but not for T-cell activation, *Nature* 405:91–4, 2000.

Dong C, Yang DD, Wysk M, Whitmarsh AJ, Davis RJ et al: Defective T cell differentiation in the absence of Jnk1, *Science* 282:2092–5, 1998.

Gupta S, Barrett T, Whitmarsh AJ, Cavanagh J, Sluss HK et al: Selective interaction of JNK protein kinase isoforms with transcription factors, *EMBO J* 15:2760–70, 1996.

Hess P, Pihan G, Sawyers CL, Flavell RA, Davis RJ: Survival signaling mediated by c-Jun NH2-terminal kinase in transformed B lymphoblasts, *Nature Genet* 32:201–5, 2002.

Hirosumi J, Tuncman G, Chang L, Gorgun CZ, Uysal KT et al: A central role for JNK in obesity and insulin resistance, *Nature* 420:333–6, 2002.

Huang C, Rajfur Z, Borchers C, Schaller MD, Jacobson K: JNK phosphorylates paxillin and regulates cell migration, *Nature* 424:219–23, 2003.

Kuan CY, Yang DD, Roy DRS, Davis RJ, Rakic P et al: The Jnk1 and Jnk2 protein kinases are required for regional specific apoptosis during early brain development, *Neuron* 22:667–76, 1999.

Kyriakis JM, Banerjee P, Nikolakaki E, Dai T, Rubie EA et al: The stress-activated protein kinase subfamily of c-Jun kinases, *Nature* 369:156–60, 1994.

Tang G, Minemoto Y, Dibling B, Purcell NH, Li Z et al: Inhibition of JNK activation through NF-kappa-B target genes, *Nature* 414:313–7, 2001.

Tournier C, Hess P, Yang DD, Xu J, Turner TK et al: Requirement of JNK for stress-induced activation of the cytochrome c-mediated death pathway, *Science* 288:870–4, 2000.

Urano F, Wang X, Bertolotti A, Zhang Y, Chung P et al: Coupling of stress in the ER to activation of JNK protein kinases by transmembrane protein kinase IRE1, *Science* 287:664–6, 2000.

Weston CR, Wong A, Hall JP, Goad MEP, Flavell RA et al: The c-Jun NH2-terminal kinase is essential for epidermal growth factor expression during epidermal morphogenesis, *Proc Natl Acad Sci USA* 101:14114–9, 2004.

c-Jun

Behrens A, Sibilia M, Wagner EF: Amino-terminal phosphorylation of c-Jun regulates stress-induced apoptosis and cellular proliferation, *Nature Genet* 21:326–9, 1999.

Bohmann D, Bos TJ, Admon A, Nishimura T, Vogt PK et al: Human proto-oncogene c-jun encodes a DNA binding protein with structural and functional properties of transcription factor AP-1, *Science* 238:1386–92, 1987.

Bos TJ, Bohmann D, Tsuchie H, Tjian R, Vogt PK: v-jun encodes a nuclear protein with enhancer binding properties of AP-1, *Cell* 52:705–12, 1988.

Eferl R, Ricci R, Kenner L, Zenz R, David JP et al: Liver tumor development: c-Jun antagonizes the proapoptotic activity of p53, *Cell* 112:181–92, 2003.

Gao M, Labuda T, Xia Y, Gallagher E, Fang D et al: Jun turnover is controlled through JNK-dependent phosphorylation of the E3 ligase Itch, *Science* 306:271–5, 2004.

Haluska FG, Huebner K, Isobe M, Nishimura T, Croce CM et al: Localization of the human JUN protooncogene to chromosome region 1p31–32, *Proc Natl Acad Sci USA* 85:2215–8, 1988.

Hilberg F, Aguzzi A, Howells N, Wagner EF: c-Jun is essential for normal mouse development and hepatogenesis, *Nature* 365:179–81, 1993.

Lamph WW, Wamsley P, Sassone-Corsi P, Verma IM: Induction of proto-oncogene JUN/AP-1 by serum and TPA, *Nature* 334:629–31, 1988.

Mathas S, Hinz M, Anagnostopoulos I, Krappmann D, Lietz A et al: Aberrantly expressed c-Jun and JunB are a hallmark of Hodgkin lymphoma cells, stimulate proliferation and synergize with NF-kappa-B, *EMBO J* 21:4104–13, 2002.

Mattei MG, Simon-Chazottes D, Hirai SI, Ryseck RP et al: Chromosomal localization of the three members of the jun proto-oncogene family in mouse and man, *Oncogene* 5:151–6, 1990.

Nateri AS, Riera-Sans L, Da Costa C, Behrens A: The ubiquitin ligase SCF(Fbw7) antagonizes apoptotic JNK signaling, *Science* 303:1374–8, 2004.

Shaulian E, Karin M: AP-1 as a regulator of cell life and death, *Nature Cell Biol* 4:E131–6, 2002.

Shaulian E, Schreiber M, Piu F, Beeche M, Wagner EF et al: The mammalian UV response: c-Jun induction is required for exit from p53-imposed growth arrest, *Cell* 103:897–907, 2000.

Wertz IE, O'Rourke KM, Zhang Z, Dornan D, Arnott D et al: Human de-etiolated-1 regulates c-Jun by assembling a CUL4A ubiquitin ligase, *Science* 303:1371–4, 2004.

Whitfield J, Neame SJ, Paquet L, Bernard O, Ham J: Dominant-negative c-Jun promotes neuronal survival by reducing BIM expression and inhibiting mitochondrial cytochrome c release, *Neuron* 29:629–43, 2001.

ATF2

Bhoumik A, Takahashi S, Breitweiser W, Shiloh Y, Jones N et al: ATM-dependent phosphorylation of ATF2 is required for the DNA damage response, *Mol Cell* 18:577–87, 2005.

Diep A, Li C, Klisak I, Mohandas T, Sparkes RS et al: Assignment of the gene for cyclic AMP-response element binding protein 2 (CREB2) to human chromosome 2q24.1-q32, *Genomics* 11:1161–3, 1991.

Kawasaki H, Schiltz L, Chiu R, Itakura K, Taira K et al: ATF-2 has intrinsic histone acetyltransferase activity which is modulated by phosphorylation, *Nature* 405:195–200, 2000.

Maekawa T, Sakura H, Kanei-Ishii C, Sudo T, Yoshimura T et al: Leucine zipper structure of the protein CRE-BP1 binding to the cyclic AMP response element in brain, *EMBO J* 8:2023–8, 1989.

Ozawa K, Sudo T, Soeda EI, Yoshida MC, Ishii S: Assignment of the human CREB2 (CRE-BP1) gene to 2q32, *Genomics* 10:1103–4, 1991.

Human protein reference data base, Johns Hopkins University and the Institute of Bioinformatics, at http://www.hprd.org/protein.

NONRECEPTOR TYROSINE KINASE-MEDIATED CELL SIGNALING

Structure and Function [5.3]

Nonreceptor tyrosine kinases belong to a group of cytoplasmic tyrosine kinases that are attached to the cell membrane but are not intrinsic kinases of any cell membrane receptors. The Src family protein kinases are typical nonreceptor tyrosine kinases that have been well characterized. The Src protein was originally discovered in the Rous sarcoma retrovirus by Dr. Peyton Rous in 1911, for which Dr. Rous won the Nobel Prize in 1966. The viral Src is referred to as *v-Src*, which is responsible for the induction of mesodermal cancers. Further work has demonstrated that a gene similar to the v-Src gene exists in chickens and mammals. The protein encoded by this gene is defined as *c-Src* (cellular Src) in chicken and mammalian cells. In normal cells, the c-Src protein has been shown to regulate cell proliferation and differentiation, contributing to the control of morphogenesis during development. Investigations with molecular approaches have revealed a number of protein tyrosine kinases that are similar to Src in structure and function. These include Yes, Fgr, Lck, Fyn, Yrk, Hck, Lyn, and Blk, which are defined as members of the Src family. Among these proteins, Src, Fyn, Yes, and Yrk are expressed in most cell types, whereas others are expressed primarily in hematopoietic cells. (See list in Table 5.4.)

The Src family proteins are characterized by the presence of several distinct domains, including an *N*-terminal domain with one or more acylation sites, a Src homology (SH)3 domain, a SH2 domain, a catalytic kinase domain, and a *C*-terminal regulatory domain with a tyrosine at location 527 (Tyr527). The *N*-terminal acylation sites anchor the protein to the cell membrane via a myristoyl group (Fig. 5.6), the SH3 and SH2 domains are capable of binding to praline-rich peptides and phosphorylated tyrosine residues, respectively, and the *C*-terminal tyrosine residue regulates the activity of the Src kinases.

Signaling Mechanisms [5.4]

In the quiescent state, the activity of the Src family tyrosine kinases is suppressed mostly as a result of phosphorylation of the *C*-terminal tyrosine residue Tyr527. The phosphorylation of Tyr527 is induced by the Csk kinase. The phosphorylated tyrosine (pTyr527) interacts with the SH2 domain of the same molecule. This interaction renders the SH2 domain of the Src kinase incapable of binding to a phosphorylated tyrosine from a stimulatory factor (Fig. 5.6).

A Src tyrosine kinase can be activated under several conditions. First, dephosphorylation of the *C*-terminal tyrosine (Tyr527) by a protein tyrosine phosphatase prevents the interaction of the *C*-terminal tyrosine with the SH2 domain and allows the access of a stimulatory factor to the SH2 domain of the Src tyrosine kinase, thus inducing Src activation. However, protein tyrosine phosphatases specific to the *C*-terminal tyrosine residue have not been identified. Second, the presence of proteins with phosphorylated tyrosine residues may competitively bind the SH2 domain of the Src tyrosine kinase, thus facilitating Src activation. A number of growth factor receptors, such insulin-like growth factor receptor, platelet-derived growth factor receptor, and epidermal growth factor receptor, contain tyrosine residues in their intracellular domains. Once activated by the binding of growth factors, these tyrosine residues can be autophosphorylated and can serve as docking sites for SH2 domain-containing Src kinases, recruiting the Src kinases to the phosphorylated tyrosine residues. The recruitment process activates Src kinases (Fig. 5.7).

TABLE 5.4. Characteristics of Selected Members of the Src Family*

Proteins	Alternative Names	Amino Acids	Molecular Weight (kDa)	Expression	Functions
Src	Avian sarcoma virus protein (ASV), p60-Src, tyrosine kinase pp60c-src, tyrosine protein kinase SRC-1, protooncogene SRC	536	60	Ubiquitous	A tyrosine kinase (the cellular homolog of the avian sarcoma virus protein) that regulates cell survival, proliferation, migration, and cell–cell communication
Yes	cellular yes-1 protein, Yamaguchi sarcoma oncogene, v-yes-1 Yamaguchi sarcoma viral oncogene homolog 1, protooncogene tyrosine-protein kinase YES, and P61-YES	543	61	Nervous system, lung, liver, intestine, kidney, skeletal muscle, skin	Acting as a tyrosine kinase that is cellular homolog of Yamaguchi sarcoma virus oncogene product, and regulating cell survival, proliferation, and differentiation
Fgr	c-fgr, p55c-fgr, Gardner–Rasheed feline sarcoma viral oncogene homolog, oncogene FGR, and SRC2	529	59	Brain, lung, liver, kidney, leukocytes	Regulating cell survival, proliferation, and differentiation
Lck	T-cell-specific protein tyrosine kinase, lymphocyte-specific protein tyrosine kinase, P56-LCK, and oncogene LCK	509	58	Lymphocytes, lung, thymus, intestine, bone marrow	Regulating lymphocyte development and proliferation, mediating immune reactions, and contributing to development of lymphomas
Fyn	Src-like kinase, c-syn protooncogene, src/yes-related novel gene, and protooncogene tyrosine-protein kinase fyn	537	61	Nervous system, skin, lymphocytes	Binding to p85 subunit of phosphatidylinositol 3-kinase, mediating axon outgrowth, and regulating cell development, proliferation, and differentiation
Hck	Hemopoietic cell kinase, tyrosine protein kinase HCK	526	60	Bone marrow, blood cells, lung	A hematopoietic cell protein tyrosine kinase that regulates blood cell development, survival, proliferation, and apoptosis, and mediates leukocyte activation and migration
Lyn	v-yes-1 Yamaguchi sarcoma viral-related oncogene homolog, and oncogene LYN	512	59	Bone marrow, leukocytes, platelet, macrophage	Regulating cytoskeletal organization and remodeling in platelets, mediating lymphocyte proliferation and activation, and regulating immune reactions
Blk	B-lymphocyte tyrosine kinase, p55-BLK, and BLK nonreceptor tyrosine kinase	505	58	B lymphocytes, thymus, liver	Regulating development and activity of B lymphocytes

*Based on bibliography 5.3.

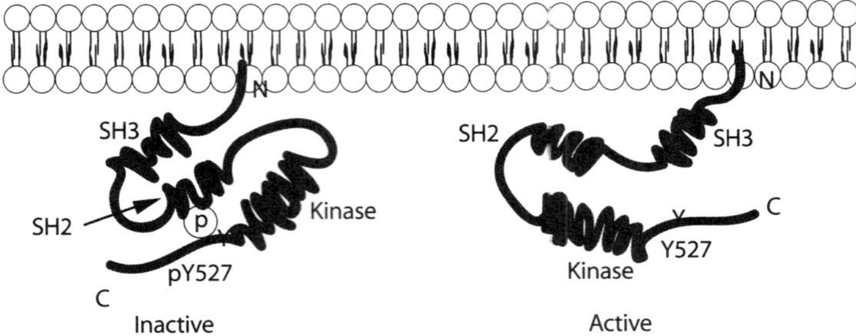

Figure 5.6. Schematic representation of the Src protein tyrosine kinase anchored to the cytoplasmic side of the cell membrane (based on bibliography 5.4).

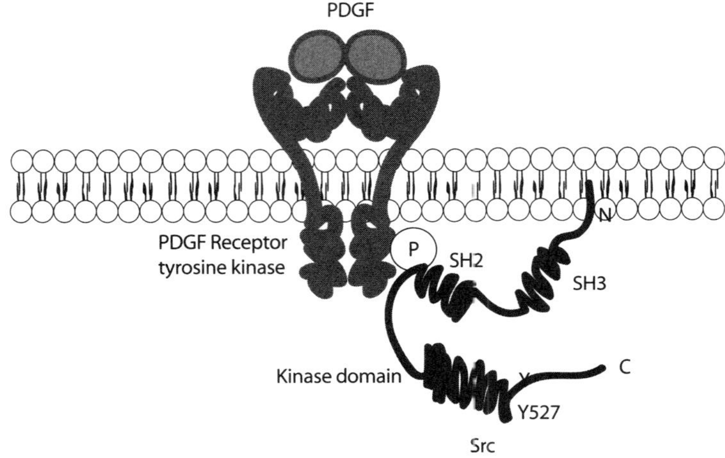

Figure 5.7. Activation of the Src protein tyrosine kinase (based on bibliography 5.4).

Previous investigations have shown that a treatment of quiescent fibroblasts with PDGF induces the activation of the Src tyrosine kinase, suggesting that the PDGF receptor may stimulate the phosphorylation of the Src kinase. Further investigations have demonstrated that several growth factor receptors, including the PDGF receptor, EGF receptor, basic FGF receptor, and colony-stimulating factor 1 receptor, are capable of interacting with the Src family tyrosine kinases, including Src, Fyn, and Yes. A Src kinase can interact with a growth factor receptor at specific sites of phosphotyrosine residues. For example, activated PDGF receptor can recruit Fyn to phosphotyrosines 579 and 581, which are located in the juxtamembrane region of the receptor. It is now clear that the formation of receptor–Src kinase complexes is a critical process for the activation of PDGF-induced cellular activities. Phosphorylated Src tyrosine kinases can activate a number of substrate proteins, including GTPase activating protein, focal adhesion kinase (FAK) (Table 5.5), and the adaptor protein Shc. These Src target proteins play critical roles in the regulation of mitogenic cellular activities.

The Src tyrosine kinases also play a role in signal transduction initiated from the extracellular matrix. Cells can interact with extracellular matrix components via integrins, a family of heterodimeric transmembrane receptors. Integrins can cluster with a number

TABLE 5.5. Characteristics of FAK*

Proteins	Alternative Names	Amino Acids	Molecular Weight (kDa)	Expression	Functions
Focal adhesion kinase	FAK, focal adhesion kinase 1 (FAK1), protein tyrosine kinase 2, and pp125FAK	1074	122	Ubiquitous	Found at cell focal adhesion contacts, interacting with membrane integrins, mediating cell adhesion to extracellular matrix, and regulating cell polarity, motility, and proliferation

*Based on bibliography 5.4.

of molecules, including talin, vinculin, actinin, paxillin, and focal adhesion kinase, forming a cell membrane structure known as the *focal adhesion contact*. Among the focal contact molecules, focal adhesion kinase plays a key role in regulating cell adhesion, migration, and proliferation. On interaction of integrins with extracellular matrix, focal adhesion kinase becomes autophosphorylated on a tyrosine residue, establishing a docking site for SH2 domain-containing proteins. Src and Fyn, which contain a SH2 domain, can bind to the phosphorylated tyrosine residue of focal adhesion kinase. Bound Src kinases can phosphorylate additional tyrosine residues in the focal adhesion kinase molecule, further activating focal adhesion kinase and creating docking sites for other SH2 domain-containing molecules. An important molecule that binds to focal adhesion kinase is the Grb2 adaptor protein. This molecule forms a complex with Sos, which activates the Ras–MAPK signaling pathway (see page 151). Thus, the Src tyrosine kinase signaling pathway is linked to the MAPK signaling pathway via focal adhesion kinase. These observations demonstrate that, through interactions with growth factor receptors and integrins, the Src family tyrosine kinases participate in the regulation of cell proliferation, adhesion, and migration.

SERINE/THREONINE KINASE-MEDIATED CELL SIGNALING

Serine/threonine protein kinases are enzymes that catalyze the transfer of the γ-phosphate of an ATP molecule to the OH group of the serine and threonine residues of substrate proteins. Protein serine/threonine kinases exist in inactive and active forms. In unstimulated cells, most protein kinases are inactive. On stimulation by specific signals, protein kinases can be activated by several mechanisms. These include activator binding, phosphorylation of activation sites, and dephosphorylation of inhibitory phosphates. There are a large number of serine/threonine protein kinases, among which receptor serine/threonine kinases, protein kinase A (cAMP-dependent protein kinase or PKA), and protein kinase C (Ca^{2+}-dependent protein kinase or PKC) have been extensively studied and well characterized. These kinases are used as examples in this section to demonstrate the mechanisms of serine/threonine kinase-mediated cell signaling.

Serine/Threonine Kinase Receptors [5.5]

Serine/threonine kinase receptors are transmembrane receptors that contain an intrinsic serine/threonine-specific kinase, which is located in the cytoplasmic domain of the receptors. These receptors can interact with several growth factors, including transforming growth factor (TGF)β, bone morphogenetic proteins, and activins. A typical serine/threonine receptor contains two subunits: receptor type I and type II. Both subunits are required for the activation of the receptor. An activated receptor serine/threonine kinase can lead to activation of a cascade of intracellular signaling molecules, including Smad 2, 3, and 4, and adaptor proteins. Smads are a family of proteins homologous to the products of the *Drosophila* gene *mothers against decapentaplegic* (*Mad*) and the C. elegans gene *Sma*. These proteins form complexes and serve as transcriptional factors and regulate the expression of target genes.

The signaling mechanisms of serine/threonine kinase receptors are similar among transforming growth factor (TGF)β, bone morphogenetic proteins, and activins. Here, the signal transduction pathway for TGFβ is used to demonstrate the signaling mechanisms (Fig. 5.8). TGFβ is capable of binding to the TGFβ receptor type II (serine/threonine

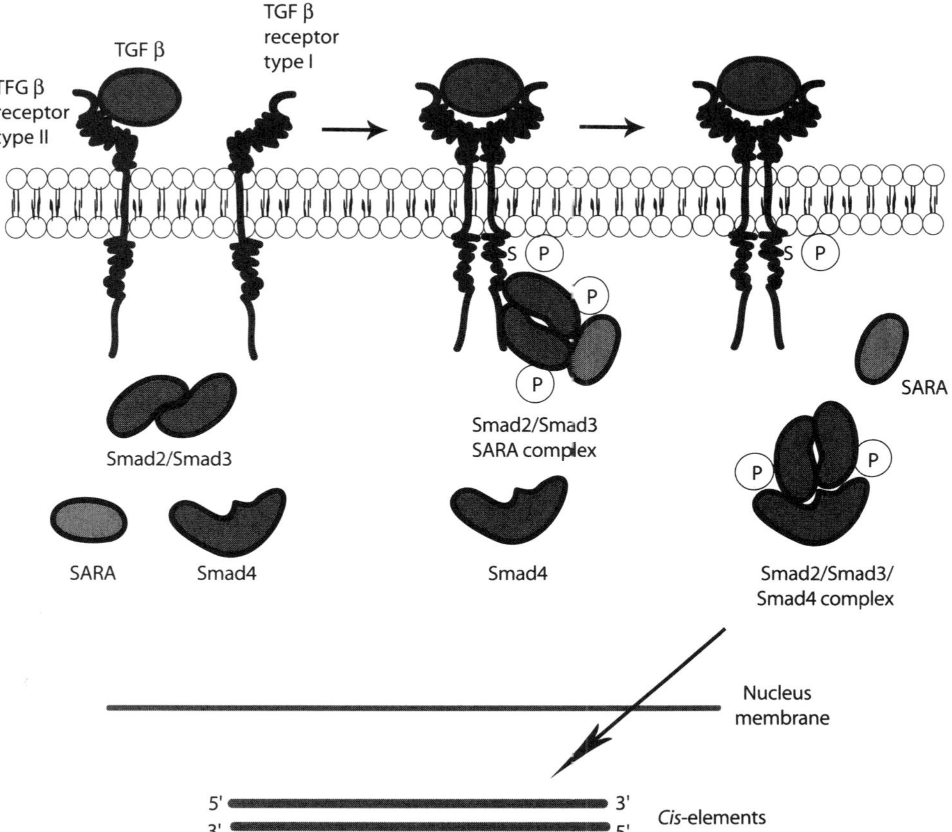

Figure 5.8. Schematic representation of transforming growth factor (TGF)-β-induced cell signaling (based on bibliography 5.5).

kinase receptor type II). The binding of TGFβ induces the association of the TGFβ receptor type I (serine/threonine kinase receptor type I) with the type II receptor, allowing the phosphorylation of the type I receptor by the type II receptor kinase. A complex of Smads, including Smad2 and Smad3, is activated in response to the activation of the TGFβ receptors and initiate interaction with an adapter protein, known as *SMAD anchor for receptor activation* (*SARA*), forming a protein complex. The complex of Smad2, Smad3, and SARA is recruited to TGFβ receptor type I, which in turn phosphorylates the Smad complex. Phosphorylated Smad complex then dissociates from the type I receptor and SARA, and binds to another Smad molecule, known as *Smad4*, which is a critical molecule mediating the translocation of the Smad complex. The Smad2/Smad3/Smad4 complex serves as a transcriptional factor, translocates from the cytoplasm to the cell nucleus, interacts with target genes, and induces gene expression. The TGFβ-activated Smad signaling pathways (see list in Table 5.6) negatively regulate cell proliferation and differentiation by activating inhibitors of cyclin-dependent kinases, in resulting cell cycle arrest in the Gl phase. Activation of the Smad signaling pathways also induces cell apoptosis.

Protein Kinase A [5.6]

Protein kinase A (PKA) (see list of isoforms in Table 5.7) is the primary target of cAMP and is involved in the regulation of sugar and lipid metabolism, ion channel activities, and nerve synaptic transduction. In unstimulated cells, protein kinase A exists in the form of tetramer, composed of two regulatory R subunits and two catalytic C subunits. The catalytic C subunits are masked by the R subunits. In response to stimulation, activated cAMP can bind to the R subunits, resulting in the dissociation of the tetramer into an R—R dimer, which is bound to four cAMP molecules, and two free C monomers. The C monomers are catalytically active and can phosphorylate transcription factors, such as CREB (cyclic AMP response element-binding protein). Phosphorylated transcriptional factors can translocate from the cytoplasm to the cell nucleus, bind to target gene promoters, and stimulate the transcription of target genes (Fig. 5.9).

In mammalian cells, there are four types of R-subunit isoforms, including RIα, RIβ, RIIα, and RIIβ, and three types of C isoforms, including Cα, Cβ, and Cγ. These isoforms mediate distinct biochemical activities, contributing to cell- and tissue-specific functions. The C isoforms contain serine/threonine phosphorylation sites. These sites can be autophosphorylated on stimulation. For all protein kinase A isoforms, there exist a consensus sequence, RRXSX, which catalyzes the phosphorylation of substrate proteins.

The activity of protein kinase A is regulated by several mechanisms. These include alterations in cAMP concentration, phosphorylation on the serine/threonine residues of protein kinase A, and binding of inhibitor proteins. It is important to note that the cAMP concentration is a primary factor that controls the activity of protein kinase A. An increase in cAMP concentration induces activation of protein kinase A. Phosphorylation of protein kinase A enhances the activity of the kinase, whereas the binding of inhibitor proteins exerts an opposite effect.

Protein Kinase C [5.7]

Protein kinase C (calcium-dependent protein kinase or PKC) (see list of isoforms in Table 5.8) is a critical signaling molecule, which is involved in the regulation of cellular activities, including cell proliferation, migration, apoptosis, and secretion. Protein kinase C can

TABLE 5.6. Characteristics of Selected Molecules for the TGFβ Receptor–Smad Signaling Pathway*

Proteins	Alternative Names	Amino Acids	Molecular Weight (kDa)	Expression	Functions
TGF β	TGFB1, transforming growth factor β1	390	44	Lung, kidney, liver	Regulating cell proliferation, differentiation, transformation, and apoptosis; mediating inflammatory reactions, and acting as a negative regulator for certain cell types including vascular smooth muscle cells
TGF β receptor type I	TGFBR1, TGF β receptor 1, TGF-β type I receptor, activin receptor-like kinase 5, serine-threonine protein kinase receptor R4	503	56	Ubiquitous	Forming a heterogenic complex with TGF β receptor type II in response to the binding of TGF β, relaying the TGF β signal from cell surface to cytoplasm, and regulating cell proliferation, differentiation, and apoptosis
TGF β receptor type II	TGF β receptor 2, TGFBR2, and TGFR-2	567	65	Ubiquitous	Forming a heterogenic complex with TGF β receptor type I in response to TGF β binding, relaying the TGF β signal from cell surface to cytoplasm, and regulating cell proliferation, differentiation, and apoptosis
Smad2	SMAD mothers against decapentaplegic homolog 2 (*Drosophila*)	467	52	Fetus, bone, pancreas, ovary, intestine, prostate gland, and skin	Forming complexes with Smad3 and Smad4, serving as a transcriptional factor, mediating signal transduction initiated by TGF β, activin, and bone morphogenetic factors, and regulating cell proliferation, differentiation, and apoptosis
Smad3	SMAD mothers against decapentaplegic homolog 3 (*Drosophila*), SMA- and MAD-related protein 3	425	48	Fetus, bone, pancreas, ovary, intestine, prostate gland, skin	Similar to function of Smad2
Smad4	SMAD mothers against decapentaplegic homolog 4 (*Drosophila*), SMA- and MAD-related protein 4	552	60	Fetus, bone, pancreas, ovary, intestine, prostate gland, skin	Similar to function of Smad2

*Based on bibliography 5.5.

TABLE 5.7. Characteristics of Selected Protein Kinase A Isoforms*

Proteins	Alternative Names	Amino Acids	Molecular Weight (kDa)	Expression	Functions
Protein kinase A RIα	cAMP-dependent protein kinase regulatory type I α, CAMP-dependent protein kinase type I-α regulatory chain, and tissue-specific extinguisher 1	381	43	Heart, liver, lymphocytes	Forming a tetramer with other protein kinase A subunits, transducing the cAMP signal to target proteins by phosphorylation, and regulating cell metabolism and activities
Protein kinase A RIβ	cAMP-dependent protein kinase regulatory type I β, cAMP-dependent protein kinase type I β regulatory chain	381	43	T lymphocytes	Similar to functions of protein kinase A RIα̃
Protein kinase A RIIα	cAMP-dependent protein kinase regulatory type II α	404	46	T lymphocytes, testis	Similar to functions of protein kinase A RIα
Protein kinase A RIIβ	cAMP-dependent protein kinase regulatory type II β, RIIβ	418	46	T lymphocytes, testis	Similar to functions of protein kinase A RIα
Protein kinase A Cα	cAMP-dependent protein kinase α catalytic subunit, PKA C α, PKACA, and A kinase α	351	41	T lymphocytes, testis	Forming a tetramer with other protein kinase A subunits, transducing cAMP signal to target proteins by phosphorylation, and regulating cell metabolism and activities
Protein kinase A Cβ	cAMP-dependent protein kinase β catalytic subunit, PKA Cβ	398	46	Brain	Similar to functions of protein kinase A Cα
Protein kinase A Cγ	cAMP-dependent protein kinase γ catalytic subunit, PKA Cγ	351	40	Testis	Similar to functions of protein kinase A Cα

*Based on bibliography 5.6.

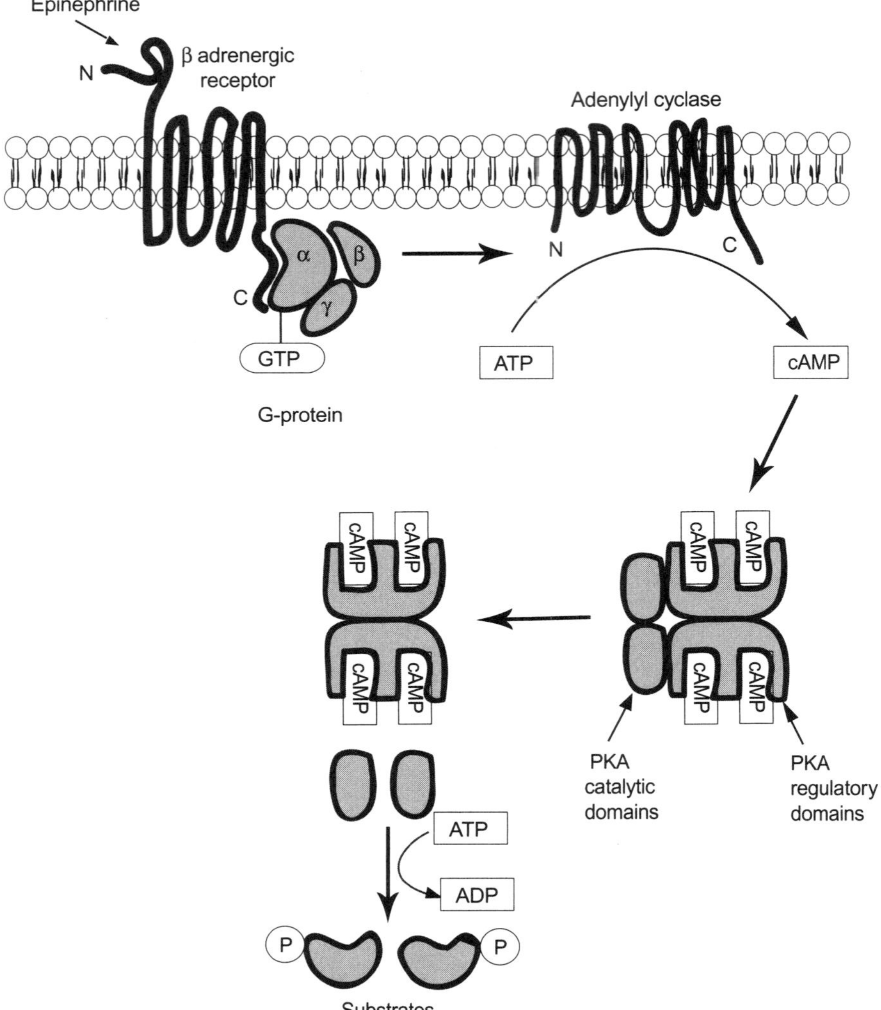

Figure 5.9. Schematic representation of activation of protein kinase A (PKA). Based on bibliography 5.6.

be activated by signals from the receptor protein tyrosine kinase pathways and the G-protein-linked receptor pathways. Protein kinase C represents a family of at least 12 subtypes of protein kinase, including PKCα, β_I, β_{II}, γ, δ, ε, η, θ, ζ, ι, μ, and ν. These subtypes exhibit different distributions in tissues. The α, δ, and ζ subtypes are widely distributed among almost all tissues, whereas others are found in specialized tissues.

Various subtypes of protein kinase C exhibit different characteristics in regulatory mechanisms. Some subtypes, such as PKCα, β_I, β_{II}, and γ, can be activated by diacylglycerol and Ca^{2+}, whereas others such as PKCζ and PKCι cannot be activated by these factors. A common feature for most PKC subtypes is the affinity and responsiveness to phorbol esters, which induce PKC activation. It is thought that tumor promoting esters, such as tetradecanoyl phorbol acetate (TPA), stimulate cell proliferation via the activation of PKC.

TABLE 5.8. Characteristics of Selected Protein Kinase C Isoforms*

Proteins	Alternative Names	Amino Acids	Molecular Weight (kDa)	Expression	Functions
Protein kinase Cα	PKCα	672	77	Pancreas, intestine, skeletal muscle	A serine/threonine-specific protein kinase that can be activated by calcium and diacylglycerol, inducing phosphorylation of target proteins, regulating cell proliferation, migration, and transformation; also mediating cardiac contractility
Protein kinase Cβ	PKCβ, PKCB, PKC beta	673	77	Widely expressed	Similar to functions of PKCα
Protein kinase Cγ	PKCγ	697	78	Nervous system	Similar to functions of PKCα

*Based on bibliography 5.7.

The activity of PKC is mediated by several mechanisms, including phosphorylation, cell membrane association, and Ca^{2+} and diacylglycerol binding (see page 217). Protein kinase C can be phosphorylated on the serine/threonine and tyrosine residues, a critical process inducing PKC activation. Such a process can be catalyzed by phosphoinositide-dependent protein kinases. PKC association with cell membrane is another approach for inducing PKC activation. Such a process is mediated by Ca^{2+} and diacylglycerol. Binding of Ca^{2+} and diacylglycerol to PKC promotes PKC association to cell membrane and thus activates PKC.

BIBLIOGRAPHY

5.3. Structure and Function

Src

Avizienyte E, Wyke AW, Jones RJ, McLean GW, Westhoff MA et al: Src-induced de-regulation of E-cadherin in colon cancer cells requires integrin signaling, *Nature Cell Biol* 4:632–8, 2002.

Azarnia R, Reddy S, Kmiecik TE, Shalloway D, Loewenstein WR: The cellular src gene product regulates junctional cell-to-cell communication, *Science* 239:398–401, 1988.

Czernilofsky AP, Levinson AD, Varmus HE, Bishop JM et al: Correction to the nucleotide sequence of the src gene of Rous sarcoma virus, *Nature* 301:736–8, 1983.

Gibbs CP, Tanaka A, Anderson SK, Radul J, Baar J et al: Isolation and structural mapping of a human c-src gene homologous to the transforming gene (v-src) of Rous sarcoma virus, *J Virol* 53:19–24, 1985.

Irby RB, Mao W, Coppola D, Kang J, Loubeau JM et al: Activating SRC mutation in a subset of advanced human colon cancers, *Nature Genet* 21:187–90, 1999.

Le Beau MM, Westbrook CA, Diaz MO, Rowley JD: c-src is consistently conserved in the chromosomal deletion (20q) observed in myeloid disorders, *Proc Natl Acad Sci USA* 82:6692–6, 1985.

Le Beau MM, Westbrook CA, Diaz MO, Rowley JD: Evidence for two distinct c-src loci on human chromosomes 1 and 20, *Nature* 312:70–1, 1984.

Lowe C, Yoneda T, Boyce BF, Chen H, Mundy GR et al: Osteopetrosis in Src-deficient mice is due to an autonomous defect of osteoclasts, *Proc Natl Acad Sci USA* 90:4485–9, 1993.

Parker RC, Mardon G, Lebo RV, Varmus HE, Bishop JM: Isolation of duplicated human c-src genes located on chromosomes 1 and 20, *Mol Cell Biol* 5:831–8, 1985.

Sakaguchi AY, Mohandas T, Naylor SL: A human c-src gene resides on the proximal long arm of chromosome 20 (cen-q13.1), *Cancer Genet Cytogenet* 18:123–9, 1985.

Soriano P, Montgomery C, Geske R, Bradley A: Targeted disruption of the c-src proto-oncogene leads to osteopetrosis in mice, *Cell* 64:693–702, 1991.

Xing L, Venegas AM, Chen A, Garrett-Beal L, Boyce BF et al: Genetic evidence for a role for Src family kinases in TNF family receptor signaling and cell survival, *Genes Dev* 15:241–53, 2001.

Yes

Overhauser J, Mewar R, Rojas K, Lia K, Kline AD et al: STS map of genes and anonymous DNA fragments on human chromosome 18 using a panel of somatic cell hybrids, *Genomics* 15:387–91, 1993.

Semba K, Yamanashi Y, Nishizawa M, Sukegawa J, Yoshida M et al: Location of the c-yes gene on the human chromosome and its expression in various tissues, *Science* 227:1038–40, 1985.

Silverman GA, Kuo, WL, Taillon-Miller P, Gray JW: Chromosomal reassignment: YACs containing both YES1 and thymidylate synthase map to the short arm of chromosome 18, *Genomics* 15:442–5, 1993.

Sukegawa J, Semba K, Yamanashi Y, Nishizawa M, Miyajima N et al: Characterization of cDNA clones for the human c-yes gene, *Mol Cell Biol* 7:41–7, 1987.

Lck

Anderson SJ, Levin SD, Perlmutter RM: Protein tyrosine kinase p56(lck) controls allelic exclusion of T-cell receptor beta-chain genes, *Nature* 365:552–4, 1993.

Goldman FD, Ballas ZK, Schutte BC, Kemp J, Hollenback C et al: Defective expression of p56lck in an infant with severe combined immunodeficiency, *J Clin Invest* 102:421–9, 1998.

Levin SD, Anderson SJ, Forbush KA, Perlmutter RM: A dominant-negative transgene defines a role for p56lck in thymopoiesis, *EMBO J* 12:1671–80, 1993.

Marth JD, Disteche C, Pravtcheva D, Ruddle F, Krebs EG et al: Localization of a lymphocyte-specific protein tyrosine kinase gene (lck) at a site of frequent chromosomal abnormalities in human lymphomas, *Proc Natl Acad Sci USA* 83:7400–4, 1986.

Marth JD, Overell RW, Meier KE, Krebs EG, Perlmutter RM: Translational activation of the lck proto-oncogene, *Nature* 332:171–3, 1988.

Molina TJ, Kishihara K, Siderovski DP, van Ewijk W, Narendran A et al: Profound block in thymocyte development in mice lacking p56‘lck’, *Nature* 357:161–4, 1992.

Seddon B, Legname G, Tomlinson P, Zamoyska R: Long-term survival but impaired homeostatic proliferation of naive T cells in the absence of p56(lck), *Science* 290:127–31, 2000.

Volpi EV, Romani M, Siniscalco M: Subregional mapping of the human lymphocyte-specific protein tyrosine kinase gene (LCK) to 1p35-p34.3 and its position relative to the 1p marker D1S57, *Cytogenet Cell Genet* 67:187–9, 1994.

Welte T, Leitenberg D, Dittel BN, al-Ramadi BK, Xie B et al: STAT5 interaction with the T cell receptor complex and stimulation of T cell proliferation, *Science* 283:222–5, 1999.

Fyn

Grant SGN, O'Dell TJ, Karl KA, Stein PL, Soriano P et al: Impaired long-term potentiation, spatial learning, and hippocampal development in fyn mutant mice, *Science* 258:1903–10, 1992.

Liu G, Beggs H, Jurgensen C, Park HT, Tang H et al: Netrin requires focal adhesion kinase and Src family kinases for axon outgrowth and attraction, *Nature Neurosci* 7:1222–32, 2004.

Popescu NC, Kawakami T, Matsui T, Robbins KC: Chromosomal localization of the human FYN gene, *Oncogene* 1:449–51, 1987.

Saijo K, Schmedt C, Su I, Karasuyama H, Lowell CA et al: Essential role of Src-family protein tyrosine kinases in NF-kappa-B activation during B cell development, *Nature Immun* 4:274–9, 2003.

Semba K, Nishizawa M, Miyajima N, Yoshida MC, Sukegawa J et al: yes-related protooncogene, syn, belongs to the protein-tyrosine kinase family, *Proc Natl Acad Sci USA* 83:5459–63, 1986.

Yagi T, Aizawa S, Tokunaga T, Shigetani Y, Takeda N et al: A role for Fyn tyrosine kinase in the suckling behaviour of neonatal mice, *Nature* 366:742–5, 1993.

Fgr

Dracopoli NC, Stanger BZ, Lager M, Housman DE: Localization of the FGR protooncogene on the genetic linkage map of human chromosome 1p, *Genomics* 3:124–8, 1988.

Le Beau MM, Westbrook CA, Diaz MO, Rowley JD: Evidence for two distinct c-src loci on human chromosomes 1 and 20, *Nature* 312:70–1, 1984.

Nishizawa M, Semba K, Yoshida MC, Yamamoto T, Sasaki M et al: Structure, expression, and chromosomal location of the human c-fgr gene, *Mol Cell Biol* 6:511–7, 1986.

Parker RC, Mardon G, Lebo RV, Varmus HE, Bishop JM: Isolation of duplicated human c-src genes located on chromosomes 1 and 20, *Mol Cell Biol* 5:831–8, 1985.

Tronick SR, Popescu NC, Cheah MSC, Swan DC, Amsbaugh SC et al: Isolation and chromosomal localization of the human fgr protooncogene, a distinct member of the tyrosine kinase gene family, *Proc Natl Acad Sci USA* 82:6595–9, 1985.

Hck

Ziegler SF, Marth JD, Lewis DB, Perlmutter RM: Novel protein-tyrosine kinase gene (hck) preferentially expressed in cells of hematopoietic origin, *Mol Cell Biol* 7:2276–85, 1987.

Podar K, Mostoslavsky G, Sattler M, Tai YT, Hayashi T et al: Critical role for hematopoietic cell kinase (Hck)-mediated phosphorylation of Gab1 and Gab2 docking proteins in interleukin 6-induced proliferation and survival of multiple myeloma cells, *J Biol Chem* 279:21658–65, 2004.

Shivakrupa R, Radha V, Sudhakar Ch, Swarup G: Physical and functional interaction between Hck tyrosine kinase and guanine nucleotide exchange factor C3G results in apoptosis, which is independent of C3G catalytic domain, *J Biol Chem* 278:52188–94, 2003.

Lyn

Brunati AM, Deana R, Folda A, Massimino ML, Marin O et al: Thrombin-induced tyrosine phosphorylation of HS1 in human platelets is sequentially catalyzed by Syk and Lyn tyrosine kinases and associated with the cellular migration of the protein, *J Biol Chem* 280:21029–35, 2005.

Baran CP, Tridandapani S, Helgason CD, Humphries RK, Krystal G et al: The inositol 5′-phosphatase SHIP-1 and the Src kinase Lyn negatively regulate macrophage colony-stimulating factor-induced Akt activity, *J Biol Chem* 278:38628–36, 2003.

Suzuki-Inoue K, Tulasne D, Shen Y, Bori-Sanz T, Inoue O et al: Association of Fyn and Lyn with the proline-rich domain of glycoprotein VI regulates intracellular signaling, *J Biol Chem* 277:21561–6, 2002.

Harder KW, Parsons LM, Armes J, Evans N, Kountouri N et al: Gain- and loss-of-function Lyn mutant mice define a critical inhibitory role for Lyn in the myeloid lineage, *Immunity* 15:603–15, 2001.

Hibbs ML, Tarlinton DM, Armes J, Grail D, Hodgson G et al: Multiple defects in the immune system of Lyn-deficient mice, culminating in autoimmune disease, *Cell* 83:301–11, 1995.

Parravicini V, Gadina M, Kovarova M, Odom S, Gonzalez-Espinosa C et al: Fyn kinase initiates complementary signals required for IgE-dependent mast cell degranulation, *Nature Immun* 3:741–8, 2002.

Saijo K, Schmedt C, Su I, Karasuyama H, Lowell CA et al: Essential role of Src-family protein tyrosine kinases in NF-kappa-B activation during B cell development, *Nature Immun* 4:274–9, 2003.

Blk

Drebin JA, Hartzell SW, Griffin C, Campbell MJ, Niederhuber JE: Molecular cloning and chromosomal localization of the human homologue of a B-lymphocyte specific protein tyrosine kinase (blk), *Oncogene* 10:477–86, 1995.

Dymecki S, Niederhuber J, Desiderio S: Specific expression of a novel tyrosine kinase gene, Blk, in B lymphoid cells, *Science* 247:332–6, 1990.

Kozak CA, Dymecki SM, Niederhuber JE, Desiderio SV: Genetic mapping of the gene for a novel tyrosine kinase, Blk, to mouse chromosome 14, *Genomics* 9:762–4, 1991.

Saijo K, Schmedt C, Su I, Karasuyama H, Lowell CA et al: Essential role of Src-family protein tyrosine kinases in NF-kappa-B activation during B cell development, *Nature Immun* 4:274–9, 2003.

Human protein reference data base, Johns Hopkins University and the Institute of Bioinformatics, at http://www.hprd.org/protein.

5.4. Signaling Mechanisms

Pelicci G, Lanfrancone L, Grignani F, McGlade J, Cavallo F et al: A novel transforming protein (SHC) with an SH2 domain is implicated in mitogenic signal transduction, *Cell* 70:93–104, 1992.

Chang JH, Wilson LK, Moyers JS, Zhang K, Parsons SJ: Increased level of p21 ras-GTP and enhanced DNA synthesis accompany elevated tyrosal phosphorylation of GAP-associated proeins, p190 and p62 in c-src overexpressors, *Oncogene* 8:959–67, 1993.

Sicheri F, Moarefi I, Kuriyan J: Crystal structure of the Src-family tyrosine kinase Hck, *Nature* 385:602–9, 1997.

Moarefi I, LaFevre-Bernt M, Sicheri F, Huse M, Lee CH et al: Activation of the Src family tyrosine kinase Hck by SH3 domain displacement, *Nature* 385:650–3, 1997.

Sicheri F, Kuriyan J: Structures of Src-family tyrosine kinases, *Curr Opin Struct Biol* 7:777–85, 1997.

Martin GS: The hunting of the Src, *Nature Rev Mol Cell Biol* 2:467–75, 2001.

Beggs HE, Schahin-Reed D, Zang K, Goebbels S, Nave KA et al: FAK deficiency in cells contributing to the basal lamina results in cortical abnormalities resembling congenital muscular dystrophies, *Neuron* 40:501–14, 2003.

Fiedorek FT Jr, Kay ES: Mapping of the focal adhesion kinase (Fadk) gene to mouse chromosome 15 and human chromosome 8, *Mam Genome* 6:123–6, 1995.

Hsia DA, Mitra SK, Hauck CR, Streblow DN, Nelson JA et al: Differential regulation of cell motility and invasion by FAK, *J Cell Biol* 160:753–67, 2003.

Li W, Lee J, Vikis HG, Lee SH, Liu G et al: Activation of FAK and Src are receptor-proximal events required for netrin signaling, *Nature Neurosci* 7:1213–21, 2004.

Liu G, Beggs H, Jurgensen C, Park HT, Tang H et al: Netrin requires focal adhesion kinase and Src family kinases for axon outgrowth and attraction, *Nature Neurosci* 7:1222–32, 2004.

Polte TR, Hanks SK: Interaction between focal adhesion kinase and Crk-associated tyrosine kinase substrate p130-Cas, *Proc Natl Acad Sci USA* 92:10678–82, 1995.

Ren X, Ming G, Xie Y, Hong Y, Sun D et al: Focal adhesion kinase in netrin-1 signaling, *Nature Neurosci* 7:1204–12, 2004.

Schaller MD, Borgman C, Cobb BS, Vines RR, Reynolds AB et al: pp125(FAK), a structurally distinctive protein-tyrosine kinase associated with focal adhesions, *Proc Natl Acad Sci USA* 89:5192–6, 1992.

Parsons JT, Parsons SJ: Src family tyrosine kinases: Cooperating with growth factor and adhesion signaling pathways, *Curr Opin Cell Biol* 9:187–92, 1997.

Erpel T, Courtneidge SA: Src family protein tyrosine kinases and cellular signal transduction pathways, *Curr Opin Cell Biol* 7:176–82, 1995.

Krauss G: *Biochemistry of Signal Transduction and Regulation*, 3rd ed, Wiley-VCH GmbH, 2003.

Human protein reference data base, Johns Hopkins University and the Institute of Bioinformatics, at http://www.hprd.org/protein.

5.5. Serine/Threonine Kinase Receptors

TGFβ

Krauss G: *Biochemistry of Signal Transduction and Regulation*, 3rd ed, Wiley-VCH GmbH, 2003.

Blobe GC, Schiemann WP, Lodish HF: Role of transforming growth factor beta in human disease, *New Engl J Med* 342:1350–8, 2000.

Border WA, Noble NA: Transforming growth factor beta in tissue fibrosis, *New Engl J Med* 331:1286–92, 1994.

Brionne TC, Tesseur I, Masliah E, Wyss-Coray T: Loss of TGF-beta-1 leads to increased neuronal cell death and microgliosis in mouse brain, *Neuron* 40:1133–45, 2003.

Derynck R, Jarrett JA, Chen EY, Eaton DH, Bell JR et al: Human transforming growth factor-beta complementary DNA sequence and expression in normal and transformed cells, *Nature* 316:701–5, 1985.

Han G, Lu SL, Li AG, He W, Corless CL et al: Distinct mechanisms of TGF-beta-1-mediated epithelial-to-mesenchymal transition and metastasis during skin carcinogenesis, *J Clin Invest* 115:1714–23, 2005.

Heldin CH, Miyazono K, ten Dijke P: TGF-beta signalling from cell membrane to nucleus through SMAD proteins, *Nature* 390:465–71, 1997.

Jang CW, Chen CH, Chen CC, Chen J, Su YH et al: TGF-beta induces apoptosis through Smad-mediated expression of DAP-kinase, *Nature Cell Biol* 4:51–8, 2001.

Munger JS, Huang X, Kawakatsu H, Griffiths MJ, Dalton SL et al: The integrin alpha v beta 6 binds and activates latent TGF beta 1: A mechanism for regulating pulmonary inflammation and fibrosis, *Cell* 96:319–28, 1999.

Pittet JF, Griffiths MJ, Geiser T, Kaminski N, Dalton SL et al: TGF-beta is a critical mediator of acute lung injury, *J Clin Invest* 107:1537–44, 2001.

Scandura JM, Boccuni P, Massague J, Nimer SD: Transforming growth factor beta-induced cell cycle arrest of human hematopoietic cells requires p57KIP2 upregulation, *Proc Natl Acad Sci USA* 101:15231–6, 2004.

Shehata M, Schwarzmeier JD, Hilgarth M, Hubmann R, Duechler M et al: TGF-beta-1 induces bone marrow reticulin fibrosis in hairy cell leukemia, *J Clin Invest* 113:676–85, 2004.

Sporn MB, Roberts AB, Wakefield LM, Assoian RK: Transforming growth factor-beta: Biological function and chemical structure, *Science* 233:532–4, 1986.

Valderrama-Carvajal H, Cocolakis E, Lacerte A, Lee EH, Krystal G et al: Activin/TGF-beta induce apoptosis through Smad-dependent expression of the lipid phosphatase SHIP, *Nature Cell Biol* 4:963–9, 2002.

TGFβ Receptor Type I

Ebner R, Chen RH, Shum L, Lawler S, Zioncheck TF et al: Cloning of a type I TGF-beta receptor and its effect on TGF-beta binding to the type II receptor, *Science* 260:1344–8, 1993.

Franzen P, ten Dijke P, Ichijo H, Yamashita H, Schulz P et al: Cloning of a TGF-beta type I receptor that forms a heteromeric complex with the TGF-beta type II receptor, *Cell* 75:681–92, 1993.

Johnson DW, Qumsiyeh M, Benkhalifa M, Marchuk DA: Assignment of human transforming growth factor-beta type I and type III receptor genes (TGFBR1 and TGFBR3) to 9q33-q34 and 1p32-p33, respectively, *Genomics* 28:356–7, 1995.

Larsson J, Goumans MJ, Sjostrand LJ, van Rooijen MA, Ward D et al: Abnormal angiogenesis but intact hematopoietic potential in TGF-beta type I receptor-deficient mice, *EMBO J* 20:1663–73, 2001.

Loeys BL, Chen J, Neptune ER, Judge DP, Podowski M et al: A syndrome of altered cardiovascular, craniofacial, neurocognitive and skeletal development caused by mutations in TGFBR1 or TGFBR2, *Nature Genet* 37:275–81, 2005.

Vellucci VF, Reiss, M: Cloning and genomic organization of the human transforming growth factor-beta type I receptor gene, *Genomics* 46:278–83, 1997.

TGFβ Receptor Type II

Chen RH, Ebner R, Derynck R: Inactivation of the type II receptor reveals two receptor pathways for the diverse TGF-beta activities, *Science* 260:1335–8, 1993.

Han G, Lu SL, Li AG, He W, Corless CL et al: Distinct mechanisms of TGF-beta-1-mediated epithelial-to-mesenchymal transition and metastasis during skin carcinogenesis, *J Clin Invest* 115:1714–23, 2005.

Ito Y, Yeo JY, Chytil A, Han J, Bringas P Jr et al: Conditional inactivation of Tgfbr2 in cranial neural crest causes cleft palate and calvaria defects, *Development* 130:5269–80, 2003.

Lin HY, Wang XF, Ng-Eaton E, Weinberg RA, Lodish HF: Expression cloning of the TGF-beta type II receptor, a functional transmembrane serine/threonine kinase, *Cell* 68:775–85, 1992.

Loeys BL, Chen J, Neptune ER, Judge DP, Podowski M et al: A syndrome of altered cardiovascular, craniofacial, neurocognitive and skeletal development caused by mutations in TGFBR1 or TGFBR2, *Nature Genet* 37:275–81, 2005.

Markowitz S, Wang J, Myeroff L, Parsons R, Sun L et al: Inactivation of the type II TGF-beta receptor in colon cancer cells with microsatellite instability, *Science* 268:1336–8, 1995.

Mathew S, Murty VVVS, Cheifetz S, George D, Massague J et al: Transforming growth factor receptor gene TGFBR2 maps to human chromosome band 3p22, *Genomics* 20:114–5, 1994.

Neptune ER, Frischmeyer PA, Arking DE, Myers L, Bunton TE et al: Dysregulation of TGF-beta activation contributes to pathogenesis in Marfan syndrome, *Nature Genet* 33:407–11, 2003.

Ozdamar B, Bose R, Barrios-Rodiles M, Wang HR, Zhang Y et al: Regulation of the polarity protein Par6 by TGF-beta receptors controls epithelial cell plasticity, *Science* 307:1603–9, 2005.

Sanford LP, Ormsby I, Gittenberger-de Groot AC, Sariola H, Friedman R et al: TGF-beta-2 knockout mice have multiple developmental defects that are non-overlapping with other TGF-beta knockout phenotypes, *Development* 124:2659–70, 1997.

Smad2

Nakao A, Roijer E, Imamura T, Souchelnytskyi S, Stenman G et al: Identification of Smad2, a human Mad-related protein in the transforming growth factor-beta signaling pathway, *J Biol Chem* 272:2896–900, 1997.

Oft M, Akhurst RJ, Balmain A: Metastasis is driven by sequential elevation of H-ras and Smad2 levels, *Nature Cell Biol* 4:487–94, 2002.

Riggins GJ, Thiagalingam S, Rozenblum E, Weinstein CL, Kern SE et al: Mad-related genes in the human, *Nature Genet* 13:347–9, 1996.

Takenoshita S, Mogi A, Nagashima M, Yang K, Yagi K et al: Characterization of the MADH2/Smad2 gene, a human Mad homolog responsible for the transforming growth factor-beta and activin signal transduction pathway, *Genomics* 48:1–11, 1998.

Waldrip WR, Bikoff EK, Hoodless PA, Wrana JL, Robertson EJ: Smad2 signaling in extraembryonic tissues determines anterior-posterior polarity of the early mouse embryo, *Cell* 92:797–808, 1998.

Wu G, Chen YG, Ozdamar B, Gyuricza CA, Chong PA et al: Structural basis of Smad2 recognition by the Smad anchor for receptor activation, *Science* 287:92–7, 2000.

Wu JW, Hu M, Chai J, Seoane J, Huse M et al: Crystal structure of a phosphorylated Smad2: recognition of phosphoserine by the MH2 domain and insights on Smad function in TGF-beta signaling, *Mol Cell* 8:1277–89, 2001.

Smad3

Matsuura I, Denissova NG, Wang G, He D, Long J et al: Cyclin-dependent kinases regulate the antiproliferative function of Smads, *Nature* 430:226–31, 2004.

Riggins GJ, Thiagalingam S, Rozenblum E, Weinstein CL, Kern SE et al: Mad-related genes in the human, *Nature Genet* 13:347–9, 1996.

Sato M, Muragaki Y, Saika S, Roberts AB, Ooshima A: Targeted disruption of TGF-beta-1/Smad3 signaling protects against renal tubulointerstitial fibrosis induced by unilateral ureteral obstruction, *J Clin Invest* 112:1486–94, 2003.

Wolfraim LA, Fernandez TM, Mamura M, Fuller WL, Kumar R et al: Loss of Smad3 in acute T-cell lymphoblastic leukemia, *New Engl J Med* 351:552–9, 2004.

Yang X, Letterio JJ, Lechleider RJ, Chen L, Hayman R et al: Targeted disruption of SMAD3 results in impaired mucosal immunity and diminished T cell responsiveness to TGF-beta, *EMBO J* 18:1280–91, 1999.

Zawel L, Dai JL, Buckhaults P, Zhou S, Kinzler KW et al: Human Smad3 and Smad4 are sequence-specific transcription activators, *Mol Cell* 1:611–7, 1998.

Zhang Y, Feng XH, Wu RY, Derynck R: Receptor-associated Mad homologues synergize as effectors of the TGF-beta response, *Nature* 383:168–72, 1996.

Zhu Y, Richardson JA, Parada LF, Graff JM: Smad3 mutant mice develop metastatic colorectal cancer, *Cell* 94:703–14, 1998.

Smad4

Inman GJ, Nicolas FJ, Hill CS: Nucleocytoplasmic shuttling of Smads 2, 3, and 4 permits sensing of TGF-beta receptor activity, *Mol Cell* 10:283–94, 2002.

MacGrogan D et al: Comparative mutational analysis of DPC4 (Smad4) in prostatic and colorectal carcinomas, *Oncogene* 15:1111–4, 1997.

Shioda T, Lechleider RJ, Dunwoodie SL, Li H, Yahata T et al: Transcriptional activating activity of Smad4: Roles of SMAD hetero-oligomerization and enhancement by an associating transactivator, *Proc Natl Acad Sci USA* 95:9785–90, 1998.

Sirard C, de la Pompa JL, Elia A, Itie A, Mirtsos C et al: The tumor suppressor gene Smad4/Dpc4 is required for gastrulation and later for anterior development of the mouse embryo, *Genes Dev* 12:107–19, 1998.

Takaku K, Oshima M, Miyoshi H, Matsui M, Seldin MF et al: Intestinal tumorigenesis in compound mutant mice of both Dpc4 (Smad4) and Apc genes, *Cell* 92:645–56, 1998.

Thiagalingam S, Lengauer C, Leach FS, Schutte M, Hahn SA et al: Evaluation of candidate tumour suppressor genes on chromosome 18 in colorectal cancers, *Nature Genet* 13:343–6, 1996.

Zawel L, Dai JL, Buckhaults P, Zhou S, Kinzler KW et al: Human Smad3 and Smad4 are sequence-specific transcription activators, *Mol Cell* 1:611–7, 1998.

Zhou S, Buckhaults P, Zawel L, Bunz F, Riggins G et al: Targeted deletion of Smad4 shows it is required for transforming growth factor beta and activin signaling in colorectal cancer cells, *Proc Natl Acad Sci USA* 95:2412–6, 1998.

Human protein reference data base, Johns Hopkins University and the Institute of Bioinformatics, at http://www.hprd.org/protein.

5.6. Protein Kinase A

PKA RI α

Bertherat J, Groussin L, Sandrini F, Matyakhina L, Bei T et al: Molecular and functional analysis of PRKAR1A and its locus (17q22–24) in sporadic adrenocortical tumors: 17q losses, somatic mutations, and protein kinase A expression and activity, *Cancer Res* 63:5308–19, 2003.

Bongarzone I, Monzini N, Borrello MG, Carcano C, Ferraresi G et al: Molecular characterization of a thyroid tumor-specific transforming sequence formed by the fusion of ret tyrosine kinase and the regulatory subunit RI alpha of cyclic AMP-dependent protein kinase A, *Mol Cell Biol* 13:358–66, 1993.

Boshart M, Weih F, Nichols M, Schutz G: The tissue-specific extinguisher locus TSE1 encodes a regulatory subunit of cAMP-dependent protein kinase, *Cell* 66:849–59, 1991.

Chen AE, Ginty DD, Fan CM: Protein kinase A signaling via CREB controls myogenesis induced by Wnt proteins, *Nature* 433:317–22, 2005.

Dodge-Kafka KL, Soughayer J, Pare GC, Michel JJC, Langeberg LK et al: The protein kinase A anchoring protein mAKAP coordinates two integrated cAMP effector pathways, *Nature* 437:574–8, 2005.

Jia J, Tong C, Wang B, Luo L, Jiang J: Hedgehog signalling activity of Smoothened requires phosphorylation by protein kinase A and casein kinase I, *Nature* 432:1045–50, 2004.

Jones KW, Shapero MH, Chevrette M, Fournier REK: Subtractive hybridization cloning of a tissue-specific extinguisher: TSE1 encodes a regulatory subunit of protein kinase A, *Cell* 66:861–72, 1991.

Kim C, Xuong NH, Taylor SS: Crystal structure of a complex between the catalytic and regulatory (RI-alpha) subunits of PKA, *Science* 307:690–6, 2005.

Zhang J, Hupfeld CJ, Taylor SS, Olefsky JM, Tsien RY: Insulin disrupts beta-adrenergic signalling to protein kinase A in adipocytes, *Nature* 437:569–73, 2005.

PKA RI β

Solberg R, Sistonen P, Traskelin AL, Berube D, Simard J et al: Mapping of the regulatory subunits RI-beta and RII-beta of cAMP-dependent protein kinase genes on human chromosome 7, *Genomics* 14:63–9, 1992.

PKA RII α

Oyen O, Myklebust F, Scott JD, Hansson V, Jahnsen T: Human testis cDNA for the regulatory subunit RII alpha of cAMP-dependent protein kinase encodes an alternate amino-terminal region, *FEBS Lett* 246:57–64, 1989.

Tasken K, Naylor SL, Solberg R, Jahnsen T: Mapping of the gene encoding the regulatory subunit RII-alpha of cAMP-dependent protein kinase (locus PRKAR2A) to human chromosome region 3p21.3-p21.2, *Genomics* 50:378–81, 1998.

PKA RII β

Adams MR, Brandon EP, Chartoff EH, Idzerda RL, Dorsa DM et al: Loss of haloperidol induced gene expression and catalepsy in protein kinase A-deficient mice, *Proc Natl Acad Sci* 94:12157–61, 1997.

Cummings DE, Brandon EP, Planas JV, Motamed K, Idzerda RL et al: Genetically lean mice result from targeted disruption of the RII-beta subunit of protein kinase A, *Nature* 382:622–6, 1996.

Solberg R, Sistonen P, Traskelin AL, Berube D, Simard J et al: Mapping of the regulatory subunits RI-beta and RII-beta of cAMP-dependent protein kinase genes on human chromosome 7, *Genomics* 14:63–9, 1992.

Wainwright B, Lench N, Davies K, Scambler P, Kruyer H et al: A human regulatory subunit of type II cAMP-dependent protein kinase localized by its linkage relationship to several cloned chromosome 7q markers, *Cytogenet Cell Genet* 45:237–9, 1987.

PKA Cα

Kim C, Xuong NH, Taylor SS: Crystal structure of a complex between the catalytic and regulatory (RI-alpha) subunits of PKA, *Science* 307:690–6, 2005.

Skalhegg BS, Huang Y, Su T, Idzerda RL, McKnight GS et al: Mutation of the C-alpha subunit of PKA leads to growth retardation and sperm dysfunction, *Mol Endocrinol* 16:630–9, 2002.

Tasken K, Solberg R, Zhao Y, Hansson V, Jahnsen T et al: The gene encoding the catalytic subunit C-alpha of cAMP-dependent protein kinase (locus PRKACA) localizes to human chromosome region 19p13.1, *Genomics* 36:535–8, 1996.

PKA Cβ

Simard J, Berube D, Sandberg M, Grzeschik KH, Gagne R et al: Assignment of the gene encoding the catalytic subunit C-beta of cAMP-dependent protein kinase to the p36 band on chromosome 1, *Hum Genet* 88:653–7, 1992.

PKA Cγ

Beebe SJ, Oyen O, Sandberg M, Froysa A, Hansson V et al: Molecular cloning of a tissue-specific protein kinase (C gamma) from human testis—representing a third isoform for the catalytic subunit of cAMP-dependent protein kinase, *Mol Endocrinol* 4:465–75, 1990.

Foss KB, Simard J, Berube D, Beebe SJ, Sandberg M et al: Localization of the catalytic subunit C-gamma of the cAMP-dependent protein kinase gene (PRKACG) to human chromosome region 9q13, *Cytogenet Cell Genet* 60:22–5, 1992.

Reinton N, Haugen TB, Orstavik S, Skalhegg BS, Hansson V et al: The gene encoding the C gamma catalytic subunit of cAMP-dependent protein kinase is a transcribed retroposon, *Genomics* 49:290–7, 1998.

Human protein reference data base, Johns Hopkins University and the Institute of Bioinformatics, at http://www.hprd.org/protein.

5.7. Protein Kinase C

PKCα

Braz JC, Gregory K, Pathak A, Zhao W, Sahin B et al: PKC-alpha regulates cardiac contractility and propensity toward heart failure, *Nature Med* 10:248–54, 2004.

Coussens L, Parker PJ, Rhee L, Yang-Feng TL, Chen E et al: Multiple, distinct forms of bovine and human protein kinase C suggest diversity in cellular signaling pathways, *Science* 233:859–66, 1986.

Linnenbach AJ, Huebner K, Reddy EP, Herlyn M, Parmiter AH et al: Structural alteration in the MYB protooncogene and deletion within the gene encoding alpha-type protein kinase C in human melanoma cell lines, *Proc Natl Acad Sci* 85:74–8, 1988.

Lorenz K, Lohse MJ, Quitterer U: Protein kinase C switches the Raf kinase inhibitor from Raf-1 to GRK-2, *Nature* 426:574–9, 2003.

Parker PJ, Coussens L, Totty N, Rhee L, Young S et al: The complete primary structure of protein kinase C—the major phorbol ester receptor, *Science* 233:853–9, 1986.

PKCβ

Coussens L, Parker PJ, Rhee L, Yang-Feng TL, Chen E et al: Multiple, distinct forms of bovine and human protein kinase C suggest diversity in cellular signaling pathways, *Science* 233:859–66, 1986.

Francke U, Darras BT, Zander NF, Kilimann MW: Assignment of human genes for phosphorylase kinase subunits alpha (PHKA) to Xq12-q13 and beta (PHKB) to 16q12–q13, *Am J Hum Genet* 45:276–82, 1989.

Greenham J, Adams M, Doggett N, Mole S: Elucidation of the exon-intron structure and size of the human protein kinase C beta gene (PRKCB), *Hum Genet* 103:483–7, 1998.

Leitges M, Schmedt C, Guinamard R, Davoust J, Schaal S et al: Immunodeficiency in protein kinase C-beta-deficient mice, *Science* 273:788–91, 1996.

Su TT, Guo B, Kawakami Y, Sommer K, Chae K et al: PKC-beta controls I-kappa-B kinase lipid raft recruitment and activation in response to BCR signaling, *Nature Immun* 3:780–6, 2002.

PKAγ

Brkanac Z, Bylenok L, Fernandez M, Matsushita M, Lipe H et al: A new dominant spinocerebellar ataxia linked to chromosome 19q13.4-qter, *Arch Neurol* 59:1291–5, 2002.

Coussens L, Parker PJ, Rhee L, Yang-Feng TL, Chen E et al: Multiple, distinct forms of bovine and human protein kinase C suggest diversity in cellular signaling pathways, *Science* 233:859–66, 1986.

Saunders AM, Seldin MF: The syntenic relationship of proximal mouse chromosome 7 and the myotonic dystrophy gene region on human chromosome 19q, *Genomics* 6:324–32, 1990.

Vithana EN, Abu-Safieh L, Allen MJ, Carey A, Papaioannou M et al: A human homolog of yeast pre-mRNA splicing gene, PRP31, underlies autosomal dominant retinitis pigmentosa on chromosome 19q13.4 (RP11), *Mol Cell* 8:375–81, 2001.

Human protein reference data base, Johns Hopkins University and the Institute of Bioinformatics, at http://www.hprd.org/protein.

PROTEIN PHOSPHATASE-MEDIATED CELL SIGNALING

Cellular activities are often regulated by counterbalanced stimulatory and inhibitory mechanisms. Protein phosphorylation and dephosphorylation are two typical mechanisms that control the activation and suppression of signaling molecules. These two processes are regulated by protein kinases and protein phosphatases, respectively. As discussed on page 151–190, the phosphorylation of mitogenic signaling proteins by protein kinases initiates and enhances cellular activities such as proliferation and migration. In contrast, the dephosphorylation of mitogenic signaling proteins by protein phosphatases elicits inhibitory effects on cellular activities. The stimulatory effects mediated by protein kinases and the inhibitory effect mediated by protein phosphatases coordinately control the mitogenic activities of cells.

Protein phosphatases are classified into three groups based on the type and specificity of substrates: protein serine/threonine phosphatases, protein tyrosine phosphatases, and dual-specificity protein phosphatases. All these enzymes reverse the action of protein kinases. *Protein serine/threonine phosphatases* induce hydrolysis of phosphate esters or dephosphorylation on the serine/threonine residues of substrate proteins. *Protein tyrosine phosphatases* hydrolyze phosphate esters or dephosphorylation on tyrosine residues. The *dual-specificity phosphatases* catalyze phosphate ester hydrolysis on the serine/threonine as well as tyrosine residues. These protein phosphatases are briefly discussed here.

Protein Serine/Threonine Phosphatase-Mediated Cell Signaling [5.8]

Structure and Function. *Protein serine/threonine phosphatases* (see Table 5.9) are enzymes that remove the phosphate esters from the serine and threonine residues of substrate proteins via hydrolysis. Several families of protein serine/threonine phosphatases have been identified in mammalian cells. Among these families, four have been extensively studied and well characterized. These include protein phosphatases (PP) 1, 2A, 2B, and 2C. In structure, a typical protein serine/threonine phosphatase is composed of one or more regulatory domains and a catalytic domain. The phosphatase 1, 2A, and 2B families possess a similar structure in the catalytic domain, which contains several unique motifs, including the -GDxHG-, -GDxVDRG-, and -GNHE- motifs, in the *N*-terminal half of the molecule. The activity of these phosphatases is dependent on two metal ions: Fe^{2+} and Zn^{2+}. The unique motifs listed above play a critical role in the binding of these

TABLE 5.9. Characteristics of Selected Protein Serine/Threonine Phosphatases*

Proteins	Alternative Names	Amino Acids	Molecular Weight (kDa)	Expression	Functions
Protein phosphatase 1 catalytic subunit α	Protein phosphatase 1 α, serine/threonine protein phosphatase PP1-α catalytic subunit, PP-1A	341	39	Liver, heart, brain	A catalytic subunit for protein serine/threonine phosphatase 1, which dephosphorylates target proteins on serine/threonine residues and regulates cell proliferation and differentiation, glycogen metabolism, muscle contractility, and learning and memory
Protein phosphatase 1 catalytic subunit β	Protein phosphatase 1 β, PP 1B, serine/threonine protein phosphatase PP1 β catalytic subunit	327	37	Widely expressed, mostly in skeletal muscle	Similar to functions of protein phosphatase 1 catalytic subunit α
Protein phosphatase 1 catalytic subunit γ	PP-1G, protein phosphatase 1 γ subunit, protein phosphatase 1C catalytic subunit, serine/threonine protein phosphatase PP1-γ catalytic subunit	323	37	Widely expressed	Similar to functions of protein phosphatase 1 catalytic subunit α
Protein phosphatase 2A catalytic subunit α isoform	PP2CA, PP2A-α, serine/threonine protein phosphatase 2A catalytic subunit α isoform	309	36	Brain, heart, lung, liver, intestine, pancreas, kidney	Forming a complex with one or more regulatory subunits, dephosphorylating target proteins, and negatively controlling cell proliferation and differentiation
Protein phosphatase 2A catalytic subunit β isoform	Serine/threonine protein phosphatase 2A catalytic subunit β isoform, PP2A β	309	36	Heart	Similar to functions of protein phosphatase 2A catalytic subunit α isoform
Protein phosphatase 2A regulatory subunit A α isoform	Protein phosphatase 2A regulatory subunit A α, PP2A subunit A R1-α isoform, PP2A subunit A PR65α isoform	589	65	Kidney, lung, intestine	Forming a complex with a catalytic subunit, regulating dephosphorylation of target proteins, and negatively regulating cell proliferation and differentiation
Protein phosphatase 2A regulatory subunit B		358	441	Heart, retina, placenta, leukocytes	Similar to functions of protein phosphatase 2A regulatory subunit A α isoform

*Based on bibliography 5.8.

metal ions. The PP2C family is characterized by the presence of several motifs, including the ED- and DG-rich motifs. The action of this phosphatase family is dependent on the metal ions Mg^{2+} and Mn^{2+}.

Protein phosphatases exert inhibitory effects on cellular activities induced by protein kinases. When a protein kinase initiates a mitogenic stimulatory effect, which promotes cell proliferation and migration, the activation of a corresponding protein phosphatase results in the suppression of the mitogenic effect. On the other hand, when protein phosphorylation elicits an inhibitory effect on a cellular activity, dephosphorylation by a protein phosphatase exerts an opposite effect. Thus, protein phosphatases and protein kinases coordinately regulate cellular activities, ensuring appropriate initiation and termination of cellular activities.

Signaling Mechanisms. The activity of protein serine/threonine phosphatases is generally regulated via several processes, including phosphorylation, activator binding, and inhibitor binding. The phosphorylation of phosphatases is a major mechanism of phosphatase activation. The regulatory and catalytic domains of a phosphatase can be phosphorylated by specific protein kinases. Such a process induces changes in the localization and catalytic activity of the phosphatase. The binding of activators to a phosphatase induces phosphatase activation, whereas the binding of inhibitors elicits an opposite effect. The phosphorylation and binding of activators and inhibitors are common regulatory mechanisms for protein serine/threonine phosphatases. Each phosphatase may possess distinct features in action and regulation.

Protein Tyrosine Phosphatase-Mediated Cell Signaling

Structure and Function [5.9]. *Protein tyrosine phosphatases* (PTPs) (see Table 5.10) are enzymes that catalyze dephosphorylation on the tyrosine residues of substrate proteins. The effect of PTPs counterbalances that of protein tyrosine kinases, which induces tyrosine phosphorylation. Tyrosine phosphorylation and dephosphorylation are two critical processes that coordinately regulate cell survival, proliferation, differentiation, migration, and adhesion. In human cells, there exists a family of about 100 protein tyrosine phosphatases. These phosphatases are characterized by the presence of a signature motif, HCxxGxxR[S/T], where H, C, G, R, S, and T are histidine, cysteine, glycine, arginine, serine, and threonine, respectively, and x represents any amino acids. This signature motif constitutes the center of the catalytic domain of PTPs. The cysteine residues play a critical role in the catalytic activity of PTPs. The arginine residues are responsible for interaction with phosphate groups. Both cysteine and arginine residues are well preserved among PTPs and are essential for the function of the enzymes. In addition to these two amino acids, there is another invariant amino acid, aspartic acid, which is located in a conformationally flexible loop and plays a critical role in regulating the catalytic activity of PTPs.

Protein tyrosine phosphatases possess distinct molecular structures and can act on a variety of substrate proteins. Based on target amino acid residues, PTPs can be classified into two subfamilies: *classical tyrosine-specific PTPs*, which recognize and act on phosphotyrosine residues in substrate proteins, and *dual-specificity phosphatases* (DSPs), which recognize and act on phosphotyrosine as well as phosphoserine and phosphothreonine residues.

TABLE 5.10. Characteristics of Selected Protein Tyrosine Phosphatases and Related Molecules*

Proteins	Alternative Names	Amino Acids	Molecular Weight (kDa)	Expression	Functions
Protein tyrosine phosphatase receptor type α	Protein tyrosine phosphatase α, tyrosine phosphatase α, leukocyte common antigen-related peptide	802	91	Brain, heart, blood vessel, liver, skeletal muscle, kidney, placenta	Dephosphorylating and activating the Src family tyrosine kinases; also regulating cell adhesion, proliferation, and migration
Protein tyrosine phosphatase nonreceptor type 1	Protein tyrosine phosphatase 1B, PTP1B, PTPN1	435	50	Lymphocytes, placenta, and skeletal muscle	Dephosphorylating a variety of protein tyrosine kinases, including insulin receptor kinase, epidermal growth factor receptor kinase, JAK2 and TYK2 kinases, and negatively regulating cell proliferation, differentiation, and migration
Dual-specificity phosphatase 1	Protein tyrosine phosphatase nonreceptor type10 (PTPN10), MAP kinase phosphatase 1 (MKP 1)	367	39	Brain, skin	Inactivating mitogen-activated protein (MAP) kinase by dephosphorylating the phosphothreonine and phosphotyrosine residues, and negatively regulating cell proliferation and differentiation

SH2 containing protein tyrosine phosphatase 1	SHP-1, SHPTP1, protein tyrosine phosphatase 1C (PTP1C), protein tyrosine phosphatase nonreceptor type 6, tyrosine phosphatase SHP1, hematopoietic cell phosphatase (HCP)	595	68	Myeloid cells, bone marrow	Inactivating a variety of protein tyrosine kinases by dephosphorylating the phosphotyrosine residues, and negatively regulating cell adhesion, proliferation, and differentiation
SH2 containing protein tyrosine phosphatase 2	SHP-2, protein tyrosine phosphatase nonreceptor type 11 (PTPN11), protein tyrosine phosphatase 2C (PTP2C), tyrosine phosphatase SHP2, PTP-1D, SHPTP2	593	68	Ubiquitous	Regulating cell mitogenic activation, metabolism, and migration
Signal regulatory protein α	Signal regulatory protein α type 1, protein tyrosine phosphatase nonreceptor type substrate 1, SHP substrate 1 (SHPS1), tyrosine phosphatase SHP substrate 1, inhibitory receptor SHPS1	504	55	Brain, heart, blood vessel, lung, liver, kidney, spleen, thymus, bone marrow, leukocytes	A receptor-type transmembrane glycoprotein belonging to the immunoglobulin superfamily, recruiting SHP1, serving as a substrate of SHP1, activating SHP1 when SHP1 is recruited, negatively regulating receptor tyrosine kinase-mediated signaling events, and suppressing cell adhesion, proliferation, and migration

*Based on bibliography 5.9.

The subfamily of *classical tyrosine-specific PTPs* is composed of 17 known members. Within this subfamily, 9 PTPs are present in the cytoplasm, which are defined as nontransmembrane PTP subtypes, whereas the remaining 8 PTPs are transmembrane molecules that are similar in structure to cell membrane receptors. Most classical nontransmembrane PTPs contain the signature motif as described above within the catalytic domain located near the *C*-terminus, whereas two PTP subtypes, including PTP1B and BDP1, possess the signature motif in the catalytic domain near the *N*-terminus. In addition to the signature motif, each subtype of PTP contains a characteristic domain. For instance, SH2 domain-containing protein tyrosine phosphatase-1 (SHP1) and SH2 domain-containing protein tyrosine phosphatase-2 (SHP2) contain two Src homology 2 domains. The type 9 nonreceptor protein tyrosine phosphatase (MEG2) contains a cellular retinaldehyde binding protein-like domain. For the transmembrane receptor-like PTPs, the catalytic domain is located near the *C*-terminus on the intracellular side. The extracellular region of the transmembrane PTPs contain various domains such as fibronectin III-like repeats, carbonic anhydrase-like domains, RGDS adhesion recognition motifs, and glycosylated domains, depending on the subtypes of PTPs.

All nontransmembrane and transmembrane PTPs can specifically recognize phosphotyrosine residues in substrate proteins. The specificity of PTPs is determined by the structure and conformation of the active-site cleft and the signature motif of the enzyme. Tyrosine-specific PTPs possess a ~9X deep active-site cleft. Such a structure allows only a substrate phosphotyrosine to reach the cysteine nucleophile at the base of the active-site cleft of a PTP, initiating dephosphorylation of the substrate phosphotyrosine, while phosphoserine/phosphothreonine cannot reach the cysteine nucleophile because of a structural mismatch and thus cannot be dephosphorylated. These observations demonstrate how PTPs selectively dephosphorylate substrate proteins.

The subfamily of *dual-specificity phosphatases* is composed of a large number of members, which exhibit a greater level of structural diversity compared with the classical PTPs. While the signature motif is similar between the two subfamilies of PTPs, other structures are significantly different. In particular, the active-site cleft of the dual-specificity phosphatases is more widely open and shallower than that of the classical PTPS, rendering dual-specificity phosphatases accessible by not only phosphotyrosine but also phosphserine/threonine residues. This feature suggests a mechanism for the selection of both phosphserine/threonine and phosphotyrosine by dual-specificity phosphatases. According to their molecular structure, dual-specificity phosphatases can be divided into three groups: VH1-like dual-specificity phosphatases, myotubularins and cdc25 phosphatases. These phosphatases recognize specific target proteins. For instance, mitogen-activated protein kinase phosphatases (MKPs), which belong to the group of VH1-like dual-specificity phosphatases, dephosphorylate mitogen-activated protein kinases, which are critical signaling molecules regulating cell survival, proliferation, and migration. The cdc25 phosphatases induce dephosphorylation of cyclin-dependent kinases, which regulate cell mitosis.

As discussed above, PTPs exhibit high specificity to substrate proteins. A "substrate trapping" approach has been developed and used for identifying individual substrate proteins for PTPs. A mutant form of a PTP can be generated to suppress the catalytic activity of the enzyme, but keep the substrate-binding domain functional. When expressed in the cell, the mutant PTP can still bind to a specific phosphotyrosine-containing substrate, but cannot initiate dephosphorylation. The complex of the mutant PTP and substrate can be immunoprecipitated, and the associated substrate can be identified by immunoblotting or amino acid sequence analysis. The activity of the substrate can be assessed by immunoblotting with an antiphosphotyrosine antibody. A number of PTP substrates have been identified by the "substrate trapping" approach. Examples include p130cas and VCP (p97/CDC48) as substrates for PTP-PEST and PTPH1, respectively.

Signaling Mechanisms [5.10]. Structural studies have suggested potential mechanisms for the action of PTPs. Once a PTP is engaged with a protein substrate, the cysteine residue (serving as a nucleophile) in the active site of the enzyme interacts with the phosphate group of a substrate phosphotyrosine, forming an intermediate complex of cysteine and phosphate. The ester bond between the phosphate group and the substrate tyrosine residue is cleaved and an aspartic acid residue donates a proton (H) to the cleaved substrate tyrosine residue, resulting in tyrosine dephosphorylation. At the same time, the aspartic acid residue in the conformationally flexible loop, together with a glutamine residue, activates a water molecule and initiates hydrolysis of the cysteine–phosphate intermediate, resulting in the dissociation of the phosphate group from the enzyme. This is a general mechanism of catalytic action for most members of the PTP family.

The catalytic activity of PTPs is regulated by a variety of factors, depending on the structure of PTPs and signaling context. *Transmembrane PTPs*, which are similar to cell membrane receptors in structure, may be directly activated by extracellular ligand binding. For example, soluble pleiotrophin can interact with and activate the transmembrane protein tyrosine phosphatase PTPζ/β, and heparan sulfate proteoglycans can activate PTPσ. Some of the transmembrane PTPs are similar in structure to cell membrane adhesion molecules, suggesting that these PTPs may be activated via cell–cell interactions. In contrast to transmembrane PTPs, *nontransmembrane PTPs* are activated through the mediation of cell membrane receptors and intracellular signaling molecules. In general, there are three known mechanisms for the regulation of PTP activities: (1) phosphorylation of PTPs, (2) conformational changes in the three-dimensional (3D) structure of PTPs, and (3) oxidation of the catalytic cysteine residue of PTPs. These mechanisms are briefly described here.

Phosphorylation is an essential process that induces PTP activation. A typical example is the activation of the protein tyrosine phosphatase Src homology (SH)2 domain-containing tyrosine phosphatase (SHP)1 (Fig. 5.10). SHP1 can interact with the inhibitory receptor signal regulatory protein (SIRP)α, also classified as Src homology 2 domain-containing tyrosine phosphatase substrate (SHPS)1, which is expressed primarily in myeloid cells. SIRP α is a transmembrane glycoprotein that transmits inhibitory signals through tyrosine phosphorylation of its intracellular immunoreceptor tyrosine-based inhibitory motif (ITIM). The phosphorylation of the ITIM, on ligand binding to SIRPα, initiates the recruitment of SHP1 to SIRPα, which is known as a *substrate* of SHP1. The recruitment of SHP1 induces phosphorylation of SHP1, which in turn dephosphorylates protein kinases, possibly including receptor tyrosine kinases, Src family protein tyrosine kinases, phosphatidylinositol 3-kinase, and the Janus family tyrosine kinases. These activities potentially suppress inflammatory and mitogenic cellular activities.

A conformational change in the 3D structure of PTP molecule is another effective mechanism that regulates the activity of PTPs. Here, SHP2 is used to demonstrate this mechanism (Fig. 5.11). A SHP2 molecule contains two SH2 domains. The *N*-terminal SH2 domain serves as a switch. In the absence of a substrate protein with phosphotyrosine residues, the *N*-terminal SH2 domain blocks the active site of SHP2, inhibiting the activity of the enzyme. The binding of SHP2 to a substrate protein induces a conformational

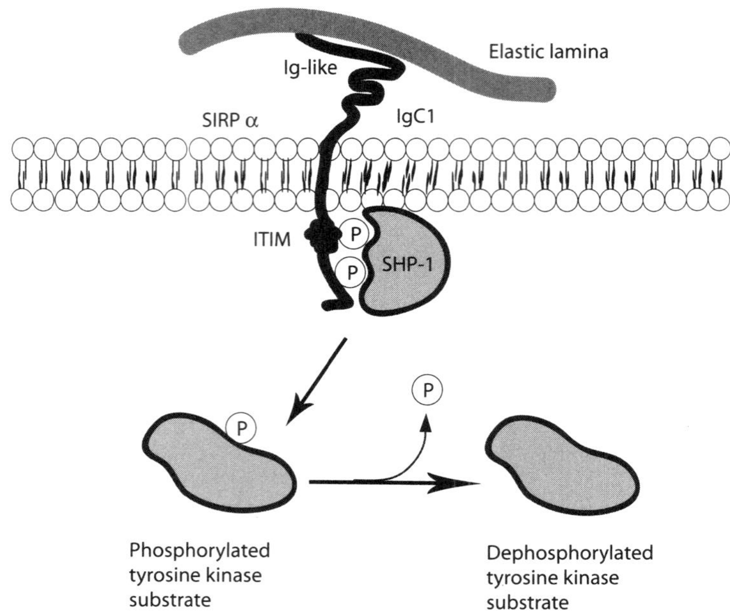

Figure 5.10. Signal regulatory protein (SIRP) α-mediated activation of SH2 domain-containing protein tyrosine phosphatase (SHP)-1 (based on bibliography 5.10).

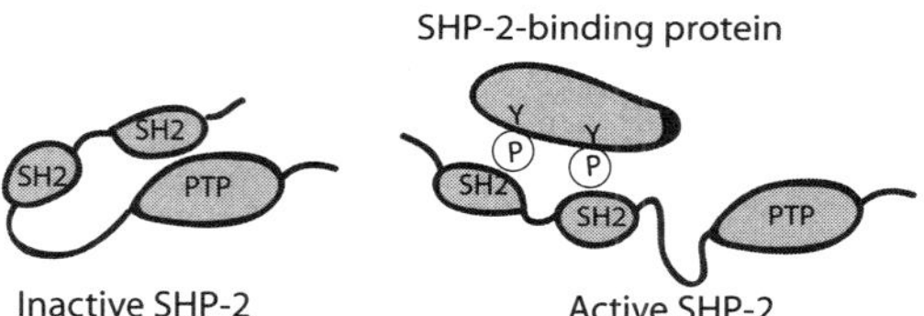

Figure 5.11. Mechanisms of the activation of SH2 domain-containing protein tyrosine phosphatase (SHP)-2. (Reprinted from Hof P et al: Crystal structure of the tyrosine phosphatase, *Cell* 92:441–50, copyright 1998, with permission from Elsevier.)

change, which removes the inhibitory effect of the *N*-terminal SH2 domain. This mechanism also applies to SHP1.

Oxidation of cysteine is a mechanism involved in the inhibition of PTPs. The activity of PTPs is rapidly suppressed when the mission of the PTPs is accomplished. As discussed above, the invariant cysteine residue at the catalytic site of PTPs is critical for the catalytic activity. Cysteine is present as a thiolate anion, which renders the residue susceptible to oxidation, and can be oxidized to a sulfenic acid form (Cys-SOH) or a disulfide form (Cys-S—S—), which inhibit the catalytic activity of PTPs. This process is reversible and is an effective mechanism for temporarily inhibiting the activity of PTPs. Oxidation of PTPs can be initiated in response to the binding of growth factors and hormones, such as epidermal growth factor and insulin, to corresponding receptors, which activate tyrosine phosphorylation-dependent signaling pathways. The suppression of PTP activities further enhances the activation of protein tyrosine kinases. These observations suggest that the activity of PTPs is controlled in coordination with the activity of protein tyrosine kinases.

CYTOKINE-JAK-STAT-MEDIATED CELL SIGNALING [5.11]

Structure and Function

The JAK–STAT, or Janus tyrosine kinase–signal transducers and activators of transduction, signaling pathways are molecular cascades that transduce cytokine signals from the extracellular space to the cell nucleus. This type of signaling pathway is composed of cell membrane and intracellular components, including cytokine receptors, JAKs, and STATs. Cytokines are small proteins (15–30 kDa), including interleukins, interferons (α, β, and γ), erythropoietin, thrombopoietin, leukemia inhibitor factor, cardiotrophin, oncostatin, granulocyte colony-stimulating factors, granulocyte macrophage colony-stimulating factors, and macrophage colony-stimulating factors. Cytokines are produced and secreted primarily by leukocytes. These molecules participate in the regulation of blood cell differentiation and proliferation as well as immune reactions. A cytokine molecule can interact with a specific cytokine receptor, inducing the activation of the JAK-STAT signaling pathways.

Cytokine receptors are classified into three types based on molecular structure: (1) cytokine receptors containing a glycoprotein (gp) 130, such as the interleukin (IL) 6 receptor; (2) cytokine receptors containing a β subunit, such as the receptor for granulocyte-macrophage colony-stimulating factor; and (3) cytokine receptors containing a γ subunit, such as the IL2 receptor. There are four common features for cytokine receptors:

1. The extracellular region is similar for most cytokine receptors.
2. There are no intracellular catalytic domains.
3. Most receptors are arranged into complexes from dimers to tetramers.
4. The intracellular region of a cytokine receptor is associated with JAKs, which relay signals from cytokine receptors and elicit catalytic activities.

The binding of cytokine to cytokine receptor induces the activation of JAKs, which in turn phosphorylate STATs. Phosphorylated STATs serve as transcriptional factors and can translocate to the nucleus, inducing gene expression.

There are four types of JAKs in mammalian cells, including JAK1, JAK2, JAK3, and Tyk2. The molecular weights of these molecules range from 120 to 140-kDa. Each JAK molecule contains 7 functional regions, defined as JAK homology (JH) domains 1–7 starting from the *C*-terminus. The first JAK homology domain, JH1, is a kinase and is able to phosphorylate substrates. JH2 is located upstream to JH1. Although JH2 is similar to a kinase in structure, it does not possess catalytic function because of the lack of several critical amino acid residues. JH2 is thus referred to as a pseudo-kinase domain. JH3 and JH4, located upstream to JH2, are similar in structure to the Src homology (SH) 2 domain, which can bind to phosphotyrosine residues. Thus, these domains are defined as SH2 domains. All remaining domains, including the N-terminal part of the JH4 domain and domains JH5 to JH7, are referred to as four-point-one, ezrin, radixin, moesin homology (FERM) domains. These domains mediate the attachment of JAKs to cytokine receptors in the cell membrane.

Signaling Mechanisms

JAKs are constitutively associated with cytokine receptors. These molecules are arranged in different configurations between an inactive (unliganded) state and active (liganded) state. As shown by crystallography, the receptor of the cytokine erythropoietin is composed of two identical molecules, each of which is associated with a JAK2 molecule. In an unliganded state, the two erythropoietin receptors of each homodimer are separated by approximately 70 Å. In response to the binding of an erythropoietin molecule, the two receptors undergo a conformational change, reducing the gap between the two receptors from 70 to ~30 Å. This change also brings together the two JAK2 molecules associated with the erythropoietin receptor homodimer, allowing reciprocal phosphorylation between the two JAK2 molecules. Phosphorylated JAK2 further activates STATs, leading to transcriptional activities.

STATs are a group of molecules that serve as transcriptional (*trans*-acting) factors for the JAK-STAT signaling pathways. Each STAT molecule contains several domains, including a *C*-terminal transcriptional activation domain, a SH2 domain, a linker domain, a DNA-binding domain, a coiled-coil domain, and an *N*-terminal domain. The SH2 domain can be recruited to the phosphotyrosine docking site of a phosphorylated JAK and can be phosphorylated by the JAK on a tyrosine residue, inducing STAT activation.

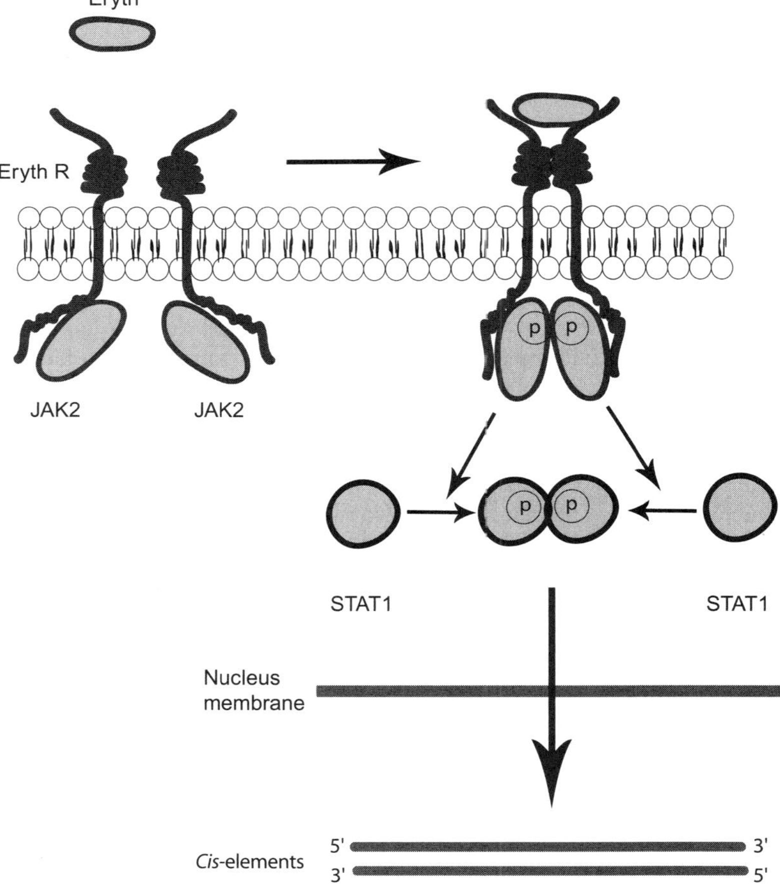

Figure 5.12. Schematic representation of the erythropoietin-mediated JAK-STAT signaling pathway (based on bibliography 5.11).

TABLE 5.11. Characteristics of Selected Molecules for the JAK-STAT Signaling Pathway*

Proteins	Alternative Names	Amino Acids	Molecular Weight (kDa)	Expression	Functions
Janus kinase 1	JAK1	1142	132	Ubiquitous	Regulating signaling activities induced by cytokines, such as interferon α, β, γ, and activating transcription factors known as signal transducers and activators of transcription (STATs)
Janus kinase 2	JAK2	1132	131	Ubiquitous	Similar to functions of JAK1
Janus kinase 2					Regulating cytokine-induced signal transduction, activating STATs, and especially regulating the development and activity of lymphocytes and the immune system
Tyk2	Protein tyrosine kinase 2, nonreceptor tyrosine protein kinase TYK2	1187	134	Leukocytes, bone marrow, macrophages, liver, thymus, skin	Regulating cytokine-induced signal transduction and activating STATs
STAT1	Signal transducer and activator of transcription 1	750	87	Ubiquitous	Forming homodimers or heterodimers with other STAT proteins, and serving as a downstream transcription factor for various cytokine ligands, such as interferon-α, interferon-γ, PDGF, and IL6
STAT2	Signal transducer and activator of transcription 2, and Interferon α-induced transcriptional activator	851	98	Thymus, skin	Forming a complex with STAT1 and interferon regulatory factor p48, acting as a coactivator without the capability of binding to DNA directly, and mediating STAT1 activity

*Based on bibliography 5.11.

Two activated STATs then interact between each other at the phosphotyrosine sites of the SH2 domains, inducing the formation of a STAT dimer, which serves as a *trans*-acting factor and translocates to the cell nucleus. A STAT dimer can bind to the promoter region of a target gene together with other transcription–regulatory factors, initiating gene transcription. An example of STAT activation is shown in Fig. 5.12.

Following gene transcription, the JAK-STAT signaling pathways are deactivated by several types of protein, including protein tyrosine phosphatases, suppressors of cytokine signaling family proteins, and inhibitors of STAT family proteins. *Protein tyrosine phosphatases* induce dephosphorylation of JAKs, diminishing their activities. A typical example of this class is the SH2 domain-containing protein tyrosine phosphatase (SHP)1, which can bind and dephosphorylate JAK2. In addition, protein tyrosine phosphatases may directly act on STATs, inducing STAT dephosphorylation and deactivation. The *suppressors of cytokine signaling family proteins* are able to bind to JAKs and block ATP binding to the kinases, inhibiting kinase activities. Some proteins of this family, such as cytokine-inducible SH2 domain-containing proteins, can directly bind to phosphorylated tyrosine residues of JAKs, inhibiting the recruitment and phosphorylation of STATs. The third type of JAK-STAT inhibitors, *inhibitors of STAT family proteins*, can bind to phosphorylated STATs and block their interaction with gene promoters, thereby inhibiting STAT-induced gene expression. These inhibitory mechanisms counterbalance the cytokine-induced activation of the JAK-STAT signaling pathways, ensuring rapid deactivation of these pathways on the accomplishment of the signaling events. (See Table 5.11.)

BIBLIOGRAPHY

5.8. Protein Serine/Threonine Phosphatase-Mediated Cell Signaling

PP1α Catalytic Subunit

Barker HM, Jones TA, da Cruz e Silva EF, Spurr NK, Sheer D et al: Localization of the gene encoding a type I protein phosphatase catalytic subunit to human chromosome band 11q13, *Genomics* 7:159–66, 1990.

Carr AN, Schmidt AG, Suzuki Y, del Monte F, Sato Y et al: Type 1 phosphatase, a negative regulator of cardiac function, *Mol Cell Biol* 22:4124–35, 2002.

Genoux D, Haditsch U, Knobloch M, Michalon A, Storm D et al: Protein phosphatase 1 is a molecular constraint on learning and memory, *Nature* 418:970–5, 2002.

Kabashima T, Kawaguchi T, Wadzinski BE, Uyeda K: Xylulose 5-phosphate mediates glucose-induced lipogenesis by xylulose 5-phosphate-activated protein phosphatase in rat liver, *Proc Natl Acad Sci USA* 100:5107–12, 2003.

Saadat M, Mizuno Y, Kikuchi K, Yoshida MC: Comparative mapping of the gene encoding the catalytic subunit of protein phosphatase type 1-alpha (PPP1CA) to human, rat, and mouse chromosomes, *Cytogenet Cell Genet* 70:55–7, 1995.

Song Q, Khanna KK, Lu H, Lavin MF: Cloning and characterization of a human protein phosphatase 1-encoding cDNA, *Gene* 129:291–5, 1993.

Terrak M, Kerff F, Langsetmo K, Tao T, Dominguez R: Structural basis of protein phosphatase 1 regulation, *Nature* 429:780–4, 2004.

PP1Cβ

Barker HM, Brewis ND, Street AJ, Spurr NK, Cohen PTW: Three genes for protein phosphatase 1 map to different human chromosomes: Sequence, expression and gene localisation of protein serine/threonine phosphatase 1 beta (PPP1CB), *Biochim Biophys Acta* 1220:212–8, 1994.

PP1Cγ

Norman SA, Mott DM: Molecular cloning and chromosomal localization of a human skeletal muscle PP-1-gamma-1 cDNA, *Mam Genome* 5:41–5, 1994.

Saadat M, Nomoto K, Mizuno Y, Kikuchi K, Yoshida MC: Assignment of the gene encoding type 1-gamma protein phosphatase catalytic subunit (PPP1CC) on human, rat, and mouse chromosomes, *Jpn J Hum Genet* 41:159–65, 1996.

PP2A Catalytic α

Gergs U, Boknik P, Buchwalow I, Fabritz L, Matus M et al: Overexpression of the catalytic subunit of protein phosphatase 2A impairs cardiac function, *J Biol Chem* 279:40827–34, 2004.

Jones TA, Barker HM, da Cruz e Silva EF, Mayer-Jaekel RE, Hemmings BA et al: Localization of the genes encoding the catalytic subunits of protein phosphatase 2A to human chromosome bands 5q23-q31 and 8p12-p11.2, respectively, *Cytogenet Cell Genet* 63:35–41, 1993.

PP2A Catalytic β

Hemmings BA, Wernet W, Mayer R, Maurer F, Hofsteenge J et al: The nucleotide sequence of the cDNA encoding the human lung protein phosphatase 2A-beta catalytic subunit, *Nucleic Acids Res* 16:11366 (only), 1988.

Imbert A, Chaffanet M, Essioux L, Noguchi T, Adelaide J et al: Integrated map of the chromosome 8p12-p21 region, a region involved in human cancers and Werner syndrome, *Genomics* 32:29–38, 1996.

Jones TA, Barker HM, da Cruz e Silva EF, Mayer-Jaekel RE, Hemmings BA et al: Localization of the genes encoding the catalytic subunits of protein phosphatase 2A to human chromosome bands 5q23-q31 and 8p12-p11.2, respectively, *Cytogenet Cell Genet* 63:35–41, 1993.

PP2A Regulatory Unit A

Everett AD, Xue C, Stoops T: Developmental expression of protein phosphatase 2A in the kidney, *J Am Soc Nephrol* 10:1737–45, 1999.

Groves MR, Hanlon N, Turowski P, Hemmings BA, Barford D: The structure of the protein phosphatase 2A PR65/A subunit reveals the conformation of its 15 tandemly repeated HEAT motifs, *Cell* 96:99–110, 1999.

Hemmings BA, Adams-Pearson C, Maurer F, Muller P, Goris J et al: Alpha- and beta-forms of the 65-kDa subunit of protein phosphatase 2A have a similar 39 amino acid repeating structure, *Biochemistry* 29:3166–73, 1990.

Hendrix P, Turowski P, Mayer-Jaekel RE, Goris J, Hofsteenge J et al: Analysis of subunit isoforms in protein phosphatase 2A holoenzymes from rabbit and Xenopus, *J Biol Chem* 268:7330–7, 1993.

PP2A Regulatory Unit B

McCright B, Rivers AM, Audlin S, Virshup DM: The B56 family of protein phosphatase 2A (PP2A) regulatory subunits encodes differentiation-induced phosphoproteins that target PP2A to both nucleus and cytoplasm, *J Biol Chem* 271:22081–9, 1996.

Van Hoof C, Aly MS, Garcia A, Cayla X, Cassiman JJ et al: Structure and chromosomal localization of the human gene of the phosphotyrosyl phosphatase activator (PTPA) of protein phosphatase 2A, *Genomics* 28:261–72, 1995.

Human protein reference data base, Johns Hopkins University and the Institute of Bioinformatics, at http://www.hprd.org/protein.

5.9. Structure and Function of Protein Tyrosine Phosphatases

PTP Receptor α

Jia Z, Barford D, Flint AJ, Tonks NK: Structural basis for phosphotyrosine peptide recognition by protein tyrosine phosphatase 1B, *Science* 268:1754–8, 1995.

Garton AJ, Flint AJ, Tonks NK: Identification of p130cas as a substrate for the cytosolic protein tyrosine phosphatase PTP-PEST, *Mol Cell Biol* 16:6408–18, 1996.

Zhang SH, Liu J, Kobayashi R, Tonks NK: Identification of the cell cycle regulator VCP (p97/CDC48) as a substrate of the band 4.1-related protein-tyrosine phosphatase PTPH1, *J Biol Chem* 274:17806–12, 1999.

Jirik FR, Anderson LL, Duncan AMV: The human protein-tyrosine phosphatase PTP-alpha/LRP gene (PTPA) is assigned to chromosome 20p13, *Cytogenet Cell Genet* 60:117–8, 1992.

Jirik FR, Janzen NM, Melhado IG, Harder KW: Cloning and chromosomal assignment of a widely expressed human receptor-like protein-tyrosine phosphatase, *FEBS Lett* 273:239–42, 1990.

Kaplan R, Morse B, Huebner K, Croce C, Howk R et al: Cloning of three human tyrosine phosphatases reveals a multigene family of receptor-linked protein-tyrosine-phosphatases expressed in brain, *Proc Natl Acad Sci USA* 87:7000–4, 1990.

Rao VVNG, Loffler C, Sap J, Schlessinger J, Hansmann I: The gene for receptor-linked protein-tyrosine-phosphatase (PTPA) is assigned to human chromosome 20p12-pter by in situ hybridization (ISH and FISH), *Genomics* 13:906–7, 1992.

PTPN1

Ahmad F, Azevedo JL Jr, Cortright R, Dohm GL, Goldstein BJ: Alternations in skeletal muscle protein-tyrosine phosphatase activity and expression in insulin-resistant human obesity and diabetes, *J Clin Invest* 100:449–58, 1997.

Brown-Shimer S, Johnson KA, Lawrence JB, Johnson C, Bruskin A et al: Molecular cloning and chromosome mapping of the human gene encoding protein phosphotyrosyl phosphatase 1B, *Proc Natl Acad Sci USA* 87:5148–52, 1990.

Charbonneau H, Tonks NK, Kumar S, Diltz CD, Harrylock M et al: Human placenta protein-tyrosine-phosphatase: Amino acid sequence and relationship to a family of receptor-like proteins, *Proc Natl Acad Sci USA* 86:5252–6, 1989.

Chernoff J, Schievella AR, Jost CA, Erikson RL, Neel BG: Cloning of a cDNA for a major human protein-tyrosine-phosphatase, *Proc Natl Acad Sci USA* 87:2735–9, 1990.

Elchebly M, Payette P, Michaliszyn E, Cromlish W, Collins S et al: Increased insulin sensitivity and obesity resistance in mice lacking the protein tyrosine phosphatase-1B gene, *Science* 283:1544–8, 1999.

Forsell PKAL, Boie Y, Montalibet J, Collins S, Kennedy BP: Genomic characterization of the human and mouse protein tyrosine phosphatase-1B genes, *Gene* 260:145–53, 2000.

Haj FG, Verveer PJ, Squire A, Neel BG, Bastiaens PIH: Imaging sites of receptor dephosphorylation by PTP1B on the surface of the endoplasmic reticulum, *Science* 295:1708–11, 2002.

Jia Z, Barford D, Flint AJ, Tonks NK: Structural basis for phosphotyrosine peptide recognition by protein tyrosine phosphatase 1B, *Science* 268:1754–8, 1995.

Kaplan R, Morse B, Huebner K, Croce C, Howk R et al: Cloning of three human tyrosine phosphatases reveals a multigene family of receptor-linked protein-tyrosine-phosphatases expressed in brain, *Proc Natl Acad Sci USA* 87:7000–4, 1990.

Klaman LD, Boss O, Peroni OD, Kim JK, Martino JL et al: Increased energy expenditure, decreased adiposity, and tissue-specific insulin sensitivity in protein-tyrosine phosphatase 1B-deficient mice, *Mol Cell Biol* 20:5479–89, 2000.

Salmeen A, Andersen JN, Myers MP, Meng TC, Hinks JA et al: Redox regulation of protein tyrosine phosphatase 1B involves a sulphenyl-amide intermediate, *Nature* 423:769–73, 2003.

Tonks NK, Diltz CD, Fischer EH: Purification of the major protein-tyrosine-phosphatases of human placenta, *J Biol Chem* 263:6722–30, 1988.

van Montfort RLM, Congreve M, Tisi D, Carr R, Jhoti H: Oxidation state of the active-site cysteine in protein tyrosine phosphatase 1B, *Nature* 423:773–7, 2003.

Dual-Specificity PTP

Alessi DR, Smythe C, Keyse SM: The human CL100 gene encodes a tyr/thr-protein phosphatase which potently and specifically inactivates MAP kinase and suppresses its activation by oncogenic ras in Xenopus oocyte extracts, *Oncogene* 8:2015–20, 1993.

Brondello JM, Pouyssegur J, McKenzie FR: Reduced MAP kinase phosphatase-1 degradation after p42/p44(MAPK)-dependent phosphorylation, *Science* 286:2514–7, 1999.

Emslie EA, Jones TA, Sheer D, Keyse SM: The CL100 gene, which encodes a dual specificity (tyr/thr) MAP kinase phosphatase, is highly conserved and maps to human chromosome 5q34, *Hum Genet* 93:513–6, 1994.

Keyse SM, Emslie EA: Oxidative stress and heat shock induce a human gene encoding a protein-tyrosine phosphatase, *Nature* 359:644–7, 1992.

SHP1

Banville D, Stocco R, Shen SH: Human protein tyrosine phosphatase 1C (PTPN6) gene structure: Alternate promoter usage and exon skipping generate multiple transcripts, *Genomics* 27:165–73, 1995.

Kamata T, Yamashita M, Kimura M, Murata K, Inami M et al: Src homology 2 domain-containing tyrosine phosphatase SHP-1 controls the development of allergic airway inflammation, *J Clin Invest* 111:109–19, 2003.

Liu SQ, Alkema PK, Tieche C, Tefft BJ, Liu DZ et al: Negative regulation of monocyte adhesion to arterial elastic laminae by signal-regulatory protein alpha and SH2 domain-containing protein tyrosine phosphatase-1, *J Biol Chem* 280:39294–301, 2005.

Plutzky J, Neel BG, Rosenberg RD, Eddy RL, Byers MG et al: Chromosomal localization of an SH2-containing tyrosine phosphatase (PTPN6), *Genomics* 13:869–72, 1992.

Shen SH, Bastien L, Posner BI, Chretien P: A protein-tyrosine phosphatase with sequence similarity to the SH2 domain of the protein-tyrosine kinases, *Nature* 352:736–9, 1991.

Shultz LD, Schweitzer PA, Rajan TV, Yi T, Ihle JN et al: Mutations at the murine motheaten locus are within the hematopoietic cell protein-tyrosine phosphatase (Hcph) gene, *Cell* 73:1445–54, 1993.

Tsui HW, Siminovitch KA, de Souza L, Tsui FWL: Motheaten and viable motheaten mice have mutations in the haematopoietic cell phosphatase gene, *Nature Genet* 4:124–9, 1993.

Yi T, Cleveland JL, Ihle JN: Protein tyrosine phosphatase containing SH2 domains: Characterization, preferential expression in hematopoietic cells, and localization to human chromosome 12p12-p13, *Mol Cell Biol* 12:836–46, 1992.

SHP2

Ahmad S, Banville D, Zhao Z, Fischer EH, Shen SH: A widely expressed human protein-tyrosine phosphatase containing src homology 2 domains, *Proc Natl Acad Sci USA* 90:2197–201, 1993.

Chen B, Bronson RT, Klaman LD, Hampton TG, Wang J et al: Mice mutant for Egfr and Shp2 have defective cardiac semilunar valvulogenesis, *Nature Genet* 24:296–9, 2000.

Higashi H, Tsutsumi R, Muto S, Sugiyama T, Azuma T et al: SHP-2 tyrosine phosphatase as an intracellular target of Helicobacter pylori CagA protein, *Science* 295:683–6, 2002.

Hof P, Pluskey S, Dhe-Paganon S, Eck MJ, Shoelson SE: Crystal structure of the tyrosine phosphatase SHP-2, *Cell* 92:441–50, 1998.

Isobe M, Hinoda Y, Imai K, Adachi M: Chromosomal localization of an SH2 containing tyrosine phosphatase (SH-PTP3) gene to chromosome 12q24.1, *Oncogene* 9:1751–3, 1994.

Qu CK, Yu WM, Azzarelli B, Cooper S, Broxmeyer HE et al: Biased suppression of hematopoiesis and multiple developmental defects in chimeric mice containing Shp-2 mutant cells, *Mol Cell Biol* 18:6075–82, 1998.

Saxton TM, Ciruna BG, Holmyard D, Kulkarni S, Harpal K et al: The SH2 tyrosine phosphatase Shp2 is required for mammalian limb development, *Nature Genet* 24:420–3, 2000.

Saxton TM, Henkemeyer M, Gasca S, Shen R, Rossi DJ et al: Abnormal mesoderm patterning in mouse embryos mutant for the SH2 tyrosine phosphatase Shp-2, *EMBO J* 16:2352–64, 1997.

Tartaglia M, Niemeyer CM, Fragale A, Song X, Buechner J et al: Somatic mutations in PTPN11 in juvenile myelomonocytic leukemia, myelodysplastic syndromes and acute myeloid leukemia, *Nature Genet* 34:148–50, 2003.

Zhang EE, Chapeau E, Hagihara K, Feng GS: Neuronal Shp2 tyrosine phosphatase controls energy balance and metabolism, *Proc Natl Acad Sci USA* 101:16064–9, 2004.

SIRPα

Eckert C, Olinsky S, Cummins J, Stephan D, Narayanan V: Mapping of the human P84 gene to the subtelomeric region of chromosome 20p, *Somat Cell Mol Genet* 23:297–301, 1997.

Fujioka Y, Matozaki T, Noguchi T, Iwamatsu A, Yamao T et al: A novel membrane glycoprotein, SHPS-1, that binds the SH2-domain-containing protein tyrosine phosphatase SHP-2 in response to mitogens and cell adhesion, *Mol Cell Biol* 16:6887–99, 1996.

Kharitonenkov A, Chen Z, Sures I, Wang H, Schilling J et al: A family of proteins that inhibit signalling through tyrosine kinase receptors, *Nature* 386:181–6, 1997.

Oldenborg PA, Zheleznyak A, Fang YF, Lagenaur CF, Gresham HD et al: Role of CD47 as a marker of self on red blood cells, *Science* 288:2051–4, 2000.

Yamao T, Matozaki T, Amano K, Matsuda Y, Takahashi N et al: Mouse and human SHPS-1: Molecular cloning of cDNAs and chromosomal localization of genes, *Biochem Biophys Res Commun* 231:61–7, 1997.

Hof P, Pluskey S, Dhe-Paganon S, Eck MJ, Shoelson SE: Crystal structure of the tyrosine phosphatase SHP-2, *Cell* 92:441–50, 1998.

Human protein reference data base, Johns Hopkins University and the Institute of Bioinformatics, at http://www.hprd.org/protein.

5.10. Signaling Mechanisms of Protein Tyrosine Phosphatases

Lee SR, Kwon KS, Rhee SG: Reversible inactivation of protein-tyrosine phosphatase 1B in A431 cells stimulated with epidermal growth factor, *J Biol Chem* 273:15366–72, 1998.

Mahadev K, Zilbering A, Zhu L, Goldstein BJ: Insulin-stimulated hydrogen peroxide reversibly inhibits protein-tyrosine phosphatase 1b in vivo and enhances the early insulin action cascade, *J Biol Chem* 276:21938–42, 2001.

Krauss G: *Biochemistry of Signal Transduction and Regulation*, 3rd ed, Wiley-VCH GmbH, 2003.

Neel B: in *Handbook of Cell Signaling, Vol 1*, Bradshaw RA, Dennis EA, eds, Academic Press, Amsterdam, 2004.

Tonks NK: Overview of protein tyrosine phosphatases, in *Handbook of Cell Signaling*, Vol 1, Bradshaw RA, Dennis EA, Academic Press, Amsterdam, 2004, pp 641–51.

Anderson JN, Mortensen OH, Peters GH, Drake PG, Iverson LF et al: Structural and evolutionary relationships among protein tyrosine phosphatase domains, *Mol Cell Biol* 21:7117–36, 2001.

5.11. Cytokine-JAK-STAT-Mediated Cell Signaling

JAK1

Ihle JN: Cytokine receptor signaling, *Nature* 377:591–4, 1995.

Modi WS, Farrar WL, Howard OMZ: Confirmed assignment of a novel human tyrosine kinase gene (JAK1A) to 1p32.3-p31.3 by nonisotopic in situ hybridization, *Cytogenet Cell Genet* 69:232–4, 1995.

Muller M, Briscoe J, Laxton C, Guschin D, Ziemiecki A et al: The protein tyrosine kinase JAK1 complements defects in interferon-alpha/beta and -gamma signal transduction, *Nature* 366:129–35, 1993.

Pritchard MA, Baker E, Callen DF, Sutherland GR, Wilks AF: Two members of the JAK family of protein tyrosine kinases map to chromosomes 1p31.3 and 9p24, *Mam Genome* 3:36–8, 1992.

Rodig SJ, Meraz MA, White JM, Lampe PA, Riley JK et al: Disruption of the Jak1 gene demonstrates obligatory and nonredundant roles of the Jaks in cytokine-induced biologic responses, *Cell* 93:373–83, 1998.

Wilks AF, Harpur AG, Kurban RR, Ralph SJ, Zurcher G et al: Two novel protein-tyrosine kinases, each with a second phosphotransferase-related catalytic domain, define a new class of protein kinase, *Mol Cell Biol* 11:2057–65, 1991.

Yang CH, Murti A, Valentine WJ, Du Z, Pfeffer LM: Interferon alpha activates NF-kappaB in JAK1-deficient cells through a TYK2-dependent pathway, *J Biol Chem* 280:25849–53, 2005.

Radtke S, Haan S, Jorissen A, Hermanns HM, Diefenbach S, Smyczek T et al: The Jak1 SH2 domain does not fulfill a classical SH2 function in Jak/STAT signaling but plays a structural role for receptor interaction and up-regulation of receptor surface expression, *J Biol Chem* 280:25760–8, 2005.

Proietti C, Salatino M, Rosemblit C, Carnevale R, Pecci A et al: Progestins induce transcriptional activation of signal transducer and activator of transcription 3 (Stat3) via a Jak- and Src-dependent mechanism in breast cancer cells, *Mol Cell Biol* 25:4826–40, 2005.

JAK2

Xie S, Lin H, Sun T, Arlinghaus RB: Jak2 is involved in c-Myc induction by Bcr-Abl, *Oncogene* 21:7137–46, 2002.

Campbell GS, Argetsinger LS, Ihle JN, Kelly PA, Rillema JA et al: Activation of JAK2 tyrosine kinase by prolactin receptors in Nb2 cells and mouse mammary gland explants, *Proc Natl Acad Sci USA* 91:5232–6, 1994.

Digicaylioglu M, Lipton SA: Erythropoietin-mediated neuroprotection involves cross-talk between Jak2 and NF-kappa-B signalling cascades, *Nature* 412:641–7, 2001.

James C, Ugo V, Le Couedic JP, Staerk J, Delhommeau F et al: A unique clonal JAK2 mutation leading to constitutive signalling causes polycythaemia vera, *Nature* 434:1144–8, 2005.

Kralovics R, Passamonti F, Buser AS, Teo SS, Tiedt R et al: A gain-of-function mutation of JAK2 in myeloproliferative disorders, *New Engl J Med* 352:1779–90, 2005.

Neubauer H, Cumano A, Muller M, Wu H, Huffstadt U et al: Jak2 deficiency defines an essential developmental checkpoint in definitive hematopoiesis, *Cell* 93:397–409, 1998.

Parganas E, Wang D, Stravopodis D, Topham DJ, Marine JC et al: Jak2 is essential for signaling through a variety of cytokine receptors, *Cell* 93:385–95, 1998.

Pritchard MA, Baker E, Callen DF, Sutherland GR, Wilks AF: Two members of the JAK family of protein tyrosine kinases map to chromosomes 1p31.3 and 9p24, *Mam Genome* 3:36–8, 1992.

Watling D, Guschin D, Muller M, Silvennoinen O, Witthuhn BA et al: Complementation by the protein tyrosine kinase JAK2 of a mutant cell line defective in the interferon-gamma signal transduction pathway, *Nature* 366:166–70, 1993.

JAK3

Changelian PS, Flanagan ME, Ball DJ, Kent CR, Magnuson KS et al (and 54 others): Prevention of organ allograft rejection by a specific Janus kinase 3 inhibitor, *Science* 302:875–8, 2003.

Gires O, Kohlhuber F, Kilger E, Baumann M, Kieser A et al: Latent membrane protein 1 of Epstein-Barr virus interacts with JAK3 and activates STAT proteins, *EMBO J* 18:3064–73, 1999.

Johnston JA, Kawamura M, Kirken RA, Chen YQ, Blake TB et al: Phosphorylation and activation of the Jak-3 Janus kinase in response to interleukin-2, *Nature* 370:151–2, 1994.

Macchi P, Villa A, Giliani S, Sacco MG, Frattini A et al: Mutations of Jak-3 gene in patients with autosomal severe combined immune deficiency, *Nature* 377:65–8, 1995.

Nosaka T, van Deursen JMA, Tripp RA, Thierfelder WE, Witthuhn BA et al: Defective lymphoid development in mice lacking Jak3, *Science* 270:799–801, 1995.

Rane SG, Reddy EP: JAK3: A novel JAK kinase associated with terminal differentiation of hematopoietic cells, *Oncogene* 9:2415–23, 1994.

Riedy MC, Dutra AS, Blake TB, Modi W, Lal BK et al: Genomic sequence, organization, and chromosomal localization of human JAK3, *Genomics* 37:57–61, 1996.

Russell SM, Johnston JA, Noguchi M, Kawamura M, Bacon CM et al: Interaction of IL-2R-beta and gamma(c) chains with Jak1 and Jak3: Implications for XSCID and XCID, *Science* 266:1042–5, 1994.

Russell SM, Tayebi N, Nakajima H, Riedy MC, Roberts JL et al: Mutation of Jak3 in a patient with SCID: essential role of Jak3 in lymphoid development, *Science* 270:797–9, 1995.

Thomis DC, Gurniak CB, Tivol E, Sharpe AH, Berg LJ: Defects in B lymphocyte maturation and T lymphocyte activation in mice lacking Jak3, *Science* 270:794–7, 1995.

TYK2

Firmbach-Kraft I, Byers M, Shows T, Dalla-Favera R, Krolewski JJ: tyk2, prototype of a novel class of non-receptor tyrosine kinase genes, *Oncogene* 5:1329–36, 1990.

Lukashova V, Asselin C, Krolewski JJ, Rola-Pleszczynski M, Stankova J: G-protein-independent activation of Tyk2 by the platelet-activating factor receptor, *J Biol Chem* 276:24113–21, 2001.

Ragimbeau J, Dondi E, Alcover A, Eid P, Uze G et al: The tyrosine kinase Tyk2 controls IFNAR1 cell surface expression, *EMBO J* 22:537–47, 2003.

Stoiber D, Kovacic B, Schuster C, Schellack C, Karaghiosoff M et al: TYK2 is a key regulator of the surveillance of B lymphoid tumors, *J Clin Invest* 114:1650–8, 2004.

Trask B, Fertitta A, Christensen M, Youngblom J, Bergmann A et al: Fluorescence in situ hybridization mapping of human chromosome 19: Cytogenetic band location of 540 cosmids and 70 genes or DNA markers, *Genomics* 15:133–45, 1993.

STAT1

Chen X, Vinkemeier U, Zhao Y, Jeruzalmi D, Darnell JE Jr et al: Crystal structure of a tyrosine phosphorylated STAT-1 dimer bound to DNA, *Cell* 93:827–39, 1998.

Copeland NG, Gilbert DJ, Schindler C, Zhong Z, Wen Z et al: Distribution of the mammalian Stat gene family in mouse chromosomes, *Genomics* 29:225–8, 1995.

Darnell JE Jr, Kerr IM, Stark GM: Jak-STAT pathways and transcriptional activation in response to IFNs and other extracellular signaling proteins, *Science* 264:1415–21, 1994.

Dupuis S, Dargemont C, Fieschi C, Thomassin N, Rosenzweig S et al: Impairment of mycobacterial but not viral immunity by a germline human STAT1 mutation, *Science* 293:300–3, 2001.

Dupuis S, Jouanguy E, Al-Hajjar S, Fieschi C, Al-Mohsen IZ et al: Impaired response to interferon-gamma/beta and lethal viral disease in human STAT1 deficiency, *Nature Genet* 33:388–91, 2003.

Durbin JE, Hackenmiller R, Simon MC, Levy DE: Targeted disruption of the mouse Stat1 gene results in compromised innate immunity to viral disease, *Cell* 84:443–50, 1996.

Haddad B, Pabon-Pena CR, Young H, Sun WH: Assignment of STAT1 to human chromosome 2q32 by FISH and radiation hybrids, *Cytogenet Cell Genet* 83:58–9, 1998.

Ihle JN: Cytokine receptor signaling, *Nature* 377:591–4, 1995.

Ihle JN: STATs: Signal transducers and activators of transcription, *Cell* 84:331–4, 1996.

Kaplan DH, Shankaran V, Dighe AS, Stockert E, Aguet M et al: Demonstration of an interferon gamma-dependent tumor surveillance system in immunocompetent mice, *Proc Natl Acad Sci USA* 95:7556–61, 1998.

Meraz MA, White JM, Sheehan KCF, Bach EA, Rodig SJ et al: Targeted disruption of the Statl gene in mice reveals unexpected physiologic specificity in the JAK-STAT signaling pathway, *Cell* 84:431–42, 1996.

Ramana CV, Chatterjee-Kishore M, Nguyen H, Stark GR: Complex roles of Statl in regulating gene expression, *Oncogene* 19:2619–27, 2000.

Shankaran V, Ikeda H, Bruce AT, White JM, Swanson PE et al: IFN-gamma and lymphocytes prevent primary tumour development and shape tumour immunogenicity, *Nature* 410:1107–11, 2001.

Wang J, Schreiber RD, Campbell IL: STAT1 deficiency unexpectedly and markedly exacerbates the pathophysiological actions of IFN-alpha in the central nervous system, *Proc Natl Acad Sci USA* 99:16209–14, 2002.

Wesemann DR, Qin H, Kokorina N, Benveniste EN: TRADD interacts with STAT1-alpha and influences interferon-gamma signaling, *Nature Immun* 5:199–207, 2004.

Yamamoto K, Kobayashi H, Arai A, Miura O, Hirosawa S et al: cDNA cloning, expression and chromosome mapping of the human STAT4 gene: Both STAT4 and STAT1 genes are mapped to 2q32.2-q32.3, *Cytogenet Cell Genet* 77:207–10, 1997.

STAT2

Bluyssen HAR, Levy DE: Stat2 is a transcriptional activator that requires sequence-specific contacts provided by Statl and p48 for stable interaction with DNA, *J Biol Chem* 272:4600–5, 1997.

Yan R, Qureshi S, Zhong Z, Wen Z, Darnell JE Jr: The genomic structure of the STAT genes: multiple exons in coincident sites in Statl and Stat2, *Nucleic Acids Res* 23:459–63, 1995.

Bhandari R, Kuriyan J: JAK-STAT signaling, in *Handbook of Cell Signaling*, Vol 1, Bradshaw RA, Dennis EA, eds, Academic Press, Amsterdam, 2004, pp 343–348.

Krauss G: *Biochemistry of Signal Transduction and Regulation*, 3rd ed, Wiley-VCH GmbH, 2003.

Human protein reference data base, Johns Hopkins University and the Institute of Bioinformatics, at http://www.hprd.org/protein.

G-PROTEIN RECEPTOR-MEDIATED CELL SIGNALING

Structure and Function [5.12]

Cell signaling events regulated by G proteins and G-protein-coupled receptors are referred to as *G-protein receptor-mediated cell signaling* (Table 5.12). *G-proteins* are guanine nucleotide-binding proteins, each composed of three subunits: α, β, and γ (see below). G proteins are localized to the cytoplasmic side of the cell membrane and coupled to a class of cell membrane receptors, defined as *G-protein-coupled receptors*. These receptors pass through the cell membrane back and forth for 7 times, and are thereby called *7-pass transmembrane receptors* (Fig. 5.13). Several types of extracellular ligands, including hormones, growth-related factors, small peptides, and neurotransmitters, can bind to G-protein receptors, initiating activation of the receptors as well as G proteins. In addition, photons and odorants can activate G-protein receptors in vision and olfactory cells, respectively.

G proteins exist in two states, inactive and active, and serve as switches for the regulation of several intracellular signaling pathways. Among the three subunits of the G protein,

TABLE 5.12. Characteristics of Selected Molecules for G-Protein Receptor-Mediated Signaling Pathways*

Proteins	Alternative Names	Amino Acids	Molecular Weight (kDa)	Expression	Functions
Guanine nucleotide binding protein G(S), α subunit	Guanine nucleotide-binding protein, α-stimulating activity polypeptide 1, Gs α subunit, adenylate cyclase stimulatory protein α subunit, adenylate cyclase-stimulating G α protein, stimulatory G protein, and GNAS	909	98	Ubiquitous	Activating adenylyl cyclase; also regulating metabolic processes and cell proliferation and differentiation
Guanine nucleotide-binding protein G(q) α subunit	G α q and guanine nucleotide-binding protein Q polypeptide	359	42	Ubiquitous	Inducing activation of phospholipase Cβ, regulating cell proliferation and differentiation, and controlling contractile activity of smooth muscle cells
Guanine nucleotide-binding protein β1 subunit	G protein β1 subunit, GNB1	340	37	Heart, red blood cells	Forming complexes with G protein α, γ subunits, and regulating the activity of α subunit

Guanine nucleotide-binding protein γ3 subunit	G protein γ3	75	8	Nervous system	Forming complexes with G protein α, β subunits, and regulating the activity of α subunit
Adrenergic receptor β1	ADRB1	477	51	Heart, adipocyte	Mediating the action of epinephrine and norepinephrine, and enhancing cardiac contractility
Adenylate cyclase 1	Adenylate cyclase type I	1119	123	Brian, heart, lung, liver, kidney, pancreas, spleen, ovary, skeletal muscle	Catalyzing the formation of cAMP
Phospholipase Cβ1	Phospholipase C β1, 1 phosphatidylinositol 4,5-bisphosphate phosphodiesterase β1, phosphoinositide-specific phospholipase C β1, PLC β1, PLC1	1216	139	Brain, lung, blood vessel	Catalyzing the formation of inositol 1,4,5-trisphosphate and diacylglycerol from phosphatidylinositol 4,5-bisphosphate, and participating in regulation of cellular activities such as cell proliferation and migration

*Based on bibliography 5.12.

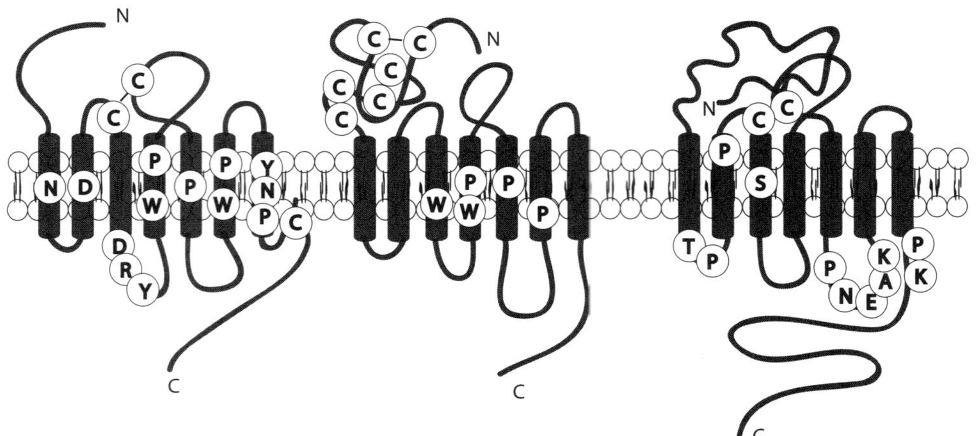

Figure 5.13. Schematic representation of the types of G-protein-coupled receptors. (Based on Gether U: *Endocr Rev* 21:90–113, 2000.)

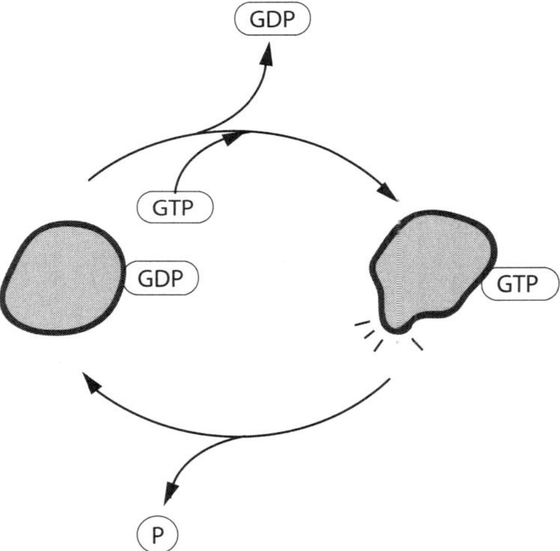

Figure 5.14. Schematic representation of G-protein activation and deactivation. (Based on bibliography 5.12).

the α subunit is responsible primarily for regulating the activity of the G-protein. Each α subunit is composed of a GTPase, which hydrolyses GTP, and a helical domain, which holds a guanine nucleotide, also known as a *guanine nucleotide pocket*. In an inactive state, a GDP molecule is bound to the guanine nucleotide pocket of the α subunit. When a G-protein is activated in response to the binding of a ligand to the G-protein receptor, the GDP molecule is released, and a GTP is bound to the α subunit (Fig. 5.14). Simultaneously, the GTP-bound α subunit is dissociated from the trimeric G protein, forming two distinct modules: the GTP-bound α subunit and βγ complex. These modules are able to activate downstream signaling pathways. On the completion of signal transduction, the GTP molecule with the α subunit is rapidly hydrolyzed by the GTPase associated with the α subunit, diminishing the activity of signaling pathways regulated by G proteins.

There exist various types of α, β, and γ subunits. To date, about 20 α, 5 β, and 12 γ subunits have been found. These subunits can be assembled into a variety of G-protein isoforms. Based on the structure of the α subunits, G proteins have been classified into four groups: G_s, G_q, $G_{i/o}$, and $G_{12/13}$. Among these G proteins, the first three types have been well studied and characterized. The G_s group includes the G_s-proteins and olfactory specific G_{olf} proteins. These G proteins are referred to as *stimulatory G proteins* for their role in the activation of adenylyl cyclase. The G_q group is composed of G_q, G_{11}, G_{14}, and G_{15} proteins. These proteins activate phospholipase Cβ, a key enzyme that induces the activation of protein kinase C (PKC) and phosphatidyl inositol triphosphate signaling pathways (described later in this section). The $G_{i/o}$ group includes a number of members, including G_{i1}, G_{i2}, G_{i3}, G_{oA}, G_{oB}, G_z, $G_{t\text{-rod}}$, $G_{t\text{-cone}}$, and G_{gust}. These $G_{i/o}$ members elicit different effects. Activated G_{i1}, G_{i2}, G_{i3}, G_{oA}, G_{oB}, and G_z inhibit the activity of adenylyl cyclase. For such a function, these G proteins are referred to as *inhibitory G proteins.* However, G_{i1}, G_{i2}, G_{i3} can stimulate the activation of phosphatidylinositol 3-OH kinase γ (PI3Kγ). $G_{t\text{-rod}}$ and $G_{t\text{-cone}}$ can activate cGMP-phosphodiesterase (PDE) and are responsible for regulating the function of rod and cone photoreceptors, respectively, in retinal vision cells. G_{gust} can activate cGMP-phosphodiesterase and regulate the function of gustatory cells. $G_{i/o}$ proteins are usually sensitive to pertussis toxin, a substance possessing the activity of ADP-ribosyl transferase and inducing ADP-ribosylation in $G_{i/o}$ proteins. Such a process suppresses the activity of the $G_{i/o}$ proteins.

Signaling Mechanisms [5.13]

The discussion above indicates that each type of G protein mediates distinct signaling pathway(s). While it is beyond the scope of this book to cover all G-protein-related mechanisms, examples from three major G-protein signaling pathways, including the G_s, G_i, and G_q pathways, are discussed briefly.

G_s proteins are often found in cells undergoing metabolic processes for the breakdown of energy-storing molecules, such as triglyceride and glycogen. A typical example is the G_s protein coupled to the β-adrenergic receptor (Fig. 5.9). The binding of ligands, such as epinephrine and norepinephrine, to the β-adrenergic receptor induces the ejection of the bound GDP molecule from the G_s α subunit and the recruitment of a GTP molecule to the guanine nucleotide pocket, while the α subunit dissociates from the βγ subunits. The free α subunit–GTP complex binds to and activates adenylyl cyclase. Activated adenylyl cyclase acts on an ATP molecule, inducing the formation cyclic AMP or cAMP. cAMP serves as a second messenger and interacts with cAMP-dependent protein kinase, or PKA, inducing the activation of this kinase. PKA is a serine/threonine kinase and can transfer the terminal phosphate group from an ATP molecule to serine and threonine residues of substrate proteins, a process known as *serine/threonine phosphorylation.* A typical substrate of PKA is phosphorylase kinase. Phosphorylated phosphorylase kinase induces the phosphorylation of glycogen phosphorylase, which catalyzes the breakdown of glycogen into glucose 1-phosphates, a process referred to as *glycolysis.* At the same time, activated PKA phosphorylates another enzyme called *glycogen synthase*, which catalyzes the synthesis of glycogen from glucose. The phosphorylation of glycogen synthase negatively regulates the activity of the enzyme, inducing suppression of glycogen synthesis. Thus, the activation of G_s-protein signaling pathway leads to the release of glucose, increasing the blood concentration of glucose and enhancing energy production.

In contrast to the G_s proteins, the *G_i proteins* elicit an inhibitory effect on the activity of adenylyl cyclase. G_i proteins are coupled to α_2-adrenergic receptors. Among the three subunits of a G protein, the β and γ subunits are identical between the G_i and G_s proteins, but the α subunit is different, which determines the distinct functions of the G_i protein. The binding of epinephrine or norepinephrine to α_2-adrenergic receptors activates the G_i proteins, resulting in the substitution of GTP for GDP in the α subunit and the dissociation of the α subunit from the $\beta\gamma$ subunits. The free α subunit acts to suppress the activity of adenylyl cyclase, resulting in inhibition of glucose metabolism. This signaling mechanism counterbalances that of the G_s proteins.

The *G_q-proteins* belong to another group of stimulatory signaling pathways. These G proteins regulate the transport of Ca^{2+} and the activity of several intracellular molecules via the mediation of cell membrane lipids. G_q proteins are coupled to cell membrane receptors, which interact with several extracellular ligands, including vasopressin, acetylcholine, thrombin, and angiotensins I and II. The binding of a ligand to a G_q-protein-coupled receptor stimulates the G_q protein, which activates phospholipase Cβ, a phosphoinositol-specific enzyme. Activated phospholipase Cβ hydrolyzes phosphatidylinositol 4,5-biphosphate (PIP_2) into inositol 1,4,5-triphosphate (IP_3) and diacylglycerol. These lipid molecules excert distinct functions. The IP_3 molecule can diffuse through the cytoplasm and acts on Ca^{2+} channels in the endoplasmic reticulum (ER), resulting in the opening of these channels and Ca^{2+} release from the ER to the cytoplasm (Fig. 5.15). Ca^{2+} mediates many cellular processes, including actin–myosin interaction in muscular cells, secretion of neurotransmitters, and signal transduction. The other molecule, diacylglycerol, can activate Ca^{2+}-dependent protein kinase, or protein kinase C (PKC), a serine/threonine protein kinase. Activated PKC can phosphorylate two known substrate proteins: mitogen-activated protein kinase (MAPK) and IκB. The phosphorylation of MAPK initiates a cascade of mitogenic activities, which is discussed on page 151. IκB forms a complex with NFκB, a *trans*-acting factor that regulates the expression of mitogenic genes. The phosphorylation of IκB induces the release of NFκB, which translocates to the cell nucleus and induces gene expression.

The G-protein signaling pathways can communicate by crosstalk with other signaling pathways, resulting in diverse biological consequences. For example, the G_q protein α subunit can interact with various molecules of the Ras-MAPK signaling pathways, activating mitogenic cellular activities. Recent studies has suggest that both $G_{\alpha i}$ and $G_{\alpha s}$ may be able to interact with and activate Src, which in turn activate STAT3, a key transcription factor that can be activated by cytokines (see page 207). Thus, the G-protein signaling pathways can impose effects on diverse cellular activities via interactions with various signaling pathways.

NFκB-MEDIATED CELL SIGNALING

Structure and Function [5.14]

Nuclear factor κB (NFκB) belongs to a family of *trans*-acting factors, which stimulate the expression of genes encoding proteins for regulating cell survival and proliferation as well as inflammatory and immune processes. There are five known members of the NFκB family in mammalian cells, including RelA (p65), RelB, c-Rel, NFκB1 (p50), and NFκB2 (p52). These proteins possess a highly conserved domain near the *N*-terminus, known as

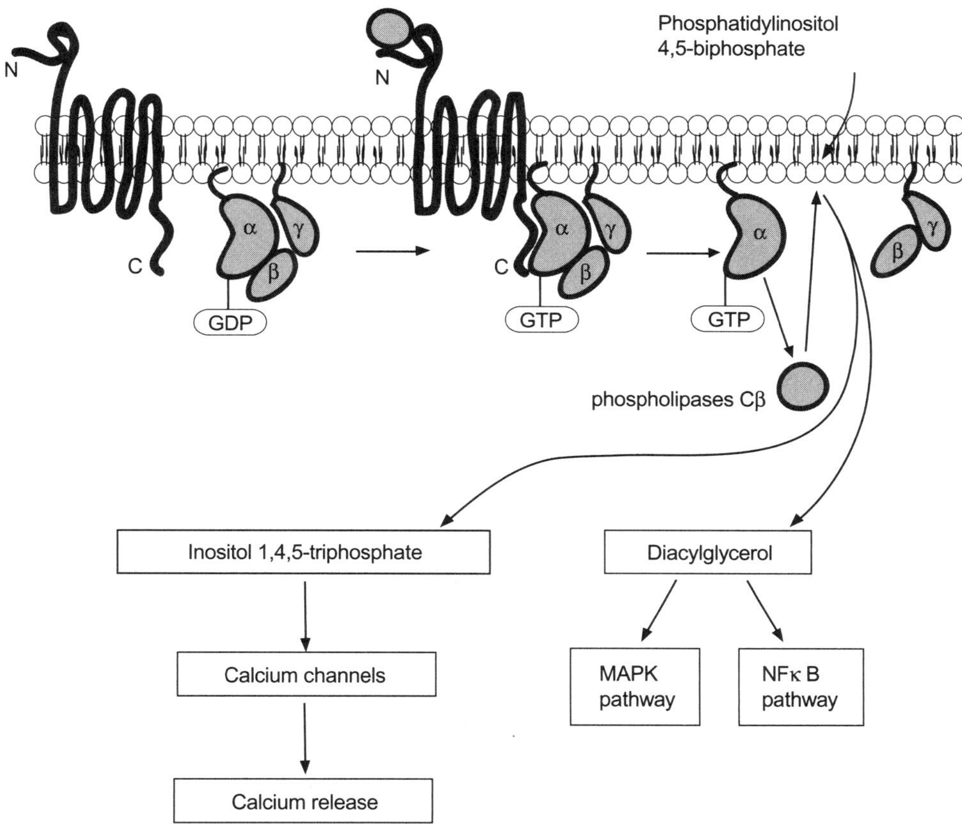

Figure 5.15. Schematic representation of the mechanisms of Gq-protein signaling (based on bibliography 5.13).

the *Rel homology domain* (RHD). This domain is responsible for NFκB binding to target DNA, dimerization, and association with inhibitory κB (IκB), which binds to NFκB and inhibits the activity of NFκB in unstimulated cells. The NFκB family proteins usually exist in the form of homodimer or heterodimer, such as NFκB1/RelA, NFκB1/ NFκB1, and NFκB2/ NFκB2. These NFκB dimeric complexes may impose apparently different effects on target genes. For instance, the NFκB1/RelA heterodimer induces the expression of target genes, whereas the NFκB1/ NFκB1 and NFκB2/ NFκB2 homodimers elicit an opposite effect.

In unstimulated cells, the NFκB family proteins, existing as heterodimers or homodimers, are associated with a protein of the IκB family, which is composed of seven members, including IκBα, IκBβ, IκBε, IκBγ, Bcl3, NFκB1 precursor (p105), and NFκB2 precursor (p100). These IκB family members can bind to and mask the nuclear localization signals of a NFκB molecule. This action renders the NFκB molecule incapable of binding to and interacting with target DNA, thus inhibiting the transcriptional activity of NFκB. The NFκB family proteins can only be activated when the inhibitory IκB protein is degraded or removed. (See Table 5.13.)

TABLE 5.13. Characteristics of Selected Molecules of NFκB-Mediated Signaling Pathways*

Proteins	Alternative Names	Amino Acids	Molecular Weight (kDa)	Expression	Functions
Nuclear factor κB subunit 1	Nuclear factor of κ light polypeptide gene enhancer in B cells 1, NFκB, NFKB1, transcription factor NFKB1, NFKB p105	968	105	Ubiquitous	Forming the protein complex of nuclear factor κB (NFκB) and regulating the activity of NFκB
Nuclear factor κB subunit 2	Nuclear factor of κ light-chain gene enhancer in B cells 2, transcription factor NFKB2, NFKB p52/p100 subunit	933	100	Ubiquitous	Forming the protein complex of nuclear factor κB (NFκB) and regulating the activity of NFκB
IκBα	Nuclear factor κB inhibitor, IKBA, nuclear factor of κ light-chain gene enhancer in B cells inhibitor α, nuclear factor of κ light-chain gene enhancer in B cells inhibitor (NFKBI), inhibitor of κ light-chain gene enhancer in B cells α	317	36	Leukocytes	Forming complex proteins with NFκB1 or NFκB2 and inhibiting the activity of these NFκB subunits
IκBβ	NFκB inhibitor β, IKB β, nuclear factor of κ light polypeptide gene enhancer in B-cell inhibitor β, Inhibitor of κ light-chain gene enhancer in B cells β, IKBB	356	38	Ubiquitous	Forming complex proteins with NFκB1 or NFκB2 and inhibiting the activity of these NFκB subunits

IκB kinase α	I κB kinase α, IKKA, I κB kinase 1, IKK1	745	85	Ubiquitous	A serine/threonine protein kinase that dephosphorylates IκB, inducing degradation of IκB via ubiquitination and activation of NFκB
IκB kinase β	Nuclear factor of κ light-chain gene enhancer in B-cell inhibitor kinase β, NFKBIKB, I-κB kinase β, IKKB, I-κB kinase 2, IKK2, inhibitor of nuclear factor κB kinase β subunit, nuclear factor NFκB inhibitor kinase β	756	87	Ubiquitous	A serine/threonine protein kinase that dephosphorylates IκB, inducing degradation of IκB via ubiquitination and activation of NFκB
IκB kinase γ	Inhibitor of κ light polypeptide gene enhancer in B cells, kinase γ, inhibitor of nuclear factor κB kinase γ subunit, I-κB kinase γ, IkB kinase γ subunit, IKKγ	419	48	Ubiquitous	A serine/threonine protein kinase that dephosphorylates IκB, inducing degradation of IκB via ubiquitination and activation of NFκB
Atk	Agammaglobulinemia tyrosine kinase, Bruton's tyrosine kinase, tyrosine-protein kinase BTK, bruton agammaglobulinemia tyrosine kinase, B-cell progenitor kinase	659	76	Lymphocytes, bone marrow	Regulating B-cell development and causing agammaglobulinemia[a] when detected

[a] Agammaglobulinemia is an X-linked immunodeficiency with impaired generation of mature B lymphocytes and failure of Ig heavy-chain rearrangement.

*Based on bibliography 5.14.

Signaling Mechanisms [5.15]

The NFκB signaling pathway can be activated by several factors, including physical stress, oxidative stress, chemical toxins, and bacterial and viral infection. These factors can interact with cells and activate NFκB, which in turn stimulates the expression of many inflammatory cytokines, chemokines, immune receptors, and cell surface adhesion molecules. Thus, NFκB has been considered as a central mediator for immune responses as well as stress responses induced by physical stress, oxidative stress, and chemical substances. For instance, NFκB can be activated in response to the stimulation of IL1. The binding of IL1 to the IL1 receptor activates a molecule known as *NFκB-inducing kinase* (NIK or serine/threonine protein kinase NIK), which in turn phosphorylates the IκB kinase. Activated IκB kinase can phosphorylate the IκB module of the IκB-NFκB complex, inducing the separation of NFκB from IκB. NFκB is a complex of p50 and p65 proetins, which serves as a transcriptional factor and regulates gene transcription. Figure 5.16 shows the mechanisms of NFκB activation in response to the stimulation of cytokines.

In addition, extracellular mitogens and growth factors can activate NFκB, which in turn regulates cell survival and proliferation. The activation of NFκB is mediated by a protein kinase known *as IκB kinase* (IKK), which is composed of at least three subunits: two catalytic subunits (IKKα and IKKβ) and a regulatory subunit (IKKγ). Extracellular mitogens and growth factors can activate IKK via various signaling pathways. For instance, protein kinases MEKK1, MEKK2, and MEKK3, which are activated through receptor protein tyrosine kinase signaling pathways, are capable of phosphorylating IKK. Activated IKK in turn phosphorylates two specific serine residuals on IκB. The phosphorylation of IκB leads to the ubiquitination and degradation of IκB by proteasome. The degradation of IκB liberates NFκB, which translocates to the cell nucleus, binds to target genes, and regulates gene transcription, resulting in the activation of inflammatory and mitogenic activities.

In addition to the degradation and removal of IκB, serine phosphorylation of NFκB may be required for certain NFκB family members for efficient binding to transcriptional activators and interaction with target genes. For instance, the catalytic domain of protein kinase A can bind to an inactive NFκB p65 protein in unstimulated cells. On IκB degradation, protein kinase A phosphorylates p65 on serine 276, resulting in a conformational change in the p65 protein and consequent interaction with a transcriptional activator, namely CBP (CREB-binding protein or cAMP response element-binding protein-binding protein), which increases the transcriptional activity of NFκB. Another example involves the activation of protein kinases PI3K and Atk (agammaglobulinemia tyrosine kinase), which mediate IL1- and TNFα-initiated NFκB activation. In cultured cells, IL1 and TNFα induce the activation of PI3K and Atk. These protein kinases in turn phosphorylate NFκB p65, enhancing the DNA-binding capacity of NFκB.

UBIQUITIN AND PROTEASOME-MEDIATED CELL SIGNALING

Structure and Function [5.16]

Ubiquitin (see Table 5.14) and *proteasome* are two critical protein structures of a protease system that recognizes and degrades damaged, unfolded, and nonfunctional proteins. Ubiquitin is a 76-amino acid protein that can attach to a target protein via an isopeptide bond linking the terminal carboxyl group of the ubiquitin molecule to a ε-amino group

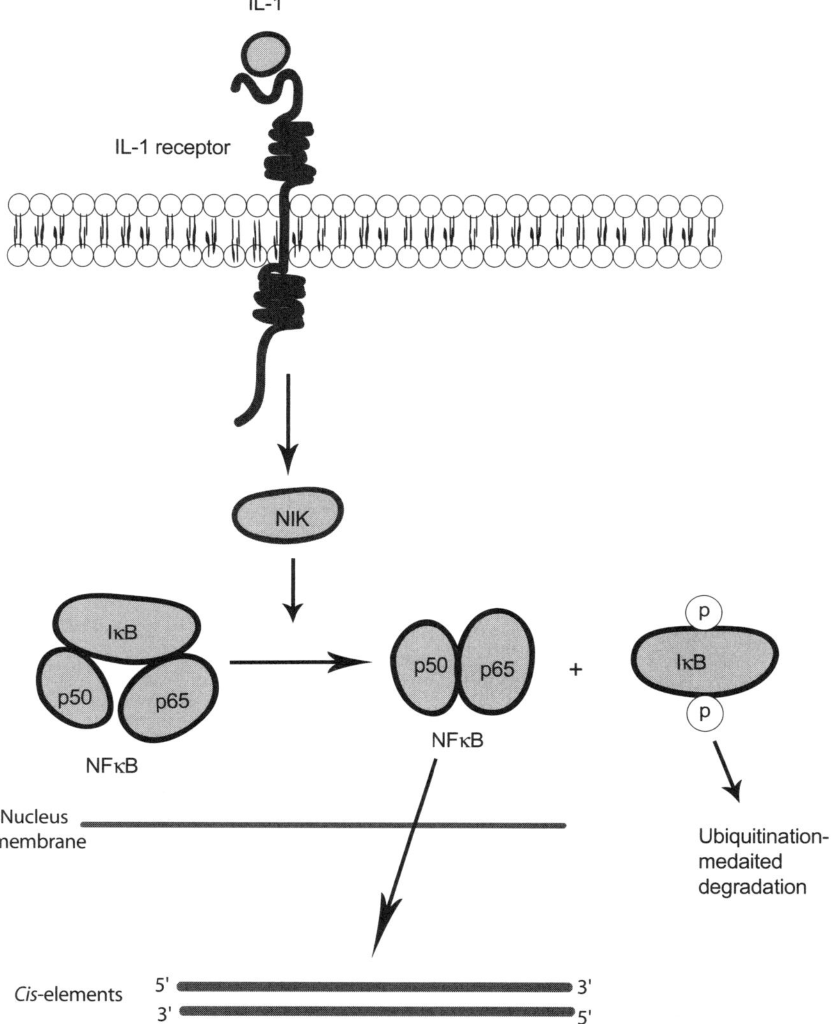

Figure 5.16. Mechanisms of nuclear factor (NF)κB activation in response to the stimulation of cytokines IL6 and tumor necrosis factor (TNF)α (based on bibliography 5.15).

of a lysine of the target protein. Additional ubiquitins can be added to the linked ubiquitin, forming a polyubiquitin chain. Such a process is known as *ubiquitination*. The polyubiquitin chain serves as a tag for the recognition of the targeted protein by a proteasome, a multicatalytic protease that destroys the ubiquitin-tagged protein. Ubiquitination and proteasome activation are an effective means for the destruction of useless proteins, an important process for protein metabolism and recycling.

In addition to ubiquitins and proteasomes, several mediating factors are required for ubiquitination. These include an ubiquitin-activating enzyme, an ubiquitin-conjugating enzyme, and an ubiquitin/protein ligase. The *ubiquitin-activating enzyme* can bind to and activate ubiquitin via a thiolester bond, a process that requires ATP. The *ubiquitin-*

TABLE 5.14. Characteristics of Selected Ubiquitin Molecules*

Proteins	Alternative Names	Amino Acids	Molecular Weight (kDa)	Expression	Functions
Ubiquitin B	Polyubiquitin B, UBA52	229	26	Brain, dependent pancreas	Regulating ATP-degradation of abnormal and nonfunctional proteins; also mediating gente expression by binding to histone H2A (note that ubiquitins do not cause histone H2A degradation)
Ubiquitin C	Polyubiquitin, PolyUB	685	77	Kidney	Similar to functions of ubiquitin B

*Based on bibliography 5.16.

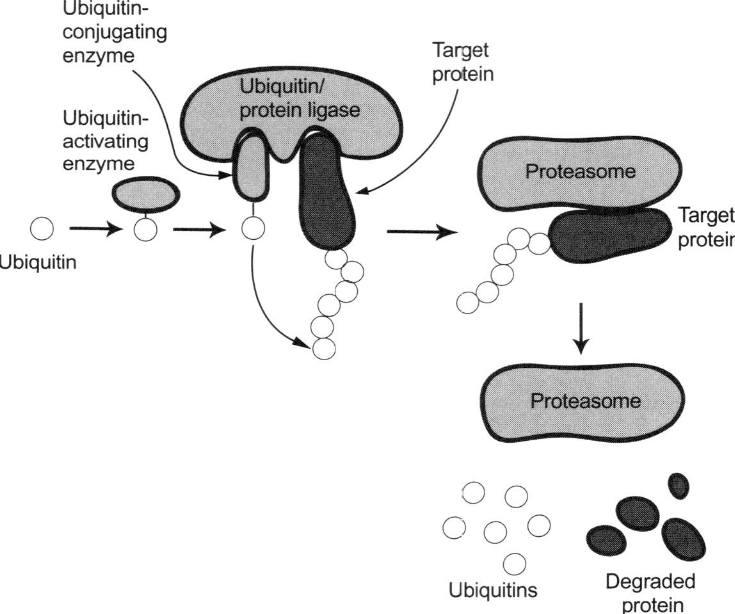

Figure 5.17. Schematic representation of the components of the ubiquitination system. (From Nakayama KI et al: Regulation of the cell cycle at the G1-S transition by proteolysis of cyclin E and p27Kip1, *Biochem Biophys Res Commun* 282:853–60, copyright 2001, with permission from Elsevier.)

conjugating enzyme can transfer an activated ubiquitin to a target protein with the help of the *ubiquitin/protein ligase*, which recognizes target proteins and promotes ubiquitin transfer and ubiquitination (Fig. 5.17). Thus, these factors play a critical role in the initiation and regulation of ubiquitination.

Signaling Mechanisms [5.17]

Protein ubiquitination is regulated at the level of substrate proteins. Two processes are often involved in the regulation of ubiquitination: substrate phosphorylation and hydroxylation. *Substrate phosphorylation* may induce activation or inhibition of ubiquitination, depending on the nature of substrate proteins and ubiquitin ligases, whereas *substrate hydroxylation* activates ubiquitination. Several examples are given here, to demonstrate these regulatory mechanisms.

Stimulation of Ubiquitination by Substrate Phosphorylation. Substrate phosphorylation is required for the activation of several ubiquitin ligases, including S-phase kinase-associated protein 1 (Skpl), Cdc53, and F-box protein, which are designated as SCF ubiquitin ligases. These ubiquitin ligases form complexes with an identical Skpl and Cdc53 protein, but a varying F-box protein. A specified F-box protein is responsible for the recognition of a specific substrate protein and the affinity of the ligase complex to the substrate. The phosphorylation of a substrate protein can activate a SCF ligase complex, inducing ubiquitination and destruction of the substrate. Several transcriptional factors, including IκB and β-catenin, are known substrates for the SCF ligase complex.

IκB is an inhibitory molecule that binds to and sequesters NFκB in unstimulated cells. On the stimulation of NFκB-related signaling pathways, the IκB molecule becomes phosphorylated at two serine residues, which stimulate the activation of a SCF ubiquitin ligase complex, namely, $SCF^{\beta\text{-TrCP}}$, where β-TrCP is a specific F-box protein. Activated $SCF^{\beta\text{-TrCP}}$ induces ubiquitination and degradation of IκB, a necessary step for the release and activation of NFκB. Activated NFκB serves as a transcriptional factor, migrates to target genes, and initiates gene transcription. In this case, ubiquitination is not only responsible for the degradation of IκB, but also contributes to the activation of NFκB.

β-Catenin is a coactivator of transcriptional factors, participating in the regulation of gene transcription. β-Catenin can be phosphorylated by glycogen synthase kinase 3β. The phosphorylation of β-catenin activates $SCF^{\beta\text{-TrCP}}$, which in turn induces β-catenin ubiquitination and destruction. This is an effective approach for reducing transcriptional activities mediated by β-catenin. (See Table 5.15.)

Inhibition of Ubiquitination by Substrate Phosphorylation. The phosphorylation of certain proteins may reduce the activity of ubiquitin ligases, thus suppressing ubiquitination of the substrate proteins. A typical example is p53, a tumor suppressor protein (see Table 5.18, later in this chapter for p53). Toxic stress can lead to p53 phosphorylation. Phosphorylated p53 in turn induces cell arrest during cell mitosis. Excessive p53 is usually degraded by ubiquitination, which is mediated by an ubiquitin ligase mdm2. Phosphorylation of p53 on the serine residues has been shown to prevent interaction of p53 with the ubiquitin ligase mdm2, suppressing ubiquitination. Such a process stabilizes p53 and enhances the function of p53.

Stimulation of Ubiquitination by Substrate Hydroxylation. In addition to substrate phosphorylation, substrate hydroxylation plays a role in regulating substrate ubiquitination. An example is the ubiquitination of the hypoxic response transcriptional regulator hypoxia inducible factor 1 α (HIF1α) (see Table 5.16). This factor is stabilized and activated under a hypoxic condition but degraded under a normoxic condition. Ubiquitination of HIF1α is a critical step in the degradation of the HIF1α protein. Following recovery from a

TABLE 5.15. Characteristics of S-Phase Kinase-Associated Protein 1A and β-Catenin*

Proteins	Alternative Names	Amino Acids	Molecular Weight (kDa)	Expression	Functions
S-phase kinase-associated protein 1A	SKP1, SKP1A	163	19	Ubiquitous	Serving as a substrate recognition component of the SCF ubiquitin ligase complex, binding to regulatory proteins involved in ubiquitin proteolysis (e.g., cyclin F and S-phase kinase-associated protein 2); also serving as an RNA polymerase II elongation factor
β-Catenin	Catenin β 1, cadherin-associated protein β	781	86	Ubiquitous	Serving as an adherens junction protein, mediating cell–cell adhesion and interaction, regulating cell attachment to matrix, cell proliferation, and differentiation during embryogenesis, wound healing, and tumor cell metastasis; also serving as a coactivator of transcriptional factors

*Based on bibliography 5.17.

TABLE 5.16. Characteristics of Hypoxia Inducible Factor 1α Subunit*

Proteins	Alternative Names	Amino Acids	Molecular Weight (kDa)	Expression	Functions
Hypoxia inducible factor 1α subunit	HIF1α	826	93	Ubiquitous	Acting as a transcriptional factor, regulating cellular responses to reduced oxygen concentration or hypoxia, mediating inflammatory reactions, enhancing angiogenesis, and mediating neural development

*Based on bibliography 5.17.

hypoxic condition, increased oxygen concentration stimulates the activation of a proline hydroxylase, which catalyzes the hydroxylation of a proline residue in the HIF1α protein. The hydroxylated proline mediates the interaction of HIF1α with an ubiquitin ligase complex known as the *VBC complex* (Von Hippel–lindau protein–elongin B–Elongin C), which is similar to the SCF complex in assembly and function. Activated VBC complex in turn induces HIF1α ubiquitination and degradation.

NUCLEAR RECEPTOR-MEDIATED CELL SIGNALING

Structure and Function [5.18]

Nuclear receptors are a family of intracellular proteins that interact with steroid, thyroid, and retinoid hormones, which are lipid-soluble molecules and can diffuse through the cell membrane. These receptors are found in the cytoplasm and cell nucleus. Among the nuclear receptors, the *steroid hormone receptor subfamily* has been studied extensively. Common nuclear receptors include estrogen receptor (ER) α and β, glucocorticoid receptor (GR), mineralocorticoid receptor (MR), progesterone receptor (PR), androgen receptor (AR), and vitamin D receptor (VDR). These receptors interact with corresponding steroid hormones (Fig. 5.18). In addition, there are several nuclear receptors that are similar to estrogen receptors in function. These are defined as estrogen-related receptors α, β, and γ, which are also referred to as "orphan" nuclear receptors.

A typical nuclear receptor is composed of several functional domains, including a DNA-binding domain (DBD), a ligand-binding domain (LBD), an activation function 1 (AF1) domain, and an activation function 2 (AF2) domain. The *DNA-binding domain* is located in the central region of the receptor and composed of two zinc finger motifs that are responsible for protein–DNA interaction. The DNA-binding domain is well reserved among different nuclear receptors. The *ligand-binding domain* is located at the *C*-terminus and is composed of sequences responsible for ligand interactions, activation of nuclear transcription, binding to chaperone proteins, and dimerization with other

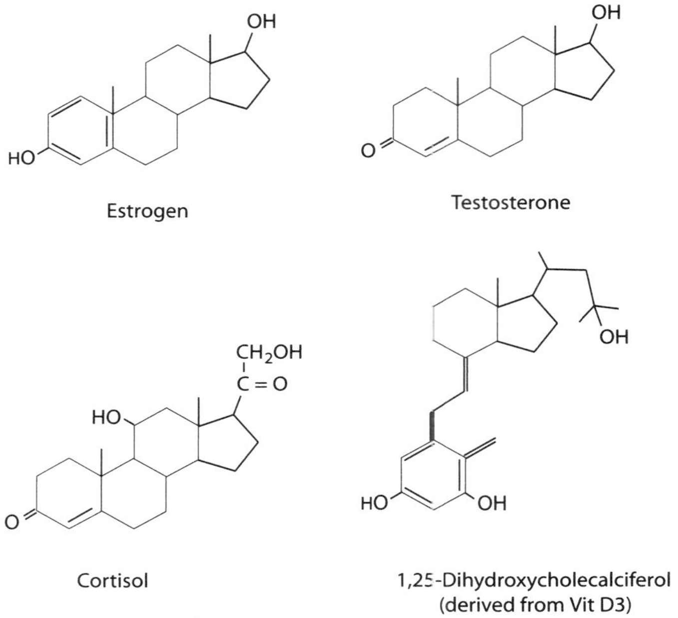

Figure 5.18. Steroid hormones for nuclear receptors. (Based on bibliography 5.18).

receptors. The ligand-binding domain is moderately conserved compared with the DNA-binding domain. The AF1 and AF2 *domains* regulate the activation of the receptor. The difference between these two domains is that the activity of the AF1 domain is independent of ligand binding, whereas the activity of the AF2 domain is dependent on ligand binding. On the interaction with hormone ligands, the hormone–receptor complexes are activated, serve as transcriptional factors, and directly bind to corresponding *cis* elements in target genes, initiating gene transcription. The nuclear receptors participate in the regulation of several physiological processes, including salt balance, glucose metabolism, reproduction, and responses to environmental stress impacts. (See Table 5.17.)

Signaling Mechanisms [5.18]

In unstimulated cells, nuclear receptors are associated with "chaperones," which suppress the activity of the receptors. On the binding of ligands to the nuclear receptors, the chaperones are dissociated from the receptors, resulting in receptor conformational changes and the exposure of the nuclear localization signals. The ligand–receptor complexes often form homodimers or heterodimers and are translocated from the cytoplasm to the nucleus, initiating gene transcription (Fig. 5.19). The activation of the nuclear receptor transcriptional factors is regulated to a large degree by the AF1 and AF2 domains of the nuclear receptor. These domains recruit and activate nuclear receptor cofactors, which are enzymes including acetylases, deacetylases, methylases, kinases, and ubiquitinases. These cofactors play a critical role in regulating the formation of the transcription-initiating complexes, the conformation of target DNA *cis*-acting elements, and the degradation and recycling of the nuclear receptors.

TABLE 5.17. Characteristics of Selected Molecules for Nuclear Receptor-Mediated Signaling Pathways*

Proteins	Alternative Names	Amino Acids	Molecular Weight (kDa)	Expression	Functions
Estrogen receptor α	Estrogen receptor 1, ESR, ER, oestrogen receptor α	595	66	Uterus, ovary, blood vessel, kidney, bone	Acting as a transcriptional factor, regulating gene expression and mediating the development of female sex characters, promoting fertilization and implantation, and mediating the formation of bone matrix
Glucocorticoid receptor	GCR, GCCR, GR	777	86	Ubiquitous	Serving as a transcriptional factor, forming a homodimer or heterodimer with another protein (e.g., retinoid receptor and a heatshock protein), regulating the expression of glucocorticoid-responsive genes, promoting gluconeogenesis and elevation of blood glucose concentration, and inhibiting inflammatory and immune responses

TABLE 5.17. *Continued*

Proteins	Alternative Names	Amino Acids	Molecular Weight (kDa)	Expression	Functions
Mineralocorticoid receptor	MR, aldosterone receptor	984	107	Kidney, brain, heart, liver, intestine	Acting as a transcriptional factor, stimulating the expression of mineralocorticoid responsive genes, and mediating electrolyte and water balance via controlling ion transport in the renal tubular system, resulting in retention of sodium and water and loss of potassium
Progesterone receptor	PR	933	99	Uterus, ovary, placenta, brain, blood vessel, leukocytes	Acting as a transcriptional factor in response to binding of progesterone; also regulating the establishment and maintenance of pregnancy
Androgen receptor	AR, dihydrotestosterone receptor (DHTR)	920	99	Prostate, adrenal gland, brain	Serving as a transcriptional factor, inducing transcription of androgen responsive genes, and regulating masculinization or the development of male sex characteristics
Vitamin D receptor	1,25-Dihydroxyvitamin D_3 receptor, VDR	427	48	Ubiquitous	Serving as a transcriptional factor and regulating calcium metabolism and the formation of bone matrix

*Based on bibliography 5.18.

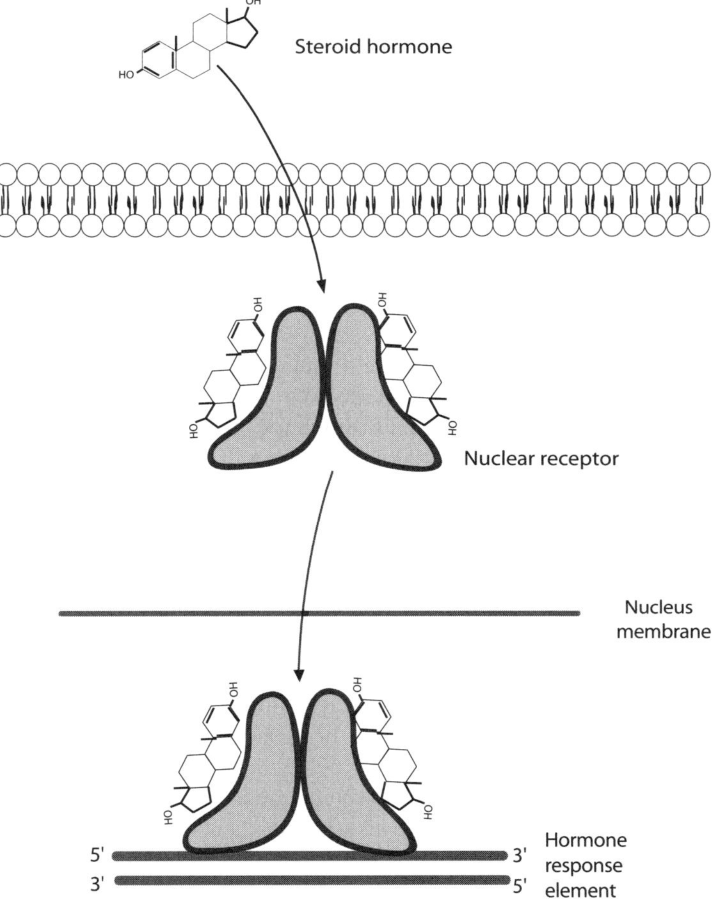

Figure 5.19. Mechanisms of nuclear receptor activation (based on bibliography 5.18).

When activated, the nuclear receptor transcriptional factors interact with specific DNA *cis*-acting elements defined as *hormone response elements* (HREs) with the assistance of cofactors. The HREs are located within the promoter and enhancer regions of target genes and are composed of unique sequences for different steroid ligands. For instance, the hormone response element for estrogen receptors and estrogen-related receptors contains repeating AGGTCA sequences, whereas that for glucocorticoid receptor, mineralocorticoid receptor, progesterone receptor, and androgen receptor contains AGAACA. Thus, the structure of these *cis* elements determines the specificity of the binding of nuclear receptor transcriptional factors to DNA. These transcriptional factors bind DNA in the form of homodimers or heterodimer.

In addition to steroid hormones, other types of molecule can activate nuclear receptors. A typical example is the activation of nuclear receptors by protein kinases, including protein kinase A, mitogen-activated protein kinases, cyclin-dependent kinases, and glycogen synthase kinases. While the exact mechanisms remain poorly understood, it appears that growth factor-induced activation of intracellular protein kinases play a role in the activation of nuclear receptors. Several ligands, including epidermal growth factor and insulin growth factor, can induce activation of protein kinases, leading to the phosphorylation of serine residues within the AF1 domain of the nuclear receptors. Such an action facilitates the recruitment of coactivators to nuclear receptors, enhancing the activity of the nuclear receptor. These activities have been demonstrated in estrogen receptor-mediated signaling events.

p53-MEDIATED CELL SIGNALING

Structure and Function [5.19]

p53 (Table 5.18) is a 393-amino acid protein that serves as a transcriptional factor. The activation of p53 leads to cell-cycle arrest, growth inhibition, and cell apoptosis. During the period of cell arrest, p53 often repairs impaired genes. These functions are implemented by regulating the expression of specific genes. These genes encode proteins that control cell cycle progression and cell apoptosis. Because of its inhibitory effect on cell growth and stimulatory effect on cell apoptosis, p53 is considered a tumor suppressor protein.

Based on the amino acid sequence and function, the p53 protein is divided into several domains, including the *N*-terminal domain from amino acids 1–101, a central DNA-binding domain from 102 to 292, and a *C*-terminal domain from 293 to 393. The *N*-terminal domain is capable of interacting with regulatory proteins, which activate or suppress the activity of the p53 protein. Several proteins, including TFIID, TFIIH, TAFs, PCAF, and the MDM2 ubiquitin ligase, have been shown to interact with the *N*-terminal domain. In this region, amino acid residues 1–31 and 80–101 are highly reserved among mammals. The central domain of the p53 protein contains DNA-specific binding sites. A consensus DNA-binding site includes two identical segments, each composed of a sequence of RRRCWWGYYY. The *C*-terminal domain contains several sequences, including a nuclear localization signal, a tetramerization sequence, and DNA-binding sequence. Tetramerization of the p53 protein is required for the activation of the protein as a transcriptional factor.

TABLE 5.18. Characteristics of p53*

Proteins	Alternative Names	Amino Acids	Molecular Weight (kDa)	Expression	Functions
p53	Tumor protein p53, transformation-related protein 53, TRP53	393	44	Ubiquitous	A transcriptional factor that binds to target genes, regulates gene transcription, inhibits cell proliferation and differentiation, and suppresses tumor growth

*Based on bibliography 5.19.

Signaling Mechanisms [5.20]

The p53 protein exists in a latent form and does not induce gene expression in unstimulated cells. It can be activated in response to stimulations induced by ionizing radiation, UV light, chemicals, hypoxia, ribonucleotide depletion, microtubule disruption, and oncogene activation. The activation of p53 requires two conditions: a critical concentration of the p53 protein and posttranscriptional modification.

The level of p53 is determined by the relative activity of protein production and degradation. In unstimulated cells, p53 degradation is more predominant than p53 production, resulting in a relatively low level of stable p53. Thus, the p53 activity is suppressed in unstimulated cells. p53 can be rapidly degraded by the ubiquitin–proteasome system (see page 226). Several mechanisms have been discovered for the ubiquitination and degradation of the p53 protein. In the G_0 phase of a cell cycle, the c-JUN *N*-terminal kinase can bind to p53, forming a complex. This complex serves as a target for an ubiquitin ligase, which induces p53 ubiquitination. Similarly, the COP9 signalosome can binds to p53, inducing p53 ubiquitination and degradation. The p53 protein can also be degraded directly by ubiquitination. One of the ubiquitin ligases, MDM2, can bind to p53 and induce p53 ubiquitination. In response to various stimulations as outlined above, the p53 production system can be activated, resulting in an increase in the level of p53. To activate p53, it is necessary to suppress p53 ubiquitination. The expression and activation of p53 are regulated by several processes as discussed below.

Posttranscriptional modification is a process that modulates the structure and function of a protein after mRNA translation. Several types of modification have been found for p53, including p53 phosphorylation and acetylation. These modifications are required for p53 activation. p53 phosphorylation occurs on the serine and threonine residues, whereas acetylation occurs on the lysine residues. Stimulating factors, such ionizing radiation, UV light, chemicals, hypoxia, microtubule disruption, and oncogene activation, may induce p53 phosphorylation and/or acetylation, although each factor may activate distinct signaling pathways. Here, the mechanisms of p53 activation in response to the stimulation of several common factors are briefly discussed.

Stimulatory factors for p53 can be divided into two groups, based on the influence on gene expression: genotoxic and nongenotoxic. Typical genotoxic factors include ionizing radiation, UV light, anticancer drugs (e.g., adriamycin, camptothecin, actinomycin D, and mitomycin C), and toxic chemicals (e.g., arsenite, cadmium, and chromate). These factors often induce gene damage. Nongenotoxic factors include hypoxia, ribonucleotide depletion, microtubule disruption, oncogene activation, and senescence. These factors may not induce significant gene damage.

A treatment with ionizing radiation induces DNA disruption and activation of protein kinase ATM (ataxia–telangiectasia M), a member of the phosphatidylinositol-3-kinase (PI3K) family. Activated ATM can phosphorylate p53 on serine 15 and activates other protein kinases such as Chk1 and Chk2, which further phosphorylate p53. Other protein kinases, such as PKA, PKC, and CDK, can also phosphorylate p53. Phosphorylated p53 serves as a transcriptional factor, stimulates the expression of selected genes, and induces apoptosis and the arrest of cell cycle (Fig. 5.20). Although DNA disruption is considered a factor for activating the protein kinases that phosphorylate p53, the exact mechanisms of protein kinase activation remains poorly understood.

A treatment with UV light induces DNA damage. Damaged DNA triggers the binding of a protein kinase ATR (ataxia–telangiectasia Rad3-related) to the damaged site, leading

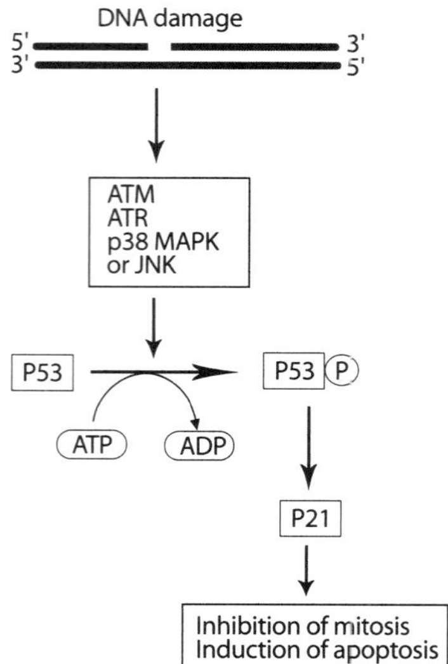

Figure 5.20. Mechanisms of p53 activation (based on bibliography 5.20).

to the activation of ATR. Activated ATR in turn phosphorylates p53 on serine 15 and serine 37. UV exposure can also activate other protein kinases, such as p38 MAPK, HIPK2, JNK, and cdc2/Cdk2 via DNA damage. These protein kinases are capable of phosphorylating p53 on multiple serine residues, leading to cell cycle arrest and apoptosis.

Both ionizing radiation and UV light can induce p53 acetylation on multiple lysine residues at the *C*-terminus, including lysines 320, 373, and 382. It is interesting to note that the *C*-terminal lysine acetylation is enhanced by *N*-terminal serine/threonine phosphorylation. p53 acetylation enhances the stability of the molecule, activates p53, and facilitates p53 binding to target genes, thus enhancing apoptosis and cell cycle arrest.

BIBLIOGRAPHY

5.12. Structure and Function of G-Protein Receptor-Related Signaling Molecules

G(s) Protein α

Bastepe M, Weinstein LS, Ogata N, Kawaguchi H, Juppner H et al: Stimulatory G protein directly regulates hypertrophic differentiation of growth plate cartilage in vivo, *Proc Natl Acad Sci USA* 101:14794–9, 2004.

Blatt C, Eversole-Cire P, Cohn VH, Zollman S, Fournier REK et al: Chromosomal localization of genes encoding guanine nucleotide-binding protein subunits in mouse and human, *Proc Natl Acad Sci USA* 85:7642–6, 1988.

Chen M, Gavrilova O, Liu J, Xie T, Deng C et al: Alternative Gnas gene products have opposite effects on glucose and lipid metabolism, *Proc Natl Acad Sci USA* 102:7386–91, 2005.

Harris BA, Robishaw JD, Mumby SM, Gilman AG: Molecular cloning of complementary DNA for the alpha subunit of the G protein that stimulates adenylate cyclase, *Science* 229:1274–77, 1985.

Harrison T, Samuel BU, Akompong T, Hamm H, Mohandas N et al: Erythrocyte G protein-coupled receptor signaling in malarial infection, *Science* 301:1734–6, 2003.

Iiri T, Herzmark P, Nakamoto JM, Van Dop C, Bourne HR: Rapid GDP release from Gs-alpha in patients with gain and loss of endocrine function, *Nature* 371:164–8, 1994.

Kozasa T, Itoh H, Tsukamoto T, Kaziro Y: Isolation and characterization of the human Gs-alpha gene, *Proc Natl Acad Sci USA* 85:2081–5, 1988.

Landis CA, Masters SB, Spada A, Pace AM, Bourne HR et al: GTPase inhibiting mutations activate the alpha chain of Gs and stimulate adenylyl cyclase in human pituitary tumours, *Nature* 340:692–6, 1989.

Levine MA, Modi WS, O'Brien SJ: Mapping of the gene encoding the alpha subunit of the stimulatory G protein of adenylyl cyclase (GNAS1) to 20q13.2-q13.3 in human by in situ hybridization, *Genomics* 11:478–9, 1991.

Mehlmann LM, Jones TLZ, Jaffe LA: Meiotic arrest in the mouse follicle maintained by a GS protein in the oocyte, *Science* 297:1343–5, 2002.

Vallar L, Spada A, Giannattasio G: Altered Gs and adenylate cyclase activity in human GH-secreting pituitary adenomas, *Nature* 330:566–8, 1987.

Warner DR, Weinstein LS: A mutation in the heterotrimeric stimulatory guanine nucleotide binding protein alpha-subunit with impaired receptor-mediated activation because of elevated GTPase activity, *Proc Natl Acad Sci USA* 96:4268–72, 1999.

Williamson CM, Ball ST, Nottingham WT, Skinner JA, Plagge A et al: A cis-acting control region is required exclusively for the tissue-specific imprinting of Gnas, *Nature Genet* 36:894–9, 2004.

Wroe SF, Kelsey G, Skinner JA, Bodle D, Ball ST et al: An imprinted transcript, antisense to Nesp, adds complexity to the cluster of imprinted genes at the mouse Gnas locus, *Proc Natl Acad Sci USA* 97:3342–6, 2000.

Gq Protein

Adams JW, Sakata Y, Davis MG, Sah VP, Wang Y et al: Enhanced G-alpha-q signaling: A common pathway mediates cardiac hypertrophy and apoptotic heart failure, *Proc Natl Acad Sci USA* 95:10140–5, 1998.

Dong Q, Shenker A, Way J, Haddad BR, Lin K et al: Molecular cloning of human G-alpha(q) cDNA and chromosomal localization of the G-alpha(q) gene (GNAQ) and a processed pseudogene, *Genomics* 30:470–5, 1995.

Offermanns S, Toombs CF, Hu YH, Simon MI: Defective platelet activation in G-alpha-q-deficient mice, *Nature* 389:183–6, 1997.

Santagata S, Boggon TJ, Baird CL, Gomez CA, Zhao J et al: G-protein signaling through tubby proteins, *Science* 292:2041–50, 2001.

Van Raamsdonk CD, Fitch KR, Fuchs H, Hrabe de Angelis M, Barsh GS: Effects of G-protein mutations on skin color, *Nature Genet* 36:961–8, 2004.

Deo DD, Bazan NG, Hunt JD: Activation of platelet-activating factor receptor-coupled G alpha q leads to stimulation of Src and focal adhesion kinase via two separate pathways in human umbilical vein endothelial cells, *J Biol Chem* 279:3497–508, 2004.

G Protein β

Danciger M, Farber DB, Peyser M, Kozak CA: The gene for the beta-subunit of retinal transducin (Gnb-1) maps to distal mouse chromosome 4, and related sequences map to mouse chromosomes 5 and 8, *Genomics* 6:428–35, 1990.

Hurowitz EH, Melnyk JM, Chen YJ, Kouros-Mehr H, Simon MI et al: Genomic characterization of the human heterotrimeric G protein alpha, beta, and gamma subunit genes, *DNA Res* 7:111–20, 2000.

Levine MA, Modi WS, O'Brien SJ: Chromosomal localization of the genes encoding two forms of the G-protein beta polypeptide, beta-1 and beta-3, in man, *Genomics* 8:380–6, 1990.

G Protein γ

Evanko DS, Thiyagarajan MM, Siderovski DP, Wedegaertner PB: Gbeta gamma isoforms selectively rescue plasma membrane localization and palmitoylation of mutant Galphas and Galphaq, *J Biol Chem* 276:23945–53, 2001.

Hurowitz EH, Melnyk JM, Chen YJ, Kouros-Mehr H, Simon MI et al: Genomic characterization of the human heterotrimeric G protein alpha, beta, and gamma subunit genes, *DNA Res* 7:111–20, 2000.

Schwindinger WF, Giger KE, Betz KS, Stauffer AM, Sunderlin EM et al: Mice with deficiency of G protein gamma-3 are lean and have seizures, *Mol Cell Biol* 24:7758–68, 2004.

β Adrenergic Receptor

Bachman ES, Dhillon H, Zhang CY, Cinti S, Bianco AC et al: Beta-AR signaling required for diet-induced thermogenesis and obesity resistance, *Science* 297:843–5, 2002.

Frielle T, Collins S, Daniel KW, Caron MG, Lefkowitz RJ et al: Cloning of the cDNA for the human beta-1-adrenergic receptor, *Proc Natl Acad Sci USA* 84:7920–4, 1987.

Frielle T, Kobilka B, Lefkowitz RJ, Caron MG: Human beta-1- and beta-2-adrenergic receptors: Structurally and functionally related receptors derived from distinct genes, *Trends Neurosci* 11:321–4, 1988.

Mason DA, Moore JD, Green SA, Liggett SB: A gain-of-function polymorphism in a G-protein coupling domain of the human beta-1-adrenergic receptor, *J Biol Chem* 274:12670–4, 1999.

Mialet Perez J, Rathz DA, Petrashevskaya NN, Hahn HS, Wagoner LE et al: Beta-1-adrenergic receptor polymorphisms confer differential function and predisposition to heart failure, *Nature Med* 9:1300–5, 2003.

Rohrer DK, Chruscinski A, Schauble EH, Bernstein D, Kobilka BK: Cardiovascular and metabolic alterations in mice lacking both beta-1- and beta-2-adrenergic receptors, *J Biol Chem* 274:16701–8, 1999.

Xu J, He J, Castleberry AM, Balasubramanian S, Lau AG et al: Heterodimerization of alpha-2A- and beta-1-adrenergic receptors, *J Biol Chem* 278:10770–7, 2003.

Yang-Feng TL, Xue F, Zhong W, Cotecchia S, Frielle T et al: Chromosomal organization of adrenergic receptor genes, *Proc Natl Acad Sci USA* 87:1516–20, 1990.

Adenylyl Cyclase

Abdel-Majid RM, Leong WL, Schalkwyk LC, Smallman DS, Wong ST et al: Loss of adenylyl cyclase I activity disrupts patterning of mouse somatosensory cortex, *Nature Genet* 19:289–91, 1998.

Edelhoff S, Villacres EC, Storm DR, Disteche CM: Mapping of adenylyl cyclase genes type I, II, III, IV, V, and VI in mouse, *Mam Genome* 6:111–3, 1995.

Lu HC, She WC, Plas DT, Neumann PE, Janz R et al: Adenylyl cyclase I regulates AMPA receptor trafficking during mouse cortical "barrel" map development, *Nature Neurosci* 6:939–47, 2003.

Villacres EC, Xia Z, Bookbinder LH, Edelhoff S, Disteche CM et al: Cloning, chromosomal mapping, and expression of human fetal brain type I adenylyl cyclase, *Genomics* 16:473–8, 1993.

Wang H, Ferguson GD, Pineda VV, Cundiff PE, Storm DR: Overexpression of type-1 adenylyl cyclase in mouse forebrain enhances recognition memory and LTP, *Nature Neurosci* 7:635–42, 2004.

Welker E, Armstrong-James M, Bronchti G, Ourednik W, Gheorghita-Baechler F et al: Altered sensory processing in the somatosensory cortex of the mutant mouse barrelless, *Science* 271:1864–7, 1996.

Phospholipase C β

Caricasole A, Sala C, Roncarati R, Formenti E, Terstappen GC: Cloning and characterization of the human phosphoinositide-specific phospholipase C-beta 1 (PLC-beta 1), *Biochim Biophys Acta* 1517:63–72, 2000.

Peruzzi D, Aluigi M, Manzoli L, Billi AM, Di Giorgio FP et al: Molecular characterization of the human PLC beta-1 gene, *Biochim Biophys Acta* 1584:46–54, 2002.

Peruzzi D, Calabrese G, Faenza I, Manzoli L, Matteucci A et al: Identification and chromosomal localisation by fluorescence in situ hybridisation of human gene of phosphoinositide-specific phospholipase C beta-1, *Biochim Biophys Acta* 1484:175–82, 2000.

Human protein reference data base, Johns Hopkins University and the Institute of Bioinformatics, at http://www.hprd.org/protein.

5.13. Signaling Mechanisms of G-Protein Receptor-Mediated Pathways

Ram PT, Iyengar R: G protein coupled receptor signaling through the Src and Stat3 pathway: Role in proliferation and transformation, *Oncogene* 20:1601–6, 2001.

Mazzoni MR, Hamm H: G-protein organization and signaling, in *Handbook of Cell Signaling*, Vol. 1, Bradshaw RA, Dennis EA, eds, Academic Press, Amsterdam, 2004, pp 335–41.

Birnbaumer L: Signal transduction by G proteins—basic principles, molecular diversity, and structural basis of their actions, in *Handbook of Cell Signaling*, Vol 2, Bradshaw RA, Dennis EA, eds, Academic Press, Amsterdam, 2004, pp 557–569.

Krauss G: *Biochemistry of Signal Transduction and Regulation*, 3rd ed, Wiley-VCH GmbH, 2003.

5.14. Structure and Function of NFκB-Related Signaling Molecules

NFκB1

Anest V, Hanson JL, Cogswell PC, Steinbrecher KA, Strahl BD et al: A nucleosomal function for I-kappa-B kinase-alpha in NF-kappa-B-dependent gene expression, *Nature* 423:659–63, 2003.

Baeuerle PA: I-kappa-B–NF-kappa-B structures: At the interface of inflammation control, *Cell* 95:729–31, 1998.

Baldwin AS Jr: The NF-kappa-B and I-kappa-B proteins: New discoveries and insights, *Annu Rev Immun* 14:649–83, 1996.

Barnes PJ, Karin M: Nuclear factor-kappa-B—a pivotal transcription factor in chronic inflammatory diseases, *New Engl J Med* 336:1066–71, 1997.

Bouwmeester T, Bauch A, Ruffner H, Angrand PO, Bergamini G et al: A physical and functional map of the human TNF-alpha/NF-kappa-B signal transduction pathway, *Nature Cell Biol* 6:97–105, 2004.

Brummelkamp TR, Nijman SMB, Dirac AMG, Bernards R: Loss of the cylindromatosis tumour suppressor inhibits apoptosis by activating NF-kappa-B, *Nature* 424:797–801, 2003.

Cai D, Yuan M, Frantz DF, Melendez PA, Hansen L et al: Local and systemic insulin resistance resulting from hepatic activation of IKK-beta and NF-kappa-B, *Nature Med* 11:183–90, 2005.

Covert MW, Leung TH, Gaston JE, Baltimore D: Achieving stability of lipopolysaccharide-induced NF-kappa-B activation, *Science* 309:1854–7, 2005.

Digicaylioglu M, Lipton SA: Erythropoietin-mediated neuroprotection involves cross-talk between Jak2 and NF-kappa-B signalling cascades, *Nature* 412:641–7, 2001.

Heron E, Deloukas P, van Loon APGM: The complete exon-intron structure of the 156-kb human gene NFKB1, which encodes the p105 and p50 proteins of transcription factors NF-kappa-B and I-kappa-B-gamma: Implications for NF-kappa-B-mediated signal transduction, *Genomics* 30:493–505, 1995.

Huxford T, Huang DB, Malek S, Ghosh G: The crystal structure of the I-kappa-B-alpha/NF-kappa-B complex reveals mechanisms of NF-kappa-B inactivation, *Cell* 95:759–70, 1998.

Iotsova V, Caamano J, Loy J, Yang Y, Lewin A et al: Osteopetrosis in mice lacking NF-kappa-B1 and NF-kappa-B2, *Nature Med* 3:1285–9, 1997.

Jacobs MD, Harrison SC: Structure of an I-kappa-B-alpha/NF-kappa-B complex, *Cell* 95:749–58, 1998.

Lawrence T, Bebien M, Liu GY, Nizet V, Karin M: IKK-alpha limits macrophage NF-kappa-B activation and contributes to the resolution of inflammation, *Nature* 434:1138–43, 2005.

Le Beau MM, Ito C, Cogswell P, Espinosa R III, Fernald AA et al: Chromosomal localization of the genes encoding the p50/p105 subunits of NF-kappa-B (NFKB2) and the I-kappa-B/MAD-3 (NFKBI) inhibitor of NF-kappa-B to 4q24 and 14q13, respectively, *Genomics* 14:529–31, 1992.

Lin L, DeMartino GN, Greene WC: Cotranslational biogenesis of NF-kappaB p50 by the 26S proteasome, *Cell* 92:819–28, 1998.

Meffert MK, Chang JM, Wiltgen BJ, Fanselow MS, Baltimore D: NF-kappa-B functions in synaptic signaling and behavior, *Nature Neurosci* 6:1072–8, 2003.

Meyer R, Hatada EN, Hohmann HP, Haiker M, Bartsch C et al: Cloning of the DNA-binding subunit of human nuclear factor kappa-B: the level of its mRNA is strongly regulated by phorbol ester or tumor necrosis factor alpha, *Proc Natl Acad Sci USA* 88:966–70, 1991.

Ozes ON, Mayo LD, Gustin JA, Pfeffer SR, Pfeffer LM et al: NF-kappa-B activation by tumour necrosis factor requires the Akt serine-threonine kinase, *Nature* 401:82–5, 1999.

Pikarsky E, Porat RM, Stein I, Abramovitch R, Amit S et al: NF-kappa-B functions as a tumour promoter in inflammation-associated cancer, *Nature* 431:461–6, 2004.

Romashkova JA, Makarov SS: NF-kappa-B is a target of AKT in anti-apoptotic PDGF signalling, *Nature* 401:86–90, 1999.

Ryan KM, Ernst MK, Rice NR, Vousden KH: Role of NF-kappa-B in p53-mediated programmed cell death, *Nature* 404:892–7, 2000.

Sha WC, Liou HC, Tuomanen EI, Baltimore D: Targeted disruption of the p50 subunit of NF-kappa-B leads to multifocal defects in immune responses, *Cell* 80:321–30, 1995.

Sigala JLD, Bottero V, Young DB, Shevchenko A, Mercurio F et al: Activation of transcription factor NF-kappa-B requires ELKS, an I-kappa-B kinase regulatory subunit, *Science* 304:1963–7, 2004.

Wertz IE, O'Rourke KM, Zhou H, Eby M, Aravind L et al: De-ubiquitination and ubiquitin ligase domains of A20 downregulate NF-kappa-B signalling, *Nature* 430:694–9, 2004.

Yamamoto Y, Verma UN, Prajapati S, Kwak YT, Gaynor RB: Histone H3 phosphorylation by IKK-alpha is critical for cytokine-induced gene expression, *Nature* 423:655–9, 2003.

NFκB2

Baldwin AS Jr: The NF-kappa-B and I-kappa-B proteins: New discoveries and insights, *Annu Rev Immun* 14:649–83, 1996.

Claudio E, Brown K, Park S, Wang H, Siebenlist U: BAFF-induced NEMO-independent processing of NF-kappa-B2 in maturing B cells, *Nature Immun* 3:958–65, 2002.

Iotsova V, Caamano J, Loy J, Yang Y, Lewin A et al: Osteopetrosis in mice lacking NF-kappa-B1 and NF-kappa-B2, *Nature Med* 3:1285–9, 1997.

Liptay S, Schmid RM, Perkins ND, Meltzer P, Altherr MR et al: Related subunits of NF-kappa-B map to two distinct loci associated with translocations in leukemia, NFKB1 and NFKB2, *Genomics* 13:287–92, 1992.

Mathew S, Murty VVVS, Dalla-Favera R, Chaganti RSK: Chromosomal localization of genes encoding the transcription factors, c-rel, NF-kappa-Bp50, NF-kappa-Bp65, and lyt-10 by fluorescence in situ hybridization, *Oncogene* 8:191–3, 1993.

Neri A, Chang CC, Lombardi L, Salina M, Corradini P et al: B cell lymphoma-associated chromosomal translocation involves candidate oncogene lyt-10, homologous to NF-kappa-B p50, *Cell* 67:1075–87, 1991.

Smahi A, Courtois G, Rabia SH, Doffinger R, Bodemer C et al: The NF-kappa-B signalling pathway in human diseases: from incontinentia pigmenti to ectodermal dysplasias and immune-deficiency syndromes, *Hum Mol Genet* 11:2371–5, 2002.

IκBα

Auphan N, DiDonato JA, Rosette C, Helmberg A, Karin M: Immunosuppression by glucocorticoids: inhibition of NF-kappa-B activity through induction of I-kappa-B synthesis, *Science* 270:286–90, 1995.

Baeuerle PA: I-kappa-B–NF-kappa-B structures: at the interface of inflammation control, *Cell* 95:729–31, 1998.

Cai D, Frantz JD, Tawa NE Jr, Melendez PA, Oh BC et al: IKK-beta/NF-kappa-B activation causes severe muscle wasting in mice, *Cell* 119:285–98, 2004.

Dajee M, Lazarov M, Zhang JY, Cai T, Green CL et al: NF-kappa-B blockade and oncogenic Ras trigger invasive human epidermal neoplasia, *Nature* 421:639–43, 2003.

Haskill S, Beg AA, Tompkins SM, Morris JS, Yurochko AD et al: Characterization of an immediate-early gene induced in adherent monocytes that encodes I-kappa-B-like activity, *Cell* 65:1281–9, 1991.

Hoffmann A, Levchenko A, Scott ML, Baltimore D: The I-kappa-B-NF-kappa-B signaling module: Temporal control and selective gene activation, *Science* 298:1241–5, 2002.

Huxford T, Huang DB, Malek S, Ghosh G: The crystal structure of the I-kappa-B-alpha/NF-kappa-B complex reveals mechanisms of NF-kappa-B inactivation, *Cell* 95:759–70, 1998.

Ito CY, Adey N, Bautch VL, Baldwin AS Jr: Structure and evolution of the human IKBA gene, *Genomics* 29:490–5, 1995.

Jacobs MD, Harrison SC: Structure of an I-kappa-B-alpha/NF-kappa-B complex, *Cell* 95:749–58, 1998.

Jung M, Zhang Y, Lee S, Dritschilo A: Correction of radiation sensitivity in ataxia telangiectasia cells by a truncated I-kappa-B-alpha, *Science* 268:1619–21, 1995.

Le Beau MM, Ito C, Cogswell P, Espinosa R III, Fernald AA et al: Chromosomal localization of the genes encoding the p50/p105 subunits of NF-kappa-B (NFKB2) and the I-kappa-B/MAD-3 (NFKBI) inhibitor of NF-kappa-B to 4q24 and 14q13, respectively, *Genomics* 14:529–31, 1992.

Neish AS, Gewirtz AT, Zeng H, Young AN, Hobert ME et al: Prokaryotic regulation of epithelial responses by inhibition of I-kappa-B-alpha ubiquitination, *Science* 289:1560–3, 2000.

Scheinman RI, Cogswell PC, Lofquist AK, Baldwin AS Jr: Role of transcriptional activation of I-kappa-B-alpha in mediation of immunosuppression by glucocorticoids, *Science* 270:283–6, 1995.

Sigala JLD, Bottero V, Young DB, Shevchenko A, Mercurio F et al: Activation of transcription factor NF-kappa-B requires ELKS, an I-kappa-B kinase regulatory subunit, *Science* 304:1963–7, 2004.

IκBβ

Fenwick C, Na SY, Voll RE, Zhong H, Im SY et al: A subclass of Ras proteins that regulate the degradation of I-kappa-B, *Science* 287:869–73, 2000.

Hoffmann A, Levchenko A, Scott ML, Baltimore D: The I-kappa-B-NF-kappa-B signaling module: Temporal control and selective gene activation, *Science* 298:1241–5, 2002.

Okamoto T, Ono T, Hori M, Yang JP, Tetsuka T et al: Assignment of the I-kappa-B-beta gene NFKBIB to human chromosome band 19q13.1 by in situ hybridization, *Cytogenet Cell Genet* 82:105–6, 1998.

IκB Kinase α

Anest V, Hanson JL, Cogswell PC, Steinbrecher KA, Strahl BD et al: A nucleosomal function for I-kappa-B kinase-alpha in NF-kappa-B-dependent gene expression, *Nature* 423:659–63, 2003.

Cao Y, Bonizzi G, Seagroves TN, Greten FR, Johnson R et al: IKK-alpha provides an essential link between RANK signaling and cyclin D1 expression during mammary gland development, *Cell* 107:763–75, 2001.

Delhase M, Hayakawa M, Chen Y, Karin M: Positive and negative regulation of I-kappa-B kinase activity through IKK-beta subunit phosphorylation, *Science* 284:309–12, 1999.

DiDonato JA, Hayakawa M, Rothwarf DM, Zandi E, Karin M: A cytokine-responsive IkappaB kinase that activates the transcription factor NF-kappaB, *Nature* 388:548–54, 1997.

Hu Y, Baud V, Delhase M, Zhang P, Deerinck T et al: Abnormal morphogenesis but intact IKK activation in mice lacking the IKK-alpha subunit of I-kappa-B kinase, *Science* 284:316–20, 1999.

Hu Y, Baud V, Oga T, Kim KI, Yoshida K et al: IKK-alpha controls formation of the epidermis independently of NF-kappa-B, *Nature* 410:710–4, 2001.

Lawrence T, Bebien M, Liu GY, Nizet V, Karin M: IKK-alpha limits macrophage NF-kappa-B activation and contributes to the resolution of inflammation, *Nature* 434:1138–43, 2005.

Matsushima A, Kaisho T, Rennert PD, Nakano H, Kurosawa K et al: Essential role of nuclear factor (NF)-kappa-B-inducing kinase and inhibitor of kappa-B (I-kappa-B) kinase alpha in NF-kappa-B activation through lymphotoxin beta receptor, but not through tumor necrosis factor receptor I, *J Exp Med* 193:631–6, 2001.

May MJ, D'Acquisto F, Madge LA, Glockner J, Pober JS et al: Selective inhibition of NF-kappa-B activation by a peptide that blocks the interaction of NEMO with the I-kappa-B kinase complex, *Science* 289:1550–4, 2000.

Mercurio F, Zhu H, Murray BW, Shevchenko A, Bennett BL et al: IKK-1 and IKK-2: Cytokine-activated I-kappa-B kinases essential for NF-kappa-B activation, *Science* 278:860–6, 1997.

Mock BA, Connelly MA, McBride OW, Kozak CA, Marcu KB: CHUK, a conserved helix-loop-helix ubiquitous kinase, maps to human chromosome 10 and mouse chromosome 19, *Genomics* 27:348–51, 1995.

Ozes ON, Mayo LD, Gustin JA, Pfeffer SR, Pfeffer LM et al: NF-kappa-B activation by tumour necrosis factor requires the Akt serine-threonine kinase, *Nature* 401:82–5, 1999.

Regnier CH, Song HY, Gao X, Goeddel DV, Cao Z et al: Identification and characterization of an I-kappa-B kinase, *Cell* 90:373–83, 1997.

Senftleben U, Cao Y, Xiao G, Greten FR, Krahn G et al: Activation by IKK-alpha of a second, evolutionarily conserved, NF-kappa-B signaling pathway, *Science* 293:1495–9, 2001.

Sil AK, Maeda S, Sano Y, Roop DR, Karin M: I-kappa-B kinase-alpha acts in the epidermis to control skeletal and craniofacial morphogenesis, *Nature* 428:660–4, 2004.

Sizemore N, Agarwal A, Das K, Lerner N, Sulak M et al: Inhibitor of kappa-B kinase is required to activate a subset of interferon gamma-stimulated genes, *Proc Natl Acad Sci USA* 101:7994–8, 2004.

Takeda K, Takeuchi O, Tsujimura T, Itami S, Adachi O et al: Limb and skin abnormalities in mice lacking IKK-alpha, *Science* 284:313–16, 1999.

Werner SL, Barken D, Hoffmann A: Stimulus specificity of gene expression programs determined by temporal control of IKK activity, *Science* 309:1857–61, 2005.

Yamamoto Y, Verma UN, Prajapati S, Kwak YT, Gaynor RB: Histone H3 phosphorylation by IKK-alpha is critical for cytokine-induced gene expression, *Nature* 423:655–9, 2003.

IκB Kinase β

Ambros PF, Schmid J, Rumpler S, Binder BR, de Martin R: Localization of the human I-kappa-B kinase-beta (IKBKB) to chromosome 8p11.2 by fluorescence in situ hybridization and radiation hybrid mapping, *Genomics* 54:575–6, 1998.

Arkan MC, Hevener AL, Greten FR, Maeda S, Li ZW et al: IKK-beta links inflammation to obesity-induced insulin resistance, *Nature Med* 11:191–8, 2005.

Cai D, Frantz JD, Tawa NE Jr, Melendez PA, Oh BC et al: IKK-beta/NF-kappa-B activation causes severe muscle wasting in mice, *Cell* 119:285–98, 2004.

Cai D, Yuan M, Frantz DF, Melendez PA, Hansen L et al: Local and systemic insulin resistance resulting from hepatic activation of IKK-beta and NF-kappa-B, *Nature Med* 11:183–90, 2005.

Delhase M, Hayakawa M, Chen Y, Karin M: Positive and negative regulation of I-kappa-B kinase activity through IKK-beta subunit phosphorylation, *Science* 284:309–12, 1999.

DiDonato JA, Hayakawa M, Rothwarf DM, Zandi E, Karin M: A cytokine-responsive IkappaB kinase that activates the transcription factor NF-kappaB, *Nature* 388:548–54, 1997.

Egan LJ, Eckmann L, Greten FR, Chae S, Li ZW et al: I-kappa-B-kinase-beta-dependent NF-kappa-B activation provides radioprotection to the intestinal epithelium, *Proc Natl Acad Sci USA* 101:2452–7, 2004.

Greten FR, Eckmann L, Greten TF, Park JM, Li ZW et al: IKK-beta links inflammation and tumorigenesis in a mouse model of colitis-associated cancer, *Cell* 118:285–96, 2004.

Hu MCT, Lee DF, Xia W, Golfman LS, Ou-Yang F et al: I-kappa-B kinase promotes tumorigenesis through inhibition of forkhead FOXO3a, *Cell* 117:225–37, 2004.

Hu MCT, Wang Y: I-kappa-B kinase-alpha and -beta genes are coexpressed in adult and embryonic tissues but localized to different human chromosomes, *Gene* 222:31–40, 1998.

Li Q, Van Antwerp D, Mercurio F, Lee KF, Verma IM: Severe liver degeneration in mice lacking the I-kappa-B kinase 2 gene, *Science* 284:321–5, 1999.

May MJ, D'Acquisto F, Madge LA, Glockner J, Pober JS et al: Selective inhibition of NF-kappa-B activation by a peptide that blocks the interaction of NEMO with the I-kappa-B kinase complex, *Science* 289:1550–4, 2000.

Mercurio F, Zhu H, Murray BW, Shevchenko A, Bennett BL et al: IKK-1 and IKK-2: Cytokine-activated I-kappa-B kinases essential for NF-kappa-B activation, *Science* 278:860–6, 1997.

Ozes ON, Mayo LD, Gustin JA, Pfeffer SR, Pfeffer LM et al: NF-kappa-B activation by tumour necrosis factor requires the Akt serine-threonine kinase, *Nature* 401:82–5, 1999.

Pasparakis M, Courtois G, Hafner M, Schmidt-Supprian M, Nenci A et al: TNF-mediated inflammatory skin disease in mice with epidermis-specific deletion of IKK2, *Nature* 417:861–6, 2002.

Rossi A, Kapahi P, Natoli G, Takahashi T, Chen Y et al: Anti-inflammatory cyclopentenone prostaglandins are direct inhibitors of I-kappa-B kinase, *Nature* 403:103–8, 2000.

Shindo M, Nakano H, Sakon S, Yagita H, Mihara M et al: Assignment of I-kappa-B kinase beta (IKBKB) to human chromosome band 8p12-p11 by in situ hybridization, *Cytogenet Cell Genet* 82:32–3, 1998.

Sizemore N, Agarwal A, Das K, Lerner N, Sulak M et al: Inhibitor of kappa-B kinase is required to activate a subset of interferon gamma-stimulated genes, *Proc Natl Acad Sci USA* 101:7994–8, 2004.

Woronicz JD, Gao X, Cao Z, Rothe M, Goeddel DV: IkappaB kinase-beta: NF-kappa-B activation and complex formation with IkappaB kinase-alpha and NIK, *Science* 278:866–9, 1997.

Yin MJ, Yamamoto Y, Gaynor RB: The anti-inflammatory agents aspirin and salicylate inhibit the activity of I-kappa-B kinase-beta, *Nature* 396:77–80, 1998.

Yuan M, Konstantopoulos N, Lee J, Hansen L, Li ZW et al: Reversal of obesity- and diet-induced insulin resistance with salicylates or targeted disruption of Ikk-beta, *Science* 293:1673–7, 2001.

Zandi E, Rothwarf DM, Delhase M, Hayakawa M, Karin M: The IkappaB kinase complex (IKK) contains two kinase subunits, IKKalpha and IKKbeta, necessary for IkappaB phosphorylation and NF-kappa-B activation, *Cell* 91:243–52, 1997.

IκB Kinase γ

Brummelkamp TR, Nijman SMB, Dirac AMG, Bernards R: Loss of the cylindromatosis tumour suppressor inhibits apoptosis by activating NF-kappa-B, *Nature* 424:797–801, 2003.

Doffinger R, Smahi A, Bessia C, Geissmann F, Feinberg J et al: X-linked anhidrotic ectodermal dysplasia with immunodeficiency is caused by impaired NF-kappa-B signaling, *Nature Genet* 27:277–85, 2001.

Jain A, Ma CA, Liu S, Brown M, Cohen J et al: Specific missense mutations in NEMO result in hyper-IgM syndrome with hypohydrotic [sic] ectodermal dysplasia, *Nature Immun* 2:223–8, 2001.

Kovalenko A, Chable-Bessia C, Cantarella G, Israel A, Wallach D et al: The tumour suppressor CYLD negatively regulates NF-kappa-B signalling by deubiquitination, *Nature* 424:801–5, 2003.

Li Q, Van Antwerp D, Mercurio F, Lee KF, Verma IM: Severe liver degeneration in mice lacking the I-kappa-B kinase 2 gene, *Science* 284:321–5, 1999.

May MJ, D'Acquisto F, Madge LA, Glockner J, Pober JS et al: Selective inhibition of NF-kappa-B activation by a peptide that blocks the interaction of NEMO with the I-kappa-B kinase complex, *Science* 289:1550–4, 2000.

Rothwarf DM, Zandi E, Natoli G, Karin M: IKK-gamma is an essential regulatory subunit of the I-kappa-B kinase complex, *Nature* 395:297–300, 1998.

The International Incontinentia Pigmenti Consortium: Genomic rearrangement in NEMO impairs NF-kappa-B activation and is a cause of incontinentia pigmenti, *Nature* 405:466–72, 2000.

Yamaoka S, Courtois G, Bessia C, Whiteside ST, Weil R et al: Complementation cloning of NEMO, a component of the I-kappa-B kinase complex essential for NF-kappa-B activation, *Cell* 93:1231–40, 1998.

Mao C, Zhou M, Uckun FM: Crystal structure of Bruton's tyrosine kinase domain suggests a novel pathway for activation and provides insights into the molecular basis of X-linked agammaglobulinemia, *J Biol Chem* 276:41435–43, 2001.

Parolini O, Hejtmancik JF, Allen RC, Belmont JW, Lassiter GL et al: Linkage analysis and physical mapping near the gene for X-linked agammaglobulinemia at Xq22, *Genomics* 15:342–9, 1993.

Rawlings DJ, Saffran DC, Tsukada S, Largaespada DA, Grimaldi JC et al: Mutation of unique region of Bruton's tyrosine kinase in immunodeficient xid mice, *Science* 261:358–61, 1993.

Saffran DC, Parolini O, Fitch-Hilgenberg ME, Rawlings DJ, Afar DEH et al: A point mutation in the SH2 domain of Bruton's tyrosine kinase in atypical X-linked agammaglobulinemia, *New Engl J Med* 330:1488–91, 1994.

Thomas JD, Sideras P, Smith CIE, Vorechovsky I, Chapman V et al: Colocalization of X-linked agammaglobulinemia and X-linked immunodeficiency genes, *Science* 261:355–8, 1993.

Tsukada S, Saffran DC, Rawlings DJ, Parolini O, Allen RC et al: Deficient expression of a B cell cytoplasmic tyrosine kinase in human X-linked agammaglobulinemia, *Cell* 72:279–90, 1993.

Uckun FM, Waddick KG, Mahajan S, Jun X, Takata M et al: BTK as a mediator of radiation-induced apoptosis in DT-40 lymphoma B cells, *Science* 273:1096–9, 1996.

Vetrie D, Vorechovsky I, Sideras P, Holland J, Davies A et al: The gene involved in X-linked agammaglobulinaemia is a member of the src family of protein-tyrosine kinases, *Nature* 361:226–33, 1993.

Yel L, Minegishi Y, Coustan-Smith E, Buckley RH, Trubel H et al: Mutations in the mu heavy-chain gene in patients with agammaglobulinemia, *New Engl J Med* 335:1486–93, 1996.

Human protein reference data base, Johns Hopkins University and the Institute of Bioinformatics, at http://www.hprd.org/protein.

5.15. Signaling Mechanisms of NFκB-Mediated Pathways

Krauss G: *Biochemistry of Signal Transduction and Regulation*, 3rd ed, Wiley-VCH GmbH, 2003.

Li X, Stark GR: NFkappaB-dependent signaling pathways, *Exp Hematol* 30:285–96, 2002.

5.16. Structure and Function of Ubiquitination-Related Signaling Molecules

Ubiquitin B

Conaway RC, Brower CS, Conaway JW: Emerging roles of ubiquitin in transcription regulation, *Science* 296:1254–8, 2002.

Sutovsky P, Moreno RD, Ramalho-Santos J, Dominko T, Simerly C et al: Ubiquitin tag for sperm mitochondria, *Nature* 402:371–2, 1999.

van Leeuwen FW, de Kleijn DPV, van den Hurk HH, Neubauer A, Sonnemans MAF et al: Frameshift mutants of beta-amyloid precursor protein and ubiquitin-B in Alzheimer's and Down patients, *Science* 279:242–7, 1998.

Webb GC, Baker RT, Fagan K, Board PG: Localization of the human UbB polyubiquitin gene to chromosome band 17p11.1–17p12, *Am J Hum Genet* 46:308–15, 1990.

Ubiquitin C

Baker RT, Board PG: Unequal crossover generates variation in ubiquitin coding unit number at the human UbC polyubiquitin locus, *Am J Hum Genet* 44:534–42, 1989.

Board PG, Coggan M, Baker RT, Vuust J, Webb GC: Localization of the human UBC polyubiquitin gene to chromosome band 12q24.3, *Genomics* 12:639–42, 1992.

Human protein reference data base, Johns Hopkins University and the Institute of Bioinformatics, at http://www.hprd.org/protein.

5.17. Signaling Mechanisms of Ubiquitination

SKP1

Bai C, Sen P, Hofmann K, Ma L, Goebl M et al: SKP1 connects cell cycle regulators to the ubiquitin proteolysis machinery through a novel motif, the F-box, *Cell* 86:263–74, 1996.

Demetrick DJ, Zhang H, Beach DH: Chromosomal mapping of the genes for the human CDK2/cyclin A-associated proteins p19 (SKP1A and SKP1B) and p45 (SKP2), *Cytogenet Cell Genet* 73:104–7, 1996.

Dias DC, Dolios G, Wang R, Pan ZQ: CUL7: A DOC domain-containing cullin selectively binds Skp1-Fbx29 to form an SCF-like complex, *Proc Natl Acad Sci USA* 99:16601–6, 2002.

Liang Y, Chen H, Asher JH Jr, Chang CC, Friedman TB: Human inner ear OCP2 cDNA maps to 5q22-5q35.2 with related sequences on chromosomes 4p16.2-4p14, 5p13-5q22, 7pter-q22, 10 and 12p13-12qter, *Gene* 184:163–7, 1997.

Maniatis T: A ubiquitin ligase complex essential for the NF-kappa-B, Wnt/Wingless, and Hedgehog signaling pathways, *Genes Dev* 13:505–10, 1999.

Schulman BA, Carrano AC, Jeffrey PD, Bowen Z, Kinnucan ERE et al: Insights into the SCF ubiquitin ligases from the structure of the Skpl-Skp2 complex, *Nature* 408:381–6, 2000.

Sowden J, Morrison K, Schofield J, Putt W, Edwards Y: A novel cDNA with homology to an RNA polymerase II elongation factors maps to human chromosome 5q31 (TCEB1L) and to mouse chromosome 11 (Tceb1l), *Genomics* 29:145–51, 1995.

Winston JT, Strack P, Beer-Romero P, Chu CY, Elledge SJ et al: The SCF(beta-TRCP)-ubiquitin ligase complex associates specifically with phosphorylated destruction motifs in I-kappa-B-alpha and beta-catenin and stimulates I-kappa-B-alpha ubiquitination in vitro, *Genes Dev* 13:270–83, 1999.

Zheng N, Schulman BA, Song L, Miller JJ, Jeffrey PD et al: Structure of the Cul1-Rbx1-Skp1-F box(Skp2) SCF ubiquitin ligase complex, *Nature* 416:703–9, 2002.

β-Catenin

Batlle E, Henderson JT, Beghtel H, van den Born MMW, Sancho E et al: Beta-catenin and TCF mediate cell positioning in the intestinal epithelium by controlling the expression of EphB/EphrinB, *Cell* 111:251–63, 2002.

Chan EF, Gat U, McNiff JM, Fuchs E: A common human skin tumour is caused by activating mutations in beta-catenin, *Nature Genet* 21:410–13, 1999.

Chenn A, Walsh CA: Regulation of cerebral cortical size by control of cell cycle exit in neural precursors, *Science* 297:365–9, 2002.

Essers MAG, de Vries-Smits LMM, Barker N, Polderman PE, Burgering BMT et al: Functional interaction between beta-catenin and FOXO in oxidative stress signaling, *Science* 308:1181–4, 2005.

Huelsken J, Vogel R, Erdmann B, Cotsarelis G, Birchmeier W: Beta-catenin controls hair follicle morphogenesis and stem cell differentiation in the skin, *Cell* 105:533–45, 2001.

Jamora C, DasGupta R, Kocieniewski P, Fuchs E: Links between signal transduction, transcription and adhesion in epithelial bud development, *Nature* 422:317–22, 2003.

Kang DE, Soriano S, Xia X, Eberhart CG, De Strooper B et al: Presenilin couples the paired phosphorylation of beta-catenin independent of Axin: Implications for beta-catenin activation in tumorigenesis, *Cell* 110:751–62, 2002.

Kaplan DD, Meigs TE, Kelly P, Casey PJ: Identification of a role for beta-catenin in the establishment of a bipolar mitotic spindle, *J Biol Chem* 279:10829–32, 2004.

Kim JH, Kim B, Cai L, Choi HJ, Ohgi KA et al: Transcriptional regulation of a metastasis suppressor gene by Tip60 and beta-catenin complexes, *Nature* 434:921–6, 2005.

Korinek V, Barker N, Morin PJ, van Wichen D, de Weger R et al: Constitutive transcriptional activation by a beta-catenin-Tcf complex in APC-/- colon carcinoma, *Science* 275:1784–7, 1997.

Kraus C, Liehr T, Hulsken J, Behrens J, Birchmeier W et al: Localization of the human beta-catenin gene (CTNNB1) to 3p21: A region implicated in tumor development, *Genomics* 23:272–4, 1994.

Lee HY, Kleber M, Hari L, Brault V, Suter U et al: Instructive role of Wnt/beta-catenin in sensory fate specification in neural crest stem cells, *Science* 303:1020–3, 2004.

Morin PJ, Sparks AB, Korinek V, Barker N, Clevers H et al: Activation of beta-catenin-Tcf signaling in colon cancer by mutations in beta-catenin or APC, *Science* 275:1787–90, 1997.

Peifer M: Cancer, catenins, and cuticle pattern: A complex connection, *Science* 262:1667–8, 1993.

Roose J, Huls G, van Beest M, Moerer P, van der Horn K et al: Synergy between tumor suppressor APC and the beta-catenin-Tcf4 target Tcf1, *Science* 285:1923–6, 1999.

Rubinfeld B, Robbins P, El-Gamil M, Albert I, Porfiri E et al: Stabilization of beta-catenin by genetic defects in melanoma cell lines, *Science* 275:1790–2, 1997.

Tetsu O, McCormick F: Beta-catenin regulates expression of cyclin D1 in colon carcinoma cells, *Nature* 398:422–6, 1999.

Trent JM, Wiltshire R, Su LK, Nicolaides NC, Vogelstein B et al: The gene for the APC-binding protein beta-catenin (CTNNB1) maps to chromosome 3p22, a region frequently altered in human malignancies, *Cytogenet Cell Genet* 71:343–4, 1995.

Widlund HR, Horstmann MA, Price ER, Cui J, Lessnick SL et al: Beta-catenin-induced melanoma growth requires the downstream target Microphthalmia-associated transcription factor, *J Cell Biol* 158:1079–87, 2002.

Wikramanayake AH, Hong M, Lee PN, Pang K, Byrum CA et al: An ancient role for nuclear beta-catenin in the evolution of axial polarity and germ layer segregation, *Nature* 426:446–50, 2003.

Xu Y, Banerjee D, Huelsken J, Birchmeier W, Sen JM: Deletion of beta-catenin impairs T cell development, *Nature Immun* 4:1177–82, 2003.

Yu X, Malenka RC: Beta-catenin is critical for dendritic morphogenesis, *Nature Neurosci* 6:1169–77, 2003.

HIF-1

Cramer T, Yamanishi Y, Clausen BE, Forster I, Pawlinski R et al: HIF-1-alpha is essential for myeloid cell-mediated inflammation, *Cell* 112:645–57, 2003.

Hon WC, Wilson MI, Harlos K, Claridge TDW, Schofield CJ et al: Structural basis for the recognition of hydroxyproline in HIF-1-alpha by pVHL, *Nature* 417:975–8, 2002.

Ivan M, Kondo K, Yang H, Kim W, Valiando J et al: HIF-alpha targeted for VHL-mediated destruction by proline hydroxylation: implications for O(2) sensing, *Science* 292:464–8, 2001.

Jaakkola P, Mole DR, Tian YM, Wilson MI, Gielbert J et al: Targeting of HIF-alpha to the von Hippel-Lindau ubiquitylation complex by O(2)-regulated prolyl hydroxylation, *Science* 292:468–72, 2001.

Jeong JW, Bae MK, Ahn MY, Kim SH, Sohn TK et al: Regulation and destabilization of HIF-1-alpha by ARD1-mediated acetylation, *Cell* 111:709–20, 2002.

Lando D, Peet DJ, Whelan DA, Gorman JJ, Whitelaw ML: Asparagine hydroxylation of the HIF transactivation domain: a hypoxic switch, *Science* 295:858–61, 2002.

Min JH, Yang H, Ivan M, Gertler F, Kaelin WG Jr et al: Structure of an HIF-1-alpha-pVHL complex: Hydroxyproline recognition in signaling, *Science* 296:1886–9, 2002.

Nakayama K, Frew IJ, Hagensen M, Skals M, Habelhah H et al: Siah2 regulates stability of prolyl-hydroxylases, controls HIF1-alpha abundance, and modulates physiologic responses to hypoxia, *Cell* 117:941–52, 2004.

Peyssonnaux C, Datta V, Cramer T, Doedens A, Theodorakis EA et al: HIF-1alpha expression regulates the bactericidal capacity of phagocytes, *J Clin Invest* 115:1806–15, 2005.

Semenza GL, Rue EA, Iyer NV, Pang MG, Kearns WG: Assignment of the hypoxia-inducible factor 1-alpha gene to a region of conserved synteny on mouse chromosome 12 and human chromosome 14q, *Genomics* 34:437–9, 1996.

Sutter CH, Laughner E, Semenza GL: Hypoxia-inducible factor 1-alpha protein expression is controlled by oxygen-regulated ubiquitination that is disrupted by deletions and missense mutations, *Proc Natl Acad Sci USA* 97:4748–53, 2000.

Krauss G: *Biochemistry of Signal Transduction and Regulation*, 3rd ed, Wiley-VCH GmbH, 2003.

Ubiquitin

Conaway RC, Brower CS, Conaway JW: Emerging roles of ubiquitin in transcription regulation, *Science* 296:1254–8. 2002.

Coux O, Tanaka K, Goldberg AL: Structure and functions of the 20S and 26S proteasomes, *Annu Rev Biochem* 65:801–47, 1996.

Hershko A, Ciechanover A: The ubiquitin system, *Annu Rev Biochem* 67:425–79, 1998.

Laney JD, Hochstrasser M: Substrate targeting in the ubiquitin system, *Cell* 97:427–30, 1999.

Pickart CM: Ubiquitin enters the new millennium, *Mol Cell* 8:499–504, 2001.

Pickart CM: Mechanisms underlying ubiquitination, *Annu Rev Biochem* 70:503–33, 2001.

Human protein reference data base, Johns Hopkins University and the Institute of Bioinformatics, at http://www.hprd.org/protein.

5.18. Structure and Function of Nuclear Receptor-Related Signaling Molecules

Estrogen Receptor α

Auboeuf D, Honig A, Berget SM, O'Malley BW: Coordinate regulation of transcription and splicing by steroid receptor coregulators, *Science* 298:416–19, 2002.

Couse JF, Hewitt SC, Bunch DO, Sar M, Walker VR et al: Postnatal sex reversal of the ovaries in mice lacking estrogen receptors alpha and beta, *Science* 286:2328–31, 1999.

Fan S, Wang JA, Yuan R, Ma Y, Meng Q et al: BRCA1 inhibition of estrogen receptor signaling in transfected cells, *Science* 284:1354–6, 1999.

Gosden JR, Middleton PG, Rout D: Localization of the human oestrogen receptor gene to chromosome 6q24-q27 by in situ hybridization, *Cytogenet Cell Genet* 43:218–20, 1986.

Green S, Walter P, Kumar V, Krust A, Bornert JM et al: Human oestrogen receptor cDNA: Sequence, expression and homology to v-erb-A, *Nature* 320:134–9, 1986.

Greene GL, Gilna P, Waterfield M, Baker A, Hort Y et al: Sequence and expression of human estrogen receptor complementary DNA, *Science* 231:1150–4, 1986.

Heine PA, Taylor JA, Iwamoto GA, Lubahn DB, Cooke PS: Increased adipose tissue in male and female estrogen receptor-alpha knockout mice, *Proc Natl Acad Sci USA* 97:12729–34, 2000.

Herrington DM, Howard TD, Hawkins GA, Reboussin DM, Xu J et al: Estrogen-receptor polymorphisms and effects of estrogen replacement on high-density lipoprotein cholesterol in women with coronary disease, *New Engl J Med* 346:967–74, 2002.

Issa JPJ, Ottaviano YL, Celano P, Hamilton SR, Davidson NE et al: Methylation of the oestrogen receptor CpG island links ageing and neoplasia in human colon, *Nature Genet* 7:536–40, 1994.

Johnson MD, Kenney N, Stoica A, Hilakivi-Clarke L, Singh B et al: Cadmium mimics the in vivo effects of estrogen in the uterus and mammary gland, *Nature Med* 9:1081–4, 2003.

Korach KS: Insights from the study of animals lacking functional estrogen receptor, *Science* 266:1524–7, 1994.

Kumar R, Wang RA, Mazumdar A, Talukder AH, Mandal M et al: A naturally occurring MTA1 variant sequesters oestrogen receptor-alpha in the cytoplasm, *Nature* 418:654–7, 2002.

Ohtake F, Takeyama K, Matsumoto T, Kitagawa H, Yamamoto Y et al: Modulation of oestrogen receptor signalling by association with the activated dioxin receptor, *Nature* 423:545–50, 2003.

Shiau AK, Barstad D, Loria PM, Cheng L, Kushner PJ et al: The structural basis of estrogen receptor/coactivator recognition and the antagonism of this interaction by tamoxifen, *Cell* 95:927–37, 1998.

Shim GJ, Kis LL, Warner M, Gustafsson JA: Autoimmune glomerulonephritis with spontaneous formation of splenic germinal centers in mice lacking the estrogen receptor alpha gene, *Proc Natl Acad Sci USA* 101:1720–4, 2004.

Simoncini T, Hafezi-Moghadam A, Brazil DP, Ley K, Chin WW et al: Interaction of oestrogen receptor with the regulatory subunit of phosphatidylinositol-3-OH kinase, *Nature* 407:538–41, 2000.

Smith EP, Boyd J, Frank GR, Takahashi H, Cohen RM et al: Estrogen resistance caused by a mutation in the estrogen-receptor gene in a man, *New Engl J Med* 331:1056–61, 1994.

Glucocorticoid Receptor

Bledsoe RK, Montana VG, Stanley TB, Delves CJ, Apolito CJ et al: Crystal structure of the glucocorticoid receptor ligand binding domain reveals a novel mode of receptor dimerization and coactivator recognition, *Cell* 110:93–105, 2002.

Brewer JA, Khor B, Vogt SK, Muglia LM, Fujiwara H et al: T-cell glucocorticoid receptor is required to suppress COX-2-mediated lethal immune activation, *Nature Med* 9:1318–22, 2003.

Encio IJ, Detera-Wadleigh SD: The genomic structure of the human glucocorticoid receptor, *J Biol Chem* 266:7182–8, 1991.

Francke U, Foellmer BE: The glucocorticoid receptor gene is in 5q31-q32, *Genomics* 4:610–12, 1989.

Hollenberg SM, Weinberger C, Ong ES, Cerelli G, Oro A et al: Primary structure and expression of a functional human glucocorticoid receptor cDNA, *Nature* 318:635–41, 1985.

McNally JG, Muller WG, Walker D, Wolford R, Hager GL: The glucocorticoid receptor: Rapid exchange with regulatory sites in living cells, *Science* 287:1262–5, 2000.

Oakley RH, Sar M, Cidlowski JA: The human glucocorticoid receptor isoform: Expression, biochemical properties, and putative function, *J Biol Chem* 271:9550–9, 1996.

Pepin MC, Pothier F, Barden N: Impaired type II glucocorticoid-receptor function in mice bearing antisense RNA transgene, *Nature* 355:725–8, 1992.

Reichardt HM, Kaestner KH, Tuckermann J, Kretz O, Wessely O et al: DNA binding of the glucocorticoid receptor is not essential for survival, *Cell* 93:531–41, 1998.

Tronche F, Kellendonk C, Kretz O, Gass P, Anlag K et al: Disruption of the glucocorticoid receptor gene in the nervous system results in reduced anxiety, *Nature Genet* 23:99–103, 1999.

Mineralocorticoid Receptor

Arriza JL, Weinberger C, Cerelli G, Glaser TM, Handelin BL et al: Cloning of human mineralocorticoid receptor complementary DNA: Structural and functional kinship with the glucocorticoid receptor, *Science* 237:268–75, 1987.

Fan YS, Eddy RL, Byers MG, Haley LL, Henry WM et al: The human mineralocorticoid receptor gene (MLR) is located on chromosome 4 at q31.2, *Cytogenet Cell Genet* 52:83–4, 1989.

Geller DS, Farhi A, Pinkerton N, Fradley M, Moritz M et al: Activating mineralocorticoid receptor mutation in hypertension exacerbated by pregnancy, *Science* 289:119–23, 2000.

Geller DS, Rodriguez-Soriano J, Vallo Boado A, Schifter S, Bayer M et al: Mutations in the mineralocorticoid receptor gene cause autosomal dominant pseudohypoaldosteronism type I, *Nature Genet* 19:279–81, 1998.

Le Menuet D, Isnard R, Bichara M, Viengchareun S, Muffat-Joly M et al: Alteration of cardiac and renal functions in transgenic mice overexpressing human mineralocorticoid receptor, *J Biol Chem* 276:38911–20, 2001.

Morrison N, Harrap SB, Arriza JL, Boyd E, Connor JM: Regional chromosomal assignment of the human mineralocorticoid receptor gene to 4q31.1, *Hum Genet* 85:130–2, 1990.

Zennaro MC, Keightley MC, Kotelevtsev Y, Conway GS, Soubrier F et al: Human mineralocorticoid receptor genomic structure and identification of expressed isoforms, *J Biol Chem* 270:21016–20, 1995.

Progesterone Receptor

Auboeuf D, Honig A, Berget SM, O'Malley BW: Coordinate regulation of transcription and splicing by steroid receptor coregulators, *Science* 298:416–19, 2002.

De Vivo I, Huggins GS, Hankinson SE, Lescault PJ, Boezen M et al: A functional polymorphism in the promoter of the progesterone receptor gene associated with endometrial cancer risk, *Proc Natl Acad Sci USA* 99:12263–8, 2002.

Giangrande PH, Pollio G, McDonnell DP: Mapping and characterization of the functional domains responsible for the differential activity of the A and B isoforms of the human progesterone receptor, *J Biol Chem* 272:32889–900, 1997.

Mulac-Jericevic B, Mullinax RA, DeMayo FJ, Lydon JP, Conneely OM: Subgroup of reproductive functions of progesterone mediated by progesterone receptor-B isoform, *Science* 289:1751–4, 2000.

Robker RL, Russell DL, Espey LL, Lydon JP, O'Malley BW et al: Progesterone-regulated genes in the ovulation process: ADAMTS-1 and cathepsin L proteases, *Proc Natl Acad Sci USA* 97:4689–94, 2000.

Rousseau-Merck MF, Misrahi M, Loosfelt H, Milgrom E, Berger R: Localization of the human progesterone receptor gene to chromosome 11q22-q23, *Hum Genet* 77:280–2, 1987.

Tibbetts TA, DeMayo F, Rich S, Conneely OM, O'Malley BW: Progesterone receptors in the thymus are required for thymic involution during pregnancy and for normal fertility, *Proc Natl Acad Sci USA* 96:12021–6, 1999.

Androgen Receptor

Brown CJ, Goss SJ, Lubahn DB, Joseph DR, Wilson EM et al: Androgen receptor locus on the human X chromosome: regional localization to Xq11-12 and description of a DNA polymorphism, *Am J Hum Genet* 44:264–9, 1989.

Chang C, Kokontis J, Liao S: Molecular cloning of human and rat complementary DNA encoding androgen receptors, *Science* 240:324–6, 1988.

Griffin JE: Androgen resistance—the clinical and molecular spectrum, *New Engl J Med* 326:611–18, 1992.

Ikeda Y, Aihara K, Sato T, Akaike M, Yoshizumi M et al: Androgen receptor gene knockout male mice exhibit impaired cardiac growth and exacerbation of angiotensin II-induced cardiac fibrosis, *J Biol Chem* 280:29661–6, 2005.

La Spada AR, Wilson EM, Lubahn DB, Harding AE, Fischbeck KH: Androgen receptor gene mutations in X-linked spinal and bulbar muscular atrophy, *Nature* 352:77–9, 1991.

Lubahn DB, Joseph DR, Sullivan PM, Willard HF, French FS et al: Cloning of human androgen receptor complementary DNA and localization to the X chromosome, *Science* 240:327–30, 1988.

Nantermet PV, Xu J, Yu Y, Hodor P, Holder D et al: Identification of genetic pathways activated by the androgen receptor during the induction of proliferation in the ventral prostate gland, *J Biol Chem* 279:1310–22, 2004.

Taplin ME, Bubley GJ, Shuster TD, Frantz ME, Spooner AE et al: Mutation of the androgen-receptor gene in metastatic androgen-independent prostate cancer, *New Engl J Med* 332:1393–8, 1995.

Tilley WD, Marcelli M, Wilson JD, McPhaul MJ: Characterization and expression of a cDNA encoding the human androgen receptor, *Proc Natl Acad Sci USA* 86:327–31, 1989.

Visakorpi T, Hyytinen E, Koivisto P, Tanner M, Keinanen R et al: In vivo amplification of the androgen receptor gene and progression of human prostate cancer, *Nature Genet* 9:401–6, 1995.

Yeh S, Tsai MY, Xu Q, Mu XM, Lardy H et al: Generation and characterization of androgen receptor knockout (ARKO) mice: An in vivo model for the study of androgen functions in selective tissues, *Proc Natl Acad Sci USA* 99:13498–503, 2002.

Zhang L, Leeflang EP, Yu J, Arnheim N: Studying human mutations by sperm typing: Instability of CAG trinucleotide repeats in the human androgen receptor gene, *Nature Genet* 7:531–5, 1994.

Zoppi S, Wilson CM, Harbison MD, Griffin JE, Wilson JD et al: Complete testicular feminization caused by an amino-terminal truncation of the androgen receptor with downstream initiation, *J Clin Invest* 91:1105–12, 1993.

Vitamin D Receptor

Baker AR, McDonnell DP, Hughes M, Crisp TM, Mangelsdorf DJ et al: Cloning and expression of full-length cDNA encoding human vitamin D receptor, *Proc Natl Acad Sci USA* 85:3294–8, 1988.

Li YC, Wei M, Kong J, Chen ZF, Liu SQ et al: 1,25-Dihydroxyvitamin D3 is a negative endocrine regulator of the renin-angiotensin system, *J Clin Invest* 110:229–38, 2002.

Malloy PJ, Eccleshall TR, Gross C, Van Maldergem L, Bouillon R et al: Hereditary vitamin D resistant rickets caused by a novel mutation in the vitamin D receptor that results in decreased affinity for hormone and cellular hyporesponsiveness, *J Clin Invest* 99:297–304, 1997.

Morrison NA, Qi JC, Tokita A, Kelly PJ, Crofts L et al: Prediction of bone density from vitamin D receptor alleles, *Nature* 367:284–7, 1994.

Palmer HG, Larriba MJ, Garcia JM, Ordonez-Moran P, Pena C et al: The transcription factor SNAIL represses vitamin D receptor expression and responsiveness in human colon cancer, *Nature Med* 10:917–19, 2004.

Szpirer J, Szpirer C, Riviere M, Levan G, Marynen P et al: The Spl transcription factor gene (SP1) and the 1,25-dihydroxyvitamin D(3) receptor gene (VDR) are colocalized on human chromosome arm 12q and rat chromosome 7, *Genomics* 11:168–73, 1991.

Yoshizawa T, Handa Y, Uematsu Y, Takeda S, Sekine K et al: Mice lacking the vitamin D receptor exhibit impaired bone formation, uterine hypoplasia and growth retardation after weaning, *Nature Genet* 16:391–6, 1997.

Giguere V: Steroid hormone receptor signaling, in *Handbook of Cell Signaling*, Vol 3, Bradshaw RA, Dennis EA, eds, Academic Press, Amsterdam, 2004, pp 35–8.

Human protein reference data base, Johns Hopkins University and the Institute of Bioinformatics, at http://www.hprd.org/protein.

5.19. Structure and Function of p53-Related Signaling Molecules

An W, Kim J, Roeder RG: Ordered cooperative functions of PRMT1, p300, and CARM1 in transcriptional activation by p53, *Cell* 117:735–48, 2004.

Bernal JA, Luna R, Espina A, Lazaro I, Ramos-Morales F et al: Human securin interacts with p53 and modulates p53-mediated transcriptional activity and apoptosis, *Nature Genet* 32:306–11, 2002.

Bunz F, Dutriaux A, Lengauer C, Waldman T, Zhou S et al: Requirement for p53 and p21 to sustain G2 arrest after DNA damage, *Science* 282:1497–501, 1998.

Caelles C, Helmberg A, Karin M: p53-dependent apoptosis in the absence of transcriptional activation of p53-target genes, *Nature* 370:220–3, 1994.

Chen PL, Chen Y, Bookstein R, Lee WH: Genetic mechanisms of tumor suppression by the human p53 gene, *Science* 250:1576–80, 1990.

Chen Z, Trotman LC, Shaffer D, Lin HK, Dotan ZA et al: Crucial role of p53-dependent cellular senescence in suppression of Pten-deficient tumorigenesis, *Nature* 436:725–30, 2005.

Chipuk JE, Bouchier-Hayes L, Kuwana T, Newmeyer DD, Green DR: PUMA couples the nuclear and cytoplasmic proapoptotic function of p53, *Science* 309:1732–5, 2005.

Chipuk JE, Kuwana T, Bouchier-Hayes L, Droin NM, Newmeyer DD et al: Direct activation of Bax by p53 mediates mitochondrial membrane permeabilization and apoptosis, *Science* 303:1010–4, 2004.

Cho Y, Gorina S, Jeffrey PD, Pavletich NP: Crystal structure of a p53 tumor suppressor-DNA complex: Understanding tumorigenic mutations, *Science* 265:346–55, 1994.

Christophorou MA, Martin-Zanca D, Soucek L, Lawlor ER, Brown-Swigart L et al: Temporal dissection of p53 function in vitro and in vivo, *Nature Genet* 37:718–26, 2005.

Chuikov S, Kurash JK, Wilson JR, Xiao B, Justin N et al: Regulation of p53 activity through lysine methylation, *Nature* 432:353–60, 2004.

Dumont P, Leu JIJ, Pietra AC D, III, George DL, Murphy M: The codon 72 polymorphic variants of p53 have markedly different apoptotic potential, *Nature Genet* 33:357–65, 2003.

Fujiwara T, Bandi M, Nitta M, Ivanova EV, Bronson RT et al: Cytokinesis failure generating tetraploids promotes tumorigenesis in p53-null cells, *Nature* 437:1043–7, 2005.

Gao Y, Ferguson DO, Xie W, Manis JP, Sekiguchi J et al: Interplay of p53 and DNA-repair protein XRCC4 in tumorigenesis, genomic stability and development, *Nature* 404:897–900, 2000.

Gorgoulis VG, Vassiliou LVF, Karakaidos P, Zacharatos P, Kotsinas A et al: Activation of the DNA damage checkpoint and genomic instability in human precancerous lesions, *Nature* 434:907–13, 2005.

Grossman SR, Deato ME, Brignone C, Chan HM, Kung AL et al: Polyubiquitination of p53 by a ubiquitin ligase activity of p300, *Science* 300:342–4, 2003.

Harris CC, Hollstein M: Clinical implications of the p53 tumor-suppressor gene, *New Engl J Med* 329:1318–27, 1993.

Hemann MT, Bric A, Teruya-Feldstein J, Herbst A, Nilsson JA et al: Evasion of the p53 tumour surveillance network by tumour-derived MYC mutants, *Nature* 436:807–11, 2005.

Hirao A, Kong YY, Matsuoka S, Wakeham A, Ruland J et al: DNA damage-induced activation of p53 by the checkpoint kinase Chk2, *Science* 287:1824–7, 2000.

Hollstein M, Sidransky D, Vogelstein B, Harris CC: p53 mutations in human cancers, *Science* 253:49–53, 1991.

Hsu IC, Metcalf RA, Sun T, Welsh JA, Wang NJ et al: Mutational hotspot in the p53 gene in human hepatocellular carcinomas, *Nature* 350:427–8, 1991.

Johnson TM, Hammond EM, Giaccia A, Attardi LD: The p53(QS) transactivation-deficient mutant shows stress-specific apoptotic activity and induces embryonic lethality, *Nature Genet* 37:145–52, 2005.

Jonkers J, Meuwissen R, van der Gulden H, Peterse H, van der Valk M et al: Synergistic tumor suppressor activity of BRCA2 and p53 in a conditional mouse model for breast cancer, *Nature Genet* 29:418–25, 2001.

Kemp CJ, Wheldon T, Balmain A: p53-deficient mice are extremely susceptible to radiation-induced tumorigenesis, *Nature Genet* 8:66–9, 1994.

Leu JIJ, Dumont P, Hafey M, Murphy ME, George DL: Mitochondrial p53 activates Bak and causes disruption of a Bak-Mcl1 complex, *Nature Cell Biol* 6:443–50, 2004.

Levine AJ: p53, the cellular gatekeeper for growth and division, *Cell* 88:323–31, 1997.

Levine AJ, Momand J, Finlay CA: The p53 tumour suppressor gene, *Nature* 351:453–6, 1991.

Miller C, Mohandas T, Wolf D, Prokocimer M, Rotter V et al: Human p53 gene localized to short arm of chromosome 17, *Nature* 319:783–4, 1986.

Olive KP, Tuveson DA, Ruhe ZC, Yin B, Willis NA et al: Mutant p53 gain of function in two mouse models of Li-Fraumeni syndrome, *Cell* 119:847–60, 2004.

Pearson M, Carbone R, Sebastiani C, Cioce M, Fagioli M et al: PML regulates p53 acetylation and premature senescence induced by oncogenic Ras, *Nature* 406:207–10, 2000.

Polyak K, Xia Y, Zweler JL, Kinzler KW, Vogelstein B: A model for p53-induced apoptosis, *Nature* 389:300–5, 1997.

Raj K, Ogston P, Beard P: Virus-mediated killing of cells that lack p53 activity, *Nature* 412:914–17, 2001.

Raman V, Martensen SA, Reisman D, Evron E, Odenwald WF et al: Compromised HOXA5 function can limit p53 expression in human breast tumours, *Nature* 405:974–8, 2000.

Ryan KM, Ernst MK, Rice NR, Vousden KH: Role of NF-kappa-B in p53-mediated programmed cell death, *Nature* 404:892–7, 2000.

Sax JK, Fei P, Murphy ME, Bernhard E, Korsmeyer SJ et al: BID regulation by p53 contributes to chemosensitivity, *Nature Cell Biol* 4:842–9, 2002.

Shieh SY, Ikeda M, Taya Y, Prives C: DNA damage-induced phosphorylation of p53 alleviates inhibition by MDM2, *Cell* 91:325–34, 1997.

Srivastava S, Zou Z, Pirollo K, Blattner W, Chang EH: Germ-line transmission of a mutated p53 gene in a cancer-prone family with Li-Fraumeni syndrome, *Nature* 348:747–9, 1990.

Storey A, Thomas M, Kalita A, Harwood C, Gardiol D et al: Role of a p53 polymorphism in the development of human papilloma-virus-associated cancer, *Nature* 393:229–34, 1998.

Takaoka A, Hayakawa S, Yanai H, Stolber D, Negishi H et al: Integration of interferon-alpha/beta signalling to p53 responses in tumour suppression and antiviral defence, *Nature* 424:516–23, 2003.

Tyner SD, Venkatachalam S, Choi J, Jones S, Ghebranious N et al: p53 mutant mice that display early ageing-associated phenotypes, *Nature* 415:45–53, 2002.

Ueda H, Ullrich SJ, Gangemi JD, Kappel CA, Ngo L et al: Functional inactivation but not structural mutation of p53 causes liver cancer, *Nature Genet* 9:41–7, 1995.

Vogelstein B, Kinzler KW: p53 function and dysfunction, *Cell* 70:523–6, 1992.

Yin Y, Liu YX, Jin YJ, Hall EJ, Barrett JC: PAC1 phosphatase is a transcription target of p53 in signalling apoptosis and growth suppression, *Nature* 422:527–31, 2003.

Yu JL, Rak JW, Coomber BL, Hicklin DJ, Kerbel RS: Effect of p53 status on tumor response to antiangiogenic therapy, *Science* 295:1526–8, 2002.

Zhang Y, Xiong Y: A p53 amino-terminal nuclear export signal inhibited by DNA damage-induced phosphorylation, *Science* 292:1910–5, 2001.

Zhu C, Mills KD, Ferguson DO, Lee C, Manis J et al: Unrepaired DNA breaks in p53-deficient cells lead to oncogenic gene amplification subsequent to translocations, *Cell* 109:811–21, 2002.

Human protein reference data base, Johns Hopkins University and the Institute of Bioinformatics, at http://www.hprd.org/protein.

5.20. Mechanisms of p53-Mediated Signaling

Anderson CW, Appella E: Signaling to the p53 tumor suppressor through pathways activated by genotoxic and nongenotoxic stresses, in *Handbook of Cell Signaling*, Bradshaw RA, Dennis EA, eds, Academic Press, Amsterdam, 2004, pp 237–47.

Krauss G: *Biochemistry of Signal Transduction and Regulation*, 3rd ed, Wiley-VCH GmbH, 2003.

Lewin B: *Genes VIII*, Pearson Prentice-Hall, Upper Saddle River, NJ, 2004.

6

FUNDAMENTAL CELLULAR FUNCTIONS

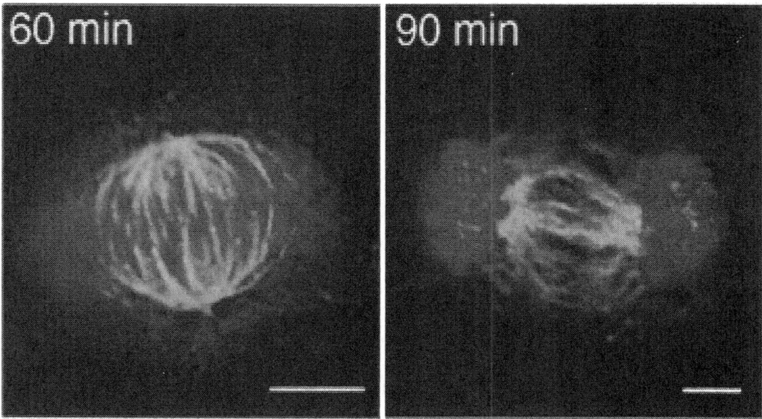

Dynamic rearrangement of chromosomes and microtubules during mitosis. Hela cells are cultured and arrested by treatment with nocodazole, which results in the synchronization of the cells in prometaphase of mitosis. Following the removal of nocodazole, cells start to enter mitosis. Microtubules (green) and chromosomes (blue) undergo dynamic rearrangement during mitosis as shown at 60 and 90 min after the removal of nocodazole. [Reprinted by permission from Macmillan Publishers Ltd: Yasuda S et al: *Nature* 428:767–71, copyright 2004.] See color insert.

Bioregenerative Engineering: Principles and Applications, by Shu Q. Liu

The cell is capable of conducting a number of basic activities, including cell division (cell proliferation and differentiation), migration, adhesion, and apoptosis. These activities play critical roles in the development, morphogenesis, and remodeling of tissues and organs. Cell differentiation, proliferation, and migration are essential processes that contribute to the formation of specialized tissues and organs during development and to the regeneration of malfunctioned and lost cells in injury. Cell-to-cell and cell-to-matrix adhesion are critical processes for the formation of coherent tissue and organ systems. Cell apoptosis is a process that eliminates unnecessary cells, ensuring appropriate morphogenesis and development of tissues and organs. These basic activities are precisely controlled to achieve specialized functions for a living tissue, organ, and system. In a multicellular organism, various cellular activities may occur at different locations and levels. These activities must be precisely coordinated between different systems at the cellular, tissue, and organ levels, which is an essential mechanism for the survival and function of the entire organism. In this chapter, these fundamental cellular activities are briefly reviewed.

CELL DIVISION

Cell division is a process of cell reproduction, which generates progeny with identical genotypes. In mammalian cells, there are two types of division: mitosis and meiosis. *Mitosis* is a cell division process for nongerm cells. This process transmits identical copies of all genes from parent cells to daughter cells. *Meiosis* is a cell division process for germ cells, which produces gametes with half or 23 of the chromosomes.

Mitosis [6.1]

Cell division via mitosis may result in two consequences: cell proliferation and differentiation. *Cell proliferation* is a process of cell division, which results in the reproduction of progeny with identical phenotypes, that is, physical, chemical, and physiological characteristics. *Cell differentiation* is a process of cell division, which results in the production of specialized cells with phenotypes different from the mother cells. These are two fundamental processes, which determine the morphogenesis of tissues and organs during embryonic development and the structure and function of tissues and organs during physiological and pathological remodeling.

Cell proliferation and differentiation may coexist at a given time. The relative activity of cell proliferation and differentiation is dependent on the stage of development. During the embryonic and fetal stages, the two processes occur simultaneously. The differentiation of embryonic stem cells gives rise to various types of specialized cells for various tissue and organ systems, whereas the proliferation of a specialized cell type contributes to cell multiplication and the construction of a specialized tissue and organ. After birth, the activity of cell differentiation is relatively reduced, while cell proliferation remains prominent for tissue and organ growth. After reaching maturation, both cell differentiation and proliferation are reduced significantly. However, a basal level of cell proliferation and differentiation remains for the replacement of malfunctioned, lost, or aging cells. Although cell proliferation and differentiation result in different cell fates, both undergo a common cell division cycle.

Cycle of Mitotic Cell Division

A *cell division cycle* is defined as the period between two mitotic cell divisions. A cell cycle is composed of two major periods: the interphase and mitotic phase (M). The *interphase* is defined as the period from the end of the prior M phase to the beginning of the next M phase and is traditionally divided into the G1 (gap 1), S (synthetic), and G2 (gap 2) phases. The M phase is the period during which cells divide, whereas the interphase is the period during which cells prepare for cell division. These different phases are corresponding to highly ordered and discrete molecular processes that regulate cell division.

The length of a cell cycle is dependent on the stage of development. During the early embryonic stage, the G1 and G2 phases are short or even missing. Embryonic cells primarily undergo alternations of the S and M phases. The duration of a cell cycle is <1 h. The short cell cycle is in coordination with the rapid growth of the early embryo. The length of a cell cycle increases with maturation. In a fully developed mammal, a typical cell cycle may last for 12–24 h with four discrete phases. Each phase of a standard cell cycle is briefly discussed here.

G1 Phase. The G1 phase is the interval during which the cell is prepared for DNA synthesis. It is 6–12 h in length, starting from the end of the M phase to the beginning of the S phase. During this phase, the cell size increases, due to the synthesis of proteins and lipids.

S Phase. The S phase is the period of DNA replication. It lasts for 7–8 hrs, starting from the end of G1 phase to the beginning of the G2 phase. The total DNA content and the number of chromatids are doubled by the end of the S phase. The nucleus increases in size apparently. In addition to DNA synthesis, other cellular components, including RNA and proteins, are also synthesized. However, DNA synthesis occurs only during the S phase, other components are synthesized continuously through the cell cycle. By the beginning of mitosis, the mass of the mother cell is doubled with cellular components equally apportioned for the two daughter cells.

G2 Phase. The G2 phase is the interval of cell preparation for cell mitosis. It is defined as the period (3–4 h) from the end of the S phase to the beginning of the M phase. During this phase, the cell possesses two complete diploid sets of chromosomes and is prepared for entering the mitotic phase. One of the major cellular activities during the G2 phase is the proofreading of the synthesized DNA. The detection of abnormal, damaged, or unreplicated DNA fragments can activate a protein kinase cascade of the G2 DNA damage checkpoint. The consequence of such activity is the inhibition of cyclin-dependent kinases, which are required for the initiation of mitosis, leading to a delay or blockade of entering the M phase.

M Phase. The M phase is the period of chromosome separation (~1 h), staring from the end of the G2 phase to the beginning of the G1 phase. During this phase, the chromosomes are separated into two equal parts for the two daughter cells. The cytoplasm is separated via cytokinesis. Chromosome separation is accomplished through several stages, including the prophase, prometaphase, metaphase, anaphase, and telophase (Fig. 6.1).

During the *prophase*, chromatins, which are composed of DNA and associated proteins and are dispersed in the nucleus during the non-M phases, are condensed into discrete

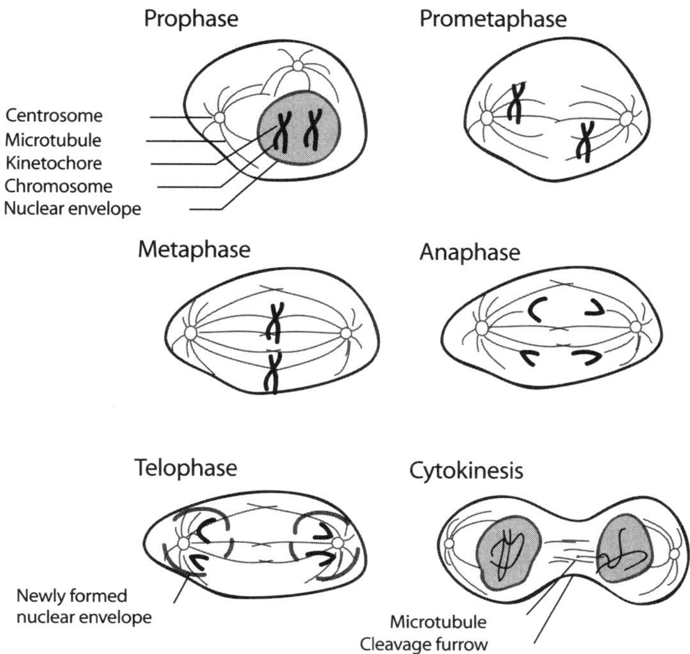

Figure 6.1. Schematic representation of cell mitosis. Mitosis is a process of cell division and consists of several defined phases, including prophase, prometaphase, metaphase, anaphase, telophase, and cytokinesis. Before a cell enters the prophase, the content of DNA is doubled. During the prophase, the chromosomes are organized from dispersed chromatins, the nuclear envelope is disassembled, the two centrosomes are deployed to the two mitotic poles, and microtubules are organized into a spindle-like network. During prometaphase, the chromosomes attach to the spindle microtubules and move toward the cell equator. During metaphase, the chromosomes are aligned along the cell equator and are ready for separation. During anaphase, the two chromatids of each chromosome complex are separated and move toward the opposite mitotic poles. During telophase, the chromosomes approach the mitotic poles, the nuclear envelope of the two daughter nuclei starts to form, and the cell is ready for cytokinesis. Cytokinesis is a process by which the cytoplasm is divided into two equal parts, each with a daughter nucleus. The formation of the daughter cells indicates the end of the cell division cycle. Based on bibliography 6.1.

chromosomes, each containing a pair of chromatids that are connected at the centromere. At the same time, the cytoskeletal microtubules start to reassemble into a mitotic spindle between two centrosomes outside the nucleus.

During the *prometaphase*, the nucleus envelope is disrupted, and the spindle microtubules are redistributed in the nucleus. The kinetochore protein, which forms a complex with the centromere of each chromatid, attaches to a selected spindle microtubule, establishing a kinetochore microtubule. Since each chromosome is composed of a pair of chromatids that are joined by two centromeres, each chromosome contains two kinetochore complexes. The kinetochore-free spindle microtubules are known as *polar microtubules.*

During the *metaphase*, with the help of the kinetochore microtubules, the chromosomes are aligned midway between the two spindle poles on a plane perpendicular to the spindle microtubules, which are connected to the spindle poles. This chromosome plane is defined as the *metaphase plate*.

During the *anaphase*, the kinetochore–centromere complex is separated and the two chromatids of each chromosome are pulled toward opposite spindle poles at a speed of ~1 μm/min. These activities are induced by the dynamic shortening of the kinetochore microtubules, which connect the chromatids. At the same time, the two spindle poles move away from each other and the polar microtubules, which are not connected to chromosomes, elongate in the spindle direction.

During the *telophase*, separated chromatids, now referred to as *chromosomes*, approach the two spindle poles symmetrically, the kinetochore microtubules gradually disappear, and the nucleus envelope appears around each group of chromosomes surrounding each spindle pole. The chromatin is gradually dispersed in the nucleus and the typical interphase nucleus starts to form, indicating the end of the M phase and beginning of the G1 phase.

Cytokinesis [6.2]

Cytokinesis is the process of cytoplasmic cleavage or segregation, which takes place immediately following chromosomal separation. During the late telophase, the cytoskeletal actin filaments and myosin are deployed along the cell equator, forming a contractile ring or cleavage furrow. The contractile ring gradually constricts the cell until the two daughter nuclei are completely separated with equal cytoplasm contents. This indicates the completion of the cell division cycle.

The formation of the cleavage furrow has always fascinated scientists. This is a complex process, which involves several structures, including the microtubule spindle, actin filaments and myosin, and cell membrane. It is now understood that the assembly of the contractile actin–myosin ring is regulated by several types of molecule. These include microtubule spindle-associated molecules, the RhoA guanosine triphosphatase (GTPase), myosin II, actin and actin-associated factors, and molecules mediating the fusion of membrane structures. The coordinated actions of all these structures and regulatory molecules contribute to the formation and performance of the contractile ring during cytokinesis.

Cell membrane fusion is an important part of cytokinesis. When the cytoplasm is separated by the constriction of the contractile ring, the cytoplasm of each daughter cell must be completely covered with the cell membrane. Two mechanisms are involved in the formation of the membrane of the daughter cells: (1) the daughter cells are able to generate additional cell membrane, which is deployed to the cleavage furrow to cover the open cytoplasm; and (2) membrane vesicles can be produced in the daughter cells. These vesicles can be transported to the cleavage furrow to bridge the open surface of the cytoplasm. Membrane fusion is required for the formation of a complete cell membrane.

Control of Cell Division [6.3]

The mitotic events described above are precisely controlled in a highly coordinated manner. The order and completeness of these events, which are critical to successful cell division, are controlled by two mechanisms: constitutive and extracellular. The constitutive mechanism is intrinsic in nature and ensures the initiation, progression, and completion of cell division. The extracellular mechanism is dependent on the stimulation of extracellular factors and ensures that cell division takes place in coordination with the global function and physiological status of a specialized tissue or organ. In other words, the extracellular mechanism controls whether, when, and to what degree cell division occurs. Once a cell cycle is triggered by an extracellular cue, the constitutive mechanism controls the progression of cell division.

Both constitutive and extracellular mechanisms are implemented via the mediation of cell cycle regulatory molecules, including cyclins, cyclin-dependent kinases (CDKs), and inhibitors of cyclin-dependent kinases (CKIs). A common feature of cell cycle regulation is the occurrence of oscillatory changes in the activity of regulatory factors, which determine the cyclic events of cell division. In particular, periodic phosphorylation/dephosphorylation of CDKs and cyclic increase and decrease in the level of cyclins play key roles in the initiation, progression, and cessation of cell division.

Constitutive Control of Cell Division. The constitutive control mechanism is dependent on several "checkpoints." These checkpoints inspect and control the progression of cell division and eliminate errors, if any. Checkpoints have been identified in all phases of the cell cycle. Several proteins, including p53, and p21, have been found to play a role in these checkpoints.

In the Gl phase, a major checkpoint is known as the *restriction point*, which controls the initiation of cell division. A cell can pass the Gl phase only if the prior mitosis is complete, DNA is undamaged, and the cell reaches a critical size. DNA damage, if any, must be repaired before a new cell cycle is initiated. Unrepaired DNA may result in the halt of the cell cycle. Also, a critical size of cell mass is required for cell division. The Gl phase is the longest period with the largest variation in length among all cell cycle phases. Thus, the Gl phase is timely flexible to adjust the rate of cell division in response to various stimulations.

The *S-phase checkpoints* assess the integrity of DNA. The checkpoints prevent the cell from synthesizing DNA, if the DNA is damaged or disrupted. All DNA molecules in the genome must be completely replicated, which is a prerequisite for the continuation of cell division to the G2 phase. The cell is arrested in the S phase if damaged DNA is found. Once a DNA replication process is initiated, it must be completed, or the cell cycle is stopped.

During the G2 phase, the newly replicated DNA is examined by the *G2-phase checkpoints* to ensure that the DNA molecules have been correctly duplicated. DNA repair factors can be activated to correct replication errors. These checkpoints determine whether the cell can move to the M phase. If the DNA molecules are not completely replicated, the cell is arrested before the M phase.

The *M-phase checkpoints* are responsible for detecting problems that potentially influence cell division. The checkpoints assess the completeness of proceeding preparatory events as well as the mitotic events. For example, unattached kinetochore can be detected by the checkpoints, leading to the halt of cell cycle progression. The completeness of the entire mitosis is also assessed by the M-phase checkpoints. Any incomplete events may result in cell arrest.

Some cells, especially terminally differentiated cells, are not committed to cell division and enter a phase known as the *G0 phase*, which resembles the Gl phase in certain aspects. These calls cannot pass the restriction points and cannot proceed to the S phase. However, under appropriate extracellular stimulations, some of these cells can be stimulated to pass the restriction points, initiating cell division.

The discovery of the cell cycle checkpoints has led to a significant advance in the understanding of the control mechanisms of cell division. All key processes of cell division are assessed by the checkpoints. The presence of continuous checkpoints through the cell cycle ensures the initiation, progression, and completion of cell division and the accuracy of cell reproduction.

Extracellular Control of Cell Division. There are a variety of extracellular factors that regulate the initiation and progression of cell division. These include growth factors, nutrient supplies, cell–cell interactions, and mechanical forces. Growth factors, such as platelet-derived growth factor, epidermal growth factor, fibroblast growth factor, activate cyclins and CDKs and promote the cell to enter the S phase and reduce the length of the G1 phase. In contrast, a growth inhibitor, such as transforming growth factor β, activates cell division inhibitors, such as p21, p27, and p57, and induces cell arrest. A reduction or depletion of nutrient supplies may promote the cell to enter the G0 phase. An increase in cell density or cell–cell contact may result in cell arrest. Mechanical forces have also been found to mediate cell mitosis. A decrease in bloodflow or fluid shear stress may activate the division cycle of vascular smooth muscle cells. In contrast, an increase in mechanical stretch or tensile stress in the wall of blood vessels induces mitosis of vascular smooth muscle cells. These mechanical factors may directly regulate cell mitosis or influence cell mitosis via the mediation of mitogenic factors.

Signaling Events of Cell Cycle Control. The progression of the cell cycle is regulated by a cascade of signaling molecules (Fig. 6.2). In the G1 phase, cells either enter the G0 phase or are committed to enter the S phase, initiating the cell division cycle. An increase in growth factors stimulates the cell to enter the S phase by inducing the expression of cyclin D. The level of cyclin D remains high through the G1, S, and G2 phases and rapidly reduces during the M phase via ubiquitination-mediated degradation (see Chapter 5). Increased cyclin D in the G1 phase promotes the formation of the cyclin D-CDK4/6 complex. Activated cyclin D-CDK4/6 complex stimulates cell growth, which is required for the passing of the G1 restriction point.

The cyclin D-CDK4/6 complex can phosphorylate retinoblastoma tumor suppressor (Rb), which contains the transcriptional factor E2F (elongation factor). The phosphorylation of Rb induces the release of E2F, which stimulates the expression of the cyclin E gene, resulting in a transient increase in cyclin E during the transition period from the G1 to S phase. Cyclin E forms a complex with CDK2 and the cyclin E/CDK2 complex is in turn activated by CDC25A-mediated dephosphorylation of CDK2 on the Thr14 and Tyr15 residues. Activated cyclin E-CDK2 complex promotes the cell to pass the G1 restriction point and enter the S phase. (See Table 6.1.)

During the early S phase, the transcriptional factor E2F stimulates the expression of cyclin A, which forms a complex with CDK2. The cyclin A-CDK2 and cyclin E-CDK2 complexes are capable of phosphorylating critical components that initiate and regulate DNA replication. The cyclin E-CDK2 complex is dissociated due to the degradation of cyclin E by the ubiquitin–proteasome system (see Chapter 5) during the early S phase, whereas the cyclin A-CDK2 complex remains active through the S and G2 phases.

During the S phase, another cyclin molecule, cyclin B, is gradually accumulated and forms a complex with cell division cycle protein (CDC)2. The cyclin B/CDC2 complex is known as the *M-phase-promoting factor* (MPF). MPF can be activated by CDC25B/C-mediated dephosphorylation at the G2/M transition. Activated MPF can phosphorylate a number of substrate proteins, including lamin, vimentin, and caldelsom, leading to cell mitosis. The phosphorylation of lamin is thought to induce the disruption of the cell

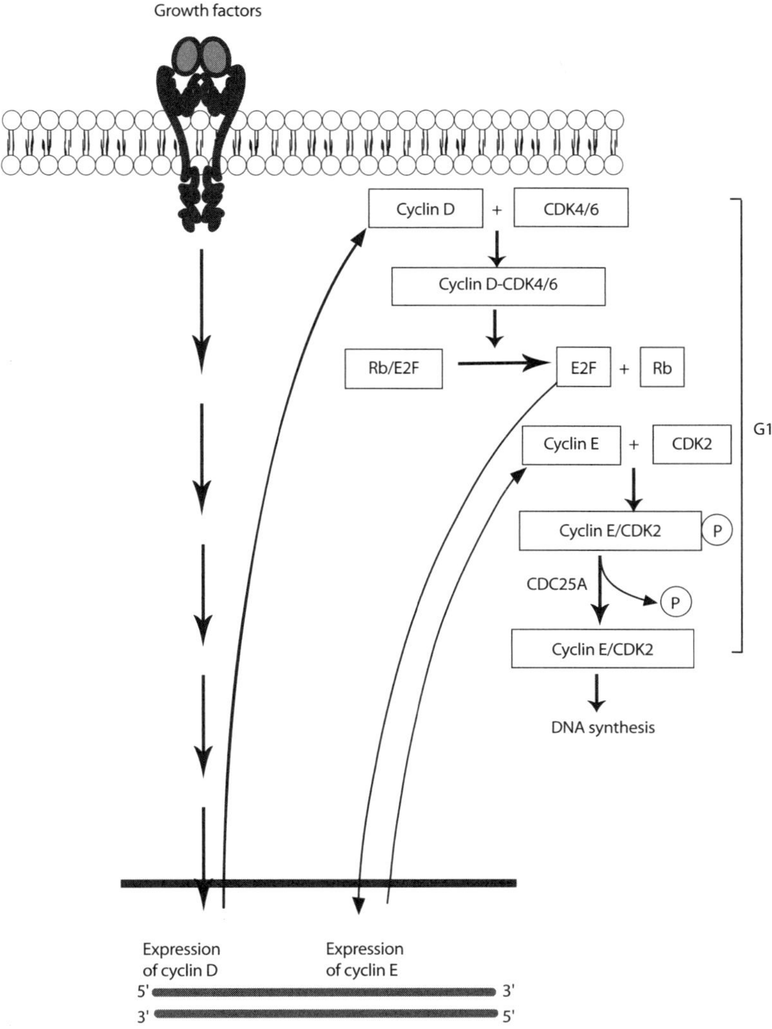

Figure 6.2. Schematic representation of the regulatory mechanisms of cell mitosis. Based on bibliography 6.3.

nucleus. The phosphorylation of vimentin is responsible for dynamic changes in microtubules and the formation of the spindle. The phosphorylation of caldelsom induces the interaction of actin filaments with myosin molecules, leading to the formation of the cleavage furrow and cell cytokinesis. MPF is deactivated by ubi-quitination of cyclin B during the M phase, which indicates the end of cell mitosis. (See Table 6.2.)

Inhibition of Cell Division Cycle. A family of proteins, including p15(INK4B), p16(INK4), p18(INK4C), and p19(INK4D), exerts an inhibitory effect on the activity of the cyclin D/CDK4/CDK 6 complex, and induces cell arrest during the G1 phase. The suppression of these inhibiting molecules leads to uncontrolled cell proliferation.

TABLE 6.1. Characteristics of Selected Cell Cycle Regulatory Molecules*

Proteins	Alternative Names	Amino Acids	Molecular Weight (kDa)	Expression	Functions
Cyclin D1	B-cell leukemia 1, BCL1	295	34	Liver, kidney, intestine, uterus, pancreas, placenta	Binding to CDK4 and CDK6, mediating the activity of these CDKs, regulating the transition of the cell division cycle from the G1 to S phase, and promoting cell division and tumorigenesis
Cyclin E1	Cyclin E, G1/S-specific cyclin E1	410	47	Lung	Binding to CDK2, mediating the activity of this CDK, regulating the transition of cell division cycle from the G1 to S phases, and promoting cell division and tumorigenesis
CDK2	Cyclin-dependent kinase 2, cell division kinase 2, p33 protein kinase, cell division protein kinase 2	298	34	Kidney, prostate, lymphocytes, skin	A serine/threonine protein kinase that regulates the transition from G1 to S phases
CDK4	Cyclin-dependent kinase 4, cell division kinase 4	303	34	Ubiquitous	A serine/threonine protein kinase that phosphorylates the retinoblastoma protein and regulates the transition from G1 to S phase
CDK6	Cyclin-dependent kinase 6, cell division protein kinase 6	326	37	Ubiquitous	A serine/threonine protein kinase that forms a complex with CDK4 and regulates the transition from the G1 to S phases
E2F1	E2F transcription factor 1, transcription factor E2F, retinoblastoma associated protein 1, and retinoblastoma-binding protein 3	437	47	Pancreas, connective tissue, skin	Serving as a transcriptional factor, stimulating gene transcription, and regulating cell division, proliferation, differentiation, and apoptosis
CDC25	Cell division cycle 25A, M-phase inducer phosphatase 1, dual-specificity phosphatase Cdc25A	523	59	Brain	A phosphatase that activates CDC2 by dephosphorylation and regulates the progression of cell division cycle from the G1 to S phases

*Based on bibliography 6.3.

The p15 protein may be activated in response to TGF-β, which suppresses the proliferation of several cell types, including epithelial cell and smooth muscle cell. Another group of proteins, including p21, p27, and p57, may inhibit the activity of the cyclin D/CDK4/CDK6 and cyclin A/CDK2 complexes, and induce cell arrest in the G1 phase. The tumor suppressor protein p53 induces activation of p27, leading to cell arrest. (See Table 6.3.)

Meiosis [6.4]

Meiosis is the process of gametogenesis or germ cell division. During such a process, a diploid germ progenitor cell undergoes DNA synthesis and two division events to produce four daughter cells with a haploid set of chromosomes. Each germ progenitor cell contains homologous pairs of chromosomes. Each homologous pair of chromosomes is composed of a maternal chromosome and a paternal chromosome, which can be identical or allelic (not completely identical). In response to the stimulation of signals for gametogenesis, the germ progenitor cell enters the S phase and initiates DNA synthesis, yielding two identical chromatids for each chromosome. The cell then enters two consecutive division processes, designated as meiosis I and meiosis II, to produce daughter germ cells. The meiosis I process is usually divided into several stages, including prophase I, metaphase I, anaphase I, and telophase I. The meiosis II process is divided into metaphase II, anaphase II, and telophase II. By the end of telophase II, four haploid daughter cells are produced from a single diploid germ progenitor cell (Fig. 6.3).

During meiosis prophase I, the nucleus envelope is reorganized, degraded, and disappeared. Centrosomes and microtubule spindles start to form. Scattered chromosomes are organized into apparently double-chromatid structures. The chromosomes can be clearly recognized under an optical microscope. Crossing over of chromatid fragments may occur between the maternal and paternal chromosomes, resulting in homologous recombination (Fig. 6.4). During meiosis metaphase I, a complete spindle and centrosome system is established. The chromosomes are aligned in the equator region. The centromeres of the chromosomes are connected to the spindles. During meiosis anaphase I, the chromosomes remain paired and are pulled toward the poles of the cell. Note that the chromosomal segregation process of meiosis is different from that of mitosis. In mitosis, the pair of chromatids for each chromosome is separated during the anaphase. In meiosis, the paired chromatids of each chromosome are not separated during anaphase I. During meiosis telophase I, the chromosomes are moved to the poles and rearranged. The germ progenitor cell is divided into two cells. However, no nucleus envelope is developed.

Meiosis II is the process by which the two daughter cells are further divided to produce haploid germ cells. The chromosomal separation in meiosis II is similar to that in mitosis. During meiosis metaphase II, the chromosomes in each daughter cell are aligned in the equator region. The centromeres are connected to the microtubule spindles. During meiosis anaphase II, the two identical chromatids of each chromosome are separated and pulled to the cell poles in opposite directions. During meiosis telophase II, each daughter cell is further divided into two granddaughter cells with a haploid set of chromosomes (a single copy of each chromosome from either the mother or the father). The chromosomes are rearranged and enveloped within the nucleus.

Experimental Assessment of Cell Division [6.5]

Cell division can be assessed by detecting DNA synthesis, which occurs only during cell division. There are two basic approaches for the detection of DNA synthesis: measuring

TABLE 6.2. Characteristics of Selected Cell Cycle Regulatory Molecules*

Proteins	Alternative Names	Amino Acids	Molecular Weight (kDa)	Expression	Functions
Cyclin A	CCNA1	465	52	Brain, testis, leukocytes	Binding to CDK2 and CDC2 kinases, mediating the activity of these kinases, regulating the transition of cell division cycle from the S to G2 phases
Cyclin B	G2/mitotic specific cyclin B1, CCNB	433	48	Liver, lung, bone marrow	Binding to CDC2 to form the M-phase-promoting factor and regulating the transition of cell cycle from G2 to M phases
CDC2	Cell division cycle 2, cell cycle controller CDC2, p34(CDC2), cyclin-dependent kinase 1 (CDK1), p34 protein kinase, Cdc2 kinase	297	34	Skin	A Ser/Thr protein kinase that serves as the catalytic subunit of the M-phase-promoting factor (MPF); phosphorylates proteins such as lamin, vimentin, and caldelsom; and regulates the transition of cell cycle from G1 to S phases and from G2 to M phases

*Based on bibliography 6.3.

TABLE 6.3. Characteristics of Selected Cell Cycle Inhibitory Molecules*

Proteins	Alternative Names	Amino Acids	Molecular Weight (kDa)	Expression	Functions
p15(INK4B)	p15 inhibitsCDK4, cyclin-dependent kinase inhibitor2B, CDKN2B, p15 INK4b, CDK4B inhibitor, cyclin-dependent kinase 4 inhibitor B, multiple-tumor suppressor 2 (MTS2)	138	15	Skin, placenta	Acting as a cyclin-dependent kinase inhibitor, forming a complex with CDK4 or CDK6, suppressing the activation of these kinases, and inducing the arrest of cell division cycle in G1 phase
p16(INK4A)	p16 inhibits CDK4, CDKN2A, cyclin-dependent kinase inhibitor 2A, CDKN2, CDK4 inhibitor, multiple-tumor suppressor 1 (MTS1)	173	18	Ubiquitous	Serving as an inhibitor for the CDK4 kinase and inducing cell arrest in G1 phase
p18(INK4C)	p18, inhibits CDK4, cyclin-dependent kinase inhibitor 2C, CDKN2C, cyclin-dependent kinase 6 inhibitor, cyclin-dependent kinase 4 inhibitor C, cyclin-dependent inhibitor, CDK6 inhibitor p18, p18-INK6, p18-INK4c, CDKN6	168	18	Brain, liver, testis, thymus	Serving as a cyclin-dependent kinase inhibitor, suppressing the activity of CDK4 and CDK6, and inducing cell arrest in G1 phase
p19(INK4D)	Cyclin-dependent kinase inhibitor 2D, p19 inhibits CDK4, INK4D, p19 INK4D, cyclin-dependent kinase 4 inhibitor D, CDK inhibitor p19 INK4D	166	18	Ubiquitous	Serving as a cyclin-dependent kinase inhibitor, and inducing cell arrest in the G1 phase.
p21	Cyclin-dependent kinase inhibitor 1A, cyclin-dependent kinase inhibitor 1, DNA synthesis inhibitor, CDK-interaction protein 1, wildtype p53-activated fragment 1, melanoma-differentiation-associated protein 6	164	18	Heart, bone	Acting as a cyclin-dependent kinase inhibitor, inhibiting the activity of cyclin-CDK2 and CDK4, and mediating p53-dependent cell arrest in G1 and G2 phases
p27	Cyclin-dependent kinase inhibitor 1B, cyclin-dependent kinase inhibitor p27, p27Kip1, KIP1, CDKN4	198	22	Ubiquitous	Serving as a cyclin-dependent kinase inhibitor, binding to and suppressing the cyclin E-CDK2 and cyclin D-CDK4 complexes, and inducing cell arrest in G1 phase
p57	Cyclin-dependent kinase inhibitor 1C, p57KIP2, KIP2	316	32	Heart, skeletal muscle, B cells, placenta	Acting as an inhibitor for G1 cyclin/CDK complexes and inducing cell arrest in G1 phase

*Based on bibliography 6.3.

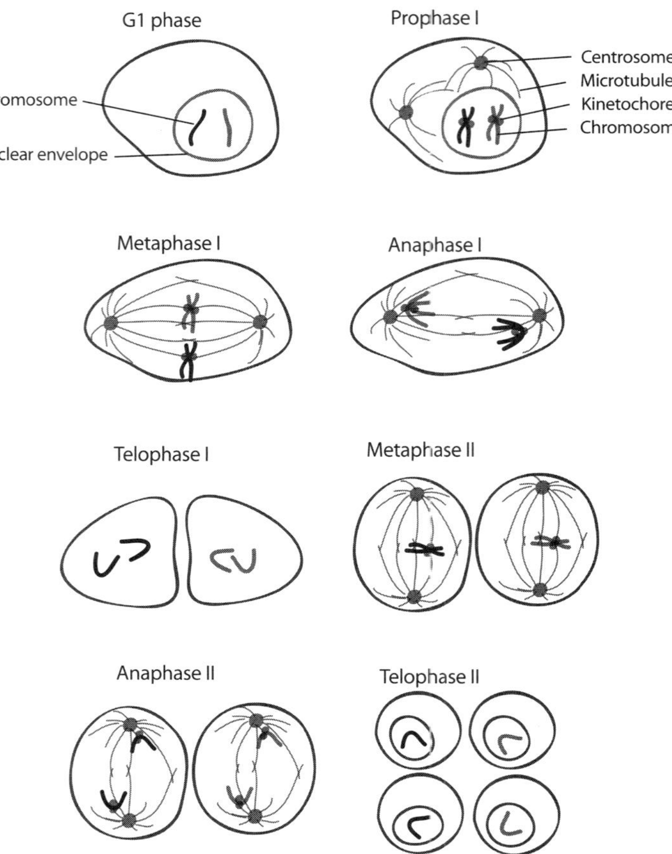

Figure 6.3. Schematic representation of cell meiosis. Meiosis is a process of germ cell division and is composed of two phases, including meiosis I and meiosis II. The phase meiosis I is consists of several stages, including prophase I, metaphase I, anaphase I, and telophase I. The phase meiosis II consists of metaphase II, anaphase II, and telophase II. During prophase I, the nucleus envelope is degraded, centrosomes and microtubule spindles start to form, and scattered chromosomes are organized into apparently double-chromatid structures. During metaphase I, a complete spindle network forms, the two centrosomes are deployed to the mitotic poles, and the chromosomes are aligned in the equator region. During anaphase I, the chromosomes remain paired and are pulled to the mitotic pole of the cell. During telophase I, the chromosomes are completely separated and rearranged near the two poles. The germ progenitor cell is divided into two cells. However, no nucleus envelope is developed. During metaphase II, the chromosomes in each daughter cell are aligned along the equator. The centromeres are connected to the microtubule spindles. During anaphase II, the two identical chromatids for each chromosome are separated and pulled to the mitotic poles. During telophase II, each daughter cell is further divided into two granddaughter cells with a haploid set of chromosomes (a single copy of each chromosome from either the mother or the father). The chromosomes are rearranged and enveloped within the nucleus. Based on bibliography 6.4.

DNA content and measuring the density of cells that undergo DNA synthesis. For DNA content measurement, DNA can be extracted from a given volume of tissue or given area of cultured cells and the DNA content can be measured by DNA extraction and spectrophotometry. Although this method is easy to use, it does not directly give the density of dividing cells.

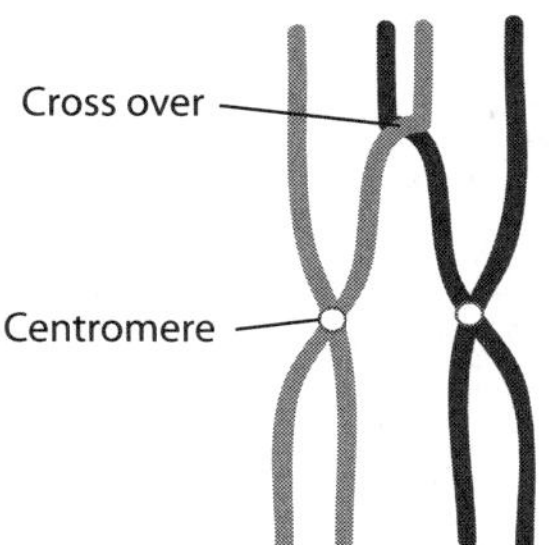

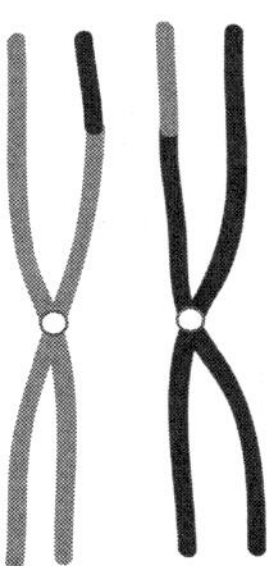

Figure 6.4. Schematic representation of chromosomal cross over. During chromosome segregation, chromosomal segments may exchange location between two chromatids of different chromosome complexes. Based on bibliography 6.4.

To directly measure the density of cells that synthesize DNA, a selected type of deoxynucleotides can be tagged with a marker and delivered to cells. Since only dividing cells take up deoxynucleotides, any cells exhibiting the tagged deoxynucleotides can be considered dividing cells. There are two types of tagging markers—radioactive isotopes and molecules—that can be detected by immunohistochemistry. A common radioactive material used for detecting cell division is [^{3}H]-thymidine. This isotope can be delivered to animal models or cultured cells. Tissue specimens or cultured cells can be collected after 24 h, fixed with 4% formaldehyde in phosphate-buffered saline (PBS), and processed for detecting [^{3}H]-thymidine incorporation. Specimens are exposed to X-ray films and cells with positive [^{3}H]-thymidine signals are considered dividing cells. Such a method is referred to as *autoradiography*. The specimen can be counterstained with hematoxylin and eosin for measuring the total number of cells. The ratio of the number of [^{3}H]-thymidine-labeled cells to that of the total cells can be used as an index for assessing cell division.

Alternatively, 5′-bromodeoxyuridine (BrdU) can be used to detect cell division instead of [^{3}H]-thymidine. BrdU can be directly injected into an animal or delivered to cultured cells. As [^{3}H]-thymidine, BrdU can be taken up only by dividing cells. BrdU can be detected by immunohistochemistry with a BrdU-specific antibody (Fig. 6.5). It is important to address several technical points for the BrdU assay. First, cultured cells or intact tissue specimens without histological sectioning should be treated with a detergent (e.g., 0.5% Triton X-100) to permeabilize cell membrane so that antibody can diffuse through the cell membrane and reach the cell nucleus. For histological tissue sections, since cells are cut open and cell nuclei are exposed, it is not necessary to treat specimens with a detergent. Second, incorporated BrdU is embedded within the cell chromatin, which prevents the anti-BrdU antibody from accessing BrdU. Thus, cultured cells or tissue specimens should be treated with pepsin to digest nucleus proteins and expose DNA, so that the anti-BrdU antibody can access the incorporated BrdU. A DNA-binding fluorochrome, e.g., Hoechst 33258, can be used to counter-staining DNA nonspecifically, allowing the measurement of the total number of cells within a selected specimen. Comparing to the [^{3}H]-thymidine incorporation method, the BrdU method is more advantageous for its simplicity and nonradioactivity.

CELL MIGRATION [6.6]

Cell migration is a fundamental cellular activity observed during development and pathogenic remodeling. During development, cell migration plays a critical role in the initiation

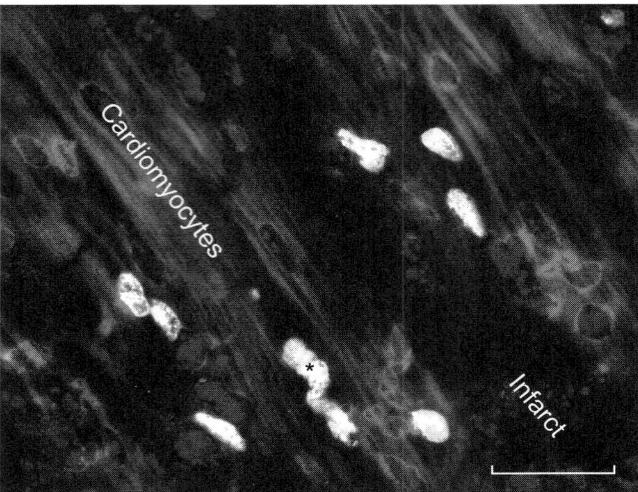

Figure 6.5. 5′-Bromodeoxyuridine (BrdU)-positive cells in injured cardiac tissue at day 5 after ischemic injury. In this preparation, cardiac injury was induced by ligating the left anterior descending coronary artery in a mouse model. BrdU was injected into the skeletal muscle of a mouse 24 hrs before observation. Cardiac specimens were fixed in 4% formaldehyde in phosphate-buffered saline (PBS), cut into cryosections, treated subsequently with 0.5% pepsin and 1.5 N HCl, incubated subsequently with an anti-BrdU antibody and a fluorescein-conjugated secondary antibody, and observed by using a fluorescence microscope. Scale bar: 10 μm.

and formation of tissues and organs, such as the nerve and cardiovascular systems. During pathogenic remodeling, cell migration contributes to the initiation and progression of pathogenic disorders, such as atherogenesis (e.g., smooth muscle cell migration from the arterial media to the arterial intima or arterial substitutes), tumorigenesis (e.g., cancer cell migration and metastasis), and inflammation (leukocyte migration to inflammatory sites). Cell migration is a mechanical event that involves a variety of molecular processes and is controlled by a number of known signaling pathways. In this chapter, the mechanics and regulatory mechanisms of cell migration are briefly reviewed.

Mechanics of Cell Migration

Cell migration is accomplished by a number of mechanical processes at the molecular and subcellular levels. Theses processes include protrusion or extension of cell membrane at the cell leading edge, attachment of protruded cell membrane to a substrate via adhesion receptors, contraction and movement of the cell body, retraction at the cell trailing edge, and recycle of adhesion receptors (Fig. 6.6). A variety of regulatory and contractile proteins are involved in the initiation and progression of cell migration. It is important to note that the five processes outlined above are arbitrarily defined. All these processes take place simultaneously and continuously in a cyclic manner. It is difficult to identify the beginning and end of a migration cycle.

Protrusion of Cell Membrane. When a cell is stimulated by a migration-activating factor, such as a chemoattractant, the cell initiates directed membrane protrusion. The direction of membrane protrusion is often determined by the stimulus. For instance, in the presence

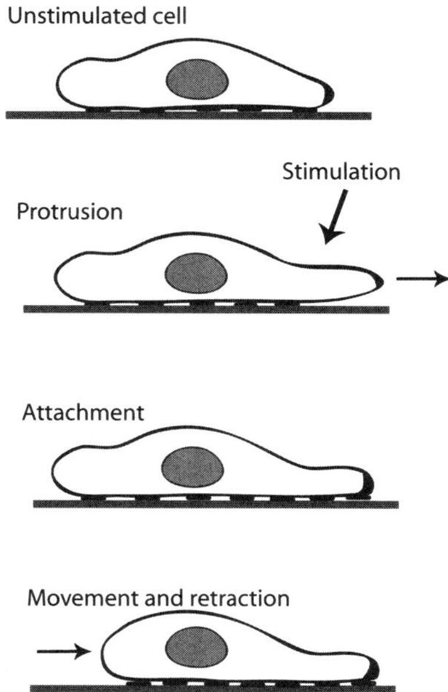

Figure 6.6. Schematic representation of cell migration. Based on bibliography 6.6.

of a chemoattractant, the cell membrane extends toward the chemoattractant. The forces that drive the membrane protrusion are generated by the actin assembly. Although the mechanisms of actin assembly is under debate, it is thought that controlled sequential extension of actin filaments toward the cell leading edge may provide a propelling force for membrane protrusion. In this model, actin subunits are added to the barbed end of the actin filaments, which point at the leading edge of cell migration. Such a process is controlled by regulatory proteins, including the Arp2/3 complex (see Chapter 3).

Attachment of Cell Membrane to Substrate Matrix at the Leading Edge. Following the protrusion of cell membrane due to directed actin assembly, the next step is the attachment of the extended membrane lamellipodia to substrate. Such a process stabilizes the leading edge of the migrating cell and allows the cell to exert forces on the substrate, a necessary condition for cell migration. The attachment of cell leading edge is mediated by integrins. Unstimulated integrins are freely suspended in the cell membrane. In response to the stimulation of extracellular matrix, integrins are activated and bind to the actin cytoskeleton. Cell membrane protrusion enhances the binding of integrins to extracellular matrix in newly formed lamellipodia. Integrin–matrix interaction further stimulates the binding of integrins to the actin cytoskeleton, facilitating the formation of focal adhesion contacts between the actin cytoskeleton and extracellular matrix. Thus, integrins play a critical role for the attachment of cell lamellipodia to substrate. Since focal adhesion contacts link the actin cytoskeleton to extracellular matrix, forces generated by the actin cytoskeleton can be transmitted to the extracellular matrix, which is critical to cell migration.

Cell Traction and Movement. Cell movement is propelled by traction forces generated by the actin cytoskeleton and exerted on the extracellular matrix substrate. In a fibroblast, for example, the actin cytoskeleton can generate traction forces about 1 nN/μm^2 on a substrate. The generation of traction forces is dependent on the interaction of actin filaments with myosin II in mammalian cells. Actin filaments are distributed around the cell periphery. The barbed ends of the actin filaments are oriented toward the cell periphery in regions near the leading and trailing edges, while those in the middle region are oriented more randomly. The myosin II molecules are distributed more heavily in the middle region than in regions near the leading and trailing edges. With such molecular distributions, the interaction of actin filaments with myosin II results in predominantly peripheral movements. The direction of cell migration may be dependent on the asymmetric distribution of actin filaments and the relationship between actin filaments and the extracellular matrix, which determine the balance of the traction forces between the cell leading and trailing edges. Directed migration can occur only if the traction force at the leading edge exceeds that at the trailing edge.

Retraction of Cell Membrane at the Trailing Edge. To initiate cell migration, the forward movement at the cell leading edge must be accompanied with a retraction at the cell trailing edge. The dynamic interaction of integrins with extracellular matrix may mediate the coordinated leading-edge movement and trailing-edge retraction. The distribution of integrin-containing focal adhesion contacts changes dynamically from the cell leading edge to the trailing edge. The density of focal adhesion contacts is relatively lower at the trailing edge than that at the leading edge. Such a distribution of focal adhesion contacts results in reduced cell membrane adhesion to the extracellular matrix at the cell trailing edge and is in favor of the dissociation of cell membrane from the extracellular matrix. Furthermore, the traction forces generated at the cell leading edge are counterbalanced by those at the trailing edge. A reduction in the density of focal adhesion contacts at the trailing edge likely results in an increase in the traction force per focal adhesion contact, which enhances the disruption of integrin–matrix bonds and thus facilitates the retraction of the cell trailing edge.

In addition to the influence of the physical factors described above, the disruption of the integrin–matrix bonds at the cell trailing edge may be regulated by biochemical processes. For instance, the disruption of $\alpha v\beta 3$–vitronectin interaction in migrating neutrophils requires the presence of calcium. The suppression of the calcium-dependent phosphatase calcineurin prevents the disruption of the $\alpha v\beta 3$–vitronectin interaction. This observation suggests that calcineurin plays a role in regulating the detachment of cell membrane from matrix substrate at the trailing edage. However, the mechanisms of chemically mediated retraction remain to be investigated.

Replenishment of Integrins. Integrins play a critical role in the mediation of membrane attachment, cell traction, and cell retraction during cell migration. The distribution and activity of integrins vary from the cell leading to trailing edges. This suggests that the cell must replenish active integrins at the cell leading edge. There are two possible ways for the replenishment of integrins: integrin synthesis and recycling. New integrins are continuously synthesized and deployed to the cell membrane. Cells are also able to endocytose and reuse the integrin molecules left behind on the substrate during cell trailing-edge retraction. In addition, cells may actively transport membrane integrins from the cell trailing to leading edges. With these approaches, the cell is able to maintain an appropriate distribution of integrins, which is necessary for the conduction of cell migration.

Regulation of Cell Migration

Role of the Rho family of GTPases. As discussed above, cell migration is accomplished by a number of complex molecular processes. These processes are regulated by a variety of signaling molecules. Among the signaling molecules, the Rho family of small GTPases, including Rho, Rac, and Cdc42, plays a critical role in the regulation of cell migration.

The GTPases of the Rho family are GTP-binding proteins with molecular weight of ~21 kDa. These proteins belong to the Ras protein superfamily, which include, in addition to the Rho family GTPases, the Rab, the ADP-ribosylation factor (ARF), and the Ran families. The Rab proteins participate in the regulation of vesicle transport, the ARF proteins mediate signal transduction and vesicle transport, whereas the Ran proteins mediate protein transport to the cell nucleus. All proteins of the Ras superfamily are able to bind GTP.

For the Rho family of small GTPases, 11 different isoforms have been identified in mammalian cells, including RhoA, RhoB, RhoC, RhoD, RhoE, RhoG, Rac1, Rac2, Cdc42, TC10, and TTF. Among these proteins, the role of RhoA (see Table 6.4.), Rac1, and Cdc42 has been extensively studies (see Chapter 3 for characteristics of these molecules). RhoA has been shown to regulate the formation and organization of actin filaments, Rac1 mediates the formation of cell lamellipodia and membrane ruffles, whereas Cdc42 is responsible for the formation of filopodia. All these processes are related to cell migration.

All GTPases of the Rho family can bind GTP or GDP. A GTPase is active when GTP is bound, whereas it is inactive when GDP is bound. Nucleotide exchange factors can stimulate the binding of GTP to GTPases, activating the GTPases. A variety of extracellular signals can activate the nucleotide exchange factors and thus the GTPases. While the exact mechanisms of GTPase activation remains poorly understood, GTPase translocation to the cell membrane or cytoskeleton may play a role. For example, the nucleotide exchange factor for Cdc42 is associated with the cell membrane. Activated nucleotide

TABLE 6.4. Characteristics of RhoA*

Proteins	Alternative Names	Amino Acids	Molecular Weight (kDa)	Expression	Functions
RhoA	Ras homolog gene family member A, ARHA, aplysia ras related homologl 2 (ARH12), oncogene rho H12, RHOH12, RHO12, RHOA, transforming protein RhoA, Ras homolog gene family member A	193	22	Ubiquitous	Regulating the organization and remodeling of actin cytoskeleton during cell morphogenesis and migration and mediating cell proliferation and differentiation

*Based on bibliography 6.6.

exchange factor can induce translocation of Cdc42 from the cytoplasm to the cell membrane, which facilitates the activation of Cdc42.

Activated RhoA, Rac1, and Cdc42 can interact with and stimulate downstream signaling molecules, including protein kinases, adapter proteins, and phosphoinositide kinases. For instance, Rho can activate Rho-associated kinase, which in turn phosphorylates myosin II light-chain kinase in smooth muscle cells. Myosin II light-chain kinase can activate myosin light chain and facilitate myosin–actin interaction. The Rho family of small GTPases can also interact with molecules that link to the actin cytoskeleton. An example is the interaction of Rho with p140mDia, which induces the activation of p140mDia. Activated p140mDia can interact with profilin, an actin-binding protein.

Rho, Rac1, and Cdc42 are involved in the regulation of actin polymerization and the formation of stress fibers, which are myosin II-associated contractile actin filamentous bundles. In cultured Swiss 3T3 fibroblasts, cell transfection with active Rho and Rac mutants stimulates the formation of stress fibers and lamellipodia, or wide cell membrane protrusions. Cells transfected with a Rho inhibitor C3 transferase or a dominant-negative mutant for Rac exhibit reduced formation of stress fibers. In addition, Rho, Rac, and Cdc42 promote the formation of focal adhesion contacts. Since actin polymerization, cell membrane protrusion (formation of lamellipodia and filopodia), and formation of focal adhesion contacts are essential processes of cell migration, the small GTPases Rho, Rac, and Cdc42 contribute to the regulation of cell migration.

Role of MAPKs. As discussed in Chapter 5, mitogen-activated protein kinases (MAPKs) are key elements for signaling pathways that respond to the stimulation of growth factors, including platelet-derived growth factor, epidermal growth factor, fibroblast growth factor, and vascular endothelial growth factor. The binding of these growth factors to cognate growth factor receptors induces autophosphorylation of the receptor tyrosine kinase located in the cytoplasmic domain of the receptor. Such a process leads to the activation of a cascade of signaling molecules, including the Ras protein, ERK kinase, and ERK1/2, which belongs to the MAPK family. Activated ERK1/2 can directly phosphorylate myosin light-chain kinase, which activates myosin light chain and promotes myosin–actin interaction. These MAPK-involved processes influence cell migration via mediating the contractility of actin filaments.

CELL ADHESION

Cell adhesion is a molecular process that mediates cell–cell (intercell) and cell–matrix interactions and is involved in the regulation of developmental morphogenesis, physiological adaptation, and pathogenic remodeling. Cells can selectively bind to other cells and extracellular matrix by activating cell adhesion mechanisms. Cell adhesion is related to a variety of molecular and cellular processes, such as cytoskeletal reorganization, alterations in cell geometry, signaling activation, gene expression, and cell mitogenic responses. It is now understood that cell–cell adhesion and cell–matrix adhesion are regulated by several classes of cell adhesion molecules. These include immunoglobulin-like cell adhesion molecules, selectins, cadherins, cell surface heparan sulfate proteoglycans, protein tyrosine phosphatases, and integrins. Cell adhesion is a process that requires coordinated interactions between various cell adhesion molecules as well as between adhesion molecules and the actin cytoskeleton. In this section, the structural and functional characteristics of major classes of adhesion molecules are discussed.

Immunoglobulin-Like Domain-Containing Cell Adhesion Molecules [6.7]

Classification and Structure. Immunoglobulin (Ig)-like domain-containing cell adhesion molecules (IgCAMs) (see Table 6.5) are cell surface adhesion molecules that belong to the immunoglobulin superfamily, which contains about 100 members. IgCAMs mediate cell–cell and cell–matrix adhesion and play a role in regulating cell signaling. These molecules contribute to the regulation of embryonic development and pathogenic remodeling. In particular, IgCAMs play a critical role in mediating polarized migration of neurons and the development of the nerve system.

A typical IgCAM is composed of one or more Ig-like domains, which are located in the extracellular region of the molecule (Fig. 6.7). Each Ig-like domain is composed of about 100 amino acids, which constitute two opposed β sheets linked by disulfide bonds between cysteine residues. In addition, a typical IgCAM is composed of fibronectin (FN)-like repeats in the extracellular region. Each repeat consists of two β sheets of about 90 amino acids. Some IgCAMs are widely distributed in almost all tissues and organs, while others are expressed in limited tissues and organs. The expression pattern of IgCAMs is regulated to suit the function of various types of tissues and organs during development and remodeling.

Several types of IgCAMs have been identified and characterized. These include neural cell adhesion molecules (NCAMs), vascular cell adhesion molecules (VCAMs), intercellular adhesion molecules (ICAMs), L1-like IgCAMs, and receptor protein tyrosine phosphatases. A typical NCAM is composed of five Ig-like domains and two FN-like repeats in the extracellular region, a transmembrane domain, and a cytoplasmic domain. NCAMs are primarily expressed in neural cells and are involved in the regulation of neural cell adhesion and development. VCAM is composed of seven extracellular Ig-like domains, a transmem-

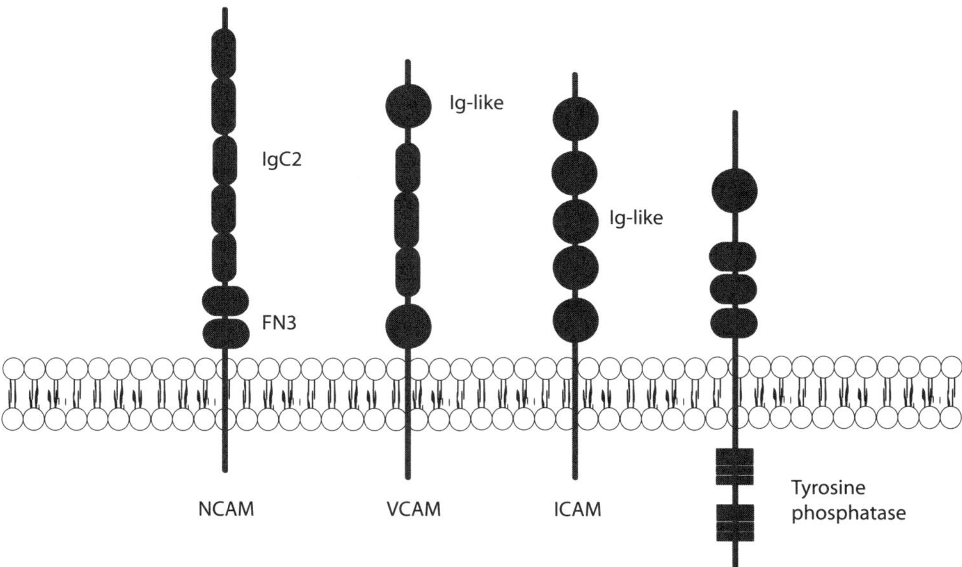

Figure 6.7. Schematic representation of the structure of immunoglobulin (Ig)-like domain-containing cell adhesion molecules. IgC2: Immunoglobulin C2-type domain. FN3: fibronectin type 3 domain. Based on bibliography 6.7.

TABLE 6.5. Characteristics of Selected Immunoglobulin-Like Domain-Containing Cell Adhesion Molecules*

Proteins	Alternative Names	Amino Acids	Molecular Weight (kDa)	Expression	Functions
Neural cell adhesion molecule 1	CD56, NCAM, NCAM1, NCA1, NCAM140	848	93	Nervous system	Regulating cell adhesion and signaling
Vascular cell adhesion molecule	VCAM, INCAM100, L1CAM	739	81	Vascular endothelial cells, heart, lung	Mediating the interaction of leukocytes with vascular endothelial cells and regulating cell signal transduction
Intercellular adhesion molecule1	ICAM1, CD54, CD54 antigen, surface antigen of activated B cells BB2	532	58	Leukocytes, vascular endothelial cell, brain, kidney, intestine, skin	Binding to integrins CD11a, CD11b, and CD18, and regulating cell adhesion
Protein tyrosine phosphatase receptor type δ	Phosphotyrosine phosphatase receptor δ, phosphotyrosine phosphatase receptor D, RPTP δ, protein tyrosine phosphatase δ	1912	215	Brain, heart, kidney, placenta	Consisting of three Ig-like and eight fibronectin type III-like domains in extracellular region, promoting neurite growth and axon extension, and mediating cell adhesion and signal transduction
Protein tyrosine phosphatase receptor type κ	Protein tyrosine phosphatase receptor type K, RPTP κ, PTPK, protein tyrosine phosphatase κ	1440	162	Ubiquitous	Regulating intercellular adhesion via interaction with β, γ-catenin at adherens junctions, and inhibiting the proliferation of certain cell types such as keratinocyte
Protein tyrosine phosphatase receptor type μ	Phosphotyrosine phosphatase receptor μ, RPTPμ, protein tyrosine phosphatase receptor-like 1, PTPRL1, protein tyrosine phosphatase μ	1452	164	Blood vessel	Regulating intercellular adhesion

*Based on bibliography 6.7.

brane domain, and a cytoplasmic domain and can be found in vascular endothelial cells. It is involved in the regulation of endothelial cell adhesion. L1-like IgCAMs consists of six Ig-like domains and five FN-like repeats, a transmembrane domain, and a cytoplasmic domain. These IgCAMs can be found in the nerve tissue and play an important role in regulating neuron–neuron and neuron–glial cell interactions. Several receptor protein tyrosine phosphatases (RPTP), including RPTPδ, RPTPκ, and RPTPμ, possess the function of IgCAMs. These molecules are composed of Ig-like domains and FN-like repeats in the extracellular region and two cytoplasmic phosphatase domains. While RPTPs catalyze dephosphorylation of protein tyrosine kinases, they also mediate cell–cell adhesion.

Functions

Role in Mediating Cell–Cell Adhesion. IgCAMs are cell membrane receptors and mediate cell adhesion via interaction between their extracellular domains and target molecules. Various forms of IgCAM interaction have been identified. Under appropriate conditions, some IgCAMs, such as NCAMs and L1-like IgCAMs, can initiate homophilic binding, specifically, interaction between identical IgCAM molecules of different cells. However, most IgCAMs undergo heterophilic interaction between different IgCAMs or between IgCAMs and non-IgCAM molecules, such as laminin and tenascin.

IgCAMs are involved in the mediation of neuron–neuron interaction. For instance, L1-like IgCAMs mediate interaction between different neurons homophilically as well as heterophilically. Such a process is critical to neuron–neuron interaction, which is the basis for neuronal communication. L1-like IgCAM interaction also facilitates neurite extension and outgrowth. In addition to the regulation of neuron–neuron interaction, IgCAMs are also involved in mediating neuron–glial cell interaction. Neural NCAMs are capable of interacting with receptor protein tyrosine phosphatase β in the membrane of glial cells. In this process, receptor protein tyrosine phosphatase β may serve as a substrate for neuron migration and outgrowth. Thus, IgCAM-mediated neuron–neuron and neuron–glial cell interactions contribute to the development of the nerve system.

Role in Mediating Cell–Matrix Adhesion. The interaction of neurons and extracellular matrix is a process that regulates neurite outgrowth and the morphogenesis of the nerve system. IgCAMs are involved in the mediation of neuron–matrix interaction. For instance, L1-like IgCAMs can interact with extracellular matrix-derived tenascin R. Such a process plays an important role in regulating neurite outgrowth. IgCAMs interact not only with extracellular matrix components but also with intracellular actin cytoskeleton. The intracellular interaction is mediated by an actin-binding molecule known as *ankyrin.* L1-like IgCAMs are capable of binding to ankyrin, which links IgCAMs to actin filaments via interaction with spectrin. The IgCAM linkage with actin cytoskeleton may enhance the interaction of the cell with extracellular matrix.

Role in Cell Signaling. IgCAMs are involved in the regulation of cell signal transduction. IgCAMs have been shown to interact with a number of signaling molecules, including the Src family nonreceptor tyrosine kinases, growth factor receptor tyrosine kinases, receptor protein tyrosine phosphatases, and serine/threonine protein kinases. For instance, IgCAM-dependent neurite outgrowth is reduced in the absence of Src and Fyn, suggesting that these nonreceptor tyrosine kinases may relay signals from IgCAMs.

As described above, certain types of receptor protein tyrosine phosphatases (RPTPs) are also IgCAMs. These RPTPs possess dual functions; the extracellular Ig-like region mediates cell adhesion, while the cytoplasmic phosphatase domain transmits adhesion-related signals to intracellular signaling pathways. Although the signaling mechanisms of the cytoplasmic domain remain poorly understood, it is possible that phosphatase-induced dephosphorylation of substrate proteins may play a role.

IgCAMs are also involved in fibroblast growth factor (FGF)-related cell signaling. L1-like IgCAMs and NCAMs can induce phosphorylation of the FGF receptor protein tyrosine kinase, which is independent of FGF ligand stimulation. The phosphorylation of the FGF protein receptor tyrosine kinase induces the activation of a cascade of signaling molecules, leading to mitogenic cellular activities.

BIBLIOGRAPHY

6.1. Mitosis

Baker DJ, Chen J, van Deursen JM: The mitotic checkpoint in cancer and aging: What have mice taught us? *Curr Opin Cell Biol* 17:583–9, 2005.

Kops GJ, Weaver BA, Cleveland DW: On the road to cancer: Aneuploidy and the mitotic checkpoint, *Nat Rev Cancer* 5:773–85, 2005.

de Gramont A, Cohen-Fix O: The many phases of anaphase, *Trends Biochem Sci* 30:559–68, 2005.

Bloom K: Chromosome segregation: Seeing is believing, *Curr Biol* 15:R500–3, 2005.

Doxsey S, Zimmerman W, Mikule K: Centrosome control of the cell cycle, *Trends Cell Biol* 15:303–11, 2005.

Blow JJ, Dutta A: Preventing re-replication of chromosomal DNA, *Nat Rev Mol Cell Biol* 6:476–86, 2005.

Bosl WJ, Li R: Mitotic-exit control as an evolved complex system, *Cell* 121:325–33, 2005.

Stillman B: Origin recognition and the chromosome cycle, *FEBS Lett* 579:877–84, 2005.

Hauf S, Watanabe Y: Kinetochore orientation in mitosis and meiosis, *Cell* 119:317–27, 2004.

6.2. Cytokenesis

Burgess DR, Chang F: Site selection for the cleavage furrow at cytokinesis, *Trends Cell Biol* 15:156–62, 2005.

Piekny A, Werner M, Glotzer M: Cytokinesis: Welcome to the Rho zone, *Trends Cell Biol* 15:651–8, 2005.

Doxsey S, McCollum D, Theurkauf W: Centrosomes in cellular regulation, *Annu Rev Cell Dev Biol* 21:411–34, 2005.

Matsumura F: Regulation of myosin II during cytokinesis in higher eukaryotes, *Trends Cell Biol* 15:371–7, 2005.

Burgess DR: Cytokinesis: New roles for myosin, *Curr Biol* 15:R310–1, 2005.

Reichl EM, Effler JC, Robinson DN: The stress and strain of cytokinesis, *Trends Cell Biol* 15:200–6, 2005.

Glotzer M: The molecular requirements for cytokinesis, *Science* 307:1735–9, 2005.

Burgess DR, Chang F: Site selection for the cleavage furrow at cytokinesis, *Trends Cell Biol* 15:156–62, 2005.

Albertson R, Riggs B, Sullivan W: Membrane traffic: A driving force in cytokinesis, *Trends Cell Biol* 15:92–101, 2005.

6.3. Control of Cell Division

Cyclin D1

Geng Y, Whoriskey W, Park MY, Bronson RT, Medema RH et al: Rescue of cyclin D1 deficiency by knockin cyclin E, *Cell* 97:767–77, 1999.

Geng Y, Yu Q, Sicinska E, Das M, Bronson RT et al: Deletion of the p27(Kip1) gene restores normal development in cyclin D1-deficient mice, *Proc Natl Acad Sci USA* 98:194–99, 2001.

Hinds PW, Dowdy SF, Eaton EN, Arnold A, Weinberg RA: Function of a human cyclin gene as an oncogene, *Proc Natl Acad Sci USA* 91:709–13, 1994.

Inaba T, Matsushime H, Valentine M, Roussel MF, Sherr CJ et al: Genomic organization, chromosomal localization, and independent expression of human cyclin D genes, *Genomics* 13:565–74, 1992.

Kozar K, Ciemerych MA, Rebel VI, Shigematsu H, Zagozdzon A et al: Mouse development and cell proliferation in the absence of D-cyclins, *Cell* 118:477–91, 2004.

Ma C, Papermaster D, Cepko CL: A unique pattern of photoreceptor degeneration in cyclin D1 mutant mice, *Proc Natl Acad Sci USA* 95:9938–43, 1998.

Muller H, Lukas J, Schneider A, Warthoe P, Bartek J et al: Cyclin D1 expression is regulated by the retinoblastoma protein, *Proc Natl Acad Sci USA* 91:2945–9, 1994.

Sicinski P, Donaher JL, Parker SB, Li T, Fazeli A et al: Cyclin D1 provides a link between development and oncogenesis in the retina and breast, *Cell* 82:621–30, 1995.

Szepetowski P, Simon MP, Grosgeorge J, Huebner K, Bastard C et al: Localization of 11q13 loci with respect to regional chromosomal breakpoints, *Genomics* 12:738–44, 1992.

Tetsu O, McCormick F: Beta-catenin regulates expression of cyclin D1 in colon carcinoma cells, *Nature* 398:422–6, 1999.

Tsikitis M, Zhang Z, Edelman W, Zagzag D, Kalpana GV: Genetic ablation of cyclin D1 abrogates genesis of rhabdoid tumors resulting from Ini1 loss, *Proc Natl Acad Sci USA* 102:12129–34, 2005.

Wang C, Pattabiraman N, Zhou JN, Fu M, Sakamaki T et al: Cyclin D1 repression of peroxisome proliferator-activated receptor gamma expression and transactivation, *Mol Cell Biol* 23:6159–73, 2003.

Wang TC, Cardiff RD, Zukerberg L, Lees E, Arnold A et al: Mammary hyperplasia and carcinoma in MMTV-cyclin D1 transgenic mice, *Nature* 369:669–71, 1994.

Xiong Y, Connelly T, Futcher B, Beach D: Human D-type cyclin, *Cell* 65:691–9, 1991.

Yu Q, Geng Y, Sicinski P: Specific protection against breast cancers by cyclin D1 ablation, *Nature* 411:1017–21, 2001.

Cyclin E

Demetrick DJ, Matsumoto S, Hannon GJ, Okamoto K, Xiong Y et al: Chromosomal mapping of the genes for the human cell cycle proteins cyclin C (CCNC), cyclin E (CCNE), p21 (CDKN1) and KAP (CDKN3), *Cytogenet Cell Genet* 69:190–2, 1995.

Geng Y, Whoriskey W, Park MY, Bronson RT, Medema RH et al: Rescue of cyclin D1 deficiency by knockin cyclin E, *Cell* 97:767–77, 1999.

Keyomarsi K, Tucker SL, Buchholz TA, Callister M, Ding Y et al: Cyclin E and survival in patients with breast cancer, *New Engl J Med* 347:1566–75, 2002.

Koff A, Cross F, Fisher A, Schumacher J, Leguellec K et al: Human cyclin E, a new cyclin that interacts with two members of the CDC2 gene family, *Cell* 66:1217–28, 1991.

Li H, Lahti JM, Valentine M, Saito M, Reed SI et al: Molecular cloning and chromosomal localization of the human cyclin C (CCNC) and cyclin E (CCNE) genes: deletion of the CCNC gene in human tumors, *Genomics* 32:253–9, 1996.

Matsumoto Y, Maller JL: A centrosomal localization signal in cyclin E required for Cdk2-independent S phase entry, *Science* 306:885–8, 2004.

Spruck CH, Won KA, Reed SI: Deregulated cyclin E induces chromosome instability, *Nature* 401:297–300, 1999.

CDK2

Bourne Y, Watson MH, Hickey MJ, Holmes W, Rocque W et al: Crystal structure and mutational analysis of the human CDK2 kinase complex with cell cycle-regulatory protein CksHs1, *Cell* 84:863–74, 1996.

De Bondt HL, Rosenblatt J, Jancarik J, Jones HD, Morgan DO et al: Crystal structure of cyclin-dependent kinase 2, *Nature* 363:595–602, 1993.

Demetrick DJ, Zhang H, Beach DH: Chromosomal mapping of human CDK2, CDK4, and CDK5 cell cycle kinase genes, *Cytogenet Cell Genet* 66:72–74, 1994.

Jeffrey PD, Russo AA, Polyak K, Gibbs E, Hurwitz J et al: Mechanism of CDK activation revealed by the structure of a cyclinA-CDK2 complex, *Nature* 376:313–20, 1995.

Matsumoto Y, Maller JL: A centrosomal localization signal in cyclin E required for Cdk2-independent S phase entry, *Science* 306:885–8, 2004.

Matsuura I, Denissova NG, Wang G, He D, Long J, Liu F: Cyclin-dependent kinases regulate the antiproliferative function of Smads, *Nature* 430:226–31, 2004.

Ninomiya-Tsuji J, Nomoto S, Yasuda H, Reed SI, Matsumoto K: Cloning of a human cDNA encoding a CDC2-related kinase by complementation of a budding yeast cdc28 mutation, *Proc Natl Acad Sci USA* 88:9006–10, 1991.

Ortega S, Prieto I, Odajima J, Martin A, Dubus P et al: Cyclin-dependent kinase 2 is essential for meiosis but not for mitotic cell division in mice, *Nature Genet* 35:25–31, 2003.

Shiffman D, Brooks EE, Brooks AR, Chan CS, Milner PG: Characterization of the human cyclin-dependent kinase 2 gene: promoter analysis and gene structure, *J Biol Chem* 271:12199–204, 1996.

Tsai L-H, Harlow E, Meyerson M: Isolation of the human cdk2 gene that encodes the cyclin A- and adenovirus E1A-associated p33 kinase, *Nature* 353:174–7, 1991.

CDK4

Demetrick DJ, Zhang H, Beach DH: Chromosomal mapping of human CDK2, CDK4, and CDK5 cell cycle kinase genes, *Cytogenet Cell Genet* 66:72–4, 1994.

Lazarov M, Kubo Y, Cai T, Dajee M, Tarutani M et al: CDK4 coexpression with Ras generates malignant human epidermal tumorigenesis, *Nature Med* 8:1105–114, 2002.

Malumbres M, Sotillo R, Santamaria D, Galan J, Cerezo A et al: Mammalian cells cycle without the D-type cyclin-dependent kinases Cdk4 and Cdk6, *Cell* 118:493–504, 2004.

Matsuura I, Denissova NG, Wang G, He D, Long J et al: Cyclin-dependent kinases regulate the antiproliferative function of Smads, *Nature* 430:226–31, 2004.

Wolfel T, Hauer M, Schneider J, Serrano M, Wolfel C et al: A p16(INK4a)-insensitive CDK4 mutant targeted by cytolytic T lymphocytes in a human melanoma, *Science* 269:1281–4, 1995.

Zuo L, Weger J, Yang Q, Goldstein AM, Tucker MA et al: Germline mutations in the p16(INK4a) binding domain of CDK4 in familial melanoma, *Nature Genet* 12:97–9, 1996.

CDK6

Harbour JW, Luo RX, Dei Santi A, Postigo AA, Dean DC: Cdk phosphorylation triggers sequential intramolecular interactions that progressively block Rb functions as cells move through G1, *Cell* 98:859–69, 1999.

Malumbres M, Sotillo R, Santamaria D, Galan J, Cerezo A et al: Mammalian cells cycle without the D-type cyclin-dependent kinases Cdk4 and Cdk6, *Cell* 118:493–504, 2004.

Meyerson M, Harlow E: Identification of G1 kinase activity for cdk6, a novel cyclin D partner, *Mol Cell Biol* 14:2077–86, 1994.

Serrano M, Hannon GJ, Beach D: A new regulatory motif in cell-cycle control causing specific inhibition of cyclin D/CDK4, *Nature* 366:704–7, 1993.

Veiga-Fernandes H, Rocha B: High expression of active CDK6 in the cytoplasm of CD8 memory cells favors rapid division, *Nature Immun* 5:31–7, 2003.

E2F

Fajas L, Annicotte JS, Miard S, Sarruf D, Watanabe M et al: Impaired pancreatic growth, beta cell mass, and beta cell function in E2F1 -/- mice, *J Clin Invest* 113:1288–95, 2004.

Field SJ, Tsai FY, Kuo F, Zubiaga AM, Kaelin WG Jr et al: E2F-1 functions in mice to promote apoptosis and suppress proliferation, *Cell* 85:549–61, 1996.

Furukawa Y, Nishimura N, Furukawa Y, Satoh M, Endo H et al: Apaf-1 is a mediator of E2F-1-induced apoptosis, *J Biol Chem* 277:39760–8, 2002.

Helin K, Lees JA, Vidal M, Dyson N, Harlow E et al: A cDNA encoding a pRB-binding protein with properties of the transcription factor E2F, *Cell* 70:337–50, 1992.

Irwin M, Marin MC, Phillips AC, Seelan RS, Smith DI et al: Role for the p53 homologue p73 in E2F-1-induced apoptosis, *Nature* 407:645–8, 2000.

Lissy NA, Davis PK, Irwin M, Kaelin WG, Dowdy SF: A common E2F-1 and p73 pathway mediates cell death induced by TCR activation, *Nature* 407:642–5, 2000.

Nevins JR: E2F: A link between the Rb tumor suppressor protein and viral oncoproteins, *Science* 258:424–9, 1992.

Ohtani K, DeGregori J, Nevins JR: Regulation of the cyclin E gene by transcription factor E2F1, *Proc Natl Acad Sci USA* 92:12146–50, 1995.

Weinberg RA: E2F and cell proliferation: A world turned upside down, *Cell* 85:457–9, 1996.

Wu L, Timmers C, Maiti B, Saavedra HI, Sang L et al: The E2F1-3 transcription factors are essential for cellular proliferation, *Nature* 414:457–62, 2001.

Yamasaki L, Jacks T, Bronson R, Goillot E, Harlow E et al: Tumor induction and tissue atrophy in mice lacking E2F-1, *Cell* 85:537–48, 1996.

Zhang HS, Postigo AA, Dean DC: Active transcriptional repression by the Rb-E2F complex mediates G1 arrest triggered by p16(INK4a), TGF-beta, and contact inhibition, *Cell* 97:53–61, 1999.

Cdc25

Demetrick DJ, Beach DH: Chromosome mapping of human CDC25A and CDC25B phosphatases, *Genomics* 18:144–7, 1993.

Falck J, Mailand N, Syljuasen RG, Bartek J, Lukas J: The ATM-Chk2-Cdc25A checkpoint pathway guards against radioresistant DNA synthesis, *Nature* 410:842–7, 2001.

Falck J, Petrini JHJ, Williams BR, Lukas J, Bartek J: The DNA damage-dependent intra-S phase checkpoint is regulated by parallel pathways, *Nature Genet* 30:290–4, 2002.

Fauman EB, Cogswell JP, Lovejoy B, Rocque WJ, Holmes W et al: Crystal structure of the catalytic domain of the human cell cycle control phosphatase, Cdc25A, *Cell* 93:617–25, 1998.

Galaktionov K, Beach D: Specific activation of cdc25 tyrosine phosphatases by B-type cyclins: evidence for multiple roles of mitotic cyclins, *Cell* 67:1181–94, 1991.

Galaktionov K, Lee AK, Eckstein J, Draetta G, Meckler J et al: CDC25 phosphatases as potential human oncogenes, *Science* 269:1575–7, 1995.

Mailand N, Falck J, Lukas C, Syljuasen RG, Welcker M et al: Rapid destruction of human Cdc25A in response to DNA damage, *Science* 288:1425–9, 2000.

Cyclin A

Liao C, Wang XY, Wei HQ, Li SQ, Merghoub T et al: Altered myelopoiesis and the development of acute myeloid leukemia in transgenic mice overexpressing cyclin A1, *Proc Natl Acad Sci USA* 98:6853–8, 2001.

Liu D, Matzuk MM, Sung WK, Guo Q, Wang P et al: Cyclin A1 is required for meiosis in the male mouse, *Nature Genet* 20:377–80, 1998.

Muller C, Yang R, Beck-von-Peccoz L, Idos G, Verbeek W et al: Cloning of the cyclin A1 genomic structure and characterization of the promoter region: GC boxes are essential for cell cycle-regulated transcription of the cyclin A1 gene, *J Biol Chem* 274:11220–8, 1999.

Rape M, Kirschner MW: Autonomous regulation of the anaphase-promoting complex couples mitosis to S-phase entry, *Nature* 432:588–95, 2004.

Cyclin B

Brandeis M, Rosewell I, Carrington M, Crompton T, Jacobs MA et al: Cyclin B2-null mice develop normally and are fertile whereas cyclin B1-null mice die in utero, *Proc Natl Acad Sci USA* 95:4344–9, 1998.

Moore JD, Kirk JA, Hunt T: Unmasking the S-phase-promoting potential of cyclin B1, *Science* 300:987–90, 2003.

Pines J, Hunter T: Isolation of a human cyclin cDNA: Evidence for cyclin mRNA and protein regulation in the cell cycle and for interaction with p34(cdc2), *Cell* 58:833–46, 1989.

Sartor H, Ehlert F, Grzeschik KH, Muller R, Adolph S: Assignment of two human cell cycle genes, CDC25C and CCNB1, to 5q31 and 5q12, respectively, *Genomics* 13:911–2, 1992.

CDC2

Draetta G, Piwnica-Worms H, Morrison D, Druker B, Roberts T et al: Human CDC2 protein kinase is a major cell-cycle regulated tyrosine kinase substrate, *Nature* 336:738–44, 1988.

Ira G, Pellicioli A, Balijja A, Wang X, Fiorani S et al: DNA end resection, homologous recombination and DNA damage checkpoint activation require CDK1, *Nature* 431:1011–7, 2004.

Lee MG, Norbury CJ, Spurr NK, Nurse P: Regulated expression and phosphorylation of a possible mammalian cell-cycle control protein, *Nature* 333:676–9, 1988.

Matsuo T, Yamaguchi S, Mitsui S, Emi A, Shimoda F et al: Control mechanism of the circadian clock for timing of cell division in vivo, *Science* 302:255–9, 2003.

Moore JD, Kirk JA, Hunt T: Unmasking the S-phase-promoting potential of cyclin B1, *Science* 300:987–90, 2003.

Nazarenko SA, Ostroverhova NV, Spurr NK: Regional assignment of the human cell cycle control gene CDC2 to chromosome 10q21 by in situ hybridization, *Hum Genet* 87:621–2, 1991.

P15

Hannon GJ, Beach D: p15(INK4B) is a potential effector of TGF-beta-induced cell cycle arrest, *Nature* 371:257–61, 1994.

Kamb A, Gruis NA, Weaver-Feldhaus J, Liu Q, Harshman K et al: A cell cycle regulator potentially involved in genesis of many tumor types, *Science* 264:436–40, 1994.

Nobori T, Miura K, Wu DJ, Lois A, Takabayashi K et al: Deletions of the cyclin-dependent kinase-4 inhibitor gene in multiple human cancers, *Nature* 368:753–6, 1994.

Quelle DE, Ashmun RA, Hannon GJ, Rehberger PA, Trono D et al: Cloning and characterization of murine p16(INK4a) and p15(INK4b) genes, *Oncogene* 11:635–45, 1995.

Stone S, Dayananth P, Jiang P, Weaver-Feldhaus JM, Tavtigian SV et al: Genomic structure, expression and mutational analysis of the P15 (MTS2) gene, *Oncogene* 11:987–91, 1995.

P16

Cairns J, Mao L, Merlo A, Lee DJ, Schwab D et al: Rates of p16 (MTS1) mutations in primary tumors with 9p loss, *Science* 265:415–7, 1994.

Goldstein AM, Fraser MC, Struewing JP, Hussussian CJ, Ranade K et al: Increased risk of pancreatic cancer in melanoma-prone kindreds with p16 (INK4) mutations, *New Engl J Med* 333:970–4, 1995.

Gruis NA, van der Velden PA, Sandkuijl LA, Prins DE, Weaver-Feldhaus J et al: Homozygotes for CDKN2 (p16) germline mutation in Dutch familial melanoma kindreds, *Nature Genet* 10:351–3, 1995.

Hussussian CJ, Struewing JP, Goldstein AM, Higgins PAT, Ally DS et al: Germline p16 mutations in familial melanoma, *Nature Genet* 8:15–21, 1994.

Kamb A, Gruis NA, Weaver-Feldhaus J, Liu Q, Harshman K et al: A cell cycle regulator potentially involved in genesis of many tumor types, *Science* 264:436–40, 1994.

Kamijo T, Zindy F, Roussel MF, Quelle DE, Downing JR et al: Tumor suppression at the mouse INK4a locus mediated by the alternative reading frame product p19(ARF), *Cell* 91:649–59, 1997.

Koh J, Enders GH, Dynlacht BD, Harlow E: Tumour-derived p16 alleles encoding proteins defective in cell-cycle inhibition, *Nature* 375:506–10, 1995.

Krimpenfort P, Quon KC, Mooi WJ, Loonstra A, Berns A: Loss of p16(Ink4a) confers susceptibility to metastatic melanoma in mice, *Nature* 413:83–5, 2001.

Linggi B, Muller-Tidow C, van de Locht L, Hu M, Nip J et al: The t(8;21) fusion protein, AML1-ETO, specifically represses the transcription of the p14(ARF) tumor suppressor in acute myeloid leukemia, *Nature Med* 8:743–50, 2002.

Liu L, Dilworth D, Gao L, Monzon J, Summers A et al: Mutation of the CDKN2A 5-prime UTR creates an aberrant initiation codon and predisposes to melanoma, *Nature Genet* 21:128–32, 1999.

Lukas J, Parry D, Aagaard L, Mann DJ, Bartkova J et al: Retinoblastoma-protein-dependent cell-cycle inhibition by the tumour suppressor p16, *Nature* 375:503–6, 1995.

Lund AH, Turner G, Trubetskoy A, Verhoeven E, Wientjens E et al: Genome-wide retroviral insertional tagging of genes involved in cancer in Cdkn2a-deficient mice, *Nature Genet* 32:160–5, 2002.

Nobori T, Miura K, Wu DJ, Lois A, Takabayashi K et al: Deletions of the cyclin-dependent kinase-4 inhibitor gene in multiple human cancers, *Nature* 368:753–6, 1994.

Ohtani N, Zebedee Z, Huot TJG, Stinson JA, Sugimoto M et al: Opposing effects of Ets and Id proteins on p16(INK4A) expression during cellular senescence, *Nature* 409:1067–70, 2001.

Qi Y, Gregory MA, Li Z, Brousal JP, West K et al: p19(ARF) directly and differentially controls the functions of c-Myc independently of p53, *Nature* 431:712–7, 2004.

Schmitt CA, Fridman JS, Yang M, Lee S, Baranov E et al: A senescence program controlled by p53 and p16-INK4a contributes to the outcome of cancer therapy, *Cell* 109:335–46, 2002.

Serrano M, Hannon GJ, Beach D: A new regulatory motif in cell-cycle control causing specific inhibition of cyclin D/CDK4, *Nature* 366:704–7, 1993.

Sharpless NE, Bardeesy N, Lee KH, Carrasco D, Castrillon DH et al: Loss of p16(Ink4a) with retention of p19(Arf) predisposes mice to tumorigenesis, *Nature* 413:86–91, 2001.

Zhang Y, Xiong Y, Yarbrough WG: ARF promotes MDM2 degradation and stabilizes p53: ARF-INK4a locus deletion impairs both the Rb and p53 tumor suppression pathways, *Cell* 92:725–34, 1998.

P18

Bai F, Pei XH, Godfrey VL, Xiong Y: Haploinsufficiency of p18(INK4c) sensitizes mice to carcinogen-induced tumorigenesis, *Mol Cell Biol* 23:1269–77, 2003.

Guan KL, Jenkins CW, Li Y, Nichols MA, Wu X et al: Growth suppression by p18, a p16(INK4/MTS1)- and p14(INK4B/MTS2)-related CDK6 inhibitor, correlates with wild-type pRb function, *Genes Dev* 8:2939–52, 1994.

Lapointe J, Lachance Y, Labrie Y, Labrie C: A p18 mutant defective in CDK6 binding in human breast cancer cells, *Cancer Res* 56:4586–9, 1996.

Yuan Y, Shen H, Franklin DS, Scadden DT, Cheng T: In vivo self-renewing divisions of haematopoietic stem cells are increased in the absence of the early G1-phase inhibitor, p18(INK4C), *Nature Cell Biol* 6:436–42, 2004.

Zindy F, den Besten W, Chen B, Rehg JE, Latres E et al: Control of spermatogenesis in mice by the cyclin D-dependent kinase inhibitors p18(Ink4c) and p19(Ink4d), *Mol Cell Biol* 21:3244–55, 2001.

P19

Chen P, Zindy F, Abdala C, Liu F, Li X et al: Progressive hearing loss in mice lacking the cyclin-dependent kinase inhibitor Ink4d, *Nature Cell Biol* 5:422–6, 2003.

Hirai H, Roussel MF, Kato J-Y, Ashmun RA, Sherr CJ: Novel INK4 proteins, p19 and p18, are specific inhibitors of cyclin D-dependent kinases CDK4 and CDK6, *Mol Cell Biol* 15:2672–81, 1995.

Matsuzaki Y, Miyazawa K, Yokota T, Hitomi T, Yamagishi H et al: Molecular cloning and characterization of the human p19(INK4d) gene promoter, *FEBS Lett* 517:272–6, 2002.

Okuda T, Hirai H, Valentine VA, Shurtleff SA, Kidd VJ et al: Molecular cloning, expression pattern, and chromosomal localization of human CDKN2D/INK4d, an inhibitor of cyclin D-dependent kinases, *Genomics* 29:623–30, 1995.

Zindy F, Cunningham JJ, Sherr CJ, Jogal S, Smeyne RJ et al: Postnatal neuronal proliferation in mice lacking Ink4d and Kip1 inhibitors of cyclin-dependent kinases, *Proc Natl Acad Sci USA* 96:13462–7, 1999.

Zindy F, den Besten W, Chen B, Rehg JE, Latres E et al: Control of spermatogenesis in mice by the cyclin D-dependent kinase inhibitors p18(Ink4c) and p19(Ink4d), *Mol Cell Biol* 21:3244–55, 2001.

Zindy F, van Deursen J, Grosveld G, Sherr CJ, Roussel MF: INK4d-deficient mice are fertile despite testicular atrophy, *Mol Cell Biol* 20:372–8, 2000.

P21

Bunz F, Dutriaux A, Lengauer C, Waldman T, Zhou S et al: Requirement for p53 and p21 to sustain G2 arrest after DNA damage, *Science* 282:1497–1501, 1998.

Carreira S, Goodall J, Aksan I, La Rocca SA, Galibert MD et al: Mitf cooperates with Rb1 and activates p21(Cip1) expression to regulate cell cycle progression, *Nature* 433:764–9, 2005.

Cheng T, Rodrigues N, Shen H, Yang Y, Dombkowski D et al: Hematopoietic stem cell quiescence maintained by p21(cip1/waf1), *Science* 287:1804–8, 2000.

Demetrick DJ, Matsumoto S, Hannon GJ, Okamoto K et al: Chromosomal mapping of the genes for the human cell cycle proteins cyclin C (CCNC), cyclin E (CCNE), p21 (CDKN1) and KAP (CDKN3), *Cytogenet Cell Genet* 69:190–2, 1995.

Megyesi J, Price PM, Tamayo E, Safirstein RL: The lack of a functional p21(WAF1/CIP1) gene ameliorates progression to chronic renal failure, *Proc Natl Acad Sci USA* 96:10830–5, 1999.

Nakatani F, Tanaka K, Sakimura R, Matsumoto Y, Matsunobu T et al: Identification of p21(WAF1/CIP1) as a direct target of EWS-Fli1 oncogenic fusion protein, *J Biol Chem* 278:15105–15, 2003.

Raj K, Ogston P, Beard P: Virus-mediated killing of cells that lack p53 activity, *Nature* 412:914–7, 2001.

Seoane J, Le HV, Massague J: Myc suppression of the p21(Cip1) Cdk inhibitor influences the outcome of the p53 response to DNA damage, *Nature* 419:729–34, 2002.

Wang YA, Elson A, Leder P: Loss of p21 increases sensitivity to ionizing radiation and delays the onset of lymphoma in atm-deficient mice, *Proc Natl Acad Sci USA* 94:14590–5, 1997.

P27

Carrano AC, Eytan E, Hershko A, Pagano M: SKP2 is required for ubiquitin-mediated degradation of the CDK inhibitor p27, *Nature Cell Biol* 1:193–9, 1999.

Di Cristofano A, De Acetis M, Koff A, Cordon-Cardo C, Pandolfi PP: Pten and p27(KIP1) cooperate in prostate cancer tumor suppression in the mouse, *Nature Genet* 27:222–4, 2001.

Fero ML, Rivkin M, Tasch M, Porter P, Carow CE et al: A syndrome of multiorgan hyperplasia with features of gigantism, tumorigenesis, and female sterility in p27(Kip1)-deficient mice, *Cell* 85:733–44, 1996.

Liang J, Zubovitz J, Petrocelli T, Kotchetkov R, Connor MK et al: PKB/Akt phosphorylates p27, impairs nuclear import of p27 and opposes p27-mediated G1 arrest, *Nature Med* 8:1153–60, 2002.

Malek NP, Sundberg H, McGrew S, Nakayama K, Kyriakidis TR et al: A mouse knock-in model exposes sequential proteolytic pathways that regulate p27(Kip1) in G1 and S phase, *Nature* 413:323–7, 2001.

Medema RH, Kops GJPL, Bos JL, Burgering BMT: AFX-like forkhead transcription factors mediate cell-cycle regulation by Ras and PKB through p27(kip1), *Nature* 404:782–7, 2000.

Mitsuhashi T, Aoki Y, Eksioglu YZ, Takahashi T, Bhide PG et al: Overexpression of p27(Kip1) lengthens the G1 phase in a mouse model that targets inducible gene expression to central nervous system progenitor cells, *Proc Natl Acad Sci USA* 98:6435–40, 2001.

Polyak K, Lee MH, Erdjument-Bromage H, Koff A, Roberts JM et al: Cloning of p27(Kip1), a cyclin-dependent kinase inhibitor and a potential mediator of extracellular antimitogenic signals, *Cell* 78:59–66, 1994.

Shin I, Yakes FM, Rojo F, Shin NY, Bakin AV et al: PKB/Akt mediates cell-cycle progression by phosphorylation of p27(Kip1) at threonine 157 and modulation of its cellular localization, *Nature Med* 8:1145–52, 2002.

Toyoshima H, Hunter T: p27, a novel inhibitor of G1 cyclin-Cdk protein kinase activity, is related to p21, *Cell* 78:67–74, 1994.

Uchida T, Nakamura T, Hashimoto N, Matsuda T, Kotani K et al: Deletion of Cdkn1b ameliorates hyperglycemia by maintaining compensatory hyperinsulinemia in diabetic mice, *Nature Med* 11:175–82, 2005.

Viglietto G, Motti ML, Bruni P, Melillo RM, D'Alessio A et al: Cytoplasmic relocalization and inhibition of the cyclin-dependent kinase inhibitor p27(Kip1) by PKB/Akt-mediated phosphorylation in breast cancer, *Nature Med* 8:1136–44, 2002.

Zindy F, Cunningham JJ, Sherr CJ, Jogal S, Smeyne RJ et al: Postnatal neuronal proliferation in mice lacking Ink4d and Kip1 inhibitors of cyclin-dependent kinases, *Proc Natl Acad Sci USA* 96:13462–7, 1999.

P57

Fitzpatrick GV, Soloway PD, Higgins MJ: Regional loss of imprinting and growth deficiency in mice with a targeted deletion of KvDMR1, *Nature Genet* 32:426–31, 2002.

Hatada I, Mukai T: Genomic imprinting of p57(KIP2), a cyclin-dependent kinase inhibitor, in mouse, *Nature Genet* 11:204–6, 1995.

Hatada I, Ohashi H, Fukushima Y, Kaneko Y, Inoue M et al: An imprinted gene p57(KIP2) is mutated in Beckwith-Wiedemann syndrome, *Nature Genet* 14:171–3, 1996.

Matsuoka S, Thompson JS, Edwards MC, Barletta JM, Grundy P et al: Imprinting of the gene encoding a human cyclin-dependent kinase inhibitor, p57(KIP2), on chromosome 11p15, *Proc Natl Acad Sci USA* 93:3026–30, 1996.

Scandura JM, Boccuni P, Massague J, Nimer SD: Transforming growth factor beta-induced cell cycle arrest of human hematopoietic cells requires p57KIP2 upregulation, *Proc Natl Acad Sci USA* 101:15231–6, 2004.

Human protein reference data base, Johns Hopkins University and the Institute of Bioinformatics, at http://www.hprd.org/protein.

6.4. Meiosis

Gilbert Scott F: *Developmental Biology*, 7th ed, Sinauer Associates, Sunderland, UK, 2003, pp 183–219.

Alberts B, Bray D, Lewis J, Raff M, Roberts K, Watson JD: *Molecular Biology of the Cell*, 3rd ed, Garland Publishing, New York, 1994.

6.5. Experimental Assessment of Cell Division

Gratzner HG: Monoclonal antibody to 5-bromo- and 5-iododeoxyuridine: A new reagent for determination of DNA replication, *Science* 218:474–5, 1982.

Krauss G: *Biochemistry of Signal Transduction and Regulation*, 3rd ed, Wiley-VCH GmbH, 2003.

Liu SQ: Focal activation of angiotensin II type 1 receptor and smooth muscle cell proliferation in the neointima of experimental vein grafts: Relation to eddy blood flow, *Arterioscl Thromb Vasc Biol* 19:2630–9, 1999.

Liu SQ, Tieche C, Alkema PK: Neointima formation on elastic lamina and collagen matrix scaffolds implanted in the rat aorta, *Biomaterials* 25:1869–82, 2004.

6.6. Cell Migration

Jaffe AB, Hall A: Rho GTPases: Biochemistry and biology, *Annu Rev Cell Dev Biol* 21:247–69, 2005.

Raftopoulou M, Hall A: Cell migration: Rho GTPases lead the way, *Dev Biol* 265:23–32, 2004.

Etienne-Manneville S, Hall A: Rho GTPases in cell biology, *Nature* 420:629–35, 2002.

Hall A: Rho GTPases and the actin cytoskeleton, *Science* 279:509–14, 1998.

Even-Ram S, Yamada KM: Cell migration in 3D matrix, *Curr Opin Cell Biol* 17:524–32, 2005.

Keller R: Cell migration during gastrulation, *Curr Opin Cell Biol* 17:533–41, 2005.

Small JV, Resch GP: The comings and goings of actin: coupling protrusion and retraction in cell motility, *Curr Opin Cell Biol* 17:517–23, 2005.

Yamaguchi H, Wyckoff J, Condeelis J: Cell migration in tumors, *Curr Opin Cell Biol* 17:559–64, 2005.

Li S, Guan JL, Chien S: Biochemistry and biomechanics of cell motility, *Annu Rev Biomed Eng* 7:105–50, 2005.

Craig SW, Johnson RP: Assembly of focal adhesions: Progress, paradigms, and proteins, *Curr Opin Cell Biol* 8:74–85, 1996.

Lauffenburger DA, Horwitz AF: Cell migration: A physically integrated molecular process, *Cell* 84:359–69, 1996.

Mitchison TJ, Cramer LP: Actin-based cell motility and cell locomotion, *Cell* 84:371–9, 1996.

Kiosses WB, Shattil SJ, Pampori N, Schwartz MA: Rac recruits high-affinity integrin alphavbeta3 to lamellipodia in endothelial cell migration, *Nature Cell Biol* 3(3):316–20, 2001.

Ridley AJ, Schwartz MA, Burridge K, Firtel RA, Ginsberg MH et al: Cell migration: Integrating signals from front to back, *Science* 302:1704–9, 2003.

Raftopoulou M, Hall A: Cell migration: Rho GTPases lead the way, *Dev Biol* 265:23–32, 2004.

Schwartz MA: Integrin signaling revisited, *Trends Cell Biol* 11:466–70, 2001.

Sieg DJ, Hauck CR, Ilic D, Klingbeil CK, Schlaefer E et al: FAK integrates growth-factor and integrin signals to promote cell migration, *Nature Cell Biol* 2:249–56, 2000.

Klemke RL, Cai S, Giannini AL, Gallagher PJ, de Lanerolle P et al: Regulation of cell motility by mitogen-activated protein kinase, *J Cell Biol* 137:481–92, 1997.

Cramer LP, Mitchison TJ: Myosin is involved in postmitotic cell spreading, *J Cell Biol* 131:179–89, 1995.

RhoA

Cannizzaro LA, Madaule P, Hecht F, Axel R, Croce CM et al: Chromosome localization of human ARH genes, a ras-related gene family, *Genomics* 6:197–203, 1990.

Kiss C, Li J, Szeles A, Gizatullin RZ, Kashuba VI et al: Assignment of the ARHA and GPX1 genes to human chromosome bands 3p21.3 by in situ hybridization and with somatic cell hybrids, *Cytogenet Cell Genet* 79:228–30, 1997.

Maekawa M, Ishizaki T, Boku S, Watanabe N, Fujita A et al: Signaling from Rho to the actin cytoskeleton through protein kinases ROCK and LIM-kinase, *Science* 285:895–8, 1999.

Wang HR, Zhang Y, Ozdamar B, Ogunjimi AA, Alexandrova E et al: Regulation of cell polarity and protrusion formation by targeting RhoA for degradation, *Science* 302:1775–9, 2003.

Wu KY, Hengst U, Cox LJ, Macosko EZ, Jeromin A et al: Local translation of RhoA regulates growth cone collapse, *Nature* 436:1020–4, 2005.

Zhou Y, Su Y, Li B, Liu F, Ryder JW et al: Nonsteroidal anti-inflammatory drugs can lower amyloidogenic A-beta(42) by inhibiting Rho, *Science* 302:1215–7, 2003.

Ridley AJ, Allen WE, Peppelenbosch M, Jones GE: Rho family proteins and cell migration, *Biochem Soc Symp* 65:111–24, 1999.

Michiels F, Collard JG: Rho-like GTPases: their role in cell adhesion and invasion, *Biochem Soc Symp* 65:125–146, 1999.

Sheetz MP, Felsenfeld D, Galbraith CG, Choquet D: Cell migration as a five-step cycle, *Biochem Soc Symp* 65:233–44, 1999.

Human protein reference data base, Johns Hopkins University and the Institute of Bioinformatics, at http://www.hprd.org/protein.

6.7. Immunoglobulin-Like Domain-Containing Cell Adhesion Molecules

NCAM

Bello MJ, Salagnon N, Rey JA, Guichaoua MR, Berge-Lefranc JL et al: Precise in situ localization of NCAM, ETS1, and D11S29 on human meiotic chromosomes, *Cytogenet Cell Genet* 52:7–10, 1989.

Cunningham BA, Hemperly JJ, Murray BA, Prediger EA, Brackenbury R et al: Neural cell adhesion molecule: Structure, immunoglobulin-like domains, cell surface modulation, and alternative RNA splicing, *Science* 236:799–806, 1987.

D'Eustachio P, Owens GC, Edelman GM, Cunningham BA: Chromosomal location of the gene encoding the neural cell adhesion molecule (N-CAM) in the mouse, *Proc Natl Acad Sci USA* 82:7631–5, 1985.

Nguyen C, Mattei MG, Mattei JF, Santoni MJ, Goridis C et al: Localization of the human NCAM gene to band q23 of chromosome 11: The third gene coding for a cell interaction molecule mapped to the distal portion of the long arm of chromosome 11, *J Cell Biol* 102:711–5, 1986.

Rabinowitz JE, Rutishauser U, Magnuson T: Targeted mutation of Ncam to produce a secreted molecule results in a dominant embryonic lethality, *Proc Natl Acad Sci USA* 93:6421–4, 1996.

Rutishauser U, Acheson A, Hall AK, Mann DM, Sunshine J: The neural cell adhesion molecule (NCAM) as a regulator of cell-cell interactions, *Science* 240:53–7, 1988.

Rutishauser U, Goridis C: NCAM: The molecule and its genetics, *Trends Genet* 2:72–6, 1986.

Telatar M, Lange E, Uhrhammer N, Gatti RA: New localization of NCAM, proximal to DRD2 at chromosome 11q23, *Mam Genome* 6:59–60, 1995.

VCAM

Besemer J, Harant H, Wang S, Oberhauser B, Marquardt K et al: Selective inhibition of cotranslational translocation of vascular cell adhesion molecule 1, *Nature* 436:290–3, 2005.

Cybulsky MI, Fries JWU, Williams AJ, Sultan P, Eddy R et al: Gene structure, chromosomal location, and basis for alternative mRNA splicing of the human VCAM1 gene, *Proc Natl Acad Sci USA* 88:7859–63, 1991.

Garmy-Susini B, Jin H, Zhu Y, Sung RJ, Hwang R et al: Integrin alpha-4-beta-1–VCAM-1–mediated adhesion between endothelial and mural cells is required for blood vessel maturation, *J Clin Invest* 115:1542–51, 2005.

Kumar AG, Dai XY, Kozak CA, Mims MP, Gotto AM et al: Murine VCAM-1: Molecular cloning, mapping, and analysis of a truncated form, *J Immun* 153:4088–98, 1994.

Lu TT, Cyster JG: Integrin-mediated long-term B cell retention in the splenic marginal zone, *Science* 297:409–12, 2002.

Minn AJ, Gupta GP, Siegel PM, Bos PD, Shu W et al: Genes that mediate breast cancer metastasis to lung, *Nature* 436:518–24, 2005.

Olson E, Srivastava D: Molecular pathways controlling heart development, *Science* 272:671–6, 1996.

ICAM

Bella J, Kolatkar PR, Marlor CW, Greve JM, Rossmann MG: The structure of the two amino-terminal domains of human ICAM-1 suggests how it functions as a rhinovirus receptor and as an LFA-1 integrin ligand, *Proc Natl Acad Sci USA* 95:4140–5, 1998.

Greve JM, Davis G, Meyer AM, Forte CP, Yost SC et al: The major human rhinovirus receptor is ICAM-1, *Cell* 56:839–47, 1989.

Le Beau MM, Ryan D Jr, Pericak-Vance MA: Report of the committee on the genetic constitution of chromosomes 18 and 19, *Cytogenet Cell Genet* 51:338–57, 1989.

Lu TT, Cyster JG: Integrin-mediated long-term B cell retention in the splenic marginal zone, *Science* 297:409–12, 2002.

Simmons D, Makgoba MW, Seed B: ICAM, an adhesion ligand of LFA-1, is homologous to the neural cell adhesion molecule NCAM, *Nature* 331:624–7, 1988.

Sligh JE Jr, Ballantyne CM, Rich SS, Hawkins HK, Smith CW et al: Inflammatory and immune responses are impaired in mice deficient in intercellular adhesion molecule 1, *Proc Natl Acad Sci USA* 90:8529–33, 1993.

PTPRδ

Mizuno K, Hasegawa K, Katagiri T, Ogimoto M, Ichikawa T et al: MPTP-delta, a putative murine homolog of HPTP-delta, is expressed in specialized regions of the brain and in the B-cell lineage, *Mol Cell Biol* 13:5513–23, 1993.

Schaapveld RQJ, van den Maagdenberg AMJM, Schepens JTG, Olde Weghuis D, Geurts van Kessel A et al: The mouse gene Ptprf encoding the leukocyte common antigen-related molecule LAR: Cloning, characterization, and chromosomal localization, *Genomics* 27:124–30, 1995.

Uetani N, Kato K, Ogura H, Mizuno K, Kawano K et al: Impaired learning with enhanced hippocampal long-term potentiation in PTP-delta-deficient mice, *EMBO J* 19:2775–85, 2000.

PTPRκ

Fuchs M, Muller T, Lerch MM, Ullrich A: Association of human protein-tyrosine phosphatase kappa with members of the armadillo family, *J Biol Chem* 271:16712–9, 1996.

Yang Y, Gil MC, Choi EY, Park SH, Pyun KH et al: Molecular cloning and chromosomal localization of a human gene homologous to the murine R-PTP-kappa, a receptor-type protein tyrosine phosphatase, *Gene* 186:77–82, 1997.

Zhang Y, Siebert R, Matthiesen P, Yang Y, Ha H et al: Cytogenetical assignment and physical mapping of the human R-PTP-kappa gene (PTPRK) to the putative tumor suppressor gene region 6q22.2-q22.3, *Genomics* 51:309–11, 1998.

PTPRμ

Gebbink MFBG, van Etten I, Hateboer G, Suijkerbuijk R, Beijersbergen RL et al: Cloning, expression and chromosomal localization of a new putative receptor-like protein tyrosine phosphatase, *FEBS Lett* 290:123–30, 1991.

Suijkerbuijk RF, Gebbink MFGB, Moolenaar WH, Geurts van Kessel A: Fine mapping of the human receptor-like protein tyrosine phosphatase gene (PTPRM) to 18p11.2 by fluorescence in situ hybridization, *Cytogenet Cell Genet* 64:245–6, 1993.

Human protein reference data base, Johns Hopkins University and the Institute of Bioinformatics, at http://www.hprd.org/protein.

Selectins

Classification and Structure [6.8]. *Selectins* (Table 6.6) are lectin-type adhesion molecules expressed in the membrane of several cell types, including vascular endothelial cells, leukocytes, and platelets. Selectins are classified into several groups: E-selectin, L-selectin, and P-selectin. E-selectin is found in endothelial cells, and its function is to mediate the interaction of endothelial cells with leukocytes via binding to corresponding ligands. L-selectin is expressed in leukocytes and is responsible for binding to ligands on endothelial cells and other leukocytes. P-selectin is expressed in platelets and endothelial cells, and is responsible for binding to ligands on leukocytes and endothelial cells. Selectins are involved in the regulation of several basic leukocyte activities, including leukocyte adhesion to, rolling on, and migration through the endothelium.

A typical selectin is composed of several domains: an *N*-terminal lectin-like domain, an epidermal growth factor (EGF)-like domain, several consensus repeats, a transmembrane domain, and a cytoplasmic domain (Fig. 6.8). The lectin-like and EGF-like domains are similar in amino acid sequence among different selectins, while other domains differ between different selectins. The *N*-terminal lectin domain is responsible for the adhesion properties of selectins in a Ca^{2+}-dependent manner.

Function [6.9]. The primary function of selectins is to mediate interaction between leukocytes, platelets, and endothelial cells. Selectins can selectively bind to the oligosaccharides of glycoproteins in the membrane of a target cell. Endothelial cells express various selectin ligands, including glycosylation cell adhesion molecule-1, CD34, mucosal addressin cell adhesion molecule-1, and podocalyxin. Leukocytes express primarily E-selectin glycoprotein ligand-1 and P-selectin glycoprotein ligand-1. E-selectin and

TABLE 6.6. Characteristics of Selected Selectins*

Proteins	Alternative Names	Amino Acids	Molecular Weight (kDa)	Expression	Functions
E-Selectin	Selectin-E, endothelial leukocyte adhesion molecule 1, ELAM1, ELAM, leukocyte endothelial cell adhesion molecule 2, LECAM2, CD62E	610	67	Vascular endothelial cells	Regulating leukocyte adhesion to endothelial cells and mediating inflammatory reactions
L-Selectin	Lymphocyte adhesion molecule 1, LYAM1, LAM1, CD62 antigen ligand, CD62L, leukocyte adhesion molecule 1, leukocyte endothelial cell adhesion molecule 1, LECAM1, SELL	385	44	Leukocytes, bone marrow cells	Regulating leukocyte adhesion to endothelial cells
P-selectin	Platelet α granule membrane protein, SELP, CD62, granulocyte membrane protein, GRMP, PSGL1	830	91	Platelets, endothelial cells	Regulating platelet adhesion to endothelial cells

*Based on bibliography 6.8.

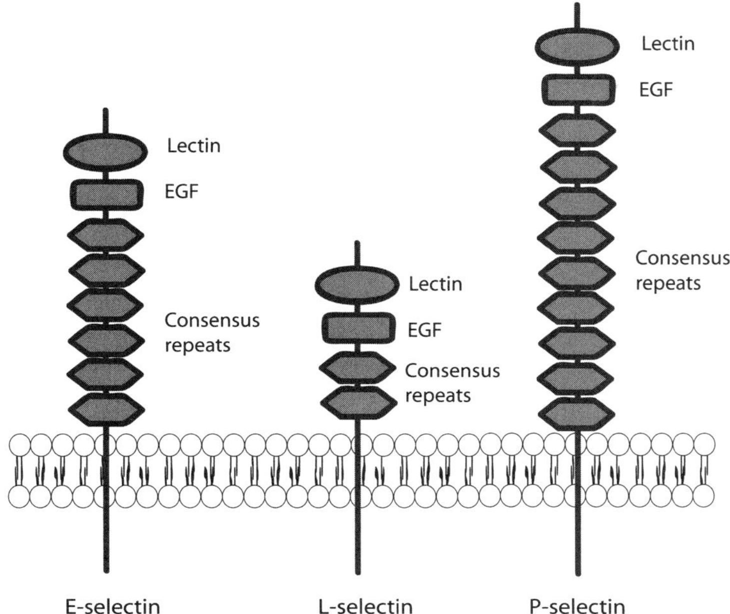

Figure 6.8. Schematic representation of the structure of selectins. Based on bibliography 6.8.

P-selectin of endothelial cells can bind to E-selectin glycoprotein ligand-1 and P-selectin glycoprotein ligand-1 of leukocytes, respectively. L-selectin of leukocytes can bind to glycosylation cell adhesion molecule-1, CD34, mucosal addressin cell adhesion molecule-1, and podocalyxin of endothelial cells. When leukocyte-leukocyte interaction takes place, the L-selectin of one cell can bind to the P-selectin glycoprotein ligand-1 of another cell. Similarly, the P-selectin of platelets can bind to the P-selectin glycoprotein ligand-1 of leukocytes and to the glycosylation cell adhesion molecule-1, CD34, and mucosal addressin cell adhesion molecule-1 of endothelial cells. The binding of selectins with corresponding ligands is the basis for leukocyte adhesion to and rolling on the endothelium.

Under physiological conditions, the constitutive level of selectins in the cell membrane is considerably low, and thus leukocytes and platelets rarely adhere to endothelial cells. Endothelial cells and platelets can synthesize and maintain a constitutive pool of selectins (primarily P-selectin). The synthesized selectin molecules are not deployed to the cell membrane, but stored in the α-granules of platelets and the Weibel–Palade bodies of endothelial cells. These selectin molecules can be redistributed to the cell membrane rapidly in response to inflammatory stimulation. The expression of selectins and ligands are also upregulated in inflammatory reactions. Increased selectin level enhances selectin–ligand interaction and thus facilitates leukocyte and platelet adhesion to the endothelium.

Since leukocytes and endothelial cells are subject to bloodflow, the formation of selectin and ligand bonds must be rapid and the bonds must be sufficiently strong to resist shearing forces imposed by the bloodflow. Under certain shearing conditions, adhered leukocytes may roll on the endothelium. Such a process requires coordination between shear stress and the adhesion bond dynamics, so that the formation of adhesion bonds at the cell leading edge due to selectin–ligand interaction is associated with an equal level of disruption of adhesion bonds at the cell trailing edge due to shear stress. Leukocyte adhesion to and rolling on the endothelium are critical processes in inflammatory responses. These processes prepare leukocytes for transmigration into the interstitial space, where inflammatory reactions take place.

Selectin–ligand interaction may contribute to signal transduction in leukocytes and endothelial cells. Although a complete mechanism is not yet demonstrated, preliminary studies have shown that leukocyte adhesion to L-selectin ligands induces calcium redistribution and activation of mitogen-activated protein kinases. The level of activation is related to the density of the selectin ligands. There is also evidence that selectin–ligand interaction induces activation of integrins. Further investigations are needed to clarify selectin-related signaling pathways.

Cadherins

Classification and Structure [6.10]. *Cadherins* (Table 6.7) are a family of calcium-dependent cell adhesion molecules, which are characterized by the presence of cadherin-specific repeats in the extracellular region of the molecule. Cadherins are traditionally classified into several subfamilies: classical cadherins, protocadherins, and desmosomal cadherins. Cadherins are usually associated with a class of molecules known as *catenins.* These adhesion molecules are involved in the regulation cell–cell interaction, tissue morphogenesis, as well as mitogenic activities such as cell proliferation and migration.

The classical cadherin subfamily includes E-, P-, and N-cadherins (Fig. 6.9). These molecules are localized to the zonula adherens or adherens junctions, which are intercellular contacts required for cell adhesion, cell–cell communication, and tissue formation and organization. These cadherins share similar amino acid sequences and mediate Ca^{2+}-dependent cell–cell interaction and connection. A typical classical cadherin is composed of an *N*-terminal precursor sequence, which contains a proteolytic processing signal sequence K/RRXKR, four characteristic cadherin repeats immediately following the *N*-terminal precursor sequence, a transmembrane domain, and a well-conserved cytoplasmic domain. Cleavage of the *N*-terminal precursor sequence is required for the activation of cadherin. Each cadherin repeat in the extracellular region contains consensus Ca^{2+}-binding sites. The binding of Ca^{2+} induces dimerization of cadherins and protection of the molecule from degradation. The cytoplasmic domain of cadherins interacts with the actin cytoskeleton via cadherin-associated proteins known as catenins.

Protocadherins constitute another cadherin subfamily. Compared with the classical cadherins, these adhesion molecules are characterized by the lack of the proteolytic precursor sequence and the presence of more than four cadherin repeats in the extracellular region (Fig. 6.10). In addition, unlike the classical cadherins, the cytoplasmic domain of protocadherins is considerably heterogeneous in structure. The structural difference suggests different mechanisms in regulating cell adhesion between protocadherins and classical cadherins. It appears that cell adhesion mediated by protocadherins is not as strong as that mediated by classical cadherins.

The third subfamily of cadherins is found in desmosomes and is defined as desmosomal cadherins. Desmosomes are intercellular structures identified in epithelial and cardiac muscular cells (Fig. 6.11) and are responsible for cell–cell interaction and connection, which play a critical role in regulating the formation and integrity of tissues and organs. A typical desmosome appears under an electron microscope as a complex with two parallel plaques (one from each cell) and a narrow gap (~30 nm in width) between two cell

TABLE 6.7. Characteristics of Selected Cadherins*

Proteins	Alternative Names	Amino Acids	Molecular Weight (kDa)	Expression	Functions
E-Cadherin	Calcium-dependent adhesion protein, cadherin 1, epithelial E-cadherin, ECAD, calcium-dependent adhesion protein, epithelial liver cell adhesion molecule, LCAM	901	100	Ubiquitous	A glycoprotein that regulates cell adhesion, which is a calcium-dependent process
P-Cadherin	Placental P cadherin, cadherin 3, PCAD, placental calcium-dependent adhesion protein, CDHP	829	91	Ubiquitous	A glycoprotein that regulates cell–cell adhesion (note that its action is dependent on calcium)
N-Cadherin	Cadherin 2, neuronal cadherin, N cadherin, NCAD, CDHN.	906	100	Nervous system, ovary, testis	A glycoprotein that regulates calcium-dependent cell–cell adhesion
Protocadherin 1	Protocadherin 42, PCDH42, PC42, cadherin-like protein 1	1237	134	Brain	A membrane protein found at cell–cell junctions, regulating neural cell adhesion and development
Desmoglein 1	Desmosomal glycoprotein 1, DG1, pemphigus foliaceus antigen, desmoglein 1 preproprotein	1049	114	Skin, esophagus	Serving as a calcium-binding glycoprotein component of desmosomes in epithelial cells, and regulating cell–cell adhesion and interaction
Desmocollin	Desmocollin 1A/1B, type I, desmocollins desmosomal glycoprotein 2/3, DG2/DG3	894	100	Skin, thymus	Serving as a calcium-binding glycoprotein component of desmosomes in epithelial cells and regulating cell–cell adhesion and interaction

*Based on bibliography 6.10.

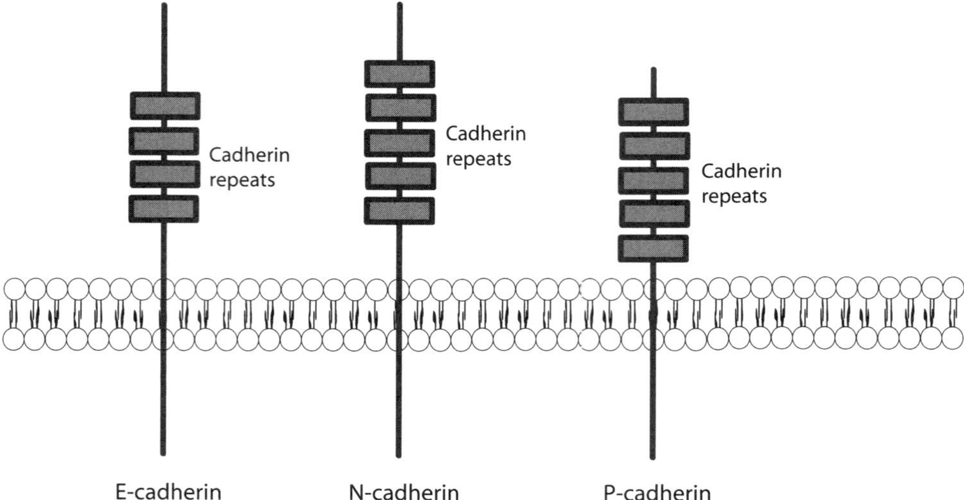

Figure 6.9. Schematic representation of the structure of cadherins. Based on bibliography 6.10.

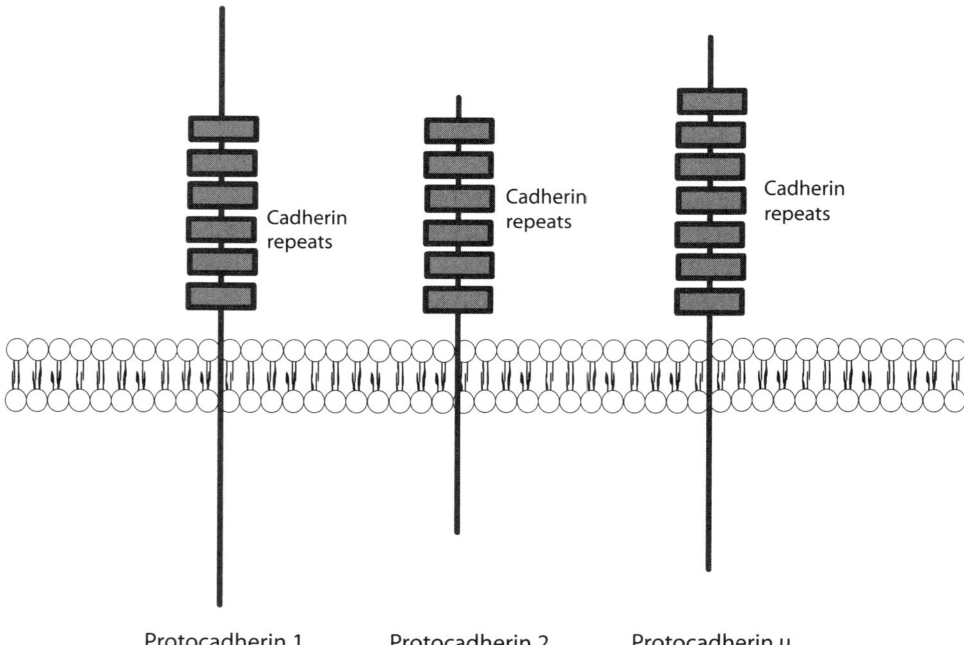

Figure 6.10. Schematic representation of the structure of Protocadherins. Based on bibliography 6.10.

membranes. Such a structure contains several components, including desmosomal cadherins, plakoglobin, plakins, and plakophilins.

There are two types of desmosomal cadherin in desmosomes: desmogleins and desmocollins. These are glycoproteins containing amino acid sequences that are similar to

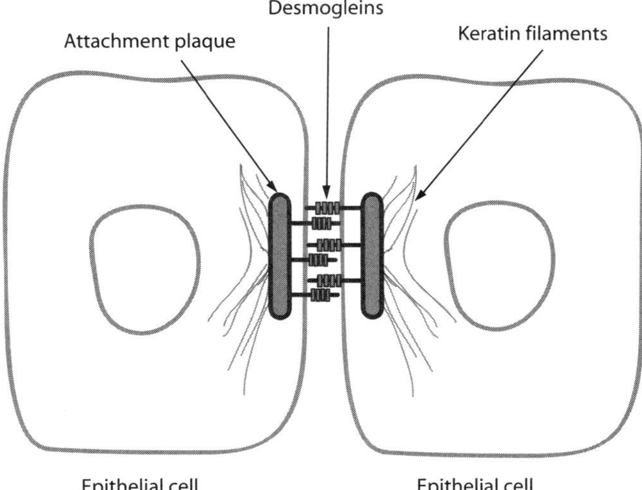

Figure 6.11. Schematic representation of the structure of desmosomes. Based on bibliography 6.10.

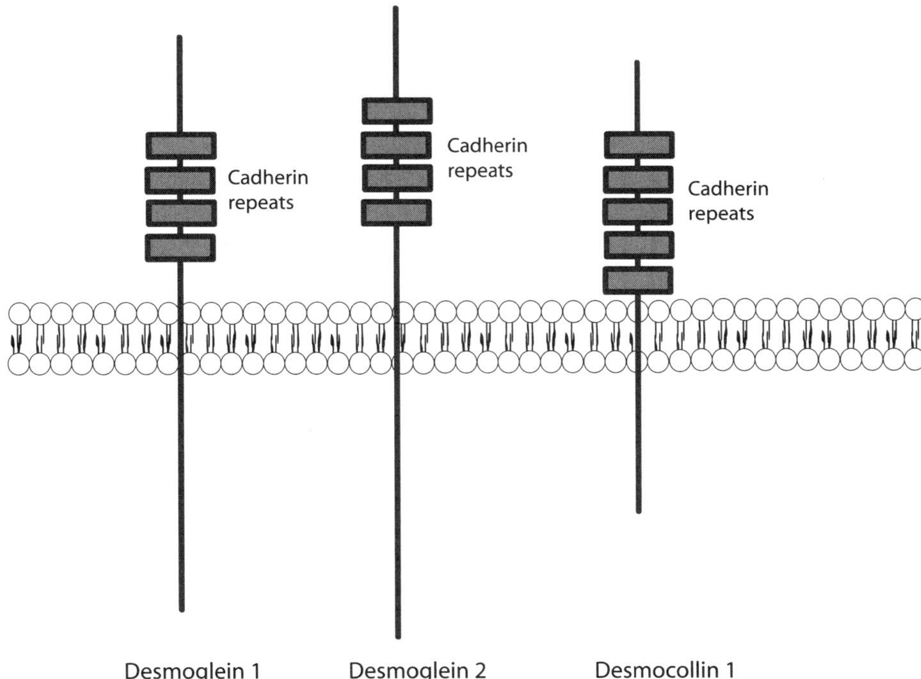

Figure 6.12. Schematic representation of the structure of desmogleins. Based on bibliography 6.10.

the classical cadherins described above. However, desmogleins contain three additional domains in their cytoplasmic tail, including a proline-rich linker, a repeating unit domain, and a terminal domain (Fig. 6.12). Each type of desmosomal cadherin exists in three isoforms. The isoforms for desmogleins are desmoglein-1,2,3, and those for desmocollins are desmocollin-1,2,3. The distribution of each desmosomal cadherin varies among tissues and organs. For instance, desmoglein-2 and desmocollin-2 are ubiquitously expressed in tissues with desmosomes, whereas desmoglein-3 and desmocollin-3 are found mostly in the basal layer of stratified epithelia.

Function [6.11]. Cadherin-mediated cell adhesion plays an important role in the regulation of tissue and organ morphogenesis and development. A change in the expression pattern of major cadherins is often associated with altered morphogenetic processes. For instance, E-cadherin is highly expressed and activated in the oocyte after fertilization, while N-cadherin is not. At gastrulation, the expression and activation pattern of the E- and N-cadherins is switched; specifically, the E-cadherin is downregulated whereas the N-cadherin is upregulated. Such a switch is associated with epithelial–mesenchymal transition. Furthermore, the overexpression of N-cadherins blocks the segregation of neural crest cells from the neural tube. These observations suggest that the expression and activation of appropriate cadherins are critical to the regulation of tissue and organ morphogenesis during embryonic development.

Cadherins are involved in the regulation of cell differentiation. This function has been demonstrated in E-cadherin-null embryonic stem cells. While wildtype embryonic stem cells can differentiate into well-organized forms of specialized tissues, E-cadherin-null stem cells develop only into tissues without specialized forms. The transfer of E-cadherin gene into E-cadherin-null stem cells restores the differentiation function of the embryonic stem cells. These observations suggest that E-cadherins play a critical role in regulating the differentiation of embryonic stem cells.

Cell adhesion mediated by cadherins plays a role in the regulation of cell survival. This is supported by several lines of evidence. The lack of E-cadherins is associated with the induction of endothelial cell apoptosis. Caspases can cleave cadherins and induce cell dissociation, contributing to cell apoptosis. Thus, the presence of cadherins is essential to cell survival.

Cell Surface Heparan Sulfate Proteoglycans

Classification and Structure [6.12]. Cell surface heparan sulfate proteoglycans are composed of heparan sulfate glycosaminoglycan (GAG) chains, which are attached to core proteins in the extracellular region. These proteoglycans serve as adhesion receptors and participate in the regulation of cell–cell and cell–matrix interactions. Heparan sulfate can form various types of heparan sulfate proteoglycan, including perlecan and agrin in extracellular matrix, serglycin in the cytoplasm, and syndecans and glypicans in the cell membrane. The function of heparan sulfate proteoglycans is determined by the heparan sulfate glycosaminoglycan group.

A heparan sulfate molecule is composed of highly sulfated heparin-like domains and poorly sulfated domains with rich glucuronic acids. These domains alternate with intermediate sulfated domains. The total length and the number of each domain of a heparan sulfate vary considerably between different cell types. The differences between cell types may be due to the existence of multiple isoforms of modification enzymes. A major function of the heparan sulfate chains in a cell surface heparan sulfate proteoglycan is to bind ligands. A large number of proteins can bind to heparan sulfate. Although the amino acid sequences of these binding proteins vary widely, the binding proteins are rich in basic amino acids such as lysine and arginine.

TABLE 6.8. Characteristics of Selected Heparan Sulfate Proteoglycans*

Proteins	Alternative Names	Amino Acids	Molecular Weight (kDa)	Expression	Functions
Syndecan 1	Syndecan, SYND1, SDC1, CD138 antigen, CD138.	310	32	Skin, kidney, placenta, lymphocytes	A transmembrane heparan sulfate that mediates cell binding, cell signaling, cytoskeletal organization, cell proliferation, cell differentiation, and HIV transmission to lymphocytes
Glypican 1	GPC1	558	62	Pancreas, intestine, bone marrow, placenta	A heparan sulfate proteoglycan that attaches to external surface of cell membrane and regulates cell–cell interaction

*Based on bibliography 6.12.

Syndecans and glypicans (see Table 6.8) are the most abundant heparan sulfate proteoglycans at the cell surface. Syndecans are transmembrane receptors with heparan sulfate chains attached to the extracellular region at the N-terminus (Fig. 6.13). Syndecans are a family of several heparan sulfate proteoglycans, including syndecan-1,2,3,4 with molecular weights 33, 22, 46, and 22 kDa, respectively. The extracellular domain of the four syndecan molecules differs considerably. Syndecan-2,-3 contain in their extracellular region primarily heparan sulfate chains, whereas syndecan-1,-4 contain chondroitin sulfates in addition to heparan sulfates. The extracellular domain of syndecans is composed of heparan sulfate attachment sites, signal peptide sites, and proteolytic cleavage sites. Syndecans can be shed from the cell membrane via protein cleavage at sites near the cell membrane. Such a process converts receptor-type to soluble syndecans. Both receptor and soluble syndecans can bind to the same type of ligands.

The transmembrane domain of syndecans is well conserved among different types of syndecan. This domain plays a role in mediating the dimerization of the syndecan molecules and localizing syndecans to appropriate membrane compartments. The cytoplasmic tail of all syndecans is highly conserved. This tail is composed of phosphorylation sites and binding sites for cytoskeletal proteins and signaling molecules.

Glypicans are globular molecules with molecular weight ~60 kDa. Six types of glypican have been identified. Each glypican is composed of an *N*-terminal cysteine-rich domain and heparan sulfate attachment motifs. Glypicans are attached to the external surface of the cell membrane via a glycosyl phosphatidylinositol anchor, which is localized to the membrane microdomains with rich glycosphingolipids. The heparan sulfate chains are attached to the *C*-terminus of the core proteins near the cell membrane. Glypicans do not pass through the cell membrane.

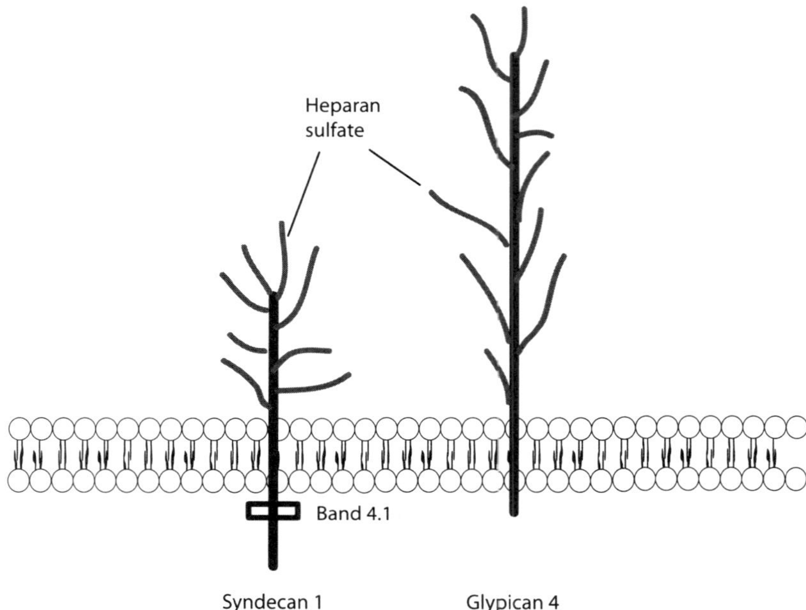

Figure 6.13. Schematic representation of the structure of syndecans. Based on bibliography 6.12.

Function [6.13]. Cell surface heparan sulfate proteoglycans participate in the regulation of cell–cell adhesion and interaction. Syndecans are localized to adherens junctions and can interact with several adhesion molecules, such as L-selectin, N-CAM, and PE-CAM. The lack of syndecan-1 is associated with reduced cell aggregation, and the transfection of the syndecan-1 gene restores cell aggregation. These observations suggest that the presence of syndecans is essential to cell–cell adhesion. However, the exact mechanisms remain to be investigated.

Cell surface heparan sulfate proteoglycans play a role in the regulation of cell–matrix adhesion. Extracellular matrix contains a number of components, including collagen, elastin, fibronectin, laminin, tenascin, vitronectin, and thrombospondin, which are capable of interacting with syndecans. During development, syndecans are colocalized with extracellular matrix components. Certain types of syndecans, such as syndecan-1 and -4 are localized to the focal adhesion contacts. In heparan sulfate-deficient cells, the formation of focal adhesion contacts is impaired. These observations demonstrate the importance of cell surface heparan sulfate proteoglycans in the control of cell–matrix interaction. The role of glypicans in regulating cell–matrix adhesion remains to be determined, although glypicans have been found to bind to collagen and fibronectin.

Integrins

Classification and Structure [6.14]. Integrins are transmembrane glycoproteins that mediate cell–matrix adhesion, providing a linkage between the cytoskeleton and extracellular matrix (Fig. 6.14). Integrins also mediate cell–cell interactions. Each integrin is a heterodimer composed of a variable α subunit and a relatively conserved β subunit. To date, at least 18 α subunits and 8 β subunits have been identified. The binding ligands of

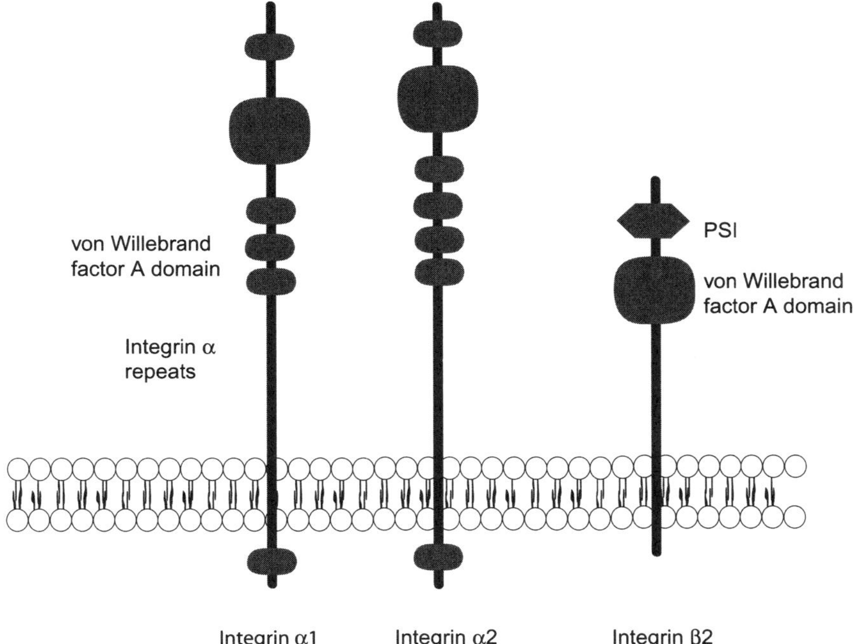

Figure 6.14. Schematic representation of the structure of integrins. PSI: domain found in plexins, semaphorins, and integrins. Based on bibliography 6.14.

integrins are determined by the combination of specific α and β subunits. Several α integrins and a common β integrin are presented in Table 6.9; some of these molecules are used for treatment of muscular dystrophy.

Both integrin α and β subunits contain an extracellular domain, a transmembrane domain, and a cytoplasmic tail. The α and β subunits contain distinct amino acid sequences. The β subunit is about 760–790 amino acids in length, whereas the α subunit is about 1000–1200 amino acids. The majority of amino acids are distributed in the extracellular region for both α and β subunits. The extracellular region of the β subunit is composed of an I-domain, which is responsible for ligand binding, and several cysteine-rich repeats. The extracellular region of the α subunit is composed of an I-domain (in at least 9 α subunits) and several repeats with a consensus sequence DxDxDGxxD (x = any amino acid). These repeats are responsible for integrin binding.

Although the cytoplasmic tail of integrins is considerably short, this region plays an important role in mediating interactions between integrins and intracellular signaling molecules, between integrins and cytoskeletal components, and between the integrin α and β subunits. The cytoplasmic tail is composed of a large number of binding sites for these interactions. The binding of integrins to the actin cytoskeleton is the basis for the formation of cell focal adhesion contacts, essential structures for the control of cell attachment to extracellular matrix and cell migration.

Each cell contains a number of different types of integrins. The β1 subunit forms a dominant subfamily of integrins with the α subunits. The β1-containing integrins are defined as β1 integrins. The ligands of β1 integrins are determined primarily by the specificities of the α subunits. For example, α1β1 and α2β1 bind to collagen and laminin, α4β1 and α5β1 bind to fibronectin, whereas α6β1 binds to laminin.

TABLE 6.9. Characteristics of Selected Integrins*

Proteins	Alternative Names	Amino Acids	Molecular Weight (kDa)	Expression	Functions
Integrin $\alpha 1$	Laminin and collagen receptor, very late activation protein 1, VLA1, CD49a	1179	131	Ubiquitous	Forming a heterodimer with integrin $\beta 1$, serving as a receptor for collagen and laminin, regulating cell–matrix interaction, and mediating cell survival, migration, and proliferation
Integrin $\alpha 2$	ITGA2, very late activation protein 2 receptor $\alpha 2$ subunit, VLA2 receptor $\alpha 2$ subunit, VLA2 α chain, VLAA2, CD49B, platelet glycoprotein Ia/IIa, platelet membrane glycoprotein Ia, GPIa, collagen receptor	1181	129	Ubiquitous	Forming a heterodimer with integrin $\beta 1$ and serving as a receptor for collagen
Integrin $\alpha 3$	ITGA3, CD49C, very late activation antigen 3 (VLA-3), very common antigen 2 (VCA-2), extracellular matrix receptor 1 (ECMR1), and galactoprotein b3 (GAPB3), VLA3 α chain	1066	119	Ubiquitous	Forming a heterodimer with integrin $\beta 1$ and and mediating cell interaction with extracellular matrix components, such as collagen, fibronectin, and laminin
Integrin $\alpha 4$	ITGA4, CD49D, very late activation protein 4 receptor $\alpha 4$ subunit, VLA4 receptor $\alpha 4$ subunit	1038	115	Blood vessels, blood cells, skin.	Forming a heterodimer with integrin $\beta 1$ regulating cell interaction with fibronectin, and mediating vascular cell–cell interaction

Integrin α5	ITGA5, CD49e, fibronectin receptor α subunit, very late activation protein 5α subunit	1049	115	Ubiquitous	Forming a heterodimer with integrin β1, regulating cell interaction with fibronectin
Integrin α6	ITGA6, CD49f, very late activation protein 6α subunit, VLA-6	1073	120	Skin, lung, intestine, stomach, pancreas.	Forming a heterodimer with integrin β1, regulating cell interaction with laminin
Integrin α7	ITGA7	1137	124	Heart, skeletal muscle, nervous system, lung, intestine, overy, prostate gland	Joining with integrin β1 to form an integrin complex, which is a major integrin complex expressed in differentiated muscle cells, binding to the extracellular matrix protein laminin-1, and regulating cell attachment to extracellular matrix
Integrin αv	ITGAV, vitrionectin receptor α polypeptide, CD51	1048	116	Ubiquitous	Forming heterodimers with several β integrins, such as integrin β1, 3, and 5; mediating cell adhesion to extracellular matrix (note that the integrin complex αvβ3 regulates cell adhesion to vitronectin), and regulating TGFβ-related signal transduction
Integrin β1	ITGB1, CD29, very late activation protein, β polypeptide, VLAβ	825	92	Ubiquitous	Joining with an integrin α subunit to form integrin complexes, regulating cell adhesion to extracellular matrix, regulating various cellular activities, including embryogenesis, cell proliferation and migration, immune response, and metastasis of tumor cells

*Based on bibliography 6.14.

Function [6.15]. Integrins play a critical role in the regulation of cell migration. As discussed on page 270 of this chapter, in order to induce migration, a cell must attach to a matrix substrate at the leading edge while detaching from the substrate at the trailing edge at a given time. Although the exact mechanisms are poorly understood, integrins are likely involved in the regulation of these processes. During cell migration, integrins at the leading edge must be engaged to form strong adhesion bonds between focal adhesion contacts and the matrix substrate (Fig. 6.15), whereas integrins at the trailing edge must dissociate from the matrix substrate. The location-dependent integrin activation and deactivation within a single cell remains a subject of research.

In addition to the regulation of cell membrane attachment to matrix substrate, integrins are involved in regulating the contractile activity of the actin cytoskeleton. Integrins can be activated by exposure to extracellular matrix. Activated integrins can lead to phosphorylation of the mitogen-activated protein kinases (MAPKs) via the mediation of focal adhesion kinase and adaptor proteins. MAPKs can directly phosphorylate the myosin light-chain kinase (MLCK), which in turn activates myosin light chain, inducing and enhancing actin–myosin interaction. Actin–myosin interaction generates forces necessary for cell protrusion and traction during migration.

Integrins are involved in regulating the assembly of extracellular matrix. One example is the control of basement membrane formation by β1 integrins in epithelial tissues. Embryonic stem cells derived from β1 integrin-null mice are not able to form a basement membrane in epidermal tissue. Another example is integrin-related fibronectin fibrillogenesis. The assembly of fibronectin fibrils can be initiated on the binding of fibronectin to the α5β1 integrin. The loss of this type of integrin is associated with impairment of fibronectin fibrillogenesis. Other types of integrin, such as α4β1 and αvβ3, also contribute to the regulation of fibronectin fibrillogenesis.

Integrins play a critical role in the regulation of cell differentiation and proliferation. In several experimental models, integrins have been shown to mediate the pattern of gene expression and cell differentiation. For instance, salivary gland cells can differentiate into duct and acinar epithelial cells in response to the interaction of integrins and extracellular matrix, whereas these cells cannot differentiate in the absence of extracellular matrix. Furthermore, a treatment with antibodies specific to collagen IV and integrin α6 and β1

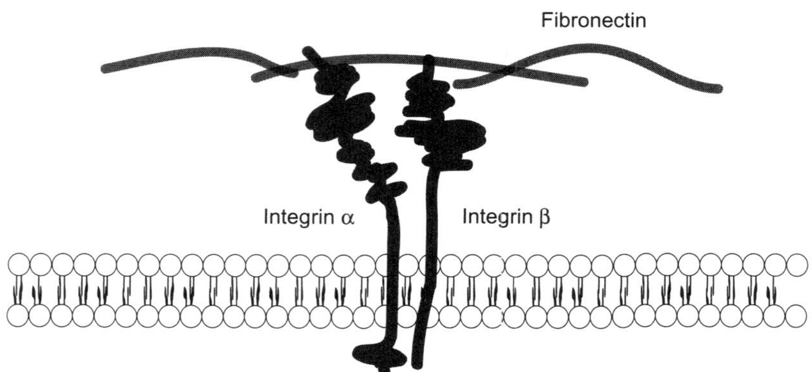

Figure 6.15. Schematic demonstration of interaction of integrins with fibronectin. Based on bibliography 6.15.

reduces the capability of cell differentiation. The attachment of cells to fibronectin- and collagen-containing matrix is often associated with extensive changes in gene expression. Altered gene activities may likely contribute to integrin-initiated cell differentiation and proliferation. The loss of β1 integrins in keratinocytes is associated with a reduction in cell differentiation and proliferation. These observations suggest a critical role for integrin-matrix interaction in the regulation of cell differentiation and proliferation.

Mechanisms of Integrin-Related Activities [6.15]. The primary function of integrins is to regulate the adhesion of cells to extracellular matrix. Integrins can be activated by exposure to extracellular matrix, which induces conformational changes and clustering of integrins. The activation of integrins is associated with an increase in integrin binding affinity. Integrins are major constituents of focal adhesion contacts, structures regulating cell adhesion to extracellular matrix. Focal adhesion contacts were original identified in cultured fibroblasts by electron microscopy, which demonstrates the presence of electron dense plaques with filamentous structures. These plaques were later found to contain a number of molecules, including integrins, actin filaments, talin, filamin, α-actinin, vinculin, profilin, paxillin, tensin, focal adhesion kinase, the Src family kinases, protein tyrosine phosphatases, the Grb2 adaptor protein, and phosphoinositide 3-kinase. Integrins serve as links between the intracellular actin cytoskeleton and extracellular matrix, and transmit signals from the extracellular matrix to the actin cytoskeleton and intracellular signaling pathways.

Integrins do not possess intrinsic catalytic activity. However, the β subunit of integrins can transmit a variety of extracellular signals via their connection with intracellular signaling molecules and actin cytoskeleton-associated molecules as described above. The importance of the β subunit can be tested by selected sequence deletion and restoration. The deletion of the β subunit is associated with diminished interaction of integrins with intracellular signaling molecules. Compared with the β subunit, the α subunit binds to fewer molecules. Identified molecules that bind to the α subunit include calreticulin, guanine nucleotide exchange factor Mss4, and calcium-binding protein. The physiological function of the α subunit-binding proteins remains to be determined.

The interaction of extracellular matrix with integrins often initiates intracellular signaling events, leading to molecular activities such as phosphorylation of protein tyrosine kinases, changes in the level of cAMP and calcium, and expression of mitogenic genes. In particular, integrins can transmit signals to two nonreceptor protein tyrosine kinases: focal adhesion kinase (FAK) and Src protein tyrosine kinase. These protein tyrosine kinases are localized to focal adhesion contacts. FAK can be phosphorylated in response to interaction of integrins with fibronectin, although the mechanisms of FAK activation remain poorly understood. Phosphorylated FAK induces the recruitment of Src to FAK. Recruited Src in turn phosphorylates FAK at various sites, enhancing the activity of FAK. These activities lead to the recruitment of adaptor proteins, including Grb2 and $pp130^{Cas}$, to the focal adhesion contacts. These adapter proteins link the integrin-FAK pathway to other signaling pathways, including the PI3-kinase, Ras, and MAPK pathways, which play critical roles in regulating cell proliferation and migration.

Integrin-related signaling molecules can communicate with the Rho family of small GTPases, including Rho, Cdc42, and Rac, which regulate the assembly and function of the actin cytoskeleton. The interaction of integrins with extracellular matrix can induce activation of Rho, Cdc42, and Rac. Activated Rho, Cdc42, and Rac enhance the assembly of focal adhesion contacts and the activity of related signaling molecules, including FAK.

However, the exact regulatory mechanisms remain to be investigated. These investigations suggest that integrin-dependent signaling pathways can "crosstalk" to other signaling pathways. Such interaction provides a synergistic mechanism for the regulation of cell adhesion, migration, proliferation, and differentiation.

APOPTOSIS [6.16]

Apoptosis is a process of naturally occurring cell death, which eliminates malfunctioned and undesired cells during development and remodeling. During development, apoptosis and cell division together contribute to the morphogenesis of tissues and organs. While cell division contributes to the growth of cell and tissue mass, apoptosis contributes to the removal of excessive cells. These processes are unnecessary for the formation of tissues and organs. After reaching maturity, apoptosis continues to play an important role in the maintenance of the homeostasis. Cells with damaged or mutant DNA are eliminated by apoptosis. Without apoptotic elimination, these cells may develop into tumor cells. Under a physiological condition, the cell density is kept at a relatively constant level through coordinated cell proliferation and apoptosis. An increase in apoptotic activity results in tissue degeneration, whereas a decrease in apoptotic activity results in hyperplasia, both of which contribute to pathogenic disorders of tissues and organs. Thus, it is important to maintain a physiological level of apoptotic activity.

Morphological Characteristics of Apoptosis [6.16]

In biological research, it is important to identify apoptotic cells, which helps to understand the mechanisms of apoptosis. Apoptotic cells undergo several stages of morphological change. On the stimulation of apoptotic signals, a cell usually starts to round up and becomes spherical in shape. These morphological changes are usually associated with cell membrane budding. The next noticeable change is DNA condensation, resulting in an increase in the nucleus density and reduction in the nucleus size, which can be seen under an optical and electron microscope (Fig. 6.16). DNA condensation is followed by DNA fragmentation and nucleus disruption. The entire cell eventually disintegrates into small pieces, which are phagocytosed by macrophages or neighboring cells. These morphological features can be used to identify apoptotic cells.

Apoptosis-Inducing Factors [6.16]. Apoptosis can be induced by a variety of extracellular factors, such as depletion of growth factors and nutrients, hypoxia, UV irradiation, mechanical stress, and binding of apoptotic ligands. Intrinsic changes, such as DNA damage and disruption, and immunoreactions, such as T-lymphocyte activation, can also trigger apoptosis. In addition, cancer cells can initiate apoptosis, an important mechanism for the elimination of cancer cells. Suppression of the apoptotic function increases the possibility of tumorigenesis.

Regulation of Apoptosis [6.16]. Apoptosis can be induced and regulated by two known signaling pathways: Fas ligand (FasL)- and tumor necrosis factor (TNF)-activated pathways. FasL and TNF are apoptosis-inducing proteins. These proteins are generated in the ER, deployed to the cell membrane, and cleaved from the cell membrane to form soluble ligands. The forms of the ligands determine the effectiveness of the ligands. For the Fas ligand, the membrane-bound form is more effective than the soluble form. In

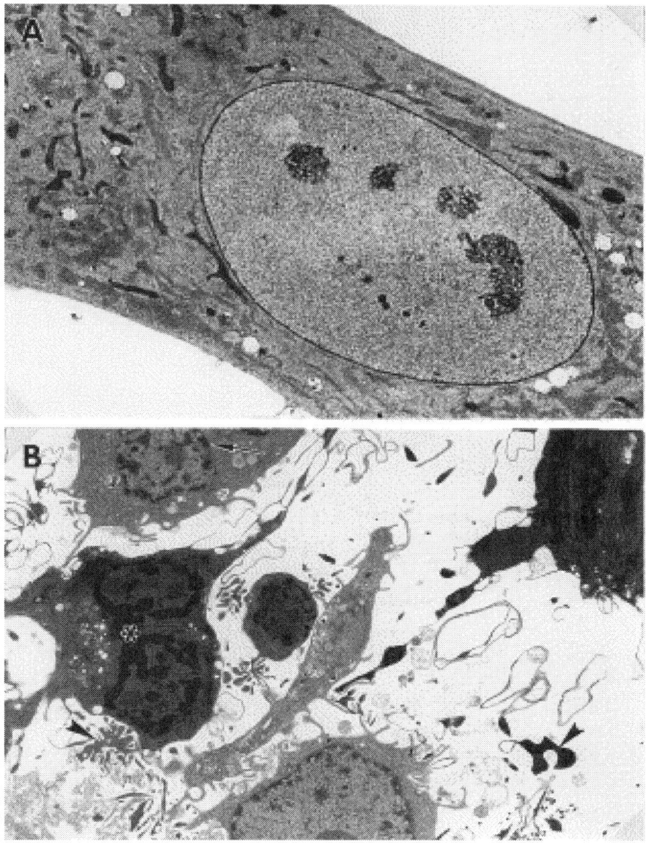

Figure 6.16. Electron micrographs of apoptotic human fibroblasts. (A) A control cell and (B) cells exposed to 0.5-μM naphthazarin (5,8-dihydroxy-1,4-naphthoquinone), an apoptosis inducer, for 8 h are shown. Different stages of apoptosis can be discerned in the treated cells: reduced cell size, condensed chromatin (stars), fragmented nuclei (arrow), and apoptotic bodies (arrow heads). Reprinted from Roberg K et al., Lysosomal release of cathepsin D precedes relocation of cytochrome c and loss of mitochondrial transmembrane potential during apoptosis induced by oxidative stress, *Free Radical Biol Med* 27:1228–37, 1999, with permission from Elsevier.

contrast, soluble TNF is more active than the membrane-bound form of TNF. The mechanism of Fas ligand-induced apoptosis is similar to that induced by TNF. Here, the FasL signaling pathway is used as an example to demonstrate the mechanisms of apoptosis (Fig. 6.17).

A Fas ligand can interact with the Fas ligand receptor, resulting in oligomerization (often trimerization) and activation of the receptor. Activated Fas ligand receptor in turn stimulates the Fas ligand-associated death domain or FADD (note that TNF can interact with and activate TNF receptor, which stimulates the TNFR-associated death domain or TRADD). Activated FADD (or TRADD) binds to a downstream protein known as *caspase 8* (cysteine aspartate protease 8), which belongs to the caspase family and possesses protease catalytic activity. In unstimulated cells, caspase 8 exists in an inactive form known as *procaspase 8*. In response to the stimulation of FADD (or TRADD), caspase 8 can be

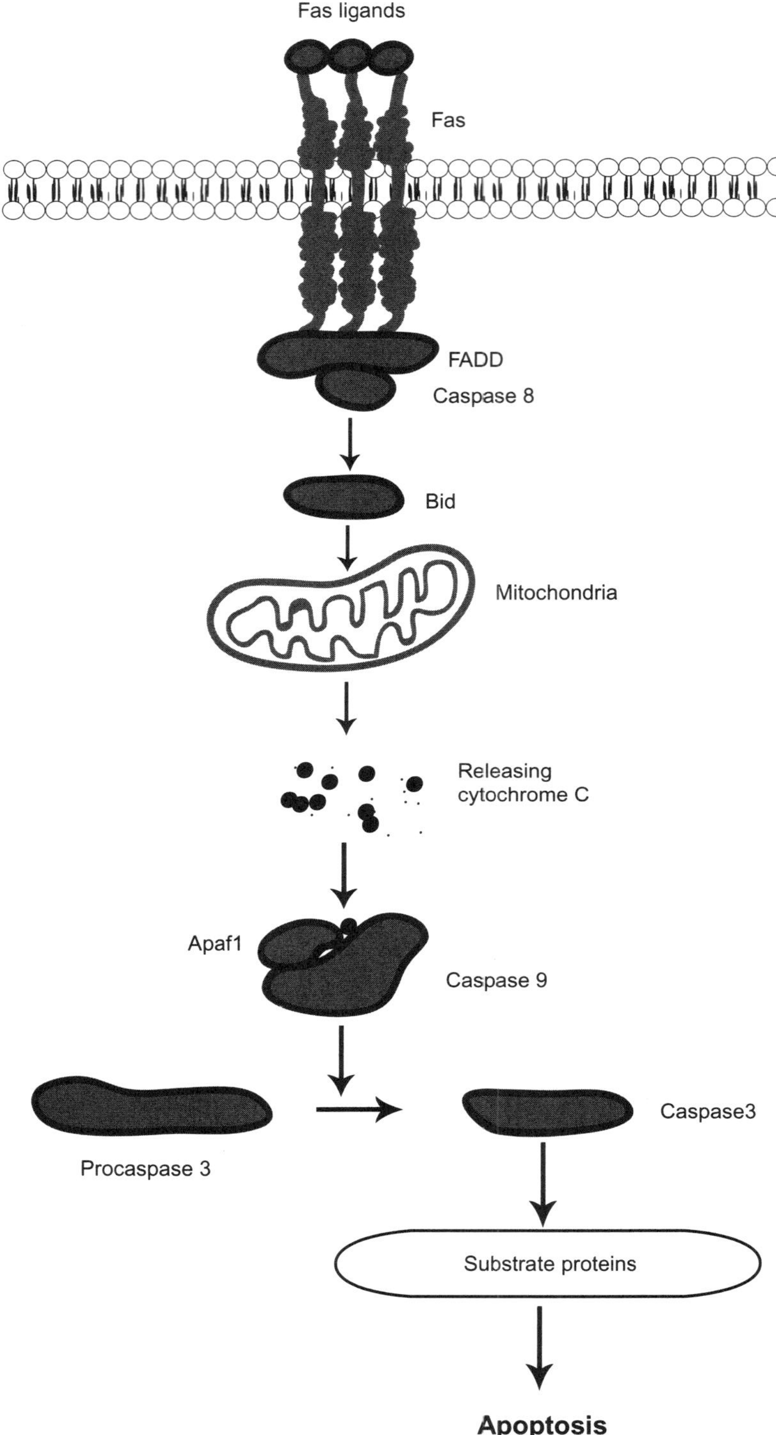

Figure 6.17. Schematic demonstration of the mechanisms of Fas-induced apoptosis. Based on bibliography 6.16.

activated via autocatalytic cleavage. Caspase 8 can further activate a downstream protein known as *Bid*, a member of the Bcl2 family, by cleaving the C-terminal domain of the substrate. Following the cleavage by caspase 8, Bid can be translocated to the mitochondrial membrane. A major action of Bid is to release mitochondrial cytochrome *c* into the cytoplasm. Cytochrome *c* in turn activates a downstream protein named apoptosis *protein-activating factor-1* (Apaf-1), which binds ATP and interacts with procaspase 9, resulting in the formation of the active form of caspase 9 via autocatalytic cleavage. Caspase 9 can cleave procaspase 3, releasing the active form caspase 3. Caspase 3 is a terminal-stage protease, which cleaves and degrades a variety of signaling and structural proteins, including protein kinases, poly[A]polymerse, and actin filaments. Caspase 3 can also cleave DNA fragmentation factor (DFF), releasing a DFF subunit. This subunit can activate nucleases, which induces DNA degradation. These activities eventually lead to DNA fragmentation and cell degeneration. (See Table 6.10.)

Assessment of Cell Apoptosis [6.17]

There are several methods that can be used for assessing cell apoptosis. These methods have been developed on the basis of cell morphological and molecular changes in apoptotic cells and are classified into several groups: (1) methods based on changes in the structure of cell membrane, (2) methods based on changes in cell morphology, (3) methods based on DNA fragmentation, (4) methods based on cytochrome *c* translocation, and (5) methods based on caspase activities. These methods are briefly discussed here.

Assessing Changes in Cell Membrane Structure. A cell membrane contains asymmetrically distributed phospholipid species in the membrane bilayer. The cytoplasmic layer of the cell membrane is composed of phosphatidylserine, phosphatidylinositol, phosphatidylethanolamine, and phosphatidylcholine. The extracellular layer is composed of sphingomyelin, phosphatidylcholine, and phosphatidylethanolamine, but not phosphatidylserine. The asymmetrical distribution of phosphatidylserine is created and maintained by the activity of aminophospholipid translocase, which transports phosphatidylserine from the extracellular layer to the cytoplasmic layer of the cell membrane. In apoptosis, the activity of the aminophospholipid translocase is inhibited, and the content of phosphatidylserine in the extracellular layer of the cell membrane increases, even though the integrity of the cell membrane is uncompromised. Thus, the appearance of phosphatidylserine in the extracellular layer of the cell membrane is indicative of early cell apoptosis. Phosphatidylserine in the extracellular layer can be detected by using an assay for Annexin A5, which is a phosphatidylserine-binding protein. To detect phosphatidylserine, Annexin A5 can be tagged with a marker (e.g., biotin or fluorochrome) and incubated with cell samples. Positive labeling of cells with Annexin A5 suggests the translocation of phosphatidylserine to the extracellular layer of the cell membrane, which indicates the occurrence of apoptosis in the labeled cells.

In addition, the permeability of cell membrane is often increased in cell apoptosis because of the disorganization of phospholipids. In such a case, the cell membrane is permeable to certain types of fluorescent dyes, such as merocyanine (MC) 540 and 7-aminoactinomycin D (7-AAD), which can not pass through the plasma membrane of normal cells. These dyes can be incubated with cell samples and detected by fluorescence microscopy. The appearance of the dye within the cell suggests the occurrence of cell apoptosis. However, these fluorescent dyes are not specific to cell apoptosis. Any factors that cause an increase in cell membrane permeability can induce positive cell labeling.

TABLE 6.10. Characteristics of Selected Apoptosis Regulatory Proteins*

Proteins	Alternative Names	Amino Acids	Molecular Weight (kDa)	Expression	Functions
Fas ligand	CD95 ligand, CD178, FAS, tumor necrosis factor ligand superfamily member 6, TNFSF6, FASL, apoptosis antigen ligand 1, apoptosis (APO 1) antigen ligand 1	281	31	Lymphocyte, testis	A protein that interacts with the Fas receptor and induces apoptosis
Fas receptor	Tumor necrosis factor receptor superfamily member 6, CD95, FAS1, apoptosis-mediating surface antigen FAS, apoptosis antigen 1	335	38	Ubiquitous	A transmembrane receptor that interacts with the Fas ligand and induces apoptosis
FADD	FAS-associated protein with death domain	208	23	Ubiquitous	An adaptor protein that interacts with the Fas receptor and TNF receptor; also mediates cell apoptosis
TNFα	Tumor necrosis factor, tumor necrosis factor α, TNFA, cachectin	233	26	Monocyte, macrophage	Interacting with TNF receptor and regulating cell apoptosis, proliferation, differentiation, and inflammatory reactions
TNF receptor	TNFR1α, TNFR1, tumor necrosis factor receptor superfamily member 1A, p55 TNFR	455	50	Heart, blood vessel, leukocytes	Interacting with TNFα, and mediating apoptosis and inflammatory reactions
TRADD	TNFR1-associated death-domain protein, tumor necrosisfactor receptor 1-associated death-domain protein, tumor necrosis factor receptor 1-associated protein	312	34	Ubiquitous	Interacting with TNF receptor 1 and regulating apoptosis and inflammation

Caspase 8	FADD homologous ICE/CED-3-ICE/CED-3-like protease, FADD-like ICE, MACH, MCH5, FLICE	496	58	Ubiquitous	A cysteine–aspartic acid protease that interacts with FADD and mediates the transduction of apoptotic signals
Bid	BH3 interacting death-domain agonist, BID	195	22	Ubiquitous	An apoptotic agonist activated by caspase 8, stimulating the release of cytochrome *c*, and regulating cell apoptosis
Cytochrome *c*	CYC	105	12	Ubiquitous	A mitochondrial electron transport chain component that mediates electron transfer and regulates apoptosis
Apaf-1	Apoptotic protease-activating factor 1	1248	142	Ubiquitous	Interacting with cytochrome *c* and forming an apoptosome, a structure that cleaves the preproprotein of caspase 9 and generates active caspase 9, resulting in apoptosis
Caspase 9	CASP9, apoptotic protease MCH6, MCH6, ICE-like apoptotic protease 6, apoptotic protease-activating factor 3, APAF3	416	46	Ubiquitous	A cysteine–aspartic acid protease that cleaves procaspase 3, inducing the formation of caspase 3 and apoptosis
Caspase 3	CASP3, cysteine protease CPP32, CPP32, apoptosis-related cysteine protease, APOPAIN	277	32	Ubiquitous	A downstream cysteine–aspartic acid protease that cleaves a variety of cytoplasmic proteins, including protein kinases, nuclear proteins, and cytoskeletal proteins, and induces apoptosis

*Based on bibliography 6.16.

Assessing Changes in Cell Morphology. Apoptotic cells are associated with morphological changes, including cell membrane blebbing, DNA condensation with increased nucleus density, and nucleus disruption. The entire cell is eventually disintegrated into small pieces. Optical and electron microscopic approaches can be used to examine these morphological features. At the optical level, hematoxylin can be used to examine the morphology of cell nuclei. In addition, cell nucleus-binding fluorescent dyes, such as DAPI and Hoechst 33258, can be used for the same purpose. Usually, the fluorescent approach provides better images. At the electron microscopic level, cell membrane blebbing and DNA condensation (Fig. 6.16) can be observed with a much better resolution compared with the optical approach. Morphological examination is a key method for the identification of cell apoptosis and is often used as a standard for the confirmation of cell apoptosis detected by using other methods.

Assessing DNA Fragmentation. DNA fragmentation is a hallmark of cell apoptosis. Thus, apoptotic cells can be identified by assessing DNA fragmentation. Two approaches can be used for such purpose: DNA electrophoresis and terminal deoxynucleotidyl transferase-mediated dUTP nick end-labeling (TUNEL). For the DNA electrophoresis method, the pattern of DNA bands can be analyzed by comparing to DNA samples from normal control cells. An increase in the number of DNA bands in a large range of molecular size suggests the occurrence of cell apoptosis.

TUNEL is a method used for visualizing DNA fragments in situ. A key enzyme used for this assay is the terminal deoxynucleotidyl transferase, which catalyzes DNA synthesis at the ends of DNA fragments in the presence of deoxynucleotides (dNUPs). When a dNTP is tagged with a marker, such as a fluorescent molecule, the DNA fragments with the added dNTP can be visualized by fluorescence microscopy. This method can be used at the single cell level. However, the method is not apoptosis-specific. DNA fragmentation induced by other factors, such as cell injury, can be detected. Thus, the identification of morphological changes in apoptotic cells is often conducted together with the TUNEL method to confirm the results by TUNEL.

Assessing the Translocation of Cytochrome c. Cytochrome *c* is a protein component of the respiratory chain in the mitochondria. It is localized to the surface of the internal membrane of the mitochondria. Cytochrome *c* can be translocated from the mitochondria to the cytoplasm and contributes to the activation of apoptotic signaling pathways. The translocation of cytochrome *c* is a critical step in apoptosis. Thus, the detection of cytochrome *c* translocation from the mitochondria to the cytoplasm is indicative of apoptosis. An antibody can be used for examining the distribution of cytochrome *c*. Typical images of cytochrome *c* translocation are shown in Fig. 6.18.

Assessing the Activity of Caspases. Caspases are a group of proteinases that degrade proteins, ranging from signaling protein kinases to structural proteins, and play critical roles for the induction of cell apoptosis. The activation of caspases indicates the occurrence of apoptosis. Caspases are expressed in the form of inactive precursors, which can be activated by proteolytic cleavage at specific sites induced by proteinases. Thus, caspases cleavage is a sign of caspase activation. Immunoblotting is an effective method for the detection of caspase cleavage. The presence of reduced caspase subunits is indicative of caspase activation and the occurrence of apoptosis.

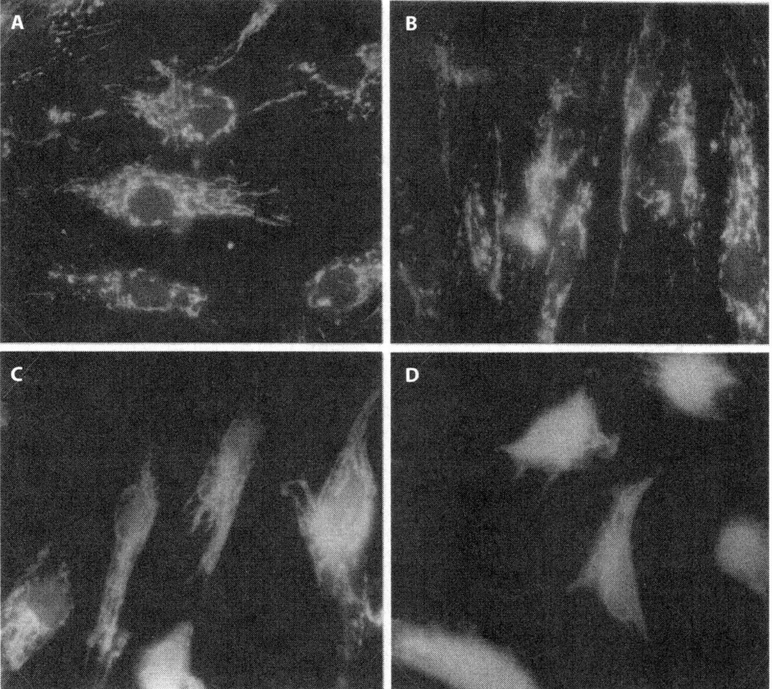

Figure 6.18. Immunofluorescence detection of cytochrome *c* in fibroblasts. The micrographs show: (A) control cells; (B–D) cells exposed to 0.5 μM naphthazarin (5,8-dihydroxy-1,4-naphthoquinone, an apoptosis inducer) for 1, 2, and 3 h, respectively. Note the translocation of cytochrome c from the mitochondria to the cytoplasm in panels B, C, and D. (Reprinted from Roberg K et al: Lysosomal release of cathepsin D precedes relocation of cytochrome c and loss of mitochondrial transmembrane potential during apoptosis induced by oxidative stress, *Free Radical Biol Med* 27:1228–37, 1999, with permission from Elsevier.)

BIBLIOGRAPHY

6.8. Classification and Structure of Selectins

E-Selectin

Bevilacqua MP, Stengelin S, Gimbrone MA Jr, Seed B: Endothelial leukocyte adhesion molecule 1: An inducible receptor for neutrophils related to complement regulatory proteins and lectins, *Science* 243:1160–5, 1989.

Collins T, Williams A, Johnston GI, Kim J, Eddy R et al: Structure and chromosomal location of the gene for endothelial-leukocyte adhesion molecule 1, *J Biol Chem* 266:2466–73, 1991.

Hidalgo A, Weiss LA, Frenette PS: Functional selectin ligands mediating human CD34+ cell interactions with bone marrow endothelium are enhanced postnatally, *J Clin Invest* 110:559–69, 2002.

Iida A, Nakamura Y: High-resolution SNP map in the 55-kb region containing the selectin gene family on chromosome 1q24-q25, *J Hum Genet* 48:150–4, 2003.

Wang N, Chintala SK, Fini ME, Schuman JS: Activation of a tissue-specific stress response in the aqueous outflow pathway of the eye defines the glaucoma disease phenotype, *Nature Med* 7:304–9, 2001.

Watson ML, Kingsmore SF, Johnston GI, Siegelman MH, Le Beau MM et al: Genomic organization of the selectin family of leukocyte adhesion molecules on human and mouse chromosome 1, *J Exp Med* 172:263–72, 1990.

L-Selectin

Genbacev OD, Prakobphol A, Foulk RA, Krtolica AR, Ilic D et al: Trophoblast L-selectin-mediated adhesion at the maternal-fetal interface, *Science* 299:405–8, 2003.

Iida A, Nakamura Y: High-resolution SNP map in the 55-kb region containing the selectin gene family on chromosome 1q24-q25, *J Hum Genet* 48:150–4, 2003.

Lasky LA, Singer MS, Dowbenko D, Imai Y, Henzel WJ et al: An endothelial ligand for L-selectin is a novel mucin-like molecule, *Cell* 69:927–38, 1992.

Lasky LA, Singer MS, Yednock TA, Dowbenko D, Fennie C et al: Cloning of a lymphocyte homing receptor reveals a lectin domain, *Cell* 56:1045–55, 1989.

Ord DC, Ernst TJ, Zhou LJ, Rambaldi A, Spertini O et al: Structure of the gene encoding the human leukocyte adhesion molecule-1 (TQ1, Leu-8) of lymphocytes and neutrophils, *J Biol Chem* 265:7760–7, 1990.

Watson ML, Kingsmore SF, Johnston GI, Siegelman MH, Le Beau MM et al: Genomic organization of the selectin family of leukocyte adhesion molecules on human and mouse chromosome 1, *J Exp Med* 172:263–72, 1990.

P-Selectin

Burger PC, Wagner DD: Platelet P-selectin facilitates atherosclerotic lesion development, *Blood* 101:2661–6, 2003.

Hidalgo A, Weiss LA, Frenette PS: Functional selectin ligands mediating human CD34+ cell interactions with bone marrow endothelium are enhanced postnatally, *J Clin Invest* 110:559–69, 2002.

Hrachovinova I, Cambien B, Hafezi-Moghadam A, Kappelmayer J, Camphausen RT et al: Interaction of P-selectin and PSGL-1 generates microparticles that correct hemostasis in a mouse model of hemophilia A, *Nature Med* 9:1020–5, 2003.

Johnston GI, Bliss GA, Newman PJ, McEver RP: Structure of the human gene encoding granule membrane protein-140, a member of the selectin family of adhesion receptors for leukocytes, *J Biol Chem* 265:21381–5, 1990.

Johnston GI, Cook RG, McEver RP: Cloning of GMP-140, a granule membrane protein of platelets and endothelium: sequence similarity to proteins involved in cell adhesion and inflammation, *Cell* 56:1033–44, 1989.

Koyama H, Maeno T, Fukumoto S, Shoji T, Yamane T et al: Platelet P-selectin expression is associated with atherosclerotic wall thickness in carotid artery in humans, *Circulation* 108:524–9, 2003.

Marshall BT, Long M, Piper JW, Yago T, McEver RP et al: Direct observation of catch bonds involving cell-adhesion molecules, *Nature* 423:190–3, 2003.

Mayadas TN, Johnson RC, Rayburn H, Hynes RO, Wagner DD: Leukocyte rolling and extravasation are severely compromised in P selectin-deficient mice, *Cell* 74:541–54, 1993.

Rosenkranz AR, Mendrick DL, Cotran RS, Mayadas TN: P-selectin deficiency exacerbates experimental glomerulonephritis: A protective role for endothelial P-selectin in inflammation, *J Clin Invest* 103:649–59, 1999.

Human protein reference data base, Johns Hopkins University and the Institute of Bioinformatics, at http://www.hprd.org/protein.

6.9. Function of Selectins

Afshar-Kharghan V, Thiagarajan P: Leukocyte adhesion and thrombosis, *Curr Opin Hematol* 13:34–9, 2006.

Simon SI, Green CE: Molecular mechanics and dynamics of leukocyte recruitment during inflammation, *Annu Rev Biomed Eng* 7:151–185, 2005.

Ley K, Kansas GS: Selectins in T-cell recruitment to non-lymphoid tissues and sites of inflammation, *Nat Rev Immunol* 4:325–35, 2004.

Kakkar AK, Lefer DJ: Leukocyte and endothelial adhesion molecule studies in knockout mice, *Curr Opin Pharmacol* 4(2):154–8, April 2004.

Rosen SD: Ligands for L-selectin: Homing, inflammation, and beyond, *Annu Rev Immunol* 22:129–56, 2004.

Grunewald M, Griesshammer M: P-selectin modulation in haemostasis: One size fits all? *Trends Mol Med* 10:9–12, 2004.

Ulbrich H, Eriksson EE, Lindbom L: Leukocyte and endothelial cell adhesion molecules as targets for therapeutic interventions in inflammatory disease, *Trends Pharmacol Sci* 24:640–7, 2003.

6.10. Classification and Structure of Cadherins

E-Cadherin

Batlle E, Sancho E, Franci C, Dominguez D, Monfar M et al: The transition factor Snail is a repressor of E-cadherin gene expression in epithelial tumour cells, *Nature Cell Biol* 2:84–9, 2000.

Berx G, Staes K, van Hengel J, Molemans F, Bussemakers MJG et al: Cloning and characterization of the human invasion suppressor gene E-cadherin (CDH1), *Genomics* 26:281–9, 1995.

Boggon TJ, Murray J, Chappuis-Flament S, Wong E, Gumbiner BM et al: C-cadherin ectodomain structure and implications for cell adhesion mechanisms, *Science* 296:1308–3, 2002.

Cano A, Perez-Moreno MA, Rodrigo I, Locascio A, Blanco MJ et al: The transcription factor Snail controls epithelial-mesenchymal transitions by repressing E-cadherin expression, *Nature Cell Biol* 2:76–83, 2000.

Ceccherini I, Romeo G, Lawrence S, Breuning MH, Harris PC et al: Construction of a map of chromosome 16 by using radiation hybrids, *Proc Natl Acad Sci USA* 89:104–8, 1992.

Grady WM, Willis J, Guilford PJ, Dunbier AK, Toro TT et al: Methylation of the CDH1 promoter as the second genetic hit in hereditary diffuse gastric cancer, *Nature Genet* 26:16–7, 2000.

Guilford P, Hopkins J, Harraway J, McLeod M, McLeod N et al: E-cadherin germline mutations in familial gastric cancer, *Nature* 392:402–5, 1998.

Hayashi T, Carthew RW: Surface mechanics mediate pattern formation in the developing retina, *Nature* 431:647–52, 2004.

Huntsman DG, Carneiro F, Lewis FR, MacLeod PM, Hayashi A et al: Early gastric cancer in young, asymptomatic carriers of germ-line E-cadherin mutations, *New Engl J Med* 344:1904–9, 2001.

Jamora C, DasGupta R, Kocieniewski P, Fuchs E: Links between signal transduction, transcription and adhesion in epithelial bud development, *Nature* 422:317–22, 2003.

Kawasaki H, Taira K: Induction of DNA methylation and gene silencing by short interfering RNAs in human cells, *Nature* 431:211–7, 2004.

Maretzky T, Reiss K, Ludwig A, Buchholz J, Scholz F et al: ADAM10 mediates E-cadherin shedding and regulates epithelial cell–cell adhesion, migration, and beta-catenin translocation, *Proc Natl Acad Sci USA* 102:9182–7, 2005.

Natt E, Magenis RE, Zimmer J, Mansouri A, Scherer G: Regional assignment of the human loci for uvomorulin (UVO) and chymotrypsinogen B (CTRB) with the help of two overlapping deletions on the long arm of chromosome 16, *Cytogenet Cell Genet* 50:145–8, 1989.

Overduin M, Harvey TS, Bagby S, Tong KI, Yau P et al: Solution structure of the epithelial cadherin domain responsible for selective cell adhesion, *Science* 267:386–9, 1995.

Palmer HG, Larriba MJ, Garcia JM, Ordonez-Moran P, Pena C et al: The transcription factor SNAIL represses vitamin D receptor expression and responsiveness in human colon cancer, *Nature Med* 10:917–9, 2004.

Perl AK, Wilgenbus P, Dahl U, Semb H, Christofori G: A causal role for E-cadherin in the transition from adenoma to carcinoma, *Nature* 392:190–3, 1998.

Risinger JI, Berchuck A, Kohler MF, Boyd J: Mutations of the E-cadherin gene in human gynecologic cancers, *Nature Genet* 7:98–102, 1994.

P-Cadherin

Hatta M, Miyatani S, Copeland NG, Gilbert DJ, Jenkins NA et al: Genomic organization and chromosomal mapping of the mouse P-cadherin gene, *Nucleic Acids Res* 19:4437–41, 1991.

Kremmidiotis G, Baker E, Crawford J, Eyre HJ, Nahmias J et al: Localization of human cadherin genes to chromosome regions exhibiting cancer-related loss of heterozygosity, *Genomics* 49:467–71, 1998.

Shimoyama Y, Yoshida T, Terada M, Shimosato Y, Abe O, Hirohashi S: Molecular cloning of a human Ca(2+)-dependent cell–cell adhesion molecule homologous to mouse placental cadherin: its low expression in human placental tissues, *J Cell Biol* 109:1787–94, 1989.

Sprecher E, Bergman R, Richard G, Lurie R, Shalev S et al: Hypotrichosis with juvenile macular dystrophy is caused by a mutation in CDH3, encoding P-cadherin, *Nature Genet* 29:134–6, 2001.

N-Cadherin

Garcia-Castro MI, Vielmetter E, Bronner-Fraser M: N-cadherin, a cell adhesion molecule involved in establishment of embryonic left-right asymmetry, *Science* 288:1047–51, 2000.

Hayashi T, Carthew RW: Surface mechanics mediate pattern formation in the developing retina, *Nature* 431:647–52, 2004.

Hermiston ML, Gordon JI: Inflammatory bowel disease and adenomas in mice expressing a dominant negative N-cadherin, *Science* 270:1203–6, 1995.

Miyatani S, Copeland NG, Gilbert DJ, Jenkins NA, Takeichi M: Genomic structure and chromosomal mapping of the mouse N-cadherin gene, *Proc Natl Acad Sci USA* 89:8443–7, 1992.

Reid RA, Hemperly JJ: Human N-cadherin: Nucleotide and deduced amino acid sequence, *Nucleic Acids Res* 18:5896 (only), 1990.

Tanaka H, Shan W, Phillips GR, Arndt K, Bozdagi O et al: Molecular modification of N-cadherin in response to synaptic activity, *Neuron* 25:93–107, 2000.

Wallis J, Fox MF, Walsh FS: Structure of the human N-cadherin gene: YAC analysis and fine chromosomal mapping to 18q11.2, *Genomics* 22:172–9, 1994.

Protocadherin 1

Del Mastro RG, Wang L, Simmons AD, Gallardo TD, Clines GA et al: Human chromosome-specific cDNA libraries: New tools for gene identification and genome annotation, *Genome Res* 5:185–94, 1995.

Sago H, Kitagawa M, Obata S, Mori N, Taketani S et al: Cloning, expression, and chromosomal localization of a novel cadherin-related protein, protocadherin-3, *Genomics* 29:631–40, 1995.

Sano K, Tanihara H, Heimark RL, Obata S, Davidson M et al: Protocadherins: A large family of cadherin-related molecules in central nervous system, *EMBO J* 12:2249–56, 1993.

Desmoglein 1

Allen E, Yu QC, Fuchs E: Mice expressing a mutant desmosomal cadherin exhibit abnormalities in desmosomes, proliferation, and epidermal differentiation, *J Cell Biol* 133:1367–82, 1996.

Amagai M, Klaus-Kovtun V, Stanley JR: Autoantibodies against a novel epithelial cadherin in pemphigus vulgaris, a disease of cell adhesion, *Cell* 67:869–77, 1991.

Arnemann J, Spurr NK, Wheeler GN, Parker AE, Buxton RS: Chromosomal assignment of the human genes coding for the major proteins of the desmosome junction, desmoglein DGI (DSG), desmocollins DGII/III (DSC), desmoplakins DPI/II (DSP), and plakoglobin DPIII (JUP), *Genomics* 10:640–5, 1991.

Buxton RS, Wheeler GN, Pidsley SC, Marsden MD, Adams MJ et al: Mouse desmocollin (Dsc3) and desmoglein (Dsgl) genes are closely linked in the proximal region of chromosome 18, *Genomics* 21:510–16, 1994.

Simrak D, Cowley CME, Buxton RS, Arnemann J: Tandem arrangement of the closely linked desmoglein genes on human chromosome 18, *Genomics* 25:591–4, 1995.

Wang Y, Amagai M, Minoshima S, Sakai K, Green KJ et al: The human genes for desmogleins (DSG1 and DSG3) are located in a small region on chromosome 18q12, *Genomics* 20:492–5, 1994.

Desmocollin

King IA, Arnemann J, Spurr NK, Buxton RS: Cloning of the cDNA (DSC1) coding for human type 1 desmocollin and its assignment to chromosome 18, *Genomics* 18:185–94, 1993.

Troyanovsky SM, Eshkind LG, Troyanovsky RB, Leube RE, Franke WW: Contributions of cytoplasmic domains of desmosomal cadherins to desmosome assembly and intermediate filament anchorage, *Cell* 72:561–74, 1993.

6.11. Function of Cadherins

Braga VM, Yap AS: The challenges of abundance: Epithelial junctions and small GTPase signaling, *Curr Opin Cell Biol* 17:466–74, 2005.

Cowin P, Rowlands TM, Hatsell SJ: Cadherins and catenins in breast cancer, *Curr Opin Cell Biol* 17:499–508, 2005.

Junghans D, Haas IG, Kemler R: Mammalian cadherins and protocadherins: About cell death, synapses and processing, *Curr Opin Cell Biol* 17:446–52, 2005.

Lilien J, Balsamo J: The regulation of cadherin-mediated adhesion by tyrosine phosphorylation/dephosphorylation of beta-catenin, *Curr Opin Cell Biol* 17:459–65, 2005.

Gumbiner BM: Regulation of cadherin-mediated adhesion in morphogenesis, *Nat Rev Mol Cell Biol* 6:622–34, 2005.

Carthew RW: Adhesion proteins and the control of cell shape, *Curr Opin Genet Dev* 15:358–63, 2005.

Takeichi M, Abe K: Synaptic contact dynamics controlled by cadherin and catenins, *Trends Cell Biol* 15:216–21, 2005.

Payne AS, Hanakawa Y, Amagai M, Stanley JR: Desmosomes and disease: Pemphigus and bullous impetigo, *Curr Opin Cell Biol* 16(5):536–43, Oct 2004.

Getsios S, Huen AC, Green KJ: Working out the strength and flexibility of desmosomes, *Nat Rev Mol Cell Biol* 5:271–81, 2004.

Nelson WJ, Nusse R: Convergence of Wnt, beta-catenin, and cadherin pathways, *Science* 303:1483–7, 2004.

Cavallaro U, Christofori G: Cell adhesion and signalling by cadherins and Ig-CAMs in cancer, *Nat Rev Cancer* 4:118–32, 2004.

Human protein reference data base, Johns Hopkins University and the Institute of Bioinformatics, at http://www.hprd.org/protein.

6.12. Classification and Structure of Cell Surface Heparan Sulfate Proteoglycans

Syndecan 1

Alexander CM, Reichsman F, Hinkes MT, Lincecum J, Becker KA et al: Syndecan-1 is required for Wnt-1-induced mammary tumorigenesis in mice, *Nature Genet* 25:329–32, 2000.

Li Q, Park PW, Wilson CL, Parks WC: Matrilysin shedding of syndecan-1 regulates chemokine mobilization and transepithelial efflux of neutrophils in acute lung injury, *Cell* 111:635–46, 2002.

Mali M, Jaakkola P, Arvilommi AM, Jalkanen M: Sequence of human syndecan indicates a novel gene family of integral membrane proteoglycans, *J Biol Chem* 265:6884–9, 1990.

Oettinger HF, Streeter H, Lose E, Copeland NG, Gilbert DJ et al: Chromosome mapping of the murine syndecan gene, *Genomics* 11:334–8, 1991.

Reizes O, Lincecum J, Wang Z, Goldberger O, Huang L et al: Transgenic expression of syndecan-1 uncovers a physiological control of feeding behavior by syndecan-3, *Cell* 106:105–16, 2001.

Glypican

Karumanchi SA, Jha V, Ramchandran R, Karihaloo A, Tsiokas L et al: Cell surface glypicans are low-affinity endostatin receptors, *Mol Cell* 7:811–22, 2001.

Vermeesch JR, Mertens G, David G, Marynen P: Assignment of the human glypican gene (GPC1) to 2q35-q37 by fluorescence in situ hybridization, *Genomics* 25:327–9, 1995.

Human protein reference data base, Johns Hopkins University and the Institute of Bioinformatics, at http://www.hprd.org/protein.

6.13. Function of Cell Surface Heparan Sulfate Proteoglycans

Turnbull J, Powell A, Guimond S: Heparan sulfate: Decoding a dynamic multifunctional cell regulator, *Trends Cell Biol* 11:75–82, 2001.

Rosenberg RD, Shworak NW, Liu J, Schwartz JJ, Zhang L: Heparan sulfate proteoglycans of the cardiovascular system. Specific structures emerge but how is synthesis regulated? *J Clin Invest* 100(Suppl 11):S67–75, 1997.

6.14. Classification and Structure Integrins

Integrin α1

Briesewitz R, Epstein MR, Marcantonio EE: Expression of native and truncated forms of the human integrin alpha-1 subunit, *J Biol Chem* 268:2989–96, 1993.

Douville P, Seldin MF, Carbonetto S: Genetic mapping of the integrin alpha-1 gene (Vla1) to mouse chromosome 13, *Genomics* 14:503–5, 1992.

Ekholm E, Hankenson KD, Uusitalo H, Hiltunen A, Gardner H et al: Diminished callus size and cartilage synthesis in alpha-1 beta-1 integrin-deficient mice during bone fracture healing, *Am J Pathol* 160:1779–85, 2002.

Zemmyo M, Meharra EJ, Kuhn K, Creighton-Achermann L, Lotz M: Accelerated, aging-dependent development of osteoarthritis in alpha-1 integrin-deficient mice, *Arthritis Rheum* 48:2873–80, 2003.

Integrin α2

Nieuwenhuis HK, Akkerman JWN, Houdijk WP, Sixma JJ: Human blood platelets showing no response to collagen fail to express surface glycoprotein Ia, *Nature* 318:470–2, 1985.

Santoso S, Kalb R, Walka M, Kiefel V, Mueller-Eckhardt C et al: The human platelet alloantigens Br(a) and Br(b) are associated with a single amino acid polymorphism on glycoprotein Ia (integrin subunit alpha-2), *J Clin Invest* 92:2427–32, 1993.

Santoso S, Kunicki TJ, Kroll H, Haberbosch W, Gardemann A: Association of the platelet glycoprotein Ia C807T gene polymorphism with nonfatal myocardial infarction in younger patients, *Blood* 93:2449–53, 1999.

Takada Y, Hemler ME: The primary structure of the VLA-2/collagen receptor alpha-2 subunit (platelet GPIa): Homology to other integrins and the presence of a possible collagen-binding domain, *J Cell Biol* 109:397–407, 1989.

von Beckerath N, Koch W, Mehilli J, Bottiger C, Schomig A et al: Glycoprotein Ia gene C807T polymorphism and risk for major adverse cardiac events within the first 30 days after coronary artery stenting, *Blood* 95:3297–301, 2000.

Woods VL Jr, Pischel KD, Avery ED, Bluestein HG: Antigenic polymorphism of human very late activation protein-2 (platelet glycoprotein Ia-IIa): Platelet alloantigen Hc(a), *J Clin Invest* 83:978–85, 1989.

Integrin α3

Dulabon L, Olson EC, Taglienti MG, Eisenhuth S, McGrath B et al: Reelin binds alpha-3-beta-1 integrin and inhibits neuronal migration, *Neuron* 27:33–44, 2000.

Jones SD, van der Flier A, Sonnenberg A: Genomic organization of the human alpha-3 integrin subunit gene, *Biochem Biophys Res Commun* 248:896–8, 1998.

Takada Y, Murphy E, Pil P, Chen C, Ginsberg MH et al: Molecular cloning and expression of the cDNA for alpha-3 subunit of human alpha-3/beta-1 (VLA-3), an integrin receptor for fibronectin, laminin, and collagen, *J Cell Biol* 115:257–66, 1991.

Tsuji T, Hakomori S, Osawa T: Identification of human galactoprotein b3, an oncogenic transformation-induced membrane glycoprotein, as VLA-3 alpha subunit: the primary structure of human integrin alpha-3, *J Biochem* 109:659–65, 1991.

Integrin α4

Fernandez-Ruiz E, Pardo-Manuel de Villena F, Rubio MA, Corbi AL, Rodriguez de Cordoba S et al: Mapping of the human VLA-alpha-4 gene to chromosome 2q31-q32, *Eur J Immun* 22: 587–90, 1992.

Garmy-Susini B, Jin H, Zhu Y, Sung RJ, Hwang R, et al: Integrin alpha-4-beta-1–VCAM-1–mediated adhesion between endothelial and mural cells is required for blood vessel maturation, *J Clin Invest* 115:1542–51, 2005.

Lu TT, Cyster JG: Integrin-mediated long-term B cell retention in the splenic marginal zone, *Science* 297:409–12, 2002.

Matsunaga T, Takemoto N, Sato T, Takimoto R, Tanaka I et al: Interaction between leukemic-cell VLA-4 and stromal fibronectin is a decisive factor for minimal residual disease of acute myelogenous leukemia, *Nature Med* 9:1158–65, 2003.

Rosemblatt M, Vuillet-Gaugler MH, Leroy C, Coulombel L: Coexpression of two fibronectin receptors, VLA-4 and VLA-5, by immature human erythroblastic precursor cells, *J Clin Invest* 87:6–11, 1991.

Zhang Z, Vekemans S, Aly MS, Jaspers M, Marynen P et al: The gene for the alpha-4 subunit of the VLA-4 integrin maps to chromosome 2q31-32, *Blood* 78:2396–9, 1991.

Integrin α5

Argraves WS, Pytela R, Suzuki S, Millan JL, Pierschbacher MD et al: cDNA sequences from the alpha subunit of the fibronectin receptor predict a transmembrane domain and a short cytoplasmic peptide, *J Biol Chem* 261:12922–4, 1986.

Argraves WS, Suzuki S, Arai H, Thompson K, Pierschbacher MD et al: Amino acid sequence of the human fibronectin receptor, *J Cell Biol* 105:1183–90, 1987.

McCarty JH, Monahan-Earley RA, Brown LF, Keller M, Gerhardt H et al: Defective associations between blood vessels and brain parenchyma lead to cerebral hemorrhage in mice lacking alpha-v integrins, *Mol Cell Biol* 22:7667–77, 2002.

Sosnoski D, Emanuel BS, Hawkins AL, van Tuinen P, Ledbetter DH, et al: Chromosomal localization of the genes for the vitronectin and fibronectin receptors alpha-subunits and for platelet glycoproteins IIb and IIIa, *J Clin Invest* 81:1993–8, 1988.

Spurr NK, Rooke L: Confirmation of the assignment of the vitronectin (VNRA) and fibronectin (FNRA) receptor alpha-subunits, *Ann Hum Genet* 55:217–23, 1991.

Integrin α6

Georges-Labouesse E, Messaddeq N, Yehia G, Cadalbert L, Dierich A et al: Absence of integrin alpha-6 leads to epidermolysis bullosa and neonatal death in mice, *Nature Genet* 13:370–3, 1996.

Hogervorst F, Kuikman I, Geurts van Kessel A, Sonnenberg A: Molecular cloning of the human alpha-6 integrin subunit: alternative splicing of alpha-6 mRNA and chromosomal localization of the alpha-6 and beta-4 genes, *Eur J Biochem* 199:425–33, 1991.

Ruzzi L, Gagnoux-Palacios L, Pinola M, Belli S, Meneguzzi G et al: A homozygous mutation in the integrin alpha-6 gene in junctional epidermolysis bullosa with pyloric atresia, *J Clin Invest* 99:2826–31, 1997.

Tamura RN, Rozzo C, Starr L, Chambers J, Reichardt LF et al: Epithelial integrin alpha-6/beta-4: complete primary structure of alpha-6 and variant forms of beta-4, *J Cell Biol* 111:1593–604, 1990.

Integrin α7

Hayashi YK, Chou FL, Engvall E, Ogawa M, Matsuda C et al: Mutations in the integrin alpha-7 gene cause congenital myopathy, *Nature Genet* 19:94–7, 1998.

Mayer U, Saher G, Fassler R, Bornemann A, Echtermeyer F et al: Absence of integrin alpha-7 causes a novel form of muscular dystrophy, *Nature Genet* 17:318–23, 1997.

Wang W, Wu W, Desai T, Ward DC, Kaufman SJ: Localization of the alpha-7 integrin gene (ITGA7) on human chromosome 12q13: Clustering of integrin and Hox genes implies parallel evolution of these gene families, *Genomics* 26:563–70, 1995.

Cunningham SA, Rodriguez JM, Arrate MP, Tran TM, Brock TA: JAM2 interacts with alpha-4/beta-1: Facilitation by JAM3, *J Biol Chem* 277:27589–92, 2002.

Garmy-Susini B, Jin H, Zhu Y, Sung RJ, Hwang R et al: Integrin alpha-4-beta-1–VCAM-1–mediated adhesion between endothelial and mural cells is required for blood vessel maturation, *J Clin Invest* 115:1542–51, 2005.

Goodfellow PJ, Nevanlinna HA, Gorman P, Sheer D, Lam G et al: Assignment of the gene encoding the beta-subunit of the human fibronectin receptor (beta-FNR) to chromosome 10p11.2, *Ann Hum Genet* 53:15–22, 1989.

Hynes RO: Integrins: a family of cell surface receptors, *Cell* 48:549–54, 1987.

Johansson S, Forsberg E, Lundgren B: Comparison of fibronectin receptors from rat hepatocytes and fibroblasts, *J Biol Chem* 262:7819–24, 1987.

Lu TT, Cyster JG: Integrin-mediated long-term B cell retention in the splenic marginal zone, *Science* 297:409–12, 2002.

McDowall A, Inwald D, Leitinger B, Jones A, Liesner R et al: A novel form of integrin dysfunction involving beta-1, beta-2, and beta-3 integrins, *J Clin Invest* 111:51–60, 2003.

Tadokoro S, Shattil SJ, Eto K, Tai V, Liddington RC et al: Talin binding to integrin beta tails: A final common step in integrin activation, *Science* 302:103–6, 2003.

Integrin αv

Bader BL, Rayburn H, Crowley D, Hynes RO: Extensive vasculogenesis, angiogenesis, and organogenesis precede lethality in mice lacking all alpha-V integrins, *Cell* 95:507–19, 1998.

Faccio R, Takeshita S, Zallone A, Ross FP, Teitelbaum SL: c-Fms and the alpha-V-beta-3 integrin collaborate during osteoclast differentiation, *J Clin Invest* 111:749–58, 2003.

Fernandez-Ruiz E, de Villena FPM, Rodriguez de Cordoba S, Sanchez-Madrid F: Regional localization of the human vitronectin receptor alpha-subunit gene (VNRA) to chromosome 2q31-q32, *Cytogenet Cell Genet* 62:26–8, 1993.

Morris DG, Huang X, Kaminski N, Wang Y, Shapiro SD et al: Loss of integrin alpha-v-beta-6-mediated TGF-beta activation causes Mmp12-dependent emphysema, *Nature* 422:169–73, 2003.

Munger JS, Huang X, Kawakatsu H, Griffiths MJ, Dalton SL et al: The integrin alpha v beta 6 binds and activates latent TGF beta 1: A mechanism for regulating pulmonary inflammation and fibrosis, *Cell* 96:319–28, 1999.

Sims MA, Field SD, Barnes MR, Shaikh N, Ellington K et al: Cloning and characterisation of ITGAV, the genomic sequence for human cell adhesion protein (vitronectin) receptor alpha subunit, CD51, *Cytogenet Cell Genet* 89:268–71, 2000.

Wang X, Huang DY, Huong SM, Huang ES: Integrin alpha-v-beta-3 is a coreceptor for human cytomegalovirus, *Nature Med* 11:515–21, 2005.

Xiong JP, Stehle T, Diefenbach B, Zhang R, Dunker R et al: Crystal structure of the extracellular segment of integrin alpha-V-beta-3, *Science* 294:339–45, 2001.

Xiong JP, Stehle T, Zhang R, Joachimiak A, Frech M et al: Crystal structure of the extracellular segment of integrin alpha-V-beta-3 in complex with an Arg-Gly-Asp ligand, *Science* 296:151–5, 2002.

Integrin β1

Arregui C, Pathre P, Lilien J, Balsamo J: The nonreceptor tyrosine kinase Fer mediates cross-talk between N-cadherin and beta-1-integrins, *J Cell Biol* 149:1263–73, 2000.

Aszodi A, Hunziker EB, Brakebusch C, Fassler R: Beta-1 integrins regulate chondrocyte rotation, G1 progression, and cytokinesis, *Genes Dev* 17:2465–79, 2003.

Garmy-Susini B, Jin H, Zhu Y, Sung RJ, Hwang R et al: Integrin alpha-4-beta-1–VCAM-1–mediated adhesion between endothelial and mural cells is required for blood vessel maturation, *J Clin Invest* 115:1542–51, 2005.

Giuffra LA, Lichter P, Wu J, Kennedy JL, Pakstis AJ et al: Genetic and physical mapping and population studies of a fibronectin receptor beta-subunit-like sequence on human chromosome 19, *Genomics* 8:340–6, 1990.

Goodfellow PJ, Nevanlinna HA, Gorman P, Sheer D, Lam G et al: Assignment of the gene encoding the beta-subunit of the human fibronectin receptor (beta-FNR) to chromosome 10p11.2, *Ann Hum Genet* 53:15–22, 1989.

Lu TT, Cyster JG: Integrin-mediated long-term B cell retention in the splenic marginal zone, *Science* 297:409–12, 2002.

McDowall A, Inwald D, Leitinger B, Jones A, Liesner R et al: A novel form of integrin dysfunction involving beta-1, beta-2, and beta-3 integrins, *J Clin Invest* 111:51–60, 2003.

Tadokoro S, Shattil SJ, Eto K, Tai V, Liddington RC et al: Talin binding to integrin beta tails: A final common step in integrin activation, *Science* 302:103–6, 2003.

Human protein reference data base, Johns Hopkins University and the Institute of Bioinformatics, at http://www.hprd.org/protein.

6.15. Function of Integrins

Beckerle MC, ed, *Cell Adhesion* (in series of Frontiers in Molecular Biology, Hames BD, Glover DM, series eds), Oxford University Press, New York, 2001.

Bennett JS: Structure and function of the platelet integrin alphaIIb beta3, *J Clin Invest* 115:3363–9, 2005.

Watt FM: Role of integrins in regulating epidermal adhesion, growth and differentiation, *EMBO J* 21:3919–26, 2002.

Arnaout MA, Mahalingam B, Xiong JP: Integrin structure, allostery, and bidirectional signaling, *Annu Rev Cell Dev Biol* 21:381–410, 2005.

Ginsberg MH, Partridge A, Shattil SJ: Integrin regulation, *Curr Opin Cell Biol* 17:509–16, 2005.

Millward-Sadler SJ, Salter DM: ntegrin-dependent signal cascades in chondrocyte mechanotransduction, *Ann Biomed Eng* 32:435–46, 2004.

Friedl P: Prespecification and plasticity: Shifting mechanisms of cell migration, *Curr Opin Cell Biol* 16:14–23, 2004.

Calderwood DA: Integrin activation, *J Cell Sci* 117(Pt 5):657–66, 2004.

Carman CV, Springer TA: Integrin avidity regulation: Are changes in affinity and conformation underemphasized? *Curr Opin Cell Biol* 15:547–56, 2003.

Humphries MJ, McEwan PA, Barton SJ, Buckley PA, Bella J et al: Integrin structure: Heady advances in ligand binding, but activation still makes the knees wobble, *Trends Biochem Sci* 28:313–20, 2003.

Humphries MJ, Symonds EJ, Mould AP: Mapping functional residues onto integrin crystal structures, *Curr Opin Struct Biol* 13:236–43, 2003.

Hynes RO: Integrins: Bidirectional, allosteric signaling machines, *Cell* 110:673–87, 2002.

Takagi J, Springer TA: Integrin activation and structural rearrangement, *Immunol Rev* 186:141–63, 2002.

Liddington RC, Ginsberg MH: Integrin activation takes shape, *J Cell Biol* 158:833–9, 2002.

Shimaoka M, Takagi J, Springer TA: Conformational regulation of integrin structure and function, *Annu Rev Biophys Biomol Struct* 31:485–516, 2002.

6.16. Apoptosis

Fas Ligand

Kuwana T, Newmeyer DD: Bcl-2-family proteins and the role of mitochondria in apoptosis, *Curr Opin Cell Biol* 15:691–9, 2003.

Bellgrau D, Gold D, Selawry H, Moore J, Franzusoff A et al: A role for CD95 ligand in preventing graft rejection, *Nature* 377:630–2, 1995.

Borges VM, Falcao H, Leite-Junior JH, Alvim L, Teixeira GP et al: Fas ligand triggers pulmonary silicosis, *J Exp Med* 194:155–63, 2001.

D'Alessio A, Riccioli A, Lauretti P, Padula F, Muciaccia B et al: Testicular FasL is expressed by sperm cells, *Proc Natl Acad Sci USA* 98:3316–21, 2001.

Demjen D, Klussmann S, Kleber S, Zuliani C, Stieltjes B et al: Neutralization of CD95 ligand promotes regeneration and functional recovery after spinal cord injury, *Nature Med* 10:389–95, 2004.

Hahne M, Rimoldi D, Schroter M, Romero P, Schreier M et al: Melanoma cell expression of Fas (Apo-1/CD95) ligand: Implications for tumor immune escape, *Science* 274:1363–6, 1996.

Hill LL, Ouhtit A, Loughlin SM, Kripke ML, Ananthaswamy, HN et al: Fas ligand: A sensor for DNA damage critical in skin cancer etiology, *Science* 285:898–900, 1999.

Ma Y, Liu H, Tu-Rapp H, Thiesen HJ, Ibrahim SM et al: Fas ligation on macrophages enhances IL-1R1-Toll-like receptor 4 signaling and promotes chronic inflammation, *Nature Immun* 5:380–7, 2004.

Pestano GA, Zhou Y, Trimble LA, Daley J, Weber GF et al: Inactivation of misselected CD8 T cells by CD8 gene methylation and cell death, *Science* 284:1187–91, 1999.

Stuart PM, Griffith TS, Usui N, Pepose J, Yu X et al: CD95 ligand (FasL)-induced apoptosis is necessary for corneal allograft survival, *J Clin Invest* 99:396–402, 1997.

Suda T, Takahashi T, Golstein P, Nagata S: Molecular cloning and expression of the Fas ligand, a novel member of the tumor necrosis factor family, *Cell* 75:1169–78, 1993.

Takahashi T, Tanaka M, Brannan CI, Jenkins NA, Copeland NG et al: Generalized lymphoproliferative disease in mice, caused by a point mutation in the Fas ligand, *Cell* 76:969–76, 1994.

Viard I, Wehrli P, Bullani R, Schneider P, Holler N et al: Inhibition of toxic epidermal necrolysis by blockade of CD95 with human intravenous immunoglobulin, *Science* 282:490–3, 1998.

Volpert OV, Zaichuk T, Zhou W, Reiher F, Ferguson TA et al: Inducer-stimulated Fas targets activated endothelium for destruction by anti-angiogenic thrombospondin-1 and pigment epithelium-derived factor, *Nature Med* 8:349–57, 2002.

Wu J, Wilson J, He J, Xiang L, Schur PH et al: Fas ligand mutation in a patient with systemic lupus erythematosus and lymphoproliferative disease, *J Clin Invest* 98:1107–13, 1996.

FAS Receptor

Brunner T, Mogil RJ, LaFace D, Yoo NJ, Mahboubi A et al: Cell-autonomous Fas (CD95)/Fas-ligand interaction mediates activation-induced apoptosis in T-cell hybridomas, *Nature* 373: 441–4, 1995.

Clementi R, Dagna L, Dianzani U, Dupre L, Dianzani I et al: Inherited perforin and Fas mutations in a patient with autoimmune lymphoproliferative syndrome and lymphoma, *New Engl J Med* 351:1419–24, 2004.

Desbarats J, Birge RB, Mimouni-Rongy M, Weinstein DE, Palerme JS et al: Fas engagement induces neurite growth through ERK activation and p35 upregulation, *Nature Cell Biol* 5:118–25, 2003.

Dhein J, Walczak H, Baumler C, Debatin KM, Krammer PH: Autocrine T-cell suicide mediated by APO-1/(Fas/CD95), *Nature* 373:438–41, 1995.

Fisher GH, Rosenberg FJ, Straus SE, Dale JK, Middelton LA et al: Dominant interfering Fas gene mutations impair apoptosis in a human autoimmune lymphoproliferative syndrome, *Cell* 81:935–46, 1995.

Grassme H, Kirschnek S, Riethmueller J, Riehle A, von Kurthy G et al: CD95/CD95 ligand interactions on epithelial cells in host defense to Pseudomonas aeruginosa, *Science* 290:527–30, 2000.

Hueber AO: CD95: More than just a death factor? *Nature Cell Biol* 2: E23–5, 2000.

Hueber AO, Zornig M, Lyon D, Suda T, Nagata S et al: Requirement for the CD95 receptor-ligand pathway in c-Myc-induced apoptosis, *Science* 278:1305–9, 1997.

Inazawa J, Itoh N, Abe T, Nagata S: Assignment of the human Fas antigen gene (FAS) to 10q24.1, *Genomics* 14:821–2, 1992.

Itoh N, Yonehara S, Ishii A, Yonehara M, Mizushima SI et al: The polypeptide encoded by the cDNA for human cell surface antigen Fas can mediate apoptosis, *Cell* 66:233–43, 1991.

Ju ST, Panka DJ, Cui H, Ettinger R, El-Khatib M et al: Fas(CD95)/FasL interactions required for programmed cell death after T-cell activation, *Nature* 373:444–8, 1995.

Lepple-Wienhues A, Belka C, Laun T, Jekle A, Walter B et al: Stimulation of CD95 (Fas) blocks T lymphocyte calcium channels through sphingomyelinase and sphingolipids, *Proc Natl Acad Sci USA* 96:13795–800, 1999.

Ma Y, Liu H, Tu-Rapp H, Thiesen HJ, Ibrahim SM et al: Fas ligation on macrophages enhances IL-1R1-Toll-like receptor 4 signaling and promotes chronic inflammation, *Nature Immun* 5:380–7, 2004.

Mannick JB, Hausladen A, Liu L, Hess DT, Zeng M et al: Fas-induced caspase denitrosylation, *Science* 284:651–4, 1999.

Pestano GA, Zhou Y, Trimble LA, Daley J, Weber GF et al: Inactivation of misselected CD8 T cells by CD8 gene methylation and cell death, *Science* 284:1187–91, 1999.

Rieux-Laucat F, Le Deist F, Hivroz C, Roberts IAG, Debatin KM et al: Mutations in Fas associated with human lymphoproliferative syndrome and autoimmunity, *Science* 268:1347–9, 1995.

Savinov AY, Tcherepanov A, Green EA, Flavell RA, Chervonsky AV: Contribution of Fas to diabetes development, *Proc Natl Acad Sci USA* 100:628–32, 2003.

Siegel RM, Frederiksen JK, Zacharias DA, Chan FKM, Johnson M et al: Fas preassociation required for apoptosis signaling and dominant inhibition by pathogenic mutations, *Science* 288:2354–7, 2000.

Viard I, Wehrli P, Bullani R, Schneider P, Holler N et al: Inhibition of toxic epidermal necrolysis by blockade of CD95 with human intravenous immunoglobulin, *Science* 282:490–3, 1998.

Volpert OV, Zaichuk T, Zhou W, Reiher F, Ferguson TA et al: Inducer-stimulated Fas targets activated endothelium for destruction by anti-angiogenic thrombospondin-1 and pigment epithelium-derived factor, *Nature Med* 8:349–57, 2002.

Watanabe-Fukunaga R, Brannan CI, Copeland NG, Jenkins NA, Nagata S: Lymphoproliferation disorder in mice explained by defects in Fas antigen that mediates apoptosis, *Nature* 356:314–17, 1992.

Watanabe-Fukunaga R, Brannan CI, Itoh N, Yonehara S, Copeland NG, et al: The cDNA structure, expression, and chromosomal assignment of the mouse Fas antigen, *J Immun* 148:1274–9, 1992.

FADD

Balachandran S, Thomas E, Barber GN: A FADD-dependent innate immune mechanism in mammalian cells, *Nature* 432:401–5, 2004.

Chinnaiyan AM, O'Rourke K, Tewari M, Dixit VM: FADD, a novel death domain-containing protein, interacts with the death domain of Fas and initiates apoptosis, *Cell* 81:505–12, 1995.

Kabra NH, Kang C, Hsing LC, Zhang J, Winoto A: T cell-specific FADD-deficient mice: FADD is required for early T cell development, *Proc Natl Acad Sci USA* 98:6307–12, 2001.

Kim PKM, Dutra AS, Chandrasekharappa SC, Puck JM: Genomic structure and mapping of human FADD, an intracellular mediator of lymphocyte apoptosis, *J Immun* 157:5461–6, 1996.

Yeh WC, de la Pompa JL, McCurrach ME, Shu HB, Elia AJ, et al: FADD: Essential for embryo development and signaling from some, but not all, inducers of apoptosis, *Science* 279:1954–8, 1998.

Zhang J, Cado D, Chen A, Kabra NH, Winoto A: Fas-mediated apoptosis and activation-induced T-cell proliferation are defective in mice lacking FADD/Mort1, *Nature* 392:296–300, 1998.

TNF α

Aggarwal BB, Eessalu TE, Hass PE: Characterization of receptors for human tumour necrosis factor and their regulation by gamma-interferon, *Nature* 318:665–7, 1985.

Beattie EC, Stellwagen D, Morishita W, Bresnahan JC, Ha BK et al: Control of synaptic strength by glial TNF-alpha, *Science* 295:2282–5, 2002.

Beutler B, Krochin N, Milsark IW, Luedke C, Cerami A: Control of cachectin (tumor necrosis factor) synthesis: Mechanisms of endotoxin resistance, *Science* 232:977–80, 1986.

Bouwmeester T, Bauch A, Ruffner H, Angrand PO, Bergamini G et al: A physical and functional map of the human TNF-alpha/NF-kappa-B signal transduction pathway, *Nature Cell Biol* 6:97–105, 2004.

Brenner DA, O'Hara M, Angel P, Chojkier M, Karin M: Prolonged activation of JUN and collagenase genes by tumour necrosis factor-alpha, *Nature* 337:661–3, 1989.

Bruce AJ, Boling W, Kindy MS, Peschon J, Kraemer PJ et al: Altered neuronal and microglial responses to excitotoxic and ischemic brain injury in mice lacking TNF receptors, *Nature Med* 2:788–94, 1996.

Kamata H, Honda S, Maeda S, Chang L, Hirata H et al: Reactive oxygen species promote TNF-alpha-induced death and sustained JNK activation by inhibiting MAP kinase phosphatases, *Cell* 120:649–61, 2005.

Knight JC, Udalova I, Hill AVS, Greenwood BM, Peshu N et al: A polymorphism that affects OCT-1 binding to the TNF promoter region is associated with severe malaria, *Nature Genet* 22:145–50, 1999.

Marino MW, Dunn A, Grail D, Inglese M, Noguchi Y et al: Characterization of tumor necrosis factor-deficient mice, *Proc Natl Acad Sci USA* 94:8093–8, 1997.

McGuire W, Hill AVS, Allsopp CEM, Greenwood BM, Kwiatkowski D: Variation in the TNF-alpha promoter region associated with susceptibility to cerebral malaria, *Nature* 371:508–11, 1994.

Obeid LM, Linardic CM, Karolak LA, Hannun YA: Programmed cell death induced by ceramide, *Science* 259:1769–71, 1993.

Old LJ: Tumor necrosis factor (TNF), *Science* 230:630–2, 1985.

Pennica D, Nedwin GE, Hayflick JS, Seeburg PH, Derynck R et al: Human tumour necrosis factor: Precursor structure, expression and homology to lymphotoxin, *Nature* 312:724–9, 1984.

Steed PM, Tansey MG, Zalevsky J, Zhukovsky EA, Desjarlais JR et al: Inactivation of TNF signaling by rationally designed dominant-negative TNF variants, *Science* 301:1895–8, 2003.

Uysal KT, Wiesbrock SM, Marino MW, Hotamisligil GS: Protection from obesity-induced insulin resistance in mice lacking TNF-alpha function, *Nature* 389:610–14, 1997.

Van Ostade X, Vandenabeele P, Everaerdt B, Loetscher H, Gentz R et al: Human TNF mutants with selective activity on the p55 receptor, *Nature* 361:266–9, 1993.

Wang AM, Creasey AA, Ladner MB, Lin LS, Strickler J et al: Molecular cloning of the complementary DNA for human tumor necrosis factor, *Science* 228:149–54, 1985.

TNF Receptor

Baker E, Chen LZ, Smith CA, Callen DF, Goodwin R et al: Chromosomal location of the human tumor necrosis factor receptor genes, *Cytogenet Cell Genet* 57:117–8, 1991.

Brockhaus M, Schoenfeld HJ, Schlaeger EJ, Hunziker W, Lesslauer W et al: Identification of two types of tumor necrosis factor receptors on human cell lines by monoclonal antibodies, *Proc Natl Acad Sci USA* 87:3127–31, 1990.

Bruce AJ, Boling W, Kindy MS, Peschon J, Kraemer PJ et al: Altered neuronal and microglial responses to excitotoxic and ischemic brain injury in mice lacking TNF receptors, *Nature Med* 2:788–94, 1996.

Castellino AM, Parker GJ, Boronenkov IV, Anderson RA, Chao MV: A novel interaction between the juxtamembrane region of the p55 tumor necrosis factor receptor and phosphatidylinositol-4-phosphate 5-kinase, *J Biol Chem* 272:5861–70, 1997.

Chan FKM, Chun HJ, Zheng L, Siegel RM, Bui KL et al: A domain in TNF receptors that mediates ligand-independent receptor assembly and signaling, *Science* 288:2351–4, 2000.

Fuchs P, Strehl S, Dworzak M, Himmler A, Ambros PF: Structure of the human TNF receptor 1 (p60) gene (TNFR1) and localization to chromosome 12p13, *Genomics* 13:219–24, 1992.

Gray PW, Barrett K, Chantry D, Turner M, Feldmann M: Cloning of human tumor necrosis factor (TNF) receptor cDNA and expression of recombinant soluble TNF-binding protein, *Proc Natl Acad Sci USA* 87:7380–4, 1990.

Loetscher H, Pan YCE, Lahm HW, Gentz R, Brockhaus M et al: Molecular cloning and expression of the human 55 kd tumor necrosis factor receptor, *Cell* 61:351–9, 1990.

McDermott MF, Aksentijevich I, Galon J, McDermott EM, Ogunkolade BW et al: Germline mutations in the extracellular domains of the 55 kDa TNF receptor, TNFR1, define a family of dominantly inherited autoinflammatory syndromes, *Cell* 97:133–44, 1999.

Micheau O, Tschopp J: Induction of TNF receptor I-mediated apoptosis via two sequential signaling complexes, *Cell* 114:181–90, 2003.

Rothe J, Lesslauer W, Lotscher H, Lang Y, Koebel P et al: Mice lacking the tumour necrosis factor receptor 1 are resistant to TNF-mediated toxicity but highly susceptible to infection by Listeria monocytogenes, *Nature* 364:798–802, 1993.

Schall TJ, Lewis M, Koller KJ, Lee A, Rice GC et al: Molecular cloning and expression of a receptor for human tumor necrosis factor, *Cell* 61:361–70, 1990.

Smith CA, Davis T, Anderson D, Solam L, Beckmann MP et al: A receptor for tumor necrosis factor defines an unusual family of cellular and viral proteins, *Science* 248:1019–23, 1990.

Tartaglia LA, Ayres TM, Wong GHW, Goeddel DV: A novel domain within the 55 kd TNF receptor signals cell death, *Cell* 74:845–53, 1993.

TRADD

Hsu H, Xiong J, Goeddel DV: The TNF receptor 1-associated protein TRADD signals cell death and NF-kappa-B activation, *Cell* 81:495–504, 1995.

Park YC, Ye H, Hsia C, Segal D, Rich RL et al: A novel mechanism of TRAF signaling revealed by structural and functional analyses of the TRADD-TRAF2 interaction, *Cell* 101:777–87, 2000.

Scheuerpflug CG, Lichter P, Debatin KM, Mincheva A: Assignment of TRADD to human chromosome band 16q22 by in situ hybridization, *Cytogenet Cell Genet* 92:347–8, 2001.

Wesemann DR, Qin H, Kokorina N, Benveniste EN: TRADD interacts with STAT1-alpha and influences interferon-gamma signaling, *Nature Immun* 5:199–207, 2004.

Caspase 8

Boldin MP, Goncharov TM, Goltsev YV, Wallach D: Involvement of MACH, a novel MORT1/FADD-interacting protease, in Fas/APO-1- and TNF receptor-induced cell death, *Cell* 85:803–15, 1996.

Chun HJ, Zheng L, Ahmad M, Wang J, Speirs CK et al: Pleiotropic defects in lymphocyte activation caused by caspase-8 mutations lead to human immunodeficiency, *Nature* 419:395–9, 2002.

Fernandes-Alnemri T, Armstrong RC, Krebs J, Srinivasula SM, Wang L et al: In vitro activation of CPP32 and Mch3 by Mch4, a novel human apoptotic cysteine protease containing two FADD-like domains, *Proc Natl Acad Sci USA* 93:7464–9, 1996.

Gervais FG, Singaraja R, Xanthoudakis S, Gutekunst CA, Leavitt BR et al: Recruitment and activation of caspase-8 by the huntingtin-interacting protein Hip-1 and a novel partner Hippi, *Nature Cell Biol* 4:95–105, 2002.

Grenet J, Teitz T, Wei T, Valentine V, Kidd VJ: Structure and chromosome localization of the human CASP8 gene, *Gene* 226:225–32, 1999.

Hadano S, Yanagisawa Y, Skaug J, Fichter K, Nasir J et al: Cloning and characterization of three novel genes, ALS2CR1, ALS2CR2, and ALS2CR3, in the juvenile amyotrophic lateral sclerosis (ALS2) critical region at chromosome 2q33-q34: Candidate genes for ALS2, *Genomics* 71:200–13, 2001.

Ikeda H, Yamaguchi M, Sugai S, Aze Y, Narumiya S et al: Expanded polyglutamine in the Machado-Joseph disease protein induces cell death in vitro and in vivo, *Nature Genet* 13:196–202, 1996.

Kischkel FC, Kioschis P, Weitz S, Poustka A, Lichter P et al: Assignment of CASP8 to human chromosome band 2q33-q34 and Casp8 to the murine syntenic region on chromosome 1B-proximal C by in situ hybridization, *Cytogenet Cell Genet* 82:95–6, 1998.

Muzio M, Chinnaiyan AM, Kischkel FC, O'Rourke K, Shevchenko A et al: FLICE, a novel FADD-homologous ICE/CED-3-like protease, is recruited to the CD95 (Fas/APO-1) death-inducing signaling complex, *Cell* 85:817–27, 1996.

Su H, Bidere N, Zheng L, Cubre A, Sakai K et al: Requirement for caspase-8 in NF-kappa-B activation by antigen receptor, *Science* 307:1465–8, 2005.

Yu L, Alva A, Su H, Dutt P, Freundt E et al: Regulation of an ATG7-beclin 1 program of autophagic cell death by caspase-8, *Science* 304:1500–2, 2004.

Zender L, Hutker S, Liedtke C, Tillmann HL, Zender S, et al: Caspase 8 small interfering RNA prevents acute liver failure in mice, *Proc Natl Acad Sci USA* 100:7797–802, 2003.

BID

Chou JJ, Li H, Salvesen GS, Yuan J, Wagner G: Solution structure of BID, an intracellular amplifier of apoptotic signaling, *Cell* 95:615–24, 1999.

Footz TK, Birren B, Minoshima S, Asakawa S, Shimizu N et al: The gene for death agonist BID maps to the region of human 22q11.2 duplicated in cat eye syndrome chromosomes and to mouse chromosome 6, *Genomics* 51:472–5, 1998.

Li H, Zhu H, Xu C, Yuan J: Cleavage of BID by caspase 8 mediates the mitochondrial damage in the Fas pathway of apoptosis, *Cell* 94:491–501, 1998.

Luo X, Budihardjo I, Zou H, Slaughter C, Wang X: Bid, a Bcl2 interacting protein, mediates cytochrome c release from mitochondria in response to activation of cell surface death receptors, *Cell* 94:481–90, 1998.

McDonnell JM, Fushman D, Milliman CL, Korsmeyer SJ, Cowburn D: Solution structure of the proapoptotic molecule BID: A structural basis for apoptotic agonists and antagonists, *Cell* 96:625–34, 1999.

Sax JK, Fei P, Murphy ME, Bernhard E, Korsmeyer SJ et al: BID regulation by p53 contributes to chemosensitivity, *Nature Cell Biol* 4:842–9, 2002.

Wang K, Yin XM, Chao DT, Milliman CL, Korsmeyer SJ: BID: A novel BH3 domain-only death agonist, *Genes Dev* 10:2859–69, 1996.

Wang K, Yin XM, Copeland NG, Gilbert DJ, Jenkins NA et al: BID, a proapoptotic BCL-2 family member, is localized to mouse chromosome 6 and human chromosome 22q11, *Genomics* 53:235–8, 1998.

Wei MC, Zong WX, Cheng EHY, Lindsten T, Panoutsakopoulou V, et al: Proapoptotic BAX or BAK: A requisite gateway to mitochondrial dysfunction and death, *Science* 292:727–30, 2001.

Yin XM, Wang K, Gross A, Zhao Y, Zinkel S et al: Bid-deficient mice are resistant to Fas-induced hepatocellular apoptosis, *Nature* 400:886–91, 1999.

Zha J, Weiler S, Oh KJ, Wei MC, Korsmeyer SJ: Posttranslational N-myristoylation of BID as a molecular switch for targeting mitochondria and apoptosis, *Science* 290:1761–5, 2000.

Cytochrome c

Boehning D, Patterson RL, Sedaghat L, Glebova NO, Kurosaki T et al: Cytochrome c binds to inositol (1,4,5) trisphosphate receptors, amplifying calcium-dependent apoptosis, *Nature Cell Biol* 5:1051–61, 2003.

Evans MJ, Scarpulla RC: The human somatic cytochrome c gene: Two classes of processed pseudogenes demarcate a period of rapid molecular evolution, *Proc Natl Acad Sci USA* 85:9625–9, 1988.

Li K, Li Y, Shelton JM, Richardson JA, Spencer E: et al: Cytochrome c deficiency causes embryonic lethality and attenuates stress-induced apoptosis, *Cell* 101:389–99, 2000.

Li P, Nijhawan D, Budihardjo I, Srinivasula SM, Ahmad M et al: Cytochrome c and dATP-dependent formation of Apaf-1/caspase-9 complex initiates an apoptotic protease cascade, *Cell* 91:479–89, 1997.

Liu X, Kim CN, Yang J, Jemmerson R, Wang X: Induction of apoptotic program in cell-free extracts: requirement for dATP and cytochrome c, *Cell* 86:147–57, 1996.

Apaf1

Cecconi F, Alvarez-Bolado G, Meyer BI, Roth KA, Gruss P: Apaf1 (CED-4 homolog) regulates programmed cell death in mammalian development, *Cell* 94:727–37, 1998.

Fortin A, Cregan SP, MacLaurin JG, Kushwaha N, Hickman ES et al: APAF1 is a key transcriptional target for p53 in the regulation of neuronal cell death, *J Cell Biol* 155:207–16, 2001.

Furukawa Y, Nishimura N, Furukawa Y, Satoh M, Endo H et al: Apaf-1 is a mediator of E2F-1-induced apoptosis, *J Biol Chem* 277:39760–8, 2002.

Honarpour N, Gilbert SL, Lahn BT, Wang X, Herz J: Apaf-1 deficiency and neural tube closure defects are found in fog mice, *Proc Natl Acad Sci USA* 98:9683–7, 2001.

Kim H, Jung YK, Kwon YK, Park SH: Assignment of apoptotic protease activating factor-1 gene (APAF1) to human chromosome band 12q23 by fluorescence in situ hybridization, *Cytogenet Cell Genet* 87:252–3, 1999.

Marsden VS, O'Connor L, O'Reilly LA, Silke J, Metcalf D et al: Apoptosis initiated by Bcl-2-regulated caspase activation independently of the cytochrome c/Apaf-1/caspase-9 apoptosome, *Nature* 419:634–7, 2002.

Riedl SJ, Li W, Chao Y, Schwarzenbacher R, Shi Y: Structure of the apoptotic protease-activating factor 1 bound to ADP, *Nature* 434:926–33, 2005.

Soengas MS, Capodieci P, Polsky D, Mora J, Esteller M et al: Inactivation of the apoptosis effector Apaf-1 in malignant melanoma, *Nature* 409:207–11, 2001.

Srinivasula SM, Ahmad M, Fernandes-Alnemri T, Alnemri ES: Autoactivation of procaspase-9 by Apaf-1-mediated oligomerization, *Mol Cell* 1:949–57, 1998.

Yoshida H, Kong YY, Yoshida R, Elia AJ, Hakem A et al: Apaf1 is required for mitochondrial pathways of apoptosis and brain development, *Cell* 94:739–50, 1998.

Zou H, Henzel WJ, Liu X, Lutschg A, Wang X: APAF-1, a human protein homologous to C. elegans CED-4, participates in cytochrome c-dependent activation of caspase-3, *Cell* 90:405–13, 1997.

Caspase 9

Hadano S, Nasir J, Nichol K, Rasper DM, Vaillancourt JP et al: Genomic organization of the human caspase-9 gene on chromosome 1p36.1-p36.3, *Mam Genome* 10:757–60, 1999.

Hakem R, Hakem A, Duncan GS, Henderson JT, Woo M et al: Differential requirement for caspase 9 in apoptotic pathways in vivo, *Cell* 94:339–52, 1998.

Kuida K, Haydar TF, Kuan CY, Gu Y, Taya C et al: Reduced apoptosis and cytochrome c-mediated caspase activation in mice lacking caspase 9, *Cell* 94:325–7, 1998.

Li P, Nijhawan D, Budihardjo I, Srinivasula SM, Ahmad M et al: Cytochrome c and dATP-dependent formation of Apaf-1/caspase-9 complex initiates an apoptotic protease cascade, *Cell* 91:479–89, 1997.

Marsden VS, O'Connor L, O'Reilly LA, Silke J, Metcalf D et al: Apoptosis initiated by Bcl-2-regulated caspase activation independently of the cytochrome c/Apaf-1/caspase-9 apoptosome, *Nature* 419:634–7, 2002.

Srinivasula SM, Hegde R, Saleh A, Datta P, Shiozaki E et al: A conserved XIAP-interaction motif in caspase-9 and Smac/DIABLO regulates caspase activity and apoptosis, *Nature* 410:112–16, 2001.

Caspase 3

Fernandes-Alnemri T, Armstrong RC, Krebs J, Srinivasula SM, Wang L et al: In vitro activation of CPP32 and Mch3 by Mch4, a novel human apoptotic cysteine protease containing two FADD-like domains, *Proc Natl Acad Sci USA* 93:7464–9, 1996.

Fernandes-Alnemri T, Litwack G, Alnemri ES: CPP32, a novel human apoptotic protein with homology to Caenorhabditis elegans cell death protein Ced-3 and mammalian interleukin-1 beta-converting enzyme, *J Biol Chem* 269:30761–4, 1994.

Fernando P, Kelly JF, Balazsi K, Slack RS, Megeney LA: Caspase 3 activity is required for skeletal muscle differentiation, *Proc Natl Acad Sci USA* 99:11025–30, 2002.

Gervais FG, Xu D, Robertson GS, Vaillancourt JP, Zhu Y et al: Involvement of caspases in proteolytic cleavage of Alzheimer's amyloid-beta precursor protein and amyloidogenic A-beta peptide formation, *Cell* 97:395–406, 1999.

Jiang X, Kim HE, Shu H, Zhao Y, Zhang H et al: Distinctive roles of PHAP proteins and prothymosin-alpha in a death regulatory pathway, *Science* 299:223–6, 2003.

Kuida K, Zheng TS, Na S, Kuan C, Yang D et al: Decreased apoptosis in the brain and premature lethality in CPP32-deficient mice, *Nature* 384:368–72, 1996.

Mannick JB, Hausladen A, Liu L, Hess DT, Zeng M et al: Fas-induced caspase denitrosylation, *Science* 284:651–4, 1999.

Miura M, Chen XD, Allen MR, Bi Y, Gronthos S et al: A crucial role of caspase-3 in osteogenic differentiation of bone marrow stromal stem cells, *J Clin Invest* 114:1704–13, 2004.

Nasir J, Theilmann JL, Chopra V, Jones AM, Walker D et al: Localization of the cell death genes CPP32 and Mch-2 to human chromosome 4q, *Mam Genome* 8:56–9, 1997.

Nicholson DW, Ali A, Thornberry NA, Vaillancourt JP, Ding CK et al: Identification and inhibition of the ICE/CED-3 protease necessary for mammalian apoptosis, *Nature* 376:37–43, 1995.

Woo M, Hakem R, Furlonger C, Hakem A, Duncan GS et al: Caspase-3 regulates cell cycle in B cells: A consequence of substrate specificity, *Nature Immun* 4:1016–22, 2003.

Human protein reference data base, Johns Hopkins University and the Institute of Bioinformatics, at http://www.hprd.org/protein.

6.17. Assessment of Cell Apoptosis

Yan N, Shi Y: Mechanisms of apoptosis through structural biology, *Annu Rev Cell Dev Biol* 21:35–56, 2005.

Nagata S: DNA degradation in development and programmed cell death, *Annu Rev Immunol* 23:853–75, 2005.

Jiang X, Wang X: Cytochrome C-mediated apoptosis, *Annu Rev Biochem* 73:87–106, 2004.

Sancar A, Lindsey-Boltz LA, Unsal-Kacmaz K, Linn S: Molecular mechanisms of mammalian DNA repair and the DNA damage checkpoints, *Annu Rev Biochem* 73:39–85, 2004.

Bard JB: Growth and death in the developing mammalian kidney: Signals, receptors and conversations, *Bioessays* 24:72–82, 2002.

Tilly JL: Commuting the death sentence: How oocytes strive to survive, *Nat Rev Mol Cell Biol* 2:838–48, 2001.

Kuan CY, Roth KA, Flavell RA, Rakic P: Mechanisms of programmed cell death in the developing brain, *Trends Neurosci* 23:291–7, 2000.

Nijhawan D, Honarpour N, Wang X: Apoptosis in neural development and disease, *Annu Rev Neurosci* 23:73–87, 2000.

Hsu SY, Hsueh AJ: Tissue-specific Bcl-2 protein partners in apoptosis: An ovarian paradigm, *Physiol Rev* 80:593–614, 2000.

Fraser A, McCarthy N, Evan GI: Biochemistry of cell death, *Curr Opin Neurobiol* 6:71–8, 1996.

SECTION 3

DEVELOPMENTAL ASPECTS OF BIOREGENERATIVE ENGINEERING

7

FERTILIZATION AND EARLY EMBRYONIC DEVELOPMENT

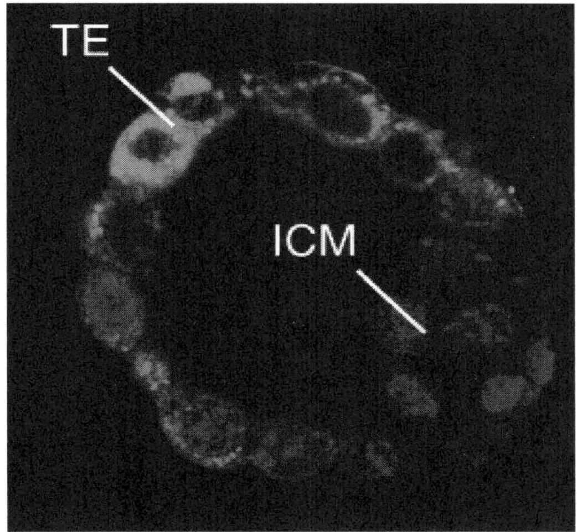

Fluorescent micrograph showing a mouse blastocyst. TE: trophoblast. ICM: inner cell mass. Cells were labeled for mitochondria (green) and nuclei (blue). (Reprinted from Houghton FD: Energy metabolism of the inner cell mass and trophectoderm of the mouse blastocyst, *Differentiation* 74:11–8, 2006 by permission of Blackwell Publishing.) See color insert.

Bioregenerative Engineering: Principles and Applications, by Shu Q. Liu

An animal undergoes a developmental cycle composed of a series of biological processes: the initiation, development, maturation, and reproduction of the animal. The repetition of these processes is the foundation for the continuation of the life. The initiation and development occur during the embryonic period and are collectively referred to as *embryogenesis*. It is in this period that a new generation of animals forms. In tradition, the embryonic development of an animal is divided into two stages: the embryonic and fetal stages based on distinct markers of functional anatomy.

The *embryonic stage* is defined in the human as the period from the conception to the formation of primary organs at about the eighth week. This stage is divided into two substages: germinal and embryonic development. Germinal development takes place in the human from the conception to the formation of the three germinal layers, including the ectoderm, mesoderm, and endoderm, at the end of the second week. During this period, a male gamete (sperm) fuses into a female gamete (oocyte) to form a zygote, a process known as *fertilization*. The fusion of the two gametes allows the integration of the genomes from both parents, an essential process for transmitting genetic information from the parents to the progeny and for initiating the development of a new individual. Fertilization triggers an early mitotic segmentation process, known as *cleavage*, by which a fertilized egg is divided continuously into smaller cells. When reaching a certain cell density, the cells are organized into various patterns, which undergo dynamic changes through different stages. The embryonic cells are subsequently committed to *gastrulation*, a process leading to the formation of a three-layered structure known as *gastrula*, composed of the ectoderm, mesoderm, and endoderm. The formation of a gastrula takes place during the first two weeks and is indicative of the ending of the germinal period.

Embryonic development takes place from the second to the eighth week. During this period, the basic forms of major organs are developed from the three germ layers: ectoderm, mesoderm, and endoderm. The *ectoderm* gives rise to the central nervous system (brain and spinal cord), peripheral nerve structures, and the epidermis of the skin, teeth, nose, and external ear. The *mesoderm* is the origin of the heart, vascular system, blood, muscle, bone, cartilage, and connective tissue. The *endoderm* develops into the gastrointestinal tract, liver, pancreas, bladder, and lung. By the end of the eighth week, tissues and organs are assembled into the primary form of a fetus. The embryonic stage is the period during which most dynamic changes take place in morphogenesis.

The *fetal stage* is the remaining period from the formation of the fetus to the parturition or birth of a new individual. During this stage, the fetus gains size rapidly from several centimeters to about half a meter (~0.5 m), but the form of the tissues and organs does not change as vigorously as that during the embryonic stage. The entire embryonic period from conception to parturition is about 40 weeks.

In this chapter, the developmental processes during embryogenesis will be introduced with emphasis on the morphogenesis of embryonic structures and related regulatory mechanisms during each stage. We will see that, although bioregeneration is defined as a process occurring during the adulthood for the repair and reconstruction of lost tissues and organs, it is similar to embryonic development in many aspects.

THE SPERM [7.1]

The male and female gametes are essential components for the initiation of embryogenesis and the development of a new individual. A male gamete, known as *sperm* or *spermato-*

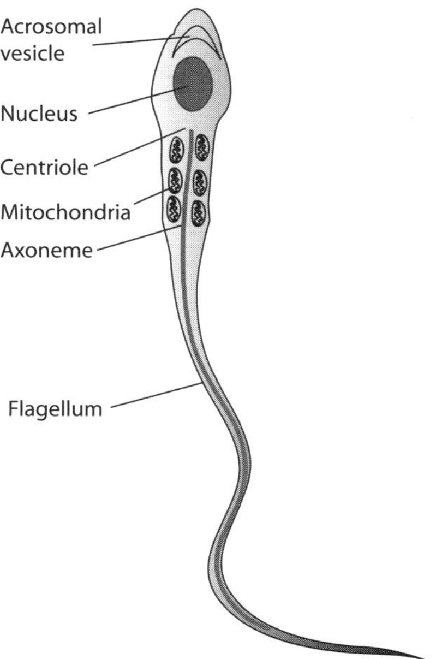

Figure 7.1. Schematic representation of a sperm cell. Based on bibliography 7.1.

zoon, is composed of several systems, including a haploid nucleus, a propulsion apparatus, and protein enzymes that are required for its interaction and fusion with a female gamete. During sperm maturation, the volume of the cytoplasm is minimized to reduce the size of the sperm, but the structures that regulate the sperm–egg interaction are evolved. These include the acrosomal vesicle and the flagellum (Fig. 7.1). The *acrosomal vesicle* is originated from the Golgi apparatus and located in front of the nucleus. This structure contains enzymes that are necessary for degrading the external layer of the egg during sperm–egg fusion. The *flagellum* is developed on the basis of the centrioles and is responsible for sperm movement, which is driven by a motile apparatus known as *axoneme*. The motile apparatus is composed of three-dimensionally organized microtubules, consisting of dimeric tubulins, and a type of motor protein known as *dynein*. A dynein molecule is attached to the microtubule and serves as an enzyme that hydrolyzes ATP. The hydrolysis of ATP provides energy necessary for the motile activity of dynein molecules and the flagellar propulsion, which induces sperm movement.

Sperms are generated from a cell type called *primordial germ cell* in the testes, the male reproductive organs. The process of sperm generation is referred to as *spermatogenesis* (Fig. 7.2). The primordial germ cells are specified during the early embryonic cell cleavage stage. A fraction of cells at the 8/16-cell stages develop into primordial germ cells. These cells usually contain mRNAs and proteins that are necessary for the development of gametes. The cytoplasm of these cells is called *germ plasm*. Sperms are formed via meiosis of the primordial germ cells. These cells are formed in the epiblast, an early embryonic structure that develops from the inner cell mass of a blastocyst and gives rise to the ectoderm, mesoderm, and endoderm. When the epiblast is developed into the three germ layers, the primordial germ cells are localized to the endoderm. With further development, these germ cells migrate into the genital ridges, where gonads (ovaries and testes)

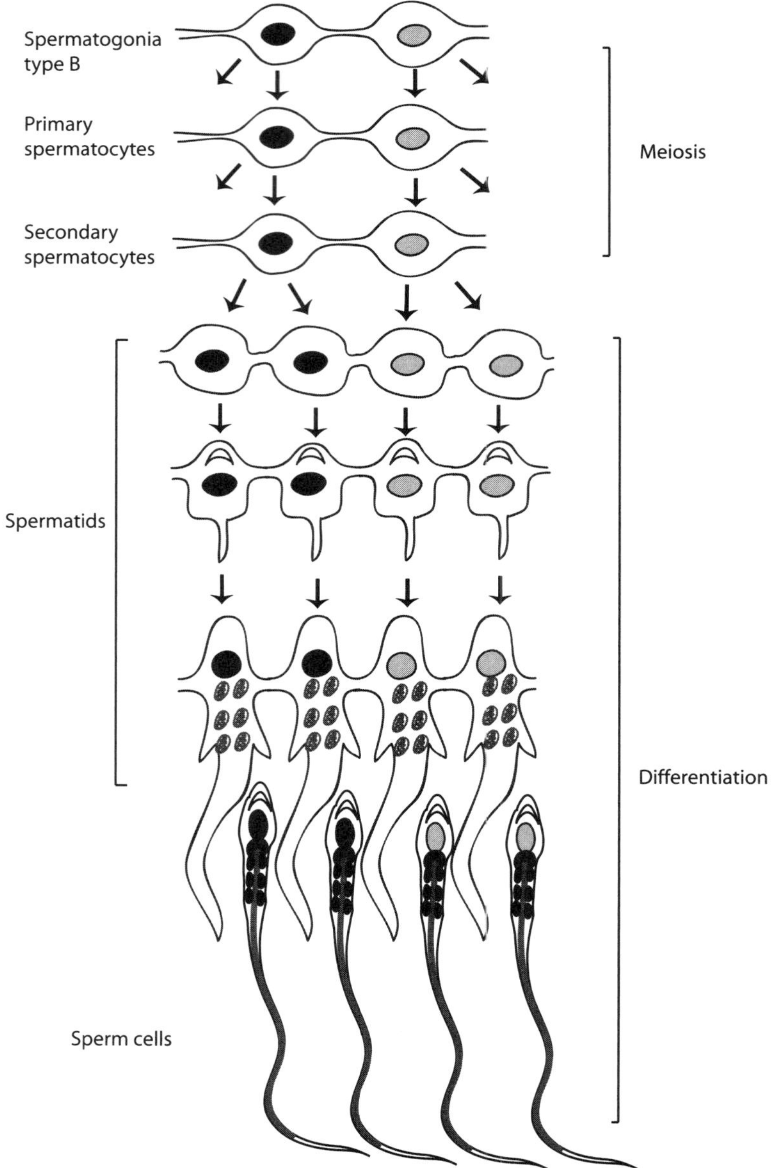

Figure 7.2. Schematic demonstration of spermatogenesis. Based on bibliography 7.1.

are developed, and remain relatively quiescent in a structure called sex cord until reaching maturity.

In the sex cord, which develops into the seminiferous tubules of the testis at puberty, the primordial germ cells differentiate into type A1 spermatogonia, which can subsequently differentiate into several levels of spermatogonia, including types A2, A3, and A4 spermatogonia (type A2 differentiates into A3, and A3 into A4). These type A spermatogonia are stem cells in nature and can self-renew themselves as well as differentiate into specified cell types. Type 4 spermatogonia can further differentiate into a hierarchy of several cell types, including intermediate spermatogonia, spermatozoa, type B spermatogonia, and primary spermatocytes. Up to this point, mitosis is the basic form of cell division. The primary spermatocytes can divide to form another hierarchy of sperm progenitor cells via meiosis, including secondary spermatocytes and spermatids. The spermatids give rise to sperms (Fig. 7.2).

THE EGG [7.1]

The *egg* is the female gamete and is also known as the *ovum*. A developing egg before its complete meiotic division or formation of a haploid nucleus is called an *oocyte*. An egg is composed of a nucleus and an enormous volume of cytoplasm. The egg stores a large amount of food and energy-producing materials in its cytoplasm for the growth and development of the embryo. In addition, the egg cytoplasm contains a variety of proteins (structural, regulatory, morphogenetic, and protective factors), rRNAs, tRNAs, and mRNAs. The egg cytoplasm is enclosed within the cell membrane. The egg cell membrane is surrounded by an extracellular matrix layer known as *zona pellucida*. Outside the zona pellucida, there exists a thick cellular structure called *cumulus*, composed of a large number of *ovarian follicular cells* (Fig. 7.3). The follicular cells provide soluble factors and mechanical protection to the egg cell.

The process of egg formation is referred to as *oogenesis*. This process is different from spermatogenesis. Whereas spermatogenesis produces a nucleus-predominant sperm with strong motility, oogenesis gives rise to eggs with materials and factors necessary for embryonic development. In the human, there exist a limited number of female primitive germ cells (about thousand) called *oogonia*. These cells can divide into several millions

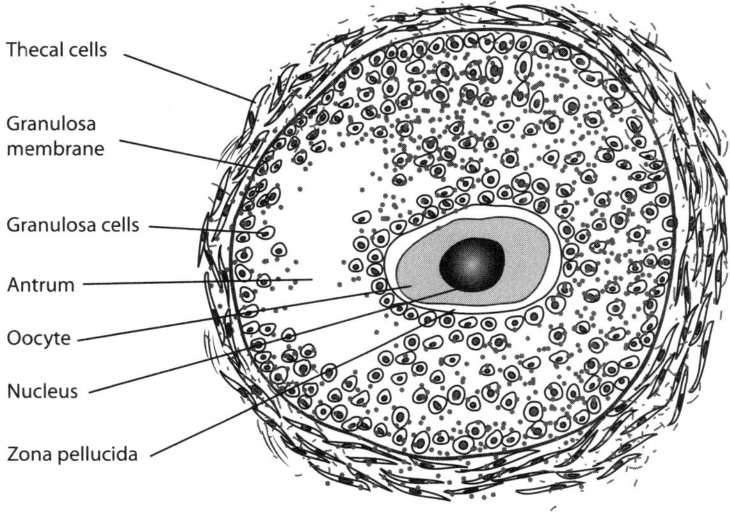

Figure 7.3. Schematic representation of an ovarian follicle. Based on bibliography 7.1.

of secondary germ cells via mitosis from the second to the seventh month of gestation. A large number of germ cells are committed to apoptosis afterward. The surviving germ cells give rise to primary oocytes via the first meiotic cycle. The primary oocytes are prompted to enter the prophase and metaphase of meiosis and maintain a quiescent state until puberty. When a human female individual reaches maturity, oocytes are periodically committed to meiosis and formation of mature eggs. Such an activity usually begins at the age of ~13 years and disappears at the age of ~50 years; these events are termed *menarche* and *menopause*, respectively. About 400 eggs can be formed through the lifespan of a female individual.

FERTILIZATION [7.1]

Fertilization is a process by which a sperm fuses into an oocyte to form a zygote, which develops into a new individual through embryogenesis. Note that the term *oocyte* is used here instead of *egg* because a sperm usually fuses with an oocyte or developing egg, which is arrested in the metaphase of meiosis and has not completed meiosis. Fertilization takes place via several steps, including: (1) attraction of sperm to an oocyte by chemotactic factors released from the oocyte, (2) interaction between the sperm and the oocyte, (3) penetration of the sperm through the extracellular layers of the oocyte, (4) the entrance of the sperm into the oocyte, and (5) fusion of the sperm with the oocyte. These steps are briefly outlined here.

Attraction of Sperm Cells to the Oocyte

In mammals, oocytes are produced and released from the ovary and moved to the oviduct. The interaction of the oocyte with sperm occurs in the ampulla region of the oviduct, which is near the ovary (Fig. 7.4). Sperm are capable of migrating toward an oocyte. Such an activity is induced and controlled by chemotaxis, or chemical gradient-directed cell movement. An oocyte can produce and release sperm-attracting proteins, which can act on specific membrane receptors in the sperm. An example of such sperm-attracting protein is resact, discovered in the sea urchin oocytes. This molecule can induce strong chemotactic activity of the sperm in vitro (Fig. 7.5). It is important to address that the chemotactic activity of sperm is dependent on animal species. Sperm can be activated and committed to directed migration only in response to a chemoattractant released from the oocytes of the same species. Furthermore, the sperm attraction activity is dependent on the developmental stage. Sperm rarely move toward oocytes that have not yet committed to the second meiosis, since these oocytes do not release sufficient chemoattractants.

Sperm–Oocyte Interaction

When a sperm approaches an oocyte, the physical interaction of the sperm with the extracellular layer of the oocyte induces the activation of the acrosomal vesicle of the sperm. In mammals, the sperm first passes through the follicular cell layer. When reaching the zona pellucida, the acrosomal vesicle undergoes exocytosis, a process that releases proteolytic enzymes. These enzymes degrade the matrix of the zona pellucida and thus help the sperm approach the oocyte membrane. The exocytosis of the acrosomal vesicle is induced by the binding of a zona pellucida protein, zona protein 3 (ZP3), to a specific

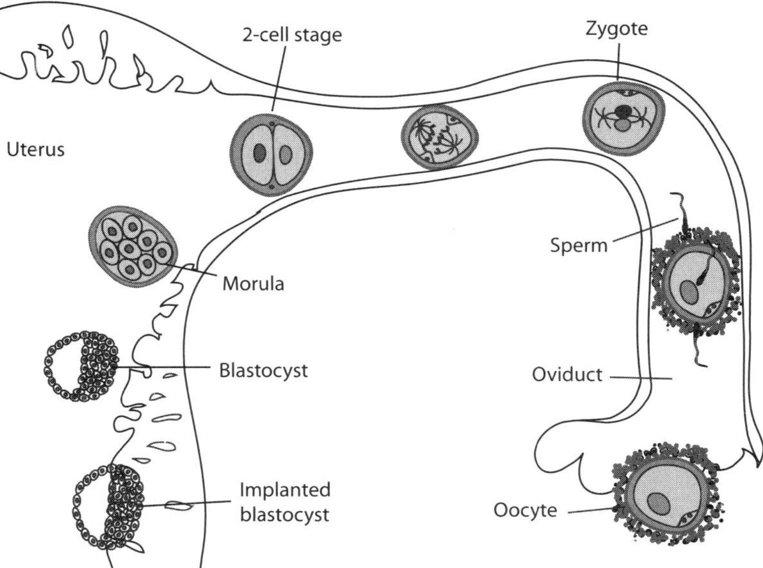

Figure 7.4. Schematic representation of the locations for oocyte fertilization, morula formation, and blastocyst formation and implantation in the human oviduct and uterus. Based on bibliography 7.1.

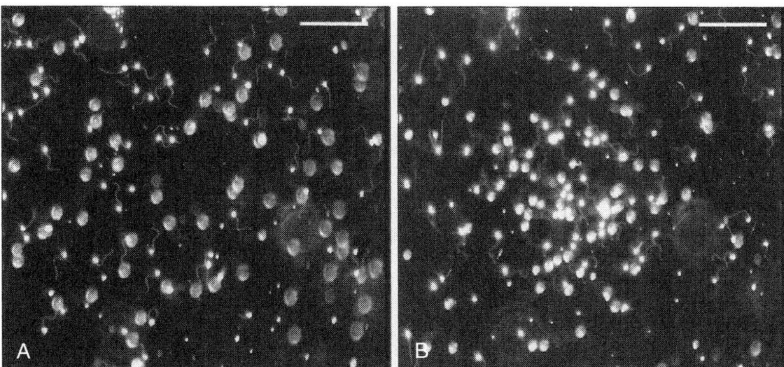

Figure 7.5. Resact-mediated movement of sperm cells. (A) Dispersed control sperm cells without the release of resact. (B) Chemotactic accumulation of sperm cells in response to the release of resact. Scale bar: 100 μm. (Reproduced from Solzin J et al: *J General Physiol* 124:115–24, 2004 copyright by permission of The Rockefeller University Press.)

sperm surface receptor, galactosyltransferase-I. The binding activity stimulates a G-protein-mediated signaling pathway and results in the release of calcium, which in turn induces the release of proteolytic enzymes from the acrosomal vesicle. Since the structure of the ligand and receptor is specific to an animal species, sperm can interact only with the egg zona pellucida of the same species.

Sperm–Oocyte Fusion

After a sperm passes through the zona pellucida, it approaches and attaches to the cell membrane of the oocyte. The cell membrane of the sperm and that of the oocyte can fuse together, bringing the sperm into the oocyte cytoplasm (Fig. 7.6). Gamete membrane fusion is a process regulated by fusion proteins. While the regulatory mechanisms of fusion are not completely understood, an oocyte membrane protein, CD9 (Table 7.1), has been suggested to play a role in the mediation of gamete fusion. The knockout of the CD9 gene is associated with impaired interaction between sperm and oocyte, resulting in infertility. When CD9 mRNA is delivered into the egg with CD9 gene knockout, infertility can be reversed. In addition, a molecule found in the membrane of sperm cells, known as the *Izumo immunoglobulin superfamily protein*, plays a critical role in regulating sperm fusion into the oocyte. In Izumo –/– mice, the sperm cells can develop to maturity and

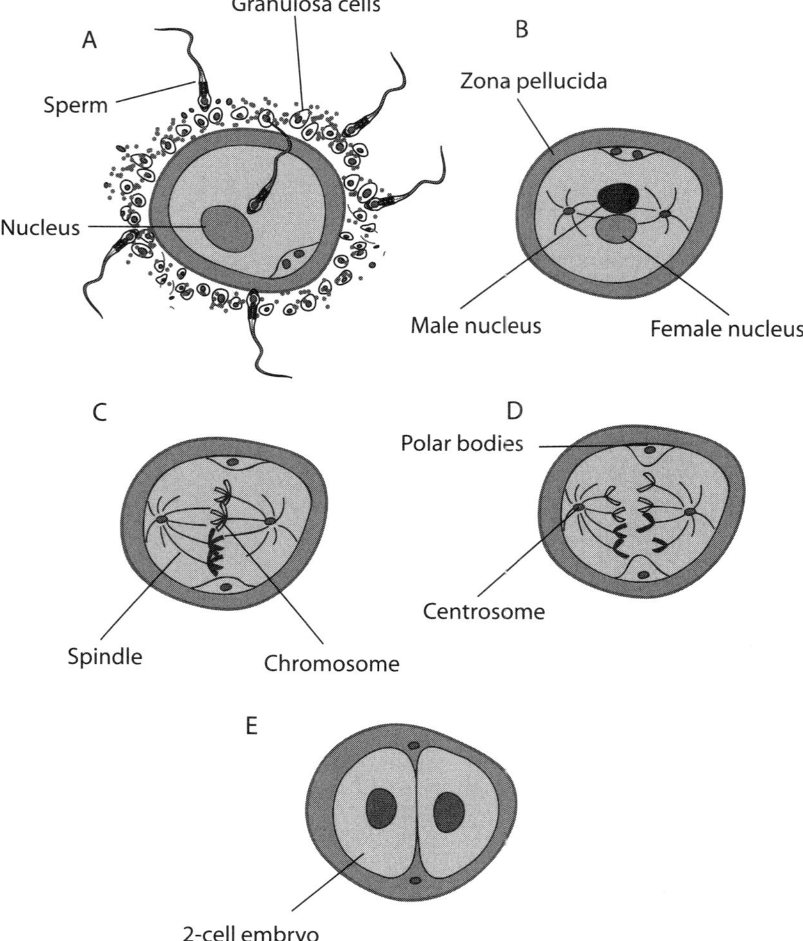

Figure 7.6. Schematic representation of sperm fusion into an oocyte. Based on bibliography 7.1.

can penetrate the zona pellucida, but cannot fuse into the oocyte (Fig. 7.7). However, the exact mechanisms of Izumo action remains poorly understood.

When a sperm enters an oocyte, the oocyte immediately becomes resistant to further sperm fusion, a mechanism that prevents polyspermy. The antipolyspermy activity is controlled by a rapid shift of cell membrane potential. A sperm can fuse with an oocyte at the normal resting membrane potential (about −90 mV). Immediately after a sperm fuses into the oocyte, the normally negative resting membrane potential is reversed to a positive value (about 20 mV), induced by the opening of the sodium channels and sodium flux into the oocyte. Such a rapid change in membrane potential prevents the interaction of other sperm cells with the oocyte during the early period after a sperm fuses with the oocyte. The shift of membrane potential is a transient process that lasts only about one minute. There exists another mechanism, by which a sperm-fused oocyte prevents further sperm fusion. The oocyte cell contains a large number of cortical granules, which contain proteolytic enzymes (Fig. 7.8). The fusion of a sperm to the oocyte triggers the release of calcium, which in turn induces the exocytosis of the cortical granules, releasing the enzymes into the zona pellucida. These enzymes cleave regulatory proteins, such as zona pellucida glycoprotein 3 (ZP3; Table 7.2), which mediate sperm–oocyte interaction. Thus, sperm can no longer bind to the zona pellucida and enter the sperm-fused oocyte.

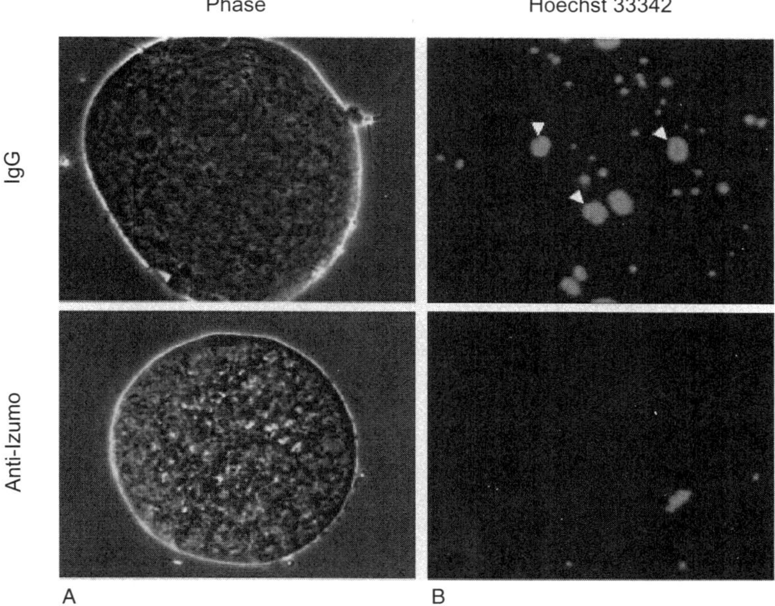

Figure 7.7. Role of the the Izumo protein in the regulation of sperm fusion to oocytes. The Izumo protein is expressed in the membrane of the sperm cell and regulates sperm fusion into eggs. (A) Izumo$^{-/-}$ mouse sperm cells are not able to fuse into a mouse egg compared to Izumo$^{+/-}$ mouse sperm cells. Note that fused sperm cells show enlarged cell nuclei. (B) Anti-human Izumo antibody (anti-hIzumo) significantly suppressed human sperm fusion into a human egg. The blue color represents cell nucei labeled with Hoechst 33342. (Reprinted by permission from Macmillan Publishers Ltd: Inoue N et al: *Nature* 434:234–8, copyright 2005.)

TABLE 7.1. Characteristics of CD9*

Proteins	Alternative Names	Amino Acids	Molecular Weight (kDa)	Expression	Functions
CD9	CD9 antigen, leukocyte antigen MIC3, p24 antigen	228	25	Sperm, oocyte, leukocyte, skin, intestine	A membrane glycoprotein that interacts with integrins and regulates cell adhesion, migration, and fusion

*Based on bibliography 7.1.

TABLE 7.2. Characteristics of ZP3*

Proteins	Alternative Names	Amino Acids	Molecular Weight (kDa)	Gene Locus	Expression	Functions
ZP3	Zona pellucida glycoprotein 3A, sperm receptor, zona pellucida protein C, zona pellucida sperm-binding protein 3	424	47	7q11.23	Oocyte	A zona pellucida component that regulates sperm–oocyte interaction and activation of fertilization processes

*Based on bibliography 7.1.

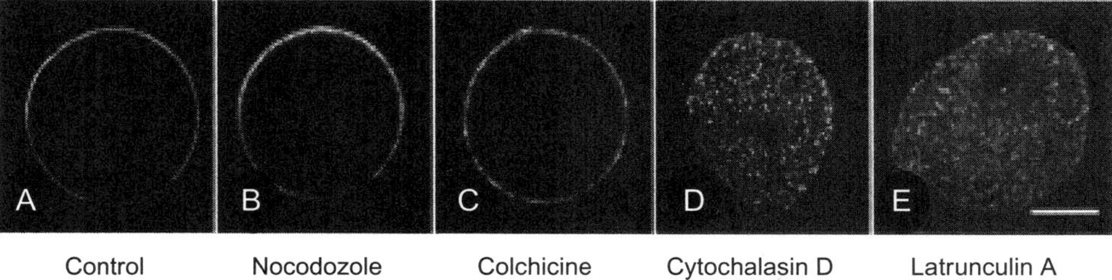

Figure 7.8. Translocation of cortical granules in oocytes. The cortical granules of oocytes are translocated to the cortical layer following germinal vesicle breakdown. This process is dependent on the organization and function of the actin filaments. A. Distribution of cortical granules in a control egg. B and C. Distribution of cortical granules in eggs treated with microtubule inhibitors nocodozole and colchicine, respectively. Note that these inhibitors did not influence the distribution of the cortical granules, suggesting that microtubules do not play a significant role in regulating the translocation of cortical granules. D and E. Distribution of cortical granules in eggs treated with actin filament inhibitors cytochalasin D and latrunculin A, respectively. Note that these inhibitors significantly influenced the distribution of the cortical granules, suggesting that actin filaments play a role in regulating the translocation of cortical granules. Reprinted from Wessel GM et al. Development 129:4315–4325, 2002 by permission of The Company of Biologists Ltd.

Activation of Embryonic Development

The fusion of a sperm cell into an oocyte triggers the activation of several important cellular activities, including DNA synthesis, restoration of mitosis, protein synthesis, and membrane synthesis. All these activities are directly or indirectly related to changes in the intracellular concentration of calcium. Since sea urchin oocytes are used extensively in the study of sperm–oocyte fusion, these cells are used here as an example. The fusion of a sperm with an oocyte, especially the binding of ZP3 to its receptor on the sperm, triggers the activation of G proteins, which further activate phospholipase C. Activated phospholipase C can cleave phosphatidylinositol biphosphate (PIP_2) to form inositol triphosphate (IP_3) and diacylglycerol. IP_3 can act on calcium channels to induce calcium release and an increase in the level of intracellular calcium. Calcium is involved in the regulation of mitosis, DNA synthesis, protein synthesis, and membrane synthesis. Diacylglycerol can further activate protein kinase C, which is involved in the regulation of DNA and protein synthesis. These activities are essential for initiating the developmental processes of the zygote or the fertilized oocyte.

Integration of Gamete Genomes

In mammals, a sperm usually fuses with an oocyte that is arrested in the metaphase of meiosis (note that an oocyte is a developing egg before completing meiosis). The interaction of the sperm with the oocytes triggers a number of events: (1) separation of the sperm nucleus and centrioles from the flagellum, (2) degradation of the sperm mitochondria and flagellum, (3) migration of the sperm nucleus to the oocyte nucleus, (4) completion of oocyte meiosis, and (5) fusion of the sperm and oocyte genetic materials (Fig. 7.6). These processes are critical to the embryonic development. Since the sperm mitochondria and microtubules are mostly degraded, the mitochondrial and microtubular systems in each individual are derived from the oocyte.

The sperm chromatids are organized and packed within the nucleus by DNA-binding proteins through disulfide bonds. When the sperm enters the oocyte cytoplasm, the oocyte is able to reduce the disulfide bonds and loose the packed chromatids. The loose sperm DNA then migrates toward the oocyte nucleus and is prepared for fusion with the oocyte DNA. At the same time, the oocyte activates its signaling pathways that stimulate the reactivation and completion of meiosis, resulting in the formation of the haploid egg nucleus (note that the oocyte has now matured to an egg). The microtubules of the egg are connected with the sperm and the egg nucleus. This process allows the migration of both nuclei toward each other. The interaction of the nuclei induces the integration of the genetic materials.

CLEAVAGE [7.2]

Cleavage is a process of zygote separation, which is accomplished through two events: mitotic division of the nucleus (karyokinesis) and cytoplasmic division (cytokinesis). The mitotic division takes place immediately following fertilization. Cytokinesis occurs subsequently. These events generate individual nucleated cells, known as *blastomeres*. The zygote is cleaved into two cells, which are subsequently cleaved into four cells, and so on, although the mammalian blastomeres may not cleave symmetrically and some cells may skip a round of cleavage (Fig. 7.9). The zygote moves along the oviduct during cleavage. The first cleavage process occurs approximately at the end portion of the oviduct. With further cleavage, the embryo moves to the uterus and is ready for implantation (Fig. 7.4).

Blastomere cleavage represents the most rapid process of cell division through the entire lifespan, including the embryonic period and adulthood. A *Drosophila* egg, for example, can be cleaved to generate about 50,000 blastomeres within about 12 h. The cleavage of the egg is initiated by a molecule known as *mitosis-promoting factor* (MPF). The fertilization process induces the activation of MPF, which stimulates the fertilized egg to enter the mitosis cycle. The mitotic process during the early cleavage undergoes only two phases: DNA synthesis (S) and mitosis (M). During the periodic cleavage cycles, the level of MPF changes cyclically: it reaches the maximal level during the M phase and reduces to the minimal level during the S phase. The periodic change in the MPF level controls the cyclically procession of cell division.

FORMATION OF THE BLASTOCYST [7.2]

Blastocyst is an embryonic structure composed of an *inner cell mass* of about 30 stem cells, which can differentiate into all specified cell types, and an external sac known as *trophoblast*, which develops into an embryo-supporting tissue called *chorion* (the external layer of the extraembryonic membrane and the embryonic part of the placenta) (see chapter-opening figure). Since the inner cell mass and the trophoblast serve as the origins of distinct embryonic and extraembryonic tissues, the formation of blastocyst is considered a milestone of embryogenesis.

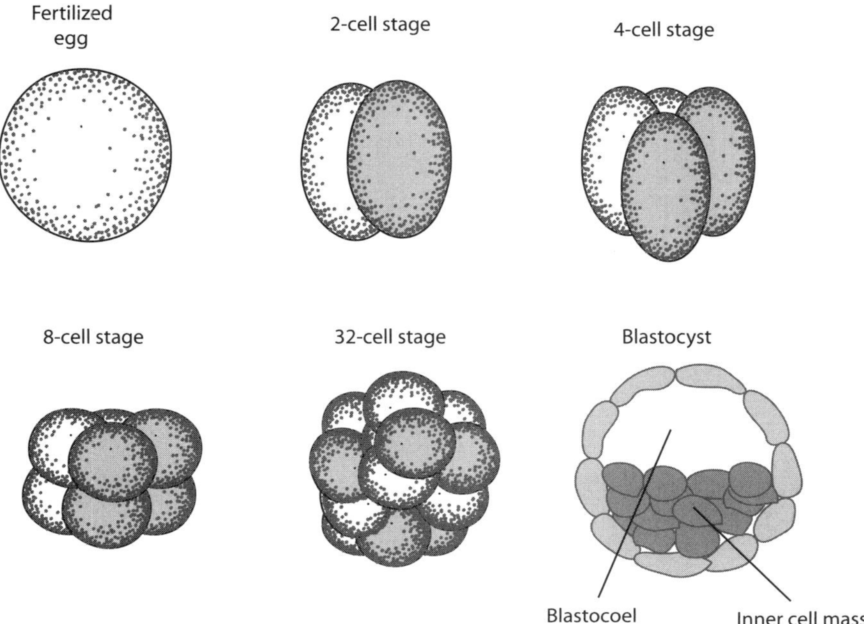

Figure 7.9. Schematic demonstration of developmental stages. Based on bibliography 7.1.

There are several key steps for the formation of the blastocyst: zygote cleavage, formation of morula, and cavitation of the morula (Fig. 7.9). Zygote cleavage has been discussed above. When cells are cleaved to a stage of 16 cells at about day 4–5, all cells are highly compacted into a solid ball-like structure. This structure is defined as a *morula*. With further cell divisions, the external cells of the morula transform into trophoblast cells at about day 6–7, while the internal cells transform into cells of the inner cell mass (Fig. 7.10). The inner cell mass further develops into two layers: the epiblast and the hypoblast. These layers eventually develop into embryonic and extraembryonic tissues as discussed below.

GASTRULATION [7.2]

Gastrulation is a process by which early embryonic cells are organized into three distinct layers: ectoderm, mesoderm, and endoderm (Fig. 7.11). These layers eventually develop into specified tissues and organs for the new individual (see page 346 for details). When cell cleavage generates a sufficient number of cells, the embryonic cells are organized into a blastocyst, composed of the *inner cell mass* and *trophoblast*. The inner cell mass gives rise to *epiblast* and *hypoblast*. The former develops into the *embryonic epiblast* and *amnionic ectoderm*, and the latter develops into *extraembryonic endoderm*. The embryonic epiblast develops into the embryonic *ectoderm* and *primitive streak*. The primitive streak gives rise to *embryonic endoderm* and *mesoderm*. The amnionic ectoderm gives rise to the *amnionic sac*. The hypoblast-derived extraembryonic endoderm develops into the *yolk sac*. The trophoblast develops into extraembryonic supporting structures.

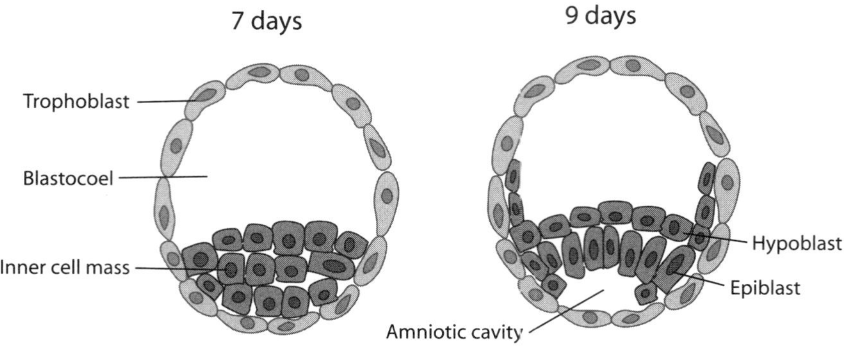

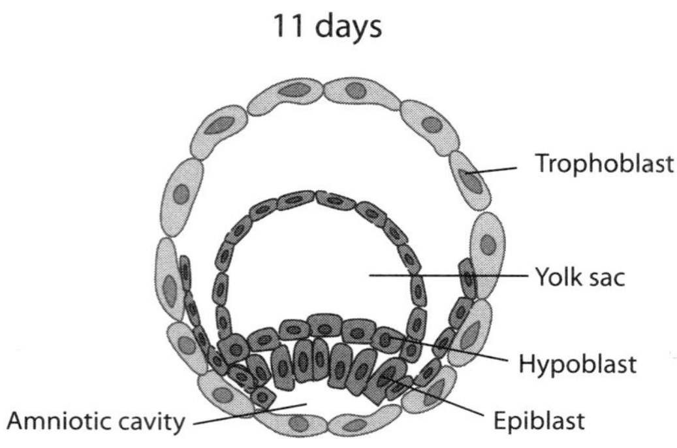

Figure 7.10. Schematic demonstration of the formation of the inner cell mass and the epiblast and hypoblast layers. Based on bibliography 7.2.

The extraembryonic supporting structures include the chorion, amnionic sac or amnion, and yolk sac. The *chorion* is the outmost extraembryonic membrane developed from the trophoblast of the blastocyst. The chorionic membrane is composed of chorionic cells, a vascular system, and connective tissue. A region of the chorion gives rise to the embryonic placenta, which is composed of two types of cellular structure called *cytotrophoblast* and *synctiotrophoblast*. The cells of these structures can interact with the maternal uterine epithelial cells via adhesion molecules, which assist in the attachment and implantation of the blastocyst into the uterine wall. Furthermore, these cells produce and release enzymes that degrade the uterine tissue, facilitating the invasion of the blastocyst into the uterine wall. Such a uterine degradation process also reorganizes the uterine vascular system in favor of the transport of nutrients and oxygen from the maternal blood to the embryo and fetus. The cytotrophoblast and synctiotrophoblast eventually develop into placenta structures including chorionic villi and intervillus space. The chorionic villi contain embryonic/fetal arteries, capillaries, and veins, which absorb and transport nutrients and oxygen from the maternal placenta to the embryo/fetus.

The *amnionic sac* is a tough extraembryonic membrane that lines the internal surface of the chorion and, when fully developed, encloses the fetus. The amnionic sac is developed from the inner cell mass-derived epiblast. During the early embryonic stage, the amnionic sac is located between the chorionic embryonic placenta and the ectoderm. It

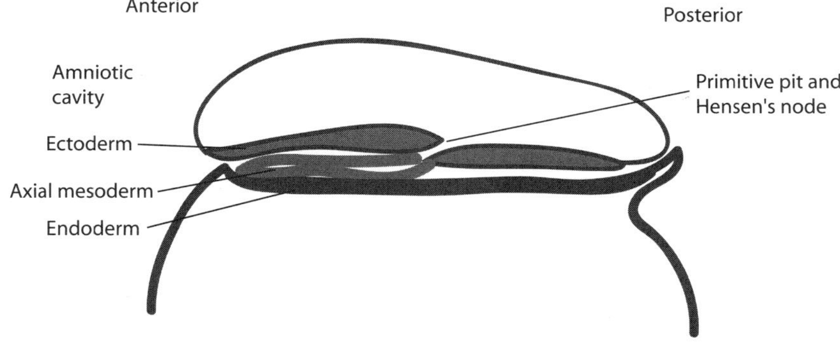

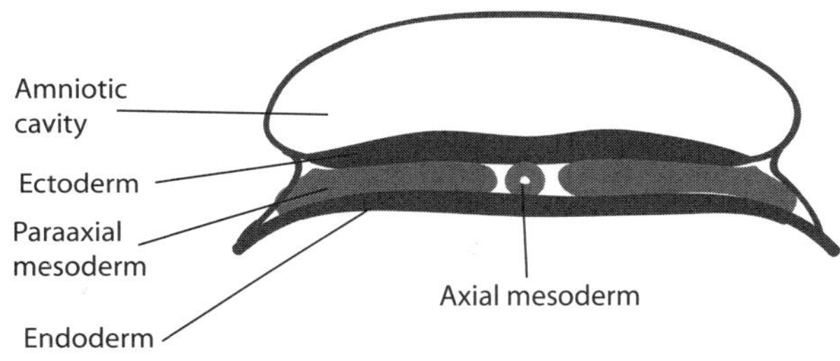

Figure 7.11. Formation of the three embryonic layers. Based on bibliography 7.2.

folds and extends gradually around the embryo, and eventually lines the chorionic membrane and encloses the embryo/fetus. The amnionic cavity is filled with a fluid called the *amnionic fluid*, in which the embryo/fetus resides. Together with the chorionic membrane, the amnionic membrane mediates the interaction of the embryo/fetus with the mother and protects the embryo/fetus from harmful environment. The *yolk sac* is a membrane structure that stores nutrients in the yolk cavity for the early development of the embryo. This structure is developed from the inner cell mass-derived hypoblast, and is located near the endoderm. The yolk cavity is relative large during the early embryonic stage, but is reduced in size and is gradually diminished during embryogenesis.

BIBLIOGRAPHY

7.1. Sperm, Egg, and Fertilization

Gilbert Scott F: *Developmental Biology*, 7th edn, Sinauer Associates, Sunderland, UK, 2003.

Inoue N, Isotani A, Okabe M: The immunoglobulin superfamily protein Izumo is required for sperm to fuse with eggs, *Nature* 434:234–8, 2005.

Gutierrez-Lopez MD, Ovalle S, Yanez-Mo M, Sanchez-Sanchez N, Rubinstein E et al: A functionally relevant conformational epitope on the CD9 tetraspanin depends on the association with activated beta1 integrin, *J Biol Chem* 278:208–18, 2003.

Zhu GZ, Miller BJ, Boucheix C, Rubinstein E, Liu CC et al: Residues SFQ (173–175) in the large extracellular loop of CD9 are required for gamete fusion, *Development* 129:1995–2002, 2002.

Benoit P, Gross MS, Frachet P, Frezal J, Uzan G et al: Assignment of the human CD9 gene to chromosome 12 (region p13) by use of human specific DNA probes, *Hum Genet* 86:268–72, 1991.

Kaji K, Oda S, Shikano T, Ohnuki T, Uematsu Y et al: The gamete fusion process is defective in eggs of Cd9-deficient mice, *Nature Genet* 24:279–82, 2000.

Le Naour F, Rubinstein E, Jasmin C, Prenant M, Boucheix C: Severely reduced female fertility in CD9-deficient mice, *Science* 287:319–21, 2000.

Miyado K, Yamada G, Yamada S, Hasuwa H, Nakamura Y et al: Requirement of CD9 on the egg plasma membrane for fertilization, *Science* 287:321–4, 2000.

Tachibana I, Hemler ME: Role of transmembrane 4 superfamily (TM4SF) proteins CD9 and CD81 in muscle cell fusion and myotube maintenance, *J Cell Biol* 146:893–904, 1999.

Waterhouse R, Ha C, Dveksler GS: Murine CD9 is the receptor for pregnancy-specific glycoprotein 17, *J Exp Med* 195:277–82, 2002.

Dean J: Biology of mammalian fertilization: Role of the zona pellucida, *J Clin Invest* 89:1055–9, 1992.

Lunsford RD, Jenkins NA, Kozak CA, Liang LF, Silan CM et al: Genomic mapping of murine Zp-2 and Zp-3, two oocyte-specific loci encoding zona pellucida proteins, *Genomics* 6:184–7, 1990.

van Duin M, Polman JEM, Suikerbuijk RF, Geurts van Kessel AHM, Olijve W: The human gene for the zona pellucida glycoprotein ZP3 and a second polymorphic locus are located on chromosome 7, *Cytogenet Cell Genet* 63:111–3, 1993.

van Duin M, Polman JEM, Verkoelen CCEH, Bunschoten H, Meyerink JH et al: Cloning and characterization of the human sperm receptor ligand ZP3: Evidence for a second polymorphic allele with a different frequency in the Caucasian and Japanese populations, *Genomics* 14:1064–70, 1992.

Zhao M, Gold L, Ginsberg AM, Liang LF, Dean J: Conserved furin cleavage site not essential for secretion and integration of ZP3 into the extracellular egg coat of transgenic mice, *Mol Cell Biol* 22:3111–20, 2002.

7.2. Cleavage, Formation of Blastocyst, and Gastrulation

Wassarman PM: Contribution of mouse egg zona pellucida glycoproteins to gamete recognition during fertilization, *J Cell Physiol* 204:388–91, 2005.

Shur BD, Ensslin MA, Rodeheffer C: SED1 function during mammalian sperm-egg adhesion, *Curr Opin Cell Biol* 16:477–85, 2004.

Santella L, Lim D, Moccia F: Calcium and fertilization: The beginning of life, *Trends Biochem Sci* 29:400–8, 2004.

Shur BD, Rodeheffer C, Ensslin MA: Mammalian fertilization, *Curr Biol* 14:R691–2, 2004.

Klonisch T, Fowler PA, Hombach-Klonisch S: Molecular and genetic regulation of testis descent and external genitalia development, *Dev Biol* 270:1–18, 2004.

Dean J: Reassessing the molecular biology of sperm-egg recognition with mouse genetics, *Bioessays* 26:29–38, 2004.

Lenart P, Ellenberg J: Nuclear envelope dynamics in oocytes: From germinal vesicle breakdown to mitosis, *Curr Opin Cell Biol* 15:88–95, 2003.

Evans JP, Florman HM: The state of the union: The cell biology of fertilization, *Nat Cell Biol* 4 (Suppl):s57–63, 2002.

Primakoff P, Myles DG: Penetration, adhesion, and fusion in mammalian sperm-egg interaction, *Science* 296:2183–5, 2002.

Matzuk MM, Burns KH, Viveiros MM, Eppig JJ: Intercellular communication in the mammalian ovary: oocytes carry the conversation, *Science* 296:2178–80, 2002.

Runft LL, Jaffe LA, Mehlmann LM: Egg activation at fertilization: Where it all begins, *Dev Biol* 245:237–54, 2002.

Wassarman PM, Jovine L, Litscher ES: A profile of fertilization in mammals, *Nat Cell Biol* 3: E59–64, 2001.

Wassarman PM: Mammalian fertilization: molecular aspects of gamete adhesion, exocytosis, and fusion, *Cell* 96:175–83, 1999.

Snell WJ, White JM: The molecules of mammalian fertilization, *Cell* 85:629–37, 1996.

Wassarman PM: Towards molecular mechanisms for gamete adhesion and fusion during mammalian fertilization, *Curr Opin Cell Biol* 7:658–64, 1995.

Nothias JY, Majumder S, Kaneko KJ, DePamphilis ML: Regulation of gene expression at the beginning of mammalian development, *J Biol Chem* 270:22077–80, 1995.

Schultz RM, Kopf GS: Molecular basis of mammalian egg activation, *Curr Top Dev Biol* 30:21–62, 1995.

8

EMBRYONIC ORGAN DEVELOPMENT

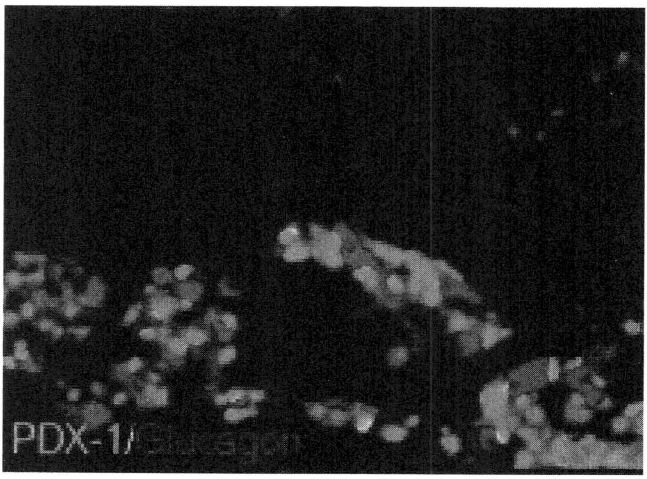

Early development of the pancreatic islets. By embryonic day 19 (E19), Pdx1-positive (green) and glucagon-positive (red) cell aggregates can be detected in the pancreas. Pdx1 is a marker for the pancreatic β cells, whereas glucagon is expressed in the α cells. The α cells are found primarily around the β cells. (From Jensen J: *Dev Dyn* 229:176–200, 2004, reprinted by permission of Wiley-Liss, Inc, a subsidiary of John Wiley & Sons, Inc.) See color insert.

Bioregenerative Engineering: Principles and Applications, by Shu Q. Liu

DEVELOPMENT OF ECTODERM-DERIVED ORGANS

The *ectoderm* is the outmost layer of the three embryonic germ layers. During the early period of embryogenesis, the ectoderm develops into a structure composed of three functional components: surface ectoderm, neural crest, and neural tube. The *surface ectoderm* is the origin of the epidermis, epidermal appendages (the nails, hair, and sebaceous glands), mucous membrane of the mouth and anus, tooth enamel, lens and cornea, and anterior pituitary. The *neural crest* gives rise to the peripheral nervous system (the peripheral ganglia, Schwann cells, and sympathetic and parasympathetic nervous systems), adrenal medulla, melanocytes, and tooth dentine. The *neural tube* develops into the brain, spinal cord, retina, and inner ear. In this chapter, we will focus on the development of the epidermis and the nervous system.

Epidermal Development [8.1]

The surface ectoderm is originally a structure with a single cell layer, which further develops into a double-layered structure. The external layer forms a temporary tissue that protects the internal layer. The internal layer contains epidermal stem cells and is referred to as the *basal layer.* The basal layer cells can differentiate into all epidermal cell types. As a first step, basal layer cells differentiate into granular cells, which contain keratin granules in the cytoplasm. The granular cells can further differentiate into epidermal cells, also known as *keratinocytes.* These cells can produce a large amount of keratin. Mature keratinocytes move to the external layer of the epidermis and eventually die, forming a keratin-rich layer (Fig. 8.1). This keratin layer is about 10 cells thick and is a tough structure that protects the internal tissues and organs from physical and chemical injury. The dead, keratin-containing keratinocytes on the epidermal surface shed constantly throughout the lifespan. The keratin layer is replenished by newly generated keratinocytes from the granular cells.

The basal cells of the epidermis reside on an extracellular matrix membrane known as the *basal lamina* (Fig. 8.1). Underneath the basal lamina lies the *dermis*, a soft connective tissue. It is important to note that the dermis is not derived from the ectoderm, but from the mesoderm. A major cell type in the dermis is fibroblast. This cell type plays a role in regulating the differentiation and proliferation of the epidermal cells by releasing growth factors, such as fibroblast growth factor, transforming growth factor α, and epidermal growth factor. Keratinocytes can also produce these growth factors. The level and timing of growth factor release may determine the rate of differentiation and proliferation of the basal cells and granular cells.

The *epidermal appendages*, including the hair, nails, and sebaceous glands, are also derived from the basal layer stem cells at designated sites. The formation of epidermal appendages is regulated by epidermal and dermal factors. For hair formation, it is necessary to establish a hair follicle primordium, the early form of the hair follicle. The follicle primordium is derived from the basal layer. At a specific site where dermal fibroblasts are activated by autocrine regulatory factors, the basal layer cells form cell clusters, undergo shape changes, and invade the dermal layer, forming scattered cell nodes. The dermal fibroblasts release growth factors, which stimulate the node cells to divide and differentiate into hair follicle cells and keratin hair shaft. The hair shaft grows out of the dermal layer and forms hair. There are follicle stem cells that can self-renew and differentiate into hair cells and regenerate the hair shaft when hair is damaged, removed, or shed.

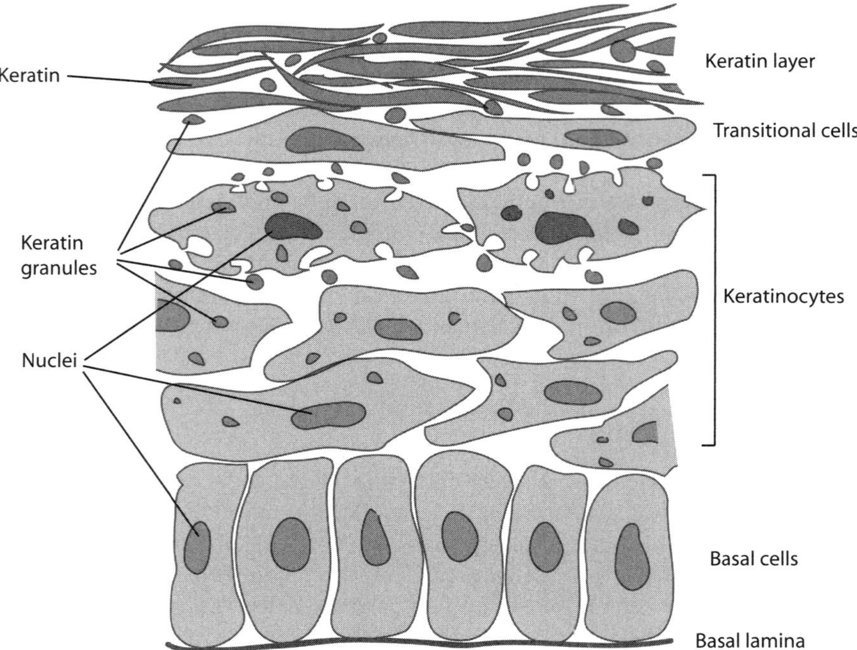

Figure 8.1. Schematic representation of the epidermis. Based on bibliography 8.1.

When the basal layer cells migrate into the dermal tissue for the formation of the hair follicle, a group of cells branches off and forms the *sebaceous glands*, which generate sebum, a lubricant that covers and protects the skin. When the sebaceous gland cells are injured, the follicle stem cells can be stimulated to differentiate into sebaceous cells and replenish the sebaceous glands.

Development of Neural Crest-Derived Systems [8.2]

The Neural Crest. The *neural crest* is an embryonic structure derived from the ectoderm. It contains stem cells that can differentiate into a number of specified cell types and tissues. These include the peripheral ganglionic neurons and glial cells, Schwann cells, sympathetic and parasympathetic neurons, medulla of the adrenal gland, epidermal pigment cells, facial bones and cartilages, corneal endothelial cells, tooth papillae, connective tissue cells and smooth muscle cells in the aortic arch, and connective tissue cells in the salivary and thyroid glands. Since these cells and structures are distributed through the entire body, the neural crest cells must migrate from the ectodermal neural crest to the periphery for a long distance. The control of the migration pattern and destination of neural crest cells as well as the specification and differentiation of these cells are major topics of developmental research.

Neurulation and Formation of Neural Crest Cells. During the stage of blastocyst formation and gastrulation, certain cells in the ectodermal region are prompted to differentiate into *neuroblasts*, cells that develop into nerve cells. This differentiation process is induced

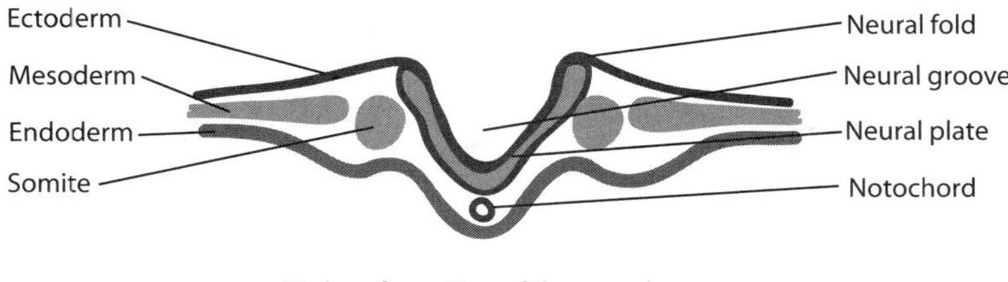

~20 days, formation of the neural groove

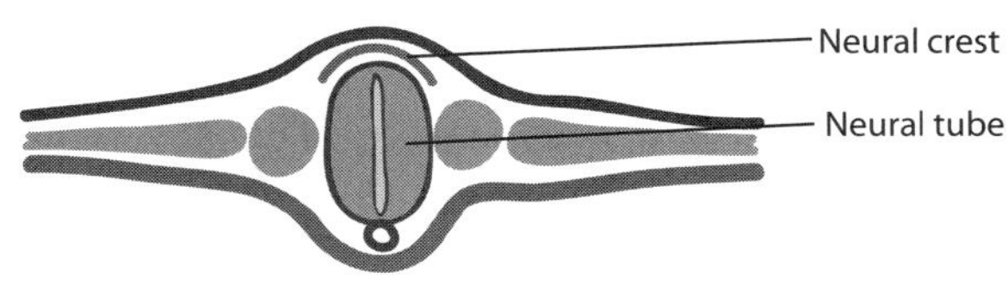

~23 days, completion of the neural tube

Figure 8.2. Schematic representation of the formation of the neural plate and neural tube, a process known as *neurulation*. Based on bibliography 8.2.

by exposing blastomeres to specific signaling molecules, such as fibroblast growth factor, noggin, and chordin. Soon following gastrulation, the narrow dorsal region of the ectoderm along the anterior–posterior (anteroposterior) axis (superior-inferior axis in the human) is transformed into a neuroblast-containing structure called *neural plate*. The neural plate undergoes a process called *neurulation*, by which the neural plate forms a *neural tube* (Fig. 8.2). There are two forms of neurulation: primary and secondary. *Primary neurulation* is the formation of the neural tube based on the ectoderm neuroblasts. This process takes place in the anterior neural plate, which develops into the brain and anterior spinal cord. In contrast, *secondary neurulation* is the formation of the neural tube from the mesenchymal cells of the mesoderm. The secondary process occurs near the posterior end of the neural plate, which develops into the posterior portion of the spinal cord.

Primary neurulation divides the ectoderm into three layers: the external presumptive epidermis, the middle neural crest cells, and the internal neural tube. At the beginning of the primary neurulation, cells in the neural plate elongate in the direction perpendicular to the neural plate and proliferate symmetrically with respect to the anterior–posterior axis. The neural plate thickens along its edges, bends its edges upward around the anterior–posterior axis, forming the *neural folds*. The groove between the two edges is defined as the *neural groove*, which is the centerline of the left and right central nervous system. The two neural folds bend further, meet each other at the ends, and fuse together to form a tube-shaped structure known as the *neural tube*. The neural tube pinches off from the ectoderm, which becomes an independent structure. At the same time, the fused ectoderm transforms to the epidermal layer at the external surface and the *neural crest* between the epidermal layer and the neural tube.

Secondary neurulation in the posterior portion originates from the mesenchymal cells, which develop into a tube-shaped structure underneath the ectoderm. This tube is

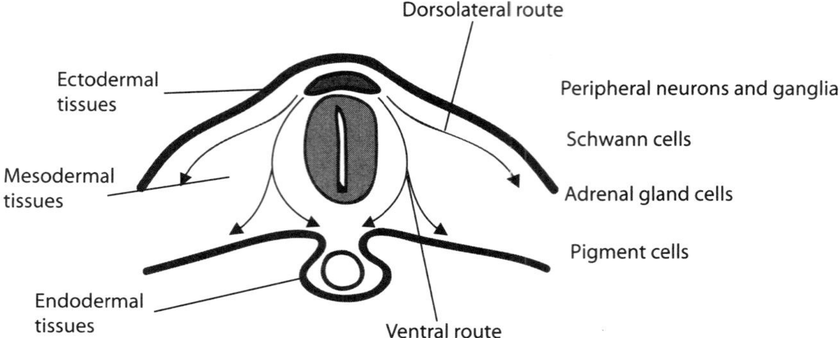

Figure 8.3. Migration routes of the trunk neural crest cells. Based on bibliography 8.2.

connected to the anterior neural tube established by primary neurulation. Secondary neurulation occurs in the region of the lumbar vertebrae. This process gives rise to the posterior spinal cord.

Migration of Neural Crest Cells. During neurulation, the neural crest cells are organized into four groups: cranial neural crest, trunk neural crest, cardiac neural crest, and vagal neural crest. Cells in the *cranial neural crest* migrate to the craniofacial region, forming cranial neurons and glial cells, craniofacial bones, cartilages, and connective tissue. The structures of the head, except for the brain, are developed mostly from the cranial neural crest cells. The *trunk neural crest* cells migrate to the peripheral systems to differentiate into several cell types, including the sensory neurons, medulla of the adrenal gland, sympathetic ganglia, and melanocytes (Fig. 8.3). The *cardiac neural crest* cells migrate to the neck region, the ascending aorta and aortic arch, and the septum to transform into nerve cells, connective tissue cells, the medial and adventitial cells of the ascending aorta. The *vagal neural crest* cells migrate to peripheral systems, primarily the gastrointestinal tract, to form parasympathetic ganglia.

The migration and differentiation of neural crest cells are fundamental processes involved in the formation of almost all body systems. How neural crest cells migrate and differentiate has become a major research topic. While these processes remain poorly understood, previous investigations have provided limited information into the mechanistic aspect. For the migration of the neural crest cells, it is necessary to initiate the following actions: activation of the cell contractile apparatus, dissociation of the cell junctions, and development of extracellular matrix pathways that lead cell migration. Cell movement is initiated by activation of the actin-dependent contractile system. Soon after the neural crest is established, the cells undergo enhanced actin polymerization, resulting in the formation of contractile actin filaments. These filaments are attached to the cell membrane. Their interaction with the myosin molecules induces cell movement. In addition, actin polymerization occurs at the leading edge of migrating cells. Such a process pushes the cell leading edge forward, enhancing cell migration. At the same time, a cell–cell dissociation process is initiated by degrading adhesion molecules such as N-cadherin, which links cells together at the cell–cell junctions. Such a process frees the cells, allowing them to migrate away from the neural crest.

The establishment of appropriate extracellular matrix pathways is another critical factor for directing the migration of the neural crest cells. There are two types of extracellular matrix involved: stimulatory and inhibitory matrix components, which coordinately navigate the migration of neural crest cells. Stimulatory matrix components include collagen, fibronectin, laminin, tenascin, and certain proteoglycan molecules. When encountering cells, these matrix components interact with corresponding integrins on the cell membrane, activating intracellular signaling pathways and promoting cell migration. In a complex biological system composed of a variety of cellular and extracellular components, directed cell migration can only be achieved in the presence of both stimulatory and inhibitory matrix, which confines cells to the stimulatory matrix pathway. Ephrins are a family of inhibitory matrix proteins. These proteins, when interacting with their receptors on the neural crest cells, prevent cell migration. In all tissue and organ systems, stimulatory and inhibitory extracellular matrix components are coordinately organized, ensuring directed migration of neural crest cells.

Differentiation of Neural Crest Cells. *Neural crest cells* are pluripotent cells that can differentiate into a variety of specified cell types essentially in all organ systems. There are several factors that are known to regulate the differentiation of the neural crest cells, including (1) the local environment within the neural crest and (2) the environment of peripheral tissue to which the neural crest cells migrate. Within the neural crest, there exist heterogeneous populations of pluripotent cells. The specification of these populations may be dependent on the local presence of different soluble factors (e.g., bone morphogenetic protein and Wnts) in the neural crest. Different combinations of these factors may determine the fate of the neural crest cells. For instance, when selected trunk neural crest cells (developing into sympathetic neurons) are removed and transplanted to the location of the vagal crest cells (differentiating into parasympathetic neurons) in the chick, the trunk crest cells can generate neurons that produce parasympathetic neurotransmitters. Thus, the initial specification of the neural crest cells is predetermined not by factors within the cells, but by extracellular factors. However, it remains poorly understood how early blastomeres elect to secrete different factors at different locations. (See Table 8.1.)

When neural crest cells migrate out of the neural crest, the fate of the cells is not finally specified. Neural crest cells committed to each pathway, such as the cranial, trunk, vagal, or cardiac pathway, have the potential to develop into different cell types. The local environment of the destination tissue determines the final fate of the neural crest cells. Soluble factors secreted by local cells may play a critical role in cell specification. These factors may include bone morphogenetic factors, glial growth factor, fibroblast growth factor, epidermal growth factors, and platelet-derived growth factor. The ratio of different factors as well as the timing of factor secretion may all contribute to the regulation of cell specification and differentiation.

Development of Neural Tube-Derived Systems [8.3]

Fate of the Neural Tube. The *neural tube* is a straight cylindrical structure located along the anterior–posterior axis of vertebrates (superior–inferior axis in the human) and is formed during neurulation as discussed in the previous section. The neural tube gives rise to the brain and spinal cord as well as their cavities. The anterior portion of the neural tube forms three major parts of the brain: the prosencephalon (forebrain), mesencephalon

TABLE 8.1. Characteristics of Selected Molecules that Regulate the Development of Neural Cells*

Proteins	Alternative Names	Amino Acids	Molecular Weight (kDa)	Expression	Functions
Wnt 2	Wingless-type MMTV integration site family member 2, Wnt 2 protein	360	40	Embryo, fetus, thalamus	Regulating embryonic development and promoting oncogenesis
Bone morphogenetic protein 2	BMP2	396	45	Embryo, fetus, bone, cartilage	A member of the transforming growth factor β superfamily, regulating the formation of bone and cartilage
Bone morphogenetic protein 4	BMP4	408	47	Embryo, fetus, cartilage, bone, heart, lung, intestine, ovary	Regulating embryonic development and endochondral bone formation after birth
Bone morphogenetic protein 6	BMP6, VGR1	513	57	Embryo, fetus, cartilage, bone	Regulating embryonic development and inducing bone formation after birth

*Based on bibliography 8.2.

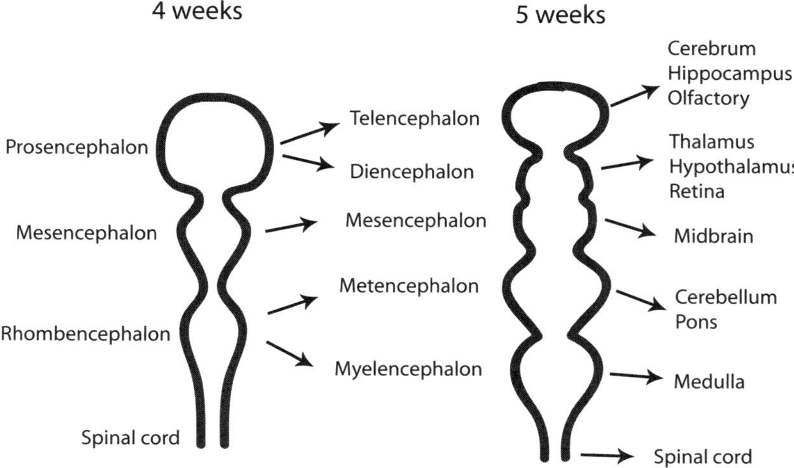

Figure 8.4. Schematic representation of brain development. Based on bibliography 8.3.

(midbrain), and rhombencephalon (hindbrain) (Fig. 8.4). The *prosencephalon* develops into the cerebrum, thalamus, hypothalamus, and retina. The *mesencephalon* gives rise to the cerebral aqueduct. The *rhombencephalon* forms the medulla oblongata and cerebellum. The posterior portion of the neural tube develops into the spinal cord.

Formation of Neurons and Glial Cells. There are two types of cell in the central nervous system: neurons and neuroglial cells. *Neurons* are the cells that can learn, memorize, sense signals from peripheral cells, and control the activity of other cell types. Each neuron is composed of a long axon, which physically connects the central and the peripheral nervous systems and is responsible for signal transduction and information transmission between the central and peripheral systems. In addition, each neuron consists of a large number of short processes known as *dendrites*, which contact other neurons within the central nervous system and are responsible for neuron–neuron communication.

Neuroglial cells serve to support and protect the neurons and to assist in neuronal development. There are three types of neuroglial cell: oligodendrocyte, astrocyte, and microglial cell. The *oligodendrocyte* can form large membrane processes called myelin sheaths, which wrap around the neuronal axons. The myelin sheaths have several functions: axon protection, molecular transport into and from the axon, maintenance of the ionic environment of the neuron, and assistance in the transmission of the action potential. The *astrocyte* also contributes to myelin formation and assists in neuronal function. The *microglial cell* behaves as a phagocyte type that degrades and removes debris from apoptotic cells.

The neuron, oligodendrocyte, and astrocyte are derived from the neural stem cells, defined as *germinal neuroepithelial cells*, in the neural tube. The local environment, to which a neural stem cell is migrating, is a critical factor that regulates the differentiation of the neural stem cells. It is believed that a neural tube cell can develop into either a neuron or glial cell depending on the local stimulation. However, the mechanisms remain poorly understood. Although the microglial cell appears as an interstitial cell type in the central nervous system, it is not derived from the ectodermal neural tube, but from the mesoderm. Mesodermal cells can migrate into the central nervous system and form microglial cells during development.

In the neural tube, neural stem cells are organized into a single-cell layer and are aligned in the direction perpendicular to the surface of the neural tube. These cells often conduct nonsymmetric division along their axial direction. The daughter cells adjacent to the luminal surface of the neural tube remain to be stem cells, whereas the daughter cells adjacent to the external surface differentiate into committed neural progenitor cells. With continuous development, the neural tube is organized into a structure with three zones: ventricular, intermediate, and marginal zones. The *ventricular zone* contains neural stem cells, the *intermediate zone* contains committed neural progenitor cells, and the *marginal zone* contain neurons and neuronal axons, depending on the subsystems of the brain such as spinal cord, cerebellum, or cerebrum. From the ventricular zone, the committed neural progenitor cells migrate outward into the intermediate zone. In the intermediate zone, different parts of the central nervous system are established. The patterns of cell migration, division, and organization vary considerably between the spinal cord, cerebellum, and cerebrum. The formation of these structures is briefly discussed here.

For the *formation of the spinal cord*, cells in the intermediate zone form axons, which extend from the intermediate zone toward the marginal zone, where the density of neurons reduces compared with the intermediate zone. With further development, established neuroglial cells migrate into the marginal zone and form myelin sheaths that enclose the axons. The neuron-containing mantle zone eventually develops into the butterfly-shaped *gray matter* of the spinal cord, whereas the axon-containing marginal zone forms the peripheral *white matter.* The terms of gray matter and white matter are defined based on the color of the structures under an optical microscope. In the spinal cord, neurons localized to different regions possess distinct functions. Neurons in the dorsal half of the spinal cord are responsible for receiving and processing sensory signals from peripheral neuronal sensors, whereas the ventral neurons are for the motor control function.

For the *formation of the cerebellum*, the pattern of neural cell migration and organization differs from that for the spinal cord. The neuronal stem cells for the development of the cerebellum are evolved to form several types of cerebellar cells, including neurons that form cerebellar nuclei, granule neurons, and Purkinje neurons. For the formation of neuronal nuclei, the neuronal progenitor cells from the neural tube migrate out of the neural tube, establish the intermediate zone, and continue to migrate forward to enter the marginal zone, where they form *neuronal nuclei*, serving as relay units between different systems of the cerebellum and between the cerebellum and other systems of the brain. Another group of neuronal progenitor cells migrates through the intermediate zone, enters the marginal zone, and forms a new layer called the *external granule layer* adjacent to the external surface. Cells in the external granule layer make a turn and migrate inward. At the interface between the marginal and intermediate zones, these cells form the granule neurons. In addition, the neural tube stem cells differentiate into progenitor cells for the Purkinje neurons. These cells migrate to the marginal zone, forming a Purkinje cell layer. The Purkinje neurons are responsible for communications between different systems of the cerebellum.

The *cerebrum* is originated from the anterior portion of the neural tube and organized into a layered structure. During cerebral formation, cerebral progenitor cells are generated from the stem cells located in the ventricular zone, pass through the intermediate zone,

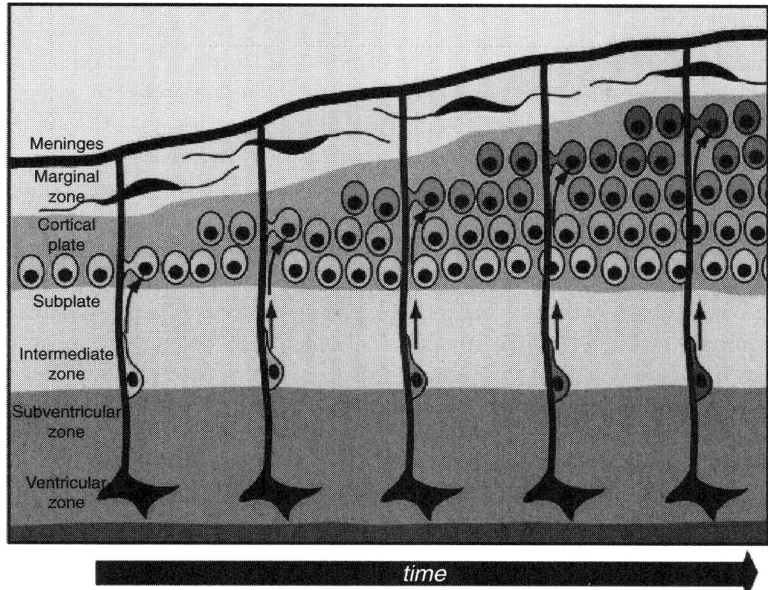

Figure 8.5. Migration and formation of cortical neurons. During development, neurons are generated in the ventricular zone, migrate along radially aligned glial cells through the subventricular zone, intermediate zone, and subplate, and are deployed to the cortical plate. (Reprinted by permission from Bielas S et al: *Annu Rev Cell Dev Biol* 20:593–618, copyright 2004 by *Annual Reviews*, www.annualreviews.org.)

and enter the marginal zone, where they form the cortical plate or the neocortex (Fig. 8.5). The neural progenitor cells in the cortical plate proliferate and differentiate into several layers of neurons in the vertical direction (perpendicular to the surface of the neural tube). Cells in different layers are responsible for the control of different peripheral systems. The cerebral neurons are also organized into a large umber of regions in the horizontal direction. Each region controls a corresponding peripheral system. In the cerebrum, neurons are concentrated in the cortical zone. This zone is referred to as the gray matter based on the color of the neuron-rich brain tissue. The intermediate zone is primarily composed of axons, which appear whitish under an optical microscope, and is referred to as the white matter.

DEVELOPMENT OF MESODERM-DERIVED ORGANS

The *mesoderm* is the middle layer of the three embryonic germ layers established during gastrulation and is the origin of a number of systems, including the cardiovascular system, lymphatic system, skeletal muscle system, bone, cartilage, connective tissue, kidney, and gonads. The mesoderm is composed of four major structures: the notochord (chordamesoderm), paraxial mesoderm, intermediate mesoderm, and lateral plate mesoderm. The organization, development, and fate of these structures are outlined here.

The Notochord [8.4]

The *notochord* is an embryonic structure established during gastrulation and located underneath the neural tube. The notochord is formed from an early embryonic structure known as the *primitive streak*, which is initiated at the posterior end of the embryo in the blastocyst stage with two blastoderm layers (epiblast and hypoblast) and plays a critical role in the formation of the mesoderm and the endoderm. The primitive streak progresses toward the anterior end along the anterior-posterior axis. At the same time, the centerline of the primitive streak depresses, forming the *primitive groove*. When the primitive streak reaches the anterior end, there forms a node structure called *Hensen's node* at the anterior end (superior or cranial end in humans) of the primitive streak. The mesoderm and endoderm are formed by the primitive streak cells, which pass through the primitive streak layer, move from the dorsal to the ventral direction, and spread on the inner surface of the epiblast, forming the mesoderm and the endoderm. These cells form two layers ventral to the epiblast: the middle layer and the internal layer. The former develops into the mesoderm and the later endoderm. A group of primitive streak cells migrates through the Hensen's node and forms the notochord along the anterior–posterior axis. During such a process, the epithelial type of primitive streak cells is transformed into mesenchymal cells. The notochord is a transient structure, which mediates the formation of the neural tube and defines the anterior–posterior axis of the body. In mammals, the notochord cells eventually form part of the endoderm and contribute to the formation of the primitive gut.

The Paraxial Mesoderm

Formation of the Somites [8.5]. The *paraxial mesoderm*, also known as the somatic dorsal mesoderm, is formed by cells from the primitive streak as described above and organized into two parallel columns that are aligned along the neural tube one at each side. The paraxial mesoderm forms segmented blocks called *somites*, also known as *mesodermal segments*. These somites develop into the vertebrae, ribs, skeletal muscles, and dermis. These structures serve as pathways for the extension of the spinal nerve axons and the migration of neural crest cells, which develop into the sympathetic and parasympathetic neurons.

The formation of the somites begins at the anterior portion of the primitive streak. As a first step, paraxial mesodermal cells are organized into two unsegmented presomitic cell clusters called *somitomeres*, which are aligned along the neural tube one at each side. The somitomere at the anterior end is first separated into individual somites. The tip portion of the somitomere adjacent to an established somite is subsequently separated from the main body of the somitomere. The segmentation process continues from the anterior to the posterior directions (superior to inferior direction in humans) until the entire somitomere is segmented. The segmentation of the somitomere is regulated by several molecules, including hairy 1, ephrin, and ephrin tyrosine kinase receptor. Hairyl encodes a transcription factor, which can activate the expression of the ephrin and ephrin tyrosine kinase receptor genes. In the somitomere, hairy 1 is expressed in the posterior region of a premature somite, leading to local activation of ephrin and ephrin tyrosine kinase receptor. Ephrin can interact with the ephrin tyrosine kinase receptor, triggering the activation of inhibitory signaling pathways. Such an activity results in local suppression of cell interaction with substrate and enhancement of somite separation from the somitomere. This pattern of molecular activities repeats along the somitomere from the anterior to the posterior direction until the completion of the whole set of somites. The primitive streak is gradually diminished during the segmentation of the somitomere.

Each pair of somites alongside the neural tube is the origin of the muscular and connective tissue specified for a designated location. Each somite, once established and matured, is committed to the specification of a tissue and its developmental course cannot be altered. This phenomenon has been demonstrated by transplantation of a somite from one location to another. Tissues developed from the transplanted somite will appear in the peripheral location of the tissues that are supposed to form from the original somite. For example, when a somite from the thoracic region is transplanted to the cervical region of a chick, ribs will form in the cervical region on the transplantation side. Overall, somites develop into several types of tissues and structures, including the vertebra, rib, limb, skeletal muscle, connective tissue, and back dermis.

At each level, the somite is organized into distinct regions in the transverse direction, including the ventral region, two lateral regions, and dorsal region. The *ventral region* develops into a group of mesenchymal stem cells collectively called the *sclerotome*. Cells in this structure give rise to chondrocytes and osteoblasts of the vertebrae, ribs, and other bones. The two *lateral regions*, one adjacent to and the other farthest from the neural tube, develop into *myotomes*, which form myoblasts, precursors of skeletal muscle cells. The lateral myotome adjacent to the neural tube develops into the back skeletal muscles, whereas the lateral myotome farthest from the neural tube form skeletal muscles in the limbs and other body parts. The *dorsal region* of a somite develops into the *dermatome*, which is the origin of the dermal tissue in the back area. The formation of these distinct somite regions is mediated by the activation of paracrine molecules, including sonic hedgehog, Wnt1, Wnt3a, neurotrophin 3, and bone morphogenetic protein 4 and 5. These factors are expressed and released from the structures adjacent to the somites, such as the neural tube and the lateral plate mesoderm. The location- and time-dependent activation of these factors may determine the specification and commitment of the cells in the sclerotome, myotomes, and dermatome.

Formation of the Skeletal Muscle System [8.6]. As discussed above, the cell lineages in the lateral myotomes develop into skeletal muscles. The myotome contains muscular progenitor cells called *myoblasts*. These cells are prompted to enter a process known as *myogenesis*, the generation of muscular cells. Myogenesis is initiated and enhanced by transcriptional factors MyoD and Myf5, which are myogenic regulatory factors. These factors activate the expression of necessary genes that encode proteins for initiating myogenesis.

During myogenesis, one of the genes that are upregulated is the fibroblast growth factor (FGF) gene. In the presence of increased FGF, myoblasts expand extensively via proliferation. These cells also produce and release extracellular matrix components such as collagen and fibronectin. At a certain time, the FGF level is suppressed, leading to a reduction in the rate of myoblast proliferation. The myoblasts attach to the extracellular matrix through the interaction of the cell membrane integrins, such as $\alpha5\beta1$, with selected extracellular matrix components, such as fibronectin. All myoblasts within a muscular bundle are aligned in the same direction by the mediation of cell membrane proteins such as cadherins and cell adhesion molecules (CAMs). The interaction between different myoblasts and between myoblasts and extracellular matrix induce an increase in the level of calcium, which initiates and regulates myoblast fusion, the formation of multinucleated muscular cells from individual myoblasts. This process is also mediated by a metalloproteinase, known as meltrin α, which degrades extracellular matrix components and enhances myoblast fusion.

Formation of the Skeleton [8.7]. The skeleton is composed of bones and cartilages. These structures are developed from two mesodermal systems, including the somites and lateral plate mesoderm, and an ectodermal system, the cranial neural crest. The somites give rise to the trunk skeleton (vertebrae and ribs), the lateral plate mesoderm gives rise to the limb skeleton, and the cranial neural crest gives rise to the craniofacial skeleton. For the ectodermal source of the skeleton, the neural crest cells are transformed from an ectodermal cell type to a mesenchymal type, which forms skeletal tissues.

There are two mechanisms for the formation of the bone: *direct osteogenesis* from soft mesenchymal tissue and *ossification of cartilage*. The craniofacial bones are formed through direct osteogenesis. In such a process, the cranial neural crest cells migrate from the neural crest to the skull, transform from ectodermal cells to mesenchymal cells, undergo extensive proliferation, and form a mesenchymal tissue. A fraction of the neural crest-derived mesenchymal cells differentiates into *osteoblasts* or bone progenitor cells. The osteoblasts can produce and release collagen and specific proteoglycans, which serve as a matrix for the formation of the bone. This matrix can bind and retain calcium, which results in the calcification or *ossification* of the matrix. All osteoblasts within the calcified matrix are considered mature bone cells and are defined as *osteocytes*. Cells adjacent to the periosteum or the surface membrane of the calcified matrix structure retain the features of osteoblasts and can differentiate into osteocytes and produce bone matrix. Several bone morphogenetic proteins, including BMP2, BMP4, and BMP7, play a critical role in osteogenesis by stimulating the differentiation of osteoblasts and the production of bone matrix components.

Most bones, including the vertebrae, ribs, and limb bones, originate from the somites and lateral plate mesoderm, are developed via cartilage formation and ossification of cartilage. There are several steps for bone morphogenesis: (1) formation of chondrocytes, (2) formation of cartilages, (3) initiation of osteogenesis, and (4) mineralization of bone matrix and formation of bone. *Chondrocytes* are formed from mesenchymal progenitor cells derived from the somites and lateral plate mesoderm under the stimulation of local osteogenic factors. The newly formed chondrocytes expand extensively via proliferation and generate extracellular matrix components that are necessary for the formation of cartilage. Extracellular matrix and chondrocytes are integrated and organized to form *cartilage*. Within the cartilage, established chondrocytes are transformed into hypertrophic chondrocytes characterized with an increase in cell volume. These chondrocytes can produce collagen type X and fibronectin. These matrix components absorb calcium and phosphate into the matrix, initiating *matrix mineralization* or *osteogenesis*. The hypertrophic chondrocytes also produce and release vascular endothelial growth factor (VEGF), which stimulates angiogenesis or blood vessel formation within the cartilage. The vascular system supplies oxygen and nutrients to the newly generated bone. The hypertrophic chondrocytes undergo apoptosis shortly after these initiating processes for bone formation. As a last step, mesenchymal progenitor cells are transformed into osteoblasts, which generate and release bone-forming extracellular matrix components and stimulate matrix mineralization. A bone forms when a sufficient amount of calcium and phosphate is deposited to the bone matrix.

The Intermediate Mesoderm: Formation of the Kidney [8.8]

The intermediate mesoderm gives rise to the *urogenital system*, including the kidney and associated duct systems as well as gonads. For kidney development, there are two critical stages: the formation of *transient kidneys* and the formation of *permanent kidneys*. For the formation of the transient kidneys, the intermediate mesoderm initiates the formation of two *pronephric ducts* at the anterior portion of the mesoderm during the early embryonic stage (about 20 days in humans). This process is symmetric with respect to the neural tube from both parts of the intermediate mesoderm. The formation of the pronephric ducts progress continuously toward the posterior end. Established pronephric ducts interact with adjacent mesodermal tissue to form *pronephroi*, which are side branches attached to the anterior portion of the pronephric ducts. When the formation of the pronephric ducts progresses to the middle portion of the mesoderm, the pronephric ducts stimulate the development of *mesonephroi*, which constitute the major *excretory system* of the embryo. The pronephric ducts further progress to the posterior end, where the two pronephric ducts meet to form the *cloaca*, a presumptive structure for the formation of the bladder, the rectum, and the genital system. During the development of the mesonephroi, the anterior end of the pronephric system undergoes progressive degeneration. This degenerative process continues when the posterior end of the pronephric ducts form new mesonephroi. Shortly after the cessation of mesonephros generation, the mesonephroi are degenerated via cell apoptosis. Thus, the mesonephroi serve only as a temporary excretory system.

While the anterior and middle portions of the pronephric system are degenerated, the posterior portion of the system remains. A side branch, known as the *ureteric bud*, forms from each of the two pronephric ducts in the posterior portion. The ureteric bud serves as the presumptive structure for the formation of the two *permanent kidneys*. The ureteric bud interacts with and stimulates surrounding mesenchymal tissue, inducing the formation of the *metanephrogenic mesenchyme* (Fig. 8.6). At the same time, the metanephrogenic mesenchyme stimulates the ureteric bud to extend and branch into the mesenchyme. The ureteric bud-derived branching system serves as a kidney rudiment, which eventually develops into the kidney. At the end of each established ureteric branch, epithelial cells from the metanephrogenic mesenchyme form a cell cluster and undergo proliferation and differentiation, resulting in the formation of specified kidney cell types, including the glomerular capsule cells, endothelial cells, and tubular cells. These cells are assembled into functional kidney units known as the *nephrons*. The cells in each nephron are integrated into the structure at the tip of each ureteric branch, establishing a connection between the nephron and the ureteric bud branch, which eventually forms the renal collecting duct. Urine forms in the nephrons and is conveyed to the ureters via the collecting ducts.

The Lateral Plate Mesoderm

The lateral plate mesoderm is located farthest from the central neural tube and is composed of two layers: the somatic mesoderm and the splanchnic mesoderm. The *somatic mesoderm* is the layer adjacent to the ectoderm, and the *splanchnic mesoderm* is the layer next to the endoderm. These mesodermal structures develop into the heart, blood vessels, and blood cells. There is a gap between the two mesodermal layers, which is known as the *coelom*. The coelom forms the three body cavities, including the pleural (thoracic), pericardial (cardiac), and peritoneal (abdominal) cavities.

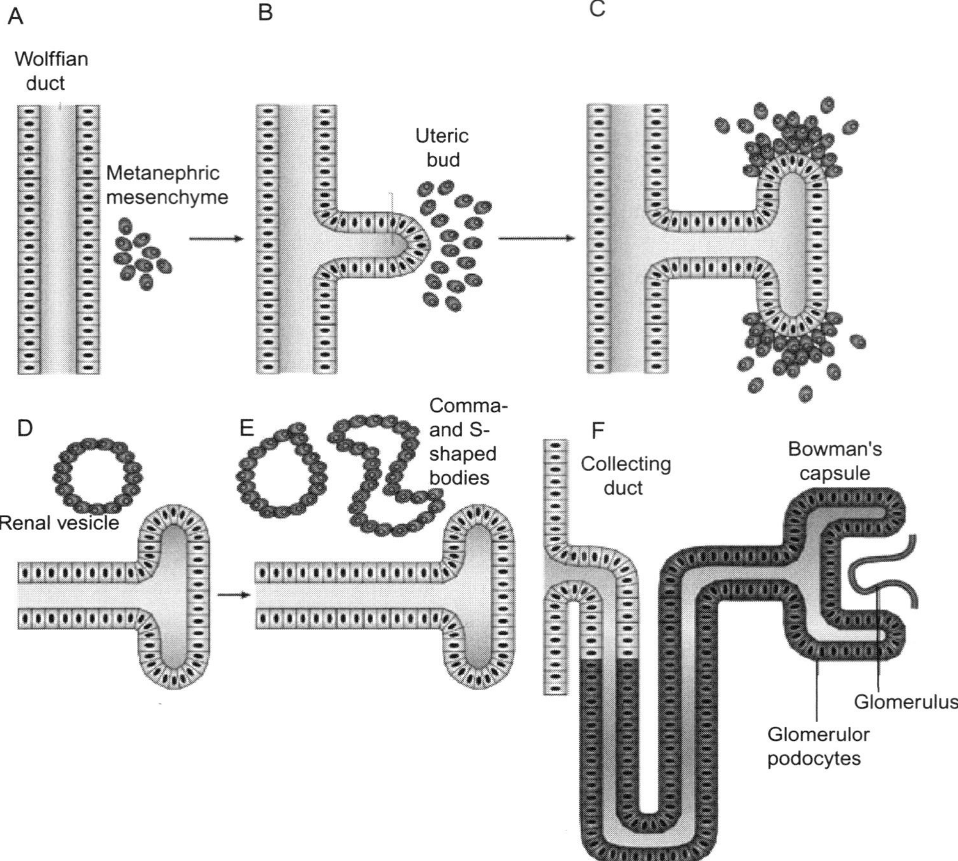

Figure 8.6. Processes of nephron formation. (A) The metanephric mesenchyme interacts with the Wolffian duct of the intermediate mesoderm. (B) At about embryonic day 10.5 (E10.5), the ureteric bud forms from the Wolffian duct and invades the metanephric mesenchyme and induces differentiation. (C) At about E11.5, the mesenchymal cells condense to form aggregates near the ureteric bud. Branches form from the ureteric bud. (D) At about E12.5, the condensed mesenchyme undergoes a mesenchymal-to-epithelial transformation and forms epithelial structures (renal vesicles). (E) At about E13.5, complex epithelial structures, comma- and S-shaped bodies, are formed. (F) From E14.5 to E16.5, the S-shaped bodies fuse into the ureteric bud derivatives, inducing the formation of mature nephron. The ureteric bud develops into collecting ducts. Endothelial cells migrate into the nephron and form the glomerulus, which is surrounded by glomerular podocytes (Bowman's capsule). Selected transcription factors and signaling molecules expressed in the uretic bud and metanephic mesenchyme are shown. WT1 (Wilms' tumor 1), PAX2 (paired box 2), and EYA1 (eyes absent 1) are transcription factors expressed in the metanephric mesenchyme and regulate ureteric bud induction. GDNF (glial-cell-line-derived neurotrophic factor) is also expressed in the metanephric mesenchyme and stimulates ureteric bud induction and branching by interacting with its receptor RET. WT1 and bone morphogenetic protein 7 (BMP7) play a role in regulating the survival of the metanephric mesenchyme. WNT4 is essential for the mesenchymal-to-epithelial transformation of the metanephric mesenchyme and might be induced by WNT6 signals from the ureteric bud. BMP4 is involved in ureteric bud growth, and WNT11 regulates branching of the ureteric bud. GPC3 modulates signals between the metanephric mesenchyme and the ureteric bud, including bone morphogenic proteins. (Reprinted by permission from Macmillan Publishers Ltd.: Rivera MN, Haber DA: *Nature Rev Cancer* 5:699–712, copyright 2005.)

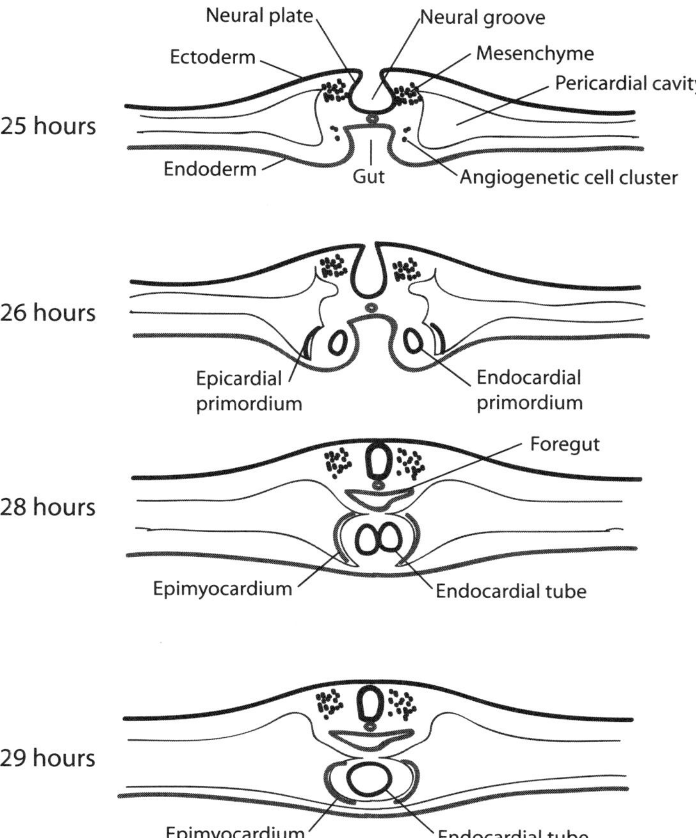

Figure 8.7. Early developmental stages of the chick heart. (After Calson BM: *Pattern Formation of Embryology*, McGraw-Hill, New York, 1981.)

Formation of the Heart [8.9]. The *presumptive cardiac cells* are specified during the early embryonic stage when the primitive streak forms. These cells migrate through the primitive streak and form two *cardiogenic mesodermal clusters*, which are symmetrical with respect to the primitive streak, in the splanchnic mesoderm at the level of the Hensen's node (Fig. 8.7). The two presumptive cardiac cell clusters move toward each other and form two *endocardial primordia*, the initial form of the heart. At the same time, the adjacent mesodermal cells are stimulated to transform into epithelial cells, forming the *pericardium*, which encloses the endocardial primordia, and the endocardium, which lines the internal surface of the heart. Within about 3 weeks in the human, the two endocardial primordia meet and fuse into each other to form a single *myocardial tube*. Cells with cardiomyocyte phenotypes are established during endocardial primordial fusion and initiate the contractile activity, causing pulsations of the endocardial primordia.

Within about 5 weeks, the myocardial tube is developed into a tube with two chambers: the atrium and ventricle (Fig. 8.8). A fraction of cells from the endocardium forms a structure called *endocardial cushion*, which separates the two-chambered tube into the left and right cardiac canals. At about the same time, the atrial septum develops from the

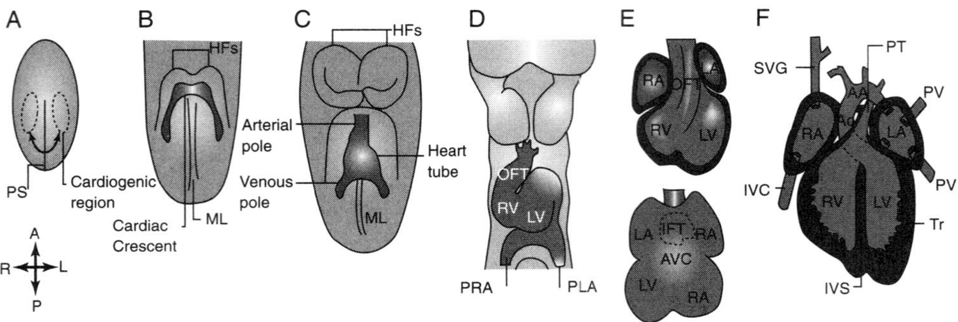

Figure 8.8. Development of the heart. (A) Myocardial progenitor cells originate in the primitive streak (PS) and migrate to the anterior of the embryo at about embryonic day E6.5. (B) The myocardial progenitor cells settle under the head folds (HF) and form the cardiac crescent at about E7.5. (C) The early cardiac tube forms through fusion of the cardiac crescent at the midline (ML) at about E8. (D) The cardiac tube forms a loop at about E8.5. (E) The cardiac tube further develop into chamber structures by about E10.5, but the chambers are still connected. (F) By about E14.5, the cardiac chambers are separated. The left and right ventricles are connected to the pulmonary trunk (PT) and aorta (Ao), respectively. (*Abbreviations*: A, Anterior; P, posterior; R, right; L, left; AA, aortic arch; AVC, atrioventricular canal; IFT, inflow tract; IVC, inferior vena cava; IVS, interventricular septum; OFT, outflow tract; PLA primitive left atrium; PRA, primitive right atrium; PV, pulmonary vein; SVC, superior vena cava; Tr, trabeculae.) Reprinted by permission from Macmillan Publishers Ltd.: Buckinghamm et al: *Nature Rev Genet* 6:826–37, copyright 2005.)

atrial portion of the myocardial tube, while the ventricular septum grows from the ventricular portion. Both septa grow toward the endocardial cushion and separate the myocardial tube into a four-chamber structure—the heart. With further development, the heart is connected to the aorta, pulmonary arterial trunk, and vena cava, which are developed simultaneously with the heart.

The *embryonic circulation* is established when the heart and blood vessels form. The embryonic circulation is different from the *postnatal circulation*. The difference is caused by distinct functional anatomy of the cardiovascular, pulmonary, and intestinal systems between the prenatal and postnatal stages. During the prenatal stage, the embryo or fetus does not have a functional lung and intestine. Oxygen and nutrients are supplied from the mother's blood through the placental circulation. During the stage of the two-chamber heart (about 5 weeks in humans), the umbilical venous blood obtains oxygen and nutrients from the placenta, and the veins conduct oxygenated blood to the heart. The heart pumps blood into the arterial system, which supplies blood to the peripheral systems. The arterial system also conveys blood to the placenta, where metabolic wastes are transported to the mother's circulation.

During the stage of the four-chamber heart when the basic structures of the vascular and other systems are formed, the umbilical veins convey oxygenated blood to the vena cava and subsequently to the right atrium. Unlike the anatomy of the postnatal heart, the right and left atria are connected through the foramen ovale, which is open during the prenatal stage. In addition, the ductus arteriosus between the aorta and pulmonary arterial trunk is also open in the embryo or fetus. Oxygenated blood from the umbilical veins can directly enter the left atrium and left ventricle through the open foramen ovale. The left ventricle pumps oxygenated blood into the arterial system, which supplies blood to the peripheral systems. Oxygenated blood enters the pulmonary circulation from the right ventricle. A fraction of blood is diverted into the aorta through the ductus arteriosus. Deoxygenated blood with metabolic wastes from the lung returns directly to the left heart, while deoxygenated blood from peripheral organs returns to the right heart. Thus there is a significant degree of mixing between oxygenated and deoxygenated blood in the prenatal circulatory system. Deoxygenated blood with metabolic wastes is conveyed to the placenta for excretion via the umbilical arteries, which bifurcate from the common iliac arteries.

When a baby is born, the umbilical circulation no longer supplies oxygenated blood. A number of anatomical changes take place immediately during the early postnatal period: (1) the lung starts to expand and function, ensuring a transition of gas exchange from the placenta to the pulmonary system; (2) the foramen ovale closes to stop blood diversion from the right to the left atrium; (3) the ductus arteriosus closes to prevent blood diversion from the pulmonary arterial trunk to the aorta; and (4) the umbilical arteries and veins are committed to degeneration. These drastic anatomical changes ensure the establishment of the postnatal circulation.

Formation of the Vascular System [8.10]. Blood vessels are formed from the splanchnic mesoderm, occurring simultaneously with heart generation. As a first step, the mesodermal cells differentiate into *hemangioblasts*, which are progenitor cells for blood vessels as well as blood cells. The hemangioblasts migrate to designated tissues and organs, where they form cell clusters known as *blood islands* (Fig. 8.9). A blood island is composed of two types of cell: *hematopoietic stem cells* (inner cells of the blood island) and *angioblasts* (outer cells). The hematopoietic stem cells develop into blood cell types, including erythrocytes, leukocytes, and platelets, whereas the angioblasts give rise to vascular cells, including endothelial cells, smooth muscle cells, and fibroblasts. There are two basic processes for the formation of blood vessels: vasculogenesis and angiogenesis. *Vasculogenesis* is the formation of primary endothelial cells and capillaries, which occurs during the embryonic stage. *Angiogenesis* is the formation of arteries, veins, and capillaries based on the established capillaries (Fig. 8.10). Angiogenesis occurs not only during the embryonic stage but also during the adulthood in response to injury and pathological disorders, such as cancer and hypertrophy. For vasculogenesis, the angioblasts differentiate into endothelial cells, a cell type that lines the internal surface of a blood vessel. The endothelial cells subsequently form tube-shaped capillaries, which are linked together into a capillary network known as the *primary capillary plexus*. These capillaries are further developed into arteries, veins, and additional capillaries via angiogenesis. During angiogenesis, endothelial cells undergo proliferation and sprout to form new blood vessels. Pericytes are recruited to newly formed capillaries and transformed into smooth muscle cells, eventually leading to the formation arteries and veins.

Several growth factors, including fibroblast growth factor (FGF), vascular endothelial growth factor (VEGF), platelet-derived growth factor (PDGF), and angiopoietin, play important roles in the regulation of vasculogenesis and angiogenesis. Fibroblast growth factor is produced by mesodermal cells and released into the interstitial space. This growth factor can stimulate the differentiation of mesodermal cells into hemangioblasts. Vascular endothelial growth factor is also produced by mesodermal cells. This factor directly stimulates angioblasts to differentiate into endothelial cells and stimulates endothelial cells to form capillaries. The effect of VEGF is dependent on the type of its receptor on

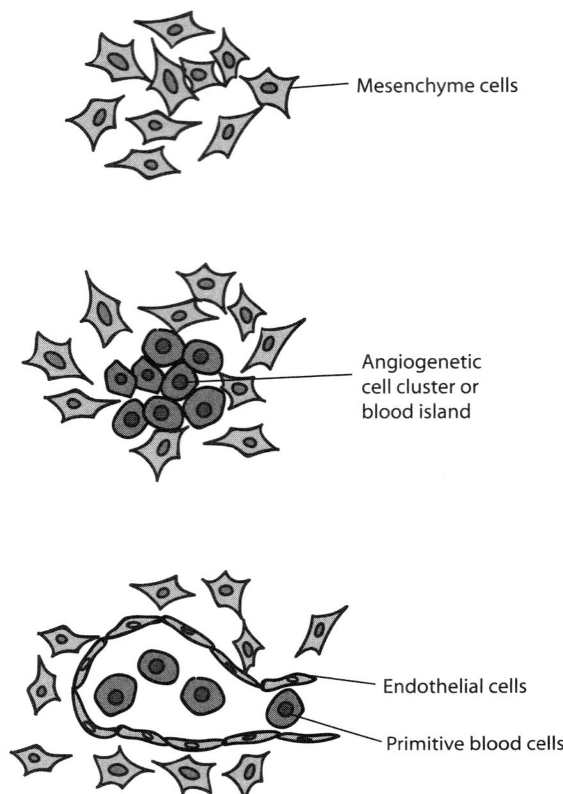

Figure 8.9. Schematic representation of embryonic blood formation and vasculogenesis. (After Langman J: *Medical Embryology*, 4th ed., Williams & Wilkins, Baltimore, 1981.)

the cell surface. There are two types of VEGF receptor: flk1 (VEGF receptor 2) and flt1 (VEGF receptor 1). The activation of flk1 induces the differentiation of angioblasts into endothelial cells. The activation of flt1 results in the formation of endothelial tubes. Platelet-derived growth factor stimulates the recruitment of pericytes to newly formed blood vessels and the transformation of these cells to smooth muscle cells. Another factor, angiopoietin, interacts with its receptor Tie2 and regulates the recruitment of smooth muscle cells and the formation of blood vessels. Thus, these growth factors are essential for both vasculogenesis and angiogenesis.

Formation of Blood Cells [8.11]. *Blood cells* are a family of circulating cells, including erythrocytes, leukocytes (monocytes, neutrophils, basophils, eosinophils, and T and B lymphocytes), and platelets. Blood cells are developed from the hematopoietic stem cells derived from the mesodermal hemangioblasts. The process of blood cell formation is defined as *hematopoiesis*, which is a continuous process through the lifespan because of constant death of blood cells. There are two types of hematopoiesis: embryonic and adult hematopoiesis, which occur during and after the embryonic stage, respectively. During the *embryonic* stage, *hematopoiesis* initially occurs in the blood islands derived from the lateral plate mesoderm (Fig. 8.9). The blood islands, however, are not permanent sites for

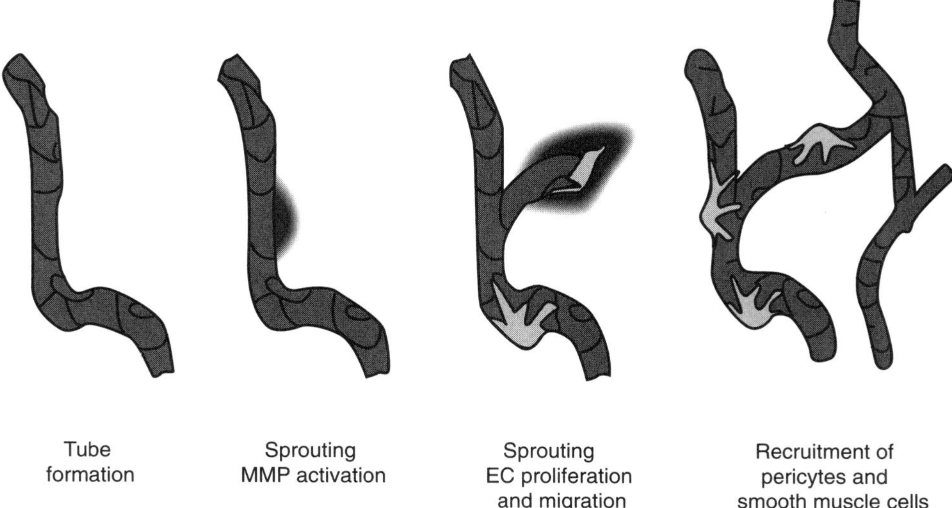

Figure 8.10. Schematic representation of the blood vessel formation, which involves endothelial cell tube formation, endothelial cell proliferation and migration, sprouting angiogenesis, and pericyte recruitment. (Reprinted from Hallmann R et al: *Physiol Rev* 85:979–1000, 2005 by permission of the American Physiological Society.)

hematopoiesis. The site of hematopoiesis changes for several times during embryogenesis. The first change occurs when the blood islands are transformed into blood vessels. The site of hematopoiesis is shifted from the blood islands to structures near the aorta known as the *aorta–gonad–mesonephroi*. With the degeneration and disappearance of the mesonephroi, the site of hematopoiesis is moved to the liver. During the late embryonic stage, hematopoiesis takes place in the bone marrow. During the adulthood, the bone marrow is a permanent site for hematopoiesis. Blood cells die constantly during the lifespan and dead cells are replaced with newly generated cells from the bone marrow.

In the bone marrow, there exist pluripotent *hematopoietic stem cells*, which can self-renew, proliferate, and differentiate into all mature blood cell types during the embryonic stage as well as the adulthood. The percentage of hematopoietic stem cells in the bone marrow is about 0.01%. Such a small fraction of cells can replenish the entire family of blood cells. Hematopoietic stem cells can differentiate into lineage progenitor cells, including B- and T-lymphocyte progenitor cells and myeloid progenitor cells. The *B-lymphocyte progenitor cells* can differentiate into B-lymphocytes and plasma cells through several intermediate cell lineages, including the pro-B cell and pre-B cell lineages. The *T-lymphocyte progenitor cells* can differentiate into T-lymphocytes through cell lineages, including the pro-T cell, pre-T cell, and immature thymocyte lineages. The *myeloid progenitor cells* can differentiate into at least five types of lineages: erythroid precursor (CFU-E), platelet precursor (CFU-MK), monocyte precursor (CFU-M), neutrophils precursor (CFU-G), basophil precursor (CFU-BUO), and eosinophil precursor cells (CFU-Eo). The *erythroid precursor cells* can differentiate subsequently into proerythroblasts, erythroblasts, reticulocytes (with nuclei removed), and erythrocytes. The *platelet precursor cells* can develop into immature megakaryocytes and megakaryocytes, which split into platelets. The *monocyte precursor cells* can give rise to monocytes, which transform into macrophages when migrating into the blood vessel wall and interstitial space. The *neutrophil*, *basophil*, and *eosinophil precursor cells* can differentiate into neutrophils, basophils, and eosinophils, respectively. Mature blood cells are generated in the bone marrow and transported to the vascular system.

DEVELOPMENT OF ENDODERM-DERIVED ORGANS

The *endoderm* is the innermost layer of the three embryonic germ layers and gives rise to the lining and gland structures of the digestive and respiratory systems, including the gastrointestinal tract, liver, pancreas for the digestive system and the lung and airways for the respiratory system (Fig. 8.11). In addition, the endoderm plays an important role in regulating the formation of the mesodermal tissues and organs, such as the heart and blood vessels, via interaction with mesodermal cells and release of soluble mediating factors. Here, we focus on the formation of the digestive and respiratory systems.

Formation of the Digestive Tract [8.12]

The digestive system develops from an early endodermal structure known as the *primitive gut*. In the human, the primitive gut forms at about 16 days following the conception and is composed of three parts in the early stage: the foregut, the midgut, and the hindgut. At about 22 days, the *liver bud* forms from the foregut and is the presumptive structure for the formation of the liver. At about 28 days, the anterior end of the foregut opens up to form the *oral opening*, which is the presumptive structure of the mouth. The foregut also gives rise to the pharynx, esophagus, thyroid bud, and lung bud, and stomach at about the

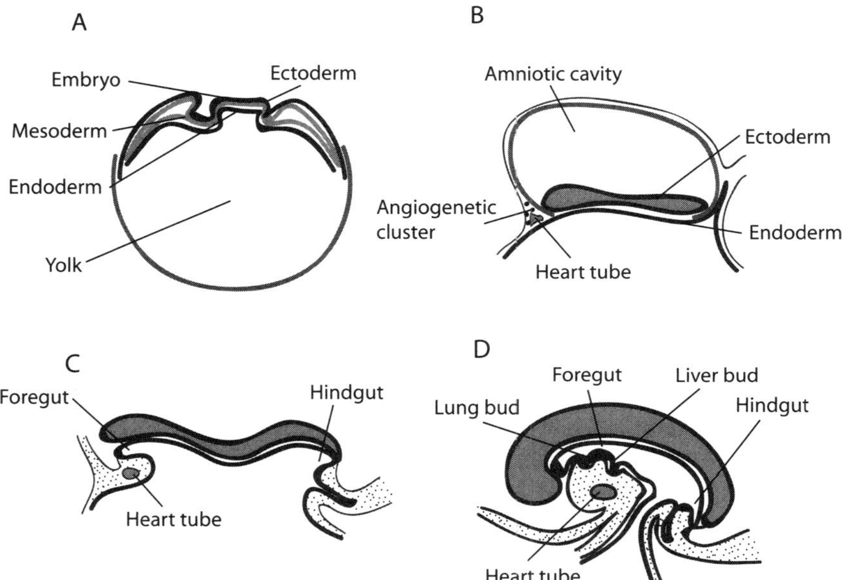

Figure 8.11. Schematic representation of the development of the endodermal organs. Based on bibliography 8.12.

same time. The midgut and hindgut are the presumptive structures for the small and large intestines.

In the anterior region of the primitive gut, there forms the *pharynx primordium*, a structure composed of several clusters of cells known as the *pharyngeal pouches*. These pouches give rise to a number of tissues and organs, including the middle ear, tonsil, thymus, thyroid gland, parathyroid gland, the respiratory tube, and pharynx. Following the establishment of the pharynx, the primitive gut forms subsequently the esophagus, stomach, and small and large intestines (Fig. 8.11). It is important to note that cells in the primitive gut only give rise to the internal epithelial cells and gland cells of the gastrointestinal system. The smooth muscle cells and fibroblasts in the submucosa, muscular, and adventitial layers are derived from the lateral plate mesoderm. The mesodermal cells not only directly contribute to the construction of the digestive tract, but also participate in regulating the specification of the endodermal cells. The interaction of endodermal cells with local mesodermal cells stimulates the specification of the endodermal cells to designated endodermal tissues and organs. Regional expression of different regulatory factors from the mesoderm and endoderm may influence the specification of the endodermal cells.

Formation of the Liver [8.13]

The primitive structure of the liver, known as the *liver bud*, sprouts from the foregut of endodermal primitive gut (Fig. 8.12). The liver bud grows into the surrounding mesodermal tissue, which plays a critical role in the formation of the liver. The mesodermal

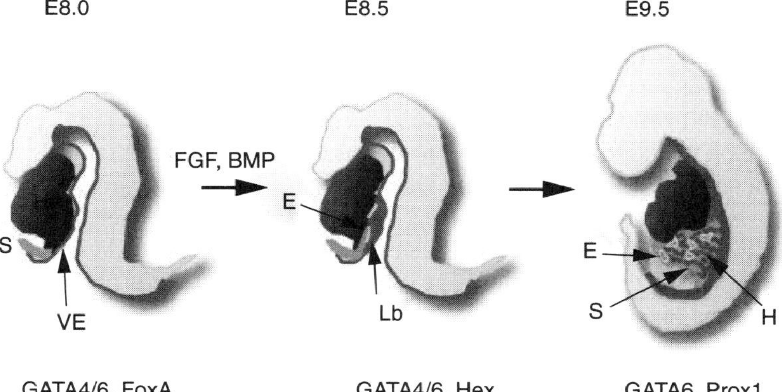

Figure 8.12. Early development of the mouse liver. By about embryonic day 8 (E8.0), a ventral portion of the endoderm near the developing heart (He) is stimulated to initiate the formation of the liver bud (Lb) in response to bone morphogenetic protein (BMPs) and fibroblast growth factor (FGFs). By embryonic day 8.5, the specified hepatic endoderm forms the liver bud, which expresses several liver-specific mRNAs, including that for albumin. Endothelial cells (E) are formed around the liver bud and are necessary for the liver bud development. By embryonic day 9.5, the nascent hepatoblasts (H) delaminate from the ventral endoderm (VE) and invade the septum transversum (S) mesenchyme, which is the source of stellate cells as well as sinusoidal endothelial cells. (Reprinted from Zhao R, Duncan SA: *Hepatology* 41:956–67, 2005 by permission of Wiley-Liss, Inc., a subsidiary of John Wiley & Sons, Inc.)

cells produce and release regulatory factors that stimulate the liver bud cells to proliferate and differentiate into various hepatic cells, such as hepatocytes and duct epithelial cells. In particular, the presence of vascular endothelial cells, which form from the mesoderm, is necessary for the formation of the liver. The removal of vascular endothelial cells results in the failure of liver formation. Gallbladder is an affiliated structure of the liver and is formed based on a branch from the hepatic drainage duct.

Formation of the Pancreas [8.14]

The *pancreas* is originated from the endodermal foregut near the liver bud. As a first step, two pancreatic rudiments form at about 30 days in the human: one from the foregut called the *dorsal pancreatic bud* and the other from the hepatic duct called the *ventral pancreatic bud*. These buds subsequently develop into the dorsal and ventral pancreas, respectively, at about 35 days. The two pancreatic structures fuse together to form the pancreas. The formation of the pancreas is mediated by mesodermal cells. Cells from the mesodermal notochord can produce and release soluble regulatory factors, such as fibroblast growth factor 2 and activin, which regulate the differentiation of endodermal cells into pancreatic cells. In addition, vascular endothelial cells are involved in the formation of pancreas. In the absence of vascular endothelial cells, the endodermal cells are unable to differentiate into pancreatic progenitor cells. In particular, endothelial cells can release regulatory factors that mediate the formation of insulin-secreting endocrine cells.

Formation of the Lung [8.15]

The lung is derived from the lung rudiment sprouted from the digestive foregut. The lung rudiment first grows into the trachea, bifurcates into the left and right bronchi, and subsequently establishes the left and right lungs (Fig. 8.13). The mesenchymal cells of the mesoderm interact with the epithelial cells derived from the endoderm and play an important role in the formation of the respiratory system. For instance, when embryonic rat tracheal epithelial cells are cultured in the absence of mesenchyme, the epithelial cells will not develop into airway structures. In contrast, when lung epithelial cells are cultured in the presence of lung mesenchyme, airway-like structures form at the distal end (Fig. 8.14). A regional difference in the characteristics of the mesenchymal cells may determine the specification of different cell types of the respiratory system. The lung is not functional during the embryonic stage, when the embryo or fetus obtains oxygen from the placenta. The lung initiates gas ventilation and exchange immediately after birth.

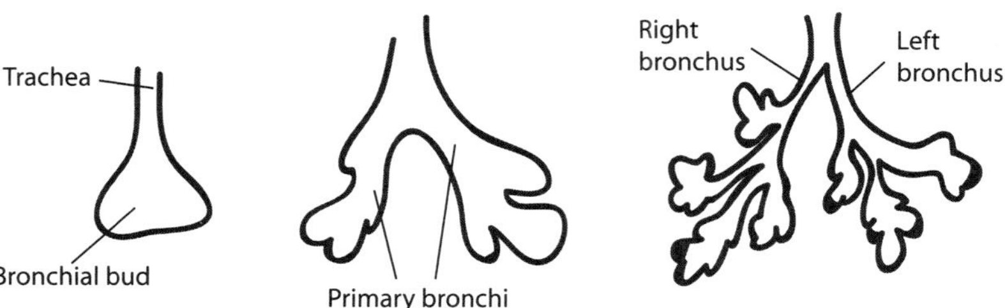

Figure 8.13. Schematic representation of the lung development. Based on bibliography 8.15.

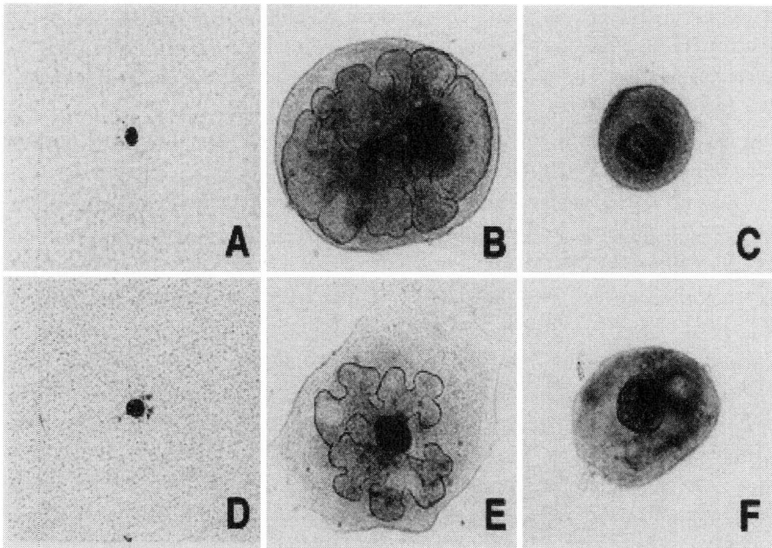

Figure 8.14. The development of the lung is mediated by the interaction of epithelial cells with mesenchymal cells. When embryonic rat lung (A) or tracheal epithelial cells (D) are cultured in Matrigel in the absence of mesenchyme, the epithelial cells will not develop into pulmonary structures. In contrast, when lung epithelial cells are cultured in the presence of lung mesenchyme (B), branching structures form at the distal end. Interestingly, when lung epithelial cells are cultured in the presence of tracheal mesenchyme (C), epithelial cell growth, but not branching, was found. Tracheal epithelial cells also exhibit similar activities (F). When tracheal epithelial cells were cultured in the presence of lung mesenchyme (E), the cells grow and form branches, lading to the development of a lung-like pattern. (Reprinted with permission from Shannon JM, Hyatt BA: *Annu Rev Physiol* 66:625–45, copyright 2004 by *Annual Reviews*, www.annualreviews.org.)

BIBLIOGRAPHY

8.1. Epidermal Development

Gilbert Scott F: Developmental Biology, 7th ed, Sinauer Associates, Sunderland, UK, 2003, pp 183–219.

Han M, Yang X, Taylor G, Burdsal CA, Anderson RA et al: Limb regeneration in higher vertebrates: Developing a roadmap, *Anat Rec B New Anat* 287:14–24, 2005.

Wolpert L: Pattern formation in epithelial development: the vertebrate limb and feather bud spacing, *Phil Trans R Soc Lond B Biol Sci* 353:871–5, 1998.

8.2. Development of Neural Crest-Derived Systems

Gilbert Scott F: *Developmental Biology*, 7th ed, Sinauer Associates, Sunderland, UK, 2003.

Garcia-Castro M, Bronner-Fraser M: Induction and differentiation of the neural crest, *Curr Opin Cell Biol* 11:695–8, 1999.

Balmer CW, LaMantia AS: Noses and neurons: Induction, morphogenesis, and neuronal differentiation in the peripheral olfactory pathway, *Dev Dyn* 234:464–81, 2005.

Jessen KR, Mirsky R: The origin and development of glial cells in peripheral nerves, *Nat Rev Neurosci* 6:671–82, 2005.

Raible DW, Ragland JW: Reiterated Wnt and BMP signals in neural crest development, *Semin Cell Dev Biol* 16:673–82, 2005.

Cornell RA, Eisen JS: Notch in the pathway: the roles of Notch signaling in neural crest development, *Semin Cell Dev Biol* 16:663–72, 2005.

Stoller JZ, Epstein JA: Cardiac neural crest, *Semin Cell Dev Biol* 16(6):704–15, 2005.

Trainor PA: Specification of neural crest cell formation and migration in mouse embryos, *Semin Cell Dev Biol* 16:683–93, 2005.

Martinsen BJ: Reference guide to the stages of chick heart embryology, *Dev Dyn* 233:1217–37, 2005.

Helms JA, Cordero D, Tapadia MD: New insights into craniofacial morphogenesis, *Development* 132:851–61, 2005.

Howard MJ: Mechanisms and perspectives on differentiation of autonomic neurons, *Dev Biol* 277:271–86, 2005.

Streit A: Early development of the cranial sensory nervous system: from a common field to individual placodes, *Dev Biol* 276:1–15, 2004.

Eisenberg LM, Markwald RR: Cellular recruitment and the development of the myocardium, *Dev Biol* 274:225–232, 2004.

Chizhikov VV, Millen KJ: Mechanisms of roof plate formation in the vertebrate CNS, *Nat Rev Neurosci* 5:808–12, 2004, 2004.

Santagati F, Rijli FM: Cranial neural crest and the building of the vertebrate head, *Nat Rev Neurosci* 4:806–18, 2003.

Gammill LS, Bronner-Fraser M: Neural crest specification: Migrating into genomics, *Nat Rev Neurosci* 4(10):795–805, Oct 2003.

Hoch RV, Soriano P: Roles of PDGF in animal development, *Development* 130:4769–84, 2003.

Thiery JP: Cell adhesion in development: A complex signaling network, *Curr Opin Genet Dev* 13:365–71, 2003.

Unsicker K, Finotto S, Krieglstein K: Generation of cell diversity in the peripheral autonomic nervous system: The sympathoadrenal cell lineage revisited, *Ann Anat* 179:495–500, 1997.

Le Douarin NM, Creuzet S, Couly G, Dupin E: Neural crest cell plasticity and its limits, *Development* 131:4637–50, 2004.

Kulesa P, Ellies DL, Trainor PA: Comparative analysis of neural crest cell death, migration, and function during vertebrate embryogenesis, *Dev Dyn* 229(1):14–29, Jan 2004.

Graham A, Begbie J, McGonnell I: Significance of the cranial neural crest, *Dev Dyn* 229:5–13, 2004.

Wnt2

Huguet EL, McMahon JA, McMahon AP, Bicknell R, Harris AL: Differential expression of human Wnt genes 2, 3, 4, and 7B in human breast cell lines and normal and disease states of human breast tissue, *Cancer Res* 54:2615–21, 1994.

Nusse R, Brown A, Papkoff J, Scambler P, Shackleford G, McMahon A et al: A new nomenclature for int-1 and related genes: The Wnt gene family, *Cell* 64:231–2, 1991.

Wainwright BJ, Scambler PJ, Stanier P, Watson EK, Bell G et al: Isolation of a human gene with protein sequence similarity to human and murine int-1 and the Drosophila segment polarity mutant wingless, *EMBO J* 7:1743–8, 1988.

Wassink TH, Piven J, Vieland VJ, Huang J, Swiderski RE et al: Evidence supporting WNT2 as an autism susceptibility gene, *Am J Med Genet* 105:406–13, 2001.

BMP2

Dickinson ME, Kobrin MS, Silan CM, Kingsley DM, Justice MJ et al: Chromosomal localization of seven members of the murine TGF-beta superfamily suggests close linkage to several morphogenetic mutant loci, *Genomics* 6:505–20, 1990.

Gopal Rao VVN, Loffler C, Wozney JM, Hansmann I: The gene for bone morphogenetic protein 2A (BMP2A) is localized to human chromosome 20p12 by radioactive and nonradioactive in situ hybridization, *Hum Genet* 90:299–302, 1992.

Hallahan AR, Pritchard JI, Chandraratna RAS, Ellenbogen RG, Geyer JR et al: BMP-2 mediates retinoid-induced apoptosis in medulloblastoma cells through a paracrine effect, *Nature Med* 9:1033–8, 2003.

Johnson RL, Tabin CJ: Molecular models for vertebrate limb development, *Cell* 90:979–90, 1997.

Kleber M, Lee HY, Wurdak H, Buchstaller J, Riccomagno MM et al: Neural crest stem cell maintenance by combinatorial Wnt and BMP signaling, *J Cell Biol* 169:309–20, 2005.

Tabas JA, Zasloff M, Wasmuth JJ, Emanuel BS, Altherr MR et al: Bone morphogenetic protein: chromosomal localization of human genes for BMP1, BMP2A, and BMP3, *Genomics* 9:283–9, 1991.

Wang EA, Rosen V, D'Alessandro JS, Bauduy M, Cordes P et al: Recombinant human bone morphogenetic protein induces bone formation, *Proc Natl Acad Sci USA* 87:2220–4, 1990.

Wozney JM, Rosen V, Celeste AJ, Mitsock LM, Whitters MJ et al: Novel regulators of bone formation: Molecular clones and activities, *Science* 242:1528–34, 1988.

BMP4

Haramis APG, Begthel H, van den Born M, van Es J, Jonkheer S et al: De novo crypt formation and juvenile polyposis on BMP inhibition in mouse intestine, *Science* 303:1684–6, 2004.

Jiao K, Kulessa H, Tompkins K, Zhou Y, Batts L et al: An essential role of Bmp4 in the atrioventricular septation of the mouse heart, *Genes Dev* 17:2362–7, 2003.

Paez-Pereda M, Giacomini D, Refojo D, Nagashima AC, Hopfner U et al: Involvement of bone morphogenetic protein 4 (BMP-4) in pituitary prolactinoma pathogenesis through a Smad/estrogen receptor crosstalk, *Proc Natl Acad Sci USA* 100:1034–9, 2003.

Shafritz AB, Shore EM, Gannon FH, Zasloff MA, Taub R et al: Overexpression of an osteogenic morphogen in fibrodysplasia ossificans progressiva, *New Engl J Med* 335:555–61, 1996.

Thomas BL, Liu JK, Rubenstein JLR, Sharpe PT: Independent regulation of Dlx2 expression in the epithelium and mesenchyme of the first branchial arch, *Development* 127:217–24, 2000.

Tucker AS, Matthews KL, Sharpe PT: Transformation of tooth type induced by inhibition of BMP signaling, *Science* 282:1136–8, 1998.

van den Wijngaard A, Olde Weghuis D, Boersma CJC, van Zoelen EJJ, Geurts van Kessel A et al: Fine mapping of the human bone morphogenetic protein-4 gene (BMP4) to chromosome 14q22-q23 by in situ hybridization, *Genomics* 27:559–60, 1995.

Zhu NL, Li C, Xiao J, Minoo P: NKX2.1 regulates transcription of the gene for human bone morphogenetic protein-4 in lung epithelial cells, *Gene* 327:25–36, 2004.

BMP 6

Dickinson ME, Kobrin MS, Silan CM, Kingsley DM, Justice MJ et al: Chromosomal localization of seven members of the murine TGF-beta superfamily suggests close linkage to several morphogenetic mutant loci, *Genomics* 6:505–20, 1990.

Hahn GV, Cohen RB, Wozney JM, Levitz CL, Shore EM et al: A bone morphogenetic protein subfamily: Chromosomal localization of human genes for BMP5, BMP6, and BMP7, *Genomics* 14:759–62, 1992.

Lyons K, Graycar JL, Lee A, Hashmi S, Lindquist PB et al: Vgr-1, a mammalian gene related to Xenopus Vg-1, is a member of the transforming growth factor beta gene superfamily, *Proc Natl Acad Sci USA* 86:4554–8, 1989.

Olavesen MG, Bentley E, Mason RVF, Stephens RJ, Ragoussis J: Fine mapping of 39 ESTs on human chromosome 6p23-p25, *Genomics* 46:303–6, 1997.

Rickard DJ, Hofbauer LC, Bonde SK, Gori F, Spelsberg TC et al: Bone morphogenetic protein-6 production in human osteoblastic cell lines: Selective regulation by estrogen, *J Clin Invest* 101:413–22, 1998.

8.3. Development of Neural Tube-Derived Systems

Gilbert Scott F: *Developmental Biology*, 7th ed, Sinauer Associates, Sunderland, UK, 2003.

Rooke JE, Xu T: Positive and negative signals between interacting cells for establishing neural fate, *Bioessays* 20:209–14, 1998.

Ma DK, Ming GL, Song H: Glial influences on neural stem cell development: Cellular niches for adult neurogenesis, *Curr Opin Neurobiol* 15:514–20, 2005.

Abrous DN, Koehl M, Le Moal M: Adult neurogenesis: From precursors to network and physiology, *Physiol Rev* 85:523–69, 2005.

Chklovskii DB, Koulakov AA: Maps in the brain: What can we learn from them? *Annu Rev Neurosci* 27:369–92, 2004.

Kiehn O, Kullander K: Central pattern generators deciphered by molecular genetics, *Neuron* 41:317–21, 2004.

Hatten ME: New directions in neuronal migration, *Science* 297:1660–3, 2002.

Mendell LM, Munson JB, Arvanian VL: Neurotrophins and synaptic plasticity in the mammalian spinal cord, *J Physiol* 533(Pt 1):91–7, 2001.

Katz LC, Shatz CJ: Synaptic activity and the construction of cortical circuits, *Science* 274:1133–8, 1996.

Aubert I, Ridet JL, Gage FH: Regeneration in the adult mammalian CNS: Guided by development, *Curr Opin Neurobiol* 5:625–35, 1995.

Jessen KR, Mirsky R: Neural development. Fate diverted, *Curr Biol* 4:824–7, 1994.

Shatz CJ, Sretavan DW: Interactions between retinal ganglion cells during the development of the mammalian visual system, *Annu Rev Neurosci* 9:171–207, 1986.

Tanabe Y, Jessell TM: Diversity and pattern in the developing spinal cord, *Science* 274:1115–23, 1996.

8.4. The Notochord

Stemple DL: Structure and function of the notochord: An essential organ for chordate development, *Development* 132:2503–12, 2005.

Cleaver O, Krieg PA: Notochord patterning of the endoderm, *Dev Biol* 234:1–12, 2001.

Narasimha M, Leptin M: Cell movements during gastrulation: Come in and be induced, *Trends Cell Biol* 10:169–7, 2000.

Ng JK, Tamura K, Buscher D, Izpisua-Belmonte JC: Molecular and cellular basis of pattern formation during vertebrate limb development, *Curr Top Dev Biol* 41:37–66, 1999.

Di Gregorio A, Levine M: Ascidian embryogenesis and the origins of the chordate body plan, *Curr Opin Genet Dev* 8:457–63, 1998.

Granato M, Nusslein-Volhard C: Fishing for genes controlling development, *Curr Opin Genet Dev* 6:461–8, 1996.

Satoh N, Jeffery WR: Chasing tails in ascidians: Developmental insights into the origin and evolution of chordates, *Trends Genet* 11:354–9, 1995.

Yamada T: Caudalization by the amphibian organizer: Brachyury, convergent extension and retinoic acid, *Development* 120:3051–62, 1994.

8.5. Formation of Somites

Lassar AB, Munsterberg AE: The role of positive and negative signals in somite patterning, *Curr Opin Neurobiol* 6:57–63, 1996.

Bumcrot DA, McMahon AP: Somite differentiation. Sonic signals somites, *Curr Biol* 5:612–4, 1995.

Jouve C, Palmeirim I, Henrique D, Beckers J, Gossler A et al: Notch signaling is required for cyclic expression of the hairy-like gene HES1 in the presomite mesoderm, *Development* 127:1421–9, 2000.

Jouve C, Imura T, Pourquie O: Onset of sementation clock in the chick embryo: evidence for oscillations in the somite precursors in the primitive streak, *Development* 129:1107–17, 2002.

Kulesa PM, Fraser SE: Cell dynamics during somite boundary formation revealed by time-lapse analysis, *Science* 298:991–5, 2002.

Palmeirim I, Henrique D, Ish-Horowicz D, Pourquie O: Avian hairy gene expression identifies a molecular clock linked to vertebrate segmentation and somitogenesis, *Cell* 91:639–48, 1997.

Sulik K, Dehart DB, Carson JL, Vrablic T, Gesteland K et al: Morphogenesis of the murine node and notochordal plate, *Dev Dyn* 201:260–78, 1994.

Brill G, Kahane N, Carmeli C, von Schack D, Barde YA et al: Epithelial-mesenchymal conversion of dermatome progenitors requires neural tube-derived signals: Characterization of the role of neurotrophin-3, *Development* 121:2583–94, 1995.

Kato N, Aoyama H: Dermamyomal origin of the ribs as revealed by extirpation and transplantation experiments in chick and quail embryos, *Development* 125:3437–43, 1998.

Chen CM, Kraut N, Groudine M, Weintraub H: I-mf, a novel myogenic repressor, interacts with members of the MyoD Family, *Cell* 86:731–41, 1996.

Christ B, Ohrdahl CP: Early stages of chick somite development, *Anat Embryol* 191:381–96, 1995.

Cossu G, Kelly R, Tajbakhsh S, Di Donna S, Vivarelli E et al: Activation of different myogenic pathways: myf-5 is induced by the neural tube and MyoD by the dorsal ectoderm in mouse paraxial mesoderm, *Development* 122:429–37, 1996.

Dietrich S, Schubert FR, Healy C, Sharpe PT, Lumsden A: Specification of hypaxial musculature, *Development* 125:2235–49, 1998.

Fan CM, Tessier-Lavigne M: Patterning of mammalian somites by surface ectoderm and notochord: Evidence for sclerotome induction by a hedgehog homolog, *Cell* 79:1175–86, 1994.

Ikeya M, Takada S: Wnt signaling from the dorsal neural tube is required for the formation of the medial dermomyotome, *Development* 125:4969–76, 1998.

Johnson R, Laufer LE, Riddle RD, Tabin C: Ectopic expression of sonic hedgehog alters dorsal-ventral patterning of somites, *Cell* 79:1165–73, 1994.

Marcelle C, Stark MR, Bronner-Fraser: Coordinate actions of BMPs, Wnts, Shh, and Noggin mediate patterning of the dorsal somite, *Development* 124:3955–63, 1997.

Olivera-Martinez I, Missier S, Fraboulet, Thelu J, Dhouaily D: Differential regulation of the chick dorsal thoracic dermal progenitors from medial dermamyotome, *Development* 129:4763–72, 2002.

Pourquie O et al: Lateral and axial signals involved in somite patterning: A role for BMP4, *Cell* 84:461–71, 1996.

Stern HM, Brown AMC, Hauschka SD: Myogenesis in paraxial mesoderm: Preferential induction by dorsal neural tube and by cells expressing Wnt-1, *Development* 121:3675–86, 1995.

Takahashi Y, Monsoro-Burq AH, Vincent C, Le Douarin N: Two domains in vertebral development: Antagonistic regulation by SHH and BMP4 proteins, *Development* 125:2631–9, 1998.

Venters SJ, Thorsteindottir S, Duxton MJ: Early development of the myotome in the mouse, *Dev Dyn* 216:219–32, 1999.

Watanabe Y, Duprez D, Monsoro-Burq AH, Vincent C, Le Douarin N: Two domains in vertebral development: Antagonistic regulation by SHH and BMP4 proteins, *Development* 125:2631–9, 1998.

8.6. Formation of the Skeletal Muscle System

Knudsen KA, McElwee SA, Myers L: A role for the neural cell adhesion molecule, N-CAM, in myoblast interaction during myogenesis, *Dev Biol* 138:159–68, 1990.

Yagami-Hiromasa T, Sato T, Kurisaki T, Kamijo K, Nabeshima YI et al: A metalloprotease-disintegrin participating in myoblast fusion, *Nature* 377:652–6, 1995.

Sartorelli V, Caretti G: Mechanisms underlying the transcriptional regulation of skeletal myogenesis, *Curr Opin Genet Dev* 15:528–35, 2005.

Tajbakhsh S: Stem cells to tissue: Molecular, cellular and anatomical heterogeneity in skeletal muscle, *Curr Opin Genet Dev* 13:413–22, 2003.

Snider L, Tapscott SJ: Emerging parallels in the generation and regeneration of skeletal muscle, *Cell* 113:811–2, 2003.

Buckingham M, Bajard L, Chang T, Daubas P, Hadchouel J et al: The formation of skeletal muscle: From somite to limb, *J Anat* 202:59–68, 2003.

Brent AE, Tabin CJ: Developmental regulation of somite derivatives: Muscle, cartilage and tendon, *Curr Opin Genet Dev* 12:548–57, 2002.

Pownall ME, Gustafsson MK, Emerson CP Jr: Myogenic regulatory factors and the specification of muscle progenitors in vertebrate embryos, *Annu Rev Cell Dev Biol* 18:747–83, 2002.

Arber S, Burden SJ, Harris AJ: Patterning of skeletal muscle, *Curr Opin Neurobiol* 12:100–3, 2002.

8.7. Formation of the Skeleton

Aoyama H, Asamoto K: The development fate of the rostral/caudal half of a somite for vertebra and rib formation: Experimental confirmation of the resegmentation theory using chick-quail chimeras, *Mech Dev* 99:71–82, 2000.

Ducy P, Zhang R, Geoffroy V, Ridall AL, Karsenty G: Osf2/Cbal: A transcriptional activator of osteoblast differentiation, *Cell* 89:747–54, 1997.

Gerber HP, Vu TH, Ryan AM, Kowalski J, Werb Z: VEGF couples hypertrophic cartilage remodeling, ossification, and angiogenesis during endochondral bone formation, *Nature Med* 5:623–8, 1999.

Haigh JJ, Gerber HP, Ferrara N, Wagner EF: Conditional inactivation of VEGF-A in areas of collagen 2a1 expression results in embryonic lethality in the heterozygous state, *Development* 127:1445–53, 2000.

Karsenti G, Wagner EF: Reaching a genetic and molecular understanding of skeletal development, *Dev Cell* 2:389–406, 2002.

Komori T et al: Targeted disruption of Cbal results in a complete lack of bone formation owning to maturational arrest of osteoblasts, *Cell* 89:755–64, 1997.

Morin-Kenmsicki EM, Melancon E, Eisen JS: Segmental relationship between somites and vertebral column in zebrafish, *Development* 129:3851–60, 2002.

Mundlos S et al: Mutations involving the transcription factor CBFA1 cause cleiocranial dysplasia, *Cell* 89:773–9, 1997.

Otto F et al: Cbal, a candidate gene for cleidocranial dysplasia syndrome, is essential for osteoblast differentiation and bone development, *Cell* 89:765–71, 1997.

Rajpurohit R, Mansfield K, Ohyama K, Ewert D, Shapiro IM: Chondrocyte death is linked to development of a mitochondrial membrane permeability transition in the growth plate, *J Cell Physiol* 179:287–96, 1999.

Yin M, Koyama E, Zasloff M, Pacifici M: Antiangiogenic treatment delays chondrocyte maturation and bone formation during limb morphogenesis, *J Bone Miner Res* 17:56–65, 2002.

Helms JA, Cordero D, Tapadia MD: New insights into craniofacial morphogenesis, *Development* 132:851–61, 2005.

Helms JA, Schneider RA: Cranial skeletal biology, *Nature* 423:326–31, 2003.

8.8. The Intermediate Mesoderm: Formation of the Kidney

Bard JBL, Davis JA, Karavanova I, Lehtonen E, Sariola H et al: Kidney development; the inductive interactions, *Semin Cell Dev Biol* 7:195–202, 1996.

Bard JBL, Gordon A, Sharp L, Sellers W: The early stages of nephron formation in the developing mouse kidney, *J Anat* 199:385–92, 2001.

Bard JBL: Growth and death in the developing mammalian kidney: Signals, receptors, and conversations, *Bioessays* 24:72–82, 2002.

Sariola H: Nephron induction revisited: From caps to condensates, *Curr Opin Nephrol Hypertens* 11:17–21, 2002.

Schejter ED, Shilo BZ: Modular tubes: common principles of renal development, *Curr Biol* 13: R511–3, 2003.

8.9. Formation of the Heart

Biben C, Harvey RP: Homeodomain factor Nkx2–5 controls left-right asymmetric expression of bHLH gene eHand during murine heart development, *Genes Dev* 11:1357–69, 1997.

Imanaka-Yoshida K, Knudsen KA, Linask KK: N-cadherin is required for the differentiation and initial myofibrillogenesis of chick cardiomyocytes, *Cell Motil Cytoskel* 39:52–62, 1998.

Linask KK, Yu X, Chen Y, Han MD: Directionality of heart looping: Effects of Pitx2c misexpression on flectin asymmetry and midline structures, *Dev Biol* 246:407–17, 2002.

Nguyen MT et al: The prostaglandin receptor EP4 triggers remodeling of the ccardiovascular system at birth, *Nature* 390:78–81, 1997.

Redkar A, Montgomery M, Litvin J: Fate map of the early avian cardiac progenitor cells, *Development* 128:2269–79, 2001.

Srivastava D, Olson EN: A genetic blueprint for cardiac development, *Nature* 407:221–32, 2000.

Srivastava D, Cserjesi P, Olson EN: A subclass of bHLH proteins required for cardiac morphogenesis, *Science* 270:1995–9, 1995.

Wakimoto K et al: Targeted disruption of Na+/Ca2+ exchanger gene leads to cardiomyocyte apoptosis and defects in heart-beat, *J Biol Chem* 275:36991–8, 2000.

Moorman AF, Christoffels VM: Cardiac chamber formation: Development, genes, and evolution, *Physiol Rev* 83:1223–67, 2003.

8.10. Formation of the Vascular System

Choi H, Kennedy M, Kazarov A, Padaimitrious JC, Keller G: A common precursor for hematopoietic and endothelial cells, *Development* 125:725–32, 1998.

Davis S et al: Isolation of angiopoietin-1, a ligand for the TIE2 rfeceptor, by secretion trap expression cloning, *Cell* 87:1161–9, 1996.

Ferrara N, Alitalo K: Clinical applications of angiogenic growth factors and their inhibitors, *Nature Med* 5:1359–64, 1999.

Ferrara N et al: Heterozygous embryonic lethality induced by targeted inactivation of the VEGF gene, *Nature* 380:439–42, 1996.

Flamme I, Risau W: Induction of vasculogenesis and hematchenesis in vitro, *Development* 116:435–9, 1992.

Fong GH, Rossant J, Gertenstein M, Breitman ML: Role of the flt-1 receptor tyrosine kinase in regulating the assembly of vascular endothelium, *Nature* 376:66–70, 1995.

Hanai JI et al: Endostatin causes G1 arrest of endothelial cells through inhibition of cyclin D1, *J Biol Chem* 277:16464–9, 2002.

Kim YM et al: Endostatin blocks VEGF-mediated signaling via direct interaction with KDR/flk-1, *J Biol Chem* 277:27872–9, 2002.

Ribatti D, Urbinati C, Nico B, Rusnati M, Roncali L et al: Endogenous basic fibroblast growth factor is implicated in the vascularization of the chick embryo chorioallantoic membrane, *Dev Biol* 170:39–49, 1995.

Hanahan D: Signaling vascular morphogenesis and maintenance, *Science* 277:48–50, 1997.

LaBarbera M: Principle of design of fluid transport systems in zoology, *Science* 249:992–1000, 1990.

LeCouter J et al: Identification of an angiogenic mitogen selective for endocrine gland endothelium, *Nature* 412:877–84, 2001.

Lindahl P, Johansson BR, Leveen P, Betscholtz C: Pericyte loss and microaneurysm formation in PDGF-B-deficient mice, *Science* 277:242–5, 1997.

Millauer B, Wizigmann-Voos S, Schnurch H, Martinez R, Muller NPH et al: High-affinity VEGF binding and developmental expression suggest flk-1 as a major regulator of vasculogenesis and angiogenesis, *Cell* 72:835–46, 1993.

Oh J et al: The membrane-anchored MMP inhibitor RECK is a key regulator of extracellular matrix integrity and angiogenesis, *Cell* 107:789–800, 2001.

Risau W: Differentiation of the endothelium, *FASEB J* 9:926–33, 1995.

Risau W: Mechanisms of angiogenesis, *Nature* 386:671–4, 1997.

Shalaby F, Rossant J, Yamaguchi TP, Gertenstein M, Wu XF et al: Failure of blood-island formation and vasculogenesis in flk-1-deficient mice, *Nature* 376:62–6, 1995.

Shalaby F et al: A requirement for Flk1 in primitive and definitive hematopoiesis and vasculogenesis, *Cell* 89:981–90, 1997.

Suri C et al: Requisite role of angiopoietin-1, a ligand for the TIE2 receptor, during embryonic angiogenesis, *Cell* 87:1171–80, 1996.

Vikkula M et al: Vascular dysmorphogenesis caused by an activating mutation in the receptor tyrosine kinase TIE2, *Cell* 87:1181–90, 1996.

Hallmann R, Horn N, Selg M, Wendler O, Pausch F et al: Expression and function of laminins in the embryonic and mature vasculature, *Physiol Rev* 85:979–1000, 2005.

Weinstein BM: Vessels and nerves: Marching to the same tune, *Cell* 120:299–302, 2005.

8.11. Formation of Blood Cells

Berardi AC, Wang A, Levine JD, Lopez P, Scadden DT: Functional characterization of human hematopoietic stem cells, *Science* 267:104–8, 1995.

Godlin IE, Garcia-Porrero JA, Coutinho A, Dieterlen-Lievre F, Marcos MAR: Para-aortic splanchnopleura from early mouse embryos contain B1a cell progenitors, *Nature* 364:67–70, 1993.

Heissig B et al: Recruitment of stem and progenitor cells from bone marrow niche requires MMP-9 mediated release of Kit-ligand, *Cell* 109:625–37, 2002.

Medvinsky AL, Samoylina NL, Muller AM, Dzierzak EA: An early pre-liver intraembryonic source of CFU-S in the development mouse, *Nature* 364:64–7, 1993.

Orkin SH, Zorn LI: Hematopoiesis and stem cells: Plasticity versus developmental heterogeneity, *Nature Immun* 3:323–8, 2002.

Potten CS, Loeffler M: Stem cells: Attributes, spirals, pitfalls, and uncertainties. Lessons for and from the crypt, *Development* 110:1001–20, 1990.

Gordon S, Taylor PR: Monocyte and macrophage heterogeneity, *Nat Rev Immunol* 5:953–64, 2005.

Murre C: Helix-loop-helix proteins and lymphocyte development, *Nat Immunol* 6:1079–86, 2005.

Baron MH, Fraser ST: The specification of early hematopoiesis in the mammal, *Curr Opin Hematol* 12:217–21, 2005.

Medina KL, Singh H: Genetic networks that regulate B lymphopoiesis, *Curr Opin Hematol* 12:203–9, 2005.

Dzierzak E: The emergence of definitive hematopoietic stem cells in the mammal, *Curr Opin Hematol* 12:197–202, 2005.

Maillard I, Fang T, Pear WS: Regulation of lymphoid development, differentiation, and function by the Notch pathway, *Annu Rev Immunol* 23:945–74, 2005.

Lacaud G, Keller G, Kouskoff V: Tracking mesoderm formation and specification to the hemangioblast in vitro, *Trends Cardiovasc Med* 14:314–7, 2004.

8.12. Formation of the Digestive Tract

Ishii Y, Fukuda K, Saiga H, Matsushita S, Yasugi S: Early specification of intestinal epithelium in the chicken embryo: A study on the localization of and regulation of CdxA expression, *Dev Growth Differ* 39:643–53, 1997.

Horb ME, Slack JM: Endoderm specification and differentiation in Xenopus embryos, *Dev Biol* 236:330–43, 2001.

Matsushita S, Ishii Y, Scotting PJ, Kuroiwa A, Yasugi S: Pre-gut endoderm of chick embryos is regionalized by 1.5 days of development, *Dev Dyn* 223:33–47, 2002.

Roberts DJ, Johnson RL, Burke A, Nelson CE, Morgan BA et al: Sonic hedgehog is an endodermal signal inducing Bmp-4 and Hox genes during induction and regionalization of the chick hindgut, *Development* 121:3163–74, 1995.

Roberts DJ, Smith DM, Goff DJ, Tabin CJ: Epithelial-mesodermal signaling during the regionalization of the chick gut, *Development* 125:2791–801, 1998.

Yokouchi Y, Sakiyama J, Kuroiwa A: Coordinate expression of Abd-B subfamily genes of the HoxA cluster in developing digestive tract of the chick embryo, *Dev Biol* 169:76–89, 1995.

Sancho E, Batlle E, Clevers H: Signaling pathways in intestinal development and cancer, *Annu Rev Cell Dev Biol* 20:695–723, 2004.

Pacha J: Development of intestinal transport function in mammals, *Physiol Rev* 80:1633–67, 2000.

8.13. Formation of the Liver

Gualdi R, Bossard P, Zheng M, Hamada Y, Coleman JR et al: Hepatic specification of the gut endoderm in vitro: Cell signaling and transcriptional control, *Genes Dev* 10:1670–82, 1996.

Jung J, Goldfarb M, Zaret KS: Initiation of mammalian liver development from endoderm by fibroblast growth factors, *Science* 284:1998–2003, 1999.

Matsumoto K, Yoshitomi H, Rossant J, Zaret KS: Liver organogenesis promoted by endothelial cells prior to vascular function, *Science* 294:559–63, 2001.

Lemaigre F, Zaret KS: Liver development update: New embryo models, cell lineage control, and morphogenesis, *Curr Opin Genet Dev* 14:582–90, 2004.

Newsome PN, Hussain MA, Theise ND: Hepatic oval cells: Helping redefine a paradigm in stem cell biology, *Curr Top Dev Biol* 61:1–28, 2004.

Duncan SA: Mechanisms controlling early development of the liver, *Mech Dev* 120:19–33, 2003.

Lemaigre FP: Development of the biliary tract, *Mech Dev* 120:81–7, 2003.

Lammert E, Cleaver O, Melton D: Role of endothelial cells in early pancreas and liver development, *Mech Dev* 120:59–64, 2003.

8.14. Formation of the Pancreas

Ahlgren U, Johnson J, Edlund H: The morphogenesis of the pancreatic mesenchyme is uncoupled from that of the pancreatic epithelium in IPF/PDX1-deficient mice, *Development* 122:1409–16, 1996.

Johnson J, Carlsson L, Edlund T, Edlund H: Insulin-promote-factor is required for pancreas development in mice, *Nature* 371:606–8, 1994.

Offield MF et al: PDX-1 is required for pancreatic outgrowth and differentiation of the rostral duodenum, *Development* 122:983–95, 1996.

St-Onge L, Wehr R, Gruss P: Pancreas development and diabetes, *Curr Opin Genet Dev* 9:295–300, 1999.

Bouwens L, Rooman I: Regulation of pancreatic beta-cell mass, *Physiol Rev* 85:1255–70, 2005.

Nir T, Dor Y: How to make pancreatic beta cells—prospects for cell therapy in diabetes, *Curr Opin Biotechnol* 16:524–9, 2005.

Homo-Delarche F, Drexhage HA: Immune cells, pancreas development, regeneration and type 1 diabetes, *Trends Immunol* 25:222–9, 2004.

Murtaugh LC, Melton DA: Genes, signals, and lineages in pancreas development, *Annu Rev Cell Dev Biol* 19:71–89, 2003.

Kumar M, Melton D: Pancreas specification: A budding question, *Curr Opin Genet Dev* 13:401–7, 2003.

8.15. Formation of the Lung

Sakiyama J, Yamagishi A, Kuroiwa A: Tbx4-Fgf10 system controls lung bud formation during chick embryonic development, *Development* 130:1225–34, 2003.

Pepicelli CV, Lewis PM, McMahon AP: Sonic hedgehog regulates branching morphogenesis in the mammalian lung, *Curr Biol* 8:1083–6, 1998.

Chinoy MR: Lung growth and development, *Front Biosci* 8:d392–415, 2003.

Metzger RJ, Krasnow MA: Genetic control of branching morphogenesis, *Science* 284:1635–9, 1999.

Warburton D, Zhao J, Berberich MA, Bernfield M: Molecular embryology of the lung: Then, now, and in the future, *Am J Physiol* 276:L697–704, 1999.

9

REGENERATION OF ADULT CELLS, TISSUES, AND ORGANS

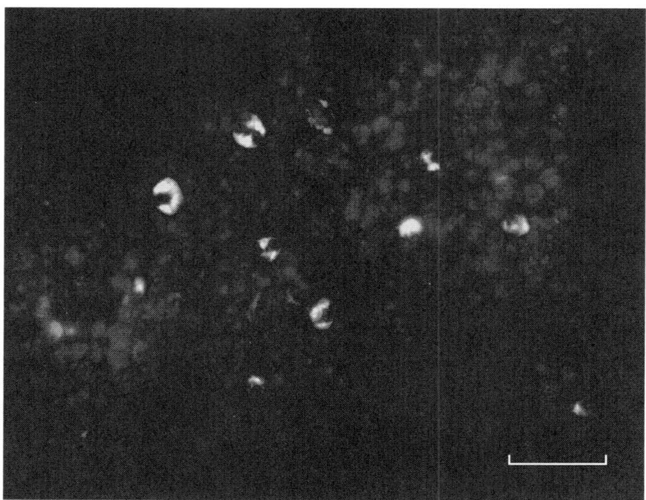

Presence of smooth muscle α-actin-positive (green) and CD34-positive (red) cells in the bone marrow. When cultured on elastic lamina-dominant arterial matrix scaffolds, these cells can transform to smooth-muscle-like cells with α-actin filaments. These cells may serve as progenitor cells for the regeneration of vascular smooth muscle cells. Blue: cell nuclei. Scale: 10 μm. See color insert.

Bioregenerative Engineering: Principles and Applications, by Shu Q. Liu

THE STEM CELL CONCEPT

Stem cells are cell types that can self-renew indefinitely and can differentiate into specified cell types. These features are critical to sustained generation and regeneration of cells, tissues, and organs during development and remodeling. Based on the function, stem cells can be classified into two subtypes: uncommitted and committed stem cells. *Uncommitted stem cells* can differentiate into all specified cell types. These cells are also known as *pluripotent stem cells* or *embryonic stem cells*, and are found in the embryo before the formation of the endodermal, mesodermal, and ectodermal layers. Uncommitted stem cells are usually obtained from the inner cell mass of the blastocyst, primordial germ cells, and epiblast. *Committed cells* are cells found in later stages of development compared to the blastocyst stage and are committed to form cells in a specified tissue, organ, or system. All committed stem cells are originated from the inner cell mass of the blastocyst. Typical examples for committed stem cells include the stem cells in the endoderm, the mesoderm, and the ectoderm. Stem cells in the fetal tissues and organs are also considered committed stem cells. In each tissue, organ, or system, there are cell types that can differentiate into specified cells, but cannot self-renew. These cell types are referred to as *progenitor* or *precursor cells*.

Embryonic Stem Cells [9.1]

Embryonic stem cells are pluripotent embryonic cells that can self-renew and self-expand indefinitely and can differentiate into all specified cell types under appropriate growth conditions. Even under cell culture conditions in vitro, embryonic stem cells exhibit these features. There are several potential sources for embryonic stem cells: the inner cell mass, primordial germ cells, and epiblast (also known as *primitive ectoderm*). The *inner cell mass* is a structure of the early embryo within the blastocyst and consists of about 30 cells. The inner cell mass cells can give rise to all cell types in the body. These cells can be collected, expanded in culture, and used for the repair and replacement of malfunctioned adult cells (Fig. 9.1). The collection and preparation of embryonic stem cells are demonstrated in Fig. 9.2. Expanded stem cells can be then delivered to a target organ, where the stem cells can differentiate into specified cell types under appropriate local environment. It should be noted that a special condition for culturing embryonic stem cells is the presence of feeder cells. Fibroblasts can serve as a feeder cell type. The feeder cells provide structural support as well as soluble factors necessary for the survival and expansion of the embryonic stem cells.

The *primordial germ cells* are a group of embryonic cells that develop into reproductive cells, including sperm and ovum (see Chapter 7). These cells arise from the proximal epiblast, a structure developed from the inner cell mass, pass through the primitive streak, migrate to the genital ridge, and differentiate into germ progenitor cells and extraembryonic mesodermal cells. The germ cells are collectively called the *embryonic germ cells* during the early embryonic stages. These cells exhibit phenotypes that are similar to those found in the inner cell mass cells and can serve as embryonic stem cells that give rise to specified cell types in all ectodermal, mesodermal, and endodermal layers.

The *epiblast* is a structure directly derived from the inner cell mass of the embryo. This structure gives rise to the three embryonic layers (ectoderm, mesoderm, and endoderm) and the amnionic ectoderm, which forms the amnionic sac. Isolated epiblast cells can be cultured and expanded in vitro. These cells demonstrate features of embryonic

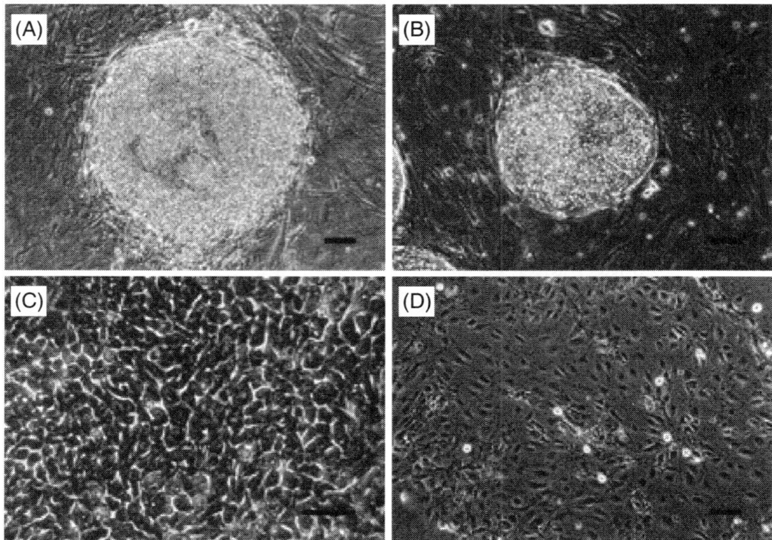

Figure 9.1. Human embryonic stem cells (cell line H9) in culture. (A) Human inner cell mass-derived cells attached to mouse embryonic fibroblast feeder layer after 8 days of culture, 24 h before first dissociation. Scale bar: 100 μm. (B) H9 colony. Scale bar: 100 μm. (C) H9 cells. Scale bar: 50 μm. (D) Differentiated H9 cells, cultured for 5 days in the absence of mouse embryonic fibroblasts, but in the presence of human LIF (20 ng/mL). Scale bar: 100 μm. (Reprinted with permission from Thomson JA et al: Embryonic stem cell lines derived from human blastocysts. *Science* 282:1145–7, copyright 1998 AAAS.)

stem cells and can differentiate into specified cell types found in all tissues and organs. Thus, epiblast cells are potential stem cells that can be used for stem cell-based therapeutic purposes.

In addition to the sources described above, stem cells can be generated by using the *somatic cell nuclear transfer* (SCNT) technology. For this technology, the nucleus of an adult cell, such as a fibroblast, from a patient is collected and transferred into an unfertilized egg. The transferred egg can be induced to give rise to stem cells with the genetic characteristics of the donor somatic cell. Because the somatic nucleus carries the genome of the patient, the derived stem cells are partially autologous to the patient, thus reducing immune rejections. The transferred stem cells may retain their pluripotent nature and can be stored for multiple transplantations. This is a potential approach for the generation of therapeutic stem cells.

To date, the developmental features of embryonic stem cells have been investigated in several mammalian species including the mouse, primate, and human. In particular, the successful preparation, expansion, and induction of differentiation of human embryonic stem cells have greatly promoted research activities in stem cell-based regenerative medicine. It is now well established that human embryonic stem cells can differentiate into numbers of functional adult cell types, including neurons, glial cells, cardiomyocytes, liver cells, vascular smooth muscle cells, pancreatic β cells, blood cells, skeletal muscle cells, and osteoblasts. Typical examples are shown in Figs. 9.3–9.5. These preliminary studies have demonstrated the clinical potential of using stem cells for regenerative therapies.

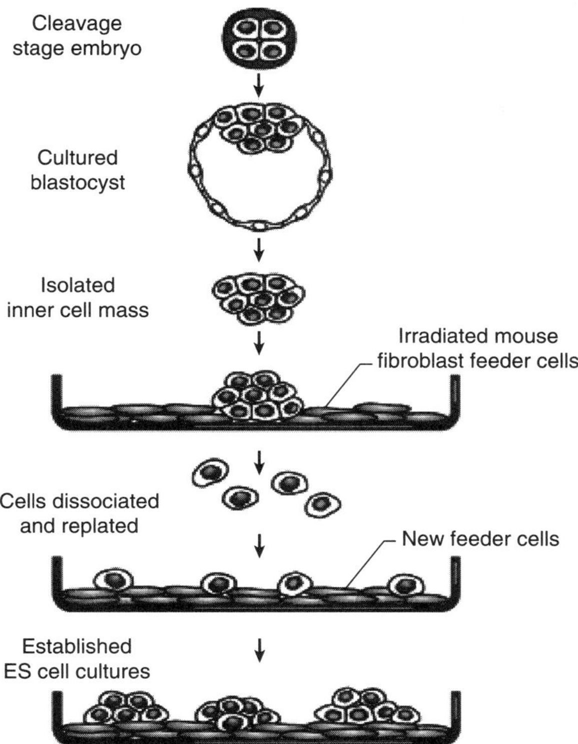

Figure 9.2. Preparation of human embryonic stem cells. Human blastocysts were grown from cleavage-stage embryos produced by in vitro fertilization. The inner cell mass (ICM) cells were collected and plated onto a fibroblast feeder substratum in medium containing fetal calf serum. Colonies were sequentially expanded and cloned. (Reprinted from Odorico JS et al: *Stem Cells* 19:193–204, 2001 by permission).

Fetal Stem and Progenitor Cells [9.1]

Fetal stem and progenitor cells are those found in the fetus. Investigations by using animal models have demonstrated the presence of various stem and progenitor cells in the fetus. Certain fetal stem cells can differentiate into specified cell types across differentiation barriers between the three developmental layers: ectoderm, mesoderm, and endoderm. However, because of ethical concerns, the therapeutic potential of human fetal stem cells has not been investigated extensively. One type of fetal stem/progenitor cells, the blood-borne fetal mesenchymal stem/progenitor cells, may be considered for clinical application. The fetal blood has been shown to contain circulating fetal mesenchymal stem/progenitor cells. Blood samples can be collected from the fetus, and circulating stem and progenitor cells can be enriched, expanded, and stored for treating disorders, if any, of the same fetus. Since these cells are from the fetal stage of development, they can potentially differentiate into various types of specialized cells.

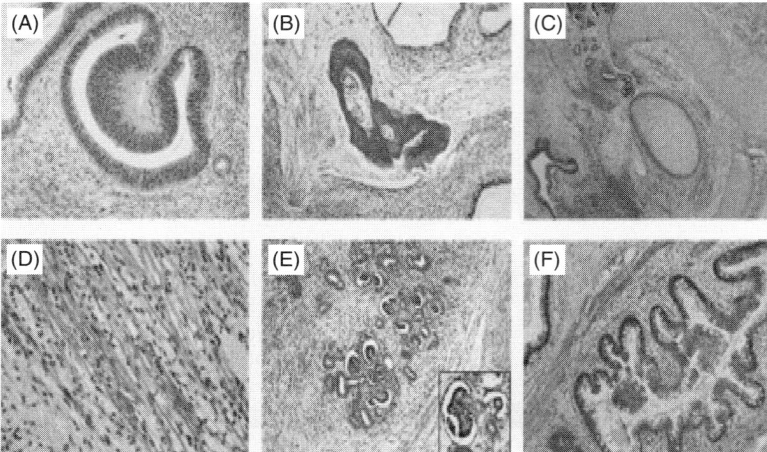

Figure 9.3. Specified cell types derived from human embryonic stem cells transplanted into the mouse. Human embryonic stem cells were prepared and transplanted into immunocompromised mice. The implanted human embryonic stem cells can form benign teratomas. Various cell and tissue types can be found from the teratomas, including neural epithelium (panel A, 100×), bone (panel B, 100×), cartilage (panel C, 40×), striated muscle (panel D, 200×), and glomeruli and renal tubules (panel E, 100×; inset 200×), and gut (panel F, 40×). All photomicrographs are of hematoxylin- and eosin-stained sections. (Reprinted from Odorico JS et al: *Stem Cells* 19:193–204, 2001 by permission.)

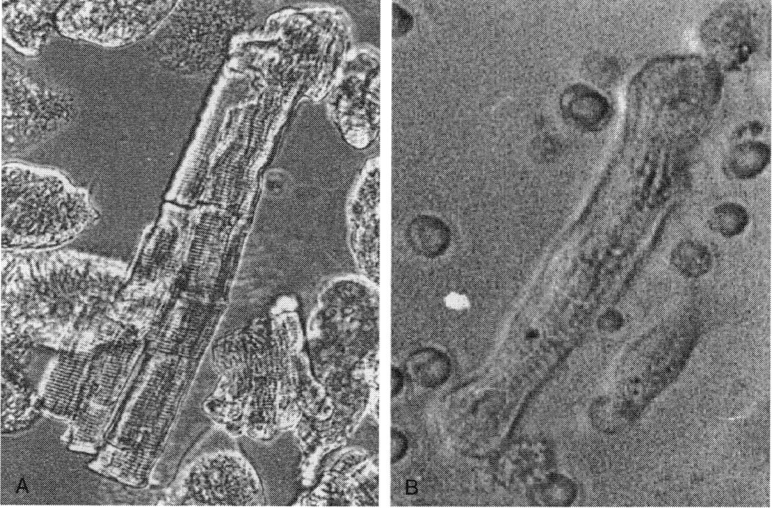

Figure 9.4. Cardiomyocytes derived from mouse embryonic stem cells. (A) Phase contrast micrograph showing a cluster of 2-day cardiomyocytes cells after isolation and fixation. (B) Micrograph showing a single cell, digitally magnified 2× compared with panel A. (Reprinted from Mummery C et al: *J Anat* 200:233–42, 2002 by permission of Blackwell Publishing.)

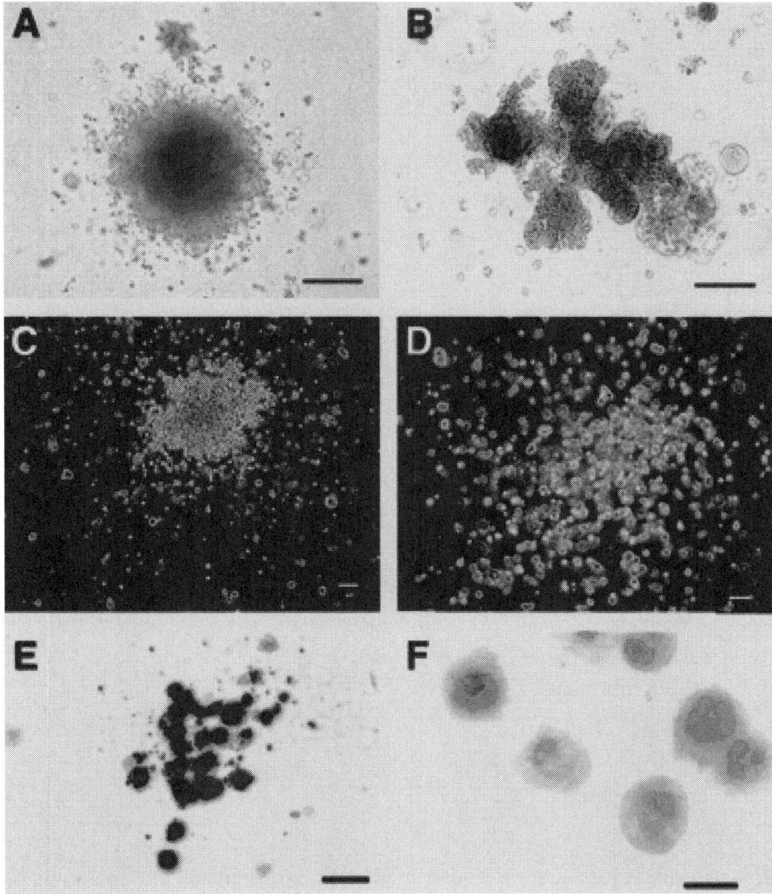

Figure 9.5. Hematopoietic cells derived from human embryonic stem cells. H1 human embryonic stem cells were allowed to differentiate on S17 cells for ~17 days. The H1 cells were harvested and cultures in semisolid media for 14 days before scoring colony phenotypes. (A) Colony of mixed erythroid and myeloid cells, inclduing CFU–granulocyte, erythroid, macrophage, and megakaryocyte (CFU-GEMM). (B) Large unstained erythroid colony (red, hemoglobin). (C) Unstained myeloid colony, CFU-GM. (D) Unstained myeloid colony, CFU-M, which was less dense than the CFU-GM colony. (E) Colony of CFU-Mk cells stained with platelet/megakaryocyte-specific antibody against CD41 (GPIIb/IIIa) with alkaline phosphatase-conjugated secondary antibody and Fast Red/naphthol reagent to provide red stain. (F) Cytospin of CFU-GM cells demonstrating granulocytes with esterase-positive red granules. Scale bars: A–D, 100 μm; E, 40 μm; F, 20 μM. (Reprinted by permission from Kaufman DS et al: *Proc Natl Acad Sci USA* 98:10716–21, copyright 2001, National Academy of Science USA.)

Adult Stem Cells

There exist various types of stem cells in adult tissues and organs. These cells can self-renew and differentiate into specified cell types. Adult stem cells can be found in the bone marrow (e.g., hematopoietic stem cells) (see chapter-opening figure), nervous system (e.g., neuroepithelial cells and olfactory cells) (Fig. 9.6), epidermis (e.g., basal epidermal

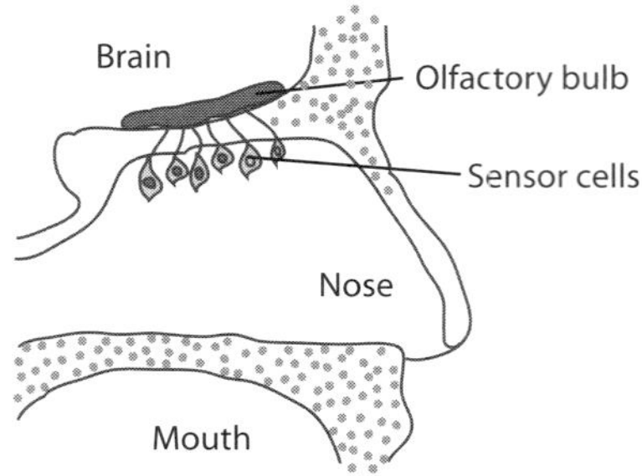

Figure 9.6. Schematic representation of the olfactory cells. Based on bibliography 9.1.

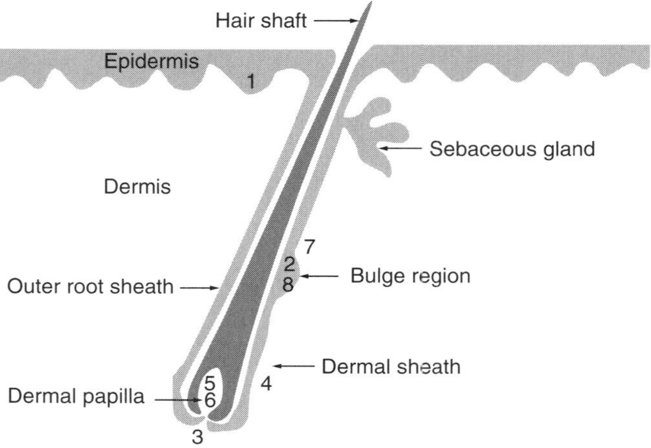

Figure 9.7. Schematic representation of skin stem cells in the hair follicle, epidermis, and dermis: (1) epidermal stem cell; (2) follicle multipotent stem cell; (3) mesenchymal stem cell; (4) dermal sheath stem cells; (5) neural crest stem cell; (6) hematopoietic stem cell; (7) endothelial stem cell; (8) melanocyte stem cell. (Reprinted from Shi C et al: *Trends Biotechnol* 24:48–52, copyright 2006 with permission from Elsevier.)

cells) (Fig. 9.7), dermis (Fig. 9.8), and intestines (e.g., epithelial stem cells) (Fig. 9.9). The presence of stem cells is the basis for cell regeneration in response to cell injury and death in these tissues. The adult stem cells have been traditionally considered committed tissue-specific stem cells, which can differentiate only into cells specific to the tissues and organs in which the stem cells reside. However, recent investigations have demonstrated that certain types of adult stem cells from one tissue type can break the developmental restriction and differentiate into specified cell types found in another tissue type. For instance, hematopoietic stem cells from the bone marrow can differentiate into not only hematopoietic cells but also mature cells in tissues other than the hematopoietic system. When hematopoietic stem cells are transplanted into the brain, heart, and liver, these cells can transform into neurons, cardiomyocytes, and hepatocytes, respectively. The property of the cross-lineage or cross-tissue transformation of adult stem cells is referred to as *plasticity*. In this section, the types and features of adult stem cells are briefly discussed.

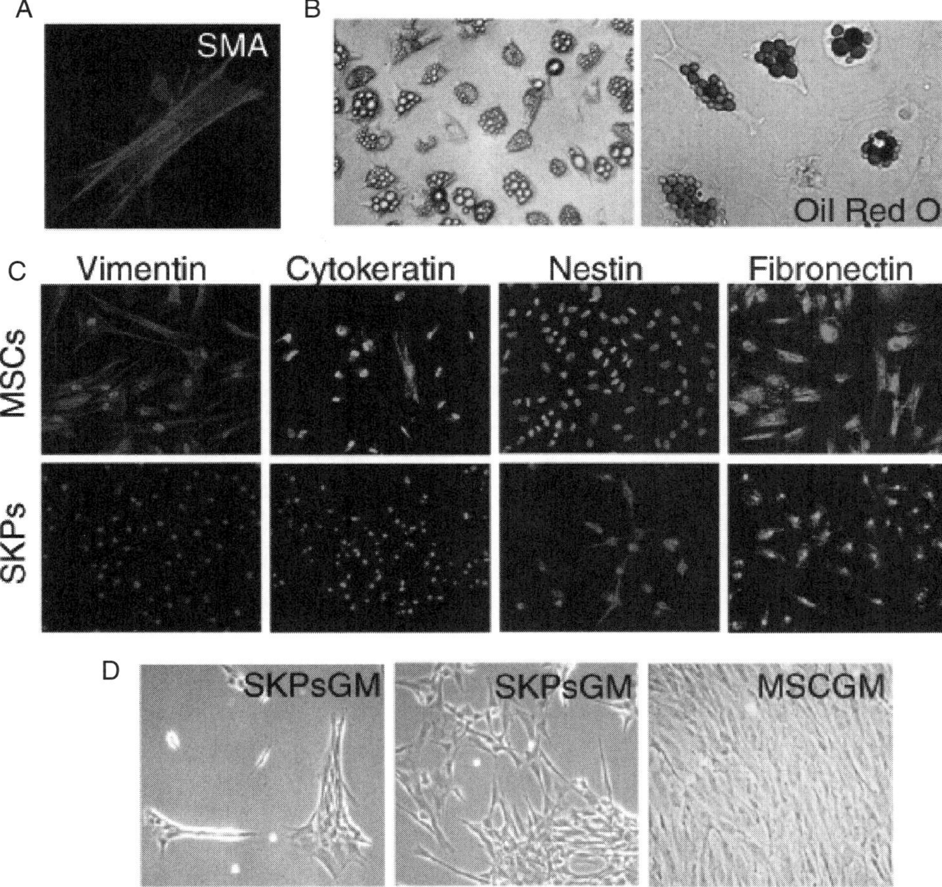

Figure 9.8. Dermis-derived precursors (SKPs) can differentiate into smooth muscle cells and adipocytes. (a) A SMA-positive smooth muscle cell differentiated from juvenile SKPs that were passaged for 3 months. (b) In culture medium with 10% FBS without added growth factors, adult SKPs differentiate into cells with the morphological characteristics of adipocytes. Left, phase-contrast micrograph; right, brightfield picture of a culture stained with Oil Red O, which stains lipid droplets. (c) Immunocytochemical analysis of mesenchymal stem cells (MSCs; top) and SKPs (bottom) for vimentin, cytokeratin, nestin, and fibronectin. SKPs differ from mesenchymal stem cells in their production of the intermediate filament proteins vimentin and nestin. In both cases, cells were dissociated and plated onto poly-D-lysine/laminin-coated slides overnight before immunocytochemistry. Mesenchymal stem cells produce high levels of vimentin and no nestin, whereas SKPs produce nestin but not vimentin. A subpopulation of mesenchymal stem cells also produce cytokeratin. Note the difference in morphology of the two cell types: SKPs are clearly much smaller and less flattened than mesenchymal stem cells. (d) Mesenchymal stem cells do not proliferate in suspension when grown under SKP conditions. Phase micrographs of mesenchymal stem cells grown on uncoated tissue culture plastic for 2 weeks in SKPs medium and growth factors (SKPsGM; left and center) or in mesenchymal stem cell medium (MSCGM; right). Note that the mesenchymal stem cells adhere to uncoated plastic and survive but do not proliferate in the SKPs medium, whereas they rapidly proliferate to reach confluence in the mesenchymal stem cell medium. (Reprinted by permission from Macmillan Publishers Ltd.: Toma JG et al: Isolation of multipotent adult stem cells from the dermis of mammalian skin, *Nature Cell Biol* 3:778–84, copyright 2001.)

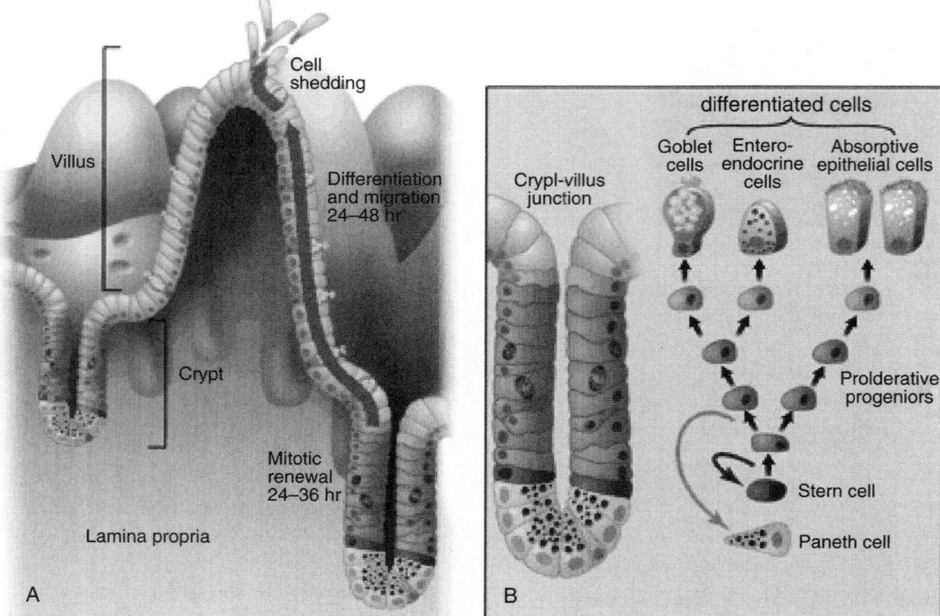

Figure 9.9. Distribution and differentiation of stem cells in the intestine. (A) Distribution of intestinal mature and stem cells. (B) Differentiation pathways of intestinal stem cells. (Reprinted with permission from Radtke F, Clevers H: *Science* 307:1904–9, copyright 2005 AAAS.)

Bone Marrow-Derived Stem Cells [9.2]. The bone marrow contains adult stem cells, including the hematopoietic stem cells and marrow stromal stem cells, which can self-renew and differentiate into specified cell types. The *hematopoietic stem cells* can differentiate into all types of mature hematopoietic cell, including erythrocytes, leukocytes, and platelets. The hematopoietic stem cells can be identified and purified based on protein makers expressed on the cell surface. Protein markers found in these cells include CD34, c-Kit, Sca-1, and Thy1.1. These markers are relatively unique to the bone marrow-derived hematopoietic stem cells. Thus, $CD34^+$c-kt^+Sca-1^+Thy-1^+ bone marrow cells are often considered hematopoietic stem cells. These markers can be identified by immunohistochemistry with specific antibodies.

The bone marrow also contains specified hematopoietic progenitor cells and mature blood cells. Since the presence of these cell lineages obscures the identification of the hematopoietic stem cells, these cells are usually identified and depleted by using lineage-specific antibodies before the hematopoietic stem cells are isolated. For instance, T cells and their immediate progenitor cells express uniquely CD3, CD4, and CD8; B cells express B220; monocytes and granulocytes express CD11b; and erythrocytes express Ter119. Blood cells with these surface proteins can be removed by using magnetic beads coated with antibodies specific to these surface proteins. Alternatively, fluorochrome-conjugated antibodies can be used to label specified cell surface markers. Antibody-labeled cells can be identified and removed by fluorescence-activated cell sorting (FACS). The remaining cells after immune depletion are referred to as *lineage-depleted cells* or *Lin^- cells.*

Based on the features of cell self-renewal and differentiation, hematopoietic stem cells can be divided into three groups: long-term and short-term hematopoietic stem cells, and progenitor cells. The long-term hematopoietic stem cells can self-renew indefinitely, the short-term hematopoietic stem cells can self-renew for 1–2 months, and the progenitor cells cannot self-renew.

The hematopoietic stem cells demonstrate several features that are useful for the application of these cells to the regeneration of nonhematopoietic cells and tissues:

1. A small number of hematopoietic stem cells can repopulate the entire hematopoietic system after a complete ablation of the blood cells. Indeed, this is the most convincing evidence that bone marrow contains hematopoietic stem cells.
2. Hematopoietic stem cells, when delivered to the vascular system, can engraft to the intima of blood vessels, pass through the vessel wall, and engraft to various organs, including the brain, liver, intestines, kidney, and connective tissue. Under organ-specific conditions, the bone marrow stem cells can differentiate into specified cell types. For instance, bone marrow stem cells can transform to hepatocytes in the liver (Fig. 9.10) and renal cells in the kidney (Fig. 9.11).
3. Hematopoietic stem cells can be used to transplant into allogenic recipients with or without myeloablation, and can home to the bone marrow and replenish the hematopoietic system permanently without significant immune rejection responses.
4. The transplantation of hematopoietic stem cells derived from healthy donors into allogenic recipients with autoimmune disorders (e.g., type I diabetes) can potentially reduce the autoimmune response of the recipients.

These features provide a foundation for the use of hematopoietic stem cells for treating human disorders.

Another population of bone marrow cells that include stem cells is the *marrow stromal cells*. These cells provide a stroma that supports the growth and development of the hematopoietic cells. Furthermore, the marrow stromal cells produce and release soluble mediators that are necessary for the differentiation and specification of the hematopoietic stem cells. The marrow stromal cell population contains mesenchymal stem cells that can differentiate into osteoblasts, chondrocytes, fibroblasts, adipocytes, endothelial cells, and smooth muscle-like cells. The marrow stromal cells are characterized by a unique property: the ability to adhere to a glass, plastic, or extracellular matrix substrate. Hematopoietic cells do not possess such a property. This property has been used for the identification and isolation of marrow stromal cells.

The marrow stromal cells undergo a dynamic change in phenotype and cell surface markers when cultured in vitro. During the early phase of culture when the marrow stromal cells are engaged only in adhesion, the adherent cell population contains $CD34^+$, $c\text{-}Kit^+$, $Sca\text{-}1^+$, $Thy1.1^+$, $CD11b^+$, $CD45^+$, smooth muscle α-$actin^+$, $calponin^+$, and smooth muscle $myosin^+$ (SM1). Certain markers are coexpressed in the adherent cells. For instance, CD34 and c-Kit are found in the majority of smooth muscle α-$actin^+$ bone marrow cells (see chapter-opening figure). Calponin and SM1, which are markers for mature smooth muscle cells, are found in a large fraction of smooth muscle α actin cells in culture (Fig. 9.12). With the progression of cell culture, stem cell markers including CD34, c-Kit, Sca-1, and Thy1.1 are gradually diminished, while the expression of CD11b, CD45, smooth muscle α-actin, calponin, and SM1 remains. Up to this stage, cells can be divided into two major groups: smooth muscle α-$actin^+$ and smooth muscle α-$actin^-$ cells. The smooth

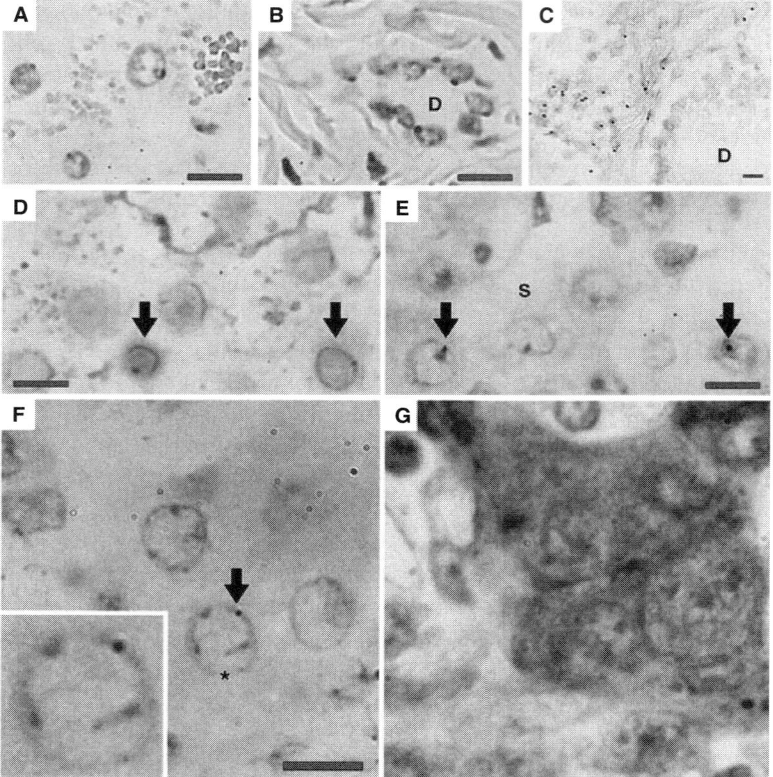

Figure 9.10. Hepatocytes derived from bone marrow cells. Adult bone marrow cells prepared from male human donors were transplanted into female patients. The presence of the Y chromosome in the female liver indicates the engraftment of the transplanted male bone marrow cells. The Y chromosome can be detected by immunolocalization of a fluorescein isothiocyanate (FITC)-labeled Y-chromosome probe using an anti-fluorescein antibody conjugated to horseradish peroxidase, and visualized using diaminobenzidine as a brown chromogen. For this preparation, the presence of the Y chromosome is indicated by a distinct brown dot, typically located at the nuclear periphery. (A) Hepatocytes were identified by their large round nuclei and cytoplasmic granules and were Y-chromosome-positive in male controls. (B) The Y chromosome was also detected in the bile ducts (D) in male controls. (C) A Y-chromosome-negative bile duct surrounded by numerous Y-positive inflammatory cells in a female liver transplanted into a male. (D) Immunodetection of two Y-chromosome-positive cells (arrows) located in a hepatocyte plate delineated by cytokeratin-8 immunostaining from a female patient who had received male bone marrow. (E–G) Y-chromosome detection in female livers transplanted into male recipients; (E) two Y-chromosome-positive hepatocytes (arrows) in a hepatocyte plate bordering a sinusoid (S); note the brown dot, demonstrating that the Y chromosome is readily distinguishable from the blue nucleolus; (F, G) consecutive sections of a group of four hepatocyte nuclei, showing (F) one Y-chromosome-positive hepatocyte (arrow, and inset at 2-fold magnification), and (G) their cytokeratin-8 immunoreactivity. Scale bars: 10 μm. (Reprinted by permission from Macmillan Publishers Ltd.: Alison MR et al: *Nature* 406:257, copyright 2000.)

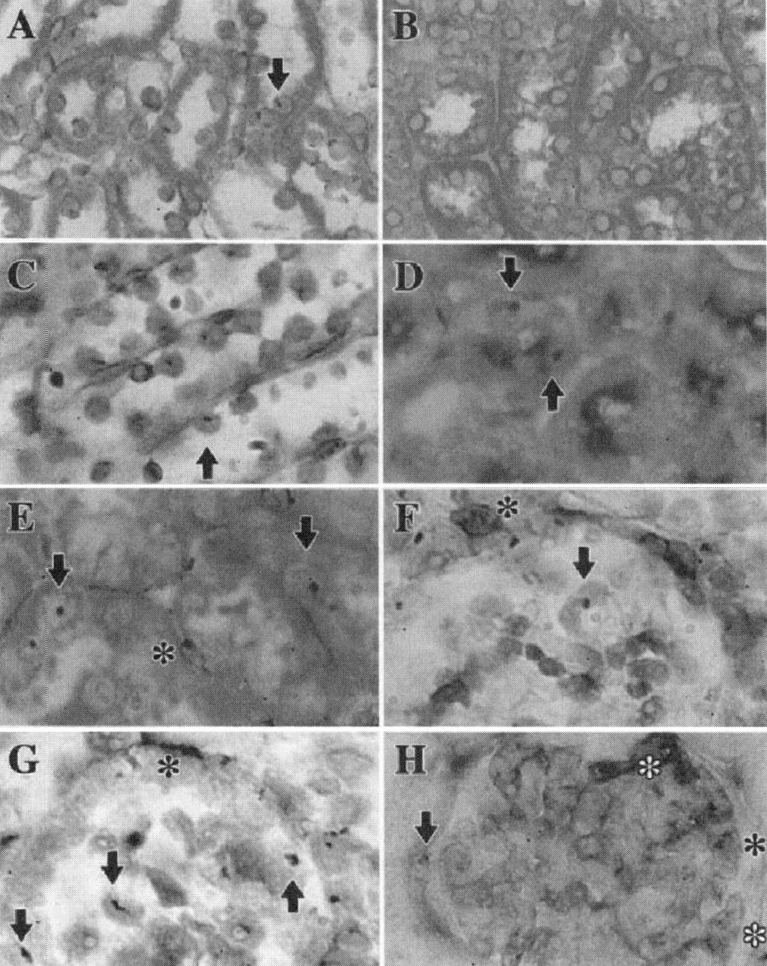

Figure 9.11. Engraftment of bone marrow cells into the kidney. Bone marrow cells were prepared from male donor mice and transplanted into female recipient mice. The Y chromosomes in the female kidney were detected by immunolabeling with the anti-CYP1A2 antibody. (A) A male control showing positive Y chromosome (arrow) in tubular epithelial cells immunostained with the anti-CYP1A2 antibody. (B) A female control showing the lack of Y chromosome within tubular epithelial cells immunostained with the anti-CYP1A2 antibody. Micrographs C–H were prepared from a female mouse 13 weeks following male whole bone marrow transplantation. (C) Y-chromosome-positive proximal tubular epithelial cells (arrow) adherent to a PAS-positive basement membrane. (D) Y-chromosome-positive tubular epithelial cells (arrows) that are reactive to RCA lectin; (E) Y-chromosome-positive proximal tubular epithelial cells (arrows) that are CYP1A2-immunoreactive. A Y-chromosome-positive interstitial cell can also be seen (*). (F) Y-chromosome-positive cells associated with the glomerulus: Y-chromosome-positive/CD45-negative cells within a glomerulus (arrow) and Y-chromosome-positive/CD45-positive cells apposed closely to the renal corpuscle (*). (G) Y-chromosome-positive F4/80-negative cells (arrows) within a glomerulus and a Y-chromosome-positive F4/80-positive cell apposed closely to the renal corpuscle (*). (H) A Y-chromosome-positive/vimentin-positive cell (black asterisk) within a glomerulus. A Y-chromosome-positive/vimentin-positive cell lining the renal corpuscle (arrow) is apparent along with several other Y-chromosome-positive cells (white asterisks) of undetermined phenotype. (Reprinted with permission from Poulsom R et al: *J Pathol* 195:229–35, 2001. Copyright Pathological Society of Great Britain and Ireland. Permission is granted by John Wiley & Sons Ltd on behalf of PathSoc.)

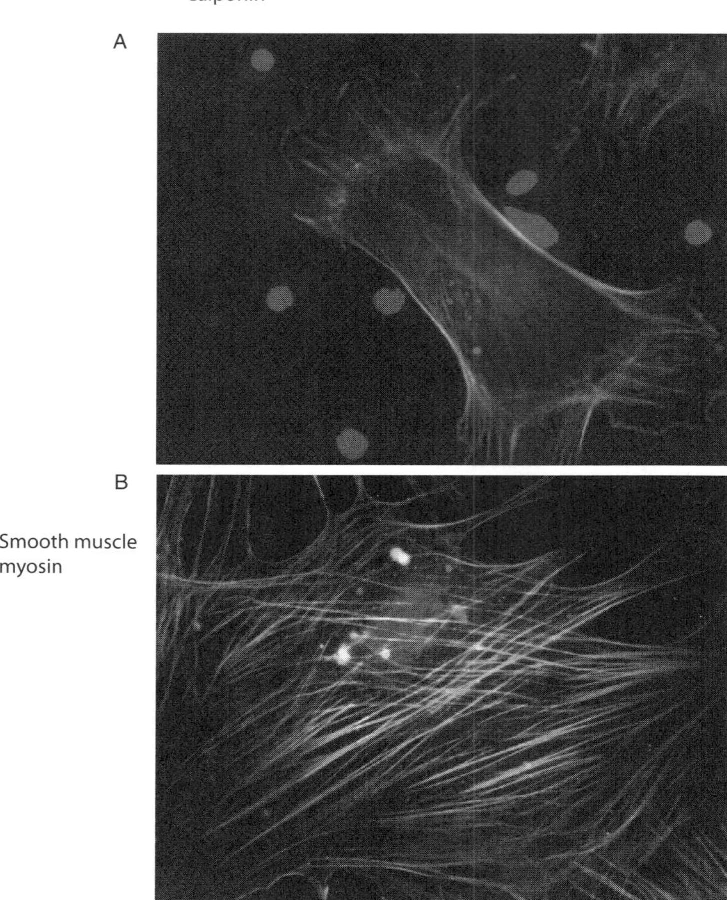

Figure 9.12. Transformation of CD34-positive cells derived from the mouse bone marrow to calponin- and smooth muscle myosin heavy chain-cells in vitro.

muscle α-actin$^+$ cells seldom express CD11b, whereas the smooth muscle α-actin$^-$ express CD11b. Both smooth muscle α-actin$^+$ and smooth muscle α-actin$^-$ cells express CD45. The smooth muscle α-actin$^+$ cells can form smooth muscle-like cells in elastic lamina-containing matrix.

Bone marrow-derived stem cells have been used to regenerate nonhematopoietic cells, including neurons, cardiomyocytes, endothelial cells, vascular smooth muscle cells, and hepatocytes. One of the major discoveries in recent studies is the transformation of hematopoietic stem cells to *hepatocytes* in an injured or transplanted liver in the mouse, rat, and human. The hematopoietic stem cells are originated from the mesoderm, whereas the hepatocytes are formed from the endoderm. The formation of hepatocytes from the hematopoietic stem cells suggests that adult stem cells can transdifferentiate across lineage barriers. Often, regional injury or disorder is necessary for the induction of stem cell transformation into a specified cell type. For instance, in a transgenic mouse model lacking the fumarylacetoacetate hydrolase, the animal experiences progressive liver injury, dysfunction, and failure due to the accumulation of tyrosine metabolites. This injury demands and stimulates the regeneration of liver cells. When allogenic $Lin^-c\text{-}kit^+Sca\text{-}1^+Thy1^+$ hematopoietic stem cells are transplanted into myeloablated mice with liver injury, the transplanted cells can differentiate into functional hepatocytes and rescue the injured liver. The hematopoietic stem cells possess a high capacity of cell regeneration. A number of ~50 $Lin^-c\text{-}kit^+Sca\text{-}1^+Thy1^+$ hematopoietic stem cells, when transplanted into a mouse, can repopulate the hepatocyte family that is necessary for the maintenance of the liver function.

The hematopoietic stem cells can be induced to regenerate cardiomyocytes in experimental cardiac injury and in human cardiac infarction. In an experimental model, $CD34^{low}c\text{-}Kit^+Sca\text{-}1^+$ bone marrow cells, which are also known as *side population cells* (so defined because these cells are found outside the main population of bone marrow cells in a flow cytometry test), can be selected and transplanted into the heart of myeloablated mice with cardiac infarction induced by coronary arterial ligation. Following cardiac injury, the transplanted bone marrow cells can engraft to the injured cardiac tissue and regenerate cardiomyocytes and vascular endothelial cells. The newly regenerated cardiomyocytes can partially compensate for the lost cardiac function, and the regenerated endothelial cells can initiate angiogenesis, a process that improves blood circulation in the injured area and facilitates the recovery from cardiac ischemia. Furthermore, the regenerated cells can produce and release various cytokines and growth factors that initiate and enhance cardiac repair. Bone marrow-derived stem cells can also be injected directly into the injury sites of the animal and human heart, ensuring a focused delivery of therapeutic cells.

Bone marrow-derived stem cells can be used to regenerate nervous cells. Bone marrow cells originate from the mesoderm, whereas the nervous system is from the ectoderm. The regeneration of ectodermal cells from mesodermal stem cells indicates that adult stem cells can differentiate into specified cells across developmental barriers, a phenomenon previously considered impossible. This discovery provides fundamental information for the clinical application of adult stem cells to the treatment of degenerative disorders in the nervous system, which possesses a relatively low capacity of regeneration. Several investigations have demonstrated that donor bone marrow cells, when injected into the central venous system of myeloablated recipient animals, can form cells that express neuronal protein markers such as NeuN and class 3b-tubulin in the brain. These observations suggest that adult bone marrow stem cells can be potentially used to treat degenerative neural disorders such as Alzheimer's and Parkinson's diseases. Bone marrow cells can also differentiate into glial cells. When bone marrow cells are delivered to a demyelinated spinal cord, these cells can transform into oligodendrocytes that express myelin protein and induce remyelination. Bone marrow cells have been used to treat spinal cord injury, resulting in functional improvement of the injured spinal cord. However, several aspects remain to be determined. First, the morphology of the neural cells derived from bone marrow cells has not been thoroughly studied: it remains poorly understood whether bone marrow-derived cells can form axon. Second, the function of bone marrow-derived cells has not been systematically characterized. It is not clear whether bone marrow cells can develop into fully functional neurons or glial cells.

Hematopoietic stem cells have also been used for the treatment of cancer in humans following radiation-induced myeloablation. For instance, $CD34^+Thy\text{-}1^+$ autologous hematopoietic stem cells can be collected from the bone marrow and delivered to the vascular system of patients with breast cancer and multiple myeloma. Such a treatment was effective in about 41% cancer patients. The minimal dose required to achieve an effective treatment (judged with respect to absolute neutrophil count) is 2–4 × 10^5 cells. These studies have demonstrated the potential of using hematopoietic stem cells for the treatment of cancers.

The pluripotent features of the hematopoietic stem cells have been described by using two models: the plasticity and heterogeneity models. The *plasticity model* suggests that there exist one or more types of pluripotent hematopoietic stem cells. These cells can cross the mesodermal lineage barrier to transdifferentiate (note that transdifferentiation is the differentiation of a stem or progenitor cell across lineage barriers) into ectodermal and endodermal cells under appropriate local conditions. These cells either possess features of the embryonic stem cells or can dedifferentiate (note that dedifferentiation is a process by which a cell changes from a mature differentiation state to a more primitive differentiation state so that the cell regains stem cell features and can differentiate into different cell types) and return to a more primitive state of specification and differentiation. In contrast, the *heterogeneity model* suggests that in the bone marrow there exist different types of stem cells, including not only the hematopoietic stem cells but also other types of committed stem cells that can develop into specified cells within a system. For instance, the bone marrow stromal cell population may contain mesenchymal stem cells that can differentiate into vascular endothelial and smooth muscle cells. Even within the hematopoietic stem cell population, there may be different subpopulations that are committed to the differentiation into specified cell types. It is important to point out that the two models are hypothetical in nature. Further investigations are necessary to test the hypotheses proposed in these models and explore cell surface markers that specify unique types of stem cells.

While bone marrow stem cells can maintain their capabilities of self-replication and differentiation throughout their lifespans in vivo, in vitro culture and expansion usually diminish these capabilities. For instance, the proliferation rate of bone marrow stem cells is reduced significantly after the first confluence in culture. The formation of specialized mesenchymal cells from in vitro expanded bone marrow stromal cells is reduced compared to freshly harvested bone marrow stromal cells. In fact, the differentiation ability of bone marrow stem cells is gradually lost when the cells are cultured for a number of passages. Such phenotypic changes are possibly due to the lack of physiological environment and necessary mediators in culture. These features should be taken into account when bone marrow stem cells are used for cellular therapies.

In vitro culture and expansion of bone marrow stem cells require the presence of growth factors, such as fibroblast growth factor, epidermal growth factor, and platelet-derived growth factor. The capability of cell proliferation is impaired in the absence of growth factors. Among the common growth factors, fibroblast growth factor 2 has been shown to effectively enhance the proliferation of bone marrow stem cells and maintains their immature stem cell phenotypes. Thus, appropriate growth factors should be used for in vitro culture and expansion of bone marrow stem cells.

In summary, bone marrow-derived stem cells can be used to regenerate a variety of cell types originally derived from the ectoderm (e.g., epidermal cells and neurons), mesoderm (e.g., skeletal muscle and kidney cells), and endoderm (e.g., pulmonary, gastrointestinal, and pancreatic cells). During the past decade, extensive investigations have been conducted to study the differentiation capacity of the adult stem cells. These investigations have established a foundation for understanding the mechanisms of adult stem cell differentiation and applying adult stem cells to the treatment of human degenerative disorders.

Neural Stem Cells [9.3]. The brain and the spinal cord have long been considered a system containing only terminally differentiated cells that cannot differentiate and regenerate in the event of cell loss. However, recent studies have demonstrated that the adult nerve system contains neural progenitor and stem cells, which can proliferate and differentiate into neurons and glial cells in response to nerve injury and cell death. This discovery challenges the previous terminal differentiation theory and suggests possibilities for the establishment of stem cell-based therapies for degenerative nerve disorders.

In the adult brain, neurogenesis has been found since the 1960s in two major regions: the olfactory bulb and the hippocampal dentate gyrus. The olfactory bulb is the bulb-like region of the olfactory tract where the olfactory nerves enter (Fig. 9.6). It is located on the undersurface of the frontal lobe of each cerebral hemisphere. The hippocampal dentate gyrus is an archicortex that develops along the edge of the hippocampal fissure. These areas demonstrate cell division activities in the adult brain as detected by autoradiography. However, because of difficulties in identifying newborn neurons during the 1960s, these discoveries were not well accepted. The phenomenon of neurogenesis was not widely recognized until neurons were generated in vitro from adult rodent brain in 1992. The generation of neuronal cells has led to extensive investigations on neurogenesis during the past decade.

It is well known now that several regions in the adult brain contain neural stem and progenitor cells. These include the caudal portion of the subventricular zone, olfactory bulb, hippocampus, striatum, optic nerve, corpus callosum, spinal cord, cortex, retina, and hypothalamus. In vitro investigations have shown that cells collected from these regions can be induced to differentiate to neurons, astrocytes, and oligodendrocytes. Stem and progenitor cells can be identified from these regions based on immunocytochemical markers on the cell surface. For instance, the NeuN protein has been used as an identification marker for neurons, and GFAP and S100β are markers for glial cells. Neural stem cells can be confirmed by the presence of common stem cell features, such as capability of self-renewal, expansion, and differentiation into specified mature cells. Cell renewal and expansion can be tested by using the BrdU incorporation assay, designed on the basis of the fact that BrdU can be taken up only by dividing cells. As for other tissues, neural stem and progenitor cells can be expanded in vitro and transplanted into a target region in the brain to regenerate injured and lost cells in degenerative nerve disorders.

There are several features for neural stem and progenitor cells in culture. First, these cells can adhere to the base of culture dishes. The proliferation and differentiation of neural stem and progenitor cells can be modulated by altering the substrate of cell culture. Second, neural stem and progenitor cells can form aggregates in cell suspension. Such a property has been used as a criterion for identifying neural stem and progenitor cells. The culture of neural stem and progenitor cells requires the presence of growth factors. In particular, epidermal growth factor and fibroblast growth factor 2 play a critical role in regulating the survival and differentiation of neural stem and progenitor cells in vitro. The lack of these growth factors may significantly influence the expansion and differentiation of neural stem and progenitor cells. These growth factors have also been shown to participate in the regulation of neurogenesis in vivo.

Several issues remain to be clarified in neurogenesis. One of the issues is that the neural stem and progenitor cells have not been characterized in terms of structure and function.

These cells may be highly heterogeneous in function and capacity of regeneration. The formation of functional adult neurons from the adult stem or progenitor cells may be dependent on specific locations and environment. Furthermore, while the in vitro formation of neural cells provides useful information for neurogenesis, the mechanisms and the final products of cell transformation in vitro may be different from those in vivo. It is necessary to establish in vivo experimental models for the investigation of neurogenesis.

Other Adult Stem Cells [9.4]. There are other types of stem and progenitor cells in mammalian tissues and organs, including stem cells from the epidermal tissue (basal cells), intestinal tissue (crypt epithelial cells), connective tissue (adipocytes), and skeletal muscle (myoblasts). These cells can be identified and characterized by using methods and criteria described for the bone marrow and neural stem and progenitor cells. Recent studies have demonstrated that stem cells from these systems can differentiate into specified mature cells within the system and can also cross the developmental barriers to form specified cells for other systems. It seems that most systems in a mature adult body contain stem and progenitor cells. These cells can regenerate specified cells in the event of cell loss due to injury and pathological disorders. Here, the epidermal and intestinal stem cells are used as examples for discussion.

There exist several types of stem and progenitor cells in the epidermal tissue. These include epidermal stem cells, follicle multipotent stem cells, melanocyte stem cells, dermal sheath stem cells, neural crest stem cells, and endothelial stem cells (Fig. 9.7). The epidermal stem cells are found in the basal layer of the epidermis and are also known as *basal cells*. These cells can primarily differentiate to epidermal cells. The follicle multipotent stem cells are found in the hair follicle bulge region and can differentiate to hair follicle epithelial cells, sebaceous gland cells, and epidermal cells. The melanocyte stem cells are also found in the hair follicle bulge and can transform into melanocytes. The dermal sheath stem cells are found in the hair follicle dermal sheath and can differentiate to dermal papilla cells and wound healing fibroblasts. The neural crest stem cells are present in the hair follicle dermal papillae and can transform into neural cells and mesenchymal cells. The endothelial stem cells are present in the dermal tissue and can differentiate to vascular endothelial cells. The dermis underneath the epidermal layer also contains stem cells. These cells can differentiate various cell types including smooth muscle-like cells and adipocytes under appropriate culture conditions (Fig. 9.8).

The intestinal system contains epidermal stem and progenitor cells. The epithelial stem cells are found mostly in the crypt region of the intestine (Fig. 9.9). These cells can differentiate to intestinal epithelial cells in response to epithelial injury. Compared to stem cell types from other organs, the epidermal and intestinal stem cells can be easily accessed and collected. These stem cells can be used as candidate cells for regenerative therapies.

REGENERATION OF ADULT TISSUES AND ORGANS

Cell regeneration occurs in selected adult tissues and organs of certain species. Typical examples include the limbs of salamander and crayfish as well as the liver of mammals. Salamander and crayfish can regrow their limbs after amputation. The regenerated limbs are identical to the original limbs in morphology and are fully functional. In mammals, the liver is the only organ that can regenerate completely after partial hepatectomy or liver removal. The regenerated hepatic tissue possesses natural structure and physiological function. Based on the principles of developmental biology, these organs can reactivate their developmental processes that take place during the embryonic stage and thus regenerate missing or damaged tissues.

There are two general mechanisms for the regeneration of adult tissues and organs in salamander and mammals: epimorphogenesis and compensatory regeneration. *Epimorphogenesis* is a process by which undifferentiated cells (found in the embryonic stage) are reestablished from adult cells via dedifferentiation. The undifferentiated cells can be respecified into adult cells. This is a mechanism for the regeneration of the salamander and crayfish limbs. *Compensatory regeneration* is a process by which existing cells reproduce themselves via division. These cells are not able to transform themselves to undifferentiated cells. Liver regeneration is an example of compensatory regeneration. The investigation of these regenerative processes can provide essential information for understanding the underlining mechanisms of regeneration, for controlling the regenerative processes, and for enhancing the restoration of impaired tissues and organs. Here, the salamander limbs and mammalian liver are used as examples to demonstrate the principle of adult organ regeneration.

Regeneration of Salamander Limbs [9.5]

Salamander is a tailed amphibian of the order Caudata. It has long been observed that, when a limb of the salamander is severed, the remaining part of the limb can grow back to the original form. The limb system can precisely control the growth process. The growth is ceased when a complete limb is reconstructed. It has been a mystery how exactly a salamander reconstructs a severed limb and how it controls the morphology and distribution of different cells and tissues. During the past several decades, extensive investigations have been conducted on the regenerative mechanisms of the salamander limbs. These investigations have demonstrated a series of biological processes that are involved in the regeneration of the amputated limb.

Following limb amputation, the first step of repair is the formation of blood clots, which seal damaged blood vessels and prevent bleeding. The second noticeable step is the migration of epidermal cells, together with proliferation, from the remaining limb to the wound area. The newly generated epidermal cells form wound epidermis that covers the area of damage. A unique feature for the regenerative response of the salamander limb is the lack of scar formation. There are few fibroblasts and little fibrotic connective tissue in the layer of wound epidermis. Such a feature is critical to the reconstruction of the severed limb. The third step is the formation of the *regeneration blastema* underneath the wound epidermis. Several mesodermal cell types, including osteoblasts, chondrocytes, muscular cells, and fibroblasts, undergo dedifferentiation and regain the phenotypes of embryonic stem cells, capable of transforming to the cell types and producing extracellular matrix components necessary for limb regeneration. An example of muscular cell regeneration is shown in Fig. 9.13. The dedifferentiated mesodermal cells are clustered beneath the wound epidermis, forming a structure known as the regeneration blastema. This structure is the basis for the regeneration of a new limb.

Next step is the proliferation of cells in the regeneration blastema and the formation of a primary structure for the limb. Several growth factors, including fibroblast growth factor and glial growth factor, play an important role for cell proliferation. To form a limb, the regenerated cells must be organized into the right pattern with each cell type deployed to

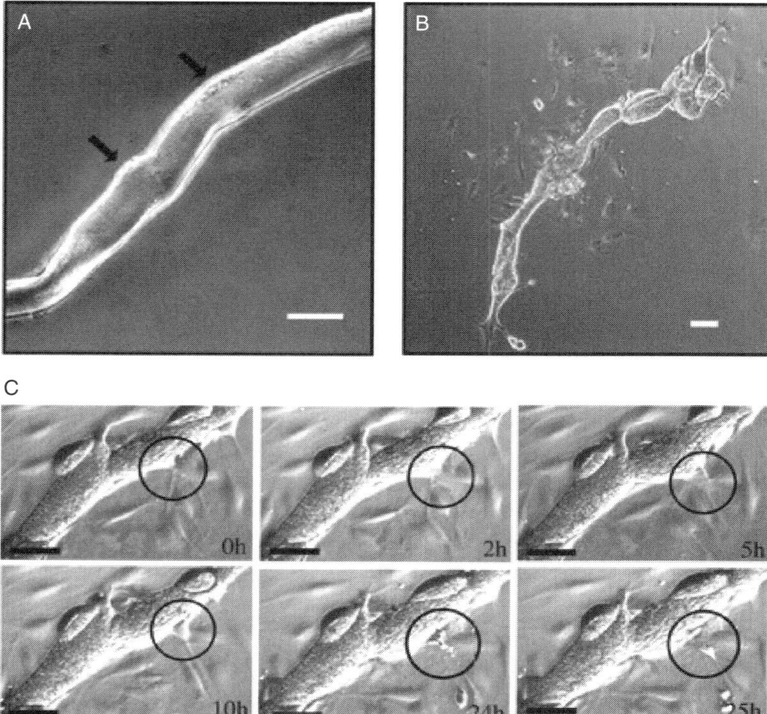

Figure 9.13. Budding of progeny cells from newt skeletal muscle cells. (A) Photomicrograph showing a freshly isolated single newt skeletal muscle cell. Arrows point to two visible nuclei. (B) Photomicrograph showing the same newt skeletal muscle cell at 15 days of culture. The myofiber morphology has changed and several lobular structures are seen while mononucleate progeny has been produced. (C) Time-lapse photomicrographs showing a sequence of a representative budding event, which leads to the derivation of a mononucleate cell. Note the protrusion of the myofiber in the circled area, which is concomitant with the appearance of a mononucleate progeny. Timepoints indicate the duration of the one specific budding event. Scale bars: 50 μm. (Reprinted from Morrison JI et al: *J Cell Biol* 172:433–40, 2006, copyright by permission of the Rockefeller University Press.)

the right location. Increasing evidence suggests that injury-induced limb regeneration resembles the process of embryonic limb formation. Molecules that regulate limb development during the embryonic stage, such as retinoic acid, sonic hedgehog, HoxA, and HoxD, can also be found in a regenerating limb in salamanders. These molecules participate in the regulation of the pattern formation of a regenerated limb. In particular, retinoic acid, produced by wound epidermal cells, is present with graded concentrations in the proximal–distal direction of the blastema. The concentration gradient of retinoic acid induces location-dependent activation of the HoxA gene. HoxA regulates the pattern formation of regenerated limb cells. Within about 2–3 months after limb amputation, a complete new limb can be regenerated with full function (Fig. 9.14).

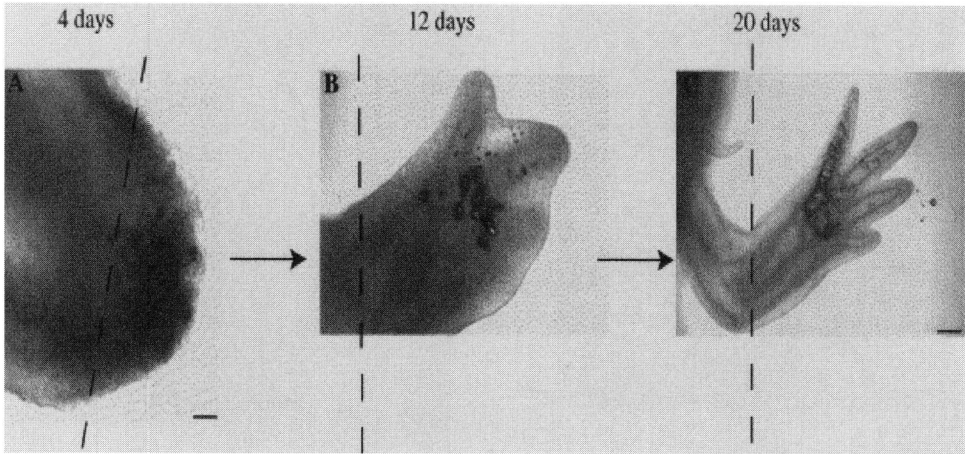

Figure 9.14. Salamander limb regeneration. (A) The blastema of a salamander limb at day 4 after amputation. The blastema was transfected with the CMV-DsRed fluorescent protein gene via the mediation of electroporation. The dark marker indicates the expression of the transfected gene. (B) By 12 days postamputation, the marked distal cells divided and remained in the most distal part. (C) At 20 days postamputation, the limb regenerated, and the marked cells contribute to the formation of the digits. The dashed line marks the plane of amputation. Scale bars: A,B = 100 µm; C = 500 µm. (Reprinted from Echeverri K, Tanaka EM: *Dev Biol* 279:391–401, copyright 2005 with permission from Elsevier.)

Regeneration of the Mammalian Liver [9.6]

The liver is an organ that can fully regenerate after partial hepatectomy or injury in mammals. Under physiological conditions, the liver is a quiescent organ with a turnover rate $<0.01\%$. In response to injury, the liver can activate its mitotic mechanisms, initiating rapid cell proliferation and regenerating functional liver cells. A fundamental experiment that demonstrates liver regeneration is liver resection or partial hepatectomy. Such a procedure stimulates rapid liver regeneration (Fig. 9.15). Following partial hepatectomy (up to 60% of the liver size), the remaining liver can grow back to the original size within 10 days in the mouse and rat! Cell proliferation and liver regeneration cease once the liver returns to its original size. Few other cell types in mammals can grow or regenerate with such a rate.

Further experimental investigations have shown that liver regeneration is a precisely controlled process, depending on the ratio of the liver mass to the entire body mass. This phenomenon is demonstrated in several experimental models, including the parabiotic animal model and liver transplantation. In the parabiotic animal model, in which the circulatory systems of two animals are anastomosed, liver removal from one animal induces rapid liver regeneration of the liver of the other animal. The regenerative process continues until the liver mass doubles and reaches the original liver : body mass ratio. In contrast, the separation of the two animals induces rapid hepatocyte apoptosis and liver shrinkage in the animal survived (note that the animal without liver will die). Liver shrinkage continues until the liver mass returns to the original size of the individual animal. In the model of liver transplantation, a liver transplant derived from a smaller donor animal or

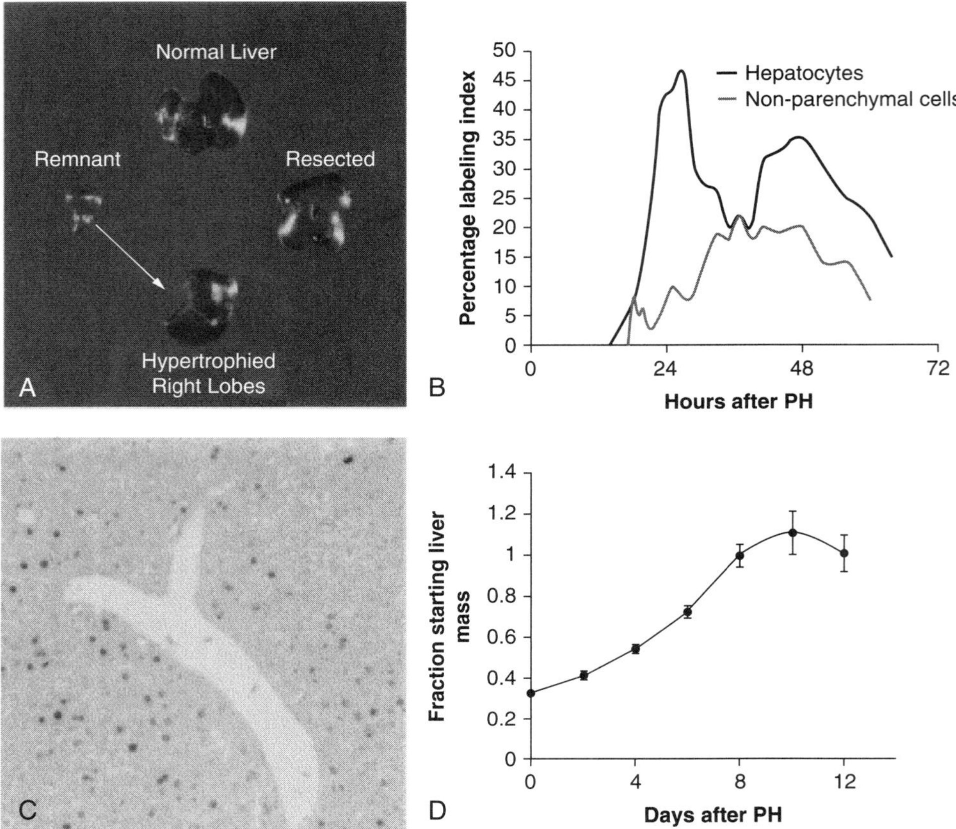

Figure 9.15. Liver regeneration after two-thirds partial hepatectomy (PH). (A) Mouse livers before and after partial hepatectomy and hypertrophied remnant 6 days after partial hepatectomy. (B) Percentage of ^{3}H-thymidine-labeled liver cells at timepoints after partial hepatectomy. (C) BrdU incorporation into proliferating rat hepatocytes after carbon tetrachloride treatment. (D) Time course of liver regeneration by mass after partial hepatectomy. (Reprinted from Koniaris LG et al: Liver regeneration, *J Am Coll Surg* 197:634–59, copyright 2003 by permission of the American College of Surgeons.)

human compared to the host will grow until reaching the size of the original liver of the larger host. In contrast, a liver transplant from a larger donor will shrink until reaching the size of the original liver of the smaller host. These investigations demonstrate that the mammalian liver can regenerate precisely in response to degree of liver loss and liver regeneration is a metabolism-controlled process. The investigations of liver regeneration have provided insights into the understanding of organ regeneration and the development of therapeutic approaches for degenerative diseases.

Biological Processes of Liver Regeneration [9.6]. There are a series of regenerative processes following partial hepatectomy. One of earliest known processes is the activation of mitogenic factors, such as hepatocyte growth factor (HGF) and interleukin (IL)6. Hepatocyte growth factor can be activated within the first hour following partial hepatectomy and plays a critical role in the initiation and regulation of hepatocyte proliferation and liver regeneration. Under the stimulation of mitogenic factors, quiescent hepatocytes are activated to enter the cell division cycle, initiating DNA synthesis and cell division. DNA synthesis can be observed within 1–2 days following partial hepatectomy in small animals such as mice and rats. In larger animals, such as dogs and primates, DNA synthesis is often observed within 3–10 days. The DNA synthesis stage is followed with rapid cell proliferation, as detected by nucleotide incorporation assays. The rate of cell proliferation is usually proportional to the degree of liver resection. Cell proliferation slows down and ceases at about 10 days in mouse and rat hepatectomy models when the liver grows back to its original size.

Features of Liver Regeneration [9.6]. There are several features for liver regeneration:

1. Each liver cell type exhibits a distinct rate of regeneration. The hepatocytes can respond rapidly (within about 1 h) to stimulation induced by liver hepatectomy and has the highest rate of proliferation compared to other liver cell types. However, the duration of hepatocyte regeneration (about 2 days) is shorter than that of other cell types. The liver endothelial cells respond more slowly to stimulation (within about 2 h), but have a longer duration of proliferation (about 10 days) compared to the hepatocytes. The bile duct epithelial, Küpffer, and Ito cells fall in a range between the hepatocytes and endothelial cells in both rate and duration of proliferation (Fig. 9.16). Hepatocyte regeneration is dependent on the animal species. Large animals such as dogs, pigs, and primates usually exhibit delayed activation of cell prolifera-

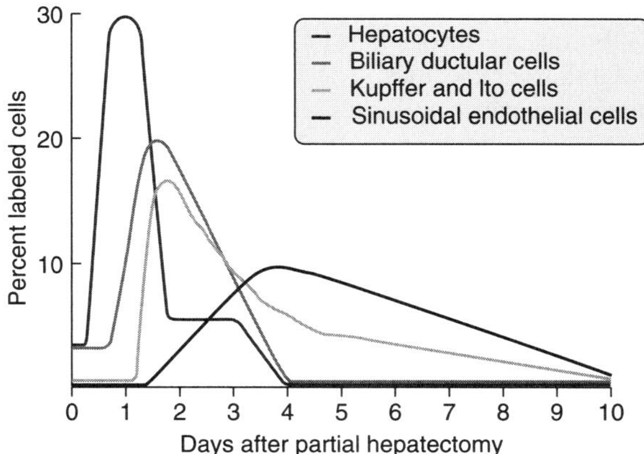

Figure 9.16. Time course of the proliferation of different liver cell types during liver regeneration after partial hepatectomy. The four major types of liver cells undergo DNA synthesis at different times. Hepatocyte proliferate peaks at 24 h, whereas the other cell types proliferate later. Regenerating hepatocytes produce growth factors that can function as mitogens for these cells. This has suggested that hepatocytes stimulate proliferation of the other cells by a paracrine mechanism. The figure was generated by graphic adaptation of the data presented in two publications (Grisham JW: *Cancer Res* 22:842, 1962; Widmann JJ, Fahimi HD: in *Liver Regeneration after Experimental Injury*, Lesch R, Reutter W, eds, Lesch R, Reutter W, eds, Stratton Intercontinental Medical Book Corp., New York, 1975, pp 89–98). (Reprinted with permission from Michalopoulos GK, DeFrances MC: Liver regeneration, *Science* 276:60–6, copyright 1997 AAAS.)

tion compared to small animals such as rats and mice. These observations suggest that liver cells contribute differently to liver regeneration following partial hepatectomy.

2. The rate of liver regeneration is proportional to the amount of liver removed. The larger is the removed portion, the faster is the rate of liver regeneration. Even a small resection (<10%) is followed by liver restoration to its full size.
3. Liver regeneration ceaseds when the injured liver grows back to its original size, suggesting that the functional demand for metabolism is a factor that initiates and controls the process of liver regeneration.
4. The liver possesses a large capacity of regeneration. A single rat hepatocyte can undergo at least 34 division cycles, resulting a clone of about 1.7×10^{10} cells (note that a normal rat liver has about 3×10^{8} cells). A rat liver can regenerate to full size after 12 sequential operations of partial hepatectomy.
5. Liver cells undergo a clonogenic growth process. Each cell can proliferate to form a cell clone, according to which a liver nodule develops.
6. Mature hepatocytes are not terminally differentiated cells and can differentiate to different types of liver cells.

Liver regeneration is a compensatory process that is dependent not on the differentiation of stem cells, but on the proliferation and differentiation of liver cells, including hepatocytes, biliary epithelial cells, fenestrated endothelial cells, Küpffer cells, and Ito cells. All these cells can thus be considered liver progenitor cells. Among these cell types, the hepatocytes are the majority of liver cells and are responsible for basic liver functions, including metabolic processing of carbohydrates, fats, and proteins, degradation of toxic compounds, production of necessary proteins (e.g., albumin), and secretion of bile. The biliary epithelial cells line the internal surface of the bile ducts and participate in the transport process of biliary molecules. The fenestrated endothelial cells are found at the internal surface of the hepatic capillaries and the hepatic sinusoids and are responsible for molecular transport across blood vessels. The Küpffer cells are hepatic macrophages found in the hepatic sinusoids and are responsible for the destruction and clearance of bacteria and cell debris. The Ito cells can synthesize extracellular matrix components. In the case of liver injury or hepatectomy, all these cell types are stimulated to proliferate, differentiate, and participate in liver regeneration.

Experimental Models of Liver Regeneration [9.7]. Liver regeneration has been studied by using experimental models. Rats and mice are often used for such models. There are three common types of experimental models for liver regeneration: chemical ingestion, ischemia, and partial hepatectomy or liver resection. Chemical ingestion can induce cell injury and death. Common chemicals used for livery injury include alcohol, carbon tetrachloride, and D-galactosamine. These chemicals may target different regions of the liver. For instance, alcohol primarily induces hepatocyte injury near the portal veins, D-galactosamine induces overall hepatocyte injury, whereas carbon tetrachloride causes central hepatocyte injury. These models are often used to simulate toxin-induced human liver injury and failure. Liver ischemia can be induced by temporary occlusion of arteries. The ischemic area can be controlled by selectively occluding different generations of hepatic arteries. This model can be used to simulate human liver injury due to atherosclerosis and embolism.

Among the three types of liver injury models, partial hepatectomy is most commonly used in experimental investigation of liver regeneration. Selected liver lobes can be ligated and excised at the lobe base. Compared to the other types of liver injury, hepatectomy reduces the liver size, but does not induce significant hepatocyte injury or death. Thus, this model can be used to study liver regeneration with respect to mechanisms of physiological adaptation. The rate of liver regeneration is usually dependent on the relative resection size. The maximal size of resection without influencing the metabolism of the body is about 60% of the original liver size. Larger resection induces metabolic disorders and acute liver failure. In this chapter, partial hepatectomy is used as an example to demonstrate the processes and mechanisms of liver regeneration.

Regulation of Liver Regeneration [9.8]. Liver regeneration is initiated and regulated by growth regulatory factors produced in the liver as well as other organs. These factors include hepatocyte growth factor (HGF), interleukin (IL)6, epidermal growth factor (EGF), insulin, Wnts, transforming growth factor (TGF)β, and activin. Among these factors, hepatocyte growth factor, interleukin-6, epidermal growth factor, insulin, and Wnts exert mostly a stimulatory effect on liver regeneration, whereas transforming growth factor-β and activin serve as inhibitory factors for liver regeneration (Fig. 9.17). Both growth stimulatory and growth inhibitory factors are upregulated following partial hepatectomy with a time-dependent manner. The early dominant expression of growth stimulatory factors promotes hepatocyte proliferation and liver regeneration, whereas the following expression of growth inhibitory factors contributes to the cessation of liver regeneration when the liver mass returns to the original level. Thus, the control of the activities of these growth regulatory factors is critical to the regulation of liver regeneration. The redundancy of the growth regulatory factors is a mechanism that ensures effective, synergistic control of liver regeneration in response to injury. The role of these factors in liver regeneration is briefly discussed in this section.

Hepatocyte growth factor is one of the most important factors that regulate liver regeneration. This growth factor is expressed in the liver cells. A major function of hepatocyte growth factor is to stimulate hepatocyte growth and regulate hepatocyte survival. This function has been confirmed in experiments in vitro and in vivo. Hepatocyte growth factor can interact with and activate the c-met protein tyrosine kinase receptor, which induces activation of the Ras—mitogen-activated protein kinase signaling pathway. When the hepatocyte growth factor gene is deleted or modulated in the embryo, hepatic development is arrested or reduced, leading to death of the embryo. The administration of hepatocyte growth factor to animals with liver injury stimulates liver regeneration. After hepatectomy, the blood level of hepatocyte growth factor can be increased by 20 folds within 1 h. Hepatocyte growth factor serves as a signal that initiates early proliferation of liver cells. Other growth factors, such as *epidermal growth factor* and *fibroblast growth factor*, also activate protein tyrosine kinase receptor-mediated signaling pathways. These growth factors have been shown to serve as potent stimulators for hepatocyte proliferation following partial hepatectomy.

Interleukin (IL)6 is a cytokine that induces inflammatory reactions. Interleukin-6 exerts diverse effects on target cells, such as growth-stimulatory effect and growth-inhibitory effect, depending on cell types. For hepatocytes, IL6 serves as a growth stimulatory factor (Fig. 9.18). Hepatocytes express interleukin-6 following partial hepatectomy. Interleukin-6 can interact with the IL6 receptor α chain, also known as *gp80*, in the hepatocyte membrane and induce dimerization with another IL6 receptor chain known as

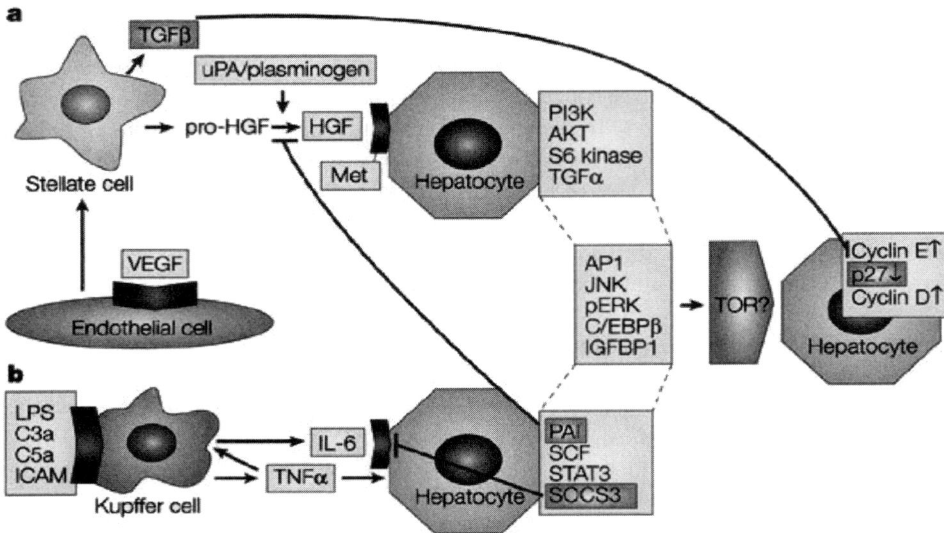

Figure 9.17. Growth factors and cytokines involved in liver regeneration. (a) Liver regeneration is regulated by several growth factors. Vascular endothelial growth factor (VEGF) binds to endothelial cells, which triggers the release of the hepatocyte growth factor (HGF) precursor, pro-HGF, from stellate cells. The urokinase-type plasminogen activator (uPA) and plasminogen proteases cleave pro-HGF, which releases HGF. HGF binds to the Met receptor on hepatocytes to activate the phosphatidylinositol 3-kinase (PI3K), AKT, and S6 kinase signal-transduction pathways. HGF signaling activates transforming growth factor (TGF)α and other downstream signals that are shared with the cytokine-mediated pathway, such as AP1, Jun *N*-terminal kinase (JNK), phosphorylated extracellular signal-regulated kinases (pERKs), CCAAT-enhancer-binding protein (C/EBP) β and insulin-like-growth-factor-binding protein (IGFBP)1. These factors are proposed to activate target of rapamycin (TOR), although this remains to be established, and this leads to cell cycle transition by increasing the expression of cyclins D and E and reducing p27. (b) Liver regeneration is also regulated by cytokines. Molecules factors, including lipopolysaccharide (LPS), complement factors C3a and C5a and intercellular adhesion molecules (ICAMs), can activate Kupffer cells, which produce tumor necrosis factor (TNF)α. TNFα in turn upregulates the expression of interleukin (IL)6 in Kupffer cells. Both TNFα and IL6 can activate the signal transducer and activator of transcription (STAT)3 and induce the expression of stem cell factor (SCF) and several proteins that are shared with the growth-factor-mediated pathway, resulting in hepatocyte activation and proliferation. During liver regeneration, various inhibitory proteins are also activated (shown in orange), including TGFβ (which is produced by stellate cells), plasminogen activator inhibitor (PAI), suppressor of cytokine signaling-3 (SOCS3) and p27 and other cyclin-dependent-kinase inhibitors. The effects of these inhibitors on liver regeneration are shown. (Reprinted by permission from Macmillan Publishers Ltd.: Taub R: *Nature Rev Mol Cell Biol* 5:836–47, copyright 2004.)

gp130. These activities further induce activation of the signaling cascade involving the Janus tyrosine kinase (JAK) and signal transducers and activators of transduction (STAT), resulting in hepatocyte proliferation. In mice with defect interleukin-6, liver regeneration is significantly impaired following partial hepatectomy. In transgenic mice with upregulated IL6, enhanced hepatocyte proliferation has been observed. These observations that IL6 serves as a potent stimulator for liver regeneration following liver injury.

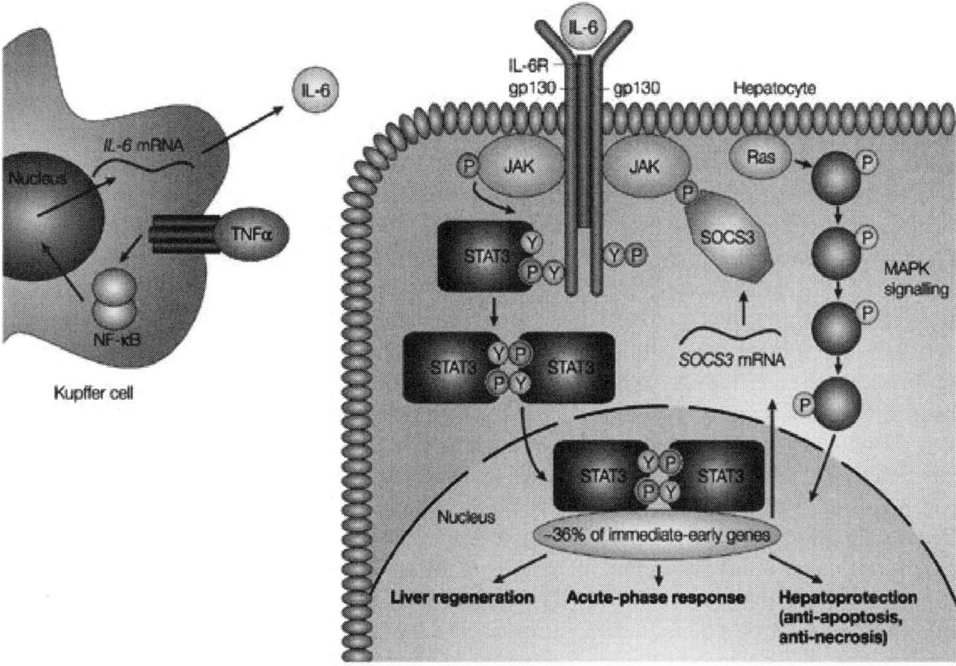

Figure 9.18. Role of the interleukin-6–STAT3-signaling pathway in regulating liver regeneration. Tumor necrosis factor (TNF)α binds to its receptor on Kupffer cells, resulting in the upregulation of interleukin-6 (IL6) transcription by the nuclear factor (NF)κB pathway. IL6 binds to the IL6 receptor (IL6R) on hepatocytes. The IL6 receptor interacts with two subunits of gp130, and activates Janus kinase (JAK). Activated JAK triggers the activation of two pathways: (1) the mitogen-activated protein kinase (MAPK) pathway, which is activated by SHP2–GRB2-SOS–Ras signal transduction (where SHP2 is SH2-domain-containing protein tyrosine phosphatase-2; GRB2 is growth-factor-receptor-bound protein-2; and SOS is "son of sevenless"); and (2) the signal transducer and activator of transcription (STAT)3 pathway, which is activated through JAK-mediated tyrosine (Y) phosphorylation. The STAT3 transcription factor dimerizes and translocates to the nucleus, where it activates transcription of ~36% of immediate–early target genes. In the liver, this process promotes liver regeneration, the acute-phase response, and hepatoprotection against Fas and toxic damage. Suppressor of cytokine signalling (SOCS)3 transcription is also regulated by IL6 signaling. SOCS3 interacts with JAK and blocks cytokine signaling. P, phosphate. (Reprinted by permission from Macmillan Publishers Ltd.: Taub R: *Nature Rev Mol Cell Biol* 5:836–47, copyright 2004.)

In addition, other biochemical factors, such as chemokines (transforming growth factor α, interferon-inducible protein 10, and macrophage inflammatory protein 1α), Wnts, norepinephrine, insulin, prostaglandins (prostaglandin E2, prostacyclin, and thromboxane), and steroid hormones (estradiol), are involved in the regulation of liver regeneration in response to liver injury. These factors, together with hepatocyte growth factor, epidermal growth factor, fibroblast growth factor, and IL6, control the initiation and progression of liver regeneration. Liver injury and hepatectomy can also activate matrix metalloproteinases. These enzymes can digest extracellular matrix, enhancing the proliferation and migration of hepatic cells.

Soluble biochemical factors may not only control the initiation, but also the cessation of liver regeneration. There are several factors, including *transforming growth factor (TGF)*β and *activin*, which serve as inhibitory factors for liver regeneration and may play a critical role in regulating the termination of liver regeneration. Transforming growth factor β is a member of the transforming growth factor β superfamily, which contains more than 30 members. This factor is upregulated in response to liver injury. Its expression is associated with suppression of hepatocyte proliferation in liver injury. Administration of exogenous transforming growth factor β to mice induces a significant reduction in liver regeneration following partial hepatectomy. In contrast, administration of inhibitors for transforming growth factor β elicits an opposite effect. The inhibitory effect of transforming growth factor β has been observed in experiments in vitro as well as in vivo. Activin is also a member of the TGFβ superfamily. This factor exerts a potent inhibitory effect on hepatocyte proliferation following liver injury. The effect of activin can be suppressed by the application of follistatin, a ligand that binds activin, inhibits the interaction of activin with its receptor, and thus suppresses the activity of activin. These observations suggest that TGFβ and activin serve as negative regulators for the regulation of liver regeneration in response to liver injury. This is a critical mechanism for the cessation of hepatocyte proliferation and control of liver regeneration.

The processes and regulatory mechanisms of liver regeneration discussed above are for normal livers or injured livers with regenerative capability. When a liver is composed of mostly damaged hepatocytes in disorders, such as liver failure, chronic hepatitis, and liver cirrhosis, the majority of hepatocytes are no longer capable of regenerating in response to additional liver injury due to exposure to toxic chemicals and partial hepatectomy. In these cases, hepatocyte or liver transplantation is necessary to restore the liver function.

BIBLIOGRAPHY

9.1. Embryonic Stem Cells

Brook FA, Gardner RL: The origin and efficient derivation of embryonic stem cells in the mouse, *Proc Natl Acad Sci USA* 94:5709–12, 1997.

Buehr M, Smith A: Genesis of embryonic stem cells, *Phil Trans R Soc Lond B Biol Sci* 358:1397–402, 2003.

Donovan PJ, de Miguel MP: Turning germ cells into stem cells, *Curr Opin Genet Dev* 13:463–71, 2003.

Extavour CG, Akam M: Mechanisms of germ cell specification across the metazoans: Epigenesist and preformation, *Development* 130:5869–84, 2003.

Geijsen N, Horoschak M, Kim K, Gribnau J, Eggan K et al: Derivation of embryonic germ cells and male gametes from embryonic stem cells, *Nature* 427:148–54, 2004.

Matsui Y, Zsebo K, Hogan BL: Derivation of pluripotential embryonic stem cells from murine primordial germ cells in culture, *Cell* 70:841–7, 1992.

Rossant J: Stem cells from the mammalian blastocyst, *Stem Cells* 19:477–82, 2001.

Saitou M, Barton SC, Surani MA: A molecular programme for the specification of germ cell fate in mice, *Nature* 418:293–300, 2002.

Shamblott MJ, Axelman J, Wang S, Bugg EM, Littlefield JW et al: Derivation of pluripotent stem cells from cultured human primordial germ cells, *Proc Natl Acad Sci USA* 95:13726–31, 1998.

Smith AG: Embryo-derived stem cells: Of mice and men, *Annu Rev Cell Dev Biol* 17:435–62, 2001.

Thomson JA, Itskovitz-Eldor J, Shapiro SS, Waknitz MA, Swiergiel JJ et al: Embryonic stem cell lines derived from human blastocysts, *Science* 282:1145–7, 1998.

Thomson JA, Kalishman J, Golos TG et al: Isolation of a primate embryonic stem cell line, *Proc Natl Acad Sci USA* 92:7844–8, 1995.

Zwaka TP, Thomson JA: A germ cell origin of embryonic stem cells? *Development* 132:227–33, 2005.

Toyooka Y, Tsunekawa N, Akasu R, Noce T: Embryonic stem cells can form germ cells in vitro, *Proc Natl Acad Sci USA* 100:11457–62, 2003.

Estes BT, Gimble JM, Guilak F: Mechanical signals as regulators of stem cell fate, *Curr Top Dev Biol* 60:91–126, 2004.

Evans MJ, Kaufman MH: Establishment in culture of pluripotential cells from mouse embryos, *Nature* 292:154–6, 1981.

Martin GR: Isolation of a pluripotent cell line from early mouse embryos cultured in medium conditioned by teratocarcinoma stem cells, *Proc Natl Acad Sci USA* 78:7634–8, 1981.

9.2. Bone Marrow-Derived Stem Cells

Jackson KA, Majka SM, Wang H, Pocius J, Hartley CJ et al: Regernation of ischemic cardiac muscle and vascular endothelium by adult stem cells, *J Clin Invest* 107:1395–402, 2001.

Quito FL, Beh J, Bashayan O et al: Effects of fibroblast growth factor-4 (k-FGF) on long-term cultures of human bone marrow cells, *Blood* 87:1282–91, 1996.

Martin I, Muraglia A, Campanile G et al: Fibroblast growth factor-2 supports ex vivo expansion and maintenance of osteogenic precursors from human bone marrow, *Endocrinology* 138:4456–62, 1997.

Walsh S, Jefferiss C, Stewart K et al: Expression of the developmental markers STRO-1 and alkaline phosphatase in cultures of human marrow stromal cells: regulation by fibroblast growth factor (FGF)-2 and relationship to the expression of FGF receptors 1–4, *Bone* 27:185–195, 2000.

Bryder D, Jackson EW: Interleukin-3 supports expansion of long-term multilineage repopulating activity after multiple stem cell division in vitro, *Blood* 96:1748–55, 2000.

Herzog E, Chai L, Krause DS: Plasticity of marrow-derived stem cells, *Blood* 102:3483–92, 2003.

Krause DS, Theise ND, Collector MI, Henegariu O, Hwang S et al: Multi-organ, multi-lineage engraftment by a single bone marrow-derived stem cell, *Cell* 105:369–77, 2001.

Lagasse E, Shizuru JA, Uchida N, Tsukamoto A, Weissman IL: Toward regeneration medicine, *Immunity* 14:425–36, 2001.

Orkin SH, Zon LI: Hematopoiesis and stem cells: Plasticity versus developmental heterogeneity, *Nat Immun* 3:323–8, 2002.

9.3. Neural Stem Cells

Altman J: Autoradiographic and histological studies of postnatal neurogenesis. IV. Cell proliferation and migration in the anterior forebrain, with special reference to persisting neurogenesis in the olfactory bulb, *J Compar Neurol* 137:433–57, 1969.

Altman J, Das GD: Autoradiographic and histological evidence of postnatal hippocampal neurogenesis in rats, *J Compar Neurol* 124:319–35, 1965.

Arvidsson A, Collin T, Kirik D, Kokaia Z, Lindvall O: Neuronal replacement from endogenous precursors in the adult brain after stroke, *Nat Med* 8:963–70, 2002.

Bauer S, Hay M, Amilhon B, Jean A, Moyse E: In vivo neurogenesis in the dorsal vagal complex of the adult rat brainstem, *Neuroscience* 130:75–90, 2005.

Emsley JG, Mitchell BD, Kempermann G, Macklis JD: Adult neurogenesis and repair of the adult CNS with neural progenitors, precursors, and stem cells, *Prog Neurobiol* 75:321–41, 1995.

Palmer TD, Willhoite AR, Gage FH: Vascular niche for adult hippocampal neurogenesis, *J Compar Neurol* 425:479–94, 2000.

Weiss S, Dunne C, Hewson J, Wohl C, Wheatley M et al: Multipotent CNS stem cells are present in the adult mammalian spinal cord and ventricular neuroaxis, *J Neurosci* 16:7599–609, 1996.

Shihabuddin LS, Horner PJ, Ray J, Gage FH: Adult spinal cord stem cells generate neurons after transplantation in the adult dentate gyrus, *J Neurosci* 20:8727–35, 2000.

Tropepe V, Coles BL, Chiasson BJ, Horsford DJ, Elia AJ et al: Retinal stem cells in the adult mammalian eye, *Science* 287:2032–6, 2000.

Lie DC, Dziewczapolski G, Willhoite AR, Kaspar BK, Shults CW et al: The adult substantia nigra contains progenitor cells with neurogenic potential, *J Neurosci* 22:6639–49, 2002.

Lie DC, Song H, Colamarino SA, Ming GL, Gage FH: Neurogenesis in the adult brain: New strategies for central nervous system diseases, *Annu Rev Parmacol Toxicol* 44:399–421, 2004.

Reynolds BA, Weiss S: Generation of neurons and astrocytes from isolated cells of the adult mammalian central nerve system, *Science* 255:1707–10, 1992.

Richards LJ, Kilpatrick TJ, Bartlett PF: De novo generation of neuronal cells from the adult mouse brain, *Proc Natl Acad Sci USA* 89:8591–5, 1002.

9.4. Other Stem Cells

Weiss MC, Strick-Marchand H: Isolation and characterization of mouse hepatic stem cells in vitro, *Semin Liver Dis* 23:313–24, 2003.

Leri A, Kajstura J, Anversa P: Cardiac stem cells and mechanisms of myocardial regeneration, *Physiol Rev* 85:1373–416, 2005.

Overturf K, al-Dhalimy M, Ou CN, Finegold M, Grompe M: Serial transplantation reveals the stem-cell-like regenerative potential of adult mouse hepatocytes, *Am J Pathol* 151:1273–80, 1997.

Alison MR, Liver stem cells: a two compartment system, *Curr Opin Cell Biol* 10:710–15, 1998.

Overturf K, al-Dhalimy M, Ou CN, Finegold M, Grompe M: Serial transplantation reveals the stem-cell-like regenerative potential of adult mouse hepatocytes, *Am J Pathol* 151:1273–80, 1997.

Bonner-Weir S, Sharma A: Pancreatic stem cells, *J Pathol* 197:519–26, 2002.

Vats A, Bielby RC, Tolley NS, Nerem R, Polak JM: Stem cells, *Lancet* 366(9485):592–602, Aug 2005 (article on cells for tissue regeneration).

9.5. Regeneration of Salamander Limbs

Brockes JP: Introduction of a retinoid reporter gene into the urodele limb blastema, *Proc Natl Acad Sci USA* 89:11386–90, 1992.

Brockes JP, Kumar A: Plasticity and reprogramming of differentiated cells in amphibian regeneration, *Nat Rev Mol Cell Biol* 3:566–74, 2002.

Chernoff EAG, Stocum D: Developmental aspects of spinal cord and limb regeneration, *Dev Growth Differ* 37:133–47, 1995.

Gardiner DM, Endo T, Bryant S: The molecular basis of amphibian limb regeneration: Integrating the old with the new, *Semin Cell Dev Biol* 13:345–52, 2002.

Imokawa Y, Yoshizato K: Exrpression of sonic hedgehog gene in regenerating newt limb blastema recapitulates that in developing limb buds, *Proc Natl Acad Sci USA* 94:9159–64, 1997.

Pecorino LT, Entwistle A, Brockes JP: Activation of a single retinoic acid receptor isoforms mediates proximo-distal respecification, *Curr Biol* 6:563–9, 1996.

Torok MA, Gardiner DM, Shubin NH, Bryant SV: Expression of HoxD genes in developing and regenerating axolotl limbs, *Dev Bol* 200:225–33, 1998.

Wang L, Marchionni MA, Tassava RA: Cloning and neuronal expression of a type III newt neuregulin and rescue of degenerated, nerve-dependent newt limb blastemas by rhGGF2, *J Neurobiol* 43:150–8, 2000.

9.6. Experimental Observations and Features of Liver Regeneration

Michalopoulos GK, DeFrances MC: Liver regeneration, *Science* 276:60–6, 1997.

Lindroos PM, Zarnegar R, Michalopoulos GK: Hepatocyte growth factor (hepatopoietin A) rapidly increases in plasma before DNA synthesis and liver regeneration stimulated by partial hepatectomy and carbon tetrachloride administration, *Hepatology* 13:743–50, 1991.

Mars WM, Liu ML, Kitson RP, Goldfarb RH, Gabauer MK et al: Immediate early detection of urokinase receptor after partial hepatectomy and its implications for initiation of liver regeneration, *Hepatology* 21:1695–701, 1995.

Taub R: Liver regeneration: From myth to mechanism, *Nat Rev Mol Cell Biol* 5:836–47, 2004.

Diehl AM: Liver regeneration, *Front Biosci* 7:e301–14, 2002.

Diehl AM: Cytokine regulation of liver injury and repair, *Immunol Rev* 174:160–71, 2000.

Alison M, Golding M, Lalani el-N, Sarraf C: Wound healing in the liver with particular reference to stem cells, *Phil Trans R Soc Lond B Biol Sci* 353:877–94, 1998.

Michalopoulos GK, DeFrances MC: Liver regeneration, *Science* 276:60–6, 1997.

Columbano A, Shinozuka H: Liver regeneration versus direct hyperplasia, *FASEB J* 10:1118–28, 1996.

Pistoi S, Morello D: Liver regeneration 7. Prometheus' myth revisited: Transgenic mice as a powerful tool to study liver regeneration, *FASEB J* 10:819–28, 1996.

Ponder KP: Analysis of liver development, regeneration, and carcinogenesis by genetic marking studies, *FASEB J* 10:673–82, 1996.

Kren BT, Steer CJ: Posttranscriptional regulation of gene expression in liver regeneration: Role of mRNA stability, *FASEB J* 10:559–73, 1996.

Taub R: Liver regeneration 4: Transcriptional control of liver regeneration, *FASEB J* 10:413–27, 1996.

Diehl AM, Rai RM: Liver regeneration 3: Regulation of signal transduction during liver regeneration, *FASEB J* 10:215–27, 1996.

Fausto N, Laird AD, Webber EM: Liver regeneration. 2. Role of growth factors and cytokines in hepatic regeneration, *FASEB J* 9:1527–36, 1995.

Martinez-Hernandez A, Amenta PS: The extracellular matrix in hepatic regeneration, *FASEB J* 9:1401–10, 1995.

Steer CJ: Liver regeneration, *FASEB J* 9:1396–400, 1995.

Koniaris LG, McKillop IH, Schwartz SI, Zimmers TA: Liver regeneration, *J Am Coll Surg* 197:634–59, 2003.

Diehl AM, Rai R: Review: Regulation of liver regeneration by pro-inflammatory cytokines, *J Gastroenterol Hepatol* 11:466–70, 1996.

Holt DR, Thiel DV, Edelstein S, Brems JJ: Hepatic resections, *Arch Surg* 135:1353–8, 2000.

Debonera F, Aldeguer X, Shen X et al: Activation of interleukin-6/STAT3 and liver regeneration following transplantation, *J Surg Res* 96:289–95, 2001.

Higgins GM, Anderson RM: Experimental pathology of the liver. I. Restoration of the liver of the white rat following surgical removal, *Arch Pathol* 12:186–206, 1931.

Moolten FL, Bucher NL: Regeneration of rat liver: transfer of humoral agent by cross circulation, *Science* 158:272–4, 1967.

Fisher B, Szuch P, Levine M, Fisher ER: A portal blood factor as the humoral agent in liver regeneration, *Science* 171:575–7, 1971.

Francavilla A, Zeng Q, Polimeno L et al: Small-for-size liver transplanted into larger recipient: A model of hepatic regeneration, *Hepatology* 19:210–16, 1994.

Shiota G, Wang TC, Nakamura T, Schmidt EV: Hepatocyte growth factor in transgenic mice: Effects on hepatocyte growth, liver regeneration and gene expression, *Hepatology* 19:962–72, 1994.

Shimada M, Matsumata T, Taketomi A et al: The role of interleukin-6, interleukin-16, tumor necrosis factor-alpha and endotoxin in hepatic resection, *Hepatogastroenterology* 42:691–7, 1995.

Weglarz TC, Sandgren EP: Timing of hepatocyte entry into DNA synthesis after partial hepatectomy is cell autonomous, *Proc Natl Acad Sci USA* 97:12595–600, 2000.

Rai RM, Lee FY, Rosen A et al: Impaired liver regeneration in inducible nitric oxide synthase deficient mice, *Proc Natl Acad Sci USA* 95:13829–34, 1998.

Cressman DE, Greenbaum LE, DeAngelis RA et al: Liver failure and defective hepatocyte regeneration in interleukin-6-deficient mice, *Science* 274:1379–83, 1996.

Zimmers TA, McKillop IH, Pierce RH et al: Massive liver growth in mice induced by systemic interleukin-6 administration, *Hepatology* 38:326–34, 2003.

Fausto N: Liver regeneration, *J Hepatol* 32:19–31, 2000.

Zarnegar R, Michalopoulos GK: The many faces of hepatocyte growth factor: From hepatopoiesis to hematopoiesis, *J Cell Biol* 129:1177–80, 1995.

Russell WE: Transforming growth factor beta (TGF-beta) inhibits hepatocyte DNA synthesis independently of EGF binding and EGF receptor autophosphorylation, *J Cell Physiol* 135:253–61, 1988.

Carr BI, Hayashi I, Branum EL, Moses HL: Inhibition of DNA synthesis in rat hepatocytes by platelet-derived type beta transforming growth factor, *Cancer Res* 46:2330–4, 1986.

Yasuda H, Mine T, Shibata H et al: Activin A: An autocrine inhibitor of initiation of DNA synthesis in rat hepatocytes, *J Clin Invest* 92:1491–6, 1993.

Schwall RH, Robbins K, Jardieu P et al: Activin induces cell death in hepatocytes in vivo and in vitro, *Hepatology* 18:347–56, 1993.

Yamada Y, Kirillova I, Peschon JJ, Fausto N: Initiation of liver growth by tumor necrosis factor: Deficient liver regeneration in mice lacking type I tumor necrosis factor receptor, *Proc Natl Acad Sci USA* 94:1441–6, 1997.

Kirillova I, Chaisson M, Fausto N: Tumor necrosis factor induces DNA replication in hepatic cells through nuclear factor kappaB activation, *Cell Growth Differ* 10:819–28, 1999.

Hirano T, Nakajima K, Hibi M: Signaling mechanisms through gp130: A model of the cytokine system, *Cytokine Growth Factor Rev* 8:241–52, 1997.

Kovalovich K, DeAngelis RA, Li W et al: Increased toxin-induced liver injury and fibrosis in interleukin-6-deficient mice, *Hepatology* 31:149–59, 2000.

Sakamoto T, Liu Z, Murase N et al: Mitosis and apoptosis in the liver of interleukin-6-deficient mice after partial hepatectomy, *Hepatology* 29:403–11, 1999.

Maione D, Di Carlo E, Li W et al: Coexpression of IL-6 and soluble IL-6R causes nodular regenerative hyperplasia and adenomas of the liver, *EMBO J* 17:5588–97, 1998.

Roos F, Ryan AM, Chamow SM et al: Induction of liver growth in normal mice by infusion of hepatocyte growth factor/scatter factor, *Am J Physiol* 268:G380–6, 1995.

Fujiwara K, Nagoshi S, Ohno A et al: Stimulation of liver growth by exogenous human hepatocyte growth factor in normal and partially hepatectomized rats, *Hepatology* 18:1443–9, 1993.

Schlessinger J: Cell signaling by receptor tyrosine kinases, *Cell* 103:211–25, 2000.

Noguchi S, Ohba Y, Oka T: Influence of epidermal growth factor on liver regeneration after partial hepatectomy in mice, *J Endocrinol* 128:425–31, 1991.

Jones DE, Jr, Tran-Patterson R, Cui DM et al: Epidermal growth factor secreted from the salivary gland is necessary for liver regeneration, *Am J Physiol* 268:G872–8, 1995.

Webber EM, Wu JC, Wang L et al: Overexpression of transforming growth factor-alpha causes liver enlargement and increased hepatocyte proliferation in transgenic mice, *Am J Pathol* 145:398–408, 1994.

Bucher NLR, Swaffield MN: Regulation of hepatic regeneration in rats by synergistic action of insulin and glucagons, *Proc Natl Acad Sci USA* 72:1157–60, 1975.

Kogure K, Zhang YQ, Kanzaki M et al: Intravenous administration of follistatin: Delivery to the liver and effect on liver regeneration after partial hepatectomy, *Hepatology* 24:361–6, 1996.

Koniaris LG, Zimmers-Koniaris T, Hsiao EC et al: Cytokine-responsive gene-2/IFN-inducible protein-10 expression in multiple models of liver and bile duct injury suggests a role in tissue regeneration, *J Immunol* 167:399–406, 2001.

Hogaboam CM, Bone-Larson CL, Steinhauser ML et al: Novel CXCR2-dependent liver regenerative qualities of ELR-containing CXC chemokines, *FASEB J* 13:1565–74, 1999.

Dierick H, Bejsovec A: Cellular mechanisms of wingless/Wnt signal transduction, *Curr Top Dev Biol* 43:153–90, 1999.

Nusse R: WNT targets. Repression and activation, *Trends Genet* 15:1–3, 1999.

Tsujii H, Okamoto Y, Kikuchi E et al: Prostaglandin E2 and rat liver regeneration, *Gastroenterology* 105:495–9, 1993.

Rudnick DA, Perlmutter DH, Muglia LJ: Prostaglandins are required for CREB activation and cellular proliferation during liver regeneration, *Proc Natl Acad Sci USA* 98:8885–90, 2001.

Kuebler JF, Jarrar D, Wang P et al: Dehydroepiandrosterone restores hepatocellular function and prevents liver damage in estrogen-deficient females following trauma and hemorrhage, *J Surg Res* 97:196–201, 2001.

Francavilla A, Polimeno L, DiLeo A et al: The effect of estrogen and tamoxifen on hepatocyte proliferation in vivo and in vitro, *Hepatology* 9:614–20, 1989.

Hirano T, Ishihara K, Hibi M: Roles of STAT3 in mediating the cell growth, differentiation and survival signals relayed through the IL-6 family of cytokine receptors, *Oncogene* 19:2548–56, 2000.

Li W, Liang X, Kellendonk C et al: STAT3 Contributes to the mitogenic response of hepatocytes during liver regeneration, *J Biol Chem* 277:28411–17, 2002.

Macias-Silva M, Li W, Leu JI et al: Up-regulated transcriptional repressors SnoN and Ski bind Smad proteins to antagonize transforming growth factor-beta signals during liver regeneration, *J Biol Chem* 28483–90, 2002.

Ichikawa T, Zhang YQ, Kogure K et al: Transforming growth factor beta and activin tonically inhibit DNA synthesis in the rat liver, *Hepatology* 34:918–25, 2001.

Hully JR, Chang L, Schwall RH et al: Induction of apoptosis in the murine liver with recombinant human activin A, *Hepatology* 20:854–62, 1994.

Kovalovich K, Li W, DeAngelis R et al: Interleukin-6 protects against Fas-mediated death by establishing a critical level of anti-apoptotic hepatic proteins FLIP, Bcl-2, and Bcl-xL, *J Biol Chem* 276:26605–13, 2001.

Selzner M, Rudiger HA, Sindram D et al: Mechanisms of ischemic injury are different in the steatotic and normal rat liver, *Hepatology* 32:1280–8, 2000.

Yang SQ, Lin HZ, Yin M et al: Effects of chronic ethanol consumption on cytokine regulation of liver regeneration, *Am J Physiol* 275:G696–704, 1998.

LeSage GD, Benedetti A, Glaser S et al: Acute carbon tetrachloride feeding selectively damages large, but not small, cholangiocytes from normal rat liver, *Hepatology* 29:307–19, 1999.

Dabeva MD, Shafritz DA: Activation, proliferation, and differentiation of progenitor cells into hepatocytes in the D-galactosamine model of liver regeneration, *Am J Pathol* 143:1606–20, 1993.

Lemire JM, Shiojiri N, Fausto N: Oval cell proliferation and the origin of small hepatocytes in liver injury induced by D-galactosamine, *Am J Pathol* 139:535–52, 1991.

Lentsch AB, Kato A, Yoshidome H et al: Inflammatory mechanisms and therapeutic strategies for warm hepatic ischemia/reperfusion injury, *Hepatology* 32:169–73, 2000.

9.7. Experimental Models of Liver Regeneration

Palmes D, Spiegel HU: Animal models of liver regeneration, *Biomaterials* 25:1601–11, 2004.

Lindroos PM, Zarnegar R, Michalopoulos GK: Hepatocyte growth factor (hepatopoietin A) rapidly increases in plasma before DNA synthesis and liver regeneration stimulated by partial hepatectomy and carbon tetrachloride administration, *Hepatology* 13:743–50, 1991.

Rahman TM, Hodgson HJ: Animal models of acute hepatic failure, *Int J Exp Pathol* 81:145–57, 2000.

Bolesta S, Haber SL: Hepatotoxicity associated with chronic acetaminophen administration in patients without risk factors, *Ann Pharmacother* 36:331–3, 2002.

Liu KX, Kato Y, Yamazaki M, Higuchi O, Nakamura T, Sugiyama Y: Decrease in the hepatic clearance of hepatocyte growth factor in carbon tetrachloride-intoxicated rats, *Hepatology* 17:651–60, 1993.

9.8. Regulation of Liver Regeneration

Schrem H, Klempnauer J, Borlak J: Liver-enriched transcription factors in liver function and development. Part II: the C/EBPs and D site-binding protein in cell cycle control, carcinogenesis, circadian gene regulation, liver regeneration, apoptosis, and liver-specific gene regulation, *Pharmacol Rev* 56(2):291–330, 2004.

Kren BT, Trembley JH, Fan G, Steer CJ: Molecular regulation of liver regeneration, *Ann NY Acad Sci* 831:361–81, 1997.

Kren BT, Steer CJ: Posttranscriptional regulation of gene expression in liver regeneration: role of mRNA stability, *FASEB J* 10(5):559–73, 1996.

Taub R: Liver regeneration 4: transcriptional control of liver regeneration, *FASEB J* 10(4):413–27, 1996.

Yamada Y, Kirillova I, Peschon JJ, Fausto N: Initiation of liver growth by tumor necrosis factor: Deficient liver regeneration in mice lacking type I tumor necrosis factor receptor, *Proc Natl Acad Sci USA* 94:1441–6, 1997.

Kirillova I, Chaisson M, Fausto N: Tumor necrosis factor induces DNA replication in hepatic cells through nuclear factor kappaB activation, *Cell Growth Differ* 10:819–28, 1999.

Zimmers TA, McKillop IH, Pierce RH et al: Massive liver growth in mice induced by systemic interleukin-6 administration, *Hepatology* 38:326–34, 2003.

Kuma S, Inaba M, Ogata H et al: Effect of human recombinant interleukin-6 on the proliferation of mouse hepatocytes in the primary culture, *Immunobiology* 180:235–42, 1990.

Satoh M, Yamazaki M: Tumor necrosis factor stimulates DNA synthesis of mouse hepatocytes in primary culture and is suppressed by transforming growth factor β and interleukin 6, *J Cell Physiol* 150:134–9, 1992.

Kordula T, Rokita H, Koj A et al: Effects of interleukin-6 and leukemia inhibitory factor on the acute phase response and DNA synthesis in cultured rat hepatocytes, *Lymphokine Cytokine Res* 10:23–6, 1991.

Hirano T, Nakajima K, Hibi M: Signaling mechanisms through gp130: A model of the cytokine system, *Cytokine Growth Factor Rev* 8:241–52, 1997.

Miki C, Iriyama K, Gunson BK et al: Influence of intraoperative blood loss on plasma levels of cytokines and endotoxin and subsequent graft liver function, *Arch Surg* 132:136–41, 1997.

Kim YI, Hwang YJ, Song KE et al: Hepatocyte protection by a protease inhibitor against ichemia/reperfusion injury of human liver, *J Am Coll Surg* 195:41–50, 2002.

Kovalovich K, DeAngelis RA, Li W et al: Increased toxin-induced liver injury and fibrosis in interleukin-6-deficient mice, *Hepatology* 31:149–59, 2000.

Sakamoto T, Liu Z, Murase N et al: Mitosis and apoptosis in the liver of interleukin-6-deficient mice after partial hepatectomy, *Hepatology* 29:403–11, 1999.

Fattori E, Della Rocca C, Costa P et al: Development of progressive kidney damage and myeloma kidney in interleukin-6 transgenic mice, *Blood* 83:2570–9, 1994.

Maione D, Di Carlo E, Li W et al: Coexpression of IL-6 and soluble IL-6R causes nodular regenerative hyperplasia and adenomas of the liver, *EMBO J* 17:5588–97, 1998.

Koga M, Ogasawara H: Induction of hepatocyte mitosis in intact adult rat by interleukin-1 alpha and interleukin-6, *Life Sci* 49:1263–70, 1991.

Schlessinger J: Cell signaling by receptor tyrosine kinases, *Cell* 103:211–25, 2000.

Noguchi S, Ohba Y, Oka T: Influence of epidermal growth factor on liver regeneration after partial hepatectomy in mice, *J Endocrinol* 128:425–31, 1991.

Jones DE, Jr, Tran-Patterson R, Cui DM et al: Epidermal growth factor secreted from the salivary gland is necessary for liver regeneration, *Am J Physiol* 268:G872–8, 1995.

Mann GB, Fowler KJ, Gabriel A et al: Mice with a null mutation of the TGF alpha gene have abnormal skin architecture, wavy hair, and curly whiskers and often develop corneal inflammation, *Cell* 73:249–61, 1993.

Webber EM, Wu JC, Wang L et al: Overexpression of transforming growth factor-alpha causes liver enlargement and increased hepatocyte proliferation in transgenic mice, *Am J Pathol* 145:398–408, 1994.

Fausto N: Liver regeneration, *J Hepatol* 32:19–31, 2000.

Sakata H, Takayama H, Sharp R et al: Hepatocyte growth factor/scatter factor overexpression induces growth, abnormal development, and tumor formation in transgenic mouse livers, *Cell Growth Differ* 7:1513–23, 1996.

Roos F, Ryan AM, Chamow SM et al: Induction of liver growth in normal mice by infusion of hepatocyte growth factor/scatter factor, *Am J Physiol* 268:G380–6, 1995.

Fujiwara K, Nagoshi S, Ohno A et al: Stimulation of liver growth by exogenous human hepatocyte growth factor in normal and partially hepatectomized rats, *Hepatology* 18:1443–9, 1993.

Yasuda H, Mine T, Shibata H et al: Activin A: An autocrine inhibitor of initiation of DNA synthesis in rat hepatocytes, *J Clin Invest* 92:1491–6, 1993.

Schwall RH, Robbins K, Jardieu P et al: Activin induces cell death in hepatocytes in vivo and in vitro, *Hepatology* 18:347–56, 1993.

Hillier SG, Miro F: Inhibin, activin, and follistatin, *Ann NY Acad Sci* 687:29–38, 1993.

Kogure K, Zhang YQ, Kanzaki M et al: Intravenous administration of follistatin: Delivery to the liver and effect on liver regeneration after partial hepatectomy, *Hepatology* 24:361–6, 1996.

Bottinger EP, Factor VM, Tsang ML et al: The recombinant proregion of transforming growth factor $beta_1$ (latency-associated peptide) inhibits active transforming growth factor beta1 in transgenic mice, *Proc Natl Acad Sci USA* 93:5877–82, 1996.

Esquela AF, Zimmers TA, Koniaris LG et al: Transient down-regulation of inhibin-betaC expression following partial hepatectomy, *Biochem Biophys Res Commun* 235:553–6, 1997.

Hsiao EC, Koniaris LG, Zimmers-Koniaris T et al: Characterization of growth-differentiation factor 15, a transforming growth factor beta superfamily member induced following liver injury, *Mol Cell Biol* 20:3742–51, 2000.

Zarnegar R, Michalopoulos GK: The many faces of hepatocyte growth factor: From hepatopoiesis to hematopoiesis, *J Cell Biol* 129:1177–80, 1995.

Yasuda H, Mine T, Shibata H et al: Activin A: An autocrine inhibitor of initiation of DNA synthesis in rat hepatocytes, *J Clin Invest* 92:1491–6, 1993.

Schwall RH, Robbins K, Jardieu P et al: Activin induces cell death in hepatocytes in vivo and in vitro, *Hepatology* 18:347–56, 1993.

Yamada Y, Kirillova I, Peschon JJ, Fausto N: Initiation of liver growth by tumor necrosis factor: deficient liver regeneration in mice lacking type I tumor necrosis factor receptor, *Proc Natl Acad Sci USA* 94:1441–6, 1997.

Diehl AM, Rai R: Review: Regulation of liver regeneration by pro-inflammatory cytokines, *J Gastroenterol Hepatol* 11:466–70, 1996.

Michalopoulos GK, DeFrances MC: Liver regeneration, *Science* 276:60–6, 1997.

Holt DR, Thiel DV, Edelstein S, Brems JJ: Hepatic resections, *Arch Surg* 135:1353–8, 2000.

TGFα

Tricoli JV, Nakai H, Byers MG, Rall LB, Bell GI et al: The gene for human transforming growth factor alpha is on the short arm of chromosome 2, *Cytogenet Cell Genet* 42:94–8, 1986.

Ellis DL, Kafka SP, Chow JC, Nanney LB, Inman WH et al: Melanoma, growth factors, acanthosis nigricans, the sign of Leser-Trelat, and multiple acrochordons: A possible role for alpha-transforming growth factor in cutaneous paraneoplastic syndromes, *New Engl J Med* 317:1582–7, 1987.

Fernandez-Larrea J, Merlos-Suarez A, Urena JM, Baselga J, Arribas J: A role for a PDZ protein in the early secretory pathway for the targeting of proTGF-alpha to the cell surface, *Mole Cell* 3:423–33, 1999.

Kramer A, Yang FC, Snodgrass P, Li X, Scammell TE et al: Regulation of daily locomotor activity and sleep by hypothalamic EGF receptor signaling, *Science* 294:2511–15, 2001.

TGFβ1

Awad MR, El-Gamel A, Hasleton P, Turner DM, Sinnott PJ et al: Genotypic variation in the transforming growth factor-beta-1 gene: Association with transforming growth factor-beta-1 production, fibrotic lung disease, and graft fibrosis after lung transplantation, *Transplantation* 66:1014–20, 1998.

Bernasconi P, Torchiana E, Confalonieri P, Brugnoni R, Barresi R et al: Expression of transforming growth factor-beta-1 in dystrophic patient muscles correlates with fibrosis: Pathogenetic role of a fibrogenic cytokine, *J Clin Invest* 96:1137–44, 1995.

Blobe GC, Schiemann WP, Lodish HF: Role of transforming growth factor beta in human disease, *New Engl J Med* 342:1350–8, 2000.

Border WA, Noble NA: Transforming growth factor beta in tissue fibrosis, *New Engl J Med* 331:1286–92, 1994.

Derynck R, Akhurst RJ, Balmain A: TGF-beta signaling in tumor suppression and cancer progression, *Nature Genet* 29:117–29, 2001.

Derynck R, Jarrett JA, Chen EY, Eaton DH, Bell JR et al: Human transforming growth factor-beta complementary DNA sequence and expression in normal and transformed cells, *Nature* 316:701–5, 1985.

Han G, Lu SL, Li AG, He W, Corless CL et al: Distinct mechanisms of TGF-beta-1-mediated epithelial-to-mesenchymal transition and metastasis during skin carcinogenesis, *J Clin Invest* 115:1714–23, 2005.

Heldin CH, Miyazono K, ten Dijke P: TGF-beta signalling from cell membrane to nucleus through SMAD proteins, *Nature* 390:465–71, 1997.

Jang CW, Chen CH, Chen CC, Chen J, Su YH et al: TGF-beta induces apoptosis through Smad-mediated expression of DAP-kinase, *Nature Cell Biol* 4:51–8, 2001.

Jobling AI, Nguyen M, Gentle A, McBrien NA: Isoform-specific changes in scleral transforming growth factor-beta expression and the regulation of collagen synthesis during myopia progression, *J Biol Chem* 279:18121–6, 2004.

Lin HK, Bergmann S, Pandolfi PP: Cytoplasmic PML function in TGF-beta signaling, *Nature* 431: 205–11, 2004.

Morris DG, Huang X, Kaminski N, Wang Y, Shapiro SD et al: Loss of integrin alpha-v-beta-6-mediated TGF-beta activation causes Mmp12-dependent emphysema, *Nature* 422:169–73, 2003.

Pittet JF, Griffiths MJ, Geiser T, Kaminski N, Dalton SL et al: TGF-beta is a critical mediator of acute lung injury, *J Clin Invest* 107:1537–44, 2001.

Shehata M, Schwarzmeier JD, Hilgarth M, Hubmann R, Duechler M et al: TGF-beta-1 induces bone marrow reticulin fibrosis in hairy cell leukemia, *J Clin Invest* 113:676–85, 2004.

Stroschein SL, Wang W, Zhou S, Zhou Q, Luo K: Negative feedback regulation of TGF-beta signaling by the SnoN oncoprotein, *Science* 286:771–4, 1999.

Valderrama-Carvajal H, Cocolakis E, Lacerte A, Lee EH, Krystal G et al: Activin/TGF-beta induce apoptosis through Smad-dependent expression of the lipid phosphatase SHIP, *Nature Cell Biol* 4:963–9, 2002.

TGFβ2

Barton DE, Foellmer BE, Du J, Tamm J, Derynck R et al: Chromosomal mapping of genes for transforming growth factors beta-2 and beta-3 in man and mouse: dispersion of TGF-beta gene family, *Oncogene Res* 3:323–31, 1988.

Fillistatin

Matzuk MM, Lu N, Vogel H, Sellheyer K, Roop DR et al: Multiple defects and perinatal death in mice deficient in follistatin, *Nature* 374:360–3, 1995.

Ueno N, Ling N, Ying SY, Esch F, Shimasaki S et al: Isolation and partial characterization of follistatin: a single-chain M(r) 35,000 monomeric protein that inhibits the release of follicle-stimulating hormone, *Proc Nat Acad Sci USA* 84:8282–6, 1987.

Wang XP, Suomalainen M, Jorgez CJ, Matzuk MM, Werner S et al: Follistatin regulates enamel patterning in mouse incisors by asymmetrically inhibiting BMP signaling and ameloblast differentiation, *Dev Cell* 7:719–30, 2004.

PART II

PRINCIPLES AND APPLICATIONS OF BIOREGENERATIVE ENGINEERING TO ORGAN SYSTEMS

SECTION 4

PRINCIPLES OF BIOREGENERATIVE ENGINEERING

10

MOLECULAR ASPECTS OF BIOREGENERATIVE ENGINEERING

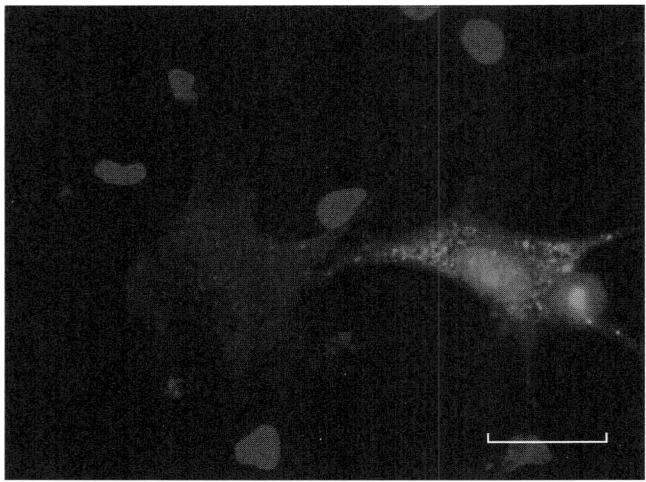

Expression of green fluorescent protein (GFP) in cultured bone marrow stromal cells transfected with a GFP gene. Green: GFP. Blue: cell nuclei. Scale: 10 μm. See color insert.

The molecular aspects of bioregenerative engineering address the principles, technologies, and applications of bioregenerative engineering at the molecular level. Cellular activities, including cell proliferation, differentiation, and regeneration, are regulated by coordinated gene expression and protein activation. Thus, the regeneration of cells, tissues, and organs can be controlled by the modulation of gene expression and/or protein activities. The process of molecular modulation is often referred to as *molecular engineering*. Since protein modulation depends on, to a large extent, DNA modulation, this book will focus on DNA engineering.

Bioregenerative Engineering: Principles and Applications, by Shu Q. Liu

DNA engineering stems from the principles of natural genetic processes, such as gene replication, transcription, recombination, and mutation. Nature has designed and created one of the most elegant biological systems, the genome, for the storage and processing of hereditary information. The human genome is composed of about 3,000,000 nucleotide pairs and more than 50,000 genes, which are the fundamental units of genetic processing. A gene encodes the sequence of a polypeptide chain or protein. Each gene can be replicated to pass genetic information to the next generation and transcribed into messenger RNA (mRNA) to produce a protein. For the past half century, genetic research has demonstrated that a gene can be identified, removed, purified, modulated, replaced, reproduced, and transferred between cells or organisms. Methods for these preparations have provided a foundation for the establishment of the most influential technology in biomedical history: the recombinant DNA technology. Such a technology has revolutionized modern biomedical research, enabling the detection of gene mutants, the identification of genetic causes for pathological disorders, the reproduction and modeling of gene mutation-induced diseases and, more importantly, the potential improvement of cell regeneration and treatment of human diseases by using genetic approaches. Biomedical research with genetic engineering may hold the key to the understanding and treatment of deadly diseases, such as degenerative disorders, cancer, and atherosclerosis.

DNA engineering approaches are established on the basis of DNA recombination and transfection technologies. The ultimate goal of DNA engineering is to repair or replace defective or mutant genes, which cause pathological disorders and cannot be naturally repaired, and thus to enhance the regeneration of disordered cells, tissues, and organs. To achieve such a goal, it is necessary to understand the structure and function of genes and the role of gene mutation in the induction pathogenic disorders, and to establish engineering technologies for constructing therapeutic genes, which are used for replacing defect or mutant genes. It is also necessary to detect the function and effectiveness of the therapeutic genes delivered into target cells. In this chapter, the principles of molecular engineering are outlined.

DNA ENGINEERING

Gene Mutation [10.1]

In a broad sense, human diseases can be induced by gene mutation, environmental stimulation, or a combination of both. *Gene mutation* is defined as a heritable alteration in the structure of DNA. Gene mutation can occur spontaneously during evolution, which may play a role for evolutionary development of plant and animal species, and can also be induced by environmental factors, such as chemical (carcinogens), physical (radiation), and biological (viruses) factors. For instance, cytosine can be deaminated to form uracil spontaneously (Fig. 10.1) or under the action of certain chemicals, such as nitrous acid. There are various types of gene mutation. These types are summarized in Fig. 10.2. During the entire lifespan of an organism, each base pair may be mutated at a rate of 10^{-9}–10^{-10}. The rate of mutation for a gene is dependent on the number of the base pairs.

There are two forms of gene mutation: alterations in chromosomal structure and alterations in specific gene structure. Chromosomal alterations usually include apparent deletion or translocation of a segment of the involved chromosome(s). Each altered segment may

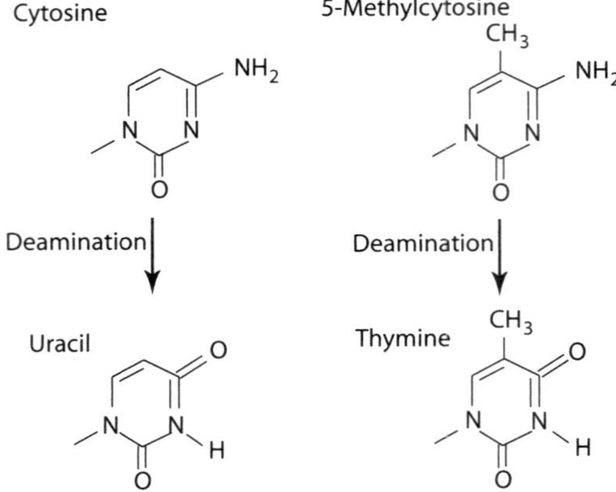

Figure 10.1. Schematic representation of cytosine mutation.

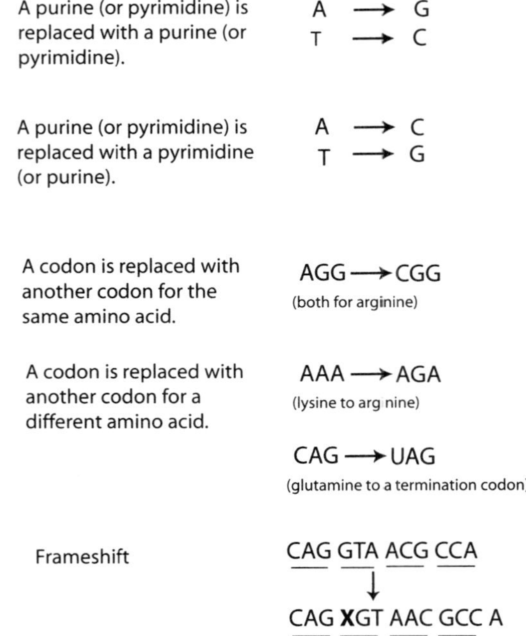

Figure 10.2. Common types of gene mutation found in the mammalian genomes. Based on bibliography 10.1.

contain multiple gene loci. In contrast, alterations in gene structure include mutation of single or multiple base pairs, which are often difficult to notice. In certain cases, a change in a single base pair, known as point mutation, may result in the translation of defective proteins, leading to severe clinical consequences. Sickle ell anemia is a typical example. This disorder is induced by the substitution of a single base C (cystine) with T (thymine) in the hemoglobin β chain gene. This substitution changes the codon CTC to TTC, resulting in the replacement of glutamic acid with lysine at the sixth residue of the hemoglobin β chain. As a result, the erythrocytes exhibit an irregular shape and increased rigidity of the cytoskeleton, rendering the erythrocytes difficult to pass through capillaries. Capillary obstruction and oxygen deficiency are often found in peripheral tissues in sickle cell anemia. These pathological alterations are associated with clinical manifestations such as hemolysis and anemia, impaired system development and growth, increased susceptibility to infection, and micro- and macroinfarction.

Gene mutation may also induce premature termination of protein synthesis, resulting in the formation of incomplete or defective proteins. The translation of proteins is controlled by the sequence of mRNA, which contains the coding information for a specified protein. Three codons, including UAA, UAG, and UGA, serve as signals that induce the termination of protein translation. When a gene mutation occurs in a gene to change an amino acid codon to one of the termination codons, protein translation will be terminated prematurely, producing a defective protein. For instance, the codon for the amino acid tyrosine is UAU. The substitution of the last base U with A changes the tyrosine codon to a termination codon UAA, which signals the protein synthesis machinery to terminate protein translation. Since proteins are major components that participate in the constitution of cell structures and in the regulation of cellular activities, defective proteins often cause pathogenic alterations in cell structure and disorder in cell function.

Disorders Due to Gene Mutation [10.1]

Genetic disorders can be classified into three types: chromosomal defects, simple inherited genetic defects, and multifactorial defects. *Chromosomal defects* are disorders induced by deletion, duplication, or abnormal arrangement of partial or entire chromosome(s). Usually, a large amount of DNA or a large number of genes may be involved in chromosomal defects. Common chromosomal defects include trisomy (the presence of 47 chromosomes), monosomy (45 chromosomes), triplody (69 chromosomes), segmental duplication, and segmental deficiency. These chromosomal defects are often associated with abnormal anatomy of selected tissues and organs, mental retardation, behavioral disorders, and disorder of growth and development. Chromosomal defects can be detected by staining chromosomes with DNA dyes and optical microscopy. In normal cells, each of the 23 pairs of chromosomes exhibit distinct characteristics in terms of size, banding pattern, and the location of the centromere, which divides a chromosome into two arms. The sex chromosomes also show these characteristics. Women contain two XX sex chromosomes, while men contain one X and one Y sex chromosome. Any changes from the standard number and form of chromosome are considered chromosomal defects.

Simple inherited gene defects are disorders induced usually by a single mutant gene. On the basis of inheritance patterns, these defects have been classified into three categories: autosomal dominant, autosomal recessive, and X chromosome (or X)-linked gene defect or mutation. *Autosomal dominant gene defects* are those that cause clinical manifestations when one copy or allele of a gene in a genetic locus is mutated and the other copy or allele is normal, a condition referred to as *heterozygous gene mutation*. The gene defect is found in the autosomes, but not in the X and Y chromosomes. Because of the heterozygous nature of the gene defect (i.e., one chromosome of each pair contains the defective gene), 50% of the offspring will inherit the defective gene. The defective gene will be transmitted continuously through all generations by the individuals who carry the defective gene. Since the defect does not involve the sex chromosomes, males and females are equally affected. Autosomal dominant gene disorders rarely cause infertility. Examples of autosomal dominant disorders include Huntington's chorea, Marfan's syndrome, familial hypercholesterolemia, myotonic dystrophy, familial hyperlipidemia, familial breast cancer, and hypertrophic obstructive cardiomyopathy.

Autosomal recessive gene defects are those that cause clinical manifestations only when both copies of a gene are defective at a gene locus. Individuals who carry only one copy of the defective gene do not exhibit clinical manifestations. While the defective gene is transmitted to offspring, the offspring do not show clinical symptoms. The condition that causes clinical manifestations is when the two parents are carriers of the same defective gene and the two defective copies of the affected gene are transmitted to the same individual. In that case, 25% offspring are not defective gene carriers, 50% are carrying one copy of the defective gene (heterozygotes), and the remaining 25% are homozygous carriers who express clinical manifestations. When a homozygote is married to a normal individual, the offspring are all heterozygous defective gene carriers, but do not expression clinical manifestations. The recessive type of gene defect occurs in the autosomes. Thus, males and females are both affected. Examples of common autosomal recessive disorders include sickle cell anemia, deafness, cystic fibrosis, hereditary emphysema, congenital adrenal hyperplasia, and familial Mediterranean fever.

X-linked gene defects are those that occur in the X chromosome. Although both males and females may have X-linked gene defects, the clinical expression of the disorder differs between the two genders. A female possesses two X chromosomes and may carry one or two copies of the defective gene. Thus, dominant and recessive forms of gene disorder can be found in females. In contrast, a male carries only one X chromosome. All male X-linked disorders are dominant. The X-linked defective gene will not be transmitted from the father to the son, because the son inherits only the Y chromosome from the father. A female can receive X chromosomes from both parents. When a male is a defective gene carrier, all his daughters will be defective gene carriers. When a female is a homozygous defective gene carrier and the father is normal, 50% of her offspring, including males and females, will become heterozygotes. Examples of common X-linked gene disorders include hemophilia A and B (factor VIII and IX deficiency, respectively), colorblindness, Duchenne muscular dystrophy, glucose 6-phosphate dehydrogenase deficiency, testicular feminization, and ocular albinism.

The third type of genetically related disorder is *multifactorial genetic disorders*. This is a group of disorders that are induced or enhanced by activation of more than one gene. The involved genes may act synergistically in the initiation and development of a disorder. Often, environmental factors are necessary components that stimulate the expression of the involved genes. When the genetic effect is accumulated to a critical level, one or more environmental factors may trigger the onset of the disorder. The environmental factors may also determine the severity of the disorder. These disorders do not express the inheritance features of the dominant or recessive gene defect-induced disorders. It is even difficult to clearly identify the genes that are involved in the pathogenesis of the disorder. Even for the same type of disorder, different genes may be involved in different individuals. Thus, it is difficult to predict the risk of inheritance for the offspring. A judgment is often made based on the basis of a family history and statistics. However, certain multifactorial genetic disorders may be dependent on other types of genetic disorder. For instance, familial hypercholesterolemia, a disorder induced by autosomal dominant gene defect, contributes significantly to the development of atherosclerosis, which is considered a multifactorial genetic disorder. Examples of common multifactorial genetic disorders include atherosclerosis, diabetes mellitus, essential hypertension, cancer, epilepsy, and congenital heart disease, rheumatoid arthritis, and Parkinson's and Alzheimer's diseases.

Principles of DNA Engineering [10.2]

DNA engineering is applied to restore or modulate the structure of defective genes and thus to restore or improve the function of these genes. DNA engineering is established to study and treat disorders due to genetic defects. To achieve such goals, it is necessary to accomplish the following tasks: (1) identification of a cell type that is involved in the disorder of interest, (2) identification of a mutant gene that is responsible for or contributes to the disorder, (3) construction of a therapeutic gene, (4) transfection of target cells with the therapeutic gene for the treatment of the disorder; and (5) test of the effectiveness of gene transfection. These approaches are briefly discussed here.

Identification of Cell Types Involved in a Disorder. A pathological process inevitably involves cells, tissues, and organs, influencing the structure and functions of these biological systems. It may be fairly straightforward to identify the cell types involved in certain diseases, but may be difficult in others, depending on the nature of the disease and the accessibility of the involved organs. For instance, hepatoma may likely involve hepatocytes and other cell types in the liver, and atherosclerosis may involve blood and vascular cells, such as monocytes, macrophages, endothelial cells, and smooth muscle cells. For these diseases, cell types involved can be easily identified since the clinical signs and the anatomical locations of the pathological disorders can be easily recognized. However, it may not be that easy to identify a cell type involved in a neurological or psychological disorder since the functional anatomy of the brain is not well understood and the brain is not accessible. In any case, the first task in molecular engineering is to identify cell types that are involved in the disorder of interest. Although the genome is identical for all cell types, the pattern and level of gene expression can be vastly different. A disease is often associated with changes in gene expression. Thus, it is a critical step to identify cell types with altered gene expression. These cell types can be collected and prepared for the identification of genes involved in the disorder of interest.

Identification of Mutant Genes. One of the most important tasks in DNA engineering is the identification of a mutant gene that is responsible for the disorder of interest. Once a mutant gene is identified, a therapeutic gene can be established for replacement of the mutant gene and treatment of the disorder. It is important to note that a large number of genes have been identified, cloned, and analyzed in terms of their structure, function, and contribution to the pathogenesis of many diseases. It is strongly advised that, before starting an investigation, one should conduct a thorough literature search for information relevant to the genes potentially involved in the disorder of interest. It can save a tremendous amount of time and effort if the investigation starts with a known gene. However, if the cause of the disorder and the contribution of potential genes are not known, one may have to start with gene identification. A number of procedures are necessary for the identification of a mutant gene, including: (1) assessment of mRNA transcription or gene expression, (2) extraction of DNA from a selected cell sample and digestion DNA with restriction enzymes, (3) construction of recombinant DNA, (4) establishment of a DNA library, (5) selection of the gene of interest, (6) amplification of the selected gene, (7) gene sequencing and analysis, and (8) test of the function of the selected gene.

Assessing mRNA Transcription. The identification of mutant genes may start with assessing the level of gene transcription. In general, a gene defect or mutation may result in an alteration in the level of mRNA transcription. The alteration may be a complete deficiency, increase, or decrease in mRNA transcription. Since a change in the regulatory activity of gene expression may also induce an increase or decrease in mRNA transcription, an alteration in gene expression may not be indicative of the presence of a mutant gene. However, such a change may serve as a clue for identifying potential mutant genes. Genes with altered expression can be selected for further analyses, such as DNA sequencing and functional test, which provide conclusive information for identifying the mutant gene.

To assess the level of mRNA transcription or gene expression, it is necessary to prepare mRNA and measure the level of mRNA transcription. Total mRNA can be extracted from target cells, and the transcription of mRNA can be assessed by analytical approaches such as Northern blotting and gene microarray analyses. Since the gene microarray analysis provides a profile of mRNA transcription covering hundreds and thousands of genes, this method is more efficient than Northern blotting analysis. Here, the principle of gene microarray analysis is briefly discussed.

Gene microarray analysis is an approach established to detect quantitatively the transcription levels of multiple mRNAs according to the rule of complementary hybridization with prearranged DNA or cDNA (complementary DNA) samples with known sequences. In this analysis, fluorochrome-conjugated mRNA or reverse-transcribed cDNA samples, known as *probes*, from a specified cell type are applied to arrays of microspots coated with selected DNA samples of known sequences, inducing hybridization reactions. The fluorescent levels of the spots with mRNA-hybridized DNA can be measured and used to represent the relative level of mRNA transcription in the selected cell type (Fig. 10.3). Because the DNA samples coated on the microarrays are selected with known structure, the hybridized mRNA or cDNA samples can be identified. With the information accumulated in the gene bank for the past decades, genes involved in most known diseases have been studied, identified, and cloned. Thus, it is not difficult to establish gene microarrays that can be used for the identification of genes involved in common diseases. A major advantage for the gene microarray analysis is that a single experiment can provide an expression profile for a large number of genes.

To conduct a gene microarray analysis, it is necessary to follow several procedures, including the preparation of gene microarrays, preparation of fluorochrome-conjugated mRNA or cDNA probes, probe-DNA hybridization, detection of the relative levels of mRNA or cDNA probes hybridized to DNA targets, and data analysis. Gene microarrays can be prepared by coating denatured genomic DNA, cDNA (DNA prepared by reverse-transcribing mRNA), or synthesized oligonucleotide samples to arrays of DNA-binding spots created by photolithography on a glass slide of a centimeter size. Each spot can be as small as 50–350 μm in dimension. Thus, a large number of different DNA samples can be arranged in a single slide (>5000 spots/cm^2), which is also known as a gene "chip." In practice, it is considerably easier to prepare microarrays with oligonucleotides compared to those with DNA and cDNA. Oligonucleotides can be synthesized in single-strand format. An appropriate size of oligonucleotides for gene microarray analysis should be 20–120 bases.

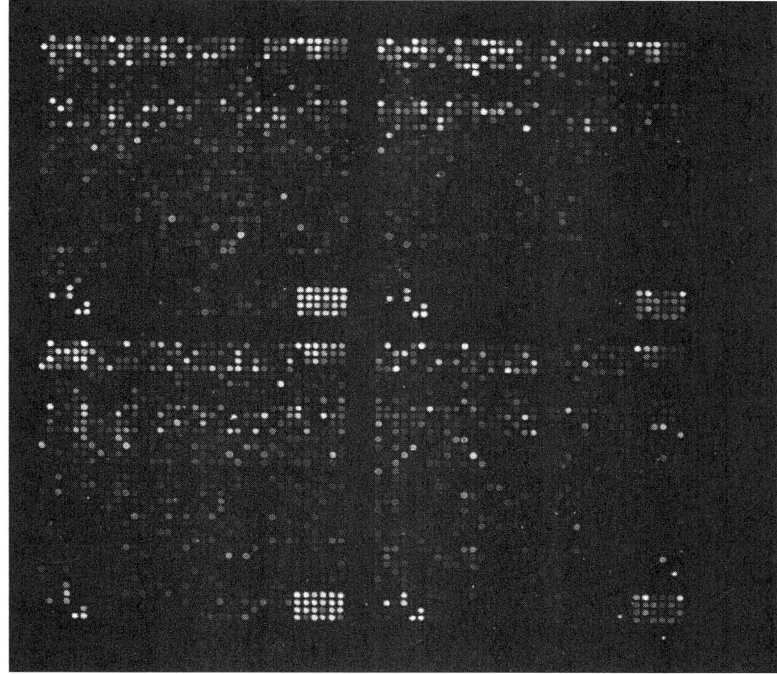

Figure 10.3. An image of a microarray containing >5000 genes. Each spot features a pool of identical single-stranded DNA molecules representing a single gene. The brightness of the spot is proportional to the amount of fluorescent mRNA hybridized to the DNA of the spot. The fluorescence spots can be identified by automated image analysis. The fluorescence intensity from each spot can be measured and compared to the background fluorescence. The images are further compared to images obtained from control measurements and transformed into a gene expression matrix, which can be analyzed by numerical methods. (Reprinted by permission of the Federation of the European Biochemical Society from Brazma A, Vilo J: Gene expression data analysis, *FEBS Lett* 480:17–24, copyright 2000.)

To detect the level of mRNA transcription from a cell sample, two types of probe can be synthesized from extracted mRNA: cDNA and RNA. cDNA probes can be synthesized by reverse transcription using extracted mRNAs as templates and the reverse transcriptase as a catalytic enzyme. Such a procedure produces single-stranded cDNA molecules. The cDNA probes can be conjugated with fluorochromes for probe identification. Alternatively, RNA probes can be synthesized from cDNA templates reverse-transcribed from mRNAs. In this preparation, T7 RNA polymerase is used to synthesize RNA molecules from double-stranded cDNA templates in the presence of fluorochrome-labeled nucleotides. The synthesized RNAs can be used as probes for gene microarray analysis.

The next step is to apply the probe, either single-stranded cDNAs or RNAs, to gene microarrays. The probing cDNA or RNA molecules hybridize with target DNA molecules via hydrogen bonds on the basis of the DNA complementary rule. Excessive and unhybridized probes can be removed by washing. The fluorescence of hybridized probes can be imaged and recorded from the microarrays by a laser scanner, and the intensity of fluorescence can be measured and analyzed. Such intensity represents the relative level of mRNA transcription. It is important to note that positive and negative controls should be introduced to the analysis. A positive control can be a selected cDNA probe that is known to hybridize to a target DNA, whereas a negative control is a cDNA probe that does not hybridize to a given target DNA.

The profile of mRNA transcription derived from a gene microarray analysis can be analyzed and compared to the profile of mRNA transcription from a normal or specified control cell type. Thus, genes with altered transcription can be identified from the microarray analysis. Although changes in the level of mRNA transcription may not indicate a gene defect or mutation, an increased, decreased, or null gene transcription suggests a candidate gene for further investigation. Once a gene with an altered transcription level is identified, the gene can be cloned and analyzed for the identification of structural mutation. If no mutation is found, the alterations in gene expression reflect changes in the regulatory activity for gene expression.

Extracting and Digesting DNA. Once a gene with altered mRNA transcription is identified as described above, DNA can be isolated and collected for gene analyses, including gene sequencing and identification of gene mutation. To obtain DNA, a selected cell sample can be lysed (for cultured cells) or homogenized (for a tissue sample), and DNA can be extracted and purified with established methods. Since DNA is a very large molecule, it should be digested into short fragments for DNA manipulation. DNA digestion can be accomplished by using restriction enzymes, which are found in bacteria and are capable of cleaving DNA at specific sequences. Although restriction enzymes are originated from bacteria, they can cleave DNA molecules from all known species as long as the DNA contains digestion target sequences. Each DNA molecule contains a large number of restriction sites for various types of restriction enzyme. These restriction sites are randomly distributed. DNA fragments with desired lengths can be generated by selecting appropriate restriction enzymes based on the locations of restriction sites. The digestion of target DNA at specified sites is one of the most important genetic approaches used for DNA manipulation. It is impossible to clone genes without the assistance of the restriction enzymes. It should be noted that restriction enzymes are not present in mammals. Thus, DNA molecules are not digested in the body systems under physiological conditions.

Restriction enzymes can cleave DNA into fragments with two different forms of end—sticky end and blunt end—depending on the type of restriction enzyme, but not the sequence of DNA. Some restriction enzymes, such as EcoRI, EcoRII, and HindIII, can cut the two strands of a DNA molecule unevenly, generating DNA ends with uneven strand length (the 3′ end is longer than the 5′ end or vise versa). The uneven end is also known as "sticky" end, so named because two such ends can hybridize to each other based on the complementary rule. Other restriction enzymes, such as HindII, HaeIII, and SmaI, cut the two strands of a DNA molecule evenly, leaving DNA ends with even strand length. The even DNA ends are also called "blunt" ends. In most cases, the substrate sequence for a restriction enzyme is identical for the two DNA strands at a given restriction site, but oriented in antiparallel directions. Thus, a "sticky" cut can produce two identical ends at a restriction site. This is a very useful feature for the creation of recombinant DNA. A "blunt" DNA end can be remodeled into a "sticky" end by an enzymatic treatment. Once DNA fragments are prepared, these fragments can be integrated into DNA vectors and used to create recombinant DNAs.

Constructing Recombinant DNA. The extracted and digested DNA fragments need to be further processed to generate DNA structures that can be cloned, screened for the desired gene, analyzed, and tested. An effective approach for such purposes is the construction of *recombinant DNA*. Recombinant DNA is a complex of DNA constructed by integrating a DNA fragment selected from a donor organism into a DNA vector collected from a different organism. The process of constructing a recombinant DNA is defined as *DNA recombination*. The purpose of DNA recombination is to generate functional genes in vitro, so that the gene of interest can be cloned, amplified, tested, used for gene transfection, and expressed in the transfected cells and tissues.

Gene cloning is a process of gene replication, generating multiple copies of the same gene. For cloning a gene, it is necessary to prepare a *gene cloning vector*. A vector is a bacterial DNA structure that contains restriction sites, and can accept foreign gene inserts and replicate in living bacteria. A vector is often engineered to add multiple restriction sites for the insertion of foreign gene fragments and to add functional genes for the selection, identification, and collection of the vector. A number of vector types, such as plasmids, λ phages, and cosmids, have been established and used in gene cloning. A typical plasmid vector carrying a LacZ gene (β-galactosidase gene) is shown in Fig. 10.4. The LacZ gene is often used as a reporter gene for gene selection or gene transfection.

To make recombinant genes, a selected donor DNA or cDNA fragment and a selected vector are treated with an identical type of restriction enzyme to generate the same type of "sticky" DNA ends. The enzyme-treated DNA fragment and vector are then incubated under a desired condition to hybridize the DNA fragment with the vector. At this stage, although the donor DNA and vector are joined with hydrogen bonds, the sugar–phosphate backbone is not linked together with phosphodiester bonds. An enzyme known as DNA ligase is needed to link the backbone, completing the recombinant process.

There are potential problems for DNA recombination. One problem is that the open ends of a vector may rejoin together without the insertion of donor DNA. Such vectors should not be included in the analysis. A gene selection method is usually used to remove the vectors that do not contain the inserted gene of interest. For this method, a selection gene, such as the β-galactosidase gene, is inserted into a selected vector in a cloning region with multiple restriction sites, also known as a *polylinker* or a *multiple cloning site*. The selection mechanism is that the β-galactosidase gene can be expressed when transfected into bacterial cells, only if the β-galactosidase gene is not interrupted by another gene. The insertion of a donor DNA fragment into the polylinker interrupts the structure of the β-galactosidase gene and renders the β-galactosidase gene unfunctional. When vectors without the donor DNA fragment are transfected into bacteria, the β-galactosidase gene

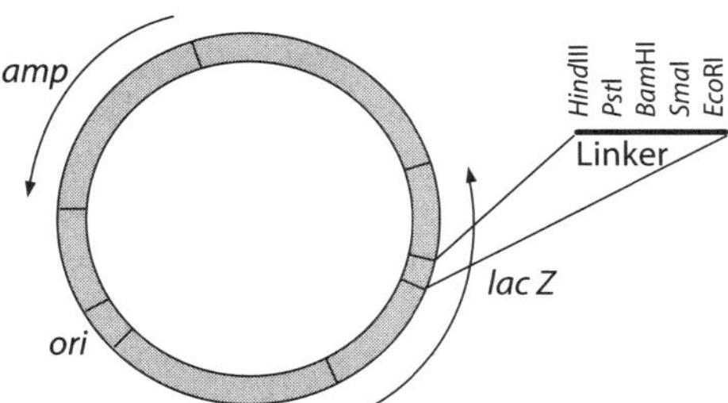

Figure 10.4. Schematic representation of the pUC18 plasmid vector containing a Lac Z gene.

can be expressed. Expressed β-galactosidase can react with a reagent known as X-gal to produce a blue-colored substance, which can be visualized under an optical microscope. However, the β-galactosidase gene in the vector with the donor DNA insert cannot be expressed and no blue-colored substance is generated when reacted with X-gal. Thus, clones with inserted donor genes can be selected on the basis of the color expression.

Another problem is that dominant "blunt" ends may be generated for the donor DNA, reducing the efficiency of donor DNA integration into the cloning vectors. To resolve such a problem, the donor DNA can be treated with a specific exonuclease that degrades the DNA at only the 5′ end, thus creating uneven "sticky" ends. However, the "sticky" ends of the donor DNA may not be complementary to the ends of a vector. To resolve such a problem, the open ends of the donor DNA fragment and the vector can be modified by synthesizing a short sequence at the ends. Such a process can be induced by treating both the donor DNA and the vectors separately with terminal transferase in the presence of different yet complementary nucleotides. For instance, the donor DNA can be treated with terminal transferase in the presence of dTTP, whereas the vector can be treated with terminal transferase in the presence of dATP. The added poly-T ends of the donor DNA are thus complementary to the poly-A ends of the vector.

Establishing a DNA Library. To identify and isolate the gene of interest from the gene pool containing all genes from a genome, it is necessary to establish a *DNA library*, which is defined as a collection of all gene clones from a cell sample of interest. A gene library can be established by using the procedures as described above, namely, extracting all DNA molecules, digesting the DNA with selected restriction enzymes into fragments of appropriate lengths, and inserting the DNA fragments into gene-carrying vectors. The vectors are then transfected into *Escherrichia coli* to establish *E. coli* colonies. Each *E. coli* colony is supposed to contain a specific DNA fragment. A DNA library is composed a large number of DNA fragments (Fig. 10.5). It is expected that one of these fragments contains the gene of interest. Such a gene can be identified and isolated by using a specific DNA or RNA probe as discussed on page 433.

For establishing a DNA library, two issues should be taken into account: vector type used in the cloning process and the DNA source. Vectors can accommodate DNA fragments with different sizes. Thus, vectors should be selected on the basis of the DNA fragment size. Plasmids and λ-phages are suitable for small DNA fragments, whereas cosmids and yeast artificial chromosomes (YACs) are used for large DNA fragments. Various vectors give different forms of DNA library. A plasmid or cosmid library is a collection of *E. coli* bacteria containing the cloned genes, a λ-phage library is a collection of λ-phages, whereas a YAC library is a collection of yeasts. The choice of vectors does not affect the structure of the cloned genes.

Another factor to consider in the construction of a DNA library is the source of DNA. DNA from two sources can be used for the construction: genomic DNA and complementary DNA (cDNA). The later is DNA synthesized from mRNA by reverse transcription. There are distinct features for the two types of DNA library. Genomic DNA contains introns and exons for all genes from a genome. A genomic DNA library is suitable for investigating the regulation of gene transcription, which requires the identification and understanding of the regulatory *cis* elements located in the introns. In contrast, a cDNA library contains only exons that encode proteins. In addition, cDNA does not contain genes that are not transcribed into mRNA at a given time and state. Thus, a cDNA library is suitable for the identification and isolation of expressed genes. Since most pathological

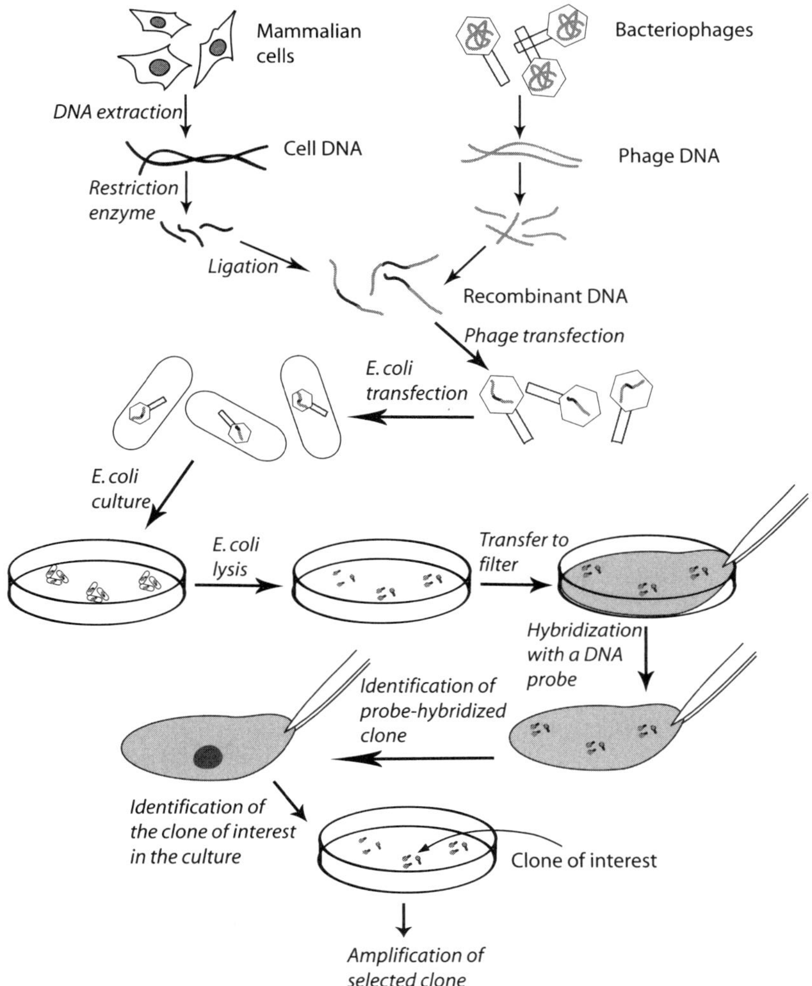

Figure 10.5. Preparation of a gene library. Based on bibliography 10.2.

disorders involve gene expression, creating a cDNA library is a practical approach for identifying genes that contribute to pathological disorders.

Selecting a Gene of Interest from a DNA Library. A DNA library may contain thousands of gene fragments. To find and collect the gene of interest, it is necessary to screen the DNA library. There are two approaches that can be used for DNA screening: directly probing the gene of interest or probing the protein encoded by the gene integrated into the expression vector, which is designed to express proteins in cultured *E. coli* cells. For the direct DNA approach, it is necessary to establish a DNA probe, which is a single-stranded DNA fragment or oligodeoxynucleotide, capable of hybridizing to its complimentary strand. With the tag of an identification marker to the probe, the target gene strand can be identified and isolated. For the protein-probing approach, an antibody specific to the protein produced by the gene of interest can be used to identify the protein, which leads to the identification of the encoding gene.

As discussed on page 426, gene selection can start from the detection of mRNA transcription in a cell sample selected from a disordered organ by gene microarray analysis. An altered transcription level (increase, decrease, or null) may help to identify a potential gene involved in the pathogenesis of a disorder. It is important to point out that a gene microarray analysis does not provide information about alterations in the gene structure. It merely demonstrates the level of mRNA transcription, suggesting potential genes for further analyses. It is the DNA library screening and probing that provide gene clones, which can be used for sequencing and functional analyses.

For direct gene probing, a key issue is how to design a DNA probe. A probe can be constructed according to a selected sequence within the gene of interest (identified by gene microarray analysis), if the gene has been identified. Usually, information for most human genes is available in the GenBank database, from which a gene sequence can be selected. A DNA probe or oligodeoxynucleotide can be synthesized based on the selected sequence. A DNA probe can also based on the structure of a protein, which may potentially contribute to a pathological disorder of interest and is identified via protein analysis. Once the identity of the protein is known (via sequencing or mass spectrometry), an oligodeoxynucleotide can be synthesized based on the amino acid sequence of the protein. It is important to note that most amino acids are coded by more than one gene codon and there may exist a number of DNA sequences for each given amino acid sequence. For instance, the amino acid tyrosine (Tyr or Y) is encoded by gene codons TAT and TAC, and serine (Ser or S) is encoded by TCT, TCC, TCA, and TCG. Thus, it may be difficult to define the exact oligonucleotide sequence on the basis of a known protein sequence. To resolve such a problem, a fragment of protein with the minimal gene codon redundancy should be used. Furthermore, all possible DNA sequences for a selected protein fragment should be synthesized and mixed together. For a peptide sequence of 5 amino acids, including Tyr, Ser, Asp (asparagines), Cys (cysteine), and His (histidine), which are encoded by 2, 4, 2, 2, and 2 gene codons, respectively, 64 oligonucleotide strands need to be synthesized. The mixture of all these oligonucleotide strands should be applied to a DNA library for hybridization. The correct oligonucleotide fragment will hybridize to the gene of interest.

In addition to DNA probes, an antibody can be used as a probe to identify the protein product of a gene, thus identifying the gene indirectly. A key procedure for this approach is to induce protein expression. Such a task can be accomplished by using an expression vector. A cDNA fragment can be inserted into a selected expression vector in frame with the gene of a bacterial protein to generate a fusion protein. When *E. coli* cells are transformed with the vector, the cells will produce the fusion protein that contains the protein encoded by the inserted cDNA and the bacterial protein. In the DNA library, the fusion protein is expressed within or near the gene clone. Since the structure of the bacterial protein is known, an antibody can be developed and used to screen the DNA library. Any clones marked with the antibody should contain the gene of interest. Such a clone can be excised and collected for further analyses.

Amplification of the Selected Gene. The procedures described above can only produce a small amount of recombinant DNA. To carry out further DNA analyses such as DNA sequencing and transfection, the recombinant DNA must be amplified to generate a sufficient amount of DNA. An effective approach for DNA amplification is to transfect *E. coli* cells with the identified and isolated recombinant gene from the DNA library. Each recombinant gene can replicate into multiple copies once transfected into *E. coli* cells. These recombinant gene copies can be rapidly amplified when *E. coli* cells grow. After cell culture for a day or two, billions of copies of each recombinant gene can be produced. The recombinant genes can be purified and isolated. The copies of the gene are referred to as gene clones.

DNA can also be amplified by using polymerase chain reaction (PCR), a method for in vitro DNA synthesis in the presence of a DNA template, DNA polymerase, DNA primers, and deoxynucleotides (dNTPs). This method can be used to select and amplify a DNA fragment of interest from genomic DNA by using an appropriate set of primers. In practice, a template DNA/cDNA fragment or a plasmid that contains a DNA fragment of interest is incubated in a PCR buffer supplemented with a temperature-resistant DNA polymerase (e.g., Teq DNA polymerase), DNA primers (specific to each selected DNA fragment), and dNTPs. The reaction mix is subject to about 30–35 thermal cycles with three alternating temperature levels for each cycle, including 95°C, 60°C, and 70°C. The temperature 95°C is for denaturing DNA, 60°C is for primer annealing, and 70°C is for DNA synthesis. After the PCR reaction, a million-fold of amplification can be achieved. It is important to note that PCR is suitable for the amplification of small DNA fragments (hundreds of base pairs) before the gene is inserted into a plasmid and for the selection of a gene fragment from a plasmid. PCR is not capable of amplifying large DNA molecules, such as plasmids or cosmids.

DNA Sequencing and Analysis. Once a gene of interest is identified, cloned, and isolated, the next steps are to confirm the isolated gene, detect the sequence of the gene, and assess possible structural changes or gene mutation. A common approach used for the confirmation of the isolated gene is Southern blotting analysis, named after E. M. Southern, who initially developed the technique. To prepare for Southern blotting analysis, the isolated gene-containing vector should be amplified as described on page 432 if the samples are not sufficient, and the gene of interest should be removed from the cloning vector by digestion with restriction enzymes that cut and release precisely a selected DNA fragment. The removed DNA fragment can be fractionated by gel electrophoresis, transferred to a filter membrane, and detected with the probe used for DNA library screening or a different probe designed on the basis of the same gene. The probe-reacted band of DNA on the filter membrane is the gene of interest. The identified band can be excised from the gel and used for gene sequencing and analyses.

To determine whether altered expression of a selected gene is due to gene defect or mutation, or due to merely a change in the regulatory activity of gene transcription, it is necessary to detect the sequence of the gene and compare it with that of a control gene. The sequence of a gene can be determined by using two methods: *base destruction sequencing* and *dideoxy sequencing*. For the first method, DNA strands are labeled at the 5′ ends with ^{32}P, the double-stranded DNA is denatured (separated), and one strand is discarded. Copies from the remaining strand are separated into four groups and used for sequencing. Each group is treated with a chemical reagent that selectively breaks one or two of the four bases. The four groups are treated with four different chemicals. For instance, group 1 is treated with a chemical that degrades the bases A and G, group 2 with a chemical for the base G, group 3 with a chemical for the base C, and group 4 with a chemical for the bases C and T. For each group, the degradation of the selected base results in the disruption of the DNA strand at all locations with the selectively degraded base. When the degrading chemical is prepared in an appropriate concentration and reaction is carried out for an appropriate period, only a fraction of the target bases is degraded for each copy of DNA. Since the chemical reaction is completely random and a DNA strand contains a large number of the same base, DNA strands can be disrupted at different locations of the target base. Since a large number of DNA copies are present for each reaction with a selected degrading chemical, multiple DNA copies with an identical disruption location can be generated. For a group of 6 identical DNA copies, each contains 3 Gs, a treatment with a degrading chemical specific to the Gs under a well-controlled chemical concentration may induce only one disruption for each copy of DNA with an equal degradation probability for the three Gs, possibly resulting in an equal number of DNA copies disrupted at each G location. On the basis of such a principle, the four groups of single-stranded DNA samples can be disrupted randomly at distinct locations. Each group is composed of DNA fragments disrupted at a given base type with an equal disruption probability for each location of the same base. Resulting DNA fragments from each group can be run in a designated lane in gel electrophoresis. All DNA bands in each lane lost the same base at the 3′ ends, which is the target of a selected degrading chemical. The 5′ ends of the DNA fragments are tagged with ^{32}P for the purpose of identification. By comparing electrophoretic results from all four groups, all ends of the DNA fragments can be read and analyzed, giving a sequence of the entire DNA strand.

For the dideoxy sequencing method, a single stranded DNA is prepared and used as a template for synthesizing DNA in the presence of dideoxy nucleotides (ddNTPs). A ddNTP contains a deoxyribose sugar that lacks the 3′-hydroxyl group. A ddNTP can be incorporated into a DNA chain, but DNA synthesis is terminated whenever a ddNTP is incorporated because of the lack of the 3′-hydroxyl group, which is necessary for DNA extension. In practice, single-stranded template DNA is divided into four groups, and DNA synthesis is carried out in the presence of DNA polymerase, primers labeled with a radioisotope or fluorescent marker for identification, and the four types of standard deoxynucleotides (dATP, dTTP, dCTP, and dGTP). In addition, each reaction group is mixed with a distinct type of ddNTPs. Group 1 is mixed with ddATP, group 2 with ddTTP, group 3 with ddCTP, and group 4 with ddGTP. For each reaction group, the ddNTP is mixed in such a percentage with respect to the dNTP that the ddNTP molecules are incorporated into only a fraction of growing DNA chains, while the remaining DNA chains incorporate the standard dNTPs at any given time. This gives an equal probability of ddNTP incorporation into each site of a corresponding dNTP. DNA synthesis continues when a dNTP is incorporated to the respective site, but stops when a ddNTP is incorporated. As a result, various sizes of DNA chains can be synthesized in each reaction group with all chains ending at the same ddNTP site. Synthesized DNA fragments of different sizes from each reaction group can be fractionated by gel electrophoresis in a distinct lane. Each DNA band indicates the relative location of a designated ddNTP. By comparing all four reaction groups (each reacted with a different ddNTP), the locations of all four ddNTPs can be determined. By reading along the electrophoretic lanes, the sequence of the DNA fragment can be resolved.

Testing the Function of the Selected Gene. Based on the gene sequence, gene mutation, if any, can be identified by comparing with the sequence of a respective control gene from a gene database or from a specified control cell sample. The next step is to design a physiological test and assess whether the identified gene mutation influences the physiological function of the gene. A test may be prepared with the following procedures:

(1) constructing a recombinant vector containing the gene of interest; (2) establishing a physiological testing system, which is usually cultured cells or an in vivo animal model; (3) transferring the vector into the physiological testing system; and (4) assessing the influence of the transferred gene on the function of corresponding proteins and related cellular activities.

A recombinant vector with a mutant gene can be constructed as shown in Fig. 10.6. Since gene expression is a critical process for the test, it is necessary to integrate into the vector a gene promoter, which drives the gene expression. Commonly used promoters include cytomegalovirus (CMV) promoter and simian virus (SV)40 promoter, but any promoter that drives the expression of the selected gene can be used. In addition, one may integrate a reporter gene, such as a β-galactosidase gene and a green or red fluorescent protein gene, into the recombinant gene. The protein products of these reporter genes can serve as markers for the identification of the cells with positive gene transfection and expression. The constructed vector can be amplified by transforming and growing *E. coli* cells as described on page 432.

A functional test can be carried out in an in vitro cell culture system or an in vivo organ system. The in vitro system is easier to use and usually provides reliable information. It should be kept in mind that the endogenous wildtype gene in the testing cells may likely interfere with the transferred exogenous gene, rendering it difficult to identify the influence of the transferred exogenous gene. To resolve such a problem, a cell line with null gene mutation (complete loss of the function of a wildtype gene, also referred to as *gene knockout*) should be used. Such a cell line can be obtained from a transgenic animal

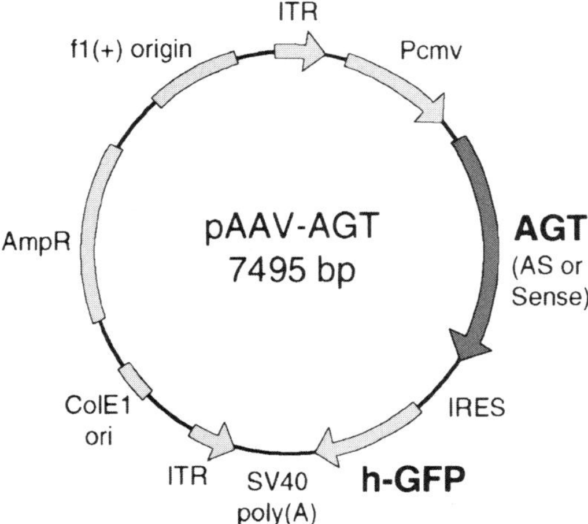

Figure 10.6. Schematic representation of the rat angiotensinogen (AGT) gene construct based on the adeno-associated virus (AAV)-derived plasmid. Plasmid pAAV-AGT-AS and pAAV-AGT-S were constructed by inserting a full-length cDNA (1.65 kb) of AGT into the unique *Hind* III site of the AAV-derived plasmid pTR-UF3 in antisense (AS) or sense (S) direction, respectively. AGT cDNA, indicated by the shaded arrow, is driven by a cytomegalovirus (CMV) early promoter (Pcmv). ITR, AAV inverted terminal repeat; IRES, polio virus type 1 internal ribosomal entry site; h-GFP, "humanized" victoria green fluorescent protein gene. Other open bars and arrows represent other basic elements of the vector. (Reprinted from Tang X et al: *Am J Physiol* 277:H2392–9, copyright 1999 by permission of the American Physiological Society.)

model without the gene of interest. Since the function of the selected wildtype gene is completely deficient, the function of the transferred recombinant gene can be assessed by comparing to a control with the function of the corresponding wildtype gene.

The next step for testing the gene function is to transfer the gene construct into cultured cells. It is a natural property that cells can endocytose large molecules, including DNA, into the cytoplasm. A fraction of endocytosed DNA can go through the nucleus membrane and integrate into the genome. Several methods, including virus-, salt-, liposome-, receptor-, and electroporation-mediated gene transfer, have been established and used to facilitate gene transfection into mammalian cells. The exogenous gene can be expressed transiently, producing encoded protein. The function of the transferred gene can be tested in several aspects, including the expression and function of the encoded protein as well as the activity of the host cells. Tests can be conducted at selected times after 12 hs from gene transfer. Protein expression can be detected by Western blotting or immunoblotting, which is a sensitive method for detecting a small amount of protein. A comparison to a sample from control cells with the corresponding wildtype gene demonstrates alterations in the mutant protein in terms of molecular weight and expression level. A functional test can be designed according to the nature of the protein. If the protein of interest is an enzyme, the catalytic activity of the protein can be assessed by detecting changes in the substrate protein. In vitro tests can be carried out by using an isolated enzyme and substrates. The influence of the transferred gene on the host cells can be assessed by measuring changes in a cellular activity known to be related to the transferred gene.

Constructing a Recombinant Therapeutic Gene. Once the pathological role of a mutant gene is identified, the next step is to design and construct a therapeutic gene that can be used to restore the physiological function of the gene (Fig. 10.7). Methods described on page 429 can be used for this purpose. Recall that there exist several types of genetic disorders, including chromosomal defects, simple inherited genetic defects, and multifactorial defects. Different strategies should be used for these types of genetic disorder. For chromosomal defects, since a large amount of DNA is involved, it is difficult to restore the structure and function of the defect chromosomes. For a simple inherited genetic defect, sine a single gene is usually involved, a corresponding wildtype gene may be isolated, cloned, and used to construct a therapeutic gene, which can be used for correcting the defect gene. For multifactorial defects, it is necessary to identify the genes involved and determine the pathogenic mechanisms of the disorder, based on which therapeutic genes can be constructed to modulate the activity of the mutant genes. For instance, the upregulation of multiple growth factor genes may contribute to the initiation and development of atherosclerosis. Growth inhibitor genes can be constructed and transferred into vascular cells to inhibit the activity of growth factors and thus to suppress atherogenesis.

Transfection of Target Cells with a Therapeutic Gene. The primary goal of molecular engineering is to correct pathological disorders due to gene mutation or changes in gene activities. An effective approach is to transfer therapeutic genes, as constructed on page 429, into target cells to restore the structure and function of the mutant or altered genes. Such an approach is referred to as *gene therapy*. The transferred genes, once in the cytoplasm, can be transported to the cell nucleus, integrated into the genome, and expressed to produce corresponding proteins. It is expected that the newly expressed proteins can function in place of the corresponding mutant proteins, thus reducing or suppressing

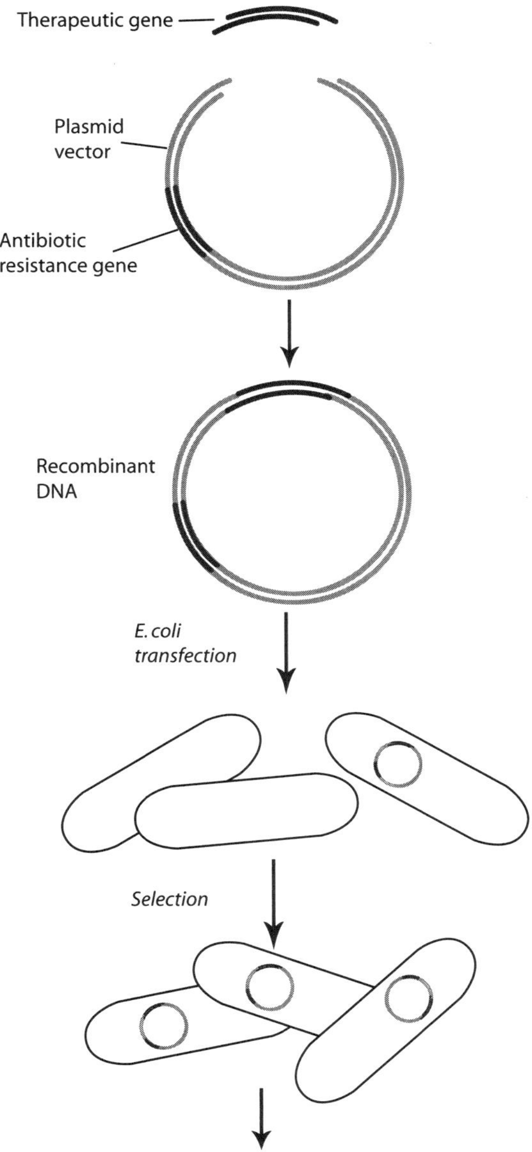

Figure 10.7. Construction of a therapeutic gene vector. Based on bibliography 10.2.

mutant protein-induced pathological disorders. Investigations for the past decade have provided tremendous evidence for the potential application of gene therapy to human disorders.

There are two general approaches that can be potentially applied to human gene therapy: embryonic gene transfer and somatic gene transfer. For *embryonic gene transfer*, a blastocyst is collected from a pregnant animal, and a therapeutic gene is injected into the cells of the blastocyst to induce site-specific gene integration or homologous gene recombination. Embryonic cells from the blastocyst are stem cells. Once these stem cells carry the therapeutic gene, the gene can be passed to differentiated cells. Gametes may also carry the therapeutic gene and pass it to next generations. The transferred gene may function in place of the mutant gene. This is potentially an effective approach for the correction of pathological disorders due to hereditary gene defect or mutation. However, procedures for blastocyst collection induce cell injury and death. A safe, reliable, and effective technique has not been established for the treatment of hereditary human disorders. A technically similar technique has been used to modulate or remove selected genes in the embryonic stem cells of animals for the establishment of transgenic models. These models have been used extensively in scientific research.

Somatic gene transfer is an approach used for transferring therapeutic genes into the somatic cells of humans and animals after birth. Obviously, it is impossible to transfer a gene into all somatic cells in the body. Thus, this approach is used for gene transfer into selected target cells. Although mammalian cells are capable of taking up DNA, a process known as *endocytosis* or "naked gene transfer," the rate of DNA taking up is very low. A number of gene transfer-mediation methods, including virus-, liposome-, receptor-, electroporation-, salt-mediated gene transfer, have been established to facilitate gene transfer into mammalian cells. Each of these gene transfer strategies has strengths and weaknesses. The therapeutic efficacy and effectiveness differ between these mediation methods. The viral approach usually results in a higher rate of gene transfection and longer period of gene expression than do the nonviral approaches. However, viral carriers may cause infectious disorders and gene mutation in transfected cells. In contrast, the nonviral approaches are safe and easy to carry out, and exhibit low immunogenicity. However, their efficacy is relatively low and the expression duration of the transferred gene is short compared to the viral approach.

The choice of a gene transfer approach may also be dependent on the structure, function, and accessibility of the target tissue or organ. For instance, electroporation may be applied to the skin and skeletal muscle cells for gene transfer, but not to the heart and brain, because these organs are difficult to access and electroporation causes cell injury and death. In contrast, viral gene carriers and liposome can be used for gene transfer into target cells in most types of tissues and organs. Furthermore, various cell types may possess different capabilities of accepting foreign genes, and the cell state may also influence the rate of gene transfer for a given type of gene transfer mediation approach. For instance, retroviral gene carriers can infect dividing cells, but not nondividing cells, whereas adenoviral gene carriers can infect both dividing and nondividing cells. However, the retroviral approach may induce more stable gene expression than the adenoviral approach. Thus, all possible factors should be taken into account for the choice of a gene transfer approach. Here, several common gene transfer mediating approaches are briefly discussed.

Virus-Mediated Gene Transfer. Several types of virus, including retrovirus, adenovirus, adeno-associated virus, herpes simplex virus, have been used as gene carriers for mediating gene transfer into mammalian cells. The basis for virus-mediated gene transfer is that viruses are capable of infecting mammalian cells, carrying therapeutic genes, and integrating the carried genes into the host cell genome. Viruses can be modified or genetically engineered to remove harmful components and accommodate desired therapeutic gene fragments. Once transferred into the cell, viruses are able to integrate their genome into the host genome and use the synthetic machineries of the host cells to produce viral DNA, RNA, and proteins. Thus, viruses are natural gene transfer carries. To be used for gene transfer mediation, a viral gene carrier ought to be replication-deficient, nonimmunogenic, and nontoxic, but are able to facilitate gene transfer to the genome of a target cell. The choice of a viral carrier for gene transfer is dependent on the nature of the virus and the state of the target cells, as we will see in the following sections.

RETROVIRUS-MEDIATED GENE TRANSFER. Retroviruses belong to a family of RNA viruses. A typical retrovirus contains two parts: the viral core and the envelope. The viral core is composed of two identical RNA strands and several enzymes, including reverse transcriptase, protease, and integrase. The envelope, composed of a membrane and glycoproteins, encloses the viral core. Common retroviruses used for gene transfer include Moloney virus and lentivirus. The Moloney virus is a murine leukemia virus that causes lymphoid leukemia in mice. The lentivirus is a retroviral strain that causes maedi (a chronic pulmonary disease found in Iceland) and visna (a disease affecting the central nervous system and causing paralysis) in sheep.

Retroviruses can interact with mammalian cells through cell membrane receptors, and enter the cell through receptor mediation. In the cytoplasm, the viral RNA genome can be converted into DNA by the reverse transcriptase. The converted DNA can be integrated into the host genome and replicate together with the host DNA during cell division. Thus, genetic information of the virus can be transmitted to the genome of the host cell and carried to the next generation. Any foreign genes that are inserted into the retroviral genome can be integrated into the host genome. The virus itself can produce viral proteins based on the viral mRNA and reproduce the same type of virus.

The RNA of the retrovirus can be engineered through a packaging process, generating viral vectors. Briefly, retroviral particles can be modulated to remove a selected RNA sequence known as the *packaging signal*, which is responsible for packaging the RNA core into the viral envelope. When transfected into a host cell line, the viral particles can express functional genes, such as env, gag, and pol. The env and pol genes encode viral envelope proteins, whereas the gag gene encodes the reverse transcriptase. While the virus is able to convert its RNA to DNA and produce necessary proteins for the construction of the viral envelope, it cannot pack the core RNA into the envelope. At the same time, another group of viral particles is selected and engineered to remove selectively the harmful sequences without modulating the packaging signal. A therapeutic gene fragment can be inserted into the viral genome to generate a recombinant viral gene carrier. When the recombinant viral gene carrier is transfected into the same host cell line containing the empty viral envelopes, the recombinant gene carrier can be packed into the empty envelopes under the action of the packaging signal, thus generating a complete viral structure that contains the therapeutic gene without the harmful gene sequences. This engineered viral structure is harmless and nonreplicable, and can be used to for gene transfer. Retrovirus-derived gene carriers have been used in experimental models and clinical trials.

There are several advantages for retrovirus-derived gene carriers. These carriers can induce a high rate of gene integration into the host cell genome with relatively stable and long-term expression (several months). However, retroviral gene carriers exhibit several disadvantages, including toxicity and limitation of effectiveness to only dividing cells. In particular, retroviral vectors may be potentially tumorigenic. The integration of a viral gene into the host cell genome may induce mutation of tumor suppressor genes or oncogenes, potentially leading to tumorigenesis. Furthermore, viral genes can be expressed to generate viral proteins, which potentially cause host cell immune reactions, a potential factor that reduces the stability of the transfected gene and the duration of gene expression.

ADNOVIRUS-MEDIATED GENE TRANSFER. Adenoviruses are double-stranded DNA viruses and belong to a family composed of more than 30 serotypes. Many of the serotypes invade human cells and cause infectious disorders of the upper respiratory tract, known as *cold* or *flu*, and conjunctivae. Others infect the simian, bovine, canine, avian, and murine species. The mechanisms by which the adenovirus enters mammalian cells and causes infection remain poorly understood. It has been hypothesized that cell membrane receptors specific to adenoviral proteins may mediate the invasion of viruses (Fig. 10.8).

The genome of adenoviruses can be modified to remove replication control sequences and thus to prevent adenoviral replication and to eliminate toxic influence. The DNA genome of a typical adenovirus contains several regions, including El to E4. The El region is responsible for viral replication. The replacement of the El region with a foreign gene results in the generation of replication-defective viruses. Such engineered adenoviruses can be used as a gene transfer carrier. There are several advantages for the use of adenoviruses in gene transfer. First, the adenoviral gene carrier can induce a relatively high efficiency of gene transfection compared to a retroviral carrier. Second, an adenoviral carrier can infect both dividing and non-dividing cells. Third, an adenoviral vector can carry large DNA inserts [≤8 kb (kilobases)]. However, gene transfer mediated by the adenoviral vector often results in a relatively low rate of gene integration into the host cell genome and a short duration of gene expression (about several weeks). Although the toxicity of adenoviruses is lower than that of retroviruses, the viral particles can still induce infectious disorders in the host cells. Furthermore, the expressed viral proteins can induce host immune reactions, a potential factor that reduces the stability of the transfected gene and the duration of gene expression.

ADENO-ASSOCIATED VIRUS AS A GENE CARRIER. Adeno-associated virus belongs to a family of nonpathogenic human parvoviruses with a single-stranded DNA genome. This viral type is smaller than other types of viruses and cannot replicate unless associated with helper viruses, usually adenoviruses or herpesviruses. In the absence of helper viruses, the adeno-associated virus usually integrates its genome into the host chromosomal DNA and enters a latent state. Adeno-associated viruses can infect dividing as well as nondividing cells, and cause stable and long-term gene expression. Compared to the

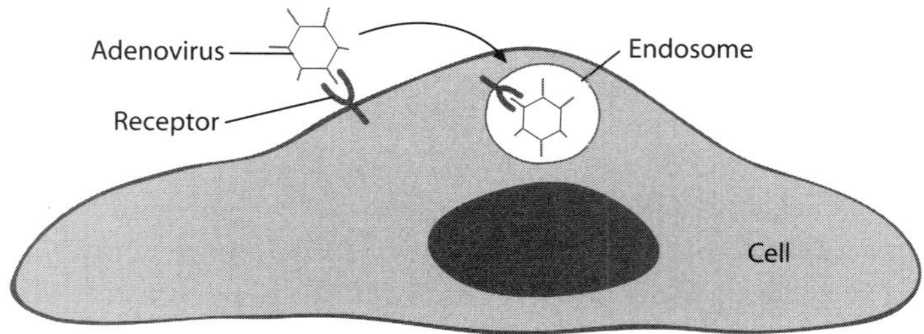

Figure 10.8. Schematic representation of adenovirus-mediated gene transfection.

adenovirus, the adeno-associated virus exhibits reduced immunogenicity and toxicity, since this virus consists of fewer genes that encode proteins harmful to the host cells. In a typical adeno-associated viral vector, the only wildtype sequences left are the inverted terminal repeats (ITRs) that flank the inserted gene cassette. These sequences are necessary for the packaging of the viral genome. Because of the lack of the replication capability, it is not necessary to modulate the genome of the adeno-associated virus for gene transfer. A desired therapeutic gene can be directly inserted into the viral genome and the viral vector can be used for gene transfer (Fig. 10.9). This type of virus has been extensively used for gene transfer in experimental models. However, adeno-associated viral vectors can accommodate relatively small gene inserts because of its limited genome size.

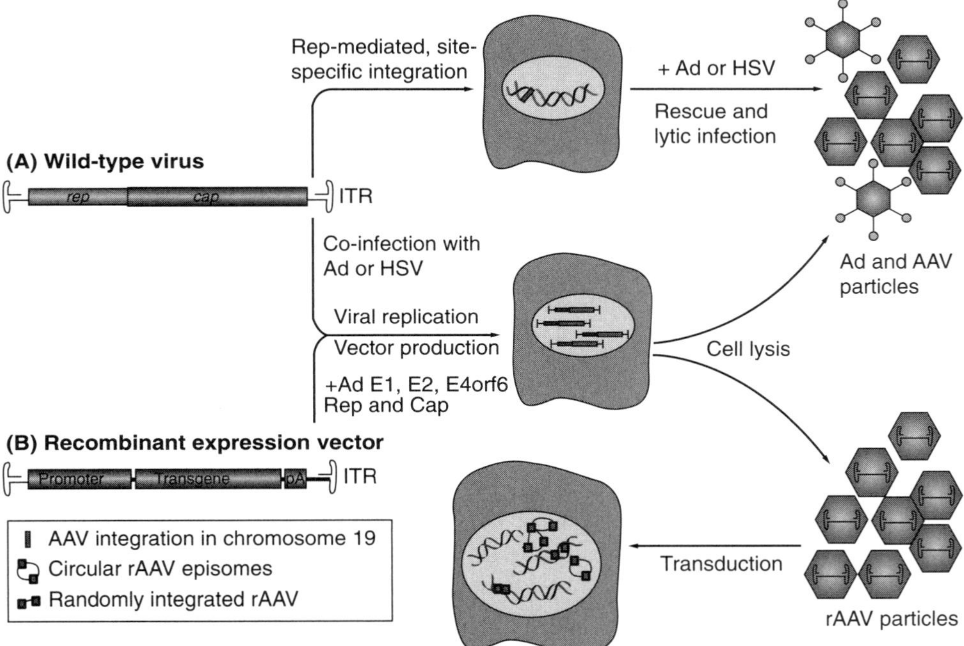

Figure 10.9. Adeno-associated virus as a vector for gene transfection. (A) The structure of the wildtype adeno-associated virus (AAV) is shown. A single-stranded DNA genome is encompassed by palindromic inverted terminal repeats (ITRs). The *rep* open reading frame (ORF) encodes proteins that are involved in viral replication, and the *cap* ORF encodes proteins that are necessary for viral packaging. AAV integrates into the human genome at a specific locus on chromosome 19 (red) and persists in a latent form. It can exit this stage only if the cell is co- or superinfected with helper virus such as adenovirus (Ad) or herpes simplex virus (HSV), which provide factors necessary for active AAV replication. (B) The generic genedelivery vector based on AAV is depicted. The viral genome is replaced by an expression cassette, which usually consists of a promoter, transgene, and polyA (pA) tail. For production of the recombinant virus (rAAV), Rep and Cap proteins as well as Ad or HSV elements (Ad E1, E2 and E4, or f6) have to be provided in *trans*. Examples of intracellular forms of the delivery vector that are responsible for transgene expression following transduction with rAAV (double-stranded circular episomes and randomly integrated vector genomes) are depicted in red. (Reprinted by permission from Macmillan Publishers Ltd.: Vasileva A, Jessberger R: *Nature Rev Microbiol* 3:837–47, copyright 2005.)

The efficiency of gene transfer mediated by adeno-associated viruses is usually lower than that mediated by adenoviruses.

HERPES SIMPLEX VIRUS-MEDIATED GENE TRANSFER. Herpes simplex virus belongs to a family of double-stranded DNA viruses that cause herpes simplex in humans. The virus can also be found in other mammalian species, such as canine, bovine, and simian. The virus can invade mammalian cells, integrate its DNA into the host genome, and mature in the host nucleus. This type of virus has been used for mediating gene transfer in experimental models. Herpes simplex virus-derived gene vectors can carry large gene fragments, infect dividing as well as non-dividing cells, and induce efficient gene transfection. However, the duration of gene expression mediated by herpes simplex viruses is often shorter (about one week) than that mediated by retroviral and adenoviral vectors. A possible reason is that herpes simplex viruses induce cytotoxic and/or immune reactions in the host cells, which shut down the expression of the transfected gene. A herpes simplex viral vector lacking the harmful genes, such as the immediately early genes, exhibits much reduced toxicity, improved gene transfection efficacy, and prolonged gene expression, as demonstrated in experimental models.

Receptor-Mediated Gene Transfer. Certain types of cell membrane receptors can be used to mediate gene transfer. A typical example is the transferrin receptor, which interacts with transferrin. The transferrin molecules can spontaneously link to a polymeric linker, such as polylysine, which can also link to DNA fragments at different sites, forming DNA–polylysine–transferrin complexes. These complexes can adhere to the cell membrane through the interaction of transferrin with the transferrin receptor (Fig. 10.10). Such an interaction activates the endocytosis mechanism of the target cells, which in turn take up the triplet DNA complexes. The transferred DNA complexes can be transported from the cytoplasm to the nucleus via endosomes.

Although the receptor-mediated gene transfer approach has been successfully used to transfer genes into mammalian cells, the transfer efficiency is low because most DNA molecules are trapped in the endosomes and cannot be released before degradation. To reduce DNA degradation and facilitate gene transport, a codelivery of replication-defective adenoviruses can improve the efficiency of receptor-mediated gene transfer. The mechanism for this enhancement is that endocytosed adenoviruses can disrupt the endosome membrane. Thus, the cotransferred triplet DNA complexes can be released. In this preparation, the DNA fragments are not inserted into the genome of the adenovirus. The virus is merely used as a gene transfer helper.

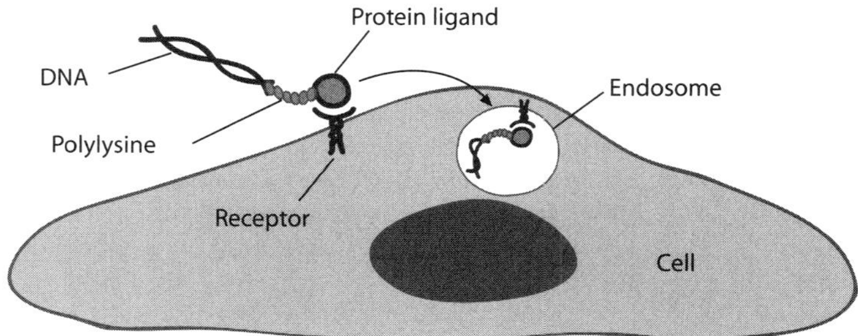

Figure 10.10. Schematic representation of receptor-mediated gene transfection.

Liposome-Mediated Gene Transfer. Liposomes are lipid vesicles of ~200 nm in diameter. Lipid molecules can spontaneously form multilaminar vesicles, which can be converted to smaller uni-laminar vesicles by sonication. Each liposome contains about 2500 lipid molecules. Each lipid molecule is composed of a hydrophilic and hydrophobic end. When liposomes are mixed with a water-based fluid, the hydrophilic end of the lipid molecule interacts with the water molecules and the hydrophobic end is enclosed inside. When liposomes and DNA molecules are mixed, liposome–DNA complexes can form spontaneously. The interaction between positive and negative charges is responsible for the complex formation. Some types of liposome, such as lipofectin (*N*-[1-(2,3-dioleyloxy)propyl]-*N,N,N*-trimethylammonium chloride), derivatives of cholesterol and diacyl glycerol, and lipopolyamines, are positively charged at the hydrophilic heads on the external surface of the vesicle. DNA molecules are negatively charged at the phosphate groups. The positively charged liposomes can form complexes with the negatively charged DNA molecules. There exist also negatively charged liposomes. These liposomes may repel negatively charged DNA molecules and are not suitable for mediating gene transfer. However, some negatively charged liposomes may entrap DNA molecules within the lipid vesicles to form DNA–liposome complexes. The negatively charged liposomes are less efficient in gene transfer than positively charged liposomes.

Liposomes have long been used for mediating gene transfer into mammalian cells. Although the transfection mediated by liposomes is not as effective as that by viral vectors, the efficiency of gene transfection is acceptable. There are several mechanisms underlying liposome-mediated gene transfer. First, liposome–DNA complexes can bind to the lipid cell membrane. Cationic liposomes attach to the cell membrane more efficiently than anionic liposomes, because the cell membrane is more negatively charged. Attached liposome–DNA complexes can be either endocytosed or fused into cells (Fig. 10.11). The endocytosed liposome–DNA complexes are transported and released into cytoplasm by endosomes. Some liposomes may be degraded by lysosomal enzymes before being released into cytoplasm. Second, liposome-DNA complexes can diffuse directly through the nucleus membrane and enter the cell nucleus.

Calcium Phosphate-Mediated Gene Transfer. Calcium phosphate can bind to DNA to form complexes that facilitate precipitation of DNA molecules to the cell surface. The interaction of calcium phosphate–DNA complexes with the cell membrane triggers cell endocytosis, by which DNA molecules are taken up and transported into the cytoplasm. A fraction of the endocytosed DNA can reach the nucleus, while the remains are degraded in the cytoplasm. There are several advantages for the use of calcium phosphate: (1) calcium phosphate is a physiological ingredient and is harmless to the host cells—calcium

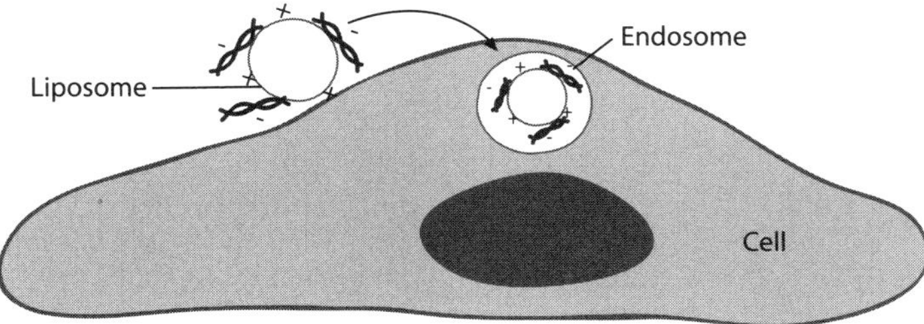

Figure 10.11. Schematic representation of liposome-mediated gene transfection.

phosphate is probably the safest substance among those used for mediating gene transfer and (2) calcium phosphate is easy to prepare and use. The only procedures involved in gene transfer are preparation of a calcium phosphate buffer supplemented with DNA and delivery of the DNA mix to target cells. However, the efficacy of calcium phosphate-mediated gene transfer is generally lower than that mediated by viral vectors and liposomes.

Electroporation-Mediated Gene Transfer. Electroporation is electrophysical process that induces the formation of hydrophilic pores in the cell membrane under the influence of an electric field. When cells are subject to an electric field between two electrodes, an electric pulse of appropriate voltage can induce the formation of pores in the cell lipid membrane. The rate of pore formation and the size of the pores are dependent on the maximal voltage across the cell membrane. The higher is the voltage, the higher is the rate of pore formation and the larger is the pore size. The duration of the pore opening is proportional to the duration of the electric pulse delivered to cells. The duration of electrical pulse may also affect the pore size. The longer is the duration of the electrical pulse, the larger is the pore size. Pores about 10 nm in diameter can be achieved in mammalian cell membranes by selecting an appropriate voltage. Short pulses usually result in a uniform population of small pores with a short duration of pore opening.

The process of electroporation is transient and reversible. After the termination of an electric pulse, the cell membrane pores shrink gradually and sealed eventually within several minutes. During the period of pore opening, DNA fragments present in the buffer can enter the cytoplasm and migrate into the cell nucleus. There are several possible mechanisms for the entrance of DNA fragments into the cytoplasm: (1) DNA fragments may directly move into the cytoplasm through the cell membrane pores, (2) DNA may be transported into the cytoplasm by osmotic forces established with respect to the concentration gradient of the DNA molecules, (3) electroporation may enhance endocytosis of the DNA fragments attached to the cell membrane, and electroporation may impose electrophoretic influence on the DNA fragments, enhancing DNA transport into the cytoplasm.

Electroporation is a highly efficient approach for gene transfer and has been extensively used in gene transfer for cultured cells. It has also been used for in vivo gene transfer into cells in accessible tissues and organs, such as the skin and skeletal muscle (Fig. 10.12). The efficacy of gene transfer mediated by electroporation is usually higher than that mediated by liposome and calcium phosphate, but lower than that mediated by viral vectors. A significant drawback for electroporation is cell injury induced by electrical pulses. In cell culture models, a standard electrical pulse (1000 V/cm in voltage), which causes effective gene transfer, may induce injury or death in about 30–50% of cells. Electroporation also causes cell injury and death for in vivo gene transfer. Thus, the voltage level for electroporation should be optimized to minimize cell injury through trial and error experiments for each specific type of cell, tissue, and organ.

Gene Gun-Mediated Gene Transfer. This is approach mediated by gene particle bombardment. Microparticles coated with the gene of interest can be accelerated by a force (e.g., a pressure) to penetrate through the cell membrane and thus to deliver the gene into the cytoplasm or the cell nucleus. The force used for gene transfer can be controlled to achieve an appropriate distance of penetration. This technique is easy to use and is efficient compared to other gene transfer-mediating approaches. However, bombardment induces cell injury and death.

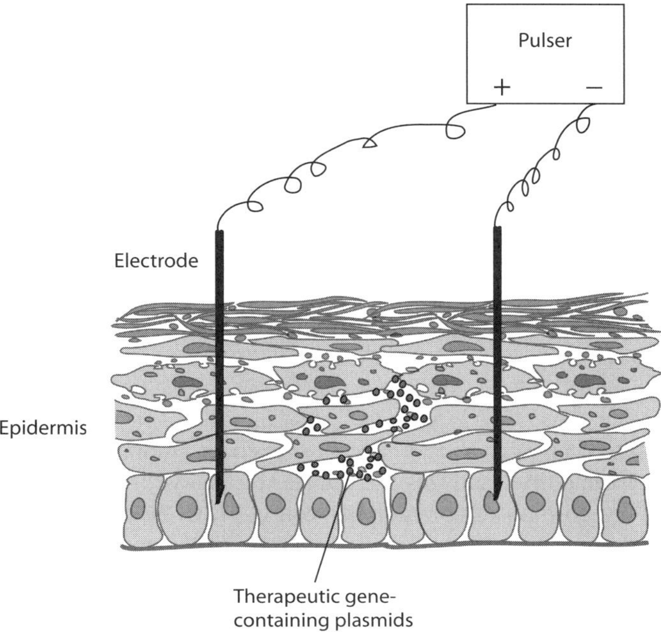

Figure 10.12. Schematic representation of electroporation-mediated gene transfection.

Assessing the Expression of the Transfected Gene. One of the most important steps in molecular engineering is to assess the expression of the transferred gene. Reporter genes are often used for such a purpose. Reporter genes are recombinant genes that encode proteins exhibiting special signs for visualization and thus can be used as markers for assessing the level of gene expression. Major criteria for the construction of a reporter gene are that the genes or expressed proteins are nontoxic, do not influence the function of the host cell, can be identified, and are not expressed in the target cell. Several types of reporter genes have been established and used for gene transfer. These include the β-galactosidase (gal) gene, luciferase gene, chloramphenicol acetyltransferase (CAT) gene, and fluorescent protein genes.

The *β-galactosidase gene* encodes a protein enzyme known as galactosidase and is found in *E. coli* cells. β-Galactosidase is a hydrolase that catalyzes the hydrolysis of the terminal residues of β-galactoside, forming d-galactose. d-galactose can react with x-gal, a chemical compound, to form a blue-colored substance. This feature has been used for detecting the efficiency of gene transfer. The presence of the blue color in cells transfected with the β-gal gene indicates the expression of this gene. β-galactosidase is not expressed in mammalian cells and nontoxic to the host cells. It is commonly used as a reporter gene in experiments of gene transfer. The β-galactosidase gene can be cotransferred with a therapeutic gene into target cells. It is assumed that the expression of the β-galactosidase gene indicates the expression of the cotransferred therapeutic gene. Alternatively, the β-galactosidase gene can be inserted into a recombinant vector that carries a therapeutic gene. The expression of the β-galactosidase gene is usually associated with the expression of the therapeutic gene. The second approach is preferable for detecting the efficiency of gene transfer.

The *luciferase gene* encodes a protein enzyme that oxidizes luciferin and is found in certain types of fish and insects, such as firefly bugs. The oxidization reduces luciferin to a compound that emits fluorescence. The emitted fluorescence can be observed by using a fluorescence microscope. Alternatively, the intensity of the emitted fluorescence can be detected by spectrophotometry. Since the luciferase gene is not expressed in mammalian cells, this gene can be used as a reporter gene for detecting the efficiency of gene transfer in terms of the fluorescence emitted by the luciferase substrate. In a cell sample transfected with the luciferase gene, positive fluorescent emission from the cell sample in the presence of luciferin indicates the expression of the transferred luciferase gene. In contrast, the absence of fluorescence in the presence of luciferin indicates the failure of gene transfer. Alternatively, an antiluciferase antibody developed by using luciferase as an antigen can be used to detect the expressed enzyme by immunohistochemistry. Luciferase is harmless to mammalian cells and is commonly used for the assessment of gene transfer efficiency.

The *chloramphenicol acetyltransferase gene* encodes chloramphenicol acetyltransferase (CAT), an enzyme that catalyzes the transfer of a catyl group from acetylcoenzyme A to the 3′-hydoxy position of chloramphenicol, $C_{11}H_{12}Cl_2N_2O_5$, which is a broad-spectrum antibiotic derived from *Streptomyces venezuelae* (fungus-like bacteria). The gene is not present in mammalian cells and can be used as a reporter gene for detecting the efficiency of gene transfer. The CAT catalytic activity can be monitored by a liquid scintillation enzyme assay and used for assessing gene expression. For the liquid scintillation enzyme assay, cell extracts can be incubated in a reaction mix containing ^{14}C- or ^{3}H-labeled chloramphenicol and *n*-butyryl coenzyme A. When the CAT gene is expressed, the CAT transfers the *n*-butyryl group of the *n*-butyryl coenzyme A molecule to chloramphenicol, forming *n*-butyryl chloramphenicol. The reaction products can be extracted with a small volume of xylene. The *n*-butyryl chloramphenicol compound can be partitioned into the xylene phase, while unmodified chloramphenicol remains predominantly in the aqueous phase. The xylene phase is mixed with scintillant and counted with a scintillation counter for the radioactivity of ^{14}C or ^{3}H, which is conjugated to chloramphenicol. The presence of radioactivity indicates the expression of the transferred CAT gene.

There are natural *fluorescent protein genes* that encode proteins capable of emitting fluorescence. These genes are found in certain types of florescent fish and insects. One of the most commonly seen natural fluorescent proteins is the green fluorescent protein, which is encoded by the green fluorescence protein gene. Recombinant fluorescent genes that encode red and cyanine proteins have been artificially constructed by modulating the gene sequences of natural fluorescence protein genes. These genes can be used as reporter genes for assessing the efficiency of gene transfer. Two methods can be used for such a purpose: (1) a fluorescence protein gene can be cotransferred with a therapeutic gene into target cells and used to assess the expression of the therapeutic gene or (2) a fluorescent protein gene can be inserted into a recombinant vector that contains a functional or therapeutic gene (Fig. 10.6). The fluorescent gene can be expressed together with the therapeutic gene when transfected into target cells, thus indicating the efficiency of gene transfer. It is important to note that the fluorescent protein gene should be inserted into an appropriate site that does not influence the transcription of the therapeutic gene. Such a site is usually located at the end of the therapeutic gene. However, trial-and-error experiments should be carried out to search for a correct insertion site. An advantage by using the fluorescent protein gene is that fluorescent signals can be examined in living cells, allowing the observation of dynamic changes in molecular action. Thus, fluorescent protein gene has been extensively used in cell biology research.

Assessing the Effectiveness of Gene Transfer. An important aspect of gene therapy is to test the effectiveness and efficacy of gene therapy. One of the critical issues is whether gene transfer restores the physiological function of the target gene, cell, tissue, and organ, and corrects pathological alterations due to genetic disorders. The design of a test is largely dependent on the function of the target gene as well as the structure and function of the target cells. For instance, for a gene that encodes an enzyme, such as a matrix metalloproteinase, the level and activity of the enzyme can be detected before and after the gene transfection. The level of the enzyme can be assessed by immunoblotting analysis, and the function of the enzyme can be evaluated by detecting the level of the substrate modification. For a gene that encodes a protein responsible for regulating a specific cell activity, such as smooth muscle cell contractility, the level of the protein and the degree of cell activity can be assessed before and after the gene transfection. Similarly, the level of cell proliferation and apoptosis can be measured to assess the effectiveness of delivered genes that encode proteins for the regulation of cell mitogenic and apoptotic activities, respectively. Regardless of the gene type, a significant phenotype change in response to gene transfection suggests the effectiveness of the therapeutic gene.

Potential Negative Effects of Gene Transfer. Gene transfer may not be always beneficial and may potentially induce harmful effects:

1. Gene transfer requires the mediation of gene carriers, which are often toxic (liposomes), traumatic (electroporation and bombard), or infectious (adenovirus and retrovirus). These mediating approaches may negatively influence the function of host cells.
2. It is difficult to control the gene insertion sites in the target genome. Gene transfer by any mediating means may potentially cause gene insertion into incorrect sites in the genome, resulting in host or guest gene mutation, upregulation, or downregulation.
3. A therapeutic gene may be suitable for certain physiological functions, but may negatively influence other functions. A typical example is the genetic manipulation of the *ras* gene, which encodes the Ras protein, a critical molecule for the transduction of mitogenic signals and for the regulation of cell survival and proliferation. Because of its proproliferative effect, the *ras* gene has been considered a potential gene target for the treatment of atherosclerosis and cardiovascular hypertrophy. However, the suppression of the *ras* gene, while potentially reduces the proliferative activity, inevitably causes a negative effect on other functions mediated by this protein such as cell survival. Thus, the potential negative effects of gene transfer should be taken into account in the design of a molecular engineering approach.

HOMOLOGOUS RECOMBINATION [10.3]

Homologous recombination is a natural process by which DNA damages, such as doublestrand breaks (DSBs) and interstrand crosslinks, are repaired by generating a functional DNA copy with the homologous DNA sequence of the intact sister chromatin as a template and replacing precisely the damaged DNA sequence. This process can completely restore the structure and function of the damaged DNA. Thus, homologous recombination is an ideal genetic approach for the correction of a mutant gene and can be utilized for therapeutic purposes.

Homologous recombination is originally discovered in yeast and *E. coli*. In mammalian cells, homologous recombination is rare, but is significantly increased in frequency when DNA double-strand breaks occur. Thus, one strategy to induce homologous recombination in mammalian cells is to introduce double strand breaks to desired target genes, which are responsible for the disorder of interest. The natural homologous recombination machinery in the cell is capable of recognizing the double-strand breaks on the target gene and initiating homologous gene recombination or site-specific gene replacement, resulting in the restoration of the structure and function of the mutant gene.

An effective approach for introducing DNA double strand breaks to a target gene is to deliver specific zinc finger nucleases to cells or tissues. Zinc finger nucleases are proteins that can recognize and cut target DNA at specific sites, induce double-strand breaks, and enhance site-specific homologous recombination. A typical zinc finger nuclease is composed of characteristic zinc finger domains. Each zinc finger domain consists of about 30 amino acids and can bind a zinc ion, which is critical to the stability and function of the zinc finger protein. A zinc finger protein can recognize and bind to a specific target gene by interacting with the major groove of the DNA double helix. The catalytic domain of the zinc finger nuclease can then cut the double strand of the bound DNA, resulting in double strand breaks.

Zinc finger nucleases can be designed and constructed with high specificity to desired target DNA sequences. To date, several types of three-finger zinc finger nucleases have been constructed with specificity to unique triplet DNA sequences, including 5′-ANN-3′, 5′-CNN-3′, 5′-GNN-3′, and 5′-TNN-3′, where N represents any nucleotide. Determination of the specificity of a zinc finger nuclease is based on the structure of the protein enzyme and the target triplet nucleotides.

For therapeutic purposes, a triplet DNA sequence, which is unique to a mutant gene and serves as a specific binding site for a zinc finger nuclease, can be identified from a cell type. A zinc finger nuclease can be designed, constructed, and delivered to the target cells. The zinc finger nuclease can recognize the specific binding site and induce double-strand breaks in the target mutant gene. The double strand breaks can initiate site-specific homologous recombination that replaces the mutant gene with a corresponding functional gene. Thus, the induction of the artificial double strand breaks and homologous recombination represent potential therapeutic approaches for the molecular treatment of genetic disorders.

ANTISENSE OLIGONUCLEOTIDE-BASED THERAPY [10.4]

Antisense oligonucleotide therapy is to design and use a DNA or RNA oligonucleotide sequence to hybridize complementary target mRNA, which is considered the "sense" sequence, and thus to block the translation of a specific protein, which is involved in the initiation and development of the disorder of interest. An example is the use of antisense oligonucleotides to downregulate the expression of angiotensinogen, thus to reduce the level of angiotensin I and II, for the treatment of hypertension (see page 699 for mechanisms). Furthermore, the hybridization of the target mRNA with an antisense oligonucleotide may induce the activation of an enzyme known as *ribonuclease* H (RNase H), which cleaves the hybridized mRNA. These approaches can be used to effectively downregulate protein translation, providing a therapeutic means for the treatment of disorders. An antisense DNA or RNA sequence specific to a selected mRNA molecule can be designed and constructed. The constructed antisense oligonucleotides can be delivered to target cells on the basis of endocytosis, a natural process that takes up small particles on the cell surface. Several gene transfer-mediating methods, such as liposome- and electroporation-mediated transfer, can be used to enhance the delivery of the antisense oligonucleotides. The antisense approach can be potentially used for therapeutic purposes by repressing the translation of specific proteins, which are involved in the pathogenesis of disorders.

SMALL INTERFERING RNA-BASED THERAPY [10.5]

Small interfering RNA (siRNA), also known as *short interfering RNA*, is a short RNA sequence, which is specific to and can hybridize to a target mRNA, induce the degradation of the target mRNA, and thus knock down the expression of the encoded protein. The process of siRNA-induced mRNA degradation is referred to as RNA interference, post-transcriptional gene silencing, or transgene silencing. This mechanism was originally discovered in petunia plant cells and *Caenorhabditis elegans*. Further investigations have demonstrated that mammalian cells also exhibit RNA interference. Since RNA interference can be effectively used to suppress the translation of specific proteins, siRNA can be potentially used as therapeutic agents for pathological disorders that involve abnormal expression of specific proteins.

A siRNA molecule is generated under the action of an enzyme, known as "dicer," which is an RNAse III molecule composed of a helicase domain, two RNAse III domains, a double-stranded RNA (dsRNA)-binding domain, and a PAZ domain (Fig. 10.13). The dicer enzyme can splice dsRNA into short siRNA. A typical siRNA is a double-stranded RNA about 21 nucleotide in length, and consists of a 5′ phosphate overhang at one end and a 3′ hydroxyl overhang at the other end. A siRNA fragment can form a complex with a multiprotein complex known as the *RNA-induced silencing complex* (RISC), which binds to a specific sequence of the siRNA and unwind the siRNA into single strands. The single-stranded siRNA can hybridize to complementary mRNA in the cytoplasm. The RISC complex is activated and can degrade the substrate mRNA bound by the siRNA, thus suppressing protein translation.

In mammalian cells, there is a RNA interference machinery, which has been evolved as a defense mechanism against the invasion of retroviruses. The RNA interference machinery can recognize and separate double-stranded RNA molecules into two single strands to form siRNAs. The siRNA molecules can recognize and recruit RNases, which degrade mRNA transcripts complementary to the siRNA molecules. On the invasion of retroviruses, siRNA sequences are generated and utilized for the destruction of the invaded viruses. This mechanism is also used for posttranscriptional gene regulation in mammalian cells. During gene transcription, certain RNA transcripts can fold to form hairpin-like double-stranded RNA structures, sometimes referred to as *microRNA*, which cannot be translated into proteins. The RNA interference machinery can detect and destroy these microRNA structures by generating siRNAs. Since the RNA interference machinery may not be able to completely remove target mRNAs, siRNA-mediated mRNA degradation is also referred to as gene knockdown.

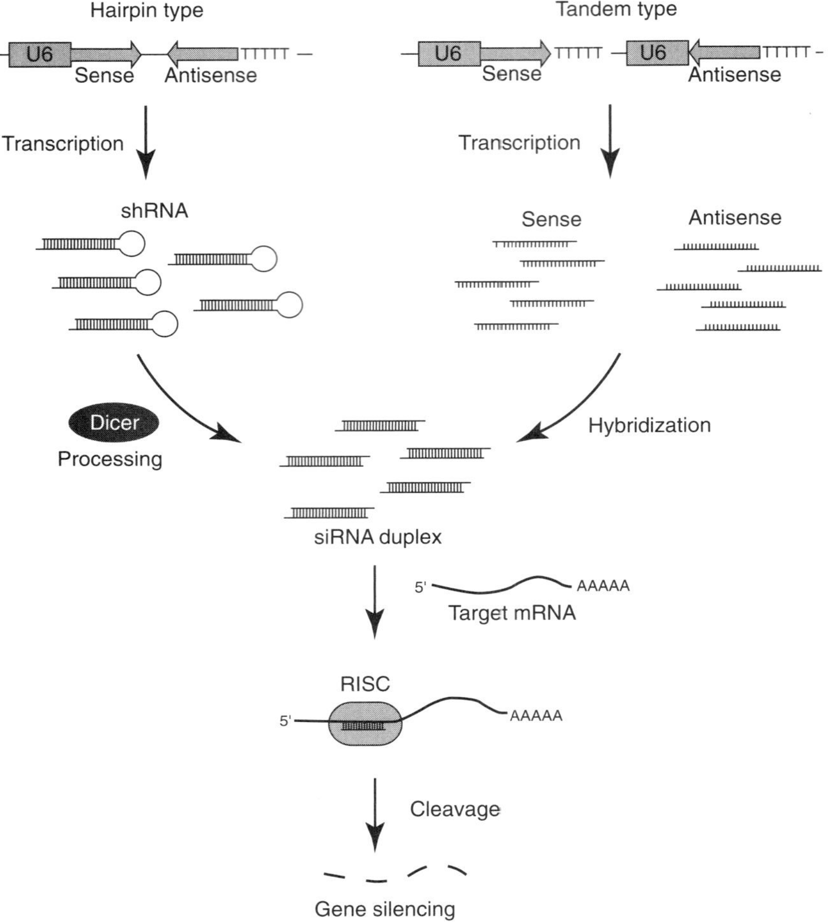

Figure 10.13. Schematic representation of a transcription system for production of siRNA. (Reprinted by permission of the Federation of the European Biochemical Societies from Itoa M et al: *FEBS Lett* 579:5988–95, copyright 2005.)

For therapeutic purposes, specific siRNA molecules can be synthesized in vitro and applied to target cells or tissues for inducing degradation of target mRNA. To synthesize a siRNA sequence, it is necessary to determine the target mRNA sequences, on which the siRNA acts, based on the sequence of a specific gene. The effective target mRNA sequences can be selected by using siRNA design programs provided by commercial carriers, such as Promega. Desired siRNA sequences can be synthesized by using an oligonucleotide synthesizer. The synthesized siRNA can be used for cell transfection. A liposome-mediated transfection method can be used to facilitate siRNA transfection. For experimental purposes, it is necessary to construct and use a siRNA control, known as "scrambled" siRNA. Such a siRNA is usually a 21-nucleotide sequence, which is nonspecific to any known mRNA. The control siRNA should be transfected to control cells simultaneously with the specific siRNA transfection. While siRNA transfection is effective in knocking down gene expression, the effect is usually transient. It is necessary to conduct multiple transfection if long-term effect is required for experimental or therapeutic purposes.

To induce a long-term effect of RNA interference, a siRNA cloning vector can be constructed and used to transfect cells (Fig. 10.14). The transfected siRNA cloning vector contains a nucleotide sequence, which encodes a desired specific siRNA, and a gene expression vector, which induces gene expression when transfected into a cell. To construct a siRNA cloning vector, it is necessary to establish siRNA encoding gene sequence, which can be done by using a siRNA design program as described above. The siRNA-encoding gene sequence can be inserted into a cloning vector, such as the neomycin-resistant gene-containing psiSTRIKE cloning vector from Promega. The established siRNA cloning vector can be amplified by transfecting and growing E. coli cells. The amplified siRNA cloning vector can be purified and used for mammalian cell transfection. The following is an example of the siRNA cloning vector for the mRNA of the protein tyrosine phosphatase SH2 domain-containing protein tyrosine phosphatase 1 (SHP1).

The sequences for the SHP1 siRNA are as follows: 5′-ACC**GAAAGGCCGGAACA AAT GTGT***TTCAAGAGAACACATTTGTTCC GGCCTTTC*TTTTTC-3′ and 5′-TGCAGA AAA***AGAAAGGCCGGAACAAATGTGT*** *TCTCTTGAA***ACACATTTGT TCCGGCC TTT**-3′. In these sequences, the boldfaced fragments represent the target

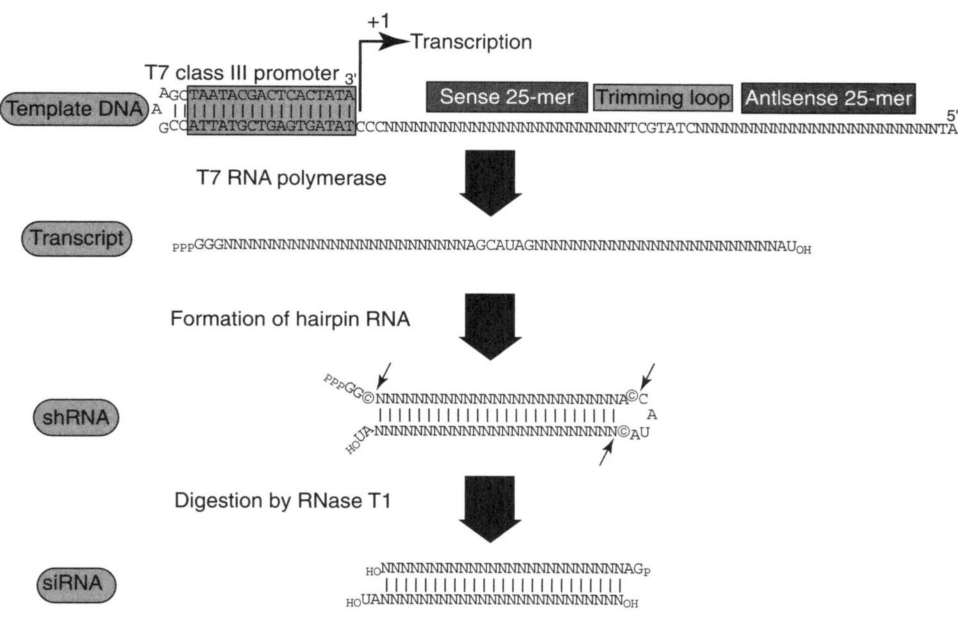

Walking across the entire gene, shifting one base at a time

Figure 10.14. Schematic representation of gene silencing by a shRNA expression vector. The shRNA is transcribed from a shRNA cloning vector and processed by Dicer to produce siRNA. The processed siRNA enters the RNA-induced silencing complex (RISC), where it targets mRNA for degradation. (Reprinted by permission of the Federation of the European Biochemical Societies from Itoa M et al: *FEBS Lett* 579:5988–95, copyright 2005.)

sequence, the italic boldfaced fragments represent the target reverse complement, the underlined fragments are for the mRNA loop, and the remainder are the overhang fragment (5′-ACC) and the U6 termination sequence (TTTTC-3′) for cloning purpose. The sequences for the control scrambled siRNA for SHP-1 are as follows: 5′-ACC**GAAGATGCGAAGGGATAC TAC***TTCAAGAGA****GTAGTATCCC T TCGCATCTTC***TTTTTC-3′ and 5′-TGCAGAAAA***AGAAGATGCGAAGGGATACTA*** *TCTCTTGAA* **GTAGTATCCC TTCGCATCTT**-3′. The SHP1 specific and scrambled siRNA sequences can be synthesized by a commercial carrier such as Proligo.

BIBLIOGRAPHY

10.1. Disorders Due to Gene Mutation

Kopp P, Jameson JL: Transmission of human genetic disease, in *Principles of Molecular Medicine*, Jameson JL, Collins FS, eds, Humana Press, Totowa, NJ, 1998, pp 43–58.

Schneider AS, Szanto PA: *Pathology*, 3rd ed, Lippincott Williams & Wilkins, Philadelphia, 2006.

Pasternak JJ: *An Introduction to Human Molecular Genetics: Mechanisms of Inherited Diseases*, 2nd ed, Wiley-Liss, Hoboken, NJ, 2005.

Killeen AA: *Principles of Molecular Pathology*, Humana Press, Totowa, NJ, 2004.

Watson JD, Berry A: *DNA: The Secret of Life*, Knopf, New York, 2003.

Pasternak JJ: *An Introduction to Human Molecular Genetics: Mechanisms of Inherited Diseases*, Fitzgerald Science Press, Bethesda, MD, 1999.

10.2. Principles of DNA Engineering

Schena M: *Microarray Analysis*, Wiley-Liss, Hoboken, NJ, 2003.

Grompe M, Johnson W, Jameson JL: Recombinant DNA and genetic techniques, in *Principles of Molecular Medicine*, Jameson JL, Collins FS, eds., Humana Press, Totowa, NJ, 1998, pp 9–24.

Watson JD, Gilman M, Witkowski J, Zoller M: *Recombinant DNA*, 2nd ed, Scientific American Books/Freeman, New York, 1992.

Kmiec EB: Gene targeting protocols, in *Methods in Molecular Biology*, Vol 133, Humana Press, Totowa, NJ, 2000.

Speicher DW, Reim DF: Microsequence analysis of electroblotted proteins, *Anal Biochem* 207:19–23, 1992.

Jackson PJ: N-terminal protein sequencing for special application, in *Methods in Molecular Biology*, Vol 211: *Protein Sequencing Protocols*, 2nd ed, Smith BJ, ed, Humana Press, Totowa, NJ, 2003, pp 287–300.

Steinke L, Cook RG: Identification of phosphorylation sites by Edman degradation, in *Methods in Molecular Biology*, Vol 211: *Protein Sequencing Protocols*, 2nd ed, Smith BJ, ed, Humana Press, Totowa, NJ, 2003, pp 301–8.

Jackson PJ et al: Validation of protein sequencing in a regulated laboratory, in *Methods in Molecular Biology*, Vol 211: *Protein Sequencing Protocols*, 2nd ed, Smith BJ, ed, Humana Press, Totowa, NJ, 2003, pp 309–18.

Dupont DR, Yuen SW, Graham KS: Automated C-terminal protein sequence analysis using the alkylated-thiohydantoin method, in *Methods in Molecular Biology*, Vol 211: *Protein Sequencing Protocols*, 2nd ed, Smith BJ, ed, Humana Press, Totowa, NJ, 2003, pp 319–31.

Harvey DJ: Identification of sites of glycosylation, in *Methods in Molecular Biology*, Vol 211: *Protein Sequencing Protocols*, 2nd ed, Smith B.J. ed, Humana Press, Totowa, NJ, 2003, pp 371–83.

Aitken A, Learmonth M: Analysis of sites of protein phosphorylation, in *Methods in Molecular Biology*, Vol 211: *Protein Sequencing Protocols*, 2nd ed, Smith BJ, ed, Humana Press, Totowa, NJ, 2003, pp 385–98.

Copley RR, Russell RB: Getting the most from your protein sequence, in *Methods in Molecular Biology*, Vol 211: *Protein Sequencing Protocols*, 2nd ed, Smith BJ, ed, Humana Press, Totowa, NJ, 2003, pp 411–30.

Shannon JD, Fox JW: Identification of phosphorylation sites by edman degradation, *Tech Protein Chem* VI:117–23, 1995.

Edman P: Method for determination of the amino acid sequence in peptides, *Acta Chem Scand* 4:283–93, 1950.

Burke DT, Carle GF, Olson MV: Cloning of large segments of exogenous DNA into yeast by means of artificial chromosome vectors, *Science* 236:806–12, 1987.

Cohen SN, Chang ACY, Boyer HW, Helling RB: Construction of biologically functional bacterial plasmids in vitro, *Proc Natl Acad Sci USA* 70:3240–4, 1973.

Nathans D, Smith HO: Restriction endonucleases in the analysis and restructuring of DNA molecules, *Annu Rev Biochem* 44:273–93, 1975.

Saiki RK, Gelfand DH, Stoffel S, Scharf SJ, Higuchi R, Horn GT et al: Primer-directed enzymatic amplification of DNA with a thermostable DNA polymerase, *Science* 239:487–91, 1988.

Sanger F, Nicklen S, Coulson AR: DNA sequencing with chain-terminating inhibitors, *Proc Natl Acad Sci USA* 74:5463–7, 1977.

Benton WD, Davis RW: Screening lambdagt recombinant clones by hybridization to single plaques in situ, *Science* 196:180–2, 1977.

Broome S, Gilbert W: Immunological screening method to detect specific translation products. *Proc Natl Acad Sci USA* 75:2746–9, 1978.

Brown PO, Botstein D: Exploring the new world of the genome with DNA microarrays, *Nat Genet* 21(Suppl 1):33–7, 1999.

Gerhold DL, Jensen RV, Gullans SR: Better therapeutics through microarrays, *Nat Genet* 32(Suppl):547–51, 2002.

Grunstein M, Hogness DS: Colony hybridization: A method for the isolation of cloned DNAs that contain a specific gene, *Proc Natl Acad Sci USA* 72:3961–5, 1975.

Maniatis T, Hardison RC, Lacy E, Lauer J, O'Connell C, et al: The isolation of structural genes from libraries of eucaryotic DNA, *Cell* 15:687–701, 1978.

Southern EM: Detection of specific sequences among DNA fragments separated by gel electrophoresis, *J Mol Biol* 98:503–17, 1975.

Schildkraut CL, Marmur J, Doty P: The formation of hybrid DNA molecules and their use in studies of DNA homologies, *J Mol Biol* 3:595–617, 1961.

Ruitenberg MJ, Eggers R, Boer GJ, Verhaagen J: Adeno-associated viral vectors as agents for gene delivery: application in disorders and trauma of the central nervous system, *Methods* 28:182–194, 2002.

Chan L, Fujimiya M, Kojima H: In vivo gene therapy for diabetes mellitus, *Trends Mol Med* 9:430–5, 2003.

Zhdanov R, Bogdanenko E, Moskovtsev A, Podobed O, Duzgunes N: Liposome-mediated gene delivery: Dependence on lipid structure, glycolipid-mediated targeting, and immunological properties, *Meth Enzymol* 373:433–65, 2003.

Bessis N, Doucet C, Cottard V, Douar AM, Firat H et al: Gene therapy for rheumatoid arthritis, *J Gene Med* 4:581–91, 2002.

Botstein D, Shortle D: Strategies and applications of in vitro mutagenesis, *Science* 229:1193–201, 1985.

Herskowitz I: Functional inactivation of genes by dominant negative mutations, *Nature* 329:219–22, 1987.

Robertson E, Bradley A, Kuehn M, Evans M: Germ-line transmission of genes introduced into cultured pluripotential cells by retroviral vector, *Nature* 323:445–8, 1986.

Jaenisch R: Transgenic animals, *Science* 240:1468–74, 1988.

10.3. Homologous Recombination

Bronson SK, Smithies O: Altering mice by homologous recombination using embryonic stem cells, *J Biol Chem* 269:27155–8, 1994.

Capecchi MR: Altering the genome by homologous recombination, *Science* 244:1288–92, 1989.

Gordon JW, Ruddle FH: Integration and stable germ line transmission of genes injected into mouse pronuclei, *Science* 214:1244–6, 1981.

Durai S, Mani M, Kandavelou K, Wu J, Porteus MH et al: Zinc finger nucleases: Custom-designed molecular scissors for genome engineering of plant and mammalian cells, *Nucleic Acids Res* 33:5978–90, 2005.

Gommans WM, Haisma HJ, Rots MG: Engineering zinc finger protein transcription factors: The therapeutic relevance of switching endogenous gene expression on or off at command, *J Mol Biol* 354:507–19, 2005.

Durai S, Mani M, Kandavelou K, Wu J, Porteus MH et al: Zinc finger nucleases: Custom-designed molecular scissors for genome engineering of plant and mammalian cells, *Nucleic Acids Res* 33:5978–90, 2005.

Vasileva A, Jessberger R. Precise hit: Adeno-associated virus in gene targeting, *Nat Rev Microbiol* 3:837–47, 2005.

Aplan PD: Causes of oncogenic chromosomal translocation, *Trends Genet* 22:46–55, 2006.

Glaser S, Anastassiadis K, Stewart AF: Current issues in mouse genome engineering, *Nat Genet* 37:1187–93, 2005.

Coghlan A, Eichler EE, Oliver SG, Paterson AH, Stein L: Chromosome evolution in eukaryotes: A multi-kingdom perspective, *Trends Genet* 21:673–82, 2005.

Porteus MH, Carroll D: Gene targeting using zinc finger nucleases, *Nat Biotechnol* 23:967–73, 2005.

Gerton JL, Hawley RS: Homologous chromosome interactions in meiosis: Diversity amidst conservation, *Nat Rev Genet* 6:477–87, 2005.

Klug A: Towards therapeutic applications of engineered zinc finger proteins, *FEBS Lett* 579:892–4, 2005.

Park Y, Gerson SL: DNA repair defects in stem cell function and aging, *Annu Rev Med* 56:495–508, 2005.

Pellegrini L, Venkitaraman A: Emerging functions of BRCA2 in DNA recombination, *Trends Biochem Sci* 29:310–6, 2004.

10.4. Antisense Oligonucleotide-Based Therapies

Izant JG, Weintraub H: Constitutive and conditional suppression of exogenous and endogenous genes by anti-sense RNA, *Science* 229:345–52, 1985.

Izant JG, Weintraub H: Inhibition of thymidine kinase gene expression by anti-sense RNA: A molecular approach to genetic analysis, *Cell* 36:1007–15, 1984.

Wagner RW: Gene inhibition using antisense oligodeoxynucleotides, *Nature* 372:333–5, 1994.

Poulaki V, Mitsiades CS, Kotoula V, Tseleni-Balafouta S, Ashkenazi A et al: Regulation of Apo2L/ tumor necrosis factor-related apoptosis-inducing ligand-induced apoptosis in thyroid carcinoma cells, *Am J Pathol* 161:643–54, 2002.

Kawai H, Sato W, Yuzawa Y, Kosugi T, Matsuo S et al: Lack of the growth factor midkine enhances survival against cisplatin-induced renal damage, *Am J Pathol* 165(5):1603–12, Nov 2004.

10.5. Small Interfering RNA-Based Therapies

Hutvagner G, McLachlan J, Pasquinelli AE, Balint E, Tuschl T et al: A cellular function for the RNA-interference enzyme Dicer in the maturation of the let-7 small temporal RNA, *Science* 293:834–8, 2001.

Hutvagner G, Zamore PD: RNAi: Nature abhors a double-strand, *Curr Opin Genet Dev* 12:225–32, 2002.

Novina CD, Sharp PA: The RNAi revolution, *Nature* 430:161–4, 2004.

Sharp PA: RNA interference—2001, *Genes Dev* 15:485–90, 2001.

Hutvagner G, Zamore PD: A microRNA in a multiple-turnover RNAi enzyme complex, *Science* 297:2056–60, 2002.

Napoli C, Lemieux C, Jorgensen R: Introduction of a chalcone synthase gene into Petunia results in reversible co-suppression of homologous genes in trans, *Plant Cell* 2:279–89, 1990.

Dehio C, Schell J: Identification of plant genetic loci involved in a post transcriptional mechanism for meiotically reversible transgene silencing, *Proc Acad Sci USA* 91:5538–42, 1994.

Fire A, Xu S, Montgomery MK, Kostas SA, Driver SE, Mello CC: Potent and specific genetic interference by double–stranded RNA in Caenorhabditis elegans, *Nature* 391:806–11, 1998.

Aronin N: Target selectivity in mRNA silencing, *Gene Ther* 13:509–16, 2006.

Rodriguez-Lebron E, Paulson HL: Allele-specific RNA interference for neurological disease, *Gene Ther* 13:576–81, 2006.

Bonini NM, La Spada AR: Silencing polyglutamine degeneration with RNAi, *Neuron* 48:715–8, 2005.

Uprichard SL: The therapeutic potential of RNA interference, *FEBS Lett* 579:5996–6007, 2005.

Stecca B, Mas C, Ruiz i Altaba A: Interference with HH-GLI signaling inhibits prostate cancer, *Trends Mol Med* 11199–203, 2005.

Bennink JR, Palmore TN: The promise of siRNAs for the treatment of influenza, *Trends Mol Med* 10:571–4, 2004.

Ryther RC, Flynt AS, Phillips JA 3rd, Patton JG: siRNA therapeutics: Big potential from small RNAs, *Gene Ther* 12:5–11, 2005.

Hannon GJ, Rossi JJ: Unlocking the potential of the human genome with RNA interference, *Nature* 431:371–8, 2004.

Caplen NJ: Gene therapy progress and prospects. Downregulating gene expression: The impact of RNA interference, *Gene Ther* 11:1241–8, 2004.

Stevenson M: Dissecting HIV-1 through RNA interference, *Nat Rev Immunol* 3:851–8, 2003.

11

CELL AND TISSUE REGENERATIVE ENGINEERING

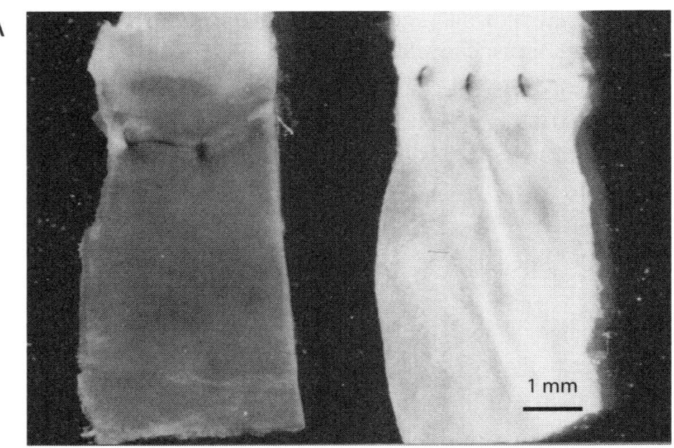

Vein graft cells transfected with a β-galactosidase (β-gal) gene vector by the mediation of electroporation. The left jugular vein of a rat was harvested, incubated in culture medium supplemented with 5 μg β-gal gene vector/100 μL, and subjected to electroporation. The transfected vein graft was incubated in culture medium for 30 min and grafted into the abdominal aorta of the rat. At day 10 after the grafting surgery, vein graft specimens were collected, fixed in 4% formaldehyde, incubated in the presence of X-gal, and observed by using an optical microscope. The blue color is indicative of the expression of the β-gal gene. (A) Vein graft specimens with (left) and without (right) β-gal gene transfection, showing the luminal surface. (B) Transverse section of a vein graft transfected with the β-gal gene. Dark blue: β-gal. Light blue: cell nuclei labeled with Hoechst 33258. Yellow: smooth muscle α-actin labeled with an anti-smooth muscle α-actin antibody. (C) En face observation of β-gal-positive cells (blue) on the luminal surface of a vein graft specimen transfected with the β-gal gene. See color insert.

Bioregenerative Engineering: Principles and Applications, by Shu Q. Liu

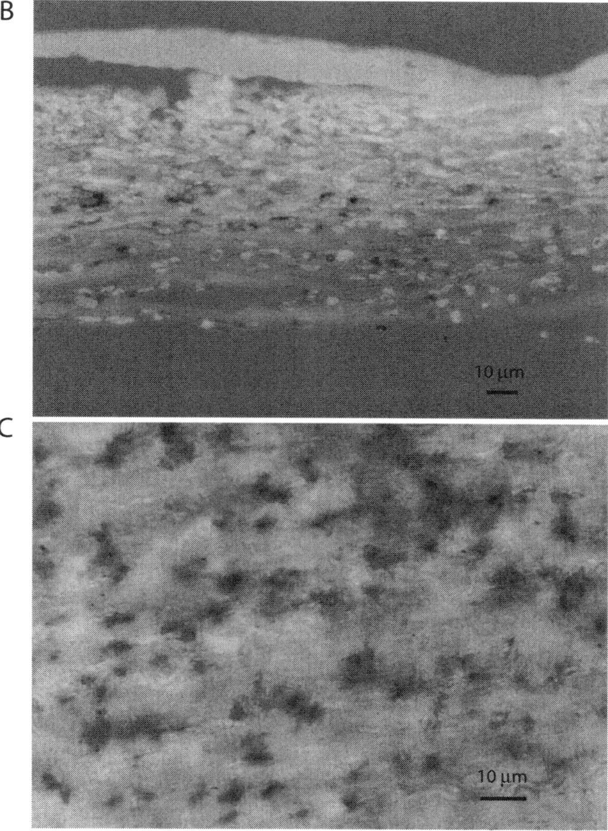

Continued

An important aspect of bioregenerative engineering is to modulate, regenerate, repair, or replace disordered cells and tissues. The work at the cellular and tissue level may be considered cell and tissue regenerative engineering, respectively. However, it is often difficult to distinguish the level of the engineering work. In fact, for any given disorder, it is necessary to carry out engineering manipulations at all levels, including the cellular and tissue levels as well as the molecular level as discussed in Chapter 10. The classification of bioregenerative engineering into different levels is mostly for the convenience of description and text organization. This chapter focuses on the cellular and tissue aspects of bioregenerative engineering.

CELL REGENERATIVE ENGINEERING

Cell regenerative engineering is to regenerate, repair, or replace injured, disordered, and lost cells with functional somatic cells of the same type or stem cells that can differentiate into a desired cell type. For the majority of tissues and organs, such as the skin, intestine, stomach, blood vessels, liver, and lung, certain types of cells including stem and progenitor cells are capable of proliferating and differentiating to replace injured and lost cells to a certain extent. However, when a pathogen is too strong and an injury is too severe, a tissue or organ may not be able to completely self-repair and regenerate. Furthermore, certain types of tissues and organs, such as the brain, spinal cord, and heart, have little capability of self-renewing and regeneration. For these systems, it is necessary to enhance the repair and regeneration processes by using engineering approaches. Cell regenerative engineering is established to achieve such a goal. To successfully conduct cell regenerative engineering, it is necessary to understand the function, organization, and development of cells and tissues, to identify and cultivate therapeutic cell lines, and to establish technologies for cell manipulation, transplantation, and functional tests.

Candidate Cell Types for Cell Regenerative Engineering [11.1]

Candidate cell types for cell regenerative engineering include somatic, stem, and progenitor cells. *Somatic cells* are defined as cells other than germ cells, including the egg and sperm, in a mature animal. Most somatic cells are specified cells that constitute tissues and organs. These cells have limited capability of proliferation, differentiation, and regeneration. Certain types of somatic cell, such as hepatocytes, smooth muscle cells, and epithelial cells, are capable of differentiating and/or proliferating within specified tissues. These cells may be used for cell regenerative engineering. For instance, hepatocytes can be used for liver repair and regeneration. Smooth muscle cells may be used for the construction of artificial blood vessels. Other types of somatic cell, such as the neurons and cardiomyocytes, exhibit minimal capacity for differentiation and proliferation. These cells are not suitable for cellular engineering. Overall, somatic cells are not considered preferred candidates for cell regenerative engineering.

Stem cells are undifferentiated cells that are capable of self-renewing and differentiating into specialized cells types. Thus, stem cells are ideal cells for the repair and regeneration of disordered or lost cells. Stem cells can be classified into three types, based on the stage of development: embryonic, fetal, and adult stem cells. *Embryonic stem cells* are cells derived primarily from an early embryonic structure, known as the inner cell mass of the blastocyst, and from embryonic germ cells. These cells are pluripotent cells that can develop into all specified cell types for peripheral tissues and organs.

Fetal stem cells are cells found in various tissues of the fetus and are committed to differentiation into specified cell types within a given tissue. Note that in humans the fetal stage is defined from the formation of the tissue and organ systems (at about the end of the second month) to the birth (at about the end of the ninth month) and the period before the fetal stage is known as the *embryonic stage*. Because of ethical concerns, fetal stem cells have not been extensively used for regenerative medicine and engineering.

Adult stem cells are committed stem cells found in a mature tissue or organ. These cells are capable of renewing and differentiating into specialized cells. Adult stem cells have been found in a number of tissues, including the bone marrow, blood, brain, liver, skin, intestine, stomach, and pancreas. While certain adult stem cells from a given tissue may be multipotent, that is, capable of differentiating into cells for a different type of tissue, most adult stem cells are committed to give rise to specialized cells in the same type of tissue. Compared to embryonic stem cells, adult stem cells are scarce and difficult to identify. Nevertheless, because of ethical concerns about using embryonic and fetal stem cells, adult stem cells are still valuable candidates for cell regenerative engineering. Detailed descriptions for stem cells are presented on page 381.

Cell Expansion

Cell expansion via cell culture is a critical step for cell regenerative engineering. This is specially important for the preparation of stem cells. It is often difficult to collect a sufficient number of stem cells for cell repair and regeneration. For instance, each blastocyst can only provide about 30 stem cells. Stem cells are also scarce in adult tissues and organs. Thus, it is necessary to culture and expand stem cells before they are used for cellular engineering. For the three types of stem cell, including embryonic, fetal, and adult stem cells, embryonic stem cells from the human and mouse have been successfully cultured. Under an appropriate culture environment, cultured embryonic stem cells are able to grow for more than 2 years with a stable complement of chromosomes. Culture conditions may significantly influence the renewal and differentiation of the embryonic stem cells. For instance, to maintain the undifferentiated state of the embryonic stem cells, it is necessary to provide a layer of embryonic fibroblasts as feeder cells for the stem cells. When cultured in suspension without a feeder layer, the embryonic stem cells form aggregates with various cell types resembling those derived from the ectoderm, mesoderm, and endoderm.

In contrast to embryonic and fetal stem cells, adult stem cells are difficult to identify, because of their scarcity and the lack of identification markers. Adult stem cells are often dispersed in tissues. The purification of adult stem cells is a major challenge in cell regenerative engineering. Furthermore, it is difficult to maintain the undifferentiated state and to expand the number of adult stem cells in culture. These technical difficulties should be resolved before adult stem cells can be used for therapeutic purposes.

Genetic Modulation of Cells

It is often desired to have cells expressing an augmented phenotype that enhances the repair and regeneration of disordered and lost cells. Genetic modulation is an effective approach for achieving such a goal. There are two potential approaches that can be used to modulate the phenotypes of candidate cells for cell regenerative engineering: (1) enhancing or reducing the expression of desired proteins and (2) generating stem cell-like cells through somatic nuclear transfer.

As discussed in Chapter 10, the expression of a desired protein can be achieved by transferring the gene that encodes the protein. This approach can be potentially used to treat pathogenic disorders due to protein deficiency. In fact, protein deficiency is a cause for a large number of diseases. For instance, the deficiency of brain-derived growth factor in the brain is known as a cause for Alzheimer's disease. The lack of insulin causes diabetes. A reduction in elastin contributes to the development of intimal hyperplasia. By applying the principles of molecular regenerative engineering, it is possible to construct a recombinant gene encoding a desired protein and transfer it into candidate stem or progenitor cells to enhance the expression of the selected gene. Such an approach may be used to treat disorders due to protein deficiency. Furthermore, with the understanding of the regulatory mechanisms of stem cell differentiation, necessary genes can be transferred to control the differentiation of stem cells into desired cell types.

Somatic nuclear transfer is an approach established to generate cells with stem cell features in vitro by transferring a somatic cell nucleus from a patient into an embryonic stem cell derived from a blastocyst. Because the somatic nucleus carries the genome of the patient, the derived stem cells are compatible to the patient in terms of histocompatibility factors, thus reducing immune rejections. The transferred stem cells may retain their pluripotent nature. This is a potential approach for the generation of therapeutic stem cells. It is important to note that, although some techniques used for such an approach are similar to those used in reproductive animal cloning, the goal is different. Animal cloning is to reproduce an animal that is identical to the nucleus donor in genotype. An enucleated egg is used for somatic nuclear injection. The egg is implanted into the uterus and allowed to develop into progeny. In contrast, somatic nuclear transfer is for the generation of stem cells used to repair or regenerate disordered or lost cells.

Cell Transplantation

After selected stem or progenitor cells are expanded to a sufficient number, the cells can be delivered to a target tissue to replace disordered somatic cells. There are several approaches for cell delivery, including direct injection, implantation of polymer capsules with enclosed cells, and implantation of polymer scaffolds with seeded cells. The choice of delivery methods is dependent on the anatomy and function of the target tissue and organ. For cell delivery into a tissue that is difficult to access and is enclosed within a tight space, such as the brain, direct cell injection may be the method of choice. For cell delivery into the abdominal organs, such as the liver and pancreas, which reside within a relatively large space and are easy to access, scaffold implantation may be an effective approach. If immune rejection is a problem, cells may be enclosed within capsules made of porous membranes that prevent the infiltration of immune cells, but allow the transport of oxygen and nutrients into the capsules and produced proteins by the therapeutic cells out of the capsules. General criteria for cell transplantation are that the approach should not induce injury of the transplanted cells and the target tissue, and can be used to effectively deliver cells into the target tissue.

Identification of Transplanted Cells

Following cell transplantation, an important step is to identify the transplanted cells, ensuring the presence of the cells in the target tissue. The transplanted cells can be identified based on markers specific to the cells of interest. Certain cell types express protein markers unique to the cells themselves. For instance, CD34, c-Kit, Sca-1, and Thy1.1 are expressed in bone marrow-derived hematopoietic stem cells. Antibodies to these proteins can be established and used to identify hematopoietic stem cells. It is important to note that the expression of stem cell markers gradually diminishes when the stem cells are specified and differentiated. The stem cell markers may only be used within a short period after cell transplantation. Protein markers for long-term expression can be established by transfecting cells with reporter genes, such as the β-galactosidase gene, luciferase gene, CAT gene, and green or red fluorescent protein genes. These genes are not expressed in mammalian cells, and the proteins encoded by these genes can be detected for assessing cell transplantation. Methods for gene transfer are discussed on page 436.

In addition, cell morphological parameters, such as the shape and the intracellular structure of the implanted cells, the overall organization of implanted cells, and the relationship of implanted cells with neighbor cells and with extracellular matrix, can be measured and used to assess the transformation of stem cells to functional cells. For instance, the formation of cardiomyocytes can be judged by the appearance of sarcomeres and contractile filaments. Morphological information can be achieved by optical, fluorescence, and electron microscopy. Immunohistochemical methods can be used to identify specific structure and components the transplanted cells.

Functional Tests

In addition to morphological tests, another important aspect is the assessment of the function of transplanted cells. Various forms of test may be designed and performed, depending on the type and function of implanted cells. In general, functional tests may include biochemical, molecular, and physiological tests. For cells that produce necessary hormones and regulatory factors, such as hepatocytes and pancreatic β cells, biochemical assays may be designed and conducted to assess the level of a specified factor. For cells that generate forces, such as cardiac and skeletal muscle cells, mechanical tests should be carried out to assess the cell contractibility. Testing methods for specific tissues and organs are discussed in chapters addressing the organ systems through the book.

TISSUE REGENERATIVE ENGINEERING [11.2]

Tissue regenerative engineering is applied to regenerate, repair, or replace disordered and malfunctioned tissues or organs by establishing and using artificial tissue constructs based on biological or synthetic materials. Certain types of tissue construct may serve as tissue replacements that possess partial or complete function of the original tissues. These tissues are often supporting and mechanical structures. Examples of such constructs include artificial joints, bones, skin, blood vessels, and heart valves. Other types of tissue constructs may serve as scaffolds or frameworks that guide the regeneration of injured or lost tissues, which are composed of cells and glands responsible for producing and secreting hormones, enzymes, and necessary biochemical components. Examples of such constructs include liver and pancreatic scaffolds.

For most pathological disorders, tissue destruction occurs as a result of cell injury and death, which is associated with partial or complete loss of the function of the involved organ. Examples of such disorders include tissue infarction due to ischemia, severe viral and bacterial infections, and cirrhosis. When a tissue or organ system is unable to rebuild its supporting structure and framework, it is necessary to replace the malfunctioned tissue with an artificial tissue construct. For certain types of tissue, such as the hepatic, cardiac, pancreatic, and nervous tissues, it is necessary to incorporate functional cells into the tissue construct. Thus, tissue regenerative engineering is dependent on cell regenerative engineering. On the other hand, for some types of tissue, such as the bone, joint, blood vessels, and heart valves, it is not necessary to incorporate living cells into the tissue construct, although cell-based tissue constructs may improve the efficacy of tissue replacement.

On of the basis of the engineering approaches, tissue regenerative engineering can be classified into two categories: cell-based and cell-free tissue regenerative engineering. *Cell-based tissue regenerative engineering* involves the integration of living cells into artificial tissue constructs. *Cell-free tissue regenerative engineering* does not require the use of living cells for tissue construction. When implanted into a target tissue in vivo, host cells can migrate into the tissue constructs and gradually transform the implanted construct into a functional tissue. For either cell-based or cell-free type, several procedures are necessary for the successful replacement of a malfunctioned tissue. These include tissue construction, functional test of tissue constructs in vitro, construct implantation, cell viability test in vivo for cell-based tissue constructs, as well as morphological and functional tests of tissue constructs in vivo. These aspects are briefly discussed in this section.

Tissue Construction

The construction of a specified tissue is dependent on the structure and function of the target tissue. A general standard for tissue construction is that an artificial tissue should possess the structure and function of a natural tissue. The necessity of incorporating living cells into tissue constructs is dependent on the type of target tissue. For tissues that provide the organ system with mechanical support, protection, strength, and elasticity, such as the bone, joint, and blood vessel, cell-free tissue replacements can be constructed and used. The geometry and mechanical properties of these replacements are essential factors that should be taken into consideration during construction. In contrast, for tissues that produce hormones and biochemical factors (liver and pancreas), generate forces (heart and skeletal muscle), and process electrical and chemical signals (neurons), it is necessary to construct cell-based tissue replacements. For these tissue replacements, the survival of cells and the maintenance of cell phenotypes are important issues. Various approaches can be used to achieve these goals, depending on the type of cells. Theses approaches are described in chapters corresponding to different systems through the book (e.g., for nerve regenerative engineering, see Chapter 13).

Functional Tests of Tissue Replacement

A tissue replacement should be tested for functionality before it is used for repairing or replacing a target tissue. The form of functional tests is dependent on the type of tissues and required functionality. In general, several forms of test can be performed for functional assessment, including mechanical, contractility, and biochemical tests.

Mechanical tests are designed for tissue constructs that are used to replace structures for fluid conduction and mechanical support and performance, including blood vessels, cardiac valves, bones, and joints. Common mechanical tests include the assessment of mechanical properties and maximal strength. The mechanical properties of a material can be represented by the relationship between stress and strain. *Stress* is the force per unit area applied to the tissue construct or produced by cells implanted into the construct. *Strain* is deformation induced by the applied or produced force. At a given stress level, distinct strains indicate different levels of stiffness or compliance of the tested material. The stress–strain relationship can be described by mathematical expressions known as *constitutive equations*. These equations can be used to model the mechanical properties of tissue constructs. The coefficients of the constitutive equations can be used to represent the stiffness or compliance of the material for the tissue construct. The maximal strength of a tissue construct can be assessed by testing the yielding stress of the material used for the tissue construct. A yielding stress is the stress level at which the material breaks when subject to a continuous increasing force. The higher is the yielding stress, the stronger is the material.

Contractility tests are designed for assessing the function of contractile muscular tissues. Muscular tissue replacements can be constructed by integrating muscular stem or progenitor cells into engineered tissue scaffolds. The contractility of the constructed muscular tissue replacements can be assessed by measuring the forces generated by the integrated muscle cells and/or the deformation of the muscular replacements. The generated forces can be tested by using a force transducer and the deformation can be assessed by measuring changes in the dimensions of the muscular replacements. In practice, the contraction of the muscular cells can be induced by electrical or chemical stimulation. At a given level of stimulation, the higher is the generated force or deformation, the higher is the contractility.

Biochemical assays can be designed and conducted for testing the functions of constructed tissue replacements, such as the production of hormones, enzymes, and regulatory factors. Various assays may be established for such a purpose, depending on the functions of a specified cell type. For instance, for an artificially constructed pancreatic tissue replacement, it is necessary to test the level of insulin. For an engineered liver tissue, the level of albumin is a critical parameter indicative of the function of the tissue replacement. Various tests are described in chapters corresponding to the organ systems through the book.

Tissue Implantation

Following functional tests in vitro, a constructed tissue can be implanted into a target tissue, organ, or body cavity. Various strategies may be designed for tissue implantation, depending on the type and function of the target tissue. For tissues that provide mechanical support, strength, and function, such as the bone, joint, blood vessel, and cardiac valve, it is necessary to conduct site-specific tissue replacement, or replacing a target tissue at its original natural site. For tissues that produce hormones, enzymes, and regulatory molecules, such as the hepatic, pancreatic, and endocrine gland tissue, a tissue construct should be connected to the vascular system, so that produced hormones and biochemical factors can be directly released into the bloodstream. It is not necessary to implant these tissue constructs to the original site of the target tissue. Specific issues will be discussed in detail in chapters corresponding to various tissue and organ systems.

Morphological Tests of Implanted Tissue Constructs

After the implantation of a tissue construct, it is necessary to test its morphology and structural relationship to neighbor tissues, which provides essential information for assessing the performance and the integration of the implanted tissue construct to the host tissue. For the first step, the implanted tissue construct should be identified based on the structural properties of the material used for the tissue construct. For cell-based tissue constructs, the implanted cells can be identified by establishing specific markers, such as green fluorescent protein (GFP) and β-galactosidase as discussed on page 445. In the second step, the anatomy of the implanted tissue construct, the microstructure of implanted cells, and associated extracellular matrix should be examined. A histological approach can be used for assessing the global morphology of the tissue construct, while an electron microscopic approach can be used for assessing the microstructure of cells and extracellular matrix. In addition, the structure and organization of inflammatory tissues and newly generated blood vessels, in response to tissue implantation, should be assessed by using similar approaches.

Test of Cell Viability and Growth

For cell-based tissue constructs, it is necessary to test the viability and growth of the seeded cells. Tissue samples can be collected at specified times for such tests. Identification of the seeded cells can be based on specific markers generated by transfected genes, such GFP and β-galactosidase genes. Several approaches can be used for testing cell viability and growth, including the test of cell density, proliferation, and apoptosis. Cell density can be assessed by labeling and counting the nuclei of mononucleated cells per unit area. Cell nuclei can be labeled with DNA-specific fluorescent dyes, such as Hoechst 33258, and histological dyes, such as hematoxylin. Cell density is easy to measure and is a reliable index for assessing the cell viability. A cell proliferation test, such as the BrdU assay, can be conducted to assess whether the cells in the implanted tissue construct undergo cell division and growth. A cell apoptosis test, such as the TUNEL assay, can be used to estimate the rate of cell death. These methods are described on page 307.

Functional Tests for Implanted Tissue Constructs

The most important task of all is to test the function of the implanted tissue constructs. Various testing strategies can be designed and conducted, depending on the type and function of the target tissue. For tissues that provide mechanical performance and support, such as the bone and joint, the mechanical performance of the implanted tissue construct should be examined under a given condition at appropriate observation times following tissue implantation. For tissue constructs that conduct fluids, such as vascular substitutes, it is necessary to test the patency of the implanted vascular constructs. For tissue constructs that assist the contraction of the heart, it is important to examine the cardiac ejection function. For tissue constructs that generate hormone and regulatory factors, biochemical and molecular approaches should be used to detect the level of a specified hormone or factor.

BIBLIOGRAPHY

11.1. Cellular Regenerative Engineering

Caplan AI: Mesenchymal stem cells, *J Orth Res* 9:641–50, 1991.

Friedenstein AJ, Petrakova KV, Kurolesova AI et al: Heterotopic of bone marrow. Analysis of precursor cells for osteogenic and hematopoietic tissues, *Transplantation* 6:230–47, 1968.

Muraglia A, Cancedda R, Quarto R: Clonal mesenchymal progenitors from human bone marrow differentiate in vitro according to a hierarchical model, *J Cell Sci* 113:1161–6, 2000.

Quito FL, Beh J, Bashayan O et al: Effects of fibroblast growth factor-4 (k-FGF) on long-term cultures of human bone marrow cells, *Blood* 87:1282–91, 1996.

Martin I, Muraglia A, Campanile G et al: Fibroblast growth factor-2 supports ex vivo expansion and maintenance of osteogenic precursors from human bone marrow, *Endocrinology* 138:4456–62, 1997.

Walsh S, Jefferiss C, Stewart K et al: Expression of the developmental markers STRO-1 and alkaline phosphatase in cultures of human marrow stromal cells: Regulation by fibroblast growth factor (FGF)-2 and relationship to the expression of FGF receptors 1-4, *Bone* 27:185–95, 2000.

Rafil S, Lyden D: Therapeutic stem and progenitor cell transplantation for organ vascularization and regeneration, *Nature Med* 9:702–12, 2003.

Wilson A, Trumpp A: Bone-marrow haematopoietic-stem-cell niches, *Nat Rev Immunol* 6:93–106, 2006.

Brockes JP, Kumar A: Appendage regeneration in adult vertebrates and implications for regenerative medicine, *Science* 310:1919–23, 2005.

Wobus AM, Boheler KR: Embryonic stem cells: Prospects for developmental biology and cell therapy, *Physiol Rev* 85:635–78, 2005.

Heng BC, Cao T, Lee EH: Directing stem cell differentiation into the chondrogenic lineage in vitro, *Stem Cells* 22(7):1152–67, 2004.

Masson S, Harrison DJ, Plevris JN, Newsome PN: Potential of hematopoietic stem cell therapy in hepatology: A critical review, *Stem Cells* 22(6):897–907, 2004.

Halban PA: Cellular sources of new pancreatic beta cells and therapeutic implications for regenerative medicine, *Nat Cell Biol* 6(11):1021–5, Nov 2004.

Pomerantz J, Blau HM: Nuclear reprogramming: A key to stem cell function in regenerative medicine, *Nat Cell Biol* 6(9):810–6, Sept 2004.

Fuchs E, Tumbar T, Guasch G: Socializing with the neighbors: Stem cells and their niche, *Cell* 116:769–78, 2004.

Down JD, White-Scharf ME: Reprogramming immune responses: Enabling cellular therapies and regenerative medicine, *Stem Cells* 21:21–32, 2003.

Orlic D, Hill JM, Arai AE: Stem cells for myocardial regeneration, *Circ Res* 91:1092–102, 2002.

Reya T, Morrison SJ, Clarke MF, Weissman IL: Stem cells, cancer, and cancer stem cells, *Nature* 414:105–11, 2001.

Lagasse E, Shizuru JA, Uchida N, Tsukamoto A, Weissman IL: Toward regenerative medicine, *Immunity* 14(4):425–36, April 2001.

Agrawal S, Schaffer DV: In situ stem cell therapy: Novel targets, familiar challenges, *Trends Biotechnol* 23:78–83, 2005.

11.2. Tissue Regenerative Engineering

Maskarinec SA, Tirrell DA: Protein engineering approaches to biomaterials design, *Curr Opin Biotechnol* 16:422–6, 2005.

Kelm JM, Fussenegger M: Microscale tissue engineering using gravity-enforced cell assembly, *Trends Biotechnol* 22:195–202, 2004.

Allen JW, Bhatia SN: Engineering liver therapies for the future, *Tissue Eng* 8:725–37, 2002.

Layer PG, Robitzki A, Rothermel A, Willbold E: Of layers and spheres: The reaggregate approach in tissue engineering, *Trends Neurosci* 25:131–4, 2002.

Lysaght MJ, Reyes J: The growth of tissue engineering, *Tissue Eng* 7(5):485–93, Oct 2001.

Fernandez P, Daculsi R, Remy-Zolghadri M, Bareille R, Bordenave L: Review: Endothelial cells cultured on engineered vascular grafts are able to transduce shear stress, *Tissue Eng* 12:1–7, 2006.

Khademhosseini A, Langer R, Borenstein J, Vacanti JP: Microscale technologies for tissue engineering and biology, *Proc Natl Acad Sci USA* 103:2480–2487, 2006.

Falconnet D, Csucs G, Michelle Grandin H, Textor M: Surface engineering approaches to micropattern surfaces for cell-based assays, *Biomaterials* 27(16):3044–63, June 2006.

Isenberg BC, Williams C, Tranquillo RT: Small-diameter artificial arteries engineered in vitro, *Circ Res* 98:25–35, 2006.

Eschenhagen T, Zimmermann WH: Engineering myocardial tissue, *Circ Res* 97:1220–31, 2005.

Lee CC, MacKay JA, Frechet JM, Szoka FC: Designing dendrimers for biological applications, *Nat Biotechnol* 23:1517–26, 2005.

Stevens MM, George JH: Exploring and engineering the cell surface interface, *Science* 310:1135–8, 2005.

Woo SL, Abramowitch SD, Kilger R, Liang R: Biomechanics of knee ligaments: Injury, healing, and repair, *J Biomech* 39:1–20, 2006.

Riha GM, Lin PH, Lumsden AB, Yao Q, Chen C: Review: Application of stem cells for vascular tissue engineering, *Tissue Eng* 11:1535–52, 2005.

Lohfeld S, Barron V, McHugh PE: Biomodels of bone: A review, *Ann Biomed Eng* 33(10):1295–311, Oct 2005.

Vesely I: Heart valve tissue engineering, *Circ Res* 97(8):743–55, Oct 2005.

Cho CS, Seo SJ, Park IK, Kim SH, Kim TH et al: Galactose-carrying polymers as extracellular matrices for liver tissue engineering, *Biomaterials* 27:576–85, 2006.

Santerre JP, Woodhouse K, Laroche G, Labow RS: Understanding the biodegradation of polyurethanes: From classical implants to tissue engineering materials, *Biomaterials* 26:7457–70, 2005.

Martin Y, Vermette P: Bioreactors for tissue mass culture: Design, characterization, and recent advances, *Biomaterials* 26:7481–503, 2005.

Hollister SJ: Porous scaffold design for tissue engineering, *Nat Mater* 4:518–24, 2005.

Laflamme MA, Murry CE: Regenerating the heart, *Nat Biotechnol* 23:845–56, 2005.

Sales KM, Salacinski HJ, Alobaid N, Mikhail M, Balakrishnan V et al: Advancing vascular tissue engineering: The role of stem cell technology, *Trends Biotechnol* 23:461–7, 2005.

Stegemann JP, Hong H, Nerem RM: Mechanical, biochemical, and extracellular matrix effects on vascular smooth muscle cell phenotype, *J Appl Physiol* 98:2321–7, 2005.

Torsney E, Hu Y, Xu Q: Adventitial progenitor cells contribute to arteriosclerosis, *Trends Cardiovasc Med* 15(2):64–8, Feb 2005.

Brey EM, Uriel S, Greisler HP, McIntire LV: Therapeutic neovascularization: contributions from bioengineering, *Tissue Eng* 11:567–84, 2005.

Kakisis JD, Liapis CD, Breuer C, Sumpio BE: Artificial blood vessel: The Holy Grail of peripheral vascular surgery, *J Vasc Surg* 41:349–54, 2005.

Ma Z, Kotaki M, Inai R, Ramakrishna S: Potential of nanofiber matrix as tissue-engineering scaffolds, *Tissue Eng* 11:101–9, 2005.

Hoenig MR, Campbell GR, Rolfe BE, Campbell JH: Tissue-engineered blood vessels: Alternative to autologous grafts? *Arterioscler Thromb Vasc Biol* 25:1128–34, 2005.

Breuer CK, Mettler BA, Anthony T, Sales VL, Schoen FJ et al: Application of tissue-engineering principles toward the development of a semilunar heart valve substitute, *Tissue Eng* 10:1725–36, 2004.

Lutolf MP, Hubbell JA: Synthetic biomaterials as instructive extracellular microenvironments for morphogenesis in tissue engineering, *Nat Biotechnol* 23:47–55, 2005.

Vorp DA, Maul T, Nieponice A: Molecular aspects of vascular tissue engineering, *Front Biosci* 10:768–89, 2005.

Vunjak-Novakovic G, Altman G, Horan R, Kaplan DL: Tissue engineering of ligaments, *Annu Rev Biomed Eng* 6:131–56, 2004.

Patrick CW: Breast tissue engineering, *Annu Rev Biomed Eng* 6:109–30, 2004.

Cowin SC: Tissue growth and remodeling, *Annu Rev Biomed Eng* 6:77–107, 2004.

Ratner BD, Bryant SJ: Biomaterials: Where we have been and where we are going, *Annu Rev Biomed Eng* 6:41–75, 2004.

Atala A, Koh CJ: Tissue engineering applications of therapeutic cloning, *Annu Rev Biomed Eng* 6:27–40, 2004.

Vogel V, Baneyx G: The tissue engineeting puzzle: A molecular perspective, *Annu Rev Biomed Eng* 5:441–63, 2003.

Schmidt CE, Leach JB: Neural tissue engineering: Strategies for repair and regeneration, *Annu Rev Biomed Eng* 5:293–347, 2003.

Tzanakakis ES, Hess DJ, Sielaff TD, Hu WS: Extracorporeal tissue engineered liver-assist devices, *Annu Rev Biomed Eng* 2:607–32, 2000.

Koffas M, Roberge C, Lee K, Stephanopoulos G: Metabolic engineering, *Annu Rev Biomed Eng* 1:535–57, 1999.

Chaikof EL: Engineering and material considerations in islet cell transplantation, *Annu Rev Biomed Eng* 1:103–27, 1999.

Laurencin CT, Ambrosio AM, Borden MD, Cooper JA Jr: Tissue engineering: Orthopedic applications, *Annu Rev Biomed Eng* 1:19–46, 1999.

Zandstra PW, Nagy A: Stem cell bioengineering, *Annu Rev Biomed Eng* 3:275–305, 2001.

Nerem RM, Seliktar D: Vascular tissue engineering, *Annu Rev Biomed Eng* 3:225–43, 2001.

Liu SQ: Prevention of focal intimal hyperplasia in rat vein grafts by using a tissue engineering approach, *Atherosclerosis* 140:365–377, 1998.

Liu SQ: Biomechanical basis of vascular tissue engineering, *Crit Rev Biomed Eng* 27:75–148, 1999.

12

BIOMATERIAL ASPECTS OF BIOREGENERATIVE ENGINEERING

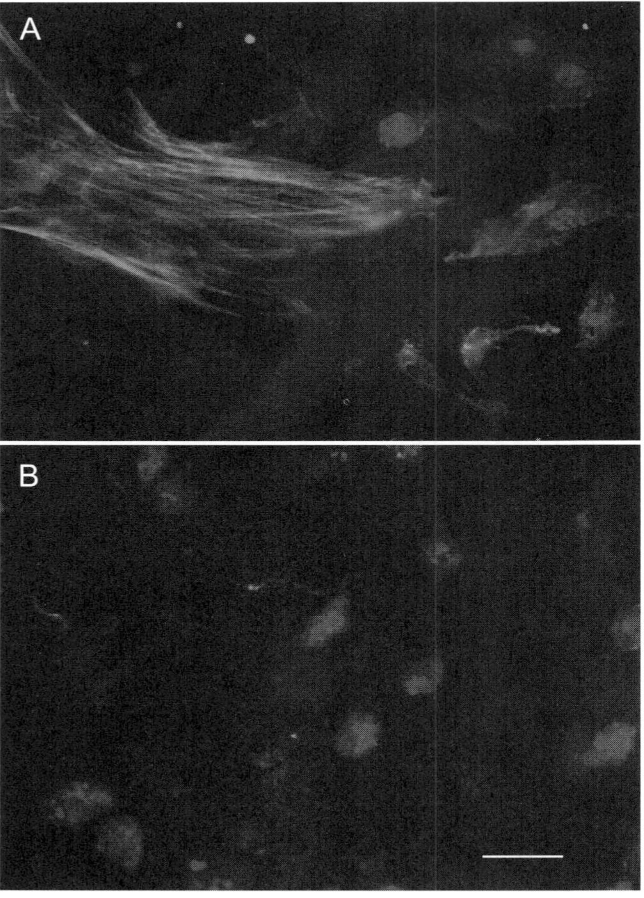

Influence of extracellular matrix on the formation of smooth muscle α-actin filaments in CD34-positive bone marrow cells in the mouse. CD34-positive bone marrow cells form smooth muscle α actin filaments (green in color) when cultured on the arterial elastic lamina matrix (A), but not on the arterial collagen (type III) matrix (B). Blue: cell nuclei. Scale: 100 μm. See color insert.

Bioregenerative Engineering: Principles and Applications, by Shu Q. Liu

Biomaterials are natural or synthetic materials that can be used to replace partially or completely a disordered tissue and thus to restore or improve the function of the disordered tissue. Several types of material, including synthetic polymers, metals, and ceramic materials, have been developed and used as biomaterials. In addition, natural polymeric materials generated by biological systems, such as collagen fibers, elastic fibers and laminae, and polysaccharides, have been used for the repair, replacement, and regeneration of disordered tissues. These materials are referred to as *biological materials*. Both biomaterials and biological materials can be used for constructing tissue scaffolds, which can be used for facilitating the regeneration of disordered tissues.

Biomaterials and biological materials have been used in a number organ systems for regenerative engineering purposes, including the skeletal, cardiovascular, gastrointestinal, epidermal, and urinary systems. Successful examples include replacement of hip joints and bones with metallic prostheses, replacement of blood vessels with biological and polymeric materials, replacement of the heart valves with biological and metallic materials, repair and regeneration of skin with biological materials, replacement of the heart with artificial cardiac pumps, argumentation of the respiratory function with polymeric artificial lungs, and improvement of the renal function with kidney dialysis devices. With the development of new technologies in biomaterial synthesis and analysis, it is expected that new biomaterials will be developed with improved performance and biocompatibility. In this chapter, several fundamental concepts, including the types, properties, and biocompatibility of biomaterials, are briefly reviewed.

SYNTHETIC POLYMERS AS BIOMATERIALS [12.1]

A synthetic polymer is a long-chain compound derived by bonding together many single-unit molecules. The single-unit molecules that constitute a polymer are known as *monomers*. Examples of synthetic polymers include nonbiodegradable polymers, such as polyethylene, poly(vinyl chloride), Dacron, nylon, and Teflon, and biodegradable polymers, such as polyglycolides, polylactides, polyanhydrides, and polysaccharides. Synthetic polymers possess unique features suitable for biological applications. These materials can be modulated to change their chemical and mechanical properties, and can be tailored into various shapes. In addition, these materials are light, strong, and inert. Thus, synthetic polymeric materials have been widely used for the repair, replacement, and regeneration of injured and disordered tissues. In this section, the classification, structure, and properties of synthetic polymers as well as their application to tissue engineering are discussed.

Classification

Polymers can be classified into two groups according to on the mechanisms of synthesis: addition polymers and condensation polymers. An *addition polymer* is synthesized by connecting monomer units via rearranging chemical bonds. The formation of poly(vinyl chloride) is a typical example of addition polymer. In this case, the double bond of a single vinyl chloride opens up in the presence of an initiator such as a peroxide molecule. The initiator can be activated by increasing temperature or exposure to ultraviolet. The initiator can activate monomers to form free bonds, which connect the monomers together. This process is known as propagation. Common addition polymers include polyethylene, polypropylene, polystyrene, and poly(vinyl chloride).

A *condensation polymer* is synthesized by joining two molecules together with the elimination of a molecule such as water and methanol. Two different types of monomers are usually participating in the polymerization reaction. The resulting product is known as copolymer. Nylon is a typical copolymer. During the polymerization process of nylon, a dicarboxylic acid molecule reacts with a diamine. A water molecule is removed when the two molecules are bonded. Examples of condensation polymers include polysaccharides, proteins, polyesters, polyamides, polyurea, polyurethane, and cellulose.

Based on the degradability of polymeric materials in a biological system, polymers can be classified into non-biodegradable and biodegradable polymers. Nonbiodegradable polymers cannot be degraded, whereas biodegradable polymers can be degraded in a biological system. Examples of nonbiodegradable polymers used as biomaterials include polyethylene, polytetrafluoroethylene, poly(vinyl chloride), and polypropylene. Examples of biodegradable polymers include polyglycolides, polylactides, polysaccharides, and poly(α-hydroxyl acids).

In terms of the type of monomer in a polymer structure, polymers can be classified into homopolymers and heteropolymers (or copolymers). Homopolymers are compounds constituted with one type of repeated monomers. Examples of homopolymers include polyethylene, polytetrafluoroethylene, poly(vinyl chloride), and polypropylene. Copolymers are polymeric compounds composed of two or more types of monomer. The different monomers may be randomly distributed or may alternate in a pattern. Examples of copolymers include poly(glycolide lactide), polyurethane, and poly(glycolide trimethylene carbonate).

General Properties

Polymeric materials exist in several forms, including liquid, elastomer, and plastic. The form of polymeric materials is determined by a number of factors, including the nature of monomers, the size and form of polymer molecules, the type and concentration of catalysts used for polymer synthesis, and curing temperature and duration. By selectively altering these factors, a desired form of polymeric materials can be generated.

An important factor that influences the structure and mechanical properties, such as flexibility and strength, of polymeric materials is the composition of polymer molecules. A substitution of a key atom in a monomer may induce a significant change in the mechanical properties. For instance, the replacement of the carbon atom with oxygen in a polyethylene molecule reduces the rigidity of the polymer.

The molecular size or polymeric chain length is another critical factor that determines the structure and mechanical properties of polymeric materials. Polymers are composed of various numbers of monomers. For the same type of monomer, a longer polymer is more flexible and tangled more easily than a shorter one. For instance, polyethylene and paraffin can be synthesized based on the same type of monomer CH_2CH_2, but with different chain length. The length of a paraffin molecule is much shorter than that of a polyethylene molecule. The long-chained polyethylene molecules are more difficult to crystallize, thus exhibiting higher flexibility, compared to paraffin. The chain length of polymers is one of the factors that determines the rigidity and strength of polymer materials. An increase in the chain length reduces the mobility of polymer molecules, and thus enhances the rigidity and strength.

The form of polymer molecules is a major factor that influences the structure and organization of polymeric materials. Polymer molecules exit in several forms: linear, branched, and crosslinked forms. A linear polymer molecule is composed of a single chain of monomers with various lengths and molecular weights. Examples of linear polymers include polyvinyls and polyesters. These molecules can be partially crystallized to form so-called semicrystalline polymers. It is usually difficult to completely crystallize these polymers.

A branched polymer molecule is composed of a mainchain and various densities of sidechains. Copolymerization may enhance the formation of branches. Since the sidebranches influence the interaction between the main polymer chains, the introduction of sidebranches may reduce polymer crystallization, yielding more flexible polymers. An increase in the length of the sidechain reduces the melting temperature of the polymer in association with a reduction in crystallization.

Polymer molecules can be crosslinked to form networks. Long-chain polymer molecules are usually linked together with sidechains. Crosslinked polymers are usually difficult to crystallize. An increase in the degree of crosslinking reduces the crystallization capability of polymeric materials. Up to a certain degree of crosslinking, polymer crystallization may be completely prevented. During the crosslinking process, the form of monomers and reaction conditions may influence the structure and mechanical properties of the polymeric material. For instance, the presence of tortuous and curved polymer chains may result in a polymeric material that is elastic and flexible. This type of material can be stretched to a large extent, and can return back to the undeformed length upon the release of the stretching force. Natural rubber (*cis*-polyisoprene) is a typical example of such materials.

In addition to the intrinsic factors of polymeric molecules, environmental factors influence the structure and mechanical properties of polymeric materials. A typical environmental factor is temperature. An increase in temperature often reduces the rigidity of polymeric materials and induces the transformation of polymer from plastic to elastomer and liquid.

Nonbiodegradable Polymers [12.2]

Polytetrafluoroethylene. Polytetrafluoroethylene (PTFE) is a fluorocarbon polymer, also known as Teflon, which is commonly used as a biomaterial for the replacement of soft tissues. Teflon is characterized by high crystallization (about 90% molecules crystallized), a relatively high density (~2 g/cm^3), and low surface tension and friction compared to other types of polymer. Teflon is used primarily for the construction of vascular substitutes. Teflon vascular grafts are sufficiently strong for withstanding stretching forces induced by arterial blood pressure. The mechanical properties of Teflon grafts can be stable for years following implantation. Because Teflon stimulates inflammatory reactions and thrombogenesis, which lead to the development of intimal hyperplasia, it can only be used for the replacement of arteries larger than 4 mm in diameter. Teflon can also be used to construct supporting sheaths for vein grafts, an approach for modulating the diameter of the vein graft, reducing diameter mismatch-induced disturbance of blood flow, and suppressing flow disturbance-induced intimal hyperplasia.

Poly(ethylene terephthalate). Poly(ethylene terephthalate) (PET), known as Dacron, is a polyester material that is characterized by hydrophobicity, resistance to hydrolysis, and high strength and toughness. Dacron has primarily been used for the construction of vascular grafts. Porous grafts can be constructed by weaving PET fibers into mesh-like materials. This type of graft usually facilitates cell integration into the graft wall when anastomosed into a host artery. Dacron grafts are often used to replace malfunctioned thoracic and abdominal aortae. As other polymeric materials, Dacron induces inflammatory reactions and thrombogenesis. Thus, this type of graft can only be applied to arteries with diameter exceeding 4 mm.

Polyethylene. Polyethylene is a polymer composed of ethylene monomers and can be synthesized into polymeric materials of various densities. A low-density polyethylene material (0.91–0.93 g/cm^3) can be synthesized by using peroxide catalysts at pressure 1000–3000 kg/cm^2 and temperature 300–500°C. Polyethylene materials generated under such conditions are composed of branched polymers that are difficult to crystallize. This type of material is tough and flexible, and is often used to fabricate thin membranes for food packaging and also for manufacturing biomedical supplies, such as tubing and containers.

A high-density polyethylene material (0.945–0.96 g/cm^3) can be synthesized by using metal catalysts at pressure ~10 kg/cm^2 and temperature 60–80°C. This polymer material is composed primarily linear polymers and is highly crystallized with strong bonds between ethylene monomers. The high-density polyethylene material is stronger than the low-density polyethylene material, and has been used for fabricating orthopedic implants, such as load-bearing caps for artificial joints.

Biodegradable Polymers [12.3]

A number of biodegradable polymeric materials have been synthesized and used as biomaterials. These materials belong to several polymer families, including linear aliphatic polyesters, polyorthoesters, polyphosphate esters, poly(ester–ether), polyanhydrides, polyamides, polysaccharides, polyamino acids, and inorganic polyphosphazenes. Some of these polymers have been increasingly used for the regeneration, repair, and replacement of tissues and organs. Because these materials can be gradually degraded and removed through various organ systems such as the liver and kidneys, harmful influences, if any, imposed by these materials can be eliminated. Furthermore, when used for constructing tissue scaffolds, biodegradable polymers with desired shapes can serve as a guidance for cell migration and pattern formation. With the degradation of the polymer scaffold, natural tissues can be gradually established and strengthened, eventually integrating into the host system.

Biodegradable polymeric materials have been applied to biomedical research in primarily two areas: wound healing and drug delivery. Linear aliphatic polyesters, such as polyglycolides, poly(glycolide-L-lactide), poly(ester–ether), and poly(glycolide–trimethylene carbonate), have been successfully used for enhancing wound closure and healing. These polymers have been extensively studied for their structure, material properties, and biological compatibility. Several types of biodegradable polymer have been used for constructing drug delivery carriers. Examples include polyanhydrides and poly(ester–ether). By controlling the rate of degradation, the rate of drug delivery can be regulated. In addition, biodegradable polymers have been investigated for their potential use in tissue regeneration and repair. For the last decade (since the mid-1990s), the synthesis and characterization of biodegradable polymers are among the most active research areas in biomedical engineering. Here, several common types of biodegradable polymers are discussed with a focus on the structure, material properties, and potential applications to bioregenerative engineering.

Linear Aliphatic Polyesters. *Linear aliphatic polyesters* are straight-chain polymers in which monomers are joined by the ester bond —COO—. Commonly used linear aliphatic polyesters include polyglycolide (PG) or poly(glycolic acid) (PGA), polylactide (PL) or poly(lactic acid) (PLA), polycaprolactone (PCL), poly(β-hydroxybutyrate), and poly(glycolide–trimethylene carbonate). These polymers have been successfully used in biomedical research and application. Among these aliphatic polymers, polyglycolide and polylactide are basic molecules that can be used to form copolymers and derive different forms of polymers. These polymers have been frequently used in biomedical research.

Polyglycolides and Polylactides [12.4]. *Polyglycolides* and *polylactides* can be synthesized from glycolic acids and lactic acids, respectively. The direct condensation method can be used to synthesize low-molecular-weight polyglycolides or polylactides (<3000 kDa). Common catalysts for such synthesis include phosphoric acids, *p*-toluene sulfonic acid, and antimony trifluoride. For a polymer larger than 3000 kDa, a process known as *ring-opening polymerization* is necessary for polymer synthesis. Several catalysts, such as stannous chloride dehydrate and aluminum alkoxide, has been used for the synthesis of large polyglycolides and polylactides.

The thermal properties of polymers are described by the melting and glass transition temperature, while the mechanical properties of polymers are expressed by the elastic modulus, tensile strength, and maximal distensibility or strain. A typical polyglycolide material has a melting temperature of ~210°C and a glass transition temperature of ~36°C, whereas a polylactide material possesses a melting temperature of ~170°C and a glass transition temperature of ~56°C. Mechanically, polylactide-based materials possess elastic moduli ranging from 1200 to 3000 MPa and tensile strength of 28–50 MPa depending on the molecular weight, and can be extended to 2–6% of their original length before reaching the breakpoint.

The rate of polymer biodegradation is an important parameter considered in the design and synthesis of biodegradable polymers. Polymers are usually degraded by hydrolysis of the ester bonds, resulting in a decrease in molecular weight. Several factors are known to influence the rate of polymer degradation. These include the chemical structure and molecular weight of the polymer material as well as environmental conditions. For instance, amorphous polymers can be degrade more easily than crystalline polymers. Polymers with a higher molecular weight may be degraded more slowly than those with a lower molecular weight. Branched polymer molecules may be degraded faster than linear molecules. An increase in temperature facilitates polymer degradation. For a typical semicrystalline polylactide material, weight loss can be detected after 30 weeks in a phosphate buffer at 37°C.

Glycolides and lactides can form copolymers with each other as well as with other types of esters. The copolymer formed on the basis of glycolides and L-lactides is known as poly(glycolide-L-lactide). The relative contents of various esters in a copolymer may influence the rate of polymer degradation. For instance, the degradation rate of poly(glycolide-L-lactide) is dependent on the relative concentration of lactides. It is interesting to note that the relationship is not linear. In the range of 0–25% of L-lactides, the rate of degradation is inversely proportional to the concentration of L-lactides, whereas in the range of 75–100% of L-lactides, the rate of degradation is directly proportional to the concentration of lactides. Interestingly, in the range of 25–75% of L-lactides, the rate of polymer degradation does not change significantly with an increase in the concentration of lactides. Thus, for the design of a copolymer, the degradation rate should be taken into account. Ideally, a polymer material should have a degradation rate that is comparable to the rate of native tissue formation.

Polyglycolides, polylactides, and their copolymers have been used for constructing various forms of matrix for biomedical applications such as drug delivery and tissue repair. These polymers can be also injected into target tissues. The biocompatibility and toxicity of these polymers have been tested extensively. The degradation product of polylactides is lactic acid, which is a natural metabolite in mammals and can be removed via physiological metabolism. The metabolite of polyglycolides, glycolic acid, has been shown to be a low-toxicity substance. An increase in acidity near a polymer implant may occur, but such a change can be mitigated by a local pH treatment. In general, these polymers exhibit low toxicity and are safe for in vivo implantation and injection.

Polycaprolactones [12.5]. Polycaprolactone (PCL) are another type of polyester that is used in biomedical research and application. Polycaprolactones can be synthesized by polymerization with anionic, cationic, and coordination catalysts. For the anionic catalytic system, tertiary amines, alkali metal alkoxides, and carboxylates are necessary for the polymerization. For the cationic polymerization system, several catalysts, including protic acids, Lewis acids, acylating agents, and alkylating agents, are often used. The coordination polymerization system is used for synthesizing high-molecular-weight polymers. Catalysts for this type of polymerization include stannous octoate, alikoxides, and metallic elements such as Al, Sn, Mg, and Ti. Polycaprolactone materials usually have the following thermal properties: melting temperature ~60°C and glass transition temperature ~−60°C. Mechanically, polycaprolactone materials can be stretched to a strain about 0.3 (~30% elongation) at a yielding stress about 11 MPa. The elastic modulus is about 0.3 GPa. Polycaprolactone materials and polycaprolactone hybrids with other materials, such as hydroxyapatite, poly-L-lactides, and silica, have been used in a number of biomedical applications, such as scaffolding and repairing soft and bone tissues (Fig. 12.1). Polycaprolactone materials can be degraded by hydrolysis. In vivo tests have shown that it takes about 2–4 years to completely degrade a polycaprolactone implant. Copolymerization or blending with glycolides and/or lactides increases the rate of degradation. In vivo animal tests have shown that polycaprolactone materials exhibit low toxicity and do not significantly influence the function of host cells and tissues.

Poly(β-hydroxybutyrate) [12.6]. Poly(β-hydroxybutyrate), or PHB, is a polymer of β-hydroxybutyrate. Poly(β-hydroxybutyrate) can be generated by natural fermentation of the bacterium *Alcaligenes* and can also be synthesized by the plant flax or linum usitatissimum, which is genetically transfected with gene constructs required for the synthesis of poly(β-hydroxybutyrate), or other types of plants such as *Arabidopsis thaliana* and *Brassica napus*. Poly(β-hydroxybutyrate) is an amorphous material with a glass transition temperature of about −40°C and a melting temperature of ~160°C. The failure strain (strain when a material is extended to the failure point) of poly(β-hydroxybutyrate) is about 0.15 or 15%. When poly(β-hydroxybutyrate) is blended with other types of polymers, such as hydroxyhexanoate, the failure strain can be increased depending on the relative contents of the polymer components. The poly(β-hydroxybutyrate)-containing copolymer can be fabricated into various forms and used for constructing tissue replacements (Fig. 12.2). The degradation of poly(β-hydroxybutyrate) is induced by hydrolysis. In vitro tests have shown that high-molecular-weight poly(β-hydroxybutyrate) films can be completely degraded at 25°C in freshwater within 3 weeks.

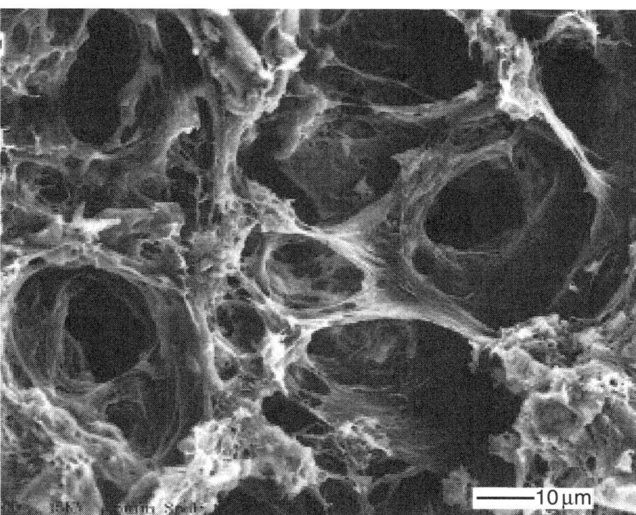

Figure 12.1. Scanning electron microscopic images of freeze fracture hydroxyapatite (HAP)/polycaprolactone (PCL) composite materials. (Reprinted with permission of John Wiley & Sons, Inc. from Verma D et al: Experimental investigation of interfaces in hydroxyapatite/polyacrylic acid/polycaprolactone composites using photoacoustic FTIR spectroscopy, *J Biomed Mater Res A* 77:59–66, copyright 2006.)

Polycarbonates [12.7]. *Polycarbonates* are a group of polyesters, including poly(urethane carbonate), poly(ethylene carbonate), and poly(propylene carbonate). These polymers are synthesized with dihydroxy compounds and carbonyl chloride. For aliphatic poly(urethane carbonate), the glass transition temperature is about −18°C. The tensile failure stress of this polymer is about 50 MPa, and the failure strain is about 416%. Clearly, this is a highly extendable material. The degradation rate varies among different polycarbonate materials. For instance, poly(ethylene carbonate) tablets implanted in the rat can be degraded within about 21 days, poly(propylene carbonate) may last for 60 days, whereas poly(urethane carbonate) is stable for a much longer time. This class of polymeric materials are biocompatible and has been used for cell culture (Fig. 12.3). These materials have also been used as biomaterials for constructing cardiovascular implants (intraaortic balloons, cardiac valves, vascular prostheses, pacemaker leads, ventricular assist devices and artificial heart diaphragms, heart valves, vascular grafts, and urethral catheters) as well as reconstructive implants (wound dressings, mammary prostheses, maxillofacial prostheses).

Polyamides [12.8]. A *polyamide* is a polymer that is formed by joining monomers with an amide bond —CONH—. Natural proteins are amide-based polymers. Amino acids can be used to synthesize artificial polyamides. Typical examples are polyglutamic acid and polylysine. Glutamic acid and lysine can also form copolymers with other types of amino acids. Unlike other synthetic polymers that have been tested in biomedical research, polymers based on amino acids are composed of naturally occurring molecules and thus possess low toxicity. Such a feature renders these polymers promising candidates as biomaterials for tissue repair and regeneration as well as drug delivery. Here, polyglutamic

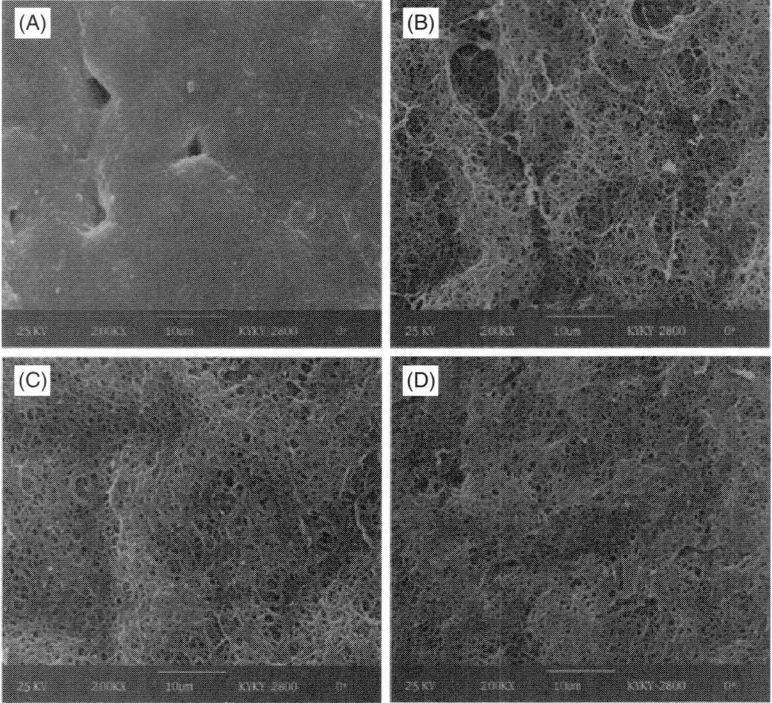

Figure 12.2. Scanning electron microscopic images of poly(3-hydroxybutyrate-*co*-3-hydroxyhexanoate)-based biomaterials: (A) poly(3-hydroxybutyrate-*co*-3-hydroxyhexanoate) materials; (B) poly(3-hydroxybutyrate-*co*-3-hydroxyhexanoate) with 5% gelatin blend; (C) poly(3-hydroxybutyrate-*co*-3-hydroxyhexanoate) with 10% gelatin blend; (D) poly(3-hydroxybutyrate-*co*-3-hydroxyhexanoate) with 30% gelatin blend. (Reprinted with permission of the American Chemical Society from Wang YW, Wu Q, Chen GQ: Gelatin blending improves the performance of poly(3-hydroxybutyrate-*co*-3-hydroxyhexanoate) films for biomedical application, *Biomacromolecules* 6:566–71, copyright 2005.)

acid is used as an example to demonstrate the principles of the synthesis, material properties, and potential biomedical application of polyamides.

Poly(glutamic acid) can be synthesized from poly(γ-benzyl-L-glutamate) by eliminating the benzyl group by using hydrogen bromide. Poly(glutamic acid) (PGA) can be degraded by enzymatic hydrolysis. In particular, cystein proteases play a critical role in the degradation of PGA. The time course of PGA degradation ranges from several hours to months, depending on the concentration of proteinase, temperature, and the composition of the polymer (with or without additional components). Poly(glutamic acid) exhibits little toxicity when implanted into an animal tissue. Animals can tolerate a single dose of $\leq$800 mg/kg and an accumulated dose of $\leq$1.8 g/kg. Polyglutamic acid exhibits little immunogenicity in animal models. This type of materials can be used in various biomedical applications, such as tissue repair and regeneration as well as drug delivery.

Polyphosphazenes [12.9]. Polyphosphazenes are inorganic biodegradable polymers that are constituted with a nitrogen–phosphorus (N=P) backbone. This type of polymer is

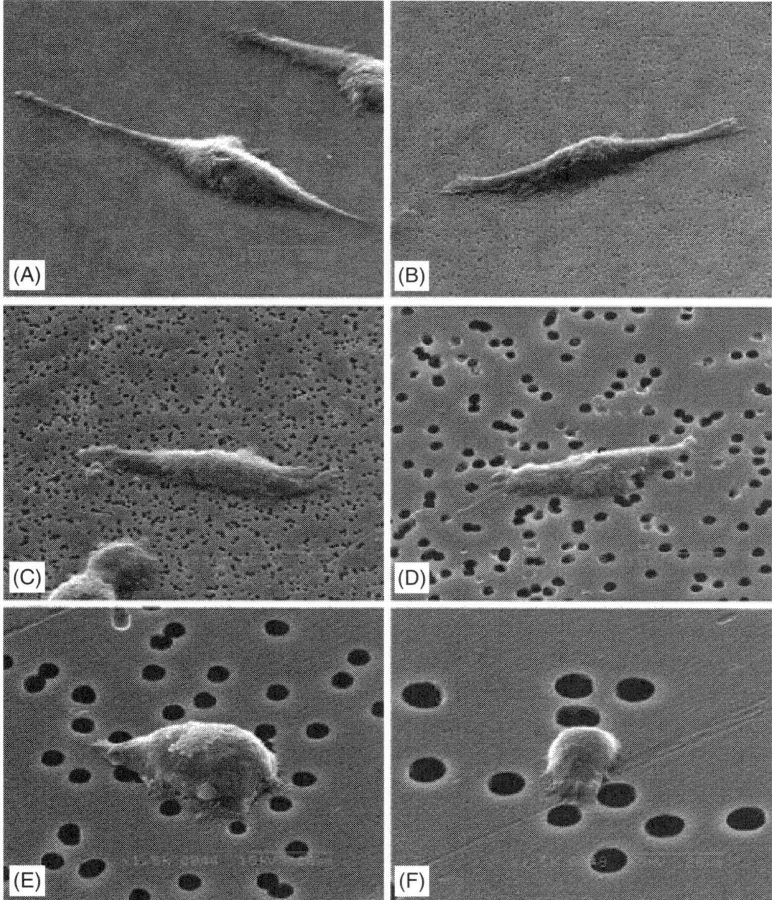

Figure 12.3. Scanning electron microscopic images of MG63 osteoblast-like cells (2-day culture) attached to the polycarbonate membrane surfaces with different micropore sizes: (A) 0.2, (B) 0.4, (C) 1.0, (D) 3.0, (E) 5.0, and (F) 8.0μm in diameter of micropore sizes. (Reprinted from Lee SJ et al: Response of MG63 osteoblast-like cells onto polycarbonate membrane surfaces with different micropore sizes, *Biomaterials* 25:4699–707, copyright 2004 with permission from Elsevier.)

synthesized via reactions of polydichlorophosphazene with amines or alkoxides in tetrahydrofuran or aromatic hydrocarbone solutions. Polymers with various sidegroups can be synthesized by mixing different components. The material properties of polyphosphazenes are dependent on the composition of the polymer. Biodegradable polyphosphazenes can be generated when amino acid derivatives are used as sidegroups. For instance, the ethylglycinato-derived polymers can be degraded into ammonia, phosphate, ethanol, and glycine. Experimental tests in vitro have demonstrated that amino acid-based polyphosphazenes can be degraded within several months. The rate of degradation is dependent on the type of amino acid selected for the sidegroups.

Polyphosphazenes have been used for constructing various structures, such as matrix scaffolds (Fig. 12.4), hydrogels, and microspheres, for drug delivery. The degradation of

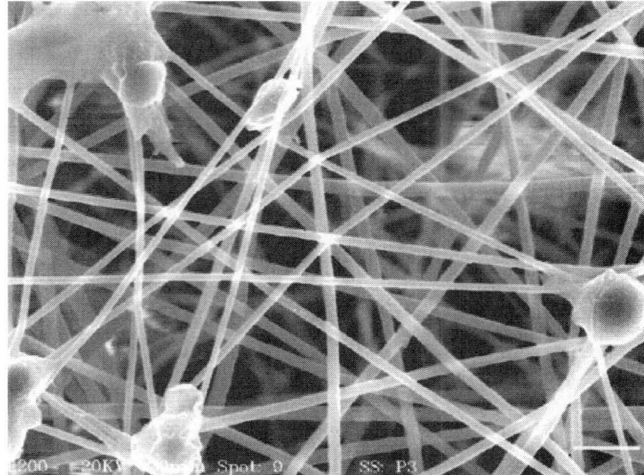

Figure 12.4. Scanning electron micrograph showing electrospun poly[bis(*p*-methylphenoxy) phosphazene fiber matrix with arterial endothelial cells after 24 h of culture. (Reprinted with permission of the American Chemical Society from Lakshmi S et al: Fabrication and optimization of methylphenoxy substituted polyphosphazene nanofibers for biomedical applications, *Biomacromolecules* 5:2212–20, copyright 2004.)

polyphosphazenes is sensitive to changes in temperature. Thus, the degradation rate of these polymers can be regulated by controlling environmental temperature. Such polymers can be used for constructing drug delivery carriers for temperature-related diseases. Other environmental factors, such as pH, may also influence the rate of degradation of polyphosphazenes. For instance, oxybenzoate-containing polyphosphazenes are pH-sensitive. The degradation of this type of polymer can be controlled by altering the content of oxybenzoate within a specified range of pH. Such polymers can be used to deliver drug for diseases that result in a change in pH. In vivo animal tests with subcutaneous implantation have shown that polyphosphazenes materials exhibit low toxicity, induce little inflammatory reactions in host tissues, and are relatively safe for implantation and drug delivery.

Polyanhydrides [12.10]. Polyanhydrides are polymers formed with anhydrides, compounds in which two carbonyl groups are joined with an oxygen atom, RCO—O—COR′, where R and R′ are any organic groups. The polymers are synthesized by reactions of diacids with anhydrides to form acetyl anhydride prepolymers. High-molecular-weight polyanhydrides can be formed from the prepolymers by melt condensation (180°C for 90 min in vacuo). The addition of coordination catalysts, such as cadmium acetate and metal oxides, can facilitate the polymerization process, increasing the molecular weight of the polymer. Polyanhydrides can be dissolved in organic solvents such as chloroform and dichloromethane. Various components can be copolymerized with anhydrides to alter the solubility. Homopolymers of anhydrides usually exhibit a high level of crystallinity. Copolymerization with different components may reduce the crystallinity, producing more amorphous materials. Copolymerization also influences the mechanical properties of polyanhydrides.

Polyanhydrides are degraded by hydrolytic erosion. A number of factors influence the rate of polyanhydrides degradation. These include pH and copolymerization with different compounds. An increase in pH facilitates polyanhydrides degradation. The incorporation of different aliphatic monomers may facilitate the degradation of the polymer, whereas the addition of methylene groups into the polymer backbone reduces the rate of degradation. Thus, polyanhydrides with various levels of degradation can be synthesized by copolymerization with various compounds. Polyanhydrides can be used to construct various forms of matrix, such as disks and pellets, and can also be injected into target tissues. The injection of mixed polyanhydrides and therapeutic substances is a promising technique for controlled drug delivery. A number of studies have shown that polyanhydrides do not significantly influence the growth of cultured cells and exhibit little toxicity when implanted into target tissues in animal tests.

BIOLOGICAL MATERIALS

Collagen Matrix [12.11]

Collagen matrix is found in mesenchymal and connective tissues, such as the subcutaneous tissue and bone, and the adventitia of tubular organs (blood vessels, airways, esophagus, stomach, and intestines). In mammalian tissues, there exist about 15 types of collagen matrix, namely, collagen types I–XV. Among these types of collagen, types I, II, III, IV, V, IX, XI, and XII are commonly found in connective tissues. Collagen types I, II, III, V, and XI are organized into filamentous structures, known as *collagen fibrils*, with a diameter of ~10–100 nm. These fibrils usually form large collagen bundles as found in subcutaneous tissues and the adventitia of tubular organs. Collagen types I and V are often found in the bone, skin, cornea, tendon, ligament, and internal organs such as the lung, liver, pancreas, and kidney. Collagen types II and XI are found in the cartilage, notochord, and intervertebral disks. Collagen type III is found in blood vessels, skin, and internal organs. Collagen types IX and XII are molecules that link other types of collagen fibril and are known as *fibril-associated collagens*. These types are found in the cartilage, tendons, and ligaments. In contrast to the filamentous collagen molecules, collagen type IV participates in the construction of a membrane-like structure, known as the *basal lamina*, which underlies epithelial and endothelial cells. The structure and biochemical features of collagen molecules are discussed on page 103. These aspects will not be repeated here.

Collagen molecules play important roles in the constitution of mammalian tissues or organs. The collagen matrix serves as a structural framework that supports cells, helps organize cells into various forms of tissues and organs, and protects cells from mechanical injury. In addition, collagen matrix participates in the regulation of cellular activities such as cell survival, adhesion, proliferation, and migration. Collagen molecules can directly interact with cells via the cell membrane collagen receptors, or indirectly via the mediation of fibronectin, a matrix component that binds collagen molecules at one side and cell membrane matrix receptors, known as *integrins*, at the other side. The binding of collagen and fibronectin molecules to the integrin receptors initiate the activation of intracellular signaling pathways that stimulate or activate mitogenic processes, including cell survival, proliferation, and migration.

Given the structural and functional features, collagen matrix has long been used for constructing drug-delivery devices and scaffolds for tissue regeneration. Collagen matrix

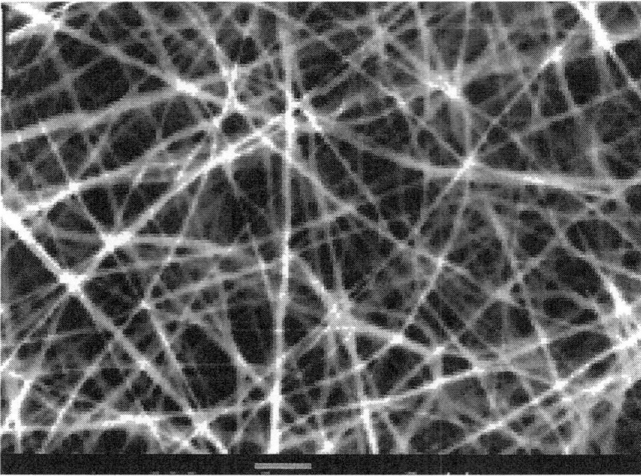

Figure 12.5. Scanning electron micrograph of electrospun collagen matrix. (Reprinted with permission of the American Chemical Society from Zhang YZ et al: Characterization of the surface biocompatibility of the electrospun PCL-collagen nanofibers using fibroblasts, *Biomacromolecules* 6:2583–9, copyright 2005.)

has been used in several forms: collagen gels, meshes, composites with different molecules, and decellularized natural matrix. Collagen gels and meshes are suitable for drug delivery, whereas cell-free natural collagen matrix can be used as scaffolds or grafts for the repair or regeneration of various tissues and organs, such as blood vessels, airways, intestines, and bladder. Collagen gel can be spun into collagen fibers, which can be used to construct collagen scaffolds for tissue repair and regeneration (Fig. 12.5).

Native collagen matrix is a suitable material for the construction of tissue scaffolds. Such a material maintains the natural biological and mechanical characteristics and exhibits superior biocompatibility compared to in vitro crosslinked collagen gels or matrix. To prepare a native collagen matrix, mammalian tissue specimens can be collected from the submucosa of intestines, the adventitia of blood vessels, and the subcutaneous tissue. Cells in these specimens can be removed by various enzymatic and hydrolytic methods. Such treatments eliminate the cellular immunogenicity of allogenic tissues (note that extracellular matrix molecules exhibit neglible immunogenicity). The resulting cell-free collagen matrix can be tailored into a scaffold with a desired form and used for tissue repair or regeneration.

Elastic Fibers and Laminae [12.12]

Elastic fibers and laminae are major extracellular matrix components found in mesenchymal and connective tissues. Elastic fibers are present in the lung, connective tissue, the submucosa of intestines, and the wall of veins, whereas elastic laminae are found primarily in the media of large and medium arteries. Elastic fibers and laminae are composed of several proteins, including elastin, microfibrils, and microfibril-associated proteins. Elastin is the most abundant protein in elastic fibers and laminae. Mature elastin is a highly insoluble and hydrophobic protein, and is formed by crosslinking the 72-kDa elastin precursor, known as *tropoelastin*. Tropoelastin is produced by several cell types, including the smooth muscle cell, endothelial cell, and fibroblast, and is released into the extracellular space where crosslinking and elastin formation take place. The structure and biochemical features of elastic fibers are discussed on page 109.

Elastic fibers and laminae play an important role in the constitution of tissues and organs as well as in the maintenance of the stability of tissues and organs. For instance, multiple layers of elastic laminae are found in large arteries. These laminae have long been known to contribute to the structural stability and mechanical strength of the arterial wall (44,45). Arteries are subject to extensive mechanical stress induced by arterial blood pressure. Without the support of the elastic laminae, vascular cells may be overstretched under arterial blood pressure. Elastic laminae also contribute to the elasticity of soft tissues, such as connective tissues and arteries. The recoil of the arterial wall is a critical mechanism for the continuation of bloodflow during diastole when cardiac ejection is ceased. Elastic laminae have also been shown to serve as a signaling structure and play a role in regulating arterial morphogenesis and pathogenesis. An important contribution of elastic laminae is to confine smooth muscle cells to the arterial media by inhibiting smooth muscle cell proliferation and migration, thus preventing intimal hyperplasia under physiological conditions. In addition, elastic laminae exhibit antiinflammatory effects and inhibit leukocyte adhesion, activation, and transmigration relative to collagen matrix. These features render elastic laminae a potential material for vascular reconstruction. Furthermore, elastic laminae and elastin-containing structures can be used to prevent inflammatory reactions after surgery.

Polysaccharides [12.13]

Polysaccharides are polymers composed of many monosaccharides bonded together by glycosidic bonds. There are a number of forms of natural polysaccharides, including glycogen, cellulose, alginate, chitosan, starch, and glycosaminoglycan. These polysaccharides are found in animals and plants, and play an important role for the survival and function of animals and plants. Glycogen is a polymer composed of glucose monomers and synthesized in animals for the storage of energy. Alginates are linear polysaccharides composed of β-mannuronic acid and α-guluronic acid, and are found in brown seaweed and in certain bacteria. Starch is a polymer found in plants and synthesized for the storage of energy. Cellulose is found in plants and bacteria. Chitosan is found in the shell of crabs and shrimps. One of the important properties of polysaccharides is their ability to form hydrogel. This property is the basis for polysaccharide-mediated drug delivery. Several types of polysaccharide, such as cellulose, chitosan, and starch, have been used as materials for tissue engineering and drug delivery.

Cellulose. *Cellulose* is a linear polysaccharide composed of D-glucose units jointed together by 1,4-β-glucosidic bonds. In plants, cellulose participates in the constitution of plant skeleton and cell wall. Cotton is a well-known cellulose-containing material. Cellulose molecules are often arranged in parallel, giving cellulose fibers high mechanical strength. Humans cannot use cellulose as an energy source because of the lack of β-glycosidase, which catalyzes the hydrolysis of β-glycosidic bonds (note that mammals have α-glycosidase that catalyzes the hydrolysis of glycogen and starch). Cellulose can also be produced by bacteria. Bacterial cellulose has been often used for tissue engineering and will be the focus here.

Several types of microorganism, including algae (*Vallonia*), fungi (*Saprolegnia*, *Dictystelium discoideum*), and bacteria (*Acetobacter*, *Achromobacter*, *Aerobacter*, *Agrobacterium*, *Pseudomonas*, *Rhizobium*, *Sarcina*, *Alcaligenes*, *Zoogloea*) can synthesize cellulose. Among these microorganisms, the bacterium Acetobacter xylinum, which is usually found in fruits, vegetables, and alcoholic beverages, has been used to generate cellulose for tissue engineering applications. In a culture medium, this bacterium can produce a network of cellulose fibers. The cellulose fibers can be collected, fabricated into desired forms, and used to construct scaffolds for tissue engineering applications.

The bacterial cellulose synthesized by *Acetobacter xylinum* is similar to the plant cellulose in molecular composition. Both types of cellulose contain D-glucose. However, bacterial cellulose exhibits a higher crystallinity, higher water absorption capacity or lower hydrophobicity, higher mechanical strength, and finer molecular arrangement compared to the plant cellulose. Cellulose and its derivatives, such as cellulose nitrate, cellulose acetate, and cellulose xanthate, can be easily fabricated into desired forms. Unlike other polysaccharides, such as glycogen and starch, cellulose exhibits low water solubility and, therefore, a low rate of degradation when implanted into an animal tissue. A decrease in the crystallinity and hydrophobicity of cellulose usually results in an increase in the biodegradability of cellulose. Given the chemical composition, bacterial cellulose is highly biocompatible and nontoxic to the host. Furthermore, bacterial cellulose is a highly moldable material and can be used to fabricate scaffolds with desired forms. Cellulose-based materials have been used in a number of biomedical applications. These include construction of cellulose membranes for hemodialysis, construction of enzyme carriers for biosensors, drug delivery, construction of scaffolds for the regeneration of various tissue types, such as the bone, cartilage, liver, skin, and blood vessels. These investigations have consistently demonstrated that cellulose-based materials elicit little inflammatory and toxic reactions. Cellulose has been proven a promising material for the construction of tissue regenerating scaffolds.

Alginates [12.14]. *Alginates* are linear polysaccharides composed of β-mannuronic acid and α-guluronic acid. Alginates are found in brown seaweed and in certain bacteria. The content of β-mannuronic acid and α-guluronic acid may vary depending on the plant or bacterial species from which alginates are obtained. Alginates can be used to form hydrogel and matrix. Divalent cations, such as Ca^{2+} and Mn^{2+}, can initiate alginate gelation by linking α-guluronic acid units between different polymer chains. The feature of gelation renders alginates a potential material for tissue engineering applications, such as cell seeding and transplantation, tissue repair, and drug delivery.

Alginate gels with various mechanical properties can be generated under different gelling conditions and by using different crosslinkers. Numerous studies have been conducted to test the elastic and shearing mechanical properties. Under compressive forces, alginate matrices exhibit elastic modulus ranging from 1 to 1000 kPa, depending on gelling and experimental conditions. Similarly, the shear modulus of alginate matrices spreads widely from 0.02 to 40 kPa under different experimental conditions. Under tensile forces, the maximal tensile strength or failure stress of alginate gels ranges from 3 to 35 kPa and the maximal or failure strain is from 0.3 to 1.25, depending on the composition of alginates and the strain rate applied.

Alginate gels crosslinked by Ca^{2+} have been used for a number of biomedical applications. One of the applications is alginate-mediated gene delivery. Alginate microspheres have been fabricated to carry genes of interest. The alginate microspheres can be delivered to target tissues, where the gene is released. Because of the biodegradability of alginates, genes can be released in a controlled manner with the releasing rate depending on the rate of alginate degradation. Similarly, an alginate-based gel or matrix can be used to mediate controlled protein and drug delivery. In addition, alginate-based materials have been fabricated and used to mediate wound healing. Alginates can form a thin layer of gel when crosslinked by Ca^{2+}. Such a gel layer can be used to cover skin wound to prevent the loss of body fluids and bacterial infection. Alginate-based materials can be used to construct various forms of matrix scaffolds for the repair or regeneration of various tissue types such as the cartilage, liver, and bone. Alginate materials have also been used to construct capsules for cell transplantation. Cells can be encapsulated within alginate capsules and delivered to target tissues (Fig. 12.6). The alginate capsules can partially protect the enclosed cells from inflammation-induced injury.

Chitosan [12.15]. Chitosan is a linear polysaccharide composed of D-glucosamine units jointed by β-1,4-glycosidic bonds with randomly inserted *N*-acetylglucosamine units. Chitosan is a partially deacetylated derivative of chitin, which is a copolymer of randomly distributed *N*-acetylglucosamine and *N*-glucosamine units. A polymer molecule with more than 50% *N*-acetylglucosamine units is known as *chitin*, and that with more than 50% *N*-glucosamine units is called *chitosan*.

Chitosan and chitin are found in the shell of crabs and shrimps and are similar to cellulose in structure. Chitosan and chitin can be collected from these shellfish sources. Chitosan is a semicrystalline molecule and is usually stable. It is insoluble in water, but soluble in acidic solutions (pH ~5). The use of chitosan for tissue engineering relies partially on its gelation ability. Chitosan solutions can be gelled in methanol and under a high pH condition. A dried chitosan structure can be mechanically very strong. Chitosan molecules are usually positively charged and can bind to molecules with negative charges, such as glycosaminoglycans and alginates. A unique feature is that the charge density of chitosan is dependent on pH. Such a feature renders chitosan a candidate material for pH-controlled drug delivery.

When implanted in vivo, chitosan is degraded by lysozyme-catalyzed hydrolysis. Chitosan is disintegrated into oligosaccharides. The rate of chitosan degradation is inversely proportional to the degree of crystallinity. The crystallization of chitosan is regulated by deacetylation. Chitosan molecules with increased deacetylation on the *N*-acetylglucosamine units exhibit increased crystallinity and reduced degradation. In a highly crystal form, it takes several months to degrade chitosan scaffolds in vivo. Amorphous chitosan exhibits more rapid degradation.

Chitosan is a polysaccharide that can be fabricated into various forms of porous matrix. To produce a chitosan matrix, chitosan can be dissolved in acetic acid. The chitosan–acetic acid solution can be frozen and lyophilized to produce chitosan matrix. The freezing process induces the formation of ice crystals. The following lyophilizing process removes the ice crystals, allowing the formation of a porous matrix. The size of the pores can be controlled by altering the rate of ice crystal formation.

Chitosan can be used to make materials with various mechanical properties. A pure chitosan material without apparent pores exhibits elastic modulus ranging from 5 to 7 MPa. However, the introduction of pores reduces the elastic modulus and mechanical strength. Porous chitosan materials could have an elastic modulus as low as 0.1 MPa. The failure strain or maximal strain of chitosan is also dependent on the porosity of the material. Nonporous chitosan materials can be stretched to a strain about 0.3, whereas a porous

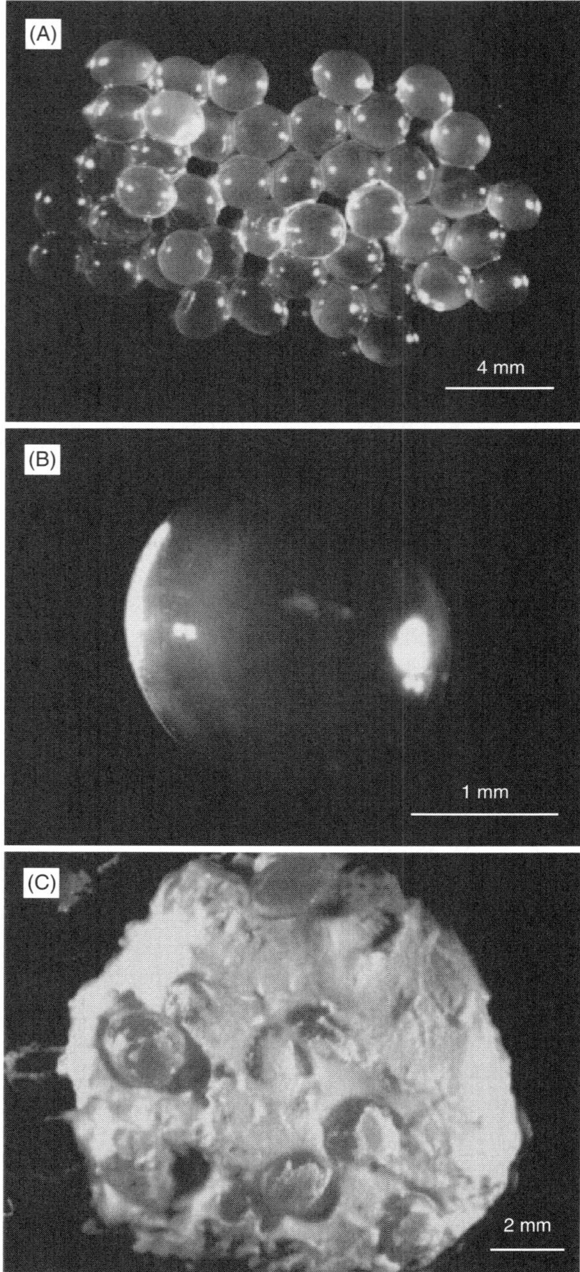

Figure 12.6. Cell-containing alginate beads for cell transplantation. Cells were encapsulated with alginate beads by dropping the cell–alginate mixture into an agitated bath of calcium chloride using a syringe. (A) A-low magnification image of the beads. (B) A higher-magnification image. (C) Cell–alginate beads were mixed into a calcium phosphate cement paste at a 54% volume fraction of alginate beads. (Reprinted with permission of John Wiley & Sons, Inc. from Weir MD et al: Strong calcium phosphate cement-chitosan-mesh construct containing cell-encapsulating hydrogel beads for bone tissue engineering, *J Biomed Mater Res Pt A*, published online Feb 15, 2006.)

chitosan material can be stretched to a strain about 1. Porous chitosan exhibits a nonlinear mechanical behavior, i.e., the mechanical behavior is dependent on the level of strain and stress. The material gains stiffness (with increased elastic modulus) when strain and stress are elevated.

Chemical modifications significantly influence the mechanical properties of chitosan materials. For instance, the coating of a chitosan material with hyaluronic acid significantly increases the tensile strength of the chitosan material. This mechanical reinforcement is due to the formation of tight bonds between positively charged chitosan molecules and negatively charged hyaluronic acid molecules. Such reinforced chitosan materials are suitable for repairing tissues with high mechanical loads such as cartilage. Furthermore, the incorporation of hydroxyapatite or other calcium containing materials into chitosan or chitin can generate composite materials with increased mechanical strength. Such materials can be used for bone repair or regeneration.

Given the molecular structure, mechanical properties, biocompatibility, and the capability of forming various matrix structures, chitosan and chitosan derivatives have been considered candidate materials for the engineering and regeneration of injured tissues and organs. Chitosan materials have been used to construct matrix scaffolds for seeding, culturing, and transplanting cells into target tissues. These materials have also been used as carriers for drug delivery. Several studies have shown that chitosan can serve as a gene transfer carrier. Genes mixed with chitosan-based materials have been successfully delivered to target cells in the knee joints in animal models. Chitosan-mediated gene transfer can also be carried out together with cell transplantation, enhancing therapeutic effects on target diseases.

Numerous investigations have shown consistently that chitosan and chitosan derivatives are relatively nontoxic and biocompatible. In particular, chitosan-based materials do not induce significant fibrous encapsulation around the implants. Although chitosan implantation induces leukocyte infiltration during the early period (within days), chronic inflammation does not occur significantly. The application of chitosan to cartilage repair and regeneration has demonstrated a beneficial effect on the recovery of injured cartilage tissues, such as stimulation of chondrocyte growth and expression of structural proteins. These observations have demonstrated the feasibility of using chitosan and chitosan derivatives as biomaterials for tissue regenerative engineering.

Starch [12.16]. Starch is composed of D-glucose and is a form of polysaccharide for the storage of energy in plants. It can be found in all plant seeds and tubers. There are two forms of starch: amylase and amylopectin. Amylose is a linear polymer with the D-glucose units joined by the α1,4-glycosidic bonds, whereas amylopectin contains branching polymer chains, in which the D-glucose units are joined by the α1,4-glycosidic bonds in the linear portion and those at branching points are joined by the α1,6-glycosidic bonds.

Cornstarch is usually used in biomedical research. Starch can be blended with chemical compounds such as ethylene vinyl alcohol and cellulose acetate to make matrices that can serve as engineering scaffolds or cell seeding/culture substrates. The fabricated matrix can be reinforced by mixing with hydroxyapatite to form a composite material. Polymer matrices of various forms can be prepared by injection molding. Starch and its composites have been used as substrates for cell culture and carriers for cell transplantation. Starch-based materials do not significantly influence the growth and function of cultured cells. These materials have also been used in vivo for several biomedical applications, including drug delivery and tissue repair and regeneration. While starch may not be mechanically strong, the addition of reinforcement compounds may enhance the mechanical strength. As starch is composed of natural D-glucose, starch-based materials are usually nontoxic and biocompatible. Such features render starch a promising material for tissue regenerative engineering.

Glycosaminoglycans [12.17]. Glycosaminoglycans (GAGs) are linear polysaccharides composed of repeated disaccharide units. Each unit contains an uronic acid and amino sugar molecule. According to the type of the disaccharide unit, GAGs can be classified into several groups, including chondroitin sulfate, hyaluronate, keratan sulfate, and heparan sulfate. A chondroitin sulfate molecule is composed of a glucuronic acid and an *N*-acetylgalactosamine unit with a SO_4^- group on the 4 or 6 carbon position. A hyaluronate molecule contains a glucuronic acid and an *N*-acetylglucosamine unit. A keratan sulfate molecule contains a galactose and an *N*-acetylglucosamine unit with a SO_4^- on carbon position 6. A heparan sulfate molecule contains a D-glucuronic acid and an *N*-acetyl-D-glucosamine unit.

Glycosaminoglycans are found in mammalian connective tissues, such as the subcutaneous tissue, cartilage, and blood vessels. These molecules attach to core proteins and form proteoglycans, major extracellular matrix molecules known as *ground substances*. Heparan sulfate is found on the surface of vascular endothelial cells and is similar in structure and function to heparin, which is a potent anticoagulant.

Glycosaminoglycans are characterized by several general features, including the presence of a high density of negative charges, high hydrophilicity and water solubility, and low crystallinity. However, there are differences in material properties between various GAGs molecules. For instance, hyaluronate is a large molecule and has a high gel-forming capability. These molecules absorb a large amount of water, constituting a major part of the extracellular matrix. Because of the gel-forming capability, hyaluronate is often used as a media for drug-delivery or a material for tissue repair and regeneration. The composition of hyaluronate may be modified to construct materials with various properties. For example, partial esterification of the carboxyl groups of hyaluronate molecules reduces the water solubility of the polymer and increases its viscosity. Extensive esterification generates materials that form water-insoluble films or gels. Thus, hyaluronate gels with desired properties can be prepared by altering chemical compositions.

Compared to hyaluronate, other types of GAGs exhibit poor gel-forming capability in vitro. These GAGs alone have not been used extensively as biomaterials. However, negatively charged GAGs can bind tightly to positively charged molecules, such as chitin and chitosan, and form composite polymeric materials. Such composite materials can be used to form gels with various material properties by altering the relative contents of GAG and/or chitosan.

Glycosaminoglycans and glycosaminoglycan-based composite polymers have been used to construct hydrogels and matrices for various biomedical applications, such as drug delivery, cell seeding and transplantation, tissue repair, and tissue regeneration. Since GAGs are natural molecules, they are biocompatible and do not cause significant toxic and inflammatory reactions. These molecules can be degraded at different rates, depending on the compositions of the materials. For fully esterified hyaluronate membranes, the lifetime is several months. A reduction in esterification increases the rate of degradation.

METALLIC MATERIALS AS BIOMATERIALS [12.18]

Several types of metallic material have been used as biomaterials. These include iron (Fe), chromium (Cr), cobalt (Co), nickel (Ni), titanium (Ti), molybdenum (Mo), and tungsten (W). These materials have also been used to create alloys, providing favorable properties for the fabrication and performance of biomaterials. Typical examples of alloys include Co–Cr and Ti alloys. In addition, stainless steels have been developed and used as biomaterials. Because of their superior strength, elasticity, and endurance, metallic materials are often used for the repair and replacement of bones and joints. For the past several decades, these alloys have been well accepted for their performance, biocompatibility, and stability.

Stainless Steels as Biomaterials

Steels are artificially modified forms of iron with various carbon contents and are characterized by mechanical hardness, elasticity, and strength. Thus, steels are considered candidate materials for the repair of the skeletal system. The mechanical features of steels are dependent on the content of carbon and temperature. The crystal structure of iron, which determines the mechanical characteristics of iron, can be modulated by altering the treatment temperature and carbon concentration. At a relatively cold temperature, say 20°C, iron atoms are organized into a unit structure with a *body-centered cubic* form. In each unit, eight neighboring atoms are symmetrically localized to the corners of an imaginary cube with one atom at the cube center. An increase in temperature to a certain degree can induce a transformation of the atomic structure from the body-centered cubic form into a unit structure with a *face-centered cubic* form, in which the atoms are localized to the faces of an imaginary cube. Carbon atoms can be integrated more easily into the iron unit structure with the face-centered cubic form than that with the body-centered cubic form. Thus, an appropriate alteration in temperature facilitates the integration of carbon into the iron. Carbon integration enhances the stability, hardness, and strength of the iron. However, the solubility of carbon in iron is relatively low. An excessive level of carbon induces carbon precipitation, a problem influencing the endurance and mechanical properties of the steel. An appropriate concentration of carbon is about 0.03%.

A common problem for using steels as biomaterials is corrosion. To resolve such a problem, chromium has been added to steels, rendering the steels stainless. For the manufacturing of biomaterials, chromium is used at a concentration ranging from 17 to 20%. However, the use of chromium introduces a problem: the mixing of chromium and carbon can form carbides, which enhances carbon precipitation. An approach used to mitigating carbide formation is to add nickel to the steel. Nickel can stabilize the iron structure, prevent carbide formation, and enhance corrosion resistance. Nickel is used at a concentration of 12–14% in steels as biomaterials. The chromium–nickel stainless steel has a high yielding strength (>170 MPa) and is considerably corrosion-resistant. However, corrosion can still occur when steel materials are implanted into the body. Thus, this material is often used to fabricate temporary implants such as fracture plates and nails.

Co–Cr Alloys as Biomaterials

Co–Cr alloys are metallic mixtures containing primarily Co and Cr as well as various amounts of other elements, such as Ni, Mo, Fe, C, Si, Mn, and Ti. There are two types of

Co–Cr alloy that have been fabricated and used as biomaterials: CoCrMo and CoNiCrMo alloys. The CoCrMo alloys are composed of Co 63–68%, Cr 27–30%, and Mo 5–7%. The CoNiCrMo alloys consist of Co 31.5–39%, Ni 33–37%, Cr 19–21%, and Mo 9–10.5%. These alloys possess high yielding strength (~450 MPa for the CoCrMo alloys, and 240–655 MPa for the CoNiCrMo alloys). The CoCrMo alloys can be cast, while the CoNiCrMo alloys can be forged, into implants of desired shapes. Both types of alloy are highly corrosion-resistant. These alloys are suitable materials for the fabrication of artificial bones and joints.

Titanium and Titanium Alloys as Biomaterials

Titanium is a metallic material that is characterized by superior hardness, corrosion resistance, and lightness (4.5 g/cm^3 compared to 7.9 g/cm^3 for iron). Given such features, titanium has been used as a biomaterial for the replacement of bones and joints. Titanium materials usually contain several elements, such as nitrogen, carbon, hydrogen, oxygen, and iron. The contents of these elements are very low, with nitrogen ranging from 0.03–0.05%, carbon about 0.1%, hydrogen about 0.015%, oxygen 0.18–0.40%, and iron 0.2–0.5%.

Titanium has been used to make alloys. A typical titanium alloy is $Ti_6A_{14}V$, which contains ~6% aluminum, ~4% vanadium, ~90% titanium and low contents of nitrogen, carbon, hydrogen, oxygen, and iron. The addition of aluminum and vanadium increases the strength and corrosion resistance of the titanium alloy. For instance, pure titanium possesses yielding strength ranging from 170 to 485 MPa, whereas $Ti_6A_{14}V$ exhibits yielding strength ~795 MPa. Other types of titanium alloys have also been created to provide more features suitable for the performance of titanium alloys as biomaterials. Examples include $Ti_{13}V_{11}Cr_3Al$ and $Ti_{13}Nb_{13}Zr$. The $Ti_{13}V_{11}Cr_3Al$ alloy contains ~13% vanadium, ~11% chromium, ~3% aluminum, and ~73% titanium. The addition of these elements enhances the strength of the titanium alloy. The $Ti_{13}Nb_{13}Zr$ alloy is composed of ~13% niobium, ~13% zirconium, and ~74% titanium. The addition of these elements enhances the corrosion resistance of the titanium alloy.

Potential Problems with Metallic Materials

There are two potential problems with the use metallic materials as biomaterials. These are corrosion and bioincompatibility. These problems potentially influence the performance and endurance of metallic biomaterials, especially when these materials are used to replace bones and joints that are subject considerably high mechanical loads.

Corrosion is a process of metal degradation induced by chemical reactions, primarily oxidization. When subject to water-based solutions containing dissolved oxygen and ions such as chloride and hydroxide, metal atoms react with these species and form oxide or hydroxide compounds. These compounds detach from the metal surface and dissolve in the solution. The metal is degraded gradually.

Since the physiological fluids in the human body contain oxidative chemical species, providing a harsh environment for metallic implants, corrosive degradation of metals occur at various rates, depending on the type of the metal and the local environment. Iron and steels can be corroded easily in the presence of water and oxygen, whereas chromium, nickel, and titanium are considerably corrosion-resistant. The concentration of oxygen in the interstitial fluids varies considerably in the different compartments of the body. Such variations significantly influence the rate of metal corrosion. An increase in oxygen concentration facilitates metal corrosion.

In addition to chemical factors, physical factors such as mechanical loads and friction accelerate metal corrosion. For instance, repetitive deformation of metal implants can induce mechanical fatigue, which facilitates chemical corrosion, a phenomenon known as *fatigue corrosion*. Shearing motions between two implants induces damage of the protective passivation layer, contributing to corrosion, which is known as fretting corrosion. These mechanical factors should be taken into account in the design of metallic implants.

There are several methods that can be used to measure the rate of corrosion. These include the estimation of the number of ions liberated from a metal per unit time, the measurement of the depth of the metal corroded away, and the measurement of the loss of the metal weight due to corrosion per unit time. These are fairly straightforward methods and can be applied to in vitro tests and in vivo tests in animal models.

Several approaches can be used to reduce the rate of metal corrosion. For steel-based implants, the addition of chromium can significantly reduce the rate of implant corrosion, since chromium can form a stable passive chromium oxide film on the steel surface. Modulation of carbon contents may also influence the rate of steel corrosion. Since excessive carbon content induces carbon precipitation, which may facilitate steel corrosion, lowering the carbon content is an effective approach to reduce the rate of steel corrosion. Other metallic materials, such as cobalt and titanium, are considerably resistant to corrosion, since these metals are inert in physiological fluids and can form a passivating oxide film. Alloys based on these metallic materials exhibit improved resistance to corrosion.

Although corrosion-resistant alloys are used, corrosion still occur in artificial metallic bones and joints. Metallic corrosion often causes local swelling and pain, and influences the function of the artificial implants. Corrosion can be detected by x-ray examination. At surgery, inflammatory reactions and metal debris can be found in tissue surrounding the metallic implants. Because corrosion accelerates metal wear and fatigue failure, metallic implants with severe corrosion should be replaced.

CERAMICS AS BIOMATERIALS [12.19]

Ceramics are a group of inorganic, polycrystalline, and refractory materials, including metallic oxides, carbides, silicates, hydrides, and sulfides. Ceramics are characterized by several physical properties, including the hardness, inertness to physiological ionic fluids, and resistance to high compressive stress. Given such properties, ceramics have been used as biomaterials for the replacement of bones and teeth. In terms of their interaction with biological tissues, ceramics can be classified into several types: bioactive, bioinert, and biodegradable ceramics. The characteristics and applications of these ceramics are discussed here.

Bioactive Ceramics

Bioactive ceramics are ceramics that can interact and form bonds with surrounding tissues. Such ceramics can be used as "adhesives" for prostheses, enhancing the attachment of prostheses to adjacent tissue. Given the mechanical strength and hardness, this type of ceramic is often used as adhesive for orthopedic applications such as the repair and replacement of bones and joints. A major type of bioactive ceramic is glass ceramics. This type of ceramics is constructed with SiO_2, CaO, Na_2O, and P_2O_5. The adhesive properties of glass ceramics are dependent on the formation of a surface layer composed of calcium phosphate and silicon oxide (SiO_2).

Bioactive glass ceramics may not only serve as structural materials, but also play a role in regulating the function of host cells. For instance, silicon–calcium glass ceramics, once implanted into the skeletal system, can release silicon and calcium ions, which stimulate osteoblast growth and differentiation. Such a process involves genes that encode proteins responsible for the regulation of cell mitosis and differentiation. In addition, a controlled release of soluble calcium and silicon from a composite material composed of bioactive glass and resorbable polymer has been shown to enhance the generation of vascularized soft tissues. Thus, by controlling the compositions and releasing rate of silicon and calcium, bioactive glass ceramics can be used to mediate the growth of bone tissues.

Bioactive glass ceramics have been used not only for bone replacement but also for soft tissue regeneration. Recent studies have shown that bioglass-coated polystyrene scaffolds stimulate the proliferation of cultured fibroblasts. Such an influence is dependent on the concentration of the coating bioactive ceramics. An excessive concentration of bioglass induces a reduction in the rate of cell proliferation, in association with a change in cell shape. A limited number of experiments have demonstrated that low concentration of bioglass (0.01%) may stimulate the expression and release of vascular endothelial growth factor. In vivo experiments have shown that bioglass-coated scaffolds can be well tolerated for up to 42 days when implanted subcutaneously in the rat. While fibroblasts actively adhere to poly(glycolic acid) (PGA) meshes, they rarely adhere to the bioglass particles. These observations demonstrate that bioactive glass ceramics can be used as compounds for the fabrication of composite scaffolds for soft tissue engineering.

Bioinert Ceramics

Bioinert ceramics, including alumina, zirconia, and carbons, have been used as biomaterials. These ceramics are generally corrosion-resistant and wear-resistant. They do not cause significant toxic, inflammatory, and allergic reactions and are relatively biocompatible. These ceramics possess common ceramic characteristics such as hardness, low friction, and resistance to compressive stress. Because of these characteristics, bioinert ceramics are often used to fabricate bone plates, screws, femoral heads, and middle ear ossicles.

Alumina, or aluminum oxide (Al_2O_3), is a typical type of bioinert ceramics. Alumina exists in nature as crystal corundum. A crystal form of alumina can be synthesized by applying fine alumina powder to a flame of mixed oxygen and hydrogen. The mechanical strength of synthetic alumina is dependent on the grain size and porosity. Alumina with small grains and low porosity has high strength. A minimum of flexural strength 400 MPa and elastic modulus 380 GPa is required for using alumina as an orthopedic biomaterial. In general, alumina is a material that is characterized by hardness, low friction, inertness to physiological fluid environment, low toxicity, and low immunogenicity. These properties render alumina a suitable orthopedic biomaterial. Alumina has been used to fabricate artificial joints and total hip prostheses.

Biodegradable Ceramics

Biodegradable ceramics are ceramics that can be degraded and absorbed in a biological system. A number of biodegradable ceramics have been developed and used as biomaterials. These include calcium phosphate, aluminum calcium phosphate, coralline, zinc-calcium-phosphorous oxide, and zinc sulfate–calcium phosphate. Most biodegradable ceramics contain calcium. Biodegradable ceramics are often used for constructing artificial bones and drug delivery carriers, as well as to repair bone damages due to trauma, tumor removal, and pathological disorders. A biodegradable ceramic implant may serve as a temporary frame that guides the formation of the shape of remodeling tissues. The absorbed ceramic material can be replaced by growing tissue, eventually restoring the natural structure and function of the damaged tissue. Thus, biodegradable ceramics are suitable materials for orthopedic tissue regeneration.

Calcium phosphate is a typical biodegradable ceramic and has been used to fabricate artificial bones. Calcium phosphate can be crystallized into a form known as hydroxyapatite. Crystallized calcium phosphate can be very stiff and strong with an elastic modulus up to ~100 GPa. Note that the hardest tissue in our body, such as compact bones, dentin, and dental enamel, is composed of the crystal form of calcium phosphate with structure similar to hydroxyapatite. Thus, calcium phosphate is commonly used in orthopedic regenerative engineering for the replacement and repair of malfunctioned bones. Calcium phosphate-based biomaterials are usually nontoxic and biocompatible.

The biocompatibility of calcium phosphate-based bioceramics has been a topic of research in orthopedic regenerative engineering. Extensive investigations have shown that osteoblasts exhibit normal growth patterns when cultured on calcium phosphate materials. In addition, calcium phosphate biomaterials exert a stimulatory effect on the expression of osteogenic proteins and the proliferation of osteoblasts. These observations demonstrate the suitability of using calcium phosphate compounds as biomaterials for orthopedic regenerative engineering.

Calcium phosphate ceramics can also be used to fabricate drug delivery devices. Drugs, hormones, or growth factors can be packed into biodegradable calcium phosphate ceramics for implantation and delivery into target tissues. With the degradation of the ceramic, drugs or proteins can be gradually released. By controlling the density or compounds of the drug delivery material, the rate of substance release can be regulated. Biodegradable ceramics can be mixed with biodegradable polymers, such as poly *d,l,*-lactic acid-polyethyleneglycol copolymer (PLA-PEG) to form composite materials. Such an approach enhances the capability of controlling the rate of substance release. A composite ceramic material can also serve as a scaffold for tissue regeneration. Biological active substances can be integrated into the scaffold for controlled substance release, which enhances the regeneration of injured tissues.

BIBLIOGRAPHY

12.1. Classification and Properties of Synthetic Biomaterials

Lee HB, Kim SS, Khang G: Polymeric biomaterials, in *The Biomedical Engineering Handbook*, Bronzino J, ed, 1995, Chap 42.

Cooke FW: Bulk properties of materials, in *An Introduction to Materials in Medicine*, Ratner BD, Hoffman AS, Schoen FJ, Lemons JE, eds, Academic Press, San Diego, 1996, pp 11–20.

Visser SA, Hergenrother RW, Cooper SL: Polymers, in *An Introduction to Materials in Medicine*, Ratner BD, Hoffman AS, Schoen FJ, Lemons JE, eds, Academic Press, San Diego, 1996, pp 50–59.

12.2. Nonbiodegradable Polymers

Lee HB, Kim SS, Khang G: Polymeric biomaterials, in *The Biomedical Engineering Handbook*, Bronzino J, ed, 1995, Chap 42.

Visser SA, Hergenrother RW, Cooper SL: Polymers, in *Biomaterials Science. An Introduction to Materials in Medicine*, Ratner BD, Hoffman AS, Schoen FJ, Lemons JE, eds, Academic Press, San Diego, 1996.

12.3. Biodegradable Polymers

Kohn J, Langer R: Bioresorbable and bioerodible materials, in *An Introduction to Materials in Medicine*, Ratner BD, Hoffman AS, Schoen FJ, Lemons JE, eds, Academic Press, San Diego, 1996, pp 64–72.

Domb AJ, Kumar N, Sheskin T, Bentolila A, Slager J et al: Biodegradable polymers as drug carrier system, in *Polymeric Biomaterials*, 2nd ed, Dumitriu S, ed, Marcel Dekker, New York, 2002, pp 91–121.

12.4. Polyglycolides and Polylactides

Domb AJ, Kumar N, Sheskin T, Bentolila A, Slager J et al: Biodegradable polymers as drug carrier system, in *Polymeric Biomaterials*, 2nd ed, Dumitrius, ed, Marcel Dekker, New York, 2002, pp 91–121.

Kissel T, Brich Z, Bantle S, Lancranjan I, Nimmerfall VP: Parenteral depot-systems on the basis of biodegradable polyesters, *J Controll Release* 16:27, 1991.

Vert M, Li SM, Garreau H: More about the degradation of LA/GA-derived matrices in aqueous media, *J Controll Release* 16:15–26, 1991.

Lindhardt R: in *Biodegradable Polymers for Controlled Release of Drugs*, Springer-Verlag, New York, 1988, Chap 2.

Miller RA, Brady JM, Cutright DE: Degradation rates of oral resorbable implants (polylactates and polyglycolates): Rate modification with changes in PLA/PGA copolymer ratios, *J Biomed Mater Res* 11:711, 1977.

12.5. Polycaprolactones

Broz ME, VanderHart DL, Washburn NR: Structure and mechanical properties of poly(D,L-lactic acid)/poly(ε-caprolactone) blends, *Biomaterials* 24:4181–4190, 2003.

Rhee SH: Bone-like apatite-forming ability and mechanical properties of poly(β-caprolactone)/silica hybrid as a function of poly(β-caprolactone) content, *Biomaterials* 25:1167–75, 2004.

Domb AJ, Kumar N, Sheskin T, Bentolila A, Slager J et al: Biodegradable polymers as drug carrier system, in *Polymeric Biomaterials*, 2nd ed., Dumitriu S, ed, Marcel Dekker, New York, 2002, pp 91–121.

12.6. Poly(β-hydroxybutyrate)

Kusaka S, Iwata T, Doi Y: Properties and biodegradability of ultra-high-molecular-weight poly[(R)-hydroxybutyrate] produced by a recombinant Escherichia coli, *Int J Biol Macromol* 25:87–94, 1999.

Wrobel M, Zebrowski J, Szopa J: Polyhydroxybutyrate synthesis in transgenic flax, *J Biotechnol* 107:41–54, 2004.

Zhao K, Deng Y, Chun Chen J, Chen GQ: Polyhydroxyalkanoate (PHA) scaffolds with good mechanical properties and biocompatibility, *Biomaterials* 24:1041–5, 2003.

Poirier Y: Production of polyesters in transgenic plants, *Adv Biochem Eng Biotechnol* 71:209–40, 2001.

Holmes PA: Application of PHB—a microbially produced biodegradable thermoplastic, *Phys Technol* 16:32–6, 1985.

Miyake M, Miyamoto C, Schnackenberg J, Kurane R, Asada Y: Phosphotransacetylase as a key factor in biological production of polyhydroxybutyrate, *Appl Biochem Biotechnol* 84–86:1039–44, 2000.

12.7. Polycarbonates

Khan I, Smith N, Jones E, Finch DS, Cameron RE: Analysis and evaluation of a biomedical polycarbonate urethane tested in an in vitro study and an ovine arthroplasty model. Part I: Materials selection and evaluation, *Biomaterials* 26:621–31, 2005.

Stoll GH, Nimmerfall F, Acemoglu M, Bodmer D, Bantle S et al: Poly(ethylene carbonate)s, part II[1]: degradation mechanisms and parenteral delivery of bioactive agents, *J Controll Release* 76:209–25, 2001.

12.8. Polyamides

Li C: Poly(L-glutamic acid)–anticancer drug conjugates, *Adv Drug Delivery Rev* 54:695–713, 2002.

Li C, Yu DF, Newman RA, Cabral F, Stephens LC et al: Complete regression of well-established tumors using novel water-soluble poly(L-glutamic acid)–paclitaxel conjugates, *Cancer Res* 58p: 2404–9, 1998.

Li C, Price JE, Milas L, Hunter NR, Ke S et al: *Antitumor activity* of poly(L-glutamic acid)–paclitaxel on syngeneic and xenografted tumors, *Clin Cancer Res* 5:891–7, 1999.

12.9. Polyphosphazenes

Ambrosio AM, Allcock HR, Katti DS, Laurencin CT: Degradable polyphosphazene/poly(alpha-hydroxyester) blends: Degradation studies, *Biomaterials* 23:1667–72, 2002.

Allcock HR, Pucher SR, Scopelianos AG: Poly[(amino acid ester) phosphazenes]: Synthesis, crystallinity and hydrolytic sensitivity in solution and the solid state, *Macromolecules* 27:1071–5, 1994.

Allcock HR, Pucher SR, Scopelianos AG: Poly[(amino acid ester) phosphazenes] as substrates for the controlled release of small molecules, *Biomaterials* 15:563–9, 1994.

12.10. Polyanhydrides

Domb AJ: Synthesis and characterization of bioerodible aromatic anhydrides copolymers, *Macromolecules* 25:12, 1993.

Domb AJ, Nudelman R: In vivo and in vitro elimination of aliphatic polyanhydrides, *Biomaterials* 16:319–23, 1995.

12.11. Collagen Matrix

Hafemann B, Ensslen S, Erdmann C, Niedballa R, Zuhlke A et al: Use of a collagen/elastin-membrane for the engineering of dermis, *Burns* 25:373–84, 1999.

Tay BK, Le AX, Heilman M, Lotz J, Bradford DS: Use of a collagen-hydroxyapatite matrix in spinal fusion. A rabbit model, *Spine* 23:2276–81, 1998.

Hakim S, Merguerian PA, Chavez DR: Use of biodegradadble mesh as a transport for a cultured uroepitheial graft: An improved method using collagen gel, *Urology* 44:139–42, 1994.

Kang HW, Tabata Y, Ikada Y: Fabrication of porous gelatin scaffolds for tissue engineering, *Biomaterials* 20:1339–44, 1999.

Gloechner DC, Sacks MS, Billiar KL, Bachrach N: Mechanical evaluation and design of a multilayered collagenous repair biomaterial, *J Biomed Mater Res* 52:365–73, 2000.

Girton TS, Oegema TR, Tranquillo RT: Exploiting glycation to stiffen and strengthen tissue equivalents for tissue engineering, *J Biomed Mater Res* 46:87–92, 1999.

12.12. Elastic Fibers and Laminae

Curran ME, Atkinson DL, Ewart AK, Morris CA, Leppert MF, Keating MT: The elastin gene is disrupted by a translocation associated with supravalvular aortic stenosis, *Cell* 73:159–68, 1993.

Emanuel BS, Cannizzaro L, Ornstein-Goldstein N, Indik ZK, Yoon K et al: Chromosomal localization of the human elastin gene, *Am J Hum Genet* 37:873–82, 1985.

Ewart AK, Jin W, Atkinson D, Morris CA, Keating MT: Supravalvular aortic stenosis associated with a deletion disrupting the elastin gene, *J Clin Invest* 93:1071–7, 1994.

Ewart AK, Morris CA, Atkinson D, Jin W, Sternes K et al: Hemizygosity at the elastin locus in a developmental disorder, Williams syndrome, *Nature Genet* 5:11–6, 1993.

Faury G, Pezet M, Knutsen RH, Boyle WA, Heximer SP et al: Developmental adaptation of the mouse cardiovascular system to elastin haploinsufficiency, *J Clin Invest* 112:1419–28, 2003.

Fazio MJ, Mattei MG, Passage E, Chu ML, Black D et al: Human elastin gene: New evidence for localization to the long arm of chromosome 7, *Am J Hum Genet* 48:696–703, 1991.

Li DY, Brooke B, Davis EC, Mecham RP, Sorensen LK et al: Elastin is an essential determinant of arterial morphogenesis, *Nature* 393:276–80, 1998.

Olson TM, Michels VV, Urban Z, Csiszar K, Christiano AM et al: A 30kb deletion within the elastin gene results in familial supravalvular aortic stenosis, *Hum Mol Genet* 4:1677–9, 1995.

Urban Z, Riazi S, Seidl TL, Katahira J, Smoot LB et al: Connection between elastin haploinsufficiency and increased cell proliferation in patients with supravalvular aortic stenosis and Williams-Beuren syndrome, *Am J Hum Genet* 71:30–44, 2002.

Urban Z, Zhang J, Davis EC, Maeda GK, Kumar A et al: Supravalvular aortic stenosis: genetic and molecular dissection of a complex mutation in the elastin gene, *Hum Genet* 109:512–20, 2001.

Urry DW, Hugel T, Seitz M, Gaub HE, Sheiba L et al: Elastin: A representative ideal protein elastomer, *Phil Trans R Soc Lond B Biol Sci* 357:169–84, 2002.

Gosline J, Lillie M, Carrington E, Guerette P, Ortlepp C et al: Elastic proteins: Biological roles and mechanical properties, *Phil Trans R Soc Lond B Biol Sci* 357:121–32, 2002.

Tatham AS, Shewry PR: Elastomeric proteins: Biological roles, structures and mechanisms, *Trends Biochem Sci* 25:567–71, 2000.

Dietz HC, Mecham RP: Mouse models of genetic diseases resulting from mutations in elastic fiber proteins, *Matrix Biol* 19:481–8, 2000.

Liu SQ, Alkema PK, Tieche C, Tefft BJ, Liu DZ et al: Negative regulation of monocyte adhesion to arterial elastic laminae by signal-regulatory protein alpha and SH2 domain-containing protein tyrosine phosphatase-1, *J Biol Chem* 280:39294–301, 2005.

Milewicz DM, Urban Z, Boyd C: Genetic disorders of the elastic fiber system, *Matrix Biol* 19:471–80, 2000.

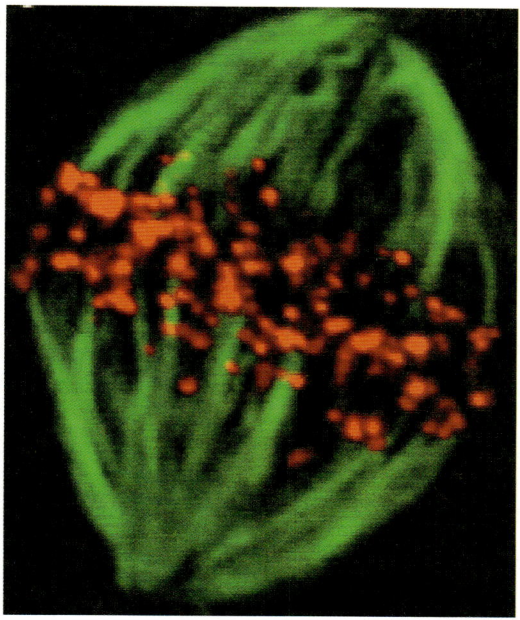

Chapter 1 opening figure. Organization of chromosomes and microtubules in an epithelial cell in the metaphase. Green and red: immunochemically labeled tubulin and kinetochores, respectively. (Reprinted by permission from Kapoor TM et al: Chromosomes can congress to the metaphase plate before biorientation, *Science* 311:388–91, 2006. Copyright 2006, AAAS.)

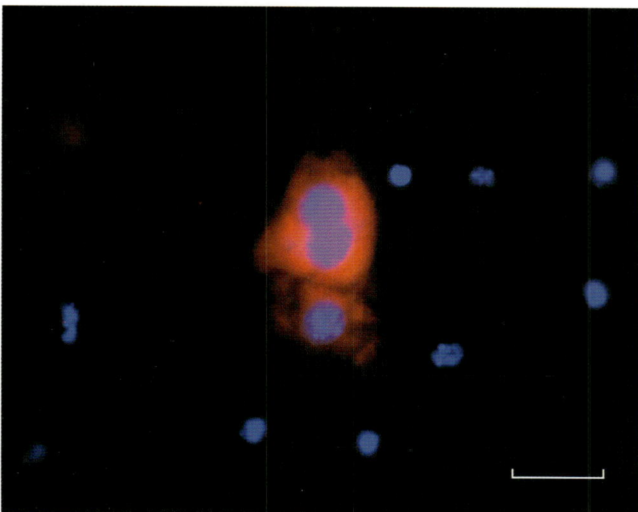

Chapter 2 opening figure. Mobilization of albumin-positive hepatocyte-like cells into the blood in response to the upregulation of interleukin-6 in ischemic cardiac injury. The mobilized hepatocyte-like cells can be recruited to injured cardiac tissue, promoting the proliferation of cardiomyocytes in the zone of cardiac injury. Red: albumin. Blue: cell nuclei. Sclae: 10 μm.

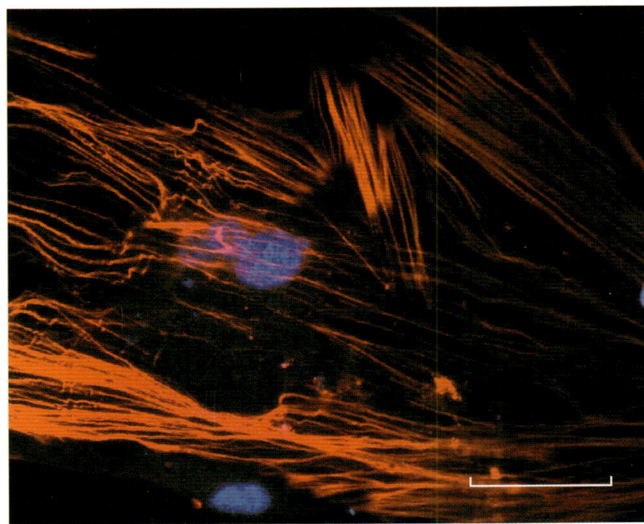

Chapter 3 opening figure. α-Actin filaments in vascular smooth muscle cells derived from the mouse aorta. Smooth muscle cells were collected from the medial layer of the mouse aorta and cultured for 10 days. The α-actin filaments were labeled with an anti-smooth-muscle α actin antibody (red in color) and observed by fluorescence microscopy. Cell nuclei were labeled with Hoechst 33258 (blue in color). Scale bar: 5 μm.

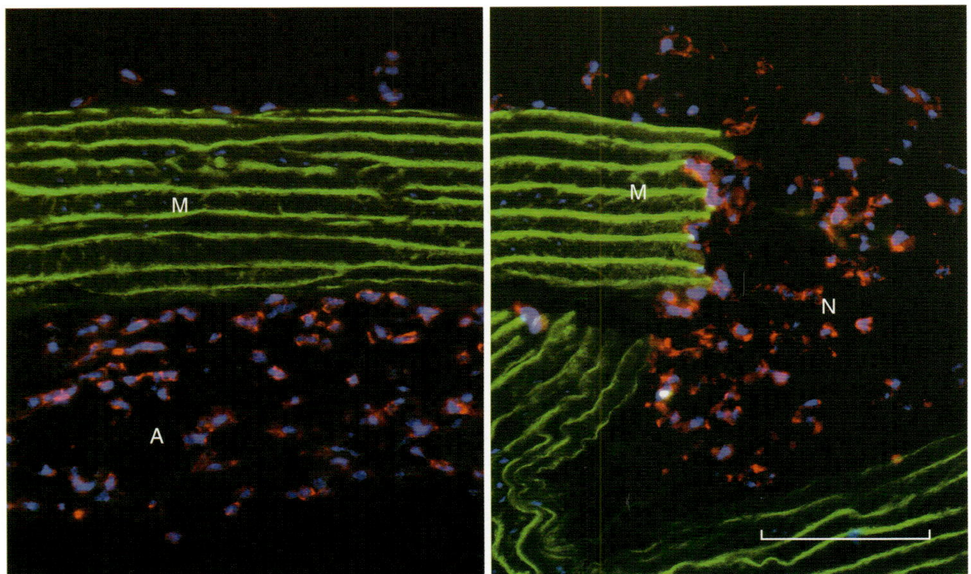

Chapter 4 opening figure. Transverse fluorescent micrographs showing the distribution of CD 11b/c-positive leukocytes in the media and adventitia of matrix-based aortic substitutes. The density of leukocytes in the elastic lamina-containing media was significantly lower than that in the collagen-containing adventitia. Note that leukocytes did not migrate into the gaps between the elastic laminae at the end of the aortic matrix substitutes (right). Red: antibody-labeled CD 11 b/c. Green: elastic laminae. Blue: Hoechst 33258-labeled cell nuclei. M, media. A, adventitia. N, neointima. Scale: 100 μm. (Reprinted from Liu SQ et al: *J Biolo Chem* 280:39294–301, 2005 with permission from the American Society for Biochemistry and Molecular Biology).

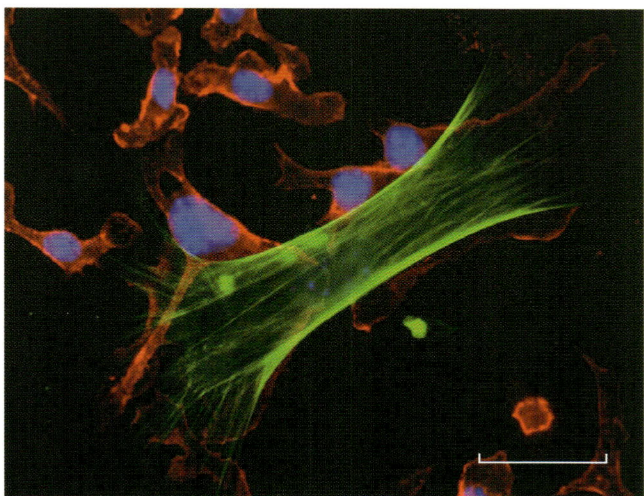

Chapter 5 opening figure. Interaction of bone marrow-derived smooth muscle α actin+ cells with CD11b+ cells in culture. Mouse bone marrow cells were collected and cultured in DMEM with 10% FBS. Smooth muscle α actin+ cells (green) were often found on top of CD11b+ cells (red). When the CD11b+ cells were selectively removed, the number of smooth muscle α actin+ cells was reduced, suggesting that the CD11b+ bone marrow cells play a role in regulating the development of smooth muscle cells from bone marrow progenitor cells. Blue: cell nuclei. Scale: 10 μm.

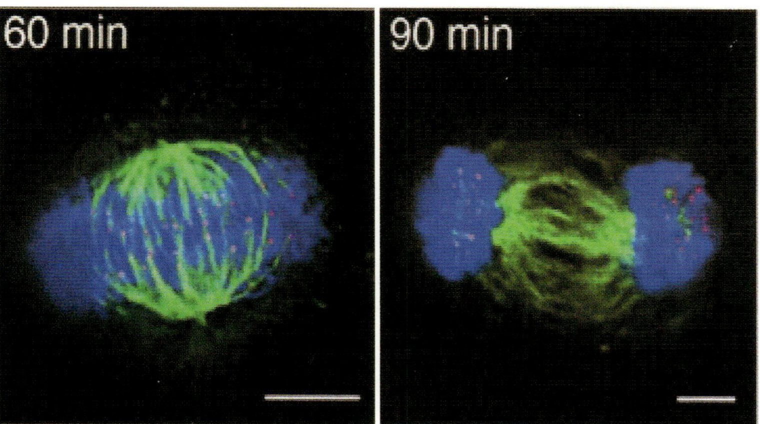

Chapter 6 opening figure. Dynamic rearrangement of chromosomes and microtubules during mitosis. Hela cells are cultured and arrested by treatment with nocodazole, which results in the synchronization of the cells in prometaphase of mitosis. Following the removal of nocodazole, cells start to enter mitosis. Microtubules (green) and chromosomes (blue) undergo dynamic rearrangement during mitosis as shown at 60 and 90 min after the removal of nocodazole. [Reprinted by permission from Macmillan Publishers Ltd: Yasuda S et al: *Nature* 428:767–71, copyright 2004.]

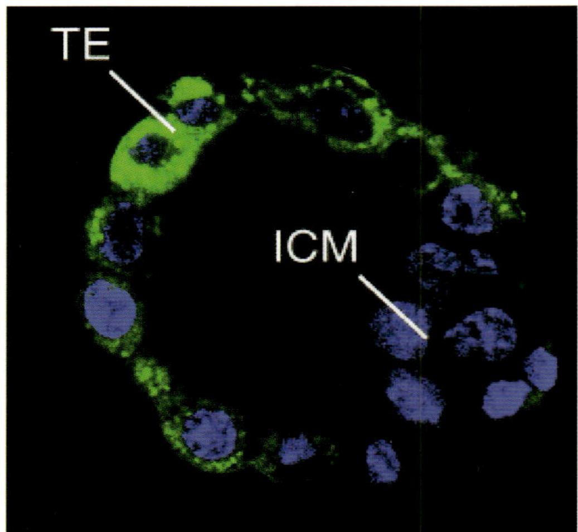

Chapter 7 opening figure. Fluorescent micrograph showing a mouse blastocyst. TE: trophoblast. ICM: inner cell mass. Cells were labeled for mitochondria (green) and nuclei (blue). (Reprinted from Houghton FD: Energy metabolism of the inner cell mass and trophectoderm of the mouse blastocyst, *Differentiation* 74:11–8, 2006 by permission of Blackwell Publishing.)

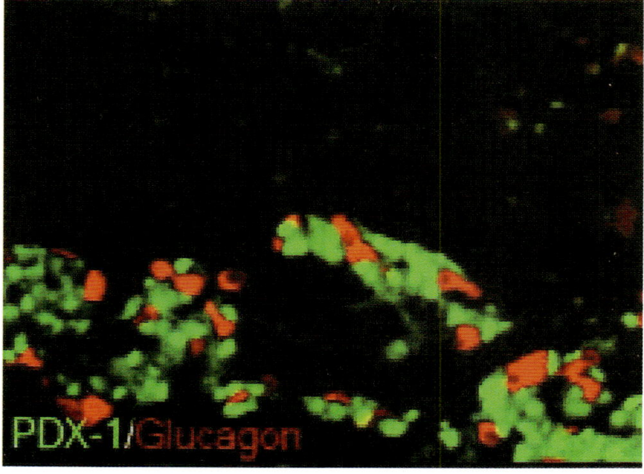

Chapter 8 opening figure. Early development of the pancreatic islets. By embryonic day 19 (E19), Pdx1-positive (green) and glucagon-positive (red) cell aggregates can be detected in the pancreas. Pdx1 is a marker for the pancreatic β cells, whereas glucagon is expressed in the α cells. The α cells are found primarily around the β cells. (From Jensen J: *Dev Dyn* 229:176–200, 2004, reprinted by permission of Wiley-Liss, Inc, a subsidiary of John Wiley & Sons, Inc.)

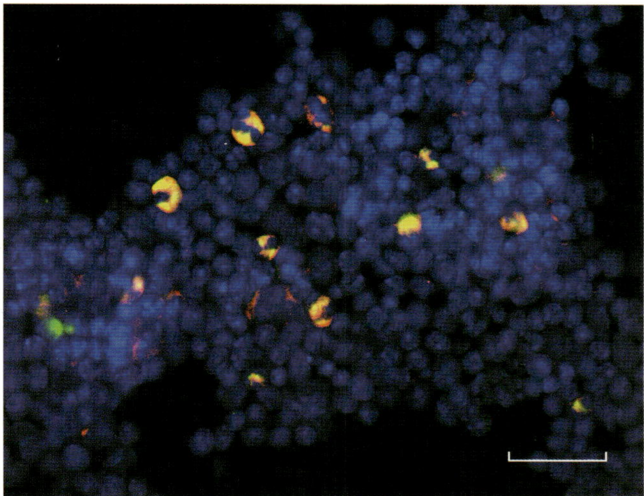

Chapter 9 opening figure. Presence of smooth muscle α-actin-positive (green) and CD34-positive (red) cells in the bone marrow. When cultured on elastic lamina-dominant arterial matrix scaffolds, these cells can transform to smooth-muscle-like cells with α-actin filaments. These cells may serve as progenitor cells for the regeneration of vascular smooth muscle cells. Blue: cell nuclei. Scale: 10 μm.

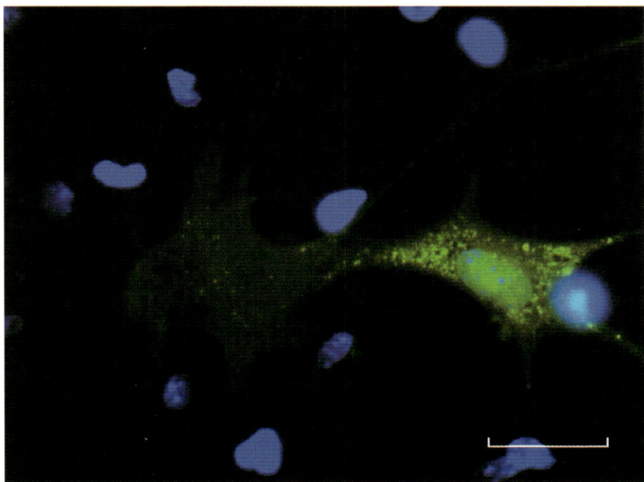

Chapter 10 opening figure. Expression of green fluorescent protein (GFP) in cultured bone marrow stromal cells transfected with a GFP gene. Green: GFP. Blue: cell nuclei. Scale: 10 μm.

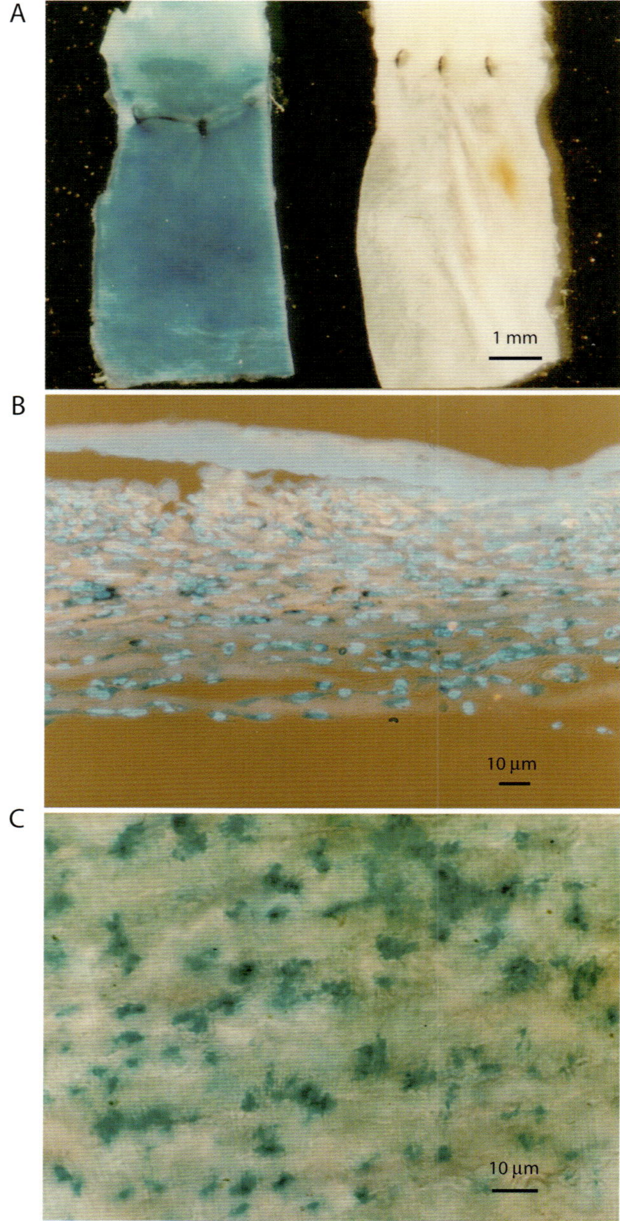

Chapter 11 opening figure. Vein graft cells transfected with a β-galactosidase (β-gal) gene vector by the mediation of electroporation. The left jugular vein of a rat was harvested, incubated in culture medium supplemented with 5 μg β-gal gene vector/100 μL, and subjected to electroporation. The transfected vein graft was incubated in culture medium for 30 min and grafted into the abdominal aorta of the rat. At day 10 after the grafting surgery, vein graft specimens were collected, fixed in 4% formaldehyde, incubated in the presence of X-gal, and observed by using an optical microscope. The blue color is indicative of the expression of the β-gal gene. (A) Vein graft specimens with (blue) and without (white) β-gal gene transfection, showing the luminal surface. (B) Transverse section of a vein graft transfected with the β-gal gene. Dark blue: β-gal. Light blue: cell nuclei labeled with Hoechst 33258. Yellow: smooth muscle α-actin labeled with an anti-smooth muscle α-actin antibody. (C) En face observation of β-gal-positive cells (blue) on the luminal surface of a vein graft specimen transfected with the β-gal gene.

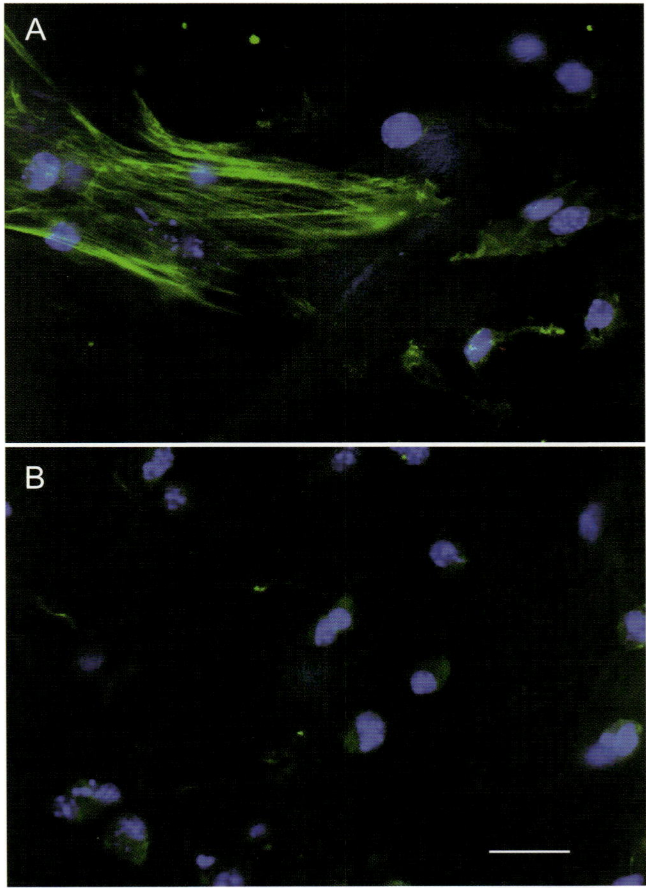

Chapter 12 opening figure. Influence of extracellular matrix on the formation of smooth muscle α-actin filaments in CD34-positive bone marrow cells in the mouse. CD34-positive bone marrow cells form smooth muscle α actin filaments (green in color) when cultured on the arterial elastic lamina matrix (A), but not on the arterial collagen (type III) matrix (B). Blue: cell nuclei. Scale: 100 μm.

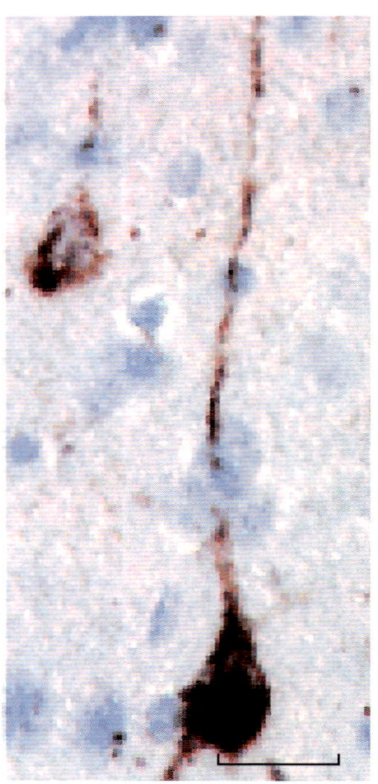

Chapter 13 opening figure. τ-protein-based tangles found in the inferior temporal gyrus of the human brain in Alzheimer's disease, stained with an antibody against the *N*-terminal sequences of the τ protein. Scale bar: 20 μm (Reprinted with permission from Horowitz PM et al: Early N-terminal changes and caspase-6 cleavage of τ in Alzheimer's disease, *J Neurosc* 24:7895–902, copyright 2004.)

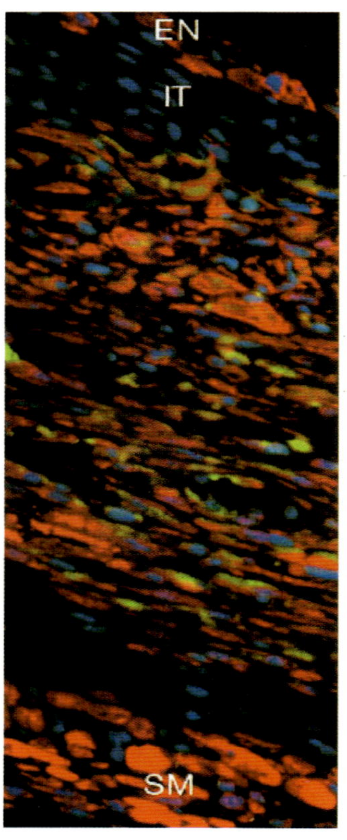

Chapter 14 opening figure. Transformation of bone marrow cells to infarcted heart. Myocardial infarction was induced and EGFP-Lin-c-*kit*^{POS} bone marrow cells were injected into the infarcted heart. Infarcted tissue (IT) can be seen in the subendocardium; spared myocytes (SM) can be seen in the subepicardium. Original magnification, ×250. EN: endocardium. Red: cardiac myosin. Green: EGFP-positive cells. (Reprinted by permission from Macmillan Publishers Ltd.: Orlic D et al: Bone marrow cells regenerate infarcted myocardium, *Nature* 410:701–5, copyright 2001.)

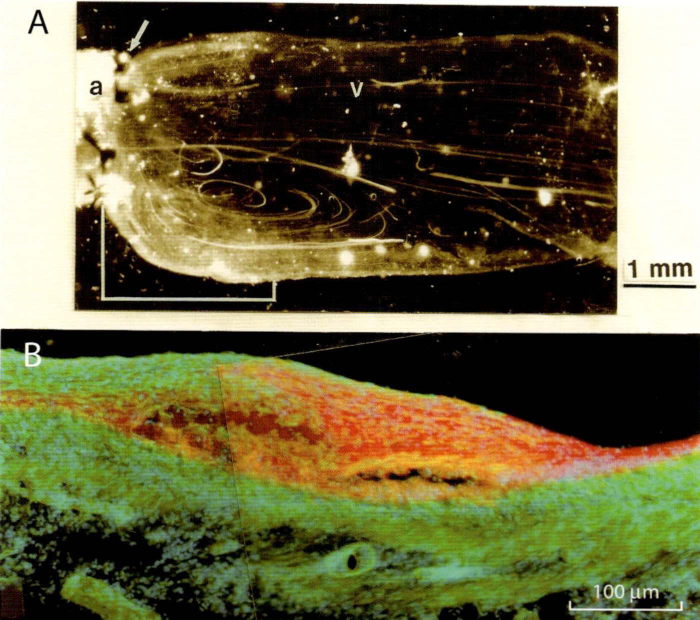

Chapter 15 opening figure. Influence of vortex bloodflow on intimal hyperplasia in experimental vein grafts. Panel A shows the bloodflow pattern in a rat vein graft. Vortex flow develops in the proximal anastomotic region due to divergent geometry from the host artery to the vein graft. Panel B is a fluorescent micrograph produced from an axial specimen section from the proximal host–vein graft junction of a rat vein graft at day 10, showing intimal hyperplasia at the site of vortex bloodflow. Red: mRNA of the angiotensin II type 1 receptor. Green: smooth muscle α-actin. Blue: cell nuclei. Note that the expression of the angiotensin II type 1 receptor, which mediates proliferative cell activities, was consistent with the presence of vortex bloodflow. When the divergent geometry and vortex bloodflow were eliminated by using a mechanical engineering approach, the expression of the angiotensin II type 1 receptor, smooth muscle cell proliferation, as well as intimal hyperplasia were significantly reduced.

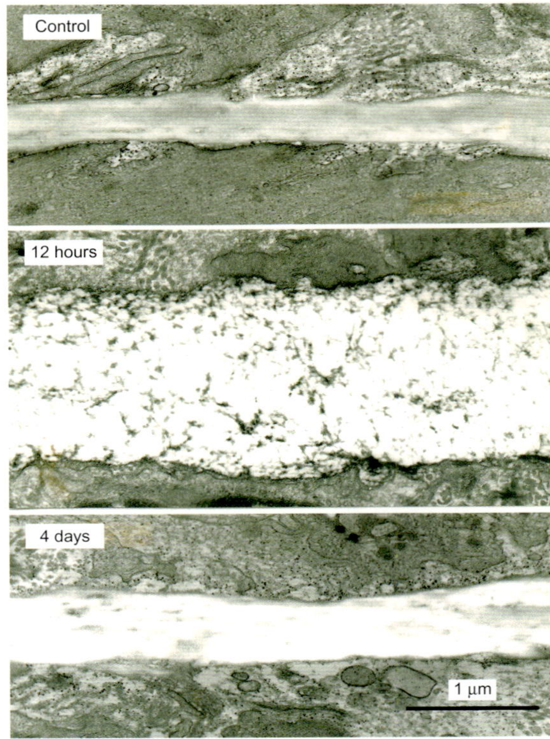

Chapter 16 opening figure. Influence of hypoxic pulmonary hypertension on the structure of the medial elastic laminae of the pulmonary arteries in the rat. Pulmonary arterial hypertension was induced by exposing rats to 10% oxygen and 90% nitrogen. The structure of the pulmonary arteries was examined by electron microscopy at 0, 0.5, and 4 days of exposure to hypoxia. An increase in the pulmonary arterial blood pressure induced rapid swelling and disorganization of the pulmonary arterial elastic laminae within 12 h of exposure to hypoxia. These pathological changes were transient. The pulmonary arterial elastic laminae regained their physiological morphology and appearance after 4 days of exposure to hypoxia without further noticeable changes even under a continuous identical hypoxic condition.

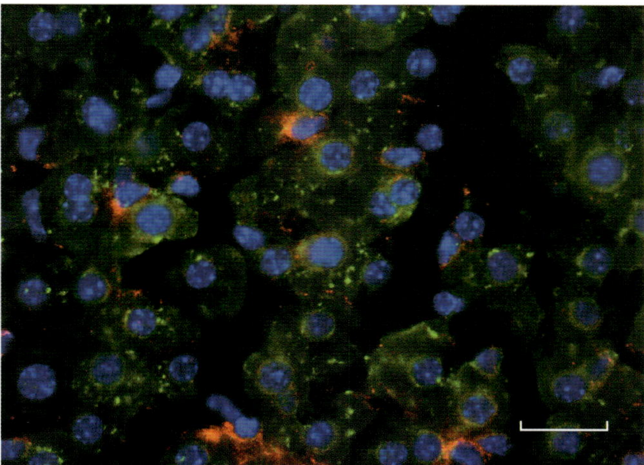

Chapter 17 opening figure. The presence of c-Kit-positive cells in the mouse liver. C-Kit is a cell membrane protein that is expressed in stem cells found in the embryo, fetus, and adult bone marrow. Red: c-Kit. Green: albumin. Blue: cell nuclei. Scale: 10 μm.

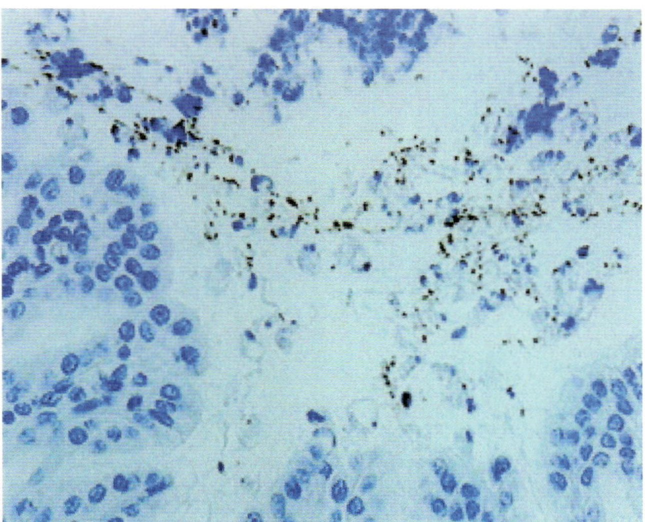

Chapter 18 opening figure. Histological micrograph showing the distribution of *Helicobacter pylori* (*H. pylori*), a type of bacteria that cause peptic ulcers in mammalian animals and humans, in the stomach of Mongolian gerbil. The *Helicobacter pylori* bacteria were detected by immunohistochemistry and appears black in color. (Reprinted with permission from Ikeno T et al: *Am J Pathol* 154:951–60, copyright 1999.)

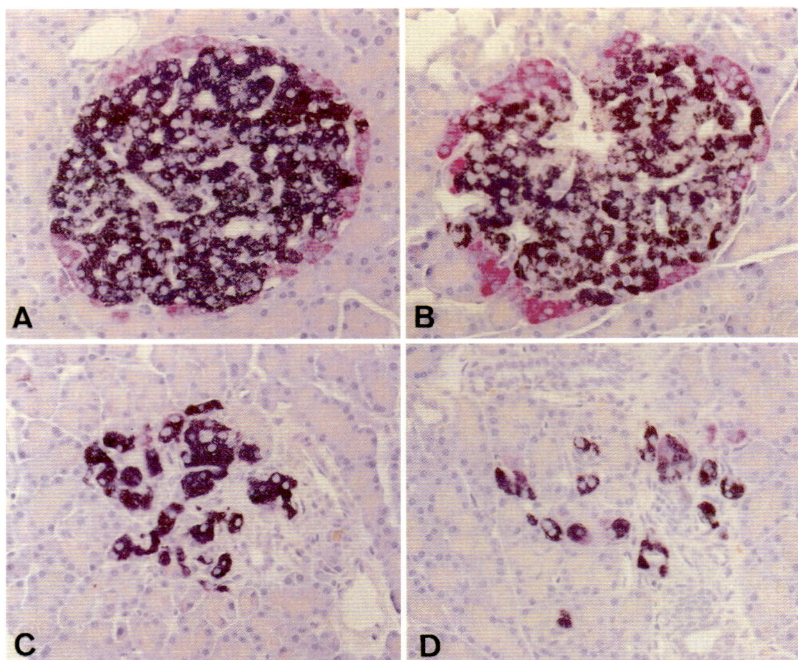

Chapter 19 opening figure. Islet structures and distribution of insulin-positive cells (black) and glucagon-positive cells (red) in Wistar rat (A), sucrose-fed Wistar rat (B), Goto–Kakizaki (GK) rat (C), and sucrose-fed GK rat (D) at 12 weeks of age. Note that the GK rat is a spontaneously diabetic animal model of non-insulin-dependent diabetes mellitus, which is characterized by progressive loss of β cells in the pancreatic islets with fibrosis. There is marked islet fibrosis with β-cell depletion in GK and sucrose-fed GK rats, in which the latter showed more severe changes. Double immunostaining for insulin and glucagon. Magnification ×300 (A–D). (Reprinted with permission from Koyama M et al: Accelerated loss of islet β cells in sucrose-fed Goto-Kakizaki rats, a genetic model of non-insulin-dependent diabetes mellitus, *Am J Pathol* 153:537–45, copyright 1998.)

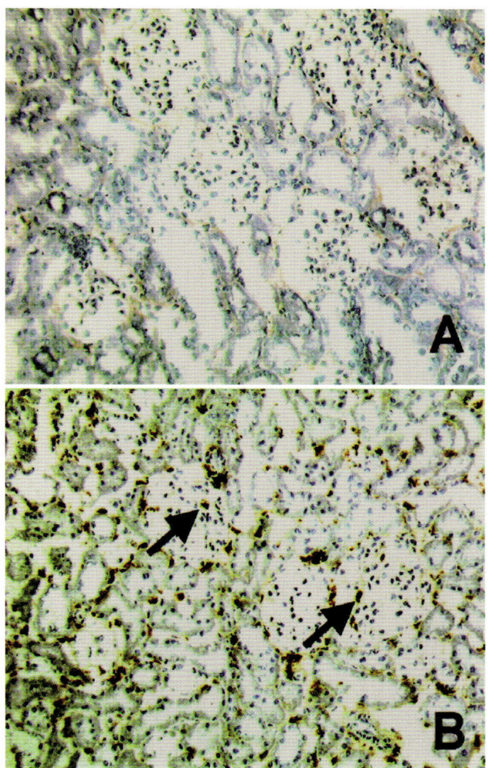

Chapter 20 opening figure. Formation of glomerular and tubular cells in the kidney of a Brown Norway/RijHsd (BN) rat from transplanted allogenic bone marrow cells derived from a WAG/RijHsd (WR) rat. The donor WR bone marrow cells wre injected into the BN rat intravenously and were detected in the recipient kidney by immunohistochemistry with an anti-WR-major histocompatibility complex class-I (MHC-I) antibody (U9F4), which reacts only with the WR MHC-I. (A) Immunoperoxidase staining of cryo-kidney sections from a control BN rat (no WR bone marrow transplantation), showing negative U9F4 staining. (B) Presence of donor WR bone marrow-derived U9F4-positive cells in the glomeruli (arrows) as well as tubulo-interstitium of the BN rat kidney 2 months after bone marrow cell transplantation (×200). Reprinted with permission from Rookmaaker MB et al., Bone-marrow-derived cells contribute to glomerular endothelial repair in experimental glomerulonephritis, Am J Pathol 163:533–562, copyright 2003.

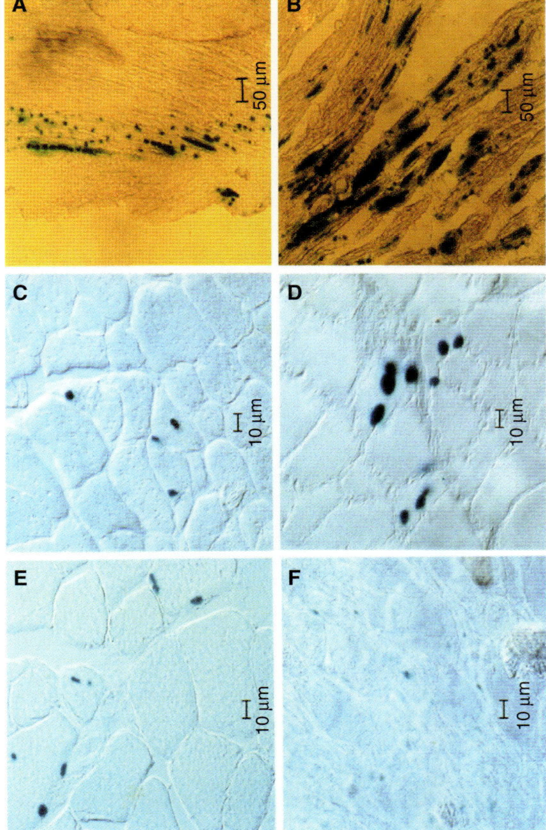

Chapter 21 opening figure. Formation of muscular cells from transplanted bone marrow cells. The images show nuclear lacZ expression in whole-mount dissected muscle fibers (A,B) or cryostat sections (C–F) of regenerating muscles of scid/bg mice. Mice were injected with unfractionated (A,C), adherent (E), or nonadherent (F) bone marrow cells, or with control satellite cells (B,D), from C57/MlacZ transgenic mice. (A,B) Brightfield; scale bars: 50 µm. (C–F) Nomarski optics; scale bars: 10 µm. (Reprinted with permission from Giuliana Ferrari G et al: Muscle regeneration by bone marrow-derived myogenic progenitors, *Science* 279:1528–30, copyright 1998 AAAS.)

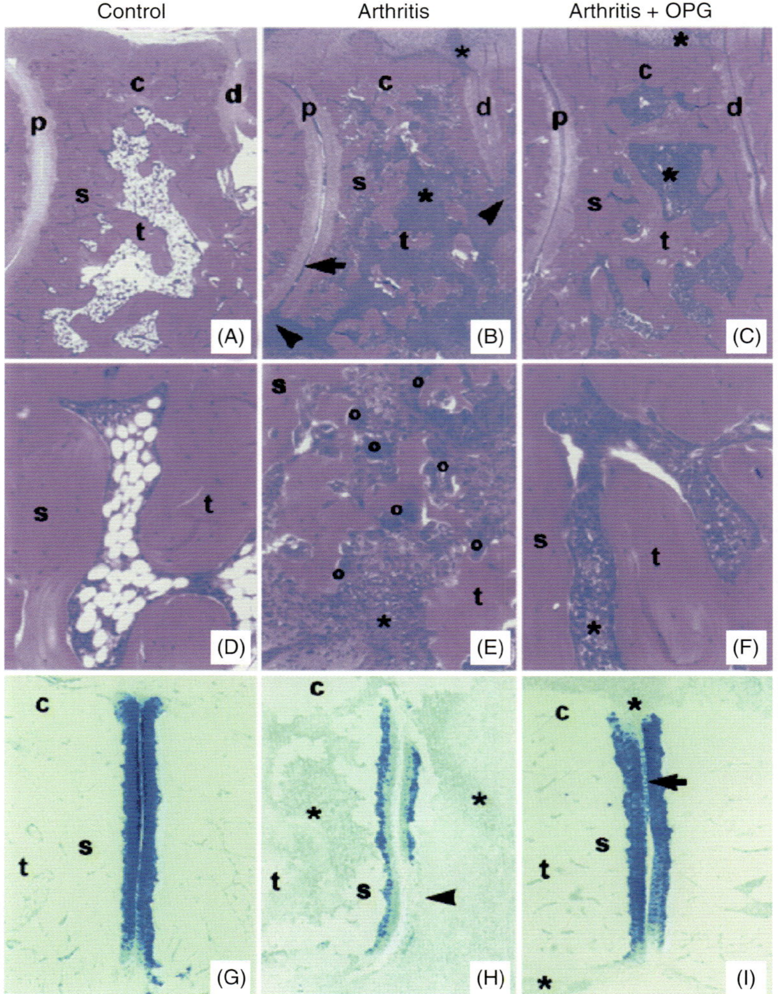

Chapter 22 opening figure. Osteoprotegerin (OPG, the decoy receptor for the tumor necrosis factor family molecule osteoprotegerin ligand or OPGL) prevents bone and cartilage destruction in the presence of severe inflammation. (A, D), Bone and joint structure in the normal hind paw showing dense bony plates, intact articular cartilages and marrow cavities containing scattered hematopoietic precursor cells. Proximal (p) distal (d) intertarsal joints. (B, E), Disrupted bone and joint structure in AdA rats (day 16) with severe mononuclear infiltration in the bone marrow (asterisk), and pannus (arrow); advanced destruction (arrowheads) of cortical (c), subchondral (s), and trabecular (t) bone; and erosion of the articular cartilages. The marrow cavity contains a marked mononuclear cell infiltration (asterisk) containing numerous osteoclasts (o). These rats show severe clinical crippling. (C, F), Preserved bone and joint structure of AdA rats (day 16) treated with OPG ($1\,mg\,kg^{-1}\,day^{-1}$) on days 9–15. OPG-treated rats exhibit extensive mononuclear cell infiltration of bone marrow (asterisk) and pannus formation (arrow), but cortical (c), subchondral (s), and trabecular (t) bone and articular cartilages are intact. Note the absence of osteoclasts as compared with non-OPG-treated rats (E). These rats do not show any clinical signs of crippling. (G) Normal cartilage integrity in control rats as determined by toluidine blue staining. The matrix of normal articular cartilage is uniformly stained, and the underlying bony plates are dense. (H) Cartilage matrix degeneration in AdA rats (day 16). Uniform pallor (small arrow) in the upper half of articular cartilages denotes extensive loss of matrix proteoglycans. Subchondral (s), cortical (c), and trabecular (t) bone is extensively eroded, and the marrow cavity is filled with inflammatory cells (asterisks). (I) OPG treatment preserves matrix proteoglycans of arthritic rats (day 16) treated with OPG ($1\,mg\,kg^{-1}\,day^{-1}$) on days 9–15. Note modest reduction of toluidine blue staining in peripheral cartilage regions that are in direct contact with inflamed synovial tissue (asterisks) or pannus (arrows). Subchondral (s), cortical (c) and trabecular (t) bone is intact. Staining: H&E (A–F); toluidine blue (G–I). Magnifications: 50× (A–C); 250× (D–F); 75× (G–I). (Reprinted by permission from Macmillan Publishers Ltd.: Kong YY et al: Activated T cells regulate bone loss and joint destruction in adjuvant arthritis through osteoprotegerin ligand, *Nature* 402:304–9, copyright 1999.)

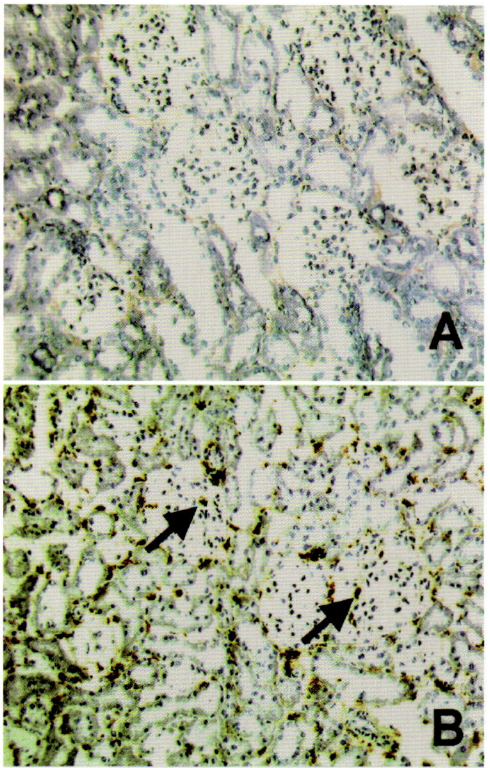

Chapter 20 opening figure. Formation of glomerular and tubular cells in the kidney of a Brown Norway/RijHsd (BN) rat from transplanted allogenic bone marrow cells derived from a WAG/RijHsd (WR) rat. The donor WR bone marrow cells wre injected into the BN rat intravenously and were detected in the recipient kidney by immunohistochemistry with an anti-WR-major histocompatibility complex class-I (MHC-I) antibody (U9F4), which reacts only with the WR MHC-I. (A) Immunoperoxidase staining of cryo-kidney sections from a control BN rat (no WR bone marrow transplantation), showing negative U9F4 staining. (B) Presence of donor WR bone marrow-derived U9F4-positive cells in the glomeruli (arrows) as well as tubulo-interstitium of the BN rat kidney 2 months after bone marrow cell transplantation (×200). Reprinted with permission from Rookmaaker MB et al., Bone-marrow-derived cells contribute to glomerular endothelial repair in experimental glomerulonephritis, Am J Pathol 163:533–562, copyright 2003.

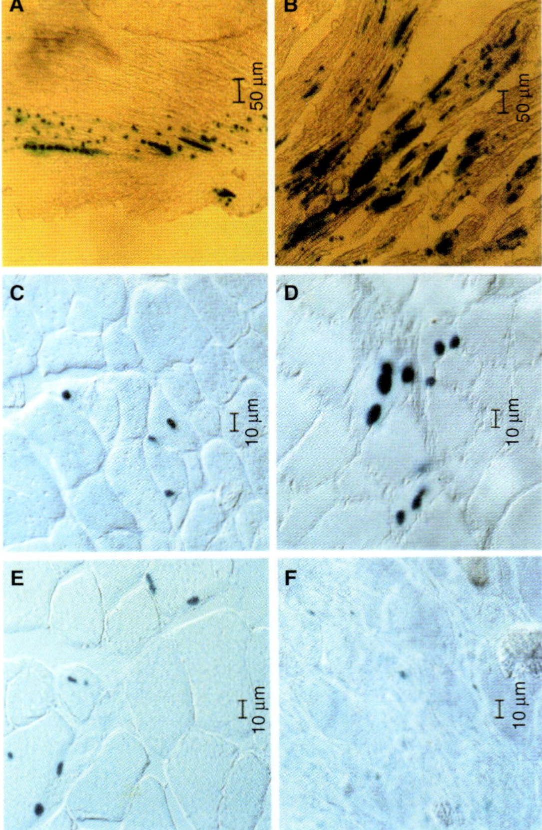

Chapter 21 opening figure. Formation of muscular cells from transplanted bone marrow cells. The images show nuclear lacZ expression in whole-mount dissected muscle fibers (A,B) or cryostat sections (C–F) of regenerating muscles of scid/bg mice. Mice were injected with unfractionated (A,C), adherent (E), or nonadherent (F) bone marrow cells, or with control satellite cells (B,D), from C57/MlacZ transgenic mice. (A,B) Brightfield; scale bars: 50 µm. (C–F) Nomarski optics; scale bars: 10 µm. (Reprinted with permission from Giuliana Ferrari G et al: Muscle regeneration by bone marrow-derived myogenic progenitors, *Science* 279:1528–30, copyright 1998 AAAS.)

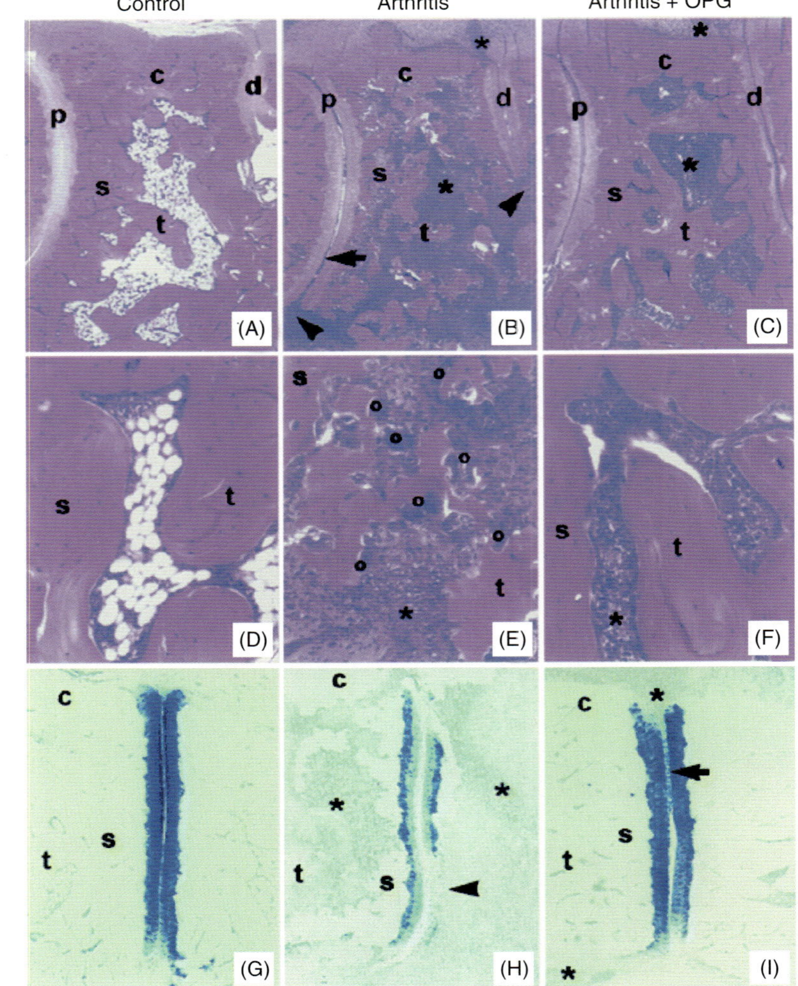

Chapter 22 opening figure. Osteoprotegerin (OPG, the decoy receptor for the tumor necrosis factor family molecule osteoprotegerin ligand or OPGL) prevents bone and cartilage destruction in the presence of severe inflammation. (A, D), Bone and joint structure in the normal hind paw showing dense bony plates, intact articular cartilages and marrow cavities containing scattered hematopoietic precursor cells. Proximal (p) distal (d) intertarsal joints. (B, E), Disrupted bone and joint structure in AdA rats (day 16) with severe mononuclear infiltration in the bone marrow (asterisk), and pannus (arrow); advanced destruction (arrowheads) of cortical (c), subchondral (s), and trabecular (t) bone; and erosion of the articular cartilages. The marrow cavity contains a marked mononuclear cell infiltration (asterisk) containing numerous osteoclasts (o). These rats show severe clinical crippling. (C, F), Preserved bone and joint structure of AdA rats (day 16) treated with OPG ($1\,mg\,kg^{-1}\,day^{-1}$) on days 9–15. OPG-treated rats exhibit extensive mononuclear cell infiltration of bone marrow (asterisk) and pannus formation (arrow), but cortical (c), subchondral (s), and trabecular (t) bone and articular cartilages are intact. Note the absence of osteoclasts as compared with non-OPG-treated rats (E). These rats do not show any clinical signs of crippling. (G) Normal cartilage integrity in control rats as determined by toluidine blue staining. The matrix of normal articular cartilage is uniformly stained, and the underlying bony plates are dense. (H) Cartilage matrix degeneration in AdA rats (day 16). Uniform pallor (small arrow) in the upper half of articular cartilages denotes extensive loss of matrix proteoglycans. Subchondral (s), cortical (c), and trabecular (t) bone is extensively eroded, and the marrow cavity is filled with inflammatory cells (asterisks). (I) OPG treatment preserves matrix proteoglycans of arthritic rats (day 16) treated with OPG ($1\,mg\,kg^{-1}\,day^{-1}$) on days 9–15. Note modest reduction of toluidine blue staining in peripheral cartilage regions that are in direct contact with inflamed synovial tissue (asterisks) or pannus (arrows). Subchondral (s), cortical (c) and trabecular (t) bone is intact. Staining: H&E (A–F); toluidine blue (G–I). Magnifications: 50× (A–C); 250× (D–F); 75× (G–I). (Reprinted by permission from Macmillan Publishers Ltd.: Kong YY et al: Activated T cells regulate bone loss and joint destruction in adjuvant arthritis through osteoprotegerin ligand, *Nature* 402:304–9, copyright 1999.)

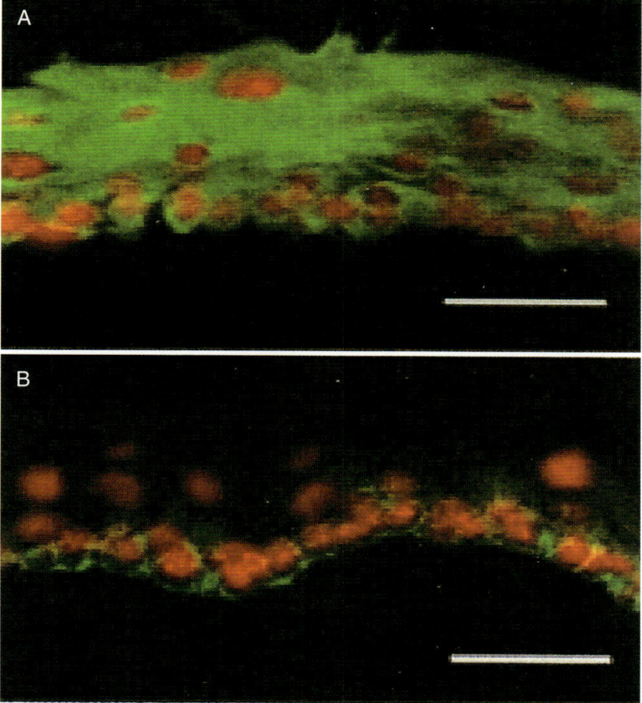

Chapter 23 opening figure. Preparation of autologous tissue-engineered epithelial cell sheets fabricated from oral mucosal epithelium. Oral mucosal tissue (3 × 3 mm) was removed from a patient's cheek. Isolated epithelial cells were seeded onto temperature-responsive cell culture inserts. After two weeks of culture at 37°C, the cells grow to form multilayered sheets of epithelial cells. These sheets were used to cover and repair injured corneal epithelium. Specimens were collected for testing the expression of keratin 3 (A) and anti-β1 integrin (B) by immunohistochemistry. Red: cell nuclei. Scale bars: 50 μm. (Reprinted with permission from Nishida K et al: Corneal reconstruction with tissue-engineered cell sheets composed of autologous oral mucosal epithelium, *New Engl J Med* 351:1187–96, copyright 2004 Massachusetts Medical Society. All rights reserved.)

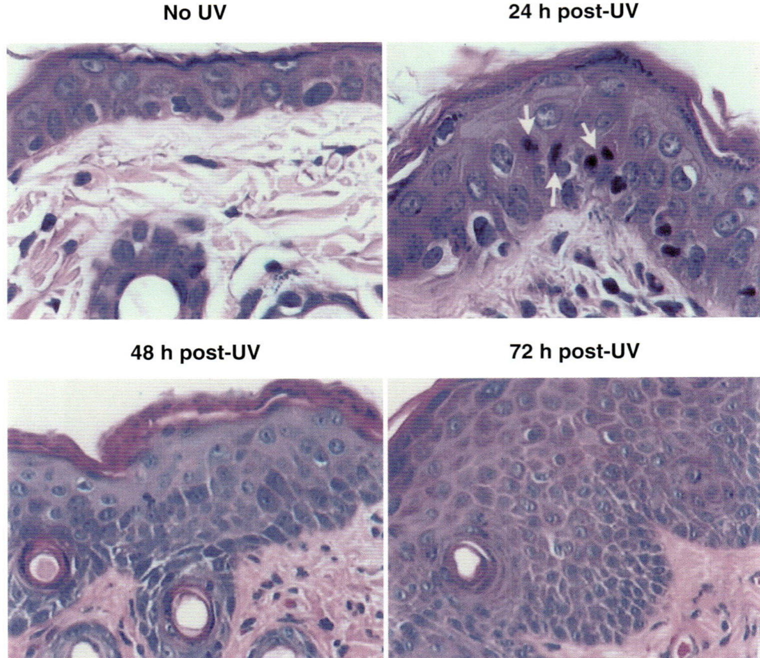

Chapter 24 opening figure. Induction of sunburn cells and epidermal hyperplasia in unirradiated and UV-irradiated mouse skin at various times after UV irradiation. Arrows indicate representative sunburn cells. Original magnification, ×400 for unirradiated and 24 h post-UV and ×200 for 48 and 72 h post-UV. (Reprinted with permission from Ouhtit A et al: Temporal events in skin injury and the early adaptive responses in ultraviolet-irradiated mouse skin, *Am J Pathol* 156:201–7, copyright 2000.)

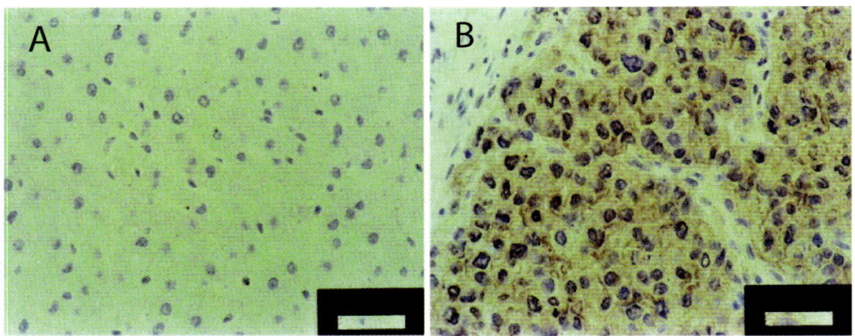

Chapter 25 opening figure. Cell nucleus density and size in control rat liver and hepatoma: (A) control rat liver; (B) hepatoma. Scale bar: 50 μm. (Reprinted with permission from la Cour JM et al: Up-regulation of ALG-2 in hepatomas and lung cancer tissue, *Am J Pathol* 163:81–9, copyright 2003.)

Ramirez F: Pathophysiology of the microfibril/elastic fiber system: introduction, *Matrix Biol* 19:455–6, 2000.

Robert L: Interaction between cells and elastin, the elastin-receptor, *Connect Tissue Res* 40:75–82, 1999.

12.13. Polysaccharides

Suh JK, Matthew HW: Application of chitosan-based polysaccharide biomaterials in cartilage tissue engineering: A review, *Biomaterials* 21:2589–98, 2000.

Klemm D, Schumann D, Udhardt U, Marsch S: Bacterial synthesized cellulose—artificial blood vessels for microsurgery, *Prog Polym Sci* 26:91561–603, 2001.

Entcheva E, Bien H, Yin L, Chiung-Yin Chung CY, Farrell F et al: Functional cardiac cell constructs on cellulose-based scaffolding, *Biomaterials* 25:5753–62, 2004.

Svensson A, Nicklasson E, Harrah T, Panilaitis B, Kaplan DL et al: Bacterial cellulose as a potential scaffold for tissue engineering of cartilage, *Biomaterials* 26(4):419–31, Feb 2005.

De Bartolo L, Morelli S, Bader A, Drioli E: Evaluation of cell behaviour related to physico-chemical properties of polymeric membranes to be used in bioartificial organs, *Biomaterials* 23:2485–97, 2002.

Entcheva EG, Yotova LK: Analytical application of membranes with covalently bound glucose-oxidase, *Anal Chim Acta* 299:171–7, 1994.

Doheny JG, Jervis EJ, Guarna MM, Humphries RK, Warren RAJ et al: Cellulose as an inert matrix for presenting cytokines to target cells: Production and properties of a stem cell factor-cellulose-binding domain fusion protein, *Biochem J* 339:429–34, 1999.

Martson M, Viljanto J, Hurme T, Saukko P: Biocompatibility of cellulose sponge with bone, *Eur Surg Res* 30:426–32, 1998.

Risbud MV, Bhonde RR, Suitability of cellulose molecular dialysis membrane for bioartificial pancreas: in vitro biocompatibility studies, *J Biomed Mater Res* 54:436–44, 2001.

Cullen B, Watt PW, Lundqvist C, Silcock D, Schmidt RJ et al: The role of oxidised regenerated cellulose/collagen in chronic wound repair and its potential mechanism of action, *Int J Biochem Cell Biol* 34:1544–56, 2006.

12.14. Alginates

Smidsrød O, Skjåk-Bræk G: Alginate as immobilization matrix for cells, *Trends Biotechnol* 8:71–8, 1990.

Drury JL, Dennis RG, Mooney DJ: The tensile properties of alginate hydrogels, *Biomaterials* 25:3187–99, 2004.

Klöck G, Pfeffermann A, Ryser C, Gröhn P, Kuttler B et al: Biocompatibility of mannuronic acid-rich alginate, *Biomaterials* 18:707–13, 1997.

Johnson FA, Craig DQM, Mercer AD: Characterization of the block structure and molecular weight of sodium alginates, *J Pharm Pharmacol* 49:639–43, 1997.

Eiselt P, Lee KY, Mooney DJ: Rigidity of two-component hydrogels prepared from alginate and poly(ethylene glycol)–diamines, *Macromolecules* 32:5561–6, 1999.

Kim BS, Nikolovski J, Bonadio J, Mooney DJ: Cylic mechanical strain regulates the development of engineered smooth muscle tissue, *Nat Biotechnol* 17:979–83, 1999.

de Chalain T, Phillips JH, Hinek A: Bioengineering of elastic cartilage with aggregated porcine and human auricular chondrocytes and hydrogels containing alginate, collagen, and *k*-elastin. *J Biomed Mater Res* 44:280–8, 1999.

Chang SCN, Rowley JA, Tobias G, Genes NG, Roy AK et al: Injection molding of chondrocyte/alginate constructs in the shape of facial implants, *J Biomed Mater Res* 55:503–11, 2001.

Suzuki Y, Tanihara M, Suzuki K, Saitou A, Sufan W et al: Alginate hydrogel linked with synthetic oligiopeptide derived from BMP-2 allows ectopic osteoinduction in vivo, *J Biomed Mater Res* 50:405–9, 2000.

Lee KY, Alsberg E, Mooney DJ: Degradable and injectable poly(aldehyde guluronate) hydrogels for bone tissue engineering, *J Biomed Mater Res* 56:228–33, 2001.

Rowley JA, Mooney DJ: Alginate type and RGD density control myoblast phenotype, *J Biomed Mater Res* 60:217–23, 2002.

Rowley JA, Sun Z, Goldman D, Mooney DJ: Biomaterials to spatially regulate cell fate, *Adv Mater* 14:886–9, 2002.

12.15. Chitosan

Suh JK, Matthew HW: Application of chitosan-based polysaccharide biomaterials in cartilage tissue engineering: A review, Biomaterials 21:2589–98, 2000.

Khor E, Lim LY: Implantable applications of chitin and chitosan, *Biomaterials* 24:2339–49, 2003.

Yamane S, Iwasaki N, Majima T, Funakoshi T, Masuko T et al: Feasibility of chitosan-based hyaluronic acid hybrid biomaterial for a novel scaffold in cartilage tissue engineering, *Biomaterials* 26:611–9, 2005.

Madihally SV, Matthew HW: Porous chitosan scaffolds for tissue engineering, *Biomaterials* 20:1133–42, 1999.

Sechriest VF, Miao YJ, Niyibizi C, Westerhausen-Larson A, Matthew HW et al: GAG-augmented polysaccharide hydrogel: A novel biocompatible and biodegradable material to support chondrogenesis, *J Biomed Mater Res* 49:534–41, 2000.

Nettles DL, Elder SH, Gilbert JA: Potential use of chitosan as a cell scaffold material for cartilage tissue engineering, *Tissue Eng* 8:1009–16, 2002.

Khor E, Lim LY: Implantable applications of chitin and chitosan, *Biomaterials* 24:2339–49, 2003.

Hirano S, Midorikawa T: Novel method for the preparation of N-acylchitosan fiber and N-acylchitosan–cellulose fiber, *Biomaterials* 19:293–7, 1998.

Gaserod O, Smidsrod O, Skjak-Braek G: Microcapsules of alginate–chitosan—I. A quantitative study of the interaction between alginate and chitosan, *Biomaterials* 19:1815–25, 1998.

Denuziere A, Ferrier D, Damour O, Domard A: Chitosan–chondroitin sulfate and chitosan–hyaluronate polyelectrolyte complexes: biological properties, *Biomaterials* 19:1275–85, 1998.

MacLaughlin FC, Mumper RJ, Wang J, Tagliaferri JM, Gill I et al: Chitosan and depolymerized chitosan oligomers as condensing carriers for in vivo plasmid delivery, *J Control Release* 56:259–72, 1998.

Roy K, Mao HQ, Huang SK, Leong KW: Oral gene delivery with chitosan–DNA nanoparticles generates immunologic protection in a murine model of peanut allergy, *Nature Med* 5:387–91, 1999.

Usami Y, Okamoto Y, Takayama T, Shigemasa Y, Minami S: Chitin and chitosan stimulate canine polymorphonuclear cells to release leukotriene B4 and prostaglandin E2, *J Biomed Mater Res* 42:517–22, 1998.

Onishi H, Machida Y: Biodegradation and distribution of water-soluble chitosan in mice, *Biomaterials* 20:175–82, 1999.

Madihally SV, Matthew HW: Porous chitosan scaffolds for tissue engineering, *Biomaterials* 20:1133–42, 1999.

Lu JX, Prudhommeaux F, Meunier A, Sedel L, Guillemin G: Effects of chitosan on rat knee cartilages, *Biomaterials* 20:1937–44, 1999.

12.16. Starch

Marques AP, Reis RL, Hunt JA: The biocompatibility of novel starch-based polymers and composites: in vitro studies, *Biomaterials* 23:1471–8, 2002.

Reis RL, Cunha AM: Characterisation of two biodegradable polymers of potential application within the biomedical field, *J Mater Sci Mater Med* 6:786–92, 1995.

Reis RL, Mendes SC, Cunha AM, Bevis M: Processing and in-vitro degradation of starch/EVOH thermoplastic blends, *Polym Int* 43:347–53, 1997.

Gomes ME, Ribeiro AS, Malafaya PB, Reis RL, Cunha AM: A new approach based on injection moulding to produce biodegradable starch-based polymeric scaffolds: Morphology, mechanical and degradation behaviour, *Biomaterials* 22:883–9, 2001.

Reis RL, Cunha AM, Bevis MJ: Structure development and control of injection moulded hydroxylapatite reinforced starch/EVOH composites, *Adv Polym Technol* 16:263–77, 1997.

12.17. Glycosaminoglycans

Detamore MS, Athanasiou KA: Motivation, characterization, and strategy for tissue engineering the temporomandibular joint disc, *Tissue Eng* 9:1065–87, 2003.

Hubbell JA: Materials as morphogenetic guides in tissue engineering, *Curr Opin Biotechnol* 14:551–8, 2003.

Spector M: Novel cell-scaffold interactions encountered in tissue engineering: Contractile behavior of musculoskeletal connective tissue cells, *Tissue Eng* 8:351–7, 2002.

Caplan AI: Tissue engineering designs for the future: new logics, old molecules, *Tissue Eng* 6:1–8, 2000.

Vercruysse KP, Prestwich GD: Hyaluronate derivatives in drug delivery, *Crit Rev Ther Drug Carrier Syst* 15:513–55, 1998.

12.18. Metallic Materials as Biomaterials

Enderle J, Blanchard S, Bronzino J: in *Introduction to Biomedical Engineering*, Academic Press, San Diego, 2000, Chap 11.

Bronzino JD: *The Biomedical Engineering Handbook*, CRC Press and IEEE Press, 1995, Chap 40.

American Society for Testing and Materials: *Annual Book of ASTM Standards*, Vol 13: *Medical Devices and Services, American Society for Testing and Materials, Philadelphia*, 1992.

12.19. Ceramics as Biomaterials

Park JB, Lakes RS: *Biomaterials: An Introduction*, 2nd ed, Plenum Press, New York, 1992.

Ducheyne P: Bioglass coating and bioglass composites as implant materials, *J Biomed Mater Res* 19:273, 1985.

Hench LL: Bioceramics: From concept to clinic, *J Am Ceramic Soc* 74:1487, 1991.

Hench LL, Xynos ID, Polak JM: Bioactive glasses for in situ tissue regeneration, *J Biomater Sci Polym Ed* 15:543–62, 2004.

Day RM, Boccaccini AR, Shurey S, Roether JA, Forbes A et al: Assessment of polyglycolic acid mesh and bioactive glass for soft-tissue engineering scaffolds, *Biomaterials* 25:5857–66. 2004.

Oonishi H: Bioceramic in orthopedic surgery—our clinical experiences, in *Bioceramics* Vol 3, Hulbert JE, Hulbert SF eds., Rose Hulman Inst Technology, Terre Haute, IN, 1992, pp 31–42.

Knabe C, Driessens FC, Planell JA, Gildenhaar R, Berger G et al: Evaluation of calcium phosphates and experimental calcium phosphate bone cements using osteogenic cultures, *J Biomed Mater Res* 52:498–508, 2000.

Knabe C, Berger G, Gildenhaar R, Howlett CR, Markovic B et al: The functional expression of human bone-derived cells grown on rapidly resorbable calcium phosphate ceramics, *Biomaterials* 25:335–44, 2004.

Bajpai PK, Billotte WG: Ceramic biomaterials, in *The Biomedical Engineering Handbook*, Bronzino JD, ed, CRC Press and IEEE Press, 1995, Chap 41.

Kaito T, Myoui A, Takaoka K, Saito N, Nishikawa M et al: Potentiation of the activity of bone morphogenetic protein-2 in bone regeneration by a PLA-PEG/hydroxyapatite composite, *Biomaterials* 26:73–9, 2005.

SECTION 5

APPLICATION OF BIOREGENERATIVE ENGINEERING TO ORGAN SYSTEMS

13

NERVOUS REGENERATIVE ENGINEERING

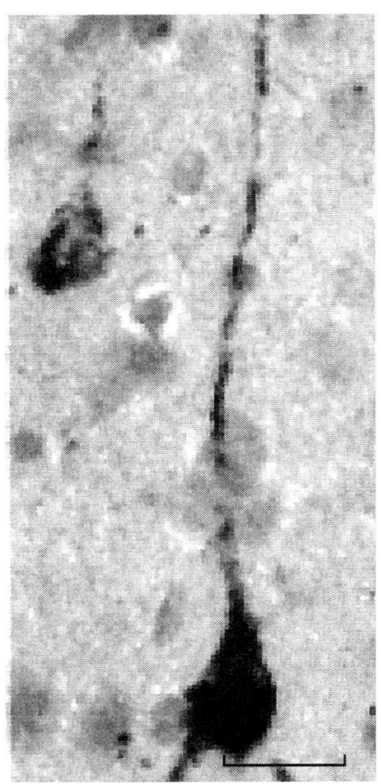

τ-protein-based tangles found in the inferior temporal gyrus of the human brain in Alzheimer's disease, stained with an antibody against the *N*-terminal sequences of the τ protein. Scale bar: 20 μm (Reprinted with permission from Horowitz PM et al: Early N-terminal changes and caspase-6 cleavage of τ in Alzheimer's disease, *J Neurosc* 24:7895–902, copyright 2004.) See color insert.

Bioregenerative Engineering: Principles and Applications, by Shu Q. Liu

The nerve system controls the activity and function of peripheral tissues and organs. The disorder of the nerve system often leads to the malfunction of not only the nervous system but also the peripheral tissues and organs. Neural disorders can be induced by a variety of factors, ranging from mechanical injury to gene mutation. In spite of extensive investigations, the mechanisms of most neural disorders remain poorly understood. Few effective approaches have been established for the treatment of neural diseases. Recent work in neural regenerative medicine and engineering has provided a basis for the establishment of molecular and cell regenerative engineering approaches for the treatment of neural disorders. Although most of these approaches have not been tested in clinical trials, preliminary investigations have demonstrated the potential of using these approaches for therapeutic purposes. In this chapter, we will focus on the organization and function of the nerve system, the pathogenic mechanisms and conventional treatment of common neural disorders, as well as recent development in nerve regenerative engineering.

ANATOMY AND PHYSIOLOGY OF THE CENTRAL AND PERIPHERAL NERVOUS SYSTEMS [13.1]

Neural Cells

The nerve system is composed of neurons and nonneuronal cells. *Neurons* are cells that receive and process signals and emit signals to control the activity and function of effector cell types. *Nonneuronal cells* are known as glial cells, which provide support and protection to the neurons.

Neurons. A neuron is composed of several parts: a cell body, short processes known as dendrites, and a long process known as the *axon*, which is also referred to as a *nerve fiber* (Fig. 13.1). The cell body contains typical cellular components such as the nucleus,

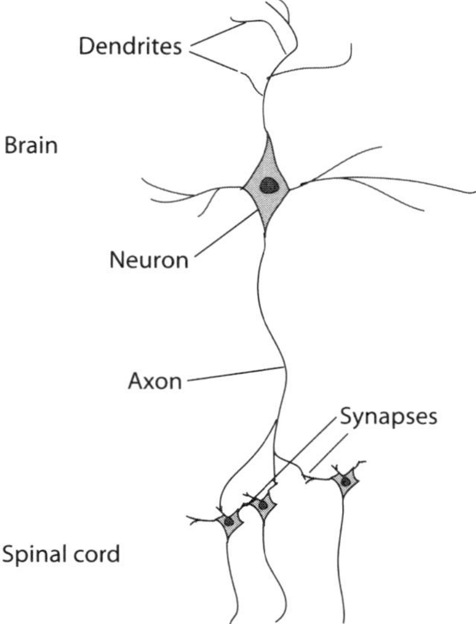

Figure 13.1. Schematic representation of the structure of neurons.

endoplasmic reticulum, Golgi apparatus, endosomes, mitochondria, and cytoskeletal filaments, including actin filaments, neurofilaments, and microtubules. Note that the neurofilaments are intermediate filaments in the neuron. *Dendrites* are short processes that are extensions from the cell body and are responsible for communication with other neurons and transmission of signals from and to the cell body. The *axon* is a long process extended from the cell body, and transmits signals (action potentials) from the cell body to peripheral effector cells via synapses (Fig. 13.2). Chemical substances such as acetylcholine and norepinephrine (Fig. 13.3), known as neurotransmitters, are required for signal transmission through the synapses. Action potentials from the presynaptic terminal stimulate the release of a neurotransmitter, which in turn activates the postsynaptic cell membrane to elicit action potentials. In the central nervous system, synapses are found between different neurons. In the peripheral systems, synapses can be found between peripheral neurons as well as between neurons and skeletal muscle cells. An axon is enclosed within a myelin sheath formed by a glial cell type known as *Schwann cells* in the peripheral systems.

On the basis of function, neurons are classified into sensory neurons, motor neurons, and interneurons. The *sensory neurons*, also known as *afferent neurons*, receive and transmit signals from the peripheral systems to the central nervous system. The *motor neurons*, known as *efferent neurons*, process received signals and emit signals that control peripheral cells such as skeletal muscle cells and gland cells. The *interneurons* transmit signals between different neurons.

Based on the structure of the cell, neurons can be classified into three types: unipolar, bipolar, and multipolar neurons. *Unipolar neurons* exhibit only a single process from the

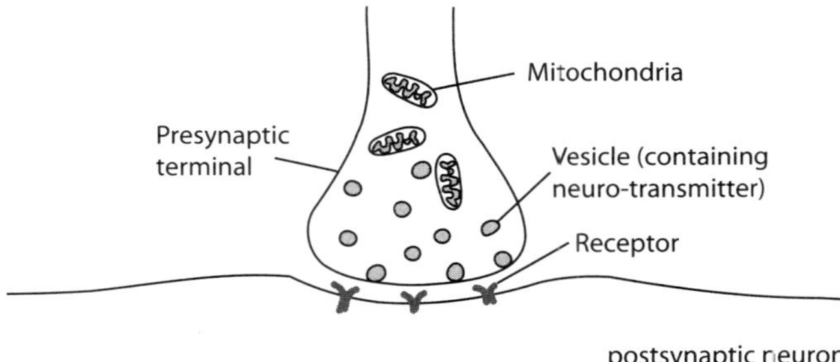

Figure 13.2. Schematic representation of the neuronal synapse. Based on bibliography 13.1.

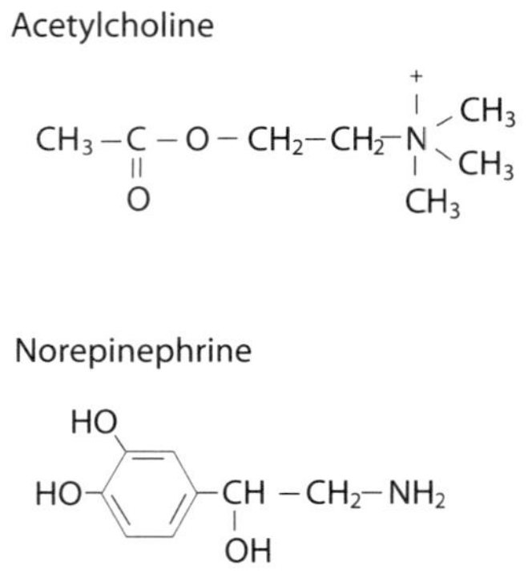

Acetylcholine

$CH_3-C(=O)-O-CH_2-CH_2-N^+(CH_3)_3$

Norepinephrine

$(HO)_2C_6H_3-CH(OH)-CH_2-NH_2$

Figure 13.3. Chemical structure of acetylcholine and norepinephrine.

cell body. The process is divided into two branches; one reaches a peripheral tissue with its end structures serving as sensory receptors and the other branch transmits sensory signals to the central nervous system. These neurons often serve as sensory neurons. *Bipolar neurons* possess an axon and a dendrite. The dendrite receives signals from a peripheral tissue and the axon transmits the signals to the central nerve system. This type of neuron also serves as sensory neuron. *Multipolar neurons* possess one axon and more than one dendrites. These are the most popular neurons and are found in the brain and spinal cord.

Glial Cells. Glial cells support and protect the neurons. There are four types of glial cells: astrocyte, microglia, oligodendrocyte, and ependymal cell. *Astrocytes* are glial cells with multiple processes and are often distributed around the neurons and capillaries. These cells participate in the regulation of the brain fluid composition. *Microglial cells* are macrophages of the central nervous system and are capable of phagocytosing and degrading microorganisms, cell debris, and harmful substances in the brain. *Oligodendrocytes* are cells that wrap axons and form axon myelin sheaths in the central nervous system (note that the cell type that wraps peripheral nerve axons is known as the *Schwann cell*). *Ependymal cells* are cells that line the ventricle cavities of the brain and the central canal of the spinal cord. These cells secrete cerebrospinal fluid in the brain ventricles.

Organization of the Nervous System

The nervous system is composed of two subdivisions: the central and peripheral nervous systems. The *central nervous system* includes the brain and spinal cord, whereas the *peripheral nerve system* contains neural cells and structures external to the central nervous system, including peripheral sensory receptors, ganglia, and nerve fibers.

Brain. The brain is composed of four major parts: cerebrum, diencephalon, cerebellum, and brainstem and is the center of the nervous system.

Cerebrum. The *cerebrum* includes the left and right hemispheres. Each hemisphere is composed of a number of lobes, including the *frontal lobe* (controlling thinking, motivation, mood, judgment, and the voluntary motor function), *parietal lobe* (controlling the integration of sensory information except for smelling, hearing, and vision), *occipital lobe* (controlling the integration of visual information), and *temporal lobe* (controlling memory and the integration of hearing and smelling).

Each lobe is composed of three distinct structures: the cortex, cerebral medulla, and cerebral nuclei. The *cortex* is the external layer of cerebrum known as the *gray matter*, which contains primarily neurons. The *cerebral medulla* is the structure between the cortex and the cerebral nuclei, known as *white matter*, and is composed of nerve fibers that link various parts of the central nervous system. The nerve fibers that connect neurons within the same hemisphere are defined as *association fibers*, those that connect the two hemispheres are *commissural fibers*, and those that connect the cerebrum and other parts of the brain are *projection fibers*. Each hemisphere contains a set of *basal nuclei*, which are neuron-containing gray matter, located laterally in regions inferior to the cerebrum and diencephalon. These nuclei include the lentiform nucleus, caudate nucleus, subthalamic nucleus, and substantia nigra. The lentiform and caudate nuclei together are known as the *corpus striatum*. The main function of the basal nuclei is to control motor activities.

The cerebrum also includes a limbic system. The *limbic system* is composed of certain cortical regions, certain nuclei such as the anterior nucleus of the thalamus and the habenular nucleus of the epithalamus, certain basal nuclei, the hypothalamus, the olfactory cortex, and nerve fibers that link various cortical regions and nuclei. The limbic system controls survival functions such as reproduction, memory, and nutrition-related activities.

Diencephalon. The *diencephalon* is a structure located between the cerebrum and the brainstem. It is composed of four parts: thalamus, subthalamus, epithalamus, and hypothalamus. The *thalamus* includes a group of nuclei: the medial geniculate, lateral geniculate, ventral posterior, ventral anterior, ventral lateral, lateral dorsal, and lateral posterior nucleus. The overall function of thalamus is to receive and process sensory information from the peripheral nervous system. Auditory information goes to the medial geniculate nucleus and visual information goes to the lateral geniculate nucleus. The ventral anterior and lateral nuclei participate in the regulation of motor activities. The lateral dorsal nucleus is linked to the cerebral cortex and other thalamus regions and is related to the control of emotion. The lateral posterior nucleus is responsible for the integration of sensory inputs from the peripheral nervous system.

The *subthalamus* is a structure below the thalamus and is composed of subthalamus nuclei and several nerve fiber bundles. The subthalamus participates in the regulation of motor activities. The *epithalamus* is a structure located above the thalamus and is composed of habenular nuclei and the pineal body. The habenular nuclei participate in the processing of smell and odor inputs. The pineal body may be involved in the regulation of the sleep cycle.

The *hypothalamus* is composed of several nuclei and nerve fiber bundles. The nuclei include the anterior, supraoptic, paraventricular, dorsomedial, posterior, and ventromedial nucleus, and the mammillary body. These nuclei play critical roles in the regulation of several functions, including cardiac contractility, bloodflow control, urine release, ion balance, sexual activity, sweat production, food and water intake, feeling and emotion, and sleep cycle. One of the most important functions for hypothalamus is to control hormone secretion from the *pituitary gland*, a structure located below the hypothalamus and connected to the hypothalamus via the infundibulum. The hypothalamus produces and releases a number of hormones, including growth hormone-releasing hormone, growth hormone-inhibiting hormone, gonadotropin-releasing hormone, corticotropin-releasing hormone, and thyroid-releasing hormone. These hormones can act on hormone-generating cells in the pituitary gland and control the secretion of the pituitary gland. The growth hormone-releasing hormone stimulates the secretion of growth hormone from the pituitary gland, whereas the growth hormone-inhibiting hormone exerts an opposite effect. The gonadotropin-releasing, corticotropin-releasing, and thyroid-releasing hormones stimulate the release of corresponding hormones from the pituitary gland. The pituitary hormones play a critical role in regulating a variety of functions, ranging from growth, water absorption, thyroid hormone secretion from the thyroid gland, glucocorticoid hormone secretion from the cortex of the adrenal gland, and follicle maturation and estrogen secretion in ovaries and sperm cell production in testes.

Cerebellum. The *cerebellum* is a structure located below the cerebrum and behind the brainstem. As the cerebrum, the cerebellum contains gray cortex, nuclei, and white medulla. The cerebellum also contains nerve fiber bundles that connect the cerebellum to other parts of the brain, including the cerebrum, brainstem, and diencephalon. The cerebellum is divided into three parts: the flocculonodular lobe, vermis, and two symmetric lateral hemispheres. The flocculonodular lobe is involved in the regulation of body balance and eye movement. The vermis and part of the lateral hemispheres are responsible for the control of the posture, movement, and motion coordination. The remaining lateral hemispheres, together with the cerebrum, control the process of learning complex skills.

Brainstem. The *brainstem* is a structure that connects the spinal cord to the brain and is composed of the medulla oblongata, pons, and midbrain (Fig. 13.4). The brainstem is responsible for the regulation of physiological activities that are critical to the survival of the body, such as cardiac contraction, the maintenance of bloodflow and pressure, respiration, swallowing, and vomiting. The *medulla oblongata* is the lower part of the brainstem and contains a number of medullary nuclei. These nuclei serve as central nerve regulatory centers that are composed of neuron-containing gray matter and control the activity of the heart and blood vessels. The medulla oblongata also contains ascending and descending nerve fiber bundles, which connect the spinal cord to the brain. It is important to note that the medullary descending nerve fibers originated from the left and right hemispheres are projected to the opposite side and cross at the lower part of the medulla oblongata. Thus, nerve fibers from the left hemisphere control the activities of the right body, whereas those from the right hemisphere control the activities of the left body. The brainstem also contains nuclei that control the body balance and motion coordination.

The *pons* is a structure located in the middle of the brainstem and contains a number of nuclei and ascending and descending nerve bundles. Some of the nuclei serve as the central nerve centers that control respiration and the sleep cycle. Other nuclei mediate the communication between the cerebrum and cerebellum. A number of cranial nerve nuclei, including those for nerve VI (abducens), VII (facial), VIII (vestibulocochlear), and IX (glossopharyngeal), are located in the pons (see below).

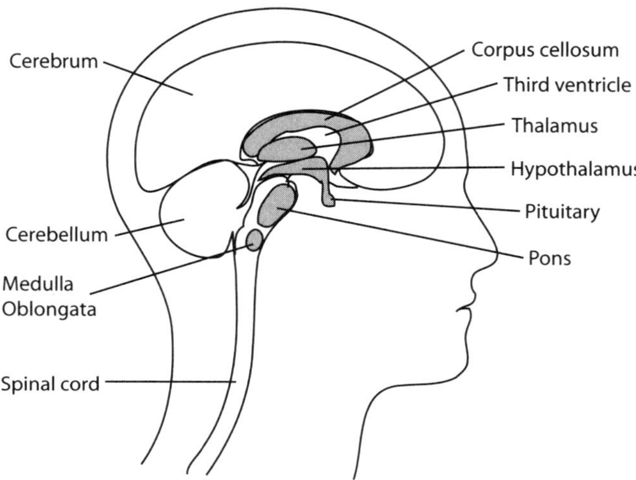

Figure 13.4. Schematic representation of the midbrain. Based on bibliography 13.1.

The *midbrain* is a structure located above the pons and contains the nuclei of cranial nerve III (oculomotor), IV (trochlear), and V (trigeminal). These nerve fibers are described in the following section. The midbrain also contains nuclei that receive and process hearing information and serve as the control center for the auditory system. The ascending and descending nerve fiber bundles pass through the midbrain. In addition, there are scattered neurons throughout the brainstem. These neurons constitute a structure called reticular formation or system. This system controls the consciousness and sleep–wake cycles.

Cranial Nerves. Cranial nerves are nerve fiber bundles that arise directly from the brain, instead of the spinal cord. There are 12 pairs of cranial nerves (Fig. 13.5). Some of these nerves are responsible for the transmission of sensory and somatic motor information, while others transmit information for the parasympathetic system, which regulates the function of the heart, large blood vessels, and glands.

Cranial nerves I and II arise from the lower part of the cerebrum and are responsible for receiving and processing smell and visual information, respectively. These nerves are known as olfactory and optic nerves. The remaining 10 nerves are from the brainstem. *Cranial nerve III* is known as oculomotor nerve and is responsible for regulating the activities of the eye. This nerve contains parasympathetic nerve fibers, which regulate the contractility of smooth muscle cells in the eye and therefore control the size of the pupil and the shape of the eye lens. *Cranial nerve IV* (trochlear nerve) controls the contraction of eyeball muscles and thus the eyeball movement. *Cranial nerve V* (trigeminal nerve) controls the contraction of ear and throat muscles and also controls the proprioceptive and cutaneous sensory function. *Cranial nerve VI* (abducens), together with nerve IV, controls the movement of the eyeball. *Cranial nerve VII* (facial nerve) controls the contraction of facial, ear, and throat muscles. *Cranial nerve VIII* (vestibulocochlear nerve) is responsible for receiving and processing hearing and balancing information from the ear. *Cranial nerve IX* (glossopharyngeal nerve) controls the contraction of pharynx muscles and the secretion of salivary glands. *Cranial nerve X* (vagus nerve) is a parasympathetic, motor, sensory nerve, controls the muscle contraction of the pharynx/larynx, regulates the cardiac activity, and mediates the contractility of smooth muscle cells in the blood vessel wall and the gastrointestinal wall. *Cranial nerve XI* (accessory nerve) is composed of cranial

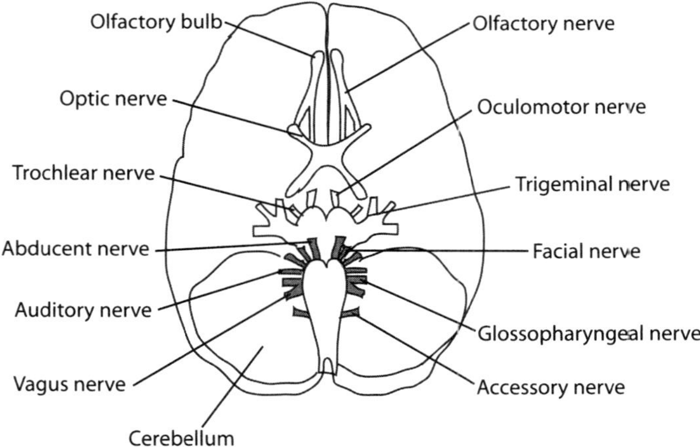

Figure 13.5. Schematic representation of the cranial nerves. Based on bibliography 13.1.

and spinal nerve fibers. The cranial nerve fibers contribute to the vagus nerve and regulate the contractility of smooth muscle cells. The spinal fibers control the movement of the neck and shoulder muscles. *Cranial nerve XII* (hypoglossal nerve) controls the contraction of the tongue and the geniohyoid muscles.

Cerebral Circulation. There is a rich network of blood vessels in the brain. The main arteries that supply blood to the brain are the left and right *common carotid arteries*, which originate from the aortic arch, and the left and right *vertebral arteries*, which originate from the subclavian arteries. Each common carotid artery is divided into the internal and external arteries before entering the cranial cavity. The *internal carotid arteries* enter the cranial cavity and supply blood to the brain, whereas the *external carotid arteries* supply blood to the facial tissues. The vertebral arteries enter the cranial cavity and join together to form the *basilar artery*. The internal carotid arteries and the basilar artery join together to form the *cerebral arterial circle*. Branches from the cerebral arterial circle supply blood to the brain. The cortex of each hemisphere is supplied by the anterior, middle, and posterior cerebral arteries from the cerebral arterial circle.

Arteries are eventually divided into numerous capillaries. The endothelium of the capillaries in the brain possesses well-developed tight junctions, constituting the *blood–brain barrier*. This barrier allows only lipid-soluble substances, such as CO_2, CO, ethanol, and nicotine, to freely pass. Small water-soluble substances, such as amino acids and glucose, can be transported across the blood–brain barrier via active energy-consuming processes. Large molecules and particles, such as viruses and bacteria, cannot pass the blood–brain barrier under physiological conditions.

Blood from the brain is collected by small veins from the capillary network. The small veins conduct blood to a number of cranial venous sinuses. There are several small and medium-sized sinuses, including the superior sagittal, inferior sagittal, straight, cavernous, and occipital sinuses. Venous blood from these sinuses converges to larger sinuses, including the *transverse* and *sigmoid sinuses*, which drain blood into the *internal jugular vein*. Venous blood eventually returns to the right atrium via the subclavian and brachiocephalic veins and the superior vena cava.

Meninges and Cerebral Ventricles. Meninges are membranes of connective tissue, which cover and protect the brain and spinal cord. There are three meningeal layers: the dura mater/periosteum, arachnoid mater, and the pia mater. The *dura mater* is the outmost layer, which folds and extends into the brain fissues at three locations: the falx cerebri (between the two cerebral hemispheres), tentorium cerebelli (between the cerebrum and cerebellum), and falx cerebelli (between the two cerebellar hemispheres). The dura mater is attached to the periosteum of the cranial cavity, forming an integrated functional layer. The *arachnoid mater* is a thin membrane, which is separated from the dura mater by the subdural space filled with serous fluid. The *pia mater* is a membrane that is tightly attached to the brain. The pia mater is separated from the arachnoid mater by the subarachnoid space, which is filled with fluid and contains blood vessels.

The brain contains a number of cavities. These include the two *lateral ventricles* (one in each cerebral hemisphere), the *third ventricle*, and the *fourth ventricle*. The four ventricles are connected via two channels. The two lateral ventricles and the third ventricle are connected by a short channel known as the *interventricular foramina*. The third and fourth ventricles are connected by a long channel called the *cerebral aqueduct*. The fourth ventricle is continuous with the subarachnoid space and the spinal cord.

The brain ventricles, spinal cord, and subarachnoid space are filled with *cerebrospinal fluid,* which is similar in composition to the serum except that the cerebrospinal fluid contains a very low concentration of proteins. The cerebrospinal fluid is produced in the ventricles by a structure known as the *choroid plexus*, which contain capillaries and connective tissue covered by ependymal cells. The cerebrospinal fluid flows through the ventricles, leaves the fourth ventricle via several apertures, including the median aperture and two lateral apertures, flows through the subarachnoid space, and enters the dural venous sinuses, where cerebrospinal fluid joins the blood.

Spinal Cord. The *spinal cord* is a part of the central nervous system and extends from the brainstem to the vertebral column. At each transverse level, the spinal cord is composed of central gray matter and peripheral white matter. The *gray matter* consists of neurons, glial cells, and nerve axons, whereas the *white matter* consists of nerve axons. *Spinal nerves* arise from a large number of rootlets of the gray matter symmetrically at two dorsal and two ventral sites. Nerve fibers from several rootlets combine to form a nerve bundle, which leaves the spinal cord through the vertebral column. The spinal cord is vertically composed of several segments, known as the *cervical*, *thoracic*, *lumbar*, and *sacral segments*. The spinal cord gives rise to 31 pairs of spinal nerve bundles. Each pair of nerve bundles leaves the vertebral column at a corresponding vertebra. These nerves innervate peripheral tissues and organs.

At each level, the gray matter of the spinal cord is divided into two functional units: sensory and somatic motor control units. The sensory unit is located in the two dorsal sites symmetrically, whereas the somatic motor control unit is located in the two ventral sites. The dorsal nerve bundles contain input nerve fibers that transmit sensory signals from the peripheral receptors to the dorsal sensory ganglia, whereas the ventral nerve bundles contain output nerve fibers that transmit signals from the ventral motor control centers to the peripheral muscles. While the somatic motor control neurons are located within the gray matter, the sensory receiving and processing neurons are located in the dorsal root ganglia, which is located outside the spinal cord. The sensory neurons transmit signals from the ganglia to the dorsal gray matter of the spinal cord.

The Peripheral Nervous System. The peripheral nerve system is composed of neurons and nerve bundles outside the brain and spinal cord. A peripheral nerve bundle contains several structures: Schwann cell-enclosed nerve axons, blood vessels, and connective tissue. Each Schwann cell/axon bundle is enclosed with a connective tissue membrane known as endoneurium. A group of axons is enclosed within a perineurium membrane, forming a nerve fascicle. A number of axon fascicles are enclosed within an epineurium sheath, forming a nerve bundle. The connective tissue membranes protect the nerve fibers from injury.

The peripheral nerve fibers from the spinal cord are divided into various groups, based on the origin of the nerves and region of innervation. Each group is defined as a plexus, which contains several nerve bundles from a number of vertebrae. Major nerve plexuses include the cervical, brachial, thoracic, lumbosacral, coccygeal plexuses. The cervical plexus contains nerves from cervical vertebrae 1–4 and innervates the head and neck skin, muscle, and tissue. The brachial plexus originates from cervical vertebra 5 and thoracic vertebra 1, and innervates the upper limbs and part of the head. The thoracic plexus originates from thoracic vertebrae 1–12, and innervates the chest skin, muscle, and tissue. The lumbosacral plexus is from lumbar vertebrae 1–4 and sacral vertebrae 1–4, and innervates the abdomen and lower limbs. The coccygeal plexus is from sacral vertebrae 4 and 5, and innervates the muscles of the pelvic floor and the coccyx.

Functional Integration of the Nervous System

Sensory Input. The nervous system has a number of sensory functions: hearing, visualizing, touching, smelling, tasting, sensing temperature, and sensing pain. All sensory activities are triggered by environmental stimulations. To sense and react to a stimulus, it is necessary to have a complete sensory system, including the *sensory receptors* that sense the environmental stimulus and covert the stimulus to action potentials, *sensory input nerves* that transmit the action potentials to the central nerve system, *sensory centers* that receive and process the input signals, and *neuronal centers* that respond to the sensory centers and generate signals for the awareness of the environmental stimulation and for corresponding responsive actions (Fig. 13.6).

Sensory receptors are the ending structures of the axon and dendrites. There are various types of sensory receptors. Based on the location, sensory receptors can be divided into three groups: exteroreceptors, visceroreceptors, and proprioceptors. The *exteroreceptors* are distributed in the superficial tissues such as the epidermis and dermis, and are responsible for sensing external stimulations. The *visceroreceptors* are found in the viscera and internal organs, and are responsible for sensing internal stimulations. The *proprioceptors* are found in connective tissues, such as the joint, tendon, and bone, and provide information about the body position and movement.

Based on the function, sensory receptors can be classified into mechanoreceptors, chemoreceptors, thermoreceptors, photoreceptors, and pain receptors. The

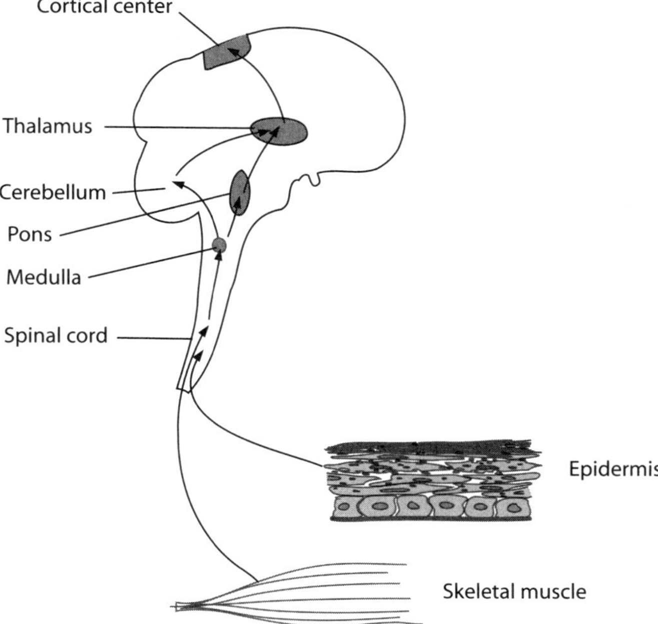

Figure 13.6. Schematic representation of the pathways of sensory signals. Based on bibliography 13.1.

mechanoreceptors respond to mechanical stimuli, such as tension, compression, and shearing, and are found in the epidermal tissue, muscle, inner ear, and internal organs. These receptors are responsible for the sensing of mechanical contacts and air vibrations (sounds). The *chemoreceptors* sense chemical stimuli, are responsible for the activities of smelling and tasting, and are found in the epidermal tissue. The *thermoreceptors* are responsible for the sensing of temperature and are found in the epidermal tissue. The *photoreceptors* sense light and are found in the retina. The *pain receptors* sense mechanical, chemical, and thermal stimuli and are found in almost all types of tissues.

The sensory signals sensed by the receptors are transmitted to the central nerve system via sensory nerves. There are various sensory pathways in the spinal cord and brainstem. These pathways are composed of neurons that are specific to sensory receptors. Each type of neuron is responsible for the transmission of a specific sensory signal. There are several ascending sensory pathways that transmit signals from the spinal cord to various parts of the brain. These include the spinothalamic pathway (transmitting pain, temperature, light, and mechanical signals from the spinal cord to the thalamus), the dorsal-column/medial–lemniscal pathway (transmitting proprioception, pressure, and vibration signals from the spinal cord to the brainstem), the spinocerebellar pathway (transmitting proprioception signals from the spinal cord to the cerebellum), and the spinoreticular pathway (transmitting tactile signals from the spinal cord to the reticular formation of the brainstem).

Sensory signals from the peripheral nerve system and the spinal cord are eventually converged to the primary sensory areas of the cerebral cortex. Each type of sensory input is projected to a designated cortical area. For instance, the somatic cortical area is located in the middle region of the cerebral cortex. Other regions are illustrated in Fig. 13.7. This area is organized in relation to the distribution of the peripheral tissues and organs. The sensory inputs from the head and face are transmitted to the inferior cortical area, whereas the sensory inputs from the lower limbs are projected to the superior cortical area. The sensory inputs from other organs are projected to corresponding areas between the head and the lower limbs.

Motor Control. The nervous motor system controls the contractile activities of the skeletal muscle cells by emitting action potentials and thus regulates the movement, posture, and balance of the body. There are two types of movements: involuntary and voluntary. An *involuntary movement* is controlled by the spinal cord motor centers and includes

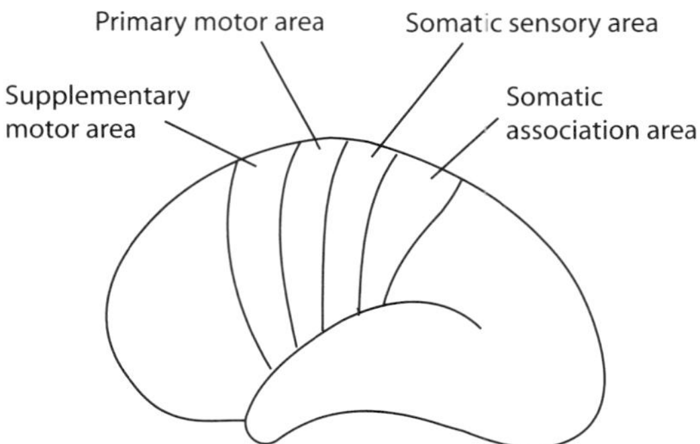

Figure 13.7. Schematic representation of the distribution of the motor and sensory control regions in the cortex. Based on bibliography 13.1.

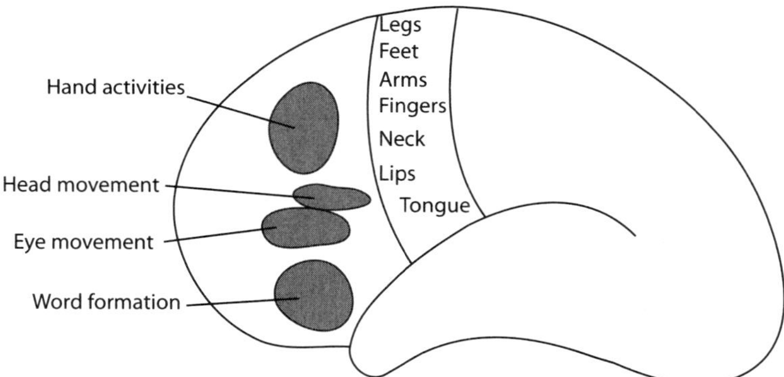

Figure 13.8. Schematic representation of the distribution of the cortical motor control regions. Based on bibliography 13.1.

primarily reflexes of the skeletal muscle system. The reflex movement is not initiated through conscious judgment by the cortex. It should be noted that the heart beating, gastrointestinal motion, and blood vessel contraction are also involuntary movements, but these are controlled by the parasympathetic nerve system, not by the motor system. In contrast, a *voluntary movement* is consciously initiated and controlled by the cortex to achieve specified goals, such as running, walking, eating, writing, and talking.

Voluntary movements are induced in response to the action potentials initiated from the *primary motor area* located anterior to the primary sensory region of the cortex (Fig. 13.8). The primary motor area is organized in relation to the distribution of the peripheral tissues and organs. The action potentials initiated from the superior region of the primary motor area controls the muscles of the lower limbs, whereas the action potentials from the inferior area of the cortex controls the muscles of the head, face, and upper limbs. There is another cortical area, called *premotor area*, which is located anterior to the primary motor area and regulates the coordination of various muscles, ensuring accurate movements. The third important motor-control area is the *prefrontal area*, which is responsible for the regulation of mood- and emotion-related movements.

The motor-control signals or action potentials are transmitted from the cortical motor control centers to the brainstem or spinal cord, and then to the peripheral skeletal muscles via motor nerves (Fig. 13.9). The motor nerves are divided into direct and indirect descending nerves. The direct nerves are responsible for the control of accurate muscle contraction, muscle tone, skilled movements, and emotion-related facial movements. The indirect nerves control less accurate muscle activities related to overall body movements. The direct nerves exist only in mammals.

Autonomic Control of Vital Activities. Vital activities include the heartbeat, regulation of bloodflow and pressure, and temperature regulation. These activities are primarily controlled by two nervous systems: the sympathetic and parasympathetic nervous systems.

Sympathetic Nervous System. The *sympathetic nerve system* is composed of a number of control units. A typical sympathetic control unit consists of preganglionic neurons and postganglionic neurons (Fig. 13.10). The *preganglionic neurons* are located in the gray matter of the spinal cord in the region from thoracic vertebra 1 to lumbar vertebra 2. The *postganglionic neurons* are located in either a structure near the vertebrate column known

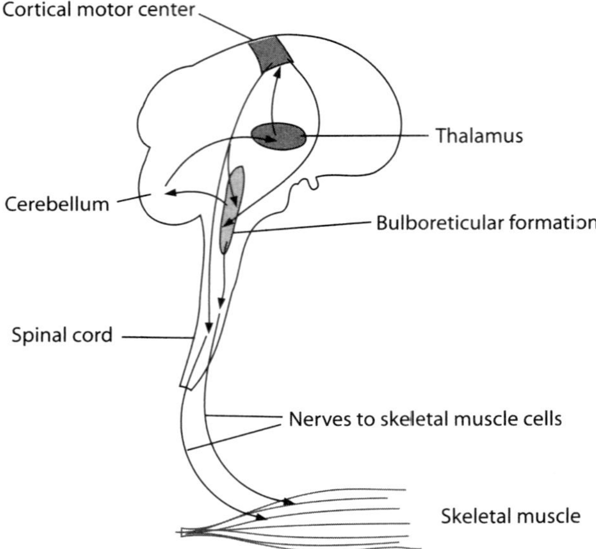

Figure 13.9. Schematic representation of the pathways of motor control signals. Based on bibliography 13.1.

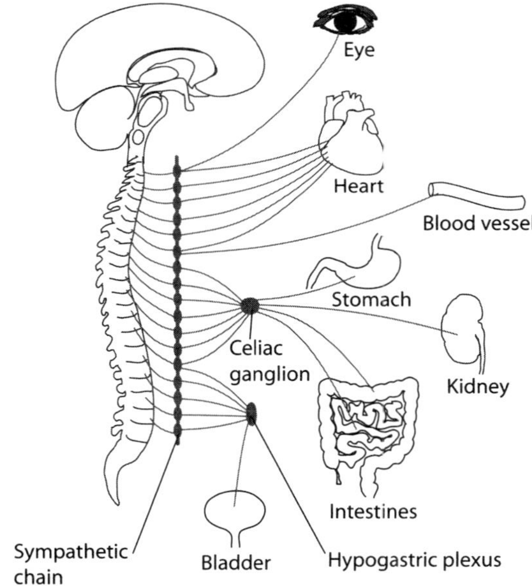

Figure 13.10. Schematic representation of the sympathetic nervous system. Based on bibliography 13.1.

as the *sympathetic chain ganglia* or a structure near the target tissues known as the *collateral ganglia*. The axons of preganglionic neurons join the spinal nerves, leave the spinal cord, and extend to the *sympathetic chain ganglia*. In the sympathetic chain ganglia, the sympathetic nerves are organized into four different groups: the spinal, sympathetic, splanchnic, and adrenal gland nerves.

For the *spinal group*, the preganglionic axons synapse with the postganglionic neurons in the sympathetic chain ganglia. The postganglionic axons leave the sympathetic chain ganglia, join the spinal nerves, and innervate peripheral tissues and organs. For the *sympathetic group*, the preganglionic axons synapse with the postganglionic neurons in the sympathetic chain ganglia. The postganglionic axons leave the sympathetic chain ganglia, form sympathetic nerves, and innervate peripheral tissues and organs. The heart is a major target of the sympathetic nerves. Activation of the sympathetic nerves induces an increase in cardiac contractility and the heartbeat. For the *splanchnic group*, the preganglionic axons enter the sympathetic chain ganglia, but do not form synapses there. They leave the sympathetic chain ganglia, form the splanchnic nerves, extend to the peripheral organs, and enter another sympathetic structure known as the *collateral ganglia*, where they synapse with postganglionic neurons. The postganglionic axons innervate target tissues, including the stomach and intestines. The *adrenal gland nerves* are different form others. The preganglionic axons do not synapse with postganglionic neurons, but directly extend to the adrenal gland and synapse with the adrenal medullar cells, stimulating the secretion of epinephrine and norepinephrine.

Parasympathetic Nervous System. This is another nervous system that controls the vital activities. As the sympathetic nervous system, the parasympathetic nervous system is composed of a number of control units. Each unit consists of preganglionic and postganglionic neurons (Fig. 13.11). The *preganglionic neurons* are located in the brainstem, and the postganglionic neurons are located in a structure near or within the target tissues or organs, known as the *terminal ganglia*. The preganglionic axons leave the brainstem, extend to and enter the terminal ganglia, where they synapse with the postganglionic

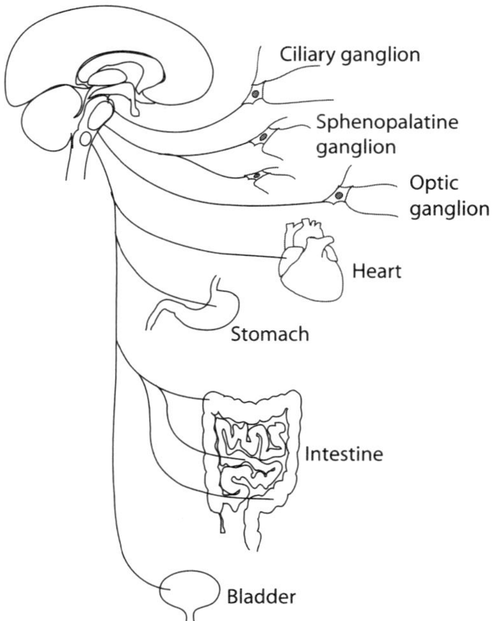

Figure 13.11. Schematic representation of the parasympathetic nervous system. Based on bibliography 13.1.

neurons. The postganglionic axons leave the terminal ganglia and innervate target tissues and organs.

Autonomic Regulation of Cardiovascular Functions. The activities of the heart and blood vessels are controlled by the sympathetic and parasympathetic nervous systems in a coordinated manner. The heart is innervated with both sympathetic and parasympathetic (vagus) nerves. Arteries and veins are primarily innervated with sympathetic nerves. Certain structures, such as the baroreceptors, which attach to the carotid arteries, are innervated with parasympathetic vagus nerves and play a critical role in regulating the cardiovascular activities.

The regulation of the cardiovascular functions is accomplished via autonomic reflexes, or nerve-controlled actions in response to peripheral inputs. In general, the sympathetic nerve system activates, whereas the parasympathetic system inhibits the cardiovascular activities. The two systems regulate the cardiovascular activities in coordination and counterbalance each other's effects. An increase in the activity of the sympathetic system will activate the parasympathetic system via a feedback mechanism, and vice versa. Thus the cardiovascular activities are maintained within a relatively stable range, which is defined as the *physiological range*.

A typical example is baroreceptor-mediated regulation of arterial blood pressure. The baroreceptors detect changes in blood pressure. An increase in blood pressure beyond the physiological level stimulates the baroreceptors. Signals from the baroreceptors are transmitted to the brainstem via the vagus nerve and activate the parasympathetic cardiovascular control neurons. The parasympathetic signals are transmitted to the heart, inducing a decrease in the heartbeat and contractility. As a result, arterial blood pressure decreases. In contrast, a decrease in arterial blood pressure to a level below the physiological level will reduce stimulation to the baroreceptors as well as to the parasympathetic control neurons, resulting in a situation with relatively dominant sympathetic activities. Thus, heartbeat and cardiac contractility both increase simultaneously, leading to an increase in blood pressure.

NERVOUS DISORDERS

Nerve Injury

Etiology, Pathology, and Clinical Features [13.2]

Brain Injury. Brain injuries can be caused by head trauma and skull fractures resulting from physical impacts or penetrations. An injury beyond the tolerant level can cause changes in the structure and function of the brain at the injury site. Open trauma can cause hemorrhage, which significantly compromises the brain function at the hemorrhage site, often leading to dysfunction of the corresponding peripheral systems. Another common type of brain injury is concussion, which is defined as mild reversible brain injury with transient amnesia and loss of consciousness. This type of injury is often caused by blunt impacts on or sudden deceleration of the brain. Usually, concussion is diagnosed when no visible brain damage and hemorrhage are observed. The pathogenic mechanisms for the loss of consciousness are related to the dysfunction of the reticular system in the brainstem due to sudden rotation or movement of the brain. The mechanisms for amnesia are related to the injury of the cerebrum.

Severe head impact or deceleration causes brain contusion, which is often associated with various degrees of apparent brain injury, ranging from small superficial hemorrhage to large area necrotic destruction of the brain structure. The site and degree of brain injury can be detected by CT scan or MRI. This type of injury is often associated with prolonged loss of consciousness and amnesia. Clinical signs are dependent on the injury location and size. In certain cases, diffusive edema may occur within a short period following the trauma due to alterations in blood pressure (hypertension) and microcirculatory function. In addition, glial cells and leukocytes migrate to the injury and hemorrhage sites within hours to days and generate extracellular matrix components, eventually resulting in the formation of fibrous tissue or scars. The scars in the brain are often the causes of post-traumatic epilepsy.

Cranial Nerve Injury. Head trauma, especially basilar skull fractures, is often associated with injury of the cranial nerves, such as the optic, trochlear, olfactory, trigeminal, facial, and auditory nerves. Nerve injuries are often induced by shearing deformation of the skull. In severe cases, the nerve fibers can be completely severed. Nerve injury often induces degradation of the distal nerve fibers in association with biochemical changes, such as reduction in the expression of synaptophysin, which is an axonal marker (Fig. 13.12). Clinical signs are dependent on the type of nerve that is injured and the degree of the injury. For instance, minor injury of the optic nerves may cause blurring of vision, whereas bruising and transecting of the optic nerves will result in partial or complete blindness.

Spinal Cord Injury. Spinal cord injury can be induced by damage to, fracture, or dislocation of the vertebral column. The thoracic spinal cord is often injured by vertical compression. The cervical spinal cord can be injured by flexion. Spinal cord injury can be detected by imaging approaches such as X-ray and MRI. Spinal cord injuries result in functional changes in peripheral sensation and motor control. In severe cases, paralysis occurs in areas below the injured spinal cord.

Immediately following spinal cord injury, hemorrhage may occur, which significantly affect the function of the spinal cord neurons. Other changes include regional spinal cord edema and ischemia during the early period (about 4 h). Peripheral signs, such as demyelination of the peripheral nerves (Fig. 13.13) and regional loss of sensory input and movement, may be observed during this period. Spinal cord injury is usually reversible during the early period. Global infarction or necrosis occurs at the injury site about 8 h after spinal cord injury. Such injury is associated with complete paralysis below the injury site. Complete spinal cord transection or global spinal cord infarction is usually irreversible. At the injury site, glial cells and leukocytes are often activated. These cells can migrate to the damaged tissue and generate extracellular matrix, eventually forming fibrous tissue or scars.

EXPERIMENTAL MODELS OF SPINAL CORD INJURY. Spinal cord injury is often created in small rodents and used to investigate the mechanisms of nerve injury and regeneration. Several types of experimental models have been developed and reported in the literature. These include complete and partial spinal cord transection, blunt contusion, and compression. In the spinal cord *transection model*, the spinal cord is completely transected at a selected location, or selected nerve tracts in the spinal cord are transected. Neuronal growth and axonal regeneration can be observed after transection. In the blunt *contusion model*, a weight impactor can be dropped from a desired height to a segment of exposed

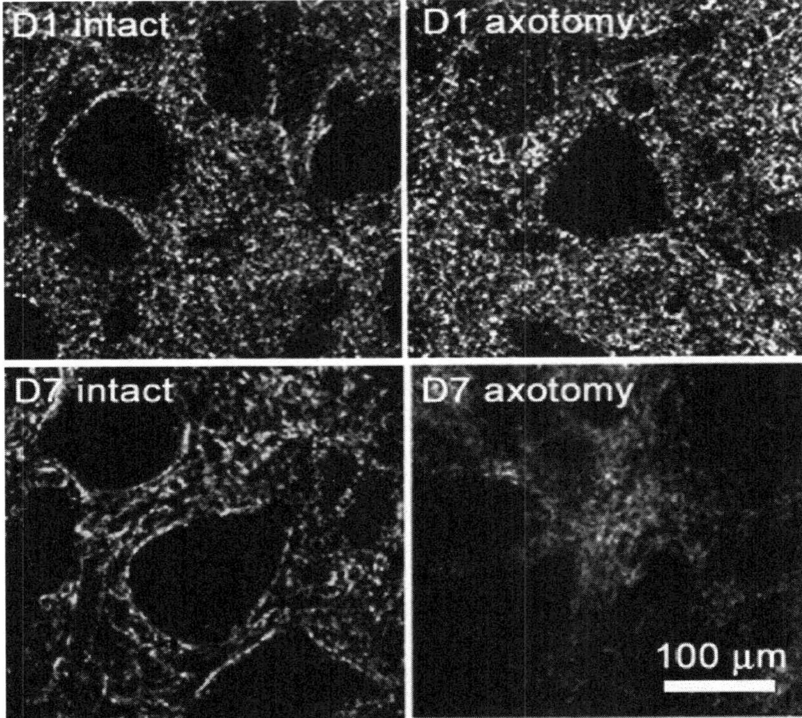

Figure 13.12. Influence of axotomy on the structure of neurons. Confocal microscopy images of facial nuclei at the intact (left) and axotomy (right) sides. D1 and D7 are days 1 and 7, respectively, after axotomy. The specimens were labeled with an antibody against mouse synaptophysin, an axonal marker. The expression level of synaptophysin is apparently reduced at the axotomy side. (Reprinted from Ikeda R, Kato F: Early and transient increase in spontaneous synaptic inputs to the rat facial motoneurons after axotomy in isolated brainstem slices of rats. *Neuroscience* 134:889–99, 2005, copyright 2005, with permission from Elsevier.)

spinal cord, causing blunt injury. The degree of injury can be controlled by altering the weight and height of the impactor. Electronically controlled mechanical impactors can be developed and used to cause blunt injury based on a similar principle. In the *compression model*, a mechanical clip or band can be applied to an exposed spinal cord to induce circumferential compression. The degree of injury can be controlled by altering the level of compression. This model simulates spinal cord injury due to the compression of vertebral columns.

An important issue in spinal cord injury modeling is to estimate the degree of injury or whether the nerve tracts are completely severed. Axonal tracers are often used to achieve such a goal. The nerve axons are capable of conducting retrograde transport of substances toward the cell body. A tracer can be applied to a site distal to the nerve injury. The completeness of transection can be judged by observing the appearance of the tracer in the cell body. Among the three types of model described above, the contusion and compression models usually result in incomplete injury. Because of the complexity of the spinal cord, it is often difficult to estimate the degree of nerve regeneration by using these incomplete injury models.

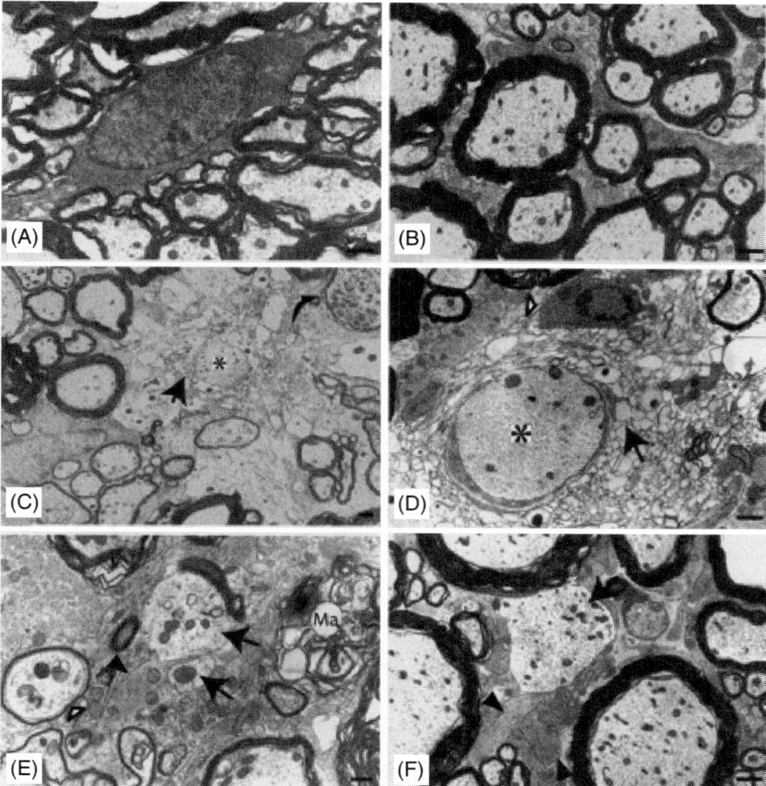

Figure 13.13. Histological micrographs showing demyelination after contusive spinal cord injury. (A,B) Normal myelin sheaths in the ventrolateral funiculus of the adult spinal cord. These sheaths surround axons. (C,D) Demyelination 1 week after contusive spinal cord injury in the ventrolateral funiculus. Many myelin sheaths degenerated and numerous intramyelinic vacuoles were seen (arrows). A macrophage was observed in close relation to degenerating myelin (open arrowhead in panel D). The axons surrounded by degenerating myelin appeared morphologically normal (asterisks). (E,F) Some demyelinated axons survived for at least 1 month after injury (arrows). (E,F) Healthy-appearing demyelinated axons were observed in the ventrolateral funiculus of spinal cord at the injury epicenter (E) and also a few millimeters away from the epicenter (panel F, arrow). Scale bars: 2.5 μm. (Reprinted with permission from Cao Q et al: *J Neurosci* 25:6947–57, copyright 2005.)

Peripheral Nerve Injury. Peripheral nerve injury is largely induced by trauma resulting from accidents and surgery. The peripheral nerves are capable of recovering from injury when they are not completely severed. However, when the nerves are completely severed, the distal axons are usually disintegrated into segments within several days, followed by further degradation of the nerve segments. At the same time, the myelin sheaths of the Schwann cells that enclose the axons are also degraded. The degraded axon and myelin are phagocytosed by macrophages.

The proximal neurons are able to regenerate and extend their axons and restore the lost distal axonal segment (Fig. 13.14). However, the outcome depends on the severity of the injury, especially the distance between the two ends of the severed axon. When the two

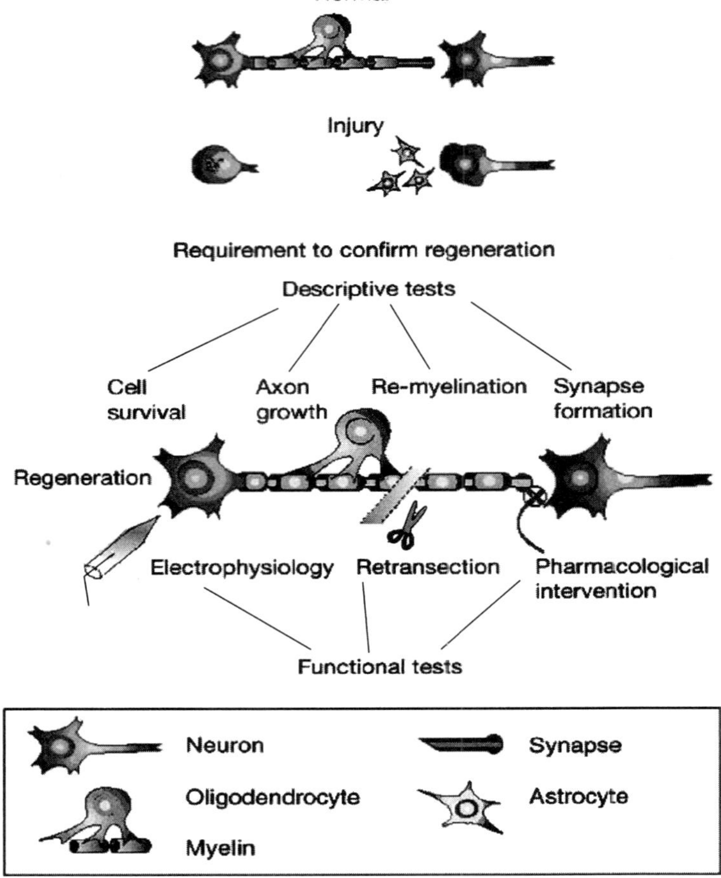

Figure 13.14. Nerve regeneration after nerve injury. A number of descriptive and functional tests have been designed and used for evaluating the outcome of nerve regeneration after nerve injury. Descriptive tests, as shown in the figure, can be used to determine the survival and integrity of the injured system, whether axonal regeneration is present, and if appropriate synaptic connections and remyelination have occurred. Functional tests, incldung electrophysiological and pharmacological intervention, can be used to assess the function and specificity of the regenerated pathway. (Reprinted by permission from Macmillan Publishers Ltd.: Horner PJ, Gage FH: *Nature* 407:963–70, copyright 2000.)

ends are close to each other and well aligned, nerve regeneration and reconnection can occur. In contrast, when the two ends of a severed nerve are separated and not aligned, the injured nerve segment may not be restored. The residual Schwann cells play a critical role in nerve regeneration. These cells start to enlarge and undergo mitosis when the proximal axon starts to regenerate. The Schwann cells are aligned along the injured axons, preparing for the regeneration and reconnection between the proximal and distal axons. The proximal axons that encounter the Schwann cells are able to restore the injured distal axons. However, the proximal axons that are not escorted by Schwann cells cannot reach the severed distal axons.

Nerve regeneration is a slow process. It usually takes about several weeks for the proximal regenerating nerves to reach the injured distal nerves. The end of each proximal nerve may form several sprouts. The sprouts that encounter the Schwann cells develop into axons and replace the severed distal nerves. The remaining sprouts are gradually disintegrated. Following the growth of the proximal axons into the Schwann cells, the regenerating axons are enclosed by newly generated myelin sheaths.

Conventional Treatment of Nerve Injury. For the injury of the central nerve system due to craniocerebral trauma, several approaches can be used for the treatment depending on the severity of the injury. For patients with concussion (transient unconsciousness), no special treatment is needed. Patients are usually given analgesics such as aspirin or acetaminophen for headache, and antidepressants for anxious depression, if any. For severe head injury, especially when patients are unconscious, several treatments should be carried out: (1) check to ensure that the airways are functional and are able to supply adequate ventilation (artificial respiration should be applied, if necessary); (2) stop bleeding, if any; (3) restore blood pressure, if hypotension exists (note that hypertension may exist with severe head injury); (4) restore the cardiac function, if necessary; (5) measure and reduce intracranial pressure when intracranial pressure is too high; (6) supply nutrients via a nasogastric tube, if coma persists; and (7) resort to surgical intervention to stop hemorrhage, relieve intracranial pressure, repair fractured or distorted bones, and/or remove coagulated blood, if necessary. For severed peripheral nerves due to trauma, simple surgical reconnection or grafting of the injured nerves with an autologous nerve segment, if necessary, can be performed. Such an approach usually gives a satisfactory result.

Molecular Nerve Regenerative Engineering [13.3]

Strategies of Molecular Nerve Regenerative Engineering. Principal strategies for the treatment of nerve injury are to promote neuronal survival and regeneration, prevent neuronal death, protect the neurons from secondary injury, stimulate axonal adhesion, extension and reconnection, reduce fibrosis and scar formation, enhance synaptic formation. In conventional medicine, there are few available methods that can be used to effectively achieve these strategies. The recovery of injured neurons and nerve fibers is very much dependent on the self-regeneration capability. For the past decade, regenerative engineering approaches have been developed at the molecular, cellular, and tissue levels to facilitate the regeneration of injured nerves. At the molecular level, mitogenic proteins and genes can be used to promote the survival and regeneration of injured neurons and to prevent neuronal death. At the cellular level, various cell types can be used and transplanted to the injury site to assist the recovery of injured neurons. At the tissue level, various biological and mechanical devices have been constructed to serve as guidance for the regeneration of injured nerve axons. Although most of these engineering approaches have not been used in clinical therapy, experimental investigations have demonstrated potential for clinical applications.

For the central nerve system, because the function of many structures has not been completely understood and the system is difficult to access, it is often a challenge to "engineer" the brain and spinal cord. Thus, engineering treatment for injured brain and spinal cord is limited. In contrast, the peripheral nerve fibers are relatively easy to access, it is not technically difficult to manipulate these fibers. Here, the principle of molecular and cell regenerative engineering for central and peripheral nerve injuries is discussed.

Molecular regenerative engineering approaches can be established and used for facilitating the regeneration of injured neurons and nerve fibers. There are several strategies for the application of molecular engineering approaches to nerve regeneration: preventing cell death, promoting cell survival, and enhancing the differentiation of stem and progenitor cells to neurons and supporting cells. There exist several types of growth factors that are known to promote cell survival and growth, and prevent cell death. Thus, growth factors and their genes are candidate therapeutic factors for the treatment of nerve injury. At the protein level, growth factors can be applied directly to the site of nerve injury to stimulate nerve cell regeneration. At the gene level, selected growth factor genes can be manipulated to enhance the level of gene expression, resulting in an increase in the production of encoded growth factors.

There are several challenges for the application of molecular regenerative engineering approaches to nerve regeneration. These include the selection of therapeutic growth factors and/or genes, the establishment of appropriate techniques for the delivery of the therapeutic factors, and precise delivery of the therapeutic factors to the injury site. The selection of therapeutic factors and genes is dependent on the nature of therapeutic molecules, the degree of nerve injury, and the type of the injured cells. While certain types of growth factors promote nerve growth and regeneration, other types may exert an opposite effect. Fr instance, neurotrophic factors and receptors stimulate nerve regeneration, whereas epidermal growth factor receptor (EGFR) inhibits nerve regeneration. EGFR can be activated or phosphorylated in response to the stimulation of chondroitin sulfate proteoglycans and activates signaling mechanisms that suppress axonal regeneration. Other therapeutic factors and genes are discussed for different cases in the following sections.

There are several approaches for the delivery of therapeutic factors and genes to the site of nerve injury:

1. Therapeutic proteins can be delivered to the nerve system via direct injection to the injury site of the central nervous system. This method gives a local delivery of highly concentrated therapeutic factors. However, it is difficult to find the precise site of injury for injection. Furthermore, direct injection induces nerve injury.
2. Therapeutic factors and genes can be injected into the spinal cavity. This cavity is connected to the four cerebral ventricles and the arachnoid space that cover the cerebrum (see page 507). The delivered factors can reach the cranial cavity via diffusion, where they are taken up by epithelial cells and transported to the nearby nerve cells. This approach is suitable for delivering therapeutic factors to the brain and spinal cord. Compared to the direct injection method, the spinal cavity delivery is a safe and reliable method, although it may require a high dose of growth factors or genes to reach an effective concentration at the injury site.
3. Selected cell types, such as glial cells or fibroblasts, can be cultured and transfected with a therapeutic gene, rendering the cells to express a high level of a desired factor. These cells can be delivered to the site of nerve injury to release the selected therapeutic factor. Direct injection is an effective method for cell delivery.

It should be noted that the bloodstream may be considered an alternative delivery route. Therapeutic factors and genes can be injected into the blood, taken up by endothelial cells, and transported to the target tissues. However, due to the presence of the blood–brain barrier, it may be difficult to deliver therapeutic proteins and genes to the brain system with high efficiency.

Enhancement of Neuron Survival. There are several growth factors that can be used for the enhancement of neuronal cell survival and regeneration. These factors include brain-derived neurotrophic factor (BDNF), neurotrophin (NT)3, neurotrophin 4, nerve growth factor (NGF), and glial cell-derived growth factor (GDNF) (Table 13.1). The functions of these growth factors are briefly discussed as follows.

BRAIN-DERIVED NEUROTROPHIC FACTOR [13.4]. Brain-derived neurotrophic factor is a 247 amino acid protein and a member of the nerve growth factor family. The brain-derived neurotrophic factor gene is mapped to chromosome 11 and locus 11p13 in humans. This factor shares a nucleotide sequence considerably similar to the nerve growth factor gene. The brain-derived neurotrophic factor gene is similar among different mammals. Brain-derived neurotrophic factor is expressed primarily in the cortical neurons. These neurons produce the precursor of brain-derived neurotrophic factor. The precursor is released into the extracellular matrix and cleaved by several enzymes, including serine protease plasmin, matrix metalloproteinases (MMP)7 and MMP3, resulting in the formation of mature brain-derived neurotrophic factor.

Brain-derived neurotrophic factor is necessary for the survival of neurons in the central nervous system. During embryonic development, brain-derived neurotrophic factor plays a critical role in the differentiation and proliferation of neuronal cells and contributes to the development of germ cells. In response to nerve injury, the expression of the BDNF gene is often upregulated, an important mechanism for the self-repair of injured nerves (Fig. 13.15). A treatment with BDNF in cultured dorsal root ganglion explants enhances neurite outgrowth (Fig. 13.16). In neural degenerative disorders, such as Alzheimer's and Huntington's diseases, the expression of brain-derived neurotrophic factor is downregulated, which contributes to the development of these disorders. In transgenic model of brain-derived neurotrophic factor deficiency induced by homologous recombination or site-specific gene targeting in embryonic stem cells, sensory neurons are lost in various areas of the central nerve system. However, the motor neurons are not significantly affected by the deficiency of the brain-derived neurotrophic factor gene.

Brain-derived neurotrophic factor can be released on neuron depolarization, resulting in the activation of intracellular signaling pathways. In dopamine neurons, brain-derived neurotrophic factor stimulates the expression of dopamine D3 receptor during development and adulthood, contributing to the control of animal behavior. In serotonergic neurons, brain-derived neurotrophic factor regulates the expression of several postsynaptic serotonin receptors in the frontal cortex, hippocampus, and hypothalamus. The deficiency of brain-derived neurotrophic factor induces a reduction in the expression of these serotonin receptors, which is associated with behavioral abnormalities such as increased aggressiveness and hyperphagia. These observations suggest that brain-derived neurotrophic factor plays a critical role in the regulation of the development, function, and regeneration of dopaminergic and serotonergic neurons.

NEUROTROPHIN 3 [13.5]. Neurotrophin 3, also known as neurotrophic factor 3, is a growth factor that is similar in structure to nerve growth factor and brain-derived neurotrophic factor. All these factors are members of the nerve growth factor family. Neurotrophin 3 is composed of 258 amino acids and is expressed in the human brain and placenta. The gene of neurotrophin 3 is localized to chromosome 12 and locus 12p13. The gene sequence of neurotrophin 3 is well conserved among mammals. The primary function of

TABLE 13.1. Characteristics of Selected Nerve Growth-Related Factors*

Proteins	Alternative Names	Amino Acids	Molecular Weight (kDa)	Expression	Functions
Brain-derived neurotrophic factor	BDNF	247	28	Central nervous system (cortex, retina, and spinal cord), fetal testis	Regulating the survival of neurons, stimulating embryonic development
Ciliary neurotrophic factor	CNTF	200	23	Brain	Promoting neurotransmitter synthesis and neurite outgrowth and regulating the survival of neurons and oligodendrocytes
Glial cell line-derived neurotrophic factor	Astrocyte-derived trophic factor 1 and glial cell-derived neurotrophic factor	211	24	Nervous system, kidney	Promoting the survival and differentiation of dopaminergic neurons and preventing apoptosis of motor neurons

*Based on bibliography 13.4 and 13.8.

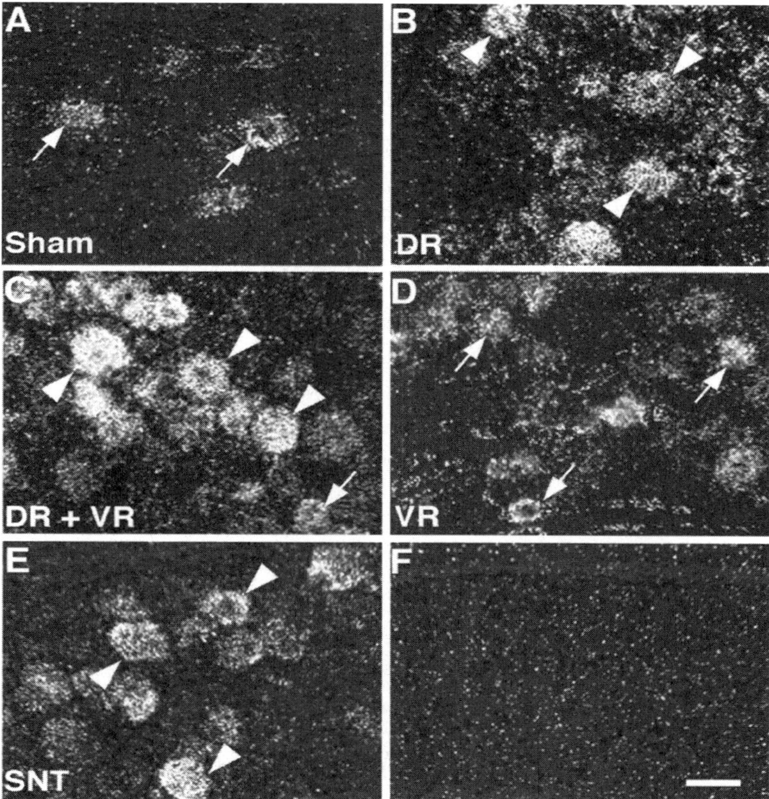

Figure 13.15. Expression of brain-derived neurotrophic factor (BDNF) in injured neurons. (A–E) Expression of BDNF mRNA in the ipsilateral L5 dorsal root ganglion (DRG) by in situ hybridization histochemistry in the sham control (A), dorsal rhizotomy only (B), dorsal rhizotomy + ventral rhizotomy (C), ventral rhizotomy only (D), and spinal nervous transection (E) groups at 7 days after surgery. Arrowheads indicate large-size sensory neurons positive for BDNF, whereas arrows indicate small-to-medium-diameter neurons. These observations show that nerve and spinal cord injury induces upregulation of the BDNF gene. (F) Hybridization with a sense probe for BDNF mRNA showed no positive signals. Scale bar: 50 μm. (Reprinted from Obata K et al: The effect of site and type of nerve injury on the expression of brain-derived neurotrophic factor in the dorsal root ganglion and on neuropathic pain behavior, *Neuroscience* 137:961–70, copyright 2006, with permission from Elsevier.)

neurotrophin 3 is to promote neuronal proliferation, differentiation, and survival as well as the neurite outgrowth during development.

Neurotrophin 3 can interact with the neurotrophin receptor p75 in neurons, exerting a stimulatory effect on neuronal mitogenic activities. In a rat model of spinal cord injury, the transfer of the neurotrophin 3 gene into the spinal cord stimulates the regeneration of injured axons (Fig. 13.17). Neurotrophin 3 (NT3) can also promote axonal growth of injured neurons into neural stem cell (NSC) grafts in injured spinal cord (Fig. 13.18). Furthermore, neurotrophin 3 plays a critical role, together with brain-derived neurotrophic factor, in the regulation of myelin formation by developing and regenerating Schwann

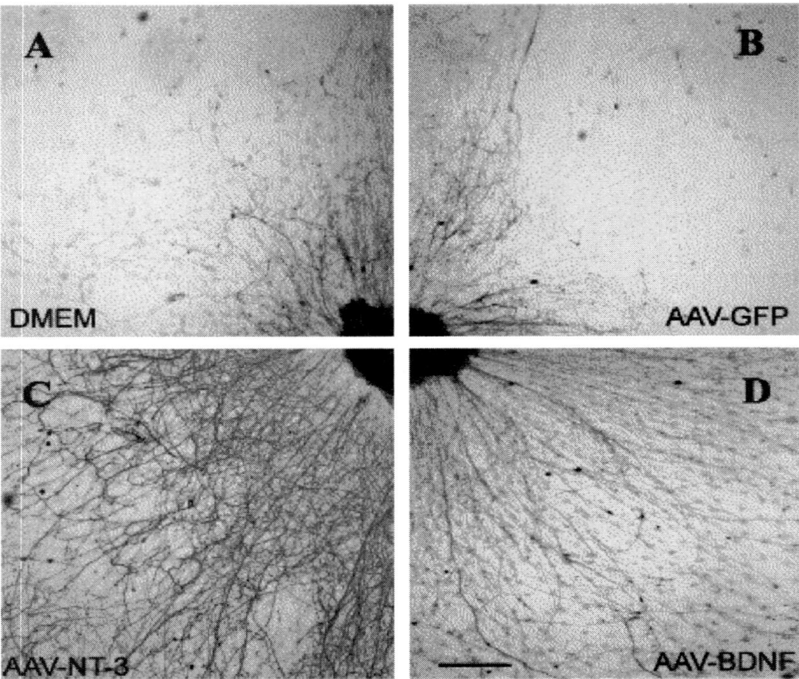

Figure 13.16. Enhancement of neurite outgrowth by neurotrophin-3 (NT3) and brain-derived neurotrophic factor (BDNF). Dorsal root ganglion explants were cultured with conditioned media collected from four groups of cells: (A) mock-infected 293 cells; (B) 293 cells transfected with adeno-associated viral vector (AAV) containing the GFP gene; (C) 293 cells transfected with AAV containing the NT3 gene; (D) 293 cells transfected with AAV containing the BDNF gene. Explant specimens were prepared and stained for neurofilaments using the RT97 antibody. In the controls (A and B), a low level of neurite outgrowth was observed. With AAV–NT3- and AAV–BDNF-conditioned medium, the rate of neurite outgrowth is significantly higher than the controls. Scale bar: 1 mm. (Reprinted from Blits B et al: Adeno-associated viral vector-mediated neurotrophin gene transfer in the injured adult rat spinal cord improves hind-limb function, *Neuroscience* 118:271–81, copyright 2003, with permission from Elsevier.)

cells. Neurotrophin 3 is also required for the outgrowth and innervation of sympathetic axons. In transgenic mice, the deficiency of neurotrophin 3 results in the loss of peripheral sensory and sympathetic neurons, while the motor neurons are not significantly affected. These defects are associated with the loss of peripheral muscular sensors, movement disorder, and death shortly after birth. Thus, neurotrophin 3 is a candidate factor for the treatment of nerve injury. It is interesting to note that neurotrophin 3 regulates not only the development and regeneration of the central and peripheral nervous system, but also the development of the heart. Neurotrophin 3 is required for the development of the cardiac atria and ventricles. The deficiency of neurotrophin 3 induces several major cardiac disorders, including atrial and ventricular septal defects and Fallot tetralogy, common forms of congenital cardiac defects. These disorders often result in perinatal death of transgenic mice.

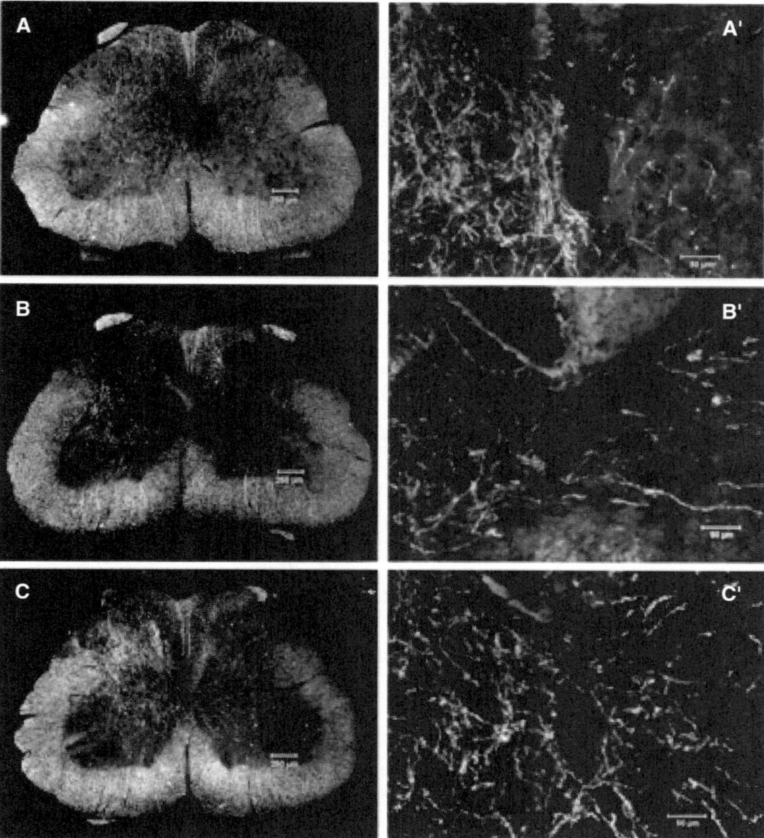

Figure 13.17. Enhancement of axonal growth by neurotrophin-3 (NT3) gene transfer to injured spinal cord. Rats with unilateral lesion of the corticospinal tract (CST) were transfected with Adv.EF α-NT3 or Adv.EF α-*LacZ*. The unlesioned corticospinal tract was labeled with BDA. (A) Section from a sham control rat. (B) Section from an Adv.EF α-*LacZ*-transfected rat. (C) Section from an Adv.EF α-NT3-transfected rat. (A′–C′) Higher-magnification photomicrographs of the regions around the central canal. Note that in panel C, BDA-labeled corticospinal tract neurites can be seen arising from the intact corticospinal tract, traversing the midline, and growing into the gray matter of the lesioned side of the spinal cord. (Reprinted with permission from Zhou L et al: *J Neurosci* 23:1424–31, copyright 2003.)

NEUROTROPHIN 4 [13.6]. Neurotrophin 4 is a 210-amino acid neurotrophic factor (13–14 kDa) expressed in the nervous system. It is also known as neurotrophin 5 or neurotrophin 4/5. The term neurotrophin 4 was derived based on the fact that the *Xenopus* neurotrophin 4 gene was used to isolate the human counterpart gene. However, the human neurotrophin 4 exhibits more diverse features and activities than the *Xenopus* neurotrophin 4. Thus, this growth factor was also termed *neurotrophin 5*. Some investigators prefer the term neurotrophin 4/5 to reflect the similarities and differences between the mammalian and *Xenopus* forms. The gene of neurotrophin 4 is localized to chromosome 19q13.3. Neurotrophin 4 plays a critical role in regulating the survival, proliferation, and differen-

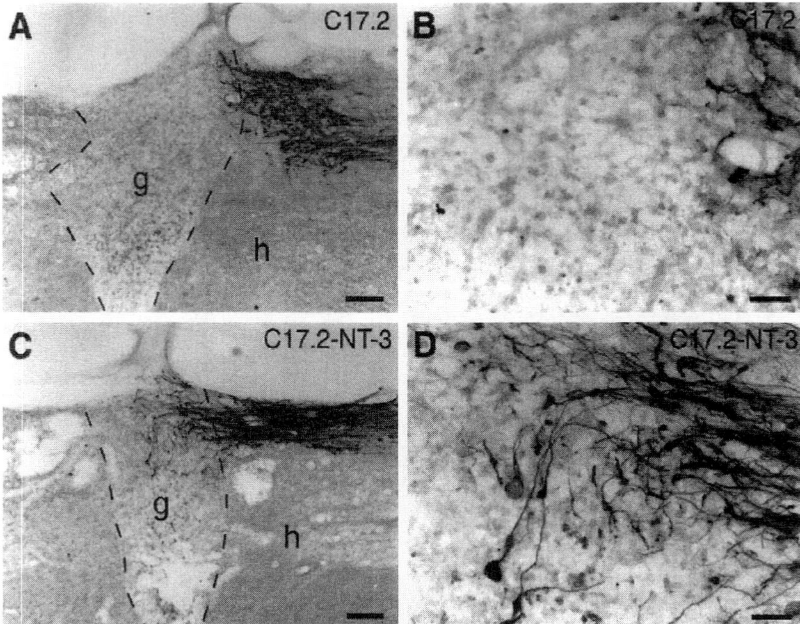

Figure 13.18. Neurotrophin-3 (NT3) enhances axonal growth of injured neurons into neural stem cell (NSC) grafts in injured spinal cord. Neural stem cell (mouse clone C17.2) grafts were transplanted to the lesion site of injured rat spinal cord. The growth of the dorsal column ascending sensory axons to the NSC graft was measured. (A) The ascending sensory axons rarely grow into the grafts of C17.2 NSCs without NT3 gene transfer. (B) Image from panel A with a higher magnification. (C) The ascending sensory axons could penetrate the graft of NSCs with NT3 gene transfer. (D) Image from panel C with a higher magnification. g: graft; h: host; dashed lines indicate host–graft interface. Scale bars: 177 μm in A,C; 44 μm in B,D. (E) Measurements of axon density within C17.2 grafts and C17.2–NT3 grafts (*$P < 0.05$). (Reprinted from Lu P et al: Neural stem cells constitutively secrete neurotrophic factors and promote extensive host axonal growth after spinal cord injury, *Exp Neurol* 181:115–29, copyright 2003, with permission from Elsevier.)

tiation of neurons. This factor also acts on the hippocampal neurons and contributes to the development of long-term memory. Furthermore, neurotrophin 4 is expressed in the fetal testis, contributing to the regulation of germ cell development and morphogenesis.

NERVE GROWTH FACTOR [13.7]. Nerve growth factor is a protein complex composed of four subunits, including one α unit, two β units, and one γ unit. The molecular weight of the complex is about 130 kDa. The genes that encode these subunits are mapped to chromosome 1 and locus 1p13.1. The gene sequence is similar between the human and mouse. Nerve growth factor subunits are produced at first as precursor proteins. The precursors are released into the extracellular matrix and cleaved by matrix metalloproteinases (MMP)7, MMP3, and serine protease plasmin. The cleavage activates the protein subunits. It is interesting to note that the precursor of nerve growth factor may exert an effect that is opposite to that of the mature form. In the rat and mouse models of brain injury, the precursor of nerve growth factor, produced and released by injured neurons, can bind to the neurotrophin receptor p75 and induce apoptosis. The blockade of the interaction of the nerve growth factor precursor with the neurotrophin receptor reduces apoptosis.

The expression of nerve growth factor is regulated by several factors, including endothelin-1 (EDN1), endothelin receptor, protein kinase C, the Src family protein tyrosine kinases, mitogen-activated protein kinases, and activator protein 1 (AP1). Upregulated endothelin 1 can interact with and activate its receptor, which in turn induces the activation of the intracellular signaling protein kinases. The deficiency of endothelin 1 in transgenic mice results in a reduction in the expression of nerve growth factor, which is associated with the loss of neurons during the late embryonic stage and a reduction in the level of cardiac norepinephrine and sympathetic innervation. The overexpression of nerve growth factor in endothelin-1-deficient mice enhances the production of norepinephrine and sympathetic innervation.

Nerve growth factor is produced and released by neurons and plays a critical role in the regulation of growth and differentiation of sympathetic and sensory neurons of the nerve system. For instance, in a rat model of spinal cord dorsal root injury, a treatment with nerve growth factor induces the regeneration of injured dorsal axons and functional reconnection of injured synapses.

GLIAL CELL LINE-DERIVED NEUROTROPHIC FACTOR [13.8]. Glial cell line-derived neurotrophic factor (GDNF) is a member of the transforming growth factor-β superfamily. This is a 211 amino acid protein produced by the neural glial cells. The gene of glial cell line-derived neurotrophic factor is localized to chromosome 5p13.1–p12. Glial cell line-derived neurotrophic factor exists in the form of disulfide-bonded homodimer, which is usually glycosylated.

The primary function of glial cell line-derived neurotrophic factor is to promote the survival, proliferation, and differentiation of neurons in the central nervous system. In particular, this neurotrophic factor is present in the midbrain and enhances the survival and proliferation of dopaminergic neurons. In the embryonic midbrain, glial cell line-derived neurotrophic factor plays a critical role in regulating the differentiation and morphogenesis of neuronal cells as well as the uptake of dopamine. It is important to note that glial cell line-derived neurotrophic factor exerts a specific effect on the dopaminergic neuron. It does not significantly influence the mitogenic activity of other types of neurons such as γ-aminobutyric-containing and serotonergic neurons. The dopaminergic neuron-specific features render this factor a potential therapeutic agent for the treatment of Parkinson's disease, which is possibly caused by progressive degeneration of the midbrain dopaminergic neurons.

Glial cell line-derived neurotrophic factor has been shown to mediate the survival and innervation of the motor neurons during development. The deficiency of glial cell line-derived neurotrophic factor in transgenic mice induces a reduction in motor axonal innervation to skeletal muscle cells. The overexpression of glial cell line-derived neurotrophic factor enhances the innervation of the motor axons. Furthermore, glial cell line-derived neurotrophic factor may be involved in the regulation of motor neuron pattern formation. In the deficiency of glial cell line-derived neurotrophic factor, motor neurons are mispositioned within the spinal cord in association with reduced innervation into the skeletal muscle tissue. These observations suggest the importance of glial cell line-derived neurotrophic factor in regulating the development of the peripheral nervous system.

Glial cell line-derived neurotrophic factor also plays a role in regulating the development of other systems. For instance, during the embryonic development, glial cell line-derived neurotrophic factor is expressed in the metanephric mesenchyme and acts on the ureteric bud epithelium, contributing to the regulation of ureter outgrowth and epithelial branching of the excretory system of the embryo (see Chapter 8 for the excretory system of the embryo).

Prevention of Cell Death. Nerve cell death is often induced by injury and injury-associated complications, such as ischemia and hypoxia, and is the most critical pathological event that deteriorates the function of the nervous system. In central nerve injury, most cells die in the injury site because of injury-associated complications. As reported in several studies, about 25–50% of the neurons die near the site of spinal cord axotomy within 4 weeks following the occurrence of the injury. The survival cells are highly atrophic with reduced capability of survival and proliferation.

Conventional approaches, as described on page 519 of this chapter, can be used to improve blood and oxygen supply in nerve injury, the most important step for the prevention of continuous cell death. However, nerve cells, especially injured nerve cells, may continue to commit to death even after the reestablishment of bloodflow. Thus, it is necessary to apply therapeutic approaches to injured nerve cells. Recent work has demonstrated that molecular engineering approaches can be used for such a purpose. It is now well established that cell death is mediated by extracellular ligands, including tumor necrosis factor α and Fas ligand, and intracellular death signaling molecules and caspases (see page 304). Extracellular stimuli, such as ischemia and hypoxia, may activate these factors and induce cell death. Molecular engineering approaches can be used to reduce cell death by suppressing the activity of cell death-promoting factors and by enhancing the activity of cell death-inhibiting factors. For instance, antibodies may be developed and used to block the extracellular cell death inducers TNF-α and Fas ligand. Genes encoding cell death inhibitors, such as Bcl2, can be used to transfect injured neuronal cells to prevent cell death. Furthermore, the nerve growth-stimulating factors, such as brain-derived neurotrophic factor, neurotrophin-4/5, nerve growth factor, and glial-derived neurotrophic factor, can prevent cell injury and death. These factors can be directly delivered to the site of nerve injury. The genes that encode these neurotrophic factors can be used to transfect injured nerve cells. These approaches have been shown to be effective for the prevention of neuronal death.

Prevention of Secondary Nerve Injury. Secondary injury is defined as disorders induced by pathological alterations initiated by the original nerve injury. Injury-induced pathophysiological alterations include, but are not limited to, hemorrhage (open wound), hypoxia (reduction in oxygen supply), inflammation (bacterial or nonbacterial), release of harmful enzymes (proteinases), and disorder of ion fluxes. These alterations significantly worsen the original injury and induce secondary injury of the nerve system, resulting in cell death and tissue disintegration. Thus, immediately after the initial nerve injury, it is critical to control these pathophysiological changes and protect the nerve system from secondary injury. Conventional approaches for the treatment of secondary injury have been described on page 519. Several molecular engineering approaches have been developed and tested for the prevention of secondary nerve injury. Strategies for preventing secondary injury include prevention of the disorder of ion transport and excitotoxicity.

PREVENTION OF ION FLUX [13.9]. The flux of ions, including sodium and calcium, is increased in nerve injury and contributes to secondary nerve injury (Fig. 13.19). Under physiological conditions, sodium is concentrated in the extracellular fluid. The concentration of sodium in the extracellular matrix is significantly higher than that in the cytoplasm. The concentration gradient for the sodium across the cell membrane is maintained by the

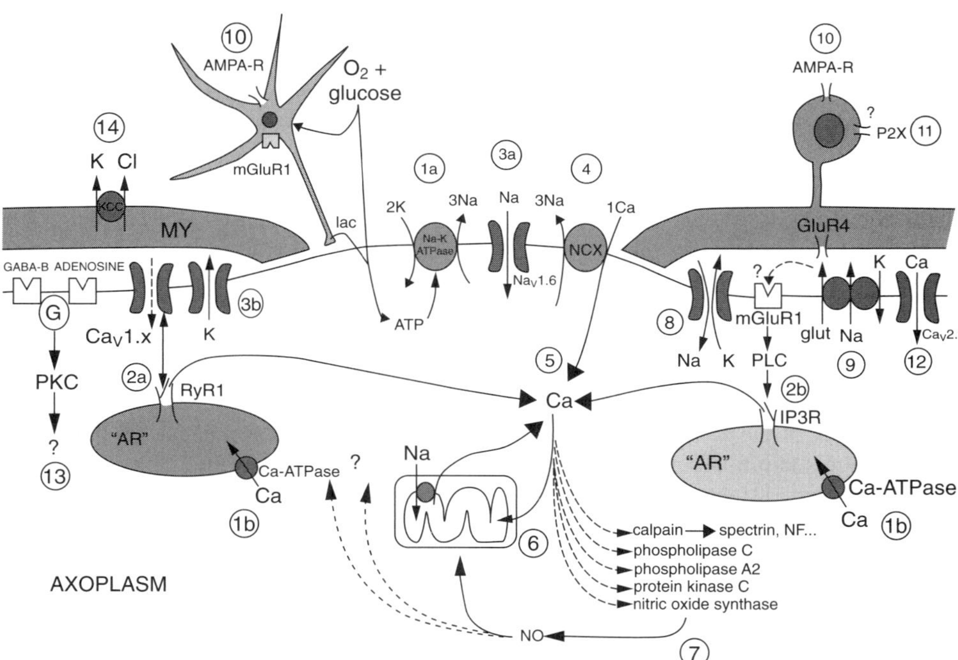

Figure 13.19. Mechanisms of ion-induced nerve injury. An energy deficit and/or excess demand impairs ATP-dependent pumps such as the Na–K ATPase (1a) and Ca ATPase (1b). Internal stores of Ca may contribute significantly to axonal Ca accumulation, triggered by depolarization via L-type Ca channels (2a) and/or generation of IP3 (2b). The rise in flux through noninactivating Na^+ channels (3a) will increase $[Na]_i$ and, together with depolarization caused by K efflux through a variety of K channels (3b), will stimulate the Na–Ca exchanger to operate in the reverse Ca import mode (4). This Ca accumulation (5) promotes destructive events including mitochondrial Ca overload (especially during reoxygenation) (6), and overactivation of several Ca-dependent enzyme systems (7). NO will inhibit mitochondrial respiration and alter other cellular proteins. Some Na influx may occur through Na/K-permeable inward rectifier channels (8). Glutamate is also released through reversal of Na-dependent glutamate transport (9), causing cellular injury from activation of ionotropic glutamate receptors (10). ATP-activated P2X purinergic receptors may cause Ca-dependent oligodendroglial injury (11). A component of Ca influx into damaged axons directly through voltage-gated Ca channels is also likely (12). GABA and adenosine release may play an "autoprotective" role (13). Anion transporters such as the K—Cl cotransporter participate in volume dysregulation in glia and the myelin sheath, contributing to conduction abnormalities (14). The locations of the various channels and transporters are drawn for convenience and do not necessarily reflect actual distributions. (Reprinted from Stys PK: General mechanisms of axonal damage and its prevention, *J Neurol Sci* 233:3–13, copyright 2005, with permission from Elsevier.)

continuous action of the sodium pumps in the cell membrane. Calcium is stored in the endoplasmic reticulum (ER) system in normal cells. The concentration of calcium in the cytoplasm is much lower than that within the ER. The calcium concentration gradient across the ER membrane is established and maintained by the calcium pumps in the ER membrane.

In nerve injury, especially in the presence of ischemia, cell membrane exhibits enhanced permeability to ions, resulting in an increase in ion flux. Such a change influences the level of membrane potential as well as the initiation and transmission of the action potential of the neurons, thus reducing the function of the nerve system. Ion channel blockers, including sodium channel blockers (QX-314, tetrodotoxin, and procaine) and calcium channel blockers (diltiazem, verapamil, and ω-conotoxin GVIA), have been developed and used to reduce ion flux. Such an approach has been shown to be effective for the prevention of secondary nerve injury. The ion channel blockers can be directly delivered into the spinal fluid. Alternatively, the blockers can be injected into the blood. Since the blockers are relatively small molecules, they can pass the blood–brain barrier and reach the injured nerve cells and axons.

PREVENTION OF EXCITOTOXICITY [13.10]. Central nerve injury is associated with excitotoxic activities mediated by glutamate, which contributes to secondary nerve injury. Glutamate-mediated secondary injury is often found in the nerve axon-concentrated white matter of the spinal cord. As discussed above, nerve injury induces disorder of sodium influx, which in turn influences the efflux of potassium. These alterations induce the disruption of the concentration gradient for sodium and potassium. Such a disruption activates the sodium-dependent glutamate transporter, resulting in the release of endogenous glutamate. Released glutamate can interact with and activate the α-amino-3-hydroxy-5-methyl-isoxazole-4-propionate (AMPA)/kainate receptor. This receptor mediates calcium release and transport, resulting in cytosolic accumulation of calcium. Calcium is known to activate various enzymes, such as calpains and phospholipases. These enzymes can cause irreversible injury of the axons. A treatment with a glutamate antagonist (e.g., kynurenic acid) or a selective AMPA antagonist (e.g., GYKI52466) significantly reduces secondary axonal injury. The inhibition of the sodium-dependent glutamate transport with L-*trans*-pyrrolidine-2,4-dicarboxylic acid can also protect the axons from injury. These substances may be potentially used for the prevention of secondary nerve injury.

Stimulation of Stem Cell Differentiation [13.11]. One of the strategies for inducing neuronal regeneration is to enhance the differentiation of the neural stem cells to neurons. Neurons have long been considered well-differentiated cells that exhibit a very low level of regenerative activity. However, recent work has demonstrated that there exist neuronal stem cells in a number of locations in the adult central nerve system, including the caudal portion of the subventricular zone, olfactory bulb, hippocampus, striatum, optic nerve, corpus callosum, spinal cord, cortex, retina, and hypothalamus. A typical type of stem cell is the neuroepithelial cells at these locations. The neuroepithelial cells are able to differentiate into neurons or glial cells to replace injured and dead cells. Thus, a conceivable therapeutic approach in nerve regenerative engineering is to promote the differentiation of the stem cells. Furthermore, neurons can extend their axons and replace injured and severed axonal segments. The enhancement of axonal growth is another approach for the treatment of nerve injury.

An effective method for stimulating the differentiation of neural stem cells is to deliver neurotrophic factors to injured nerve tissues. Neurotrophic factors have long been known to stimulate neural stem cell proliferation and differentiation. In the nerve system, there are a number of growth factors, also known as neurotrophic factors. These include nerve growth factor (NGF), brain-derived neurotrophic factor (BDNF), neurotrophin 3 (NT3), and neurotrophin-4/5 (NT4/5). These neurotrophic factors are required for the expression of the nerve regeneration-associated genes and stimulate neuronal regeneration. The neurotrophic factors also directly stimulate axonal regeneration and enhance structural and functional reconnection between injured neurons. The functions of these regeneration-stimulating factors are discussed on page 521 of this chapter.

It is important to address that neuronal regeneration is regulated not only by stimulatory neurotrophic factors but also by inhibitory factors. The inhibitory factors suppress the regeneration of neuronal cells. The understanding of the inhibitory effect helps to clarify the mechanisms of controlling neuronal regeneration. A typical inhibitory factor is growth and differentiation factor 11 (GDF11), a member of the transforming growth factor (TGF)β protein superfamily. This factor is expressed in the olfactory neuroepithelial cells, which are thought to include neural stem cells. When upregulated, GDF11 exerts an inhibitory effect on olfactory neurogenesis by activating an inhibitory factor p27, which is implicated in the induction of cell cycle arrest. The knockout of the GDF11 gene results in an increase in neurogenesis. The inhibitory effect of GDF11 represents a mechanism by which the density of neurons is checked and maintained at a certain level. GDF11 may be activated when excessive neurogenesis is induced, whereas GDF11 is suppressed when neurogenesis is reduced. This is a typical example of the feedback regulatory mechanism.

Neuronal regeneration is also inhibited by bone morphogenetic protein 2 (BMP2), a member of the BMP family, which are involved in the development of the nervous system. A treatment with BMP2 in cultured neural stem cells induces a significant reduction in the formation of a certain type of nerve cell that expresses the neuronal marker microtubule-associated protein 2, but enhances the formation of other type of nerve cells that express a glial cell marker S100β. These observations suggest that BMP2 may stimulate the transformation of neural stem cells to glial cells, while inhibiting the formation of neuronal cells. Further investigations have demonstrated that BMP7, another member of the BMP family, is upregulated in spinal cord injury. This factor may stimulate the regeneration of glial cells in injured spinal cord. Taken together, these investigations suggest that, in nerve regenerative engineering, both the stimulatory and inhibitory mechanisms should be taken into account.

Enhancement of Axonal Extension, Adhesion, and Reconnection [13.12]. Under appropriate conditions, the neuron is capable of regenerating its axon when the axon is severed. It is interesting to note that axon transection can serve as a stimulus for the initiation of axon regeneration. Such an injury can induce the expression of regeneration-associated genes, including c-jun and c-fos. The protein products of these genes enhance the expression of several factors that are involved in axon extension, including the Tα1-tubulin and nerve cell adhesion molecule (NCAM). The Tα1-tubulin is a cytoplasmic growth cone protein and is involved in growth cone guidance. The nerve cell adhesion molecule is a factor that regulates cell attachment and migration. These molecules promote axonal regeneration. Nerve regeneration fails in the absence of these factors. For therapeutic purposes, the genes of these factors can be used to construct recombinant genes, which can be transferred to the injury site of the central nerve system by using methods described on page 436. The enhancement of gene expression by gene transfer facilitates the extension of injured axon.

Cell adhesion and attachment to substrate are critical processes for nerve axonal extension and regeneration. A number of adhesion molecules, such as growth-associated protein 43 (GAP43), cell adhesion molecule L1, N-cadherin, have been found to modulate neuronal adhesion. For instance, the overexpression of growth-associated protein 43 or cell adhesion molecule L1 in the Purkinje neurons of transgenic mice stimulates axonal sprouting of the Purkinje neurons into nerve grafts compared to wildtype mice, which do not significantly express growth-associated protein 43 and cell adhesion molecule L1 in the Purkinje neurons. It is interesting to note that growth-associated protein 43 and cell adhesion molecule L1 promote neuronal growth and axonal sprouting in a synergistic manner. When both molecules are overexpressed in the Purkinje neurons by genetic modulation, the rate of neuronal growth and axonal extension into nerve grafts is significantly enhanced compared to the transgenic model with overexpression of only a single molecule (either growth-associated protein 43 or cell adhesion molecule L1). These observations suggest that genetic upregulation of neuronal adhesion molecules via gene transfer can enhance neuronal regeneration and axonal growth in nerve injury. It is important to address that axonal outgrowth is regulated by opposing stimulatory and inhibitory factors. For instance, a protein known as netrin 1 significantly enhanced the axonal outgrowth of dopaminergic neurons in the embryonic ventral midbrain. In contrast, a protein known as slit-2 suppresses the axonal outgrowth of the dopaminergic neurons. These observations suggest that opposing regulatory factors may coordinately control the neuronal outgrowth in nerve injury. Thus, the inhibitory factors ought to be considered in the treatment of nerve injury.

Prevention of Fibrous Scar Formation [13.13]. Nerve injury is associated with glial cell proliferation, fibrosis, and scar formation. The scar tissue is composed of microglial cells, astrocytes, oligodendrocytes, and extracellular matrix. The scar tissue obviously imposes a physical barrier to the regeneration of axons. In addition, the cells in the scar tissue can secret a number of molecules, including neurocan (a major chondroitin sulfate proteoglycan or CSPG in the nerve system), proteoglycan phosphocan, brevican, tenascin, semaphorins, and ephrins. These molecules exist in the form of either membrane-bound molecules or soluble molecules and exert an inhibitory effect on axonal regeneration. Antagonistic molecules and pharmacological substances that inhibit the activity of these molecules can be used to enhance axonal regeneration in nerve injury. For example, urokinase has been used to degrade proteoglycans in injured nerve tissues and shown to promote axonal regeneration. Antibodies can be developed and used to neutralize the activity of the inhibitory molecules. Such a strategy can be potentially applied to the treatment of nerve injury.

Recent investigations have shown that nuclear factor (NF)κB plays a critical role in secondary inflammatory reactions and the formation of scar tissue after nerve injury. NFκB is a transcription factor, which activates the expression of genes encoding proinflammatory factors, such as CXCL10, CCL2, and transforming growth factor β2. Furthermore, NFκB stimulates the formation of chondroitin sulfate proteoglycans. When NFκB is selectively suppressed in a transgenic mouse model, inflammatory reactions and scar formation are significantly inhibited after contusive spinal cord injury. These observations show that selective inhibition of NFκB in glial cells exerts a protective effect on injured nerve axons.

There exist a number of central nerve myelin-associated inhibitory factors, including NI35, NI250, and several glycoproteins. These factors directly inhibit neuronal axon regeneration. The neutralization of these inhibitory factors with antibodies results in enhanced sprouting of injured spinal neurons. These experiments confirm the inhibitory role of the myelin-associated factors and suggest therapeutic applications of the antibodies.

Enhancement of Synaptic Formation [13.14]. Nerve injuries are often associated with the discontinuation of nerve synapses. An important strategy in nerve regeneration is to restore the nerve synapses. Several molecules, such as neurotrophins (NT), agrin, and s-laminin, have been shown to promote synaptic formation. For instance, neurotrophins, especially neurotrophin 3, stimulate not only spinal axonal outgrowth, but also the reconnection of transected axons to terminal muscular cells via the regeneration of synapses. Most neurotrophic factors exert a promoting effect on axonal outgrowth after injury. Thus, synapse-stimulating proteins and their genes can be used to treat nerve injury to enhance the restoration of injured synapses.

Nerve Cell Regenerative Engineering. Cell regenerative engineering is to identify, collect, modulate, and transplant functional cells to injury or lesion sites of the nerve system to replace lost cells and improve nerve regeneration. As discussed in the last section, molecular regenerative engineering can be potentially applied to the central nervous system to facilitate the regeneration of injured neurons and axons. However, in most cases, neuronal regeneration and axon outgrowth are hindered by pathological changes, such as glial proliferation, fibroblast infiltration, and fibrosis. These lesions encapsulate injured neurons and axons, preventing neuronal regeneration and axonal outgrowth. Thus, it is necessary to develop strategies to overcome these detrimental conditions. A potential strategy is to transplant stem and progenitor cells as well as supporting cells to enhance neuronal regeneration and axonal outgrowth. Candidate cell types may include embryonic and fetal stem cells, Schwann cells, and olfactory ensheathing cells. These cells may promote neuronal and axonal growth and regeneration, while protecting them from detrimental effects. It is expected that cell regenerative engineering, together with molecular regenerative engineering, may enhance the regeneration of injured neurons and the outgrowth of injured axons.

Embryonic and Fetal Stem Cells [13.15]. Embryonic and fetal stem cells are the primary cell candidates for the treatment of nerve injury. As discussed on page 381, embryonic stem cells are capable of differentiating into all specified cell types. Under the stimulation of appropriate environmental cues for nerve cell development, these stem cells may be induced to differentiate into neurons or glial cells. Fetal neural stem cells may also be used to generate neurons. In experimental models, stem cell transplantation has been shown to reduce neuronal cell death, enhance neuronal survival, and promote axonal outgrowth after nerve injury. Clinical trials for treating posttraumatic syringomyelia with fetal spinal cord grafts have demonstrated the feasibility of cell regenerative engineering for treating nerve injury. Although investigations with stem cell transplantation have provided promising results, embryonic and fetal cells may not be used until related ethical issues are resolved.

Adult Neural Stem Cells [13.16]. There exist adult stem and progenitor cells that can differentiate to neuronal and glial cells. These cells are neuroepithelial cells and can be

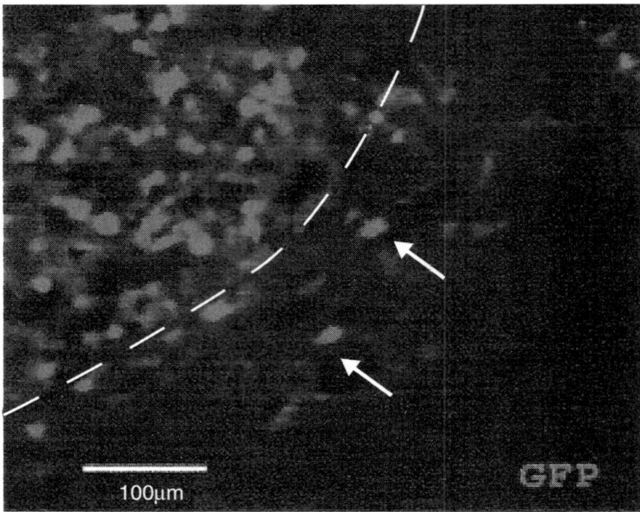

Figure 13.20. Fluorescent micrograph showing transplantation and engraftment of olfactory progenitor cells to the spinal cord of the rat. Neurosphere-forming progenitor cells were prepared from human adult olfactory epithelium, transfected with a GFP gene, and transplanted into traumatized rat spinal cord. At 1 week, specimens were collected from the injured spinal cord and prepared for observing GFP-labeled neurosphere-forming progenitor cells at the lesion site. The dashed line indicates the host–graft interface. Several transplanted GFP-positive cells appear with the host spinal cord bridging the injury site (arrows). (Reprinted from Xiao M et al: Human adult olfactory neural progenitors rescue axotomized rodent rubrospinal neurons and promote functional recovery, *Exp Neurol* 194:12–30, copyright 2005, with permission from Elsevier.)

found in the central nerve system as well as in the bone marrow. In the adult brain, neural stem and progenitor cells are present in several regions, including the caudal portion of the subventricular zone, olfactory bulb, hippocampus, striatum, optic nerve, corpus callosum, spinal cord, cortex, retina, and hypothalamus. Cells collected from these regions can be induced to differentiate to neurons, astrocytes, and oligodendrocytes. Figure 13.20 shows that olfactory cells transplanted into the rat spinal cord can engraft and integrate into the host tissue.

Stem and progenitor cells have been shown to express specific proteins. For instance, neuronal progenitor cells express the NeuN protein, whereas glial progenitor cells express GFAP and S100β. These proteins can be used as markers to identify neuronal and glial progenitor cells. For therapeutic purposes, neural stem and progenitor cells can be identified, collected, expanded in vitro, and transplanted to the site of nerve injury to enhance nerve regeneration. See page 395 for characteristics of neural stem and progenitor cells.

Bone Marrow Cells [13.17]. The bone marrow contains stem and progenitor cells for a variety of specified tissue types, including the nerve system. Several investigations have demonstrated that bone marrow-derived stem cells, when cultured in the presence of EGF or BDNF, can transform to cells that express the neural progenitor cell marker nestin, neuron-specific nuclear protein (NeuN), and glial fibrillary acidic protein (GFAP). When injected into the venous system of myeloablated animals, bone marrow cells can engraft to the brain and form cells that express neuronal protein markers such as NeuN, 200-

kilodalton neurofilaments, and class III β-tubulin in the brain. In human studies, transplanted bone marrow cells can transform to neurons in the cerebellum and cerebrum. Bone marrow-derived neurons account for about 1% of all neurons.

In a chicken spinal cord injury model, the implantation of human hematopoietic CD34+ stem cells into the injured spinal cord induces the formation of neuron-like cells that express the neuronal markers NeuN and MAP2. These neuron-like cells can extend their axons into the white matter of the spinal cord with synaptic terminals. Furthermore, these cells demonstrate spontaneous synaptic action potentials characteristic of functional neurons. These observations suggest that bone marrow-derived cells can transform to neuron-like cells and can be potentially used to treat spinal cord injury.

Bone marrow cells can also differentiate into glial cells. When bone marrow cells are injected to a demyelinated spinal cord, these cells can transform to oligodendrocytes that express myelin protein and induce remyelination. Bone marrow cells have been used to treat spinal cord injury, resulting in functional improvement of the spinal cord. However, there are several aspects that remain to be investigated. First, the morphology of the neural cells derived from bone marrow cells has not been thoroughly studied. It remains poorly understood whether bone marrow-derived cells can form axons. Second, the function of bone marrow-derived cells has not been systematically characterized. It is not clear whether bone marrow cells can develop into fully functional neurons or glial cells. Furthermore, the use of bone marrow cells for nerve regeneration remains a controversial topic. A recent study has demonstrated that the bone marrow hematopoietic stem cells do not contribute significantly to neuronal regeneration. Further investigations are necessary to clarify this issue.

Neuronal Supporting Cells [13.18]. Neuronal supporting cells include glial and Schwann cells, which can be used to serve as substrate for neuronal regeneration and axonal outgrowth. Schwann cells can be easily collected and cultured. These cells have been used to transplant to the site of nerve injury and to serve as bridges for axonal outgrowth and regeneration. To enhance the growth of transplanted cells, neurotrophic factors and/or their genes can be delivered, together with cell transplantation, to nerve injury sites. Schwann cells have been transplanted together with brain-derived neurotrophic factor (BDNF) and neurotrophin (NT)-3, in experimental spinal cord injury. Such a combination has been shown to promote axonal outgrowth. Furthermore, transplanted Schwann cells can guide axonal extension, a critical process for the reinnervation of axons to peripheral skeletal muscle cells.

Another type of neuronal supporting cells is the olfactory ensheathing cells. These cells are glial cells and are found in the nerve system in association with the olfactory neurons and axons. The primary function of olfactory ensheathing cells is to escort axons from the peripheral to the central nerve system. These cells are also capable of promoting axonal outgrowth. Such a property may be utilized to bridge the gap of nerve injury and enhance the myelination of newly generated axons. The role of olfactory ensheathing cells in regulating axonal regeneration has been observed in numbers of investigations. These observations demonstrate the potential of using olfactory glial cells for neural cell regenerative engineering.

Transgenic Cell Lines [13.19]. For the treatment of nerve injury, it is critical to promote nerve regeneration as rapidly as possible. To achieve such a goal, transgenic cell lines have been established by transfecting selected cell types with neurotrophic factor genes, such as nerve growth factor, neurotrophin 3, brain-derived neurotrophic factor, and basic fibroblast growth factor genes. The transgenic cells can be used for transplantation into injured nerve tissues. These cells can express and release the transfected neurotrophic genes at a high rate, rapidly stimulating the regeneration of injured neurons and axons. Candidate cell types for such a purpose include embryonic stem cells, glial cells, Schwann cells, olfactory ensheathing cells, and fibroblasts. Preliminary investigations have demonstrated promising results for the use of genetically modified cells for the regeneration of injured neurons and axons.

Tissue Regenerative Engineering for Nerve Injury. Tissue regenerative engineering is to construct and provide nerve matrix scaffolds, which serve to enhance and guide neuronal regeneration and axonal outgrowth. Compared to molecular and cell regenerative engineering, which can be applied to both the central and peripheral nerve systems, tissue regenerative engineering has been primarily used for the treatment of peripheral nerve injury. This is attributed to the difficulty of accessing the central nerve system and the possibility of inducing nerve injury by tissue scaffold implantation. For peripheral nerve engineering, several strategies have been developed and used, including guidance of axonal regeneration and outgrowth, stimulation of neuronal proliferation by graft-based assistance, and improvement of neuronal function.

Stimulation of Neuronal Regeneration and Guidance of Axonal Outgrowth by Graft-Based Assistance [13.20]. As discussed on page 517 of this Chapter, severed nerve axons can regenerate and establish reconnection between the severed segments. However, such reconnection requires appropriate guidance from the Schwann cells, which form a sheath along the severed axons. When the severed proximal axon is relocated away from the distal axon, it is difficult for the proximal axon to reconnect to the severed distal axon. In conventional treatment, severed nerve bundles are reconnected surgically. However, when a segment of a nerve bundle is severely damaged or removed, it is impossible to reconnect the severed proximal to the distal nerve bundle. Under such as circumstance, it is necessary to provide guidance to the injured axons, allowing axonal extension along a desired path. For the past decade, a number of biological and synthetic materials have been developed and used to construct nerve scaffolds for the stimulation of axonal sprouting and the guidance of axonal extension. These materials include autologous nerve tissue grafts, allogenic (from the same animal species) and xenogeneic nerve tissue grafts (from different animal species), extracellular matrix components (laminin, fibronectin, collagen, and fibrin), and synthetic polymeric materials [poly (L-lactic acid), poly-L-lactide-ε-caprolactone, polyurethane, poly(2-hydroxyethyl methacrylate-co-methyl methacrylate), and polypyrrole-hyaluronic acid]. Nerve grafts based on these materials are briefly discussed here.

AUTOLOGOUS NERVE TISSUE GRAFTS [13.21]. Autologous nerve tissue specimens can be collected from the patient who receives grafts. Autologous nerve tissue is an ideal material type for constructing nerve grafts. Certain subcutaneous nerves, such as the branches of the saphenous nerve, can be used as nerve grafts. A nerve branch can be isolated and grafted into a severed nerve bundle. Such a graft usually provides effective support and guidance to the injured nerve and significantly enhances neuronal regeneration and axonal outgrowth. Autologous nerve tissue is usually considered the gold standard for constructing nerve grafts. Other types of natural tissue, such as autologous blood vessels, skeletal muscle, and soft connective tissues, have also been used as nerve grafts. These tissues can be tailored into the shape of nerve bundles and grafted into severed nerves. However, the effectiveness of these tissues is limited compared to the autologous nerve grafts.

ALLOGENIC AND XENOGENEIC TISSUE GRAFTS [13.22]. Allogenic grafts are those collected from different individuals of the same animal species. Xenogeneic grafts are from individuals of different animal species. Various types of allogenic and xenogeneic tissue, such as nerve tissue, soft connective tissue, intestinal submucosa, and amnion, have been used as nerve grafts. However, the cellular components of allogenic and xenogeneic tissues, especially living cells, induce acute immune rejection reactions. Thus, the cellular components must be removed from the graft tissue. Decellularized matrix scaffolds can be produced by appropriate thermal, detergent, or NaOH (or KOH) treatment. The cell-free grafts can be used as scaffolds that guide neuronal regeneration and axonal outgrowth. These grafts, however, are not as effective as the autologous nerve grafts.

EXTRACELLULAR MATRIX COMPONENTS [13.23]. Extracellular components, including collagen, fibronectin, laminin, and fibrin, have been used to construct scaffolds for stimulating and guiding neuronal regeneration and axonal outgrowth. These molecules play an important role in the regulation of neuronal proliferation and migration as well as axonal extension and innervation during development. Some of these matrix components, such as collagen, fibronectin, and laminin, serve as substrates for cell attachment and migration, which are critical processes for the morphogenesis of the nerve system during development and regeneration of injured nerve axons. Other extracellular matrix molecules, such as fibrinogen, fibrin gels, peptide scaffolds, alginate, agarose, and chitosan, have been applied to injured nerves for enhancing axonal outgrowth in experimental models. It is important to point out that not all extracellular matrix components are neuronal growth promoters. Certain molecules of the proteoglycan family, such as chondroitin sulfate proteoglycan (CSPG) and proteoglycan phosphocan, are typical inhibitory factors for neuronal regeneration and axonal outgrowth. The inhibitory effect of these molecules may play a role in the prevention of glial scar formation, a hindrance for the regeneration and extension of nerve axons.

SYNTHETIC MATERIALS [13.24]. A number of synthetic polymeric materials have been developed and used for the regeneration of injured nerves in experimental models. These materials include poly(glycolic acid) (PGA), poly(lactic acid) (PLA), poly(lactic-co-glycolic acid) (PLGA), poly(caprolactones), biodegradable poly(urethane), polyorganophosphazene, methacrylate-based hydrogels, and poly(3-hydroxybutyrate). These polymeric materials are biodegradable, relatively low in toxicity, and easy to process. Scaffolds made of these materials have been tested in animal models for nerve regeneration. Results from these investigations are promising. Nonbiodegradable synthetic materials, such as polytetrafluoroethylene (PTFE), have also been used as scaffolds for nerve regeneration with satisfactory results.

Polymeric materials can be used in combination with nerve regeneration stimulators, such as neurotrophic factors and antiapoptotic factors or their genes. A polymeric scaffold may serve not only as a guide for nerve regeneration but also as a drug delivery device. The polymeric scaffold and the nerve regeneration stimulators may synergistically enhance the regeneration of injured nerves. However, compared to autogenous nerve grafts, the effectiveness of the synthetic polymeric materials remains limited. A critical problem with the synthetic materials is their properties of inducing host inflammatory reactions. Improving biocompatibility of polymers is a major challenge in nerve regeneration with synthetic materials.

ROLE OF GEOMETRIC FACTORS [13.25]. While the material properties are critical to the modulation of neuronal regeneration and axonal outgrowth, geometric factors, such as the shape, orientation, curvature, and dimension of matrix substrate and scaffolds, may influence the extension and direction of regenerating axons. For instance, when neurons are cultured on a polymeric matrix surface, the direction of axonal outgrowth is dependent on the matrix structure and the orientation of the matrix fibers (Fig. 13.21). The curvature of

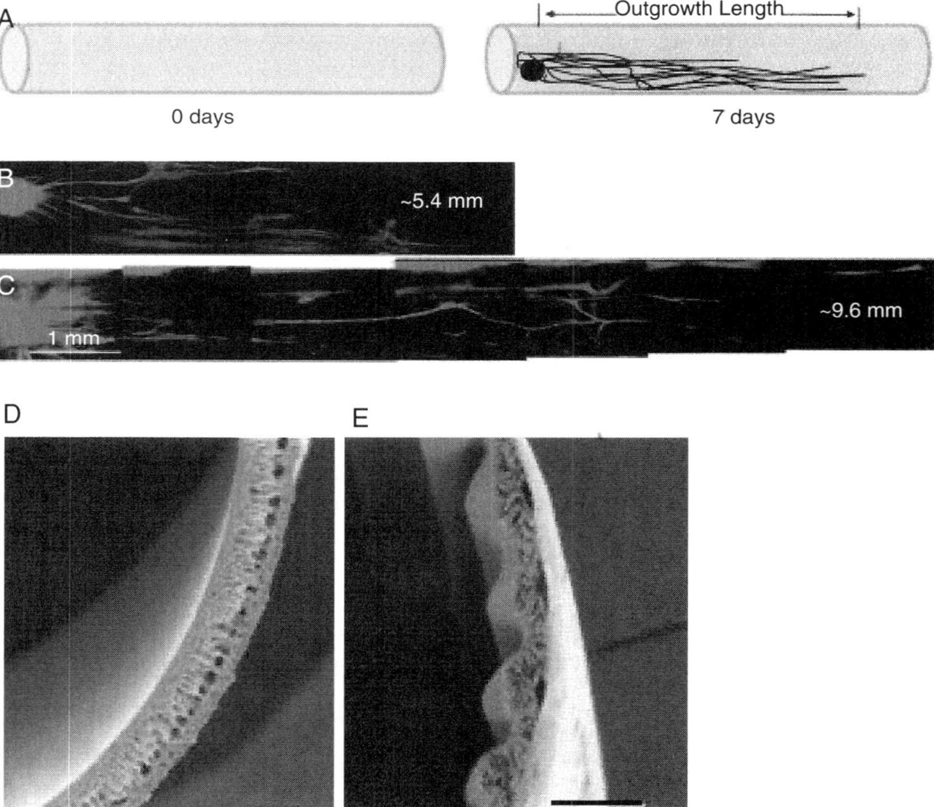

Figure 13.21. Biomaterial implantation for guiding axonal outgrowth. (A) Schematic illustration of neurite outgrowth evaluation in a dorsal root ganglion (DRG) explant model. DRG explant was seeded at one end of a cylinder of polyurethane hollow fiber membrane (HFM) and allowed to grow for 7 days. Neurite outgrowth was assessed based on neurite length measurement. (B) Neurite outgrowth on a smooth HFM surface for 7 days. (C) Neurite outgrowth on a HFM surface with aligned grooves for 7 days. (D,E) Scanning electron micrographs showing the morphology of HFM membrane with smooth and grooved surfaces, respectively. (Reprinted with permission of John Wiley & Sons, Inc. from Zhang N et al: Fabrication of semipermeable hollow fiber membranes with highly aligned texture for nervous guidance, *J Biomed Mater Rese Pt A* 75:941–9, copyright 2005.)

the matrix fibers influences the direction of axonal extension. Furthermore, the geometry of the matrix substrate regulates the rate of neuronal regeneration. As shown in several studies, magnetically aligned collagen fibers elicit a more significant effect on neuronal regeneration than randomly aligned collagen fibers. These observations demonstrate that geometric factors may serve as cues for the neuronal regeneration and axonal outgrowth.

IMPROVEMENT OF NERVE FUNCTION [13.26]. For the treatment of nerve injury, in addition to the promotion of nerve regeneration, another important strategy is to improve the function of regenerated axons. A major approach for such a purpose is to electrically stimulate impaired and regenerated neurons and axons. Several types of devices have been developed and tested. These include cochlear, retinal, spinal cord, and brain stimulators. These devices can be used to improve nerve functions such as hearing, vision, and muscular movement control. It becomes now clear that a successful treatment of nerve injury requires both anatomical nerve regeneration and functional recovery.

BIBLIOGRAPHY

13.1. Anatomy and Physiology of the Nervous System

Lai C, Peripheral glia: Schwann cells in motion, *Curr Biol* 15(9):R332–4, May 2005.

Guyton AC, Hall JE: *Textbook of Medical Physiology*, 11th ed, Saunders, Philadelphia, 2006.

McArdle WD, Katch FI, Katch VL: *Essentials of Exercise Physiology*, 3rd ed, Lippincott Williams & Wilkins, Baltimore, 2006.

Germann WJ, Stanfield CL (with contributors Niles MJ, Cannon JG): *Principles of Human Physiology*, 2nd ed, Pearson Benjamin Cummings, San Francisco, 2005.

Thibodeau GA, Patton KT: *Anatomy & Physiology*, 5th ed, Mosby, St Louis, 2003.

Boron WF, Boulpaep EL: *Medical Physiology: A Cellular and Molecular Approach*, Saunders, Philadelphia, 2003.

Ganong WF: *Review of Medical Physiology*, 21st ed, McGraw-Hill, New York, 2003.

13.2. Nerve Injury

Schneider AS, Szanto PA: *Pathology*, 3rd ed, Lippincott Williams & Wilkins, Philadelphia, 2006.

McCance KL, Huether SE: *Pathophysiology: The Biologic Basis for Disease in Adults & Children*, 5th ed, Elsevier Mosby, St Louis, 2006.

Porth CM: *Pathophysiology: Concepts of Altered Health States*, 7th ed, Lippincott Williams & Wilkins, Philadelphia, 2005.

Frazier MS, Drzymkowski JW: *Essentials of Human Diseases and Conditions*, 3rd ed, Elsevier Saunders, St Louis, 2004.

Kwon BK, Tetzlaff W: Spinal cord regeneration: From gene to transplants, *Spine* 26(Suppl):S13–22, 2001.

Feringa ER, McBride RL, Pruitt JN: Loss of neurons in the red nucleus after spinal cord transection, *Exp Neurol* 100:112–20, 1988.

Goshgarian HG, Koistinen JM, Schmidt ER: Cell death and changes in the retrograde transport of horseradish peroxidase in rubrospinal neurons following spinal cord hemisection in the adult rat, *J Compar Neurol* 214:251–7, 1983.

McDonald JW, Sadowsky C: Spinal cord injury, *Lancet* 359:417–25, 2002.

Schwab ME: Repairing the injured spinal cord, *Science* 295:1029–31, 2002.

13.3. Molecular Regenerative Engineering for the Nervous System

Koprivica V, Cho KS, Park JB, Yiu G, Atwal J et al: EGFR activation mediates inhibition of axon regeneration by myelin and chondroitin sulfate proteoglycans, *Science* 310:106–10, 2005.

Houle JD, Ye JH: Survival of chronically-injured neurons can be prolonged by treatment with neurotrophic factors, *Neuroscience* 94:929–36, 1999.

Yamada M, Natsume A, Mata M, Oligino T, Goss J et al: Herpes simplex virus vector-mediated expression of Bcl-2 protects spinal motor neurons from degeneration following root avulsion, *Exp Neurol* 168:225–30, 2001.

Chen DF, Schneider GE, Martinou JC, Tonegawa S: Bcl-2 promotes regeneration of severed axons in mammalian CNS, *Nature* 385:434–39, 1997.

Chierzi S, Strettoi E, Cenni MC, Maffei L: Optic nerve crush: axonal responses in wild-type and bcl-2 transgenic mice, *J Neurosci* 19:8367–76, 1999.

Holm K, Isacson O: Factors intrinsic to the neuron can induce and maintain its ability to promote axonal outgrowth: A role for BCL2? *Trends Neurosci* 22:269–73, 1999.

Kobayashi NR, Fan DP, Giehl KM et al: BDNF and NT-4/5 prevent atrophy of rat rubrospinal neurons after cervical axotomy, stimulate GAP-43 and Talpha1-tubulin mRNA expression, and promote axonal regeneration, *J Neurosci* 17:9583–95, 1997.

Schumacher PA, Siman RG, Fehlings MG: Pretreatment with calpain inhibitor CEP-4143 inhibits calpain I activation and cytoskeletal degradation, improves neurological function, and enhances axonal survival after traumatic spinal cord injury, *J Neurochem* 74:1646–55, 2000.

Beattie MS, Farooqui AA, Bresnahan JC: Review of current evidence for apoptosis after spinal cord injury, *J Neurotrauma* 17:915–25, 2000.

Crowe MJ, Bresnahan JC, Shuman SL et al: Apoptosis and delayed degeneration after spinal cord injury in rats and monkeys, *Nat Med* 3:73–6, 1997.

Emery E, Aldana P, Bunge MB et al: Apoptosis after traumatic human spinal cord injury, *J Neurosurg* 89:911–20, 1998.

Martin LJ, Al Abdulla NA, Brambrink AM et al: Neurodegeneration in excitotoxicity, global cerebral ischemia, and target deprivation: A perspective on the contributions of apoptosis and necrosis, *Brain Res Bull* 46:281–309, 1998.

Liu XZ, Xu XM, Hu R et al: Neuronal and glial apoptosis after traumatic spinal cord injury, *J Neurosci* 17:5395–406, 1997.

13.4. Brain-Derived Neurotrophic Factor (BDNF)

Jones KR, Reichardt LF: Molecular cloning of a human gene that is a member of the nerve growth factor family, *Proc Natl Acad Sci USA* 87:8060–4, 1990.

Maisonpierre PC, Le Beau MM, Espinosa R III, Ip NY, Belluscio L et al: Human and rat brain-derived neurotrophic factor and neurotrophin-3: Gene structures, distributions and chromosomal localizations, *Genomics* 10:558–68, 1991.

Lee R, Kermani P, Teng KK, Hempstead BL: Regulation of cell survival by secreted proneurotrophins, *Science* 294:1945–8, 2001.

Pang PT, Teng HK, Zaitsev E, Woo NT, Sakata K et al: Cleavage of proBDNF by tPA/plasmin is essential for long-term hippocampal plasticity, *Science* 306:487–91, 2004.

Chen H, Weber AJ: BDNF enhances retinal ganglion cell survival in cats with optic nerve damage, *Invest Ophthalm Vis Sci* 42:966–74, 2001.

Kawamura K, Kawamura N, Mulders SM, Sollewijn Gelpke MD et al: Ovarian brain-derived neurotrophic factor (BDNF) promotes the development of oocytes into preimplantation embryos, *Proc Natl Acad Sci USA* 102:9206–11, 2005.

Conover JC, Erickson JT, Katz DM, Bianchi LM, Poueymirou WT et al: Neuronal deficits, not involving motor neurons, in mice lacking BDNF and/or NT4, *Nature* 375:235–8, 1995.

Liu X, Ernfors P, Wu H, Jaenisch R: Sensory but not motor neuron deficits in mice lacking NT4 and BDNF, *Nature* 375:238–41, 1995.

Guillin O, Diaz J, Carroll P, Griffon N, Schwartz J-C et al: BDNF controls dopamine D3 receptor expression and triggers behavioural sensitization, *Nature* 411:86–9, 2001.

Lyons WE, Mamounas LA, Ricaurte GA, Coppola V, Reid SW et al: Brain-derived neurotrophic factor-deficient mice develop aggressiveness and hyperphagia in conjunction with brain serotonergic abnormalities, *Proc Natl Acad Sci USA* 96:15239–44, 1999.

Peng S, Wuu J, Mufson EJ, Fahnestock M: Precursor form of brain-derived neurotrophic factor and mature brain-derived neurotrophic factor are decreased in the pre-clinical stages of Alzheimer's disease, *J Neurochem* 93(6):1412–21, 2005.

Jones KR, Reichardt LF: Molecular cloning of a human gene that is a member of the nerve growth factor family, *Proc Natl Acad Sci USA* 87(20):8060–4, 1990.

Baker-Herman TL, Fuller DD, Bavis RW, Zabka AG, Golder FJ et al: BDNF is necessary and sufficient for spinal respiratory plasticity following intermittent hypoxia, *Nature Neurosci* 7:48–55, 2004.

Chen H, Weber AJ: BDNF enhances retinal ganglion cell survival in cats with optic nerve damage, *Invest Ophthalm Vis Sci* 42:966–74, 2001.

Conover JC, Erickson JT, Katz DM, Bianchi LM, Poueymirou WT et al: Neuronal deficits, not involving motor neurons, in mice lacking BDNF and/or NT4, *Nature* 375:235–8, 1995.

Du J, Poo M: Rapid BDNF-induced retrograde synaptic modification in a developing retinotectal system, *Nature* 429:878–83, 2004.

Guillin O, Diaz J, Carroll P, Griffon N, Schwartz J-C et al: BDNF controls dopamine D3 receptor expression and triggers behavioural sensitization, *Nature* 411:86–9, 2001.

Hofer MM, Barde Y-A: Brain-derived neurotrophic factor prevents neuronal death in vivo, *Nature* 331:261–2, 1988.

Huang ZJ, Kirkwood A, Pizzorusso T, Porciatti V, Morales B et al: BDNF regulates the maturation of inhibition and the critical period of plasticity in mouse visual cortex, *Cell* 98:739–55, 1999.

Jones KR, Reichardt LF: Molecular cloning of a human gene that is a member of the nerve growth factor family, *Proc Natl Acad Sci USA* 87:8060–4, 1990.

Kawamura K, Kawamura N, Mulders SM, Sollewijn Gelpke MD, Hsueh AJW: Ovarian brain-derived neurotrophic factor (BDNF) promotes the development of oocytes into preimplantation embryos, *Proc Natl Acad Sci USA* 102:9206–11, 2005.

Kernie SG, Liebl DJ, Parada LF: BDNF regulates eating behavior and locomotor activity in mice, *EMBO J* 19:1290–1300, 2000.

Kovalchuk Y, Hanse E, Kafitz KW, Konnerth A: Postsynaptic induction of BDNF-mediated long-term potentiation, *Science* 295:1729–34, 2002.

Lawrence JM, Keegan DJ, Muir EM, Coffey PJ, Rogers JH et al: Transplantation of Schwann cell line clones secreting GDNF or BDNF into the retinas of dystrophic Royal College of Surgeons rats, *Invest Ophthalm Vis Sci* 45:267–74, 2004.

Lee R, Kermani P, Teng KK, Hempstead BL: Regulation of cell survival by secreted proneurotrophins, *Science* 294:1945–8, 2001.

Li Y, Jian J-C, Cui K, Li N, Zheng Z-Y et al: Essential role of TRPC channels in the guidance of nerve growth cones by brain-derived neurotrophic factor, *Nature* 434:894–8, 2005.

Liu X, Ernfors P, Wu H, Jaenisch R: Sensory but not motor neuron deficits in mice lacking NT4 and BDNF, *Nature* 375:238–41, 1995.

Lyons WE, Mamounas LA, Ricaurte GA, Coppola V, Reid SW et al: Brain-derived neurotrophic factor-deficient mice develop aggressiveness and hyperphagia in conjunction with brain serotonergic abnormalities, *Proc Natl Acad Sci USA* 96:15239–44, 1999.

Ming G, Wong ST, Henley J, Yuan X, Song H et al: Adaptation in the chemotactic guidance of nerve growth cones, *Nature* 417:411–8, 2002.

Pang PT, Teng HK, Zaitsev E, Woo NT, Sakata K et al: Cleavage of proBDNF by tPA/plasmin is essential for long-term hippocampal plasticity, *Science* 306:487–91, 2004.

Human protein reference data base, Johns Hopkins University and the Institute of Bioinformatics, at http://www.hprd.org/protein.

13.5. Neurotrophin 3

Jones KR, Reichardt LF: Molecular cloning of a human gene that is a member of the nerve growth factor family, *Proc Natl Acad Sci USA* 87:8060–4, 1990.

Maisonpierre PC, Le Beau MM, Espinosa R III, Ip NY, Belluscio L et al: Human and rat brain-derived neurotrophic factor and neurotrophin-3: Gene structures, distributions, and chromosomal localizations, *Genomics* 10:558–68, 1991.

Kalcheim C, Carmeli C, Rosenthal A: Neurotrophin 3 is a mitogen for cultured neural crest cells, *Proc Natl Acad Sci USA* 89:1661–5, 1992.

Ramer MS, Priestley JV, McMahon SB: Functional regeneration of sensory axons into the adult spinal cord, *Nature* 403:312–6, 2000.

Cosgaya JM, Chan JR, Shooter EM: The neurotrophin receptor p75(NTR) as a positive modulator of myelination, *Science* 298:1245–8, 2002.

Kuruvilla R, Zweifel LS, Glebova NO, Lonze BE, Valdez G et al: A neurotrophin signaling cascade coordinates sympathetic neuron development through differential control of TrkA trafficking and retrograde signaling, *Cell* 118:243–55, 2004.

Ernfors P, Lee K-F, Kucera J, Jaenisch R: Lack of neurotrophin-3 leads to deficiencies in the peripheral nervous system and loss of limb proprioceptive afferents, *Cell* 77:503–12, 1994.

Tessarollo L, Vogel KS, Palko ME, Reid SW, Parada LF: Targeted mutation in the neurotrophin-3 gene results in loss of muscle sensory neurons, *Proc Natl Acad Sci USA* 91:11844–8, 1994.

Donovan MJ, Hahn R, Tessarollo L, Hempstead BL: Identification of an essential nonneuronal function of neurotrophin 3 in mammalian cardiac development, *Nature Genet* 14:210–3, 1996.

13.6. Neurotrophin (NT)4

Ip NY, Ibanez CF, Nye SH, McClain J, Jones PF et al: Mammalian neurotrophin-4: structure, chromosomal localization, tissue distribution, and receptor specificity, *Proc Natl Acad Sci USA* 89:3060–4, 1992.

Ibanez CF: Neurotrophin-4: the odd one out in the neurotrophin family, *Neurochem Res* 21:787–93, 1996.

Robinson LLL, Townsend J, Anderson RA: The human fetal testis is a site of expression of neurotrophins and their receptors: regulation of the germ cell and peritubular cell population, *J Clin Endocr Metab* 88:3943–51, 2003.

Xie C-W, Sayah D, Chen Q-S, Wei W-Z, Smith D et al: Deficient long-term memory and long-lasting long-term potentiation in mice with a targeted deletion of neurotrophin-4 gene, *Proc Natl Acad Sci USA* 97:8116–21, 2000.

13.7. Nerve Growth Factor (NGF)

Ramer MS, Priestley JV, McMahon SB: Functional regeneration of sensory axons into the adult spinal cord, *Nature* 403:312–6, 2000.

Ieda M, Fukuda K, Hisaka Y, Kimura K, Kawaguchi H et al: Endothelin-1 regulates cardiac sympathetic innervation in the rodent heart by controlling nerve growth factor expression, *J Clin Invest* 113:876–84, 2004.

Zabel BU, Eddy RL, Scott J, Shows TB: The human nerve growth factor gene (NGF) is located on the short arm of chromosome 1 [abstract], *Cytogenet Cell Genet* 37:614, 1984.

Ullrich A, Gray A, Berman C, Dull TJ: Human beta-nerve growth factor gene sequence highly homologous to that of mouse, *Nature* 303:821–5, 1983.

Lee R, Kermani P, Teng KK, Hempstead BL: Regulation of cell survival by secreted proneurotrophins, *Science* 294:1945–8, 2001.

Harrington AW, Leiner B, Blechschmitt C, Arevalo JC, Lee R et al: Secreted proNGF is a pathophysiological death-inducing ligand after adult CNS injury, *Proc Natl Acad Sci USA* 101:6226–30, 2004.

13.8. Glial Cell-Derived Neurotrophic Factor

Lin LF, Doherty DH, Lile JD, Bektesh S, Collins F: GDNF: A glial cell line-derived neurotrophic factor for midbrain dopaminergic neurons, *Science* 260(5111):1130–2, May 1993.

Quintero EM, Willis LM, Zaman V, Lee J, Boger HA et al: Glial cell line-derived neurotrophic factor is essential for neuronal survival in the locus coeruleus-hippocampal noradrenergic pathway, *Neuroscience* 124(1):137–46, 2004.

Gianino S, Grider JR, Cresswell J, Enomoto H, Heuckeroth RO: GDNF availability determines enteric neuron number by controlling precursor proliferation, *Development* 130(10):2187–98, May 2003.

Haase G, Dessaud E, Garces A, de Bovis B, Birling M et al: GDNF acts through PEA3 to regulate cell body positioning and muscle innervation of specific motor neuron pools, *Neuron* 35(5):893–905, Aug 2002.

Shakya R, Jho EH, Kotka P, Wu Z, Kholodilov N et al: The role of GDNF in patterning the excretory system, *Dev Biol* 283(1):70–84, July 2005.

Whitehead J, Keller-Peck C, Kucera J, Tourtellotte WG: Glial cell-line derived neurotrophic factor-dependent fusimotor neuron survival during development, *Mech Dev* 122(1):27–41, Jan 2005.

Beck KD, Valverde J, Alexi T, Poulsen K, Moffat B et al: Mesencephalic dopaminergic neurons protected by GDNF from axotomy-induced degeneration in the adult brain, *Nature* 373:339–41, 1995.

Boucher TJ, Okuse K, Bennett DLH, Munson JB, Wood JN et al: Potent analgesic effects of GDNF in neuropathic pain states, *Science* 290:124–7, 2000.

Durbec P, Marcos-Gutierrez CV, Kilkenny C, Grigoriou M, Wartiowaara K et al: GDNF signalling through the Ret receptor tyrosine kinase, *Nature* 381:789–93, 1996.

Gash DM, Zhang Z, Ovadia A, Cass WA, Yi A et al: Functional recovery in parkinsonian monkeys treated with GDNF, *Nature* 380:252–5, 1996.

Gill SS, Patel NK, Hotton GR, O'Sullivan K, McCarter R et al: Direct brain infusion of glial cell line-derived neurotrophic factor in Parkinson disease, *Nature Med* 9:589–95, 2003.

Kordower JH, Emborg ME, Bloch J, Ma SY, Chu Y et al: Neurodegeneration prevented by lentiviral vector delivery of GDNF in primate models of Parkinson's disease, *Science* 290:767–73, 2000.

Lin L-FH, Doherty DH, Lile JD, Bektesh S, Collins F: GDNF: A glial cell line-derived neurotrophic factor for midbrain dopaminergic neurons, *Science* 260:1130–2, 1993.

Meng X, Lindahl M, Hyvonen ME, Parvinen M, de Rooij DG et al: Regulation of cell fate decision of undifferentiated spermatogonia by GDNF, *Science* 287:1489–93, 2000.

Moore MW, Klein RD, Farinas I, Sauer H, Armanini M et al: Renal and neuronal abnormalities in mice lacking GDNF, *Nature* 382:76–9, 1996.

Nguyen QT, Parsadanian AS, Snider WD, Lichtman JW: Hyperinnervation of neuromuscular junctions caused by GDNF overexpression in muscle, *Science* 279:1725–9, 1998.

Oppenheim RW, Houenou LJ, Johnson JE, Lin L-FH, Li L et al: Developing motor neurons rescued from programmed and axotomy-induced cell death by GDNF, *Nature* 373:344–6, 1995.

Pichel JG, Shen L, Shang HZ, Granholm AC, Drago J et al: Defects in enteric innervation and kidney development in mice lacking GDNF, *Nature* 382:73–6, 1996.

Sanchez MP, Silos-Santiago I, Frisen J, He B, Lira SA et al: Renal agenesis and the absence of enteric neurons in mice lacking GDNF, *Nature* 382:70–3, 1996.

Tomac A, Lindqvist E, Lin L-FH, Ogren SO, Young D et al: Protection and repair of the nigrostriatal dopaminergic system by GDNF in vivo, *Nature* 373:335–9, 1995.

Treanor JJS, Goodman L, de Sauvage F, Stone DM, Poulsen KT et al: Characterization of a multicomponent receptor for GDNF, *Nature* 382:80–3, 1996.

Human protein reference data base, Johns Hopkins University and the Institute of Bioinformatics, at http://www.hprd.org/protein.

13.9. Prevention of Ion Flux

Agrawal SK, Fehlings MG: Mechanisms of secondary injury to spinal cord axons in vitro: Role of Na+, Na(+)-K(+)-ATPase, the Na(+)-H+ exchanger, and the Na(+)-Ca2+ exchanger, *J Neurosci* 16:545–52, 1996.

Agrawal SK, Fehlings MG: Role of NMDA and non-NMDA ionotropic glutamate receptors in traumatic spinal cord axonal injury, *J Neurosci* 17:1055–63, 1997.

Agrawal SK, Fehlings MG: The effect of the sodium channel blocker QX-314 on recovery after acute spinal cord injury, *J Neurotrauma* 14:81–8, 1997.

Agrawal SK, Nashmi R, Fehlings MG: Role of L- and N-type calcium channels in the pathophysiology of traumatic spinal cord white matter injury, *Neuroscience* 99:179–88, 2000.

Li S, Jiang Q, Stys PK: Important role of reverse Na(+)-Ca(2+) exchange in spinal cord white matter injury at physiological temperature, *J Neurophysiol* 84:1116–9, 2000.

Nashmi R, Jones OT, Fehlings MG: Abnormal axonal physiology is associated with altered expression and distribution of Kv1.1 and Kv1.2 K+ channels after chronic spinal cord injury, *Eur J Neurosci* 12:491–506, 2000.

Agrawal SK, Fehlings MG: Mechanisms of secondary injury to spinal cord axons in vitro: Role of Na+, Na(+)-K(+)-ATPase, the Na(+)-H+ exchanger, and the Na(+)-Ca2+ exchanger, *J Neurosci* 16(2):545–52, Jan 1996.

Rosenberg LJ, Teng YD, Wrathall JR: 2,3-Dihydroxy-6-nitro-7-sulfamoyl-benzo(f)quinoxaline reduces glial loss and acute white matter pathology after experimental spinal cord contusion, *J Neurosci* 19:464–75, 1999.

Shi R, Blight AR: Differential effects of low and high concentrations of 4-aminopyridine on axonal conduction in normal and injured spinal cord, *Neuroscience* 77:553–62, 1997.

Stys PK, Steffensen I: Na(+)-Ca2+ exchange in anoxic/ischemic injury of CNS myelinated axons, *Ann NY Acad Sci* 779:366–78, 1996.

Stys PK: General mechanisms of axonal damage and its prevention, *J Neurol Sci* 233:3–13, 2005.

13.10. Prevention of Excitotoxicity

Li S, Stys PK: Mechanisms of ionotropic glutamate receptor-mediated excitotoxicity in isolated spinal cord white matter, *J Neurosci* 20:1190–8, 2000.

Wada S, Yone K, Ishidou Y et al: Apoptosis following spinal cord injury in rats and preventative effect of N-methyl-D-aspartate receptor antagonist, *J Neurosurg* 91:98–104, 1999.

Li S, Stys PK: Na(+)-K(+)-ATPase inhibition and depolarization induce glutamate release via reverse Na(+)-dependent transport in spinal cord white matter, *Neuroscience* 107(4):675–83, 2001.

Stys PK: General mechanisms of axonal damage and its prevention, *J Neurol Sci* 233:3–13, 2005.

Stys PK: White matter injury mechanisms, *Curr Mol Med* 4(2):113–30, March 2004.

13.11. Stimulation of Stem Cell Differentiation

Kobayashi NR, Fan DP, Giehl KM et al: BDNF and NT-4/5 prevent atrophy of rat rubrospinal neurons after cervical axotomy, stimulate GAP-43 and Talpha1-tubulin mRNA expression, and promote axonal regeneration, *J Neurosci* 17:9583–95, 1997.

Ramer MS, Priestley JV, McMahon SB: Functional regeneration of sensory axons into the adult spinal cord, *Nature* 403:312–6, 2000.

Houweling DA, Bar PR, Gispen WH, Joosten EA: Spinal cord injury: Bridging the lesion and the role of neurotrophic factors in repair, *Prog Brain Res* 117:455–71, 1998.

Terenghi G: Peripheral nerve regeneration and neurotrophic factors, *J Anat* 194:1–14, 1999.

Blesch A, Lu P, Tuszynski MH: Neurotrophic factors, gene therapy, and neural stem cells for spinal cord repair, *Brain Res Bull* 57:833–8, 2002.

Jones LL, Oudega M, Bunge MB, Tuszynski MH: Neurotrophic factors, cellular bridges and gene therapy for spinal cord injury, *J Physiol* 533:83–9, 2001.

Murakami Y, Furukawa S, Nitta A, Furukawa Y: Accumulation of nerve growth factor protein at both rostral and caudal stumps in the transected rat spinal cord, *J Neurol Sci* 198:63–9, 2002.

Whitworth IH, Brown RA, Dore CJ, Anand P, Green CJ et al: Nerve growth factor enhances nerve regeneration through fibronectin grafts, *J Hand Surg* 21:514–22, 1996.

Bloch J, Fine EG, Bouche N, Zurn AD, Aebischer P: Nerve growth factor- and neurotrophin-3-releasing guidance channels promote regeneration of the transected rat dorsal root, *Exp Neurol* 172:425–32, 2001.

Ramer MS, Priestley JV, McMahon SB: Functional regeneration of sensory axons into the adult spinal cord, *Nature* 403:312–16, 2000.

Sendtner M, Holtmann B, Kolbeck R, Thoenen H, Barde YA: Brain-derived neurotrophic factor prevents the death of motoneurons in newborn rats after nerve section, *Nature* 360:757–9, 1992.

Yan Q, Elliott J, Snider WD: Brain-derived neurotrophic factor rescues spinal motor neurons from axotomy-induced cell death, *Nature* 360:753–5, 1992.

Henderson CE, Camu W, Mettling C, Gouin A, Poulsen K et al: Neurotrophins promote motor neuron survival and are present in embryonic limb bud, *Nature* 363:266–70, 1993.

Oudega M, Hagg T: Neurotrophins promote regeneration of sensory axons in the adult rat spinal cord, *Brain Res* 818:431–8, 1999.

Bloch J, Fine EG, Bouche N, Zurn AD, Aebischer P: Nerve growth factor- and neurotrophin-3-releasing guidance channels promote regeneration of the transected rat dorsal root, *Exp Neurol* 172:425–32, 2001.

Dijkhuizen PA, Hermens WT, Teunis MA, Verhaagen J: Adenoviral vector-directed expression of neurotrophin-3 in rat dorsal root ganglion explants results in a robust neurite outgrowth response, *J Neurobiol* 33:172–84, 1997.

Schmalbruch H, Rosenthal A: Neurotrophin-4/5 postpones the death of injured spinal motoneurons in newborn rats, *Brain Res* 700:254–60, 1995.

Stucky C, Shin JB, Lewin GR: Neurotrophin-4: A survival factor for adult sensory neurons, *Curr Biol* 12:1401–4, 2002.

Sendtner M, Kreutzberg GW, Thoenen H: Ciliary neurotrophic factor prevents the degeneration of motor neurons after axotomy, *Nature* 345:440–1, 1990.

Gravel C, Gotz R, Lorrain A, Sendtner M: Adenoviral gene transfer of ciliary neurotrophic factor and brain-derived neurotrophic factor leads to long-term survival of axotomized motor neurons, *Nat Med* 3:765–70, 1997.

Siegel SG, Patton B, English AW: Ciliary neurotrophic factor is required for motoneuron sprouting, *Exp Neurol* 166:205–12, 2000.

Levison SW, Ducceschi MH, Young GM, Wood TL: Acute exposure to CNTF in vivo induces multiple components of reactive gliosis, *Exp Neurol* 141:256–68, 1996.

Henderson CE, Phillips HS, Pollock RA, Davies AM, Lemeulle C et al: GDNF: A potent survival factor for motoneurons present in peripheral nerve and muscle, *Science* 266:1062–4, 1994.

Buj-Bello A, Buchman VL, Horton A, Rosenthal A, Davies AM: GDNF is an age-specific survival factor for sensory and autonomic neurons, *Neuron* 15:821–8, 1995.

Friesel RE, Maciag T: Molecular mechanisms of angiogenesis: Fibroblast growth factor signal transduction, *FASEB J* 9:919–25, 1995.

Romero MI, Rangappa N, Li L et al: Extensive sprouting of sensory afferents and hyperalgesia induced by conditional expression of nerve growth factor in the adult spinal cord, *J Neurosci* 20:4435–45, 2000.

Zhang Y, Dijkhuizen PA, Anderson PN et al: NT-3 delivered by an adenoviral vector induces injured dorsal root axons to regenerate into the spinal cord of adult rats, *J Neurosci Res* 54:554–62, 1998.

Hastings NB, Gould E: Neurons inhibit neurogenesis, *Nat Med* 9:264–6, 2003.

Wu HH, Ivkovic S, Murray RC, Jaramillo S, Lyons KM et al: Autoregulation of neurogenesis by GDF11, *Neuron* 37:197–207, 2003.

Nakashima K, Takizawa T, Ochiai W, Yanagisawa M, Hisatsune T et al: BMP2-mediated alteration in the developmental pathway of fetal mouse brain cells from neurogenesis to astrocytogenesis, *Proc Natl Acad Sci USA* 98:5868–73, 2001.

Setoguchi T, Yone K, Matsuoka E, Takenouchi H, Nakashima K et al: Traumatic injury-induced BMP7 expression in the adult rat spinal cord, *Brain Res* 921:219–25, 2001.

13.12. Enhancement of Axonal Extension, Adhesion, and Reconnection

Morrow DR, Campbell G, Lieberman AR et al: Differential regenerative growth of CNS axons into tibial and peroneal nervous grafts in the thalamus of adult rats, *Exp Neurol* 120:60–9, 1993.

Fernandes KJ, Fan DP, Tsui BJ et al: Influence of the axotomy to cell body distance in rat rubrospinal and spinal motoneurons: Differential regulation of GAP-43, tubulins, and neurofilament-M, *J Compar Neurol* 414:495–510, 1999.

Woolhead CL, Zhang Y, Lieberman AR et al: Differential effects of autologous peripheral nervous grafts to the corpus striatum of adult rats on the regeneration of axons of striatal and nigral neurons and on the expression of GAP-43 and the cell adhesion molecules N-CAM and L1, *J Compar Neurol* 391:259–73, 1998.

Richardson PM, Issa VM: Peripheral injury enhances central regeneration of primary sensory neurones, *Nature* 309:791–3, 1984.

Herdegen T, Skene P, Bahr M: The c-Jun transcription factor: Bipotential mediator of neuronal death, survival and regeneration, *Trends Neurosci* 20:227–31, 1997.

Jenkins R, Tetzlaff W, Hunt SP: Differential expression of immediate early genes in rubrospinal neurons following axotomy in rat, *Eur J Neurosci* 5:203–9, 1993.

Fernandes KJ, Fan DP, Tsui BJ et al: Influence of the axotomy to cell body distance in rat rubrospinal and spinal motoneurons: Differential regulation of GAP-43, tubulins, and neurofilament-M, *J Compar Neurol* 414:495–510, 1999.

Becker T, Bernhardt RR, Reinhard E et al: Readiness of zebrafish brain neurons to regenerate a spinal axon correlates with differential expression of specific cell recognition molecules, *J Neurosci* 18:5789–803, 1998.

Jung M, Petrausch B, Stuermer CA: Axon-regenerating retinal ganglion cells in adult rats synthesize the cell adhesion molecule L1 but not TAG-1 or SC-1, *Mol Cell Neurosci* 9:116–31, 1997.

Woolhead CL, Zhang Y, Lieberman AR et al: Differential effects of autologous peripheral nervous grafts to the corpus striatum of adult rats on the regeneration of axons of striatal and nigral neurons and on the expression of GAP-43 and the cell adhesion molecules N-CAM and L1, *J Compar Neurol* 391:259–73, 1998.

Schreyer DJ, Skene JH: Injury-associated induction of GAP-43 expression displays axon branch specificity in rat dorsal root ganglion neurons, *J Neurobiol* 24:959–70, 1993.

Zhang Y, Bo X, Schoepfer R, Holtmaat AJ, Verhaagen J et al: Growth-associated protein GAP-43 and L1 act synergistically to promote regenerative growth of Purkinje cell axons in vivo, *Proc Natl Acad Sci USA* 102:14883–14888, 2005.

Carulli D, Buffo A, Strata P: Reparative mechanisms in the cerebellar cortex, *Prog Neurobiol* 72:373–98, 2004.

Smith GM, Romero MI: Adenoviral-mediated gene transfer to enhance neuronal survival, growth, and regeneration, *J Neurosci Res* 55:147–57, 1999.

Thanos PK, Okajima S, Terzis JK: Ultrastructure and cellular biology of nerve regeneration, *J Reconstr Microsurg* 14:423–36, 1998.

Mohajeri MH, Bartsch U, van der Putten H, Sansig G, Mucke L et al: Neurite outgrowth on non-permissive substrates in vitro is enhanced by ectopic expression of the neural adhesion molecule L1 by mouse astrocytes, *Eur J Neurosci* 8:1085–97, 1996.

Kobayashi S, Miura M, Asou H, Inoue HK, Ohye C et al: Grafts of genetically modified fibroblasts expressing neural cell adhesion molecule L1 into transected spinal cord of adult rats, *Neurosci Lett* 188:191–4, 1995.

Lin L, Rao Y, Isacson O: Netrin-1 and slit-2 regulate and direct neurite growth of ventral midbrain dopaminergic neurons, *Mol Cell Neurosci* 28:547–55, 2005.

13.13. Prevention of Fibrous Scar Formation

Beggah AT, Dours-Zimmermann MT, Barras FM, Brosius A, Zimmermann DR et al: Lesion-induced differential expression and cell association of Neurocan, Brevican, Versican V1 and V2 in the mouse dorsal root entry zone, *Neuroscience* 133(3):749–62, 2005.

Moon LD, Brecknell JE, Franklin RJ et al: Robust regeneration of CNS axons through a track depleted of CNS glia, *Exp Neurol* 161:49–66, 2000.

Brambilla R, Bracchi-Ricard V, Hu WH, Frydel B, Bramwell A et al: Inhibition of astroglial nuclear factor kappaB reduces inflammation and improves functional recovery after spinal cord injury, *J Exp Med* 202(1):145–56, July 2005.

Caroni P, Schwab ME: Two membrane protein fractions from rat central myelin with inhibitory properties for neurite growth and fibroblast spreading, *J Cell Biol* 106:1281–8, 1988.

Schwab ME, Caroni P: Oligodendrocytes and CNS myelin are nonpermissive substrates for neurite growth and fibroblast spreading in vitro, *J Neurosci* 8:2381–93, 1988.

Cai D, Shen Y, De Bellard M et al: Prior exposure to neurotrophins blocks inhibition of axonal regeneration by MAG and myelin via a cAMP-dependent mechanism, *Neuron* 22:89–101, 1999.

McKerracher L, David S, Jackson DL et al: Identification of myelin-associated glycoprotein as a major myelin-derived inhibitor of neurite growth, *Neuron* 13:805–11, 1994.

Mukhopadhyay G, Doherty P, Walsh FS et al: A novel role for myelin-associated glycoprotein as an inhibitor of axonal regeneration, *Neuron* 13:757–67, 1994.

Brosamle C, Huber AB, Fiedler M et al: Regeneration of lesioned corticospinal tract fibers in the adult rat induced by a recombinant, humanized IN-1 antibody fragment, *J Neurosci* 20:8061–8, 2000.

Caroni P, Schwab ME: Antibody against myelin-associated inhibitor of neurite growth neutralizes nonpermissive substrate properties of CNS white matter, *Neuron* 1:85–96, 1988.

13.14. Enhancement of Synaptic Formation

Mendell LM, Munson JB, Arvanian VL: Neurotrophins and synaptic plasticity in the mammalian spinal cord, *J Physiol* 533:91–7, 2001.

Jones LL, Oudega M, Bunge MB, Tuszynski MH: Neurotrophic factors, cellular bridges and gene therapy for spinal cord injury, *J Physiol* 533:83–9, 2001.

Ruegg MA: Agrin, laminin beta 2 (s-laminin) and ARIA: Their role in neuromuscular development, *Curr Opin Neurobiol* 6:97–103, 1996.

13.15. Embryonic and Fetal Stem Cells

McDonald JW, Liu XZ, Qu Y et al: Transplanted embryonic stem cells survive, differentiate and promote recovery in injured rat spinal cord, *Nat Med* 5:1410–2, 1999.

Lakatos A, Franklin RJ: Transplant mediated repair of the central nervous system: An imminent solution? *Curr Opin Neurol* 15:701–5, 2002.

Kwon BK, Tetzlaff W: Spinal cord regeneration: From gene to transplants, *Spine* 26(Suppl 24): S13–22, 2001.

Akesson E, Kjaeldgaard A, Seiger A: Human embryonic spinal cord grafts in adult rat spinal cord cavities: Survival, growth, and interactions with the host, *Exp Neurol* 149:262–76, 1998.

Bernstein-Goral H, Bregman BS: Axotomized rubrospinal neurons rescued by fetal spinal cord transplants maintain axon collaterals to rostral CNS targets, *Exp Neurol* 148:13–25, 1997.

Bregman BS, Broude E, McAtee M et al: Transplants and neurotrophic factors prevent atrophy of mature CNS neurons after spinal cord injury, *Exp Neurol* 149:13–27, 1998.

Mori F, Himes BT, Kowada M et al: Fetal spinal cord transplants rescue some axotomized rubrospinal neurons from retrograde cell death in adult rats, *Exp Neurol* 143:45–60, 1997.

Gokhan S, Mehler MF: Basic and clinical neuroscience applications of embryonic stem cells, *Anat Rec* 265:142–56, 2001.

13.16. Adult Nerve Stem Cells

Song HJ, Stevens CF, Gage FH: Neural stem cells from adult hippocampus develop essential properties of functional CNS neurons, *Nat Neurosci* 5:438–45, 2002.

Arvidsson A, Collin T, Kirik D, Kokaia Z, Lindvall O: Neuronal replacement from endogenous precursors in the adult brain after stroke, *Nat Med* 8:963–70, 2002.

Bauer S, Hay M, Amilhon B, Jean A, Moyse E: In vivo neurogenesis in the dorsal vagal complex of the adult rat brainstem, *Neuroscience* 130:75–90, 2005.

Magnus T, Rao MS: Neural stem cells in inflammatory CNS diseases: mechanisms and therapy, *J Cell Mol Med* 9:303–19, 2005.

Palmer TD, Willhoite AR, Gage FH: Vascular niche for adult hippocampal neurogenesis, *J Compar Neurol* 425:479–94, 2000.

Shihabuddin LS, Horner PJ, Ray J, Gage FH: Adult spinal cord stem cells generate neurons after transplantation in the adult dentate gyrus, *J Neurosci* 20:8727–35, 2000.

Tropepe V, Coles BL, Chiasson BJ, Horsford DJ, Elia AJ et al: Retinal stem cells in the adult mammalian eye, *Science* 287:2032–6, 2000.

Lie DC, Dziewczapolski G, Willhoite AR, Kaspar BK, Shults CW et al: The adult substantia nigra contains progenitor cells with neurogenic potential, *J Neurosci* 22:6639–49, 2002.

Lie DC, Song H, Colamarino SA, Ming GL, Gage FH: Neurogenesis in the adult brain: New strategies for central nervous system diseases, *Annu Rev Pharmacol Toxicol* 44:399–421, 2004.

Reynolds BA, Weiss S: Generation of neurons and astrocytes from isolated cells of the adult mammalian central nerve system, *Science* 255:1707–10, 1992.

Richards LJ, Kilpatrick TJ, Bartlett PF: De novo generation of neuronal cells from the adult mouse brain, *Proc Natl Acad Sci USA* 89:8591–5, 2002.

Park KI, Ourednik J, Ourednik V, Taylor RM, Aboody KS et al: Global gene and cell replacement strategies via stem cells, *Gene Ther* 9:613–24, 2002.

13.17. Bone Marrow Cells

Sanchez-Ramos J, Song S, Cardozo-Pelaez F, Hazzi C, Stedeford T et al: Adult bone marrow stromal cells differentiate into neural cells in vitro, *Exp Neurol* 164:247–56, 2000.

Rice CM, Scolding NJ: Adult stem cells—reprogramming neurological repair? *Lancet* 364:193–9, 2004.

Brazelton TR, Rossi FMV, Keshet GI, Blau HM: From Marrow to brain: Expression of neuronal phenotypes in adult mice, *Science* 290:1775–9, 2000.

Mezey E, Chandross KJ, Harta G, Maki RA, McKercher SR: Turning blood into brain: Cells bearing neuronal antigens generated in vivo from bone marrow, *Science* 290:1779–82, 2000.

Sanchez-Ramos J, Song S, Cardozo-Pelaez F, Hazzi C, Stedeford T, Willing A et al: Adult bone marrow stromal cells differentiate into neural cells in vitro, *Exp Neurol* 164:247–56, 2000.

Weimann JM, Charlton CA, Brazelton TR, Hackman RC, Blau HM: Contribution of transplanted bone marrow cells to Purkinje neurons in human adult brains, *Proc Natl Acad Sci USA* 100:2088–93, 2003.

Cogle CR, Yachnis AT, Laywell ED, Zander DS, Wingard JR et al: Bone marrow transdifferentiation in brain after transplantation: A retrospective study, *Lancet* 363:1432–7, 2004.

Sigurjonsson OE, Perreault MC, Egeland T, Glover JC: Adult human hematopoietic stem cells produce neurons efficiently in the regenerating chicken embryo spinal cord, *Proc Natl Acad Sci USA* 102:5227–32, 2005.

Wagers AJ, Sherwood RI, Christensen JL, Weissman IL: Little evidence for developmental plasticity of adult hematopoietic stem cells, *Science* 297:2256–9, 2002.

13.18. Neuronal Supporting Cells

Imaizumi T, Lankford KL, Kocsis JD: Transplantation of olfactory ensheathing cells or Schwann cells restores rapid and secure conduction across the transected spinal cord, *Brain Res* 854:70–8, 2000.

Cheng H, Cao Y, Olson L: Spinal cord repair in adult paraplegic rats: Partial restoration of hind limb function, *Science* 273:510–3, 1996.

Xu XM, Guenard V, Kleitman N et al: A combination of BDNF and NT-3 promotes supraspinal axonal regeneration into Schwann cell grafts in adult rat thoracic spinal cord, *Exp Neurol* 134:261–72, 1995.

Jones LL, Oudega M, Bunge MB, Tuszynski MH: Neurotrophic factors, cellular bridges and gene therapy for spinal cord injury, *J Physiol* 533:83–9, 2001.

Devon R, Doucette R: Olfactory ensheathing cells myelinate dorsal root ganglion neurites, *Brain Res* 589:175–9, 1992.

Ramon-Cueto A, Valverde F: Olfactory bulb ensheathing glia: A unique cell type with axonal growth-promoting properties, *Glia* 14:163–73, 1995.

Navarro X, Valero A, Gudino G et al: Ensheathing glia transplants promote dorsal root regeneration and spinal reflex restitution after multiple lumbar rhizotomy, *Ann Neurol* 45:207–15, 1999.

Ramon-Cueto A, Cordero MI, Santos-Benito FF et al: Functional recovery of paraplegic rats and motor axon regeneration in their spinal cords by olfactory ensheathing glia, *Neuron* 25:425–35, 2000.

Li Y, Field PM, Raisman G: Repair of adult rat corticospinal tract by transplants of olfactory ensheathing cells, *Science* 277:2000–2, 1997.

Ramon-Cueto A, Nieto-Sampedro M: Regeneration into the spinal cord of transected dorsal root axons is promoted by ensheathing glia transplants, *Exp Neurol* 127:232–44, 1994.

13.19. Transgenic Cell Lines

Blesch A, Tuszynski MH: Robust growth of chronically injured spinal cord axons induced by grafts of genetically modified NGF-secreting cells, *Exp Neurol* 148:444–52, 1997.

Grill R, Murai K, Blesch A et al: Cellular delivery of neurotrophin-3 promotes corticospinal axonal growth and partial functional recovery after spinal cord injury, *J Neurosci* 17:5560–72, 1997.

Liu Y, Kim D, Himes BT et al: Transplants of fibroblasts genetically modified to express BDNF promote regeneration of adult rat rubrospinal axons and recovery of forelimb function, *J Neurosci* 19:4370–87, 1999.

Tuszynski MH, Peterson DA, Ray J et al: Fibroblasts genetically modified to produce nervous growth factor induce robust neuritic ingrowth after grafting to the spinal cord, *Exp Neurol* 126:1–14, 1994.

Menei P, Montero-Menei C, Whittemore SR et al: Schwann cells genetically modified to secrete human BDNF promote enhanced axonal regrowth across transected adult rat spinal cord, *Eur J Neurosci* 10:607–21, 1998.

Tuszynski MH, Weidner N, McCormack M et al: Grafts of genetically modified Schwann cells to the spinal cord: Survival, axon growth, and myelination, *Cell Transplant* 197:187–96, 1998.

Weidner N, Blesch A, Grill RJ et al: Nerve growth factor-hypersecreting Schwann cell grafts augment and guide spinal cord axonal growth and remyelinate central nervous system axons in a phenotypically appropriate manner that correlates with expression of L1, *J Compar Neurol* 413:495–506, 1999.

Hendriks WT, Ruitenberg MJ, Blits B, Boer GJ, Verhaagen J: Viral vector-mediated gene transfer of neurotrophins to promote regeneration of the injured spinal cord, *Prog Brain Res* 146:451–76, 2004.

Blits B, Boer GJ, Verhaagen J: Pharmacological, cell, and gene therapy strategies to promote spinal cord regeneration, *Cell Transplant* 11:593–613, 2002.

Kwon BK, Tetzlaff W: Spinal cord regeneration: From gene to transplants, *Spine* 26(Suppl 24): S13–22, 2001.

Cheng H, Cao Y, Olson L: Spinal cord repair in adult paraplegic rats: Partial restoration of hind limb function, *Science* 273:510–13, 1996.

Davies SJ, Fitch MT, Memberg SP, Hall AK, Raisman G et al: Regeneration of adult axons in white matter tracts of the central nervous system, *Nature* 390:680–3, 1997.

Lakatos A, Franklin RJ: Transplant mediated repair of the central nervous system: An imminent solution? *Curr Opin Neurol* 15:701–5, 2002.

13.20. Stimulation of Neuronal Regeneration and Guidance of Axonal Outgrowth by Graft-Based Assistance

Schmidt CE, Leach JB: Neural tissue engineering: strategies for repair and regeneration, *Annu Rev Biomed Eng* 5:293–347, 2003.

Ide C, Tohyama K, Tajima K, Endoh K, Sano K et al: Long acellular nervous transplants for allogeneic grafting and the effects of basic fibroblast growth factor on the growth of regenerating axons in dogs: A preliminary report, *Exp Neurol* 154:99–112, 1998.

Sondell M, Lundborg G, Kanje M: Regeneration of the rat sciatic nerve into allografts made acellular through chemical extraction, *Brain Res* 795:44–54, 1998.

Sondell M, Lundborg G, Kanje M: Vascular endothelial growth factor stimulates Schwann cell invasion and neovascularization of acellular nerve grafts, *Brain Res* 846:219–28, 1999.

Kauppila T, Jyvasjarvi E, Huopaniemi T, Hujanen E, Liesi P: A laminin graft replaces neurorrhaphy in the restorative surgery of the rat sciatic nerve, *Exp Neurol* 123:181–91, 1993.

Tong XJ, Hirai K, Shimada H, Mizutani Y, Izumi T et al: Sciatic nervous regeneration navigated by laminin-fibronectin double coated biodegradable collagen grafts in rats, *Brain Res* 663:155–62, 1994.

Chen YS, Hsieh CL, Tsai CC, Chen TH, Cheng WC et al: Peripheral nerve regeneration using silicone rubber chambers filled with collagen, laminin and fibronectin, *Biomaterials* 21:1541–7, 2000.

Yoshii S, Oka M: Peripheral nerve regeneration along collagen filaments, *Brain Res* 888:158–62, 2001.

Itoh S, Takakuda K, Kawabata S, Aso Y, Kasai K et al: Evaluation of cross-linking procedures of collagen tubes used in peripheral nerve repair, *Biomaterials* 23:4475–81, 2002.

Yoshii S, Oka M, Shima M, Taniguchi A, Akagi M: 30 mm regeneration of rat sciatic nerve along collagen filaments, *Brain Res* 949:202–8, 2002.

Evans GR, Brandt K, Katz S, Chauvin P, Otto L et al: Bioactive poly(L-lactic acid) conduits seeded with Schwann cells for peripheral nervous regeneration, *Biomaterials* 23:841–48, 2002.

Valero-Cabre A, Tsironis K, Skouras E, Perego G, Navarro X et al: Superior muscle reinnervation after autologous nerve graft or poly-L-lactide-epsilon-caprolactone (PLC) tube implantation in comparison to silicone tube repair, *J Neurosci Res* 63:214–23, 2001.

Soldani G, Varelli G, Minnocci A, Dario P: Manufacturing and microscopical characterisation of polyurethane nerve guidance channel featuring a highly smooth internal surface, *Biomaterials* 19:1919–24, 1998.

Dalton PD, Flynn L, Shoichet MS: Manufacture of poly(2-hydroxyethyl methacrylate-co-methyl methacrylate) hydrogel tubes for use as nerve guidance channels, *Biomaterials* 23:3843–51, 2002.

Schmidt CE, Shastri VR, Vacanti JP, Langer R: Stimulation of neurite outgrowth using an electrically conducting polymer, *Proc Natl Acad Sci USA* 94:8948–53, 1997.

Collier JH, Camp JP, Hudson TW, Schmidt CE: Synthesis and characterization of polypyrrole-hyaluronic acid composite biomaterials for tissue engineering applications, *J Biomed Mater Res* 50:574–84, 2000.

Lore AB, Hubbell JA, Bobb DS Jr, Ballinger ML, Loftin KL et al: Rapid induction of functional and morphological continuity between severed ends of mammalian or earthworm myelinated axons, *J Neurosci* 19:2442–54, 1999.

Novikova LN, Novikov LN, Kellerth JO: Biopolymers and biodegradable smart implants for tissue regeneration after spinal cord injury, *Curr Opin Neurol* 16:711–15, 2003.

Bellamkonda R, Aebischer P: Review: Tissue engineering in the nervous system, *Biotechnol Bioeng* 43:543–54, 1994.

Valentini RF: Nerve guidance channels, in *The Biomedical Engineering Handbook*, 2nd ed, Bronzino JD, ed. CRC Press, Boca Raton, FL, 2000, pp 2:135-1–12.

Furnish EJ, Schmidt CE: Tissue engineering of the peripheral nervous system, in *Frontiers in Tissue Engineering*, Patrick CW Jr, Mikos AG, McIntire LV, eds, Elsevier, New York, 1998, pp 514–35.

Thompson DM, Buettner HM: Neurite outgrowth is directed by Schwann cell alignment in the absence of other guidance cues, *Ann Biomed Eng* (in press).

13.21. Autologous Nerve Tissue Grafts

Sunderland S: *Nervous Injuries and Their Repair*, Churchill Livingston, London, 1991.

Chiu DT, Janecka I, Krizek TJ, Wolff M, Lovelace RE: Autogenous vein graft as a conduit for nervous regeneration, *Surgery* 91:226–33, 1982.

Walton RL, Brown RE, Matory WE Jr, Borah GL, Dolph JL: Autogenous vein graft repair of digital nervous defects in the finger: a retrospective clinical study, *Plast Reconstr Surg* 84:944–9, 1989.

Risitano G, Cavallaro G, Merrino T, Coppolino S, Ruggeri F: Clinical results and thoughts on sensory nerve repair by autologous vein graft in emergency hand reconstruction, *Chir Main* 21:194–7, 2002.

Battiston B, Tos P, Cushway TR, Geuna S: Nerve repair by means of vein filled with muscle grafts I. Clinical results, *Microsurgery* 20:32–6, 2000.

Meek MF, Varejao AS, Geuna S: Muscle grafts and alternatives for nerve repair, *J Oral Maxillofac Surg* 60:1095–96, 2002.

Brandt J, Dahlin LB, Lundborg G: Autologous tendons used as grafts for bridging peripheral nerve defects, *J Hand Surg* 24:284–90, 1999.

Hudson TW, Zawko S, Deister C, Lundy S, Hu CY et al: Optimized acellular nerve graft is immunologically tolerated and supports regeneration, *Tissue Eng* 10:1641–51, 2004.

13.22. Allogenic and Xenogeneic Tissue Grafts

Hadlock TA, Sundback CA, Hunter DA, Vacanti JP, Cheney ML: A new artificial nerve graft containing rolled Schwann cell monolayers, *Microsurgery* 21:96–101, 2001.

Mohammad JA, Warnke PH, Pan YC, Shenaq S: Increased axonal regeneration through a biodegradable amnionic tube nerve conduit: Effect of local delivery and incorporation of nerve growth factor/hyaluronic acid media, *Ann Plast Surg* 44:59–64, 2000.

Meek MF, Coert JH, Nicolai JP: Amnion tube for nerv regeneration, *Plast Reconstr Surg* 107:622–23, 2001.

Mligiliche N, Endo K, Okamoto K, Fujimoto E, Ide C: Extracellular matrix of human amnion manufactured into tubes as conduits for peripheral nerve regeneration, *J Biomed Mater Res* 63:591–600, 2002.

13.23. Extracellular Matrix Components

Grimpe B, Silver J: The extracellular matrix in axon regeneration, *Prog Brain Res* 137:333–49, 2002.

Bovolenta P, Fernaud-Espinosa I: Nervous system proteoglycans as modulators of neurite outgrowth, *Prog Neurobiol* 61:113–32, 2000.

Asher RA, Morgenstern DA, Moon LD, Fawcett JW: Chondroitin sulphate proteoglycans: Inhibitory components of the glial scar, *Prog Brain Res* 132:611–19, 2001.

Chamberlain LJ, Yannas IV, Hsu HP, Strichartz GR, Spector M: Near-terminus axonal structure and function following rat sciatic nerve regeneration through a collagen-GAG matrix in a ten-millimeter gap, *J Neurosci Res* 60:666–77, 2000.

Herbert CB, Nagaswami C, Bittner GD, Hubbell JA, Weisel JW: Effects of fibrin micromorphology on neurite growth from dorsal root ganglia cultured in three-dimensional fibrin gels, *J Biomed Mater Res* 40:551–9, 1998.

Holmes TC, de Lacalle S, Su X, Liu G, Rich A et al: Extensive neurite outgrowth and active synapse formation on self-assembling peptide scaffolds, *Proc Natl Acad Sci USA* 97:6728–33, 2000.

Hashimoto T, Suzuki Y, Kitada M, Kataoka K, Wu S et al: Peripheral nerve regeneration through alginate gel: Analysis of early outgrowth and late increase in diameter of regenerating axons, *Exp Brain Res* 146:356–68, 2002.

Balgude AP, Yu X, Szymanski A, Bellamkonda RV: Agarose gel stiffness determines rate of DRG neurite extension in 3D cultures, *Biomaterials* 22:1077–84, 2001.

Mosahebi A, Simon M, Wiberg M, Terenghi G: A novel use of alginate hydrogel as Schwann cell matrix, *Tissue Eng* 7:525–34, 2001.

Haipeng G, Yinghui Z, Jianchun L, Yandao G, Nanming Z et al: Studies on nerve cell affinity of chitosan-derived materials, *J Biomed Mater Res* 52:285–95, 2000.

Rosner BI, Siegel RA, Grosberg A, Tranquillo RT: Rational design of contact guiding, neurotrophic matrices for peripheral nerve regeneration, *Ann Biomed Eng* 31:1383–401, 2003.

13.24. Synthetic Materials

Molander H, Olsson Y, Engkvist O, Bowald S, Eriksson I: Regeneration of peripheral nerve through a polyglactin tube, *Muscle Nervous* 5:54–7, 1982.

Nyilas E, Chiu TH, Sidman RL, Henry EW, Brushart TM et al: Peripheral nerve repair with bioresorbable prosthesis, *Trans Am Soc Artif Intern Organs* 29:307–13, 1983.

13.25. Role of Geometric Factors

Smeal RM, Rabbitt R, Biran R, Tresco PA: Substrate curvature influences the direction of nerve outgrowth, *Ann Biomed Eng* 33:376–82, 2005.

Dubey N, Letourneau PC, Tranquillo RT: Guided neurite elongation and Schwann cell invasion into magnetically aligned collagen in simulated peripheral nerve regeneration, *Exp Neurol* 158:338–50, 1999.

Ceballos D, Navarro X, Dubey N, Wendelschafer-Crabb G, Kennedy WR et al: Magnetically aligned collagen gel filling a collagen nerve guide improves peripheral nerve regeneration, *Exp Neurol* 158:290–300, 1999.

13.26. Improvement of Nerve Function

Prochazka A, Mushahwar VK, McCreery DB: Neural prostheses, *J Physiol* 533:99–109, 2001.

Grill WM, McDonald JW, Peckham PH, Heetderks W, Kocsis J et al: At the interface: Convergence of neural regeneration and neural prostheses for restoration of function, *J Rehab Res Dev* 38:633–9, 2001.

Condic ML, Lemons ML: Extracellular matrix in spinal cord regeneration: Getting beyond attraction and inhibition, *NeuroReport* 13:A37–48, 2002.

Degenerative Neural Diseases

Degenerative neural diseases are a group of disorders characterized by progressive atrophic changes in the brain and spinal cord, resulting in the disorder of intellectual ability, motor control, and/or sensation, depending on the types of neuron involved. While there are numbers of degenerative neural disorders, we will focus on several common types, including Alzheimer's disease (progressive dementia), Huntington's chorea (progressive dementia complicated by other neurological disorders), Parkinson's disease (disorder of the motor control system), and muscular atrophy (progressive spinal muscular atrophy). The clinical symptoms and signs, etiology, pathology, pathogenesis, and treatment of these disorders are discussed in the following sections.

Alzheimer's Disease

Etiology, Pathology, and Clinical Manifestations [13.27]. Alzheimer's disease, also known as *presenile dementia*, is a neural degenerative disorder characterized by progressive loss of intellectual capability involving the impairment of memory, judgment, and logical thinking as well as changes in personality within a course of 5–10 years. The primary cause of Alzheimer's disease is apoptosis of the cholinergic neurons in the basal forebrain. The disorder may occur at any age, but is traditionally diagnosed as Alzheimer's disease when it is initiated before age 65. Similar neurological changes after 65 are considered consequences of aging, known as *senile dementia*. About 1% of patients with Alzheimer's disease have a family history of the disease, suggesting genetic influence on the occurrence of the disease.

Alzheimer's disease is a progressive disorder associated with sequential clinical features. The onset of the disease is insidious and difficult to notice. There may be a long asymptomatic, preclinical period (7 years or more) before the disease is diagnosed. It is usually found at a certain level of severity. There are several loosely defined neurological deficits, which may occur in sequence. These include amnesia, dysnomia, disorientation, and paranoia. Amnesia may occur in the early stage of Alzheimer's disease. The patient may exhibit gradual loss of short- or long-term retentive memory. The disorders may be intermittent. With the progression of the disease, the patient may show dysnomia. Major signs include inappropriate use of names, forgetting common words, and impairment of speech. Eventually, the patient may be unable to read and understand. Visuospatial disorientation is another common sign of Alzheimer's disease. The patient may find it difficult to park a car, put arms into sleeves, put books in order, or find the way home. Paranoia or changes in personality may occur in a large fraction of patients. Eventually the patient completely loses intellectual abilities, including memory and judgment.

Alzheimer's disease is associated with a number of pathological changes. These include diffuse atrophy of the cerebral cortex with an apparent decrease in brain size, especially the size of the hippocampus, and an increase in the volume of cerebral ventricles (Fig. 13.22). At the microscopic level, the loss of neurons or neuronal apoptosis is noticeable in the hippocampus and cerebral cortex (Fig. 13.23). Neuronal degeneration is associated with proliferation of astrocytes as a compensatory mechanism. There exist scattered pathological lesions in the extracellular space known as *senile plaques*, which are composed of fragmented axons and dendrites surrounding an amyloid core structure. In the cytoplasm of the neurons, fiber-like structures, such as coils and knots, known as neurofibrillary *tangles* or *clumps*, are often observed. The neurofibrillary tangles are composed of a highly phosphorylated form of a microtubule-associated protein τ. These changes are associated with progressive degeneration of neurons primarily in the cerebral cortex.

Alzheimer's disease is characterized by the death of cholinergic neurons (Fig. 13.24). A number of factors contribute to the pathogenesis of cholinergic neuron death. These include the accumulation of amyloid plaques in the extracellular matrix, the accumulation of neurofibrillary tangles in the neuronal cytoplasm (Fig. 13.25), and progressive loss of choline acetyltransferase and acetylcholine in the cerebral cortex. In particular, the formation of amyloid plaques and neurofibrillary tangles is considered a major pathological change that contributes to the death of cholinergic neurons and the progression of Alzheimer's disease. However, the mechanisms remain poorly understood.

Formation of amyloid plaques is due to the accumulation of proteins, inclduing a small protein known as *amyloid* β (Aβ) *peptide*. This peptide is a product of the cleavage of the amyloid β protein precursor (APP), a protein normally bound to the membrane of the neurons. The amyloid β protein precursor can be cleaved by β- and γ-secretases and metalloproteases, generating 40- and 42-amino acid amyloid β peptides. The 42-amino acid peptide is shed from the neuronal membrane, and aggregates into Alzheimer plaques (Fig. 13.26).

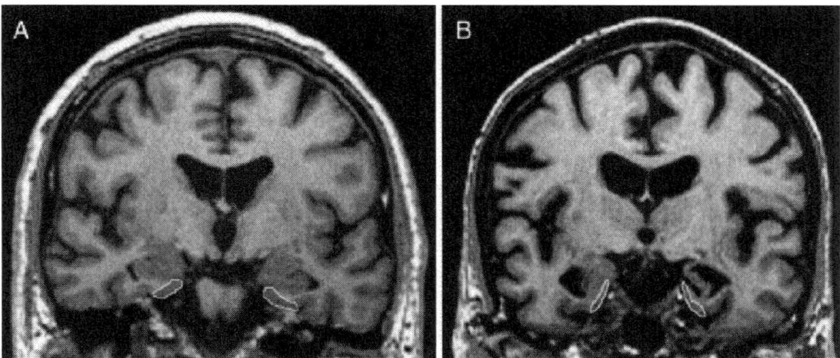

Figure 13.22. Coronal MRI at the level of the mammillary bodies. (A) A normal control; (B) a person with mild cognitive impairment. (Reprinted from Jose C, Masdeu JC et al: Neuroimaging as a marker of the onset and progression of Alzheimer's disease, *J Neurol Sci* 236:55–64, copyright 2005, with permission from Elsevier.)

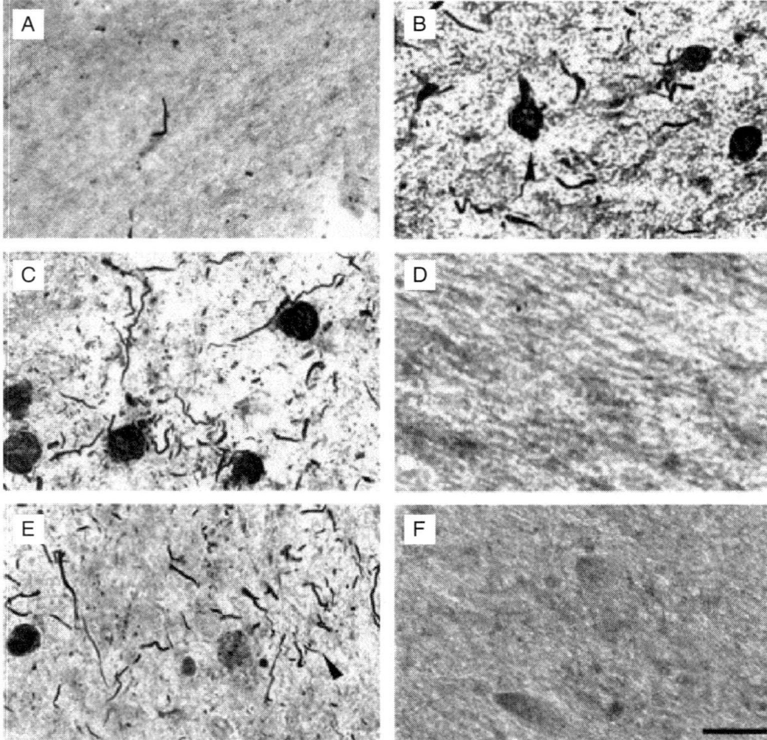

Figure 13.23. In the nucleus basalis of Meynert (NBM) of normal control individuals, rare Fas-associated death-domain (FADD)-positive neurites were observed (A). However, many FADD-positive tangle-like structures (arrowhead in panel B) and neurites (arrowhead in panel E) were observed within the NBM of all the Alzheimer's cases. The Alzheimer's brains with either Braak stage V (B) or Braak stage VI (C) displayed the same pattern of FADD-positive structures within the NBM. Preabsorption of the antibody with excess FADD peptide resulted in complete absence of FADD immunoreactivity (D). Immunoreactivity for caspase-3 (F) was completely absent from the NBM of AD. Scale bar: 100 μm for all panels. (Reprinted from Wu CK et al: Apoptotic signals within the basal forebrain cholinergic neurons in Alzheimer's disease, *Exp Neurol* 195:484–96, copyright 2005, with permission from Elsevier.)

→

Figure 13.25. Formation of amyloid plaques and neurofibrillary tangles in Alzheimer's disease. Amyloid plaques contain the Aβ peptide, which is produced from the amyloid precursor protein (APP). Cleavage of APP at one point (not shown) generates a carboxy-terminal fragment. This is then severed by the γ-secretase complex (which consists of at least four proteins: presenilin-1, APH1, PEN2, and nicastrin), producing Aβ. Tangles are produced from hyperphosphorylated τ protein, possibly by means of the enzyme glycogen synthase kinase-3 (GSK3). GSK3 might also regulate the cleavage of APP carboxy-terminal fragments. Lithium and kenpaullone, two GSK3 inhibitors, as well as small interfering RNAs that knock down GSK3 expression, inhibit Aβ production. The GSK3 inhibitors might interfere with the generation of both amyloid plaques and tangles in Alzheimer's disease. The precise roles of the two forms of GSK3, α and β, are unknown. (Reprinted by perission from Macmillan Publishers Ltd.: De Strooper B, Woodgett J: Alzheimer's disease: Mental plaque removal, *Nature* 423:392–3, copyright 2003.)

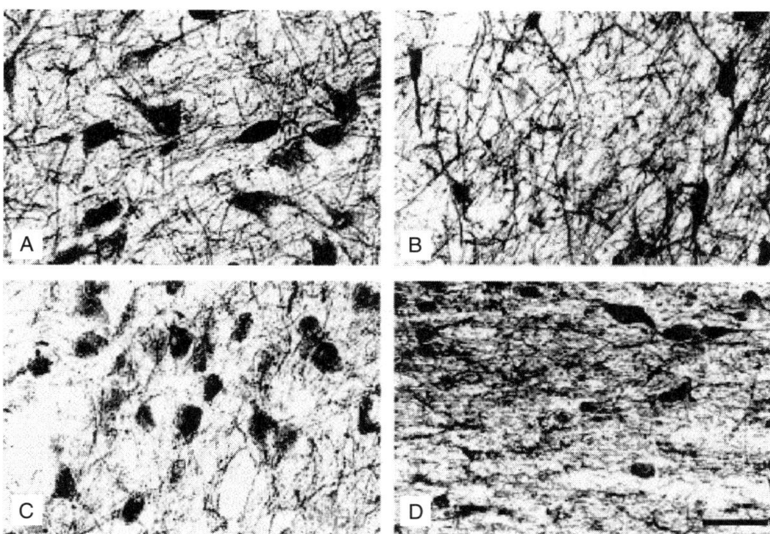

Figure 13.24. Choline acetyltransferase (ChAT) neurons in normal and Alzheimer's brains. In the basal forebrain of normal cases, many ChAT- (A) or p75-positive (C) cholinergic neurons were present. In contrast, very few ChAT- (B) or p75-positive (D) cholinergic neurons were found within the nucleus basalis of Meynert (NBM) of Alzheimer's cases. Scale bar: 100 μm, in all panels. (Reprinted from Wu CK et al: Apoptotic signals within the basal forebrain cholinergic neurons in Alzheimer's disease, *Exp Neurol* 195:484–96, copyright 2005, with permission from Elsevier.)

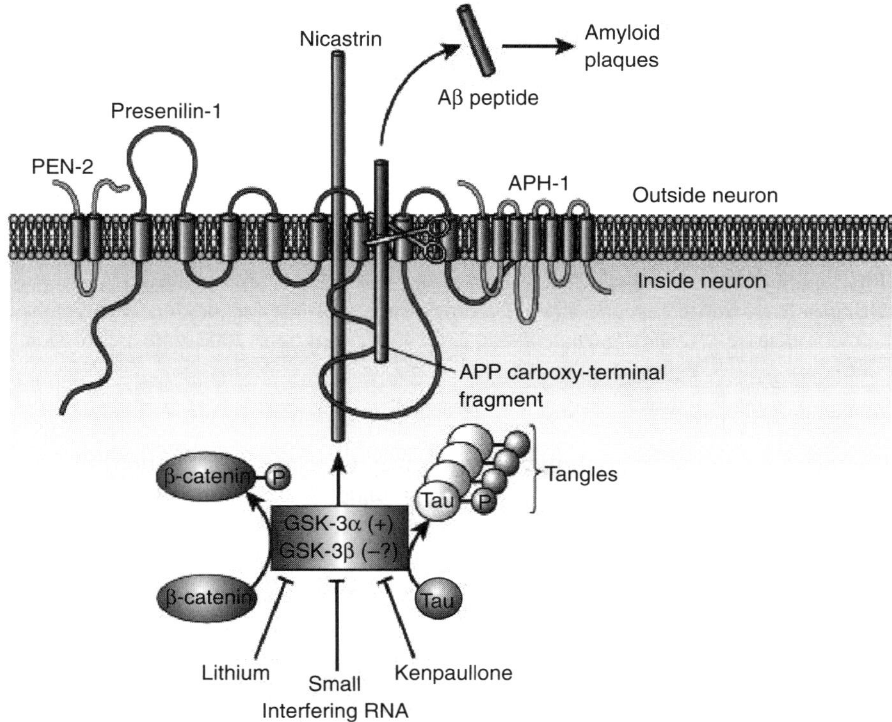

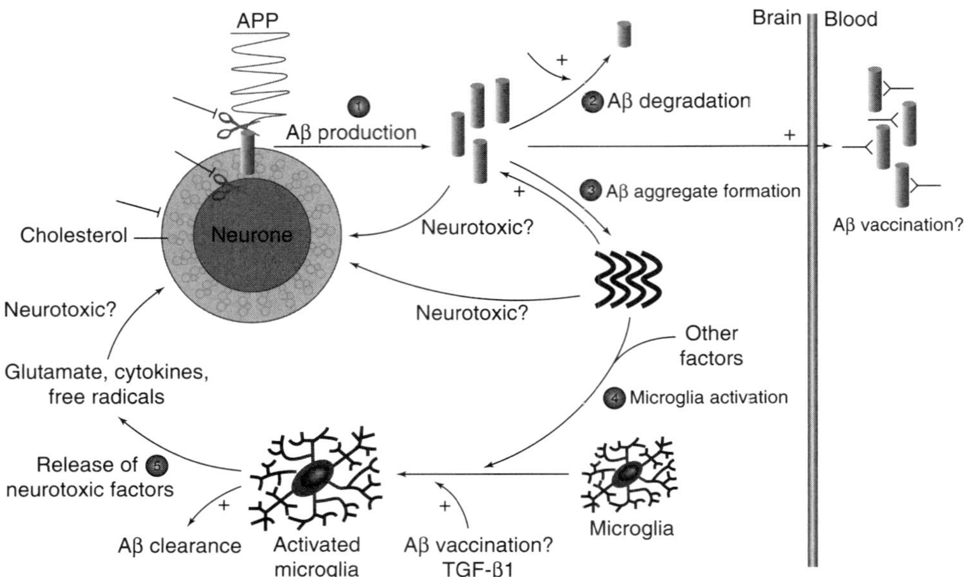

Figure 13.26. The amyloid hypothesis in Alzheimer's disease (AD) and candidate targets for therapeutic intervention. AD is characterized by two types of protein aggregates, neurofibrillary tangles and β-amyloid (Aβ) plaques, distributed in regions of the central nervous system involved in learning and memory. Soluble Aβ is formed following the cleavage of Aβ precursor protein (APP) by enzymatic activities known as β-secretase and γ-secretase (1). Aβ formed is then degraded by enzymes (2). The balance between Aβ production and degradation can be disrupted, leading to Aβ accumulation beyond pathological levels and, in turn, increased levels of Aβ aggregates (3) and deposits in the brain. Aggregates and Aβ fibrils could themselves be neurotoxic or could activate microglia (4), which can release neurotoxic factors as part of an inflammatory response (5). Several steps could be targeted pharmacologically to treat AD. For example, inhibitors of β-secretase and γ-secretase or cholesterol-lowering drugs (blunt arrows) could be used to decrease the production of Aβ. It is possible that activators of Aβ-degrading enzymes could be developed to reduce Aβ levels and metal chelators could be used to dissolve amyloid plaques. Furthermore, Aβ vaccination is proposed to sequester Aβ in the blood, which in turn would induce a rapid efflux of Aβ from the brain. Microglia can also be activated by Aβ vaccination or by transforming growth factor β1 (TGFβ1), leading to increased Aβ clearance and neuroprotection. (Reprinted from Dominguez DI, De Strooper B: Novel therapeutic strategies provide the real test for the amyloid hypothesis of Alzheimer's disease, *Trends Pharmacol Sci* 23:324–30, copyright 2002 with permission from Elsevier.)

The formation of the neurofibrillary tangles may be related to alterations in the phosphoralation of the τ protein. This protein may undergo several pathological changes, including hyperphosphorylation, enzymatic cleavage, and conformational alterations, which contribute to the formation of Alzheimer's tangles and the pathogenesis of Alzheimer's disease. The role of the protein τ in initiating Alzheimer's disease has been established with the observation that hyperphosphorylated τ is present in the neuronal tangles in Alzheimer's disease (chapter-opening figure). However, the exact mechanisms of τ-involved pathogenesis remains poorly understood. A serine/threonine kinase that induces τ hyperphosphorylation is the enzyme glycogen synthase kinase 3 (GSK3) (Fig.

13.25). This enzyme may also mediate the cleavage of the amyloid β protein precursor, contributing to the formation of the amyloid plaques. The inhibition of the GSK3 by specific pharmarcological inhibitors, such as lithium and kenpaullone, and small interfering RNA prevents the formation of neurofibrillary tangles and amyloid plaques.

Another factor that contributes to the initiation and development of Alzheimer's disease is the progressive loss of choline acetyltransferase. This enzyme catalyzes the transfer of an acetyl group to choline to form acetylcholine, which is the primary neurotransmitter of the cholinergic neurons. The loss of this enzyme reduces the production of acetylcholine, thus inducing the impairment of the function and survival of cholinergic neurons. The defect of the choline acetyltransferase has been considered a major pathogenic factor for Alzheimer's disease.

Conventional Treatment of Alzheimer's Disease [13.28]. Alzheimer's disease has been traditionally treated with several approaches, including: (1) improvement of the cerebral bloodflow, (2) supplementation of cholinergic neurotransmitter precursors, (3) reduction of the degradation of the neurotransmitter acetylcholine, and (4) treatment of abnormal behaviors. The cerebral bloodflow can be enhanced by administrating vasodilators. The improvement of blood supply to the brain may reduce the degeneration of neurons, thus slowing down the progression of Alzheimer's disease. Cholinergic neurotransmitter precursors, such as choline, lecithin, and acetyl-L-carnitine, are often used to enhance the production of acetylcholine, the neurotransmitter of cholinergic neurons. However, the concentration of a precursor may not be the rate-limiting factor for the production of acetylcholine. Many studies in which these substances were used failed to show substantial beneficial effects. An alternative approach for reducing the degradation of acetylcholine is to suppress the enzyme acetylcholinesterase, which breaks down acetylcholine in synapses. Several pharmacological inhibitors, including tacrine, physostigmine, donepezil, and rivastigmine, have been used to suppress acetylcholinesterase, thus reducing the degradation of acetylcholine. The use of acetylcholinesterase inhibitors has shown mild beneficial effects on the progression of Alzheimer's disease. When patients express significant signs of depression and hallucinations, tranquilizers or antidepressants can be administrated to control the patients.

A new approach has been established and used to treat Alzheimer's disease by modulating the activity of the nicotinic and muscarinic receptors in the neuron. These receptors, which can interact with acetylcholine and participate in the regulation of cognitive function, are decreased in the brain of Alzheimer's patients. A potential treatment is to administrate muscarinic receptor agonists, such as AF102B [(+/−)-cis-2-methyl-spiro (1,3-oxathiolane-5,3′)quinuclidine] and AF150(S) [1-methylpiperidine-4-spiro-(2′-methylthiazoline)], which are cholinomimetics or analogs of acetylcholine. These substances can enhance the activity of the muscarinic receptor, act synergistically with neurotrophic factors to stimulate neuronal regeneration, protect neurons from oxidative stress-induced apoptosis, and reduce the level of β-amyloid, which contributes to the development of Alzheimer's disease. Preliminary investigations have shown that these substances exert a beneficial effect on the cognitive function of patients with Alzheimer's disease.

Similarly, agonists to the nicotinic acetylcholine receptor, such as SIB-1553A and GTS-21, have been used to enhance the function of the receptor in animal models of Alzheimer's disease. SIB-1553A is a nicotinic acetylcholine receptor ligand, also known as (+/−)-4-[[2-(1-methyl-2-pyrrolidinyl)ethyl]thio]phenol hydrochloride. This substance has been tested in aged rodents and nonhuman primates for its effect on the cognitive function of these animals. A treatment with SIB-1553A (subcutaneous, intramuscular, or oral administration) improves choice accuracy as well as spatial and nonspatial working memory in aged rhesus monkeys. Furthermore, a treatment with SIB-1553A induces the release of acetylcholine in the hippocampus of aged rats, suggesting that the SIB-1553A-induced cognitive improvement may be dependent on the release acetylcholine. Thus SIB-1553A may serve as a therapeutic substance for Alzheimer's disease.

GTS-21, also known as *DMXBA*, is a benzylidene anabaseine derivative [3-(2,4-dimet hoxybenzylidene)anabaseine] based on a marine worm toxin named *anabaseine*. This substance nonselectively activates both muscle-type and neuronal nicotinic acetylcholine receptors. GTS-21 has been shown to exert a therapeutic effect on Alzheimer's disease. GTS-21 exhibits neuroprotective activity in cultured neurons deprived of nervous growth factors or exposed to β-amyloid. A treatment with this substance enhances cognitive function in animal models such as the mouse, monkey, rat, and rabbit. This substance exhibits lower toxicity than nicotine. At a dose that enhances the cognitive behavior of animals, it does not significantly affect the activities of the nervous and skeletal muscle systems. Clinical trials have demonstrated promising results for the treatment of Alzheimer's disease. Oral administration of GTS-21 improves the cognitive behavior of patients with Alzheimer's disease.

Molecular Regenerative Engineering for Alzheimer's Disease. Based on the pathogenic mechanisms as discussed on page 554, several strategies have been established for the molecular treatment of Alzheimer's disease, including prevention of neuronal degeneration and inhibition of glycogen synthase kinase 3. These strategies are briefly discussed as follows.

PREVENTION OF NEURONAL DEGENERATION BY DELIVERING NEUROTROPHIC FACTORS [13.29]. As discussed above, Alzheimer's disease is caused at least in part by the degeneration of cerebral cholinergic neurons resulting from the reduction in the neurotransmitter acetylcholine. Thus, the principle of molecular treatment for Alzheimer's disease is to reduce the degradation of acetylcholine and promote the generation of this neurotransmitter. A common molecular approach for the treatment of Alzheimer's disease is to deliver locally neurotrophic factors, including brain-derived neurotrophic factor, neurotrophins, and nervous growth factor, or the genes encoding these growth factors, into target sites such as the forebrain cortex and the hippocampus, which are involved in Alzheimer's disease. These growth factors play a critical role in controlling neuronal survival (see page 521 for the characteristics of neurotrophic factors). In particular, neurotrophins and nervous growth factor can stimulate neuronal release of acetylcholine. These growth factors or their genes can be delivered by using methods described on page 436.

Experimental investigations have produced promising results for the treatment of Alzheimer's disease with neurotrphins. For instance, the transfection of the forebrain neurons with the nervous growth factor and brain-derived neurotrophic factor genes can significantly boost the production of acetylcholine and enhances the survival of impaired cholinergic neurons in animal models. It should be noted that the selection of the delivery route for cerebral gene transfer is an important issue. The bloodstream delivery of growth factors or their genes is a relatively simple method, but these molecules may not be able to reach the target sites because of the blockade of the blood–brain barrier. The direct injection of nervous growth factors or their genes into the brain has been shown to be an effective method. However, it is difficult to deliver the therapeutic factors to a precise site.

Furthermore, this method causes neural injury. Spinal cavity delivery of growth factors or their genes is a relatively safe and effective approach. Delivered proteins or genes can be transported to the central nervous system via the cerebrospinal fluid. However, a large amount of protein or gene is required to reach an effective level of therapy.

Another approach for neuronal survival is to promote the production and release of nervous growth-related factors by using pharmacological substances. The synthetic purine AIT-082 or Neotrofin {4-[[3-(1,6 dihydro-6-oxo-9-purin-9-yl)-1-oxypropyl] amino] benzoic acid} has been shown to stimulate neurotrophin secretion from astrocytes. This substance also enhances several other neural activities, such as neuronal regeneration, axonal outgrowth, cognitive activity of animals and humans, and the restoration of age-induced memory deficits in animals. AIT-082 has been applied to patients with mild to moderate Alzheimer's disease. A treatment with AIT-082 induces an increase in the glucose metabolic rate, indicative of enhanced bloodflow, in the cerebellum and the prefrontal cortical area as detected by PET scanning. Furthermore, AIT-082 improves the memory, executive functioning, and attention of patients with Alzheimer's disease. Several other pharmacological substances, such as propentofylline (a synthetic xanthine derivative that inhibits phosphodiesterase), idebenone (a quinone derivative that serves as an antioxidant), and pyrroloquinoline quinone (a vitamin and redox cofactor of quinoprotein dehydrogenases), have been shown to stimulate the production and release of nervous growth factor and exert indirectly a therapeutic effect on Alzheimer's disease.

INHIBITION OF GLYCOGEN SYNTHASE KINASE 3 [13.30]. As discussed on page 556, the enzyme glycogen synthase kinase 3 induces hyperphosphorylation of the protein tau, contributing to the formation of the neurofibrillary tangles, the production of amyloid β peptides, and the development of Alzheimer's disease. Thus, one of the strategies for treating Alzheimer's disease is to suppress the activity of glycogen synthase kinase 3 (GSK3). An effective means of inhibiting GSK3 is to construct and deliver small interfering RNA (siRNA) specific to the GSK3 mRNA. This approach has been shown to significantly suppress the expression of GSK3 and reduce the formation of the neurofibrillary tangles and amyloid plaques. In addition, there are two pharmarcological inhibitors, lithium and kenpaullone, specific to the enzyme glycogen synthase kinase-3. These inhibitors can be used to suppress the activity of GSK3 and are effective for the treatment of Alzheimer's disease. However, lithium is toxic to the kidney at a high blood concentration. It should be used with caution.

Cell Regenerative Engineering for Alzheimer's Disease [13.31]. For the treatment of Alzheimer's disease, several cell types have been used for transplantation into the brain to improve the production and release of the cholinergic neurotransmitter in the basal forebrain area. These cells include fetal basal forebrain cells, embryonic acetylcholine-enriched cells, and gene-transfected stem cells. The transplantation of embryonic and fetal stem cells into the brain has long been known to enhance the production of acetylcholine and the survival of cholinergic neurons. The transplanted stem cells may differentiate into cholinergic neurons, release acetylcholine, and prevent apoptosis of neurons. Furthermore, genetic modulation of the stem cells can greatly enhance the cell's ability to stimulate neuronal cell survival. For example, the transfection of stem or progenitor cells with the nervous growth factor gene significantly promotes the cell's ability to secrete nervous growth factor. Such a cell line, when transplanted into a target tissue, can release excessive nervous growth factor, thus enhancing cell survival. In addition to embryonic and fetal stem cells, other cell types, such as bone marrow cells and umbilical cord blood cells, have been used and tested in animal Alzheimer's models. The transplantation of these cell types has been shown to exert a beneficial effect on the treatment of Alzheimer's disease. For cell transplantation into the brain, direct injection is the method of choice (see page 460 for method).

Huntington's Disease

Etiology, Pathology, and Clinical Features [13.32]. Huntington's disease, also known as *Huntington's chorea*, is a hereditary autosomal dominant neural disorder characterized by chronic ceaseless occurrence of rapid, involuntary, complex jerky movements, mental deterioration, and dementia. This is one of the most frequently observed hereditary diseases in the United States. The average occurrence is about 4–5 per million. The disease is often found at the age of 40–50, but also occurs in children and young adults.

There are several distinct clinical features for Huntington's disease. In the early phase, patients may express annoying behaviors, such as constant complaining and reduced self-control. With the progression of the disease, patients may show mood disturbance and depression signs, reduced communication and fine manual skills, reduced work performance and responsibility, difficulties in maintaining attention, changes in personality, and deteriorating intellectual abilities. These mental changes are associated with abnormalities in muscular movement. Early motor-related signs include reduced movement of fingers and hands and increased rate of blinking. In the advanced stage, patients exhibit constant and rapid involuntary jerky movements and are not able to speak well because of inadequate control of the tongue muscles. The speed of voluntary movements is significantly reduced.

Huntington's disease is associated with a number of pathological changes. These include atrophy of the central nervous system, loss of neuron dendrites, disappearance of myelinated axons, and compensatory astrocyte proliferation. The pathogenesis of Huntington's disease is thought to relate to several biochemical factors. An increase in the production of dopamine and/or in the sensitivity of striatal dopamine receptors may promote the dopamine effect, inducing involuntary movement. Other neurotransmitters, such as acetylcholine and norepinephrine, are disturbed in Huntington's disease, contributing to the pathogenesis of the disease. However, the exact mechanisms remain poorly understood.

The pathogenesis of Huntington's disease is related to several genetic mechanisms, including the mutation of the CAG-repeating sequence, mutation of the huntingtin gene, and the impairment of histone acetylation. The CAG-repeating sequence is found in genes that encode multiple glutamine or polyglutamine residues. The number of the CAG groups ranges from 3 to 34 in a normal gene. In Huntington's disease, however, the number of the CAG groups can increase significantly, ranging from 36 to 121, resulting in a significant increase in the length of the polyglutamine sequence in involved proteins. Proteins with expanded polyglutamine sequences exhibit reduced solubility and form aggregates with amyloid or other polyglutamine proteins. These protein aggregates exert a toxic effect on the cerebral neurons by abnormal interaction with essential cellular signaling proteins. Such toxicity may influence a number of cellular processes, including protein folding, proteasomal degradation, mitochondrial energy metabolism, and gene transcription. These abnormalities may contribute to the initiation and development of Huntington's disease. A typical example is the polyglutamine expansion in the TATA-binding protein (TBP), which binds the TATA box of genes and regulates gene transcription. Polyglutamine expansion in this protein disturbs the binding of the protein to the TATA box of target genes involved in the regulation of neuronal activities, thus contributing to the development of the symptoms of Huntington's disease.

The mutation of the huntingtin gene is another cause of Huntington's disease. *Huntingtin*, also known as *Huntington's disease protein*, is a 348-kDa protein expressed ubiquitously and plays a critical role in the regulation of development. Huntingtin exerts several functions, including the protection of neural and nonneural cells from apoptosis and excitotoxicity, enhancement of brain-derived neurotrophic factor gene expression, and regulation of fast axonal trasport and synaptic transmission. Huntingtin mutation induces a decrease in the transcription of the brain-derived neurotrophic factor gene, alterations in axonal trafficking and synaptic transmission, and impairment of neuronal functions. Experimental knockout of the huntingtin gene is assocaited with pathological changes seen in Huntington's disease. Overexpression of the huntington gene improves the function of neurons. These observations suggest a role for huntingtin in the pathogenesis of Huntington's disease.

Another potential pathogenic mechanism for Huntington's chorea is the impairment of histone acetylation and gene transcription regulated by cAMP responsive element binding protein (CREB)-binding protein (CBP). CBP is a 250-kDa transcriptional coactivator that binds to and mediates the activity of phosphorylated CREB, which is a transcriptional factor activated by protein kinase A and responsible for activating genes involved in learning and memory in the central nervous system. The CREB coactivator CBP is a histone acetyltransferase that catalyzes histone acetylation. This protein has been found to aggregate with proteins containing expanded polyglutamines in cultured cells. In transgenic mouse models with polyglutamine-expanded huntingtin, CBP can form aggregates with huntingtin, resulting in a reduction in the level of CBP. Such a change ultimately impairs the function of CREB and possibly contributes to the pathogenesis of Huntington's disease.

In the brain of Huntington's patients, aggregates of CBP with polyglutamine-expanded proteins can be found by electron microscopy. The expression of glutamine-expanded huntingtin can be observed by immunohistochemistry. These changes are often associated with a decrease in the level of soluble CBP, which affects CREB-mediated transcription activities, resulting in cell death in tissue culture models of Huntington disease. These changes can be mitigated by overexpression of CBP. Furthermore, the aggregates of CBP with polyglutamine-expanded proteins are greatly reduced when the polyglutamine-containing domain is deleted. These observations suggest that polyglutamine-related sequestration of CBP may contribute to the impairment of histone acetylation and disorder of gene transcription, thus leading to the development of Huntington's disease. However, how these alterations influence the structure and function of cerebral neurons remains poorly understood.

Conventional Treatment of Huntington's Disease [13.33]. There are few methods available for the treatment of Huntington's disease. Dopamine antagonists are used to suppress movement disorders and behavioral abnormalities. One of such antagonists is haloperidol, which is also used as a tranquilizer for the management of psychoses. Although this drug is effective, it often causes dyskinesia or impairment of voluntary movements. Substances that enhance the degradation of dopamine or block the dopamine receptor are also used for treating the symptoms of Huntington's disease. These substances include reserpine, clozapine, and tetrabenazine, which are effective for the control of Huntington's symptoms. However, these drugs often cause severe side effects such as drowsiness, dyskinesia, and akathisia. It is important to note that these drugs are used only to treat the symptoms of the disease without influencing the progression of the disease. Unfortunately, drugs that prevent and suppress Huntington's disease are not available.

Molecular Regenerative Engineering for Huntington's Disease [13.34]. Since the mechanisms of Huntington's disease are not well understood, there are few therapeutic approaches that can be used for the treatment of the disease. Recent investigations have begun to reveal the genetics of Huntington's disease, which suggest potential therapeutic approaches. As discussed on page 562, the sequestration of CREB-binding protein by polyglutamine-expanded proteins induces a significant decrease in the level of the CREB-binding protein and may contribute to the pathogenesis of Huntington's disease. CREB-binding protein is a signaling molecule that participates in the regulation of a variety of cellular processes such as cell survival and proliferation. A potential therapeutic approach is to enhance the expression of the CREB-binding protein. Investigations with cell culture models have demonstrated that overexpression of the CREB-binding protein gene can reduce the influence of polyglutamine-expanded proteins. The CREB-binding protein gene can be potentially used for the treatment of Huntington's disease.

The impairment of histone acetylation contributes to the pathogenesis of Huntington's disease. A typical example is histone deacetylation, which is catalyzed by histone deacetylase. The prevention of histone deacetylation may be a potential therapeutic approach for treating Huntington's disease. Histone deacetylase inhibitors, such as butyrate and suberoylanilide hydroxamic acid (SAHA), have been used and tested in models of neuronal degeneration induced by polyglutamine-expanded proteins. In a transgenic *Drosophila* model of neurodegeneration induced by expression of expanded polyglutamines, histone deacetylase inhibitors (suberoylanilide hydroxamic acid) significantly reduced the progression of neurodegeneration. In another study with a cell culture model, polyglutamine expansion in the androgen receptor caused cell death, which was reversed by a treatment with the histone deacetylase inhibitor suberoylanilide hydroxamic acid. These investigations demonstrate that histone deacetylase inhibitors may be potentially used as therapeutic agents for Huntington's disease. To date, histone deacetylase inhibitors have only been tested in cell culture and *Drosophila* models. Further experimental tests are needed in mammalian models.

Another potential therapeutic approach is to modulate the activity of huntingtin (Fig. 13.27). Huntingtin is a soluble cytoplasmic protein expressed widely in the central nervous system and found in somatodendritic regions and axons. Huntingtin can interact with numbers of signaling molecules and plays a critical role in regulating cellular processes, such as channel control, protein processing, and pre-mRNA splicing. The mutation of huntingtin may disrupt these cellular processes and contribute to the development of Huntington's disease. Other functions of huntingtin are to stimulate the expression of brain-derived neurotrophic factor in the cortical area and facilitate vesicular transport of brain-derived neurotrophic factor along microtubules in neurons. Huntingtin mutation is associated with a reduction in brain-derived neurotrophic factor, resulting in insufficient neurotrophic support and apoptosis of cortical neurons. The expression and release of brain-derived neurotrophic factor are attenuated when the level of wildtype huntingtin is reduced. The restoration of wildtype huntingtin via gene transfer enhances the expression of brain-derived neurotrophic factor and the survival of cortical neurons. This may be a potential approach for the treatment of Huntington's disease.

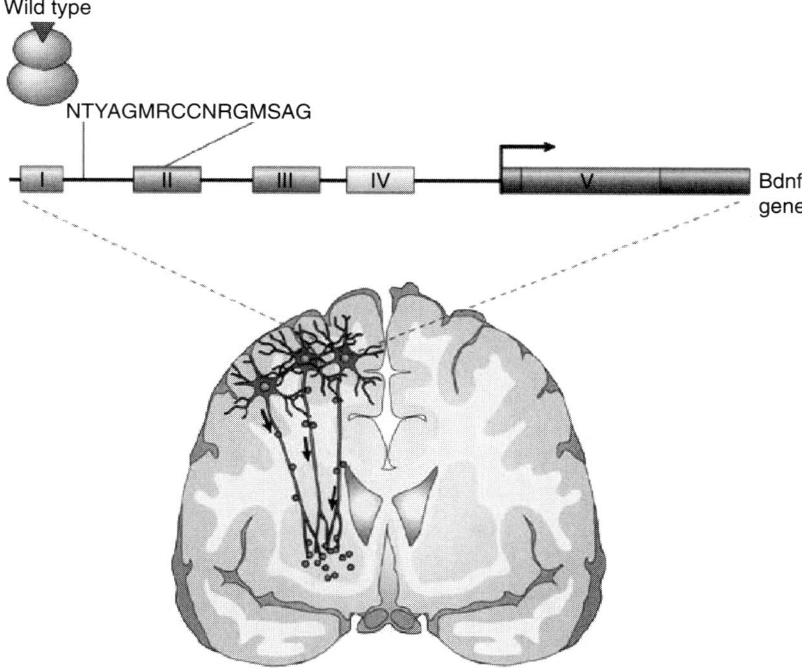

Figure 13.27. Treatment of Huntington's disease with huntingtin. Wildtype but not mutant huntingtin facilitates production of brain-derived neurotrophic factor (BDNF) in the cortical neurons, which project to the striatum, by inhibiting the repressor element 1/neuron-restrictive silencer element (RE1/NRSE) that is located in BDNF promoter exon II. I–IV indicate BDNF promoter exons in the rodent BDNF gene; V indicates the coding region. The RE1/NRSE consensus sequence is shown. Inactivation of the RE1/NRSE in the BDNF gene leads to increased mRNA transcription and protein production in the cortex. BDNF, which is also produced through translation from exons III and IV is then made available to the striatal targets via the corticostriatal afferents. Wildtype huntingtin might also facilitate vesicular BDNF transport from the cortex to the striatum. (Reprinted by permission from Macmillan Publishers Ltd.: Cattaneo E, Zuccato C, Tartari M: Normal huntingtin function: an alternative approach to Huntington's disease, *Nat Rev Neurosci* 6:919–30, copyright 2005.)

Parkinson's disease

Etiology, Pathology, and Clinical Manifestation [13.35]. Parkinson's disease is a neural disorder that is characterized by tremor of resting muscles, slow voluntary movements, festinating gait, peculiar posture, and muscle weakness. Patients also exhibit excessive sweating and feeling of heat. This disease often occurs at the age of 40–70. The incidence of Parkinson's disease is very low in the population under 30 years of age, but is fairly high (about 1%) in the elder population (over 65) in the United States. Several factors such as cranial injury, overwork, emotional shock, extreme stress, and exposure to unusual environments may facilitate the deterioration of the disease. Indeed, the disease may often be found after such incidents.

Parkinson's disease is associated with several distinct pathological changes. These include the progressive loss of pigmented neurons in the substantia nigra and dorsal motor nucleus, the presence of Lewy bodies (eosinophilic cytoplasmic contents with halos) (Fig. 13.28), and reduction in nonpigmented neurons. The pathogenesis of the disease may be related to the disorder of the dopamine metabolic system, resulting in a reduction in the production of dopamine. It has been observed that the level of tyrosine hydroxylase, a key enzyme for the production of dopamine, is reduced significantly in patients with Parkinson's disease. This is consistent with the reduction in the level of dopamine. Thus, reduced dopamine in the central motor control system may contribute to the degeneration of the pigmented neurons. Although it is not certain whether hereditary factors play a role, there are patients with a family history of Parkinson's disease.

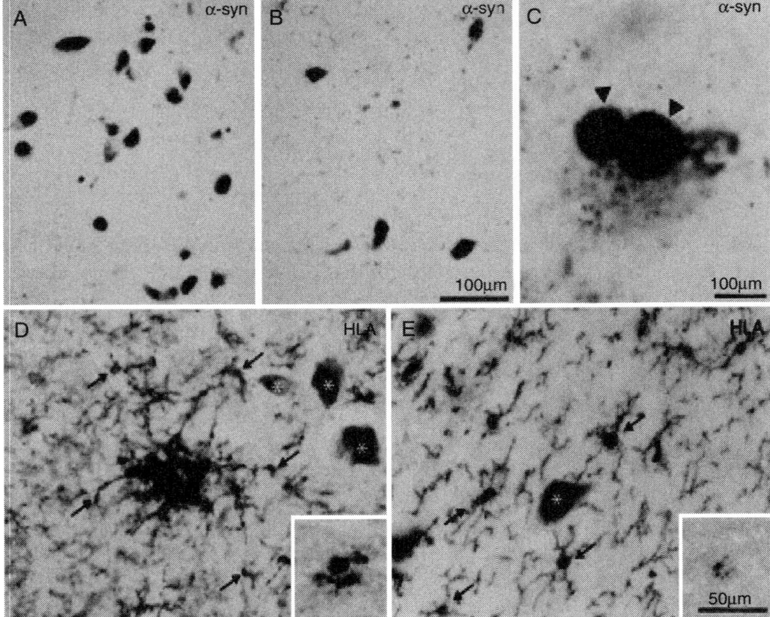

Figure 13.28. Neuron loss and Lewy body formation in Parkinson's disease. (A) Control. (B–E) Parkinson's disease. (A–C) Neuromelanin pigmented neurons immunohistochemically stained with an anti-α synuclein antibody. Scale in B is equivalent to that for A. There is an obvious loss of pigmented dopamine neurons in Parkinson's disease (B) compared with the control (A). Some remaining pigmented dopamine neurons in Parkinson's disease contain α-synuclein-immunoreactive Lewy bodies (arrowheads in C). (D,E) Specimens immunohistochemically stained with antibodies to HLA-DP/DQ/DR (HLA), a marker for the major histocompatibility complex class II protein. Scale in E is equivalent to that for D. HLA-immunoreactive upregulated microglia (arrows) near nonimmunoreactive pigmented neurons (asterisks) in thick (D) and thin (E) midbrain sections from patients with Parkinson's disease (D and E). The inserts in D and E were from Parkinson's disease patients with α-synuclein and parkin gene mutation, respectively. (Reprinted from Orr CF et al: A possible role for humoral immunity in the pathogenesis of Parkinson's disease, *Brain* 128:2665–74, copyright 2005 by permission of Oxford University Press.)

Conventional Treatment [13.35]. Parkinson's disease is induced predominantly by reduction or depletion of cerebral dopamine, the neurotransmitter of dopaminergic neurons. Thus, the principle of treating Parkinson's disease is to maintain or restore the level of dopamine in the central nervous system. Several types of pharmacological substances have bee used to achieve such a goal. These include dopamine precursors, such as L-dihydroxyphenylalanine (L-dopa; also known as *levodapa*), dopamine agonists, such as bromocriptine, pergolide, lisuride, ropinirole, and pramipexole, and substances that inhibit the degradation of dopamine, such as selegiline. L-Dopa is the most effective drug for the treatment of Parkinson's disease. L-Dopa can be used alone or in combination with other agents. One type of such agents is decarboxylase inhibitors such as carbidopa and benserizide. Decarboxylase is an enzyme that catalyzes L-dopa breakdown and the conversion of L-dopa to dopamine. Carbidopa reduced the activity of decarboxylase. Because carbidopa cannot pass through the blood–brain barrier, the administration of carbidopa together with L-dopa reduces the rate of L-dopa breakdown in the peripheral system, allowing a large fraction of L-dopa to reach the brain. This strategy also reduces the side effect of dopamine, such as nausea and hypotension, in the peripheral systems.

Dopamine agonists, such as bromocriptine, pergolide, lisuride, ropinirole, and pramipexole, can interact with the dopamine receptor, mimicking the activity of dopamine. These agents can be administrated before L-dopa is used or as a supplement to L-dopa treatment. Selegiline is an inhibitor for the monoamine oxidase, which degrades dopamine, and thus can be used to reduce the degradation of dopamine in the brain. It should be noted that treatment with all these agents may only be effective for treating the symptoms of Parkinson's disease, but not be effective for preventing the progression of the disease.

Molecular Regenerative Engineering for Parkinson's Disease [13.36]. Given the pathogenic mechanisms of Parkinson's disease as discussed on page 565, strategies for molecular engineering treatment of Parkinson's disease include the promotion of the survival of dopaminergic neurons, maintenance of the dopamine level in the brain, and protection of dopaminergic neurons from apoptosis. These strategies are tested primarily in rodent and nonhuman primate models of Parkinson's disease. These models are induced by administration of chemical toxins, such as 6-hydroxydopamine (6-OHDA) and 1-methyl-4-phenyl-1,2,3,6-tetrahydropyridine (MPTP), which reduce the production of dopamine in neurons and promote neuronal apoptosis. It is important to note that these animal models may not completely assemble the human Parkinson's disease. Results form these models should be interpreted with caution.

PROMOTION OF THE SURVIVAL OF DOPAMINERGIC NEURONS. For promoting the survival of dopaminergic neurons, neurotrophic factors, such as nervous growth factor (NGF), brain-derived neurotrophic factor (BDNF), glial cell line-derived neurotrophic factor (GDNF), and platelet-derived growth factor (PDGF) or their genes can be delivered to the brain by using approaches of direct injection or spinal cavity injection (see page 521 of this chapter for these neurotrophic factors) (Fig. 13.29). In animal models of Parkinson's disease, the overexpression of neurotrophic factors can exert several effects that benefit the treatment of the disease, including an increase in the synthesis of dopamine, protection of dopaminergic neurons from toxic injury and apoptosis, and enhancement of the sprouting of dopaminergic neurons. A large number of investigations have demonstrated the effectiveness of neurotrophic factor gene transfer in mitigating the symptoms of Parkinson's

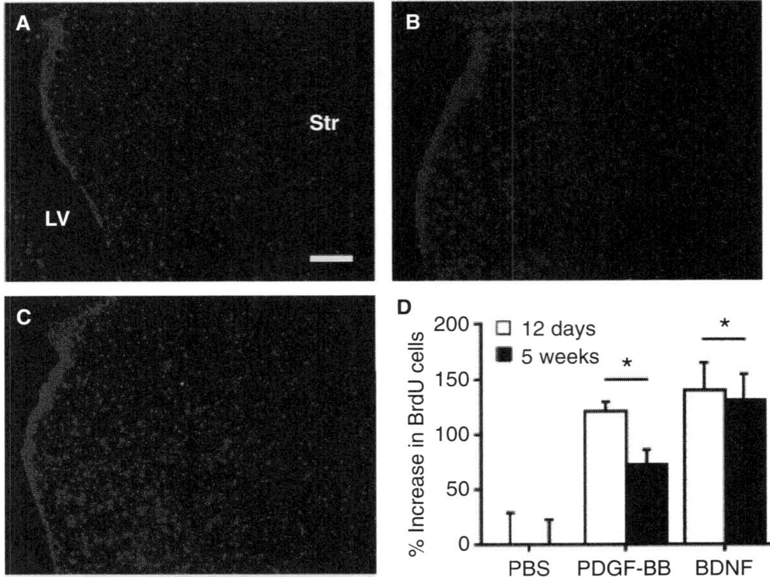

Figure 13.29. Infusion of platelet-derived growth factor BB (PDGF-BB) and brain-derived neurotrophic factor (BDNF) into the right lateral ventricle (LV) enhances the density of BrdU-positive cells in the denervated striatum (Str) of the rat brain with unilateral 6-OHDA lesions. Red BrdU-labeled cells in PBS- (A), PDGF-BB- (B), and BDNF-treated (C) rats at 12 days after growth factor infusion (5 weeks after 6-OHDA lesions). (D) Measurements of BrdU-positive cells in the Str at 12 days and 5 weeks after growth factor infusion. Both growth factors enhanced BrdU incorporation compared to the PBS control, with no differences between PDGF-BB and BDNF (* $P < 0.05$). Scale bar: 100 µm. (Reprinted from Mohapel P et al: Platelet-derived growth factor (PDGF-BB) and brain-derived neurotrophic factor (BDNF) induce striatal neurogenesis in adult rats with 6-hydroxydopamine lesions, *Neuroscience* 132:767–76, copyright 2005, with permission from Elsevier.)

disease. For instance, in the rodent model of 6-OHDA-induced Parkinson's disease, the transfer of the glial cell line-derived neurotrophic factor gene into the striatum results in a significant increase in the dopamine level and a reduction in the animal rotary behavior, a typical sign of Parkinson's animals. In the nonhuman primate model of Parkinson's disease, striatal transfer of the glial cell line-derived neurotrophic factor gene induces increased gene expression of tyrosine hydroxylase, an enzyme participating in the synthesis of dopamine.

MAINTENANCE OF THE DOPAMINE LEVEL IN THE BRAIN. A critical issue for promoting the survival of dopaminergic neurons is to maintain the level of dopamine in the central nervous system. Since tyrosine hydroxylase catalyzes the synthesis of L-dopa, one of the approaches for achieving such a goal is to transfer the tyrosine hydroxylase gene into the striatum. A number of investigations have shown that such an approach results in a significant increase in the dopamine level, and a decrease in the symptoms of Parkinson's disease, such as rotational movements. There are other enzymes that regulate the synthesis of dopamine. One of such enzymes is the amino acid decarboxylase (AADC), which converts L-dopa to dopamine. The deficiency of this enzyme may contribute to the reduction in the dopamine level and the pathogenesis of Parkinson's disease. The transfer of the amino acid decarboxylase gene into the striatum of nonhuman primates with induced Parkinson's disease results in increased conversion of L-dopa to dopamine and reduction in Parkinson's symptoms. The cotransfer of the genes encoding multiple enzymes for the synthesis of dopamine has been demonstrated to provide a synergistic effect and exhibit significant improvement in the dopamine level and motor control in Parkinson's animal models. Similar results have also been found in studies with rodent models. The tyrosine hydroxylase and amino acid decarboxylase genes are considered potential candidate genes for the treatment of human Parkinson's disease.

PROTECTION OF DOPAMINERGIC NEURONS FROM APOPTOSIS. Another molecular approach for treating Parkinson's disease is to protect neurons from apoptosis. Parkinson's disease is associated with an increase in neuronal apoptosis, which is thought to contribute to the pathogenesis of Parkinson's disease. Several factors, including oxidative stress and free radicals resulting from dopamine metabolism and ion toxicity, may induce neuronal apoptosis. An enzyme, known as Cu/Zn superoxide dismutase, plays a critical role in detoxifying superoxide, which forms toxic peroxynitrite with nitric oxide. The overexpression of the Cu/Zn superoxide dismutase gene in cultured neurons by gene transfection has been shown to protect the cells from 6-OHDA-induced toxic effects. Animals with overexpressed Cu/Zn superoxide dismutase gene exhibit increased resistance to oxidative stress and apoptosis. These observations suggest that the Cu/Zn superoxide dismutase gene can potentially serve as a therapeutic gene for the treatment of Parkinson's disease.

Apoptosis is regulated by the Bcl2 family of proteins. While some of these proteins are proapoptotic, the Bcl2 protein exerts an antiapoptotic effect (see page 304 for signaling mechanisms of apoptosis). Experimental investigations have shown that overexpression of the Bcl2 protein in the striatum by gene transfection protects neurons from apoptosis and induces an increase in the number of tyrosine hydroxylse-positive neurons. In transgenic mice with overexpressed Bcl2 gene, neurons in the central nervous system are protected from the toxic effect of 1-methyl-4-phenyl-1,2,3,6-tetrahydropyridine (MPTP), which induces a reduction in dopamine and degeneration of dopaminergic neurons. Cotransfer of the glial cell line-derived neurotrophic factor gene with the Bcl2 gene significantly augments the beneficial effect of glial cell line-derived neurotrophic factor gene. Both factors synergistically prevent neurons from apoptosis. These investigations suggest that the antiapoptosis protein genes can be used to as therapeutic genes for enhancing the survival of dopaminergic neurons.

Cell apoptosis is regulated by a protein kinase known as C-Jun *N*-terminal kinase (JNK). The phosphorylation of C-Jun *N*-terminal kinase has been shown to activate c-Jun and promote cell apoptosis. A protein called *JNK-interacting protein 1* (JIP1) can inhibit the phosphorylation of C-Jun *N*-terminal kinase. The transfer of the JIP1 gene into the striatum induces the overexpression of the gene, which is associated with reduced phosphorylation of C-Jun *N*-terminal *kinase and* reduced activity of caspase 3, a proteinase that induces cell apoptosis. Furthermore, the transfer of a dominant negative c-jun gene protects neurons from the toxic effect of 1-methyl-4-phenyl-1,2,3,6-tetrahydropyridine (MPTP) and promotes cell survival. Thus, the negative regulators of JNK may serve as therapeutic factors for the molecular treatment of Parkinson's disease.

Caspase 3 is a critical downstream proteinase that cleaves a variety of intracellular proteins, such as actin filaments, signaling protein kinases, and nuclear proteins, and induces cell apoptosis. Caspase inhibitor genes can be used and transferred into the brain to suppress cell apoptosis. Such inhibitor genes include the p35 gene and the family of the inhibitor of apoptosis protein genes. The transfer of these genes into the striatum in the model of 6-OHDA-induced Parkinson's disease significantly inhibits the activity of caspases and delays the occurrence of cell apoptosis. Transgenic mice with p35 overexpression exhibit increased neuronal tolerance to the toxic effect of 1-methyl-4-phenyl-1,2,3,6-tetrahydropyridine (MPTP). Caspase inhibitors also augment the mitogenic effect of neurotrophic factors. In general, all caspase inhibitor genes can be used for molecular treatment of Parkinson's disease.

Cell Regenerative Engineering for Parkinson's Disease [13.37]. In cell regenerative engineering for Parkinson's disease, the primary goal is to transplant stem or other types of cell into the brain to restore the structure and function of impaired dopaminergic neurons. Typical cell types for such a purpose include embryonic stem cells, fetal neuronal stem and progenitor cells, adult neuronal stem cells, and cell lines with transferred genes that enhance desired functions. To date, cell regenerative engineering approaches have been tested primarily in animal models. Since these models may not completely assemble the human Parkinson's disease, results from animal models should be interpreted with caution.

As discussed on page 381, embryonic stem cells from the blastocyst are capable of differentiating into all specified cell types, including neurons, under appropriate conditions. Fetal neuronal precursor cells can differentiate into dopamine-synthesizing neurons. These stem cells can be used and transplanted into the central nervous system for neuronal regeneration. It is expected that the embryonic and fetal stem cells can differentiate into dopaminergic neurons, thus improving the motor control capability. A number of investigations have demonstrated that dopamine-synthesizing neurons from embryonic and fetal stem cells can be used for transplantation into the central nervous system. The transplanted cells can be integrated into the host brain (Fig. 13.30). Such a treatment significantly improves the level of dopamine production and the survival of dopaminergic neurons in association with reduced Parkinson's symptoms. However, the application of embryonic or fetal stem cells to human patients remains an issue of ethical debate. To date, most investigations in this area are conducted by using animal models.

Given the controversial issues for the use of embryonic and fetal stem cells, many investigators have searched for neuronal stem cells in the central nervous system. Indeed, the adult brain contains cells that exhibit stem cell characteristics such as self-renewing and differentiation. These cells are considered adult neural stem cells and can differentiate into neurons, astrocytes, and oligodendrocytes. The adult neural stem cells can be found in the wall of the ventricular system, the olfactory system, and the hippocampus of the brain (see page 395 for details). These cells can be potentially used for the treatment of Parkinson's disease.

Although the transplantation of stem and progenitor cells demonstrates beneficial effects on the treatment of Parkinson's disease, natural wildtype cells may not be able to sufficiently boost the production and release of the neurotransmitter dopamine and neurotrophic factors necessary for the survival of neurons. To resolve such an issue, stem and progenitor cells have been transfected with genes encoding proteins that promote the differentiation of stem and progenitor cells to dopaminergic neurons. Two types of gene, including the Nurrl and von Hippel–Lindau (VHL) protein genes, have been used for such a purpose.

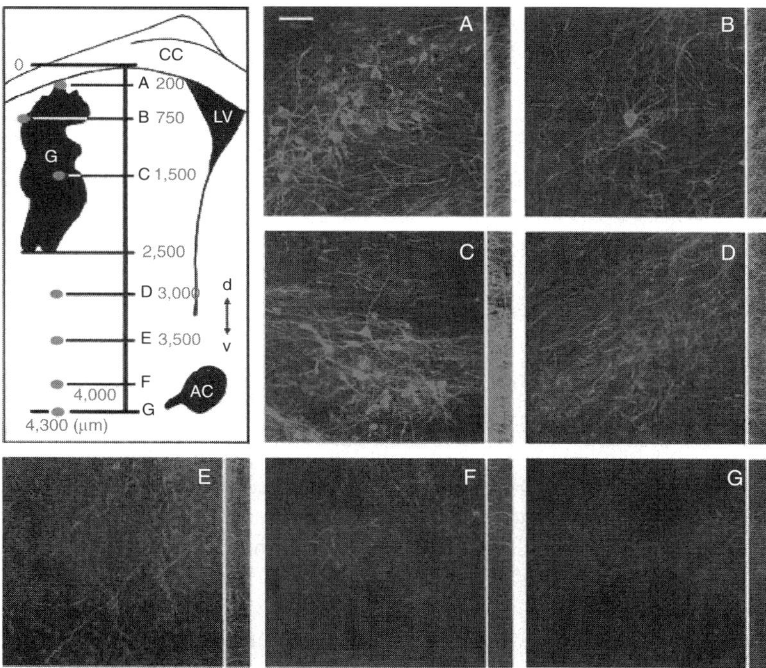

Figure 13.30. Integration of Nurrl embryonic stem cells into the striatum of hemiparkinsonian rats. The diagram shows a drawing of a single section through a graft (G) in the striatum (LV, lateral ventricle; AC, anterior commissure). Single confocal images after immunohistochemistry for tyrosine hydroxylase (TH) are shown (A–G) from regions marked by red dots in the diagram. The distribution of cells and processes through the thickness of the section (35 μm) is shown by the *z* series displayed in green on the right. Note the many TH^+ processes that extend away from the graft into the parenchyma of the host striatum (D–F). Scale bar: 50 μm. (Reprinted by permission from Macmillan Publishers Ltd.: Kim JH et al: Dopamine neurons derived from embryonic stem cells function in an animal model of Parkinson's disease, *Nature* 418:50–6, copyright 2002.)

Nurrl is a protein that belongs to the nuclear receptor superfamily of transcription factors. This protein is expressed predominantly in the developing and mature dopaminergic neurons of the central nervous system, and is essential for the differentiation of the mesencephalic precursors to dopaminergic neurons and for the survival of mature dopaminergic neurons. The von Hippel–Lindau protein is a tumor suppressor that downregulates the transcriptional activity of mitogenic genes. This protein is primarily expressed in the central nervous system and plays a role in regulating the differentiation of the mesencephalic precursors to dopaminergic neurons. Thus, genes encoding the Nurrl and von Hippel–Lindau proteins can be potentially used to boost the differentiation of stem and progenitor cells into dopamine-synthesizing neurons. Several investigations have demonstrated that the transfer of these genes into stem or progenitor cells enhances the synthesis of dopamine, increases the density of tyrosine hydroxylase-positive neurons, improves the survival of dopaminergic cells, and reestablishes dopamine-dependent motor behaviors in animal models of Parkinson's disease. Other potential genes that can be used to boost the function of transplanted neuronal stem cells include neurotrophic factor genes, dopamine synthesis-promoting genes, and Cu/Zn superoxide dismutase genes.

Nonneuronal cell types have also been used to create dopamine-synthesizing cell lines by gene transfer. These cells include fibroblasts, astrocytes, Schwann cells, myoblasts, and marrow stromal cells. These cell types are readily available and easy to collect in comparison to embryonic and fetal stem cells. Although these cell types may not be able to differentiate into neurons, they can carry and deliver necessary proteins such as neurotrophic factors and dopamine synthesis enzymes for enhancing the survival of neurons.

Multiple Sclerosis

Etiology, Pathogenesis, and Clinical Manifestations [13.38]. *Multiple sclerosis* is a neural disorder that is characterized by myelin degradation, demyelination, and the generation of insoluble precipitates or plaques in various locations, in association with dysfunction of multiple cerebral regions and the spinal cord. Clinical signs include motor weakness, partial paralysis, abnormal sensation, abnormal vision, tremor, and dysarthria. The insoluble plaques can be detected by X-ray and magnetic resonance imaging. The disorder progresses slowly with a latent period of 1–10 years starting with a minor symptom. One of the distinct features of the disorder is the intermittent occurrence of the symptoms and signs. Multiple sclerosis is found in about 0.001% of the population. The disorder is often detected in patients at the age of 20–40. Hereditary factors may influence the occurrence of the disease. About 15% of the patients with multiple sclerosis have relatives with the same disorder. Siblings of patients with multiple sclerosis have a risk factor significantly greater than that of the general population.

Multiple sclerosis is associated with the destruction of myelin sheaths. However, this process does not apparently affect the structure of neurons and axons. A pathological examination often reveals an uneven surface of the spinal cord, while the brain surface may appear normal. Scattered lesions or insoluble plaques may be found in the nervous system with a size ranging from millimeters to centimeters with pink/gray-colored white matter due to demyelination. The lesion is found primarily in the white matter near the cerebral ventricles, in the brainstem, and in the spinal cord, but not in peripheral nerves. The optic nerves are often affected by the disorder. Structural alterations are dependent on the progression of the disorder. Fresh lesions exhibit partial destruction of myelin sheaths and infiltration of lymphocytes and mononuclear cells. With the progression of the disorder, there appears increased infiltration of microglial cells (macrophages) and increased size of sclerotic lesions. In the late stage, few myelin sheaths can be found. Affected regions often exhibit increased fibrous tissue with reduced lymphocytes and macrophages.

A number of factors may contribute to the pathogenesis of multiple sclerosis. These factors include viral infection, autoimmune reactions, and synergistic effects of both viral and autoimmune factors. Patients with multiple sclerosis exhibit immune reactions against viral products. Herpes viruses have been implicated in the initiation and development of multiple sclerosis, since DNA from these viruses is found in multiple sclerosis plaques. However, little direct evidence has been obtained for the role of viruses. Another factor that potentially contributes to the induction and development is autoimmune reaction. In such a case, the host immune cells may not recognize myelin proteins as the body's own components and may initiate immune reactions to attack and destroy the myelin structure and sometimes the nerve axon. A major line of supporting evidence for the autoimmune mechanism is the existence of antibodies against the myelin components in the serum of patients with multiple sclerosis. Viral infection and autoimmune reactions may coordinately influence the process of demyelination. Several types of virus, such as rubella and rubeola, may contain protein components that are similar to some proteins in the myelin sheath of nerve axons. The exposure of T and B lymphocytes to these viruses may induce lymphocyte immunization. Immunized lymphocytes in turn recognize and attack the host myelin.

Further investigations have demonstrated that multiple sclerosis is associated with the infiltration of mononuclear cells in the area of demyelination, axonal loss, and glial fibrosis. Given the involvement of T lymphocytes, it has been hypothesized that activated antigen-specific T cells initiate specific immune reactions and induce the infiltration of non-antigen-specific mononuclear cells into the brain. The mononuclear cells in turn interact with and destroy oligodendrocytes by releasing toxic substances in association with the degradation of myelin sheaths. In the early stage of the disease, oligodendrocytes are capable of surviving and remyelinating axons. However, in the late stage, these cells are gradually committed to apoptosis.

Multiple sclerosis causes changes in the physiological function of nervous axons. A major change is delayed transmission of action potentials. This is due to the destruction of the axon myelin sheath. Under physiological conditions, the transmission of action potentials in the myelinated axons (as high as 100 m/s) is much faster than that in unmyelinated axons (as low as 0.25 m/s). The destruction of the myelin sheath influences the electrical conduction of the axons. In severe cases, the transmission of electrical signals can be completely blocked. The impairment and blockade of electrical signals ultimately influence the function of peripheral tissues and organs.

Conventional Treatment [13.39]. Multiple sclerosis is a disease possibly induced by viral infection and autoimmune reactions. The principle of treating multiple sclerosis is administration of antiinflammatory and immunosuppressor agents. Corticosteroids are commonly used as antiinflammatory agents. These agents usually give noticeable results within about 2 weeks. Immunosuppressor agents, such as azathioprine and cyclophosphamide, have been used for the treatment of multiple sclerosis with some positive results. However, these substances compromise with the normal immune function and impose toxic effects. Such harmful influences may preclude the widespread use of the immunosuppressor agents. In addition, appropriate physical exercise is necessary to stimulate impaired motor control systems. Bacterial infection should be prevented and treated promptly, if any. Other disorders associated with multiple sclerosis should be treated properly.

Molecular Regenerative Engineering for Multiple Sclerosis [13.40]. As the pathogenesis of multiple sclerosis is attributed to the destruction of the myelin sheath and oligodendrocytes by inflammatory reactions involving activated mononuclear cells, a potential approach for this disease is to suppress inflammatory reactions and inhibit the infiltration of bloodborne mononuclear cells into the brain. The enhancement of oligodendrocyte proliferation and migration into the area of demyelinated axons may also provide therapeutic effects. Gene therapeutic approaches have been developed to achieve these goals. These approaches have been tested primarily in the animal model of multiple sclerosis: autoimmune encephalomyelitis (EAE). It should be noted that, since the animal model may not completely assemble the human disease, information from experimental observations may not be directly applied to the human disease.

Several therapeutic strategies have been developed and used to suppress inflammatory reactions. These include the enhancement of antiinflammatory cytokines and induction of B-lymphocyte tolerance to myelin-related antigens. Since mononuclear cells and B-lymphocytes are bloodborne cells, genes encoding antiinflammatory proteins can be delivered into the bloodstream. Potential genes for antiinflammatory purposes include the interleukin (IL)1β, IL2, IL4, IL6, IL10, tumor necrosis factor (TNF)α, and transforming growth factor (TGF)β1 genes. The transfer of these genes into animals with autoimmune encephalomyelitis reduces pathological signs and clinical symptoms of the disorder, although controversial results are observed for some IL molecules, such as IL4 and IL10. These antiinflammatory protein genes can be delivered to target tissues by three approaches: injection into the bloodstream, directly into the brain, and into the cerebrospinal fluid cavities.

The enhancement of B-lymphocyte tolerance to myelin-related antigens is another potential approach for the treatment of multiple sclerosis. Certain viruses may contain components that partially assemble the structure of myelin proteins. Viral infection may expose B lymphocytes to the myelin-like viral components and induce immunization of the lymphocytes, producing antimyelin protein antibodies. These antibodies may interact with host myelin proteins and contribute to autoimmune processes, potentially inducing multiple sclerosis. In an experimental study, a recombinant IgG–myelin basic protein (MBP) gene is inserted into a retroviral vector and transferred into B lymphocytes. These B cells were introduced into the bloodstream of mice with autoimmune encephalomyelitis-induced multiple sclerosis. Compared to control mice without B cell transplantation, the B cell-transplanted mice exhibit reduced pathological signs and clinical symptoms of multiple sclerosis. It is thought that the presence of myelin basic protein in the B cells increases the tolerance of these cells to the myelin basic protein, thus reducing the production and secretion of antimyelin protein antibodies and mitigating autoimmune reactions.

Gene therapeutic approaches have also been developed to enhance the proliferation and migration of oligodendrocytes and to promote the remyelination of impaired axons, as demyelination results in axonal loss and neurological impairment. Candidate genes for such a purpose include neurotrophic factor and nerve growth factor genes. Since the therapeutic targets of these genes are the oligodendrocytes in the central nervous system, direct brain gene delivery usually gives satisfactory results. Although neurotrophic factors and their genes can be injected into the bloodstream, the therapeutic efficiency is usually low, as it is difficult for protein and DNA molecules to pass through the blood–brain barrier. Another delivery route is the cerebrospinal fluid. In an experimental model of autoimmune encephalomyelitis-induced multiple sclerosis, a herpes virus-derived vector containing the fibroblast growth factor gene was transferred into the cerebrospinal fluid. Fibroblast growth factor is known to promote the proliferation and differentiation of oligodendrocytes. The introduction of this growth factor to the cerebrospinal fluid enhances remyelination of impaired axons and reduces pathological signs and clinical symptoms of multiple sclerosis.

Cell Regenerative Engineering [13.41]. Cellular engineering approaches have been developed for the treatment of experimental multiple sclerosis induced by autoimmune encephalomyelitis. These include the transplantation of oligodendrocytes, oligodendrocyte precursors, or genetically modified memory T lymphocytes with enhanced secretion of growth factors or antiinflammatory factors. Oligodendrocytes or their precursors can be directly delivered to the lesion sites of multiple sclerosis. A fraction of these cells can integrate into the native system and generate myelin proteins and sheaths. These cells can also be transfected with growth factor genes to enhance their capability of proliferation and migration. T lymphocytes can be transfected with antiinflammatory protein genes, such as the interleukin (IL)1β, IL2, IL4, IL6, IL10, tumor necrosis factor (TNF)α genes, enhancing the production and secretion of antiinflammatory factors. These cells can be transplanted to the bloodstream, from where they can migrate into the lesion sites of the brain. Alternatively, T cells can be directly delivered to the central nervous system. T lymphocytes can also be transfected with growth factor genes to promote their capability of producing growth factors.

BIBLIOGRAPHY

13.27. Etiology, Pathology, and Clinical Manifestations for Alzheimer's Disease

Wilquet V, De Strooper B: Amyloid-beta precursor protein processing in neurodegeneration, *Curr Opin Neurobiol* 14:582–8, 2004.

Reinhard C, Hebert SS, De Strooper B: The amyloid-beta precursor protein: Integrating structure with biological function, *EMBO J* 24:3996–4006, 2005.

Annaert W, De Strooper B: A cell biological perspective on Alzheimer's disease, *Annu Rev Cell Dev Biol* 18:25–51, 2002.

Dominguez DI, De Strooper B: Novel therapeutic strategies provide the real test for the amyloid hypothesis of Alzheimer's disease, *Trends Pharmacol Sci* 23:324–30, 2002.

Lu PJ, Wulf G, Zhou XZ, Davies P, Lu KP: The prolyl isomerase Pin1 restores the function of Alzheimer-associated phosphorylated tau protein, *Nature* 399:784–8, 1999.

Panda D, Samuel JC, Massie M, Feinstein SC, Wilson L: Differential regulation of microtubule dynamics by three- and four-repeat tau: Implications for the onset of neurodegenerative disease, *Proc Natl Acad Sci USA* 100:9548–53, 2003.

Rapoport M, Dawson HN, Binder LI, Vitek MP, Ferreira A: Tau is essential to beta-amyloid-induced neurotoxicity, *Proc Natl Acad Sci USA* 99:6364–9, 2002.

Ros R, Thobois S, Streichenberger N, Kopp N, Sanchez MP et al: A new mutation of the tau gene, G303V, in early-onset familial progressive supranuclear palsy, *Arch Neurol* 62:1444–50, 2005.

SantaCruz K, Lewis J, Spires T, Paulson J, Kotilinek L et al: Tau suppression in a neurodegenerative mouse model improves memory function, *Science* 309:476–81, 2005.

Skipper L, Wilkes K, Toft M, Baker M, Lincoln S et al: Linkage disequilibrium and association of MAPT H1 in Parkinson disease, *Am J Hum Genet* 75:669–77, 2004.

Spillantini MG, Murrell JR, Goedert M, Farlow MR, Klug A et al: Mutation in the tau gene in familial multiple system tauopathy with presenile dementia, *Proc Natl Acad Sci USA* 95:7737–41, 1998.

Spittaels K, van den Haute C, van Dorpe J, Geerts H, Mercken M et al: Glycogen synthase kinase-3-beta phosphorylates protein tau and rescues the axonopathy in the central nervous system of human four-repeat tau transgenic mice, *J Biol Chem* 275:41340–9, 2000.

Stamer K, Vogel R, Thies E, Mandelkow E, Mandelkow EM: Tau blocks traffic of organelles, neurofilaments, and APP vesicles in neurons and enhances oxidative stress, *J Cell Biol* 156:1051–63, 2002.

Tatebayashi Y, Miyasaka T, Chui DH, Akagi T, Mishima K et al: Tau filament formation and associative memory deficit in aged mice expressing mutant (R406W) human tau, *Proc Natl Acad Sci USA* 99:13896–901, 2002.

Tesseur I, van Dorpe J, Spittaels K, van den Haute C, Moechars D et al: Expression of human apolipoprotein E4 in neurons causes hyperphosphorylation of protein tau in the brains of transgenic mice, *Am J Pathol* 156:951–64, 2000.

13.28. Conventional Treatment of Alzheimer's Disease

Victor M, Ropper AH: *Principles of Neurology*, 7th ed, McGraw-Hill, New York, 2001 pp 1106–74.

Etienne P, Robitaille Y, Wood P, Gauthier S: Nair NP et al: Nucleus basalis neuronal loss, neuritic plaques and choline acetyltransferase activity in advanced Alzheimer's disease, *Neuroscience* 19:1279–91, 1986.

Tucek S: Short-term control of the synthesis of acetylcholine, *Prog Biophys Mol Biol* 60:59–69, 1993.

Bontempi B, Whelan KT, Risbrough VB, Rao TS, Buccafusco JJ et al: SIB-1553A, (+/−)-4-[[2-(1-methyl-2-pyrrolidinyl)ethyl]thio]phenol hydrochloride, a subtype-selective ligand for nicotinic acetylcholine receptors with putative cognitive-enhancing properties: effects on working and reference memory performances in aged rodents and nonhuman primates, *J Pharmacol Exp Ther* 299:297–306, 2001.

Woodruff-Pak DS, Santos IS: Nicotinic modulation in an animal model of a form of associative learning impaired in Alzheimer's disease, *Behav Brain Res* 113:11–9, 2000.

Fisher A, Pittel Z, Haring R, Bar-Ner N, Kliger-Spatz M et al: Ml muscarinic agonists can modulate some of the hallmarks in Alzheimer's disease: implications in future therapy, *J Mol Neurosci* 20:349–56, 2003.

Auld DS, Kornecook TJ, Bastianetto S, Quirion R: Alzheimer's disease and the basal forebrain cholinergic system: Relations to beta-amyloid peptides, cognition, and treatment strategies, *Prog Neurobiol* 68:209–45, 2002.

Ruske AC, White KG: Facilitation of memory performance by a novel muscarinic agonist in young and old rats, *Pharmacol Biochem Behav* 63:663–7, 1999.

Papke RL, Meyer EM, Lavieri S, Bollampally SR, Papke TA et al: Effects at a distance in alpha 7 nAChR selective agonists: Benzylidene substitutions that regulate potency and efficacy, *Neuropharmacology* 46:1023–38, 2004.

Kem WR: The brain alpha7 nicotinic receptor may be an important therapeutic target for the treatment of Alzheimer's disease: Studies with DMXBA (GTS-21), *Behav Brain Res* 113:169–81, 2000.

13.29. Prevention of Neuronal Degeneration by Delivering Neurotrophic Factors

Auld DS, Kornecook TJ, Bastianetto S, Quirion R: Alzheimer's disease and the basal forebrain cholinergic system: Relations to beta-amyloid peptides, cognition, and treatment strategies, *Prog Neurobiol* 68:209–45, 2002.

Gustilo MC, Markowska AL, Breckler SJ, Fleischman CA, Price DL et al: Evidence that nerve growth factor influences recent memory through structural changes in septohippocampal cholinergic neurons, *J Comp Neurol* 405:491–507, 1999.

Fischer W, Bjorklund A, Chen K, Gage FH: NGF improves spatial memory in aged rodents as a function of age, *J Neurosci* 11:1889–906, 1991.

Markowska AL, Koliatsos VE, Breckler SJ, Price DL, Olton DS: Human nervous growth factor improves spatial memory in aged but not in young rats, *J Neurosci* 14:4815–24, 1994.

Scali C, Casamenti F, Pazzagli M, Bartolini L, Pepeu G: Nerve growth factor increases extracellular acetylcholine levels in the parietal cortex and hippocampus of aged rats and restores object recognition, *Neurosci Lett* 170:117–20, 1994.

Auld DS, Mennicken F, Day JC, Quirion R: Neurotrophins differentially enhance acetylcholine release, acetylcholine content and choline acetyltransferase activity in basal forebrain neurons, *J Neurochem* 77:253–62, 2001.

Auld DS, Mennicken F, Quirion R: Nervous growth factor rapidly induces prolonged acetylcholine release from cultured basal forebrain neurons: Differentiation between neuromodulatory and neurotrophic influences, *J Neurosci* 21:3375–82, 2001.

Klein RL, Muir D, King MA, Peel AL, Zolotukhin S et al: Long-term actions of vector-derived nervous growth factor or brain-derived neurotrophic factor on choline acetyltransferase and Trk receptor levels in the adult rat basal forebrain, *Neuroscience* 90:815–21, 1999.

Mandel RJ, Gage FH, Clevenger DG, Spratt SK, Snyder RO et al: Nerve growth factor expressed in the medial septum following in vivo gene delivery using a recombinant adeno-associated viral vector protects cholinergic neurons from fimbria–fornix lesion-induced degeneration, *Exp Neurol* 155:59–64, 1999.

Mahoney MJ, Saltzman WM: Millimeter-scale positioning of a nervous-growth-factor source and biological activity in the brain, *Proc Natl Acad Sci USA* 96:4536–9, 1999.

Lahiri DK, Ge YW, Farlow MR: Effect of a memory-enhancing drug, AIT-082, on the level of synaptophysin, *Ann NY Acad Sci* 903:387–93, 2000.

Keator D, Carreon D, Fleming K, Fallon JH: Brain metabolic effects of Neotrofin in patients with Alzheimer's disease, *Brain Res* 951:87–95, 2002.

Arenas E, Persson H: Neurotrophin-3 prevents the death of adult central noradrenergic neurons in vivo, *Nature* 367:368–71, 1994.

13.30. Inhibition of Glycogen Synthase Kinase 3 (GSK3)

Phiel CJ, Wilson CA, Lee VM, Klein PS: GSK-3alpha regulates production of Alzheimer's disease amyloid-beta peptides, *Nature* 423:435–9, 2003.

Dominguez DI, De Strooper B: Novel therapeutic strategies provide the real test for the amyloid hypothesis of Alzheimer's disease, *Trends Pharmacol Sci* 23:324–30, 2002.

De Strooper B, Woodgett J: Alzheimer's disease: Mental plaque removal, *Nature* 423:392–3, 2003.

13.31. Cell Regenerative Engineering for Alzheimer's Disease

Fine A, Dunnett SB, Bjorklund A, Iversen SD: Cholinergic ventral forebrain grafts into the neocortex improve passive avoidance memory in a rat model of Alzheimer's disease, *Proc Natl Acad Sci USA* 82:5227–30, 1985.

Dunnett SB: Neural transplants as a treatment for Alzheimer's disease, *Psychol Med* 21:825–30, 1991.

Turnpenny L, Cameron IT, Spalluto CM, Hanley KP, Wilson DI et al: Human embryonic germ cells for future neuronal replacement therapy, *Brain Res Bull* 68:76–82, 2005.

Holden C: Neuroscience. Versatile cells against intractable diseases, *Science* 297:500–2, 2002.

Stahl T, Goldammer A, Luschekina E, Beck M, Schliebs R, Bigl V: Long-term basal forebrain cholinergic-rich grafts derived from trisomy 16 mice do not develop beta-amyloid pathology and neurodegeneration but demonstrate neuroinflammatory responses, *Int J Dev Neurosci* 16:763–75, 1998.

Baekelandt V, De Strooper B, Nuttin B, Debyser Z: Gene therapeutic strategies for neurodegenerative diseases, *Curr Opin Mol Ther* 2:540–54, 2000.

Barkats M, Bilang-Bleuel A, Buc-Caron MH, Castel-Barthe MN, Corti O et al: Adenovirus in the brain: recent advances of gene therapy for neurodegenerative diseases, *Prog Neurobiol* 55:333–41, 1998.

Fischer MSW, Bjorklund A: Reversal of age-dependent cognitive impairments and cholinergic neuron atrophy by NGF-secreting neural progenitors grafted to the basal forebrain, *Neuron* 15:473–84, 1995.

Wyman T, Rohrer D, Kirigiti P, Nichols H, Pilcher K et al: Promoter-activated expression of nerve growth factor for treatment of neurodegenerative diseases, *Gene Ther* 6:1648–60, 1999.

Ende N, Chen R, Ende-Harris D: Human umbilical cord blood cells ameliorate Alzheimer's disease in transgenic mice, *J Med* 32:241–7, 2001.

13.32. Etiology, Pathology, and Clinical Manifestations of Huntington's Disease

Shiwach RS, Norbury CG: A controlled psychiatric study of individuals at risk for Huntington's disease, *Br J Psychiat* 165:500–5, 1994.

Lovestone S, Hodgson S, Sham P, Differ AM, Levy R: Familial psychiatric presentation of Huntington's disease, *J Med Genet* 33:128–31, 1996.

Giordani B, Berent S, Boivin MJ, Penney JB, Lehtinen S et al: Longitudinal neuropsychological and genetic linkage analysis of persons at risk for Huntington's disease, *Arch Neurol* 52:59–64, 1995.

Zoghbi HY, Orr HT: Glutamine repeats and neurodegeneration, *Annu Rev Neurosci* 23:217–47, 2000.

Hughes RE: Polyglutamine disease: Acetyltransferases awry, *Curr Biol* 12:R141–3, 2002.

Perez MK, Paulson HL, Pendse SJ, Saionz SJ, Bonini NM, Pittman RN: Recruitment and the role of nuclear localization in polyglutamine-mediated aggregation, *J Cell Biol* 143:1457–70, 1998.

Fujigasaki H, Martin JJ, De Deyn PP, Camuzat A, Deffond D et al: CAG repeat expansion in the TATA box-binding protein gene causes autosomal dominant cerebellar ataxia, *Brain* 124:1939–47, 2001.

Carlezon WA Jr, Duman RS, Nestler EJ: The many faces of CREB, *Trends Neurosci* 28:436–45, 2005.

Steffan JS, Kazantsev A, Spasic-Boskovic O, Greenwald M, Zhu YZ et al: The Huntington's disease protein interacts with p53 and CREB-binding protein and represses transcription, *Proc Natl Acad Sci USA* 97:6763–8, 2000.

Steffan JS, Bodal L, Pallos J, Poelman M, McCampbell A et al: Histone deacetylase inhibitors arrest polyglutamine-dependent neurodegeneration in Drosophila, *Nature* 413:739–43, 2001.

Kazantsev A, Preisinger E, Dranovsky A, Goldgaber D, Housman D: Insoluble detergent-resistant aggregates form between pathological and nonpathological lengths of polyglutamine in mammalian cells, *Proc Natl Acad Sci USA* 96:11404–9, 1999.

Nucifora FC Jr, Sasaki M, Peters MF, Huang H, Cooper JK et al: Interference by huntingtin and atrophin-1 with cbp-mediated transcription leading to cellular toxicity, *Science* 291:2423–8, 2001.

Cattaneo E, Zuccato C, Tartari M: Normal huntingtin function: An alternative approach to Huntington's disease, *Nat Rev Neurosci* 6:919–30, 2005.

13.33. Conventional Treatment for Huntington's Disease

Victor M, Ropper AH: *Principles of Nerology*, 7th ed, McGraw-Hill, New York, 2001, pp 1106–74.

13.34. Molecular Regenerative Engineering for Huntington's Disease

Nucifora FC Jr, Sasaki M, Peters MF, Huang H, Cooper JK et al: Interference by huntingtin and atrophin-1 with cbp-mediated transcription leading to cellular toxicity, *Science* 29:2423–8, 2001.

Steffan JS, Bodai L, Pallos J, Poelman M, McCampbell A et al: Histone deacetylase inhibitors arrest polyglutamine-dependent neurodegeneration in Drosophila, *Nature* 413:739–43, 2001.

McCampbell AA, Taye L, Whitty EP, Steffan JS, Fischbeck KH: Histone deacetylase inhibitors reduce polyglutamine toxicity, *Proc Natl Acad Sci USA* 98:15179–84, 2001.

Zoghbi HY, Orr HT: Glutamine repeats and neurodegeneration, *Annu Rev Neurosci* 23:217–47, 2000.

Zuccato C, Ciammola A, Rigamonti D, Leavitt BR, Goffredo D et al: Loss of huntingtin-mediated BDNF gene transcription in Huntington's disease, *Science* 293:493–8, 2001.

Gauthier LR, Charrin BC, Borrell-Pages M, Dompierre JP, Rangone H et al: Huntingtin controls neurotrophic support and survival of neurons by enhancing BDNF vesicular transport along microtubules, *Cell* 118:127–38, 2004.

Cattaneo E, Zuccato C, Tartari M: Normal huntingtin function: An alternative approach to Huntington's disease, *Nat Rev Neurosci* 6:919–30, 2005.

13.35. Etiology, Pathology, Clinical Manifestation, and Conventional Treatment of Parkinson's Disease

Victor M, Ropper AH: *Principles of Nerology*, 7th ed, McGraw-Hill, New York, 2001, pp 1106–74.

Anderson KE, Weiner WL, Lang AE: *Behavioral Neurology of Movement Disorders*, 2nd ed, Lippincott Williams & Wilkins, Philadelphia, 2005.

Schapira AHV, Warren Olanow C: *Principles of Treatment in Parkinson's Disease*, Butterworth-Heinemann/Elsevier, Philadelphia, 2005.

Silverstein A, Silverstein V, Silverstein Nunn L: *Parkinson's Disease*, Enslow Publishers, Berkeley Heights, NJ, 2002.

13.36. Molecular Regenerative Engineering for Parkinson's Disease

Mamah CE, Lesnick TG, Lincoln SJ, Strain KJ, de Andrade M et al: Interaction of alpha-synuclein and tau genotypes in Parkinson's disease, *Ann Neurol* 57:439–43, 2005.

Martin ER, Scott WK, Nance MA, Watts RL, Hubble JP et al: Association of single-nucleotide polymorphisms of the Tau gene with late onset Parkinson disease, *JAMA* 286:2245–50, 2001.

Burton EA, Glorioso JC, Fink DJ: Gene therapy progress and prospects: Parkinson's disease, *Gene Ther* 10:1721–7, 2003.

Dunnett SB, Björklund A: Prospects for new restorative and neuroprotective treatments in Parkinson's disease, *Nature* 399(Suppl):A32–9, 1999.

Kordower JH: In vivo gene delivery of glial cell line–derived neurotrophic factor for Parkinson's disease, *Ann Neurol* 53(Suppl 3):S120–32, 2003.

Choi-Lundberg DL, Lin Q, Chang YN, Chiang YL, Hay CM et al: Dopaminergic neurons protected from degereration by GDNF gene therapy, *Science* 275:838–41, 1997.

During MJ, Naegele JR, O'Malley KL, Geller AI: Long-term behavioral recovery in parkinsonian rats by an HSV vector expressing tyrosine hydroxylase, *Science* 266:1399–403, 1994.

Horellou P, Vigne E, Castel MN, Barneoud P, Colin P et al: Direct intracerebral gene transfer of an adenoviral vector expressing tyrosine hydroxylase in a rat model of Parkinson's disease, *Neuroreport* 6:49–53, 1994.

Bankiewicz KS, Eberling JL, Kohutnicka M et al: Convection-enhanced delivery of AAV vector in parkinsonian monkeys; in vivo detection of gene expression and restoration of dopaminergic function using pro-drug approach, *Exp Neurol* 164:2–14, 2000.

Lee WY, Chang JW, Nemeth NL, Kang UJ: Vesicular monoamine transporter-2 and aromatic L-amino acid decarboxylase enhance dopamine delivery after L-3, 4-dihydroxyphenylalanine administration in parkinsonian rats, *J Neurosci* 19:3266–74, 1999.

Leff SE, Spratt SK, Snyder RO, Mandel RJ: Long-term restoration of striatal l-aromatic amino acid decarboxylase activity using recombinant adeno-associated viral vector gene transfer in a rodent model of Parkinson's disease, *Neuroscience* 92:185–96, 1999.

Azzouz M, Martin-Rendon E, Barber RD, Mitrophanous KA, Carter EE et al: Multicistronic lentiviral vector-mediated striatal gene transfer of aromatic L-amino acid decarboxylase, tyrosine hydroxylase, and GTP cyclohydrolase I induces sustained transgene expression, dopamine production, and functional improvement in a rat model of Parkinson's disease, *J Neurosci* 22:10302–12, 2002.

Shen Y, Muramatsu SI, Ikeguchi K, Fujimoto KI, Fan DS et al: Triple transduction with adeno-associated virus vectors expressing tyrosine hydroxylase, aromatic-L-amino-acid decarboxylase, and GTP cyclohydrolase I for gene therapy of Parkinson's disease, *Hum Gene Ther* 11:1509–19, 2000.

Muramatsu S, Fujimoto K, Ikeguchi K, Shizuma N, Kawasaki K et al: Behavioral recovery in a primate model of Parkinson's disease by triple transduction of striatal cells with adeno-associated viral vectors expressing dopamine-synthesizing enzymes, *Hum Gene Ther* 13:345–54, 2002.

Sun M, Zhang GR, Kong L, Holmes C, Wang X et al: Correction of a rat model of Parkinson's disease by coexpression of tyrosine hydroxylase and aromatic amino acid decarboxylase from a helper virus-free Herpes simplex virus type 1 vector, *Hum Gene Ther* 14:415–24, 2003.

Asanuma M, Hirata H, Cadet JL: Attenuation of 6-hydroxydopamine-induced dopaminergic nigrostriatal lesions in superoxide dismutase transgenic mice, *Neuroscience* 85:907–17, 1998.

Yamada M, Oligino T, Mata M, Goss JR, Glorioso JC et al: Herpes simplex virus vector-mediated expression of Bcl-2 prevents 6-hydroxydopamine-induced degeneration of neurons in the substantia nigra in vivo, *Proc Natl Acad Sci USA* 96:4078–83, 1999.

Natsume A, Mata M, Goss J, Huang S, Wolfe D et al: Bcl-2 and GDNF delivered by HSV-mediated gene transfer act additively to protect dopaminergic neurons from 6-OHDA-induced degeneration, *Exp Neurol* 169:231–8, 2001.

Xia XG, Harding T, Weller M, Bieneman A, Uney JB et al: Gene transfer of the JNK interacting protein-1 protects dopaminergic neurons in the MPTP model of Parkinson's disease, *Proc Natl Acad Sci USA* 98:10433–8, 2001.

Hayley S, Crocker SJ, Smith PD, Shree T, Jackson-Lewis V et al: Regulation of dopaminergic loss by Fas in a 1-methyl-4-phenyl-1,2,3,6-tetrahydropyridine model of Parkinson's disease, *J Neurosci* 24:2045–53, 2004.

13.37. Cellular Regenerative Engineering for Parkinson's Disease

Dunnett SB, Björklund A: Prospects for new restorative and neuroprotective treatments in Parkinson's disease, *Nature* 399 (Suppl):A32–9, 1999.

Kim JH, Auerbach JM, Rodriguez-Gomez JA, Velasco I, Gavin D et al: Dopamine neurons derived from embryonic stem cells function in an animal model of Parkinson's disease, *Nature* 418:50–6, 2002.

Lee SH, Lumelsky N, Auerbach JM, McKay RD: Efficient generation of midbrain and hindbrain neurons from mouse embryonic stem cells, *Nature Biotechnol* 18:675–9, 2000.

Studer L, Tabar V, McKay RD: Transplantation of expanded mesencephalic precursors leads to recovery in parkinsonian rats, *Nature Neurosci* 1:290–5, 1998.

McKay R: Stem cells in the central nervous system, *Science* 276:66–71, 1997.

Sanchez-Pernaute R, Studer L, Bankiewicz KS, Major EO, McKay RD: In vitro generation and transplantation of precursor-derived human dopamine neurons, *J Neurosci Res* 65:284–8, 2001.

Thomson JA et al: Embryonic stem cell lines derived from human blastocysts, *Science* 282:1145–7, 1998.

Lee SH, Lumelsky N, Auerbach JM, McKay RD: Efficient generation of midbrain and hindbrain neurons from mouse embryonic stem cells, *Nature Biotechnol* 18:675–9, 2000.

Okabe S, Forsberg-Nilsson K, Spiro AC, Segal M, McKay RD: Development of neuronal precursor cells and functional postmitotic neurons from embryonic stem cells in vitro, *Mech Dev* 59:89–102, 1996.

Lumelsky N et al: Differentiation of embryonic stem cells to insulin-secreting structures similar to pancreatic islets, *Science* 292:1389–94, 2001.

Studer L, Tabar V, McKay RD: Transplantation of expanded mesencephalic precursors leads to recovery in parkinsonian rats, *Nature Neurosci* 1:290–5, 1998.

Kim JH, Auerbach JM, Rodriguez-Gomez JA, Velasco I, Gavin D et al: Dopamine neurons derived from embryonic stem cells function in an animal model of Parkinson's disease, *Nature* 418:50–6, 2002.

Reynolds BA, Weiss S: Clonal and population analyses demonstrate that an EGF-responsive mammalian embryonic CNS precursor is a stem cell, *Dev Biol* 175:1–13, 1996.

Lois C, Alvarez-Buylla A: Proliferating subventricular zone cells in the adult mammalian forebrain can differentiate into neurons and glia, *Proc Natl Acad Sci USA* 90:2074–7, 1993.

Morshead CM, Reynolds BA, Craig CG et al: Neural stem cells in the adult mammalian forebrain: A relatively quiescent subpopulation of subependymal cells, *Neuron* 13:1071–82, 1994.

Weiss S, Dunne C, Hewson J et al: Multipotent CNS stem cells are present in the adult mammalian spinal cord and ventricular neuroaxis, *J Neurosci* 16:7599–609, 1996.

Palmer TD, Takahashi J, Gage H: The adult rat hippocampus contains primordial neural stem cells, *Mol Cell Neurosci* 8:389–404, 1997.

Sabate O, Horellou P, Vigne E, Colin P, Perricaudet M et al: Transplantation to the rat brain of human neural progenitors that were genetically modified using adenoviruses, *Nat Genet* 9:256–60, 1995.

Yamada H, Dezawa M, Shimazu S, Baba M, Sawada H et al: Transfer of the von Hippel–Lindau gene to neuronal progenitor cells in treatment for Parkinson's disease, *Ann Neurol* 54:352–9, 2003.

Saucedo-Cardenas O et al: Nurrl is essential for the induction of the dopaminergic phenotype and the survival of ventral mesencephalic late dopaminergic precursor neurons, *Proc Natl Acad Sci USA* 95:4013–8, 1998.

Le W et al: Selective agenesis of mesencephalic dopaminergic neurons in Nurrl-deficient mice, *Exp Neurol* 159:451–8, 1999.

Kanno H, Saljooque F, Yamamoto I et al: Role of the von Hippel–Lindau tumor suppressor protein during neuronal differentiation, *Cancer Res* 60:2820–4, 2000.

Shingo T, Date I, Yoshida H, Ohmoto T: Neuroprotective and restorative effects of intrastriatal grafting of encapsulated GDNF-producing cells in a rat model of Parkinson's disease, *J Neurosci Res* 69:946–54, 2002.

Barkats M, Nakao N, Grasbon-Frodl EM, Bilang-Bleuel A, Revah F et al: Intrastriatal grafts of embryonic mesencephalic rat neurons genetically modified using an adenovirus encoding human Cu/Zn superoxide dismutase, *Neuroscience* 78:703–13, 1997.

Wyman T, Rohrer D, Kirigiti P, Nichols H, Pilcher K et al: Promoter-activated expression of nervous growth factor for treatment of neurodegenerative diseases, *Gene Ther* 6:1648–60, 1999.

Hermens WT, Verhaagen J: Viral vectors, tools for gene transfer in the nervous system, *Prog Neurobiol* 55:399–432, 1998.

Ericson C, Wictorin K, Lundberg C: Ex vivo and in vitro studies of transgene expression in rat astrocytes transduced with lentiviral vectors, *Exp Neurol* 173:22–30, 2002.

Cortez N, Trejo F, Vergara P, Segovia J: Primary astrocytes retrovirally transduced with a tyrosine hydroxylase transgene driven by a glial-specific promoter elicit behavioral recovery in experimental parkinsonism, *J Neurosci Res* 59:39–46, 2000.

Cunningham LA, Su C: Astrocyte delivery of glial cell line-derived neurotrophic factor in a mouse model of Parkinson's disease, *Exp Neurol* 174:230–42, 2002.

Wang ZH, Ji Y, Shan W, Zeng B, Raksadawan N et al: Therapeutic effects of astrocytes expressing both tyrosine hydroxylase and brain-derived neurotrophic factor on a rat model of Parkinson's disease, *Neuroscience* 113:629–40, 2002.

Cao L, Zhao YC, Jiang ZH, Xu DH, Liu ZG et al: Long-term phenotypic correction of rodent hemiparkinsonism by gene therapy using genetically modified myoblasts, *Gene Ther* 7:445–9, 2000.

Schwarz EJ, Alexander GM, Prockop DJ, Azizi SA: Multipotential marrow stromal cells transduced to produce L-DOPA: Engraftment in a rat model of Parkinson disease, *Hum Gene Ther* 10:2539–49, 1999.

13.38. Etiology, Pathogenesis, and Clinical Features of Multiple Sclerosis

Prat A, Antel J: Pathogenesis of multiple sclerosis, *Curr Opin Neurol* 18:225–30, 2005.

Franklin RJ: Why does remyelination fail in multiple sclerosis? *Nat Rev Neurosci* 3:705–14, 2002.

Furlan R, Pluchino S, Martino G: The therapeutic use of gene therapy in inflammatory demyelinating diseases of the central nervous system, *Curr Opin Neurol* 16:385–92, 2003.

Martino G, Hartung HP: Immunopathogenesis of multiple sclerosis: The role of T cells, *Curr Opin Neurol* 12:309–21, 1999.

Kieseier BC, Storch MK, Archelos JJ et al: Effector pathways in immune mediated central nervous system demyelination, *Curr Opin Neurol* 12:323–36, 1999.

Martino G, Adorini L, Rieckmann P et al: Inflammation in multiple sclerosis: The good, the bad, and the complex, *Lancet Neurol* 1:499–509, 2002.

Franklin RJ: Why does remyelination fail in multiple sclerosis? *Nat Rev Neurosci* 3:705–14, 2002.

13.39. Molecular Regenerative Engineering for Multiple Sclerosis

Shaw MK, Lorens JB, Dhawan A, DalCanto R, Tse HY et al: Local delivery of interleukin 4 by retrovirus-transduced T lymphocytes ameliorates experimental autoimmune encephalomyelitis, *J Exp Med* 185:1711–4, 1997.

Mathisen PM, Yu M, Johnson JM et al: Treatment of experimental autoimmune encephalomyelitis with genetically modified memory T cells, *J Exp Med* 186:159–64, 1997.

Chen LZ, Hochwald GM, Huang C et al: Gene therapy in allergic encephalomyelitis using myelin basic protein-specific T cells engineered to express latent transforming growth factor-beta1, *Proc Natl Acad Sci USA* 95:12516–21, 1998.

Piccirillo CA, Prud'homme GJ: Prevention of experimental allergic encephalomyelitis by intramuscular gene transfer with cytokine-encoding plasmid vectors, *Hum Gene Ther* 10:1915–22, 1999.

Garren H, Ruiz PJ, Watkins TA et al: Combination of gene delivery and DNA vaccination to protect from and reverse Th1 autoimmune disease via deviation to the Th2 pathway, *Immunity* 15:15–22, 2001.

Croxford JL, Triantaphyllopoulos K, Podhajcer OL et al: Cytokine gene therapy in experimental allergic encephalomyelitis by injection of plasmid DNA-cationic liposome complex into the central nervous system, *J Immunol* 160:5181–7, 1998.

Croxford JL, Triantaphyllopoulos KA, Neve RM et al: Gene therapy for chronic relapsing experimental allergic encephalomyelitis using cells expressing a novel soluble p75 dimeric TNF receptor, *J Immunol* 164:2776–81, 2000.

Triantaphyllopoulos K, Croxford J, Baker D, Chernajovsky Y: Cloning and expression of murine IFN beta and a TNF antagonist for gene therapy of experimental allergic encephalomyelitis, *Gene Ther* 5:253–63, 1998.

Weiner HL: Immunosuppressive treatment in multiple sclerosis, *J Neurol Sci* 223:1–11, 2004.

Melo ME, Qian J, El-Amine M et al: Gene transfer of Ig-fusion proteins into B cells prevents and treats autoimmune diseases, *J Immunol* 168:4788–95, 2002.

Ruffini F, Furlan R, Poliani PL et al: Fibroblast growth factor-II gene therapy reverts the clinical course and the pathological signs of chronic experimental autoimmune encephalomyelitis in C57BL/6 mice, *Gene Ther* 8:1207–13, 2001.

13.40. Cellular Regenerative Engineering or Multiple Sclerosis

Franklin RJ: Why does remyelination fail in multiple sclerosis? *Nat Rev Neurosci* 3:705–14, 2002.

Croxford JL, Triantaphyllopoulos KA, Neve RM et al: Gene therapy for chronic relapsing experimental allergic encephalomyelitis using cells expressing a novel soluble p75 dimeric TNF receptor, *J Immunol* 164:2776–81, 2000.

Ohashi T, Watabe K, Uehara K et al: Adenovirus-mediated gene transfer and expression of human [beta]-glucuronidase gene in the liver, spleen, and central nervous system in mucopolysaccharidosis type VII mice, *Proc Natl Acad Sci USA* 94:1287–92, 1997.

Furlan R, Pluchino S, Martino G: The therapeutic use of gene therapy in inflammatory demyelinating diseases of the central nervous system, *Curr Opin Neurol* 16:385–92, 2003.

Shaw MK, Lorens JB, Dhawan A et al: Local delivery of interleukin 4 by retrovirus-transduced T lymphocytes ameliorates experimental autoimmune encephalomyelitis, *J Exp Med* 185:1711–4, 1997.

Dal Canto RA, Shaw MK, Nolan GP et al: Local delivery of TNF by retrovirus-transduced T lymphocytes exacerbates experimental autoimmune encephalomyelitis, *Clin Immunol* 90:10–14, 1999.

Mathisen PM, Yu M, Johnson JM et al: Treatment of experimental autoimmune encephalomyelitis with genetically modified memory T cells, *J Exp Med* 186:159–64, 1997.

Chen LZ, Hochwald GM, Huang C et al: Gene therapy in allergic encephalomyelitis using myelin basic protein-specific T cells engineered to express latent transforming growth factor-beta1, *Proc Natl Acad Sci USA* 95:12516–21, 1998.

14

CARDIAC REGENERATIVE ENGINEERING

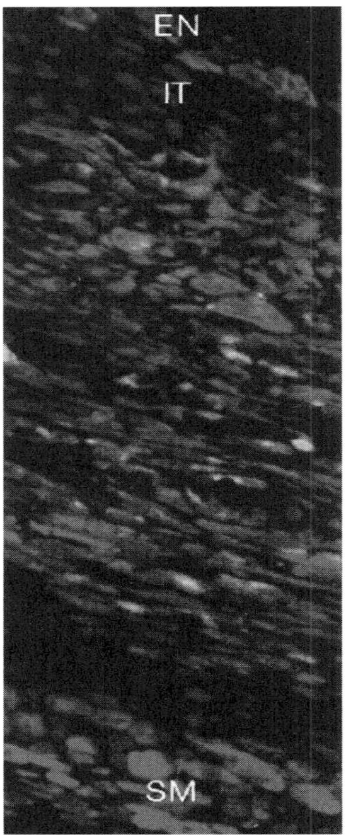

Transformation of bone marrow cells to infarcted heart. Myocardial infarction was induced and EGFP-Lin-c-*kit*^{POS} bone marrow cells were injected into the infarcted heart. Infarcted tissue (IT) can be seen in the subendocardium; spared myocytes (SM) can be seen in the subepicardium. Original magnification, ×250. EN: endocardium. Red: cardiac myosin. Green: EGFP-positive cells. (Reprinted by permission from Macmillan Publishers Ltd.: Orlic D et al: Bone marrow cells regenerate infarcted myocardium, *Nature* 410:701–5, copyright 2001.) See color insert.

Bioregenerative Engineering: Principles and Applications, by Shu Q. Liu

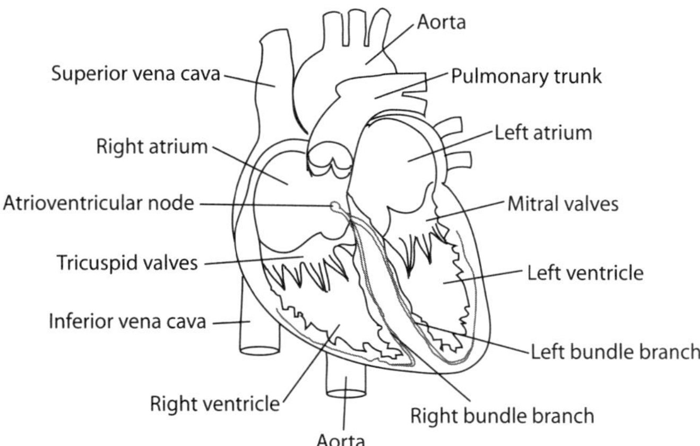

Figure 14.1. Schematic representation of cardiac structure. Based on bibliography 14.1.

ANATOMY AND PHYSIOLOGY OF THE HEART [14.1]

Cardiac Structure

The heart is a muscular organ that provides driving forces for blood circulation. The heart is located within the thoracic cavity, together with the left and right lungs, and is composed of the left atrium and ventricle, the right atrium and ventricle, and the pericardium (Fig. 14.1). The left atrium is a muscular chamber that receives oxygenated blood from the lung via the pulmonary veins and conducts blood to the left ventricle. The left ventricle pumps oxygenated blood to the aorta, the largest artery in the body. Oxygenated blood is delivered to the peripheral tissues and organs via various generations of arteries and capillaries. After oxygen is released and used by peripheral cells, deoxygenated blood is converged into various generations of veins and returned to the right atrium. The right atrium conducts blood to the right ventricle, which pumps deoxygenated blood to the lung via the pulmonary arteries for oxygenation. The pericardium is a double-layered thin sac that encloses the heart. The external layer of the pericardium is composed of tough connective tissue, known as the *fibrous pericardium*, whereas the internal layer consists of epithelial cells, known as the *serous pericardium*.

The structure and geometry differ considerably among the four atrial and ventricular chambers. In a human adult, the left ventricle consists of a chamber about 125 mL in volume and a muscular wall about 1 cm in thickness, which is the highest wall thickness in the heart because the left ventricle pumps blood against the highest blood pressure in the vascular system (80–120 mm Hg under physiological conditions). The right ventricle is about the same size in volume as the left ventricle and its wall is about half of the thickness of the left ventricular wall. The thinner wall of the right ventricle is attributed to the lower work load (pulmonary arterial blood pressure ~20–30 mm Hg) compared to the left ventricle. The left and right atria are smaller chambers compared with the ventricles, and their walls are about half the thickness of the right ventricular wall. The atria encounter lower blood pressure (about 0–5 mm Hg) compared to the ventricles.

The cardiac chambers and large blood vessels, including the aorta and pulmonary artery, are separated by four sets of fibrous valves: the mitral, tricuspid, aortic, and pulmonary valves. The mitral valves divide the left atrium and the left ventricle. They are open during the diastole, allowing blood entering the left ventricle. They are closed during the systole, preventing blood from regurgitation into the left atrium. The tricuspid valves separate the right atrium from the right ventricle, and have function similar to that of the mitral valves. Both mitral and tricuspid valves are referred to as *atrioventricular valves*. The aortic valves separate the left ventricle from the ascending aorta. These valves are open during systole, allowing blood to enter the aorta. They are closed during the diastole, preventing blood from flowing back to the left ventricle. The pulmonary valves separate the right ventricle from the pulmonary arterial trunk, and have function similar to that of the aortic valves. The aortic and pulmonary valves are referred to as *arterial valves*. The driving forces for the opening and closure of the cardiac valves are blood pressure gradients across the valves.

The heart is composed of a blood circulatory system, known as the *coronary circulation*. This system consists of coronary arteries, capillaries, and veins. Two major coronary arteries, the left and right coronary arteries, stem from the ascending aorta and supply blood to the left and right heart, respectively. The left coronary artery is divided into two major branches: the anterior descending artery and the left circumflex artery. The anterior descending artery supplies blood to the anterior wall of the left heart. The left circumflex artery supplies blood to the lateral and posterior wall of the left heart. The right coronary artery is divided into two major branches: the posterior descending artery and the right circumflex artery, which supply blood to the posterior and lateral wall of the right heart, respectively.

The heart consists of a rich capillary network to accommodate its high oxygen consumption activities. Capillaries are distributed within the cardiac muscular bundles. Deoxygenated blood is drained to the coronary venous system. Blood from the left heart is drained to a coronary vein known as the *great cardiac vein*, whereas blood from the right heart is drained to the *small cardiac vein*. These veins converge to the largest coronary vein, known as the coronary sinus, which empties blood into the right atrium.

Cardiac Cells

The heart is composed of a number of cell types, including cardiomyocytes, conducting cells, fibroblasts, epithelial cells, and vascular cells. Cardiomyocytes are elongated, multinucleated, striated, and involuntary contractile muscle cells. These cells consist of actin–myosin filament-containing sarcomeres, the fundamental units of cardiac contraction. Actin and myosin filaments are contractile structures, and their interactions induce sliding motions between the two types of filament, resulting in the contraction of the cardiomyocytes. The ordered arrangement of actin and myosin filaments gives the cardiomyocytes a striated pattern. Other subcellular structures of the cardiomyocytes are similar to those found in other cell types. Cardiomyocytes are organized into muscular bundles, which are further arranged with optimal alignments for the generation of contractile forces.

Cardiac conducting cells are specially differentiated cardiac muscle cells and are found in three structures: the sinoatrial (SA) node, the atrioventricular (AV) node, and the conducting bundles. The sinoatrial node is located at the right atrium near the root of the superior vena cava. The atrioventricular node is located at the right atrium near the intercept of the atrial and ventricular septa. The conducting bundle, known as the atrioventricular bundle, relays action potentials from the atrium to the ventricle. This bundle is divided into the left and right bundle branches. The left bundle branch is further divided into various generations of branches, which innervate the left ventricle. Similarly, the branches of the right bundle branch innervate the right ventricle. The terminal branches of the bundle branches are called *Purkinje fibers*, which conduct action potentials to the ventricular muscle cells. The conducting cells in the sinoatrial node, atrioventricular node, and conducting branches undergo automatic depolarization and initiate periodic action potentials, which induce cyclic cardiac contraction and relaxation. The conducting cells in these structures emit cyclic action potentials at different frequencies with the highest frequency found in the sinoatrial node (70–80 per minute). Thus, the action potentials from the sinoatrial node override those from other conducting structures and control the frequency of cardiac muscle contraction and heartbeat.

The heart also consists of fibroblasts. These cells are found in the connective tissue, which resides in extramuscular space, and are capable of producing and secreting extracellular matrix components, including collagen, elastin, and proteoglycans. Other cell types found in the heart include epithelial cells and vascular cells. Epithelial cells are the cells that line the internal and external surfaces of the atrial and ventricular wall and the surface of the valves. Vascular cells include endothelial cells and smooth muscle cells, which are found in the coronary blood vessels. Endothelial cells line the internal surface of blood vessels, including arteries, capillaries, and veins. Smooth muscle cells are found in the media of the arteries and veins. The structure and function of vascular cells are discussed further on page 660.

Cardiac Performance and Cycle

The heart conducts cyclic contractile activities and pumps blood into the arterial system. The cyclic activities include two mechanical events: contraction and relaxation (Fig. 14.2). Contraction occurs in a phase called *systole*, which lasts about 0.3–0.4 s in healthy humans, and relaxation occurs in another phase called *diastole*, which lasts about 0.4–0.5 s. Systole is defined as the period from the beginning to the end of ventricular contraction, and the diastole is from the beginning to the end of ventricular relaxation. Systole and diastole constitute a cardiac cycle. A cardiac cycle is defined from the beginning of ventricular contraction to the ending of ventricular relaxation or the beginning of next ventricular contraction. Since action potential waves recorded by electrocardiography are consistent with cardiac mechanical events, these waves are usually used to identify cardiac cycles. The QRS wave complex, which reflects the spreading of action potentials within the ventricles and initiates ventricular contraction, is used to indicate the beginning of the systole and the ending of the diastole, although the beginning of the QRS wave complex is slightly earlier than the beginning of ventricular contraction.

A sequence of mechanical events occurs during the systole. When the ventricular muscle cells are excited by action potentials from the sinoatrial node, the ventricular muscle cells start contraction almost at the same time (owning to the fast-conducting activity of the Purkinje fibers). The contraction of the left ventricle rapidly increases ventricular blood pressure from about 5 to 120 mm Hg. The increased ventricular blood pressure rapidly closes the mitral valves due to the pressure gradient across the valves. There is a short period when the mitral valves are closed, but the aortic valves are not yet open. This period is known as the period of isometric contraction, since blood is not flowing from or to the ventricle. A continuous increase in the ventricular blood pressure results

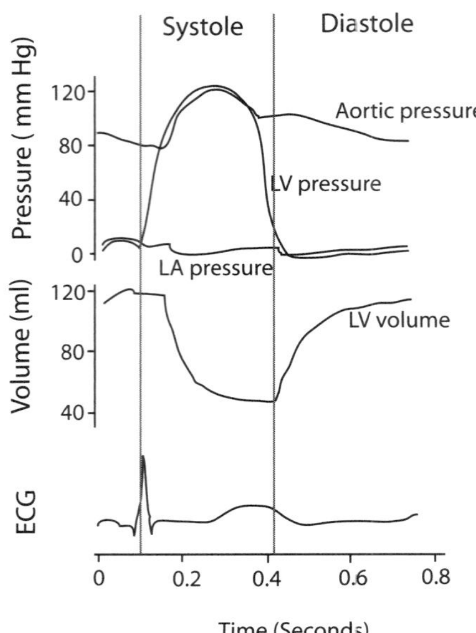

Figure 14.2. Electrical and mechanical events during a cardiac cycle. Based on bibliography 14.1.

in a pressure gradient across the aortic valves, leading to the opening of these valves and blood ejection from the left ventricle to the aorta. A similar sequence of mechanical events occurs in the right ventricle, except that blood pressure in the right ventricle is lower than that in the left ventricle. At the same time, the left and right atria are in the relaxing mode. The two atria collect blood from the pulmonary veins and vena cava, respectively. Approaching the end of ventricular contraction, the ventricles reduce contractile strength in association with a significant decrease in the ventricular blood pressure and volume. At the point when the ventricular blood pressure reaches a level just below the aortic blood pressure (for the left ventricle) and the pulmonary arterial blood pressure (for the right ventricle), the aortic and pulmonary valves are closed owning to the reversed pressure gradients across these valves. The closure of the aortic and pulmonary arterial valves indicates the ending of the systole and the beginning of the diastole.

During early diastole following the closure of the arterial valves, there is a period when the ventricular blood pressure is higher than that in the atria and the atrioventricular valves (mitral and tricuspid valves) are still closed. This period is called the *period of isometric relaxation*, during which blood is not flowing from or to the ventricles. Continuous ventricular relaxation results in a rapid decrease in the ventricular blood pressure. Soon, when the ventricular blood pressure drops below the atrial blood pressure, the atrioventricular valves are open, owing to the reversed pressure gradients across these valves, and blood flows from the left and right atria to the left and right ventricles, respectively. Two factors facilitate ventricular blood filling: continuous expansion of the ventricles and atrial contraction, which occurs in the late stage of diastole. The ventricles are completely filled before the next QRS-exciting wave complex arrives at the ventricle and initiates another cardiac cycle.

Regulation of Cardiac Performance

Cardiac activities undergo constant changes in response to various demands of bloodflow and oxygen consumption in the peripheral tissues and organs. A typical example is the increase in cardiac contraction and heartbeat in response to increased physical activities such as running and other forms of exercise. The change in cardiac activity is regulated by two mechanisms: intrinsic and extrinsic. The *intrinsic* mechanism is controlled by the natural properties of the cardiomyocytes, such as the length of the sarcomere or the fraction of the actin filaments that overlap with the myosin filaments. The contractile strength is proportional to the sarcomere length within a limited range. The muscle generates maximal forces at an optimal sarcomere length. With a further stretch or lengthening of the sarcomere, the contractile strength reduces. Such properties are described by *Starling's law*, which is represented by a sarcomere length–tensile force curve. Under physiological conditions, the heart works in the ascending range of the sarcomere length–tensile force curve; that is, an increase in sarcomere length results in enhanced cardiac contraction. In a beating heart, the sarcomere length of the ventricular muscles is controlled by the diastolic blood volume of the ventricle. An increase in the diastolic blood volume results in an increase in the sarcomere length within a certain range.

The *extrinsic* mechanism is controlled by neural and hormonal factors. The heart is innervated with parasympathetic and sympathetic nerves. Activation of the parasympathetic nervous system suppresses the cardiac activity or contractility, whereas activation of the sympathetic nervous system exerts an opposite effect. The two nerve systems coordinately control the performance of the heart. In addition, the performance of the heart is regulated by hormones, including epinephrine and norepinephrine. Epinephrine and norepinephrine are produced by the adrenal medulla and released into the blood. The secretion of these hormones are controlled by the sympathetic nervous system and elevated in response to increased physical activities and emotional changes. The function of these hormones, especially epinephrine, is to stimulate cardiac contraction. These hormones can bind to and activate the β-adrenergic receptors in the membrane of the cardiomyocytes, inducing the activation of G-protein signaling pathways (see page 217). As a result, calcium is released into the cytoplasm. As calcium is required for the initiation of actin–myosin interactions, an increase in calcium concentration enhances actin–myosin interactions and the muscular contractile strength.

CARDIAC DISORDERS

Heart Failure and Cardiomyopathy

Pathogenesis, Pathology, and Clinical Features of Heart Failure [14.2]. Heart failure is a pathophysiological state in which the heart is unable to pump sufficient blood to the arterial system, resulting in the dysfunction of tissues and organs due to the lack of blood supply. Heart failure is an end event of cardiac and pulmonary disorders, which are among the most popular diseases. Each year about 960,000 patients are admitted to hospital because of heart failure in the United States. About 20% patients are associated with life-threatening heart failure. Heart failure can be induced by a variety of disorders, including cardiac infarction, congenital heart disease, cardiomyopathy, myocarditis, anemia, hemorrhage, cardiac trauma, cardiac valve rupture, cardiac valvular diseases, arrhythmia, systemic hypertension, and excessive physical exercise. Although these disorders involve

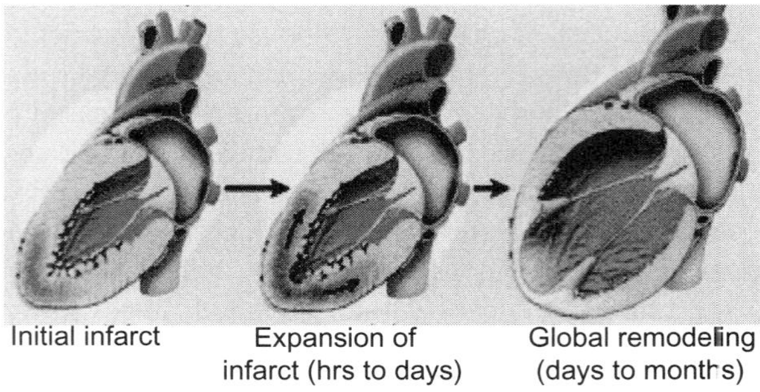

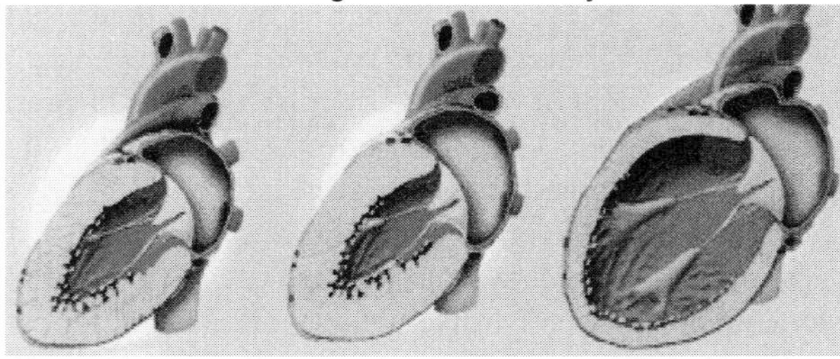

Figure 14.3. Ventricular remodeling after infarction (panel A) and in diastolic and systolic heart failure (panel B). At the time of an acute myocardial infarction, there is no clinically significant change in overall ventricular geometry (panel A). Within hours to days, the area of myocardium affected by the infarction begins to expand and become thinner. Within days to months, global remodeling can occur, resulting in overall ventricular dilatation, decreased systolic function, mitral valve dysfunction, and the formation of an aneurysm. The classic ventricular remodeling that occurs with hypertensive heart disease (middle of panel B) results in a normal-sized left ventricular cavity with thickened ventricular walls (concentric left ventricular hypertrophy) and preserved systolic function. There may be some thickening of the mitral valve apparatus. In contrast, the classic remodeling that occurs with dilated cardiomyopathy (right side of panel B) results in a globular shape of the heart, a thinning of the left ventricular walls, an overall decrease in systolic function, and distortion of the mitral valve apparatus, leading to mitral regurgitation. (Reprinted with permission from Jessup M, Brozena S: *New Engl J Med* 348:2007–18, copyright 2003 Massachusetts Medical Society. All rights reserved.)

different systems and are caused by different pathological factors, the consequence of these disorders is the impairment of cardiac contractility, eventually leading to cardiac failure (Fig. 14.3).

At the molecular and cellular levels, cardiac disorders may induce apoptosis of cardiomyocytes and/or reduction in cardiac contractility due to the impairment of calcium transport and handling. Cell apoptosis results in reduction in the density of cardiomyocytes, which impairs cardiac performance. Cardiac ischemia and infarction are typical cardiac disorders that induce cell apoptosis. The impairment of calcium transport is a primary cause for nonapoptotic heart failure. Calcium plays a critical role in the regulation of muscular contraction. Calcium is primarily stored in a subcellular structure known as sarcoplasmic reticulum (SR). In patients with heart failure and experimental heart failure models, the load of calcium in the sarcoplasmic reticulum is often reduced. Thus, calcium released to the cytoplasm, in response to an action potential, is insufficient to induce optimal actin–myosin interaction, resulting in impaired cardiac performance. The reduction in the load of sarcoplasmic reticulum calcium is related to malfunction of several molecules, including sarcoplasmic reticulum calcium ATPase (SERCA2a), phospholamban (PLB), and/or the Na^+–Ca^{2+}-exchanger (NCX), which participate in the regulation of calcium handling and transport.

In response to the stimulation of an action potential, cardiomyocytes undergo depolarization, which stimulates the cell membrane voltage-dependent calcium channels to induce calcium influx. The increase in the cytoplasmic calcium concentration activates the ryanodine receptor and induces the transport of additional calcium from the sarcoplasmic reticulum to the cytoplasm. Released calcium binds to troponin C, which in turn initiates actin–myosin interaction and muscle cell contraction. Immediately after the contraction process, calcium must be removed from the cytoplasm, allowing the relaxation of the muscular cells. The transport of calcium from the cytoplasm to the sarcoplasmic reticulum is controlled primarily by the sarcoplasmic reticulum Ca^{2+} ATPase, while calcium transport from the cytoplasm to the extracellular space is controlled by the Na^+–Ca^{2+} exchanger. A reduction in the activity of sarcoplasmic reticulum calcium ATPase is thought to contribute to the reduction in the sarcoplasmic reticulum calcium load and the impairment of cardiac contractility, leading to the development of heart failure.

Cardiac contractility and performance are also regulated by the β-adrenergic receptor signaling system. These receptors interact with catecholamines, including epinephrine and norepinephrine, and stimulate the contraction of cardiac muscle cells. The binding of catecholamine ligands to the β-adrenergic receptor activates the G_s proteins, which induce the activation of a cascade of signaling molecules, including adenylyl cyclase, cAMP, and protein kinase A (PKA). Activated protein kinase A can phosphorylate downstream proteins that regulate the contractile activity of the cardiac muscle cells. One of the substrate molecules for protein kinase A is the cardiac ryanodine receptor, which controls the gating the calcium channels in the sarcoplasmic reticulum. The phosphorylation of the ryanodine receptor promotes diastolic calcium release from the sarcoplasmic reticulum, initiating cardiomyocyte contraction.

Furthermore, activated protein kinase A by the β-adrenergic receptor can phosphorylate a molecule known as *phospholamban*. In its dephosphorylated form, phospholamban inhibits the activity of sarcoplasmic reticulum Ca^{2+} ATPase. The PKA-induced phosphorylation of phospholamban results in the deinhibition of the sarcoplasmic reticulum Ca^{2+} ATPase activity. Thus, protein kinase A-induced phosphorylation of phospholamban removes the inhibitory effect of phospholamban on sarcoplasmic reticulum Ca^{2+} ATPase and enhances the cardiac contractility. The accumulation of dephosphorylated phosholamban or a decrease in the phosphorylation of the molecule reduces the cardiac contractility and promotes the development of heart failure. The knockout of phospholamban in mice with transgenic heart failure can reduce ventricular dilation and cardiac fibrosis, and improves the ventricular performance (Fig. 14.4). Thus, phospholamban is a potential target molecule for the treatment of cardiac failure.

There are various types of heart failure. In terms of the rate of progression, heart failure can be classified into acute and chronic heart failure. Acute heart failure occurs suddenly, often due to large cardiac infarction, cardiac trauma, sudden cardiac valve rupture, or severe hemorrhage, resulting in a rapid reduction in cardiac output and arterial blood

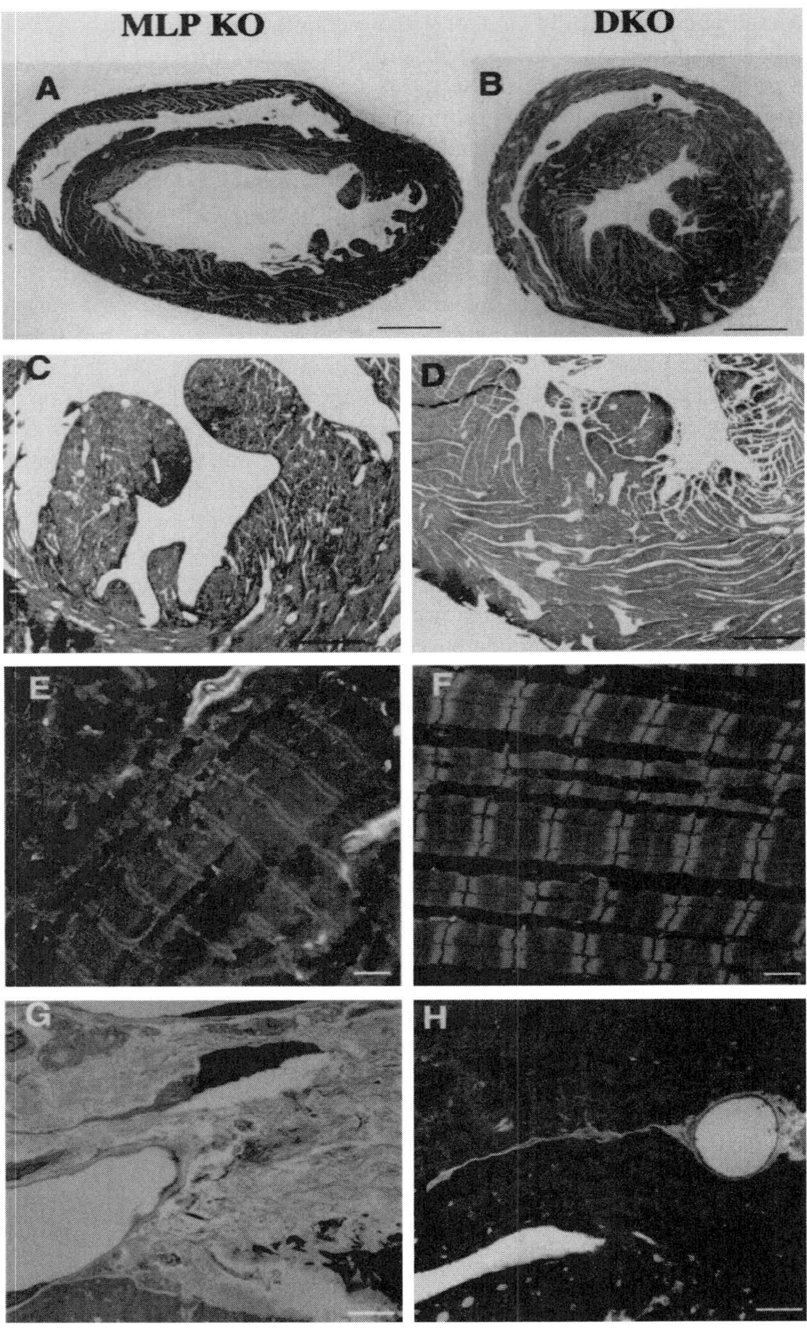

Figure 14.4. Rescue of ultrastructural defects and fibrosis in DKO hearts. Histological micrographs of transverse sections with hematoxylin-eosin (A, B) and trichrome (C, D) stain from cardiac specimens of muscle-specific LIM protein (MLP) knockout mice (MLPKO) and MLP/phospholamban (PLB) knockout mice (DKO). Marked chamber dilation (A) and massive fibrosis [red color, (C)] in MLPKO hearts were rescued in DKO hearts (B, D). Electron microscopic analysis documented that myofibrillar disarray, pronounced increases in nonmyofibrillar space (E), and massive fibrosis (G) in MLPKO hearts were rescued in DKO hearts (F and H). Bars: 1 mm (A, B), 0.5 mm (C, D), 1.0 µm (E, F), and 2.5 µm (G, H). (Reprinted from Minamisawa S et al: *Cell* 99:313–22, copyright 1999, with permission from Elsevier.)

pressure. Acute heart failure is followed with the failure of all peripheral organs within a short period. Changes in the function of peripheral organs depend on the degree of heart failure. A complete loss of arterial blood pressure due to acute heart failure may result in cell death within minutes in vital organs such as the brain, heart, and kidney. Chronic heart failure progresses slowly over months, years, or decades, and are often a result of hypertension, cardiomyopathy, and valvular diseases, such as aortic valve stenosis and mitral valve regurgitation. Hypertension and valvular disorders induce a gradual increase in the cardiac workload. The heart may initially adapt to the workload by increasing its muscle size, a process known as *hypertrophy*. Within a certain period (months, years, or even decades), the heart is still capable of providing sufficient bloodflow to the peripheral systems. However, when the heart reaches the limitation of its adaptive capacity, additional workload may lead to heart failure. Cardiomyopathy is a pathological disorder that gradually deteriorates the structure and function of the cardiac muscle cells, resulting in cardiac malfunction and eventually cardiac failure.

According to the structure of the heart, heart failure can be classified into left and right heart failure. Left heart failure is defined as the failure of the left ventricle, whereas right heart failure is the failure of the right ventricle. Left heart failure can be induced by left coronary arteriosclerosis and left ventricular infarction, systemic hypertension, mitral and aortic valvular diseases, and cardiomyopathy. Right heart failure is often a result of pulmonary hypertension, hypoxia, and tricuspid and pulmonary valvular diseases.

Pathological changes in heart failure often include enlargement of the heart, left and right ventricular dilation for left and right heart failure, respectively, and ventricular wall hypertrophy. Usually, before the onset of heart failure, the heart intends to compensate for the loss of its function by increasing the mass of cardiac muscle via cellular hypertrophy. Such a compensatory remodeling process results in cardiac hypertrophy. Other pathological changes may be dependent on the disease that causes heart failure.

Patients with heart failure often express a number of clinical symptoms and signs. These include dyspnea, orthopnea, paroxysmal dyspnea, cyclic respiration, fatigue, headache, anxiety, insomnia, memory impairment, an increase in heartbeat, and a reduction in blood pressure. All these symptoms and signs are due to the lack of blood supply in various organs such as the lung (dyspnea, orthopnea, paroxysmal dyspnea, and cyclic respiration) and brain (headache, anxiety, insomnia, memory impairment). The increase in heartbeat is a sign of cardiac adaptation. A decrease in blood pressure or blood supply in the peripheral tissue stimulates the sympathetic nerve system, while it suppresses the parasympathetic system, resulting in an increase in the cardiac activity.

Experimental Models of Heart Failure [14.3]. Experimental heart failure can be established by using three methods: cardiac infarction, aortic constriction, and gene knockout. Cardiac infarction can be induced by coronary arterial ligation. In an open-chest surgery, a selected coronary artery can be ligated with a suture, inducing acute myocardial infarction distal to the ligated artery. The severity of heart failure is dependent on the size or generation of the ligated coronary artery. Usually, the ligation of a major branch of the left coronary artery can induce a severe left ventricular failure. Left heart failure can also be induced by increasing the afterload (or arterial blood pressure) of the left ventricle. Such a condition can be achieved by partial constriction of the ascending aorta. The degree of change in aortic blood pressure proximal to the constriction is dependent on the level of constriction. Following the constriction surgery, left ventricular hypertrophy can be developed within days. Severe hypertension and ventricular hypertrophy can lead to heart failure. Another method for inducing heart failure is to knock out genes involved in the regulation of cardiac performance. A typical gene is the sarcoplasmic reticulum Ca^{2+} ATPase gene. As discussed above, sarcoplasmic reticulum Ca^{2+} ATPase plays a critical role in calcium handling and cardiac muscle contraction. A decrease in the expression of the sarcoplasmic reticulum Ca^{2+} ATPase gene induces heart failure. For these models, the degree of heart failure can be assessed by measuring the movement of the ventricular wall and the ejection fraction of the left ventricle by magnetic resonance imaging or echocardiography.

Pathogenesis, Pathology, and Clinical Features of Cardiomyopathy [14.4]. *Cardiomyopathy* is a disorder that originates from the intrinsic properties of the cardiomyocytes, but not as a result of other diseases, such as coronary arteriosclerosis, congenital heart disease, valvular disease, or hypertension. This disorder is characterized by ventricular dilation and/or hypertrophy and a progressive reduction in cardiac contractility and performance, eventually resulting in heart failure. Cardiomyopathy can be divided into two different types on the basis of etiology: primary and secondary cardiomyopathy. Primary cardiomyopathy is a myocardial disorder of unknown cause. Major forms of primary cardiomyopathy include the idiopathic and hereditary cardiomyopathy. Secondary cardiomyopathy is a myocardial disorder of known cause, often induced by diseases involving other organs. Typical examples include cardiomyopathy resulting from viral, bacterial, and fungal myocarditis, metabolic disorders, electrolyte and nutritional deficiency, systemic lupus erythematosus, rheumatoid arthritis, amyloidosis, muscular dystrophy, and toxic reactions resulting from alcohol and drugs. In either primary or secondary cardiomyopathy, molecular changes described in the section on heart failure may also occur.

Cardiomyopathy is associated with a number of characteristic pathological changes. These include enlargement of the heart and dilation of the ventricles, hypertrophy of the ventricular wall, interstitial fibrosis, leukocyte infiltration, and occasionally myocardial cell death. In dilated ventricles, thrombi may be found on the wall near the apex. Ventricular dilation is induced by a decrease in the cardiomyocyte contractility, which renders the ventricle unable to pump sufficient blood into the arterial system. Excessive residual blood contributes to the dilation of the ventricles. Ventricular dilation induces extension of the sarcomeres, which enhances the cardiac performance within a certain range (see page 589 of this chapter). Cell death may be attributed to the effects of toxic, metabolic wastes and infectious factors. Interstitial fibrosis develops as a result of enhanced fibroblast proliferation and excess extracellular matrix generation in response to cell death. Ventricular hypertrophy is often a result of adaptation to reduction in cardiac contractility and performance. Cardiomyocytes increase their size to compensate the loss of cardiac performance. Leukocyte infiltration is induced by infectious factors, if any, or cell death.

The clinical manifestations of cardiomyopathy are dependent on the level of changes in cardiac function, regardless the causes of the disorder. In the early stage, the patient may not experience any noticeable symptoms. With progressive deterioration of the cardiac function, patients may show heart failure symptoms and signs, such as dyspnea, fatigue, orthopnea, lower-limb edema, and palpitation. Patients may also feel chest pain. Cardiomyopathy eventually develops to heart failure. Strategies for the treatment of cardiomyopathy include the removal of primary causes, protection of cardiomyocytes from injury, and enhancement of myocardial contractility. In the end stage, the treatment of cardiomyopathy is similar to that of heart failure.

Experimental Model of Cardiomyopathy [14.4]. Experimental cardiomyopathy can be established by modulating or knocking out genes encoding proteins that regulate the integrity and contractility of cardiomyocytes. Candidate genes for such a purpose include the muscle-specific lim protein (MLP), dystrophin, sarcoglycans, desmin, myosin heavy-chain, myosin-binding protein C genes. These genes are involved in the regulation of cardiac contractility. The modulation or knockout of these genes exhibits the phenotype of cardiomyopathy. These models are usually established in the mouse. Transgenic mice with a desired disorder may be acquired from investigators who developed the model or from commercial carriers such as the Jackson Laboratory.

Conventional Treatment of Cardiac Failure and Cardiomyopathy [14.5]. Heart failure is usually treated with several approaches, including the removal of primary causes, reduction of cardiac workload, enhancement of cardiac contractility, and control of salt and water intake. The first approach (removal of primary causes) is dependent on the original disease. Each disease should be treated with distinct methods. The reduction of the cardiac workload is often a priority for the treatment of heart failure. There are several methods that can be used for such a purpose. These include reduction in blood volume and arterial blood pressure as well as reduction in physical activities. Blood volume can be reduced by administration of diuretics, and arterial blood pressure can be reduced by administration of vasodilators. In addition, salt and water intake should be carefully controlled to reduce circulating blood volume and the workload of the heart. The enhancement of cardiac contractility can be achieved by the administration of digitalis glycosides. These compounds can act on cardiac muscle cells, increase muscular contractility, and reduce the conduction of action potentials, thus slowing down the heartbeat. A commonly used compound is digoxin. Chemical compounds that augment the sympathetic nerve function by acting on the β-adrenergic receptors, such as epinephrine and dopamine, can be used to improve the cardiac contractility.

Molecular Therapy for Cardiac Failure and Cardiomyopathy [14.6]. Although numbers of conventional approaches have been established for the treatment of heart failure, these approaches are primarily used for the relief of cardiac symptoms, but not for long-term improvement of the cardiac function. Molecular and cellular studies have demonstrated that cardiac contractile dysfunction in heart failure may be related to intrinsic defects in molecular signaling mechanisms that control the cardiac contractility and/or cell apoptosis. Thus, therapeutic modulation of the contractility-controlling molecular mechanisms and prevention of cell apoptosis may provide new approaches for the treatment of heart failure. Available experimental evidence indicates that cardiac contractility is controlled by calcium handling and transport, which are regulated by several molecules localized to the cell membrane and sarcoplasmic reticulum, including sarcoplasmic reticulum Ca^{2+} ATPase, phospholamban, and Na^+–Ca^{2+} exchanger. The engineering manipulation of the genes that encode these proteins may exert therapeutic effects on heart failure.

In cardiac failure, the expression and activity of sarcoplasmic reticulum Ca^{2+} ATPase are often reduced. In animal models of heart failure, the overexpression of the sarcoplasmic reticulum Ca^{2+} ATPase gene by gene transfer enhances the contractility of cardiomyocytes and improves the cardiac performance of failing hearts. As discussed above, phospholamban is a molecule that regulates the activity of sarcoplasmic reticulum Ca^{2+} ATPase. In the dephosphorylated form, phospholamban inhibits the activity of sarcoplasmic reticulum Ca^{2+} ATPase, suppressing the sarcoplasmic reticulum calcium pump function. The phosphorylation of phospholamban, induced by the β-adrenergic receptor-activated protein kinase A, results in the deinhibition of the sarcoplasmic reticulum Ca^{2+} ATPase activity. In mouse models of heart failure, the knockout of the phospholamban gene enhances the cardiac contractility and improves cardiac performance, suggesting that dephosphorylated phospholamban may be dominant in failing hearts. In vitro investigations have shown that the application of antisense mRNA for phospholamban to rat and human cardiomyocytes reduces the translation of the phospholamban protein and improves the contractile performance of these cells. The transfer of a dominant-negative mutant phospholamban gene to hamsters with experimental cardiomyopathy enhances the performance of failing hearts. These observations suggest that the suppression of phospholamban expression may be a potential approach for the treatment of heart failure. Since the activation of the β-adrenergic receptor–protein kinase A signaling pathway induces the phosphorylation of phospholamban, which removes the inhibitory effect of phospholamban on sarcoplasmic reticulum Ca^{2+} ATPase, enhancement of the β-adrenergic receptor may serve as an alternative approach (see below). The Na^+–Ca^{2+} exchanger regulates the transport of calcium from the cytoplasm to the extracellular space. The overexpression of the Na^+–Ca^{2+} exchanger gene reduces the contractility of cardiac myocytes, suggesting that the suppression of the Na^+–Ca^{2+} exchanger may exert therapeutic effects on a failing heart.

Myocardial β-adrenergic receptors play an important role in the regulation of cardiac contractility. Myocytes contain two dominant types of adrenergic receptor: β_1-adrenergic receptor (accounting for 75–80%) and β_2-adrenergic receptor (20–25%). The two types of receptor possess different functions. The activation of the β_1-adrenergic receptors may induce cell apoptosis, whereas that of the β_2-adrenergic receptors exerts a protective effect against cell apoptosis, leading to enhanced cardiac function. The different roles of the two types of receptor have been demonstrated in transgenic mice with targeted overexpression of β_1- and β_2-adrenergic receptors. In these mice, the overexpression of the β_2-adrenergic receptor enhances the contractile performance of the cardiac muscle and prevents heart failure (Figs. 14.5 and 14.6), whereas the overexpression of the β_1-adrenergic receptor facilitates heart failure. It has been suggested that the selective involvement of G_i proteins may be responsible for the different functions of the two receptors. In vivo transfer of the β_2-adrenergic receptor to the animal heart has demonstrated an increase in the contractile activity of cardiac muscle cells in response to catecholamine and enhances the left ventricular function. Based on these investigations, it seems that the β_2-adrenergic receptor gene may be used for the treatment of heart failure.

Tissue Regenerative Engineering for Cardiac Failure and Cardiomyopathy [14.7]. Based on the pathogenic mechanisms and clinical manifestations of cardiac failure, tissue regenerative engineering approaches have been developed for treating cardiac failure. The principle of cardiac tissue regenerative engineering is to enhance the cardiac performance. One method for such a purpose is to construct cardiomyocyte-containing tissue scaffolds, implant the constructed cardiac scaffold around the heart, and induce synchronized cardiomyocyte beating between the heart and the cardiac scaffold, thus enhancing the performance of the heart. Alternatively, the constructed cardiomyocyte-containing scaffold can be implanted around the abdominal aorta. The cyclic contractile activity of the cardiac scaffold can enhance the circulation by dynamically altering the diameter of the aorta.

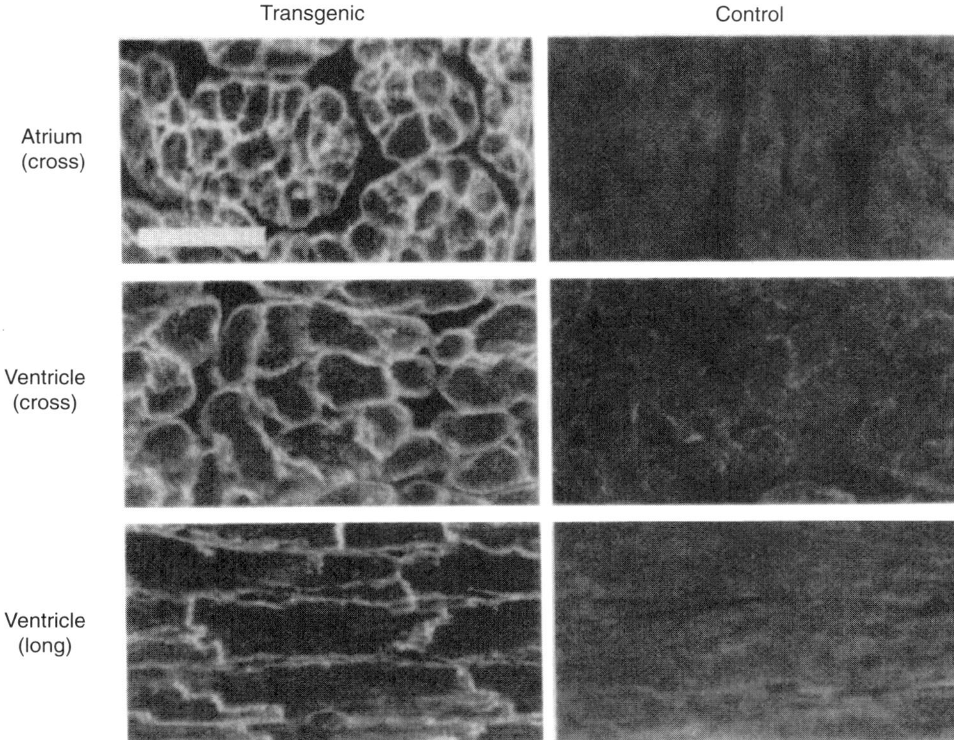

Figure 14.5. In situ demonstration of β2-adrenergic receptor transgene expression. Immunohistochemical labeling of the human β2-adrenergic receptor in TG4 (with transgenic overexpression of the β2-adrenergic receptor) and control myocardial frozen sections was done with rabbit antiserum to the COOH terminus of the human β2-adrenergic receptor. Atrium (cross), cross-sectioned atria; ventricle (cross), cross-sectioned ventricle; ventricle (long), longitudinal sectioned ventricle. Scale bar: 50 μm. (Reprinted with permission from Milano CA et al: *Science* 264:582–6, copyright 1994 AAAS.)

In an experimental model in the rat, researchers have constructed cardiac tissue scaffolds by seeding neonatal cardiomyocytes into a gel mixture containing collagen type I, matrigel, serum, and cell culture medium. The seeded cardiomyocytes can exhibit contractile activities. The matrix gel can be tailored into a desired shape according to the size and geometry of the host heart. The constructed cardiac tissue scaffold can be implanted around to host heart to cover a desired area. Following the implantation surgery, the implanted cardiac tissue scaffold can develop into a structure with organized cardiac muscle cells, which express mature cardiac protein markers such as actinin, connexin 43, and cadherins. More importantly, the implanted cardiac tissue scaffold demonstrates periodic contractile activities. Such activities can enhance the contractile performance of the host heart and reduce the pathological effect of heart failure. These investigations demonstrate that cell-based tissue regenerative engineering approaches can be used to improve the cardiac function in heart failure.

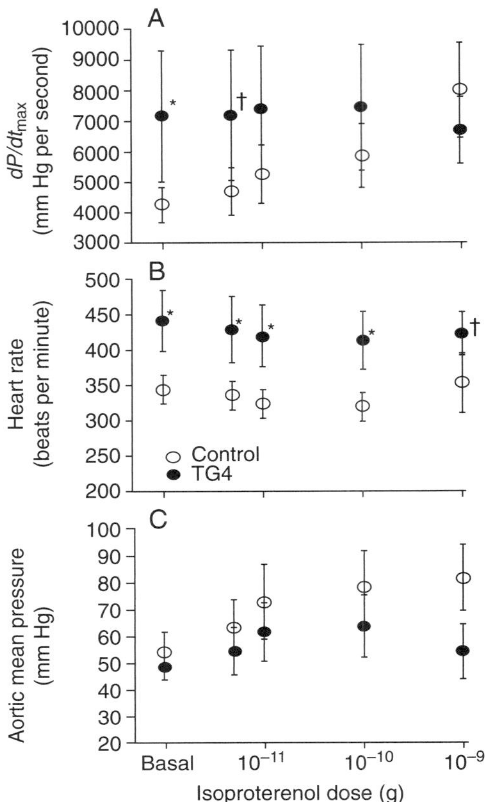

Figure 14.6. In vivo assessment of left ventricular function. Seven TG4 (transgenic) animals and seven controls were anesthetized and aortic and left ventricular catheters placed. Three measured parameters are shown at baseline and after doses of isoproterenol: (A) LV dP/dt_{max}; (B) heart rate; (C) mean aortic pressure. Data are reported as means plus minus SD and were analyzed with a two-way repeated measures analysis of variance with post hoc tests based on *t* test with Bonferroni correction for five comparisons (*, $P < 0.005$; dagger, $P < 0.05$). (Reprinted with permission from Milano CA et al: *Science* 264:582–6, copyright 1994 AAAS.)

Ischemic Heart Disease

Pathogenesis, Pathology, and Clinical Features [14.8]. Ischemic heart disease is a disorder induced by inadequate blood supply to the cardiac tissue, a condition known as *ischemia*. This disorder is characterized by various levels of impairment of cardiomyocytes, ranging from reversible injury to acute death or infarction, depending on the degree of bloodflow deficiency. A complete lack of bloodflow or oxygen in the cardiac muscle induces acute cardiac infarction within 3–5 min. Cardiac infarction is seldom reversible. Dead cardiomyocytes are usually replaced with noncontractile fibrous tissue produced by fibroblasts (Fig. 14.3). The clinical consequence of cardiac infarction is the impairment of cardiac function. Severe cardiac infarction with a large volume of infracted cardiac tissue may result in acute heart failure or cardiac arrest. Cardiac ischemia and infarction are often found in the elder population and are among the leading causes of human death.

There are a number of vascular disorders that contribute to the pathogenesis of cardiac ischemia and infarction. These vascular disorders include coronary arteriosclerosis, thrombosis, and embolism. All these disorders block the coronary arterial bloodflow, resulting in blood and oxygen deficiency. The most common cause is coronary arteriosclerosis, a vascular disorder characterized by the presence of intimal atheroma, which grows continuously and partially or completely blocks bloodflow (see page 674 for details). Coronary arterial thrombosis is an acute, complex process involving endothelial cell injury, activation of the blood coagulation system, adhesion of platelets and leukocytes to injured endothelium, and formation of thrombi, which partially or completely block the lumen of an artery. Embolism is a condition with the arterial lumen blocked with loose thrombi detached from an upstream artery. All these vascular disorders cause similar changes in the heart.

Cardiac infarction is associated with apparent structural changes in the heart, while cardiac ischemia may not exhibit noticeable structural changes. When specimens are collected from an infracted area, one may find characteristic structural changes at different stages. In the acute stage, major structural changes include local edema, massive death of cardiac muscle cells, and leukocyte infiltration, in association with locally reduced contractile activities. If the patient survives, the dead cardiac muscles cannot self-regenerate, but are gradually replaced with proliferating fibroblasts and fibrous tissue composed of primarily collagen matrix and proteoglycans. In the late stage of cardiac infarction, a pathological examination often reveals changes seen in a typical scar tissue, including scattered fibroblasts and fibrous extracellular matrix.

The clinical manifestations of ischemic heart disease are dependent on the degree and location of arterial obstruction. When a coronary artery is partially blocked, the distal cardiac muscles may experience transient ischemia, but are not completely devoid of bloodflow. Such a condition often gives rise to a clinical syndrome known as *angina pectoris*. A major symptom of this syndrome is transient chest pain or discomfort. Often, pain is radiated to the left shoulder and both arms or to the back, neck, jaw, and teeth. Angina usually occurs after physical activities or emotional changes. An electrocardiographic examination often reveals ST-segment depression. In particular, a change in the ST segment after a defined level of exercise in a "stress test" provides further evidence for cardiac ischemia. A coronary angiographic test can provide convincing evidence ensuring the presence of partial arterial obstruction and cardiac ischemia.

In the case of acute cardiac infarction, patients often experience deep, heavy, crushing pain in the chest. Although the pain occurs at locations similar to those in angina pectoris, it is often more severe and lasts longer. The pain may radiate to the shoulder, neck, back, and jaw. The pain is usually associated with other symptoms and signs, such as sweating, weakness, nausea, vomiting, and sudden drop in arterial blood pressure. However, about 15% of patients with cardiac infarction may not experience any pain. Such infarction is referred to as *silent infarction*. This is a more dangerous situation, because patients with severe cardiac infarction can be easily ignored without prompt medical attention.

Conventional Treatment of Ischemic Heart Disease [14.8]. Acute cardiac infarction is often associated with two types of life-threatening disorder: electrical rhythmic disorder (arrhythmia) and mechanical pump failure. The principle of treating acute cardiac infarction is thus to prevent these disorders and minimize the size of cardiac infarction. Patients are always given additional oxygen to maintain an optimal level of blood oxygen to minimize the spread of cardiac infarction, in association with the administration of analgesics, which keeps the patients calm, lowers physical activities and emotional stress, and reduces the heartbeat and oxygen consumption.

An electrocardiographic examination may reveal various forms of atrial and ventricular arrhythmia. The most serous life-threatening arrhythmias are ventricular tachycardia (heart rate >100 beats/min) and ventricular fibrillation. These forms of arrhythmia occur during the first 24 h following the onset of cardiac infarction. Often, patients are given a preventive treatment with an antiarrhythmia drug such as lidocaine. Ventricular tachycardia and fibrillation, if any, should be treated immediately by defibrillation. Another form of severe arrhythmia is sinus bradycardia (heart rate <45 beats/min). Patients may be administrated with atropine (note that atropine increases heart rate and should be used with caution). In a life-threatening case, patients should be given electrical pacing when blood pressure drops rapidly.

The impairment of the mechanical or contractile performance of an infracted heart is dependent on the size of the infract. Cardiac infarcts with a size larger than a critical level may result in heart failure. In the absence of heart failure, patients are often associated with tachycardia and an increase in arterial blood pressure as the heart intends to compensate for the lost function due to infarction. In such a case, α-adrenergic blocker should be administrated to lower the heart rate and arterial blood pressure. In the presence of heart failure, inotropic agents such as digitalis glycosides or catecholamines may be administrated to raise arterial blood pressure. However, these agents are not given for preventive purposes because they increase heart rate, cardiac contractility, and oxygen consumption, which may facilitate the spread of cardiac infarction. Other treatments described above should be applied in the presence of acute heart failure.

In a large fraction of patients with cardiac infarction, thrombus formation is a major cause of acute arterial obstruction near an atheromatic lesion. In such a case, thrombus dissolution should be carried out promptly with thrombolytic agents to reduce the arterial obstruction, introduce reperfusion, and minimize the size of cardiac infarcts. Common thrombolytic agents include streptokinase and tissue plasminogen activator. These agents can be used to effectively lyse freshly formed thrombi.

Molecular Regenerative Engineering for Ischemic Heart Disease. Molecular engineering approaches can be used to treat cardiac infarction and to improve cardiac function. The principle of cardiac molecular engineering for cardiac infarction is to prevent acute cell death, promote cell survival, protect cells from reperfusion injury, enhance angiogenesis, and improve the contractility of impaired myocardial cells. A number of molecules can be used to prevent cell death and promote cell survival. These include growth factors, such as fibroblast growth factor (FGF), platelet-derived growth factor (PDGF), epidermal growth factor (EGF), and vascular endothelial growth factor or (VEGF), antiapoptotic proteins, such as Bcl2, protein kinase B, and Akt, and inhibitors of proinflammatory cytokines. Experimental investigations have demonstrated that these factors can be used effectively to protect impaired cells from death.

Growth Factors as Therapeutic Agents for Cardiac Regenerative Engineering

FIBROBLAST GROWTH FACTORS. Fibroblast growth factors (FGFs) are a family of growth factors, including about 22 known members, including FGF1–22. Among these members, FGF1 and FGF2, also known as acidic and basic FGF, respectively, have been intensively studied and characterized. Other members are discovered more recently and have become the targets of increasing investigations. The members of the FGF family share about 30–70% identical amino acid sequence. Fibroblast growth factors play a critical role in the regulation of multiple biological processes, including cell proliferation, cell differentiation, cell migration, tissue morphogenesis, and organ formation during development and remodeling. Different FGFs exhibit distinct functions, depending on the structure of the individual members and the types of target cells. In this section, the two well-characterized FGFs, including FGF1 and FGF2, are discussed.

Fibroblast Growth Factor 1 (FGF1, Acidic FGF, or aFGF) [14.9]. Fibroblast growth factor 1 is a 155-amino acid protein with molecular weight of ~17 kDa. This growth factor is also known as *heparin-binding growth factor 1* (HBGF1), since heparin sulfate glycosaminoglycans can bind to FGF1 and facilitate the interaction of FGF1 with FGF receptor (FGFR). Fibroblast growth factor 1 shares the same gene with two growth factors known as endothelial cell growth factors α and β (ECGF α and ECGF β, respectively). The distinct forms of FGF1, ECGFα, and ECGFβ, result from different processes of posttranslational splicing. Fibroblast growth factor 1 is 14 amino acids shorter and ECGFα is 20 amino acid shorter at the *N*-terminus than ECGFβ. The gene for these growth factors is localized to chromosome 5 at gene locus 5q31. The amino acid sequence is highly conserved among mammals. Fibroblast growth factor 1 is expressed in several tissue types, including the central nervous system (primarily the cortex), kidney, pancreas, spleen, and skeletal muscle. This growth factor can interact with all four types of FGF receptors, including FGFR1, FGFR2, FGFR3, and FGFR4.

Fibroblast growth factor 1 exerts a mitogenic effect via the interaction with the fibroblast growth factor receptor (FGFR, see section on fibroblast growth factor receptors below for characterization of FGFR). Investigations by crystallography have demonstrated that FGF1 can interact with the extracellular immunoglobulin-like ligand-binding domain 2 and the linker between domains 2 and 3 of the FGFR. Indeed, this domain and the linker are general binding sites for all FGFs. The binding specificity of distinct FGFs is achieved via the interaction of the *N*-terminus of the FGFs and the immunoglobulin-like ligand-binding domain 3. This structural analysis provides a basis for understanding the interaction of FGF1 with its receptor.

Fibroblast growth factor 1 plays an important role in the regulation of cell proliferation, differentiation, and morphogenesis in a variety of tissue types during embryonic development and adult remodeling. FGF1 is expressed in the neuronal cells of the central nerve system and is responsible for the regeneration of neuronal cells in nerve injury. There exist three of the four types of FGF receptors in the central nerve system. These receptors can interact with FGF1 to induce mitogenic responses. Selective heparan proteoglycans (HSPGs) may mediate the interaction of FGF1 with its receptor. FGF1 also contributes to the formation of the liver from the foregut endoderm. During the embryonic stage, FGF1 is expressed in the mesoderm, which is close to the foregut endoderm, and released to the endoderm, where it stimulates the differentiation of stem and progenitor cells into liver cells. A treatment with FGF1 can induce the foregut endoderm to express genes that are required for liver generation.

Fibroblast Growth Factor 2 (FGF2, Basic FGF, or bFGF) [14.10]. Fibroblast growth factor 2 is protein of 288 amino acids with molecular weight approximately 23 kDa. This growth factor is expressed widely and participates in the regulation of tissue and organ development, angiogenesis, wound healing, cell regeneration, and tumorigenesis. The level of FGF2 expression in the central nervous system is considerably higher than that in other systems, suggesting a critical role for FGF2 in regulating the development and survival of the neurons. Fibroblast growth factor 2 is encoded by a single gene, which is localized to chromosome 4 at locus 4q25-q27. The length of the human FGF2 is 288 amino acids, whereas that of the mouse FGF2 is 154 amino acids.

Fibroblast growth factor 2 can interact with all four types of FGF receptor, initiating intracellular signaling events. Studies with crystallography have demonstrated that FGF2 can bind to the extracellular immunoglobulin-like domains 2 and 3 of the FGF receptors. The binding of FGF2 induces dimerization of the receptors. The interaction of FGF2 with the immunoglobulin domain 2 is critical for stabilizing the dimerization of the FGF receptor. The interaction of FGF2's *N*-terminus with the immunoglobulin domain 3 of the FGF receptor plays a critical role in the specificity of the ligand–receptor interaction. These observations provide a structural basis for understanding the function of FGF2.

Fibroblast growth factor 2 plays a critical role in the regulation of multiple biological processes, including cell proliferation, differentiation, and morphogenesis during embryonic development and adult remodeling. This growth factor is highly expressed in the central nervous system during embryonic development. The interaction of FGF2 with FGF receptors activates intracellular signaling events, which control the differentiation of neuronal stem cells and the pattern formation of neurons. In the transgenic mouse model of FGF2 knockout induced by homologous recombination, histological abnormalities can be found in the frontal motor sensory area of the cortex with a significant reduction in the neuronal density, although the genetic modulation does not significantly influence the lifespan and fertility of the mouse. These observations suggest a role for FGF2 in regulating neurogenesis. Fibroblast growth factor 2 also participates in regulating the formation of other systems, including the limb, liver, and the Langerhans islets of the pancreas by controlling cell differentiation and proliferation.

Fibroblast growth factor 2 is known as a factor that regulates angiogenesis. In animal models of arterial ligation, a treatment with FGF2 significantly promotes arteriogenesis in regions distal to the ligation site. The application of platelet-derived growth factor (PDGF), together with FGF2, to the arterial ligation model exerts a synergistic effect on arteriogenesis. These growth factors upregulate the expression of PDGF receptors α and β, and activate intracellular mitogenic signaling pathways, leading to enhanced activity of angiogenesis. Furthermore, FGF2 has been shown to induce lymphangiogenesis in the mouse cornea.

Another function of FGF2 is the regulation of wound healing and cell regeneration in tissue and organ injury. For instance, in animal models of bone fracture, a treatment with FGF2 can significantly enhance the union of fractured bones. Such a treatment can also stimulate the mineralization and enhance the mechanical stiffness of the fractured bones. When the FGF2 gene is disrupted by genetic modulation, the rate of mineral deposition and bone formation are significantly reduced. Furthermore, FGF2 promotes cell survival and protects cells from apoptosis. In particular, FGF2 enhances the differentiation of transplanted cardiomyocyte progenitor cells into mature cardiomyocytes. FGF2-deficient progenitor cells exhibit reduced capability of differentiating to cardiomyocytes. These observations suggest that FGF2 is a critical factor for the regulation of wound healing and cell regeneration.

FIBROBLAST GROWTH FACTOR RECEPTORS. Fibroblast growth factor receptors (FGFRs) are a group of single-pass membrane proteins that interact with FGFs and can activate corresponding intracellular signaling pathways, resulting in the activation of mitogenic cellular processes such as cell proliferation, differentiation, migration, and pattern formation. There exist four known types of FGFR, designated as FGFR1, FGFR2, FGFR3, and FGFR4. These receptor types exhibit differential ligand binding affinity and tissue-specific expression. In general, each FGFR is composed of an extracellular region, a transmembrane region, and a cytoplasmic region. In the extracellular region, there are three distinct domains known as *immunoglobulin-like domains*, which are responsible for the binding of FGFs. A catalytic tyrosine kinase domain is found in the cytoplasmic region. This domain serves as a protein tyrosine kinase that phosphorylates downstream signaling molecules. Upon the binding of a FGF ligand, which is mediated by heparan sulfate glycosaminoglycans, the FGFRs are stimulated to form hetero- or homodimers, which induce autophosphorylation of the cytoplasmic domains of the receptors. This process induces the recruitment of adaptor and linker proteins to the receptor cytoplasmic domains, resulting in the activation of corresponding signaling pathways, such as the ras–mitogen-activated protein kinase pathway (see page 151), and activation of mitogenic cellular processes, such as cell proliferation and migration.

Fibroblast Growth Factor Receptor 1 [14.11]. Fibroblast growth factor receptor 1 (FGFR1) is also known as *FMS-like tyrosine kinase 2* or FLT2 (note that FMS is macrophage colony-stimulating factor receptor encoded by the *fms* oncogene). The human FGFR1 is composed of 820 amino acids with molecular weight ~92 kDa. The FGFR1 gene is localized to chromosome 8 at locus 8p11.1 and 8p11.2. Fibroblast growth factor receptor 1 is expressed in a variety of tissue types, including the central nerve system, kidney, lung, mammary gland, blood vessels, stomach, pancreas, thymus, uterus, and cornea. This growth factor receptor primarily interacts with FGF1, FGF2, and FGF5, resulting in the activation of the receptor. Activated FGFR1 can interact with adapter proteins, including growth factor receptor-bound protein 2 (Grb2) and Sos, which in turn activate intracellular signaling pathways and regulate mitogenic processes such as cell proliferation, differentiation, migration, and pattern formation. Mutation of FGFR1 contributes to the development of several disorders, including Pfeiffer syndrome (an anautosomal dominant craniosynostosis syndrome characterized by craniofacial anomalies and broad thumbs and large toes) and Kallmann syndrome (characterized by craniosynostosis).

Fibroblast Growth Factor Receptor 2 [14.12]. Fibroblast growth factor receptor 2 (FGFR2) is a 758-amino acid transmembrane receptor protein tyrosine kinase with molecular weight of approximately 92 kDa. It is also known as a *keratinocyte growth factor receptor*, *fibroblast growth factor receptor BEK*, *protein tyrosine kinase receptor-like 14* (TK14), or *BEK*. The receptor is composed of several domains, including three extracellular immunoglobulin-like domains, a transmembrane domain, and a tyrosine kinase domain. The FGFR2 gene is localized to chromosome 10 at locus 10q26. This receptor is expressed in a variety of tissues, including the brain, thymus, cornea, skin, pancreas, stomach, prostate gland and uterus. This receptor can interact with numbers of ligands, including FGF1, FGF2, FGF5, FGF7, FGF9, FGF10, and phospholipase C/γ1. The primary function of the FGFR2 is to regulate cell proliferation, differentiation, and pattern formation during embryonic development and adult remodeling. Mutation of the FGFR2 gene may induce several disorders, such as Pfeiffer syndrome, colorectal carcinoma, and gastric cancer.

Fibroblast Growth Factor Receptor 3 [14.13]. Fibroblast growth factor receptor 3 (FGFR3) is an 806 amino acid transmembrane protein tyrosine kinase with molecular weight ~88 kDa. This receptor is also known as *human tyrosine kinase JTK4*. It is expressed in the central nerve system, liver, intestine, cartilage, lung, and thymus. Fibroblast growth factor receptor 3 can interact with FGF1, FGF8, and FGF9, inducing the activation of the receptor. FGFR3 can activate intracellular adapter proteins, including Grb2 and SH2 domain-containing protein tyrosine phosphatase 2 (SHP2). These adaptor proteins induce activation of intracellular mitogenic signaling pathways. The mutation of FGFR3 induces several disorders, including thanatophoric dwarfism (sporadic lethal skeletal dysplasia with limb shortening, macrocephaly, platyspondyly, and reduced thoracic cavity), gastric and colorectal cancers, and hypochondroplasia (chondrodystrophy or abnormal development of cartilage).

Fibroblast Growth Factor Receptor 4 [14.14]. Fibroblast growth factor receptor 4 (FGFR4), also known as *TKF*, is a 802-amino acid transmembrane protein tyrosine kinase with molecular weight ~88 kDa. This receptor is expressed in the central nervous system, heart, lung, liver, intestine, adrenal gland, pancreas, spleen, thymus, retina, and cornea. Fibroblast growth factor receptor 4 can interact with FGF1, FGF2, FGF6, FGF8, and FGF19, inducing mitogenic cellular activities. The mutation of FGFR4 induces breast cancer and ovarian cancer.

EPIDERMAL GROWTH FACTOR [14.15]. Epidermal growth factor (EGF), also known as *urogastrone*, is synthesized first as a precursor of 1168 amino acids with molecular weight ~128 kDa. The EGF precursor os converted to mature EGF by protein cleavage. The mature EGF is a 53 polypeptide with molecular weight ~6 kDa. The EGF gene is localized to chromosome 4 at locus 4q25. Epidermal growth factor is expressed in the skin, intestine, ovary, pancreas, prostate gland, uterus, and blood vessels. Epidermal growth factor interacts with and activates EGF receptor (EGFR), leading to the activation of intracellular signaling pathways involving Grb2, PI3 kinase, Ras, and mitogen-activated protein kinase (see page 151). The activation of these signaling molecules results in cellular activities, such as cell proliferation and migration. The mutation of EGF gene may contribute to the development sporadic malignant melanoma.

EPIDERMAL GROWTH FACTOR RECEPTOR [14.16]. Epidermal growth factor receptor (EGFR) is a transmembrane receptor protein kinase of 1210 amino acids with molecular weight ~134 kDa. The gene of EGFR is localized to chromosome 7 at locus 7p12.3–p12.1. This receptor is expressed ubiquitously. The EGFR is composed of two cheY-homologous receiver (REC) domains and three furin-like repeats in the extracellular region, a transmembrane domain, and a cytoplasmic protein tyrosine kinase. Note that the REC domain is a sequence homologous to that of the cheY protein, which regulates the rotation of *E. coli* flagellate motors, and the furin-like repeat is a cysteine-rich sequence found in receptors involved in signal transduction mediated by receptor tyrosine kinases. Epidermal growth factor receptor can interact with EGF, inducing autophosphorylation of the receptor. Phosphorylated EGFR can activate intracellular signaling molecules, including protein kinase A, Ras, focal adhesion kinase, integrins, Sos, protein kinase Cα, STAT, and SH2 domain-containing protein tyrosine kinase 2 (SHP2). The activation of these signaling molecules leads to mitogenic cellular processes such as cell proliferation, migration, and adhesion.

PLATELET-DERIVED GROWTH FACTOR A CHAIN [14.17]. Platelet-derived growth factor A chain (PDGF A), also known as *platelet-derived growth factor α peptide*, is a protein of 211 amino acids with molecular weight ~24 kDa. The PDGF A gene is localized to gene locus 7p22. Platelet-derived growth factor A is expressed in several tissues, including the uterus, lung, and blood vessels. Platelet-derived growth factor A chain can form a homodimer, known as *PDGF AA*, with another PDGF A molecule. PDGF A chain can interact with PDGF receptor α to induce mitogenic cellular activities, such as cell proliferation and migration. Platelet-derived growth factor A can bind to several extracellular matrix proteins, including collagen, laminin 1, and perlecan. The binding of PDGF A to collagen molecules requires the presence of calcium, whereas the binding to perlecan is not calcium-dependent. The formation of PDGF A–matrix protein complexes is an effective mechanism for the storage of PDGF A in the extracellular space. Platelet-derived growth factor A can be rapidly released in the case of cell injury, facilitating cell regeneration. The expression of PDGF A is upregulated in the vascular smooth muscle cells and endothelial cells in vascular disorders such as atherosclerosis, inducing smooth muscle cell proliferation and migration from the media to the intima, critical processes that contribute to atherogenesis. Furthermore, PDGF A plays a role in regulating spermatogenesis. The deficiency of PDGF A in a transgenic mouse model induces the arrest of spermatogenesis.

PLATELET-DERIVED GROWTH FACTOR B CHAIN [14.18]. Platelet-derived growth factor B chain (PDGF B) is also known as *PDGFβ polypeptide*, *V-SIS platelet-derived growth factor β polypeptide*, and *Simian sarcoma viral oncogene homolog*. It is a 241-amino acid protein with molecular weight ~27 kDa. The PDGF B gene, also known as the *sis* oncogene, is localized to chromosome 22 at locus 22q12.3–q13.1. Platelet-derived growth factor B is expressed in the heart, blood vessels, testis, kidney, eye, and ovary. This growth factor often forms a dimer with another PDGF B chain, known as *PDGF BB*. The dimeric complex can interact with PDGF receptor α and PDGF receptor β. The binding of PDGF BB to PDGFR induces dimerization of the receptors and autophosphorylation of the receptor cytoplasmic tyrosine kinase domains, a critical process for the action of mitogenic signaling pathways involving Ras, Raf1, and mitogen-activated protein kinases. The activation of these signaling molecules enhances cell proliferation and migration. PDGF B can bind to extracellular matrix proteins, including collagen and perlecan, a process for the storage of PDGF B.

PLATELET-DERIVED GROWTH FACTOR RECEPTOR α [14.19]. Platelet-derived growth factor receptor α (PDGFRα) is a transmembrane receptor protein tyrosine kinase (1089 amino acids, ~123 kDa). It is also known as *PDGFR A* or *PDGFR2*. The gene of PDGFRα is localized to chromosome 4 at locus 4q12. The receptor is composed of an extracellular region, a transmembrane region, and a cytoplasmic region. The extracellular region contains three immunoglobulin (Ig)-like domains and an immunoglobulin constant 2-type (IGC2) domain. PDGFRα is expressed in the brain, liver, pancreas, bone, platelets, and B cells. The extracellular region of the PDGFRα can interact with PDGF A and PDGF B, inducing dimerization of the receptors and autophosphorylation of the receptor cytoplasmic tyrosine kinase domains. Phosphorylated PDGFRα can activate several intracellular signaling molecules, including guanine nucleotide-releasing factor 2, growth factor receptor-bound protein 2 (Grb2), phospholipase Cγ and SH2 domain-containing protein tyrosine phosphatase 2 (SHP2). The activation of these signaling molecules results in mitogenic cellular activities such as cell proliferation, adhesion, and migration. PDGFRα can also interact with integrins, including integrin αV and integrin 3, to mediate synergistically the transduction of signals from extracellular matrix proteins and thus augment mitogenic cellular processes.

PLATELET-DERIVED GROWTH FACTOR RECEPTOR β [14.20]. Platelet-derived growth factor receptor β (PDGFRβ), also known as *PDGFR1*, is a receptor protein tyrosine kinase of 1106 amino acids with molecular weight 124 kDa. The receptor is composed of an extracellular region, a transmembrane region, and a cytoplasmic region. The extracellular region contains several domains, including two Ig-like domains and an Ig constant 2 domain. The cytoplasmic region contains a protein tyrosine kinase domain. The gene of PDGFRβ is localized to chromosome 5 at locus 5q21–q32. PDGFRβ is expressed in the lung, kidney, blood vessels, bone, and prostate gland.

The extracellular region of PDGFR β can interact with primarily the PDGF B chain, inducing dimerization of the receptors and autophosphorylation of the receptor cytoplasmic domains. Phosphorylated PDGFR β can in turn activate a number of intracellular signaling molecules, including SH2 domain-containing protein tyrosine phosphatase (SHP)2, Raf1, focal adhesion kinase, growth factor receptor-bound protein 2 (Grb2), Grb4, and the Ras protein. These signaling molecules further induce activation of mitogenic processes such as cell proliferation and migration. PDGFRβ can also interact with integrins, including integrin αV and β3. This interaction synergistically enhances mitogenic activities induced by extracellular ligands and matrix components.

VASCULAR ENDOTHELIAL GROWTH FACTOR A [14.21]. Vascular endothelial growth factor (VEGF) A, also known as the *vascular permeability factor*, is a 189 amino acid protein of molecular weight approximately 22 kDa. This growth factor is encoded by the VEGF A gene, which contains 8 exons. Alternative splicing of the VEGF A gene may give rise to different sizes of the VEGF protein. The missing of the exon 6-encoded residues results in the formation of a VEGF variant of 165 amino acids, whereas the missing of exons 6 and 7 results in the formation of a VEGF variant of 121 amino acids. All these forms possess the normal biological activity of the growth factor. This growth factor may exist in the form of homodimer with molecular weight ~45 kDa.

The VEGF gene is localized to chromosome 6 at locus 6p12. VEGF A is expressed in a number of tissues, including the heart, blood vessels, kidney, adrenal gland, lung, liver, stomach, pancreas, uterus, retina, and skin. VEGF A can interact with VEGF receptor 1 (VEGFR1) and VEGFR2. Such interaction activates these receptors, leading to mitogenic cellular responses such as cell proliferation, differentiation, and migration. In particular, VEGF A is involved in the regulation of endothelial cell differentiation and angiogenesis, or the formation of blood vessels on the basis of an established vascular network. Another mitogenic factor, angiopoietin 2, acts synergistically with VEGF A in the regulation of angiogenesis. Under conditions with reduced oxygen and nutrient supply, VEGF plays a critical role for the survival of vascular endothelial cells and the formation of blood vessels, which is an adaptive process in response to hypoxia and nutrient depletion.

VASCULAR ENDOTHELIAL GROWTH FACTOR B [14.22]. Vascular endothelial growth factor B (VEGF B), also known as *vascular endothelial growth factor-related factor* (VRF), is a 22 kDa protein, which exists in two different isoforms: one with 186 amino acids and the other with 167 amino acids. The two VEGF B isoforms differ at their carboxyl ends due to a shift in the open reading frame. Both VEGF B isoforms possess similar function. VEGF B exhibits strong homology to VEGF A. The VEGF B gene is localized to chromosome 11 at locus 11q13. The human VEGF B gene shares about 88% amino acid sequence identity with the mouse VEGF B gene. VEGF B is expressed in the heart, skeletal muscle, pancreas, and prostate gland. VEGF B can bind to VEGF receptor (VEGFR) 1 to induce phosphorylation of the receptor protein tyrosine kinase, leading to activation of intracellular mitogenic signaling cascades, involving Grb2, Ras, mitogen-activated protein kinases, and mitogenic cellular activities, such as cell proliferation, migration, and angiogenesis. The deficiency of VEGF B in transgenic mice is associated with abnormal vascular structure and impaired recovery from cardiac injury.

VASCULAR ENDOTHELIAL GROWTH FACTOR RECEPTOR 1 [14.23]. Vascular endothelial growth factor 1 (VEGF1) is a protein tyrosine kinase-containing transmembrane receptor. It is also known as vascular permeability factor receptor, FLT, and FLT1. The length of the receptor is 1338 amino acids and the molecular weight is approximately 151 kDa. The receptor is composed of five immunoglobulin-like domains and two immunoglobulin constant 2 domains in the extracellular region, a transmembrane region, and a cytoplasmic region, which contains a protein tyrosine kinase domain. VEGFR1 is encoded by the oncogene flt1, which belongs to the src oncogene family and is localized to chromosome 13 at locus 13q12. VEGFR1 is expressed primarily in vascular endothelial cells in several tissues and organs, including the bone marrow, testis, intestine, pancreas, ovary, prostate gland, and placenta. VEGFR1 can interact with VEGF A and VEGF B at the extracellular domains 2 and 3, inducing autophosphorylation of the cytoplasmic protein tyrosine kinase domain. Phosphorylated VEGFR1 activates intracellular signaling molecules, including phospholipase Cγ2 and SHC. An important function of VEGFR1 is to regulate endothelial cell proliferation and angiogenesis. Furthermore, VEGFR1 can interact with placental growth factor (PGF), which mediates the crosstalk between VEGFR1 and VEGFR2. VEGF and PGF can form heterodimers, which stimulate the formation of VEGFR1 and VEGFR2 heterodimers. The VEGFR1/PGF combination synergistically enhances the activity of VEGFR1 and VEGFR2. A soluble form of VEGFR1 can be generated by alternative splicing of the VEGFR1 pre-mRNA and is termed *sFLT1.* This form of VEGFR1 can bind to VEGF with high affinity. Thus, sFLT1 sequesters VEGF and competitively inhibits the activity of VEGF.

VASCULAR ENDOTHELIAL GROWTH FACTOR RECEPTOR 2 [14.24]. Vascular endothelial growth factor receptor 2 (VEGFR2) is a transmembrane receptor with a cytoplasmic protein tyrosine kinase domain. VEGFR2 is also known as *fetal liver kinase 1* (Flk1), *kinase insert domain receptor* (KDR), and *tyrosine kinase growth factor receptor.* The length of the receptor is 1356 amino acids, and the molecular weight is approximately 151 kDa. This receptor is encoded by the oncogene kdr, which is localized to the gene locus 4q12. VEGFR2 is composed of five immunoglobulin-like domains in the extracellular region, a transmembrane region, and a cytoplasmic region with a catalytic protein tyrosine kinase domain. The structure of the protein tyrosine kinase domain is similar to that of the platelet-derived growth factor receptor, colony-stimulating factor 1 receptor, fibroblast growth factor receptor, and KIT. The receptor is primarily expressed in the vascular endothelial cells and is often used as a marker for the identification of the endothelial cells. VEGFR2 can interact with VEGF A. The binding of VEGF A to VEGFR2 induces autophosphorylation of the cytoplasmic protein tyrosine kinase domain, which further activates intracellular signaling molecules, including Grb2, Grb10, phospholipase Cγ2, Src-homology collagen protein (Shc), and Shc-like protein (Sck). These signaling molecules participate in the regulation of endothelial cell division and angiogenesis. VEGFR2 and associated signaling molecules also regulate the differentiation and development of vascular endothelial cells as well as vasculogenesis during the embryonic stage.

VEGFR2 is found in embryonic hematopoietic stem cells. About 0.1–0.5% of $CD34^+$ embryonic hematopoietic stem cells co-express VEGFR2. The $CD34^+VEGFR2^+$ cell population contains pluripotent hematopoietic stem cells that can differentiate into hematopoietic progenitor cells and vascular endothelial cells. The adult bone marrow also contains $CD34^+VEGFR2^+$ cells. These cells can potentially differentiate into vascular endothelial cells. VEGFR2 is considered a marker for the identification of endothelial cell progenitor cells. Furthermore, $VEGFR2^+$ progenitor cells contribute to the formation of blood vessels and regulate the organization of the vasculature. It is interesting to note that the growth factor VEGF stimulates the transformation of $VEGFR2^+$ progenitor cells to endothelial cells, whereas PDGF induces the formation of the vascular smooth muscle cells.

HEPATOCYTE GROWTH FACTOR [14.25]. Hepatocyte growth factor (HGF) is a protein of 728 amino acids with molecular weight ~83 kDa. HGF is also known as hepatopoeitin A, lung fibroblast-derived mitogen, and scatter factor (SF). This growth factor is composed of a heavy and light chain. The heavy chain is about 50–60-kDa in molecular weight, and the light chain is about 30–35 kDa. The two chains are linked by disulfide bonds. The HGF gene is localized to the gene locus 7q21.1 of chromosome 7. The structure of HGF is apparently different from that of other growth factors as described above. HGF is composed of an apple-like domain, four Kringle domains, and a trypsin-like serine protease domain. The apple-like domain is a fourfold repeat apple-like structure found in plasma kallikrein and coagulation factor XI. This domain mediates the binding of factor XI to platelets. The Kringle domain is a triple-loop, three-disulphide bridged structure found in several serine proteases such as prothrombin and urokinase-type plasminogen activator. This domain mediates the binding of molecules. HGF is expressed in a number of tissues, including the liver, blood vessels, brain, bone marrow, and placenta. This growth factor can interact with HGF receptor (HGFR), inducing mitogenic responses. HGF plays a critical role in regulating the development and morphogenesis of embryonic tissues and organs. The deficiency of HGF in transgenic mice is associated with incomplete liver and placental development, resulting in premature death of the animal.

HGF exerts a protective effect on injured cardiomyocytes (Fig. 14.7). HGF is upregulated in cardiac injury and stimulates the survival and proliferation of injured cardiomyocytes. In particular, this growth factor protects injured cardiac tissue from fibrosis (note that fibrosis reduces cardiac contractility). These effects render HGF a potential therapeutic factor for the treatment of cardiac injury.

HEPATOCYTE GROWTH FACTOR RECEPTOR [14.26]. Hepatocyte growth factor receptor (HGFR), also known as *MET* and *RCCP2*, is a transmembrane receptor with a cytoplasmic protein tyrosine kinase. The length of the receptor is 1400 amino acids, and the molecular

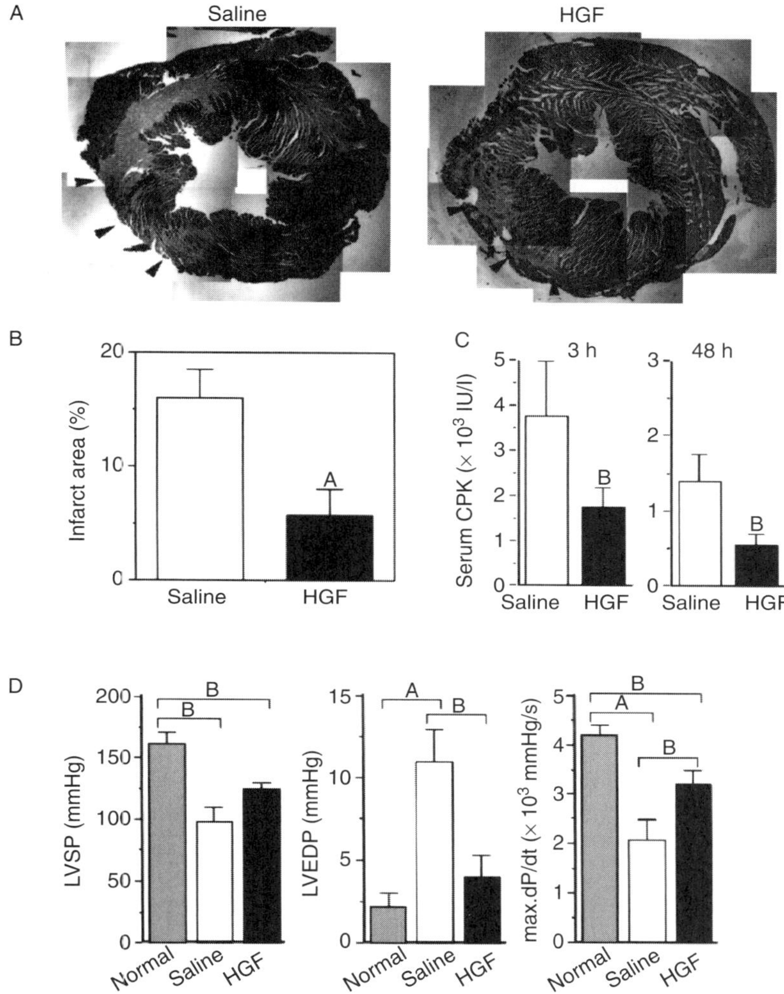

Figure 14.7. Amelioration of ischemia/reperfusion injury by hepatocyte growth factor (HGF). Recombinant human HGF ($n = 8$) or saline ($n = 8$) was injected immediately after and every 12 h after reperfusion. After 48 h, specimens were collected for observation. (A) α-Sarcomeric actin staining done to depict the infarct area (original magnification, ×40). Arrowheads indicate the alpha-actin–negative infarct area. (B, C) Changes in infarct area (B) and serum CPK activity (C). $^{A}P < 0.01$, $^{B}P < 0.05$. (D) Change in cardiac functions after ischemia/reperfusion injury. $^{A}P < 0.01$, $^{B}P < 0.05$. LVSP, left ventricular systolic pressure; LVEDP, left ventricular end-diastolic pressure; max *dP*/*dt*, maximal rate of left ventricular pressure rise. (Reprinted with permission from Nakamura T et al: Myocardial protection from ischemia/reperfusion injury by endogenous and exogenous HGF, *J Clin Invest* 106:1511–9, copyright 2000.)

weight is 157 kDa. This receptor is composed of a semaphorin (SEMA) domain, a PSI domain, and four IPT domains in the extracellular region, a transmembrane domain, and cytoplasmic protein tyrosine kinase domain. The SEMA domain is a structure found in semaphorins (proteins that induce the collapse and paralysis of neuronal growth cones during development) and is characterized by the presence of conserved cysteine residues. These residues form disulfide bonds and stabilize the protein structure. The PSI domain is found in plexins (proteins involved in the development of neural and epithelial tissues), semaphorins, and integrins. The IPT domain is an immunoglobulin-like structure found in plexins and certain transcription factors. HGFR is encoded by the oncogene *met*, which is localized to the gene locus 7q31. The protein product of the *met* oncogene is a dimer composed of an α subunit and a β subunit linked by disulfide bonds. The α subunit contains only an extracellular region, whereas the β subunit consists of extracellular, transmembrane, and cytoplasmic regions. The β subunit of the MET protein is the primary subunit that interacts with HGF. HGFR is expressed in the liver, brain, placenta, and skeletal muscle. The deficiency of the HGFR gene in transgenic mouse models induces liver and limb muscle defects, resulting in embryonic death of the animal. HGFR is often upregulated in human cancer cells. The expression of HGFR is enhanced during metastasis. This receptor may contribute to the metastatic properties to nontumorigenic and tumorigenic cells.

BIBLIOGRAPHY

14.1. Anatomy and Physiology

Guyton AC, Hall JE: *Textbook of Medical Physiology*, 11th ed, Saunders, Philadelphia, 2006.

McArdle WD, Katch FI, Katch VL: *Essentials of Exercise Physiology*, 3rd ed, Lippincott Williams & Wilkins, Baltimore, 2006.

Germann WJ, Stanfield CL: (with contributors Niles MJ, Cannon JG): *Principles of Human Physiology*, 2nd ed, Pearson Benjamin Cummings, San Francisco, 2005.

Thibodeau GA, Patton KT: *Anatomy & Physiology*, 5th ed, Mosby, St Louis, 2003.

Boron WF, Boulpaep EL: *Medical Physiology: A Cellular and Molecular Approach*, Saunders, Philadelphia, 2003.

Ganong WF: *Review of Medical Physiology*, 21st ed, McGraw-Hill, New York, 2003.

14.2. Pathogenesis, Pathology, and Clinical Features of Heart Failure

Schneider AS, Szanto PA: *Pathology*, 3rd ed, Lippincott Williams & Wilkins, Philadelphia, 2006.

McCance KL, Huether SE: *Pathophysiology: The Biologic Basis for Disease in Adults & Children*, 5th ed, Elsevier Mosby, St Louis, 2006.

Porth CL: *Pathophysiology: Concepts of Altered Health States*, 7th ed, Lippincott Williams & Wilkins, Philadelphia, 2005.

Frazier MS, Drzymkowski JW: *Essentials of Human Diseases and Conditions*, 3rd ed, Elsevier Saunders, St Louis, 2004.

Lloyd-Jones DM, Larson MG, Leip EP, Beiser A, D'Agostino RB et al: Lifetime risk for developing congestive heart failure: The Framingham Heart Study, *Circulation* 106:3068–72, 2002.

Davidson MJ, Koch WJ: Genetic manipulation of b-adrenergic signalling in heart failure, *Acta Physiol Scand* 173:145, 2001.

Birkeland JA, Sejersted OM, Taraldsen T, Sjaastad I: EC-coupling in normal and failing hearts, *Scand Cardiovasc J* 39:13–23, 2005.

Bers DM: Cardiac excitation-contraction coupling, *Nature* 415:198–205, 2002.

Hoshijima M: Gene therapy targeted at calcium handling as an approach to the treatment of heart failure, *Pharmacol Ther* 105:211–28, 2005.

Brodde OE: Beta-adrenoceptors in cardiac disease, *Pharmacol Ther* 60:405–30, 1993.

Rockman HA, Koch WJ, Lefkowitz RJ: Seven-transmembrane-spanning receptors and heart function, *Nature* 415:206–212, 2002.

Brodde OE: Beta-adrenoceptors in cardiac disease, *Pharmacol Ther* 60:405–30, 1993.

Wehrens XH, Marks AR: Molecular determinants of altered contractility in heart failure, *Ann Med* 36(Suppl 1):70–80, 2002.

Movsesian MA: cAMP-mediated signal transduction and sarcoplasmic reticulum function in heart failure, *Ann NY Acad Sci* 853:231–9, 1998.

Jessup M, Brozena S: Heart failure, *New Engl J Med* 348:2007–18, 2003.

14.3. Experimental Models of Heart Failure

Yue P et al: Post-infarction heart failure in the rat is associated with distinct alterations in cardiac myocyte molecular phenotype, *J Mol Cell Cardiol* 30:1615–30, 1998.

Swynghedauw B: Molecular mechanisms of myocardial remodeling *Physiol Rev* 79:215–62, 1999.

Gomez AM et al: Heart failure after myocardial infarction: Altered excitation-contraction coupling, *Circulation* 104:688–93, 2001.

Loennechen JP et al: Cardiomyocyte contractility and calcium handling partially recover after early deterioration during post-infarction failure in rat, *Acta Physiol Scand* 176:17–26, 2002.

Lorell BH: Transition from hypertrophy to failure, *Circulation* 96:3824–7, 1997.

Ito K et al: Contractile reserve and intracellular calcium regulation in mouse myocytes from normal and hypertrophied failing hearts, *Circ Res* 87:588–95, 2000.

Babu GJ, Periasamy M: Transgenic mouse models for cardiac dysfunction by a specific gene manipulation, *Meth Mol Med* 112:365–77, 2005.

Ho KKL, Anderson KM, Kannel WB, Grossman W, Levy D: Survival after the onset of congestive heart failure in Framingham Heart Study subjects, *Circulation* 88:107–15, 1993.

14.4. Cardiomyopathy

Hoshijima M et al: The MLP family of cytoskeletal Z disc proteins and dilated cardiomyopathy: A stress pathway model for heart failure progression, *Cold Spring Harb Symp Quant Biol* 67:399–408, 2002.

Pashmforoush M et al: Adult mice deficient in actinin-associated LIM-domain protein reveal a developmental pathway for right ventricular cardiomyopathy, *Nat Med* 7:591–7, 2001.

Zhou Q et al: Ablation of cypher, a PDZ-LIM domain Z-line protein, causes a severe form of congenital myopathy. *J Cell Biol* 155:605–12, 2001.

Megeney LA et al: Severe cardiomyopathy in mice lacking dystrophin and MyoD, *Proc Natl Acad Sci USA* 96:220–5, 1999.

Grady RM et al: Skeletal and cardiac myopathies in mice lacking utrophin and dystrophin: A model for Duchenne muscular dystrophy, *Cell* 90:729–38, 1997.

Hunter EG, Hughes V, White J: Cardiomyopathic hamsters, CHF 146 and CHF 147: A preliminary study, *Can J Physiol Pharmacol* 62:1423–8, 1984.

Durbeej M, Campbell KP: Muscular dystrophies involving the dystrophin-glycoprotein complex: An overview of current mouse models, *Curr Opin Genet Dev* 12:349–61, 2002.

Milner DJ et al: Disruption of muscle architecture and myocardial degeneration in mice lacking desmin, *J Cell Biol* 134:1255–70, 1996.

Fatkin D et al: Neonatal cardiomyopathy in mice homozygous for the Arg403Gln mutation in the alpha cardiac myosin heavy chain gene, *J Clin Invest* 103:147–53, 1999.

McConnell BK et al: Dilated cardiomyopathy in homozygous myosin-binding protein-C mutant mice, *J Clin Invest* 104:1235–44, 1999.

Hoshijima M: Models of dilated cardiomyopathy in small animals and novel positive inotropic therapies, *Ann NY Acad Sci* 1015:320–31, 2004.

14.5. Conventional Treatment of Cardiac Failure

Williams JF Jr, Bristow MR, Fowler MB, Francis GR, Garson A et al: Guidelines for the evaluation and management of heart failure: Report of the American College of Cardiology/American Heart Association Task Force on Practice Guidelines (Committee on Evaluation and Management of Heart Failure), *J Am Coll Cardiol* 26:1376–98, 1995.

Cohn JN, Franciosa JA: Vasodilator therapy of cardiac failure, *New Engl J Med* 297:27–31, 1977.

Packer M: The development of positive inotropic agents for chronic heart failure: How have we gone astray? *J Am Coll Cardiol* 22(Suppl A):119A–26A, 1993.

Cohn JN: The management of chronic heart failure, *New Engl J Med* 335:490–8, 1996.

Eichhorn EJ, Bristow MR: Medical therapy can improve the biological properties of the chronically failing heart: A new era in the treatment of heart failure, *Circulation* 94:2285–96, 1996.

Cohn JN, Archibald DG, Ziesche S, Franciosa JA, Harston WE et al: Effect of vasodilator therapy on mortality in chronic congestive heart failure: Results of a Veterans Administration Cooperative Study, *New Engl J Med* 314:1547–52, 1986.

The CONSENSUS Trial Study Group: Effects of enalapril on mortality in severe congestive heart failure: Results of the Cooperative North Scandinavian Enalapril Survival Study (CONSENSUS), *New Engl J Med* 316:1429–35, 1987.

DiBianco R, Shabetai R, Kostuk W, Moran J, Schlant RC et al: for the Milrinone Multicenter Trial Group: A comparison of oral milrinone, digoxin, and their combination in the treatment of patients with chronic heart failure, *New Engl J Med* 320:677–83, 1989.

The SOLVD Investigators: Effect of enalapril on survival in patients with reduced left ventricular ejection fractions and congestive heart failure, *New Engl J Med* 325:293–302, 1991.

Cohn JN, Johnson G, Ziesche S, Cobb F, Francis G et al: A comparison of enalapril with hydralazine-isosorbide dinitrate in the treatment of chronic congestive heart failure, *New Engl J Med* 325:303–10, 1991.

Packer M, Gheorghiade M, Young JB, Constantini PJ, Adams KF et al: for the RADIANCE Study Group: Withdrawal of digoxin from patients with chronic heart failure treated with angiotensin-converting-enzyme inhibitors, *New Engl J Med* 329:1–7, 1993.

Packer M, Bristow MR, Cohn JN, Colucci WS, Fowler MB et al: for the U.S. Carvedilol Heart Failure Study Group: The effect of carvedilol on morbidity and mortality in patients with chronic heart failure, *New Engl J Med* 334:1349–55, 1996.

The Digitalis Investigation Group: The effect of digoxin on mortality and morbidity in patients with heart failure, *New Engl J Med* 336:525–33, 1997.

Pfeffer MA, Braunwald E, Moye LA, Basta L, Brown EJ et al: on behalf of the SAVE Investigators: Effect of captopril on mortality and morbidity in patients with left ventricular dysfunction after myocardial infarction: Results of the Survival and Ventricular Enlargement Trial, *New Engl J Med* 327:669–77, 1992.

The Acute Infarction Ramipril Efficacy (AIRE) Study Investigators: Effect of ramipril on mortality and morbidity of survivors of acute myocardial infarction with clinical evidence of heart failure, *Lancet* 342:821–8, 1993.

Køber L, Torp-Pedersen C, Carlsen JE, Bagger H, Eliasen P et al: for the Trandolapril Cardiac Evaluation (TRACE) Study Group: A clinical trial of the angiotensin-converting-enzyme inhibitor trandolapril in patients with left ventricular dysfunction after myocardial infarction, *New Engl J Med* 333:1670–6, 1995.

Pitt B, Segal R, Martinez FA, Meurers G, Cowley AJ et al: on behalf of the ELITE Study Investigators: Randomised trial of Losartan versus captopril in patients over 65 with heart failure (Evaluation of Losartan in the Elderly Study, ELITE), *Lancet* 349:747–52, 1997.

Cohn JN, Ziesche S, Smith R, Anand I, Dunkman B et al: for the Vasodilator Heart Failure Trial (VHeFT) Study Group: Effect of the calcium antagonist felodipine as supplementary vasodilatory therapy in patients with chronic heart failure treated with enalapril: VHeFT III, *Circulation* 96:856–63, 1997.

Elkayam U, Amin J, Mehra A, Vasquez J, Weber L et al: A prospective, randomized, double-blind, crossover study to compare the efficacy and safety of chronic nifedipine therapy with that of isosorbide dinitrate and their combination in the treatment of chronic congestive heart failure, *Circulation* 82:1954–61, 1990.

Hampton JR, van Veldhuisen DJ, Kleber FX, Cowley AJ, Ardia A et al: for the Second Prospective Randomised Study of Ibopamine on Mortality, and Efficacy (PRIME II) Investigators: Randomised study of effect of ibopamine on survival in patients with advanced severe heart failure, *Lancet* 349:971–7, 1997.

The Cardiac Arrhythmia Suppression Trial II Investigators: Effect of antiarrhythmic agent moricize on survival after myocardial infarction, *New Engl J Med* 327:227–33, 1992.

Levine GN, Keaney JF, Vita JA: Cholesterol reduction in cardiovascular disease: Clinical benefits and possible mechanisms, *New Engl J Med* 332:512–21, 1995.

Pedersen TR, Kjekshus J, Berg K, Olsson AG, Wilhelmsen L et al: for the Scandinavian Simvastatin Survival Study Group: Cholesterol lowering and the use of healthcare resources: Results of the Scandinavian Simvastatin Survival Study, *Circulation* 93:1796–802, 1996.

14.6. Molecular Therapy for Cardiac Failure

Williams ML, Koch WJ: Viral-based myocardial gene therapy approaches to alter cardiac function, *Annu Rev Physiol* 66:49–75, 2004.

Houser SR, Margulies KB: Is depressed myocyte contractility centrally involved in heart failure? *Circ Res* 92:350–8, 2003.

Miyamoto MI, del Monte F, Schmidt U, DiSalvo TS, Kang ZB et al: Adenoviral gene transfer of SERCA2a improves left-ventricular function in aortic-banded rats in transition to heart failure, *Proc Natl Acad Sci USA* 97:793–8, 2000.

del Monte F, Williams E, Lebeche D, Schmidt U, Rosenzweig A et al: Improvement in survival and cardiac metabolism after gene transfer of sarcoplasmic reticulum Ca^{2+}-ATPase in a rat model of heart failure, *Circulation* 104:1424–9, 2001.

Hagemann D, Xiao RP: Dual site phospholamban phosphorylation and its physiological relevance in the heart, *Trends Cardiovasc Med* 12:51–6, 2002.

Movsesian MA: cAMP-mediated signal transduction and sarcoplasmic reticulum function in heart failure, *Ann NY Acad Sci* 853:231–9, 1998.

Minamisawa S, Hoshijima M, Chu G, Ward CA, Frank K et al: Chronic phospholamban-sarcoplasmic reticulum calcium ATPase interaction is the critical calcium cycling defect in dilated cardiomyopathy, *Cell* 99:313–22, 1999.

Freeman K, Lerman I, Kranias EG, Bohlmeyer T, Bristow MR et al: Alterations in cardiac adrenergic signaling and calcium cycling differentially affect the progression of cardiomyopathy, *J Clin Invest* 107:967–74, 2001.

Sato Y, Kiriazis H, Yatani A, Schmidt AG, Hahn H et al: Rescue of contractile parameters and myocyte hypertrophy in calsequestrin overexpressing myocardium by phospholamban ablation, *J Biol Chem* 276:9392–9, 2001.

Eizema K, Fechner H, Bezstarosti K, Schneider-Rasp S, van der Laarse A et al: Adenovirus-based phospholamban antisense expression as a novel approach to improve cardiac contractile dysfunction: Comparison of a constitutive viral versus an endothelin-1-responsive cardiac promoter, *Circulation* 101:2193–9, 2000.

del Monte F, Harding SE, Dec GW, Gwathmey JK, Hajjar RJ: Targeting phospholamban by gene transfer in human heart failure, *Circulation* 105:904–7, 2002.

Hoshijima M, Ikeda Y, Iwanaga Y, Minamisawa S, Date MO et al: Chronic suppression of heart-failure progression by a pseudophosphorylated mutant of phospholamban via in vivo cardiac rAAV gene delivery, *Nat Med* 8:864–71, 2002.

Haghighi K, Gregory KN, Kranias EG: Sarcoplasmic reticulum Ca-ATPase-phospholamban interactions and dilated cardiomyopathy, *Biochem Biophys Res Commun* 322:1214–22, 2004.

Movsesian MA: cAMP-mediated signal transduction and sarcoplasmic reticulum function in heart failure, *Ann NY Acad Sci* 853:231–9, 1998.

Schillinger W, Janssen PM, Enami S, Henderson SA, Ross RS et al: Impaired contractile performance of cultured rabbit ventricular myocytes after adenoviral gene transfer of Na^+-Ca^{2+} exchanger, *Circ Res* 87:581–7, 2000.

Communal C, Singh K, Sawyer DB, Colucci WS: Opposing effects of B1 and B2 adrenergic receptors on cardiac myocyte apoptosis: Role of pertussis toxin-sensitive G protein, *Circulation* 100:2210–2, 1999.

Milano CA, Allen LF, Rockman HA, Dolber PC, McMinn TR et al: Enhanced myocardial function in transgenic mice overexpressing the β_2-adrenergic receptor, *Science* 264:582–6, 1994.

Engelhardt S, Hein L, Wiesmann F, Lohse MJ: Progressive hypertrophy and heart failure in β_1-adrenergic receptor transgenic mice, *Proc Natl Acad Sci USA* 96:7059–64, 1999.

Milano CA, Allen LF, Rockman HA, Dolber PC, McMinn TR et al: Enhanced myocardial function in transgenic mice overexpressing the β_2-adrenergic receptor, *Science* 264:582–6, 1994.

Dorn GW, Tepe NM, Lorenz JN, Davis MG, Koch WJ et al: Low and high transgenic expression of β_2-adrenergic receptors differentially affects cardiac hypertrophy and function in Gαq overexpressing mice, *Proc Natl Acad Sci USA* 96:6400–5, 1999.

Maurice JP, Hata JA, Shah AS, White DC, McDonald PH et al: Enhancement of cardiac function after adenoviral-mediated in vivo intracoronary beta-2-adrenergic receptor gene delivery, *J Clin Invest* 104:21–9, 1999.

Engelhardt S, Hein L, Wiesmann F, Lohse MJ: Progressive hypertrophy and heart failure in β_1-adrenergic receptor transgenic mice, *Proc Natl Acad Sci USA* 96:7059–64, 1999.

Brodde OE: Beta-adrenoceptors in cardiac disease, *Pharmacol Ther* 60:405–30, 1993.

Steinberg SF: The molecular basis for distinct β-adrenergic receptor subtype actions in cardiomyocytes, *Circ Res* 85:1101–11, 1999.

Maurice JP, Hata JA, Shah AS, White DC, McDonald PH et al: Enhancement of cardiac function after adenoviral-mediated in vivo intracoronary β_2-adrenergic receptor gene delivery, *J Clin Invest* 104:21–9, 1999.

Kawahira Y, Sawa Y, Nishimura M, Sakakida S, Ueda H et al: Gene transfection of β_2-adrenergic receptor into the normal rat heart enhances cardiac response to β-adrenergic agonist, *J Thorac Cardiovasc Surg* 118:446–51, 1999.

Kawahira Y, Sawa Y, Nishimura M, Sakakida S, Ueda H et al: In vivo transfer of a β_2-adrenergic receptor gene into the pressure-overloaded rat heart enhances cardiac response to β-adrenergic agonist, *Circulation* 98:II262–7, 1998.

Shah AS, Lilly RE, Kypson AP, Tai O, Hata JA et al: Intracoronary adenovirus-mediated delivery and overexpression of the β_2-adrenergic receptor in the heart: Prospects for molecular ventricular assistance, *Circulation* 101:408–14, 2000.

Kreusser MM, Haass M, Buss SJ, Hardt SE, Gerber SH et al: Injection of nerve growth factor into stellate ganglia improves norepinephrine reuptake into failing hearts, *Hypertension* 47:209–15, 2006.

14.7. Tissue Engineering Therapy for Cardiac Failure

Zimmermann WH, Didie M, Wasmeier GH, Nixdorff U, Hess A et al: Cardiac grafting of engineered heart tissue in syngenic rats, *Circulation* 106(Suppl 1):151–7, 2002.

14.8. Pathogenesis, Pathology, and Clinical Features of Ischemic Heart Disease

Gheorghiade M, Bonow RO: Chronic heart failure in the United States: A manifestation of coronary artery disease, *Circulation* 97:282–9, 1998.

Sytkowski PA, Kannel WB, D'Agostino RB: Changes in risk factors and the decline in mortality from cardiovascular disease: The Framingham Heart Study, *New Engl J Med* 322:1635–41, 1990.

Andrews TC, Raby K, Barry J, Naimi CL, Allred E et al: Effect of cholesterol reduction on myocardial ischemia in patients with coronary disease, *Circulation* 95:324–8, 1997.

Frye RL: Clinical reality of lowering total and LDL cholesterol, *Circulation* 95:306–7, 1997.

Mancini GBJ, Henry GC, Macaya C, O'Neill BR, Pucillo AL et al: Angiotensin-converting enzyme inhibition with quinapril improves endothelial vasomotor dysfunction in patients with coronary artery disease: The TREND (Trial on Reversing Endothelial Dysfunction) Study, *Circulation* 94:258–65, 1996.

Daemen MJAP, Lombardi DM, Bosman FT, Schwatz SM: Angiotensin II induces smooth muscle cell proliferation in the normal and injured rat arterial wall, *Circ Res* 68:450–6, 1991.

Falk E, Shah PK, Fuster V: Coronary plaque disruption, *Circulation* 92:657–71, 1995.

Latini R, Maggioni AP, Flather M, Sleight P, Tognoni G: ACE inhibitor use in patients with myocardial infarction: Summary of evidence from clinical trials, *Circulation* 92:3132–7, 1995.

Rapaport E, Gheorghiade M: Pharmacologic therapies after myocardial infarction, *Am J Med* 101(Suppl 4A):61S–70S, 1996.

Ambrosio G, Betocchi S, Pace L, Losi MA, Perrone-Filardi P et al: Prolonged impairment of regional contractile function after resolution of exercise-induced angina: Evidence of myocardial stunning in patients with coronary artery disease, *Circulation* 94:2455–64, 1996.

14.8. Molecular Regenerative Engineering for Ischemic Heart Disease

Melo LG, Pachori AS, Kong D, Gnecchi M, Wang K et al: Gene and cell-based therapies for heart disease, *FASEB J* 18:648–63, 2004.

Brocheriou V, Hagege AA, Oubenaissa A, Lambert M, Mallet VO et al: Cardiac functional improvement by a human Bcl-2 transgene in a mouse model of ischemia/reperfusion injury, *J Gene Med* 2:326–33, 2000.

Miao W, Luo Z, Kitsis RN, Walsh K: Intracoronary, adenovirus-mediated Akt gene transfer in heart limits infarct size following ischemia-reperfusion injury in vivo, *J Mol Cell Cardiol* 32:2397–402, 2000.

Brauner R, Nonoyama M, Laks H, Drinkwater DC, McCaffery S et al: Intracoronary adenovirus-mediated transfer of immunosuppressive cytokine genes prolongs allograft survival, *J Thorac Cardiovasc Surg* 114:923–33, 1997.

14.9. Fibroblast Growth Factor 1 (Acidic)

Mergia A, Eddy R, Abraham JA, Fiddes JC, Shows TB: The genes for basic and acidic fibroblast growth factors are on different human chromosomes, *Biochem Biophys Res Commun* 138:644–51, 1986.

Burgess WH, Mehlman T, Marshak DR, Fraser BA, Maciag T: Structural evidence that endothelial cell growth factor beta is the precursor of both endothelial cell growth factor alpha and acidic fibroblast growth factor, *Proc Natl Acad Sci USA* 83:7216–20, 1986.

Jaye M, Howk R, Burgess W, Ricca GA, Chiu IM et al: Human endothelial cell growth factor: Cloning, nucleotide sequence, and chromosome localization, *Science* 233:541–5, 1986.

Le Beau MM, Espinosa R III, Neuman WL, Stock W, Roulston D et al: Cytogenetic and molecular delineation of the smallest commonly deleted region of chromosome 5 in malignant myeloid diseases, *Proc Natl Acad Sci USA* 90:5484–8, 1993.

Plotnikov AN, Hubbard SR, Schlessinger J, Mohammadi M: Crystal structures of two FGF-FGFR complexes reveal the determinants of ligand-receptor specificity, *Cell* 101:413–24, 2000.

Pellegrini L, Burke DF, von Delft F, Mulloy B, Blundell TL: Crystal structure of fibroblast growth factor receptor ectodomain bound to ligand and heparin, *Nature* 407:1029–34, 2000.

Eckenstein FP: Fibroblast growth factors in the nervous system, *J Neurobiol* 25(11):1467–80, Nov 1994.

Jung J, Zheng M, Goldfarb M, Zaret KS: Initiation of mammalian liver development from endoderm by fibroblast growth factors, *Science* 284:1998–2003, 1999.

14.10. Fibroblast Growth Factor 2 (Basic)

Abraham JA, Mergia A, Whang JL, Tumolo A, Friedman J et al: Nucleotide sequence of a bovine clone encoding the angiogenic protein, basic fibroblast growth factor, *Science* 233:545–8, 1986.

Fukushima Y, Byers MG, Fiddes JC, Shows TB: The human basic fibroblast growth factor gene (FGFB) is assigned to chromosome 4q25, *Cytogenet Cell Genet* 54:159–60, 1990.

Plotnikov AN, Hubbard SR, Schlessinger J, Mohammadi M: Crystal structures of two FGF-FGFR complexes reveal the determinants of ligand-receptor specificity, *Cell* 101:413–24, 2000.

Plotnikov AN, Schlessinger J, Hubbard SR, Mohammadi M: Structural basis for FGF receptor dimerization and activation, *Cell* 98:641–50, 1999.

Gritti A, Parati EA, Cova L, Frolichsthal P, Galli R et al: Multipotential stem cells from the adult mouse brain proliferate and self-renew in response to basic fibroblast growth factor, *J Neurosci* 16:1091–100, 1996.

Ortega S, Ittmann M, Tsang SH, Ehrlich M, Basilico C: Neuronal defects and delayed wound healing in mice lacking fibroblast growth factor 2, *Proc Natl Acad Sci USA* 95:5672–7, 1998.

Johnson RL, Tabin CJ: Molecular models for vertebrate limb development, *Cell* 90:979–90, 1997.

Jung J, Zheng M, Goldfarb M, Zaret KS: Initiation of mammalian liver development from endoderm by fibroblast growth factors, *Science* 284:1998–2003, 1999.

Hardikar AA, Marcus-Samuels B, Geras-Raaka E, Raaka BM, Gershengorn MC: Human pancreatic precursor cells secrete FGF2 to stimulate clustering into hormone-expressing islet-like cell aggregates, *Proc Natl Acad Sci USA* 100:7117–22, 2003.

Cao R, Brakenhielm E, Pawliuk R, Wariaro D, Post MJ et al: Angiogenic synergism, vascular stability and improvement of hind-limb ischemia by a combination of PDGF-BB and FGF-2, *Nature Med* 9:604–13, 2003.

Chang LK, Garcia-Cardena G, Farnebo F, Fannon M, Chen EJ et al: Dose-dependent response of FGF-2 for lymphangiogenesis, *Proc Natl Acad Sci USA* 101:11658–63, 2004.

Montero A, Okada Y, Tomita M, Ito M, Tsurukami H et al: Disruption of the fibroblast growth factor-2 gene results in decreased bone mass and bone formation, *J Clin Invest* 105:1085–93, 2000.

Rosenblatt-Velin N, Lepore MG, Cartoni C, Beermann F, Pedrazzini T: FGF-2 controls the differentiation of resident cardiac precursors into functional cardiomyocytes, *J Clin Invest* 115:1724–33, 2005.

Alavi A, Hood JD, Frausto R, Stupack DG, Cheresh DA: Role of Raf in vascular protection from distinct apoptotic stimuli, *Science* 301:94–6, 2003.

14.11. Fibroblast Growth Factor Receptor 1

Ruta M, Howk R, Ricca G, Drohan W, Zabelshansky M et al: A novel protein tyrosine kinase gene whose expression is modulated during endothelial cell differentiation, *Oncogene* 3:9–15, 1988.

Ruta M, Burgess W, Givol D, Epstein J, Neiger N et al: Receptor for acidic fibroblast growth factor is related to the tyrosine kinase encoded by the FMS-like gene (FLG), *Proc Natl Acad Sci USA* 86:8722–6, 1989.

Lee PL, Johnson DE, Cousens LS, Fried VA, Williams LT: Purification and complementary DNA cloning of a receptor for basic fibroblast growth factor, *Science* 245:57–60, 1989.

Pirvola U, Ylikoski J, Trokovic R, Hebert JM, McConnell SK et al: FGFR1 is required for the development of the auditory sensory epithelium, *Neuron* 35:671–80, 2002.

Muenke M, Schell U, Hehr A, Robin NH, Losken HW et al: A common mutation in the fibroblast growth factor receptor 1 gene in Pfeiffer syndrome, *Nat Genet* 8:269–74, 1994.

Dode C, Levilliers J, Dupont JM, De Paepe A, Le Du N et al: Loss-of-function mutations in FGFR1 cause autosomal dominant Kallmann syndrome, *Nat Genet* 33:463–5, 2003.

14.12. Fibroblast Growth Factor Receptor 2

Mattei M-G, Moreau A, Gesnel M-C, Houssaint E, Breathnach R: Assignment by in situ hybridization of a fibroblast growth factor receptor gene to human chromosome band 10q26, *Hum Genet* 87:84–6, 1991.

Dionne CA, Crumley G, Bellot F, Kaplow JM, Searfoss G et al: Cloning and expression of two distinct high-affinity receptors cross-reacting with acidic and basic fibroblast growth factors, *EMBO J* 9:2685–92, 1990.

Cornejo-Roldan LR, Roessler E, Muenke M: Analysis of the mutational spectrum of the FGFR2 gene in Pfeiffer syndrome, *Hum Genet* 104:425–31, 1999.

Jang JH, Shin KH, Park JG: Mutations in fibroblast growth factor receptor 2 and fibroblast growth factor receptor 3 genes associated with human gastric and colorectal cancers, *Cancer Res* 61:3541–3, 2001.

14.13. Fibroblast Growth Factor Receptor 3

Rousseau F, el Ghouzzi V, Delezoide AL, Legeai-Mallet L, Le Merrer M et al: Missense FGFR3 mutations create cysteine residues in thanatophoric dwarfism type I (TD1), *Hum Mol Genet* 5:509–12, 1996.

Jang JH, Shin KH, Park JG: Mutations in fibroblast growth factor receptor 2 and fibroblast growth factor receptor 3 genes associated with human gastric and colorectal cancers, *Cancer Res* 61:3541–3, 2001.

Mortier G, Nuytinck L, Craen M, Renard JP, Leroy JG et al: Clinical and radiographic features of a family with hypochondroplasia owing to a novel Asn540Ser mutation in the fibroblast growth factor receptor 3 gene, *J Med Genet* 37:220–4, 2000.

14.14. Fibroblast Growth Factor Receptor 4

Jaakkola S, Salmikangas P, Nylund S, Partanen J, Armstrong E et al: Amplification of fgfr4 gene in human breast and gynecological cancers, *Int J Cancer* 54:378–82, 1993.

14.15. Epidermal Growth Factor

Gray A, Dull TJ, Ullrich A: Nucleotide sequence of epidermal growth factor cDNA predicts a 128,000-molecular weight protein precursor, *Nature* 303:722–5, 1983.

Urdea MS, Merryweather JP, Mullenbach GT, Coit D, Heberlein U et al: Chemical synthesis of a gene for human epidermal growth factor urogastrone and its expression in yeast, *Proc Natl Acad Sci USA* 80:7461–5, 1983.

Lei ZM, Rao CV: Expression of epidermal growth factor (EGF) receptor and its ligands, EGF and transforming growth factor-alpha, in human fallopian tubes, *Endocrinology* 131:947–57, 1992.

Barnard JA, Graves-Deal R, Pittelkow MR, DuBois R, Cook P et al: Auto- and cross-induction within the mammalian epidermal growth factor-related peptide family, *J Biol Chem* 269:22817–22, 1994.

McClellan M, Kievit P, Auersperg N, Rodland K: Regulation of proliferation and apoptosis by epidermal growth factor and protein kinase C in human ovarian surface epithelial cells, *Exp Cell Res* 246:471–9, 1999.

Gansauge S, Gansauge F, Yang Y, Muller J, Seufferlein T et al: Interleukin 1beta-converting enzyme (caspase-1) is overexpressed in adenocarcinoma of the pancreas, *Cancer Res* 58:2703–6, 1998.

Adam RM, Borer JG, Williams J, Eastham JA, Loughlin KR et al: Amphiregulin is coordinately expressed with heparin-binding epidermal growth factor-like growth factor in the interstitial smooth muscle of the human prostate, *Endocrinology* 140:5866–75, 1999.

Edmondson SR, Murashita MM, Russo VC, Wraight CJ, Werther GA: Expression of insulin-like growth factor binding protein-3 (IGFBP-3) in human keratinocytes is regulated by EGF and TGFbeta1, *J Cell Physiol* 179:201–7, 1999.

Shahbazi M, Pravica V, Nasreen N, Fakhoury H, Fryer AA et al: Association between functional polymorphism in EGF gene and malignant melanoma, *Lancet* 359:397–401, 2002.

14.16. Epidermal Growth Factor Receptor

Carpenter G: The EGF receptor: A nexus for trafficking and signaling, *Bioessays* 22:697–707, 2000.

Lowenstein EJ, Daly RJ, Batzer AG, Li W, Margolis B et al: The SH2 and SH3 domain-containing protein GRB2 links receptor tyrosine kinases to ras signaling, *Cell* 70:431–42, 1992.

Tortora G, Damiano V, Bianco C, Baldassarre G, Bianco AR et al: The RIalpha subunit of protein kinase A (PKA) binds to Grb2 and allows PKA interaction with the activated EGF-receptor, *Oncogene* 14:923–8, 1997.

Sieg DJ, Hauck CR, Ilic D, Klingbeil CK, Schaefer E et al: FAK integrates growth-factor and integrin signals to promote cell migration, *Nat Cell Biol* 2:249–56, 2000.

Kuwada SK, Li X: Integrin alpha5/beta1 mediates fibronectin-dependent epithelial cell proliferation through epidermal growth factor receptor activation, *Mol Biol Cell* 11:2485–96, 2000.

Gauthier ML, Torretto C, Ly J, Francescutti V, O'Day DH: Protein kinase Calpha negatively regulates cell spreading and motility in MDA-MB-231 human breast cancer cells downstream of epidermal growth factor receptor, *Biochem Biophys Res Commun* 307:839–46, 2003.

Qian X, Esteban L, Vass WC, Upadhyaya C, Papageorge AG et al: The Sos1 and Sos2 Ras-specific exchange factors: Differences in placental expression and signaling properties, *EMBO J* 19:642–54, 2000.

Tomic S, Greiser U, Lammers R, Kharitonenkov A, Imyanitov E et al: Association of SH2 domain protein tyrosine phosphatases with the epidermal growth factor receptor in human tumor cells. Phosphatidic acid activates receptor dephosphorylation by PTP1C, *J Biol Chem* 270:21277–84, 1995.

14.17. Platelet-Derived Growth Factor A Chain

Bonthron D, Collins T, Grzeschik KH, van Roy N, Speleman F: Platelet-derived growth factor A chain: Confirmation of localization of PDGFA to chromosome 7p22 and description of an unusual minisatellite, *Genomics* 13:257–63, 1992.

Mendoza AE, Young R, Orkin SH, Collins T: Increased platelet-derived growth factor A-chain expression in human uterine smooth muscle cells during the physiologic hypertrophy of pregnancy, *Proc Natl Acad Sci USA* 87:2177–81, 1990.

Betsholtz C, Johnsson A, Heldin CH, Westermark B, Lind P et al: cDNA sequence and chromosomal localization of human platelet-derived growth factor A-chain and its expression in tumour cell lines, *Nature* 320:695–9, 1986.

Claesson-Welsh L, Eriksson A, Westermark B, Heldin CH: cDNA cloning and expression of the human A-type platelet-derived growth factor (PDGF) receptor establishes structural similarity to the B-type PDGF receptor, *Proc Natl Acad Sci USA* 86:4917–21, 1989.

Wilcox JN, Smith KM, Williams LT, Schwartz SM, Gordon D: Platelet-derived growth factor mRNA detection in human atherosclerotic plaques by in situ hybridization, *J Clin Invest* 82:1134–43, 1988.

Gnessi L, Basciani S, Mariani S, Arizzi M, Spera G et al: Leydig cell loss and spermatogenic arrest in platelet-derived growth factor (PDGF)-A-deficient mice, *J Cell Biol* 149:1019–25, 2000.

14.18. Platelet-Derived Growth Factor B Chain

Hermansson M, Nister M, Betsholtz C, Heldin CH, Westermark B et al: Endothelial cell hyperplasia in human glioblastoma: Coexpression of mRNA for platelet-derived growth factor (PDGF) B chain and PDGF receptor suggests autocrine growth stimulation, *Proc Natl Acad Sci USA* 85:7748–52, 1988.

Matsuda M, Shikata K, Makino H, Sugimoto H, Ota K et al: Gene expression of PDGF and PDGF receptor in various forms of glomerulonephritis, *Am J Nephrol* 17:25–31, 1997.

Versnel MA, Haarbrink M, Langerak AW, de Laat PA, Hagemeijer A et al: Human ovarian tumors of epithelial origin express PDGF in vitro and in vivo, *Cancer Genet Cytogenet* 73:60–4, 1994.

Li DQ, Tseng SC: Three patterns of cytokine expression potentially involved in epithelial-fibroblast interactions of human ocular surface, *J Cell Physiol* 163:61–79, 1995.

Claesson-Welsh L, Eriksson A, Westermark B, Heldin CH: cDNA cloning and expression of the human A-type platelet-derived growth factor (PDGF) receptor establishes structural similarity to the B-type PDGF receptor, *Proc Natl Acad Sci USA* 86:4917–21, 1989.

Cartel NJ, Wang J, Post M: Platelet-derived growth factor-BB-mediated glycosaminoglycan synthesis is transduced through Akt, *Biochem J* 363:19–28, 2002.

Gohring W, Sasaki T, Heldin CH, Timpl R: Mapping of the binding of platelet-derived growth factor to distinct domains of the basement membrane proteins BM-40 and perlecan and distinction from the BM-40 collagen-binding epitope, *Eur J Biochem* 255:60–6, 1998.

14.19. Platelet-Derived Growth Factor Receptor α

Claesson-Welsh L, Eriksson A, Westermark B, Heldin CH: cDNA cloning and expression of the human A-type platelet-derived growth factor (PDGF) receptor establishes structural similarity to the B-type PDGF receptor, *Proc Natl Acad Sci USA* 86:4917–21, 1989.

Bazenet CE, Gelderloos JA, Kazlauskas A: Phosphorylation of tyrosine 720 in the platelet-derived growth factor alpha receptor is required for binding of Grb2 and SHP-2 but not for activation of Ras or cell proliferation, *Mol Cell Biol* 16:6926–36, 1996.

Yokote K, Hellman U, Ekman S, Saito Y, Ronnstrand L et al: Mori SIdentification of Tyr-762 in the platelet-derived growth factor alpha-receptor as the binding site for Crk proteins, *Oncogene* 16:1229–39, 1998.

Eriksson A, Nanberg E, Ronnstrand L, Engstrom U, Hellman U et al: Demonstration of functionally different interactions between phospholipase C-gamma and the two types of platelet-derived growth factor receptors, *J Biol Chem* 270:7773–81, 1995.

Kawagishi J, Kumabe T, Yoshimoto T, Yamamoto T: Structure, organization, and transcription units of the human alpha-platelet-derived growth factor receptor gene, PDGFRA, *Genomics* 30:224–32, 1995.

Muschen M, Lee S, Zhou G, Feldhahn N, Barath VS et al: Molecular portraits of B cell lineage commitment, *Proc Natl Acad Sci USA* 99:10014–9, 2002.

Gogate N, Verma L, Zhou JM, Milward E, Rusten R et al: Plasticity in the adult human oligodendrocyte lineage, *J Neurosci* 14:4571–87, 1994.

Vassbotn FS, Havnen OK, Heldin CH, Holmsen H: Negative feedback regulation of human platelets via autocrine activation of the platelet-derived growth factor alpha-receptor, *J Biol Chem* 269:13874–9, 1994.

Baron W, Shattil SJ, ffrench-Constant C: The oligodendrocyte precursor mitogen PDGF stimulates proliferation by activation of alpha(v)beta3 integrins, *EMBO J* 21:1957–66, 2002.

14.20. Platelet-Derived Growth Factor Receptor Beta

Gronwald RGK, Grant FJ, Haldeman BA, Hart CE, O'Hara PJ et al: Cloning and expression of a cDNA coding for the human platelet-derived growth factor receptor: Evidence for more than one receptor class, *Proc Natl Acad Sci USA* 85:3435–9, 1988.

Claesson-Welsh L, Eriksson A, Moren A, Severinsson L, Ek B et al: cDNA cloning and expression of a human platelet-derived growth factor (PDGF) receptor specific for B-chain-containing PDGF molecules, *Molec Cell Biol* 8:3476–86, 1988.

Holtrich U, Brauninger A, Strebhardt K, Rubsamen-Waigmann H: Two additional protein-tyrosine kinases expressed in human lung: Fourth member of the fibroblast growth factor receptor family and an intracellular protein-tyrosine kinase, *Proc Natl Acad Sci USA* 88:10411–5, 1991.

Satomura K, Derubeis AR, Fedarko NS, Ibaraki-O'Connor K, Kuznetsov SA et al: Receptor tyrosine kinase expression in human bone marrow stromal cells, *J Cell Physiol* 177:426–38, 1998.

Liu SQ, Tieche C, Tang D, Alkema P: Pattern formation of vascular smooth muscle cells subject to non-uniform fluid shear stress: Role of platelet-derived growth factor β receptor and Src, *Am J Physiol* 285:H1081–90, 2003.

Cartel NJ, Wang J, Post M: Platelet-derived growth factor-BB-mediated glycosaminoglycan synthesis is transduced through Akt, *Biochem J* 363:19–28, 2002.

Lechleider RJ, Sugimoto S, Bennett AM, Kashishian AS, Cooper JA et al: Activation of the SH2-containing phosphotyrosine phosphatase SH-PTP2 by its binding site, phosphotyrosine 1009, on the human platelet-derived growth factor receptor, *J Biol Chem* 268:21478–81, 1993.

Morrison DK, Kaplan DR, Escobedo JA, Rapp UR, Roberts TM et al: Direct activation of the serine/threonine kinase activity of Raf-1 through tyrosine phosphorylation by the PDGF beta-receptor, *Cell* 58:649–57, 1989.

Sieg DJ, Hauck CR, Ilic D, Klingbeil CK, Schaefer E et al: FAK integrates growth-factor and integrin signals to promote cell migration, *Nat Cell Biol* 2:249–56, 2000.

Arvidsson AK, Rupp E, Nanberg E, Downward J, Ronnstrand L et al: Tyr-716 in the platelet-derived growth factor beta-receptor kinase insert is involved in GRB2 binding and Ras activation, *Mol Cell Biol* 14:6715–26, 1994.

Chen M, She H, Kim A, Woodley DT, Li W: Nckbeta adapter regulates actin polymerization in NIH 3T3 fibroblasts in response to platelet-derived growth factor bb, *Mol Cell Biol* 20:7867–80, 2000.

Farooqui T, Kelley T, Coggeshall KM, Rampersaud AA, Yates AJ: GM1 inhibits early signaling events mediated by PDGF receptor in cultured human glioma cells, *Anticancer Res* 19:5007–13, 1999.

Ekman S, Kallin A, Engstrom U, Heldin CH, Ronnstrand L: SHP-2 is involved in heterodimer specific loss of phosphorylation of Tyr771 in the PDGF beta-receptor, *Oncogene* 21:1870–5, 2002.

Borges E, Jan Y, Ruoslahti E: Platelet-derived growth factor receptor beta and vascular endothelial growth factor receptor 2 bind to the beta 3 integrin through its extracellular domain, *J Biol Chem* 275:39867–73, 2000.

14.21. Vascular Endothelial Growth Factor A

Tischer E, Mitchell R, Hartman T, Silva M, Gospodarowicz D et al: The human gene for vascular endothelial growth factor: Multiple protein forms are encoded through alternative exon splicing, *J Biol Chem* 266:11947–54, 1991.

Ferrara N, Carver-Moore K, Chen H, Dowd M, Lu L et al: Heterozygous embryonic lethality induced by targeted inactivation of the VEGF gene, *Nature* 380:439–42, 1996.

Leung DW, Cachianes G, Kuang WJ, Goeddel DV, Ferrara N: Vascular endothelial growth factor is a secreted angiogenic mitogen, *Science* 246:1306–9, 1989.

Berse B, Brown LF, Van de Water L, Dvorak HF, Senger DR: Vascular permeability factor (vascular endothelial growth factor) gene is expressed differentially in normal tissues, macrophages, and tumors, *Mol Biol Cell* 3:211–20, 1992.

Takahashi A, Sasaki H, Kim SJ, Tobisu K, Kakizoe T et al: Markedly increased amounts of messenger RNAs for vascular endothelial growth factor and placenta growth factor in renal cell carcinoma associated with angiogenesis, *Cancer Res* 54:4233–7, 1994.

von Marschall Z, Cramer T, Hocker M, Finkenzeller G, Wiedenmann B et al: Dual mechanism of vascular endothelial growth factor upregulation by hypoxia in human hepatocellular carcinoma, *Gut* 48:87–96, 2001.

Yonekura H, Sakurai S, Liu X, Migita H, Wang H et al: Placenta growth factor and vascular endothelial growth factor B and C expression in microvascular endothelial cells and pericytes. Implication in autocrine and paracrine regulation of angiogenesis, *J Biol Chem* 274:35172–8, 1999.

Frank S, Hubner G, Breier G, Longaker MT, Greenhalgh DG et al: Regulation of vascular endothelial growth factor expression in cultured keratinocytes. Implications for normal and impaired wound healing, *J Biol Chem* 270:12607–13, 1995.

Gerber HP, Malik AK, Solar GP, Sherman D, Liang XH et al: VEGF regulates haematopoietic stem cell survival by an internal autocrine loop mechanism, *Nature* 417:954–8, 2002.

Keyt BA, Nguyen HV, Berleau LT, Duarte CM, Park J et al: Identification of vascular endothelial growth factor determinants for binding KDR and FLT-1 receptors. Generation of receptor-selective VEGF variants by site-directed mutagenesis, *J Biol Chem* 271:5638–46, 1996.

Watanabe D, Suzuma K, Suzuma I, Ohashi H, Ojima T et al: Vitreous levels of angiopoietin 2 and vascular endothelial growth factor in patients with proliferative diabetic retinopathy, *Am J Ophthal* 139:476–81, 2005.

Helmlinger G, Endo M, Ferrara N, Hlatky L, Jain RK: Formation of endothelial cell networks, *Nature* 405:139–41, 2000.

14.22. Vascular Endothelial Growth Factor B

Olofsson B, Pajusola K, Kaipainen A, von Euler G, Joukov V et al: Vascular endothelial growth factor B, a novel growth factor for endothelial cells, *Proc Natl Acad Sci USA* 93:2576–81, 1996.

Olofsson B, Pajusola K, von Euler G, Chilov D, Alitalo K et al: Genomic organization of the mouse and human genes for vascular endothelial growth factor B (VEGF-B) and characterization of a second splice isoform, *J Biol Chem* 271:19310–7, 1996.

Silvestre JS, Tamarat R, Ebrahimian TG, Le-Roux A, Clergue M et al: Vascular endothelial growth factor-B promotes in vivo angiogenesis, *Circ Res* 93:114–23, 2003.

Bellomo D, Headrick JP, Silins GU, Paterson CA, Thomas PS et al: Mice lacking the vascular endothelial growth factor-B gene (Vegfb) have smaller hearts, dysfunctional coronary vasculature, and impaired recovery from cardiac ischemia, *Circ Res* 86:e29–35, 2000.

Olofsson B, Pajusola K, Kaipainen A, von Euler G, Joukov V et al: Vascular endothelial growth factor B, a novel growth factor for endothelial cells, *Proc Natl Acad Sci USA* 93:2576–81, 1996.

14.23. Vascular Endothelial Growth Factor Receptor 1

Ergun S, Kilic N, Fiedler W, Mukhopadhyay AK: Vascular endothelial growth factor and its receptors in normal human testicular tissue, *Mol Cell Endocrinol* 131:9–20, 1997.

Bellamy WT, Richter L, Frutiger Y, Grogan TM: Expression of vascular endothelial growth factor and its receptors in hematopoietic malignancies, *Cancer Res* 59:728–33, 1999.

Wiesmann C, Fuh G, Christinger HW, Eigenbrot C, Wells JA et al: Crystal structure at 1.7 Angstrom resolution of VEGF in complex with domain 2 of the Flt-1 receptor, *Cell* 91:695–704, 1997.

Keyt BA, Nguyen HV, Berleau LT, Duarte CM, Park J et al: Identification of vascular endothelial growth factor determinants for binding KDR and FLT-1 receptors. Generation of receptor-selective VEGF variants by site-directed mutagenesis, *J Biol Chem* 271:5638–46, 1996.

Olofsson B, Korpelainen E, Pepper MS, Mandriota SJ, Aase K et al: Vascular endothelial growth factor B (VEGF-B) binds to VEGF receptor-1 and regulates plasminogen activator activity in endothelial cells, *Proc Natl Acad Sci USA* 95:11709–14, 1998.

Autiero M, Waltenberger J, Communi D, Kranz A, Moons L et al: Role of PlGF in the intra- and intermolecular cross talk between the VEGF receptors Flt1 and Flk1, *Nature Med* 9:936–43, 2003.

Kendall RL, Wang G, Thomas KA: Identification of a natural soluble form of the vascular endothelial growth factor receptor, FLT-1, and its heterodimerization with KDR, *Biochem Biophys Res Commun* 226:324–8, 1996.

He Y, Smith SK, Day KA, Clark DE, Licence DR et al: Alternative splicing of vascular endothelial growth factor (VEGF)-R1 (FLT-1) pre-mRNA is important for the regulation of VEGF activity, *Mol Endocr* 13:537–45, 1999.

14.24. Vascular Endothelial Growth Factor Receptor 2

Ziegler BL, Valtieri M, Porada GA, De Maria R, Muller R et al: KDR receptor: A key marker defining hematopoietic stem cells, *Science* 285:1553–8, 1999.

Yamashita J, Itoh H, Hirashima M, Ogawa M, Nishikawa S et al: Flk1-positive cells derived from embryonic stem cells serve as vascular progenitors, *Nature* 408:92–6, 2000.

Terman BI, Carrion ME, Kovacs E, Rasmussen BA, Eddy RL et al: Identification of a new endothelial cell growth factor receptor tyrosine kinase, *Oncogene* 6:1677–83, 1991.

Plate KH, Breier G, Millauer B, Ullrich A, Risau W: Up-regulation of vascular endothelial growth factor and its cognate receptors in a rat glioma model of tumor angiogenesis, *Cancer Res* 53:5822–7, 1993.

Millauer B, Shawver LK, Plate KH, Risau W, Ullrich A: Glioblastoma growth inhibited in vivo by a dominant-negative Flk-1 mutant, *Nature* 367:576–9, 1994.

Millauer B, Wizigmann-Voos S, Schnurch H, Martinez R, Moller NPH et al: High affinity VEGF binding and developmental expression suggest Flk-1 as a major regulator of vasculogenesis and angiogenesis, *Cell* 72:835–46, 1993.

Giorgetti-Peraldi S, Murdaca J, Mas JC, Van Obberghen E: The adapter protein, Grb10, is a positive regulator of vascular endothelial growth factor signaling, *Oncogene* 20:3959–68, 2001.

Warner AJ, Lopez-Dee J, Knight EL, Feramisco JR, Prigent SA: The Shc-related adaptor protein, Sck, forms a complex with the vascular-endothelial-growth-factor receptor KDR in transfected cells, *Biochem J* 347:501–9, 2000.

Cunningham SA, Arrate MP, Brock TA, Waxham MN: Interactions of FLT-1 and KDR with phospholipase C gamma: Identification of the phosphotyrosine binding sites, *Biochem Biophys Res Commun* 240:635–9, 1997.

Holmqvist K, Cross MJ, Rolny C, Hagerkvist R, Rahimi N et al: The adaptor protein shb binds to tyrosine 1175 in vascular endothelial growth factor (VEGF) receptor-2 and regulates VEGF-dependent cellular migration, *J Biol Chem* 279:22267–75, 2004.

Quinn TP, Peters KG, De Vries C, Ferrara N, Williams LT: Fetal liver kinase 1 is a receptor for vascular endothelial growth factor and is selectively expressed in vascular endothelium, *Proc Natl Acad Sci USA* 90:7533–7, 1993.

14.25. Hepatocyte Growth Factor

Gohda E, Tsubouchi H, Nakayama H, Hirono S, Sakiyama O et al: Purification and partial characterization of hepatocyte growth factor from plasma of a patient with fulminant hepatic failure, *J Clin Invest* 81:414–9, 1988.

Szpirer C, Riviere M, Cortese R, Nakamura T, Islam MQ et al: Chromosomal localization in man and rat of the genes encoding the liver-enriched transcription factors C/EBP, DBP, and HNF1/

LFB-1 (CEBP, DBP, and transcription factor 1, TCF1, respectively) and of the hepatocyte growth factor/scatter factor gene (HGF), *Genomics* 13:293–300, 1992.

Naka D, Ishii T, Yoshiyama Y, Miyazawa K, Hara H et al: Activation of hepatocyte growth factor by proteolytic conversion of a single chain form to a heterodimer, *J Biol Chem* 267:20114–9, 1992.

Sakaguchi H, Seki S, Tsubouchi H, Daikuhara Y, Niitani Y et al: Ultrastructural location of human hepatocyte growth factor in human liver, *Hepatology* 19:1157–63, 1994.

Nakamura Y, Morishita R, Higaki J, Kida I, Aoki M et al: Expression of local hepatocyte growth factor system in vascular tissues, *Biochem Biophys Res Commun* 215:483–8, 1995.

Achim CL, Katyal S, Wiley CA, Shiratori M, Wang G et al: Expression of HGF and cMet in the developing and adult brain, *Brain Res Dev Brain Res* 102:299–303, 1997.

Naldini L, Weidner KM, Vigna E, Gaudino G, Bardelli A et al: Scatter factor and hepatocyte growth factor are indistinguishable ligands for the MET receptor, *EMBO J* 10:2867–78, 1991.

Schmidt C, Bladt F, Goedecke S, Brinkmann V, Zschiesche W et al: Scatter factor/hepatocyte growth factor is essential for liver development, *Nature* 373:699–702, 1995.

Uehara Y, Minowa O, Mori C, Shiota K, Kuno et al: Placental defect and embryonic lethality in mice lacking hepatocyte growth factor/scatter factor, *Nature* 373:702–5, 1995.

14.26. Hepatocyte Growth Factor Receptor

Dean M, Kozak C, Robbins J, Callahan R, O'Brien S et al: Chromosomal localization of the met proto-oncogene in the mouse and cat genome, *Genomics* 1:167–73, 1987.

Park M, Dean M, Kaul K, Braun MJ, Gonda MA et al: Sequence of MET protooncogene cDNA has features characteristic of the tyrosine kinase family of growth-factor receptors, *Proc Natl Acad Sci USA* 84:6379–83, 1987.

Bottaro DP, Rubin JS, Faletto DL, Chan AML, Kmiecik TE et al: Identification of the hepatocyte growth factor receptor as the c-met proto-oncogene product, *Science* 251:802–4, 1991.

Powell EM, Mars WM, Levitt P: Hepatocyte growth factor/scatter factor is a motogen for interneurons migrating from the ventral to dorsal telencephalon, *Neuron* 30:79–89, 2001.

Maina F, Casagranda F, Audero E, Simeone A, Comoglio PM et al: Uncoupling of Grb2 from the Met receptor in vivo reveals complex roles in muscle development, *Cell* 87:531–42, 1996.

Giordano S, Bardelli A, Zhen Z, Menard S, Ponzetto C et al: A point mutation in the MET oncogene abrogates metastasis without affecting transformation, *Proc Natl Acad Sci USA* 94:13868–72, 1997.

Giordano S, Corso S, Conrotto P, Artigiani S, Gilestro G et al: The Semaphorin 4D receptor controls invasive growth by coupling with Met, *Nature Cell Biol* 4:720–4, 2002.

Prevention of Cardiac Injury [14.27]. Blood reperfusion following cardiac infarction may introduce additional harmful stress to the heart, exacerbating the damage induced by the original ischemia and infarction. The reoxygenation of ischemic myocardial cells may induce an in increase in the formation of reactive-oxygen species (ROS). The reactive oxygen species may react with endogenous antioxidant factors, potentially depleting the buffering capacity of the endogenous antioxidant system. Excessive reactive oxygen species may activate inflammatory reactions, inducing cell membrane injury, contractile dysfunction, and cell death, which increase the propensity of heart failure. Thus, the prevention of reperfusion injury is a critical step for the treatment of cardiac infarction.

There exist a number of antioxidant enzymes, such as superoxide dismutase (an enzyme dismutating the oxygen free radical superoxide O_2^-) and heme oxygenase (an enzyme involved in the catabolism of heme), which reduce the activities of reactive oxygen species. These protein enzymes or their genes can be delivered to the infarct area of the heart for therapeutic purposes. The overexpression of heme oxygenase and superoxide dismutase in cardiac infarction has been shown to significantly reduce the size of infarcts, in association with a decrease in oxidative stress and inflammatory reactions. These changes are associated with improvement of the contractile function of impaired cardiac tissue. Other proteins that exert antioxidant effects include glutathione peroxidase, stress-induced heat-shock proteins, survival genes (Bcl2, Akt), immunosuppressive cytokines, adenosine A_1 and A_3 receptors, and hepatocyte growth factor. These molecules can be used to protect the heart from reperfusion injury.

Antioxidant Molecules as Therapeutic Agents

SUPEROXIDE DISMUTASES [14.28]. Superoxide dismutases (SODs) are a family of enzymes, including SOD1 (CuZn-SOD), SOD2 (Mn-SOD), and SOD3 (EC-SOD), which remove the reactive superoxide radical O_2^-. The superoxide radical is produced by oxygen reduction and exerts a highly toxic effect on molecular and cellular functions and activities. This radical has been implicated in degenerative processes including amyotrophic lateral sclerosis, ischemic heart disease, Alzheimer's disease, Parkinson's disease, and aging. Superoxide dismutase can catalyze the dismutation of the toxic superoxide anion O_2^- to O_2 and H_2O_2, thus reducing the toxicity of the superoxide radical. Although the three superoxide dismutases exert a similar function, the structure and gene locus of these enzymes are different. SOD1 is a 154-amino acid protein with molecular weight about 16-kDa. This enzyme is also known as *soluble superoxide dismutase*, *copper–zinc superoxide dismutase*, and *indophenoloxidase A* (IPOA). SOD1 is a copper- and zinc-containing homodimer found primarily in the cytoplasm. The SOD1 gene is localized to gene locus 21q22.1. SOD1 is expressed ubiquitously. The mutation of SOD1 can induce amyotrophic lateral sclerosis, a degenerative motor neuron disorder.

Superoxide dismutase 2 (SOD2) is a manganese-containing enzyme of 222 amino acids with molecular weight ~25 kDa. This enzyme is also known as *mitochondrial superoxide dismutase*, *indophenoloxidase B* (IPO-B), and *manganese superoxide dismutase*. It exists in the form of tetramer and is found primarily in the mitochondria. The SOD2 gene is localized to the gene locus 6q25.3. SOD2 is expressed ubiquitously. The function of SOD2 is similar to that of SOD1 as described above. SOD2 helps to stabilize the activity of mitochondrial enzymes, which are susceptible to superoxide-induced toxicity. The knockout of the SOD2 gene in mice results in cardiomyopathy, lipid accumulation in the liver and skeletal muscle, metabolic acidosis, and premature animal death. At the molecular level, the deficiency of the SOD2 gene induces the suppression of the respiratory chain enzymes NADH-dehydrogenase (complex I) and succinate dehydrogenase (complex II), the inhibition of the tricarboxylic acid cycle enzyme aconitase, functional defect in the 3-hydroxy-3-methylglutaryl-CoA lyase in association with aciduria, and oxidative damage of DNA. Mutation in the SOD2 gene causes idiopathic dilated cardiomyopathy.

Superoxide dismutase 3 (SOD3) is a copper- and zinc-containing enzyme (240 amino acids, 26-kDa) and exists in the form of tetramer. It is also known as extracellular *superoxide dismutase* [Cu–Zn] and extracellular superoxide dismutase. The gene is localized to gene locus 4p15.3-p15.1. SOD3 is found primarily in the extracellular space of several systems, including the heart, lung, liver, kidney, pancreas, thyroid, uterus, and skeletal muscle.

HEME OXYGENASE [14.29]. Heme oxygenase, also known as HO and HMOX, is an enzyme that cleaves heme to form biliverdin, a molecule subsequently converted to bilirubin by biliverdin reductase. This process consumes oxygen and electrons donated by the NADPH-cytochrome p450 reductase. Activated HO exerts an antioxidative effect. Such an effect is exerted through the production of bilirubin. Whereas heme is known as a prooxidant, bilirubin is an antioxidant. The accumulation of bilirubin contributes to the antioxidative effect. There are two isoforms for the heme oxygenase: HO1 and HO2. HO1 is the inducible form of HO. This enzyme is 288 amino acids in length and about 33 kDa in molecular weight. The gene of HO1 is localized to the gene locus 22q12. HO is expressed in a number of tissue and organ systems, including the central nervous system (cerebral cortex and hippocampus), lung, blood vessels, kidney, and prostate gland. HO1 plays a critical role for the prevention of oxidant-induced cell injury and death. For instance, the expression of HO1 is upregulated in skin wounds in response to heme release. Enhanced HO1 exerts an antagonistic effect on oxidative stresses and inflammation induced by heme.

Heme oxygenase 2, or HO2, is the constitutive form of heme oxygenase (316 amino acids, 36 kDa). HO2 exerts an antioxidative effect similar to that of HO1. The HO2 gene is localized to the gene locus 16p13.3. HO2 is expressed in the brain, skin, and placenta. In HO2-null mice, exposure to hyperoxia induces more severe oxidative injury and cell death compared to control animals. Furthermore, HO2 is expressed primarily in the endothelial cells of blood vessels and the neurons of several autonomic ganglia, including the petrosal, superior cervical, and nodose ganglia. HO2 catalyzes the formation of carbon monoxide (CO), which potentially induces endothelial relaxation. Thus, HO1 is thought to serve as a vasodilator.

GLUTATHIONE PEROXIDASE [14.30]. Glutathione peroxidase (GPX) is an enzyme that reduces hydrogen peroxide H_2O_2. The enzyme reduces H_2O_2 to H_2O by oxidizing glutathione ($H_2O_2 + 2\,GSH \rightarrow GSSG + 2\,H_2O$, where GSH is glutathione and GSSG is oxidized glutathione). A further reduction of oxidized glutathione is catalyzed by glutathione reductase ($GSSG + NADPH + H^+ \rightarrow 2\,GSH + NADP^+$). Since H_2O_2 induces cell injury and death, GPX plays a critical role for protecting cells from injury and death. The action of glutathione peroxidase requires a metal cofactor selenium. Transgenic mice with GPX deficiency exhibit increased sensitivity to oxidative stress. Cells derived from GPX-deficient mice are committed to apoptosis when exposed to hydrogen peroxide, whereas cells from wildtype control mice exhibit a significant decrease in susceptibility to hydrogen peroxide. There exist at least six GPX isoforms: GPX1–6, which are encoded by different genes and expressed in different tissues and organs. The characteristics of these isoforms are presented in Table 14.1.

STRESS-INDUCED HEATSHOCK PROTEINS [14.31–14.35]. Heatshock proteins (HSPs) are a family of cytoplasmic proteins that are upregulated in response to environmental stress conditions, such as an alteration in temperature, inflammation, viral and bacterial infection, hypoxia, starvation, and exposure to toxins and ultraviolet light. These proteins are also called *stress proteins*. The HSP family contains a number of members with various protein structure and molecular weight. These members are grouped on the basis of their moelcular weights. For example, HSPs with molecular weight ~10 kDa are classified as HSP10, and those with molecualr weight ~40 kDa HSP40, and so on. The primary functions of HSPs are to protect cells from stress-induced injury and present signals of infection and inflammation to the immune defense system in resposne to altered environmetnal stress condictions. Under physiological conditions, HSPs are expressed at a basal level. These proteins serve as chaperones that mediate protein folding and assembly, intracellular protein sorting and transport, protein–protein interactions, and disposal of diordered and aged proteins. Release HSPs from necrotic or apoptotic cells, such as infarcted cardiac cells, may serve as signals which activate the defense system to remove disintegrated cells and repair injured cells. These functions are well conserved among most organisms and mammals ranging from bacteria to humans. Several common HSPs are listed in Table 14.2.

BCL2 [14.36]. Bcl2 (B-cell lymphoma 2) is a protein of 239 amino acids with molecular weight ~26 kDa. This protein is also known as *B-cell lymphoma protein 2α* and *apoptosis regulator Bcl2*. The molecule is composed of a Bcl2 homology (BH)1, BH2, and BH4 domain as well as a transmembrane domain. The Bcl2 gene is localized to gene locus 19q21.3. Bcl2 is expressed ubiquitously and is primarily found in the mitochondria.

The primary function of Bcl2 is to inhibit cell apoptosis. The anti-apoptotic effect of Bcl2 was initially found in studies with pro-B-lymphocyte cells. Overexpression of Bcl2 blocks the apoptosis of these cells. Overexpressed Bcl2 can also protect neurons from apoptosis by enhacing the expression of choline acetyltransferase during development, resulting in hypertrophy of the central nervous system. Transgenic mice with Bcl2 overexpression exhibit enhanced resistance to ischemic injury induced by cerebral artery occlusion as well as increased proliferation of monocytes. In contrast, the deficiency of Bcl2 in transgenic mice induces a marked decrease in the number of $CD4^+$ and $CD8^+$ lymphocytes shortly after birth, retardation of tissue and organ growth, and animal death within several weeks after birth. These observations suggest that Bcl2 plays a role in the regulation of not only cell survival but also organ tissue and organ development.

There are a number of proteins, including Bcl2, Bcl-x(L), Bcl-w, Bcl-x(S), Bad, and Bak, which exhibit a structure similar to that of Bcl2 and participate in the regulation of cell apoptosis. These proteins are defined as members of the Bcl2 family. A unique feature is that all Bcl2 family proteins contain BH1 and BH2 domains. Among the Bcl2 family members, some exert an antiapoptotic effect, such as Bcl2, Bcl-x(L), and Bcl-w, whereas others exerts a proapoptotic effect. It is interesting to note that the antiaoptotic proteins usually contain an N-terminal BH4 domain. In contrast, proapoptotic proteins usually contain a BH3 domain except for Bad.

AKT1 [14.37]. Akt1 is a serine/threonine protein kinase of 480 amino acids with molecular weight ~56 kDa and participates in the regulation of inflammatory and mitogenic activities. Akt1 is also classified as v-Akt murine thymoma viral oncogene homolog 1, oncogene Akt1, protein kinase B-alpha, and Rac serine/threonine protein kinase. The Akt1 protein is characterized by the presence of a pleckstrin homology (PH) domain, which is a sequence about 100 residues found in a wide range of proteins involved in intracellular signaling or as a constituent of the cytoskeleton, and a serine/threonine protein kinase (ST kinase) domain, which is a signature of serine/threonine kinases. Akt1 resides in the cytoplasm in quiescent cells. On stimulation by mitogenic factors, such as platelet-derived growth factor and insulin-like growth factor, Akt1 is activated and recruited to cell membrane by the phosphoinositide 3 kinase (PI3K), which is activated by platelet-derived growth factor receptor β and insulin-like growth factor receptor. The Akt1 gene is localized to the gene locus 14q32.3. Akt1 is expressed in the brain, heart, lung, thymus, liver, pancreas, kidney, intestine, placenta, ovary, prostate gland, spleen, and testis. The Akt1

TABLE 14.1. Characteristics of Selected Glutathione Peroxidase Isoforms*

Isoforms	Amino Acids	Molecular Weight (kDa)	Expression
GPX1	203	22	Sperm, alveolar macrophage, monocyte, red blood cells
GPX2	189	22	Liver, intestine, mammary gland
GPX3	225	25	Heart, lung, liver, kidney, eye, placenta, mammary gland
GPX4	196	22	Sperm
GPX5	221	25	Testis
GPX6	221	25	

*Based on bibliography 14.30.

TABLE 14.2. Characteristics of Selected Heatshock Proteins*

Proteins	Alternative Names	Amino Acids	Molecular Weight (kDa)	Expression
Heatshock 10-kDa protein (HSP10)	HSPE1, chaperonin 10 homolog, CPN10, GroES	102	11	Heart, uterus
Heatshock 27-kDa protein (HSP27)	HSPB2	182	20	Heart, skeletal muscle, skin
Heatshock protein (HSP40)	Dna J homolog subfamily B member 1, HSPF1, HDJ1, DNAJ1	340	38	Brain, heart, lung, liver, kidney, pancreas, skeletal muscle, placenta
Heatshock 60-kDa protein (HSP60)	Chaperonin, chaperonin 60 homolog, cpn60 homolog, CPN60, 60-kDa chaperonin, mitochondrial 60-kDa heatshock protein, GroEL homolog, mitochondrial matrix protein P1, P60 lymphocyte protein, and HuCHA60	573	61	Heart, adrenal gland, liver, kidney, pancreas, intestine, placenta, thyroid gland, retina, skeletal muscle, blood cells, macrophages
Heatshock 70-kDa protein 2 (HSP70-2)	Heatshock-related 70-kDa protein 2	633	70	Ubiquitous

*Based on bibliography 14.31–14.35.

protein kinase exhibits about 70% similarity in amino acid sequence to protein kinase A (PKA) and protein kinase C (PKC).

Akt1 is capable of regulating the activity of downstream molecules through phosphorylation. An example is Akt1-mediated activation of nuclear factor κB (NFκB). NFκB can be activated by inflammatory cytokines via the mediation of the IκB-kinase (IKK) complex composed of IKKα and IKKβ. Akt1 can phosphorylate IKKα, which in turn phosphorylates IκB and activates NFκB (see page 222). NFκB mediates cytokine-initiated immune and inflammatory responses. Akt1-induced activation of IκB-kinase has also been implicated in antiapoptotic events initiated by platelet-derived growth factor. Akt1 plays a critical role in the protection of cells from apoptosis. Furthermore, Akt1 can directly phosphorylate and activate nitric oxide synthase (NOS), which catalyzes the synthesis of nitric oxide (NO). Activated Akt1 induces an increase in basal NO release. Mutation of the NOS gene at Akt1 phosphorylation site results in a reduction in the responsiveness of NOS to Akt1 stimulation and a decrease in NO production. PIK3 is involved in the activation of Akt1 and NOS. In transgenic mouse models, the deficiency of Akt1 gene induces an increase in spontaneous apoptosis in the testis and thymus, in association with reduced spermatogenesis and enhanced thymocyte susceptibility to cell injury in response to irradiation, toxins, apoptotic ligands, and serum withdrawal. In contrast, the transfection of cells with an active form of Akt1 gene promotes cell proliferation and tumor development. These observations suggest that Akt1 is a mitogenic factor that mediates the signaling process of cell survival and proliferation.

CYTOKINES [14.38–14.52]. Cytokines are proteins produced and secreted by lymphocytes, monocytes, macrophages, and mast cells in response to immune and inflammatory stimuli. Fibroblasts and endothelial cells can also produce cytokines. These molecules are responsible for the regulation of immune responses, inflammatory reactions, and/or hematopoiesis via interacting with corresponding cell membrane receptors. There are several alternative names for cytokines. Cytokines produced by lymphocytes and monocytes are referred to as *lymphokines* and *monokines*, respectively. Cytokines that mediate cell chemotactic activities are known as *chemokines*. Cytokines produced by one cell type and acting on other cell types are known as interleukins.

There are several common features for cytokines:

1. A cytokine may be produced by multiple cell types. For instance, tumor necrosis factor (TNF)α can be produced by macrophages, mast cells, and natural killer (NK) cells. This mechanism ensures the production of sufficient cytokines for the regulation of immune or inflammatory reactions.
2. Different cytokines may exert a similar function. For instance, granulocyte–monocyte colony-stimulating factor (GM-CSF), interleukin (IL)1, IL2, IL3, IL4, and IL5 all initiate and promote cell proliferation and differentiation. This functional redundancy is another mechanism to provide sufficient cytokine activity in response to immune and inflammatory stimuli.
3. One type of cytokine, once secreted and activated, may stimulate the production of another cytokine, an effective mechanism for amplifying cytokine production and cell activation in response to the stimulation of immune and inflammatory factors.
4. Certain types of cytokine can act synergistically. For instance, IL2 and IL4 can coordinately regulate the proliferation of lymphocytes. On the other hand, certain

types can exert opposing effects. For instance, tumor necrosis factor induces cell apoptosis, whereas IL3 promote cell survival and proliferation. The characteristics of common cytokines are presented in Table 14.3.

Cytokines exert their effects via interacting with corresponding cell membrane receptors, which are functionally linked to the Janus tyrosine kinase (JAK)—signal transducers and activators of transduction (STAT) signaling pathways. The binding of a cytokine molecule to a cytokine receptor induces the activation of JAKs, which in turn phosphorylate STATs. Phosphorylated STATs serve as transcriptional factors and translocate to the nucleus, inducing gene expression and cellular activities. See Chapter 5 for signaling pathways and mechanisms of cytokines.

Enhancement of Angiogenesis [14.53]. The enhancement of angiogenesis is another major task for the treatment of cardiac infarction. Several proangiogenic factors such as vascular endothelial growth factor (VEGF), fibroblast growth factor (FGF), and hepatocyte growth factor (HGF) have been used for such a purpose (see page 600–610 for these factors). These factors have been shown to effectively promote neovascularization and improve the function of ischemic myocardium. In addition, it is beneficial to enhance the contractility of impaired myocardial cells in the infarct area. Methods described on page 595 can be used for such a purpose.

The therapeutic proteins described above can be directly delivered to the infarct area of the heart. Consecutive deliveries are usually required because growth factors are degraded rapidly. Alternatively, genes encoding selected growth factors can be delivered to the target sites of an infracted heart. The effectiveness of the transferred genes can last longer than that of injected proteins. For the past several years, a number of clinical trials have been carried out for treatment of cardiac infarction with gene transfer. Growth factor genes, including VEGF and FGF genes, have been used in these trials. Promising results have been obtained from these clinical trials. It is important to note that, for acute cardiac infarction, direct protein delivery may be more effective than gene transfer, because proteins can exert therapeutic effect immediately following delivery, whereas it requires 1–2 days for gene expression.

There are potential problems for the application of molecular engineering approaches to the heart. One of the problems is the promotion of cell proliferation and migration in atherosclerotic coronary arteries following the delivery of growth factor genes. Such an influence may enhance the progression of atherosclerosis, a major cause of cardiac infarction. A potential approach for solving such a problem is to use local condition-sensitive gene promoters, such as a promoter that can be controlled by local hypoxia or reduced oxygen concentration, a condition often associated with cardiac infarction. As shown in previous studies, the erythropoietin-responsive element (HRE) is a hypoxia-sensitive gene promoter. The incorporation of this promoter to the VEGF gene, carried by an adenovirus-associated viral vector, renders the VEGF gene hypoxia-sensitive; that is, it can be activated only under a reduced oxygen condition as seen in cardiac infarction. With such a controlling approach, the VEGF gene is not expressed in normal cardiac tissue with a physiological level of oxygen.

Cell Regenerative Engineering for Ischemic Heart Disease [14.54]. Cardiac infarction can be potentially treated with cellular engineering approaches, specifically, transplanting appropriate cell types into an infracted heart to replace dead or injured cells. Cellular

TABLE 14.3. Characteristics of Selected Cytokines*

Cytokines	Alternative Names	Amino Acids	Molecular Weight (kDa)	Expression	Functions
Colony-stimulating factor 1	CSF1, macrophage colony-stimulating factor 1, MCSF1, macrophage granulocyte inducer IM, MGI-IM	554	60	Lymphocytes, osteoclasts, microglia, astrocytes, bone marrow stromal cells, liver, and placenta	Promoting proliferation and differentiation
Erythropoietin	EPO	193	21	Bone marrow, spleen, liver, kidney, nervous system	Promoting erythroid differentiation and hemoglobin synthesis, protecting nervous system from injuries, and preventing apoptosis
IL1	IL1α, hematopoietin 1, IL1F1	271	31	Monocytes, macrophages, brain, skin, lung	Regulating inflammatory and immune processes, mediating hematopoiesis, inducing apoptosis, mediating osteoclastogenesis and gene polymorphisms, and inducing rheumatoid arthritis and Alzheimer's disease
IL2	T cell growth factor (TCGF), and aldesleukin	153	18	Leukocytes, bone marrow, brain, colon, thymus, kidney	Regulating T- and B-cell proliferation and mediating immune and inflammatory reactions
IL3	Multipotential colony-stimulating factor (multiCSF), hematopoietic growth factor, mast cell growth factor (MCGF), P-cell-stimulating factor	151	17	Eosinophils, brain, intestine	Promoting proliferation of hematopoietic cells, regulating cell differentiaon, preventing apoptosis, and mediating nerve development

TABLE 14.3. *Continued*

Cytokines	Alternative Names	Amino Acids	Molecular Weight (kDa)	Expression	Functions
IL4	B-cell stimulatory factor 1 (BSF1), lymphocyte stimulatory factor 1, B-cell growth factor (BCGF), B-cell differentiation factor γ (BCDF γ)	153	17	T cells, basophils, eosinophils, mast cells	Regulating immune and inflammatory reactions and recruiting myoblasts during mammalian muscle growth
IL5	T-cell-replacing factor (TRF), eosinophil differentiation factor (EDF), B-cell differentiation factor 1	134	15	T cells	Regulating growth and differentiation of B cells and eosinophils, stimulating eosinophil maturation and activation, and contributing to asthma or hypereosinophilic syndrome
IL6	Interferon β2, IFNB2, B-cell differentiation factor, BSF2, hepatocyte stimulatory factor (HSF), hybridoma growth factor (HGF), B-cell stimulatory factor 2, 26-kDa protein	212	24	Monocytes, fibroblasts, B cells, brain, kidney, placenta, thymus, adipocytes	Mediating inflammatory reactions, regulating T-cell development and activity, inducing neuroendocrine cell proliferation and differentiation, and mediating hepatocyte proliferation and regeneration

IL7	Lymphopoietin 1, LP1, pre-B-cell factor	177	20	Thymus, bone marrow, intestine, skin, dendritic cells	Regulating B/T-cell development, promoting lymphocyte survival
IL8	Small inducible cytokine subfamily B member 8 (SCYB8), monocyte-derived neutrophil chemotactic factor (MDNCF), monocyte-derived neutrophil-activating protein, neutrophil-activating peptide 1 (NAP1), Granulocyte chemotactic protein 1 (GCP1), CXC chemokine ligand 8 (CXCL8), T-cell chemotactic factor, lymphocyte-derived neutrophil-activating factor (LYNAP), protein 3-10C, neutrophil-activating factor (NAF), emoctakin	99	11	Monocytes, macrophages, lymphocytes, bone marrow, intestine, kidney, placenta	Mediating inflammatory response, stimulating angiogenesis, and inducing chemotaxis

TABLE 14.3. *Continued*

Cytokines	Alternative Names	Amino Acids	Molecular Weight (kDa)	Expression	Functions
IL9	T-cell/mast cell growth factor P40, P40 T-cell and mast cell growth factor, P40 cytokine, TCGF III, MEA	144	16	T cells	Stimulating cell proliferation, preventing apoptosis, inducing inflammatory reactions, and contributing to pathogenesis of asthma
IL10	Cytokine synthesis inhibitory factor (CSIF), mast cell growth factor III, B-cell-derived T-cell growth factor (B TCGF)	178	21	Manocytes, T cells, macrophages, epidermal cells.	Downregulation of cytokine expression in Th1 cells and macrophages, enhancement of B-cell survival, proliferation, and antibody production; also inhibition of NFκB activity
IFNα	α interferon, Interferon α1, IFN, interferon leukocytic, IFN leukocytic, interferon, αD, LeIF D	189	22	Leukocytes	Activating natural killer cells, exerting antiviral inflammatory reactions, and suppressing tumor cell growth
IFNβ	IFN β1, interferon fibroblast (IFF), β interferon, IFB, IFNB, fibroblast interferon	187	22	Leukocytes	Exerting antiviral and antiproliferative activites, regulating immune and inflammatory reactions, and suppressing tumor cell proliferation

*Based on bibliography 14.38–14.52.

engineering approaches are used to achieve three goals: replacement of malfunctioned myocardiocytes, induction or stimulation of angiogenesis, and delivery of therapeutic agents for the prevention of myocardiocyte death.

Replacement of Malfunctioned Cardiomyocytes. A number of cell types have been tested and used for the replacement of dead and severely injured myocardial cells. These cell types include embryonic stem cells, fetal cadiomyocyte progenitor cells, bone marrow myocyte progenitor cells (chapter-opening figure), and adult cardiomyocyte stem cells. Embryonic stem cells from blastocyst, as described on page 381, are capable of differentiating into all specialized cell types, including cardiomyocytes, under appropriate developmental conditions. These cells are primary candidates for cardiac cell transplantation. Fetal cadiomyocyte progenitor cells can be collected from the mesoderm, where the heart is generated. These are committed cells that are to be differentiated into cardiomyocytes and are excellent candidates for cardiac cell transplantation. The bone marrow contains various types of progenitor cell, including mesenchymal stem cells, which contain cardiomyocyte progenitor cells. These cells can be enriched based on cell surface markers, such as Sca1 and c-Kit, by fluorescence-activated cell sorting (FACS) or magnetic bead-mediated cell sorting. The bone marrow progenitor cells can differentiate into cardiomyocytes in vitro and in vivo (Fig. 14.8). It is interesting to note that, under appropriate culture conditions, mesenchymal stem cells can transform into beating cardiomyocytes. The transplantation of bone marrow cells into models of cardiac infarction significantly improves the performance of the injured heart. A number of clinical trials have been conducted for the treatment of myocardial infarction by transplantation of bone marrow cells. These studies have shown that such a cellular therapy results in a reduction in the infarct size and improvement of cardiac perfusion and function in patients with late-stage ischemic heart disease.

The adult heart contains undifferentiated adult stem cells and progenitor cells, which are capable of proliferating and differentiating into cardiomyocytes, endothelial cells, and

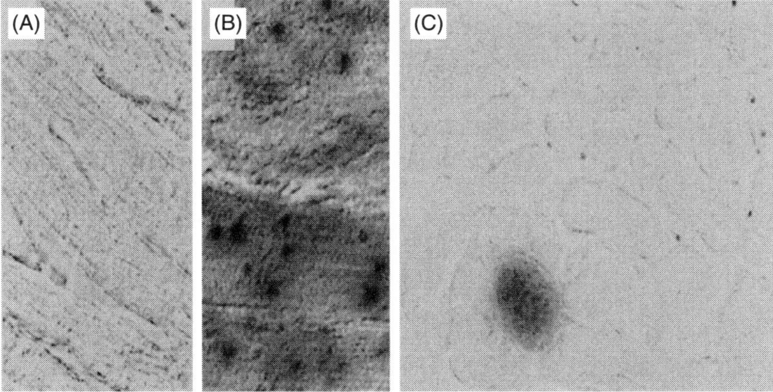

Figure 14.8. Incorporation of bone marrow side population (SP) cells into cardiomyocytes. (A) Negative control: C57Bl/6 cardiac tissue stained for lacZ expression. (B) Positive control: C57Bl/6-Rosa26 cardiac tissue stained for lacZ expression. (C) Cross section of a heart from an SP cell transplant recipient, which received an infarct. (Reprinted by permission from Macmillan Publishers Ltd.: Orlic D et al: Bone marrow cells regenerate infarcted myocardium, *Nature* 410:701–5, copyright 2001.)

vascular smooth muscle cells, although the density of these cells is extremely low. These cells can be enriched by FACS or magnetic bead-mediated cell sorting based on cell surface markers as described above for bone marrow cell enrichment. The transplantation of adult cardiomyocyte stem and progeniotor cells into infarcted heart in animal models significantly improves the cardiac performance. The discovery of adult cardiac stem cells has changed the traditional view that the heart contains finally differentiated cells, which cannot regenerate, and will greatly facilitate the development of cardiac cell regenerative engineering.

Although cardiac cell regenerative engineering is successful in experimental investigations and certain clinical applications, there are still problems. A major problem is immune rejection reactions in response to the transplantation of allogenic embryonic stem cells, fetal progenitor cells, and adult stem cells. This problem may be overcome by using autogenous bone marrow-derived progenitor cells. Another problem is cell death after transplantation. Many transplanted cells die within a short period. The maintenance of cell survival remains a challenge in cell regenerative engineering. A potential approach is to transfer growth factor genes into cell candidates for transplantation. Such an approach has been shown to reduce the rate of cell death.

Enhancement of Angiogenesis. Another strategy for treating myocardial infarction is to stimulate angiogenesis in the ischemic areas by transplant angiogenic cells. Potential cell types include the bone marrow endothelial progenitor cells. These cells originate from the bone marrow and are characterized by the expression of a number of endothelial lineage markers, including von Willebrand factor, VE-cadherin, Flk-1 (vascular endothelial growth factor receptor 2), PECAM1, CD34, and E-selectin. These cells can be isolated from the bone morrow or peripheral blood and transplanted into an ischemic heart via direct injection or intravenous delivery for therapeutic purposes. The transplanted cells may participate in the process of angiogenesis in injured and infarcted areas, contributing to the neovascularization and recovery from cardiac infarction (Fig. 14.9).

As a general strategy, cells selected for therapeutic purposes can be engineered for augmentation of specified properties by gene transfer. Cytoprotective genes (e.g., Bcl2, Akt1, and growth factor genes) and angiogenic genes (e.g., VEGF and Flk-1) can be transferred into candidate cells in vitro. At the time of maximal gene expression (usually 2–3 days), the cells can be transplanted to the target tissues. For myocardial infarction, such an approach may be more important than direct delivery of therapeutic genes into injured cardiomyocytes. The reason is that cardiomyocytes are either injured or dead in cardiac infarction. Even though therapeutic genes are delivered to the infarction site, the injured cardiomyocytes may not be able to express the delivered genes efficiently. Genetically engineered cells can release proteins encoded by the transferred genes, such as anti-apoptotic factors and angiogenic factors. These factors in turn promote the survival of cardiomyocytes and angiogenesis, respectively.

Tissue Regenerative Engineering for Ischemic Heart Disease [14.55]. Ischemic heart disease is characterized by the presence of regional cardiac injury and/or infarction. In severe cases, cardiomyocyte death occurs in large areas, often resulting in acute heart failure. To reduce the effect of cardiac injury and infarction, a tissue engineering strategy is to construct a cardiac tissue scaffold and implant the scaffold to the inury site of the heart. The cardiac scaffold can be constructed with the integration of various cell types, such as cardiac stem cells, neonatal cardiomyocytes, fibroblasts, or vascualr smooth

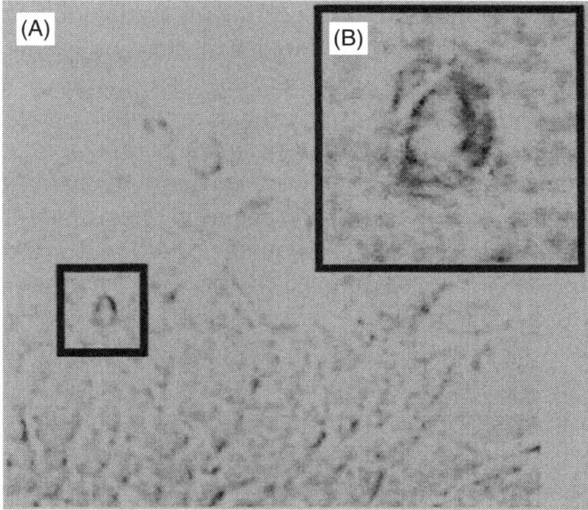

Figure 14.9. Incorporation of bone marrow side population (SP) cells into vascular endothelial cells. (A) X-gal-stained section of cardiac tissue from an infarcted SP cell transplant recipient. Panel B shows magnification of the indicated capillaries from panel A. (Reprinted with permission from Jackson KA et al: Regeneration of ischemic cardiac muscle and vascular endothelium by adult stem cells, *J Clin Invest* 107:1395–402, copyright 2001.)

muscle cells. The cardiac stem and progenitor cells can potentially transform to cardiomyocytes, whereas the fibroblasts and smooth muscle cells can produce mitogenic factors, which enhance the survival of injured cardiac cells and stimulate the transformation of stem/progenitor cells into cardiomyocytes. Furtherore, various mitogenic factors, such as fibroblast growth factor and vascular endothelial growth factor, can be integrated into the constructed cardiac scaffold. These mitogenic factors can directly promote the survivcal of injured cardiomyocytes and enhance the differentiation of cardiac stem and progector cells into functional cardiomyocytes. Alternatively, the integrated cells can be transfected with growth factor genes to enhance the expression of growth factors. These approaches have been tested in experimental models. These investigations have demonstrated the potential of applying tissue regenerative engineering approaches to cardiac therapy.

Valvular Diseases

Pathogenesis, Pathological Changes, and Clinical Features. Valvular diseases are a group of disorders that occur in the mitral, tricuspid, aortic, and pulmonary valves, are often caused by inflammatory reactions, and are characterized by structural distortion, calcification, and mechanical stiffening of the valves in association with altered hemodynamics in the atrial and ventricular chambers and remodeling of the atrial and ventricular structure and function. The pathogenesis of the valvular diseases is closely related to rheumatic fever, which is possibly a result of group A streptococcal infection. Rheumatic fever often occurs several days after acute streptococcal infection and involves the heart, joints, and nervous system. Antibodies developed in response to the stimulation of antigens from streptococci have been shown to cross-react with components of cardiac valves. Thus, autoimmune reactions are a potential cause for the inflammatory responses in the cardiac valves. Although all valves are susceptible to rheumatic fever, the mitral valves are most frequently involved. Several pathological changes can be found in involved valves, including fibrosis, thickening, calcification, fusion, distortion, and shortening of the valves.

Valvular diseases can be classified into two types according to the form of the disease: valvular stenosis and regurgitation. Each type may occur in either of the valves. The most common types are mitral valve stenosis and regurgitation. While all types of valvular disease influence the performance and hemodynamics of the heart, the type and severity of cardiac malfunction and associated clinical consequences are dependent on the location of the involved valves and the degree of the structural and geometric changes.

Mitral stenosis is a form of valvular disease characterized by the narrowing of the mitral orifice resulting from mitral valve fibrosis, fusion, and stiffening. Mitral stenosis exhibits a sequence of pathophysiological changes. The resistance of the mitral orifice to bloodflow increases and the rate of bloodflow from the left atrium to the left ventricle decreases during the diastole. When bloodflow is reduced to a rate below the critical level required for filling the left ventricle, the left ventricle receives insufficient blood volume during the diastole and is thus unable to pump sufficient blood into the arterial system. The physical activities of patients are often limited due to the lack of arterial bloodflow. In severe cases, patients often exhibit orthopnea and paroxysmal dyspnea. In regions above the mitral valves, an excessive amount of blood accumulates in the left atrium and the pulmonary veins, raising pulmonary venous and capillary pressure. To a certain level, pulmonary edema occurs due to an increase in the capillary transmural pressure. Pathophysiological changes in tricuspid stenosis are similar to those described above except that the changes occur in the right heart and blood accumulation occurs in the right atrium and the systemic veins.

Mitral regurgitation is a valvular disease characterized by incomplete closure of the mitral valves during systole due to fibrosis, distortion, and stiffening of the mitral valves. As a result, blood flows back from the left ventricle to the left atrium during the systole, leading to a reduction in blood ejection into the arterial system and excessive expansion of the left atrium. The reduction in systemic arterial bloodflow results in limited physical activities. To compensate the reduction in arterial bloodflow, the sympathetic system is usually activated, inducing an increase in the heart rate and contractility. The heart also undergoes adaptive remodeling in structure, resulting in an increase in its myocardial mass, a process referred to as cardiac hypertrophy. Long-term hypertrophy may lead to left heart failure. The increase in atrial blood volume induces an elevation in pulmonary capillary blood pressure, which is a common cause of pulmonary edema. Similar changes can be found in tricuspid valve regurgitation in the right heart.

Aortic valve stenosis is characterized by the narrowing of the aortic orifice and is induced by the fibrosis, fusion, and stiffening of the aortic valves. These changes result in an increase in the resistance of the aortic orifice to bloodflow, a decrease in aortic bloodflow, and thus an increase in workload for the left ventricle during the systole. The left ventricle adapts to these changes by increasing its contractile strength and mass, which often leads to ventricular hypertrophy. When the left ventricle is capable of compensating for the decrease in arterial bloodflow, there may be no apparent clinical symptoms. However, in a severe case of aortic stenosis, the left ventricle is not capable of overcoming the aortic orifice resistance. Left heart failure occurs due to left ventricular hypertrophy and/or excessive workload. Similar changes are observed in pulmonary valve stenosis in the right heart.

Aortic regurgitation is characterized by incomplete closure of the aortic valves during the diastole due to fibrosis, distortion, shortening, and stiffening of the aortic valves. As a result, there is backward bloodflow from the aorta to the left ventricle during the diastole. Pathophysiological changes include a reduction in arterial bloodflow toward the peripheral systems during the diastole, an increase in diastolic filling volume in the left ventricle, and an increase in workload for the left ventricle. The left ventricle adapts to these changes by increasing its contractile strength, cardiac muscle mass, and chamber size, resulting in daptive ventricular hypertrophy and dilation. During the end-stage, left heart failure and shortage of arterial bloodflow are often observed. Pulmonary regurgitation exhibits similar changes in the right heart.

Treatment. Surgical correction and replacement of malfunctioned cardiac valves are the most effective approaches for the treatment of valvular diseases. For mitral stenosis, the narrowed mitral orifice can be widened surgically via a procedure known as *mitral valvotomy*. Stenosis in the tricuspid, aortic, and pulmonary valves can be treated with a similar approach. For valvular regurgitation, including mitral, tricuspid, aortic, and pulmonary regurgitation, valve replacement with a prosthesis is usually required, especially when malfunction occurs in the left or right ventricle. In special cases with ruptured chordae or flail leaflets, it is usually necessary to conduct valve reconstruction.

Artificial Cardiac Valves [14.56]. Cardiac valve prostheses have been developed for the replacement of malfunctioned mitral, tricuspid, aortic, and pulmonary valves. Since the first case of aortic valve replacement in 1952 by Dr. Charles Hufnagel and colleagues, cardiac valve replacement has become a common treatment for severe valvular diseases. There are two major types of cardiac valve prostheses: mechanical and tissue valves. A *mechanical valve* is composed a frame and a ball- or disk-shaped valve. The ball or disk opens and closes depending on the pressure gradient across the device during a cardiac period (systole or diastole). According to the shape of the valve, mechanical valves are classified into several subtypes: ball-and-cage valves (such as the Star-Edwards ball-and-cage valve, Smeloff–Cutter valve, and Magovern valve), caged disk valves (the Kay–Shiley and Beall valves), tilting disk valves (the Bjork–Shiley tilting disk valve, Medtronic-Hall valves), and bileaflet valves (the St. Jude bileaflet valve, CarboMedics bileaflet valve, and parallel bileaflet valve). The ball, tilting disk, and bileaflet valves have been used for replacing all types of valves, and the disk valves have been used primarily for replacing the mitral and tricuspid valves.

The mechanical valves, although strong in material, are problematic in several aspects: (1) these devices stimulate blood coagulation and thrombogenesis—it is necessary to administrate anticoagulants following valve replacement, (2) mechanical valves induce blood flow disturbance, and (3) these devices may induce blood cell damage. To overcome these problems, natural tissue-based cardiac valves have been developed and tested. Major types include autogenous valves (using the pulmonary valves for replacing the aortic valves of the same patient), allogenic valves (collected from human cadavers), and glutaraldehyde-treated zenogeneic tissue valves (porcine valves and calf pericardium-based valves). Overall, these valves exhibit improved performance and hemodynamics compared with mechanical valves. However, there are potential problems. While the autogenous valves are ideal, the source of the valves is limited and the removal of the pulmonary valves obviously influences the function of the right heart. Allogenic valves often cause immune responses and undergo progressive degradation and leaf wear, reducing the lifespan of the valves. These valves also cause blood coagulation and thrombosis. Glutaraldehyde-treated zenogeneic valves, although exhibiting improved material strength, cause blood coagulation, thrombosis, and valve calcification. These problems remain to be resolved.

Tissue-Engineered Cardiac Valves [14.57]. Tissue-engineered valves are valves constructed with synthetic polymers or extracellular matrix constituents with seeded stem or somatic cells. Because of the presence of living cells, it is expected that these valves may adapt to the physiological environment, integrate into the host tissue, and maintain functions for a longer time than mechanical and tissue-based valves. Because cardiac valves are subject to dynamic movements and fluid shear stress, there are several issues that ought to be considered specially: (1) the material for constructing the valve frame must be mechanically strong, flexible, and durable; (2) the construction material should be antiinflammatory and thrombosis-resistant; (3) the cell type selected for seeding should be able to survive under dynamically moving conditions and withstand the influence of fluid shear stress; and (4) the material and cells should be non-immunogenic.

Several tissue-engineered cardiac valve designs have been developed and tested in experimental models. These include biodegradable polymer valves based on polyglycolic acid and polyhydroxyoctanoates, as well as zenogeneic valve matrix seeded with cells. Preliminary studies have provided promising results. Cells seeded in the valve frame are able to grow under flow conditions in vitro. When grafted into the heart of host animals, these valves can carry out normal valve functions and maintain pressure gradients across the valves. However, results from a small clinical trial are disappointing. Tissue-engineered porcine heart valves grafted into the human heart induce inflammatory reactions and thrombosis. Valve failure occurs due to valve rupture and degeneration. These human trial studies provide insights into future design and improvement of cardiac valves.

BIBLIOGRAPHY

14.27. Prevention of Cardiac Injury

Laham RJ, Simons M, Sellke F: Gene transfer for angiogenesis in coronary artery disease, *Annu Rev Med* 52:485–502, 2001.

Okudo S, Wildner O, Shah MR, Chelliah JC, Hess ML et al: Gene transfer of heat shock protein 70 reduces infarct size in vivo after ischemia/reperfusion in the rabbit heart, *Circulation* 103:877–81, 2001.

Li Q, Bolli R, Qiu Y, Tang XL, Guo Y et al: Gene therapy with extracellular superoxide dismutase protects conscious rabbits against myocardial infarction, *Circulation* 103:1893–8, 2001.

Melo LG, Agrawal R, Zhang L, Rezvani M, Mangi AA et al: Gene therapy strategy for long term myocardial protection using adeno-associated virus-mediated delivery of heme oxygenase gene, *Circulation* 105:602–7, 2002.

Lefer DJ, Granger DN: Oxidative stress and cardiac disease, *Am J Med* 109:315–23, 2000.

Woo YZ, Zhang JC, Vijayasarathy C, Zwacka RM, Engehardt JF et al: Recombinant adenovirus-mediated cardiac gene transfer of superoxide dismutase and catalase attenuates postischemic contractile dysfunction, *Circulation* 98(Suppl):II255–60, 1998.

Suzuki K, Sawa Y, Kaneda Y: In vivo gene transfer of heat shock protein 70 enhances myocardial tolerance to ischemia-reperfusion injury in rat, *J Clin Invest* 99:1645–50, 1997.

Yoshida T, Watanabe M, Engelman DT, Engelman RM, Schley JA et al: Transgenic mice overexpressing glutathione peroxidase are resistant to myocardial reperfusion injury, *J Mol Cell Cardiol* 28:1759–67, 1996.

Chatterjee S, Stewart AS, Bish LT, Jayasankar V, Kim EM et al: Viral gene transfer of the anti-apoptotic factor Bcl-2 protects against chronic ischemic heart failure, *Circulation* 106(Suppl): I212–17, 2002.

Brauner R, Nonoyama M, Laks H, Drinkwater DC, McCaffery S et al: Intracoronary adenovirus-mediated transfer of immunosuppressive cytokine genes prolongs allograft survival, *J Thor Cardiovasc Surg* 114:923–33, 1997.

Yang Z, Cerniway RJ, Byford AM: Cardiac overexpression of A1-adenosine receptor protects intact mice against myocardial infarction, *Am J Physiol* 282:H949–55, 2002.

Ueda H, Sawa Y, Matsumoto K, Kitagawa-Sakakida S, Kawahira Y, et al: Gene transfection of hepatocyte growth factor attenuates reperfusion injury in the heart, *Ann Thor Surg* 67:1726–31, 1999.

14.28. Superoxide Dismutases

Huret JL, Delabar JM, Marlhens F, Aurias A, Nicole A et al: Down syndrome with duplication of a region of chromosome 21 containing the CuZn superoxide dismutase gene without detectable karyotypic abnormality, *Hum Genet* 75:251–7, 1987.

Zelko IN, Mariani TJ, Folz RJ: Superoxide dismutase multigene family: A comparison of the CuZn-SOD (SOD1), Mn-SOD (SOD2), and EC-SOD (SOD3) gene structures, evolution, and expression, *Free Radic Biol Med* 33:337–49, 2002.

Rosen DR, Siddique T, Patterson D, Figlewicz DA, Sapp P et al: Mutations in Cu/Zn superoxide dismutase gene are associated with familial amyotrophic lateral sclerosis, *Nature* 362:59–62, 1993.

Borgstahl GE, Parge HE, Hickey MJ, Beyer WF Jr et al: The structure of human mitochondrial manganese superoxide dismutase reveals a novel tetrameric interface of two 4-helix bundles, *Cell* 71:107–18, 1992.

Rosenblum JS, Gilula NB, Lerner RA: On signal sequence polymorphisms and diseases of distribution, *Proc Natl Acad Sci USA* 93:4471–3, 1996.

Church SL, Grant JW, Meese EU, Trent JM: Sublocalization of the gene encoding manganese superoxide dismutase (MnSOD/SOD2) to 6q25 by fluorescence in situ hybridization and somatic cell hybrid mapping, *Genomics* 14:823–5, 1992.

Li Y, Huang TT, Carlson EJ, Melov S, Ursell PC et al: Dilated cardiomyopathy and neonatal lethality in mutant mice lacking manganese superoxide dismutase, *Nature Genet* 11:376–81, 1995.

Melov S, Coskun P, Patel M, Tuinstra R, Cottrell B et al: Mitochondrial disease in superoxide dismutase 2 mutant mice, *Proc Natl Acad Sci USA* 96:846–51, 1999.

Stern LF, Chapman NH, Wijsman EM, Altherr MR, Rosen DR: Assignment of SOD3 to human chromosome band 4p15.3-p15.1 with somatic cell and radiation hybrid mapping, linkage mapping, and fluorescent in-situ hybridization, *Cytogenet Genome Res* 101:178, 2003.

Hjalmarsson K, Marklund SL, Engstrom A, Edlund T: Isolation and sequence of complementary DNA encoding human extracellular superoxide dismutase, *Proc Natl Acad Sci USA* 84:6340–4, 1987.

Marklund SL: Extracellular superoxide dismutase in human tissues and human cell lines, *J Clin Invest* 74:1398–403, 1984.

Folz RJ, Crapo JD: Extracellular superoxide dismutase (SOD3): Tissue-specific expression, genomic characterization, and computer-assisted sequence analysis of the human EC SOD gene, *Genomics* 22:162–71, 1994.

14.29. Heme Oxygenase

Kutty RK, Kutty G, Rodriguez IR, Chader GJ, Wiggert B: Chromosomal localization of the human oxygenase genes: heme oxygenase-1 (HMOX1) maps to chromosome 22q12 and heme oxygenase-2 (HMOX2) maps to chromosome 16p13.3, *Genomics* 20:513–6, 1994.

Schipper HM, Cisse S: Stopa EGExpression of heme oxygenase-1 in the senescent and Alzheimer-diseased brain, *Ann Neurol* 37:758–68, 1995.

Duckers HJ, Boehm M, True AL, Yet SF, San H et al: Heme oxygenase-1 protects against vascular constriction and proliferation, *Nat Med* 7:693–8, 2001.

Yamada N, Yamaya M, Okinaga S, Nakayama K, Sekizawa K et al: Microsatellite polymorphism in the heme oxygenase-1 gene promoter is associated with susceptibility to emphysema, *Am J Hum Genet* 66:187–95, 2000.

Yang Y, Ohta K, Shimizu M, Morimoto K, Goto C et al: Selective protection of renal tubular epithelial cells by heme oxygenase (HO)-1 during stress-induced injury, *Kidney Int* 64:1302–9, 2003.

Maines MD, Abrahamsson PA: Expression of heme oxygenase-1 (HSP32) in human prostate: normal, hyperplastic, and tumor tissue distribution, *Urology* 47:727–33, 1996.

Vile GF, Basu-Modak S, Waltner C, Tyrrell RM: Heme oxygenase 1 mediates an adaptive response to oxidative stress in human skin fibroblasts, *Proc Natl Acad Sci USA* 91:2607–10, 1994.

Wagener FADTG, van Beurden HE, von den Hoff JW, Adema GJ, Figdor CG: The heme-heme oxygenase system: A molecular switch in wound healing, *Blood* 102:521–8, 2003.

Takahashi K, Hara E, Suzuki H, Sasano H, Shibahara S: Expression of heme oxygenase isozyme mRNAs in the human brain and induction of heme oxygenase-1 by nitric oxide donors, *J Neurochem* 67:482–9, 1996.

Dennery PA, Spitz DR, Yang G, Tatarov A, Lee CS et al: Oxygen toxicity and iron accumulation in the lungs of mice lacking heme oxygenase-2, *J Clin Invest* 101:1001–11, 1998.

Zakhary R, Gaine SP, Dinerman JL, Ruat M, Flavahan NA et al: Heme oxygenase 2: endothelial and neuronal localization and role in endothelium-dependent relaxation, *Proc Natl Acad Sci USA* 93:795–8, 1996.

14.30. Glutathione Peroxidase

Kryukov GV, Castellano S, Novoselov SV, Lobanov AV, Zehtab O et al: Characterization of mammalian selenoproteomes, *Science* 300:1439–43, 2003.

de Haan JB, Bladier C, Griffiths P, Kelner M, O'Shea RD et al: Mice with a homozygous null mutation for the most abundant glutathione peroxidase, Gpx1, show increased susceptibility to the oxidative stress-inducing agents paraquat and hydrogen peroxide, *J Biol Chem* 273:22528–36, 1998.

Kiss C, Li J, Szeles A, Gizatullin RZ, Kashuba VI, Lushnikova T et al: Assignment of the ARHA and GPX1 genes to human chromosome bands 3p21.3 by in situ hybridization and with somatic cell hybrids, *Cytogenet Cell Genet* 79:228–30, 1997.

Carter AB, Tephly LA, Venkataraman S, Oberley LW, Zhang Y et al: High levels of catalase and glutathione peroxidase activity dampen H2O2 signaling in human alveolar macrophages, *Am J Res Cell Mol Biol* 31:43–53, 2004.

Chen X, Scholl TO, Leskiw MJ, Donaldson MR, Stein TP: Association of glutathione peroxidase activity with insulin resistance and dietary fat intake during normal pregnancy, *J Clin Endocrinol Metab* 88:5963–8, 2003.

Chu FF, Doroshow JH, Esworthy RS: Expression, characterization, and tissue distribution of a new cellular selenium-dependent glutathione peroxidase, GSHPx-GI, *J Biol Chem* 268:2571–6, 1993.

Chu FF, Esworthy RS, Doroshow JH, Doan K, Liu XF: Expression of plasma glutathione peroxidase in human liver in addition to kidney, heart, lung, and breast in humans and rodents, *Blood* 79(12):3233–8, 1992.

Human protein reference data base, Johns Hopkins University and the Institute of Bioinformatics, at http://www.hprd.org/protein.

14.31. Heatshock 10-kDa Protein

Shan YX, Yang TL, Mestril R, Wang PH: Hsp10 and Hsp60 suppress ubiquitination of insulin-like growth factor-1 receptor and augment insulin-like growth factor-1 receptor signaling in cardiac muscle: Implications on decreased myocardial protection in diabetic cardiomyopathy, *J Biol Chem* 278(46):45492–8, 2003.

Zeilstra-Ryalls J, Fayet O, Georgopoulos C: The universally conserved GroE (Hsp60) chaperonins, *Annu Rev Microbiol* 45:301–25, 1991.

Georgopoulos C, Welch WJ: Role of the major heat shock proteins as molecular chaperones, *Annu Rev Cell Biol* 9:601–34, 1993.

Hansen JJ, Bross P, Westergaard M, Nielsen MN, Eiberg H et al: Genomic structure of the human mitochondrial chaperonin genes: HSP60 and HSP10 are localised head to head on chromosome 2 separated by a bidirectional promoter, *Hum Genet* 112:71–7, 2003.

14.32. Heatshock 27-kDa Protein

Boxman IL, Kempenaar J, de Haas E, Ponec M: Induction of HSP27 nuclear immunoreactivity during stress is modulated by vitamin C, *Exp Dermatol* 11(6):509–17, 2002.

Nagano T, Nakagawa M, Iwaki T, Fukumaki Y: Identification and characterization of the gene encoding a new member of the alpha-crystallin/Small hsp family, closely linked to the alpha-B-crystallin gene in a head-to-head manner, *Genomics* 45:386–94, 1997.

14.33. Heatshock 40-kDa Protein

Ohtsuka K: Cloning of a cDNA for heat-shock protein hsp40, a human homologue of bacterial DnaJ, *Biochem Biophys Res Commun* 197:235–40, 1993.

Terada K, Mori M: Human DnaJ homologs dj2 and dj3, and bag-1 are positive cochaperones of hsc70, *J Biol Chem* 275:24728–34, 2000.

Hata M, Okumura K, Seto M, Ohtsuka K: Genomic cloning of a human heat shock protein 40 (Hsp40) gene (HSPF1) and its chromosomal localization to 19p13.2, *Genomics* 38:446–9, 1996.

14.34. Heatshock 60-kDa Protein

Hansen JJ, Bross P, Westergaard M, Nielsen MN, Eiberg H et al: Genomic structure of the human mitochondrial chaperonin genes: HSP60 and HSP10 are localised head to head on chromosome 2 separated by a bidirectional promoter, *Hum Genet* 112:71–7, 2003.

Schmidt M, Rutkat K, Rachel R, Pfeifer G, Jaenicke R et al: Symmetric complexes of GroE chaperonins as part of the functional cycle, *Science* 265:656–9, 1994.

Cheng MY, Hartl FU, Martin J, Pollock RA, Kalousek F et al: Mitochondrial heat-shock protein hsp60 is essential for assembly of proteins imported into yeast mitochondria, *Nature* 337:620–5, 1989.

Schett G, Metzler B, Mayr M, Amberger A, Niederwieser D et al: Macrophage-lysis mediated by autoantibodies to heat shock protein 65/60, *Atherosclerosis* 128(1):27–38, 1997.

14.35. Heatshock 70-kDa Protein 2

Bonnycastle LLC, Yu CE, Hunt CR, Trask BJ, Clancy KP et al: Cloning, sequencing, and mapping of the human chromosome 14 heat shock protein gene (HSPA2), *Genomics* 23:85–93, 1994.

Human protein reference data base, Johns Hopkins University and the Institute of Bioinformatics, at http://www.hprd.org/protein.

14.36. Bcl2

Kataoka T, Holler N, Micheau O, Martinon F, Tinel A et al: Bcl-rambo, a novel Bcl-2 homologue that induces apoptosis via its unique C-terminal extension, *J Biol Chem* 276(22):19548–54, 2001.

Hockenbery D, Nunez G, Milliman C, Schreiber RD, Korsmeyer SJ: Bcl-2 is an inner mitochondrial membrane protein that blocks programmed cell death, *Nature* 348:334–6, 1990.

Williams GT: Programmed cell death: Apoptosis and oncogenesis, *Cell* 65:1097–8, 1991.

Jacobson MD, Burne JF, King MP, Miyashita T, Reed JC et al: Bcl-2 blocks apoptosis in cells lacking mitochondrial DNA, *Nature* 361:365–9, 1993.

Sagot Y, Dubois-Dauphin M, Tan SA, de Bilbao F, Aebischer P et al: Bcl-2 overexpression prevents motoneuron cell body loss but not axonal degeneration in a mouse model of a neurodegenerative disease, *J Neurosci* 15:7727–33, 1995.

Farlie PG, Dringen R, Rees SM, Kannourakis G, Bernard O: Bcl-2 transgene expression can protect neurons against developmental and induced cell death, *Proc Natl Acad Sci USA* 92:4397–401, 1995.

Chen J, Flannery JG, LaVail MM, Steinberg RH, Xu J et al: Bcl-2 overexpression reduces apoptotic photoreceptor cell death in three different retinal degenerations, *Proc Natl Acad Sci USA* 93:7042–7, 1996.

Martinou JC, Dubois-Dauphin M, Staple JK, Rodriguez I, Frankowski H et al: Overexpression of Bcl-2 in transgenic mice protects neurons from naturally occurring cell death and experimental ischemia, *Neuron* 13:1017–30, 1994.

Limana F, Urbanek K, Chimenti S, Quaini F, Leri A et al: Bcl-2 overexpression promotes myocyte proliferation, *Proc Natl Acad Sci USA* 99:6257–62, 2002.

Nakayama K, Nakayama K, Negishi I, Kuida K, Sawa H et al: Targeted disruption of Bcl-2-alpha-beta in mice: Occurrence of gray hair, polycystic kidney disease, and lymphocytopenia, *Proc Natl Acad Sci USA* 91:3700–4, 1994.

14.37. Akt1

Franke TF, Kaplan DR, Cantley LC, Toker A: Direct regulation of the Akt proto-oncogene product by phosphatidylinositol-3,4-bisphosphate, *Science* 275:665–7, 1997.

Franke TF, Yang SI, Chan TO, Datta K et al: The protein kinase encoded by the Akt proto-oncogene is a target of the PDGF-activated phosphatidylinositol 3-kinase, *Cell* 81:727–36, 1995.

Dudek H, Datta SR, Franke TF, Birnbaum MJ, Yao R et al: Regulation of neuronal survival by the serine-threonine protein kinase Akt, *Science* 275:661–3, 1997.

Staal SP, Huebner K, Croce CM, Parsa NZ, Testa JR: The AKT1 proto-oncogene maps to human chromosome 14, band q32, *Genomics* 2:96–8, 1988.

Jones PF, Jakubowicz T, Pitossi FJ, Maurer F, Hemmings BA: Molecular cloning and identification of a serine/threonine protein kinase of the second-messenger subfamily, *Proc Natl Acad Sci USA* 88:4171–5, 1991.

Ozes ON, Mayo LD, Gustin JA, Pfeffer SR, Pfeffer LM et al: NF-kappa-B activation by tumour necrosis factor requires the Akt serine-threonine kinase, *Nature* 401:82–5, 1999.

Romashkova JA, Makarov SS: NF-kappa-B is a target of AKT in anti-apoptotic PDGF signalling, *Nature* 401:86–90, 1999.

Fulton D, Gratton JP, McCabe TJ, Fontana J, Fujio Y et al: Regulation of endothelium-derived nitric oxide production by the protein kinase Akt, *Nature* 399:597–601, 1999.

Dimmeler S, Fleming I, Fisslthaler B, Hermann C, Busse R et al: Activation of nitric oxide synthase in endothelial cells by Akt-dependent phosphorylation, *Nature* 399:601–5, 1999.

Chen WS, Xu PZ, Gottlob K, Chen ML, Sokol K et al: Growth retardation and increased apoptosis in mice with homozygous disruption of the akt1 gene, *Genes Dev* 15:2203–8, 2001.

Holland EC, Celestino J, Dai C, Schaefer L, Sawaya RE et al: Combined activation of Ras and Akt in neural progenitors induces glioblastoma formation in mice, *Nature Genet* 25:55–7, 2000.

14.38. CSF1

Mitrasinovic OM, Perez GV, Zhao F, Lee YL, Poon C et al: Overexpression of macrophage colony-stimulating factor receptor on microglial cells induces an inflammatory response, *J Biol Chem* 276(32):30142–9, 2001.

Praloran V, Chevalier S, Gascan H: Macrophage colony-stimulating factor is produced by activated T lymphocytes in vitro and is detected in vivo in T cells from reactive lymph nodes, *Blood* 79(9):2500–1, 1992.

Besse A, Trimoreau F, Praloran V, Denizot Y: Effect of cytokines and growth factors on the macrophage colony-stimulating factor secretion by human bone marrow stromal cells, *Cytokine* 12(5):522–5, 2000.

Pollard JW, Bartocci A, Arceci R, Orlofsky A, Ladner MB et al: Apparent role of the macrophage growth factor, CSF-1, in placental development, *Nature* 330:484–6, 1987.

Wong GG, Temple PA, Leary AC, Witek-Giannotti JS, Yang YC et al: Human CSF-1: Molecular cloning and expression of 4-kb cDNA encoding the human urinary protein, *Science* 235:1504–8, 1987.

14.39. IL1

Bensen JT, Langefeld CD, Hawkins GA, Green LE, Mychaleckyj JC et al: Nucleotide variation, haplotype structure, and association with end-stage renal disease of the human interleukin-1 gene cluster, *Genomics* 82:194–217, 2003.

Cox A, Camp NJ, Cannings C, di Giovine FS, Dale M et al: Combined sib-TDT and TDT provide evidence for linkage of the interleukin-1 gene cluster to erosive rheumatoid arthritis, *Hum Mol Genet* 8:1707–13, 1999.

Du Y, Dodel RC, Eastwood BJ, Bales KR, Gao F et al: Association of an interleukin 1-alpha polymorphism with Alzheimer's disease, *Neurology* 55:480–4, 2000.

Furutani Y, Notake M, Fukui T, Ohue M, Nomura H et al: Complete nucleotide sequence of the gene for human interleukin 1 alpha, *Nucleic Acids Res* 14:3167–79, 1986.

Green EK, Harris JM, Lemmon H, Lambert JC, Chartier-Harlin MC et al: Are interleukin-1 gene polymorphisms risk factors or disease modifiers in AD? *Neurology* 58:1566–8, 2002.

Grimaldi LME, Casadei VM, Ferri C, Veglia F, Licastro F et al: Association of early-onset Alzheimer's disease with an interleukin-1-alpha gene polymorphism, *Ann Neurol* 47:361–5, 2000.

Hogquist KA, Nett MA, Unanue ER, Chaplin DD: Interleukin 1 is processed and released during apoptosis, *Proc Natl Acad Sci USA* 88:8485–9, 1991.

Lord PCW, Wilmoth LMG, Mizel SB, McCall CE: Expression of interleukin-1 alpha and beta genes by human blood polymorphonuclear leukocytes, *J Clin Invest* 87:1312–21, 1991.

Nicoll JAR, Mrak RE, Graham DI, Stewart J, Wilcock G et al: Association of interleukin-1 gene polymorphisms with Alzheimer's disease, *Ann Neurol* 47:365–8, 2000.

Wei S, Kitaura H, Zhou P, Ross FP, Teitelbaum SL: IL-1 mediates TNF-induced osteoclastogenesis, *J Clin Invest* 115:282–90, 2005.

Werman A, Werman-Venkert R, White R, Lee JK, Werman B et al: The precursor form of IL-1-alpha is an intracine proinflammatory activator of transcription, *Proc Natl Acad Sci USA* 101:2434–9, 2004.

14.40. IL2

Clark SC, Arya SK, Wong-Staal F, Matsumoto-Kobayashi M, Kay RM et al: Human T-cell growth factor: Partial amino acid sequence, cDNA cloning, and organization and expression in normal and leukemic cells, *Proc Natl Acad Sci USA* 81:2543–7, 1984.

Ferlazzo G, Thomas D, Lin SL, Goodman K, Morandi B et al: The abundant NK cells in human secondary lymphoid tissues require activation to express killer cell Ig-like receptors and become cytolytic, *J Immun* 172:1455–62, 2004.

Cantrell DA, Smith KA: The interleukin-2 T-cell system: A new cell growth model, *Science* 224:1312–6, 1984.

Hollander GA, Zuklys S, Morel C, Mizoguchi E, Mobisson K et al: Monoallelic expression of the interleukin-2 locus, *Science* 279:2118–21, 1998.

Ku CC, Murakami M, Sakamoto A, Kappler J, Marrack P: Control of homeostasis of CD8+ memory T cells by opposing cytokines, *Science* 288:675–8, 2000.

Lowenthal JW, Zubler RH, Nabholz M, MacDonald HR: Similarities between interleukin-2 receptor number and affinity on activated B and T lymphocytes, *Nature* 315:669–72, 1985.

Rickert M, Wang X, Boulanger MJ, Goriatcheva N et al: The structure of interleukin-2 complexed with its alpha receptor, *Science* 308:1477–80, 2005.

Rosenberg SA, Grimm EA, McGrogan M, Doyle M, Kawasaki E et al: Biological activity of recombinant human interleukin-2 produced in Escherichia coli, *Science* 223:1412–5, 1984.

Seigel LJ, Harper ME, Wong-Staal F, Gallo RC, Nash WG et al: Gene for T-cell growth factor: Location on human chromosome 4q and feline chromosome B1, *Science* 223:175–8, 1984.

Weinberg K, Parkman R: Severe combined immunodeficiency due to a specific defect in the production of interleukin-2, *New Engl J Med* 322:1718–23, 1990.

14.41. IL3

Celestin J, Rotschke O, Falk K, Ramesh N, Jabara H et al: IL-3 induces B7.2 (CD86) expression and costimulatory activity in human eosinophils, *J Immun* 167:6097–104, 2001.

Chiang CS, Powell HC, Gold LH, Samimi A, Campbell IL: Macrophage/microglial-mediated primary demyelination and motor disease induced by the central nervous system production of interleukin-3 in transgenic mice, *J Clin Invest* 97:1512–24, 1996.

Clark-Lewis I, Aebersold R, Ziltener H, Schrader JW, Hood LE et al: Automated chemical synthesis of a protein growth factor for hemopoietic cells, interleukin-3, *Science* 231:134–9, 1986.

Devireddy LR, Teodoro JG, Richard FA, Green MR: Induction of apoptosis by a secreted lipocalin that is transcriptionally regulated by IL-3 deprivation, *Science* 293:829–34, 2001.

Donahue RE, Seehra J, Metzger M, Lefebvre D, Rock B et al: Human IL-3 and GM-CSF act synergistically in stimulating hematopoiesis in primates, *Science* 241:1820–3, 1988.

Fung MC, Hapel AJ, Ymer S, Cohen DR, Johnson RM et al: Molecular cloning of cDNA for murine interleukin-3, *Nature* 307:233–7, 1984.

14.42. IL4

Carvalho LH, Sano G, Hafalla JCR, Morrot A, Curotto de Lafaille MA et al: IL-4-secreting CD4+ T cells are crucial to the development of CD8+ T-cell responses against malaria liver stages, *Nature Med* 8:166–70, 2002.

Dickensheets HL, Venkataraman C, Schindler U, Donnelly RP: Interferons inhibit activation of STAT6 by interleukin 4 in human monocytes by inducing SOCS-1 gene expression, *Proc Natl Acad Sci USA* 96:10800–5, 1999.

Ghoreschi K, Thomas P, Breit S, Dugas M, Mailhammer R et al: Interleukin-4 therapy of psoriasis induces Th2 responses and improves human autoimmune disease, *Nature Med* 9:40–6, 2003.

Horsley V, Jansen KM, Mills ST, Pavlath GK: IL-4 acts as a myoblast recruitment factor during mammalian muscle growth, *Cell* 113:483–94, 2003.

Kelly-Welch AE, Hanson EM, Boothby MR, Keegan AD: Interleukin-4 and interleukin-13 signaling connections maps, *Science* 300:1527–8, 2003.

Kotanides H, Reich NC: Interleukin-4-induced STAT6 recognizes and activates a target site in the promoter of the interleukin-4 receptor gene, *J Biol Chem* 271:25555–61, 1996.

Lacy DA, Wang ZE, Symula DJ, McArthur CJ, Rubin EM et al: Faithful expression of the human 5q31 cytokine cluster in transgenic mice, *J Immun* 164:4569–74, 2000.

Loots GG, Locksley RM, Blankespoor CM, Wang ZE, Miller W et al: Identification of a coordinate regulator of interleukins 4, 13, and 5 by cross-species sequence comparisons, *Science* 288:136–40, 2000.

Messi M, Giacchetto I, Nagata K, Lanzavecchia A, Natoli G et al: Memory and flexibility of cytokine gene expression as separable properties of human T(H)1 and T(H)2 lymphocytes, *Nature Immun* 4:78–86, 2003.

Spilianakis CG, Lalioti MD, Town T, Lee GR, Flavell RA: Interchromosomal associations between alternatively expressed loci, *Nature* 435:637–45, 2005.

14.43. IL5

Broide DH, Hoffman H, Sriramarao P: Genes that regulate eosinophilic inflammation, *Am J Hum Genet* 65:302–7, 1999.

Campbell HD, Tucker WQJ, Hort Y, Martinson ME, Mayo G et al: Molecular cloning, nucleotide sequence, and expression of the gene encoding human eosinophil differentiation factor (interleukin-5), *Proc Natl Acad Sci USA* 84:6629–33, 1987.

Chandrasekharappa SC, Rebelsky MS, Firak TA, Le Beau MM, Westbrook CA: A long-range restriction map of the interleukin-4 and interleukin-5 linkage group on chromosome 5, *Genomics* 6:94–9, 1990.

Coffman RL, Seymour BWP, Hudak S, Jackson J, Rennick D: Antibody to interleukin-5 inhibits helminth-induced eosinophilia in mice, *Science* 245:308–10, 1989.

Cogan E, Schandene L, Crusiaux A, Cochaux P, Velu T et al: Clonal proliferation of type 2 helper T cells in a man with the hypereosinophilic syndrome, *New Engl J Med* 330:535–8, 1994.

Loots GG, Locksley RM, Blankespoor CM, Wang ZE, Miller W et al: Identification of a coordinate regulator of interleukins 4, 13, and 5 by cross-species sequence comparisons, *Science* 288:136–40, 2000.

Simon HU, Plotz SG, Dummer R, Blaser K: Abnormal clones of T cells producing interleukin-5 in idiopathic eosinophilia, *New Engl J Med* 341:1112–20, 1999.

Tanabe T, Konishi M, Mizuta T, Noma T, Honjo T: Molecular cloning and structure of the human interleukin-5 gene, *J Biol Chem* 262:16580–4, 1987.

Willman CL, Sever CE, Pallavicini MG, Harada H, Tanaka N et al: Deletion of IRF-1, mapping to chromosome 5q31.1, in human leukemia and preleukemic myelodysplasia, *Science* 259:968–71, 1993.

Yokota T, Coffman RL, Hagiwara H, Rennick DM, Takebe Y et al: Isolation and characterization of lymphokine cDNA clones encoding mouse and human IgA-enhancing factor and eosinophil colony-stimulating factor activities: relationship to interleukin 5, *Proc Natl Acad Sci USA* 84:7388–92, 1987.

14.44. IL10

Fickenscher H, Hor S, Kupers H, Knappe A, Wittmann S et al: The interleukin-10 family of cytokines, *Trends Immunol* 23(2):89–96, Feb 2002.

Spittler A, Reissner CM, Oehler R, Gornikiewicz A, Gruenberger T et al: Immunomodulatory effects of glycine on LPS-treated monocytes: Reduced TNF-alpha production and accelerated IL-10 expression, *FASEB J* 13(3):563–71, March 1999.

Foey AD, Feldmann M, Brennan FM: CD40 ligation induces macrophage IL-10 and TNF-alpha production: differential use of the PI3K and p42/44 MAPK-pathways, *Cytokine* 16(4):131–42, Nov 2001.

Crawley JB, Williams LM, Mander T, Brennan FM, Foxwell BM: Interleukin-10 stimulation of phosphatidylinositol 3-kinase and p70 S6 kinase is required for the proliferative but not the antiinflammatory effects of the cytokine, *J Biol Chem* 271:16357–62, 1996.

Eskdale J, Gallagher G, Verweij CL, Keijsers V, Westendorp RGJ et al: Interleukin 10 secretion in relation to human IL-10 locus haplotypes, *Proc Natl Acad Sci USA* 95:9465–70, 1998.

Gesser B, Leffers H, Jinquan T, Vestergaard C, Kirstein N et al: Identification of functional domains on human interleukin 10, *Proc Natl Acad Sci USA* 94:14620–5, 1997.

Lee TS, Chau LY: Heme oxygenase-1 mediates the anti-inflammatory effect of interleukin-10 in mice, *Nature Med* 8:240–6, 2002.

14.45. Erythropoietin

Jacobs K, Shoemaker C, Rudersdorf R, Neill SD, Kaufman RJ et al: Isolation and characterization of genomic and cDNA clones of human erythropoietin, *Nature* 313(6005):806–10, Feb–March 1985.

Becerra SP, Amaral J: Erythropoietin—an endogenous retinal survival factor, *New Engl J Med* 347:1968–70, 2002.

Bianchi R, Buyukakilli B, Brines M, Savino C, Cavaletti G et al: Erythropoietin both protects from and reverses experimental diabetic neuropathy, *Proc Natl Acad Sci USA* 101:823–8, 2004.

Celik M, Gokmen N, Erbayraktar S, Akhisaroglu M, Konakc S et al: Erythropoietin prevents motor neuron apoptosis and neurologic disability in experimental spinal cord ischemic injury, *Proc Natl Acad Sci USA* 99:2258–63, 2002.

Digicaylioglu M, Lipton SA: Erythropoietin-mediated neuroprotection involves cross-talk between Jak2 and NF-kappa-B signalling cascades, *Nature* 412:641–7, 2001.

Gorio A, Gokmen N, Erbayraktar S, Yilmaz O, Madaschi L et al: Recombinant human erythropoietin counteracts secondary injury and markedly enhances neurological recovery from experimental spinal cord trauma, *Proc Natl Acad Sci USA* 99:9450–5, 2002.

Grimm C, Wenzel A, Groszer M, Mayser H, Seeliger M et al: HIF-1-induced erythropoietin in the hypoxic retina protects against light-induced retinal degeneration, *Nature Med* 8:718–24, 2002.

Junk AK, Mammis A, Savitz SI, Singh M, Roth S et al: Erythropoietin administration protects retinal neurons from acute ischemia-reperfusion injury, *Proc Natl Acad Sci USA* 99:10659–64, 2002.

Leist M, Ghezzi P, Grasso G, Bianchi R, Villa P et al: Derivatives of erythropoietin that are tissue protective but not erythropoietic, *Science* 305:239–42, 2004.

14.46. IL7

Carvalho TL, Mota-Santos T, Cumano A, Demengeot J, Vieira P: Arrested B lymphopoiesis and persistence of activated B cells in adult interleukin 7-/- mice, *J Exp Med* 194:1141–50, 2001.

Goodwin RG, Lupton S, Schmierer A, Hjerrild KJ, Jerzy R et al: Human interleukin 7: molecular cloning and growth factor activity on human and murine B-lineage cells, *Proc Natl Acad Sci USA* 86:302–6, 1989.

Lai L, Goldschneider I: Cutting edge: identification of a hybrid cytokine consisting of IL-7 and the beta-chain of the hepatocyte growth factor/scatter factor, *J Immun* 167:3550–4, 2001.

Namen AE, Lupton S, Hjerrild K, Wignall J, Mochizuki DY et al: Stimulation of B-cell progenitors by cloned murine interleukin-7, *Nature* 333:571–3, 1988.

Seddon B, Tomlinson P, Zamoyska R: Interleukin 7 and T cell receptor signals regulate homeostasis of CD4 memory cells, *Nature Immun* 4:680–6, 2003.

Vobhenrich CAJ, Cumano A, Muller W, Di Santo JP, Vieira P: Thymic stromal-derived lymphopoietin distinguishes fetal from adult B cell development, *Nature Immun* 4:773–9, 2003.

Watanabe M, Ueno Y, Yajima T, Iwao Y, Tsuchiya M et al: Interleukin 7 is produced by human intestinal epithelial cells and regulates the proliferation of intestinal mucosal lymphocytes, *J Clin Invest* 95:2945–53, 1995.

14.47. IL8

Baggiolini M, Walz A, Kunkel SL: Neutrophil-activating peptide-1/interleukin 8, a novel cytokine that activates neutrophils, *J Clin Invest* 84:1045–9, 1989.

Bruun JM, Pedersen SB, Richelsen B: Regulation of interleukin 8 production and gene expression in human adipose tissue in vitro, *J Clin Endocr Metab* 86:1267–73, 2001.

Emi M, Keicho N, Tokunaga K, Katsumata H, Souma S et al: Association of diffuse panbronchiolitis with microsatellite polymorphism of the human interleukin 8 (IL-8) gene, *J Hum Genet* 44:169–72, 1999.

Hull J, Ackerman H, Isles K, Usen S, Pinder M et al: Unusual haplotypic structure of IL8, a susceptibility locus for a common respiratory virus, *Am J Hum Genet* 69:413–9, 2001.

Modi WS, Chen ZQ: Localization of the human CXC chemokine subfamily on the long arm of chromosome 4 using radiation hybrids, *Genomics* 47:136–9, 1998.

Modi WS, Dean M, Seuanez HN, Mukaida N, Matsushima K et al: Monocyte-derived neutrophil chemotactic factor (MDNCF/IL-8) resides in a gene cluster along with several other members of the platelet factor 4 gene superfamily, *Hum Genet* 84:185–7, 1990.

Pruijt JFM, Verzaal P, van Os R, de Kruijf EJFM, van Schie MLJ et al: Neutrophils are indispensable for hematopoietic stem cell mobilization induced by interleukin-8 in mice, *Proc Natl Acad Sci USA* 99:6228–33, 2002.

14.48. IL9

Collaborative Study on the Genetics of Asthma: A genome-wide search for asthma susceptibility loci in ethnically diverse populations, *Nature Genet* 15:389–92, 1997.

Gounni AS, Hamid Q, Rahman SM, Hoeck J, Yang J et al: IL-9-mediated induction of eotaxin1/CCL11 in human airway smooth muscle cells, *J Immun* 173:2771–9, 2004.

Holbrook ST, Ohls RK, Schibler KR, Yang YC, Christensen RD: Effect of interleukin-9 on clonogenic maturation and cell-cycle status of fetal and adult hematopoietic progenitors, *Blood* 77:2129–34, 1991.

Kelleher K, Bean K, Clark SC, Leung WY, Yang-Feng TL et al: Human interleukin-9: genomic sequence, chromosomal location, and sequences essential for its expression in human T-cell leukemia virus (HTLV)-I-transformed human T cells, *Blood* 77:1436–41, 1991.

Le Beau MM, Espinosa R III, Neuman WL, Stock W, Roulston D et al: Cytogenetic and molecular delineation of the smallest commonly deleted region of chromosome 5 in malignant myeloid diseases, *Proc Natl Acad Sci USA* 90:5484–8, 1993.

Merz H, Houssiau FA, Orscheschek K, Renauld JC, Fliedner A et al: Interleukin-9 expression in human malignant lymphomas: unique association with Hodgkin's disease and large cell anaplastic lymphoma, *Blood* 78:1311–7, 1991.

Mock BA, Krall M, Kozak CA, Nesbitt MN, McBride OW et al: IL9 maps to mouse chromosome 13 and human chromosome 5, *Immunogenetics* 31:265–70, 1990.

Modi WS, Pollock DD, Mock BA, Banner C, Renauld JC et al: Regional localization of the human glutaminase (GLS) and interleukin-9 (IL9) genes by in situ hybridization, *Cytogenet Cell Genet* 57:114–6, 1991.

Nicolaides NC, Holroyd KJ, Ewart SL, Eleff SM, Kiser MB et al: Interleukin 9: a candidate gene for asthma, *Proc Natl Acad Sci USA* 94:13175–80, 1997.

Postma DS, Bleecker ER, Amelung PJ, Holroyd KJ, Xu J et al: Genetic susceptibility to asthma: bronchial hyperresponsiveness coinherited with a major gene for atopy, *New Engl J Med* 333:894–900, 1995.

Van Snick J, Goethals A, Renauld JC, Van Roost E, Uyttenhove C et al: Cloning and characterization of a cDNA for a new mouse T cell growth factor (P40), *J Exp Med* 169:363–8, 1989.

Yang YC, Ricciardi S, Ciarletta A, Calvetti J, Kelleher K et al: Expression cloning of a cDNA encoding a novel human hematopoietic growth factor: Human homologue of murine T-cell growth factor P40, *Blood* 74:1880–4, 1989.

14.49. IFNγ

Pyo JO, Jang MH, Kwon YK, Lee HJ, Jun JI et al: Essential roles of Atg5 and FADD in autophagic cell death: Dissection of autophagic cell death into vacuole formation and cell death, *J Biol Chem* 280(21):20722–9, May 2005.

Badovinac VP, Tvinnereim AR, Harty JT: Regulation of antigen-specific CD8(+) T cell homeostasis by perforin and interferon-gamma, *Science* 290:1354–7, 2000.

Ben-Asouli Y, Banai Y, Pel-Or Y, Shir A, Kaempfer R: Human interferon-gamma mRNA autoregulates its translation through a pseudoknot that activates the interferon-inducible protein kinase PKR, *Cell* 108:221–32, 2002.

Binder GK, Griffin DE: Interferon-gamma-mediated site-specific clearance of alphavirus from CNS neurons, *Science* 293:303–6, 2001.

Cavet J, Dickinson AM, Norden J, Taylor PR, Jackson GH et al: Interferon-gamma and interleukin-6 gene polymorphisms associate with graft-versus-host disease in HLA-matched sibling bone marrow transplantation, *Blood* 98:1594–600, 2001.

Khani-Hanjani A, Lacaille D, Hoar D, Chalmers A, Horsman D et al: Association between dinucleotide repeat in non-coding region of interferon-gamma gene and susceptibility to, and severity of, rheumatoid arthritis, *Lancet* 356:820–5, 2000.

Koh KP, Wang Y, Yi T, Shiao SL, Lorber MI et al: T cell-mediated vascular dysfunction of human allografts results from IFN-gamma dysregulation of NO synthase, *J Clin Invest* 114:846–56, 2004.

Messi M, Giacchetto I, Nagata K, Lanzavecchia A, Natoli G et al: Memory and flexibility of cytokine gene expression as separable properties of human T(H)1 and T(H)2 lymphocytes, *Nature Immun* 4:78–86, 2003.

Szabo SJ, Sullivan BM, Stemmann C, Satoskar AR, Sleckman BP et al: Distinct effects of T-bet in T(H)1 lineage commitment and IFN-gamma production in CD4 and CD8 T cells, *Science* 295:338–42, 2002.

Takayanagi H, Ogasawara K, Hida S, Chiba T, Murata S et al: T-cell-mediated regulation of osteoclastogenesis by signalling cross-talk between RANKL and IFN-gamma, *Nature* 408:600–5, 2000.

14.50. IL6

Boulanger MJ, Chow D, Brevnova EE, Garcia KC: Hexameric structure and assembly of the interleukin-6/IL-6 alpha-receptor/gp130 complex, *Science* 300:2101–4, 2003.

Cressman DE, Greenbaum LE, DeAngelis RA, Ciliberto G, Furth EE et al: Liver failure and defective hepatocyte regeneration in interleukin-6-deficient mice, *Science* 274:1379–82, 1996.

Ferguson-Smith AC, Chen YF, Newman MS, May LT, Sehgal PB et al: Regional localization of the interferon beta-2/B-cell stimulatory factor 2/hepatocyte stimulating factor gene to human chromosome 7p15-p21, *Genomics* 2:203–8, 1988.

Kawano M, Hirano T, Matsuda T, Taga T, Horii Y et al: Autocrine generation and requirement of BSF-2/IL-6 for human multiple myelomas, *Nature* 332:83–5, 1988.

Kiecolt-Glaser JK, Preacher KJ, MacCullum RC, Atkinson C, Malarkey WB et al: Chronic stress and age-related increases in the proinflammatory cytokine IL-6, *Proc Natl Acad Sci USA* 100:9090–5, 2003.

McLoughlin RM, Jenkins BJ, Grail D, Williams AS, Fielding CA et al: IL-6 trans-signaling via STAT3 directs T cell infiltration in acute inflammation, *Proc Natl Acad Sci USA* 102:9589–94, 2005.

Nemeth E, Rivera S, Gabayan V, Keller C, Taudorf S et al: IL-6 mediates hypoferremia of inflammation by inducing the synthesis of the iron regulatory hormone hepcidin, *J Clin Invest* 113:1271–6, 2004.

Sawczenko A, Azooz O, Paraszczuk J, Idestrom M, Croft NM et al: Intestinal inflammation-induced growth retardation acts through IL-6 in rats and depends on the -174 IL-6 G/C polymorphism in children, *Proc Natl Acad Sci USA* 102:13260–5, 2005.

Venihaki M, Dikkes P, Carrigan A, Karalis KP: Corticotropin-releasing hormone regulates IL-6 expression during inflammation, *J Clin Invest* 108:1159–66, 2001.

14.51. IFNα

Jinushi M, Takehara T, Kanto T, Tatsumi T, Groh V et al: Critical role of MHC class I-related chain A and B expression on IFN-alpha-stimulated dendritic cells in NK cell activation: Impairment in chronic hepatitis C virus infection, *J Immunol* 170(3):1249–56, 2003.

Nguyen KB, Watford WT, Salomon R, Hofmann SR, Pien GC et al: Critical role for STAT4 activation by type 1 interferons in the interferon-gamma response to viral infection, *Science* 297(5589):2063–6, 2002.

Liang S, Wei H, Sun R, Tian Z: IFNalpha regulates NK cell cytotoxicity through STAT1 pathway, *Cytokine* 23(6):190–9, 2003.

Diaz MO, Le Beau MM, Pitha P, Rowley JD: Interferon and c-ets-1 genes in the translocation (9;11)(p22;q23) in human acute monocytic leukemia, *Science* 231:265–7, 1986.

Diaz MO, Pomykala HM, Bohlander SK, Maltepe E, Malik K et al: Structure of the human type-I interferon gene cluster determined from a YAC clone contig, *Genomics* 22:540–52, 1994.

Edge MD, Green AR, Heathcliffe GR, Meacock PA, Schuch W et al: Total synthesis of a human leukocyte interferon gene, *Nature* 292:756–62, 1981.

Fountain JW, Karayiorgou M, Taruscio D, Graw SL, Buckler AJ et al: Genetic and physical map of the interferon region on chromosome 9p, *Genomics* 14:105–12, 1992.

Hitzeman RA, Hagie FE, Levine HL, Goeddel DV, Ammerer G et al: Expression of a human gene for interferon in yeast, *Nature* 293:717–22, 1981.

Siegal FP, Kadowaki N, Shodell M, Fitzgerald-Bocarsly PA, Shah K et al: The nature of the principal type 1 interferon-producing cells in human blood, *Science* 284:1835–7, 1999.

Takaoka A, Hayakawa S, Yanai H, Stolber D, Negishi H et al: Integration of interferon-alpha/beta signalling to p53 responses in tumour suppression and antiviral defence, *Nature* 424:516–23, 2003.

14.52. IFNβ

Bekisz J, Schmeisser H, Hernandez J, Goldman ND, Zoon KC: Human interferons alpha, beta and omega, *Growth Factors* 22(4):243–51, 2004.

Satomi H, Wang B, Fujisawa H, Otsuka F: Interferon-beta from melanoma cells suppresses the proliferations of melanoma cells in an autocrine manner, *Cytokine* 18(2):108–15, 2002.

Diaz MO, Le Beau MM, Pitha P, Rowley JD: Interferon and c-ets-1 genes in the translocation (9;11)(p22;q23) in human acute monocytic leukemia, *Science* 231:265–7, 1986.

Knight E Jr: Human fibroblast interferon: Amino acid analysis and amino terminal amino acid sequence, *Science* 207:525–7, 1980.

Siegal FP, Kadowaki N, Shodell M, Fitzgerald-Bocarsly PA, Shah K et al: The nature of the principal type 1 interferon-producing cells in human blood, *Science* 284:1835–7, 1999.

Takaoka A, Hayakawa S, Yanai H, Stolber D, Negishi H et al: Integration of interferon-alpha/beta signalling to p53 responses in tumour suppression and antiviral defence, *Nature* 424:516–23, 2003.

Human protein reference data base, Johns Hopkins University and the Institute of Bioinformatics, at http://www.hprd.org/protein.

14.53. Enhancement of Angiogenesis

Ueda H, Sawa Y, Matsumoto K, Kitagawa-Sakakida S, Kawahira Y et al: Gene transfection of hepatocyte growth factor attenuates reperfusion injury in the heart, *Ann Thorac Surg* 67:1726–31, 2002.

Losordo DW, Vale PR, Symes JF, Dunnington CH, Esakof DD et al: Gene therapy for myocardial angiogenesis. Initial clinical results with direct myocardial injection of $phVEGF_{165}$ as sole therapy for myocardial ischemia, *Circulation* 98:2800–4, 1998.

Rosengart TK, Lee LY, Patel SR, Sanborn TA, Parikh M et al: Angiogenesis gene therapy, phase I assessment of direct intramyocardial administration of an adenovirus vector expressing $VEGF_{121}$ cDNA to individuals with clinically significant severe coronary artery disease, *Circulation* 100:468–74, 1999.

Vale PR, Losordo DW, Milliken CE, Maysky M, Esakof DD et al: Randomized, single-blind, placebo-controlled pilot study of catheter-based myocardial gene transfer for therapeutic angiogenesis using left ventricular electromechanical mapping in patients with chronic myocardial ischemia, *Circulation* 103:2138–43, 2001.

Grines CL, Watkins MW, Helmer G, Penny W, Brinker J et al: Angiogenic gene therapy (AGENT) trial in patients with stable angina pectoris, *Circulation* 105:1291–7, 2002.

Laham RJ, Simons M, Sellke F: Gene transfer for angiogenesis in coronary artery disease, *Annu Rev Med* 52:485–502, 2001.

Isner JM: Myocardial gene therapy, *Nature* 415:234–9, 2002.

Giordano FJ, Ping P, McKirnan MD, Nozaki S, DeMaria AN et al: Intracoronary gene transfer of fibroblast growth factor-5 increases blood flow and contractile function in an ischemic region of the heart, *Nat Med* 2:534–9, 1996.

Celletti FL, Waugh JM, Amabile PG, Brendolan A, Hilfiker PR et al: Vascular endothelial growth factor enhances atherosclerotic plaque progression, *Nat Med* 7:425–9, 2001.

Su H, Arakawa-Hoyt J, Kan YW: Adeno-associated viral vector-mediated hypoxia response element-regulated gene expression in mouse ischemic heart model, *Proc Natl Acad Sci USA* 99:9480–5, 2002.

14.54. Cellular Regenerative Engineering for Ischemic Heart Disease

Leor J, Aboulafia-Etzion S, Dar A, Shapiro L, Barbash IM et al: Bioengineered cardiac grafts. A new approach to repair the infarcted myocardium? *Circulation* 102(Suppl. III):56–61, 2000.

Shimizu T, Yamato M, Isoi Y, Akutsu T, Setomaru T et al: Fabrication of pulsatile cardiac tissue grafts using a novel 3-dimensional cell sheet manipulation technique and temperature-responsive cell culture surfaces, *Circ Res* 90(3):e40, Feb 2002.

Kucia M, Dawn B, Hunt G, Guo Y, Wysoczynski M et al: Cells expressing early cardiac markers reside in the bone marrow and are mobilized into the peripheral blood after myocardial infarction, *Circ Res* 95(12):1191–9, 2004.

Menasche P, Hagege AA, Vilquin JT, Desnos M, Abergel E et al: Autologous skeletal myoblast transplantation for severe postinfarction left ventricular dysfunction, *J Am Coll Cardiol* 41:1078–83, 2003.

Kofidis T, de Bruin JL, Yamane T, Tanaka M, Lebl DR et al: Stimulation of paracrine pathways with growth factors enhances embryonic stem cell engraftment and host-specific differentiation in the heart after ischemic myocardial injury, *Circulation* 111:2486–93, 2005.

Orlic D, Kajstura J, Chimenti S, Jakoniuk I, Anderson SM et al: Bone marrow cells regenerate infracted myocardium, *Nature* 410:701–5, 2001.

Balsam LB, Wagers AJ, Christensen JL, Kofidis T, Weissman IL et al: Haematopoietic stem cells adopt mature haematopoietic fates in ischaemic myocardium, *Nature* 428:668–73, 2004.

Murry CE, Soonpaa MH, Reinecke H, Nakajima H, Nakajima HO et al: Haematopoietic stem cells do not transdifferentiate into cardiac myocytes in myocardial infarcts, *Nature* 428:664–8, 2004.

Nygren JM, Jovinge S, Breitbach M, Säwén P, Röll W et al: Bone marrow-derived hematopoietic cells generate cardiomyocytes at a low frequency through cell fusion, but not transdifferentiation, *Nat Med* 10:494–501, 2004.

Harada M, Qin Y, Takano H, Minamino T, Zou Y et al: G-CSF prevents cardiac remodeling after myocardial infarction by activating the Jak-Stat pathway in cardiomyocytes, *Nat Med* 11:305–11, 2005.

Beltrami AP, Barlucchi L, Torella D, Baker M, Limana F et al: Adult cardiac stem cells are multipotent and support myocardial regeneration, *Cell* 114:763–76, 2003.

Oh H, Bradfute SB, Gallardo TD, Nakamura T, Gaussin V et al: Cardiac progenitor cells from adult myocardium: Homing, differentiation, and fusion after infarction, *Proc Natl Acad Sci USA* 100:12313–8, 2003.

Matsuura K, Nagai T, Nishigaki N, Oyama T, Nishi J et al: Adult cardiac Sca-1-positive cells differentiate into beating cardiomyocytes, *J Biol Chem* 279:11384–91, 2004.

Laugwitz KL, Moretti A, Lam J, Gruber P, Chen Y et al: Postnatal isl1+ cardioblasts enter fully differentiated cardiomyocyte lineages, *Nature* 433:647–53, 2005.

Martin CM, Meeson AP, Robertson SM, Hawke TJ, Richardson JA et al: Persistent expression of the ATP-binding cassette transporter, Abcg2, identifies cardiac SP cells in the developing and adult heart, *Dev Biol* 265:262–75, 2004.

Urbanek K, Rota M, Cascapera S, Bearzi C, Nascimbene A et al: Cardiac stem cell possesses the growth factor-receptor system that following activation the regenerative the infarcted myocardium improving ventricular function and long term survival, *Circ Res* 97:663–73, 2005.

Chimenti C, Kajstura J, Torella D, Urbanek K, Heleniak H et al: Senescence and death of primitive cells and myocytes lead to premature cardiac aging and heart failure, *Circ Res* 93:604–13, 2003.

Messina E, Angelis LD, Frati G, Morrone S, Chimenti S et al: Isolation and expansion of adult cardiac stem cells from human and murine heart, *Circ Res* 95:911–21, 2004.

Arai F, Hirao A, Ohmura M, Sato H, Matsuoka S et al: Tie2/angiopoietin-1 signaling regulates hematopoietic stem cell quiescence in the bone marrow niche, *Cell* 118:149–61, 2004.

Khwaja A, Lehmann K, Marte BM, Downward J: Phosphoinositide 3-kinase induces scattering and tubulogenesis in epithelial cells through a novel pathway, *J Biol Chem* 273:18793–801, 1998.

Royal I, Lamarche-Vane N, Lamorte L, Kaibuchi K, Park M: Activation of cdc42, rac, PAK, and rho-kinase in response to hepatocyte growth factor differentially regulates epithelial cell colony spreading and dissociation, *Mol Biol Cell* 11:1709–25, 2000.

Kawamura K, Takano K, Suetsugu S, Kurisu S, Yamazaki D et al: N-WASP and WAVE2 acting downstream of phosphatidylinositol 3-kinase are required for myogenic cell migration induced by hepatocyte growth factor, *J Biol Chem* 279:54862–71, 2004.

Soonpaa MH, Koh GY, Klug MG, Field LJ: Formation of nascent intercalated disks between grafted fetal cardiomyocytes and host myocardium, *Science* 264:98–101, 1994.

Torella D, Ellison GM, Mendez-Ferrer S, Ibanez B, Nadal-Ginard B: Resident human cardiac stem cells: Role in cardiac cellular homeostasis and potential for myocardial regeneration, *Nat Clin Pract Cardiovasc Med* 3(Suppl 1):S8–13, 2006.

Laflamme MA, Murry CE: Regenerating the heart, *Nat Biotechnol* 23(7):845–56, July 2005.

Rafii S, Lyden D: Therapeutic stem and progenitor cell transplantation for organ vascularization and regeneration, *Nat Med* 9:702–12, 2003.

Reinlib L, Field L: Cell transplantation as future therapy for cardiovascular disease? *Circulation* 101:e192–7, 2000.

Orlic D, Kajstura J, Chimenti S, Jakoniuk I, Anderson SM et al: Bone marrow cells regenerate infarcted myocardium, *Nature* 410:710–5, 2001.

Jackson K, Majka SM, Wang H, Pocious J, Hartley CJ et al: Regeneration of ischemic cardiac muscle and vascular endothelium by adult stem cells, *J Clin Invest* 107:1395–402, 2001.

Reinlib L, Field L: Cell transplantation as future therapy for cardiovascular disease? *Circulation* 101:e192–7, 2000.

Jiang Y, Jahagirdar BN, Reinhardt RL, Schwartz RE, Keene CD et al: Pluripotency of mesenchymal stem cells derived from adult marrow, *Nature* 418:41–9, 2002.

Hakuno D, Fukuda K, Makino S, Konishi F, Tomita Y et al: Bone marrow-derived regenerated cardiomyocytes (CMG cells) express functional adrenergic and muscarinic receptors, *Circulation* 105:380–6, 2002.

Makino S, Fukuda K, Miyoshi S, Konishi F, Kodama H et al: Cardiomyocytes can be generated from marrow stromal cells in vitro, *J Clin Invest* 103:697–705, 1999.

Tomita S, Li RK, Weisel RD, Mickle DAG, Kim EJ et al: Autologous transplantation of bone marrow cells improves damaged heart function, *Circulation* 100(Suppl):II-247–56, 1999.

Toma C, Pittenger MF, Cahill KS, Byrne BJ, Kessler PD: Human mesenchymal stem cells differentiate to a cardiomyocyte phenotype in the adult murine heart, *Circulation* 105:93–8, 2002.

Makino S, Fukuda K, Miyoshi S, Konishi F, Kodama H et al: Cardiomyocytes can be generated from marrow stromal cells in vitro, *J Clin Invest* 103:697–705, 1999.

Strauer BE, Brehm M, Zeus T, Kostering M, Hernandez A et al: Repair of infarcted myocardium by autologous intracoronary mononuclear bone marrow cell transplantation in humans, *Circulation* 106:1913–8, 2002.

Stamm C, Westphal B, Kleine, HD, Petzsch M, Kittner C et al: Autologous bone marrow stem cell transplantation for myocardial regeneration, *Lancet* 361:45–6, 2003.

Assmus B, Schachinger V, Teupe C, Britten M, Lehmann R et al: Transplantation of progenitor cells and regeneration enhancement in acute myocardial infarction (TOPCARE-AMI), *Circulation* 106:3009–17, 2002.

Perin EC, Dohmann HF, Borojevic R, Silva SA, Sousa AL et al: Transendocardial, autologous bone marrow cell transplantation for severe, chronic ischemic heart failure, *Circulation* 107:2294–302, 2003.

Tse HF, Kwong YL, Chan JK, Lo G, Ho CL et al: Angiogenesis in ischemic myocardium by intramyocardial autologous bone marrow mononuclear cell implantation, *Lancet* 361:47–9, 2003.

Orlic D, Hill JM, Arai AE: Stem cells for myocardial regeneration, *Circ Res* 91:1092–102, 2002.

Beltrami AP, Beltrami AP, Urbanek K, Kajstura J, Yan SM et al: Evidence that human cardiac myocytes divide after myocardial infarction, *New Engl J Med* 344:175–7, 2001.

Beltrami AP, Barlucchi L, Torella D, Baker M, Limana F et al: Adult cardiac stem cells are multipotent and support myocardial regeneration, *Cell* 114:763–76, 2003.

Urbanek K, Quaini F, Tasca G, Torella D, Castaldo C et al: Intense myocyte formation from cardiac stem cells in human cardiac hypertrophy, *Proc Natl Acad Sci USA* 100:10440–5, 2003.

Kajstura J, Leri A, Finato N, Di Loreto C, Beltrami CA et al: Myocyte proliferation in end-stage cardiac failure in humans, *Proc Natl Acad Sci USA* 95:8801–5, 1998.

Mangi AA, Nouseaux N, Kong D, He H, Rezvani M et al: Mesenchymal stem cells modified with Akt prevent remodeling and restore performance of infarcted hearts, *Nat Med* 9:1195–201, 2003.

Rafii S, Lyden D: Therapeutic stem and progenitor cell transplantation for organ vascularization and regeneration, *Nat Med* 9:702–12, 2003.

Hristov M, Erl W, Weber PC: Endothelial progenitor cells, mobilization, differentiation, and homing, *Arterioscler Thromb Vasc Biol* 23:1185–9, 2003.

Shintani S, Murohara T, Ikeda H, Ueno T, Sasaki K et al: Augmentation of postnatal neovascularization with autologous bone marrow transplantation, *Circulation* 103:897–903, 2001.

Ikenaga S, Hamano K, Nishida M, Kobayashi T, Li TS et al: Autologous bone marrow implantation induced angiogenesis and improved deteriorated exercise capacity in a rat ischemic hindlimb model, *J Surg Res* 96:277–83, 2001.

Orlic D, Kajstura J, Chimenti S, Jakoniuk I, Anderson SM et al: Bone marrow cells regenerate infarcted myocardium, *Nature* 410:710–5, 2001.

Jackson K, Majka SM, Wang H, Pocious J, Hartley CJ et al: Regeneration of ischemic cardiac muscle and vascular endothelium by adult stem cells, *J Clin Invest* 107:1395–402, 2001.

14.55. Tissue Regenerative Engineering for Ischemic Heart Disease

Kochupura PV, Azeloglu EU, Kelly DJ, Doronin SV, Badylak SF et al: Tissue-engineered myocardial patch derived from extracellular matrix provides regional mechanical function, *Circulation* 112:I-144–9, 2005.

Matsubayashi K, Fedak PW, Mickle DA, Weisel RD, Ozawa T et al: Improved left ventricular aneurysm repair with bioengineered vascular smooth muscle grafts, *Circulation* 108(Suppl 1): II219–25, Sept 2003.

Robinson KA, Li J, Mathison M, Redkar A, Cui J et al: Extracellular matrix scaffold for cardiac repair, *Circulation* 112(Suppl 9):I135–43, Aug 2005.

14.56. Artificial Cardiac Valves

Yoganathan AP: Cardiac valve prostheses, in *Biomedical Engineering Handbook*, Bronzino JD, ed, CRC Press/IEEE Press, Boca Raton, FL, 1995, pp 1847–70.

Vesely I: Heart valve tissue engineering, *Circ Res* 97(8):743–55, 2005.

14.57. Tissue-Engineered Cardiac Valves

Vesely I: Heart valve tissue engineering, *Circ Res* 97:743–55, 2005.

Stock UA, Nagashima M, Khalil PN, Nollert GD, Herden T et al: Tissue-engineered valved conduits in the pulmonary circulation, *J Thor Cardiovasc Surg* 119(4 Pt 1):732–40, 2000.

Sodian R, Hoerstrup SP, Sperling JS, Daebritz S, Martin DP et al: Early in vivo experience with tissue-engineered trileaflet heart valves, *Circulation* 102(Suppl III):22–9, 2000.

Rezai N, Podor TJ, McManus BM: Bone marrow cells in the repair and modulation of heart and blood vessels: Emerging opportunities in native and engineered tissue and biomechanical materials, *Artif Organs* 28(2):142–51, 2004.

Leyh RG, Wilhelmi M, Walles T, Kallenbach K, Rebe P et al: Acellularized porcine heart valve scaffolds for heart valve tissue engineering and the risk of cross-species transmission of porcine endogenous retrovirus, *J Thor Cardiovasc Surg* 126(4):1000–4, 2003.

Sarraf CE, Harris AB, McCulloch AD, Eastwood M: Heart valve and arterial tissue engineering, *Cell Prolif* 36(5):241–54, 2003.

Simon P, Kasimir MT, Seebacher G, Weigel G, Ullrich R et al: Early failure of the tissue engineered porcine heart valve SYNERGRAFT in pediatric patients, *Eur J Cardiothor Surg* 23(6):1002–6, 2003.

Mol A, Bouten CV, Zund G, Gunter CI, Visjager JF et al: The relevance of large strains in functional tissue engineering of heart valves, *Thor Cardiovasc Surg* 51(2):78–83, 2003.

Dohmen PM, Ozaki S, Nitsch R, Yperman J, Flameng W et al: Tissue engineered heart valve implanted in a juvenile sheep model, *Med Sci Monit* 9(4):BR97–104, 2003.

Barron V, Lyons E, Stenson-Cox C, McHugh PE, Pandit A: Bioreactors for cardiovascular cell and tissue growth: A review, *Ann Biomed Eng* 31:1017–30, 2003.

Breuer CK, Mettler BA, Anthony T, Sales VL, Schoen FJ et al: Application of tissue-engineering principles toward the development of a semilunar heart valve substitute, *Tissue Eng* 10:1725–36, 2004.

Elkins RC: Tissue-engineered valves, *Ann Thor Surg* 74:1434, 2002.

Hopkins RA: Tissue engineering of heart valves: Decellularized valve scaffolds, *Circulation* 111:2712–4, 2005.

Korossis SA, Fisher J, Ingham E: Cardiac valve replacement: A bioengineering approach, *Biomed Mater Eng* 10:83–124, 2000.

Mayer JE Jr, Shin'oka T, Shum-Tim D: Tissue engineering of cardiovascular structures, *Curr Opin Cardiol* 12:528–32, 1997.

Mol A, Bouten CV, Baaijens FP, Zund G, Turina MI et al: Review article: Tissue engineering of semilunar heart valves: current status and future developments, *J Heart Valve Dis* 13:272–80, 2004.

Nugent HM, Edelman ER: Tissue engineering therapy for cardiovascular disease, *Circ Res* 92:1068–78, 2003.

Rabkin-Aikawa E, Mayer JE Jr, Schoen FJ: Heart valve regeneration, *Adv Biochem Eng Biotechnol* 94:141–79, 2005.

Rashid ST, Salacinski HJ, Hamilton G, Seifalian AM: The use of animal models in developing the discipline of cardiovascular tissue engineering: A review, *Biomaterials* 25:1627–37, 2004.

Stock UA, Vacanti JP: Tissue engineering: Current state and prospects, *Annu Rev Med* 52:443–51, 2001.

Stock UA, Vacanti JP, Mayer JE Jr, Wahlers T: Tissue engineering of heart valves–current aspects, *Thor Cardiovasc Surg* 50:184–93, 2002.

Vesely I, Noseworthy R, Pringle G: The hybrid xenograft/autograft bioprosthetic heart valve: in vivo evaluation of tissue extraction, *Ann Thor Surg* 60:S359–64, 1995.

Zimmermann WH, Eschenhagen T: Tissue engineering of aortic heart valves, *Cardiovasc Res* 60:460–2, 2003.

Berthiaume F, Yarmush ML: Tissue engineering, in: *The Biomedical Engineering Handbook*, Bronzino JD ed, CRC Press, Boca Raton, FL, 1995, pp 1556–66.

Schoen FJ, Levy RJ: Founder's Award, 25th Annual Meeting of the Society for Biomaterials, perspectives, Providence, RI, April 28–May 2, 1999. Tissue heart valves: Current challenges and future research perspectives, *J Biomed Mater Res* 47:439–65, 1999.

Courtman DW, Pereira CA, Omar S, Langdon SE, Lee JM et al: Biomechanical and ultrastructural comparison of cryopreservation and a novel cellular extraction of porcine aortic valve leaflets, *J Biomed Mater Res* 29:1507–16, 1995.

Schenke-Layland K, Riemann I, Opitz F, Konig K et al: Comparative study of cellular and extracellular matrix composition of native and tissue engineered heart valves, *Matrix Biol* 23:113–25, 2004.

Sodian R, Sperling JS, Martin DP, Egozy A, Stock U et al: Fabrication of a trileaflet heart valve scaffold from a polyhydroxyalkanoate biopolyester for use in tissue engineering, *Tissue Eng* 6:183–8, 2000.

Shi Y, Vesely I: Fabrication of mitral valve chordae by directed collagen gel shrinkage, *Tissue Eng* 9:1233–42, 2003.

Girton TS, Oegema TR, Tranquillo RT: Exploiting glycation to stiffen and strengthen tissue equivalents for tissue engineering, *J Biomed Mater Res* 46:87–92, 1999.

Grinnell F: Fibroblast biology in three-dimensional collagen matrices, *Trends Cell Biol* 13:264–9, 2003.

Cuy JL, Beckstead BL, Brown CD, Hoffman AS, Giachelli CM: Adhesive protein interactions with chitosan: Consequences for valve endothelial cell growth on tissue-engineering materials, *J Biomed Mater Res A* 67:538–47, 2003.

Masters KS, Shah DN, Leinwand LA, Anseth KS: Crosslinked hyaluronan scaffolds as a biologically active carrier for valvular interstitial cells, *Biomaterials* 26:2517–25, 2005.

Rothenburger M, Volker W, Vischer P, Glasmacher B, Scheld HH et al: Ultrastructure of proteoglycans in tissue-engineered cardiovascular structures, *Tissue Eng* 8:1049–56, 2002.

Rothenburger M, Volker W, Vischer JP, Berendes E, Glasmacher B et al: Tissue engineering of heart valves: formation of a three-dimensional tissue using porcine heart valve cells, *ASAIO J* 48:586–91, 2002.

Vesely I, Boughner DR, Leeson-Dietrich J: Bioprosthetic valve tissue viscoelasticity: Implications on accelerated pulse duplicator testing, *Ann Thor Surg* 60: S379–82, 1995.

Liao J, Vesely I: Relationship between collagen fibrils, glycosaminoglycans, and stress relaxation in mitral valve chordae tendineae, *Ann Biomed Eng* 32:977–83, 2004.

Sacks MS, Smith DB, Hiester ED: The aortic valve microstructure: effects of transvalvular pressure, *J Biomed Mater Res* 41:131–41, 1998.

15

VASCULAR REGENERATIVE ENGINEERING

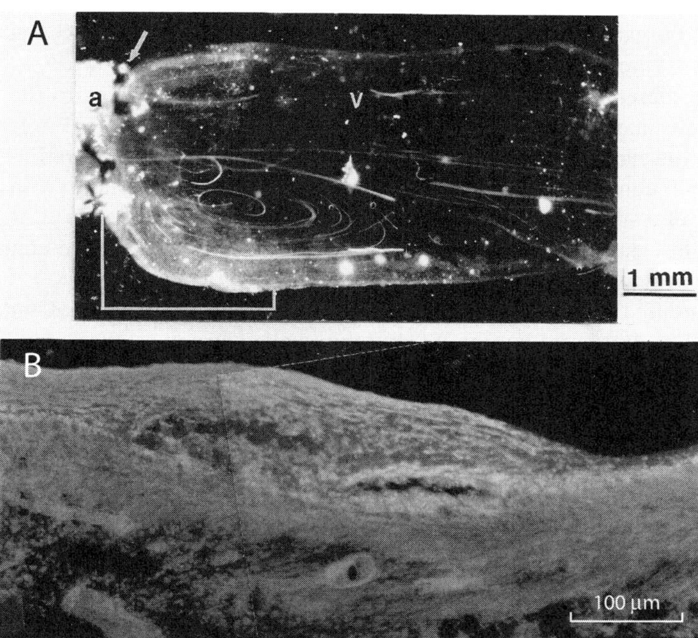

Influence of vortex bloodflow on intimal hyperplasia in experimental vein grafts. Panel A shows the bloodflow pattern in a rat vein graft. Vortex flow develops in the proximal anastomotic region due to divergent geometry from the host artery to the vein graft. Panel B is a fluorescent micrograph produced from an axial specimen section from the proximal host–vein graft junction of a rat vein graft at day 10, showing intimal hyperplasia at the site of vortex bloodflow. Red: mRNA of the angiotensin II type 1 receptor. Green: smooth muscle α-actin. Blue: cell nuclei. Note that the expression of the angiotensin II type 1 receptor, which mediates proliferative cell activities, was consistent with the presence of vortex bloodflow. When the divergent geometry and vortex bloodflow were eliminated by using a mechanical engineering approach, the expression of the angiotensin II type 1 receptor, smooth muscle cell proliferation, as well as intimal hyperplasia were significantly reduced. See color insert.

Bioregenerative Engineering: Principles and Applications, by Shu Q. Liu

ANATOMY AND PHYSIOLOGY OF THE VASCULAR SYSTEM

Structure and Organization of Blood Vessels [15.1]

The vascular system is composed of three subsystems: arteries, capillaries, and veins. The arterial subsystem conducts oxygenated blood from the heart to peripheral tissues and consists of various generations of arteries, including elastic arteries, muscular arteries, and arterioles. *Elastic arteries* are defined as arteries characterized by the presence of three distinct layers, including the tunica intima, tunica media, and tunica adventitia and the presence of multiple elastic laminae in the tunica media. Aorta is the largest elastic artery, which includes two segments: the thoracic and abdominal aorta. The thoracic aorta bifurcates into the brachiocephalic artery (a short artery extending from the aorta to the right common carotid and right subclavian arteries), left common carotid artery, left subclavian artery, and intercostal arteries. The abdominal aorta bifurcates into celiac, superior and inferior mesenteric, renal, and common iliac arteries. These arteries supply blood to the peripheral organs: the common carotid arteries to the head, the subclavian arteries to the upper arms, the intercostal arteries to the chest wall, the celiac artery to the liver and stomach, the mesenteric arteries to the intestines, the renal arteries to the kidneys and adrenal glands, and the common iliac arteries to the lower abdominal organs and lower limbs. The large elastic arteries are further bifurcated into various generations of *muscular arteries*, which contain the three tunica layers described above, but are characterized by the presence of a single internal elastic lamina and rich smooth muscle cells in the tunica media. Smallest muscular arteries are defined as *arterioles*, which are connected to the capillary network.

Capillaries are the smallest blood vessels that contain a single layer of endothelial cells and a subendothelial basal lamina. Such a thin-walled structure is designed for effective transport of oxygen and nutrients from the blood to the peripheral tissue as well as transport of carbon dioxide and waste products from the peripheral tissue to the blood. Capillaries are connected to the venous system at the distal end. The *venous system* includes venules and small, medium-sized, and large veins. Capillary blood is converged to the venules, various generations of veins, and eventually to the largest vein, the vena cava, which conducts blood to the right atrium. The vena cava is divided into two portions: the superior and inferior vena cava. The superior vena cava collects blood from the head and upper arms, whereas the inferior vena cava collects blood from the remaining parts of the body. Each vein is composed of three layers: the tunica intima, tunica media, and tunica adventitia. While the tunica intima and adventitia are similar in structure to those of the arteries, the tunica media is considerably different. The venous tunica media contains loosely organized elastic fibers instead of multilayered elastic laminae. In addition, a vein contains a single layer of smooth muscle cells, whereas an artery contains multiple layers of smooth muscle cells.

Types and Functions of Vascular Cells [15.1]

The vascular system consists of several cell types, including the endothelial cell (EC), smooth muscle cell (SMC), and fibroblast, which reside in the extracellular matrix of the blood vessel wall. The vascular cells possess distinct structure and function. The structural and functional characteristics of each cell type are discussed as follows.

Endothelial Cells. Endothelial cells are monolayered squamous epithelial cells, which line the blood-contacting surface of blood vessels. These cells are primarily aligned along the direction of bloodflow (Fig. 15.1). Endothelial cells express several surface molecules specific to endothelial cells, including von Willebrand factor, vascular endothelial growth factor receptor 1 (VEGFR1 or Flt-1), vascular endothelial growth factor receptor 2 (VEGFR2 or Flk-1), and factor VIII. These molecules can be used to identify endothelial cells. Endothelial cells possess a number of functions, including selective transport of plasma substances and molecules, regulation of anti- and pro-anticoagulation activities, regulation of leukocyte and platelet adhesion and trans-migration, regulation of vascular contractility, and regulation of vascular cell proliferation and migration. These functions are mediated by various molecular processes and are essential for maintenance of the homeostatic state.

Endothelial Barrier Function. Endothelial cells serve as a barrier that allows free transport of lipid-soluble molecules and regulated transport of ions, glucose, amino acids, and

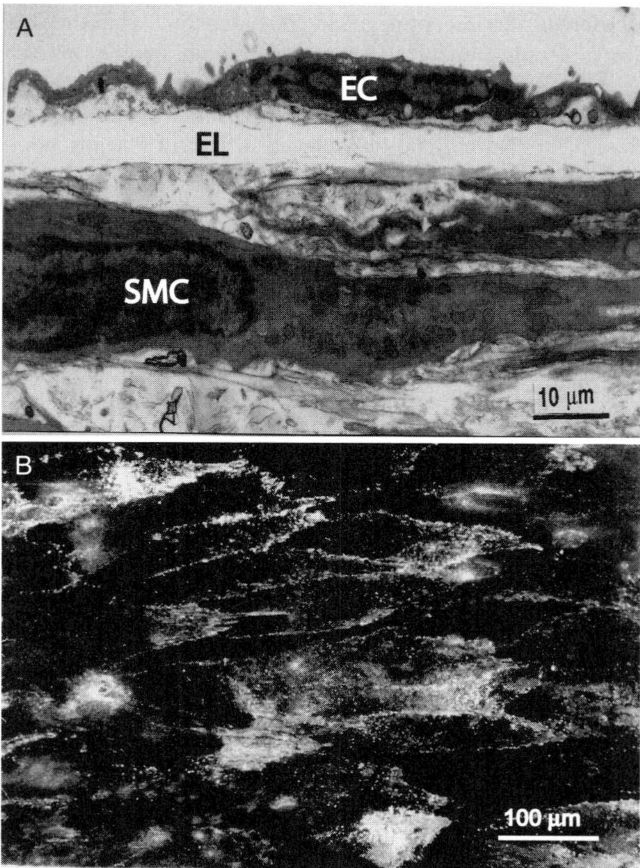

Figure 15.1. Structure and organization of vascular endothelial cells in the rat aorta. (A) Electron micrograph of vascular endothelial and smooth muscle cells. (B) En face fluorescent micrograph of vascular endothelial cells labeled with an anti-ICAM1 antibody. EC: endothelial cell. EL: elastic lamina. SMC: smooth muscle cell.

proteins. Endothelial cells control the exchange of substances and fluids between blood and tissue. This function is the basis for maintaining the stability of a physiological environment, which is essential to cell survival and function. It is known that the endothelium is permeable to lipid-soluble molecules, water and ionic solutes, and is impermeable to plasma proteins, although a small amount of proteins are able to escape from the capillary network to the interstitial space. This selective barrier function has been studied by using radioactive and fluorescent tracers, such as ^{3}H-water, ^{125}I-albumin, rhodamine-albumin, and ^{14}C-sucrose. The flux rate of each selected tracer across an endothelial membrane can be measured and used to estimate the permeability of the endothelial cells. For a selected substance and a specified tissue type, the flux rate across the endothelium depends on the concentration gradient, molecular weight and size, electrical charges, and chemical properties of the substance. Among commonly used tracers such as water, sucrose and albumin, water has highest flux rate, followed by sucrose and albumin, as shown in studies by using cultured endothelial cells.

Several endothelial transport pathways have been identified. These include the cell membrane, the plasmalemmal vesicles, the transendothelial channels, the intercellular junction pores, the receptor-mediated pathways, and the open and closed fenestrae. Certain molecules, such as water, small lipid-soluble substances, and ionic solutes, may be passively or actively transported across the endothelial membrane, intercellular junctions, transendothelial channels and other openings. Certain molecules, such as albumin and dextran, may be transported through the transendothelial channels and plasmalemmal vesicles. Other molecules, such as low-density lipoproteins, insulin, and transferrin, may be transported through the receptor-mediated pathways and endocytosis. The open fenestrae of the endothelium in some tissues may allow large molecular weight plasma proteins, except for cellular components, to escape from the capillaries.

Endothelial permeability to plasma substances depends on the state of the endothelial cells. Injured endothelial cells due to exposure to mechanical stress, such as excessive tensile stress, and chemical toxins may exhibit reduced barrier function and increase permeability. Endothelial mitosis is associated with endothelial junction leakage and an increase in endothelial permeability. An exposure to histamine in inflammatory reactions may induce endothelial cell injury or endothelial junction dilatation, leading to an increase in endothelial permeability. Biochemical regulatory factors, such as atrial natriuretic peptide, can induce an increase in endothelial permeability.

The permeability of endothelial cells varies from one organ to another. For instance, the cerebral capillaries, which form a blood–brain barrier, are less permeable to plasma substances than are capillaries in other organs. This feature is partially attributed to the presence of tighter intercellular junctions and the scarcity of transport vesicles, and is crucial to the protection of central nervous cells from exposure to harmful substances and microbiological organisms. In contrast, the renal glomerular endothelium is more permeable than that in other organs, which is attributed to the presence of fenestrae and is essential to the renal functions such as fluid filtration and clearance. Other organs, which exhibit high endothelial permeability, include the liver, glands, and bone marrow. In these organs, high endothelial permeability is necessary for facilitating the transport of enzymes and hormones. From the developmental point of view, a specified structure of endothelial cells in each organ is a result of cell differentiation based on the functional necessity. However, the identification of the determinants of cell differentiation remains a difficult task in developmental research.

Regulation of Anti- and Procoagulation Activities [15.2]. Endothelial cells participate in the regulation of blood coagulation and ular fibrinolysis, two important functions for the maintenance of the homeostatic internal environment in multicelled mammalian systems. These functions are regulated coordinately in the vascular system. The coagulation function allows rapid formation of insoluble blood plug at locations of vascular injury so that hemorrhage can be rapidly stopped. On the other hand, the anticoagulation and fibrinolytic function not only assures blood fluidity and vascular patency under physiological conditions but also prevents the spreading of coagulation activities from injured to normal blood vessels. (Characteristics of anti/procoagulation molecules are listed in Table 15.1.)

Endothelial cells regulate *anticoagulation* activities by expressing and secreting heparan sulfate proteoglycan (a heparin-like proteoglycan expressed in the membrane of endothelial cells), plasminogen activators (specific serine proteases that convert plasminogen into plasmin, a hydrolase capable of converting coagulant fibrin into a soluble form), and protein C (an enhancer of fibrinolysis). These factors play critical roles in the maintenance of the blood fluidity and the prevention of blood coagulation and thrombogenesis under physiological conditions.

Endothelial cells participate in the regulation of *procoagulation* activities in the event of vascular trauma and injury. Injured endothelial cells can express and release procoagulants such as von Willebrand factor (mediating platelet adhesion), tissue thromboplastin (a cell surface prothrombin activator that converts prothrombin to thrombin, a pivotal enzyme capable of converting soluble fibrinogen (a bloodborne glycoprotein composed of three pairs of polypeptide chains, α, β, and γ) to insoluble gel-like coagulant fibrin), leading to the activation of the coagulation cascade and the formation of solid blood plugs and thrombi. In such a coagulation process, intact endothelial cells near the injury site are able to prevent the spreading of coagulation and platelet aggregation through several mechanisms, including the release of prostaglandins, plasminogen activators, and protein C. Several factors, such as endotoxin, tumor necrosis factor, and interleukin 1, can induce a shift of the balance of coagulation activities in favor of coagulation activation. It is important to note that blood coagulation and thrombosis are considered the initial steps in the development of atherosclerosis.

Regulation of Vascular Contractility [15.3]. Endothelial cells regulate vascular contractility by secreting regulatory factors, such as endothelial cell-derived vasorelaxants and endothelial cell-derived vasoconstrictors (see Table 15.2). Endothelial cell-derived vasorelaxants, such as nitric oxide, induce rapid smooth muscle relaxation and blood vessel dilation, leading to an increase in local bloodflow. In contrast, endothelial cell-derived vasoconstrictors, such as endothelin, induce smooth muscle contraction and blood vessel constriction, leading to a decrease in local bloodflow. In a homeostatic state, vasorelaxants and vaso-constrictors are released in a coordinated manner to maintain a basal smooth muscle activity and a basal vascular tone. Endothelial cell injury or denudation may influence the release of these vasoactive molecules and, thus, influence the activity of smooth muscle cells. In general, the vasoconstrictor system is activated in response to endothelial or vascular injury. Such a reaction often contributes to proliferation of blood and vascular cells as well as to intimal hyperplasia and atherogenesis. In contrast, the activation of the vasorelaxant system results in the suppression of these pathogenic processes.

Regulation of Leukocyte and Platelet Adhesion [15.4]. Endothelial cells participate in the regulation of leukocyte activities by expressing and releasing several cytokines, such

TABLE 15.1. Characteristics of Selected Anti- and Procoagulation Molecules*

Proteins	Alternative Names	Amino Acids	Molecular Weight (kDa)	Expression	Functions
Plasminogen activator	Tissue type plasminogen activator (TPA), Alteplase, Reteplase, tPU	562	63	Endothelial cells, skin, lung, uterus	A serine protease that converts the proenzyme plasminogen to plasmin, a fibrinolytic enzyme
Plasminogen	Microplasmin, angiostatin	810	91	Brain, liver, blood cells, kidney, testis, cornea	Forming plasmin, a hydrolase capable of converting coagulant fibrin into a soluble form
Protein C	PROC, PC, blood coagulation factor XIV, anticoagulant protein C, autoprothrombin IIA, vitamin K-dependent protein C	461	52	Blood cells, liver, blood vessels, skeletal muscle	Enhancing fibrinolysis and inhibiting coagulation
von Willebrand factor	Coagulation factor VIII VWF	2813	309	Endothelial cells, platelets, bone marrow, lung, eye, kidney, skin	Serving as an antihemophilic factor and mediating platelet–endothelial interaction
Tissue thromboplastin	Tissue factor (TF), thromboplastin tissue factor, CD142 antigen, coagulation factor III, F3	295	33	Monocytes, lymphocytes, epidermis, brain, kidney, heart, lung	Interacting with coagulation factor VII and activating coagulation protease cascades

Prothrombin	Coagulation factor II, factor II, F2	622	70	Ovary, liver	Forming thrombin (via proteolytic cleavage), which converts fibrinogen to fibrin
Fibrinogen α	FGA	866	95	Leukocytes, platelets, bone marrow, brain, liver, lung, ovary	Forming fibrin (via thrombin-mediated cleavage), which participates in coagulation, regulates cell adhesion, spreading, and proliferation via cleavage products and mediates vasoconstriction and chemotaxis via cleavage products
Fibrinogen β	FGB	491	56	Leukocytes, platelets, bone marrow, brain, liver, lung, and ovary	Forming complex with fibrinogen α and γ and possessing identical function as fibrinogen α
Fibrinogen γ	FGG	453	51	Leukocytes, platelets, bone marrow, brain, liver, lung, and ovary	Forming complexes with fibrinogen β and γ and possessing function identical to that of fibrinogen α

*Based on bibliography 15.2.

TABLE 15.2. Characteristics of Selected Vasoconstrictors*

Proteins	Alternative Names	Amino Acids	Molecular Weight (kDa)	Expression	Functions
Endothelin 1	ET1, EDN1	212	24	Ubiquitous	Stimulating smooth muscle contraction, inducing cell proliferation, causing tumorigenesis, and mediating matrix production
Endothelin 2	ET2, EDN2	178	20	Ubiquitous	Inducing vasoconstriction (most potent vasoconstrictor among endothelin 1,2,3), promoting tumor cell survival, and serving as a macrophage chemoattractant
Endothelin 3	ET3, EDN3	238	25	Lung, spleen, pancreas	Inducing smooth muscle contraction, regulating the development of neural crest-derived cells, and stimulating cell division
Endothelin receptor type A	ETA, ETRA	427	49	Blood vessels, placenta, skeletal muscle	A G-protein-coupled receptor for endothelin 1,3
Endothelin receptor type B	ETB, ETRB, endothelin receptor, nonselective type	442	50	Ubiquitous	A G-protein-coupled receptor for endothelin 1,3, regulating neural development

*Based on bibliography 15.3.

as monocyte chemotactic protein (MCP)-1 (Table 15.3) and interleukin (IL)1, and by expressing adhesion molecules, such as intercellular adhesion molecules (ICAMs) (Fig. 15.1) and vascular cell adhesion molecules (VCAMs). In response to mechanical, chemical, or biological stimuli, endothelial cells are able to release cytokines, which in turn attract leukocytes to the endothelium, a critical process initiating leukocyte adhesion and thrombus formation. Injured endothelial cells can also upregulate the adhesion molecules ICAMs and VCAMs, which mediate leukocyte attachment to the endothelium (see page 274 for the function of adhesion molecules). Activated leukocytes can release additional cytokines and growth factors, contributing to leukocyte activation as well as smooth muscle cell proliferation and migration.

Regulation of Cell Proliferation and Migration [15.1]. Endothelial cells participate in regulating the growth of several cell types, including endothelial cells themselves, smooth muscle cells, and fibroblasts. Endothelial cells can synthesize and secret a number of cytokines and growth factors, including acidic and basic fibroblast growth factors (aFGF and bFGF), platelet-derived growth factor (PDGF), vascular endothelial cell growth factor (VEGF), insulin-like growth factor (ILGF), interleukin-1 (IL1), and endothelin (see pages 600, 631, 666 for characteristics of these factors). These growth factors can activate the tyrosine kinase receptor signaling system and induce cell proliferation and migration (see pages 151, 207, for regulatory mechanisms). Activated smooth muscle cells can migrate from the media into the intima, and play a critical role in the formation of focal intimal hyperplasia and atherosclerosis.

TABLE 15.3. Characteristics of MCP1*

Proteins	Alternative Names	Amino Acids	Molecular Weight (kDa)	Expression	Functions
Monocyte chemotactic protein 1	MCP-1, chemokine (C–C motif) ligand 2 (CCL2), small inducible cytokine A2, monocyte chemoattractant protein 1, monocyte chemotactic and activating factor (MCAF), monocyte secretory protein JE	99	11	Leukocytes, endothelial cells	Regulating immune and inflammatory processes, exerting a chemotactic effect on monocytes and basophils (but not for neutrophils or eosinophils), and contributing to atherogenesis

*Based on bibliography 15.4.

Endothelial cells are also able to synthesize and release growth inhibitors, including transforming growth factor β (TGFβ), heparan sulfate, and nitric oxide synthase. TGFβ has been shown to inhibit the proliferation of endothelial and smooth muscle cells. Heparan sulfate is a potent inhibitor of smooth muscle proliferation as well as an inhibitor of smooth muscle phenotype modulation. Nitric oxide synthase catalyzes the formation of nitric oxide, which exerts an inhibitory effect on the proliferation of endothelial and smooth muscle cells. In a homeostatic state, a coordinated regulation of the synthesis and secretion of growth promoters and inhibitors is critical to the control of cell proliferation and vascular remodeling.

Production of Extracellular Matrix [15.1]. Endothelial cells can synthesize and release several extracellular components, including collagen type IV, fibronectin, and laminin. These components participate in the construction of the vascular basal lamina. In addition, endothelial cells can produce proteoglycans, which constitute the ground substance compartment of the intima.

Smooth Muscle Cells [15.5]. Smooth muscle cells are fiber-like muscular cells found in the tunica media of arteries and veins. These cells are primarily aligned in the circumferential direction of blood vessels. Vascular smooth muscle cells can be identified by their location; all cells localized to the media of arteries and veins are defined as smooth muscle cells. Smooth muscle cells express several distinct molecular markers, including smooth muscle α actin (see Chapter 15 opening figure), smooth myosin 1, and calponin. These markers are used in biological research to identify vascular smooth muscle cells.

Although smooth muscle cells are present in the wall of arteries and veins, the distribution of smooth muscle cells differs between the two type of blood vessels. Arteries contain multiple layers of smooth muscle cells, which are organized between elastic laminae. In contrast, veins contain a single layer of smooth muscle cells, which are found underneath the intimal layer of the venous wall. For the same generation of blood vessels, the number of smooth muscle cells in the arterial wall is considerably larger than that in the venous wall.

Smooth muscle cells possess a number of functions, including contraction and relaxation, regulation of vascular cell proliferation and migration, and synthesis of extracellular matrix constituents. These functions are discussed as follows.

Contractility [15.5]. Smooth muscle contraction and relaxation are processes induced by interaction between α-actin filaments and myosin molecules. The sliding of myosin molecules along actin filaments drives the shortening of the cells, whereas the dislodgment of the myosin molecules from the actin filaments results in cell relaxation. The interaction of myosin molecules with actin filaments is initiated by an increase in the concentration of cytoplasmic calcium, which is stored in and released from the endoplasmic reticulum. Calcium is an critical ion that participates in regulating the activities of contractility-mediating signaling molecules. While a complete signaling pathway remains to be determined, it has been demonstrated that myosin light chain kinase (MLCK) is a key downstream regulatory factor for the activation of myosin molecules. MLCK can be activated in response to an increase in the cytoplasmic calcium concentration. Activated MLCK induces phosphorylation of the myosin light chain, resulting in the activation of the myosin ATPase and the interaction of myosin molecules with actin filaments. Several signaling molecules, including Rho kinase and mitogen-activated protein kinase, have been shown to directly phosphorylate and activate MLCK.

There are several hormones, including angiotensin I, angiotensin II, and norepinephrine, which can act on smooth muscle cells and induce calcium release from the endoplasmic reticulum. These molecules serve as ligands of G-protein receptors. The binding of a ligand to a G-protein receptor induces activation of a specified G-protein signaling cascade, resulting in calcium release from the endoplasmic reticulum and activation of the mitogen-activated protein kinase pathway. The regulatory mechanisms of the G-protein signaling cascade are discussed on page 217.

Smooth muscle contraction and relaxation are also regulated by the calponin signaling system. Calponin is a smooth muscle-specific protein which is associated with α-actin filaments. This protein exerts inhibitory effects on the interaction between myosin and α-actin. Activated calponin suppresses the activity of the myosin ATPase and the sliding movement of actin filaments on immobilized myosin molecules in vitro, and induces reduction in shorting velocity and force development. While the exact regulatory mechanisms remain a research topic, the activation of the type 2A phosphatase may mediate the activity of calponin. The activation of protein kinase C (PKC) suppresses the activity of calponin. Thus, calponin serves to to counterbalance the stimulatory effect of vasoconstricting molecules and promote smooth muscle relaxation.

Smooth muscle contraction is also regulated by mechanical factors, including tensile stress and fluid shear stress. *Tensile stress* is the force per unit area within the wall of a blood vessel, and is induced by the internal blood pressure. Tensile stress is present in two directions of a blood vessel: the circumferential and longitudinal directions. The average circumferential tensile stress (σ_θ) is dependent on the internal radius (r) and the wall thickness (h) of a blood vessel and the internal blood pressure (p) and can be estimated with the following mathematical formula:

$$\sigma_\theta = \frac{pr}{h}$$

The average longitudinal tensile stress (σ_z) is dependent on the internal and external radii (r_i and r_o, respectively) of a blood vessel and the internal blood pressure and can be estimated with the following formula:

$$\sigma_z = \frac{p r_i^2}{r_o^2 - r_i^2}$$

An increase in blood pressure induces elevation in tensile stress, which leads to elongation of smooth muscle cells. Such a mechanical effect usually activates the contractile mechanism, initiating smooth muscle contraction, a process referred to as myogenic response. The physiological significance of myogenic response is that an increase in local blood pressure, which is associated with an elevation in local bloodflow, may trigger smooth muscle contraction and vascular constriction, resulting in the restoration of the physiological level of bloodflow. Thus, myogenic response is a mechanism which contributes to the regulation of regional bloodflow.

Fluid shear stress is another mechanical factor, which regulates smooth muscle activity. Fluid shear stress is a flow-induced shearing force per unit area at the endothelial surface of blood vessels. The average fluid shear stress (τ) of a blood vessel is dependent on the blood viscosity (μ), bloodflow rate (Q), and the radius of the vessel (*r*) and can be estimated by using the following formula:

$$\tau = \frac{4\mu Q}{\pi r^3}$$

An increase in fluid shear stress, due to elevated bloodflow or reduced vessel diameter, may induce deformation or conformational changes of molecules at the cell surface. Such mechanical stimulation may activate signaling pathways, resulting activation of the vasorelaxant mechanisms, including the activation of nitric oxide synthase and the release of nitric oxide. The consequence is smooth muscle relaxation and vessel dilation. The physiological significance of these activities is that an increase in fluid shear stress, often resulting from elevated demand of regional bloodflow, stimulates the dilation of blood vessels, leading to the enhancement of regional blood supply.

Regulation of Cell Proliferation and Migration [15.5]. Smooth muscle cells play an important role in regulating the proliferation and migration of vascular cells, including themselves and fibroblasts. Smooth muscle cells exert such an effect via synthesizing and releasing growth factors, including platelet-derived growth factor (PDGF), insulin-like growth factor (ILGF), and fibroblast growth factor (FGF). These growth factors can act on vascular cells via interaction with corresponding growth factor receptor protein tyrosine kinases and activate cytoplasmic mitogenic signaling pathways, resulting in cell proliferation and migration. In particular, smooth muscle cell proliferation and migration from the media to the intima contribute to intimal hyperplasia and atherogenesis. The characteristics and activation mechanisms of the growth factors are discussed on page 151, respectively.

Regulation of Extracellular Matrix Production [15.5]. Smooth muscle cells can synthesize and release extracellular matrix constituents, including primarily collagen type III, tropoelastin, and proteoglycans. Collagen is assembled into collagen fibers, tropoelastin is organized into elastic fibers, which constitute elastic laminae in the wall of arteries, and proteoglycans constitute the ground substance compartment of blood vessels.

Fibroblasts [15.1]. Fibroblasts are cells found in the adventitia of arteries and veins. These cells are randomly organized through the adventitia. The density of fibroblasts is considerably lower than that of the medial smooth muscle cells. Fibroblasts possess two major functions: production of extracellular matrix constituents and regulation of vascular cell proliferation. Fibroblasts can synthesize and release collagen type III and proteoglycans, which constitute the extracellular matrix of the adventitia. Fibroblasts regulate the proliferation of vascular cells via production of growth factors, such as fibroblast growth factor (FGF) and epidermal growth factor (EGF). These growth factor can interact with corresponding growth factor receptors and induce proliferative activities.

Extracellular Matrix of Blood Vessels [15.1]

The extracellular matrix of blood vessels is composed of the basal lamina, collagen fibers, elastic fibers and laminae, and proteoglycans. These matrix components provide structural support, mechanical strength, and elasticity to the wall of blood vessels, and play critical roles in regulating the development, morphogenesis, and pathogenic remodeling of blood vessels. The basal lamina is a thin membrane found underneath the endothelium and serves as a substrate for the endothelial cells. This membrane consists of collagen type IV, fibronectin, and laminin. In the case of blood vessel injury and endothelial detachment, the basal lamina initiates platelet and leukocyte adhesion and triggers blood coagulation, contributing to the repair of damaged blood vessels.

The vascular collagen matrix is composed of primarily collagen type III and determines the strength of blood vessels, especially at a high blood pressure level. This matrix also serves as a substrate for the attachment of smooth muscle cells and fibroblasts, stimulates the proliferation and migration of vascular cells, and controls the pattern formation of smooth muscle cells. Elastic fibers and laminae are primary matrix components that are responsible for the elasticity of blood vessels. In addition, these components exert an inhibitory effect on cell proliferation and migration via activation of a signaling pathway involving the inhibitory receptor signal-regulatory protein α and SH2 domain-containing protein tyrosine phosphatase (see page 201 for details). Proteoglycans are the components of ground substances, which serve as a structure for the assembly and organization of vascular cells. A detailed discussion of extracellular matrix can be found on page 102.

Regulation of Bloodflow [15.1]

The primary function of the vascular system is to conduct oxygenated bloodflow from the beating heart to peripheral tissues and return deoxygenated blood from the peripheral tissues to the heart, maintaining the circulation of bloodflow. At a given state and a given time, the rate of blood supply differs between different organ systems because of different metabolic activities. For instance, the cardiac and skeletal muscular systems demand a considerable increase in bloodflow during physical exercise, while bloodflow to other less active systems, such as the gastrointestinal system, is accordingly reduced. Such a coordinated control of regional bloodflow is achieved by regulating the contractility of local arterial smooth muscle cells. The arterial smooth muscle cells in the cardiac and skeletal muscle systems undergo relaxation, whereas the intestinal arterial smooth muscle cells undergo enhanced contraction. Thus, bloodflow is redistributed from the intestinal system to the cardiac and skeletal muscle systems during physical exercises.

The redistribution of bloodflow is regulated via several mechanisms, including regulation by local metabolites, the nervous system, and hormones. Local metabolites, including carbon dioxide, H^+, lactic acid, adenosine, adenosine monophosphate, and adenosine diphosphate, play a critical role in regulating regional bloodflow. These metabolites are generated during metabolic processes and stimulate the relaxation of vascular smooth muscle cells. An increase in metabolism during physical exercises results in the elevation of these metabolites and induces the dilation of local arteries, leading to an increase in the rate of regional bloodflow. This mechanism contributes to the enhancement of bloodflow to the heart and skeletal muscle system during exercises.

Arteries are innervated with sympathetic nerves, which secret norepinephrine. The function of norepinephrine is dependent on the type of adrenergic receptors in the smooth muscle cells. There are two major types of adrenergic receptors: α- and β-adrenergic receptors. The binding of norepinephrine to the α-adrenergic receptor induces primarily smooth muscle cell contraction, whereas the binding to the β-adrenergic receptor induces smooth muscle cell relaxation. The distribution of the two types of adrenergic receptor in arteries varies between different organs. Arteries in the brain, heart, and skeletal muscle system are dominantly innervated with the β-adrenergic receptor, whereas the gastrointestines, kidney, and skin are primarily innervated with the α-adrenergic receptor. Under conditions of increased physical activities, the activation of the sympathetic nerve system induces arterial dilation in the heart, brain, and skeletal muscular systems, whereas such a change exerts an opposite effect on the arterial system in the stomach, intestines, kidneys, and skin. This mechanism contributes to blood flow redistribution to the brain, heart, and skeletal muscle system during physical exercises. The effect of bloodborne hormones, such as epinephrine and norepinephrine produced in the adrenal gland, elicit a similar effect on the contractility of arterial smooth muscle cells. These hormones are upregulated during exercises or under stress conditions (e.g., escaping from dangers).

Regulation of Blood Pressure [15.1]

The circulation of blood flow is driven by the arterial blood pressure, which is generated by the cardiac pumping activity. Arterial blood pressure varies periodically during a cardiac cycle. It is about 120 mm Hg during the systole and about 80 mm Hg during the diastole under physiological conditions. An increase or decrease in arterial blood pressure, defined as *hypertension* or *hypotension*, respectively, may influence the function of a variety of tissues and organs. For instance, hypertension may cause cardiac hypertrophy and failure, and extreme hypotension may result in deficiency of blood and oxygen supply. Arterial blood pressure is controlled within a narrow range under physiological conditions.

There are a number of mechanisms which control the arterial blood pressure. These mechanisms include regulation by the baroreceptors, chemoreceptors, central nerve control centers, rennin–angiotensin–aldosterone system, and vasopressin system. These mechanisms are briefly discussed as follows.

Regulation by Baroreceptors [15.1]. Baroreceptors are sensory cells localized to the carotid sinus near the carotid bifurcation and to the aortic arch, and are capable of sensing mechanical stretching signals due to arterial blood pressure. An increase in arterial blood pressure stimulates the baroreceptors, enhancing the generation of action potentials. The action potentials are transmitted to the central cardiovascular control centers via the glossopharyngeal and vagus nerves. The central control centers in turn reduce the activity of the sympathetic nerve system, leading to a decrease in arterial blood pressure. A decrease in arterial blood pressure exerts an opposite effect on the central nerve control centers, resulting in an increase in arterial blood pressure. Through such a feedback system, the baroreceptors control the arterial blood pressure within a narrow range under physiological conditions.

Regulation by Chemoreceptors [15.1]. Chemoreceptors are sensory cells found in structures known as *carotid* and *aortic bodies* localized to the bifurcation of carotid arteries and aortic arch. These receptors sense chemical signals such as oxygen, carbon dioxide, and hydrogen ion and regulate arterial blood pressure based on the blood concentration of these substances. The carotid and aortic bodies receive rich blood supply. When arterial blood pressure decreases, the concentration of oxygen decreases in the carotid and aortic bodies, whereas the concentration of hydrogen and carbon dioxide increases. These changes activate the chemoreceptors, increasing the frequency of action potentials in the receptor cells. The action potentials are transmitted to the central vasomotor nerve control centers, which in turn activate the sympathetic nervous system. Activated sympathetic nerves enhance cardiac contractility and arterial contraction, resulting in an increase in arterial blood pressure. It should be noted that the chemoreceptors are usually activated in response to a rapid decrease in arterial blood pressure under emergency conditions, such as hemorrhage and shock. They are not activated under physiological conditions.

Regulation by Central Nerve Control Centers [15.1]. There are neurons which sense changes in blood pressure in the medulla oblongata of the central nervous system. A sudden decrease in arterial blood pressure stimulates these cells to activate the sympathetic nervous system, leading to an increase in arterial blood pressure by a mechanism as described above.

Regulation by the Renin–Angiotensin–Aldosterone System [15.6]. This is a hormonal system that controls the basal tone and contractility of the vascular system. Renin is an enzyme secreted by a specialized type of arterial smooth muscle cell in a structure known as the *juxtaglomerular apparatus* located in the wall of renal arteries. These specialized cells are capable of sensing mechanical stretching signals. A decrease in arterial blood pressure can stimulate these cells to secret renin. Renin is released into the blood and can cleave a plasma protein called *angiotensinogen* to produce angiotensin I. Angiotensin I can be cleaved again and converted into angiotensin II by another enzyme called *angiotensin-converting enzyme*, which is produced primarily in the lungs. Angiotensin II can interact with the angiotensin II type 1 receptor in the membrane of vascular smooth muscle cells, activate a G-protein signaling pathway, induce the release of calcium and activation of protein kinases (see page 217 for G-protein-related regulatory mechanisms), and enhance actin–myosin interactions and cell contraction, resulting in an increase in arterial blood pressure (see page 668 of this chapter for the regulation of actin–myosin interaction). At a resting state, angiotensin I and II are kept at a relatively constant level, which is critical to the maintenance of the physiological arterial blood pressure.

Angiotensin II can stimulate the adrenal cortical cells to secret a hormone known as *aldosterone*. This hormone can act on the tubular system of the kidney to increase the absorption and retention of sodium and water, resulting in an increase in blood volume. The activation of the renin–angiotensin–aldosterone system is a major mechanism that regulates the basal level of the arterial blood pressure. A disorder of this system contributes to the pathogenesis of arterial hypertension (see page 699 of this chapter for pathogenesis of arterial hypertension).

The production of renin is regulated by hormones. A typical hormone is vitamin D. This molecule can difuse through the cell membrane and interact with the nuclear receptor vitamin D receptor. The complex of vitamin D and vitamin D receptor serves as a transcriptional factor and binds to the renin gene, suppressing its expression. Such a mechanism results in a decrease in the level of angitesin II. In experimental models, knockout of the vitamin D receptor is associated with a significant increase in the lelve of serum renin and angiotensin II, which is associated with an increase in the arterial blood pressure (Fig. 15.2).

Characteristics of renin–angiotensin system molecules are presented in Table 15.4.

Regulation by the Vasopressin System [15.1]. Vasopressin is also known as antidiuretic hormone (ADH) and is produced in the posterior pituitary. The release of vasopressin is regulated by baroreceptors and the central vasomotor control system. A decrease in arterial blood pressure stimulates the baroreceptors and central vasomotor system, which in turn activates the posterior pituitary to release vasopressin. Released vasopressin can enhance the contractility of vascular smooth muscle cells and thus increase the arterial blood pressure.

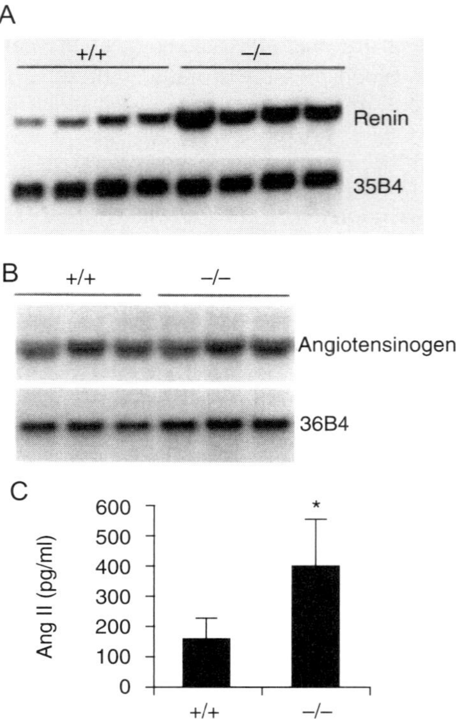

Figure 15.2. Effect of vitamin D receptor (VDR) inactivation on renin expression and plasma Angiotensin II (Ang II) production. (A) Renin mRNA expression in the kidney. Kidney total RNAs were isolated from wild-type (+/+) and $VDR^{-/-}$ (–/–) mice and analyzed by Northern blot. The same membrane was sequentially hybridized with mouse renin and 36B4 (control) cDNA probes. Each lane represents an individual animal. (B) Liver angiotensinogen mRNA expression in wild-type and $VDR^{-/-}$ mice, determined by Northern blot. The membrane was sequentially hybridized with mouse angiotensinogen and 36B4 cDNA probes. Each lane represents an individual mouse. (C) Plasma Ang II concentrations in wild-type and $VDR^{-/-}$ mice. $*P < 0.001$ vs. wild-type mice; $n = 15$ in each group. Reprinted with permission from Li Y C, et al. 1,25-Dihydroxyvitamin D_3 is a negative endocrine regulator of the renin-angiotensin system, J Clin Invest 110:229–238, copyright 2002.

VASCULAR DISORDERS

Atherosclerosis

Pathogenesis, Pathology, and Clinical Features [15.7]. *Atherosclerosis* is a disorder found in the lumen of primarily large and medium-sized arteries and characterized by the progressive development of focal atheroma, a structure containing extracellular matrix, cholesterol, calcium, smooth muscle cells, and macrophages, which are covered with a layer of endothelial cells (Fig. 15.3). Atherosclerosis may be initiated at any age. The formation of atheroma may begin during the childhood. As the development of atheroma

TABLE 15.4. Characteristics of Selected Renin–Angiotensin System Proteins*

Proteins	Alternative Names	Amino Acids	Molecular Weight (kDa)	Expression	Functions
Renin	Angiotensin forming enzyme, and angiotensinogenase	406	45	Kidney, lung, brain	An aspartyl protease-cleaving angiotensinogen to form angiotensin I
Angiotensin I	AngI	485	53	Blood vessels, heart, brain, adrenal gland, liver, ovary	Inducing contraction of vascular smooth muscle cells
Angiotensin-converting enzyme	ACE, ACE1, dipeptidyl carboxypeptidase 1, kininase II	1306	150	Ubiquitous	Catalyzing the conversion of angiotensin I into a physiologically active peptide angiotensin II
Angiotensin II type 1 receptor	AGTR1, angiotensin II receptor vascular type 1, AT2R1, angiotensin receptor 1A, AGTR1A, angiotensin receptor 1B, AGTR1B	359	41	Vascular smooth muscle cells, macrophages, kidney, eye, lung, skeletal muscle, adrenal gland, uterus, ovary, placenta	Interacting with angiotensin II to induce smooth muscle contraction and enhancing cell proliferation

*Based on bibliography 15.6.

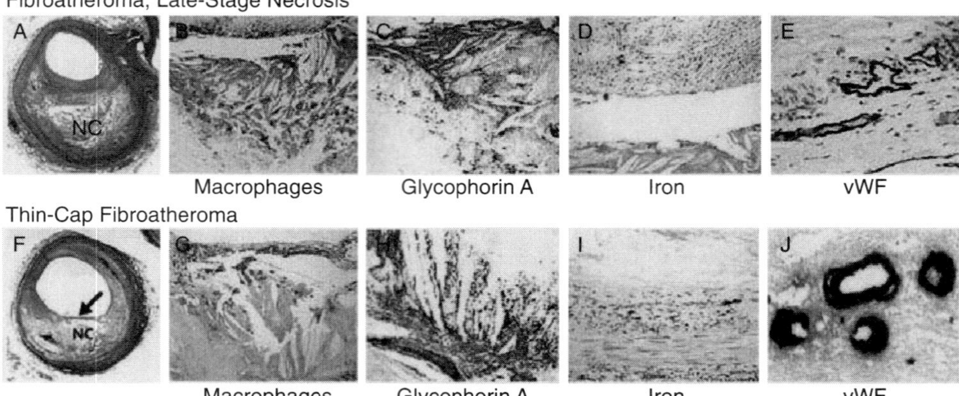

Figure 15.3. Pathological changes in human late-stage necrotic fibroatheroma (panels A–E) and thin-cap fibroatheroma (panels F–J). Panel A shows a low-power view of a fibroatheroma with a late-stage necrotic core (NC) (Movat pentachrome, ×20). Panel B shows intense staining of CD68-positive macrophages within the necrotic core (×200). Panel C shows extensive staining for glycophorin A in erythrocyte membranes localized with numerous cholesterol clefts within the necrotic core (×200). Panel D shows iron deposits (blue pigment) within foam cells (Mallory's stain ×200). Panel E shows microvessels bordering the necrotic core with perivascular deposition of von Willebrand factor (vWF) (×400). Panel F shows a low-power view of a fibroatheroma with a thin fibrous cap (arrow) overlying a relatively large necrotic core (Movat pentachrome, ×20). The fibrous cap is devoid of smooth muscle cells (not shown) and is heavily infiltrated by CD68-positive macrophages (panel G, ×200). Panel H shows intense staining for glycophorin A in erythrocyte membranes within the necrotic core, together with cholesterol clefts (×100). Panel I shows an adjacent coronary segment with iron deposits (blue pigment) in a macrophage-rich region deep within the plaque (Mallory's stain, ×200). Panel J shows diffuse, perivascular deposits of von Willebrand factor in microvessels, indicating that leaky vessels border the necrotic core (×400). (Reprinted with permission from Kolodgie FD et al: Intraplaque hemorrhage and progression of coronary atheroma, *New Engl J Med* 349:2316–25, copyright 2003 Massachusetts Medical Society. All rights reserved.)

takes a considerably long time, ranging from years to decades, atherosclerosis is often detected in aged population (> 40 years old). Grown atheroma can partially or completely block the lumen of arteries, resulting in ischemia or infarction of distal tissues or organs. Atherosclerosis often involves the brain (causing stroke), the heart (cardiac ischemia and infarction), kidney (renal infarction), and lower limbs (ischemia). Atherosclerosis is the leading cause of human death.

The pathogenesis of atherosclerosis is related to various types of endothelial injury. A number of factors have been found to induce endothelial injury, including bacterial and viral infection, physical trauma (surgery and catheterization), chemical toxins, bloodflow-related shear and tensile forces, and physical trauma. These factors induce endothelial cell injury or detachment and inflammatory reactions in the wall of blood vessels (Fig. 15.4). Endothelial cell injury may result in changes in endothelial permeability and adhesiveness to blood cells, and the release of growth factors and cytokines. Altered endothelial permeability may facilitate the transport of cholesterol and low-density lipoproteins across the endothelium. Altered endothelial adhesiveness may promote leukocyte attachment to and migrate through the endothelium. Released growth factors and cytokines may attract more

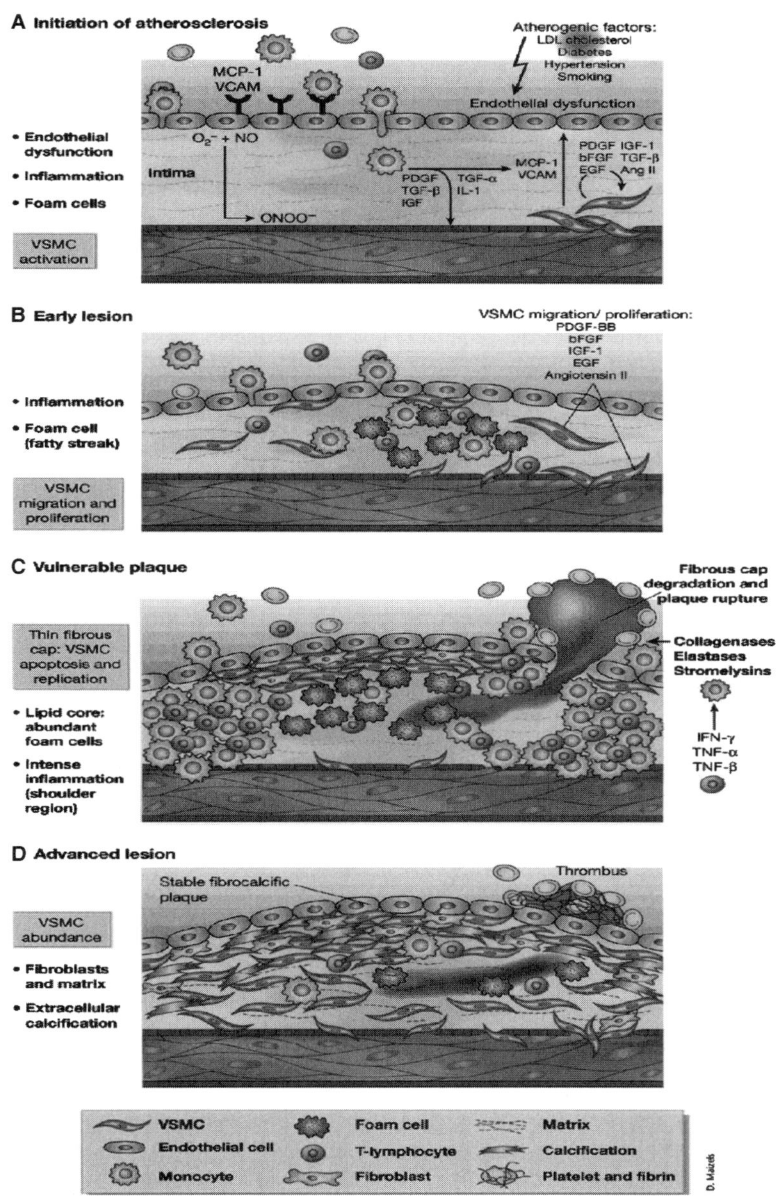

Figure 15.4. Function of vascular smooth muscle cells (VSMCs) during different stages of atherosclerosis. Cardiovascular risk factors alter the vascular endothelium (EC), which triggers a cascade of events, including the recruitment of leukocytes. Cytokines and growth factors are released by leukocytes and vascular cells, generating a highly mitogenic milieu. VSMCs migrate, proliferate, and synthesize extracellular matrix components on the luminal side of the vessel wall, forming the fibrous cap of the atherosclerotic lesion. Inflammatory mediators ultimately induce thinning of the fibrous cap by expression of proteases, rendering the plaque weak and susceptible to rupture and thrombus formation. In advanced disease, fibroblasts and VSMCs with extracellular calcification give rise to fibrocalcific lesions. LDL, low-density lipoprotein; MCP, monocyte chemoattractant protein; VCAM, vascular cell adhesion molecule; PDGF-BB, platelet-derived growth factor (BB, β-chain homodimer); TNF, tumor necrosis factor; TGF, transforming growth factor; IL, interleukin-1; IGF, insulin-like growth factor; bFGF, basic fibroblast growth factor; AngII, angiotensin II; EGF, epidermal growth factor; IFN, interferon. (Reprinted by permission from Macmillan Publishers Ltd.: Dzau VJ et al: Vascular proliferation and atherosclerosis: New perspectives and therapeutic strategies, *Nature Med* 8:1249–56, copyright 2002.)

leukocytes to the endothelium, promote smooth muscle proliferation and migration from the media to the intima (Fig. 15.5), and induce excessive production of extracellular matrix. In particular, monocytes can migrate into the intima and transform to macrophages, which may take up low-density lipoproteins (LDLs) and convert to "foam" cells, a characteristic cell type found in atheroma. Low-density lipoproteins undergo oxidation in macrophages. Oxidized low-density lipoproteins may facilitate the conversion of macrophages to foam cells, serve as monocyte chemoattractants, and confine macrophages to the intima. All these processes contribute to the formation of atheroma during different stages of atherogenesis. During the late stage, accelerated SMC proliferation, excessive production of extracellular matrix, and accumulation of cholesterol and low-density lipoproteins cause rapid growth of the atheroma lesion, leading to clinical consequences such as vascular stenosis and occlusion.

The process of atherogenesis is influenced by mechanical factors, including bloodflow-associated shear stress at the luminal surface of blood vessels and blood pressure-associated tensile stress in the wall of blood vessels. Atherosclerosis is often initiated at arterial bifurcation regions, where vortex bloodflow and reduced fluid shear stress are present. In experimental modles of intimal hyperplasia, a pathological event similar to atherosclerosis, the presence of vortex bloodflow and a reduction in blood shear stress enhance thrombosis, leukocyte adhesion, smooth muscle cell migration, and intimal hyperplasia (Chapter 15 opening figure and Fig. 15.6). These observations suggest that a

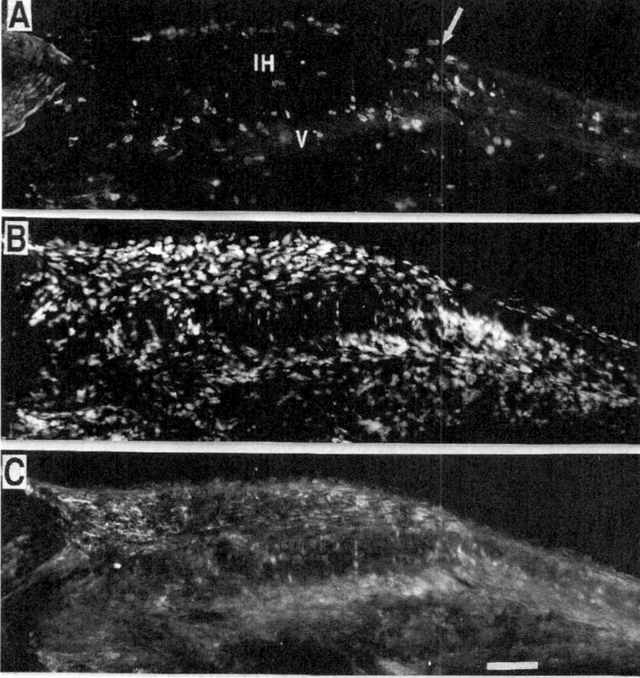

Figure 15.5. BrdU-positive cells in the neointima of an experimental vein graft. (A) BrdU-positive cells detected by using an anti-BrdU antibody in the neointima; (B) Hoechst 33258-labeled cell nuclei in the same specimen as shown in panel A; (C) Smooth muscle cells detected by using an anti-smooth muscle antibody in the same specimen as shown in panels A and B. Scale: 100 μm.

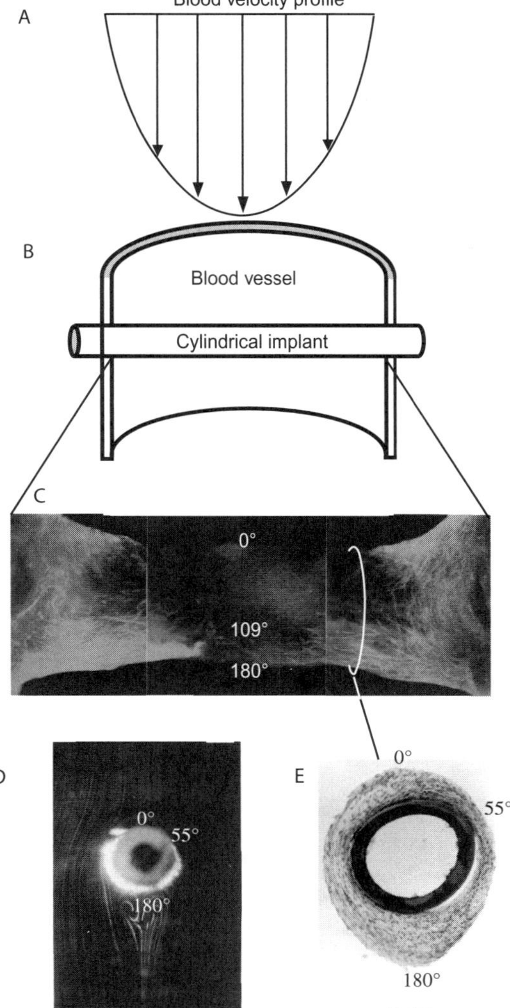

Figure 15.6. Influence of blood flow and shear stress on thrombosis, intimal hyperplasia, and the migration of vascular smooth muscle cells on a cylindrical polymer tube implanted into the rat aorta. (A) Parabolic profile of bloodflow. (B) Schematic representation of the polymer implant in the aorta. (C) Thrombotic tissue on the implanted polymer tube with smooth muscle cells (green in color, labeled with an anti-smooth muscle α-actin antibody), which migrated from the wall of the host aorta. Note that the maximal bloodflow velocity is associated with a minimal level of thrombosis in the central region of the implanted polymer tube. In this model, the implanted cylindrical tube is subject to a laminar bloodflow with the tube axis perpendicular to the bloodflow. The interaction of the implanted cylindrical tube with the bloodflow results in a unique distribution of fluid shear stress on the surface of the implanted tube in its circumferential direction: a leading stagnation point at 0° (with zero shear stress), an increasing shear stress region from 0° to 55° with a maximal shear stress point at 55°, a decreasing shear stress region from 55° to 109°, a flow separation point at 109° (with zero shear stress), a vortex flow region from 109° to 180°, and a trailing stagnation point at 180° (with zero shear stress). The flow pattern is symmetric from 0° to –180°. The distribution of fluid shear stress on the surface of the implanted cylindrical tube can be estimated by using an fluid dynamic model. The level of fluid shear stress is inversely correlated to the migration of the vascular smooth muscle cells. As shown in panel C, the maximal shear stress point at 55° is associated with minimal migration of smooth muscle cells. (D) Flow pattern around the implanted cylindrical polymer tube detected in an flow simulation model with a physiological Reynolds number in vitro. Note the formation of vortex flow behind the polymer tube. (E) A histological micrograph derived from a transverse section prepared from the implanted polymer tube with encapsulating thrombotic tissue as indicated in panel C. Note that the minimal shear stress at the leading and trailing stagnation points (0° and 180°) is associated with maximal thrombosis and intimal hyperplasia whereas the maximal shear stress at 55° is associated with minimal thrombosis and intimal hyperplasia. Scale: 100 μm.

reduction in the bloodflow velocity or bloos shear stress promote thrombogenesis, intimal hyperplasia, and atherogenesis. The tensile stretch stress in the wall of blood vessels is another mechanical factor that contributes to intimal hyperplasia and atherogenesis. In experimental renovascular hypertension induced by aortic constriction, increased blood pressure induces profound arterial hypertrophy, which is associated with intimal hyperplasia in large arteries such as the common iliac arteries (Fig. 15.7).

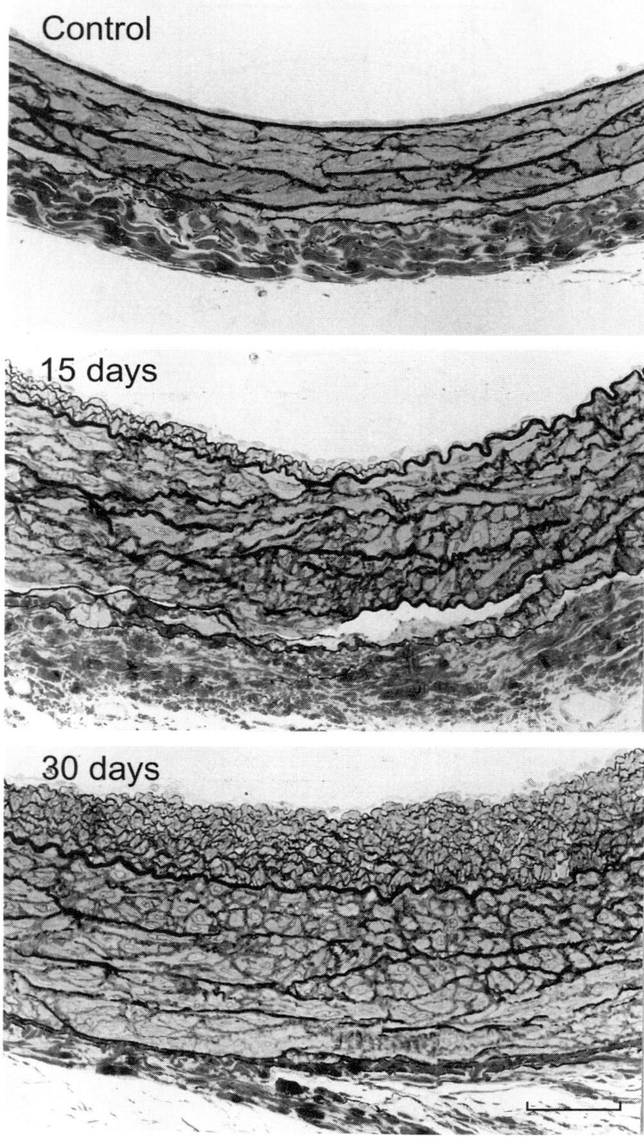

Figure 15.7. Influence of renovascular arterial hypertension on the intimal structure of the rat common iliac artery. Hypertension was induced by aortic constriction. The arterial specimens were stained with toluidine blue O and examined by optical microscopy. Scale: 100 μm.

Several metabolic disorders contribute to the initiation and development of atherosclerosis. These include disbetes and hyperlipidemia. Diabetes is a disorder due to insulin deficiency and characterized by an increase in the serum level of glucose. Insulin deficiency is induced by the malfunction and apoptosis of the pancreatic β cells (see chapter 19 for detailed mechanisms). Diabetes is often associated with endothelial cell injury, leukocyte activation, intimal hyperplasia, and atherosclerosis. Figure 15.8 shows an example of endothelial cell injury and leukocyte adhesion to injured endothelial cells in a rat experimental diabetes model. Indeed, atherosclerosis is a major cause of death in diabetic patients. However, the mechanisms of diabetes-induced atherogenesis remains poorly understood.

Hyperlipidemia is a disorder characterized by an increase in the plasma concentration of lipids, including cholesterol and triglyceride. Clinical observations have demonstrated that an increase in cholesterol and triglyceride accelerates atherogenesis, whereas a reduction in these components diminishes the risk of atherogenesis. In the United States, hyperlipidemia is found in about half of the adult population. Thus, suppression of hyperlipidemia is considered an effective approach for the prevention of atherogenesis.

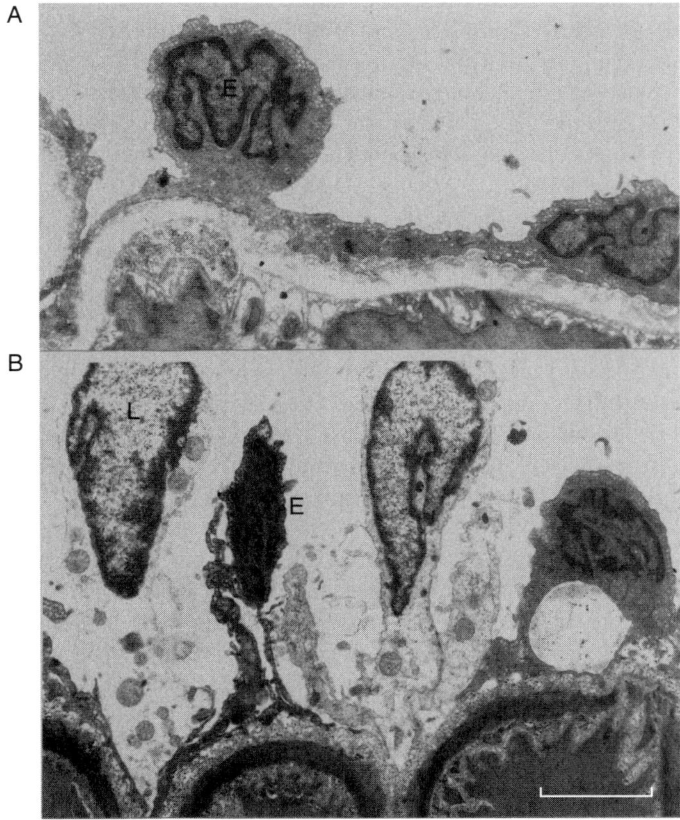

Figure 15.8. Interaction of leukocytes with endothelial cells in the femoral artery of experimental diabetic rats. A. Control. B. Diabetic. Diabetes was induced by administration of Streptozocin. The arterial specimens were examined by electron microscopy. E: endothelial cell. L: Leukocyte. Scale: 5 μm.

Hyperlipidemia is induced by two mechanisms: excessive intake of lipids (fatty acids and cholesterol) and reduced clearance of plasma low-density lipoproteins, which carry lipids. The later is considered a disorder. Hyperlipidemia is divided into two groups, based on the pathogenesis of the disorder: primary and secondary. The primary form is a hereditary disorder and is related to genetic defects. In some patients, this is due to a single gene defect, which is called monogenic hyperlipidemia. This type of disorder can be predicted on the basis of the Mendelian genetic trait. In others, it is caused by a combination of multiple gene mutations and environmental insults, which is called *polygenic hyperlipidemia*. In this case, the plasma cholesterol level of the majority members of a family may go up some time during the lifespan, if the intake of saturated fats and cholesterol is high. Usually, the polygenic type prevails in the human population. Secondary hyperlipidemia is a complication of metabolic disorders, such as diabetes. A common sign of hyperlipidemia is the elevation of plasma cholesterol and triglyceride.

Atherosclerosis is characterized by several stages of progressive pathological lesions, including the fatty streak, the intermediate, and the fibrous plaque stages. During the *fatty streak stage*, monocytes and T lymphocytes migrate into and accumulate in the intima. The monocytes are transformed into macrophages, which convert to foam cells. The fatty streaks are often found at locations such as curved arteries and bifurcations. During the *intermediate stage*, atherosclerotic lesions often contain several types of cell, including foam cells, T lymphocytes, and smooth muscle cells, and loosely organized extracellular matrix components, including collagen and elastin fibers. During the *fibrous plaque stage*, the density of smooth muscle cells increases together with excessive production of extracellular matrix. Foam cells and T lymphocytes as well as lipid components are found in atheroma. These components are usually capped with a layer of connective tissue. Since the lesion increases in size during this stage, clinical consequences such as vascular stenosis and occlusion often occur. Furthermore, atherosclerotic lesions may detach from the wall of blood vessels, resulting in arterial embolism.

Atherosclerosis is associated with cell apoptosis in humans and animal models. Apoptotic cells are often found at the base of atheroma. Several death regulatory factors, including Bcl-2, caspase 1 and 3, and Fas receptor are upregulated in atheroma, suggesting a role of these death factors in the regulation of cell apoptosis. The activation of apoptosis in atherosclerotic lesions may be a self-defense mechanism that is initiated in response to increased smooth muscle cell proliferation. However, this process may reduce the stability of atherosclerotic plaques, leading to formation of emboli.

Animal Models of Atherosclerosis [15.8]. Atherosclerosis models can be established in the mouse, rat, guinea pig, rabbit, dog, pig, and nonhuman primate. Several approaches can be used to create an atherosclerosis model, including high-cholesterol diet feeding, genetic modulation, and vascular injury. High-cholesterol-diet feeding induces hypercholesterolemia and fatty streak-like lesions in the arterial system. Although fatty streaks do not represent human atherosclerosis, they are similar in structure to early atherosclerotic lesions in humans. Rabbits are often used for this model since these animals are susceptible to hypercholesterolemia and develop fatty streak-like lesions within a short period.

Genetic modulation of proteins that regulate lipid metabolism can render animals vulnerable to atherosclerosis. A typical model is experimental atherosclerosis due to apolipoprotein E (apoE)-deficiency. ApoE is a protein that plays a role in the clearance of blood lipid molecules, which contribute to atherogenesis. The deficiency of apoE enhances the

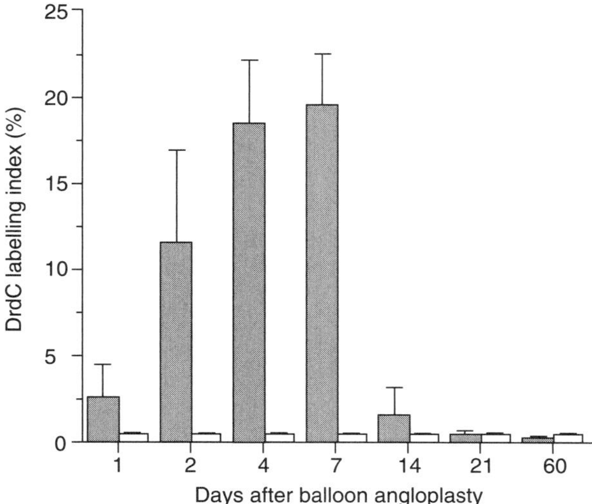

Figure 15.9. Effect of balloon injury on smooth muscle cell proliferation in injured porcine arteries. Proliferation of intimal cells was measured in injured porcine iliofemoral arteries with BrdC labeling 1–60 days after balloon angioplasty ($n = 4$ arteries at each timepoint, 2 arteries per animal). Intimal cell proliferation was measured by counting the number of labeled and unlabeled nuclei in four quadrant cross sections of balloon-injured and uninjured arteries with a microscope-based video image analysis system. Injured iliofemoral arteries (solid bars) and uninjured carotid arteries (open bars) were examined in the same animal. Standard error bars are shown. Additional immunohistochemical studies with an antibody to smooth muscle α-actin identified proliferating intimal cells as smooth muscle in origin. (Reprinted with permission from Ohno T et al: *Science* 265:781–4, copyright 1994 AAAS.)

development of hyperlipidemia and atherosclerosis. ApoE-deficient mice are available from commercial sources such as Jackson Labs.

Another approach for the induction of experimental atherosclerosis is mechanical injury of arterial endothelial and smooth muscle cells (Fig. 15.9). The injury of these cells can be induced by inserting a balloon-tipped catheter into and distending an artery to a desired dimension. A back-and-forth movement of the balloon can cause endothelial cell denudation or detachment, resulting in the exposure of the subendothelial basal lamina. The collagen matrix in the basal lamina initiates rapid platelet and leukocyte adhesion, blood coagulation, and thrombosis, establishing a basis for the development of atherosclerosis. The injury of smooth muscle cells, together with mitogenic factors released by injured vascular cells and activated blood cells, stimulates the proliferation and migration of smooth muscle cells from the arterial media to the intima, contributing to intimal hyperplasia and the formation of atheroma. This model can be used in combination with other models such as apoE deficiency and/or high-cholesterol diet feeding to facilitate atherogenesis.

Prevention of Atherosclerosis [15.7]. The development of symptomatic atherosclerosis usually takes a long time, ranging from years to decades. Once developed, it is difficult to remove the atherosclerotic lesions. Thus, an ideal approach is to prevent atherosclerosis at the first place, especially, for those who have a familial history of hyperlipidemia, hypertension, diabetes, and obesity. These disorders facilitate the initiation and development of atherosclerosis. A number of methods have been established and used for the prevention of atherosclerosis. These include diet control, appropriate daily exercise, and administration of antioxidant agents. For diet control, Low-fat, low-cholesterol, and soybean-rich foods are beneficial. Daily exercise is also effective for the prevention of atherogenesis. While the mechanisms are poorly understood, exercise-enhanced bloodflow and increased fluid shear stress may reduce the rate of atherogenesis. Experimental investigations have demonstrated that an increase in blood shear stress with respect to the physiological level exerts an inhibitory effect on the activity of growth factors and the proliferation and migration of vascular smooth muscle cells. These mechanisms may contribute to shear stress-related suppression of atherogenesis. Furthermore, the administration of antioxidants, such as vitamin E, vitamin C, vitamin A, and β-carotene (a precursor of vitamin A), may help to prevent atherogenesis. The oxidation of low-density lipoproteins (LDL) facilitates macrophage endocytosis of LDL, enhances macrophage conversion to foam cells, and contributes to atherogenesis. A treatment with antioxidant agents can reduce the level of LDL oxidation, thus reducing atherogenic activities.

Conventional Treatment of Atherosclerosis [15.9]. When atherosclerosis is diagnosed, it can be treated with several approaches, including the administration of antihyperlipidemia agents, anti-SMC proliferation agents, and vasodilators, reduction in cardiac workload, angioplasty, vascular stenting, and vascular reconstruction. These approaches are described as follows.

Antihyperlipidemia Agents. Common anti-hyperlipidemia agents include 3-hydroxy-3-methylglutaryl coenzyme A (HMG CoA) reductase inhibitor, fibric acids, and bile acid-binding resins. 3-Hydroxy-3-methylglutaryl coenzyme A (HMG CoA) reductase inhibitor is a fungus-derived compound that inhibits HMG CoA reductase, a rate-controlling enzyme for the synthesis of cholesterol. Fibric acids are substances that reduce the plasma level of very-low-density lipoproteins, low density lipoproteins, and intermediate density lipoproteins. Bile acid-binding resins are anion exchange polymeric resins that facilitate the removal of cholesterol from the intestinal system. All these agents can be used to lower the level of lipids.

Antiproliferative Agents. Several types of agents have been developed and used for suppressing smooth muscle cell proliferation, including inhibitors for angiotensin converting enzyme (ACE), angiotensin II type 1 receptor, and growth factors. Angiotensin converting enzyme catalyzes the cleavage of angiotensin I and converts angiotensin I to angiotensin II, which not only stimulates smooth muscle contraction but also promotes smooth muscle proliferation. The suppression of ACE reduces smooth muscle proliferation. Common ACE inhibitors include captopril, enalapril, and lisinopril.

The effect of angiotensin II on the proliferative activity of vascular smooth muscle cells depends on its interaction with the angiotensin II type 1 receptor on the cell membrane. The blockage of this type of receptor can significantly suppress angiotensin II-induced smooth muscle proliferation. Most angiotensin II antagonists are peptide-based angiotensin II analogues, which have a structure similar to that of angiotensin II with a modified amino acid sequence. These peptides compete with angiotensin II for receptor binding, but do not induce smooth muscle activation and, therefore, can reduce the mitogenic effect of angiotensin II.

Growth factors, such as platelet-derived growth factor, stimulate smooth muscle proliferation and migration. The suppression of growth factors reduces smooth muscle mitogenic activities. A typical antagonist for platelet-derived growth factor is terbinafine. This agent has been shown to inhibit the activity of platelet-derived growth factor and thus suppress cell proliferation in experimental models of atherosclerosis. Other antiproliferative agents include heparin, rapamycin, and Taxol. Heparin is an anticoagulant and also exerts an inhibitory effect on smooth muscle proliferation. Rpamycin is a natural antibiotic found in a microorganism *Streptomyxes hygroscopius*. This agent inhibits the progression of cell cycle and thus suppresses smooth muscle proliferation. Taxol is a compound extracted from the bark of *Taxus brevifolia*, a Pacific yew. Taxol has been found to stabilize microtubules, suppress cell division, and induce cell apoptosis. When applied to blood vessels, Taxol inhibits SMC proliferation and prevent intimal hyperplasia.

Treatment of Angina Due to Ischemia. When angina occurs due to ischemia, vasodilators can be used to increase bloodflow to the heart, thus reducing ischemia. Two types of drug are commonly used for such a purpose: organic nitrates and calcium channel blockers. Organic nitrates induce relaxation of coronary arterial and venous smooth muscle cells, resulting dilation of these blood vessels and an increase in bloodflow. Commonly used nitrates include amyl nitrite, nitroglycerin, and isosorbide dinitrate. Calcium channel blockers can be used to reduce the release of calcium. Calcium is necessary for the initiation of vascular smooth muscle contraction and is released into the cytoplasm through its channels. The blockade of calcium channels leads to a reduction in the contractility of smooth muscle cells. Commonly used agents in this class include nicardipine, nifedipine, nimodipine, and verapamil.

Reduction in Cardiac Workload. A decrease in cardiac workload is beneficial to the treatment of ischemia and prevention of cardiac infarction. β-adrenergic antagonists are effective agents for such a purpose. Adrenergic agents, including epinephrine and norepinephrine, enhance myocardial contractility and, therefore, induce an increase in the workload and oxygen consumption of the heart. β-adrenergic antagonists, such as atenolol, acebutolol, metoprolol, nadolol, pindolol, propranolol, and timolol, can block the β-adrenergic receptors, thus reducing cardiac contractility and workload.

Treatment of Malfunctioned Arteries with Angioplasty. For advanced atherosclerosis with clinical consequences, such as vascular stenosis and occlusion, surgical interventions are necessary. One surgical approach is angioplasty, by which stenotic arteries are reopened by inserting a balloon-tipped catheter into and expand the stenotic arteries by balloon inflation. This approach is relatively simple compared to other types of surgical interventions, such as stenting and grafting. However, mechanical expansion induces stretch and injury of the arterial wall, which initiates blood coagulation, thrombogenesis, and intimal hyperplasia. Restenosis occurs after angioplasty in a large fraction of patients. Thus, angioplasty is primarily used for temporary relief of arterial occlusive disorders.

Treatment of Malfunctioned Arteries with Stents. Arterial stenting is a procedure by which a cylindrical wire-frame is inserted into a stenotic artery with the aid of a balloon-tipped catheter. The wire-frame is expanded by inflating the balloon. The balloon and catheter are then retreated, leaving the opened wire-frame within the lumen so that the artery is kept open. Although this is a relatively simple procedure, stenting often induces vascular injury and cell death. These changes facilitate blood coagulation, thrombogenesis, and intimal hyperplasia, resulting in arterial restenosis.

To reduce stent-induced vascular disorders, materials with β and γ radiation or antiproliferative drug coating, have been used to construct vascular stents. Stents with radiation materials can induce DNA damage and cell arrest in the G1 phase. Such stents have been shown to inhibit smooth muscle cell proliferation and prevent intimal hyperplasia after stenting surgery. However, radiation may induce cell death and vascular aneurysm. The control of radiation dosage is a critical issue. Antiproliferative drug-coated stents are effective for preventing smooth muscle proliferation, thus reducing intimal hyperplasia and stent failure.

Arterial Reconstruction. Vascular grafts are often used to replace malfunctioned arteries due to atherosclerosis. A number of materials have been developed and used for clinical or experimental arterial reconstruction. These materials include autogenous arteries and veins, allogenic arteries, non-biodegradable polymers (polytetrafluoroethylene and Dacron), biodegradable polymers (polyglycolides and polylactides), and extracellular matrix constituents (collagen matrix, fibrin matrix, and elastic laminae). Among these grafting materials, the arterial and vein grafts are considered the gold standard grafts because of their high patency rate and long lifespan following arterial reconstruction. However, not all patients possess arteries and veins available for arterial reconstruction. Allogenic vascular grafts and synthetic polymers have been used for arterial reconstruction. Although these materials exhibit characteristics suitable for arterial reconstruction, they induce inflammation and thrombogenesis, contributing to arterial restenosis. Further research is necessary to improve the anti-inflammatory and antithrombotic properties of these materials. Compared to angioplasty and stenting, arterial reconstruction has been proven a more effective approach for the treatment of atherosclerosis.

Arterial reconstruction models are often used in vascular tissue regenerative engineering to test new materials and assess the interaction of materials with host cells and tissues. This type of model can be created in a number of animals, including nonhuman primates, pigs, dogs, rabbits, rats, and mice. To establish such a model, an animal can be anesthetized by peritoneal injection of sodium pentobarbital (50 mg/kg). The artery of interest can be exposed and a segment of the artery is surgically separated from surrounding tissue and clamped with vascular clamps to stop local blood flow at two locations: one proximal and the other distal to the reconstruction site. The selected artery is transected between the two clamps. The lumen of the transected artery is immediately treated with 100 U/mL heparin. A segment of the host artery can be removed with its length depending on the length of the arterial graft. The arterial graft can be anastomosed to the host artery by using disrupted suture stitches. Note that continuous stitches often cause arterial constriction at the anastomosis. After anastomotic procedures, the vascular clamps can be released slowly. When bleeding occurs at the anastomoses, the artery can be reclamped and the bleeding sites can be sealed with additional suture stitches. After vascular clamps are removed, the artery ought to be inspected for blood flow by observing pulse activities. The surgical wounds can be closed in the order of the muscular tissue, soft connective tissue, and skin. Continuous suture stitches can be used to close the muscular and connective tissues, but discontinuous stitches are necessary for skin closure, since animals, especially rodents, intend to chew the suture stitches.

Molecular Treatment of Atherosclerosis. The strategies of treating atherosclerosis by molecular engineering are to prevent blood cell adhesion and thrombogenesis, to inhibit the proliferation and migration of smooth muscle cells and macrophages, and to suppress intimal hyperplasia. These strategies are different from those for treating disorders in other organ systems, such as the heart, brain, and liver, which often require the enhancement of cell proliferation and differentiation. A number of candidate DNA molecules have been selected and tested for their roles in the suppression of atherogenesis and intimal hyperplasia. These molecules are discussed as follows.

Antisense Oligonucleotides for Mitogenic Factors [15.10]. As described on page 448, an antisense oligonucleotide is a DNA fragment that is complimentary to and can hybridize with a specific mRNA in the cytoplasm, rendering the mRNA incapable of translating proteins. Cyclin-dependent kinases (CDKs) are proteins responsible for the regulation of cell mitosis (see page 262 for characteristics and regulatory mechanisms of CDKs). The blockade of the translation of these proteins in blood vessels exerts an inhibitory effect on cell proliferation, thus preventing atherogenesis and intimal hyperplasia. Similarly, the blockade of other mitogenic factors, such as growth factors and angiotensinogen, by antisense oligonucleotides can suppress the translation of the mitogenic factors (Fig. 15.10) and potentially inhibit the proliferation of vascular smooth muscle cells. Antisense oligonucleotides for these mRNAs have been successfully used in animal models.

Transcription factors are proteins that initiate and control gene transcription. The blockade of the mRNA of mitogenic transcription factors inhibits the expression of mitogenic genes and thus suppresses cell proliferation. A typical transcription factor is E2F (elongation factor), which control the expression of mitogenic genes. The application of antisense oligonucleotides for E2F mRNA has been shown to inhibit the proliferation of

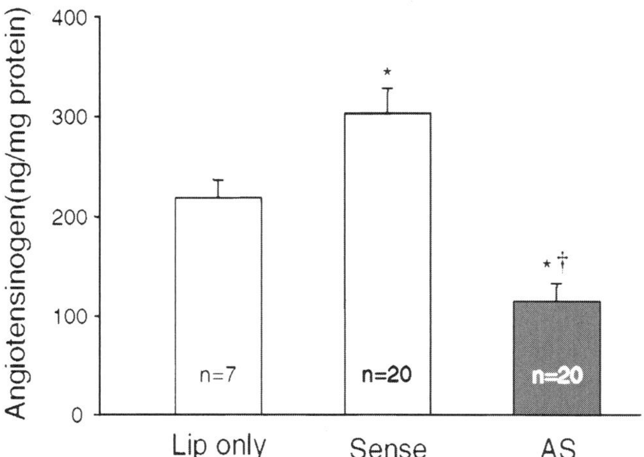

Figure 15.10. Bar graph of secreted AGT levels in medium of H4-II-E cells treated with sense or antisense (AS) plasmid. Cells treated with Lipofectamine (Lip) served only as the control. After transfections (48 h), medium samples were harvested. AGT levels were measured by RIA. $^{*}P <$ 0.05, compared with control group; $^{*}P < 0.05$ compared with sense-treated group. (Reprinted from Tang X et al: Intravenous angiotensinogen antisense in AAV-based vector decreases hypertension, *Am J Physiol* 277:H2392–9, copyright 1999 by permission of American Physiological Society.)

vascular smooth muscle cells and prevent intimal hyperplasia in vascular injury and graft models. Antisense oligonucleotides for E2F mRNA have also been used in clinical trials of human vein grafts. Preliminary studies have shown promising results.

Cell Cycle Inhibiting Genes [15.11]. The progression of cell cycle is negatively regulated by a number of inhibiting proteins, including the p53, $p21^{CIP1}$, $p27^{KIP1}$ (see page 265 for characteristics and regulation of these proteins). The genes encoding these inhibitory proteins can be prepared and used for transferring into arterial lesion sites. The overexpression of these genes has been shown to inhibit vascular cell proliferation and intimal hyperplasia in animal models.

Nitric Oxide Synthase Gene [15.12]. Nitric oxide synthase is an enzyme that catalyzes the deamination of L-arginine to produce nitric oxide. Nitric oxide can readily diffuse through the cell membrane, reacts with guanylyl cyclase to produce cyclic GMP (cGMP), and potentially inhibits the activity of cyclin A, a molecule that regulates cell mitosis. In experimental investigations, nitric oxide has been shown to effectively inhibit the proliferation of vascular smooth muscle cells (Fig. 15.11). Furthermore, nitric oxide enhances the expression and activity of the cell cycle-inhibiting protein $p21^{CIP1}$, inhibits cell proliferation, and suppresses platelet aggregation and leukocyte adhesion. The enhancement of nitric oxide production in vascular cells leads to the suppression of cell proliferation. Since nitric oxide synthase is a limiting factor for the production of nitric oxide, the overexpression of this gene via gene transfer enhances the production of nitric oxide.

Characteristics of several nitric oxide synthases are listed in Table 15.5.

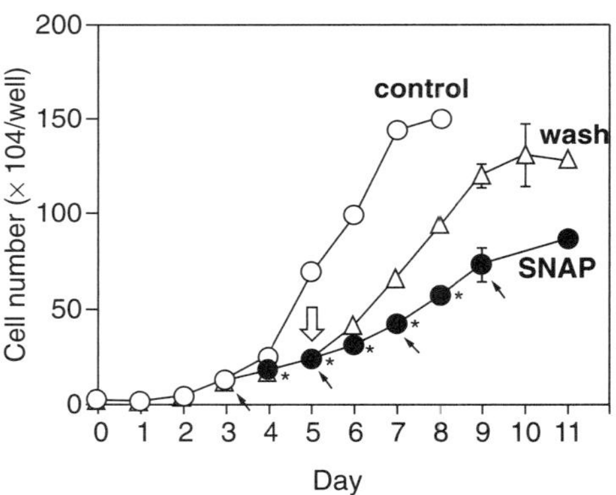

Figure 15.11. Influence of nitric oxide on vascular smooth muscle cell (VSMC) proliferation. The effect of *S*-nitroso-*N*-acetylpenicillamine (SNAP, a nitric oxide-releasing substance) on VSMC proliferation. VSMCs seeded in 6-well plates (3×10^4/well) were cultured in growth medium (open circle). The medium was changed every other day and the cells were counted every day. Two-thirds of the cultures (solid circle) were exposed to SNAP (100 μM) added every other day simultaneously with the changing medium (filled arrows). On day 5 (open arrow), one-half of the cultures treated with SNAP were placed in SNAP-free growth medium after several washes (triangle), and the other half were continuously exposed to SNAP. Data represent the means ± SD ($n = 3$). $^*p < 0.01$ versus control. (Reprinted with permission from Ishida A et al: Induction of the cyclin-dependent kinase inhibitor p21(Sdi1/Cip1/Waf1) by nitric oxide-generating vasodilator in vascular smooth muscle cells, *J Biol Chem* 272:10050–7, copyright 1997.)

TABLE 15.5. Characteristics of Selected Nitric Oxide Synthases*

Proteins	Alternative Names	Amino Acids	Molecular Weight (kDa)	Expression	Functions
Nitric oxide synthase 1	NOS1, PNNOS, nitric oxide synthase penile neuronal, nitric oxide synthase neuronal	1434	161	Brain, skeletal muscle, heart, kidney	Catalyzing the synthesis of nitric oxide
Nitric oxide synthase 2A	NOS2A, NOS2, nitric oxide synthase, macrophage, NOS2A inducible, hepatocyte	1153	131	Brain, blood vessels, retina, liver, lung, monocytes	An inducible synthase that catalyzes the synthesis of nitric oxide
Nitric oxide synthase 3	NOS3, NOS type III, NOSIII, constitutive NOS, endothelial nitric oxide synthase (ENOS, eNOS, or EC-NOS), endothelial NOS, cNOS	1203	133	Blood vessels, nervous system, heart, eye, lung, liver, placenta	A constitutive synthase that catalyzes the synthesis of nitric oxide

*Based on bibliography 15.12.

Herpes Virus Thymidine Kinase Gene [15.13]. Viral thymidine kinase can phosphorylate nucleosides and nucleoside substitutes. When viral thymidine kinase is present together with a nucleoside substitute, such as ganciclovir, the thymidine kinase can phosphorylate the nucleoside substitute, forming nucleoside triphosphate. The substitute nucleotide can be incorporated into DNA during DNA replication. Once the substitute nucleosite is integrated into a DNA molecule, DNA replication is terminated because the substitute nucleosite can no longer bind to additional nucleotides. Thus, a treatment with Herpes virus thymidine kinase gene and ganciclovir inhibits DNA replication and cell proliferation (Fig. 15.12).

Dominant Negative Mutant Mitogenic Genes [15.14]. Mitogenic genes encode proteins which regulate cell proliferative activities. Proteins encoded by dominant negative mutant genes may lose their mitogenic function, but keep their binding capability. When such dominant negative genes are expressed, their protein products can compete with the physiological form of the same proteins for substrate proteins that regulate cell mitogenic activities and thus partially block the substrate proteins, resulting in reduced mitogenic activities. A typical example of such dominant negative mutant genes is rasN17. This mutant gene is generated by replacing the gene codon for a serine residue at the 17th amino acid with a codon for asparagine. The RasN17 protein can competitively bind to substrate proteins, such as mitogen activated protein kinase kinase kinases, which are critical protein kinases for the regulation of cell mitogenic activities. The binding of RasN17 does not activate the substrate proteins, but block their interactions with the physiological form of Ras, thus reducing mitogenic activities (Fig. 15.13).

Vascular Tissue Regenerative Engineering. Vascular tissue regenerative engineering is to reconstruct blood vessels with cell-integrated vascular substitutes and to improve

Figure 15.12. Effect of ganciclovir (GC) on intimal and medial areas in arteries after balloon injury and infection with ADV-tk vector. Representative cross sections from iliofemoral arteries of pigs (A) injured for 1 min, then infected with adenoviral (ADV)- thymidine kinase (tk) vector and treated with saline (left) or ganciclovir (right); (B) injured for 5 min, then infected with ADV-tk vector and treated with saline (left) or ganciclovir (right); (C) injured for 5 min, then infected with ADV-δ E1 and treated with saline (left) or ganciclovir (right). These arteries were examined 3 weeks after injury and gene transfer (hematoxylin and eosin stain magnification 87×). Measurements of I/M area ratios (D) are from arteries infected after a 1-min injury and 3 weeks posttransfection with ADV-tk vector and treated with saline (0.445 +/− 0.047, $n = 4$ arteries) or ganciclovir (0.057 +/− 0.027, $n = 4$ arteries); infected after a 5-min injury and 3 weeks posttransfection with ADV-tk vector and treated with saline (0.512 +/− 0.047, $n = 8$ arteries) or ganciclovir (0.205 +/− 0.065, $n = 8$ arteries); infected after a 5-min injury and 6 weeks posttransfection with ADV-tk vector and treated with saline (0.639 +/− 0.099, $n = 4$ arteries) or ganciclovir (0.334 +/− 0.024, $n = 4$ arteries); infected after a 5-min injury and 3 weeks posttransfection with ADV-δ E1 vector and treated with saline (0.481 +/− 0.020, $n = 8$ arteries), or ganciclovir (0.445 +/− 0.027, $n = 8$ arteries). A statistically significant reduction in I/M area ratios was observed in the ADV-tk/positive GC group compared with ADV-tk/negative GC (1 min, 3 weeks, two-tailed unpaired t test, $P < 0.05$); ADV-tk/positive GC compared with ADV-tk/negative GC, ADV-δ E1/negative GC, and ADV-δ E1/positive GC (5 min, 3 weeks, ANOVA with Dunnett's t test, $P < 0.05$); and ADV-tk/positive GC compared with ADV-tk/negative GC (5 min, 6 weeks, two-tailed unpaired t test, $P < 0.05$). I, intima; M, media; GC, ganciclovir. (Reprinted with permission from Ohno T et al: *Science* 265:781–4, copyright 1994 AAAS.)

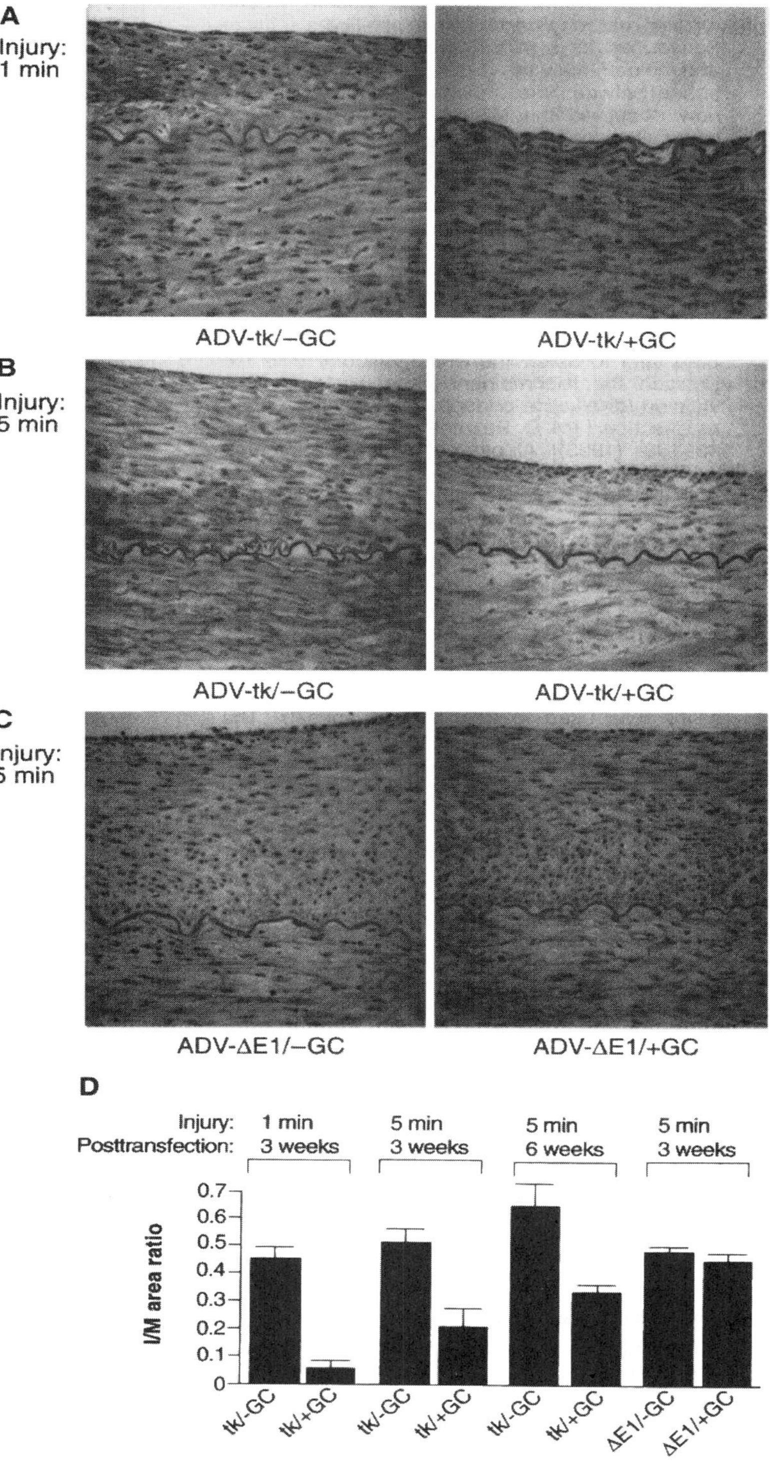
A
Injury:
1 min
ADV-tk/−GC
ADV-tk/+GC
B
Injury:
5 min
ADV-tk/−GC
ADV-tk/+GC
C
Injury:
5 min
ADV-ΔE1/−GC
ADV-ΔE1/+GC
D
Injury: 1 min 5 min 5 min 5 min
Posttransfection: 3 weeks 3 weeks 6 weeks 3 weeks
I/M area ratio
0.7
0.6
0.5
0.4
0.3
0.2
0.1
0
tk/-GC
tk/+GC
tk/-GC
tk/+GC
tk/-GC
tk/+GC
ΔE1/-GC
ΔE1/+GC

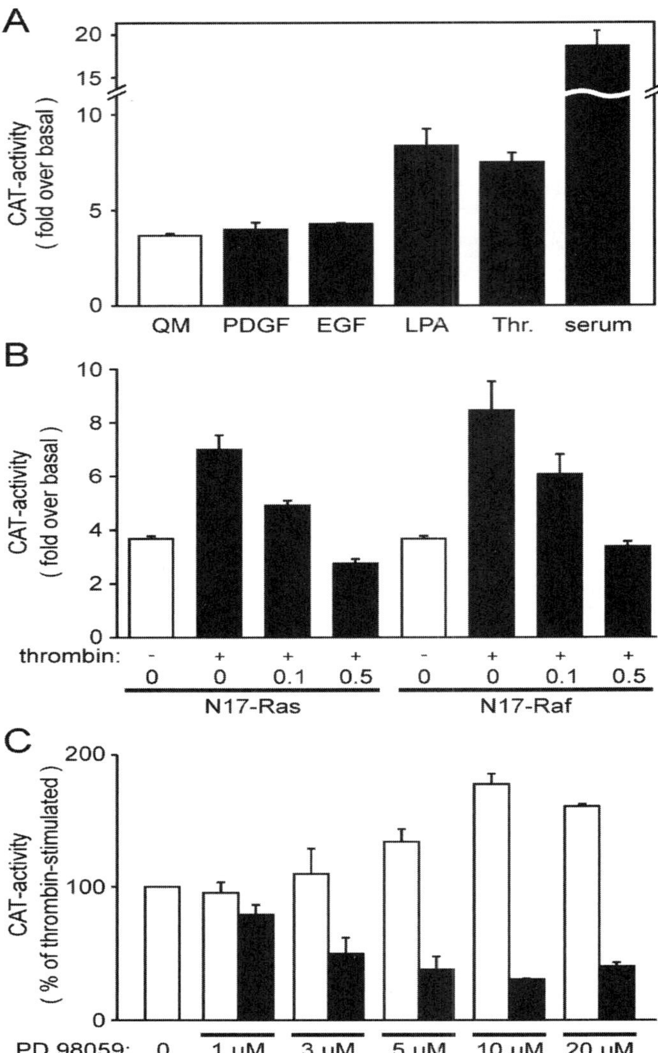

Figure 15.13. Negative regulation of extracellular signal-regulated kinase (ERK)1/2 phosphorylation by N17-Ras and N17-Raf. Receptor-mediated activation of the smooth muscle (SM)–myosin heavy-chain (MHC) promoter requires the Ras/Raf/ERK signaling cascade. Panel A, Vascular smooth muscle cells (VSM) cells, serum-starved for 48 h, were transfected with a 1346-nucleotide SM-MHC promoter-CAT fusion construct (pCAT-1346) and then incubated in the presence of serum free medium (QM) or in QM supplemented with 10 ng/mL PDGF, 10 ng/mL EGF, 10 μM LPA, 1 unit/mL thrombin (Thr.), or 10% serum for another 48 h. Cells were lysed and assayed for CAT activity. Depicted CAT activities were normalized for protein concentrations and compared with the CAT activity of cells transfected with a reporter gene construct lacking the SM-MHC promoter. Bars represent the means ± SE of at least five independent transfection experiments. Panel B, to test for participation of Ras/Raf in the thrombin-mediated SM-MHC promoter induction, VSM cells were cotransfected with 0.5 μg/well pCAT-1346 and the indicated amounts (in μg) of dominant negative N17-Ras or N17-Raf expression constructs. The total amount of plasmid DNA was kept constant (1 μg/well) with promoterless pCAT-basic. Panel C, VSM cells transfected with 1 μg/ml pCAT-1346 were preincubated for 30 min with different concentrations of PD98059, an inhibitor for mitogen activated protein kinase/ERK kinase (MEK black bars), and then stimulated with 1 unit/mL thrombin. Because dimethyl sulfoxide (the solvent of PD98059) further enhanced thrombin-stimulated SM-MHC promoter activities, controls were incubated in equivalent concentrations of the solvent (open bars). All values were normalized to the thrombin-mediated CAT activity without PD98059. (Reprinted with permission from Reusch HP et al: Gbeta gamma mediate differentiation of vascular smooth muscle cells, *J Biol Chem* 276:19540–7, 2001 with permission.)

the performance of reconstructed blood vessels. A number of approaches have been developed and used for engineering blood vessels. These approaches are discussed as follows.

Construction of Polymeric Arterial Substitutes with Endothelial Cell Seeding [15.15]. Polymeric materials have long been used for constructing arterial substitutes. Common polymeric materials used for vascular reconstruction include polytetrafluoroethylene (PTFE, Teflon) and polyethylene terephthalate (PET, Dacron). However, these materials induce inflammatory reactions and thrombogenesis, resulting in arterial restenosis and occlusion. Thus, polymeric materials can be used only for the reconstruction of blood vessels with diameter larger than 4 mm. To resolve such a problem, endothelial cells have been used to seed polymeric arterial substitutes on the luminal surface. While the approach seems reasonable, it is difficult to retain endothelial cells on the substitute during and after the reconstruction surgery. Major factors that influence the retention of endothelial cells include surgical trauma and bloodflow-induced shear stress. Further investigations are necessary to improve the polymeric surface properties and to enhance the retention of endothelial cells.

Construction of Cell-Integrated Arterial Substitutes with Biodegradable Polymers [15.16]. Another strategy in vascular tissue regenerative engineering is to construct biodegradable polymeric vascular substitutes with integrated vascular endothelial cells on the luminal surface and smooth muscle cells within the wall of the substitute. Several types of biodegradable polymeric materials, including polyglactin, poly-L-lactic-co-epsilon-caprolactone, poly(glycolide-co-caprolactone, polydioxanone, polyglycolic acid, and polyhydroxyoctanoates, have been used and tested for such a purpose. It is expected that, with the gradual degradation of polymeric materials, implanted cells within the materials can proliferate and produce extracellular matrix, self-generating a functional arterial substitute. While experimental results are promising, there are potential problems. First, implanted cells may not be able to generate an arterial substitute with the structural and mechanical properties of natural arteries. With the degradation of the polymeric materials, the arterial substitutes may exhibit reduced mechanical strength, inducing arterial aneurysm and rupture. Second, it is difficult to retain implanted endothelial cells. Third, polymeric materials, once subject to blood, induce blood coagulation and thrombosis, resulting in intimal hyperplasia and arterial restenosis. It is necessary to resolve these problems before biodegradable polymeric materials can be used for arterial reconstruction.

Construction of Arterial Substitutes with Extracellular Matrix [15.17]. Extracellular matrix components serve as substrate for cell attachment and growth, and determine the strength and elasticity of blood vessels. Thus, these components have long been considered materials for arterial reconstruction. Allogenic extracellular matrix is readily available and is often used for such a purpose. Because allogenic cells induce immune rejection reactions, matrix tissues are usually decellularized before being applied to arterial reconstruction. The extracellular matrix components exhibit significantly lower immunogenicity compared to cellular components. The implantation of matrix-based vascular constructs does not cause severe acute immune rejection responses.

Several types of extracellular matrix have been used for the construction of arterial substitutes. These include allogenic arterial matrix scaffolds, intestinal submucosa, and dermal collagen matrix. These are mostly collagen-based connective tissues characterized by natural matrix structure and suitable mechanical properties such as stiffness, elasticity, and strength. Experimental investigations have shown that collagen matrix-based arterial substitutes can serve as substrates for the attachment and migration of vascular endothelial and smooth muscle cells. These substitutes can eventually integrate into the host arteries, forming naturalized arteries. However, collagen matrix components stimulate blood cell adhesion, coagulation, thrombogenesis, smooth muscle cell proliferation and migration, and intimal hyperplasia. These are pathological changes that cause substitute restenosis and failure.

Elastin-containing structures, including elastic fibers and laminae, are major extracellular matrix constituents. These structures contribute to the elasticity, stiffness, and strength of blood vessels. In particular, large elastic arteries contain multiple elastic laminae, which are essential for the maintenance of the stability of arteries. Furthermore, elastin-containing matrix exerts an inhibitory effect on inflammatory reactions, leukocyte adhesion and activation, thrombosis, smooth muscle cell proliferation, and intimal hyperplasia. In vitro studies have demonstrated that arterial elastic lamina specimens induce significantly lower monocyte adhesion and activation compared to arterial adventitial collagen matrix. In vivo studies in animal models have shown that arterial grafts with an elastic lamina-based blood-contacting surface exhibit significantly lower thrombosis and intimal hyperplasia compared to arterial grafts with a collagen matrix-based blood contacting surface (Fig. 15.14). These inhibitory features render elastic laminae a potential material for constructing the blood contacting surface of arterial substitutes.

The inhibitory role of arterial elastic laminae is mediated by a cell signaling pathway, which involves the inhibitory receptor signal-regulatory protein α (SIRPα) and SH2 domain-containing protein tyrosine phosphatase 1 (SHP1). The interaction of monocytes with elastic laminae induces autophosphorylation of SIRPα and recruitment of SHP1 to the cytoplasmic domain of SIRPα. This process activates SHP1, which in turn dephosphorylates several potential protein tyrosine kinases, including receptor tyrosine kinases, the Src family protein tyrosine kinases, phosphatidylinositol 3-kinase, and the Janus family tyrosine kinases. The dephosphorylation of these protein tyrosine kinases results in the suppression of inflammatory and mitogenic responses. These preliminary studies have provided promising results for the use of elastic laminae as a blood-contacting material for arterial reconstruction.

Construction of Arterial Substitutes with Decellularized Arterial Matrix Scaffolds Seeded with Endothelial Progenitor Cells [15.17] Bone marrow endothelial progenitor cells can be mobilized to the blood and are present in the blood at a basal level under physiological conditions. These cells can be isolated from blood samples based on their adherent properties and expression of endothelial progenitor markers. Isolated endothelial progenitor cells can be expanded by cell culture *in vitro*, and used to seed decellularized arterial matrix scaffolds, which can be used for arterial reconstruction. As shown in an experimental investigation, while the matrix scaffold-based arterial substitutes without endothelial progenitor cell seeding failed within 15 days, the arterial substitutes with endothelial progenitor cell seeding remained patent for 130 days. Furthermore, the arterial substitutes with endothelial progenitor cell seeding exhibited contractile function compared to the control arterial substitutes without endothelial progenitor cell seeding. These observations suggest that bone marrow-derived endothelial progenitor cells can be potentially used for improving the performance of reconstructed arteries.

Construction of Arterial Substitutes in vivo [15.18]. Cells and tissues are capable of initiating inflammatory reactions and producing extracellular matrix in response to the

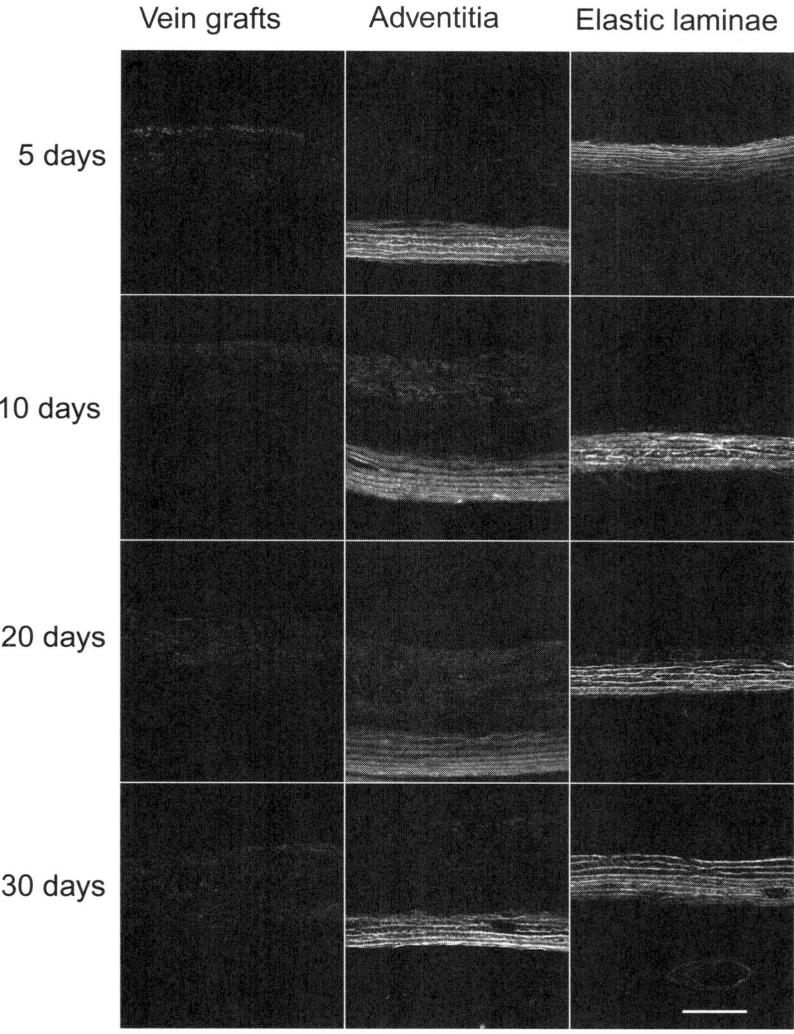

Figure 15.14. Matrix-based arterial reconstruction with an elastic lamina blood-contacting surface. Fluorescent micrographs showing anti-smooth muscle (SM) α actin antibody-labeled SM cells in matrix-based rat aortic constructs. Note that the neointimal layer on the elastic lamina blood-contacting surface is apparently thinner than that on matrix-based constructs with a collagen surface and vein grafts. Arrow: blood-contacting surface. Scale: 100 µm.

stimulation of foreign implants of various materials, including polymers, metals, and allogenic and xenogeneic biological tissues. Such features can be used to generate arterial substitutes in vivo. When a cylindrical tube or rod is implanted into a soft connective tissue (e.g., subcutaneous tissue), muscular tissue, or the abdominal cavity, the implant can stimulate inflammatory reactions, resulting in the formation of a layer of encapsulating tissue, containing various types of cells, such as macrophages, fibroblasts, and epithelial cells, and extracellular matrix composed of collagen fibers and proteoglycans. The encapsulating tissue can be removed from the cylindrical implant and used as an arterial substitute. With a sufficient time of implantation and matrix accumulation in vivo, the arterial substitute can gain mechanical strength and withstand arterial blood pressure. When used for arterial reconstruction in animal models, the encapsulating tissue can be integrated into the host artery and arterialized within a short period. A major advantage of this type of arterial substitute is that the substitute is generated in the host tissue and does not induce immune rejection reactions. However, as other types of matrix-based arterial substitute, the encapsulating tissue induces inflammatory reactions, thrombogenesis, and intimal hyperplasia when anastomosed into an artery.

Construction of Arterial Substitutes with Vascular Cells and Matrix Components in vitro [15.19]. Researchers have intended to construct arterial substitutes in vitro by integrating vascular progenitor cells into scaffolds constituted with extracellular components such as collagen and fibrin. These matrix components can be polymerized in a test tube and used as substrates for cell seeding and culture. Cell-integrated matrix scaffolds can be used to construct arterial substitutes. Experimental investigations in cell culture models have demonstrated promising results for the integration of vascular cells into matrix scaffolds based on collagen and fibrin. However, it is difficult to construct a matrix scaffold that is sufficiently strong and can withstand arterial blood pressure.

Modulation of the Structure and Function of Arterial Substitutes [15.20]. In addition to the selection of materials, there are other tasks for arterial reconstruction. These include the control of cell growth and the modulation of the structure and function of arterial substitutes. These tasks can be achieved by using two general approaches: biological and mechanical manipulations. Biologically, cell growth regulators can be applied locally to the arterial substitute by using polymer-mediated or osmotic pump-mediated delivery. Biological substances or molecules can be integrated into the polymer material of an arterial substitute or integrated into a polymer gel that is placed to the exterior of the arterial substitute. The integrated biological substances and molecules can reach cells via diffusion. Alternatively, biological substances and molecules can be delivered by using an osmotic pump, which can be implanted near the arterial substitute. An osmotic pump is a double-layered sac system. The exterior layer of the sac is made of a semipermeable polymer and can absorb water. The interior bag is made of a watertight polymer and contains an aqueous solution of a selected substance or molecule. When implanted, the exterior layer of the osmotic pump absorbs water and squeezes the interior sac, resulting in controlled release the interior content. The rate of substance release can be regulated by controlling the water absorption rate of the exterior sac.

The proliferation of vascular cells and the rate of matrix production are regulated by mechanical factors such as stretching tensile stress in the arterial wall and fluid shear stress at the endothelial surface. For instance, an increase in mechanical stretch, compared to the physiological level, stimulates cell proliferation and matrix production, whereas an increase in fluid shear stress exerts an opposite effect. Three factors usually influence the level of mechanical stretch: blood pressure, vessel lumen diameter, and vessel wall thickness. An increase in blood pressure and lumen diameter induces an elevation in mechanical stretch, whereas an increase in the wall thickness elicits an opposite effect. Thus, the level of mechanical stretch in the wall of an arterial substitute can be manipulated by engineering the vessel lumen diameter and wall thickness. For instance, experimental vein grafts are associated with a significantly increased tensile stress in the wall of the vein grafts deue to exposure to arterial blood pressure. Such a stretching effect induces detachment of endothelial cells (Fig. 15.15) and disruption of smooth muscle cells (Fig. 15.16)

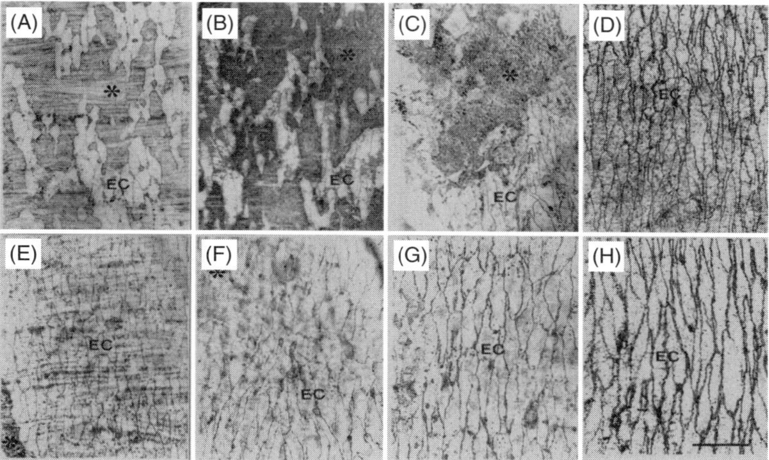

Figure 15.15. Influence of mechanical stretch on the integrity of endothelial cells in experimental vein grafts in the rat. (A–D) Endothelial cells labeled with silver nitrate in experimental vein grafts with increased mechanical stretching stress due to exposure to arterial blood pressure at 1, 5, 10, and 20 days, respectively. Note the detachment of endothelial cells at day 1 after the grafting surgery. The luminal surface of the vein graft was partially covered with regenerated endothelial cells within 10 days. (E–H) Prevention of endothelial detachment in engineered experimental vein grafts with significantly reduced mechanical stretching stress, which was achieved by reinforcing the vein grafts by using an external supporting sheath, at 1, 5, 10, and 20 days, respectively. Scale: 10 μm (Reprinted from Liu SQ et al., J. Biomech. Eng. 122:31–38, 2000 by permission).

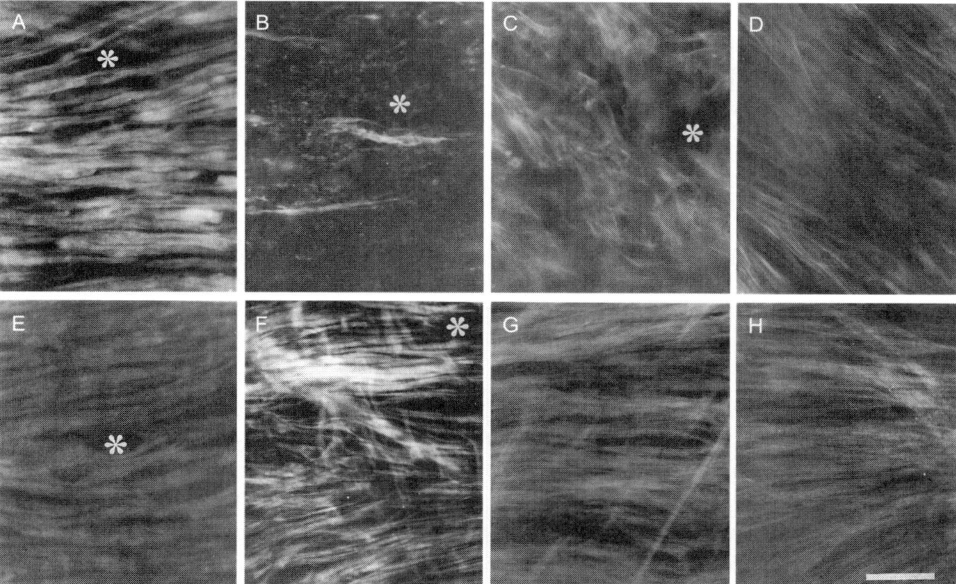

Figure 15.16. Influence of mechanical stretch on the integrity of smooth muscle cells in experimental vein grafts in the rat. (A–D) Smooth muscle cells labeled with anti-smooth muscle α-actin antibody in experimental vein grafts with increased mechanical stretching stress due to exposure to arterial blood pressure at 1, 5, 10, and 20 days, respectively. Note the disruption of the smooth muscle cells in the vein grafts. (E–H) Prevention of smooth muscle cell disruption in engineered experimental vein grafts with significantly reduced mechanical stretching stress which was achieved by reinforcing the vein grafts by using an external supporting sheath at 1, 5, 10, and 20 days, respectively. Scale: 10 μm (Reprinted from Liu SQ et al., J. Biomech. Eng. 122:31–38, 2000 by permission).

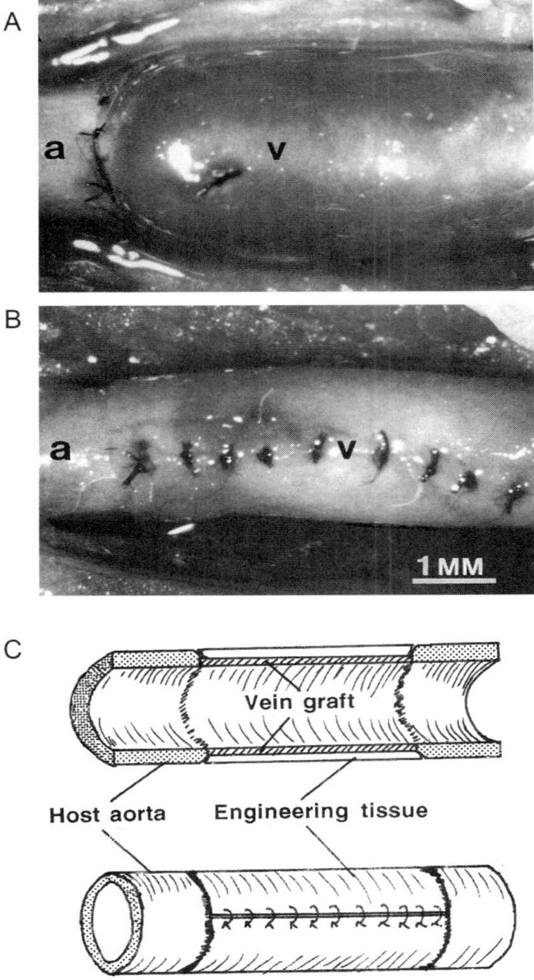

Figure 15.17. Vein graft engineering for the reduction of mechanical stretch stress in the wall as well as the elimination of vortex bloodflow in the lumen of vein grafts by reinforcing the vein grafts with an external supporting sheath, which was constructed by using the submucosal layer of the rat small intestine.

during the early period. These early changes are followed by profound smooth muscle proliferation and wall hypertrophy. One approach for mitigating these pathological changes is to eliminate the stretching tensile stress by reducing the lumen diameter and increasing the wall thickness of the vein grafts. Such a goal can be achieved by reinforcing the vein graft with a tube-shaped sheath (Fig. 15.17). The sheath can be constructed with a polymeric material or extracellular matrix scaffold. Such an approach has been shown to effectively reduce the detachment of endothelial cells (Fig. 15.15) and the injury of smooth muscle cells (Fig. 15.16), and control the rate of vascular hypertrophy.

Fluid shear stress at the endothelial surface is influenced by the rate and pattern of bloodflow. An increase in the bloodflow rate results in an elevation in fluid shear stress. Laminar bloodflow in a straight blood vessel is associated with spatially constant shear stress (note that shear stress is not constant with respect to time), whereas vortex bloodflow

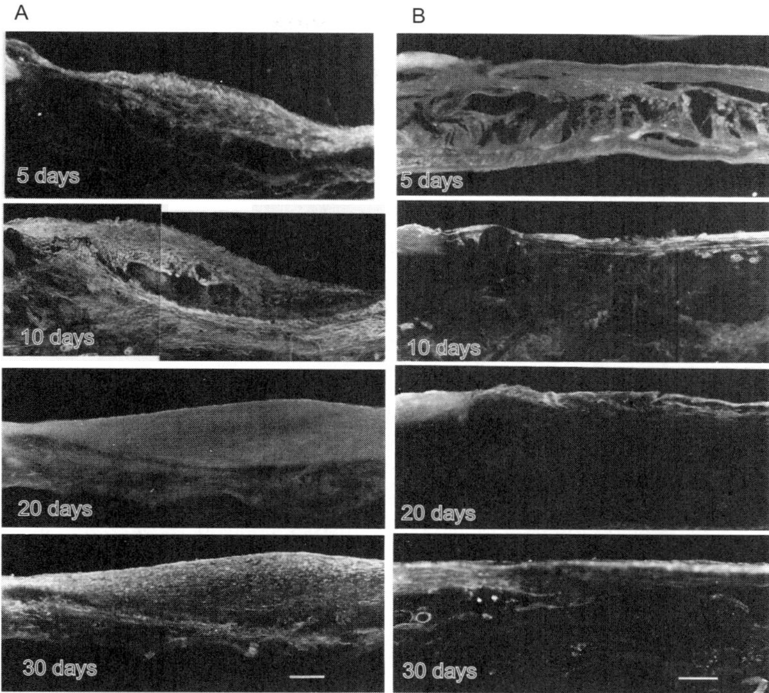

Figure 15.18. Formation of neointima in the vortex bloodflow region in experimental vein grafts and prevention of neointimal formation by eliminating vortex bloodflow with an engineering method as shown in Fig. 15.18. (A) Neointimal formation in experimental vein grafts with vortex bloodflow at 10, 20, and 30 days after the grafting surgery. (B) Prevention of neointimal formation in engineered experimental vein grafts without vortex bloodflow. Scale: 100 μm (Reprinted from Liu SQ. Arterioscl. Thromb. Vasc. Biol. 19:2630–2639, 1999 by permission).

is associated with spatial shear gradients, stagnation points (with zero-shear stress), and reduced fluid shear stress. Since zero shear stress and reduced shear stress, compared to the physiological level, enhance cell proliferation, vortex bloodflow contributes to the initiation of thrombogenesis and intimal hyperplasia. In the fluid dynamic perspective, vortex bloodflow forms due to adverse blood pressure gradients at divergent arterial bifurcations or at locations of divergent geometry of arterial substitutes. A typical example of divergent flow is the bloodflow from a smaller tube to a larger tube (Chapter 15 opening figure). Such a case often exists in arterial substitutes because of the diameter mismatch of the substitute with the host artery. The removal of divergent geometry is an ideal approach for eliminating vortex bloodflow, thus preventing shear stress-induced thrombosis and intimal hyperplasia. A suitable method for removing divergent geometry is to match the diameter of the substitute to that of the host artery by sheathing the substitute into an exterior tube with a desired diameter (Fig. 15.17). Such an approach has been shown to effectively control the pattern of bloodflow and reduce thrombogenesis and intimal hyperplasia in experimental arterial reconstruction (Fig. 15.18).

Hypertension

Pathogenesis, Pathology, and Clinical Features [15.21]. *Hypertension* is a vascular disorder defined as an increase in arterial blood pressure: diastolic pressure >85 mm Hg and/or systolic pressure >140 mm Hg. Hypertension is a common vascular disorder, which affects more than 15% of the human adult population. There exist several types of hypertension. These include essential hypertension, renal hypertension, and endocrine hypertension.

Essential Hypertension. *Essential hypertension* is a form of hypertension without known causes and is the primary form of hypertension in humans. It is detected primarily in the population over the age of 40–50. Several factors have been thought to contribute to the development of essential hypertension. These include genetic factors, environmental factors, and activation of the renin-angiotensin system. Genetic alterations have long been considered factors that contribute to the pathogenesis of hypertension. Although no hypertension-specific genes have been found, the regulatory disorder of multiple genes, which encode blood pressure-controlling proteins, such as renin, angiotensinogen, and angiotensin II receptor, may contribute to the development of hypertension. In particular, the disorder of the angiotensinogen gene expression has been proven a critical factor contributing to the initiation and development of hypertension. Angiotensinogen is synthesized in hepatocytes and vascular endothelial cells, released into the blood, and cleaved by renin to generate angiotensin I, which is further cleaved by angiotensin-converting enzyme into angiotensin II. Angiotensin II acts on vascular smooth muscle cells, activates specific G-protein signaling pathways, induces the release of calcium, and stimulates smooth muscle cell contraction, resulting in arterial hypertension.

The role of altered angiotensinogen gene in the pathogenesis of hypertension has been demonstrated by several lines of evidence: (1) hypertensive sibling and relatives often exhibit an increase in the level of serum angiotensinogen, (2) hypertension is associated with upregulation of angiotensinogen gene expression and mutation of the angiotensinogen gene, and (3) in transgenic animal models, the level of the angiotensinogen gene expression, which is controlled by manipulating the angiotensinogen gene, is proportional to the level of arterial blood pressure. These observations strongly support the role of angiotensinogen and its derivatives in the pathogenesis of hypertension.

In addition to the genetic influence, environmental factors, such as salt intake, occupation, smoking, and mental stress have been implicated in the development of hypertension. The most important factor is probably salt intake. Related research has suggested that excessive salt loading may lead to an increase in circulating natriuretic factors, which promote intracellular calcium accumulation and therefore enhance the contractility of vascular smooth muscle cells, resulting arterial constriction and hypertension. Persistent influence of mental stress has been suggested to contribute to the pathogenesis of hypertension. However, the mechanisms by which environmental factors influence arterial blood pressure remain poorly understood.

Renovascular Hypertension. *Renovascular hypertension* is a form of hypertension induced by renal arterial stenosis. Renal arterial stenosis is often a result of congenital defects, scar formation after trauma, tumor compression, or renal arterial atherosclerosis. The hemodynamic consequence of renal arterial stenosis is a decrease in renal arterial blood pressure and flow. The renal arteries contain a specially differentiated type of smooth muscle cells, which produce and secret renin. The rate of renin production and secretion is regulated by the renal arterial bloodflow and blood pressure. A decrease in the renal arterial blood flow and pressure due to arterial stenosis activates the renin-producing cells, resulting in an increase in renin production and secretion. The consequence of these activities is an increase in the blood level of angiotensin I/II, which stimulate smooth muscle contraction and induce an increase in arterial blood pressure. Angiorensin also stimulates the expression of mitogenic factors and promotes smooth muscle cell proliferation and migration. Furthermore, increased blood pressure enhances mechanical stretch of the arterial wall, stimulating the expression of mitogenic factors, such as platelet-derived growth factor and vascular endothelial growth factor. These growth factors in turn stimulate the proliferation of vascular smooth muscle cells (Fig. 15.19). The consequence of cell proliferation is vascular hypertrophy (Fig. 15.20). An effective treatment for renal arterial hypertension is to surgically remove the cause of renal arterial stenosis. Once stenosis is removed, the level of arterial blood pressure returns to the physiological level.

Endocrine Hypertension. Endocrine hypertension is a form of hypertension due to the disorder of the endocrine system. There are several types of endocrine hypertension, including glucocorticoid-remediable aldosteronism (GRA) and syndrome of apparent

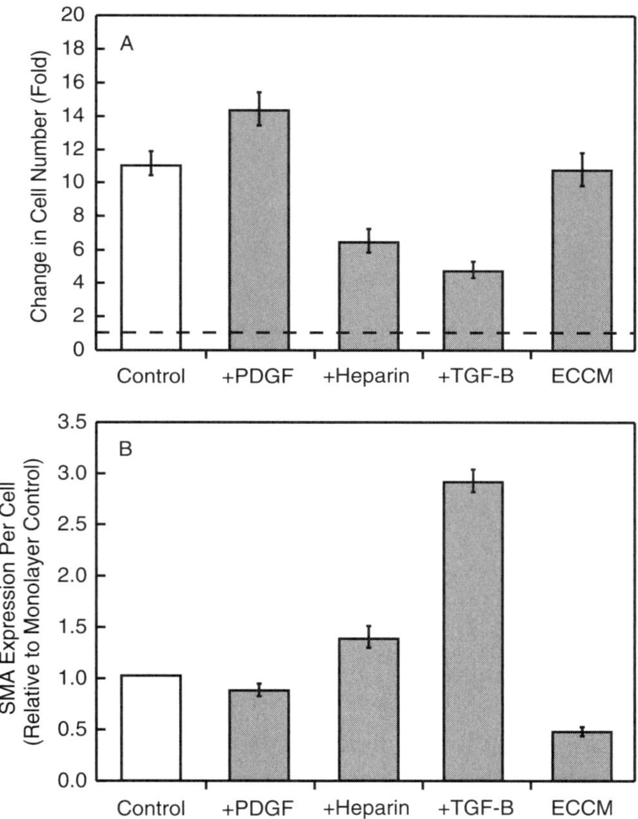

Figure 15.19. Cell proliferation (A) and smooth muscle (SM) actin expression (B) in two-dimensional (monolayer) culture in response to exogenous biochemical stimulation. RASMC were cultured on tissue culture polystyrene for 6 days in control medium and in the presence of platelet-derived growth factor BB (PDGF), heparin transforming growth factor β1 (TGFβ), or endothelial cell-conditioned medium (ECCM). Cells were then removed for cell counting and flow cytometric analysis of SM actin expression ($n = 6$, ± SEM). (Reprinted from Stegemann JP, Nerem RM: *Exp Cell Res* 283:146–55, copyright 2003, with permission from Elsevier.)

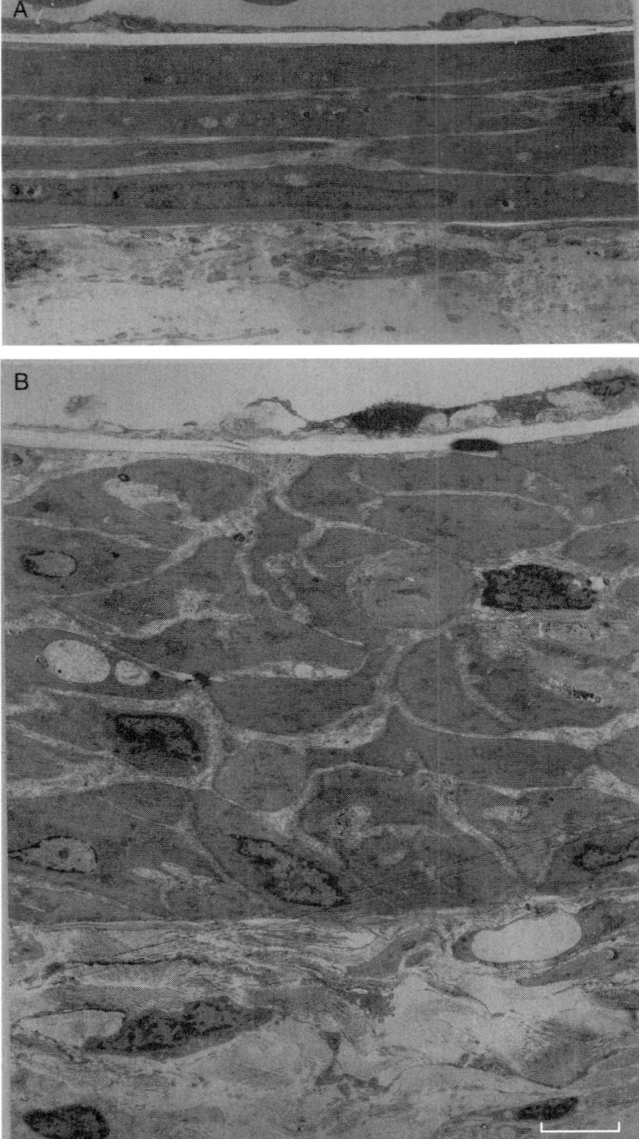

Figure 15.20. Induction of arterial hypertrophy in renovascular hypertension. (A) Electron micrograph of an ileal artery derived from a control rat; (B) Electron micrograph of an ileal artery from a hypertensive rat. Note the change in thickness of the arterial wall and the density of the medial smooth muscle cells. Scale: 5 μm.

mineralocorticoid excess (AME). *Glucocorticoid-remediable aldosteronism* is an autosomal dominant genetic disorder (see page 423 for forms of genetic disorders). This disorder is characterized by an increase in the level of aldosterone, a steroid hormone secreted by the adrenal cortex. Under physiological conditions, the production and secretion of aldosterone is regulated by angiotensin II. In glucocorticoid-remediable aldosteronism, a mechanism involving the adrenocorticotropic hormone (ACTH) is activated. This hormone directly stimulates the production and secretion of aldosterone. Aldosterone can interact with the mineralocorticoid receptor in the renal tubular cells and stimulate sodium and water reabsorption. The consequence of aldosterone activation is an increase in water retention, which leads to expansion of blood volume, enhancement of cardiac activity, and hypertension. This is a major form of endocrine hypertension.

ACTH-induced glucocorticoid-remediable aldosteronism is probably to the disorder of the aldosterone synthase gene and the steroid 11-β-hydroxylase gene. These genes encode proteins that regulate the synthesis of aldosterone. In glucocorticoid-remediable aldosteronism, the aldosterone synthase and steroid 11-β-hydroxylase genes undergo a mutation process, resulting in the formation of new chimeric genes. In the chimeric genes, the 5 prime regulatory sequences from the steroid 11-β-hydroxylase gene can be fused to the aldosterone synthase gene. With the newly integrated 5 prime regulatory sequences, the chimeric aldosterone synthase gene is now subject to the control of ACTH (note that adrenocorticotropic hormone does not significantly influence the expression of the aldosterone synthase gene under physiological conditions without gene mutation). The synthesis of aldosterone synthase is significantly enhanced in response to the stimulation of ACTH, thus resulting in water retention and hypertension. It is important to note that, in glucocorticoid-remediable aldosteronism, the rennin–angiotensin II system is not activated and does not exert significantly influence on the production and secretion of aldosterone. Indeed, renin secretion may be suppressed due to water retention and bloodflow elevation.

Syndrome of apparent mineralocorticoid excess is an autosomal recessive genetic disorder (see page 423 for the definition of genetic disorders). This disorder is induced by the mutation of the 11-β-hydroxysteroid dehydrogenase gene. Under physiological conditions, 11-β-hydroxysteroid dehydrogenase can convert the glucocorticoid hormone cortisol (11-β, 17-α, 21-trihydroxy-4-pregnene-3, 20-dione) to cortisone (17-α, 21-dihydroxy-4-pregnene-3, 11, 20-trione). For the two hormones, cortisol can activate the renal mineralocorticoid receptor and induce renal water retention and thus hypertension, whereas cortisone does not possess such a function. In the presence of a physiological level of 11-β-hydroxysteroid dehydrogenase, most cortisol is converted to cortisone, so that the renal mineralocorticoid receptor is not activated. However, in the absence of 11-β-hydroxysteroid dehydrogenase due to gene mutation, cortisol is accumulated in the blood, resulting in the activation of the renal mineralocorticoid receptor and induction of renal water retention. The ratio of cortisol to cortisone can be tested and used for the diagnosis of syndrome of apparent mineralocorticoid.

There are several common pathological changes in the arterial system in all types of hypertension. These changes include an increase in the density of smooth muscle cells and fibroblasts, an increase in the density of collagen fibers, and thickening of the arterial wall. These changes result from arterial adaptation to hypertension. An increase in the arterial blood pressure enhances mechanical stretch in the wall of arteries. Increased stretch stimulates arterial cells to secret growth factors, which in turn promote cell proliferation and production of extracellular matrix components, resulting in vascular hypertrophy.

Animal Models of Renovascular Hypertension [15.22]. Renovascular hypertension is a form of hypertension caused by a reduction in the renal blood pressure and flow (see section for mechanisms). This type of hypertension can be induced by physical constriction of the renal arteries or the abdominal aorta at a location immediately above the renal arteries (Fig. 15.21). Arterial constriction can be introduced by clamping a selected artery

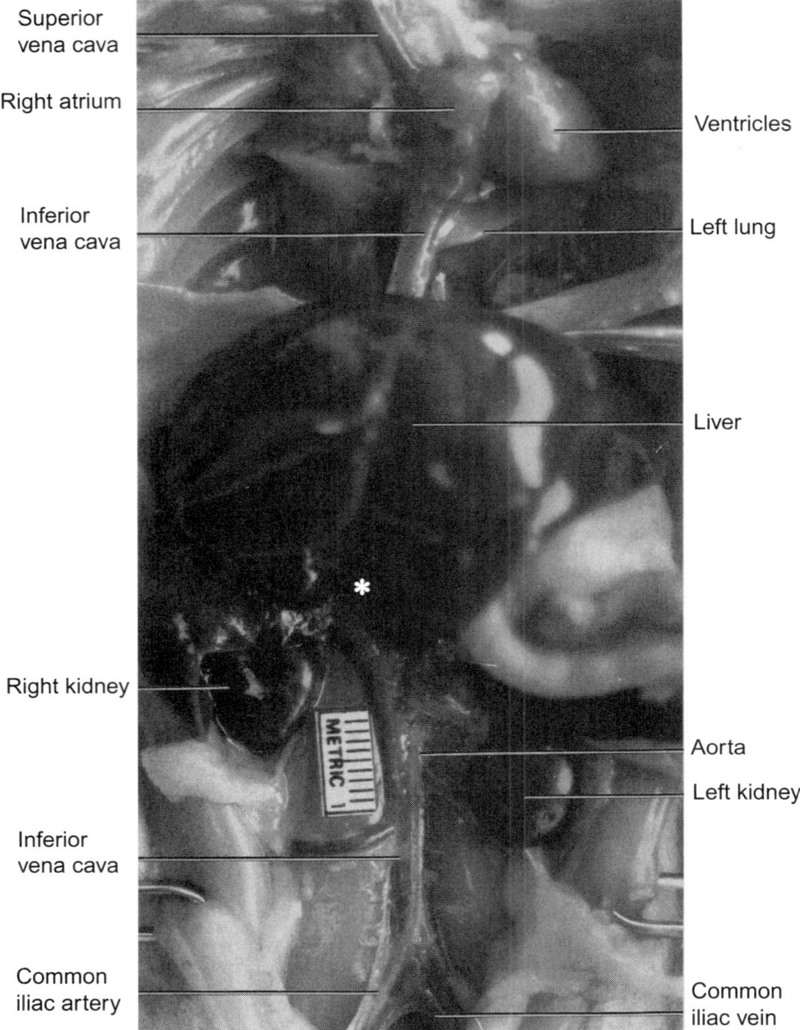

Figure 15.21. Photograph showing the anatomy of the thoracic and abdominal organs of a rat.

with a metal band (Fig. 15.22). A reduction of ~90% of the original arterial lumen area can induce a significant decrease in the renal arterial blood pressure and flow and induces noticeable hypertension within 5–10 days (Fig. 15.23). The cross-sectional area of the constricted artery can be measured after surgery by using a histological method. The measured lumen area should be used to calibrate the constriction level of the metal band.

Another method to introduce arterial constriction is to ligate a selected artery together with a polymer tubing placed in parallel with the artery. Following the ligation with a surgical suture, the polymer tube can be pulled out, leaving the artery narrowed to a dimension comparable to the diameter of the polymer tubing. The degree of arterial constriction can be controlled by selecting a polymer tubing with a desired diameter. This

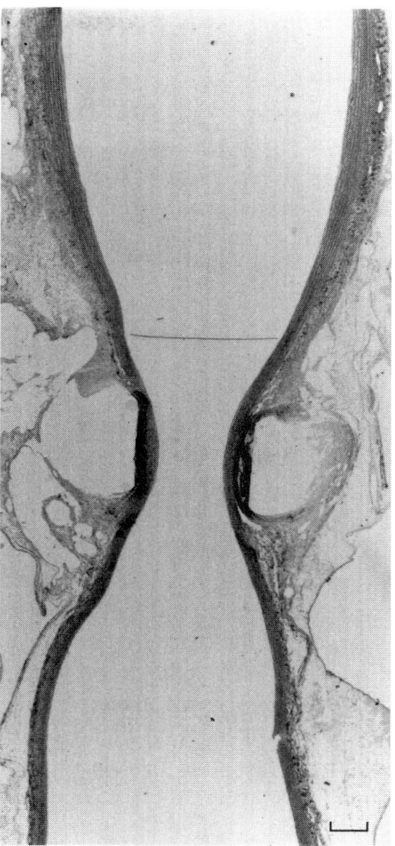

Figure 15.22. Histological micrograph from an axial section of the rat abdominal aorta showing aortic constriction for the induction of experimental hypertension. Scale: 100 μm.

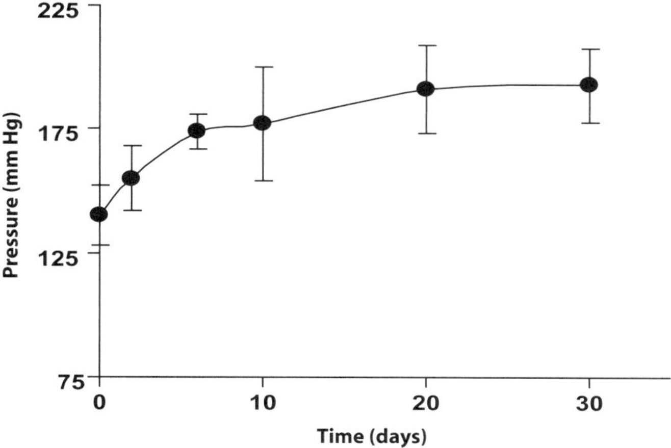

Figure 15.23. Changes in arterial blood pressure measured from the carotid artery in experimental arterial hypertension in the rat.

method is superior to that with a metal band described above in terms of controlling the accuracy of arterial constriction.

Conventional Treatment of Hypertension [15.23]. The principle of hypertension treatment is to remove factors that cause hypertension and to reduce arterial blood pressure. As discussed on page 700 of this chapter, renovascular hypertension is caused by renal arterial stenosis. Surgical correction of stenosed renal arteries is usually an effective treatment. Endocrine hypertension may be caused by adrenal gland tumors, which induce excessive secretion of aldosterone (in the case of a cortical tumor) or epinephrine and norepinephrine (in the case of a medulla tumor). Surgical removal of the tumor usually results in a decrease in arterial blood pressure. For essential hypertension, since the causes are poorly understood, there are no effective approaches for the removal of the pathogenic factors.

An increase in arterial blood pressure usually stimulates mitogenic responses of vascular endothelial and smooth muscle cells and enhances atherogenesis. Thus, excessive hypertension should be controlled appropriately. There are two approaches that can be used to reduce arterial blood pressure: diuresis and vasodilation. Diuresis can be induced by agents known as diuretics, which enhance the excretion of water in the kidney. Typical diuretic agents are thiazide derivatives. These agents inhibit the reabsorption of sodium and potassium in the proximal renal tubules and stimulate the secretion of chloride, enhancing water excretion and reducing arterial blood pressure.

Vasodilators are agents that induce the relaxation of vascular smooth muscle cells and thus the reduction of blood pressure. There are several types of vasodilators: blockers of the α adrenergic receptor, suppressors of sympathetic vasomotor centers, blockers of the angiotensin II type 1 receptor, and blockers of calcium channels. The α-adrenergic receptor mediates norepinephrine-induced smooth muscle cell contraction and its activation induces an elevation in arterial blood pressure. The inhibition of the α adrenergic receptor can effectively reduce arterial blood pressure. Typical agents of α adrenergic receptor blockers include phentolamine and phenoxybenzamine. The sympathetic vasomotor centers control the basal tone of small arteries. An increase in the basal tone is a critical factor that contributes to hypertension. Several agents, such as clonidine and methyldopa, can be used to suppress the activity of the sympathetic vasomotor centers and reduce arterial blood pressure. The angiotensin II type 1 receptor mediates angiotensin II-induced contraction of smooth muscle cells. The suppression of the activity of this receptor reduces arterial blood pressure. A typical agent for such a purpose is losartan. Calcium is necessary for vascular smooth muscle contraction. The blockade of calcium channels leads to a reduction in the contractility of smooth muscle cells. Commonly used calcium blockers include nicardipine, nifedipine, nimodipine, and verapamil.

Molecular Engineering [15.24]. While hypertension is commonly treated with agents as described above, these agents are short-lived and often elicit side effects. Since the pathogenesis of hypertension, especially, essential hypertension, is related to the disorder of vascular control genes, the modulation of these genes provides a means for the treatment of hypertension. Strategies for molecular manipulation of hypertension are to enhance the expression of vasodilator genes and suppress the expression of vasoconstrictor genes. There are several basic approaches for achieving the therapeutic goals, including application of full-length DNA and antisense oligonucleotides. These genetic materials can be directly delivered into the blood. Potential genes and oligodeoxynucleotides for these purposes are described as follows.

Nitric Oxide Synthase Gene [15.25]. Nitric oxide synthase catalyzes the formation of nitric oxide, a potent smooth muscle cell relaxant (see page 689). The transfer of the nitric oxide synthase gene enhances the expression of nitric oxide synthase and the production of nitric oxide. This gene has been shown to reduce arterial blood pressure (Fig. 15.24).

Atrial Natriuretic Peptide Gene [15.26]. Atrial natriuretic peptides are a group of hormones generated by atrial muscle cells. The functions of these hormones are to relax vascular smooth muscle cells and stimulate sodium and water excretion in the kidney, resulting in a reduction in arterial blood pressure. The overexpression of the atrial natriuretic peptide gene by gene transfer enhances the production of these peptides. Experimental investigations have demonstrated that the transfer of the atrial natriuretic peptide gene into animal hypertension model can effectively reduce the arterial blood pressure.

Characteristics of vasodilating molecules are presented in Table 15.6.

Kallikrein Genes [15.27]. Kallikreins are serine proteases present in the blood plasma and are capable of cleaving a group of plasma proteins, known as *kininogens*, to produce kinins, including bradykinin and lysyl-bradykinin. Bradykinin and lysyl-bradykinin are potent vasodilators and can induce smooth muscle cell relaxation, thus reducing the arterial blood pressure. Experimental investigations have provided promising results for the effectiveness of kallikrein gene delivery for the treatment of experimental hypertension.

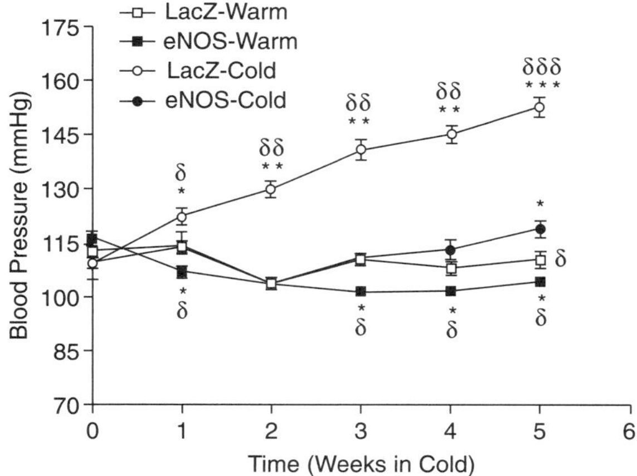

Figure 15.24. Effects of eNOS gene delivery on cold-induced elevation of blood pressure (values are means ± SE; $n = 6$ rats). Intravenous injections of rAdv.heNOS and rAdv.LacZ were carried out immediately before exposure to cold. Four groups included LacZ-cold, rats treated with rAdv. LacZ and exposed to cold; eNOS-cold rats, treated with rAdv.heNOS and exposed to cold; eNOS-warm, rats treated with rAdv.heNOS and kept at room temperature; and LacZ-warm, rats treated with rAdv.LacZ and kept at room temperature. $*P < 0.05$; $**P < 0.01$; $***P < 0.001$ compared with LacZ-warm group; $dP < 0.05$; $ddP < 0.01$; $dddP < 0.001$ compared with eNOS-cold group. (Reprinted from Wang X, Cade R, Sun Z: Human eNOS gene delivery attenuates cold-induced elevation of blood pressure in rats, *Am J Physiol Heart Circ Physiol* 289:H1161–8, 2005, with permission from American Physiological Society.)

TABLE 15.6. Characteristics of Selected Vasodilating Molecules*

Proteins	Alternative Names	Amino Acids	Molecular Weight (kDa)	Expression	Functions
Atrial natriuretic peptide	Natriuretic peptide precursor A (NPPA), atrial natriuretic polypeptide (ANP), prepronatriodilatin, cardionatrin, atrionatriuretic factor (ANF), pronatriodilatin (PND), atriopeptin	151	16	Atrium, prostate gland	Inducing smooth muscle relaxation, stimulating water and sodium excretion, enhancing endothelial regeneration, and regulating the activity of NFκB
Kallikrein	KLK1, renaL/pancreatic/salivary kallikrein, KLKR, kallikrein serine protease 1	262	29	Prostate gland, uterus	One of the 15 kallikrein subfamily members that cleaves kininogens to produce bradykinin and lysylbradykinin
Kininogen	KNG, α2-thiol proteinase inhibitor	427	48	Liver, blood vessels, platelets, kidney, placenta	Precursor of kinins (bradykinin and lysylbradykinin)

*Based on bibliography 15.26.

TABLE 15.7. Characteristics of Selected Adrenergic Receptors*

Proteins	Alternative Names	Amino Acids	Molecular Weight (kDa)	Expression	Functions
Adrenergic receptor α1A	α1A adrenoceptor, ADRA1A, α1C adrenergic receptor, ADRA1C, α1A adrenergic receptor, isoform 2	499	55	Blood vessels, heart, brain, urinary bladder, prostate gland	A G-protein-coupled receptor that interacts with epinephrine and norepinephrine, inducing smooth muscle contraction
Adrenergic receptor α1B	α1 adrenergic receptor, ADRA1, α1B adrenoceptor, ADRA1B	520	57	Liver, blood vessels, kidney, lung and spleen	Function similar to that of α1A-adrenergic receptor
Adrenergic receptor α1D	α1D adrenoceptor, ADRA1D, α1A adrenergic receptor, α-adrenergic receptor 1a	572	60	Blood vessels, prostate gland, heart, hippocampus	Function similar to that of α1A-adrenergic receptor
Adrenergic receptor β1	ADRB1R, B1AR, ADRB1	477	51	Heart, blood vessels, adipocytes	A G-protein-coupled receptor that interacts with epinephrine and norepinephrine, inducing smooth muscle relaxation and an increase in cardiac activity, regulating diet-induced thermogenesis, preventing obesity, and inducing cardiaomyocyte apoptosis
Adrenergic receptor β2	B2AR, ADRB2, ADRB2R	413	47	Heart, lymphocytes, lung, thyroid gland	Function similar to that of adrenergic receptor β1, except that this receptor prevents cardiomyocyte apoptosis

*Based on bibliography 15.30.

Antisense Oligonucleotides for Angiotensinogen mRNA [15.28]. As described on page 700 of this chapter, angiotensinogen is a precursor for angiotensin I and can be converted to angiotensin I under the action of renin, which is produced by specialized renal arterial smooth muscle cells. Angiotensin I can be converted to angiotensin II by angiotensin-converting enzyme (ACE). Angiotensin II stimulates smooth muscle contraction and induces an increase in arterial blood pressure. Suppression of the expression of angiotensinogen is an effective approach for reducing the level of angiotensin II. The application of anti-sense oligonucleotides for angiotensinogen mRNA has been shown to reduce the translation of angiotensinogen, resulting in a decrease in arterial blood pressure.

Antisense Oligonucleotides for Angiotensin II type 1 (AT_1) Receptor mRNA [15.29]. Angiotensin II type 1 (AT_1) receptor interacts with angiotensin II and mediates angiotensin II-induced smooth muscle cell contraction, resulting in an increase in the arterial blood pressure (see page 700 of this chapter for mechanisms). The hybridization of antisense oligonucleotides to the angiotensin II type 1 receptor mRNA reduces the translation of angiotensin II type 1 receptor and thus reduces the effect of angiotensin II. Since angiotensin II is a potent vasoconstrictor, the blockade of the angiotensin II signaling pathway can significantly reduce the arterial blood pressure.

Antisense Oligonucleotides for Adrenergic Receptor mRNA [15.30]. The β_1-adrenergic receptor interacts with epinephrine and enhances the contractility of cardiomyocytes, resulting in an increase in the arterial blood pressure. A decrease in the activity of the β_1-adrenergic receptor can reduce the arterial blood pressure. The hybridization of antisense oligodeoxynucleotides to the β_1-adrenergic receptor mRNA can reduce the translation of the β_1-adrenergic receptor. This is a potential approach for the treatment of hypertension.

Characteristics of adrenergic receptors are listed in Table 15.7.

BIBLIOGRAPHY

15.1. Anatomy and Physiology

Guyton AL, Hall JE: *Textbook of Medical Physiology*, 11th ed, Saunders, Philadelphia, 2006.

McArdle WD, Katch KI, Katch VL: *Essentials of Exercise Physiology*, 3rd ed, Lippincott Williams & Wilkins, Baltimore, 2006.

Germann WL, Stanfield CL (with contributors Niles MJ, Cannon JG), *Principles of Human Physiology*, 2nd ed, Pearson Benjamin Cummings, San Francisco, 2005.

Thibodeau GA, Patton KT: *Anatomy & Physiology*, 5th ed, Mosby, St Louis, 2003.

Boron WF, Boulpaep EL: *Medical Physiology: A Cellular and Molecular Approach*, Saunders, Philadelphia, 2003.

Ganong WF: *Review of Medical Physiology*, 21st ed, McGraw-Hill, New York, 2003.

15.2. Regulation of Anti- and Procoagulation Factors

Plasminogen Activator

Bell LD, Smith JC, Derbyshire R, Finlay M, Johnson I et al: Chemical synthesis, cloning and expression in mammalian cells of a gene coding for human tissue-type plasminogen activator, *Gene* 63:155–63, 1988.

Ladenvall P, Nilsson S, Jood K, Rosengren A, Blomstrand C et al: Genetic variation at the human tissue-type plasminogen activator (tPA) locus: haplotypes and analysis of association to plasma levels of tPA, *Eur J Hum Genet* 11:603–10, 2003.

Loscalzo J, Braunwald E: Tissue plasminogen activator, *New Engl J Med* 319:925–31, 1988.

Ludwig M, Wohn KD, Schleuning WD, Olek K: Allelic dimorphism in the human tissue-type plasminogen activator (TPA) gene as a result of an Alu insertion/deletion event, *Hum Genet* 88:388–92, 1992.

Pang PT, Teng HK, Zaitsev E, Woo NT, Sakata K et al: Cleavage of proBDNF by tPA/plasmin is essential for long-term hippocampal plasticity, *Science* 306:487–91, 2004.

Pennica D, Holmes WE, Kohr WJ, Harkins RN, Vehar GA et al: Cloning and expression of human tissue-type plasminogen activator cDNA in E. coli, *Nature* 301:214–21, 1983.

Seeds NW, Basham ME, Haffke SP: Neuronal migration is retarded in mice lacking the tissue plasminogen activator gene, *Proc Natl Acad Sci USA* 96:14118–23, 1999.

Seeds NW, Williams BL, Bickford PC: Tissue plasminogen activator induction in Purkinje neurons after cerebellar motor learning, *Science* 270:1992–4, 1996.

Tishkoff SA, Pakstis AJ, Stoneking M, Kidd JR, Destro-Bisol G et al: Short tandem-repeat polymorphism/Alu haplotype variation at the PLAT locus: Implications for modern human origins, *Am J Hum Genet* 67:901–25, 2000.

Wang X, Lee SR, Arai K, Lee SR, Tsuji K et al: Lipoprotein receptor-mediated induction of matrix metalloproteinase by tissue plasminogen activator, *Nature Med* 9:1313–7, 2003.

Plasminogen

Azuma H, Uno Y, Shigekiyo T, Saito S: Congenital plasminogen deficiency caused by a ser572-to-pro mutation, *Blood* 82:475–80, 1993.

Bugge TH, Flick MJ, Daugherty CC, Degen JL: Plasminogen deficiency causes severe thrombosis but is compatible with development and reproduction, *Genes Dev* 9:794–807, 1995.

Cao Y, Ji RW, Davidson D, Schaller J, Marti D et al: Kringle domains of human angiostatin: Characterization of the anti-proliferative activity on endothelial cells *J Biol Chem* 271:29461–7, 1996.

Cao Y, O'Reilly MS, Marshall B, Flynn E, Ji RW et al: Expression of angiostatin cDNA in a murine fibrosarcoma suppresses primary tumor growth and produces long-term dormancy of metastases, *J Clin Invest* 101:1055–63, 1998.

Drew AF, Kaufman AH, Kombrinck KW, Danton MJ, Daugherty CC et al: Ligneous conjunctivitis in plasminogen-deficient mice, *Blood* 91:1616–24, 1998.

Fischer MB, Roeckl C, Parizek P, Schwarz HP, Aguzzi A: Binding of disease-associated prion protein to plasminogen, *Nature* 408:479–83, 2000.

Ichinose A, Espling ES, Takamatsu J, Saito H, Shinmyozu K et al: Two types of abnormal genes for plasminogen in families with a predisposition for thrombosis, *Proc Natl Acad Sci USA* 88:115–9, 1991.

O'Reilly MS, Holmgren L, Shing Y, Chen C, Rosenthal RA et al: Angiostatin: A novel angiogenesis inhibitor that mediates the suppression of metastases by a Lewis lung carcinoma, *Cell* 79:315–28, 1994.

Petersen TE, Martzen MR, Ichinose A, Davie EW: Characterization of the gene for human plasminogen, a key proenzyme in the fibrinolytic system, *J Biol Chem* 265:6104–11, 1990.

Romer J, Bugge TH, Pyke C, Lund LR, Flick MJ et al: Impaired wound healing in mice with a disrupted plasminogen gene, *Nature Med* 2:287–92, 1995.

Schott D, Dempfle CE, Beck P, Liermann A, Mohr-Pennert A et al: Therapy with a purified plasminogen concentrate in an infant with ligneous conjunctivitis and homozygous plasminogen deficiency, *New Engl J Med* 339:1679–86, 1998.

Swaisgood CM, Schmitt D, Eaton D, Plow EF: In vivo regulation of plasminogen function by plasma carboxypeptidase, *Br J Clin Invest* 110:1275–82, 2002.

Protein C

Allaart CF, Poort SR, Rosendaal FR, Reitsma PH, Bertina RM et al: Increased risk of venous thrombosis in carriers of hereditary protein C deficiency defect, *Lancet* 341:134–8, 1993.

Beckmann RJ, Schmidt RJ, Santerre RF, Plutzky J, Crabtree GR et al: The structure and evolution of a 461 amino acid human protein C precursor and its messenger RNA, based upon the DNA sequence of cloned human liver cDNAs, *Nucleic Acids Res* 13:5233–47, 1985.

Bernard GR, Vincent JL, Laterre PF, LaRosa SP, Dhainaut JF et al: Recombinant Human Activated Protein C Worldwide Evaluation in Severe Sepsis (PROWESS) Study Group Efficacy and safety of recombinant human activated protein C for severe sepsis, *New Engl J Med* 344:699–709, 2001.

Conard J, Horellou MH, van Dreden P, Samama M, Reitsma PH et al: Homozygous protein C deficiency with late onset and recurrent coumarin-induced skin necrosis, *Lancet* 339:743–4, 1992.

Drews R, Paleyanda RK, Lee TK, Chang RR, Rehemtulla A et al: Proteolytic maturation of protein C upon engineering the mouse mammary gland to express furin, *Proc Natl Acad Sci USA* 92:10462–6, 1995.

Dreyfus M, Magny JF, Bridey F, Schwarz HP, Planche C et al: Treatment of homozygous protein C deficiency and neonatal purpura fulminans with a purified protein C concentrate, *New Engl J Med* 325:1565–8, 1991.

Faust SN, Levin M, Harrison OB, Goldin RD, Lockhart MS et al: Dysfunction of endothelial protein C activation in severe meningococcal sepsis, *New Engl J Med* 345:408–16, 2001.

Hasstedt SJ, Bovill EG, Callas PW, Long GL: An unknown genetic defect increases venous thrombosis risk, through interaction with protein C deficiency, *Am J Hum Genet* 63:569–76, 1998.

Lay AJ, Liang Z, Rosen ED, Castellino FJ: Mice with a severe deficiency in protein C display prothrombotic and proinflammatory phenotypes and compromised maternal reproductive capabilities, *J Clin Invest* 115:1552–61, 2005.

Manns BJ, Lee H, Doig CJ, Johnson D, Donaldson C: An economic evaluation of activated protein C treatment for severe sepsis, *New Engl J Med* 347:993–1000, 2002.

Riewald M, Petrovan RJ, Donner A, Mueller BM, Ruf W: Activation of endothelial cell protease activated receptor 1 by the protein C pathway, *Science* 296:1880–2, 2002.

Siegel JP: Assessing the use of activated protein C in the treatment of severe sepsis, *New Engl J Med* 347:1030–4, 2002.

Warren HS, Suffredini AF, Eichacker PQ, Munford RS: Risks and benefits of activated protein C treatment for severe sepsis, *New Engl J Med* 347:1027–30, 2002.

Von Willebrand Factor

Mancuso DJ, Tuley EA, Westfield LA, Worrall NK, Shelton-Inloes BB et al: Structure of the gene for human von Willebrand factor, *J Biol Chem* 264:19514–27, 1989.

Sadler JE, Shelton-Inloes BB, Sorace JM, Harlan JM, Titani K et al: Cloning and characterization of two cDNAs coding for human von Willebrand factor, *Proc Natl Acad Sci USA* 82:6394–8, 1985.

Mancuso DJ, Tuley EA, Westfield LA, Lester-Mancuso TL, Le Beau MM et al: Human von Willebrand factor gene and pseudogene: structural analysis and differentiation by polymerase chain reaction, *Biochemistry* 30:253–69, 1991.

Ginsburg D, Handin RI, Bonthron DT, Donlon TA, Bruns GAP et al: Human von Willebrand factor (vWF): Isolation of complementary DNA (cDNA) clones and chromosomal localization, *Science* 228:1401–6, 1985.

Lynch DC, Zimmerman TS, Collins CJ, Brown M, Morin MJ et al: Molecular cloning of cDNA for human von Willebrand factor: Authentication by a new method, *Cell* 41:49–56, 1985.

Emsley J, Cruz M, Handin R, Liddington R: Crystal structure of the von Willebrand factor A1 domain and implications for the binding of platelet glycoprotein Ib, *J Biol Chem* 273:10396–401, 1998.

Tuley EA, Gaucher C, Jorieux S, Worrall NK, Sadler JE et al: Expression of von Willebrand factor "Normandy": an autosomal mutation that mimics hemophilia A, *Proc Natl Acad Sci USA* 88:6377–81, 1991.

Ribba AS, Voorberg J, Meyer D, Pannekoek H, Pietu G: Characterization of recombinant von Willebrand factor corresponding to mutations in type IIA and type IIB von Willebrand disease, *J Biol Chem* 267:23209–15, 1992.

Tissue Thromboplastin

Bogdanov VY, Balasubramanian V, Hathcock J, Vele O, Lieb M et al: Alternatively spliced human tissue factor: a circulating, soluble, thrombogenic protein, *Nature Med* 9:458–62, 2003.

Carson SD, Henry WM, Shows TB: Tissue factor gene localized to human chromosome 1 (1pter-1p21), *Science* 229:991–3, 1985.

Davie EW, Fujikawa K, Kisiel W: The coagulation cascade: Initiation, maintenance, and regulation, *Biochemistry* 30:10363–70, 1991.

Erlich J, Parry GCN, Fearns C, Muller M, Carmeliet P et al: Tissue factor is required for uterine hemostasis and maintenance of the placental labyrinth during gestation, *Proc Natl Acad Sci USA* 96:8138–43, 1999.

Isermann B, Sood R, Pawlinski R, Zogg M, Kalloway S et al: The thrombomodulin-protein C system is essential for the maintenance of pregnancy, *Nature Med* 9:331–7, 2003.

Pawlinski R, Fernandes A, Kehrle B, Pedersen B, Parry G et al: Tissue factor deficiency causes cardiac fibrosis and left ventricular dysfunction, *Proc Natl Acad Sci USA* 99:15333–8, 2002.

Toomey JR, Kratzer KE, Lasky NM, Broze GJ, Jr: Effect of tissue factor deficiency on mouse and tumor development, *Proc Natl Acad Sci USA* 94:6922–6, 1997.

Prothrombin

Banfield DK, MacGillivray RTA: Partial characterization of vertebrate prothrombin cDNAs: Amplification and sequence analysis of the B chain of thrombin from nine different species, *Proc Natl Acad Sci USA* 89:2779–83, 1992.

Celikel R, McClintock RA, Roberts JR, Mendolicchio GL, Ware J et al: Modulation of alpha-thrombin function by distinct interactions with platelet glycoprotein Ib-alpha, *Science* 301:218–21, 2003.

Degen SJF, Davie EW: Nucleotide sequence of the gene for human prothrombin, *Biochemistry* 26:6165–77, 1987.

De Stefano V, Martinelli I, Mannucci PM, Paciaroni K, Chiusolo P et al: The risk of recurrent deep venous thrombosis among heterozygous carriers of both factor V Leiden and the G20210A prothrombin mutation, *New Engl J Med* 341:801–6, 1999.

Dumas JJ, Kumar R, Seehra J, Somers WS, Mosyak L: Crystal structure of the Gplb-alpha-thrombin complex essential for platelet aggregation, *Science* 301:222–6, 2003.

Gehring NH, Frede U, Neu-Yilik G, Hundsdoerfer P, Vetter B et al: Increased efficiency of mRNA 3-prime end formation: a new genetic mechanism contributing to hereditary thrombophilia, *Nature Genet* 28:389–92, 2001.

Martinelli I, Sacchi E, Landi G, Taioli E, Duca F et al: High risk of cerebral-vein thrombosis in carriers of a prothrombin-gene mutation and in users of oral contraceptives, *New Engl J Med* 338:1793–7, 1998.

Sun WY, Witte DP, Degen JL, Colbert MC, Burkart MC et al: Prothrombin deficiency results in embryonic and neonatal lethality in mice, *Proc Natl Acad Sci USA* 95:7597–602, 1998.

Fibrinogen

Collen A, Maas A, Kooistra T, Lupu F, Grimbergen J et al: Aberrant fibrin formation and cross-linking of fibrinogen Nieuwegein, a variant with a shortened A-alpha-chain, alters endothelial capillary tube formation, *Blood* 97:973–80, 2001.

Crabtree GR, Kant JA: Molecular cloning of cDNA for the alpha, beta, and gamma chains of rat fibrinogen: A family of coordinately regulated genes, *J Biol Chem* 256:9718–23, 1981.

Crabtree GR, Kant JA: Coordinate accumulation of the mRNAs for the alpha, beta, and gamma chains of rat fibrinogen following defibrination, *J Biol Chem* 257:7277–9, 1982.

Drew AF, Liu H, Davidson JM, Daugherty CC, Degen JL: Wound-healing defects in mice lacking fibrinogen, *Blood* 97:3691–8, 2001.

Fu Y, Cao Y, Hertzberg KM, Grieninger G: Fibrinogen alpha genes: Conservation of bipartite transcripts and carboxy-terminal-extended alpha subunits in vertebrates, *Genomics* 30:71–6, 1995.

Kant JA, Fornace AJ Jr, Saxe D, Simon MI, McBride OW et al: Evolution and organization of the fibrinogen locus on chromosome 4: gene duplication accompanied by transposition and inversion, *Proc Natl Acad Sci USA* 82:2344–8, 1985.

Neerman-Arbez M, de Moerloose P, Bridel C, Honsberger A, Schonborner A et al: Mutations in the fibrinogen A-alpha gene account for the majority of cases of congenital afibrinogenemia, *Blood* 96:149–52, 2000.

Neerman-Arbez M, Honsberger A, Antonarakis SE, Morris MA: Deletion of the fibrogen (sic) alpha-chain gene (FGA) causes congenital afibrogenemia [sic], *J Clin Invest* 103:215–8, 1999.

15.3. Regulation of Vascular Contractility

Endothelin 1

Kim TH, Xiong H, Zhang Z, Ren B: beta-Catenin activates the growth factor endothelin-1 in colon cancer cells, *Oncogene* 24(4):597–604, 2005.

Jin JJ, Nakura J, Wu Z, Yamamoto M, Abe M et al: Association of endothelin-1 gene variant with hypertension, *Hypertension* 41(1):163–7, 2003.

Arinami T, Ishikawa M, Inoue A, Yanagisawa M, Masaki T et al: Chromosomal assignments of the human endothelin family genes: the endothelin-1 gene (EDN1) to 6p23–p24, the endothelin-2 gene (EDN2) to 1p34, and the endothelin-3 gene (EDN3) to 20q13.2–q13.3, *Am J Hum Genet* 48:990–6, 1991.

Bloch KD, Friedrich SP, Lee ME, Eddy RL, Shows TB et al: Structural organization and chromosomal assignment of the gene encoding endothelin, *J Biol Chem* 264:10851–7, 1989.

Bourgeois C, Robert B, Rebourcet R, Mondon F, Mignot TM et al: Endothelin-1 and ET(A) receptor expression in vascular smooth muscle cells from human placenta: A new ET(A) receptor messenger ribonucleic acid is generated by alternative splicing of exon 3, *J Clin Endocr Metab* 82:3116–23, 1997.

Campia U, Cardillo C, Panza JA: Ethnic differences in the vasoconstrictor activity of endogenous endothelin-1 in hypertensive patients, *Circulation* 109:3191–5, 2004.

Ieda M, Fukuda K, Hisaka Y, Kimura K, Kawaguchi H et al: Endothelin-1 regulates cardiac sympathetic innervation in the rodent heart by controlling nerve growth factor expression, *J Clin Invest* 113:876–84, 2004.

Inoue A, Yanagisawa M, Takuwa Y, Mitsui Y, Kobayashi M et al: The human preproendothelin-1 gene: Complete nucleotide sequence and regulation of expression, *J Biol Chem* 264:14954–9, 1989.

Khodorova A, Navarro B, Jouaville LS, Murphy JE, Rice FI et al: Endothelin-B receptor activation triggers an endogenous analgesic cascade at sites of peripheral injury, *Nature Med* 9:1055–61, 2003.

Kurihara Y, Kurihara H, Oda H, Maemura K, Nagai R et al: Aortic arch malformations and ventricular septal defect in mice deficient in endothelin-1, *J Clin Invest* 96:293–300, 1995.

Kurihara Y, Kurihara H, Suzuki H, Kodama T, Maemura K et al: Elevated blood pressure and craniofacial abnormalities in mice deficient in endothelin-1, *Nature* 368:703–10, 1994.

Remuzzi G, Perico N, Benigni A: New therapeutics that antagonize endothelin: Promises and frustrations, *Nature Rev* 1:986–1001, 2002.

Shohet RV, Kisanuki YY, Zhao XS, Siddiquee Z, Franco F et al: Mice with cardiomyocyte-specific disruption of the endothelin-1 gene are resistant to hyperthyroid cardiac hypertrophy, *Proc Natl Acad Sci USA* 101:2088–93, 2004.

Yanagisawa H, Hammer RE, Richardson JA, Williams SC, Clouthier DE et al: Role of endothelin-1/endothelin-A receptor-mediated signaling pathway in the aortic arch patterning in mice, *J Clin Invest* 102:22–33, 1998.

Yanagisawa H, Yanagisawa M, Kapur RP, Richardson JA, Williams SC et al: Dual genetic pathways of endothelin-mediated intercellular signaling revealed by targeted disruption of endothelin converting enzyme-1 gene, *Development* 125:825–36, 1998.

Yang LL, Gros R, Kabir MG, Sadi A, Gotlieb AI et al: Conditional cardiac overexpression of endothelin-1 induces inflammation and dilated cardiomyopathy in mice, *Circulation* 109:255–61, 2004.

Yin JJ, Mohammad KS, Kakonen SM, Harris S, Wu-Wong JR et al: A causal role for endothelin-1 in the pathogenesis of osteoblastic bone metastases, *Proc Natl Acad Sci USA* 100:10954–9, 2003.

Xu SW, Howat SL, Renzoni EA, Holmes A, Pearson JD et al: Endothelin-1 induces expression of matrix-associated genes in lung fibroblasts through MEK/ERK, *J Biol Chem* 279(22):23098–103, 2004.

Endothelin 2

Arinami T, Ishikawa M, Inoue A, Yanagisawa M, Masaki T et al: Chromosomal assignments of the human endothelin family genes: The endothelin-1 gene (EDN1) to 6p23-p24, the endothelin-2 gene (EDN2) to 1p34, and the endothelin-3 gene (EDN3) to 20q13.2-q13.3, *Am J Hum Genet* 48:990–6, 1991.

Bloch KD, Hong CC, Eddy RL, Shows TB, Quertermous T: cDNA cloning and chromosomal assignment of the endothelin 2 gene: Vasoactive intestinal contractor peptide is rat endothelin 2, *Genomics* 10:236–42, 1991.

Deng AY, Dene H, Pravenec M, Rapp JP: Genetic mapping of two new blood pressure quantitative trait loci in the rat by genotyping endothelin system genes, *J Clin Invest* 93:2701–9, 1994.

Ohkubo S, Ogi K, Hosoya M, Matsumoto H, Suzuki N et al: Specific expression of human endothelin-2 (ET-2) gene in a renal adenocarcinoma cell line: Molecular cloning of cDNA encoding the precursor of ET-2 and its characterization, *FEBS Lett* 274:136–40, 1990.

Endothelin 3

Arinami T, Ishikawa M, Inoue A, Yanagisawa M, Masaki T et al: Chromosomal assignments of the human endothelin family genes: the endothelin-1 gene (EDN1) to 6p23–p24, the endothelin-2 gene (EDN2) to 1p34, and the endothelin-3 gene (EDN3) to 20q13.2–q13.3, *Am J Hum Genet* 48:990–6, 1991.

Baynash AG, Hosoda K, Giaid A, Richardson JA, Emoto N et al: Interaction of endothelin-3 with endothelin-B receptor is essential for development of epidermal melanocytes and enteric neurons, *Cell* 79:1277–85, 1994.

Bloch K, Eddy RL, Shows TB, Quertermous T: cDNA cloning and chromosomal assignment of the gene encoding endothelin 3, *J Biol Chem* 264:18156–61, 1989.

Bolk S, Angrist M, Xie J, Yanagisawa M, Silvestri JM et al: Endothelin-3 frameshift mutation in congenital central hypoventilation syndrome, *Nature Genet* 13:395–6, 1996.

Dupin E, Glavieux C, Vaigot P, Le Douarin NM: Endothelin 3 induces the reversion of melanocytes to glia through a neural crest-derived glial-melanocytic progenitor, *Proc Natl Acad Sci USA* 97:7882–7, 2000.

Gopal Rao VVN, Loffler C, Hansmann I: The gene for the novel vasoactive peptide endothelin 3 (EDN3) is localized to human chromosome 20q13.2-qter, *Genomics* 10:840–1, 1991.

Lee S, Lin M, Mele A, Cao Y, Farmar J et al: Proteolytic processing of big endothelin-3 by the Kell blood group protein, *Blood* 94:1440–50, 1999.

Endothelin Receptor A

Bourgeois C, Robert B, Rebourcet R, Mondon F, Mignot TM et al: Endothelin-1 and ET(A) receptor expression in vascular smooth muscle cells from human placenta: A new ET(A) receptor messenger ribonucleic acid is generated by alternative splicing of exon 3, *J Clin Endocr Metab* 82:3116–23, 1997.

Campia U, Cardillo C, Panza JA: Ethnic differences in the vasoconstrictor activity of endogenous endothelin-1 in hypertensive patients, *Circulation* 109:3191–5, 2004.

Hosoda K, Nakao K, Tamura N, Arai H, Ogawa Y et al: Organization structure chromosomal assignment and expression of the gene encoding the human endothelin-A receptor, *J Biol Chem* 267:18797–804, 1992.

Maggi M, Barni T, Fantoni G, Mancina R, Pupilli C et al: Expression and biological effects of endothelin-1 in human gonadotropin-releasing hormone-secreting neurons, *J Clin Endocr Metab* 85:1658–65, 2000.

Maurer M, Wedemeyer J, Metz M, Piliponsky AM, Weller K et al: Mast cells promote homeostasis by limiting endothelin-1-induced toxicity, *Nature* 432:512–6, 2004.

Tzourio C, El Amrani M, Poirier O, Nicaud V, Bousser MG et al: Association between migraine and endothelin type A receptor (ETA -231 A/G) gene polymorphism, *Neurology* 56:1273–7, 2001.

Endothtelin Receptor B

Arai H, Nakao K, Takaya K, Hosoda K, Ogawa Y et al: The human endothelin-B receptor gene: Structural organization and chromosomal assignment, *J Biol Chem* 268:3463–70, 1993.

Baynash AG, Hosoda K, Giaid A, Richardson JA, Emoto N et al: Interaction of endothelin-3 with endothelin-B receptor is essential for development of epidermal melanocytes and enteric neurons, *Cell* 79:1277–85, 1994.

Carpenter T, Schomberg S, Steudel W, Ozimek J, Colvin K et al: Endothelin B receptor deficiency predisposes to pulmonary edema formation via increased lung vascular endothelial cell growth factor expression, *Circ Res* 93:456–63, 2003.

Ceccherini I, Zhang AL, Matera I, Yang G, Devoto M et al: Interstitial deletion of the endothelin-B receptor gene in the spotting lethal (sl) rat. *Hum Mol Genet* 4:2089–96, 1995.

Elshourbagy NA, Adamou JE, Gagnon AW, Wu HL, Pullen M et al: Molecular characterization of a novel human endothelin receptor splice variant, *J Biol Chem* 271:25300–7, 1996.

Gariepy CE, Cass DT, Yanagisawa M: Null mutation of endothelin receptor type B gene in spotting lethal rats causes aganglionic megacolon and white coat color, *Proc Natl Acad Sci USA* 93:867–72, 1996.

Gariepy CE, Ohuchi T, Williams SC, Richardson JA, Yanagisawa M: Salt-sensitive hypertension in endothelin-B receptor-deficient rats, *J Clin Invest* 105:925–33, 2000.

Gariepy CE, Williams SC, Richardson JA, Hammer RE, Yanagisawa M: Transgenic expression of the endothelin-B receptor prevents congenital intestinal aganglionosis in a rat model of Hirschsprung disease, *J Clin Invest* 102:1092–101, 1998.

Hosoda K, Hammer RE, Richardson JA, Baynash AG, Cheung JC et al: Targeted and natural (piebald-lethal) mutations of endothelin-B receptor gene produce megacolon associated with spotted coat color in mice, *Cell* 79:1267–76, 1994.

Iwashita T, Kruger GM, Pardal R, Kiel MJ, Morrison SJ: Hirschsprung disease is linked to defects in neural crest stem cell function, *Science* 301:972–6, 2003.

Khodorova A, Navarro B, Jouaville LS, Murphy JE, Rice FI et al: Endothelin-B receptor activation triggers an endogenous analgesic cascade at sites of peripheral injury, *Nature Med* 9:1055–61, 2003.

Shin MK, Levorse JM, Ingram RS, Tilghman SM: The temporal requirement for endothelin receptor-B signalling during neural crest development, *Nature* 402:496–501, 1999.

Shin MK, Russell LB, Tilghman SM: Molecular characterization of four induced alleles at the Ednrb locus, *Proc Natl Acad Sci USA* 94:13105–10, 1997.

Nishiyama M, Masaki T, Yanagisawa M, Sekiya S, Kimura S: Novel mutations of the endothelin B receptor gene in patients with Hirschsprung's disease and their characterization, *J Biol Chem* 273:11378–83, 1998.

Yang LL, Gros R, Kabir MG, Sadi A, Gotlieb AI et al: Conditional cardiac overexpression of endothelin-1 induces inflammation and dilated cardiomyopathy in mice, *Circulation* 109:255–61, 2004.

Zhu L, Lee HO, Jordan CS, Cantrell VA, Southard-Smith EM et al: Spatiotemporal regulation of endothelin receptor-B by SOX10 in neural crest-derived enteric neuron precursors, *Nature Genet* 36:732–7, 2004.

Human protein reference data base, Johns Hopkins University and the Institute of Bioinformatics, at http://www.hprd.org/protein.

15.4. Regulation of Leukocytes and Platelet Adhesion

MCP1

Ambati J, Anand A, Fernandez S, Sakurai E, Lynn BC et al: An animal model of age-related macular degeneration in senescent Ccl-2- or Ccr-2-deficient mice, *Nature Med* 9:1390–7, 2003.

Corrigall VM, Arastu M, Khan S, Shah C, Fife M et al: Functional IL-2 receptor beta (CD122) and gamma (CD132) chains are expressed by fibroblast-like synoviocytes: Activation by IL-2 stimulates monocyte chemoattractant protein-1 production, *J Immun* 166:4141–7, 2001.

Galindo M, Santiago B, Rivero M, Rullas J, Alcami J et al: Chemokine expression by systemic sclerosis fibroblasts: Abnormal regulation of monocyte chemoattractant protein 1 expression, *Arthritis Rheum* 44:1382–6, 2001.

Gosling J, Slaymaker S, Gu L, Tseng S, Zlot CH et al: MCP-1 deficiency reduces susceptibility to atherosclerosis in mice that overexpress human apolipoprotein B, *J Clin Invest* 103:773–8, 1999.

Gu L, Tseng S, Horner RM, Tam C, Loda M et al: Control of TH2 polarization by the chemokine monocyte chemoattractant protein-1, *Nature* 404:407–11, 2000.

Lu B, Rutledge BJ, Gu L, Fiorillo J, Lukacs NW et al: Abnormalities in monocyte recruitment and cytokine expression in monocyte chemoattractant protein 1-deficient mice, *J Exp Med* 187:601–8, 1998.

Sartipy P, Loskutoff DJ: Monocyte chemoattractant protein 1 in obesity and insulin resistance, *Proc Natl Acad Sci USA* 100:7265–70, 2003.

15.5. Smooth Muscle Cells

Arner A, Malmqvist U: Cross-bridge cycling in smooth muscle: A short review, *Acta Physiol Scand* 164(4):363–72, 1998.

Kohama K, Ye LH, Hayakawa K, Okagaki T: Myosin light chain kinase: An actin-binding protein that regulates an ATP-dependent interaction with myosin, *Trends Pharmacol Sci* 8:284–7, 1996.

Winder SJ, Allen BG, Clement-Chomienne O, Walsh MP: Regulation of smooth muscle actin-myosin interaction and force by calponin, *Acta Physiol Scand* 164(4):415–26, 1998.

Liu SQ, Fung YC: Relationship between hypertension hypertrophy and opening angle of zero-stress state of arteries following aortic constriction, *J Biomech Eng* 111:325–35, 1989.

Liu SQ, Fung YC: Influence of STZ-diabetes on zero-stress state of rat pulmonary and systemic arteries, *Diabetes* 41:136–46, 1992.

Liu SQ, Fung YC: Material coefficients of the strain energy function of pulmonary arteries in normal and cigarette smoke-exposed rats, *J Biomech* 26:1261–9, 1993.

Liu SQ, Fung YC: Changes in the structure and mechanical properties of pulmonary arteries of rats exposed to cigarette smoke, *Am Rev Resp Dis* 148:768–77, 1993.

Fung YC, Liu SQ: Determination of the mechanical properties of the different layers of blood vessels in vivo, *Proc Natl Acad Sci USA* 92:2169–73, 1995.

Liu SQ, Fung YC: Indicial functions of arterial remodeling in response to locally altered mechanical stress, *Am J Physiol* 270: H1323–33, 1996.

Liu SQ: Alterations in structure of elastic laminae of rat pulmonary arteries in hypoxic hypertension, *J Appl Physiol* 81:2147–55, 1996.

Liu SQ: Regression of hypoxic hypertension-induced changes in the elastic laminae of rat pulmonary arteries, *J Appl Physiol* 82:1677–84, 1997.

Liu SQ: Influence of tensile strain on smooth muscle cell orientation in rat blood vessels, *J Biomech Eng* 120:313–20, 1998.

Liu SQ: Biomechanical basis of vascular tissue engineering, *Cri Rev Biomed Eng* 27:75–148, 1999.

Liu SQ, Moore MM, Glucksberg MR, Mockros LF, Grotberg JB et al: Partial prevention of monocyte and granulocyte activation in experimental vein grafts using a biomechanical engineering approach, *J Biomech* 32:1165–75, 1999.

Liu SQ, Moore MM, Yap C: Prevention of tensile stress-induced endothelial and smooth muscle cell injury in experimental vein grafts, *J Biomech Eng* 122:31–8, 2000.

Moore MM, Goldman J, Patel A, Chien S, Liu SQ: Role of mechanical stretch in the induction of vascular cell death, *J Biomech* 34:289–97, 2001.

Goldman J, Zhong L, Liu SQ: Degradation of α actin filaments in vascular smooth muscle cells in response to mechanical stretch, *Am J Physiol* 284:H1839–47, 2003.

15.6. Regulation by the Renin–Angiotensin–Aldosterone System

Renin

Caron KMI, James LR, Kim HS, Knowles J, Uhlir R et al: Cardiac hypertrophy and sudden death in mice with a genetically clamped renin transgene, *Proc Natl Acad Sci USA* 101:3106–11, 2004.

Frossard PM, Lestringant GG, Malloy MJ, Kane JP: Human renin gene BglI dimorphism associated with hypertension in two independent populations, *Clin Genet* 56:428–33, 1999.

Gribouval O, Gonzales M, Neuhaus T, Aziza J, Bieth E et al: Mutations in genes in the renin-angiotensin system are associated with autosomal recessive renal tubular dysgenesis, *Nature Genet* 37:964–8, 2005.

Nguyen G, Delarue F, Burckle C, Bouzhir L, Giller T et al: Pivotal role of the renin/prorenin receptor in angiotensin II production and cellular responses to renin, *J Clin Invest* 109:1417–27, 2002.

Qin H, Chen YH, Yip MY, Lam-Po-Tang PRL, Morris BJ: Reassignment of human renin gene to chromosome 1q32 in studies of a (1;4)(q42;p16) translocation, *Hum Hered* 43:261–4, 1993.

van Hooft IMS, Grobbee DE, Derkx FHM, de Leeuw PW, Schalekamp MADH et al: Renal hemodynamics and the renin-angiotensin-aldosterone system in normotensive subjects with hypertensive and normotensive parents, *New Engl J Med* 324:1305–11, 1991.

Villard E, Lalau JD, van Hooft IS, Derkx FHM, Houot A-M et al: A mutant renin gene in familial elevation of prorenin, *J Biol Chem* 269:30307–12, 1994.

Angiotensin I

Bloem LJ, Manatunga AK, Tewksbury DA, Pratt JH: The serum angiotensinogen concentration and variants of the angiotensinogen gene in white and black children, *J Clin Invest* 95:948–53, 199.

Caulfield M, Lavender P, Farrall M, Munroe P, Lawson M et al: Linkage of the angiotensinogen gene to essential hypertension, *New Engl J Med* 330:1629–33, 1994.

Caulfield M, Lavender P, Newell-Price J, Farrall M, Kamdar S et al: Linkage of the angiotensinogen gene locus to human essential hypertension in African Caribbeans, *J Clin Invest* 96:687–92, 1995.

Davisson RL, Kim HS, Krege JH, Lager DJ, Smithies O et al: Complementation of reduced survival hypotension, and renal abnormalities in angiotensinogen-deficient mice by the human renin and human angiotensinogen genes, *J Clin Invest* 99:1258–64, 1997.

Ding Y, Stec DE, Sigmund CD: Genetic evidence that lethality in angiotensinogen-deficient mice is due to loss of systemic but not renal angiotensinogen, *J Biol Chem* 276:7431–6, 2001.

Gribouval O, Gonzales M, Neuhaus T, Aziza J, Bieth E et al: Mutations in genes in the renin-angiotensin system are associated with autosomal recessive renal tubular dysgenesis, *Nature Genet* 37:964–8, 2005.

Hata A, Namikawa C, Sasaki M, Sato K, Nakamura T et al: Angiotensinogen as a risk factor for essential hypertension in Japan, *J Clin Invest* 93:1285–7, 1994.

Inoue I, Nakajima T, Williams CS, Quackenbush J, Puryear R et al: A nucleotide substitution in the promoter of human angiotensinogen is associated with essential hypertension and affects basal transcription in vitro, *J Clin Invest* 99:1786–97, 1997.

Jeunemaitre X, Soubrier F, Kotelevtsev YV, Lifton RP, Williams CS et al: Molecular basis of human hypertension: Role of angiotensinogen, *Cell* 71:7–20, 1992.

Katsuya T, Koike G, Yee TW, Sharpe N, Jackson R et al: Association of angiotensinogen gene T235 variant with increased risk of coronary heart disease, *Lancet* 345:1600–3, 1995.

Sadoshima J, Xu Y, Slayter HS, Izumo S: Autocrine release of angiotensin II mediates stretch-induced hypertrophy of cardiac myocytes in vitro, *Cell* 75:977–84, 1993.

Tanimoto K, Sugiyama F, Goto Y, Ishida J, Takimoto E et al: Angiotensinogen-deficient mice with hypotension, *J Biol Chem* 269:31334–7, 1994.

Angiotensin-Converting Enzyme

Cambien F, Poirier O, Lecerf L, Evans A, Cambou JP et al: Deletion polymorphism in the gene for angiotensin-converting enzyme is a potent risk factor for myocardial infarction, *Nature* 359:641–4, 1992.

Doria A, Warram JH, Krolewski AS: Genetic predisposition to diabetic nephropathy: Evidence for a role of the angiotensin I-converting enzyme gene, *Diabetes* 43:690–5, 1994.

Esther CR Jr, Marino EM, Howard TE, Machaud A, Corvol P et al: The critical role of tissue angiotensin-converting enzyme as revealed by gene targeting in mice, *J Clin Invest* 99:2375–85, 1997.

Jeunemaitre X, Lifton RP, Hunt SC, Williams RR, Lalouel JM: Absence of linkage between the angiotensin converting enzyme locus and human essential hypertension, *Nature Genet* 1:72–5, 1992.

Keavney B, McKenzie C, Parish S, Palmer A, Clark S et al: Large-scale test of hypothesized associations between the angiotensin-converting-enzyme insertion/deletion polymorphism and myocardial infarction in about 5000 cases and 6000 controls, *Lancet* 355:434–42, 2000.

Kessler SP, deS Senanayake P, Scheidemantel TS, Gomos JB, Rowe TM et al: Maintenance of normal blood pressure and renal functions are independent effects of angiotensin-converting enzyme, *J Biol Chem* 278:21105–12, 2003.

Kramers C, Danilov SM, Deinum J, Balyasnikova IV, Scharenborg N et al: Point mutation in the stalk of angiotensin-converting enzyme causes a dramatic increase in serum angiotensin-converting enzyme but no cardiovascular disease, *Circulation* 104:1236–40, 2001.

Lindpaintner K, Lee M, Larson MG, Rao VS, Pfeffer MA et al: Absence of association or genetic linkage between the angiotensin-converting-enzyme gene and left ventricular mass, *New Engl J Med* 334:1023–8, 1996.

Lindpaintner K, Pfeffer MA, Kreutz R, Stampfer MJ, Grodstein F et al: A prospective evaluation of an angiotensin-converting-enzyme gene polymorphism and the risk of ischemic heart disease, *New Engl J Med* 332:706–11, 1995.

Marre M, Jeunemaitre X, Gallois Y, Rodier M, Chatellier G et al: Contribution of genetic polymorphism in the renin-angiotensin system to the development of renal complications in insulin-dependent diabetes, *J Clin Invest* 99:1585–95, 1997.

Montgomery HE, Clarkson P, Dollery CM, Prasad K, Losi MA et al: Association of angiotensin-converting enzyme gene I/D polymorphism with change in left ventricular mass in response to physical training, *Circulation* 96:741–7, 1997.

Natesh R, Schwager SLU, Sturrock ED, Acharya KR: Crystal structure of the human angiotensin-converting enzyme-lisinopril complex, *Nature* 421:551–4, 2003.

Ohishi M, Fujii K, Minamino T, Hagaki J, Kamitani A et al: A potent genetic risk factor for restenosis [Letter], *Nature Genet* 5:324–5, 1993.

Oike Y, Hata A, Ogata Y, Numata Y, Shido K et al: Angiotensin converting enzyme as a genetic risk factor for coronary artery spasm: implication in the pathogenesis of myocardial infarction, *J Clin Invest* 96:2975–9, 1995.

Ramaraj P, Kessler SP, Colmenares C, Sen GC: Selective restoration of male fertility in mice lacking angiotensin-converting enzymes by sperm-specific expression of the testicular isozyme, *J Clin Invest* 102:371–8, 1998.

AT1

AbdAlla S, Lother H, Langer A, el Faramawy Y, Quitterer U: Factor XIIIA transglutaminase crosslinks AT1 receptor dimers of monocytes at the onset of atherosclerosis, *Cell* 119:343–54, 2004.

AbdAlla S, Lother H, Quitterer U: AT(1)-receptor heterodimers show enhanced G-protein activation and altered receptor sequestration, *Nature* 407:94–8, 2000.

Bonnardeaux A, Davies E, Jeunemaitre X, Fery I, Charru A et al: Angiotensin II type 1 receptor gene polymorphisms in human essential hypertension, *Hypertension* 24:63–9, 1994.

Haywood GA, Gullestad L, Katsuya T, Hutchinson HG, Pratt RE et al: AT(1) and AT(2) angiotensin receptor gene expression in human heart failure, *Circulation* 95:1201–6, 1997.

Herzig TC, Jobe SM, Aoki H, Molkentin JD, Cowley AW Jr et al: Angiotensin II type-1a receptor gene expression in the heart: AP-1 and GATA-4 participate in the response to pressure overload, *Proc Natl Acad Sci USA* 94:7543–8, 1997.

Kambayashi Y, Bardhan S, Takahashi K, Tsuzuki S, Inui H et al: Molecular cloning of a novel angiotensin II receptor isoform involved in phosphotyrosine phosphatase inhibition, *J Biol Chem* 268:24543–6, 1993.

Murphy TJ, Alexander RW, Griendling KK, Runge MS, Bernstein KE: Isolation of a cDNA encoding the vascular type-1 angiotensin II receptor, *Nature* 351:233–6, 1991.

Sasaki K, Murohara T, Ikeda H, Sugaya T, Shimada T et al: Evidence for the importance of angiotensin II type 1 receptor in ischemia-induced angiogenesis, *J Clin Invest* 109:603–11, 2002.

Sasaki K, Yamano Y, Bardhan S, Iwai N, Murray JJ et al: Cloning and expression of a complementary DNA encoding a bovine adrenal angiotensin II type-1 receptor, *Nature* 351:230–3, 1991.

Human protein reference data base, Johns Hopkins University and the Institute of Bioinformatics, at http://www.hprd.org/protein.

15.7. Pathogenesis, Pathology, and Clinical Features of Atherosclerosis

Ross R: Atherosclerosis—an inflammatory disease, *New Engl J Med* 340(2):115–26, 1999.

Ross R: The pathogenesis of atherosclerosis: A perspective for the 1990s, *Nature* 362(6423):801–9, 1993.

Schneider AS, Szanto PA: *Pathology*, 3rd ed, Lippincott Williams & Wilkins Philadelphia, 2006.

McCance KL, Huether SE: *Pathophysiology: The Biologic Basis for Disease in Adults & Children*, 5th ed, Elsevier Mosby, St Louis, 2006.

Porth CM: *Pathophysiology: Concepts of Altered Health States*, 7th ed, Lippincott Williams & Wilkins, Philadelphia, 2005.

Frazier MS, Drzymkowski JW: *Essentials of Human Diseases and Conditions*, 3rd ed, Elsevier Saunders, St Louis, 2004.

15.8. Animal Models of Atherosclerosis

Xu Q: Mouse models of arteriosclerosis: from arterial injuries to vascular grafts, *Am J Pathol* 165:1–10, 2004.

Meir KS, Leitersdorf E: Atherosclerosis in the apolipoprotein E-deficient mouse: A decade of progress, *Arterioscler Thromb Vasc Biol* 24:1006–14, 2004.

Leidenfrost JE, Khan MF, Boc KP, Villa BR, Collins ET et al: A model of primary atherosclerosis and post-angioplasty restenosis in mice, *Am J Pathol* 163(2):773–8, 2003.

15.9. Conventional Treatment of Atherosclerosis

Gershlick AH: Treating atherosclerosis: Local drug delivery from laboratory studies to clinical trials, *Atherosclerosis* 160:259–71, 2002.

Jordan MA, Toso RJ, Thrower D, Wilson L: Mechanism of mitotic block and inhibition of cell proliferation by taxol at low concentrations, *Proc Natl Acad Sci USA* 90:9552–6, 1993.

Axel DI et al: Paclitaxel inhibits arterial smooth muscle cell proliferation and migration in vitro and in vivo using local drug delivery, *Circulation* 96:636–45, 1997.

Liistro F et al: First clinical experience with a paclitaxel derivate-eluting polymer stent system implantation for in-stent restenosis: Immediate and long-term clinical and angiographic outcome, *Circulation* 105:1883–6, 2002.

Dzau VJ, Braun-Dullaeus RC, Sedding DG: Vascular proliferation and atherosclerosis: New perspectives and therapeutic strategies, *Nat Med* 8(11):1249–56, 2002.

Sehgal SN, Baker H, Vezina C: Rapamycin (AY-22,989), a new antifungal antibiotic. II. Fermentation, isolation and characterization, *J Antibiot* (Tokyo) 28:727–32, 1975.

Gregory CR, Huie P, Billingham ME, Morris RE: Rapamycin inhibits arterial intimal thickening caused by both alloimmune and mechanical injury. Its effect on cellular, growth factor, and cytokine response in injured vessels, *Transplantation* 55:1409–18, 1993.

Braun-Dullaeus RC et al: Cell cycle protein expression in vascular smooth muscle cells in vitro and in vivo is regulated through phosphatidylinositol 3-kinase and mammalian target of rapamycin, *Arterioscler Thromb Vasc Biol* 21:1152–8, 2001.

Gallo R et al: Inhibition of intimal thickening after balloon angioplasty in porcine coronary arteries by targeting regulators of the cell cycle, *Circulation* 99:2164–70, 1999.

Grise MA et al: Five-year clinical follow-up after intracoronary radiation: results of a randomized clinical trial, *Circulation* 105:2737–40, 2002.

Waksman R et al: Intracoronary gamma-radiation therapy after angioplasty inhibits recurrence in patients with in-stent restenosis, *Circulation* 101:2165–71, 2000.

Scott S, O'Sullivan M, Hafizi S, Shapiro LM, Bennett MR: Human vascular smooth muscle cells from restenosis or in-stent stenosis sites demonstrate enhanced responses to p53: Implications for brachytherapy and drug treatment for restenosis, *Circ Res* 90:398–404, 2002.

Liu SQ: Prevention of focal intimal hyperplasia in rat vein grafts by using a tissue engineering approach, *Atherosclerosis* 140:365–77, 1998.

Liu SQ: Biomechanical basis of vascular tissue engineering, *Crit Rev Biomed Eng* 27:75–148, 1999.

Liu SQ: Focal activation of angiotensin II type 1 receptor and smooth muscle cell proliferation in the neointima of experimental vein grafts: Relation to eddy blood flow, *Arterioscl Thromb Vasc Biol* 19:2630–9, 1999.

15.10. Antisense Oligodeoxynucleotides for CDKs

Dzau VJ, Braun-Dullaeus RC, Sedding DG: Vascular proliferation and atherosclerosis: New perspectives and therapeutic strategies, *Nat Med* 8(11):1249–56, 2002.

Braun-Dullaeus RC, Mann MJ, Dzau VJ: Cell cycle progression. New therapeutic target for vascular proliferative disease, *Circulation* 98:82–89, 1998.

Morishita R et al: Single intraluminal delivery of antisense cdc2 kinase and proliferating-cell nuclear antigen oligonucleotides results in chronic inhibition of neointimal hyperplasia, *Proc Natl Acad Sci USA* 90:8474–8, 1993.

Simons M, Edelman ER, Rosenberg RD: Antisense proliferating cell nuclear antigen oligonucleotides inhibit intimal hyperplasia in a rat carotid artery injury model, *J Clin Invest* 93:2351–6, 1994.

Morishita R et al: A gene therapy strategy using a transcription factor decoy of the E2F binding site inhibits smooth muscle proliferation in vivo, *Proc Natl Acad Sci USA* 92:5855–9, 1995.

Mann MJ et al: Cell cycle inhibition preserves endothelial function in genetically engineered rabbit vein grafts, *J Clin Invest* 99:1295–301, 1997.

Mann MJ et al: Ex vivo gene therapy of human vascular bypass grafts with E2F decoy: The PREVENT single-centre, randomised, controlled trial, *Lancet* 354:1493–8, 1999.

Mangi AA, Dzau VJ: Gene therapy for human bypass grafts, *Ann Med* 33:153–5, 2001.

15.11. Cell Cycle-Inhibiting Genes

Smith RC et al: p21CIP1-mediated inhibition of cell proliferation by overexpression of the gax homeodomain gene, *Genes Dev* 11:1674–89, 1997.

Yang ZY et al: Role of the p21 cyclin-dependent kinase inhibitor in limiting intimal cell proliferation in response to arterial injury, *Proc Natl Acad Sci USA* 93:7905–10, 1996.

Yonemitsu Y, Kaneda Y, Hataa Y, Nakashima Y, Sueishi K: Wild-type p53 gene transfer: A novel therapeutic strategy for neointimal hyperplasia after arterial injury, *Ann NY Acad Sci* 811:395–9, 1997.

15.12. Nitric Oxide Synthase

Nitric Oxide Synthase 1 Gene

Guo K, Andres V, Walsh K: Nitric oxide-induced downregulation of Cdk2 activity and cyclin A gene transcription in vascular smooth muscle cells, *Circulation* 97:2066–72, 1998.

Ishida A, Sasaguri T, Kosaka C, Nojima H, Ogata J: Induction of the cyclin-dependent kinase inhibitor p21(Sdi1/Cip1/Waf1) by nitric oxide-generating vasodilator in vascular smooth muscle cells, *J Biol Chem* 272:10050–7, 1997.

von der Leyen HE et al: Gene therapy inhibiting neointimal vascular lesion: in vivo transfer of endothelial cell nitric oxide synthase gene, *Proc Natl Acad Sci USA* 92:1137–41, 1995.

Qian H, Neplioueva V, Shetty GA, Channon KM, George SE: Nitric oxide synthase gene therapy rapidly reduces adhesion molecule expression and inflammatory cell infiltration in carotid arteries of cholesterol-fed rabbits, *Circulation* 99:2979–82, 1999.

von der Leyen HE, Dzau VJ: Therapeutic potential of nitric oxide synthase gene manipulation, *Circulation* 103:2760–5, 2001.

Kong D, Melo LG, Mangi AA, Zhang L, Lopez-Ilasaca M et al: Enhanced inhibition of neointimal hyperplasia by genetically engineered endothelial progenitor cells, *Circulation* 109(14):1769–75, 2004.

Barouch LA, Harrison RW, Skaf MW, Rosas GO, Cappola TP et al: Nitric oxide regulates the heart by spatial confinement of nitric oxide synthase isoforms, *Nature* 416:337–40, 2002.

Bredt DS, Hwang PM, Glatt CE, Lowenstein C, Reed RR et al: Cloned and expressed nitric oxide synthase structurally resembles cytochrome P-450 reductase, *Nature* 351:714–8, 1991.

Brenman JE, Chao DS, Xia H, Aldape K, Bredt DS: Nitric oxide synthase complexed with dystrophin and absent from skeletal muscle sarcolemma in Duchenne muscular dystrophy, *Cell* 82:743–52, 1995.

Deans Z, Dawson SJ, Xie J, Young AP, Wallace D et al: Differential regulation of the two neuronal nitric-oxide synthase gene promoters by the Oct-2 transcription factor, *J Biol Chem* 271:32153–8, 1996.

Burnett AL, Lowenstein CJ, Bredt DS, Chang TSK, Snyder SH: Nitric oxide: A physiologic mediator of penile erection, *Science* 257:401–3, 1992.

Kuo RC, Baxter GT, Thompson SH, Stricker SA, Patton C et al: NO is necessary and sufficient for egg activation at fertilization, *Nature* 406:633–6, 2000.

Marsden PA, Heng HHQ, Scherer SW, Stewart RJ, Hall AV et al: Structure and chromosomal localization of the human constitutive endothelial nitric oxide synthase gene, *J Biol Chem* 268:17478–88, 1993.

Nelson RJ, Demas GE, Huang PL, Fishman MC, Dawson VL et al: Behavioural abnormalities in male mice lacking neuronal nitric oxide synthase, *Nature* 378:383–6, 1995.

Newton DC, Bevan SC, Choi S, Robb GB, Millar A et al: Translational regulation of human neuronal nitric-oxide synthase by an alternatively spliced 5-prime untranslated region leader exon, *J Biol Chem* 278:636–44, 2003.

Sander M, Chavoshan B, Harris SA, Iannaccone ST, Stull JT et al: Functional muscle ischemia in neuronal nitric oxide synthase-deficient skeletal muscle of children with Duchenne muscular dystrophy, *Proc Natl Acad Sci USA* 97:13818–23, 2000.

Nitric Oxide Synthase 2A

Bloch KD, Wolfram JR, Brown DM, Roberts JD Jr, Zapol DG et al: Three members of the nitric oxide synthase II gene family (NOS2A, NOS2B, and NOS2C) colocalize to human chromosome 17, *Genomics* 27:526–30, 1995.

Chartrain NA, Geller DA, Koty PP, Sitrin NF, Nussler AK et al: Molecular cloning, structure, and chromosomal localization of the human inducible nitric oxide synthase gene, *J Biol Chem* 269:6765–72, 1994.

Diefenbach A, Schindler H, Rollinghoff M, Yokoyama WM, Bogdan C: Requirement for type 2 NO synthase for IL-12 signaling in innate immunity, *Science* 284:951–5, 1999.

Koh KP, Wang Y, Yi T, Shiao SL, Lorber MI, Sessa WC et al: T cell-mediated vascular dysfunction of human allografts results from IFN-gamma dysregulation of NO synthase, *J Clin Invest* 114:846–56, 2004.

Kolodziejski PJ, Musial A, Koo JS, Eissa NT: Ubiquitination of inducible nitric oxide synthase is required for its degradation, *Proc Natl Acad Sci USA* 99:12315–20, 2002.

Nitric Oxide Synthase 3

Aicher A, Heeschen C, Mildner-Rihm C, Urbich C, Ihling C et al: Essential role of endothelial nitric oxide synthase for mobilization of stem and progenitor cells, *Nature Med* 9:1370–6, 2003.

Barouch LA, Harrison RW, Skaf MW, Rosas GO, Cappola TP et al: Nitric oxide regulates the heart by spatial confinement of nitric oxide synthase isoforms, *Nature* 416:337–40, 2002.

Casas JP, Bautista LE, Humphries SE, Hingorani AD: Endothelial nitric oxide synthase genotype and ischemic heart disease: Meta-analysis of 26 studies involving 23028 subjects, *Circulation* 109:1359–65, 2004.

Connelly L, Madhani M, Hobbs AJ: Resistance to endotoxic shock in endothelial nitric-oxide synthase (eNOS) knock-out mice: A pro-inflammatory role for eNOS-derived NO in vivo, *J Biol Chem* 280:10040–6, 2005.

Dimmeler S, Fleming I, Fisslthaler B, Hermann C, Busse R et al: Activation of nitric oxide synthase in endothelial cells by Akt-dependent phosphorylation, *Nature* 399:601–5, 1999.

Fulton D, Gratton JP, McCabe TJ, Fontana J, Fujio Y et al: Regulation of endothelium-derived nitric oxide production by the protein kinase Akt, *Nature* 399:597–601, 1999.

Huang PL, Huang Z, Mashimo H, Bloch KD, Moskowitz MA et al: Hypertension in mice lacking the gene for endothelial nitric oxide synthase, *Nature* 377:239–42, 1995.

Janssens SP, Shimouchi A, Quertermous T, Bloch DB, Bloch KD: Cloning and expression of a cDNA encoding human endothelium-derived relaxing factor/nitric oxide synthase, *J Biol Chem* 267:14519–22, 1992.

Liu S, Premont RT, Kontos CD, Zhu S, Rockey DC: A crucial role for GRK2 in regulation of endothelial cell nitric oxide synthase function in portal hypertension, *Nature Med* 11:952–8, 2005.

Marsden PA, Heng HHQ, Scherer SW, Stewart RJ, Hall AV et al: Structure and chromosomal localization of the human constitutive endothelial nitric oxide synthase gene, *J Biol Chem* 268:17478–88, 1993.

Nisoli E, Tonello C, Cardile A, Cozzi V, Bracale R et al: Calorie restriction promotes mitochondrial biogenesis by inducing the expression of eNOS, *Science* 310:314–7, 2005.

Ohashi Y, Kawashima S, Hirata K, Yamashita T, Ishida T et al: Hypotension and reduced nitric oxide-elicited vasorelaxation in transgenic mice overexpressing endothelial nitric oxide synthase, *J Clin Invest* 102:2061–71, 1998.

Raman CS, Li H, Martasak P, Kral V, Masters BSS et al: Crystal structure of constitutive endothelial nitric oxide synthase: A paradigm for pterin function involving a novel metal center, *Cell* 95:939–50, 1998.

Human protein reference data base, Johns Hopkins University and the Institute of Bioinformatics, at http://www.hprd.org/protein.

15.13. Herpes Virus Thymidine Kinase Gene

Ohno T, Gordon D, San H, Pompili VJ, Imperiale MJ et al: Gene therapy for vascular smooth muscle cell proliferation after arterial injury, *Science* 265(5173):781–4, 1994.

Chang MW, Ohno T, Gordon D, Lu MM, Nabel GJ et al: Adenovirus-mediated transfer of the herpes simplex virus thymidine kinase gene inhibits vascular smooth muscle cell proliferation and neointima formation following balloon angioplasty of the rat carotid artery, *Mol Med* 1(2):172–81, 1995.

15.14. Dominant-Negative Mutant Mitogenic Genes

Reusch HP, Schaefer M, Plum C, Schultz G, Paul M: Gbeta gamma mediate differentiation of vascular smooth muscle cells, *J Biol Chem* 276(22):19540–7, 2001.

Ginnan R, Pfleiderer PJ, Pumiglia K, Singer HA: PKC-delta and CaMKII-delta 2 mediate ATP-dependent activation of ERK1/2 in vascular smooth muscle, *Am J Physiol Cell Physiol* 286(6): C128–9, 2004.

Miura S, Matsuo Y, Kawamura A, Saku K: JTT-705 blocks cell proliferation and angiogenesis through p38 kinase/p27(kip1) and Ras/p21(waf1) pathways, *Atherosclerosis* 182(2):267–75, 2005.

15.15. Construction of Polymeric Arterial Substitutes with Endothelial Cell Seeding

Brewster DC, Rutherford RB: Prosthetic grafts, in: *Vascular Surgery*, Rutherford RB, ed, Saunders, Philadelphia, 2000, pp 492–521.

O'Donnell TF: Correlation of operative findings with angiographic and noninvasive hemodynamic factors associated with failure of PTFE grafts, *J Vasc Surg* 1:136–48, 1984.

Bowlin GL, Meyer A, Fields C, Cassano A, Makhoul RG et al: The persistence of electrostatically seeded endothelial cells lining a small diameter expanded polytetrafluoroethylene vascular graft, *J Biomater Appl* 16:157–73, 2001.

Braddon LG, Karoyli D, Harrison DG, Nerem RM: Maintenance of a functional endothelial cell monolayer on a fibroblast/polymer substrate under physiologically relevant shear stress conditions, *Tissue Eng* 8:695–708, 2002.

Deutsch M, Meinhart J, Fischlein T, Preiss P, Zilla P: Clinical autologous in vitro endothelialization of infrainguinal ePTFE grafts in 100 patients: A 9-year experience, *Surgery* 126:847–55, 1999.

Herring MB, Dilley R, Jersild RA Jr, Boxer L, Gardner A et al: Seeding arterial prostheses with vascular endothelium. The nature of the lining, *Ann Surg* 190:84–90, 1979.

Herring MB, Compton RS, LeGrand DR, Gardner AL, Madison DL et al: Endothelial seeding of polytetrafluoroethylene popliteal bypasses. A preliminary report, *J Vasc Surg* 6:114–8, 1987.

Herring MB, LeGrand DR: The histology of seeded PTFE grafts in humans, *Ann Vasc Surg* 3:96–103, 1989.

Herring MB, Compton R, Legrand DR, Gardner AL: Endothelial cell seeding in the management of vascular thrombosis, *Semin Thromb Hemost* 15:200–5, 1989.

Herring MB: The use of endothelial seeding of prosthetic arterial bypass grafts, *Surg Ann* 23(Pt 2):157–71, 1991.

Herring MB: Endothelial cell seeding, *J Vasc Surg* 13:731–2, 1991.

Kesler KA, Herring MB, Arnold MP, Park HM, Baughman S et al: Short-term in vivo stability of endothelial-lined polyester elastomer and polytetrafluoroethylene grafts, *Ann Vasc Surg* 1:60–5, 1986.

Ratcliffe A: Tissue engineering of vascular grafts, *Matrix Biol* 19:353–7, 2000.

15.16. Construction of Cell-Integrated Arterial Substitutes with Biodegradable Polymers

Motlagh D, Yang J, Lui KY, Webb AR, Ameer GA: Hemocompatibility evaluation of poly(glycerol-sebacate) in vitro for vascular tissue engineering, *Biomaterials* 27(24):4315–24, Aug 2006.

Yang J, Motlagh D, Webb AR, Ameer GA: Novel biphasic elastomeric scaffold for small-diameter blood vessel tissue engineering, *Tissue Eng* 11(11–12):1876–86, Nov–Dec 2005.

Jeremy JY, Bulbulia R, Johnson JL, Gadsdon P, Vijayan V et al: A bioabsorbable (polyglactin), nonrestrictive, external sheath inhibits porcine saphenous vein graft thickening, *J Thor Cardiovasc Surg* 127(6):1766–72, June 2004.

Perego G, Preda P, Pasquinelli G, Curti T, Freyrie A et al: Functionalization of poly-L-lactic-co-epsilon-caprolactone: Effects of surface modification on endothelial cell proliferation and hemocompatibility, *J Biomater Sci Polym Ed* 14(10):1057–75, 2003.

Xu CY, Inai R, Kotaki M, Ramakrishna S: Aligned biodegradable nanofibrous structure: A potential scaffold for blood vessel engineering, *Biomaterials* 25(5):877–86, Feb 2004.

Lee SH, Kim BS, Kim SH, Choi SW, Jeong SI et al: Elastic biodegradable poly(glycolide-co-caprolactone) scaffold for tissue engineering, *J Biomed Mater Res* 66A(1):29–37, July 2003.

Valente M, Pettenazzo E, Di Filippo L, Laborde F, Rinaldi S et al: Biodegradable polymer (D,L-lactide-epsilon-caprolactone) in aortic vascular prosthesis: morphological evaluation in an animal model, *Int J Artif Organs* 25(8):777–82, Aug 2002.

Teebken OE, Pichlmaier AM, Haverich A: Cell seeded decellularised allogeneic matrix grafts and biodegradable polydioxanone-prostheses compared with arterial autografts in a porcine model, *Eur J Vasc Endovasc Surg* 22(2):139–45, Aug 2001.

Langer R, Vacanti JP: Tissue engineering, *Science* 260:920–6, 1993.

Mayer JE Jr, Shinoka T, Shum-Tim D: Tissue engineering of cardiovascular structures, *Cur Opin Cardiol* 12:528–32, 1997.

Niklason LE, Gao J, Abbott WM, Hirschi KK, Houser S et al: Functional arteries grown in vitro, *Science* 284:489–93, 1999.

Hoerstrup SP, Kadner A, Breymann C, Maurus CF, Guenter CI et al: Living, autologous pulmonary artery conduits tissue engineered from human umbilical cord cells, *Ann Thor Surg* 74:46–52, 2002.

15.17. Construction of Arterial Substitutes with Decellularized Collagen Matrix

Huynh T, Abraham G, Murray J, Brockbank K, Hagen PO et al: Remodeling of an acellular collagen graft into a physiologically responsive neovessel, *Nat Biotech* 17:1083–6, 1999.

Liu SQ, Tieche C, Alkema PK: Neointima formation on elastic lamina and collagen matrix scaffolds implanted in the rat aorta, *Biomaterials* 25:1869–82, 2004.

Liu SQ, Alkema PK, Tieche C, Tefft BJ, Liu DZ et al: Negative regulation of monocyte adhesion to arterial elastic laminae by signal-regulatory protein alpha and SH2 domain-containing protein tyrosine phosphatase-1, *J Biol Chem* 280:39294–301, 2005.

Jernigan TW, Croce MA, Cagiannos C, Shell DH, Handorf CR et al: Small intestinal submucosa for vascular reconstruction in the presence of gastrointestinal contamination, *Ann Surg* 239(5): 733–8, May 2004.

Tieché C, Alkema PK, Liu SQ: Vascular elastic laminae: Anti-inflammatory properties and potential applications to arterial reconstruction, *Front Biosci* 9:2205–17, Sept 2004.

Hinek A: Biological roles of the non-integrin elastin/laminin receptor, *J Biol Chem* 377(7–8):471–80, 1996.

Hinek A, Wrenn DS, Mecham RP, Barondes SH: The elastin receptor: A galactoside-binding protein, *Science* 239(4847):1539–41, 1988.

Mecham RP, Hinek A, Entwistle R, Wrenn DS, Griffin GL et al: Elastin binds to a multifunctional 67-kilodalton peripheral membrane protein, *Biochemistry* 28(9):3716–22, May 1989.

Barocas VH, Girton TS, Tranquillo RT: Engineered alignment in media equivalents: Magnetic prealignment and mandrel compaction, *J Biomech Eng* 120:660–6, 1998.

Huynh T, Abraham G, Murray J, Brockbank K, Hagen PO et al: Remodeling of an acellular collagen graft into a physiologically responsive neovessel, *Natl Biotechnol* 17:1083–6, 1999.

Inoue Y, Anthony JP, Lleon P, Young DM: Acellular human dermal matrix as a small vessel substitute, *J Reconstr Microsurg* 12:307–11, 1996.

Kanda K, Matsuda T: Mechanical stress induced cellular orientation and phenotypic modulations of 3-D cultured smooth muscle cells, *ASAIO J* 39:M86–90, 1994.

Lantz GC, Badylak SF, Hiles MC, Coffey AC, Geddes LA et al: Small intestinal submucosa as a vascular graft: A review, *J Invest Surg* 6:297–310, 1993.

Sandusky GE, Lantz GC, Badylak SF: Healing comparison of small intestine submucosa and ePTFE grafts in the canine carotid artery, *J Surg Res* 58:415–20, 1995.

Seliktar D, Nerem RM, Galis ZS: The role of matrix metalloproteinase-2 in the remodeling of cell-seeded vascular constructs subjected to cyclic strain, *Ann Biomed Eng* 29:923–34, 2001.

Tranquillo RT, Girton TS, Bromberek BA, Triebes TG, Mooradian DL: Magnetically orientated tissue-equivalent tubes: application to a circumferentially orientated media-equivalent, *Biomaterials* 17:349–57, 1996.

Wilson GJ, Yeger H, Klement P, Lee JM, Courtman DW: Acellular matrix allograft small caliber vascular prostheses, *ASAIO Trans* 36:M340–3, 1990.

Wilson GJ, Courtman DW, Klement P, Lee JM, Yeger H: Acellular matrix: A biomaterials approach for coronary artery bypass and heart valve replacement, *Ann Thor Surg* 60(Suppl 2):S353–8, 1995.

Adams S, van der Laan LJ, Vernon-Wilson E, Renardel de Lavalette C, Dopp EA et al: Signal-regulatory protein is selectively expressed by myeloid and neuronal cells, *J Immunol* 161:1853–9, 1998.

Berg KL, Carlberg K, Rohrschneider LR, Siminovitch KA, Stanley ER: The major SHP-1-binding tyrosine phosphorylated protein in macrophages is a member of the KIR/LIR family and an SHP-1 substrate, *Oncogene* 17:2535–41, 1998.

Kharitonenkov A, Chen Z, Sures I, Wang H, Schilling J et al: A family of proteins that inhibit signaling through tyrosine kinase receptors, *Nature* 386:181–6, 1997.

Gardai SJ, Xiao YO, Dickinson M, Nick JA, Voelker DR et al: By binding SIRP α or calreticulin/CD91, lung collectins act as dual function surveillance molecules to suppress or enhance inflammation, *Cell* 115:13–23, 2003.

Lienard H, Bruhns P, Malbec O, Fridman WH, Daeron M: Signal regulatory proteins negatively regulate immunoreceptor-dependent cell activation, *J Biol Chem* 274:32493–9, 1999.

Stofega MR, Argetsinger LS, Wang H, Ullrich A, Carter-Su C: Negative regulation of growth hormone receptor/JAK2 signaling by signal regulatory protein alpha, *J Biol Chem* 275:28222–9, 2000.

Oldenborg PA, Gresham HD, Lindberg FP: CD47-Signal regulatory protein α SIRP) regulates Fc7 and complement receptor-mediated phagocytosis, *J Exp Med* 193:855–62, 2001.

Yamao T, Noguchi T, Takeuchi O, Nishiyama U, Morita H et al: Negative regulation of platelet clearance and of the macrophage phagocytic response by the transmembrane glycoprotein SHPS-1, *J Biol Chem* 277:39833–9, 2002.

Oshima K, Ruhul Amin AR, Suzuki A, Hamaguchi M, Matsuda S: SHPS-1, a multifunctional transmembrane glycoprotein, *FEBS Lett* 519:1–7, 2002.

Veillette A, Thibaudeau E, Latour S: High expression of inhibitory receptor SHPS-1 and its association with protein-tyrosine phosphatase SHP-1 in macrophages, *J Biol Chem* 273:22719–28, 1998.

Neel BG, Gu H, Pao L: SH2-domain-containing protein-tyrosine phosphatases, in *Handbook of Cell Signaling*, Vol 1 Brandshaw RA, Dennis EA, eds, Academic Press, Amsterdam, 2004, pp 707–28.

Timms JF, Carlberg K, Gu H, Chen H, Kamatkar S et al: Identification of major binding proteins and substrates for the SH2-containing protein tyrosine phosphatase SHP-1 in macrophages, *Mol Cell Biol* 18:3838–50, 1998.

Tran KT, Rusu SO, Satish L, Wells A: Aging-related attenuation of EGF receptor signaling is mediated in part by increased protein tyrosine phosphatase activity, *Exp Cell Res* 289:359–67, 2003.

Roach TI, Slater SE, White LS, Zhang X, Majerus PW et al: The protein tyrosine phosphatase SHP-1 regulates integrin-mediated adhesion of macrophages, *Curr Biol* 8:1035–8, 1998.

David M, Chen HE, Goelz S, Larner AC, Neel BG: Differential regulation of the / interferon-stimulated Jak/Stat pathway by the SH2 domain-contaning tyrosine phosphatase SHPTP-1, *Mol Cell Biol* 15:7050–8, 1995.

Klingmuller U, Lorenx U, Cantley LC, Neel BG, Lodish HF: Specific recruitment of SH-PTP1 to the erythropoietin receptor causes inactivation of JAK2 and termination of proliferative signals, *Cell* 80:729–38, 1995.

Zhang J, Somani AK, Siminovitch KA: Roles of the SHP-1 tyrosine phosphatase in the negative regulation of cell signaling, *Semin Immunol* 12:361–78, 2000.

Dong Q, Siminovitch KA, Fialkow L, Fukushima T, Downey GP: Negative regulation of myeloid cell proliferation and function by the SH2 domain-containing tyrosine phosphatase-1, *J Immunol* 162:3220–30, 1999.

Kamata T, Yamashita M, Kimura M, Murata K, Inami M et al: Src homology 2 domain-containing tyrosine phosphatase SHP-1 controls the development of allergic airway inflammation, *J Clin Invest* 111:109–19, 2003.

Daigle I, Yousefi S, Colonna M, Green DR, Simon HU: Death receptors bind SHP-1 and block cytokine-induced anti-apoptotic signaling in neutrophils, *Nat Med* 8:61–7, 2002.

Kaushal S, Amiel GE, Guleserian KJ, Shapira OM, Perry T, Sutherland FW, Rabkin E, Moran AM, Schoen FJ, Atala A, Soker S, Bischoff J, Mayer JE, Jr. Functional small-diameter neovessels created using endothelial progenitor cells expanded ex vivo, *Nat Med* 7:1035–1040, 2001.

Urbich C, Dimmeler S. Endothelial progenitor cells: characterization and role in vascular biology. *Circ Res.* 95:343–353, 2004.

15.18. Construction of Arterial Substitutes in vivo

Hoenig MR, Campbell GR, Rolfe BE, Campbell JH: Tissue-engineered blood vessels: Alternative to autologous grafts? *Arterioscler Thromb Vasc Biol* 25(6):1128–34, 2005.

Daly CD, Campbell GR, Walker PJ, Campbell JH: In vivo engineering of blood vessels, *Front Biosci* 9:1915–24, 2004.

Chue WL, Campbell GR, Caplice N, Muhammed A, Berry CL et al: Dog peritoneal and pleural cavities as bioreactors to grow autologous vascular grafts, *J Vasc Surg* 39(4):859–67, 2004.

15.19. Construction of Arterial Substitutes with Vascualr Cells and Matrix Components in vitro

Barocas VH, Girton TS, Tranquillo RT: Engineered alignment in media equivalents: Magnetic prealignment and mandrel compaction, *J Biomech Eng* 120:660–6, 1998.

L'Heureux N, Paquet S, Labbe R, Germain L, Auger FA: A completely biological tissue-engineered human blood vessel, *FASEB J* 12:47–56, 1998.

L'Heureux N, Stoclet JC, Auger FA, Lagaud GJ, Germain L et al: A human tissue-engineered vascular media: A new model for pharmacological studies of contractile responses, *FASEB J* 15:515–24, 2001.

Matsuda T, Miwa H: A hybrid vascular model biomimicking the hierarchic structure of arterial wall: Neointimal stability and neoarterial regeneration process under arterial circulation, *J Thor Cardiovasc Surg* 110:988–97, 1995.

Nerem RM, Seliktar D: Vascular tissue engineering, *Annu Rev Biomed Eng* 3:225–43, 2001.

Grassl ED, Oegema TR, Tranquillo RT: Fibrin as an alternative biopolymer to type-I collagen for the fabrication of a media equivalent, *J Biomed Mater Res* 60:607–12, 2002.

Grassl ED, Oegema TR, Tranquillo RT: A fibrin-based arterial media equivalent, *J Biomed Mater Res A* 66:550–61, 2003.

Ross JJ, Tranquillo RT: ECM gene expression correlates with in vitro tissue growth and development in fibrin gel remodeled by neonatal smooth muscle cells, *Matrix Biol* 22:477–90, 2003.

15.20. Modulation of Growth and Function of Arterial Substitutes

McCloskey KE, Gilroy ME, Nerem RM: Use of embryonic stem cell-derived endothelial cells as a cell source to generate vessel structures in vitro, *Tissue Eng* 11(3–4):497–505, 2005.

Berglund JD, Nerem RM, Sambanis A: Incorporation of intact elastin scaffolds in tissue-engineered collagen-based vascular grafts, *Tissue Eng* 10(9–10):1526–35, 2004.

Butcher JT, Nerem RM: Porcine aortic valve interstitial cells in three-dimensional culture: Comparison of phenotype with aortic smooth muscle cells, *J Heart Valve Dis* 13(3):478–85, 2004.

Kladakis SM, Nerem RM: Endothelial cell monolayer formation: Effect of substrate and fluid shear stress, *Endothelium* 11(1):29–44, 2004.

Butcher JT, Penrod AM, Garcia AJ, Nerem RM: Unique morphology and focal adhesion development of valvular endothelial cells in static and fluid flow environments, *Arterioscler Thromb Vasc Biol* 24(8):1429–34, 2004.

Cummings CL, Gawlitta D, Nerem RM, Stegemann JP: Properties of engineered vascular constructs made from collagen, fibrin, and collagen-fibrin mixtures, *Biomaterials* 25(17):3699–706, 2004.

Stegemann JP, Dey NB, Lincoln TM, Nerem RM: Genetic modification of smooth muscle cells to control phenotype and function in vascular tissue engineering, *Tissue Eng* 10(1–2):189–99, 2004.

Nerem RM, Ensley AE: The tissue engineering of blood vessels and the heart, *Am J Transplant* 4(Suppl 6):36–42, 2004.

McCloskey KE, Lyons I, Rao RR, Stice SL, Nerem RM: Purified and proliferating endothelial cells derived and expanded in vitro from embryonic stem cells, *Endothelium* 10(6):329–36, 2003.

Stegemann JP, Nerem RM: Phenotype modulation in vascular tissue engineering using biochemical and mechanical stimulation, *Ann Biomed Eng* 31(4):391–402, 2003.

Stegemann JP, Nerem RM: Altered response of vascular smooth muscle cells to exogenous biochemical stimulation in two- and three-dimensional culture, *Exp Cell Res* 283(2):146–55, 2003.

Nerem RM: Role of mechanics in vascular tissue engineering, *Biorheology* 40(1–3):281–7, 2003.

Braddon LG, Karoyli D, Harrison DG, Nerem RM: Maintenance of a functional endothelial cell monolayer on a fibroblast/polymer substrate under physiologically relevant shear stress conditions, *Tissue Eng* 8(4):695–708, Aug 2002.

Nerem RM, Seliktar D: Vascular tissue engineering, *Annu Rev Biomed Eng* 3:225–43, 2001.

Weinberg CB, Bell E: A blood vessel model constructed from collagen and cultured vascular cells, *Science* 231(4736):397–400, 1986.

Ilveskero S, Siljander P, Lassila R: Procoagulant activity on platelets adhered to collagen or plasma clot, *Arterioscler Thromb Vasc Biolo* 21(4):628–35, 2001.

Nerem RM, Seliktar D: Vascular tissue engineering, *Annu Rev Biomed Eng* 3:225–43, 2001.

15.21. Pathogenesis, Pathology, and Clinical Features of Hypertension

Lifton RP: Molecular genetics of human blood pressure variation, *Science* 272(5262):676–80, 1996.

Kaplan NM: *Kaplan's Clinical Hypertension*, 9th ed, Lippincott Williams & Wilkins, Philadelphia, 2006.

Prisant ML: *Hypertension in the Elderly*, Humana Press, Totowa, NJ, 2005.

15.22. Animal Models of Renovascular Hypertension

Fung YC, Liu SQ: Change of residual strains in arteries due to hypertrophy caused by aortic constriction, *Circ Res* 65:1340–9, 1989.

Liu SQ, Fung YC: Relationship between hypertension, hypertrophy, and opening angle of zero-stress state of arteries following aortic constriction, *J Biomech Eng* 111:325–35, 1989.

Liu SQ, Fung YC: Indicial functions of arterial remodeling in response to locally altered mechanical stress, *Am J Physiol* 270:H1323–33, 1996.

15.23. Conventional Treatment of Hypertension

Brunton LL, Lazo JS, Parker KL: *Goodman & Gilman's Pharmacological Basis of Therapeutics*, 11th ed, McGraw-Hill New York, 2006.

Pacak K, Eisenhofer G: *Endocrine Hypertension*, Academy of Sciences, New York, 2002.

15.24. Molecular Engineering

Lifton RP: Molecular genetics of human blood pressure variation, *Science* 272:676–80, 1996.

Jeunemaitre X, Soubrier F, Kotelevetsev YV, Lifton RP, Williams CS et al: Molecular basis of human hypertension: Role of angiotensinogen, *Cell* 71:169–80, 1992.

Phillips MI: Gene therapy for hypertension: The preclinical data, *Hypertension* 38(3 Pt 2):543–8, 2001.

15.25. Nitric Oxide Synthase Gene

Lin KF, Chao L, Chao J: Prolonged reduction of high blood pressure with human nitric oxide synthase gene delivery, *Hypertension* 30:307–13, 1997.

Wang X, Cade R, Sun Z: Human eNOS gene delivery attenuates cold-induced elevation of blood pressure in rats, *Am J Physiol Heart Circ Physiol* 289(3):H1161–8, 2005.

Li L, Crockett E, Wang DH, Galligan JJ, Fink GD et al: Gene transfer of endothelial NO synthase and manganese superoxide dismutase on arterial vascular cell adhesion molecule-1 expression and superoxide production in deoxycorticosterone acetate-salt hypertension, *Arterioscler Thromb Vasc Biol* 22(2):249–55, 2002.

Pachori AS, Huentelman MJ, Francis SC, Gelband CH, Katovich MJ et al: The future of hypertension therapy: Sense, antisense, or nonsense? *Hypertension* 37(2 Pt 2):357–64, 2001.

Yu Q, Shao R, Qian HS, George SE, Rockey DC: Gene transfer of the neuronal NO synthase isoform to cirrhotic rat liver ameliorates portal hypertension, *J Clin Invest* 105(6):741–8, 2000.

Dominiczak AF, Negrin DC, Clark JS, Brosnan MJ, McBride MW et al: Genes and hypertension: From gene mapping in experimental models to vascular gene transfer strategies, *Hypertension* 35(1 Pt 2):164–72, 2000.

Lin KF, Chao L, Chao J: Prolonged reduction of high blood pressure with human nitric oxide synthase gene delivery, *Hypertension* 30(3 Pt 1):307–13, 1997.

Cuevas P, Garcia-Calvo M, Carceller F, Reimers D, Zazo M et al: Correction of hypertension by normalization of endothelial levels of fibroblast growth factor and nitric oxide synthase in spontaneously hypertensive rats, *Proc Natl Acad Sci USA* 93(21):11996–2001, 1996.

15.26. Atrial Natriuretic Peptide Gene

Lin KF, Chao J, Chao L: Atrial natriuretic peptide gene delivery attenuates hypertension, cardiac hypertrophy, and renal injury in salt-sensitive rats, *Hum Gene Ther* 9:1429–38, 1998.

Kook H, Itoh H, Choi BS, Sawada N, Doi K et al: Physiological concentration of atrial natriuretic peptide induces endothelial regeneration in vitro, *Am J Physiol Heart Circ Physiol* 284(4): H1388–97, 2003.

Kiemer AK, Weber NC, Vollmar AM: Induction of IkappaB: Atrial natriuretic peptide as a regulator of the NF-kappaB pathway, *Biochem Biophys Res Commun* 295(5):1068–76, 2002.

de Bold AJ: Atrial natriuretic factor: A hormone produced by the heart, *Science* 230:767–70, 1985.

Greenberg BD, Bencen GH, Seilhamer JJ, Lewicki JA, Fiddes JC: Nucleotide sequence of the gene encoding human atrial natriuretic factor precursor, *Nature* 312:656–8, 1984.

John SWM, Krege JH, Oliver PM, Hagaman JR, Hodgin JB et al: Genetic decreases in atrial natriuretic peptide and salt-sensitive hypertension, *Science* 267:679–81, 1995.

Lang RE, Tholken H, Ganten D, Luft FC, Ruskoaho H et al: Atrial natriuretic factor—a circulating hormone simulated by volume loading, *Nature* 314:264–6, 1985.

Needleman P, Greenwald JE: Atriopeptin: A cardiac hormone intimately involved in fluid, electrolyte, and blood-pressure homeostasis, *New Engl J Med* 314:828–34, 1986.

Nemer M, Chamberland M, Sirois D, Argentin S, Drouin J et al: Gene structure of human cardiac hormone precursor, pronatriodilatin, *Nature* 312:654–6, 1984.

Human protein reference data base, Johns Hopkins University and the Institute of Bioinformatics, at http://www.hprd.org/protein.

15.27. Kallikrein

Azizi M, Boutouyrie P, Bissery A, Agharazii M, Verbeke F et al: Arterial and renal consequences of partial genetic deficiency in tissue kallikrein activity in humans, *J Clin Invest* 115:780–7, 2005.

Meneton P, Bloch-Faure M, Hagege AA, Ruetten H, Huang W et al: Cardiovascular abnormalities with normal blood pressure in tissue kallikrein-deficient mice, *Proc Natl Acad Sci USA* 98:2634–9, 2001.

Riegman PHJ, Vlietstra RJ, Suurmeijer L, Cleutjens CBJM, Trapman J: Characterization of the human kallikrein locus, *Genomics* 14:6–11, 1992.

Chao J, Chao L: Kallikrein gene therapy: A new strategy for hypertensive diseases, *Immunopharmacology* 36:229–36, 1997.

Dobrzynski E, Yoshida H, Chao J, Chao L: Adenovirus-mediated kallikrein gene delivery attenuates hypertension and protects against renal injury in deoxycorticosterone-salt rats, *Immunopharmacology* 44:57–65, 1999.

15.28. Antisense Oligodeoxynucleotides for Angiotensinogen mRNA

Wielbo D, Simon A, Phillips MI, Toffolo S: Inhibition of hypertension by peripheral administration of antisense oligodeoxynucleotides, *Hypertension* 28:147–51, 1996.

Tomita N, Morishita R, Higaki J, Aoki M, Nakamura Y et al: Transient decrease in high blood pressure by in vivo transfer of antisense oligodeoxynucleotides against rat angiotensinogen, *Hypertension* 26:131–6, 1995.

Makino K, Sugano M, Ohtsuka S, Sawada S: Intravenous injection with antisense oligodeoxynucleotides against angiotensinogen decreases blood pressure in spontaneously hypertensive rats, *Hypertension* 31:1166–70, 1998.

Tang X, Mohuczy D, Zhang CY, Kimura B, Galli SM et al: Intravenous angiotensinogen antisense in AAV-based vector decreases hypertension, *Am J Physiol* 277(6 Pt 2):H2392–9, 1999.

15.29. Antisense Oligodeoxynucleotides for Angiotensin II type 1 (AT_1) Receptor mRNA

Phillips MI: Gene therapy for hypertension: The preclinical data, *Hypertension* 38(3 Pt 2):543–8, 2001.

Tang X, Mohuczy D, Zhang YC, Kimura B, Galli SM et al: Intravenous angiotensinogen antisense in AAV-based vector decreases hypertension, *Am J Physiol* 277(6 Pt 2):H2392–9, 1999.

Kagiyama S, Varela A, Phillips MI, Galli SM: Antisense inhibition of brain renin angiotensin system decreased blood pressure in chronic 2-kidney, 1-clip hypertensive rats, *Hypertension* 32:371–5, 2001.

Peng JF, Kimura B, Fregly MJ, Phillips MI: Reduction of cold-induced hypertension by antisense oligodeoxynucleotides to angiotensinogen mRNA and AT_1 receptor mRNA in brain and blood, *Hypertension* 31:1317–23, 1998.

Gyurko R, Tran D, Phillips MI: Time course of inhibition of hypertension by antisense oligonucleotides targeted to AT_1 angiotensin receptor mRNA in spontaneously hypertensive rats, *Am J Hypertens* 10:56S–62S, 1997.

Galli SM, Phillips MI: Angiotensin II AT_{1A} receptor antisense lowers blood pressure in acute 2-kidney, 1-clip hypertension, *Hypertension* 38:674–8, 2001.

Iyer SN, Lu D, Katovich M, Raizada MK: Chronic control of high blood pressure in the spontaneously hypertensive rats by delivery of angiotensin type 1 receptor antisense, *Proc Natl Acad Sci USA* 93:9960–5, 1996.

Reaves PY, Gelband CH, Wang H, Yang H, Lu D et al: Permanent prevention of hypertension by angiotensin II type 1 receptor antisense gene therapy, *Circ Res* 86:E44–50, 1999.

15.30. Antisense Oligonucleotides for Adrenergic Receptor mRNA

α1A-Adrenergic Receptors

Chang DJ, Chang TK, Yamanishi SS, Salazar FHR, Kosaka AH et al: Molecular cloning, genomic characterization and expression of novel human alpha-1A-adrenoceptor isoforms, *FEBS Lett* 422:279–83, 1998.

Forray C, Bard JA, Wetzel JM, Chiu G, Shapiro E et al: The alpha-1-adrenergic receptor that mediates smooth muscle contraction in human prostate has the pharmacological properties of the cloned human alpha-1C subtype, *Mol Pharm* 45:703–8, 1994.

Hirasawa A, Horie K, Tanaka T, Takagaki K, Murai M et al: Cloning, functional expression and tissue distribution of human cDNA for the alpha-1C-adrenergic receptor, *Biochem Biophys Res Commun* 195:902–9, 1993.

Hirasawa A, Shibata K, Horie K, Takei Y, Obika K et al: Cloning, functional expression and tissue distribution of human alpha-1C-adrenoceptor splice variants, *FEBS Lett* 363:256–60, 1995.

Schwinn DA, Johnston GI, Page SO, Mosley MJ, Wilson KH et al: Cloning and pharmacological characterization of human alpha-1 adrenergic receptors: Sequence corrections and direct comparison with other species homologues, *J Pharm Exp Ther* 272:134–42, 1995.

Tseng-Crank J, Kost T, Goetz A, Hazum S, Roberson KM et al: The alpha-1C-adrenoceptor in human prostate: Cloning, functional expression and localization to specific prostatic cell types, *Br J Pharm* 115:1475–85, 1995.

Weinberg DH, Trivedi P, Tan CP, Mitra S, Perkins-Barrow A et al: Cloning, expression and characterization of human alpha adrenergic receptors alpha-1A, alpha-1B, and alpha-1C, *Biochem Biophys Res Commun* 201:1296–304, 1994.

α1B-Adrenergic Receptors

Allen LF, Lefkowitz RJ, Caron MG, Cotecchia S: G-protein-coupled receptor genes as protooncogenes: Constitutively activating mutation of the alpha-1B-adrenergic receptor enhances mitogenesis and tumorigenicity, *Proc Natl Acad Sci USA* 88:11354–8, 1991.

Cavalli A, Lattion AL, Hummler E, Nenniger M, Pedrazzini T et al: Decreased blood pressure response in mice deficient of the alpha(1b)-adrenergic receptor, *Proc Natl Acad Sci USA* 94:11589–94, 1997.

Krushkal J, Xiong M, Ferrell R, Sing CF, Turner ST et al: Linkage and association of adrenergic and dopamine receptor genes in the distal portion of the long arm of chromosome 5 with systolic blood pressure variation, *Hum Mol Genet* 7:1379–83, 1998.

Loftus SK, Shiang R, Warrington JA, Bengtsson U, McPherson JD et al: Genes encoding adrenergic receptors are not clustered on the long arm of human chromosome 5, *Cytogenet Cell Genet* 67:69–74, 1994.

Lomasney JW, Cotecchia S, Lorenz W, Leung WY, Schwinn DA et al: Molecular cloning and expression of the cDNA for the alpha-1A-adrenergic receptor: The gene for which is located on human chromosome 5, *J Biol Chem* 266:6365–9, 1991.

Oakey RJ, Caron MG, Lefkowitz RJ, Seldin MF: Genomic organization of adrenergic and serotonin receptors in the mouse: Linkage mapping of sequence-related genes provides a method for examining mammalian chromosome evolution, *Genomics* 10:338–44, 1991.

Ramarao CS, Kincade Denker JM, Perez DM, Gaivin RJ, Riek RP et al: Genomic organization and expression of the human alpha-1B-adrenergic receptor, *J Biol Chem* 267:21936–45, 1992.

Yang-Feng TL, Xue F, Zhong W, Cotecchia S, Frielle T et al: Chromosomal organization of adrenergic receptor genes, *Proc Natl Acad Sci USA* 87:1516–20, 1990.

α1D-Adrenergic Receptors

Bruno JF, Whittaker J, Song J, Berelowitz M: Molecular cloning and sequencing of a cDNA encoding a human alpha-1A-adrenergic receptor, *Biochem Biophys Res Commun* 179:1485–90, 1991.

Esbenshade TA, Hirasawa A, Tsujimoto G, Tanaka T, Yano J et al: Cloning of the human alpha-1D-adrenergic receptor and inducible expression of three subtypes in SK-N-MC cells, *Mol Pharm* 47:977–85, 1995.

Lomasney JW, Cotecchia S, Lorenz W, Leung WY, Schwinn DA et al: Molecular cloning and expression of the cDNA for the alpha-1A-adrenergic receptor: The gene for which is located on human chromosome 5, *J Biol Chem* 266:6365–9, 1991.

Weinberg DH, Trivedi P, Tan CP, Mitra S, Perkins-Barrow A et al: Cloning, expression and characterization of human alpha adrenergic receptors alpha-1A, alpha-1B, and alpha-1C, *Biochem Biophys Res Commun* 201:1296–304, 1994.

Wilkie TM, Chen Y, Gilbert DJ, Moore KJ, Yu L et al: Identification chromosomal location and genome organization of mammalian G-protein-coupled receptors, *Genomics* 18:175–84, 1993.

Adrenergic Receptors β_1, β_2, β_3

Bachman ES, Dhillon H, Zhang CY, Cinti S, Bianco AC et al: Beta-AR signaling required for diet-induced thermogenesis and obesity resistance, *Science* 297:843–5, 2002.

Frielle T, Collins S, Daniel KW, Caron MG, Lefkowitz RJ et al: Cloning of the cDNA for the human beta-1-adrenergic receptor, *Proc Natl Acad Sci USA* 84:7920–4, 1987.

Frielle T, Kobilka B, Lefkowitz RJ, Caron MG: Human beta-1- and beta-2-adrenergic receptors: Structurally and functionally related receptors derived from distinct genes, *Trends Neurosci* 11:321–4, 1988.

Jahns R, Boivin V, Hein L, Triebel S, Angermann CE et al: Direct evidence for a beta-1-adrenergic receptor-directed autoimmune attack as a cause of idiopathic dilated cardiomyopathy, *J Clin Invest* 113:1419–29, 2004.

Magnusson Y, Marullo S, Hoyer S, Waagstein F, Andersson B et al: Mapping of a functional autoimmune epitope on the beta(1)-adrenergic receptor in patients with idiopathic dilated cardiomyopathy, *J Clin Invest* 86:1658–63, 1990.

Mason DA, Moore JD, Green SA, Liggett SB: A gain-of-function polymorphism in a G-protein coupling domain of the human beta-1-adrenergic receptor, *J Biol Chem* 274:12670–4, 1999.

Mialet Perez J, Rathz DA, Petrashevskaya NN, Hahn HS, Wagoner LE et al: Beta-1-adrenergic receptor polymorphisms confer differential function and predisposition to heart failure, *Nature Med* 9:1300–5, 2003.

Rohrer DK, Chruscinski A, Schauble EH, Bernstein D, Kobilka BK: Cardiovascular and metabolic alterations in mice lacking both beta-1- and beta-2-adrenergic receptors, *J Biol Chem* 274:16701–8, 1999.

Small KM, Wagoner LE, Levin AM, Kardia SLR, Liggett SB: Synergistic polymorphisms of beta-1- and alpha-2C-adrenergic receptors and the risk of congestive heart failure, *New Engl J Med* 347:1135–42, 2002.

Xu J, He J, Castleberry AM, Balasubramanian S, Lau AG et al: Heterodimerization of alpha-2A- and beta-1-adrenergic receptors, *J Biol Chem* 278:10770–7, 2003.

Large V, Hellstrom L, Reynisdottir S, Lonnqvist F, Eriksson P et al: Human beta-2 adrenoceptor gene polymorphisms are highly frequent in obesity and associate with altered adipocyte beta-2 adrenoceptor function, *J Clin Invest* 100:3005–13, 1997.

Luttrell LM, Ferguson SSG, Daaka Y, Miller WE, Maudsley S et al: Beta-arrestin-dependent formation of beta-2 adrenergic receptor-Src protein kinase complexes, *Science* 283:655–61, 1999.

Maurice JP, Hata JA, Shah AS, White DC, McDonald PH et al: Enhancement of cardiac function after adenoviral-mediated in vivo intracoronary beta-2-adrenergic receptor gene delivery, *J Clin Invest* 104:21–9, 1999.

Communal C, Singh K, Sawyer DB, Colucci WS: Opposing effects of β_1- and β_2-adrenergic receptors on cardiac myocyte apoptosis: role of pertussis toxin-sensitive G protein, *Circulation* 100:2210–12, 1999.

Milano CA, Allen LF, Rockman HA, Dolber PC, McMinn TR et al: Enhanced myocardial function in transgenic mice overexpressing the β_2-adrenergic receptor, *Science* 264:582–6, 1994.

Engelhardt S, Hein L, Wiesmann F, Lohse MJ: Progressive hypertrophy and heart failure in β_1-adrenergic receptor transgenic mice, *Proc Natl Acad Sci USA* 96:7059–64, 1999.

Milano CA, Allen LF, Rockman HA, Dolber PC, McMinn TR et al: Enhanced myocardial function in transgenic mice overexpressing the β_2-adrenergic receptor, *Science* 264:582–6, 1994.

Dorn GW, Tepe NM, Lorenz JN, Davis MG, Koch WJ et al: Low and high transgenic expression of β_2-adrenergic receptors differentially affects cardiac hypertrophy and function in $G\alpha q$ overexpressing mice, *Proc Natl Acad Sci USA* 96:6400–5, 1999.

Maurice JP, Hata JA, Shah AS, White DC, McDonald PH et al: Enhancement of cardiac function after adenoviral-mediated in vivo intracoronary beta-2-adrenergic receptor gene delivery, *J Clin Invest* 104:21–9, 1999.

Steinberg SF: The molecular basis for distinct β-adrenergic receptor subtype actions in cardiomyocytes, *Circ Res* 85:1101–11, 1999.

Maurice JP, Hata JA, Shah AS, White DC, McDonald PH et al: Enhancement of cardiac function after adenoviral-mediated in vivo intracoronary β_2-adrenergic receptor gene delivery, *J Clin Invest* 104:21–9, 1999.

Kawahira Y, Sawa Y, Nishimura M, Sakakida S, Ueda H et al: Gene transfection of β_2-adrenergic receptor into the normal rat heart enhances cardiac response to β-adrenergic agonist, *J Thor Cardiovasc Surg* 118:446–51, 1999.

Kawahira Y, Sawa Y, Nishimura M, Sakakida S, Ueda H et al: In vivo transfer of a β_2-adrenergic receptor gene into the pressure-overloaded rat heart enhances cardiac response to β-adrenergic agonist, *Circulation* 98:II262–7, 1998.

Shah AS, Lilly RE, Kypson AP, Tai O, Hata JA et al: Intracoronary adenovirus-mediated delivery and overexpression of the β_2-adrenergic receptor in the heart: Prospects for molecular ventricular assistance, *Circulation* 101:408–14, 2000.

Human protein reference data base, Johns Hopkins University and the Institute of Bioinformatics, at http://www.hprd.org/protein.

16

PULMONARY REGENERATIVE ENGINEERING

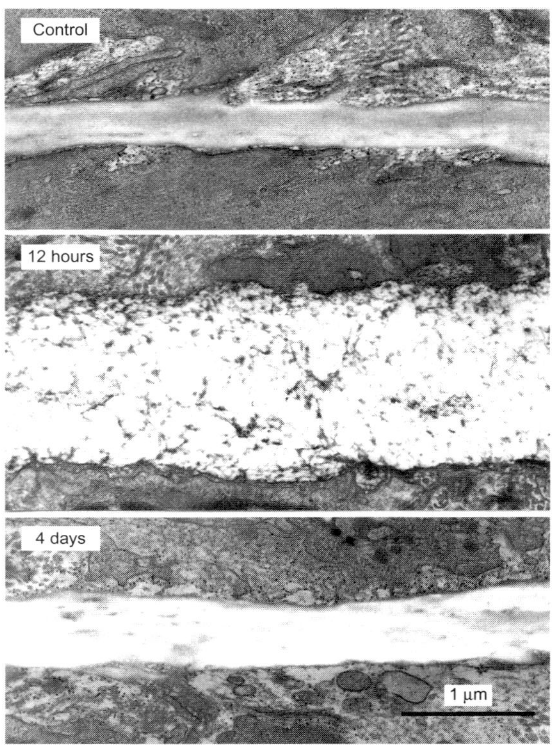

Influence of hypoxic pulmonary hypertension on the structure of the medial elastic laminae of the pulmonary arteries in the rat. Pulmonary arterial hypertension was induced by exposing rats to 10% oxygen and 90% nitrogen. The structure of the pulmonary arteries was examined by electron microscopy at 0, 0.5, and 4 days of exposure to hypoxia. An increase in the pulmonary arterial blood pressure induced rapid swelling and disorganization of the pulmonary arterial elastic laminae within 12 h of exposure to hypoxia. These pathological changes were transient. The pulmonary arterial elastic laminae regained their physiological morphology and appearance after 4 days of exposure to hypoxia without further noticeable changes even under a continuous identical hypoxic condition. See color insert.

Bioregenerative Engineering: Principles and Applications, by Shu Q. Liu

ANATOMY AND PHYSIOLOGY OF THE RESPIRATORY SYSTEM

Pulmonary Structure [16.1]

The pulmonary or respiratory system consists of the rib cage, respiratory skeletal muscles, diaphragm, pleura, and lung. The rib cage is the wall of the thoracic cavity. Respiratory muscles are attached to the ribs and other thoracic bones such as the clavicle bone. The diaphragm is a muscular and connective tissue membrane and constitutes the inferior border of the thoracic cavity. The pleura is a membrane system that covers the external surface of the lung, known as the visceral pleura, and internal surface of the rib cage, known as the *parietal pleura*. The two pleural membranes are continuous, forming a narrow, closed pleural cavity between the exterior surface of the lung and the internal surface of the rib cage (Fig. 16.1). As we will see below, this closed cavity plays a critical role in air ventilation of the lung.

The lung is composed of the airway, alveolar, vascular, and lymphatic systems. The lung is divided into the left and right lung in the thoracic cavity. The primary functions of the lung are blood oxygenation and carbon dioxide removal via gas ventilation and exchange. The airways are a tubular system responsible for gas ventilation. The alveoli constitute a membrane system responsible for gas exchange, specifically, oxygen transport from the air to the blood and carbon dioxide transport from the blood to the air. The vascular system carries deoxygenated blood from the systemic veins and right heart to the lung and delivers oxygenated blood from the lung to the left heart and systemic arteries. The lymphatic system collects and removes excessive fluid from the lung parenchyma. The anatomy and functions of these systems are described as follows.

The Airway and Alveolar Systems. The airway system consists of the larynx and various generations of airway. The *larynx* is a cartilage structure located between the pharynx and trachea and its function is to keep the trachea (the main airway) closed during swallowing, thus preventing water and foods from entering the airways and lung. The larynx also contains the vocal cords, which are structures for sound generation. The airways are a tubular system, consisting of tree-like cylindrical structures of various diameters. The largest airway is the trachea, which is composed of three layers: the mucosal, cartilage/smooth muscle, and connective tissue layers. The mucosa is constituted with two types of columnar epithelial cells: goblet and ciliated epithelial cells. The goblet epithelial cells secret mucus

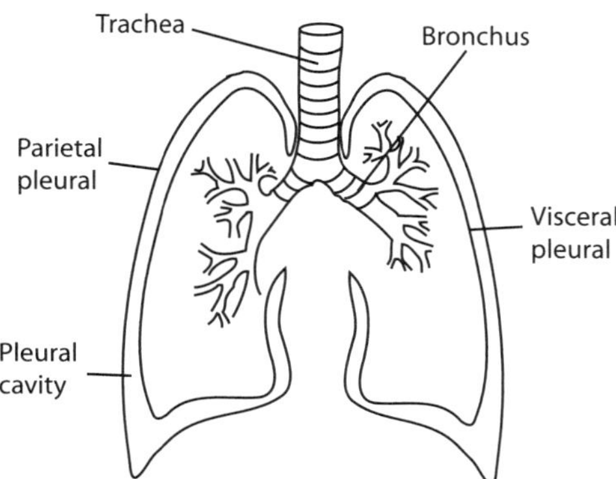

Figure 16.1. Schematic representation of the structure of the lung.

in response to the stimulation of inhaled particles. The ciliated epithelial cells are characterized by the presence of cilia at the cell surface. The cilia conduct periodic movements and are responsible for propelling and removing mucus and inhaled particles. In the middle layer, there is a series of C-shaped cartilage rings which alternate with connective/muscular tissue. The cartilage appears in the anterior and side wall, but not in the posterior wall of the trachea. The cartilage structures protect the trachea from collapsing.

The trachea is divided into the left and right bronchi, which enter the left and right lungs, respectively. The structure and cellular components of the left and right bronchi are similar to those of the trachea. Each bronchus is further divided into about 19 generations of bronchi with a graded decrease in diameter. The smallest airways are defined as bronchioles and the last generation is called terminal bronchioles. There are a large number of terminal bronchioles in the lung. The function of the bronchial tree is to conduct gases into and out of the lung. The structure of the small bronchi is different from that of the trachea and major bronchi. A major difference is that bronchi after the second generation do not contain cartilage rings. In addition, the density of goblet and ciliated epithelial cells reduces gradually as the airway diameter decreases. All bronchi contain smooth muscle cells, which control the diameter of the airways by contraction and relaxation.

The terminal bronchi are connected to the alveolar ducts, a structure mixed with tubular terminal bronchioles and alveoli. Alveoli are clusters of thin-walled, connected membrane sacs with an average size about 200 μm. There are about 300–400 million alveoli in the lung. The primary function of the alveoli is gas exchange between the blood and alveolar air. The wall of each alveolus is composed of a monolayer of alveolar epithelial cells on each side of the wall and a dense network of capillaries constituted with a monolayer of endothelial cells. There are two types of alveolar epithelial cells: type I and II epithelial cells. *Type I cells* are squamous thin epithelial cells that covers about 90% of the alveolar surface, where gas exchange takes places. *Type II cells* are specially differentiated epithelial cells which produce and secret surfactant, a mixture of lipids and proteins that spreads over the alveolar surface. The function of the surfactant is to reduce the surface tension of the alveolar wall at the interface between the air and the cell surface, ensuring even expansion and reduction of the alveoli through the entire lung during inspiration and expiration, respectively.

The Vascular System. The vascular system is composed of pulmonary arteries, capillaries, and veins. The pulmonary arteries originate from the right ventricle and are organized into a tree-like tubular system with a graded decrease in vessel diameter. Pulmonary arteries at each generation are arranged together with airways and pulmonary veins of the same generation. These arteries conduct deoxygenated blood from the right ventricle to the alveolar capillaries. A pulmonary capillary is a tube-like structure composed of a monolayer of endothelial cells about 6–10 μm in diameter and a basement membrane around the endothelial cells. The wall of alveoli contains a dense network of capillaries, where gas exchange occurs: oxygen diffuses from the alveolar air into the blood, and carbon dioxide diffuses from the blood to the alveolar air. The pulmonary veins are a tree-like tubular system, composed of multiple generations of veins and organized in parallel to the arterial system. The pulmonary veins conduct oxygenated blood from the capillaries to the left atrium.

The Lymphatic System. The pulmonary lymphatic system is a network of multiple generations of lymphatic vessels distributed around the airways and within the parenchymal tissue. The alveolar wall does not contain lymphatic vessels. The primary function of the lymphatic system is to collect and remove excessive fluid from the parenchymal tissue. Because of the lack of lymphatic vessels at the alveolar level, the lung is susceptible to pulmonary edema, a disorder with excessive fluid in the parenchymal tissue and alveolar space. Pulmonary edema reduces the rate of gas exchange across the alveolar wall and is often found in patients with left heart failure, which induces an increase in blood pressure in the alveolar capillaries and excessive fluid transport from the blood to the alveolar space.

Pulmonary Function [16.1]

Gas Ventilation and Exchange. *Gas ventilation* is a cyclic process that includes an inspiration and an expiration phase. During the inspiration phase, fresh air is moved into the lung, whereas during the expiration phase exhaust air is removed from the lung. Gas ventilation is accomplished by coordinated action of a number of pulmonary constituents, including the airways, alveoli, pleural cavity, chest wall, skeletal muscles, and diaphragm. During the inspiration phase, the inward air flow is driven by a pressure gradient from the nasal cavity to the alveoli. The pressure gradient is generated by the contraction of inspiratory skeletal muscles, including the scalenes, pectoralis minor, external intercostals, and diaphragm. The contraction of these muscles increases the volume of the thoracic cavity. Because the pleural cavity is a closed negative system, the expansion of the thoracic cavity induces simultaneous expansion of the lung, resulting in a pressure gradient from the nasal cavity to the alveoli. During the expiration phase, the inspiratory muscles relax and the rib cage recoiled back, resulting in a decrease in the thoracic cavity and the volume of the lung. The volumetric change establishes an adverse pressure gradient from the alveoli to the nasal cavity, which removes the exhaust air from the lung.

Gas exchange is a process of gas diffusion across the alveolar epithelial and endothelial cells, involving oxygen and carbon dioxide. Oxygen diffuses from the alveolar air space into the blood, whereas carbon dioxide diffuses from the blood to the alveolar space. The driving force for gas diffusion is the gradient of partial gas pressure, which is defined as the pressure contributed by a specified type of gas in a mixture of multiple gases. The total air pressure is about 760 mm Hg at the sea level. The *oxygen partial pressure* (Po_2) is about 155 mm Hg in the air, about 105 mm Hg in the alveolar space, and about 40 mm Hg in the deoxygenated venous blood in the alveolar capillary network (Fig. 16.2). The difference of Po_2 is about 65 mm Hg across the alveolar wall. This partial pressure difference drives oxygen diffusion from the alveolar space to the blood. Since oxygen diffusion is very efficient across the alveolar wall and the capillaries are fairly long, the capillary blood can be saturated with oxygen within the first half of the capillary length. Oxygenated blood is conducted to the pulmonary veins, left heart, and the arterial system.

At the same time, carbon dioxide diffuses from the blood to the alveolar space based on a gradient of partial pressure. The *partial pressure of* CO_2 is about 45 mm Hg in the deoxygenated venous blood and about 40 mm Hg in the alveolar space. The difference of carbon dioxide partial pressure across the alveolar wall drives CO_2 diffusion from the blood to the alveolar space. During expiration, the exhaust air containing a higher concentration of CO_2 is removed from the lung.

Ratio of Air Ventilation to Blood Perfusion. The rate of air ventilation in a region of the lung or in the entire lung is proportional to the rate of blood perfusion within the same region. The ratio of air ventilation to blood perfusion is about 0.75 under physiological conditions. The pulmonary system intends to maintain this ratio within a narrow range around 0.75 under pathological conditions. For instance, when a bronchus is partially

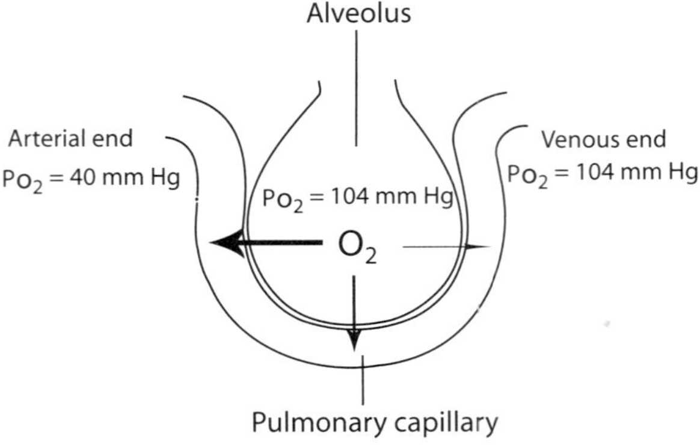

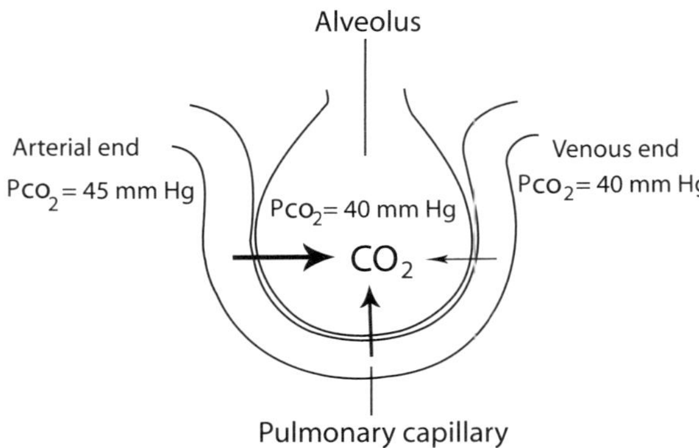

Figure 16.2. Schematic representation of the structure of the airways and alveoli. Based on bibliography 16.1.

blocked with a tumor, the rate of air ventilation is accordingly reduced in the alveolar system distal to the blocked bronchus, resulting in a reduction in the ventilation-to-perfusion ratio. Reduced oxygen or hypoxia in the under-ventilated region enhances the contractility of the arterial smooth muscle cells and induces arterial constriction, reducing the rate of blood perfusion into the region. Such a response brings up the ventilation-to-perfusion ratio toward the physiological level. With such a mechanism, the ventilation-to-perfusion ratio is maintained at a relatively constant range. The physiological significance of such a mechanism is to redistribute blood volume to well-ventilated regions and to ensure a sufficient volume of fully oxygenated blood.

Control of Gas Ventilation. There are two types of respiratory movement: involuntary and voluntary. The *involuntary movement* is rhythmic and responsible for the maintenance of the basal level of air ventilation. Such a movement is controlled by the respiratory center located in the medulla oblongata of the brainstem. In the case of increased physical exercise, which demands more oxygen in the skeletal muscle system and the heart, the respiratory center increases the contraction frequency and depth of the respiratory muscles, resulting in an increase in air ventilation and oxygen supply to the alveoli. At the same time, the heart beating rate and cardiac contractility are also increased under the influence of activated cardiovascular control center, resulting in an increase in blood supply to the lung. The increased blood supply can be fully oxygenated because of enhanced air ventilation. The *voluntary movement* causes forced respiration. Such an activity is controlled by the cortex neurons for regulating the respiratory muscles. Under physiological conditions, it is not necessary to carry out voluntary respiration.

PULMONARY DISORDERS

Asthma

Pathogenesis, Pathology, and Clinical Features [16.2]. *Asthma* is defined as a chronic disorder of the airways characterized by increased responsiveness of the airways to a multiplicity of stimuli and inflammatory reactions in the airway wall. It is manifested by widespread narrowing of the air passages in association with clinical symptoms, such as paroxysms of dyspnea, cough, and wheezing. The clinical symptoms may be relieved spontaneously or as a result of therapy. Asthma is an episodic disease with acute exacerbations interspersed with symptom-free periods. Typically, most attacks are short-lived, lasting minutes to hours. The patient can recover completely after an attack. However, there can be a phase in which the patient experiences some degree of airway obstruction daily. This phase can be mild, with or without superimposed severe episodes, or much more serious, with severe obstruction persisting for days or weeks. The latter condition is known as *status asthmaticus*. In unusual circumstances, acute episodes can cause death.

Asthma is a chronic airway disorder characterized by transient airway constriction or obstruction (Fig. 16.3), airway inflammation and remodeling (Fig. 16.4), airway cell

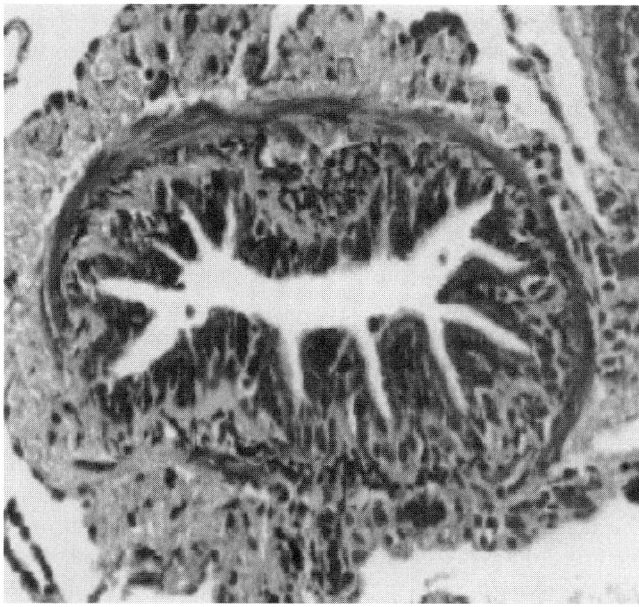

Figure 16.3. Airway remodeling in asthma. Remodeling of small airway in chronic asthma. Note connective tissue deposition in subepithelial and adventitial compartments of the airway wall. Connective tissue deposition seems to encase the airway smooth muscle bundles. (Reprinted from Chung KF: The role of airway smooth muscle in the pathogenesis of airway wall remodeling in chronic obstructive pulmonary disease, *Proc Am Thorac Soc*) 2:347–54, copyright 2005, with permission from American Thoracic Society.)

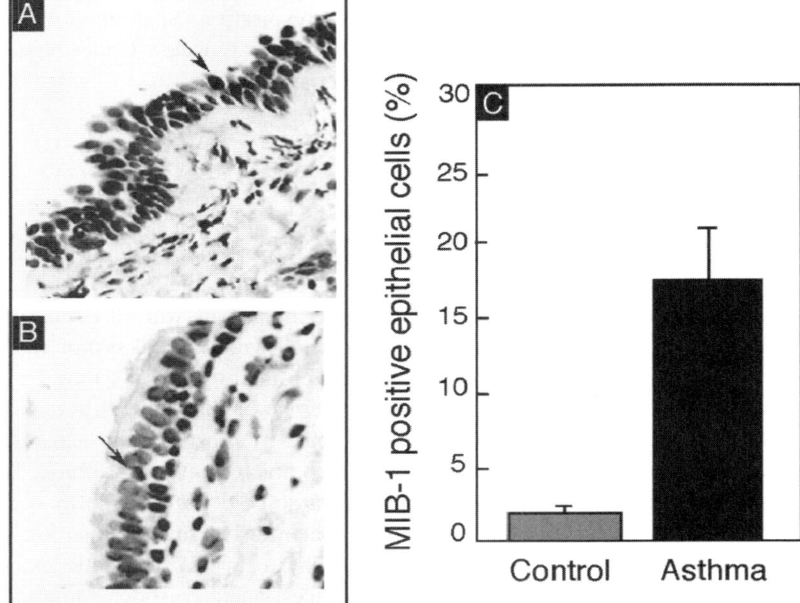

Figure 16.4. Airway cell proliferation in control and asthma. Cell proliferation was detected by an antibody (MIB1) directed against the Ki-67 antigen, a marker for proliferating cells. The black color indicates positive MIB1 staining in the asthmatic epithelial cells (A) and healthy controls (B). (C) The graph shows MIB1-positive cells (mean ± SD) of three healthy controls and four asthmatics. Note that some fields in asthmatic airways show more than 80% MIB1-positive cells. Arrows show positive cells. (Reprinted with permission from Comhair SA et al: Superoxide dismutase inactivation in pathophysiology of asthmatic airway remodeling and reactivity, *Am J Pathol* 166:663–74, copyright 2005.)

proliferation (Fig. 16.5), and persistently increased responsiveness and contractility of airway smooth muscle cells in response to the stimulation of allergens, environmental factors, and pharmacological agents. Typical asthmatic allergens include air-borne particles derived from plants (e.g., grass pollens, ragweed pollens, birch pollens, mountain cedar pollens, and peanuts), animals (cat and dog fur dusts), and microorganisms (bacteria and viruses). Examples of environmental factors include air pollutants, chemicals, metal particles, dusts, and polymer particles. Pharmacological agents that cause asthma include antibiotics, aspirin, and β-adrenergic antagonists. In addition, viral infection of the lung, physical exercise, and emotional stress can initiate asthmatic attacks. All these factors can cause an increase in the reactivity of the airway smooth muscle cells, resulting in airway constriction and a reduction in the rate of ventilation. Asthmatic attacks are usually transient, lasting for a period from minutes to hours. The clinical consequence of asthma is dependent on the degree of airway constriction and oxygen deficiency. Severe airway constriction is often life threatening. Asthma occurs frequently in children and young adults. Asthmatic patients often have a family history of allergic diseases.

The pathogenic mechanisms of asthma vary, depending on the factors that cause asthma. Allergic asthma, one of the most common types, is related to enhanced immune

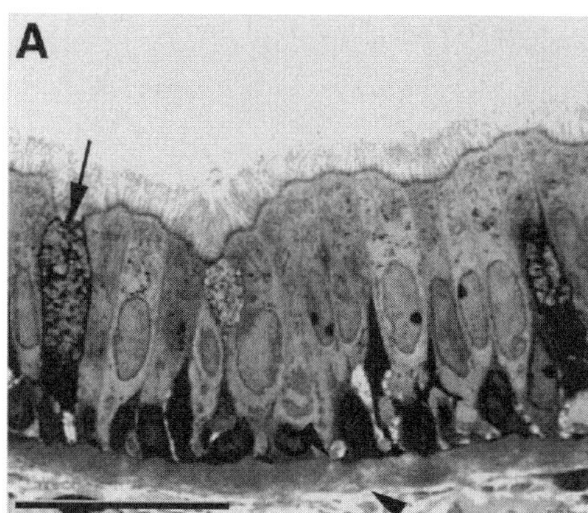

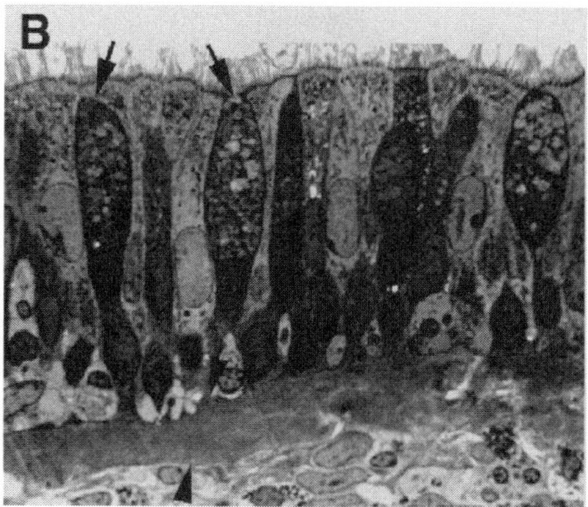

Figure 16.5. Inflammatory changes in asthmatic airways in rhesus monkeys. Histopathological comparison of epithelial morphology in the intrapulmonary bronchi of sensitized (B) and nonsensitized (A) rhesus monkeys with house dust mite (*Dermatophagoides farinae*). Note that the asthmatic airway was associated with inflammatory cell infiltration, epithelial hypertrophy, and mucous goblet cell hyperplasia (arrows). Scale bar: 20 μm. (Reprinted with permission from Schelegle ES et al: Allergic asthma induced in rhesus monkeys by house dust mite (Dermatophagoides farinae), *Am J Pathol* 158:333–41, copyright 2001.)

responses of lymphocytes exposed to allergens. In patients genetically susceptible to asthma, a primary exposure to an allergen in the airway system leads to the activation of the antigen-presenting dentritic cells (Fig. 16.6). These cells present the inhaled antigen to $CD4^+$ T-helper cell precursors in the bronchial lymph nodes, potentially inducing the differentiation of the precursor cells into two types of cells: type 1 and type 2 T-helper cells. The fate of the differentiation is dependent on the presence of dominant cytokines. When interleukin (IL)12 is dominant, the precursor cells are induced to differentiate into type 1 T-helper cells. In contrast, in the presence of dominant IL4, the precursor cells are differentiated into type 2 T-helper cells. The type 1 T-helper cells can produce interferon-β, IL2, and Tumor necrosis factor (TNF)α. These factors exert inhibitory effects on asthmatic activities. In contrast, the type 2 T-helper cells produce IL4, IL5, and IL9, which promote asthmatic activities.

The type 2 T-helper cell cytokines can stimulate B lymphocytes to produce IgE antibodies against the allergen. These antibodies can attach to the surface of mast cells. A secondary exposure to the same allergen induces reaction of the allergen with the IgE antibodies attached to the mast cells, resulting in antibody activation. The activated antibodies in turn stimulate mast cells to release inflammatory factors, such as histamine, prostaglandins, and bradykinin, causing the contraction of airway smooth muscle cells. These inflammatory factors also cause mucosal edema, an increase in mucus secretion, and infiltration of leukocytes, especially eosinophils. IgE and allergens can bind to eosinophils, inducing the release of major basic protein (MBP) from the eosinophils. MBP can cause airway injury and inflammation. All these pathological changes contribute to the reduction in the luminal area of the bronchi. As asthma attacks continue, chronic inflammatory reactions may occur, inducing persistent leukocyte infiltration, mucosal thickening, and airway constriction.

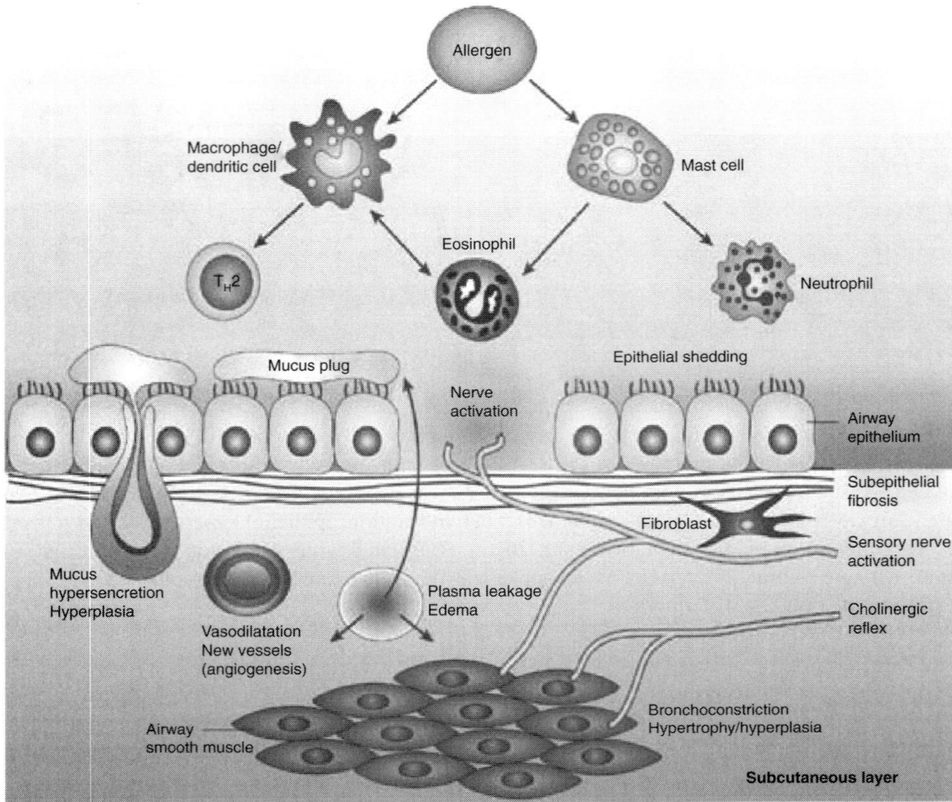

Figure 16.6. Pathogenesis of asthma. Several inflammatory cell types=s are recruited to and/or activated in the airways, releasing a variety of inflammatory mediators that have acute effects on the airway, such as bronchoconstriction, plasma leakage, vasodilatation, mucus secretion, sensory nerve activation, and cholinergic reflex-induced bronchoconstriction. These acute changes are followed with structural remodeling, resulting in subepithelial fibrosis, increased numbers of blood vessels and mucus-secreting cells, and increased thickness of airway smooth muscle, and airway hyperplasia and hypertrophy. (Reprinted by permission from Macmillan Publishers Ltd.: Barnes PJ: New drugs for asthma, *Nature Revs Drug Discov* 3:831–44, copyright 2004.)

Asthma is associated with several pathological changes. At the cellular and tissue level, there often exist epithelial cell detachment, airway mucosal edema, eosinophil infiltration, subepithelial fibrosis, basal lamina thickening, and smooth muscle hypertrophy and hyperplasia (Fig. 16.4). In addition, goblet cell hyperplasia occurs (Fig. 16.5), inducing mucus overproduction and secretion. All these changes contribute to the reduction in the airway diameter. In asthma, the sensory nerve endings are sensitized to a certain extent, contributing to the elevation of the smooth muscle tone. In severe cases, an apparent change is lung overexpansion due to airway obstruction and air retention in the alveoli.

Experimental models of asthma [16.3]. Experimental asthma can be induced by sensitizing rats or mice by administration of allergens. A common allergen used in asthma induction is ovalbumin. For primary allergen sensitization, a mixture of ovalbumin (10–100 μg) and aluminum hydroxide 1 mg in 0.2–1 ml saline can be prepared and injected into the peritoneal cavity of an animal three times at day 0, 7, and 14. The animal can be rechallenged by exposure to ovalbumin aerosol later. The presence of asthma can be assessed by measuring pathological changes, such as airway mucosal edema, smooth muscle hypertrophy, epithelial cell detachment, eosinophil infiltration, mucus overproduction, basal lamina thickening, and airway constriction.

Conventional Treatment of Asthma [16.4]. Several approaches have been developed and used to treat asthma, including the removal of the causative factors, immunotherapy, and drug therapy. The removal of the causative factors is an effective approach for the prevention of asthma. However, it is usually difficult to find the cause of asthma. Even though the causative factors are known, it is impossible to remove them completely from the environment.

Allergen immunotherapy is an effective method for the treatment of asthma. The hypothetical basis for immunotherapy is that the introduction of a selected allergen via a parenteral route may activate regulatory T lymphocytes, which inhibit Th2 lymphocytes and thus reduces the production of IgE antibodies by B lymphocytes. These changes lead to reduced activity of the immune system to subsequently inhaled allergens. To carry out immunotherapy, allergens specific to a patient should be identified, prepared, and used for therapeutic purposes.

Several types of drugs have been used to treat asthma. These include bronchodilators and glucocorticoids. Bronchodilators include β-adrenergic agonists and anticholinergic agents. β-adrenergic agonists, such as epinephrine, isoproterenol, and resorcinols, activate the β-adrenergic receptor of the airway smooth muscle cells, inducing airway dilation and relieving the symptoms of asthma. Anticholinergic agents, such as atropine sulfate, atropine methylnitrate, and ipratropium bromide, can be used to suppress acetylcholine, a substance that stimulates airway smooth muscle contraction, and thus to induce airway dilation. Airway inhalation is an effective method for the delivery of these agents. However, most bronchodilators stimulate cardiac activities and should be used with caution for patients with cardiac diseases. Glucocorticoids are hormones produced in the cortex of the adrenal gland and can be used to suppress inflammatory reactions. Since inflammation occurs in asthma and contributes to the obstruction of bronchi, glucocorticoids are often used to reduce inflammation.

Molecular Therapies for Asthma [16.5]. Based on the pathogenic mechanisms of asthma, two molecular therapeutic strategies have been developed: (1) suppressing inflammatory reactions and mucus secretion, and (2) inducing airway dilation. The suppression of airway inflammation can be achieved by the transfection of target cells with antiinflammatory cytokine genes and the glucocorticoid receptor gene, administration with inhibitors to inflammatory transcription factors and kinases, and the blockade of inflammatory cytokines and cytokine receptors (Fig. 16.7). Airway dilation can be achieved by transferring the β_2-adrenergic receptor gene. These approaches are described as follows.

Suppression of Asthmatic Changes by Administration of Antiinflammatory Cytokine Genes, Antibodies, and Inhibitors [16.6]. There are several approaches for inhibiting allergic responses and asthmatic changes. These include the transfection of target cells with antiinflammatory cytokines or their genes, administration with antibodies against

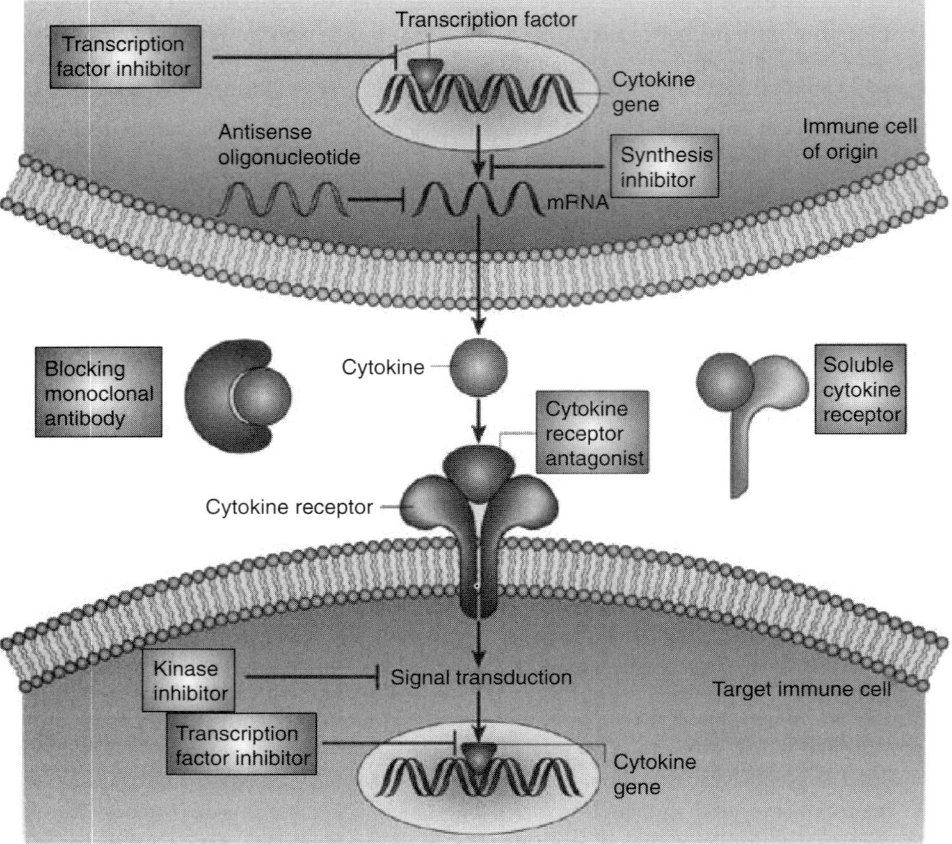

Figure 16.7. Strategies for inhibiting proinflammatory cytokines in asthma. These include inhibition of cytokine synthesis (e.g., corticosteroids), inhibition of transcription factors regulating cytokine expression (e.g., calcineurin inhibitors or decoy oligonucleotides), inhibition of secreted cytokines with blocking antibodies [e.g., antiinterleukin (IL)5 antibody] or soluble receptors (e.g., soluble IL4 receptors), blocking cytokine receptors (e.g., chemokine receptor antagonists), blocking signal-transduction pathways (e.g., p38 mitogen-activated protein kinase inhibitors) or transcription factors activated by cytokines (e.g., STAT6 inhibitors). (Reprinted by permission from Macmillan Publishers Ltd.: Barnes PJ: New drugs for asthma, *Nature Revs Drug Discov* 3:831–44, copyright 2004.)

inflammatory factors, and administration with inflammation inhibitors (Fig. 16.8). There are a number of antiinflammatory cytokines, including interleukin (IL)10, interleukin-12, and interferon-γ. Interleukin-10 is a cytokine that suppresses inflammatory reactions in the airway. In patients with asthma, the expression of the interleukin-10 gene is reduced compared to the normal population. In experimental models, the administration of interleukin-10 significantly suppresses inflammatory reactions in response to the stimulation of allergens. Immunotherapeutic approaches often induce upregulation of the interleukin-10 gene. These observations suggest that interleukin-10 or its gene can be used as therapeutic agents for the treatment of asthma.

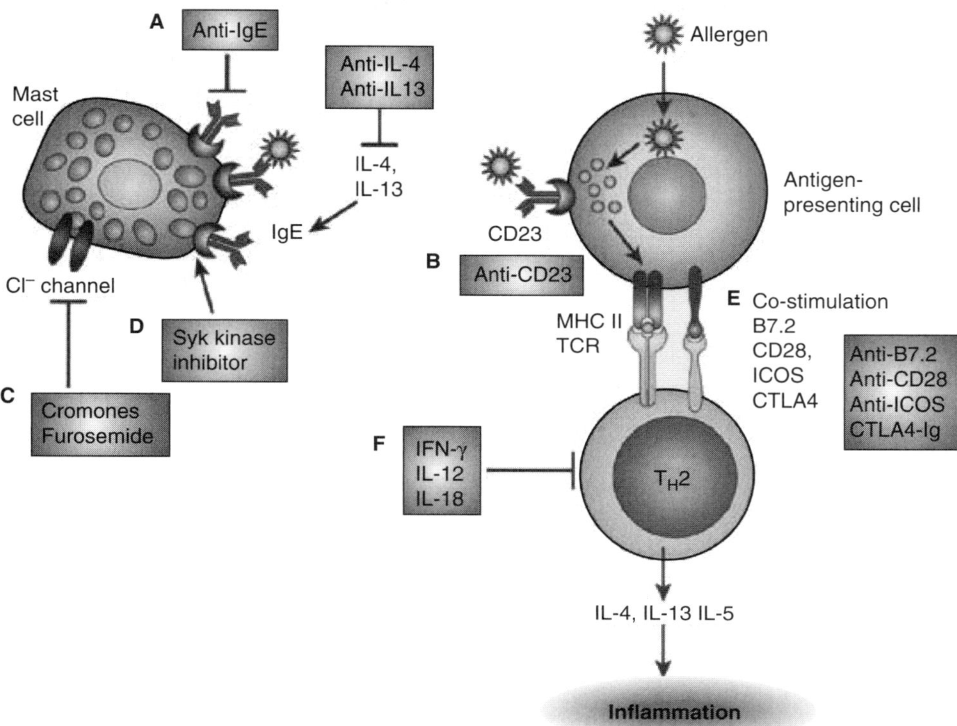

Figure 16.8. Strategies for inhibiting allergic responses underlying asthma. Immunoglobulin E (IgE) can be inhibited by the antibody omalizumab (A) and low-affinity IgE receptors by anti-CD23 (B). Mast cells can also be blocked by cromones and furosemide (C), probably acting on a chloride channel and by inhibitors of Syk kinase, which inhibit the signal-transduction pathways activated by IgE receptors (D). Antigen presentation can be blocked by inhibitors of costimulatory molecules (E), including B7.2, CD28, inducible costimulatory molecule (ICOS), and cytotoxic T-lymphocyte antigen-4 (CTLA4). T_H2 cells can also be directly inhibited by interferon-γ (IFNγ), interleukin (IL)12 and IL18 (F). (Reprinted by permission from Macmillan Publishers Ltd.: Barnes PJ: New drugs for asthma, *Nature Revs Drug Discov* 3:831–44, copyright 2004.)

IL12 is a cytokine that suppresses the production of asthma-promoting cytokines from the type 2 T-helper cells, thus exerting an inhibitory effect on asthmatic activities, such as mucus secretion, inflammation, and eosinophil infiltration. Experimental investigations have demonstrated that the transfer of the IL12 gene into the lung of ovalbumin- or dust mite-sensitized animals results in an increase in the level of interferon-γ, an anti-asthmatic cytokine, and a reduction in the activity of the type 2 T-helper cells, eosinophil infiltration, and inflammatory reactions. The cotransfer of IL12 with IL18, another anti-asthmatic cytokine, elicits a synergistic inhibitory effect on asthmatic changes. The IL12 and IL18 genes are potential candidates for the treatment of human asthma. Furthermore, the activation of nitric oxide synthase type 2 is required for the activity of IL12. Thus a cotransfer of the nitric oxide synthase type 2 gene with the IL12 gene may enhance the anti-asthmatic effect of IL12. The characteristics of IL12 and IL18 are presented in the following table.

Interferon-γ is a cytokine that suppresses inflammatory reactions and inhibits the production and secretion of asthma-promoting cytokines, including IL4 and IL5 (note that these cytokines stimulate the activation and migration of eosinophils and secretion of inflammatory agents such as IgE immunoglobulins), from the type 2 T-helper cells, and thus reduce asthmatic activities (see Table 16.1 for characteristics of interferon-γ). In several experimental investigations, interferon-γ gene has been transferred into the lung of asthmatic animals by using virus-mediated gene transfer approaches. These investigations have demonstrated that the overexpression of the interferon-γ gene is associated with a reduction in the density of eosinophils and mucus secretion, resulting in a decrease in asthmatic symptoms. Clinical investigations have demonstrated similar results. Interferon-γ and its gene have been proven effective antiinflammatory molecules that can be used for the treatment of asthma.

In addition to the administration of antiinflammatory cytokines and their genes, the inflammatory cytokines can be suppressed by local delivery of antisense oligonucleotides or small interfering RNA specific to these cytokines. As discussed on pages 448 and 449, antisense oligonucleotides or small interfering RNA (siRNA) are short nucleotide sequences, can be synthesized based on the sequence of target mRNA, and can be used as therapeutic agents. A selected agent can be delivered via inhalation to the airways. These short nucleotide sequences can be taken up by airway epithelial cells. Experimental investigations have demonstrated that either oligonucleotides or small siRNA can effectively suppress inflammatory reactions in the airways. Other potential therapeutic approaches are presented in Fig. 16.8.

Characteristics of asthma-related interleukins are listed in Table 16.1.

Suppression of Inflammatory Reactions by Transferring the Glucocorticoid Receptor Gene [16.7]. Glucocorticoid is a hormone produced in the cortex of the adrenal glands. It interacts with the glucocorticoid receptor and suppresses inflammatory reactions, such as leukocyte infiltration, the production of asthmatic cytokines, and mucus secretion. The glucocorticoid receptor, also known as GCR and nuclear receptor subfamily 3 group C member 1 (NR3C1), is a nuclear receptor of 777 amino acids and about 86 kDa. This receptor is expressed in a variety of cell and tissue types, including leukocytes, cardiomyocytes, brain, lung, kidney, skeletal muscle, liver, pancreas, intestine, eye, skin, and osteoblasts. The primary functions of this receptor are to interact with glucocorticoid and act as a transcription factor to induce the expression of antiinflammatory genes. Thus the activation of glucocorticoid receptor results in the suppression of inflammatory reactions. In vitro investigations by using human airway cells have demonstrated that the overexpression of the glucocorticoid receptor gene results in the suppression of several transcriptional factors, such as activator protein 1 and nuclear factor κB, which induce inflammatory reactions. Further investigations are necessary to verify these antiinflammatory activities in vivo.

Inducing Bronchodilation by Transferring Bronchodilator Genes and Proteins [16.8]. There are several types of bronchodilator proteins, including the β_2-adrenergic receptor and atrial natriuretic peptide (ANP). The β_2-adrenergic receptor of the airway smooth muscle cells interacts with epinephrine and induces smooth muscle relaxation and airway dilation, an effective approach for the relief of asthmatic symptoms. The transfection of the β_2-adrenergic receptor gene into the airway smooth muscle cells can up-regulate

TABLE 16.1. Characteristics of Selected Asthma-Related Interleukins*

Proteins	Alternative Names	Amino Acids	Molecular Weight (kDa)	Expression	Functions
IL12α	IL12A, IL-12A, interleukin 12 α chain, IL12 subunit p35, cytotoxic lymphocyte maturation factor (CLMF), cytotoxic lymphocyte maturation factor 35-kDa subunit (CLMF p35), natural killer cell stimulatory factor 35-kDa subunit (NKSF1), NK cell stimulatory factor chain 1, T-cell-stimulating factor (TSF)	219	25	Dendritic cells	Stimulating T-cell-independen induction of interferon-γ, inducing the differentiation of Th1 cells, suppressing the production of asthma-promoting cytokines from the Th2 cells, and sustaining memory/effector Th1 cells to mediate long-term immune protection
IL12β	IL12B, IL12 subunit p40, interleukin 12 β chain, cytotoxic lymphocyte maturation factor 40-kDa subunit (CLMF p40), NK cell stimulatory factor chain 2 (NKSF2), IL23 subunit p40, cytotoxic lymphocyte maturation factor 2 (CLMF2), natural killer cell stimulatory factor 40-kDa subunit, T-cell-stimulating factor (TSF)	328	37	Dendritic cells, B cells, macrophages	Forming heterodimer via disulfide bonds
IL13	P600	132	14	Th2 cells, skin	Regulating B- cell maturation and differentiation, promoting IgE isotype development of B cells, and promoting asthmatic activities

*Based on bibliography 16.6.

**Note that other asthma-related interleukins and interferons are listed on page 631.

the expression of the receptor and enhance the activity of the bronchodilator epinephrine. Experimental investigations have demonstrated the effectiveness of this approach for the treatment of asthma. Atrial natriuretic peptide (ANP) serves as a potent bronchodilator. In experimental investigations, this peptide can protect the airways from chemical-induced bronchoconstriction. Blood delivery and local application of atrial natriuretic peptide are effective therapeutic approaches for the treatment of asthma.

Cystic Fibrosis

Pathogenesis, Pathology, and Clinical Features [16.9]. *Cystic fibrosis* is a fatal, autosomal, inherited disease which involves multiple systems, including the pulmonary, cardiovascular, gastrointestinal, skeletal, and reproductive systems. Cystic fibrosis in the pulmonary system is characterized by the presence of profound inflammation, bronchitis, bronchopneumonia, lung abscesses, atelectasis, pulmonary hypertension, cor pulmonale, and congestive heart failure. The pulmonary involvement is a major cause of death in cystic fibrosis. Cystic fibrosis is a common disease in the Caucasian population and the occurrence is about 1 in 2500. This disorder is often found in children.

The pathogenesis of cystic fibrosis is related to the mutation of a gene that encodes the cystic fibrosis transmembrane conductance regulator (CFTR), also known as cystic fibrosis transmembrane conductance regulator ATP binding cassette, MRP7, ABC35, and ABCC7 (1480 amino acids and 169 kDa). CFTR is a member of the superfamily of ATP-binding cassette (ABC) transporters and is found in the epithelial membrane of the lung and intestine. It serves as a cAMP-regulated chloride channel. CFTR mediates the transport of chloride as well as other ions such as sodium and HCO_3^- (Fig. 16.9). The mutation of the CFTR gene induces the disorder of ion transport across the epithelial cells, resulting in a reduction in chloride secretion and an increase in sodium absorption. These changes influence the function of epithelial cells and contribute to the development of cystic fibrosis. There are several types of genetic alterations, which may influence the function of the CFTR protein: null expression, mutation, and regulatory disorder of CFTR. CFTR gene mutation is the most common cause of cystic fibrosis. A typical mutation is the deletion of the 508 phenylalanine. In the lung of cystic fibrosis, the concentration of interleukin-10, an antiinflammatory cytokine, is reduced compared to that in normal people. This change also contributes to inflammatory reactions in cystic fibrosis.

In cystic fibrosis, pathological examinations often exhibit excessive growth of bronchial goblet cells and increased secretion of mucus. Inflammatory reactions are found around bronchi with increased leukocyte infiltration. These changes usually lead to partial or complete obstruction of the bronchi. Patients with cystic fibrosis are often associated with infection by bacteria, such as *Pseudomonas aeruginosa* and *E. coli,* viruses, and fungi. It is often difficult to remove these micro-organisms from the airways because of the functional impairment of the ciliated epithelial cells. As the disorder develops, extensive fibrosis in the parenchyma and bronchi occurs (Fig. 16.10), which is associated with irreversible bronchial distortion and obstruction, increased stiffness of parenchyma, and reduced ventilation. Fibrous disorders also occur in the alveolar wall, resulting in the impairment of oxygen exchange. Patients usually die of oxygen deficiency due to impaired gas ventilation and exchange.

Experimental Models of Cystic Fibrosis [16.10]. Experimental cystic fibrosis can be induced by introducing *P. aeruginosa* into the airways of the rat, mouse, or other animals.

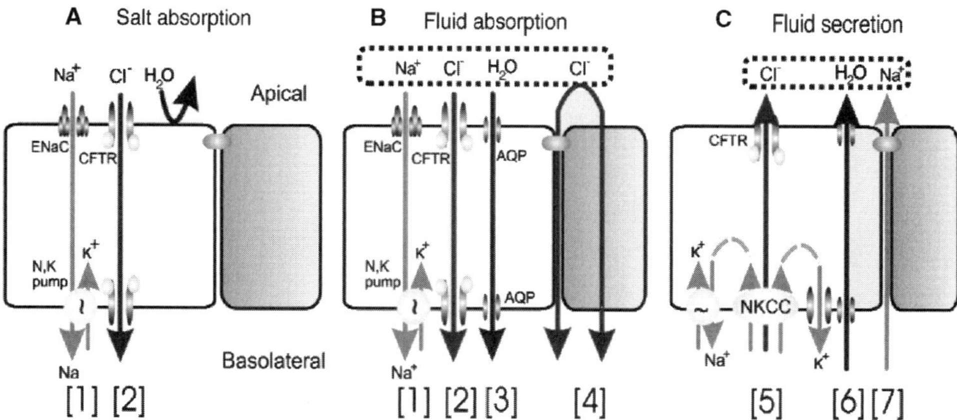

Figure 16.9. CFTR's multiple roles in fluid and electrolyte transport. (A) Salt absorption. In the sweat duct, high apical conductance for Na^+ [1] and Cl^- [2] and relatively low water conductance allows salt to be reabsorbed in excess of water (hypertonic absorption), leaving a hypotonic luminal fluid. In the sweat duct CFTR is the only available anion conductance pathway, and when it is lost in CF the lumen quickly becomes highly electronegative and transport virtually ceases, resulting in high (similar to plasma) luminal salt (B). Fluid absorption. In epithelia with high water permeability (3) relative to electrolyte permeability water will absorbed osmotically with salt to decrease the volume of luminal fluid. If no other osmolytes or forces are present, the salt concentration will remain unchanged. If water-retaining forces are present, permeant electrolytes can be reduced preferentially. The consequences of eliminating CFTR depend on the magnitude of such forces, the relative magnitude of alternate pathways for transepithelial anion flow (4), and how CFTR affects other ion channels. The high-salt and low-volume hypotheses differ on each of these points. (C) Anion-mediated fluid secretion. Secreting epithelia lack a significant apical Na^+ conductance. Basolateral transporters such as NKCC move Cl^- uphill into the cell; it then flows passively into the lumen via CFTR [5], K^+ exits basolaterally, Na^+ flows paracellularly and water follows transcellularly. Elimination of CFTR eliminates secretion. (Reprinted with permission from JJ: Wine The genesis of cystic fibrosis lung disease, *J Clin Invest* 103:309–12, 1999.)

Agarose beads can be used as carriers for bacterial delivery. To prepare *P. aeruginosa*-containing agarose beads, *P. aeruginosa* can be isolated from the airway of human cystic fibrosis patients, mixed with an agarose solution, and grown to near saturation. The bacterial and agarose solution is added to a bath of mineral oil at temperature 50°C, stirred for several minutes, and placed on ice for several minutes. Agarose beads form in the oil bath when temperature reduces. The agarose beads can be washed in 0.5% deoxycholic acid in phosphate buffered saline (PBS, pH 7.4) and subsequently in PBS. The diameter of the beads can be controlled by altering the stirring speed and can be measured by microscopy.

To introduce the constructed *P. aeruginosa*-agorose beads into the airways of an animal, the animal can be anesthetized by intraperitoneal injection of sodium pentobarbital (50 mg/kg body weight). After sterilization, the trachea is cut open, and a volume of 50 μL bead slurry is instilled into the left and right bronchi. The tracheal wound is closed with one to two suture stitches (10-O nylon suture for mice or rats). At a scheduled time (e.g., 3 days after surgery), samples can be collected from the airways for detecting the presence of the bacterium. Mucus and lung tissue samples can be also collected for the

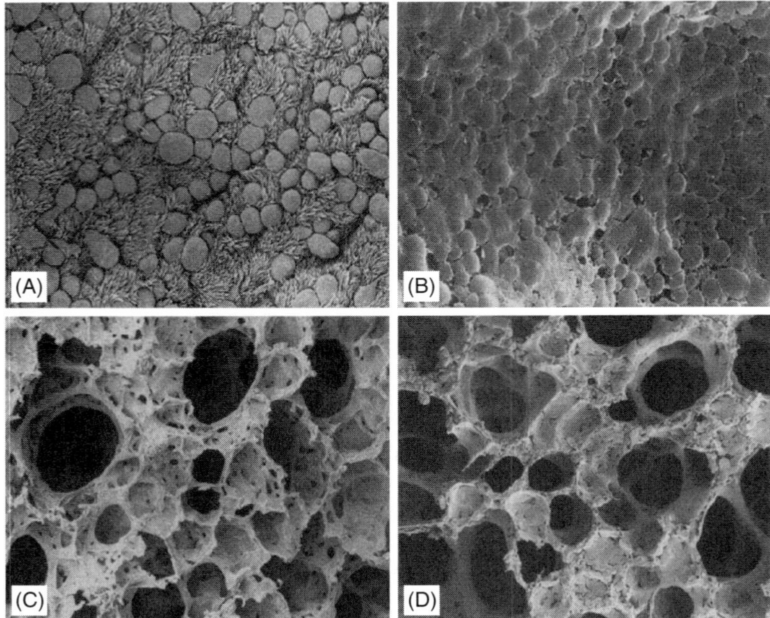

Figure 16.10. Morphological changes in the lung of cystic fibrosis. (A) Scanning electron micrographs of the surface of the respiratory epithelium from a terminal bronchiole in an 11-month-old wildtype mouse. Note the numerous ciliated and nonciliated cells. (B) Terminal bronchiole from a *Cftr*$^{-/-}$ littermate. Respiratory epithelium is encrusted in mucus-like material. (C) Alveoli from the wildtype animal. (D) Alveoli from the affected animal. Distal airways were caked with mucus-like material. Original magnifications: ×1000 (A,B); ×650 (C,D). (Reprinted with permission from Durie PR et al: Characteristic multiorgan pathology of cystic fibrosis in a long-living cystic fibrosis transmembrane regulator knockout murine model, *Am J Pathol* 164:1481–93, copyright 2004.)

detection and analysis of cytokines by biochemistry. Structural changes in the airways can be assessed by microscopy. It is important to note that this model is merely a model of *P. aeruginosa* infection, which often occurs in human patients with cystic fibrosis and exhibits certain features of cystic fibrosis. However, it is not a realistic cystic fibrosis model induced by the mutation of the cystic fibrosis transmembrane regulator gene.

Conventional Treatment of Cystic Fibrosis [16.11]. Cystic fibrosis can be treated with approaches which enhance the function of CFTR and relieve the symptoms of the disorder. Several types of chemical compound, including phenylbutyrate, 8-cyclopentyl-1,3-dipropylxanthine (CPX), and genistein have been tested in previous investigations. These agents may serve as chaperones that promote the deployment of CFTR to the cell membrane. Agents that stimulate chloride secretion (e.g., uridine triphosphate) and inhibit sodium absorption (e.g., amiloride) have also been considered and tested for the treatment of cystic fibrosis. However, few conventional approaches are available for the removal of the causative factors of the disorder.

Cystic fibrosis is associated with life-threatening alterations, such as bronchial constriction and obstruction by inflammatory cells and mucus. A critical issue for treating cystic fibrosis is to relieve bronchial obstruction. Bronchodilators, such as epinephrine, are often used to achieve such as goal. As discussed on page 750 of this chapter, cystic fibrosis is often associated with bacterial infection. In such a case, antibiotics should be given to control infection. In addition, glucocorticoids are often used to control inflammatory reactions in cystic fibrosis. In the end-stage of cystic fibrosis, it is usually necessary to carry out lung transplantation. However, the long-term survival of lung transplants is often disappointing.

Molecular Engineering [16.12]. Cystic fibrosis can be genetically treated with two approaches: to remove the causative factors and reduce symptoms. Since CFTR gene mutation and deletion are major causes of cystic fibrosis, one strategy for the treatment of cystic fibrosis is to restore the structure and function of the CFTR gene. For the last two decades, numbers of investigations have been carried out to test the effectiveness of CFTR gene transfer into the respiratory system of cystic fibrosis. Limited clinical trials have been carried out. These investigations have shown promising results. However, it is often difficult to achieve long-term therapeutic effects by gene transfer. Furthermore, the effectiveness of the CFTR gene transfer, which requires viral gene-carriers, is limited by local immune reactions to the viral factors.

Another approach is to reduce inflammatory reactions, which often occur in cystic fibrosis. Inflammation is a major factor that contributes to the obstruction of bronchi in cystic fibrosis. As for the treatment of asthma, the overexpression of the glucocorticoid receptor gene by gene transfer is a potential therapeutic approach for reducing inflammatory reactions in cystic fibrosis. Since bronchial constriction and obstruction are major causes of cystic fibrosis-associated symptoms, the transfer of genes encoding bronchodilator proteins is a potential approach for the treatment of cystic fibrosis. A candidate gene is the β_2-adrenergic receptor gene. The upregulation of the β_2-adrenergic receptor induces bronchial dilation and relieves airway obstruction.

Pulmonary Hypertension

Pathogenesis, Pathology, and Clinical Features

Primary Pulmonary Hypertension [16.13]. Primary pulmonary hypertension, also known as idiopathic pulmonary hypertension, is a disorder characterized by an increase in pulmonary arterial blood pressure. The term "primary" or "idiopathic" is used because the pathogenic causes of pulmonary hypertension are poorly understood. There are other types of pulmonary hypertension, such as hypoxia- and pulmonary embolism-induced hypertension. Primary pulmonary hypertension is diagnosed only if these types of pulmonary hypertension are excluded.

Although the cause of primary pulmonary hypertension remains unknown, there are several hypothetical mechanisms for the pathogenesis of the disorder. One of the possible causes is disordered regulation of the arterial basal tone or contractility. In pulmonary hypertension, the basal tone of the pulmonary arteries is increased compared to that in the general population. The contribution of altered arterial contractility is supported by the observation that a treatment with vasodilators reduces pulmonary arterial blood pressure. However, the exact cause of disordered arterial contractility remains poorly understood. Another possible cause of primary pulmonary hypertension is the obstruction of pulmonary arteries by unrecognized thromboemboli. Such an obstruction induces elevation in the resistance to the right ventricle, stimulating the right ventricle to increase its contractility. As a result, pulmonary hypertension occurs.

In pulmonary hypertension, pathological examinations exhibit several changes in the structure of pulmonary arteries. These include intimal hyperplasia and wall hypertrophy of pulmonary arteries, a reduction in the density of small pulmonary arteries and capillaries, thickening of endothelial cells and basement membrane of capillaries, and occasionally atherosclerotic plaques in large pulmonary arteries. Pulmonary hypertension is often associated with enlarged right atrium and hypertrophied right ventricle. These changes are attributed to the increase in the workload of the right ventricle resulting from pulmonary hypertension. When the right ventricle is unable to overcome the resistance of the pulmonary arteries and eject sufficient blood into the pulmonary arteries, excessive blood is retained in the right atrium, resulting in the enlargement of the right atrium. The treatment for primary pulmonary hypertension is similar to that for hypoxic pulmonary hypertension (see next section).

Hypoxic Pulmonary Hypertension [16.14]. Hypoxia can induce rapid contraction of smooth muscle cells in the pulmonary arteries of mammals, resulting in acute pulmonary hypertension. Hypoxia is defined as a reduction in the oxygen concentration (<21%). The level of atmospheric oxygen changes in proportion to the altitude. An elevation of 1000 m in altitude is associated with a reduction of about 2% in oxygen concentration. The oxygen level at the altitude 5000 m is about 10%. The exposure of mammals to such an altitude induces acute pulmonary hypertension (Fig. 16.11). The severity of hypertension is proportional to the decrease in the oxygen concentration. Mammals may not survive in an environment with an oxygen level lower than 5%.

It is important to note that, while hypoxia directly induces pulmonary arterial constriction, it exerts an opposite effect on the contractility of the systemic arteries. In the pulmonary arterial system, a reduction in the oxygen level results in a decrease in the air ventilation-to-blood perfusion ratio. As discussed on page 739, the pulmonary system 739 intends to maintain a constant air ventilation-to-blood perfusion ratio and thus to reduce blood perfusion by activating smooth muscle contraction. In the systemic arterial system,

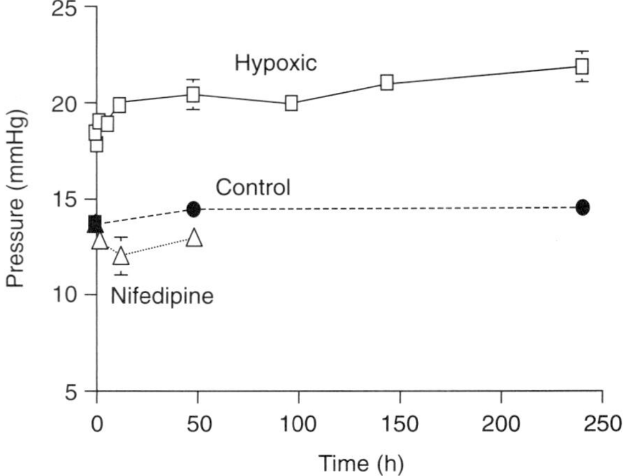

Figure 16.11. Changes in pulmonary arterial blood pressure in hypoxic pulmonary hypertension.

hypoxia is often associated with an increase in metabolic wastes, such as CO_2, H^+, and adenosine. These metabolites induce smooth muscle relaxation in the systemic arteries. However, the effect of these metabolites cannot overcome the constrictive effect of hypoxia in the pulmonary arteries.

Hypoxic pulmonary hypertension can induce rapid pathological changes in the pulmonary arteries. Primary changes include early bleb formation in the endothelial cells (Fig. 16.12), transient edematous swelling and disorganization of the medial elastic laminae (Chapter 16 opening Figure), and progressive hyperplasia of the pulmonary arterial smooth muscle cells and hypertrophy of the pulmonary arterial wall (Fig. 16.13). These are considered adaptive alterations in response to a rapid increase in the pulmonary arterial blood pressure and the stretching tensile stress in the vessel wall. In the presence of hypertension suppressors, such as nifedipine, these pathological changes were significantly inhibited in association with reduced pulmonary arterial blood pressure. These observations support the role of hypertension, but not hypoxia, in the induction of pulmonary arterial pathological changes. When hypoxia is removed, the pathological changes can be gradually recovered. However, the time required for recovery is significantly longer than the time required for the development of the pathological changes.

Rats and mice are often used to induce experimental hypoxic pulmonary hypertension. A hypoxic environment can be created by using a hypoxic chamber with a controlled flow of oxygen and nitrogen. It is necessary to install an oxygen sensor and a CO_2 sensor to monitor the concentrations of oxygen and CO_2, respectively. Animals can be placed in the chamber for a desired period. Pulmonary arterial blood pressure usually starts to increase within 5 min, continues to increase for about 5 days, and reaches a plateau after 5 days. The pulmonary arterial blood pressure can be monitored in living animals by implanting a catheter into the pulmonary arterial trunk or the right ventricle via the jugular vein and the right atrium.

Conventional Treatment of Pulmonary Hypertension [16.13, 16.14]. The principle of treating pulmonary hypertension is to induce arterial dilation and reduce the pulmonary arterial blood pressure. Vascular smooth muscle relaxants are often used to achieve such a goal. Common smooth muscle relaxants include direct smooth muscle relaxants (e.g., nitroprusside and nitroglycerine), β-adrenergic agonists (e.g., isoproterenol), α-adrenergic blockers (e.g., phentolamine and phenoxybenzamine), and calcium channel blockers (e.g., nifedipine). These agents are effective for the temporary relief of pulmonary hypertension. When pulmonary arterial embolism is the cause of pulmonary arterial hypertension, patients should be treated with anticoagulants.

Molecular Therapies for Pulmonary Hypertension. Primary pulmonary hypertension is a result of pulmonary arterial constriction due to hyperactivity of smooth muscle cells. Thus, the principle of treating pulmonary hypertension by molecular engineering is to relax smooth muscle cells, dilate pulmonary arteries, and reduce pulmonary arterial blood pressure. Because the genes responsible for the hyperactivity of smooth muscle cells remain poorly understood, there are few molecular engineering approaches available for removing the causative factors. Several genes encoding vasodilator proteins have been used in experimental models for the treatment of pulmonary hypertension. These include the nitric oxide synthase gene, the prostaglandin synthase gene, and the prepro-calcitonin gene-related peptide gene. These genes are briefly described as follows.

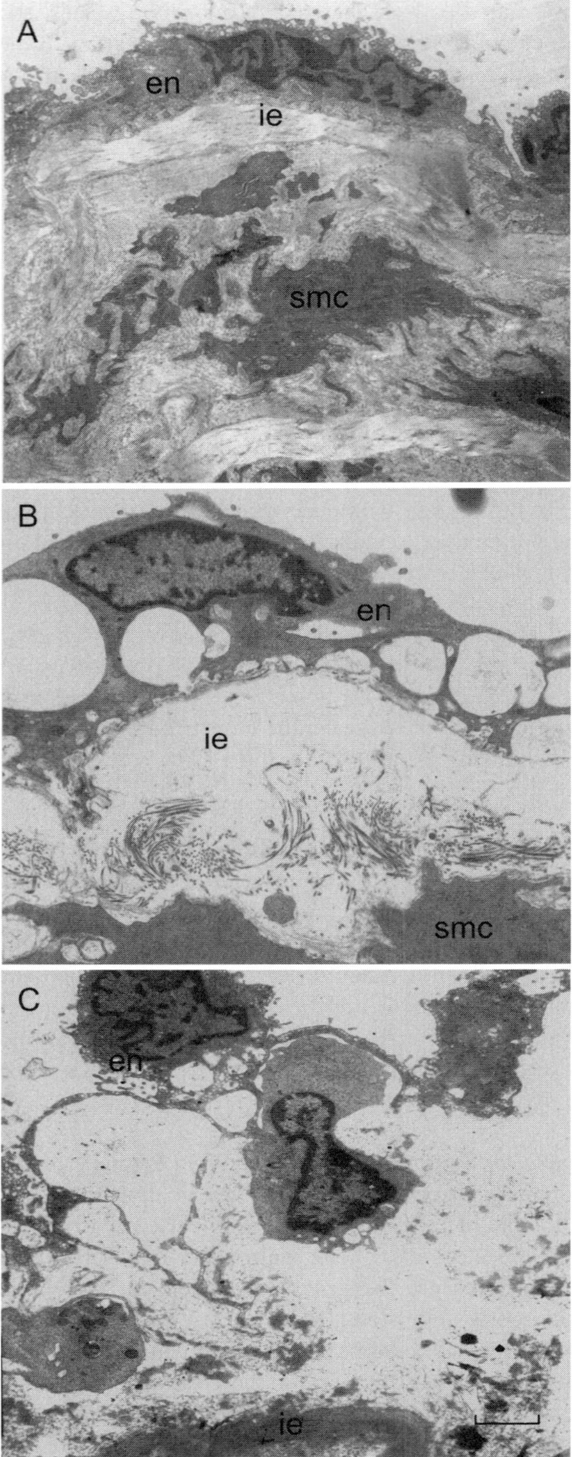

Figure 16.12. Electron micrographs of pulmonary arteries in hypoxia-induced pulmonary hypertension. (A) Control. (B) 12 hours of hypoxia. (C) 30 days of hypoxia. en: endothelial cell. ie: internal elastic lamina. smc: smooth muscle cell. Scale: 1 μm.

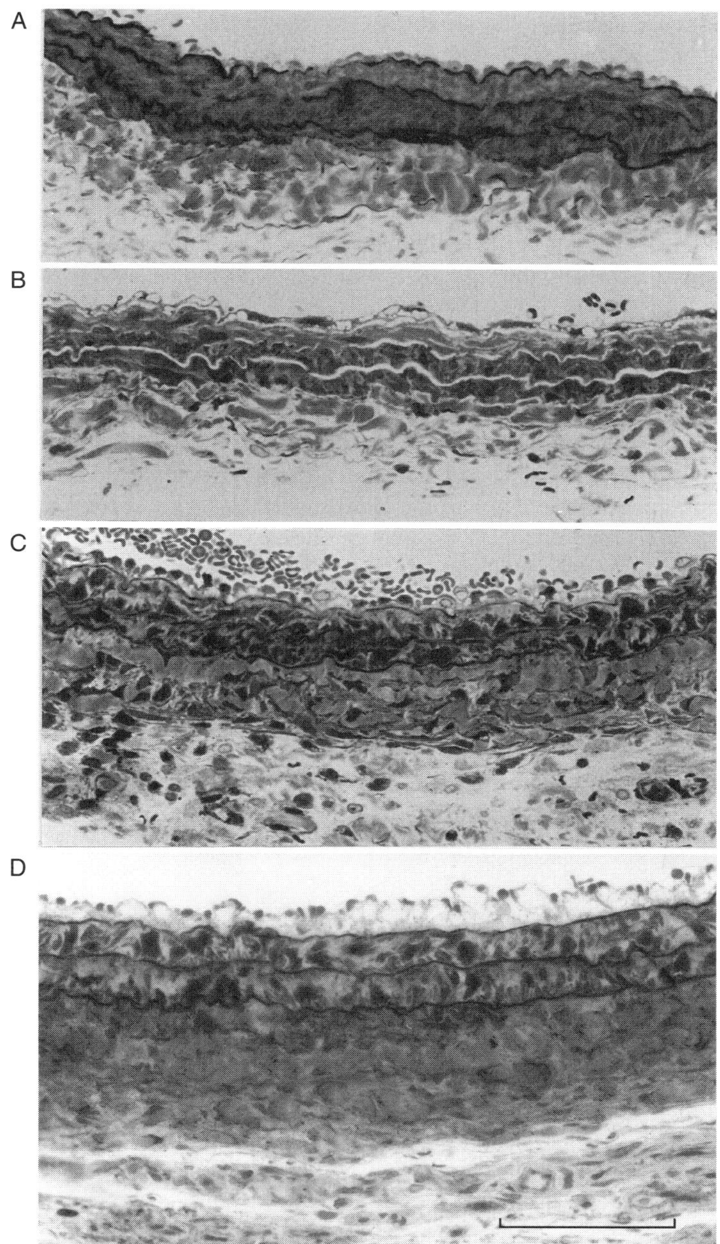

Figure 16.13. Pulmonary arterial structure in hypoxia-induced pulmonary hypertension. (A) Control. (B) 12 hours of hypoxia. (C) 10 days of hypoxia. (D) 30 days of hypoxia. Scale: 100 μm.

Nitric Oxide Synthase Gene [16.15]. As discussed in Chapter 15, nitric oxide synthase catalyzes the formation of nitric oxide from L-arginine. Nitric oxide induces rapid relaxation of smooth muscle cells, dilation of arteries, and reduction in arterial blood pressure. The overexpression of the nitric oxide synthase gene enhances the translation of nitric oxide synthase and therefore the production of nitric oxide. Experimental investigations have shown that adenovirus-mediated transfer of the nitric oxide synthase gene into the lung of the rat with pulmonary hypertension results in a decrease in pulmonary arterial blood pressure and a reduction in the responsiveness arterial smooth muscle cells to vasoconstrictors, such as angiotensin II and endothelin-1. In contrast, the deletion of the nitric oxide synthase gene in mice is associated with profound pulmonary hypertension. The transfer of the nitric oxide synthase gene into the lung of nitric oxide synthase-deficient mice results in a reduction in the pulmonary arterial blood pressure. These observations have demonstrated that the nitric oxide synthase gene is a potential therapeutic gene for the treatment of pulmonary hypertension.

Prostaglandin I2 Synthase Gene [16.16]. Prostaglandins are fatty acid-derived molecules that act on cell membrane receptors and induce relaxation of vascular smooth muscle cells and dilation of arteries. Prostaglandin I2 synthase, also known as prostacyclin synthase, PGI2 synthase, and PGIS, is an enzyme of 500 amino acids and about 57 kDa. This enzyme is a member of the cytochrome P450 superfamily of enzymes and is expressed in the blood vessels (endothelial cells, smooth muscle cells), lung, ovary, brain, intestine, and testis. Prostaglandin I2 synthase catalyzes the conversion of prostglandin H2 to prostacyclin (prostaglandin I2), which is a potent vasodilator and inhibitor of platelet aggregation and atherogenesis. The overexpression of the prostaglandin I2 synthase gene enhances the formation of the prostaglandin I2 synthase protein and prostaglandins as well. Experimental studies have demonstrated that mice with genetically induced overexpression of the prostaglandin I2 synthase gene exhibit increased production of pulmonary prostaglandins. When exposed to hypoxia, which induces pulmonary hypertension, the transgenic mice are more resistant to hypoxic hypertension than control mice without overexpressing the prostaglandin I2 synthase gene. These observations suggest that the prostaglandin I2 synthase gene can be considered a potential gene for the treatment of human pulmonary hypertension.

Preprocalcitonin-Related Peptide Gene [16.17]. The preprocalcitonin gene-related peptide is a precursor for calcitonin-related peptide (CGRP) and is expressed in the nerve and endocrine cells of the airways. CGRP, also known as *calcitonin gene-related peptide 2*, CGRP2, CGRP II, β-type CGRP, is a protein of 127 amino acids and about 14-kDa. This protein can act on its receptors in the vascular smooth muscle cell and induce dilation of the pulmonary arteries. In pulmonary hypertension induced by hypoxia, the level of CGRP is reduced, an alteration potentially contributing to the development of pulmonary hypertension. Thus, the enhancement of CGRP production and activation is a potential approach for the treatment of pulmonary hypertension. Experimental investigations have demonstrated that, in the animal model of hypoxia-induced pulmonary hypertension, the overexpression of the prepro-CGRP gene results in an increase in the production of CGRP and a reduction in the pulmonary arterial blood pressure in association with an attenuation of pulmonary arterial and right ventricular hypertrophy. These preliminary investigations suggest that the prepro-CGRP gene may be considered a potential gene for the treatment of pulmonary hypertension. Although hypoxia-induced hypertension does not assemble human primary pulmonary hypertension, vasodilation is a general approach for relieving the symptoms of pulmonary hypertension regardless the causative factors.

BIBLIOGRAPHY

16.1. Anatomy and Physiology of the Respiratory System

Guyton AC, Hall JE: *Textbook of Medical Physiology*, 11th ed, Saunders, Philadelphia, 2006.

McArdle WD, Katch FI, Katch VL: *Essentials of Exercise Physiology*, 3rd ed, Lippincott Williams & Wilkins, Baltimore, 2006.

Germann WJ, Stanfield CL; (with contributors Niles MJ, Cannon JG), *Principles of Human Physiology*, 2nd ed, Pearson Benjamin Cummings, San Francisco, 2005.

Thibodeau GA, Patton KT: *Anatomy & Physiology*, 5th ed, Mosby, St Louis, 2003.

Boron WF, Boulpaep EL: *Medical Physiology: A Cellular and Molecular Approach*, Saunders, Philadelphia, 2003.

Ganong WF: *Review of Medical Physiology*, 21st ed, McGraw-Hill, New York, 2003.

16.2. Pathogenesis, Pathology, and Clinical Features of Asthma

Schneider AS, Szanto PA: *Pathology*, 3rd ed, Lippincott Williams & Wilkins, Philadelphia, 2006.

McCance KL, Huether SE: *Pathophysiology: The Biologic Basis for Disease in Adults & Children*, 5th ed, Elsevier Mosby, St Louis, 2006.

Porth CM: *Pathophysiology: Concepts of Altered Health States*, 7th ed, Lippincott Williams & Wilkins, Philadelphia, 2005.

Frazier MS, Drzymkowski JW: *Essentials of Human Diseases and Conditions*, 3rd ed, Elsevier Saunders, St Louis, 2004.

Maddox L, Schwartz DA: The pathophysiology of asthma, *Annu Rev Med* 53:477–98, 2002.

Elias JA, Lee CG, Zheng T, Ma B, Homer RJ et al: New insights into the pathogenesis of asthma, *J Clin Invest* 111:291–7, 2003.

Lemanske RF Jr, Busse WW: Asthma, *J Allergy Clin Immunol* 111(Suppl):502–19, 2003.

Herrick CA, Bottomly K: To respond or not to respond: T cells in allergic asthma, *Nat Rev Immunol* 3:405–12, 2003.

Romagnani S: The role of lymphocytes in allergic disease, *J Allergy Clin Immunol* 105:399–408, 2000.

Factor P: Gene therapy for asthma, *Mol Ther* 7(2):148–52, 2003.

Corry DB, Kheradmand F: Induction and regulation of the IgE response, *Nature* 402:B18–23, 1999.

16.3. Experimental Models of Asthma

Kitagaki K, Jain VV, Businga TR, Hussain I, Kline JN: Immunomodulatory effects of CpG oligodeoxynucleotides on established Th2 responses, *Clin Diagn Lab Immunol* 9:1260–9, 2002.

16.4. Conventional Treatment of Asthma

Norman PS, Immunotherapy, *J Allergy Clin Immunol* 113:1013–23, 2004.

Cools M, Van Bever HP, Weyler JJ, Stevens WJ: Long term effects of specific immunotherapy, administered during childhood, in asthmatic patients allergic to either house-dust mite or to both house-dust mite and grass pollen, *Allergy* 55:69–73, 2000.

Moller C, Dreborg S, Ferdousi HA, Halken S, Host A et al: Pollen immunotherapy reduces the development of asthma in children with seasonal rhinoconjunctivitis (the PAT-study), *J Allergy Clin Immunol* 109:251–6, 2002.

Corry DB, Kheradmand F: Induction and regulation of the IgE response, *Nature* 402:B18–23, 1999.

16.5. Molecular Engineering Approaches

Barnes PJ: New drugs for asthma, *Nature Rev Drug Discov* 3:831–44, 2004.

Factor P: Gene therapy for asthma, *Mol Ther* 7(2):148–52, Feb 2003.

16.6. Suppression of Asthmatic Changes by Transferring Antiinflammatory Cytokine Genes

Behera AK, Kumar M, Lockey RF, Mohapatra SS: Adenovirus-mediated interferon gamma gene therapy for allergic asthma: Involvement of interleukin 12 and STAT4 signaling, *Hum Gene Ther* 13:1697–1709, 2002.

Li XM et al: Mucosal IFN-gamma gene transfer inhibits pulmonary allergic responses in mice, *J Immunol* 157:3216–9, 1996.

Dow SW, Schwarze J, Heath TD, Potter TA, Gelfand EW: Systemic and local interferon gamma gene delivery to the lungs for treatment of allergen-induced airway hyperresponsiveness in mice, *Hum Gene Ther* 10:1905–14, 1999.

Hogan SP, Foster PS, Tan X, Ramsay AJ: Mucosal IL-12 gene delivery inhibits allergic airways disease and restores local antiviral immunity, *Eur J Immunol* 28:413–23, 1998.

Lee YL et al: Construction of single-chain interleukin-12 DNA plasmid to treat airway hyperresponsiveness in an animal model of asthma, *Hum Gene Ther* 12:2065–79, 2001.

Stampfli MR et al: Regulation of allergic mucosal sensitization by interleukin-12 gene transfer to the airway, *Am J Resp Cell Mol Biol* 21:317–26, 1999.

del Pozo V et al: Gene therapy with galectin-3 inhibits bronchial obstruction and inflammation in antigen-challenged rats through interleukin-5 gene downregulation, *Am J Respir Crit Care Med* 166:732–7, 2002.

Walter DM et al: Il-18 gene transfer by adenovirus prevents the development of and reverses established allergen-induced airway hyperreactivity, *J Immunol* 166:6392–8, 2001.

Diefenbach A, Schindler H, Rollinghoff M, Yokoyama WM, Bogdan C: Requirement for type 2 NO synthase for IL-12 signaling in innate immunity, *Science* 284:951–5, 1999.

IL12α and IL12β

Cooper AM, Kipnis A, Turner J, Magram J, Ferrante J et al: Mice lacking bioactive IL-12 can generate protective, antigen-specific cellular responses to mycobacterial infection only if the IL-12 p40 subunit is present, *J Immun* 168:1322–7, 2002.

Cua DJ, Sherlock J, Chen Y, Murphy CA, Joyce B et al: Interleukin-23 rather than interleukin-12 is the critical cytokine for autoimmune inflammation of the brain, *Nature* 421:744–8, 2003.

Diefenbach A, Schindler H, Rollinghoff M, Yokoyama WM, Bogdan C: Requirement for type 2 NO synthase for IL-12 signaling in innate immunity, *Science* 284:951-5, 1999.

Ferlazzo G, Pack M, Thomas D, Paludan C, Schmid D et al: Distinct roles of IL-12 and IL-15 in human natural killer cell activation by dendritic cells from secondary lymphoid organs, *Proc Nat USA Acad Sci USA* 101:16606–11, 2004.

Grabie N, Delfs MW, Westrich JR, Love VA, Stavrakis G et al: IL-12 is required for differentiation of pathogenic CD8-positive T cell effectors that cause myocarditis, *J Clin Invest* 111:671–80, 2003.

Schwarz A, StSnder S, Berneburg M, Bshm M, Kulms D et al: Interleukin-12 suppresses ultraviolet radiation-induced apoptosis by inducing DNA repair, *Nature Cell Biol* 4:26–31, 2002.

Wolf SF, Sieburth D, Sypek J: Interleukin 12: A key modulator of immune function, *Stem Cells* 12:154–68, 1994.

Altare F, Durandy A, Lammas D, Emile JF, Lamhamedi S et al: Impairment of mycobacterial immunity in human interleukin-12 receptor deficiency, *Science* 280:1432–35, 1998.

Brightbill HD, Libraty DH, Krutzik SR, Yang RB, Bellsie JT et al: Host defense mechanisms triggered by microbial lipoproteins through toll-like receptors, *Science* 285:732–6, 1999.

de Jong R, Altare F, Haagen IA, Elferink DG, de Boer T et al: Severe mycobacterial and Salmonella infections in interleukin-12 receptor-deficient patients, *Science* 280:1435–8, 1998.

Ferlazzo G, Pack M, Thomas D, Paludan C, Schmid D et al: Distinct roles of IL-12 and IL-15 in human natural killer cell activation by dendritic cells from secondary lymphoid organs, *Proc Nat Acad Sci USA* 101:16606–11, 2004.

Ferlazzo G, Thomas D, Lin SL, Goodman K, Morandi B et al: The abundant NK cells in human secondary lymphoid tissues require activation to express killer cell Ig-like receptors and become cytolytic, *J Immun* 172:1455–62, 2004.

Haraguchi S, Day NK, Nelson RP Jr, Emmanuel P, Duplantier JE et al: Interleukin 12 deficiency associated with recurrent infections, *Proc Nat Acad Sci USA* 95:13125–9, 1998.

Jankovic D, Kullberg MC, Hieny S, Caspar P, Collazo CM et al: In the absence of IL-12, CD4+ T cell responses to intracellular pathogens fail to default to a Th2 pattern and are host protective in an IL-10-/- setting, *Immunity* 16:429–39, 2002.

Oppmann B, Lesley R, Blom B, Timans JC, Xu Y et al: Novel p19 protein engages IL-12p40 to form a cytokine, IL-23, with biological activities similar as well as distinct from IL-12, *Immunity* 13:715–25, 2000.

Sugimoto K, Ohata M, Miyoshi J, Ishizaki H, Tsuboi N et al: A serine/threonine kinase, Cot/Tpl2, modulates bacterial DNA-induced IL-12 production and Th cell differentiation, *J Clin Invest* 114:857–66, 2004.

IL13

Blackburn MR, Lee CG, Young HWJ, Zhu Z, Chunn JL et al: Adenosine mediates IL-13-induced inflammation and remodeling in the lung and interacts in an IL-13-adenosine amplification pathway, *J Clin Invest* 112:332–44, 2003.

Grunig G, Warnock M, Wakil AE, Venkayya R, Brombacher F et al: Requirement for IL-13 independently of IL-4 in experimental asthma, *Science* 282:2261–3, 1998.

Howard TD, Whittaker PA, Zaiman AL, Koppelman GH, Xu J et al: Identification and association of polymorphisms in the interleukin-13 gene with asthma and atopy in a Dutch population, *Am J Resp Cell Mol Biol* 25:377–84, 2001.

Kelly-Welch AE, Hanson EM, Boothby MR, Keegan AD: Interleukin-4 and interleukin-13 signaling connections maps, *Science* 300:1527–8, 2003.

Kuperman DA, Huang X, Koth LL, Chang GH, Dolganov GM et al: Direct effects of interleukin-13 on epithelial cells cause airway hyperreactivity and mucus overproduction in asthma, *Nature Med* 8:885–9, 2002.

Lacy DA, Wang ZE, Symula DJ, McArthur CJ, Rubin EM et al: Faithful expression of the human 5q31 cytokine cluster in transgenic mice, *J Immun* 164:4569–74, 2000.

Loots GG, Locksley RM, Blankespoor CM, Wang ZE, Miller W et al: Identification of a coordinate regulator of interleukins 4, 13, and 5 by cross-species sequence comparisons, *Science* 288:136–40, 2000.

Minty A, Chalon P, Derocq JM, Dumont X, Guillemot JC et al: Interleukin-13 is a new human lymphokine regulating inflammatory and immune responses, *Nature* 362:248–50, 1993.

Vladich FD, Brazille SM, Stern D, Peck ML, Ghittoni R et al: IL-13 R130Q, a common variant associated with allergy and asthma, enhances effector mechanisms essential for human allergic inflammation, *J Clin Invest* 115:747–54, 2005.

Wills-Karp M, Luyimbazi J, Xu X, Schofield B, Neben TY et al: Interleukin-13: Central mediator of allergic asthma, *Science* 282:2258–61, 1998.

Zheng T, Zhu Z, Wang Z, Homer RJ, Ma B et al: Inducible targeting of IL-13 to the adult lung causes matrix metalloproteinase- and cathepsin-dependent emphysema, *J Clin Invest* 106:1081–93, 2000.

Zhu Z, Homer RJ, Wang Z, Chen Q, Geba GP et al: Pulmonary expression of interleukin-13 causes inflammation, mucus hypersecretion, subepithelial fibrosis, physiologic abnormalities, and eotaxin production, *J Clin Invest* 103:779–88, 1999.

Zhu Z, Zheng T, Homer RJ, Kim YK, Chen NY et al: Acidic mammalian chitinase in asthmatic Th2 inflammation and IL-13 pathway activation, *Science* 304:1678–82, 2004.

Human protein reference data base, Johns Hopkins University and the Institute of Bioinformatics, at http://www.hprd.org/protein.

16.7. Suppression of Inflammatory Reactions by Transferring the Glucocorticoid Receptor Gene

Mathieu M et al: The glucocorticoid receptor gene as a candidate for gene therapy in asthma, *Gene Ther* 6:245–52, 1999.

Strickland I, Kisich K, Hauk PJ, Vottero A, Chrousos GP et al: High constitutive glucocorticoid receptor beta in human neutrophils enables them to reduce their spontaneous rate of cell death in response to corticosteroids, *J Exp Med* 193(5):585–93, 2001.

Oakley RH, Sar M, Cidlowski JA: The human glucocorticoid receptor beta isoform. Expression, biochemical properties, and putative function, *J Biol Chem* 271(16):9550–9, 1996.

Bamberger CM, Bamberger AM, de Castro M, Chrousos GP: Glucocorticoid receptor beta, a potential endogenous inhibitor of glucocorticoid action in humans, *J Clin Invest* 95:2435–41, 1995.

Bledsoe RK, Montana VG, Stanley TB, Delves CJ, Apolito CJ et al: Crystal structure of the glucocorticoid receptor ligand binding domain reveals a novel mode of receptor dimerization and coactivator recognition, *Cell* 110:93–105, 2002.

Brewer JA, Khor B, Vogt SK, Muglia LM, Fujiwara H et al: T-cell glucocorticoid receptor is required to suppress COX-2-mediated lethal immune activation, *Nature Med* 9:1318–22, 2003.

Huizenga NATM, de Lange P, Koper JW, Clayton RN, Farrell WE et al: Human adrenocorticotropin-secreting pituitary adenomas show frequent loss of heterozygosity at the glucocorticoid receptor gene locus, *J Clin Endocr Metab* 83:917–21, 1998.

McNally JG, Muller WG, Walker D, Wolford R, Hager GL: The glucocorticoid receptor: Rapid exchange with regulatory sites in living cells, *Science* 287:1262–5, 2000.

Reichardt HM, Kaestner KH, Tuckermann J, Kretz O, Wessely O et al: DNA binding of the glucocorticoid receptor is not essential for survival, *Cell* 93:531–41, 1998.

Strickland I, Kisich K, Hauk PJ, Vottero A, Chrousos GP et al: High constitutive glucocorticoid receptor beta in human neutrophils enables them to reduce their spontaneous rate of cell death in response to corticosteroids, *J Exp Med* 193:585–93, 2001.

Weinberger C, Hollenberg SM, Ong ES, Harmon JM, Brower ST et al: Identification of human glucocorticoid receptor complementary DNA clones by epitope selection, *Science* 228:740–2, 1985.

16.8. Inducing Bronchodilation by Transferring Bronchodilator Genes and Proteins

McGraw DW et al: Targeted transgenic expression of beta(2)-adrenergic receptors to type II cells increases alveolar fluid clearance, *Am J Physiol Lung Cell Mol Physiol* 281:L895–903, 2001.

Small KM, Brown KM, Forbes SL, Liggett SB: Modification of the beta 2-adrenergic receptor to engineer a receptor-effector complex for gene therapy, *J Biol Chem* 276:31596–601, 2001.

Barnes PJ: New drugs for asthma, *Nature Rev Drug Discov* 3:831–44, 2004.

16.9. Pathogenesis, Pathology, and Clinical Features

Ratjen F, Döring G: Cystic fibrosis, *Lancet* 361(9358):681–9, 2003.

Lohi H, Makela S, Pulkkinen K, Hoglund P, Karjalainen-Lindsberg ML et al: Upregulation of CFTR expression but not SLC26A3 and SLC9A3 in ulcerative colitis, *Am J Physiol Gastrointest Liver Physiol* 283(3):G567–75, 2002.

Bonfield TL, Konstan MW, Burfeind P et al: Normal bronchial epithelial cells constitutively produce the anti-inflammatory cytokine interleukin-10, which is downregulated in cystic fibrosis, *Am J Resp Cell Mol Biol* 13:257–61, 1995.

Chmiel JF, Konstan MW, Knesebeck JE et al: IL-10 attenuates excessive inflammation in chronic Pseudomonas infection in mice, *Am J Resp Crit Care Med* 160:2040–7, 1999.

Riordan JR, Rommens JM, Kerem BS et al: Identification of the cystic fibrosis gene: Cloning and characterization of complementary DNA, *Science* 245:1066–73, 1989.

Rommens JM, Iannuzzi MC, Kerem BS et al: Identification of the cystic fibrosis gene: Chromosome walking and jumping, *Science* 245:1059–65, 1989.

Zielinski J, Rozmahel R, Bozon D et al: Genomic DNA sequence of the cystic fibrosis transmembrane conductance regulator, *Genomics* 10:241–8, 1991.

Collins FS: Cystic fibrosis: Molecular biology and therapeutic implications, *Science* 256:774–9, 1992.

Schwiebert EM, Morales MM, Devidas S et al: Chloride channel and chloride conductance regulator domains of CFTR, the cystic fibrosis transmembrane conductance regulator, *Proc Natl Acad Sci USA* 95:2674–89, 1998.

Reddy MM, Light MJ, Quinton PM: Activation of the epithelial Na+ channel (ENaC) requires CFTR Cl– channel function, *Nature* 402:301–4, 1999.

Knowles MR, Clarke LL, Boucher RC: Activation by extracellular nucleotides of chloride secretion in the airway epithelia of patients with cystic fibrosis, *New Engl J Med* 325:533–8, 1991.

Morral N, Bertranpetit J, Estivill X et al: The origin of the major cystic fibrosis mutation (delta F508) in European populations, *Nat Genet* 7:169–75, 1994.

Bedwell DM, Kaenjak A, Benos DJ, Bebok Z, Bubien JK et al: Suppression of a CFTR premature stop mutation in a bronchial epithelial cell line, *Nature Med* 3:1280–4, 1997.

Bronsveld I, Mekus F, Bijman J, Ballmann M, de Jonge HR et al: Chloride conductance and genetic background modulate the cystic fibrosis phenotype of delta-F508 homozygous twins and siblings, *J Clin Invest* 108:1705–15, 2001.

Chanson M, Scerri I, Suter S: Defective regulation of gap junctional coupling in cystic fibrosis pancreatic duct cells, *J Clin Invest* 103:1677–84, 1999.

Cheng SH, Gregory RJ, Marshall J, Paul S, Souza DW et al: Defective intracellular transport and processing of CFTR is the molecular basis of most cystic fibrosis, *Cell* 63:827–34, 1990.

Chillon M, Casals T, Mercier B, Bassas L, Lissens W et al: Mutations in the cystic fibrosis gene in patients with congenital absence of the vas deferens, *New Eng J Med* 332:1475–80, 1995.

Choi JY, Muallem D, Kiselyov K, Lee MG, Thomas PJ et al: Aberrant CFTR-dependent HCO(-3) transport in mutations associated with cystic fibrosis, *Nature* 410:94–7, 2001.

Cutting GR, Kasch LM, Rosenstein BJ, Zielenski J, Tsui LC et al: A cluster of cystic fibrosis mutations in the first nucleotide-binding fold of the cystic fibrosis conductance regulator protein, *Nature* 346:366–9, 1990.

Dean M, White MB, Amos J, Gerrard B, Stewart C et al: Multiple mutations in highly conserved residues are found in mildly affected cystic fibrosis patients, *Cell* 61:863–70, 1990.

Denning GM, Anderson MP, Amara JF, Marshall J, Smith AE et al: Processing of mutant cystic fibrosis transmembrane conductance regulator is temperature-sensitive, *Nature* 358:761–4, 1992.

Devor DC, Schultz BD: Ibuprofen inhibits cystic fibrosis transmembrane conductance regulator-mediated CI(-) secretion, *J Clin Invest* 102:679–87, 1998.

Dorin JR, Dickinson P, Alton EWFW, Smith SN, Geddes DM et al: Cystic fibrosis in the mouse by targeted insertional mutagenesis, *Nature* 359:211–5, 1992.

Ellsworth RE, Jamison DC, Touchman JW, Chissoe SL, Maduro VVB et al: Comparative genomic sequence analysis of the human and mouse cystic fibrosis transmembrane conductance regulator genes, *Proc Nat USA Acad Sci* 97:1172–7, 2000.

Konstan MW, Byard PJ, Hoppel CL, Davis PB: Effect of high-dose ibuprofen in patients with cystic fibrosis, *New Eng J Med* 332:848–54, 1995.

Logan J, Hiestand D, Daram P, Huang Z, Muccio DD et al: Cystic fibrosis transmembrane conductance regulator mutations that disrupt nucleotide binding, *J Clin Invest* 94:228–36, 1994.

Reddy MM, Quinton PM: Control of dynamic CFTR selectivity by glutamate and ATP in epithelial cells, *Nature* 423:756–60, 2003.

Rich DP, Anderson MP, Gregory RJ, Cheng SH, Paul S et al: Expression of cystic fibrosis transmembrane conductance regulator corrects defective chloride channel regulation in cystic fibrosis airway epithelial cells, *Nature* 347:358–63, 1990.

Riordan JR, Rommens JM, Kerem B, Alon N, Rozmahel R et al: Identification of the cystic fibrosis gene: Cloning and characterization of complementary DNA, *Science* 245:1066–73, 1989.

Rosenfeld MA, Yoshimura K, Trapnell BC, Yoneyama K, Rosenthal ER et al: In vivo transfer of the human cystic fibrosis transmembrane conductance regulator gene to the airway epithelium, *Cell* 68:143–55, 1992.

Schwiebert EM, Egan ME, Hwang TH, Fulmer SB, Allen SS et al: CFTR regulates outwardly rectifying chloride channels through an autocrine mechanism involving ATP, *Cell* 81:1063–73, 1995.

Snouwaert JN, Brigman KK, Latour AM, Malouf NN, Boucher RC et al: An animal model for cystic fibrosis made by gene targeting, *Science* 257:1083–8, 1992.

16.10. Experimental Models of Cystic Fibrosis

van Heeckeren AM, Scaria A, Schluchter MD, Ferkol TW, Wadsworth S et al: Delivery of CFTR by adenoviral vector to cystic fibrosis mouse lung in a model of chronic Pseudomonas aeruginosa lung infection, *Am J Physiol Lung Cell Mol Physiol* 286:L717–26, 2004.

Cash HA, Woods DE, McCullough B, Johanson WG Jr, Bass JA: A rat model of chronic respiratory infection with Pseudomonas aeruginosa, *Am Rev Resp Dis* 119:453–9, 1979.

Starke JR, Edwards MS, Langston C, Baker CJ: A mouse model of chronic pulmonary infection with Pseudomonas aeruginosa and Pseudomonas cepacia, *Pediatr Res* 22:698–702, 1987.

16.11. Conventional Treatment of Cystic Fibrosis

Rubenstein RC, Zeitlin PL: A pilot clinical trial of oral sodium 4-phenylbutyrate (Buphenyl) in deltaF508-homozygous cystic fibrosis patients: Partial restoration of nasal epithelial CFTR function, *Am J Resp Crit Care Med* 157:484–90, 1998.

Galietta LV, Jayaraman S, Verkman AS: Cell-based assay for high-throughput quantitative screening of CFTR chloride transport agonists, *Am J Physiol Cell Physiol* 281:C1734–42, 2001.

Wang F, Zeltwanger S, Hu S et al: Deletion of phenylalanine 508 causes attenuated phosphorylation-dependent activation of CFTR chloride channels, *J Physiol* 524:637–48, 2000.

Knowles MR, Church NL, Waltner WE et al: A pilot study of aerosolized amiloride for the treatment of lung disease in cystic fibrosis, *New Engl J Med* 322:1189–94, 1990.

Knowles MR, Clarke LL, Boucher RC: Activation by extracellular nucleotides of chloride secretion in the airway epithelia of patients with cystic fibrosis, *New Engl J Med* 325:533–8, 1991.

Rodgers HC, Knox AJ: The effect of topical benzamil and amiloride on nasal potential difference in cystic fibrosis, *Eur Resp J* 14:693–6, 1999.

Liou TG, Adler FR, Cahill BC et al: Survival effect of lung transplantation among patients with cystic fibrosis, *JAMA* 286:2683–9, 2001.

Aurora P, Whitehead B, Wade A et al: Lung transplantation and life extension in children with cystic fibrosis, *Lancet* 354:1591–3, 1999.

Aris RM, Routh JC, LiPuma JJ et al: Lung transplantation for cystic fibrosis patients with Burkholderia cepacia complex: Survival linked to genomovar type, *Am J Resp Crit Care Med* 164:2102–6, 2001.

Kerem E, Reisman J, Corey M et al: Prediction of mortality in patients with cystic fibrosis, *New Engl J Med* 326:1187–91, 1992.

Ratjen F, Döring G: Cystic fibrosis, *Lancet* 361:681–9, 2003.

Guggino WB: Cystic fibrosis and the salt controversy, *Cell* 96(5):607–10, 1999.

16.12. Molecular Engineering Approaches

Driskell RA, Engelhardt JF: Current status of gene therapy for inherited lung diseases, *Annu Rev Physiol* 65:585–612, 2003.

Schwiebert LM: Cystic fibrosis, gene therapy, and lung inflammation: for better or worse? *Am J Physiol Lung Cell Mol Physiol* 286: L715–6, 2004.

Flotte TR, Laube BL: Gene therapy in cystic fibrosis, *Chest* 120:124S–31S, 2001.

Knowles MR, Hohneker KW, Zhou Z et al: A controlled study of adenoviral-vector-mediated gene transfer in the nasal epithelium of patients with cystic fibrosis, *New Engl J Med* 333:823–31, 1995.

Alton EW, Stern M, Farley R et al: Cationic lipid-mediated CFTR gene transfer to the lungs and nose of patients with cystic fibrosis: A double-blind placebo-controlled trial, *Lancet* 353:947–54, 1999.

Zabner J, Ramsey BW, Meeker DP et al: Repeat administration of an adenovirus vector encoding cystic fibrosis transmembrane conductance regulator to the nasal epithelium of patients with cystic fibrosis, *J Clin Invest* 97:1504–11, 1996.

Rosenfeld MA, Yoshimura K, Trapnell BC, Yoneyama K, Rosenthal ER et al: In vivo transfer of the human cystic fibrosis transmembrane conductance regulator gene to the airway epithelium, *Cell* 68: 143–55, 1992.

16.13. Pathogenesis, Pathology, and Clinical Features of Primary Pulmonary Hypertension

Archer S, Rich S: Primary, pulmonary hypertension: A vascular biology and translational research "Work in progress," *Circulation* 102(22):2781–91, 2000.

Moser KM, Fedullo PF, Finkbeiner WE, Golden J: Do patients with primary pulmonary hypertension develop extensive central thrombi? *Circulation* 91(3):741–5, 1995.

Rubin LJ: Pathology and pathophysiology of primary pulmonary hypertension, *Am J Cardiol* 75(3):51A–54A, Jan 1995.

Klinger JR, Hill NS: Right ventricular dysfunction in chronic obstructive pulmonary disease. Evaluation and management, *Chest* 99(3):715–23, 1991.

16.14. Pathogenesis, Pathology, and Clinical Features of Hypoxic Pulmonary Hypertension

Fung YC, Liu SQ: Change of zero-stress state of rat pulmonary arteries in hypoxic pulmonary hypertension, *J Appl Physiol* 70:2455–70, 1991.

Liu SQ: Alterations in structure of elastic laminae of rat pulmonary arteries in hypoxic hypertension, *J Appl Physiol* 81:2147–55, 1996.

Liu SQ: Regression of hypoxic hypertension-induced changes in the elastic laminae of rat pulmonary arteries, *J Appl Physiol* 82:1677–84, 1997.

16.15. Nitric Oxide Synthase Gene

Gelband CH, Katovich MJ, Raizada MK: Current perspectives on the use of gene therapy for hypertension, *Cir Res* 87:1118, 2000.

Champion HC, Bivalacqua TJ, D'Souza FM, Ortiz LA, Jeter JR et al: Gene transfer of endothelial nitric oxide synthase to the lung of the mouse in vivo: Effect on agonist-induced and flow-mediated vascular responses, *Circ Res* 84:1422–32, 1999.

Champion HC, Bivalacqua TJ, Greenberg SS, Giles TD, Heistad DD et al: Gene transfer of endothelial nitric oxide synthase to the lung of the mouse in vivo: selective rescue of pulmonary hypertension in eNOS-deficient mice, *Circulation* 100:1–28, 1999.

Janssens SP, Bloch KD, Nong Z, Gerard RD, Zoldhelyi P et al: Adenoviral-mediated transfer of the human endothelial nitric oxide synthase gene reduces acute hypoxic pulmonary vasoconstriction in rats, *J Clin Invest* 98:317–24, 1996.

16.16. Prostaglandin I2 Synthase Gene

Geraci MW, Gao B, Shepherd DC, Moore MD, Westcott JY et al: Pulmonary prostacyclin synthase overexpression in transgenic mice protects against development of hypoxic pulmonary hypertension, *J Clin Invest* 103:1509–15, 1999.

Chevalier D, Cauffiez C, Bernard C, Lo-Guidice JM, Allorge D et al: Characterization of new mutations in the coding sequence and 5-prime-untranslated region of the human prostacyclin synthase gene (CYP8A1), *Hum. Genet* 108:148–55, 2001.

Miyata A, Hara S, Yokoyama C, Inoue H, Ullrich V et al: Molecular cloning and expression of human prostacyclin synthase, *Biochem Biophys Res Commun* 200:1728–34, 1994.

Nakayama T, Soma M, Watanabe Y, Hasimu B, Sato M et al: Splicing mutation of the prostacyclin synthase gene in a family associated with hypertension, *Biochem Biophys Res Commun* 297:1135–9, 2002.

Wang LH, Chen L: Organization of the gene encoding human prostacyclin synthase, *Biochem Biophys Res Commun* 226:631–7, 1996.

Yokoyama C, Yabuki T, Inoue H, Tone Y, Hara S et al: Human gene encoding prostacyclin synthase (PTGIS): Genomic organization, chromosomal localization, and promoter activity, *Genomics* 36:296–304, 1996.

16.17. Preprocalcitonin-Related Peptide Gene

Champion HC, Bivalacqua TJ, Toyoda K, Heistad DD, Hyman AL et al: In vivo gene transfer of prepro-calcitonin gene-related peptide to the lung attenuates chronic hypoxia-induced pulmonary hypertension in the mouse, *Circulation* 101(8):923–30, 2000.

Kwan YW, Wadsworth RM, Kane KA: Effects of neuropeptide Y and calcitonin gene-related peptide on sheep coronary artery rings under oxygenated, hypoxic and simulated myocardial ischaemic conditions, *Br J Pharmacol* 99:774–8, 1990.

Tjen ALS, Ekman R, Lippton H, Cary J, Keith I: CGRP and somatostatin modulate chronic hypoxic pulmonary hypertension, *Am J Physiol* 263:H681–90, 1992.

Shimosegawa T, Said SI: Pulmonary calcitonin gene-related peptide immunoreactivity: Nerve-endocrine cell interrelationships, *Am J Resp Cell Mol Biol* 4:126–34, 1991.

Springall DR, Polak JM: Calcitonin gene-related peptide and pulmonary hypertension in experimental hypoxia, *Anat Rec* 236:96–104, 1993.

Stevens TP, McBride JT, Peake JL, Pinkerton KE, Stripp BR: Cell proliferation contributes to PNEC hyperplasia after acute airway injury, *Am J Physiol* 272:L486–93, 1997.

Kusakabe T, Kawakami T, Powell FL, Ellisman MH, Sawada H et al: Distribution of substance P and calcitonin gene-related peptide immunoreactive nerve fibers in the trachea of chronically hypoxic rats, *Brain Res Bull* 39:335–9, 1996.

Champion HC, Bivalacqua TJ, Toyoda K, Heistad DD, Hyman AL et al: In vivo gene transfer of prepro-calcitonin gene-related peptide to the lung attenuates chronic hypoxia-induced pulmonary hypertension in the mouse, *Circulation* 101:923–30, 2000.

Amara SG, Arriza JL, Leff SE, Swanson LW, Evans RM et al: Expression in brain of a messenger RNA encoding a novel neuropeptide homologous to calcitonin gene-related peptide, *Science* 229:1094–7, 1985.

Hoovers JMN, Redeker E, Speleman F, Hoppener JWM, Bhola S et al: High-resolution chromosomal localization of the human calcitonin/CGRP/IAPP gene family members, *Genomics* 15:525–9, 1993.

17

LIVER REGENERATIVE ENGINEERING

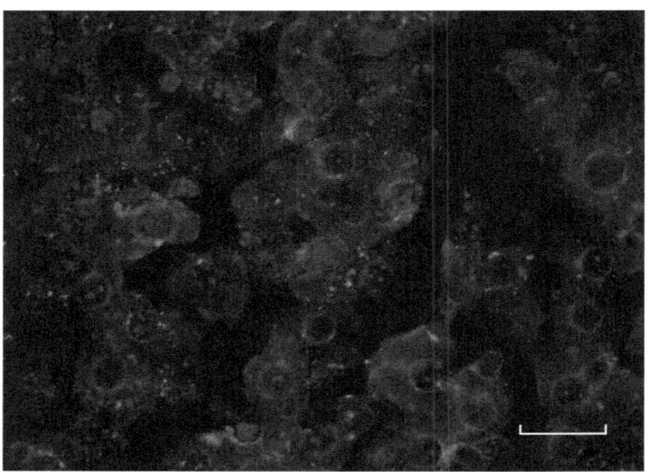

The presence of c-Kit-positive cells in the mouse liver. C-Kit is a cell membrane protein that is expressed in stem cells found in the embryo, fetus, and adult bone marrow. Red: c-Kit. Green: albumin. Blue: cell nuclei. Scale: 10 μm. See color insert.

Bioregenerative Engineering: Principles and Applications, by Shu Q. Liu

ANATOMY AND PHYSIOLOGY OF THE LIVER

Structure [17.1]

The liver is located in the upper right abdominal cavity. This organ is composed of a large number of functional units, known as hepatic lobules. Each *hepatic lobule* is a cylindrical structure of several millimeters in length and about 1 mm in diameter, and is composed of hepatic cell plates and several tubular systems, including the central vein, portal vein, hepatic artery, lymphatic vessels, and bile ducts. The *portal vein* in each unit is a branch of the major hepatic portal vein and conducts blood from the gastrointestinal tracts into the central vein through a structure between the hepatic cell plates known as *hepatic sinusoid*. The *central vein* in each unit converges blood into the hepatic vein and then into the inferior vena cava. The *hepatic artery* in each unit is a branch of the major hepatic artery and supplies oxygenated blood to the hepatic unit. The *lymphatic vessels* collect and conduct excessive fluids from the interstitial tissue to the large lymphatic vessels and then to the vena cava. The *bile ducts* collect and conduct bile to larger bile ducts and then to the duodenum.

The liver consists of several types of cell, including the hepatocytes, Küpffer cells, Ito cells, epithelial cells, and endothelial cells. *Hepatocytes* are the largest cell population in the liver, are major constituents for the functional units of the liver, and are characterized by the expression of albumin (Chapter 17 opening figure). *Küpffer cells* are macrophages and are found in the hepatic sinusoids. These cells are responsible for the destruction and clearance of microorganisms present in the blood. *Ito cells* are found in the hepatic sinusoids and are responsible for the generation of extracellular matrix components. *Epithelial cells* are constituents of the bile ducts. *Endothelial cells* are found in blood vessels and lymphatic vessels and are responsible for the transport of molecules and electrolytes.

Functions [17.1]

The liver is a vital organ that conducts several essential functions, including nutrient metabolism, detoxification, blood filtration and storage, and bile excretion. The liver possesses a large capacity of functional reserve that is not used under physiological conditions. About one-third of the total liver is sufficient for the maintenance of metabolic homeostasis. The reserve capacity is developed during evolution for sudden changes in metabolic demand under unusual conditions, such as ingestion of a large amount of toxins, liver trauma, and liver infection. Here, the basic functions of the liver are briefly discussed.

Metabolism. The liver is responsible for the metabolism of the three major types of nutrient: carbohydrates, lipids, and proteins. *Carbohydrates* include glucose, galactose, and fructose. These substances are absorbed from the small intestine to the blood, and transported to and processed in the liver. Glucose is the most important carbohydrate for energy production. The liver is responsible for the control of a stable level of blood glucose at ~80 mg/dL. When the level of blood glucose is increased (for instance, immediately after a meal), the hepatocytes are able to synthesize glycogen, a glucose polymer, from glucose molecules via a process known as *glycogenesis*, thus reducing the blood glucose level. In contrast, when the level of blood glucose is decreased in the fasting state, the hepatocytes can hydrolyze glycogen to produce glucose via a process called glycogenolysis, bringing back the blood glucose level. Other types of carbohydrate, such as galactose and fructose, can be converted to glucose in the liver, contributing to the accumulation of blood glucose. In addition, under conditions with a very low level of blood glucose, the liver is capable of converting amino acids and glycerol into glucose, a process called gluconeogenesis. The maintenance of the blood glucose level is essential for the function of vital organs and tissues, including the brain, heart, and skeletal muscles.

The liver conducts several functions related to lipid metabolism, including energy generation from fatty acids, synthesis of phospholipids and lipoproteins, and conversion of amino acids and carbohydrates to fatty acids. *Fatty acids* can be oxidized to form acetylcoenzyme A, which is further oxidized to generate energy. Hepatocytes are responsible for the synthesis of *cholesterols*, which are released to the blood, transported to cells in other organs, and used for the construction of cell membrane. Cholesterols are also used to constitute bile, which is released into the duodenum via the bile ducts. Hepatocytes can synthesize phospholipids, which are utilized to construct cell membranes. The liver is a major organ that synthesizes and processes lipoproteins, which are responsible for the transport of lipids and cholesterols between the blood, liver, and other organs. The liver is also responsible for the conversion of amino acids and carbohydrates to fatty acids, which are stored in the fat cells or adipocytes.

The liver is the most important organ in the body for protein metabolism. Hepatocytes conduct several protein-related metabolic functions, including the formation of urea, synthesis of proteins, formation of amino acids, and transamination and deamination of amino acids. Protein metabolism generates highly toxic ammonia, which is converted to urea in the hepatocytes. Urea is released into the blood and removed from the kidney. Hepatocytes can synthesize a number of plasma proteins, including albumin, fibrinogen, heparin, globulins, and blood coagulation factors. These factors play critical roles in many important cellular activities. Hepatocytes are capable of forming a number of amino acids, known as nonessential amino acids (meaning that it is not necessary to ingest these amino acids from diets). In addition, hepatocytes are able to carry out transamination and deamination of amino acids. Transamination is a process by which an amino group is transferred from an amino acid to a keto acid by an aminotransferase. After such a process, amino acids can participate in the metabolism of the citric acid cycle and generate energy. Deamination is a process by which the amino group of amino acids is removed by deaminases. Such a process generates ammonia, a waste product of protein metabolism.

Detoxification. The processes of metabolism as described above generate waste substances, most of which are toxic. In addition, various types of toxic substances, such as medicines and chemicals, may be ingested on a daily basis. Hepatocytes are able to chemically modify toxic substances, reducing their toxicity and rendering the substances less toxic and removable. For instance, biological agents, such as penicillin and ampicillin, are processed in the liver, released into the bile, and transported to the intestinal system, where the substances are removed. Excessive hormones, such as cortisol and estrogen, are modified and detoxified in the liver.

Blood Filtration and Storage. In the small intestine, bacterial and toxic particles can be transported into the blood during nutrient absorption. These toxic particles are transported to the liver through the portal vein. The Küpffer cells (specialized reticuloendothelial cells) can endocytose bacterial and toxic particles so that portal blood can be cleansed. In addition, the liver can serve as an organ for blood storage. Under physiological conditions, about 0.5 L of blood can be stored in the hepatic sinusoids and blood vessels. The blood storage function is important when heart failure occurs. In such a case, excessive blood can be stored in the liver to prevent systemic edema.

Bile Excretion. Bile is produced by hepatocytes, released into the bile ducts, and transported into the duodenum. A human liver can produce about 1,000 ml bile per day. Bile is composed of bile salts, bilirubin, cholesterol, electrolytes, and water. The bile salts are produced from cholesterol and possess two important functions: (1) emulsifying fat diets in the small intestine into minute particles that can be digested by pancreatic enzymes and (2) facilitating the transport and absorption of fat particles through the intestinal epithelial cells. Bilirubin is a metabolic product of hemoglobin, is produced in the hepatocytes, and is transported to the bile ducts and then to the duodenum, where the substance is removed. The accumulation of bilirubin, in the case of liver dysfunction, induces jaundice. Excessive cholesterol molecules are also removed through bile formation and excretion.

Liver Regeneration [17.2]

Organ regeneration is a process of organ self-reconstruction by cell differentiation and proliferation as well as matrix production following organ injury or partial removal. The capacity of regeneration varies among different organs. For instance, the brain and heart have very a low capacity of regeneration, whereas digestive organs are able to regenerate extensively following organ injury. In particular, the liver has a very high capacity of regeneration. It can completely regenerate itself when it is partially removed. No other organs in the human body have the regeneration capacity of the liver. Such a phenomenon has been documented since year the 1890. For more than a century, liver regeneration has fascinated physicians and scientists. For the past several decades, extensive investigations have been carried out for the mechanisms of liver regeneration. These investigations have provided a foundation for today's liver regenerative engineering and reconstruction. See page 399 for detailed discussion about liver regeneration.

HEPATIC DISORDERS

Acute Viral Hepatitis and Liver Failure

Pathogenesis, Pathology, and Clinical Features [17.3]. *Acute viral hepatitis* is a liver infectious disorder induced predominantly by hepatitis A virus and hepatitis B virus. Hepatitis is one of the most popular diseases in the world. About 33% of the human population are infected by hepatitis viruses. This disorder is characterized by inflammatory reactions, edema, cell necrosis, and hyperplasia of Küpffer cells in the liver. The clinical manifestations and consequences of hepatitis vary widely, ranging from mild asymptomatic infections without any noticeable pathological changes in the liver to fatal acute liver failure with massive hepatocyte necrosis. Some patients can completely recover from acute hepatitis, whereas others exhibit persistent infections, which eventually develop into chronic hepatitis and cirrhosis. The outcome of the disease is largely dependent on the responsiveness or sensitivity of the immune system as well as the age of individual patients. The occurrence of chronic hepatitis from the population with acute hepatitis decreases with age. While chronic hepatitis may be found in about 30% of children at the age of <5, the disorder may be found in only about 2% of the adult population with acute hepatitis. The mechanisms for the age related epidemiology remain poorly understood.

Hepatitis A virus is a virus (~26 nm in diameter) that invades the digestive system and blood of humans. This virus is often transmitted via the oral route. Poor hygiene and overpopulation are factors that facilitate the transmission of the virus. Hepatitis A virus can be found in the liver, bile, and stools from patients who carry the virus. Antibodies against hepatitis virus A (IgM class) can be detected in the serum of patients and is often used for the diagnosis of the disease. *Hepatitis B virus* is transmitted via several routes, including oral ingestion, blood infusion, intimate contact, and perinatal transmission. In patients with hepatitis B, viral antigens and corresponding antibodies can be detected in almost all body fluids, including the saliva, tears, serum, gastric fluid, and urine. The presence of hepatitis B antigens and antibodies indicates the infection of the virus.

The pathogenic mechanisms of hepatocyte injury and necrosis in response to the stimulation of hepatitis virus A and B remain poorly understood. It has been thought that the host cell-initiated immune responses may play a role in the development of infectious reactions. The invasion of hepatitis viruses stimulates the host immune system to produce antibodies, which form complexes with the viral antigens. These complexes may sensitize cytotoxic T cells, which recognize hepatitis antigens as well as certain host hepatic molecules that are similar to the viral antigens in structure. The cytotoxic T cells may in turn attack the host liver cells, inducing cell injury and necrosis.

In acute infection of hepatitis A and B viruses, *pathological examinations* often reveal several changes. These include infiltration of mononuclear cells into the parenchyma of the liver, hepatocyte degeneration and necrosis, and edema, which are often associated with hepatocyte proliferation. Immunohistochemical examinations demonstrate the presence of hepatitis viral antigens in the cytoplasm and plasma membrane of hepatocytes. In severe cases, massive hepatocyte necrosis occurs, resulting in acute hepatic atrophy and failure. Acute liver failure is accompanied with rapid jaundice, imbalance of fluid electrolytes, and accumulation of toxins, which cause symptoms such as anorexia, vomiting, fever, fatigue, and headache. Hepatitis induced by hepatitis viruses may contribute to the development of hepatoma. The incidence of hepatoma in patients with hepatitis is considerably higher than that of the general population.

Conventional Treatment [17.4]. Viral hepatitis can be effectively prevented by vaccination with specific vaccines. The effectiveness of vaccination can usually reach about 95%. However, once hepatitis occurs, there are few effective approaches for the treatment of the disorder. For patients with symptoms such as nausea and vomiting, bed rest may help to relieve the symptoms. Patients should avoid taking drugs that are metabolized and reduced in the liver. Hypoglycemia and imbalance of fluids and electrolytes, if any, should be corrected immediately via venous infusion of glucose and physiological fluids. Protein-rich diets should be limited to reduce the workload for the liver. Most patients can be self-cured without clinical consequences.

Certain drugs have been developed and used for suppressing the activities of hepatitis viruses and treat hepatitis. These drugs are primarily nucleoside analogs, including lamivudine and adefovir dipivoxil, which are analogues for deoxycytidine and deoxyadenosine, respectively. These nucleoside analogues can integrate into the viral genome during DNA replication, stop viral DNA elongation, and suppress viral amplification. Thus, a treatment with these nucleoside analogues reduces pathological changes in hepatitis.

In the case of acute liver failure or liver atrophy with complete loss of liver function, it is necessary to conduct allogenic liver transplantation. Since allogenic liver cells induce immune rejection responses, it is necessary to administrate immune suppressors for protecting the transplanted liver from acute rejection. Although liver cells have high capacity of regeneration, it is impossible to generate a new liver within a short period. Furthermore, in acute liver failure, almost all hepatocytes are necrotic or injured. It is difficult to collect sufficient healthy hepatocytes that can be used for liver regeneration.

Molecular Regenerative Therapies. The strategies of molecular treatment are similar among acute and chronic hepatitis as well as cirrhosis, including suppressing viral activities, inhibiting inflammatory reactions, preventing fibrosis, and enhancing hepatocyte proliferation. Several approaches have been developed and used to achieve these goals. These include gene transfer, antisense oligonucleotide delivery, and genetic vaccination.

Suppression of Viral Activities [17.5]. Certain types of cytokines exert antiviral effects. A typical cytokine that inhibits the activity of hepatitis viruses is interferon α. As described above, a treatment with interferon α results in reduced activities of hepatitis viruses and improved hepatic function in chronic hepatitis. However, interferon α protein undergoes rapid degradation. It is difficult to induce long-term effects by protein delivery. The transfer of the interferon α gene represents a potential approach for overcoming such a problem. Experimental investigations have demonstrated that the transfer of the interferon α gene into the liver can induce a sustained increase in the level of interferon α. Such a gene transfer approach results in a reduction in fibrogenesis, improvement of hepatic function, and prevention of the development of hepatoma.

Another approach used for the suppression of the viral activities is to deliver antisense oligodeoxynucleotides or genes that encode oligodeoxynucleotides specific to viral mRNA. The delivered or expressed oligodeoxynucleotides bind to viral mRNAs and render the mRNAs incapable of translating necessary proteins, thus suppressing viral activities and replication. During the end stage of cirrhosis, few functional hepatocytes may be found in the liver. In such a case, the therapeutic oligodeoxynucleotides or genes can be used to transfect functional hepatocytes in vitro. The transfected hepatocytes can then be transplanted into the liver. These cells may become virus-resistant cells and may proliferate and repopulate the liver.

Genetic vaccination is an approach by which viral antigen genes are constructed and delivered into the host cells. The gene products, once expressed, may sensitize the host immune cells and induce antibody generation, rendering the cells prepared for further virus invasion. Sensitized cytotoxic T cells can suppress viral activities. A typical viral antigen gene is the gene that encodes the nucleocapsid protein of the woodchuck hepatitis virus. This gene has been used for immunization in animal models. Interestingly, the cotransfer of certain types of cytokine, such as interleukin-12, is required for the activation of the transferred antigen gene. The expression of the viral antigen gene alone does not effectively protect the hepatocytes from virus attacks.

Enhancement of Hepatocyte Proliferation [17.6]. Chronic hepatitis and cirrhosis are both associated with the loss of hepatocytes, a critical pathological change that causes functional deficiency of the liver. Thus, the enhancement of hepatocyte proliferation is an essential strategy for the treatment of these liver disorders. Several growth factors, including hepatocyte growth factor (HGF), interleukin-6 (IL6), epidermal growth factor (EGF), transforming growth factor (TGF)α, and keratinocyte growth factor (KGF) have been shown to stimulate hepatocyte proliferation. Several other factors, such as insulin, insulin-like growth factors, vasopressin, angiotensin, norepinephrine, and glucagon, can enhance the activity of the growth factors listed above, thus enhancing hepatocyte proliferation. Among these factors, HGF is one of the most potent growth factors that stimulates hepatocyte proliferation and inhibits hepatocyte apoptosis (Fig. 17.1). In addition, HGF exerts inhibitory effects on the activation of Ito cells, the expression of procollagen genes, and the expression of TGFβ, and thus suppressing hepatic fibrogenesis and the progression of cirrhosis. These observations suggest that the HGF gene is a potential therapeutic candidate gene for the treatment of chronic hepatitis and cirrhosis. Experimental investigations have shown that the transfer of the HGF gene into the liver results in an increase in the expression of HGF. The expression of HGF is associated with a number of changes, including (1) tyrosine phosphorylation of the HGF receptor, (2) an elevation in the expression of proliferative cellular nuclear antigen (PCNA), (3) enhancement of hepatocyte proliferation, (4) facilitation of angiogenesis, (5) a decrease in the expression of TGF-β, (6) an reduction in hepatocyte apoptosis; and (7) improvement of hepatic structure and function.

Another hepatic growth promoter is hepatopoietin or augmenter of liver regeneration (ALR), which is a 30-kDa homodimeric protein (see Table 17.1 for characteristics of

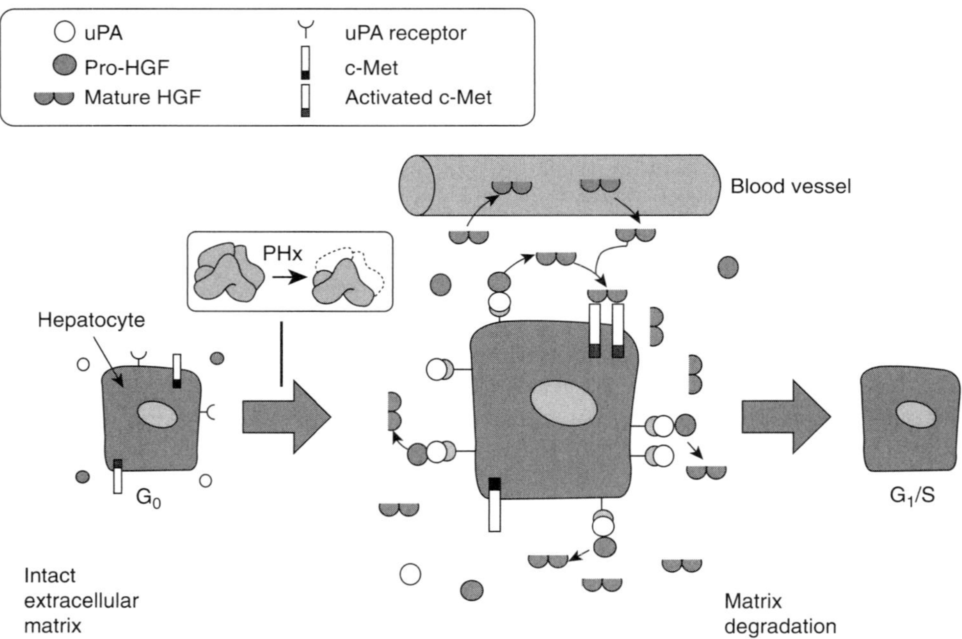

Figure 17.1. Proposed model for the role of HGF in liver regeneration. Rapid upregulation of the uPA receptor leads to activation of uPA within 5 min after PHx. This initiates a protease cascade causing degradation of the scant extracellular matrix surrounding hepatocytes and releasing, among others, matrix-bound inactive pro-HGF. uPA activates pro-HGF into the mature active form. Active HGF is released in the blood and stimulates hepatocyte DNA synthesis by an endocrine or paracrine mechanism by binding to the c-Met receptor. (Reprinted with permission from Michalopoulos GK, DeFrances MC; Liver regeneration, *Science* 276:60–6, copyright 1997 AAAS.)

TABLE 17.1. Characteristics of Selected Therapeutic Proteins for Chronic Hepatitis and Cirrhosis*

Proteins	Alternative Names	Amino Acids	Molecular Weight (kDa)	Expression	Functions
Keratinocyte growth factor	KGF, fibroblast growth factor 7 (FGF7)	194	23	Skin, blood vessels, pancreas, uterus, intestine, ovary, skeletal muscle, cornea	A member of the fibroblast growth factor (FGF) family, stimulating the survival and proliferation of epithelial cells and keratinocytes, regulating embryonic development and morphogenesis, promoting tissue repair, and enhancing tumor growth and invasion
Insulin	INS	110	12	Pancreas	Stimulating glucose uptake via interaction with insulin receptor (INSR) and augmenting the activity of growth factors
Insulin-like growth factor	IGFI, somatomedin C	195	22	Brain, blood cells, adrenal gland, bone marrow, intestine, skeletal muscle, uterus	Enhancing the effect of growth hormone (HG) during development, stimulating cell proliferation, and protecting neuronal cells from injury and apoptosis

TABLE 17.1. *Continued*

Proteins	Alternative Names	Amino Acids	Molecular Weight (kDa)	Expression	Functions
Vasopressin	VP, antidiuretic hormone (ADH), vasopressin neurophysin II, arginine vasopressin neurophysin II, vasopressin neurophysin II copeptin	164	17	Brain	A posterior pituitary hormone synthesized in the supraoptic nucleus and paraventricular nucleus of the hypothalamus, stimulating cell growth, inducing vasoconstriction by acting on smooth muscle cells, and exerting an antidiuretic effect on the kidney
Glucagon	Glucagon-like peptide 1 (GLP1), glucagon-like peptide 2 (GLP2), glicentin-related polypeptide (GRPP),glucagon preproprotein	180	21	Intestine, pancreas	Increasing the glucose level by stimulating glycogenolysi and gluconeogenesis, enhancing cell proliferation via the G-protein-coupled receptor signaling pathways, and inducing transformation of intestinal epithelial cells to insulin-producing cells
Hepatopoietin	HERV1, augmenter of liver regeneration (ALR), growth factor ERV1-like, hepatic regenerative stimulation substance (HSS)	125	15 (monomer)	Liver, testis	Serving as a hepatotrophic factor that stimulates hepatocyte proliferation and liver regeneration

*Based on bibliography 17.6.

hepatopoietin). This protein has been shown to stimulate hepatocyte proliferation and protect liver cells from injury and apoptosis. The transfer of the hepatopoietin gene into animal model of cirrhosis results in a reduction in the progression of hepatic fibrosis and improvement of hepatic structure and function. Genes that encode other hepatic growth promoters can also be used for therapeutic purposes.

Suppression of Inflammatory Reactions [17.7]. Inflammatory reactions in chronic hepatitis and cirrhosis often induce fibrogenesis, which prevents hepatocyte regeneration and deteriorates the function of the liver. Thus, one of the strategies for the treatment of chronic hepatitis and cirrhosis is to suppress inflammation. Interleukin (IL)10 is known as an anti-inflammatory cytokine, which suppresses the activity of proinflammatory cytokines. This factor is produced in several types of cells, including lymphocytes, monocytes, and macrophages (see page 634 for the characteristics of IL10). Genetically induced deficiency of IL10 is associated with enhanced hepatic fibrosis and monocyte infiltration in animal models. The transfer of IL10 gene into the liver cells induces a reduction in the activity of the collagen gene promoter, leukocyte infiltration, and fibrosis in experimental models of liver cirrhosis. Thus, the IL10 gene can be considered a potential gene for the treatment of human chronic hepatitis and cirrhosis.

Inhibition of Fibrosis [17.8]. Hepatic fibrosis is often promoted by certain factors. Transforming growth factor (TGF)β is one of such factors. TGFβ stimulates the transformation of Ito cells to fibroblast-like cells, which produce excessive extracellular matrix components, including collagen and fibronectin, and contribute to hepatic fibrosis. Thus, the blockade of the TGFβ signal transduction pathway may exert an inhibitory effect on hepatic fibrosis. One approach to reduce the activity of TGFβ is to modify the structure of the TGFβ receptor. Experimental investigations have demonstrated that the transfer of a truncated dominant-negative TGFβ receptor gene into the liver of experimental cirrhosis results in the inhibition of TGFβ activity, which is associated with a reduction in the production of collagen and fibronectin, monocyte infiltration, activity of the Ito and Küpffer cells, and hepatic fibrosis (Fig. 17.2). The hepatic function is improved accordingly. The truncated TGFβ receptor can compete for the TGFβ ligands with the wild-type TGFβ receptor present in the cells, but cannot transmit the TGFβ signal to the intracellular signaling pathways, thus reducing the effect of TGFβ. A large quantity of the truncated TGFβ receptor gene is usually required to achieve therapeutic effectiveness.

Enhancing the Activity of Telomerase [17.9]. Telomerase is a complex enzyme composed of RNA and two protein subunits and is responsible for the maintenance of telomere integrity and function. Telomere is a cap structure for eukaryotic chromosomes, and consists of a TTAGGG-rich DNA sequence and catalytic protein enzymes. This structure plays a critical role in the maintenance of the stability and function of chromosomes and in the regulation of DNA synthesis and cell mitosis. Hepatic cirrhosis is associated with reduced length and altered function of telomere, which is thought to contribute to the progression of hepatic cell apoptosis and fibrosis. These changes are attributed to alterations in the activity of the telomerase. Thus, the enhancement of the telomerase activity by telomerase gene transfer into the liver may improve the integrity of telomere and reduce hepatocyte apoptosis and hepatic fibrosis. Experimental investigations have provided evidence that supports such a possibility.

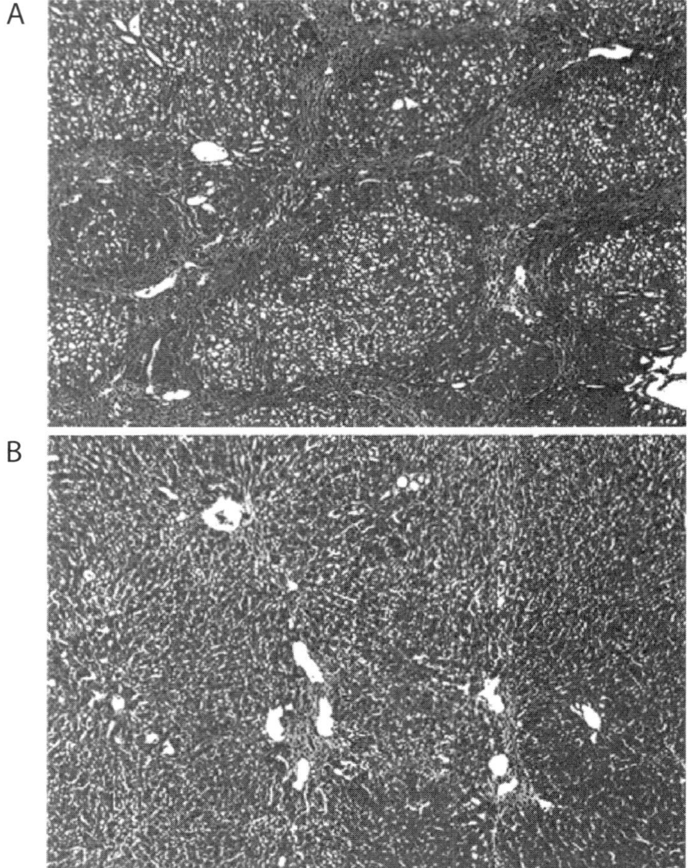

Figure 17.2. Histological micrographs of the rat liver treated with dimethylnitrosamine (DMN), a substance causing persistent liver fibrosis, and a dominant-negative type II TGFβ receptor gene. Rats were infused once via the portal vein with saline (A), or adenoviruses containing a dominant-negative type II TGFβ receptor gene (AdCAT βTR) (B). Both groups of rats were then treated with DMN for 3 weeks. Liver sections were examined histologically by Masson trichrome staining (×200). Note the formation of fibrotic structure in saline-treated liver, but not in the liver transfected with the dominant-negative type II TGFβ receptor gene. (Reprinted with permission from Qi Z et al: *Proc Natl Acad Sci USA* 96:2345–9, copyright 1999 National Academy of Science USA.)

Cell and Tissue Regenerative Engineering [17.10]. The goal of hepatic cell and tissue regenerative engineering is to replace malfunctioned hepatocytes or augment the function of an injured liver by using an engineered liver construct containing necessary liver cells. In principle, a liver construct can be established by assembling liver cells into a liver scaffold, which provides an appropriate environment for liver cell survival, proliferation, and differentiation. Essential criteria are that a liver construct should be implantable and able to conduct essential liver functions. There are a number of issues that should be taken into account for liver reconstruction. These include (1) selection, culture, and manipulation of liver cells; (2) fabrication of liver scaffolds; (3) maintenance of cell viability and functions; (4) implantation of liver constructs; and (5) test of liver functions. These issues are discussed as follows.

Selection, Culture, and Manipulation of Liver Cells [17.11]. A critical issue for liver reconstruction is the selection of cell types. Ideally, a liver construct should contain all necessary hepatic cells. The liver construct should be assembled into the form of the natural liver. However, it is difficult to construct a realistic liver with available technologies. The current liver constructs are mostly extracellular matrix- or polymer-based scaffolds containing selected liver or stem cells. Several types of liver cells have been used for such a purpose, including adult hepatocytes, genetically modulated hepatocytes, and hepatoma cells. Stem cells and hepatic progenitor cells derived from the embryo, fetus, and adult bone marrow are also candidates for liver regeneration.

Hepatocytes are the primary choice for liver regeneration since these cells can proliferate rapidly and conduct necessary functions immediately following implantation. Healthy autogenous hepatocytes from the host patient are ideal candidates for liver construction. Such cells do not induce acute immune rejection, which is the most serious problem in cell and tissue transplantation. However, patients who need liver reconstruction may not possess sufficient functional hepatocytes. In such a case, hepatocytes from a close relative may be considered. Other choices of hepatocyte sources may include allogenic and xenogenic livers. Allogenic cells are those collected from different individuals of the same species. Xenogenic cells are from different species. Hepatocytes from pigs are often used for constructing artificial livers in experimental models. Obviously, allogenic and xenogeneic cells induce acute immune rejection responses. Transplanted hepatocytes will be attacked by host immune cells, resulting in cell apoptosis and rejection. Immune suppressor agents should always be used to prevent immune reactions. Since hepatocytes have a high capacity of regeneration, a small biopsy sample of hepatocytes may generate a sufficient number of cells within relatively short period.

Genetically modified immortal hepatic cell lines have been used for liver reconstruction in experimental models. A major feature of these cell lines is that cells are immortalized and can survive in an engineering system, which is difficult to achieve by using primary hepatocytes or stem cells. Immortalized cell lines can be established by viral transformation. For instance, the introduction of simian virus 40 into cultured hepatocytes can transform the cells into an immortal form, which still exhibits certain characteristics of the hepatocytes, such as the generation of albumin and process of bilirubin. Another approach for cell immortalization is to coculture and transform hepatocytes with a different cell type from a different species. An example is the transformation of human hepatocytes by coculturing with the rat liver epithelial cells. The transformed hepatocytes exhibit not only immortal properties, but also certain hepatic characteristics such as the generation of albumin and α-fetoprotein. Hepatoma cell lines may also be considered for the construction of an artificial liver, since these cells are able to survive and keep certain hepatic characteristics. An example is the human hepatoblastoma C3A cell line. Experimental studies have demonstrated that this type of cells can survive in animal transplantation models for a longer time compared to primary cell lines and can produce hepatic proteins. Overall, cell lines are potential cell candidates for liver reconstruction. However, there is a risk of introducing cancers to the host system. In addition, the transformed cells may lose hepatic functions. Further investigations are necessary to clarify these issues.

Stem and progenitor cells derived from the embryo, fetus, and adult bone marrow are potential sources for regenerating functional liver cells, repairing an injured liver, and reconstructing a malfunctioned liver. Embryonic stem cells are capable of differentiating to all specified cell types. Under an appropriate condition, these cells can differentiate to liver cells and thus can be used for liver reconstruction. Fetal stem and progenitor cells can also be used to regenerate liver cells and reconstruct malfunctioned liver. However, the use of embryonic and fetal cells remains an ethically debating issue, which will likely last for a long time. Alternatively, the adult bone marrow stem and progenitor cells can be used to regenerate liver cells. Several investigations have demonstrated that bone marrow cells can transform to liver cells when the bone marrow cells are delivered to the liver (Fig. 17.3). While the mechanisms of cell transformation remains a research topic, the fusion of bone marrow cells into liver cells has been considered a potential mechanisms for bone marrow cell-based liver regeneration.

Fabrication of Liver Scaffolds and Maintenance of Cell Viability and Function [17.12]. To construct an artificial liver, it is necessary to assemble liver cells in a scaffold. Biological extracellular matrix and synthetic polymers have been used for the construction of such a scaffold. Extracellular matrix components, such as various types of collagens, are natural polymeric materials, which serve as cell substrates and participate in the regulation of cell activities, including cell adhesion, proliferation, and migration. Collagen matrix has been used extensively for the construction of tissue scaffolds. Biodegradable polymers have also been synthesized and used for such a purpose. A unique feature for this type of material is that the scaffold can be gradually degraded in the host system and replaced with cells and natural extracellular matrix.

Several issues should be considered for the construction of hepatic scaffolds:

1. The selected material should be compatible with seeded cells and should not influence the survival and function of the cells. When synthetic polymers are used, biological molecules can be used to coat the surface to which cells attach. A polymeric material should be always tested for toxic effects before being used for constructing tissue scaffolds.
2. A scaffold should be constructed with an appropriate form and structure, factors that may influence the cell function and performance.
3. It is necessary to establish a circulatory system, which introduces blood to the cells seeded in a tissue scaffold.
4. Transplanted cells are subject to an environment that is not natural in a reconstructed liver. Cell apoptosis often occurs within a short period of cell transplantation. It is always a challenge to maintain cell viability when cells are transplanted into a host system in vivo. Thus, cell survival stimulators, such as hepatic growth factor and insulin or their genes, should be applied to liver constructs. These factors play an important role in regulating the survival and proliferation of transplanted liver cells.

Various forms of liver constructs have been established in experimental models. These include scaffolds based on extracellular matrix and polymeric materials, hollow fibers, and microcarriers. These constructs can be used to encapsulate liver cells for transplantation. Experimental investigations and clinical trials have demonstrated the feasibility of using the liver constructs for the treatment of liver failure. It should be noted that all forms of liver constructs established to date are relatively small with respect to the natural liver

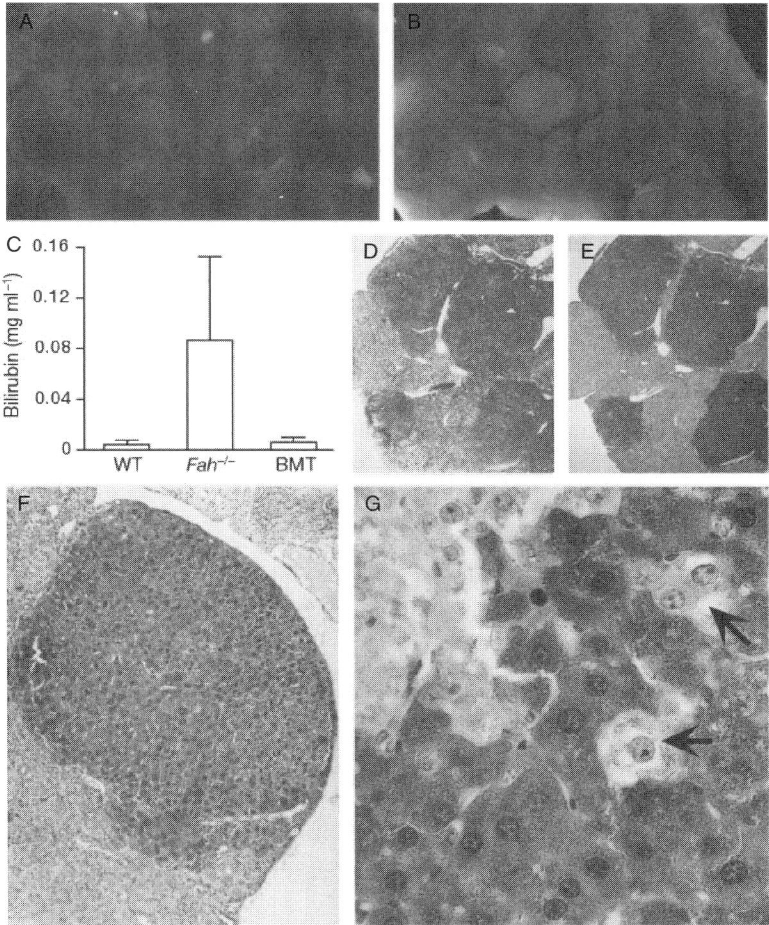

Figure 17.3. Formation of functional fumarylacetoacetate hydrolase-positive (Fah^+) liver nodules in $Fah^{-/-}$ mice by transplantation of $Fah^{+/+}$ wildtype bone marrow cells. (A, B) Gross liver specimens from transplant recipients showing embedded (A) and protruding (B) nodules photographed on a dissecting microscope. (C) Total serum bilirubin levels (mean +/–s.d., $n >= 4$) from wildtype (WT) mice, $Fah^{-/-}$ mice maintained without NTBC for >4 weeks, and $Fah^{-/-}$ mice after wildtype bone marrow transplantation (BMT). (D, E) Serial liver sections from transplant recipients stained with haematoxylin and eosin (D) or with an anti-Fah antibody to stain expressing cells brown (E–G). Images were photographed with ×2.5 (D, E), ×10 (F), or ×40 (G) objectives. Arrows indicate the locations of nonexpressing cells at the edge of an Fah^+ nodule. These observations show that transplanted $Fah^{+/+}$ wildtype bone marrow cells can engraft to the liver and differentiate into hepatocytes with fah function in transgenic $Fah^{-/-}$ mice. (Reprinted by permission from Macmillan Publishers Ltd.: Vassilopoulos G, Wang PR, Russell DW: *Nature* 422:901–4, copyright 2003.)

because of difficulties in the construction of a vascular system. These models are briefly discussed as follows.

For the *scaffold model*, hepatocytes can be seeded in an extracellular matrix or polymeric scaffold, which provides a substrate for cell attachment and assembly. The cell-seeded scaffold can be then enclosed within a semipermeable membrane system

(Fig. 17.4). Several types of material, including polysaccharide hydrogels, hydroxyethyl methacrylate-methyl methacrylate matrix, calcium alginate, and collagen matrix, have been used for constructing liver scaffolds. The semipermeable membrane can be constructed with polymeric materials, such as cellulose and polysulfone. This membrane separates the enclosed hepatocytes from the host tissue and thus prevents leukocyte infiltration and immune rejection responses, when allogenic or xenogenic cells are used. At the same time, the semipermeable membrane allows the release of proteins produced by the enclosed liver cells from the scaffold to the surrounding tissue. The semipermeable membrane can also restrain the transplanted cells within the scaffold, preventing potentially harmful effects, such as carcinogenesis when immortal cells are used.

A liver construct can be implanted into the abdominal cavity of the host. Ideally, the liver construct should be connected to the host circulatory system so that liver-produced proteins can be release into the blood (see next section). However, it is often difficult to establish blood circulation in an artificial liver. Alternatively, multiple small liver scaffolds can be constructed and implanted into the abdominal cavity without connecting to the host circulatory system. When the scaffolds are sufficient small, oxygen and nutrients can diffuse to the cells seeded in the scaffolds.

A. Matrix-based implant

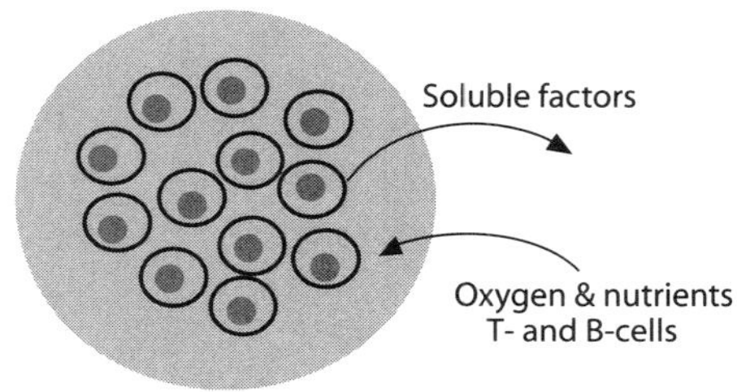

B. Encapsulated implant

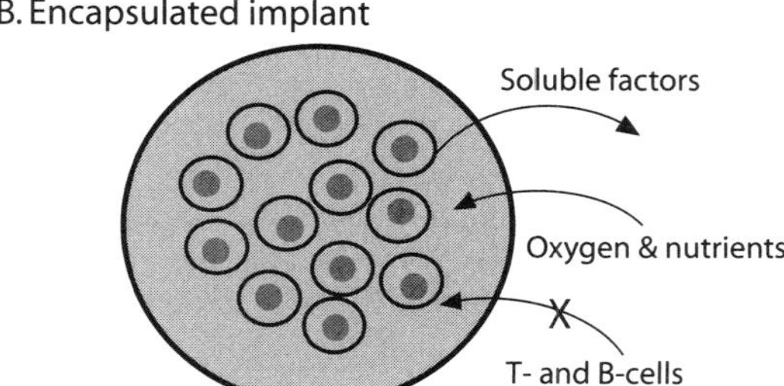

Figure 17.4. Types of implantable devices with hepatocytes. (A) Cells are in an open matrix, which is often biodegradable. Surrounding tissue, including blood vessels, can grow into the implanted matrix. This provides no protection from the immune system from the host. (B) Cells are encapsulated so that they are protected by a barrier that prevents immune cells and factors from reaching the cells while allowing small (usually <50 kDa) metabolites to transport from the capsule to surrounding tissue. Metabolite transport to and from the nearest vascular bed is chiefly by diffusion and may be adversely affected by the presence of a fibrotic layer, which often develops around such implants. (Reprinted from Chan C et al: Hepatic tissue engineering for adjunct and temporary liver support: Critical technologies, *Liver Transplant* 10:1331–42, copyright 2004 by permission of John Wiley & Sons, Inc.)

For the *microcarrier model*, polymeric materials can be used to construct bead-like carriers. The carriers can be coated with extracellular matrix molecules, such as collagen and fibronectin, which enhance cell attachment and survival. Hepatocytes can be collected and cultured on the carrier beads. When cells reach confluence, the carriers can be implanted into the abdominal cavity. In this model, the transplanted cells are directly exposed to the host system. Immune rejection reactions will occur when allogenic and xenogeneic cells are used.

For the *hollow fiber model*, polymeric hollow fibers can be filled with a gel of extracellular matrix component, such as collagen, mixed with hepatocytes (Fig. 17.5). The collagen gel serves as a matrix for the attachment and assembly of the seeded hepatocytes.

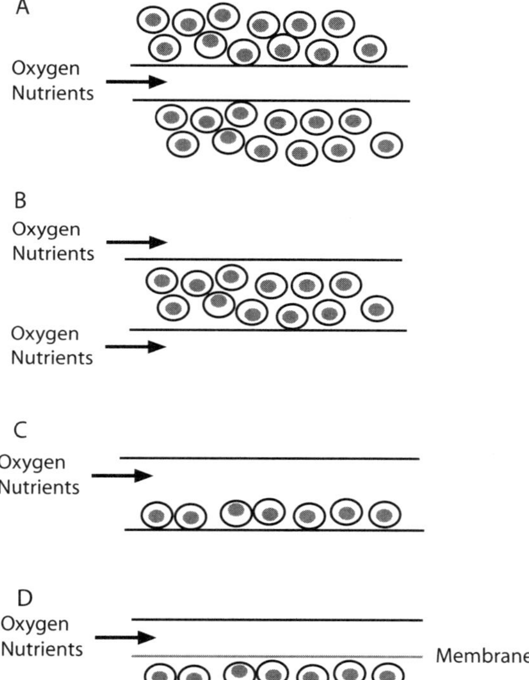

Figure 17.5. Common bioreactor designs for bioartificial livers. (A) Hepatocyte aggregates on microcarriers are placed on the outside of hollow fibers. Oxygenated plasma is flown through the hollow fibers. (B) Hepatocyte aggregates in a supporting matrix are inside hollow fibers and oxygenated plasma is flown outside the hollow fibers. (C) Similar to panel A, although separate hollow fibers are used to deliver hepatocyte culture medium, plasma, and oxygen into the system. Circle with O_2 is a hollow fiber perpendicular to the plane of the paper. (D) Hepatocyte aggregates are in a supporting matrix next to hollow fibers that deliver oxygen. Oxygenated plasma is flown in the space outside of the hollow fibers and percolates through the matrix–hepatocyte network. (E) Hepatocytes are seeded as a monolayer on the bottom surface of a flat plate and placed within a parallel-plate flow chamber. Oxygenated plasma is flown directly above the cells. (F) System is similar to that shown in panel E, except that oxygen is delivered through a permeable membrane directly above the flow channel with the hepatocytes. (Reprinted from Chan C et al: Hepatic tissue engineering for adjunct and temporary liver support: Critical technologies, *Liver Transplant* 10:1331–42, copyright 2004 by permission of John Wiley & Sons, Inc.)

When the fibers are small, oxygen and nutrients can easily diffuse to the enclosed hepatocytes. By using porous polymeric materials with an appropriate pore size, the fiber wall can prevent the invasion of host immune cells and immune rejection responses. Multiple hollow fibers can be grouped and assembled into a large tubular device, which can be used for implantation. Host blood can be introduced to the interfiber spaces via vascular anastomoses (see next section), ensuring sufficient oxygen and nutrient supplies. Alternatively, hepatocytes can be cultured on the exterior surface of semipermeable hollow fibers. Multiple fibers can be assembled within a larger tubular device. Upon implantation, host blood can be introduced into the lumens of the hollow fibers via vascular anastomoses. The hollow fiber model gives a large surface area for molecular diffusion, ensuring efficient release of proteins produced by transplanted hepatocytes. The hollow fiber device can be implanted into the abdominal cavity of the host. Experimental studies have demonstrated the feasibility and usefulness of this model.

Implantation of Liver Constructs [17.13]. Various methods can be used to implant liver constructs, depending on the form of the construct. Microcarriers and capsules can be directly implanted into the abdominal cavity. Since these devices are small, oxygen and nutrients can diffuse from the abdominal serous fluid into the device. For large devices that require blood supply, such as the hollow fiber liver construct, an artery and a vein should be selected and anastomosed to the blood circulatory system of the liver construct. It is important to note that the artery and vein selected for such a purpose should be rich in collateral circulation, so that the use of the blood vessels for the liver construct does not influence blood supply to the distal tissues of the host system. In the abdominal cavity, the small and large intestines are supplied by the superior and inferior mesenteric arteries, which are connected by collateral arteries through the intestinal system. There are also collateral blood vessels for the mesenteric veins. The blockade of the inferior mesenteric artery and vein does not significantly influence the blood supply to the intestines. Thus, the inferior mesenteric artery and vein can be used as a blood supplying system for an implanted liver construct. A common problem for anastomoses with a nonvascular structure is blood coagulation within the liver construct as well as thrombosis and intimal hyperplasia within the anastomotic blood vessels. These pathological changes often result in obstruction of the blood circulation within the implanted liver construct. A persistent administration of anticoagulants and anti-proliferative agents is necessary for preventing thrombosis and intimal hyperplasia.

Testing Liver Function [17.13]. It is important to test the function and durability of the implanted liver construct. There are a number of parameters that are used for testing the liver function. These include the blood concentration of albumin, aminotransferases (aspartate aminotransferases and alanine aminotransferases), clotting factors, and ammonia. *Albumin* is produced by hepatocytes and its blood concentration is a useful index for the assessment of the liver function. The normal serum level of albumin is 3.5–5 g/dL. A significant decrease in the albumin concentration compared to normal controls suggests insufficient function or malfunction of the liver construct. Aspartate aminotransferases and alanine aminotransferases are two enzymes that catalyze the transfer of the γ-amino group from aspartate and alanine to the γ-keto group of ketoglutarate, forming oxaloacetic acid and pyruvic acid. The normal blood level of these enzymes is about 40 IU. Under physiological conditions, these enzymes are degraded in the liver. An increase in the blood level of these enzymes suggests insufficient hepatic function. The liver synthesizes several blood coagulation factors, including coagulation factor I (fibrinogen), II (prothrombin), V, VII, IX, and X. A reduction in the blood concentrations of these factors indicates insufficient hepatic function. *Ammonia* is a waste product generated by protein metabolism and is transformed into urea in the liver. An elevation in the blood concentration of ammonia strongly suggests deficiency of the hepatic function. Thus, these proteins and substances can be measured and used for assessing the function of an implanted liver construct.

Chronic Hepatitis and Cirrhosis

Pathogenesis, Pathology, and Clinical Features [17.3]. *Chronic hepatitis* is a disorder caused by various pathogens, including hepatitis viruses and chemical toxins, and characterized by continuous inflammatory reactions, hepatocyte necrosis, and fibrosis in the liver. Persistent pathological changes may eventually lead to cirrhosis and liver failure. About 30% of patients with chronic hepatitis have a history of hepatitis B infection. These patients often exhibit positive hepatitis B antigens in their serum. The clinical manifestations of chronic hepatitis vary from mild asymptomatic illness to liver failure. The mechanisms for such a wide range of changes remain poorly understood.

The pathogenesis of chronic hepatitis has been hypothetically related to immune responses in the liver. In many cases, hepatic lesions are associated with T-cell infiltration and activation. Antibodies against host components have been detected in some patients. Other types of autoimmune disorders, such as diabetes and thyroiditis, can be found in some patients with chronic hepatitis. The administration of corticosteroids, which are used to treat autoimmune disorders, is effective for the treatment of chronic hepatitis. These observations support the hypothesis that chronic hepatitis may be induced by autoimmune reactions in the liver.

Chronic hepatitis is usually diagnosed on the basis of biopsy examinations. Typical pathological changes include inflammatory reactions characterized by the presence of dense mononuclear cells, hepatocyte necrosis in peripheral regions of liver lobules, excessive formation of fibrous extracellular matrix, and regeneration of hepatic lobules. A large fraction of patients (up to 50%) are associated with cirrhosis (see next paragraph for details). These pathological changes usually develop within 1–2 years.

Cirrhosis is a hepatic disorder characterized by massive fibrosis, structural distortion, formation of regenerative nodules, and deterioration in the function of the liver. When most hepatocytes are replaced with fibrous tissue, the liver loses its function and dies. A number of factors contribute to the development of cirrhosis. These include alcohol toxicity, hepatitis viral infection, obstruction of biliary ducts, and right heart failure. These factors induce inflammatory reactions in the liver in association with excessive production of extracellular matrix components. Several types of liver cells, including the Kupffer cells, Ito cells, and natural killer cells, can release various cytokines and inflammatory mediators, which may further enhance inflammatory reactions. A typical inflammatory mediator is transforming growth factor β (TGFβ), which stimulates hepatic fibrogenesis and accelerates cirrhosis. Here, the role of alcohol toxicity, hepatitis viral infection, obstruction of biliary ducts, and right heart failure in regulating the development of cirrhosis is briefly discussed.

Alcohol toxicity is a common cause of cirrhosis. Long-term exposure to alcohol induces various changes in the liver. In certain cases, extensive fatty acid deposition occurs in hepatocytes due to the impairment of fatty acid oxidation, leading to fat accumulation in the liver (Fig. 17.6). This change is associated with enlarged liver, uniform loss of hepatocytes, formation of regenerative nodules, and deposition of extracellular matrix or fibrosis. Alcohol toxicity is often associated with hepatitis, characterized by hepatocyte necrosis, infiltration of leukocytes, and deposition of collagen matrix. During the end-stage, hepatocytes are committed to apoptosis and are replaced with fibroblasts, which produce excessive collagen matrix, leading to the formation of a fibrous network. The liver exhibits structural distortion, extensive fibrosis, and regenerative nodules, in association with progressive shrinkage of the liver. The end-stage disorder is often referred to as alcoholic cirrhosis.

Hepatitis viral infection is a major cause of cirrhosis. In the majority of patients, viral hepatitis can be completely self-cured. However, in a small fraction of patients, such a disorder may develop into chronic hepatitis and eventually into cirrhosis. Pathological

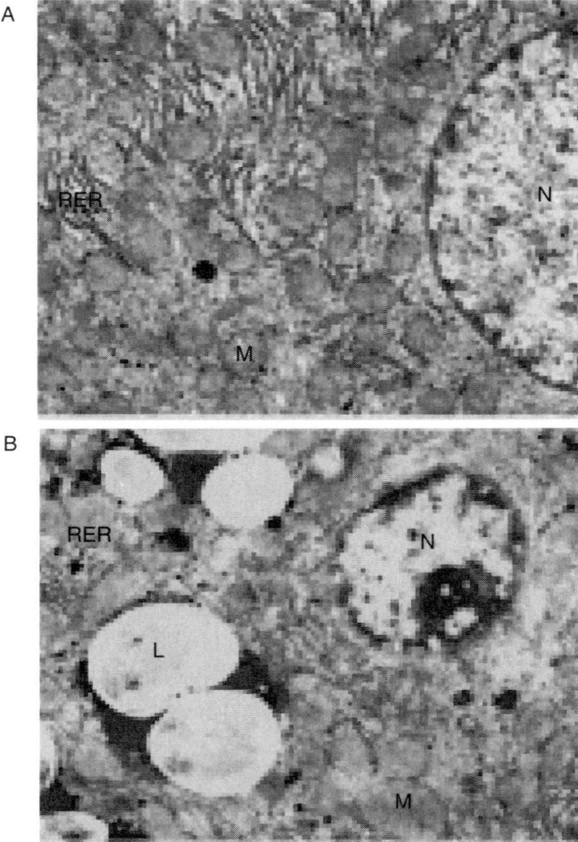

Figure 17.6. Ethanol-induced ultrastructural changes in the hepatocytes of wildtype mice. (A) control; (B) ethanol. Note that presence of lipid droplets in ethanol-treated liver. N, nucleus; M, mitochondria; RER, rough endoplasmic reticulum; L, lipid droplets; FCD, focal cytoplasmic degeneration. Magnification, ×7500. (Reprinted with permission from Zhou Z et al: Metallothionein-independent zinc protection from alcoholic liver injury, *Am J Pathol* 160:2267–74, copyright 2002.)

features in viral hepatitis-induced cirrhosis are similar to those found in alcoholic cirrhosis, except that few fatty acid-filled cells are found. Hepatitis viral infection often causes the obstruction of the biliary ducts. Such an alteration causes bile accumulation, leading to inflammatory reactions and fibrosis in the liver (see below for more details).

Biliary obstruction is another major cause of cirrhosis. There are two types of biliary obstruction: primary and secondary obstruction. Primary obstruction occurs in small intrahepatic biliary ducts. The mechanisms of primary obstruction remain poorly understood. Secondary obstruction is found in large extrahepatic bile ducts and is caused by the compression of gallstones, tumors, and surgical scars. Both types induce similar pathological changes, which include destruction of biliary ducts, infiltration of leukocytes, fibrosis, and progressive loss of hepatocytes, eventually leading to the development of cirrhosis.

Right heart failure is a disorder characterized by reduced performance of the right ventricle and reduced cardiac output. As a result, venous blood pressure increases due to elevated resistance in the right heart, resulting in blood accumulation in the venous system and congestion of the liver. Furthermore, liver ischemia may occur because of heart failure. These changes cause chronic hepatocyte injury and necrosis. Necrotic hepatocytes are often replaced with fibroblasts and fibrous matrix, leading to fibrosis and scar formation. These changes continue unless the right ventricular function is improved. The ultimate consequence is cirrhosis and the death of the liver.

Treatment of Chronic Hepatitis and Cirrhosis [17.4]. Chronic hepatitis and cirrhosis often exhibit similar changes in hepatic structure and function, except that the degree of changes may differ. Chronic hepatitis often develops into cirrhosis, and cirrhosis is considered the final stage of chronic hepatitis. Thus, both disorders are treated with similar approaches. Since chronic viral hepatitis and subsequent cirrhosis are the most common forms of hepatic disorder, we use these disorders as examples for discussing therapeutic strategies. General strategies for treating chronic hepatitis and cirrhosis include the suppression of inflammatory reactions, prevention of fibrosis, and inhibition of viral replication.

Corticosteroids can be used to suppress inflammatory reactions in chronic hepatitis. In about 70% of patients, corticosteroid treatment significantly reduces clinical symptoms, slows down pathological changes, and improves the liver function in chronic hepatitis. Typical signs of improvement include disappearance of fatigue and anorexia, an increase in serum albumin, and a fall in serum bilirubin. Pathological examinations may find a decrease in the infiltration of mononuclear cells and hepatocyte necrosis. However, pathological changes often resume when corticosteroid administration is ceased. Corticosteroid administration may not be effective for the treatment of cirrhosis, since structural changes are often irreversible.

Patients may be administrated with interferon (IFN) α, a cytokine that exerts antiviral, antifibrogenic, and anti-tumoral effects. Such a treatment has been shown to reduce pathological changes and improve hepatic function in chronic hepatitis. Recently, antiviral nucleoside analogues, such as lamivudine and famciclovir, have been used to treat chronic viral hepatitis. These nucleoside analogues can be incorporated into the viral genome and can terminate viral replication. A treatment with these agents results in the suppression of the activity of hepatitis B virus and improvement of hepatic function. The molecular, cell, and tissue regenerative engineering approaches described above for hepatitis and liver failure also apply to the treatment for chronic hepatitis and cirrhosis.

Liver Cancers

Pathogenesis, Pathology, and Clinical Features. The liver may develop two types of cancer: primary carcinoma and metastatic cancer. Primary hepatic carcinoma, also known as *hepatoma*, can initiate from hepatocytes or bile duct epithelial cells. Hepatocyte carcinoma accounts for about 90% of liver carcinomas. Several etiologic factors may contribute to the initiation and development of liver carcinoma, including chronic hepatitis, cirrhosis, viral hepatitis, and exposure to certain carcinogens. The incidence of primary hepatic carcinoma is significantly higher in patients with chronic hepatitis, cirrhosis, and viral hepatitis than in the general population. Such a statistical analysis suggests a causative effect of these disorders on the carcinogenesis of the liver. Carcinogens, such as aflatoxins and formaldehydes, may induce gene mutation or changes in the DNA sequence, contributing to carcinogenesis. Metastatic cancer is a cancer that originated from a different organ and spread to the liver. The liver is susceptible to the invasion of metastatic cancer cells because of its large volume and function as a blood reservoir, which enhance cancer cell accumulation in the liver. The pathogenic mechanisms and pathological changes in cancers will be discussed in detail on page in Chapter 25. These aspects are similar in all types of cancers.

Treatment of Liver Cancers. Liver cancers, including primary and secondary carcinomas, exhibit genetic, biochemical, pathological, and clinical characteristics that are similar to those of other types of cancer. Thus similar therapeutic strategies can be applied to all types of cancer with modifications with regard to anatomical differences. The therapeutic aspect of cancers is discussed in Chapter 25.

BIBLIOGRAPHY

17.1. Anatomy and Physiology of the Liver

Guyton AC, Hall JE: *Textbook of Medical Physiology*, 11th ed, Saunders, Philadelphia, 2006.

McArdle WA, Katch FI, Katch VL: *Essentials of Exercise Physiology*, 3rd ed, Lippincott Williams & Wilkins, Baltimore, 2006.

Germann WL, Stanfield CL (with contributors Niles MJ, JG), *Cannon: Principles of Human Physiology*, 2nd ed, Pearson Benjamin Cummings, San Francisco, 2005.

Thibodeau GA, Patton KT: *Anatomy & Physiology*, 5th ed, Mosby, St Louis, 2003.

Boron WF, Boulpaep EL: *Medical Physiology: a Cellular and Molecular Approach*, Saunders, Philadelphia, 2003.

Ganong WF: *Review of Medical Physiology*, 21th ed, McGraw-Hill, New York, 2003.

17.2. Liver Regeneration

Bucher NLR: Regeneration of mammalian liver, *Int Rev Cytol* 15:245–300, 1963.

Sigel B, Acevedo FJ, Dunn MR: The effect of partial hepatectomy on autotransplanted liver tissue, *Surg Gynecol Obstet* 117:29–36, 1963.

Virolainen M: Mitotic response in liver autograft after partial hepatectomy in the rat, *Exp Cell Res* 33:588–91, 1964.

Koniaris LG, McKillop IH, Schwartz SI, Zimmers TA: Liver regeneration, *J Am Col Surg* 197: 634–59, 2003.

Michalopoulos GK, Bowen WC, Mule K, Stolz DB: Histological organization in hepatocyte organoid cultures, *Am J Pathol* 159:1877–87, 2001.

Sauer IM, Obermeyer N, Kardassis D et al: Development of a hybrid liver support system, *Ann NY Acad Sci* 944:308–19, 2001.

Bhandari RN, Riccalton LA, Lewis AL et al: Liver tissue engineering: A role for co-culture systems in modifying hepatocyte function and viability, *Tissue Eng* 7:345–57, 2001.

Miyoshi H, Ehashi T, Ema H et al: Long-term culture of fetal liver cells using a three-dimensional porous polymer substrate, *ASAIO J* 46:397–402, 2000.

Heckel JL, Sandgren EP, Degen JL et al: Neonatal bleeding in transgenic mice expressing urokinase-type plasminogen activator, *Cell* 62:447–56, 1990.

Sandgren EP, Palmiter RD, Heckel JL et al: Complete hepatic regeneration after somatic deletion of an albumin-plasminogen activator transgene, *Cell* 66:245–56, 1991.

Rhim JA, Sandgren EP, Degen JL et al: Replacement of diseased mouse liver by hepatic cell transplantation, *Science* 263:1149–52, 1994.

Petrowsky H, Gonen M, Jarnagin W et al: Second liver resections are safe and effective treatment for recurrent hepatic metastases from colorectal cancer: A bi-institutional analysis, *Ann Surg* 235:863–71, 2002.

Little SA, Jarnagin WR, DeMatteo RP et al: Diabetes is associated with increased perioperative mortality but equivalent long-term outcome after hepatic resection for colorectal cancer, *J Gastrointest Surg* 6:88–94, 2002.

Kawasaki S, Makuuchi M, Matsunami H et al: Living related liver transplantation in adults, *Ann Surg* 227:269–74, 1998.

Nishizaki T, Ikegami T, Hiroshige S et al: Small graft for living donor liver transplantation, *Ann Surg* 233:575–80, 2001.

Palmes D, Spiegel HU: Animal models of liver regeneration, *Biomaterials* 25:1601–11, 2004.

Lindroos PM, Zarnegar R, Michalopoulos GK: Hepatocyte growth factor (hepatopoietin A) rapidly increases in plasma before DNA synthesis and liver regeneration stimulated by partial hepatectomy and carbon tetrachloride administration, *Hepatology* 13:743–50, 1991.

Rahman TM, Hodgson HJ: Animal models of acute hepatic failure, *Int J Exp Pathol* 81:145–57, 2000.

Bolesta S, Haber SL: Hepatotoxicity associated with chronic acetaminophen administration in patients without risk factors, *Ann Pharmacother* 36:331–3, 2002.

Liu KX, Kato Y, Yamazaki M, Higuchi O, Nakamura T, Sugiyama Y: Decrease in the hepatic clearance of hepatocyte growth factor in carbon tetrachloride-intoxicated rats, *Hepatology* 17:651–60, 1993.

17.3. Pathogenesis, Pathology, and Clinical Features of Hepatitis and Liver Failure

Schneider AS, Szanto PA: *Pathology*, 3rd ed, Lippincott Williams & Wilkins, Philadelphia, 2006.

McCance KL, Huether SE: *Pathophysiology: the Biologic Basis for Disease in Adults & Children*, 5th ed, Elsevier Mosby, St Louis, 2006.

Porth CM: *Pathophysiology: Concepts of Altered Health States*, 7th ed, Lippincott Williams & Wilkins, Philadelphia, 2005.

Lai CL, Ratziu V, Yuen MF, Poynard T: Viral hepatitis *B. Lancet* 362(9401):2089–94, Dec 2003.

Frazier MS, Drzymkowski JW: *Essentials of Human Diseases and Conditions*, 3rd ed, Elsevier Saunders, St Louis, 2004.

Friedman SL: The cellular basis of hepatic fibrosis: Mechanisms andtreatment strategies, *New Engl J Med* 328:1828–35, 1993.

Friedman SL: Molecular regulation of hepatic fibrosis, an integrated cellular response to tissue injury, *J Biol Chem* 275:2247–50, 2000.

Bissell DM: Hepatic fibrosis as wound repair: A progress report, *J Gastroenterol* 33:295–302, 1998.

Border WA, Noble NA: Transforming growth factor B in tissue fibrosis, *New Engl J Med* 331:1286–91, 1994.

Shetty K, Wu GY, Wu CH: Gene therapy of hepatic diseases: prospects for the new millennium, *Gut* 46:136–9, 2000.

17.4. Conventional Treatment

Hoofnagle JH, Lau D: New therapies for chronic hepatitis B, *J Viral Hepat* 4(Suppl 1):41–50, 1997.

Davis GL, Esteban-Mur R, Rustgi V et al: Interferon alfa-2b alone or in combination with ribavirin for the treatment of relapse of chronic hepatitis C. International Hepatitis Interventional Therapy Group, *New Engl J Med* 339:1493–9, 1998.

McHutchison JG, Gordon SC, Schiff ER et al: Interferon alfa-2b alone or in combination with ribavirin as initial treatment for chronic hepatitis C. Hepatitis Interventional Therapy Group, *New Engl J Med* 339:1485–92, 1998.

Zeuzem S, Feinman SV, Rasenack J et al: Peginterferon alfa-2a in patients with chronic hepatitis C, *New Engl J Med* 343:1666–72, 2000.

Sobesky R, Mathurin P, Charlotte F et al: Modeling the impact of interferon alfa treatment on liver fibrosis progression in chronic hepatitis C: A dynamic view. The Multivirc Group, *Gastroenterology* 116:378–86, 1999.

17.5. Suppression of Viral Activities

Sobesky R, Mathurin P, Charlotte F et al: Modeling the impact of interferon alfa treatment on liver fibrosis progression in chronic hepatitis C: A dynamic view. The Multivirc Group, *Gastroenterology* 116:378–86, 1999.

Yoshida H, Shiratori Y, Moriyama M et al: Interferon therapy reduces the risk for hepatocellular carcinoma: National surveillance program of cirrhotic and noncirrhotic patients with chronic hepatitis C in Japan. IHIT Study Group. Inhibition of Hepatocarcinogenesis by Interferon Therapy, *Ann Intern Med* 131:174–81, 1999.

Aurisicchio L, Delmastro P, Salucci V et al: Liver-specific alpha 2 interferon gene expression results in protection from induced hepatitis, *J Virol* 74:4816–23, 2000.

Alt M, Renz R, Hofschneider PH et al: Specific inhibition of hepatitis C viral gene expression by antisense phosphorothioate oligodeoxynucleotides, *Hepatology* 22:707–17, 1995.

Blum HE, Galun E, Weizsacker FV et al: Inhibition of hepatitis B virus by antisense oligodeoxynucleotides, *Lancet* 337:1230, 1991.

Welch PJ, Yei S, Barber JR: Ribozyme gene therapy for hepatitis C virus infection, *Clin Diagn Virol* 10:163–71, 1998.

Garcia-Navarro R, Blanco-Urgoiti B, Berraondo P et al: Protection against woodchuck hepatitis virus (WHV) infection by gene gun coimmunization with WHV core and interleukin-12, *J Virol* 75:9068–76, 2001.

Lasarte JJ, Corrales FJ, Casares N et al: Different doses of adenoviral vector expressing IL-12 enhance or depress the immune response to a coadministered antigen: The role of nitric oxide, *J Immunol* 162:5270–7, 1999.

Prieto J, Herraiz M, Sangro B, Qian C, Mazzolini G, Melero I, Ruiz J: The promise of gene therapy in gastrointestinal and liver diseases, *Gut* 52:ii49, 2003.

Wands JR, Geissler M, Putlitz JZ, Blum H, von Weizsacker F et al: Nucleic acid-based antiviral and gene therapy of chronic hepatitis B infection, *J Gastroenterol Hepatol* 12(9–10):S354–69, 1997.

Hoofnagle JH, Lau D: New therapies for chronic hepatitis B, *J Viral Hepat* 4(Suppl 1):41–50, 1997.

17.6. Enhancement of Hepatocyte Proliferation

Steer CJ: Liver regeneration, *FASEB J* 9:1396–400, 1995.

Michalopoulos GK, DeFrances MC: Liver regeneration, *Science* 276:60–6, 1997.

Zarnegar R, Michalopoulos GK: The many faces of hepatocyte growth factor: From hepatopoiesis to hematopoiesis, *J Cell Biol* 1995;129:1177–80, 1995.

Kopp JB: Hepatocyte growth factor: Mesenchymal signal for epithelial homeostasis, *Kidney Int* 54:1392–3, 1998.

Yamaguchi K, Nalesnik MA, Michalopoulos GK: Hepatocyte growth factor mRNA in human liver cirrhosis as evidenced by in situ hybridization, *Scand J Gastroenterol* 31:921–7, 1996.

Itakura A, Kurauchi O, Morikawa S, Okamura M, Furugori K et al: Involvement of hepatocyte growth factor in formation of bronchoalveolar structures in embryonic rat lung in primary culture, *Biochem Biophys Res Commun* 241:98–103, 1997.

Shiota G, Wang TC, Nakamura T, Schmidt EV: Hepatocyte growth factor intransgenic mice: Effects on hepatocyte growth, liver regeneration and gene expression, *Hepatology* 19:962–72, 1994.

Yasuda H, Imai E, Shiota A, Fujise N, Morinaga T et al: Antifibrogenic effect of a deletion variant of hepatocyte growth factor on liver fibrosis in rats, *Hepatology* 24:636–42, 1996.

Ignotz RA, Endo T, Massague J: Regulation of fibronectin and type I collagen mRNA levels by transforming growth factor-B, *J Biol Chem* 262:6443–6, 1987.

Burr AW, Toole K, Chapman C, Hines JE, Burt AD: Anti-hepatocyte growth factor antibody inhibits hepatocyte proliferation during liver regeneration, *J Pathol* 185:298–302, 1998.

Ueki T, Kaneda Y, Tsutsui H, Nakanishi K, Sawa Y et al: Hepatocyte growth factor gene therapy of liver cirrhosis in rats, *Nature Med* 5:226–30, 1999.

Fujimoto J: Reversing liver cirrhosis: impact of gene therapy for liver cirrhosis, *Gene Ther* 6:305–6, 1999.

Liu Y, Michalopoulos GK, Zarnegar R: Structural and functional characterization of the mouse hepatocyte growth factor gene promoter, *J Biol Chem* 269:4152–60, 1994.

He FC, Wu C, Tu Q, Xing G: Human hepatic stimulator substance: A product of gene expression of human fetal liver tissue, *Hepatology* 17:225–9, 1993.

Hagiya M, Francavilla A, Polimeno L, Ihara I, Sakai H et al: Cloning and sequence analysis of the rat augmenter of liver regeneration (ALR) gene: expression of biologically active recombinant ALR and demonstration of tissue distribution, *Proc Natl Acad Sci USA* 91:8142–6, 1994.

Gandhi CR, Kuddus R, Subbotin VM, Prelich J, Murase N et al: A fresh look at augmenter of liver regeneration in rats, *Hepatology* 29:1435–45, 1999.

Francavilla A, Vujanovic NL, Polimeno L, Azzarone A, Iacobellis A et al: The in vivo effect of hepatotrophic factors augmenter of liver regeneration, hepatocyte growth factor, and insulin-like growth factor-IIon liver natural killer cell functions, *Hepatology* 25:411–5, 1997.

Polimeno L, Margiotta M, Marangi L, Lisowsky T, Azzarone A et al: Molecular mechanisms of augmenter of liver regeneration as immunoregulator: Its effect on interferon-B expression in rat liver, *Digest Liver Dis* 32:217–25, 2000.

KGF

Ray P, Devaux Y, Stolz DB, Yarlagadda M, Watkins SC et al: Inducible expression of keratinocyte growth factor (KGF) in mice inhibits lung epithelial cell death induced by hyperoxia, *Proc Natl Acad Sci* 100:6098–103, 2003.

Rubin, JS, Osada H, Finch PW, Taylor WG, Rudikoff S et al: Purification and characterization of a newly identified growth factor specific for epithelial cells, *Proc Natl Acad Sci* 86:802–6, 1989.

Werner S, Smola H, Liao X, Longaker MT, Krieg T et al: The function of KGF in morphogenesis of epithelium and reepithelialization of wounds, *Science* 266:819–22, 1994.

Zimonjic DB, Kelley MJ, Rubin JS, Aaronson SA et al: Fluorescence in situ hybridization analysis of keratinocyte growth factor gene amplification and dispersion in evolution of great apes and humans, *Proc Nat Acad Sci* 94:11461–5, 1997.

Hepatopoietin

Francavilla A, Hagiya M, Porter KA, Polimeno L, Ihara I et al: Augmenter of liver regeneration: Its place in the universe of hepatic growth factors, *Hepatology* 20:747–57, 1994.

Giorda R, Hagiya M, Seki T, Shimonishi M, Sakai H et al: Analysis of the structure and expression of the augmenter of liver regeneration (ALR) gene, *Mol Med* 2:97–108, 1996.

Hagiya M, Francavilla A, Polimeno L, Ihara I, Sakai H et al: Cloning and sequence analysis of the rat augmenter of liver regeneration (ALR) gene: Expression of biologically active recombinant ALR and demonstration of tissue distribution, *Proc Nat Acad Sci* 91:8142–6, 1994. *Note*: Erratum *Proc Nat Acad Sci USA* 92:3076 (only), 1995.

Lisowsky T, Weinstat-Saslow DL, Barton N, Reeders ST, Schneider MC: A new human gene located in the PKD1 region of chromosome 16 is a functional homologue to ERV1 of yeast, *Genomics* 29:690–7, 1995.

Li Y, Liu W, Xing G, Tian C, Zhu Y et al: Direct association of hepatopoietin with thioredoxin constitutes a redox signal transduction in activation of AP-1/NF-kappaB, *Cell Signal* 17(8):985–96, 2005.

Shen L, Hu J, Lu H, Wu M, Qin W et al: The apoptosis-associated protein BNIPL interacts with two cell proliferation-related proteins, MIF and GFER, *FEBS Lett* 540(1–3):86–90, 2003.

Insulin

Bell GI, Pictet RL, Rutter WJ, Cordell B, Tischer E et al: Sequence of the human insulin gene, *Nature* 284:26–32, 1980.

Owerbach D, Bell GI, Rutter WJ, Shows TB: The insulin gene is located on chromosome 11 in human, *Nature* 286:82–4, 1980.

IGF1

Baker J, Liu JP, Robertson EJ, Efstratiadis A: Role of insulin-like growth factors in embryonic and postnatal growth, *Cell* 75:73-82, 1993.

Barton ER, Morris L, Musaro A, Rosenthal N, Sweeney HL: Muscle-specific expression of insulin-like growth factor I counters muscle decline in mdx mice, *J Cell Biol* 157:137–47, 2002.

Brissenden JE, Ullrich A, Francke U: Human chromosomal mapping of genes for insulin-like growth factors I and II and epidermal growth factor, *Nature* 310:781–4, 1984.

Carro E, Trejo JL, Gomez-Isla T, LeRoith D, Torres-Aleman I: Serum insulin-like growth factor I regulates brain amyloid-beta levels, *Nature Med* 8:1390–7, 2002.

Hankinson SE, Willett WC, Colditz GA, Hunter DJ, Michaud DS et al: Circulating concentrations of insulin-like growth factor-I and risk of breast cancer, *Lancet* 351:1393–6, 1998.

Holly J: Insulin-like growth factor-I and new opportunities for cancer prevention, *Lancet* 351:1373–5, 1998.

Humbert S, Bryson EA, Cordelieres FP, Connors NC, Datta SR et al: The IGF-1/Akt pathway is neuroprotective in Huntington's disease and involves huntingtin phosphorylation by Akt, *Dev Cell* 2:831–7, 2002.

Liu JP, Baker J, Perkins AS, Robertson EJ, Efstratiadis A: Mice carrying null mutations of the genes encoding insulin-like growth factor I (Igf-1) and type 1 IGF receptor (Igf1r), *Cell* 75:59–72, 1993.

Musaro A, McCullagh K, Paul A, Houghton L, Dobrowolny G et al: Localized Igf-1 transgene expression sustains hypertrophy and regeneration in senescent skeletal muscle, *Nature Genet* 27:195–200, 2001.

Rosenfeld RG: Insulin-like growth factors and the basis of growth, *New Eng J Med* 349:2184–6, 2003.

Ruberte J, Ayuso E, Navarro M, Carretero A, Nacher V et al: Increased ocular levels of IGF-1 in transgenic mice lead to diabetes-like eye disease, *J Clin Invest* 113:1149–57, 2004.

Tricoli JV, Rall LB, Scott J, Bell GI, Shows TB: Localization of insulin-like growth factor genes to human chromosomes 11 and 12, *Nature* 310:784–6, 1984.

Woods KA, Camacho-Hubner C, Savage MO, Clark AJL: Intrauterine growth retardation and postnatal growth failure associated with deletion of the insulin-like growth factor I gene, *New Eng J Med* 335:1363–7, 1996.

Vasopressin

Brownstein MJ, Russell JT, Gainer H: Synthesis, transport, and release of posterior pituitary hormones, *Science* 207:373–8, 1980.

Korbonits M, Kaltsas G, Perry LA, Putignano P, Grossman AB et al: The growth hormone secretagogue hexarelin stimulates the hypothalamo-pituitary-adrenal axis via arginine vasopressin, *J Clin Endocr Metab* 84:2489–95, 1999.

Land H, Schutz G, Schmale H, Richter D: Nucleotide sequence of cloned cDNA encoding bovine arginine vasopressin-neurophysin II precursor, *Nature* 295:299–303, 1982.

Russell TA, Ito M, Ito M, Yu RN, Martinson FA et al: A murine model of autosomal dominant neurohypophyseal diabetes insipidus reveals progressive loss of vasopressin-producing neurons, *J Clin Invest* 112:1697–706, 2003.

Glucagon

Bell GI, Sanchez-Pescador R, Laybourn PJ, Najarian RC: Exon duplication and divergence in the human preproglucagon gene, *Nature* 304:368–71, 1983.

Drucker DJ: Glucagon-like peptide 2, *J Clin Endocr Metab* 86:1759–64, 2001.

Hirasawa A, Tsumaya K, Awaji T, Katsuma S, Adachi T et al: Free fatty acids regulate gut incretin glucagon-like peptide-1 secretion through GPR120, *Nature Med* 11:90–4, 2005.

Holz GG IV, Kuhtreiber WM, Habener JF: Pancreatic beta-cells are rendered glucose-competent by the insulinotropic hormone glucagon-like peptide-1 (7-37), *Nature* 361:362–5, 1993.

Suzuki A, Nakauchi H, Taniguchi H: Glucagon-like peptide 1 (1-37) converts intestinal epithelial cells into insulin-producing cells, *Proc Nat Acad Sci* 100:5034–9, 2003.

White JW, Saunders GF: Structure of the human glucagon gene, *Nucleic Acids Res* 14:4719–30, 1986.

Human protein reference data base, Johns Hopkins University and the Institute of Bioinformatics, at http://www.hprd.org/protein.

17.7. Suppression of Inflammatory Reactions

Moore KW, Vieira P, Fiorentino DF, Trounstine ML, Khan TA et al: Homology of cytokine synthesis inhibitory factor (IL-10) to the Epstein-Barr virus gene BCRF1, *Science* 248:1230–4, 1990.

Thompson K, Maltby J, Fallowfield J, McAulay M, Millwarde-Sadler H et al: Interleukin-10 expression and function in experimental murine liver inflammation and fibrosis, *Hepatology* 28:1597–606, 1998.

Louis H, Van Laethem JL, Wu W, Quertinmont E, Degraef C et al: Interleukin-10 controls neutrophilic infiltration, hepatocyte proliferation, and liver fibrosis induced by carbon tetrachloride in mice, *Hepatology* 28:1607–15, 1998.

Wang SC, Ohata M, Schrum L, Rippe RA, Tsukamoto H: Expression of interleukin-10 by in vitro and in vivo activated hepatic stellate cells, *J Biol Chem* 273:302–8, 1998.

Thompson KC, Trowern A, Fowell A, Marathe M, Haycock C et al: Primary rat and mouse hepatic stellate cells express the macrophage inhibitor cytokine interleukin-10 during the course of activation in vitro, *Hepatology* 28:1518–24, 1998.

Rai RM, Loffreda S, Karp CL, Yang SQ, Lin HZ et al: Kupffer cell depletion abolishes induction of interleukin-10 and permits sustained overexpression of tumor necrosis factor alpha messenger RNA in the regenerating rat liver, *Hepatology* 25:889–95, 1997.

Le Moine O, Marchant A, De Groote D, Azar C, Goldman M et al: Role of defective monocyte interleukin-10 release in tumor necrosis factor-alpha overproduction in alcoholic cirrhosis, *Hepatology* 22:1436–9, 1995.

Reitamo S, Remitz A, Tamai K, Uitto J: Interleukin-10 modulates type I collagen and matrix metalloprotease gene expression in cultured human skin fibroblasts, *J Clin Invest*, 94:2489–92, 1994.

Lacraz S, Nicod LP, Chicheportiche R, Welgus HG, Dayer JM: IL-10 inhibits metalloproteinase and stimulates TIMP-1 production in human mononuclear phagocytes, *J Clin Invest* 96:2304–10, 1995.

Van Vlasselaer P, Borremans B, Van Gorp U, Dasch JR, Malefyt R et al: Interleukin 10 inhibits transforming growth factor-B (TGF-B) synthesis required for osteogenic commitment of mouse bone marrow cells, *J Cell Biol* 124:569–77, 1994.

Ertel W, Keel M, Steckholzer U, Ungethom U, Trentz O: Interleukin-10 attenuates the release of proinflammatory cytokines but depresses splenocyte functions in murine endotoxemia, *Arch Surg* 131:51–6, 1996.

Louis H, Le Moine O, Peny MO, Quertinmont E, Fokan D et al: Production and role of interleukin-10 in concanavalin A-induced hepatitis in mice, *Hepatology* 25:1382–9, 1997.

Louis H, Le Moine A, Quertinmont E, Peny MO, Geerts A et al: Repeated concanavalin A challenge in mice induces an interleukin 10 producing phenotype and liver fibrosis, *Hepatology* 31:381–90, 2000.

Tsukamoto H: Is interleukin-10 antifibrogenic in chronic liver injury, *Hepatology* 28:1707–9, 1998.

Nelson DR, Lauwers GY, Lau JYN, Davis GL: Interleukin 10 treatment reduces fibrosis in patients with chronic hepatitis C: A pilot trial of interferon nonresponders, *Gastroenterology* 118:655–60, 2000.

17.8. Inhibition of Fibrosis

Wrana JL: Transforming growth factor signaling and cirrhosis, *Hepatology* 29:1909–10, 1999.

Armendariza-Borunda J, Katayama K, Seyer JM: Transcriptional mechanisms of type collagen gene expression are differentially regulated by interleukin-1, tumor necrosis factor alpha, and transforming growth factor beta in Ito cells, *J Biol Chem* 267:14316–21, 1992.

Casini A, Pinzani M, Milani S, Grappone C, Galli G et al: Regulation of extracellular matrix synthesis by transforming growth factor Ba-1 in human fat-storing cells, *Gastroenterology* 105:245–53, 1993.

Ramadori G, Knittel T, Odenthal M, Schwigler S, Neubauer K et al: Synthesis of cellular fibronectin by rat liver fat-storing (Ito) cells: Regulation by cytokines, *Gastroenterology* 103:1313–21, 1992.

Qi Z, Atsuchi N, Ooshima A, Takeshita A, Ueno H: Blockade of type beta transforming growth factor signaling prevents liver fibrosis and dysfunction in the rat, *Proc Natl Acad Sci USA* 96:2345–9, 1999.

17.9. Enhancing the Activity of Telomerase

Counter CM, Gupta J, Harley CB, Leber B, Bacchetti S: Telomerase activity in normal leukocytes and in hematologic malignancies, *Blood* 85:2315–20, 1995.

Komine F, Shimojima M, Moriyama M, Amaki S, Uchida T et al: Telomerase activity of needlea-biopsied liver samples: Its usefulness for diagnosis and judgment of efficacy of treatment of small hepatocellular carcinoma, *J Hepatol* 32:235–41, 2000.

Ogami M, Ikura Y, Nishiguchi S, Kuroki T, Ueda M et al: Quantitative analysis and in situ localization of human telomerase RNA in chronic liver disease and hepatocellular carcinoma, *Lab Invest* 79:15–26, 1999.

Kojima H, Yokosuka O, Kato N, Shiina S, Imazeki F et al: Quantitative evaluation of telomerase activity in small liver tumors: Analysis of ultrasonographya-guided liver biopsy specimens *J Hepatol* 31:514–20, 1999.

Hytiroglou P, Kotoula V, Thung SN, Tsokos M, Fiel MI et al: Telomerase activity in precancerous hepatic nodules, *Cancer*, 82:1831–8, 1998.

Bodnar AG, Ouellette M, Frolkis M, Holt SE, Chiu CP et al: Extension of life-span by introduction of telomerase into normal human cells, *Science* 279:349–52, 1998.

Kiyono T, Foster SA, Koop JI, McDougall JK, Galloway DA et al: Both Rb/p16aa?INK4aaa inactivation and telomerase activity are required to immortalize human epithelial cells, *Nature* 396:84–8, 1998.

Lee HW, Blasco MA, Gottlieb GJ, Horner JW, Greider CW et al: Essential role of mouse telomerase in highly proliferative organs, *Nature* 392:569–74, 1998.

Blasco MA, Lee HW, Hande MD, Samper E, Lansdorp PM et al: Telomere shortening and tumor formation by mouse cells lacking telomerase RNA, *Cell* 91:25–34, 1997.

Rudolph KL, Chang S, Millard M, Schreibera-Agus N, DePinho RA: Inhibition of experimental liver cirrhosis in mice by telomerase gene delivery, *Science* 287:1253–8, 2000.

Hagmann M: New genetic tricks to rejuvenate ailing livers, *Science* 287:1185–7, 2000.

17.10. Cell and Tissue Regenerative Engineering

Tzanakakis ES, Hess DJ, Sielaff TD, Wei-Shou Hu WS: Extracorporeal tissue engineered liver-assist devices, *Ann Rev Biomed Eng* 2:607–32, 2000.

Allen JW, Bhatia SN: Engineering liver therapies for the future, *Tissue Eng* 8(5):725–37, 2002.

Davis MW, Vacanti JP: Toward development of an implantable tissue engineered liver, *Biomaterials* 17(3):365–72, 1996.

17.11. Selection, Culture, and Manipulation of Liver Cells

Sielaff TD, Hu MY, Rao S, Groehler K, Olson D et al: A technique for porcine hepatocyte harvest and description of differentiated metabolic functions in static culture, *Transplantation* 59:145–63, 1995.

Dixit V, Gitnick G: Artificial liver support: State of the art, *Scand J Gastroenterol Suppl* 220:101–14, 1996.

Jauregui HO, Mullon CJP, Solomon BA: Extracorporeal artificial liver support, in *Principles of Tissue Engineering*, ed. R Lanza R, Langer W, Landes, Austin, TX, 1997, pp 463–79.

Sielaff TD, Chari RS, Cattral MC: Extracorporeal support of the failing liver, *Curr Opin Crit Care* 4:125–30, 1998.

Roberts EA, Letarte M, Squire J, Yang SY: Characterization of human hepatocyte lines derived from normal liver tissue, *Hepatology* 19:1390–99, 1994.

Schumacher IK, Okamoto T, Kim BH, Chowdhury NR, Chowdhury JR et al: Transplantation of conditionally immortalized hepatocytes to treat hepatic encephalopathy, *Hepatology* 24:337–43, 1996.

Pfeifer AMA, Cole KE, Smoot DT, Weston A, Groopman JD et al: Simian virus 40 large tumor antigen-immortalized normal human liver epithelial cells express hepatocyte characteristics and metabolize chemical carcinogens, *Proc Natl Acad Sci USA* 90:5123–7, 1993.

Woodworth C, Secott T, Isom H: Transformation of rat hepatocytes by transfection with simian virus 40 DNA to yield proliferating differentiated cells, *Cancer Res* 46:4018–26, 1986.

Wilson JM, Jefferson DM, Chowdhury R, Movikoff PM, Johnston DE et al: Retrovirus-mediated transduction of adult hepatocytes, *Proc Natl Acad Sci USA* 85:3014–18, 1988.

Nyberg SL, Remmel RP, Mann HJ, Peshwa MV, Hu WS et al: Primary hepatocytes outperform Hep G2 cells as the source of biotransformation functions in a bioartificial liver, *Ann Surg* 220:59–67, 1994.

Sussman NL, Chong MG, Koussayer T, He DE, Shang TA et al: Reversal of fulminant hepatic failure using an extracorporeal liver assist device, *Hepatology* 16:60–5, 1992.

Sussman NL, Gislason GT, Conlin CA, Kelly JH: The hepatic extracorporeal liver assist device: Initial clinical experience, *Artif Organs* 18:390–6, 1994.

Sussman NL, Kelly JH: Improved liver function following treatment with an extracorporeal liver assist device, *Artif Organs* 17:27–30, 1993.

Wang X, Montini E, Al-Dhalimy M, Lagasse E, Finegold M et al: Kinetics of liver repopulation after bone marrow transplantation, *Am J Pathol* 161(2):565–74, 2002.

Alvarez-Dolada M, Pardal R, Garcia-Verdugo JM, Kike JR, Lee HO et al: Fusion of bone marrow-derived cells with Purkinje neurons, cardiomyocytes and hepatocytes, *Nature* 425:968–73, 2003.

Vassilopoulos G, Russell DW: Cell fusion: An alternative to stem cell plasticity and its therapeutic implications, *Curr Opin Genet Dev* 13:480–5, 2003.

Vassilopoulos G, Wang PR, Russell DW: Transplanted bone marrow regenerates liver by cell fusion, *Nature* 422:901–4, 2003.

Petersen BE, Bowen WC, Patrene KD, Mars WM, Sullivan AK et al: Bone marrow as a potential source of hepatic oval cells, *Science* 284:1168–70, 1999.

Lagasse E, Connors H, Al-Dhalimy M, Reitsma M, Dohse M et al: Purified hematopoietic stem cells can differentiate into hepatocytes in vivo, *Nat Med* 6:1229–34, 2000.

17.12. Fabrication of Liver Scaffolds and Maintenance of Cell Viability and Function

Sauer IM, Obermeyer N, Kardassis D et al: Development of a hybrid liver support system, *Ann NY Acad Sci* 944:308–19, 2001.

Bhandari RN, Riccalton LA, Lewis AL et al: Liver tissue engineering: a role for co-culture systems in modifying hepatocyte function and viability, *Tissue Eng* 7:345–57, 2001.

Miyoshi H, Ehashi T, Ema H et al: Long-term culture of fetal liver cells using a three-dimensional porous polymer substrate, *ASAIO J* 46:397–402, 2000.

Rhim JA, Sandgren EP, Degen JL et al: Replacement of diseased mouse liver by hepatic cell transplantation, *Science* 263:1149–52, 1994.

Sussman NL, Gislason GT, Conlin CA, Kelly JH: The hepatic extracorporeal liver assist device: Initial clinical experience, *Artif Organs* 18:390–6, 1994.

Demetriou AA, Rozga J, Podesta L, Lepage E, Morsiani E et al: Early clinical experience with a hybrid bioartificial liver, *Scand J Gastroenterol* 108:111–7, 1995.

Rozga J, Williams F, Ro MS, Neuzil DF, Giorgio TD et al: Development of a bioartificial liver: Properties and function of a hollow-fiber module inoculated with liver cells, *Hepatology* 17:258–65, 1993.

Rozga J, Holzman MD, Ro MS, Griffin DW, Neuzil DF et al: Development of a hybrid bioartificial liver, *Ann Surg* 217:502–11, 1993.

Wells GD, Fisher MM, Sefton MV: Microencapsulation of viable hepatocytes in HEMA-MMA microcapsules: A preliminary study, *Biomaterials* 14:615–20, 1993.

Eisa T, Sefton MV: Towards the preparation of a MMA-PEO block copolymer for the microencapsulation of mammalian cells, *Biomaterials* 14:755–61, 1993.

Koebe HG, Pahernik SA, Thasler WE, Schildberg FW: Porcine hepatocytes for biohybrid artificial liver devices: A comparison of hypothermic storage techniques, *Artif Organs* 20:1181–90, 1996.

Dixit V, Darvasi R, Arthur M, Lewin K, Gitnick G: Cryopreserved microencapsulated hepatocytes: Transplantation studies in Gunn rats, *Transplantation* 55:616–22, 1993.

Landry J, Bernier D, Ouellet C, Goyette R, Marceau N: Spheroidal aggregate culture of rat liver cells: histotypic reorganization, biomatrix deposition, and maintenance of functional activities, *J Cell Biol* 101:914–23, 1985.

Peshwa MV, Wu FJ, Sharp HL, Cerra FB, Hu WS: Mechanistics of formation and ultrastructural evaluation of hepatocyte spheroids, *In Vitro Cell Dev Biol* 32:197–203, 1996.

Wu FJ, Peshwa MV, Cerra FB, Hu WS: Entrapment of hepatocyte spheroids in a hollow fiber bioreactor as a potential bioartificial liver, *Tissue Eng* 1:29–40, 1995.

Wu FJ, Friend JR, Mann HJ, Remmel RP, Cerra FB et al: Hollow fiber bioartificial liver utilizing collagen entrapped porcine hepatocyte spheroids, *Biotechnol Bioeng* 52:34–44, 1996.

Catapano G, Di Lorenzo MC, Della Volpe C, De Bartolo L, Migliaresi C: Polymeric membranes for hybrid liver support devices: the effect of membrane surface wettability on hepatocyte viability and functions, *J Biomater Sci Polym Ed* 7:1017–27, 1996.

Lin P, Chan WC, Badylak SF, Bhatia SN: Assessing porcine liver-derived biomatrix for hepatic tissue engineering, *Tissue Eng* 10(7–8):1046–53, 2004.

17.13. Implantation of Liver Constructs and Test of Liver Functions

Rahaman MN, Mao JJ: Stem cell-based composite tissue constructs for regenerative medicine, *Biotechnol Bioeng* 91(3):261–84, 2005.

Kulig KM, Vacanti JP: Hepatic tissue engineering, *Transpl Immunol* 12(3–4):303–10, 2004.

Selden C, Hodgson H: Cellular therapies for liver replacement, *Transpl Immunol* 12(3–4):273–88, 2004.

Allen JW, Bhatia SN: Engineering liver therapies for the future, *Tissue Eng* 8(5):725–37, 2002.

Balis UJ, Yarmush ML, Toner M: Bioartificial liver process monitoring and control systems with integrated systems capability, *Tissue Eng* 8(3):483–98, 2002.

Yarmush ML, Toner M, Dunn JC, Rotem A, Hubel A et al: Hepatic tissue engineering. Development of critical technologies, *Ann NY Acad Sci* 13(665):238–52, 1992.

Bhatia SN, Yarmush ML, Toner M: Controlling cell interactions by micropatterning in co-cultures: Hepatocytes and 3T3 fibroblasts, *J Biomed Mater Res* 34(2):189–99, 1997.

18

GASTROINTESTINAL REGENERATIVE ENGINEERING

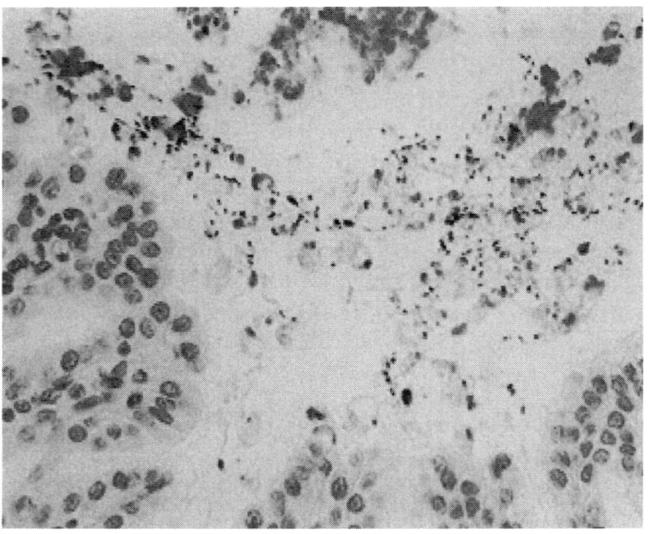

Histological micrograph showing the distribution of *Helicobacter pylori* (*H. pylori*), a type of bacteria that cause peptic ulcers in mammalian animals and humans, in the stomach of Mongolian gerbil. The *Helicobacter pylori* bacteria were detected by immunohistochemistry and appears black in color. (Reprinted with permission from Ikeno T et al: *Am J Pathol* 154:951–60, copyright 1999.) See color insert.

Bioregenerative Engineering: Principles and Applications, by Shu Q. Liu

ANATOMY AND PHYSIOLOGY OF THE GASTROINTESTINAL SYSTEM

Structure [18.1]

The gastrointestinal system includes the esophagus, stomach, small intestine, and large intestine. The primary functions of the gastrointestinal system are food ingestion, digestion, and absorption, as well as waste product elimination. The *esophagus* is a tubular organ that is extended from the pharynx to the stomach. The function of the esophagus is food transport from the mouth to the stomach. The esophagus is composed of four distinct tissue layers: mucosa, submucosa, muscular layer, and adventitia. The *mucosa* consists of stratified epithelial cells and a subendothelial connective tissue layer. The *submucosa* is a connective tissue layer, composed of fibroblasts, extracellular matrix (collagen fibers, elastic fibers, and proteoglycans), blood vessels, nerves, and small glands. The gland cells can secret mucus, which serves as a lubricant for food ingestion. The *muscular layer* is composed of striated muscular cells in the upper esophagus and smooth muscle cells near the stomach. There are two layers of muscular cells: the inner layer with circumferentially aligned muscular cells and the outer layer with longitudinally aligned muscular cells. In addition, the muscular layer contains extracellular matrix and nerves. The nerves control the movement of the esophagus. The *adventitia* is a thin layer composed of fibroblasts, extracellular matrix, and a squamous epithelium on the outer surface. The epithelium is also known as visceral peritoneum in gastrointestinal organs.

The stomach is located in the upper left abdominal cavity and is responsible for the primary digestion of ingested foods. As other digestive tracts, the stomach is composed of four layers: mucosa, submucosa, muscular layer, and adventitia. The structure of these layers is similar to that of the esophagus with several exceptions. First, the inner surface of the stomach is lined with a monolayer of columnar epithelial cells, instead of stratified cells. The epithelial cells of the stomach are specially differentiated cells. Some of these cells line the stomach surface and produce mucus. Others form gastric glands, which produce hydrochloric acid, pepsinogen, and mucus. The acidic environment (pH 1–3) helps to digest foods. Pepsinogen is the precursor of the enzyme pepsin, which cleaves proteins into peptides for further digestion. Pepsinogen can be converted into pepsin when it is released into the stomach under the action of hydrochloride acid and pepsin. Second, the muscular layer contains three sublayers of smooth muscle cells: oblique, circumferential, and longitudinal muscular sublayers. The reinforced muscular structure enhances gastric movement and sufficient food digestion. The stomach undergoes constant movements, which help to mix and digest foods.

The small intestine is a tubular organ that is located in the abdominal cavity and is extended from the stomach to the large intestine. The small intestine is composed of three segments: the duodenum, jejunum, and ileum. At the tissue level, the small intestine consists of the four standard layers: the mucosa, submucosa, muscular layer, and adventitia. The mucosa and partial submucosa form numerous folds about 1 mm in length, also known as villi, on the surface of the small intestine. The villi are covered with columnar epithelial cells. These cells form cell membrane projections called *microvilli*. The formation of villi and microvilli increases the total surface area of the small intestine, which facilitates nutrient absorption. There are epithelial goblet cells and submucosa gland cells. These cells produce mucus. The muscular layer of the small intestine contains circumferential and longitudinal smooth muscle cells. These cells conduct constant movements to move and mix ingested foods. The structure is similar among the duodenum, jejunum, and ileum. The functions of the small intestine are food digestion and absorption. A number of digestive enzymes, including those for digesting proteins, polysaccharides, and lipids, are secreted from the pancreas into the small intestine. Enzymes for protein digestion include trypsin, chymotrypsin, and carboxypeptidase. Polysaccharides are digested by amylases, and lipids are digested by lipases. In addition, the pancreas secrets deoxyribonucleases and ribonucleases, which break down DNA and RNA, respectively.

The large intestine is a digestive tract that is extended from the ileum to the anus. The large intestine is usually divided into several segments, including the colon, rectum, and anal canal. The general structure of the large intestine is similar to that of the other digestive tracts. The large intestine possesses a large number of goblet cells which secret mucus. The primary function of the large intestine is to remove food wastes. Water is absorbed in the large intestine. However, little nutrient absorption takes place in the large intestine. The large intestine undergoes constant movements, which help to move the food remains.

Nutrient Digestion and Absorption [18.1]

The ingested foods contain three major types of nutrient: protein, carbohydrate, and lipid. All these nutrients are digested in the stomach and small intestine, and absorbed in the small intestine. Proteins from ingested meats and plants are first digested in the stomach by pepsin, which cleaves proteins into polypeptides. Remaining proteins and digested polypeptides are moved into the small intestine, where they are further digested into tripeptides, dipeptides, and amino acids by peptidases. Amino acids and short peptides are absorbed into the epithelial cells by an energy-consuming process involving sodium cotransport. The absorbed short peptides are digested into amino acids in the epithelial cells. The amino acids are transported into the blood of the portal vein and used for protein synthesis in different types of cell.

Ingested foods contains polysaccharides (starches and glycogen), disaccharides (sucrose and lactose), and monosaccharides (glucose). Polysaccharides are digested by salivary and pancreatic amylases into disaccharides and monosaccharides in the stomach and small intestine. Disaccharides are further digested into monosaccharides at the epithelial surface. Monosaccharides are absorbed into the epithelial cells of the small intestine by an active transport mechanism involving the co-transport of sodium and released into the intestinal capillaries.

Lipids include cholesterol, triglycerides, phospholipids, and steroids. Ingested lipids are emulsified in the small intestine by bile compounds produced by the liver and are digested by pancreatic lipases. Digested dietary lipids form small particles known as *micelles*. The lipid molecules diffuse into the epithelial cells according to the gradient across the cell membrane. Within the epithelial cell, lipids form chylomicrons (80–500 nm in diameter) by conjugating with apoproteins. The chylomicrons are transported into the blood and are further digested by bloodborne lipases, releasing chylomicron remnants and free fatty acids (triglycerides). The chylomicron remnants are taken up by hepatocytes in the liver through receptor-mediated endocytosis, digested in the lysosomes of the hepatocytes. Free cholesterol molecules are released from the chylomicron remnants and stored in the hepatocytes as cholesteryl esters. The cholesterol molecules are either excreted into the bile or used to form very-low-density-lipoproteins (VLDL, diameter 30–80 nm) in the hepatocytes. The VLDL molecules are released into the blood and digested by lipoprotein lipases, resulting in the formation of intermediate-density-lipoproteins (IDL) after releasing a fraction of fatty acids. The IDL is digested by lipoprotein lipases into low-density-lipoproteins (LDL), which usually circulate in the blood for 1–2 days and constitute the major reserve of plasma cholesterol (60–70% of the total cholesterol pool). The cholesterol molecules of LDL can be taken up by cells and used for the construction of cell membranes. The promotion of cell intake of cholesterol reduces the plasma LDL and cholesterol level, a beneficial process for reducing the risk of atherogenesis. With the release of the majority of cholesterol molecules, LDL is converted to high-density-lipoproteins (HDL), which consist of apoproteins and residual cholesterol molecules and can form complexes with cholesterols and fatty acids, contributing to the clearance of plasma lipids.

GASTROINTESTINAL DISORDERS

Peptic Ulcer

Pathogenesis, Pathology, and Clinical Features [18.2]. *Peptic ulcer* is a disorder that occurs primarily in the duodenum and stomach and is characterized by the presence of round or oval ulcerative lesions with various sizes. The lesions include injury and detachment of epithelial cells, necrosis and fibrosis of mucosa and submucosa, and infiltration of inflammatory cells. These pathological changes are possibly induced by the corrosive effect of acid and pepsin. *Duodenal ulcer* is often found in the proximal segment of duodenum near the stomach. About 10% of the human population is affected by duodenal ulcers some times during the lifespan. The incidence of duodenal ulcer is higher than that of gastric ulcer. Duodenal ulcer is a chronic disorder with a high rate of recurrence and is often found in patients about 50 years old. A large fraction of patients experience reoccurring duodenal ulcers within a period of 2–3 years. *Gastric ulcer* occurs in the stomach and exhibit similar pathological changes as found in the duodenal ulcer. This type of ulcer is often found in patients about 60 years old.

The pathogenesis of duodenal and gastric ulcers is related to several factors:

1. A type of bacteria known as *Helicobacter pylori* (*H. pylori*) can cause inflammatory reactions and peptic ulcers in the stomach of mammalian animals and humans (Chapter 18 opening figure).
2. Acid and pepsin secreted by the stomach epithelial cells may exert a corrosive effect on the duodenal and gastric epithelial cells. Some ulcer patients are associated with increased secretion of acid and pepsin.
3. A reduction in the mucosal resistance to the corrosion of acid and pepsin contributes to the development of duodenal and gastric ulcers.
4. An increase in the release of gastrin, a molecule that stimulates the secretion of hydrochloric acid in the stomach, is another potential factor that contributes to the development of duodenal and gastric ulcers. Even though the level of gastrin may be normal, the gastric epithelial cells in ulcer patients often exhibit increased responsiveness to gastrin stimulation, enhancing the secretion of hydrochloric acid. Fourth, hereditary factors may also play a role in the development of duodenal and gastric ulcers. Ulcer patients often have a familial history of ulcer. In addition, cigarette smoking is associated with increased incidence of duodenal and gastric ulcer, although smoking does not alter acid secretion. The mechanisms of smoking-related ulcers remain poorly understood.

Patients with duodenal and gastric ulcers often experience epigastric burning pain. When a blood vessel is corroded and damaged, hemorrhage may occur. The severity of hemorrhage is dependent on the size of the damaged blood vessel. In general, ulcers do not significantly influence food digestion and absorption, since ulcers are usually small.

Experimental Models of Gastrointestinal Ulcers [18.3]. Gastric and duodenal ulcers can be induced by application of acid and pepsin to the stomach or duodenum of animal models. Here, the duodenum is used to demonstrate the experimental procedures. An animal can be anesthetized by peritoneal injection of sodium pentobarbital at a dose of 50 mg/kg body weight. The upper abdominal skin can be sterilized with 75% alcohol, Betadine (povidone-iodine), and 75% alcohol again. The abdominal cavity can be opened at a location in the upper middle area and the duodenum is identified. A segment of the duodenum can be isolated by applying a pair of intestinal clamps. A mixture of hydrochloric acid and pepsin can be injected into and kept in the duodenum for a desired period. The concentration of acid and pepsin and the treatment time should be determined by conducting a series of experiments with graded concentrations and different treatment times. The clamps can be released and the abdominal wounds closed after the treatment. At scheduled times following the surgery, specimens can be collected from the duodenum and used for pathological examinations.

Conventional Treatment [18.2]. There are several general approaches for the treatment of gastrointestinal ulcers (see Table 18.1 for a list of therapeatic proteins used to treat peptic ulcer). These include the administration of acid-reducing agents, protective coating agents, and diet control. Commonly used acid-reducing agents include antacids, anticholinergic agents, antagonists to histamine H2 receptor, and prostaglandins. The administration of antacid agents, such as sodium bicarbonate and calcium carbonate, results in direct neutralization of the acidic content in the stomach and duodenum, suppressing ulcer formation and progression. Acetylcholine is a neurotransmitter that promotes the production and secretion of gastric acid. A treatment with anticholinergic agents, such as atropine and atropine derivatives, can reduce the secretion of gastric acid. The H2 receptor of histamine mediates histamine-induced secretion of gastric acid. Antagonists of the histamine H2 receptor, such as cimetidine, suppress the activity of the receptor and reduce the secretion of gastric acid. Prostaglandins exert an inhibitory effect on gastric acid secretion in the stomach and enhance the capability of gastric mucosal resistance to acid corrosion. Stomach mucosal coating agents, such as sucralfate (polyaluminum hydroxide salt of sucrose sulfate), can adhere to the mucosal surface and protect the mucosa from acid and pepsin corrosion. In addition, the control of acidic diet intake reduces the progression of gastrointestinal ulcers. In the case of severe hemorrhage, it is necessary to remove the ulcerative tissue via surgery.

Molecular Regenerative Engineering. The principle of molecular engineering or therapy for peptic ulcer is to promote tissue recovery from ulcerative injury. Although the reduction of acid secretion in the stomach is the most important approach for treating peptic ulcer, there are no proteins or genes available for such a purpose. Several genes, including the serum response factor, platelet-derived growth factor, vascular endothelial growth factor, and angiopoietin-1 genes, have been identified and used in animal models for enhancing recovery from peptic ulcer-induced injury. These genes encode mitogenic factors that stimulate cell proliferation and migration, thus enhancing the recovery from

TABLE 18.1. Characteristics of Selected Therapeutic Proteins for Peptic Ulcer*

Proteins	Alternative Names	Amino Acids	Molecular Weight (kDa)	Expression	Functions
Serum response factor	SRF, c-fos serum response-element-binding transcription factor	508	52	Ubiquitous	A downstream target of the mitogen-activated protein kinase pathway, binding directly to the serum response element (SRE) in promoter region of immediate–early genes (e.g., c-fos), and stimulating cell proliferation and differentiation
Angiopoietin-1	ANG1, ANGPT1	498	58	Blood vessels (endothelial cells), heart, lung, placenta, bone	Regulating the early development of the heart, and promoting vascular development and angiogenesis

*Based on bibliography 18.4 and 18.7.

ulcerative injury. A selected gene can be transferred into the stomach and small intestine by catheter-mediated luminal instillation.

Serum Response Factor (SRF) [18.4]. Serum response factor is a transcriptional factor that mediates the transcription of immediate early genes (see Table 18.1 for the characteristics of SRF). Proteins encoded by the immediate early genes participate in the regulation of mitogenic responses, including cell proliferation and migration. When the SRF gene is transferred into the stomach of animal models with gastric ulcer, the expression level of SRF increases, which is associated with enhanced proliferation and migration of epithelial and smooth muscle cells in the ulcerative tissue. These activities enhance the recovery from gastric ulcer compared to a control ulcer model without SRF gene transfer. These observations suggest that the SRF gene may serve as a potential candidate gene for the molecular therapy of human peptic ulcer.

Platelet-Derived Growth Factor (PDGF) [18.5]. Platelet-derived growth factor is a protein that stimulates mitogenic activities, including cell proliferation and migration (see page 600 for the characteristics of PDGF). In particular, this growth factor enhances the survival and proliferation of smooth muscle cells. Experimental investigations have demonstrated that the overexpression of the PDGF gene by gene transfer in rat models of duodenal ulcer enhances cell proliferation in ulcerative tissue and facilitates the healing process of ulcer (Fig. 18.1).

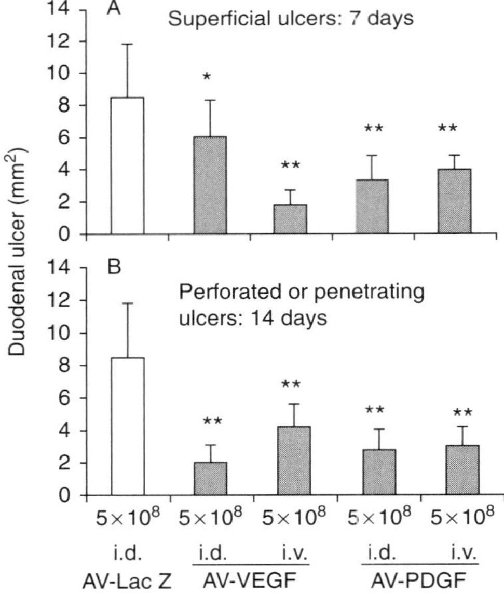

Figure 18.1. The size of duodenal ulcers at 7 and 14 days after transfection with adenoviral vector (AV) containing the vascular endothelial growth factor (VEGF) gene or the platelet-derived growth factor (PDGF) gene compared with a control vector containing the Lac Z gene (β-galactosidase gene). *, $P < 0.05$; **, $P < 0.01$. $n = 6$–12. (Reprinted with permission from Deng X et al: Gene therapy with adenoviral plasmids or naked DNA of vascular endothelial growth factor and platelet-derived growth factor accelerates healing of duodenal ulcer in rats, *J Pharmacol Exp Ther* 311:982–8, copyright 2004.)

Vascular Endothelial Growth Factor (VEGF) [18.6]. As platelet-derived growth factor, VEGF stimulates cell proliferation and migration (see page 600 for the characteristics of VEGF). This growth factor has been shown to particularly regulate the development, survival, and proliferation of the vascular endothelial cells. In experimental models of peptic ulcer, this growth factor enhances angiogenesis as well as the proliferation and migration gastrointestinal epithelial cells. Thus, the overexpression of the VEGF gene may potentially improve the healing of peptic ulcer (Fig. 18.1).

Angiopoietin-1 [18.7]. Angiopoietin-1 is a protein that regulates angiogenesis (see Table 18.1 for the characteristics of angiopoietin-1). Local delivery of angiopoietin-1 expression vectors results in up-regulation of the angiopoietin-1 gene and an increase in angiogenic activities. Since angiogenesis is required for the regeneration of ulcerative tissue, the overexpression of the angiopoietin-1 gene can enhance ulcer healing.

Gastrointestinal Cancers

Pathogenesis, Pathology, and Clinical Features [18.8]. *Gastrointestinal cancers* are malignant tumors found in the mucosal, submucosal, and muscular layers of the gastrointestinal tracts. Gastrointestinal cancers are divided into several types, including carcinoma, leiomyosarcoma, and lymphoma based on the cellular origin of the tumor. As cancers in other organs, these cancers are highly metastatic.

Gastrointestinal carcinoma is a form of cancer originated from the mucosal epithelial cells. This form of cancer is more common in men than in women. It is often found in people about 60 years old or older. The etiology of gastrointestinal carcinoma remains poorly understood. Hereditary and dietary factors may play a role in the development of the disease. In addition, the presence of several disorders, such as atrophic gastritis and intestinal metaplasia, may enhance the development of gastrointestinal carcinoma. Pathological examinations usually reveal several forms of carcinoma, including ulcerative, superficial, and infiltrative carcinomas. Ulcerative carcinomas are cancers that appear similar to ulcers at the surface. Superficial carcinomas are those found within the epithelial layer. Infiltrative carcinomas are those that invade the deep layers. At the cellular level, carcinoma cells are characterized by the enlargement of cell nuclei and an increase in cell density. During the end-stage, carcinoma cell metastasis or infiltration into deeper layers can be found. A cell proliferation assay often demonstrates an increase in the rate of cell proliferation.

Gastrointestinal leiomyosarcoma is a form of cancer originated from the submucosal smooth muscle cells. This type of cancer is not as common as gastrointestinal carcinoma and accounts for about 3% of total gastrointestinal cancers. The tumor is usually spherical in shape with necrosis at the center. *Gastrointestinal lymphoma* is a type of cancer originated from the lymphoid tissue. This type of cancer accounts for about 5% of gastrointestinal cancers. The etiological and pathological features of gastrointestinal leiomyosarcoma and lymphoma are similar to those of gastrointestinal carcinomas.

Conventional Treatment. Several conventional approaches, including surgical removal, chemotherapy, and radiotherapy, have been developed and used for the treatment of cancer. These approaches will be discussed in Chapter 25 in detail.

Molecular Therapy [18.9]. A number of molecular strategies have been developed and used for the treatment of cancers. These include the up-regulation of tumor suppressor genes, correction of mutant tumor suppressor genes, enhancement of anti-cancer immune responses, activation of tumor suppressor drugs, introduction of oncolytic viruses, and application of antisense oligodeoxynucleotides. Since different types of cancers exhibit common pathogenic mechanisms and features, these therapeutic approaches can be applied to all types of cancers. These approaches will be discussed in detail in Chapter 25.

Tissue Regenerative Engineering. *Gastrointestinal tissue engineering* is to repair or reconstruct malfunctioned esophagus, stomach, or intestines with cell-containing tissue constructs. Cancer is a major disorder that requires tissue and organ repair or reconstruction. Several other disorders, including inflammatory bowel disease, intestinal infarction, and short bowel syndrome, may also require intestinal repair or reconstruction. Since the esophagus, stomach, and intestines possess distinct functions, disorders in these organs should be treated with different approaches. The function of the esophagus is to conduct foods from the mouth to the stomach. A simple structural reconstruction may be sufficient to restore the function of a disordered esophagus. Other gastrointestinal functions such as nutrient absorption and transport may not be a critical issue for esophagus tissue engineering. Since the length of the esophagus is limited (no excessive esophagus available), it is necessary to replace a malfunctioned esophagus with a substitute (see following sections for details).

Compared to the esophagus, the stomach and intestines are responsible not only for food conduction, but also for food digestion and absorption (absorption occurring primarily in the small intestine). Thus in stomach and intestinal tissue engineering, these functional aspects should be taken into account. Another difference from esophagus engineering is that it is not necessary to replace the injured regions of the stomach and intestines when the lesions are small and do not involve the entire organ. The reason is that the stomach and intestines possess a large capacity of reserve that is not used under physiological conditions. A malfunctioned region of the stomach or intestines can be simply removed with the remaining organ reanastomosed. In general, the removal of 50% of the stomach or the intestine may not significantly influence food digestion and absorption. When the lesion involves a large area or the entire organ, which impairs the digestion and absorption of nutrients, fluids, and electrolytes, engineering replacement of the stomach or intestine is necessary. However, it remains difficult to construct a gastrointestinal substitute with the natural functions of molecular absorption and transport. The construction of such a substitute is a critical issue in gastrointestinal tissue regenerative engineering.

Several approaches have been established and used for the construction of gastrointestinal substitutes. These include organ transplantation, substitution with autogenous pedicle with blood supply, expansion of existing intestinal tracts, intestinal regeneration with the peritoneal membrane, substitution based on biodegradable and nonbiodegradable polymeric materials, substitution based on allogenic intestinal submucosa, and substitution based on extracellular matrix components.

Gastrointestinal Transplantation [18.10]. Organ transplantation is an effective approach for the replacement of malfunctioned esophagus, stomach, and intestines. A fresh, viable, allogenic organ can be harvested and used to substitute for a host equivalent. The transplant can usually maintain the functions such as food digestion and nutrient absorption. However, allogenic cells induce immune reactions, resulting in acute transplant rejection. A common approach to reduce immune reactions is to administrate immunosuppressing agents. Patients with organ transplantation usually need to take immune suppressing agents for the entire lifespan. While these agents inhibit immune rejection responses, they also suppress normal immune functions, resulting in increased susceptibility to infectious diseases.

Gastrointestinal Reconstruction Based on Autogenous Pedicles [18.11]. Given the difficulties in the maintenance of allogenic gastrointestinal transplants, scientists have established approaches for gastrointestinal regeneration based on autogenous connective and muscular tissue pedicles with blood supply. Abdominal wall pedicles with functional blood vessels can be prepared and used for such a purpose. A layer of the abdominal wall can be isolated at one end. The other end remains connected to the abdominal wall, ensuring sufficient blood supply. The free end of the pedicle can be tailored into a tubular structure and anastomosed to a malfunctioned gastrointestinal organ. In experimental models of intestinal reconstruction, the host intestinal mucosa can extend to the graft surface, forming a mucosa-like structure, known as neomucosa, within 1–2 months following the reconstruction surgery. Examinations by optical and electron microscopy have demonstrated that the neomucosa is similar in structure to the natural mucosa. Furthermore, the neomucosa exhibits certain intestinal functions such as absorption of glucose and electrolytes. These observations suggest that connective and muscular tissues can be transformed into intestinal tissue in an appropriate intestinal environment. Autogenous pedicle-based gastrointestinal reconstruction represents one of the most effective approaches for intestinal regeneration.

Expansion of Intestines [18.12]. Gastrointestinal cells, especially the mucosal epithelial cells, undergo constant proliferation, a mechanism for the replacement of apoptotic cells and for tissue expansion. Intestinal tissue expansion occurs in response to an increased demand for nutrient absorption when the absorptive surface is reduced due to various disorders, such as cancers and inflammatory bowel disease. Based on such a feature, scientists have developed an approach for intestinal self-regeneration or expansion. A segment of intact intestine can be removed and split longitudinally into two equal parts. Each part can be constructed into a tubular structure. Both tubes can be anastomosed to the intestine in a series so that the length can be doubled. The narrowed lumen can be expanded in the radial direction through adaptation to luminal inflation and enhanced wall tension. It is the nature of biological systems that tissues and organs can grow in response to mechanical stretch. This adaptive mechanism has also been found in the cardiovascular and pulmonary systems. Intestinal expansion is an effective approach for the regeneration of intestines. The regenerated intestine can be used for the repair or replacement of a malfunctioned intestine.

Regeneration with Peritoneum [18.13]. Peritoneum is an epithelium-covered connective tissue membrane found on the surface of the abdominal organs. A loose peritoneal membrane can be harvested and used to construct tubular structures. When these tubular structures are anastomosed into the intestine, host mucosal cells can migrate to the graft, forming neomucosa. The neomucosa is similar in structure to the host mucosa and exhibits intestinal functions, such as absorption of fluids and electrolytes. However, the source of the peritoneum is limited. It is difficult to collect sufficient peritoneal membrane for the replacement of a large area of malfunctioned intestine.

Gastrointestinal Reconstruction Based on Polymeric Materials [18.14]. While autogenous tissues are ideal materials for the regeneration of malfunctioned gastrointestinal tracts, it is often difficult to collect sufficient amount of tissue for large lesions. To resolve such a problem, scientists have been searching for synthetic materials that can be used for gastrointestinal reconstruction. In early studies, nonbiodegradable polymers, such as Dacron and polytetrafluoroethylene (PTFE), have been used to repair malfunctioned gastrointestinal tracts. When grafted into the host esophagus or intestine, the polymeric materials can serve as a bridge for the generation of neomucosa. These investigations suggest that polymeric materials can be used as gastrointestinal conduits. However, polymeric materials cannot be integrated into the host systems and the regenerated tissue does not assemble the native gastrointestinal system.

Biodegradable polymeric materials have also been used for gastrointestinal reconstruction. Several types of such materials, including polyglycolic acid and polylactic acid, have been used in animal models of esophagus, stomach, and intestinal reconstruction. These polymers can be absorbed gradually in the host system. The rate of degradation can be controlled by altering the composition of the materials. Biodegradable polymers are often used to construct scaffolds for cell seeding and growth. It has been demonstrated that, with controlled degradation of the polymeric materials, the gastrointestinal cells are able to regenerate a gastrointestinal substitute with the native structure of the system. To enhance the regenerative process, disintegrated intestinal tissues can be seeded in the polymeric scaffold. Since intestinal tissues contain epithelial progenitor cells, the seeding of intestinal tissues facilitate intestinal regeneration. Preliminary investigations based on animal models have demonstrated that regenerated gastrointestinal tissues in biodegradable polymer scaffolds are similar in structure to the native gastrointestinal tissues and exhibit physiological functions such as absorption of fluids and electrolytes.

Extracellular Matrix-Based Gastrointestinal Reconstruction [18.15]. Extracellular matrix is a structure constituted with collagen fibers, elastic fibers, and proteoglycans and serves as a frame for the attachment, support, organization, and communication of cells and for the formation and assembly of tissues. The submucosa of the gastrointestinal tracts is an extracellular matrix-rich structure. Thus, extracellular matrix is an ideal structure for constructing gastrointestinal scaffolds, which can be used for tissue regeneration. Since collagen fibers provide structural support and mechanical strength to tissues and organs, collagen-rich matrix is often used for tissue reconstruction. Collagen-rich matrix materials can be collected from soft connective tissues, such as the dermis and the submucosa of intestines. The collected matrix specimens can be used to reconstruct the gastrointestinal tracts.

Allogenic decellularized extracellular matrix components, such as collagen, have long been used for constructing scaffolds for tissue regeneration. It is usually necessary to remove cellular components, because these components cause host immune reactions, resulting in acute transplant rejection. Unlike cellular components, allogenic matrix components exhibit low immunogenicity and do not cause acute immune rejection responses. Several methods can be used for extracting collagen-matrix, including treatment of connective tissues with alkalines, acids, and detergents, which lyse and remove cells, leaving a collagen-rich matrix. Such a matrix can be used to construct a scaffold with desired shape and dimensions and to seed gastrointestinal cells or stem cells for tissue regeneration. Collagen-based scaffolds usually stimulate cell proliferation and migration, enhancing tissue regeneration.

Experimental investigations based on animal models have demonstrated that collagen-based matrix scaffolds can be used for the regeneration of esophagus and small intestines.

When grafted into a host esophagus or intestine, host cells are able to migrate into the matrix scaffold, forming an esophagus- or intestine-like structure, respectively. The matrix scaffold can be gradually integrated into the host system. These preliminary investigations have demonstrated that extracellular matrix is a suitable material for the regeneration of gastrointestinal tracts.

Experimental Models of Gastrointestinal Reconstruction [18.16]. Malfunctioned esophagus, stomach, and intestines can be reconstructed in animal models by using allogenic tissue equivalents, extracellular matrix specimens, autogenous connective or muscular tissue pedicles. An animal can be anesthetized by peritoneal injection of sodium pentobarbital at a dose of 50 mg/kg body weight. To reconstruct a segment of the small intestine, the abdominal skin is sterilized, the abdominal cavity is opened, and the small intestine is identified. A segment of the small intestine can be isolated by applying a pair of intestinal clamps and removed between the clamps. An intestinal substitute with diameter similar to that of the host intestine can be anastomosed to the host intestine at the two ends by using continuous stitches. The two clamps are then released and the anastomotic sites are inspected for leakage. Leaking sites, if any, can be sealed by additional suture stitches. The abdominal wound can be closed by using continuous suture stitches for the muscular layer and disrupted suture stitches or surgical staples for the skin.

Inflammatory Bowel Disease

Pathogenesis, Pathology, and Clinical Features [18.17]. *Inflammatory bowel disease* is a gastrointestinal disorder characterized by the presence of continuous and uniform inflammatory reactions, ulcerative lesions, and hemorrhage in the mucosa of the esophagus, stomach, small intestine, and/or large intestine. The etiology of this disorder remains poorly understood. The incidence of the disease is 2–6 per 100,000 in the United States. The disease is often found in the population at the age of 15–35.

The pathogenesis of inflammatory bowel disease is related to several factors, including genetics, infection, and immune responses. Patients with inflammatory bowel disease usually show a familial history of the disease, suggesting a hereditary predisposition to the disease. Bacterial and viral infection induces acute inflammatory reactions in the gastrointestinal tracts. These inflammatory reactions may contribute to the pathogenesis of inflammatory bowel disease. However, bacteria or viruses specific to inflammatory bowel disease remain unidentified. Inflammatory bowel disease is sometime associated with autoimmune diseases, such as arthritis and pericholangitis. Antiinflammation agents, such as corticosteroids, often reduce the symptoms of the disease. Patients with inflammatory bowel disease may generate antibodies against bacterial antigens, which are similar to certain cellular components of the hosts. These observations suggest an autoimmune mechanism for the development of inflammatory bowel disease.

Inflammatory bowel disease is associated a number of pathological changes. These include the loss of epithelial cells, ulcerative lesions, leukocyte infiltration into the mucosa and submucosa, submucosal edema, hemorrhage, the presence of red blood cells in the submucosa, submucosal fibrosis, and thickening of the intestinal wall. Submucosal fibrosis can induce regional distortion of the intestine. In severe cases, inflammatory and fibrous changes can result in intestinal obstruction. These changes may significantly influence nutrient absorption in the small intestine, resulting in nutritional deficiency. Intestinal perforation may occur as a result of altered structure and mechanical stiffness of the intestinal wall. Severe hemorrhage may occur occasionally.

Conventional Treatment [18.17]. There are two primary conventional treatments for inflammatory bowel disease: control of inflammation and surgical removal of malfunctioned intestine. Inflammatory reactions can be controlled by using hormones, including glucocorticoids and adrenocorticotropic hormone (ACTH). Glucocorticoids are produced in the cortex of the adrenal gland. ACTH is produced in the anterior pituitary and stimulates adrenal cortex to produce glucocorticoids. These hormones can be used to effectively suppress inflammatory reactions in inflammatory bowel disease. Nutritional treatment is necessary if there is a nutrient loss due to impaired absorption in the small intestine. Immunosuppressive agents, such as azathioprine may be used to suppress autoimmune reactions, which potentially contribute to the development of inflammatory bowel disease. In severe cases, such as those with potential perforation of intestinal wall and severe hemorrhage, it is necessary to remove the malfunctioned intestine by surgery.

Molecular Therapy. Inflammatory bowel disease is a disorder with chronic inflammation and possible autoimmune reactions. Molecular therapeutic approaches have been developed to suppress these activities. Two genes have been targeted for the potential treatment of inflammatory bowel disease: interleukin (IL)10 and IL18.

Interleukin-10 [18.18]. Interleukin (IL)10 is a cytokine that inhibits the release of proinflammatory cytokines and suppresses inflammatory reactions. Genetically induced deficiency of interleukin-10 is associated with accelerated inflammatory activities, which simulate inflammatory reactions found in inflammatory bowel disease. A treatment with the interleukin-10 protein reduces the degree of intestinal inflammation (Fig. 18.2). The delivery of the interleukin-10 gene into animal models of inflammatory bowel disease results in local upregulation of interleukin-10 expression, leading to a reduction in inflammatory reactions, such as leukocyte infiltration, edema, and fibrosis. These observations suggest that the interleukin-10 gene is a potential candidate gene for the molecular therapy of inflammatory bowel disease.

Interleukin-18 [18.19]. Interleukin (IL)18, also known as *interferon gamma inducing factor* (IGIF) and interleukin-1γ (IL1γ), is a cytokine of 193 amino acid residues and about 22 kDa in molecular weight. This cytokine is expressed in the gastrointestinal tract, lung, liver, kidney, and skeletal muscle. Interleukin-18 can induce interferon-γ production in T cells and accelerate inflammatory responses. The expression of interleukin-18 is upregulated in inflammatory bowel disease, contributing to the pathogenesis of the disease. A strategy for the treatment of inflammatory bowel disease is to suppress the expression of the IL18 gene. Local delivery of antisense oligodeoxynucleotides specific to the interleukin-18 mRNA is a potential method for such a purpose. Another approach is to transfer gene vectors that encode interleukin-18 antisense oligodeoxynucleotides. This approach has been used to suppress the translation of interleukin-18 in intestinal cells and reduce inflammatory reactions in experimental models. Alternatively, small interfering RNA (siRNA) specific to interleukin-18 can be delivered to target cells to degrade Interleukin-18 mRNA and thus reduce the translation of interleukin-18.

Tissue Regenerative Engineering. When inflammatory bowel disease involves a large fraction of the intestine, it is necessary to remove the malfunctioned intestinal segment

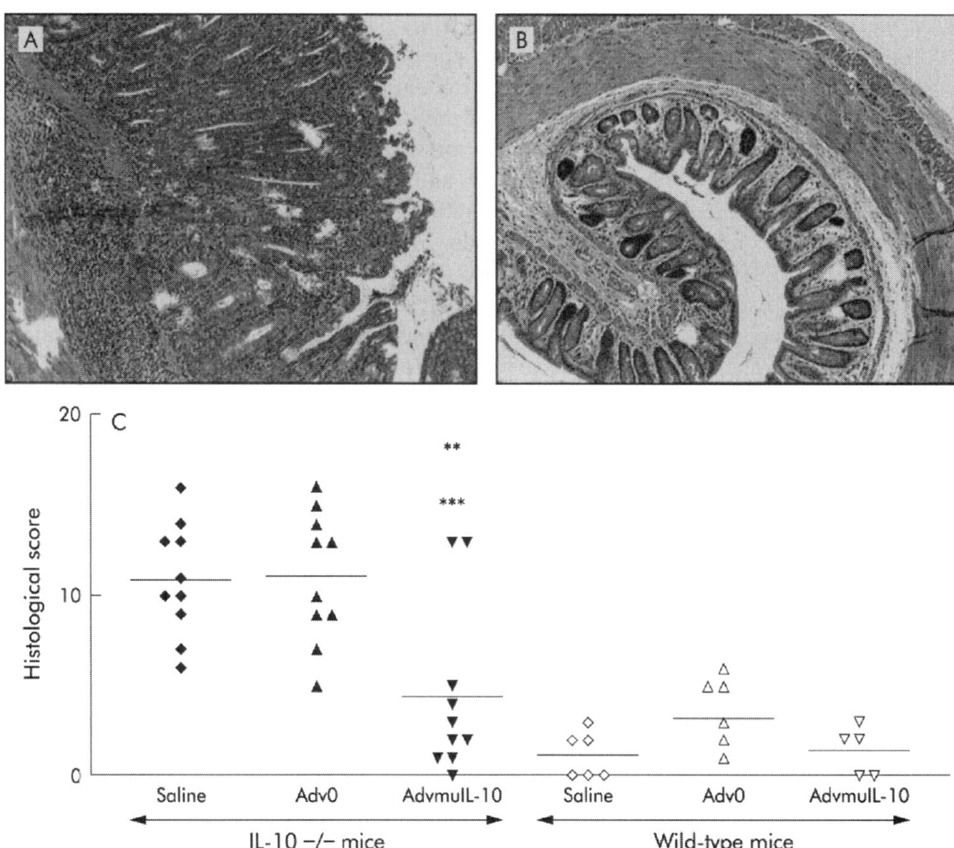

Figure 18.2. Local adenoviral vector encoding murine interleukin-10 (AdvmuIL10) therapy reduced histological colitis scores in IL10−/− mice with established disease. Four weeks after therapy, intestinal specimens were harvested for observation. Representative samples from an IL10−/− mouse treated with empty cassette adenoviral vector (Adv0) and AdvmuIL10 are shown in panels A and B, respectively. Panel C shows the histological colitis scores for different groups. (Reprinted with permission from Lindsay JO et al: *Gut* 52:363–9, copyright 2003.)

and reconstruct the intestine. There are two approaches for intestinal reconstruction: intestinal transplantation and substitution with engineered tissue substitutes. These approaches have been discussed on page 806 of this chapter.

Intestinal Ischemia and Infarction [18.20]

Pathogenesis, Pathology, and Clinical Features. *Intestinal ischemia and infarction* are disorders due to the reduction and obstruction of blood supply to the intestine, respectively. Common causes are arterial thrombosis and embolism. Arterial thrombosis occurs in response to arterial injury and bloodflow reduction due to low cardiac output. Arterial embolism is often found in patients with atrial fibrillation, artificial cardiac valves, and valvular heart diseases. Acute intestinal ischemia or infarction is associated with severe abdominal pain, vomiting, anorexia, and diarrhea. Pathological examinations often exhibit necrotic changes, including edema, cell death, hemorrhage, and tissue degradation.

Treatment of Intestinal Ischemia and Infarction. The treatment of intestinal ischemia and infarction is dependent on the nature and stage of the disorder as well as the condition of the involved intestine. In embolus-induced ischemia, embolectomy (surgical removal of emboli) is sufficient to restore the function of the intestine when the involved intestine is still viable. When infarction occurs in association with intestinal necrosis, intestinal resection is usually required. For intestinal ischemia and infarction due to severe arterial thrombosis, it is usually difficult to remove the thrombi. Instead, arterial bypass surgery may be carried out to reconstruct the occluded artery if the involved intestine is viable. For infarcted intestines, it is necessary to remove the involved intestinal segment. Intestinal reconstruction is usually necessary when a large fraction of intestine (>50%) is removed. Intestinal transplantation and substitution with engineered tissue equivalents are options for the treatment. These approaches have been discussed on page 806 of this chapter.

Molecular therapy can be applied to ischemic or partially occluded intestinal arteries. Since these vascular disorders are primarily induced by thrombosis and atherosclerosis, molecular therapies established for treating atherosclerosis can be used for treating intestinal ischemic disorders (see Chapter 15).

Short Bowel Syndrome [18.21]

Pathogenesis, Pathology, and Clinical Features. *Short bowel syndrome* is an intestinal disorder characterized by inadequate absorption of nutrients, fluids, and electrolytes resulting from massive surgical resection of the small intestine. Intestinal resection is often necessary for the following diseases: metastatic intestinal cancers, large area of inflammatory bowel disease, and large areas of intestinal infarction. These disorders result in an inadequate absorptive surface area. When more than half of the small intestine is removed, insufficient absorption of nutrients, fluids, and electrolytes often occurs.

Treatment of Short Bowel Syndrome. Several conventional approaches have been established and used for treating short bowel syndrome. These include: (1) supply of low fat diets but with high carbohydrates and proteins; (2) administration of vitamins and mineral supplements; (3) control of the motility of the intestine by using smooth muscle relaxants, allowing a sufficient amount of time for the absorption of nutrients, fluids, and electrolytes; and (4) parenteral hyperalimentation via the jugular vein and superior vena cava in severe cases. While these treatments are effective for a short-term relief of the symptoms, nutritional disorders often occur after long-term treatments because of insufficient nutrient absorption. An alternative approach for the treatment of short bowel syndrome is to reconstruct the disordered intestine with functional intestinal substitutes. This aspect has been discussed on page 806 of this chapter.

BIBLIOGRAPHY

18.1. Anatomy and Physiology of the Gastrointestinal System

Guyton AC, Hall JE: *Textbook of Medical Physiology*, 11th ed, Saunders, Philadelphia, 2006.

McArdle WD, Katch FI, Katch VL. *Essentials of Exercise Physiology*, 3rd ed, Lippincott Williams & Wilkins, Baltimore, 2006.

Germann WJ, Stanfield CL: (with contributors Niles MJ, Cannon JG), *Principles of Human Physiology*, 2nd ed, Pearson Benjamin Cummings, San Francisco, 2005.

Thibodeau GA, Patton KT: *Anatomy & Physiology*, 5th ed, Mosby, St Louis, 2003.

Boron WF, Boulpaep EL: *Medical Physiology: A Cellular and Molecular Approach*, Saunders, Philadelphia, 2003.

Ganong WF: *Review of Medical Physiology*, 21st ed, McGraw-Hill, New York, 2003.

18.2. Pathogenesis, Pathology, and Clinical Features of Gastrointestinal Ulcer

Schneider AS, Szanto PA: *Pathology*, 3rd ed, Lippincott Williams & Wilkins, Philadelphia, 2006.

McCance KL, Huether SE: *Pathophysiology: The Biologic Basis for Disease in Adults & Children*, 5th ed, Elsevier Mosby, St Louis, 2006.

Porth CM: *Pathophysiology: Concepts of Altered Health States*, 7th ed, Lippincott Williams & Wilkins, Philadelphia, 2005.

Frazier MS, Drzymkowski JW: *Essentials of Human Diseases and Conditions*, 3rd ed, Elsevier Saunders, St Louis, 2004.

Ford AC, Delaney BC, Forman D, Moayyedi P: Eradication therapy in Helicobacter pylori positive peptic ulcer disease: Systematic review and economic analysis, *Am J Gastroenterol* 99(9):1833–55, 2004.

Dore MP, Graham DY: Ulcers and gastritis, *Endoscopy* 36(1):42–7, 2004.

Kelley JR, Duggan JM: Gastric cancer epidemiology and risk factors, *J Clin Epidemiol* 56(1):1–9, 2003.

Berstad AE, Berstad K, Berstad A: PH-activated phospholipase A2: An important mucosal barrier breaker in peptic ulcer disease, *Scand J Gastroenterol* 37(6):738–42, 2002.

18.3. Experimental Model of Gastrointestinal Ulcer

Okabe S, Amagase K: An overview of acetic acid ulcer models—the history and state of the art of peptic ulcer research, *Biol Pharm Bull* 28(8):1321–41, 2005.

18.4. Serum Response Factor (SRF)

Chang J, Wei L, Otani T, Youker KA, Entman ML et al: Inhibitory cardiac transcription factor, SRF-N, is generated by caspase 3 cleavage in human heart failure and attenuated by ventricular unloading, *Circulation* 108:407–13, 2003.

Norman C, Runswick M, Pollock R, Treisman R: Isolation and properties of cDNA clones encoding SRF, a transcription factor that binds to the c-fos serum response element, *Cell* 55:989–1003, 1988.

Wang Z, Wang DZ, Hockemeyer D, McAnally J, Nordheim A et al: Myocardin and ternary complex factors compete for SRF to control smooth muscle gene expression, *Nature* 428:185–9, 2004.

Chai J, Baatar D, Tarnawski A: Serum response factor promotes re-epithelialization and muscular structure restoration during gastric ulcer healing, *Gastroenterology* 126(7):1809–18, June 2004.

Human protein reference data base, Johns Hopkins University and the Institute of Bioinformatics, at http://www.hprd.org/protein.

18.5. Platelet-Derived Growth Factor (PDGF)

Deng X, Szabo S, Khomenko T, Jadus MR, Yoshida M: Gene therapy with adenoviral plasmids or naked DNA of VEGF and PDGF accelerates healing of duodenal ulcer in rats, *J Pharmacol Exp Ther* 311:982–988, 2004.

18.6. Vascular Endothelial Growth Factor (VEGF)

Deng X, Szabo S, Khomenko T, Jadus MR, Yoshida M: Gene therapy with adenoviral plasmids or naked DNA of VEGF and PDGF accelerates healing of duodenal ulcer in rats, *J Pharmacol Exp Ther* 311:982–988, 2004.

18.7. Angiopoietin-1

Baatar D, Jones MK, Tsugawa K, Pai R, Moon WS et al: Esophageal ulceration triggers expression of hypoxia-inducible factor-1 alpha and activates vascular endothelial growth factor gene: Implications for angiogenesis and ulcer healing, *Am J Pathol* 161(4):1449–57, Oct 2002.

Jones MK, Kawanaka H, Baatar D, Szabo IL, Tsugawa K et al: Gene therapy for gastric ulcers with single local injection of naked DNA encoding VEGF and angiopoietin-1, *Gastroenterology* 121(5):1040–7, Nov 2001.

Arai F, Hirao A, Ohmura M, Sato H, Matsuoka S et al: Tie2/Angiopoietin-1 signaling regulates hematopoietic stem cell quiescence in the bone marrow niche, *Cell* 118:149–61, 2004.

Davis S, Aldrich TH, Jones PF, Acheson A, Compton DL et al: Isolation of angiopoietin-1, a ligand for the TIE2 receptor, by secretion-trap expression cloning, *Cell* 87:1161–9, 1996.

Folkman J, D'Amore PA: Blood vessel formation: What is its molecular basis? *Cell* 87:1153–5, 1996.

Hanahan D: Signaling vascular morphogenesis and maintenance, *Science* 277:48–50, 1997.

Holash J, Maisonpierre PC, Compton D, Boland P, Alexander CR et al: Vessel cooption, regression, and growth in tumors mediated by angiopoietins and VEGF, *Science* 284:1994–8, 1999.

Loughna S, Sato TN: A combinatorial role of angiopoietin-1 and orphan receptor TIE1 pathways in establishing vascular polarity during angiogenesis, *Mol Cell* 7:233–9, 2001.

Sato TN, Tozawa Y, Deutsch U, Wolburg-Buchholz K, Fujiwara Y et al: Distinct roles of the receptor tyrosine kinases Tie-1 and Tie-2 in blood vessel formation, *Nature* 376:70–3, 1995.

Suri C, Jones PF, Patan S, Bartunkova S, Maisonpierre PC et al: Requisite role of angiopoietin-1, a ligand for the TIE2 receptor, during embryonic angiogenesis, *Cell* 87:1171–80, 1996.

Suri C, McClain J, Thurston G, McDonald DM, Zhou H et al: Increased vascularization in mice overexpressing angiopoietin-1, *Science* 282:468–71, 1998.

Thurston G, Suri C, Smith K, McClain J, Sato TN et al: Leakage-resistant blood vessels in mice transgenically overexpressing angiopoietin-1, *Science* 286:2511–14, 1999.

Vikkula M, Boon LM, Carraway KL III, Calvert JT, Diamonti AJ et al: Vascular dysmorphogenesis caused by an activating mutation in the receptor tyrosine kinase TIE2, *Cell* 87:1181–90, 1996.

18.8. Pathogenesis, Pathology, and Clinical Features of Gastrointestinal Cancers

Jankowski JA, Wright NA, Meltzer SJ, Triadafilopoulos G, Geboes K et al: Molecular evolution of the metaplasia-dysplasia-adenocarcinoma sequence in the esophagus, *Am J Pathol* 154(4):965–73, 1999.

Oliveira C, Seruca R, Seixas M, Sobrinho-Simoes M: The clinicopathological features of gastric carcinomas with microsatellite instability may be mediated by mutations of different "target genes." A study of the TGFbeta RII, IGFII R, and BAX genes, *Am J Pathol* 153(4):1211–19, 1998.

Smyrk TC: Colon cancer connections. Cancer syndrome meets molecular biology meets histopathology, *Am J Pathol* 145(1):1–6, 1994.

18.9. Molecular Engineering

Shimada H, Shimizu T, Ochiai T, Liu TL, Sashiyama H et al: Preclinical study of adenoviral p53 gene therapy for esophageal cancer, *Surg Today* 31(7):597–604, 2001.

Prieto J, Herraiz M, Sangro B, Qian C, Mazzolini G et al: The promise of gene therapy in gastrointestinal and liver diseases, *Gut* 252(Suppl 2):ii49–54, 2003.

Schmid RM, Weidenbach H, Draenert GF, Liptay S, Luhrs H et al: Liposome mediated gene transfer into the rat oesophagus, *Gut* 41(4):549–56, 1997.

18.10. Gastrointestinal Transplantation

Chen MK, Beierle EA: Animal models for intestinal tissue engineering, *Biomaterials* 25(9):1675–81, 2004.

Abu-Elmagd K, Reyes J, Bond G, Mazariegos G, Wu T et al: Clinical intestinal transplantation: A decade of experience at a single center, *Ann Surg* 234:404–17, 2001.

Reyes J, Mazariegos GV, Bond GM, Green M, Dvorchik I et al: Pediatric intestinal transplantation: Historical notes, principles and controversies, *Pediatr Transplant* 6:193–207, 2002.

Kato T, Nishida S, Mittal N, Levi D, Nery J et al: Intestinal transplantation at the University of Miami, *Transplant Proc* 34:868, 2002.

18.11. Gastrointestinal Reconstruction Based on Autogenous Pedicles

Lillemoe KD, Berry WR, Harmon JW, Tai YH, Weichbrod RH et al: Use of vascularized abdominal wall pedicle flaps to grow small bowel neomucosa, *Surgery* 91:293–300, 1982.

Thompson JS, Vanderhoof JA, Antonson DL, Newland JR, Hodgson PE: Comparison of techniques for growing small bowel neomucosa, *J Surg Res* 36:401–6, 1984.

18.12. Expansion of Intestines

Saday C, Mir E: A surgical model to increase the intestinal absorptive surface: Intestinal lengthening and growing neomucosa in the same approach, *J Surg Res* 62:184–91, 1996.

Weber TR: Isoperistaltic bowel lengthening for short bowel syndrome in children, *Am J Surg* 178:600–4, 1999.

Bianchi A: Experience with longitudinal intestinal lengthening and tailoring, *Eur J Pediatr Surg* 9:256–9, 1999.

18.13. Regeneration with Peritoneum

Ring-Mrozik E: Experimental studies of the small intestine mucosa, *Z Kinderchir* 44:363–9, 1989.

Erez I, Rode H, Cywes S: Enteroperitoneal anastomosis for short bowel syndrome, *Harefuah* 123:5–8, 1992.

18.14. Gastrointestinal Reconstruction Based on Polymeric Materials

Watson LC, Friedman HI, Griffin DG, Norton LW, Mellick PW: Small bowel neomucosa, *J Surg Res* 28:280–91, 1980.

Kim SS, Kaihara S, Benvenuto MS, Choi RS, Kim BS et al: Effects of anastomosis of tissue engineered neointestine to native small bowel, *J Surg Res* 87:6–13, 1999.

Kim SS, Kaihara S, Benvenuto MS, Choi RS, Kim BS et al: Regenerative signals for intestinal epithelial organoid units transplanted on biodegradable polymer scaffolds for tissue engineering of small intestine, *Transplantation* 67(2):227–33, 1999.

Fukushima M, Kako N, Chiba K, Kawaguchi T, Kimura Y et al: Seven year follow-up study after the replacement of the esophagus with an artificial esophagus in the dog, *Surgery* 93:70–7, 1983.

Thompson JS, Kampfe PW, Newland JR, Vanderhoof JA: Growth of intestinal neomucosa on prosthetic materials, *J Surg Res* 41:484–92, 1986.

Choi RS, Riegler M, Pothoulakis C, Kim BS, Mooney D et al: Studies of brush border enzymes, basement membrane components, and electrophysiology of tissue-engineered neointestine, *J Pediatr Surg* 33:991–7, 1988.

Kim SS, Kaihara S, Benvenuto MS, Choi RS, Kim BS et al: Effects of anastomosis of tissue-engineered neointestine to native small bowel, *J Surg Res* 87:6–13, 1999.

Kaihara S, Kim SS, Kim BS, Mooney D, Tanaka K et al: Long-term follow-up of tissue-engineered intestine after anastomosis to native small bowel, *Transplantation* 69:1927–32, 2000.

Grikscheit TC, Ogilvie JB, Ochoa ER, Alsberg E, Mooney D et al: Tissue-engineered colon exhibits function in vivo, *Surgery* 132:200–4, 2002.

Grikscheit TC, Vacanti JP: Tissue-engineered stomach from autologous and syngeneic tissue, *J Surg Res* 107:277–8, 2002.

Organ GM, Mooney DJ, Hansen LK, Schloo B, Vacanti JP: Transplantation of enterocytes utilizing polymer-cell constructs to produce a neointestine, *Transplant Proc* 24:3009–11, 1992.

Duxbury MS, Grikscheit TC, Gardner-Thorpe J, Rocha FG, Ito H et al: Lymphangiogenesis in tissue-engineered small intestine, *Transplantation* 77(8):1162–6, 2004.

18.15. Extracellular Matrix-Based Gastrointestinal Reconstruction

Badylak S, Meurling S, Chen M, Spievack A, Simmons-Byrd A: Resorbable bioscaffold for esophageal repair in a dog model, *J Pediatr Surg* 35:1097–103, 2000.

Chen MK, Badylak S: Small bowel tissue engineering using small intestinal submucosa as a scaffold, *J Surg Res* 99:352–8, 2001.

Hori Y, Nakamura T, Kimura D, Kaino K, Kurokawa Y et al: Functional analysis of the tissue-engineered stomach wall, *Artif Organs* 26:868–72, 2002.

Hori Y, Nakamura T, Matsumoto K, Kurokawa Y, Satomi S et al: Tissue engineering of the small intestine by acellular collagen sponge scaffold grafting, *Int J Artif Organs* 24:50–4, 2001.

Isch JA, Engum SA, Ruble CA, Davis MM, Grosfeld JL: Patch esophagoplasty using AlloDerm as a tissue scaffold, *J Pediatr Surg* 36266–8, 2001.

18.16. Experimental Model of Gastrointestinal Reconstruction

Chen MK, Beierle EA: Animal models for intestinal tissue engineering, *Biomaterials* 25(9): 1675–81, 2004.

18.17. Pathogenesis, Pathology, and Clinical Features of Inflammatory Bowel Disease

Caprilli R, Gassull MA, Escher JC, Moser G, Munkholm P et al: European Crohn's and Colitis Organisation. European evidence based consensus on the diagnosis and management of Crohn's disease: Special situations, *Gut* 55(Suppl 1):i36–58, 2006.

Harpaz N, Schiano T, Ruf AE, Shukla D, Tao Y et al: Early and frequent histological recurrence of Crohn's disease in small intestinal allografts. Transplantation. 27;80(12):1667–70, 2005.

Blonski W, Lichtenstein GR: Complications of biological therapy for inflammatory bowel diseases, *Curr Opin Gastroenterol* 22(1):30–43, 2006.

Naito Y, Takano H, Yoshikawa T: Oxidative stress-related molecules as a therapeutic target for inflammatory and allergic diseases, *Curr Drug Targets Inflamm Allergy* 4(4):511–15, Aug 2005.

Ullman TA: Colonoscopic surveillance in inflammatory bowel disease, *Curr Opin Gastroenterol* 21(5):585–8, 2005.

18.18. Interleukin-10

Duchmann R, Schmitt E, Knolle P et al: Tolerance towards resident intestinal flora in mice is abrogated in experimental colitis and restored by treatment with interleukin-10 or antibodies to interleukin-12, *Eur J Immunol* 26:934–8, 1996.

Kuhn R, Lohler J, Rennick D et al: Interleukin-10-deficient mice develop chronic enterocolitis, *Cell* 75:263–74, 1993.

Sellon RK, Tonkonogy S, Schultz M et al: Resident enteric bacteria are necessary for development of spontaneous colitis and immune system activation in interleukin-10-deficient mice, *Infect Immun* 66:5224–31, 1998.

Lindsay JO, Ciesielski CJ, Scheinin T et al: The prevention and treatment of murine colitis using gene therapy with adenoviral vectors encoding IL-10, *J Immunol* 166:7625–33, 2001.

Rogy MA, Beinhauer BG, Reinisch W et al: Transfer of interleukin-4 and interleukin-10 in patients with severe inflammatory bowel disease of the rectum, *Hum Gene Ther* 11:1731–41, 2000.

18.19. Interleukin-18

Ushio S, Namba M, Okura T, Hattori K, Nukada Y et al: Cloning of the cDNA for human IFN-gamma-inducing factor, expression in Escherichia coli, and studies on the biologic activities of the protein, *J Immunol* 156(11):4274–9, 1996.

Konishi H, Tsutsui H, Murakami T, Yumikura-Futatsugi S, Yamanaka K et al: IL-18 contributes to the spontaneous development of atopic dermatitis-like inflammatory skin lesion independently of IgE/stat6 under specific pathogen-free conditions, *Proc Nat Acad Sci USA* 99:11340–5, 2002.

Okamura H, Tsutsui H, Komatsu T, Yutsudo M, Hakura A et al: Cloning of a new cytokine that induces IFN-gamma production by T cells, *Nature* 378:88–91, 1995.

Pizarro TT, Michie MH, Bentz M, Woraratanadharm J, Smith MF Jr et al: IL-18, a novel immunoregulatory cytokine, is up-regulated in Crohn's disease: Expression and localization in intestinal mucosal cells, *J Immunol* 162:6829–35, 1999.

Corbaz A, ten Hove T, Herren S, Graber P, Schwartsburd B et al: IL-18-binding protein expression by endothelial cells and macrophages is up-regulated during active Crohn's disease, *J Immunol* 168:3608–16, 2002.

Sugawara S, Uehara A, Nochi T, Yamaguchi T, Ueda H et al: Neutrophil proteinase 3-mediated induction of bioactive IL-18 secretion by human oral epithelial cells, *J Immunol* 167:6568–75, 2001.

ten Hove T, Corbaz A, Amitai H, Aloni S, Belzer I et al: Blockade of endogenous IL-18 ameliorates TNBS-induced colitis by decreasing local TNF-alpha production in mice, *Gastroenterology* 121:1372–9, 2001.

Wirtz S, Becker C, Blumberg R et al: Treatment of T cell-dependent experimental colitis in SCID mice by local administration of an adenovirus expressing IL-18 antisense mRNA, *J Immunol* 168:411–20, 2002.

Prieto J, Herraiz M, Sangro B, Qian C, Mazzolini G et al: The promise of gene therapy in gastrointestinal and liver diseases, *Gut* 2(Suppl 2):ii49–54, 2003.

18.20. Intestinal Ischemia and Infarction

Sreenarasimhaiah J: Chronic mesenteric ischemia, *Best Pract Res Clin Gastroenterol* 19(2):283–95, 2005.

Kolkman JJ, Mensink PB, van Petersen AS, Huisman AB, Geelkerken RH: Clinical approach to chronic gastrointestinal ischaemia: from 'intestinal angina' to the spectrum of chronic splanchnic disease, *Scand J Gastroenterol Suppl* 241:9–16, 2004.

18.21. Short Bowel Syndrome

Scolapio JS: Short bowel syndrome: Recent clinical outcomes with growth hormone, *Gastroenterology* 130(2 Suppl 1):S122–6, 2006.

Jackson C, Buchman AL: Advances in the management of short bowel syndrome, *Curr Gastroenterol Rep* 7(5):373–8, 2005.

Scolapio JS: Current update of short-bowel syndrome, *Curr Opin Gastroenterol* 20(2):143–5, 2004.

19

PANCREATIC REGENERATIVE ENGINEERING

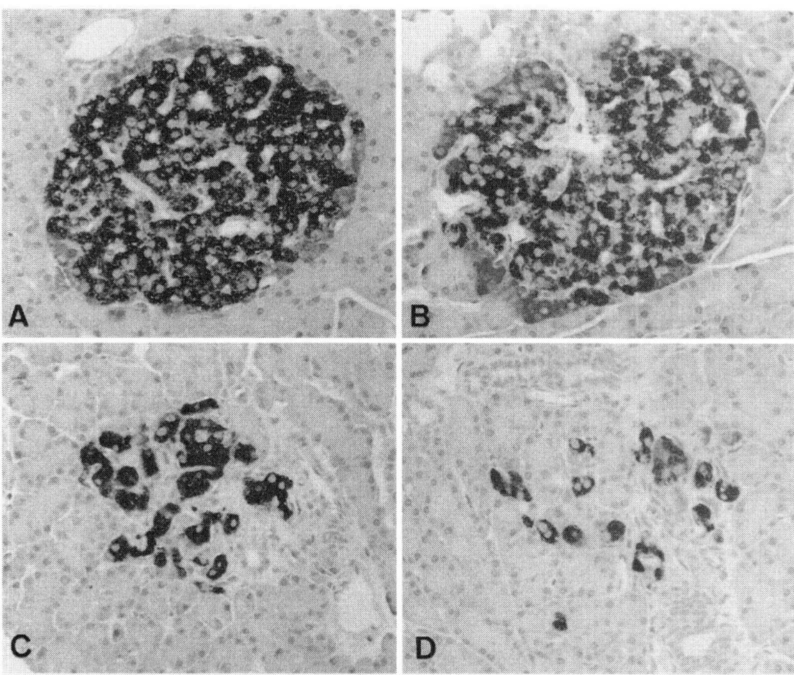

Islet structures and distribution of insulin-positive cells (black) and glucagon-positive cells (red) in Wistar rat (A), sucrose-fed Wistar rat (B), Goto–Kakizaki (GK) rat (C), and sucrose-fed GK rat (D) at 12 weeks of age. Note that the GK rat is a spontaneously diabetic animal model of non-insulin-dependent diabetes mellitus, which is characterized by progressive loss of β cells in the pancreatic islets with fibrosis. There is marked islet fibrosis with β-cell depletion in GK and sucrose-fed GK rats, in which the latter showed more severe changes. Double immunostaining for insulin and glucagon. Magnification ×300 (A–D). (Reprinted with permission from Koyama M et al: Accelerated loss of islet β cells in sucrose-fed Goto-Kakizaki rats, a genetic model of non-insulin-dependent diabetes mellitus, *Am J Pathol* 153:537–45, copyright 1998.) See color insert.

Bioregenerative Engineering: Principles and Applications, by Shu Q. Liu

ANATOMY AND PHYSIOLOGY OF THE PANCREAS

Structure [19.1]

The pancreas is an organ located in the upper abdominal cavity, behind the stomach, and between the spleen and duodenum. The pancreas consists of two functional systems: the endocrine and exocrine systems. The *endocrine system* is composed of the islets of Langerhans, which produce and secret insulin, glucagon, and somatostatin into the blood (Fig. 19.1). In the human pancreas, there are more than 1 million of Langerhans islets. These islets contain several types of cell, including the α, β, and δ cells. The *exocrine system* is composed of secretary units known as *pancreatic acini*, which contain exocrine cells. These cells produce and secrete enzymes for the digestion of proteins, carbohydrates, and fats. The pancreatic acini are connected to small ductules, which converge to larger ducts and eventually to the pancreatic duct. The pancreatic duct system conducts pancreatic juice to the duodenum, where proteins, carbohydrates, and fats are digested.

Functions of the Pancreatic Endocrine System [19.1]

The endocrine α, β, and δ cells of the Langerhans islets produce three hormones, including glucagon, insulin (chapter 19 cover page), and somatostatin, respectively. *Insulin* is a hormone that participates in regulating the metabolism of carbohydrates, fats, and proteins. Insulin is initially produced in the form of preproinsulin, which is cleaved into

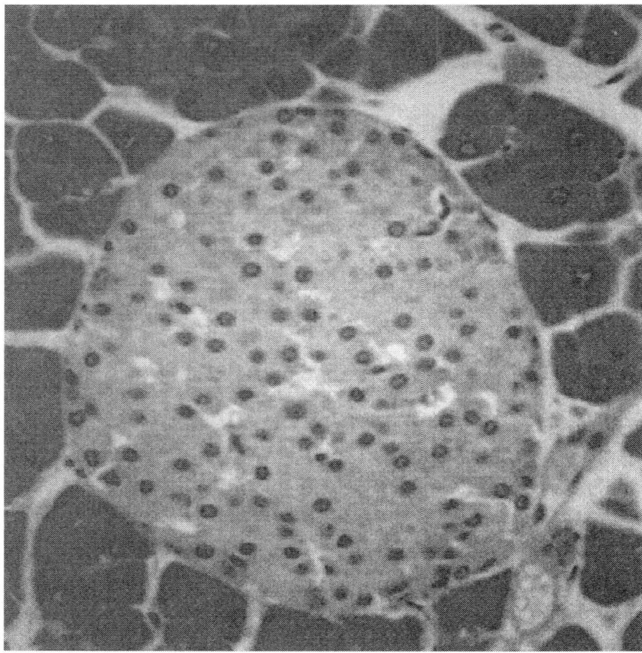

Figure 19.1. Histological micrograph of pancreatic islet from a wildtype mouse (C57BL/6J). Magnification ×200. (Reprinted with permission from Barlow SC et al: *Am J Pathol* 165:1849–52, copyright 2004.)

pro-insulin and then into insulin. Insulin is stored in the secretory granules. The release of insulin is triggered in response to an increase in the level of blood glucose after food ingestion. Once in the blood, insulin stimulates glucose uptake, metabolism, and storage by almost all cell types, thus reducing the level of blood glucose. There are several mechanisms for these processes. In the skeletal muscular system, insulin enhances glucose transport from the blood into the muscular cells. Excessive glucose is converted to glycogen for glucose storage. In addition, insulin stimulates the liver cells to take up glucose, which is converted to glycogen in the liver. This is a rapid process when the level of blood insulin is increased following food ingestion. When insulin is degraded and blood insulin is decreased, the stored glycogen is converted to glucose, an important mechanism for the maintenance of the blood glucose level. The lack of insulin results in a persistent increase in blood glucose, a pathological disorder known as diabetes mellitus (see page 822 of this chapter). After released into the blood, insulin can be degraded within about 15 min, a mechanism by which glucose metabolism can be effectively controlled.

Insulin plays a role in the regulation of fat metabolism. Insulin stimulates the synthesis of fatty acids from glucose. When the glycogen level reaches a critical level, excessive glucose is used to synthesize fatty acids and triglycerides in the liver cells. Insulin mediates the storage of fatty acids and triglycerides in adipocytes. In the absence of insulin, fatty acids are metabolized and used for energy production. Insulin also participates in the regulation of protein metabolism. Insulin stimulates protein synthesis and storage, which occur primarily following food ingestion. There are several mechanisms for these processes. Insulin enhances the transport of amino acids into the cell, stimulates mRNA transcription and protein translation, and inhibits protein degradation. Thus, the lack of insulin not only influences carbohydrate metabolism, but also fat and protein metabolism.

Glucagon is a hormone produced by the α-cells of the Langerhans islets. Its functions are opposite to those of the insulin. Glucagon is capable of degrading glycogen to produce glucose and enhancing gluconeogenesis from amino acids, thus increasing the level of blood glucose. Glucagon is activated when the blood glucose level reduces to a critical level, and suppressed when blood glucose is increased to a critical level. *Somatostatin* (SST or SMST) is a hormone produced by the δ cells of the Langerhans islets. It is first generated as a preproprotein (116 amino acids, about 13 kDa in molecular weight). The preproprotein is cleaved into two active forms of somatostatin: 14- and 28-amino acid peptides. The primary function of somatostatin is to inhibit the secretion of insulin and glucagon via interacting with the G-protein-coupled somatostatin receptors. This hormone is activated by an increase in blood glucose, amino acids, and fatty acids. Somatostatin also interacts with pituitary growth hormone and hormones produced by the gastrointestinal tracts and enhance the function of these hormones. Taken together, insulin, glucagon, and somatostatin coordinately control the level of blood glucose, amino acids, and fatty acids, ensuring appropriate metabolism of these nutrients.

Functions of the Pancreatic Exocrine System [19.1]

The acini of the pancreatic exocrine system produce a number of enzymes, including trypsin, chymotrypsin, polypeptidase, amylase, lipase, cholesterol esterase, phospholipase, and nuclease. Trypsin and chymotrypsin are responsible for the digestion of proteins into peptides. Polypeptidase can digest peptides into amino acids. Pancreatic amylase hydrolyzes glycogens and starches into disaccharides and trisaccharides. Pancreatic lipase can break down fats into fatty acids and glycerides. Cholesterol esterase and phospholipase are responsible for the digestion of cholesterol and phospholipids, respectively. Deoxyribonuclease and ribonuclease can digest DNA and RNA, respectively. It is important to note that the exocrine enzymes are initially produced in inactive forms, which are activated by enzyme cleavage in the small intestine. For example, the inactive forms of trypsin and chymotrypsin are trypsinogen and chymotrypsinogen, respectively. These inactive proenzymes are produced in the pancreas and secreted into the intestine, where the proenzymes are cleaved by trypsin and enterokinase, resulting in the activation of these enzymes.

The epithelial cells of the pancreatic ductules and ducts can produce and secret sodium bicarbonate, which plays a critical role in neutralizing the acidic solution secreted by the stomach. The formation of sodium bicarbonate involves mechanisms of ion transport across the epithelial cells of the pancreatic ductules and ducts. The source of bicarbonate is carbon dioxide, which diffuses from the blood to the epithelial cells. Under the action of an enzyme called *carbonic anhydrase*, carbon dioxide reacts with water to form carbonic acid, which is further dissociated into bicarbonates and hydrogen ions. The bicarbonate ions are transported into the pancreatic ductules and ducts. The hydrogen ions are transported into the blood in exchange with sodium transport in the opposite direction. The sodium ions are secreted into the pancreatic ductules and ducts and react with bicarbonate to form sodium bicarbonate.

The secretion of pancreatic exocrine enzymes and sodium bicarbonate is regulated by several substances, including acetylcholine, cholecystokinin, and secretin. *Acetylcholine* is a neurotransmitter for the parasympathetic nervous system. The ingested foods, when entering the small intestine, can activate the parasympathetic nerves to release acetylcholine, which stimulates the pancreatic acinar cells to secret exocrine enzymes. Similarly, the ingested foods stimulate the release of *cholecystokinin* by the epithelial cells of the duodenum. This substance stimulates the acinar cells of the pancreas to secret exocrine enzymes. *Secretin* is released by the epithelial cells of the duodenum when acidic foods enter the duodenum. This substance stimulates the secretion of sodium bicarbonate by the pancreatic ductule cells.

PANCREATIC DISORDERS

Diabetes Mellitus

Pathogenesis, Pathology, and Clinical Features [19.2]. *Diabetes mellitus* is a metabolic disorder induced by decreased or abolished secretion of insulin from the pancreatic β cells and characterized by an increase in the level of blood glucose. Diabetes is also associated with abnormalities in the metabolism of glucose, fatty acids, and proteins as well as the formation of pathological lesions in blood vessels (arteriosclerosis). During the end stage, acidosis and diabetic coma may occur. Diabetes is a relatively common disorder. The prevalence of the disorder is about 1%. Diabetes is often associated with obesity, hypertension, hyperlipidemia, and atherosclerosis. However, the cause-and-effect relationship between these disorders remains poorly understood.

Diabetes is classified in to two types: primary and secondary diabetes. *Primary diabetes* is defined as diabetes that is not induced by other diseases and is further divided into two subtypes: insulin-dependent and non-insulin-dependent diabetes. Primary diabetes is also defined on the basis of pathogenic mechanisms related to the involvement of immune reactions. Immune reaction-mediated diabetes is defined as type I diabetes, whereas non-immune-reaction-mediated diabetes is defined as type II diabetes. Type I diabetes may include insulin- and non-insulin-dependent diabetes. Type II diabetes is usually non-insulin-dependent. *Secondary diabetes* is defined as diabetes induced by other diseases and is further divided into several subtypes: diabetes due to pancreatic diseases, such as pancreatitis and cancers, chemical toxicity, hormonal abnormalities, and genetic mutation. Chronic pancreatitis is often associated with diabetes due to the involvement of the β cells. Certain drugs and chemicals may impair the function of the β cells and reduce the production and release of insulin. Hormonal abnormalities such as Cushing's syndrome, pheochromocytoma, and administration of steroid hormones, may cause malfunction of the β cells, leading to a reduction in insulin release. Several genetic disorders, including myotonic dystrophy, lipodystrophy, and ataxia–telangiectasia, are often associated with the impairment of the β cells, reducing the secretion of insulin.

The pathogenic mechanisms vary among different types of diabetes. Primary type I insulin-dependent diabetes is a disorder potentially induced by autoimmune reactions. In patients genetically susceptible to diabetes, viral infection activates the host immune system. Certain viruses may contain antigens that are similar in structure to the membrane components of the pancreatic β cells. Virus-activated T cells may infiltrate into the islets of Langerhans and attack the β cells that contain proteins similar to the viral antigens, resulting in β-cell destruction and insulin deficiency. The pathogenic mechanisms of non-insulin-dependent primary diabetes remains poorly understood. It has been thought that this type of diabetes may be a result of genetic disorders. Gene mutation may play a role in the initiation and development of non-insulin-dependent primary diabetes. In patients with this type of diabetes, abnormal insulin secretion is often found. In addition, cells are usually resistant to insulin, meaning that insulin is no longer effective in the regulation of glucose metabolism. A reduction in the density of insulin receptor and intracellular disorders of glucose metabolism may be responsible for the pathogenesis of non-insulin-dependent primary diabetes. In *secondary diabetes,* the destruction of the β cells due to pancreatic diseases as listed above is responsible for the deficiency of insulin and the pathogenesis of diabetes.

In diabetes, there are several common pathophysiological changes regardless the types and causes of the disorder. The β cells are often committed to apoptosis, a major cause for insulin deficiency (Fig. 19.2). Because of insulin deficiency, the level of blood glucose increases, a change known as *hyperglycemia.* When the blood glucose content exceeds a critical level (about 180 mg/dL), glucose is excreted from the kidneys. Because the presence of glucose increases the osmotic pressure in the renal tubules, which reduces tubular reabsorption and enhances urea formation, osmotic diuresis often occurs, resulting in increased urination. In severe cases, diabetes may induce two acute complications: ketoacidosis and hyperosmotic coma. Ketoacidosis is usually found in insulin-dependent diabetes and results from cessation of insulin administration. Because of insulin deficiency, the utilization of glucose is reduced. As a compensating mechanism, fatty acids are mobilized from fat-storage tissue (liver and adipose tissue) and used for energy production. The metabolism of fatty acids generates several acidic substances, including keto acids and acetoacetic acid, resulting in an increase in the serum concentration of hydrogen ions, a condition known as acidosis. The acidic environment induces impairment of cell functions. When the serum pH reduces to a critical level, the central nerve cells are injured,

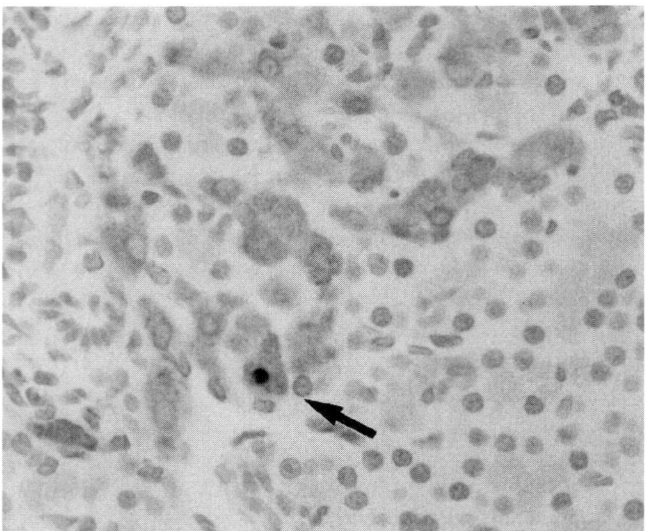

Figure 19.2. Apoptotic β cells (arrow) (black nucleus) in sucrose-fed Goto–Kakizaki (GK) rat, a spontaneously diabetic animal model of non-insulin-dependent diabetes mellitus, at 12 weeks of age detected by the TUNEL method. Apoptotic β cells were found only in sucrose-fed GK rats, not in GK and Wistar rats. Double staining is shown for β cells positive for insulin (red) and apoptosis (black). Magnification ×480. (Reprinted with permission from Koyama M et al: Accelerated loss, of islet beta cells in sucrose-fed Goto-Kakizaki rats, a genetic model of non-insulin-dependent diabetes mellitus, *Am J Pathol* 153:537–45, copyright 1998.)

resulting in acidosis coma. In addition, hyperglycemia is associated with an increase in the osmotic pressure in the serum. This osmotic change exerts a dehydration effect on the cells, mobilizing water from the interior to the exterior of the cells. Hyperosmotic coma occurs when the central nerve cells are injured due to overdehydration.

Long-term diabetes is associated with chronic complications in the vascular system. A major complication is arteriosclerosis. The incidence of arteriosclerosis in diabetics is much higher than that in the general population. Arteriosclerotic lesions are often found in the arteries of the heart, brain, kidney, and extremities of patients with diabetes. Pathological changes of the atherosclerotic lesions are similar to those described in Chapter 15. Clinical consequences of these changes include cardiac ischemia and infarction, stroke, renal ischemia, and extremity ulcers and gangrene. In addition to pathological lesions in large arteries, other blood vessels including small arteries, arterioles, and capillaries also undergo pathological changes in diabetes. A typical example is retinopathy. In this case, diabetes induces an increase in the permeability of the retinal capillaries, which is followed by gradual destruction and occlusion of the capillaries. The capillary lesions are associated with scattered hemorrhages in the retina. These pathological changes induce proliferative reactions, fibrosis, and scar formation, leading to retinal detachment and blindness. However, the pathogenic mechanisms of diabetic retinopathy remain poorly understood. Another example is nephropathy, one of the leading causes of death due to diabetes. In this disorder, diabetes induces thickening of the basement membrane of the glomerular blood vessels, and hyalinization and occlusion of glomerular arterioles. These lesions are collectively defined as *glomerulosclerosis*. The consequences of this disorder are renal dysfunction and failure.

Experimental Models of Diabetes Mellitus [19.2]. Experimental diabetes can be established in rodents by intravenous injection of an antineoplastic biotic streptozocin (2-deoxy-2-[(methylnitrosoamino)carbonyl]amino-D-glucopyranose), which is derived from *Streptomyces achromogenes*. This substance induces the destruction of pancreatic β cells, inducing experimental diabetes. In the rat and mouse models, the substance can be injected into the femoral vein. One injection is sufficient for the induction of diabetes. Blood glucose can be measured with a glucose sensor at desired timepoints. An increase in the blood glucose level can be seen within 5 days. It is important to note that streptozocin is only effective in rodents and does not induce diabetes in large animals and humans.

Another model is pancreatectomy or the removal of the pancreas. To create a pancreatectomy model, an animal is anesthetized by peritoneal injection of sodium pentobarbital at a dose of 50 mg/kg body weight. The upper abdominal skin is sterilized with 75% alcohol, Betadine, and 75% alcohol again. The abdominal cavity is opened at a location in the upper middle area and the pancreas is identified. The pancreatic blood vessels are tied off with surgical sutures, the pancreas is removed, and the abdominal wound is closed. At scheduled times following the surgery, blood glucose level can be measured with a glucose sensor.

Conventional Treatment of Diabetes [19.3]. There are several strategies for the treatment of diabetes. These include dietary control, insulin administration, and managements of complications, if any. The most important treatment is to control diet and to prevent obesity. The total calories necessary for each patient should be estimated on the basis of accepted standards, which are about 40 kcal (1000 calories) per kg body weight per day for youths and about 35 kcal per kg body weight per day for adults. Insulin administration is necessary for all type I diabetics and is recommended for type II diabetics. Such a treatment significantly reduces or prevents the occurrence of diabetic complications such as arteriosclerosis, retinopathy, and nephropathy. Insulin can be administrated via muscular and subcutaneous injections or mechanical pump-mediated subcutaneous insulin infusion. The insulin infusion method provides a sustained injection of insulin at a constant rate, an effective approach for the achievement of a relatively stable concentration of blood glucose. Treatments for common diabetic complications, including arteriosclerosis, retinopathy, and nephropathy are discussed in Chapter 15, 20, and 23, respectively. In the case of gangrene, it is often necessary to carry out amputations.

Molecular Regenerative Engineering. As discussed above, the administration of insulin is effective in the treatment of diabetes. However, it is difficult to achieve a rate of insulin delivery that matches or simulates the physiological insulin profile. Often, blood insulin concentration overshoots the physiological level immediately following insulin injection, resulting in rapid development of hypoglycemia. Since insulin is rapidly degraded in the blood, hyperglycemia occurs when the concentration of blood insulin is below the physiological level. Even though the pump-mediated insulin infusion method provides insulin delivery at a constant rate, the insulin level is often insufficient for the metabolism of glucose immediately following food ingestion, but exceeds the physiological level in the fasting state. To date, a method that controls insulin release in response to the blood glucose level is not available.

Molecular regenerative engineering or therapy has been proven a potential approach to control glucose metabolism. Several strategies have been established for the molecular treatment of diabetes, including the enhancement of glucose uptake and storage, inhibition of glucose production, facilitation of insulin synthesis, promotion of the survival and proliferation of the pancreatic β cells, and suppression of autoimmune processes.

Enhancement of Glucose Uptake and Storage and Inhibition of Glucose Production [19.4]. Glucose is stored in the form of glycogen. Glycogen formation is a process mediated by enzymes. A critical enzyme is glucokinase, which catalyzes the phosphorylation of glucose in the presence of ATP, forming glucose-6-phosphate (see Table 19.1). Glucose-6-phosphate can be catalyzed to form glucose-1-phosphate, which is further converted to uridine diphosphate glucose. Uridine diphosphate glucose is the final form used for the synthesis of glycogen. Thus an increase in the glucokinase activity enhances glycogen synthesis and reduces the level of blood glucose. Mutations of the glucokinase gene have been shown to induce non-insulin-dependent diabetes mellitus (NIDDM), also known as type 2 maturity onset diabetes of the young (MODY2). The transfer of the glucokinase gene into the liver and skeletal muscles results in over-expression of the glucokinase gene and enhances glycogen synthesis. The enhancement of glycogen synthesis is associated with an increase in glucose uptake and a decrease in glucose production from the stored glycogen in the cells, thus lowering the blood glucose concentration. In general, genes encoding proteins that facilitate glycogen synthesis can all be used for the molecular treatment of diabetes.

Facilitation of Insulin Synthesis and Activation [19.5]. Insulin is produced in the pancreatic β cells by several processes. The translation of insulin mRNA generates preproinsulin (~12 kDa), which is cleaved in the β cells to form proinsulin (~9 kDa). A large fraction of proinsulin (~80%) is further cleaved in the β cells to form insulin (~6 kDa), while the remaining proinsulin is released into the blood. One of the molecular approaches for treating diabetes is to facilitate the expression of insulin gene in diabetes. The transfer of the insulin gene into the pancreatic β cells enhances the expression of preproinsulin, leading to an increase in the production of insulin. An important aspect in molecular treatment of diabetes is to enhance the responsiveness of the β cells for releasing insulin upon an increase in the blood glucose level. In experimental investigations, several

TABLE 19.1. Characteristics of Glucokinase*

Proteins	Alternative Names	Amino Acids	Molecular Weight (kDa)	Expression	Functions
Glucokinase	GK, GCK, GLK, hexokinase 4 (HK4)	466	52	Liver, skeletal muscle, pancreas	Phosphorylating glucose to produce glucose-6-phosphate and promoting the formation of glycogen

*Based on bibliography 19.4.

glucose-responsive gene promoters, including those from the glucose-6-phosphatase gene and L-pyruvate kinase gene (see list of proteins in Table 19.2), have been used for such a purpose. These investigations have demonstrated that the overexpression of these gene promoters by gene transfer enhances the responsiveness of the insulin gene upon the stimulation of increased blood glucose. However, there is always a lag of several hours in insulin release following the stimulation. This period is necessary for gene transcription and protein translation.

Another approach to enhance insulin activity is to activate proinsulin in the liver and other tissues. As discussed above, about 20% insulin exists in the form of proinsulin, which does not have insulin activity. The activation of the proinsulin may significantly increase the activity of insulin. In pancreatic β cells, the proinsulin is converted to insulin by proinsulin convertases. However, other types of cell are not able to conduct such a function because of the lack of these conversion enzymes. To solve such a problem, researchers have engineered the structure of the insulin gene by adding proteolytic target sites for proteases. An example of such proteases is furin, also known as *paired basic amino acid cleaving* enzyme and proprotein convertase subtilisin/kexin type 3 (794 amino acids and 87 kDa in molecular weight). Furin is expressed in the liver cells and can convert precursor proteins to their active forms by cleavage at their paired basic amino acid sites. When a target encoding site for furin is inserted into the insulin gene at an appropriate location and the modified insulin gene is transferred into the liver cells, the proinsulin proteins can be cleaved by furin to form insulin. This is a potential method that can be used to activate proinsulin, thus enhancing the total activity of insulin. In addition to proinsulin, furin can cleave other protein precursors such as proparathyroid hormone, transforming growth factor β1 precursor, proalbumin, pro-β-secretase, membrane type 1 matrix metalloproteinase, the β subunit of pronerve growth factor and von Willebrand factor.

Promotion of Survival and Prevention of Apoptosis of β-Cells [19.6]. The apoptosis of pancreatic β cells is a major cause for type I and type II diabetes. Thus it is essential to prevent apoptosis and promote the survival and proliferation of the β cells. Adult pancreatic β cells can regenerate through two mechanisms: β cell proliferation and differentiation of stem and progenitor cells into β-cells. The latter is referred to as *β-cell neogenesis*. β-cell proliferation and neogenesis are regulated by a number of growth factors, including insulin-like growth factor (IGF)1, growth hormone (GH), epithelial growth factor (EGF), fibroblast growth factor (FGF), hepatocyte growth factor (HGF), keratinocyte growth factor (KGF), nerve growth factor (NGF), platelet derived growth factor (PDGF), and vascular endothelial growth factor (VEGF). These growth factors also play a role in the prevention of cell apoptosis. The genes encoding these growth factors can be used for molecular treatment of diabetes by gene transfer. The characteristics of these growth factors are described in Chapter 15.

Among the growth factors listed above, the role of insulin-like growth factor 1 in the regulation of β-cell survival and proliferation have been investigated extensively. Insulin-like growth factor 1 interacts with its receptor, inducing activation of the receptor tyrosine kinase in the cytoplasmic domain. The receptor induces tyrosine phosphorylation of insulin receptor substrate (IRS) family members (primarily IRS2) and a Src family member Shc. The phosphorylated tyrosine residues serve as docking sites for the recruitment of downstream signaling molecules, including Grb2 and PI3 kinase. Grb2 is coupled to the Ras-MAPK signaling pathway. The Ras-MAPK and PI3 kinase signaling pathways

TABLE 19.2. Characteristics of Glucose-6-Phosphatase and L-Pyruvate Kinase*

Protein	Alternative Names	Amino Acids	Molecular Weight (kDa)	Expression	Functions
Glucose-6-phosphatase	G6PC, G6Pase, G-6-Pase	357	41	Liver, skeletal muscle, brain, kidney	Catalyzing the hydrolysis of D-glucose 6-phosphate to D-glucose and orthophosphate
L-Pyruvate kinase	Pyruvate kinase liver and RBC, pyruvate kinase type L, pyruvate kinase liver and blood cell	574	62	Liver, red blood cells	Catalyzing the formation of phohsphoenolpyruvate from pyruvate and ATP and causing chronic hereditary nonspherocytic hemolytic anemia (CNSHA) when mutated

*Based on bibliography 19.5.

are described in Chapter 5. The activation of these pathways promotes the transcription of mitogenic genes, prevents cell apoptosis, and enhances cell survival and proliferation. Investigations with transgenic and gene transfer models have demonstrated the role of insulin-like growth factor-1 in promoting the survival of the β cells (Fig. 19.3). The overexpression of the insulin-like growth factor 1 gene in a transgenic mouse model suppresses hyperglycemia when the mouse was administrated with streptozocin, a substance causing β-cell death and diabetes. The administration of the same dose of streptozocin to wildtype mice induced a significantly higher level of hyperglycemia (Fig. 19.4). The overexpression of the insulin-like growth factor 1 gene also extended the lifespan of the mice with streptozocin-induced diabetes (Fig. 19.4).

The gene of insulin-like growth factor 1 is a potential candidate for the molecular treatment of human diabetes. In addition, genes that encode signaling factors, such as

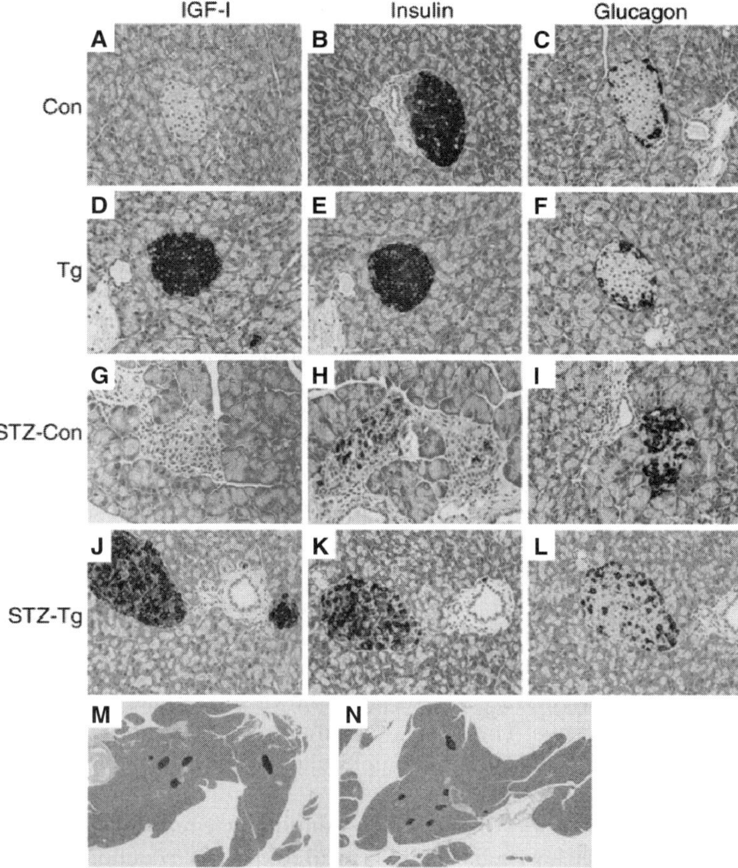

Figure 19.3. Immunohistochemical analysis of insulin growth factor I (IGF-I), glucagon, and insulin expression in pancreatic islets. IGF-I (A,D,G,J), insulin (B,E,H,K,M,N), and glucagon (C,F,I,L) staining of representative sections of pancreas before (A–F) and 3 months after (G–N) STZ treatment. Wildtype mice (Con): A–C, G–I (×400) and M (×40); C57BL/6–SJL transgenic mice (Tg, overexpressing IGF-I): D–F, J–L (×400) and N (×40). (Reprinted with permission from George M et al: *J Clin Invest* 109:1153–63, copyright 2002.)

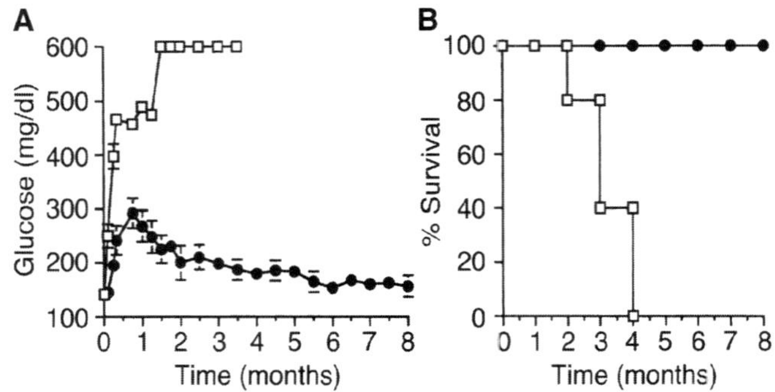

Figure 19.4. (A) Blood glucose levels in wildtype mice and C57BL/6–SJL transgenic mice, which overexpress insulin growth factor-I (IGF-I), after STZ treatment. Squares, wildtype mice ($n = 15$); circles, transgenic mice ($n = 15$). (B) Percent survival of control (squares; $n = 20$) and transgenic (circles; $n = 20$) mice after STZ treatment. Results are mean ± SEM of the indicated mice. (Reprinted with permission from George M et al: *J Clin Invest* 109:1153–63, copyright 2002.)

MAPK and protein kinase B (PKB; note that PBK is a molecule downstream to insulin receptor substrate 2), can be potentially used for molecular therapy.

There are several genes encoding proteins that regulate the differentiation of stem and progenitor cells to pancreatic β cells. One of such genes is the insulin promoter factor 1 gene (IPF1), also known as *pancreatic duodenal homeobox gene 1* (Pdx1) and somatostatin transcription factor 1 (STF1). This gene encodes the insulin promoter factor 1 protein (283 amino acids, 31-kDa), which is expressed in the pancreas, brain, and intestine. This protein promotes the differentiation of stem and progenitor cells to insulin-producing β cells, stimulates pancreatic development, and activates the transcription of the insulin and somatostatin genes. When this gene is transferred into the liver cells in animal models, its protein product stimulates the formation of β cells and the expression of insulin in these cells.

Another gene is the neurogenic differentiation factor 1 (NeuroD1; see Table 19.3) gene that encodes a protein for the regulation of β-cell differentiation. In humans and mice, the deficiency of this gene results in pathological changes found in diabetes. In experimental models of diabetes, the transfer of the NeuroD gene into the mouse liver induces the formation of pancreatic islets. The islet cells can produce insulin, glucagon, and somatostatin. As a result, the level of blood glucose is restored and diabetic changes are reduced. Fibroblast growth factor (FGF) has also been shown to regulate the development and survival of pancreatic β cells. The genes of insulin promoter factor 1, NeuroD, and FGF can be considered potential genes for the molecular treatment of human diabetes.

Suppression of Autoimmune Processes [19.7]. Autoimmune reactions, which are immune processes directed against host cells, play a critical role in the induction of type I diabetes. Thus, an important strategy in molecular treatment for diabetes is to suppress autoimmune reactions. Cell types that are directly involved in autoimmune reactions are antigen presenting dendritic cells and antigen-specific T cells. These cell types are the targets of molecular therapy for autoimmune disorders. Antiautoimmune cytokine genes can be prepared and transferred into these cells, reducing the cell immune activities. Potential antiautoimmune cytokines include transforming growth factor (TGF)β, interleukin (IL)4,

TABLE 19.3. Characteristics of Selected Proteins that Stimulate the Differentiation of Insulin-Producing β Cells*

Protein	Alternative Names	Amino Acids	Molecular Weight (kDa)	Expression	Functions
Insulin promoter factor 1	Pancreatic duodenal homeobox gene 1 (Pdx1) and somatostatin transcription factor 1 (STF1)	283	31	Pancreas, brain, intestine	Promoting the differentiation of insulin-producing β cells, stimulating pancreatic development, and activating the transcription of the insulin and somatostatin genes
Neurogenic differentiation factor 1	NeuroD1, NEUROD, β-cell E box transactivator 2	356	40	Pancreas, retina, placenta	Stimulating the differentiation of nerve cells and pancreatic β cells

*Based on bibliography 19.6.

and IL10, which suppress the activity of T cells. Genes encoding these factors can be transferred into T cells and antigen-presenting cells. The transferred cells can be transplanted into the host. Preliminary investigations have demonstrated that the transfer of these genes significantly prevents the damage of pancreatic islet cells and reduces diabetic changes in animal models of diabetes.

Cell and Tissue Regenerative Engineering. Diabetes is induced primarily by β-cell malfunction and apoptosis. There are about 10^9 β cells in the pancreatic islets of Langerhans. At a given time, certain β cells are committed to apoptosis. The apoptotic cells are replaced with new cells generated from progenitor cells or existing β cells (Fig. 19.5). Thus, the total number of β cells is maintained at a relatively constant level. Under the influence of genetic and environmental factors, such as viral infection, the rate of cell apoptosis may increase and exceed that of cell proliferation. When the majority of cells are destroyed, diabetes occurs. Thus, the principle of cell and tissue engineering for treating diabetes is to restore and maintain the β-cell population. In the section on molecular regenerative engineering, various methods have been discussed for the promotion of cell regeneration and prevention of cell apoptosis. In this section, strategies and methods are introduced for the restoration of β cells by cell transplantation.

There are several steps for the restoration of β cells, including the identification and collection of candidate cells, manipulation of collected cells, if necessary, packaging of cells into appropriate devices, and transplantation of prepared cells into a desired organ or tissue. There are several types of candidate cells for the treatment of diabetes, including autogenous β cells, autogenous stem and progenitor cells, embryonic and fetal stem cells,

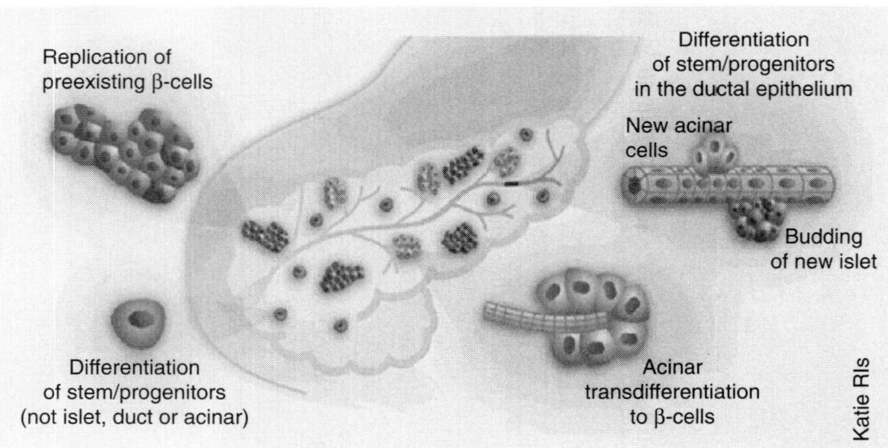

Figure 19.5. The pancreas as a source of new β cells. The pancreas itself is likely to be the main source of new insulin-producing β cells and of cells that can regenerate the acini and ducts. Several processes have been proposed: replication of preexisting β cells, possibly thorough epithelial–mesenchymal transition, including dedifferentiation, expansion, and redifferentiation; differentiation of progenitors within the ductal epithelium; transdifferentiation of acinar cells; and differentiation of pancreatic stem/progenitors that are not of β-cell, duct, or acinar origin. (Reprinted by permission from Macmillan Publishers Ltd.: Bonner-Weir S, Weir GC: New sources of pancreatic beta-cells, *Nature Biotechnol* 23:857–61, copyright 2005.)

allogenic β cells, allogenic stem and progenitor cells, xenogenic β cells, and xenogenic stem and progenitor cells. It is important to note that the cell transplantation therapy for diabetes is only in its infant stage. At present, cell transplantation may have not offered advantages over insulin injection. It is hoped, however, that with a better understanding of the differentiation control of the β cells the function of the Langerhans islets can be restored permanently with molecular and cellular regenerative approaches.

Candidate Cell Types

AUTOGENOUS PANCREATIC β CELLS [19.8]. Autogenous β cells derived from the host pancreas are an ideal cell type for the treatment of diabetes by cell transplantation (Fig. 19.5). However, a large number of cells are usually needed for cell transplantation. Diabetic patients rarely possess a sufficient number of functional β cells when the disease is identified. Even though the disease can be identified in the early stage, it is difficult to collect a large number of β cells. Whereas host β-cell regeneration and transplantation are not a suitable approach at present, it is possible to generate semi-autogenous β cells by transferring the host β-cell nuclei into donor oocytes. Such an approach may produce functional β cells with reduced immune rejection reactions. Further research is necessary to achieve such a goal.

ALLOGENIC β CELLS [19.8]. Allogenic β cells can be harvested from donor subjects, cultured for expansion, and used for cell transplantation. To successfully restore the function of the islets of Langerhans, a sufficient number of β cells are needed for each recipient. While allogenic cell transplantation is potentially an appropriate treatment for diabetes, three problems hinder the application of this approach: (1) there is a shortage of organ donors—in the United States, the incidence of type I diabetes is about 1 million per year, whereas, the number of suitable organ donors is only about 10,000 per year; (2) allogenic cell transplantation always induces acute immune rejection—patients with allogenic β-cell transplantation will have to receive lifetime immunosuppression therapy; and (3) transplanted β cells may undergo cell apoptosis even under immunosuppression therapy. It is usually difficult to maintain the survival of the transplanted cells. To overcome these difficulties, allogenic β cells can be transfected with oncogenes or growth factor genes, promoting cell survival and preventing cell apoptosis. Transformed β cells usually become immortal with enhanced cell proliferation. Such a manipulation has been shown to improve the survival rate of transplanted β cells. However, oncogene transformation may impair the physiological function of the β cells and introduce the risk of tumorigenesis to the cell recipient.

PANCREATIC STEM AND PROGENITOR CELLS [19.9]. The pancreas contains stem and progenitor cells, which can differentiate into the insulin-producing β cells. The epithelial cells of the pancreatic ductules and ducts are potential stem cells. In preliminary studies conducted in humans and mice, the pancreatic epithelial cells could be collected, cultured, and induced to form insulin-producing cells in cell culture models. These cells are suitable candidates for the cellular treatment of diabetes. However, a practical difficulty is that diabetic patients, when diabetes is diagnosed, do not have a sufficient number of functional stem and progenitor cells. It is often necessary to collect stem and progenitor cells from allogenic donors.

EMBRYONIC STEM CELLS [19.10]. Embryonic stem cells can be induced to differentiate into various specialized cells types, including the insulin-producing β cells. Thus, embryonic cells are candidate cells for the cellular treatment of diabetes. To use embryonic cells, it is necessary to carry out several steps: (1) collecting embryonic cells, (2) inducing the differentiation of stem cells to insulin-producing β cells, (3) identifying and selecting β cells, (4) engineering selected cells to express desired features such as enhanced survival and antiautoimmune capabilities, (5) expanding selected cells to a sufficient number for cell transplantation, and (6) transplanting the insulin-producing cells directly to a target tissue or packaging the cells into a desired device followed by device transplantation.

Embryonic stem cells can be collected from the embryonic blastocyst as described in Chapter 9. The collected stem cells can be cultured under pancreatic conditions to induce differentiation into insulin-producing β cells. Preliminary investigations have demonstrated the possibility of forming insulin-producing β cells from human and mouse embryonic stem cells, although the fraction of insulin-producing β cells is small. To identify insulin-producing cells, it is necessary to introduce a marker specific to these cells. The promoter of the insulin gene can be turned on by factors that stimulate insulin gene transcription and thus can be used as a specific marker. A green fluorescent protein (GFP) gene can be inserted into the insulin gene to form a recombinant gene so that the GFP gene can be driven by the insulin gene promoter. The recombinant gene can be transferred into embryonic stem cells. Any cells that express the green fluorescent protein are cells with activated insulin gene. These cells are considered insulin-producing β cells. The identified cells can be collected by fluorescence-activated cell sorting and further expanded in culture. The collected cells can be engineered by transferring genes for enhancing desired features. For example, the insulin-producing β cells can be transferred with oncogenes or growth factor genes to enhance their survival and proliferative capabilities. Certain cell membrane molecules may serve as antigens for autoimmune reactions. Such antigens can be identified and the genes of the antigens can be removed, resulting in a reduction in autoimmune responses. When the cell number reaches a sufficient level, the cells can be used for transplantation.

ADULT STEM CELLS [19.11]. There exist stem cells in various types of adult tissue and organ, including the bone marrow, liver, intestine, and the nerve system. These cells are responsible for the regeneration of adult cells when cell injury and death occur. While most adult stem cells are committed to the formation of specialized cells within a defined developmental system, certain types of adult stem cells are capable of differentiating into cells for different systems. A typical example is the bone marrow stromal cells. These cells have been shown to differentiate into various specialized cell types including muscular cells and neurons, depending on the local environment of a tissue or organ, in humans and rodents. The transplantation of bone marrow stem cells into the pancreas may induce differentiation of the stem cells into insulin-producing β cells.

XENOGENEIC β CELLS, STEM CELLS, AND PROGENITOR CELLS [19.11]. Xenogeneic β cells, stem cells, and progenitor cells are considered only when no other cell sources are available. The identification, collection, culture, manipulation, and transplantation of these cells are similar to those described above. A major concern is that, because of their xenogenic nature, these cells cause severe acute immune rejection reactions. It is necessary to establish devices for the isolation of the transplanted cells from the host system. This issue is discussed in the section on "transplantation of β-cell-protecting devices."

Prevention of Immune Reactions and β-Cell Injury [19.12]. While the approaches discussed above show potential for the treatment of diabetes, there is a common problem for most approaches: acute immune rejection. Furthermore, autoimmune reactions may occur as these reactions are the original cause of diabetes. Immunity-suppressing agents have long been used for suppressing immune responses. The requirement of lifelong administration of immune suppressing agents renders molecular and cellular engineering approaches less favorable compared to insulin injection. To overcome such a problem, selected cells for transplantation can be transfected with genes that encode immune suppressing cytokines, such as interleukin (IL)4, IL10, and transforming growth factor (TGF)β. These cytokines inhibit the function of T cells, which are responsible for immune rejection and autoimmune responses. Preliminary investigations have demonstrated the effectiveness of these suppressing cytokines. In addition, calls can be transfected with genes encoding protective antioxidant proteins, such as copper/zinc and manganese superoxide dismutases, catalase, and thioredoxin, and antiapoptotic proteins, such as Bcl2 and A20. Experimental investigations have demonstrated that the transfer of these genes protects insulin-producing β cells from injury and apoptosis.

Transplantation of β-Cell-Protecting Devices [19.13]. Because of the susceptibility of transplanted cells to immune attacks, several types of protective devices have been developed and used for β-cell transplantation. These devices include microcapsules and biohybrid pancreas-mimicking apparatuses with blood circulation. These devices are manufactured with semipermeable membranes, which prevent the entrance of immune cells into the device and protect the β cells from immune attack. Various polymeric materials can be used to fabricate semipermeable membranes with a desired pore size (see Chapter 12 for polymeric materials). Microcapsules can be used to enclose β cells and transplant into a desired tissue or cavity of the recipients. Since the capsules are usually small, allowing oxygen and nutrient diffusion into the enclosed cells, it is not necessary to introduce blood circulation into the capsules.

When a large pancreas-mimicking device is used, it is often necessary to establish a blood circulation system. A double-tube system can be fabricated with a semipermeable membrane. One of the tubes is used for packaging the β cells and the other is used for introducing bloodflow to the β cells (Fig. 19.6). The device can be implanted into the abdominal cavity of the recipients. The inlet and outlet ports of the blood circulatory system can be anastomosed to a selected artery and vein, such as the inferior mesentery

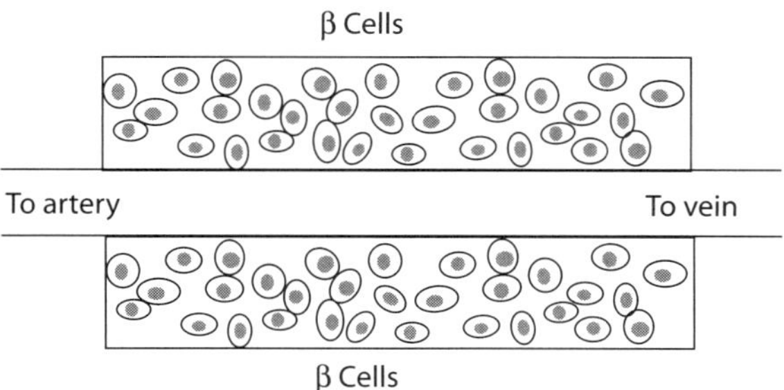

Figure 19.6. Schematic representation of an artificial pancreas containing functional β cells. Based on bibliography 19.13.

artery and vein. The blood circulation provides oxygen and nutrients to the cells, and insulin and other molecules from the β cells can diffuse across the semipermeable membrane into the blood compartment. Such a device can be potentially used for the transplantation of β cells into diabetic patients. A problem for such a device is blood coagulation and thrombosis. Blood coagulation occurs within the device, and thrombosis occurs at the anastomoses. It is necessary to administrate anticoagulants when such a device is transplanted.

Pancreatic Cancer

Pathogenesis, Pathology, and Clinical Features. Pancreatic cancer is one of the four most frequent types of cancer in humans, which include lung, colon, breast, and pancreatic cancers. Pancreatic cancer is often originated from the epithelial cells of the pancreatic ductules and ducts. Pancreatic cancer progresses rapidly. When pancreatic cancer is diagnosed, more than half of the patients are associated with cancer metastasis. As for other types of cancer, the pathogenesis of pancreatic cancer remains poorly understood. Epidemiologic studies have suggested that cigarette smoking, fat diets, and coffee intake may serve as risking factors for the development of pancreatic cancer. Pathological changes in pancreatic cancer are similar to those found in other types of cancer as described in Chapter 25. There are several unique clinical features for pancreatic cancer. These include rapid and extensive loss of body weight (due to the loss of digestion enzymes), severe upper abdominal pain, anorexia, nausea, vomiting, hyperglycemia (due to the destruction of the β cells) and jaundice (due to the compression of the bile duct by tumors in the head region of the pancreas).

Treatment of Pancreatic Cancer. Pancreatic cancer exhibits genetic, pathological, and clinical characteristics similar to those described in Chapter 25. Thus, similar therapeutic strategies can be used for the treatment of pancreatic cancer. In general, surgical removal of pancreatic cancer is the most effective treatment, provided that the cancer is identified in the early stage before the occurrence of metastasis. In the presence of metastasis, chemotherapy and radiotherapy are the methods of choice. Molecular therapy can be potentially used. The principles and methods of molecular therapy for cancers are discussed in Chapter 25.

BIBLIOGRAPHY

19.1. Anatomy and Physiology of the Pancreas

Guyton AC, Hall JE: *Textbook of Medical Physiology*, 11th ed, Saunders, Philadelphia, 2006.

McArdle WD, Katch FI, Katch VL: *Essentials of Exercise Physiology*, 3rd ed, Lippincott Williams & Wilkins, Baltimore, 2006.

Germann WJ, Stanfield CL (with contributors Niles MJ, Cannon JG): *Principles of Human Physiology*, 2nd ed, Pearson Benjamin Cummings, San Francisco, 2005.

Thibodeau GA, Patton KT: *Anatomy & Physiology*, 5th ed, Mosby, St Louis, 2003.

Boron WF, Boulpaep EL: *Medical Physiology: A Cellular and Molecular Approach*, Saunders, Philadelphia, 2003.

Ganong WF: *Review of Medical Physiology*, 21st ed, McGraw-Hill, New York, 2003.

19.2. Pathogenesis, Pathology, and Clinical Features of Diabetes

Schneider AS, Szanto PA: *Pathology*, 3rd ed, Lippincott Williams & Wilkins, Philadelphia, 2006.

McCance KL, Huether SE: *Pathophysiology: The Biologic Basis for Disease in Adults & Children*, 5th ed, Elsevier Mosby, St Louis, 2006.

Porth CM: *Pathophysiology: Concepts of Altered Health States*, 7th ed, Lippincott Williams & Wilkins, Philadelphia, 2005.

Frazier MS, Drzymkowski JW: *Essentials of Human Diseases and Conditions*, 3rd ed, Elsevier Saunders, St Louis, 2004.

Kahn SE: The relative contributions of insulin resistance and β-cell dysfunction to the pathophysiology of type 2 diabetes, *Diabetologia* 46:3–19, 2003.

Remuzzi G, Benigni A, Remuzzi A: Mechanisms of progression and regression of renal lesions of chronic nephropathies and diabetes, *J Clin Invest* 116(2):288–96, 2006.

Kelkar P: Diabetic neuropathy, *Semin Neurol* 25(2):168–73, 2005.

Wolf G, Chen S, Ziyadeh FN: From the periphery of the glomerular capillary wall toward the center of disease: Podocyte injury comes of age in diabetic nephropathy, *Diabetes* 54(6):1626–34, 2005.

Van den Berghe G: How does blood glucose control with insulin save lives in intensive care? *J Clin Invest* 114(9):1187–95, 2004.

Donath MY, Halban PA: Decreased beta-cell mass in diabetes: Significance, mechanisms and therapeutic implications, *Diabetologia* 47(3):581–9, 2004.

Mandrup-Poulsen T: Beta cell death and protection, *Ann NY Acad Sci* 1005:32–42, 2003.

Wiernsperger NF, Bouskela E: Microcirculation in insulin resistance and diabetes: More than just a complication, *Diabetes Metab* 29(4 Pt 2):6S77–87, 2003.

Rorsman P, Renstrom E: Insulin granule dynamics in pancreatic beta cells, *Diabetologia* 46(8):1029–45, 2003.

Tsilibary EC: Microvascular basement membranes in diabetes mellitus, *J Pathol* 200(4):537–46, 2003.

Caramori ML, Mauer M: Diabetes and nephropathy, *Curr Opin Nephrol Hypertens* 12(3):273–82, 2003.

Sesti G: Apoptosis in the beta cells: cause or consequence of insulin secretion defect in diabetes? *Ann Med* 34(6):444–50, 2002.

Bloomgarden ZT: Developments in diabetes and insulin resistance, *Diabetes Care* 29(1):161–7, 2006.

Verges B: New insight into the pathophysiology of lipid abnormalities in type 2 diabetes, *Diabetes Metab* 31(5):429–39, 2005.

Liu SQ, Fung YC: Influence of STZ diabetes on zero-stress state of rat pulmonary and systemic arteries, *Diabetes* 41:136–146, 1992.

19.3. Conventional Treatment of Diabetes

Cryer PE: Banting lecture. Hypoglycemia: The limiting factor in the management of IDDM, *Diabetes* 43:1378–89, 1994.

Cryer PE: Hypoglycaemia: The limiting factor in the glycaemic management of type I and type II diabetes, *Diabetologia* 45:937–48, 2002.

Chan L, Fujimiya M, Kojima H: In vivo gene therapy for diabetes mellitus, *Trends Mol Med* 9:430–5, 2003.

McKinney JL, Cao H, Robinson JF, Metzger DL, Cummings E et al: Spectrum of HNF1A and GCK mutations in Canadian families with maturity-onset diabetes of the young (MODY), *Clin Invest Med* 27(3):135–41, 2004.

Johansen A, Ek J, Mortensen HB, Pedersen O, Hansen T: Half of clinically defined maturity-onset diabetes of the young patients in Denmark do not have mutations in HNF4A, GCK, and TCF1, *J Clin Endocr Metab* 90(8):4607–14, 2005.

Gloyn AL, Odili S, Zelent D, Buettger C, Castleden HA et al: Insights into the structure and regulation of glucokinase from a novel mutation (V62M), which causes maturity-onset diabetes of the young, *J Biol Chem* 280(14):14105–13, 2005.

Weedon MN, Frayling TM, Shields B, Knight B, Turner T et al: Genetic regulation of birth weight and fasting glucose by a common polymorphism in the islet cell promoter of the glucokinase gene, *Diabetes* 54(2):576–81, 2005.

Barrio R, Bellanne-Chantelot C, Moreno JC, Morel V, Calle H et al: Nine novel mutations in maturity-onset diabetes of the young (MODY) candidate genes in 22 Spanish families, *J Clin Endocr Metab* 87(6):2532–9, 2002.

19.4. Enhancement of Glucose Uptake and Storage and Inhibition of Glucose Production

Brocklehurst KJ, Davies RA, Agius L: Differences in regulatory properties between human and rat glucokinase regulatory protein, *Biochem J* 378(Pt 2):693–7, March 2004.

Farrelly D, Brown KS, Tieman A, Ren J, Lira SA et al: Mice mutant for glucokinase regulatory protein exhibit decreased liver glucokinase: A sequestration mechanism in metabolic regulation, *Proc Natl Acad Sci USA* 96:14511–6, 1999.

Grimsby J, Coffey JW, Dvorozniak MT, Magram J, Li G et al: Characterization of glucokinase regulatory protein-deficient mice, *J Biol Chem* 275:7826–31, 2000.

Hayward BE, Dunlop N, Intody S, Leek JP, Markham AF et al: Organization of the human glucokinase regulator gene GCKR, *Genomics* 49:137–42, 1998.

Vaxillaire M, Vionnet N, Vigouroux C, Sun F, Espinosa R III et al: Search for a third susceptibility gene for maturity-onset diabetes of the young: Studies with eleven candidate genes, *Diabetes* 43:389–95, 1994.

Bali D, Svetlanov A, Lee HW, Fusco-DeMane D, Leiser M et al: Animal model for maturity-onset diabetes of the young generated by disruption of the mouse glucokinase gene, *J Biol Chem* 270:21464–7, 1995.

Byrne MM, Sturis J, Clement K, Vionnet N, Pueyo ME et al: Insulin secretory abnormalities in subjects with hyperglycemia due to glucokinase mutations, *J Clin Invest* 93:1120–30, 1994.

Danial NN, Gramm CF, Scorrano L, Zhang CY, Krauss S et al: BAD and glucokinase reside in a mitochondrial complex that integrates glycolysis and apoptosis, *Nature* 424:952–6, 2003.

Froguel P, Vaxillaire M, Sun F, Velho G, Zouali H et al: Close linkage of glucokinase locus on chromosome 7p to early-onset non-insulin-dependent diabetes mellitus, *Nature* 356:162–4, 1992.

Froguel P, Zouali H, Vionnet N, Velho G, Vaxillaire M et al: Familial hyperglycemia due to mutations in glucokinase: Definition of a subtype of diabetes mellitus, *New Eng J Med* 328:697–702, 1993.

Gloyn AL, Noordam K, Willemsen MAAP, Ellard S, Lam WWK et al: Insights into the biochemical and genetic basis of glucokinase activation from naturally occurring hypoglycemia mutations, *Diabetes* 52:2433–40, 2003.

Grimsby J, Sarabu R, Corbett WL, Haynes NE, Bizzarro FT et al: Allosteric activators of glucokinase: Potential role in diabetes therapy, *Science* 301:370–3, 2003.

Grupe A, Hultgren B, Ryan A, Ma YH, Bauer M et al: Transgenic knockouts reveal a critical requirement for pancreatic beta cell glucokinase in maintaining glucose homeostasis, *Cell* 83:69–78, 1995.

Matschinsky F, Liang Y, Kesavan P, Wang L, Froguel P et al: Glucokinase as pancreatic beta-cell glucose sensor and diabetes gene, *J Clin Invest* 92:2092–8, 1993.

Njolstad PR, Sovik O, Cuesta-Munoz A, Bjorkhaug L, Massa O et al: Neonatal diabetes mellitus due to complete glucokinase deficiency, *New Eng J Med* 344:1588–92, 2001.

Rowe RE, Wapelhorst B, Bell GI, Risch N, Spielman RS et al: Linkage and association between insulin-dependent diabetes mellitus (IDDM) susceptibility and markers near the glucokinase gene on chromosome 7, *Nature Genet* 10:240–5, 1995.

Sun F, Knebelmann B, Pueyo ME, Zouali H, Lesage S et al: Deletion of the donor splice site of intron 4 in the glucokinase gene causes maturity-onset diabetes of the young, *J Clin Invest* 92:1174–80, 1993.

Velho G, Petersen KF, Perseghin G, Hwang JH, Rothman DL et al: Impaired hepatic glycogen synthesis in glucokinase-deficient (MODY-2) subjects, *J Clin Invest* 98:1755–61, 1996.

Vionnet N, Stoffel M, Takeda J, Yasuda K, Bell GI et al: Nonsense mutation in the glucokinase gene causes early-onset non-insulin-dependent diabetes mellitus, *Nature* 356:721–2, 1992.

Ferre T et al: Correction of diabetic alterations by glucokinase, *Proc Natl Acad Sci USA* 93:7225–30, 1996.

O'Doherty RM et al: Metabolic impact of glucokinase overexpression in liver, *Diabetes* 48:2022–7, 1999.

Shiota M et al: Glucokinase gene locus transgenic mice are resistant to the development of obesity-induced type 2 diabetes, *Diabetes* 50:622–9, 2001.

Desai UJ et al: Phenotypic correction of diabetic mice by adenovirus-mediated glucokinase expression, *Diabetes* 50:2287–95, 2001.

Morral N et al: Adenovirus-mediated expression of glucokinase in the liver as an adjuvant treatment for type 1 diabetes, *Hum Gene Ther* 13:1561–70, 2002.

Morral N: Novel targets and therapeutic strategies for type 2 diabetes, *Trends Endocr Metab* 14:169–75, 2003.

Otaegui PJ et al: Expression of glucokinase in skeletal muscle: A new approach to counteract diabetic hyperglycemia, *Hum Gene Ther* 11:1543–52, 2000.

Slosberg ED et al: Treatment of type 2 diabetes by adenoviral-mediated overexpression of the glucokinase regulatory protein, *Diabetes* 50:1813–20, 2001.

Wu C et al: Increasing fructose 2,6-bisphosphate overcomes hepatic insulin resistance of type 2 diabetes, *Am J Physiol Endocr Metab* 282:E38–45, 2002.

O'Doherty RM et al: Activation of direct and indirect pathways of glycogen synthesis by hepatic overexpression of protein targeting to glycogen, *J Clin Invest* 105:479–88, 2000.

Newgard CB et al: Organizing glucose disposal. Emerging roles of the glycogen targeting subunits of protein phosphatase-1, *Diabetes* 49:1967–77, 2000.

Human protein reference data base, Johns Hopkins University and the Institute of Bioinformatics, at http://www.hprd.org/protein.

19.5. Facilitation of Insulin Synthesis and Activation

Blanchette F, Day R, Dong W, Laprise MH, Dubois CM: TGF-beta-1 regulates gene expression of its own converting enzyme furin, *J Clin Invest* 99:1974–83, 1997.

Dubois CM, Laprise MH, Blanchette F, Gentry LE, Leduc R: Processing of transforming growth factor beta-1 precursor by human furin convertase, *J Biol Chem* 270:10618–24, 1995.

Hendy GN, Bennett HPJ, Gibbs BF, Lazure C, Day R et al: Proparathyroid hormone is preferentially cleaved to parathyroid hormone by the prohormone convertase furin: A mass spectrometric study, *J Biol Chem* 270:9517–25, 1995.

Short DK et al: Adenovirus-mediated transfer of a modified human proinsulin gene reverses hyperglycemia in diabetic mice, *Am J Physiol* 275:E748–56, 1998.

Muzzin P et al: Hepatic insulin gene expression as treatment for type 1 diabetes mellitus in rats, *Mol Endocr* 11:833–7, 1997.

Auricchio A et al: Constitutive and regulated expression of processed insulin following in vivo hepatic gene transfer, *Gene Ther* 9:963–71, 2002.

19.6. Promotion of the Survival and Prevention of β-cell Apoptosis

Dickson LM, Rhodes CL: Pancreatic β-cell growth and survival in the onset of type 2 diabetes: A role for protein kinase B in the Akt? *Am J Physiol Endocr Metab* 287:E192–8, 2004.

Pick A et al: Role of apoptosis in failure of β-cell mass compensation for insulin resistance and β-cell defects in the male Zucker diabetic fatty rat, *Diabetes* 47:358–64, 1998.

Withers DJ et al: Disruption of IRS-2 causes type 2 diabetes in mice, *Nature* 391:900–4, 1998.

Bruning JC et al: Development of a novel polygenic model of NIDDM in mice heterozygous for IR and IRS-1 null alleles, *Cell* 88:561–72, 1997.

Bonner-Weir S: Islet growth and development in the adult, *J Mol Endocr* 24:297–302, 2000.

Kahn SE: Clinical review 135: The importance of β-cell failure in the development and progression of type 2 diabetes, *J Clin Endocr Metab* 86:4047–58, 2001.

Lingohr MK, Buettner R, Rhodes CJ: Pancreatic β-cell growth and survival—a role in obesity-linked type 2 diabetes? *Trends Mol Med* 8:375–84, 2002.

Nielsen JH et al: Regulation of β-cell mass by hormones and growth factors, *Diabetes* 50(Suppl 1):S25–9, 2001.

Argetsinger LS, Carter-Su C: Mechanism of signaling by growth hormone receptor, *Physiol Rev* 76:1089–107, 1996.

Benito M et al: IGF-I: A mitogen also involved in differentiation processes in mammalian cells, *Int J Biochem Cell Biol* 28:499–510, 1996.

Pearson G et al: Mitogen-activated protein (MAP) kinase pathways: Regulation and physiological functions, *Endocr Rev* 22:153–83, 2001.

Chan TO et al: AKT/PKB and other D3 phosphoinositide-regulated kinases: Kinase activation by phosphoinositide-dependent phosphorylation, *Annu Rev Biochem* 68:965–1014, 1999.

Withers DJ et al: IRS-2 coordinates IGF-1 receptor-mediated β-cell development and peripheral insulin signaling, *Nat Genet* 23:32–40, 1999.

Lingohr MK et al: Activation of IRS-2 mediated signal transduction by IGF-1, but not TGF-β or EGF, augments pancreatic β-cell proliferation, *Diabetes* 51:966–76, 2002.

Tuttle RL et al: Regulation of pancreatic β-cell growth and survival by the serine/threonine protein kinase Akt1/PKBβ, *Nat Med* 7:1133–7, 2001.

Dickson LM et al: Differential activation of protein kinase B and p70(S6)K by glucose and insulin-like growth factor 1 in pancreatic β-cells (INS-1), *J Biol Chem* 276:21110–20, 2001.

Sesti G: Apoptosis in the beta cells: Cause or consequence of insulin secretion defect in diabetes? *Ann Med* 34(6):444–50, 2002.

NeuroD

Fajans SS, Bell GI, Polonsky KS: Molecular mechanisms and clinical pathophysiology of maturity-onset diabetes of the young, *New Eng J Med* 345:971–80, 2001.

Lee JE, Hollenberg SM, Snider L, Turner DL, Lipnick N et al: Conversion of Xenopus ectoderm into neurons by NeuroD, a basic helix-loop-helix protein, *Science* 268:836–44, 1995.

Malecki MT, Jhala US, Antonellis A, Fields L, Doria A et al: Mutations in NEUROD1 are associated with the development of type 2 diabetes mellitus, *Nature Genet* 23:323–8, 1999.

Naya FJ, Huang HP, Qiu Y, Mutoh H, DeMayo FJ et al: Diabetes, defective pancreatic morphogenesis, and abnormal enteroendocrine differentiation in BETA2/neurod-deficient mice, *Genes Dev* 11:2323–4, 1997.

Yan RT, Wang SZ: Requirement of neuroD for photoreceptor formation in the chick retina, *Invest Ophthalm Vis Sci* 45:48–58, 2004.

Insulin Promoter Factor 1

Cockburn BN, Bermano G, Boodram LLG, Teelucksingh S, Tsuchiya T et al: Insulin promoter factor-1 mutations and diabetes in Trinidad: Identification of a novel diabetes-associated mutation (E224K) in an Indo-Trinidadian family, *J Clin Endocr Metab* 89:971–8, 2004.

Hani EH, Stoffers DA, Chevre JC, Durand E, Stanojevic V et al: Defective mutations in the insulin promoter factor-1 (IPF-1) gene in late-onset type 2 diabetes mellitus, *J Clin Invest* 104:R41–8, 1999.

Hart AW, Baeza N, Apelqvist A, Edlund H: Attenuation of FGF signalling in mouse beta-cells leads to diabetes, *Nature* 408:864–8, 2000.

Jonnson J, Carlsson L, Edlund T, Edlund H: Insulin-promoter-factor 1 is required for pancreas development in mice, *Nature* 371:606–9, 1994.

Kim SK, Selleri L, Lee JS, Zhang AY, Gu X et al: Pbx1 inactivation disrupts pancreas development and in Ipf1-deficient mice promotes diabetes mellitus, *Nature Genet* 30:430–5, 2002.

Macfarlane WM, Frayling TM, Ellard S, Evans JC, Allen LIS et al: Missense mutations in the insulin promoter factor-1 gene predispose to type 2 diabetes, *J Clin Invest* 104:R33–9, 1999.

Sharma S, Jhala US, Johnson T, Ferreri K, Leonard J et al: Hormonal regulation of an islet-specific enhancer in the pancreatic homeobox gene STF-1, *M Cell Biol* 17:2598–604, 1997.

Stoffers DA, Stanojevic V, Habener JF: Insulin promoter factor-1 gene mutation linked to early-onset type 2 diabetes mellitus directs expression of a dominant negative isoprotein, *J Clin Invest* 102:232–41, 1998.

Ferber S et al: Pancreatic and duodenal homeobox gene 1 induces expression of insulin genes in liver and ameliorates streptozotocin-induced hyperglycemia, *Nat Med* 6:568–72, 2000.

Kojima H et al: NeuroD/betacellulin gene therapy induces islet neogenesis in the liver and reverses diabetes in mice, *Nat Med* 9:596–603, 2003.

Zalzman M et al: Reversal of hyperglycemia in mice by using human expandable insulin-producing cells differentiated from fetal liver progenitor cells, *Proc Natl Acad Sci USA* 100:7253–8, 2003.

Newgard CB: While tinkering with the β-cell: metabolic regulatory mechanisms and new therapeutic strategies. American Diabetes Association Lilly Lecture, 2001, *Diabetes* 51:3141–50, 2002.

Bailey CJ et al: Prospects for insulin delivery by ex vivo somatic cell gene therapy, *J Mol Med* 77:244–9, 1999.

Levine F: Gene therapy for diabetes: strategies for β-cell modification and replacement, *Diabetes Metab Rev* 13:209–46, 1997.

Human protein reference data base, Johns Hopkins University and the Institute of Bioinformatics, at http://www.hprd.org/protein.

19.7. Suppression of Autoimmune Processes

Gallichman WS et al: Lentivirus-mediated transduction of islet grafts with interleukin 4 results in sustained gene expression and protection from insulitis, *Hum Gene Ther* 9:2717–26, 1998.

Moritani M et al: Abrogation of autoimmune diabetes in nonobese diabetic mice and protection against effector lymphocytes by transgenic TGF-beta1, *J Clin Invest* 102:499–506, 1998.

Goudy K et al: Adeno-associated virus vector-mediated IL-10 gene delivery prevents type 1 diabetes in NOD mice, *Proc Natl Acad Sci USA* 98:13913–8, 2001.

Tominaga Y et al: Administration of IL-4 prevents autoimmune diabetes, but enhances pancreatic insulitis in NOD mice, *Clin Immunol Immunopathol* 86:209–18, 1998.

Pennline KJ, Roque-Gaffney E, Monahan M: Recombinant human IL-10 prevents the onset of diabetes in the nonobese diabetic mouse, *Clin Immunol Immunopathol* 71:169–75, 1994.

Koh JJ et al: Degradable polymeric carrier for the delivery of IL-10 plasmid DNA to prevent autoimmune insulitis of NOD mice, *Gene Ther* 7:2099–104, 2000.

Tarner IH, Slavin AJ, McBride J, Levicnik A, Smith R et al: Treatment of autoimmune disease by adoptive cellular gene therapy, *Ann NY Acad Sci* 998:512–9, 2003.

19.8. Pancreatic β Cells

Kubota C, Yama Kuchi H, Todoroki J et al. Six cloned calves produced from adult fibroblast cells after long-term culture, *Proc Natl Acad Sci USA* 97:990–5, 2000.

Shapiro AM et al: Islet transplantation in seven patients with type 1 diabetes mellitus using a glucocorticoid-free immunosuppressive regimen, *New Engl J Med* 343:230–8, 2000.

Efrat S et al: Conditional transformation of a pancreatic beta-cell line derived from transgenic mice expressing a tetracycline-regulated oncogene, *Proc Natl Acad Sci USA* 92:3576–80, 1995.

Fleischer N et al: Functional analysis of a conditionally-transformed pancreatic beta-cell line, *Diabetes* 47:1419–25, 1998.

Milo-Landesman D et al: Correction of hyperglycemia in diabetic mice transplanted with reversibly-immortalized pancreatic beta cells controlled by the tet-on regulatory system, *Cell Transplant* 10:645–50, 2001.

19.9. Pancreatic Stem and Progenitor Cells

Bonner-Weir S et al: In vitro cultivation of human islets from expanded ductal tissue, *Proc Natl Acad Sci USA* 97:7999–8004, 2000.

Ramiya VK et al: Reversal of insulin-dependent diabetes using islets generated in vitro from pancreatic stem cells, *Nat Med* 6:278–82, 2000.

19.10. Embryonic Stem Cells

Thomson JA et al: Embryonic stem cell lines derived from human blastocysts, *Science* 282:1145–7, 1998.

Soria B et al: Insulin-secreting cells derived from embryonic stem cells normalize glycemia in streptozotocin-induced diabetic mice, *Diabetes* 49:157–62, 2000.

Assady S et al: Insulin production by human embryonic stem cells, *Diabetes* 50:1691–7, 2001.

Lumelsky N et al: Differentiation of embryonic stem cells to insulin-secreting structures similar to pancreatic islets, *Science* 292:1389–94, 2001.

Halban PA et al: Gene and cell-replacement therapy in the treatment of type 1 diabetes: How high must the standards be set? *Diabetes* 50:2181–91, 2001.

Edlund H: Pancreas: How to get there from the gut? *Curr Opin Cell Biol* 11:663–8, 1999.

Shih DQ, Stoffel M: Dissecting the transcriptional network of pancreatic islets during development and differentiation, *Proc Natl Acad Sci USA* 98:14189–91, 2001.

Tiedge M et al: Relation between antioxidant enzyme gene expression and antioxidative defense status of insulin-producing cells, *Diabetes* 46:1733–42, 1997.

19.11. Adult Stem Cells

Colter DC et al: Identification of a subpopulation of rapidly self-renewing and multipotential adult stem cells in colonies of human marrow stromal cells, *Proc Natl Acad Sci USA* 98:7841–5, 2001.

Woodbury D et al: Adult rat and human bone marrow stromal cells differentiate into neurons, *J Neurosci Res* 61:364–70, 2000.

Ferber S et al: Pancreatic and duodenal homeobox gene 1 induces expression of insulin genes in liver and ameliorates streptozotocin-induced hyperglycemia, *Nat Med* 6:568–72, 2000.

19.12. Prevention of β-Cell Injury and Death

Gallichman WS et al: Lentivirus-mediated transduction of islet grafts with interleukin 4 results in sustained gene expression and protection from insulitis, *Hum Gene Ther* 9:2717–26, 1998.

Moritani M et al: Abrogation of autoimmune diabetes in nonobese diabetic mice and protection against effector lymphocytes by transgenic TGF-beta1, *J Clin Invest* 102:499–506, 1998.

Goudy K et al: Adeno-associated virus vector-mediated IL-10 gene delivery prevents type 1 diabetes in NOD mice, *Proc Natl Acad Sci USA* 98:13913–8, 2001.

Kubisch HM et al: Transgenic copper/zinc superoxide dismutase modulates susceptibility to type I diabetes, *Proc Natl Acad Sci USA* 91:9956–9, 1994.

Hohmeier HE et al: Stable expression of manganese superoxide dismutase (MnSOD) in insulinoma cells prevents IL-1 beta-induced cytotoxicity and reduces nitric oxide production, *J Clin Invest* 101:1811–20, 1998.

Benhamou PY et al: Adenovirus-mediated catalase gene transfer reduces oxidant stress in human, porcine, and rat pancreatic islets, *Diabetologia* 41:1093–100, 1998.

Hotta M et al: Pancreatic betacell-specific expression of thioredoxin, an antioxidative and antiapoptotic protein, prevents autoimmune and streptozotocin-induced diabetes, *J Exp Med* 188:1445–51, 1998.

Rabinovitch A et al: Transfection of human pancreatic islets with an anti-apoptotic gene (bcl-2) protects beta-cells from cytokine-induced destruction, *Diabetes* 48:1223–9, 1999.

Dupraz P et al: Lentivirus-mediated Bcl-2 expression in beta TC-tet cells improves resistance to hypoxia and cytokine-induced apoptosis while preserving in vitro and in vivo control of insulin secretion, *Gene Ther* 6:1160–9, 1999.

Grey ST et al: A20 inhibits cytokine-induced apoptosis and nuclear factor kappa B-dependent gene activation in islets, *J Exp Med* 190:1135–45, 1999.

Efrat S et al: Adenovirus early region 3 (E3) immunomodulatory genes decrease the incidence of autoimmune diabetes in nonobese diabetic (NOD) mice, *Diabetes* 50:980–4, 2001.

19.13. Transplantation of β-Cell-Protecting Devices

Sullivan SJ, Maki T, Borland KM, Mahoney MD, Solomon BA et al: Biohybrid artificial pancreas: Long-term implantation studies in diabetic, pancreatectomized dogs, *Science* 252:718–21, 1991.

Sun Y et al: Normalization of diabetes in spontaneously diabetic cynomolgus monkeys by xenografts of microencapsulated porcine islets without immunosuppression, *J Clin Invest* 98:1417–22, 1996.

Duvivier-Kali VF et al: Complete protection of islets against allorejection and autoimmunity by a simple barium–alginate membrane, *Diabetes* 50:1698–705, 2001.

Shapiro AMJ et al: Islet transplantation in seven patients with type 1 diabetes mellitus using a glucocorticoid-free immunosuppressive regimen, *New Engl J Med* 343:230–8, 2000.

Paty BW et al: Intrahepatic islet transplantation in type 1 diabetic patients does not restore hypoglycemic hormonal counterregulation or symptom recognition after insulin independence, *Diabetes* 51:3428–34, 2002.

Ryan EA et al: Clinical outcomes and insulin secretion after islet transplantation with the Edmonton protocol, *Diabetes* 50:710–9, 2001.

Giannoukakis N, Pietropaolo M, Trucco M: Genes and engineered cells as drugs for type I and type II diabetes mellitus therapy and prevention, *Curr Opin Invest Drugs* 3(5):735–51, 2002.

Holland AM, Gonez LJ, Harrison LC: Progenitor cells in the adult pancreas, *Diabetes Metab Res Rev* 20(1):13–27, 2004.

20

URINARY REGENERATIVE ENGINEERING

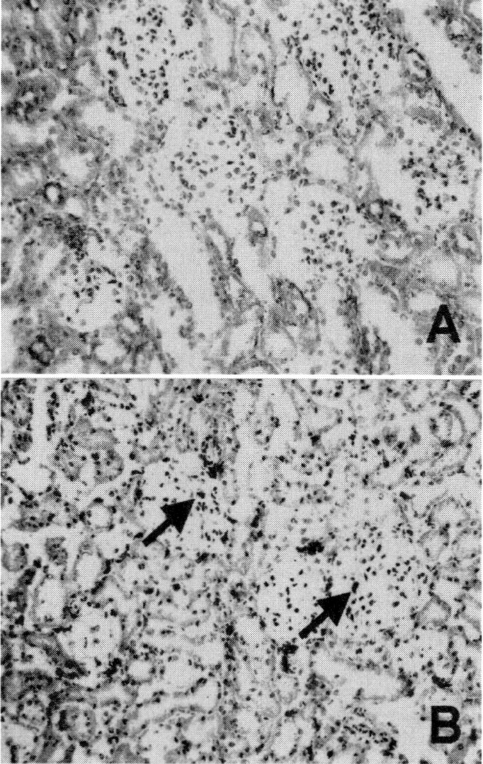

Formation of glomerular and tubular cells in the kidney of a Brown Norway/RijHsd (BN) rat from transplanted allogenic bone marrow cells derived from a WAG/RijHsd (WR) rat. The donor WR bone marrow cells wre injected into the BN rat intravenously and were detected in the recipient kidney by immunohistochemistry with an anti-WR-major histocompatibility complex class-I (MHC-I) antibody (U9F4), which reacts only with the WR MHC-I. (A) Immunoperoxidase staining of cryo-kidney sections from a control BN rat (no WR bone marrow transplantation), showing negative U9F4 staining. (B) Presence of donor WR bone marrow-derived U9F4-positive cells in the glomeruli (arrows) as well as tubulo-interstitium of the BN rat kidney 2 months after bone marrow cell transplantation (×200). Reprinted with permission from Rookmaaker MB et al., Bone-marrow-derived cells contribute to glomerular endothelial repair in experimental glomerulonephritis, Am J Pathol 163:533–562, copyright 2003. See color insert.

Bioregenerative Engineering: Principles and Applications, by Shu Q. Liu

ANATOMY AND PHYSIOLOGY OF THE URINARY SYSTEM

The urinary system is composed of the left and right kidneys, left and right ureters, urinary bladder, and urethra. The urinary system is responsible for the filtration of blood, formation of urine, excretion of waste and toxic substances via urination, control of blood volume and ionic composition, and regulation of the pH of the blood and body fluids. The structure and function of major urinary organs are discussed as follows.

Structure and Function of the Kidneys [20.1]

The kidneys are two oval organs with dimensions about 1, 2, and 4 inches in thickness, width, and length, respectively. They are located in the middle of the abdominal cavity, against the posterior wall, behind the peritoneum, and below the liver. The right kidney is slightly lower than the left kidney. Each kidney contains a concaved notch known as the *hilum*, where the renal artery, vein, ureter, and nerves are connected to the kidney. An adrenal gland is found on the top of each kidney. When a kidney is cut vertically in a plane parallel to the front abdominal surface, two distinct structures can be found: the external cortex and internal medulla. The cortex hosts a large number of renal corpuscles (see below). The medulla is composed of 10 to 15 pyramid structures, with their bases originated from the cortex–medulla border and their apices (called papillae) projected to the renal calices (funnel-shaped chambers), which converge to the renal pelvis and then to the ureter.

Each kidney is composed of about 1 million nephrons, which are the functional units of the kidney. A *nephron* is composed of the renal corpuscle, proximal convoluted tubule, Henle's loop, distal convoluted tubule, connecting tubule, and collecting tubule (Fig. 20.1). The *renal corpuscle* consists of the Bowman's capsule and glomerulus and is located in the cortex. *Bowman's capsule* is a double-membrane sac that encloses the glomerulus. The internal membrane of Bowman's capsule, known as the *visceral layer*, is constituted with podocytes and is attached to the glomerulus. The *podocytes* form processes on the external surface of the glomerulus. There are open slits between these processes, allowing

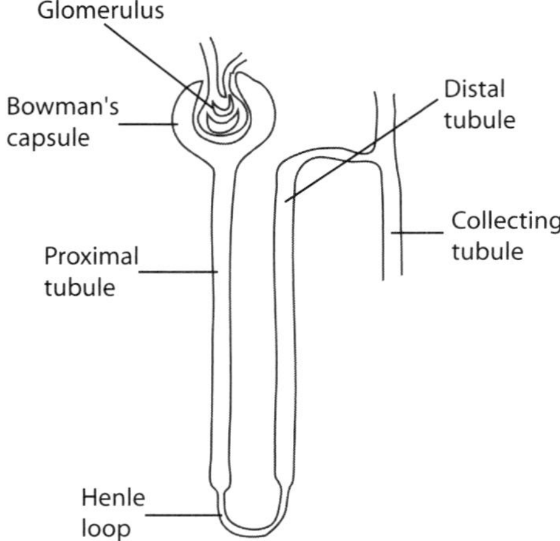

Figure 20.1. Schematic representation of a nephron unit.

fluid filtration from the glomerulus capillaries to Bowman's capsule. The external membrane of the Bowman's capsule, also called the *parietal layer*, is composed of epithelial cells, which are extended to the proximal tubule. The *glomerulus* is a capillary network with an afferent and an efferent arteriole. The afferent arteriole supplies blood to and the efferent arteriole drains blood from the glomerulus. The primary function of glomerulus is fluid filtration from the capillaries to Bowman's capsule. There exist fenestrae in the capillary endothelial cells, allowing efficient fluid filtration.

Fluid generated in the renal corpuscle is processed through the renal tubular system, eventually forming urine. All renal tubules are composed of epithelial cells, which reside on a basement membrane. The first segment of the tubular system is the *proximal convoluted tubule*, which is located in the renal cortex. The proximal tubule is extended to the *Henle's loop*, which is located in the renal medulla and consists of a descending and an ascending tubule. The ascending tubule is extended to the distal tubule, which is located in the renal cortex. The *distal tubule* is connected to the connecting tubule and then to cortical collecting tubules. A number of *cortical collecting tubules* join together to form a medullary collecting tubule, which merges into a larger collecting duct. The *collecting duct* is connected to the renal pelvis. The proximal, Henle loop, distal, and connecting tubules are associated with a rich network of capillaries, which are originated from the glomerular efferent arterioles.

The formation of urine is accomplished via several processes: glomerular filtration, renal tubular reabsorption, and secretion from peritubular capillaries. Fluid and substances from the blood, except for cells and proteins, can be freely filtered through the capillaries of the glomerulus. Major barriers for filtration are the fenestrated endothelial cells, the basement membrane, and the podocytes. Under physiological conditions, about 20% of bloodflow is filtered through the glomerular capillaries. Several factors determine the rate of glomerular filtration rate. These include the hydrostatic and colloid pressure within the glomerular capillaries, and the total surface area and conductivity of the glomerular capillaries. An increase in the hydrostatic pressure in the capillaries enhances filtration, whereas an increase in the hydrostatic pressure in Bowman's capsule reduces filtration. An increase in colloid pressure (depending on the concentration of proteins) in the capillaries reduces filtration. An increase in the total area and conductivity of capillaries enhances filtration.

Filtered fluid and substances, collectively called *filtrate*, in Bowman's capsule are not completely excreted from the kidney. Instead, they are reabsorbed and processed via coordinated work of the renal tubules and peritubular capillaries. Different substances are processed in different ways. For instance, certain types of substances, such as toxic chemicals and drugs, are filtered through the glomerular capillaries and completely excreted without reabsorption. Some substances, such as electrolytes (Na^+, K^+, Ca^{2+}, Cl^-, and HCO^-), can be partially reabsorbed in the renal tubules, resulting in a decrease in the concentration of the substances in the excreted urine. Water can be partially reabsorbed. There are also substances that are completely reabsorbed after filtration. These include amino acids and glucose. Through various processing mechanisms, a large fraction of water and useful substances are reabsorbed, whereas waste and toxic substances are excreted via the formation of urine.

The reabsorption of water and electrolytes is regulated by hormones including the antidiuretic hormone (ADH) and proteins involved in the renin-angiotensin-aldosterone system. Antidiuretic hormone is produced in the posterior pituitary. The production of antidiuretic hormone is controlled by the osmolarity-sensitive cells located in the supraoptic nuclei of the hypothalamus. These cells can sense changes in blood osmolarity. An increase in blood osmolarity induces the activation of the osmolarity-sensitive cells, which stimulate the ADH-producing cells to release ADH. A reduction in arterial blood pressure also stimulates the production and release of ADH. This hormone acts on the renal tubules, increasing water reabsorption. Such a process results in an increase in blood volume and a decrease in blood osmolarity.

Another mechanism that regulates water reabsorption involves the rennin–angiotensin–aldosterone system. As discussed in Chapter 15, cells in the renal juxtaglomerular apparatus produce and secret a proteolytic enzyme known as renin. A decrease in blood pressure and flow enhances renin secretion. Renin is an enzyme that converts angiotensinogen to angiotensin I. Another proteolytic enzyme, known as angiotensin converting enzyme (ACE), can covert angiotensin I to angiotensin II. Angiotensin II directly induces vascular smooth muscle contraction and stimulates the adrenal cortex to secret a steroid hormone called *aldosterone*. This hormone can act on the epithelial cells of the renal distal and collecting tubules, increasing Na^+ transport and water reabsorption. Such a process results in an increase in blood volume.

Structure and Function of the Urinary Tract [20.1]

The *urinary tract* includes three parts: a pair of ureters (left and right), the urinary bladder, and the urethra. The left and right *ureters* are tubular structures originated from the left and right renal pelvis, respectively, and extended to the urinary bladder. The function of the ureters is to conduct urine from the kidneys to the urinary bladder. The *urinary bladder* is a muscular sac located in the lower abdominal cavity and anterior to the rectum. The function of the urinary bladder is for the storage of urine. The *urethra* is a muscular tube that removes urine from the urinary bladder.

The structure and cellular organization are similar for the ureters, urinary bladder, and urethra. There are four layers in the wall of these structures, including the epithelium, the lamina propria, muscular layer, and adventitia. The epithelium is composed of multilayered epithelial cells. The lamina propria is a connective tissue layer, consisting of fibroblasts and extracellular matrix. The muscular layer is thicker than other layers and is composed of smooth muscle cells. The adventitia is another connective tissue layer composed of similar components as the lamina propria.

DISORDERS OF THE URINARY SYSTEM

Acute Renal Failure

Pathogenesis, Pathology, and Clinical Features [20.2]. *Acute renal failure* is a disorder characterized by rapid deterioration of the renal function, resulting in the accumulation of blood urea and metabolic wastes. These metabolic wastes, when reaching a critical level, induce cell injury and death, resulting in the failure of the brain, heart, liver, and other organs. Pathogenic factors that induce acute renal failure include renal ischemia due to trauma and atherosclerosis, acute glomerulonephritis, toxicity of chemicals and substances, such as organic solvents, heavy metals, glycols, and radiographic contrast agents, and an increase in the blood concentration of myoglobin due to massive crush trauma, muscular ischemia, and massive infection. Renal ischemia causes glomerular and tubular cell injury and death. Acute glomerulonephritis is a disorder characterized by rapid injury of glomerular capillaries and renal tubules, which results in a reduction in glomerular filtration, water and salt retention, proteinuria, and hematuria. Renal failure occurs in severe cases. Toxins impair the function of tubular epithelial cells. A sudden increase in myoglobin induces the obstruction of the renal tubules. All these changes cause a rapid reduction in the glomerular filtration and tubular processing capability, resulting in acute renal failure. When the causative factors are removed, acute renal failure is generally reversible, if there is no massive renal necrosis.

Acute renal failure is associated with a number of pathophysiological changes in the kidneys. Ischemia and toxins often result in three abnormalities: (1) reduction in the renal arterial perfusion pressure, leading to a decrease in the glomerular hydrostatic pressure and filtration; (2) reduction in the glomerular permeability and effective area, leading to a reduction in glomerular filtration; and (3) obstruction of renal tubules and reduction in tubular transport function, leading to a decrease in the capability of excreting toxic wastes, chemicals, and acidic substances. All these changes contribute to the accumulation of toxins and wastes in the blood. The accumulation of acidic substances results in a disorder known as *metabolic acidosis*. The accumulation of urea is referred to as uremia. In these cases, the toxic and acidic substances impair the function of the cells in almost all systems. The nerve cells and cardiomyocytes are especially sensitive to these substances. When the toxic and acidic substances are accumulated to a critical level, cells are injured and committed to death. Massive cell death ultimately results in the failure of tissues and organs.

The kidneys with acute renal failure exhibit a number of pathological abnormalities. Examinations by optical and electron microscopy often demonstrate disruption and necrosis of tubular epithelial cells, tubular obstruction, tubular dilation or collapse, interstitial edema, and leukocyte infiltration. In severe renal ischemia, large areas of renal infarction and tissue necrosis can be found. When recovered from acute renal failure, the kidneys usually regain normal structure. Acute renal failure may be associated with complications in the nervous, cardiovascular, and gastrointestinal systems. These complications are mostly due to cell injury or death resulting from uremia and acidosis. Nerve cell injury induces clinical symptoms including somnolence, disorientation, agitation, asterixis, and seizures. Cardiovascular complications induce arrhythmias, hypertension, circulatory congestion, and heart failure. Arrhythmias are due to electrolyte abnormalities and injury of the cardiac conductive system. Hypertension and circulatory congestion are induced by sodium and water retention resulting from renal dysfunction. Heart failure is a result of cardiac cell injury and death. In the gastrointestinal system, hemorrhage often occurs in association with symptoms such as nausea and vomiting.

Experimental Models of Acute Renal Failure [20.2]. *Renal ischemia* is a major cause of acute renal failure and is commonly used as a model of experimental renal failure. Mice and rats are often used for this model. To create renal ischemia, an animal can be anesthetized by peritoneal injection of sodium pentobarbital at a dose of 50 mg/kg body weight. The abdominal skin is sterilized, the abdominal cavity is opened, and the left or the right kidney is identified. The renal artery of the identified kidney can be isolated and tied off completely with a surgical suture for a period from 10 to 30 min. The severity of acute renal failure is dependent on the duration of ischemia. It is often desired to test a number of times (e.g., 10, 20, and 30 min) to determine an appropriate level of acute renal failure. The other kidney can be used as a control. The abdominal cavity is then closed and the animal is allowed to recover.

Acute renal failure can also be induced by administration of chemical toxins. A typical toxin is folic acid, which induces injury of the renal tubular epithelial cells and thus acute renal failure. Folic acid (250 mg/kg body weight in 150 mM sodium bicarbonate) can be introduced to the animal via a single intravenous injection. Animals with vehicle injection (150 mM sodium bicarbonate) can serve as controls. Assays and functional tests can be carried out at desired times following the injection.

Conventional Treatment [20.2]. Acute renal failure can be treated with several strategies, including (1) identification and treatment of causative diseases, such as renal ischemia, acute glomerulonephritis, and chemical toxicity; (2) establishment of urine output; (3) control of the intake of proteins, water, and electrolytes; (4) control of the intake of any drugs and chemicals that are excreted from the kidneys; and (5) treatment of nervous and cardiovascular disorders, if any. When the causative factors are removed, acute renal failure can self-recover in most cases.

In severe cases when urea and acidic substances reach a critical level, renal dialysis ought to be conducted to remove the toxic substances. Renal dialysis is a process by which toxic substances are removed via diffusion from the blood to a dialysate across a semipermeable polymeric membrane based on the gradient of substance concentration. There are two types of dialysis: hemodialysis and peritoneal dialysis. Hemodialysis is a process by perfusing blood through a semipermeable polymeric membrane system. The semipermeable membrane is usually fabricated with cellophane, cellulose acetate, or polyacrylnitrile. A typical dialyzer is composed of two systems: blood and dialysate perfusion systems. The blood is separated from the dialysate by a dialysis membrane. There are two forms of dialyzer: flat plate and hollow fiber forms. In the plate form, semipermeable membranes are arranged to form narrow slits. Blood and dialysate are introduced to separated slit systems. In the hollow fiber form, blood is perfused through the lumen of the semipermeable hollow fibers, while dialysate flow is introduced to the outside the hollow fibers. The flow rate of blood and dialysate can be controlled based on the conditions of the patient. It is usually necessary to conduct 10–15-hr dialysis for patients with acute and chronic renal failure to bring down the toxic substances to an acceptable level. In hemodialysis, there are several common clinical problems, including blood coagulation, pulmonary embolism, and intimal hyperplasia in blood vessels for dialysis. These problems remain to be resolved.

Peritoneal dialysis is a process by which urea and toxic substances are removed by diffusion from the capillaries to the dialysate introduced into the peritoneal cavity. The peritoneal membrane on the surface of the abdominal organs are ideal semipermeable membranes and suitable for dialysis. Dialysate can be introduced into and removed from the peritoneal cavity via a catheter. A period of 20–70 hr dialysis is usually sufficient for most patients with acute and chronic renal failure. Compared to hemodialysis, peritoneal dialysis does not cause vascular complications and is relatively safe. However, peritoneal dialysis is not as efficient as hemodialysis.

When massive renal infarction occurs and cannot be rescued, it is necessary to carry out renal transplantation. Allogenic kidneys can be harvested and transplanted into the host patient by renal arterial and venous anastomosis. Since allogenic tissues cause acute immune rejection, it is necessary to administer immunity-suppressing agents.

Molecular Regenerative Engineering. Acute renal failure is a devastating disease with a rate of mortality rate of ~20%. The primary cause of renal failure is the injury and death of glomerular endothelial cells and tubular epithelial cells. In spite of tremendous efforts, few conventional approaches are available for the treatment of the disease. Given the fact that mitogenic molecules protect cells from injury and death, it is conceivable that the genes encoding these proteins can be used as therapeutic agents for the treatment of acute renal failure. There are several types of genes that can be potentially used for such a purpose, including growth factor, mitogenic signaling molecule, and cell death inhibitor genes. Representative genes for each group are discussed as follows.

Growth Factor Genes [20.3]. Growth factors not only promote cell proliferation and differentiation, but also protect cells from injury and death. Thus, growth factors are potential therapeutic agents for the treatment of acute renal failure. Hepatocyte growth factor (HGF) is one of such agents. This growth factor is upregulated in the kidneys in response to renal injury, exerts renotropic, tubulogenic, antiapoptotic, and antifibrotic activities in glomerular endothelial cells and renal tubular epithelial cells (Fig. 20.2). The gene encoding hepatocyte growth factor can be transferred into the kidney to enhance the expression of HGF. In experimental models of renal injury induced by administration of nephrotoxic agents, such as cyclosporin A, the transfer of the HGF gene into the impaired kidney can induce significant upregulation of HGF expression. Upregulated HGF activates protein kinase B/Akt kinase and stimulates the expression of antiapoptotic factors such

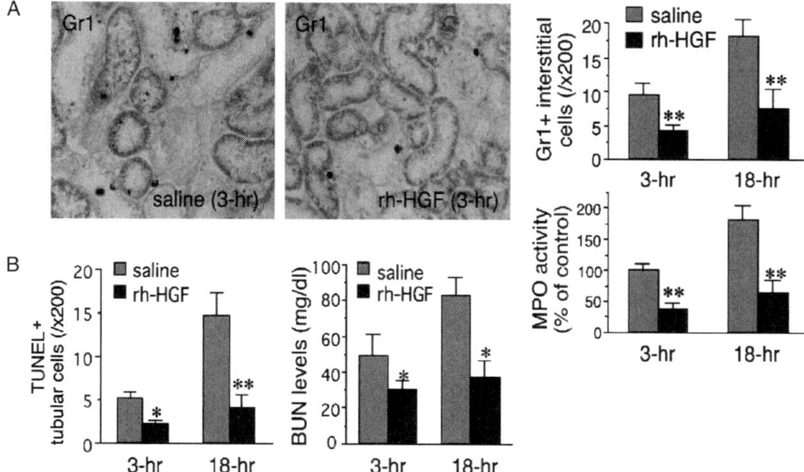

Figure 20.2. Preventive effects of recombinant human hepatocyte growth factor (rh-HGF) on neutrophil infiltration, tubular apoptosis, and renal dysfunction in mice that underwent renal ischemia and reperfusion (I/R). (A) Changes in neutrophil accumulations in interstitial spaces of ischemic kidneys, treated with or without rh-HGF therapy (Grl staining, ×300). Renal tissues were collected at 3 and 18 hour after I/R challenge and then subjected to immunohistochemical procedures. Furthermore, granulocyte-specific myeloperoxidase (MPO) activities were measured in a part of samples used in immunohistochemistry. In this assay, the saline control level (3 h post-ischemia) was defined as 100%. (B) Decrease in tubular apoptosis levels and blood urea nitrogen (BUN) levels in the mouse model of renal I/R injury by rh-HGF treatment. Data are shown as means ± SD ($n = 6$). *$P < 0.05$; **$P < 0.01$ compared with the saline-injected control group. (Reprinted with permission from Mizuno S, Nakamura T: *Am J Pathol* 166:1895–1905, copyright 2005.)

as Bcl2. The expression of these factors is associated with a reduction in tubular epithelial cell injury and death, a decrease in interstitial leukocyte infiltration, and amelioration of renal impairment. Furthermore, the overexpression of HGF induces a decrease in the production of collagen, leading to a reduction in renal fibrosis. All these effects potentially improve the function of the kidneys.

Other growth factor genes such as the epidermal growth factor (EGF), fibroblast growth factor (FGF), platelet-derived growth factor (PDGF) genes can also be used for the treatment of acute renal failure (see Chapter 15 for the characteristics of these growth factors). It should be noted that growth factors can be directly used to treat acute renal failure. However, most proteins are rapidly degraded by the liver, and have a short half-life in the blood. It is a challenging task to maintain a sustained level of exogenously delivered protein. Thus, gene transfer is a more effective approach for the delivery of growth factors.

Genes Encoding Mitogenic Signaling Proteins [20.4]. Mitogenic signaling molecules represent a group of intracellular molecules that mediate the transduction of growth-promoting signals and the induction of cell proliferation and migration. Several signaling molecule genes have been tested in experimental models for the treatment of acute renal failure. Typical examples include protein kinase B, protein kinase C, mitogen-activated protein kinase (MAPK) genes. These genes encode proteins that mediate the transduction of growth factor-initiated signals and, thus, can be potentially used for the treatment of acute renal failure.

Genes Encoding Cell Death Inhibitors [20.4]. As described above, growth factor genes, such as platelet-derived growth factor, hepatocyte growth factor, and epidermal growth factor genes, can be used for the prevention of renal cell injury and death. In addition, there exist anti-apoptotic factors in the cells. A typical example is Bcl2, which inhibits the activity of apoptotic signaling molecules. Thus, the gene encoding Bcl2 can be used for the molecular treatment of acute renal failure.

Cell and Tissue Regenerative Engineering. Renal cell and tissue regenerative engineering is to restore the structure and function of disordered kidneys by inducing renal cell and tissue regeneration. There are several strategies for such a purpose, including: (1) kidney regeneration based on stem cells, (2) kidney regeneration based on embryonic renal tissues, and (3) construction of artificial kidneys based on adult renal tubular epithelial cells. The first two strategies have been tested in experimental models, and the third strategy has been tested in both experimental studies and clinical trials. These investigations have provided promising results for kidney regeneration and reconstruction.

Stem Cell-Based Kidney Regeneration [20.5]. As discussed on page 381, embryonic and fetal stem cells are multipotent cells that are capable of differentiating into specialized cell types. Under appropriate conditions, stem cells can differentiate into renal cells. A typical example is the regeneration of kidney-like structures by using stem cells from the embryonic metanephric mesenchyme and the ureteric bud, which are intermediate mesodermal tissues for kidney formation (see page 359). During the embryonic and fetal stage, the kidney is developed via reciprocal interactions between the metanephric mesenchyme and the ureteric bud. The metanephric mesenchyme forms within about 1 month in the intermediate mesoderm. Its formation stimulates the generation of branch-like ureteric buds from an epithelial structure known as the nephric duct. The ureteric buds in turn induce mesenchyme cells to form epithelial nodules, which proliferate and differentiate into various renal cell types, including capsule cells, podocytes, and renal tubular cells, necessary for the constitution of the renal nephrons. At the same time, the ureteric buds give rise to the renal collecting ducts and the ureter. The newly generated nephrons and collecting ducts fuse with each other, forming the kidney.

In light of the fact that the metanephric mesenchyme and the ureteric bud are required for the development of the kidney, it is conceivable that stem cells from both structures are necessary for the regeneration of the kidney. In cell culture models, when stem cells from the metanephric mesenchyme and the ureteric bud are seeded in an extracellular matrix gel with mature renal epithelial cells, the stem cells from both embryonic tissues can proliferate and migrate. Interestingly, the stem cells can form branched and polarized tubules with internal lumens. A number of growth factors, including pleiotrophin, hepatocyte growth factor (HGF), fibroblast growth factor (FGF), and insulin-like growth factor (IGF)1, promote tubule formation and branching. In contrast, transforming growth factor (TGF)β inhibits branching morphogenesis. The presence of extracellular matrix is also required for the generation of the renal structures. These investigations suggest that the formation of the kidney is a complex process, requiring coordinated regulatory processes that involve multiple cells and mediators. Further investigations are needed to identify additional regulatory factors and clarify the regulatory mechanisms.

There are other types of stem cell that can be potentially used for kidney regeneration. The embryonic stem cells of the blastocyst are pluripotent stem cells that can differentiate into all types of specialized cells. Under appropriate conditions, embryonic stem cells can be induced to form renal glomerular endothelial and tubular epithelial cells. In addition, the adult bone marrow stem cells have the capability of differentiate in specialized cells. Experimental studies have shown that bone marrow stem cells can engraft to the kidney and differentiate into renal glomerular and tubular cells (Chapter 20 opening figure). These investigations have provided promising results, demonstrating the possibility of generating a functional kidney or a kidney-like structure.

Embryonic Tissue-Based Kidney Regeneration [20.5]. A kidney-like structure can be generated from the embryonic metanephros, a progenitor organ for the kidney. In experimental models, the embryonic metanephros can be collected and cultured in vitro for the generation of the kidney. Alternatively, the embryonic metanephric mesenchyme and the ureteric bud can be isolated from the metanephros and cultured together. The later model allows to understand the contributions of individual parts of the kidney progenitors. When the ureteric bud is placed in the proximity of the metanephric mesenchyme, the ureteric bud can form polarized branches with internal lumens and extend into the metanephric mesenchyme. At the same time, the metanephric mesenchyme can transform to epithelial structures and generate tubular nephrons. Interestingly, the mesenchyme tubular nephrons can form connections with the collecting tubules derived from the ureteric bud.

As for regeneration of the kidney based on stem cells, the formation of the renal structures from the embryonic metanephros are dependent on various growth factors and extracellular matrix. For example, pleiotrophin (see Table 20.1), fibroblast growth factor-1, and glial cell line-derived neurotrophic factor can induce branch formation of the ureteric bud in vitro. Fibroblast growth factor 2 enhances the survival of mesenchyme. In contrast, transforming growth factor β has been shown to inhibit the branch formation of the ureteric bud. Endostatin, a cleavage product of collagen XVIII present in the basement

TABLE 20.1. Characteristics of Selected Growth Regulatory Factors*

Proteins	Alternative Names	Amino Acids	Molecular Weight (kDa)	Expression	Functions
Pleiotrophin	PTN, heparin-binding neurite outgrowth promoting factor, neurite growth-promoting factor 1 (NEGF1), heparin-binding growth factor 8 (HBGF8), heparin-binding growth-associated molecule (HB-GAM), osteoblast-specific factor 1 (OSF1)	168	19	Brain, pancreas, kidney	Stimulating cell proliferation and migration, inducing angiogenesis, and enhancing tumor growth
Glial cell-derived neurotrophic factor	Astrocyte-derived trophic factor 1, glial cell-line-derived neurotrophic factor	211	24	Nervous system, kidney, testis, lung	Existing as a homodimer, promoting survival and differentiation of dopaminergic neurons, protecting neurons from apoptosis, stimulating the proliferation of kidney cells, and regulating kidney development
Collagen XVIII α chain	COL18A1, type XVIII collagen	1516	154	Heart, brain, liver, kidney, overy, skeletal muscle, intestine, prostate gland	Generating endostatin, which serves as a growth inhibitor and an antiangiogenic factor

*Based on bibliography 20.5.

membrane of the ureteric bud, also inhibits branch formation of the ureteric bud. These investigations show that kidney generation is a process regulated by a variety of signaling factors and extracellular matrix components. With an appropriate selection of the regulatory factors, it is possible to generate a kidney-like structure or "neokidney" based on embryonic kidney progenitor tissues. The kidney progenitor tissues can also be directly transplanted into an adult animal to generate a functional kidney-like structure.

Nuclear Transfer-Based Kidney Regeneration [20.6]. Nuclear transfer is a genetic technique that can be used to transfer adult or somatic cell nuclei into unfertilized oocytes and generate desired clones of tissues, organs, or animals with genetic characteristics of the donor somatic cells. Scientists have used such a technique to generate renal tissues for therapeutic purposes (Fig. 20.3). In experimental models, animal fetuses can be cloned by transferring adult cell nuclei into selected oocytes. At a selected stage, renal progenitor cells can be collected from the cloned fetus. The collected progenitor cells can be seeded in an extracellular matrix structure and implanted into a tissue of the cell nucleus donor. The implanted renal progenitor cells can differentiate into renal epithelial cells, which form tubule-like structures. Because half of the genome is from the donor, immune rejection response is reduced. These investigations have demonstrated the potential of the "cloning" technology for the regeneration of the kidney.

Adult Tubular Cell-Based Kidney Regeneration [20.7]. A cell-based artificial kidney can be constructed with two essential components: a dialyzer and a bioreactor with renal tubular epithelial cells (known as a *renal tubule assist device*). The dialyzer is similar to a conventional hemodialyzer, composed of semipermeable hollow fibers surrounded by an exterior chamber. The hollow fibers are for blood perfusion, and the exterior chamber is for dialysis filtrates. The bioreactor is also composed of semipermeable hollow fibers and an exterior chamber. The hollow fibers are for seeding and culturing renal tubular epithelial cells, and the exterior chamber is for blood perfusion. The dialyzer and the bioreactor are connected into a series. The dialyzer is connected to a selected vein from an animal model or a patient. Blood from the dialyzer is directed into the exterior chamber of the bioreactor, and the filtrate from the dialyzer is directed into the hollow fibers with renal tubular cells. Toxic substances and metabolic wastes can be dialyzed from the blood to the filtrate, whereas sodium, glucose, amino acids, and water are reabsorbed through the tubular epithelial cells in the bioreactor. Blood from the bioreactor is returned to host circulation and the waste-containing filtrate is discarded. Compared to conventional dialysis, the cell-based bioreactor not only serves as a blood filter but also provides necessary renal functions, including the reabsorption of nutrients and electrolytes, production of hormones, and synthesis of 1,25-dihydroxyvitamin D_3 and glutathione. This type of artificial kidney has been used successfully in dog and pig models. In limited clinical trials, the renal tubule assist device can be functional for up to 24h. These trials have provided fundamental information for the application of cell-based renal assist devices to renal failure.

Chronic Renal Failure

Pathogenesis, Pathology, and Clinical Features [20.8]. *Chronic renal failure* is a disorder characterized by a progressive reduction in the density of nephrons and deterioration

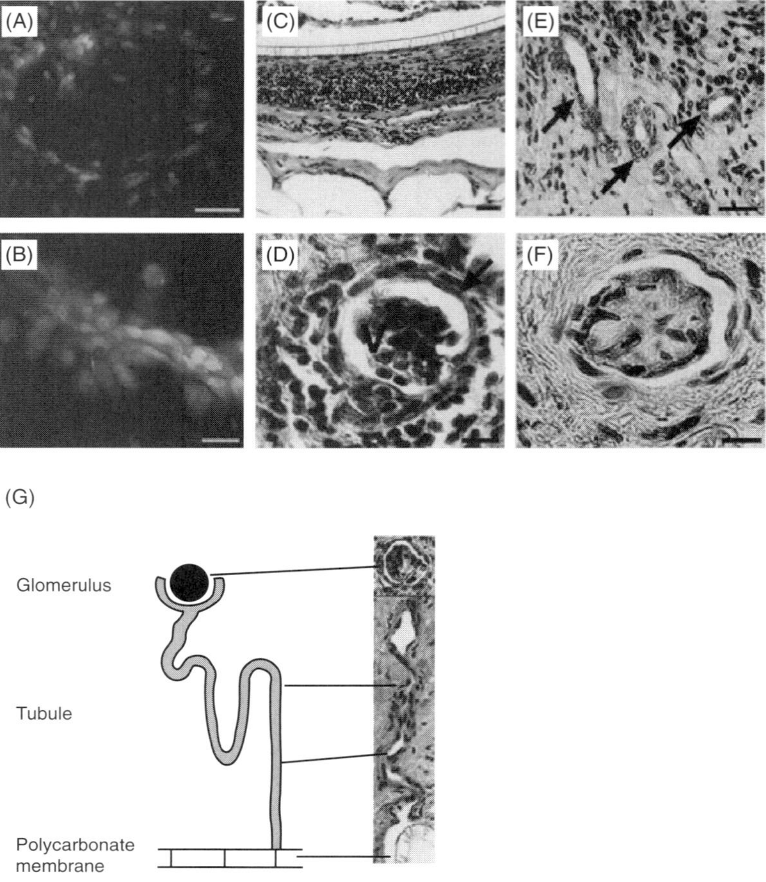

Figure 20.3. Characterization of renal explants. (A,B) Cloned cells stained positively with synaptopodin antibody (A) and AQP1 antibody (B). (C) The allogeneic controls displayed a foreign-body reaction with necrosis. (D) Cloned explant shows organized glomeruli-like structures. Vascular tufts (v); visceral epithelium (arrow). H&E stain. (E) Organized tubules (arrows) were shown in the retrieved cloned explant. (F) Immunohistochemical analysis using factor VIII antibodies (black) identifies vascular structures. (G) There was a clear unidirectional continuity between the mature glomeruli, their tubules, and the polycarbonate membrane. Scale bars: 100 μm (B,D–F); 200 μm (A); 800 μm (C). (Reprinted by permission from Macmillan Publishers Ltd.: Robert P et al: Generation of histocompatible tissues using nuclear transplantation, *Nature Biotechnol* 20: 689–96, copyright 2002.)

of renal function, eventually leading to the loss of renal function and the accumulation of toxic and acidic substances in the blood. Chronic renal failure is associated with impaired metabolism and deteriorating functions of virtually all organ systems, resulting from cell injury by uremia or the retention of blood urea. Common diseases that induce chronic renal failure include glomerulonephritis, glomerulosclerosis, tubulointerstitial fibrosis, diabetes mellitus, and hypertension.

Glomerulonephritis is a glomerular and tubular disorder characterized by the injury of glomerular capillary endothelial cells and tubular epithelial cells, impaired glomerular

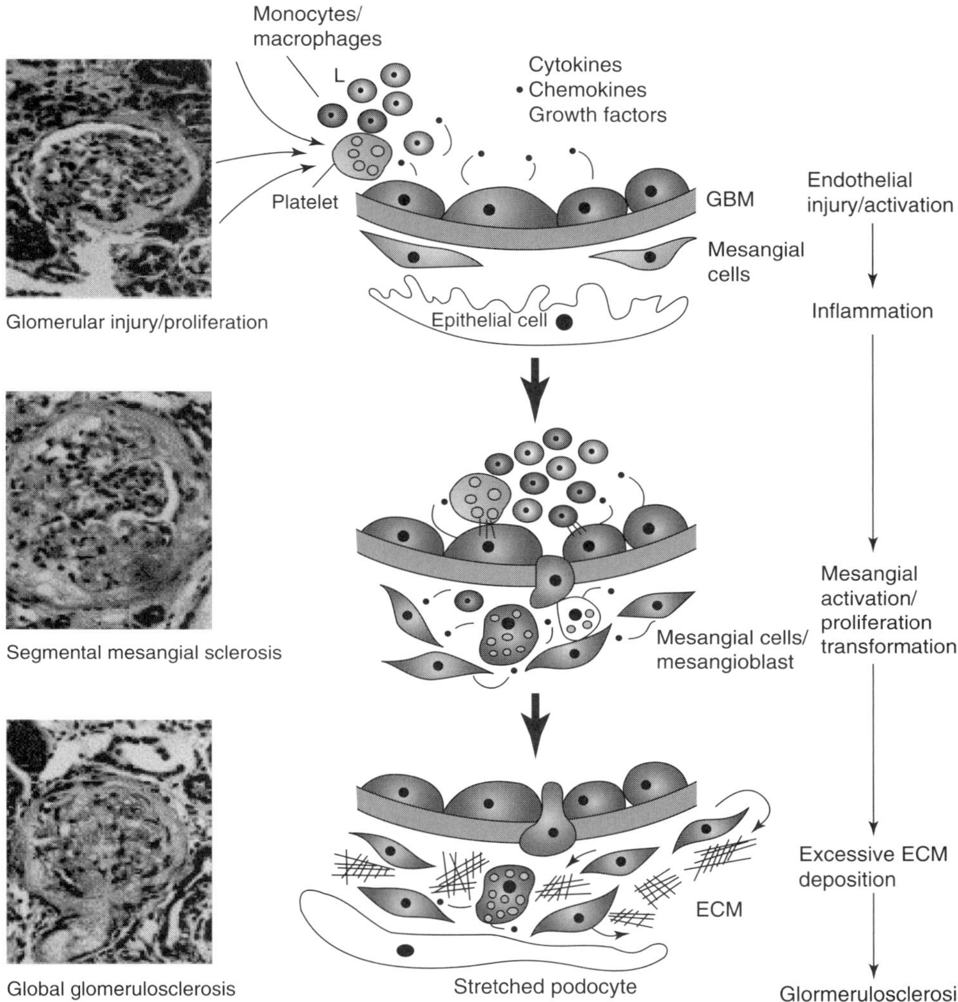

Figure 20.4. Schematic representation of the stages of glomerulosclerosis. GBM = glomerular basement membrane; ECM = extracellular matrix. (Reprinted from Meguid El, Nahas A, Bello AK: Chronic kidney disease: The global challenge, *Lancet* 365:331–40, copyright 2005, with permission from Elsevier.)

filtration, and malfunctioned tubular processing of water, electrolytes, and waste substances. Glomerulosclerosis is a glomerular disorder characterized by glomerular endothelial injury, glomerular inflammatory reactions, mesangial cell proliferation, extracellular matrix deposition, and impairment of glomerular filtration (Fig. 20.4). Tubulointerstitial fibrosis is primarily a tubular disorder, characterized by tubular epithelial injury, epithelial cell apoptosis, inflammatory reactions, fibroblast proliferation, extracellular matrix deposition and fibrosis, tubular atrophy, and impairment of the tubular functions (Fig. 20.5). Diabetes mellitus, as discussed in Chapter 19, induces arteriosclerotic changes in the renal arteries, which influences glomerular filtration. Hypertension causes renal

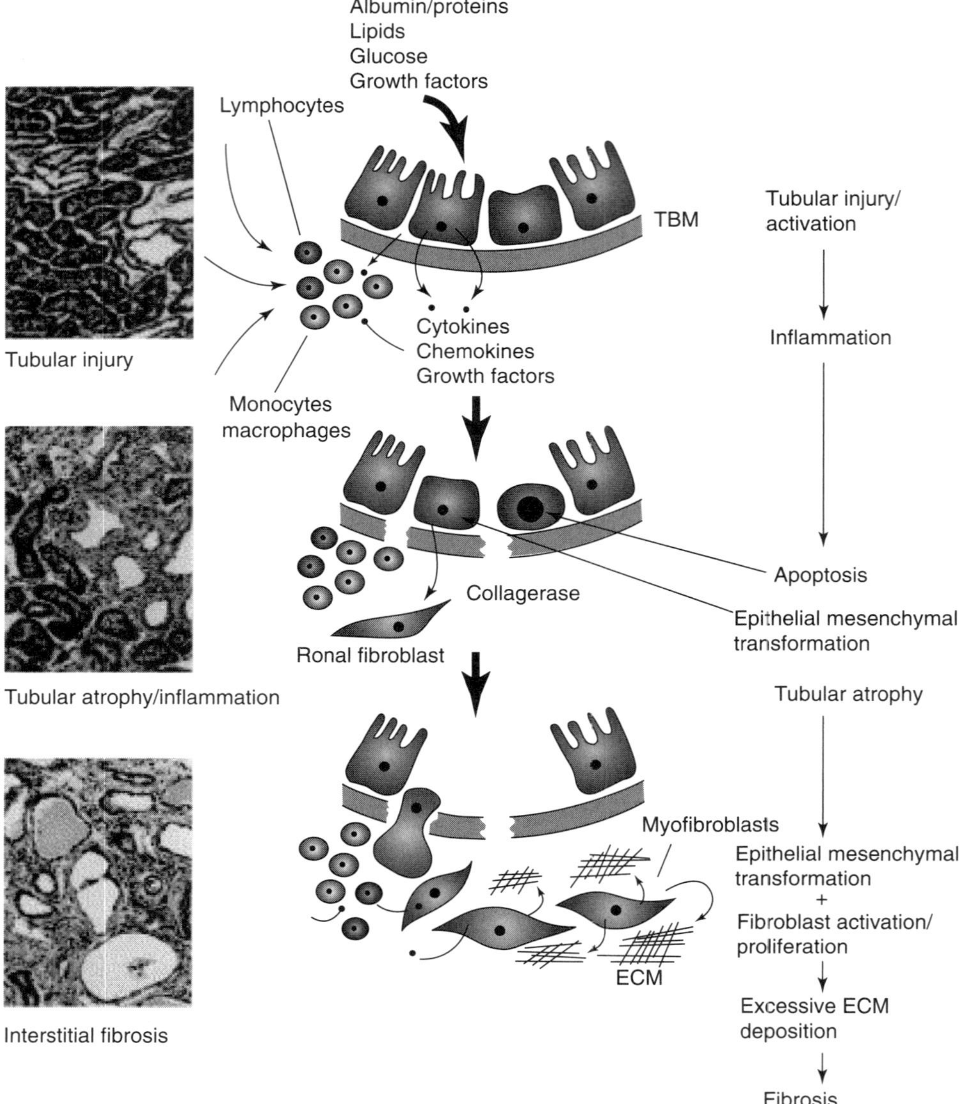

Figure 20.5. Diagrammatic representation of the stages of tubulointerstitial fibrosis. TBM = tubular basement membrane; ECM = extracellular matrix. (Reprinted from Meguid El, Nahas A, Bello AK: Chronic kidney disease: The global challenge, *Lancet* 365:331–40, copyright 2005, with permission from Elsevier.)

arterial hypertrophy and functional alterations, resulting in a reduction in glomerular filtration. All these disorders can induce chronic renal failure. Once chronic renal failure occurs, it is usually irreversible.

There are several pathophysiological changes in chronic renal failure. These changes include the imbalance of electrolytes, cellular dysfunction in almost all tissues and organs, and metabolic disorders. The pathophysiological changes are mostly due to accumulated urea or uremia. Urea is a waste product of protein metabolism and is removed from the kidneys. Chronic renal failure is always associated with uremia. Increased urea in the blood suppresses the activity of ATPases and the transport of electrolytes across the cell membrane, resulting in the imbalance of Na^+ and K^+. Such a disorder influences the resting and action potentials across the cell membrane and thus impairs cell functions. Uremia often induces a reduction in the responsiveness of cells to insulin, resulting in a decrease in the utilization of glucose by peripheral tissues and organs. The toxicity of urea also influences lipid metabolism. Uremia is commonly associated with hyperlipidemia and a decrease in high-density lipoproteins. The impairment of cellular metabolic functions in uremia is responsible for these changes. It is important to note that the kidneys have a high capacity of functional reserve. For a healthy person, one normal kidney is sufficient for the removal of toxins and the maintenance of the homeostasis. Thus, the clinical symptoms described above appear only when the majority of nephrons are destroyed in both kidneys.

Pathological examinations may demonstrate a number of structural changes in chronic renal failure, including a decrease in the density of nephrons, distortion of renal tubules, necrosis of glomerular endothelial cells and tubular epithelial cells, and formation of massive fibrous tissue. Chronic renal failure is often associated with complications in other systems, such as the cardiovascular, pulmonary, hematopoietic, nervous, and gastrointestinal systems. As in acute renal failure, fluid retention occurs commonly in chronic renal failure. Such a disorder can induce congestion, which further causes congestive heart failure and pulmonary edema. Uremia exerts toxic effects on the cardiomyocytes and pulmonary cells, facilitating the development of cardiac and pulmonary disorders. In the hematopoietic system, the toxic effect of urea on the bone marrow results in a reduction in the generation of hematopoietic cells. Uremia also induces hemolysis and a decrease in the number of red blood cells. In the nervous system, uremia induces neuronal injury, resulting in clinical symptoms such as drowsiness, insomnia, misjudgment, and loss of memory. In the gastrointestinal system, uremia causes epithelial cell injury and mucosal ulceration, resulting in disorders of digestion and absorption. Patients with chronic renal failure often express clinical symptoms such as anorexia, nausea, vomiting, and hiccups.

Treatment of Chronic Renal Failure [20.8]. Chronic renal failure is often associated with uremia and accumulation of metabolic wastes in the blood. Thus, dialysis or renal transplantation is necessary to control the blood concentration of urea and metabolic wastes. In addition, patients with chronic renal failure should be managed with the following approaches: (1) treatment of diseases that cause chronic renal failure; (2) management of complications in other systems, such as circulatory congestion, hypertension, heart failure, and pulmonary edema; (3) restriction of salt and water intake; (4) control of protein intake; and (5) control of the intake of drugs and chemicals that are excreted from the kidney. Molecular and cellular approaches described on page 850 can be used for the treatment of chronic renal failure. Once chronic renal failure reaches its end stage, dialysis and renal transplantation are the approaches of choice.

Acute Glomerulonephritis

Pathogenesis, Pathology, and Clinical Features [20.9]. *Acute glomerulonephritis* is a disorder characterized by rapid injury of glomerular capillaries and renal tubules, which results in a reduction in glomerular filtration, water and salt retention, proteinuria, and hematuria. Acute glomerulonephritis is caused by several pathogenic factors, including bacterial infection (streptococci, pneumococci, syphilis, and meningococcemia), viral infection (hepatitis B, mumps, measles, varicella, and coxsackievirus), parasitic infection (malaria and toxoplasmosis), systemic lupus erythematosus, and vasculitis. Salt and water retention is due to reduced glomerular filtration and tubular dysfunction and can induce circulatory congestion, pulmonary edema, and hypertension. Proteinuria is due to the damage of glomerular endothelial cells and increased capillary permeability, resulting in protein transport from the blood to the renal tubules. Similarly, hematuria is a result of the damage of the glomerular endothelial cells, allowing red blood cell migration to the renal tubules.

Pathological examinations or biopsies often demonstrate several structural changes, including injury of glomerular endothelial cells and tubular epithelial cells, infiltration of leukocytes, and the presence of red blood cells in the renal tubules. A considerable fraction of patients may experience acute renal failure. In most cases, acute glomerulonephritis can be recovered when the causative factors are removed. However, acute glomerulonephritis may develop into chronic glomerulonephritis in a certain fraction of patients (see next section).

Treatment of Acute Glomerulonephritis [20.9]. Acute glomerulonephritis is often treated with several approaches, including: (1) administration of antibiotics for causative bacterial infection, (2) appropriate rest until clinical symptoms disappear, (3) restriction of water and salt intake, (4) restriction of protein intake, (5) administration of diuretics, and (6) administration of vasodilators in the presence of hypertension. In about half of the cases, the disorder can be self-cured without long-term clinical consequences.

It is important to point out that an ideal treatment for acute glomerulonephritis is to prevent the development of acute renal failure. It is obvious that the conventional approaches described above are merely supportive and cannot be used to achieve such a goal. Molecular regenerative engineering is thought an effective approach that prevents acute renal failure. The strategies of molecular therapy for acute glomerulonephritis are identical to those described on page 850 for acute renal failure. When acute renal failure occurs, the cellular approach described on page 852 can be used to reduce uremia and toxic effects.

Chronic Glomerulonephritis

Pathogenesis, Pathology, and Clinical Features [20.10]. *Chronic glomerulonephritis* is a disorder with slow, persistent, and progressive impairment of renal function, which is manifested by the presence of long-term proteinuria and hematuria, eventually leading to chronic renal failure. Acute renal disorders, such as acute glomerulonephritis and nephrotic syndrome, can lead to chronic glomerulonephritis. Chronic glomerulonephritis is associated with a number of pathological changes, including the proliferation of glomerular endothelial cells and tubular epithelial cells, excessive formation of extracellular matrix and fibrosis, glomerulosclerosis, and tubulointerstitial fibrosis. During the early stage, patients with chronic glomerulonephritis may not demonstrate apparent clinical symptoms. During the late stage, chronic glomerulonephritis inevitably develop into chronic renal failure, exhibiting typical symptoms of renal failure such as uremia, water and salt retention, and complications in the cardiovascular, pulmonary, nervous, and gastrointestinal systems, as described on page 855.

Treatment of Chronic Glomerulonephritis [20.10]. Chronic glomerulonephritis is treated primarily with supportive approaches. A treatment with steroids may slow the progress of the disorder. Hypertension, if present, should be treated with vasodilators. It is necessary to administrate diuretics to reduce salt and water retention. Furthermore, the intake of water, salt, and protein should be controlled. In most cases of chronic glomerulonephritis, there exist various degrees of chronic renal failure. The treatment of chronic glomerulonephritis is similar to that of chronic renal failure. Molecular and cellular regenerative approaches described on page 850 and 852 can be used for the treatment of chronic renal glomerulonephritis.

Urinary Tract Obstruction

Pathogenesis, Pathology, and Clinical Features [20.11]. *Urinary tract obstruction* is the mechanical blockade of the urinary drainage system at a location from the renal calices to the urethra, often resulting in acute and chronic renal failure. A number of pathogenic factors can induce urinary tract obstruction. These factors include congenital abnormalities (ureteropelvic narrowing or obstruction, bladder neck narrowing and obstruction, and urethral valve stricture), inflammation, trauma, tumor, retroperitoneal fibrosis, aortic aneurysm, and kidney stone (inducing urine blockade within the kidney). Urinary tract obstruction can be effectively treated by surgical removal of the mechanical blockade and urinary tract reconstruction. Acute renal failure due to urinary tract obstruction can be recovered once the blockade is removed. However, chronic renal failure is usually irreversible. The clinical consequences of urinary tract obstruction are dependent on the degree of the mechanical blockade. Partial obstruction may not exert any influence on the renal functions. In contrast, complete urinary obstruction may induce acute renal failure within several days.

Conventional Treatment of Urinary Tract Obstruction [20.11]. The principle of treating urinary tract obstruction is to remove the factors that cause the obstruction. As discussed above, common causes for urinary obstruction are urinary tract tumors, congenital urinary tract constriction, compression by aortic aneurysms, trauma-induced scar formation, and inflammation-induced fibrosis. Surgical removal of tumors, congenital constriction, aortic aneurysm, and scars usually results in the restoration of renal functions within months if chronic renal failure is absent. When the urinary tract is severely damaged or a segment is removed due to tumors or other factors, it is necessary to reconstruct the urinary tract. This aspect is discussed in the following section. In addition, bacterial infection, if any, should be immediately controlled with antibiotics.

Cellular and Tissue Engineering. Severe urinary tract damage or loss can occur in several disorders, such as cancer, trauma, infection, and congenital abnormalities. In such cases, urinary tract reconstruction is often required to restore the function of the urinary tracts. There are two basic strategies for the reconstruction of the urinary tracts: (1) reconstruction with polymeric or extracellular matrix scaffolds and (2) reconstruction with cell-seeded matrix scaffolds. For the first strategy, a tubular scaffold can be used directly for the reconstruction of a malfunctioned urinary tract. The host urinary tract is capable of regenerating necessary cells to cover the reconstructed segment. For the second strategy, selected cells can be seeded in a scaffold of extracellular matrix or synthetic materials. The seeded cells can proliferate and produce extracellular matrix, thus facilitating the formation of the urinary tract. A fundamental issue in urinary tract reconstruction is to find a biocompatible material that can be used to construct a suitable scaffold. Furthermore, the materials should be chemically stable in urine and resistant to inflammation. Several types of biomaterials exhibit characteristics suitable for the repair and reconstruction of the urinary tracts, including the ureter, bladder, and urethra. These materials are briefly described as follows.

Polymeric Biomaterials for Urinary Tract Reconstruction [20.12]. A number of synthetic polymers, including polyurethane, silicone, and their copolymers, have been used for the reconstruction of the urinary tracts. Polyurethane has been particularly used for the construction of the ureter stents. Although this material is mechanically flexible and suitable for the reconstruction of the urinary tracts, it often causes inflammatory reactions and ulceration. Silicone is an inert material relatively compatible with biological tissues. Silicone and its copolymers have been widely used for the repair and reconstruction of the urinary tracts.

Biodegradable polymeric materials have been recently used for the repair and reconstruction of the urinary tracts. Commonly used biodegradable materials include polyglycolic polymers and poly-L-lactic and poly-L-glycolic copolymers. By altering the concentrations of various ingredients, the rate of degradation can be controlled to suite specific types of tissue. The advantage of using biodegradable materials is that, with the gradual degradation of the materials, native cells and tissue can grow into the polymeric matrix and self-reconstruct the lost structures. In experimental studies, poly-L-lactic and poly-L-glycolic copolymers have been used to fabricate urethral stents for the restoration of the urethral function. For a copolymer composed of 80% L-lactide and 20% glycolic acid, the copolymer stents are biocompatible and mechanically flexible, and can degrade within several months. In preliminary clinical studies, the same type of copolymeric urethral stent has been used in human patients with prostate disorders. The stents can be used to maintain the function of the urethra and can be degraded within 6 months.

Polymeric materials have been used for constructing the urinary bladder in animal experiments. Autologous epithelial and smooth muscle cells can be collected from the urinary bladder and seeded in a bladder-shaped polymeric scaffold to construct a neobladder. Such a construct can be used to replace the bladder of the animal that donates the epithelial and smooth muscle cells (Fig. 20.6). In preliminary tests, the reconstructed bladder can function for nearly a year and demonstrates the capability of retaining urine (Fig. 20.7). Furthermore, the neobladder can integrate to the native bladder tissue and exhibits natural structure and mechanical properties. These investigations suggest a potential for reconstructing a disordered urinary bladder by using cell and tissue regenerative engineering approaches.

Metallic Materials for Urinary Tract Reconstruction [20.13]. Several types of metallic materials, including stainless steel and titanium and nickel alloys, have been used for the reconstruction of the ureter and urethra. Given the mechanical strength and durability, metallic materials are often used for long-term implantation. The material surface can be gradually covered with epithelial cells following implantation, which enhances the com-

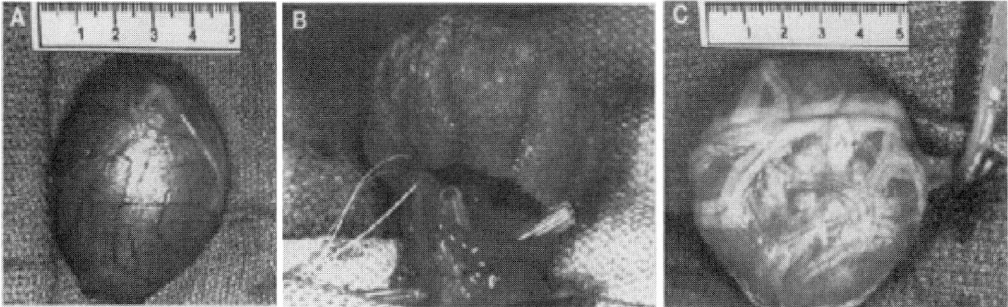

Figure 20.6. Surgical technique for cystectomy and preparation of implants: (A) the native canine bladder prior to trigone-sparing cystectomy; (B) the engineered neoorgan is anastomosed to the trigone; (C) the implant, decompressed by a transurethral and suprapubic catheter, is wrapped with omentum. (Reprinted by permission from Macmillan Publishers Ltd.: Oberpenning F et al: De novo reconstitution of a functional mammalian urinary bladder by tissue engineering, *Nature Biotechnol* 17:149–55, copyright 1999.)

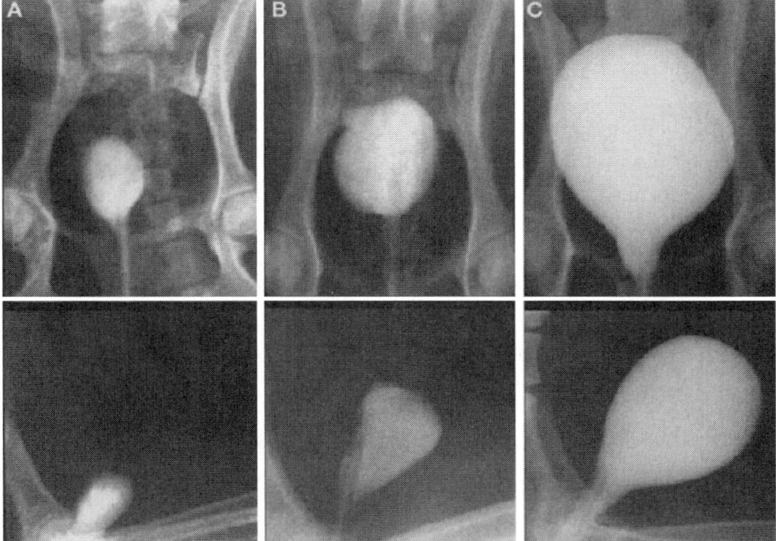

Figure 20.7. Radiographic cystograms 11 months after subtotal cystectomy: (A) subtotal cystectomy without reconstruction; (B) polymer-only implant; (C) tissue-engineered neoorgan. (Reprinted by permission from Macmillan Publishers Ltd.: Oberpenning F et al: De novo reconstitution of a functional mammalian urinary bladder by tissue engineering, *Nature Biotechnol* 17:149–55, copyright 1999.)

patibility of the metal implant with the host system. Although the success rate is acceptable in some experimental and clinical studies, metal devices usually stimulate host inflammatory responses, cell proliferation, matrix production, and tissue fibrosis, resulting in implant occlusion. The administration of antiinflammatory agents and growth inhibitors may reduce the rate of occlusion.

Biological Materials for Urinary Tract Reconstruction [20.14]. Biological tissues, such as intestinal submucosa and bladder submucosa, have been used as materials for urinary tract reconstruction. Given the natural structure and mechanical properties, these tissues are preferred compared to synthetic polymers and metallic materials. In the lack of autogenous tissues, allogenic tissues can be used. A problem with allogenic tissues is acute immune rejection. To reduce such a problem, it is necessary to remove the cellular components, which stimulate host immune responses. Decellularized matrix tissues exhibit significantly reduced immunogenicity and rarely cause acute immune rejection. Allogenic intestinal or bladder specimens can be decellularized by treatment with NaOH, KOH, or detergents. The mucosa and muscular layers can be removed and the submucosa matrix can be used for constructing ureteric, urethral, and bladder substitutes. Experimental and clinical studies have demonstrated promising results, although further investigations are necessary to improve the performance of the substitutes.

BIBLIOGRAPHY

20.1. Anatomy and Physiology of the Urinary System

Guyton AC, Hall JE: *Textbook of Medical Physiology*, 11th ed, Saunders, Philadelphia, 2006.

McArdle WD, Katch FI, Katch VL: *Essentials of Exercise Physiology*, 3rd ed, Lippincott Williams & Wilkins, Baltimore, 2006.

Germann WL, Stanfield CL (with contributors Niles MJ, Cannon JG), *Principles of Human Physiology*, 2nd ed, Pearson Benjamin Cummings, San Francisco, 2005.

Thibodeau GA, Patton KT: *Anatomy & Physiology*, 5th ed, Mosby, St Louis, 2003.

Boron WF, Boulpaep EL: *Medical Physiology: A Cellular and Molecular Approach*, Saunders, Philadelphia, 2003.

Ganong WF: *Review of Medical Physiology*, 21st ed., McGraw-Hill, New York, 2003.

20.2. Pathogenesis, Pathology, and Clinical Features

Schneider AS, Szanto PA: Pathology. 3rd ed., Lippincott Williams & Wilkins, Philadelphia, 2006.

McCance KL, Huether SE: Pathophysiology: the biologic basis for disease in adults & children. 5th ed. Elsevier Mosby, St. Louis, 2006.

Porth CM: Pathophysiology: concepts of altered health states, 7th ed., Lippincott Williams & Wilkins, Philadelphia, 2005.

Frazier MS, Drzymkowski JW: Essentials of human diseases and conditions, 3rd ed., Elsevier Saunders, St. Louis, 2004.

Dai C, Yang J, Liu Y: Single injection of naked plasmid encoding hepatocyte growth factor prevents cell death and ameliorates acute renal failure in mice, *J Am Soc Nephrol* 13:411–422, 2002.

Liu Y, Tolbert EM, Lin L, Thursby MA, Sun AM et al: Up-regulation of hepatocyte growth factor receptor: An amplification and targeting mechanism for hepatocyte growth factor action in acute renal failure, *Kidney Int* 55:442–53, 1999.

20.3. Growth Factor Genes

Yang J, Dai C, Liu Y: Systemic administration of naked plasmid encoding hepatocyte growth factor ameliorates chronic renal fibrosis in mice, *Gene Ther* 8(19):1470–9, 2001.

Mizuno S, Nakamura T: Prevention of neutrophil extravasation by hepatocyte growth factor leads to attenuations of tubular apoptosis and renal dysfunction in mouse ischemic kidneys, *Am J Pathol* 166(6):1895–905, 2005.

Bonventre JV, Siegel N, Rosen S, Portilla D, Venkatachalam M: Acute renal failure. I. Relative importance of proximal vs. distal tubular injury, *Am J Physiol Renal Physiol* 275:F623–32, 1998.

Edelstein CL, Ling H, Wangsiripaisan A, Schrier RW: Emerging therapies for acute renal failure, *Am J Kidney Dis* 30:S89–95, 1997.

Kelly KJ, Molitoris BA: Acute renal failure in the new millennium: Time to consider combination therapy, *Semin Nephrol* 20:4–19, 2000.

Hammerman MR, Miller SB: Therapeutic use of growth factors in renal failure, *J Am Soc Nephrol* 5:1–11, 1994.

Humes HD, MacKay SM, Funke AJ, Buffington DA: Acute renal failure: Growth factors, cell therapy, and gene therapy, *Proc Assoc Am Physicians* 109:547–57, 1997.

Hammerman MR, Safirstein R, Harris RC, Toback FG, Humes HD: Acute renal failure. III. The role of growth factors in the process of renal regeneration and repair, *Am J Physiol Renal Physiol* 279:F3–11, 2000.

Zarnegar R, Michalopoulos GK: The many faces of hepatocyte growth factor: From hepatopoiesis to hematopoiesis, *J Cell Biol* 129:1177–80, 1995.

Matsumoto K, Nakamura T: Emerging multipotent aspects of hepatocyte growth factor, *J Biochem* 119:591–600, 1996.

Vargas GA, Hoeflich A, Jehle PM: Hepatocyte growth factor in renal failure: Promise and reality, *Kidney Int* 57:1426–36, 2000.

Matsumoto K, Mizuno S, Nakamura T: Hepatocyte growth factor in renal regeneration, renal disease and potential therapeutics, *Curr Opin Nephrol Hypertens* 9:395–402, 2000.

Miller SB, Martin DR, Kissane J, Hammerman MR: Hepatocyte growth factor accelerates recovery from acute ischemic renal injury in rats, *Am J Physiol* 266:F129–34, 1994.

Kawaida K, Matsumoto K, Shimazu H, Nakamura T: Hepatocyte growth factor prevents acute renal failure and accelerates renal regeneration in mice, *Proc Natl Acad Sci USA* 91:4357–61, 1994.

Ueda T, Takeyama Y, Hori Y, Shinkai M, Takase K et al: Hepatocyte growth factor increases in injured organs and functions as an organotrophic factor in rats with experimental acute pancreatitis, *Pancreas* 20:84–93, 2000.

Liu Y, Tolbert EM, Lin L, Thursby MA, Sun AM et al: Up-regulation of hepatocyte growth factor receptor: An amplification and targeting mechanism for hepatocyte growth factor action in acute renal failure, *Kidney Int* 55:442–53, 1999.

Liu ML, Mars WM, Zarnegar R, Michalopoulos GK: Uptake and distribution of hepatocyte growth factor in normal and regenerating adult rat liver, *Am J Pathol* 144:129–40, 1994.

Kelley VR, Suhatme VP: Gene transfer in the kidney, *Am J Physiol* 276:F1–9, 1999.

Imai E, Isaka Y: New paradigm of gene therapy: Skeletal-muscle-targeting gene therapy for kidney disease, *Nephron* 83:296–300, 1999.

Yang J, Chen S, Huang L, Michalopoulos GK, Liu Y: Sustained expression of naked plasmid DNA encoding hepatocyte growth factor in mice promotes liver and overall body growth, *Hepatology* 33:848–59, 2001.

Liu Y, Centracchio JN, Lin L, Sun AM, Dworkin LD: Constitutive expression of HGF modulates renal epithelial cell phenotype and induces c-met and fibronectin expression, *Exp Cell Res* 242:174–85, 1998.

Dai C, Yang J, Liu Y: Single injection of naked plasmid encoding hepatocyte growth factor prevents cell death and ameliorates acute renal failure in mice, *J Am Soc Nephrol* 13:411–22, 2002.

20.4. Genes Encoding Mitogenic Signaling Proteins

Dai C, Yang J, Liu Y: Single injection of naked plasmid encoding hepatocyte growth factor prevents cell death and ameliorates acute renal failure in mice, *J Am Soc Nephrol* 13:411–22, 2002.

Cao CC, Ding XQ, Ou ZL, Liu CF, Li P et al: In vivo transfection of NF-kappaB decoy oligodeoxynucleotides attenuate renal ischemia/reperfusion injury in rats, *Kidney Int* 65(3):834–45, March 2004.

Lippin Y, Dranitzki-Elhalel M, Brill-Almon E, Mei-Zahav C, Mizrachi S et al: Human erythropoietin gene therapy for patients with chronic renal failure, *Blood* 106(7):2280–6, Oct 2005.

20.5. Stem Cell-Based Regeneration of Renal Tissue

Lange C, Togel F, Ittrich H, Clayton F, Nolte-Ernsting C et al: Administered mesenchymal stem cells enhance recovery from ischemia/reperfusion-induced acute renal failure in rats, *Kidney Int* 68(4):1613–7, 2005.

Bard JBL, Gordon A, Sharp L, Sellers W: The early stages of nephron formation in the developing mouse kidney, *J Anat* 199:385–92, 2001.

Steer DL, Nigam SK: Developmental approaches to kidney tissue engineering, *Am J Physiol Renal Physiol* 286:F1–7, 2004.

Pohl M, Bhatnagar V, Mendoza SA, Nigam SK: Toward an etiological classification of developmental disorders of the kidney and upper urinary tract, *Kidney Int* 61:10–19, 2002.

Pohl M, Stuart RO, Sakurai H, Nigam SK: Branching morphogenesis during kidney development, *Annu Rev Physiol* 62:595–620, 2000.

Sakurai H, Barros EJ, Tsukamoto T, Barasch J, Nigam SK: An in vitro tubulogenesis system using cell lines derived from the embryonic kidney shows dependence on multiple soluble growth factors, *Proc Natl Acad Sci USA* 94:6279–84, 1997.

Sakurai H, Bush KT, Nigam SK: Identification of pleiotrophin as a mesenchymal factor involved in ureteric bud branching morphogenesis, *Development* 128:3283–93, 2001.

Santos OF, Barros EJ, Yang XM, Matsumoto K, Nakamura T et al: Involvement of hepatocyte growth factor in kidney development, *Dev Biol* 163:525–9, 1994.

Barros EJ, Santos OF, Matsumoto K, Nakamura T, Nigam SK: Differential tubulogenic and branching morphogenetic activities of growth factors: Implications for epithelial tissue development, *Proc Natl Acad Sci USA* 92:4412–6, 1995.

Sakurai H, Nigam SK: Transforming growth factor-β selectively inhibits branching morphogenesis but not tubulogenesis, *Am J Physiol Renal Physiol* 272:F139–46, 1997.

Nigam S, Lieberthal W: Acute renal failure. III. The role of growth factors in the process of renal regeneration and repair, *Am J Physiol Renal Physiol* 279:F3–11, 2000.

Poulsom R, Forbes SJ, Hodivala-Dilke K, Ryan E, Wyles S et al: Bone marrow contributes to renal parenchymal turnover and regeneration, *J Pathol* 195:229–35, 2001.

Gupta S, Verfaillie C, Chmielewski D, Kim Y, Rosenberg ME: A role for extrarenal cells in the regeneration following acute renal failure, *Kidney Int* 62:1285–90, 2002.

Pleiotrophin

Florin L, Maas-Szabowski N, Werner S, Szabowski A, Angel P: Increased keratinocyte proliferation by JUN-dependent expression of PTN and SDF-1 in fibroblasts, *J Cell Sci* 118(Pt 9):1981–9, 2005.

Lu KV, Jong KA, Kim GY, Singh J, Dia EQ et al: Differential induction of glioblastoma migration and growth by two forms of pleiotrophin, *J Biol Chem* 280(29):26953–64, 2005.

Christman KL, Fang Q, Kim AJ, Sievers RE, Fok HH et al: Pleiotrophin induces formation of functional neovasculature in vivo, *Biochem Biophys Res Commun* 332(4):1146–52, 2005.

Powers C, Aigner A, Stoica GE, McDonnell K, Wellstein A: Pleiotrophin signaling through anaplastic lymphoma kinase is rate-limiting for glioblastoma growth, *J Biol Chem* 277(16): 14153–8, 2002.

Lai S, Czubayko F, Riegel AT, Wellstein A: Structure of the human heparin-binding growth factor gene pleiotrophin, *Biochem Biophys Res Commun* 187:1113–22, 1992.

Li YS, Hoffman RM, Le Beau MM, Espinosa R III, Jenkins NA et al: Characterization of the human pleiotrophin gene: Promoter region and chromosomal localization, *J Biol Chem* 267:26011–16, 1992.

Li YS, Milner PG, Chauhan AK, Watson MA, Hoffman RM et al: Cloning and expression of a developmentally regulated protein that induces mitogenic and neurite outgrowth activity, *Science* 250:1690–4, 1990.

Souttou B, Juhl H, Hackenbruck J, Rockseisen M, Klomp HJ et al: Relationship between serum concentrations of the growth factor pleiotrophin and pleiotrophin-positive tumors, *J Natl Cancer Inst* 90:1468–73, 1998.

Zhang N, Yeh HJ, Zhong R, Li YS, Deuel TF: A dominant-negative pleiotrophin mutant introduced by homologous recombination leads to germ-cell apoptosis in male mice, *Proc Natl Acad Sci USA* 96:6734–8, 1999.

Glial Cell-Derived Neurotrophic Factor

Beck KD, Valverde J, Alexi T, Poulsen K, Moffat B et al: Mesencephalic dopaminergic neurons protected by GDNF from axotomy-induced degeneration in the adult brain, *Nature* 373:339–41, 1995.

Bermingham N, Hillermann R, Gilmour F, Martin JE, Fisher EM: Human glial cell line-derived neurotrophic factor (GDNF) maps to chromosome 5, *Hum Genet* 96:671–3, 1995.

Durbec P, Marcos-Gutierrez CV, Kilkenny C, Grigoriou M, Wartiowaara K et al: GDNF signalling through the Ret receptor tyrosine kinase, *Nature* 381:789–93, 1996.

Gash DM, Zhang Z, Ovadia A, Cass WA, Yi A et al: Functional recovery in parkinsonian monkeys treated with GDNF, *Nature* 380:252–5, 1996.

Gill SS, Patel NK, Hotton GR, O'Sullivan K, McCarter R et al: Direct brain infusion of glial cell line-derived neurotrophic factor in Parkinson disease, *Nature Med* 9:589–95, 2003.

Kordower JH, Emborg ME, Bloch J, Ma SY, Chu Y et al: Neurodegeneration prevented by lentiviral vector delivery of GDNF in primate models of Parkinson's disease, *Science* 290:767–73, 2000.

Lin LFH, Doherty DH, Lile JD, Bektesh S, Collins F: GDNF: A glial cell line-derived neurotrophic factor for midbrain dopaminergic neurons, *Science* 260:1130–2, 1993.

Meng X, Lindahl M, Hyvonen ME, Parvinen M, de Rooij DG et al: Regulation of cell fate decision of undifferentiated spermatogonia by GDNF, *Science* 287:1489–93, 2000.

Moore MW, Klein RD, Farinas I, Sauer H, Armanini M et al: Renal and neuronal abnormalities in mice lacking GDNF, *Nature* 382:76–9, 1996.

Nguyen QT, Parsadanian AS, Snider WD, Lichtman JW: Hyperinnervation of neuromuscular junctions caused by GDNF overexpression in muscle, *Science* 279:1725–9, 1998.

Oppenheim RW, Houenou LJ, Johnson JE, Lin LFH, Li L et al: Developing motor neurons rescued from programmed and axotomy-induced cell death by GDNF, *Nature* 373:344–6, 1995.

Pichel JG, Shen L, Shang HZ, Granholm AC, Drago J et al: Defects in enteric innervation and kidney development in mice lacking GDNF, *Nature* 382:73–6, 1996.

Pichel JG, Shen L, Sheng HZ, Granholm AC, Drago J et al: GDNF is required for kidney development and enteric innervation, *Cold Spring Harbor Symp Quant Biol* 61:445–57, 1996.

Ramer MS, Priestley JV, McMahon SB: Functional regeneration of sensory axons into the adult spinal cord, *Nature* 403:312–16, 2000.

Sanchez MP, Silos-Santiago I, Frisen J, He B, Lira SA et al: Renal agenesis and the absence of enteric neurons in mice lacking GDNF, *Nature* 382:70–3, 1996.

Tomac A, Lindqvist E, Lin LFH, Ogren SO, Young D et al: Protection and repair of the nigrostriatal dopaminergic system by GDNF in vivo, *Nature* 373:335–9, 1995.

Treanor JJS, Goodman L, de Sauvage F, Stone DM, Poulsen KT et al: Characterization of a multicomponent receptor for GDNF, *Nature* 382:80–3, 1996.

Collagen Type XVIII

O'Reilly MS, Boehm T, Shing Y, Fukai N, Vasios G et al: Endostatin: An endogenous inhibitor of angiogenesis and tumor growth, *Cell* 88:277–85, 1997.

O'Reilly MS, Holmgren L, Shing Y, Chen C, Rosenthal RA et al: Angiostatin: A novel angiogenesis inhibitor that mediates the suppression of metastases by a Lewis lung carcinoma, *Cell* 79: 315–28, 1994.

Oh SP, Kamagata Y, Muragaki Y, Timmons S, Ooshima A et al: Isolation and sequencing of cDNAs for proteins with multiple domains of Gly-X-Y repeats identify a novel family of collagenous proteins, *Proc Natl Acad Sci* 91:4229–33, 1994.

Oh SP, Warman ML, Seldin MF, Cheng SD, Knoll JHM et al: Cloning of cDNA and genomic DNA encoding human type XVIII collagen and localization of the alpha-1(XVIII) collagen gene to mouse chromosome 10 and human chromosome 21, *Genomics* 19:494–9, 1994.

Rehn M, Hintikka E, Pihlajaniemi T: Characterization of the mouse gene for the alpha-1 chain of type XVIII collagen (Col18a1) reveals that the three variant N-terminal polypeptide forms are transcribed from two widely separated promoters, *Genomics* 32:436–46, 1996.

Sertie AL, Sossi V, Camargo AA, Zatz M, Brahe C et al: Collagen XVIII, containing an endogenous inhibitor of angiogenesis and tumor growth, plays a critical role in the maintenance of retinal structure and in neural tube closure (Knobloch syndrome), *Hum Mol Genet* 9:2051–8, 2000.

Sudhakar A, Sugimoto H, Yang C, Lively J, Zeisberg M et al: Human tumstatin and human endostatin exhibit distinct antiangiogenic activities mediated by alpha-V-beta-3 and alpha-5-beta-1 integrins, *Proc Natl Acad Sci* 100:4766–71, 2003.

Qiao J, Sakurai H, Nigam SK: Branching morphogenesis independent of mesenchymal-epithelial contact in the developing kidney, *Proc Natl Acad Sci USA* 96:7330–5, 1999.

Sakurai H, Bush KT, Nigam SK: Identification of pleiotrophin as a mesenchymal factor involved in ureteric bud branching morphogenesis, *Development* 128:3283–93, 2001.

Karihaloo A, Karumanchi SA, Barasch J, Jha V, Nickel CH et al: Endostatin regulates branching morphogenesis of renal epithelial cells and ureteric bud, *Proc Natl Acad Sci USA* 98:12509–14, 2001.

Plisov SY, Yoshino K, Dove LF, Higinbotham KG, Rubin JS et al: TGF b_2, LIF and FGF2 cooperate to induce nephrogenesis, *Development* 128:1045–57, 2001.

Barasch J, Yang J, Ware CB, Taga T, Yoshida K et al: Mesenchymal to epithelial conversion in rat metanephros is induced by LIF, *Cell* 99:377–86, 1999.

Steer DL, Bush KT, Meyer TN, Schwesinger C, Nigam SK: A strategy for in vitro propagation of rat nephrons, *Kidney Int* 62:1958–65, 2002.

Rogers SA, Hammerman MR: Transplantation of rat metanephroi into mice, *Am J Physiol Regul Integr Comp Physiol* 280:R1865–9, 2001.

Rogers SA, Lowell JA, Hammerman NA, Hammerman MR: Transplantation of developing metanephroi into adult rats, *Kidney Int* 54:27–37, 1998.

Dekel B, Amariglio N, Kaminski N, Schwartz A, Goshen E et al: Engraftment and differentiation of human metanephroi into functional mature nephrons after transplantation into mice is accompanied by a profile of gene expression similar to normal human kidney development, *J Am Soc Nephrol* 13:977–90, 2002.

Dekel B, Burakova T, Arditti FD, Reich-Zeliger S, Milstein O et al: Human and porcine early kidney precursors as a new source for transplantation, *Nat Med* 9:53–60, 2003.

Human protein reference data base, Johns Hopkins University and the Institute of Bioinformatics, at http://www.hprd.org/protein.

20.6. Nuclear Transfer-Based Generation of Renal Tissues

Lanza RP, Chung HY, Yoo JJ, Wettstein PJ, Blackwell C et al: Generation of histocompatible tissues using nuclear transplantation, *Nat Biotechnol* 20:689–96, 2002.

Atala A, Koh CJ: Tissue engineering applications of therapeutic cloning, *Annu Rev Biomed Eng* 6:27–40, 2004.

20.7. Adult Tubular Cell-Based Kidney Generation

Thomas MK, Lloyd-Jones DM, Thadhani RI et al: Hypovitaminosis D in medical inpatients, *New Engl J Med* 338:777, 1998.

Humes HD, MacKay SM, Funke AJ et al: Tissue engineering of a bioartificial renal tubule assist device: in vitro transport and metabolic, *Kidney Int* 55:2502, 1999.

Humes HD, Buffington DA, MacKay SM et al: Replacement of renal function in uremic animals with a tissue-engineered, *Nature Biotechnol* 17:451, 1999.

Weitzel WF, Fissell WH, Humes HD: Initial clinical experience with a human proximal tubule cell renal assist device (RAD), *J Am Soc Nephrol* 12:279A, 2001.

Humes HD: Bioartificial kidney for full renal replacement therapy, *Semin Nephrol* 20:71–82, 2000.

Fisselland WH, Humes HD: Cell therapy of renal failure, *Transplant Proc* 35:2837–42, 2003.

Atala A: Tissue engineering in urology, *Curr Urol Rep* 2(1):83–92, 2001.

20.8. Pathogenesis, Pathology, and Clinical Features of Chronic Real Failure

Garcia-Donaire JA, Alcazar JM: Ischemic nephropathy: Detection and therapeutic intervention, *Kidney Int Suppl* 99:S131–6, 2005.

Ruggenenti P, Schieppati A, Remuzzi G: Progression, remission, regression of chronic renal diseases, *Lancet* 357(9268):1601–8, 2001.

Alcazar JM, Rodicio JL: Ischemic nephropathy: Clinical characteristics and treatment, *Am J Kidney Dis* 36:883–93, 2000.

Morgan DB, Will EJ: Selection, presentation, and interpretation of biochemical data in renal failure, *Kidney Int* 24:438–45, 1983.

20.9. Pathogenesis, Pathology, and Clinical Features of Acute and Chronic Glomerulonephritis

Hughes J, Savill JS: Apoptosis in glomerulonephritis, *Curr Opin Nephrol Hypertens* 14:389–95, 2005.

D'Amico G, Bazzi C: Urinary protein and enzyme excretion as markers of tubular damage, *Curr Opin Nephrol Hypertens* 12:639–43, 2003.

Hoschek JC, Dreyer P, Dahal S, Walker PD: Rapidly progressive renal failure in childhood, *Am J Kidney Dis* 40:1342–7, 2002.

Andreucci M, Federico S, Andreucci VE: Edema and acute renal failure, *Semin Nephrol* 21:251–6, 2001.

Kunis CL, Teng SN: Treatment of glomerulonephritis in the elderly, *Semin Nephrol* 20:256–64, 2000.

Pan CG: Glomerulonephritis in childhood, *Curr Opin Pediatr* 9:154–9, 1997.

20.10. Pathogenesis, Pathology, and Clinical Features of Chronic Glomerulonephritis

Liu Y: Renal fibrosis: New insights into the pathogenesis and therapeutics, *Kidney Int* 69:213–7, 2006.

Joosten SA, van Kooten C, Sijpkens YW, de Fijter JW, Paul LC: The pathobiology of chronic allograft nephropathy: Immune-mediated damage and accelerated aging, *Kidney Int* 65:1556–9, 2004.

Imai E, Takabatake Y, Mizui M, and Isaka Y. Gene therapy in renal diseases, *Kidney Int* 65:1551–5, 2004.

Frasca GM, Onetti-Muda A, Renieri A: Thin glomerular basement membrane disease, *J Nephrol* 13:15–9, 2000.

Tomson CR: Recent advances: Nephrology, *Br Med J* 320:98–101, 2000.

20.11. Pathogenesis, Pathology, and Clinical Features of Urinary Tract Obstruction

Duvdevani M, Chew BH, Denstedt JD: Minimizing symptoms in patients with ureteric stents, *Curr Opin Urol* 16:77–82, 2006.

Woderich R, Fowler CJ: Management of lower urinary tract symptoms in men with progressive neurological disease, *Curr Opin Urol* 16:30–6, 2006.

Suri A, Srivastava A, Singh KJ, Dubey D, Mandhani A et al: Endoscopic incision for functional bladder neck obstruction in men: Long-term outcome, *Urology* 66:323–6, 2005.

Andersson KE, Arner A: Urinary bladder contraction and relaxation: Physiology and pathophysiology, *Physiol Rev* 84:935–86, 2004.

Tan BJ, Smith AD: Ureteropelvic junction obstruction repair: When, how, what? *Curr Opin Urol* 14:55–9, 2004.

20.12. Polymeric Biomaterials for Urinary Tract Repair and Reconstruction

Laaksovirta S, Laurila M, Isotalo T, Valimaa T, Tammela TL et al: Rabbit muscle and urethral in situ biocompatibility properties of the self-reinforced L-lactide-glycolic acid copolymer 80:20 spiral stent, *J Urol* 167:1527, 2002.

Laaksovirta S, Isotalo T, Talja M, Valimaa T, Tormala P et al: Interstitial laser coagulation and biodegradable self-expandable, self-reinforced poly-L-lactic and poly-L-glycolic copolymer spiral stent in the treatment of benign prostatic enlargement, *J Endourol* 16:311, 2002.

Marx M, Bettmann MA, Bridge S, Brodsky G, Boxt LM et al: The effects of various indwelling ureteral catheter materials on the normal canine ureter, *J Urol* 139:180, 1988.

Denstedt JD, Wollin TA, Reid G: Biomaterials used in urology: Current issues of biocompatibility, infection, and encrustation, *J Endourol* 12:493, 1998.

Beiko DT, Knudsen BE, Watterson JD, Cadieux PA, Reid G et al: Urinary tract biomaterials, *J Urol* 171(6 Pt 1):2438–44, June 2004.

Olweny EO, Landman J, Andreoni C, Collyer W, Kerbl K et al: Evaluation of the use of a biodegradable ureteral stent after retrograde endopyelotomy in a porcine model, *J Urol* 167:2198, 2002.

Auge BK, Ferraro RF, Madenjian AR, Preminger GM: Evaluation of a dissolvable ureteral drainage stent in a Swine model, *J Urol* 168:808, 2002.

Lingeman JE, Preminger GM, Berger Y, Denstedt JD, Goldstone L et al: Use of a temporary ureteral drainage stent after uncomplicated ureteroscopy: Results from a phase II clinical trial, *J Urol* 169:1682, 2003.

Moriya K, Kakizaki H, Murakumo M, Watanabe S, Chen Q et al: Creation of luminal tissue covered with urothelium by implantation of cultured urothelial cells into the peritoneal cavity, *J Urol* 170(6 Pt 1):2480–5, Dec 2003.

20.13. Metallic Materials for Urinary Tract Repair and Reconstruction

Lam JS, Volpe MA, Kaplan SA: Use of prostatic stents for the treatment of benign prostatic hyperplasia in high-risk patients, *Curr Urol Rep* 2:277, 2001.

Pauer W, Eckerstorfer GM: Use of self-expanding permanent endoluminal stents for benign ureteral strictures: mid-term results, *J Urol* 162:319, 1999.

Ahmed M, Bishop MC, Bates CP, Manhire AR: Metal mesh stents for ureteral obstruction caused by hormone-resistant carcinoma of prostate, *J Endourol* 13:221, 1999.

Kulkarni R, Bellamy E: Nickel-titanium shape memory alloy Memokath 051 ureteral stent for managing long-term ureteral obstruction: 4-year experience, *J Urol* 166:1750, 2001.

Barbalias GA, Siablis D, Liatsikos EN, Karnabatidis D, Yarmenitis S et al: Metal stents: A new treatment of malignant ureteral obstruction, *J Urol* 158:54, 1997.

Braf Z, Chen J, Sofer M, Matzkin H: Intraprostatic metal stents (Prostakath and Urospiral): More than 6 years' clinical experience with 110 patients, *J Endourol* 10:555, 1996.

Barbalias GA, Liatsikos EN, Kalogeropoulou C, Karnabatidis D, Zabakis P et al: Externally coated ureteral metallic stents: An unfavorable clinical experience, *Eur Urol* 42:276, 2002.

20.14. Biological Materials for Urinary Tract Repair and Reconstruction

El-Kassaby AW, Retik AB, Yoo JJ, Atala A: Urethral stricture repair with an off-the-shelf collagen matrix, 169(1):170–3, 2003.

De Filippo RE, Yoo JJ, Atala A: Urethral replacement using cell seeded tubularized collagen matrices, *J Urol* 168(4 Pt 2):1789–92, 2002.

Grossklaus DJ, Shappell SB, Adams MC, Brock JW III, Pope JC IV: Small intestinal submucosa as a urethral coverage layer, *J Urol* 166:636, 2001.

Rotariu P, Yohannes P, Alexianu M, Gershbaum D, Pinkashov D et al: Reconstruction of rabbit urethra with surgisis small intestinal submucosa, *J Endourol* 16:617, 2002.

Liatsikos EN, Dinlenc CZ, Kapoor R, Bernardo NO, Pikhasov D et al: Ureteral reconstruction: Small intestine submucosa for the management of strictures and defects of the upper third of the ureter, *J Urol* 165:1719, 2001.

Sofer M, Rowe E, Forder DM, Denstedt JD: Ureteral segmental replacement using multilayer porcine small-intestinal submucosa, *J Endourol* 16:27, 2002.

O'Connor RC, Harding JN 3rd, Steinberg GD: Novel modification of partial nephrectomy technique using porcine small intestine submucosa, *Urology* 60:906, 2002.

Atala A: Tissue engineering in urology, *Curr Urol Rep* 2(1):83–92, 2001.

Nuininga JE, van Moerkerk H, Hanssen A, Hulsbergen CA, Oosterwijk-Wakka J et al: A rabbit model to tissue engineer the bladder, *Biomaterials* 25(9):1657–61, 2004.

Beiko DT, Knudsen BE, Watterson JD, Cadieux PA, Reid G et al: Urinary tract biomaterials, *J Urol* 171(6 Pt 1):2438–44, 2004.

Atala A: Tissue engineering in urology, *Curr Urol Rep* 2(1):83–92, 2001.

21

SKELETAL MUSCLE REGENERATIVE ENGINEERING

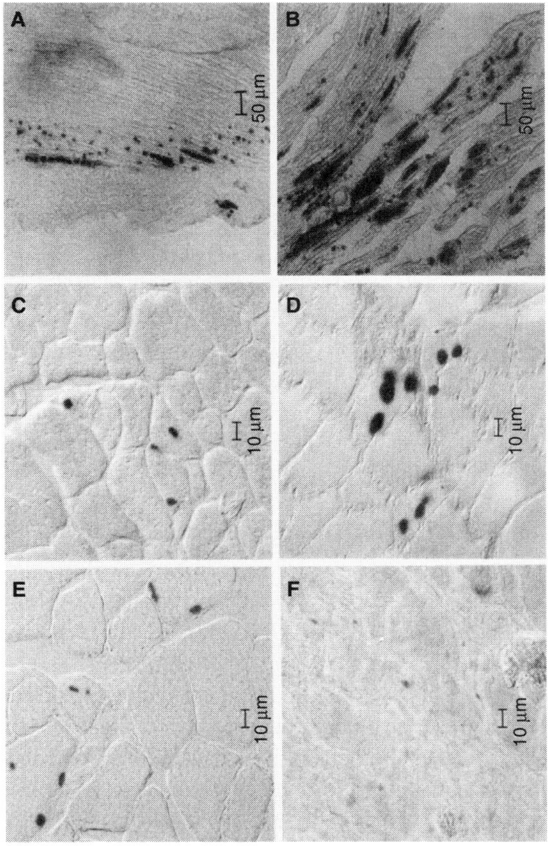

Formation of muscular cells from transplanted bone marrow cells. The images show nuclear lacZ expression in whole-mount dissected muscle fibers (A,B) or cryostat sections (C–F) of regenerating muscles of scid/bg mice. Mice were injected with unfractionated (A,C), adherent (E), or nonadherent (F) bone marrow cells, or with control satellite cells (B,D), from C57/MlacZ transgenic mice. (A,B) Brightfield; scale bars: 50 µm. (C–F) Nomarski optics; scale bars: 10 µm. (Reprinted with permission from Giuliana Ferrari G et al: Muscle regeneration by bone marrow-derived myogenic progenitors, *Science* 279:1528–30, copyright 1998 AAAS.) See color insert.

Bioregenerative Engineering: Principles and Applications, by Shu Q. Liu

ANATOMY AND PHYSIOLOGY OF THE SKELETAL MUSCLE SYSTEM

Structure [21.1]

The skeletal muscle system is composed of muscle cells, connective tissue, blood vessels, and nerve fibers. Skeletal muscle cells are organized into bundles, which are attached to the bone via the tendon. The primary functions of the skeletal muscle system are contraction, force generation, and induction of skeletal movement. A skeletal muscle cell is a fiber-like cell and is also called a muscle fiber. Each muscle cell contains a number of nuclei, contractile myofibrils, sarcoplasmic reticulum, and common subcellular organelles, including mitochondria, glycogen granules, endosomes, and Golgi apparatus. The myofibrils are thread-like structures that are aligned in the axial direction of the cell and are responsible for cell contraction and force generation. Each muscle cell contains a large number of myofibrils. Each myofibril is composed of contractile filaments, regulatory proteins, and supporting structures. There are two types of contractile filament: actin filaments and myosin filaments. Since the actin filaments (~8 nm in diameter) appear thinner than the myosin filaments (~12 nm), the actin filaments are referred to as *thin filaments*, while the myosin filaments are referred to as *thick filaments*.

An *actin filament* is composed of two filamentous actin strands, known as F actin, which are organized into a double helix (Fig. 21.1). Each F actin strand is polymerized from a large number of globular actin monomers, known as the G actin. Each G actin monomer is capable of interacting with a myosin molecule at an active myosin-binding site. The double helical actin strands are associated with several proteins, including tropomyosin and troponin. *Tropomyosin* is a fiber-like protein that is aligned along the groove of the helical F actin filaments. *Troponin* is a protein complex, which is responsible for binding to actin, tropomyosin, and calcium ions. Tropomyosin and troponin play critical roles in regulating the interaction between the active sites of actin and myosin filaments, an essential process for muscular contraction.

A *myosin filament* consists of a large number of rod-shaped myosin molecules. Each myosin molecule is composed of two heavy chains, each of which consists of a head and a tail. The tails of the two heavy chains are organized into a helical myosin rod. Each heavy chain is associated with two light chains, which are located at the hinge region between the myosin head and tail. The myosin light chains regulate the activity of the myosin heavy-chains and the interaction of myosin with actin. The myosin molecules are organized into myosin filaments with the myosin heavy-chain heads arranged at both ends of each myosin filament. The heavy-chain head plays a critical role in mediating myosin–actin interaction. It can bind to the active site of actin and can bend at the hinge region between the head and the tail, causing the actin filament to slide along myosin filaments and thus inducing muscular contraction. The heavy-chain head possesses ATPase activity, providing energy for contractile activities.

The actin and myosin filaments are assembled into a structure with highly ordered organization. Each myosin filament is surrounded by six parallel, equally spaced actin filaments. Each actin filament is aligned with a row of myosin heavy-chain heads, which can physically interact with the active sites of the actin filaments. The actin and myosin filaments are organized into functional units called *sarcomeres*, which appear as consecutive short segments along a myofibril. Under an electron microscope, there appear several structures within each sarcomere. These include two Z-disks (located at the two ends of

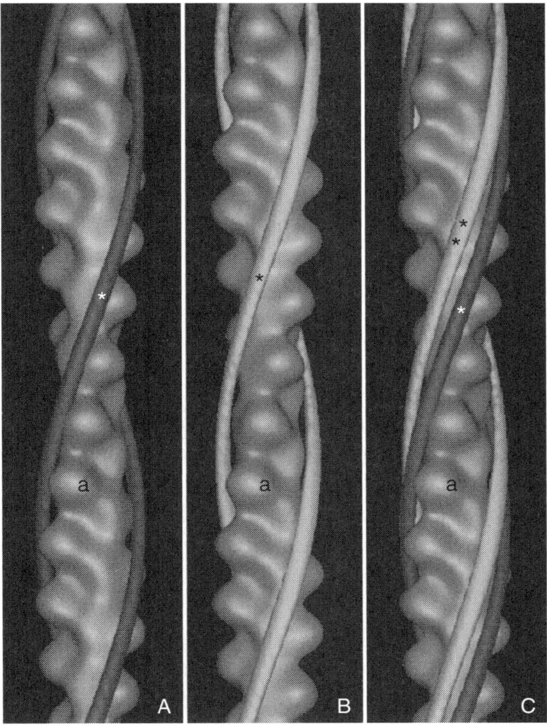

Figure 21.1. Surface views of reconstructions of thin actin filaments (a) showing the positions of tropomyosin strands (*) superimposed on actin in the presence of EGTA (A) and Ca^{2+} (B). In panel C tropomyosin strands associated with both EGTA and Ca^{2+} are superimposed on actin for comparison. Reconstructions show characteristic bilobed actin (a) and continuous tropomyosin strands. In EGTA, tropomyosin (*) occupies a position on the inner edge of the outer domain of actin, whereas in Ca^{2+}, tropomyosin (**) lies along the outer edge of the inner domain. Surface rendering was carried out by superimposing tropomyosin strand densities obtained by difference analysis on the maps of pure F-actin. (Reprinted with permission from Xu C et al: *Biophys J* 77:985–92, copyright 1999.)

a sarcomere and shared by adjacent sarcomeres), a middle band known as the A-band (anisotropic band, defined based on optical properties), a H-zone in the middle of the A-band, an M-line in the middle of the H-zone, and an I-band (isotropic band) between each end of the A band and the Z-line. The formation of these structures is based on the arrangement of actin and myosin filaments. The *Z-disk* is a filamentous network for the anchorage of the actin filaments; the *A-band* covers the entire length of the myosin filaments; the *H-zone* is the region where myosin filaments do not overlap with the actin filaments; the *M-line* is a filamentous network for the anchorage of the myosin filaments; and the *I-band* is the region with actin filaments only. Such a arrangement of actin and myosin filaments gives a striated appearance for muscular cells under an optical or electron microscope.

In addition to contractile elements, the sarcomere contains supporting protein structures. Two well known proteins are titin and nebulin. *Titin* is a filamentous protein and is anchored to the Z-disk at one end and to the M-line at the other end. This molecular structure provides anchorage to the myosin filaments, contributing to the structural integrity and stability of the myosin filaments. Furthermore, titin contributes to the elasticity of the sarcomeres and muscle cells. *Nebulin* is a filamentous protein that is distributed with the actin filaments and provides structural and mechanical supports to the actin filaments.

Mechanisms of Muscle Contraction [21.1]

Based on a hypothetical model, muscle contraction is induced by actin filament sliding against myosin filaments. The myosin filaments are organized with the heavy-chain heads localized symmetrically to both ends of the myosin filaments. The actin filaments are distributed symmetrically with respect to the myosin filaments, so that the actin filaments can interact with the myosin heavy chain heads simultaneously at both ends of the myosin filaments. Thus, the interaction of myosin heads with the actin filaments induces sliding of the actin filaments toward the center of the myosin filaments or the M-line, resulting in the shortening of the sarcomeres and the contraction of the muscle cells.

The contraction of the muscle cells is a highly regulated process, which involves complicated regulatory mechanisms and numbers of regulatory molecules. A contractile process is initiated and controlled by signals from the motor centers of the central nervous system, including the brain and spinal cord. The skeletal muscle system is innervated with nerve fibers originated from these nerve centers. The central motor-controlling neurons generate electric signals called *action potentials*, which can be transmitted from the neurons to the peripheral muscle cells via nerve axons, initiating muscle contraction. The generation of action potentials is dependent on the resting *membrane potential*, which is defined as the voltage difference across the cell membrane at the resting state (without actin potentials). In a normal cell, the cytoplasmic surface of the cell membrane is negatively charged, whereas the extracellular surface is positively charged. The resting potential difference in almost all cell types ranges from −70 to −90 mV under physiological conditions. Such a resting potential difference results from the gradients of ion concentrations across the cell membrane. The concentration of ions, such as Na^+, K^+, Ca^{2+}, and Cl^-, differs from the cytoplasm to extracellular space with the degree of difference depending on the type of the ion. Ion channels specific to each ion type in the cell membrane are responsible for the generation and maintenance of the ion difference. The formation of the resting membrane potential is often called *polarization*. A cell with resting membrane potentials is said to be polarized.

The resting membrane potential can be reversed from a cytoplasmic negative to a positive value in response to various types of stimulation. Such a process is known as *depolarization* and the resulting potential is defined as *action potential*. The stimulation of resting cells induces the opening of gated Na^+ channels, resulting in Na^+ fluxes from the extracellular space to the cytoplasm (note that the Na^+ concentration in the extracellular space is higher than that in the cytoplasm) and a reduction in the membrane potential. When the membrane potential is reduced to a threshold level, a rapid depolarization process is triggered, leading to the generation of an action potential. Shortly after the generation of the action potentials, the Na^+ channels are closed, while the K^+ channels are open, resulting in the termination of inward Na^+ flux and the beginning of outward K^+ flux (note that the K^+ concentration in the cytoplasm is higher than that in the extracellular space). The outward K^+ flux results in the reestablishment of the resting membrane potential. Such a process is defined as *repolarization*.

There are two characteristics for the action potential. First, an action potential is initiated only when the resting membrane potential is reduced to the threshold level in response to a stimulus. Any change in the resting membrane potential that does not reach the threshold will not initiate action potentials. Once an action potential is initiated, the amplitude of the action potential remains constant. The level of stimulation does not influence the amplitude of the action potential. This phenomenon is known as the *all-or-none phenomenon*. Second, an action potential can propagate through the cell membrane. A cell is initially depolarized within a small area of plasma membrane. A local action potential can stimulate the cell membrane near the depolarized area, inducing the spreading of the action potential. This is a basic process for the propagation of action potentials.

The contractile activity of the skeletal muscle system is controlled by the action potentials that are initiated in the neurons of the motor control centers and transmitted to the peripheral muscle cells via the nerve axons. Each axon develops multiple levels of branches, which project toward peripheral muscle cells. The end of each axon branch enlarges to form a terminal structure, which interacts with the plasma membrane of a muscle cell. The axon terminal and the local area of the muscle cell that interacts with the axon terminal are together called *synapse*. The axon terminal is referred to as the *presynaptic terminal*, the plasma membrane of the muscle cell is called the *postsynaptic membrane*, and the gap between the two types of membrane is known as the *synaptic cleft*. Each presynaptic terminal is composed of plasma membrane vesicles, referred to as *synaptic vesicles*. These vesicles contain a substance known as acetylcholine, a neurotransmitter that can be released into the synaptic cleft and stimulates the initiation of action potentials in the muscle cells.

When an action potential is transmitted from a motor neuron to the presynaptic terminal, the actin potential induces the opening of voltage-gated calcium channels, resulting in calcium flux into the presynaptic terminal. Increased calcium concentration in turn induces the release of acetylcholine from the synaptic vesicles to the synaptic cleft. Released acetylcholine interacts with and activates the acetylcholine receptors located in the postsynaptic membrane of the muscle cell. The activation of the acetylcholine receptors induces the opening of voltage-gated sodium channels, resulting in sodium flux into the muscle cell. The inward sodium flux causes depolarization of the muscle cell and initiation of action potentials.

The action potentials in the plasma membrane of the muscle cell can initiates muscle contraction via a process known as *excitation–contraction coupling*. In this process, the action potentials can be transmitted from the synapse to a tubular network called *T-tubules*, which are distributed continuously around the sarcoplasmic reticulum and the sarcomeres (note that the sarcoplasmic reticulum contains a high concentration of calcium). The action potentials in the T-tubules stimulate the sarcoplasmic reticulum, inducing the opening of the voltage-gated calcium channels. This action results in calcium release from the sarcoplasmic reticulum to the sarcoplasm, where the contractile myofibrils are located. Calcium ions bind to troponin, inducing a conformational change in the troponin–tropomyosin complex and the exposure of the active binding sites of the actin filaments. The active sites in turn interact with the myosin heads and induce the movement of the myosin heads. The actin–myosin interaction results in the sliding of the actin filaments against the myosin filaments and thus contraction of the muscle cells. After the sliding process, the myosin heads are released from the actin filaments and are prepared for another cycle of contraction. These activities require energy from ATPs. The myosin heads are composed of ATPases, which hydrolyze ATP molecules and generate energy. The energy is stored in the myosin heads and used for the release of the myosin heads from the actin filaments. Following the contraction process, the calcium ions are actively transported from the sarcoplasm to the sarcoplasmic reticulum. The decrease of the calcium concentration in the sarcoplasm induces muscle relaxation.

DISORDERS OF THE SKELETAL MUSCLE SYSTEM

Muscular Dystrophies

Pathogenesis, Pathology, and Clinical Features [21.2]. *Muscular dystrophies* are a group of hereditary disorders characterized by progressive loss of muscle mass and function, resulting in the disorder of the skeletal muscle system. Gene mutation-induced protein alterations and deficiency are primary causes for muscular dystrophies. Genetic studies have identified more than 30 forms of muscular dystrophy. Among these forms of muscular dystrophies, several forms are commonly seen, including the Duchenne's, Becker's, myotonic, and facioscapulohumeral muscular dystrophies. The clinical manifestations and pathological changes of these types are briefly discussed as follows.

Duchenne's muscular dystrophy is a recessive genetic disorder found in male patients. The disorder is induced by mutation of the dystrophin gene (*dys*), which encodes the dystrophin protein (see Table 21.1). Dystrophin is a constituent of the sarcolemma of muscle cells. This protein is capable of binding to actin filaments at the *N*-terminus, binding to syntrophin at the *C*-terminus, and binding to β-dystroglycan at a cysteine-rich domain. These molecular links provide a structural basis for the interaction of the actin cytoskeleton with extracellular matrix. In Duchenne's muscular dystrophy, dystrophin deficiency is often found, suggesting a role for dystrophin deficiency in the initiation and development of muscular dystrophy. Molecular analyses have demonstrated that dystrophin deficiency is often a result of dystrophin (*dys*) gene mutation. In most patients with muscular dystrophy, dystrophin gene mutation is induced by large fragment deletion, insertion, or point mutation that causes frameshifting of gene codons. These genetic alterations result in the deficiency or modulation (primarily in the cysteine-rich domain) of the dystrophin protein. The deficiency or modulation of dystrophin influences the interaction of the actin filaments with extracellular matrix, resulting in instability of the sarcolemmal structure and reducing cell adhesion and survival capabilities. All these changes promote cell degeneration, eventually leading to cell apoptosis and muscular dystrophy.

Duchenne's muscular dystrophy is the most common type of muscular dystrophy. The incidence of this disorder is about 0.01–0.03%. Signs of muscular dystrophy, such as muscle weakness and movement disorders, are usually found at the age of 5. Gait analyses often provide useful information for the diagnosis of the disorder. The patients eventually lose muscle strength and the ability of movements. The weakening of the

TABLE 21.1. Characteristics of Selected Muscular Dystrophy-Related Proteins*

Proteins	Alternative Names	Amino Acids	Molecular Weight (kDa)	Gene Locus	Expression	Functions
Dystrophin	DMD, apo-dystrophin 1	3685	427	Xp21.2	Skeletal muscle, heart, brain	Regulating the integrity and stability of skeletal and cardiac muscles
Syntrophin 1	SNT1, syntrophin α1, dystrophin-associated protein A1, Pro TGF α-cytoplasmic domain interacting protein, 159-kDa dystrophin-associated protein A1	505	54	20q11.2	Heart, skeletal muscle, brain, lung, liver, kidney, pancreas	A membrane protein associated with dystrophin, playing a role in regulating the integrity and stability of skeletal and cardiac muscles
Dystroglycan	Dystrophin-associated glycoprotein 1 (DAG), dystroglycan α, agrin receptor, dystroglycan β	895	98	3p21	Skeletal muscle, heart, brain, lung, liver, kidney	Forming complexes with dystrophin, regulating the integrity and stability of skeletal and cardiac muscles, and serving as a dual receptor for agrin and laminin-2 in the Schwann cell membrane

*Based on bibliography 21.2.

respiratory muscle system influences the function of the lung. Most patients with muscular dystrophy die of respiratory failure. Pulmonary infection often occurs because of food aspiration resulting from the loss of muscular strength responsible for swallowing. A unique feature of Duchenne's muscular dystrophy is the elevation of the serum c reatine kinase, although the level of the serum creatine kinase is normal at birth. Most patients show a significant increase in serum creatine kinase when muscular weakening occurs. Pathological examinations often demonstrate muscular necrosis, a reduction in the mass of skeletal muscle cells, and an increase in adipocytes and extracellular matrix.

Duchenne's muscular dystrophy is often associated with cardiac disorders, such as cardiomyopathy, which induces cardiac dilation and a reduction in cardiac contractility. Cardiomyopathy-induced heart failure is the second leading cause of fatality in Duchenne's muscular dystrophy (note that the first leading cause of death in this disorder is respiratory failure). Mutation or deletion of the dystrophin gene is responsible for the cardiac disorders. Cardiomyopathy is usually induced by truncated or reduced level of dystrophin, which may not cause noticeable skeletal muscle dystrophy. In dystrophin-deficient mice, the cardiomyocytes exhibit reduced compliance and enhanced cell contracture, presumably due to increased cell susceptibility to stretch-mediated calcium overload. Prolonged cardiac myocyte contracture induces cell death, a critical mechanism for the development of cardiomyopathy.

Characteristics of muscular dystrophy-related proteins are listed in Table 21.1.

Becker's muscular dystrophy is a disorder similar to Duchenne's muscular dystrophy in pathogenesis, pathology, and clinical features. Becker's muscular dystrophy is induced by mutation of the dystrophin gene. Compared to the Duchenne's muscular dystrophy, which is induced by complete dystrophin deficiency or mutation of the cysteine-rich domain of the dystrophin gene, Becker's muscular dystrophy is not associated with complete dystrophin deficiency or mutation of the cysteine-rich domain of the dystrophin gene. Thus, Becker's muscular dystrophy is not as severe as and progresses slower than the Duchenne's muscular dystrophy. The Becker's type of muscular disorder is also referred to as the "benign" form of Duchenne's muscular dystrophy. Clinical manifestations are usually found after the age of 15. Death due to muscular dystrophy occurs after the age of 40, which is much later than that due to Duchenne's muscular dystrophy.

Myotonic dystrophy is an autosomal dominant genetic disorder induced by gene mutation in chromosome 19. This disorder is characterized by progressive muscle weakening and the association of disorders in other systems, including cardiac disorders, cataract, intellectual impairment, gastrointestinal disorders, and respiratory failure. Muscle weakening is usually found in the distal extremities at the age of 10–30. Some patients may not show apparent signs or symptoms for a long time. Compared to the Duchenne and Becker muscular dystrophies, the level of serum creatine kinase may remain normal. Pathological examinations often show muscular atrophy or a reduction in the muscle mass. In the heart, conduction system is often involved with a high incidence of conduction block. In the lung, weakening of the respiratory muscles may result in ventilation disorder, hypoxia, cor pulmonale, and respiratory failure.

Facioscapulohumeral muscular dystrophy is a disorder characterized by progressive weakening of the facial, shoulder girdle, and arm muscles. This disorder is induced by autosomal dominant gene mutation and is found in both males and females at any age with a high incidence from age 30 to 40. Some patients may not show any signs of disorder.

Unlike the Duchenne and Becker muscular dystrophies, other systems are usually not involved. The serum creatine kinase level is normal or slightly increased.

Transgenic Models of Dystrophin Deficiency [21.3]. The dystrophin gene (dys) is a large gene with approximately 2.25 million base pairs, which contains 79 exons and a large fraction of introns (99.4%). In a large population of patients with muscular dystrophy (about 50%), dystrophin deficiency is caused by gene deletion- or insertion-induced frameshifting. A mouse *mdx* dystrophin-deficient model has been established for the study of human muscular dystrophy. The mouse dystrophin gene is basically homologous to that of the human. The genetic mechanisms of dystrophin gene mutations are also similar between mice and humans. Thus, the mouse *mdx* dystrophin-deficient model is commonly used for the study of the human disorder. In this mouse model, a C → T transition is induced in exon 23 of the dystrophin gene, resulting in frameshifting and dystrophin deficiency. Mice with dystrophin deficiency show increased degeneration and death of skeletal muscle cells and reduced muscle cell density (Fig. 21.2), in association with a reduction in the contractile strength (Fig. 21.3), which resemble pathological and functional alterations found in human muscular dystrophies. The overexpression of a dystrophin gene in the *mdx* model significantly reduces muscle cell degeneration and death, and improves the contractility of the skeletal muscle system. It was found from the mouse model that dystrophin is a necessary protein for maintaining the stability of the skeletal muscles and for preventing muscular dystrophy. The dystrophin deficiency model has also been established in dogs and cats, but the mouse *mdx* model is more popularly used for investigating the pathogenic mechanisms of human muscular dystrophy.

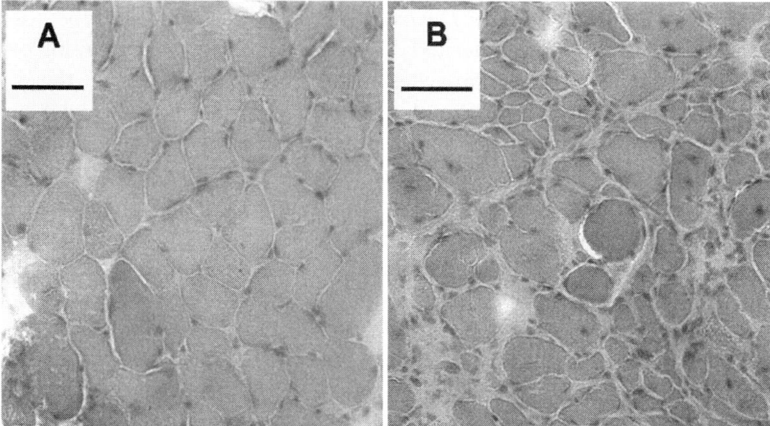

Figure 21.2. Photomicrographs of H&E-stained transverse sections of diaphragm muscles of control mice (A) and *mdx* mice, a model for Duchenne muscular dystrophy (B). Scale bars: 50 μm. (Reprinted with permission from Gregorevic P et al: Improved contractile function of the mdx dystrophic mouse diaphragm muscle after insulin-like growth factor-I administration, *Am J Pathol* 161:2263–72, copyright 2002.)

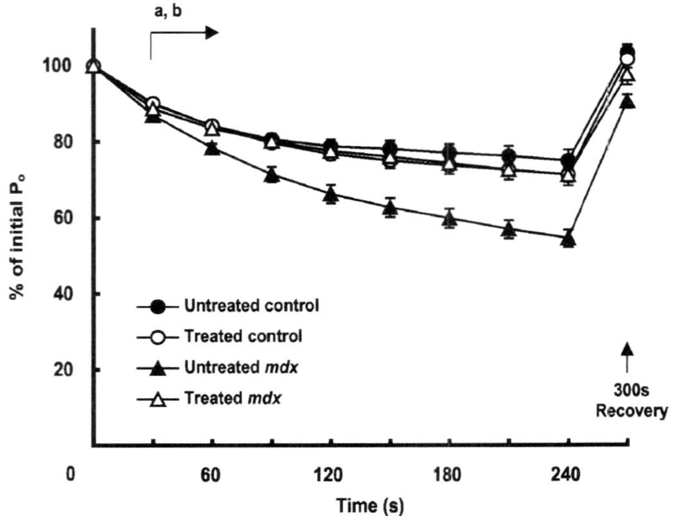

Figure 21.3. Maximum force output during and 300 s after a repeated stimulation fatigue protocol for diaphragm muscle preparations from control and *mdx* mice (an animal model for Duchenne muscular dystrophy) with and without insulin-like growth factor. Note the comparatively greater reduction in force throughout time in the diaphragm muscles of untreated dystrophic mice compared with untreated control mice (panel A, $P < 0.05$), and the enhanced resistance of muscles from treated dystrophic mice compared with the untreated dystrophic muscles (panel B, $P < 0.05$) present from 30 s after commencement of stimulation, to the cessation of stimulation. (Reprinted with permission from Gregorevic P et al: Improved contractile function of the mdx dystrophic mouse diaphragm muscle after insulin-like growth factor-I administration, *Am J Pathol* 161:2263–72, copyright 2002.)

Molecular Treatment of Muscular Dystrophy [21.4]. The various types of muscular dystrophy as described above are inherited disorders induced by gene deletion or mutation. To date, there are few effective approaches which can be used to treat these disorders. Recent studies have demonstrated that molecular regenerative therapies may be potentially used to treat muscular dystrophy and cardiomyopathy. Numerous genes that are deleted or mutated in muscular dystrophy, contributing to the development of the disorder. The repair and replacement of these genes are potential approaches for the treatment of muscular dystrophy. There are two basic molecular approaches: transfer genes that encode proteins directly responsible for the disorder, and transfer "booster" genes that encode proteins responsible for the survival of muscle cells and the prevention of cell apoptosis. In this section, Duchenne's muscular dystrophy is used as an examples to demonstrate the principles of molecular therapy for muscular dystrophy.

Transfer of Wildtype Dystrophin Gene [21.5]. For Duchenne's muscular dystrophy, the primary cause is the deletion or mutation of the dystrophin gene. Thus, direct transfer of a functional wildtype dystrophin gene into the muscular cells is a potential approach for the treatment of the disorder. This gene has been tested in a transgenic animal model, the dystrophin-deficient *mdx* mouse, which resembles the Duchenne's muscular dystrophy in humans. A number of approaches have been developed and used for the delivery of the dystrophin gene, including virus-, liposome-, and electroporation-mediated gene deliveries. Among these approaches, the virus-mediated gene delivery is the most effective approach. The dystrophin gene can be integrated into modified viral vectors and used to transfer into target muscular cells. Experimental investigations have demonstrated the effectiveness of such an approach. In particular, the transfer of the full-length dystrophin gene with a muscle-specific gene promoter (muscle creatine kinase promoter) has been shown to effectively prevent the progression of muscular dystrophy in the *mdx* mouse model of muscular dystrophy. The use of the muscle-specific gene promoter can increase the efficiency of gene transfer. However, there are potential problems. These include limited efficiency of gene transfer, immune responses provoked by the expression of the transferred gene and corresponding protein products, temporary gene expression, and poor cell survival. These problems remain to be resolved.

Given the problems with the transfer of the wildtype dystrophin gene, several alternative strategies have been developed and used for the molecular treatment of muscular dystrophy. These include the construction and delivery of truncated dystrophin gene, mutant gene correction by small fragment homologous replacement, correction of mutant genes by chimeraplasty, removal of mutant gene fragments by exon skipping, and compensation for the lost function of mutant dystrophin. These approaches are discussed as follows.

Delivery of Truncated Dystrophin Genes or Microdystrophin Gene Constructs [21.6]. The dystrophin gene is composed of a large number of base pairs, which complicate the preparation and manipulation of the dystrophin gene. Genetic and functional analyses has demonstrated that not all gene sequences are necessary for the production of functional dystrophin. Indeed, selected regions of the dystrophin gene can be deleted and the remaining regions can be recombined to generate a minidystrophin gene. A minimal requirement is that the reconstructed gene must contain critical domains responsible for regulating the structural stability and function of the sarcolemma of muscular cells. The reconstructed minidystrophin gene can be highly functional. When delivered into the mouse *mdx* muscular dystrophy model, pathological alterations of muscular dystrophy can be significantly prevented and the contractility of the skeletal muscles can be improved (Fig. 21.4). Furthermore, such an approach can be used to reduce muscular cell degeneration in the *mdx* mouse model of muscular dystrophy. However, it is still debating whether the delivery of the truncated dystrophin gene is more advantageous than that of the full-length dystrophin gene.

Mutant Gene Correction by Small Fragment Homologous Replacement (SFHR) [21.7]. Small fragment homologous replacement is a technique used for constructing and inserting PCR-generated DNA amplicons into the host genome to correct mutant genes. PCR can generate large DNA fragments up to several hundred base pairs. Designed corrective DNA fragments can be constructed and transferred into target cells. When integrated into the genome, these fragments can replace and correct mutant genes, resulting in the generation of functional genes. Such a technique has been applied to cells from the *mdx* mouse model of muscular dystrophy to correct mutant dystrophin gene. While the mechanisms of gene correction remain poorly understood, preliminary investigations have

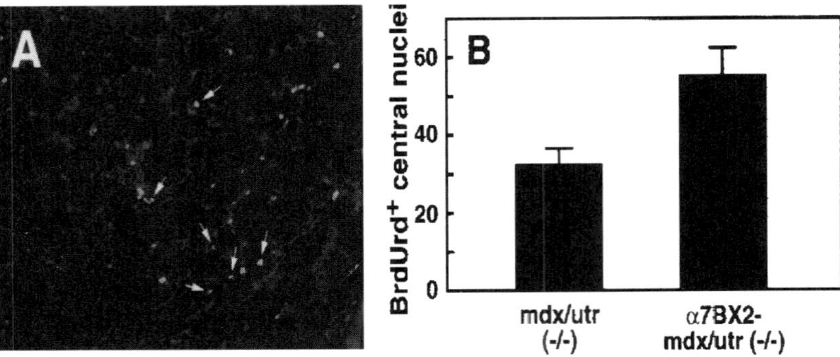

Figure 21.4. Increased integrin promotes the proliferation of satellite cells *mdx/utr*$^{-/-}$ mice (a animal model of muscular dystrophy) and *α7BX2-mdx/utr*$^{-/-}$ mice (a muscular dystrophy model with enhanced expression of the α7β1 integrin). Mice were injected with BrdU to label replicating cells. Muscle specimens were collected and analyzed by immunohistochemistry for BrdU incorporation into DNA. Nuclei are stained with DAPI. BrdUrd-labeled central nuclei (arrows in panel A) in 50 random fields were scored for each animal. Mean numbers (±SEM) are given for 11 animals for each genotype. (B) Increased integrin expression enhanced the proliferation of satellite cells and the regenerative capacity of dystrophic muscle. (Reprinted with permission from Burkin DJ et al: Transgenic expression of alpha7beta1 integrin maintains muscle integrity, increases regenerative capacity, promotes hypertrophy, and reduces cardiomyopathy in dystrophic mice, *Am J Pathol* 166:253–63, copyright 2005.)

provided promising results for this technique. As shown in a study with cultured muscular cells, the delivery of a PCR amplicon into muscular cells with dystrophin deficiency results in the correction of the dystrophin gene in about 20% cells. However, a higher efficiency may be needed to achieve therapeutic effects.

Correction of Mutant Genes by Chimeraplasty [21.8]. Chimeraplasty is a technique used for correcting mutant genes with chimeric RNA–DNA oligonucleotides, also known as chimeraplasts. Short chimeric genetic structures can be constructed by hybridizing complementary 2′-O-methyl ribonucleotide analogues to desired DNA fragments. Such a chimeric complex protects the DNA fragments from exonucleolytic digestion. When the chimeraplasts are delivered to the cell nucleus, the DNA and RNA fragments can anneal to the target site during gene transcription. The chimeraplasts can repair or replace base-pair mismatches, if any, resulting in the correction of gene mutation, although the exact mechanisms remain poorly understood. Experimental investigations with in vitro models have demonstrated a correction efficiency about 30%. The efficiency may be further improved when the mechanisms of chimeraplasty are fully understood. Chimeraplasty has been applied to the *mdx* mouse model of muscular dystrophy. The delivery of chimeraplasts for the dystrophin gene into the skeletal muscle cells results in the correction of mutant dystrophin gene in about 10% cells. Such a manipulation induces an increase in the expression of functional dystrophin gene and a reduction in the symptoms of muscular dystrophy.

Removal of Mutant Gene Fragments by Exon Skipping [21.9]. Exon skipping is a technique used to target selected mutant gene fragments and block the transcription of the targeted fragments by introducing specific antisense 2′-*O*-methyl ribonucleotide analogs to cell nuclei. The antisense ribonucleotides can bind to and block specific homologous DNA exons or sequences in the genome during transcription. Such a process induces a transcription-skip over the ribonucleotide-blocked exons. In other words, the blocked exons can no longer be transcribed. When a ribonucleotide sequence is designed and delivered to target a specific mutant gene fragment, the mutant fragment cannot be expressed and the function of the generated protein may be improved.

For the treatment of experimental muscular dystrophy in the *mdx* mouse model, a 2′-*O*-methyl ribonucleotide analogue sequence can be designed to target the exon 23 (at the junction with intron 22), which contains a mutant fragment responsible for the development of muscular dystrophy. The delivery of this ribonucleotide analogue into the *mdx* mouse model stops the transcription of the exon 23. Such a manipulation results in the generation of a dystrophin form similar to that found in Becker's muscular dystrophy, which is significantly less severe than Duchenne's muscular dystrophy. Although the approach does not provide a complete cure of the muscular dystrophy, the pathological changes are reduced and the contractility of the muscle system is improved.

Compensation for Lost Function of Dystrophin [21.10]. There exist proteins that potentially compensate for the function of dystrophin. One of such compensating factors is utrophin, also known as *dystrophin-like protein* and dystrophin related protein 1. Utrophin is a protein of 3422 amino acid residues and about 395 kDa in molecular weight. This protein is similar to dystrophin in structure and function. As dystrophin, utrophin is expressed in skeletal muscle cells and is localized to the sarcolemma and acetylcholine receptors at the neuromuscular synapses and myotendinous junctions, where it regulates the function of the postsynaptic membrane and, especially, the activity of the acetylcholine receptors. Utrophin is also expressed in the heart, brain, lung, kidney, liver, intestine, and testis. Utrophin can interact with dystrophin at the *C*-terminus. The suppression or loss of the utrophin activity exacerbates pathological changes of muscular dystrophy in the *mdx* mouse model of muscular dystrophy. The upregulation of the utrophin gene has been shown to reduce pathological alterations and compensates for functional abnormalities due to dystrophin deficiency in the *mdx* mouse model of muscular dystrophy. Growth factors, interleukin-6, l-arginine, and nitric oxide can enhance the expression of the utrophin gene promoter.

Transfer of Dystrophin "Booster" Genes [21.11]. In addition to the dystrophin gene, a number of "booster" genes have been discovered and used for the molecular treatment of muscular dystrophy. These genes encode proteins that mediate the survival and enhance the function of the of striated muscular cells. Common "booster" genes include integrin α7β1, ADAM12, calpastatin, nitric oxide synthase, insulin-like growth factor (IGF)I, myostatin, and mini-agrin.

INTEGRIN α7β1. Cell adhesion to extracellular matrix is critical to cell survival. The impairment of muscle cell adhesion to extracellular matrix may induce cell apoptosis, contributing to muscular dystrophy. In skeletal muscle cells, there are two major types of cell–matrix interaction-mediating molecules, including the dystrophin-associated glycoproteins (see Table 21.1 chapter for dystroglycan, a major dystrophin-associated glycoprotein) and integrin α7β1. These two molecules coordinately regulate muscular cell attachment to extracellular matrix. In the case of dystrophin deficiency, cell attachment mediated by the dystrophin-associated glycoproteins is impaired. The overexpression of the integrin α7β1 gene can partially rescue the functional loss of dystrophin (Fig. 21.5). Thus, the transfer of the integrin α7β1 gene into target muscular cells is a potential approach for the treatment of muscular dystrophy.

Characteristics of several therapeutic molecules for muscular dystrophy are presented in Table 21.2.

ADAM12. ADAM (A disintegrin and metalloprotease or meltrin)12 is a molecule that possesses integrin-binding and metalloproteinase activities. This molecule is expressed in skeletal muscle cells during development and regeneration, and plays a critical role in the regulation of muscular formation and morphogenesis. Furthermore, ADAM12 enhances cell attachment to extracellular matrix through interacting with syndecans and promotes cell spreading via binding to β1 integrin-containing complexes. In the *mdx* mouse model of muscular dystrophy, the overexpression of the ADAM12 gene results in a reduction in pathological changes found in muscular dystrophy and enhancement of muscular cell regeneration. Such effects may be related to the function of ADAM12 in regulating cell adhesion via interacting with integrins and syndecans.

CALPASTATIN. Calpastatin is a protein that inhibits the activity of calpain, a calcium-dependent protease that induces autoproteolysis and cell death. Calpain may participate in the regulation of cell degeneration in muscular dystrophy. In the transgenic *mdx* mouse model of muscular dystrophy, the transfer of the calpastatin gene into the target skeletal muscle cells results in the suppression of the activity of calpain in association with a reduction in muscle cell death and degeneration.

NITRIC OXIDE SYNTHASE. Nitric oxide synthase is an enzyme that catalyzes the formation of nitric oxide from L-arginine. Nitric oxide has been shown to exert an inhibitory effect on inflammatory reactions in various systems. Skeletal muscle cells express nitric oxide synthase, which is localized to the cell membrane. In muscular dystrophy, the expression of the nitric oxide synthase gene is impaired and translocation of nitric oxide synthase occurs, resulting in a reduction in the production of nitric oxide. These changes are associated with profound inflammatory reactions in the skeletal muscles, which are thought to contribute to the development of muscular dystrophy. The overexpression of the nitric oxide synthase gene in the skeletal muscle cells of the *mdx* mouse muscular dystrophy model induces an increase in the level of nitric oxide as well as a reduction in inflammatory reactions and muscle cell death (Fig. 21.6).

INSULIN-LIKE GROWTH FACTOR. Cell degeneration is a critical process that leads to muscular dystrophy. One treatment strategy for muscular dystrophy is to enhance cell regeneration. Insulin-like growth factor is a molecule that stimulates such a process. Experimental investigations have shown that the overexpression of the insulin-like growth factor gene in the mouse *mdx* muscular dystrophy model results in a reduction in muscular cell death and improvement of muscular cell survival and regeneration. The insulin-like growth

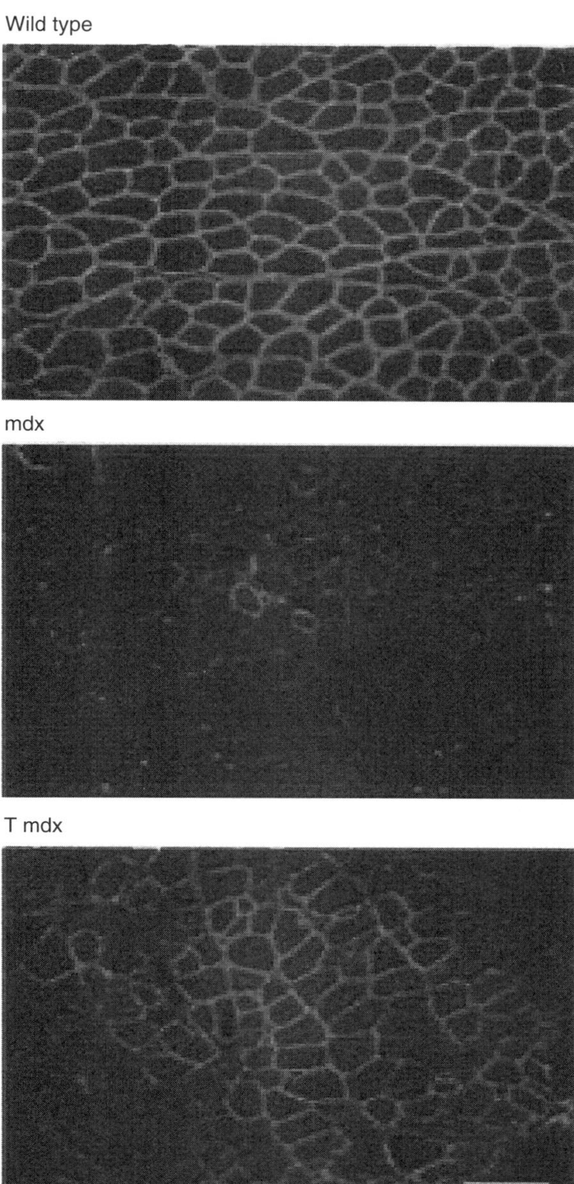

Figure 21.5. Systemic delivery of microdystrophin to dystrophic mice. (A) Antidystrophin immunofluorescence microscopy of tibialis anterior muscles from treated *mdx* mice (T*mdx*) administered 1×10^{12} vector genomes of rAAV6–CK6–microdystrophin and 10μg VEGF, compared with wild-type and untreated mdx mice, a model of muscular dystrophy. Dystrophin expression is increased in the muscles of treated compared with untreated *mdx* mice, but remains mosaic compared with wildtype mice. Scale bars: 100mm. (Reprinted by permission from Macmillan Publishers Ltd.: Gregorevic P et al: Systemic delivery of genes to striated muscles using adeno-associated viral vectors, *Nature Med* 10:828–34, copyright 2004.)

TABLE 21.2. Characteristics of Selected Therapeutic Proteins for Muscular Dystrophy*

Proteins	Alternative Names	Amino Acids	Molecular Weight (kDa)	Expression	Functions
Integrin α7	ITGA7	1137	124	Heart, skeletal muscle, nervous system, lung, intestine, ovary, prostate gland	Joining with integrin β1 to form an integrin complex, which is a major integrin complex expressed in differentiated muscular cells (note that all integrins are heterodimeric integral membrane proteins composed of an α chain and a β chain), binding to the extracellular matrix protein laminin-1, and regulating cell attachment to extracellular matrix
Integrin β1	ITGB1, fibronectin receptor β subunit (FNRB), fibronectin receptor β subunit-like, very late activation protein β polypeptide (VLA β)	825	92	Heart, nervous system, skeletal muscle, lymphocytes, liver, bone, cartilage, skin	Joining with an integrin α subunit to form integrin complexes, regulating cell adhesion to extracellular matrix, and regulating various cellular activities, including embryogenesis, cell proliferation and migration, immune responses, and metastasis of tumor cells

ADAM12	A disintegrin and metalloproteinase domain 12, meltrin α	909	100	Heart, skeletal muscle, placenta, intestine, stomach, uterus, urinary bladder	A membrane-anchored protein that regulates cell–cell and cell–matrix interactions, and mediates other cellular processes including fertilization, skeletal muscle development, and regeneration, as well as neurogenesis
Calpastatin	Calpain inhibitor	708	77	Red blood cells, leukocytes, intestine, skeletal muscle, testis, prostate gland, spleen, intestine	Serving as an inhibitor for endogenous calpain (calcium-dependent cysteine protease), regulating membrane fusion events, such as neural vesicle exocytosis and platelet and red cell aggregation
Myostatin	MSTN, growth differentiation factor 8 (GDF8)	375	43	Heart, skeletal muscle	A member of the bone morphogenetic protein (BMP) family and the TGFβ family, negatively regulating skeletal muscle growth
Agrin	AGRN	2045	215	Ubiquitous	Inducing the aggregation of acetylcholine receptors and other postsynaptic proteins on muscle fibers; also regulating the formation of neuromuscular junction

*Based on bibliography 21.11.

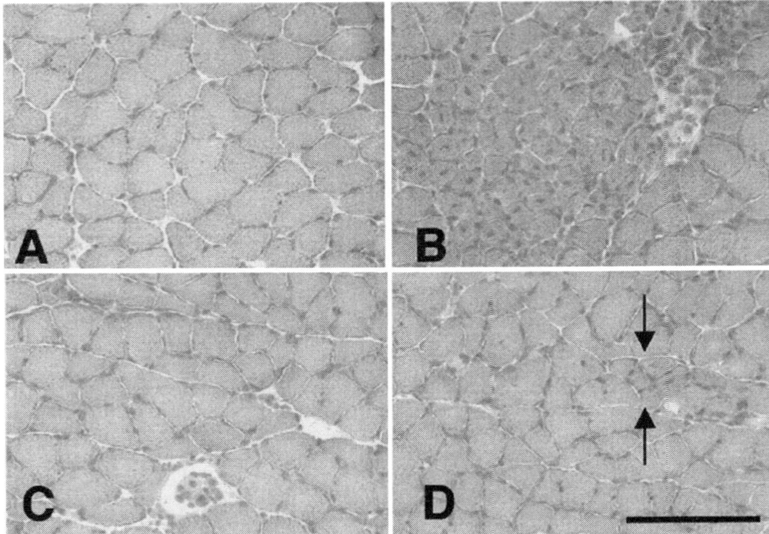

Figure 21.6. Influence of NOS expression on the morphology of skeletal muscle cells: (A) C57 control muscle showing fibers of uniform diameter, no central nucleation, and no clusters of inflammatory cells between adjacent fibers; (B) mdx muscle showing a typical focus of muscle pathology characterized by fiber populations of variable diameters and central nucleation (note that transgenic mdx mice are null mutants for dystrophin). Dark staining nuclei of inflammatory cells appear between adjacent fibers; (C) NOS Tg/mdx muscle showing typical histology, where fiber diameter is more uniform than age-matched mdx muscle, and there is little inflammation or central nucleation (note that NOS Tg/mdx is a transgenic mouse model with deficient dystrophin but with the expression of the NOS transgene); (D) NOS Tg/mdx muscle showing an example of the relatively small lesions that appear in NOS Tg/mdx muscle (between arrows) where there are small clusters of small-diameter, central-nucleated fibers. All micrographs are at the same magnification. Scale bar: 250 μm. (Reprinted with permission from Wehling M et al: Clonal isolation of muscle-derived cells capable of enhancing muscle regeneration and bone healing, *J Cell Biol* 155:123–32, copyright 2001.)

factor gene can serve as a candidate gene for the molecular treatment of muscular dystrophy.

MYOSTATIN. Myostatin is a protein that negatively regulates the development and regeneration of skeletal muscles. In animal models with myostatin gene mutation, hyperplasia and hypertrophy occur in the skeletal muscle system. The overexpression of the myostatin gene results in the degeneration of the skeletal muscle cells. Thus, it is conceivable that the suppression of the activity of the myostatin gene is beneficial for the treatment of muscular dystrophy. Such a goal can be achieved by delivering antisense oligonucleotides or small interfering RNA specific to the myostatin mRNA to compromise the translation of the myosin protein. In addition, local delivery of antimyostatin antibody and myostatin inhibitors can achieve the same goal.

MINIAGRIN. Miniagrin is a fragment of a large protein called *agrin*, which is known to induce the aggregation of acetylcholine receptors and other postsynaptic proteins in muscular cells and regulate the formation of the neuromuscular junction. Agrin also interacts with laminin and dystroglycan, enhancing cell adhesion and survival. There are two important domains within the agrin protein: the *N*-terminal domain (responsible for binding to laminin) and the *C*-terminal domain (responsible for binding to dystroglycan). A miniagrin gene has been constructed with the *N*- and *C*-terminal domains. The delivery of the constructed miniagrin gene into target skeletal muscle cells in a laminin-2-deficient model (associated with impairment of cell adhesion and muscle weakness) results in upregulation of laminin and crosslink of laminin with dystroglycan. These activities are associated with enhanced cell adhesion, reduced muscle degeneration and dystrophic symptoms, and improved muscular contractility.

Cellular Regenerative Engineering for Muscular Dystrophy. Cell transplantation is a potential approach for the treatment of muscular dystrophy. There are two potential cell types, including muscular progenitor cells and stem cells, which can be used for cell transplantation. Cell transplantation may elicit two possible therapeutic effects: (1) transplanted cells can differentiate into muscular cells, replacing cells with muscular dystrophy; and (2) transplanted cells can serve as carriers for the delivery of therapeutic genes such as the dystrophin gene and "booster" genes.

Muscular Progenitor Cells [21.12]. The skeletal muscle system contains muscular progenitor cells, also known as *skeletal myoblasts*, *myogenic cells*, and *satellite cells*, which are capable of differentiating to mature muscular cells. Experimental investigations have identified muscular progenitor cells based on stem cell- and progenitor cell-specific markers. The muscular progenitor cells express a cell surface molecule specific for stem-like cells, known as *stem cell antigen1* (Sca1). In addition, these cells may coexpress other cell surface markers, including CD34, myf5, and m-cadherin. The muscular progenitor cells can be identified by immunochemical labeling of specific surface markers in conjunction with a cell sorting approach, such as magnetic bead-assisted cell sorting (tagging iron beads with a specific antibody and enriching antibody-labeled cells by magnetization) and fluorescence-activated cell sorting or FACS (labeling cells with a fluorescent antibody and enriching antibody-labeled cells by fluorescence-based cell sorting). Once muscular progenitor cells are identified and enriched, the cells can be used for transplantation into target muscular cells. In the mouse *mdx* muscular dystrophy model, the transplanted muscular progenitor cells are capable of differentiating into mature muscular cells, replacing dystrophic cells, and reducing pathological changes and symptoms of muscular dystrophy (Fig. 21.7). Furthermore, the muscular progenitor cells can be transfected with therapeutic genes for muscular dystrophy, such as the dystrophin gene and "booster" genes, and used as gene delivery carriers. When the progenitor cells are transplanted into target muscular cells, the proteins produced by the transplanted cells can elicit a therapeutic effect on the host dystrophic cells.

Stem Cells [21.13]. As discussed on Chapter 9, there are several types of stem cells, including embryonic, fetal, and adult stem cells. These stem cell types can be potentially

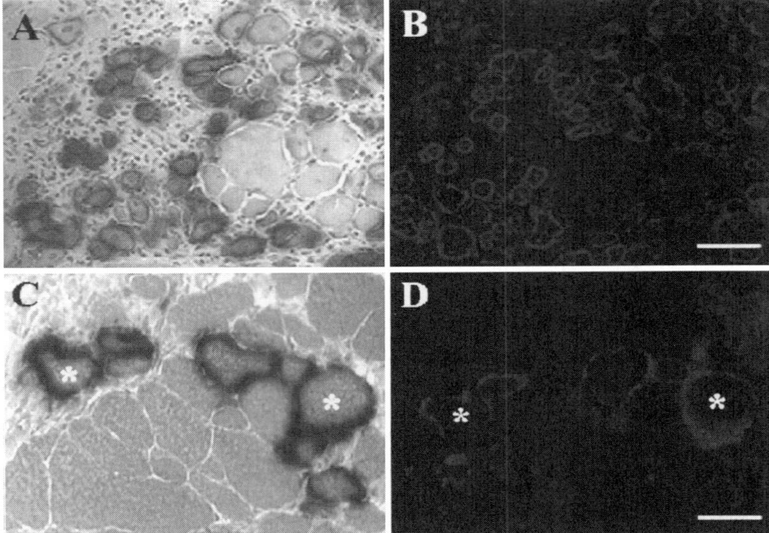

Figure 21.7. In vivo differentiation of mc13 cells into myogenic lineage after intramuscular (IM) and intravenous (IV) injection (note that mc13 cells are muscle-derived stem cells and are capable of differentiating into myogenic and osteogenic lineage in vitro and in vivo). The mc13 cells were stably transfected with a plasmid DNA construct encoding LacZ, dystrophin, and neomycin resistance genes and injected intramuscularly into hind limbs of *mdx* mice. After 7 days, and hind-limb musculature was isolated for histology. Many LacZ-positive myofibers (A) were found at the injected site that colocalized with dystrophin-positive myofibers (B). Some LacZ (C,*) and dystrophin positive myofibers (D,*) were also found in the hind limb muscle of mdx mice after IV injection of mc13. Scale bars: 100 μm (A,B); 50 μm (C,D). (Reprinted with permission from Lee JY et al: *J Cell Biol* 150:1085–100, copyright 2000.)

used for the treatment of muscular dystrophy. A desired type of stem cells can be identified, enriched, and transplanted into target muscular cells. These procedures are technically similar to those described above for the transplantation of muscular progenitor cells. In experimental investigations, multipotent embryonic stem cells have been used for the treatment of muscular dystrophy in the mouse *mdx* dystrophin-deficient model. These investigations have shown that embryonic stem cell transplantation is an effective approach for the prevention of pathological changes and the relief of the symptoms of muscular dystrophy. In addition, extensive investigations have been conducted to demonstrate the possibility of using adult stem cells for the treatment of muscular dystrophy. Bone marrow stem cells have been identified, enriched, and delivered into the circulation of the *mdx* mouse with dystrophin deficiency. The delivered cells can integrate into the skeletal muscle system and express dystrophin, improving the function of the dystrophic muscular cells. Other types of adult stem cells, such as the liver adipocytes, have also been used for transplantation into dystrophic muscular cells. Such an approach results in beneficial effects for the treatment of dystrophin deficiency-induced muscular dystrophy.

Potential Limitations [21.14]. There are several potential problems for the cellular treatment of muscular dystrophy. First, in patients with dystrophin deficiency and muscular dystrophy, the muscular progenitor cells or other types of stem cells are unlikely capable of expressing dystrophin. Thus, therapeutic cells can only be collected from an allogenic source or a donor individual. The transplantation of living allogenic cells induces acute immune rejection responses, resulting in rapid death of the transplanted cells. Second, it is impossible to deliver therapeutic cells to all dystrophic muscular cells over the entire body. The effects of cellular therapy are often limited to a small area around the cell delivery site. Although cells can be delivered through the blood circulation, the rate of cell integration into the skeletal muscle system is very low. Further investigations are needed to resolve these problems.

BIBLIOGRAPHY

21.1. Anatomy and Physiology

Guyton AC, Hall JE: *Textbook of Medical Physiology*, 11th ed, Saunders, Philadelphia, 2006.

McArdle WD, Katch FI, Katch VL: *Essentials of Exercise Physiology*, 3rd ed, Lippincott Williams & Wilkins, Baltimore, 2006.

Germann WJ, Stanfield CL (with contributors Niles MJ, Cannon JG), *Principles of Human Physiology*, 2nd ed, Pearson Benjamin Cummings, San Francisco, 2005.

Thibodeau GA, Patton KT: *Anatomy & Physiology*, 5th ed, Mosby, St Louis, 2003.

Boron WF, Boulpaep EL: Medical physiology: A cellular and molecular approach, Saunders, Philadelphia, 2003.

Ganong WF: *Review of Medical Physiology*, 21st ed, McGraw-Hill, New York, 2003.

21.2. Pathogenesis, Pathology, and Clinical Features of Muscular Dystrophies

Koenig M, Hoffman EP, Bertelson CJ, Monaco AP, Feener C et al: Complete cloning of the Duchenne muscular dystrophy (DMD) cDNA and preliminary genomic organization of the DMD gene in normal and affected individuals, *Cell* 50:509–17, 1987.

Hoffman EP, Brown RH Jr, Kunkel LM: Dystrophin: the protein product of the Duchenne muscular dystrophy locus, *Cell* 51:919–28, 1987.

Cohn RD, Campbell KP: Molecular basis of muscular dystrophies, *Muscle Nerve* 23:1456–71, 2000.

Bornemann A, Anderson LV: Diagnostic protein expression in human muscle biopsies, *Brain Pathol* 10:193–214, 2000.

Watkins SC, Hoffman EP, Slayter HS, Kunkel LM: *Immunoelectron microscopic localization of dystrophin in myofibres*, *Nature* 333:863–6, 1988.

Ohlendieck K, Campbell KP: Dystrophin constitutes 5% of membrane cytoskeleton in skeletal muscle, *FEBS Lett* 283:230–4, 1991.

Hoffman EP, Brown RH Jr, Kunkel LM: Dystrophin: The protein product of the Duchenne muscular dystrophy locus, *Cell* 51:919–28, 1987.

Koenig M, Monaco AP, Kunkel LM: The complete sequence of dystrophin predicts a rod-shaped cytoskeletal protein, *Cell* 53:219–26, 1988.

Ohlendieck K, Campbell KP: Dystrophin-associated proteins are greatly reduced in skeletal muscle from mdx mice, *J Cell Biol* 115:1685–94, 1991.

Ohlendieck K, Matsumura K, Ionasescu VV et al: Duchenne muscular dystrophy: Deficiency of dystrophin-associated proteins in the sarcolemma, *Neurology* 43:795–800, 1993.

Hoffman EP, Fischbeck KH, Brown RH et al: Characterization of dystrophin in muscle-biopsy specimens from patients with Duchenne's or Becker's muscular dystrophy, *New Engl J Med* 318:1363–8, 1988.

Blau HM, Webster C, Pavlath GK: Defective myoblasts identified in Duchenne muscular dystrophy, *Proc Natl Acad Sci USA* 80:4856–60, 1983.

Koenig M, Hoffman EP, Bertelson CJ, Monaco AP, Feener C et al: Complete cloning of the Duchenne muscular dystrophy (DMD) cDNA and preliminary genomic organization of the DMD gene in normal and affected individuals, *Cell* 50:509–17, 1987.

Hoffman EP, Monaco AP, Feener CC, Kunkel LM: Conservation of the Duchenne muscular dystrophy gene in mice and humans, *Science* 238:347–50, 1987.

Monaco AP, Bertelson CJ, Colletti-Feener C, Kunkel LM: Localization and cloning of Xp21 deletion breakpoints involved in muscular dystrophy, *Hum Genet* 75:221–7, 1987.

Nobile C, Marchi J, Nigro V, Roberts RG, Danieli GA: Exon-intron organization of the human dystrophin gene, *Genomics* 45:421–4, 1997.

Bies RD, Phelps SF, Cortez MD, Roberts R, Caskey CT, Chamberlain JS: Human and murine dystrophin mRNA transcripts are differentially expressed during skeletal muscle, heart, and brain development, *Nucleic Acids Res* 20:1725–31, 1992.

Boyce FM, Beggs AH, Feener C, Kunkel LM: Dystrophin is transcribed in brain from a distant upstream promoter, *Proc Natl Acad Sci USA* 88:1276–80, 1991.

Lederfein D, Levy Z, Augier N et al: A 71-kilodalton protein is a major product of the Duchenne muscular dystrophy gene in brain and other nonmuscle tissues, *Proc Natl Acad Sci USA* 89:5346–50, 1992.

Lidov HG, Kunkel LM: Dp140: Alternatively spliced isoforms in brain and kidney, *Genomics* 45:132–9, 1997.

Hoffman EP: Dystrophinopathies, in: *Disorders of Voluntary Muscle*, 7th edn, Karpati G, Hilton-Jones D, Griggs RC, eds, Cambridge University Press, 2001, pp 385–432.

Bulfield G, Siller WG, Wight PA, Moore KJ: X chromosome-linked muscular dystrophy (mdx) in the mouse, *Proc Natl Acad Sci USA* 81:1189–92, 1984.

Cooper BJ, Winand NJ, Stedman H et al: The homologue of the Duchenne locus is defective in X-linked muscular dystrophy of dogs, *Nature* 334:154–6, 1988.

Carpenter JL, Hoffman EP, Romanul FC et al: Feline muscular dystrophy with dystrophin deficiency, *Am J Pathol* 135:909–19, 1989.

Im WB, Phelps SF, Copen EH, Adams EG, Slightom JL, Chamberlain JS: Differential expression of dystrophin isoforms in strains of mdx mice with different mutations, *Hum Mol Genet* 5:1149–53, 1996.

Sicinski P, Geng Y, Ryder-Cook AS, Barnard EA, Darlison MG, Barnard PJ: The molecular basis of muscular dystrophy in the mdx mouse: A point mutation, *Science* 244:1578–80, 1989.

Cox GA, Cole NM, Matsumura K et al: Overexpression of dystrophin in transgenic mdx mice eliminates dystrophic symptoms without toxicity, *Nature* 364:725–9, 1993.

Lynch GS, Rafael JA, Chamberlain JS, Faulkner JA: Contraction-induced injury to single permeabilized muscle fibres from mdx, transgenic mdx, and control mice, *Am J Physiol Cell Physiol* 279:C1290–4, 2000.

Dystrophin

Ahn AH, Kunkel LM: The structural and functional diversity of dystrophin, *Nature Genet* 3:283–91, 1993.

Chelly J, Kaplan JC, Maire P, Gautron S, Kahn A: Transcription of the dystrophin gene in human muscle and non-muscle tissues, *Nature* 333:858–60, 1988.

Bies RD: X-linked dilated cardiomyopathy, *New Engl J Med* 330:368–9, 1994.

Bogdanovich S, Krag TOB, Barton ER, Morris LD, Whittemore LA et al: Functional improvement of dystrophic muscle by myostatin blockade, *Nature* 420:418–21, 2002.

Cox GA, Cole NM, Matsumura K, Phelps SF, Hauschka SD et al: Overexpression of dystrophin in transgenic mdx mice eliminates dystrophic symptoms without toxicity, *Nature* 364:725–9, 1993.

England SB, Nicholson LVB, Johnson MA, Forrest SM, Love DR et al: Very mild muscular dystrophy associated with the deletion of 46% dystrophin, *Nature* 343:180–2, 1990.

Forrest SM, Cross GS, Speer A, Gardner-Medwin D, Burn J et al: Preferential deletion of exons in Duchenne and Becker muscular dystrophies, *Nature* 329:638–40, 1987.

Goyenvalle A, Vulin A, Fougerousse F, Leturcq F, Kaplan JC et al: Rescue of dystrophic muscle through U7 snRNA-mediated exon skipping, *Science* 306:1796–9, 2004.

Gussoni E, Soneoka Y, Strickland CD, Buzney EA, Khan MK et al: Dystrophin expression in the mdx mouse restored by stem cell transplantation, *Nature* 401:390–4, 1999.

Harper SQ, Hauser MA, DelloRusso C, Duan D, Crawford RW et al: Modular flexibility of dystrophin: Implications for gene therapy of Duchenne muscular dystrophy, *Nature Med* 8:253–61, 2002.

Hoffman EP, Knudson CM, Campbell KP, Kunkel LM: Subcellular fractionation of dystrophin to the triads of skeletal muscle, *Nature* 330:754–8, 1987.

Hoffman EP, Monaco AP, Feener CC, Kunkel LM: Conservation of the Duchenne muscular dystrophy gene in mice and humans, *Science* 238:347–50, 1987.

Kronqvist P, Kawaguchi N, Albrechtsen R, Xu X, Daa Schroder H et al: ADAM12 alleviates the skeletal muscle pathology in mdx dystrophic mice, *Am J Pathol* 161:1535–40, 2002.

Kunkel LM: Analysis of deletions in DNA from patients with Becker and Duchenne muscular dystrophy, *Nature* 322:73–7, 1986.

Lee CC, Pearlman JA, Chamberlain JS, Caskey CT: Expression of recombinant dystrophin and its localization to the cell membrane, *Nature* 349:334–6, 1991.

Lu QL, Mann CJ, Lou F, Bou-Gharios G, Morris GE et al: Functional amounts of dystrophin produced by skipping the mutated exon in the mdx dystrophic mouse, *Nature Med* 9:1009–14, 2003.

Monaco AP, Neve RL, Colletti-Feener C, Bertelson CJ, Kurnit DM et al: Isolation of candidate cDNAs for portions of the Duchenne muscular dystrophy gene, *Nature* 323:646–50, 1986.

Muntoni F, Cau M, Ganau A, Congiu R, Arvedi G, Mateddu A et al: Deletion of the dystrophin muscle-promoter region associated with X-linked dilated cardiomyopathy, *New Engl J Med* 329:921–5, 1993.

Ray PN, Belfall B, Duff C, Logan C, Kean V et al: Cloning of the breakpoint of an X;21 translocation associated with Duchenne muscular dystrophy, *Nature* 318:672–5, 1985.

Scott MO, Sylvester JE, Heiman-Patterson T, Shi YJ, Fieles W et al: Duchenne muscular dystrophy gene expression in normal and diseased human muscle, *Science* 239:1418–20, 1988.

Sicinski P, Geng Y, Ryder-Cook AS, Barnard EA, Darlison MG et al: The molecular basis of muscular dystrophy in the mdx mouse: a point mutation, *Science* 244:1578–80, 1989.

Tinsley JM, Potter AC, Phelps SR, Fisher R, Trickett JI et al: Amelioration of the dystrophic phenotype of mdx mice using a truncated utrophin transgene, *Nature* 384:349–53, 1996.

Toyo-Oka T, Kawada T, Nakata J, Xie H, Urabe M et al: Translocation and cleavage of myocardial dystrophin as a common pathway to advanced heart failure: A scheme for the progression of cardiac dysfunction, *Proc Natl Acad Sci USA* 101:7381–5, 2004.

Wehling M, Spencer MJ, Tidball JG: A nitric oxide synthase transgene ameliorates muscular dystrophy in mdx mice, *J Cell Biol* 155:123–31, 2001.

Xiong D, Lee GH, Badorff C, Dorner A, Lee S et al: Dystrophin deficiency markedly increases enterovirus-induced cardiomyopathy: A genetic predisposition to viral heart disease, *Nature Med* 8:872–7, 2002.

Yasuda S, Townsend D, Michele DE, Favre EG, Day SM et al: Dystrophic heart failure blocked by membrane sealant poloaxmer, *Nature* 436:1025–9, 2005.

Zubrzycka-Gaarn EE, Bulman DE, Karpati G, Burghes AHM, Belfall B et al: The Duchenne muscular dystrophy gene product is localized in sarcolemma of human skeletal muscle, *Nature* 333:466–9, 1988.

Syntrophin 1

Adams ME, Dwyer TM, Dowler LL, White RA, Froehner SC: Mouse alpha-1- and beta-2-syntrophin gene structure, chromosome localization, and homology with a discs large domain, *J Biol Chem* 270:25859–65, 1995.

Ahn AH, Freener CA, Gussoni E, Yoshida M, Ozawa E et al: The three human syntrophin genes are expressed in diverse tissues, have distinct chromosomal locations, and each bind to dystrophin and its relatives, *J Biol Chem* 271:2724–30, 1996.

Fernandez-Larrea J, Merlos-Suarez A, Urena JM, Baselga J, Arribas J: A role for a PDZ protein in the early secretory pathway for the targeting of proTGF-alpha to the cell surface, *Mol Cell* 3:423–33, 1999.

Hosaka Y, Yokota T, Miyagoe-Suzuki Y, Yuasa K, Imamura M et al: Alpha-1-syntrophin-deficient skeletal muscle exhibits hypertrophy and aberrant formation of neuromuscular junctions during regeneration, *J Cell Biol* 158:1097–107, 2002.

Dystroglycan

Campanelli JT, Roberds SL, Campbell KP, Scheller RH: A role for dystrophin-associated glycoproteins and utrophin in agrin-induced AChR clustering, *Cell* 77:663–74, 1994.

Cohn RD, Henry MD, Michele DE, Barresi R, Saito F et al: Disruption of Dag1 in differentiated skeletal muscle reveals a role for dystroglycan in muscle regeneration, *Cell* 110:639–48, 2002.

Cote PD, Moukhles H, Lindenbaum M, Carbonetto S: Chimaeric mice deficient in dystroglycans develop muscular dystrophy and have disrupted myoneural synapses, *Nature Genet* 23:338–42, 1999.

Gee SH, Montanaro F, Lindenbaum MH, Carbonetto S: Dystroglycan-alpha, a dystrophin-associated glycoprotein, is a functional agrin receptor, *Cell* 77:675–86, 1994.

Gorecki DC, Derry JMJ, Barnard EA: Dystroglycan: Brain localisation and chromosome mapping in the mouse, *Hum Mol Genet* 3:1589–97, 1994.

Hayashi YK, Ogawa M, Tagawa K, Noguchi S, Ishihara T et al: Selective deficiency of alpha-dystroglycan in Fukuyama-type congenital muscular dystrophy, *Neurology* 57:115–21, 2001.

Henry MD, Campbell KP: A role for dystroglycan in basement membrane assembly, *Cell* 95:859–970, 1998.

Ibraghimov-Beskrovnaya O, Ervasti JM, Leveille CJ, Slaughter CA, Sernett SW et al: Primary structure of dystrophin-associated glycoproteins linking dystrophin to the extracellular matrix, *Nature* 355:696–702, 1992.

Ibraghimov-Beskrovnaya O, Milatovich A, Ozcelik T, Yang B, Koepnick K et al: Human dystroglycan: Skeletal muscle cDNA, genomic structure, origin of tissue specific isoforms and chromosomal localization, *Hum Mol Genet* 2:1651–7, 1993.

Kanagawa M, Saito F, Kunz S, Yoshida-Moriguchi T, Barresi R et al: Molecular recognition by LARGE is essential for expression of functional dystroglycan, *Cell* 117:953–64, 2004.

Ma J, Nastuk MA, McKechnie BA, Fallon JR: The agrin receptor: Localization in the postsynaptic membrane, interaction with agrin, and relationship to the acetylcholine receptor, *J Biol Chem* 268:25108–17, 1993.

Matsumura K, Nonaka I, Campbell KP: Abnormal expression of dystrophin-associated proteins in Fukuyama-type congenital muscular dystrophy, *Lancet* 341:521–2, 1993.

Matsumura K, Tome FMS, Collin H, Azibi K, Chaouch M et al: Deficiency of the 50K dystrophin-associated glycoprotein in severe childhood autosomal recessive muscular dystrophy, *Nature* 359:320–2, 1992.

Matsumura K, Tome FMS, Ionasescu V, Ervasti JM, Anderson RD et al: Deficiency of dystrophin-associated proteins in Duchenne muscular dystrophy patients lacking COOH-terminal domains of dystrophin, *J Clin Invest* 92:866–71, 1993.

Michele DE, Barresi R, Kanagawa M, Saito F, Cohn RD et al: Post-translational disruption of dystroglycan-ligand interactions in congenital muscular dystrophies, *Nature* 418:417–22, 2002.

Moore SA, Saito F, Chen J, Michele DE, Henry MD et al: Deletion of brain dystroglycan recapitulates aspects of congenital muscular dystrophy, *Nature* 418:422–5, 2002.

Rambukkana A, Yamada H, Zanazzi G, Mathus T, Salzer JL et al: Role of alpha-dystroglycan as a Schwann cell receptor for Mycobacterium leprae, *Science* 282:2076–8, 1998.

Sealock R, Froehner SC: Dystrophin-associated proteins and synapse formation: Is alpha-dystroglycan the agrin receptor? *Cell* 77:617–9, 1994.

Tinsley JM, Blake DJ, Zuellig RA, Davies KE: Increasing complexity of the dystrophin-associated protein complex, *Proc Natl Acad Sci USA* 91:8307–13, 1994.

Yamada H, Denzer AJ, Hori H, Tanaka T, Anderson LVB et al: Dystroglycan is a dual receptor for agrin and laminin-2 in Schwann cell membrane, *J Biol Chem* 271:23418–23, 1996.

Human protein reference data base, Johns Hopkins University and the Institute of Bioinformatics, at http://www.hprd.org/protein.

21.3. Transgenic Model of Dystrophin Deficiency

Bulfield G, Siller WG, Wight PA, Moore KJ: X chromosome-linked muscular dystrophy (mdx) in the mouse, *Proc Natl Acad Sci USA* 81:1189–92, 1984.

Cooper BJ, Winand NJ, Stedman H et al: The homologue of the Duchenne locus is defective in X-linked muscular dystrophy of dogs, *Nature* 334:154–6, 1988.

Carpenter JL, Hoffman EP, Romanul FC et al: Feline muscular dystrophy with dystrophin deficiency, *Am J Pathol* 135:909–19, 1989.

Sicinski P, Geng Y, Ryder-Cook AS, Barnard EA, Darlison MG et al: The molecular basis of muscular dystrophy in the mdx mouse: A point mutation, *Science* 244:1578–80, 1989.

Cox GA, Cole NM, Matsumura K et al: Overexpression of dystrophin in transgenic mdx mice eliminates dystrophic symptoms without toxicity, *Nature* 364:725–9, 1993.

Lynch GS, Rafael JA, Chamberlain JS, Faulkner JA: Contraction-induced injury to single permeabilized muscle fibres from mdx, transgenic mdx, and control mice, *Am J Physiol Cell Physiol* 279:C1290–4, 2000.

Phelps SF, Hauser MA, Cole NM et al: Expression of full-length and truncated dystrophin mini-genes in transgenic mdx mice, *Hum Mol Genet* 4:1251–8, 1995.

21.4. Molecular Treatment of Muscular Dystrophy

Koenig M, Hoffman EP, Bertelson CJ, Monaco AP, Feener C et al: Complete cloning of the Duchenne muscular dystrophy (DMD) cDNA and preliminary genomic organization of the DMD gene in normal and affected individuals, *Cell* 50:509–17, 1987.

Hoffman EP, Brown RH Jr, Kunkel LM: Dystrophin: The protein product of the Duchenne muscular dystrophy locus, *Cell* 51:919–28, 1987.

Cohn RD, Campbell KP: Molecular basis of muscular dystrophies, *Muscle Nerve* 23:1456–71, 2000.

Bornemann A, Anderson LV: Diagnostic protein expression in human muscle biopsies, *Brain Pathol* 10:193–214, 2000.

Engvall E, Wewer UM: The new frontier in muscular dystrophy research: Booster genes, *FASEB J* 17:1579–84, 2003.

21.5. Transfer of Wildtype Dystrophin Gene

Cox GA, Cole NM, Matsumura K et al: Overexpression of dystrophin in transgenic mdx mice eliminates dystrophic symptoms without toxicity, *Nature* 364:725–9, 1993.

Acsadi G, Lochmuller H, Jani A et al: Dystrophin expression in muscles of mdx mice after adenovirus-mediated in vivo gene transfer, *Hum Gene Ther* 7:129–40, 1996.

Hauser MA, Robinson A, Hartigan-O'Connor D et al: Analysis of muscle creatine kinase regulatory elements in recombinant adenoviral vectors, *Mol Ther* 2:16–25, 2000.

Hartigan-O'Connor D, Kirk CJ, Crawford R, Mule JJ, Chamberlain JS: Immune evasion by muscle-specific gene expression in dystrophic muscle, *Mol Ther* 4:525–33, 2001.

21.6. Delivery of Truncated Dystrophin Genes or Microdystrophin Gene Constructs

Harper SQ, Hauser MA, DelloRusso C, Duan D, Crawford RW et al: Modular flexibility of dystrophin: Implications for gene therapy of Duchenne muscular dystrophy, *Nat Med* 8(3):253–61, 2002.

Sakamoto M, Yuasa K, Yoshimura M et al: Micro-dystrophin cDNA ameliorates dystrophic phenotypes when introduced into mdx mice as a transgene, *Biochem Biophys Res Commun* 293:1265–72, 2002.

Decrouy A, Renaud JM, Lunde JA, Dickson G, Jasmin BJ: Mini- and full-length dystrophin gene transfer induces the recovery of nitric oxide synthase at the sarcolemma of mdx 4cv skeletal muscle fibres, *Gene Ther* 5:59–64, 1998.

Wehling M, Spencer MJ, Tidball JG: A nitric oxide synthase transgene ameliorates muscular dystrophy in mdx mice, *J Cell Biol* 155:123–31, 2001.

21.7. Mutant Gene Correction by Small Fragment Homologous Replacement (SFHR)

Dowty ME, Williams P, Zhang G, Hagstrom JE, Wolff JA: Plasmid DNA entry into postmitotic nuclei of primary rat myotubes, *Proc Natl Acad Sci USA* 92:4572–6, 1995.

Hagstrom JE, Ludtke JJ, Bassik MC, Sebestyen MG, Adam SA, Wolff JA: Nuclear import of DNA in digitonin-permeabilized cells, *J Cell Sci* 110:2323–31, 1997.

Kapsa R, Quigley A, Lynch GS et al: In vivo and in vitro correction of the mdx dystrophin gene nonsense mutation by short-fragment homologous replacement, *Hum Gene Ther* 12:629–42, 2001.

21.8. Correction of Mutant Genes by Chimeraplasty

Cole-Strauss A, Yoon K, Xiang Y et al: Correction of the mutation responsible for sickle cell anemia by an RNA-DNA oligonucleotide, *Science* 273:1386–9, 1996.

Gamper HB Jr, Cole-Strauss A, Metz R, Parekh H, Kumar R et al: A plausible mechanism for gene correction by chimeric oligonucleotides, *Biochemistry* 39:5808–16, 2000.

Gamper HB, Parekh H, Rice MC, Bruner M, Youkey H et al: The DNA strand of chimeric RNA/DNA oligonucleotides can direct gene repair/conversion activity in mammalian and plant cell-free extracts, *Nucleic Acids Res* 28:4332–9, 2000.

Yoon K, Cole-Strauss A, Kmiec EB: Targeted gene correction of episomal DNA in mammalian cells mediated by a chimeric RNA-DNA oligonucleotide, *Proc Natl Acad Sci USA* 93:2071–6, 1996.

Kren BT, Cole-Strauss A, Kmiec EB, Steer CJ: Targeted nucleotide exchange in the alkaline phosphatase gene of HuH-7 cells mediated by a chimeric RNA/DNA oligonucleotide, *Hepatology* 25:1462–8, 1997.

Rando TA, Disatnik MH, Zhou LZ: Rescue of dystrophin expression in mdx mouse muscle by RNA/DNA oligonucleotides, *Proc Natl Acad Sci USA* 97:5363–8, 2000.

Bertoni C, Rando TA: Dystrophin gene repair in mdx muscle precursor cells in vitro and in vivo mediated by RNA-DNA chimeric oligonucleotides, *Hum Gene Ther* 13:707–18, 2002.

Bartlett RJ, Stockinger S, Denis MM et al: In vivo targeted repair of a point mutation in the canine dystrophin gene by a chimeric RNA/DNA oligonucleotide, *Nat Biotechnol* 18:615–22, 2000.

Kren BT, Parashar B, Bandyopadhyay P, Chowdhury NR, Chowdhury JR et al: Correction of the UDP-glucuronosyltransferase gene defect in the gunn rat model of crigler-najjar syndrome type I with a chimeric oligonucleotide, *Proc Natl Acad Sci USA* 96:10349–54, 1999.

21.9. Removal of Mutant Gene Fragments by Exon Skipping

De Angelis FG, Sthandier O, Berarducci B, Toso S, Galluzzi G et al: Chimeric snRNA molecules carrying antisense sequences against the splice junctions of exon 51 of the dystrophin pre-mRNA induce exon skipping and restoration of a dystrophin synthesis in Delta 48–50 DMD cells, *Proc Natl Acad Sci USA* 99(14):9456–61, 2002.

Goyenvalle A, Vulin A, Fougerousse F, Leturcq F, Kaplan JC et al: Rescue of dystrophic muscle through U7 snRNA-mediated exon skipping, *Science* 306(5702):1796–9, 2004.

Dunckley MG, Manoharan M, Villiet P, Eperon IC, Dickson G: Modification of splicing in the dystrophin gene in cultured Mdx muscle cells by antisense oligoribonucleotides, *Hum Mol Genet* 7:1083–90, 1998.

Wilton SD, Lloyd F, Carville K et al: Specific removal of the nonsense mutation from the mdx dystrophin mRNA using antisense oligonucleotides, *Neuromuscul Disord* 9:330–8, 1999.

Mann CJ, Honeyman K, Cheng AJ et al: Antisense-induced exon skipping and synthesis of dystrophin in the mdx mouse, *Proc Natl Acad Sci USA* 98:42–7, 2001.

van Deutekom JC, Bremmer-Bout M, Janson AA et al: Antisense-induced exon skipping restores dystrophin expression in DMD patient derived muscle cells, *Hum Mol Genet* 10:1547–54, 2001.

21.10. Compensation for Lost Function of Dystrophin

Deconinck AE, Rafael JA, Skinner JA, Brown SC, Potter AC et al: Utrophin-dystrophin-deficient mice as a model for Duchenne muscular dystrophy, *Cell* 90:717–27, 1997.

Grady RM, Teng H, Nichol MC, Cunningham JC, Wilkinson RS et al: Skeletal and cardiac myopathies in mice lacking utrophin and dystrophin: A model for Duchenne muscular dystrophy, *Cell* 90:729–38, 1997.

Love DR, Hill DF, Dickson G, Spurr NK, Byth BC et al: An autosomal transcript in skeletal muscle with homology to dystrophin, *Nature* 339:55–8, 1989.

Matsumura K, Ervasti JM, Ohlendieck K, Kahl SD, Campbell KP: Association of dystrophin-related protein with dystrophin-associated proteins in mdx mouse muscle, *Nature* 360:588–91, 1992.

Tinsley JM, Blake DJ, Roche A, Fairbrother U, Riss J et al: Primary structure of dystrophin-related protein, *Nature* 360:591–3, 1992.

Ohlendieck K, Ervasti JM, Matsumura K, Kahl SD, Leveille CJ, Campbell KP: Dystrophin-related protein is localized to neuromuscular junctions of adult skeletal muscle, *Neuron* 7:499–508, 1991.

Buckle VJ, Guenet JL, Simon-Chazottes D, Love DR, Davies KE: Localisation of a dystrophin-related autosomal gene to 6q24 in man, and to mouse chromosome 10 in the region of the dystrophia muscularis (dy) locus, *Hum Genet* 85:324–6, 1990.

Nguyen TM, Ellis JM, Love DR et al: Localization of the DMDL gene-encoded dystrophin-related protein using a panel of nineteen monoclonal antibodies: presence at neuromuscular junctions, in the sarcolemma of dystrophic skeletal muscle, in vascular and other smooth muscles, and in proliferating brain cell lines, *J Cell Biol* 115:1695–700, 1991.

Deconinck AE, Rafael JA, Skinner JA et al: Utrophin-dystrophin-deficient mice as a model for Duchenne muscular dystrophy, *Cell* 90:717–27, 1997.

Tinsley JM, Potter AC, Phelps SR, Fisher R, Trickett JI, Davies KE: Amelioration of the dystrophic phenotype of mdx mice using a truncated utrophin transgene, *Nature* 384:349–53, 1996.

Fujimori K, Itoh Y, Yamamoto K et al: Interleukin 6 induces overexpression of the sarcolemmal utrophin in neonatal mdx skeletal muscle, *Hum Gene Ther* 13:509–18, 2002.

Chaubourt E, Fossier P, Baux G, Leprince C, Israel M, De La PS: Nitric oxide and l-arginine cause an accumulation of utrophin at the sarcolemma: A possible compensation for dystrophin loss in Duchenne muscular dystrophy, *Neurobiol Dis* 6:499–507, 1999.

van Deutekom JCT, van Ommen GJB: Advances in Duchenne muscular dystrophy gene therapy, *Nature Rev Genet* 4:774–83, 2003.

21.11. Transfer of Dystrophin "Booster" Genes

Integrin $\alpha 7\beta 1$

Burkin DJ, Wallace GQ, Nicol KJ, Kaufman DJ, Kaufman SJ: Enhanced expression of the alpha 7 beta 1 integrin reduces muscular dystrophy and restores viability in dystrophic mice, *J Cell Biol* 152:1207–18, 2001.

Hayashi YK, Chou FL, Engvall E, Ogawa M, Matsuda C et al: Mutations in the integrin alpha-7 gene cause congenital myopathy, *Nature Genet* 19:94–7, 1998.

Mayer U, Saher G, Fassler R, Bornemann A, Echtermeyer F et al: Absence of integrin alpha-7 causes a novel form of muscular dystrophy, *Nature Genet* 17:318–23, 1997.

Nawrotzki R, Willem M, Miosge N, Brinkmeier H, Mayer U: Defective integrin switch and matrix composition at alpha 7-deficient myotendinous junctions precede the onset of muscular dystrophy in mice, *Hum Molec Genet* 12:483–95, 2003.

Wang W, Wu W, Desai T, Ward DC, Kaufman SJ: Localization of the alpha-7 integrin gene (ITGA7) on human chromosome 12q13: Clustering of integrin and Hox genes implies parallel evolution of these gene families, *Genomics* 26:563–70, 1995.

Aszodi A, Hunziker EB, Brakebusch C, Fassler R: Beta-1 integrins regulate chondrocyte rotation, G1 progression, and cytokinesis, *Genes Dev* 17:2465–79, 2003.

Cunningham SA, Rodriguez JM, Arrate MP, Tran TM, Brock TA: JAM2 interacts with alpha-4/beta-1: Facilitation by JAM3, *J Biol Chem* 277:27589–92, 2002.

Garmy-Susini B, Jin H, Zhu Y, Sung RJ, Hwang R et al: Integrin alpha-4-beta-1–VCAM-1–mediated adhesion between endothelial and mural cells is required for blood vessel maturation, *J Clin Invest* 115:1542–51, 2005.

Goodfellow PJ, Nevanlinna HA, Gorman P, Sheer D, Lam G et al: Assignment of the gene encoding the beta-subunit of the human fibronectin receptor (beta-FNR) to chromosome 10p11.2, *Ann Hum Genet* 53:15–22, 1989.

Graus-Porta D, Blaess S, Senften M, Littlewood-Evans A, Damsky C et al: Beta-1-class integrins regulate the development of laminae and folia in the cerebral and cerebellar cortex, *Neuron* 31:367–79, 2001.

Hynes RO: Integrins: A family of cell surface receptors, *Cell* 48:549–54, 1987.

Johansson S, Forsberg E, Lundgren B: Comparison of fibronectin receptors from rat hepatocytes and fibroblasts, *J Biol Chem* 262:7819–24, 1987.

Lu TT, Cyster JG: Integrin-mediated long-term B cell retention in the splenic marginal zone, *Science* 297:409–12, 2002.

McDowall A, Inwald D, Leitinger B, Jones A, Liesner R et al: A novel form of integrin dysfunction involving beta-1, beta-2, and beta-3 integrins, *J Clin Invest* 111:51–60, 2003.

Tadokoro S, Shattil SJ, Eto K, Tai V, Liddington RC et al: Talin binding to integrin beta tails: A final common step in integrin activation, *Science* 302:103–6, 2003.

ADAM12

Galliano MF, Huet C, Frygelius J, Polgren A, Wewer UM et al: Binding of ADAM12, a marker of skeletal muscle regeneration, to the muscle-specific actin-binding protein, alpha-actinin-2, is required for myoblast fusion, *J Biol Chem* 275:13933–9, 2000.

Gilpin BJ, Loechel F, Mattei MG, Engvall E, Albrechtsen R et al: A novel, secreted form of human ADAM 12 (meltrin alpha) provokes myogenesis in vivo, *J Biol Chem* 273:157–66, 1998.

Kurisaki T, Masuda A, Sudo K, Sakagami J, Higashiyama S et al: Phenotypic analysis of meltrin alpha (ADAM12)-deficient mice: Involvement of meltrin alpha in adipogenesis and myogenesis, *Mol Cell Biol* 23:55–61, 2003.

Moghadaszadeh B, Albrechtsen R, Guo LT, Zaik M, Kawaguchi N et al: Compensation for dystrophin-deficiency: ADAM12 overexpression in skeletal muscle results in increased alpha-7 integrin, utrophin and associated glycoproteins, *Hum Mol Genet* 12:2467–79, 2003.

Yagami-Hiromasa T, Sato T, Kurisaki T, Kamijo K, Nabeshima Y et al: A metalloprotease-disintegrin participating in myoblast fusion, *Nature* 377:652–6, 1995.

Gilpin BJ, Loechel F, Mattei MG, Engvall E, Albrechtsen R et al: A novel secreted form of human ADAM12 (meltrin) provokes myogenesis in vivo, *J Biol Chem* 273:157–66, 1998.

Kronqvist P, Kawaguchi N, Albrechtsen R, Xu X, Schroder HD et al: ADAM12 alleviates the skeletal muscle pathology in mdx dystrophic mice, *Am J Pathol* 161:1535–40, 2002.

Yagami-Hiromasa T, Sato T, Kurisaki T, Kamijo K, Nabeshima YI et al: A metalloprotease-disintegrin participating in myoblast fusion, *Nature* (London) 377:652–6, 1995.

Galliano MF, Huet C, Frygelius J, Polgren A, Wewer UM et al: Binding of ADAM12, a marker of skeletal muscle regeneration, to the muscle-specific actin-binding protein, alpha-actinin-2, is required for myoblast fusion, *J Biol Chem* 275:13933–9, 2000.

Kronqvist P, Kawaguchi N, Albrechtsen R, Xu X, Schroder HD et al: ADAM12 alleviates the skeletal muscle pathology in mdx dystrophic mice, *Am J Pathol* 161:1535–40, 2002.

Iba K, Albrechtsen R, Gilpin B, Frohlich C, Loechel F et al: The cysteine-rich domain of human ADAM 12 supports cell adhesion through syndecans and triggers signaling events that lead to beta1 integrin-dependent cell spreading, *J Cell Biol* 149:1143–56, 2000.

Eto K, Huet C, Tarui T, Kupriyanov S, Liu HZ et al: Functional classification of ADAMs based on a conserved motif for binding to integrin alpha 9beta 1: Implications for sperm-egg binding and other cell interactions, *J Biol Chem* 277:17804–10, 2002.

Loechel F, Gilpin BJ, Engvall E, Albrechtsen R, Wewer UM: Human ADAM12 (meltrin) is an active metalloprotease, *J Biol Chem* 273:16993–7, 1998.

Calpastatin

Spencer MJ, Mellgren RL: Overexpression of a calpastatin transgene in mdx muscle reduces dystrophic pathology, *Hum Mol Genet* 11:2645–55, 2002.

Inazawa J, Nakagawa H, Misawa S, Abe T, Minoshima S et al: Assignment of the human calpastatin gene (CAST) to chromosome 5 at region q14–q22, *Cytogenet Cell Genet* 54:156–8, 1990.

Mimori T, Suganuma K, Tanami Y, Nojima T, Matsumura M et al: Autoantibodies to calpastatin (an endogenous inhibitor for calcium-dependent neutral protease, calpain) in systemic rheumatic diseases, *Proc Natl Acad Sci USA* 92:7267–71, 1995.

Pontremoli S, Salamino F, Sparatore B, De Tullio R, Pontremoli R et al: Characterization of the calpastatin defect in erythrocytes from patients with essential hypertension, *Biochem Biophys Res Commun* 157:867–74, 1988.

Spencer MJ, Mellgren RL: Overexpression of a calpastatin transgene in mdx muscle reduces dystrophic pathology, *Hum Mol Genet* 11:2645–55, 2002.

Nitric Oxide Synthase

Spencer MJ, Tidball JG: Do immune cells promote the pathology of dystrophin-deficient myopathies? *Neuromusc Disord* 11:556–64, 2001.

Wehling M, Spencer MJ, Tidball JG: A nitric oxide synthase transgene ameliorates muscular dystrophy in mdx mice, *J Cell Biol* 155:123–31, 2001.

Insulin-like Growth Factor

Barton ER, Morris L, Musaro A, Rosenthal N, Sweeney HL: Muscle-specific expression of insulin-like growth factor I counters muscle decline in mdx mice, *J Cell Biol* 157:137–48, 2002.

Myostatin

Bogdanovich S, Krag TOB, Barton ER, Morris LD, Whittemore LA et al: Functional improvement of dystrophic muscle by myostatin blockade, *Nature* 420:418–21, 2002.

Ferrell RE, Conte V, Lawrence EC, Roth SM, Hagberg JM et al: Frequent sequence variation in the human myostatin (GDF8) gene as a marker for analysis of muscle-related phenotypes, *Genomics* 62:203–7, 1999.

Lin J, Arnold HB, Della-Fera MA, Azain MJ, Hartzell DL et al: Myostatin knockout in mice increases myogenesis and decreases adipogenesis, *Biochem Biophys Res Commun* 291:701–6, 2002.

McPherron AC, Lawler AM, Lee SJ: Regulation of skeletal muscle mass in mice by a new TGF-beta superfamily member, *Nature* 387:83–90, 1997.

McPherron AC, Lee SJ: Suppression of body fat accumulation in myostatin-deficient mice, *J Clin Invest* 109:595–601, 2002.

Schuelke M, Wagner KR, Stolz LE, Huber C, Riebel T et al: Myostatin mutation associated with gross muscle hypertrophy in a child, *New Engl J Med* 350:2682–8, 2004.

Wagner KR, McPherron AC, Winik N, Lee SJ: Loss of myostatin attenuates severity of muscular dystrophy in mdx mice, *Ann Neurol* 52:832–6, 2002.

McPherron AC, Lee SJ: Suppression of body fat accumulation in myostatin-deficient mice, *J Clin Invest* 109:595–601, 2002.

Zimmers TA, Davies MV, Koniaris LG, Haynes P, Esquela AF et al: Induction of cachexia in mice by systemically administered myostatin, *Science* 296:1486–8, 2002.

Bogdanovich S, Krag TOB, Barton ER, Morris LD, Whittemore LA et al: Functional improvement of dystrophic muscle by myostatin blockade, *Nature* 420:418–21, 2002.

Wagner KR, McPherron AC, Winik N, Lee SJ: Loss of myostatin attenuates severity of muscular dystrophy in mdx mice, *Ann Neurol* 52:832–6, 2002.

Lee SJ, McPherron AC: Regulation of myostatin activity and muscle growth, *Proc Natl Acad Sci USA* 98:9306–11, 2001.

Hill JJ, Davies MV, Pearson AA, Wang JH, Hewick RM et al: The myostatin propeptide and the follistatin-related gene are inhibitory binding proteins of myostatin in normal serum, *J Biol Chem* 277:40735–41, 2002.

Mini-agrin

Moll J, Barzaghi P, Lin S, Bezakova G, Lochmuller H et al: An agrin minigene rescues dystrophic symptoms in a mouse model for congenital muscular dystrophy, *Nature* 413:302–7, 2001.

Engvall E, Wewer UM: The new frontier in muscular dystrophy research: Booster genes, *FASEB J* 17:1579–84, 2003.

Harper SQ, Hauser MA, DelloRusso C, Duan D, Crawford RW et al: Modular flexibility of dystrophin: Implications for gene therapy of Duchenne muscular dystrophy, *Nat Med* 8:253–61, 2002.

Sakamoto M, Yuasa K, Yohimura M, Yokota T, Ikemoto T et al: Micro-dystrophin cDNA ameliorates dystrophic phenotypes when introduced into mdx mice as a transgene, *Biochim Biophys Res Commun* 293:1265–72, 2002.

Campanelli JT, Hoch W, Rupp F, Kreiner T, Scheller RH: Agrin mediates cell contact-induced acetylcholine receptor clustering, *Cell* 67:909–16, 1991.

Finn AJ, Feng G, Pendergast AM: Postsynaptic requirement for Abl kinases in assembly of the neuromuscular junction, *Nature Neurosci* 6:717–23, 2003.

Gautam M, Noakes PG, Moscoso L, Rupp F, Scheller RH et al: Defective neuromuscular synaptogenesis in agrin-deficient mutant mice, *Cell* 85:525–35, 1996.

Glass DJ, Bowen DC, Stitt TN, Radziejewski C, Bruno J et al: Agrin acts via a MuSK receptor complex, *Cell* 85:513–23, 1996.

Khan AA, Bose C, Yam LS, Soloski MJ, Rupp F: Physiological regulation of the immunological synapse by agrin, *Science* 292:1681–6, 2001.

Lin W, Burgess RW, Dominguez B, Pfaff SL, Sanes JR et al: Distinct roles of nerve and muscle in postsynaptic differentiation of the neuromuscular synapse, *Nature* 410:1057–64, 2001.

McMahan UJ: The agrin hypothesis, *Cold Spring Harb Symp Quant Biol* 50:407–18, 1990.

Rupp F, Ozcelik T, Linial M, Peterson K, Francke U et al: Structure and chromosomal localization of the mammalian agrin gene, *J Neurosci* 12:3535–44, 1992.

Rupp F, Payan DG, Magill-Solc C, Cowan DM, Scheller RH: Structure and expression of a rat agrin, *Neuron* 6:811–23, 1991.

Wang J, Jing Z, Zhang L, Zhou G, Braun J et al: Regulation of acetylcholine receptor clustering by the tumor suppressor APC, *Nature Neurosci* 6:1017–8, 2003.

Human protein reference data base, Johns Hopkins University and the Institute of Bioinformatics, at http://www.hprd.org/protein.

21.12. Muscular Progenitor Cells

Rando TA, Blau HM: Primary mouse myoblast purification, characterization, and transplantation for cell-mediated gene therapy, *J Cell Biol* 125:1275–87, 1994.

Rosenblatt JD, Lunt AI, Parry DJ, Partridge TA: Culturing satellite cells from living single muscle fibre explants, *In Vitro Cell Dev Biol Anim* 31:773–9, 1995.

Jankowski RJ, Haluszczak C, Trucco M, Huard J: Flow cytometric characterization of myogenic cell populations obtained via the preplate technique: potential for rapid isolation of muscle-derived stem cells, *Hum Gene Ther* 12:619–28, 2001.

Beauchamp JR, Heslop L, Yu DS et al: Expression of CD34 and Myf5 defines the majority of quiescent adult skeletal muscle satellite cells, *J Cell Biol* 151:1221–34, 2000.

Lee JY, Qu-Petersen Z, Cao B et al: Clonal isolation of muscle-derived cells capable of enhancing muscle regeneration and bone healing, *J Cell Biol* 150:1085–100, 2000.

Beauchamp JR, Morgan JE, Pagel CN, Partridge TA: Dynamics of myoblast transplantation reveal a discrete minority of precursors with stem cell-like properties as the myogenic source, *J Cell Biol* 144:1113–22, 1999.

Jankowski RJ, Deasy BM, Huard J: Muscle-derived stem cells, *Gene Ther* 9:642–7, 2002.

Torrente Y, Tremblay JP, Pisati F et al: Intraarterial injection of muscle-derived CD34(+)Sca-1(+) stem cells restores dystrophin in mdx mice, *J Cell Biol* 152:335–48, 2001.

Heslop L, Beauchamp JR, Tajbakhsh S, Buckingham ME, Partridge TA et al: Transplanted primary neonatal myoblasts can give rise to functional satellite cells as identified using the Myf5nlacZl+ mouse, *Gene Ther* 8:778–83, 2001.

Gussoni E, Soneoka Y, Strickland CD et al: Dystrophin expression in the mdx mouse restored by stem cell transplantation, *Nature* 401:390–4, 1999.

21.13. Stem Cells

Gussoni E, Soneoka Y, Strickland CD, Buzney EA, Khan MK et al: Dystrophin expression in the mdx mouse restored by stem cell transplantation, *Nature* 401:390–4, 1999.

Bittner RE, Schofer C, Weipoltshammer K et al: Recruitment of bone-marrow-derived cells by skeletal and cardiac muscle in adult dystrophic mdx mice, *Anat Embryol* 199:391–6, 1999.

Ferrari G, Cusella-De Angelis G, Coletta M et al: Muscle regeneration by bone marrow-derived myogenic progenitors, *Science* 279:1528–30, 1998.

Malouf NN, Coleman WB, Grisham JW et al: Adult-derived stem cells from the liver become myocytes in the heart in vivo, *Am J Pathol* 158:1929–35, 2001.

Zuk PA, Zhu M, Mizuno H et al: Multilineage cells from human adipose tissue: implications for cell-based therapies, *Tissue Eng* 7:211–28, 2001.

Jiang Y, Vaessen B, Lenvik T, Blackstad M, Reyes M, Verfaillie CM: Multipotent progenitor cells can be isolated from postnatal murine bone marrow, muscle, and brain, *Exp Hematol* 30:896–904, 2002.

Sampaolesi M, Torrente Y, Innocenzi A, Tonlorenzi R, D'Antona G et al: Cell therapy of alpha-sarcoglycan null dystrophic mice through intra-arterial delivery of mesoangioblasts, *Science* 301(5632):487–92, 2003.

Gussoni E, Bennett RR, Muskiewicz KR, Meyerrose T, Nolta JA et al: Long-term persistence of donor nuclei in a Duchenne muscular dystrophy patient receiving bone marrow transplantation, *J Clin Invest* 110(6):807–14, 2002.

Chakkalakal JV, Thompson J, Parks RJ, Jasmin BJ: Molecular, cellular, and pharmacological therapies for Duchenne/Becker muscular dystrophies, *FASEB J* 19(8):880–91, 2005.

Prockop DJ: Further proof of the plasticity of adult stem cells and their role in tissue repair, *J Cell Biol* 160(6):807–9, 2003.

21.14. Potential Limitations

Huard J, Roy R, Bouchard JP, Malouin F, Richards CL et al: Human myoblast transplantation between immunohistocompatible donors and recipients produces immune reactions, *Transplant Proc* 24:3049–51, 1992.

Hodgetts SI, Beilharz MW, Scalzo AA, Grounds MD: Why do cultured transplanted myoblasts die in vivo? DNA quantification shows enhanced survival of donor male myoblasts in host mice depleted of CD4+ and CD8+ cells or Nk1.1+ cells, *Cell Transplant* 9:489–502, 2000.

Gussoni E, Soneoka Y, Strickland CD et al: Dystrophin expression in the mdx mouse restored by stem cell transplantation, *Nature* 401:390–4, 1999.

22

BONE AND CARTILAGE REGENERATIVE ENGINEERING

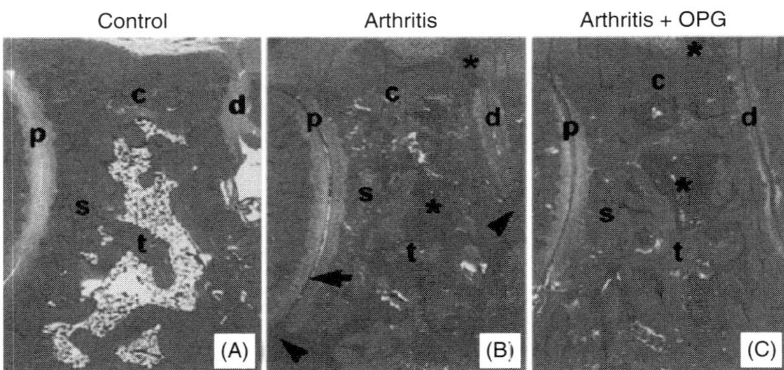

Osteoprotegerin (OPG, the decoy receptor for the tumor necrosis factor family molecule osteoprotegerin ligand or OPGL) prevents bone and cartilage destruction in the presence of severe inflammation. (A, D), Bone and joint structure in the normal hind paw showing dense bony plates, intact articular cartilages and marrow cavities containing scattered hematopoietic precursor cells. Proximal (p) distal (d) intertarsal joints. (B, E), Disrupted bone and joint structure in AdA rats (day 16) with severe mononuclear infiltration in the bone marrow (asterisk), and pannus (arrow); advanced destruction (arrowheads) of cortical (c), subchondral (s), and trabecular (t) bone; and erosion of the articular cartilages. The marrow cavity contains a marked mononuclear cell infiltration (asterisk) containing numerous osteoclasts (o). These rats show severe clinical crippling. (C, F), Preserved bone and joint structure of AdA rats (day 16) treated with OPG ($1\,mg\,kg^{-1}\,day^{-1}$) on days 9–15. OPG-treated rats exhibit extensive mononuclear cell infiltration of bone marrow (asterisk) and pannus formation (arrow), but cortical (c), subchondral (s), and trabecular (t) bone and articular cartilages are intact. Note the absence of osteoclasts as compared with non-OPG-treated rats (E). See color insert.

Bioregenerative Engineering: Principles and Applications, by Shu Q. Liu

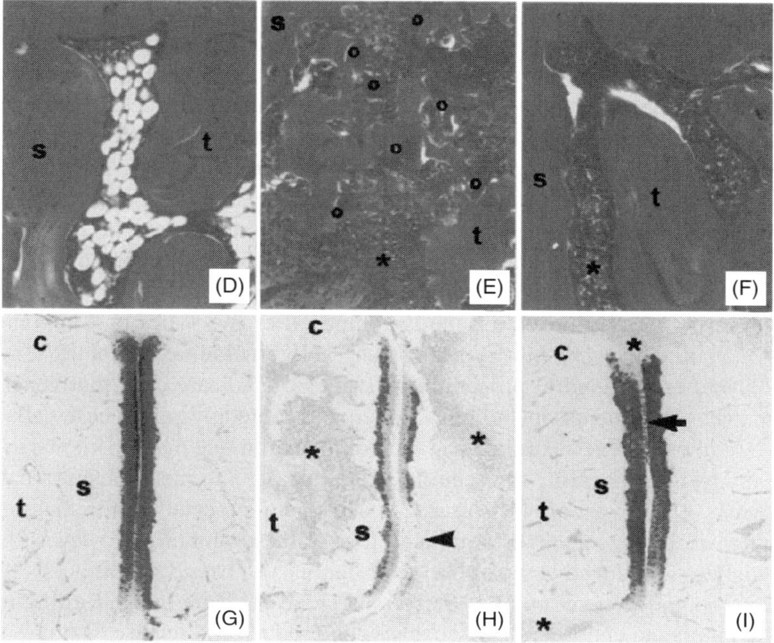

These rats do not show any clinical signs of crippling. (G) Normal cartilage integrity in control rats as determined by toluidine blue staining. The matrix of normal articular cartilage is uniformly stained, and the underlying bony plates are dense. (H) Cartilage matrix degeneration in AdA rats (day 16). Uniform pallor (small arrow) in the upper half of articular cartilages denotes extensive loss of matrix proteoglycans. Subchondral (s), cortical (c), and trabecular (t) bone is extensively eroded, and the marrow cavity is filled with inflammatory cells (asterisks). (I) OPG treatment preserves matrix proteoglycans of arthritic rats (day 16) treated with OPG ($1\,mg\,kg^{-1}\,day^{-1}$) on days 9–15. Note modest reduction of toluidine blue staining in peripheral cartilage regions that are in direct contact with inflamed synovial tissue (asterisks) or pannus (arrows). Subchondral (s), cortical (c) and trabecular (t) bone is intact. Staining: H&E (A–F); toluidine blue (G–I). Magnifications: 50× (A–C); 250× (D–F); 75× (G–I). (Reprinted by permission from Macmillan Publishers Ltd.: Kong YY et al: Activated T cells regulate bone loss and joint destruction in adjuvant arthritis through osteoprotegerin ligand, *Nature* 402:304–9, copyright 1999.) See color insert.

ANATOMY AND PHYSIOLOGY OF THE BONE

Structure [22.1]

Bones are considered hard connective tissues and are major constituents of the skeletal system. A typical bone is composed of several structures, including the periosteum, compact bone, cancellous bone, and endosteum. The *periosteum* is a soft connective tissue structure, which covers the exterior of the bone. The periosteum is composed of several layers, including an external connective tissue layer containing blood vessels and nerve fibers and an internal cellular layer containing osteoblasts, osteoclasts, and osteochondral progenitor cells. The *compact bone* is a thick layer immediately beneath the periosteum. This layer consists of primarily hard bone matrix, a component that carries mechanical loads. The compact bone consists of blood vessels, which enter the bone from the periosteum through perforating canals. The *cancellous bone* is located internal to the compact bone and is a spongy meshwork composed of trabeculae (about 100 µm in size) and small cavities between the trabeculae. The trabeculae are aligned in the direction of weight-bearing stress. The *endosteum* is the innermost layer of a bone, which is composed of cell types including osteoblasts, osteoclasts, and bone progenitor cells. In long bones, such as the femur and tibia, there is a central cavity known as the *medullary cavity*. Short and flat bones are not composed of a central cavity. The small cavities of the cancellous bones and the medullary cavity contain bone marrow, where hematopoietic stem cells and stromal cells are located.

At the micro-structural level, a bone is composed of two types of components: bone cells and matrix. There are four types of bone cells, including osteoblasts, osteocytes, osteoclasts, and osteochondral progenitor cells. *Osteoblasts* are cells that generate extracellular matrix components, including collagen and proteoglycans. These cells also play a critical role in osteogenesis or the formation of the hard bone matrix, known as hydroxyapatite crystals, by controlling the accumulation, release, and metabolism of calcium and phosphate ions. *Osteocytes* are mature osteoblasts that are located within the bone matrix. The function of the osteocytes is similar to that of the osteoblast. However, the activity of the osteocytes is reduced compared to the osteoblasts. *Osteoclasts* are cells that induce and regulate bone degeneration and resorption. These cells are located in the inner layer of the periosteum and endosteum. Osteoclasts can produce and release enzymes that break down the bone matrix. The matrix debris is endocytosed by osteoclasts. *Osteochondral progenitor cells* are cells that can give rise to osteoblasts and osteoclasts in the bone and chondroblasts in the cartilage. These cells are located in the internal layer of the periosteum and endosteum. Damaged osteoblasts and osteoclasts are replaced with newly generated cells from the osteochondral progenitor cells. A typical bone is composed of about 40% cellular and extracellular matrix components (primarily collage fibers) and about 60% hard mineral materials. The extracellular matrix and the mineral materials constitute the *bone matrix*. The mineral phase is composed of hydroxyapatite, a calcium phosphate crystal structure. The mineral materials provide mechanical strength, while the extracellular matrix provides flexibility to the bone.

Functions of the Bone [22.1]

Structural and Mechanical Support. Bones and cartilages provide support to tissues and organs, ensuring the structural integrity and stability of tissues and organs. There are a large number of bones in the body. These bones are connected by soft connective tissues known as *ligaments*. Bones participate in the movement of the body. Skeletal muscles attach to various bones by tendons. Forces generated by muscular contraction are transmitted to the bones, resulting in the movement of the involved body parts. Bones are the most important components for the protection of internal organs from mechanical injury. For instance, the skull protects the brain and the rib cage protects the heart and lungs.

Generation of Stem Cells. Bones contain bone marrow in the internal cavities. Bone marrow is composed of hematopoietic stem cells, which are capable of generating blood cells. In addition, the bone marrow contains marrow stromal cells, which support hematopoietic stem cells by providing necessary soluble factors for the development, survival, and differentiation of the hematopoietic stem cells. The marrow stromal cells also contain stem cells for a number of systems, including the connective tissue, blood vessels, heart, brain, and liver. These aspects are discussed in detail on page in Chapter 9.

Bone Formation and Resorption. The skeletal system conducts coordinated bone formation and resorption. The function of the bone is dependent on the maintenance of the dynamic balance of both processes. When bone resorption exceeds bone formation, the mass and strength of the bone reduces, resulting in a disorder known as *osteoporosis* (see page 918 of this chapter). Bone formation is a series of regulated processes, including the generation and release of collagen-dominant extracellular matrix (primarily type I collagen) by osteoblasts, the organization of collagen matrix into a framework, and the deposition of calcium and phosphorus minerals in the matrix framework, forming crystal structures of hydroxyapatite. Bone formation is influenced by a number of factors, including the activity of osteoblasts and the concentration of calcium and phosphorus. Osteoblasts control the production and release of collagen matrix, which nucleates mineral deposition and bone formation. Thus the rate of collagen production influences the rate of bone formation. A critical concentration of calcium and phosphorus is required for the formation of the mineral phase. The concentration of calcium and phosphorus is proportional to the rate of bone formation.

Osteoblasts are a major cell type that controls bone formation. These cells differentiate from the mesenchymal progenitor cells. The mesenchymal progenitor cells also give rise to myocytes, adipocytes, and chondrocytes. The specification of the mesenchymal progenitor cells to different cell types is controlled by distint proteins (Fig. 22.1). For instance, a protein known as runt-related transcription factor 2 (Runx2) promotes osteoblast formation. Another protein known as osterix (Osx) acts downstream of Runx2 to stimulate the maturation of osteoblasts. Two types of protein known as *myogenic regulatory factors* (MRFs) and *myocyte-enhancer factor 2* (MEF2) stimulate the mesenchymal progenitor cells to form myocytes. The proteins CCAAT-enhancer-binding protein (C/EBP) and peroxisome proliferator-activated receptor γ (PPAR γ) induce the differentiation of mesenchymal progenitor cells to adipocytes. The proteins Sox5, −6 and −9 and signal transducers and activators of transcription-1 (STAT1) stimulate the formation of chondrocytes from the mesenchymal progenitor cells.

A number of biochemical factors are known to regulate the activity of osteoblasts and bone development. A typical factor the Wnt protein. This protein promotes the survival, proliferation, and expansion of pre- and immature osteoblasts, thus enhancing bone formation. The Wnt-mediated signaling activities are counterregulated by several molecules, such as Dkks, Sfrps, and Wif-1. These molecules antagonize the Wnt signaling processes in osteoblasts to facilitate the death of immature cells, resulting in a reduction in bone formation (Fig. 22.2).

Bone resorption is a process by which the bone matrix is degenerated, collagen matrix is resorbed, and calcium and phosphorus ions are released from the bone to the blood. Osteoclasts are responsible for bone resorption. These cells can create a low pH environment and release proteinases. The acidic condition induces the degradation of the mineral phase of the bone. Proteinases can cleave extracellular matrix components, including collagen and proteoglycans. These activities result in bone resorption. The regulatory mechanisms of osteoclast activation are demonstrated in Fig. 22.3. The relative activities of bone formation and resorption vary during the different developmental stages. During the childhood, bone formation is dominant. During the adulthood, bone formation and resorption

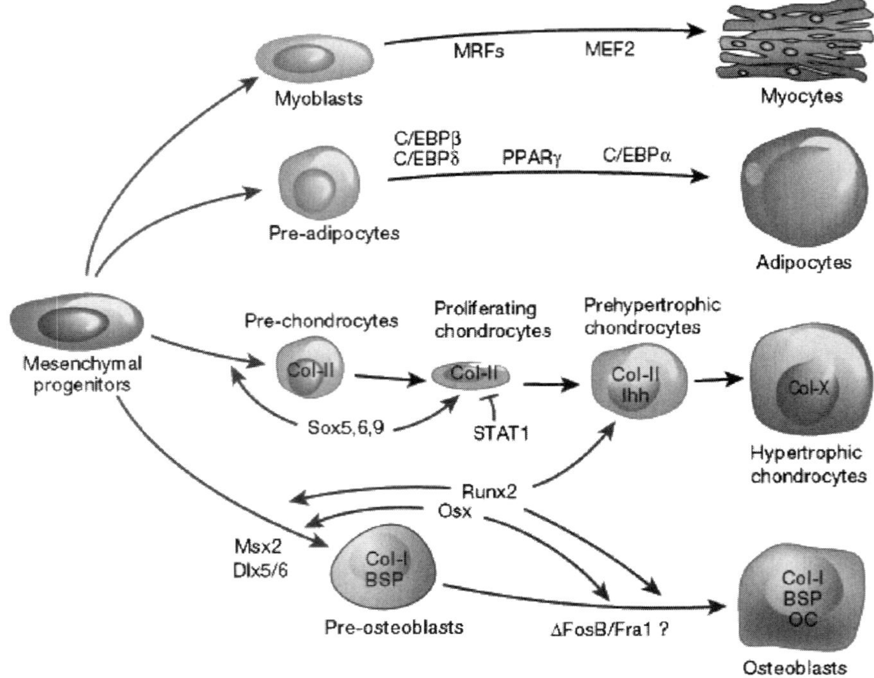

Figure 22.1. Transcriptional control of osteoblastic, chondrocytic, adipocytic and myocytic differentiation. Osteoblasts differentiate from mesenchymal progenitor cells that also give rise to myocytes, under the control of MRFs and MEF2[31], to adipocytes under the control of C/EBP α, β and δ and PPAR γ[30], and to chondrocytes under the control of Sox5, -6 and -9[33] and STAT1. Runx2 is essential for osteoblast differentiation and is also involved in chondrocyte maturation. Osterix (Osx) acts downstream of Runx2 to induce mature osteoblasts that express osteoblast markers, including osteocalcin. *Abbreviations*: MRFs, myogenic regulatory factors (including MyoD, myogenin, myogenic factor 5 and myogenic regulatory factor 4); MEF2, myocyte-enhancer factor 2; C/EBP, CCAAT-enhancer-binding protein; PPARgamma, peroxisome proliferator-activated receptor γ; STAT1, signal transducers and activators of transcription-1; Runx2, runt-related transcription factor 2; Col-I/II/X, type I/II/X collagen; Ihh, Indian hedgehog; BSP, bone sialoprotein; OC, osteocalcin. (Reprinted by permission from Macmillan Publishers Ltd.: Harada S Rodan GA: Control of osteoblast function and regulation of bone mass, *Nature* 423:349–55, copyright 2003.)

are dynamically balanced, maintaining a relatively constant density of bone matrix. During the late stage of the lifespan, the rate of bone resorption exceeds that of bone formation, often resulting in osteoporosis, a disorder with reduced bone matrix and strength (Fig. 22.4).

The process of bone resorption is controlled by the osteoclasts. These cells are differentiated from the haematopoietic precursor cells, which are present as bloodborne mononuclear cells. Osteoclastogenesis or the formation of osteoclasts is regulated by several biochemical fators, including M-CSF (CSF-1) and RANKL (Fig. 22.5). The activation of M-CSF (CSF-1) and RANKL can induce the recuitment of the haematopoietic precursor cells to the target bone. The recuited haematopoietic precursor cells can be transformed

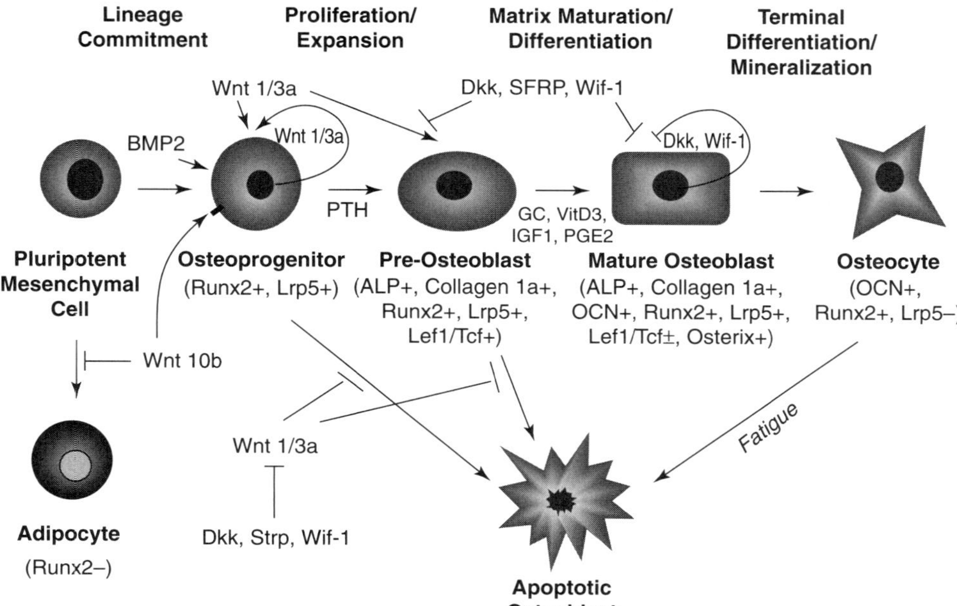

Figure 22.2. Effects of Wnt signaling on osseous cells. The canonical Wnt signaling pathway promotes the proliferation, expansion and survival of pre- and immature osteoblasts. Dkks, Sfrps, and Wif-1 antagonize Wnt signaling in osteoblasts to facilitate death of immature cells, but they may also downregulate the pathway in mature cells to induce terminal differentiation. (Reprinted from Westendorf JJ, Kahler RA, Schroeder TM: Wnt signaling in osteoblasts and bone diseases, *Gene* 341:19–39, copyright 2004, with permission from Elsevier.)

to osteoclasts. The molecular mechanisms of osteoclast formation and maturation are shown in Fig. 22.6. Another factor known as osteoprotegerin (OPG) serves as an antagonist for these osteoclast-stimulating factors. OPG can bind to and reduce the activity of RANKL and M-CSF, resulting in a reduction in osteoclastogenesis and the activity of osteoclasts.

Bone Metabolism. The bone matrix is constituted with minerals including calcium and phosphorus. The bone is a tissue for the storage, transport, and metabolism of these elements. Calcium and phosphorus are transported from the blood to the bone for storage. When the blood level of these elements is reduced, the elements are released from the bone, ensuring an appropriate level for metabolic and regulatory processes. Calcium and phosphorus are fundamental elements that participate in the regulation of cellular processes such as cell division, migration, adhesion, and contraction.

There is about 1–2 kg calcium in an adult body. More than 95% of the total calcium is stored in the bone. Calcium can be mobilized and released from the bone matrix when the calcium concentration in the extracellular space is reduced below a critical level, whereas calcium is deposited to the bone matrix when the extracellular calcium concentration is increased. The concentration of extracellular calcium is maintained within a narrow range from 9 to 10 mg/dL. The maintenance of such a calcium concentration is essential

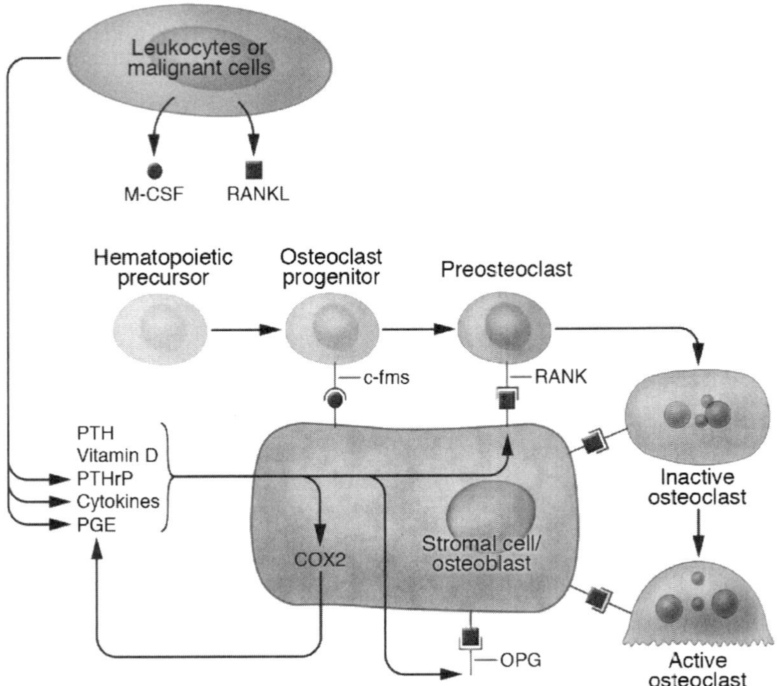

Figure 22.3. Regulation of osteoclast formation and activity. In physiologic remodeling, activation of bone resorption requires contact between cells of the osteoblast and osteoclast lineages. M-CSF, which may be either membrane bound or secreted, interacts with its receptor, c-fms, to stimulate differentiation and proliferation of hematopoietic progenitors, which then express RANK as pre-osteoclasts. Osteoclast differentiation and activity are stimulated by RANK/RANKL interaction, and this interaction can be blocked by soluble osteoprotegerin (OPG). Bone-resorbing factors can also stimulate COX2 activity, which may amplify responses to RANKL and OPG by producing prostaglandins. In pathologic conditions, inflammatory and malignant cells can increase osteoclastogenesis by producing soluble or membrane-bound M-CSF and RANKL as well as PTH-related protein (PTHrP), cytokines, and prostaglandins. (Reprinted with permission from Raisz LG, Pathogenesis of osteoporosis: concepts, conflicts, and prospects, *J Clin Invest* 115:3318–25, copyright 2005.)

to the regulation of cellular activities such as cell adhesion, migration, proliferation, and contraction.

The blood concentration of calcium is controlled by several factors, including calcium absorption from the intestinal tract, calcium mobilization from and deposition to the bone matrix, calcium excretion by the kidneys, and calcium loss via sweating. Calcium is absorbed from diets via active transport and diffusion in the small intestine. Vitamin D regulates the absorption of calcium (see section below). The urinary excretion of calcium is influenced by the concentration of blood calcium. A reduction in calcium intake and the level of blood calcium, or hypocalcemia, is associated with a decrease in urinary calcium excretion. Hypocalcemia may occur in the presence of vitamin D deficiency, intestinal disorders, and dietary calcium deprivation.

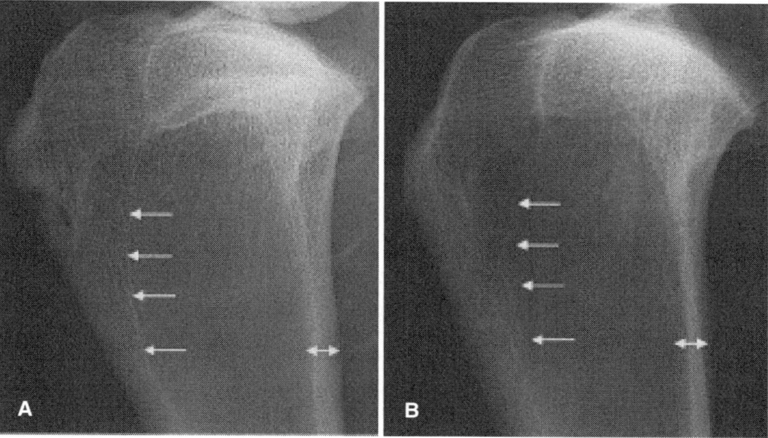

Figure 22.4. In vivo radiological survey of typical mineralization in young (18 months, panel A) and old (8 years, panel B) sheep tibia. Note in particular the overall reduced X-ray absorption in the old tibia due to lowered mineral density, the reduction in trabecular organization, the lack of a defined border towards the marrow cavity (arrows), and the thinning of the cortical bone (double arrowheads), all features pointing to significant osteoporosis in the old animal. (Reprinted from Sachse A et al: Osteointegration of hydroxyapatite-titanium implants coated with nonglycosylated recombinant human bone morphogenetic protein-2 (BMP-2) in aged sheep, *Bone* 37:699–710, copyright 2005 with permission of Elsevier.)

Phosphorus is not only a constituent of the bone matrix, but also an element participating in the regulation of molecular signaling processes, such as phosphorylation and dephosphorylation. There is about 1 kg phosphorus in an adult body. About 85% of the total phosphorus is stored in the bone matrix. When blood phosphorus is low, phosphorus can be mobilized from the bone matrix to form free phosphorus. Free phosphorus is present in the form of inorganic phosphate (HPO_4^{2-}, $H_2PO_4^-$, or $NaHPO_4^-$), which is found in the blood and extracellular space (3–4 mg/dL). Dietary phosphorus is absorbed in the small intestine. The rate of intestinal absorption controls the blood level of phosphorus. In addition to phosphorus mobilization from the bone and absorption from the intestine, the kidney also participates in the regulation of blood phosphorus. The renal tubules can efficiently reabsorb phosphorus from the glomerular filtrate. When the blood level of phosphorus is low, the renal tubules can reabsorb all phosphorus from the glomerular filtrate if necessary.

Regulation of Bone Metabolism

Role of Parathyroid Hormone. Parathyroid hormone is a protein that stimulates bone resorption and mobilizes calcium and phosphorus from the bone, resulting in an increase in the blood concentration of calcium and phosphorus. The hormone is produced in the two parathyroid glands, which are located near the thyroid glands. The parathyroid glands are composed of several cell types, including the chief cells and oxyphil cells. The chief cells produce parathyroid hormone. The function of the oxyphil cells remains poorly understood. Parathyroid hormone is synthesized in the ribosomes in the form of a preprohormone polypeptide with about 110 amino acids. The preprohormone is modified in the

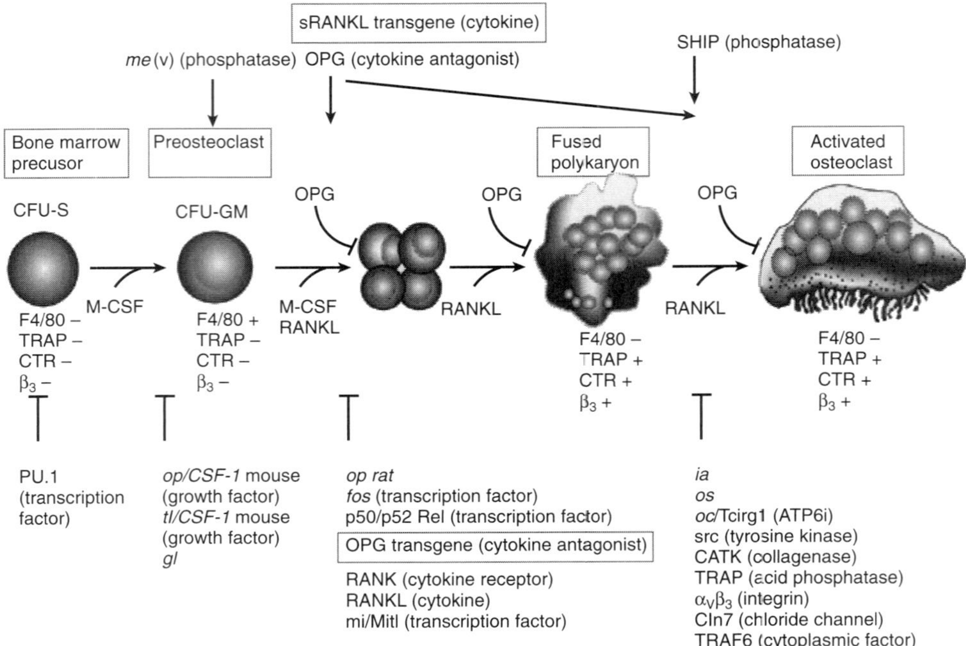

Figure 22.5. Osteoclastogenesis. Development schema of haematopoietic precursor cell differentiation into mature osteoclasts, which are fused polykaryons arising from multiple (10–20) individual cells. Maturation occurs on bone from peripheral bloodborne mononuclear cells with traits of the macrophage lineage shown below. M-CSF (CSF-1) and RANKL are essential for osteoclastogenesis, and their action during lineage allocation and maturation is shown. OPG can bind and neutralize RANKL, and can negatively regulate both osteoclastogenesis and activation of mature osteoclasts. Shown below are the single-gene mutations that block osteoclastogenesis and activation. Those indicated in italic font are naturally occurring mutations in rodents and humans, whereas the others are the result of targeted mutagenesis to generate null alleles. Shown above are the single-gene mutant alleles that increase osteoclastogenesis and/or activation and survival and result in osteoporosis. Note that all of these mutants represent null mutations with the exception of the OPG and sRANKL transgenic mouse overexpression models. (Reprinted by permission from Macmillan Publishers Ltd.: Boyle WJ, Simonet WS, Lacey DL: Osteoclast differentiation and activation, *Nature* 423:337–42, copyright 2003.)

endoplasmic reticulum and Golgi apparatus to form parathyroid hormone. The functional form of the hormone is stored in the secretory granules.

Parathyroid hormone can be released in response to a decrease in calcium concentration. The hormone can activate osteocytes and osteoblasts within a period as short as several minutes. These cells can release intermediate signals, which in turn activate osteoclasts, a type of bone-degrading cell. Note that, although the primary function of osteocytes and osteoblasts is to promote mineral deposition and bone formation, these cells mediate the activity of the bone-degrading osteoclasts in the presence of a high concentration of parathyroid hormone. Activated osteoclasts degrade bone matrix, releasing calcium and phosphorus from the bone matrix to the extracellular space and blood. In addition,

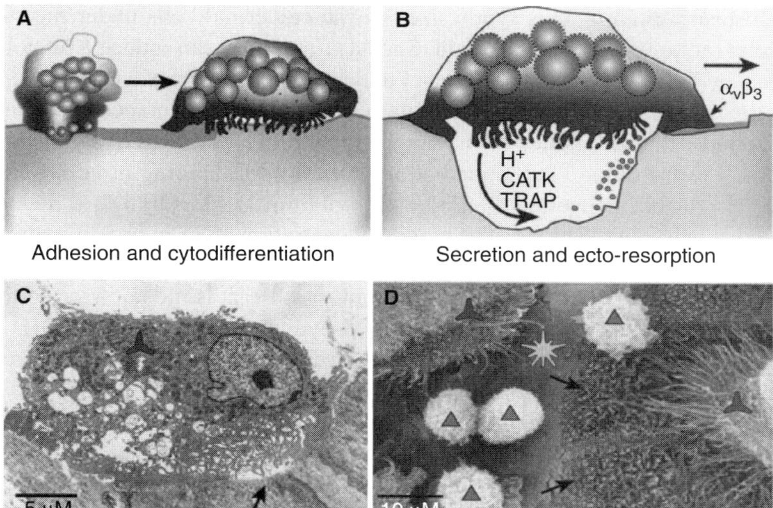

Figure 22.6. Activation of bone resorption. (A) Multinucleated polykaryons are recruited by the action of CSF1 and RANKL, which then adhere to bone and undergo cytodifferentiation into a mature osteoclast. (B) RANKL stimulates osteoclast activation by inducing secretion of protons and lytic enzymes into a sealed resorption vacuole formed between the basal surface of the osteoclast and the bone surface. Acidification of this compartment by secretion of protons leads to the activation of TRAP and CATK, which are the two main enzymes responsible for the degradation of bone mineral and collagen matrices. (C) Transmission electron micrograph of an activated mouse osteoclast with a visible ruffled border in a resorption lacunae on the periosteal femoral cortical bone surface. Red propeller, osteoclast; black arrow, a resorption pit. (D) Scanning electron micrograph of human osteoclasts generated in vitro on cortical bone slices from CSF1- and RANKL-treated peripheral blood mononuclear cells. Red propellers, osteoclasts; black arrows, a resorption pit where the normally smooth lamellar bone surface has been resorbed to expose collagen bundles; yellow star, nonresorbed bone surface; blue triangles, mononuclear cells (potential osteoclast precursors). (Reprinted by permission from Macmillan Publishers Ltd.: Boyle WJ, Simonet WS, Lacey DL: Osteoclast differentiation and activation, *Nature* 423:337–42, copyright 2003.)

parathyroid hormone can stimulate the proliferation of osteoclasts. Increased osteoclasts further enhance bone resorption and mineral release.

Parathyroid hormone can also act on the epithelial cells of the small intestine and the tubular cells of the kidney. In the intestine, parathyroid hormone enhances the absorption of calcium and phosphorus, resulting in an increase in the blood concentration of these ions. In the kidney, the effect of parathyroid hormone is more complicated than that in the intestine. The hormone stimulates the renal tubular epithelial cells to reabsorb calcium, while it reduces the reabsorption of phosphorus. Such activities result in an increase in the blood concentration of calcium and a decrease in the blood concentration of phosphorus. At the same time, the hormone also regulates the transport of magnesium, sodium, and potassium in the kidney. It enhances the reabsorption of magnesium, while it reduces the reabsorption of sodium and potassium.

Role of Vitamin D. Vitamin D is a hormone that is derived from 7-dehydrocholesterol, a precursor of cholesterol, and plays a critical role in the regulation of calcium metabolism and homeostasis. Vitamin D is synthesized in the epidermal cells under the stimulation of ultraviolet radiation from the sunlight. Such a stimulation can induce a photobiochemical process, by which 7-dehydrocholesterol undergoes a conformational change, resulting in the formation of previtamin D_3. Previtamin D_3 can be spontaneously converted to vitamin D_3 under the stimulation of the body temperature. Vitamin D_3 is transported from the epidermis to the blood via the mediation of vitamin D-binding proteins. In the liver, vitamin D_3 is further converted to 25-hydroxyvitamin D [25(OH)D], a major form of vitamin D that exists in the blood. The blood concentration of 25-hydroxyvitamin D varies considerably under physiological conditions, ranging from 5 to 80 ng/mL. However, 25-hydroxyvitamin D is not active. This form of vitamin D is metabolized in the kidney and converted to an active form known as 1,25-hydroxyvitamin D [1,25$(OH)_2$D] under the action of 25(OH)D-1α hydroxylase. 1,25-hydroxyvitamin D is an active form of vitamin D.

The level of 1,25-hydroxyvitamin D is controlled by a number of factors, including the level of sunlight exposure, the intensity of ultraviolet light, aging, the calcium concentration, and the level of parathyroid hormone. An increase in exposure to sunlight enhances vitamin D_3 synthesis and promotes the formation of 1,25-hydroxyvitamin D. Aging is associated with a progressive reduction in the rate of vitamin D_3 synthesis. A decrease in the blood concentration of calcium stimulates the conversion of 25-hydroxyvitamin D to 1,25-hydroxyvitamin D. Such a calcium change also stimulates the secretion of parathyroid hormone, which acts on renal tubule cells and enhances the formation of 1,25-hydroxyvitamin D. There exist many forms of vitamin D metabolites in the blood and extracellular matrix. These forms are mostly products of vitamin D degradation and possess vitamin D activity. However, the vitamin D metabolites are not as active as 1,25-hydroxyvitamin D.

There are a number of functions for 1,25-hydroxyvitamin D. These include the regulation of calcium and phosphorus absorption in the epithelial cells of the small intestine, the regulation of bone resorption, and the control of cell proliferation and differentiation. It is well known that 1,25-hydroxyvitamin D stimulates the absorption of calcium and phosphorus in the epithelial cells of the small intestine, resulting in an elevation in the blood concentration of these ions. The influence of vitamin D on bone resorption is dependent on the concentration of vitamin D. At a high concentration in the extracellular space, 1,25-hydroxyvitamin D enhances bone resorption. This is possibly due to the stimulatory effect of vitamin D on the bone-resorption activity of the parathyroid hormone. Such an effect results in an increase in the concentration of calcium and phosphorus in the extracellular space and blood. However, at a low level, 1,25-hydroxyvitamin D enhances bone mineralization and matrix formation. A possible mechanism for this phenomenon is that a low level of vitamin D may reduce the activity of the parathyroid hormone. 1,25-hydroxyvitamin D has also been hsown to inhibit the development and activation of osteoclasts (Fig. 22.7), thus reducing bone resorption.

Experimental investigations have demonstrated that 1,25-hydroxyvitamin D exerts inhibitory effects on the proliferation of normal and cancer cells. For instance, 1,25-hydroxyvitamin D inhibits the generation and secretion of renin in the renal arteries, resulting in a decrease in the level and activity of angiotensin II. Since angiotensin II enhances the proliferation of vascular smooth muscle cells, 1,25-hydroxyvitamin D suppresses vascular mitogenic activities via mediating the function of angiotensin II. Vitamin D is also known to promote cell differentiation A treatment with 1,25-hydroxyvitamin D can induce the differentiation of monocytes to osteoclast-like cells.

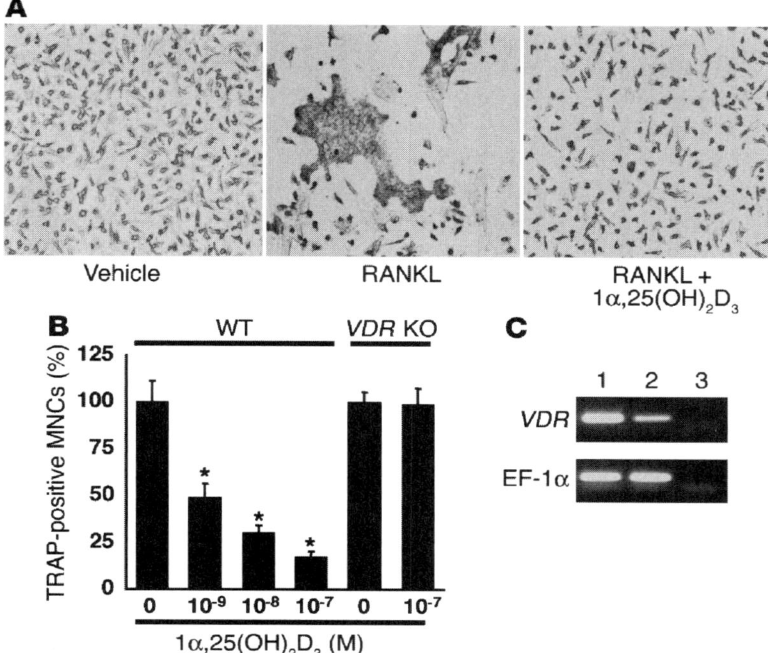

Figure 22.7. 1α-25$(OH)_2D_3$ inhibits osteoclast development through *VDR* by acting directly on osteoclast precursor cells in bone marrow. (A) Osteoclast precursor cells were isolated from the bone marrow of wildtype C57BL/6J and *VDR* Knockout mice as M-CSF–dependent adherent cells, and were further treated with RANKL (40 ng/mL) in the absence or presence of 10^{-7} M 1α-25$(OH)_2D_3$ for 3 days. Note that the development of TRAP-positive multinucleate osteoclasts induced by RANKL (receptor activator of NFκB ligand) was markedly inhibited by cotreatment with 1α-25$(OH)_2D_3$. (B) The inhibitory effect of 1α-25$(OH)_2D_3$ on the formation of TRAP-positive multinucleate cells (MNCs) was dose-dependent and was not seen in marrow cultures derived from *VDR* knockout mice, even at the highest dose of 10^{-7} M. Data are expressed as a percentage of vehicle-treated cultures. $^*P < 0.05$ versus vehicle group, $n = 6$. (C) Expression of *VDR*s in the intestine (lane 1) and osteoclast precursor cells (lane 2) as detected by RT-PCR. EF-1α mRNA served as control for PCR. Lane 3 contained water as a negative control. (Reprinted with permission from Takasu H et al: c-Fos protein as a target of antiosteoclastogenic action of vitamin D, and synthesis of new analogs, *J Clin Invest* 116:528–35, copyright 2006.)

Mammalian cells express a nuclear receptor for 1,25-hydroxyvitamin D. The interaction of 1,25-hydroxyvitamin D with this receptor induces the phosphorylation and activation of the receptor. The complex of 1,25-hydroxyvitamin D and activated nuclear receptor serves as a transcription factor and can translocate to the cell nucleus, initiating the transcription of specific target genes. The gene products in turn regulate related physiological activities.

ANATOMY AND PHYSIOLOGY OF THE CARTILAGE [22.1]

Cartilage is a connective tissue that is mostly associated with the bone except for several organs such as the large airways and ears. Cartilage is composed of two structures: perichondrium and internal cartilage. The perichondrium contains an external and internal connective tissue layer. The external layer is composed of fibroblasts, extracellular matrix (collagen fibers, elastic fibers, and proteoglycans), blood vessels, and nerves. The internal layer is composed of cells known as *chondroblasts*. These cells can generate cartilage matrix components, including collagen and proteoglycans. Chondroblasts can self-reproduce and are responsible for cartilage growth. The internal cartilage is composed of chondrocytes and cartilage matrix. *Chondrocytes* are developed from chondroblasts and are surrounded by cartilage matrix. The cartilage matrix is an amorphous structure, which contains scattered chondrocytes. It should be noted that a cartilage layer can be found on the terminal surface of the bone at a joint. This type of cartilage, called articular cartilage, does not contain perichondrium.

There are several functions for the cartilage. It provides support to adjacent bones, serves as a structural framework for several organs, including the trachea, nose, and ear, and constitutes the surface structure of the joint, ensuring smooth interaction between bone surfaces. In addition, cartilage plays a critical role in mediating bone growth. In long bones such as the femur and tibia, there exists a cartilage plate between the end and the body of the bone. This plate is responsible for the growth of the bone, which determines the height of the entire body. When bone growth is ceased, the cartilage growth plate is ossified into a thin bone matrix structure known as the epiphyseal line.

BONE AND CARTILAGE DISORDERS

Osteoporosis

Pathogenesis, Pathology, and Clinical Features [22.2]. *Osteoporosis* is a bone disorder characterized by increased degeneration or resorption of the bone matrix in the entire skeletal system, resulting in a progressive reduction in the mass of the bone, the thickness of the compact bone, and the size and number of the trabeculae of the cancellous bone. Osteoporosis often results in bone deterioration and fracture. Osteoporosis is a consequence of imbalance between bone formation and resorption. In the healthy population, the rate of bone formation is dynamically balanced with the rate of bone resorption. When the rate of bone resorption exceeds that of bone formation, the bone mass reduces and osteoporosis occurs. In such a case, the skeletal system is no longer able to resist the physiological level of mechanical loads. A slight increase in the mechanical load may result in bone fracture. In the normal population, bone resorption starts to exceed bone formation at the age of 40–50. The rate of bone resorption increases with aging. Osteoporosis is a common disorder in the population over the age of 60.

Based on etiological factors, osteoporosis is classified into several types, including idiopathic, postmenopausal, glucocorticoid, thyrotoxicotic, and inherited osteoporosis. *Idiopathic osteoporosis* is defined as osteoporosis without identified etiological factors. This disorder is often found in young people, especially premenopausal women. The disorder is associated with a decrease in blood calcium and phosphorus. Idiopathic osteoporosis can be self-cured spontaneously within several years. *Postmenopausal osteoporosis* is found in women with reduced production and release of estrogen, a hormone that suppresses bone resorption. A decrease in the level of estrogen in postmenopausal women induces increased bone resorption and osteoporosis.

An excessive increase in glucocorticoids, as seen in Cushing's syndrome, can reduce bone formation and simultaneously enhance bone resorption, often resulting in osteoporosis. Glucocorticoids have been shown to enhance the effect of parathyroid hormone, which activates the osteoclasts and mobilizes calcium from the bone matrix. Furthermore, glucocorticoids inhibit calcium absorption in the intestine. All these effects contribute to the development of osteoporosis.

Thyrotoxicity is a condition with increased secretion of the thyroid hormone, which is also referred to as hyperthyroidism. Increased thyroid hormone mobilizes calcium and phosphorus from the bone, promotes bone resorption, and enhances the excretion of calcium and phosphorus via urine and feces. These changes can lead to osteoporosis if hyperthyroidism prolongs. In postmenopausal women, hyperthyroidism may induce more severe osteoporosis than the general population.

Osteogenesis imperfecta is an inherited form of osteoporosis and is characterized by heterogeneous reduction in the mass of bone matrix. There are two types of disorder: autosomal dominant and autosomal recessive osteogenesis imperfecta. The autosomal dominant type of imperfecta is associated with relatively mild bone resorption and functional defects. In contrast, the autosomal recessive subtype is often found within a short period after birth and is associated with a severe reduction in the bone mass. Patients may experience recurrent fracture of long bones such as femurs and tibias. Pathological examinations usually demonstrate reduced synthesis of type I collagen and altered organization of collagen matrix.

Conventional Treatment [22.2]. Osteoporosis is a group of disorders induced by different etiological factors. One of the strategies for the treatment of osteoporosis is to eliminate or alleviate the primary etiological factors, if known. For instance, postmenopausal osteoporosis is caused by a reduction in the level of estrogen. Administration of estrogen is the primary choice of method for the treatment of postmenopausal osteoporosis. In osteoporosis induced by increased level of glucocorticoids and thyroid hormone, a primary approach is to treat the original diseases that enhance the production and secretion of these hormones. Since calcium is a major component of the bone matrix, calcium administration is often necessary. An increase in the level of extracellular calcium reduces bone resorption. In addition, fluorides can be incorporated into the bone matrix, enhancing the crystal formation and strength of the bone. Thus fluorides have been used to treat osteoporosis.

Molecular Engineering Therapy. Osteoporosis is induced by progressive degeneration of the bone matrix. Molecular regenerative approaches can be used to protect the bone from degeneration and promote bone matrix formation. There are a number of genes encoding proteins that are known to regulate bone matrix formation and mineralization, including vitamin D receptor, estrogen receptor alpha, type I collagen, transforming growth factor, and interleukin-6. The mutation of these genes and/or disorders in regulating the expression of these genes may predispose to bone degeneration, leading to osteoporosis. Thus, the genetic manipulation of these genes may potentially enhance bone formation and prevent bone degeneration. Here the application of these genes to osteoporosis is briefly discussed.

Vitamin D Receptor (VDR) [22.3]. The vitamin D receptor (Table 22.1) interacts with vitamin D and participates in the regulation of calcium metabolism and bone formation.

TABLE 22.1. Characteristics of Vitamin D Receptor*

Proteins	Alternative Names	Amino Acids	Molecular Weight (kDa)	Expression	Functions
Vitamin D receptor	VDR, 1,25-dihydroxyvitamin D3 receptor	427	48	Ubiquitous	Serving as a *trans*-acting transcriptional regulatory factor, which is similar to steroid and thyroid hormone receptors in structure, regulating genes involved in mineral (calcium) metabolism

*Based on bibliograhpy 22.3.

TABLE 22.2. Characteristics of Selected Collagen Molecules*

Proteins	Alternative Names	Amino Acids	Molecular Weight (kDa)	Expression	Functions
Collagen type I α1	COL1A1, collagen α1(I) chain	1464	139	Bone, skin, tendon	Constituting collagen type 1 together with collagen type I α2
Collagen type I α2	COL1A2, collagen I α2 polypeptide	1366	129	Bone, skin, tendon	Constituting collagen type 1 together with collagen type I α1

*Based on bibliograhpy 22.4.

At a relatively low level, vitamin D can activate the vitamin D receptor, enhancing calcium deposition and bone mineralization. Furthermore, vitamin D inhibits the activity of osteoclasts, suppressing bone resorption. The mutation of the vitamin D receptor gene induces disorder of calcium metabolism, influencing the mineralization process of the bone matrix. When the mutation of the vitamin D receptor is an identified cause for osteoporosis, the transfer of the wildtype vitamin D receptor gene may be considered for the treatment of the disorder.

Type I Collagen Gene [22.4]. Type I collagen is a major type of matrix molecule in the bone. This collagen is composed of collagen type I α1 and collagen type I α2 chains (both are listed in Table 22.2). The genes encoding these chains are COLIA1 and COLIA2, respectively. The mutation of these genes may influence the formation of the collagen matrix. For instance, there exists a polymorphism in the introns of the COLIA1 gene. This polymorphism induces alterations in the regulation of gene transcription, resulting in a shift in the ratio of the alpha1 to alpha2 collagen chains. These abnormalities affect the formation of type I collagen as well as the collagen matrix in the bone. Since the integrity of the collagen matrix is critical to the mineralization of the bone matrix, any disorder in the formation of collagen matrix may contribute to bone degeneration and the development of osteoporosis. When disordered formation of collagen matrix is the cause of osteoporosis, the transfer of the wildtype type I collagen gene should be considered for the treatment of the disorder. Furthermore, when gene mutation is the cause of bone disorders, targeted correction of mutant genes by transfecting cells with viral vectors may be an effective approach for the treatment of bone disorders.

Estrogen Receptor [22.5]. This receptor interacts with the hormone estrogen and enhances bone formation. Postmenopausal women are often associated with osteoporosis because of estrogen reduction or deficiency. Estrogen is known to suppress the activity of cytokines, including tumor necrosis factor α, interleukin (IL)1, IL6, IL11, macrophage-colony stimulating factor, and prostaglandin E, via interaction with the estrogen receptor. The loss of estrogen and/or estrogen receptor is often associated with activation of these cytokines, which stimulate the proliferation of osteoclasts. Activated osteoclasts induce bone resorption. Mutation of the estrogen receptor gene may also lead to enhanced bone resorption. In the case of osteoporosis due to the malfunction of the estrogen receptor gene, the transfer of a functional gene into the skeletal system may help to enhance bone formation.

Characteristics of several estrogen receptors are listed in Table 22.3.

Calcitonin [22.6]. Calcitonin (Table 22.4) is a peptide hormone that is synthesized by the parafollicular cells of the thyroid. Its function is to inhibit the activity of osteoclasts, resulting in the suppression of bone resorption and a reduction in serum calcium. Calcitonin exerts an effect opposite to that of the parathyroid hormone. The protein form of calcitonin can be delivered to disordered bones for the treatment of osteoporosis. Clinical studies have provided promising results for the use of calcitonin. Alternatively, the gene of calcitonin can be cloned, amplified, and used for therapeutic purposes. As for other therapeutic genes, the delivery of the calcitonin gene can prolong the effectiveness of the hormone compared with direct delivery of the calcitonin protein.

Osteoprotegerin (OPG) and Osteoprotegerin Ligand (OPGL) [22.7]. Osteoprotegerin is a secreted soluble receptor protein that belongs to the tumor necrosis factor receptor superfamily and exists in the extracellular matrix. Osteoprotegerin is produced by the osteoblasts and is capable of suppressing the activity of osteoclasts and thus inhibiting bone resorption. Osteoprotegerin can bind to a ligand known as osteoprotegerin ligand (OPGL), which is a member of the tumor necrosis factor cytokine family and is also known as tumor necrosis factor ligand superfamily member 11 (TNFSF11). Osteoprotegerin ligand is a protein that can interact with a cell membrane receptor in hematopoietic cells to induce the differentiation of hematopoietic stem cells to osteoclasts. The binding of osteoprotegerin, which serves as a decoy receptor in the extracellular matrix, to osteoprotegerin ligand inhibits the interaction of the osteoprotegerin ligand to cell membrane receptor and thus suppresses the formation of osteoclasts from hematopoietic stem cells.

TABLE 22.3. Characteristics of Selected Estrogen Receptors*

Proteins	Alternative Names	Amino Acids	Molecular Weight (kDa)	Expression	Functions
Estrogen receptor α	Estrogen receptor 1 (ESR1), estrogen receptor (ER)	595	66	Uterus, ovary, breast, bones, skeletal muscle, blood vessels, lung, skin	Serving as a nuclear receptor and transcription factor, regulating the development of the sex organs, and regulating cell proliferation
Estrogen receptor β	ER β, estrogen receptor 2 (ESR2), ESRB, ESR β	530	59	Uterus, ovary, breast, prostate gland, bones, skeletal muscle, blood vessels, brain, lung, skin	Similar to functions of estrogen receptor α

*Based on bibliograhpy 22.5.

TABLE 22.4. Characteristics of Calcitonin*

Proteins	Alternative Names	Amino Acids	Molecular Weight (kDa)	Expression	Functions
Calcitonin	CT, α calcitonin, katacalcin	141	15	Nervous system, bone, thyroid gland	Suppressing bone resorption and reducing the level of serum calcium

*Based on bibliograhpy 22.6.

Since osteoclasts are responsible for bone resorption, the activation of osteoprotegerin leads to a reduction in bone resorption. Animals without the osteoprotegerin gene usually develop osteoporosis-like disorders. The application of osteoprotegerin to osteoporotic bones has been shown to prevent bone degeneration. In animal models with oophorectomy (surgical removal of the ovary), the administration of osteoprotegerin reduces estrogen deficiency-induced bone degeneration. The transfer of the osteoprotegerin gene into osteoporotic bones is a potential approach for the treatment of osteoporosis.

Properties of osteoprotegerin and its ligand are listed in Table 22.5.

Integrin-Binding Proteins [22.8]. The integrin complex αvβ3 is expressed in osteoclasts, plays a role in regulating the adhesion and proliferation of osteoclasts, and thus enhances osteoclast-mediated bone resorption. A family of integrin-binding proteins, known as disintegrins, can bind to the αvβ3 integrin and inhibit the activity of the integrin, resulting in the suppression of osteoclast activation and bone resorption. Disintegrins are small proteins found in the venom of snakes. The disintegrin family includes several members, which are echistatin, kistrin, albolabrin, bitistatin, elegantin, flavoridin, halysin, and triflavin. These proteins can specifically bind to β1 and β3 integrins and block the interaction of integrins with extracellular matrix components. Disintegrins also inhibit the activity of parathyroid hormone-induced bone resorption. Thus, the proteins or genes of disintegrins can be potentially used for the treatment of osteoporosis.

Growth Factors [22.9]. Several growth factors, including insulin-like growth factor (IGFβ1), fibroblast growth factor (FGF), the bone morphogenetic proteins (BMPs), and transforming growth factor-beta (TGFβ), are known to stimulate osteoblast proliferation and promote bone formation. Mutation of the genes encoding these growth factors is associated with an increased incidence of osteoporosis. A typical example is the association of the mutation of the transforming growth factor β gene (e.g., C509→T polymorphism) with osteoporosis. Local delivery of these growth factors enhances bone regeneration and recovery from osteoporosis. Alternatively, the genes of these growth factors can be used and transferred into disordered bones. As shown in experimental investigations, the overexpression of the bone morphogenetic protein gene in osteoblasts by gene transfer enhances bone formation. The construction of the bone morphogenetic protein gene vector is shown in Fig. 22.8. The effectiveness of the bone morphogenetic protein gene in bone formation is shown in Fig. 22.9.

Cell Therapy for Bone Regeneration [22.10]. Osteoporosis is a disorder induced by reduced osteogenesis due to impaired function of the osteoblasts, which regulate calcium deposition and bone formation. A potential approach for improving the function of osteoblasts is to transplant stem or progenitor cells to target bone tissue and replace malfunctioned osteoblasts. Candidate stem and progenitor cell types include embryonic, fetal, and adult stem and progenitor cells. It is important to point out the osteoporosis is a disorder that involves the entire skeletal system. Thus a systematic approach, such as intravenous delivery of osteogenic stem or progenitor cells, is required for the treatment of the disorder. To achieve a therapeutic goal, it is necessary to carry out several steps: (1) identify and collect a stem or progenitor cell type; (2) expand the cells in vitro; (3) genetically manipulate the cells (e.g., transfection of the cells with desired genes to enhance selected functions), if necessary; and (4) deliver expanded cells to the venous system.

TABLE 22.5. Characteristics of Osteoprotegerin and Osteoprotegerin Ligand*

Proteins	Alternative Names	Amino Acids	Molecular Weight (kDa)	Expression	Functions
Osteoprotegerin	OPG, osteoclastogenesis inhibitory factor (OCIF), tumor necrosis factor receptor superfamily member 11B (TNFRSF11B)	401	46	Thyroid gland, kidney	Inhibiting the formation and activation of osteoclasts, suppressing bone resorption, and regulating lymph node organogenesis and vascular calcification
Osteoprotegerin ligand	OPGL, tumor necrosis factor ligand superfamily, member 11 (TNFSF11), receptor activator of NKκB ligand (RANKL), TNF-related activation-induced cytokine, osteoclast differentiation factor, ODF	317	35	T-cell, dendritic cell, thymus, lymph node	Stimulating osteoclast differentiation and activation, inducing bone resorption, serving as a dentritic cell survival factor, and regulating T-cell-dependent immune responses

*Based on bibliograhpy 22.7.

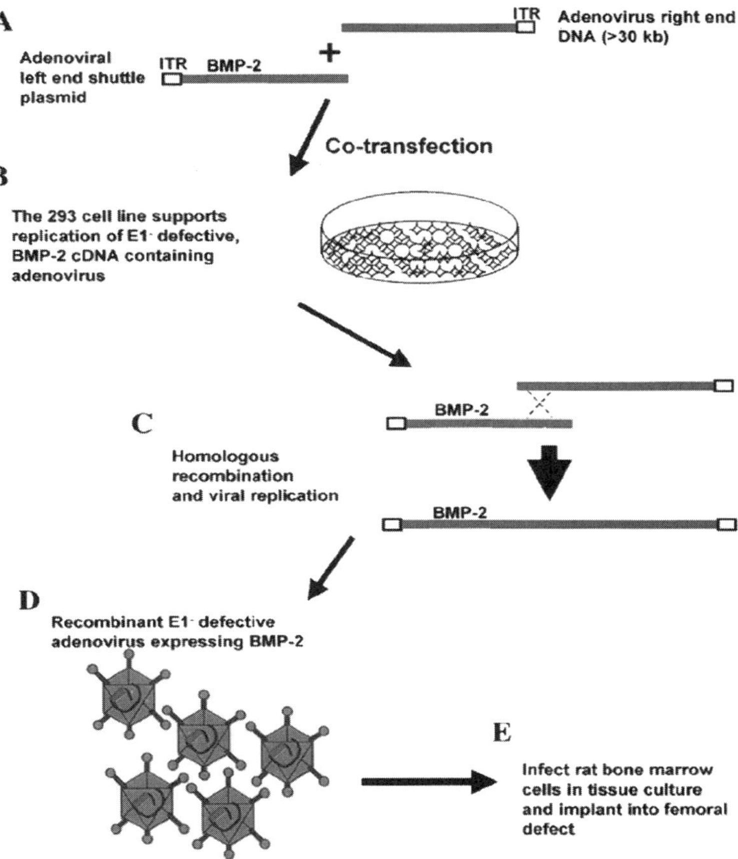

Figure 22.8. Diagram showing the construction of recombinant adenovirus containing rhBMP-2 cDNA. (A) Adenovirus El genes were deleted and replaced by BMP-2 cDNA on a plasmid (shuttle plasmid) containing the left inverted terminal repeat (ITR) required for viral replication. This BMP-2 shuttle plasmid and the large adenoviral right genome (~30 kb) were transfected into a human embryonic kidney fibroblast cell line, referred to as 293 cells. (B) The 293 cells contain integrated adenoviral El genes and express El proteins (key growth-regulatory proteins of the adenovirus) constitutively. Thus, the El-defective adenovirus (the El genes have been deleted) can be propagated only in the 293 cells. (C) The cotransfected BMP-2 shuttle plasmid DNA and the adenoviral right end DNA can undergo recombination through the shared homologous viral sequence in vivo in the 293 cells. The resultant BMP-2-expressing El-defective adenovirus will be able to replicate and form plaques on the 293 cells. (D) BMP-2 recombinant adenoviral clones are further purified and expanded from individual plaques, and their DNA structure is confirmed. (E) The purified BMP-2 recombinant adenovirus then can be used to infect the rat-bone-marrow cells that have been grown in tissue culture. (Reprinted with permission from Lieberman JR et al: The effect of regional gene therapy with bone morphogenetic protein-2-producing bone-marrow cells on the repair of segmental femoral defects in rats, *J Bone Joint Surg* 81:905–17, copyright 1999.)

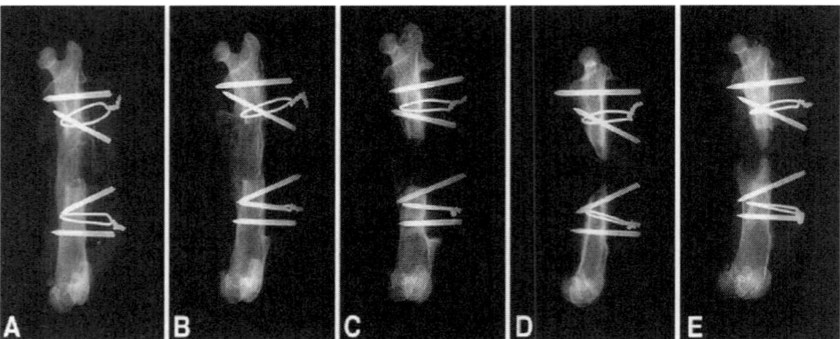

Figure 22.9. Radiographs of the specimens, made 2 months after the operation; 20 mg of guanidine hydrochloride-extracted demineralized bone matrix was used as a substrate in all defects. (A) Group I—local delivery of BMP-2-producing bone marrow cells (5×10^6), established by transferring a BMP2 gene-containing adenoviral vector. Dense, coarse trabecular bone, which was remodeling to form a new cortex, was present in these defects. (B) Group II—local delivery of rhBMP-2 (recombinant human BMP2, 20 μg). The healed defect is filled with lace-like trabecular bone. (C–E) Group III—β-galactosidase-producing rat bone marrow cells; group IV—uninfected rat bone marrow cells; group V—demineralized bone matrix alone. Minimum bone formation was noted in these three groups. (Reprinted with permission from Lieberman JR et al: The effect of regional gene therapy with bone morphogenetic protein-2-producing bone-marrow cells on the repair of segmental femoral defects in rats, *J Bone Joint Surg* 81:905–17, copyright 1999.)

Among the stem and progenitor cell types, including embryonic, fetal, and adult stem and progenitor cells, the adult bone marrow stem cells have been studied extensively, since these cells can directly differentiate to osteoblasts and autogenous cells can be used to prevent immune rejection reactions (Fig. 22.10). As discussed Chapter 9, the bone marrow contains adult stem cells, which can self-replicate and differentiate into various types of specialized cells, depending on the presence of environmental cues. Bone marrow stem cells are composed of hematopoietic and mesenchymal stem and progenitor cells. The hematopoietic stem and progenitor cells can give rise to all blood cells, including the erythrocytes, leukocytes, and platelets, whereas the mesenchymal stem and progenitor cells can differentiate into mesenchymal lineages of connective tissues, such as osteoblasts, chondroblasts, adipocytes, and fibroblasts. The bone marrow mesenchymal stem cells have been considered potential candidate cells for treatment of osteoporosis.

There are different types of mesenchymal stem cells in the bone marrow. While not all these cell types are characterized, the bone marrow stromal cells have been shown to contain mesenchymal stem cells that can transform to cell lineages of soft and hard connective tissues (Fig. 22.11). These cells show several unique features, including the capability of adhering to a substrate and forming colonies. These features can be used to isolate bone marrow stromal cells, since other types of bone marrow cells, including the hematopoietic stem cells, do not show these features. The bone marrow stromal cells have been studied and used extensively for the repair and reconstruction of impaired bones.

When expanded autogenous bone marrow stromal cells are delivered into the circulatory system, these cells can engraft and integrate into various mesenchymal tissues, including the bone, cartilage, and soft connective tissue. In the bone, engrafted osteogenic progenitor cells can transform into osteoblasts, which produce collagen matrix, promote

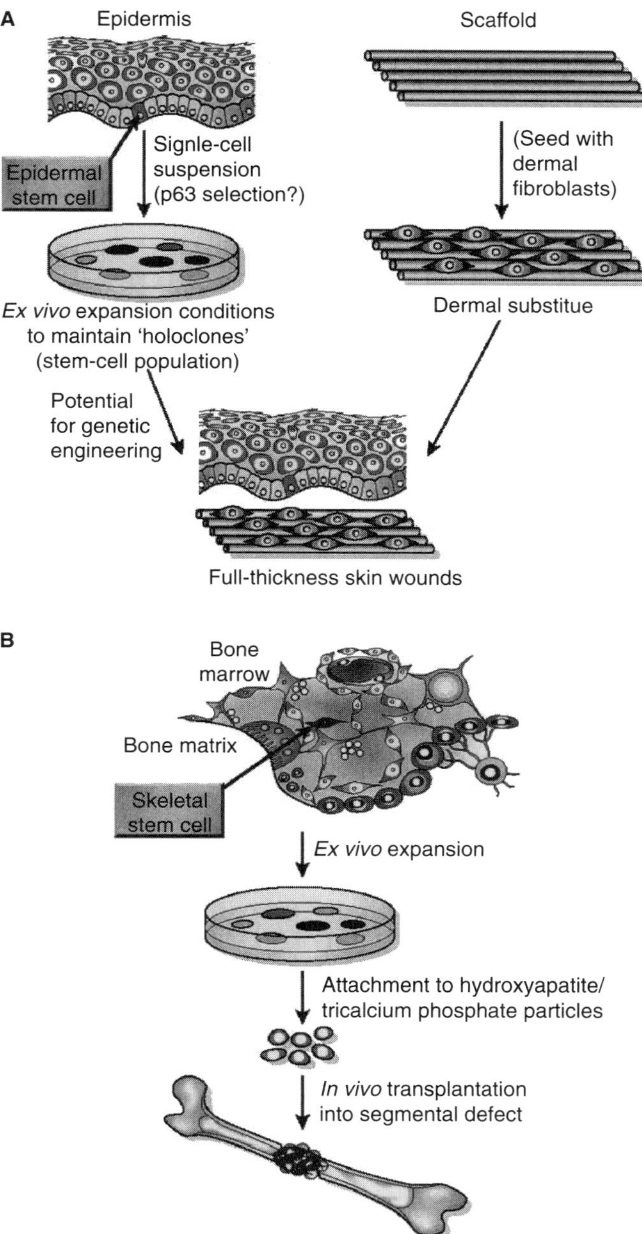

Figure 22.10. Regeneration of two-dimensional (skin) and three-dimensional (bone) tissues using stem cells. (A) Skin autografts are produced by culturing keratinocytes (which may be sorted for p63, the more recently described, epidermal stem cell marker) under appropriate conditions not only to generate an epidermal sheet but also to maintain the stem cell population (holoclones). The epidermal sheet is then placed on top of a dermal substitute comprising devitalized dermis or bioengineered dermal substitutes seeded with dermal fibroblasts. Such two-dimensional composites, generated ex vivo, completely regenerate full-thickness wounds. (B) Bone regeneration requires ex vivo expansion of bone marrow-derived skeletal stem cells and their attachment to three-dimensional scaffolds, such as particles of a hydroxyapatite/tricalcium phosphate ceramic. This composite can be transplanted into segmental defects and will subsequently regenerate an appropriate three-dimensional structure in vivo. (Reprinted by permission from Macmillan Publishers Ltd.: Bianco P, Robey PG: Stem cells in tissue engineering, *Nature* 414:118–21, copyright 2001.)

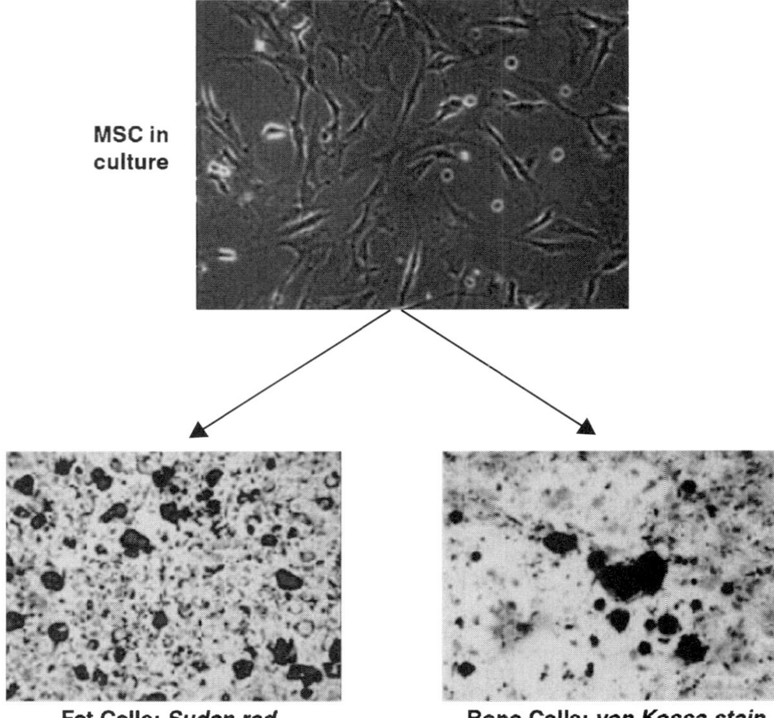

Figure 22.11. Mesenchymal stem cells (MSC) were characterized by their ability to differentiate into adipocytes and osteocytes under culture conditions. (Reprinted with permission from Togel F et al: Administered mesenchymal stem cells protect against ischemic acute renal failure through differentiation-independent mechanisms, *Am J Physiol Renal Physiol* 289:F31–42, copyright 2005.)

bone mineralization, and stimulate bone formation. Extensive investigations have demonstrated promising results for the treatment of bone osteoporosis and other disorders with cell transplantation. Allogenic osteogenic stem and progenitor cells can also be used for such a purpose. However, immune rejection is a serious problem. Immunosuppressive drugs must be administrated when allogenic cells are used.

Osteoporosis is associated with a progressive reduction in the bone strength due to the loss of bone mass and mineralization, often resulting bone fracture or functional impairment. In the case of bone fracture, a local treatment is required. Although bone injury and fracture can be treated with conventional approaches based on the self-healing nature of the bone, the regeneration process of the bone is usually slow and nonunion may occur because of poor blood circulation. In addition, bone grafts are required for the reconstruction of the lost bone. To resolve these problems, it is necessary to construct bone substitutes with living bone-forming cells and graft the substitutes to the injury sites. Several common approaches have been established and used for bone grafting and regeneration. These including: (1) identification and collection of osteogenic progenitor cells from the bone marrow or other sources, (2) expansion of the collected cells in vitro, (3) seeding and culture of expanded cells in porous bioceramic scaffolds (e.g., hydroxyapatite scaffolds) with a desired shape, and (4) reconstruction of fractured or injured bones with the cell-containing bone scaffold. Experimental investigations have consistently demonstrated that bone reconstruction with cell-containing scaffolds can significantly enhance bone regeneration and improve the healing process of injured bones. The grafted bone substitutes can be integrated into the natural bone when autogenous cells are used.

Paget's Disease

Pathogenesis, Pathology, and Clinical Features [22.11]. *Paget's disease* is a disorder characterized by excessive focal bone resorption due to increased activity of osteoclasts, subsequent fibrosis of bone marrow, and chaotic focal deposition of new bone matrix. Such activities often lead to focal bone distortion and weakening, rendering the bone fragile and prone to fracture. The incidence of Paget's disease is about 3% for the population over 40 years old. Aging is associated with an increase in the incidence of the disease. In the early stage, the disease may be asymptomatic and is often diagnosed by roentgenography for unrelated diseases. The etiology of Paget's disease is poorly understood. Since a treatment with glucocorticoids can relieve the symptoms of the disease, Paget's disease has been considered a disorder resulting from inflammatory reactions. Viral infection may contribute to the pathogenesis of the disorder. Potential viruses include respiratory syncitial, measles, and canine distemper viruses.

Pathological examinations demonstrate characteristic changes in Paget's disease, such as (1) bone mass reduction in association with enhanced angiogenic responses in the bone matrix during the early stage, (2) subsequent bone formation with irregular structure of the bone matrix, (3) an increase in the density and size of osteoclasts in association with an increase in the density of osteoblasts, (4) replacement of bone marrow with stromal tissue, and (5) reduction in the rate of bone resorption and formation during the late stage. Because of bone degeneration is associated with bone formation, which compensates for the loss of bone matrix, many patients may not show severe symptoms. During the late stage of the disease, clinical signs may occurs, including bone swelling, gait abnormalities due to unequal length of the long bones of the lower extremities (resulting from heterogeneous bone resorption and formation), and enlargement of the skull. Patients may also complain about headache, backache, and pain in other locations. The pain may be a result of nerve stimulation during bone resorption and formation. Some patients may experience hearing loss due to distortion of the ossicles and bones of the cochlea as well as the impingement of the hearing nerves by deformed bones. Brain tissue at the skull base (e.g., brainstem) may be compressed and impaired due to the distortion of the skull. The spinal cord may also be compressed by distorted vertebrae.

Conventional Therapy [22.11]. For patients without apparent symptoms, therapies are not usually necessary. When rapidly progressive manifestations are present, such as severe pains, signs of nervous compression, apparent bone distortions, and bone fractures, nonsteroid antiinflammatory drugs (such as indomethacin) and analgesics (such as aspirin) can be administrated. Although steroid antiinflammatory hormones, such as glucocorticoids, suppress the progression of Paget's disease, these hormones are not recommended because they cause osteoporosis and other disorders, such as reduction in cardiac output. In severe cases, surgical bone replacement may be recommended.

Molecular and Cellular Therapies [22.12]. Since the pathological changes in Paget's disease are similar to those of osteoporosis, all molecular and cellular therapies described for osteoporosis in Chapter 25 can be used for the treatment of Paget's disease.

Bone Tumors

Pathogenesis, Pathology, and Clinical Features [22.13]. Tumors are common disorders in the skeletal system. Bone tumors are classified into two types: benign tumors and malignant tumors. *Benign bone tumors* are those that do not invade neighboring tissues and do not metastasize. Examples of benign bone tumors include osteochondromas, endochondromas, benign giant cell tumors, and unicameral bone cysts. These tumors may arise from osteoblasts, osteoclasts, chondroblasts, and fibroblasts. Most benign bone tumors are not painful. Pathological examinations usually show restrained tumor cells, clearly identified tumor boundary, and bone enlargement and deformation. Clinical manifestations include slowly progressive enlargement of involved bones, bone deformation, and bone fractures.

Malignant bone tumors or cancers are tumors characterized by rapid invasion into neighboring tissues and metastasis. Common malignant bone tumors include multiple myeloma, osteosarcomas, chondrasarcomas, fibrosarcomas, and malignant giant cell tumors. Myeloma can arise from the bone marrow myeloid cells. Other bone cancers may be originated from osteoprogenitor cells, osteoblasts, chondroblasts, and fibroblasts. These malignant tumors are often found in the population at age of 20–40. Exposure to radiation, bone infarction, and Paget's disease may serve as predisposing factors for these tumors. Patients with malignant tumors experience severe pains, swelling, deformation, and destruction of the involved bones, which progress rapidly. Some tumors may destroy surrounding soft tissues. Laboratory tests often demonstrate an increase in the blood level of alkaline phosphatase. These tumors are usually associated with rapid metastasis. The primary target of metastasis is the lung. Bone malignant tumors often have poor prognosis.

Bone cancers often enhance the activity of osteoclasts, and stimulate the proliferation of these cells, promoting bone resorption. These events are possibly related to the production of parathyroid hormone and other mediators, such as interleukin-6 and interleukin-11, in cancer cells. These factors are known to activate osteoclasts and stimulate bone resorption. The bone-degrading activity of cancer cells facilitates cancer cell migration and metastasis.

Therapies for Bone Tumors. As for other types of tumor, conventional approaches, including surgical removal, chemotherapy, and radiotherapy, as well as molecular and cellular therapies can be used for the treatment of bone tumors. These approaches are discussed in detail in Chapter 25.

Rheumatoid Arthritis

Pathogenesis, Pathology, and Clinical Features [22.14]. *Rheumatoid arthritis* is a disorder characterized by inflammatory autoimmune reactions in the joints, resulting in chronic synovitis and progressive destruction of articular cartilage and bone. These pathological changes often lead to joint distortion, malfunction, pain, discomfort, and stiffening. The disorder is found in 0.5–1% of the human population. The incidence in the women is about 2–3 times higher than that in the men. The etiology of arthritis is poorly understood. The disorder is increased with age and may be influenced by disorders in gene regulation. For instance, most patients with rheumatoid arthritis express a protein complex known as histocompatibility antigen HLA-DR4. This protein complex may contribute to immune reactions, which are responsible for the development of arthritis.

Although the cause of arthritis is not clear, immune reactions have been thought to contribute to the pathogenesis of the disorder. In the site of the disorder, several types of immune cell are often found. These include Tlymphocytes, Blymphocytes, and macrophages. Theses cells express and release cytokines, including interleukin (IL)1, IL6, IL11, IL13, IL17, interferons, and monocyte chemotactic factors, which promote inflammatory reactions and attract additional leukocytes. The presence of these cells supports the possibility that arthritis is a disorder induced by immune reactions. Such reactions activate osteoclasts, stimulate osteoclast formation, enhance cartilage and bone resorption, and induce bone and cartilage destruction. It remains to be investigated, however, whether exogenous antigens or autoimmune factors, such as extracellular matrix components, initiate the immune reactions. The presence of autoantibodies in patients with arthritis suggests that the disorder may involve autoimmune reactions.

Pathological examinations often demonstrate hyperplasia of synovial lining cells, infiltration of mononuclear cells, and angiogenesis during the early stage. The mononuclear cells often aggregate around small blood vessels. Granulation tissue can be found in the site of arthritis. This type of tissue contains blood vessels, fibroblasts, and macrophages. During the late stage, bone and cartilage destruction can be found in peripheral joints.

Patients with arthritis may show signs of fatigue, weakness, and anorexia during the early stage. With the progression of the disorder, arthritis-specific manifestations may appear, including pain, stiffening, and swelling of the hands, feet, wrists, and knees. During the late stage, the movement of the joints may be severely limited and apparent joint distortion can be found. In addition, distortion of ligaments and tendons as well as imbalance of skeletal muscles may occur, which contribute to joint distortion and malfunction. The manifestations described above are all attributed to inflammatory reactions, cell hyperplasia and hypertrophy, excessive production of fibrous tissue, and bone and cartilage destruction.

Conventional Therapy [22.14]. Rheumatoid arthritis is treated with several types of agents, including (1) antiinflammatory drugs, (2) analgesic drugs, (3) disorder-modifying drugs, and (4) immunosuppressive drugs. For the anti-inflammatory therapy, steroidal and nonsteroidal agents can be used. Steroidal agents primarily include glucocorticoid hormones. These hormones can effectively suppress inflammatory reactions. However, steroidal hormones do not remove the causative factors and do not significantly change the prognosis of the disorder. Nonsteroidal agents, such as aspirin, fenoprofen, indomethacin, and tolmetin, exert not onlyanti-inflammatory but also analgesic effects. Among these nonsteroidal agents, aspirin is the most effective agent for the treatment of arthritis. Several drugs, including d-penicillamine, antimalarials, and gold compounds, have been used for the treatment of arthritis. These drugs slow down the progression of the disorder, although the exact mechanisms remain poorly understood. Since arthritis is a disorder related to immune reactions, it is conceivable to administrate immunosuppressive agents. Common immunosuppressive drugs include azathioprine and cyclophosphamide. These drugs can effectively reduce immune responses at the site of disorder and exert therapeutic effects.

When drug therapies are not effective and arthritic joints are severely damaged, surgical approaches can be used to reconstruct or replace the joints. Such surgical procedures can correct joint distortions, relieve pain, and improve joint function to a certain degree. The reconstruction of the hips and knees is usually more successful than that of other joints.

Molecular Therapy [22.15]. While conventional approaches can be used to reduce the symptoms of rheumatoid arthritis, these approaches may not significantly alter the progression of the disorder. Furthermore, most agents used for the treatment of rheumatoid arthritis exert adverse effects. Thus it is necessary to develop effective approaches that prevent or reduce the progression of the disorder. Recent investigations have demonstrated that genetic manipulation may represent such an approach.

Rheumatoid arthritis is possibly induced by autoimmune inflammatory responses. The suppression of immune reactions is a critical approach for the prevention and treatment of rheumatoid arthritis. A number of cytokines, including interleukin (IL)1, tumor necrosis factor (TNF)α, IL4, IL10, and IL13, have been known to regulate the activity of T and B cells, which are involved in immune reactions. IL1 and TNFα stimulate inflammatory responses, whereas IL4, IL10, and IL13 are anti-inflammatory factors. The genetic manipulation of these factors may provide a means for the suppression of autoimmune inflammatory reactions. These molecules are briefly discussed as follows.

Interleukin-1 [22.16]. Interleukin (IL)1 (see list of IL1 isoforms and receptors in Table 22.6) is a family of inflammatory mediators, including 10 known members: IL1α (IL1F1), IL1β (IL1F2), IL1Ra (IL1F3), IL18 (IL1F4), IL1F5, IL1F6, IL1F7, IL1F8, IL1F9, and IL1F10. Among these members, IL1α, IL1β, and IL1Ra have been known to play critical roles in the mediation of inflammatory responses. The activity of other members has not been clearly identified. While IL1α and IL1β serve as proinflammatory mediators that stimulate autoimmune responses, IL1Ra is an IL1 receptor agonist (which gives the name IL1Ra) that competes for the IL1 receptor and blocks the effects of IL1α and IL1β. IL1α and IL1β contribute to the development of autoimmune disorders, including rheumatoid arthritis, by initiating inflammatory responses. Bacterial infection may induce autoimmune disorders, since some bacteria may contain antigens that are similar in structure to certain proteins in the host systems. When infected with bacteria, the host monocytes and macrophages are activated to produce and release IL1. IL1 can act on endothelial cells to stimulate the production of chemotactic proteins, such as monocyte chemotactic protein 1, and adhesion molecules, such as E-selectin, intercellular adhesion molecules (ICAM), and vascular cell adhesion molecule (VCAM). These molecules promote leukocyte activation and attachment to endothelial cells, leading to leukocyte infiltration to infected areas. Furthermore, IL1 can activate monocytes to release more proinflammatory cytokines. These activities all contribute to inflammatory responses, which potentially induce autoimmune disorders and destroy host cells and structure. Thus, IL1α and IL1β are the primary targets for the treatment of autoimmune disorders.

IL1α and IL1β are two cytokines which share about 22% identity. These factors exert similar effects and induce similar cellular activities by interacting with their receptors. There are two types of IL1 receptor: type I IL1 receptor (IL1RI, ~80 kDa, 552 amino acids) and type II IL1 receptor (IL1RII, 60–68 kDa, 385 amino acids). The IL1RI *has been found in* T cells, endothelium, fibroblasts, astrocytes, chondrocytes, keratinocytes, neurons, smooth muscle cells, whereas IL1RII is found primarily in leukocytes, including

TABLE 22.6. Characteristics of IL1 Isoforms and IL1 Receptors*

Proteins	Alternative Names	Amino Acids	Molecular Weight (kDa)	Expression	Functions
IL1α	IL1α, hematopoietin 1, IL1F1	271	31	Monocytes, macrophages, brain, skin, lung	Regulating inflammatory and immune processes, mediating hematopoiesis, inducing apoptosis, mediating osteoclastogenesis, and inducing rheumatoid arthritis and Alzheimer's disease
IL1β	Catabolin, IL1F2	269	31	Macrophages, lung, skin	Similar to IL1α
Type I IL1 receptor	Interleukin-1 receptor α type 1, IL1Rα, IL1RA interleukin-1 receptor, IL1R, CD121A, antigen CD121a	569	65	T cell, monocyte, skin	Serving as a receptor for interleukin-1α (IL1α), interleukin-1β (IL1β), and mediating IL1 induced immune and inflammatory responses
Type II IL1 receptor	IL1RB, IL1Rβ, antigen CDw121b	398	45	T cell, B cell, monocytes, skin, nervous system, pancreas, testis	Serving as a receptor for interleukin-1 alpha (IL1α), interleukin-1β (IL1β), and acting as a decoy receptor that inhibits the activity of its ligands

*Based on bibliography 22.16.

B cells, T cells, monocytes, and neutrophils. Both receptors are composed of an extracellular region with three immunoglobulin-like domains, and share about 28% identity for these domains. However, the cytoplasmic domain differs considerable between the two types of receptor. The IL1RI possesses a 213-amino acid cytoplasmic region, whereas the IL1RII has only a 29-amino acid cytoplasmic region. The structural difference may influence the function of these receptors. IL1RI has been shown to interact with a transmembrane glycoprotein known as IL1 RI accessory protein (IL1 RAcP, ~66 kDa, 550 amino acids), which induces the internalization of the IL1/IL1 RI complex, transducing active signals to the intracellular signaling pathways. In contrast, IL1RII cannot interact with IL1 RAcP and serve as a decoy or dummy receptor that does not transduce active signals into the cytoplasm. Both IL1RI and IL1RII can interact with ligands IL1α and IL1β, but with different affinities. IL-1α preferentially binds to IL1RI, whereas IL1β preferentially binds to IL1RII. Because the binding of IL1 to IL1RII does not activate the intracellular signaling pathways, the role of IL1RII is to downregulate the activity of IL1.

There exist soluble forms of IL1RI (~60 kDa) and IL1RII (47 kDa and 57 kDa, representing two forms of soluble IL1RII) in the extracellular space. The soluble form of IL1RI (soluble type I IL1 receptor) can bind to IL1Ra (IL1 receptor agonist) with high affinity, thus reducing the binding of IL1Ra to the cell membrane IL1RI (membrane type I IL1 receptor) and leaving more cell membrane IL1RI available for IL1α and IL1β. These processes enhance the activity of IL1. In contrast, soluble IL1RII (soluble type II IL1 receptor) preferentially binds to IL1α and IL1β. The affinity of soluble IL1RII to IL1Ra is reduced considerably compared to its membrane-bound form, leaving more IL1Ra available for competitive binding to IL1RI, which reduces the effect of IL1α and IL1β. This is another mechanism by which IL1RII downregulates the activity of IL1α and IL1β.

Molecular therapies have been developed to reduce the inflammatory effect of the IL1 system. Major strategies include (1) enhance the expression of IL1Ra and (2) enhance the expression of IL1RII and promote the formation of soluble IL1RII. As discussed above, both IL1Ra and IL1RII exert an inhibitory effect on the activity of IL1. Thus the overexpression of the genes for these proteins potentially suppresses the activity of IL1 and reduces inflammatory responses, which are beneficial for the treatment of rheumatoid arthritis.

Tumor Necrosis Factors (TNFs) [22.17]. Tumor necrosis factors (see Table 22.7 for a list of TNF isoforms and receptors) are a family of proteins that mediate inflammatory and immune reactions. There are two types of TNFs: TNFα and TNFβ. TNF-α exists in the form of either a membrane protein (~26 kDa, 233 amino acids) or soluble protein (~17-kDa, 157 amino acids). The membrane form of TNFα is composed of extracellular, transmembrane, and cytoplasmic domains. The soluble form of TNFα is produced by cleaving the membrane form by TNFα converting enzyme (TACE) and exists in the form of homodimer. Both membrane and soluble forms are biologically active, but the soluble form is more potent. In contrast to TNFα, TNFβ (25 kDa, 171 amino acids) exists only in the form of soluble protein. TNFα and TNFβ share about 28% identity in the amino acid structure and both bind to the same types of receptor.

There are two types of TNF receptors: type I TNF receptor (TNFRI, 55 kDa, 455 amino acids) and type II TNF receptor (TNFRII, 75 kDa, 461 amino acids). TNFRI is a receptor that can interact with both TNFα and TNFβ. This receptor possesses double-sided functions. While TNFRI can induce cell apoptosis via the activation of the "death domain" in

TABLE 22.7. Characteristics of Tumor Necrosis Factor Isoforms and Receptors*

Proteins	Alternative Names	Amino Acids	Molecular Weight (kDa)	Expression	Functions
Tumor necrosis factor α	TNFα, tumor necrosis factor, TNFA, cachectin	233	26	Leukocyte, macrophage	Mediating inflammatory reactions, and regulating cellular activities including cell proliferation, differentiation, apoptosis
Tumor necrosis factor β	TNFB, TNF superfamily member 1, lymphotoxin α, lymphotoxin, lymphotoxin A, lyphotoxin α (LTα), lymphocyte-derived TNF, tumor necrosis factor ligand superfamily member 1	205	22	Lymphocyte, monocyte, dendritic cell, placenta	Mediating inflammatory and immune responses, regulating the formation of lymphoid organs, and inducing cell apoptosis
Type I tumor necrosis factor receptor	TNFR1, TNFR1α, tumor necrosis factor receptor superfamily, member 1A, p55 TNFR, TNF-R55, TNFR, 55-kDa, TNFR, 60-kDa	455	50	Lymphocytes, heart, tonsil	Interacting with tumor necrosis factor α, mediating inflammatory reactions, inducing apoptosis, and activating nuclear factor κB
Type II tumor necrosis factor receptor	TNFRSF1B, TNFR2, TNFβ receptor, TNFBR, CD120b	461	48	T cell, monocyte, dendritic cell, brain, heart, uterus, lung, thymus, intestine, kidney	Possibly serving as a decoy receptor; the soluble form reduces the activity of TNF α and β by binding to these ligands

*Based on bibliography 22.17.

its cytoplasmic region, it can also activate the nuclear factor κB mitogenic signaling pathway. However, how TNFRI selectively activates different signaling pathways remains poorly understood. TNFRI also exists in the form of soluble protein, which is generated by the cleavage of the membrane TNFRI. Soluble TNFRI can bind to and block the activity of TNFα. A number of cell types express TNFRI, including monocytes, neutrophils, endothelial cells, and hepatocytes.

Compared to TNFRI, TNFRII shows relatively low affinity to TNFα. Thus, the activity of TNFα is thought to be regulated primarily by TNFRI. TNFRII can interact with TNFβ. However, the binding of TNFβ to TNFRII does not induce activation of intracellular signaling pathways, suggesting that TNFRII may serve as a decoy receptor. There exists a soluble form of TNFRII, which is generated by cleaving the membrane form by a metalloproteinase known as TNF-receptor releasing enzyme (TRRE). The soluble form can bind to TNFα and TNFβ, which reduces the activity of these factors.

Molecular therapies can be developed on the basis of the observations described above. The overexpression of TNFRII can impose an inhibitory effect on the inflammatory effect of TNFα and TNFβ. Such a strategy can be achieved by transferring the TNFRII gene into target cells. Overexpressed TNFRII can effectively inhibit collagen-induced arthritis in mouse models. Furthermore, since soluble TNFRI binds to and block the activity of TNFα and reduces inflammatory reactions, the promotion of soluble TNFRI production helps to reduce arthritis. In particular, the cotransfer of a soluble TNF receptor-IgG1 fusion protein gene and the IL10 gene significantly suppresses the development of collagen-induced arthritis in animal models. Experimental investigations have demonstrated positive results for the genetic treatment of autoimmune disorders by manipulating the TNFR gene.

Antiinflammatory Cytokines [22.18]. There are a number of cytokines, including IL4, IL10, and IL13 (see Table 22.8 for IL13 characteristics), which exert antiinflammatory effects (see page 631 for the characteristics of these factors). These factors inhibit the release of proinflammatory factors, such as interferon-γ, IL1α, IL1β, and tumor necrosis factor, and stimulate the production and release of IL1Ra, which is known to competitively suppress the inflammatory effect of IL1α and IL1β. In particular, the structure and function

TABLE 22.8. Characteristics of IL13*

Proteins	Alternative Names	Amino Acids	Molecular Weight (kDa)	Expression	Functions
IL13	—	132	14	Lymphocyte, primarily activated Th2 cells, skin	Regulating B-cell differentiation, suppressing monocyte and macrophage activity, and inhibiting the production of proinflammatory cytokines and chemokines

*Based on bibliography 22.18.

of IL-10 have been studied extensively. Here, IL-10 is used as an example to demonstrate how the antiinflammatory cytokines inhibit the activity of proinflammatory factors.

Interleukin-10 is a protein (~18 kDa, 178 amino acids) present in the form of homodimer. IL10 is expressed in a umber of cell types, including monocytes, macrophages, T cells, B cells, natural killer cells, microglial cells, dendritic cells, eosinophils, and keratinocytes. The concentration of circulating IL10 is about 0.5 pg/mL in humans and other mammals. IL10 exerts its anti-inflammatory effect via interacting with the IL10 receptor. The IL10 receptor is a transmembrane glycoprotein complex composed of heterodimers α and β chains. The α chain is responsible for the binding of ligands, whereas the β chain transduces signals to the cytoplasmic signaling pathways. Each chain is composed of an extracellular, transmembrane, and cytoplasmic region. The binding of IL-10 to its receptor induces a number of antiinflammatory activities, including the inhibition of interferon-γ production and release from the T cells, suppression of chemokine secretion from neutrophils (e.g., MIP1α, MIP1β, and IL8), blockade of proinflammatory effects of IL1 and TNFα, reduction of inflammatory mediator generation in monocytes (e.g., IL8), and promotion of IL1Ra production. These activities results in the suppression of T-cell activation and immune responses, exerting beneficial effects for the treatment of rheumatoid arthritis. For therapeutic purposes, the genes encoding IL10 and IL10 receptor can be transferred into target cells in rheumatoid arthritis. Preliminary studies have demonstrated promising results for the treatment of arthritis by using IL10.

Dominant-Negative Mutant Ras Gene [22.19]. Ras is a protein that mediates the transduction of mitogenic factor signals. Acivated Ras stimulates cell proliferation and inflammatory reactions. Thus, Ras activation enhances the development of arthritis. The suppression of the Ras activity is an effective approach for the treatment of arthritis. One memthod for suppressing the Ras activity is to construct and transfer a dominant-negative mutant ras gene into the target cells. The dominant-negative mutant ras gene can be integrated into a gene-carrying vector such as a replication-deficient adenovirus vector. The vector can be delivered to the site of arthritis. Experimental investigations with such an approach have demonstrated that the delivered dominant-negative mutant ras gene can be expressed in target cells, resulting in a significant reduction in the level of inflammatory reactions in the joints (Fig. 22.12).

Osteoprotegerin [22.20]. Osteoprotegerin (OGP) is a protein of the cytokine tumor necrosis factor (TNF) receptor superfamily. This proetin serves as a decoy receptor for the osteoprotegerin ligand (OPGL) and can bind and inactivate OPGL. Osteoprotegerin ligand is a factor that stimulates the differentiation and proliferation of osteoclasts, which induce bone resorption and degeneration. Thus, osteoprotegerin can inhibit the activity of OPGL, suppress bone resorption, and enhance bone mineral deposition and bone formation. In the skeletal system, bone growth and resorption are controlled to a certain extent by the balance between OPGL and its decoy receptor osteoprotegerin. In transgenic animal models, the overexpression of osteoprotegerin is associated with eenhanced bone growth and reduced bone resorption. In contrast, the genetic disrutpion or knock out of the osteoprotegerin gene enhances bone resorption, resulting in osteoporosis-like alterations. Osteoprotegerin is present in the circulation and interstitial fluids of various tissues and organs. The direct delivery of osteoprotegerin or the transfer of the osteoprotegerin gene into the skeletal system prevents bone and cartilage resorption and destruction in inflammation and arthritis.

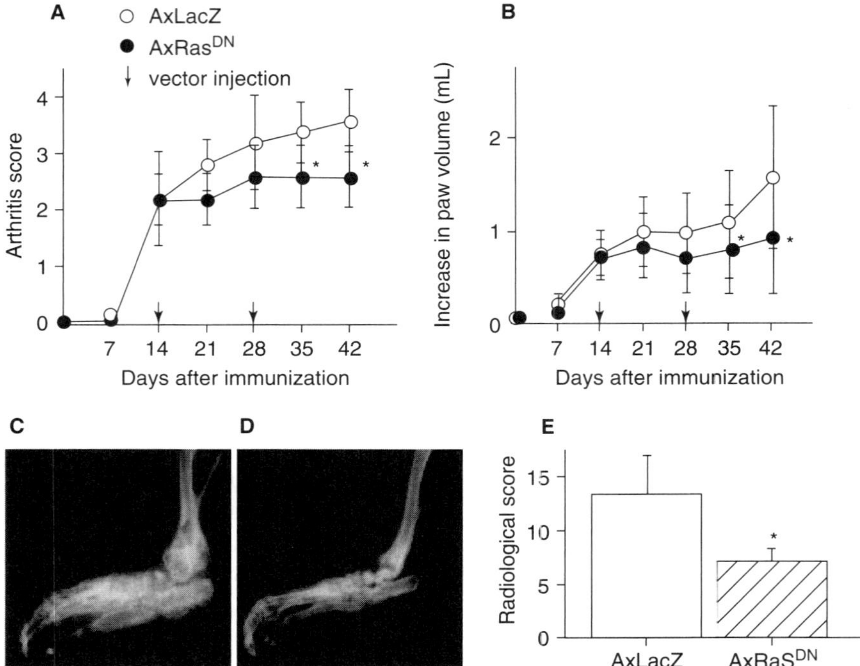

Figure 22.12. Therapeutic effects of replication-deficient adenovirus vector carrying the dominant-negative mutant of the Ras gene (AxRasDN) on rat adjuvant arthritis. All rats were immunized with a subcutaneous injection of adjuvant in the base of the tail (day 0). Viruses were then intraarticularly injected into the right ankles on days 7 and 14. Bars show the mean ± SD of 10 rats per group. (A) Effects of AxRasDN injection, evaluated by arthritis score. The arthritis score of the AxRasDN group was significantly lower than that of the control AxLacZ group on days 35 and 42. (B) Effects of AxRasDN injection, evaluated by the increase in paw volume. The increase in paw volume of the AxRasDN group was significantly less than that of the AxLacZ group on days 35 and 42. (C) The radiologic findings in the right ankles of AxLacZ-injected rats indicate severe joint destruction. (D) The radiologic findings in the right ankles of AxRasDN-injected rats show minimal destructive changes in the joint. (E) The radiologic score of the AxRasDN-injected ankles was significantly decreased in comparison with that of the AxLacZ group. * = $P < 0.01$ versus AxLacZ-injected joints. (Reprinted with permission of Wiley-Liss, Inc, a subsidiary of John Wiley & Sons, Inc., from Yamamoto A et al: Suppression of arthritic bone destruction by adenovirus-mediated dominant-negative Ras gene transfer to synoviocytes and osteoclasts, *Arthritis Rheum* 48:2682–92, copyright 2003.)

Viral Thymidine Kinase [22.21]. Rheumatoid arthritis often affects the synovium, inducing synoviocyte proliferation and synovium hypertrophy. These pathological changes often impose harmful effects on the function of the involved joints. One of the treatments for such a disorder is to remove altered synovium by surgical synovectomy. However, surgical trauma often induces inflammatory reactions and scar formation, which facilitate the distortion and destruction of the joint tissue. Alternatively, genes encoding growth inhibiting proteins can be used for the suppression of synoviocyte proliferation. One of such genes is the herpes viral thymidine kinase gene. This gene encodes a thymidine kinase protein, which can convert nucleosides to nucleotides by phosphorylation. In dividing or proliferating cells, nucleotides are taken up by cells for DNA synthesis. When nucleoside analogues (e.g., ganciclovir) are present, the viral thymidine kinase can covert the nucleoside analogues to nucleotides, which can be used for DNA synthesis. However, the incorporation of the nucleotide analogs results in the termination of DNA synthesis, because no more nucleotides can be added to the incorporated nucleotide analogues. By such a mechanism, the viral thymidine kinase gene, together with nucleoside analogues, can be used to suppress the proliferation of synoviocytes and reduce pathological changes in rheumatoid arthritis.

Bone and Cartilage Injury

Pathogenesis, Pathology, and Clinical Features [22.22]. Bone and cartilage injury occurs due to mechanical overload in sports and accidents. Various types of injury can be induced, depending on the magnitude and direction of the mechanical load. The most common injury is bone fracture. Such an injury is followed by several remodeling processes, including inflammatory, reparative, and modeling processes. The understanding of these processes is critical to the treatment of bone injury. Here, bone fracture is used as an example to discuss these remodeling processes.

Inflammatory Responses. Inflammation is a process that occurs in response to injury. Such a process is necessary for the self-healing of injured tissues or organs. While inflammation is common to all types of tissue, there are several unique features for bone injury. Bone injury induces transport and deposition of calcium and phosphorus, which do not occur in the inflammation of other tissues; also, since bone is subject to large mechanical loads, bone inflammation and recovery are influenced by mechanical forces.

Bone injury induces a series of inflammatory reactions. Blood vessel injury usually occurs with bone fracture, inducing hemorrhage and the formation of hematoma. At the same time, necrotic and injured cells can release cytokines and growth factors, which stimulate the infiltration of leukocytes and the migration of fibroblasts and other cell types to the injury site. Cytokines and growth factors also stimulate the proliferation of these cells and the production of extracellular matrix. All these reactions contribute to the formation of granulation tissue, which replaces the necrotic tissue and hematoma. Furthermore, angiogenesis occurs near the injury site, generating new blood vessels that provide oxygen and nutrients necessary for cell and tissue regeneration. Within about two weeks, calcium and phosphorus start to deposit to the injured tissue, resulting in the formation of immature membranous bone structure, which is referred to as a *callus*. A callus can be further mineralized to form a mature bone as described in the following section.

Reparative Reactions. Reparative reactions are initiated for the formation of mature bones based on calluses. Following the inflammatory phase, callus formation occurs at several locations. One is formed near the cortical surface of the bone at the injury site by the periosteum and adjacent skeletal muscle cells. In the medullary cavity at the injury site, the bone marrow cells can form a callus, which seals the fracture. Another type of callus forms between the two fracture-ends, serving to bridge the gap between the calluses at the ends. Additional calluses can be formed to join all separate calluses. Within about a month, a bone fracture can be filled with joined calluses.

Modeling Process. Modeling is a process by which the newly formed bone is further matured, organized, and aligned along the direction of the principal mechanical forces. The modeling process is thought to be regulated by mechanical stress. There are several events for the modeling phase. First, the newly generated bone structure is reshaped in response to the distribution of the mechanical stress. In regions with sufficient mechanical stress, the new bone is strengthened with additional mineralization, whereas in regions without sufficient mechanical stress new bone may be absorbed and degraded. Second, the medullary cavity and bone marrow are gradually restored. Third, the restoration of the natural form of bone structure (also referred to as *bone reconstitution*) is accomplished by coordinated bone resorption and regeneration. Bone resorption is induced by osteoclasts, whereas bone regeneration is induced by osteoblasts. Each type of cell may be activated in response to mechanical stress. A mechanical stress below a critical level may activate osteoclasts, initiating bone resorption. In contrast, a mechanical stress above a critical level may activate osteoblasts, resulting in bone regeneration. With such a stress-regulated process, the reconstituted bone can be eventually shaped to the original natural form. Bone regeneration is a long-term process. The entire reconstitution process may take about several years.

Complications of Bone Injury. There are several complications that may occur during bone healing. The most common complications include fibrous union and nonunion. Fibrous union is a form of bone reconstitution with the establishment of a fibrous tissue bridge without mineralization between the fractured bones. A major cause for the formation of the fibrous tissue bridge is the lack of blood supply to the injured bone. A poor blood supply negatively influences the formation of calluses while promoting the formation of fibrous scar tissue, resulting in the formation of fibrous tissue bridges. Fibrous union often occurs in the injury of the distal pretibia and carpal navicular bone, which are associated with scarce blood vessels and insufficient blood circulation. Nonunion is a form of incomplete bone reconstitution, leaving a boneless gap between the ends of healed bone. Several factors, including bone loss, dislocation of fractured bone, infection, and severe soft tissue damage, may contribute to the bone nonunion. Such a consequence may occur in long-bone fracture.

Conventional Therapy [22.22]. The treatment of bone fracture and other types of bone injury is primarily dependent on the ability of the bone in self-healing and self-regeneration. Several approaches can be used to assist fractured bones in the healing processes. These include the realignment and restriction of dislocated bones to their natural positions by using external fixation devices, which ensures the anatomical reconstitution of the bones, stop of hemorrhage promptly, protection of injured bones from infection by administrating antibiotics, and treatment of bone non-union, if any, by bone transplantation or grafting. These approaches are usually effective for the treatment of bone fracture.

Molecular Regenerative Engineering [22.23]. The strategies for the molecular treatment of bone injury are to enhance bone regeneration and prevent bone nonunion or fibrous union. Although injured bones can be self-healed and conventional therapies are effective for most cases, bone reconstitution may take several months or longer. Furthermore, a fraction about 10% of bone fractures may experience delayed union or form nonunion structures. Bone regeneration can be greatly enhanced by using molecular regenerative approaches. One effective approach is delivering bone formation-stimulating proteins or their genes into the cells responsible for bone regeneration. Typical factors that stimulate bone formation are bone morphogenetic proteins (BMPs), which can be used to enhance bone regeneration.

Bone morphogenetic protein 2, a member of the bone morphogenetic protein family, has been characterized and studied extensively in experimental bone injury models established in the rat, rabbit, and dog. The delivery of bone morphogenetic protein-2 or its gene to fractured bone enhances bone regeneration (Fig. 22.9). Clinical trials have demonstrated the effectiveness of locally delivered bone morphogenetic protein-2 in promoting bone regeneration. Bone morphogenetic protein 4 has also been used to enhance bone formation in experimental models of bone injury, demonstrating similar therapeutic effects as bone morphogenetic protein 2. The genes of these proteins can be used for gene therapy for improving bone regeneration. In addition to the bone morphogenetic protein genes, genes encoding angiogenic factors and vascular endothelial growth factor have been used for the promotion of bone regeneration. These genes can be delivered directly to the injury sites or indirectly delivered via the mediation of a matrix scaffold. Adenovirus vectors have been primarily used for mediating bone gene delivery.

Other growth factors, such as fibroblast growth factors (FGFs) and insulin-like growth factor (IGF), regulate the proliferation and differentiation of bone and cartilage cells (Fig. 22.13). These growth factors can be used to enhance the repair of injured bone and cartilage. In particular, fibroblast growth factors represent a family of potent growth factors for bone regeneration. The FGF family contains 23 known members. Among these members, FGF1 (~18 kDa, 155 amino acids) and FGF2 (~18 kDa, 155 amino acids) have been known to play an important role in the regulation of cell proliferation and differentiation. These growth factors are produced by many cell types, including the fibroblasts, endothelial cells, macrophages, hepatocytes, and keratinocytes. Both FGF1 and FGF2 can interact with FGF receptor tyrosine kinases and induce the proliferation of many cell

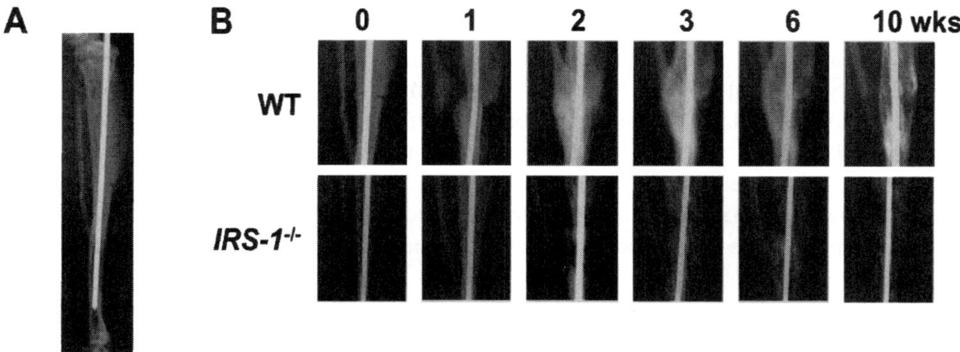

Figure 22.13. X-ray features of bone healing in WT and *IRS-1*$^{-/-}$ mice (IRS: insulin receptor substrate-1). (A) The fracture model used in this study. After exposing the right tibiae of 8-week-old mice, a transverse osteotomy was performed at the midshaft with a bone saw. The bone marrow cavity was then stabilized with an intramedullary nail. (B) Time course of the fracture healing in representative WT and *IRS-1*$^{-/-}$ mice. (Reprinted with permission from Shimoaka T et al: Impairment of bone healing by insulin receptor substrate-1 deficiency, *J Biol Chem* 279:15314–22, copyright 2004.)

types, including osteoblasts, chondrocytes, endothelial cells, and fibroblasts, thus enhacing bone and cartilage formation.

Cell Regenerative Engineering [22.24]. Bone regeneration involves the activation of osteoprogenitor cells, which differentiate to osteoblasts and other types of bone cells. It is thus conceivable to transplant osteoprogenitor cells or stem cells to promote bone formation. The bone marrow is known to contain osteoprogenitor cells. Bone marrow-derived cells have been used for the enhancement of bone regeneration in animal models as well as in humans. The transplantation of these cells to injured bones significantly facilitates bone formation and injury recovery. Other types of stem cells, such as embryonic, fetal, muscle-derived adult, and fat-derived adult stem cells, may also be used for bone regenerative engineering. These stem cells may be induced to differentiate to osteoblasts and other types of bone cells, when appropriate experimental conditions and extracellular environment are provided.

For the treatment of cartilage injury, several types of cells, including chondrocytes and osteochondrocytes, can be used for cell transplantation. The cells can be seeded in a scaffold constructed with an appropriate material (e.g., hyaluronan-based biopolymers or collagen matrix). A cell-seeded scaffold can be used to replace injured cartilage. In addition to chondrocytes, other cell types, such as embryonic stem cells and bone marrow stromal cells or mesenchymal stem cells, can be used for cartilage regeneration.

The type of matrix scaffolds may influence the differentiation of progenitor and stem cells. For instance, when bone marrow stem cells are cultured in a hyaluronan matrix with the supplement of transforming growth factor β1, the stem cells can differentiate into chondrocytes, forming a cartilage-like structure. When bone marrow stem cells are seeded in a porous calcium phosphate scaffold, the stem cells are transformed into osteoblasts, forming a bone-like structure. Bone morphogenetic proteins and growth factors can be used in cartilage constructs to facilitate cartilage formation. These approaches have been successfully used in experimental models for cartilage regeneration.

Progenitor and stem cells for bone and cartilage regeneration can be genetically transfected with genes encoding bone regeneration-promoting factors, such as the bone morphogenetic protein and vascular endothelial growth factor genes. Such cells may exhibit enhanced capability of differentiation and proliferation, thus facilitating bone regeneration and recovery. Alternatively, the bone formation-stimulating genes can be delivered by using gene carriers. Fibroblasts derived from soft connective tissue have been used as such gene carriers. Fibroblasts can be collected, cultured, and transfected with a selected gene in vitro, and used for cell transplantation in vivo. These approaches have been successfully used in experimental models. In human trials, mesenchymal stem cells have been used as gene carriers for bone-injury therapy. These trials have demonstrated encouraging results.

Tissue Regenerative Engineering [22.25]. While molecular and cellular therapies enhance bone regeneration, engineering manipulations at the tissue level is also important for the recovery from bone injury, especially in delayed bone union and nonunion. A major type of engineering manipulation is bone grafting or reconstruction with bone substitutes. Such a manipulation is necessary in the case of bone loss and destruction. There are several types of materials that can be used for such a purpose. These include autogenous cancellous bone specimens, metal prostheses, calcium phosphate ceramics, and polymeric materials. Among these materials, the autogenous cancellous bone is the gold standard material. The cancellous bone contains osteoprogenitor cells and osteoblasts, which play a critical role in bone repair and regeneration. However, in the case of large bone destruction, it is difficult to collect sufficient cancellous bone specimens. It is necessary to use other types of materials, such as synthetic materials and allogenic bones.

Synthetic materials have been used and tested for bone reconstruction. Osteoprogenitor or stem cells can be seeded in scaffolds of synthetic materials. A cell-containing scaffold can be tailored into a desired shape and used for bone grafting and reconstruction. However, synthetic materials cannon be integrated into the natural skeletal system. It is often difficult for cells and blood vessels to grow into the synthetic bone substitute. These limitations hinder the use of synthetic materials for bone reconstruction. Allogenic bone specimens are alternative materials for bone reconstruction. However, allogenic bones with living cells induce acute immune rejection reactions. It is necessary to administrate immune suppressive agents for patients with allogenic bone grafting. The removal of living cells from allogenic bone grafts can significantly reduce immune responses. Decellularized allogenic bone specimens can be used as bone substitutes for bone reconstruction.

BIBLIOGRAPHY

22.1. Anatomy and Physiology of the Bone and Cartilage

Li YC, Kong J, Wei M, Chen ZF, Liu SQ et al: 1,25-Dihydroxyvitamin D3 is a negative endocrine regulator of the renin-angiotensin system, *J Clin Invest* 110:229–38, 2002.

Suda T, Ueno Y, Fujii K, Shinki T: Vitamin D and bone, *J Cell Biochem* 88(2):259–66, 2003.

Guyton AC, Hall JE: *Textbook of Medical Physiology*, 11th ed, Saunders, Philadelphia, 2006.

McArdle WD, Katch FI, Katch VL: *Essentials of Exercise Physiology*, 3rd ed, Lippincott Williams & Wilkins, Baltimore, 2006.

Germann WJ, Stanfield CL (with contributors Niles MJ, Cannon JG), *Principles of Human Physiology*, 2nd ed, Pearson Benjamin Cummings, San Francisco, 2005.

Thibodeau GA, Patton KT: *Anatomy & Physiology*, 5th ed, Mosby, St. Louis, 2003.

Boron WF, Boulpaep EL: *Medical Physiology: A Cellular and Molecular Approach*, Saunders, Philadelphia, 2003.

Ganong WF: *Review of Medical Physiology*, 21st ed, McGraw-Hill, New York, 2003.

22.2. Pathogenesis, Pathology, and Clinical Features of Osteoporosis

Kanis JA, Melton III LJ, Christiansen C, Johnston CC, Khaltaev N: The diagnosis of osteoporosis, *J Bone Miner Res* 9:1137–41, 1994.

Downey PA, Siegel MI: Bone biology and the clinical implications for osteoporosis, *Phys Ther* 86(1):77–91, 2006.

Raisz LG: Pathogenesis of osteoporosis: Concepts, conflicts, and prospects, *J Clin Invest* 115(12):3318–25, 2005.

Rosen CJ, Brown SA: A rational approach to evidence gaps in the management of osteoporosis, *Am J Med* 118(11):1183–9, 2005.

Feng X: Regulatory roles and molecular signaling of TNF family members in osteoclasts, *Gene* 350(1):1–13, 2005.

Chien KR, Karsenty G: Longevity and lineages: Toward the integrative biology of degenerative diseases in heart, muscle, and bone. *Cell* 120(4):533–44, 2005.

Bassett JH, Williams GR: The molecular actions of thyroid hormone in bone, *Trends Endocr Metab* 14(8):356–64, 2003.

Harada S, Rodan GA: Control of osteoblast function and regulation of bone mass, *Nature* 423(6937):349–55, 2003.

Boyle WJ, Simonet WS, Lacey DL: Osteoclast differentiation and activation, *Nature* 423(6937):337–42, 2003.

Schneider AS, Szanto PA: *Pathology*, 3rd ed, Lippincott Williams & Wilkins, Philadelphia, 2006.

Frazier MS, Drzymkowski JW: *Essentials of Human Diseases and Conditions*, 3rd ed, Elsevier Saunders, St Louis, 2004.

22.3. Vitamin D Receptor (VDR)

Morrison NA, Qi JC, Tokita A, Kelly P, Crofts L et al: Prediction of bone density from vitamin D receptor alleles, *Nature* 367:284–7, 1994.

Arai H, Miyamoto KI, Taketani Y, Yamamoto H, Iemori Y et al: A vitamin D receptor gene polymorphism in the translation initiation codon: Effect on protein activity and relation to bone mineral density in Japanese women, *J Bone Miner Res* 12:915–21, 1997.

Garnero P, Munoz F, Borel O, Sornay-Rendu E, Delmas PD: Vitamin D receptor gene polymorphisms are associated with the risk of fractures in postmenopausal women, independently of bone mineral density, *J Clin Endocr Metab* 90(8):4829–35, 2005.

Carling T, Rastad J, Akerstrom G, Westin G: Vitamin D receptor (VDR) and parathyroid hormone messenger ribonucleic acid levels correspond to polymorphic VDR alleles in human parathyroid tumors, *J Clin Endocr Metab* 83:2255–9, 1998.

Colin EM, Uitterlinden AG, Meurs JBJ, Bergink AP, van de Klift M et al: Interaction between vitamin D receptor genotype and estrogen receptor alpha genotype influences vertebral fracture risk, *J Clin Endocr Metab* 88:3777–84, 2003.

Ensrud KE, Stone K, Cauley JA, White C, Zmuda JM et al: Vitamin D receptor gene polymorphisms and the risk of fractures in older women, *J Bone Miner Res* 14:1637–45, 1999.

Ferrari S, Manen D, Bonjour JP, Slosman D, Rizzoli R: Bone mineral mass and calcium and phosphate metabolism in young men: Relationships with vitamin D receptor allelic polymorphisms, *J Clin Endocr Metab* 84:2043–8, 1999.

Houston LA, Grant SFA, Reid DM, Ralston SH: Vitamin D receptor polymorphism, bone mineral density, and osteoporotic vertebral fracture: Studies in a UK population, *Bone* 18:249–52, 1996.

Hughes MR, Malloy PJ, Kieback DG, Kesterson RA, Pike JW et al: Point mutations in the human vitamin D receptor gene associated with hypocalcemic rickets, *Science* 242:1702–5, 1988.

Hustmyer FG, Peacock M, Hui S, Johnston CC, Christian J: Bone mineral density in relation to polymorphism at the vitamin D receptor gene locus, *J Clin Invest* 94:2130–4, 1994.

Kelly PJ, Hopper JL, Macaskill GT, Pocock NA, Sambrook PN et al: Genetic factors in bone turnover, *J Clin Endocr Metab* 72:808–13, 1991.

Lorentzon M, Lorentzon R, Nordstrom P: Vitamin D receptor gene polymorphism is associated with birth height, growth to adolescence, and adult stature in healthy Caucasian men: A cross-sectional and longitudinal study, *J Clin Endocr Metab* 85:1666–71, 2000.

Malloy PJ, Eccleshall TR, Gross C, Van Maldergem L, Bouillon R et al: Hereditary vitamin D resistant rickets caused by a novel mutation in the vitamin D receptor that results in decreased affinity for hormone and cellular hyporesponsiveness, *J Clin Invest* 99:297–304, 1997.

Morrison NA, Qi JC, Tokita A, Kelly PJ, Crofts L et al: Prediction of bone density from vitamin D receptor alleles, *Nature* 367:284–7, 1994. Note: Erratum: *Nature* 387: 106 only, 1997.

Riggs BL: Vitamin D-receptor genotypes and bone density, [editorial], *New Engl J Med* 337:125–126, 1997.

Sainz J, Van Tornout JM, Loro L, Sayre J, Roe TF et al: Vitamin D-receptor gene polymorphisms and bone density in prepubertal American girls of Mexican descent, *New Engl J Med* 337:77–82, 1997.

Uitterlinden AG, Weel AEAM, Burger H, Fang Y, Van Duijn CM et al: Interaction between the vitamin D receptor gene and collagen type I-alpha-1 gene in susceptibility for fracture, *J Bone Miner Res* 16:379–85, 2001.

Yoshizawa T, Handa Y, Uematsu Y, Takeda S, Sekine K et al: Mice lacking the vitamin D receptor exhibit impaired bone formation, uterine hypoplasia and growth retardation after weaning, *Nature Genet* 16:391–6, 1997.

Human protein reference data base, Johns Hopkins University and the Institute of Bioinformatics, at http://www.hprd.org/protein.

22.4. Type I Collagen Gene

Aitchison K, Ogilvie D, Honeyman M, Thompson E, Sykes B: Homozygous osteogenesis imperfecta unlinked to collagen I genes, *Hum Genet* 78:233–6, 1988.

Bateman JF, Chan D, Walker ID, Rogers JG, Cole WG: Lethal perinatal osteogenesis imperfecta due to the substitution of arginine for glycine at residue 391 of the alpha-1(I) chain of type I collagen, *J Biol Chem* 262:7021–7, 1987.

Bateman JF, Lamande SR, Dahl HHM, Chan D, Cole WG: Substitution of arginine for glycine 664 in the collagen alpha-1(I) chain in lethal perinatal osteogenesis imperfecta, *J Biol Chem* 263:11627–30, 1988.

Boedtker H, Fuller F, Tate V: The structure of collagen genes, *Int Rev Connect Tissue Res* 10:1–63, 1983.

Bonadio J, Ramirez F, Barr M: An intron mutation in the human alpha-1(I) collagen gene alters the efficiency of pre-mRNA splicing and is associated with osteogenesis imperfecta type II, *J Biol Chem* 265:2262–8, 1990.

Cabral WA, Mertts MV, Makareeva E, Colige A, Tekin M et al: Type I collagen triplet duplication mutation in lethal osteogenesis imperfecta shifts register of alpha chains throughout the helix and disrupts incorporation of mutant helices into fibrils and extracellular matrix, *J Biol Chem* 278:10006–12, 2003.

Chamberlain JR, Schwarze U, Wang PR, Hirata RK, Hankenson KD et al: Gene targeting in stem cells from individuals with osteogenesis imperfecta, *Science* 303:1198–1201, 2004.

Chu ML, de Wet W, Bernard M, Ramirez F: Fine structural analysis of the human pro-alpha-1(I) collagen gene: Promoter structure, AluI repeats, and polymorphic transcripts, *J Biol Chem* 260:2315–20, 1985.

Constantinou CD, Nielsen KB, Prockop DJ: A lethal variant of osteogenesis imperfecta has a single base mutation that substitutes cysteine for glycine 904 of the alpha-1(I) chain of type I procollagen: The asymptomatic mother has an unidentified mutation producing an overmodified and unstable type I procollagen, *J Clin Invest* 83:574–84, 1989.

Dalgleish R: The human type I collagen mutation database, *Nucleic Acids Res* 25:181–7, 1997.

Di Lullo GA, Sweeney SM, Korkko J, Ala-Kokko L, San Antonio JD: Mapping the ligand-binding sites and disease-associated mutations on the most abundant protein in the human, type I collagen, *J Biol Chem* 277:4223–31, 2002.

Grant SFA, Reid DM, Blake G, Herd R, Fogelman I et al: Reduced bone density and osteoporosis associated with a polymorphic Sp1 binding site in the collagen type I-alpha 1 gene, *Nature Genet* 14:203–5, 1996.

Long JR, Liu PY, Lu Y, Xiong DH, Zhao LJ et al: Association between COL1A1 gene polymorphisms and bone size in Caucasians, *Eur J Hum Genet* 12:383–8, 2004.

Pereira R, Khillan JS, Helminen HJ, Hume EL, Prockop DJ: Transgenic mice expressing a partially deleted gene for type I procollagen (COL1A1): A breeding line with a phenotype of spontaneous fractures and decreased bone collagen and mineral, *J Clin Invest* 91:709–16, 1993.

Uitterlinden AG, Burger H, Huang Q, Yue F, McGuigan FEA et al: Relation of alleles of the collagen type I-alpha-1 gene to bone density and the risk of osteoporotic fractures in postmenopausal women, *New Engl J Med* 338:1016–21, 1998.

Uitterlinden AG, Weel, AEAM, Burger H, Fang Y, Van Duijn CM et al: Interaction between the vitamin D receptor gene and collagen type I-alpha-1 gene in susceptibility for fracture, *J Bone Miner Res* 16:379–85, 2001.

Grant SFA, Reid DM, Blake G, Herd R, Fogelman I et al: Reduced bone density and osteoporosis associated with a polymorphic Spl site in the collagen type I alpha 1 gene, *Nat Genet* 14:203–5, 1996.

Mann V, Hobson EE, Li B, Stewart TL, Grant SF, Robins SP et al: A COL1A1 Spl binding site polymorphism predisposes to osteoporotic fracture by affecting bone density and quality, *J Clin Invest* 107:899–907, 2001.

Qureshi AM, McGuigan FEA, Seymour DG, Hutchison JD, Reid DM et al: Association between COLIA1 Spl alleles and femoral neck geometry, *Calcif Tissue Int* 69:67–72, 2001.

Garcia-Giralt N, Nogues X, Enjuanes A, Puig J, Mellibovsky L et al: Two new single nucleotide polymorphisms in the COLIA1 upstream regulatory region and their relationship with bone mineral density, *J Bone Miner Res* 17:384–93, 2002.

Niyibizi C, Wang S, Mi Z, Robbins PD: Gene therapy approaches for osteogenesis imperfecta, *Gene Ther* 11(4):408–16, 2004.

22.5. Estrogen Receptor

Rodan GA, Martin TJ: Therapeutic approaches to bone diseases, *Science* 289:1508–14, 2000.

Alves SE, Lopez V, McEwen BS, Weiland NG: Differential colocalization of estrogen receptor beta (ER-beta) with oxytocin and vasopressin in the paraventricular and supraoptic nuclei of the female rat brain: An immunocytochemical study, *Proc Natl Acad Sci USA* 95:3281–3286, 1998.

Bonnelye E, Aubin JE: Estrogen receptor-related receptor alpha: A mediator of estrogen response in bone, *J Clin Endocr Metab* 90(5):3115–21, 2005.

Rosen CJ: Clinical practice. Postmenopausal osteoporosis, *New Engl J Med* 353(6):595–603, 2005.

Kurita T, Medina R, Schabel AB, Young P, Gama P et al: The activation function-1 domain of estrogen receptor alpha in uterine stromal cells is required for mouse but not human uterine epithelial response to estrogen, *Differentiation* 73(6):313–22, 2005.

Vaskivuo TE, Maentausta M, Torn S, Oduwole O, Lonnberg A et al: Estrogen receptors and estrogen-metabolizing enzymes in human ovaries during fetal development, *J Clin Endocr Metab* 90(6):3752–6, 2005.

Bershtein LM, Poroshina TE, Zimarina TS, Tsyrlina EV, Zhil'tsova EK et al: Expression of estrogen receptors-alpha and -beta in primary breast neoplasms and tumors exposed to neoadjuvant hormonal therapy, *Bull Exp Biol Med* 138(5):494–6, 2004.

Poola I, Abraham J, Liu A: Estrogen receptor beta splice variant mRNAs are differentially altered during breast carcinogenesis, *J Steroid Biochem Mol Biol* 82(2–3):169–79, 2002.

Cartoni R, Leger B, Hock MB, Praz M, Crettenand A et al: Mitofusins 1/2 and ERRalpha expression are increased in human skeletal muscle after physical exercise, *J Physiol* 567(Pt 1):349–58, 2005.

Mansur Ade P, Nogueira CC, Strunz CM, Aldrighi JM, Ramires JA: Genetic polymorphisms of estrogen receptors in patients with premature coronary artery disease, *Arch Med Res* 36(5):511–7, 2005.

Albagha OME, Pettersson U, Stewart A, McGuigan FEA, MacDonald HM et al: Association of oestrogen receptor alpha gene polymorphisms with postmenopausal bone loss, bone mass, and quantitative ultrasound properties of bone, *J Med Genet* 42:240–6, 2005.

Bord S, Horner A, Beavan S, Compston J: Estrogen receptors alpha and beta are differentially expressed in developing human bone, *J Clin Endocr Metab* 86:2309–14, 2001.

Chu S, Mamers P, Burger HG, Fuller PJ: Estrogen receptor isoform gene expression in ovarian stromal and epithelial tumors, *J Clin Endocr Metab* 85:1200–5, 2000.

Colin EM, Uitterlinden AG, Meurs JBJ, Bergink AP, van de Klift M et al: Interaction between vitamin D receptor genotype and estrogen receptor alpha genotype influences vertebral fracture risk, *J Clin Endocr Metab* 88:3777–84, 2003.

Couse JF, Hewitt SC, Bunch DO, Sar M, Walker VR et al: Postnatal sex reversal of the ovaries in mice lacking estrogen receptors alpha and beta, *Science* 286:2328–31, 1999.

Kudwa AE, Bodo C, Gustafsson, JA, Rissman EF: A previously uncharacterized role for estrogen receptor-beta: Defeminization of male brain and behavior, *Proc Natl Acad Sci USA* 102:4608–12, 2005.

Gosden JR, Middleton PG, Rout D: Localization of the human oestrogen receptor gene to chromosome 6q24-q27 by in situ hybridization, *Cytogenet Cell Genet* 43:218–20, 1986.

Green S, Walter P, Kumar V, Krust A, Bornert, JM et al: Human oestrogen receptor cDNA: Sequence, expression and homology to v-erb-A, *Nature* 320:134–9, 1986.

Greene GL, Gilna P, Waterfield M, Baker A, Hort Y et al: Sequence and expression of human estrogen receptor complementary DNA, *Science* 231:1150–4, 1986.

Khosla S, Riggs BL, Atkinson EJ, Oberg AL, Mavilia C et al: Relationship of estrogen receptor genotypes to bone mineral density and to rates of bone loss in men, *J Clin Endocr Metab* 89:1808–16, 2004.

Korach KS: Insights from the study of animals lacking functional estrogen receptor, *Science* 266:1524–7, 1994.

Lee K, Jessop H, Suswillo R, Zaman G, Lanyon L: Bone adaptation requires oestrogen receptor-alpha, *Nature* 424: 389 (only), 2003.

Moore JT, McKee DD, Slentz-Kesler K, Moore LB, Jones SA et al: Cloning and characterization of human estrogen receptor beta isoforms, *Biochem Biophys Res Commun* 247:75–8, 1998.

Kuiper GGJM, Enmark E, Pelto-Huikko M, Nilsson S, Gustafsson JA: Cloning of a novel estrogen receptor expressed in rat prostate and ovary, *Proc Natl Acad Sci USA* 93:5925–30, 1996.

Mosselman S, Polman J, Dijkema R: ER-beta: Identification and characterization of a novel human estrogen receptor, *FEBS Lett* 392:49–53, 1996.

Rissman EF, Heck AL, Leonard JE, Shupnik MA, Gustafsson JA: Disruption of estrogen receptor beta gene impairs spatial learning in female mice, *Proc Natl Acad Sci USA* 99:3996–4001, 2002.

Shim GJ, Wang L, Andersson S, Nagy N, Kis LL et al: Disruption of the estrogen receptor beta gene in mice causes myeloproliferative disease resembling chronic myeloid leukemia with lymphoid blast crisis, *Proc Natl Acad Sci* 100:6694–9, 2003.

Wang L, Andersson S, Warner M, Gustafsson JA: Estrogen receptor (ER)-beta knockout mice reveal a role for ER-beta in migration of cortical neurons in the developing brain, *Proc Natl Acad Sci* 100:703–8, 2003.

Weihua Z, Saji S, Makinen S, Cheng G, Jensen EV et al: Estrogen receptor (ER)-beta, a modulator of ER-alpha in the uterus, *Proc Natl Acad Sci USA* 97:5936–41, 2000.

Zhu Y, Bian Z, Lu P, Karas RH, Bao L et al: Abnormal vascular function and hypertension in mice deficient in estrogen receptor beta, *Science* 295:505–8, 2002.

Weel AM, Uitterlinden AG, Burger H, Schuit SC, Hofman A et al: Estrogen receptor polymorphism predicts the onset of natural and surgical menopause, *J Clin Endocr Metab* 84:3146–50, 1999.

Albagha OM, McGuigan FEA, Reid DM, Ralston SH: Estrogen receptor alpha gene polymorphisms and bone mineral density: Haplotype analysis in women from the United Kingdom, *J Bone Miner Res* 16:128–34, 2001.

Human protein reference data base, Johns Hopkins University and the Institute of Bioinformatics, at http://www.hprd.org/protein.

22.6. Calcitonin

Rodan GA, Martin TJ: Therapeutic approaches to bone diseases, *Science* 289:1508–14, 2000.

Martin TJ et al: in *Metabolic Bone Disease*, Avioli LV, Krane SM, eds, Academic Press, New York, 1998, p 95.

Hoovers JMN, Redeker E, Speleman F, Hoppener JWM, Bhola S et al: High-resolution chromosomal localization of the human calcitonin/CGRP/IAPP gene family members, *Genomics* 15:525–9, 1993.

Mathe AA, Agren H, Lindstrom L, Theodorsson E: Increased concentration of calcitonin gene-related peptide in cerebrospinal fluid of depressed patients: A possible trait marker of major depressive disorder, *Neurosci Lett* 182:138–42, 1994.

New HV, Mudge AW: Calcitonin gene-related peptide regulates muscle acetylcholine receptor synthesis, *Nature* 323:809–11, 1986.

Olesen J, Diener HC, Husstedt IW, Goadsby PJ, Hall D et al: Calcitonin gene-related peptide receptor antagonist BIBN 4096 BS for the acute treatment of migraine, *New Engl J Med* 350:1104–10, 2004.

Saller B, Feldmann G, Haupt K, Broecker M, Janssen OE et al: RT-PCR-based detection of circulating calcitonin-producing cells in patients with advanced medullary thyroid cancer, *J Clin Endocr Metab* 87:292–6, 2002.

22.7. Osteoprotegerin (OPG)

Ulrich-Vinther M, Schwarz EM, Pedersen FS, Soballe K, Andreassen TT: Gene therapy with human osteoprotegerin decreases callus remodeling with limited effects on biomechanical properties, *Bone* 37(6):751–8, 2005.

Kostenuik PJ, Bolon B, Morony S, Daris M, Geng Z et al: Gene therapy with human recombinant osteoprotegerin reverses established osteopenia in ovariectomized mice, *Bone* 34(4):656–64, 2004.

Bolon B, Carter C, Daris M, Morony S, Capparelli C et al: Adenoviral delivery of osteoprotegerin ameliorates bone resorption in a mouse ovariectomy model of osteoporosis, *Mol Ther* 3(2):197–205, 2001.

Bucay N, Sarosi I, Dunstan CR, Morony S, Tarpley J et al: Osteoprotegerin-deficient mice develop early onset osteoporosis and arterial calcification, *Genes Dev* 12:1260–8, 1998.

Croucher PI, Shipman CM, Lippitt J, Perry M, Asosingh K et al: Osteoprotegerin inhibits the development of osteolytic bone disease in multiple myeloma, *Blood* 98:3534–40, 2001.

Cundy T, Davidson J, Rutland MD, Stewart C, DePaoli, AM: Recombinant osteoprotegerin for juvenile Paget's disease, *New Engl J Med* 353:918–23, 2005.

Kong YY, Feige U, Sarosi I, Bolon B, Tafuri A et al: Activated T cells regulate bone loss and joint destruction in adjuvant arthritis through osteoprotegerin ligand, *Nature* 402:304–9, 1999.

Krane SM: Genetic control of bone remodeling—insights from a rare disease, *New Engl J Med* 347:210–2, 2002.

Mizuno A, Amizuka N, Irie K, Murakami A, Fujise N et al: Severe osteoporosis in mice lacking osteoclastogenesis inhibitory factor/osteoprotegerin, *Biochem Biophys Res Commun* 247:610–5, 1998.

Morinaga T, Nakagawa N, Yasuda H, Tsuda E, Higashio K: Cloning and characterization of the gene encoding human osteoprotegerin/osteoclastogenesis-inhibitory factor, *Eur J Biochem* 254:685–91, 1998.

Ohmori H, Makita Y, Funamizu M, Hirooka K, Hosoi T et al: Linkage and association analyses of the osteoprotegerin gene locus with human osteoporosis, *J Hum Genet* 47:400–6, 2002.

Simonet WS, Lacey DL, Dunstan CR, Kelley M, Chang MS et al: Osteoprotegerin: A novel secreted protein involved in the regulation of bone density, *Cell* 89:309–19, 1997.

Soufi M, Schoppet M, Sattler AM, Herzum M, Maisch B et al: Osteoprotegerin gene polymorphisms in men with coronary artery disease, *J Clin Endocr Metab* 89:3764–8, 2004.

Tan KB, Harrop J, Reddy M, Young P, Terrett J et al: Characterization of a novel TNF-like ligand and recently described TNF ligand and TNF receptor superfamily genes and their constitutive and inducible expression in hematopoietic and non-hematopoietic cells, *Gene* 204:35–46, 1997.

Tsuda E, Goto M, Mochizuki S, Yano K, Kobayashi F et al: Isolation of a novel cytokine from human fibroblasts that specifically inhibits osteoclastogenesis, *Biochem Biophys Res Commun* 234:137–42, 1997.

Whyte MP, Obrecht SE, Finnegan PM, Jones JL, Podgornik MN et al: Osteoprotegerin deficiency and juvenile Paget's disease, *New Engl J Med* 347:175–84, 2002.

Yasuda H, Shima N, Nakagawa N, Mochizuki SI, Yano K et al: Identity of osteoclastogenesis inhibitory factor (OCIF) and osteoprotegerin (OPG): A mechanism by which OPG/OCIF inhibits osteoclastogenesis in vitro, *Endocrinology* 139:1329–37, 1998.

Osteoprotegerin Ligand

Anderson DM, Maraskovsky E, Billingsley WL, Dougall WC, Tometsko ME et al: A homologue of the TNF receptor and its ligand enhance T-cell growth and dendritic-cell function, *Nature* 390:175–9, 1997.

Croucher PI, Shipman CM, Lippitt J, Perry M, Asosingh K et al: Osteoprotegerin inhibits the development of osteolytic bone disease in multiple myeloma, *Blood* 98:3534–40, 2001.

Fata JE, Kong YY, Li J, Sasaki T, Irie-Sasaki J et al: The osteoclast differentiation factor osteoprotegerin-ligand is essential for mammary gland development, *Cell* 103:41–50, 2000.

Glass DA II, Patel MS, Karsenty G: A new insight into the formation of osteolytic lesions in multiple myeloma, *New Engl J Med* 349:2479–80, 2003.

Ikeda F, Nishimura R, Matsubara T, Tanaka S, Ioune J et al: Critical roles of c-Jun signaling in regulation of NFAT family and RANKL-regulated osteoclast differentiation, *J Clin Invest* 114:475–84, 2004.

Kim N, Odgren PR, Kim DK, Marks SC Jr, Choi Y: Diverse roles of the tumor necrosis factor family member TRANCE in skeletal physiology revealed by TRANCE deficiency and partial rescue by a lymphocyte-expressed TRANCE transgene, *Proc Natl Acad Sci USA* 97:10905–10, 2000.

Koga T, Inui M, Inoue K, Kim S, Suematsu A et al: Costimulatory signals mediated by the ITAM motif cooperate with RANKL for bone homeostasis, *Nature* 428:758–63, 2004.

Kong YY, Feige U, Sarosi I, Bolon B, Tafuri A et al: Activated T cells regulate bone loss and joint destruction in adjuvant arthritis through osteoprotegerin ligand, *Nature* 402:304–9, 1999.

Lacey DL, Timms E, Tan HL, Kelley MJ, Dunstan CR et al: Osteoprotegerin ligand is a cytokine that regulates osteoclast differentiation and activation, *Cell* 93:165–76, 1998.

Pearse RN, Sordillo EM, Yaccoby S, Wong BR, Liau DF et al: Multiple myeloma disrupts the TRANCE/osteoprotegerin cytokine axis to trigger bone destruction and promote tumor progression, *Proc Natl Acad Sci USA* 98:11581–6, 2001.

Sezer O, Heider U, Zavrski I, Kuhne CA, Hofbauer LC: RANK ligand and osteoprotegerin in myeloma bone disease, *Blood* 101:2094–8, 2003.

Takayanagi H, Kim S, Matsuo K, Suzuki H, Suzuki T et al: RANKL maintains bone homeostasis through c-Fos-dependent induction of interferon-B, *Nature* 416:744–9, 2002.

Wei S, Kitaura H, Zhou P, Ross FP, Teitelbaum SL: IL-1 mediates TNF-induced osteoclastogenesis, *J Clin Invest* 115:282–90, 2005.

Wong BR, Rho J, Arron J, Robinson E, Orlinick J et al: TRANCE is a novel ligand of the tumor necrosis factor receptor family that activates c-Jun N-terminal kinase in T cells, *J Biol Chem* 272:25190–4, 1997.

Yasuda H, Shima N, Nakagawa N, Yamaguchi K, Kinosaki M et al: Osteoclast differentiation factor is a ligand for osteoprotegerin/osteoclastogenesis-inhibitory factor and is identical to TRANCE/RANKL, *Proc Natl Acad Sci USA* 95:3597–602, 1998.

22.8. Integrin-Binding Proteins

McLane MA, Sanchez EE, Wong A, Paquette-Straub C, Perez JC: Disintegrins, *Curr Drug Targets Cardiovasc Haematol Disord* 4(4):327–55, 2004.

Mercer B, Markland F, Minkin C: Contortrostatin, a homodimeric snake venom disintegrin, is a potent inhibitor of osteoclast attachment, *J Bone Miner Res* 13(3):409–14, 1998.

Kumar CC, Nie H, Rogers CP, Malkowski M, Maxwell E et al: Biochemical characterization of the binding of echistatin to integrin alphavbeta3 receptor, *J Pharmacol Exp Ther* 283(2):843–53, 1997.

22.9. Growth Factors

Hiltunen MO, Ruuskanen M, Huuskonen J, Mahonen AJ, Ahonen M et al: Adenovirus-mediated VEGF-A gene transfer induces bone formation in vivo, *FASEB J* 17(9):1147–9, 2003.

Langdahl BL, Knudsen JY, Jensen HK, Gregersen N, Eriksen EF: A sequence variation: 713-8delC in the transforming growth factor-beta 1 gene has higher prevalence in osteoporotic women than in normal women and is associated with very low bone mass in osteoporotic women and increased bone turnover in both osteoporotic and normal women, *Bone* 20:289–94, 1997.

Yamada Y, Miyauchi A, Takagi Y, Tanaka M, Mizuno M, Harada A: Association of the C-509→T polymorphism, alone of in combination with the T869→C polymorphism, of the transforming growth factor-beta1 gene with bone mineral density and genetic susceptibility to osteoporosis in Japanese women, *J Mol Med* 79:149–56, 2001.

Ralston SH: Genetic determinants of susceptibility to osteoporosis, *Curr Opin Pharmacol* 3:286–90, 2003.

Niyibizi C, Wang S, Mi Z, Robbins PD: Gene therapy approaches for osteogenesis imperfecta, *Gene Ther* 11(4):408–16, 2004.

Rodan GA, Martin TJ: Therapeutic approaches to bone diseases, *Science* 289:1508–14, 2000.

22.10. Cell Therapy for Bone Regeneration

Allen TD, Dexter TM, Simmons PJ: Marrow biology and stem cells, *Immunol Ser* 49:1–38, 1990.

Caplan AI: Mesenchymal stem cells, *J Orthop Res* 9:641–50, 1991.

Bianco P, Cossu G: Uno, nessuno e centomila: Searching for the identity of mesodermal progenitors, *Exp Cell Res* 251:257–63, 1999.

Caplan AI: Mesenchymal stem cells, *J Orthop Res* 9:641–50, 1991.

Caplan AI: Review: Mesenchymal stem cells: Cell-based reconstructive therapy in orthopedics, *Tissue Eng* 11(7–8):1198–211, 2005.

Cancedda R, Bianchi G, Derubeis A, Quarto R: Cell therapy for bone disease: A review of current status, *Stem Cells* 21:610–19, 2003.

Quarto R, Mastrogiacomo M, Cancedda R et al: Repair of large bone defects with the use of autologous bone marrow stromal cells, *New Engl J Med* 344:385–6.

Goan SR, Junghahn I, Wissler M et al: Donor stromal cells from human blood engraft in NOD/SCID mice, *Blood* 96:3971–8, 2000.

Pereira RF, Halford KW, O'Hara MD et al: Cultured adherent cells from marrow can serve as long-lasting precursor cells for bone, cartilage, and lung in irradiated mice, *Proc Natl Acad Sci USA* 92:4857–61, 1995.

Horwitz EM, Prockop DJ, Fitzpatrick LA et al: Transplantability and therapeutic effects of bone marrow-derived mesenchymal cells in children with osteogenesis imperfecta, *Nat Med* 5:309–13, 1999.

Horwitz EM, Prockop DJ, Gordon PL et al: Clinical responses to bone marrow transplantation in children with severe osteogenesis imperfecta, *Blood* 97:1227–31, 2001.

Kadiyala S, Young RG, Thiede MA et al: Culture expanded canine mesenchymal stem cells possess osteochondrogenic potential in vivo and in vitro, *Cell Transplant* 6:125–34, 1997.

Krebsbach PH, Kuznetsov SA, Satomura K et al: Bone formation in vivo: Comparison of osteogenesis by transplanted mouse and human marrow stromal fibroblasts, *Transplantation* 63: 1059–69, 1997.

Ohgushi H, Goldberg VM, Caplan AI: Repair of bone defects with marrow cells and porous ceramic. Experiments in rats, *Acta Orthop Scand* 60:334–9, 1989.

Krebsbach PH, Mankani MH, Satomura K et al: Repair of craniotomy defects using bone marrow stromal cells, *Transplantation* 66:1272–8, 1998.

Casabona F, Martin I, Muraglia A et al: Prefabricated engineered bone flaps: An experimental model of tissue reconstruction in plastic surgery, *Plast Reconstr Surg* 101:577–81, 1998.

Mankani MH, Krebsbach PH, Satomura K et al: Pedicled bone flap formation using transplanted bone marrow stromal cells, *Arch Surg* 136:263–70, 2001.

Bruder SP, Kraus KH, Goldberg VM et al: The effect of implants loaded with autologous mesenchymal stem cells on the healing of canine segmental bone defects, *J Bone Joint Surg Am* 80:985–96, 1998.

Kon E, Muraglia A, Corsi A et al: Autologous bone marrow stromal cells loaded onto porous hydroxyapatite ceramic accelerate bone repair in critical-size defects of sheep long bones, *J Biomed Mater Res* 49:328–37, 2000.

Petite H, Viateau V, Bensaid W et al: Tissue-engineered bone regeneration, *Nat Biotechnol* 18:959–63, 2000.

Quarto R, Mastrogiacomo M, Cancedda R et al: Repair of large bone defects with the use of autologous bone marrow stromal cells, *New Engl J Med* 344:385–6, 2001.

Reddi AH: Role of morphogenetic proteins in skeletal tissue engineering and regeneration, *Nat Biotechnol* 16(3):247–52, 1998.

22.11. Pathogenesis, Pathology, and Clinical Features of Paget's Disease

Roodman GD, Windle JJ: Paget disease of bone, *J Clin Invest* 115(2):200–8, 2005.

Pierie JP, Choudry U, Muzikansky A, Finkelstein DM, Ott MJ: Prognosis and management of extramammary Paget's disease and the association with secondary malignancies, *J Am Coll Surg* 196(1):45–50, 2003.

Hadjipavlou AG, Gaitanis IN, Kontakis GM: Paget's disease of the bone and its management, *J Bone Joint Surg Br* 84(2):160–9, 2002.

Roodman GD: Studies in Paget's disease and their relevance to oncology, *Semin Oncol* 28(4 Suppl 11):15–21, 2001.

Dickinson CJ: The possible role of osteoclastogenic oral bacterial products in etiology of Paget's disease, *Bone* 26(2):101–2, 2000.

22.12. Molecular and Cellular Therapies

Cundy T, Davidson J, Rutland MD, Stewart C, DePaoli AM: Recombinant osteoprotegerin for juvenile Paget's disease, *New Engl J Med* 353:918–23, 2005.

22.13. Pathogenesis, Pathology, and Clinical Features of Bone Tumors

Rodan GA, Martin JT: Therapeutic approaches to bone diseases, *Science* 289:1508–14, 2000.

22.14. Pathogenesis, Pathology, and Clinical Features of Rheumatoid Arthritis

Kavanaugh ALP: in *Rheumatoid Arthritis*, 3rd ed, Gallin JI, Snyderman R, eds, Lippincott Williams and Wilkins, 1999.

Firestein GS: Immunologic mechanisms in the pathogenesis of rheumatoid arthritis, *J Clin Rheumatol* 11(3 Suppl):S39–44, 2005.

Walsh NC, Crotti TN, Goldring SR, Gravallese EM: Rheumatic diseases: The effects of inflammation on bone, *Immunol Rev* 208:228–51, 2005.

Rosenstein ED, Greenwald RA, Kushner LJ, Weissmann G: Hypothesis: The humoral immune response to oral bacteria provides a stimulus for the development of rheumatoid arthritis, *Inflammation* 28(6):311–8, 2004.

Sakaguchi S, Sakaguchi N: Animal models of arthritis caused by systemic alteration of the immune system, *Curr Opin Immunol* 17(6):589–94, 2005.

Choileain NN, Redmond HP: Regulatory T-cells and autoimmunity, *J Surg Res* 130(1):124–35, 2006.

Ma Y, Pope RM: The role of macrophages in rheumatoid arthritis, *Curr Pharm Des* 11(5):569–80, 2005.

Meinecke I, Rutkauskaite E, Gay S, Pap T: The role of synovial fibroblasts in mediating joint destruction in rheumatoid arthritis, *Curr Pharm Des* 11(5):563–8, 2005.

22.15. Molecular Therapy

Furlan R, Butti E, Pluchino S, Martino G: Gene therapy for autoimmune diseases, *Curr Opin Mol Ther* 6(5):525–36, 2004.

22.16. Interleukin-1

Wieser C, Stumpf D, Grillhosl C, Lengenfelder D, Gay S et al: A Regulated and constitutive expression of anti-inflammatory cytokines by nontransforming herpesvirus saimiri vectors, *Gene Ther* 12(5):395–406, 2005.

IL-1α

Bensen JT, Langefeld CD, Hawkins GA, Green LE, Mychaleckyj JC et al: Nucleotide variation, haplotype structure, and association with end-stage renal disease of the human interleukin-1 gene cluster, *Genomics* 82:194–217, 2003.

Cox A, Camp NJ, Cannings C, di Giovine FS, Dale M et al: Combined sib-TDT and TDT provide evidence for linkage of the interleukin-1 gene cluster to erosive rheumatoid arthritis, *Hum Mol Genet* 8:1707–13, 1999.

Du Y, Dodel RC, Eastwood BJ, Bales KR, Gao F et al: Association of an interleukin 1-alpha polymorphism with Alzheimer's disease, *Neurology* 55:480–4, 2000.

Furutani Y, Notake M, Fukui T, Ohue M, Nomura H et al: Complete nucleotide sequence of the gene for human interleukin 1 alpha, *Nucleic Acids Res* 14:3167–79, 1986.

Green EK, Harris JM, Lemmon H, Lambert JC, Chartier-Harlin MC et al: Are interleukin-1 gene polymorphisms risk factors or disease modifiers in AD? *Neurology* 58:1566–8, 2002.

Grimaldi LME, Casadei VM, Ferri C, Veglia F, Licastro F et al: Association of early-onset Alzheimer's disease with an interleukin-1-alpha gene polymorphism, *Ann Neurol* 47:361–5, 2000.

Hogquist KA, Nett MA, Unanue ER, Chaplin DD: Interleukin 1 is processed and released during apoptosis, *Proc Natl Acad Sci USA* 88:8485–89, 1991.

Lord PCW, Wilmoth LMG, Mizel SB, McCall CE: Expression of interleukin-1 alpha and beta genes by human blood polymorphonuclear leukocytes, *J Clin Invest* 87:1312–21, 1991.

Nicoll JAR, Mrak RE, Graham DI, Stewart J, Wilcock G et al: Association of interleukin-1 gene polymorphisms with Alzheimer's disease, *Ann Neurol* 47:365–8, 2000.

Wei S, Kitaura H, Zhou P, Ross FP, Teitelbaum SL: IL-1 mediates TNF-induced osteoclastogenesis, *J Clin Invest* 115:282–90, 2005.

Werman A, Werman-Venkert R, White R, Lee JK, Werman B et al: The precursor form of IL-1-alpha is an intracine proinflammatory activator of transcription, *Proc Nat Acad Sci* 101:2434–9, 2004.

IL1β

Auron PE, Webb AC, Rosenwasser LJ, Mucci SF, Rich A et al: Nucleotide sequence of human monocyte interleukin 1 precursor cDNA, *Proc Natl Acad Sci USA* 81:7907–11, 1984.

Cameron P, Limjuco G, Rodkey J, Bennett C, Schmidt JA: Amino acid sequence analysis of human interleukin 1 (IL-1): Evidence for biochemically distinct forms of IL-1, *J Exp Med* 162:790–801, 1985.

El-Omar EM, Carrington M, Chow WH, McColl KEL, Bream JH et al: Interleukin-1 polymorphisms associated with increased risk of gastric cancer, *Nature* 404:398–402, 2000.

Langdahl BL, Lokke E, Carstens M, Stenkjaer LL, Eriksen EF: Osteoporotic fractures are associated with an 86-base pair repeat polymorphism in the interleukin-1-receptor antagonist gene but not with polymorphisms in the interleukin-1 beta gene, *J Bone Miner Res* 15:402–14, 2000.

March CJ, Mosley B, Larsen A, Cerretti DP, Braedt G et al: Cloning, sequence and expression of two distinct human interleukin-1 complementary DNAs, *Nature* 315:641–7, 1985.

Nicklin MJH, Weith A, Duff GW: A physical map of the region encompassing the human interleukin-1-alpha, interleukin-1-beta, and interleukin-1 receptor antagonist genes, *Genomics* 19:382–4, 1994.

Ohmura K, Johnsen A, Ortiz-Lopez A, Desany P, Roy M et al: Variation in IL-1-beta gene expression is a major determinant of genetic differences in arthritis aggressivity in mice, *Proc Natl Acad Sci USA* 102:12489–94, 2005.

Patterson D, Jones C, Hart I, Bleskan J, Berger R et al: The human interleukin-1 receptor antagonist (IL1RN) gene is located in the chromosome 2q14 region, *Genomics* 15:173–6, 1993.

Samad TA, Moore KA, Sapirstein A, Billet S, Allchorne A et al: Interleukin-1-beta-mediated induction of Cox-2 in the CNS contributes to inflammatory pain hypersensitivity, *Nature* 410:471–5, 2001.

Voronov E, Shouval DS, Krelin Y, Cagnano E, Benharroch D et al: IL-1 is required for tumor invasiveness and angiogenesis, *Proc Natl Acad Sci USA* 100:2645–50, 2003.

Type I IL1 Receptor

Copeland NG, Silan CM, Kingsley DM, Jenkins NA, Cannizzaro LA et al: Chromosomal location of murine and human IL-1 receptor genes, *Genomics* 9:44–50, 1991.

Dale M, Nicklin MJ: Interleukin-1 receptor cluster: Gene organization of IL1R2, IL1R1, IL1RL2 (IL-1Rrp2), IL1RL1 (T1/ST2), and IL18R1 (IL-1Rrp) on human chromosome 2q, *Genomics* 57:177–9, 1999.

Dower SK, Kronheim SR, Hopp TP, Cantrell M, Deeley M et al: The cell surface receptors for interleukin-1(alpha) and interleukin-1(beta) are identical, *Nature* 324:266–8, 1986.

Sims JE, Acres RB, Grubin CE, McMahan CJ, Wignall JM et al: Cloning the interleukin 1 receptor from human T cells, *Proc Natl Acad Sci USA* 86:8946–50, 1989.

Colotta F, Re F, Muzio M et al: Interleukin-1 type II receptor: A decoy target for IL-1 that is regulated by IL-4, *Science* 261:472–5, 1993.

Burger D, Chicheportiche R, Giri JG, Dayer JM: The inhibitory activity of human interleukin-1 receptor antagonist is enhanced by type II interleukin-1 soluble receptor and hindered by type I interleukin-1 soluble receptor, *J Clin Invest* 96:38–41, 1995.

Dinarello CA: Interleukin-1, interleukin-1 receptors and interleukin-1 receptor antagonist, *Int Rev Immunol* 16(5–6):457–99, 1998.

Type II IL1 Receptor

Colotta F, Re F, Muzio M, Bertini R, Polentarutti N et al: Interleukin-1 type II receptor: A decoy target for IL-1 that is regulated by IL-4, *Science* 261:472–5, 1993.

Dale M, Nicklin MJ: Interleukin-1 receptor cluster: Gene organization of IL1R2, IL1R1, IL1RL2 (IL-1Rrp2), IL1RL1 (T1/ST2), and IL18R1 (IL-1Rrp) on human chromosome 2q, *Genomics* 57:177–9, 1999.

McMahan CJ, Slack JL, Mosley B, Cosman D, Lupton SD et al: A novel IL-1 receptor, cloned from B cells by mammalian expression, is expressed in many cell types, *EMBO J* 10:2821–32, 1991.

22.17. Tumor Necrosis Factor (TNF)

Burstein H Gene therapy for rheumatoid arthritis. Curr Opin Mol Ther. 3(4):362–74, 2001.

Kim KN, Watanabe S, Ma Y, Thornton S, Giannini EH, Hirsch R: Viral IL-10 and soluble TNF receptor act synergistically to inhibit collagen-induced arthritis following adenovirus-mediated gene transfer, *J Immunol* 164(3):1576–81, 2000.

TNF α

Aggarwal BB, Eessalu TE, Hass PE: Characterization of receptors for human tumour necrosis factor and their regulation by gamma-interferon, *Nature* 318:665–7, 1985.

Balding J, Kane D, Livingstone W, Mynett-Johnson L, Bresnihan B et al: Cytokine gene polymorphisms: Association with psoriatic arthritis susceptibility and severity, *Arthritis Rheum* 48:1408–13, 2003.

Bouwmeester T, Bauch A, Ruffner H, Angrand PO, Bergamini G et al: A physical and functional map of the human TNF-alpha/NF-kappa-B signal transduction pathway, *Nature Cell Biol* 6:97–105, 2004.

Garcia-Ruiz C, Colell A, Mari M, Morales A, Calvo M et al: Defective TNF-alpha-mediated hepatocellular apoptosis and liver damage in acidic sphingomyelinase knockout mice, *J Clin Invest* 111:197–208, 2003.

Kamata H, Honda S, Maeda S, Chang L, Hirata H et al: Reactive oxygen species promote TNF-alpha-induced death and sustained JNK activation by inhibiting MAP kinase phosphatases, *Cell* 120:649–61, 2005.

Marino MW, Dunn A, Grail D, Inglese M, Noguchi Y et al: Characterization of tumor necrosis factor-deficient mice, *Proc Natl Acad Sci USA* 94:8093–8, 1997.

Obeid LM, Linardic CM, Karolak LA, Hannun YA: Programmed cell death induced by ceramide, *Science* 259:1769–71, 1993.

Pennica D, Nedwin GE, Hayflick JS, Seeburg PH, Derynck R et al: Human tumour necrosis factor: Precursor structure, expression and homology to lymphotoxin, *Nature* 312:724–9, 1984.

Steed PM, Tansey MG, Zalevsky J, Zhukovsky EA, Desjarlais JR et al: Inactivation of TNF signaling by rationally designed dominant-negative TNF variants, *Science* 301:1895–8, 2003.

Takahashi JL, Giuliani F, Power C, Imai Y, Yong VW: Interleukin-1-beta promotes oligodendrocyte death through glutamate excitotoxicity, *Ann Neurol* 53:588–95, 2003.

Vielhauer V, Stavrakis G, Mayadas TN: Renal cell-expressed TNF receptor 2, not receptor 1, is essential for the development of glomerulonephritis, *J Clin Invest* 115:1199–1209, 2005.

Wang AM, Creasey AA, Ladner MB, Lin LS, Strickler J et al: Molecular cloning of the complementary DNA for human tumor necrosis factor, *Science* 228:149–54, 1985.

TNF β

Aggarwal BB, Eessalu TE, Hass PE: Characterization of receptors for human tumour necrosis factor and their regulation by gamma-interferon, *Nature* 318:665–7, 1985.

Balding J, Kane D, Livingstone W, Mynett-Johnson L, Bresnihan B et al: Cytokine gene polymorphisms: Association with psoriatic arthritis susceptibility and severity, *Arthritis Rheum* 48:1408–13, 2003.

Gray PW, Aggarwal BB, Benton CV, Bringman TS, Henzel WJ et al: Cloning and expression of cDNA for human lymphotoxin, a lymphokine with tumour necrosis activity, *Nature* 312:721–4, 1984.

Knight JC, Keating BJ, Kwiatkowski DP: Allele-specific repression of lymphotoxin-alpha by activated B cell factor-1, *Nature Genet* 36:394–9, 2004.

Type I TNF Receptor

Baker E, Chen LZ, Smith CA, Callen DF, Goodwin R et al: Chromosomal location of the human tumor necrosis factor receptor genes, *Cytogenet Cell Genet* 57:117–8, 1991.

Brockhaus M, Schoenfeld HJ, Schlaeger EJ, Hunziker W, Lesslauer W et al: Identification of two types of tumor necrosis factor receptors on human cell lines by monoclonal antibodies, *Proc Nat Acad Sci USA* 87:3127–31, 1990.

Castellino AM, Parker GJ, Boronenkov IV, Anderson RA, Chao, MV: A novel interaction between the juxtamembrane region of the p55 tumor necrosis factor receptor and phosphatidylinositol-4-phosphate 5-kinase, *J Biol Chem* 272:5861–70, 1997.

Chan FKM, Chun HJ, Zheng L, Siegel RM, Bui KL et al: A domain in TNF receptors that mediates ligand-independent receptor assembly and signaling, *Science* 288:2351–4, 2000.

Engelmann H, Novick D, Wallach D: Two tumor necrosis factor-binding proteins purified from human urine: Evidence for immunological cross-reactivity with cell surface tumor necrosis factor receptors, *J Biol Chem* 265:1531–6, 1990.

Fuchs P, Strehl S, Dworzak M, Himmler A, Ambros, PF: Structure of the human TNF receptor 1 (p60) gene (TNFR1) and localization to chromosome 12p13, *Genomics* 13:219–24, 1992.

Gray PW, Barrett K, Chantry D, Turner M, Feldmann M: Cloning of human tumor necrosis factor (TNF) receptor cDNA and expression of recombinant soluble TNF-binding protein, *Proc Nat Acad Sci USA* 87:7380–4, 1990.

Loetscher H, Pan YCE, Lahm HW, Gentz R, Brockhaus M et al: Molecular cloning and expression of the human 55 kd tumor necrosis factor receptor, *Cell* 61:351–9, 1990.

Micheau O, Tschopp J: Induction of TNF receptor I-mediated apoptosis via two sequential signaling complexes, *Cell* 114:181–190, 2003.

Nophar Y, Kemper O, Brakebusch C, Engelmann H, Zwang R et al: Soluble forms of tumor necrosis factor receptors (TNF-Rs): The cDNA for the type I TNF-R, cloned using amino acid sequence data of its soluble form, encodes both the cell surface and a soluble form of the receptor, *EMBO J* 9:3269–78, 1990.

Rothe J, Lesslauer W, Lotscher H, Lang Y, Koebel P et al: Mice lacking the tumour necrosis factor receptor 1 are resistant to TNF-mediated toxicity but highly susceptible to infection by Listeria monocytogenes, *Nature* 364:798–802, 1993.

Schall TJ, Lewis M, Koller KJ, Lee A, Rice GC et al: Molecular cloning and expression of a receptor for human tumor necrosis factor, *Cell* 61:361–70, 1990.

Schievella AR, Chen JH, Graham JR, Lin LL: MADD, a novel death domain protein that interacts with the type 1 tumor necrosis factor receptor and activates mitogen-activated protein kinase, *J Biol Chem* 272:12069–75, 1997.

Smith CA, Davis T, Anderson D, Solam L, Beckmann MP et al: A receptor for tumor necrosis factor defines an unusual family of cellular and viral proteins, *Science* 248:1019–23, 1990.

Zhang JY, Green CL, Tao S, Khavari, PA: NF-kappa-B RelA opposes epidermal proliferation driven by TNFR1 and JNK, *Genes Dev* 18:17–22, 2004.

Type II TNF Receptor

Baker E, Chen LZ, Smith CA, Callen DF, Goodwin R et al: Chromosomal location of the human tumor necrosis factor receptor genes, *Cytogenet Cell Genet* 57:117–118, 1991.

Beltinger CP, White PS, Maris JM, Sulman EP, Jensen SJ et al: Physical mapping and genomic structure of the human TNFR2 gene, *Genomics* 35:94–100, 1996.

Bruce AJ, Boling W, Kindy MS, Peschon J, Kraemer PJ et al: Altered neuronal and microglial responses to excitotoxic and ischemic brain injury in mice lacking TNF receptors, *Nature Med* 2:788–94, 1996.

Chan, FK-M, Chun HJ, Zheng L, Siegel RM, Bui KL et al: A domain in TNF receptors that mediates ligand-independent receptor assembly and signaling, *Science* 288:2351–4, 2000.

Kemper O, Derre J, Cherif D, Engelmann H, Wallach D et al: The gene for the type II (p75) tumor necrosis factor receptor (TNF-RII) is localized on band 1p36.2-p36.3, *Hum Genet* 87:623–4, 1991.

Li X, Yang Y, Ashwell JD: TNF-RII and c-IAP1 mediate ubiquitination and degradation of TRAF2, *Nature* 416:345–9, 2002.

Santee SM, Owen-Schaub LB: Human tumor necrosis factor receptor p75/80 (CD120b) gene structure and promoter characterization, *J Biol Chem* 271:21151–9, 1996.

Sashio H, Tamura K, Ito R, Yamamoto Y, Bamba H et al: Polymorphisms of the TNF gene and the TNF receptor superfamily member 1B gene are associated with susceptibility to ulcerative colitis and Crohn's disease, respectively, *Immunogenetics* 53:1020–7, 2002.

Schall TJ, Lewis M, Koller KJ, Lee A, Rice GC et al: Molecular cloning and expression of a receptor for human tumor necrosis factor, *Cell* 61:361–70, 1990.

Van Ostade X, Vandenabeele P, Everaerdt B, Loetscher H, Gentz R et al: Human TNF mutants with selective activity on the p55 receptor, *Nature* 361:266–9, 1993.

Vielhauer V, Stavrakis G, Mayadas, TN: Renal cell-expressed TNF receptor 2, not receptor 1, is essential for the development of glomerulonephritis, *J Clin Invest* 115:1199–1209, 2005.

Oligino T, Ghivizzani S, Wolfe D et al: Intra-articular delivery of a herpes simplex virus IL-1Ra gene vector reduces inflammation in a rabbit model of arthritis, *Gene Ther* 6:1713–20, 1999.

Chernajovsky Y, Adams G, Podhajcer OL et al: Inhibition of transfer of collagen-induced arthritis into SCID mice by ex vivo infection of spleen cells with retroviruses expressing soluble tumor necrosis factor receptor, *Gene Ther* 2:731–35, 1995.

Quattrocchi E, Walmsley M, Browne K et al: Paradoxical effects of adenovirus-mediated blockade of TNF activity in murine collagen-induced arthritis, *J Immunol* 163:1000–1009, 1999.

Le CH, Nicolson AG, Morales A, Sewell KL: Suppression of collagen-induced arthritis through adenovirus-mediated transfer of a modified tumor necrosis factor alpha receptor gene, *Arthritis Rheum* 40:1662–1669, 1997.

22.18. Antiinflammatory Cytokines

Wieser C, Stumpf D, Grillhosl C, Lengenfelder D, Gay S et al: Regulated and constitutive expression of anti-inflammatory cytokines by nontransforming herpesvirus saimiri vectors, *Gene Ther* 12(5):395–406, 2005.

Neumann E, Judex M, Kullmann F, Grifka J, Robbins PD et al: Inhibition of cartilage destruction by double gene transfer of IL-1Ra and IL-10 involves the activin pathway, *Gene Ther* 9(22): 1508–19, 2002.

Blackburn MR, Lee CG, Young HWJ, Zhu Z, Chunn JL et al: Adenosine mediates IL-13-induced inflammation and remodeling in the lung and interacts in an IL-13-adenosine amplification pathway, *J Clin Invest* 112:332–44, 2003.

Grunig G, Warnock M, Wakil AE, Venkayya R, Brombacher F et al: Requirement for IL-13 independently of IL-4 in experimental asthma, *Science* 282:2261–3, 1998.

Heinzmann H, Mao XQ, Akaiwa M, Kreomer RT, Gao PS et al: Genetic variants of IL-13 signalling and human asthma and atopy, *Hum Mol Genet* 9:549–59, 2000.

Howard TD, Koppelman GH, Xu J, Zheng SL, Postma DS et al: Gene-gene interaction in asthma: IL4RA and IL13 in a Dutch population with asthma, *Am J Hum Genet* 70:230–6, 2002.

Howard TD, Whittaker PA, Zaiman AL, Koppelman, GH Xu J et al: Identification and association of polymorphisms in the interleukin-13 gene with asthma and atopy in a Dutch population, *Am J Resp Cell Molec Biol* 25:377–84, 2001.

Kelly-Welch AE, Hanson EM, Boothby MR, Keegan AD: Interleukin-4 and interleukin-13 signaling connections maps, *Science* 300:1527–28, 2003.

Kuperman DA, Huang X, Koth LL, Chang GH, Dolganov GM et al: Direct effects of interleukin-13 on epithelial cells cause airway hyperreactivity and mucus overproduction in asthma, *Nature Med* 8:885–9, 2002.

Lacy DA, Wang, Z-E, Symula DJ, McArthur CJ, Rubin EM et al: Faithful expression of the human 5q31 cytokine cluster in transgenic mice, *J Immun* 164:4569–74, 2000.

Loots GG, Locksley RM, Blankespoor CM, Wang ZE, Miller W et al: Identification of a coordinate regulator of interleukins 4, 13, and 5 by cross-species sequence comparisons, *Science* 288: 136–40, 2000.

McKenzie ANJ, Culpepper JA, de Waal Malefyt R, Briere F, Punnonen J et al: Interleukin 13, a T-cell-derived cytokine that regulates human monocyte and B-cell function, *Proc Nat Acad Sci USA* 90:3735–9, 1993.

Minty A, Chalon P, Derocq JM, Dumont X, Guillemot JC et al: Interleukin-13 is a new human lymphokine regulating inflammatory and immune responses, *Nature* 362:248–50, 1993.

Morgan JG, Dolganov GM, Robbins SE, Hinton LM, Lovett M: The selective isolation of novel cDNAs encoded by the regions surrounding the human interleukin 4 and 5 genes, *Nucleic Acids Res* 20:5173–9, 1992.

Vladich FD, Brazille SM, Stern D, Peck ML, Ghittoni R: IL-13 R130Q, a common variant associated with allergy and asthma, enhances effector mechanisms essential for human allergic inflammation, *J Clin Invest* 115:747–54, 2005.

Wang M, Xing ZM, Lu C, Ma YX, Yu DL: A common IL-13 arg130-to-gln single nucleotide polymorphism among Chinese atopy patients with allergic rhinitis, *Hum Genet* 113:387–90, 2003.

Wills-Karp M, Luyimbazi J, Xu X, Schofield B, Neben TY et al: Interleukin-13: central mediator of allergic asthma, *Science* 282:2258–61, 1998.

Zhu Z, Homer RJ, Wang Z, Chen Q, Geba GP et al: Pulmonary expression of interleukin-13 causes inflammation, mucus hypersecretion, subepithelial fibrosis, physiologic abnormalities, and eotaxin production, *J Clin Invest* 103:779–88, 1999.

Zhu Z, Zheng T, Homer RJ, Kim YK, Chen NY et al: Acidic mammalian chitinase in asthmatic Th2 inflammation and IL-13 pathway activation, *Science* 304:1678–82, 2004.

Zurawski G, de Vries, JE: Interleukin 13 elicits a subset of the activities of its close relative interleukin 4, *Stem Cells* 12:169–74, 1994.

Bessis N, Honiger J, Damotte D et al: Encapsulation in hollow fibres of xenogeneic cells engineered to secrete IL-4 or IL-13 ameliorates murine collagen-induced arthritis (CIA), *Clin Exp Immunol* 117:376–82, 1999.

Bessis N, Chiocchia G, Kollias G et al: Modulation of proinflammatory cytokine production in tumour necrosis factor-alpha (TNF-alpha)-transgenic mice by treatment with cells engineered to secrete IL-4, IL-10 or IL-13, *Clin Exp Immunol* 111:391–6, 1998.

Bessis N, Boissier MC, Ferrara P et al: Attenuation of collagen-induced arthritis in mice by treatment with vector cells engineered to secrete interleukin-13, *Eur J Immunol* 26:2399–2403, 1996.

Bessis N, Doucet C, Cottard V, Anne-Marie Douar AM, Firat H et al: Gene therapy for rheumatoid arthritis, *J Gene Med* 4:581–91, 2002.

Human protein reference data base, Johns Hopkins University and the Institute of Bioinformatics, at http://www.hprd.org/protein.

22.19. Dominant-Negative Mutant ras Gene

Yamamoto A et al: Suppression of arthritic bone destruction by adenovirus-mediated dominant-negative Ras gene transfer to synoviocytes and osteoclasts, *Arthritis Rheum* 48:2682–92, 2003.

22.20. Osteoprotegerin

Kong YY, Feige U, Sarosi I, Bolons B, Tafuri A et al: Activated T cells regulate bone loss and joint destruction in adjuvant arthritis through osteoprotegerin ligand, *Nature* 402, 304–309, 1999.

22.21. Viral Thymidine Kinase

Goossens PH, Schouten GJ, 't Hart B et al: Feasibility of adenovirus-mediated nonsurgical synovectomy in collagen-induced arthritis-affected rhesus monkeys, *Hum Gene Ther* 10:1139–49, 1999.

Honore P, Luger NM, Sabino MA, Schwei MJ, Rogers SD et al: Osteoprotegerin blocks bone cancer-induced skeletal destruction, skeletal pain and pain-related neurochemical reorganization of the spinal cord, *Nat Med* 6(5):521–8, 2000.

Rodan GA, Martin TJ: Therapeutic approaches to bone diseases, *Science* 289:1508–14, 2000.

22.22. Pathogenesis, Pathology, and Clinical Features of Bone and Cartilage Injury

Lewis CL, Sahrmann SA: Acetabular labral tears, *Phys Ther* 86(1):110–21, Jan 2006.

Beris AE, Lykissas MG, Papageorgiou CD, Georgoulis AD: Advances in articular cartilage repair, *Injury* 36 (Suppl 4):S14–23, 2005.

Giannoudis PV, Pountos I: Tissue regeneration. The past, the present and the future, *Injury* 36 (Suppl 4):S2–5, 2005.

Beynnon BD, Johnson RJ, Abate JA, Fleming BC, Nichols CE: Treatment of anterior cruciate ligament injuries, part I, *Am J Sports Med* 33(10):1579–602, 2005.

Schachter AK, Chen AL, Reddy PD, Tejwani NC: Osteochondral lesions of the talus, *J Am Acad Orthop Surg* 13(3):152–8, 2005.

Smith GD, Knutsen G, Richardson JB: A clinical review of cartilage repair techniques, *J Bone Joint Surg Br* 87(4):445–9, 2005.

Alford JW, Cole BJ: Cartilage restoration, part 2: Techniques, outcomes, and future directions, *Am J Sports Med* 33(3):443–60, 2005.

Ritchie PK, McCarty EC: Surgical management of cartilage defects in athletes, *Clin Sports Med* 24(1):163–74, 2005.

Dirschl DR, Marsh JL, Buckwalter JA, Gelberman R, Olson SA et al: Articular fractures, *J Am Acad Orthop Surg* 12(6):416–23, 2004.

22.23. Molecular Regenerative Engineering

Mont MA, Ragland PS, Biggins B, Friedlaender G, Patel T et al: Use of bone morphogenetic proteins for musculoskeletal applications. An overview, *J Bone Joint Surg Am* 86(Suppl 2):41–55, 2004.

Kofron MD, Laurencin CT: Orthopaedic applications of gene therapy, *Curr Gene Ther* 5(1):37–61, 2005.

Lee J, Qu-Petersen Z, Cao B et al: Clonal isolation of muscle-derived cells capable of enhancing muscle regeneration and bone healing, *J Cell Biol* 150:1085–1100, 2000.

Wozney J, Rosen V, Celeste A et al: Novel regulators of bone formation: Molecular clones and activities, *Science* 242:1528–34, 1988.

Bostrom M, Lane J, Tomin E et al: Use of bone morphogenetic protein-2 in the rabbit ulnar nonunion model, *Clin Orthop* 327:272–82, 1996.

Wang E, Rosen V, D'Assandro J et al: Recombinant human bone morphogenetic protein induces bone formation, *Proc Natl Acad Sci USA* 87:2220–4, 1990.

Cook S, Baffes G, Wolfe M et al: The effects of recombinant human osteogenic protein-1 on healing of large segmental bone defects, *J Bone Joint Surg* 76:827–38, 1994.

Sandhu H, Kanim L, Kabo J et al: Effective doses of recombinant human bone morphogenetic protein-2 in experimental spinal fusion, *Spine* 21:2115–22, 1996.

Boden S, Moskovitz P, Morone M et al: Video-assisted lateral intertransverse process arthrodesis. Validation of a new minimally invasive lumbar spinal fusion technique in the rabbit and non-human primate (rhesus) models, *Spine* 21:2689–97, 1996.

Mundy G: Regulation of bone formation by morphogenetic proteins and other growth factors, *Clin Orthop* 323:24–8, 1996.

Sampath T, Maliakai J, Hauschka P et al: Recombinant human osteogenic protein-1 (hOP-1) induces new bone formation in vivo with a specific activity comparable with natural bovine osteogenic protein and stimulates osteoblast proliferation and differentiation in vitro, *J Biol Chem* 267: 20-352–20-362, 1992.

Sato M, Ochi T, Nakase T et al: Mechanical tension-stress induces expression of bone morphogenetic protein (BMP-2) and BMP-4, but not BMP-6, BMP-7, and GDF-f mRNA, during distraction osteogenesis, *J Bone Miner Res* 14:1084–95, 1999.

Fang J, Zhu YY, Smiley E et al: Stimulation of new bone formation by direct transfer of osteogenic plasmid genes, *Proc Natl Acad Sci USA* 93:5753–8, 1996.

Leong L, Brickell P: Molecules in focus: Bone morphogenetic protein-4, *Int J Biochem Cell Biol* 28:1293–6, 1996.

Sampath T, Rashka K, Doctor J et al: Drosophila transforming growth factor beta superfamily proteins induce endochondral bone formation in mammals, *Proc Natl Acad Sci USA* 90:6004–8, 1993.

Vainio S, Karavanova I, Jowett A et al: Identification of BMP-4 as a signal mediating secondary induction between epithelial and mesenchymal tissues during early tooth development, *Cell* 75:45–58, 1993.

Mori S, Yoshikawa H, Hashimoto J et al: Antiangiogenic agent (TNP-470) inhibition of ectopic bone formation induced by bone morphogenetic protein-2, *Bone* 22:99–105, 1998.

Ferguson C, Alpern E, Miclau T et al: Does adult fracture recapitulate embryonic skeletal formation? *Mech Dev* 87:57–66, 1999.

Nakagawa M, Kaneda T, Arakawa T et al: Vascular endothelial growth factor (VEGF) directly enhances osteoclastic bone resorption and survival of mature osteoclasts, *FEBS Lett* 473:161–4, 2000.

Goad D, Rubin J, Wang H et al: Enhanced expression of vascular endothelial growth factor in human SaOS-2 osteoblast-like cells and murine osteoblasts induced by insulin-like growth factor I, *Endocrinology* 137:2262–68, 1996.

Gerber H, Vu T, Ryan A et al: VEGF couples hypertrophic cartilage remodeling, ossification and angiogenesis during endochondral bone formation, *Nat Med* 5:623–8, 1999.

Hidaka M, Stanford W, Bernstein A: Conditional requirement for the Flk-1 receptor in the in vitro generation of early hematopoietic cells, *Proc Natl Acad Sci USA* 96:7370–5, 1999.

Mitola S, Sozzani S, Luini W et al: Tat-human immunodeficiency virus-1 induces human monocyte chemotaxis by activation of vascular endothelial growth factor receptor-1, *Blood* 90:1365–72, 1997.

Peng H, Wright V, Usas A et al: VEGF enhances bone formation and bone healing elicited by genetically engineered muscle derived stem cells expressing BMP-4, *J Clin Invest* 110:751–9, 2002.

Alden T, Pittman D, Hankins G et al: In vivo endochondral bone formation using a bone morphogenetic protein 2 adenoviral vector, *Hum Gene Ther* 10:2245–53, 1999.

Baltzer A, Lattermann C, Whalen J et al: Genetic enhancement of fracture repair: Healing of an experimental segmental defect by adenoviral transfer of the BMP-2 gene, *Gene Ther* 7:734–9, 2000.

Musgrave D, Bosch P, Ghivizzani S et al: Adenovirus-mediated direct gene therapy with bone morphogenetic protein-2 produces bone, *Bone* 24:541–7, 1999.

Okubo Y, Bessho K, Fujimura K et al: Osteoinduction by bone morphogenetic protein-2 via adenoviral vector under transient immunosuppression, *Biochem Biophys Res Commun* 267:382–7, 2000.

Fang J, Zhu Y, Smiley E et al: Stimulation of new bone formation by direct transfer of osteogenic plasmid genes, *Proc Natl Acad Sci USA* 93:5753–8, 1996.

Bonadio J, Smiley E, Patil P et al: Localized, direct plasmid gene delivery in vivo: Prolonged therapy results in reproducible tissue regeneration, *Nat Med* 5:753–9, 1999.

Goldstein S, Bonadio J: Potential role for direct gene transfer in the enhancement of fracture healing, *Clin Orthop* 355:S154–62, 1998.

22.24. Cell Regenerative Engineering

Steadman J, Rodkey W, Rodrigo J: Microfracture: Surgical technique and rehabilitation to treat chondral defects, *Clin Orthop* 39: S362–9, 2001.

Brittberg M, Lindahl A, Nilsson A et al: Treatment of deep cartilage defects in the knee with autologous chondrocyte transplantation, *New Engl J Med* 331:889–95, 1994.

Buckwalter J, Mankin H: Articular cartilage repair and transplantation, *Arthritis Rheum* 41: 1331–42, 1998.

Radice M, Brun P, Cortivo R et al: Hyaluronan-based biopolymers as delivery vehicles for bone-marrow-derived mesenchymal progenitors, *J Biomed Mater Res* 50:101–9, 2000.

Martin I, Padera RF, Vunjak-Novakovic G, Freed LE: In vitro differentiation of chick embryo bone marrow stromal cells into cartilaginous and bone-like tissues, *J Orthop Res* 16:181–9, 1998.

Gao J, Dennis JE, Solchaga LA, Awadallah AS, Goldberg VM et al: Tissue-engineered fabrication of an osteochondral composite graft using rat bone marrow-derived mesenchymal stem cells, *Tissue Eng* 7:363–71, 2001.

Arai Y, Kubo T, Kobayashi K et al: Adenovirus vector-mediated gene transduction to chondrocytes: In vitro evaluation of therapeutic efficacy of transforming growth factor-beta 1 and heat shock protein 70 gene transduction, *J Rheumatol* 24:1787–95, 1997.

Smith P, Shuler F, Georgescu H et al: Genetic enhancement of matrix synthesis by articular chondrocytes: Comparison of different growth factor genes in the presence and absence of interleukin-1, *Arthritis Rheum* 43:1156–64, 2000.

Wehling P, Schulitz K, Robbins P et al: Transfer of genes to chondrocytic cells of the lumbar spine. Proposal for a treatment strategy of spinal disorders by local gene therapy, *Spine* 22:1092–7, 1997.

Ikeda T, Kubo T, Nakanishi T et al: Ex vivo gene delivery using an adenovirus vector in treatment for cartilage defects, *J Rheumatol* 27:990–6, 2000.

Goomer R, Maris T, Gelberman R et al: Nonviral in vivo gene therapy for tissue engineering of articular cartilage and tendon repair, *Clin Orthop* 379: S189–200, 2000.

Mason J, Grande D, Barcia M et al: Expression of human bone morphogenic protein 7 in primary rabbit periosteal cells: Potential utility in gene therapy for osteochondral repair, *Gene Ther* 5:1098–1104, 1998.

Adachi N, Sato K, Usas A et al: Muscle derived cell based ex vivo gene therapy for the treatment of full-thickness articular cartilage defects, *J Rheumatol* 29:1920–30, 2002.

Bruder S, Kraus K, Goldberg V et al: The effect of implants loaded with autologous mesenchymal stem cells on the healing of canine segmental bone defects, *J Bone Joint Surg Am* 80:985–96, 1998.

Bruder S, Kurth A, Shea M et al: Bone regeneration by implantation of purified, culture-expanded human mesenchymal stem cells, *J Orthop Res* 16:155–62, 1998.

Lieberman J, Daluiski A, Stevenson S et al: The effect of regional gene therapy with bone morphogenetic protein-2-producing bone-marrow cells on the repair of segmental femoral defects in rats, *J Bone Joint Surg Am* 81:905–17, 1999.

Turgeman G, Pittman DD, Muller R et al: Engineered human mesenchymal stem cells: A novel platform for skeletal cell mediated gene therapy, *J Gene Med* 3:240–51, 2001.

Krebsbach P, Gu K, Franceschi R et al: Gene therapy-directed osteogenesis: BMP-7-transduced human fibroblasts form bone in vivo, *Hum Gene Ther* 11:1201–10, 2000.

Zuk P, Zhu M, Mizuno H et al: Multilineage cells from human adipose tissue: Implications for cell-based therapies, *Tissue Eng* 7:211–28, 2001.

Qu-Petersen Z, Deasy B, Jankowski R et al: Identification of a novel population of muscle stem cells in mice: Potential for muscle regeneration, *J Cell Biol* 157:851–64, 2002.

Lee J, Musgrave D, Pelinkovic D et al: Effect of bone morphogenetic protein-2-expressing muscle-derived cells on healing of critical-sized bone defects in mice, *J Bone Joint Surg Am* 83A:1032–9, 2001.

Bosch P, Musgrave D, Lee J et al: Osteoprogenitor cells within skeletal muscle, *J Orthop Res* 18:933–44, 2000.

22.25. Tissue Regenerative Engineering

Lohfeld S, Barron V, McHugh PE: Biomodels of bone: A review, *Ann Biomed Eng* 33(10):1295–311, 2005.

Yoshikawa H, Myoui A: Bone tissue engineering with porous hydroxyapatite ceramics, *J Artif Organs* 8(3):131–6, 2005.

Anderson DG, Albert TJ, Fraser JK, Risbud M, Wuisman P et al: Cellular therapy for disc degeneration, *Spine* 30(17 Suppl):S14–9, 2005.

Kang Y, Yang J, Khan S, Anissian L, Ameer GA: A new biodegradable polyester elastomer for cartilage tissue engineering, *J Biomed Mater Res A* 77(2):331–9, May 2006.

Mauney JR, Volloch V, Kaplan DL: Role of adult mesenchymal stem cells in bone tissue engineering applications: Current status and future prospects, *Tissue Eng* 11(5–6):787–802, 2005.

Mistry AS, Mikos AG: Tissue engineering strategies for bone regeneration, *Adv Biochem Eng Biotechnol* 94:1–22, 2005.

Sanchez C, Arribart H, Guille MM: Biomimetism and bioinspiration as tools for the design of innovative materials and systems, *Nat Mater* 4(4):277–88, 2005.

Hutmacher DW, Garcia AJ: Scaffold-based bone engineering by using genetically modified cells, *Gene* 347(1):1–10, 2005.

Wilson CJ, Clegg RE, Leavesley DI, Pearcy MJ: Mediation of biomaterial-cell interactions by adsorbed proteins: A review, *Tissue Eng* 11(1–2):1–18, 2005.

Lynn AK, Brooks RA, Bonfield W, Rushton N: Repair of defects in articular joints. Prospects for material-based solutions in tissue engineering, *J Bone Joint Surg Br* 86(8):1093–9, Nov 2004.

Heng BC, Cao T, Stanton LW, Robson P, Olsen B: Strategies for directing the differentiation of stem cells into the osteogenic lineage in vitro, *J Bone Miner Res* 19(9):1379–94, 2004.

Nussenbaum B, Teknos TN, Chepeha DB: Tissue engineering: The current status of this futuristic modality in head neck reconstruction, *Curr Opin Otolaryngol Head Neck Surg* 12(4):311–5, 2004.

Renner G, Lane RV: Auricular reconstruction: An update, *Curr Opin Otolaryngol Head Neck Surg* 12(4):277–80, 2004.

Liu X, Ma PX: Polymeric scaffolds for bone tissue engineering, *Ann Biomed Eng* 32(3):477–86, 2004.

Liu Y, de Groot K, Hunziker EB: Osteoinductive implants: The mise-en-scene for drug-bearing biomimetic coatings, *Ann Biomed Eng* 32(3):398–406, 2004.

Partridge KA, Oreffo RO: Gene delivery in bone tissue engineering: Progress and prospects using viral and nonviral strategies, *Tissue Eng* 10(1–2):295–307, 2004.

Derubeis AR, Cancedda R: Bone marrow stromal cells (BMSCs) in bone engineering: Limitations and recent advances, *Ann Biomed Eng* 32(1):160–5, 2004.

Sharma B, Elisseeff JH: Engineering structurally organized cartilage and bone tissues, *Ann Biomed Eng* 32(1):148–59, 2004.

Sander EA, Nauman EA: Permeability of musculoskeletal tissues and scaffolding materials: Experimental results and theoretical predictions, *Crit Rev Biomed Eng* 31(1–2):1–26, 2003.

23

OCULAR REGENERATIVE ENGINEERING

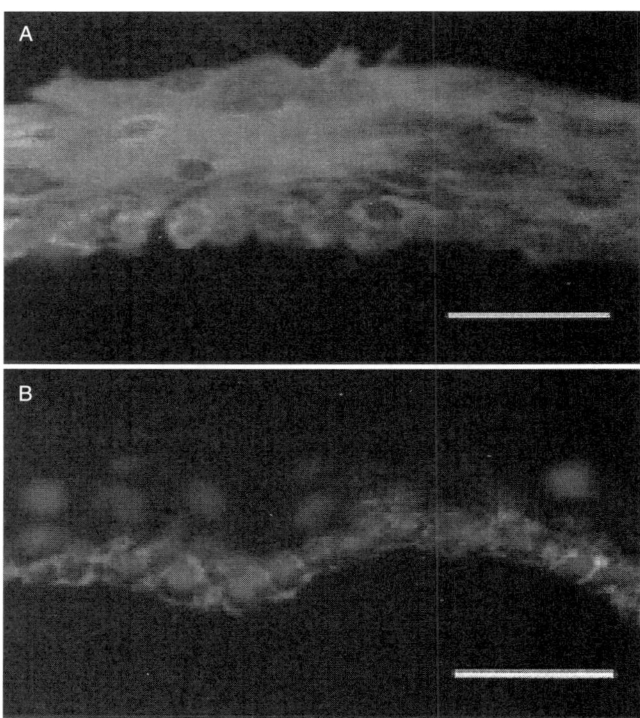

Preparation of autologous tissue-engineered epithelial cell sheets fabricated from oral mucosal epithelium. Oral mucosal tissue (3 × 3 mm) was removed from a patient's cheek. Isolated epithelial cells were seeded onto temperature-responsive cell culture inserts. After two weeks of culture at 37°C, the cells grow to form multilayered sheets of epithelial cells. These sheets were used to cover and repair injured corneal epithelium. Specimens were collected for testing the expression of keratin 3 (A) and anti-β1 integrin (B) by immunohistochemistry. Red: cell nuclei. Scale bars: 50 μm. (Reprinted with permission from Nishida K et al: Corneal reconstruction with tissue-engineered cell sheets composed of autologous oral mucosal epithelium, *New Engl J Med* 351:1187–96, copyright 2004 Massachusetts Medical Society. All rights reserved.) See color insert.

Bioregenerative Engineering: Principles and Applications, by Shu Q. Liu

ANATOMY AND PHYSIOLOGY OF THE OCULAR SYSTEM [23.1]

The *ocular system* is composed of the eyes, optic nerve, and a number of ocular accessory structures, including eyelids, lacrimal apparatus, and ocular muscles. The eye consists of several layered structures: the fibrous tunica, vascular tunica, and the nervous tunica (Fig. 23.1). The *fibrous tunica* is the outmost layer of the eye, composed of sclera and cornea. The *sclera* is a layer of collagen-rich connective tissue, which is distributed around the lateral and posterior sides of the eye and encloses the intraocular structures. It provides structural and mechanical stability to the eye, protects ocular structures from injury, and serves as an anchoring base for the ocular muscles. The *cornea* is an avascular and transparent membrane structure, which is continuous with the sclera and located in the front of the eye. It is composed of three layers: the external stratified epithelium, the middle stroma with a collagen and proteoglycan-rich matrix, and internal endothelium. The cornea is a key optical structure that refracts light to the retina. Since the cornea also serves as a protective barrier for the intraocular structures, it is heavily innervated with sensory nerve endings, which sense mechanical stimuli and temperature changes.

The *vascular tunica* is the middle layer of the eye and contains a rich network of arteries and veins, which supply oxygenated blood to and drain deoxygenated blood from the eye. The arteries of the eye are originated from the internal carotid artery and the veins drain blood to the internal jugular vein. The vascular tunica is composed of choroid, ciliary body, and iris. The *choroid* is the portion associated with the sclera and is distributed around the lateral and posterior sides of the eye. In the front of the eye, the choroid is connected to the *ciliary body*, a structure composed of the *ciliary ring* of smooth muscle cells and *ciliary processes*. The ciliary smooth muscle cells are aligned in the radial direction of the ciliary ring in the outer region, whereas aligned in the circumferential direction in the inner region. The ciliary processes are structures of epithelial cells and are connected to the suspensory ligaments. The suspensory ligaments are fibrous structures linking the ciliary body to the lens. The contraction of the ciliary smooth muscle cells stretches the lens, reducing the thickness of the lens. The lens recoils back to the original shape when the ciliary muscles relax. The contractile activity of the ciliary

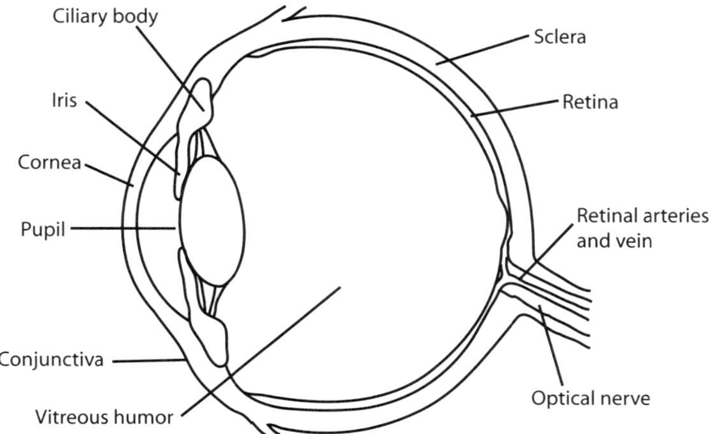

Figure 23.1. Schematic representation of the eye structures.

muscles and the shape change of the lens control the focal distance of the eye. The ciliary body is connected to the *iris*, which contains smooth muscle cells and surrounds the pupil, the central opening of the eye. The iris is composed of circumferentially and radially aligned smooth muscle cells. The circumferential smooth muscle cells are controlled by parasympathetic nerves. The contraction of these muscles reduces the size of the pupil, resulting in a decrease in the amount of light entering the eye. The radial smooth muscle cells are controlled by sympathetic nerves. Their contraction induces the dilation of the pupil, increasing the amount of light entering the eye.

The *nervous tunica* is also called the *retina*, which is the internal layer of the eye. The *retina* is composed of a surface epithelial layer, known as the pigmented retina, and a neuronal layer, known as the sensory retina. The *sensory retina* consists of a large number of rod and cone photoreceptor neurons and relay neurons. The retina covers the internal surface of the eye except for the front area with the lens and the ciliary body. In the posterior center, there are two spots that contain highly concentrated photoreceptor neurons: the macula lutea and fovea centralis. These structures are capable of identifying fine objects and images. Near these structures, there is another spot, known as the optic disc, a location where blood vessels enter and the nerve fiber bundles from photoreceptor neurons leave the eye. The optic disc is not able to sense light because of the lack of photoreceptors.

The *sensory retina* is composed of three layers: photoreceptor neurons, bipolar neurons, and ganglionic neurons. Between these layers, there exist various types of association neurons. The photoreceptor layer contains rod and cone neurons. The *rod neurons* can sense light of low intensity and are insensitive to colors. These cells contain a protein complex, known as rhodopsin, which is responsible for the sensation of dim light. In addition, rhodopsin participates in the regulation of light adaptation. When a person is suddenly exposed to bright light, rhodopsin is rapidly degraded, reducing light-initiated stimulatory signaling activities. In contrast, when a person is suddenly exposed to dim light, the production of rhodopsin increases, enhancing the sensitivity of the retina to dim light.

The *cone neurons* are responsible for the sensation of colors and ordinary light. Color identification by these cells requires the presence of a critical level of bright light. Below such a critical level, these cells lose the capability of color identification. Cone cells contain protein complexes called iodopsins. These complexes are composed of three types of opsin proteins for the sensation of red, blue, and green colors. Each type is only sensitive to a narrow spectrum of light corresponding to a specific color. The distribution of the cone neurons differs from that of the rod neurons. The fovea centralis is primarily composed of cone neurons with almost no rod neurons. Thus this structure is for the sensation of color and bright light and, especially, for accurate identification of images. In contrast, the rod neurons are spread over the remaining retina, a distribution essential for the sensation of dim light.

The *bipolar* and *ganglionic neurons* in other retinal layers play critical roles in the transmission of optic signals from the rod and cone neurons to the central visual centers. The bipolar neurons synapse with the rod and cone neurons at one side and synapse with the ganglionic neurons at the other side. In these layers, there are several types of association neurons, including horizontal neurons, amacrine neurons, and interplexiform neurons. These neurons synapse with the photoreceptor, bipolar, and ganglionic neurons, and relay, integrate, and modify signals from the photoreceptor neurons. Nerve fibers from the ganglionic neurons converge to the optic disc, where they form the *optic nerve*, exit the eye, and enter the central visual centers of the brain, including the superior colliculi, lateral geniculate nuclei of thalamus, and visual cortex.

Enclosed within the eyeball are two compartments: the anterior and posterior compartments. The *anterior compartment* is the chamber between the cornea and the lens, and is filled with aqueous humor, a fluid that is produced by the ciliary processes, released into the anterior compartment, and returned to the vein through the trabecular meshwork and the canal of Schlemm. The *aqueous humor* circulates constantly with a stable hydrostatic pressure and supplies oxygen and nutrients to the cornea and lens. The obstruction of the trabecular meshwork and the canal of Schlemm results in an increase in the intraocular pressure, a disorder known as *glaucoma* (see page 979). The *posterior compartment* is the chamber surrounded by the retina and filled with *vitreous humor*, a transparent gel-like structure. The vitreous humor plays a critical role in the maintenance of the ocular shape and transmission of light.

The *lens* is an avascular, transparent, biconvex structure that is composed of two types of epithelial cells, including the cuboidal and fiber-like epithelial cells. The cuboidal epithelial cells are found on the anterior surface, whereas the fiber-like epithelial cells are found in the remaining body of the lens. The fiber-like cells are specially differentiated cells that do not contain nuclei and cellular organelles. Instead, these cells contain a special type of protein known as crystalline. The presence of crystalline renders the lens highly transparent. The crystalline-containing cells are enclosed by an elastic layer of tissue, which is connected to the ciliary body via the suspensory ligaments.

The *ocular accessory* structures include the eyelid, conjunctiva, lacrimal apparatus, and eye muscles. The *eyelid* is composed of several layers, including the skin, areolar connective tissue, skeletal muscles, tarsal plate, and palpebral conjunctiva. The function of the eyelid is to protect the eye from injury. The *conjunctiva* is a fibrous membrane that covers the internal surface of the eyelid (the palpebral conjunctiva) and the anterior surface of the eyes (the bulbar conjunctiva). The palpebral conjunctiva directly interacts with the cornea. Because of the presence of fluids, the friction between the conjunctiva and cornea is small. The *lacrimal apparatus* is composed of the lacrimal gland, lacrimal canaliculi, lacrimal sac, and nasolacrimal duct. The lacrimal gland produces tears, which are released to the external surface of the eye. The production and release of tears are controlled by the parasympathetic nerves. Tears serve as a lubricant for the interaction of the eyeball with the eyelid. Excessive tears enter the lacrimal canaliculi through two openings called punctas, flow into the nasal cavity via the lacrimal sac and nasolacrimal duct. Each eyeball is associated with six skeletal muscle bundles. These muscles are anchored to the external surface of the sclera and control the movement of the eyeball.

OCULAR DISORDERS

Corneal Injury

Pathogenesis, Pathology, and Clinical Features [23.2]. The cornea is a structure exposed to the exterior environment and is subject to various hazards, such as mechanical injury, chemical corrosion, radiation, and infection by bacteria and viruses. Corneal injury due to mechanical trauma and chemical corrosion is commonly seen. Corneal injury often induces inflammatory reactions, followed by fibrosis and scar formation in the cornea, reducing light transmission. Since the cornea is a collagen-rich structure, disorders with collagen degradation may affect the function of the cornea. Metabolic disorders can also induce dysfunction of the cornea. For instance, hypercalcemia is associated with calcium precipitation underneath the cornea epithelial cells. Cystinosis can cause the formation of cystine crystals in the cornea. Hypercholesterolemia induces cholesterol deposition in the cornea. All these disorders influence light transmission through the cornea and induce visual impairment.

Conventional Treatment of Cornea Injury [23.2]. There are two strategies for the treatment of corneal disorders: removing the factors that cause corneal abnormalities and conducting corneal transplantation. When causative factors can be identified, these factors should be removed, if possible, to reduce or stop the progression of corneal abnormalities. For example, ocular bacterial infection should be controlled by local administration of antibiotics. When hypercalcemia is identified as a causative disorder, the blood calcium concentration should be reduced to the normal level. When severe corneal scars and opacity develop, the cornea can be replaced with an allogenic corneal specimen, a procedure known as corneal transplantation.

Molecular Regenerative Engineering. Molecular engineering approaches can be applied to corneal disorders. Corneal disorders often involve molecular activities, such as activation of pro-inflammatory factors, upregulation of proliferative genes, and production of extracellular matrix. Thus, molecular strategies for the treatment of corneal disorders are to suppress inflammation and selectively inhibit the proliferation of certain cell types such as fibroblasts. Selected genes can be prepared and used for the treatment of ocular disorders. For example, corneal haze and cloudiness after mechanical injury are due to excessive inflammatory reactions, including leukocyte infiltration, cell proliferation, and extracellular matrix deposition. Genes encoding anti-inflammatory and antiproliferative proteins can be used to suppress inflammatory reactions and fibrous changes. In addition, dominant negative genes for proinflammatory and mitogenic factors can also be used for treating corneal inflammation. Another example is the molecular treatment of primary glaucoma. This disorder is induced by the obstruction of the trabecular meshwork by excessive production of extracellular matrix. Genes encoding matrix metalloproteinases, which degrade extracellular matrix components, can be used for the treatment of glaucoma.

Given the anatomical features of the ocular system, several approaches can be used for gene delivery. For the molecular treatment of the corneal epithelial disorders, a topical gene delivery is effective. For disorders of the iris, ciliary body, and trabecular meshwork, gene injection into the anterior compartment is required. For retinal disorders, it is necessary to conduct intravitreal gene injection. As for other organs and tissues, various methods can be used to mediate gene delivery to the ocular system, depending on the anatomical features of and cell types in the target tissue. For instance, electroporation is an effective method for gene delivery to the corneal epithelial cells, but may not be a suitable method for gene delivery to the intraocular structures. Genetically modified adenoviruses and retroviruses are often used for mediating gene delivery into ocular tissues, including the cornea, trabecular meshwork, and retina. These mediating methods have been shown to be more effective than other mediating methods, such as salt- and liposome-mediated delivery, for the ocular system.

Molecular engineering approaches have been developed and used for treating several corneal disorders, including immune rejection of corneal transplants, corneal inflammation and haze, and corneal complications due to metabolic disorders such as mucopolysaccharidosis. These approaches are discussed in the following sections.

Molecular Therapies for Corneal Immune Rejection [23.3]. Allogenic corneal transplantation is an effective approach for the treatment of corneal dysfunction. However, the presence of functional epithelial cells, which are essential for successful corneal transplantation, often causes immune reactions, resulting in acute rejection. It is often necessary to administrate immune suppressor agents to patients with corneal transplantation. However, these immune suppressor agents induce side effects by inhibiting the activity of the entire immune system. Furthermore, it is required to conduct daily agent deliveries. Molecular engineering approaches can be used to overcome these problems. Genes that encode immune suppressor cytokines and antisense oligonucleotides for immune activator genes can serve as immune suppressor agents. Given the anatomical features of the cornea, it is relatively easier to deliver genes to the cornea than to the internal structures. Three approaches may be used for corneal gene delivery: application of genes to the exterior surface of the cornea, gene injection to the anterior compartment, and augmentation of gene delivery by electroporation (see page 444 for these methods). Typical genes for corneal disorders include the CD152 and interleukin (IL)10 genes, which have been used for the treatment of corneal transplant immune rejection.

The CD152 gene (CTLA-4) encodes a membrane protein in the T lymphocytes. The CD152 protein exerts an inhibitory effect on T lymphocyte-related immune reactions. When allogenic tissues are transplanted to the host, the allogenic antigens activate antigen-presenting cells (APCs), which in turn interact with the T lymphocytes, leading to activation of the T lymphocytes and initiating immune reactions. In particular, a cell membrane protein known as CD80 can interact with another membrane protein CD86 to form complexes. The CD80 and CD86 complexes on the APC surface interact with CD28 (see Table 23.1) on the T lymphocyte surface, eliciting co-stimulating signals for the activation of the T lymphocytes. The CD152 complexes on the T lymphocyte surface, when present, can bind to CD80 and CD86 in antigen-presenting cells, suppressing the activity of these cells as well as the T lymphocytes. The overexpression of the CD152 gene by gene transfer has been shown to induce the arrest of T lymphocyte division, reduce immune responses, prevent corneal immune rejection, and prolong the survival of transplanted allogenic cornea. The CD152 gene can be conjugated with an Ig gene, forming a recombinant gene complex, which can facilitate gene delivery and expression.

Another gene used for the treatment of corneal transplant immune rejection is the interleukin-10 gene (see page 634 for characteristics of IL10). This gene encodes a cytokine that suppresses the activity of T lymphocytes. Experimental investigations have demonstrated that the interleukin-10 gene can be effectively transferred into more than 70% of the epithelial cells of the cornea with a virus-mediated gene transfer approach. The transferred gene can be expressed for about 3 weeks. Such an approach has been shown to reduce immune responses in transplanted allogenic cornea and prolong corneal survival.

Molecular Therapies for Corneal Inflammation and Fibrosis [23.4]. Corneal inflammation is induced by trauma and therapeutic keratectomy. Inflammation often results in epithelial cell proliferation, extracellular matrix production, and fibrosis. Thus, the principle of molecular engineering therapy for corneal inflammation is to introduce genes that encode antiproliferative proteins. A gene encoding the dominant negative cyclin G1 (Table 23.2) protein has been constructed and used to treat corneal inflammation. Cyclin G1 plays a critical role in stimulating the progression of the cell division cycle. The

TABLE 23.1. Characteristics of Selected Molecules that Regulate Immune Responses*

Proteins	Alternative Names	Amino Acids	Molecular Weight (kDa)	Expression	Functions
CD28	Antigen CD28, T-cell antigen CD28	220	25	T cell, B cell	Regulating CD4-positive T-cell survival and proliferation, inducing interleukin-2 production from T cells, and promoting the development of T-helper type-2 (Th2) cells
CD80	CD28 antigen ligand 1 (CD28LG1), B lymphocyte activation antigen B7-1, activation B7-1 antigen, B7-1 antigen	288	33	B cell, dendritic cell, monocyte, mast cell, nervous system, blood vessels	Interacting with CD28 on T cells and regulating T-cell proliferation and activation
CD86	CD86 antigen, CD28 antigen ligand 2 (CD28LG2), B lymphocyte activation antigen B7-2, Activation B7-2 antigen, CTLA4 counterreceptor B7.2, LAB7-2, B70	329	38	T cell, B cell, monocytes, macrophages, dendritic cells, vascular endothelial cells, smooth muscle cells, bone marrow, intestinal epithelial cells	A member of the immunoglobulin superfamily expressed by antigen-presenting cells, binding to CD28 on T cells to regulate T-cell proliferation, survival, and activation, and binding to cytotoxic T-lymphocyte-associated protein 4 on T cells to negatively regulate T-cell activation and diminish T-cell-mediated immune responses
CD152	Cytotoxic T lymphocyte associated 4, cytotoxic T lymphocyte antigen 4 (CTLA4)	223	25	T cell	A member of the immunoglobulin superfamily that inhibits the activity of T cells

*Based on bibliography 23.3.

TABLE 23.2. Characteristics of Cyclin G1*

Proteins	Alternative Names	Amino Acids	Molecular Weight (kDa)	Expression	Functions
Cyclin Gl	Cyclin G, CCNG	295	34	Lymphocytes, lung, kidney, intestine, spleen, thymus, testis, ovary, prostate gland, skeletal muscle	Regulating cell division

*Based on bibliography 23.4.

negative dominant cyclin Gl protein competes with the wild-type or natural form of cyclin Gl for substrate binding, but does not activate the substrate. Thus, the introduction of this dominant negative gene to the corneal epithelial cells suppresses the activity of the natural cyclin Gl, induces the arrest of the cell division cycle, and reduces cell proliferation and matrix production. These processes are associated with a decrease in inflammatory reactions and suppression of extracellular matrix production and haze development. Experimental investigations have demonstrated promising results for the transfer of the dominant negative cyclin Gl gene into the cornea with laser keratectomy-induced injury. Dominant negative genes constructed for other types of mitogenic signaling factors, such as growth factor receptors, protein tyrosine kinases, and cell cycle regulators, can also be used to inhibit corneal inflammatory reactions.

Pharmacological inhibitors can be used to suppress the activities of inflammatory factors. One example is the use of the nuclear factor κB inhibitor SN50. Nuclear factor κB is a transcription factor that stimulate the expression of inflammatory genes. The suppression of the activity of nuclear factor κB may reduce inflammatory reactions in corneal injury. Experimental investigations have demonstrated that the topical application of SN50 can facilitate the healing process of alkali-induced corneal injury (Fig. 23.2).

Molecular Therapies for Corneal Complications Due to Mucopolysaccharidosis Type VII (MPSVII) [23.5]. Mucopolysaccharidosis type Vll is an hereditary disorder due to the deficiency of the enzyme β-glucuronidase (GUSB) (Table 23.3), which breaks down proteoglycans. This disorder is associated with the lysosomal accumulation of undegraded glycosominogycans (GAGs), resulting in corneal abnormalities and opacity. The transfer of the β-glucuronidase gene into the corneal epithelial cells induces the over-expression of β-glucuronidase, prevents the accumulation of undegraded glycosominogycans, and reduces corneal opacity.

Cellular and Tissue Engineering. Cellular and tissue engineering approaches have been established and used to treat ocular disorders in experimental models and preliminary clinical trials. As for other organ and tissue systems, a successful replacement of a malfunctioned ocular tissue requires the construction of a functional cellular and tissue structure and the integration of the constructed structure into the ocular system. To date, cellular and tissue engineering approaches have been used to repair and reconstruct disordered cornea in experimental models and clinical trials. However, the application of cellular and tissue approaches to other ocular tissues has been limited because of difficulties in the construction and assembly of functional ocular structures, such as the retina and lens.

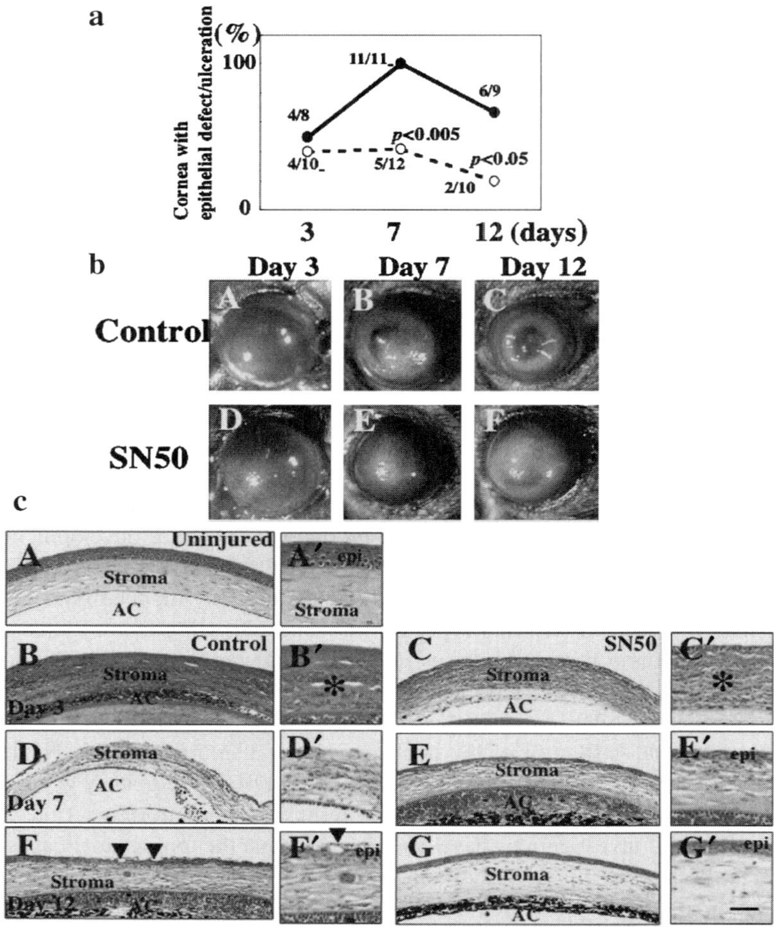

Figure 23.2. Healing of alkali-burned mouse cornea treated with topical SN50, an inhibitor of NFκB. (a) Percentage of corneas with epithelial defect (including ulceration) at each healing interval. The incidence of epithelial defect/ulceration is significantly higher in the control group than in the SN50-treated group at days 7 and 12 as judged by the χ^2 test. (b) Macroscopic observation shows similar initial resurfacing in both control (A) and SN50-treated groups (D) at day 3 after alkali burning. Recurrence of the epithelial defect with stromal opacification is observed more frequently in the control group at days 7 (B) and 12 (C) as compared with SN50-treated group (E, F). (c) Histology of burned corneas stained with H&E. (A) An uninjured cornea. Stratified epithelium and stroma are seen. There is no histological difference between central corneas in the control (B) and SN50-treated group (C) at day 3. The epithelium shows a large defect and many inflammatory cells are observed. At day 7 the burned cornea in the control (D) shows more stromal inflammation, and a large epithelial defect as compared with the SN50-treated corneas that has been resurfaced with a thin epithelium (E). At day 12 the control cornea still shows marked inflammation and hypercellularity in the stroma (F), whereas the treated cornea exhibits a well-regenerated epithelium with a less stromal inflammation (G). Regenerated epithelium in control exhibits conjunctiva-like appearance with goblet cells (arrowheads). A′–G′ are high-magnification pictures of the central area of the healing corneas in A–G, respectively. Scale bar: 100 μm (A–G), 25 μm (A′–G′). (Reprinted with permission from Saika S et al: *Am J Pathol* 166:1393–1403, copyrights 2005.)

TABLE 23.3. Characteristics of β-Glucuronidase*

Proteins	Alternative Names	Amino Acids	Molecular Weight (kDa)	Expression	Functions
β-Glucuronidase	βG1	651	75	Retinal pigmented epithelial cells, leukocyte, liver, kidney, spleen, placenta, intestine, pancreas	An enzyme that degrades proteoglycans

*Based on bibliography 23.5.

Corneal dysfunction requires corneal replacement by transplantation. Allogenic cornea is often used for such a purpose. However, there two major problems for allogenic cornea transplantation: immune rejection and shortage of cornea donors. In particular, immune responses induce injury and death of corneal epithelial and endothelial cells, leading to opacification of the transplanted cornea. Several cellular and tissue engineering strategies have been developed and used to overcome these problems, including: (1) corneal surface reconstruction with epithelial stem cells or autogenous epithelial cells, which can be applied to native or transplanted corneas with injured epithelial cells and (2) corneal reconstruction with extracellular matrix and polymeric materials.

Corneal Surface Reconstruction [23.6]. The cornea is covered at the external surface with an epithelial layer. The injury or denudation of this layer induces inflammatory reactions and fibrosis, resulting in alterations in the optical properties of the cornea and visual acuity. In such a case, it is necessary to reconstruct the corneal surface. Corneal surface reconstruction can be accomplished by using an epithelial cell layer constructed in vitro. Several types of cells can be used to construct an epithelial cell layer: limbal epithelial stem cells, adult epithelial cells, and stem cells. Epithelial stem cells are present in the limbal region or the junction of the cornea and conjunctiva. These cells can differentiate into corneal epithelial cells in corneal injury. The deficiency of limbal stem cells can cause ocular surface disorders, leading to blindness. In corneal injury, limbal stem cells can be collected and used for enhancing corneal cell regeneration by cell transplantation.

Limbal stem cells can be collected from several sources, including the cadaver eyes, the conjunctival limbal tissue from the donors, and autogenous limbal tissue. Harvested limbal epithelial stem cells by biopsy can be cultured and expanded in vitro on a suitable carrier membrane, such as a polymeric or natural matrix membrane, forming a transplantable epithelial membrane structure. The cultured epithelial cells can produce extracellular matrix, which serve as a basal meshwork for the formation of a stable epithelial membrane structure. The epithelial membrane is readily adhesive and can be used for the construction of a corneal epithelium-like structure (Fig. 23.3). Clinical investigations have shown that this approach can be used to effectively prevent corneal inflammatory reactions and improve visual acuity (Fig. 23.4).

Natural biological membranes can also be used to serve as epithelial cell carriers. A typical example is the human amniotic membrane, which can be used as a basal membrane

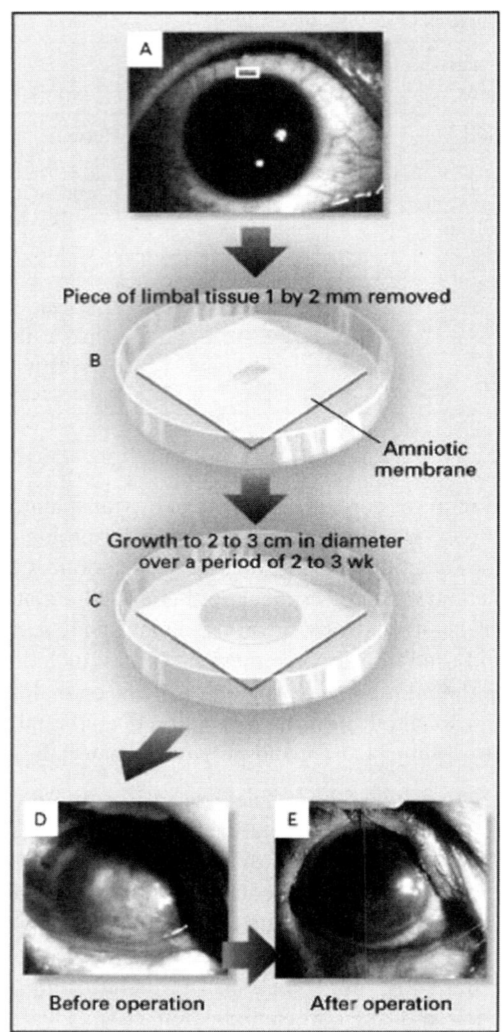

Figure 23.3. Transplantation of autologous limbal epithelial cells cultured on amniotic membrane. Limbal tissue (1 × 2 mm) was removed by lamellar keratectomy from the superior limbus of the healthy contralateral eye (panel A). The explanted tissue was placed on amniotic membrane in a 35-mm dish containing 1.5 mL of culture medium (panel B). After 2–3 weeks, the epithelial cells had grown and spread to form a circular sheet of cells with a diameter of 2–3 cm (panel C). The cultured limbal epithelial cells with amniotic membrane were then transplanted to the diseased eye (panels D and E). (Reprinted with permission from Tsai RJ, Li LM, Chen JK: *New Engl J Med* 343:86–93, copyright 2000 Massachusetts Medical Society. All rights reserved.)

for culturing epithelial cells and constructing corneal epithelial layers. The constructed epithelial layer can be directly applied to the exterior surface of the cornea. Since the epithelial cell layer is usually thin and self-adhesive, it is not necessary to fast the cell layer with suture stitches or adhesives. Compared to synthetic biomaterials, a biological membrane is compatible with cells and provides a suitable substrate for the formation of an epithelial cell layer.

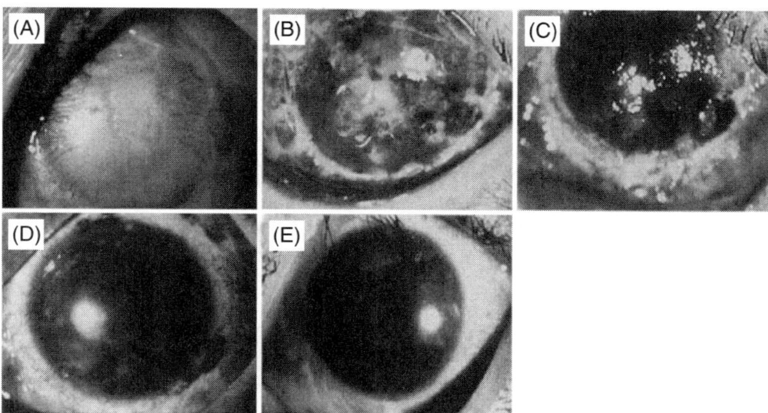

Figure 23.4. Serial photographs of the eye of patient 6 before and after transplantation initial examination. Panel A reveals corneal opacity with central corneal erosion and neovascular growth extending into the entire cornea for 4–6 mm before operation. Lamellar keratectomy was performed to remove the entire opacified limbal and corneal area to a thickness of ~⅓rd of the corneal layer (panel B). Limbal epithelial cells with the amniotic membrane substrate were transplanted onto the denuded limbal and corneal area. Photographs were taken 1 day (panel B), 7 days (panel C), 30 days (panel D), and 450 days (panel E) after the operation. (Reprinted with permission from Tsai RJ, Li LM, Chen JK: *New Engl J Med* 343:86–93, copyright 2000 Massachusetts Medical Society. All rights reserved.)

Adult epithelial cells can proliferate and can be used for constructing the corneal external surface. Ideally, autogenous corneal epithelial cells should be used for corneal reconstruction. However, in patients with corneal disorders, corneal epithelial cells are usually not available. Given the fact that most epithelial cell types on the exterior surface of the body exhibit certain common phenotypes, it is conceivable that epithelial cells from other exterior tissues may be used for constructing a corneal epithelial structure. A candidate epithelial cell type is the oral mucosal epithelial cells. These cells can be easily identified, harvested, manipulated, and expanded in culture. A major advantage of using the oral mucosal epithelial cells is that cells can be harvested from the host patients, thus avoiding immune rejection responses.

Prepared oral mucosal epithelial cells can be seeded on a membrane for the construction of an epithelial layer. Researchers have developed a temperature-sensitive synthetic polymer material, which can be used to construct membranous cell carriers (Fig. 23.5). Epithelial cells or stem cells can be seeded on the carrier for expansion and formation of an epithelial membrane. A reduction in temperature causes shrinkage of the carrier polymeric material, inducing the separation of the cells from the carrier membrane. Since this approach does not require the use of proteinases for cell separation from the membrane carrier, epithelial cells remain intact and functional.

Epithelial membranes constructed with autogenous oral epithelial cells have been shown to express corneal epithelial markers, such as keratin-3 (Chapter 23 opening figure), and have been applied to patients with complete denudation of the corneal epithelium with severe impairment of visual acuity. These investigations have demonstrated that reepithelialization of the corneal surface occurs within one week, resulting in the restoration of the corneal transparency and significant improvement of visual acuity (Fig. 23.6). In

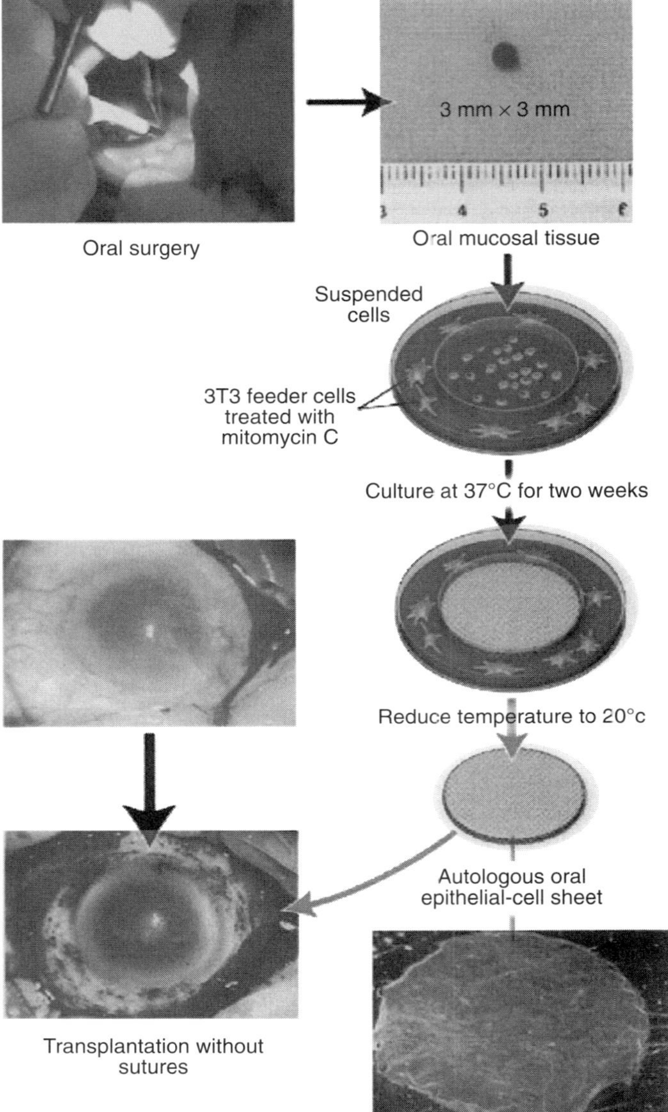

Figure 23.5. Transplantation of autologous tissue-engineered epithelial cell sheets fabricated from oral mucosal epithelium to injured cornea. Oral mucosal tissue (3 × 3 mm) was removed from a patient's cheek. Isolated epithelial cells are seeded onto temperature-responsive cell culture inserts. After 2 weeks at 37°C, these cells grow to form multilayered sheets of epithelial cells. The viable cell sheet was harvested with intact cell-to-cell junctions and extracellular matrix in a transplantable form simply by reducing the temperature of the culture to 20°C for 30 min. The cell sheet is then transplanted directly to the diseased eye without sutures. (Reprinted with permission from Nishida K et al: Corneal reconstruction with tissue-engineered cell sheets composed of autologous oral mucosal epithelium, *New Engl J Med* 351:1187–96, copyright 2004 Massachusetts Medical Society. All rights reserved.)

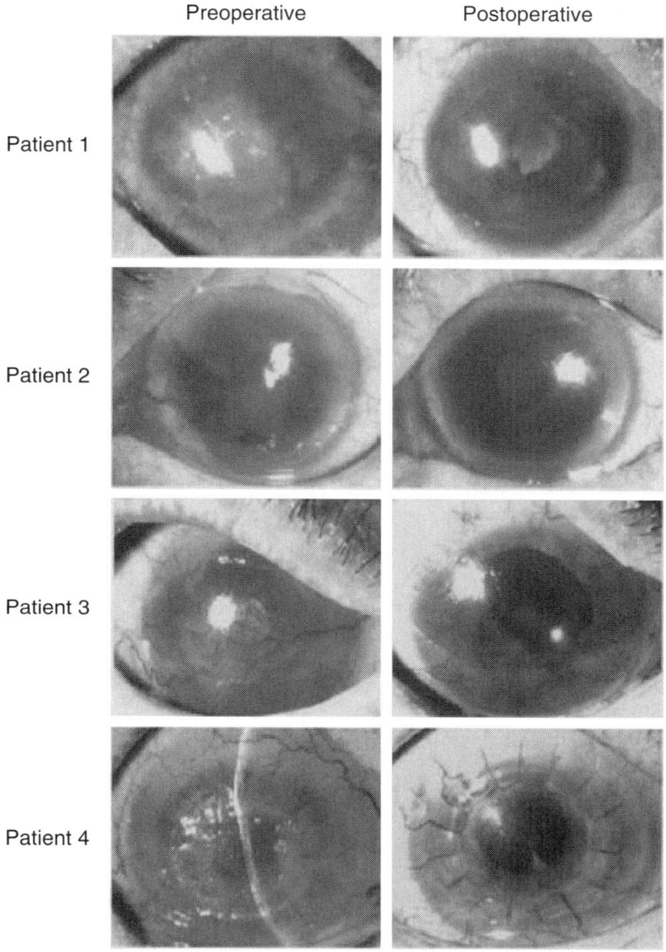

Figure 23.6. Eyes of patients before and after transplantation of sheets of tissue-engineered autologous epithelial cells. These photographs were taken just before transplantation of the cell sheets and postoperatively at 13, 14, or 15 months. (Reprinted with permission from Nishida K et al: Corneal reconstruction with tissue-engineered cell sheets composed of autologous oral mucosal epithelium, *New Engl J Med* 351:1187–96, copyright 2004 Massachusetts Medical Society. All rights reserved.)

addition to epithelial progenitor cells and mature epithelial cells, embryonic stem cells can also be used for corneal reconstruction. An example is shown in Fig. 23.7.

Corneal Reconstruction [23.7]. When severe opacity exists in the cornea, it is necessary to conduct complete corneal reconstruction. Although allogenic corneal transplantation is an option, there is a shortage of the supply of allogenic corneas. Thus, it is necessary to construct artificial corneal substitutes. Extracellular matrix components, including collagen and proteoglycans, are potential constituents for the construction of artificial corneas. Collagen type I and a type of glycosaminoglycan called chondroitin sulfate can be blended together and molded into a cornea-like structure in vitro. The presence of chondroitin

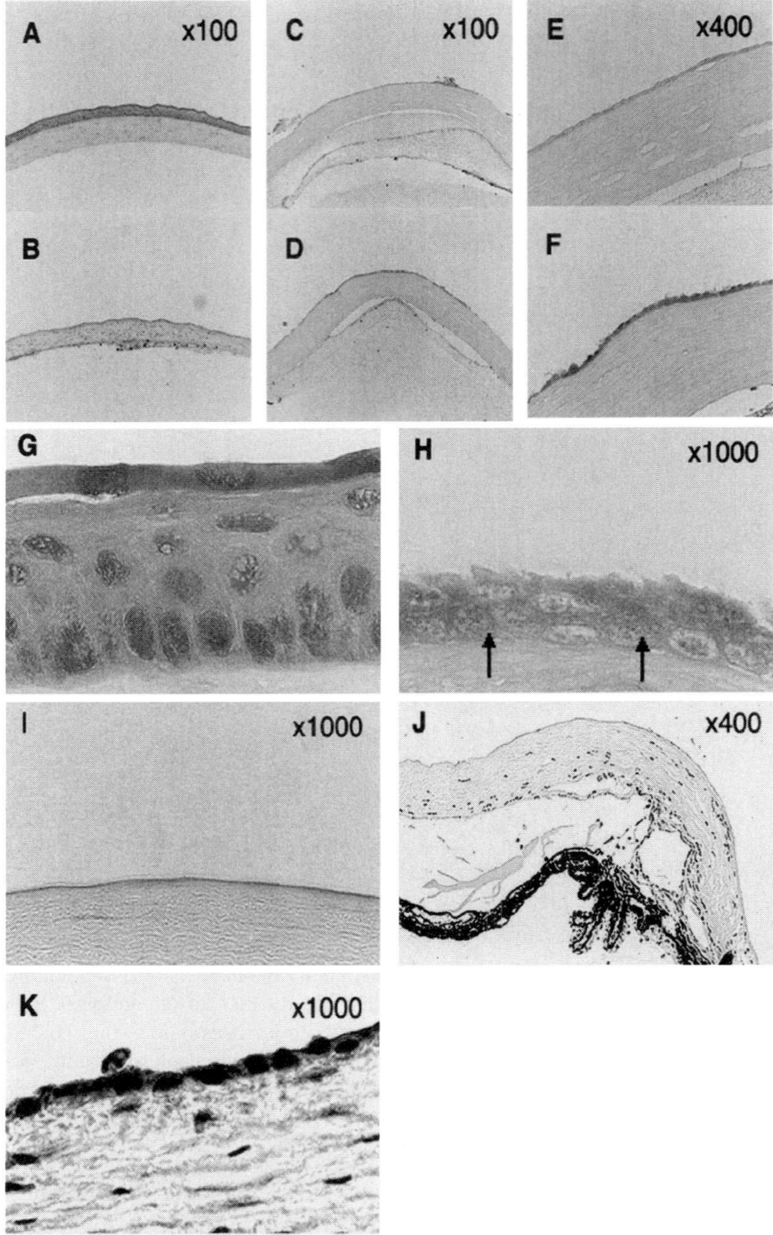

sulfate enhances the transparency of the corneal construct. However, these matrix components cannot be naturally crosslinked to form a structure with sufficient mechanical strength. It is often required to enhance the mechanical strength by adding crosslinking agents, such as glutaraldehyde and formaldehyde. The constructed matrix scaffold can be used for seeding and culturing corneal epithelial cells, keratocytes, and endothelial cells, forming a functional corneal substitute. Preliminary studies have demonstrated that this

Figure 23.7. Histologic analysis of injured cornea, with or without transplantation of the embryonic stem (ES)-cell-derived epithelial progenitor cells. The ES-cell-derived epithelial progenitor cells (day 8 culture) were transplanted to *n*-heptanol-injured cornea of mice. (A) Normal mouse cornea. (B). *n*-Heptanol-injured cornea without transplantation. (C–F). Mouse eyes were injured with *n*-heptanol. At 1 h (C, E) and 12 h (D, F) after transplantation, the eyes were enucleated. Cryostat sections were fixed with 20% formaldehyde in methanol, stained with H&E, and compared with those of normal cornea. (G) Higher magnification of the normal corneal epithelium shown in (A). (H) Higher magnification of another preparation of the ES-cell-derived epithelial progenitor cells at 12 h after transplantation. *Arrows*: the basal or wing-cell–like transplanted cells. (I) Higher magnification of *n*-heptanol-injured cornea without transplantation 12 hours after the injury. No corneal epithelial cells were observed. (J) Limbus of *n*-heptanol-injured cornea without transplantation 24 h after the injury. Migration of the host-originated progenitor cells onto the corneal surface was not observed. (K) Immunostaining for E-cadherin of the corneal epithelial cells 12 h after transplantation of ES-cell-derived graft cells. E-cadherin-positive epithelial cells are stained red. (Reprinted with permission from Homma R et al: Induction of epithelial progenitors in vitro from mouse embryonic stem cells and application for reconstruction of damaged cornea in mice, *Invest Ophthalmol Vis Sci* 45:4320–6, copyright 2004.)

type of corneal scaffold is suitable for the growth and expansion of corneal cells and can be potentially used for corneal reconstruction. However, matrix structures generated by aldehyde-induced cross-linking are not natural and may exert side effects on the ocular system.

Another approach for the construction of corneal substitutes is to integrate extracellular matrix components into a synthetic polymer material, forming a natural and synthetic copolymer material. Scientists have used hydrated collagen and a polymeric material called poly(*N*-isopropylacrylamide–coacrylic acid–coacryloxysuccinimide) to fabricate a copolymer stromal scaffold. A cornea-like structure can be generated by casting a corneal mold with mixed collagen gel and synthetic polymer. Corneal cells can be seeded and grown on the matrix scaffold to establish a functional corneal substitute. Such a matrix scaffold can also be used for nerve innervation. In addition to collagen and glycosaminoglycans, other biological matrix molecules, such as fibrin and fibronectin, have been used for the construction of corneal substitutes. These investigations have demonstrated that corneal substitutes generated with these materials can be potentially used for the reconstruction of malfunctioned cornea.

Glaucoma

Pathogenesis, Pathology, and Clinical Features [23.8]. *Glaucoma* is a disorder characterized by an increase in the pressure of the intraocular aqueous humor. Under physiological conditions, the aqueous humor maintains a narrow range of pressure about 15 mm Hg. When the outflow of the aqueous humor is obstructed, the intraocular pressure increases, resulting in *glaucoma*. Such a disorder is often induced by the abnormality and occlusion of the trabecular meshwork and the canal of Schlemm. Glaucoma is diagnosed when the aqueous humor pressure is increased over 22 mm Hg. Increased aqueous humor pressure in the anterior compartment is transmitted to the posterior compartment via the vitreous humor, resulting in the compression and impairment of the retina and optic nerves. These alterations exert harmful effects on the retinal neurons, eventually leading to blindness. In the United States, glaucoma is the second leading cause of blindness.

Based of on pathogenic mechanisms, glaucoma can be classified into two types: primary glaucoma due to the obstruction of the trabecular meshwork and secondary glaucoma as a complication of other disorders such as leukemia, rheumatoid arthritis (collagen disorder), infectious diseases (rubella and onchocerciasis), amyloidosis, cancer metastases, asthma, emphysema, renal disorders, administration of corticosteroids, chemical toxicity, Marfan's syndrome, and ocular trauma. The pathogenic mechanisms of the primary obstruction of the trabecular meshwork are not fully understood. Glaucoma may be associated with ocular pain and corneal edema. The diagnosis of glaucoma relies on the measurement of intraocular pressure and visual field tests. In severe cases, partial or complete visual loss may occur.

Conventional Treatment of Glaucoma [23.8]. Strategies for glaucoma treatment are to reduce the resistance of the trabecular meshwork to the outflow of the aqueous humor and reduce the intraocular pressure. There are two conventional approaches that can be used to achieve these goals: administration of pharmacological agents and conduction of laser trabeculoplasty. Several types of agents, including cholinergic agonists and β-adrenergic antagonists, have been used to reduce the resistance of the trabecular meshwork. Cholinergic agonists, such as pilocarpine and carbachol, stimulate the contraction of the circumferential smooth muscle cells of the iris. This action shrinks the pupil, decreases the thickness of the iris, and increases the diameter of the canal of Schlemm, thus reducing the resistance to the outflow of the aqueous humor. A treatment with β-adrenergic antagonists, such as timolol maleate, reduces the formation of aqueous humor, thus lowering the intraocular pressure. When these drugs are ineffective, it is necessary to carry out trabeculoplasty, a laser-based surgical procedure that widens the trabecular meshwork and reduces resistance to the outflow of the aqueous humor. Another surgical intervention is to create a fistula from the anterior chamber to the subconjunctival gap, a procedure known as *filtration surgery*. This approach facilitates the outflow of the aqueous humor.

Molecular Regenerative Engineering. There are several strategies for the molecular treatment of glaucoma. These include facilitation of the aqueous humor outflow through the trabecular meshwork, prevention of scar formation and occlusion of surgically created aqueous humor fistula, and protection of retinal neurons from glaucoma-induced injury and death. A number of genes can be used for these purposes. These genes are discussed as follows.

Facilitation of Aqueous Humor Outflow through the Trabecular Meshwork [23.9]. Primary glaucoma is induced by increased resistance of the trabecular meshwork to the outflow of the aqueous humor. Extracellular matrix components, including proteoglycans, in the trabecular meshwork contribute to the resistance. It has been thought that an increase in the production and a decrease in the degradation of proteoglycans may play a role in the induction of primary glaucoma. Extracellular matrix components are degraded by a class of proteinases, known as matrix metalloproteinases (MMPs). Stromelysins (Table 23.4) are a group of matrix metalloproteinases that degrade proteoglycans, several types of collagen, and other matrix components. This group of proteinases includes three known members: stromelysin 1, 2, and 3. Stromelysin 1 and 2, also known as matrix metalloproteinase 3 and 10, respectively, degrade proteoglycans, collagen types III, IV, V, and IX, fibronectin, and laminin. Stromelysin 3, known as *matrix metalloproteinase* 11, degrades primarily fibronectin and laminin. The transfer of stromelysin genes into the cells of the

TABLE 23.4. Characteristics of Stromelysin Isoforms*

Proteins	Alternative Names	Amino Acids	Molecular Weight (kDa)	Locus	Expression	Functions
Stromelysin 1	Matrix metalloproteinase-3 (MMP 3), transin, progelatinase	477	54	11q22.3	Skin, connective tissue, heart	Degrading fibronectin, laminin, collagens III, IV, IX, and X, and proteoglycans
Stromelysin 2	Matrix metalloproteinase-10, MMP10	476	54	11q22.3q23	Heart, lung, liver, kidney	Similar to those of stromelysin 1
Stromelysin 3	Matrix metalloproteinase-11 (MMP11)	488	55	22q11.2	Skin, connective tissue	Degrading fibronectin and laminin

*Based on bibliography 23.9.

trabecular meshwork may upregulate the expression of these proteinases, enhance the degradation of extracellular matrix, and reduce the resistance of the trabecular meshwork. Experimental investigations have provided promising results for the use of these proteinases. Thus, the genes encoding stromelysins are candidate genes for the molecular therapy of human primary glaucoma. Another potential gene is the interleukin-1 gene. Interleukin-1 is known to stimulate the expression of trabecular matrix metalloproteinases. The overexpression of the interleukin-1 gene in the cells of the trabecular meshwork has been shown to enhance the outflow of the aqueous humor in experimental models. For gene transfer into the eye, electroporation has been proven an effective method (Fig. 23.8).

Prevention of the Occlusion of Surgically Created Aqueous Humor Fistula [23.10]. Surgical construction of a fistula through the trabecular meshwork is an effective approach for the treatment of glaucoma. However, surgical trauma induces inflammatory reactions, cell proliferation, extracellular matrix production, and scar formation, ultimately leading to the occlusion and failure of the fistula. To resolve such a problem, it is necessary to apply anti-inflammatory and anti-proliferative agents to the fistula. While pharmacological agents have been used for such a purpose, these agents can be degraded rapidly and it is necessary to conduct daily multiple deliveries. The delivery of therapeutic genes to the surgical site can provide long-term effects.

A large number of proteins have been known to participate in the regulation of inflammatory and proliferative activities. While some proteins, such as growth factors, cell cycle regulatory proteins, and mitogenic protein kinases, enhance inflammatory and proliferative activities, others exert an opposite effect. A typical example of negative regulators is the p21 WAF-1/Cip-1 protein (Table 23.5). This protein induces cell cycle arrest, thus suppressing cell proliferation and production of extracellular matrix. The gene encoding p21 WAF-1/Cip-1 has been tested extensively in experimental glaucoma. The transfer of the p21 WAF-1/Cip-1 gene results in a reduction in inflammatory reactions, cell proliferation, and scar formation in the surgical site of ocular fistula. Genes that encode matrix metalloproteinases can also be used to prevent fistula occlusion. As discussed above,

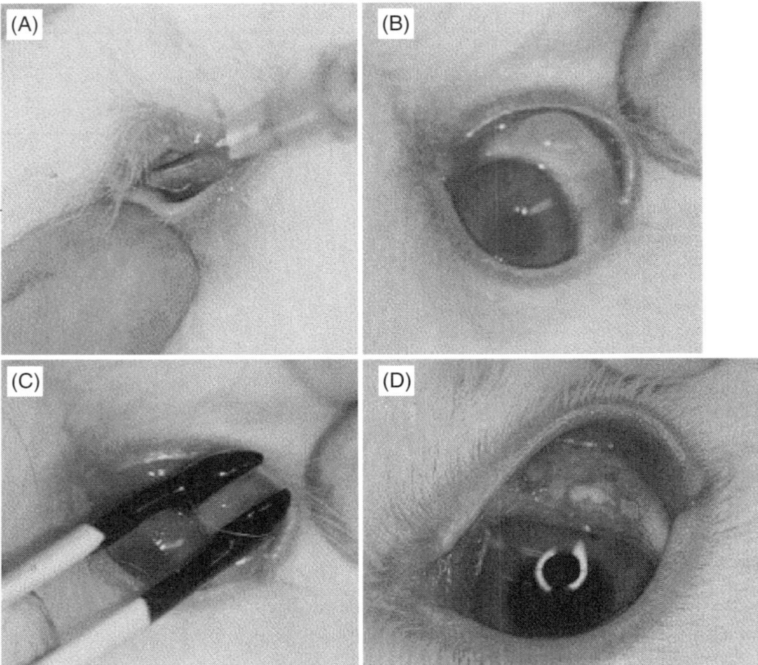

Figure 23.8. Photographic demonstration of surgical procedures of electroporation of MMP-3 cDNA into rabbit conjunctiva followed by trabeculectomy: (A) Injection of 0.1 mL of PBS containing CMV/MMP-3 vector (0.5 mg DNA/mL) into superior conjunctiva with 26 G needle; (B) bleb formation; (C) electroporation-mediated transfection using cup-shaped electrodes; (D) 7 days after trabeculectomy, which was performed 3 days after electroporation. (Reprinted from Mamiya K et al: Effects of matrix metalloproteinase-3 gene transfer by electroporation in glaucoma filter surgery, *Exp Eye Res* 79:405–10, copyright 2004, with permission from Elsevier.)

matrix metalloproteinases degrade extracellular matrix, thus reducing the rate of occlusion of surgically established ocular fistula. The stromelysin genes have been used for such a purpose in experimental models.

Protection of Retinal Neurons from Glaucoma-Induced Injury and Death [23.11]. A glaucoma-associated increase in the intraocular pressure, when reaching a certain level, often induces injury and apoptosis of the retinal neurons, a common cause for blindness. A strategy to prevent retinal neuron injury and death is to deliver genes encoding survival factors or antiapoptotic factors to the retina. Most growth factors are known to promote cell survival and their genes can be used for such a purpose. An example of cell survival factors is the brain-derived neurotrophic factor (BDNF). The transfer of the gene encoding this factor into the retina can effectively protect the retinal neurons from apoptosis and prolong the survival of these cells. In addition, genes encode ciliary neurotrophic factor (CNTF) and glial cell line-derived neurotrophic factor (GDNF) can also be used for such a purpose. These genes have been tested in experimental models, demonstrating promising results. Characteristics of several growth factors are listed in Table 23.6.

TABLE 23.5. Characteristics of p21*

Proteins	Alternative Names	Amino Acids	Molecular Weight (kDa)	Expression	Functions
P21 WAF1/ Cip-1	p21, CDK-interaction protein 1 (CIP1), wildtype p53-activated fragment 1 (WAF1), melanoma-differentiation-associated protein 6 (MDA6), DNA synthesis inhibitor, cyclin-dependent kinase inhibitor 1	164	18	Heart, eye, bone marrow, mammary gland, kidney	Binding to and inhibiting the activity of cyclin-CDK2 or cyclin-CDK4 and inducing cell cycle arrest at G1 and G2

*Based on bibliography 23.10.

Cataract

Pathogenesis, Pathology, and Clinical Features [23.12]. A major lens disorder is *cataract*, which is characterized by progressive opacification of the lens. Lens opacity may be found in the center or the peripheral region of the lens. Pathogenic factors that cause cataract include infectious diseases (such as rubella, herpes simplex, and syphilis), ocular trauma, exposure to radiation, chemical toxicity, diabetes-related metabolic disorders, and aging (senile cataract). Senile cataract occurs much earlier in diabetics than in the general population. Long-term administration of corticosteroids enhances the progression of cataract. Cataract is often associated with visual impairment, such as image blurriness and duplication, alterations in color perception, and reduction in visual acuity.

Conventional Treatment of Cataract [23.12]. When cataract is a secondary disorder due to other diseases, such as infectious diseases, metabolic disorders, and diabetes, it is necessary to treat these causative diseases. The alleviation of these diseases prevents or slows down the progression of cataract. When the transparency of the lens is significantly reduced and the visual acuity is severely impaired, surgical removal of the lens is an effective approach for the treatment of cataract.

Molecular Regenerative Engineering [23.13]. The development of cataract is related to cell proliferation in the lens, often induced by diabetes-induced pathogenic alterations or inflammatory reactions in infectious diseases. An important strategy in molecular therapy for cataract is to deliver genes that encode anti-inflammatory and antiproliferative or proapoptotic proteins. There are two approaches that have been used for the treatment of cataract: transferring genes encoding proteins that activate cell mitosis-inhibiting or pro-apoptotic mechanisms and delivering antisense oligonucleotides or siRNA that inhibit the translation of mitogenic mRNAs.

TABLE 23.6. Characteristics of Selected Growth Factors*

Proteins	Alternative Names	Amino Acids	Molecular Weight (kDa)	Gene Locus	Expression	Functions
Brain-derived neurotrophic factor	BDNF	247	28	11p13	Central nervous system (cortex, retina, and spinal cord), fetal testis	Regulating the survival of neurons and stimulating embryonic development
Ciliary neurotrophic factor	CNTF	200	23	11q12.2	Brain	Promoting neurotransmitter synthesis and neurite outgrowth and regulating the survival of neurons and oligodendrocytes
Glial cell line-derived neurotrophic factor	Astrocyte-derived trophic factor 1, and glial-cell-derived neurotrophic factor	211	24	5p13.1-p12	Nervous system, kidney	Promoting the survival and differentiation of dopaminergic neurons, and preventing the apoptosis of motor neurons

*Based on bibliography 23.11.

For the first approach, the herpes simplex virus-thymidine kinase (HSV-tk) gene is a typical example. This gene encodes a protein kinase that catalyzes the phosphorylation of deoxynucleosides. When a deoxynucleoside analogue such as ganciclovir (an analog of 2′-deoxyguanosine) is present, the analogue can be phosphorylated into a deoxynucleotide, which can be incorporated into the genome of newly formed cells during mitosis. Unlike the natural types of deoxynucleotides, the incorporation of the deoxynucleotide analogues results in the termination of DNA synthesis, an effective process suppressing cell proliferation. Thus the herpes simplex virus-thymidine kinase gene and ganciclovir can be codelivered to the lens cells for the inhibition of cell proliferation. Such an approach prevents or reduces the progression of cataract.

For the second approach, an example is the use of the antisense oligonucleotides specific to the mRNAs of cell cycle regulatory proteins, such as cyclin G1. The selected genes can be injected into the anterior compartment of the eye. Experimental investigations have demonstrated that the transfer of antisense cyclin G1 oligonucleotides into the human lens epithelial cells in vitro induces downregulation of the cyclin G1 gene, an increase in cell apoptosis, and a decrease in cell proliferation. These observations suggest that antisense oligonucleotides to cell cycle regulators or mitogenic factors may be potentially used for the treatment of cataract.

Retinopathy

Pathogenesis, Pathology, and Clinical Features [23.14]. *Retinopathy* is a retinal disorder characterized by retinal arterial stenosis and occlusion, hemorrhage, edema, neuronal injury and apoptosis, reduction in visual acuity, and blindness. Retinopathy occurs due to several diseases, including atherosclerosis in the retinal arteries, systemic hypertension, and diabetes. Atherosclerosis induces stenosis and occlusion of the retinal arteries, leading to injury and death of retinal neurons. Systemic hypertension can induce pathological alterations in the retinal structures, including arteriolar reduction, hemorrhage, retinal edema, and focal ischemia. These alterations influence the function of the retinal neurons, reducing visual acuity. Long-term hypertension may cause severe retinal damage and blindness. Diabetes often induces wall thickening of the retinal arterioles, microaneurysms, hemorrhage, and neovascularization or angiogenesis. These changes result in the injury and death of retinal neurons and thus reduce visual acuity. The ultimate consequence of retinopathy is blindness. Retinopathy is the most common cause of visual loss in the elders.

Another major type of retinal disorder is inherited retinal degeneration. This disorder is characterized by progressive apoptosis of retinal neurons, ultimately leading to blindness. Retinitis pigmentosa is a common form of retinal degeneration. The pathogenic mechanisms of this disorder remain poorly understood. Molecular research has demonstrated that the mutation of several genes may contribute to the development of retinal degeneration. Potential genes include the retinal cyclic GMP phosphodiesterase (PDE) gene, the peripherin 2 gene, and the retinal pigmented epithelium (RPE) 65 gene. In transgenic animal models, the mutation of these genes is associated with retinal degeneration and visual impairment. The retinal cyclic GMP phosphodiesterase gene encodes a protein that regulates phototransduction in the retinal neurons. This protein is composed of catalytic α and β subunits and two inhibitory γ subunits. The mutation of the γ gene results in the loss of the catalytic activity of the α and β units, in association with retinal degeneration similar to that found in human retinitis pigmentosa. The mutation of other subunits also induces retinal degenerative changes. Peripherin 2 is a membrane glycoprotein that is necessary for the formation of the external segment discs of the retinal photoreceptors. The mutation of the gene encoding this protein results in changes leading to retinal degeneration. In addition, the mutation of the RPE65 gene, encoding a protein participating in the regulation of retinoid metabolism, is associated with retinal degeneration. In general, the mutation of the genes that are involved in regulating the function of the visual system often results in visual impairment.

Conventional Treatment of Retinopathy [23.14]. The principle of treating diabetic retinopathy is to control the primary causative disease: diabetes. The alleviation of diabetes can prevent or reduce significantly the progression of retinopathy. When neovascularization is a factor causing retinopathy, photocoagulation can be carried out to reduce the progression of neovascularization. However, only a small fraction of patients are eligible for this treatment, and about half of treated patients experience recurrence of retinopathy. When the retina is severely impaired, there are few conventional approaches available for the treatment of the disorder.

Molecular Regenerative Engineering [23.15]

Molecular Therapy for Diabetic Retinopathy. Diabetic retinopathy is associated with choroid neovascularization, which reduces visual acuity and causes blindness. Such a process is induced and enhanced by the activation of angiogenic factors. Thus, a strategy in molecular therapy for retinopathy is to prevent or suppress the activity of angiogenic factors. Several genes have been used for such a purpose: the angiostatin, endostatin, and pigment epithelium-derived factor (PEDF) genes. Angiostatin is a 38-kDa fragment of plasminogen (number of amino acids 98–440), which exerts an inhibitory effect on the proliferation of vascular endothelial cells and angiogenesis. An in vivo injection of angiostatin into tumor models induces the suppression of tumor growth and angiogenesis. The angiostatin gene has been prepared and transferred into the subretinal space by using an adeno-associated virus vector in animal models with laser injury-induced neovascularization. Such a procedure induces sustained expression of the gene in the chorioretinal tissue for up to several months, and results in a reduction in the degree of neovascularization. Other factors, including endostatin (a fragment of collagen type XVIII) and pigment epithelium-derived factor, also exert an inhibitory effect on endothelial cell proliferation and angiogenesis in tumor tissues. The genes encoding these factors can be delivered into the subretinal space and used to suppress retinal neovascularization. Several angiogenesis-inhibiting proteins are listed in Table 23.7.

Molecular Therapy for Retinal Degeneration [23.16]. Retinal degeneration is induced by mutation of several genes, including the retinal cyclic GMP (see Table 23.8) phosphodiesterase (PDE), peripherin, and Bcl2 genes. Thus the correction of the mutated genes is a potential approach for the treatment of retinal degeneration. The transfer of the retinal cyclic GMP phosphodiesterase β gene into the subretinal space of animal models of retinal degeneration induces sustained expression of the gene for several months, reduces the rate of retinal apoptosis, promotes the survival of retinal neurons, and enhances the function of the retinal neurons. The transfer of the peripherin and Bcl2 genes into the subretinal space results in similar changes.

Another approach for the treatment of retinal degeneration is to transfect the retinal neurons with cell survival-stimulating genes, such as the nerve growth factor (NGF),

TABLE 23.7. Characteristics of Selected Proteins that Inhibit Angiogenesis*

Proteins	Alternative Names	Amino Acids	Molecular Weight (kDa)	Expression	Functions
Plasminogen	Angiostatin, microplasmin	810	91	Liver, kidney, brain	Forming plasmin, a hydrolase capable of converting coagulant fibrin into soluble forms
Collagen type XVIII	COL18A1, endostatin, human type XVIII collagen	1516	154	Liver, kidney, placenta, ovary, heart, skeletal muscle, intestine	Generating a *C*-terminal fragment known as endostatin, which is a potent antiangiogenic factor
Pigment-epithelium-derived factor	PEDF, serine (or cysteine) proteinase inhibitor, serpin peptidase inhibitor	418	46	Retina, cornea, brain, heart, lung, liver, kidney, intestine, ovary, pancreas, prostate gland, bone marrow	Inhibiting angiogenesis and promoting neurite growth

*Based on bibliography 23.15.

TABLE 23.8. Characteristics of Selected Proteins that Regulate the Growth of Retinal Neurons*

Proteins	Alternative Names	Amino Acids	Molecular Weight (kDa)	Expression	Functions
Retinal rod photoreceptor cGMP phosphodiesterase β subunit	Rod cGMP-specific 3′,5′-cyclic phosphodiesterase β subunit	854	98	Retina, brain	Promoting the survival of retinal neurons and preventing retinal neuronal apoptosis
Peripherin		470	54	Nervous system, retina, fetus	A type III intermediate filament protein that regulates nerve cell development, activation, growth, and motility

*Based on bibliography 23.16.

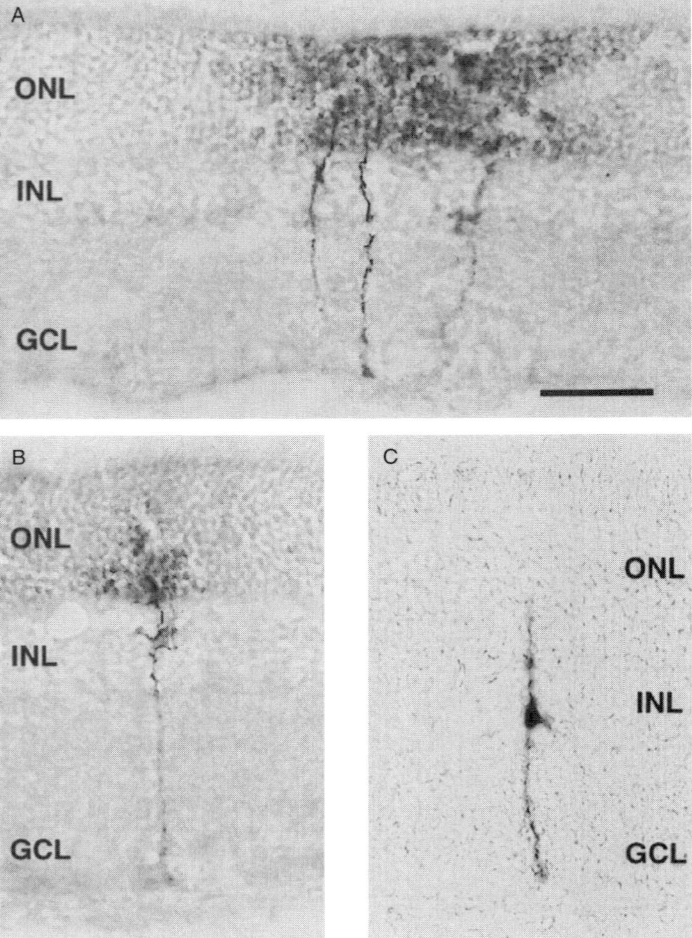

Figure 23.9. Retinal radial cryosections showing typical Müller cell transgene expression at 7 days after intravitreal administration of Ad vectors ($5\,\mu L = 10^7$ plaque-forming units/mL) in the adult rat eye. (A) A group of Müller cells and (B) a single Müller cell transduced in vivo with Ad.BDNF were visualized with an anti-c-myc antibody. (C) A single Müller cell infected with the control virus Ad.LacZ was visualized by 5-bromo-4-chloro-3-indolyl b-D-galactoside staining. Note the diffuse reaction product at the level of the ONL only in cells exposed to Ad.BDNF. INL: inner nuclear layer, GCL: ganglion cell layer. (Scale bar = $50\,\mu m$). (Reprinted with permission from Di Polo A et al: Prolonged delivery of brain-derived neurotrophic factor by adenovirus-infected Muller cells temporarily rescues injured retinal ganglion cells, *Proc Natl Acad Sci USA* 95:3978–83, copyright 1998.)

brain-derived neurotrophic factor (BDNF), and fibroblast growth factor (FGF) genes. The proteins encoded by these genes not only promote cell survival, but also prevent cell apoptosis. Figures 23.9 and 23.10 show the effectiveness of BDNF gene transfer in improving the survival of the retinal ganglion cells.

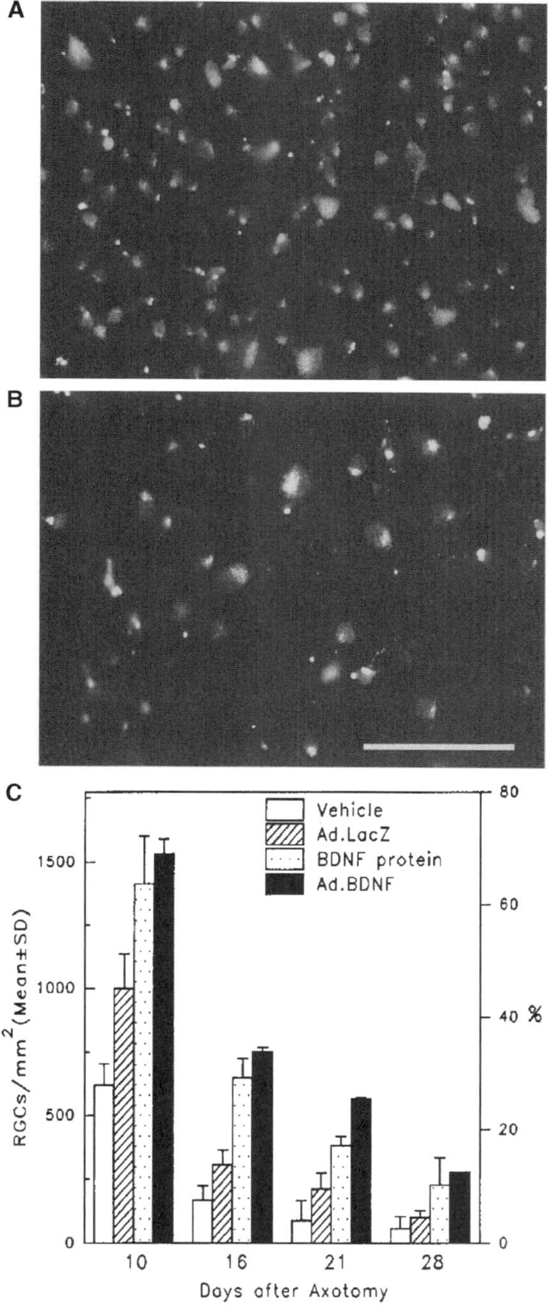

Figure 23.10. Flat-mounted retinas showing Fluorogold-labeled retinal ganglion cells (RGCs) at 10 days after optic nerve transection and intravitreal injection of Ad.BDNF (A) or vehicle (B). Note that BDNF is for brain-derived neurotrophic factor. Scale bar: 100 μm. (C) Quantitative analysis of RGC survival in vivo after axotomy and intravitreal administration of 5 μL of Ad.BDNF, recombinant BDNF, Ad.LacZ, or vehicle (n = 3–8 rats per group). At all times examined, significantly greater numbers of RGCs survived in the retinas treated with Ad.BDNF (solid bars) than in the retinas exposed to Ad.LacZ (hatched bars), or vehicle (open bars) (Student's t-test, $P < 0.001$). RGC densities were similar for the groups of retinas treated with Ad.BDNF (solid bars) or recombinant BDNF (stippled bars) but decreased in all groups at longer times after axotomy. (Reprinted with permission from Di Polo A et al: Prolonged delivery of brain-derived neurotrophic factor by adenovirus-infected Muller cells temporarily rescues injured retinal ganglion cells, *Proc Natl Acad Sci USA* 95:3978–83, copyright 1998.)

BIBLIOGRAPHY

23.1. Anatomy and Physiology of the Visual System

Guyton AC, Hall JE: *Textbook of Medical Physiology*, 11th ed, Saunders, Philadelphia, 2006.

McArdle WD, Katch FI, Katch VL: *Essentials of Exercise Physiology*, 3rd ed, Lippincott Williams & Wilkins, Baltimore, 2006.

Germann WJ, Stanfield CL (with contributors Niles MJ, Cannon JG), *Principles of Human Physiology*, 2nd ed, Pearson Benjamin Cummings, San Francisco, 2005.

Thibodeau GA, Patton KT: *Anatomy & Physiology*, 5th ed, Mosby, St Louis, 2003.

Boron WF, Boulpaep EL: *Medical Physiology: A Cellular and Molecular Approach*, Saunders, Philadelphia, 2003.

Ganong WF: *Review of Medical Physiology*, 21st ed, McGraw-Hill, New York, 2003.

Gallar J, Pozo MA, Tuckett RP, Belmonte C: Response of sensory units with unmyelinated fibres to mechanical, thermal and chemical stimulation of the cat's cornea, *J Physiol* 468:609–22, 1993.

Pflugfelder SC: Tear fluid influence on the ocular surface, *Adv Exp Med Biol* 438:611–7, 1998.

Stern ME, Beuerman RW, Fox RI, Gao J, Mircheff AK et al: A unified theory of the role of the ocular surface in dry eye, *Adv Exp Med Biol* 438:643–51, 1998.

Gilbard JP, Rossi SR: Tear film and ocular surface changes in a rabbit model of neurotrophic keratitis, *Ophthalmolog.* 97(3):308–12, 1990.

23.2. Pathogenesis, Pathology, and Clinical Features of Cornea Injury

Schneider AS, Szanto PA: *Pathology*, 3rd ed, Lippincott Williams & Wilkins, Philadelphia, 2006.

McCance KL, Huether SE: *Pathophysiology: The Biologic Basis for Disease in Adults & Children*, 5th ed, Elsevier Mosby, St Louis, 2006.

Porth CM: *Pathophysiology: Concepts of Altered Health States*, 7th ed, Lippincott Williams & Wilkins, Philadelphia, 2005.

Frazier MS, Drzymkowski JW: *Essentials of Human Diseases and Conditions*, 3rd ed, Elsevier Saunders, St Louis, 2004.

Netto MV, Mohan RR, Ambrosio R Jr, Hutcheon AE, Zieske JD et al: Wound healing in the cornea: a review of refractive surgery complications and new prospects for therapy, *Cornea* 24(5):509–22, 2005.

Kuo IC: Corneal wound healing, *Curr Opin Ophthalmol* 15(4):311–5, 2004.

Belmonte C, Acosta MC, Gallar J: Neural basis of sensation in intact and injured corneas, *Exp Eye Res* 78(3):513–25, 2004.

Khaw PT, Shah P, Elkington AR: Injury to the eye, *Br Med J* 328(7430):36–8, 2004.

Bourcier T, Thomas F, Borderie V, Chaumeil C, Laroche L: Bacterial keratitis: Predisposing factors, clinical and microbiological review of 300 cases, *Br J Ophthalmol* 87(7):834–8, 2003.

Wilson SE, Mohan RR, Hong J, Lee J, Choi R et al: Apoptosis in the cornea in response to epithelial injury: Significance to wound healing and dry eye, *Adv Exp Med Biol* 506(Pt B):821–6, 2002.

23.3. Molecular Therapies for Corneal Immune Rejection

CD28

Oral HB, Larkin DF, Fehervari Z, Byrnes AP, Rankin AM et al: Ex vivo adenovirus-mediated gene transfer and immunomodulatory protein production in human cornea, *Gene Ther* 4(7):639–47, 1997.

Andres PG, Howland KC, Nirula A, Kane LP, Barron L et al: Distinct regions in the CD28 cytoplasmic domain are required for T helper type 2 differentiation, *Nature Immun* 5:435–42, 2004.

Aruffo A, Seed B: Molecular cloning of a CD28 cDNA by a high-efficiency COS cell expression system, *Proc Natl Acad Sci USA* 84:8573–7, 1987.

Lee KP, Taylor C, Petryniak B, Turka LA, June CH et al: The genomic organization of the CD28 gene: Implications for the regulation of CD28 mRNA expression and heterogeneity, *J Immun* 145:344–52, 1990.

Okkenhaug K, Wu L, Garza KM, La Rose J, Khoo W et al: A point mutation in CD28 distinguishes proliferative signals from survival signals, *Nature Immun* 2:325–32, 2001.

Human protein reference data base, Johns Hopkins University and the Institute of Bioinformatics, at http://www.hprd.org/protein.

CD80

Freeman GJ, Disteche CM, Gribben JG, Adler DA, Freedman AS et al: The gene for B7, a costimulatory signal for T-cell activation, maps to chromosomal region 3q13.3–3q21, *Blood* 79:489–94, 1992.

Reeves RH, Patch D, Sharpe AH, Borriello F, Freeman GJ et al: The costimulatory genes Cd80 and Cd86 are linked on mouse chromosome 16 and human chromosome 3, *Mam Genome* 8:581–2, 1997.

Reiser J, von Gersdorff G, Loos M, Oh J, Asanuma K et al: Induction of B7-1 in podocytes is associated with nephrotic syndrome, *J Clin Invest* 113:1390–7, 2004.

Selvakumar A, Mohanraj BK, Eddy RL, Shows TB, White PC et al: Genomic organization and chromosomal location of the human gene encoding the B-lymphocyte activation antigen B7, *Immunogenetics* 36:175–81, 1992.

Stamper CC, Zhang Y, Tobin JF, Erbe DV, Ikemizu S et al: Crystal structure of the B7–1/CTLA-4 complex that inhibits human immune responses, *Nature* 410:608–11, 2001.

CD86

Celestin J, Rotschke O, Falk K, Ramesh N, Jabara H et al: IL-3 induces B7.2 (CD86) expression and costimulatory activity in human eosinophils, *J Immun* 167:6097–104, 2001.

Freeman GJ, Gribben JG, Boussiotis VA, Ng JW, Restivo VA et al: Cloning of B7-2:a CTLA-4 counter-receptor that costimulates human T cell proliferation. Science 262:909–11, 1993.

Jeannin P, Magistrelli G, Aubry JP, Caron G, Gauchat JF et al: Soluble CD86 is a costimulatory molecule for human T lymphocytes, *Immunity* 13:303–12, 2000.

Jellis CL, Wang SS, Rennert P, Borriello F, Sharpe AH et al: Genomic organization of the gene coding for the costimulatory human B-lymphocyte antigen B7-2 (CD86), *Immunogenetics* 42:85–9, 1995.

Schwartz JCD, Zhang X, Fedorov AA, Nathenson SG, Almo SC: Structural basis for co-stimulation by the human CTLA-4/B7-2 complex, *Nature* 410:604–8, 2001.

Shah R, Banks K, Patel A, Dogra S, Terrell R et al: Intense expression of the B7-2 antigen presentation coactivator is an unfavorable prognostic indicator for differentiated thyroid carcinoma of children and adolescents, *J Clin Endocr Metab* 87:4391–7, 2002.

CD152

Allison JP, Krummel MF: The yin and yang of T cell costimulation, *Science* 270:932–3, 1995.

Bour-Jordan H, Grogan JL, Tang Q, Auger JA, Locksley RM et al: CTLA-4 regulates the requirement for cytokine-induced signals in T(H)2 lineage commitment, *Nature Immun* 4:182–8, 2003.

Dariavach P, Mattei MG, Golstein P, Lefranc MP: Human Ig superfamily CTLA-4 gene: Chromosomal localization and identity of protein sequence between murine and human CTLA-4 cytoplasmic domains, *Eur J Immun* 18:1901–5, 1988.

Gozalo-Sanmillan S, McNally JM, Lin MY, Chambers CA, Berg LJ: Cutting edge: two distinct mechanisms lead to impaired T cell homeostasis in Janus kinase 3- and CTLA-4-deficient mice, *J Immun* 166:727–30, 2001.

Harper K, Balzano C, Rouvier E, Mattei MG, Luciani MF et al: PCTLA-4 and CD28 activated lymphocyte molecules are closely related in both mouse and human as to sequence, message expression, gene structure, and chromosomal location, *J Immun* 147:1037–44, 1991.

Ling V, Wu PW, Finnerty HF, Sharpe AH, Gray GS et al: Complete sequence determination of the mouse and human CTLA4 gene loci: Cross-species DNA sequence similarity beyond exon borders, *Genomics* 60:341–55, 1999.

Linsley PS, Nadler SG, Bajorath J, Peach R, Leung HT et al: Binding stoichiometry of the cytotoxic T lymphocyte-associated molecule-4 (CTLA-4): A disulfide-linked homodimer binds two CD86 molecules, *J Biol Chem* 270:15417–24, 1995.

Lohmueller KE, Pearce CL, Pike M, Lander ES, Hirschhorn JN: Meta-analysis of genetic association studies supports a contribution of common variants to susceptibility to common disease, *Nature Genet* 33:177–82, 2003.

Ostrov DA, Shi W, Schwartz JCD, Almo SC, Nathenson SG: Structure of murine CTLA-4 and its role in modulating T cell responsiveness, *Science* 290:816–9, 2000.

Reiser H, Stadecker MJ: Costimulatory B7 molecules in the pathogenesis of infectious and autoimmune diseases. *New Engl J Med* 335:1369–77, 1996.

Sayegh MH, Turka LA: The role of T-cell costimulatory activation pathways in transplant rejection, *New Engl J Med* 338:1813–21, 1998.

Schwartz JCD, Zhang X, Fedorov AA, Nathenson SG, Almo SC: Structural basis for co-stimulation by the human CTLA-4/B7-2 complex, *Nature* 410:604–8, 2001.

Shahinian A, Pfeffer K, Lee KP, Kundig TM, Kishihara K et al: Differential T cell costimulatory requirements in CD28-deficient mice, *Science* 261:609–12, 1993.

Stamper CC, Zhang Y, Tobin JF, Erbe DV, Ikemizu S et al: Crystal structure of the B7-1/CTLA-4 complex that inhibits human immune responses, *Nature* 410:608–11, 2001.

Ueda H, Howson JMM, Esposito L, Heward J, Snook H et al: Association of the T-cell regulatory gene CTLA4 with susceptibility to autoimmune disease, *Nature* 423:506–11, 2003.

van Belzen MJ, Mulder CJJ, Zhernakova A, Pearson PL, Houwen RHJ et al: Lymphoproliferative disorders with early lethality in mice deficient in Ctla-4, *Science* 270:985–8, 1995.

Price FW Jr, Whitson WE, Collins KS, Marks RG: Five year corneal graft survival: A large, single-center patient cohort, *Arch Ophthalmol* 111:799–805, 1993.

Williams KA, Muehlberg SM, Lewis RF, Coster DJ: How successful is corneal transplantation? A report from The Australian Corneal Graft Register, *Eye* 9:219–27, 1995.

Marrack P, Kappler J: The T cell receptor, *Science* 238:1073–9, 1987.

Linsley PS, Brady W, Grosmaire L, Aruffo A, Damle NK et al: Binding of the B cell activation antigen B7 to CD28 costimulates T cell proliferation and interleukin 2 mRNA accumulation, *J Exp Med* 173:721–30, 1991.

Walunas TL, Lenschow DJ, Bakker CY et al: CTLA-4 can function as a negative regulator of T cell activation, *Immunity* 1:405–13, 1994.

Krummel MF, Allison JP: CD28 and CTLA-4 deliver opposing signals which regulate the response of T cells to stimulation, *J Exp Med* 182:459–65, 1995.

Schwartz RH: A cell culture model for T lymphocyte clonal anergy, *Science* 248:1349–56, 1990.

Wallace PM, Rodgers JN, Leytze GM, Johnson JS, Linsley PS: Induction and reversal of long-lived specific unresponsiveness to a T-dependent antigen following CTLA4Ig treatment, *J Immunol* 154:5885–95, 1995.

Steurer W, Nickerson PW, Steele AW, Steiger J, Zheng XX, Strom TB: Ex vivo coating of islet cell allografts with murine CTLA4/Fc promotes graft tolerance, *J Immunol* 155:1165–74, 1995.

Hoffman F, Zhang EP, Wachtlin J: Inhibition of corneal allograft reaction by CTLA4-Ig Graefes, *Arch Clin Exp Ophthalmol* 235:535–40, 1997.

Gebhardt BM, Hodkin M, Varnell ED, Kaufman HE: Protection of corneal allografts by CTLA-4 Ig, *Cornea* 18:314–20, 1999.

Oshima Y, Sakamoto T, Hisatomi T, Tsutsumi C, Sassa Y et al: Targeted gene transfer to corneal stroma in vivo by electric pulses, *Exp Eye Res* 74:191–8, 2002.

Bertelmann E, Ritter T, Vogt K, Reszka R, Hartmann C et al: Efficiency of cytokine gene transfer in corneal endothelial cells and organ-cultured corneas mediated by liposomal vehicles and recombinant adenovirus, *Ophthalm Res* 35(2):117–24, 2003.

Klebe S, Sykes PJ, Coster DJ, Krishnan R, Williams KA: Prolongation of sheep corneal allograft survival by ex vivo transfer of the gene encoding interleukin-10, *Transplantation* 71(9):1214–20, 2001.

Tan PH, King WJ, Chen D, Awad HM, Mackett M et al: Transferrin receptor-mediated gene transfer to the corneal endothelium, *Transplantation* 71(4):552–60, 2001.

23.4. Molecular Therapies for Corneal Inflammation and Fibrosis

Bates S, Rowan S, Vousden KH: Characterisation of human cyclin G1 and G2: DNA damage inducible genes, *Oncogene* 13:1103–9, 1996.

Endo Y, Fujita T, Tamura K, Tsuruga H, Nojima H: Structure and chromosomal assignment of the human cyclin G gene, *Genomics* 38:92–5, 1996.

Horne MC, Goolsby GL, Donaldson KL, Tran D, Neubauer M et al: Cyclin G1 and cyclin G2 comprise a new family of cyclins with contrasting tissue-specific and cell cycle-regulated expression, *J Biol Chem* 271:6050–61, 1996.

Jensen MR, Factor VM, Zimonjic DB, Miller MJ, Keck CL et al: Chromosome localization and structure of the murine cyclin G1 gene promoter sequence, *Genomics* 45:297–303, 1997.

Behrens A, Gordon EM, Li L, Liu PX, Chen Z et al: Retroviral gene therapy vectors for prevention of excimer laser-induced corneal haze, *Invest Ophthalmol Vis Sci* 43:968–77, 2002.

Human protein reference data base, Johns Hopkins University and the Institute of Bioinformatics, at http://www.hprd.org/protein.

23.5. Molecular Therapies for Corneal Complications Due to Mucopolysaccharidosis Type VII (MPS VII)

Birkenmeier EH, Barker JE, Vogler CA, Kyle JW, Sly WS et al: Increased life span and correction of metabolic defects in murine mucopolysaccharidosis type VII after syngeneic bone marrow transplantation, *Blood* 78:3081–92, 1991.

Birkenmeier EH, Davisson MT, Beamer WG, Ganschow RE, Vogler CA et al: Murine mucopolysaccharidosis type VII: Characterization of a mouse with beta-glucuronidase deficiency, *J Clin Invest* 83:1258–66, 1989.

Fyfe JC, Kurzhals RL, Lassaline ME, Henthorn PS, Alur PRK et al: Molecular basis of feline beta-glucuronidase deficiency: An animal model of mucopolysaccharidosis VII, *Genomics* 58:121–8, 1999.

Haskins ME, Aguirre GD, Jezyk PF, Schuchman EH, Desnick RJ et al: Mucopolysaccharidosis type VII (Sly syndrome): beta-glucuronidase-deficient mucopolysaccharidosis in the dog, *Am J Pathol* 138:1553–5, 1991.

Heuer GG, Passini MA, Jiang K, Parente MK, Lee VMY et al: Selective neurodegeneration in murine mucopolysaccharidosis VII is progressive and reversible, *Ann Neurol* 52:762–70, 2002.

Kyle JW, Birkenmeier EH, Gwynn B, Vogler C, Hoppe PC et al: Correction of murine mucopolysaccharidosis VII by a human beta-glucuronidase transgene, *Proc Natl Acad Sci USA* 87:3914–8, 1990.

LeBowitz JH, Grubb JH, Maga JA, Schmiel DH, Vogler C et al: Glycosylation-independent targeting enhances enzyme delivery to lysosomes and decreases storage in mucopolysaccharidosis type VII mice, *Proc Natl Acad Sci USA* 101:3083–8, 2004.

Li T, Davidson BL: Phenotype correction in retinal pigment epithelium in murine mucopolysaccharidosis VII by adenovirus-mediated gene transfer, *Proc Natl Acad Sci USA* 92:7700–4, 1995.

Miller RD, Hoffmann JW, Powell PP, Kyle JW, Shipley JM et al: Cloning and characterization of the human beta-glucuronidase gene, *Genomics* 7:280–3, 1990.

Moullier P, Bohl D, Heard JM, Danos O: Correction of lysosomal storage in the liver and spleen of MPS VII mice by implantation of genetically modified skin fibroblasts, *Nature Genet* 4:154–9, 1993.

Ponder KP, Melniczek JR, Xu L, Weil MA, O'Malley TM et al: Therapeutic neonatal hepatic gene therapy in mucopolysaccharidosis VII dogs, *Proc Natl Acad Sci USA* 99:13102–7, 2002.

Ray J, Bouvet A, DeSanto C, Fyfe JC, Xu D et al: Cloning of the canine beta-glucuronidase cDNA mutation identification in canine MPS VII, and retroviral vector-mediated correction of MPS VII cells, *Genomics* 48:248–53, 1998.

Sands MS, Birkenmeier EH: A single-base-pair deletion in the beta-glucuronidase gene accounts for the phenotype of murine mucopolysaccharidosis type VII, *Proc Natl Acad Sci USA* 90:6567–71, 1993.

Sands MS, Vogler C, Kyle JW, Grubb JH, Levy B et al: Enzyme replacement therapy for murine mucopolysaccharidosis type VII, *J Clin Invest* 93:2324–31, 1994.

Sands MS, Vogler C, Torrey A, Levy B, Gwynn B et al: Murine mucopolysaccharidosis type VII: Long term therapeutic effects of enzyme replacement and enzyme replacement followed by bone marrow transplantation, *J Clin Invest* 99:1596–605, 1997.

Sly WS, Vogler C, Grubb JH, Zhou M, Jiang J et al: Active site mutant transgene confers tolerance to human beta-glucuronidase without affecting the phenotype of MPS VII mice, *Proc Natl Acad Sci USA* 98:2205–10, 2001.

Wu BM, Tomatsu S, Fukuda S, Sukegawa K, Orii T et al: Overexpression rescues the mutant phenotype of L176F mutation causing beta-glucuronidase deficiency mucopolysaccharidosis in two Mennonite siblings, *J Biol Chem* 269:23681–8, 1994.

Kamata Y, Okuyama T, Kosuga M, O'hira A, Kanaji A et al: Adenovirus-mediated gene therapy for corneal clouding in mice with mucopolysaccharidosis type VII, *Mol Ther* 4:307–12, 2001.

Human protein reference data base, Johns Hopkins University and the Institute of Bioinformatics, at http://www.hprd.org/protein.

23.6. Corneal Surface Reconstruction

Nishida K, Yamato M, Hayashida Y, Watanabe K, Yamamoto K et al: Corneal reconstruction with tissue-engineered cell sheets composed of autologous oral mucosal epithelium, *New Engl J Med* 351:1187–96, 2004.

23.7. Corneal Reconstruction

Nishida K, Yamato M, Hayashida Y, Watanabe K, Yamamoto K et al: Corneal reconstruction with tissue-engineered cell sheets composed of autologous oral mucosal epithelium, *New Engl J Med* 351(12):1187–96, 2004.

Homma R, Yoshikawa H, Takeno M, Kurokawa MS, Masuda C et al: Induction of epithelial progenitors in vitro from mouse embryonic stem cells and application for reconstruction of damaged cornea in mice, *Invest Ophthalmol Vis Sci* 45(12):4320–6, 2004.

Ramaesh K, Dhillon B: Ex vivo expansion of corneal limbal epithelial/stem cells for corneal surface reconstruction, *Eur J Ophthalmol* 13(6):515–24, 2003.

Suuronen EJ, Nakamura M, Watsky MA, Stys PK, Muller LJ et al: Innervated human corneal equivalents as in vitro models for nerve-target cell interactions, *FASEB J* 18:170–2, 2004.

Engelmann K, Bednarz J, Valtink M: Prospects for endothelial transplantation, *Exp Eye Res* 78:573–8, 2004.

Kinoshita S, Koizumi N, Nakamura T: Transplantable cultivated mucosal epithelial sheet for ocular surface reconstruction, *Exp Eye Res* 78:483–91, 2004.

Borene ML, Barocas VH, Hubel A: Mechanical and cellular changes during compaction of a collagen-sponge-based corneal stromal equivalent, *Ann Biomed Eng* 32:274–83, 2004.

Nishida K, Yamato M, Hayashida Y, Watanabe K, Maeda N et al: Functional bioengineered corneal epithelial sheet grafts from corneal stem cells expanded ex vivo on a temperature-responsive cell culture surface, *Transplantation* 77:379–85, 2004.

Kinoshita S, Nakamura T: Development of cultivated mucosal epithelial sheet transplantation for ocular surface reconstruction, *Artif Organs* 28:22–7, 2004.

Nishida K: Tissue engineering of the cornea, *Cornea* 22(Suppl 7):S28–34, 2003.

Doillon CJ, Watsky MA, Hakim M, Wang J, Munger R et al: A collagen-based scaffold for a tissue engineered human cornea: Physical and physiological properties, *Int J Artif Organs* 26:764–73, 2003.

Shimmura S, Tsubota K: Ocular surface reconstruction update, *Curr Opin Ophthalmol* 13:213–9, 2002.

Han B, Schwab IR, Madsen TK, Isseroff RR: A fibrin-based bioengineered ocular surface with human corneal epithelial stem cells, *Cornea* 21:505–10, 2002.

Li F, Carlsson D, Lohmann C, Suuronen E, Vascotto S et al: Cellular and nerve regeneration within a biosynthetic extracellular matrix for corneal transplantation, *PNAS* 100:15346–51, 2003.

23.8. Pathogenesis Pathology and Clinical Features of Glaucoma

Morrison JC: Elevated intraocular pressure and optic nerve injury models in the rat, *J Glaucoma* 14(4):315–7, 2005.

Schwartz M, Yoles E: Neuroprotection: A new treatment modality for glaucoma? *Curr Opin Ophthalmol* 11(2):107–11, 2000.

Osborne NN, Chidlow G, Nash MS, Wood JP: The potential of neuroprotection in glaucoma treatment, *Curr Opin Ophthalmol* 10(2):82–92, 1999.

Tan JC, Peters DM, Kaufman PL: Recent developments in understanding the pathophysiology of elevated intraocular pressure, *Curr Opin Ophthalmol* 17(2):168–74, April 2006.

Chintala SK: The emerging role of proteases in retinal ganglion cell death, *Exp Eye Res* 82(1):5–12, 2006.

Osborne NN, Chidlow G, Layton CJ, Wood JP, Casson RJ et al: Optic nerve and neuroprotection strategies, *Eye* 18(11):1075–84, 2004.

Hoh ST, Aung T, Chew PT: Medical management of angle closure glaucoma, *Semin Ophthalmol* 17(2):79–83, 2002.

Civan MM, Macknight AD: The ins and outs of aqueous humour secretion, *Exp Eye Res* 78(3):625–31, 2004.

Weisschuh N, Schiefer U: Progress in the genetics of glaucoma, *Dev Ophthalmol* 37:83–93, 2003.

Clark AF, Yorio T: Ophthalmic drug discovery, *Nat Rev Drug Discov* 2(6):448–59, 2003.

23.9. Facilitation of Aqueous Humor Outflow through the Trabecular Meshwork

D'Souza CA, Mak B, Moscarello MA: The up-regulation of stromelysin-1 (MMP-3) in a spontaneously demyelinating transgenic mouse precedes onset of disease, *J Biol Chem* 277:13589–96, 2002.

Formstone CJ, Byrd PJ, Ambrose HJ, Riley JH, Hernandez D et al: The order and orientation of a cluster of metalloproteinase genes, stromelysin 2, collagenase and stromelysin, together with D11S385, on chromosome 11q22–q23, *Genomics* 16:289–91, 1993.

Imai K, Yokohama Y, Nakanishi I, Ohuchi E, Fujii Y et al: Matrix metalloproteinase 7 (matrilysin) from human rectal carcinoma cells: Activation of the precursor, interaction with other matrix metalloproteinases and enzymic properties, *J Biol Chem* 270:6691–7, 1995.

Lu PCS, Ye H, Maeda M, Azar DT: Immunolocalization and gene expression of matrilysin during corneal wound healing, *Invest Ophthalm Vis Sci* 40:20–7, 1999.

Medley TL, Kingwell BA, Gatzka CD, Pillay P, Cole TJ: Matrix metalloproteinase-3 genotype contributes to age-related aortic stiffening through modulation of gene and protein expression, *Circ Res* 92:1254–61, 2003.

Pendas AM, Santamaria I, Alvarez MV, Pritchard M, Lopez-Otin C: Fine physical mapping of the human matrix metalloproteinase genes clustered on chromosome 11q22.3, *Genomics* 37:266–9, 1996.

Radisky DC, Levy DD, Littlepage LE, Liu H, Nelson CM et al: Rac1b and reactive oxygen species mediate MMP-3-induced EMT and genomic instability, *Nature* 436:123–7, 2005.

Saarialho-Kere UK, Chang ES, Welgus HG, Parks WC: Distinct localization of collagenase and tissue inhibitor of metalloproteinases: Expression in wound healing associated with ulcerative pyogenic granuloma, *J Clin Invest* 90:1952–7, 1992.

Saarialho-Kere UK, Pentland AP, Birkedal-Hansen H, Parks WC, Welgus HG: Distinct populations of basal keratinocytes express stromelysin-1 and stromelysin-2 in chronic wounds, *J Clin Invest* 94:79–88, 1994.

Saus J, Quinones S, Otani Y, Nagase H, Harris ED Jr, Kurkinen M: The complete primary structure of human matrix metalloproteinase-3: Identity with stromelysin, *J Biol Chem* 263:6742–5, 1988.

Sternlicht MD, Lochter A, Sympson CJ, Huey B, Rougier JP et al: The stromal proteinase MMP3/stromelysin-1 promotes mammary carcinogenesis, *Cell* 98:137–46, 1999.

Wilhelm SM, Collier IE, Kronberger A, Eisen AZ, Marmer BL et al: Human skin fibroblast stromelysin: structure glycosylation substrate specificity, and differential expression in normal and tumorigenic cells, *Proc Natl Acad Sci USA* 84:6725–9, 1987.

Yamada Y, Izawa H, Ichihara S, Takatsu F, Ishihara H et al: Prediction of the risk of myocardial infarction from polymorphisms in candidate genes, *New Engl J Med* 347:1916–23, 2002.

Yè S, Eriksson P, Hamsten A, Kurkinen M, Humphries SE et al: Progression of coronary atherosclerosis is associated with a common genetic variant of the human stromelysin-1 promoter which results in reduced gene expression, *J Biol Chem* 271:13055–60, 1996.

Levy A, Zucman J, Delattre O, Mattei MG, Rio MC et al: Assignment of the human stromelysin 3 (STMY3) gene to the q11.2 region of chromosome 22, *Genomics* 13:881–3, 1992.

Matrisian LM: Metalloproteinases and their inhibitors in matrix remodeling, *Trends Genet* 6:121–5, 1990.

Pei D, Weiss SJ: Furin-dependent intracellular activation of the human stromelysin-3 zymogen, *Nature* 375:244–7, 1995.

Kee C, Sohn S, Hwang JM: Stromelysin gene transfer into cultured human trabecular cells and rat trabecular meshwork in vivo, *Invest Ophthalm Vis Sci* 42:2856–60, 2001.

Bradley JM, Vranka J, Colvis CM, Conger DM, Alexander JP et al: Effect of matrix metalloproteinases activity on outflow in perfused human organ culture, *Invest Ophthalmolm Vis Sci* 39:2649–58, 1998.

Lütjen-Drecoll E, Rohen JW: Pathology of the trabecular meshwork in primary open-angle glaucoma, *J Glaucoma* 8:37–9, 1994.

Samples JR, Alexander JP, Acott TS: Regulation of the levels of human trabecular matrix metalloproteinases and inhibitor by interleukin-1 and dexamethasone, *Invest Ophthalmol Vis Sci* 34:3386–95, 1993.

Woessner JF Jr: Matrix metalloproteinases and the inhibitors in connective tissue remodeling, *FASEB J* 5:2145–54, 1991.

Matrisian LM: The matrix-degrading metalloproteinases, *Bioassays* 14:455–63, 1992.

Parshley DE, Bradley JBM, Fisk A et al: Laser trabeculoplasty induced stromelysin expression by trabecular juxtacanalicular cells, *Invest Ophthalm Vis Sci* 37:795–804, 1996.

Kee C, Seo K: The effect of interleukin-1α on outflow facility in rat eyes, *J Glaucoma* 6:246–9, 1997.

Human protein reference data base, Johns Hopkins University and the Institute of Bioinformatics, at http://www.hprd.org/protein.

23.10. Prevention of Occlusion of Surgically Created Aqueous Humor Fistula

Johnson KT, Rodicker F, Heise K, Heinz C, Steuhl KP et al: Adenoviral p53 gene transfer inhibits human Tenon's capsule fibroblast proliferation, *Br J Ophthalmol* 89(4):508–12, 2005.

Wen SF, Chen Z, Nery J, Faha B: Characterization of adenovirus p21 gene transfer, biodistribution, and immune response after local ocular delivery in New Zealand white rabbits, *Exp Eye Res* 77(3):355–65, 2003.

Bunz F, Dutriaux A, Lengauer C, Waldman T, Zhou S et al: Requirement for p53 and p21 to sustain G2 arrest after DNA damage, *Science* 282:1497–501, 1998.

Carreira S, Goodall J, Aksan I, La Rocca SA, Galibert MD et al: Mitf cooperates with Rb1 and activates p21(Cip1) expression to regulate cell cycle progression, *Nature* 433:764–9, 2005.

Cheng T, Rodrigues N, Shen H, Yang Y, Dombkowski D et al: Hematopoietic stem cell quiescence maintained by p21(cip1/waf1), *Science* 287:1804–8, 2000.

Demetrick DJ, Matsumoto S, Hannon GJ, Okamoto K, Xiong Y et al: Chromosomal mapping of the genes for the human cell cycle proteins cyclin C (CCNC), cyclin E (CCNE), p21 (CDKN1) and KAP (CDKN3), *Cytogenet Cell Genet* 69:190–2, 1995.

El-Deiry WS, Tokino T, Velculescu VE, Levy DB, Parsons R et al: WAF1, a potential mediator of p53 tumor suppression, *Cell* 75:817–25, 1993.

Harper JW, Adami GR, Wei N, Keyomarsi K, Elledge SJ: The p21 Cdk-interacting protein Cip1 is a potent inhibitor of G1 cyclin-dependent kinases, *Cell* 75:805–16, 1993.

Huppi K, Siwarski D, Dosik J, Michieli P, Chedid M et al: Molecular cloning, sequencing, chromosomal localization and expression of mouse p21 (Waf1), *Oncogene* 9:3017–20, 1994.

Megyesi J, Price PM, Tamayo E, Safirstein RL: The lack of a functional p21(WAF1/CIP1) gene ameliorates progression to chronic renal failure, *Proc Natl Acad Sci USA* 96:10830–5, 1999.

Seoane J, Le HV, Massague J: Myc suppression of the p21(Cip1) Cdk inhibitor influences the outcome of the p53 response to DNA damage, *Nature* 419:729–34, 2002.

Mamiya K, Ohguro H, Ohguro I, Metoki T, Ishikawa F et al: Effects of matrix metalloproteinase-3 gene transfer by electroporation in glaucoma filter surgery, *Exp Eye Res* 79:405–10, 2004.

Heatley G, Kiland J, Faha B, Seeman J, Schlamp CL et al: Gene therapy using p21WAF-1/Cip-1 to modulate wound healing after glaucoma trabeculectomy surgery in a primate model of ocular hypertension, *Gene Ther* 11:949–55, 2004.

Wen SF, Chen Z, Nery J, Faha B: Characterization of adenovirus p21 gene transfer biodistribution and immune response after local ocular delivery in New Zealand white rabbits, *Exp Eye Res* 77:355–65, 2003.

Perkins TW, Faha B, Ni M, Kiland JA, Poulsen GL et al: Adenovirus-mediated gene therapy using human p21WAF-1/Cip-1 to prevent wound healing in a rabbit model of glaucoma filtration surgery, *Arch Ophthalmol* 120(7):941–9, 2002.

Human protein reference data base, Johns Hopkins University and the Institute of Bioinformatics, at http://www.hprd.org/protein.

23.11. Protection of Retinal Neurons from Glaucoma-Induced Injury and Death

Brain-Derived Neurotrophic Factor (BDNF)

Baker-Herman TL, Fuller DD, Bavis RW, Zabka AG, Golder FJ et al: BDNF is necessary and sufficient for spinal respiratory plasticity following intermittent hypoxia, *Nature Neuros* 7:48–55, 2004.

Chen H, Weber AJ: BDNF enhances retinal ganglion cell survival in cats with optic nerve damage, *Invest Ophthalm Vis Sci* 42:966–74, 2001.

Conover JC, Erickson JT, Katz DM, Bianchi LM, Poueymirou WT et al: Neuronal deficits, not involving motor neurons, in mice lacking BDNF and/or NT4, *Nature* 375:235–8, 1995.

Du J, Poo M: Rapid BDNF-induced retrograde synaptic modification in a developing retinotectal system, *Nature* 429:878–83, 2004.

Guillin O, Diaz J, Carroll P, Griffon N, Schwartz JC et al: BDNF controls dopamine D3 receptor expression and triggers behavioural sensitization, *Nature* 411:86–9, 2001.

Hofer MM, Barde YA: Brain-derived neurotrophic factor prevents neuronal death in vivo, *Nature* 331:261–2, 1988.

Huang ZJ, Kirkwood A, Pizzorusso T, Porciatti V, Morales B et al: BDNF regulates the maturation of inhibition and the critical period of plasticity in mouse visual cortex, *Cell* 98:739–55, 1999.

Jones KR, Reichardt LF: Molecular cloning of a human gene that is a member of the nerve growth factor family, *Proc Natl Acad Sci USA* 87:8060–4, 1990.

Kawamura K, Kawamura N, Mulders SM, Sollewijn Gelpke MD, Hsueh AJW: Ovarian brain-derived neurotrophic factor (BDNF) promotes the development of oocytes into preimplantation embryos, *Proc Natl Acad Sci* 102:9206–11, 2005.

Kernie SG, Liebl DJ, Parada LF: BDNF regulates eating behavior and locomotor activity in mice, *EMBO J* 19:1290–300, 2000.

Kovalchuk Y, Hanse E, Kafitz KW, Konnerth A: Postsynaptic induction of BDNF-mediated long-term potentiation, *Science* 295:1729–34, 2002.

Lawrence JM, Keegan DJ, Muir EM, Coffey PJ, Rogers JH et al: Transplantation of Schwann cell line clones secreting GDNF or BDNF into the retinas of dystrophic Royal College of Surgeons rats, *Invest Ophthalm Vis Sci* 45:267–74, 2004.

Lee R, Kermani P, Teng KK, Hempstead BL: Regulation of cell survival by secreted proneurotrophins, *Science* 294:1945–8, 2001.

Li Y, Jian JC, Cui K, Li N, Zheng ZY et al: Essential role of TRPC channels in the guidance of nerve growth cones by brain-derived neurotrophic factor, *Nature* 434:894–8, 2005.

Liu X, Ernfors P, Wu H, Jaenisch R: Sensory but not motor neuron deficits in mice lacking NT4 and BDNF, *Nature* 375:238–41, 1995.

Lyons WE, Mamounas LA, Ricaurte GA, Coppola V, Reid SW et al: Brain-derived neurotrophic factor-deficient mice develop aggressiveness and hyperphagia in conjunction with brain serotonergic abnormalities, *Proc Natl Acad Sci USA* 96:15239–44, 1999.

Ming G, Wong ST, Henley J, Yuan X, Song H et al: Adaptation in the chemotactic guidance of nerve growth cones, *Nature* 417:411–8, 2002.

Pang PT, Teng HK, Zaitsev E, Woo NT, Sakata K et al: Cleavage of proBDNF by tPA/plasmin is essential for long-term hippocampal plasticity, *Science* 306:487–91, 2004.

Robinson LLL, Townsend J, Anderson RA: The human fetal testis is a site of expression of neurotrophins and their receptors: Regulation of the germ cell and peritubular cell population, *J Clin Endocr Metab* 88:3943–51, 2003.

Ciliary Neurotrophic Factor (CNF)

DeChiara TM, Vejsada R, Poueymirou WT, Acheson A, Suri C et al: Mice lacking the CNTF receptor, unlike mice lacking CNTF, exhibit profound motor neuron deficits at birth, *Cell* 83:313–22, 1995.

Emerich DF, Winn SR, Hantraye PM, Peschanski M, Chen EY et al: Protective effect of encapsulated cells producing neurotrophic factor CNTF in a monkey model of Huntington's disease, *Nature* 386:395–9, 1997.

Giess R, Maurer M, Linker R, Gold R, Warmuth-Metz M et al: Association of a null mutation in the CNTF gene with early onset of multiple sclerosis, *Arch Neurol* 59:407–9, 2002.

Giovannini M, Romo AJ, Evans GA: Chromosomal localization of the human ciliary neurotrophic factor gene (CNTF) to 11q12 by fluorescence in situ hybridization, *Cytogenet Cell Genet* 63:62–3, 1993.

Kokoeva MV, Yin H, Flier JS: Neurogenesis in the hypothalamus of adult mice: potential role in energy balance, *Science* 310:679–83, 2005.

Lam A, Fuller F, Miller J, Kloss J, Manthorpe M et al: Sequence and structural organization of the human gene encoding ciliary neurotrophic factor, *Gene* 102:271–6, 1991.

Masu Y, Wolf E, Holtmann B, Sendtner M, Brem G et al: Disruption of the CNTF gene results in motor neuron degeneration, *Nature* 365:27–32, 1993.

Sendtner M, Schmalbruch H, Stockli KA, Carroll P, Kreutzberg GW et al: Ciliary neurotrophic factor prevents degeneration of motor neurons in mouse mutant progressive motor neuronopathy, *Nature* 358:502–4, 1992.

Song MR, Ghosh A: FGF2-induced chromatin remodeling regulates CNTF-mediated gene expression and astrocyte differentiation, *Nature Neurosci* 7:229–35, 2004.

Yokoji H, Ariyama T, Takahashi R, Inazawa J, Misawa H et al: cDNA cloning and chromosomal localization of the human ciliary neurotrophic factor gene, *Neurosci Lett* 185:175–8, 1995.

Glial Cell Line-Derived Neurotrophic Factor (GDNF)

Beck KD, Valverde J, Alexi T, Poulsen K, Moffat B et al: Mesencephalic dopaminergic neurons protected by GDNF from axotomy-induced degeneration in the adult brain, *Nature* 373:339–41, 1995.

Boucher TJ, Okuse K, Bennett DLH, Munson JB, Wood JN et al: Potent analgesic effects of GDNF in neuropathic pain states, *Science* 290:124–7, 2000.

Durbec P, Marcos-Gutierrez CV, Kilkenny C, Grigoriou M, Wartiowaara K et al: GDNF signalling through the Ret receptor tyrosine kinase, *Nature* 381:789–93, 1996.

Gash DM, Zhang Z, Ovadia A, Cass WA, Yi A et al: Functional recovery in parkinsonian monkeys treated with GDNF, *Nature* 380:252–5, 1996.

Gill SS, Patel NK, Hotton GR, O'Sullivan K, McCarter R et al: Direct brain infusion of glial cell line-derived neurotrophic factor in Parkinson disease, *Nature Med* 9:589–95, 2003.

Japon MA, Urbano AG, Saez C, Segura DI, Cerro AL et al: Glial-derived neurotropic factor and RET gene expression in normal human anterior pituitary cell types and in pituitary tumors, *J Clin Endocr Metab* 87:1879–84, 2002.

Kordower JH, Emborg ME, Bloch J, Ma, SY, Chu Y et al: Neurodegeneration prevented by lentiviral vector delivery of GDNF in primate models of Parkinson's disease, *Science* 290:767–73, 2000.

Lawrence JM, Keegan DJ, Muir EM, Coffey PJ, Rogers JH et al: Transplantation of Schwann cell line clones secreting GDNF or BDNF into the retinas of dystrophic Royal College of Surgeons rats, *Invest Ophthalm Vis Sci* 45:267–74, 2004.

Lin LFH, Doherty DH, Lile JD, Bektesh S, Collins F: GDNF: A glial cell line-derived neurotrophic factor for midbrain dopaminergic neurons, *Science* 260:1130–2, 1993.

Meng X, Lindahl M, Hyvonen ME, Parvinen M, de Rooij DG et al: Regulation of cell fate decision of undifferentiated spermatogonia by GDNF, *Science* 287:1489–93, 2000.

Moore MW, Klein RD, Farinas I, Sauer H, Armanini M et al: Renal and neuronal abnormalities in mice lacking GDNF, *Nature* 382:76–9, 1996.

Nguyen QT, Parsadanian AS, Snider WD, Lichtman JW: Hyperinnervation of neuromuscular junctions caused by GDNF overexpression in muscle, *Science* 279:1725–9, 1998.

Oppenheim RW, Houenou LJ, Johnson JE, Lin LFH, Li L et al: Developing motor neurons rescued from programmed and axotomy-induced cell death by GDNF, *Nature* 373:344–6, 1995.

Pichel JG, Shen L, Shang HZ, Granholm AC, Drago J et al: Defects in enteric innervation and kidney development in mice lacking GDNF, *Nature* 382:73–6, 1996.

Pichel JG, Shen L, Sheng HZ, Granholm AC, Drago J et al: GDNF is required for kidney development and enteric innervation, *Cold Spring Harbor Symp Quant Biol* 61:445–57, 1996.

Ramer MS, Priestley JV, McMahon SB: Functional regeneration of sensory axons into the adult spinal cord, *Nature* 403:312–6, 2000.

Sanchez MP, Silos-Santiago I, Frisen J, He B, Lira SA et al: Renal agenesis and the absence of enteric neurons in mice lacking GDNF, *Nature* 382:70–3, 1996.

Tomac A, Lindqvist E, Lin LFH, Ogren SO, Young D et al: Protection and repair of the nigrostriatal dopaminergic system by GDNF in vivo, *Nature* 373:335–9, 1995.

Treanor JJS, Goodman L, de Sauvage F, Stone DM, Poulsen KT et al: Characterization of a multicomponent receptor for GDNF, *Nature* 382: 80–3, 1996.

Di Polo A, Aigner LJ, Dunn RJ, Bray GM, Aguayo AJ: Prolonged delivery of brain-derived neurotrophic factor by adenovirus-infected Müller cells temporarily rescues injured retinal ganglion cells, *Proc Natl Acad Sci USA* 95:3978–83, 1998.

Isenmann S, Klocker N, Gravel C, Bähr M: Short communication: protection of axotomized retinal ganglion cells by adenovirally delivered BDNF in vivo, *Eur J Neurosci* 10:2751–6, 1998.

Martin KR, Quigley HA, Zack DJ, Levkovitch-Verbin H, Kielczewski J et al: Gene therapy with brain-derived neurotrophic factor as a protection: retinal ganglion cells in a rat glaucoma model, *Invest Ophthalmol Vis Sci* 44:4357–65, 2003.

Martin KR, Quigley HA: Gene therapy for optic nerve disease, *Eye* 18:1049–55, 2004.

McKinnon SJ: Glaucoma, apoptosis, and neuroprotection, *Curr Opin Ophthalmol* 8:28–37, 1997.

Human protein reference data base, Johns Hopkins University and the Institute of Bioinformatics, at http://www.hprd.org/protein.

23.12. Pathogenesis, Pathology, and Clinical Features of Cataract

Stitt AW: The maillard reaction in eye diseases, *Ann NY Acad Sci* 1043:582–97, 2005.

Vavvas D, Azar NF, Azar DT: Mechanisms of disease: Cataracts, *Ophthalmol Clin North Am* 15(1):49–60, 2002.

Duncan G, Wormstone IM: Calcium cell signalling and cataract: Role of the endoplasmic reticulum, *Eye* 13:480–3, 1999.

McAvoy JW, Chamberlain CG, de Iongh RU, Hales AM, Lovicu FJ: Lens development, *Eye* 13 (Pt 3b):425–37, 1999.

Hutnik CM, Nichols BD: Cataracts in systemic diseases and syndromes, *Curr Opin Ophthalmol* 10(1):22–8, 1999.

Francis PJ, Berry V, Moore AT, Bhattacharya S: Lens biology: development and human cataractogenesis, *Trends Genet* 15(5):191–6, 1999.

Stevens A: The contribution of glycation to cataract formation in diabetes, *J Am Optom Assoc* 69(8):519–30, 1998.

Smith RS, Sundberg JP, Linder CC: Mouse mutations as models for studying cataracts, *Pathobiology* 65(3):146–54, 1997.

23.13. Molecular Regenerative Engineering

Malecaze F, Decha A, Serre B, Penary M, Duboue M et al: Prevention of posterior capsule opacification by the induction of therapeutic apoptosis of residual lens cells, *Gene Ther* 13(5):440–8, 2006.

Kampmeier J, Behrens A, Wang Y, Yee A, Anderson WF et al: Inhibition of rabbit keratocyte and human fetal lens epithelial cell proliferation by retrovirus-mediated transfer of antisense cyclin G1 and antisense MAT1 constructs, *Hum Gene Ther* 11:1–8, 2000.

Couderc BC, de Neuville S, Douin-Echinard V, Serres B, Manenti S et al: Retrovirus-mediated transfer of a suicide gene into lens epithelial cells in vitro and in an experimental model of posterior capsule opacification, *Curr Eye Res* 19:472–82, 1999.

Malecaze F, Couderc B, de Neuville S, Serres B, Mallet J et al: Adenovirus-mediated suicide gene transduction: Feasibility in lens epithelium and in prevention of posterior capsule opacification in rabbits, *Hum Gene Ther* 10:2365–72, 1999.

23.14. Pathogenesis, Pathology, and Clinical Features of Retinopathy

Jenkins AJ, Rowley KG, Lyons TJ, Best JD, Hill MA et al: Lipoproteins and diabetic microvascular complications, *Curr Pharm Des* 10(27):3395–418, 2004.

Jain A, Sarraf D, Fong D: Preventing diabetic retinopathy through control of systemic factors, *Curr Opin Ophthalmol* 14(6):389–94, 2003.

Pacione LR, Szego MJ, Ikeda S, Nishina PM, McInnes RR: Progress toward understanding the genetic and biochemical mechanisms of inherited photoreceptor degenerations, *Annu Rev Neurosci* 26:657–700, 2003.

Ciulla TA, Amador AG, Zinman B: Diabetic retinopathy and diabetic macular edema: Pathophysiology, screening, and novel therapies, *Diabetes Care* 26(9):2653–64, 2003.

Marc RE, Jones BW, Watt CB, Strettoi E: Neural remodeling in retinal degeneration, *Prog Retin Eye Res* 22(5):607–55, 2003.

Stitt AW: The role of advanced glycation in the pathogenesis of diabetic retinopathy, *Exp Mol Pathol* 75(1):95–108, 2003.

D'Amore PA: Mechanisms of retinal and choroidal neovascularization, *Invest Ophthalm Vis Sci* 35:3974–9, 1994.

Macular Photocoagulation Study Group: Laser photocoagulation of subfoveal neovascular lesions in age-related macular degeneration. I: Results of a randomized clinical trial, *Arch Ophthalmol* 109:1220–31, 1991.

Macular Photocoagulation Study Group: Laser photocoagulation of subfoveal recurrent neovascular lesions in age-related macular degeneration. II: Results of a randomized clinical trial, *Arch Ophthalmol* 109:1232–41, 1991.

Lopez PF, Lambert HM, Grossniklaus HE, Sternberg P Jr: Well-defined subfoveal choroidal neovascular membranes in age-related macular degeneration, *Ophthalmology* 100:415–22, 1993.

Moisseiev J, Alhael A, Masuri R, Treister G: The impact of the macular photocoagulation study results on the treatment of exudative age-related macular degeneration, *Arch Ophthalmol* 113:185–9, 1995.

Miller JW, Schmidt-Erfurth U, Sickenberg M et al: Photodynamic therapy with verteporfin for choroidal neovascularization caused by age-related macular degeneration: results of a single treatment in a phase 1 and 2 study, *Arch Ophthalmol* 117:1161–73, 1999.

23.15. Molecular Regenerative Engineering

Borras T: Recent developments in ocular gene therapy, *Exp Eye Res* 76(6):643–52, June 2003.

Igarashi T, Miyake K, Kato K, Watanabe A, Ishizaki M et al: Lentivirus-mediated expression of angiostatin efficiently inhibits neovascularization in a murine proliferative retinopathy model, *Gene Ther* 10(3):219–26, Feb 2003.

Gauthier R, Joly S, Pernet V, Lachapelle P, Di Polo A: Brain-derived neurotrophic factor gene delivery to muller glia preserves structure and function of light-damaged photoreceptors, *Invest Ophthalm Vis Sci* 46(9):3383–92, Sept 2005.

Imai D, Yoneya S, Gehlbach PL, Wei LL, Mori K: Intraocular gene transfer of pigment epithelium-derived factor rescues photoreceptors from light-induced cell death, *J Cell Physiol* 202(2):570–8, Feb 2005.

Endostatin

Marneros AG, Olsen BR: Age-dependent iris abnormalities in collagen XVIII/endostatin deficient mice with similarities to human pigment dispersion syndrome, *Invest Ophthalm Vis Sci* 44:2367–72, 2003.

O'Reilly MS, Boehm T, Shing Y, Fukai N, Vasios G et al: Endostatin: An endogenous inhibitor of angiogenesis and tumor growth, *Cell* 88:277–85, 1997.

O'Reilly MS, Holmgren L, Shing Y, Chen C, Rosenthal RA et al: Angiostatin: A novel angiogenesis inhibitor that mediates the suppression of metastases by a Lewis lung carcinoma, *Cell* 79:315–28, 1994.

Oh SP, Kamagata Y, Muragaki Y, Timmons S, Ooshima A et al: Isolation and sequencing of cDNAs for proteins with multiple domains of Gly-X-Y repeats identify a novel family of collagenous proteins, *Proc Natl Acad Sci USA* 91:4229–33, 1994.

Oh SP, Warman ML, Seldin MF, Cheng SD, Knoll JHM et al: Cloning of cDNA and genomic DNA encoding human type XVIII collagen and localization of the alpha-1(XVIII) collagen gene to mouse chromosome 10 and human chromosome 21, *Genomics* 19:494–9, 1994.

Rehn M, Hintikka E, Pihlajaniemi T: Characterization of the mouse gene for the alpha-1 chain of type XVIII collagen (Col18a1) reveals that the three variant N-terminal polypeptide forms are transcribed from two widely separated promoters, *Genomics* 32:436–46, 1996.

Rehn M, Pihlajaniemi T: Alpha-1(XVIII), a collagen chain with frequent interruptions in the collagenous sequence, a distinct tissue distribution, and homology with type XV collagen, *Proc Natl Acad Sci USA* 91:4234–8, 1994.

Saarela J, Ylikarppa R, Rehn M, Purmonen S, Pihlajaniemi T: Complete primary structure of two variant forms of human type XVIII collagen and tissue-specific differences in the expression of the corresponding transcripts, *Matrix Biol* 16:319–28, 1998.

Sudhakar A, Sugimoto H, Yang C, Lively J, Zeisberg M et al: Human tumstatin and human endostatin exhibit distinct antiangiogenic activities mediated by alpha-V-beta-3 and alpha-5-beta-1 integrins, *Proc Natl Acad Sci USA* 100:4766–71, 2003.

Pigment Epithelium-Derived Factor (PEDF)

Aymerich MS, Alberdi EM, Martinez A, Becerra SP: Evidence for pigment epithelium-derived factor receptors in the neural retina. *Invest Ophthalm Vis Sci* 42:3287–93, 2001.

Dawson DW, Volpert OV, Gillis P, Crawford SE, Xu HJ et al: Pigment epithelium-derived factor: A potent inhibitor of angiogenesis, *Science* 285:245–8, 1999.

Doll JA, Stellmach VM, Bouck NP, Bergh ARJ, Lee C et al: Pigment epithelium-derived factor regulates the vasculature and mass of the prostate and pancreas, *Nature Med* 9:774–80, 2003.

Greenberg J, Goliath R, Tombran-Tink J, Chader G, Ramesar R: Growth factors in the retina: pigment epithelium-derived factor (PEDF) now fine mapped to 17p13.3 and tightly linked to the RP13 locus, In *Degenerative Retinal Diseases*, La Vail MM, Hollyfield JG, Anderson RE, eds, Plenum Press, New York, 1997, pp 291–4.

King GL, Suzuma K: Pigment-epithelium-derived factor: A key coordinator of retinal neuronal and vascular functions, *New Engl J Med* 342:349–51, 2000.

Ogata N, Nishikawa M, Nishimura T, Mitsuma Y, Matsumura M: Unbalanced vitreous levels of pigment epithelium-derived factor and vascular endothelial growth factor in diabetic retinopathy, *Am J Ophthalm* 134:348–53, 2002.

Simonovic M, Gettins PGW, Vol K: Crystal structure of human PEDF, a potent antiangiogenic and neurite growth-promoting factor, *Proc Natl Acad Sci* 98:11131–5, 2001.

Steele FR, Chader GJ, Johnson LV, Tombran-Tink J: Pigment epithelium-derived factor: Neurotrophic activity and identification as a member of the serine protease inhibitor gene family, *Proc Natl Acad Sci* 90:1526–30, 1993.

Tombran-Tink J, Pawar H, Swaroop A, Rodriguez I, Chader GJ: Localization of the gene for pigment epithelium-derived factor (PEDF) to chromosome 17p13.1 and expression in cultured human retinoblastoma cells, *Genomics* 19:266–72, 1994.

Volpert OV, Zaichuk T, Zhou W, Reiher F, Ferguson TA et al: Inducer-stimulated Fas targets activated endothelium for destruction by anti-angiogenic thrombospondin-1 and pigment epithelium-derived factor, *Nature Med* 8:349–57, 2002.

O'Reilly MS, Holmgren L, Shring Y et al: Angiostatin: A novel angiogenesis inhibitor that mediates the suppression of metastases by a Lewis lung carcinoma [see comments] Cell 79:7315–28, 1994.

Cao Y, O'Reilly MS, Marshall B, Flynn E, Ji RW et al: Expression of angiostatin cDNA in a murine fibrosarcoma suppresses primary tumor growth and produces long-term dormancy of metastases, *J Clin Invest* 101:1055–63, 1998.

O'Reilly MS, Holmgren L, Chen C, Folkman J: Angiostatin induces and sustains dormancy of human primary tumors in mice, *Nat Med* 2:689–92, 1996.

Kvanta A, Algvere PV, Berglin L, Seregard S: Subfoveal fibrovascular membranes in age-related macular degeneration express vascular endothelial growth factor, *Invest Ophthalm Vis Sci* 37:1929–34, 1996.

Crossniklaus HE, Green WR: Histopathologic and ultrastructural findings of surgically excised choroidal neovascularization: Submacular Surgery Trials Research Group, *Arch Ophthalmol* 116:745–9, 1998.

Reddy VM, Zamora RL, Kaplan HJ: Distribution of growth factors in subfoveal neovascular membranes in age-related macular degeneration and presumed ocular histoplasmosis syndrome, *Am J Ophthalm* 120:291–301, 1995.

Amin R, Puklin JE, Frank RN: Growth factor localization in choroidal neovascular membranes of age-related macular degeneration, *Invest Ophthalm Vis Sci* 35:3178–88, 1994.

Frank RN, Amin RH, Eliott D, Puklin JE, Abrams GW: Basic fibroblast growth factor and vascular endothelial growth factor are present in epiretinal and choroidal neovascular membranes, *Am J Ophthalmol* 122:393–403, 1996.

Lopez PF, Sippy BD, Lambert HM, Thach AB, Hinton DR: Transdifferentiated retinal pigment epithelial cells are immunoreactive for vascular endothelial growth factor in surgically excised age-related macular degeneration-related choroidal neovascular membranes, *Invest Ophthalm Vis Sci* 37:855–68, 1996.

Zhang NL, Samadani EE, Frank RN: Mitogenesis and retinal pigment epithelial cell antigen expression in the rat after krypton laser photocoagulation, *Invest Ophthalm Vis Sci* 34:2412–24, 1993.

Lai CC, Wu WC, Chen SL, Xiao X, Tsai TC et al: Suppression of choroidal neovascularization by adeno-associated virus vector expressing angiostatin, *Invest Ophthalm Vis Sci* 42:2401–7, 2001.

Mori K, Duh E, Gehlbach P et al: Pigment epithelium-derived factor inhibits retinal and choroidal neovascularization, *J Cell Physiol* 188:253–63, 2001.

Mori K, Ando A, Gehlbach P et al: Inhibition of choroidal neovascularization by intravenous injection of adenoviral vectors expressing secretable endostatin, *Am J Pathol* 159:313–20, 2001.

Seo MS, Kwak N, Ozaki H et al: Dramatic inhibition of retinal and choroidal neovascularization by oral administration of a kinase inhibitor, *Am J Pathol* 154:1743–53, 1999.

Ozaki H, Seo MS, Ozaki K et al: Blockade of vascular endothelial cell growth factor receptor signaling is sufficient to completely prevent retinal neovascularization, *Am J Pathol* 156:679–707, 2000.

Kwak N, Okamoto N, Wood JM, Campochiaro PA: VEGF is an important stimulator in a model of choroidal neovascularization, *Invest Ophthalm Vis Sci* 41:3158–64, 2000.

O'Reilly MS, Holmgren S, Shing Y et al: Angiostatin: A novel angiogenesis inhibitor that mediates the suppression of metastases by a Lewis lung carcinoma, *Cell* 79:315–28 , 1994.

O'Reilly MS, Boehm T, Shing Y et al: Endostatin: An endogenous inhibitor of angiogenesis and tumor growth, *Cell* 88:277–85, 1997.

Maione TE, Gray GS, Petro J et al: Inhibition of angiogenesis by recombinant human platelet factor-4 and related peptides, *Science* 247:77–9, 1990.

Dawson DW, Volpert OV, Gillis P et al: Pigment epithelium-derived factor: a potent inhibitor of angiogenesis, *Science* 285:245–8, 1999.

Mori K, Gehlbach P, Yamamoto S, Duh E, Zack DJ et al: AAV-mediated gene transfer of pigment epithelium-derived factor inhibits choroidal neovascularization, *Invest Ophthalm Vis Sci* 43:1994–2000, 2002.

Human protein reference data base, Johns Hopkins University and the Institute of Bioinformatics, at http://www.hprd.org/protein.

23.16. Molecular Regenerative Engineering for Retinal Degeneration

Bennett J, Zeng Y, Bajwa R, Klatt L, Li Y et al: Adenovirus-mediated delivery of rhodopsin-promoted bcl-2 results in a delay in photoreceptor cell death in the rd/rd mouse, *Gene Ther* 5(9):1156–64, Sept 1998.

Retinal Phosphodiesterase

Altherr MR, Wasmuth JJ, Seldin MF, Nadeau JH, Baehr W et al: Chromosome mapping of the rod photoreceptor cGMP phosphodiesterase beta-subunit gene in mouse and human: Tight linkage to the Huntington disease region (4p16.3), *Genomics* 12:750–4, 1992.

Bateman JB, Klisak I, Kojis T, Mohandas T, Sparkes RS et al: Assignment of the beta-subunit of rod photoreceptor cGMP phosphodiesterase gene PDEB (homolog of the mouse rd gene) to human chromosome 4p16, *Genomics* 12:601–3, 1992.

Bennett J, Tanabe T, Sun D, Zeng Y, Kjeldbye H et al: Photoreceptor cell rescue in retinal degeneration (rd) mice by in vivo gene therapy, *Nature Med* 2:649, 1996.

Bowes C, Danciger M, Kozak CA, Farber DB: Isolation of a candidate cDNA for the gene causing retinal degeneration in the rd mouse, *Proc Natl Acad Sci USA* 86:9722–26, 1989. Note: Erratum: *Proc Natl Acad Sci USA* 87:1625, 1990.

Bowes C, Li T, Danciger M, Baxter LC, Applebury ML et al: Retinal degeneration in the rd mouse is caused by a defect in the beta subunit of rod cGMP-phosphodiesterase, *Nature* 347:677–80, 1990.

Collins C, Hutchinson G, Kowbel D, Riess O, Weber B et al: The human beta-subunit of rod photoreceptor cGMP phosphodiesterase: complete retinal cDNA sequence and evidence for expression in brain, *Genomics* 13:698–704, 1992.

Farber DB, Danciger JS, Aguirre G: The beta subunit of cyclic GMP phosphodiesterase mRNA is deficient in canine rod-cone dysplasia 1, *Neuron* 9:349–56, 1992.

Gal A, Orth U, Baehr W, Schwinger E, Rosenberg T: Heterozygous missense mutation in the rod cGMP phosphodiesterase beta-subunit gene in autosomal dominant stationary night blindness, *Nature Genet* 7:64–68, 1994. Note: Correction: *Nature Genet* 7:551 only, 1994.

Khramtsov NV, Feshchenko EA, Suslova VA, Shmukler BE, Terpugov BE et al: The human rod photoreceptor cGMP phosphodiesterase beta-subunit: Structural studies of its cDNA and gene, *FEBS Lett* 327:275–8, 1993.

Lem J, Flannery JG, Li T, Applebury ML, Farber DB et al: Retinal degeneration is rescued in transgenic rd mice by expression of the cGMP phosphodiesterase beta subunit, *Proc Natl Acad Sci USA* 89:4422–6, 1992.

McLaughlin ME, Ehrhart TL, Berson EL, Dryja TP: Mutation spectrum of the gene encoding the beta subunit of rod phosphodiesterase among patients with autosomal recessive retinitis pigmentosa, *Proc Natl Acad Sci USA* 92:3249–53, 1995.

McLaughlin ME, Sandberg MA, Berson EL, Dryja TP: Recessive mutations in the gene encoding the beta-subunit of rod phosphodiesterase in patients with retinitis pigmentosa, *Nature Genet* 4:130–4, 1993.

Weber B, Riess O, Hutchinson G, Collins C, Lin B et al: Genomic organization and complete sequence of the human gene encoding the beta-subunit of the cGMP phosphodiesterase and its localisation to 4p16.3, *Nucleic Acids Res* 19:6263–8, 1991.

Peripherin

Beaulieu JM, Nguyen MD, Julien JP: Late onset death of motor neurons in mice overexpressing wild-type peripherin, *J Cell Biol* 147:531–44, 1999.

Foley J, Ley CA, Parysek LM: The structure of the human peripherin gene (PRPH) and identification of potential regulatory elements, *Genomics* 22:456–61, 1994.

Gros-Louis F, Lariviere R, Gowing G, Laurent S, Camu W et al: A frameshift deletion in peripherin gene associated with amyotrophic lateral sclerosis, *J Biol Chem* 279:45951–6, 2004.

He CZ, Hays AP: Expression of peripherin in ubiquinated inclusions of amyotrophic lateral sclerosis, *J Neurol Sci* 217:47–54, 2004.

Leonard DG, Gorham JD, Cole P, Greene LA, Ziff EB: A nerve growth factor-regulated messenger RNA encodes a new intermediate filament protein, *J Cell Biol* 106:181–93, 1988.

Moncla A, Landon F, Mattei MG, Portier MM: Chromosomal localisation of the mouse and human peripherin genes, *Genet Res* 59:125–9, 1992.

Thompson MA, Ziff EB: Structure of the gene encoding peripherin, an NGF-regulated neuronal-specific type III intermediate filament protein, *Neuron* 2:1043–53, 1989.

Bennett J, Tanabe T, Sun D, Zeng Y, Kjeldbye H et al: Photoreceptor cell rescue in retinal degeneration (rd) mice by in vivo gene therapy, *Nat Med* 2:649–54, 1996.

Bennett J, Zeng Y, Bajwa R, Klatt L, Li Y et al: Adenovirus-mediated delivery of rhodopsin-promoted bcl-2 results in a delay in photoreceptor cell death in the rd/rd mouse, *Gene Ther* 5:1156–64, 1998.

Tsang SH, Gouras P, Yamashita CK, Kjeldbye H, Fisher J et al: Retinal degeneration in mice lacking the gamma subunit of the rod cGMP phosphodiesterase, *Science* 272:1026–9, 1996.

Human protein reference data base, Johns Hopkins University and the Institute of Bioinformatics, at http://www.hprd.org/protein.

24

SKIN REGENERATIVE ENGINEERING

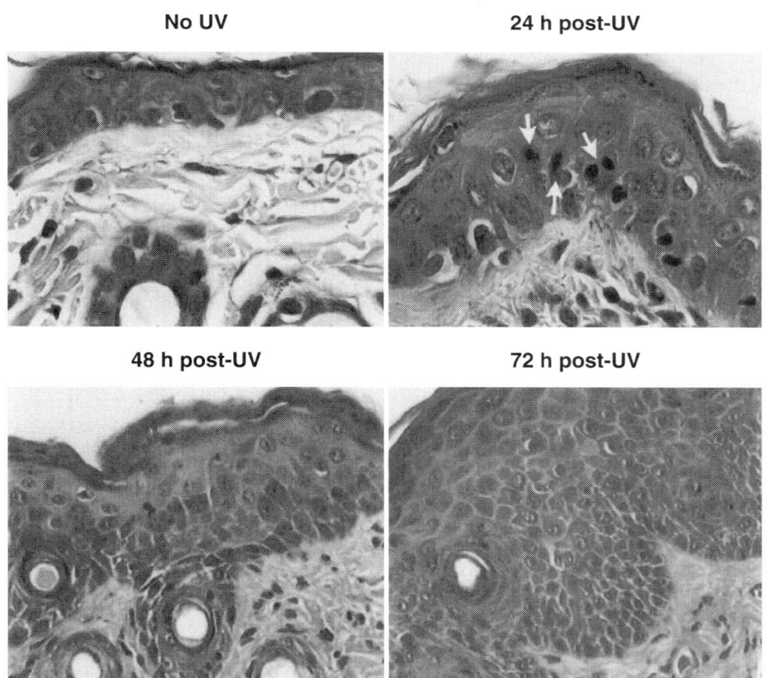

Induction of sunburn cells and epidermal hyperplasia in unirradiated and UV-irradiated mouse skin at various times after UV irradiation. Arrows indicate representative sunburn cells. Original magnification, ×400 for unirradiated and 24 h post-UV and ×200 for 48 and 72 h post-UV. (Reprinted with permission from Ouhtit A et al: Temporal events in skin injury and the early adaptive responses in ultraviolet-irradiated mouse skin, *Am J Pathol* 156:201–7, copyright 2000.) See color insert.

Bioregenerative Engineering: Principles and Applications, by Shu Q. Liu

ANATOMY AND PHYSIOLOGY OF THE SKIN [24.1]

The skin is composed of two layers: the epidermis and dermis. The *epidermis* is a layer of stratified squamous epithelial cells. There are several types of cell in the epidermal layer: keratinocytes, melanocytes, and sensory cells. *Keratinocytes* are the largest group of epithelial cells, which produce and release keratin, a protein forming the surface keratin layer of the epidermis. The *keratin* layer protects the epidermis from mechanical injury and chemical corrosion. *Melanocytes* are pigmented epithelial cells that determine the color of the skin. *Sensory cells* are specialized nerve cells. The endings of these cells are responsible for sensing mechanical contacting and temperature changes. The epidermis resides on a matrix layer known as the *basement membrane*, which separates the epidermis from the dermis.

The *dermis* is a layer of connective tissue and serves as the base for the epidermis, providing structural support and mechanical strength to the skin. The dermis is composed of several cell types, including the fibroblasts, adipocytes, and macrophages. *Fibroblasts* are responsible for the generation of extracellular matrix components such as collagen, elastin, and proteoglycans. *Adipocytes* are cells for the metabolism and storage of lipids. *Macrophages* are originated from the bloodborne monocytes, which are transformed to macrophages when migrating from the blood to tissue. These cells are responsible for the destruction of bacteria and clearance of cell debris. All cell types in the dermal layer reside in a meshwork of extracellular matrix, composed of collagen fibers, elastic fibers, and ground substances. In the dermal tissue, *collagen fibers* are constructed primarily with collagen type I. *Elastic fibers* consist of elastin, microfibrils, and microfibril-associated proteins. *Ground substances* are composed of proteoglycans (see Chapter 4 for details). These matrix components are responsible for the organization, integrity, stability, elasticity, and strength of the dermis. The dermis is also composed of several other structures, including blood vessels (arteries, capillaries, and veins), lymphatic vessels, nerve endings, hair follicles, and sweat and sebaceous glands.

The dermis resides on a deeper layer of connective tissue, known as *subcutaneous tissue*. This layer is composed of several cell types, including fibroblasts, adipocytes, and macrophages, and extracellular matrix. Large blood vessels (arteries and veins) and nerve bundles are surrounded by the subcutaneous tissue. A large fraction of adipocytes in the body is found in this type of tissue. The subcutaneous tissue provides a connection between the skin and internal structures, such as bones and skeletal muscles.

There are several functions for the skin, including protection, regulation of temperature, and sensation. Given the anatomical location and histological structure, the skin serves to protect the internal tissues and organs from injury induced by environmental factors, including chemical toxins, corrosive agents (acids and bases), mechanical forces (stretching, compressing, and shearing), microorganisms (bacteria and viruses), and radiations (UV and X-ray). The skin contains various types of sensory nerve endings, which can sense temperature changes, pressure, mechanical contacting, and chemical corrosion. These sensory structures are critical to the protection of the body from dangerous environmental factors. The skin consists of a rich network of blood vessels, which play an important role in the regulation of body temperature. An increase in the body temperature induces arteriolar dilation, leading to an increase in blood flow to the body surface and thus facilitating heat loss. A decrease in the body temperature exerts an opposite effect. The excretion of sweat is another mechanism that facilitates the loss of body heat and reduces body temperature. The skin participates in the synthesis of vitamin D, a hormone that regulates the metabolism of calcium and phosphate (stimulating the absorption of calcium and phosphate in the intestines, and increasing the level of blood calcium and phosphate). Vitamin D is synthesized from a cholesterol molecule, 7-dehydrocholesterol, which is converted into cholecalciferol under the stimulation of ultraviolet light. Cholecalciferol is released into the blood and converted into vitamin D in the liver via hydroxylation. In addition, the skin is an important structure that protects the body from water loss.

SKIN DISORDERS

Skin Injury

Pathogenesis, Pathology, and Clinical Features [24.2]. The skin directly interacts with the external environment and is often subject to hazard factors, such as heat, chemical toxins, radiation, electricity, and mechanical impacts. These factors induce various types of skin injury ranging from reversible inflammation to the destruction of the skin. The severity of skin injury is dependent on the skin layers involved and is classified into three degrees. *First-degree* skin injury is defined as injury limited to the epidermis, characterized by the presence of redness, an increase in local temperature, itching, and pain in the injured area. Examples of first-degree injury include scalds and sunburn. *Second-degree* injury involves the epidermis and dermis, and is characterized by severe local pain, swelling, and the formation of blisters containing fluids or blood, which arise from tissue cleavage within the epidermis (intraepidermal vesication) or the separation of the epidermis from the dermis (subepidermal vesication). Sweat glands and hair follicles are often damaged in the second degree of injury. *Third-degree* of skin injury involves the epidermis, dermis, and subcutaneous tissue, characterized by the destruction of all skin layers, the exposure of the subcutaneous tissue or other internal tissues, such as the skeletal muscles, tendons, ligaments, and bones, in association with massive swelling. Since sensory nerve endings are destroyed, third degree skin injury may not be associated with severe pain. The first and second degrees of skin injury can be self-healed within 1 to 3 weeks without the formation of scars. The recovery of the third degree of skin injury is often associated with scar formation. In severe cases with the exposure of skeletal muscles, tendons, ligaments, and bones, distortion and malfunction of the internal tissues may occur.

The impact of skin injury on the function of the skin and internal tissues is dependent on the area and degree of the injury. A small area of skin injury, referred to as minor injury, may not significantly influence the function, except that scars may form when the subcutaneous tissue is involved. A large area of third-degree skin injury, referred to as major injury, can cause pathological changes in not only the skin but also internal tissues and organs. Major changes in the skin include loss of the protective function and loss of body fluids. The loss of the protective function renders the skin vulnerable to bacterial and viral infection. Patients with fire- or chemical-induced injury are often associated with bacterial infection, the most common cause of death in these patients. Because of the destruction of the skin, water loss by evaporation is greatly enhanced in the area of injury, resulting in a reduction in the volume of blood and interstitial fluids. In addition, scar formation may result in distortion of the subcutaneous and internal tissues. When scars form over a joint, the flexibility of the joint will be reduced, influencing the mobility of the joint and extremities.

The pathological effects of skin injury are not limited to the skin. A large area of third-degree skin injury can result in malfunction of the internal tissues, organs, and systems. When internal tissues such as skeletal muscles and bones are injured, the functions of these tissues are impaired. Severe skin injury may also involves the vascular, lymphatic, and hormonal systems. Injured cells can release inflammatory mediators, such as histamine and prostaglandins. These mediators act on capillary endothelial cells and increase the permeability of capillaries, resulting in the transport of fluids from the blood to the interstitial tissue and a decrease in the volume of circulating blood. In severe cases, the deficit of blood volume may affect the performance of the heart and leads to oxygen deficiency in peripheral tissues and organs. In the case of bacterial infection, toxins released by the bacteria, together with inflammatory mediators released by infected and injured cells, may induce cell malfunction in the brain, heart, kidney, and liver, resulting in the impairment and failure of these organs.

Conventional Treatment of Skin Injury [24.2]. For the first and second degree of skin injury, injured skin can usually be self-cured. However, the third degree of skin injury cannot be self-cured and may result in serious clinical consequences. There are several strategies for the treatment of third-degree skin injury. These include the restoration of fluid and electrolyte balance, protection of the skin from bacterial infection, coverage of injured skin to prevent water loss and bacterial infection, and prevention of scar formation. Water loss and increased capillary permeability result in a reduction in the volume of circulating blood and imbalance of electrolytes. Thus, patients with severe skin injury should be treated to restore the water and electrolyte balance. Bacterial infection occurs in almost all cases of sever skin injury. Antibiotics should be used to prevent bacterial infection. The most important treatment for severe injury is skin transplantation. In principle, all areas of third-degree skin injury should be covered with autogenous skin specimens collected from intact areas of the patient. This is a critical treatment for the prevention of water loss and bacterial infection. In most cases of third-degree skin injury, however, it is difficult to collect sufficient skin specimens. Artificial skin substitutes can be constructed and used for the treatment of severe skin injury. When scar formation influences the function of the joints and extremities, the scar should be removed and replaced with intact skin specimens or skin substitutes.

Skin Regenerative Engineering. Skin regenerative engineering is to develop cell-based functional skin substitutes that can be used for the coverage of injured skin or replacement of the lost skin. Skin regenerative approaches are established based on the natural healing processes of skin wounds. In response to skin injury, the wound site is rapidly plugged with fibrin clots. Injured epithelial cells and fibroblasts at the wound site may release cytokines and growth factors, which attract leukocytes and fibroblasts to the wound site, inducing inflammatory reactions and formation of granulation tissue. The vascular endothelial cells are activated in response to angiogenic factors to induce the formation of new blood vessels. The epithelial cells at the edge of the wound are stimulated to proliferate and migrate over the wound area to form a new epidermal layer. The fibroblasts within the wound produce and release extracellular matrix, which contributes to fibrosis and scar formation (Fig. 24.1). When the skin wound only involves the epidermis and a small area of the dermis (first- and second-degree of wounds), the wound can be completely self-healed. However, when the wound involves the deep subcutaneous tissue (third-degree wounds), the wound cannot be self-healed, often resulting in permanent scars and distor-

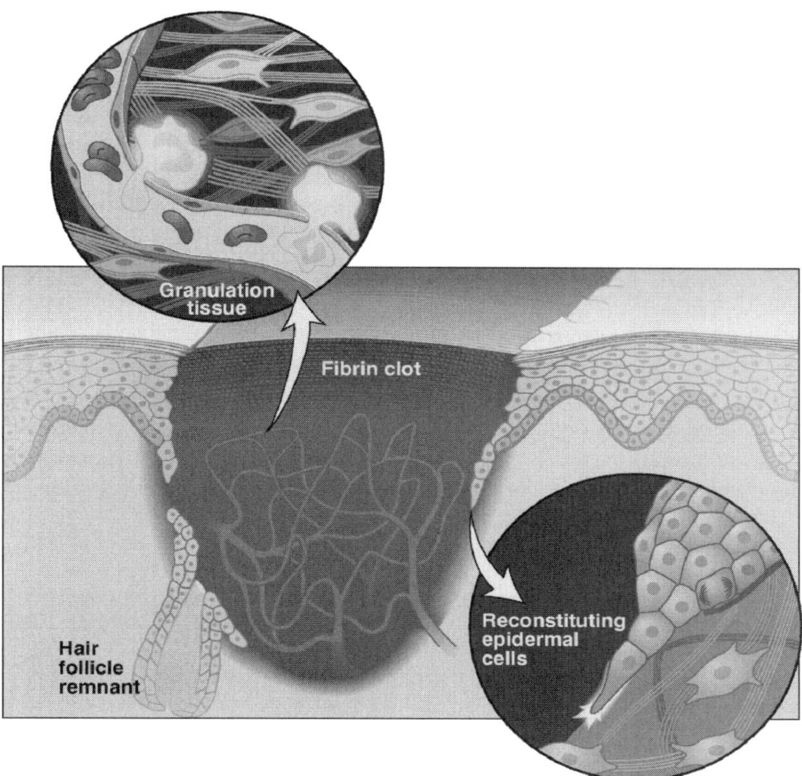

Figure 24.1. Schematic representation of the key players in the healing of a skin wound. The defect is temporarily plugged with a fibrin clot, which is infiltrated by inflammatory cells, fibroblasts, and a dense capillary plexus of new granulation tissue. An epidermal covering is reconstituted from the edges of the wound and from the cut remnants of hair follicles. At the migrating keratinocyte leading edge, cells bore a passageway enabling them to crawl beyond the cut basal lamina and over provisional matrix and healthy dermis, Cell division occurs back from the leading edge. Monocytes emigrate from wound capillaries into the granulation tissue, which contracts by means of smooth-muscle-like myofibroblasts that tug on one another and the surrounding collagen matrix. (Reprinted with permission from Martin P: Wound healing—aiming for perfect skin regeneration, *Science* 276:75–81, copyright 1997 AAAS.)

tion of the skin and involved joints. For third-degree of skin wounds, it is necessary to cover the wounds with skin substitutes and reduce inflammatory reactions, which facilitate the healing process of skin wounds. Skin regenerative engineering approaches are established for these purposes. A skin substitute can be generated by seeding and growing stem cells or epidermal cells on extracellular matrix or synthetic scaffolds. Several factors should be taken into account for the construction of skin substitutes. These include cell types, matrix types and forms, appropriate growth stimulators, and a suitable growth system. These factors are discussed as follows.

Cell Types for Constructing Skin Substitutes [24.3]. Several cell types can be used for the construction of skin substitutes. These include embryonic stem cells, multipotent fetal stem cells, adult bone marrow stem cells, adult epidermal progenitor cells, and mature epidermal and dermal cells. Embryonic stem cells, collected from the inner cell mass of the blastocyst, are pluripotent cells that can differentiate into all specialized cell types, including epidermal cells, when appropriate growth conditions are provided. Fetal ectodermal stem cells are multipotent cells that are committed to the differentiation into ectodermal cells, including epidermal cells. Adult stem cells, such as bone marrow stem cells, can also be induced to differentiate into epidermal cells. Since mature epidermal cells can proliferate, these cells can also be used for the construction of skin substitutes (Fig. 24.2).

Although embryonic and fetal stem cells are potential candidate cell types for skin regeneration, these cells are derived from allogenic sources and cause immune rejection responses. Thus, an ideal approach is to identify and use autogenous epidermal stem cells. In the adult skin, there exist potential epidermal stem and progenitor cells. The hair follicles and sebaceous glands contain multipotent epidermal stem cells. In response to skin injury, these stem cells can be activated to form epidermal cells, follicle cells, and sebaceous gland cells. The differentiation of the epidermal stem cells is a critical processes

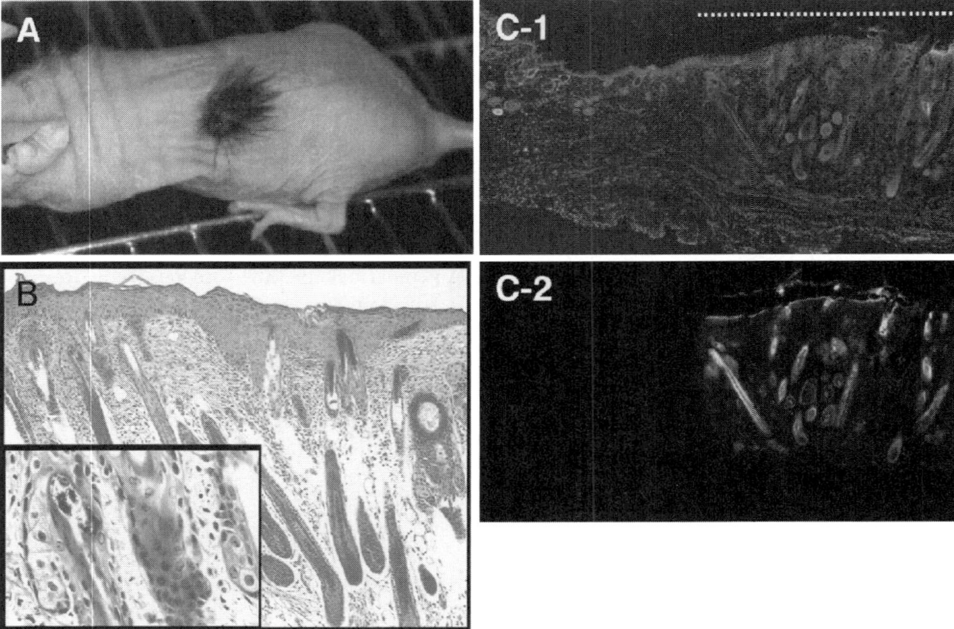

Figure 24.2. Skin tissues reconstituted by a mixture of epidermal and dermal cells. (A) Active hair growth in a region transplanted with embryonic epidermal and dermal cells observed 3 weeks after transplantation. (B) Histology of the reconstituted skin stained with hematoxilin/eosin. Epidermis and hair follicles with sebaceous glands (inset) were seen. Scale bar: 50 μm (inset, 10 μm). (C) Skin reconstituted by epidermal and dermal cells derived from GFP-transgenic mice: (C-1) DAPI staining of nuclei; (C-2) fluorescent microscopic image. The dotted line in C-1 indicates the part of reconstituted skin. Scale bar: 200 μm. (Reprinted with permission from Kataoka K et al: Participation of adult mouse bone marrow cells in reconstitution of skin, *Am J Pathol* 163:1227–31, copyright 2003.)

for the self-healing of injured skin. In addition, the basal cells of the epidermis, residing on the basement membrane, are capable of proliferating and differentiating into mature keratinocytes. These cells are considered epidermal progenitor cells and can be used for skin regeneration. Adult epidermal stem cells have a large growth capacity. Under appropriate conditions, basal epidermal and hair follicle stem cells can differentiate into keratinocytes and expand rapidly in a cell culture system. Expanded cells can be used for skin regeneration and reconstruction. Experimental investigations have shown that the transplantation of epidermal stem cells to injured skin significantly enhances skin recovery from injury (Fig. 24.3).

As for other tissue types, it is necessary to identify epidermal stem cells before they are collected and used for skin reconstruction. This can be done by detecting molecules specific to the epidermal stem cells. Compared to mature epidermal cells, the epidermal stem cells express a higher level of $\alpha 6$ and $\beta 1$ integrins, although these integrins are not specific to epidermal stem cells. The hair follicle stem cells and basal cells express a unique keratin molecule known as *keratin 15*. This molecule has been considered a marker for identifying epidermal stem cells. Antibodies specific to keratin 15 can be used to identify epidermal stem cells. The p63 protein, a homologue of the p53 tumor suppressor, is expressed in the basal cells and follicle cells. The expression level of this gene is directly proportional to the activity of cell proliferation. Thus, p63 has been considered a potential marker protein for identifying epidermal stem cells. Several other proteins, including

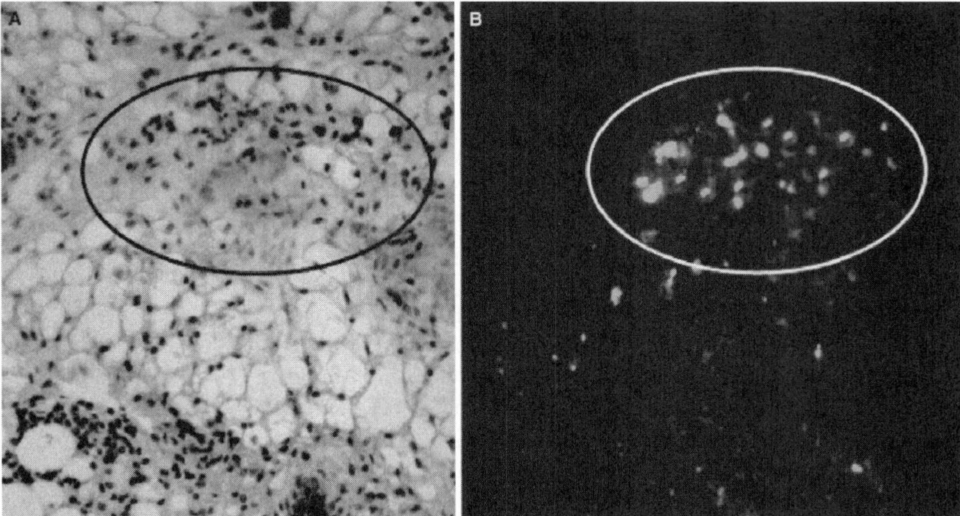

Figure 24.3. Wounds receiving epidermal stem cells (EpiSC) healed faster than did wounds receiving no cells. Full-thickness skin wounds were created in the back skin of C57BL/6 nontransgenic mice. Two days later EpiSC were isolated from the back skin of neonatal GFP transgenic mice, and injected beside and beneath the wound beds. Shown here are adjacent sections through the middle of the healed wound bed 21 days after GFP^+ EpiSC were injected. (A) H&E-stained section; (B) unstained adjacent section showing GFP fluorescence. Circle surrounds a cluster of GFP^+ cells showing varying morphology. (Reprinted with permission from Bickenbach JR, Grinnell KL: Epidermal stem cells: Interactions in developmental environments, *Differentiation* 72:371–80, copyright 2004.)

CD34 and AC133-2 (an isoform of CD133), have also been identified as potential markers for epidermal stem cells. Antibodies for these proteins can be used for the identification of epidermal stem cells. However, mature epidermal cells and follicle cells may express some of these proteins in certain developmental stages. These marker proteins should be used in caution. The use of multiple protein markers may help to identify epidermal stem cells.

Matrix Scaffolds for Constructing Skin Substitutes [24.4]. Epidermal cells reside on the dermis, a soft connective tissue containing fibroblasts and extracellular matrix that is constituted with collagen fibers, elastic fibers, and proteoglycans. These matrix components not only provide a structural basis for the integrity and mechanical strength of the skin, but also play critical roles in the regulation of cell attachment, proliferation, migration, regeneration, and wound repair. Thus, extracellular matrix is an essential component for skin regeneration and reconstruction. Collagen matrix is commonly used as a substrate for culturing epidermal cells and constructing skin substitutes. Artificially constituted collagen gel and natural collagen matrix collected from the host or allogenic patients can be used for such a purpose. Collagen stimulates cell adhesion, proliferation, and migration, thus enhancing skin regeneration. Other matrix components, such as fibronectin, proteoglycans, and fibrin, can also be used or added to the collagen matrix to construct matrix scaffolds for skin regeneration and reconstruction.

In addition to natural extracellular matrix components, synthetic polymers, such as polyethylene glycol and polyglycolide, can be used for skin regeneration and reconstruction. These polymers are often used as substrates for seeding and culturing cells or for delivering biological substances that are required for cell growth and differentiation. In particular, the use of biodegradable polymers, such as polyglycolide and poly(glycolide-L-lactide), may provide suitable conditions for the growth of epidermal cells. Biological substances can be easily incorporated into the polymer matrix and delivered to the target cells. The polymer matrix can serve as a substrate for cell attachment, growth, migration, and pattern formation. By controlling the degradation of the polymeric matrix, seeded cells can gradually form a natural structure with cell-synthesized extracellular matrix, which replaces the polymeric matrix and provide a permanent substrate for grown cells.

Growth Factors for Stimulating the Growth of Epidermal Cells [24.5]. Epidermal cells express a number of growth factors and cytokines, including epidermal growth factor (EGF), keratinocyte growth factor (KGF), fibroblast growth factor (FGF), platelet-derived growth factor (PDGF), hepatocyte growth factor (HGF), vascular endothelial growth factor (VEGF), insulin-like growth factor (IGF), macrophage colony-stimulating factor (M-CSF), granulocyte-macrophage colony-stimulating factor (GM-CSF), and interleukin-1,3,6,8,10,18. These growth factors and cytokines play important roles in regulating the proliferation and differentiation of the epidermal cells. Figure 24.4 shows the effectiveness of EGF in improving the proliferation and migration of epithelial cells.

Although the exact mechanisms of cell differentiation remain poorly understood, time-dependent dynamic activation of selected growth factors and cytokines is critical to the differentiation of epidermal stem cells. Thus, these factors can be applied to cultured epidermal stem cells to stimulate cell differentiation and growth and facilitate skin regeneration. Alternatively, the genes that encode these factors can be used and transferred into epidermal stem cells to enhance the expression of these factors and facilitate cell differentiation and growth. (See Table 24.1 for further details.)

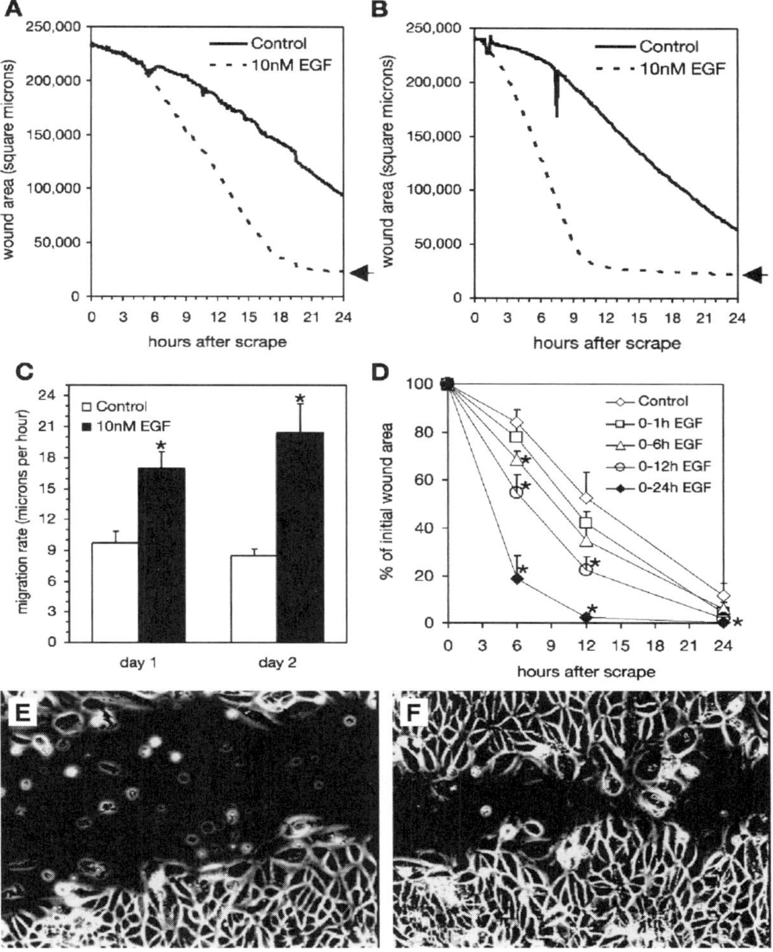

Figure 24.4. Epidermal growth factor (EGF) enhances serum-dependent closure of CV-1 cell scrape wounds: (A) parallel serum-starved CV-1 cell cultures were scrape wounded on the microscope stage, and images were collected at 6-min intervals for 24h. The area of the scrape wound was measured for each image and is plotted for cells cultured in 10% calf-serum-containing medium (control) or in 10% calf-serum-containing medium supplemented with 10nM EGF. (B) At the end of the experiment shown in panel A, the medium in both cultures was replaced with 10% calf serum and the assay was repeated. No EGF was present on day 2 of the experiment. The arrows to the *right* of the plots in A and B indicate the point of complete wound closure. (C) the sustained maximum migration rates were calculated using linear regression (means ± SE). *Significant increase ($P < 0.05$) in the monolayer migration rate compared with time-matched controls. (D) Cells cultured in calf-serum-containing medium were treated with EGF during the indicated intervals on day 1, and wound closure on day 2 was measured using a static assay. Values are means ± SE from triplicate determinations in 4 independent experiments. *Significant reduction ($P < 0.05$) in wound area compared with the time-matched control treated with calf serum only. The morphology of wound closure after 15h is shown for cells cultured in serum (E) or in serum plus 10nM EGF (F). See Video 3 for an example of time lapse. (Reprinted with permission from Kurten RC et al: Coordinating epidermal growth factor-induced motility promotes efficient wound closure, *Am J Physiol Cell Physiol* 288:C109–21, copyright 2005.)

TABLE 24.1. Characteristics of Selected Proteins that Regulate the Development and Growth of Epidermal Cells*

Proteins	Alternative Names	Amino Acids	Molecular Weight (kDa)	Expression	Functions
Keratinocyte growth factor	KGF, fibroblast growth factor 7 (FGF7)	194	23	Skin, skeletal muscle, blood vessel, pancreas, intestine, ovary, cornea	Regulating embryonic development and morphogenesis; mediating cell and tissue regeneration; and promoting cell survival, proliferation, and differentiation
Macrophage colony-stimulating factor	MCSF1, colony-stimulating factor 1 (CSF1), macrophage granulocyte inducer IM, MGI-IM	554	60	Lymphocytes, osteoclasts, microglia, astrocytes, bone marrow stromal cells, liver, and placenta	Promoting cell proliferation and differentiation
Granulocyte-macrophage colony-stimulating factor	GMCSF, colony-stimulating factor 2 (CSF2)	144	16	Monocytes, eosinophil, epithelial cells, brain, lung, retina	Existing as a homodimer, regulating the differentiation, proliferation, and activity of granulocytes and macrophages
IL18	Interleukin-1γ, Interferon-γ-inducing factor	193	22	Liver, kidney, lung, skeletal muscle	A proinflammatory cytokine that induces the IFNγ production of T cells and enhances IgG2a production of B cells

*Based on bibliography 24.5.

Construction of Skin Substitutes [24.6]. The skin is a structure that interacts with the external environment at one side and the internal connective tissue at the other side. Thus, a growth model for constructing skin substitutes should provide an environment that mimics the physiological conditions for the natural skin. Such a model can be established by using several necessary components, including epidermal stem cells or keratinocytes, fibroblasts (for the formation of dermis-like connective tissue), a matrix scaffold, essential growth-stimulating factors, and culture media. To avoid immune rejection responses, cells should be collected from the host patients, if possible.

In a skin growth apparatus, a sheet-like matrix scaffold can be constructed with either collagen gel, composite matrix components including collagen, fibronectin, and proteoglycan, or biodegradable polymers. Collected fibroblasts can be seeded in the matrix scaffold with the cells submerged under a culture medium. Epidermal cells or stem cells can be identified, collected, expanded, and seeded on the top of the fibroblast-containing matrix scaffold. Alternatively, host skin specimens can be collected and directly placed on the fibroblast-containing matrix scaffold. In the later case, the surface of the skin specimen should be exposed to the air. The culture medium may be supplemented with desired growth-stimulating factors, such as epidermal growth factor (EGF), keratinocyte growth factor (KGF), and/or fibroblast growth factor (FGF). Other necessary components may also be added, such as insulin (to promote the uptake of glucose and amino acids) and hydrocortisone (to promote cell adhesion and proliferation). The skin constructs can be cultured under standard conditions (37°C, 5% CO_2 and 95% air). The constructed skin substitutes usually exhibit a skin-like structure and express common epidermal cell markers such as keratin 6, 15, 16, and 17 (Fig. 24.5).

When a sheet-like skin structure is established, an enzyme called *dispase* can be applied to the skin construct to remove the epidermal layer from the underneath fibroblast-containing matrix. This enzyme cleaves adhesion molecules between the epidermal cell layer and the substrate without breaking the molecular adhesion bonds between the epidermal cells. The skin construct can be collected and applied to the injured skin. Alternatively, the skin construct can be stored at 4°C for a short period. When autogenous skin specimens are not available in patients with a large area of third-degree injury, allogenic epidermal cells can be used instead. In such a case, immuno-suppressors should be administrated to prevent skin-substitute rejection.

Skin Cancer

Pathogenesis, Pathology, and Clinical Features [24.7]. Skin cancers are originate primarily in the epidermal cells and belong to the carcinoma family (defined as cancers of epithelial origin). There are several types of common skin cancer, defined on the basis of the types of epidermal cells. These include basal cell carcinoma, squamous cell carcinoma, and melanoma. *Basal cell carcinoma* arises from deep basal epidermal cells, which are located on the basement membrane and can differentiate into superficial keratin-producing squamous epidermal cells for replacing lost cells. Once becoming cancer cells, the basal epidermal cells can no longer differentiate into squamous epidermal cells. Basal cell carcinoma accounts for about 70% of skin cancers. *Squamous cell carcinoma* originates from differentiated keratin-producing epidermal cells. This type of cancer can arise from a location with chronic inflammation, a burn scar, a sun-damaged area, chronic ulcers, or keratoses. It can also arise from normal skin. *Melanoma* is a type of malignant skin cancer, originating from the epidermal pigment cells, which are concentrated in the skin moles.

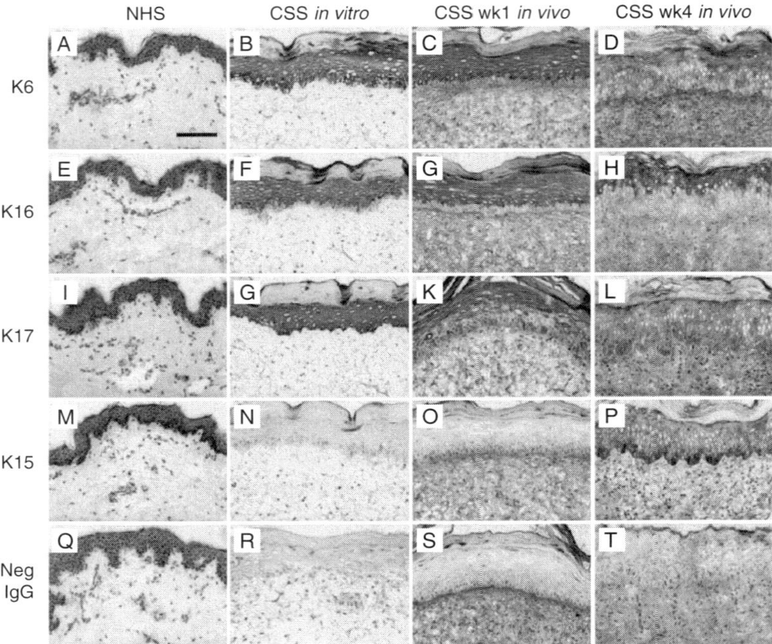

Figure 24.5. Immunohistochemical localization of keratin proteins in native human skin (NHS) and constructed skin substitutes (CSS). Shown are sections of native human skin (A,E,I,M,Q), CSS at 2 weeks incubation in vitro (B,F,J,N,R), and CSS at 1 week (C,G,K,O,S) and 4 weeks (D,H,L,P,T) after grafting to athymic mice (in vivo). For all sections, epidermis is at the top of the panel. (A–D) Keratin 6, (E–H) keratin 16, (I–L) keratin 17, (M–P) keratin 15. (Q–T) Immunohistochemistry using a nonimmune mouse IgG antibody as a negative control. Note the nonspecific background staining in the dermis of CSS sections in vivo, seen with the nonimmune negative control primary antibody (S,T) as well as the keratin-specific antibodies (C,D,G,H,K,L,O,P). This background staining, seen only in sections of CSS after grafting, resulted from the detection procedure used for mouse monoclonal antibodies on tissues excised from mice. Scale bar in (A) is same for all panels (100 μm). (Reprinted from Smiley AK et al: Keratin expression in cultured skin substitutes suggests that the hyperproliferative phenotype observed in vitro is normalized after grafting, *Burns* 32:135–8, copyright 2006, with permission from Elsevier Ltd. and the International Society for Burn Injuries.)

Melanoma develops rapidly with a high capability of invasion and metastasis. This type of cancer is characterized by progressive changes in the color, border shape, and the surface roughness of the skin moles. Melanoma is the leading cause of death among patients with skin diseases.

Several factors have been known to contribute to the pathogenesis of skin cancer. Exposure to ultraviolet light is a common cause for skin cancer, especially in patients with weak melanin pigmentation (see chapter-opening figure). People with poor tolerance to sunlight have a higher incidence of skin cancers compared to the general population. Exposure to chemical carcinogens (e.g., inorganic arsenicals and organic hydrocarbons) and ionizing radiation (e.g., X rays and γ rays) can also cause skin cancers. The pathogenesis of cancers is discussed in detail in Chapter 25.

Treatment of Skin Cancers. As for cancers in other tissues and organs, surgical removal of skin cancers prior to metastasis remains the primary treatment. Thus, early diagnosis of skin cancers is critical to the treatment of skin cancers. When cancers become metastatic, general treatment, such as chemotherapy and radiotherapy, should be used. Molecular therapy has also been used for the treatment of skin cancer in experimental investigations and clinical trials. These methods are discussed in Chapter 25.

BIBLIOGRAPHY

24.1. Anatomy and Physiology of the Skin

Guyton AC, Hall JE: *Textbook of Medical Physiology*, 11th ed, Saunders, Philadelphia, 2006.

McArdle WD, Katch FI, Katch VL: *Essentials of Exercise Physiology*, 3rd ed, Lippincott Williams & Wilkins, Baltimore, 2006.

Germann WJ, Stanfield CL (with contributors Niles MJ, Cannon JG): *Principles of Human Physiology*, 2nd ed, Pearson Benjamin Cummings, San Francisco, 2005.

Thibodeau GA, Patton KT: *Anatomy & Physiology*, 5th ed, Mosby, St Louis, 2003.

Boron WF, Boulpaep EL: *Medical Physiology: A Cellular and Molecular Approach*, Saunders, Philadelphia, 2003.

Ganong WF: *Review of Medical Physiology*, 21st ed, McGraw-Hill, New York, 2003.

24.2. Pathogenesis, Pathology, and Clinical Features of Skin Injury

Schneider AS, Szanto PA: *Pathology*, 3rd ed, Lippincott Williams & Wilkins, Philadelphia, 2006.

McCance KL, Huether SE: *Pathophysiology: The Biologic Basis for Disease in Adults & Children*, 5th ed, Elsevier Mosby, St. Louis, 2006.

Porth CM: *Pathophysiology: Concepts of Altered Health States*, 7th ed, Lippincott Williams & Wilkins, Philadelphia, 2005.

Frazier MS, Drzymkowski JW: *Essentials of Human Diseases and Conditions*, 3rd ed, Elsevier Saunders, St Louis, 2004.

Burd A, Chiu T: Allogenic skin in the treatment of burns, *Clin Dermatol* 23(4):376–87, 2005.

Andreassi A, Bilenchi R, Biagioli M, D'Aniello C: Classification and pathophysiology of skin grafts, *Clin Dermatol* 23(4):332–7, 2005.

Werner S, Grose R: Regulation of wound healing by growth factors and cytokines, *Physiol Rev* 83(3):835–70, 2003.

Ong YS, Samuel M, Song C: Meta-analysis of early excision of burns, *Burns* 32(2):145–50, 2006.

Collier M: Understanding the principles of wound management, *J Wound Care* 15(1):S7–10, 2006.

Chiu T, Burd A: "Xenograft" dressing in the treatment of burns, *Clin Dermatol* 23(4):419–23, 2005.

Shakespeare PG: The role of skin substitutes in the treatment of burn injuries, *Clin Dermatol* 23(4):413–18, 2005.

Supp DM, Boyce ST: Engineered skin substitutes: practices and potentials, *Clin Dermatol* 23(4):403–12, 2005.

Andreassi A, Bilenchi R, Biagioli M, D'Aniello C: Classification and pathophysiology of skin grafts, *Clin Dermatol* 23(4):332–7, 2005.

Rabbitts A, Alden NE, Scalabrino M, Yurt RW: Outpatient firefighter burn injuries: A 3-year review, *J Burn Care Rehab* 26(4):348–51, 2005.

Greenhalgh DG: Models of wound healing, *J Burn Care Rehab* 26(4):293–305, 2005.

24.3. Cell Types for Constructing Skin Substitutes

Owens DM, Watt FM: Contribution of stem cells and differentiated cells to epidermal tumours, *Nat Rev Cancer* 3:444–51, 2003.

Perez-Losada J, Balmain A: Stem-cell hierarchy in skin cancer, *Nat Rev Cancer* 3:434–43, 2003.

Limat A, Noser FK: Serial cultivation of single keratinocytes from the outer root sheath of human scalp hair follicles, *J Invest Dermatol* 87:485–8, 1986.

Gallico GGI, O'Connor NE, Compton CC, Kehinde O, Green H: Permanent coverage of large burn wounds with autologous cultured human epithelium, *New Engl J Med* 311:448–51, 1984.

Tong X, Coulombe PA: Mouse models of alopecia: Identifying structural genes that are baldly needed, *Trends Mol Med* 9:79–84, 2003.

Nakamura M, Sundberg JP, Paus R: Mutant laboratory mice with abnormalities in hair follicle morphogenesis, cycling, and/or structure: Annotated tables, *Exp Dermatol* 10:369–90, 2001.

Cotsarelis G, Sun TT, Lavker RM: Label-retaining cells reside in the bulge area of pilosebaceous unit: Implications for follicular stem cells, hair cycle, and skin carcinogenesis, *Cell* 61:1329–37, 1990.

Barrandon Y, Green H: Three clonal types of keratinocyte with different capacities for multiplication, *Proc Natl Acad Sci USA* 84:2302–6, 1987.

Jones PH, Watt FM: Separation of human epidermal stem cells from transit amplifying cells on the basis of differences in integrin function and expression, *Cell* 73:713–24, 1993.

Tani H, Morris RJ, Kaur P: Enrichment for murine keratinocyte stem cells based on cell surface phenotype, *Proc Natl Acad Sci USA* 97:10960–5, 2000.

Blau HM, Brazelton TR, Weimann JM: The evolving concept of a stem cell: Entity or function? *Cell* 105:829–41, 2001.

Pellegrini G, Dellambra E, Golisano O, Martinelli E, Fantozzi I et al: p63 identifies keratinocyte stem cells, *Proc Natl Acad Sci USA* 98:3156–61, 2001.

Taylor G, Lehrer MS, Jensen PJ, Sun TT, Lavker RM: Involvement of follicular stem cells in forming not only the follicle but also the epidermis, *Cell* 102:451–61, 2000.

Oshima H, Rochat A, Kedzia C, Kobayashi K, Barrandon Y: Morphogenesis and renewal of hair follicles from adult multipotent stem cells, *Cell* 104:233–45, 2001.

Rochat A, Kobayashi K, Barrandon Y: Location of stem cells of human hair follicles by clonal analysis, *Cell* 76:1063–73, 1994.

Krause DS, Theise ND, Collector MI, Henegariu O, Hwang S et al: Multi-organ, multi-lineage engraftment by a single bone-marrow-derived stem cell, *Cell* 105:369–77, 2001.

Tran SD, Pillemer SR, Dutra A, Barrett AJ, Brownstein MJ et al: Differentiation of human bone marrow-derived cells into buccal epithelial cells in vivo: A molecular analytical study, *Lancet* 361:1084–8, 2003.

Pellegrini G, Dellambra E, Golisano O, Martinelli E, Fantozzi I et al: p63 identifies keratinocyte stem cells, *Proc Natl Acad Sci USA* 98:3156–61, 2001.

Trempus CS, Morris RJ, Bortner CD, Cotsarelis G, Faircloth RS et al: Enrichment for living murine keratinocytes from the hair follicle bulge with the cell surface marker CD34, *J Invest Dermatol* 120:501–11, 2003.

Yu Y, Flint A, Dvorin EL, Bischoff J: AC133-2, a novel isoform of human AC133 stem cell antigen, *J Biol Chem* 277:20711–16, 2002.

Jones PH, Watt FM: Separation of human epidermal stem cells from transit amplifying cells on the basis of differences in integrin function and expression, *Cell* 73:713–24, 1993.

Tani H, Morris RJ, Kaur P: Enrichment for murine keratinocyte stem cells based on cell surface phenotype, *Proc Natl Acad Sci USA* 97:10960–5, 2000.

Lyle S, Christofidou-Solomidou M, Liu Y, Elder DE, Albelda S et al: The C8/144B monoclonal antibody recognizes cytokeratin 15 and defines the location of human hair follicle stem cells, *J Cell Sci* 111:3179–88, 1998.

Porter RM, Lunny DP, Ogden PH, Morley SM, McLean WH et al: K15 expression implies lateral differentiation within stratified epithelial basal cells, *Lab Invest* 80:1701–10, 2000.

Whitbread LA, Powell BC: Expression of the intermediate filament keratin gene, K15, in the basal cell layers of epithelia and the hair follicle, *Exp Cell Res* 244:448–59, 1998.

24.4. Matrix Scaffolds for Constructing Skin Substitutes

Yang S, Leong KF, Du Z, Chua CK: The design of scaffolds for use in tissue engineering: II. Rapid prototyping techniques, *Tissue Eng* 8:1–11, 2002.

Whitaker MJ, Quirk RA, Howdle SM, Shakesheff KM: Growth factor release from tissue engineering scaffolds, *J Pharm Pharmacol* 53:1427–37, 2001.

Horner PJ, Gage FH: Regenerating the damaged central nervous system, *Nature* 407:963–70, 2000.

Borgens RB: Cellular engineering: Molecular repair of membranes to rescue cells of the damaged nervous system, *Neurosurgery* 49:370–9, 2001.

Currie LJ, Sharpe JR, Martin R: The use of fibrin glue in skin grafts and tissue-engineered skin replacements: A review, *Plast Reconstr Surg* 108:1713–26, 2001.

24.5. Growth Factors for Stimulating the Growth of Epidermal Cells

Ray P, Devaux Y, Stolz DB, Yarlagadda M, Watkins SC et al: Inducible expression of keratinocyte growth factor (KGF) in mice inhibits lung epithelial cell death induced by hyperoxia, *Proc Natl Acad Sci USA* 100:6098–103, 2003.

Rubin JS, Osada H, Finch PW, Taylor WG, Rudikoff S et al: Purification and characterization of a newly identified growth factor specific for epithelial cells, *Proc Natl Acad Sci USA* 86:802–6, 1989.

Werner S, Smola H, Liao X, Longaker MT, Krieg T et al: The function of KGF in morphogenesis of epithelium and reepithelialization of wounds, *Science* 266:819–22, 1994.

Zimonjic DB, Kelley MJ, Rubin JS, Aaronson SA, Popescu NC: Fluorescence in situ hybridization analysis of keratinocyte growth factor gene amplification and dispersion in evolution of great apes and humans, *Proc Natl Acad Sci USA* 94:11461–5, 1997.

Wilgus TA, Matthies AM, Radek KA, Dovi JV, Burns AL et al: Novel function for vascular endothelial growth factor receptor-1 on epidermal keratinocytes, *Am J Pathol* 167(5):1257–66, Nov 2005.

Akita S, Akino K, Imaizumi T, Hirano A: A basic fibroblast growth factor improved the quality of skin grafting in burn patients, *Burns* 31(7):855–8, Nov 2005.

Tscharntke M, Pofahl R, Krieg T, Haase I: Ras-induced spreading and wound closure in human epidermal keratinocytes, *FASEB J* 19(13):1836–8, Nov 2005.

Gao Z, Sasaoka T, Fujimori T, Oya T, Ishii Y et al: Deletion of the PDGFR-beta gene affects key fibroblast functions important for wound healing, *J Biol Chem* 280(10):9375–89, March 2005.

Geer DJ, Swartz DD, Andreadis ST: In vivo model of wound healing based on transplanted tissue-engineered skin, *Tissue Eng* 10(7–8):1006–17, July–Aug 2004.

Colony-Stimulating Factor (CSF)

Mitrasinovic OM, Perez GV, Zhao F, Lee YL, Poon C et al: Overexpression of macrophage colony-stimulating factor receptor on microglial cells induces an inflammatory response, *J Biol Chem* 276(32):30142–9, 2001.

Praloran V, Chevalier S, Gascan H: Macrophage colony-stimulating factor is produced by activated T lymphocytes in vitro and is detected in vivo in T cells from reactive lymph nodes, *Blood* 79(9):2500–1, 1992.

Besse A, Trimoreau F, Praloran V, Denizot Y: Effect of cytokines and growth factors on the macrophage colony-stimulating factor secretion by human bone marrow stromal cells, *Cytokine* 12(5):522–5, 2000.

Muench MO, Roncarolo MG, Rosnet O, Birnbaum D, Namikawa R: Colony-forming cells expressing high levels of CD34 are the main targets for granulocyte colony-stimulating factor and macrophage colony-stimulating factor in the human fetal liver, *Exp Hematol* 25(4):277–87, 1997.

Fujikawa Y, Sabokbar A, Neale SD, Itonaga I, Torisu T et al: The effect of macrophage-colony stimulating factor and other humoral factors (interleukin-1, -3, -6, and -11, tumor necrosis factor-alpha, and granulocyte macrophage-colony stimulating factor) on human osteoclast formation from circulating cells, *Bone* 28(3):261–7, 2001.

Pollard JW, Bartocci A, Arceci R, Orlofsky A, Ladner MB et al: Apparent role of the macrophage growth factor, CSF-1, in placental development, *Nature* 330:484–6, 1987.

Wong GG, Temple PA, Leary AC, Witek-Giannotti JS, Yang YC et al: Human CSF-1: Molecular cloning and expression of 4-kb cDNA encoding the human urinary protein, *Science* 235:1504–8, 1987.

Granulocyte/Macrophage Colony-Stimulating Factor (GM-CSF)

Cantrell MA, Anderson D, Cerretti DP, Price V, McKereghan K et al: Cloning, sequence, and expression of a human granulocyte/macrophage colony-stimulating factor, *Proc Natl Acad Sci USA* 82:6250–4, 1985.

Frolova EI, Dolganov GM, Mazo IA, Smirnov DV, Copeland P et al: Linkage mapping of the human CSF2 and IL3 genes, *Proc Natl Acad Sci USA* 88:4821–4, 1991.

Grabstein KH, Urdal DL, Tushinski RJ, Mochizuki DY, Price VL et al: Induction of macrophage tumoricidal activity by granulocyte-macrophage colony-stimulating factor, *Science* 232:506–8, 1986.

Huebner K, Isobe M, Croce CM, Golde DW, Kaufman SE et al: The human gene encoding GM-CSF is at 5q21-q32, the chromosome region deleted in the 5q- anomaly, *Science* 230: 1282–5, 1985.

LeVine AM, Reed JA, Kurak KE, Cianciolo E, Whitsett JA: GM-CSF-deficient mice are susceptible to pulmonary group B streptococcal infection, *J Clin Invest* 103:563–9, 1999.

Pettenati MJ, Le Beau MM, Lemons RS, Shima EA, Kawasaki ES et al: Assignment of CSF-1 to 5q33.1: evidence for clustering of genes regulating hematopoiesis and for their involvement in the deletion of the long arm of chromosome 5 in myeloid disorders, *Proc Natl Acad Sci USA* 84:2970–4, 1987.

Thangavelu M, Neuman WL, Espinosa, R III; Nakamura Y, Westbrook CA et al: A physical and genetic linkage map of the distal long arm of human chromosome 5, *Cytogenet Cell Genet* 59:27–30, 1992.

Wong GG, Witek JS, Temple PA, Wilkens KM, Leary AC et al: Human GM-CSF: Molecular cloning of the complementary DNA and purification of the natural and recombinant proteins, *Science* 228:810–15, 1985.

IL18

Bossu P, Neumann D, Del Giudice E, Ciaramella A, Gloaguen I et al: IL-18 cDNA vaccination protects mice from spontaneous lupus-like autoimmune disease, *Proc Natl Acad Sci* 100:14181–6, 2003.

Corbaz A, ten Hove T, Herren S, Graber P, Schwartsburd B et al: IL-18-binding protein expression by endothelial cells and macrophages is up-regulated during active Crohn's disease, *J Immun* 168:3608–16, 2002.

Konishi H, Tsutsui H, Murakami T, Yumikura-Futatsugi S, Yamanaka K et al: IL-18 contributes to the spontaneous development of atopic dermatitis-like inflammatory skin lesion independently of IgE/stat6 under specific pathogen-free conditions, *Proc Natl Acad Sci USA* 99: 11340–5, 2002.

Nolan KF, Greaves DR, Waldmann H: The human interleukin 18 gene IL18 maps to 11q22.2-q22.3, closely linked to the DRD2 gene locus and distinct from mapped IDDM loci, *Genomics* 51: 161–3, 1998.

Okamura H, Tsutsui H, Komatsu T, Yutsudo M, Hakura A et al: Cloning of a new cytokine that induces IFN-gamma production by T cells, *Nature* 378:88–91, 1995.

Pizarro TT, Michie MH, Bentz M, Woraratanadharm J, Smith MF Jr. et al: IL-18, a novel immunoregulatory cytokine, is up-regulated in Crohn's disease: Expression and localization in intestinal mucosal cells, *J Immun* 162:6829–35, 1999.

Reddy P, Teshima T, Kukuruga M, Ordemann R, Liu C et al: Interleukin-18 regulates acute graft-versus-host disease by enhancing Fas-mediated donor T cell apoptosis, *J Exp Med* 194:1433–40, 2001.

Rothe H, Jenkins NA, Copeland NG, Kolb H: Active stage of autoimmune diabetes is associated with the expression of a novel cytokine, IGIF, which is located near Idd2, *J Clin Invest* 99: 469–74, 1997.

Sugawara S, Uehara A, Nochi T, Yamaguchi T, Ueda H et al: Neutrophil proteinase 3-mediated induction of bioactive IL-18 secretion by human oral epithelial cells, *J Immun* 167:6568–75, 2001.

Vidal-Vanaclocha F, Fantuzzi G, Mendoza L, Fuentes AM, Anasagasti MJ et al: IL-18 regulates IL-1-beta-dependent hepatic melanoma metastasis via vascular cell adhesion molecule-1, *Proc Natl Acad Sci USA* 97:734–9, 2000.

Stoof TJ, Boorsma DM, Nickoloff BJ: Keratinocytes and immunological cytokines, in *The Keratinocyte Handbook*, Watt FM, ed, Cambridge University Press, 1994, pp 365–99.

Werner S, Peters KG, Longaker MT, Fuller-Pace F, Banda MJ et al: Large induction of keratinocyte growth factor expression in the dermis during wound healing, *Proc Natl Acad Sci USA* 89:6896–900, 1992.

Singer AJ, Clark RA: Cutaneous wound healing, *New Engl J Med* 341:738–46, 1999.

Matsuguchi T, Lilly MB, Kraft AS: Cytoplasmic domains of the human granulocyte-macrophage colony-stimulating factor (GM-CSF) receptor beta chain (h beta c) responsible for human GM-CSF-induced myeloid cell differentiation, *J Biol Chem* 273:9411–18, 1998.

Enk AH, Katz SI: Identification and induction of keratinocyte-derived IL-10, *J Immunol* 149:92–5, 1992.

Stoll S, Muller G, Kurimoto M et al: Production of IL-18 (IFN-gamma-inducing factor) messenger RNA and functional protein by murine keratinocytes, *J Immunol* 159:298–302, 1997.

Eriksson A, Siegbahn A, Westermark B, Heldin CH, Claesson-Welsh L et al: PDGF alpha- and beta-receptors activate unique and common signal transduction pathways, *EMBO J* 11:543–50, 1992.

Brown LF, Yeo KT, Berse B et al: Expression of vascular permeability factor (vascular endothelial growth factor) by epidermal keratinocytes during wound healing, *J Exp Med* 176:1375–9, 1992.

Detmar M, Brown LF, Berse B et al: Hypoxia regulates the expression of vascular permeability factor/vascular endothelial growth factor (VPF/VEGF) and its receptors in human skin, *J Invest Dermatol* 108:263–8, 1997.

Eming SA, Medalie DA, Tompkins RG, Yarmush ML, Morgan JR: Genetically modified human keratinocytes overexpressing PDGF-A enhance the performance of a composite skin graft, *Hum Gene Ther* 9:529–39, 1998.

Liu Y, Kim D, Himes BT et al: Transplants of fibroblasts genetically modified to express BDNF promote regeneration of adult rat rubrospinal axons and recovery of forelimb function, *J Neurosci* 19:4370–87, 1999.

Eming SA, Lee J, Snow RG, Tompkins RG, Yarmush ML et al: Genetically modified human epidermis overexpressing PDGF-A directs the development of a cellular and vascular connective tissue stroma when transplanted to athymic mice—implications for the use of genetically modified keratinocytes to modulate dermal regeneration, *J Invest Dermatol* 105:756–63, 1995.

Kozarsky KF, Wilson JM: Gene therapy: Adenovirus vectors, *Curr Opin Genet Dev* 3:499–503, 1993.

Ghazizadeh S, Taichman LB: Multiple classes of stem cells in cutaneous epithelium: A lineage analysis of adult mouse skin, *EMBO J* 20:1215–22, 2001.

Bett AJ, Prevec L, Graham FL: Packaging capacity and stability of human adenovirus type 5 vectors, *J Virol* 67:5911–21, 1993.

Feng M, Cabrera G, Deshane J, Scanlon KJ, Curiel DT: Neoplastic reversion accomplished by high efficiency adenoviral-mediated delivery of an anti-ras ribozyme, *Cancer Res* 55:2024–8, 1995.

Barr E, Carroll J, Kalynych AM et al: Efficient catheter-mediated gene transfer into the heart using replication-defective adenovirus, *Gene Ther* 1:51–8, 1994.

Human protein reference data base, Johns Hopkins University and the Institute of Bioinformatics, at http://www.hprd.org/protein.

24.6. Construction of Skin Substitutes

Cha ST, Talavera D, Demir E, Nath AK, Sierra-Honigmann MR: A method of isolation and culture of microvascular endothelial cells from mouse skin, *Microvasc Res* 70(3):198–204, 2005.

Ishii K, Harada R, Matsuo I, Shirakata Y, Hashimoto K et al: In vitro keratinocyte dissociation assay for evaluation of the pathogenicity of anti-desmoglein 3 IgG autoantibodies in pemphigus vulgaris, *J Invest Dermatol* 124(5):939–46, 2005.

Navsaria HA et al: Culturing skin in vitro for wound therapy, *Trends Biotechnol* 13:91–100, 1995.

Chandler LA, Gu DL, Ma C et al: Matrix-enabled gene transfer for cutaneous wound repair, *Wound Repair Regen* 8:473–9, 2000.

Martin P: Wound healing-aiming for perfect skin regeneration, *Science* 276:75–81, 1997.

Parenteau NL, Rosenberg L, Hardin-Young J: The engineering of tissues using progenitor cells, *Curr Top Dev Biol* 64:101–39, 2004.

24.7. Pathogenesis, Pathology, and Clinical Features of Skin Cancer

Abdulla FR, Feldman SR, Williford PM, Krowchuk D, Kaur M: Tanning and skin cancer, *Pediatr Dermatol* 22:501–12, 2005.

Eide MJ, Weinstock MA: Public health challenges in sun protection, *Dermatol Clin North Am* 24:119–24, 2006.

Gandini S, Sera F, Cattaruzza MS, Pasquini P, Zanetti R et al: Meta-analysis of risk factors for cutaneous melanoma: III. Family history, actinic damage and phenotypic factors, *Eur J Cancer* 41:2040–59, 2005.

Glanz K, Mayer JA: Reducing ultraviolet radiation exposure to prevent skin cancer methodology and measurement, *Am J Prevent Med* 29:131–42, 2005.

Heenan PJ: Local recurrence of melanoma, *Pathology* 36:491–5, 2004.

Nguyen TH: Mechanisms of metastasis, *Clin Dermatol* 22:209–16, 2004.

Nathanson SD: Insights into the mechanisms of lymph node metastasis, *Cancer* 98:413–23, 2003.

Perez-Losada J, Balmain A: Stem-cell hierarchy in skin cancer, *Nat Rev Cancer* 3:434–43, June 2003.

Bode AM, Dong Z: Signal transduction pathways: Targets for chemoprevention of skin cancer, *Lancet Oncol* 1:181–8, 2000.

Halachmi S, Gilchrest BA: Update on genetic events in the pathogenesis of melanoma, *Curr Opin Oncol* 13:129–36, 2001.

Tsao H: Update on familial cancer syndromes and the skin, *J Am Acad Dermatol* 42:939–69, 2000.

25

REGENERATIVE ENGINEERING FOR CANCER

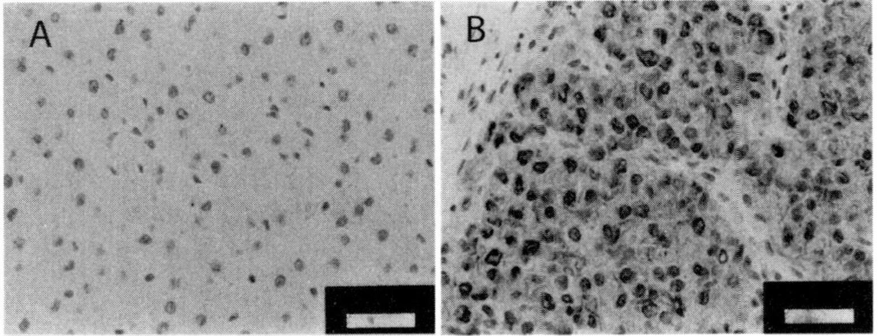

Cell nucleus density and size in control rat liver and hepatoma: (A) control rat liver; (B) hepatoma. Scale bar: 50 μm. (Reprinted with permission from la Cour JM et al: Up-regulation of ALG-2 in hepatomas and lung cancer tissue, *Am J Pathol* 163:81–9, copyright 2003.) See color insert.

CLASSIFICATION OF CANCERS [25.1]

Cancer is a disorder characterized by the formation of immortal cells that undergo uncontrolled excessive cell proliferation. Cancer cells are aberrantly differentiated cells due to gene mutation and do not possess physiological functions. These cells can aggressively invade neighboring tissues and organs, spread to remote tissues and organs through the lymphatic and vascular systems (a process known as *metastasis*), form colonies in invaded tissues and organs, and expand in the expense of normal cells. As a result, normal cells die because of deprivation of necessary nutrients, leading to the malfunction

Bioregenerative Engineering: Principles and Applications, by Shu Q. Liu

and death of involved tissues and organs. It should be noted that cancer is a malignant type of tumor. There are also tumors that grow slowly, do not invade neighboring tissues, do not metastasize, and do not significantly influence the functions of host tissues and organs. This type of tumor is defined as benign tumors and will not be covered in this book.

Based on the origin of cancer formation, cancers are classified into several types: carcinoma, sarcoma, leukemia, and neural tumors. *Carcinoma* is a type of cancer that arises from epithelial cells, including gland cells, and can be found in a variety of organs, such as the nasal and oral cavities, gastrointestinal tract, liver, pancreas, lung, breast, prostate gland, overy, uterus, bladder, and skin. Carcinoma is the most common type of cancer, accounting for about 80% of total cancers. The high incidence of carcinoma is possibly related to the high turnover or renewal rate of epithelial cells. These cells are developed for the protection of tissues and organs, and are often subject to various types of harmful environmental factors. Thus, mature epithelial cells are still able to differentiate and proliferate for the repair and replacement of injured and lost cells. Such a feature naturally increases the rate of cancer formation. Various terms have been used to describe carcinoma, depending on the type of cells, tissues, and organs where cancers arise. For instance, *hepatoma* is a term for liver carcinoma, *melanoma* is used to describe carcinoma derived from the epidermal melanocytes, which contains the pigment melanin, and *adenocarcinoma* is cancer from gland epithelial cells.

Sarcoma is a type of cancer that arises from connective and muscular tissues. This type of cancer is found in soft connective tissues, bones, cartilages, skeletal muscles, and blood vessels. Since mature connective tissue cells and muscular cells are well-differentiated cells and undergo a low rate of proliferation, the incidence of sarcoma (about 1%) is much lower than that of carcinoma. *Leukemia* is a type of cancer derived from hematopoietic or blood cells, primarily involving leukocytes. Leukemia derived from lymphocytes is called *lymphoma*. This type of cancer accounts for about 9% of the total cancers. Another major type of cancer is *neural cancers*, which are found in the nervous system. This type of cancer increases the volume of the brain within the limited skull space and induces the compression of normal brain tissue, resulting in various symptoms of neural disorders depending on the region involved. Typical nervous cancers include gliomas and retinoblastomas.

PATHOGENESIS OF CANCERS [25.1]

While the pathogenesis of cancers remains a research topic, increasing investigations have suggested that cancers are possibly a result of gene mutation or changes in DNA structure. Cancer may be originated from a single cell that undergoes cancerigenic gene mutation. It is important to note that, although various types of gene mutation may occur under physiological conditions, not all gene mutations lead to carcinogenesis. However, it remains poorly understood what types of gene mutation are carcinogenic.

Gene mutation may occur spontaneously during cell division as a natural process, which is responsible for evolutionary alternations in living organisms. In addition, gene mutation can be induced by environmental factors, including radiation, viral infection, exposure to carcinogens, and therapeutic gene transfer. Radiation (e.g., X rays and ultraviolet light) often causes DNA damage and chromosome disruption. Gene mutation may be introduced to the genome when damaged genes and chromosomes are repaired. Viruses are able to integrate their genome into the host genome, a process often inducing gene mutation. Chemical carcinogens, such as formaldehyde and peanut mold-toxin (aflatoxin), can cause changes in DNA sequences. Gene transfer is thought a therapeutic method that is used to correct mutant genes. However, the insertion of foreign genes into the genome may induce gene mutation. Cancers may be induced when patients are frequently exposed to these factors.

While many types of gene mutation can contribute to carcinogenesis, there are two types that play a major role: the activation of growth stimulatory genes, which are also known as protooncogenes, and the suppression of growth inhibitory genes, which are also called tumor suppressor genes. *Protooncogenes* are normal genes that encode mitogenic proteins and can be converted to carcinogenic oncogenes or cancer-inducing genes by mutation. Examples of proto-oncogenes include c-fos (activator protein-1 gene), c-jun (another activator protein-1 gene), c-raf (protein serine/threonine kinase gene), c-myc (gene-regulatory protein gene), c-sis (platelet-derived growth factor gene), and c-src (Src protein tyrosine kinase gene). These genes play critical roles in the regulation of physiological development and morphogenesis. These genes can be transformed into oncogenes in response to the stimulation of environmental factors. A typical cause of protooncogene transformation is retrovirus infection. Retroviruses can convert their RNA genome into DNA and insert the converted DNA into selected protooncogenes in the host genome. The protooncogenes can be structurally altered or can be subject to the control of viral gene promoters, leading to the formation of carcinogenic oncogenes that stimulate cell differentiation and proliferation.

Another type of gene mutation is the alteration or loss of tumor suppressor genes. These genes play critical roles in the inhibition of carcinogenesis. The mutation or loss of such genes contributes to the initiation and development of cancers. A typical tumor suppressor gene is the retinoblastoma tumor suppressor gene. This gene is expressed in almost all cell types and encodes a protein that controls the progression of the cell division cycle. Cancers in several organs, including the lung and breast, are associated with reduced expression of the retinoblastoma tumor suppressor gene. The mutation and loss of this gene contributes to carcinogenesis. Another tumor suppressor gene is the p53 gene (see page 236 for the characteristics of p53). Patients with mutation or loss of this gene are susceptible to carcinogenesis. Some viruses, such as papillomaviruses, exert an inhibitory effect on the tumor suppressor genes.

PATHOLOGICAL CHARACTERISTICS OF CANCERS [25.1]

In pathological examinations, cancerous changes can be found at the cellular and tissue levels. Pathological changes at the cellular level include an increase in cell proliferation (assessed by BrdU assay), cell density, and the size of cell nuclei. It is important to note that these changes should be always determined on the basis of a comparison with normal cells. Pathological changes at the tissue level include the formation of tumors without a clearly defined boundary, tumor invasion of neighboring tissues, an increase in angiogenesis, and disruption or destruction of normal tissues where cancer cells invade. The appearance of cancer cells at multiple locations suggests metastasis. Cancers are often associated with rapid functional deterioration of involved organs. For instance, stomach cancer, when involving a large area, exhibits reduced capability of food digestion and rapid loss of body weight. Lung cancer is associated with difficulties in respiration. The ultimate consequence of cancer is the death of the involved organ. During the late stage, metastasis occurs in almost all cancer patients, often resulting in rapid failure of involved organ systems.

TREATMENT OF CANCERS

Conventional Treatment [25.1]

There are three major conventional approaches for the treatment of cancer: surgical removal, chemotherapy, and radiotherapy. *Surgical removal of cancers* is the most effective treatment when cancers are limited to a local area and are not spread to the surrounding tissues and lymph nodes. Early diagnosis of cancer is critical to the success of surgical treatment. Once cancers spread to neighboring tissues or metastasize to different organs, surgery is no longer effective. Chemotherapy and/or radiotherapy can be used for spread or metastasized cancers. *Chemotherapy* is an approach by which chemical agents are used to suppress the proliferation of cells, including cancer and normal cells. Since normal cells undergo a lower rate of division than cancer cells, chemical agents primarily affect cancer cells. A typical chemotherapeutic agent is 5-fluorouracil (5-FU), a uracil derivative that interrupts DNA synthesis and therefore suppresses cell division when incorporated into the genome of dividing cells. *Radiotherapy* is an approach used to treat cancers by exposing patients to radiation. While radiation destroys cancer cells and is effective for cancer treatment, it also induces normal cell injury and death.

Molecular Engineering Therapies

A number of molecular engineering strategies have been established and used for the treatment of cancers. These include the up-regulation of tumor suppressor genes, correction of mutant tumor suppressor genes, enhancement of anti-cancer immune responses, activation of tumor suppressor drugs, introduction of oncolytic viruses, and inhibition of growth-promoting genes by using antisense and siRNA oligonucleotides. A large number of investigations have been carried out to test these strategies in experimental models. Some of the strategies have been applied to preliminary clinical trials. Selected strategies are discussed as follows.

Overexpression of Tumor Suppressor Genes and Correction of Mutant Tumor Suppressor Genes [25.2]. As discussed on page 265, several proteins, including $p16^{iNK4}$, $p15^{iNK4B}$, $p18^{INK4C}$, $p19^{INK4D}$, p21, p27, and p57, are known to exert an inhibitory effect on cell division. These proteins may suppress the activity of cyclin D/CDK 4/CDK6, an important protein complex that regulates the progression of the cell division cycle from the G1 to S phase. Another protein, p53, suppresses cell division by activating p27. The overexpression of these proteins enhances the inhibitory effect on cell division and suppresses tumor cell growth. In particular, the p53 gene has been tested extensively. A large fraction of cancer patients exhibit mutant p53 gene, a potential factor contributing to the initiation and development of cancers. Experimental investigations in animal models of cancers have shown that the transfer of the wildtype p53 gene into cancer tissues results in the suppression of cancer cell proliferation and reduction in cancer progression. Preliminary studies in human trials have demonstrated promising results for the therapeutic effect of the wildtype p53 gene. Other growth-inhibitory protein genes, as described above, can also serve as candidate genes for cancer therapy.

Enhancement of Anticancer Immune Responses [25.3]. Cancer cells express tumor antigens that can be recognized by the immune system under physiological conditions. The immune system is capable of destroying recognized cancer cells, as cancer cells are considered as foreign invaders. Such recognition and destruction activities are regulated by several signaling processes. When cancer cells form due to gene mutations, antigens expressed in the cancer cells can be recognized by antigen-presenting cells, which present the antigens to the T-helper cells. The T-helper cells produce and release cytokines, such as interleukin (IL)2, which activate cancer-specific T lymphocytes. The activated T lymphocytes can produce and release killer cytokines to suppress cancer growth and progression.

However, when the function of the immune system is suppressed, the immune system can no longer detect and destroy cancer cells, allowing cancer cells to proliferate and spread. Indeed, the suppression of the immune system may contribute to the initiation and development of cancers. Furthermore, cancer cells can produce and release cytokines that suppress the activity of host immune system and help cancer cells to escape from the immune surveillance and attack. Thus, a major approach for the treatment of cancers is to enhance the tumor-recognizing and suppressing functions of the host immune system. While it is difficult to achieve such a goal, several hypothetical strategies have been proposed and tested. These include the overexpression of T-lymphocyte activating factors, T-cell costimulating factors, and application of tumor-antigen vaccines.

The activity of the T lymphocytes can be stimulated by introducing several factors, such as human leukocyte antigen (HLA), major histocompatibility complex (MHC), cytokines (e.g., IL2 and interferon-γ) to target cells. The level of human leukocyte antigen is reduced in a large fraction of cancer patients. A molecule analogous to human leukocyte antigen is the major histocompatibility complex. The overexpression of the major histocompatibility complex gene in animal models of cancers induces activation of cytotoxic T cells and boosts anti-cancer responses. In human studies, the transfer of HLA-expressing vectors into melanoma cells induces the activation of cytotoxic T-cells and a reduction in tumor growth.

Several cytokines, including IL2, IL4, IL12α (see Table 25.1 for specifics on IL12α and CEA), granulocyte macrophage-colony stimulating factor (GM-CSF), and interferon-γ (see page 631 and the following table in this section for the characteristics of these factors), are known to promote the activation of T lymphocytes. The overexpression of these cytokines by gene transfer is a potential approach for the treatment of cancers. In particular, the IL2 gene has been tested in experimental models for anti-cancer effects. These studies have demonstrated that the overexpression of the IL2 gene activates cytotoxic T cells and reduces tumor growth in murine cancer models. Investigations in human trials have also shown promising results. Direct administration of cytokine proteins is another approach. However, cytokines undergo rapid degradation and the general administration of cytokines often induces toxic responses. Local gene transfer is an approach that can be used to overcome these problems.

TABLE 25.1. Characteristics of Carcinoembryonic Antigen and IL12α*

Proteins	Alternative Names	Amino Acids	Molecular Weight (kDa)	Expression	Functions
Carcinoembryonic antigen	CEA, carcinoembryonic antigen-related antigen-related cell adhesion molecule 5 (CEACAM5), meconium antigen 100, CD66e antigen	702	77	Leukocyte, intestine, epithelial cells	Promoting cell survival and proliferation and enhancing cancer cell growth
IL12 α	IL12A, cytotoxic lymphocyte maturation factor, natural killer cell stimulatory factor chain (NKSF1), T-cell-stimulating factor	219	25	Dendritic cells	Inducing the expression of interferon (IFN)γ and promoting the differentiation of Th1 and Th2 cells

*Based on bibliography 25.3.

TABLE 25.2. Characteristics of B7.1*

Proteins	Alternative Names	Amino Acids	Molecular Weight (kDa)	Expression	Functions
B7.1	B7-1 antigen, B lymphocyte activation antigen B7-1, activation B7-1 antigen, CD28 antigen ligand 1 (CD28LG1), CD80 antigen	288	33	Liver, dendritic cells, mast cells, brain, blood vessel	Stimulating activation of T cells

*Based on bibliography 25.3.

Since the repression of the anticancer immune responses is considered a factor that contributes to tumorigenesis, the boosting of immune activities is a potential approach for the treatment of cancers. An effective approach for boosting immune activities is to deliver cancer-associated antigens to the host systems. Such antigens include viral analogues of tumor antigens, mutated oncogene proteins, and carcinoembryonic antigens. The genes of these proteins can be used to construct plasmids, known as recombinant vaccines, which can be used for gene transfer and for producing cancer-associated antigens. The host immune system is able to recognize these antigens and generate specific antibodies, which contribute to the anticancer activities. A typical tumor-associated antigen is the carcino-embryonic antigen, a glycoprotein receptor that is upregulated in certain types of cancer, such as colon cancer. The gene of the carcino-embryonic antigen can be inserted into a viral vector, such as the *Canary pox* virus, forming a recombinant DNA molecule. The overexpression of such a recombinant gene in a transgenic mouse model results in the production of antibodies against the carcinoembryonic antigen, in association with the enhancement of T-cell activities against tumor cells that express the carcinoembryonic antigen. Clinical investigations in patients with colorectal cancer have shown promising results, including the generation of antibodies against the carcinoembryonic antigen and activation of the cytotoxic T cells.

The activation of cytotoxic T cells can be induced by exposure to cancer cell antigens. The activity of the cytotoxic T cells can be boosted by costimulating factors such as the B7.1 protein (see Table 25.2). In certain types of cancer, the expression of costimulating factors is repressed, a potential factor that reduces the anticancer immune responses. Thus, a strategy for enhancing the activity of the cytotoxic T cells is to deliver cancer cell antigens and costimulating factors together. A recombinant gene can be constructed with the carcinoembryonic antigen gene and the costimulating factor B7.1 gene. Such a gene construct has been applied to human patients with adenocarcinoma in a preliminary clinical trail. This study has demonstrated that the overexpression of the recombinant gene is associated with activation of T cells specific to the carcinoembryonic antigen. The application of the costimulating factor gene enhances the anticancer immune reactions.

Activation of Tumor-Suppressing Prodrugs [25.4]. Chemotherapy is an effective approach for the inhibition of cancer metastasis. However, most chemotherapeutic agents destroy not only cancer cells, but also normal cells, significantly repressing the activity of the immune system. It is desired to introduce chemotherapeutic agents that are effective only in targeted cancer cells without influencing the normal cells. Synthetic deoxynucleosides may serve as such chemotherapeutic agents. A typical agent is ganciclovir, an analogue of deoxyguanosine. Ganciclovir can be phosphorylated by the herpes simplex virus tyrosine kinase to form ganciclovir triphosphate, a deoxynucleotide that can be incorporated into DNA during DNA synthesis in place of dGTP. The incorporation of the ganciclovir deoxynucleotide induces the termination of DNA synthesis, thus arresting cell division in the S phase. Ganciclovir without phosphorylation is not effective. Based on such a feature, ganciclovir can be administrated through blood injection, while herpes simplex virus tyrosine kinase gene can be delivered to local target cancer cells. With the expression of the viral tyrosine kinase gene in the local cancer cells, ganciclovir can only be phosphorylated in the cancer cells, thus repressing DNA synthesis in cancers but not in normal cells. Experimental investigations have demonstrated that in animal models of colon cancer, general ganciclovir administration with local delivery of the herpes simplex viral tyrosine kinase gene results in a significant reduction in tumor growth. Clinical investigations with a similar approach have also shown promising results for the treatment of human colorectal adenocarcinoma.

Another tumor-suppressing prodrug is 5-fluorocytosine, a cytosine derivative. This agent is relatively inactive in its natural form. A cytosine-specific enzyme called cytosine deaminase, which is found in bacteria and fungi, can convert 5-fluorocytosine to 5-fluorouracil, a potent anti-cancer agent that blocks the methylation reaction of deoxyuridylic acid to thymidylic acid and thus interferes with DNA synthesis. A blood injection of 5-fluorocytosine, together with a delivery of a bacterial cytosine deaminase gene into local target cancers can induce local conversion of 5-fluorocytosine to 5-fluorouracil, resulting in the repression of cancer growth with a minimal influence on normal cell function. Such an approach has been shown to effectively inhibit cancer proliferation in animal models as well as in human patients with colorectal cancer.

Application of Oncolytic Viruses [25.5]. Several types of viruses, including adenovirus and herpes simplex virus, can be genetically modified to establish the capability of lysing cancer cells, but not normal cells. A mutant adenovirus has been created by removing the E1B segment of the viral genome, resulting in the deficiency of the E1B-55 kDa viral protein. When delivered to cancer cells, the mutant adenoviruses can replicate in cancer cells lacking p53 and then lyse these cells. Clinical investigations have shown that the delivery of mutant adenoviruses into metastatic tumors results in cancer cell death. However, controversial results have been reported regarding the target specificity of the mutant adenoviruses. Further investigations are necessary to confirm the preliminary discoveries.

Herpes simplex virus is another type of virus that can be modified to establish cancer cell-specific lysing activity. The removal of selective genes, including the tyrosine kinase gene or the ribonucleotide reductase, from the herpes simplex virus results in mutant viruses that can replicate in dividing cells. Since cancer cells undergo a much higher rate of division compared to normal cells, the transfer of the mutant herpes simplex viruses into cancer cells results in the lysis of these cells. Experimental studies have demonstrated selective oncolytic activity of herpes simplex virus in animal colon cancer models.

Application of Antisense Oligonucleotides and siRNA [25.6]. The initiation and development of cancers involve the upregulation of oncogenic genes, such as *abl* (protein tyrosine kinase), *raf* (protein serine/threonine kinase), *ras* (GTP-binding protein), *sis* (platelet-derived growth factor B chain), and *src* (Src protein tyrosine kinase). The expression of other genes that encode mitogenic proteins also contributes to carcinogenesis. The suppression of carcinogenic gene expression is an effective approach for the treatment of cancers. A typical approach for such a purpose is the administration of antisense oligonucleotides specific to a target mRNA molecule. Antisense oligonucleotides are short fragments of nucleotides with a length of 20–40 base pairs and can hybridize to complementary mRNA in the cytoplasm, blocking the translation of proteins. These fragments can be directly delivered to target cells for therapeutic purposes. An alternative method is to transfer a gene construct that encodes a fragment of antisense deoxynucleotides, which can be expressed and exerts antisense effects. Furthermore, the transfection of target cancer cells with siRNAs specific to proliferative mRNAs can effectively suppress the proliferation of cancer cells (Fig. 25.1).

Application of Combined Therapeutic Approaches [25.7]. The treatment of cancers is a challenging task. Often, the application of a single approach as described above may not be effective, especially in the presence of cancer metastasis. The combination of anticancer therapeutic methods may provide an alternative means for the treatment of cancers. Various combinations have been tested in experimental and clinical investigations. Exam-

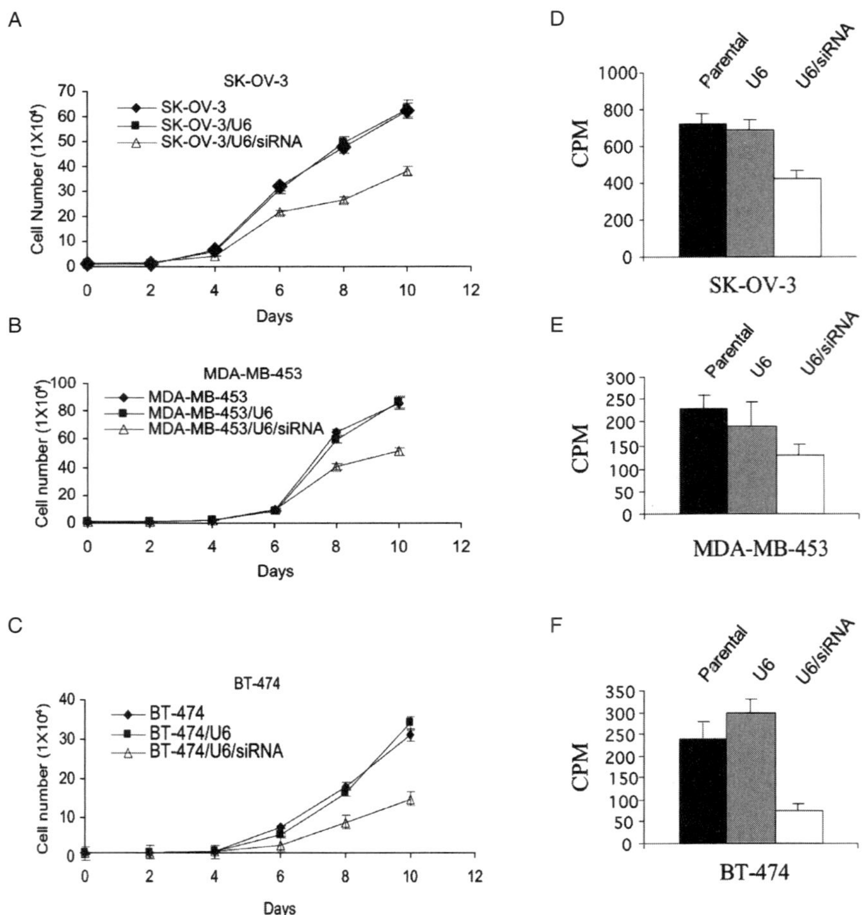

Figure 25.1. Cell proliferation, measured by cell counts (A–C) and [^{3}H]thymidine uptake (D–F) of SK-OV-3, MDA-MB-453, and BT-474 cells with and without transfection with Her-2/neu siRNA. Note that the *Her-2/neu* gene is an oncogene that is overexpressed in ~30% of breast and ovarian cancer cases. The expression of this gene often indicates a poor prognosis. CPM: counts per minute. (Reprinted with permission from Yang G et al: Inhibition of breast and ovarian tumor growth through multiple signaling pathways by using retrovirus-mediated small interfering RNA against Her-2/neu gene expression, *J Biol Chem* 279:4339–45, copyright 2004.)

ples include the combination of oncolytic viruses and anticancer prodrugs, the combination of radiotherapy with oncolytic viruses, and the combination of tumor-suppressing genes and immune response activators. These combination approaches have been shown to be more effective compared to single approaches.

BIBLIOGRAPHY

25.1. Classification, Pathogenesis, and Treatment of Cancer

Brinkman BM, Wong DT: Disease mechanism and biomarkers of oral squamous cell carcinoma, *Curr Opin Oncol* 18(3):228–33, May 2006.

Kleinsmith LJ: *Principles of Cancer Biology*, Pearson Benjamin Cummings, San Francisco, 2006.

Schiff D, O'Neill B: *Principles of Neurooncology*, McGraw-Hill, New York, 2005.

Vokes EE, Golomb HM: *Oncologic Therapies*, Springer, Berlin, 2003.

Pitot HC: *Fundamentals of Oncology*, 4th ed, Marcel Dekker, New York, 2002.

Stone RM: *Harrison's Principles of Internal Med*, 15th ed, McGraw-Hill, New York, 2001.

Schneider AS, Szanto PA: *Pathology*, 3rd ed, Lippincott Williams & Wilkins, Philadelphia, 2006.

McCance KL, Huether SE: *Pathophysiology: The Biologic Basis for Disease in Adults & Children*, 5th ed, Elsevier Mosby, St. Louis, 2006.

Porth CM: *Pathophysiology: Concepts of Altered Health States*, 7th ed, Lippincott Williams & Wilkins, Philadelphia, 2005.

Frazier MS, Drzymkowski JW: *Essentials of Human Diseases and Conditions*, 3rd ed, Elsevier Saunders, St Louis, 2004.

25.2. Overexpression of Tumor Suppressor Genes and Correction of Mutant Tumor Suppressor Genes

Harris MP, Sutjipto S, Wills KN et al: Adenovirus-mediated p53 gene transfer inhibits growth of human tumor cells expressing mutant p53 protein, *Cancer Gene Ther* 3:121–30, 1996.

Bouvet M, Ellis LM, Nishizaki M et al: Adenovirus-mediated wild-type p53 gene transfer down-regulates vascular endothelial growth factor expression and inhibits angiogenesis in human colon cancer, *Cancer Res* 58:2288–92, 1998.

Cohen AM, Kemeny NE, Kohne CH, Wils J, de Takats PG et al: Is intra-arterial chemotherapy worthwhile in the treatment of patients with unresectable hepatic colorectal cancer metastases? *Eur J Cancer* 32A:2195–205, 1996.

Nguyen DM, Spitz FR, Yen N, Cristiano RJ, Roth JA: Gene therapy for lung cancer: Enhancement of tumor suppression by a combination of sequential systemic cisplatin and adenovirus-mediated p53 gene transfer, *J Thor Cardiovasc Surg* 112:1372–6, 1376–7, 1996.

25.3. Enhancement of Anti-Cancer Immune Responses

Todryk SM, Chong H, Vile RG, Pandha H, Lemoine NR: Can immunotherapy by gene transfer tip the balance against colorectal cancer? *Gut* 43:445–9, 1998.

Bodey B, Bodey Jr B, Siegel SE, Kaiser HE: Failure of cancer vaccines: The significant limitations of this approach to immunotherapy, *Anticancer Res* 20:2665–76, 2000.

Ostrand-Rosenberg S, Thakur A, Clements V: Rejection of mouse sarcoma cells after transfection of MHC class II genes, *J Immunol* 144:4068–71, 1990.

Nabel GJ, Nabel EG, Yang ZY et al: Direct gene transfer with DNA-liposome complexes in melanoma: Expression, biologic activity, and lack of toxicity in humans, *Proc Natl Acad Sci USA* 90:11307–11, 1993.

Gonzalez R, Atkins M, Schwarzenberger P et al: Phase II trial of HLA-B7 plasmid DNA/lipid (Allovectin®) immunotherapy in patients with metastatic melanoma, *Proc Am Soc Clin Oncol* 20:1007, 2001.

Rubin J, Galanis E, Pitot HC et al: Phase I study of immunotherapy of hepatic metastases of colorectal carcinoma by direct gene transfer of an allogeneic histocompatibility antigen, HLA-B7, *Gene Ther* 4:419–25, 1997.

Carcinoembryonic Antigen (CEA)

Brandriff BF, Gordon LA, Tynan KT, Olsen AS, Mohrenweiser HW et al: Order and genomic distances among members of the carcinoembryonic antigen (CEA) gene family determined by fluorescence in situ hybridization, *Genomics* 12:773–9, 1992.

Nishi M, Inazawa J, Inoue K, Nakagawa H, Taniwaki M et al: Regional chromosomal assignment of carcinoembryonic antigen gene (CEA) to chromosome 19 at band q13.2, *Cancer Genet Cytogenet* 54:77–81, 1991.

Thompson JA, Pande H, Paxton RJ, Shively L, Padma A et al: Molecular cloning of a gene belonging to the carcinoembryonic antigen gene family and discussion of a domain model, *Proc Natl Acad Sci USA* 84:2965–9, 1987.

Willcocks TC, Craig SP, Craig IW: Assignment of the coding sequence for carcinoembryonic antigen (CEA) and normal cross-reacting antigen (NCA) to human chromosome 19q13, *Ann Hum Genet* 53:141–8, 1989.

Zimmer R, Thomas P: Mutations in the carcinoembroyonic antigen gene in colorectal cancer patients: Implications on liver metastasis, *Cancer Res* 61:2822–6, 2001.

Zimmermann W, Ortlieb B, Friedrich R, von Kleist S: Isolation and characterization of cDNA clones encoding the human carcinoembryonic antigen reveal a highly conserved repeating structure, *Proc Natl Acad Sci USA* 84:2960–4, 1987.

IL12

Kass E, Schlom J, Thompson J, Guadagni F, Graziano P et al: Induction of protective host immunity to carcinoembryonic antigen (CEA), a self-antigen in CEA transgenic mice, by immunizing with a recombinant vaccinia-CEA virus, *Cancer Res* 59:676–83, 1999.

Tsang KY, Zaremba S, Nieroda CA, Zhu MZ, Hamilton JM et al: Generation of human cytotoxic T cells specific for human carcinoembryonic antigen epitopes from patients immunized with recombinant vaccinia-CEA vaccine, *J Natl Cancer Inst* 87:982–90, 1995.

Conry RM, Khazaeli MB, Saleh MN et al: Phase I trial of a recombinant vaccinia virus encoding carcinoembryonic antigen in metastatic adenocarcinoma: Comparison of intradermal versus subcutaneous administration, *Clin Cancer Res* 5:2330–7, 1999.

Conry RM, Allen KO, Lee S, Moore SE, Shaw DR et al: Human autoantibodies to carcinoembryonic antigen (CEA) induced by a vaccinia-CEA vaccine, *Clin Cancer Res* 6:34–41, 2000.

Hodge JW, McLaughlin JP, Kantor JA, Schlom J: Diversified prime and boost protocols using recombinant vaccinia virus and recombinant non-replicating avian pox virus to enhance T-cell immunity and antitumor responses, *Vaccine* 15:759–68, 1997.

Marshall JL, Hawkins MJ, Tsang KY et al: Phase I study in cancer patients of a replication-defective avipox recombinant vaccine that expresses human carcinoembryonic antigen, *J Clin Oncol* 17:332–7, 1999.

Zhu MZ, Marshall J, Cole D, Schlom J, Tsang KY: Specific cytolytic T-cell responses to human CEA from patients immunized with recombinant avipox-CEA vaccine, *Clin Cancer Res* 6: 24–33, 2000.

B7.1

Freeman GJ, Disteche CM, Gribben JG, Adler DA, Freedman AS et al: The gene for B7, a costimulatory signal for T-cell activation, maps to chromosomal region 3q13.3-3q21, *Blood* 79:489–94, 1992.

Reiser J, von Gersdorff G, Loos M, Oh J, Asanuma K et al: Induction of B7-1 in podocytes is associated with nephrotic syndrome, *J Clin Invest* 113:1390–7, 2004.

Selvakumar A, Mohanraj BK, Eddy RL, Shows TB, White PC et al: Genomic organization and chromosomal location of the human gene encoding the B-lymphocyte activation antigen B7, *Immunogenetics* 36:175–81, 1992.

Shah R, Banks K, Patel A, Dogra S, Terrell R et al: Intense expression of the B7-2 antigen presentation coactivator is an unfavorable prognostic indicator for differentiated thyroid carcinoma of children and adolescents, *J Clin Endocr Metab* 87:4391–7, 2002.

Stamper CC, Zhang Y, Tobin JF, Erbe DV, Ikemizu S et al: Crystal structure of the B7-1/CTLA-4 complex that inhibits human immune responses, *Nature* 410:608–11, 2001.

Horig H, Lee DS, Conkright W et al: Phase I clinical trial of a recombinant canarypoxvirus (ALVAC) vaccine expressing human carcinoembryonic antigen and the B7.1 co-stimulatory molecule, *Cancer Immunol Immunother* 49:504–14, 2000.

Fakhrai H, Shawler DL, Gjerset R et al: Cytokine gene therapy with interleukin-2-transduced fibroblasts: Effects of IL-2 dose on anti-tumor immunity, *Hum Gene Ther* 6:591–601, 1995.

Sobol RE, Shawler DL, Carson C et al: Interleukin 2 gene therapy of colorectal carcinoma with autologous irradiated tumor cells and genetically engineered fibroblasts: A phase I study, *Clin Cancer Res* 5:2359–65, 1999.

Suminami Y, Elder EM, Lotze MT, Whiteside TL: In situ interleukin-4 gene expression in cancer patients treated with genetically modified tumor vaccine, *J Immunother Emphasis Tumor Immunol* 17:238–48, 1995.

Schmidt-Wolf IG, Finke S, Trojaneck B et al: Phase I clinical study applying autologous immunological effector cells transfected with the interleukin-2 gene in patients with metastatic renal cancer, colorectal cancer and lymphoma, *Br J Cancer* 81:1009–16, 1999.

Rochlitz CF, Jantscheff P, Bongartz G et al: Gene therapy with cytokine-transfected xenogeneic cells in metastatic tumors, *Adv Exp Med Biol* 451:531–7, 1998.

Galanis E, Hersh EM, Stopeck AT et al: Immunotherapy of advanced malignancy by direct gene transfer of an interleukin-2 DNA/DMRIE/DOPE lipid complex: Phase I/II experience, *J Clin Oncol* 17:3313–23, 1999.

Geutskens SB, van der Eb MM, Plomp AC et al: Recombinant adenoviral vectors have adjuvant activity and stimulate T cell responses against tumor cells, *Gene Ther* 7:1410–6, 2000.

Human protein reference data base, Johns Hopkins University and the Institute of Bioinformatics, at http://www.hprd.org/protein.

25.4. Activation of Tumor-Suppressing Pro-Drugs

Link Jr CJ, Levy JP, McCann LZ, Moorman DW: Gene therapy for colon cancer with the herpes simplex thymidine kinase gene, *J Surg Oncol* 64:289–94, 1997.

Sung MW, Yeh HC, Thung SN et al: Intratumoral adenovirus-mediated suicide gene transfer for hepatic metastases from colorectal adenocarcinoma: Results of a phase I clinical trial, *Mol Ther* 4:182–91, 2001.

Boucher PD, Im MM, Freytag SO, Shewach DS: A novel mechanism of synergistic cytotoxicity with 5-fluorocytosine and ganciclovir in double suicide gene therapy, *Cancer Res* 66(6):3230–7, 2006.

Kuriyama S, Kikukawa M, Masui K et al: Cytosine deaminase/5-fluorocytosine gene therapy can induce efficient anti-tumor effects and protective immunity in immunocompetent mice but not in athymic nude, *Int J Cancer* 81:592–7, 1999.

Huber BE, Austin EA, Richards CA, Davis ST, Good SS: Metabolism of 5-fluorocytosine to 5-fluorouracil in human colorectal tumor cells transduced with the cytosine deaminase gene: Significant antitumor effects when only a small percentage of tumor cells express cytosine deaminase, *Proc Natl Acad Sci USA* 91:8302–6, 1994.

Crystal RG, Hirschowitz E, Lieberman M et al: Phase I study of direct administration of a replication deficient adenovirus vector containing the E. coli cytosine deaminase gene to metastatic colon carcinoma of the liver in association with the oral administration of the pro-drug 5-fluorocytosine, *Hum Gene Ther* 8:985–1001, 1997.

25.5. Application of Oncolytic Viruses

Chernajovsky Y, Layward L, Lemoine N: Fighting cancer with oncolytic viruses, *Br Med J* 332(7534):170–2, 2006.

Aghi M, Martuza RL: Oncolytic viral therapies—the clinical experience, *Oncogene* 24(52): 7802–16, 2005.

Working PK, Lin A, Borellini F: Meeting product development challenges in manufacturing clinical grade oncolytic adenoviruses, *Oncogene* 24(52):7792–801, 2005.

Mathis JM, Stoff-Khalili MA, Curiel DT: Oncolytic adenoviruses—selective retargeting to tumor cells, *Oncogene* 24(52):7775–91, 2005.

Ko D, Hawkins L, Yu DC: Development of transcriptionally regulated oncolytic adenoviruses, *Oncogene* 24(52):7763–74, 2005.

Shmulevitz M, Marcato P, Lee PW: Unshackling the links between reovirus oncolysis, Ras signaling, translational control and cancer, *Oncogene* 24(52):7720–8, 2005.

Barber GN: VSV-tumor selective replication and protein translation, *Oncogene* 24(52):7710–9, 2005.

Mohr I: To replicate or not to replicate: Achieving selective oncolytic virus replication in cancer cells through translational control, *Oncogene* 24(52):7697–709, 2005.

O'Shea CC: Viruses—seeking and destroying the tumor program, *Oncogene* 24(52):7640–55, 2005.

Chou J, Roizman B: The gamma 1(34.5) gene of herpes simplex virus 1 precludes neuroblastoma cells from triggering total shutoff of protein synthesis characteristic of programmed cell death in neuronal cells, *Proc Natl Acad Sci USA* 89:3266–70, 1992.

Mineta T, Rabkin SD, Yazaki T, Hunter WD, Martuza RL: Attenuated multi-mutated herpes simplex virus-1 for the treatment of malignant gliomas, *Nat Med* 1:938–43, 1995.

Walker JR, McGeagh KG, Sundaresan P, Jorgensen TJ, Rabkin SD et al: Local and systemic therapy of human prostate adenocarcinoma with the conditionally replicating herpes simplex virus vector G207, *Hum Gene Ther* 10:2237–43, 1999.

Delman KA, Bennett JJ, Zager JS et al: Effects of preexisting immunity on the response to herpes simplex-based oncolytic viral therapy, *Hum Gene Ther* 11:2465–72, 2000.

Wildner O, Blaese RM, Morris JC: Therapy of colon cancer with oncolytic adenovirus is enhanced by the addition of herpes simplex virus-thymidine kinase, *Cancer Res* 59:410–3, 1999.

Bischoff JR, Kirn DH, Williams A et al: An adenovirus mutant that replicates selectively in p53-deficient human tumor cells, *Science* 274:373–6, 1996.

Rothmann T, Hengstermann A, Whitaker NJ, Scheffner M, zur Hausen H: Replication of ONYX-015, a potential anticancer adenovirus, is independent of p53 status in tumor cells, *J Virol* 72:9470–8, 1998.

Ries SJ, Brandts CH, Chung AS et al: Loss of p14ARF in tumor cells facilitates replication of the adenovirus mutant dl1520 (ONYX-015), *Nat Med* 6:1128–33, 2000.

Reid T, Galanis E, Abbruzzese J et al: Intra-arterial administration of a replication-selective adenovirus (dl1520) in patients with colorectal carcinoma metastatic to the liver: A phase I trial, *Gene Ther* 8:1618–26, 2001.

Habib NA, Sarraf CE, Mitry RR et al: E1B-deleted adenovirus (dl1520) gene therapy for patients with primary and secondary liver tumors, *Hum Gene Ther* 12:219–26, 2001.

Khuri FR, Nemunaitis J, Ganly I et al: A controlled trial of intratumoral ONYX-015, a selectively-replicating adenovirus, in combination with cisplatin and 5-fluorouracil in patients with recurrent head and neck cancer, *Nat Med* 6:879–85, 2000.

Heise C, Hermiston T, Johnson L et al: An adenovirus E1A mutant that demonstrates potent and selective systemic anti-tumoral efficacy, *Nat Med* 6:1134–9, 2000.

25.6. Application of Antisense Oligodeoxynucleotides

Tortora G, Caputo R, Damiano V et al: A novel MDM2 anti-sense oligonucleotide has anti-tumor activity and potentiates cytotoxic drugs acting by different mechanisms in human colon cancer, *Int J Cancer* 88:804–9, 2000.

Berg RW, Werner M, Ferguson PJ et al: Tumor growth inhibition in vivo and G2/M cell cycle arrest induced by antisense oligodeoxynucleotide targeting thymidylate synthase, *J Pharmacol Exp Ther* 298:477–84, 2001.

Nishizuka I, Ichikawa Y, Ishikawa T et al: Matrilysin stimulates DNA synthesis of cultured vascular endothelial cells and induces angiogenesis in vivo, *Cancer Lett* 173:175–82, 2001.

Tamm I, Dorken B, Hartmann G: Antisense therapy in oncology: New hope for an old idea? *Lancet* 358:489–97, 2001.

25.7. Application of Combined Therapeutic Approaches

Carroll NM, Chase M, Chiocca EA, Tanabe KK: The effect of ganciclovir on herpes simplex virus-mediated oncolysis, *J Surg Res* 69:413–7, 1997.

Rogulski KR, Wing MS, Paielli DL, Gilbert JD, Kim JH et al: Double suicide gene therapy augments the antitumor activity of a replication-competent lytic adenovirus through enhanced cytotoxicity and radiosensitization, *Hum Gene Ther* 11:67–76, 2000.

Andreansky S, He B, van Cott J et al: Treatment of intracranial gliomas in immunocompetent mice using herpes simplex viruses that express murine interleukins, *Gene Ther* 5:121–30, 1998.

Advani SJ, Sibley GS, Song PY et al: Enhancement of replication of genetically engineered herpes simplex viruses by ionizing radiation: A new paradigm for destruction of therapeutically intractable tumors, *Gene Ther* 5:160–5, 1998.

Shimada H, Shimizu T, Ochiai T, Liu TL, Sashiyama H et al: Preclinical study of adenoviral p53 gene therapy for esophageal cancer, *Surg Today* 31(7):597–604, 2001.

INDEX

Bioregenerative Engineering: Principles and Applications, by Shu Q. Liu